S0-AWR-444

HEATH

ALGEBRA 1

AN INTEGRATED APPROACH

Teacher's Edition

Roland E. Larson

Timothy D. Kanold

Lee Stiff

Evanston, Illinois Boston ◆ Dallas

Published simultaneously in Canada

Printed in the United States of America

International Standard Book Number: 0-669-31605-9

2 3 4 5 6 7 8 9 10 VHP 02 01 00 99 98 97

Authentic algebra
is algebra
that models the real world.
It's the all-new
ALGEBRA from D.C. Heath.

CONTENTS

Presenting algebra for all students

ALGEBRA and the NCTM Standards

The NCTM Standards serve as the foundation of the new ALGEBRA from D.C. Heath, with its emphasis on

- Problem Solving
- Reasoning
- Communication
- Connections.

Look for the spirit of the Standards throughout the text.

All of us learn and understand concepts better in context, and algebra is no exception. But over the last 20 years, most algebra texts have become so narrowly focused that they have lost track of the real purpose of algebra—to help us solve real-life problems. Algebra has been little more than the manipulation of symbols, producing symbolic solutions, with no larger context in sight.

ALGEBRA from D.C. Heath is designed to provide a context for the symbols of algebra. What you'll find in this new program is neither "old algebra" (however you might define that), nor some mysterious "new algebra." It is simply authentic algebra that makes sense— in the same way an architect's specifications suddenly make sense when examined next to a model of the design.

At every opportunity in this text, we demonstrate the usefulness and vitality of algebra—and in so doing we make it accessible to every student. We help each student set reasonable, reachable goals at the beginning of each lesson, and we provide continuing reasons for all students to develop confidence and a sense of accomplishment in their work.

Presenting algebra for the 21st Century—our distinguished author team

Roland E. Larson is Professor of Mathematics at the Behrend College of Pennsylvania State University in Erie. He is a member of NCTM and a highly successful author of many high school and college mathematics textbooks. Several of his D.C. Heath college titles have proven to be top choices for high school mathematics courses.

Timothy D. Kanold is Chairman of Mathematics and Science and a teacher at Adlai Stevenson High School in Prairie View, Illinois. He is the 1986 recipient of the Presidential Award for Excellence in Mathematics Teaching, as well as the 1991 recipient of the Outstanding Young Alumni Award from Illinois State University. A member of NCTM, he served on NCTM's *Professional Standards for Teaching Mathematics* Commission and is a member of the *Regional Services* Committee.

Lee Stiff is an Associate Professor of Mathematics Education in the College of Education and Psychology of North Carolina State University at Raleigh. He has taught mathematics at the high school and middle school levels. He is a member of the NCTM Board of Directors (1992-93), has served on NCTM's *Professional Standards for Teaching Mathematics* Commission, and was a founding member of the *Committee for a Comprehensive Mathematics Education for Every Child.*

What to look for in ALGEBRA from D.C. Heath

Real Life
Climatology

Real-life applications show the value of algebra. You don't have to search ALGEBRA for a subtle "orientation" to reality. You'll see it as solid grounding in every lesson, with exercise sets that blend interesting information with skill building and problem solving. *See pages T6, T18.*

Problem Solving
Draw a Graph

Problem solving is a continuing process in this program, not just an occasional topic. Students are asked to explain, justify, verify, interpret, draw and label—in short, to think critically all the time. *See page T8.*

Connections
Technology

Connections give algebra greater meaning. ALGEBRA from D.C. Heath helps you integrate such topics as geometry, statistics, trigonometry, probability, and matrix theory into your algebra course—together with topics from other disciplines, such as geography, history, economics, and the physical and biological sciences. *See page T10.*

Technology is used to investigate and verify findings. Students are shown in every chapter how they can use a calculator or a computer to evaluate expressions, graph equations, draw scatter plots and best-fitting lines, and much more. Data analysis is presented in separate lessons and is integrated throughout the text. *See pages T12, T29.*

Formal and Alternative Assessments

Assessment gives a clear picture of individual needs. In ALGEBRA from D.C. Heath, both formal and alternative forms of assessment are natural components of the program, designed to help you meet the needs of all students. *See page T14.*

Communication is central to math activity. Lesson features in ALGEBRA encourage students to verbalize math concepts and share ideas with each other—to reinforce each others' understanding of algebra and emphasize the connections between other disciplines and algebra. *See page T16.*

Communicating about ALGEBRA

The text addresses a diverse student body as a matter of course—by referring naturally in problems and examples to students of varied interests and backgrounds, by presenting selected topics in historical perspective, by highlighting current career opportunities, and by giving teachers a range of ancillary materials. *See page T30.*

Milestones ORIGINS OF ALGEBRA

Career Interview

ALGEBRA provides cooperative learning opportunities. Mathematics should not be a solitary activity. It can be done and learned with others—through discussions, projects, activities, etc.—with students of different abilities, different backgrounds, different interests. *See page T32.*

Communicating about ALGEBRA

Cooperative Learning

Extensive ancillaries help you reach all your students. You can enrich the new ALGEBRA with a variety of special materials designed to help you meet the needs of a diverse student population. *See page T33.*

COMPONENTS

To walk through a lesson from ALGEBRA 1, please turn to page T20.

A lesson that connects with life connects with students.

In developing this algebra text, the authors have kept uppermost in their minds the ways students learn—what builds interest in a topic, what makes an idea memorable.

Algebra instruction has for some time emphasized formal derivations—which ask a student to learn generalizations first, then to apply the results to particular cases. Applications, on the other hand, ask a student to learn particular, concrete cases first; then a general understanding follows. The authors have chosen to emphasize the applications of algebra in this text, feeling that moving from the need for algebra to the mechanics of algebra is what makes the most sense to students.

Derivations and proofs are provided in the Teacher's Edition, but the focus of the program is on the applications of algebra—its use in meaningful activities. Every lesson has at least one example that shows how the material in the lesson can be applied to a real-life situation.

These examples, and their corresponding exercises, comprise the most extensive, creative collection of mathematical applications ever assembled in an algebra series. The applications often use authentic data. This helps students validate what they are learning—a central theme of the NCTM Standards.

" We look at the hard data and see life in it. We're making it possible for students to do that, too."

—Rita Campanella, Executive Editor, ALGEBRA

Note: A separate Applications Handbook provides interesting additional background for applications in the text. Topics covered include astronomy, chemistry, physics, sports, economics, genetics, and music.

Look for meaning in this lesson.

While other texts discuss concepts as concepts, ALGEBRA from D.C. Heath discusses how a concept fits into the continuing story of the mathematics of real life. Compare various text treatments of the multiplication of real numbers, for instance. You'll find that the context and the logic of presentation in ALGEBRA are what makes the topic most meaningful.

Here a pattern is described to show the reasonableness of the fact that the product of two negative numbers is a positive number.

5

Multiplication of Real Numbers

ld learn:

tiply real
rties of

calculators
e in real

learn it:

itive and
solve
ich as
by time to
n also
ure to help
not make a

Goal 1 Multiplying Real Numbers

In this lesson, you will learn how to find products that have one or more negative factors. Here are two examples.

$3(-2) = (-2) + (-2) + (-2) = -6$
$3(-1) = (-1) + (-1) + (-1) = -3$

In both cases, multiplying a positive number and a negative number produces a negative number. One way to see what would happen if you multiplied two negative numbers is to look for patterns, such as:

One factor −3	One Factor −2	One Factor −1
$(3)(-3) = -9$	$(3)(-2) = -6$	$(3)(-1) = -3$
$(2)(-3) = -6$	$(2)(-2) = -4$	$(2)(-1) = -2$
$(1)(-3) = -3$	$(1)(-2) = -2$	$(1)(-1) = -1$
$(0)(-3) = 0$	$(0)(-2) = 0$	$(0)(-1) = 0$
$(-1)(-3) = ?$	$(-1)(-2) = ?$	$(-1)(-1) = ?$
$(-2)(-3) = ?$	$(-2)(-2) = ?$	$(-2)(-1) = ?$
Products increase by 3.	Products increase by 2.	Products increase by 1.

From these patterns, it makes sense that the missing numbers are positive. In other words, multiplying two negative numbers produces a positive number.

$(-1)(-3) = 3$ $\quad (-1)(-2) = 2$ $\quad (-1)(-1) = 1$
$(-2)(-3) = 6$ $\quad (-2)(-2) = 4$ $\quad (-2)(-1) = 2$

Properties of Multiplication

1. $a \cdot b = b \cdot a$ The order in which two numbers are multiplied does not change their product.
2. $1 \cdot a = a$ The product of one and a is a.
3. $0 \cdot a = 0$ The product of zero and a is 0.
4. $(-1) \cdot a = -a$ The product of -1 and a is the opposite of a.
5. $(a)(-b) = -ab$ The product of a and $-b$ is the opposite of ab.
6. $(-a)(-b) = ab$ The product of $-a$ and $-b$ is ab.

2.5 Multiplication of Real Numbers 89

Then students are given a list of the six properties of multiplication.

Nine examples of skill development follow. The three shown here involve real-life applications, calculators, and "unit analysis."

Example 3 ***Real-Life Applications of Products of Negative Numbers***

a. A football team loses 3 yards on each of 4 downs. The net gain is

$4(-3) = -12$ yards.

A gain of -12 means the same as a loss of 12.

b. A parachutist is falling with a velocity of -20 feet per second. The distance traveled in 3 seconds is

$3(-20) = -60$ feet.

The negative distance indicates a downward motion.

c. To attract customers, a grocery store runs a sale on strawberries. The strawberries are *loss leaders*, which means the store loses money on the strawberries but hopes to make it up on other sales. Suppose 3000 baskets are sold at a loss of

Always be careful when using a calculator for evaluating a product involving negative numbers—especially with the change-sign key [+/−].

Example 4 ***Using a Calculator***

	Expression	Change-Sign Key	Negation Key
a.	$(-2)(-4)$	2 [+/−] [×] 4 [+/−] [=]	[(−)] 2 [×] [(−)] 4 [=]

Your calculator should display 8 as an answer. The keystroke sequence [+/−] 2 [×] [+/−] 4 [=] is incorrect even though, in this instance, it may produce the correct answer.

	Expression	Change-Sign Key	Negation Key
b.	-3^4	3 [y^x] 4 [=] [+/−]	[(−)] 3 [^] 4 [=]

Your calculator should display -81 as an answer. The keystroke sequence [+/−] 3 [y^x] 4 [=] will not produce the correct answer. ■

Most real-life applications of products involve rates such as miles *per* hour or dollars *per* year. Example 5 shows how to decide which unit of measure to assign to a product. Notice how the units for a rate can be written in fraction form.

Example 5 ***Unit Analysis***

A car traveling 70 kilometers per hour for 2 hours will travel a distance of

$(70 \frac{\text{kilometers}}{\text{hour}})(2 \text{ hours}) = 140$ kilometers.

You can think of the "hours" as canceling each other. ■

...cating about ALGEBRA

SHARING IDEAS about the...

The Standards recommend...

Mathematics as Problem Solving

In grades 9-12, the mathematics curriculum should include the refinement and extension of methods of mathematical problem solving so that all students can—

- use, with increasing confidence, problem-solving approaches to investigate and understand mathematical content.
- apply integrated mathematical problem-solving strategies to solve problems from within and outside mathematics.
- recognize and formulate problems from situations within and outside mathematics.
- apply the process of mathematical modeling to real-world problem situations.

Increased attention to

word problems with a variety of structures.

everyday problems, applications, and open-ended questions.

investigation and discussion of patterns, relationships, and problem-solving strategies.

situations represented verbally, numerically, graphically, geometrically, or symbolically.

— from the NCTM Standards

Algebra is a way to solve problems—and to understand the world around us.

Algebra helps us solve problems efficiently. The real-life value of algebra as a problem-solving tool is a major theme stressed by Larson, Kanold, and Stiff throughout the new ALGEBRA.

Algebra is useful in real life.

Whenever students begin a new algebraic concept, the **Why you should learn it** feature in the Student Text explains how that concept will help them: that it's useful, for instance, when determining which video store offers the best deal or how much to charge for tickets to a fund-raiser. Because students encounter real-life problem solving in *every lesson*, they will come to see how valuable algebra can bc as a problem-solving tool.

"I like the way ALGEBRA uses the strategy of moving from a verbal model to an algebraic model to solve a problem. When teaching algebra, I usually try to get students to use their own words first ... but I haven't seen any [other] textbook employ this technique."

— Regina Kiczek, ALGEBRA reviewer, Ferris High School, Jersey City, NJ

Verbal modeling gets to the point of a problem.

Early in the text, ALGEBRA develops a problem-solving model. Students learn to develop a verbal model first—or "translate" a problem into their own words.

Once they have defined labels for the verbal model, they put together an equation, or algebraic model. Moving from the verbal to the algebraic helps students focus on the content of a problem.

Visual models broaden the picture.

To demonstrate the usefulness of algebra, D.C. Heath's ALGEBRA program connects linear equations—a difficult concept for many students—to real-life data. ALGEBRA also uses the power of new technology—the computer and the graphing calculator—to create meaningful visual representations of algebraic concepts. This visual approach gives students a firm grasp of concepts and strengthens reasoning skills.

Authentic data connect algebra to the real world.

Many problems and examples use real-life data and graphs to make algebraic concepts more meaningful to students—and to reinforce the role algebra plays in the real world. The use of authentic data also teaches students how to evaluate the information on graphs often found in newspapers and magazines.

Exercises build basic and critical-thinking skills.

Unusually rich and creative exercise sets blend basic skill building with critical thinking to strengthen students' problem-solving abilities. Students are continually asked to explain, justify, verify, interpret—in short, to think critically.

D.C. Heath's ALGEBRA provides...

- algebraic concepts set in real-life context.
- verbal modeling—to help students focus on what a problem is asking for.
- visual representations of algebraic concepts—for deeper understanding and stronger reasoning skills.
- exercises that strengthen students' skills *and* critical thinking.

"ALGEBRA enhances a teacher's role as guide, coach, and cheerleader. We try to help teachers nurture the problem-solving skills students already possess in order to build new problem-solving skills."

— *Roland E. Larson, ALGEBRA author*

The Standards recommend...

The mathematics curriculum should include investigation of the connections and interplay among various mathematical topics and their applications so that all students can—
- recognize equivalent representations of the same concept.
- relate procedures in one representation to procedures in an equivalent representation.
- use and value the connections among mathematical topics.
- use and value the connections between mathematics and other disciplines.

Increased attention to

connections among math topics, among math and other curriculum areas, and between math and daily life.

the relevance and value of mathematics in students' studies and lives so they view algebra as a whole rather than an isolated set of topics.

the use of real-world problems to motivate and apply theory.

—from the NCTM Standards

Connections strengthen understanding of algebra.

This all-new program helps your students view algebra not as isolated collections of symbols, but as a vital body of knowledge with relevance to other math topics, to other academic disciplines, and to their own world. Seeing algebra's importance to their own lives makes studying it more meaningful and enjoyable.

Connections to other math topics

In this new program, concepts from geometry, statistics, probability, and other branches of mathematics are integrated with algebra so that students have an opportunity to recognize and understand algebraic principles.

ALGEBRA provides **Integrated Review** exercises to show students how previously learned concepts relate to the new concepts of a lesson.

Exploring Data lessons and exercises connect algebra to statistics, data organization, and finance.

Connections to other academic disciplines

In this new program, algebra is linked to many other disciplines, such as art, biology, geography, history, music, medicine, and business. This integration will expand your students' sense of the usefulness of algebra. They will see how algebraic concepts and procedures can be applied to problems arising in diverse areas.

Connections to real life

ALGEBRA presents mathematics in a relevant, meaningful context using real-life data and applications. Students value algebra as they understand the interrelationships that connect what they find in their text to their world. Not only are the real-life examples important to understanding algebra—they are also interesting and fun!

Each lesson of the text begins with a **Why you should learn it** feature that lets your students know—right from the start—the importance of what they are about to study.

Career Interview features in alternate chapters introduce students to the broad range of careers in which math is a useful tool.

The **Independent Practice** sections at the end of each lesson contain both skill-building exercises and real-life applications.

A sampling of real-life connections in ALGEBRA 1

Chapter 1, page 39	Law Enforcement
Chapter 1, page 40	Air Traffic Control
Chapter 2, page 83	Accounting, Art
Chapter 3, page 137	Consumer Decisions, Video Games
Chapter 3, page 141	Wildlife
Chapter 3, page 143	Page Design
Chapter 3, page 169	Ecology, Chemistry, Retail, Purchasing
Chapter 4, page 178	Animal Studies
Chapter 4, page 203	Demographics
Chapter 4, page 207	Printing, Business Decisions
Chapter 5, page 246	Hockey, Cable TV, Fitness, Savings, History
Chapter 5, page 275	Geography, Games
Chapter 6, page 317	Game Show, Car Mileage, Test Scores, Track, Fireworks, Chemistry
Chapter 6, page 338	Forestry, Ecology
Chapter 7, page 355	Energy Saving
Chapter 7, page 362	History, Chemistry
Chapter 7, page 396	Airplane Navigation, Airplane Design
Chapter 8, page 404	Profit, Native American Culture
Chapter 8, page 434	Biology, Human Development, Salary Comparisons
Chapter 8, page 436	Business, Furniture Sales, Earnings, Inflation
Chapter 8, page 441	Psychology, Human Memory
Chapter 8, page 443	Demographics, Depreciation, Biology
Chapter 9, page 494	Mathematical Form in Nature
Chapter 9, page 496	Pendulum Motion, Physics
Chapter 10, page 517	Dairy Farming, Agriculture, Business
Chapter 10, page 539	Art
Chapter 11, page 564	Archaeology
Chapter 11, page 577	Airships, Physics, History
Chapter 11, page 583	Social Research, High School Students' Goals
Chapter 11, page 596	Cooking, Physics, Geography
Chapter 12, page 637	Music, Mathematical Design
Chapter 13, page 678	Geography, American History
Chapter 13, page 686	Geometry, Oceanography, Geography

D.C. Heath's ALGEBRA provides...

- **Integrated Review** exercises to relate previously learned concepts to new topics from the lesson.
- **Exploring Data** lessons and exercises to connect algebra to other disciplines.
- **Mathematical Models** and related exercises to apply materials found in each lesson to real-life situations.
- **Why You Should Learn It** features to let your students know the importance of what they are about to study.
- **Career Interview** features to acquaint students with the broad range of careers in which math is a useful tool.
- **Independent Practice** to promote understanding through skill-building exercises and real-life applications.
- **Real-life Themes** in Chapter Reviews to help students connect what they have just studied to the world around them.

"Modeling puts a mathematical overlay on the real world. It's attainable, reasonable, straightforward, and it fits with the topics at hand. Real-life connections make ALGEBRA meaningful for all students and promote success at all levels."

—Lee Stiff, ALGEBRA author

The Standards recommend...

- the integration of ideas from algebra and geometry, with graphic representation playing an important connecting role.
- the use of scientific calculators with graphing capabilities.
- the use of computers for demonstration purposes and for students to use in individual and group work.

Increased attention to

visual representation of algebraic concepts.

the use of calculators and computers as tools for learning and doing mathematics.

the use of computer utilities to develop conceptual understanding.

computer-based methods such as successive approximations and graphing utilities for solving equations and inequalities.

— from the NCTM Standards

Technology gives a visual dimension to algebra.

Seeing what the graph of an equation looks like helps students understand what that equation means. ALGEBRA emphasizes a visual approach to algebra—with the graphing calculator and the computer—because visual models of algebraic expressions make algebra more meaningful.

Technology helps students visualize algebra.

Graphing calculators and computers help students visualize statistics and data, linear equations, and other algebraic functions. When students use graphing calculators to graph equations, they are making visual models of algebraic expressions; they can readily see that an equation represents something. Graphing calculators let students create graphs quickly, too, so there's more time for "What if ... ?" questions—and for development of reasoning skills.

Technology is integrated throughout.

The graphing calculator is integrated into problem sets throughout the text, although it is not required. And in every chapter, the **Using a Graphing Calculator** feature shows students how to explore equations visually, practice data analysis techniques, and solve problems. The exercises in this feature are compatible with the newest TI, Casio and Sharp models. In the appendix, keystrokes are given for the TI-82, Casio fx-9700GE and Sharp EL-9300C along with a new application for use with these calculators.

ALGEBRA makes using technology easy.

Only the new ALGEBRA from D.C. Heath spells out the keystrokes so students learn the uses and nuances of graphing calculators quickly. The appendix pages 734–747 feature keystrokes for the TI-82, Casio fx-9700GE, and Sharp EL-9300C.

D.C. Heath's ALGEBRA provides...

- a visual approach to algebra.
- optional but fully integrated use of the graphing calculator and computer.
- useful suggestions on when and how to use the graphing calculator.

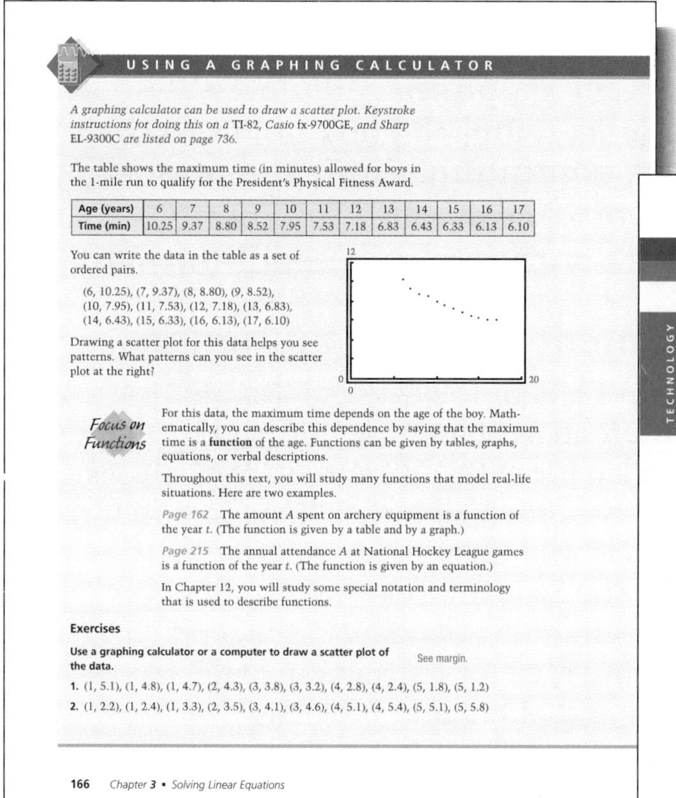

USING A GRAPHING CALCULATOR

A graphing calculator can be used to draw a scatter plot. Keystroke instructions for doing this on a TI-82, *Casio* fx-9700GE, *and Sharp* EL-9300C *are listed on page 736.*

The table shows the maximum time (in minutes) allowed for boys in the 1-mile run to qualify for the President's Physical Fitness Award.

Age (years)	6	7	8	9	10	11	12	13	14	15	16	17
Time (min)	10.25	9.37	8.80	8.52	7.95	7.53	7.18	6.83	6.43	6.33	6.13	6.10

You can write the data in the table as a set of ordered pairs.

(6, 10.25), (7, 9.37), (8, 8.80), (9, 8.52),
(10, 7.95), (11, 7.53), (12, 7.18), (13, 6.83),
(14, 6.43), (15, 6.33), (16, 6.13), (17, 6.10)

Drawing a scatter plot for this data helps you see patterns. What patterns can you see in the scatter plot at the right?

Focus on Functions

For this data, the maximum time depends on the age of the boy. Mathematically, you can describe this dependence by saying that the maximum time is a **function** of the age. Functions can be given by tables, graphs, equations, or verbal descriptions.

Throughout this text, you will study many functions that model real-life situations. Here are two examples.

Page 162 The amount A spent on archery equipment is a function of the year t. (The function is given by a table and by a graph.)

Page 215 The annual attendance A at National Hockey League games is a function of the year t. (The function is given by an equation.)

In Chapter 12, you will study some special notation and terminology that is used to describe functions.

Exercises

Use a graphing calculator or a computer to draw a scatter plot of the data. See margin.

1. (1, 5.1), (1, 4.8), (1, 4.7), (2, 4.3), (3, 3.8), (3, 3.2), (4, 2.8), (4, 2.4), (5, 1.8), (5, 1.2)
2. (1, 2.2), (1, 2.4), (1, 3.3), (2, 3.5), (3, 4.1), (3, 4.6), (4, 5.1), (4, 5.4), (5, 5.1), (5, 5.8)

166 *Chapter 3 • Solving Linear Equations*

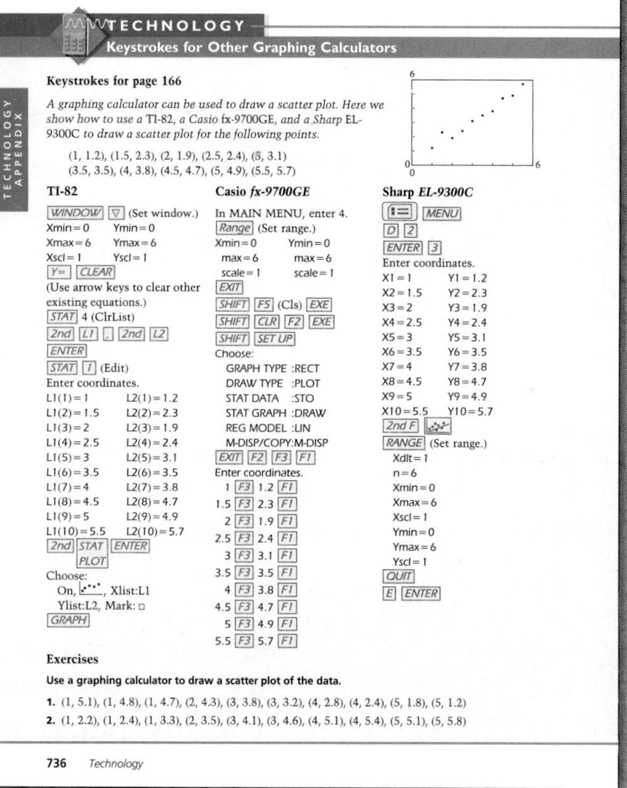

TECHNOLOGY APPENDIX

TECHNOLOGY
Keystrokes for Other Graphing Calculators

Keystrokes for page 166

A graphing calculator can be used to draw a scatter plot. Here we show how to use a TI-82, *a Casio* fx-9700GE, *and a Sharp* EL-9300C *to draw a scatter plot for the following points.*

(1, 1.2), (1.5, 2.3), (2, 1.9), (2.5, 2.4), (3, 3.1)
(3.5, 3.5), (4, 3.8), (4.5, 4.7), (5, 4.9), (5.5, 5.7)

TI-82

[WINDOW] [▽] (Set window.)
Xmin = 0 Ymin = 0
Xmax = 6 Ymax = 6
Xscl = 1 Yscl = 1
[Y=] [CLEAR]
(Use arrow keys to clear other existing equations.)
[STAT] 4 (ClrList)
[2nd] [L1] [,] [2nd] [L2]
[ENTER]
[STAT] [1] (Edit)
Enter coordinates.
L1(1) = 1 L2(1) = 1.2
L1(2) = 1.5 L2(2) = 2.3
L1(3) = 2 L2(3) = 1.9
L1(4) = 2.5 L2(4) = 2.4
L1(5) = 3 L2(5) = 3.1
L1(6) = 3.5 L2(6) = 3.5
L1(7) = 4 L2(7) = 3.8
L1(8) = 4.5 L2(8) = 4.7
L1(9) = 5 L2(9) = 4.9
L1(10) = 5.5 L2(10) = 5.7
[2nd] [STAT PLOT] [ENTER]
Choose:
On, [scatter icon], Xlist:L1
Ylist:L2, Mark: □
[GRAPH]

Casio *fx-9700GE*

In MAIN MENU, enter 4.
[Range] (Set range.)
Xmin = 0 Ymin = 0
max = 6 max = 6
scale = 1 scale = 1
[EXIT]
[SHIFT] [F5] (Cls) [EXE]
[SHIFT] [CLR] [F2] [EXE]
[SHIFT] [SET UP]
Choose:
GRAPH TYPE :RECT
DRAW TYPE :PLOT
STAT DATA :STO
STAT GRAPH :DRAW
REG MODEL :LIN
M-DISP/COPY:M-DISP
[EXIT] [F2] [F3] [F1]
Enter coordinates.
1 [F3] 1.2 [F1]
1.5 [F3] 2.3 [F1]
2 [F3] 1.9 [F1]
2.5 [F3] 2.4 [F1]
3 [F3] 3.1 [F1]
3.5 [F3] 3.5 [F1]
4 [F3] 3.8 [F1]
4.5 [F3] 4.7 [F1]
5 [F3] 4.9 [F1]
5.5 [F3] 5.7 [F1]

Sharp *EL-9300C*

[≡] [MENU]
[D] [2]
[ENTER] [3]
Enter coordinates.
X1 = 1 Y1 = 1.2
X2 = 1.5 Y2 = 2.3
X3 = 2 Y3 = 1.9
X4 = 2.5 Y4 = 2.4
X5 = 3 Y5 = 3.1
X6 = 3.5 Y6 = 3.5
X7 = 4 Y7 = 3.8
X8 = 4.5 Y8 = 4.7
X9 = 5 Y9 = 4.9
X10 = 5.5 Y10 = 5.7
[2nd F] [scatter icon]
[RANGE] (Set range.)
Xdlt = 1
n = 6
Xmin = 0
Xmax = 6
Xscl = 1
Ymin = 0
Ymax = 6
Yscl = 1
[QUIT]
[E] [ENTER]

Exercises

Use a graphing calculator to draw a scatter plot of the data.

1. (1, 5.1), (1, 4.8), (1, 4.7), (2, 4.3), (3, 3.8), (3, 3.2), (4, 2.8), (4, 2.4), (5, 1.8), (5, 1.2)
2. (1, 2.2), (1, 2.4), (1, 3.3), (2, 3.5), (3, 4.1), (3, 4.6), (4, 5.1), (4, 5.4), (5, 5.1), (5, 5.8)

736 *Technology*

Many lessons show students how to graph equations on a graphing calculator. Seeing visual representations of algebraic expressions makes algebraic concepts more meaningful to students.

> "The Using a Graphing Calculator feature was wonderful! I've never seen anything written down for students to follow; it has always been 'This is what you do.' This feature gives a lot more freedom to explore."
>
> *— Deanna Mauldin,*
> *ALGEBRA field test teacher,*
> *Liberty Bell Middle School,*
> *Johnson City, TN*

The Standards recommend...

- Students should be given tasks that are challenging and multifaceted and that allow them to perform at their maximum level of ability.
- Assessment tools should not stress only one type of task or mode of response because this does not give an accurate indication of performance, nor does it allow students to show their individual capabilities.
- Assessment programs should have opportunities for students to show how well they have integrated their math knowledge by applying what they've learned in a larger context.
- Assessment must be more than testing—it must be a continuous, dynamic, and often informal process.

Increased attention to

assessing the whole student.
ongoing assessment programs.
incorporating formal and alternative assessment programs.

— from the NCTM Standards

ALGEBRA'S assessment tools help you evaluate the whole student.

ALGEBRA provides you with options to assess your students' total progress. You'll evaluate knowledge and mathematical power, as well as problem solving, communication, and reasoning. Throughout the program, you'll find many ways to assess students both formally and informally.

Formal assessment tools are numerous.

- **Mid-Chapter Self-Tests** in the Student Text help students assess their progress before they finish the chapter. Answers are provided for every exercise in the self-tests.
- **Mid-Chapter Tests**—provided for teachers in two forms —allow you to assess mid-chapter progress.
- **Chapter Tests** in the Student Text help students evaluate their understanding of chapter concepts.
- **Chapter Tests**—provided for teachers in three forms— allow you to assess end-of-chapter progress.
- **Independent Practice** in the Student Text reviews basic concepts directly correlated to lesson goals and lesson examples. Many of these homework exercises ask for explanations, comparisons, conclusions, and other responses that promote decision making.
- **A Computer Test Bank** allows you to generate customized tests by choosing from over 2,000 items to meet your students' individual needs.

Alternative assessment allows students to demonstrate individual capabilities.

- **Communicating about Algebra**—a part of each lesson—lets you see how much your students understand as they talk about the lesson.
- **Guided Practice** helps you interact with students and judge their readiness to begin an assignment.
- **Integrated Review**, which draws on material from previous chapters relating to the current lesson, helps you measure retention and cumulative understanding. These exercises also contain *College Entrance Exam* questions that are relevant to the new material.
- **Exploration and Extension** exercises augment or extend material. Most of the exercises are appropriate for all students. In the Teacher's Edition, the exercises that present more challenge are starred.★
- **Teacher's Guide to Alternative Assessment** offers suggestions for evaluating communication skills (both oral and written), group work, and problem solving. It also includes forms for assessing student performance in these activities, and **Math Log Copymasters,** which give students opportunities to write about algebra procedures they are learning, do research, and work on open-ended problems.
- **Partner Quizzes** get students working in pairs to communicate ideas as you evaluate students' understanding.

"With ALGEBRA, teachers assess not just knowledge and mathematical power, but problem solving, communication, reasoning, and mathematical disposition—students' desire to know and do math with confidence. You assess the whole student."

—Timothy D. Kanold, ALGEBRA author

D.C. Heath's ALGEBRA provides...

an assessment program to evaluate the whole student in a continuous process with—

Formal Assessment

- Mid-Chapter Self-Tests
- Mid-Chapter Tests
- Chapter Tests
- Independent Practice
- Computer Test Bank

Alternative Assessment

- Communicating about Algebra
- Guided Practice
- Integrated Review
- Exploration and Extension
- Teacher's Guide to Alternative Assessment
- Math Log Copymasters
- Partner Quizzes

The Standards recommend...

The mathematics curriculum should include the continued development of language and symbolism to communicate mathematical ideas so that all students can—

- reflect upon and clarify their thinking about mathematical ideas and relationships.
- formulate mathematical definitions and express generalizations discovered through investigations.
- express mathematical ideas orally and in writing.
- read written presentations of mathematics with understanding.
- ask clarifying and extending questions related to mathematics they have read or heard about.

Increased attention to

experience in listening to, reading about, writing about, speaking about, reflecting on, and demonstrating mathematical ideas.

individual and small-group explorations that provide multiple opportunities for discussion, questioning, listening, and summarizing.

acknowledging the merit of students' ideas and the importance of their own language in explaining their thinking.

— from the NCTM Standards

Communicating about algebra sharpens thinking.

Throughout the text, your students are given opportunities to exercise communication skills—in talking, listening, writing, representing, modeling, and reading. When students communicate about mathematical concepts, they refine their understanding. Their ability to verbalize demonstrates higher-order thinking skills in process.

Within the text, you'll find **Communicating about Algebra** features to help your students identify and share ideas relevant to *every* lesson. These exercises—designed for classroom discussion or for working in groups—offer opportunities for communication.

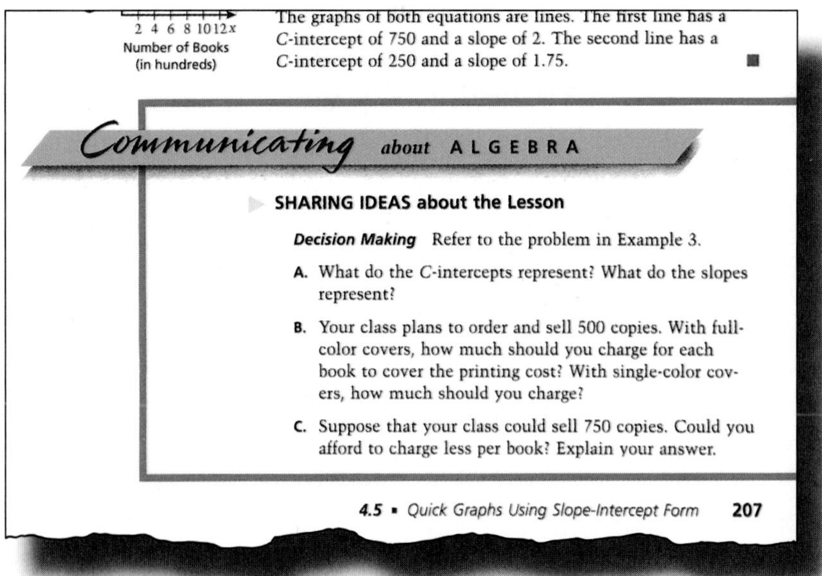
2 4 6 8 10 12 x
Number of Books (in hundreds)

The graphs of both equations are lines. The first line has a C-intercept of 750 and a slope of 2. The second line has a C-intercept of 250 and a slope of 1.75. ■

Communicating about ALGEBRA

SHARING IDEAS about the Lesson

Decision Making Refer to the problem in Example 3.

A. What do the C-intercepts represent? What do the slopes represent?

B. Your class plans to order and sell 500 copies. With full-color covers, how much should you charge for each book to cover the printing cost? With single-color covers, how much should you charge?

C. Suppose that your class could sell 750 copies. Could you afford to charge less per book? Explain your answer.

4.5 ▪ *Quick Graphs Using Slope-Intercept Form* **207**

An innovative approach to creative problem solving that promotes communication can be found in ALGEBRA's **Partner Quizzes.** Here, students work in pairs, verbalizing different ideas and listening to classmates justify their own answers. The partners confirm or reject ideas, connect others' thoughts with their own, and derive solutions based on teamwork and cooperation.

Math Logs provide opportunities for students to write their ideas about algebra. They investigate the algebra procedures they are learning, do research, and work on open-ended problem solving.

Communicating about Algebra, **Partner Quizzes,** and **Math Logs** foster active communication about algebra within your community of learners. These features also offer *you*, your students' mentor, the opportunity to monitor questions and responses.

D.C. Heath's ALGEBRA provides...

opportunities in every chapter to engage your students in meaningful communication—sharing, reflecting, and summarizing.

- **Communicating about Algebra** exercises are designed for classroom discussion or group work.
- **Partner Quizzes** make it possible for students to work together in pairs on algebra.
- **Math Logs** provide opportunities for students to write about algebra.

"Communication in a classroom gets students talking about algebra. In concert, students come up with conclusions they couldn't come up with alone. When students communicate, they learn new, different perspectives—because problems don't always have just one 'right' solution."

—Lee Stiff, ALGEBRA author

The Standards recommend...

- a core curriculum that provides a common body of mathematical ideas accessible to all students and that can be extended to meet the needs, interests, and performance levels of individual students or groups of students.
- that differences in background and ability be addressed by enrichment and extension of topics rather than by deletion.
- the use of a variety of instructional formats, such as small groups, individual exploration, peer instruction, whole-class discussion, and project work.

— from the NCTM Standards

ALGEBRA provides realistic ways to respond to diverse needs.

A basic premise of ALGEBRA is the belief that every student can succeed in algebra. It makes math concepts accessible to a wide range of students through an engaging writing style, carefully developed core concepts followed by opportunities for extension, emphasis on real-life applications of algebra, and ways to accommodate varied learning styles.

Thorough content coverage means success for all.

To reach students of all abilities, D.C. Heath's new ALGEBRA program puts the main concepts of algebra at the focus of instruction—followed by many opportunities in the exercise sets for challenge and extension. Each section of the exercise sets—Guided Practice, Independent Practice, Integrated Review, and Exploration and Extension—provides meaningful practice for every student.

Engaging writing involves students in the text.

Larson, Kanold, and Stiff write in a relaxed, encouraging style that speaks directly to the student. To get students even more involved in what they're learning, **What You Should Learn** and **Why You Should Learn It** features clearly spell out the objectives of the lesson and the usefulness of the material both in math class and in real life.

Real-life applications appeal to all.

Math problems and concepts that are based upon real-life situations give students a common basis for understanding. Because so many problems in ALGEBRA are set in real-life context, students of all abilities will find topics of interest—and will recognize the real-life value of the concepts they're learning.

ALGEBRA recognizes all learning styles.

To accommodate students of all ability levels and learning styles, ALGEBRA offers a range of activities—peer teaching, cooperative learning, and group activities. Students will find that working with others and sharing ideas are lifelong skills that go beyond algebra class.

D.C. Heath's ALGEBRA provides...

- content coverage that reaches all students and offers opportunities for extension.
- an engaging writing style, plus objectives that clarify the point of the lesson.
- real-life applications that facilitate student understanding.
- activities that recognize all abilities and learning styles.

"One of our goals in writing ALGEBRA was to show students that math is useful, interesting—and fun! Another goal was to provide enough real-life situations to give each student many opportunities to be an 'expert.' Maybe he or she has been to a location described in the text, or knows about motorcycling or music or baseball or dairy farming or raising collies—or any of the other real-life situations found in ALGEBRA"

— *Roland E. Larson, ALGEBRA author*

To walk through a lesson from ALGEBRA 1, please turn to page T20.

In this chapter, techniques for graphing linear and absolute value equations will be developed. Emphasis will be given to sketching "quick graphs" instead of plotting points. Connections between solutions of linear and absolute value equations and their graphs will be made.

Indicate to students that using graphs allows you to connect ideas in algebra with ideas in geometry. Another way of stating this is that graphing provides a way of "picturing" patterns and relationships.

Frequently, the sketch of a graph provides useful information about the real-life situation it models. More often, a comparison of graphs helps you decide among several options. Consequently, graphs are useful tools for understanding and managing the world around us.

The Chapter Summary on page 229 provides you and the students with a synopsis of the chapter. It identifies key skills and concepts. You may want to have students look at the Chapter Summary as an overview before beginning the chapter.

174

Relevance to real life

Each chapter begins with a two-page Opener, showing a table of contents for the chapter and a motivational example of the relevance of algebra to real life.

CHAPTER 4

Graphing Linear Equations

LESSONS

San Francisco traffic is a colorful reminder of the millions of automobiles in the United States.

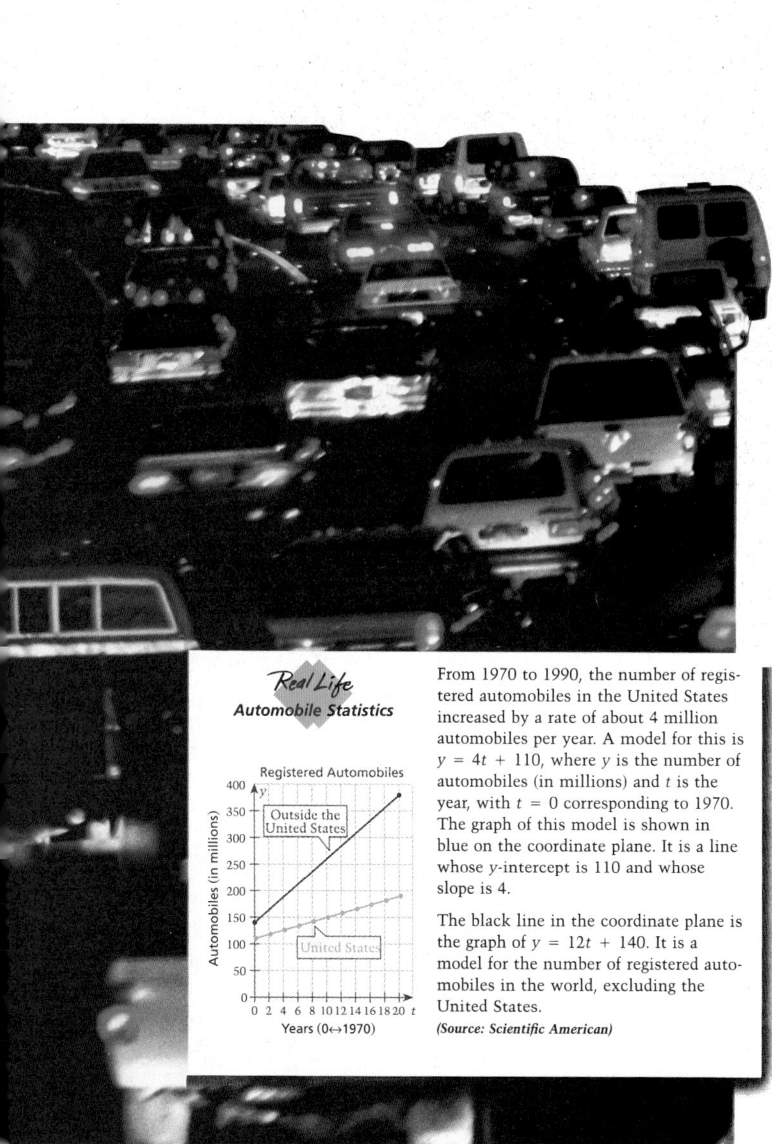

Real Life
Automobile Statistics

From 1970 to 1990, the number of registered automobiles in the United States increased by a rate of about 4 million automobiles per year. A model for this is $y = 4t + 110$, where y is the number of automobiles (in millions) and t is the year, with $t = 0$ corresponding to 1970. The graph of this model is shown in blue on the coordinate plane. It is a line whose y-intercept is 110 and whose slope is 4.

The black line in the coordinate plane is the graph of $y = 12t + 140$. It is a model for the number of registered automobiles in the world, excluding the United States.

(Source: Scientific American)

You can use the graphs of these algebraic models to easily compare the registration rates of automobiles within and outside the United States. The straight line representing the registration of cars outside the United States is steeper than the line of registered cars in the United States. What interpretation can be assigned to this observation? In the same time period, the rate of registrations outside the United States is growing faster than inside the United States. What information is provided by the y-intercepts of each line? The number of registrations in 1970.

As students learn more about the graphs of linear equations, they can use their knowledge to interpret graphs to answer questions and solve problems.

175

Real data, not contrived

Many problems and examples in the new ALGEBRA from D.C. Heath use real data and graphs to make algebraic concepts more meaningful to students. The applications reinforce the role algebra plays in the real world. Problems that use real data and newspaperlike graphics teach students how to interpret information in graphs they encounter in daily life.

What to Look For in Lesson 4.5*

- ▲ Stated objectives and purpose
- ▲ Problem of the Day
- ▲ Warm-Up Exercises
- ▲ Real-life examples
- ▲ Common-Error ALERT!
- ▲ Communicating about Algebra
- ▲ Assignment Guide
- ▲ Guided Practice
- ▲ Independent Practice
- ▲ Enrichment

*Lesson 4.5 is just a portion of Chapter 4.
The complete chapter, Graphing Linear Equations, begins on page 176 of this book.

4.5 Quick Graphs Using Slope-Intercept Form

What you should learn:

Goal 1 How to find slope and y-intercept from an equation

Goal 2 How to use the slope-intercept form to sketch a line and to solve real-life problems

Why you should learn it:

You can write an equation of a line in slope-intercept form to make it easier to sketch and use.

Goal 1 Finding the Slope and y-Intercept

In Lesson 4.4, you were given two points of a line and you learned how to find the slope of the line. Now suppose you are given an equation of a line. How could you find its slope?

One way would be to find two points on the line and use the formula for slope. Let's try this with the line $y = \frac{1}{2}x + 1$. Notice that no matter which pair of points we use to calculate the slope, the result is the same.

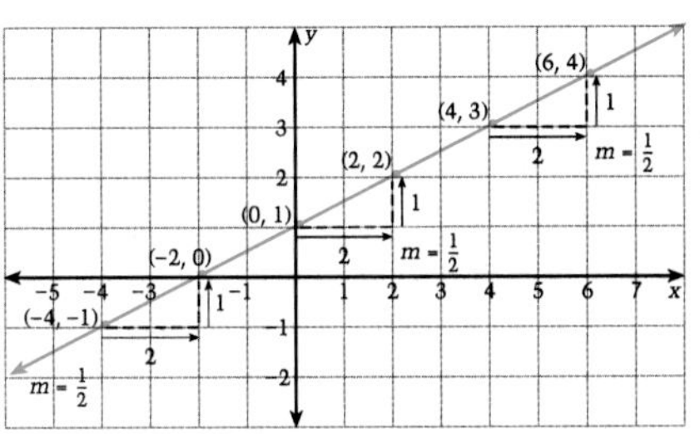

Another important observation is that the slope of the line is the coefficient of x. By substituting 0 for x, we can see that the y-intercept is the constant term in the equation.

Slope m y-intercept b

$$y = \frac{1}{2}x + 1$$

This result suggests a quick way to find the slope and the y-intercept. Just write the equation in the form $y = mx + b$.

Slope-Intercept Form of the Equation of a Line

The linear equation

$y = mx + b$

is in slope-intercept form. The slope of the line is m. The y-intercept is b.

Problem of the Day

How many different 3-digit numbers can be formed using the digits 1, 2, and 3 once in each number? How many different 4-digit numbers can be formed using the digits 1, 2, 3, and 4 once in each number? 6; 24

ORGANIZER

Warm-Up Exercises

1. For each equation, find the value of y for which the ordered pair, $(0, y)$, is a solution.
   a. $3x - 6y = 9$ $-\frac{3}{2}$
   b. $8x = 5y - 30$ 6
   c. $2y = 7x - 11$ $-\frac{11}{2}$
   d. $y = \frac{2}{3}x + 8$ 8
2. Solve each equation for y in terms of x.
   a. $3x - 6y = 9$ $y = \frac{1}{2}x - \frac{3}{2}$
   b. $8x = 5y - 30$ $y = \frac{8}{5}x + 6$
   c. $2y = 7x - 11$ $y = \frac{7}{2}x - \frac{11}{2}$

LESSON Notes

GOAL 1 Demonstrate for students how points on the line $y = \frac{1}{2}x + 1$ can be determined.

Written for understanding

ALGEBRA is written in a relaxed, encouraging style that draws students into the text and helps them understand.

Organized for success

Your Teacher's Edition helps you organize each lesson. It gives you a Problem of the Day and Warm-Up Exercises to get students oriented for the lesson, and many point-of-use teaching suggestions.

Clearly stated objectives

Each lesson begins with a set of objectives that spell out the point of the lesson and the usefulness of the material both in math class and in real life.

Visual connections

Graphing is introduced early in ALGEBRA to help students make visual connections to algebraic equations right away.

Clear explanations

The title of an example gives students a quick overview of what the example is showing. Most steps in the examples have side comments that explain the solution steps. Notice Example 1 on this page.

Example 1

Show students that the point (3, 3), at which you arrive, satisfies the equation $y = \frac{1}{3}x + 2$. This provides a check that you moved correctly from the y-intercept to another point on the line.

Example 2

You may wish to observe for students that any given line is parallel to itself! This occurs when different equations represent the same line.

Example 3

A "signature" in printing is a large sheet on which are printed a number of pages in multiples of four. This large sheet is then folded to page size and becomes one section of a book.

Extra Examples

Here are additional examples similar to **Examples 1–3.**

1. Identify three points on the line $y = \frac{3}{2}x - 3$ and verify that the slope of the line remains the same for each pair of points selected.
2. Sketch the graph of each line using its slope and y-intercept.
   a. $8x - 4y = 12$
   b. $5y + 3x = 8$
   a. $m = 2, b = -3$; b. $m = -\frac{3}{5}, b = \frac{8}{5}$

termine which lines are
'lel to each other.
the set of parallel
a coordinate plane.
1y = 12
6x − 17
8
+ 2

Lines a, b, and d are parallel.

Goal 2 Using Slope-Intercept Form

In Lesson 4.3, you learned to sketch a line quickly using the x-intercept and y-intercept. The *slope* and *y-intercept* can also be used to sketch a quick graph.

Example 1 *Sketching a Quick Graph*

Sketch the graph of $-x + 3y = 6$.

Solution Begin by solving the equation for y.

$-x + 3y = 6$ *Original equation*

$3y = x + 6$ *Add x to both sides.*

$y = \underbrace{\tfrac{1}{3}}_{m}x + \underbrace{2}_{b}$ *Divide both sides by 3.*

You can sketch the line as follows. Plot the y-intercept, (0, 2). Locate a second point by moving 3 units to the right and 1 unit up. Then draw the line through the two points.

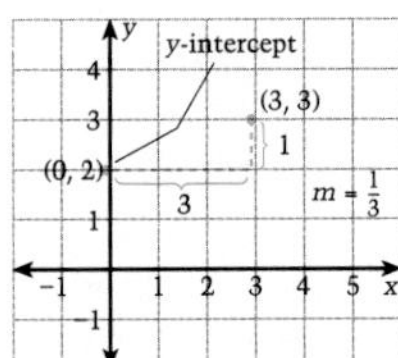

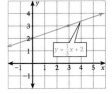

Plot the y-intercept. Then use slope to find a second point.

Draw the line that passes through the two points.

Two different lines that have the same slope are **parallel**; that is, they do not intersect.

Example 2 *Sketching a Family of Parallel Lines*

Sketch the following lines on the same coordinate plane.

a. $y = -2x + 3$ **b.** $y = -2x$ **c.** $y = -2x - 1$

Solution Each of the lines has a slope of $m = -2$. The y-intercepts of each are the values of b, as follows.

a. $y = \underset{m}{-2}x + \underset{b}{3}$ **b.** $y = \underset{m}{-2}x + \underset{b}{0}$ **c.** $y = \underset{m}{-2}x + \underset{b}{(-1)}$

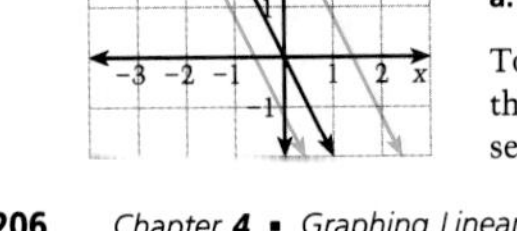

To sketch each line, plot its y-intercept. Then move 1 unit to the right and 2 units *down* (the slope is negative), and plot a second point. Finally, draw a line through the two points.

These students are planning a creative writing project that will give them some valuable insights into the realities of publishing a small book.

Problem Solving
Make a Graph

Example 3 *Printing a Book*

Your creative writing class wants to have selections of its poems and prose printed and bound into a book. The book will have 128 pages (8 *signatures* of 16 pages each). The inside pages will be printed with black ink, but the cover may have a full-color picture. A printer gives you the following prices:

With full-color cover: Cost is $750, plus $2.00 per book.
With single-color cover: Cost is $250, plus $1.75 per book.

Cost of Printing Books in the U.S.

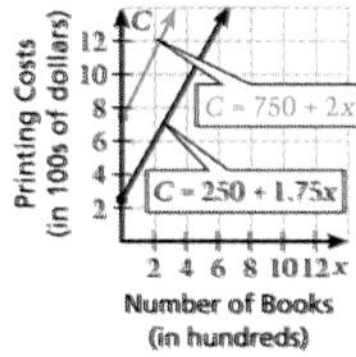

Sketch the graphs of these two costs. Let x be the number of books printed and let C be the cost of printing x books.

Solution The two equations are as follows.

With full-color cover: $C = 750 + 2x$
With single-color cover: $C = 250 + 1.75x$

The graphs of both equations are lines. The first line has a C-intercept of 750 and a slope of 2. The second line has a C-intercept of 250 and a slope of 1.75. ■

Check Understanding

1. Find the slope and y-intercept for each equation.
   a. $4x - 6y = 3$
   b. $11x = -5y + 14$
   a. $m = \frac{2}{3}, b = -\frac{1}{2}$;
   b. $m = -\frac{11}{5}, b = \frac{14}{5}$
2. Sketch the graphs of the line through the given point with the given slope.
   a. $m = -4, (-3, 6)$
   b. $m = \frac{2}{3}, (7, 1)$

Communicating about ALGEBRA

Ask students to provide a rationale about whether the cost of printing the full-color version of the book will ever be the same as the cost of printing the single-color version.

Communicating about ALGEBRA

SHARING IDEAS about the Lesson A. Cost that is independent of the number of books printed; cost per book printed

Decision Making Refer to the problem in Example 3.

A. What do the C-intercepts represent? What do the slopes represent?

B. Your class plans to order and sell 500 copies. With full-color covers, how much should you charge for each book to cover the printing cost? With single-color covers, how much should you charge? $3.50, $2.25

C. Suppose that your class could sell 750 copies. Could you afford to charge less per book? Explain your answer. See below.

Common-Error ALERT!

Remind students to label the two points selected for finding the slope of the line. This will help them avoid the mistake of mismatching the coordinates in the slope formula.

C. Yes, the cost that is independent of the number of books printed would be spread among more books.

Modeling with math

Real-life connections help students see math as a language for modeling the real world, not simply a language used to manipulate symbols.

Daily communication

Communicating about Algebra exercises in every lesson offer daily opportunities for students to share ideas about algebra.

Error prevention

In your Teacher's Edition, you'll find Common-Error ALERTs—reminders of preventive action students should take to avoid common mistakes.

Meeting individual needs

The Assignment Guide in each lesson of your Teacher's Edition helps you tailor exercises to the varying abilities of your students.

Plenty of practice

Guided Practice provides teacher-directed practice in basic algebra skills and gives teachers the opportunity to monitor students' comprehension of the material.

EXERCISE Notes

ASSIGNMENT GUIDE

Basic/Average: Ex. 1–4, 5–27 odd, 29–32, 33–43 odd, 45–54 multiples of 3, 58, 59

Above Average: Ex. 1–4, 6–27 multiples of 3, 29–32, 36, 39–40, 42–43, 45–54 multiples of 3, 57–62

Advanced: Ex. 1–4, 6–27 multiples of 3, 29–32, 37–40, 42–43, 57–62

Selected Answers
Exercises 1–4, 5–55 odd

More Difficult Exercises
Exercises 37–40, 57–62

Guided Practice

Ex. 3 Ask students to create two different equations of a given line. Discuss the reasonableness of saying that these equations represent "lines" with the same slope.

Common-Error ALERT!

In **Exercises 15–16,** remind students that the coefficient of x in an equation of a straight line represents the slope of the line only when the equation is in slope-intercept form.

208

EXERCISES

1. Yes; slope of line does not change.

Guided Practice

CRITICAL THINKING about the Lesson

1. Can any pair of points on a line be used to calculate the slope of the line? Explain.
2. In the form $y = mx + b$, what does m represent? What does b represent? Slope, y-intercept
3. What word may be used to describe two different lines that have the same slope? Parallel
4. What is the slope of the line given by $3x + 2y = 1$? What is the y-intercept? $-\frac{3}{2}, \frac{1}{2}$

Independent Practice

In Exercises 5–10, solve for y.

22. $y = \frac{3}{2}x - 1$ 23. $y = \frac{3}{4}x + \frac{1}{2}$ 24. $y = -\frac{5}{3}x + \frac{1}{2}$

5. $5x + 7y = 0$ $y = -\frac{5}{7}x$
6. $2x - 4y = 0$ $y = \frac{1}{2}x$
7. $3x + 4y = 12$ $y = -\frac{3}{4}x + 3$
8. $-4x + 5y = 8$ $y = \frac{4}{5}x + \frac{8}{5}$
9. $y + 10 = 0$ $y = -10$
10. $y - 12 = 0$ $y = 12$

In Exercises 11–16, find the slope and the y-intercept of the line.

11. $y = -2x + 1$ $-2, 1$
12. $y = 3x - 6$ $3, -6$
13. $y = -4 + (-8x)$ $-8, -4$
14. $y = 4x - 20$ $4, -20$
15. $x - y = 3x + 4$ $-2, -4$
16. $2y - x = 7x - 9$ $4, -\frac{9}{2}$

In Exercises 17–28, write in slope-intercept form. Then sketch the line. See Additional Answers.

17. $2x - y - 3 = 0$ $y = 2x - 3$
18. $x - y + 2 = 0$ $y = x + 2$
19. $x + y = 0$ $y = -x$
20. $x - y = 0$ $y = x$
21. $x + 2y - 2 = 0$ $y = -\frac{1}{2}x + 1$
22. $3x - 2y - 2 = 0$
23. $3x - 4y + 2 = 0$
24. $10x + 6y - 3 = 0$
25. $y - 3 = 0$ $y = 3$
26. $y + 5 = 0$ $y = -5$
27. $2x + 3y - 4 = x + 5$ $y = -\frac{1}{3}x + 3$
28. $-x + 4y + 3 = 2x - 7$ $y = \frac{3}{4}x - \frac{5}{2}$

In Exercises 29–32, match the equation with its graph.

29. $y = \frac{1}{2}x + 1$ d
30. $y = \frac{1}{2}x - 1$ c
31. $y = x + 2$ a
32. $y = -x - 1$ b

a. 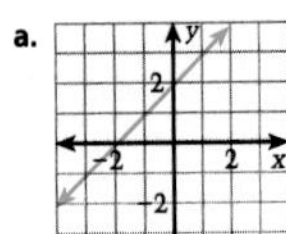
b. 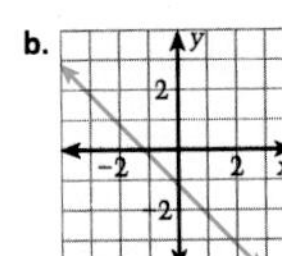
c. 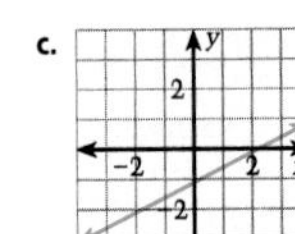
d. 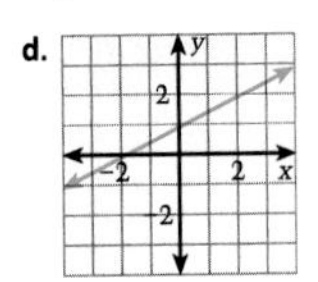

In Exercises 33–36, sketch the two lines on the same coordinate plane. See Additional Answers. **Find the slope and x- and y-intercepts of the lines.**

33. $y = -3x + 2$, $y = -3x - 2$
34. $y = -x + 6$, $y = -x + 10$ -1; (6, 0), (0, 6); (10, 0), (0, 10)
35. $y = 6x + 8$, $y = 6x - 2$ 6; $(-\frac{4}{3}, 0)$, (0, 8); $(\frac{1}{3}, 0)$, (0, −2)
36. $y = \frac{4}{3}x - 1$, $y = \frac{4}{3}x + 3$ $\frac{4}{3}$; $(\frac{3}{4}, 0)$, (0, −1); $(-\frac{9}{4}, 0)$, (0, 3)

33. −3; $(\frac{2}{3}, 0)$, (0, 2); $(-\frac{2}{3}, 0)$, (0, −2)

Varied homework assignments

Independent Practice exercises comprise the basic part of the homework assignment. They contain a balance of skill-building exercises and real-life applications. Many questions ask for explanations, comparisons, and other responses that teach decision making.

In Exercises 37–40, use the following information.

Two triangles are *similar* if their angles have the same measures. Corresponding sides of similar triangles have the same ratio. For instance, in the similar triangles at the right the ratio of AB to BC is equal to the ratio of DE to EF.

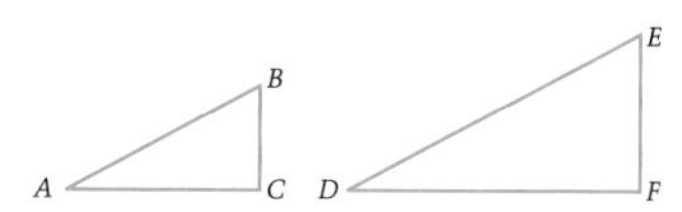

37. *Slope of a Subway Track* You are the supervisor for a construction project to build a subway track. Over one 200-foot (horizontal) portion, the track must rise 3 feet. You decide to check the elevation after your crew has covered 50 of the 200 feet. How much should the track have risen over 50 feet? $\frac{3}{4}$ ft

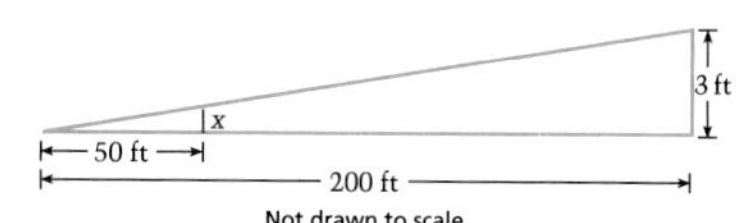

38. *Architecture* You have designed plans for a house, and are now building a scale model. For the actual house, the center of the roof is 45 feet (horizontal) from the eaves and 18 feet higher than the eaves. For the model, the center is 10 inches (horizontal) from the eaves. How high should you build the center on the model? 4 in.

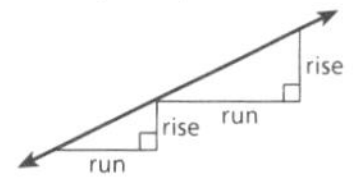

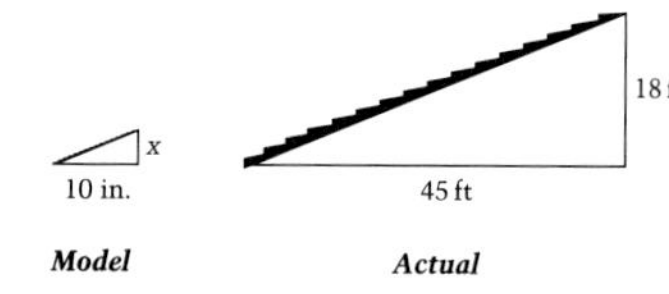

39. *King Kong* You are visiting New York City with a friend and are standing by the World Trade Center—the building from which King Kong fell in the remake of the movie *King Kong*. In the original movie, King Kong fell from the Empire State Building. Your friend claims that the Empire State Building is taller than the World Trade Center. You disagree, so to find the height of the World Trade Center, you measure the building's shadow to be 170.25 feet. Then you measure the shadow cast by a 48-inch post to be 6 inches. A few calculations give you the height. Who was correct, you or your friend? You

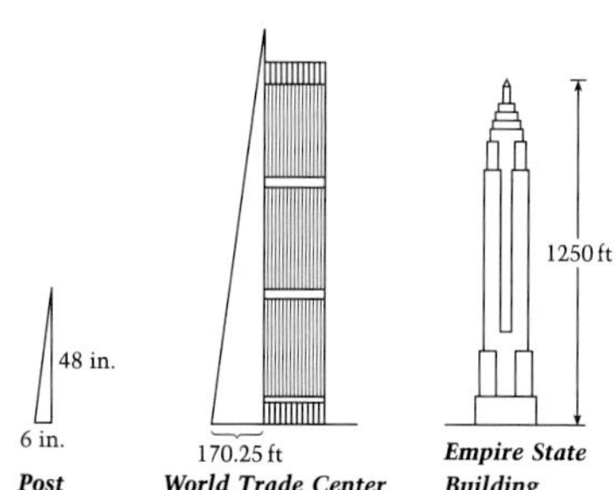

40. *Connecting Slope and Similar Triangles* Find the slope of the line passing through the hypotenuse of each triangle in Exercises 37–39. Explain how these "similar triangle" problems are related to the fact that any two points on a line can be used to calculate the slope of the line. See margin.

Independent Practice

▶ **Ex. 29–32** WRITING Ask students to provide a rationale for their matches.

▶ **Ex. 38** Eaves are the lower edge or edges of a roof, usually projecting beyond the sides of the building.

▶ **Ex. 40** COOPERATIVE LEARNING Discuss students' answers to this exercise in class. Through any two points on a line (neither vertical nor horizontal) a triangle is formed by a vertical line through one point and a horizontal line through the other. All such triangles are similar. So the $\frac{\text{rise}}{\text{run}}$ ratio is the same for any two points.

rise
run
rise
run

Enrichment Activities

WRITING Following the class discussion, students might record their personal responses in their math journals.

1. Have students write a narrative describing how a linear equation in two variables written in standard form, $Ax + By = C$, can be transformed to provide information which is then useful in graphing the equation of the line. Answers should include writing the equation in slope-intercept form.
2. Given the standard form of a linear equation in two variables, express the slope and y-intercept in terms of A, B, and C. $m = -\frac{A}{B}$, $b = \frac{C}{B}$

Ideas for cooperative learning

Your Teacher's Edition suggests opportunities for cooperative learning beyond those provided in the Communicating about Algebra feature.

Enrichment resources

A wealth of enrichment activities are provided in the margins of the ALGEBRA Teacher's Edition. Writing is one of these activities. Students may be asked to discuss their thoughts in both narrative and symbolic form and to record their observations in a journal.

Problem solving process

ALGEBRA presents problem solving as a continuing process, not as a separate or single event.

Showing algebra's value

Because students encounter real-life problems in every lesson, they will come to see how useful algebra can be as a problem-solving tool.

Continuing review

Integrated Review exercises help students fit new material into the context of the bigger picture of algebra. The emphasis is on showing how previously learned concepts are related to the new concepts in the lesson. Samples of college entrance exam questions may be included in this section.

▶ **Ex. 43–44** PROBLEM SOLVING Ask students to identify the problem-solving strategies used to solve these problems. Stress the connections between the plan and uses of tables or graphs to solve the problems.

Answers

43.

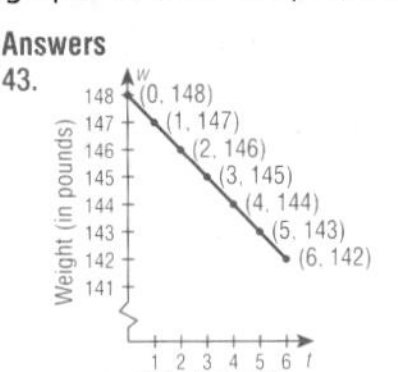

−1, rate of weight loss in pounds per week; 148, Mark's weight in pounds before he starts the weight-loss program.

44. (1, 0.46), (2, 0.77), (3, 1.08), (4, 1.39), (5, 1.7), (6, 2.01); Cost (in dollars); Time (in minutes)

0.31, cost for each additional minute

▶ **Ex. 59–62** COOPERATIVE LEARNING These are useful exercises to go over in class.

210

41. Which line is parallel to $x + 3y = 2$?
a. $y = \frac{1}{3}x + 5$ **b.** $y = -\frac{1}{3}x + 4$

42. Which line is parallel to $2x - 4y = 6$?
a. $y = \frac{1}{2}x + 1$ **b.** $y = -\frac{1}{2}x - 1$

43. *Getting Ready for the Team* Huan decides to lose weight before wrestling team practice begins. He weighs 148 pounds now and loses one pound per week for six weeks. Let w represent Huan's weight and let t represent the time in weeks. Plot points for his weights at one-week intervals, then draw a line through the points. Find the slope of the line. What does it represent? Find the w-intercept of the line. What does it represent? See margin.

44. *Making a Phone Call* The cost of a telephone call between New York City and Philadelphia is \$0.46 for the first minute and \$0.31 for each additional minute. Let C represent the total cost of making a call that lasts for t minutes. Plot points for the costs of calls that last 1, 2, 3, 4, 5, and 6 minutes, then draw a line through the points. Find the slope of the line. What does it represent? See margin.

Integrated Review

In Exercises 45–48, evaluate the expression.

45. $\frac{-3-(-4)}{5-(-1)}$ $\frac{1}{6}$ **46.** $\frac{0-3}{9-3}$ $-\frac{1}{2}$ **47.** $\frac{-2-(-2)}{8-1}$ 0 **48.** $\frac{6-10}{0-(-4)}$ −1

In Exercises 49–52, find five solutions of the equation. Solutions vary.

49. $y = -2x + 4$ (0, 4), etc. **50.** $y = 5x - 2$ (0, −2), etc. **51.** $y = -\frac{5}{7}x$ (0, 0), etc. **52.** $y = \frac{1}{2}x$ (0, 0), etc.

In Exercises 53–56, sketch the line that passes through (0, 2) with the given slope. See Additional Answers.

53. $m = 3$ **54.** $m = 0$ **55.** $m = -\frac{2}{3}$ **56.** $m = \frac{3}{4}$

Exploration and Extension

In Exercises 57 and 58, match the description with its visual model.

57. A person starts from home and rides a bike at a speed of 10 miles per hour. B

58. A person starts 25 miles from home and drives toward home at a speed of 25 miles per hour. A 59.–62. See Additional Answers.

A.

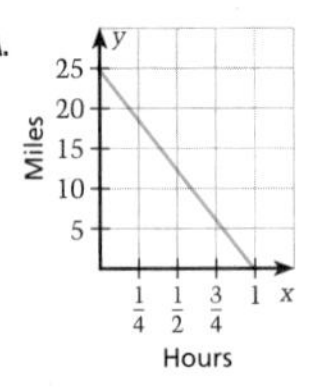

B.

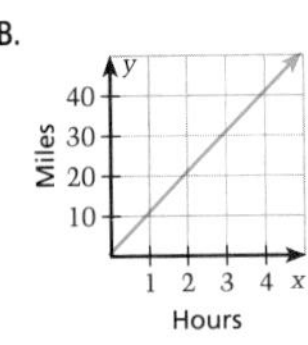

59. Sketch the line that has a y-intercept of −4 and is parallel to $y = 2x + 3$.

60. Sketch the line that has a y-intercept of −8 and is parallel to $y = -\frac{1}{2}x + 5$.

61. Sketch the line that has a y-intercept of 2 and a slope of $-\frac{1}{8}$.

62. Sketch the line that has a y-intercept of 14 and a slope of 0.

210 *Chapter 4 ▪ Graphing Linear Equations*

Extending the lesson

Exploration and Extension exercises ask all students to go beyond the lesson material in interesting ways.

ALGEBRA makes learning and using technology easier.

In every chapter of ALGEBRA 1 and 2, there is a **Using a Graphing Calculator** feature in which students are shown how to use graphing calculators to evaluate expressions, draw scatter plots and best-fitting lines, graph equations, perform matrix operations, create visual representations of polynomial addition and subtraction, and much more.

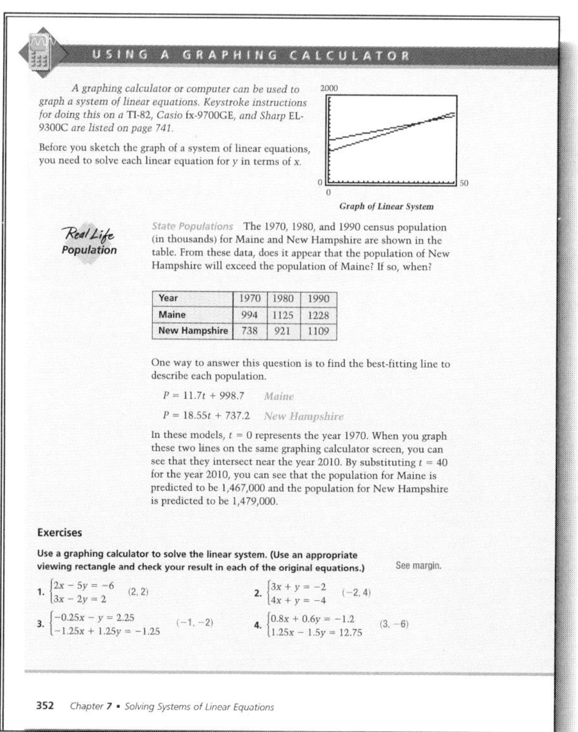

USING A GRAPHING CALCULATOR

A graphing calculator or computer can be used to graph a system of linear equations. Keystroke instructions for doing this on a TI-82, *Casio* fx-9700GE, *and Sharp* EL-9300C *are listed on page 741.*

Before you sketch the graph of a system of linear equations, you need to solve each linear equation for y in terms of x.

Graph of Linear System

Real Life Population

State Populations The 1970, 1980, and 1990 census population (in thousands) for Maine and New Hampshire are shown in the table. From these data, does it appear that the population of New Hampshire will exceed the population of Maine? If so, when?

Year	1970	1980	1990
Maine	994	1125	1228
New Hampshire	738	921	1109

One way to answer this question is to find the best-fitting line to describe each population.

$P = 11.7t + 998.7$ *Maine*

$P = 18.55t + 737.2$ *New Hampshire*

In these models, $t = 0$ represents the year 1970. When you graph these two lines on the same graphing calculator screen, you can see that they intersect near the year 2010. By substituting $t = 40$ for the year 2010, you can see that the population for Maine is predicted to be 1,467,000 and the population for New Hampshire is predicted to be 1,479,000.

Exercises

Use a graphing calculator to solve the linear system. (Use an appropriate viewing rectangle and check your result in each of the original equations.) See margin.

1. $\begin{cases} 2x - 5y = -6 \\ 3x - 2y = 2 \end{cases}$ (2, 2)
2. $\begin{cases} 3x + y = -2 \\ 4x + y = -4 \end{cases}$ (−2, 4)
3. $\begin{cases} -0.25x - y = 2.25 \\ -1.25x + 1.25y = -1.25 \end{cases}$ (−1, −2)
4. $\begin{cases} 0.8x + 0.6y = -1.2 \\ 1.25x - 1.5y = 12.75 \end{cases}$ (3, −6)

352 *Chapter 7 • Solving Systems of Linear Equations*

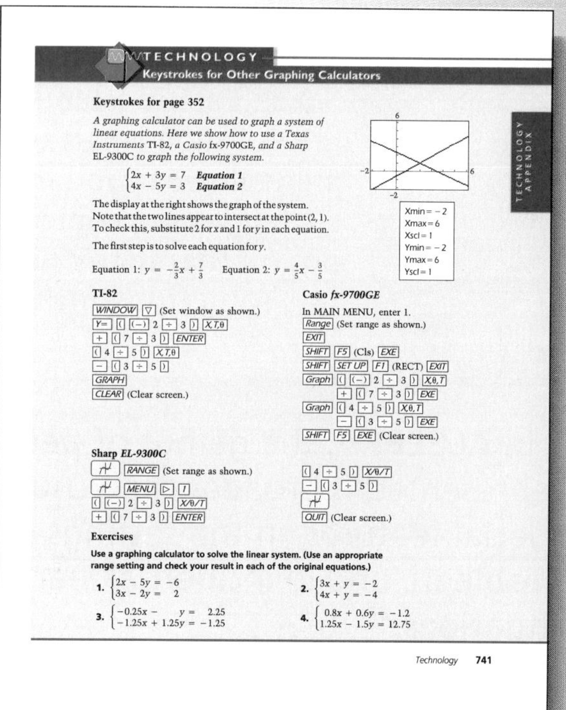

TECHNOLOGY
Keystrokes for Other Graphing Calculators

TECHNOLOGY APPENDIX

Keystrokes for page 352

A graphing calculator can be used to graph a system of linear equations. Here we show how to use a Texas Instruments TI-82, *a Casio* fx-9700GE, *and a Sharp* EL-9300C *to graph the following system.*

$\begin{cases} 2x + 3y = 7 & \textit{Equation 1} \\ 4x - 5y = 3 & \textit{Equation 2} \end{cases}$

The display at the right shows the graph of the system. Note that the two lines appear to intersect at the point (2, 1). To check this, substitute 2 for x and 1 for y in each equation.

The first step is to solve each equation for y.

Equation 1: $y = -\frac{2}{3}x + \frac{7}{3}$ Equation 2: $y = \frac{4}{5}x - \frac{3}{5}$

TI-82

[WINDOW] [▽] (Set window as shown.)
[Y=] [(] [(−)] 2 [÷] 3 [)] [X,T,θ]
[+] [(] 7 [÷] 3 [)] [ENTER]
[(] 4 [÷] 5 [)] [X,T,θ]
[−] [(] 3 [÷] 5 [)]
[GRAPH]
[CLEAR] (Clear screen.)

Casio *fx-9700GE*

In MAIN MENU, enter 1.
[Range] (Set range as shown.)
[EXIT]
[SHIFT] [F5] (Cls) [EXE]
[SHIFT] [SET UP] [F1] (RECT) [EXIT]
[Graph] [(] [(−)] 2 [÷] 3 [)] [X,θ,T]
[+] [(] 7 [÷] 3 [)] [EXE]
[Graph] [(] 4 [÷] 5 [)] [X,θ,T]
[−] [(] 3 [÷] 5 [)] [EXE]
[SHIFT] [F5] [EXE] (Clear screen.)

Sharp *EL-9300C*

[graph key] [RANGE] (Set range as shown.)
[graph key] [MENU] [▷] [1]
[(] [(−)] 2 [÷] 3 [)] [X/θ/T]
[+] [(] 7 [÷] 3 [)] [ENTER]
[(] 4 [÷] 5 [)] [X/θ/T]
[−] [(] 3 [÷] 5 [)]
[graph key]
[QUIT] (Clear screen.)

Exercises

Use a graphing calculator to solve the linear system. (Use an appropriate range setting and check your result in each of the original equations.)

1. $\begin{cases} 2x - 5y = -6 \\ 3x - 2y = 2 \end{cases}$
2. $\begin{cases} 3x + y = -2 \\ 4x + y = -4 \end{cases}$
3. $\begin{cases} -0.25x - y = 2.25 \\ -1.25x + 1.25y = -1.25 \end{cases}$
4. $\begin{cases} 0.8x + 0.6y = -1.2 \\ 1.25x - 1.5y = 12.75 \end{cases}$

Technology 741

"Calculators and computers with appropriate software transform the mathematics classroom into a laboratory much like the environment in many science classes, where students use technology to investigate, conjecture, and verify their findings. In this setting, the teacher encourages experimentation and provides opportunities for students to summarize ideas and establish connections with previously studied topics."

— from the NCTM Standards

Many lessons show students how to graph equations on a graphing calculator. Seeing visual representations of algebraic expressions makes algebraic concepts more meaningful to students. In the **Using a Graphing Calculator** feature, students learn to analyze data. In the lesson shown above, students use a graphing calculator to find the best-fitting line. Some lessons provide a BASIC program that performs equivalent operations on a computer. In the appendix, keystrokes are given for the TI-82, Casio fx-9700GE and Sharp EL-9300C along with a new application for use with these calculators.

Algebra—past, present, and future—is a language for everyone.

ALGEBRA seeks to make math appealing to students of all backgrounds and cultures—and to motivate all students to appreciate and succeed at algebra.

ALGEBRA provides ...

- representation of people of all backgrounds—not just in special features, but as an integral part of problems and applications.
- features that celebrate the contributions of both men and women, past and present, to the study of mathematics.

People of all walks of life are well represented.

People from all walks of life are an integral part of the settings of problems in ALGEBRA. Ethiopian weavers, South Pacific Islanders, and Native Americans in Cahokia are as much a part of the text as, say, the art of Africa, an archeological find in China, or the Taj Mahal. Including such diverse peoples and topics in the text reinforces the notion that many people find algebra important—and that it is something many people can do well.

"In ALGEBRA, the gamut of people represented is broader than usual—in special features, in the settings of problems, and in incidental learning situations."

— Lee Stiff, ALGEBRA author

Special features spotlight the role of math past and present.

Through anecdotes from the history of math, **Milestones** highlight the development of mathematics over the centuries, with time lines that show other important events of a period. By presenting men and women who find math a valuable tool in a range of careers, **Career Interviews** demonstrate that anyone can use and enjoy math.

In Exercises 53 and 54, give a convincing argument to explain why the property is true.

53. the Power of a Power
54. the Power of a Product

Technology **In Exercises 55–60, use a calculator to evaluate the expression. Round the results to two decimal places when appropriate.**

55. $(1.1 + 3.3)^3$
56. $5.5^3 \cdot 5.5^4$
57. $2.4^4 \cdot 2.4^2$
58. $(4.0 + 3.9)^2$
59. $(2.9^3)^5$
60. $(9.1^2)^4$

Native Americans in Cahokia **In Exercises 61 and 62, use the following information.**

Over 900 years ago, Native Americans lived in a city called Cahokia, in what is now Illinois. The city had over 30,000 inhabitants and contained several temple mounds. Some of the mounds were over 10 stories tall. The base of the Great Temple Mound covered 15 acres, 2 acres more than the base of the largest pyramid in Egypt.

Native American mounds were built as burial places and as platforms to hold temples and the houses of chiefs. Thousands of these mounds still stand in the U.S. and Canada.

61. The volume of the Great Temple Mound was $\frac{1}{3}(\frac{1}{2})(800^3 - 600^3)$ cubic feet. Evaluate this volume.
62. A cubic foot of soil weighs about 90 pounds. How many pounds

ALGEBRA provides a wide range of settings and multicultural representations that give students the confidence that everyone can succeed in algebra.

Mixed REVIEW

1. Write the reciprocal of a^4. **(2.7)**
2. Write the reciprocal of 3. **(2.7)**
3. Solve $p(1 + q) = 2 + q$ for p. **(3.6)**
4. Solve $3x - 4y = -20$ for y. **(3.6)**
5. Evaluate $3x^3 + 2x - 2$ when $x = -2$. **(1.3)**
6. Evaluate $3z^0 + 4z^{-1}$ when $z = \frac{1}{4}$. **(8.2)**
7. What is the slope of $y = 5x + 6$? **(4.4)**
8. What is the slope of $3y + 2 = 0$? **(4.4)**
9. Write 1,299,000,000 in scientific notation. **(8.4)**
10. Write 0.000000496 in scientific notation. **(8.4)**
11. Is 2 a solution of $1 + 2x < 4x$? **(1.5)**
12. Is 6 a solution of $3x - 6 = 10$? **(1.5)**
13. Write $10x - 4y = -16$ in slope-intercept form. **(4.5)**
14. Write $y = \frac{1}{3}x - 2$ in standard form. **(5.5)**
15. Simplify $3x(1 + x) - x^2 + x$. **(2.6)**
16. Simplify $(x^2y^{-3})^{-2} \cdot x^2y^{-4}$. **(8.1)**
17. Solve $|x - 3| = 6$. **(4.8)**
18. Solve $|2x + 4| \le 12$. **(4.8)**
19. Evaluate $2.2 \times 10^{-6} \cdot 3.6 \times 10^4$ **(8.4)**
20. Evaluate $7.6 \times 10^{-6} \div (2.5 \times 10^8)$. **(8.4)**

Career Interview

Printer

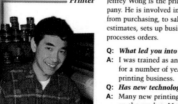

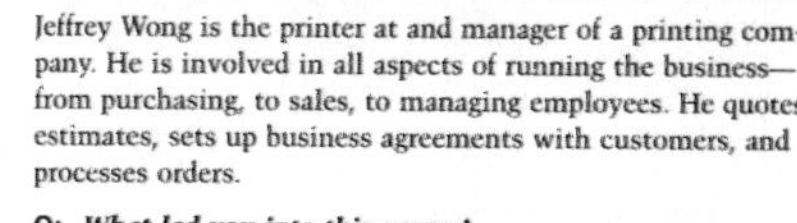

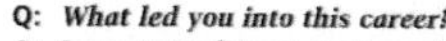

Jeffrey Wong is the printer at and manager of a printing company. He is involved in all aspects of running the business—from purchasing, to sales, to managing employees. He quotes estimates, sets up business agreements with customers, and processes orders.

Q: ***What led you into this career?***
A: I was trained as an engineer. After working in that field for a number of years, I decided to take over the family printing business.
Q: ***Has new technology changed your job experiences?***
A: Many new printing machines and equipment have come on the market. Most have built-in functions that can take care of some computations we used to do by hand.
Q: ***Does this mean you don't need to understand math?***
A: No. All the functions on a new printer don't mean a thing if I don't know how they can be used, or how they impact on what's happening during a particular process.
Q: ***What would you like to tell kids about math?***
A: Often kids think that if they are good in English, then they don't need to worry about math, and vice versa. But we really need to do well in both. English is the vehicle to express your thoughts and feelings; math is the key to meeting everyday needs.

464 *Chapter* **9** • *Quadratic Equations*

Career Interviews, which feature people of various backgrounds who use math on the job, demonstrate the utility of math in daily life.

Cooperative learning means practicing life skills.

"A variety of instructional methods should be used to cultivate students' abilities to investigate, to make sense of, and to construct meanings from new situations; to make and provide arguments for conjectures; and to use a flexible set of strategies to solve problems from both within and outside mathematics. In addition to traditional teacher demonstrations and teacher-led discussions, greater opportunities should be provided for small-group work, individual explorations, peer instruction, and whole-class discussions in which the teacher serves as moderator."

—from the NCTM Standards

There are as many different ways to engage in cooperative learning as there are ways to solve a problem. Throughout ALGEBRA, there's a continual emphasis on getting students together because they benefit from each others' discoveries and experiences.

At least one **Communicating about Algebra** feature in every chapter is an opportunity for cooperative learning. Additional suggestions for cooperative learning activities are provided in your Teacher's Edition.

Learning in pairs is encouraged with **Partner Quizzes**. Students work in tandem as they communicate mathematically and benefit from each others' thoughts.

$= \frac{500}{1 + 0.065(1)}$ *Substitute values for A, r, and t.*

$= \frac{500}{1.065}$ *Simplify.*

≈ 469.48 *Principal*

Thus, you should deposit \$469.48. ■

Communicating about ALGEBRA

Cooperative Learning

SHARING IDEAS about the Lesson

A. ***Work with a Partner*** The formula for the volume of a pyramid with a square base is $V = \frac{1}{3}hs^2$, where h is the height of the pyramid and s is the length of each side of the base. Solve this formula for h.

B. ***Apply*** The Great Pyramid in Egypt has a volume of about 90 million cubic feet and a square base with sides of about 750 feet. Use the formula for h to approximate the height of the Great Pyramid.

156 *Chapter 3 • Solving Linear Equations*

"Using ALGEBRA creates a community of learners within your classroom. Students communicate their thoughts and ideas about algebra with each other—in pairs and groups—as they engage in positive cooperative learning."

—Timothy D. Kanold, ALGEBRA author

ALGEBRA components give you a variety of options to support your teaching style.

1 The **Teacher's Edition** provides you with extensive teaching support. This easy-to-use, extended-margin text helps define essential curriculum and instructional goals.

2 **Color Transparencies for Real-Life Applications** provide 78 full-color visuals from the Student Text, such as maps, technical drawings, and statistical graphs, which facilitate discussion of examples and exercises.

3 **Applications Handbook** provides more information about specific topics in the Student Text, such as astronomy, chemistry, physics, sports, economics, genetics, and music.

4 **Cultural Diversity Extensions** offers projects for each chapter, including historical and contemporary investigations, library research, and interviews.

5 **Algebra 1 Investigations for Performance Assessment** consists of eight long-term investigations providing rich opportunities for students to exercise mathematical thinking.

6 **Technology: Using Calculators and Computers** consists of a Teacher's Guide and Copymasters for using graphing calculators, spreadsheets, and BASIC programs.

7 **Technology Update** provides keystrokes for the TI-82, Casio fx-7700GE, Casio fx-9700GE, and Sharp EL-9300C graphing calculators to be used in conjunction with the Student Text and *Technology: Using Calculators and Computers.* New activities using these calculators are also included.

8 **Interactive Real-Life Investigations** (CD-ROM and diskette versions) provides high-interest interactive investigations for solving real-world problems using rich data sets. One investigation is correlated to each chapter.

9 **Formal Assessment** provides two forms of Mid-Chapter Tests, three forms of Chapter Tests, and Practice for College Entrance Tests.

10 **Alternative Assessment** suggests ways of using portfolios, oral presentations, communication skills, models, group projects, and problem-solving situations and includes forms for assessing student performance in these activities. Also includes Math Log copymasters that provide opportunities for students to write about algebra, to engage in research, and to solve open-ended problems.

11 **Computer Test Bank** (for Macintosh, Apple, or IBM) with graphic capability allows you to create customized tests by choosing from over 2000 test items.

12 **Teaching Tools: Transparencies and Copymasters** include transparencies for number lines, coordinate grids, parabolas, rational functions, etc., plus copymasters for Problem of the Day Exercises, Warm-Up Exercises, and Answer Masters. The transparencies are in addition to the 78 full-color *Transparencies for Real-Life Applications.*

13 **Lesson Plans**

14 **Extra Practice Copymasters** offer a worksheet for every lesson with additional exercises like those found in the Student Text.

15 **Reteaching Copymasters** include teacher-directed and independent activities correlated to the Mid-Chapter Self-Tests and Chapter Tests.

16 **Complete Solutions Manual** includes step-by-step solutions for all exercises found in the Student Text.

Suggested Pacing Chart

*Full Course	*Basic Course
Chapter 1 **Connections to Algebra** 8 lessons 8–9 teaching days	Chapter 1 **Connections to Algebra** 8 lessons 8–11 teaching days
Chapter 2 **Rules of Algebra** 8 lessons 8 teaching days	Chapter 2 **Rules of Algebra** 8 lessons 8–11 teaching days
Chapter 3 **Solving Linear Equations** 7 lessons 8–9 teaching days	Chapter 3 **Solving Linear Equations** 7 lessons 9–12 teaching days
Chapter 4 **Graphing Linear Equations** 8 lessons 12–14 teaching days	Chapter 4 **Graphing Linear Equations** 5–8 lessons 10–12 teaching days
Chapter 5 **Writing Linear Equations** 7 lessons 8 teaching days	Chapter 5 **Writing Linear Equations** 7 lessons 10–14 teaching days
Chapter 6 **Solving and Graphing Linear Inequalities** 6 lessons 11–12 teaching days	Chapter 6 **Solving and Graphing Linear Inequalities** 6 lessons 12–13 teaching days
Chapter 7 **Solving Systems of Linear Equations** 7 lessons 9 teaching days	Chapter 7 **Solving Systems of Linear Equations** 6–7 lessons 9–10 teaching days
Chapter 8 **Powers and Exponents** 7 lessons 11 teaching days	Chapter 8 **Powers and Exponents** 6–7 lessons 12 teaching days
Chapter 9 **Quadratic Equations** 7 lessons 9 teaching days	Chapter 9 **Quadratic Equations** 6–7 lessons 10 teaching days
Chapter 10 **Polynomials and Factoring** 7 lessons 10–12 teaching days	Chapter 10 **Polynomials and Factoring** 6–7 lessons 11–14 teaching days
Chapter 11 **Using Proportions and Rational Functions** 8 lessons 12–13 teaching days	Chapter 11 **Using Proportions and Rational Functions** 6 lessons 9–11 teaching days
Chapter 12 **Functions** 7 lessons 7 teaching days	Chapter 12 **Functions** 4–7 lessons 6–8 teaching days
Chapter 13 **Radicals and More Connections to Geometry** 6 lessons 10 teaching days	Chapter 13 **Radicals and More Connections to Geometry** 3 lessons 5–7 teaching days
Total (average) 127 teaching days **Assessment (average)** 45 days **Cumulative Reviews** 4 days **Total Course (average)** 176 days	**Total (average)** 132 teaching days **Assessment (average)** 44 days **Cumulative Reviews** 4 days **Total Course (average)** 180 days

*See additional suggestions for pacing on page T35.

Suggestions for Pacing

Full Course:

Chapter 1 Allow 2 days for Lesson 1.6 which develops verbal and algebraic modeling, an important problem-solving theme that will be used throughout the text.

Chapter 3 Allow 2 days each for the problem solving lesson, Lesson 3.4, and for Lesson 3.5 which extends the topic of linear equations to include decimals.

Chapter 4 Allow 2 days each for developing the techniques of graphing linear equations in Lessons 4.2 and 4.5, and 3 days each for the introduction of absolute-value equations and their graphs in Lessons 4.7 and 4.8.

Chapter 5 Allow 2 days for the problem solving lesson, Lesson 5.7.

Chapter 6 Allow 3 days each for the problem-solving lesson, Lesson 6.2, and for Lesson 6.3 on compound inequalities, and 2 days each for development of Lessons 6.4 and 6.5 which extend coordinate graphing by introducing absolute-value and linear inequalities.

Chapter 7 Allow 2 days each for the development of techniques for solving and using systems of linear equations in Lessons 7.3 and 7.4.

Chapter 8 Allow 3 days each for the problem solving lesson, Lesson 8.6, and for the development of exponential growth and decay as presented in Lesson 8.7.

Chapter 9 Allow 2 days each for the problem-solving lesson, Lesson 9.5, and for Lesson 9.7 which compares all the data models used to relate real-life data.

Chapter 10 Allow 2 days each for the development of the special cases of polynomial multiplication in Lesson 10.3, and for Lessons 10.4, and 10.5 on factoring polynomials which will be used later for solving quadratic and rational equations.

Chapter 11 Allow 2 days for Lesson 11.3 on direct and inverse variation, and 3 days each for the development of techniques for simplifying and operating with rational expressions in Lessons 11.5 and 11.6.

Chapter 13 Allow 3 days each for the development of techniques for simplifying and operating with radicals as presented in Lessons 13.2 and 13.3.

Algebra 1 may be taught as a two-year course.

To subdivide the course into two years, it is suggested that Chapters 1–6 be taught in the first year. The second year can begin with several weeks of review of Chapters 1–6 before starting the six new chapters—Chapters 7–12. Chapter 13 may be omitted.

Basic Course:

Chapter 1 Allow 2 days each for the important problem solving lessons, Lessons 1.6 and 1.7.

Chapter 2 Allow 2 days for Lesson 2.5 on the multiplication of real numbers, and for Lesson 2.8 on exploring data.

Chapter 3 Allow 2 days each for Lessons 3.3–3.7 which introduce equation-solving techniques that are basic to the course.

Chapter 4 Omitting Lesson 4.6 on intercepts and Lessons 4.7, and 4.8 on absolute-value equations would allow you to spend 2 days each for Lessons 4.1–4.5 on graphing linear equations.

Chapter 5 Allow 2 days each for all lessons. Writing equations as algebraic models is an important problem-solving skill.

Chapter 6 Allow 3 days each for the problem-solving lesson, Lesson 6.2, and for Lesson 6.3 on compound inequalities, and 2 days each for Lessons 6.4 and 6.5 which solve and graph absolute-value and linear inequalities, and for Lesson 6.6 which extends the graphing techniques to include data graphs.

Chapter 7 Omitting Lesson 7.7 on linear programming would allow you 2 days each for solving and using linear systems of equations and inequalities in Lessons 7.3, 7.4, and 7.6.

Chapter 8 Omitting Lesson 8.7 on exponential growth and decay would allow you 2 days each for Lessons 8.1–8.4 which develop and use exponents, and 3 days for the problem-solving lesson, Lesson 8.6.

Chapter 9 Omitting Lesson 9.7 on data models would allow 2 days each for Lesson 9.1 on the Pythagorean Theorem and Lessons 9.3, 9.4, and 9.5 on solving and graphing quadratic equations which is the central focus of this chapter.

Chapter 10 Omitting Lesson 10.7 on completing the square would allow 2 days each for the rest of the lessons in the chapter.

Chapter 11 Omitting Lessons 11.7 on dividing polynomials and 11.8 on rational equations would allow you 2 days each for Lessons 11.3 and 11.5, which apply proportions to the real-life topics of variation and probability, and for Lesson 11.6 on multiplying rational expressions.

Chapter 12 Omitting Lessons 12.3, 12.4, and 12.5 on exponential, quadratic, and rational functions would allow you 2 days each to introduce functions in Lesson 12.1 and the real-life application on measures of central tendency in Lesson 12.7.

Chapter 13 Omitting Lessons 13.4, 13.5, and 13.6 on radical equations would allow you 2 days each for the introduction of radical expressions as presented in Lessons 13.2 and 13.3.

HEATH

ALGEBRA 1

AN INTEGRATED APPROACH

Roland E. Larson

Timothy D. Kanold

Lee Stiff

About the Cover

The cover shows the relationship of music and mathematics, one of the many real-life applications that you will work with in this text. The background is an enlargement of a Compact Disc (CD), and the lines are actually grooves in the CD. The mathematics shown on the cover represents the algebra of musical harmony. A symphony orchestra "tunes up" before a performance by having an oboist play the note "A-440," which has a frequency (pitch) of 440 vibrations per second. The frequency of every note on the piano keyboard is a multiple of 440, shown by the formula in the upper right of the cover. The exponent n represents the position of the note above or below A-440. Two notes harmonize if the ratio of their frequencies, $\frac{f_2}{f_1}$, is an integer or a simple rational number.

As you read the text, see if you can find information that is directly related to the cover.

Acknowledgments

Editorial Development Jane Bordzol, Rita Campanella, Anne M. Collier, Peter R. Devine, Tamara Trombetta Gorman, Marc Hurwitz, Albert Jacobson, Barbara M. Kelley, Savitri Kaur Khalsa, Pearl Ling, Marlys Mahajan, James O'Connell, George J. Summers, Folkert Van Karssen

Marketing/Advertising Jo DiGiustini, Phyllis D. Lindsay, Richard Ravich, Debbie R. Secrist

Design Robert Botsford, Pamela Daly, Leslie Dews, Robin Herr, Cynthia Maciel, Joan Williams

Production Sandra Easton, Lalia Nuzzolo, Mark Tricca

Author Acknowledgments

In Algebra 1, we have included numerous examples and exercises that use real-life data. This would not have been possible without the help of many people and organizations. Our wholehearted thanks goes to all for their time and effort.

We tried to use the data accurately—to give honest and unbiased portrayals of real-life situations. In the cases where models were fit to data, we used the least-squares method. In all cases, the square of the correlation coefficient, r^2, was at least 0.95. In most cases, it was 0.99 or greater.

(Acknowledgments continue following the Index.)

Published simultaneously in Canada

Printed in the United States of America

International Standard Book Number: 0-669-43359-4

1 2 3 4 5 6 7 8 9 10 VHP 02 01 00 99 98 97 96

Roland E. Larson is Professor of Mathematics at the Behrend College of Pennsylvania State University in Erie. He is a member of NCTM and a highly successful author of many high school and college mathematics textbooks. Several of his D.C. Heath College titles have proven to be top choices for senior high school mathematics courses.

Timothy D. Kanold is Chairman of Mathematics and Science and a teacher at Adlai Stevenson High School in Prairie View, Illinois. He is the 1986 recipient of the Presidential Award for Excellence in Mathematics Teaching, as well as the 1991 recipient of the Outstanding Young Alumni Award fiom Illinois State University. A member of NCTM, he served on NCTM's *Professional Standards for Teaching Mathematics* Commission and is a member of the Regional Services Committee.

Lee Stiff is an Associate Professor of Mathematics Education in the College of Education and Psychology of North Carolina State University at Raleigh. He has taught mathematics at the high school and middle school levels. He is a member of the NCTM Board of Directors, served on NCTM's *Professional Standards for Teaching Mathematics* Commission, and is a member of the *Committee for a Comprehensive Mathematics Education of Every Child.*

iii

Linda Bailey
Putnam City Schools
Oklahoma City, OK

Pat Beck
Putnam City North High School
Oklahoma City, OK

Lawrence Demetrak
San Marin High School
Novato, CA

Charleen M. DeRidder
Knox County Schools
Knoxville, TN

Melanie Donnell
Russellville High School
Russellville, AR

Joseph P. Dooley
Upper Darby High School
Upper Darby, PA

David J. Enterlin
Upper Darby High School
Upper Darby, PA

M. Jayne Fleener
University of Oklahoma
Norman, OK

Gail Gonyo
Adirondack Community College
Queensbury, NY

Wanda Heflin
Russellville High School
Russellville, AR

Karen Humble
Western Oaks Junior High School
Oklahoma City, OK

Regina D. Kiczek
Ferris High School
Jersey City, NJ

Fran Kormes
Harry S. Truman High School
Levittown, PA

Chicha Lynch
Capuchino High School
San Bruno, CA

Terry McBroom
San Marin High School
Novato, CA

Elizabeth J. Manes
Harry S. Truman High School
Levittown, PA

Deanna Mauldin
Liberty Bell Middle School
Johnson City, TN

Dominic Mendalla
Capuchino High School
San Bruno, CA

Dennis H. Noe
School District of Waukesha
Waukesha, WI

Dean Orfanedes
Capuchino High School
San Bruno, CA

Jennifer J. Penley
Liberty Bell Middle School
Johnson City, TN

Susan Shumate
Edmond Public Schools
Edmond, OK

Sharon Siegel
Dodgen Middle School
Marietta, GA

Linda Tassano
San Marin High School
Novato, CA

John Wellington
Upper Darby High School
Upper Darby, PA

iv

To the Students

Mathematics evolved over the past few thousand years in many stages. *All* the early stages were centered around the use of mathematics to answer questions about real life. This is very much like the way we wrote *Algebra 1*. We centered the concepts around the real-life use of mathematics.

As mathematics development matured, people began to collect and categorize the different rules, formulas, and properties that had been discovered. This took place independently in many different parts of the world: Africa, Asia, Europe, North America, and South America. The mathematics that we use today is a combination of the work of literally thousands of people.

As you study our algebra book, be sure you understand the value and purpose of the concepts you're learning. Knowing **why** you are learning a concept helps you master it—and understand its relevance to your own life. That's why we begin each lesson explaining what you should learn and why you should learn it.

Remember, math is not a spectator sport—it's a valuable tool you can use in everyday life!

Roland E. Larson

Timothy D. Kanold

Lee Stiff

Real-Life Applications

Look through this list for things that interest you. Then find out how they are used with algebra.

Problem Solving in Every Lesson
Chapter 1 introduces you to a general problem-solving plan (Building an Algebraic Model), relates the plan to general problem-solving strategies, and introduces some real-life algebra models.

A Variety of Problem-Solving Approaches
Throughout the program, **Algebra 1** provides many ways to look at problems in algebra: algebraically, graphically, using tables and charts, through real-life modeling, using reasoning, and using technology. Students learn to use algebra to solve problems in the real world.

CHAPTER 1 Connections to Algebra

Connections Between Pre-Algebra and Algebra
Chapters 1–3 provide you with connections between pre-algebra and algebra — reviewing many of the essential skills you have been exposed to before Algebra 1.

Emphasis on Mathematical Modeling
Algebra 1 is driven by the unifying idea of mathematical modeling and uses algebra to model data through the study of functions. Students explore the mathematical modeling of linear, exponential, quadratic, and rational functions.

CHAPTER 2 Rules of Algebra

C Integrates Coordinate Geometry   D Integrates Data Analysis
F Integrates Functions   G Integrates Geometry

Emphasis on Graphing
Algebra 1 emphasizes graphical representations of algebra. Throughout the text, students have many opportunities to make connections between algebraic, tabular, and graphical models. Bar graphs and tables in Chapters 1 and 2 help students anticipate the coordinate geometry introduced in Lesson 3.7.

Integrating Algebra and Geometry
In all chapters, algebra and geometry — both metric and coordinate geometry — are integrated together to strengthen important skills, develop concepts, and pave the way for future study of mathematics.

CHAPTER 3 Solving Linear Equations

Visualizing Concepts: A Critical Problem-Solving Tool.
Algebra 1 integrates a visual approach to the study of functions. Chapter 4 introduces students to quick graphs and develops coordinate geometry. Students explore functions visually first—by graphing them—then they solve them. This approach helps students grasp concepts and strengthens reasoning skills.

Early Emphasis on Graphing and Coordinate Geometry
Chapter 4 presents a thorough introduction to the drawing and interpreting of linear graphs — introducing you to quick graphs and the graphing techniques of coordinate geometry, a practice that continues throughout the text.

CHAPTER 4 Graphing Linear Equations

C Integrates Coordinate Geometry D Integrates Data Analysis
F Integrates Functions G Integrates Geometry

Integrating Technology
Algebra 1 integrates graphing calculator technology throughout the program in lessons, in exercises, and in the Using a Graphing Calculator feature — one per chapter — allowing you to continually explore, discover, and reinforce algebraic concepts.

CHAPTER 5 Writing Linear Equations

CHAPTER 6 Solving and Graphing Linear Inequalities

The Power of Technology
Since coordinate geometry is introduced early, so are graphing calculators. Introducing graphing calculators early (page 166) encourages students to make connections between algebraic and graphical techniques and ultimately to use technology as a problem-solving tool.

Connections Across Content Strands
Algebra 1 provides connections within mathematics—between data analysis, functions, algebra, and geometry—and to other disciplines, such as social sciences, physical sciences, music, and many more.

Integrating Data Analysis
This program integrates frequent use of authentic data to model algebra in real-life applications. Special Exploring Data lessons, like the one on linear programming in Chapter 7, also develop data analysis techniques.

CHAPTER 7

Solving Systems of Linear Equations

Modeling Algebra in Real Life
Algebra 1 uses an abundance of applications to model real life and to develop critical thinking. The use of real-life data and graphs makes algebraic concepts more meaningful to students—and reinforces the role algebra plays in the real world.

Modeling Algebra in Real Life
Algebra 1 uses an abundance of applications to model algebra in real life and to develop critical thinking. Chapter 8 explores the laws of exponents and follows with the study of exponential growth models — which ties the laws of exponents to real-life models.

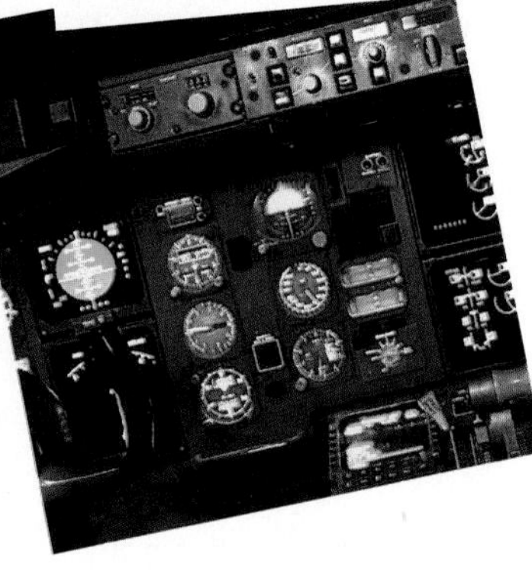

CHAPTER 8

Powers and Exponents

C Integrates Coordinate Geometry D Integrates Data Analysis
F Integrates Functions G Integrates Geometry

CHAPTER 9 Quadratic Equations

CHAPTER 10 Polynomials and Factoring

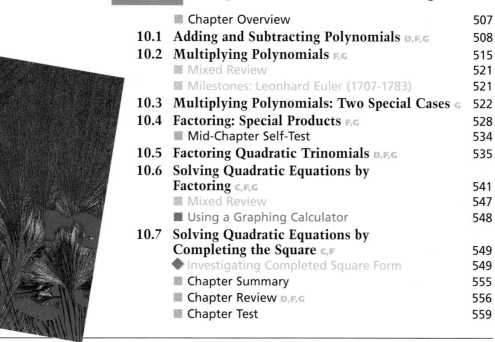

Emphasis on Communication
Communicating about Algebra—a feature found in every lesson—offers opportunities for students to share their ideas about the lesson. Other vehicles for communicating and reasoning are found in the Teacher's Edition and the Alternative Assessment package.

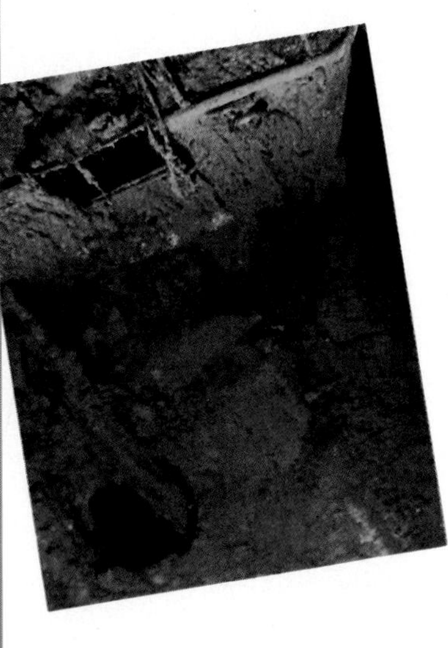

CHAPTER 11

Using Proportions and Rational Equations

Algebraic Models: A Functions Perspective
Chapter 12 revisits the four major types of algebraic models — linear, exponential, quadratic, and rational — this time from a functions point of view.

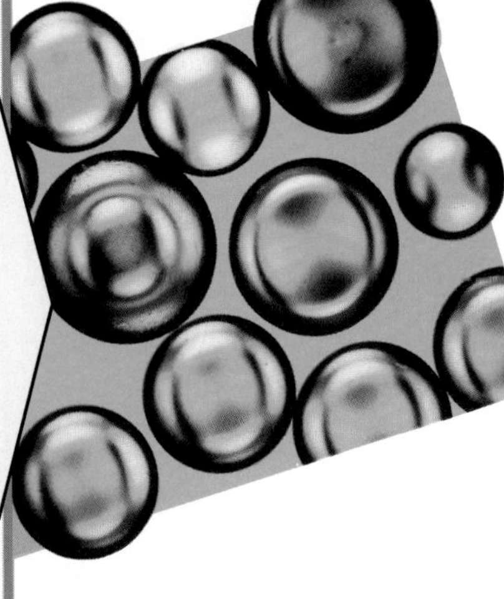

Authentic Assessment
The program's assessment strand provides a variety of options for assessing students' total progress: their knowledge; their mathematical power; their problem-solving, communication, and reasoning skills; and ultimately their mathematical confidence.

CHAPTER 12

Functions

C Integrates Coordinate Geometry D Integrates Data Analysis
F Integrates Functions G Integrates Geometry

More Connections to Geometry
Chapter 13 explores more connections to geometry, trigonometry, radicals, and formal properties of algebra.

CHAPTER 13 Radicals and More Connections to Geometry

Interactive Technology

The following program is available for use with this textbook:

■ Larson Interactive Real-Life Investigations for Algebra 1

available on CD-ROM or diskette (Macintosh or Windows)

Chapter 1	Population Growth Patterns	**Chapter 2**	Musical Sounds
Chapter 3	Sporting Goods Sales	**Chapter 4**	Medical Personnel
Chapter 5	Summer Olympics	**Chapter 6**	Chemistry and Color
Chapter 7	Calendar Years and Solar Years	**Chapter 8**	Solar Eclipses
Chapter 9	Gravitation and the Path of a Projectile	**Chapter 10**	Food Consumption
Chapter 11	Opinion Polls	**Chapter 12**	Government Budgets
Chapter 13	Capital Cities		

Technology Appendix

The **Technology Appendix** provides keystrokes for the *TI-82*, Casio *fx-9700GE*, and Sharp *EL-9300C* to be used in conjunction with the *Using a Graphing Calculator* feature in every chapter as well as in the problem sets throughout the text. The appendix also features new applications for use with these calculators.

In the Teacher's Edition:

A distinctive feature of the text is its readability. Lessons provide clear and concise instruction. You may choose to build class presentations around them. Encourage students to read the text. It has been found that when you use the textbook lessons as the basis for your own lessons, students find it easier to read and follow classroom instruction.

Chapter 1 begins by making connections between algebra and computational skills and concepts that most students already understand.

The Chapter Summary on page 55 provides you and the students with a synopsis of the chapter. It identifies key skills and concepts. You may want to have students look at the Chapter Summary as an overview before beginning the chapter.

CHAPTER 1

Connections to Algebra

LESSONS

Algebra is not just about x and y. It is also about toddlers. It models their world.

Real Life
Demographics

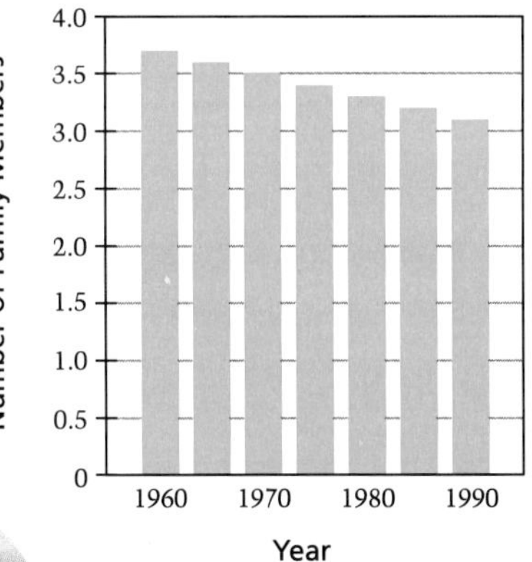

One of the goals of algebra is to recognize patterns and relationships that occur in real life. *Tables* can help you recognize patterns. For instance, the following table shows that between 1960 and 1990, the average family size in the United States decreased by about one tenth of a family member every five years.
(Source: U.S. Bureau of the Census)

Year	1960	1965	1970	1975	1980	1985	1990
Family Size	3.7	3.6	3.5	3.4	3.3	3.2	3.1

Once a pattern has been found, it can be used to answer questions about the real-life situation. For instance, from the pattern above you could estimate the 1995 average family size to be about 3.0. The *bar graph* gives a picture of the decreasing family size between 1960 and 1990.

An emphasis in this Algebra 1 text will be on making connections between mathematics and real-life situations and activities. The example given here is but one of many that will be used in developing lessons, piquing student interest, evaluating student progress, and promoting student communication in algebra, especially as an answer to the question—"Why do we have to study this?"

In Lesson 1.1, for example, Example 2 connects mathematics with basketball, something with which most students are familiar. The Guided Practice exercises of Lesson 1.1 use Little League baseball to promote critical thinking by encouraging students to share their ideas and explanations about possible solutions.

Encourage students to ask questions about the mathematics of real-life situations found in their lessons. Questions may be asked during class or written in student math journals for consideration at a later time. For example, using the table presented here, students might question if it is possible for the average number of people in a family in the United States to be less than 3.0 by the year 2000, or they might be interested in the patterns of the average family size in another country or in specific states or regions within the United States.

Connections to Algebra

PACING CHART

The daily Pacing Chart is meant to help you adjust your teaching pace. Students in the full course should finish the entire text by the end of the year. Students in the basic course may not complete the entire text in the school year. The Pacing Chart for each chapter contains suggestions for lessons that require more than one day and lessons that may be omitted for the basic course.

DAY	FULL COURSE	BASIC COURSE
1	1.1	1.1
2	1.2	1.2
3	1.3 & Programming a Calculator	1.3
4	1.4	Programming a Calculator
5	Mid-Chapter Self-Test	1.4
6	1.5	Mid-Chapter Self-Test
7	1.6	1.5
8	1.6	1.6
9	1.7	1.6
10	1.8	1.7
11	Chapter Review	1.7
12	Chapter Review	1.8
13	Chapter Test	Chapter Review
14		Chapter Review
15		Chapter Test

CHAPTER ORGANIZATION

LESSON	PAGES	GOALS	MEETING THE NCTM STANDARDS
1.1	2-7	1. Represent numbers and number operations 2. Use grouping symbols	Problem Solving, Communication, Reasoning, Connections
1.2	8-13	1. Evaluate variable expressions 2. Represent real-life quantities	Problem Solving, Communication, Reasoning, Connections, Geometry, Technology
Mixed Review	13	Review of algebra and arithmetic	
1.3	14-18	1. Evaluate expressions containing exponents 2. Use Exponents	Problem Solving, Communication, Geometry, Connections, Technology
Programming a Calculator	19	Evaluate expressions using a graphing calculator	Technology
1.4	20-25	1. Use order of operations to evaluate algebraic expressions 2. Evaluate expressions with a calculator	Problem Solving, Communication, Reasoning, Connections, Geometry, Technology
Career Interview	25	Medical Technologist	Connections
Mid-Chapter Self-Test	26	Diagnose student weaknesses and remediate with correlated Reteaching worksheets	
1.5	27-32	1. Check and solve equations 2. Check solutions of inequalities	Problem Solving, Communication, Reasoning, Connections
1.6	33-38	1. Translate verbal phrases into algebraic expressions 2. Translate verbal sentences into algebraic equations and inequalities	Problem Solving, Communication, Reasoning, Connections, Geometry
Mixed Review	38	Review of algebra and arithmetic	
1.7	39-48	1. Use algebra to solve simple real-life problems 2. Make an algebraic model for a real-life problem	Problem Solving, Communication, Reasoning, Connections
1.8	49-54	1. Use tables to organize data 2. Use graphs to model data visually	Problem Solving, Communication, Reasoning, Connections, Discrete Mathematics, Statistics
Chapter Summary	55	A restatement of what has been learned, why it has been learned, and how it fits into the structure of algebra	Structure, Connections
Chapter Review	56-58	Review of concepts and skills learned in the chapter	
Chapter Test	59	Diagnose student weaknesses and remediate with correlated Reteaching worksheets.	

LESSON RESOURCES

MEETING INDIVIDUAL NEEDS

RETEACHING For students who need to spend more time on basics:

If a mid-chapter self-test or chapter test indicates a deficiency, teachers can help students with the appropriate ***Reteaching Copymaster.***

PRACTICE For students who need more practice:

Additional exercises like those in the Pupil's Edition are provided for each lesson in ***Extra Practice Copymasters.***

ENRICHMENT For enriching and broadening students' experiences:

Problem of the Day copymasters in ***Teaching Tools*** provide a daily opportunity to use logical reasoning, looking for a pattern, writing an equation, and other routine and non-routine problem-solving strategies.

Math Log copymasters in ***Alternative Assessment*** provide opportunities to report on investigations, research, and open-ended problems.

Enriching activities with graphing and scientific calculators and computers are provided in ***Technology: Using Calculators and Computers.***

The ***Applications Handbook*** provides additional information about the cross-curriculum topics such as astronomy, chemistry, physics, sports, economics, genetics, and music that are integrated into the Pupil's Edition.

LESSON	1.1	1.2	1.3	1.4	1.5	1.6	1.7	1.8
PAGES	2-7	8-13	14-18	20-25	27-32	33-38	39-48	49-54
Teaching Tools								
Transparencies							✓	✓
Problem of the Day	✓	✓	✓	✓	✓	✓	✓	✓
Warm-up Exercises		✓	✓		✓	✓	✓	✓
Answer Masters	✓	✓	✓	✓	✓	✓	✓	✓
Extra Practice Copymasters	✓	✓	✓	✓	✓	✓	✓	✓
Reteaching Copymasters	Teacher-directed and independent activities tied to results on the Mid-Chapter Self-Tests and Chapter Tests							
Color Transparencies	✓			✓				✓
Applications Handbook	Additional background information is supplied for many real-life applications.							
Technology Handbook	Calculator and computer worksheets are supplied for appropriate lessons.							
Complete Solutions Manual	✓	✓	✓	✓	✓	✓	✓	✓
Alternative Assessment	Assess student's ability to reason, analyze, solve problems, and communicate using mathematical language.							
Formal Assessment	Mid-Chapter Self-Tests, Chapter Tests, Cumulative Tests, and Practice for College Entrance Tests							
Computer Test Bank	Customized tests can be created by choosing from over 2000 items.							

INSIGHTS

1.1 Numbers and Number Operations

Algebra builds upon students' computational skills. An understanding of the different ways in which numbers can be expressed is useful in handling real-life applications. Sharing a common way of expressing number ideas lets people communicate effectively with each other.

1.2 Variables in Algebra

Moviegoers frequently decide to see the matinee instead of the evening show because matinees usually cost less. Below is a verbal model of what each show would cost a group of five students.

Total cost = Number of tickets × Price of ticket

To translate the verbal model into an algebraic model, its parts must be labeled. One translation of the verbal model is "total cost = 5 × ticket price." Since the ticket price varies with the time of day, the letter p is used to represent the ticket price. This use of variables helps to shorten the expression of the total cost to $5 \times p$, or $5p$, where the variable p represents the ticket price.

1.3 Exponents and Powers

Ask students if they have ever wondered how far it is from the earth to the sun (about 93,000,000 miles), or how fast the typical computer can perform a single computation (0.000 000 01 second). Very large numbers and very small numbers can more easily be written and recognized if exponents are used. The study of exponents will be a major focus in Chapter 8.

1.4 Order of Operations

Students often have difficulty with the order of operations, since they are accustomed to reading left to right. Point out that an agreement about the meaning of a numerical expression is necessary. Have students with different kinds of calculators (with and without order of operations) evaluate the expression $18 \div 3 + 6 \times 5$ and discuss the differing answers.

1.5 Equations and Inequalities

This lesson begins to develop two of the most important skills in algebra—finding and checking solutions to equations and inequalities. Although most students have used these skills in simple situations in previous courses, this lesson will help them connect what they already know intuitively with formal vocabulary and algebraic procedures.

1.6 Verbal Models and Algebraic Models

Discuss with the class the meaning of the word *consultant*. Explain, if necessary, that a consultant frequently tries to help his or her clients solve problems, often with the aid of computers. In the retail business, for example, this might mean that a manager needs support in keeping financial records or monitoring financial transactions via the latest technology. To be successful, a consultant must be able to translate solutions into a language that computers will understand. This generally means translating verbal phrases into algebraic ones.

1.7 A Problem-Solving Plan

In this course, students will encounter many real-life applications and problems. Help students understand that the skills they are developing today will be useful in their future. They should understand, too, that problem solving is a process, not an event.

1.8 Exploring Data: Tables and Graphs

This lesson is the first in a series on an integrated theme—data analysis—that will occur in many chapters.

Ask students how they would decide what flavor ice cream is the favorite of most students at their school. One way is to collect information from students in the school and tabulate the results. Once the data have been collected and organized, they can be used to determine a variety of information: the girls' favorite flavor, the flavor chosen most often, and so on.

Problem of the Day

Arrange the digits 5, 6, 7, 8, and 9 in the boxes to make the greatest product.

□□□ × □□

$875 \times 96 = 84{,}000$

ORGANIZER

Warm-Up Exercises

Warm-Up Exercises are designed to get students ready quickly for the day's lesson.

Practice your listening skills. (Students should listen carefully and use mental math as you read these exercises aloud.)

1. Identify any whole number in the list: $\frac{5}{7}$, 35%, the product of 12 and $\frac{1}{2}$, π. the product of 12 and $\frac{1}{2}$
2. Identify the area formula for any rectangle: $2l + 2w$ (twice the length plus twice the width), $\frac{1}{2}bh$ (one half the base times the height), $l \times w$ (length times width), and s^2 (the square of the length of a side). $l \times w$

Record student responses. Then write the lists on the chalkboard or overhead projector. Ask if any students wish to change their answers. Discuss results.

Lesson Resources

Teaching Tools
- Transparency: 5
- Problem of the Day: 1.1
- Warm-up Exercises: 1.1
- Answer Masters: 1.1

Extra Practice: 1.1
Color Transparencies: 1, 2
Technology Handbook: p. 1, 4

Lesson Investigation Answers

a. 0.375 b. 0.333 . . .
c. 0.833 . . . d. 0.3125
e. 0.57142857 . . .
f. 0.222 . . .
Yes, Examples will vary.

1.1 Numbers and Number Operations

What you should learn:

Goal 1 How to represent numbers and number operations

Goal 2 How to use grouping symbols

Why you should learn it:

You can apply what you know about numbers and number operations to help solve problems in algebra.

Goal 1 Numbers and Number Operations

Much of what you will learn in algebra is based on numerical skills that you already have. For instance, you already know how to add, subtract, multiply, and divide numbers, and you will continue to use these skills in algebra.

Different kinds of numbers have different names. For example, the numbers 0, 1, 2, 3, and 4 are **whole numbers,** and the numbers $\frac{1}{2}, \frac{2}{3}$, and $\frac{5}{4}$ are **fractions.** What you call a number sometimes depends on the way it is written. For example, the number $\frac{1}{2}$ can be written in three forms.

Fraction: $\frac{1}{2}$ Decimal: 0.5 Percent: 50%

The decimal form of some numbers contains too many digits to write. For instance,

$$\frac{3}{7} = 0.4285714285\ldots$$

(The three dots mean that the number has more digits than are shown.) You can *approximate* this number to different degrees of accuracy. For instance, rounded to two decimal places, you could write $\frac{3}{7} \approx 0.43$. The symbol $\approx$ means *is approximately equal to.*

Study Tip

The number of decimal places that you should list when writing a decimal approximation depends on the context of the problem. For instance, with dollar amounts, you probably would not use more than two decimal places.

LESSON INVESTIGATION

Investigating Fractions and Decimals

Partner Activity The decimal form of a fraction must either have a finite number of digits or an infinite number of digits. Decide the decimal form of each of the following fractions.

a. $\frac{3}{8}$ b. $\frac{1}{3}$ c. $\frac{5}{6}$ d. $\frac{5}{16}$ e. $\frac{4}{7}$ f. $\frac{2}{9}$

If the decimal has an infinite number of digits, must the digits occur in a repeating pattern? Give examples with your answer.

There are four basic operations that can be performed with numbers: **addition, subtraction, multiplication,** and **division.**

Example 1 *Four Basic Number Operations*

a. *Addition:* The **sum** of 12 and 7 is 19.

$12 + 7 = 19$

b. *Subtraction:* The **difference** of 15 and 8 is 7.

$15 - 8 = 7$ *The difference of a and b is $a - b$.*

c. *Multiplication:* The **product** of 3 and 6 is 18.

$3 \cdot 6 = 18,\quad 3(6) = 18,\quad (3)6 = 18,\quad (3)(6) = 18$

In algebra, we usually do not use the multiplication symbol $\times$ because it is easily confused with the letter x, which is used for another purpose.

d. *Division:* The **quotient** of 24 and 6 is 4.

$24 \div 6 = 4,\quad \frac{24}{6} = 4$ *The quotient of a and b is $\frac{a}{b}$.* ■

You can perform each of the operations shown in Example 1 in your head. We call this **mental math.** For more complicated problems, you may want to use a calculator.

Real Life
Building Costs

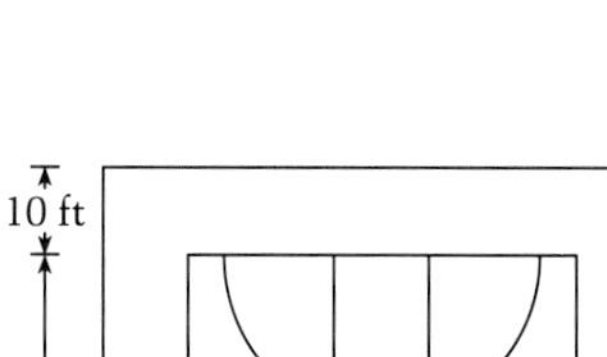

Example 2 *Building a Basketball Court*

A basketball court is 46 feet wide and 84 feet long. The court has a border that is 10 feet wide. The cost of a hardwood floor is \$9 per square foot. How much will it cost to put a new floor on both the court and its border?

Solution To find the cost of the floor, multiply the cost per square foot times the area (in square feet). Recall that the area of a rectangle is found by multiplying its length times its width.

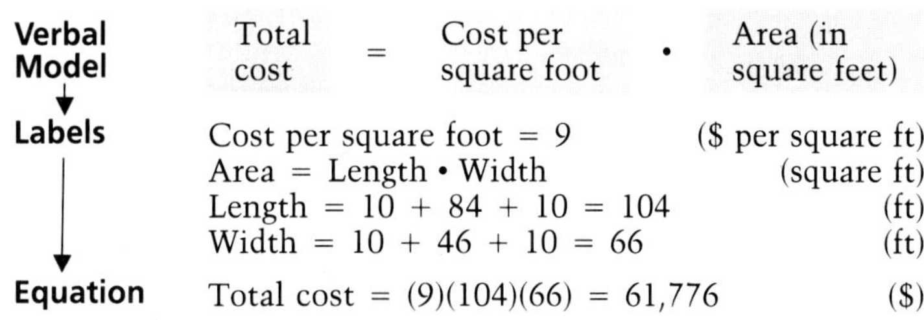

Verbal Model Total cost = Cost per square foot • Area (in square feet)

Labels
Cost per square foot = 9 (\$ per square ft)
Area = Length • Width (square ft)
Length = 10 + 84 + 10 = 104 (ft)
Width = 10 + 46 + 10 = 66 (ft)

Equation Total cost = (9)(104)(66) = 61,776 (\$)

Thus, it would cost \$61,776 to put a new floor on the basketball court and its border. ■

LESSON Notes

Example 1

Stress that in this text the phrases *the difference of* and *the quotient of* have a definite order: The difference of b and a means $b - a$, not $a - b$; the quotient of b and a means $b \div a$, not $a \div b$.

Example 2

PROBLEM SOLVING This Example introduces a problem-solving plan—Use the necessary information stated in the problem to write a *verbal model.* Assign *labels* (statements that relate the different numerical or algebraic quantities within the problem). Translate the labels into an *algebraic model,* or equation. *Solve* the problem. (This problem-solving plan will be formally introduced to students on page 40.) Stress the connection between the verbal model and the algebraic model that is solved to find a solution to the problem.

Encourage students to use their own words as much as possible in verbal models and labels. Using labels gives students a chance to make some of the intermediate calculations needed to apply the verbal model. Note also the connection between the units at the right of the labels and the units of the solution.

Verbal model: Total Cost = Cost per ft^2 • Area in ft^2

Equation:

Total Cost = (9)(104)(66)

The answer, or total cost, is given in dollars.

Example 3

An Alternate Approach
Using Technology
Encourage students to check the reasonableness of their simplifications of expressions containing grouping symbols by using calculators. For example, check [16 ÷ (2 + 6)] − 2 by following this key sequence.

[(] 16 [÷] [(] 2 [+] 6 [)] [)] [−] 2 [=]

Extra Examples

Here are additional examples similar to those in **Example 3.**

Simplify. Remember to evaluate the innermost grouping symbols first!

1. 12 + [(35 ÷ 7) x 2] 22
2. 363 ÷ [17 − (18 ÷ 3)] 33
3. [16 ÷ (2 + 6)] − 2 0

Check Understanding

Check Understanding questions can be used as a lesson quiz to assess students' understanding of the lesson goals.

State reasons why computational skills are important in understanding and solving problems.

Communicating about ALGEBRA

COOPERATIVE LEARNING
Have students work together in small groups or with partners to talk about how they would estimate the expressions. Encourage them to use their own words. Student-invented strategies are often just as valid as traditional strategies. Point out that estimating is not just rounding an answer, but an underlying skill used to solve problems and make decisions. In many real-life applications, an estimate is appropriate, since exact answers are not always possible or needed.

Goal 2 Using Grouping Symbols

For problems that have more than one operation, it is important to know which operation to do first. **Grouping symbols** such as parentheses ( ) or brackets [ ] indicate the order in which the operations should be performed. Different orders can produce different results. For instance, the number $(3 \cdot 4) + 8 = 12 + 8 = 20$ is not the same as the number $3 \cdot (4 + 8) = 3 \cdot 12 = 36$.

A collection of numbers, operations, and grouping symbols is a **numerical expression.** When we *simplify* a numerical expression to find its value, we are *evaluating* the expression.

Example 3 *Evaluating Expressions Containing Grouping Symbols*

a. $16 - [(4 \cdot 5) - 6] = 16 - [20 - 6]$ *Simplify 4 • 5.*
$= 16 - 14$ *Simplify 20 − 6.*
$= 2$ *Simplify 16 − 14.*

b. $\dfrac{210}{20 - (13 - 3)} = 210 \div (20 - 10)$ *Simplify 13 − 3.*
$= 210 \div 10$ *Simplify 20 − 10.*
$= 21$ *Simplify 210 ÷ 10.*

Connections **Technology**

c. To evaluate the expression $[413.2 \div (12.5 \cdot 6.1)] - 2.6$, use a calculator and follow these steps.

[(] 413.2 [÷] [(] 12.5 [×] 6.1 [)] [)] [−] 2.6 [=]

The calculator display of 2.819016393 can be rounded to 2.8.■

Communicating about ALGEBRA

SHARING IDEAS about the Lesson

Estimate for Reasonableness When you use a calculator to evaluate an expression, such as in Example 3, one way to check that the answer is reasonable is to estimate the value of the expression first. Using the methods of estimating you know, find three estimates of each answer and explain how you found them. See Additional Answers.

A. $[413.2 \div (12.5 - 6.1)] - 2.6$ **B.** $0.45 \cdot (1.92 - 0.84) \cdot 17.6$

EXERCISES

Guided Practice

CRITICAL THINKING about the Lesson

Little League **In Exercises 1–6, use the bar graph showing the five states with the most Little League teams.**
(Source: Little League Baseball, Inc.)

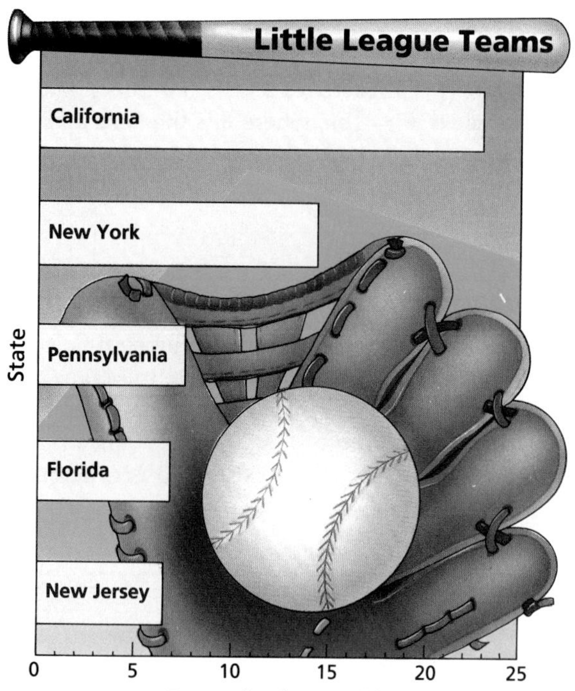

1. Which state has the most Little League teams? California
2. Why do you think California has so many more teams than New Jersey? See below.
3. Estimate the total number of teams from the five states listed. 59,000
4. Which of the following is a good estimate of the difference between the numbers of teams in New York and Florida?

   a. 14,000 **b.** 6,000 **c.** 7,000
   d. 10,000 **e.** 8,000

5. How did you arrive at the estimate in Exercise 3? In Exercise 4? See below.
6. In Exercise 4, what would be a good estimate to the nearest hundred? 7300

Independent Practice

In Exercises 7–14, use mental math.

7. What is the sum of 4 and 9? 13
8. What is the sum of 12 and 6? 18
9. What is the difference of 8 and 7? 1
10. What is the difference of 9 and 2? 7
11. What is the product of 4 and 6? 24
12. What is the product of 7 and 3? 21
13. What is the quotient of 16 and 4? 4
14. What is the quotient of 5 and 1? 5

In Exercises 15–28, evaluate the expression.

15. $[8 - 3(2)] \cdot 6$ 12
16. $12 - [(4)(2) + 2]$ 2
17. $[3(6) + 2] + 4$ 24
18. $4 - [(2 \cdot 9) - 17]$ 3
19. $24 \div [10 - (4 \div 2)]$ 3
20. $12 + [16 \div (3 + 1)]$ 16
21. $10 + [9 \div (1 + 2)]$ 13
22. $[12 \div (1 + 5)] - 2$ 0
23. $18 \div [(8)(2) - 7]$ 2
24. $[(10 \cdot 3) - 10] \div 4$ 5
25. $[12 \div (3 \cdot 2)] + 4$ 6
26. $16 - [54 \div (3)(3)]$ 10
27. ✪ $\frac{10}{(4 - 3) \cdot 2}$ 5
28. ✪ $\frac{28}{(4)(2) - 4}$ 7

2. Accept any reasonable answer such as one that involves population or climate.
5. Accept any reasonable answers that involve operating on rounded numbers.

EXERCISE Notes

ASSIGNMENT GUIDE

Basic/Average: Ex. 1–14, 15–33 odd, 35–39 odd, 41–42, 43–53 odd, 55–58

Above Average: Ex. 1–14, 19–33 odd, 35–39 odd, 41–42, 43–53 odd, 55–58

Advanced: Ex. 1–6, 7–13 odd, 19–33 odd, 35–42, 51–58

Selected Answers
Exercises 1–6, 7–53 odd

✪ More Difficult Exercises
Exercises 27, 28, 38, 40, 53, 54

Guided Practice

Guided Practice exercises require students to engage in critical thinking about the skills and concepts of the lesson. You are encouraged to complete these exercises as a whole-class activity, so that all students might benefit. The answers to all Guided Practice exercises will be found in the "Selected Answers" section of the student text.

▶ **Ex. 1–6** In these Guided Practice exercises, students use numbers and number operations to gain information about a real-life organization, Little League Baseball. Exercise 2 gives each student a chance to contribute; there is no right or wrong answer. Invite students to share how they estimated in Exercise 4.

Common-Error ALERT!

In **Exercises 7–14,** students frequently confuse the order of operands in statements involving *difference* and *quotient.* Emphasize that the "difference of 18 and 7" translates into 18 − 7 and that the "quotient of 24 and 3" becomes 24 ÷ 3.

Independent Practice

Each lesson is composed of two types of independent practice exercises based on levels of difficulty. One type of exercise presents challenging problems and extensions to the lesson Goals and lesson Examples. This type of exercise has been indicated in this Teacher's Edition by labeling each one using a star (✪). The answers to the odd-numbered exercises of the Independent Practice and Integrated Review sections will be found in the "Selected Answers" section of the student text.

▶ **Ex. 38–42** The blue titles will appear throughout the text. These titles indicate applications to real-life, connections to other mathematical topics, or technology.

▶ **Ex. 38–40** PROBLEM SOLVING Encourage students to write verbal models for the equations they intend to use. Here is a possible verbal model for Exercise 38.

Cost of fencing	=	Cost per meter	•	Perimeter (in meters)

▶ **Ex. 39** Assume in Exercise 39 that the banner is made of one layer of material.

▶ **Ex. 41–42** The labels in the parentheses refer to the set of cards and the year of issue.

Common-Error ALERT!

Using grouping symbols helps to avoid misunderstandings about the order of operations. In **Exercises 27–28**, note the use of the fraction bars as grouping symbols.

Technology **In Exercises 29–34, use a calculator to evaluate the expression. Round your result to two decimal places. Use estimation to check your answer.**

29. $[52.4 - (16.1 \div 3.6)] - 30.2$ 17.73

30. $13.9 \div [(7.4 - 6.2) + 4.9]$ 2.28

31. $42.1 \div [(3.2 \cdot 6.1) - 1.3]$ 2.31

32. $[(16.0)(7.2 \div 3.9) + 6.2] - 29.8$ 5.94

33. $[439.9 + (12.9 \div 4.1)] - (42.2)(6.1)$ 185.63

34. $(9.4 \div 2.7) + [(4.3)(16.1) - 17.7]$ 55.01

Geometry **In Exercises 35–37, find the area of the figure. (The area of a triangle is $A = \frac{1}{2}bh$, where b is the base and h is the height.)**

35. 16 ft² (square, 4 ft by 4 ft)

36. 15 m²

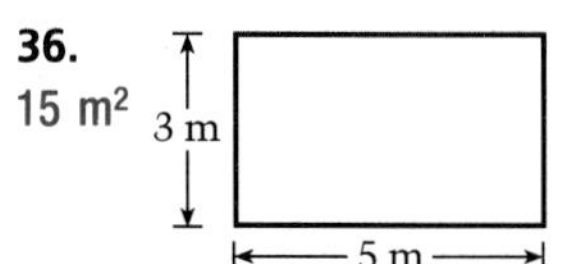

37. 20 in.²

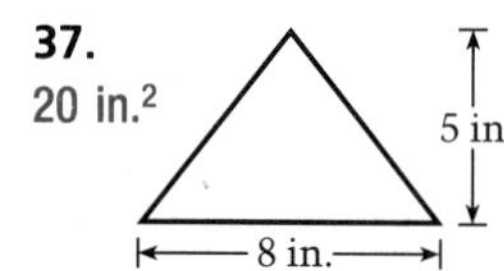

✪ **38.** *Cost of Fencing* A swimming pool is 5 meters wide and 6 meters long. A concrete walk that is 2 meters wide surrounds the pool. Draw and label a diagram of the pool and walk. Then find how much it will cost for wooden fencing around the outside of the walk if the fencing sells for \$13 per meter. \$494

39. *Airplane Banner* An airplane advertising banner has a height of $2\frac{1}{2}$ yards and a length of 15 yards. How many square yards of material are needed to make the banner? 37.5 yd²

✪ **40.** *Horse Show* A circular arena for horses has a diameter of 200 feet. How far must a horse walk to go around the edge of the arena once? (The circumference of a circle is $C = \pi d$, where $\pi \approx 3.14$ and d is the diameter.) 628 ft

Baseball Cards **In Exercises 41 and 42, use the chart showing the estimated value in 1991 of five rare baseball cards.**
(Source: Sports Collectors Digest, Baseball Card Monthly Price Guide)

	Estimated Value	Number in Existence
Honus Wagner (T206 1909)	\$110,000	50
Fred Lindstrom (U.S. Carmel 1932)	\$ 25,000	1
Napoleon Lajoie (Goudey 1933)	\$ 25,000	50
Joe Doyle (T206 1910)	\$ 17,000	2
Eddie Plank (T206 1910)	\$ 14,000	100

41. What is the total value of all the Honus Wagner cards in existence?

42. What is the total value of two Joe Doyle and three Eddie Plank cards?

41. \$5,500,000 \$76,000

Integrated Review

43. Write $\frac{6}{5}$ as a decimal. 1.2

44. Write 50% as a fraction. $\frac{1}{2}$

45. Write $\frac{3}{4}$ as a percent. 75%

46. Write 60% as a decimal. 0.60

47. Write 0.32 as a fraction. $\frac{8}{25}$

48. Write 0.07 as a percent. 7%

49. Approximate $\frac{4}{9}$ to two decimal places. 0.44

50. Approximate $\frac{7}{6}$ to two decimal places. 1.17

Geometry **In Exercises 51 and 52, find the missing portion of the *circle graph*. (The portions total 100%.)**

51. 25%

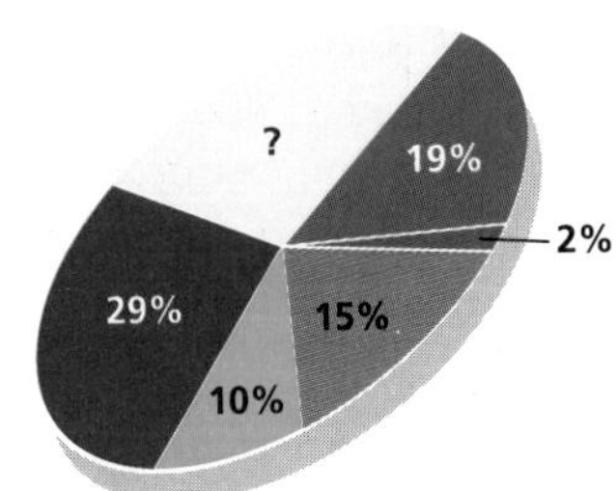

52. 23%

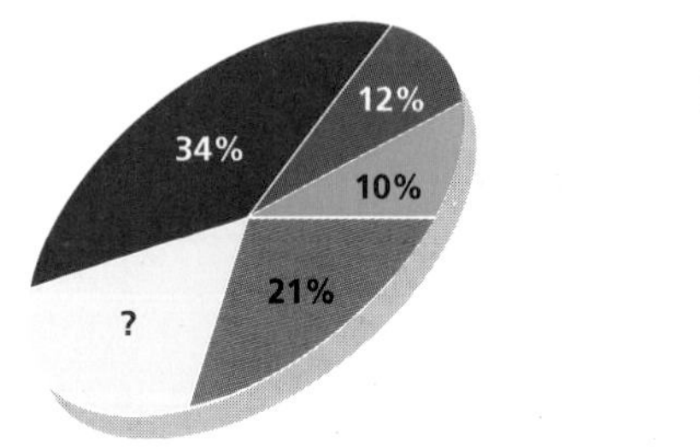

Geometry **In Exercises 53 and 54, find the area of the figure.**

53. 36

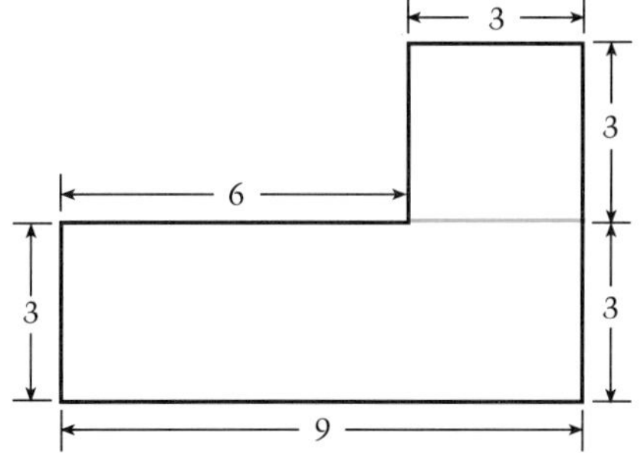

54. 128

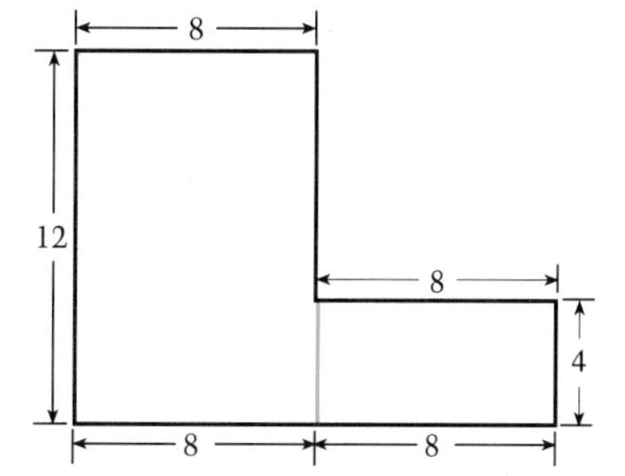

Exploration and Extension

Balance in a Savings Account **In Exercises 55–58, use the following data from a newly opened savings account.**

Date	Deposit	Withdrawal	Interest Earned	Balance
June 1	$50.00			$50.00
June 15	$50.00			
June 30			$0.37	
July 15		$30.00		
July 30			$0.58	

55. What was the balance on June 15? $100.00

56. What was the balance on June 30? $100.37

57. What was the balance on July 15? $70.37

58. What was the balance on July 30? $70.95

Ex. 53–54 There is more than one way to find the area of each figure. (Exercise 53, for example, can be solved by using either (9 x 6) – (6 x 3), in which the area of a rectangle of dimension 6 x 3 is subtracted from the area of a larger rectangle of dimension 9 x 6; or (9 x 3) + (3 x 3), in which the area of the rectangles of dimensions 9 x 3 and 3 x 3 are added.) Poll students to see what the preferred way is. You may wish to point out to students that the solution is in square units.

EXTEND Ex. 55–58 PROBLEM SOLVING Write a verbal model and an equation for the balance in the savings account after an interest period.
balance = (previous balance + deposits) + interest earned – withdrawals

Enrichment Activities

These activities require students to go beyond lesson goals.

WRITING Following the class discussion, students might record their personal responses in their math journals.

1. When is it appropriate or useful to use the fraction form, the decimal form, and the percent form of a number?
   Answers might include ease of use, calculator usage, etc.
2. How is using a verbal model for a problem related to writing an equation for that problem?
   See Lesson Notes on page 3.

Problem of the Day

Insert grouping symbols to make the statements true.

$7 - 4 \times 6 + 4 = 30$

$6 + 2 \times 3 - 4 = 20$

$9 - 4 \times 2 + 5 = 15$

$5 \times 8 - 6 \div 2 = 5$

$(7 - 4) \times (6 + 4) = 30$;
$(6 + 2) \times 3 - 4 = 20$;
$(9 - 4) \times 2 + 5 = 15$;
$5 \times (8 - 6) \div 2 = 5$

ORGANIZER

Warm-Up Exercises

Write these exercises on the chalkboard or overhead.

Simplify

1. $7(3) + 2$ — 23
2. $6(9) - 3(4)$ — 42
3. $13 + [(6 \cdot 7) - 36]$ — 19

Lesson Resources

Teaching Tools
- Problem of the Day: 1.2
- Warm-up Exercises: 1.2
- Answer Masters: 1.2

Extra Practice: 1.2

Technology Handbook: p. 7

LESSON Notes

Emphasize to students that numerical expressions are also algebraic expressions. Tell students that $8(y)$, $8 \cdot y$, and $8 \times y$ can all be written as $8y$.

Flowcharts, like the one above Example 1, can be used to help students understand a sequence of steps.

1.2 Variables in Algebra

What you should learn:

Goal 1 How to evaluate variable expressions

Goal 2 How to represent real-life quantities using variable expressions

Why you should learn it:

You can use variables in algebra to represent numbers. These variables can be used to express number relationships, such as in the formula $d = rt$ for distance, rate, and time.

Goal 1 Evaluating Variable Expressions

A **variable** is a letter that is used to represent one or more numbers. The numbers are the **values of the variable.** An **algebraic expression** is a collection of numbers, variables, operations, and grouping symbols. Here are some examples.

Algebraic Expression	Meaning
$8y$	8 times y
$\frac{16a}{b}$	16 times a, divided by b
$4r + s$	4 times r, plus s

When each variable in an algebraic expression is replaced by a number, we say that we are *evaluating* the expression, and the resulting number is the **value of the expression.** For example, when $x = 2$, the value of the expression $4x$ is found by *substituting* 2 for x. In this case, the resulting value of the expression is $4(2) = 8$.

To evaluate an algebraic expression, use this *flowchart*:

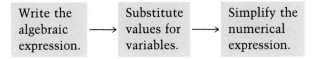

Write the algebraic expression. → Substitute values for variables. → Simplify the numerical expression.

Example 1 *Evaluating an Algebraic Expression*

Evaluate $5x + 4y$ when $x = 3$ and $y = 2$.

Solution To evaluate this expression, substitute 3 for x and 2 for y.

$5x + 4y$	*Write the expression.*
$= 5(3) + 4(2)$	*Substitute 3 for x and 2 for y.*
$= 15 + 8$	*Simplify 5(3) and 4(2).*
$= 23$	*Value of the expression*

Thus, the expression $5x + 4y$ has a value of 23 when $x = 3$ and $y = 2$. ■

When substituting a number for a variable, you must replace each occurrence of the variable in the expression. For example, if $x = 3$, then:

$x + (2x - 1)$	$= 3 + [(2 \cdot 3) - 1]$	*Substitute 3 for x twice.*
	$= 3 + [6 - 1]$	*Simplify $2 \cdot 3$.*
	$= 3 + 5$	*Simplify $6 - 1$.*
	$= 8$	*Value of the expression*

An algebraic expression can have different values when different numbers are substituted for the variable, as shown in Example 2.

Example 2 *Substituting Different Values for a Variable*

Evaluate the expression $(x+7) \div 5$ when x has the following values.

a. $x = 3$ **b.** $x = 208$

Solution

	Write the expression.		Substitute values for x.		Simplify the expression.
a.	$\frac{x+7}{5}$	=	$\frac{3+7}{5}$	=	$\frac{10}{5} = 2$
b.	$\frac{x+7}{5}$	=	$\frac{208+7}{5}$	=	$\frac{215}{5} = 43$

■

The parts of an expression formed as a sum, product, or quotient have special names.

Sum: $3x$ and 2 are **terms** of $3x + 2$.
Product: 4 and x are **factors** of $4x$.
Quotient: $5x$ is the **numerator** and 2 is the **denominator** of $\frac{5x}{2}$.

Example 3 *Using a Calculator to Evaluate an Expression*

Use a calculator to evaluate the following expression when $r = 8.6$ and $s = 4.3$. Round your result to one decimal place.

$$\frac{6.37(r + s)}{3.8}$$

Solution

$\frac{6.37(r+s)}{3.8} = \frac{6.37(8.6+4.3)}{3.8}$	*Substitute 8.6 for r and 4.3 for s.*
≈ 21.6	*Round to one decimal place.*

One possible sequence of keystrokes is as follows.

6.37 [×] [(] 8.6 [+] 4.3 [)] [÷] 3.8 [=] ■

The most frequently used variables in algebra are x and y. Of course, other letters (such as r and s) work as well, and students should understand that they have a choice when selecting a variable to use. Encourage them to select meaningful variables, such as w for width, h for height, or a for age.

Ask students if they recognize the formulas at the top of page 10. Discuss those that are unfamiliar to them. Suggest that a location in their math journals (perhaps the last five pages) be reserved for a list of formulas.

Extra Examples

Here are additional examples similar to **Examples 1, 3,** and **4.**

1. Evaluate the expression
   $5x - (2x + 4)$
   when x has the following values.
   a. $x = 2$ 2
   b. $x = 3$ 5
   c. $x = 5$ 11
   What happens to the value of the expression as the value of x increases?
   The value increases.
2. Estimate the value of
   $6.3(m - 2n) \div 1.7$
   when m and n have the following values. Then use a calculator to find the exact values. Round answers to one decimal place.
   a. $m = 2.1, n = 0.8$ 1.9
   b. $m = 15, n = 6.7$ 5.9
   c. $m = 9.8, n = 2.9$ 14.8
   How close were your estimated values?
3. In South America, standard temperatures are given in degrees Celsius. If it is 25 degrees Celsius in Lima, Peru, what temperature would that temperature be in degrees Fahrenheit?
   77 degrees Fahrenheit

Check Understanding

1. Why is evaluating expressions a useful real-life skill?
2. What applications of evaluating formulas do you know about? (Note: Students might record their responses to this question in their math journals.)

Communicating *about* ALGEBRA

COOPERATIVE LEARNING
Students should work with a partner to evaluate each statement and then compare the results.

For those statements that are false, students should work together to decide on the values of the variable that make the statement true. Encourage students to use mental math to evaluate expressions.

EXTEND *Communicating* ***WRITING***
Have students write a brief paragraph explaining why each false statement A–F on page 10 is false and how to find the values for the variables that make the statement true. (For example, Statement F, $8 + b = 7 + x$, is false when $b = 2$ and $x = 8$ because $8 + 2 = 10$ and $7 + 8 = 15$, and $10 \neq 15$. Since $8 + b = 10$, $8 + b$ is 5 less than $7 + x$, or 15. This suggests that adding 5 to the b value, 2, will give values of b and x that will make $8 + b = 7 + x$ true.)

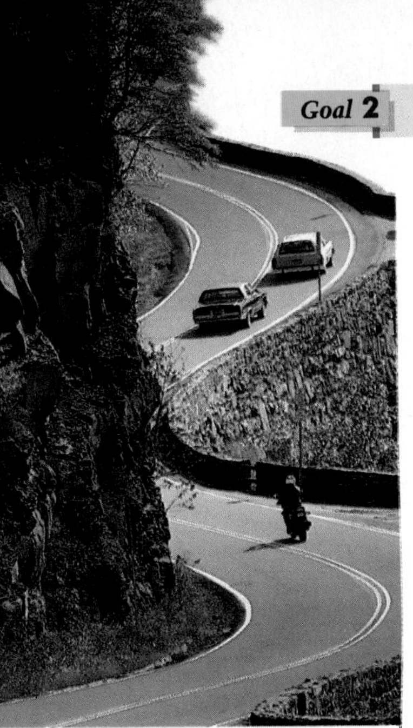

Goal 2 Representing Real-Life Quantities

Algebraic expressions are often used in real-life situations. Here are some examples.

	Labels	Formula
Area of Rectangle	A = area l = length w = width	$A = lw$
Interest Earned	I = interest P = principal r = interest rate (decimal) t = time (years)	$I = Prt$
Temperature	F = degrees Fahrenheit C = degrees Celsius	$F = \frac{9}{5}C + 32$
Distance	d = distance r = rate or speed t = time	$d = rt$

Connections **Physics**

Example 4 ***Finding the Distance Traveled***

Suppose you are traveling in a car at a constant speed of 75 kilometers per hour for one and one-half hours. How far will you travel?

Solution Use the formula for distance.

$d = rt$ *Formula for distance*

$= (75)(1.5)$ *Substitute 75 for r and 1.5 for t.*

$= 112.5$ *Simplify.*

Thus, you will travel a distance of 112.5 kilometers. ■

Communicating *about* ALGEBRA

SHARING IDEAS about the Lesson

Multiple Solutions Decide whether each statement is true or false when $a = 1$, $b = 2$, $c = 3$, and $x = 8$. If a statement is false, find values of the variables that make the statement true. (The values you find may differ from those found by other students.) Sample values given in C and F.

A. $2a = b$ True
B. $5a = b + 3$ True
C. $5c - x = 4b$ False, $x = 7$
D. $7b - x = 2c$ True
E. $7 + b = x + a$ True
F. $8 + b = 7 + x$ False, $b = 7$

EXERCISES

Guided Practice

CRITICAL THINKING about the Lesson

1. State the meaning of the variable expression $\frac{r}{3s}$ in words. *r* divided by the product of 3 and *s*
2. What is the variable expression for "6 times *x*, plus *y*"? $6x + y$
3. Evaluate the expression $8x$ when $x = 3$. 24
4. Evaluate the expression $9a - 2$ when $a = 1$. 7
5. Evaluate the expression $a - b$ when $a = 10$ and $b = 3$. 7
6. Evaluate the expression $2x + 4y$ when $x = 2$ and $y = 3$. Find other values for *x* and *y* that give the same result. 16; $x = 6$ and $y = 1$, $x = 0$ and $y = 4$, among others

Independent Practice

In Exercises 7–16, evaluate the expression.

7. $2x + 1$, when $x = 4$ 9
8. $x - 9$, when $x = 21$ 12
9. $2r - s + 4$, when $r = 4$ and $s = 5$ 7
10. $3a + b$, when $a = 2$ and $b = 1$ 7
11. $(a - 2) + 4a$, when $a = 4$ 18
12. $6a - (1 + a)$, when $a = 3$ 14
13. $3x(4 + x)$, when $x = 2$ 36
14. $5y + 4(y - 3)$, when $y = 4$ 24
15. $x \div (3y - 4)$, when $x = 6$ and $y = 2$ 3
16. $[9 - (x \div 4)] + y$, when $x = 8$ and $y = 7$ 14

In Exercises 17–20, evaluate $24 \div (x + 2)$ for the given value of *x*.

17. $x = 2$ 6
18. $x = 0$ 12
19. $x = 4$ 4
20. $x = 10$ 2

Technology **In Exercises 21–24, use a calculator to evaluate the expression. Round your result to two decimal places. Use estimation to check your answer.**

21. $4.97a + 9.21$, when $a = 6.21$ 40.07
22. $3.2(2.1a - 4.3)$, when $a = 4.8$ 18.50
23. $6.9(x - 3.2y) - 56.2$, when $x = 19.1$ and $y = 1.1$ 51.30
24. $15.9 \div (x + y)$, when $x = 3.2$ and $y = 3.4$ 2.41

In Exercises 25–28, you may need to use the formulas on page 10.

25. *Artificial Turf* A football field is 110 meters long and 49 meters wide. How many square meters of artificial turf are needed to cover the field? 5390
26. *Interest Earned* Suppose you deposit \$80 in a savings account that pays an annual interest rate of 6%. How much interest would you earn in 9 months? \$3.60
27. *Distance Traveled* A car traveled for 2 hours at a constant speed of 72 kilometers per hour. How far did the car travel? 144 km
28. *Running* If you jog 300 feet per minute, how many feet will you jog in an hour? 18,000

EXERCISE Notes

ASSIGNMENT GUIDE

Basic/Average: Ex. 1–6, 7–15 odd, 17–24, 25–45 odd, 47–50

Above Average: Ex. 1–6, 7–41 odd, 43–50

Advanced: Ex. 1–6, 7–23 odd, 25–28, 43–50

Selected Answers
Exercises 1–6, 7–45 odd

Use **Mixed Review** as needed.

✪ More Difficult Exercises
Exercises 37–42, 45–46, 48, 50

Guided Practice

▶ **Ex. 2** Call attention to the use of the comma to indicate the grouping $6x + y$ instead of $6(x + y)$. Ask students for a verbal statement that would indicate $6(x + y)$.
One such statement is "six times the sum of *x* and *y*."

Independent Practice

▶ **Ex. 25–28** PROBLEM SOLVING Encourage students to use the Draw a Diagram strategy to understand the information provided. Students will find it helpful to look back at the formulas on page 10.

Common-Error ALERT!

In **Exercises 7–24,** caution students about using the order of operations properly when evaluating these expressions. Remind students to begin each exercise by rewriting the original expression. When doing each exercise, students should show the substitution and then the evaluation of the expression.

EXTEND Ex. 41–42 GROUP ACTIVITY You might have students work in small groups to find some other values that can be made from the given expressions just by changing grouping symbols. Compare class outcomes for each expression. What is the largest value obtained? The smallest? The second most frequently obtained value? (Note: Students might include negative values.) Sample answers: 41. $-\frac{2}{3}$, 4, 24, 26; 42. 48, 158

Ex. 43–44 The algebraic expressions found in these exercises are simple models of geometric relationships.

Enrichment Activities

WRITING Following the class discussion, students might record their personal responses in their math journals.

1. What are the steps in evaluating a variable expression? See page 8.
2. What is the difference between numerical expressions and variable expressions? See page 8
3. How do factors differ from terms? See page 9.
4. Why are variables used to express number ideas? To shorten verbal expressions, ease of use, etc.

Integrated Review

In Exercises 29–32, use mental math to evaluate the expression.

29. The sum of 6 and 11 17
30. The difference of 14 and 9 5
31. The product of 6 and 8 48
32. The quotient of 27 and 3 9

Technology **In Exercises 33–36, use a calculator to evaluate the expression. Round your result to two decimal places, if appropriate. Use estimation to check your answers.**

33. $42.6 + 9.3$ 51.9
34. $21.3 - 16.8$ 4.5
35. $9.39 \div 6.12$ 1.53
36. $3.9 \cdot 6.4$ 24.96

In Exercises 37–40, evaluate the expression.

37. $10 - [8 \div (4 - 2)]$ 6
38. $6 \cdot (12 \div 4) + 9$ 27
39. $[(9 - 5) \cdot 4] \div 4$ 4
40. $(16 + 4) \div (2 \cdot 5)$ 2
41. Insert grouping symbols in the expression $30 - [(36 \div 3) + 6]$ so that its value is 12.
42. Insert grouping symbols in the expression $[(12 \cdot 2) + 12] - 10$ so that its value is 26. or $(12 \cdot 2) + (12 - 10)$
43. *Geometry* Write an algebraic expression that represents the perimeter of a triangle whose sides have lengths *a*, *b*, and *c*. (The perimeter of a triangle is the sum of the lengths of its sides.) Then use the expression to find the perimeter of each triangle. $P = a + b + c$

a. 9

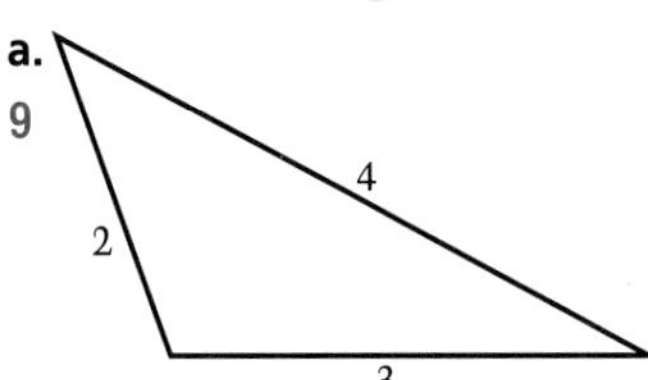

b. 12

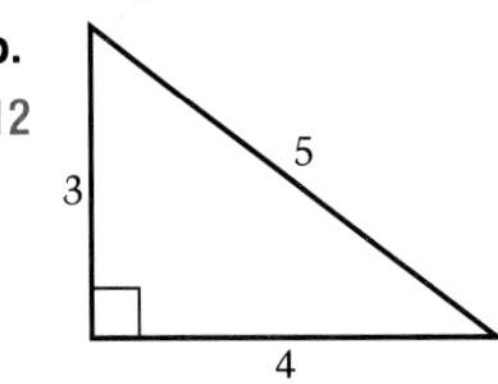

c. 15

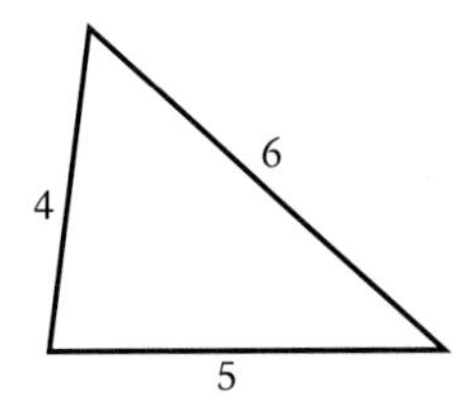

44. *Geometry* Write an algebraic expression that represents the circumference of a circle whose radius is *r*. (The circumference of a circle is π times twice its radius, where $\pi \approx 3.14$.) Then use the expression to find the circumference of each circle. $C = 2\pi r$

a. 12.56

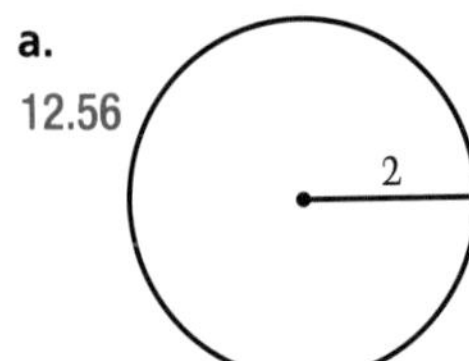

b. 18.84

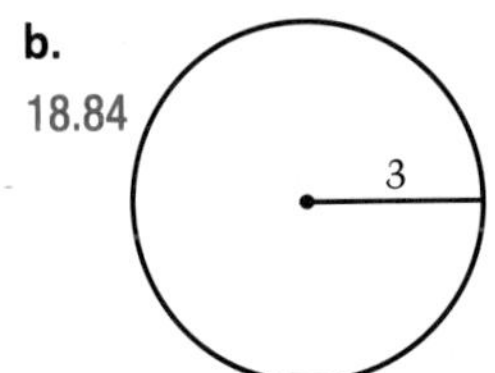

c. 25.12

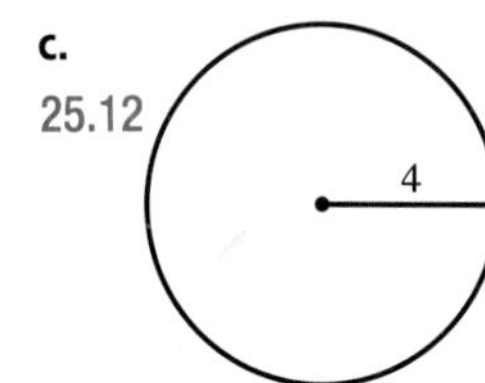

45. In Exercise 43, as the sides increase by one unit each, how does the perimeter change? Increases by 3 units
46. In Exercise 44, as each radius increases by one unit, how does the circumference change? Increases by 2π, or about 6.28

Exploration and Extension

47. *Drink-Boxes* Drink-box consumption in the United States increased in the years 1983 to 1990. An equation that approximates the millions of drink-boxes sold, D, in terms of the year, t, is

$D = 270t - 200$ 610, 880, 1150, 1420, 1690, 1960, 2230, 2500

where $t = 3$ corresponds to 1983. Use this equation to complete the table. ***(Source: Nielsen ScanTrack, Tetra Pak, Inc.)***

Year	1983	1984	1985	1986	1987	1988	1989	1990
t	3	4	5	6	7	8	9	10
D	?	?	?	?	?	?	?	?

48. Use the information in Exercise 47 to construct a bar graph showing the number of drink-boxes sold in the United States from 1983 to 1990. Describe the pattern formed by the tops of the bars.

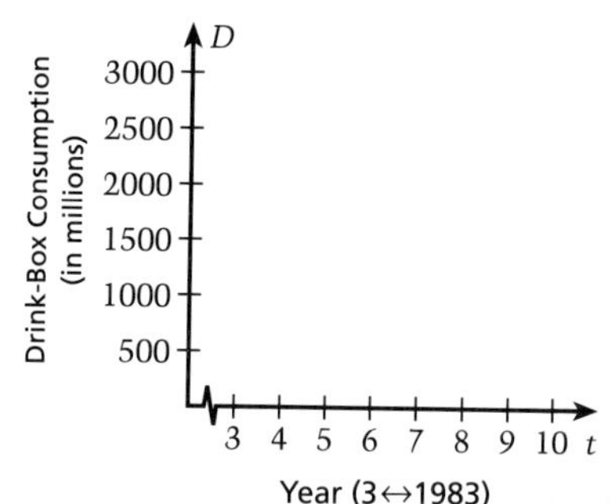

49. In 1990, the population of the United States was about 250 million. Find the average number of drink-boxes that Americans consumed *per person* in 1990. 10

50. Did the per person consumption of drink-boxes increase from 1983 to 1990? Why do you think this was so? See margin.

48. See margin. Each year the consumption increases by the same amount.

Mixed REVIEW

NOTE: Numbers in parentheses refer to lessons.
15. $x + x + x + x$, or $4x$

1. Evaluate $(17 - 9) + 28$. **(1.1)** 36
2. Write the sum as a fraction: $\frac{2}{3}$ and 3. $\frac{11}{3}$
3. What is 10% of \$245? \$24.50
4. Which is greater, $\frac{1}{2}$ or 0.75? 0.75
5. What is the product of 9 and 4? **(1.1)** 36
6. What is the difference of 13 and 5? **(1.1)** 8
7. Evaluate $9 \div [3 + (11 - 5)]$. **(1.1)** 1
8. Evaluate $[(7 \cdot 3) - 2] + (8 \div 4)$. **(1.1)** 21
9. What is the quotient of 21 and 7? **(1.1)** 3
10. Evaluate $x - 2y$ when $x = 14$ and $y = 3$. **(1.2)** 8
11. Evaluate $3x + 1$ when $x = 2$. **(1.2)** 7
12. Evaluate $3x + 2$ when $x = 2$. **(1.2)** 8
13. Evaluate $\frac{1}{2}(x - 7)$ when $x = 11$. **(1.2)** 2
14. Evaluate $\frac{x}{2} + \frac{x}{4}$ when $x = 1$. **(1.2)** $\frac{3}{4}$
15. Write an expression for the perimeter of a square with sides of length x. **(1.2)**
16. Find the perimeter of a rectangular field 103 feet long and 210 feet wide. **(1.2)** 626 ft
17. Use a calculator to evaluate $1.4x + 7.4$ when $x = 3.2$. **(1.2)** 11.88
18. What is the average height: 72 in., 68 in., 60 in., 76 in., 64 in.? 68 in.
19. What is the value of $\frac{1}{2}(2a - 3b)$ when $a = 9$ and $b = 4$? **(1.2)** 3

Ex. 47–50 In Exercise 47, the equation relates the number of drink boxes sold, D, to the year in which they are sold, t. Because t is to have values of 3, 4, 5, etc., you say that $t = 3$ corresponds to 1983. The alternative is to use the equation $D = 270(t–1980)–200$ and to let $t = 1983, 1984, 1985$, etc. This technique for representing years such as 1980, 1981, 1982, etc., by numbers such as 0, 1, 2, etc. will be used throughout the book. You may find it worthwhile to spend some time on this notation. In order to check their understanding, have students discuss the reasons for it.

Expressions like $D = 270t - 200$ are called equations and not formulas, because the relationship between D and t will probably vary for a different collection of years.

Answers

48.

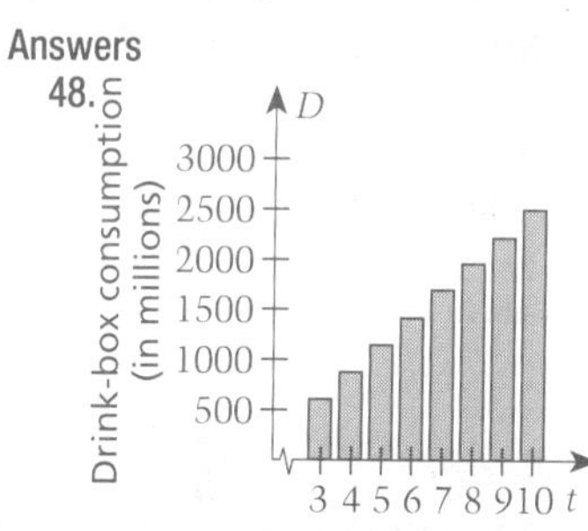

50. Increased consumption of drink boxes might not mean that the average rate of consumption *per person* increased. The per-person increase occurred because the rate of increase of drink box consumption was greater than the rate of population growth in the same time period.

Mixed Review exercises help students to check their understanding of previous lessons and prior courses. Answers to the odd-numbered exercises will be found in the "Selected Answers" section of the student text.

Problem of the Day

Draw two lines to separate this figure into three parts that have the same size and same shape.

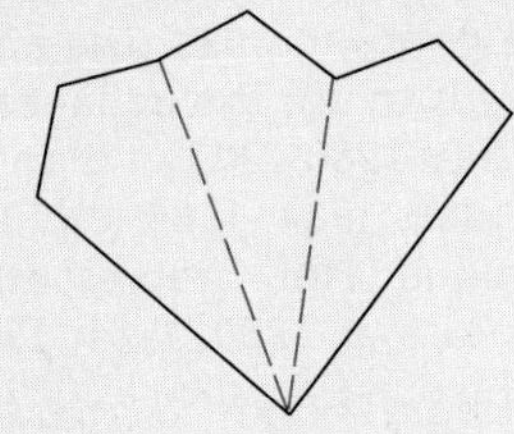

ORGANIZER

Warm-Up Exercises

1. Identify the factors in each expression.
   a. $23x$ b. $(3 + r)y$ c. $13tz$
   a. 23 and x; b. $(3 + r)$ and y; c. 13, t, and z are the three factors.
2. How many different factors are there in this expression? $4 \cdot 4 \cdot 4 \cdot 4 \cdot 7 \cdot 7$ 2
3. How many factors of 3 are there in this expression? $2 \cdot 2 \cdot 2 \cdot 3 \cdot 3 \cdot 3 \cdot 3 \cdot 3 \cdot 5$ 5
4. Evaluate the expression in Exercise 3 above. 9720

Lesson Resources

Teaching Tools
- Transparencies: 1, 2
- Problem of the Day: 1.3
- Warm-up Exercises: 1.3
- Answer Masters: 1.3

Extra Practice: 1.3

Common-Error ALERT!

In **Examples 1–2,** monitor student computations. Frequently, students evaluate 5^3 as $5 \cdot 3$. Emphasize that exponents indicate how many times to repeat a given factor.

1.3 Exponents and Powers

What you should learn:

Goal 1 How to evaluate expressions containing exponents

Goal 2 How to use exponents in real-life problems

Why you should learn it:

You can use exponents to simplify writing in algebra as in the formula for the area of a square, $A = s^2$. You can also use exponents to simplify computation by using the x^y key on your calculator.

Goal 1 Expressions Containing Exponents

Exponents can be used to represent repeated multiplication. For instance, the expression 4^6 represents the number that you obtain when 4 is used as a factor 6 times.

$$4^6 = \underbrace{4 \cdot 4 \cdot 4 \cdot 4 \cdot 4 \cdot 4}_{\text{6 factors of 4}}$$

The number 4 is the **base,** the number 6 is the **exponent,** and the number 4^6 is a **power.** The exponent represents the number of times the base is used as a factor. Exponential forms are read as follows.

7^2 is read as *seven to the second power* or as *seven squared.*
5^3 is read as *five to the third power* or as *five cubed.*
4^6 is read as *four to the sixth power.*

For a number raised to the *first power,* the exponent 1 is not usually written. For instance, simply write 5^1 as 5.

Example 1 *Using an Exponent to Represent Repeated Multiplication*

Use an exponent to represent the expression $2 \cdot 2 \cdot 2 \cdot 2 \cdot 2$.

Solution Because the number 2 is used as a factor 5 times, you know that the base is 2 and the exponent is 5. Thus, this expression can be written as

$$2 \cdot 2 \cdot 2 \cdot 2 \cdot 2 = 2^5.$$

Simplified, the value of this expression is 32. ■

Example 2 *Evaluating an Algebraic Expression Involving an Exponent*

Evaluate the expression x^3 when $x = 5$.

Solution

$x^3 = 5^3$ *Substitute 5 for x.*
$= 5 \cdot 5 \cdot 5$
$= 125$ *Value of the expression* ■

Example 3 Evaluating an Algebraic Expression Involving an Exponent

Evaluate the expression $(3a + 6b)^2$ when $a = 1$ and $b = 2$.

Solution

$$
\begin{aligned}
(3a + 6b)^2 &= [3(1) + 6(2)]^2 && \textit{Substitute 1 for a and 2 for b.}\\
&= [3 + 12]^2 && \textit{Simplify 3(1) and 6(2).}\\
&= 15^2 && \textit{Simplify 3 + 12.}\\
&= 15 \cdot 15 &&\\
&= 225 && \textit{Value of the expression}
\end{aligned}
$$

■

Example 4 Calculators and Exponents

Evaluate the expression 6^7.

Solution The expression

$$6^7 = \underbrace{6 \cdot 6 \cdot 6 \cdot 6 \cdot 6 \cdot 6 \cdot 6}_{\text{7 factors of 6}}$$

is difficult to evaluate with pencil and paper. The work is easier if you have a calculator with an exponent key. This key is often labeled as [x^y], as [y^x], or as [^]. If your calculator has one of these keys, find out how the key works. The correct result is

$$6^7 = 279{,}936.$$

■

An exponent applies only to the number, variable, or expression that is immediately to its left. In the expression $2x^3$, the base is x, not $2x$. In the expression $(2x)^3$, the base is $2x$.

Example 5 Exponents and Grouping Symbols

Evaluate $2x^3$ and $(2x)^3$ when $x = 4$.

Solution

a.
$$
\begin{aligned}
2x^3 &= 2(4^3) && \textit{Substitute 4 for x.}\\
&= 2(64) && \textit{Simplify } 4^3.\\
&= 128 && \textit{Simplify 2(64).}
\end{aligned}
$$

b.
$$
\begin{aligned}
(2x)^3 &= (2 \cdot 4)^3 && \textit{Substitute 4 for x.}\\
&= (8)^3 && \textit{Simplify } 2 \cdot 4.\\
&= 512 && \textit{Simplify } 8^3.
\end{aligned}
$$

■

LESSON Notes

Example 3

Remind students that they must simplify the expression inside parentheses before they apply the exponent.

Example 5

Later in the text, when negative integers are introduced, students will need to distinguish between -2^4 and $(-2)^4$. For now, emphasize that the exponent controls only the base value immediately to its left unless parentheses or other grouping symbols are used.

Extra Examples

Here are additional examples similar to **Examples 1–5.**

1. Express $3 \cdot 3 \cdot 3 \cdot 3 \cdot 3 \cdot 3$ and $7 \cdot 7 \cdot 7 \cdot 7$ using exponents. 3^6 and 7^4
2. Evaluate m^4 when $m = 3$ and t^5 when $t = 4$. 81; 1024
3. Evaluate $(2a + 4b)^2$ when $a = 2$ and $b = 3$. 256
4. Use a calculator to compute 3^8. 6561
5. Evaluate when $x = 2$.
   a. $4x^2$ and $(4x)^2$ 16; 64
   b. $3x^4$ and $(3x)^4$ 48; 1296

Example 6

It may be helpful for students if you display an actual cube before writing the formula for its surface area. Give students time to compute the area of each face of the cube and then add to find its area. Discuss the possibilities for the formula in class.

Check Understanding

1. Explain how using exponents is related to using multiplication.
   Exponents indicate how many times to repeat a given factor.
2. Here are two expressions.

   $$(\underset{\substack{\uparrow\\1}}{3a} \underset{\substack{\uparrow\\3}}{+} \underset{\substack{\uparrow\\2}}{2b}\underset{\substack{\uparrow\\4}}{)^3} \quad \text{and} \quad (2\underset{\substack{\uparrow\\1}}{x}\underset{\substack{\uparrow\\2}}{)^3},$$

   Indicate which operation is performed first, second, etc. Justify your responses by substituting numerical values for a, b, and x in each expression.

Communicating about ALGEBRA

Ask students the following.

1. Why might it be important to know the surface area of Epcot Center?
   Answers may relate to making repairs, painting the surface, or lighting the interior.
2. What happens to the surface area of a sphere when its radius is doubled?
   It is quadrupled.

Common-Error ALERT!

In **Exercises 11–15,** caution students against reading

$a \cdot a \cdot a \cdot a \cdot a \cdot a$

as $6a$! Remind them that six factors of a is not written $6a$, but a^6. Use a numerical example to illustrate.

$2 \cdot 2 \cdot 2 \cdot 2 \cdot 2 \cdot 2 = 64$, not $6 \cdot 2 = 12$!

Goal 2 Using Exponents

Many formulas for area and volume involve exponents. In fact, the words *squared* and *cubed* come from the formulas for the area of a square, $A = s^2$, and the volume of a cube, $V = s^3$. (In these formulas, s is the length of each side of the square or each edge of the cube.)

Connections
Geometry

Example 6 Surface Area

Use exponents to write a formula for the surface area of a cube. Then use the formula to find the surface area of a cube whose edges are each 3 inches.

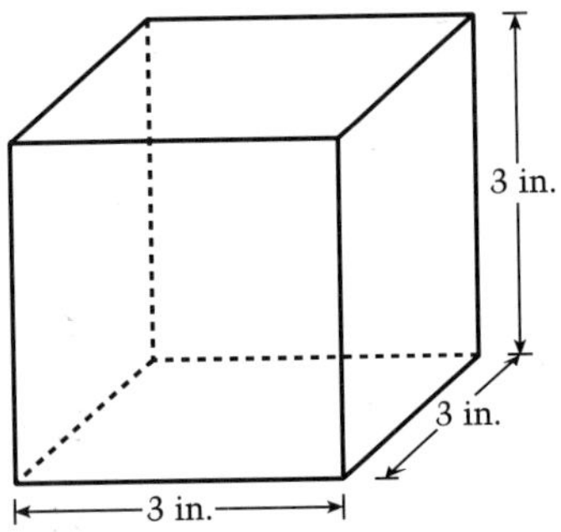

Solution Let s represent the length of each edge of the cube. Each of the six faces of the cube has an area of s^2, which means that the surface area, S, of the cube is $S = 6s^2$. Using this formula, you can find the surface area of a cube with 3-inch edges as follows.

$S = 6s^2$	*Formula for surface area of cube*
$= 6(3^2)$	*Substitute 3 for s.*
$= 6(9)$	*Simplify 3^2.*
$= 54$	*Simplify 6(9).*

Thus, the cube has a surface area of 54 square inches. (Note that this can be written as 54 in.^2) ■

Communicating about ALGEBRA

SHARING IDEAS about the Lesson

What Process to Use The surface area, S, of a sphere is

$$S = 4\pi r^2$$

where r is the radius of the sphere and $\pi \approx 3.14$.

A. A major feature of Epcot Center is called Spaceship Earth. The building is shaped like a sphere and has a diameter of 165 feet. Find the approximate surface area of the building. ***(Source: Walt Disney World)*** 85,486.5 ft^2

B. Earth has a radius of approximately 6370 kilometers. Approximate the surface area of Earth. 509,645,864 km^2

C. Land forms 29.2% of the total surface area of Earth. Approximate the area of Earth's land. 148,816,592 km^2

D. Approximate the area of Earth's oceans. 360,829,272 km^2

EXERCISES

Guided Practice

CRITICAL THINKING about the Lesson

1. Express 18^3 in words (two different ways). 1. Eighteen to the third power, eighteen cubed
2. In the expression 18^3, what is 18 called? Base
3. In the expression 18^3, what is 3 called? 3. Exponent
4. Use a calculator to evaluate 18^3. 5832
5. Explain why x^2 is called *x squared.* x^2 is the area of a square with side x.
6. Explain why y^3 is called *y cubed.* y^3 is the volume of a cube with edge y.

Independent Practice

In Exercises 7–15, write the expression in exponential form.

7. two cubed 2^3
8. x squared x^2
9. nine to the y power 9^y
10. y to the ninth power y^9
11. $3 \cdot 3 \cdot 3 \cdot 3$ 3^4
12. $y \cdot y$ y^2
13. $a \cdot a \cdot a \cdot a \cdot a \cdot a$ a^6
14. $4 \cdot 4 \cdot 4$ 4^3
15. $2x \cdot 2x \cdot 2x$ $(2x)^3$

In Exercises 16–25, evaluate the expression.

16. y^2, when $y = 9$ 81
17. a^4, when $a = 3$ 81
18. $a^2 - 3$, when $a = 3$ 6
19. $16 + x^3$, when $x = 2$ 24
20. $(x + y)^2$, when $x = 3$ and $y = 4$ 49
21. $x^2 + y^2$, when $x = 2$ and $y = 6$ 40
22. $(2a)^2 - b$, when $a = 4$ and $b = 10$ 54
23. $(6x + y)^2$, when $x = 1$ and $y = 4$ 100
24. $3r^2 + 6s$, when $r = 2$ and $s = 3$ 30
25. $4x - 3y^3$, when $x = 7$ and $y = 2$ 4

Technology **In Exercises 26–29, use a calculator to evaluate the expression. Use estimation to check your results.**

26. 9^5 59,049
27. 13^4 28,561
28. 5^9 1,953,125
29. 3^{11} 177,147
30. Find the area of a square with 8-inch sides. 64 in.^{2}
31. Find the volume of a cube with 11-centimeter edges. 1331 in.^{3}
32. *Giraffe Exhibit* The giraffe exhibit at a zoo is circular and bordered by a fence. The distance from the center of the exhibit to the fence is 60 feet. How much area do the giraffes have? (The area of a circle is $A = \pi r^2$, where $\pi \approx 3.14$ and r is the radius.) 11,304 ft^2
33. *Tennis Ball* A tennis ball has a diameter of 6.67 centimeters. What is the surface area of the ball? 139.7 cm^2
34. *Compound Interest* \$1000 is deposited in a savings account that is compounded annually at 6%. The balance, A, in the account after t years is $A = 1000(1 + 0.06)^t$. Find the balance after 1 year, 10 years, and 20 years. \$1060.00, \$1790.85, \$3207.14

EXERCISE Notes

ASSIGNMENT GUIDE

Basic/Average: Ex. 1–6, 7–15, 17–45 odd

Above Average: Ex. 1–6, 7–23 odd, 24, 25–35 odd, 41–42 odd, 43–46

Advanced: Ex. 1–6, 7–27 odd, 32–36, 41–46

Selected Answers
Exercises 1–6, 7–41 odd

More Difficult Exercises
Exercises 34–35, 41–42, 44, 46

Guided Practice

Ex. 1–6 These exercises provide a concise review of the key elements of the lesson.

Independent Practice

EXTEND Ex. 16–25 PROBLEM SOLVING Have students write verbal phrases for the expressions in Exercises 16–25. For example, $(6x + y)^2$ can be written as "the square of the sum of 6x and y."

Ex. 26–29, 37–40 TECHNOLOGY Review with students how to use the *power keys*, x^y, y^x, or ^ on their calculators.

Ex. 32 PROBLEM SOLVING Students could draw a diagram of the giraffe exhibit and fence.

Ex. 33 Refer students to the formula for the surface area of a sphere on page 16: $S = 4\pi r^2$.

Ex. 34 Round the dollar amounts up to the nearest cent.

EXTEND Ex. 35 RESEARCH Ask students if they ever heard of the Great Pyramid of Egypt in Africa, one of the seven wonders of the world. Suggest they begin with the encyclopedia to find information about this or any other Egyptian pyramid.

▶ **Ex. 44 PROBLEM SOLVING** This is a fun problem for students to do.

▶ **Ex. 45–46 PROBLEM SOLVING** Tell students that the powers of 7 are the products obtained from raising 7 to successive values of the exponents. These exercises show the importance of the problem-solving strategies—Make a Table and Look for a Pattern.

Enrichment Activities

These activities require students to go beyond lesson goals.

1. The volume of a sphere of radius r is $\frac{4}{3}\pi r^3$. To two decimal places, find the volume of the sphere when given the following radii. Use calculators if available.
   a. 4 b. 2 c. 6 d. 9
   a. 268.08; b. 33.51; c. 904.78; d. 3053.63
2. Evaluate the variable expressions when $a = 3$ and $b = 5$.
   a. $(4a - 2b)^2$ 4
   b. $(4ab)^3$ 216,000
   c. $(a^2 + b^2)^2$ 1156
3. Evaluate 1^{13} and 0^{24}. 1; 0 Explain your results.
   Any number times 1 is that number; any number times 0 is 0.

The Louvre pyramid in Paris has a height of 20.6 meters and a square base with sides of 35 meters.

✪ **35.** *Volume of a Pyramid* The volume of a pyramid is one third the height times the area of the base. What is the volume of the Louvre pyramid? ≈8411.67 m³

✪ **36.** *Volume of a Safe* A fireproof safe has a cubical storage space inside. What is the volume of the storage space if each interior dimension is 12 inches? 1728 in.³

Integrated Review

Technology **In Exercises 37–40, use a calculator to evaluate the expression. Round your result to two decimal places. Use estimation to check your answer.**

37. 1.6^5 10.49 **38.** 3.9^3 59.32 **39.** 6.9^2 47.61 **40.** 4.3^4 341.88

✪ **41.** What value of x would make the expression $x^2 + 1$ equal to 10? 3

✪ **42.** What value of x would make the expression $x^3 - 3$ equal to 5? 2

Exploration and Extension

43. *Preparation Time* In an English class, the suggested amount of time to spend on each paper triples for each new paper. If $1\frac{1}{2}$ hours was the suggested time for the first paper, how much time should be planned for the third paper? $13\frac{1}{2}$ hours

✪ **44.** *What an Allowance!* If a child's one-penny allowance doubles every day for 30 days, how much is the allowance on the 30th day? (Day 1: 1¢, Day 2: 2¢, Day 3: 4¢, and so on) $5,368,709.12

45. Copy and complete the table on the right by evaluating the powers of 9. What pattern do you see for the last digit of each product?

9^1	9^2	9^3	9^4	9^5	9^6	9^7	9^8
?	?	?	?	?	?	?	?

9; 81; 729; 6561; 59,049; 531,441; 4,782,969; 43,046,721. 9, 1.

✪ **46.** Try the experiment described in Exercise 45 again, first with powers of 8 and then with powers of 7. 8, 4, 2, 6. 7, 9, 3, 1.

PROGRAMMING A CALCULATOR

Programmable calculators and computers can be instructed to evaluate an expression. The list of instructions is a **program,** *and each separate instruction is a* **program step.** *When the calculator or computer performs the program steps, it is* executing *or* running *the program.*

Here are examples of two *flowcharts* for a program that will evaluate the algebraic expression $x^2 + 2x$ for different values of x.

Programmable Calculator

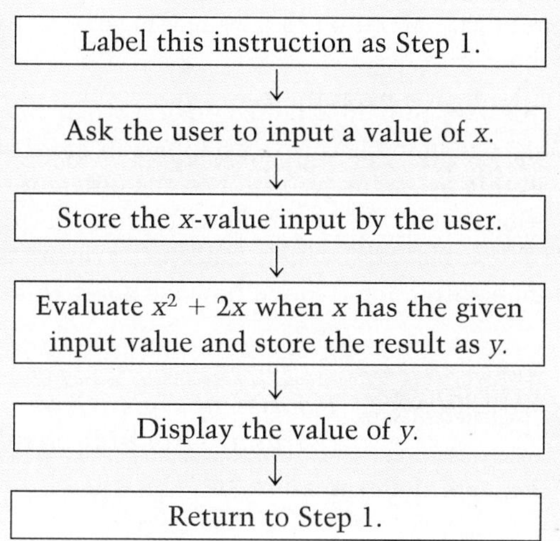
Label this instruction as Step 1.
Ask the user to input a value of x.
Store the x-value input by the user.
Evaluate x² + 2x when x has the given input value and store the result as y.
Display the value of y.
Return to Step 1.

Computer Program

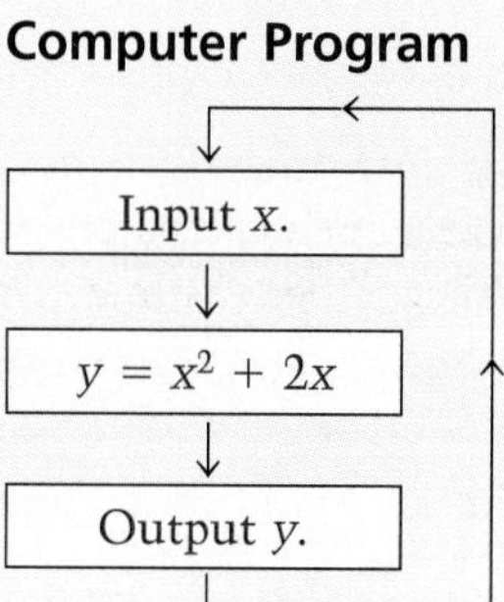
Input x.
y = x² + 2x
Output y.

Sample program steps for the *TI-82,* Casio *fx-9700GE,* and Sharp *EL-9300C* are listed on page 734. When you run the program, enter the x-value when prompted, and the calculator will display the value of the expression $x^2 + 2x$. For instance, the screen at the right shows the value of $x^2 + 2x$ when $x = 3$.

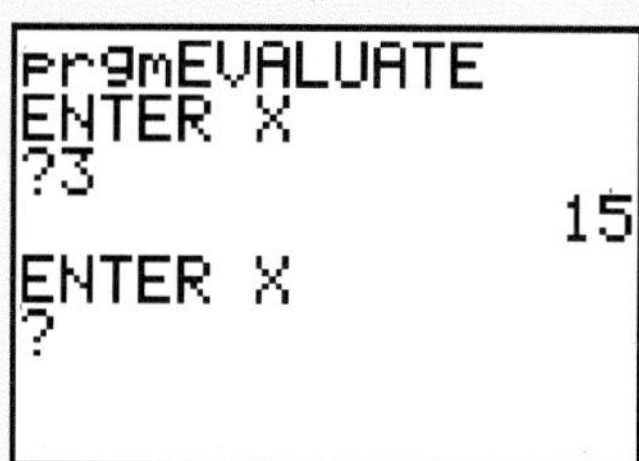
prgmEVALUATE
ENTER X
?3
15
ENTER X
?

Exercises

Use a programmable calculator or a computer to evaluate the following expressions when $x = 0$, $x = 1$, $x = 2$, and $x = 5.36$. (Round each result to two decimal places.)

1. $x^2 + 2x$
0, 3, 8, 39.45

2. $60 - 9x$
60, 51, 42, 11.76

3. $0.5x + 10$
10, 10.5, 11, 12.68

4. $\frac{1}{3}x$
0, 0.33, 0.67, 1.79

5. x^2
0, 1, 4, 28.73

6. $x^2 + 1$
1, 2, 5, 29.73

7. $3x^2 + 8x$
0, 11, 28, 129.07

8. $5x^2$
0, 5, 20, 143.65

TECHNOLOGY Four types of technology will be featured throughout the text: Texas Instrument's TI-82 graphing calculator, Casio's *fx*-9700GE graphing calculator, Sharp's EL-9300C graphing calculator, and BASIC computer programming. Technology can play an important role in learning algebra skills and concepts. In the textbook are uses of technology that enhance the study of algebra.

Students may use a programmable calculator or computer to reconsider Exercise 34 in Exercise Set 1.3. The expression $1000(1 + 0.06)^t$ can be quickly evaluated for different values of years (t) and different interest rates (r). A second program, in which the interest rate r (now at 6%) may be varied for a fixed number of years, t, would be $1000(1 + r)^t$. Students can now ask "what if" questions about the balance in an account for different interest rates and different numbers of years. Note: Compound interest will be highlighted in Chapter 8 of the text.

Problem of the Day

Which is greater? How did you know?

a. 4^3 or 2^7 — 2^7

b. 4^{15} or 2^{29} — 4^{15}

a. $4^3 = 4 \cdot 4 \cdot 4 = (2 \cdot 2)(2 \cdot 2)(2 \cdot 2) = 2^6$; $2^6 < 2^7$

b. $4^{15} = 4 \cdot 4 \cdot \ldots \cdot 4 \cdot 4 = (2 \cdot 2)(2 \cdot 2) \ldots (2 \cdot 2)(2 \cdot 2) = 2^{30}$; $2^{30} > 2^{29}$

ORGANIZER

Warm-Up Exercises

Practice your listening skills. (Students should listen carefully and perform mental computations as you read these exercises aloud.)

1. $3(5)^2$ (read, "three times the quantity of five squared") 75
2. $4(15 - 3)$ ("four times the difference of fifteen and three") 48
3. $24 + (27 \div 9)$ ("twenty-four plus the quotient of twenty-seven and nine") 27
4. $(32 - 16) + (4 - 2)^2$ ("the sum of the difference of thirty-two and sixteen and the square of four minus two") 20

Lesson Resources

Teaching Tools
- Problem of the Day: 1.4
- Warm-up Exercises: 1.4
- Answer Masters: 1.4

Extra Practice: 1.4

Color Transparency: 2

LESSON Notes

Example 1

Point out that no grouping symbols are used in the expression in this example.

1.4 Order of Operations

What you should learn:

Goal 1 How to use an established order of operations to evaluate algebraic expressions

Goal 2 How to evaluate expressions with a calculator

Why you should learn it:

You should use the order of operations that everyone agrees on so that you get the same results as everyone else. You should also know how to operate your calculator so that it performs operations in the desired order.

Goal 1 Using the Order of Operations

One of your goals as you study this book is to learn to communicate about algebra by reading and writing information about numbers.

One way to help avoid confusion when we are communicating algebraic ideas is to establish an **order of operations.** We do this by giving priorities to the different operations. First priority is given to exponents, second priority is given to multiplication and division, and third priority is given to addition and subtraction. Notice how we apply this order in Example 1.

Example 1 ***Priority of Operations (without Grouping Symbols)***

Evaluate the expression $3x^2 + 1$ when $x = 4$.

Solution

$$3x^2 + 1 = 3 \cdot 4^2 + 1 \quad \text{Substitute 4 for } x.$$
$$= 3 \cdot 16 + 1 \quad \text{Simplify } 4^2.$$
$$= 48 + 1 \quad \text{Simplify } 3 \cdot 16.$$
$$= 49 \quad \text{Simplify } 48 + 1.$$

Notice the order of the three operations: Raise to a power first, multiply second, and add third. ■

When you want to change the established order of operations for an expression, you *must* use parentheses or other grouping symbols. Notice in Example 2 how using grouping symbols in $3 \cdot 4^2 + 1$ gives expressions that have different values.

Example 2 ***Priority of Operations (with Grouping Symbols)***

a. $(3 \cdot 4)^2 + 1 = 12^2 + 1 = 144 + 1 = 145$

b. $3(4^2 + 1) = 3(16 + 1) = 3(17) = 51$

c. $3 \cdot 4^{(2+1)} = 3 \cdot 4^3 = 3 \cdot 64 = 192$

d. $(3 \cdot 4)^{(2+1)} = 12^3 = 1728$ ■

Operations that have the same priority, such as multiplication and division *or* addition and subtraction, are performed from left to right (the Left-to-Right Rule).

Example 3 *Using the Left-to-Right Rule*

a. $24 - 8 - 6 = (24 - 8) - 6$ *Work from left to right.*
$= 16 - 6$
$= 10$

b. $15 - 2 + 8 = (15 - 2) + 8$ *Work from left to right.*
$= 13 + 8$
$= 21$

c. $16 + 4 \div 2 - 3 = 16 + (4 \div 2) - 3$ *Divide before adding.*
$= 16 + 2 - 3$
$= (16 + 2) - 3$ *Work from left to right.*
$= 18 - 3$
$= 15$ ■

Remember that the Left-to-Right Rule applies only to operations having the same priority. For Part **c**, we do not perform the operations from left to right because division has a higher priority than addition. On page 4, you saw how parentheses and brackets can be used to define the order in which operations should be performed. Working from left to right would have been correct in Part **c** only if "16 + 4" had been enclosed in parentheses. If you are writing an expression and want to make sure that you will not be misunderstood, you can use grouping symbols. For instance, to make sure that $24 \div 4 \div 2$ is correctly evaluated, it would be better to write $(24 \div 4) \div 2$.

We summarize the Order of Operations in the following list.

Order of Operations

To evaluate an expression involving more than one operation, use the following order.

1. First do operations that occur within symbols of grouping.
2. Then evaluate powers.
3. Then do multiplications and divisions from left to right.
4. Finally do additions and subtractions from left to right.

Example 2

Grouping symbols can always be used in writing expressions to avoid confusing the order of operations.

Ask each student to evaluate the following expression.

$42 \div 3 + 3^2 \cdot (12 - 8) + 3$ 53

Call on students for their answers. Record student results on the overhead or chalkboard for the entire class to see. It is likely that there will be different results because students followed different orders of operations. The need to have one order of operations can now be discussed.

Example 3

Remind students that the left-to-right rule applies only to operations of the same priority. In such cases, perform the operations as they are encountered, left to right. An easy way to recall the order of operations is to use the phrase "Please Excuse My Dear Aunt Sally." The first letter of each word stands for an operation in the order of priority: ***P***arentheses and ***E***xponents, ***M***ultiplication and ***D***ivision, and ***A***ddition and ***S***ubtraction.

Extra Examples

Here are additional examples similar to **Example 2.**

Evaluate these expressions.

1. $23 - [(45 \div 15)^3 \div 9]^2$ 14
2. $(8 + [(84 \div 12) - 4]^2 - 7) + 3$ 13

Check Understanding

1. State the established order of operations. See the summary on page 21.
2. What is the acronym that helps you remember the order of operations? PEMDAS

Example 4

Have students investigate the order in which their calculators perform operations. These expressions can be used to help students in their investigation.

1. $18 + 6 \div 2 - 7 \cdot 2$
   7, established (built-in) order; 10, left-right order
2. $24 - 6 \cdot 3 + 15 \div 3$
   11, established order; 23, left-right order

Caution students to use parentheses appropriately if their calculators do not follow the established order.

Example 5

Stress that $27.17 is an unreasonable answer because sales tax is never nearly half of the total purchase! Ask students for other reasons why the sales tax seems excessive. The sales clerk must have made an error to get such an answer.

Communicating about ALGEBRA

PROBLEM SOLVING Students could use the Guess and Check strategy. Guesses should be organized in a table or list. Have students read the expressions that they found to their partners, who should then write what they hear. After each expression has been read (by each partner), students should compare what they wrote with what their partner read. Discuss any discrepancies.

Common-Error ALERT!

The error found in **Example 5** is a classic violation of the established order of operations—performing the operations from left to right in sequence.

Goal 2 Evaluating Expressions with a Calculator

Many calculators use the Order of Operations used in this book, but some do not. Try using your calculator to see whether it gives a result that is listed in the next example.

Connections
Technology

Example 4 *Using a Calculator*

When you enter 4 [+] 6 [÷] 2 [−] 1 [=] on your calculator, does it display 6 or 4? Using the established order of operations, it should display 6.

$$4 + (6 \div 2) - 1 = 4 + 3 - 1 = 6$$

If it displays 4, it would have performed the operations as follows: $[(4 + 6) \div 2] - 1$. ■

Real Life
Retail Sales

Example 5 *Calculating Sales Tax*

Suppose you live in a state that charges 6% sales tax on clothes. You find two sweaters that you want to buy. One sweater costs $24.95 and the other costs $36.95. The salesclerk uses a calculator to compute your sales tax as follows.

24.95 [+] 36.95 [×] .06 [=]

From the calculator display, the salesclerk tells you the sales tax is $27.17. You know this is too much. What went wrong?

Solution The calculator used the Order of Operations.

$$24.95 + (36.95)(0.06) \approx 24.95 + 2.22 = 27.17$$

The correct sales tax of $3.71 is obtained by using the following order of operations.

$$(24.95 + 36.95)(0.06) = (61.90)(0.06) \approx 3.71$$ ■

Communicating about ALGEBRA

SHARING IDEAS about the Lesson

Systematic Guess and Check List the different values that can be obtained from each of the following expressions by performing the operations in different orders. Which value is correct using the established order of operations?

A. $100 - 50 - 10$ 40, 60; 40 **B.** $4 + 3 \cdot 5 - 2$ 33, 17, 13, 21; 17

C. $18 - 6 + 5 - 4$ 13, 3, 11; 13 **D.** $24 \div 4 \div 2$ 3, 12; 3

EXERCISES

Guided Practice

CRITICAL THINKING about the Lesson

1. If an expression without grouping symbols includes addition and an exponent, which operation is performed first? Exponent operation

2. If an expression without grouping symbols includes multiplication and division, which operation is performed first? Leftmost operation

3. For the expression $4 + 16 \div 8 - 2$, where could you put grouping symbols to make sure the expression is correctly evaluated? $4 + (16 \div 8) - 2$

4. What are the values of $3 + (4 \cdot 9)$ and $(3 + 4) \cdot 9$? Using the Order of Operations, which of these is equal to $3 + 4 \cdot 9$? 39, 63, $3 + (4 \cdot 9)$

Independent Practice

In Exercises 5–14, evaluate the expression.

5. $4 + 2 - 1$ 5
6. $3 \cdot 1 + 4$ 7
7. $8 \div 4 + 4 \cdot 2$ 10
8. $5 + 8 \cdot 3 + 2$ 31
9. $14 \div 7 + 3^2$ 11
10. $2 \cdot 3^2 - 9$ 9
11. $10 \cdot (3 + 1) - 16$ 24
12. $13 + (3 \cdot 2)^2 - 8$ 41
13. $10 - 3 \cdot 2$ 4
14. $15 - 2 \cdot 2$ 11
15. Which is correct? b

    a. $\frac{8 \cdot 3}{4 + 3^2 - 1} = 8 \cdot 3 \div 4 + 3^2 - 1$

    b. $\frac{8 \cdot 3}{4 + 3^2 - 1} = 8 \cdot 3 \div (4 + 3^2 - 1)$

16. Which is correct? b

    a. $\frac{(4 - 2)^2 + 5}{3} = (4 - 2)^2 + 5 \div 3$

    b. $\frac{(4 - 2)^2 + 5}{3} = [(4 - 2)^2 + 5] \div 3$

In Exercises 17–24, evaluate the expression.

17. $2 + 2x^3$, when $x = 2$ 18
18. $16 - x^2 \div 12$, when $x = 6$ 13
19. $x^2 + 2x + 1$, when $x = 5$ 36
20. $x^3 - 4x + 9$, when $x = 4$ 57
21. $2l + 2w$, when $l = 3$ and $w = 19$ 44
22. $4r^2 - s$, when $r = 4$ and $s = 14$ 50
23. $\frac{2x + y}{7}$, when $x = 12$ and $y = 11$ 5
24. $6 + \frac{x}{y} \cdot 4$, when $x = 8$ and $y = 4$ 14

Technology **In Exercises 25–30, two calculators were used to evaluate the expression. They gave different results. Which calculator used the established order of operations? Rewrite the calculator steps with grouping symbols so that both calculators will give the correct result.**

25. 16 [−] [(4 [÷] 2) [×] 5] [=]
Calc. #1 Calculator 1: 6, Calculator 2: 30

26. 15 [−] (6 [÷] 3) [−] 2 [=]
Calc. #2 Calculator 1: 1, Calculator 2: 11

27. 10 [+] (15 [÷] 5) [+] 16 [=]
Calc. #2 Calculator 1: 21, Calculator 2: 29

28. (4 [×] 3) [+] (6 [÷] 2) [=]
Calc. #1 Calculator 1: 15, Calculator 2: 9

✪ 29. (12 [÷] 3) [+] (16 [÷] 4) [=]
Calc. #1 Calculator 1: 8, Calculator 2: 5

✪ 30. 8 [+] 22 [−] (15 [÷] 5) [=]
Calc. #2 Calculator 1: 3, Calculator 2: 27

EXERCISE Notes

ASSIGNMENT GUIDE

Basic/Average: Ex. 1–4, 5–39 odd, 41–45 odd

Above Average: Ex. 1–4, 17–39 odd, 40–45

Advanced: Ex. 1–4, 17–29 odd, 31–35, 39–45

Selected Answers
Exercises 1–4, 5–43 odd

✪ More Difficult Exercises
Exercises 29, 30, 40–44

Guided Practice

The Guided Practice exercises provide a concise review of the key elements of the lesson. For each of these exercises, allow students to reflect on the responses of their classmates before going to the next exercise.

▶ **Ex. 1** Which operation first? It depends! If the expression is 3^{3+1}, then the sum 3 + 1 must be added first to get 3^4 because there is an implied grouping of the exponents. However, if the addition operation is not in the exponent, then you would apply the exponent before performing addition.

Independent Practice

▶ **Ex. 15–16** TECHNOLOGY Remind students that the division bar is a grouping symbol that separates everything in the numerator from everything in the denominator. After students have computed with paper and pencil, have them compute these exercises using their calculators.

▶ **Ex. 25–30** COOPERATIVE LEARNING You might have students work in cooperative groups to solve these exercises.

▶ **Ex. 33–34 PROBLEM SOLVING** Encourage students to begin with a verbal model of the problem and to use labels. Help students move from the verbal models to the algebraic models. For example, in Exercise 33, recall that an average of a set of values is the sum of the values divided by the number of values in the set. So, the following is one possible verbal model.

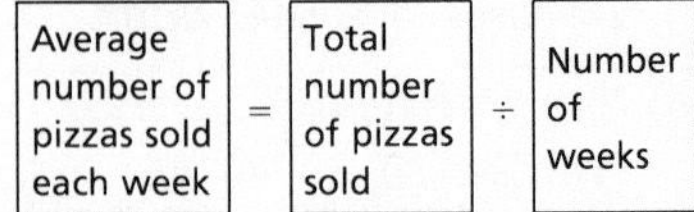

▶ **Ex. 40** If there are students who don't follow football, have other members of the class explain the game and the dimensions of the playing field. This is a good way to assert that students in the class often have the expertise needed to solve real-life applications.

▶ **Ex. 45** Point out to students that the formula for the area of a trapezoid is an expression that requires its user to know the order of operations.

31. *Brochure Error* A sports arena is 314.5 feet long and 252.5 feet wide. The public relations department wrote a brochure on the history of the arena and used a calculator to compute the perimeter of the arena as follows.

2 [×] 314.5 [+] 2 [×] 252.5 [=]

The brochure states that the perimeter of the arena is 159,327.5 feet. Is this the correct perimeter? If not, what did the department do wrong? No, the department did not follow the correct order of operations.

32. *Sales Tax Error* A family goes shopping for sneakers. One child's choice costs \$35.99 and the other child's choice costs \$42.99. If the sales tax is 7%, what would the sales tax be for both pairs of sneakers? If the cashier uses a calculator to obtain the sales tax and gets \$39, what did the cashier do wrong? Did not follow the correct order of operations.

33. *Pizza Project* As a fund-raiser, a school band sold pizzas. The number of pizzas sold for each of three weeks was 1248, 741, and 402. Write an expression that represents the average number of pizzas sold per week. Evaluate the expression. $(1248 + 741 + 402) \div 3 = 797$

34. *Average Score* Suppose your first three test scores in a math class are 87, 92, and 81. Write an expression that represents your average score. Evaluate the expression. $(87 + 92 + 81) \div 3 = 86\frac{2}{3}$

Integrated Review

35. *Geometry* Write an expression for the perimeter of the triangle. Use the expression to find the perimeter when $x = 12.9$ centimeters. $2x + 3x + 4x$, 116.1 cm

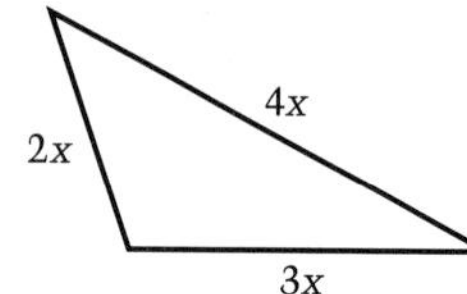

36. *Geometry* Write an expression for the perimeter of the figure. Use the expression to find the perimeter when $x = 4.2$ feet. $x + 1.5x + 2x + 3x$, 31.5 ft

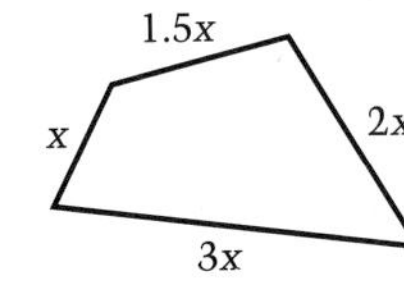

37. What is twice the area of a square with sides of length 2.9? 16.82 units^2

38. What is twice the volume of a cube with edges of length 2.9? 48.778 units^3

39. Write an expression that represents twice the area of a circle of radius r. $2\pi r^2$

40. *Football Field* Including the end zones, a football field is 360 feet by 160 feet. Use the picture at the right to approximate the total area of the floor (in green) of the Hubert Humphrey Metrodome (the home of the Minnesota Vikings). 115,200 ft^2

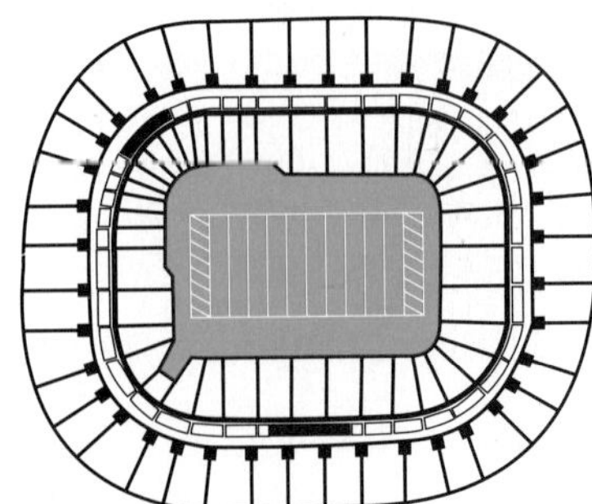

In Exercises 41–44, insert grouping symbols in the expression so that any calculator would give the same value.

41. $4.3 + (6.9 \div 2.3)$

42. $9.6 \cdot 1.3 - (3.6 \div 2.4)$

43. $6.2 \div 3.1 + (4.7 \cdot 5.3)$

44. $9.3 + (5.4 \cdot 1.1)$

Exploration and Extension

45. Geometry The area of a trapezoid with parallel bases of lengths b_1 and b_2 and height h is given by $A = \frac{1}{2}h(b_1 + b_2)$. (The variable b_1 is read "b sub 1," and the variable b_2 is read "b sub 2.") Find the area of a trapezoid whose height is 2 meters and whose bases are 6 meters and 10 meters. 16 m²

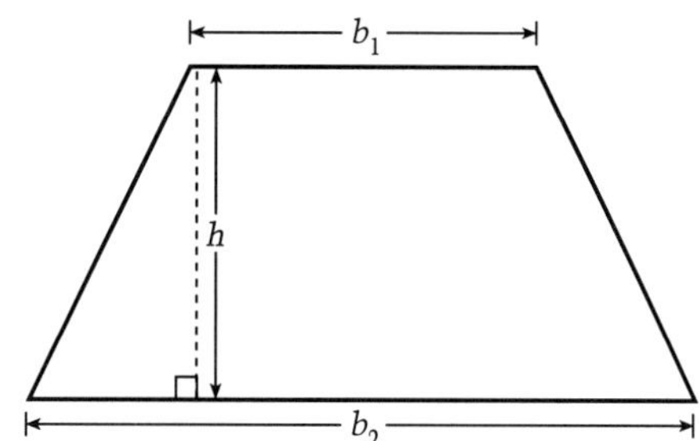

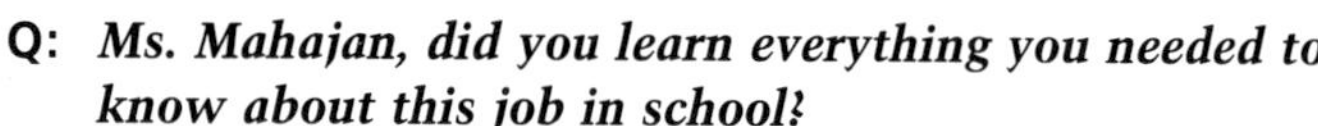

Medical Technologist

Jyoti Mahajan is a medical technologist in a large hospital. She supervises the laboratory that performs the medical tests ordered by physicians.

Q: ***Ms. Mahajan, did you learn everything you needed to know about this job in school?***

A: Not at all! We have so many new tests that we do now. I have learned at least as much on the job as I did in school.

Q: ***Can you give us an example of something you have learned on the job?***

A: One of the most exciting new technologies we are using is monoclonal antibodies to detect diseases caused by viruses. The test is relatively simple to perform using our new machine with its built-in computer, but I had to learn to calibrate the machine, to judge when I have a good sample, and to interpret the results. All of these skills require mathematical reasoning since all of the data are numeric.

Q: ***What impact has this new technology had on the practice of medicine?***

A: Because medical technologists can use these machines to obtain reliable results, doctors are now able to quickly and efficiently diagnose many diseases which they could not a few years ago.

Enrichment Activities

These activities require students to go beyond lesson goals.

1. Identify expressions (or formulas) in Lessons 1.1–1.4 for which it is important to know the order of operations. For example, the formula $A = P(1 + r)^t$ on page 17.
2. Evaluate these expressions.
   a. $144 \cdot 4 \cdot 4 \cdot 4$ 9216
   b. $144 \div 4 \div 4 \div 4$ $\frac{9}{4}$
   c. $144 + 4 + 4 + 4$ 156
   d. $144 - 4 - 4 - 4$ 132

   Which expressions are affected by the insertion of grouping symbols? b, d Which are not? a, c

Career Interview

Throughout the text, careers and individuals who use mathematics every day will be profiled. Having students interview local professionals who use mathematics at work is an excellent semester-long project.

Students could share their findings with the entire class during a Career Day that they could have the responsibility of planning. A broad representation of people in various careers about which students can read has been identified. This collection of interviews should help convince all students of the importance of mathematics to their futures.

MULTICULTURAL The magazines, *Winds of Change—A Magazine of American Indians* and *Ebony,* an African-American magazine, are two sources from which you can identify people using mathematics in their careers.

Mid-Chapter SELF-TEST

Remind students to complete this test under testlike conditions. The answers to the Mid-Chapter Self-Tests will be found in the "Selected Answers" section of the student text.

▶ **Ex. 17, 19–20** **PROBLEM SOLVING** Encourage students to use verbal models before they commit to algebraic models into which they will substitute values for the variables.

Mid-Chapter SELF-TEST

Take this test as you would take a test in class. The answers to the exercises are given in the back of the book. (1.1)

1. Write $\frac{5}{8}$ in decimal form. 0.625

2. Write $\frac{1}{5}$ in percent form. 20%

3. Find the sum of $\frac{1}{2}$ and $\frac{1}{4}$. $\frac{3}{4}$

4. Find the difference of 6 and 2. 4

5. What is the product of 1 and 8? 8

6. What is the quotient of 7 and 21? $\frac{1}{3}$

In Exercises 7–16, evaluate the expression. (1.2-1.4)

7. $2[3(4 \div 2) + 6] - 5$ 19

8. $3 + [(4 - 1) \cdot 5]$ 18

9. $3 \cdot 8 \div 2 - 9 \div 3$ 9

10. $9 - 3 \cdot 2 + 1 \div 2$ $3\frac{1}{2}$

11. $12 \div (5x - y)$, when $x = 2$ and $y = 1$ $1\frac{1}{3}$

12. $(3x + 2) \cdot y$, when $x = 1$ and $y = 4$ 20

13. $13 - 2x^2 \div 8$, when $x = 6$ 4

14. $9x^2 - x \div 2$, when $x = 6$ 321

15. $1.3x + 0.75$, when $x = 2.7$ 4.26

16. $3.47 - 2.3y$, when $y = 1.25$ 0.595

17. A rectangular region is two times longer than it is wide. The width of the region is 4 meters. What is the area? **(1.2)** 32 m^2

18. Write an expression for the volume of a cube whose sides are of length x. **(1.3)** x^3

19. Several families who live in an apartment building plan to fence in a play area in the back of the building. The fence will extend out from each side of the building, x feet along both sides of the play area and 250 feet across the back. The apartment building is 150 feet wide. Write an expression for the number of feet of fencing needed. Evaluate the expression when $x = 75$ feet. **(1.2)** $2x + 350$, 500 ft

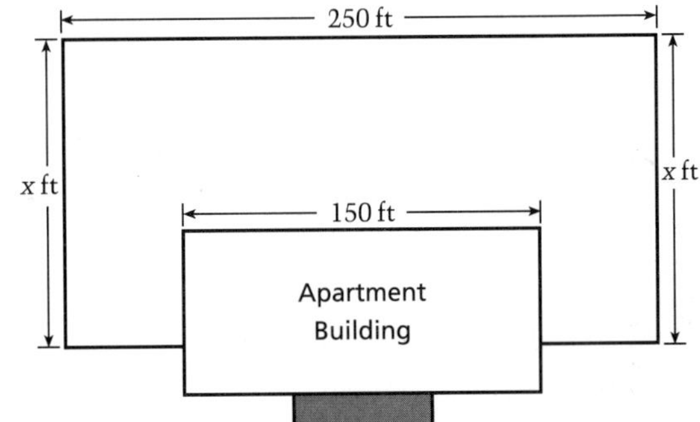

20. Bowling pins are set up in an equilateral triangle that is approximately 35 inches deep with sides that are 40.5 inches long. How much area is needed for the placement of the pins? **(1.2)** 708.75 in.^{2}

21. The surface area of a sphere is $s = 4\pi r^2$, where r is the radius of the sphere and π is approximately 3.14. A skateboard park is constructing a bowl for the skateboarders. If the radius of the bowl is 3 meters, how much surface area is available for the skateboarders? **(1.3)** 56.52 m^2

1.5 Equations and Inequalities

What you should learn:

Goal 1 How to check and solve equations

Goal 2 How to check solutions of inequalities

Why you should learn it:

You can solve real-life problems by finding solutions of equations. You can check your solutions to help assure that you did not make a mistake.

Goal 1 Checking and Solving Equations

This lesson introduces you to three very important concepts: equations, solutions, and inequalities.

An **equation** is formed when an equal sign is placed between two expressions. The two expressions form the **left side** and the **right side** of the equation. Here are some examples.

$$4 - 1 = 3, \quad 3x + 1 = 7, \quad a + 3 = 3 + a$$

Equations that contain variables are **open sentences.**

When the variable in a single-variable equation is replaced by a number, the resulting statement may be true or false. If the resulting statement is true, then the number is a **solution** of the equation. For example, the number 2 is a solution of $3x + 1 = 7$ because $3(2) + 1 = 7$ is a true statement. We also say that the number 2 *satisfies* the equation.

Substituting a number into an equation to decide whether the number is a solution is called *checking* a possible solution.

Example 1 *Checking Possible Solutions*

Check whether the numbers 2, 3, and 4 are solutions of the equation $4x - 2 = 10$.

Solution To check each number, substitute the number into the equation. If the left and right sides of the equation have the same value, then the number is a solution. If the left and right sides have different values, then the number is not a solution.

x	Substitution	Conclusion
2	$4(2) - 2 = 6$	2 is not a solution.
3	$4(3) - 2 = 10$	3 is a solution.
4	$4(4) - 2 = 14$	4 is not a solution.

Thus, 3 is a solution, and 2 and 4 are *not* solutions, of the equation $4x - 2 = 10$. ■

Problem of the Day

1	2	3
Red	Black	Mixed

The three canisters contain only red, only black, or only mixed red and black marbles. However, each canister has the wrong label. Alison claims that by just picking one marble from a canister without looking in it, she can position the labels correctly. How can Alison do this?
She picks the marble from the canister 3. If it is red, then canister 1 is black and canister 2 is mixed. If it is black, then canister 1 is mixed and canister 2 is red.

ORGANIZER

Warm-Up Exercises

1. If $a = 3$, does $2a - 5$ equal 1? How do you know? Yes; Substitute 3.
2. If $t = 7$, does $t^2 + 26$ equal 65? How do you know? No; Substitute 7.
3. Is $4 \cdot 5 - 3$ greater than $4(5 - 3)$? How do you know? Yes; Evaluate; $17 > 8$.
4. If x is 23, is x less than $3(7 - 4)^2$? How do you know? Yes; Evaluate; $23 < 27$.

Lesson Resources

Teaching Tools
- Problem of the Day: 1.5
- Warm-up Exercises: 1.5
- Answer Masters: 1.5

Extra Practice: 1.5

LESSON Notes

A good way to begin is to tell students to use the Guess and Check strategy to find what value placed in the box will make this statement true.

$\square^2 - 5 = 31$ 6

Example 1

MATH JOURNAL Emphasize the importance of using the new vocabulary to discuss the ideas of this lesson. Help students attach meaning to these terms by having them define the terms in their own words in their math journals.

Example 2

Encourage students to talk to their cooperative learning partners about possible solutions. Often mistakes can be self-corrected when you hear yourself speak them.

Example 3

Encourage students to use a verbal model to express the relationship between buying two compact disks and having $25 to spend. Verbal models may differ and still be correct. Here is one possible verbal model.

Cost of CD's	=	Amount of money I have	+	Amount of money I need

Here is another

Cost of CD's	−	Amount of money I have	=	Amount of money I need

Tell students that *tax included* means that the cost of the compact disks already includes the tax that is owed.

Example 4

Checking solutions to inequalities reinforces the meaning of a solution—namely, a number that produces a true statement when substituted for the variable in the statement. Regularly stress the concept of "solution" for both equalities and inequalities.

Finding all the solutions of an equation is called *solving* the equation. Later in the book you will study several ways to systematically solve equations. Some equations, however, are simple enough to solve in your head. For example, to solve $x + 2 = 5$, ask yourself the question "What number plus 2 gives 5?" If you can see that the answer is 3, then you have solved the equation!

Example 2 *Using Mental Math to Solve Equations*

	Equation	Mental Math
a.	$x + 2 = 6$	What number plus 2 gives 6?
b.	$x - 3 = 4$	What number minus 3 gives 4?
c.	$2x = 10$	2 times what number gives 10?
d.	$\frac{x}{3} = 1$	What number divided by 3 gives 1?
e.	$x^3 = 8$	What number cubed gives 8?

Solution

a. The solution is 4 because $4 + 2 = 6$.
b. The solution is 7 because $7 - 3 = 4$.
c. The solution is 5 because $2 \cdot 5 = 10$.
d. The solution is 3 because $\frac{3}{3} = 1$.
e. The solution is 2 because $2^3 = 8$. ■

Using mental math to solve equations is something that you already do in real life. You probably don't think of it as solving equations, but the process is the same.

Example 3 *Buying Compact Disks*

Suppose that you want to buy two compact disks that cost $14.95 each, tax included. You have only $25. How much more do you need?

Solution The mental math that you could use to solve this problem is

$$2(14.95) = 25 + \boxed{?}.$$

Since 2 times $14.95 is $29.90, you can conclude that you need $4.90 more to be able to buy both compact disks. ■

Goal 2 Checking Solutions of Inequalities

Another type of open sentence is an **inequality.** An inequality is formed when an **inequality symbol** is placed between two expressions. Here are five kinds of inequality symbols.

Inequality Symbol	Meaning
$<$	is less than
$\leq$	is less than or equal to
$>$	is greater than
$\geq$	is greater than or equal to
$\neq$	is not equal to

A **solution** of an inequality is a number that produces a true statement when it is substituted for the variable in the inequality. For example, 3 is a solution of $x < 5$ because $3 < 5$ is a true statement.

Example 4 *Checking Solutions of Inequalities*

Decide whether 4 is a solution of each inequality.

a. $2x - 1 < 8$ **b.** $x + 4 > 9$ **c.** $x - 3 \leq 1$

Solution

Inequality	Solution Check
a. $2x - 1 < 8$	4 is a solution because $2(4) - 1 < 8$ is a true statement.
b. $x + 4 > 9$	4 is not a solution because $4 + 4 > 9$ is not a true statement.
c. $x - 3 \leq 1$	4 is a solution because $4 - 3 \leq 1$ is a true statement. ■

Communicating about ALGEBRA

SHARING IDEAS about the Lesson

A. Check whether 1, 2, 3, and 4 are solutions of $x^2 + 8 - 5x = 2$. Only 2 and 3 are.

B. ***Give an Argument.*** Match each equation with the correct description. Justify your answers.

1. $x + 2 = 6$ b   **a.** has no solution

2. $x + 2 = 2 + x$ c   **b.** has only one solution

3. $x + 2 = x + 3$ a   **c.** has many solutions

Extra Examples

Here are additional examples similar to **Examples 2–4.**

1. What values of x will result in a true statement? Explain.

a. $x^2 - 3 = 6$ 3

b. $2(13 - x) = 16$ 5

c. $x^5 - 1 = 31$ 2

d. $\frac{x + 7}{8} = 6$ 41

2. Bea saw two pairs of shoes she wanted to buy. Each pair was on sale at $4 off. If she has $30 to spend, how much more money will she need to purchase shoes that originally sold for $21 a pair? $4

3. Decide whether 6 is a solution of each inequality.

a. $3x + 1 < 8$ no

b. $14 - x > 9$ no

c. $2x - 1 \leq 13$ yes

Check Understanding

WRITING Ask students to write definitions (in their own words) for the following key vocabulary terms and phrases.

1. solution

2. satisfies

3. solving an equation

4. solving an inequality

Communicating about ALGEBRA

Students could work with partners or in small groups to complete the following activities. Encourage group discussion.

EXTEND *Communicating*
COOPERATIVE LEARNING
Have students work together to generate other sets of equations that have either no solution, only one solution, or many solutions. Sample answers: $2y = 2(y + 4)$; $2y = y + 4$; $2y = 2(y + 4) - 8$

Challenge! Can you write an inequality that has no solution or many solutions? Sample answers: $x + 1 < x$, no solution; $x + 3 > 4$, many solutions.

EXERCISE Notes

ASSIGNMENT GUIDE

Basic/Average: Ex. 1–8, 9–30 multiples of 3, 32, 33–57 odd, 60–63

Above Average: Ex. 1–8, 9–30 multiples of 3, 32, 33–57 odd, 59–65 odd

Advanced: Ex. 1–8, 9–30 multiples of 3, 32, 33–36, 51–65 odd, 66

Selected Answers
Exercises 1–8, 9–49 odd

More Difficult Exercises
Exercises 35, 36, 58, 59, 66

Guided Practice

Ex. 1–6 Stress the differences among equations, inequalities, and expressions. Caution students about changing expressions to turn them into equations.

Independent Practice

Ex. 15–26 These exercises refer back to the type of questions posed in Example 2 on page 28.

Ex. 27–32 Stress the importance of the meaning of a solution to an algebraic statement–namely, that a solution is any value that makes the statement true.

Ex. 33–36 PROBLEM SOLVING Encourage students to show the verbal models that match the equations given in the text for these problems.

EXERCISES

1–6: An equation has "=," an inequality has "<," "≤," ">," or "≥," and an expression has none of these.

Guided Practice

CRITICAL THINKING about the Lesson

In Exercises 1–6, decide whether what is shown is an expression, an equation, or an inequality. Explain your decision.

1. $6x + 1 = 12$ Equation
2. $2y - 3$ Expression
3. $5(y + 4) - 7$ Expression
4. $4 + 3x = 2 + x$ Equation
5. $2x + 1 \le 4$ Inequality
6. $5x > 20$ Inequality

7. Identify the left and right sides of the equation $13 + 4x = 2x + 1$. left: $13 + 4x$; right: $2x + 1$

8. Match the equation $5 - x = 1$ with one of the questions.
a. What number can 5 be subtracted from to obtain 1?
b. What number can be subtracted from 5 to obtain 1?
c. What number can 1 be subtracted from to obtain 5?

Independent Practice

In Exercises 9–14, check whether the given number is a solution of the equation.

9. $3x + 1 = 13$, 4 Yes
10. $3 + 2x = 9$, 2 No
11. $4y + 2 = 2y + 8$, 3 Yes
12. $2y + y = 2y + 5$, 5 Yes
13. $5x - 2 = 11$, 2 No
14. $6x - 4 = 3$, 1 No

In Exercises 15–26, write a question that could be used to solve the equation. Then use mental math to solve the equation. See below for sample questions.

15. $x + 4 = 7$ 3
16. $x + 6 = 11$ 5
17. $x - 2 = 5$ 7
18. $x - 5 = 1$ 6
19. $3x = 12$ 4
20. $2x = 6$ 3
21. $\frac{x}{5} = 4$ 20
22. $\frac{x}{4} = 5$ 20
23. $2x + 1 = 7$ 3
24. $2x - 1 = 9$ 5
25. $x^3 = 27$ 3
26. $x^3 = 1$ 1

In Exercises 27–32, check whether the given number is a solution of the inequality.

27. $x - 2 < 6$, 7 It is.
28. $5 + x > 8$, 4 It is.
29. $6 + 2x \ge 10$, 2 It is.
30. $3x - 1 \ge 7$, 3 It is.
31. $4x - 3 < 0$, 1 It is not.
32. $x - 7 \ge 11 - x$, 9 It is.

In Exercises 33–36, consider the problem, then answer the question about the equation that represents the problem.

Computer Center Suppose that your school is building a new computer center. Four hundred square feet of the center will be available for computer stations. Each station requires 20 square feet. How many computer stations can be placed in the new center?

33. If this problem is represented by the equation $20x = 400$, what do the 20, the x, and the 400 represent? See Additional Answers.

15. What number plus 4 is 7? **17.** What number minus 2 is 5?
19. What number times 3 is 12? **21.** What number divided by 5 is 4?
23. What number times 2, plus 1, is 7? **25.** What number cubed is 27?

Playing a Computer Game Suppose you are playing a new computer game. You discover that for every eight screens you complete, you receive an energy bar. If you complete 96 screens, how many energy bars will you receive?

34. If this problem is represented by the equation $8x = 96$, what do the 8, the x, and the 96 represent? See Additional Answers.

Buying Gas Suppose you are going on an automobile trip with a family of a friend and you have \$65 to help pay for gas. It costs \$18 to fill the tank. How many times can you completely fill the tank? How much money will you have left?

35. If this problem is represented by the equation $18x + y = 65$, what do the 18, the x, the y, and the 65 represent? See Additional Answers.

For most car-owners, gasoline expenses form a significant part of their monthly budget.

Buying a T-Shirt Suppose you are at a shopping mall and see a T-shirt that you really like. The shirt costs \$18.75, including tax. If you can save \$5 a week, how many weeks must you save to buy the T-shirt? How much money will you have left?

36. If this problem is represented by the equation $5n - m = 18.75$, what do the 5, the n, the m, and the 18.75 represent? See Additional Answers.

Integrated Review

In Exercises 37–40, use mental math.

37. What is the sum of 6 and 5? 11

38. What is the difference of 15 and 9? 6

39. What is the product of 4 and 8? 32

40. What is the quotient of 12 and 3? 4

In Exercises 41–50, evaluate the expression.

41. $6 - [(3 \cdot 8) - 21]$ 3

42. $[18 \div (3 + 3)] + 5$ 8

43. $7 + [12 - (3)(4)]$ 7

44. $16 - (12 \div 2)$ 10

45. $x + 11$, when $x = 14$ 25

46. $2n - 3$, when $n = 9$ 15

47. $3y + 2x$, when $y = 7$ and $x = 2$ 25

48. $7a - 4b$, when $a = 2$ and $b = 1$ 10

49. $(x + y)^2$, when $x = 8$ and $y = 4$ 144

50. $3s^2 + 6r^3$, when $s = 5$ and $r = 1$ 81

▶ **Ex. 37–50** Encourage students to use mental math to answer these exercises.

▶ **Ex. 51–57** GROUP ACTIVITY You may want to use these translating exercises as oral problems in class in order to discuss them with students.

Enrichment Activities

These activities require students to go beyond lesson goals.

1. Find the values of x that satisfy each statement. Check to verify your answer.
   a. $3x + 4 = 4$ 0
   b. $4(x - 6) = 16$ 10
   c. $(3x)^3 = 216$ 2
2. Decide whether 8 is a solution of each statement.
   a. $4x - 6 = 24$ no
   b. $6(x - 5) + 3 = 21$ yes
   c. $20 - 2x \div 4 < 4$ no
   d. $(x - 5)6 + 6 > 36$ no
3. Write an open sentence that has
   a. only one solution Sample answer: $x + 3 = 5$
   b. exactly two solutions. Sample answer: $x^2 = x$

Common-Error ALERT!

In **Exercises 53–54,** remind students that the order in which differences and quotients are written is important.

▶ **Ex. 59** These "College Entrance Exam Sample" questions will be a recurring feature in the lesson exercises.

▶ **Ex. 64–66** PROBLEM SOLVING It is important that students understand the relationship between the variables x and y. Ask questions like the following to help students see the relationship clearly.

a. In terms of the variables you have, how long is the iguana's tail? y
b. How long is its body? x
c. How long is the iguana from head to tail? $x + y$
d. Which appears to be longer, the body or the tail? the tail
e. What does the equation $y = x$ say about the relationship? The iguana's body and tail are of equal length.
f. What does the equation $y = 2x$ say? The length of the iguana's tail is double its body length.
g. Does the equation $y = 0.5x$ make sense? Explain. no; This equation indicates that the tail is only half as long as its body—a contradiction.

EXTEND Ex. 64–66 GROUP ACTIVITY Students might work in groups to perform the following experiment.

Ask students if they think there is a relationship between the length of a person's forearm (elbow to wrist) and the length of a person's foot.

Tell them to work with a partner to make the necessary measurements. Record the data in a class table. Use the data to determine and express a relationship algebraically. Usually, the length of a person's forearm and foot are about equal.

Exploration and Extension

51. $x + 15 < 29$

In Exercises 51–57, write the indicated equation or inequality.

$9y = 72$

51. The sum of x and 15 is less than 29.

52. The product of 9 and y is equal to 72.

53. The difference of x and 4 is equal to 18. $x - 4 = 18$

54. The quotient of n and 3 is greater than or equal to 4. $n \div 3 \geq 4$

55. The sum of three times x and 12 is equal to 32. $3x + 12 = 32$

56. The winning score, x, was 20 points more than the losing score, y. $x = y + 20$

57. The surface temperature, V (in degrees Celsius), of the planet Venus is 380 degrees more than the temperature at which water boils (100 degrees). $V = 100 + 380$

✪ **58.** *Boeing 747* Let x represent the wingspan of a Boeing 747 jet aircraft and let y represent its length. What does the equation $1.2y = x$ say about the relationship between the length and width of a 747? Is a 747 longer than it is wide or wider than it is long? The width is 1.2 times the length; it is wider than it is long.

✪ **59.** *College Entrance Exam Sample* Using the triangle provided, $x = \boxed{?}$

a. 59°
b. 60°
c. 61°
d. 62°
e. 63°

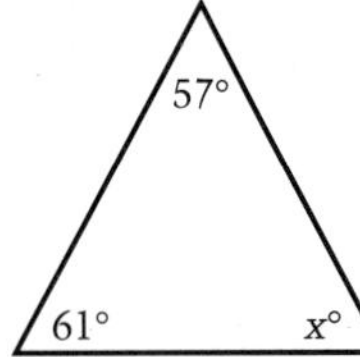

In Exercises 60–63, solve the two equations. What do you notice to be true about the pair of equations? See below.

60. $x + x = 12$ and $2x = 12$

61. $y - 8 = 10$ and $y = 18$

62. $2x + 4 = 8$ and $2x = 4$

63. $2a + 3a = 10$ and $5a = 10$

64. *The Tail of an Iguana* Let y represent the length of the tail of an iguana and x represent the length of the body of the iguana. What does the equation $y = 1.8x$ say about the relationship between the tail and body of an iguana?

65. If an iguana has a body length of 30 centimeters, what is the length of the tail? 54 cm

✪ **66.** If an iguana has a tail length of 18 inches, what is the length of the body? 10 in.

If cornered by an enemy, an iguana uses its long tail as a lash.

64. The length of the tail is 1.8 times the length of the body.

60. Both have 6 as a solution.
61. Both have 18 as a solution.
62. Both have 2 as a solution.
63. Both have 2 as a solution.

1.6 Verbal Models and Algebraic Models

What you should learn:

Goal 1 How to translate verbal phrases into algebraic expressions

Goal 2 How to translate verbal sentences into algebraic equations and inequalities

Why you should learn it:

You can use mathematics to solve real-life problems if you first can translate verbal phrases and sentences into algebraic expressions, equations, and inequalities.

Goal 1 Translating Verbal Phrases

To help you translate a verbal phrase into an algebraic expression, you should look for words that indicate a number operation. The following list gives several examples.

	Verbal Phrase	Algebraic Expression
Addition:	The *sum* of 6 and a number	$6 + x$
	Eight *more than* a number	$y + 8$
	A number *plus* 5	$n + 5$
Subtraction:	The *difference* of 5 and a number	$5 - y$
	Four *less than* a number	$x - 4$
	Seven *minus* a number	$7 - n$
Multiplication:	The *product* of 9 and a number	$9x$
	Ten *times* a number	$10n$
	A number *multiplied by* 3	$3y$
Division:	The *quotient* of a number and 4	$\frac{n}{4}$
	Seven *divided by* a number	$\frac{7}{x}$

Notice that order is important for subtraction and division, but not for addition and multiplication. For example, "four less than a number" is written as $x - 4$, not $4 - x$. On the other hand, "the sum of 6 and a number" can be written as $6 + x$ or $x + 6$.

Example 1 *Translating Verbal Phrases*

a. Three more than five times a number can be written as

$5n + 3$. *Think: 3 more than what number?*

b. Two less than the sum of six and a number can be written as

$(6 + m) - 2$. *Think: 2 less than what number?*

c. One number decreased by the sum of 10 and the square of another number can be written as

$x - (10 + y^2)$. *Think: x decreased by what number?* ■

Problem of the Day

How many triangles can be found in each drawing?

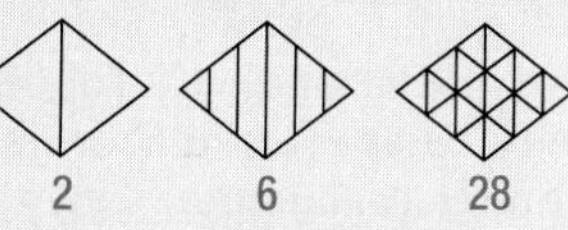

ORGANIZER

Warm-Up Exercises

Words often have more than one meaning. An extreme example is *bad,* which may mean "evil" or "rotten," or is slang for "good." Identify the different ways these words can be used.

1. *plus* can mean "add," or "a benefit"
2. *product* can mean "the result after multiplying," or "something produced or manufactured"
3. *difference* can mean "the result after subtracting," or "the condition of being unlike something else"

Lesson Resources

Teaching Tools
 Problem of the Day: 1.6
 Warm-up Exercises: 1.6
 Answer Masters: 1.6
Extra Practice: 1.6
Applications Handbook: p. 40, 56

LESSON Notes

Example 1

Be patient. Encourage questions about the translations from verbal phrases to algebraic ones. Be careful to give students an opportunity to state in their own words what they think the expressions mean.

Example 2

Students have seen these three steps for writing a mathematical model before. Emphasize the importance of following these steps when translating phrases to algebra. Many times you attempt to translate from phrases to algebra only after you have tried a problem-solving strategy. In any event, let students know that whether a verbal model is written before or after employing some problem-solving strategy, the ability to model real-life situations empowers them to understand and manage their world.

Encourage students to use their own words to describe the total amount paid for the collection of cans and bottles. Students should talk with one another to see if what they think makes sense also makes sense to someone else.

Check to see that students understand the information found in the recycling table. Have students explain how the entries in the table might have been generated.

Example 3

Encourage students by telling them that writing equations to model real-life situations is only a small step up from translating verbal phrases to algebraic ones.

Extra Examples

Here are additional examples similar to **Examples 1–3.**

Read the following problems to students and have students translate them into algebra. This will promote discussion and provide practice for students to improve their listening skills.

Writing algebraic expressions or equations that represent real-life situations is called *modeling* because the expression or equation is a **mathematical model** of the real-life situation. When you are writing a mathematical model, we suggest that you use three steps. First, write a verbal model; then, assign labels to the model; and finally, write the algebraic model.

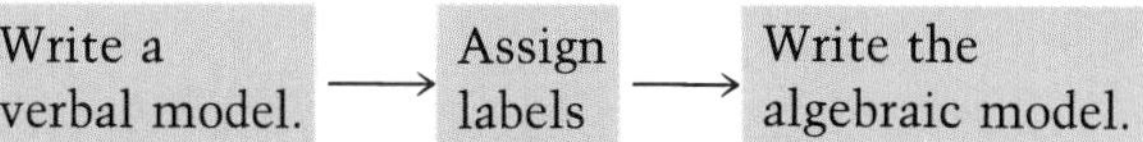

Real Life
Recycling

Example 2 *Writing an Expression as an Algebraic Model*

A recycling center pays 3¢ apiece for aluminum cans and 5¢ apiece for certain glass bottles. Write an expression representing the total amount paid for a collection of cans and bottles.

Solution

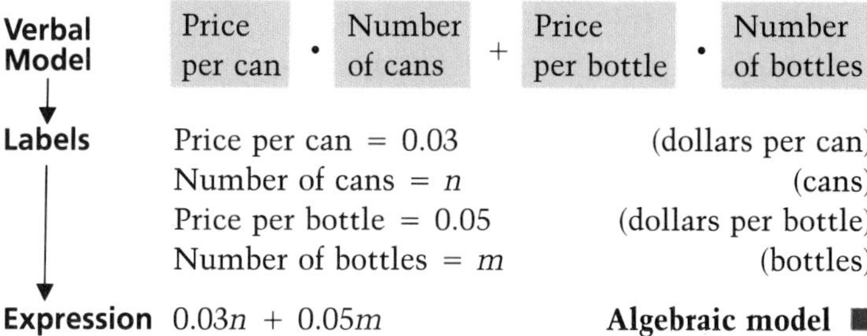

Verbal Model [Price per can] · [Number of cans] + [Price per bottle] · [Number of bottles]

Labels
Price per can $= 0.03$ (dollars per can)
Number of cans $= n$ (cans)
Price per bottle $= 0.05$ (dollars per bottle)
Number of bottles $= m$ (bottles)

Expression $0.03n + 0.05m$ **Algebraic model** ■

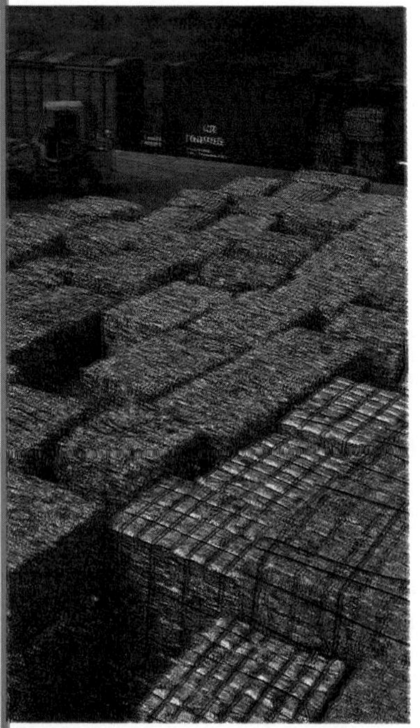

Algebraic models can help you compute and organize data about real-life situations. For instance, you could use the model in Example 2 to compute the total amounts received for various numbers of aluminum cans and glass bottles. Your results could be organized in a table like the one below.

Number of Bottles, m	Number of Cans, n: 0	5	10	15	20	25	30
0	\$0.00	\$0.15	\$0.30	\$0.45	\$0.60	\$0.75	\$0.90
5	\$0.25	\$0.40	\$0.55	\$0.70	\$0.85	\$1.00	\$1.15
10	\$0.50	\$0.65	\$0.80	\$0.95	\$1.10	\$1.25	\$1.40
15	\$0.75	\$0.90	\$1.05	\$1.20	\$1.35	\$1.50	\$1.65
20	\$1.00	\$1.15	\$1.30	\$1.45	\$1.60	\$1.75	\$1.90
25	\$1.25	\$1.40	\$1.55	\$1.70	\$1.85	\$2.00	\$2.15
30	\$1.50	\$1.65	\$1.80	\$1.95	\$2.10	\$2.25	\$2.40
35	\$1.75	\$1.90	\$2.05	\$2.20	\$2.35	\$2.50	\$2.65

Each dollar amount in the table was found by evaluating the expression $0.03n + 0.05m$ for appropriate values of n and m.

Goal 2 Translating Verbal Sentences

Verbal phrases about quantities in the English language translate to mathematical expressions. Verbal sentences translate to equations or inequalities.

Phrase	The sum of 6 and a number	$6 + x$
Sentence	The sum of 6 and a number is 14.	$6 + x = 14$
Sentence	7 times a number is less than 50.	$7x < 50$

Sentences that translate have words that tell how one quantity relates to another. For instance, in the first *sentence* above, "is" says that one quantity is equal to another.

Example 3 *Writing an Equation as an Algebraic Model*

Real Life **Nutrition**

Cheddar cheese has 30 milligrams of cholesterol per ounce, and lean ground beef has 27 milligrams of cholesterol per ounce. A cheeseburger made with 4 ounces of lean ground beef has 123 milligrams of cholesterol. Write an equation that represents this information. (The other ingredients have no cholesterol.)

Solution

Verbal Model: Cheese cholesterol + Beef cholesterol = Total cholesterol

Labels

Cheese cholesterol per ounce = 30	(mg per oz)
Ounces of cheese = x	(oz)
Beef cholesterol per ounce = 27	(mg per oz)
Ounces of beef = 4	(oz)
Total cholesterol = 123	(mg)

Equation $30x + 27(4) = 123$ **Algebraic model** ■

Communicating about ALGEBRA

SHARING IDEAS about the Lesson

Understand an Algebraic Model Refer to the equation $30x + 27(4) = 123$ in Example 3. $30x$, (mg per oz) × oz = mg

A. Which term in the equation represents the number of milligrams of cholesterol in the cheese? Explain why this term is written as a product.

B. If the cheeseburger contains one ounce of cheese, how many milligrams of cholesterol does it have? 30

1. a. The difference of a number n and 8 $n - 8$
   b. The product of 13 and a number n is 39. $13 \cdot n = 39$
   c. A number n multiplied by 15 $n \cdot 15$ or $15n$
   d. Nine more than a number n is 34. $9 + n = 34$
   e. The quotient of 51 and 17 is a number n minus 4. $\frac{51}{17} = n - 4$

Write an expression or equation for each situation.

2. The cost for a family to see a movie is \$5.00 for adults and \$3.50 for children under twelve. Sample answer: $5a + 3.5c$
3. The cost of 6 composition notebooks and 6 homework folders for school. Sample answer: $6n + 6f$ or $6(n + f)$
4. The cost to see a movie is \$5 for adults and \$3 for children under twelve. Two adults spent \$22 to take several children to the movies. Sample answer: $22 = 5(2) + 3c$

Check Understanding

MATH JOURNAL Write a paragraph in your math journal about the importance of translating verbal phrases to algebra. Use examples to make your main points. The ability to translate verbal phrases to algebra allows you to model real-life situations and to examine relationships in a systematic way.

Communicating *about* ALGEBRA

If you are watching cholesterol levels and do not want to get more than 123 milligrams per cheeseburger, write an inequality that represents this information. $30x + 27y \le 123$, where x is ounces of cheese and y is ounces of beef

EXERCISE Notes

ASSIGNMENT GUIDE

Basic/Average: Ex. 1–8, 9–33 odd, 39–53 odd

Above Average: Ex. 1–8, 9–33 odd, 39–53 odd

Advanced: Ex. 1–8, 9–25 odd, 26, 31–34, 39–55 odd, 56

Selected Answers
Exercises 1–8, 9–47 odd

Use **Mixed Review** as needed.

More Difficult Exercises
Exercises 25–26, 33–34, 55–56

Guided Practice

Ex. 4 MATH JOURNAL It is a good idea to have students record the three steps for writing an algebraic model in their math journals.

EXTEND Ex. 5–8 WRITING Have students rewrite the expressions using other verbal phrases or statements.

Independent Practice

Ex. 9–24 COOPERATIVE LEARNING These exercises may be done by students working in small groups. In any event, students should be encouraged to discuss the exercises with each other. The next day while going over homework might be an appropriate time to do this.

EXERCISES

Guided Practice

CRITICAL THINKING about the Lesson

In Exercises 1–3, consider the verbal phrase, *the difference of 8 and a number.*

1. The word *difference* indicates what operation? Subtraction
2. Translate the verbal expression into an algebraic expression. $8 - x$
3. Is order important in this expression? Yes
4. List the three steps for writing a mathematical model. Write a verbal model, assign labels, and write an algebraic model.

In Exercises 5–8, translate the sentence into an equation or an inequality.

5. The sum of a number x and 5 is 12. $x + 5 = 12$
6. The product of 4 and a number n is less than 36. $4n < 36$
7. The difference of a number m and 5 is less than 20. $m - 5 < 20$
8. The quotient of 10 and a number y is 5. $10 \div y = 5$

Independent Practice

In Exercises 9–14, translate the phrase into an algebraic expression.

9. 12 more than a number $x + 12$
10. 5 more than twice a number $2y + 5$
11. Half of a given number $\frac{1}{2}m$
12. 3 less than 4 times a given number $4n - 3$
13. 11 plus the quotient of a number and 7 $11 + r \div 7$
14. A number divided by 6 $x \div 6$

In Exercises 15–20, match each verbal phrase on the left with its corresponding algebraic expression on the right.

15. Twelve decreased by three times a number d
16. Eleven more than $\frac{1}{3}$ of a number a
17. Eleven times a number plus $\frac{1}{3}$ e
18. Three increased by 12 times a number f
19. The difference between three times a number and 12 b
20. Three times the difference of a number and twelve c

a. $11 + \frac{1}{3}s$
b. $3x - 12$
c. $3(x - 12)$
d. $12 - 3x$
e. $11s + \frac{1}{3}$
f. $12x + 3$

In Exercises 21–24, translate the sentence into an equation or an inequality.

21. Sixteen more than a number s is 36. $s + 16 = 36$
22. Ten is less than three times a number b. $10 < 3b$
23. Twenty-five decreased by a number y is 10.5. $25 - y = 10.5$
24. The product of 17 and a number n is one. $17n = 1$

25. *Geometry* Write an expression that represents the perimeter of the hexagon.
$8 + 12 + (x + 4) + 8 + 6 + 2x$

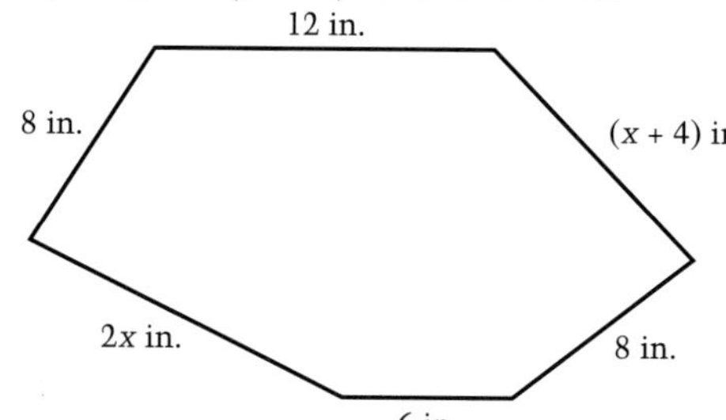

26. *Geometry* Write an expression that represents the area, A, of the shaded region. (The area of a circle is $A = \pi r^2$.)
$\pi \cdot 12^2 - \pi \cdot 4^2$

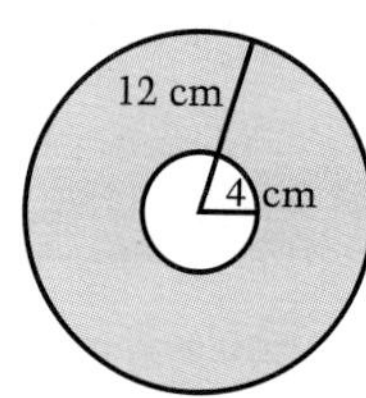

In Exercises 27–30, decide whether the equation is a correct translation of the verbal sentence.

	Sentence	Equation
27.	Half of the sum of six and a number is seven. Correct	$\frac{1}{2}(6 + n) = 7$
28.	Seven less than three times a number is eight. Incorrect	$7 - 3x = 8$
29.	Three times the difference of two and a number is four. Incorrect	$(3 \cdot 2) - y = 4$
30.	Four times the sum of a number and five is thirty-two. Correct	$4(m + 5) = 32$

In Exercises 31–34, which equation correctly models the situation?

31. *Pizza Party* You and three friends bought a pizza. You paid \$2.30 for your share ($\frac{1}{4}$ of the pizza). Let p be the cost of the pizza.

a. $4p = 2.30$ **b.** $\frac{1}{4}p = 2.30$

32. *Scale Model for a Car* A model for a new car is scaled $\frac{1}{10}$ of the actual size. The length of the model's body is 45. Let L be the actual length of the car's body.

a. $\frac{1}{10}L = 45$ **b.** $L = \frac{1}{10}(45)$

33. *Sports Cars* There are 24 sports cars in a car lot. Each car is either red or white. There are four fewer red cars than white cars. Let w be the number of white cars.

a. $w + (w - 4) = 24$ **b.** $w + 4 = 24$

34. *Coin Collection* Suppose you have a collection of rare coins. Your favorite coin will be fifty years old in only six years. Let c be the age of the coin now.

a. $c - 6 = 50$ **b.** $c + 6 = 50$

Integrated Review

In Exercises 35–38, write the expression in exponential form.

35. x raised to the seventh power x^7

36. Eight cubed 8^3

37. $y \cdot y \cdot y \cdot y \cdot y \cdot y$ y^6

38. $12 \cdot 12 \cdot 12 \cdot 12$ 12^4

In Exercises 39–42, evaluate the expression.

39. $\frac{4x + 2y}{5}$, when $x = 1$ and $y = 8$ 4

40. $6 + \frac{x + 2}{y}$, when $x = 7$ and $y = 3$ 9

41. $3y^2 - w$, when $y = 8$ and $w = 27$ 165

42. $2x^3 - 3y$, when $x = 4$ and $y = 6$ 110

Ex. 25–26 In Exercise 26, the expression for the shaded area is given in terms of π. Students might simplify the expression to $\pi(12^2 - 4^2)$, or 128π.

Ex. 39–42 Remind students that the order of operations still holds when evaluating algebraic expressions.

Ex. 31–32 PROBLEM SOLVING Translation skills are important to the development of a problem-solving plan.

Enrichment Activities

WRITING Use your own words to translate the following algebraic expressions into verbal expressions.

1. $13n - 4$ Sample answer: the difference of thirteen times a number n and four
2. $(17 + m) \cdot 7$
3. $\frac{n}{6}$
4. $(4 + a) - 9$
5. $3(n + 2) + 8$

Translate these algebraic statements into verbal ones.

6. $4 + 2x = 8$ Sample answer: The sum of four and two times a number x is eight.
7. $15(x - 3) = 12$
8. $27m < 63$
9. $8a + 14 \geq 32$

▶ **Ex. 43–48** Continue to emphasize the importance of understanding the meaning of a solution to an equation or an inequality.

▶ **Ex. 49–54** COOPERATIVE LEARNING Students might work with partners or in small groups to write a verbal phrase or sentence to represent the given expressions or equations.

Sample answers

49. 5 minus the product of 2 and a number
50. 4 subtracted from the product of 3 and a number
51. 6 times the sum of 3 and a number
52. A number times the sum of another number and 6
53. The difference of 5 times a number and 2 is 14.
54. The difference of 15 and twice a number is 7.

In Exercises 43–48, check whether the number is a solution of the equation or an inequality.

43. $7y + 2 = 4y + 8$, 2 It is.
44. $5x - 1 = 3x + 2$, 4 It is not.
45. $32 - 5y > 11$, 3 It is.
46. $7m + 2 < 12$, 1 It is.
47. $s - 7 \geq 12 - s$, 9 It is not.
48. $n + 2 \leq 2n - 2$, 4 It is.

Exploration and Extension

In Exercises 49–54, write a verbal phrase or sentence to represent the expression or equation. See margin.

49. $5 - 2x$
50. $3y - 4$
51. $6(3 + y)$
52. $r(y + 6)$
53. $5m - 2 = 14$
54. $15 - 2n = 7$

In Exercises 55 and 56, use the following information.

Super Bowl The San Francisco 49ers and the Cincinnati Bengals met in Super Bowl XVI and Super Bowl XXIII. The attendance at Super Bowl XXIII was 6141 less than the attendance at Super Bowl XVI. Let x represent the attendance at Super Bowl XVI.

55. Write an algebraic equation to represent the given information. $x - 6141 = y$

56. The attendance at Super Bowl XVI was 81,270. What was the attendance at Super Bowl XXIII? 75,129

Mixed REVIEW

1. What is the value of $17 - 5 + 3$? **(1.1)** 15
2. Evaluate $x^2 - 2$ when $x = 7$. **(1.3)** 47
3. Determine the value of $4 + 8 \cdot 4 - 1$. **(1.4)** 35
4. Determine the value of $22 - 4^2 \div 2$. **(1.4)** 14
5. What is the value of $2x^3 + x$ when $x = 3$? **(1.3, 1.4)** 57
6. Does $16 + x^2 \div 4 = 17$ when $x = 2$? **(1.3, 1.4)** Yes
7. Is 3 a solution of $x^2 - 4 = 5$? **(1.5)** Yes
8. Is 11 a solution of $2x - 9 = 13$? **(1.5)** Yes
9. Solve $x - 1 = 5$. **(1.5)** 6
10. Solve $x^3 = 1$. **(1.5)** 1
11. Is 9 a solution of $2x - 3 < 15$? **(1.5)** No
12. Is 2 a solution of $3 + 7x \geq 13$? **(1.5)** Yes
13. Which are solutions of $3x + 4 \leq 16$: $\underline{2}$, $\underline{4}$, 6? **(1.5)**
14. Use a calculator to evaluate $3.8 + 18.2 \div 7 - 1.6$. **(1.4)** 4.8
15. If your state has a 7% sales tax, what is the total cost of a keyboard marked $149.95? **(1.4)** $160.45
16. A fund-raiser generates 25¢ for each drink-box sold and 40¢ for each submarine sandwich sold. Write an expression for the total amount of money raised by both. **(1.6)** $0.25x + 0.40y$

1.7 A Problem-Solving Plan

What you should learn:

Goal 1 How to use algebra to solve simple real-life problems

Goal 2 How to make an algebraic model for a real-life problem

Why you should learn it:

You can use algebra to solve a variety of real-life problems, such as costs, rates, averages, percent, etc.

Goal 1 Using Algebra to Solve Problems

You may already know many ways to solve problems, such as guess and check, make a table, look for a pattern, and draw a diagram. In this course, you will continue to use problem-solving strategies you have already developed, and you will also learn how to use algebra to become an even better problem solver and decision maker.

In the problems in this (and every) lesson, don't be afraid to try to find the solution without using algebra. On the other hand, be sure that you also know *how* to use algebra routines to solve the problems. Then you will be able to apply these routines to more difficult problems.

Real Life
Law Enforcement

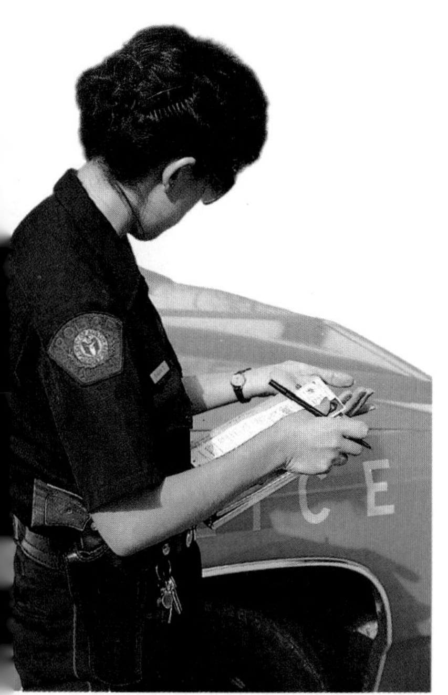

Example 1 *Getting a Speeding Ticket*

Greg lives in a state in which speeders are fined $15 for each mile per hour over the speed limit. Greg was given a ticket for $180 for speeding on a road where the speed limit is 45 miles per hour. How fast was Greg driving?

Solution

Verbal Model: [$15] • [Miles per hour over speed limit] = [Amount of speeding ticket]

Labels: Miles per hour over speed limit $= x$
Amount of speeding ticket $= \$180$

Equation: $15x = 180$ **Algebraic model**

Using mental math, you can solve this equation by finding the number that can be multiplied by 15 to obtain 180. After trying some numbers, you can see that $x = 12$. Therefore, Greg was driving 12 miles per hour over the speed limit (of 45). This means that he was driving 57 miles per hour. ■

Check your result with a table.

Speed	46	47	...	56	57
Fine	$15	$30	...	$165	$180

Problem of the Day

This dot square has 5 dots on a side. How many dots are there in all? 16

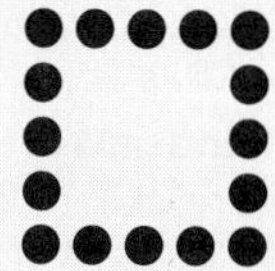

Can you write an expression for the total number of dots in a square with n dots on a side? $4(n - 1)$

ORGANIZER

TIME SCHEDULE: All levels, two days

Warm-Up Exercise

THE HANDSHAKE PROBLEM Suppose there are ten people in a room who have never met, and as they are introduced, each shakes hands with everyone else just once. How many handshakes occur?
45 handshakes

Students have a variety of ways to solve this problem. Some students might actually want to enact the event and have ten people shake hands as described; some may build a table like the one that follows; and some might even see a geometric interpretation of the problem.

Number of People	Number of Handshakes
1	0
2	1
3	3
4	6
5	10
↓	↓
10	45

Lesson Resources

Teaching Tools
- Transparency: 3
- Problem of the Day: 1.7
- Warm-up Exercises: 1.7
- Answer Masters: 1.7

Extra Practice: 1.7

The stars are people; the lines are handshakes.

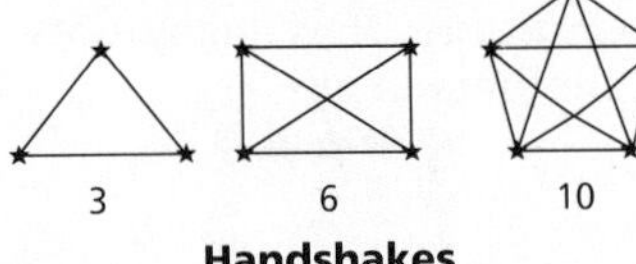

Handshakes

LESSON Notes

Examples 1–3

The first step in solving any kind of problem is to understand what the problem is about. Restating a problem in your own words is a step toward solving it. Encourage students to use their own verbal models of the problems they encounter. Support students who offer alternate approaches to solving problems. Frequently, they employ approaches that may be better understood by other students.

Example 1

Call attention to the use of a table to find the solution. Give students an opportunity to compare the verbal model they created with the one found in the text. Talk about similarities and differences in the models.

Example 2

Recall that the ground speed of an airplane is its speed relative to the ground over which it travels.

Example 3

Point out to students how the inequality in Example 3 models a real-life situation. Clearly, not all relationships that are used in algebra are about equality! Tell students that total-test-points is the sum of perfect scores on each exam.

Goal 2 Making an Algebraic Model

In Lesson 1.6, you studied the three steps for algebraic modeling. These steps can be used as part of your **general problem-solving plan.**

Real Life
Air Traffic Control

Example 2 *Piloting an Airplane*

A commercial jet aircraft is flying from Dallas to Nashville at a speed of 500 miles per hour. As the plane is passing over Little Rock, 360 miles from Nashville, an air-traffic controller at Nashville radios the pilot and says that the plane must wait two hours for clearance to land. At what speed should the aircraft fly from Little Rock to Nashville to arrive in two hours? Should the pilot reduce to this speed and fly directly to Nashville or must the pilot take some other action?

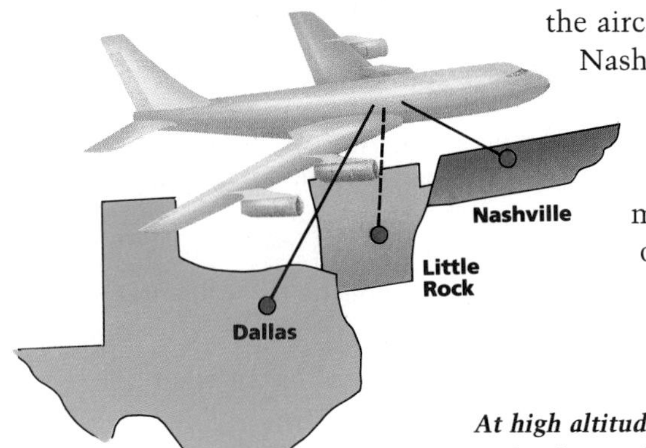

At high altitudes, a commercial jet aircraft must maintain speed of at least 350 miles per hour to avoid stalling.

A Problem-Solving Plan

Ask yourself what you need to know to solve the problem. Then **write a verbal model** that will give you what you need to know.

↓

Assign labels to each part of your verbal model.

↓

Use the labels to **write an algebraic model** based on your verbal model.

↓

Solve the algebraic model.

↓

Answer the original question. **Check** that your answer is reasonable.

Solution To solve this problem, you need to know the speed it would take to fly 360 miles in two hours. A verbal model for this can be based on the formula (rate)(time) = (distance). You would also need to know something about reasonable speeds for commercial jet aircraft.

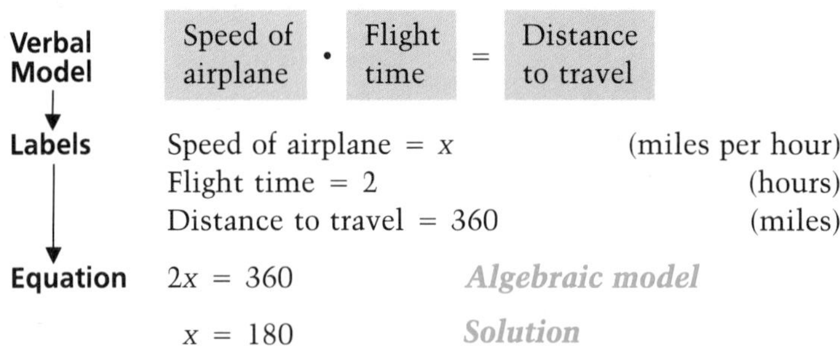

Verbal Model Speed of airplane · Flight time = Distance to travel

Labels
Speed of airplane $= x$ (miles per hour)
Flight time $= 2$ (hours)
Distance to travel $= 360$ (miles)

Equation
$2x = 360$ *Algebraic model*
$x = 180$ *Solution*

To arrive in two hours, the airplane should fly at 180 miles per hour. *Is the answer reasonable?* From the information given with the drawings, you can see that this speed is too slow. Therefore the pilot should take some other action, such as circling in a holding pattern, to use up some of the time. ■

Sometimes the solution of a real-life problem will involve an inequality, as shown in Example 3.

Real Life **Testing**

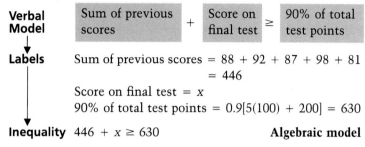

Example 3 *Getting an A in Class*

Suppose you are taking a class that has five 100-point tests and one final test worth 200 points. You need an average of 90% to get an A in the class. Your scores for the first five tests are 88, 92, 87, 98, and 81. Is it possible to score high enough on the final test to get an A?

Solution

Verbal Model

Sum of previous scores	+	Score on final test	≥	90% of total test points

Labels

Sum of previous scores $= 88 + 92 + 87 + 98 + 81 = 446$

Score on final test $= x$

90% of total test points $= 0.9[5(100) + 200] = 630$

Inequality $446 + x \geq 630$ **Algebraic model**

Since there are 200 points possible on the final test, it is possible for you to score high enough to get an A. In fact, any score of 184 or more would be enough. ■

Communicating about ALGEBRA

Cooperative Learning

SHARING IDEAS about the Lesson

Does the Answer Make Sense? When solving problems, always ask whether the answer makes sense. Work with a partner to explain why each of the answers below is suspicious, then write several similar problems.

A. A problem asked you to find the sales tax on a bottle of cologne, and your answer is \$39.56. Sales tax is too much.

B. A problem asked you to find the area of a kitchen floor, and your answer is 736 square feet. Area is too big.

C. A problem asked you to find the interest on a \$2000 deposit that is left in a savings account for one month, and your answer is \$105.75. Interest is too much.

D. A problem asked you to find the time it takes for a passenger plane to travel from New York to Los Angeles, and your answer is $1\frac{1}{2}$ hours. Time is too little.

Extra Example

Here is additional example to use in class.

BUYING A COMPUTER Rashida decided to purchase a computer that cost \$1500. Using the money she saved from baby-sitting and birthday presents, she was able to make a \$575 down payment. She borrowed the rest from her aunt and promised to repay her in 20 equal monthly payments. How much must each payment be? \$46.25

a. Write a verbal model that relates the amount of the monthly payments, the number of monthly payments, and the balance to be paid on the computer.

b. Assign labels to the verbal model.

c. Use labels to translate the verbal model into an algebraic model.

d. How much is each monthly payment?

e. What can you do to check the answer?

Check Understanding

MATH JOURNAL Students might record their responses in their math journals.

What are the steps you should use to solve problems? See page 40.

Communicating *about* **ALGEBRA**

Have students provide possible reasons why these answers were obtained. In Part D, students must understand the relationship among the different time zones in the United States. Tell them that when it's 9:00 A.M. in Los Angeles, it's 12:00 noon in New York City.

EXTEND *Communicating PROBLEM SOLVING*

Have students establish a range of acceptable answers for each situation described in Parts A-D on page 41. Students should be prepared to give reasons to support their choices.

EXERCISE Notes

ASSIGNMENT GUIDE

Day 1:
Basic/Average: Ex. 1–24
Above Average: Ex. 1–24
Advanced: Ex. 1–28

Day 2:
Basic/Average: Ex. 25–35
Above Average: Ex. 25–36
Advanced: Ex. 29–42

Selected Answers
Exercises 1–6, 7–35 odd

More Difficult Exercises
Exercises 35–42

Guided Practice

Ex. 1–6 These exercises cover the important features of using a general problem-solving plan to solve real-life problems.

Answers

1. Here is a verbal model for the given problem.

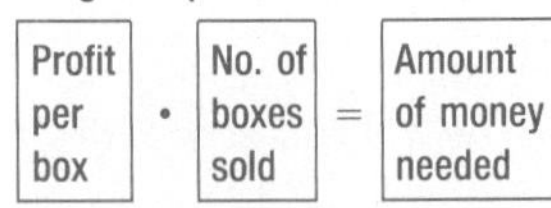

2. Profit per box = 1.50 (dollars)
Number of boxes to be sold = x
Amount of money needed = 1800 (dollars)

EXERCISES

Guided Practice

CRITICAL THINKING about the Lesson

In Exercises 1–6, use the Problem-Solving Plan to answer the questions.

Send a Student to Washington The science club at school is selling boxes of greeting cards at a profit of \$1.50 each. The club needs \$1800 more to have enough money for its trip to the Smithsonian Institution in Washington, D.C. How many boxes does it need to sell in order to earn \$1800?

Vietnam War Memorial, Washington, D.C. The Washington Monument is in the background.

1. Write a verbal model that relates the number of boxes, the profit for each box, and the amount of money the club needs to raise. See margin.
2. Assign labels to the three parts of your model. What would you let x equal? See margin.
3. Use the labels to translate your verbal model into an equation. $1.50x = 1800$
4. Use mental math to solve the equation. $x = 1200$
5. How many boxes of cards does the club need to sell? 1200
6. How could you check your answer? Multiply \$1.50 by 1200.

Independent Practice

In Exercises 7–12, consider the following question.

Walk or Take the Subway? Suppose you are one mile from your home. You are in a hurry to get home and are trying to decide whether to walk or take the subway. You can walk at a speed of 4 miles per hour. The subway schedule states that the train comes by every 15 minutes, and you heard one come by 3 minutes ago. The subway ride takes 8 minutes. Which way would get you home faster?

7. How many minutes would it take to get home by subway if you took the next train? 20
8. Write a verbal model that relates the time it would take to walk home, your walking speed, and the distance to your home. See margin.
9. Assign labels to the three parts of your model. What should you let t equal? See margin.
10. Use the labels to translate your verbal model into an equation. $t \cdot 4 = 1$ or $t \cdot \frac{1}{15} = 1$
11. Use mental math to solve the equation. $t = \frac{1}{4}$ hr or $t = 15$ min
12. Should you walk or take the subway? Explain. Walk; it takes less time.

In Exercises 13–18, consider the following question.

Disc Jockey or Live Band? Suppose you are in charge of the music for a school dance. The school budget allows only $150 for music, which is enough to hire a disc jockey for 4 hours. You would rather hire a live band, but the band charges $75 an hour, and your school does not allow students who attend the dance to be charged an admission fee. If you ask each person who attends the dance to voluntarily contribute $0.75, how many must contribute to cover the extra cost for the band?

13. How much extra money do you need to raise? $150

14. Write a verbal model that relates the extra cost, the number of contributors, and the suggested contribution amount. See margin.

15. Assign labels to the three parts of your model. See margin.

16. Use the labels to translate your verbal model into an equation. $150 = y \cdot 0.75$

17. Use mental math to solve the equation.

18. How many students must contribute to cover the cost? 200

17. $y = 200$

In Exercises 19–24, consider the following question. 20., 21., 25., 26., 30., 31. See Additional Answers.

Keeping Peace in the Family It's 9 o'clock on Saturday morning and your uncle calls. He wants you to go with him to your grandmother's home to help move furniture. He figures that it will take 1 hour to move the furniture. You already promised your mom that you would thoroughly clean your room (about 2 hours of work), *and* you have plans to go out with some friends at 6 o'clock in the evening. If your uncle can average 50 miles per hour driving to your grandmother's, what is the greatest distance she can live from you so that you can get everything done?

19. How many hours are available for driving to and from your grandmother's house? 6

20. Write a verbal model that relates the hours available, the speed of your uncle's car, and the greatest distance possible to your grandmother's home.

21. Assign labels to the three parts of your model.

22. Use the labels to translate your verbal model into an equation. $50 \cdot 6 = x$

23. Use mental math to solve the equation. $x = 300$ round trip

24. Suppose that your grandmother lives 100 miles from you. Can you get everything done? Explain. Yes; 200 mi is fewer miles than 300 mi.

Independent Practice

▶ **Ex. 7–12, 19–24, 29–34**
Use the distance formula, $d = rt$, to write verbal models for these problems. Point out to the students that each of these verbal models uses labels whose values must be found before the verbal models can be evaluated.

Answers

8. Here is one verbal model.

Time to walk home	•	Your walking speed	=	Distance from home

9. Time to walk home = t (hours)
Your walking speed = 4 (mph)
Distance from home = 1 (miles)
Alternate labels can be written using *minutes* and *miles per minute*
Time to walk home = t (minutes)
Your walking speed = $\frac{1}{15}$ (miles per minute)

▶ **Ex. 13.** Observe that the extra money needed is $300 ($75 • 4) minus the $150 from the school budget, or $150.

Answers

14. Here is one verbal model.

Extra cost	=	No. of contributors	•	Suggested contribution

15. Extra cost = 150 (dollars)
Number of contributors = y
Suggested contribution = 0.75 (dollars)

In Exercises 25–28, consider the following question.

Buying a Stereo An appliance store sells two different stereo models. The model without a compact disk player is \$350, and the model with a compact disk player is \$432. Suppose you have a summer job that lasts for 12 weeks. You are fairly sure that you can save \$30 a week, which would allow you to buy the model without the compact disk player. How much would you have to save each week to be able to buy the other model? 28. 36; you would have to save \$36 each week.

25. Write a verbal model that relates the number of weeks worked, the amount you save each week, and the price of the stereo with the compact disk player.

26. Assign labels to the three parts of your model.

25., 26. See Additional Answers.

27. Use the labels to translate your verbal model into an equation. $12m = 432$

28. Use mental math to solve the equation. Tell what its solution means. See above.

In Exercises 29–34, consider the following question.

Detective Work Suppose you are a detective and have been assigned to a robbery case. The robbery occurred at 4:30 P.M., and you have a witness who spotted a suspect at a gas station 12 miles from the robbery site at 4:48 P.M. The suspect claims that it would have been impossible to get from the robbery site to the gas station in the given amount of time. Do you agree?

29. How much time would the suspect have had to go from the robbery site to the gas station? 18 minutes

30. Write a verbal model that relates the travel time, the rate of travel, and the distance from the robbery site to the gas station.

31. Assign labels to the three parts of your model. 30., 31. See Additional Answers.

32. Use the labels to translate your verbal model into an equation. $18x = 12$ or $\frac{3}{10}x = 12$

33. Use mental math to solve the equation. $x = \frac{2}{3}$ mi per min or $x = 40$ mi per hr

34. Do you agree with the suspect? Explain. No; the suspect could have traveled by car at $\frac{2}{3}$ mile per minute or 40 mph.

Integrated Review

35. *Playing Pinball* Suppose you are playing a pinball machine that gives you five balls. In order to get an extra ball, you must have a score of at least 2500. On the first four balls, your points are 412, 613, 510, and 620. Let x represent the number of points you get on the fifth ball. Write an inequality that shows the values of x that will allow you to get an extra ball. What is the smallest value of x that is a solution of this inequality? $412 + 613 + 510 + 620 + x \geq 2500$, 345

36. *Class Election* Suppose you are running for class president. Your classmates have one week to vote for you or your opponent. Halfway through the week, you have 95 votes and your opponent has 120 votes. Forty-five more students will be voting. Let x represent the number of students (of the 45) who vote for you. Write an inequality that shows the values of x that will allow you to win the election. What is the smallest value of x that is a solution of this inequality? $95 + x > 130$, 36

Exploration and Extension

The highest averge lap speed attained on a closed-circuit motorcycle race is 160.3 miles per hour. The driver of the cycle was Yvon du Hamel, at Daytona International Speedway, Florida, March 1973. (Source:* Guinness Book of World Records, *1991)

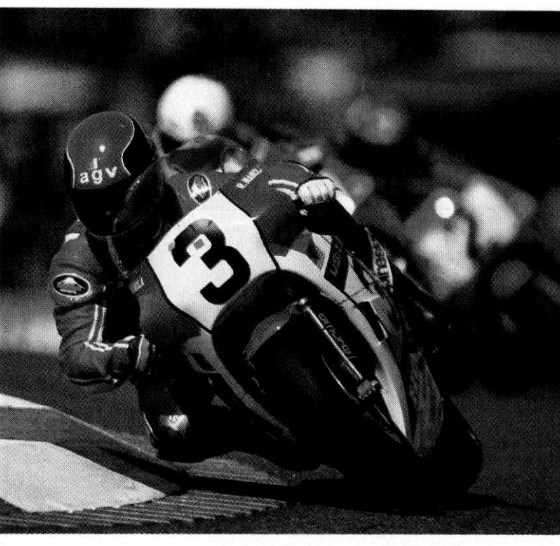

In Exercises 37–42, consider the following question.

Pit-Stop Problem Suppose you enter a closed-circuit motorcycle race. Your goal is to cross the finish line at 10:20 A.M. At 6:20 A.M., you resume racing after making an unexpected pit stop that is 640 miles from the finish line. Do you think you can reach your goal?

37. How much time do you need to reach your goal? 4 hours

38. Write a verbal model that relates the travel time you have, the distance, and the rate at which you must travel. See margin.

39. Assign labels to the three parts of your model. See margin.

40. Use the labels to translate your verbal model into an equation. $4x = 640$

41. Use mental math to solve the equation. $x = 160$

42. Do you think you will meet your goal? Explain. See margin.

Ex. 35-36 Inequalities are needed to express the relationships described in these problems. Emphasize to students that the sequence of steps used to write algebraic equations (see pages 34 and 40) are also used to write inequalities. One inequality that models the information of Exercise 35 is $x \geq 2500 - (412 + 613 + 510 + 620)$.

Here is a verbal model of the information in Exercise 36.

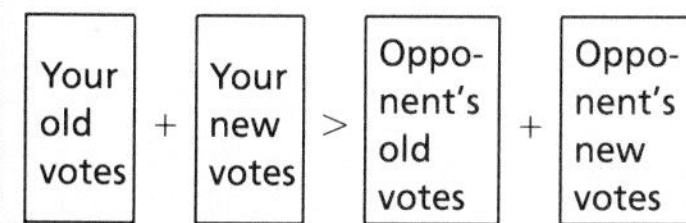

The algebraic inequality becomes $95 + x > 120 + (45 - x)$, where x is the number of votes you received and $(45 - x)$ is the number of votes received by your opponent.

Answers

38. Here is one verbal model.

Time to finish race	•	Rate you must travel	=	Distance

39. Time to finish race = 4 (hours)
Rate you must travel = x (mph)
Distance = 640 (miles)

42. No, the chances of your maintaining an average of 160 mph for 4 hours are very slight.

Enrichment Activity

WRITING PROBLEMS A good way to see how well students understand the use of algebra to solve problems is to have them create problems for their cooperative group partner(s) to solve. Students find this challenging and fun.

Problem Solving

Make an Algebraic Model

Write a verbal model.

Assign labels (values and variables).

Write an algebraic model (an equation or inequality) using the labels.

↓

Solve the equation or inequality.

↓

Answer the original question. Check that your answer is reasonable.

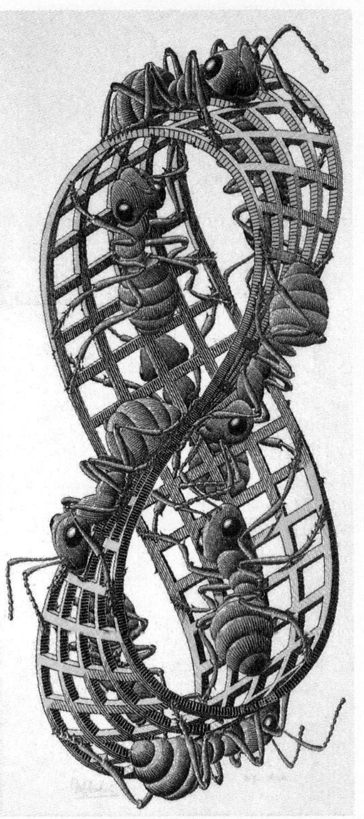

A Problem-Solving Plan

Problem solving is a process for discovering relationships; it is not simply a process for finding answers. In this text, problem solving is something you will do while you learn new skills. To help you solve problems successfully, you will have many problem-solving opportunities.

In Lesson 1.7, you were introduced to a systematic plan for solving problems algebraically, as shown in the flowchart at the left. First, write a *verbal model* for the problem. Then assign labels to the verbal model and write an *algebraic model* (an equation or inequality). Finally, solve the equation or inequality. You will be using this algebraic-model strategy throughout this algebra text.

Other problem-solving strategies can be used with and without algebraic modeling. In fact, in this chapter you have been informally introduced to the strategies *making a table, looking for a pattern, using a model,* and *drawing a diagram.* The uses of such strategies are

- to provide techniques for solving problems without using equations or inequalities.
- to provide insight and information when using the general algebraic-model plan. For example, it is often useful to draw a diagram to help you understand the algebraic model.
- to provide a second technique as a check to a problem solved algebraically. In Example 1 on page 39, *making a table* is used to check the solution provided by the algebraic model.

Some of the names for these nonalgebraic problem-solving strategies may be familiar already. On the following two pages, you will see how three such strategies can be used. Each strategy has a general "plan of attack" that may help you get started and then proceed through a problem. Each can be a powerful tool in your investigation of the problems and applications in this course.

Mobius Strip II, *1963, woodcut by M. C. Escher.*

There may be several ways to approach each of the following three problems. Since these are examples, a specific strategy is suggested at the left. However, an essential part of problem solving is making a decision about which strategy seems best to use.

Problem Solving
Guess, Check, Revise

Guess a reasonable solution based on data in the problem.

↓

Check the guess.

↓

Revise the guess and continue until a correct solution is found.

Example 1 *A Puzzle with Parentheses*

Insert parentheses so that the expression 54 − 6 • 8 − 5 has a value of 36.

Solution A strategy to try with this problem is *Guess, Check, and Revise.* Copy the expression, guess where to place the parentheses, and evaluate the expression. Always record each guess, because patterns in your results may help you with your next guess.

Original expression: 54 − 6 • 8 − 5

First guess: (54 − 6)(8 − 5) = (48)(3). Too great.

Second guess: (54 − 6)8 − 5 = (48)(8) − 5. Too great.

Third guess: 54 − 6(8 − 5) = 54 − (6)(3) = 54 − 18 = 36. √

Problem Solving
Make a Table/ Look for a Pattern

Make a table using the data from the problem.

↓

Look for number patterns.

↓

Use a pattern to complete the table and/or find a solution.

Example 2 *Mrs. Mellina's Strange Algebra Assignment*

Mrs. Mellina tells her algebra students that they will have 1 problem for homework the first day, 1 the second day, 2 the third day, 3 the fourth day, 5 the fifth day, 8 the sixth day, and 13 on the seventh day. If she continues this pattern, how many problems will be assigned on the eleventh day?

Solution A strategy to try with this problem is *Make a Table/Look for a Pattern.* Use a table that shows the number of problems assigned each day.

Day	1	2	3	4	5	6	7	8	9	10	11
Number of problems	1	1	2	3	5	8	13	?	?	?	?

Do you see a pattern in the second row? It appears that the number of problems for each day is the sum of the numbers from the previous two days. Thus, day 8 will have 21 problems, day 9 will have 34 problems, day 10 will have 55 problems, and day 11 will have 89 problems!

Answers

1. Use *Make a Table.*

Perimeter (cm)	Length (cm)	Width (cm)	Area (cm^2)
24	11	1	11
24	10	2	20
24	9	3	27
24	8	4	32
24	7	5	35
24	6	6	36

6 cm by 6 cm gives the greatest area

2. Use *Guess, Check, and Revise.*
$3 + (3 \div 3) \cdot (3 + 3) \cdot 3 - 3 = 18$
3. Use *Solve a Simpler Problem* a few times.
2 teams require 1 game, 3 teams require 2 games, 4 teams require 3 games, 5 teams require 4 games, 8 teams require 7 games. Use *Look for a Pattern:* the number of games is 1 less than the number of teams.
128 teams require 127 games.
4. Use *Draw a Diagram.*
He can get 7 pieces.

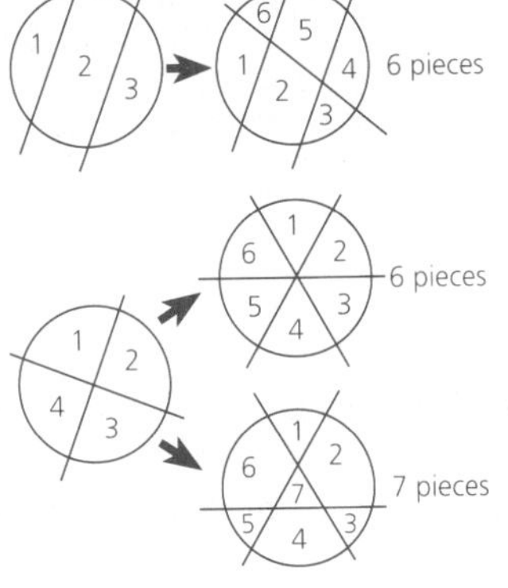

Problem Solving
Draw a Diagram

Draw a diagram that shows the facts from the problem.

↓

Use the diagram to visualize the action of the problem.

↓

Use algebra to find a solution. Then check the solution against the facts.

Liberation, *1955, lithograph by* M. C. Escher.

Example 3 ***Mr. Galloway's Quiz Bowl Budget***

Central High School is hosting an 8-team single-elimination Academic Quiz Bowl tournament. Mr. Galloway, the supervisor, must pay each moderator $25 per match. How much money should Mr. Galloway budget to pay the moderators?

Solution Mr. Galloway needs to know the total number of matches for the tournament. A strategy to try with this problem is *Draw a Diagram* of the matches for each round of the tournament.

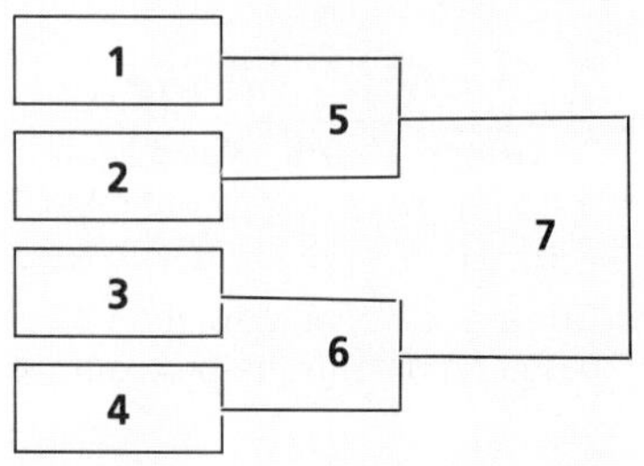

From the diagram, Mr. Galloway can see that the first round has four matches, the second round has two, and the last round has one. Thus, he has a total of $4 + 2 + 1 = 7$ matches and must budget (7)($25) = $175 for moderator fees.

Exercises

Decide which problem-solving strategy (or strategies) would be best to use and then solve the problems. See margin.

1. A rectangle has a perimeter of 24 centimeters. Its length and width are whole numbers. What are all the possibilities for the length and width? Which dimensions give the greatest area?
2. Insert parentheses so that the expression
$3 + 3 \div 3 \cdot 3 + 3 \cdot 3 - 3$
is equal to 18.
3. Example 3 above shows with a diagram that an 8-team elimination tournament requires 7 games. Now, instead of a diagram, use other strategies to find how many games would be needed in such a tournament for 128 teams.
4. Rafael is slicing a round pizza. He is challenged to use three straight cuts through the pizza to create as many pieces as possible. Draw a diagram that illustrates the most pieces possible.

1.8 Exploring Data: Tables and Graphs

What you should learn:

Goal 1 How to use tables to organize data

Goal 2 How to use graphs to model data visually

Why you should learn it:

You can use tables and graphs to help you see relationships among such data collections as survey results.

Goal 1 Using Tables to Organize Data

The world you live in is becoming more and more complex. Almost every day of your life you will be given **data** that describe real-life situations. The word *data* is plural and it means facts or numbers that describe something.

A collection of data is easier to understand when it is organized in a table or in a graph. There is no "best way" to organize data, but there are many good techniques. Often it helps to put numbers in order (either increasing order or decreasing order). It also helps to group numbers so that patterns or trends are more apparent.

Real Life **Physical Fitness**

Example 1 *Burning Calories*

The average number of Calories that you burn per minute depends on your activity. Resting burns 1.1 Calories per minute, walking burns 5.5, swimming burns 10.9, and running burns 14.7. Suppose that you are the manager of a physical fitness center and want to create a poster that conveys this information to the members of the center. How would you organize the information?

Number of Calories in Foods	
Quarter-pound cheeseburger	520
Turkey deluxe sandwich	510
Beef burrito	466
Big roast beef sandwich	418
Two pieces pizza	380

Calories are averages taken from an analysis of eleven fast food restaurant chains.

Solution One way would be to find the number of Calories in several different common foods, as shown in the table at the left. Then calculate the number of minutes of each activity that is required to burn the Calories from each food. Finally make a table that compares the times required to burn the Calories from different foods.

	Time Required to Burn Calories			
Food	**Resting**	**Walking**	**Swimming**	**Running**
Cheeseburger	473 min	95 min	48 min	35 min
Turkey deluxe	464 min	93 min	47 min	35 min
Beef burrito	424 min	85 min	43 min	32 min
Beef sandwich	380 min	76 min	38 min	28 min
Pizza	345 min	69 min	35 min	26 min

■

Problem of the Day

Look for the pattern and find the values of x and y.

$a + 45 = 101$
$b + 445 = 1001$
$c + 4445 = 10001$
$d + 44445 = 100001$
•
•
•
$x + y = 100000001$

$x = 55{,}555{,}556$; $y = 44{,}444{,}445$

ORGANIZER

Warm-Up Exercises

GROUP ACTIVITY Prepare a copy of these questions for each student (or for groups of students.)

1. How would you decide what the ten most-watched television programs are for the students at your school? Sample answer: Students should make a class list of the information by boys and by girls. Discuss how the information can be displayed and permit several representations of the data—such as tables, tallies, organized lists, etc.

Use the results of Exercise 1 to answer the following.

2. What programs do most girls watch?
3. What programs do most boys watch?
4. What time of day do most students watch television ?

Lesson Resources

Teaching Tools
- Transparency: 4
- Problem of the Day: 1.8
- Warm-up Exercises: 1.8
- Answer Masters: 1.8

Extra Practice: 1.8

Color Transparencies: 3, 4, 5, 6

Applications Handbook, p. 40, 53

Technology Handbook: p. 9

LESSON Notes

Example 1

Ask students if they can think of another way to represent the data in Example 1. (Example 2 might suggest a way.)

This lesson is the first of a series of exploring-data lessons in the text. The feature is designed to develop students' quantitative-literacy skills through the use of meaningful and relevant data sets. Students will explore data by studying tables and graphs, time-line plots, pictographs, circle graphs, measures of variance and dispersion, stem-and-leaf plots, box-and-whisker plots, and techniques in probability appropriate to the problem-solving process of algebra.

Example 2

In the bar graph in Example 2, note that each age group has a span of 10 years. For instance, everyone who has not had their tenth birthday is in the group labeled 0–9.

Common-Error ALERT!

In **Example 2,** students frequently misread the legend in bar graphs. Caution students to notice the difference in the way the lines are drawn inside each bar and how each line pattern in a bar is identical to one of the three bars (the legend) in the upper right-hand corner of the graph. Matching the line patterns between the legend and the bars helps students identify the year represented by each bar.

Goal 2 Using Graphs as Visual Models

One of the most common ways to organize data is with a **bar graph.**

Real Life **Demography**

Example 2 *Changes in Age Distribution in the United States*

The following bar graph gives the numbers of people in different age groups in the United States in 1970, 1980, and 1990. ***(Source: U.S. Bureau of the Census)***

a. How many people who were between 10 and 19 years old lived in the United States in 1970?

b. Which age groups had the most people in 1970, in 1980, and in 1990? Can you see a pattern to your three answers?

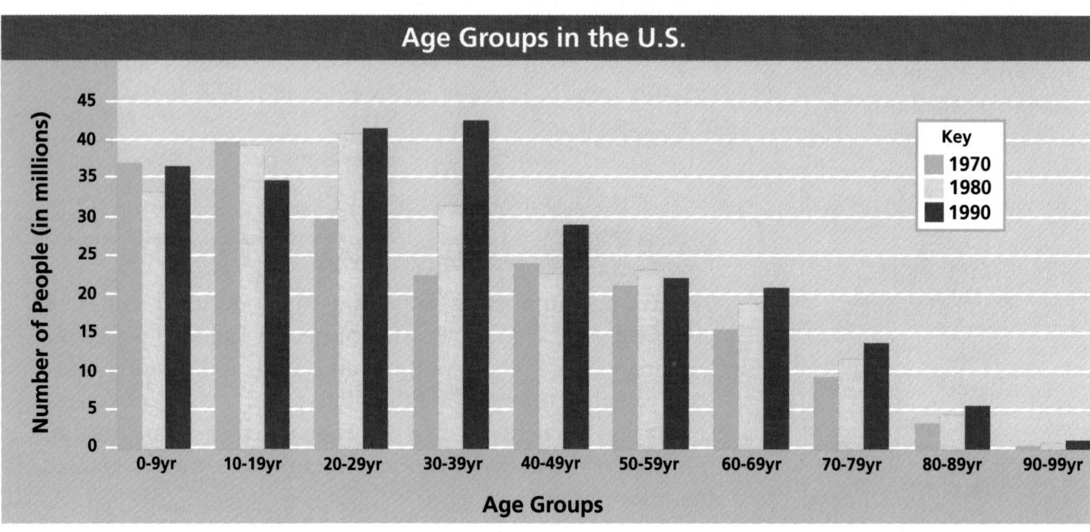

Solution

a. Look for the label *10–19* at the bottom of the graph. There are three bars above this label. The left bar represents 1970, the middle bar represents 1980, and the right bar represents 1990. From the scale given on the left side of the graph, you can read that there were about 40 million people between the ages of 10 and 19 in 1970.

b. In 1970, the age group 10–19 had the most people. In 1980, the age group 20–29 had the most people. In 1990, the group 30–39 had the most people. This makes sense because if there were more 10–19-year-olds than any other age group in 1970, then ten years later, in 1980, there should have been more 20–29-year-olds than any other age group. ■

Real Life
Leisure Activities

Example 3 *Number of Movies Watched in a Week*

Thirty students in a class were asked to keep track of the number of movies each watched during a specific week. The results were as follows.

6, 3, 0, 10, 4, 1, 7, 2, 3, 4, 4, 4, 1, 3, 4
3, 5, 5, 7, 2, 5, 4, 3, 3, 0, 2, 4, 5, 1, 5

Organize these data to help show a pattern.

Solution Three common ways to organize a collection of numbers are shown below. The first is a **frequency distribution.** Note that the possible numbers are written *in order* and tally marks are used to count how many times each number occurs in the collection. The second is a **line plot.** Each × in the graph represents a number from the collection. A third is a bar graph, as shown below.

a.

Number	Tally	Frequency
10	I	1
9		0
8		0
7	II	2
6	I	1
5	IIIII	5
4	IIIIIII	7
3	IIIIII	6
2	III	3
1	III	3
0	II	2

Frequency distribution

b.

```
          ×
        × ×
        × × ×
        × × ×
      × × × ×
  × × × × × ×   ×
  × × × × × × × ×     ×
+-+-+-+-+-+-+-+-+-+-+
0 1 2 3 4 5 6 7 8 9 10
```

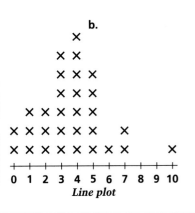

Line plot

c.

Bar graph ■

Communicating about ALGEBRA

SHARING IDEAS about the Lesson

Interpreting Data **Use Examples 1–3 of this lesson.**

A. Use the information given in Example 1 to find the number of minutes resting, walking, swimming, and running it would take to burn a sandwich that gives 350 calories. 318, 64, 32, 24

B. Use the information given in Example 2 to approximate the number of 30–39-year-olds in the United States in 1970, 1980, and 1990. See below.

C What is the average number of movies watched per student during the week discussed in Example 3? $3\frac{2}{3}$

B. 23 million, 32 million, 42 million

Example 3

Ask students to tell you what the numbers in the frequency distribution chart represent. Be prepared for lots of interesting responses.

Check Understanding

MATH JOURNAL Students might record their personal responses in their math journals.

Write about the uses and frequency of information that is presented by using tables, bar graphs, etc. Answers should include the importance of using tables and graphs to organize data to make it easier to read and interpret.

Communicating **about ALGEBRA**

Encourage students to compare the results of these activities and explain their findings to each other.

In Part C, determine the most common way students used to compute the average. Ask if the frequency distribution shown in Example 3 can be used to compute the average number of movies watched by a student. (You can obtain the total number of movies watched by using information from the frequency distribution table. The total is (10 × 1) + (9 × 0) + (8 × 0) + (7 × 2) + (6 × 1) + (5 × 5) + ... + (1 × 3) + (0 × 2), where (m × n) indicates that n students watched m movies.)

EXERCISE Notes

ASSIGNMENT GUIDE

Basic/Average: Ex. 1–4, 5–20

Above Average: Ex. 1–4, 5–23

Advanced: Ex. 1–4, 5–25

Selected Answers

Exercises 1–4, 5–21 odd

More Difficult Exercises
Exercises 11–12, 14, 21–22

Guided Practice

Ex. 1–4 The Guided Practice exercises ask general questions about data collection and data representation.

Independent Practice

Answers

5. List the records down the left-hand column and the events across the top or list events down and records across.
6. No; they are equal in readability.
8. Compact disk, no, first
9. Numbers sold are added cumulatively in the lower graph.

EXERCISES

Guided Practice

CRITICAL THINKING about the Lesson

1. What does the word *data* mean?
Facts or numbers that describe something.

2. Can a table of numbers help organize data? Explain. Yes, because a table is an organized set of data.

3. What is a frequency distribution?
It is an organized set of data that tells how often each datum occurs.

4. When drawing a bar graph, does it help to make a frequency distribution first? Explain. Sometimes; it helps when the numbers in the data are small and occur at various frequencies.

Independent Practice

5. *Track Records* Data for track records for an upcoming track meet must include the particular meet record, the school's record, the national high school record, and the world high school record for six running events: 100-meter dash, 200-meter dash, 400-meter dash, 800-meter run, 1500-meter run, and 3000-meter run. Describe two ways that these data could be organized in a table.
5., 6., 8., 9. See margin.

6. In Exercise 5, do you think a table with four rows and six columns would be easier to read than a table with six rows and four columns? Explain.

CD's, Record Albums, and Cassettes **In Exercises 7–10, use the bar graphs which show the numbers (in millions) of compact discs (CDs), record albums, and cassettes sold from 1990 to 1994.** ***(Source: Recording Industry Association of America)***

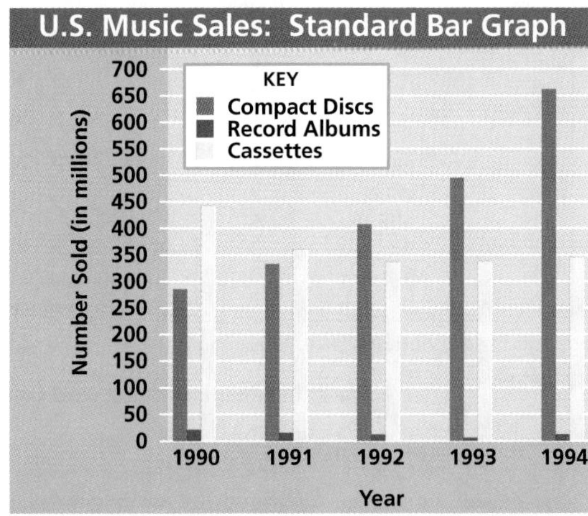

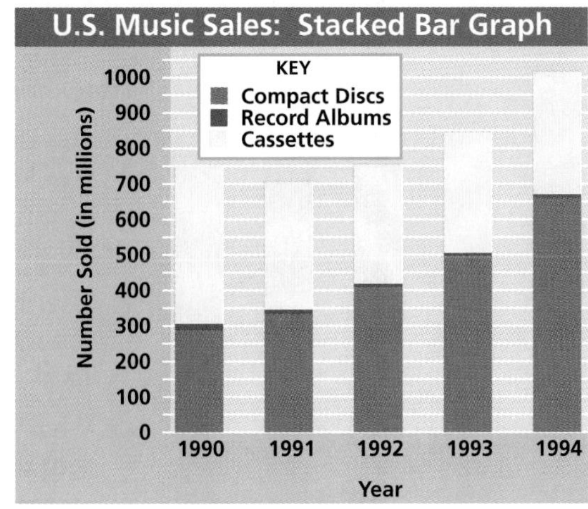

7. In which year did the compact disc sales first surpass the cassette sales? Which bar graph is easier to use to answer this question? 1992, first

8. Which type(s) of recording had increasing sales from 1990 to 1994? Did any have decreasing sales? Which bar graph is easier to use to answer these questions?

9. How did the lower **stacked bar graph** differ from the upper bar graph?

10. In which two years were total sales approximately the same? Which bar graph is easier to use to answer this question?
1990 and 1992, second

College Entrance Exam Sample **In Exercises 11 and 12, use the table showing the price of different quantities of bagels.**

Number of Bagels	1	Bag of 6	Bag of 12
Total Price	$0.40	$1.89	$3.59

11. Which is the closest approximation of the cost per bagel when you purchase a bag of six bagels?

a. $0.20 **b.** $0.30 **c.** $0.40 **d.** $0.50 **e.** $0.60

12. What would be the *least* amount of money needed to purchase exactly 21 bagels?

a. $5.88 **b.** $6.68 **c.** $7.19 **d.** $7.38 **e.** $8.40

Study Time **In Exercises 13–16, use the following data which represent the number of hours that 30 different students in a class studied during a week.**

13., 16. See Additional Answers.

7,	4,	10,	8,	5,	2
7,	12,	3,	4,	12,	7
15,	3,	6,	9,	11,	10
7,	2,	8,	7,	5,	6
18,	6,	1,	0,	17,	13

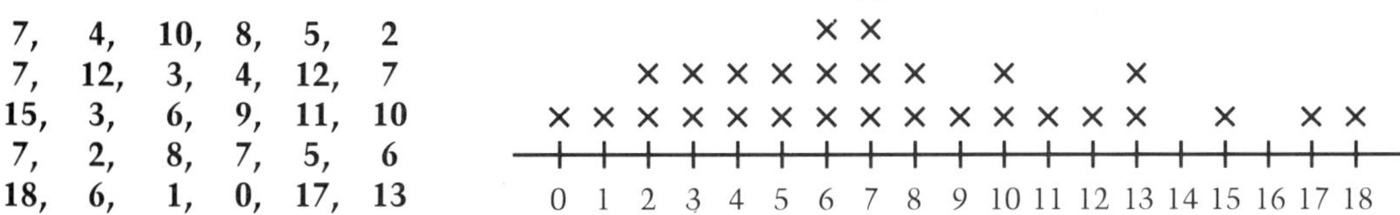

13. Construct a frequency distribution for these data.

14. What was the average number of hours studied by a student in the class? **7.5**

15. Find the error in the line plot above.
12 hours should have 2 ×'s, 13 hours should have 1 ×.

16. Sketch a bar graph for these data.

Weather Map **In Exercises 17 and 18, refer to the weather map. Note that this map uses different colors to represent different temperature ranges. The temperatures shown on this map represent the average minimum temperatures in each region.**
(Source: U.S. Department of Agriculture)

17. Approximate the average minimum temperature in Columbus, Ohio. **−23°C to −18°C**

18. Approximate the average minimum temperature in Washington, D.C. **−18°C to −12°C**

Temperature Zones

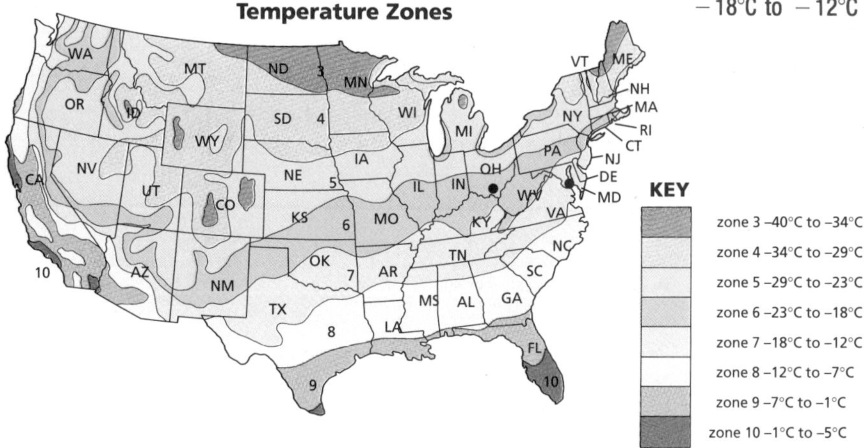

▶ **Ex. 7–9** Emphasize the importance of carefully reading the legends of the graphs. Magazines and newspapers frequently have graphs similar to the ones in these exercises. Students will find real-life examples of these graphs interesting and informative.

EXTEND Ex. 17–18 **CAREER CONNECTIONS** The lesson exercises demonstrate that many professions are concerned with the useful display of data. Can you identify other occupations in which the appropriate display of data is important? **Advertising, teaching, and sales come quickly to mind.**

▶ **Ex. 19–22** GEOMETRY These exercises connect geometry with data collection nicely. Frequently, it is important to collect data in search of patterns that helps you understand many key relationships in mathematics.

Answers

20. Length (ft) = 50, 45, 40, 35, 30, 25, 20, 15, 10

    Area(ft²) = 500, 675, 800, 875, 900, 875, 800, 675, 500

21. The closer the length is to the width, the greater the area; b, the length is equal to the width

22.

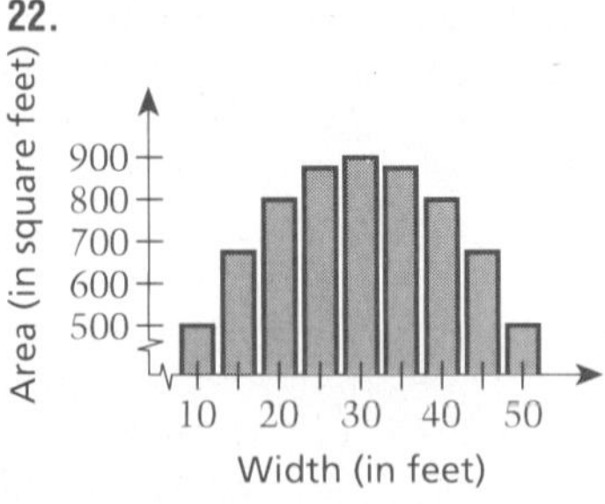

Enrichment Activities

WRITING Students might record their responses in their math journals.

1. Write a report on the use of graphs in magazines and newspapers. Include examples of the types of graphs you find. The national newspaper, *USA Today*, shows many different types of graphs in each issue.
2. Discuss how graphs can sometimes mislead the reader. One way is by representing data using a scale on which small differences appear large.
3. Are any of the graphs you found in newspapers or magazines misleading? Why?

Integrated Review

Perimeter and Area of a Rectangle **In Exercises 19–22, suppose that you have 120 feet of fencing with which to enclose a rectangular region.**

19. Let W be the width of the rectangle and let L be the length. Write an equation involving W, L, and 120. $W + L + W + L = 120$ or $2W + 2L = 120$

20. Copy and complete the following table which compares the areas of several rectangles whose perimeters are 120. 20.–22. See margin.

Width	10 ft	15 ft	20 ft	25 ft	30 ft	35 ft	40 ft	45 ft	50 ft
Length	?	?	?	?	?	?	?	?	?
Area	?	?	?	?	?	?	?	?	?

✪ **21.** If the area is to be as large as possible, which of the following rectangles should you construct? Explain.

a. A rectangle that is longer than it is wide

b. A rectangle whose width is equal to its length (a square)

✪ **22.** Construct a bar graph that compares the width of a rectangle, whose perimeter is 120 feet, with its area.

Exploration and Extension

Exercise Activities **In Exercises 23–24, use the bar graph which shows the percent of Americans in various age groups who participate in bike riding, exercise walking, or swimming at least six times a year.**
(Source: National Sporting Goods Association)

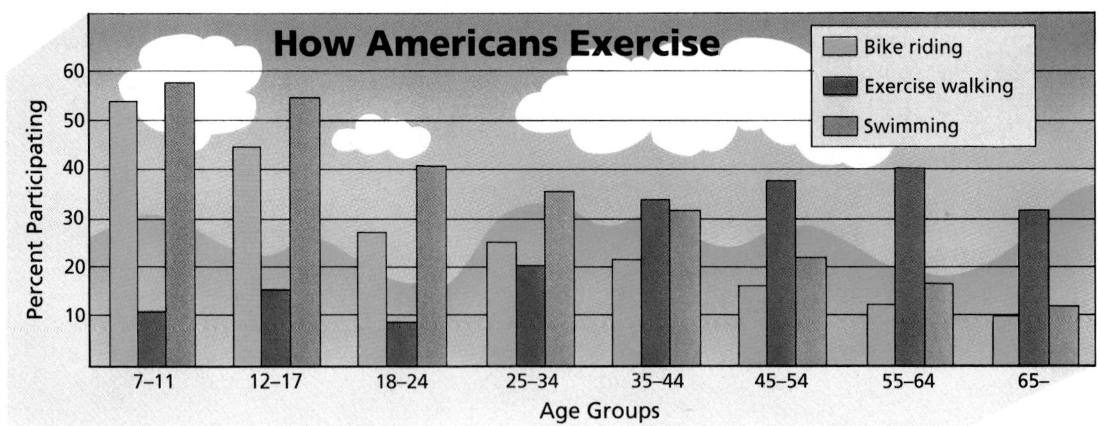

23. Which two of these three exercise activities do people tend to participate in less as they grow older? Bike riding and swimming

24. What percent of people between the ages of 12 and 17 would you expect to ride a bike at least six times a year? About 45

1 Chapter Summary

What did you learn?

Skills

1. Evaluate an expression
   - containing numbers, grouping symbols, and exponents. **(1.1, 1.3)**
   - containing variables. **(1.2)**
   - using the established order of operations. **(1.4)**
   - using a calculator. **(1.1–1.4)**
2. Check a solution of an equation or inequality. **(1.5)**
3. Solve an equation
   - by expressing the equation as a question, and then **(1.5)**
   - using mental math to answer the question. **(1.5, 1.6)**
4. Translate
   - a verbal phrase into an expression. **(1.6)**
   - a verbal sentence into an equation or inequality. **(1.6)**

Strategies

5. Create a problem-solving plan for a real-life situation
   - writing a *verbal model,* **(1.6)**
   - assigning *labels,* **(1.6)**
   - writing an *algebraic model,* **(1.6)**
   - using mental math to *solve* the model, **(1.7)**
   - answering questions about the real-life situation. **(1.7)**
6. Use algebra to enhance your problem-solving strategies. **(1.7)**

Exploring Data

7. Organize data with a table, frequency distribution, and graph. **(1.8)**

Why did you learn it?

The world you live in is becoming more and more technical. Many real-life situations involve data, numbers, and patterns that can be modeled by algebraic expressions, equations, and inequalities. By learning how to create and solve a mathematical model, you will be able to use the mathematical solution to answer questions in school, in business, and in your career.

How does it fit into the bigger picture of algebra?

In this chapter, you were introduced to many of the terms and goals of algebra. Communication is a very important part of algebra, so it is important that words such as *variable, expression, equation, solution,* and *inequality* become part of your vocabulary. Some people think of algebra as only a collection of rules. Algebra is much *more* than that. It is a language that can be used to answer questions about real-life situations.

ORGANIZER

TIME SCHEDULE: All levels, two days

The Chapter Summary helps students organize the main ideas of the chapter. It is a convenient way for you to review the chapter with students in preparation for a chapter test. It is suggested that the first day of review be a combination of the students and you working problems. The first day's homework assignment will be the basis for the second day of review. You may want to check the homework assignment to avoid working those problems during the first day of review.

Be creative. The second day of review, for example, could be set up as a "class challenge" day. Teams of students would work to solve selected problems from the homework set.

COOPERATIVE LEARNING
Encourage students to study together. Emphasize the importance of teaching a classmate how to perform a skill or how to conceptualize the ideas in the chapter. When students work together, everyone wins. The students receiving help get additional instruction, and the students giving help gain a deeper understanding of the skills and concepts involved.

Chapter SUMMARY

Review the chapter by asking students "What did you learn? Why did you learn it? How does it fit into the bigger picture of algebra?" Encourage students to answer these questions in their own way.

Chapter REVIEW

ASSIGNMENT GUIDE

Basic/Average: Ex. 1–19 odd, 25–39 odd, 43–48, 49–53 odd, 55–58

Above Average: Ex. 1–19 odd, 25–39 odd, 43–48, 49–53 odd, 55–58

Advanced: Ex. 1–19 odd, 25–39 odd, 43–48, 49–53 odd, 55–58

▶ **Ex. 1–58** Since the Chapter Review exercises parallel the list of Skills, Strategies, and Exploring Data statements on page 55, tell students to use the Chapter Summary to recall the lesson in which a skill or strategy was first taught if they are uncertain about how to do a given exercise in the Chapter Review. This forces students to categorize the task to be performed and to locate information in the chapter about the task, both beneficial activities.

The answers to the odd-numbered exercises of the Chapter Reviews will be found in the "Selected Answers" section of the student text.

Chapter REVIEW

1. Write $\frac{1}{8}$ in decimal form. **(1.1)** 0.125

2. Write $\frac{2}{5}$ in percent form. **(1.1)** 40%

3. Write 0.75 in fraction form. **(1.1)** $\frac{3}{4}$

4. Write 0.33 in percent form. **(1.1)** 33%

5. Write 10% in decimal form. **(1.1)** 0.10

6. Write 25% in fraction form. **(1.1)** $\frac{1}{4}$

7. Find the sum of 5 and $\frac{3}{2}$. **(1.1)** $\frac{13}{2}$

8. Find the product of 9 and 9. **(1.1)** 81

9. What is the difference of 23 and 7? **(1.1)** 16

10. What is the quotient of 14 and 4? **(1.1)** $\frac{7}{2}$

In Exercises 11–14, evaluate the given expression. (1.1)

11. $3 - [(4 \div 2) + 1]$ 0

12. $3(7) + [9 - (18 \div 6)]$ 27

13. $[(4 + 16) \div 5] - 2$ 2

14. $\frac{(3)(5)}{(8 - 7) \cdot 3}$ 5

15. Use a calculator to evaluate $[224.9 \div (4.3 + 16.5)] - 9.3$. **(1.1)** 1.5125

16. Use a calculator to evaluate $14.4 + [(6.8 - 1.1) \cdot 4.7]$. **(1.1)** 41.19

In Exercises 17–20, evaluate the given expression. (1.2)

17. $2s - 3$, when $s = 14$ 25

18. $(x + 4) \div 9$, when $x = 23$ 3

19. $2x - 3y + 2$, when $x = 4$ and $y = 1$ 7

20. $(3x - 5)(y + 2)$, when $x = 3$ and $y = 2$ 16

21. Use a calculator to evaluate $2.6 + (3.9 \div x)$ when $x = 1.5$. **(1.2)** 5.2

22. Use a calculator to evaluate $3.7x - 6.2y$ when $x = 5.5$ and $y = 1.9$. **(1.2)** 8.57

23. Write "x raised to the fifth power" as an algebraic expression. **(1.3)** x^5

24. Write the expression $9 \cdot 9 \cdot 9 \cdot 9 \cdot 9 \cdot 9 \cdot 9$ in exponential form. **(1.3)** 9^7

In Exercises 25–32, evaluate the given expression. (1.4)

25. $4 + 21 \div 3 - 3^2$ 2

26. $(12 - 7)^2 + 5$ 30

27. $\frac{6 + 2^2}{17 - 6 \cdot 2}$ 2

28. $\frac{9 - 1 \cdot 7}{42}$ $\frac{1}{21}$

29. $12 - 2x^2$, when $x = 2$ 4

30. $x^3 + x - 7$, when $x = 3$ 23

31. $\frac{x - 3y}{6}$, when $x = 15$ and $y = 2$ $\frac{3}{2}$

32. $\frac{x^2}{y^2} + 4$, when $x = 4$ and $y = 2$ 8

In Exercises 33–36, use mental math to solve the equation. (1.5)

33. $2x - 8 = 0$ 4

34. $\frac{4}{x} = \frac{1}{2}$ 8

35. $x^3 = 64$ 4

36. $\frac{x}{8} = 0$ 0

37. Is 7 a solution of the equation $x^2 - x + 2 = 44$? **(1.5)** Yes

38. Is 4 a solution of $2x - 3 = 2$? **(1.5)** No

39. Determine whether 3 is a solution of the inequality $9x - 3 > 24$. **(1.5)** No

40. Determine whether 6 is a solution of the inequality $13 + 2x \leq 25$. **(1.5)** Yes

In Exercises 41 and 42, consider the problem, then answer the question about the equation that represents the problem. (1.6)

41. *Buying a Compact Disk* Suppose you want to buy a compact disk that costs \$15.90, including tax. If you can save \$4.00 a week, how many weeks must you save to buy the compact disk?

If this problem is represented by the equation $4x - y = 15.90$, what do the 4, the x, the y, and the 15.90 represent? See margin.

42. *Dance Supplies* Suppose you are in charge of buying refreshments for a school dance. You are given \$65 and each bottle of juice is 35¢. How many bottles of juice can you buy?

If this problem is represented by the equation $0.35a + b = 65$, what do the 0.35, the a, the b, and the 65 represent? See margin.

In Exercises 43–48, write an algebraic expression, equation, or inequality for the phrase or sentence. 49. $180x + 200y$ 46. $9 + 6y \leq 20$

43. Six more than a number squared $x^2 + 6$

44. Half of a number, decreased by nine $\frac{1}{2}y - 9$

45. The quotient of a number and two is greater than seven $x \div 2 > 7$

46. Nine plus the product of a number and six is less than or equal to twenty.

47. A number cubed minus another number is sixty. $x^3 - y = 60$

48. The difference of a number and two times another number is ten. $m - 2n = 10$

49. *Calorie Counter* A 12-ounce can of fruit punch has 180 Calories. A 12-ounce can of orange soda has 200 Calories. Write an expression that represents the number of Calories consumed if you drink x cans of fruit punch and y cans of orange soda.

50. *Buying Fishing Gear* New line for your fishing reel will cost 2¢ a yard. Lures are on sale for \$3.50 each. Write an expression that represents the total amount you will spend if you purchase x yards of line and y lures. $0.02x + 3.50y$

51. *A Riddle* This rhyme is one of the world's oldest mathematical riddles:

As I was going to St. Ives,
I met a man with seven wives.
Each wife had seven sacks.
Each sack had seven cats.
Each cat had seven kittens.
Kittens, cats, sacks, and wives,
How many were going to St. Ives? Just 1

The number of kittens, cats, sacks, and wives is $7^4 + 7^3 + 7^2 + 7^1 = 2800$, but what is the answer to the riddle?

52. *Falcon in Flight* A peregrine falcon sights a sparrow three miles away. If the falcon's average speed is 180 miles per hour during the attack, how long will it take for the falcon to reach the sparrow?

The peregrine falcon can attain speeds over 200 miles per hour when attacking its prey.

52. $\frac{1}{60}$ hour or 1 minute

Answers

41. 4: dollar amount saved each week, x: number of weeks you must save, y: dollar amount left, 15.90: dollar cost of disk

42. 0.35: dollar cost of a bottle of juice, a: number of bottles of juice, b: dollar amount left, 65: dollar amount available

▶ **Ex. 49–55** Continue to stress the importance of beginning with a verbal model and moving toward an algebraic one. It may seem to students that time is lost when you follow these steps, but point out that understanding the problem does eventually result in gains in time and success!

53. *Computer Screen* A square computer screen has sides whose lengths are s inches each. Write an algebraic expression that represents the area of the computer screen. What is the area of the screen if $s = 9$ inches? s^2, 81 in.^{2}

54. *Perimeter of a Picture* A rectangular picture frame has sides of length $2w$ and $3w$. Write an expression for the perimeter of the picture frame. What is the perimeter if $w = 4$ inches?

$2w + 3w + 2w + 3w$, 40 in.

55. *Basketball Score* Suppose that you are the second leading scorer in your high school's basketball league. The league's leading scorer has finished the season with a game average of 20 points (for 25 games). You have a 19.5-point game average and have yet to play your 25th game. How many points do you need to score to overtake the league's leading scorer? 33

Unemployment Rate **In Exercises 56–58, use the following bar graph which shows the percentages of Americans (of working age) who were unemployed for each year from 1986 to 1991, according to the number of years of high school completed.**

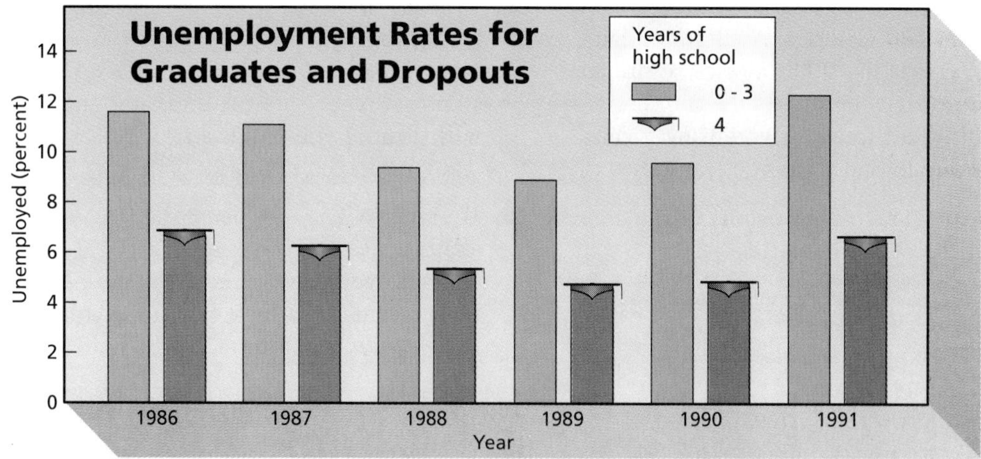

56. What percent of the people who went to three or fewer years of high school were unemployed in 1986? About 11.5%

57. Was the unemployment rate for Americans who had completed high school increasing or decreasing from 1986 to 1989? Decreasing

58. Out of 100 working-age Americans who had completed high school in 1991, how many would you expect to have been unemployed? 7

Chapter TEST

For exercises 1 and 2, evaluate the expressions.

1. Evaluate $2(x - 3) + x^2 \div 4$ when $x = 6$. (1.4) 15

2. Evaluate $9(x - y^2)$ when $x = 6$ and $y = 2$. (1.4) 18

In Exercises 3–8, decide whether the statement is true or false.

3. $(3 \cdot 4)^2 = 3 \cdot 4^2$ (1.3, 1.4) False
4. The quotient of 2 and 6 is 3. (1.1) False
5. $4 - 3 = 3 - 4$ (1.1) False
6. $5 \div (3 - 1) + 1 > 5 - 3 \cdot 4 \div 6$ (1.5) True
7. 10% of \$27 is 27¢. (1.1) False
8. $14 - x^2 \leq 7$, when $x = 3$. (1.5) True
9. Use mental math to solve $7 - x = 4$. (1.5) 3
10. Use mental math to solve $x^3 = 27$. (1.5) 3
11. Insert grouping symbols in $13 + 7 \cdot 4 \div 2$ so that its value is 40. (1.4) $(13 + 7) \cdot 4 \div 2$
12. Insert grouping symbols in $2x + 2 \cdot 4 - 3$ so that its value is 12 when $x = 5$. (1.4) $(2x + 2) \cdot (4 - 3)$
13. If you can only travel 35 miles per hour, is $2\frac{1}{2}$ hours enough time to get to a concert that is 85 miles away? (1.4) Yes
14. Write an algebraic expression for the phrase "two times a number, decreased by the sum of another number squared and 2." (1.6) $2x - (y^2 + 2)$
15. A zoo is building a glass cylindrical tank for the small sharks. The tank is 10 feet high and has a diameter of 16 feet. How much water is needed to fill the tank? (The volume of a right circular cylinder is $V = \pi r^2 h$, where r is the radius, h is the height, and $\pi \approx 3.14$.) 2009.6 ft^3
16. Members of the marching band are making their own color-guard flags. Each rectangular flag is 1.2 yards by 0.5 yards. How much fabric is needed to make 20 flags? 12 yd^2
17. The distance between consecutive bases on a baseball diamond is 90 feet. What is the perimeter of the baseball diamond? 360 ft
18. The senior class is planning a trip that will cost \$35 per student. If \$3920 has been collected from the seniors for the trip, how many have paid for their trip? 112
19. The bar graph gives the numbers of homes in the U.S. that had household pets in 1983 and 1987.
(Source: American Veterinary Medical Association)
    a. Approximately how many homes had a cat in 1983? 25 million
    b. For which types of pets was the number of homes increasing? Cats and birds

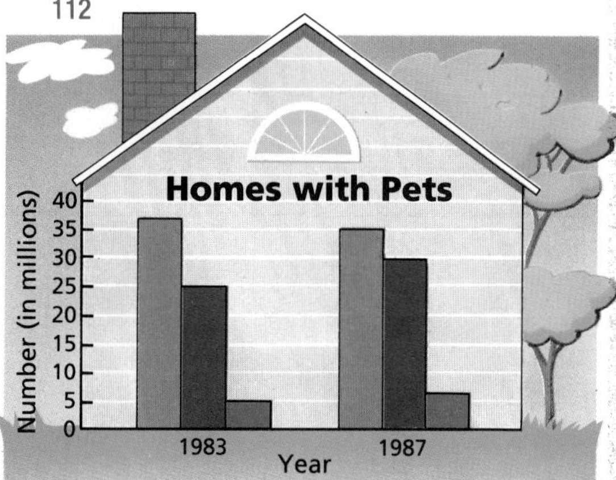

Chapter TEST

Taking the Chapter Test is a good way for students to prepare for your exam. Encourage students to take this test under testlike conditions.

In this chapter, alternate strategies for teaching the rules of algebra will be presented. These strategies emphasize the use of manipulatives. You must decide which approach (the one presented in the pupil text or the alternate approach) will better serve your students. Perhaps you may decide to present both! This will allow students who process information differently to select the presentation best-suited for them. Students seldom object to choices of presentation.

If you think that students will not respond well to the use of concrete materials, employing a pictorial representation of the alternate strategies should be considered.

The Chapter Summary on page 114 provides you and the students with a synopsis of the chapter. It identifies key skills and concepts. You may want to have students look at the Chapter Summary as an overview before beginning the chapter.

CHAPTER 2

Rules of Algebra

LESSONS

Scientists are constantly trying to understand and predict patterns in Nature, using algebra as one of their most important tools.

Real Life
Earth Science

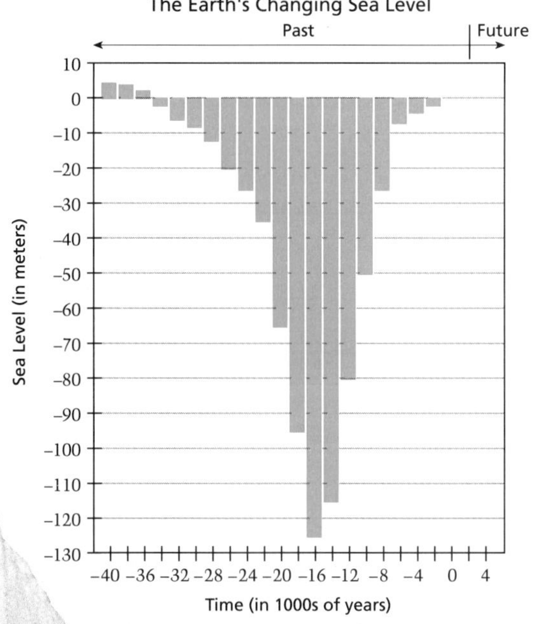

Changes in sea level occur primarily as the result of glacier activity. In an ice age, the sea level drops. In an age of ice melting, the sea level rises. At present, Earth is in an interglacial period. The graph at the left shows the variations in sea level for the eastern United States during the last 40,000 years.

On the graph, time is measured horizontally (in years), and sea level is measured vertically (in meters). On the horizontal scale, 0 corresponds to the present time, negative numbers correspond to past times, and positive numbers correspond to future times. On the vertical scale, 0 corresponds to the present sea level, negative numbers correspond to sea levels below the present level, and positive numbers correspond to sea levels above the present level.

During the last 5000 years, the sea level along the eastern United States has risen an average of one millimeter per year. (Since 1940, it has risen about three millimeters per year.)

Ask students to read several data points on the graph. Discuss why the graph represents *below* sea level by using bars extending *downward.* (One reason is that the display itself suggests "below sea level.") Ask students to identify similarities between the rectangular coordinate system and the graph presented here.

Following a general discussion of the graph, ask this series of questions.

About how much did the sea level change between

a. 12 000 and 8 000 years ago? a rise of 55 meters

b. 22 000 and 16 000 years ago? a fall of 90 meters

c. 40 000 and 28 000 years ago? a fall of 17.5 meters

An issue related to the information contained in the graph is that of global warming. Students might be interested in gathering information from the library about the effects of global warming. A class project that emphasizes the importance of mathematics in understanding the environment would be appropriate.

Rules of Algebra

PACING CHART

The daily Pacing Chart is meant to help you adjust your teaching pace. Students in the full course should finish the entire text by the end of the year. Students in the basic course may not complete the entire text in the school year. The Pacing Chart for each chapter contains suggestions for lessons that require more than one day and lessons that may be omitted for the basic course.

DAY	FULL COURSE	BASIC COURSE
1	2.1	2.1
2	2.2	2.2
3	2.3	2.3
4	2.4 & Programming a Calculator	2.4
5	Mid-Chapter Self-Test	Programming a Calculator
6	2.5	Mid-Chapter Self-Test
7	2.6	2.5
8	2.7	2.5
9	2.8	2.6
10	Chapter Review	2.7
11	Chapter Review	2.8
12	Chapter Test	2.8
13		Chapter Review
14		Chapter Review
15		Chapter Test

CHAPTER ORGANIZATION

LESSON	PAGES	GOALS	MEETING THE NCTM STANDARDS
2.1	62-67	1. Graph and compare real numbers 2. Find opposites and absolute values	Problem Solving, Communication, Reasoning, Connections, Geometry, Structure
2.2	68-73	1. Add real numbers 2. Add real numbers using a calculator	Problem Solving, Communication, Reasoning, Connections, Structure, Geometry, Technology
Mixed Review	74	Review of algebra and arithmetic	
Milestones	74	Origins of algebra	Connections
2.3	75-80	1. Subtract real numbers 2. Simplify the difference of two algebraic expressions	Problem Solving, Communication, Reasoning, Connections, Structure
2.4	81-86	1. Organize data in a matrix 2. Add and subtract matrices	Problem Solving, Communication, Connections, Discrete Mathematics, Statistics
Programming a Calculator	87	Add and subtract matrices	Technology
Mid-Chapter Self-Test	88	Diagnose student weaknesses and remediate with correlated Reteaching worksheets	
2.5	89-94	1. Multiply real numbers 2. Use calculators and units of measure in products	Problem Solving, Communication, Reasoning, Connections, Structure, Technology
2.6	95-100	1. Use the Distributive Property 2. Simplify expressions by combining like terms	Problem Solving, Communication, Reasoning, Connections, Structure, Geometry
Mixed Review	100	Review of algebra and arithmetic	
2.7	101-106	1. Express division as multiplication 2. Divide real numbers	Problem Solving, Communication, Reasoning, Connections, Structure, Technology
2.8	107-113	1. Use rates to relate quantities measured in different units 2. Use ratios to relate quantities measured in the same units	Problem Solving, Communication, Reasoning, Connections
Chapter Summary	114	A restatement of what has been learned, why it has been learned, and how it fits into the structure of algebra	Structure, Connections
Chapter Review	115-118	Review of concepts and skills learned in the chapter	
Chapter Test	119	Diagnose student weaknesses and remediate with correlated Reteaching worksheets	

LESSON RESOURCES

MEETING INDIVIDUAL NEEDS

RETEACHING For students who need to spend more time on basics:

If a mid-chapter self-test or chapter test indicates a deficiency, teachers can help students with the appropriate ***Reteaching Copymaster.***

PRACTICE For students who need more practice:

Additional exercises like those in the Pupil's Edition are provided for each lesson in ***Extra Practice Copymasters.***

ENRICHMENT For enriching and broadening students' experiences:

Problem of the Day copymasters in ***Teaching Tools*** provide a daily opportunity to use logical reasoning, looking for a pattern, writing an equation, and other routine and non-routine problem-solving strategies.

Math Log copymasters in ***Alternative Assessment*** provide opportunities to report on investigations, research, and open-ended problems.

Enriching activities with graphing and scientific calculators and computers are provided in ***Technology: Using Calculators and Computers.***

The ***Applications Handbook*** provides additional information about the cross-curriculum topics such as astronomy, chemistry, physics, sports, economics, genetics, and music that are integrated into the Pupil's Edition.

LESSON	2.1	2.2	2.3	2.4	2.5	2.6	2.7	2.8
PAGES	62-67	68-73	75-80	81-86	89-94	95-100	101-106	107-113
Teaching Tools								
Transparencies	✓	✓						
Problem of the Day	✓	✓	✓	✓	✓	✓	✓	✓
Warm-up Exercises	✓	✓	✓	✓	✓	✓	✓	✓
Answer Masters	✓	✓	✓	✓	✓	✓	✓	✓
Extra Practice Copymasters	✓	✓	✓	✓	✓	✓	✓	✓
Reteaching Copymasters	Teacher-directed and independent activities tied to results on the Mid-Chapter Self-Tests and Chapter Tests							
Color Transparencies		✓	✓				✓	✓
Applications Handbook	Additional background information is supplied for many real-life applications.							
Technology Handbook	Calculator and computer worksheets are supplied for appropriate lessons.							
Complete Solutions Manual	✓	✓	✓	✓	✓	✓	✓	✓
Alternative Assessment	Assess student's ability to reason, analyze, solve problems, and communicate using mathematical language.							
Formal Assessment	Mid-Chapter Self-Tests, Chapter Tests, Cumulative Tests, and Practice for College Entrance Tests							
Computer Test Bank	Customized tests can be created by choosing from over 2000 items.							

INSIGHTS

2-1 The Real Number Line

This lesson helps students understand the representation and use in real-life applications of negative numbers.

2-2 Addition of Real Numbers

Learning to add positive and negative real numbers is important, since real numbers are used to represent situations found in sports, chemistry, physics, business, and so on.

2-3 Subtraction of Real Numbers

Learning to subtract real numbers is important for representing and understanding real-life situations found in sports, chemistry, physics, and business.

2-4 Exploring Data: Matrices

Matrices are used to organize data. In video games, numbers that give the position of images on the screen are stored in matrices. Matrix operations can be used to locate and move these images during the play of a game.

2-5 Multiplication of Real Numbers

Since real numbers are used in many applications, students should be able to add, subtract, and now multiply real numbers.

2-6 The Distributive Property

The Distributive Property of Multiplication over Addition is used to simplify expressions in both arithmetic and algebra. It provides an alternative method (some say, more efficient method) for multiplying real numbers and real-number expressions.

2-7 Division of Real Numbers

The Division Rule for real numbers is used to simplify expressions in both arithmetic and algebra. It allows you to replace the operation of division with that of multiplication, which is often more efficient.

2-8 Exploring Data: Rates and Ratios

When numbers are compared by division, they form rates or ratios. When the numbers compared have different units of measurement, they form *rates.* Rates can be identified by comparison of the units, such as dollars per pound, miles per gallon, income per household, yards per carry (in football), and so on.

When the numbers compared have the same unit of measurement, they form *ratios*. For example, ratios compare games won to games lost, model size to actual size (of furniture, for example), an original document to its photocopy. Knowledge of rates and ratios is important, since consumer and scientific information is often presented in such form.

Problem of the Day

Suppose you are delivering mail in an office building. You leave the mailroom and enter the elevator next door. You go up four floors, down seven, and up nine to the executive offices on the top floor. Then, you go down six, up two, and down eight to the lobby on the first floor. What floor is the mailroom on?
seventh

ORGANIZER

Warm-Up Exercises

1. Use a number line to represent each number and order the following numbers from smallest to largest.
   0, 18, 10.5, $\frac{1}{2}$, 4.1, 13.7, 6

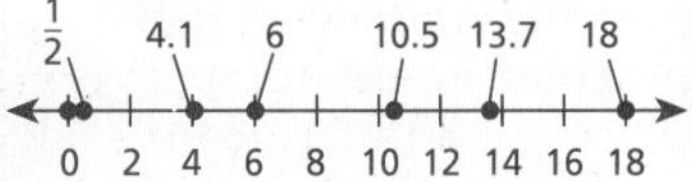

2. After you have ordered the numbers in Exercise 1, use < or > to list them in increasing order, then in decreasing order. Increasing order: $0 < \frac{1}{2} < 4.1 < 6 < \ldots < 18$; decreasing order: $18 > 13.7 > 10.5 > 6 \ldots > 0$

Lesson Resources

Teaching Tools
- Transparency: 6
- Problem of the Day: 2.1
- Warm-up Exercises: 2.1
- Answer Masters: 2.1

Extra Practice: 2.1
Color Transparency: 8
Technology Handbook: p. 11

2.1 The Real Number Line

What you should learn:

Goal 1 How to graph and compare real numbers using the real number line

Goal 2 How to find the opposite and the absolute value of a real number

Why you should learn it:

You can use the real number line as a model to help you visualize relationships among real numbers and to help you understand negative amounts such as "below zero" temperatures and negative velocities.

Goal 1 Graphing Real Numbers

The numbers used in algebra are **real numbers,** and they can be pictured as points on a horizontal line called a **real number line.** The point for 0 is the **origin.** Points to the left of 0 represent **negative numbers,** and points to the right of 0 **positive numbers.** Zero is neither positive nor negative.

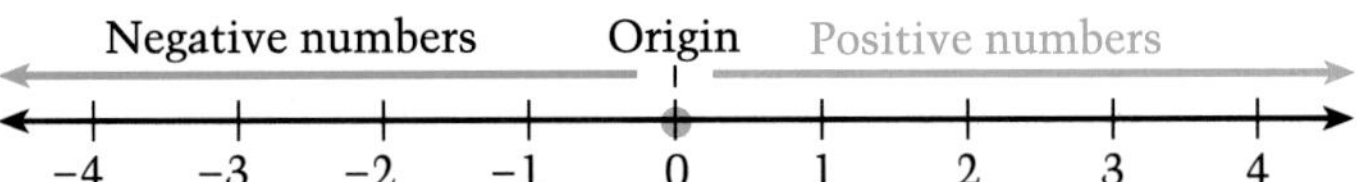

The scale marks on the real number line are equally spaced and represent **integers.** An integer is either negative, zero, or positive.

$$\underbrace{\{\ldots, -3, -2, -1,}_{\text{Negative}}\ \underbrace{0,}_{\text{Zero}}\ \underbrace{1, 2, 3, \ldots\}}_{\text{Positive}} \quad \textit{Integers}$$

By enclosing the numbers in braces { }, we are indicating that they form a **set.** The three dots on each side indicate that the list continues in both directions without end. For instance, the next integer to the left is -4, read as "negative four," and the next integer to the right is $+4$, or simply 4, read as "positive four," or simply "four."

The real number line contains points that represent fractions and decimals as well as integers. The point that corresponds to a number is the **graph** of the number, and drawing the point is called *graphing* the number or *plotting* the point.

Example 1 ***Graphing Numbers on the Real Number Line***

Graph the numbers $\frac{1}{2}$ and -2.3 on the real number line.

Solution The point that corresponds to the positive number $\frac{1}{2}$ is one-half unit to the *right* of 0. The point that corresponds to the negative number -2.3 is 2.3 units to the *left* of 0.

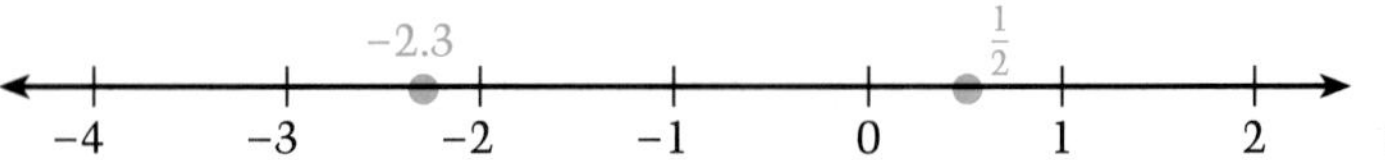

■

Having the real number line as a *visual model* of the real numbers helps answer questions about real numbers. You already know that 5 is greater than 4, but what about -5 and -4? Which is greater? Graphing these numbers on the real number line gives you the answer. Numbers to the right get larger, numbers to the left get smaller. Because -4 is to the right of -5, you know that -4 is greater than -5.

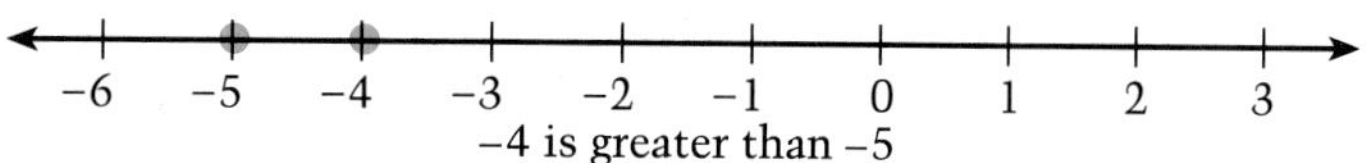

-4 is greater than -5

The inequality symbols $<$ and $>$ can be used to show the **order** of two numbers.

Verbal Statement	Inequality	Number Line Interpretation
-4 is *greater* than -5.	$-4 > -5$	-4 is to the *right* of -5.
2 is *less* than 3.	$2 < 3$	2 is to the *left* of 3.

The statements $2 < 3$ and $3 > 2$ give the same information.

Example 2 *Ordering Real Numbers*

Write the numbers -2, 4, 0, 1.5, $\frac{1}{2}$, and $-\frac{3}{2}$ in increasing order.

Solution One way to order the numbers is to graph them on the real number line.

From the picture, the order is -2, $-\frac{3}{2}$, 0, $\frac{1}{2}$, 1.5, 4. ■

Real Life
Meteorology

Example 3 *Comparing Temperatures*

A weather station in Alaska recorded the temperature (in degrees Celsius) each day at noon for one week.

a. Which noon temperature reading was the coldest?

b. Which days had noon temperatures below -10°C?

c. Which days had noon temperatures above 0°C?

Day	Date	Temp.
Sun.	Mar. 3	-17°C
Mon.	Mar. 4	-21°C
Tue.	Mar. 5	-22°C
Wed.	Mar. 6	-9°C
Thu.	Mar. 7	1°C
Fri.	Mar. 8	-3°C
Sat.	Mar. 9	-12°C

Solution

a. The coldest temperature was -22°C.

b. Days with noon temperatures below -10°C were Sunday (-17°C), Monday (-21°C), Tuesday (-22°C), and Saturday (-12°C).

c. Only Thursday's noon temperature was above 0°C. ■

LESSON Notes

Note that some students may be unfamiliar with positive and negative (+ and −) signs written at the base level instead of in the superior position to denote positive and negative integers. Indicate to students that $+5 = {}^{+}5 = 5$ and that $-13 = {}^{-}13$. Remind them also that it is usual to omit the use of the plus sign when writing positive integers.

Example 2

Help students recall how to insert $<$ and $>$ when comparing numbers. Remind them that the symbol $<$ opens toward the larger of the two numbers. So comparing -12 and -7 means you open the symbol toward -7 because -7 is greater than -12. Hence, $-12 < -7$ (read as "negative 12 is less than negative 7"). Students should also be shown that comparing -7 and -12 (in that order) means $-7 > -12$ (read as "negative 7 is greater than negative 12").

Explain to students that when ordering integers, they should compare and arrange them in a sequence that is either from smallest to largest (ascending order) or in a sequence from largest to smallest (descending order).

An Alternate Approach
Using Calculators

If students have difficulty ordering the numbers in **Example 2,** have them use a calculator to write all numbers in decimal form. For example, keying in

3 [÷] 2 [=]

will instruct the calculator to convert the fraction $\frac{3}{2}$ into a decimal. Students can now use decimal representations to decide how to order the numbers in the list.

Example 4

The use of the term "opposite" is important because the term "negative" has two meanings: opposite direction and negative value. Students more easily understand that the opposite of -2 is 2 and vice versa.

Example 5

The definition of absolute value as the distance between the origin and the point on the real number line representing the number is a key concept. Emphasize the definition using many examples. Be careful to say that distance is always a nonnegative number value. Later this definition will be useful when solving equations.

Example 6

Be certain that students understand the difference between velocity and speed. Note that traveling to one location might be considered positive and returning would then be considered negative.

Common-Error ALERT!

In **Example 5,** stress that variables represent values and that sometimes the substituted value is negative. In particular, a may represent the value -5. If so, the absolute value of a would be the opposite value, $-a$ (or 5, the opposite of -5). In algebra you write $|a| = -a$.

Goal 2 Finding Opposites and Absolute Values

Two points on the real number line that are the same distance from the origin but are on opposite sides of the origin are **opposites.** For instance, -3 and 3 are opposites of each other, but -2 and 4 are not opposites because -2 is closer to the origin than 4. As a special case, we will agree that 0 is its own opposite. In general, a and $-a$ are opposites. Thus, the opposite of $-a$ is $-(-a) = a$.

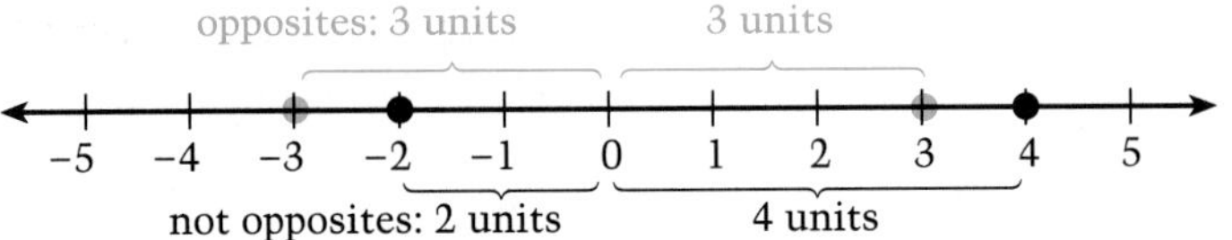

Example 4 *Finding the Opposite of a Number*

a. If $a = 5$, then $-a = -5$, the opposite of a.

b. If $a = -6$, then $-a = -(-6) = 6$, the opposite of a. ■

You should not assume that $-a$ is a negative number. For instance, in Example 4b, $-a = -(-6) = 6$, which is positive.

Properties of Opposites

1. If a is positive, then its opposite, $-a$, is negative.
2. The opposite of 0 is 0.
3. If a is negative, then its opposite, $-a$, is positive.

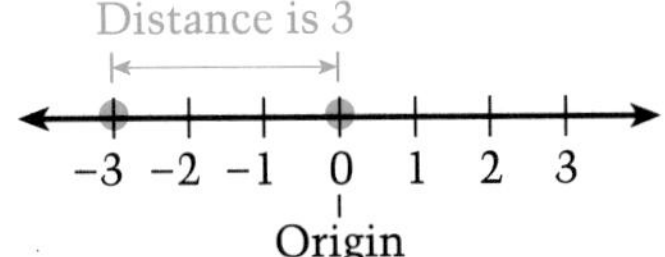

The **absolute value** of a real number is the distance between the origin and the point representing the real number. For instance, the absolute value of -3 is 3, which we write as

$|-3| = 3.$ *The absolute value of -3 is 3.*

Example 5 *Finding the Absolute Value of a Number*

a. If a is a positive number, then $|a| = a$.

$|4| = 4$ *The absolute value of 4 is 4.*

b. $|0| = 0$ *The absolute value of 0 is 0.*

c. If a is a negative, then $|a|$ is the opposite of a. Another way of saying this is that if a is negative, then $|a| = -a$.

$|-\frac{1}{2}| = -(-\frac{1}{2}) = \frac{1}{2}$ *The absolute value of $-\frac{1}{2}$ is $\frac{1}{2}$.* ■

By the time the two solid rocket boosters disengage from the space shuttle* Columbia, *it has reached a speed of about 16,000 miles per hour. The liquid hydrogen fuel has a temperature of −258°C.

The **speed** of an object is the absolute value of its **velocity.** *Velocity* indicates both speed and direction. In Example 6, positive velocities are used for motion that is straight up. Negative velocities are used for motion that is straight down.

Connections
Physical Science

Example 6 *Finding Velocity and Speed*

A parachutist is falling at a rate of 20 feet per second.

a. What is the velocity, v, of the parachutist?

b. What is the speed, s, of the parachutist?

Solution

a. Because the motion is down, the velocity is negative. Thus, $v = -20$ feet per second.

b. The speed of an object is the absolute value of its velocity. Thus, $s = |-20| = 20$ feet per second. ■

Communicating about ALGEBRA

SHARING IDEAS about the Lesson

Justify Your Response True or false? Discuss your answers.

A. Each integer is either positive or negative. False

B. The absolute value of a negative number is positive. True

C. The expression $-a$ is always negative. False

D. If a is negative and b is positive, then $a < b$. True

E. If a and b are different real numbers, then either $a < b$ or $a > b$. True. See Additional Answers for discussions.

Extra Examples

Here are additional examples similar to **Examples 2–6.**

1. Compare the following numbers. Complete, using the appropriate sign: <, >, or =

a. -20 ◆ -16 $<$
b. 23 ◆ -45 $>$
c. $-\frac{26}{3}$ ◆ -8.5 $<$
d. $\frac{35}{3}$ ◆ 11 $>$
e. 3 ◆ -17 $>$
f. -4.3 ◆ $\frac{43}{-10}$ $=$

2. Write the numbers in each list in increasing order.

a. $-3\frac{1}{2}$, 14, 4.6, $-4\frac{2}{3}$, 1.2, 2
b. 4.1, -3.9, 6, -1.5, $\frac{3}{4}$, $\frac{2}{3}$
c. $-\frac{2}{3}$, 5.4, -2.1, $\frac{3}{2}$, -1, 0

a. $-4\frac{2}{3}$, $-3\frac{1}{2}$, 1.2, 2, 4.6, 14;
b. -3.9, -1.5, $\frac{2}{3}$, $\frac{3}{4}$, 4.1, 6;
c. -2.1, -1, $-\frac{2}{3}$, 0, $\frac{3}{2}$, 5.4

3. Find the opposite of the number.

a. -4.2 4.2 **b.** 1 -1
c. -9.8 9.8 **d.** 2.7 -2.7

4. Evaluate the expression.

a. $|-3.5|$ **b.** $|-2.3|$
c. $|17|$ **d.** $|\frac{4}{5}|$
e. $|-0.1|$ **f.** $|0|$

a. 3.5, b. 2.3, c. 17, d. $\frac{4}{5}$ e. 0.1, f. 0

5. A diver jumps from her platform and falls 10 feet per second. What is the velocity of the diver? At what speed is she falling? v: -10 feet per second; s:10 feet per second

Check Understanding

Use an integer to represent each situation.

1. a debt of $25,000 $-25{,}000$
2. a gain of 18 pounds 18
3. three steps to the left -3
4. a 6-ft drop -6
5. 13,000 ft above sea level 13,000

Communicating about ALGEBRA

Ask students if there are other ways of representing positive and negative integers. (Students may have been exposed to models such as vectors, electric charges, or red and black checkers.) Encourage students to discuss how the various integer models might be used to represent positive and negative integers. **Sample answers:** Vectors are useful for showing the magnitude and direction of integers. Checkers provide concrete representations of integers.

EXERCISE Notes

ASSIGNMENT GUIDE

Basic/Average: Ex.1–6, 7–27 odd, 31–32, 35–36, 39–51 odd

Above Average: Ex.1–6, 9–27 multiples of 3, 29–32, 39–53 odd

Advanced: Ex.1–6, 29–32, 37–38, 39–47 odd, 49–56

Selected Answers
Exercises 1–6, 7–51 odd

More Difficult Exercises
Exercises 39, 40, 55, 56

Guided Practice

Ex. 1–3 Check students' understanding of inequalities.

Ex. 4–6 It is important that students translate English phrases that involve positive and negative numbers into algebraic ones. Note that left connotes negative.

EXERCISES

Guided Practice

CRITICAL THINKING about the Lesson

1. How is the inequality $5 < 6$ read? How can it be interpreted on a real number line? See below.

2. Write an inequality for this sentence: Two is greater than negative three. $2 > -3$

3. Decide whether the inequalities are true or false. Explain.

$-4 < -5$ $\quad -2 < 0$ $\quad 6 > -7$ $\quad -1 < -\frac{1}{2}$ Location of numbers on the real number line verifies that only the first one is false.

In Exercises 4–6, write an integer to represent the situation.

4. a loss of ten dollars -10

5. six feet above ground 6

6. two meters to the left -2

Independent Practice

In Exercises 7–14, graph the numbers on a number line. Which is greater? See Additional Answers for graphs.

7. 4 and -1 4

8. 3 and -3 3

9. $-\frac{1}{2}$ and 2 2

10. -1.5 and 2.5 2.5

11. 4.5 and -5.5 4.5

12. -1.3 and 2.1 2.1

13. -1.4 and $\frac{3}{4}$ $\frac{3}{4}$

14. -2 and $-\frac{2}{3}$ $-\frac{2}{3}$

15. Write the numbers in increasing order.
3.1, -1.9, 4, 0, $\frac{1}{2}$, and $-\frac{1}{2}$ $\quad -1.9, -\frac{1}{2}, 0, \frac{1}{2}, 3.1, 4$

16. Write the numbers in increasing order.
1.5, $-\frac{1}{2}$, 6, -4.6, $-\frac{3}{4}$, and $\frac{1}{3}$ $\quad -4.6, -\frac{3}{4}, -\frac{1}{2}, \frac{1}{3}, 1.5, 6$

In Exercises 17–24, find the opposite of the number.

17. 6 -6

18. -10 10

19. -2.4 2.4

20. 1.9 -1.9

21. -9 9

22. 3.6 -3.6

23. 0

24. -1

In Exercises 25–32, evaluate the expression.

25. $|5|$ 5

26. $|-3|$ 3

27. $|-\frac{4}{5}|$ $\frac{4}{5}$

28. $|\frac{1}{2}|$ $\frac{1}{2}$

29. $|3.2|$ 3.2

30. $|-4.1|$ 4.1

31. $|0|$ 0

32. $|16.3|$ 16.3

Brightness of a Star **In Exercises 33 and 34, use the following information.**

A star's brightness to a person on Earth is measured by its apparent magnitude. The less the apparent magnitude, the brighter the star. The table lists several stars and their apparent magnitudes.

Star	Apparent magnitude	Star	Apparent magnitude
Canopus	-0.73	Vega	0.04
Procyon	0.30	Sirius	-1.40
Pollux	1.16	Sun	-26.70
Deneb	1.25	Aldebaran	0.85
Beta Crucis	1.25	Alpha Centauri	-0.29

33. Which of the stars is the brightest?

34. Which are the dimmest? Deneb, Beta Crucis

1. Five is less than six, 5 is to the left of 6.

Greenwich Mean Time (GMT) **In Exercises 35–38, use the following data.**

Greenwich, England, is on the 0° meridian. Standard time elsewhere is calculated from GMT. The table lists a few countries and the number of hours each differs from GMT.

Country	Standard Time (in hours)
Chile	−4
Libya	+1
Ireland	−1
Israel	+2
Kenya	+3
Japan	+9
Mexico	−7
Pakistan	+5

37. Chile, Ireland, Mexico

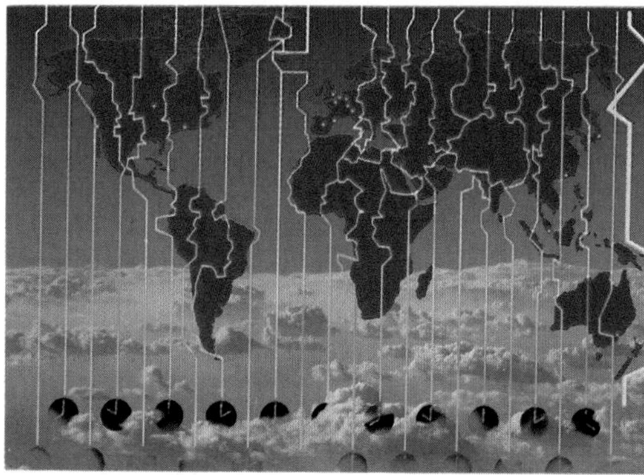

35. Which countries have standard times closest to GMT? Libya, Ireland

36. Which country has standard time farthest from GMT? Japan

37. Which countries are west of Greenwich?

38. Which countries are east of Greenwich?

39. *Velocity of a Helicopter* A helicopter is 240 feet above its landing pad. Twenty seconds later, it lands. What was its velocity? What was its speed?
−12 ft per second, 12 ft per second

38. Libya, Israel, Kenya, Japan, Pakistan

40. *Bromley Alpine Slide* The longest slide in the world is the 4600-foot Bromley Alpine Slide in Peru, Vermont. What speed must you have to complete the slide in 8 minutes? 575 feet per minute

Integrated Review

In Exercises 41–48, evaluate the expression.

41. $4x + 3y$, when $x = 2$ and $y = 1$ 11

42. $2a - b$, when $a = 6$ and $b = 3$ 9

43. $\frac{1}{4}x - y$, when $x = 9$ and $y = 5$ $-\frac{11}{4}$ or $-2\frac{3}{4}$

44. $\frac{1}{2}a + 4b$, when $a = 10$ and $b = 3$ 17

45. $|-4| + |3|$ 7

46. $|6| - |-1|$ 5

47. $|x| + |y|$, when $x = -3$ and $y = -1$ 4

48. $|y| - |x|$, when $y = -7$ and $x = 3$ 4

In Exercises 49–52, use mental math to solve the equation.

49. $2a = 8$ 4

50. $\frac{a}{3} = 3$ 9

51. $\frac{x}{2} = 4$ 8

52. $9 - y = 1$ 8

Exploration and Extension

In Exercises 53–56, use mental math to solve the equation. Explain your answers.

53. $|x| = 2$ 2, −2
See below for explanations.

54. $|x| = 0$ 0

55. $|x| = -1$ No solution

56. $|x| = |-x|$ Any real number

53. Both values are 2 units from the origin. 54. Only 0 is at the origin.
55. The absolute value of a number is always non-negative
56. A number and its opposite are the same distance from the origin.

Independent Practice

▶ **Ex. 33–34** Positive and negative numbers are also used in astronomy. As night begins to fall, the apparent magnitude of a star represents a comparison (a ratio) of one star to another. The larger the magnitude, the dimmer the star appears to the naked eye. The human eye can see only an apparent magnitude of about 6. Because the sun is so bright, its apparent magnitude is −26.70.

EXTEND Ex. 35–38 PROJECT You might have students use reference books to determine the number of hours by which certain states or Canadian provinces differ from GMT.

▶ **Ex. 38** Which countries are east of Greenwich? (It can be argued that every country is east of Greenwich, since the earth is spherical!) This should lead to some good discussion. Students might find it helpful if you provided a globe or a world map for their use.

Enrichment Activities

WRITING Following the class discussion, students might record their personal responses in their math journals.

1. Identify ways that people use negative numbers at school, at work, and even at play. Ask friends or family members if they have used negative numbers to interpret some real-life situation. thermometers, computers, rulers, battery ends, debits, back moves and penalties in games, etc.
2. Describe the way (or ways) that positive and negative numbers can be represented. using models, using + and − signs, etc.

Problem of the Day

Find the start number.

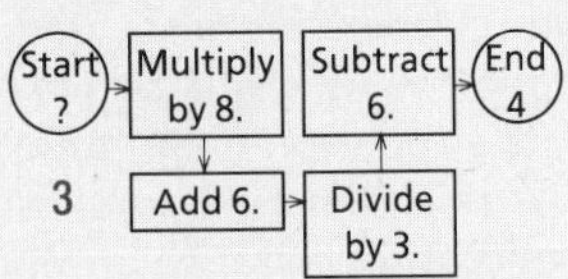

3

ORGANIZER

Warm-Up Exercises

Given a starting point (A) on the real number line, determine the distance and direction from point A to point B.

a. A: 13 B: −5 18, left
b. A: −6 B: 4 10, right
c. A: −2 B: −9 7, left
d. A: 8 B: 1 7, left
e. A: −3 B: −1 2, right

Lesson Resources

Teaching Tools
Transparency: 6
Problem of the Day: 2.2
Warm-up Exercises: 2.2
Answer Masters: 2.2
Extra Practice: 2.2
Color Transparencies: 9, 10
Applications Handbook: p. 15–17
Technology Handbook: p. 13

LESSON Notes

Example 2

The net gain can also be written as $0 + (-4) + 12$.

An Alternate Approach
Using Manipulatives
Ordinary checkers can be used to model positive and negative integers and to build an understanding of the rules of addition of real numbers. Let black checkers represent positive numbers and red checkers represent negative numbers. (See Enrichment Activities on pages 73 and 74.)

2.2 Addition of Real Numbers

What you should learn:

Goal 1 How to add real numbers using a number line or Addition Rules

Goal 2 How to add real numbers using a calculator

Why you should learn it:

You can add positive and negative numbers to solve real-life problems, such as finding the cash balance on a business balance sheet.

Goal 1 Adding Real Numbers

Addition can be modeled with movements on the real number line. You add a positive number by moving to the right. You add a negative number by moving to the left. For instance, to find the sum of -2 and 3, you can start at -2 and move 3 units to the right, arriving at 1. This can be written as $-2 + 3 = 1$. Another way to find the sum is to start at 3 and move 2 units to the left, arriving at 1. This can be written as $3 + (-2) = 1$.

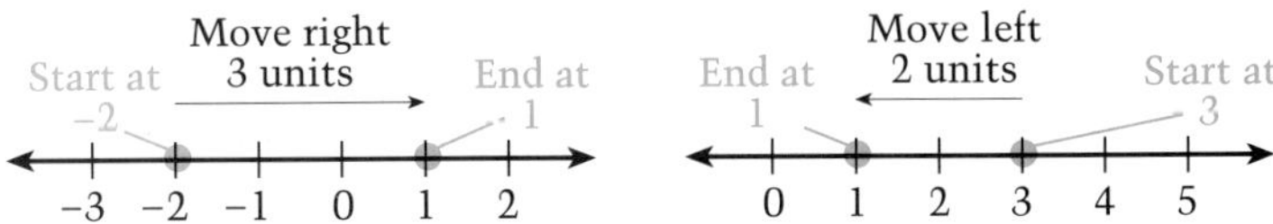

Note that adding in either order produces the same sum.

Example 1 ***Adding Real Numbers Using a Number Line***

a. To add -5 and -2, start at -5 and move left 2 units to -7. The action can be written as $-5 + (-2) = -7$.

b. To add 5, -10, and 3, start at 5, move left 10 units to -5, and move right 3 units to -2. The action can be written as $5 + (-10) + 3 = -2$. ■

Problem Solving
Using a Model

Example 2 ***Finding the Number of Yards Gained***

The Pittsburgh Steelers lose 4 yards on their first play and gain 12 yards on their second play. What is the net gain?

Solution The two plays can be represented on a real number line in which 0 represents the line of scrimmage.

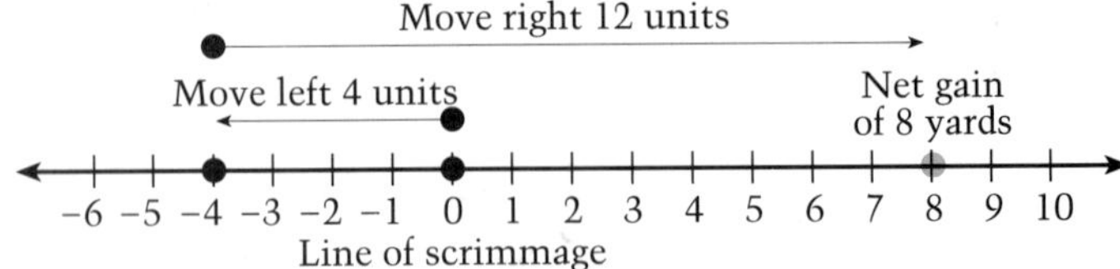

The net gain can be written as $0 + (-4) + 12 = 8$. ■

The symbol $-$, used to represent a negative number, is a **negative sign.** A positive number is considered to have a **positive sign,** $+$. The following Addition Rules show how to find the sum of two real numbers without using a number line.

1. To add two numbers with the *same sign,* add their absolute values and write the common sign.
2. To add two numbers with *opposite signs,* find the difference of their absolute values and write the sign of the number with the greater absolute value.

Example 3 *Adding Real Numbers Using Addition Rules*

a. Same sign

$$-4 + (-5) = -(4 + 5) = -9$$

b. Opposite signs

$$3 + (-9) = -(9 - 3) = -6$$

■

Properties of Addition

1. $a + b = b + a$	The order in which two numbers are added does not change their sum.
2. $a + 0 = a$	The sum of a number and 0 is the number.
3. $a + (-a) = 0$	The sum of a number and its opposite is 0.
4. $-(a + b) = -a + (-b)$	The opposite of a sum of two numbers is equal to the sum of the opposites of the numbers.

Connections

Physical Science

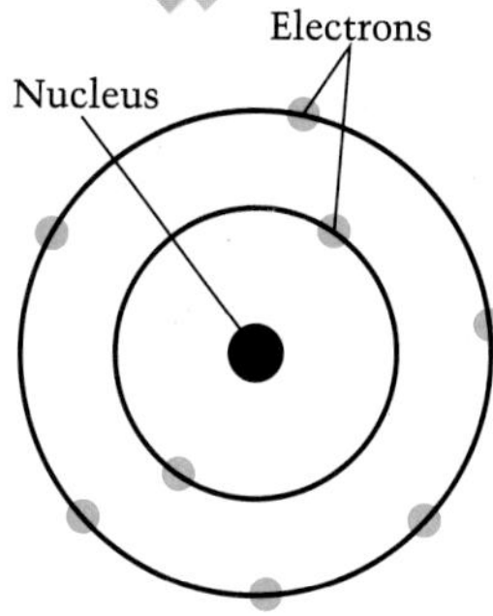

This Bohr model for an oxygen atom shows 8 electrons orbiting about a nucleus.

Example 4 *Atoms and their Ions*

Atoms and ions are composed of electrons, neutrons, and protons. Each electron has a charge of -1, and each proton has a charge of $+1$. The total charge of an atom or ion is the sum of the charges of its electrons and protons. An atom has zero charge. An ion has a positive or negative charge. Which of the following arrangements represents an atom? An ion?

a. 8 electrons, 8 protons

b. 10 electrons, 8 protons

Solution

a. The total charge is $-8 + 8 = 0$. Thus it is an atom, not an ion.

b. The total charge is $-10 + 8 = -2$. Thus it is an ion. In chemistry, this oxygen ion is written as O^{-2}. ■

Example 3

An Alternate Approach *Using Manipulatives*

The rules of addition of real numbers can be easily understood by using the checker model of addition of integers. Compare the rules of addition in Example 3 with the checker-rules:

a. Same sign: Checkers are either all red or all black. In either case, the sum is all red or all black.

b. Opposite signs: Adding more red checkers than black ones together will result in red checkers left over after opposite colors cancel each other. Because you have concrete objects, you can easily identify the smaller number of checkers in a given color (the absolute value) and count what remains after cancellations, noting the color (that is, the sign) as you do so.

Example 4

HISTORICAL NOTE The Bohr model for an oxygen atom is named for Danish scientist Niels Bohr (1885–1962). He is best known for his work in quantum theory.

Example 5

TECHNOLOGY Poll students to determine the types of calculators in the class. Discuss the merits of the different types of calculators for adding real numbers. Are change-sign keys easier to understand and use than negation keys?

Example 6

Note that the sum of the profits and losses is given in units (dollars). Some students may not wish to include the units in their computations.

Extra Examples

Here are additional examples similar to **Examples 1–6.**

1. Use a number line to find the sums.
   a. $-13 + (-12)$ — -25
   b. $17 + (-4)$ — 13
   c. $-9 + 17$ — 8
2. Use the rules of addition to find the sums.
   a. $-7 + (-6)$ — -13
   b. $-8 + 13$ — 5
   c. $18 + (-23)$ — -5
3. Estimate the sums. Then use a calculator to evaluate the sums.
   a. $2.9 + (-3.6) + (-3)$ — -3.7
   b. $-43 + 12.7 + (-26.89)$ — -57.19
   c. $109 + (-89) + (26)$ — 46
4. The balance sheet of a student-run company shows a loss the first week of $38. In the next three weeks, the company shows a loss of $39, a profit of $97, and a loss of $12. Could the students continue with their company? Yes; company has a profit of $8.
5. Diarra needs to know how much money is in her apartment rental account. At the beginning of the month, she had $1,200 in the account. During the month, she credited a rental payment of $240, debited $80 to fix a leak, paid $273 to replace a heater, and received another rental payment of $264. How much money is in her account at the end of the month? $1,351

Check Understanding

TECHNOLOGY Most calculators have a change-sign key, +/−, and a subtraction key, − (or a negation key, −, and a subtraction key, -). Note that negation keys are found on graphics calculators but seldom on four-function calculators.

Goal 2 Adding Real Numbers Using a Calculator

A calculator is useful for adding large numbers or numbers in decimal form. The key used to enter negative numbers depends on the type of calculator. Some calculators have a change-sign key [+/−], others have a negation key [(−)], and some have no key for entering negative numbers. On most calculators, the subtraction key [−] cannot be used to enter a negative sign.

Example 5 *Evaluating a Sum Using a Calculator*

Use a calculator to evaluate $-2.6 + 3.1 + (-1.4)$.

Solution

Change-sign key 2.6 [+/−] [+] 3.1 [+] 1.4 [+/−] [=]

Negation key [(−)] 2.6 [+] 3.1 [+] [(−)] 1.4 [=]

With either keystroke sequence, the calculator display should read -0.9. ■

A company has a profit when its income is greater than its expenses. It has a loss when income is less than expenses. Business losses are generally indicated by negative numbers.

Real Life **Business**

Example 6 *Finding the Total Profit*

The balance sheet of a company shows a profit of $13,142.50 in January, a loss of $6,783.16 in February, and a profit of $2,589.82 in March. What was the overall profit during the first quarter? (January, February, and March form the first quarter of the year.)

Solution The overall profit during the first quarter was

$$13{,}142.50 + (-6{,}783.16) + 2{,}589.82 = \$8{,}949.16.$$ ■

Communicating about ALGEBRA

Cooperative Learning

SHARING IDEAS about the Lesson See Additional Answers.

Work with a Partner and Use the Real Number Line

A. Show how to model the sum of -4 and 3 in two ways. Make a sketch to illustrate both ways.

B. Describe different ways the sum of -3, 2, and -1 can be modeled. Make a sketch to illustrate each way.

EXERCISES

Guided Practice

CRITICAL THINKING about the Lesson

1. **True or False?** On a number line, adding a negative number can be shown by a move to the right. False
2. **True or False?** When adding numbers with opposite signs, subtract the smaller absolute value from the larger absolute value and attach the sign of the number with the larger absolute value. True
3. Is the order in which you add two numbers important? Illustrate your answer with an example that shows the sum of -3 and 2 in two ways on a number line. No See below for graphs.
4. The number zero is called the **additive identity.** Why do you think zero was given this name?
   Any number added to 0 is that identical number.

Independent Practice

In Exercises 5–12, use a number line to help you find the sum of the numbers.

5. -7 and 11 4
6. 3 and -4 -1
7. -3 and $3\frac{1}{2}$ $\frac{1}{2}$
8. -2 and -8 -10
9. -4 and 5 1
10. -5 and 4 -1
11. -5, 8, and -2 1
12. 10, -17, and 1 -6

See Additional Answers for graphs.

In Exercises 13–20, use the Addition Rules to find the sum.

13. $-4 + 2$ -2
14. $6 + 3$ 9
15. $7 + 0$ 7
16. $0 + (-6)$ -6
17. $-11 + (-4)$ -15
18. $12 + (-9)$ 3
19. $5 + (-10) + (-3)$ -8
20. $-2 + 11 + (-7)$ 2

In Exercises 21–24, evaluate the expression in two ways. (The results should agree.)

21. $-(8 + 5)$ -13
22. $-[-(4+3)]$ 7
23. $-[12 + (-6)]$ -6
24. $-[1 - (-3)]$ -4

In Exercises 25–28, use a calculator to evaluate the sum. Use an estimate to check.

25. $-11.4 + 6.2 + (-2.0)$ -7.2
26. $5.8 + (-9.3) + 5.2$ 1.7
27. $6.4 + 3.1 + (-6.4)$ 3.1
28. $-2.9 + 5.7 + (-8.6)$ -5.8

In Exercises 29–32, simplify the expression. (Sample: $2 + x + (-3) = -1 + x$)

✪ 29. $3 + x + (-8)$ $x - 5$
✪ 30. $2x + 8 + (-8)$ $2x$
✪ 31. $-6 + x + 2$ $x - 4$
✪ 32. $3 + (-2) + p + 19$ $20 + p$

33. *Los Angeles Raiders* The Los Angeles Raiders completed a pass for a gain of 37 yards. After the play was over, a flag was thrown and a 15-yard penalty was called against the Raiders. How many yards did the Raiders actually gain? Draw a diagram to help you. 22 See margin.

3.
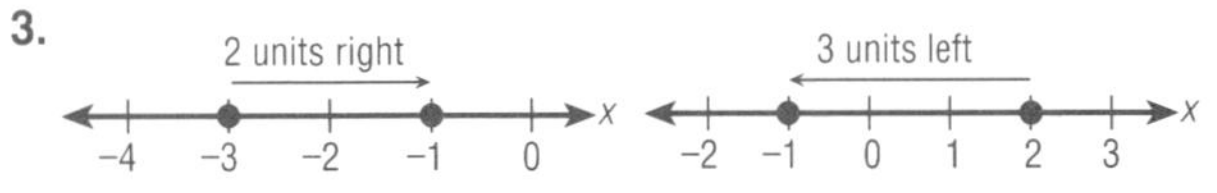

WRITING Explain why two keys are necessary and give examples in which one must not confuse the two keys. This question requires students to distinguish between the naming of numbers [the change-sign or negation key] and the operation [the subtraction key] in number expressions.

Communicating about ALGEBRA

COOPERATIVE LEARNING If you have introduced students to the checker model (or other manipulative model) for integers, have pairs of students describe how they would complete Parts A and B on page 70 using the model.

EXERCISE Notes

ASSIGNMENT GUIDE

Basic/Average: Ex. 1–4, 5–31 odd, 33–42 multiples of 3, 45–51 odd. 54–56

Above Average: Ex. 1–4, 6–30 multiples of 3, 33–39 odd, 45–55 odd

Advanced: Ex. 1–4, 6–30 multiples of 3, 33–43 odd, 51–53, 57

Selected Answers
Exercises 1–4, 5–53 odd

Use **Mixed Review** as needed.

✪ **More Difficult Exercises**
Exercises 29–32, 40, 53, 57

Guided Practice

▶ **Ex. 1–4** These exercises target several key ideas from the lesson. Emphasize the importance of using absolute values in applying the rules of addition.

Independent Practice

Ex. 5–20 Students should be encouraged to use the checker model if they are having difficulty finding the sums with paper and pencil alone.

Ex. 25–28 TECHNOLOGY Caution students to use their negation or change-sign keys properly when entering negative numbers.

Ex. 29–32 Stress the importance of using the commutative property of addition to find the sums in these exercises. For example, in simplifying Exercise 29 as $3 + x + (-8) = 3 + (-8) + x = -5 + x$, the commutative property of addition was used to rewrite $x + (-8)$ as $(-8) + x$.

Ex. 33–34 PROBLEM SOLVING Stress the importance of using a diagram to understand the problem. Pay close attention to how students set up Exercise 34.

Answers

33.

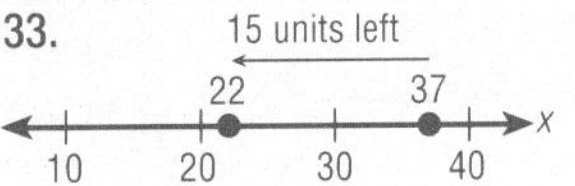

34. Using a number line model, −12 is the starting point.

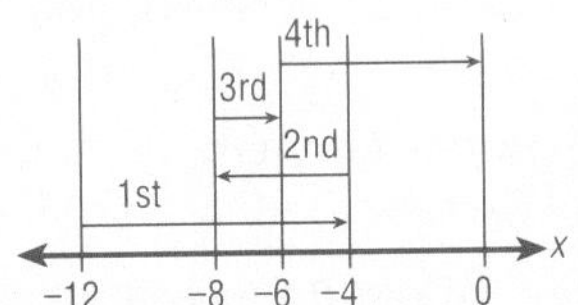

Be prepared for a discussion about whether your team won the game because the ball is advanced to point 0 (the goal line).

Ex. 35–36 Refer back to Example 4 in the lesson to determine how to use symbols to express the results of adding electrons and protons.

34. *Championship Game* Suppose it is the game that decides the high school football championship. Your team is behind by five points and needs a touchdown to win. Starting on the opponent's 12-yard line, your team's final four plays result in an 8-yard completed pass, a loss of 4 yards on a quarterback sack, 2 yards gained by the fullback, and a 6-yard gain on a quarterback sneak as time runs out. Does your team win? Draw a diagram and explain. See margin.

35. +1, ion; 0, atom; 0, atom; −1, ion
36. 2nd, 1st, 4th
38. Yes, profits and losses total a profit of $432.37.

35. *Atom or Ion?* Find the total charge. Then decide whether you have an atom or an ion.

36. *Chemical Symbol* For the elements in Exercise 35, which has a symbol of Na? Of Na^{+1}? Of F^{-1}?

	Electrons	Protons	Total charge	Atom or ion?
Sodium	10	11	?	?
Sodium	11	11	?	?
Fluorine	9	9	?	?
Fluorine	10	9	?	?

37. *Second-Quarter Profit?* A company had a profit of $3,412.53 in April, a profit of $5,784.25 in May, and a loss of $9,013.86 in June. Did the company enjoy a profit during this quarter? Explain. Yes, profits and losses total a profit of $182.92.

38. *Third-Quarter Profit or Loss?* A company had a loss of $5,519.80 in July, a profit of $2,337.06 in August, and a profit of $3,615.11 in September. Did the company enjoy a profit during the third quarter of the year? Explain.

39. *Televised Live* A soccer team from North Carolina has made it to the championship game in Oregon. The game is to be televised live and is scheduled to begin at 2:00 P.M. At what time will the game be seen in North Carolina? 5:00 P.M.

40. *What Time Is It?* Martha left California at 7:00 A.M. and flew to Disney World in Florida. The flight included $6\frac{3}{4}$ hours of airtime and two $1\frac{1}{2}$-hour stops. When she landed in Florida, what was the local time? 7:45 P.M.

The continental United States has four time zones: Pacific, Mountain, Central, and Eastern.

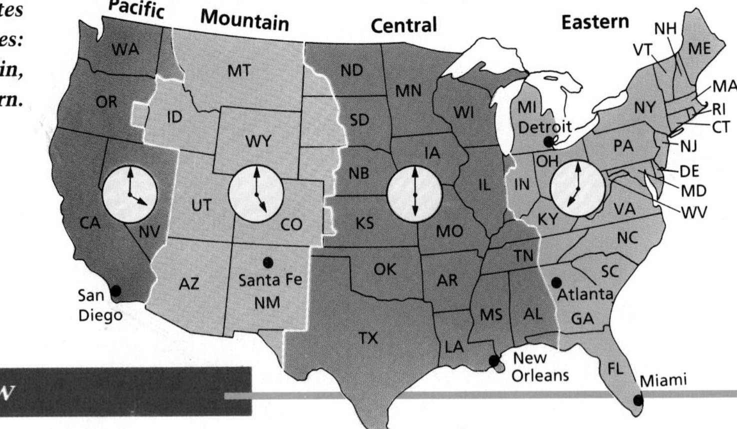

Integrated Review

41. Find the opposite of −16. 16

42. Find the opposite of 0. 0

43. Find the absolute value of 7. 7

44. Find the absolute value of −11. 11

In Exercises 45–48, evaluate the expression.

45. $a^2 - 5$, when $a = 3$ 4

46. $3x^3 - 9$, when $x = 2$ 15

47. $(6s - t)^2$, when $s = 3$ and $t = 9$ 81

48. $17 - (3m + 1)^2$, when $m = 1$ 1

In Exercises 49 and 50, use mental math to solve the equation.

49. $5x = 15$ 3

50. $\frac{21}{x} = 7$ 3

51. *Grocery Scales* Suppose you go to a grocery store to buy 5 kilograms of potatoes for a salad. You put several in a sack and find the weight is 3.8 kilograms. Let x represent the weight of the additional potatoes you need. Which of the following equations correctly model the situation?

a. $x + 3.8 = 5$ **b.** $x + 5 = 3.8$ **c.** $5 - 3.8 = x$ **d.** $5 - x = 3.8$

How many kilograms do you need? Check that this amount is a solution of each correct equation. 1.2

52. *Apartment Size* An apartment is 26 feet by 30 feet. Which of the following equations correctly describes the relationship among the areas of the rooms as shown on the floor plan? Explain.

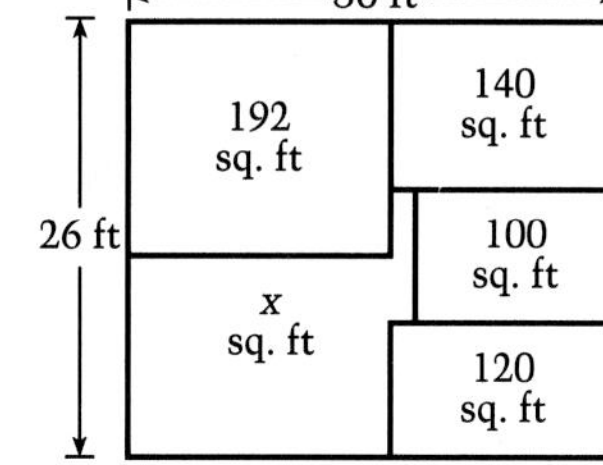

a. $x + (192 + 140 + 100 + 120) = (26)(30)$

b. $(26)(30) - (192 + 140 + 100 + 120) = x$

How many square feet are in the area labeled with x square feet? See Additional Answers.

53. *Playing Putt-Putt* Suppose that your sister challenges you to three rounds of Putt-Putt golf. In the first round, she wins by two strokes and finishes six over par. In the second round, you win by one stroke and finish three over par. In the final round, there is one hole left, and you are tied for the round. Your sister scores a two, giving her a round total of three over par. She says there is no way for you to win either that round or overall. Is she correct? Make a table to support your answer. See Additional Answers.

Exploration and Extension

In Exercises 54–56, use mental math to solve the equation.

54. $-6 = 3 + x$ −9

55. $-29 + x = 16$ 45

56. $-22 + (-14) = x + 5$ −41

57. *Entrepreneur* On Saturday, your brother decides to open a lemonade and cookie stand to earn money for a new video game that costs \$29.95. It costs \$2.00 to make a pitcher (16 glasses) of fresh lemonade and \$5.00 to make a batch of cookies (60 cookies). He charges \$0.30 for a glass of lemonade and \$0.15 for a cookie. If he makes x batches of cookies and x pitchers of lemonade, which inequality models the outcome desired by your brother?

a. $[16(0.30) - 2]x + [60(0.15) - 5]x \geq 29.95$

b. $[16(0.30) - 2]x + [60(0.15) - 5]x \leq 29.95$

What is the smallest integer value of x that will allow your brother to buy the game? 5

Ex. 39–40 Explain the relationship of different time zones to one another. Help students to understand that when it is 2:00 P.M. in Oregon, it is 5:00 P.M. in North Carolina.

Ex. 51–52 PROBLEM SOLVING Emphasize that the given equations represent possible models of the problem situations. Modeling problems is one of the ways that algebra helps in interpreting and solving real-life applications.

Enrichment Activities

These activities require students to go beyond lesson goals.

1. Suppose that red checkers represent negative integers and black checkers represent positive integers. How would you model the following integers?
   - a. −4 4 red checkers
   - b. 7 7 black checkers
   - c. −6 6 red checkers
   - d. 0 blank space or the same number of red and black checkers

Because black and red checkers cancel each other, −4 can also be modeled using 7 red and 3 black checkers.

Complete the following exercises, using checkers to model addition. Combine checkers from one set with checkers from another. The sum is what remains after red and black checkers cancel each other.

2. Use checkers to add.
   - a. $-9 + (-5)$ −14
   - b. $-12 + 7$ −5
   - c. $8 + (-14)$ −6

Using a pictorial model often works as well as using actual checkers. Complete the following exercises using pictorial models. Draw black and red circles to represent the checkers.

3. Provide a pictorial model of these integers.
   a. -2 b. -8 c. 6 d. 12
   a. 2 red circles; b. 8 red circles; c. 6 black circles; d. 12 black circles
4. Use a pictorial model to add.
   a. $7 + (-12)$ -5
   b. $-8 + (-3)$ -11
   c. $-13 + 17$ 4
   d. $5 + 9$ 14

Mixed REVIEW

1. Evaluate $2y + 14$ when $y = 5$. **(1.2)** 24
2. Evaluate $(4 \div 2) + 3(8 - 7)$. **(1.1)** 5
3. Write $\frac{3}{8}$ in percent form. **(1.1)** 37.5%
4. Which is greater, $-\frac{1}{2}$ or $-\frac{3}{4}$? **(2.1)** $-\frac{1}{2}$
5. Write $s \cdot s \cdot s \cdot s \cdot s$ in exponential form. **(1.3)** s^5
6. Evaluate $2r^2 + 5y^3$ when $r = 4$ and $y = 2$. **(1.3)** 72
7. What value of t makes $t^3 - 4$ equal to 23? **(1.5)** 3
8. Find the absolute value of -4.3. **(2.1)** 4.3
9. Is 8 a solution of $x \div 4 + 3 \geq 5$? **(1.5)** Yes
10. Is 5 a solution of $6y = 20 + 2y$? **(1.5)** Yes
11. What is the product of 8 and 3? **(1.4)** 24
12. What is the quotient of 3 and 4? **(1.4)** $\frac{3}{4}$
13. Evaluate $16.3 + 9.4 \cdot 2.9$. **(1.4)** 43.56
14. Evaluate $39.7 \div 16.2 + 5.5$. **(1.4)** 7.95
15. Write 4, -8, 3, 0, -5, -2, 1 in increasing order. **(2.1)** $-8, -5, -2, 0, 1, 3, 4$
16. What is the area of a circle with a radius of 6 centimeters? **(1.3)** 113.04 cm^2
17. What is the volume of a cube with edges of 5 inches each? **(1.3)** 125 in.^{3}
18. Write an inequality for "The difference of three times a number and 8 is less than 20." **(1.6)** $3n - 8 < 20$

Milestones ORIGINS OF ALGEBRA

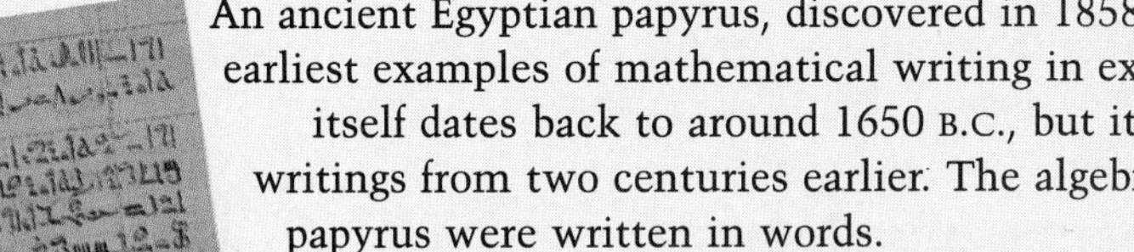

1800 1600 1400 1200 1000 800 600 400 200 B.C. A.D. 200 400 600 800 1000 1200

Rhind Papyrus, Egypt — Diophantus, Greece — al-Khowarizmi, Persia

An ancient Egyptian papyrus, discovered in 1858, contains one of the earliest examples of mathematical writing in existence. The papyrus itself dates back to around 1650 B.C., but it is actually a copy of writings from two centuries earlier. The algebraic equations on the papyrus were written in words.

The Babylonians are believed to have explored algebraic concepts before the Egyptians. They also wrote their mathematical thoughts in words. Although this was cumbersome, the Babylonians were able to show some sophisticated algebraic solutions.

Diophantus, a Greek who lived around A.D. 250, is often called the Father of Algebra. He was the first to use abbreviated word forms in equations. In India, Hindu mathematicians, such as Brahmagupta (A.D. 628), worked with negative and irrational roots. In A.D. 825, Mohammed ibn-Musa al-Khowarizmi wrote a book with a practical discussion of equations and how to solve them. The book title, *Hisab al-jabr w'al-muqabalah*, contains the word *al-jabr*, which was later translated into the Latin word *algebra*.

Rhind Papyrus

Who introduced the following: equal sign, multiplication sign (•), the use of letters as variables? In what century were they introduced?

Milestones

Language Arts: Critical Reading

1. About how many centuries have passed in the study of algebra since the writings referred to in the Egyptian papyrus? about 37
2. Write an algebraic equation using words instead of symbols. What do you notice? Students should notice that equations written out in words are considerably longer than equations that use algebraic symbols. Using algebraic symbols provides mathematicians a shorthand method of expressing the words.

Answer to Milestones: equal sign—Robert Recorde, sixteenth century; multiplication sign—Gottfried Wilhelm von Leibniz, seventeenth century; the use of letters as variables—Francis Vieta, sixteenth century

2.3 Subtraction of Real Numbers

What you should learn:

Goal 1 How to subtract real numbers using a Subtraction Rule or a calculator

Goal 2 How to simplify the difference of two algebraic expressions

Why you should learn it:

You can subtract positive and negative numbers by knowing how to rewrite a subtraction expression in terms of addition. This is useful in solving real-life problems such as finding differences in stock-market prices.

Goal 1 Subtracting Real Numbers

In Lesson 2.2, you learned that some addition expressions can be evaluated using subtraction. Here are two examples.

Addition		Corresponding Subtraction
$5 + (-3) = 2$	$\longleftrightarrow$	$5 - 3 = 2$
$9 + (-6) = 3$	$\longleftrightarrow$	$9 - 6 = 3$

In the first example, adding the opposite of 3 produces the same result as subtracting 3. In the second example, adding the opposite of 6 produces the same result as subtracting 6.

Subtraction Rule

To subtract b from a, add the opposite of b to a.

$$a - b = a + (-b)$$

The result is the difference of a and b.

In Lesson 2.2, you saw that the order in which two numbers are added doesn't affect their sum. For subtraction, however, different orders can produce different results.

$5 - 2 = 3$ *The difference of 5 and 2 is 3.*

$2 - 5 = -3$ *The difference of 2 and 5 is −3.*

Example 1 *Using the Subtraction Rule*

a. $-4 - 3 = -4 + (-3)$
$= -7$

b. $10 - 11 = 10 + (-11)$
$= -1$

c. $5 - (-4) = 5 + 4$
$= 9$

d. $-\frac{3}{2} - (-\frac{1}{2}) = -\frac{3}{2} + \frac{1}{2}$
$= -1$ ■

When an expression is written as a *sum,* the parts that form the sum are the **terms** of the expression. For instance, the expression $-9 - x$ can be written as $-9 + (-x)$ and has two terms. The first term is -9, and the second term is $-x$.

Problem of the Day

If a and b are positive one-digit integers and $5 + a + b = 17$, then what is the smallest value b can have?
3

ORGANIZER

Warm-Up Exercises

1. Evaluate the following.
   - **a.** $12 + (-23)$ -11
   - **b.** $-15 + (-32)$ -47
   - **c.** $-2.3 + (-7.4) + 8$ -1.7
   - **d.** $18.5 + (-18.5)$ 0
2. Compare the sums with the differences.
   - **a.** $13 + (-5)$ $13 - 5$
   - **b.** $-57 + (-46)$ $-57 - 46$
   - **c.** $23 + (-37)$ $23 - 37$

   Respectively, each sum and difference is the same; a. 8, b. −103, c. −14

Lesson Resources

Teaching Tools
  Problem of the Day: 2.3
  Warm-up Exercises: 2.3
  Answer Masters: 2.3
Extra Practice: 2.3
Applications Handbook: p. 58–59

LESSON Notes

Example 1

An Alternate Approach
Using Manipulatives

In **Example 1,** checkers can be used to model subtraction of integers. For example, the differences for the problems $5 - 2$ and $5 - (-8)$ can be easily found using checkers. Specifically, to find $5 - 2$: Place 5 black checkers on your desk; remove two of them. Three black checkers remain. So, $5 - 2 = +3$.

To find $5 - (-8)$: Place 5 black checkers on your desk. How can you remove 8 red checkers? Because there are no red checkers on the desk, this may seem impossible! However, since 0 can be represented in a number of different ways and because 0 is the additive identity, you may place 8 black and 8 red checkers on the desk without changing the integer value. Now 5 looks like this.

Remove 8 red checkers. That leaves 13 black checkers in all! So $5 - (-8) = 13$.

Example 3

Note that for penny stocks, closing prices may be given in 16ths, 32nds, 64ths, or 128ths. Most calculators will change the fractions in this example to decimal numbers. There are fraction calculators (such as Texas Instruments' Math Explorer) that will compute and display numbers in their fractional form.

Common-Error ALERT!

Students frequently forget to apply the distributive property correctly when multiplying by -1. Stress that each of the two terms in the two expressions $(a - b)$ and $(a + b)$ must be multiplied by -1.

Expressions containing more than one subtraction can also be evaluated by "adding opposites." Note how the Left-to-Right Rule for order of operations is used in Example 2.

Example 2 *Repeated Subtraction and Addition*

$$\begin{aligned} 3 - (-4) - 2 + 8 &= 3 + 4 + (-2) + 8 && -(-4) \\ &= 7 + (-2) + 8 && 3 + 4 \\ &= 5 + 8 && 7 + (-2) \\ &= 13 && 5 + 8 \end{aligned}$$

■

On a typical day, over 200 million shares worth over \$7.5 billion are bought and sold through the New York Stock Exchange. Stock prices often vary throughout a day. The closing price of each stock is the price of the last sale in the day. Changes in closing prices are usually given in multiples of eighths of dollars.

Real Life
Stock Market

Example 3 *Finding the Value of a Stock*

On Friday, May 10, 1991, the closing price of a share of Coca-Cola™ stock was $\$54\frac{3}{4}$. The daily changes in closing values during the following week were as follows.

Monday	*Tuesday*	*Wednesday*	*Thursday*	*Friday*
$-\frac{3}{8}$	$-\frac{1}{2}$	$-\frac{1}{8}$	$+1\frac{1}{4}$	$-\frac{1}{8}$

What was the closing price of a share on Friday, May 17, 1991?

Solution You could add and subtract fractions by hand to solve the problem, or you could use a calculator.

$$\begin{aligned} \text{Closing price} &= 54\tfrac{3}{4} + (-\tfrac{3}{8}) + (-\tfrac{1}{2}) + (-\tfrac{1}{8}) + 1\tfrac{1}{4} + (-\tfrac{1}{8}) \\ &= 54\tfrac{3}{4} - \tfrac{3}{8} - \tfrac{1}{2} - \tfrac{1}{8} + 1\tfrac{1}{4} - \tfrac{1}{8} \\ &= 54\tfrac{7}{8} \end{aligned}$$

Thus, the closing price on May 17 was $\$54\frac{7}{8}$ per share. ■

Goal 2 Simplifying Algebraic Expressions

You can *simplify* the expression $8 + (-x + 1)$ by rewriting it as $9 - x$. The second form has fewer terms and fewer grouping symbols, but it is **equivalent** to the first because both expressions have the same value for each number represented by the variable. Whenever you are performing any algebraic operation, you can *substitute* an equivalent form of an expression without changing the result.

The next properties are useful for simplifying an expression.

Property	Statement	Example
Opposite of a Difference	$-(a - b) = -a + b$	$-(5 - 3) = -5 + 3 = -2$
Opposite of a Sum	$-(a + b) = -a - b$	$-(-8 + 5) = 8 - 5 = 3$

Example 4 *Simplifying before Evaluating*

Evaluate $7 - (2 - x)$ when $x = -3$.

Solution First, you can use the "opposite of a difference" property to write an equivalent expression. Then simplify before substituting.

$7 - (2 - x) = 7 - 2 + x$	*Opposite of a Difference*
$= 5 + x$	*Simplify $7 - 2$.*
$= 5 + (-3)$	*Substitute -3 for x.*
$= 2$	*Simplify $5 + (-3)$.*

■

Communicating about ALGEBRA

SHARING IDEAS about the Lesson

Demonstrate Understanding Which of the following pairs of expressions are equivalent? For all pairs that are not equivalent, find a value of x that produces different values for the two expressions. (For instance, the expressions $x - 4$ and $4 - x$ are not equivalent because when $x = 1$, they will have values of -3 and 3, respectively.)

	Expression 1	Expression 2
A.	$r - (-3)$	$r + 3$
B.	$x - (2 - x)$	$x - 2 - x$
C.	$5 - y - 2$	$(5 - y) + 2$

A., Equivalent
B., C. Not equivalent, for any number

Example 4

Point out that the expression $7 - (2 - x)$ can be rewritten as $7 + [-(2 - x)]$ so that the Opposite of a Difference property can be applied.

Extra Examples

Here are additional examples similar to **Examples 1–4.**

1. Simplify.
a. $-4 - 7$ **b.** $12 - 19$
c. $-\frac{1}{3} - \frac{4}{5}$ **d.** $13 - (-6)$
e. $8 - (-6) - 7 + 14$
f. $-6 + (-15) + 9 - (-21)$
g. $33\frac{2}{3} + (-4) - 16\frac{1}{3} + 7$
h. $34\frac{1}{8} - \frac{5}{8} + 4\frac{1}{2} - (-3\frac{3}{8})$

a. -11, b. -7, c. $-1\frac{2}{15}$, d. 19, e. 21, f. 9, g. $20\frac{1}{3}$, h. $41\frac{3}{8}$

2. Evaluate in two ways: Use the distributive property and then simplify inside parentheses first. Compare your results.
a. $-(29 - 15)$ -14
b. $-(-7 - 9)$ 16
c. $-(31 + (-24))$ -7
d. $-(-17 + 28)$ -11

3. Simplify before evaluating.
a. $9 - (4 + x), x = 2$ 3
b. $-7 - (3 - x), x = -2$ -12
c. $8 - (5 + x), x = 3$ 0
d. $-2 - (9 - x), x = 5$ -6

Check Understanding

If adding the opposite of a number is the same as subtracting the number and if addition is commutative, is subtraction of integers commutative? Discuss your reasoning with others. No; for all real numbers a and b, $a - b \neq b - a$. However, since addition of integers is commutative, $a - b = a + (-b)$ implies that $a + (-b) = (-b) + a$.

Communicating about ALGEBRA

A pair of equivalent expressions is any pair of expressions in which each expression can be transformed by properties and rules of algebra into the other. For example, $x - (-3)$ is equivalent to $x + 3$ because $x - (-3)$ can be rewritten as $x + 3$ by the Subtraction Rule. Require students to use a property or rule to justify or explain why given pairs of expressions are equivalent.

EXTEND *Communicating COOPERATIVE LEARNING*

You might also ask pairs (or small groups) of students to create six examples of expressions that may or may not be equivalent. They should include the answers on the back of the list. Problems could be exchanged and checked, or they could be collected to be used later for review.

EXERCISE Notes

ASSIGNMENT GUIDE

Basic/Average: Ex. 1–6, 9–42 multiples of 3, 43, 45–46, 48–57 multiples of 3, 59–62

Above Average: Ex. 1–6, 9–42 multiples of 3, 45–50, 59–64

Advanced: Ex. 1–6, 24–42 multiples of 3, 44–50, 61–66

Selected Answers
Exercises 1–6, 7–57 odd

More Difficult Exercises
Exercises 43, 44, 63–66

Guided Practice

Ex. 4–6 Students need to use their critical-thinking skills to complete.

EXERCISES

Guided Practice

CRITICAL THINKING about the Lesson

1. **True or False?** Adding the opposite of a number is the same as subtracting the number. True
2. **True or False?** The difference of a and b is $a - b$. True
3. What is the first term of $-x + (-3)$? What is the second term? $-x$, -3
4. For two numbers, is the order in which subtraction is performed important? Give an example to support your answer. Yes; $4 - 3 = 1$ and $3 - 4 = -1$
5. Evaluate $-3 - (-2) - (-1)$. 0
6. Can $-(6 - 3)$ be simplified as $-6 + 3$? Explain. Yes; $-(6 - 3) = -(3) = -3$, $-6 + 3 = -3$

Independent Practice

In Exercises 7–22, find the difference.

7. $3 - 8$ -5
8. $12 - 8$ 4
9. $7 - 4$ 3
10. $4 - 7$ -3
11. $\frac{6}{2} - \frac{3}{2}$ $\frac{3}{2}$
12. $\frac{2}{3} - \frac{12}{3}$ $-\frac{10}{3}$
13. $-1 - 5$ -6
14. $-9 - 6$ -15
15. $8 - |-6|$ 2
16. $12 - |-7|$ 5
17. $-6 - |4|$ -10
18. $-3 - |5|$ -8
19. $|10| - 6.5$ 3.5
20. $|15| - 2.4$ 12.6
21. $|-\frac{3}{4}| - \frac{7}{4}$ -1
22. $|-\frac{7}{4}| - \frac{3}{4}$ 1

In Exercises 23–28, evaluate the expression.

23. $2 - (-6) - 7$ 1
24. $3 + (-4) - 2$ -3
25. $-6 + 5 - 4$ -5
26. $-11 - (-12) + 1$ 2
27. $5 - 3 + 12 - 9$ 5
28. $10 - 4 - 9 + 2$ -1

Technology **In Exercises 29–34, use a calculator to evaluate the expression. Use estimation to check your result.**

29. $4.8 - (-3.1) - 2.2$ 5.7
30. $7.1 - (-1.1) - 6.4$ 1.8
31. $-6.23 + 4.52 - (-2.75)$ 1.04
32. $-3.14 + 2.20 - (-3.21)$ 2.27
33. $-11.99 - (-11.99) + 6.8$ 6.8
34. $-17.3 + 0.0 + 4.2$ -13.1

In Exercises 35–42, evaluate the expression. Then simplify the expression and evaluate the simplified form. Check to see that the two values are equal.

35. $4 - (1 + x)$, when $x = 2$ 1
36. $2 - (4 - t)$, when $t = 1$ -1
37. $15 - (-x) + 7$, when $x = -6$ 16
38. $12 - (-x) - 5$, when $x = -2$ 5
39. $-9 - (-13) - p$, when $p = -7$ 11
40. $-8 - (-5) - r$, when $r = -3$ 0
41. $-x - (3 - 8) + 4$, when $x = 10$ -1
42. $-x - (7 + 6) + 2$, when $x = 9$ -20

43. *Monopoly* Suppose you are playing Monopoly. So far you have collected \$200 for passing **GO** and have received \$300 from another player. But now you have drawn this card. If you own four hotels and eight houses, how much does this cost? Did you gain or lose money on this trip around the board? Explain. See below.

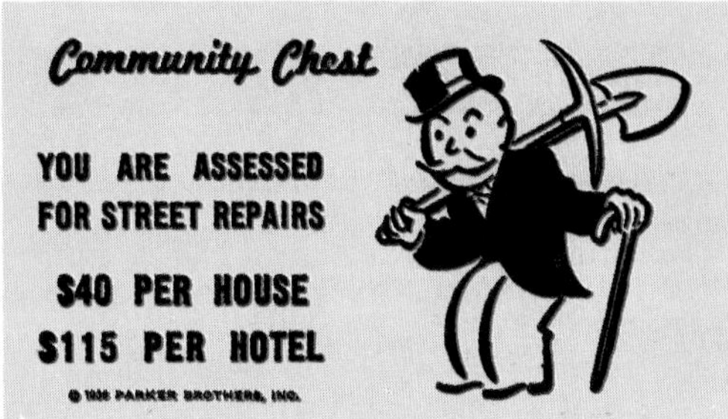

MONOPOLY is a registered trademark of Parker Brothers, division of Tonka Corporation, and used with permission.

44. *Field Trip* While on a field trip, you and a few classmates go exploring. You climb a hill and visit a cave before joining the rest of the class by the lake for lunch. You are curious to discover how high you actually climbed, so you find a map of the area. Here is a list of the various elevations.

A. Parking lot	300 feet
B. Start of trail	275 feet
C. Foot of hill	325 feet
D. Peak of hill	515 feet
E. Cave on hill	440 feet
F. Edge of lake	210 feet

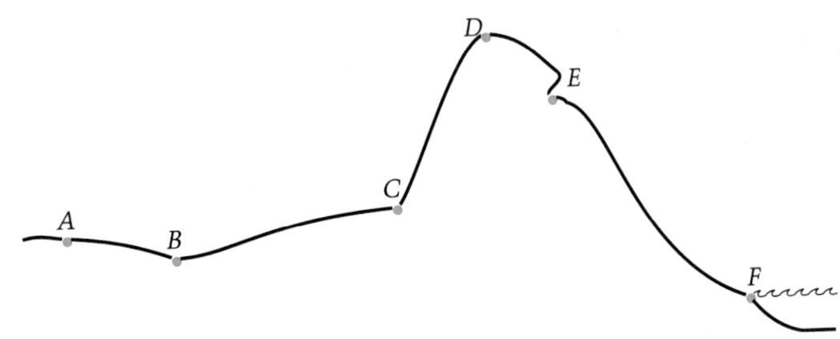

How high is the hill? Determine the change in elevation between each two points. Find the total of these changes to discover your net change in elevation. See below.

Escorted Tours **In Exercises 45 and 46, use the bar graph.**

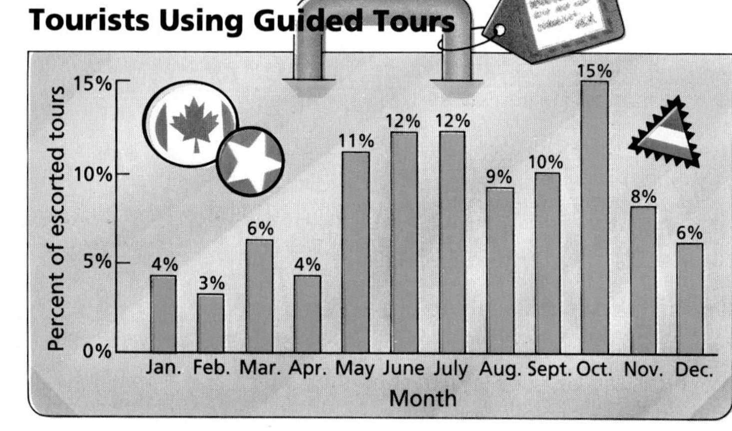

This bar graph shows the percentages of tourists in the United States who went on escorted tours in 1990. (Source: The National Tour Association)

45. Complete the table by finding the monthly changes in the percentages given in the bar graph. −2%, 7%, 1%, 0%, −3%, 1%, 5%, −7%, −2%

Month	J	F	M	A	M	J	J	A	S	O	N	D
Change in percentage	−2	−1	3	?	?	?	?	?	?	?	?	?

46. What is the overall change in percentage for the first six months? For the second six months? For the year? 6%, −6%, 0%

43. \$780, lose; income is \$500 and expenses are \$280 more than that.

44. The hill is 190 ft high; A to B: −25 ft, B to C: 50 ft, C to D: 190 ft, D to E: −75 ft, E to F: −230 ft; net change: −90 ft.

Independent Practice

▶ **Ex. 15–22** Remind students that the absolute value of a number is the number if it is positive, and its opposite if it is negative.

EXTEND Ex. 35–42 **USING PROPERTIES** Ask students to identify the rule or property that allowed them to simplify each expression.

▶ **Ex. 43** The popular game Monopoly is mentioned. How many students in class have actually played this game? For those who have not, have one or more students briefly describe how the game is played and encourage students who have not played to consider playing the game in the future.

▶ **Ex. 44–46** **PROBLEM SOLVING** Indicate that information used to solve problems is frequently contained in tables. Reading and interpreting tables correctly is an important mathematical skill.

Answers

47. $-5, -3, -2\frac{1}{2}, -1.25, 1, 2\frac{1}{2}, 4.3$

48. 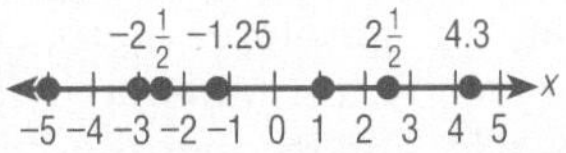

Ex. 59–60 PROBLEM SOLVING Students should use the numerical expressions to develop and test the patterns. Note also that the numbers in these expressions are the coefficients of the variables.

Ex. 63–66 A measure in music is a division of the staff (the five horizontal lines and the four spaces between them) into groups of beats, as indicated by the time signature. Encourage students in the class who read music to share their knowledge with those who do not.

Enrichment Activities

WRITING Following the class discussion, students might record their personal responses in their math journals.

Explain how the rules of addition and subtraction of integers can be shown using concrete objects or pictorial models. See Lesson Notes for Lessons 2.2 and 2.3.

Integrated Review

In Exercises 47–50, use the numbers $-3, 2\frac{1}{2}, 4.3, -1.25, -5, -2\frac{1}{2}$, and 1. 47., 48. See margin.

47. Write the numbers in increasing order.

48. Graph the numbers on the number line.

49. Which are less than 3 units from 0? $-2\frac{1}{2}, -1.25, 1, 2\frac{1}{2}$.

50. Which are more than 3 units from 0? $-5, 4.3$

In Exercises 51–54, evaluate the expression.

51. $6 + (1 - 3)$ 4

52. $-5 + 11$ 6

53. $-12 + 4$ -8

54. $8 + (-2)$ 6

In Exercises 55–58, use mental math to solve the equation.

55. $x - 8 = 10$ 18

56. $x + 45 = 30$ -15

57. $-[(-x) + 8] = 16$ 24

58. $-(x - 5) = 8$ -3

Exploration and Extension

59. Consider the sums $-1 + (-1) = -2$ and $-1 + (-1) + (-1) = -3$. Using these patterns, simplify $-x + (-x)$ and $-x + (-x) + (-x)$. $-2x$, $-3x$

60. Consider the sums $-1 + (-2) = -3$ and $-2 + (-3) + (-4) = -9$. Using these patterns, simplify $-x + (-2x)$ and $-2x + (-3x) + (-4x)$. $-3x$, $-9x$

In Exercises 61–62, use Exercises 59 and 60 to simplify the expression.

61. $4 + (-x) + (-2x)$ $4 - 3x$

62. $3 - [-x + (-x)]$ $3 + 2x$

Reading Music **In Exercises 63–66, show that each measure has the correct number of beats.** See Additional Answers.

In music, the time signature indicates the number of beats to a measure. Two common time signatures are $\frac{3}{4}$ (three beats to a measure) and $\frac{4}{4}$ (four beats to a measure).

Dotting a note means to hold it one and one-half times as long. The symbols 𝄽 and 𝄾 mean to rest, or not play anything, for one beat and one-half beat respectively.

Name	Symbol	Number of beats
Whole note	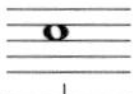	4
Half note	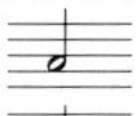	2
Quarter note		1
Eighth note		$\frac{1}{2}$
Sixteenth note		$\frac{1}{4}$

63. 

64. 

65. 

66. 

2.4 Exploring Data: Matrices

What you should learn:

Goal 1 How to organize data into a matrix

Goal 2 How to add and subtract two matrices

Why you should learn it:

You can use matrices for effective organization of many forms of real-life data, such as library statistics, political charts, and accounting spreadsheets.

Goal 1 Organizing Data in a Matrix

A **matrix** is a rectangular arrangement of numbers into rows and columns. For instance, the matrix

$$\begin{bmatrix} 3 & 1 & 0 \\ -1 & 2 & 4 \end{bmatrix}$$

has two rows and three columns. The numbers in the matrix are called **entries.** In the above matrix, the entry in the second row and third column is 4. (The plural of *matrix* is *matrices.*)

Two matrices are **equal** if the entries in corresponding positions are equal.

$$\begin{bmatrix} 3 & -2 \\ \frac{1}{2} & 0 \end{bmatrix} = \begin{bmatrix} 3 & -2 \\ 0.5 & 0 \end{bmatrix} \qquad \begin{bmatrix} -4 & 7 \\ 0 & -1 \end{bmatrix} \neq \begin{bmatrix} 7 & -4 \\ 0 & -1 \end{bmatrix}$$

You can think of a matrix as a type of table that can be used to organize data.

	Circulation	Reference
Books	10,000	3,000
Magazines	500	6,000
Videos	2,000	500

Example 1 *Writing a Table as a Matrix*

Some of the books, magazines, and videos in a library can be checked out. Others are for reference only and must be used in the library. The table at the left shows how many of each are in the Springfield High School library.

Write this table as a matrix. How many reference books are in the library's collection?

Solution The matrix associated with this table has three rows and two columns.

$$\begin{array}{r} \\ \text{Books} \\ \text{Magazines} \\ \text{Videos} \end{array} \begin{array}{c} \textbf{Circulation} \quad \textbf{Reference} \\ \begin{bmatrix} 10{,}000 & 3{,}000 \\ 500 & 6{,}000 \\ 2{,}000 & 500 \end{bmatrix} \end{array}$$

The number of reference books is 3,000 because that is the entry in the row labeled *Books* and the column labeled *Reference.* ■

Problem of the Day

A teacher entered a room and saw four students—Andrea, Beth, Carlos, and Don. On the teacher's desk was an apple. Curious, he asked each student who brought it. Andrea said "I did not bring it." Don said "Beth brought it." Beth said "Carlos brought it." Carlos said "Beth did not tell you the truth when she said I brought it."

Only one of those statements is true. So, who brought the apple?
Andrea

ORGANIZER

Warm-Up Exercises

1. Evaluate each expression.
 a. 2 − (−7) **b.** −11 − (−7)
 c. 0 − (−5) **d.** 14 + (−19)
 a. 9, b. −4, c. 5, d. −5
2. Use the table below.
 a. How many comedy videotapes does the store presently have? 55
 b. How many new sci–fi video-tapes are there? 7
 c. How many new releases in the categories identified are there? 30

	New Releases	Regular Selection
Comedy	9	46
Adventure	14	97
Sci-Fi	7	86

Lesson Resources

Teaching Tools
 Transparencies: 1, 2
 Problem of the Day: 2.4
 Warm-up Exercises: 2.4
 Answer Masters: 2.4
Extra Practice: 2.4
Technology Handbook: p. 16

LESSON Notes

Example 1

Emphasize that the rows of a matrix are the horizontal arrangements of numbers; the columns are the vertical arrangements of numbers. A matrix of 3 rows and 4 columns is said to have dimension 3×4 (read as 3 by 4).

Example 2

Corresponding entries have the same row number and the same column number.

Example 4

Make sure students understand that if the year is divided into four sets of months, then the first group of months, the first quarter, consists of January, February, and March.

Companies usually state their profits as a percent of the total income. You might ask students to find the percent of profit in Example 4. 4.8%

Extra Examples

Here are additional examples similar to **Examples 1–4.**

1. The following table shows a retail store's selection of jeans. Write the table below as a matrix.

	Denim	Polyester
Size 8	12	8
Size 10	18	19
Size 12	22	14
Size 14	13	10

   a. How many pairs of jeans does the store have in size 12? in size 10? 36, 37
   b. How many pairs of denim jeans does the store carry? 65
   c. How many pairs of jeans in size 8 are needed to equal the number of jeans in size 14? 3

Goal 2 Adding and Subtracting Matrices

To **add** or **subtract** matrices, you simply add or subtract corresponding entries, as shown in Example 2.

Example 2 *Adding and Subtracting Matrices*

a. $$\begin{bmatrix} 4 & 2 \\ 0 & -3 \\ -5 & 1 \end{bmatrix} + \begin{bmatrix} 1 & 0 \\ 2 & -1 \\ 6 & -4 \end{bmatrix} = \begin{bmatrix} 4+1 & 2+0 \\ 0+2 & -3+(-1) \\ -5+6 & 1+(-4) \end{bmatrix} = \begin{bmatrix} 5 & 2 \\ 2 & -4 \\ 1 & -3 \end{bmatrix}$$

b. $$\begin{bmatrix} 10 & -4 \\ 5 & 0 \end{bmatrix} - \begin{bmatrix} 4 & 5 \\ -3 & 2 \end{bmatrix} = \begin{bmatrix} 10-4 & -4-5 \\ 5-(-3) & 0-2 \end{bmatrix} = \begin{bmatrix} 10-4 & -4-5 \\ 5+3 & 0-2 \end{bmatrix} = \begin{bmatrix} 6 & -9 \\ 8 & -2 \end{bmatrix}$$

You cannot add or subtract matrices that have different numbers of rows or columns. For instance, you cannot add a matrix that has three rows to a matrix that has only two rows.

Real Life **Political Science**

Example 3 *Political Composition of Congress*

Find the number of Democrats, Republicans, and Independents in the Congress between 1983 and 1995.
(Source: United States Congress)

Senate	1983	1985	1987	1989	1991	1993	1995
Democrats	46	47	55	55	56	57	48
Republicans	54	53	45	45	44	43	52
Independents	0	0	0	0	0	0	0

House	1983	1985	1987	1989	1991	1993	1995
Democrats	269	252	258	259	267	258	204
Republicans	165	182	177	174	167	176	230
Independents	0	0	0	0	1	1	1

Solution The United States Congress is composed of the Senate and the House of Representatives. By adding the given matrices, you obtain a matrix that shows the political composition of the Congress.

House	1983	1985	1987	1989	1991	1993	1995
Democrats	315	299	313	314	323	315	252
Republicans	219	235	222	219	211	219	282
Independents	0	0	0	0	1	1	1

The profit that a company makes is the difference between its income (or revenue) and its expenses (or costs). For instance, if a company had income of \$125,000 and expenses of \$110,000, its profit would be \$125,000 − \$110,000 = \$15,000.

Melencholia, **with detail below. Albrecht Dürer (1471–1528)**

Example 4 *Finding the First-Quarter Profit*

A retail company has two stores. The incomes and expenses (in dollars) for the two stores for three months are shown in the matrices.

a. Write a matrix that shows the first-quarter incomes and expenses for the two stores.

b. Find the first-quarter profit for each store.

January	Income	Expenses
Store 1	200,000	180,000
Store 2	110,000	120,000

February	Income	Expenses
Store 1	220,000	205,000
Store 2	135,000	130,000

March	Income	Expenses
Store 1	230,000	225,000
Store 2	140,000	125,000

Solution

a. To find the matrix for the first-quarter incomes and expenses, add the matrices for January, February, and March.

1st Quarter	Income (\$)	Expenses (\$)
Store 1	650,000	610,000
Store 2	385,000	375,000

b. Profit for Store 1 = 650,000 − 610,000 = \$40,000

Profit for Store 2 = 385,000 − 375,000 = \$10,000 ■

Communicating about ALGEBRA

SHARING IDEAS about the Lesson

A matrix that has the same number of rows and columns is a **square matrix.** A **magic square** is a square matrix whose rows, columns, and diagonals each add to the same sum. Find the missing entries in the following magic square with entries 1, 2, 3, . . . , 15, 16.

$$\begin{bmatrix} 16 & 3 & 2 & 13 \\ 5 & 10 & 11 & 8 \\ \boxed{?} & \boxed{?} & \boxed{?} & \boxed{?} \\ \boxed{?} & \boxed{?} & \boxed{?} & \boxed{?} \end{bmatrix}$$

9, 6, 7, 12

4, 15, 14, 1

2. Perform the indicated matrix operation.

a. $\begin{bmatrix} -3 & 8 & 0 \\ 5 & 2 & -4 \end{bmatrix} + \begin{bmatrix} 2 & 11 & -7 \\ -8 & 9 & -3 \end{bmatrix}$

b. $\begin{bmatrix} 7 & -15 \\ 3 & 2 \\ 8 & 19 \\ 9 & 3 \end{bmatrix} - \begin{bmatrix} 9 & -23 \\ -5 & 12 \\ 0 & 22 \\ 1 & 4 \end{bmatrix}$

a. $\begin{bmatrix} -1 & 19 & -7 \\ -3 & 11 & -7 \end{bmatrix}$

b. $\begin{bmatrix} -2 & 8 \\ 8 & -10 \\ 8 & -3 \\ 8 & -1 \end{bmatrix}$

Check Understanding

1. Explain how the difference of two matrices $A - B$ can be expressed as a sum. Illustrate with an example. $A - B = A + (-B)$; each entry in the matrix $-B$ is the opposite of the corresponding entry in matrix B.
2. Is there a matrix, I, which when added to any other matrix, M, has a sum equal to the second matrix, M? Yes; I is the zero matrix, which has all 0 entries.

Communicating about ALGEBRA

EXTEND *Communicating*
MAGIC SQUARES

Once a magic square has been found, ask students to investigate what happens when a number value (say, 5) is added to each of the entries. Is the resulting square a magic square? Yes What happens when a number value is subtracted from each entry? Are there other operations or combinations of operations that do not affect the magic of the square? Any combination of addition, subtraction, or nonzero multiplication and division applied to each entry results in another magic square.

EXERCISE Notes

ASSIGNMENT GUIDE

Basic/Average: Ex. 1–19, 21–23 odd, 27–36

Above Average: Ex. 1–19, 21–24, 27–36

Advanced: Ex. 1–8, 9–23 odd, 25–26, 29–36

Selected Answers
Exercises 1–4, 5–31 odd

More Difficult Exercises
Exercises 24–26, 33–36

Guided Practice

Ex. 3 Students should use mental math to determine the values of *a, b, c,* and *d*. Have them explain their reasoning in solving this exercise. Similar reasoning will be needed to complete Exercises 19–20 later.

Ex. 4 Remember that the difference between two matrices depends on the order in which the matrices are given. Assume that the left-most matrix is given first.

Independent Practice

Ex. 9–18 Students' work with matrices is a good way for them to review their computational skills involving integers.

EXTEND Ex. 21–23 PROBLEM SOLVING You or your students might suggest other questions related to the information contained in these matrices. For example, in Exercise 21, you might ask, "How many new releases does the video store have? How many horror titles does the store have?" etc. 80, 205

EXERCISES

Guided Practice

CRITICAL THINKING about the Lesson

In Exercises 1 and 2, use the matrix at the right. $\begin{bmatrix} 4 & -3 \\ 6 & 1 \\ -5 & -6 \end{bmatrix}$

1. How many rows are in the matrix? How many columns? 3, 2

2. What is the entry in the first row and second column? −3

3. If
$$\begin{bmatrix} 1 & 3 \\ -1 & 0 \end{bmatrix} = \begin{bmatrix} a & b \\ c & d \end{bmatrix},$$
what are the values of a, b, c, and d? $a = 1$, $b = 3$, $c = -1$, $d = 0$

4. Find the sum and difference of these matrices. $\begin{bmatrix} -1 & 0 & -4 \\ 2 & -3 & 0 \end{bmatrix}, \begin{bmatrix} 3 & -2 & 0 \\ -2 & 1 & -4 \end{bmatrix}$ $\begin{bmatrix} 2 & -2 & -4 \\ 0 & -2 & -4 \end{bmatrix}, \begin{bmatrix} -4 & 2 & -4 \\ 4 & -4 & 4 \end{bmatrix}$

Independent Practice

In Exercises 5–8, decide whether the matrices can be added.

5. $\begin{bmatrix} 3 & -1 \\ 6 & 2 \end{bmatrix}, \begin{bmatrix} 3 & -3 \\ -2 & 1 \end{bmatrix}$ Can

6. $\begin{bmatrix} 2 & -1 & 6 \\ 3 & 0 & -9 \end{bmatrix}, \begin{bmatrix} -4 & 5 \\ 7 & 5 \end{bmatrix}$ Cannot

7. $\begin{bmatrix} 4 & 2 & 6 \\ -6 & 3 & 9 \\ -1 & -2 & 1 \end{bmatrix}, \begin{bmatrix} 6 & 4 & -3 \\ 7 & -8 & 1 \end{bmatrix}$ Cannot

8. $\begin{bmatrix} 9 & 4 & -6 \\ 6 & 1 & 3 \end{bmatrix}, \begin{bmatrix} -3 & -6 & 0 \\ 4 & -1 & 9 \end{bmatrix}$ Can

In Exercises 9–14, find the sum of the matrices. 12.–14. See Additional Answers.

9. $\begin{bmatrix} 4 & -1 \\ 6 & 0 \end{bmatrix} + \begin{bmatrix} 4 & -3 \\ -7 & 2 \end{bmatrix}$ $\begin{bmatrix} 8 & -4 \\ -1 & 2 \end{bmatrix}$

10. $\begin{bmatrix} 6 & -3 \\ -4 & -2 \end{bmatrix} + \begin{bmatrix} -5 & 3 \\ 4 & 3 \end{bmatrix}$ $\begin{bmatrix} 1 & 0 \\ 0 & 1 \end{bmatrix}$

11. $\begin{bmatrix} 2 & 1 & -2 \\ 3 & 0 & 3 \end{bmatrix} + \begin{bmatrix} 1 & -3 & 4 \\ 2 & 5 & -5 \end{bmatrix}$ $\begin{bmatrix} 3 & -2 & 2 \\ 5 & 5 & -2 \end{bmatrix}$

12. $\begin{bmatrix} 2 & 9 & -3 \\ 1 & 8 & -2 \\ -4 & 1 & 0 \end{bmatrix} + \begin{bmatrix} -1 & -6 & 4 \\ -1 & 2 & 6 \\ 2 & -4 & 2 \end{bmatrix}$

13. $\begin{bmatrix} -1.3 & 2.4 & -6.9 \\ 15.8 & 0 & -3.4 \end{bmatrix} + \begin{bmatrix} 2.6 & 2.3 & -6.9 \\ 1.7 & 3.2 & -5.8 \end{bmatrix}$

14. $\begin{bmatrix} 6.1 & -1.7 \\ -1.3 & 4.4 \\ 3.2 & -5.9 \end{bmatrix} + \begin{bmatrix} 1.5 & 9.1 \\ -6.4 & 4.9 \\ 5.4 & -3.3 \end{bmatrix}$

In Exercises 15–18, find the difference of the matrices.

15. $\begin{bmatrix} 9 & -4 \\ 2 & -4 \end{bmatrix} - \begin{bmatrix} 6 & 6 \\ -2 & -4 \end{bmatrix}$ $\begin{bmatrix} 3 & -10 \\ 4 & 0 \end{bmatrix}$

16. $\begin{bmatrix} 6 & 4 \\ 10 & -5 \end{bmatrix} - \begin{bmatrix} 7 & 1 \\ -1 & -6 \end{bmatrix}$ $\begin{bmatrix} -1 & 3 \\ 11 & 1 \end{bmatrix}$

17. $\begin{bmatrix} -3 & 1 \\ 0 & -9 \\ 1 & 7 \end{bmatrix} - \begin{bmatrix} 6 & 4 \\ -5 & 8 \\ -1 & 4 \end{bmatrix}$ $\begin{bmatrix} -9 & -3 \\ 5 & -17 \\ 2 & 3 \end{bmatrix}$

18. $\begin{bmatrix} -5 & 12 & -1 \\ -10 & 4 & 0 \end{bmatrix} - \begin{bmatrix} -3 & 0 & 1 \\ 9 & 2 & -2 \end{bmatrix}$ $\begin{bmatrix} -2 & 12 & -2 \\ -19 & 2 & 2 \end{bmatrix}$

In Exercises 19 and 20, use mental math to find *a, b, c,* and *d.*

19. $\begin{bmatrix} 2a & 3b \\ c-4 & d \end{bmatrix} = \begin{bmatrix} 10 & 9 \\ 16 & -4 \end{bmatrix}$ $a = 5, b = 3, c = 20, d = -4$

20. $\begin{bmatrix} a & b+1 \\ 4c & d-1 \end{bmatrix} = \begin{bmatrix} 4 & 9 \\ 12 & 16 \end{bmatrix}$ $a = 4, b = 8, c = 3, d = 17$

21. *Video Rentals* A video store rents comedies, dramas, and horror movies. The table shows how many of each are new releases and how many of each are regular selections in the store.

	New releases	Regular selections
Comedy	25	215
Drama	30	350
Horror	25	180

	New releases	Regular selections
Comedy	25	215
Drama	30	350
Horror	25	180

Write this table as a matrix. How many newly released comedy videos does the store have? 25

22. *On Sale* A music store is having a sale on some of its cassette tapes and compact disks. The table shows how many different titles are available at the sale price and how many are available at the regular price.

	Sale price titles	Regular price titles
Tapes	45	3000
Compact disks	20	2500

	Sale price titles	Regular price titles
Tapes	45	3000
Compact disks	20	2500

Write this table as a matrix. How many compact disk titles are on sale? 20

23. *NFL Games* Did each of the four listed NFL Western Division teams play the same number of games per year from 1991 through 1994? Explain. ***(Source: National Football League)*** Yes. All teams played 16 regular season games.

Wins	1991	1992	1993	1994
San Francisco	10	14	10	13
L.A. Rams	3	6	5	4
Atlanta	10	6	6	7
New Orleans	11	12	8	7

Losses/Ties	1991	1992	1993	1994
San Francisco	6	2	6	3
L.A. Rams	13	10	11	12
Atlanta	6	10	10	9
New Orleans	5	4	8	9

24. *Grade-Level Wage* The hourly wage rates for the clerk-typist positions at a company depend on the number of years of experience. The company decides to give $0.50 raises to all typists in grade level 4, $0.75 raises to all typists in grade level 5, and $1.00 raises to all typists in grade level 6. Find the entries for the *raise* matrix and determine the resulting wage-rate matrix. See Additional Answers.

Wage rates	0–1 yr	2–3 yr	4+ yr
Level 4	8.00	9.00	9.50
Level 5	10.00	10.50	11.00
Level 6	11.50	12.00	12.50

+

Raises	0–1 yr	2–3 yr	4+ yr
Level 4	?	?	?
Level 5	?	?	?
Level 6	?	?	?

EXTEND Ex. 24 PROBLEM SOLVING Suppose the company described in Exercise 24 on page 85 decided to give raises based on years of experience. That is, those with 0–1 year's experience would get $.50 more per hour; those with 2–3 years' experience would get $.75 more per hour; and those with 4 or more years' experience would get $1.00 more per hour. Have students show this wage-rate matrix.

	Years of Experience		
Level	**0–1**	**2–3**	**4+**
4	8.50	9.75	10.50
5	10.50	11.25	12.00
6	12.00	12.75	13.50

You or your students might suggest questions such as the following that compare the two raise matrices.

a. Which workers do better if the raises are based on experience? Level 4, 2–3 years, and Level 4, 4+ years; Level 5, 4+ years

b. Which workers do better if the raises are based on grade level? Level 5, 0–1 year; Level 6, 0–1 year, and Level 6, 2–3 years

c. For which workers are both raise plans the same? Level 4, 0–1 year; Level 5, 2–3 years; and Level 6, 4+ years

Enrichment Activities

These activities require students to go beyond lesson goals.

1. Find the sum of the matrices.

a. $\begin{bmatrix} -2 & 4 & 7 \\ 0 & 1 & -3 \\ 8 & 1 & 13 \end{bmatrix} + \begin{bmatrix} 3 & 0 & -7 \\ 6 & 6 & -8 \\ 0 & 14 & 2 \end{bmatrix} + \begin{bmatrix} 1 & -2 & 9 \\ 0 & 3 & -4 \\ 1 & 19 & 0 \end{bmatrix}$

b. $\begin{bmatrix} -3 & 8 & 12 \\ 16 & 8 & 4 \\ z & 0 & 1 \end{bmatrix} + \begin{bmatrix} x & 3 & 0 \\ 0 & -2 & y \\ 1 & 6 & 8 \end{bmatrix}$

a. $\begin{bmatrix} 2 & 2 & 9 \\ 6 & 10 & -15 \\ 9 & 34 & 15 \end{bmatrix}$

b. $\begin{bmatrix} -3+x & 11 & 12 \\ 16 & 6 & 4+y \\ z+1 & 6 & 9 \end{bmatrix}$

2. Find the values of m, n, p, q, r, and s.

$\begin{bmatrix} 8 & -4 \\ 12 & 5 \\ r & -3 \end{bmatrix} + \begin{bmatrix} 0 & -6 \\ m & 7 \\ -2 & 8 \end{bmatrix} = \begin{bmatrix} p & q \\ -8 & n \\ 10 & s \end{bmatrix}$

$m = -20$; $n = 12$; $p = 8$; $q = -10$, $r = 12$; $s = 5$

Business **In Exercises 25 and 26, use the following information.**

You decide to sell food at a fair to raise enough money to buy a bike for $200. You hold the sale on Saturday and Sunday. The income and expenses (in dollars) for the two days for each type of item are shown in these matrices.

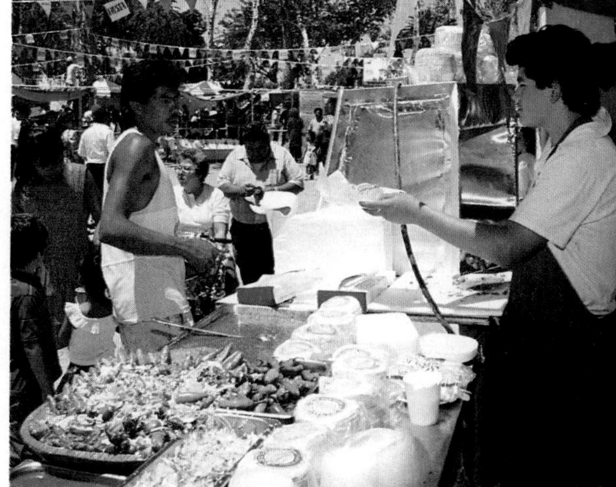

Saturday	Income	Expenses
Burritos	33	15
Tacos	38	30
Cheeseburgers	65	18

Sunday	Income	Expenses
Burritos	39	18
Tacos	29	23
Cheeseburgers	78	21

25. Find the total profit you received from each item. Burritos: $39, Tacos: $14, Cheeseburgers: $104

26. Did you raise enough money to pay for the bike? No

Integrated Review

In Exercises 27–28, evaluate the expression.

27. $\begin{bmatrix} 2 & -1 \\ 3 & 6 \end{bmatrix} + \begin{bmatrix} 6 & 10 \\ -1 & -6 \end{bmatrix}$ $\begin{bmatrix} 8 & 9 \\ 2 & 0 \end{bmatrix}$

28. $\begin{bmatrix} -4 & 5 \\ 6 & 2 \end{bmatrix} - \begin{bmatrix} 3 & -1 \\ -5 & 6 \end{bmatrix}$ $\begin{bmatrix} -7 & 6 \\ 11 & -4 \end{bmatrix}$

29. Evaluate the expression $\frac{1}{2}x + 1$ when $x = \frac{1}{2}$. $1\frac{1}{4}$

30. Evaluate the expression $6x - 1$ when $x = \frac{1}{3}$. 1

31. Use mental math to solve the equation $\frac{x}{9} = 3$. 27

32. Use mental math to solve the equation $2x + 1 = 19$. 9

Exploration and Extension

Business **In Exercises 33–36, use the following information.**

A company has two stores. For Store 1, the May and June incomes were $98,000 and $81,500, and the May and June profits were $16,500 and $10,500. For Store 2, the May and June incomes were $61,800 and $72,900, and the May and June profits were $9,600 and $10,600. See Additional Answers.

33. Set up a matrix that shows the incomes of the two stores for May and June.

34. Set up a matrix for the profits of the two stores for May and June.

35. Use the two matrices you wrote in Exercises 33 and 34 to write a matrix that shows the expenses of the two stores in May and June.

36. What were the expenses of Store 2 in May? In June? In May and June? $52,200; $62,300; $114,500

USING A GRAPHING CALCULATOR

Most graphing calculators and computers can store, add, and subtract matrices. Keystroke instructions for doing this on a TI-82, *Casio* fx-9700GE, *and Sharp* EL-9300C *are listed on page 735.*

The table shows the number of middle schools and high schools with modems, networks, and CD-ROMs from 1992 to 1995. One way to organize this data is to store it in two matrices—one showing middle schools and the other showing high schools. Try entering these matrices in a graphing calculator or computer. Then find the sum and difference of the matrices. What does the matrix $A + B$ represent? What does the matrix $B - A$ represent?

Schools with Computer Equipment

	1992		1993		1994		1995	
	Middle Schools	High Schools	Middle Schools	High Schools	Middle Schools	High Schools	Middle Schools	High Schools
Modems	2608	5001	3431	6371	4246	7402	5652	8790
Networks	776	1736	2030	4895	3220	6576	4599	8159
CD-ROMs	1231	2543	2326	4168	4261	6713	6510	9327

(Source: Quality Education Data, Inc.)

```
MATRIX[A]  3×4
[2608 3431 4246 5652]
[776  2030 3220 4599]
[1231 2326 4261 6510]
```

```
MATRIX[B]  3×4
[5001 6371 7402 8790]
[1736 4895 6576 8159]
[2543 4168 6713 9327]
```

Exercises

Use a graphing calculator or a computer to find the sum and difference of the matrices.

1. $\begin{bmatrix} 1 & 2 & 5 \\ -2 & 0 & -3 \end{bmatrix}, \begin{bmatrix} 0 & -4 & -2 \\ 2 & 0 & 3 \end{bmatrix}$

$\begin{bmatrix} 1 & -2 & 3 \\ 0 & 0 & 0 \end{bmatrix}, \begin{bmatrix} 1 & 6 & 7 \\ -4 & 0 & -6 \end{bmatrix}$

2. $\begin{bmatrix} -3 & 4 & 2 \\ 6 & 1 & -4 \\ 0 & -2 & -5 \end{bmatrix}, \begin{bmatrix} 2 & 0 & -1 \\ -4 & 5 & 0 \\ 0 & -3 & 9 \end{bmatrix}$

$\begin{bmatrix} -1 & 4 & 1 \\ 2 & 6 & -4 \\ 0 & -5 & 4 \end{bmatrix}, \begin{bmatrix} -5 & 4 & 3 \\ 10 & -4 & -4 \\ 0 & 1 & -14 \end{bmatrix}$

TECHNOLOGY Remind students that the difference of matrices depends on the order in which matrices are subtracted. In these exercises, let the leftmost matrix be the first of the two matrices given.

Answer

22. $\begin{bmatrix} -1 & -5 & 0 \\ 8 & -8 & -1 \\ 6 & 10 & 3 \end{bmatrix}$

25.

	Inc.	Exp.
Bowlathon	384	192
Car Wash	150	10
Sub Sale	400	200

$32 over

26.

	Feet	Meters
Sears	1454	443
World Trade	1350	411
Empire State	1250	381
Amoco Bldg	1136	346
John Hancock	1127	344

Mid-Chapter SELF-TEST

Take this test as you would take a test in class. The answers to the exercises are given in the back of the book.

1. Which is greater, -5 or -4? **(2.1)** -4

2. Which is less, $-\frac{1}{2}$ or $-\frac{1}{3}$? **(2.1)** $-\frac{1}{2}$

3. Find the opposite of -6. **(2.1)** 6

4. Find the opposite of $\frac{3}{2}$. **(2.1)** $-\frac{3}{2}$

5. Evaluate $|-4|$. **(2.1)** 4

6. Evaluate $-|-3\frac{1}{3}|$. **(2.1)** $-3\frac{1}{3}$

7. Evaluate $7 + (-5)$. **(2.2)** 2

8. Evaluate $-6 + 2$. **(2.2)** -4

9. Evaluate $-\frac{1}{2} + \frac{3}{2} + (-\frac{3}{4})$. **(2.2)** $\frac{1}{4}$

10. Evaluate $9 + 7 + (-6.5) + 1.1$ **(2.2)** 10.6

11. Evaluate $6.7 + (-4.5) + 1.2$ **(2.2)** 3.4

12. Evaluate $-12.5 + 8.6 + (-3.8)$. **(2.2)** -7.7

13. Evaluate $0.4 + 7.2 + (-2.2)$. **(2.2)** 5.4

14. Evaluate $9.3 + (-10.5) + 1.2$. **(2.2)** 0

In Exercises 15–20, simplify the expression. Then evaluate the expression. (2.2, 2.3)

15. $2 - (x - 3)$, when $x = 4$ $5 - x$, 1

16. $3 - (-x) - (-\frac{1}{2})$, when $x = 1$ $3\frac{1}{2} + x$, $4\frac{1}{2}$

17. $-[y + (-6)] - (y + 4)$, when $y = -1$ $-2y + 2$, 4

18. $|r| - (r + 3)$, when $r = -5$ $|r| - r - 3$, 7

19. $s + (7 - s) - 4$, when $s = 0$ 3, 3

20. $s + |-\frac{3}{2}| - (8 - s)$, when $s = -2$ $2s - \frac{13}{2}$, $-\frac{21}{2}$

In Exercises 21 and 22, find the sum or difference. (2.2, 2.3, 2.4)

21. $\begin{bmatrix} 3 & 6 & 7 \\ 4 & 0 & 8 \\ -1 & 2 & -4 \end{bmatrix} + \begin{bmatrix} -3 & 6 & 0 \\ 1 & 8 & 4 \\ 1 & 6 & 5 \end{bmatrix}$ $\begin{bmatrix} 0 & 12 & 7 \\ 5 & 8 & 12 \\ 0 & 8 & 1 \end{bmatrix}$

22. $\begin{bmatrix} -\frac{1}{2} & 3 & \frac{3}{4} \\ 10 & -7 & 4 \\ 6 & 9 & 0 \end{bmatrix} - \begin{bmatrix} \frac{1}{2} & 8 & \frac{3}{4} \\ 2 & 1 & 5 \\ 0 & -1 & -3 \end{bmatrix}$ See margin.

23. A ball is falling from the third floor of an apartment building at a rate of 15 feet per second. What is its velocity? What is its speed? **(2.1)**
-15 feet per second, 15 feet per second

24. The water temperature in an outdoor pool is 60°F. The temperature rises 2°F each day for four days. Because of a rainstorm, the temperature drops 5°F on the fifth day. What is the temperature on the fifth day after the temperature drop? **(2.2)** 63°F

25. A scout troop needs to raise $500 for a week of camping. Income and expenses for three projects are shown in the table. Write this table as a matrix. Did the troop raise enough money for its trip? How much was it short or over? **(2.4)** See margin.

	Income (in dollars)	Expenses (in dollars)
Bowlathon	384	192
Car wash	150	10
Sub sale	400	200

26. The world's five tallest skyscrapers and their heights in feet and meters are as follows. Set up a matrix that contains this information. (Sears Tower, Amoco Building, and John Hancock Center are in Chicago. The other two buildings are in New York City.) **(2.4)** See margin.

Sears Tower: 1454 feet, 443 meters
World Trade Center: 1350 feet, 411 meters
Empire State Building: 1250 feet, 381 meters
Amoco Building: 1136 feet, 346 meters
John Hancock Center: 1127 feet, 344 meters

2.5 Multiplication of Real Numbers

What you should learn:

Goal 1 How to multiply real numbers using Properties of Multiplication

Goal 2 How to use calculators and units of measure in real number products

Why you should learn it:

You can multiply positive and negative numbers to solve real-life problems, such as multiplying velocity by time to find distance. You can also check units of measure to help assure that you did not make a mistake.

Factor of -3: 3, 6
Factor of -2: 2, 4
Factor of -1: 1, 2
The sign of the product of two negative numbers is positive.

Study Tip

In the properties shown at the right, a and b don't have to be positive numbers. If either or both are negative, the properties are still true. For instance, the product of -1 and -2 is 2, which is the opposite of -2.

Goal 1 Multiplying Real Numbers

In this lesson, you will learn how to find products that have one or more negative factors. Here are two examples.

$$3(-2) = (-2) + (-2) + (-2) = -6$$
$$3(-1) = (-1) + (-1) + (-1) = -3$$

In both cases, multiplying a positive number and a negative number produces a negative number.

LESSON INVESTIGATION

■ Investigating Multiplication Patterns

Partner Activity Copy the following products. In each column, use the pattern of the first four products to write the last two products.

Factor of -3	Factor of -2	Factor of -1
$(3)(\mathbf{-3}) = -9$	$(3)(\mathbf{-2}) = -6$	$(3)(-1) = -3$
$(2)(-3) = -6$	$(2)(-2) = -4$	$(2)(\mathbf{-1}) = -2$
$(1)(-3) = -3$	$(1)(-2) = -2$	$(1)(-1) = -1$
$(0)(-3) = 0$	$(0)(-2) = 0$	$(0)(-1) = 0$
$(-1)(-3) = ?$ 3	$(-1)(-2) = ?$ 2	$(-1)(-1) = ?$ 1
$(-2)(\mathbf{-3}) = ?$ 6	$(-2)(\mathbf{-2}) = ?$ 4	$(-2)(-1) = ?$ 2

Discuss any generalizations that you can make about the sign of a product. For instance, what is the sign of the product of two negative numbers?

The sign of the product of two negative numbers is positive.

Properties of Multiplication

1. $a \cdot b = b \cdot a$	The order in which two numbers are multiplied does not change their product.
2. $1 \cdot a = a$	The product of one and a is a.
3. $0 \cdot a = 0$	The product of zero and a is 0.
4. $(-1) \cdot a = -a$	The product of -1 and a is the opposite of a.
5. $(a)(-b) = -ab$	The product of a and $-b$ is the opposite of ab.
6. $(-a)(-b) = ab$	The product of $-a$ and $-b$ is ab.

Problem of the Day

Show that $(10^2)^3$ and $(10^3)^2$ are equal.

$(10^2)^3 = (100)^3 = 100 \cdot 100 \cdot 100 = 1{,}000{,}000$

$(10^3)^2 = (1000)^2 = 1000 \cdot 1000 = 1{,}000{,}000$

ORGANIZER

Warm-Up Exercises

1. Multiply.
   a. $2 \cdot 7$ **b.** $11 \cdot 7$
   c. $0 \cdot 5$ **d.** $4 \cdot 19$
   a. 14, b. 77, c. 0, d. 76
2. One interpretation of multiplication is repeated addition. If $2 \cdot 7$ means two sets of 7 objects, then describe each problem as repeated addition and find its sum. (You could use the checker model to represent positive and negative integers.)
   a. $2 \cdot -7$ **b.** $11 \cdot -3$
   c. $0 \cdot -5$ **d.** $4 \cdot -9$
   a. 2 sets of 7 red, -14; b. 11 sets of 3 red, -33; c. 0 sets of 5 red, 0; d. 4 sets of 9 red; -36.
3. Write three more numbers to complete the patterns.
   a. 2, 4, 6, 8, ?, ?, ?
   10 ,12, 14
   b. 14, 11, 8, 5, ?, ?, ?
   2, -1, -4
   c. -12, -8, -4, 0, ?, ?, ?
   d. -10, -7, -4, ?, ?, ?
   c. 4, 8, 12; d. -1, 2, 5

Lesson Resources

Teaching Tools
  Problem of the Day: 2.5
  Warm-up Exercises: 2.5
  Answer Masters: 2.5
Extra Practice: 2.5
Applications Handbook: p. 60–61

LESSON Notes

Call students' attention to the patterns resulting from multiplying a positive and a negative integer. Identify the amount by which successive products increase. In order for each pattern to hold, the product of two negative integers must be positive.

Example 1

Example 1b follows from the 5th property of multiplication. Note that you can think of $(6)(-x)$ as $(6)(-1)(x) = (-6)(x) = -6x$.

Example 2

Many students will prefer to apply the properties of multiplication one step at a time. Such students will multiply from left to right, applying the properties of multiplication as needed. For example, $(-4)(-2)(-x) = (-4 \cdot -2)(-x) = (8)(-x) = -8x$.

Example 4

TECHNOLOGY Give students several opportunities to use their calculators to multiply real numbers. The behavior of the change-sign key and the negation key are quite different, and students may need to discuss in class how their particular calculator works. Observe that calculators with the negation key compute products consistent with the rules of order of operations. In particular, -3^4 equals -81 and $(-3)^4$ equals 81, as it should. (See *Extra Examples* on page 91.)

Example 5

Keeping track of units in a problem is a useful way to check whether the computations are correct.

Example 1 Products Involving Negative Factors

a. $(-4)(5) = -(4)(5) = -20$

b. $(6)(-x) = (6)(-1)(x) = -6x$

c. $(-\frac{1}{4})(-4) = (\frac{1}{4})(4) = 1$

d. $(-\frac{2}{5})(-\frac{3}{4}) = \frac{6}{20}$ or $\frac{3}{10}$ ■

The following rules can be helpful in simplifying products with three or more factors.

1. The product of an *even* number of negative factors is positive.
2. The product of an *odd* number of negative factors is negative.

Example 2 Products with Three or More Factors

a. $(-4)(-2)(-x) = -(4)(2)x = -8x$

b. $(-3)(5)(-y) = (3)(5)y = 15y$

c. $(-2)^4 = (-2)(-2)(-2)(-2) = 16$

d. $-2^4 = -(2 \cdot 2 \cdot 2 \cdot 2) = -16$

e. $(-x)^3 = (-x)(-x)(-x) = -x^3$ ■

Be sure you understand why $(-2)^4$ is not the same as -2^4.

Example 3 Real-Life Applications of Products of Negative Numbers

a. A football team loses 3 yards on each of 4 downs. The net gain is

$4(-3) = -12$ yards.

A gain of -12 means the same as a loss of 12.

b. A parachutist is falling with a velocity of -20 feet per second. The distance traveled in 3 seconds is

$3(-20) = -60$ feet.

The negative distance indicates a downward motion.

Real Life **Retailing**

c. To attract customers, a grocery store runs a sale on strawberries. The strawberries are *loss leaders,* which means the store loses money on the strawberries but hopes to make it up on other sales. Suppose 3000 baskets are sold at a loss of \$0.15 each.

$3000(-0.15) = -\$450$.

The negative result indicates a loss of \$450. ■

Goal 2 Using a Calculator and Units of Measure

Always be careful when using a calculator for evaluating a product involving negative numbers—especially with the change-sign key [+/−].

Example 4 ***Using a Calculator***

a. Expression | **Change-Sign Key** | **Negation Key**

$(-2)(-4)$ 2 [+/−] [×] 4 [+/−] [=]   [(−)] 2 [×] [(−)] 4 [=]

Your calculator should display 8 as an answer. The keystroke sequence [+/−] 2 [×] [+/−] 4 [=] is incorrect even though, in this instance, it may produce the correct answer.

b. Expression | **Change-Sign Key** | **Negation Key**

-3^4 3 [y^x] 4 [=] [+/−]   [(−)] 3 [∧] 4 [=]

Your calculator should display -81 as an answer. The keystroke sequence [+/−] 3 [y^x] 4 [=] will not produce the correct answer. ■

Most real-life applications of products involve rates such as miles *per* hour or dollars *per* year. Example 5 shows how to decide which unit of measure to assign to a product. Notice how the units for a rate can be written in fraction form.

Connections
Physical Science

Example 5 ***Unit Analysis***

A car traveling 70 kilometers per hour for 2 hours will travel a distance of

$$\left(70\ \frac{\text{kilometers}}{\text{hour}}\right)(2\ \text{hours}) = 140\ \text{kilometers}.$$

You can think of the "hours" as canceling each other. ■

Communicating about ALGEBRA

Connections
Unit Analysis

SHARING IDEAS about the Lesson

Checking Your Work Apply the unit analysis technique of Example 5 to these products given in Example 3.

A. $(4\ \text{downs})\left(-3\ \frac{?}{?}\right) = -12$ yds
yards, down

B. $(3000\ ?)\left(-0.15\ \frac{?}{?}\right) = -\$450$ (or -450 dollars)
baskets, dollars, basket

Extra Examples

Here are additional examples similar to **Examples 1–5.**

1. Find the product.
   a. $(-6)(3)$ b. $(12)(-x)$
   c. $\left(-\frac{2}{5}\right)(-15)$
   d. $(16)\left(-\frac{2}{3}\right)$
   e. -4^4 f. $(-5)^3$
   g. $(-4)(-2)(-6)$
   h. $(-1)(-9)$
   i. $(-m)(2m)(-m)$
   j. $(-n)(n)$
   k. $-x^4(-x)(-x)$

   a. -18, b. $-12x$, c. 6, d. $-\frac{32}{3}$, e. -256, f. -125, g. -48, h. 9, i. $2m^3$, j. $-n^2$, k. $-x^6$
2. Use a calculator to verify the numerical answers you found in Exercise 1 above.
3. A department store has a weekend sale on all of its fall hats, hoping to attract buyers for its winter merchandise. It sold 300 fall hats at a loss of \$1.35 each. How much of a loss did the store incur during its weekend sale? a loss of \$405

Check Understanding

Which of the following statements are true for both addition and multiplication? All are.

a. The order in which you perform the operation does not matter.
b. When adding or multiplying three or more real numbers, it does not matter which two values are added or multiplied first.
c. There is a real number value for each operation, which when combined with a second value by the operation, leaves the second value unchanged.

Communicating about ALGEBRA

EXTEND Communicating
COOPERATIVE LEARNING
Have pairs (or small groups) of students write three problems that require using unit analysis to solve them. Solutions should be shown on the back of the paper. For extra credit, you could collect the problems and have individual students pick problems at random. The students would then determine whether the writer solved the problem correctly.

EXERCISE Notes

ASSIGNMENT GUIDE

Basic/Average: Ex. 1–6, 7–27 odd, 31–34, 35–49 odd, 51–60, 61–65 odd

Above Average: Ex. 1–6, 8–26 even, 29–34, 36–50 even, 59–60, 62–66

Advanced: Ex. 1–6, 15–49 odd, 59–68

Selected Answers
Exercises 1–6, 7–59 odd

✪ More Difficult Exercises
Exercises 27–30, 61–64

Guided Practice

These exercises assess students' understanding of the properties of multiplication of real numbers. Have students give reasons (or identify properties) for their answers.

Answers
1. For example, $-2 \cdot 3 = -6$
2. For example, $0 \cdot 4 = 0$.
3. Commutative property of multiplication
4. Rule for simplifying products

EXERCISES

Guided Practice

CRITICAL THINKING about the Lesson

In Exercises 1–4, decide whether the statement is true or false and explain. See margin for explanations.

1. Multiplying a negative number by a positive number produces a positive number. False
2. Multiplying a number by 0 produces the original number. False
3. Multiplying a by b produces the same result as multiplying b by a. True
4. The product of an odd number of negative factors is negative. True
5. Does the order in which two numbers are multiplied affect their product? No
6. If a is measured in feet per second and b in seconds, what is the unit for the product ab? Feet

Independent Practice

In Exercises 7–30, find the product. 27. $168x^2$ 28. $330y$ 29. $-\frac{1}{10}a^3$ 30. $\frac{1}{40}b^3$

7. $(-6)(8)$ -48
8. $(-3)(3)$ -9
9. $(4)(-7)$ -28
10. $(2)(-9)$ -18
11. $(-1)(x)$ $-x$
12. $(-4)(15p)$ $-60p$
13. $(16)(-3a)$ $-48a$
14. $(5)(-2c)$ $-10c$
15. $|(-5)|(6)$ 30
16. $(6)|(-5)|$ 30
17. $|(-2)(21)|$ 42
18. $|(8)(-9)|$ 72
19. $(-3)^2$ 9
20. -4^2 -16
21. $(-2x)^3$ $-8x^3$
22. $-3(x^2)$ $-3x^2$
23. $(-2)(-3)(-5)$ -30
24. $(-10)(-4)(-2)$ -80
25. $(-4)(x)(12)$ $-48x$
26. $(-2)(-x)(10)$ $20x$
27. ✪ $(-8x)(7x)(-3)$
28. ✪ $(11)(-5y)(-6)$
29. ✪ $(\frac{1}{3}a)(-\frac{2}{5}a)(\frac{3}{4}a)$
30. ✪ $(-\frac{1}{4}b)(-\frac{1}{5}b)(\frac{1}{2}b)$

Technology **In Exercises 31–34, use a calculator to find the product. Round your result to two decimal places. Use estimation to check.**

31. $(-2.68)(4.32)(-6.1)$ 70.62
32. $(3.62)(-3.45)(3.68)$ -45.96
33. $(4.97)(-2.13)^3(-7.35)$ 353.01
34. $(-5.11)^2(-8.24)(-6.59)$ 1417.93

In Exercises 35–44, evaluate the expression.

35. $2x^2 - 4x$, when $x = -3$ 30
36. $3x + x^2$, when $x = -7$ 28
37. $-2x - 5x$, when $x = 4$ -28
38. $-6x + x$, when $x = 9$ -45
39. $-2(|x - 7|)$, when $x = -3$ -20
40. $-5(|a - 4|)$, when $a = -4$ -40
41. $(|9 - y|)(-1)$, when $y = -5$ -14
42. $(|12 - m|)(-3)$, when $m = 6$ -18
43. $-2x^2 + 3x + 1$, when $x = 5$ -34
44. $-7x - (-4x^3)$, when $x = 4$ 228

45. *Drop in Temperature* The daily high temperature is decreasing at a rate of 2°F per day. Let T represent the daily high temperature now. Write an expression for the high temperature in 5 days. The current high temperature is 82°F. Find the high temperature in 5 days. Check your answer by making a table and looking for a pattern. $T - 10$, 72°

46. *Smaller Fry* Suppose you and your family go fishing every year in Canada. You find that the average length of the fish is decreasing by 1.8 cm per year. Let x represent the average length of the fish in 1988. Write an expression for the average length in 1992. If the average length was 76 cm in 1988, what was it in 1992? Check your answer by making a table and looking for a pattern. $x - 7.2$, 68.8 cm

47. *Jumping Frogs* People from around Calaveras County, California, have been training their frogs for the annual jumping competition. A measured jump consists of three consecutive leaps by the frog. A local favorite jumps six feet on the first leap, but then jumps six inches less than the previous leap on each of its last two leaps. Which of the following correctly represents the total length of the three leaps?

a. $6 + (6 - \frac{1}{2}) + (6 - \frac{1}{2})$

b. $6 + (6 - \frac{1}{2}) + 2(6 - \frac{1}{2})$

c. $6 + (6 - \frac{1}{2}) + [6 - 2(\frac{1}{2})]$

Evaluate the correct expression. Did this jump beat the record set by Rosie the Ribeter? $16\frac{1}{2}$ ft, no

The record at the annual Calaveras Jumping Jubilee is 21 feet $5\frac{3}{4}$ inches by a bullfrog named Rosie the Ribeter on May 18, 1986. Mark Twain's first popular story was "The Celebrated Jumping Frog of Calaveras County," published in 1865.

48. *Health Bar Promotion* To promote sales, a grocery store lowered the price on its health bars from \$0.55 to \$0.45 per bar. During the first weekend of the sale, the store sold 1321 health bars. Describe in words what the expression 1321(0.55 − 0.45) represents. See below.

49. *Train Crossing* A train traveling at a speed of 80 kilometers per hour will reach a railroad crossing in 0.6 hour. Which of the following correctly represents the distance to the crossing?

a. $(80 \frac{\text{kilometers}}{\text{hour}})$ (0.6 hours) **b.** $(80 \frac{\text{kilometers}}{\text{hour}}) \div$ (0.6 hours)

How far away is the crossing? 48 km

50. *Cycling* How far behind the leaders will the trailers be after they have cycled for 90 minutes? 4.5 mi

On a trip to Yellowstone Park, the lead riders of the Cheyenne cycling team average a speed of 20 miles per hour. The trailers average a speed of 17 miles per hour.

Independent Practice

▶ **Ex. 27, 29–30** Refer students to Example 2 in the lesson. These exercises require students to combine the properties of multiplication of variables and real numbers into one problem. For example:
$(-6x)(4x)(-3x)$
$= (-6 \cdot 4 \cdot -3)(x \cdot x \cdot x)$
$= (72)(x^3) = 72x^3$

▶ **Ex. 45–46** **PROBLEM SOLVING** Remind students that making a table and looking for patterns are important problem-solving strategies.

▶ **Ex. 47** Rosie the Ribeter must have been an exceptional frog! Her name was most likely inspired by the mythical aircraft worker Rosie the Riveter, a World War II symbol for women doing factory work as part of the war effort.

48. Amount of money in dollars that the store lost due to the price reduction.

Ex. 59–60 Tell students to use mental math to complete these exercises.

Ex. 68 The inventory value of the medium T-shirts in Store 2 is the total wholesale cost of the T-shirts.

Enrichment Activity

This activity requires students to go beyond lesson goals.

COOPERATIVE LEARNING

Work with a partner to investigate models of the multiplication of real numbers.

Consider how the checker model or the vector model of real numbers might be used to represent real-number multiplication. (Hint: The checker model's representation is based upon multiplication as repeated addition.)

Integrated Review

In Exercises 51–54, find the sum.

51. $6 + (-3.5)$ 2.5

52. $-7 + (-4)$ -11

53. $-3^2 + 4$ -5

54. $-5 + (-2)^3$ -13

In Exercises 55–58, find the difference.

55. $12 - 9$ 3

56. $9 - 12$ -3

57. $|-3| - 7$ -4

58. $4 - |-8|$ -4

59. Solve for a, b, c, and d.

$$\begin{bmatrix} a & -4 \\ 0 & 1 \end{bmatrix} + \begin{bmatrix} 3 & b \\ c & 4 \end{bmatrix} = \begin{bmatrix} 10 & 6 \\ 8 & d \end{bmatrix}$$ $a = 7$, $b = 10$, $c = 8$, $d = 5$

60. Find the difference.

$$\begin{bmatrix} -6 & 2 & |4| \\ 0 & -9 & 18 \end{bmatrix} - \begin{bmatrix} -6 & (-1)^5 & 0 \\ 3 & 14 & |-5| \end{bmatrix}$$

$$\begin{bmatrix} 0 & 3 & 4 \\ -3 & -23 & 13 \end{bmatrix}$$

Exploration and Extension

In Exercises 61–64, multiply each matrix by the real number as shown in the sample.

Sample: $-2\begin{bmatrix} 2 & -1 \\ -3 & 0 \end{bmatrix} = \begin{bmatrix} -2(2) & -2(-1) \\ -2(-3) & -2(0) \end{bmatrix} = \begin{bmatrix} -4 & 2 \\ 6 & 0 \end{bmatrix}$

61. $-3\begin{bmatrix} 2 & 6 \\ -1 & -8 \end{bmatrix}$ $\begin{bmatrix} -6 & -18 \\ 3 & 24 \end{bmatrix}$

62. $-7\begin{bmatrix} -2 & 0 & 3 \\ 2^2 & -11 & -8 \end{bmatrix}$ $\begin{bmatrix} 14 & 0 & -21 \\ -28 & 77 & 56 \end{bmatrix}$

63. $-1\begin{bmatrix} 2x & -5y \\ -10m & 13n \end{bmatrix}$ $\begin{bmatrix} -2x & 5y \\ 10m & -13n \end{bmatrix}$

64. $2\begin{bmatrix} x & 2y \\ -3m & 12n \end{bmatrix}$ $\begin{bmatrix} 2x & 4y \\ -6m & 24n \end{bmatrix}$

65. *Sales Tax* The matrix at the right shows the prices of a sweatshirt, a pair of jeans, and a pair of sneakers at two stores. If the state sales tax on clothes is 5%, what does the following product represent?

Prices	Store 1	Store 2
Sweatshirt	\$14.80	\$13.95
Jeans	\$32.95	\$29.95
Sneakers	\$85.50	\$79.95

$$0.05\begin{bmatrix} 14.80 & 13.95 \\ 32.95 & 29.95 \\ 85.50 & 79.95 \end{bmatrix}$$ Amount of sales tax for each item at each store

66. What is the sales tax on a pair of jeans in Store 1? \$1.65

67. *T-Shirt Inventory* The matrix at the right shows the number of *Save the Whales* T-shirts in stock at three stores. If each T-shirt has a wholesale cost of \$10.50, what does the following product represent?

Inventory	Store 1	Store 2	Store 3
Small	25	30	14
Medium	14	10	12
Large	16	12	10
X-Large	5	6	5

$$10.50\begin{bmatrix} 25 & 30 & 14 \\ 14 & 10 & 12 \\ 16 & 12 & 10 \\ 5 & 6 & 5 \end{bmatrix}$$

Total wholesale cost of all the T-shirts at the store

68. What is the inventory value of medium-size *Save the Whale* T-shirts in Store 2? \$105.00

2.6 The Distributive Property

What you should learn:

Goal 1 How to use the Distributive Property

Goal 2 How to simplify expressions by combining like terms

Why you should learn it:

You can use the Distributive Property to simplify expressions and to make calculations easier.

Goal 1 Using the Distributive Property

The Distributive Property is one of the most important properties in algebra. Before looking at the property, let's look at an example that shows why the property is true.

Suppose you want to find the area of a rectangle whose width is 3 and whose length is $x + 2$. There are two ways to do this: as the area of a single rectangle or as the sum of the areas of two rectangles.

Area of One Rectangle

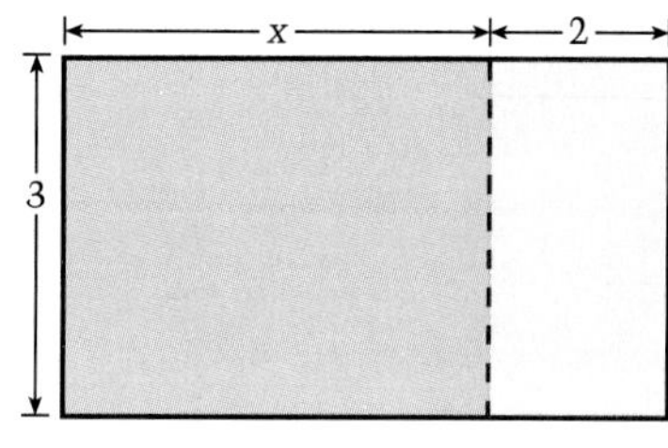

Area $= 3(x + 2)$

Area of Two Rectangles

x | 2

3

Area $= 3(x) + 3(2)$

Connections
Geometry

Because both ways produce the same area, you can conclude that

$$3(x + 2) = 3(x) + 3(2).$$

This is an example of the **Distributive Property.** To distribute means to give something to each member of a group. In this equation, the factor 3 is "distributed" to *each* term of the sum $(x + 2)$.

The Distributive Property

The product of a and $(b + c)$ is given by

$$a(b + c) = ab + ac \quad \text{or} \quad (b + c)a = ba + ca.$$

The product of a and $(b - c)$ is given by

$$a(b - c) = ab - ac \quad \text{or} \quad (b - c)a = ba - ca.$$

Problem of the Day

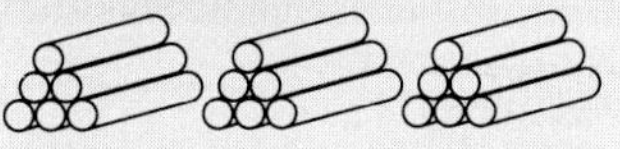

There are three piles of logs with six logs in each pile as shown. What is the least number of logs that can be moved so that the second pile contains twice as many logs as the first, and the third pile contains three times as many logs as the first?

3 logs (moved from the first pile to the third pile)

ORGANIZER

Warm-Up Exercises

1. Multiply each expression as indicated. Remember: Combine values inside parentheses first. Indicate which product is easier to find.
   a. $3(4 + 7)$ $3 \cdot 4 + 3 \cdot 7$
   b. $7(20 - 5)$ $7 \cdot 20 - 7 \cdot 5$
   c. $(17 - 7)6$ $17 \cdot 6 - 7 \cdot 6$
   The product in each case is the same. a. 33, b. 105, c. 60
2. Multiply by expanding the second factor so that the distributive property may be used. For example, $4 \cdot 23 = 4(20 + 3) = 80 + 12 = 92$.
   a. $-5 \cdot 12$ **b.** $-7 \cdot -13$
   c. $6 \cdot 18$ **d.** $9 \cdot -65$
   a. $12 = (10 + 2)$, -60;
   b. $-13 = -10 + (-3)$, 91;
   c. $18 = 10 + 8$, 108;
   d. $-65 = -60 + (-5)$, -585

Lesson Resources

Teaching Tools
  Problem of the Day: 2.6
  Warm-up Exercises: 2.6
  Answer Masters: 2.6
Extra Practice: 2.6
Technology Handbook: p. 18

LESSON Notes

Examples 1–2

Encourage students to use mental math to compute the second step in applying the distributive property.

Forgetting to distribute the factor over each term in the sum or difference is particularly evident when the factor has a negative sign or is negated. Using arrows to show the distribution is helpful to students. For example,

$-3(x + 4) =$
$-3(x) + (-3)(4) = -3x - 12$

Example 3

Bowlers refer to a score of 145 points as 145 pins. A perfect score in bowling is 300 pins. Have one or more students explain how the game of bowling is scored, using an actual scoring sheet if possible. In more modern bowling lanes, scoring is done automatically.

GOAL 2 Like terms have the same variable expression even though the coefficients may be different. Hence, $3x$ and $-5x$ are like terms because each has the variable expression x. Similarly, $12xy$ and $5xy$ are like terms because each contains xy. The terms $-6xz$ and $5x$ are not alike.

Common-Error ALERT!

In evaluating expressions such as $8x + x$, students frequently forget that the understood coefficient of the variable x is 1. (See **Example 4.**) In the beginning, encourage students to rewrite expressions to show explicitly that the coefficient is 1—that is,

$8x + x = 8x + 1x = 9x$

Example 1 Using the Distributive Property

a. $2(x + 5) = 2(x) + 2(5)$
$= 2x + 10$

b. $(x - 4)x = x(x) - 4(x)$
$= x^2 - 4x$

c. $(1 + 2x)8 = (1)8 + (2x)8$
$= 8 + 16x$

d. $y(1 - y) = (y)(1) - y(y)$
$= y - y^2$ ■

In Example 1, the middle steps are usually done as mental math, without writing the step. For instance, Part **a** would usually be written as $2(x + 5) = 2x + 10$.

The products in the next example are more difficult because the sum or difference is multiplied by a factor that has a negative sign. Study these examples carefully. Products like these are the cause of many "careless errors." For each product, remember the factor that has a negative sign must multiply *each* term.

Example 2 Using the Distributive Property

a. $-3(x + 4) = -3(x) + (-3)(4) = -3x - 12$

b. $(y + 5)(-4) = y(-4) + 5(-4) = -4y - 20$

c. $(-1)(6 - 3x) = (-1)(6) - (-1)(3x) = -6 + 3x$

d. $(x - 1)(-9x) = x(-9x) - (1)(-9x) = -9x^2 + 9x$ ■

Example 3 Bowling Handicap

Some bowling leagues are called *handicap leagues* because players are awarded extra points called a handicap. Players with low averages have larger handicaps than those with high averages. This tends to keep the competition close. Suppose your average is 145. The handicap in your league is 80% of the difference between 200 and your average. What is your handicap?

Solution

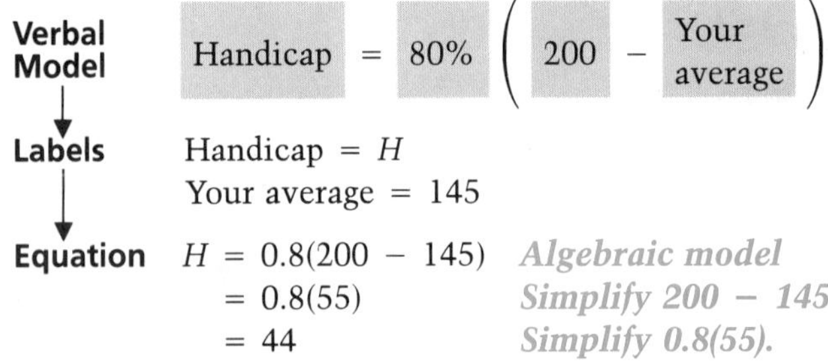

Thus, your handicap is 44 points. Using the Distributive Property to evaluate $0.8(200 - 145)$, do you get the same result? ■

Goal 2 Simplifying by Combining Like Terms

In a term that is the product of a number and a variable, the number is the **coefficient** of the variable. For instance, in the term $5x$, 5 is the coefficient of x. Remember that the terms of an expression are separated by *addition*.

$$5x - 3x = \underbrace{5x}_{\text{Term}} + \underbrace{(-3x)}_{\text{Term}}$$ *Terms are separated by addition.*

Thus, the coefficient of x on the second term is -3.

The terms $5x$ and $-3x$ are called **like terms** because the variable part of each term is the same. The Distributive Property allows you to *combine like terms* in an expression by simply adding or subtracting coefficients.

$5x - 3x = (5 - 3)x$ *Distributive Property*

$= 2x$ *Simplify.*

Example 4 *Simplifying by Combining Like Terms*

a. $8x + x = 8x + 1x$
$= (8 + 1)x$
$= 9x$

b. $4x^2 + 2 + 3x^2 = 4x^2 + 3x^2 + 2$
$= 7x^2 + 2$

c. $9 - x + 4 = 9 + 4 - x$
$= 13 - x$

d. $2(x - 4) + 5 = 2x - 8 + 5$
$= 2x - 3$ ■

Remember that simplifying an expression is not the same as solving an equation. Simplifying means rewriting in simpler form. Solving means finding values of the variables.

Communicating about ALGEBRA

Real Life Testing

SHARING IDEAS about the Lesson

Understanding a Definition The following question appeared on a College Entrance Exam. Explain how this question is related to the Distributive Property.

If $y = 2x - 3$, then $-2y = $ [?].

a. $5x + 3$ **b.** $5x - 3$ **c.** $2x + 6$

d. $-4x - 6$ **e.** $-4x + 6$

$-2y = -2(2x - 3)$
$= -4x + 6$
by the Distributive Property.

Extra Examples

Here are additional examples similar to **Examples 1–4 .**

1. Multiply using the distributive property.
   - **a.** $7(x + 6)$ — $7x + 42$
   - **b.** $-2(9 - x)$ — $-18 + 2x$
   - **c.** $(3x - 4)3$ — $9x - 12$
   - **d.** $x(8 + 4x)$ — $8x + 4x^2$
   - **e.** $(5x + 6)(-x)$ — $-5x^2 - 6x$
2. Use mental math.
   - **a.** $-5(2x - 4)$ — $-10x + 20$
   - **b.** $2(3x - y)$ — $6x - 2y$
   - **c.** $(y + 8)2$ — $2y + 16$
   - **d.** $(7x - 3)4$ — $28x - 12$
   - **e.** $-(9 - 6x)$ — $-9 + 6x$
3. Combine like terms.
   - **a.** $3x - 4 + 5x$ — $8x - 4$
   - **b.** $5y + 6x + 7y + 2x$ — $8x + 12y$
   - **c.** $13 + 5x - 9 - 8x$ — $4 - 3x$
   - **d.** $3(2x - 8) + 20$ — $6x - 4$
   - **e.** $14 - (3x + 12)$ — $2 - 3x$

Check Understanding

WRITING Describe how the associative, commutative, and distributive properties are used to combine algebraic expressions. Be sure students include the associative property of addition, which states that $(a + b) + c = a + (b + c)$ and the commutative property of addition, which states that $a + b = b + a$. In other words, in an expression with three or more terms, you may add the terms in any order.

Communicating about ALGEBRA

EXTEND *Communicating*
COOPERATIVE LEARNING
You may wish to assign pairs (or small groups) of students this activity for further study of the distributive property. Suppose you are a fourth-grade student who only knows only the multiplication facts from $1 \times 1 = 1$ through $1 \times 5 = 5$, up to $5 \times 1 = 5$ through $5 \times 5 = 25$.

Show by examples how you could use the distributive property and the multiplication facts you know to develop the rest of the basic multiplication facts from $1 \times 6 = 6$ through $9 \times 9 = 81$.

Sample answers: $6 = 1 + 5$ or $2 + 4$ or $3 + 3$; $7 = 2 + 5$ or $3 + 4$, etc. So, $5 \times 9 = 5(4 + 5) = 5(4) + 5(5) = 20 + 25 = 45$; $7 \times 8 = (5 + 2)(4 + 4) = 5(4 + 4) + 2(4 + 4) = 5(4) + 5(4) + 2(4) + 2(4) = 20 + 20 + 8 + 8 = 56$

EXERCISE Notes

ASSIGNMENT GUIDE

Basic/Average: Ex. 1–8, 9–36 multiples of 3, 37–45 odd, 47–48, 51–65 odd, 67–69

Above Average: Ex. 1–8, 9–36 multiples of 3, 37–45 odd, 49–50, 51–54, 55–65 odd, 67–69

Advanced: Ex. 1–8, 25–36, 38–46 even, 47–54, 59–64, 67–69

Selected Answers
Exercises 1–8, 9–65 odd

Use **Mixed Review** as needed.

More Difficult Exercises
Exercises 49–54, 69

Guided Practice

Ex. 6–8 Ask students to show each step in the simplifications of the expressions. This is a good way for students to demonstrate their understanding of the commutative and distributive properties. You may want to remind students that expressions such as $5x - 3$ can be rewritten, using addition, as $5x + (-3)$ in order to use the commutative property of addition.

EXERCISES

Guided Practice

CRITICAL THINKING about the Lesson

In Exercises 1–4, decide whether the equation is true. Use the Distributive Property to explain your answers. See answers below.

1. $2(4 + 6) = 2(4) + 6$
2. $(4 + 6)2 = 4(2) + 6(2)$
3. $16(3 - 2) = 16(3) - 16(2)$
4. $(3 - 2)16 = 3 - 2(16)$

5. State the property that allows us to combine like terms. Distributive Property

6. Simplify $3x - 5 - 2x$. $x - 5$ **7.** Simplify $2(x - 3) + x$. $3x - 6$ **8.** Simplify $-(x - 1) + 2x$. $x + 1$

Independent Practice

In Exercises 9–36, apply the Distributive Property.

9. $3(y + 6)$ $3y + 18$ **10.** $(x + 1)11$ $11x + 11$ **11.** $(4 - x)7$ $28 - 7x$ **12.** $-4(y - 2)$ $-4y + 8$
13. $(x + 5)(-2)$ **14.** $10(y - 6)$ $10y - 60$ **15.** $x(8 + x)$ $8x + x^2$ **16.** $(x - 14)x$
17. $-y(y - 9)$ **18.** $-x(8 + x)$ $-8x - x^2$ **19.** $a(a - 2)$ $a^2 - 2a$ **20.** $(3 + y)y$
21. $4(10 - 3x)$ **22.** $(16 + 3t)t$ $16t + 3t^2$ **23.** $(4s + 3)(-5)$ $-20s - 15$ **24.** $-13(1 - x)$
25. $(6x + 9)x$ $6x^2 + 9x$ **26.** $4x(x - 20)$ $4x^2 - 80x$ **27.** $(4 - 3x)(-6x)$ **28.** $-5s(s + 2)$
29. $2x(9 + x)$ $18x + 2x^2$ **30.** $(12 + 9y)6$ $72 + 54y$ **31.** $(12 - 5y)y$ $12y - 5y^2$ **32.** $(5 - 2x)(-10x)$
33. $-8y(2y + 7)$ **34.** $x(18 + 2x)$ $18x + 2x^2$ **35.** $-4a(3a - 4)$ $-12a^2 + 16a$ **36.** $-a(5 - 2a)$ $-5a + 2a^2$

In Exercises 37–46, simplify the expression.

37. $4 + x - 1$ $3 + x$ **38.** $x + 5 + x$ $2x + 5$ **39.** $2x^2 + 9x^2 + 4$ $11x^2 + 4$ **40.** $10x - 4 - 5x$ $5x - 4$
41. $-3(x + 1) - 2$ **42.** $(2x - 1)(2) + x$ **43.** $11x + (8 - x)3$ **44.** $7x(2 - x) - 4x$ $10x - 7x^2$
45. $-6x(x - 1) + x^2$ $-5x^2 + 6x$ **46.** $x^2 - (4 + x^2)$ -4

47. *How Much Are the Jeans?* You want to buy a pair of jeans and a \$15 T-shirt. There is a 6% sales tax. If x represents the cost of the jeans and you have \$42, then the following inequality is a model that shows what you can spend for the jeans.

$$\underbrace{15 + x}_{\text{cost of T-shirt and jeans}} + \underbrace{0.06(15 + x)}_{\text{sales tax}} \le 42$$

Simplify the left side of this inequality. $1.06x + 15.9 \le 42$

48. Suppose the jeans in Exercise 47 cost \$25. Do you have enough money to pay for both the T-shirt and jeans? What is the most that you could pay for the jeans? No, \$24.62

13. $-2x - 10$
16. $x^2 - 14x$
17. $-y^2 + 9y$
20. $3y + y^2$
21. $40 - 12x$
24. $-13 + 13x$
27. $-24x + 18x^2$
28. $-5s^2 - 10s$
32. $-50x + 20x^2$
33. $-16y^2 - 56y$
41. $-3x - 5$
42. $5x - 2$
43. $8x + 24$

1. Not true, $2(4 + 6) = 2(4) + 2(6)$. 2. True, application of distributive property.
3. True, application of distributive property 4. Not true, $(3 - 2)16 = 3(16) - 2(16)$.

49. *Bowling Handicap* A handicap in a bowling league is 80% of the difference between 200 and a bowler's average score. (Players with averages of 200 or more have zero handicap.) Let H represent the handicap for a bowler with an average of A. Which of the following equations correctly model the relationship between H and A?

a. $\underline{H = 0.8(200 - A)}$ **b.** $\underline{H = 160 - 0.8A}$
c. $A = 0.8(200 - H)$ **d.** $A = 160 - 0.8H$

50. *Changing the Handicap Formula* Suppose your bowling league wonders if there might be a better way to compute handicaps than by the formula described in Exercise 49. It considers awarding a handicap of 90% of the difference between 180 and a bowler's average. Complete the following table, then compare the handicaps given by the two formulas. **See Additional Answers.**

Average (A)	120	130	140	150	160	170	180	190	200
H (old) $= 0.8(200 - A)$	64	56	48	40	32	24	16	8	0
H (new) = ?	?	?	?	?	?	?	?	0	0

Give one advantage and one disadvantage of each handicap system.

51. *Geometry* Write an expression for the perimeter of the movie screen shown below and simplify.

$x + (x - 7) + x + (x - 7)$, $4x - 14$

52. *Geometry* Write an expression for the perimeter of the triangle shown below and simplify.

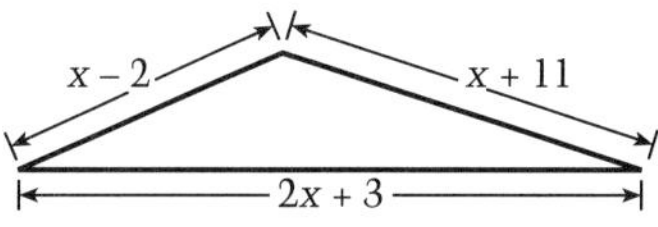

$(x - 2) + (x + 11) + (2x + 3)$, $4x + 12$

53. *Geometry* Find the area of the rectangle in two different ways. Show how the results are related to the Distributive Property. $a(b + c) = ab + ac$

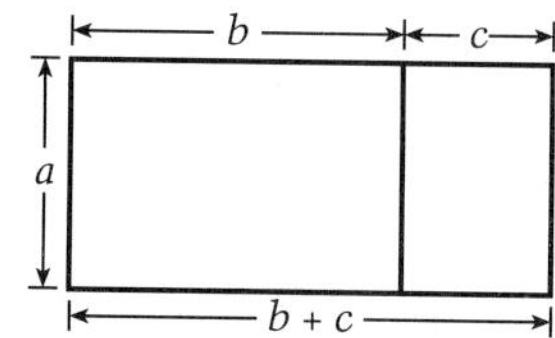

54. *Geometry* Find the area of the blue rectangle in two different ways. Show how the results are related to the Distributive Property. $a(b - c) = ab - ac$

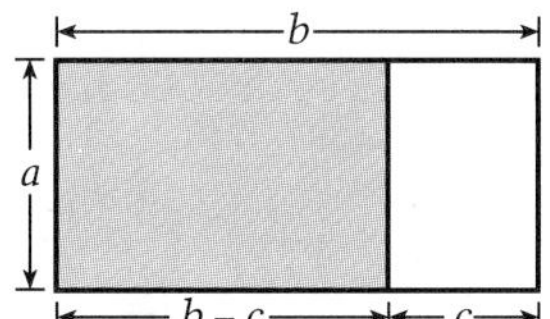

Integrated Review

Technology **In Exercises 55–60, use a calculator to help you simplify the expression.**

55. $3.6(x - 1.2)$ $3.6x - 4.32$

56. $9.1(4.3 + 1.1x)$ $39.13 + 10.01x$

57. $(6.1 - 32.3x)(-2)$ $-12.2 + 64.6x$

58. $-6.6(10.9 - 3.5x)$ $-71.94 + 23.1x$

59. $-4.1x(2.1x - 40.9)$ $-8.61x^2 + 167.69x$

60. $(3.3 - 0.9x)(-2x)$ $-6.6x + 1.8x^2$

Independent Practice

Ex. 47–48 If you simplify the given inequality in Exercise 47, then $15.90 + 1.06x \le 42$. If $x = 25$, then $15.90 + 1.06(25) = 42.40$. This is more than the \$42 you have to spend! To determine the most that you can pay, you should solve the inequality for x.

$$15.90 + 1.06x \le 42$$
$$1.06x \le 42 - 15.90$$
$$1.06x \le 26.10$$
$$\frac{1.06x}{1.06} \le \frac{26.10}{1.06}$$
$$x \le \$24.62$$

EXTEND Ex. 51 **USING PROPERTIES** Ask students what property of real numbers they can use to determine the area of the movie screen. (Using the distributive property, the rectangle has area $A = lw = x(x - 7) = x^2 - 7x$)

Ex. 53–54 **PROBLEM SOLVING** These exercises use area models to illustrate the distributive property. The first rectangle has area

$$a(b + c) = ab + ac$$

Area of the entire rectangle → $a(b + c)$; Sum of the areas of the larger and smaller rectangles → $ab + ac$

The area of the blue rectangle is this.

$$a(b - c) = ab - ac$$

Area of the blue rectangle → $a(b - c)$; Difference of the entire area and the non-blue area → $ab - ac$

Common-Error ALERT!

In **Exercise 46,** students often forget to take the opposite of both 4 and x^2 when applying the distributive property. The following may help.
$-(4 + x^2) = -1(4 + x^2)$

▶ **Ex. 67–68** To multiply a matrix by a real number means to multiply each entry of the matrix by the real number (sometimes called a scalar).

$$4 \cdot \begin{bmatrix} -3 & 4 & 0 \\ 2 & 1 & -6 \\ -8 & -3 & 7 \end{bmatrix} =$$

$$\begin{bmatrix} -12 & 16 & 0 \\ 8 & 4 & -24 \\ -32 & -12 & 28 \end{bmatrix}$$

Enrichment Activities

WRITING Following the class discussion, students might record their personal responses in their math journals.

1. The fourth property of addition states that $-(a + b) = -a + (-b)$. Use the distributive property to explain this property of addition. $-(a + b) = -1(a + b) = (-1)a + (-1)b = -a + (-b)$
2. How can the distributive property be used to simplify expressions such as $(2x + 1)(x + 3)$? Discuss the possibilities and provide examples. The distributive property can be applied twice as follows.

$$\begin{aligned}&(2x + 1)(x + 3)\\ &= (2x + 1)x + (2x + 1)3\\ &= 2x(x) + 1(x) + 2x(3) + 1(3)\\ &= 2x^2 + x + 6x + 3\\ &= 2x^2 + 7x + 3\end{aligned}$$

In Exercises 61–66, simplify the left side of the equation by combining like terms. Then use mental math to solve the equation.

61. $6x - 3x = 12$ $3x = 12$, 4

62. $8y - 4y = 48$ $4y = 48$, 12

63. $6y + 3 - y = 8$ $5y + 3 = 8$, 1

64. $4 + 2y - 4 = 16$ $2y = 16$, 8

65. $-z + 4z = 27$ $3z = 27$, 9

66. $4x - 8 - x = 1$ $3x - 8 = 1$, 3

Exploration and Extension

In Exercises 67 and 68, evaluate the two matrix expressions. Does this suggest that multiplication of a matrix by a real number distributes over matrix addition and matrix subtraction?

67. $4\left(\begin{bmatrix} 2 & -1 \\ 4 & -6 \end{bmatrix} + \begin{bmatrix} -4 & 6 \\ 2 & 9 \end{bmatrix}\right)$, $4\begin{bmatrix} 2 & -1 \\ 4 & -6 \end{bmatrix} + 4\begin{bmatrix} -4 & 6 \\ 2 & 9 \end{bmatrix}$ $\begin{bmatrix} -8 & 20 \\ 24 & 12 \end{bmatrix}, \begin{bmatrix} -8 & 20 \\ 24 & 12 \end{bmatrix}$; yes

68. $2\left(\begin{bmatrix} -1 & 0 \\ 6 & 4 \end{bmatrix} - \begin{bmatrix} 3 & 6 \\ -7 & 1 \end{bmatrix}\right)$, $2\begin{bmatrix} -1 & 0 \\ 6 & 4 \end{bmatrix} - 2\begin{bmatrix} 3 & 6 \\ -7 & 1 \end{bmatrix}$ $\begin{bmatrix} -8 & -12 \\ 26 & 6 \end{bmatrix}, \begin{bmatrix} -8 & -12 \\ 26 & 6 \end{bmatrix}$; yes

69. *Common Error* Suppose that you are tutoring a friend in algebra. After learning the Distributive Property, your friend tries to apply the property to multiplication and obtains $2(xy) = 2x \cdot 2y$. Write a convincing argument to show your friend that this is not correct.
Substitute any two numbers, except 0, for x and y, then evaluate.

Mixed REVIEW

1. Insert grouping symbols in the expression $3 \div 4 - 1$ so that its value is 1. **(1.1)** $3 \div (4 - 1)$
2. Insert grouping symbols in the expression $10 - 4 \cdot 2 \cdot 7$ so that its value is 14. **(1.1)** $[10 - (4 \cdot 2)] \cdot 7$
3. Find the absolute value of -19. **(2.1)** 19
4. Write $\frac{2}{5}$ in percent form. **(1.1)** 40%
5. Evaluate $\frac{1}{4}(-\frac{1}{2}) + 2$. **(2.5)** $1\frac{7}{8}$
6. Evaluate $11 - [54 \div (2)(9)]$. **(1.1)** 8
7. Evaluate $2x^3$ when $x = 5$. **(1.3)** 250
8. Evaluate $8x - 3y$ when $x = -1$ and $y = 3$. **(2.5)** -17
9. Evaluate $\frac{1}{5}(x + 4)$ when $x = -3$. **(2.2)** $\frac{1}{5}$
10. Evaluate $|x| - 1 + (-x)$ when $x = -3$. **(2.2)** 5
11. Is $3 \cdot (6 - 2) + 17 = 3 \cdot 6 - 2 + 17$? **(1.5)** No
12. Is -2 a solution of $4x + x^2 > 6x$? **(2.5)** Yes
13. Is -1 a solution of $3x - (-8) \le -9$? **(2.3)** No
14. Simplify $-4x + 3 - 2(1 - x)$. **(2.6)** $-2x + 1$
15. Evaluate $(-11)(-3) + 12(-2)$. **(2.5)** 9
16. Evaluate $(-1.2)(3.8)(0.2)$. **(2.5)** -0.912
17. Graph $-\frac{4}{3}, -2\frac{1}{2}, 2, \frac{5}{8}, \frac{1}{3}$ on the number line. **(2.1)** See below.
18. Use an exponent to write $3 \cdot 3 \cdot 3$. **(1.3)** 3^3

17. 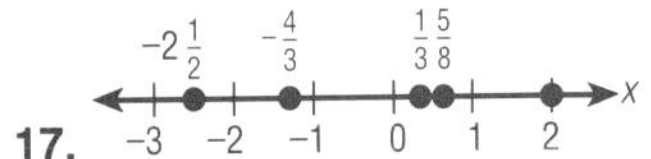

2.7 Division of Real Numbers

What you should learn:

Goal 1 How to express division as multiplication

Goal 2 How to divide real numbers using the Division Rule or a calculator

Why you should learn it:

You can divide positive and negative numbers by knowing how to express division as multiplication. This is useful in solving real-life problems, such as finding the frequency of a sound wave given its period.

Goal 1 Expressing Division as Multiplication

Just as subtraction can be expressed as addition, division can be expressed as multiplication.

Subtraction	Corresponding Addition
$8 - 3 = 5$	$8 + (-3) = 5$
Division	**Corresponding Multiplication**
$12 \div 4 = 3$	$(12)(\frac{1}{4}) = 3$

Subtracting a number is the same as adding the *opposite* of the number, and dividing by a number is the same as multiplying by the *reciprocal* of the number. If $\frac{a}{b}$ is a number that is not zero, then its **reciprocal** is $\frac{b}{a}$. In particular, if a is not zero, then its reciprocal is $\frac{1}{a}$. The product of a number and its reciprocal is 1.

$a \cdot \frac{1}{a} = 1$ *Multiplying a by $\frac{1}{a}$ produces 1.*

Division Rule

To divide a number a by a nonzero number b, multiply a by the reciprocal of b.

$$a \div b = a \cdot \frac{1}{b}$$

The result is the **quotient** of a and b.

Zero is the only real number that does not have a reciprocal. This means that you cannot divide a number by zero.

Example 1 *Expressing Division as Multiplication*

a. The reciprocal of 4 is $\frac{1}{4}$.

$20 \div 4 = (20)(\frac{1}{4}) = 5$

b. The reciprocal of $\frac{1}{3}$ is 3.

$3x \div \frac{1}{3} = 3x \cdot 3 = 9x$

c. The reciprocal of 2 is $\frac{1}{2}$.

$\frac{3}{5} \div 2 = \frac{3}{5} \cdot \frac{1}{2} = \frac{3}{10}$

d. The reciprocal of $\frac{2}{3}$ is $\frac{3}{2}$.

$\frac{12}{\frac{2}{3}} = 12 \cdot \frac{3}{2} = 18$

Problem of the Day

Suppose the 4 key on your calculator does not work. How can you use the calculator to multiply the following products?)

a. 1874×45
b. 427×444

Sample answers:
a. $(1873 + 1)(50 - 5)$
b. $(327 + 100)(222 + 222)$

ORGANIZER

Warm-Up Exercises

1. Multiply.
   a. $36\left(\frac{1}{6}\right)$ b. $45\left(\frac{1}{9}\right)$
   c. $-15\left(\frac{2}{3}\right)$ d. $\frac{1}{7}(28)$
   e. $(12x)\left(\frac{1}{4}\right)$ f. $\left(\frac{2}{5}\right)\left(\frac{3}{4}\right)$
   a. 6, b. 5, c. –10, d. 4, e. $3x$, f. $\frac{3}{10}$
2. Evaluate.
   a. $2x - 4$ when $x = -3$
   b. $3(9 - 2x)$ when $x = 6$
   c. $\frac{2x}{x + 3}$ when $x = 1$
   a. -10, b. -9, c. $\frac{1}{2}$

Lesson Resources

Teaching Tools
  Problem of the Day: 2.7
  Warm-up Exercises: 2.7
  Answer Masters: 2.7
Extra Practice: 2.7
Color Transparency: 11
Applications Handbook: p. 55–58

LESSON Notes

GOAL 1 Division by zero is not permitted because it would not be well defined. That is, if $\frac{4}{0} = b$, and b is a real number, then because multiplication and division are inverse operations, $b \cdot 0 = 4$ would have to be true!

Example 1

Point out to students that $20 \div 4$ can be rewritten as $\frac{20}{4}$ and that $\frac{20}{4}$ can be rewritten as $20 \cdot \frac{1}{4}$. Also, the fraction $\frac{12}{\frac{2}{3}}$ can be rewritten as $12 \cdot \frac{3}{2}$.

Example 2

Show students that negative fractions may have their sign in any of three locations and still be equivalent fractions. That is,

$$\frac{-3}{4} = \frac{3}{-4} = -\frac{3}{4}$$

Example 3

Indicate to students that the relationship $\frac{a-b}{c} = \frac{a}{c} - \frac{b}{c}$ is another way to express the given fraction.

Extra Examples

Here are additional examples similar to **Examples 1–4.**

1. Divide using the Division Rule for real numbers.
   a. $35 \div 5$ b. $126 \div 9$
   c. $51 \div -3$ d. $-216 \div 12$
   e. $5 \div \frac{2}{3}$ f. $\frac{4}{5} \div \frac{3}{10}$
   g. $-48 \div -9$
   h. $5x \div \frac{7}{10}$
   i. $-3x \div \frac{2}{5}$ j. $1 \div \frac{6}{5}$
   a. 7, b. 14, c. -17, d. -18, e. $\frac{15}{2}$, f. $\frac{8}{3}$, g. $\frac{16}{3}$, h. $\frac{50x}{7}$, i. $-\frac{15x}{2}$, j. $\frac{5}{6}$
2. Simplify each quotient.
   a. $\frac{51x + 34}{17}$ b. $\frac{12 - 18x}{6}$
   c. $\frac{23x - 15 + 7x}{5}$
   d. $\frac{-16 + 8x}{-4}$
   a. $3x + 2$, b. $2 - 3x$, c. $6x - 3$, d. $4 - 2x$
3. Evaluate and simplify when $a = 3$ and $b = -4$.
   a. $\frac{3b - a}{b + a}$ b. $\frac{b}{2a - b}$
   c. $\frac{a + 2b}{a}$ d. $\frac{4b + 2a}{2b}$
   a. 15, b. $-\frac{2}{5}$, c. $-\frac{5}{3}$, d. $\frac{5}{4}$

Goal 2 Dividing Real Numbers

Each division in Example 1 involves only positive numbers. Reciprocals can also be used in divisions that involve negative numbers.

Example 2 *Division Involving Negative Numbers*

a. $10 \div -2 = (10)(-\frac{1}{2}) = -5$ — *Reciprocal of -2 is $-\frac{1}{2}$.*

b. $-1 \div 3 = -1 \cdot \frac{1}{3} = -\frac{1}{3}$ — *Reciprocal of 3 is $\frac{1}{3}$.*

c. $\frac{-x}{-6} = -x \div -6 = -x(-\frac{1}{6})$ — *Reciprocal of -6 is $-\frac{1}{6}$.*
$= \frac{1}{6}x$ or $\frac{x}{6}$

d. $\frac{-\frac{1}{3}}{4} = -\frac{1}{3} \div 4 = -\frac{1}{3} \cdot \frac{1}{4} = -\frac{1}{12}$ — *Reciprocal of 4 is $\frac{1}{4}$.*

e. $\frac{1}{-\frac{3}{4}} = 1 \div -\frac{3}{4} = (1)(-\frac{4}{3}) = -\frac{4}{3}$ — *Reciprocal of $-\frac{3}{4}$ is $-\frac{4}{3}$.* ■

Example 3 *Simplifying a Quotient*

$\frac{32x - 8}{4} = (32x - 8)(\frac{1}{4})$ — *Multiply by the reciprocal of 4.*

$= (32x)(\frac{1}{4}) - (8)(\frac{1}{4})$ — *Distributive Property*

$= 8x - 2$ — *Simplify.* ■

Example 4 *Evaluating an Expression*

Evaluate the following expressions when $a = -2$ and $b = -3$.

a. $\frac{2a - b}{ab}$ b. $\frac{a + 2}{b}$ c. $\frac{a}{3 + b}$

Solution

a. $\frac{2a - b}{ab} = \frac{2(-2) - (-3)}{(-2)(-3)} = \frac{-4 + 3}{6} = \frac{-1}{6} = -\frac{1}{6}$

b. $\frac{a + 2}{b} = \frac{-2 + 2}{-3} = \frac{0}{-3} = 0$

c. $\frac{a}{3 + b} = \frac{-2}{3 + (-3)} = \frac{-2}{0}$ (Division by 0 is undefined.) ■

In Example 4b, note that 0 can be divided by a nonzero number. The result will always be 0.

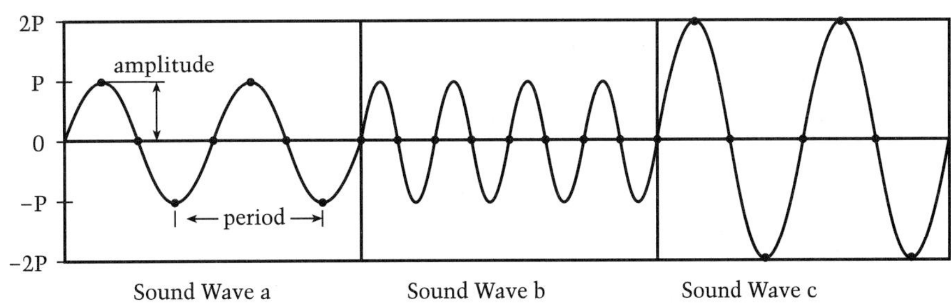

Sound waves (vibrations) are measured by amplitude, period, and frequency. Amplitude measures the "height" of the wave, the period is the distance between adjacent wave "troughs," and the frequency is the number of waves passing a fixed point per second. Waves* a *and* b *have the same amplitude but different frequencies, whereas waves* a *and* c *have the same frequency but different amplitudes.

Example 5 *Finding the Frequency of a Musical Note*

The musical note A above middle C has a period of $\frac{1}{440}$ second. What is the frequency of this note? How many sound waves does this note have per second?

Solution The frequency is the reciprocal of the period.

$$\text{frequency} = \frac{1}{\text{period}} = \frac{1}{\frac{1}{440}} = 440$$

Thus, the note has 440 waves per second. (Higher notes have more waves per second, and lower notes have fewer.) ■

Communicating about ALGEBRA

Connections **Technology**

SHARING IDEAS about the Lesson

Using the Reciprocal Key The following calculator keystrokes use the reciprocal key, labeled [1/x] or [x^{-1}], to find the reciprocal of −8. In either case, the display should be −0.125.

Using [1/x] Key	**Using [x^{-1}] Key**
8 [+/−] [1/x]	[(−)] 8 [x^{-1}] [=]

A. Use a calculator to multiply −8 by −0.125. Explain why the result is 1. Product of a number and its reciprocal is 1.

B. Use a calculator to divide 7 by −8 in two different ways: once using the division key and once using the reciprocal key. Which way do you prefer? Explain why.

Fewer keys to press

Check Understanding

1. What number is its own reciprocal? 1
2. Can every division problem be expressed as a multiplication problem? Yes, assuming there are no 0 divisors

Communicating about ALGEBRA

COOPERATIVE LEARNING
Students might work with partners or in small groups with a calculator to find the reciprocals of all the numbers between 1 and 21. Ask students to identify those numbers with terminating decimal reciprocals (that is, decimals that have a finite decimal part.) One such number is 8 since its reciprocal is 0.125. 1, 2, 4, 5, 8, 10, 16, 20

EXTEND *Communicating*
PROBLEM SOLVING
Ask: "What do the numbers with terminating decimal reciprocals have in common?" The numbers 2, 4, 5, 8, 10, 16, and 20 have only powers of 2 and 5 as non-unit factors. For example,
8 = 2 • 2 • 2.

Common-Error ALERT!
In **Example 4,** students frequently forget to substitute correctly when the replacement value involves a negative sign. Encourage them to use parentheses, ( ), in which to insert the replacement values. This helps students avoid computational errors.

EXERCISE Notes

ASSIGNMENT GUIDE

Basic/Average: Ex. 1–8, 9–36 multiples of 3, 37–40, 41–47 odd, 49–51, 57–73 odd, 75–78

Above Average: Ex. 1–8, 9–36 multiples of 3, 38–48 even, 50, 53–56, 57–72 multiples of 3, 75–81

Advanced: Ex. 1–8, 33–40, 42–48 even, 50–56, 57–72 multiples of 3, 75–81

Selected Answers
Exercises 1–8, 9–73 odd

More Difficult Exercises
Exercises 51–56, 75–80

Guided Practice

Ex. 1–8 The Guided Practice exercises summarize the main ideas of the lesson. Students should explain their rationales for the answers to Exercises 1–2 and 7–8.

Independent Practice

Ex. 37–40 The first step before using the distributive property is to express the quotient as a product.

So in Exercise 37, $\frac{9x-27}{3}$ becomes $\frac{1}{3}(9x - 27)$.

EXERCISES

Guided Practice

CRITICAL THINKING about the Lesson

1. **True or False?** Dividing a number is the same as multiplying by the opposite of the number. Explain your reasoning. False, dividing 8 by 2 is not the same as multiplying 8 by -2.
2. **True or False?** The product of a number and its reciprocal is 1. Explain your reasoning. True, by definition of reciprocal

In Exercises 3–6, find the reciprocal of the number.

3. 22 $\frac{1}{22}$
4. -17 $-\frac{1}{17}$
5. $\frac{4}{5}$ $\frac{5}{4}$
6. $-\frac{7}{2}$ $-\frac{2}{7}$
7. What is the only number that does not have a reciprocal? 0
8. Is it possible to divide 0 by a nonzero number? If yes, what will be the result? If no, explain why. Yes, 0

Independent Practice

In Exercises 9–16, express the division as multiplication.

9. $16 \div 4$ $16 \cdot \frac{1}{4}$
10. $24 \div 8$ $24 \cdot \frac{1}{8}$
11. $2x \div \frac{1}{9}$ $2x \cdot 9$
12. $\frac{1}{2}x \div 3$ $\frac{1}{2}x \cdot \frac{1}{3}$
13. $\frac{4}{9} \div 6$ $\frac{4}{9} \cdot \frac{1}{6}$
14. $12x \div \frac{2}{3}$ $12x \cdot \frac{3}{2}$
15. $\frac{48}{\frac{1}{2}}$ $48 \cdot 2$
16. $\frac{4}{\frac{3}{4}}$ $4 \cdot \frac{4}{3}$

In Exercises 17–36, perform the indicated division.

17. $6 \div 3$ 2
18. $27 \div 9$ 3
19. $9 \div \frac{1}{2}$ 18
20. $8 \div \frac{1}{5}$ 40
21. $\frac{92}{\frac{1}{2}}$ 184
22. $\frac{16}{\frac{4}{9}}$ 36
23. $-8 \div 4$ -2
24. $36 \div -6$ -6
25. $18 \div -9$ -2
26. $-64 \div 16$ -4
27. $-10 \div \frac{1}{9}$ -90
28. $-33 \div -\frac{1}{2}$ 66
29. $-\frac{1}{7} \div -3$ $\frac{1}{21}$
30. $-\frac{3}{4} \div 4$ $-\frac{3}{16}$
31. $\frac{x}{4} \div 3$ $\frac{x}{12}$
32. $\frac{2x}{3} \div \frac{1}{2}$ $\frac{4x}{3}$
33. $35x \div \frac{1}{7}$ $245x$
34. $4x \div -\frac{1}{2}$ $-8x$
35. $84x \div -\frac{4}{5}$ $-105x$
36. $-\frac{x}{4} \div 3$ $-\frac{x}{12}$

In Exercises 37–40, use the Distributive Property to simplify the quotient.

37. $\frac{9x-27}{3}$ $3x - 9$
38. $\frac{24x+18}{-2}$ $-12x - 9$
39. $\frac{-32+x}{-8}$ $4 - \frac{1}{8}x$
40. $\frac{45-9x}{9}$ $5 - x$

In Exercises 41–48, evaluate the expression.

41. $\frac{x-y}{6}$, when $x = 12$ and $y = 0$ 2
42. $\frac{r-4}{s}$, when $r = 24$ and $s = \frac{1}{2}$ 40
43. $\frac{3a-b}{a}$, when $a = \frac{1}{3}$ and $b = -2$ 9
44. $\frac{16-4x}{y}$, when $x = 2$ and $y = \frac{1}{2}$ 16
45. $\frac{15x-10}{y}$, when $x = 3$ and $y = \frac{2}{3}$ $\frac{105}{2}$
46. $\frac{3a-4b}{ab}$, when $a = -\frac{1}{3}$ and $b = \frac{1}{4}$ 24
47. $\frac{ab}{2a-8}$, when $a = 8$ and $b = 10$ 10
48. $\frac{x}{4y+6}$, when $x = \frac{3}{5}$ and $y = \frac{1}{2}$ $\frac{3}{40}$

49. *How Many Breadsticks?* Suppose you made 60 ounces of dough for breadsticks. Each breadstick is made of $\frac{5}{4}$ ounces of dough. How many breadsticks can you make? 48

50. *For the Birds* You have 126 feet of 1-inch by 8-inch wood. If you can make one birdhouse from $4\frac{1}{2}$ feet of wood, how many birdhouses can you make? 28

51. *How Much Can I Make?* Each birdhouse in Exercise 50 can be sold for $15. If the lumber and other supplies cost $32, how much profit can you make on *each* birdhouse? $13.86

52. *Is It Worth It?* Each birdhouse in Exercise 50 takes you two hours to make. How much profit can you make per hour of work? $6.93

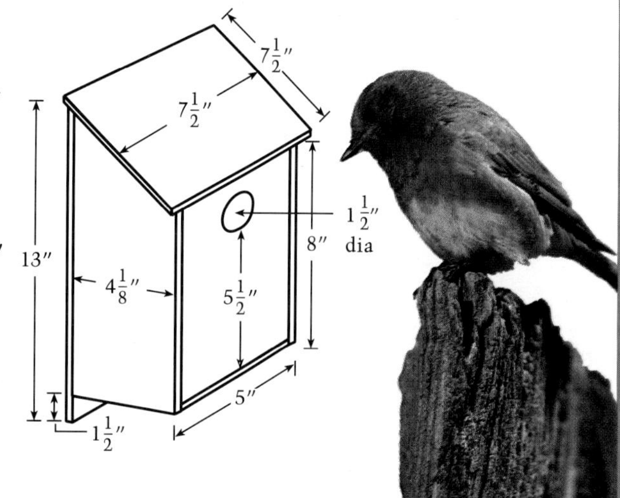

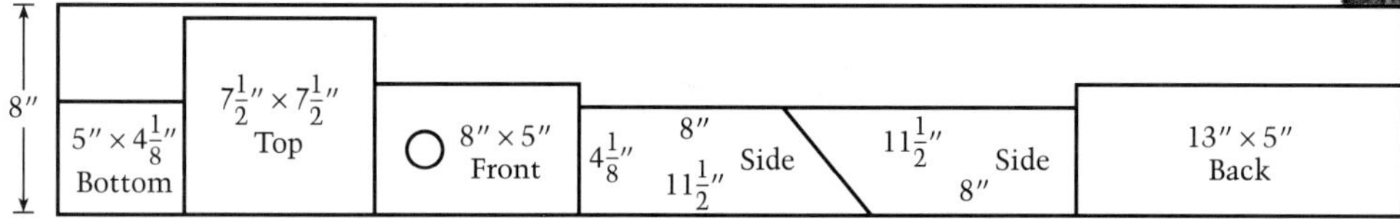

Broadcasting **In Exercises 53–56, use the following information.**

The frequency of an AM station is obtained by multiplying the station number by 1000. For example, the station 1370 AM has a frequency of 1370 • 1000 = 1,370,000 vibrations per second. The frequency of an FM station is obtained by multiplying the station number by 1,000,000. For example, the station 101.7 FM has a frequency of 101.7 • 1,000,000 = 101,700,000 vibrations per second.

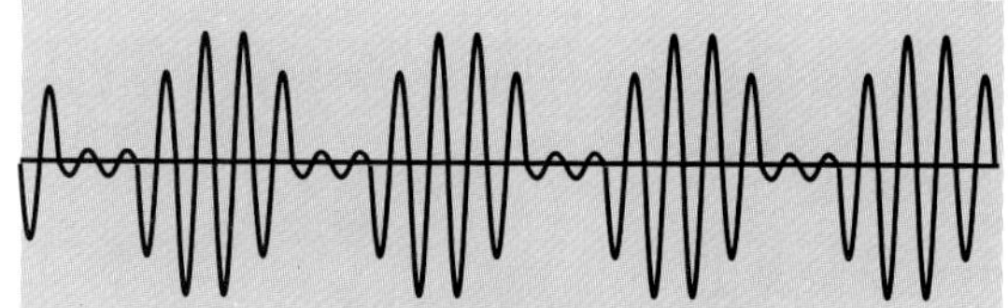

Amplitude modulation (AM)

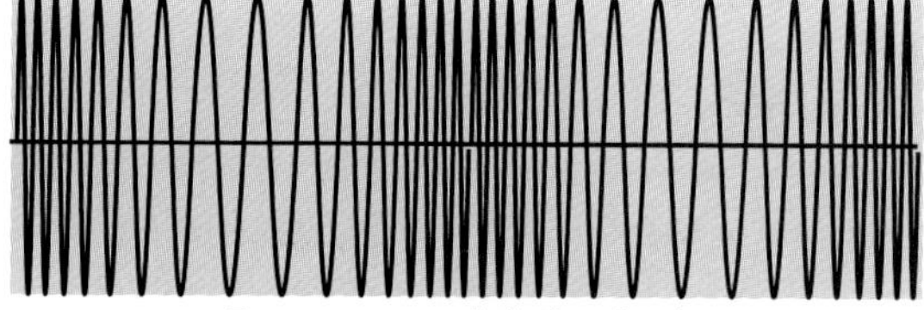

Frequency modulation (FM)

AM stations broadcast on frequencies between 535,000 and 1,705,000 vibrations per second. FM stations broadcast on frequencies between 88,000,000 and 108,000,000 vibrations per second.

53. Find the period of the radio waves transmitted by the radio station 1400 AM.

54. Find the period of the radio waves transmitted by the radio station 92.3 FM.

55. Find the radio station setting for a station transmitting waves that have a period of $\frac{1}{90,000,000}$ second. Is this an AM station or an FM station? Explain.

56. Find the radio station setting for a station transmitting waves that have a period of $\frac{1}{905,000}$ second. Is this an AM station or an FM station? Explain.

53. $\frac{1}{1,400,000}$ second 54. $\frac{1}{92,300,000}$ second

55. FM, the frequency is 90,000,000.

AM, the setting is at 905.

Ex. 53–56 Recall from Example 5 in this lesson that frequency $= \frac{1}{\text{period}}$, that is, the period is the reciprocal of the frequency. A radio station that transmits at 1400 AM has a frequency of 1,400,000, or a period of $\frac{1}{1\ 400\ 000}$.

Enrichment Activities

These activities require students to go beyond lesson goals.

1. **WRITING** Make up a problem whose solution requires the use of division. (For example, see Exercise 49 of this lesson's exercise set.)
2. **Challenge!** Explain how the distributive property and the Division Rule can be used to compute the following easily.

$$\frac{24x - 12xy + 32z - 28}{4}$$

$24x - 12xy + 32z - 28 = 4(6x - 3xy + 8z - 7)$; $\frac{4}{4}$ simplifies to 1.

▶ **Ex. 75–80** PROBLEM SOLVING Point out to students that these exercises make use of two important strategies: Make a Table and Find a Pattern.

Answers

75. $100x$; $1000x$; $10{,}000x$; $100{,}000x$ There is one zero for each decimal place in the original expression. $1{,}000{,}000x$

76. 1, −1, −1, 1, 1, −1, 1. When the sum of the exponents is even, the answer is positive; otherwise, the answer is negative.

EXTEND Ex. 81 PROBLEM SOLVING Students computed the average amount per pound the Boy Scouts kept. Ask: "Using the same data, what was the total amount donated to the National Elk Refuge?", "What amount did the Boy Scouts keep?" $88,605.60, $22,151.40

Common-Error ALERT!

In **Exercise 70,** once $\frac{2}{3}$ replaces y in $\frac{14}{y} + 2$, caution students that the order of operations tells us that the quotient of 14 and y is added to 2. That is, you compute $(14 \div y) + 2$.

Integrated Review

In Exercises 57–68, simplify the expression.

57. $16 - 22$ −6
58. $9 - 7$ 2
59. $24 \div 2$ 12
60. $64 \div 4$ 16
61. $4 \div \frac{1}{2} + 9 - 12$ 5
62. $6 + 9 \div \frac{1}{4} - 14$ 28
63. $8 - 14 \div 2 \cdot 4$ −20
64. $3 - 6^2 \div 2$ −15
65. $(-4 \div \frac{1}{3})^2 - 9$ 135
66. $14 - (2 \div -\frac{1}{4})^2$ −50
67. $|-2| - |-4|$ −2
68. $2|-1| - |-3|$ −1

In Exercises 69–74, evaluate the expression.

69. $-x \div \frac{1}{2}$, when $x = 6$ −12
70. $14 \div y + 2$, when $y = \frac{2}{3}$ 23
71. $\frac{x}{2} \div 8$, when $x = 4$ $\frac{1}{4}$
72. $2 \div \frac{x}{9}$, when $x = 3$ 6
73. $x \div 2$, when $x = \frac{1}{4}$ $\frac{1}{8}$
74. $8 \cdot x$, when $x = \frac{1}{8}$ 1

Exploration and Extension

✪ **75.** *Make a Table, Find a Pattern* Complete the table by expressing each division as a product that has no decimal factor. Describe the pattern. Use the pattern to simplify the expression $\frac{x}{0.000001}$. See margin.

$\frac{x}{1}$	$\frac{x}{0.1}$	$\frac{x}{0.01}$	$\frac{x}{0.001}$	$\frac{x}{0.0001}$	$\frac{x}{0.00001}$
x	$10x$	?	?	?	?

✪ **76.** *Make a Table, Find a Pattern* Evaluate each expression and complete the table. Describe the pattern. See margin.

$\frac{(-1)^1}{(-1)^1}$	$\frac{(-1)^1}{(-1)^2}$	$\frac{(-1)^2}{(-1)^1}$	$\frac{(-1)^2}{(-1)^2}$	$\frac{(-1)^3}{(-1)^1}$	$\frac{(-1)^3}{(-1)^2}$	$\frac{(-1)^3}{(-1)^3}$
?	?	?	?	?	?	?

In Exercises 77–80, use the result of Exercise 76 to find the value of $\frac{(-1)^n}{(-1)^m}$. (Assume n and m are positive integers.)

✪ **77.** n is even and m is odd. −1
✪ **78.** n is even and m is even. 1
✪ **79.** n is odd and m is even. −1
✪ **80.** n is odd and m is odd. 1

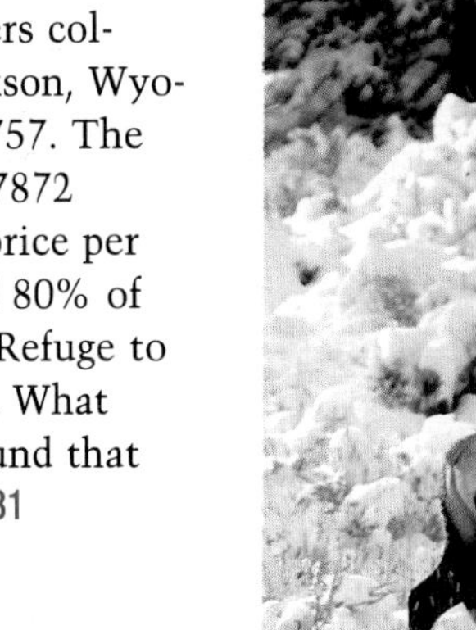

81. *Antler Sale* In 1989, the antlers collected by the Boy Scouts in Jackson, Wyoming, were auctioned for $110,757. The total weight of the antlers was 7872 pounds. What was the average price per pound? The Boy Scouts donated 80% of the money to the National Elk Refuge to purchase food for wintering elk. What was the average amount per pound that the Boy Scouts kept? $14.07, $2.81

Since 1965, the Boy Scouts of America in the Jackson District of Wyoming have been permitted to collect the freshly shed antlers of elk.

2.8 Exploring Data: Rates and Ratios

What you should learn:

Goal 1 How to use rates to relate quantities measured in different units

Goal 2 How to use ratios to relate quantities measured in the same units

Why you should learn it:

You can use rates to compare quantities with different units, such as the amount of money spent by a number of people. You can use ratios to compare quantities with like units, such as the areas of two rooms, both measured in square feet.

Goal 1 Using Rates

If a and b are two quantities that are measured in *different* units, then the **average rate** (or **average number**) of a per b is $\frac{a}{b}$. Finding a rate is something that most people do instinctively. Here are two common examples.

- You drive a distance of 250 miles and use 11 gallons of gas. The average mileage for your car is

$$\frac{250 \text{ miles}}{11 \text{ gallons}} = \frac{250}{11} \frac{\text{miles}}{\text{gallon}} \approx 22.7 \text{ miles per gallon.}$$

- You get paid \$20 for 4 hours of work. Your hourly rate is

$$\frac{20 \text{ dollars}}{4 \text{ hours}} = \frac{20}{4} \frac{\text{dollars}}{\text{hour}} = 5 \text{ dollars per hour.}$$

Example 1 *Spending in the United States*

The table shows some total amounts spent in the United States in 1990. How could you present this information to someone so that it is easier to understand?

1990 Spending ***(Source: Census of Retail Trade.)***

Clothing & Shoes	Food (Restaurants)	Food (Grocery Stores)
\$90,000,000,000	\$180,000,000,000	\$340,000,000,000

Solution One way to personalize these large numbers is to find the average rate of spending *per person.* You can do this by dividing the spending amounts by the total population. (It was about 250 million in 1990.) For instance, the average amount spent per person for clothing and shoes in 1990 was

$$\frac{90{,}000{,}000{,}000 \text{ dollars}}{250{,}000{,}000 \text{ people}} = 360 \text{ dollars per person.}$$

By changing the amount to a *per person* rate, you allow people to compare themselves to the average. All three rates are summarized in the following table.

1990 Average Rate of Spending Per Person

Clothing & Shoes	Food (Restaurants)	Food (Grocery Stores)
\$360 per person	\$720 per person	\$1360 per person

■

Problem of the Day

Suppose X, Y, and Z are negative numbers, and $X < Y$ and $Y < Z$. Are the following expressions negative or positive?

a. $\frac{Y - Z}{|X|}$ b. $\frac{Z - X}{Y^2}$

c. $\frac{Y^5}{Z^8}$ d. $\frac{(-X)(Y)}{Z}$

a. Negative, b. Positive, c. Negative, d. Positive

ORGANIZER

Warm-Up Exercises

Use the given formulas and unit analysis to solve the problems.

1. A sky diver falls at a velocity of 25 feet per second. How far does she fall in 4 seconds?
   $d = r \cdot t$
   $d = 25\frac{ft}{s} \cdot 4s$
   $d = (25 \cdot 4)\left(\frac{ft}{s} \cdot s\right)$
   $d = 100$ ft
2. A secretary can type an average of 65 words per minute. If he types for 15 minutes nonstop, about how many words will he type?
   Number of words = Typing speed • Time
   975 words
3. How far can a truck travel if its 20 gallon tank is full of gas and it get 28 miles per gallon?
   Distance = miles per gallon • Number of gallons
   560 miles

Lesson Resources

Teaching Tools
- Problem of the Day: 2.8
- Warm-up Exercises: 2.8
- Answer Masters: 2.8

Extra Practice: 2.8
Color Transparency: 12

LESSON Notes

When the numbers you compare have different units of measurement, they form rates. When the numbers you compare have the same unit of measurement, they form ratios. Ratios occur when you compare numbers with the same units.

Examples 1–2

Large numbers do not always convey information well. Data becomes personalized when per person rates are found. This smaller value gives you a basis with which to compare the given average rate with your personal behavior or life-style.

Example 3

Point out to students how unit analysis is helpful in setting up the solution to the problem of currency exchange. Observe that the exchange rate from United States funds to Canadian funds is not the same as the rate from Canadian funds to United States funds.

You might wish to update the problem by having students use the financial pages of the newspaper to find the current exchange rates for Canada or for other countries. Students might also find percent of increase (or percent of decrease) comparisons.

Example 2 *Rate of Hair Growth*

Elise has just returned from spending six months in Australia in a student exchange program. The second thing her mother said to Elise when she got off the plane was, "I can't believe how much your hair has grown." How much could Elise's hair have grown while she was gone? Why do you think Elise's mother noticed the difference in hair length?

Solution The normal rate of growth for human hair is 3 to 5 inches per year. Even at the faster rate, Elise's hair would have only grown

$$\frac{5 \text{ inches}}{1 \text{ year}} \cdot \frac{1}{2} \text{ year} = 2.5 \text{ inches.}$$

As to why Elise's mother noticed the difference, this is one of the surprising things about people. We are very sensitive to relatively small changes in the appearance of people we know well. ■

Example 3 *Currency Exchange*

Suppose you visited Canada and took \$175 to spend. The rate of currency exchange was \$1.1515 Canadian dollars per U.S. dollar. If you exchanged the entire amount, how much did you receive in Canadian dollars? While in Canada, you spent everything but \$32.50 (Canadian). How much could you exchange this for in U.S. dollars?

Solution For the exchange from U.S. dollars to Canadian, you want to end up with Canadian dollars, so you need to write the exchange rate as a fraction with Canadian dollars in the numerator.

$$\frac{1.1515 \text{ Canadian dollars}}{1 \text{ U.S. dollar}} \cdot 175 \text{ U.S. dollars} = \$201.51 \text{ (Canadian)}$$

For the exchange from Canadian dollars to U.S. dollars, you want to end up with U.S. dollars, so you need to write the exchange rate as a fraction with U.S. dollars in the numerator.

$$\frac{1 \text{ U.S. dollar}}{1.1515 \text{ Canadian dollars}} \cdot 32.50 \text{ Canadian dollars} = \$28.22 \text{ (U.S.)}$$

■

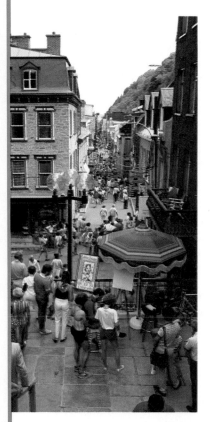

Historic Quebec City, Canada

Rates that are used to compare the prices per unit of consumer items are called **unit prices.** For instance, by comparing the unit prices of the following bottles of cologne, you can see that the larger bottle costs less per ounce.

$$\text{4-ounce bottle: } \frac{20.95 \text{ dollars}}{4 \text{ ounces}} = \$5.24 \text{ per ounce}$$

$$\text{6-ounce bottle: } \frac{29.95 \text{ dollars}}{6 \text{ ounces}} = \$4.99 \text{ per ounce}$$

Goal 2 Using Ratios

If a and b are two quantities that are measured in the *same* units, then the ratio of a to b is $\frac{a}{b}$. The difference between a rate and a ratio is that a rate compares two quantities measured in different units, and a ratio compares two quantities measured in the same unit. (In itself, a ratio has no units.)

Often ratios are left in fractional form. For instance, in 1991 the Stevenson High School girls' volleyball team won 10 of its 16 games. Thus, the win-loss ratio was

$$\text{win-loss ratio} = \frac{\text{games won}}{\text{games lost}} = \frac{10 \text{ games}}{6 \text{ games}} = \frac{5}{3}.$$

Example 4 *Comparing Two Square Rooms*

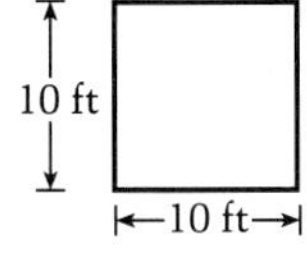

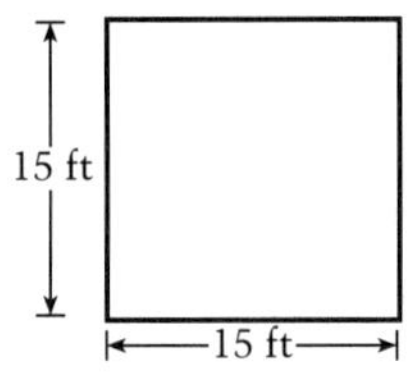

One bedroom in a house is 10 feet by 10 feet. The other bedroom is 15 feet by 15 feet. Which is a correct comparison of the sizes of the two rooms?

- The larger room has $1\frac{1}{2}$ times the area of the smaller room.
- The larger room has over twice the area of the smaller room.

Solution The ratio of the areas of the two rooms is

$$\frac{\text{area of larger}}{\text{area of smaller}} = \frac{225 \text{ square feet}}{100 \text{ square feet}} = \frac{225}{100} \text{ or } 2.25.$$

Thus, the second statement is correct. Drawing diagrams provides a visual check. ■

Communicating about ALGEBRA

SHARING IDEAS about the Lesson

Connections
Technology

Finding Needed Information For many years, mathematicians wondered whether π was a **rational number.** That is, they wondered if π could be written as the ratio of two integers. Finally, in 1882, a mathematician named Ferdinand Lindemann published a proof that π is not rational. There are, however, rational numbers whose values are very close to *pi.*

A. Use a calculator to write the following in increasing order.

π, 3.1416, $\frac{22}{7}$, $\frac{355}{113}$ π, $\frac{355}{113}$, 3.1416, $\frac{22}{7}$

B. Which of the three rational numbers is closest in value to π? $\frac{355}{113}$

Extra Examples

Here are additional examples similar to **Examples 1–4.**

1. In 1990, Americans ate about 11,750,000,000 pounds of chicken. Present this information so that it is easier to understand. Each American ate about 47 pounds of chicken in 1990.
2. At the grocery store, you notice that your favorite cereal comes in a new giant size of 27 ounces and costs \$3.25. You normally buy the 19-ounce size, which costs \$2.50. What should you do now? Buy the new giant size at about \$0.12 per oz instead of the regular size at \$0.13 per oz.
3. One bedroom in a house is 9 feet by 9 feet. The master bedroom is 27 feet by 27 feet. Which is a correct comparison of the sizes of the two rooms? c; The larger room is 9 times as large, $729 \div 81 = 9$.
   a. The larger room is 3 times as big as the smaller.
   b. The larger room is 6 times as big as the smaller.
   c. The larger room is 9 times as big as the smaller.

Check Understanding

MATH JOURNAL Find examples of real-life uses of rates and ratios. Explain the uses or purposes of the comparisons you find.

***Communicating* about ALGEBRA**

EXTEND *Communicating*
COOPERATIVE LEARNING
Have students work with a partner (or in small groups). Tell them to find some round objects in the classroom such as wastebaskets, lids, watch faces, coins, etc. Tell students to use a tape measure to find the circumference C (distance around) and diameter d of each object.

Have each pair (or group) of students compute the ratio $\frac{C}{d}$ for each object to obtain an estimate of the value of pi. Ask students to compare their computed values to the usual value for pi to determine how close their estimates were.

Make a class list of the computed values. Have students find the average value of pi. Is this value closer to the usual value of 3.1416 than the individual computed values?

EXERCISE Notes

ASSIGNMENT GUIDE

Basic/Average: Ex. 1–4, 5–31 odd, 34, 37–44, 45–54 multiples of 3, 57

Above Average: Ex. 1–4, 5–31 odd, 35–44, 49–59 odd

Advanced: Ex. 1–4, 6–34 even, 35–44, 54–59

Selected Answers
Exercises 1–4, 5–55 odd

More Difficult Exercises
Exercises 16–25, 35, 36, 57, 58

Guided Practice

Ex. 1–4 Do these exercises in class as oral exercises. Ask students to indicate other forms besides fractions that a ratio may take.

Answer

1. A rate compares two quantities measured in different units. A ratio compares two quantities measured in the same unit.

Independent Practice

EXTEND Ex. 5–6 PROBLEM SOLVING Students might compare the rates of fuel consumption. The motorcycle's mile-per-gallon rate is about 383% of the car's mile-per-gallon rate.

EXERCISES

Guided Practice

CRITICAL THINKING about the Lesson

1. Explain the difference between a rate and a ratio. See margin.
2. Would you use a rate or a ratio to compare unit prices? Rate
3. Is the quotient of the circumference and the diameter of a circle a rate or a ratio? What is the symbol for this quotient? Ratio, π
4. Is a speed of 40 kilometers per hour a rate or a ratio? Is the quotient of 4 hours to 24 hours a rate or a ratio? Rate, ratio

Independent Practice

5. *Motorcycle Mileage* A motorcycle uses 3 gallons of gasoline to travel 440 miles. Find the average number of miles per gallon. $146\frac{2}{3}$
6. *Car Mileage* A car uses 12 gallons to travel 460 miles. Find the average number of miles per gallon. $38\frac{1}{3}$
7. *Hourly Wage* You get paid \$85 for working 20 hours. Find your hourly rate of pay. \$4.25
8. *Hourly Wage* You get paid \$127.50 for working 30 hours. Find your hourly rate of pay. \$4.25 per hour
9. *Average Speed* A family drove 80 kilometers in $\frac{3}{4}$ hours. What was the average speed? $\approx$ 106.7 km per hour
10. *Average Speed* A family drove m miles in $\frac{3}{4}$ hours. What was the average speed? $\frac{4m}{3}$ mi per hour
11. *Video Rental* The regular rental fee for a video is \$1.79. The rental store runs a special on three movies for \$4.99. What is the cost per video under the special? How much do you save per video? \$1.66, \$0.13
12. *Frozen Yogurt Bars* A store sells a box of 5 frozen yogurt bars for \$1.20. The same store sells a box containing 7 bars for \$1.59. Find the unit price to determine which is the better buy. 24¢ and 22.7¢, box of 7
13. *Average Golf Score* During a golf game, you score an 83 on an 18-hole course. What was your average score per hole? Is this average better or worse than a score of 44 for 9 holes of golf? $\approx$ 4.6, better
14. *Library Space* A library has 10,513 books and 255 shelves. All the shelves are full. What is the average number of books per shelf? If the library receives 659 new books from another library that is closing, approximately how many more shelves does it need? $\approx$ 41.2, 16
15. *Compact Disk Collection* Suppose you have 127 compact disks. You want to buy storage cases that can hold 35 disks each. How many cases do you need? How many more compact disks can you buy before you need to buy another case? 4, 13

An Average Day **In Exercises 16–25, yearly rates for the United States are given. Find the rate per day.** See margin.

(Source: **On an Average Day,** ***Fawcett Press, 1989)***

16. 888,045 automobiles are stolen.

17. 352,225,000 colas are drunk at breakfast.

18. 1,572,481,430 credit card purchases are made.

19. 12,045 new consumer products are introduced.

20. 350,000,000,000 photocopies are made.

21. $450,000,000 is spent on sun-care products.

22. 88,695 patents are issued.

23. 20,440 house fires are started by arson.

24. 13,140,000,000 postage stamps are printed.

25. 24,820 animals are treated with acupuncture.

26. *Sales Tax* You buy a new pair of high-top sneakers for $50. The tax on the shoes is $2.50. What is the ratio of the sales tax to the cost of the shoes? $\frac{1}{20}$

27. *Height* Your brother is 72 inches tall. The average height for males is 70 inches. What is the ratio of your brother's height to the average height for males? $\frac{36}{35}$

28. *Who Is Faster?* It takes you $3\frac{1}{2}$ hours to mow your lawn, and it takes your sister 3 hours. Find the ratio of your time to your sister's time. Find the ratio of your sister's time to yours. Are these two ratios reciprocals or opposites? $\frac{7}{6}$, $\frac{6}{7}$, reciprocals

29. *Dog Shelter* An animal shelter has housed 225 dogs during a 12-week summer. Now it is the last week of summer and there are 15 dogs in the shelter. What is the ratio of currently sheltered dogs to the total number of summer dogs? The average stay per dog is one week. Is the shelter busier than normal? $\frac{1}{15}$, no

30. *Windbreaker Sale* A windbreaker regularly sells for $32. You save $8 during a sale. What is the percent discount? 25%

31. *Baseball Diamond* What is the ratio of the baseball diamond's area to the baseball field's area? $\frac{1}{9}$

32. *Field of Dreams* The Cleveland Indians baseball team has a season record so far of 80 wins and 40 losses. What is the ratio of games won to total games played? What is the ratio of games lost to total games played? $\frac{2}{3}$, $\frac{1}{3}$

The diamond portion is 90 feet by 90 feet. The entire field is 270 feet by 270 feet.

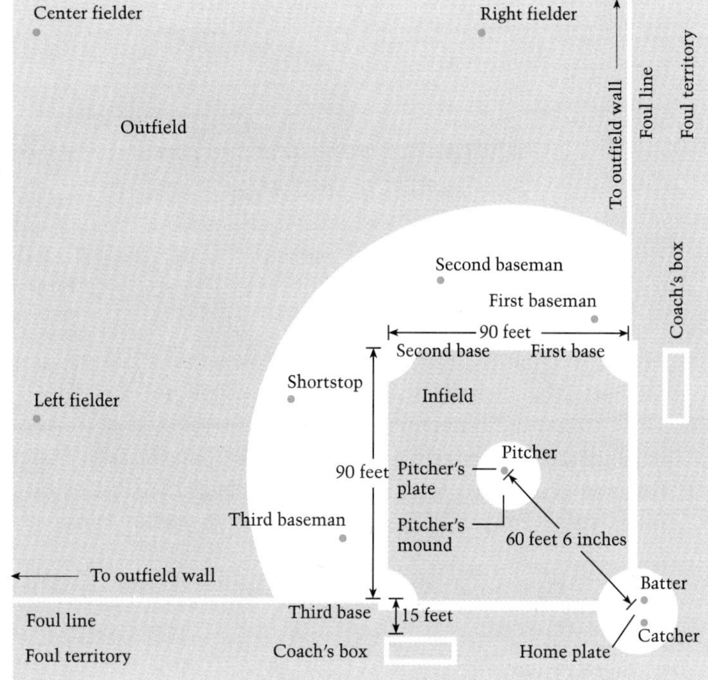

Ex. 16–25 Assume that a year has 365 days. Give results rounded to the nearest whole number.

Ex. 16–25 TECHNOLOGY Some of the numbers are larger than the 8- or 10-digit displays of most calculators. Some calculators will require students to truncate a number before entering it in the calculator. For example, to calculate

352,225,000 ÷ 365

students might calculate

352,225 ÷ 365

and then multiply the result by 1,000 by moving the decimal point in the result 3 places to the right. Students should first estimate whether their answers should be in ones, hundreds, thousands, . . ., millions, etc., in order to decide whether their answers are reasonable.

Answers

In one day, there are

16. 2433 automobiles stolen
17. 965,000 colas drunk at breakfast
18. 4,308,168 credit card purchases
19. 33 new consumer products introduced
20. 958,904,110 photocopies made
21. $1,232,877 spent on sun-care products
22. 243 patents issued
23. 56 house fires started by arson
24. 36,000,000 postage stamps printed
25. 68 animals treated with acupuncture

Ex. 30 A discount is an amount of money by which a regular sales price is reduced. In this exercise, the discount is $8. The percent of discount is a ratio, $\frac{\$8}{\$32} = 0.25$, or 25%.

Ex. 34 PROBLEM SOLVING Point out to students that a matrix is used to present the necessary information for this problem.

Ex. 35–36 PROBLEM SOLVING Point out that students must use the diagram to get the information needed to complete Exercise 35. Ask students to rely on their own experiences when deciding which gear is easiest to pedal. How do the ratios indicate the ease with which one can pedal?

Answer

36. Ist gear: $\frac{13}{7}$; 2nd gear: $\frac{13}{6}$; 3rd gear: $\frac{13}{5}$; 4th gear: $\frac{52}{17}$; 5th gear: $\frac{26}{7}$. The first gear because the lower the gear ratio, the easier it is to turn the axle.

Ex. 37–44 COOPERATIVE LEARNING These rates have not been discussed in the lesson, but most students will know what the appropriate units of measurement are. You might wish to do them as oral classroom exercises.

33. *Years of School* Maya completed eighth grade in 1992. What is the ratio of her completed number of years in school to the total number of years she will spend in school if her highest degree is the following? (Don't count kindergarten.)

- High school diploma, 1996 $\frac{2}{3}$
- Associate's Degree, 1998 $\frac{4}{7}$
- Bachelor's Degree, 2000 $\frac{1}{2}$
- Master's Degree, 2002 $\frac{4}{9}$
- Doctor's Degree, 2004 $\frac{2}{5}$

34. *Population* Using the matrix below, find the ratio of the U.S. population in 1980 to the world population in 1980. Compare this to the ratio of the U.S. population in 1989 to the world population in 1989. Is the ratio of the U.S. population to the world population increasing or decreasing? ***(Source: U.S. Bureau of Census)***

Population	1980	1989
World	4,477,000,000	5,239,000,000
U.S.	227,757,000	248,231,000

≈ 0.051 in 1980, ≈ 0.047 in 1989; decreasing

Gear Ratio **In Exercises 35 and 36, use the following information.**

The *gear ratio* of two connected gears, Gear A and Gear B, is the ratio of the number of teeth on Gear A to the number of teeth on Gear B. Gear ratios for vehicles are normally given in terms of the ratio of the "engine gear" to the "axle gear."

35. The diagram shows the gears for an experimental electric car called *Pohlmann E1.* What is the gear ratio for this car? $\frac{1}{4}$

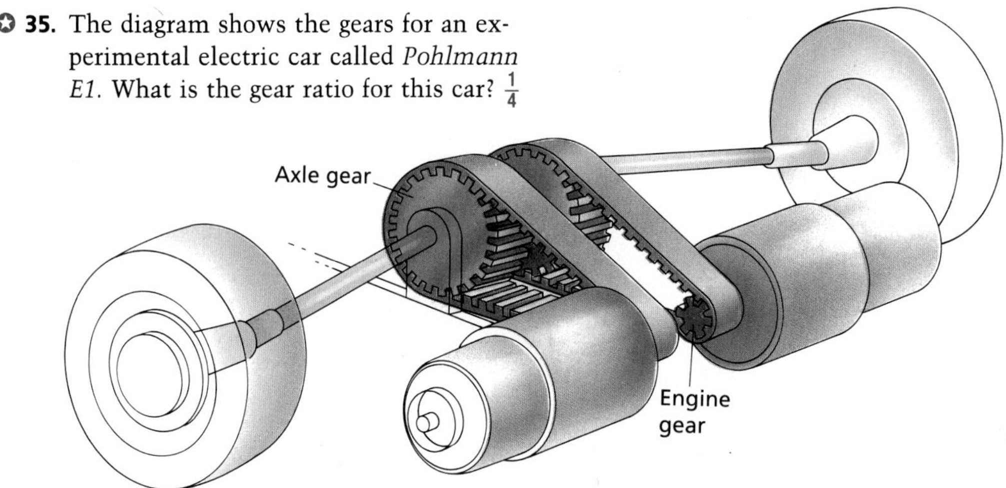

36. On a five-speed bicycle, the ratio of the pedal gear to the axle gear depends on which axle gear is engaged. Use the following table to find the gear ratios for the five different gears. (Ratios on five-speed bikes vary.) For which gear is it easiest to pedal? Why?

	1st Gear	2nd Gear	3rd Gear	4th Gear	5th Gear
Teeth on Pedal Gear	52	52	52	52	52
Teeth on Axle Gear	28	24	20	17	14

See margin.

In Exercises 37–44, match the common name for a rate to one of the measures.

37. Speed e	**38.** Wage h	**a.** Heartbeats per minute	**b.** Vibrations per second
39. Pulse a	**40.** Density f	**c.** Pounds per square inch	**d.** Dollars per ounce
41. Time (in music) g	**42.** Unit Price d	**e.** Miles per hour	**f.** Pounds per cubic inch
43. Frequency b	**44.** Air Pressure c	**g.** Beats per measure	**h.** Dollars per hour

Integrated Review

In Exercises 45–56, evaluate the expression.

45. $-3 - (-2)$ -1

46. $5 - (3 + 7)$ -5

47. $2 + |-3| - 12$ -7

48. $11 - 4 + |-8|$ 15

49. $6x - 4x$ $2x$

50. $-5y + 2y$ $-3y$

51. $4^3 - 8x$, when $x = 7$ 8

52. $\frac{1}{12}(-11 - t^2)$, when $t = 2$ 52. $-\frac{5}{4}$ or $-1\frac{1}{4}$

53. $\frac{1}{8}[m + (-5)]$, when $m = 21$ 2

54. $63x \div -3$, when $x = 7$ -147

55. $49y \div \frac{7}{3}$, when $y = -1$ -21

56. $-3x - (-6)^2$, when $x = -9$ -9

Exploration and Extension

57. *Photocopier* The page on the left was put into a photocopier and reduced to produce the copy on the right. Estimate the ratio of the size of the original to the size of the copy. About $\frac{4}{3}$

Original size

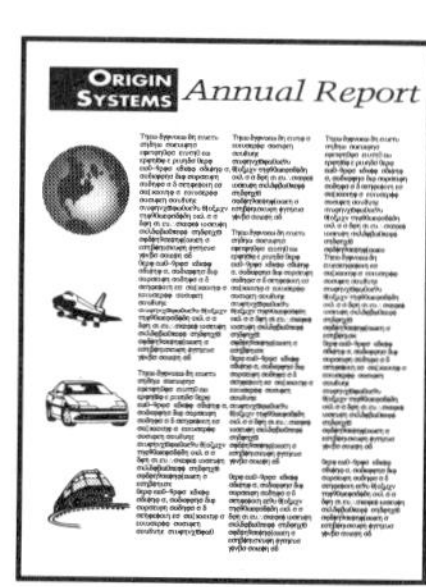

Size of Photocopy

58. *Map Scale* Use the map of Arizona to approximate the distance between Tucson and Phoenix. About 100 mi

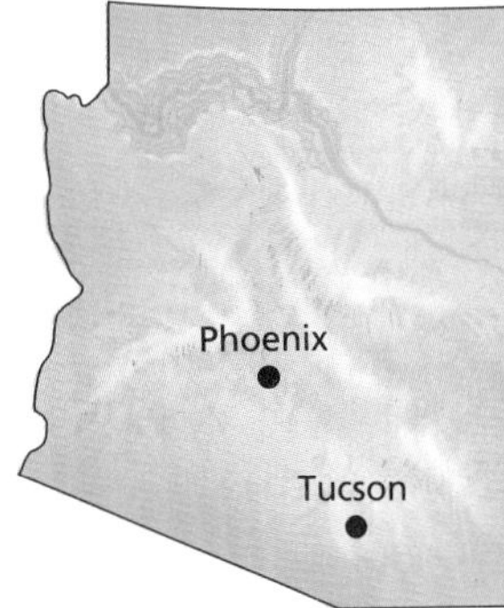

The scale of this map uses a ratio of 1 inch to 200 miles.

59. *That's a Lot of Cousins* Let y be the number of cousins you have. Let x be the number of cousins your best friend has. The ratio of your cousins to your friend's cousins is 2 to 5. Write an equation that describes the relationship between the number of cousins you have and the number of cousins your friend has. If you have 12 cousins, how many does your friend have? $\frac{y}{x} = \frac{2}{5}$, 30

Enrichment Activity

This activity requires students to go beyond lesson goals.

Before assigning the project described below, you may wish to tell students that the relationship between the circumference of circles and pi has been known since ancient times. Babylonian tablets from around 1800–1600 B.C. show two approximations for pi—3 and $3\frac{1}{8}$.

THE HISTORY OF PI A Chinese writer, Tsu Ch'ung-chih, (A.D. 470) computed pi correctly to 6 decimal places as $\frac{355}{113}$, or 3.1415929. . . . Aryabhata of India (A.D. 510) approximated pi as $\frac{62,832}{20,000}$ or 3.1416. . . .

Use resource materials to find out the ways that pi has been computed over the years, from ancient times to the modern technological age. One source of information is Howard Eves' *An Introduction to the History of Mathematics.* An encyclopedia should also have interesting facts about pi.

ORGANIZER

TIME SCHEDULE: All levels, two days

The Chapter Summary helps students organize the main ideas of the chapter. As a class review, ask students to identify the skills that were taught in this chapter. Students should then be asked to summarize the processes or steps required to perform the skill under review successfully.

It is suggested that the first day of review be a combination of you and students working problems from the Chapter Review exercises. The first day's homework assignment should be the basis for the second day of review.

COOPERATIVE LEARNING
Encourage students to study together in small groups. Emphasize the importance of teaching a classmate how to perform a skill or how to recall a strategy. When students work together, everyone wins. The students receiving help get additional instruction, and the students giving help gain a deeper understanding of the skills and concepts involved.

Chapter SUMMARY

Review the chapter by asking students, "What did you learn? Why did you learn it? How does it fit with other algebra skills and concepts?"

2

Chapter Summary

What did you learn?

Skills

1. Use a real number line
   - to graph real numbers. **(2.1)**
   - to order real numbers. **(2.1)**
   - to add real numbers. **(2.2)**
2. Find the absolute value of a real number. **(2.1)**
3. Find the opposite of a real number. **(2.1)**
4. Find the reciprocal of a real number. **(2.7)**
5. Add, subtract, multiply, and divide real numbers. **(2.2–2.7)**
6. Add and subtract matrices. **(2.4)**
7. Simplify expressions. **(2.2–2.7)**
   - Combine like terms. **(2.6)**
8. Use the Distributive Property. **(2.6)**

Strategies

9. Interpret real numbers in real-life settings. **(2.1–2.8)**
10. Find and use rates and ratios in real-life settings. **(2.8)**

Exploring Data

11. Use a matrix to organize data. **(2.4)**

Why did you learn it?

Many real-life situations can be modeled with algebraic expressions and with matrices. By learning to create, simplify, and evaluate algebraic expressions and matrices, you will be able to answer questions pertaining to the real world. Learning to create, simplify, and evaluate algebraic expressions will also help prepare you for the algebra that you will study later in this course.

How does it fit into the bigger picture of algebra?

All the numbers you worked with in Chapter 1 were either zero or positive. In this chapter, you studied rules for adding, subtracting, multiplying, and dividing with both positive and negative numbers. You should try to build an intuitive sense for the type of problems that are associated with negative numbers, such as losses, deficits, low temperatures, etc. This chapter introduces some of the rules of algebra. Your goal should be to practice using the rules so much that they become second nature as you become familiar with the bigger picture of algebra.

Chapter REVIEW

1. Graph 1 and -4 on the real number line. Which one is greater? **(2.1)** 1.–4. See margin.

2. Graph $\frac{1}{2}$, $-\frac{2}{3}$, 0, -2, and $\frac{5}{3}$ on the real number line. Then list the numbers in increasing order. **(2.1)**

3. Graph -5 and its opposite on the number line. **(2.1)**

4. Graph $\frac{3}{4}$ and its opposite on the number line. **(2.1)**

5. What is the absolute value of -1.1? **(2.1)** 1.1

6. What is the absolute value of 4.5? **(2.1)** 4.5

In Exercises 7–24, evaluate the expression. (2.1, 2.2, 2.3, 2.5, 2.7)

7. $8 + (-4)$ 4
8. $-5 + (-3) + 4$ -4
9. $|-5| - 4$ 1
10. $10 - (-2) - |5|$ 7
11. $-[(-7) + 8]$ -1
12. $-[3 + (-4)]$ 1
13. $-14.3 + (-5.5) + 6.2$ -13.6
14. $15.3 + (8.9) + (-7.4)$ 16.8
15. $(-14)(-2)$ 28
16. $(-9)(-2)(-\frac{1}{3})$ -6
17. $\frac{3}{8} \div \frac{1}{2}$ $\frac{3}{4}$
18. $-16 \div -4$ 4
19. $\frac{1}{3}m - m^2$, when $m = \frac{1}{2}$ $-\frac{1}{12}$
20. $-6x + 2|x - 3|$, when $x = -3$ 30
21. $\frac{3r - 6s}{3}$, when $r = 1$ and $s = 2$ -3
22. $\frac{xy}{3x + 4y}$, when $x = 4$ and $y = \frac{1}{2}$ $\frac{1}{7}$
23. $4 + (-x) + 10$, when $x = 5$ 9
24. $|m| - m + 3$, when $m = \frac{1}{2}$ 3

In Exercises 25–28, simplify the expression. (2.3, 2.6, 2.7)

25. $7(y - 1) + 10$ $7y + 3$
26. $3m^2 - (1 + 2m^2)$ $m^2 - 1$
27. $\frac{4x - 16}{2}$ $2x - 8$
28. $\frac{-45 + 15y}{-5}$ $9 - 3y$

In Exercises 29 and 30, add or subtract the matrices. (2.4)

29. Add: $\begin{bmatrix} 3 & -\frac{1}{2} \\ 4 & 0 \\ 1 & -5 \end{bmatrix} + \begin{bmatrix} -4 & \frac{5}{2} \\ 6 & -3 \\ 7 & 2 \end{bmatrix}$ $\begin{bmatrix} -1 & 2 \\ 10 & -3 \\ 8 & -3 \end{bmatrix}$

30. Subtract: $\begin{bmatrix} -\frac{3}{4} & \frac{1}{2} & \frac{3}{2} \\ \frac{5}{4} & \frac{1}{8} & \frac{3}{4} \end{bmatrix} - \begin{bmatrix} \frac{1}{4} & -\frac{3}{2} & 0 \\ -\frac{1}{4} & \frac{3}{8} & 1 \end{bmatrix}$. $\begin{bmatrix} -1 & 2 & \frac{3}{2} \\ \frac{3}{2} & -\frac{1}{4} & -\frac{1}{4} \end{bmatrix}$

In Exercises 31–34, use the table showing average monthly high and low temperatures and average days of rain or snow in San Francisco. (2.4)

	Jan	Feb	Mar	Apr	May	Jun	Jul	Aug	Sep	Oct	Nov	Dec
High Temperature (F)	56°	59°	61°	63°	65°	69°	69°	70°	72°	69°	64°	57°
Low Temperature (F)	40°	43°	44°	45°	48°	50°	52°	52°	52°	49°	45°	45°
Days of Rain or Snow	12	10	9	6	3	2	0	0	1	4	6	11

31. Write a matrix for the average high and low temperatures in San Francisco from November to April. See below.

32. Which six-month period is San Francisco's rainy season, May through October or November through April? November through April

33. During which months are the days getting warmer, on the average? January through September

34. In which months are the average high and low temperatures farthest apart? Closest together? September and October, December

31.

	Nov.	Dec.	Jan.	Feb.	Mar.	Apr.
high	64	57	56	59	61	63
low	45	45	40	43	44	45

ASSIGNMENT GUIDE

Basic/Average: Ex. 1–29 odd, 31–34, 35–39 odd, 46–49, 54–57

Above Average: Ex. 1–29 odd, 31–34, 35–39 odd, 50–53, 54–57

Advanced: Ex. 1–29 odd, 31–34, 35–39 odd, 50–53, 51–57

More Difficult Exercises
Exercises 50–53, 55, 57

Ex. 1–57 The lesson references in the Chapter Review direct students to the lesson in which each skill or strategy was first taught. Students should look back to this lesson if they are having difficulty attacking a set of exercises. Note also that the review exercises parallel the list of Skills, Strategies, and Exploring Data statements on page 114. Students may use this list to identify each skill or strategy, and perhaps check other chapters to see where this skill is used.

Answers

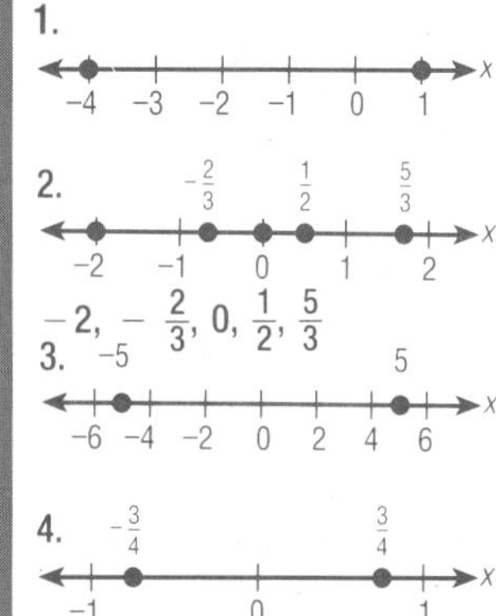

▶ Ex. 50–53 HISTORICAL NOTE The French mathematician René Descartes was the first to observe the relationship among a nautilus's chambers.

35. *Opening a Parachute* Carmen is skydiving and opens her parachute at a height of h feet. After the parachute opens, Carmen's velocity is -20 feet per second. What is her speed? Carmen lands on the ground one minute after her parachute opens. Which equation shows the height at which Carmen opened her parachute?

a. $h = \frac{60}{20}$ **b.** $60h = 20$ **c.** $\underline{h = 20 \cdot 60}$ 20 feet per second

36. *Visiting Your Dad* You live in New Mexico. It normally takes you three hours, traveling at 60 miles per hour, to get to your father's house in Texas. Because of construction, today's trip takes 4 hours. How far is it to your father's house? Which equation shows your average speed, r, on today's trip?

a. $r = \frac{180}{3}$ **b.** $3 = \frac{180}{r}$ **c.** $\underline{4r = 180}$ 180 mi

37. *Flying to San Diego* A pilot leaves Detroit at 7:00 A.M. and lands at San Diego six hours later. What time is it when the pilot lands at San Diego? (Use the map on page 72.) 10 A.M.

38. *Annual Profit or Loss?* A company had a loss of $1,002.91 in the first quarter, a loss of $565.77 in the second quarter, a profit of $14,232.01 in the third quarter, and a profit of $3,027.88 in the fourth quarter. What was its profit for the year? $15,691.21

In Exercises 39–41, decide whether the element is an atom or an ion.

39. Magnesium (Mg): 12 electrons, 12 protons Atom

40. Magnesium (Mg): 10 electrons, 12 protons Ion

41. Sodium (Na): 10 electrons, 11 protons Ion

42. Which of Exercises 39–40 has an element with Mg^{+2} as its chemical symbol? Exercise 40

Where the Buffalo Roam **In Exercises 43–45, use the table showing the average number of buffalo in four parks in the United States and Canada.**

Park	Number of Buffalo	Number of Acres
Elk Island National Park, Alberta	750	48,000
Wood Buffalo National Park, Alberta/Northwest Territories	3375	11,072,000
Wind Cave National Park, South Dakota	325	28,300
Yellowstone National Park, Idaho/Montana/Wyoming	2850	2,219,800

43. Find the average number of acres per buffalo in each park.

44. Which park has the greatest number of acres per buffalo? Wood Buffalo

45. Which park has the least number of acres per buffalo? Elk Island

43. Elk Island: 64, Wood Buffalo: 3280.6, Wind Cave: 87.1, Yellowstone: 778.9

Earning Money **In Exercises 46–49, use the information in the bar graph.**

The bar graph shows the percent of eighth graders in the United States who earn money after school. (Those who earned money in two or more jobs were asked to select their primary job.)

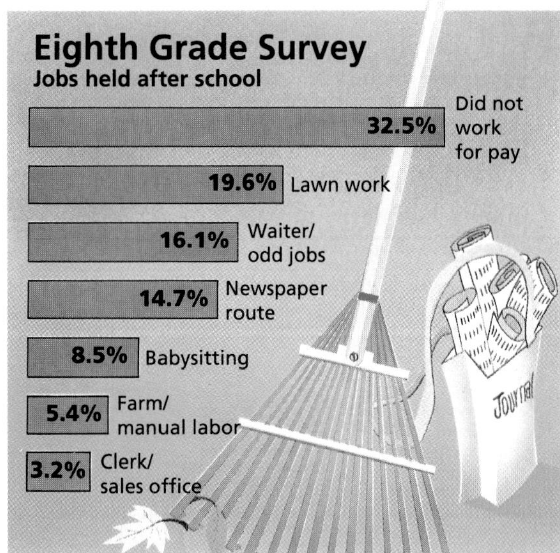

46. Out of 2000 eighth graders, how many would you expect to earn money primarily in a newspaper route? 294

47. Out of 3500 eighth graders, how many would you expect to earn money primarily by baby-sitting? 298

48. Out of 1500 eighth graders, how many *more* would you expect to do lawn work than farm work or manual labor? 213

49. Out of 1500 eighth graders, which of the following expressions tells you how many have earned money in some job? Explain your answer. See margin.

a. $1500 - 0.325(1500)$
b. $1500(1 - 0.325)$

Chambered Nautilus **In Exercises 50–53, use the following information.**

A chambered nautilus is a mollusk that builds its spiral shell by constructing new living chambers, each with the same shape. Each consecutive living chamber is 6.3% larger than the previous chamber.

50. Consider three consecutive living chambers. What percent larger is Chamber 3 than Chamber 1? $\approx$ 13.0%

51. Consider two consecutive living chambers. What percent *smaller* is Chamber 1 than Chamber 2? $\approx$ 5.9%

52. Consider four consecutive living chambers. If Chamber 1 has a volume of 3 cm^3, what is the volume of Chamber 4? ≈ 3.6 cm^3

53. After completing any given living chamber (call it Chamber 0), the nautilus must complete 18 more chambers to bring the spiral full circle. The 18th additional chamber will be about 2.85 times larger than Chamber 0. Does a. or b. describe the relative size of the 18th chamber to Chamber 0? Justify your answer.

a. $(1.06)^{18}$ **b.** $(1.06)^{19}$

$(1.06)^{18} \approx 2.85$

Answer

49. a and b; in a, the number who have not worked for pay is subtracted from 1500; in b, the fraction who have worked for pay is multiplied by 1500.

Human-Powered Submarines **In Exercises 54–57, use the information about human-powered submarines. Each of the submarines participated in the Second International Submarine Races held in June 1991, in Riviera Beach, Florida.** *(Source:* **Popular Science Magazine***)*

54. Find the ratio of each submarine's width to its length. Which ratio is closest to the ideal ratio for minimum drag? (See page 121.)

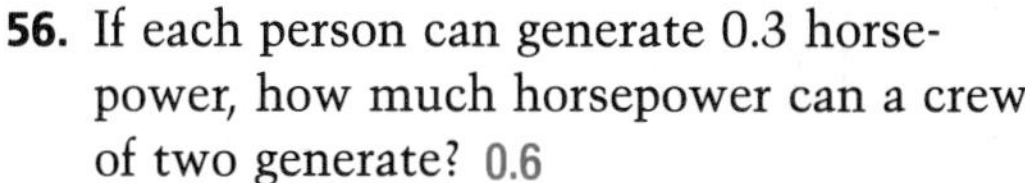

54. $\frac{37}{184}, \frac{2}{17}, \frac{4}{21}, \frac{1}{4}, \frac{24}{67}; \frac{1}{4}$

55. Suppose a submarine had a speed of 3 knots. One knot is one nautical mile per hour. One nautical mile is 6080 feet. One mile is 5280 feet. What is the submarine's speed in miles per hour? $3\frac{5}{11}$

$(3 \text{ knots})\left(\frac{1 \text{ n.mi/h}}{1 \text{ knot}}\right)\left(\frac{6080 \text{ ft}}{1 \text{ n.mi}}\right)\left(\frac{1 \text{ mi}}{5280 \text{ ft}}\right)$

56. If each person can generate 0.3 horsepower, how much horsepower can a crew of two generate? 0.6

57. If the foot pedals have a 2 to 1 gear ratio to the submarine paddle axle, how many revolutions per minute must a crew member pump to make the paddles turn at 70 revolutions per minute? 140

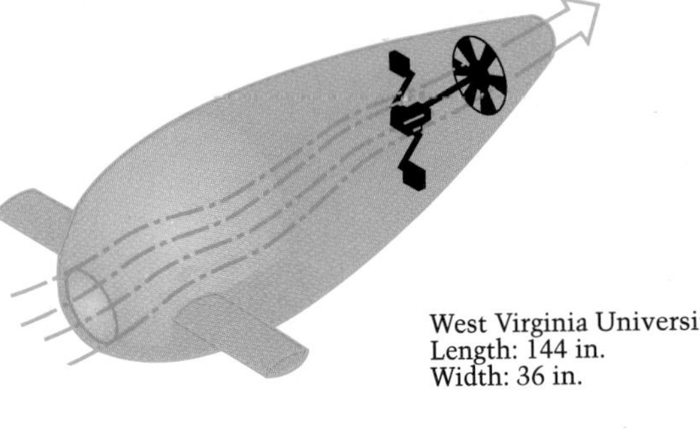

West Virginia University
Length: 144 in.
Width: 36 in.

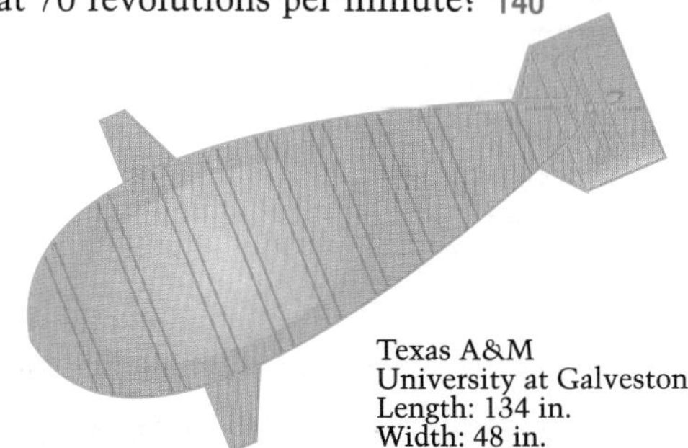

Texas A&M
University at Galveston
Length: 134 in.
Width: 48 in.

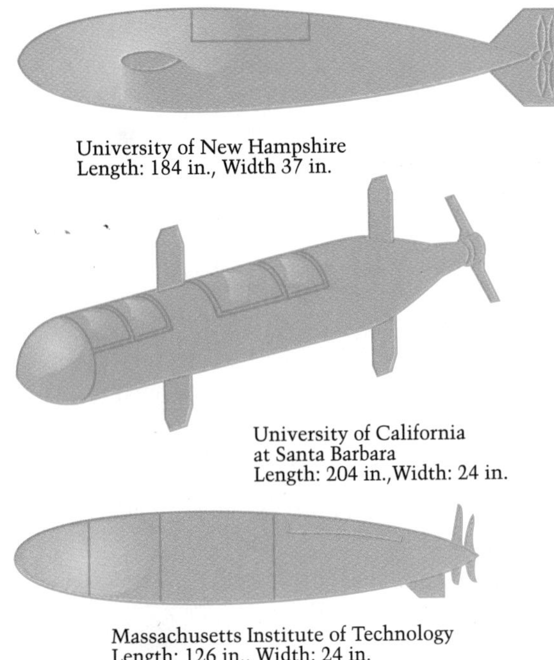

University of New Hampshire
Length: 184 in., Width 37 in.

University of California
at Santa Barbara
Length: 204 in.,Width: 24 in.

Massachusetts Institute of Technology
Length: 126 in., Width: 24 in.

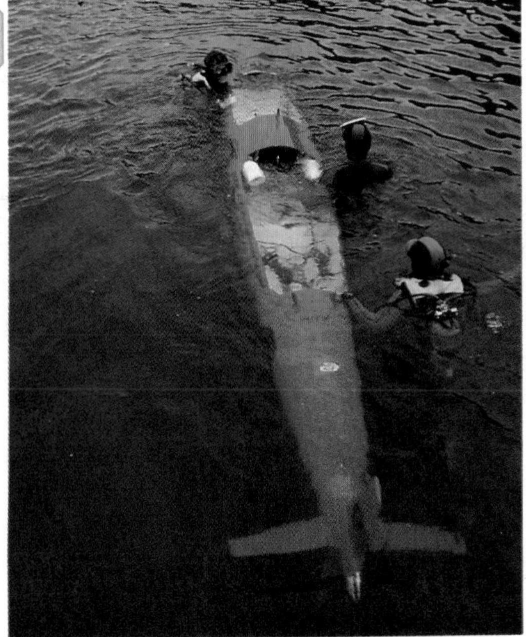

Answer
24.

	9th	10th	11th	12th
Males	307	323	290	289
Females	331	300	270	312

Chapter TEST

1. Write in increasing order. $-\frac{5}{4}, -\frac{5}{2}, \frac{3}{4}$ (2.1) $-\frac{5}{2}, -\frac{5}{4}, \frac{3}{4}$

2. Which is correct? (2.1)
   a. $-3 < -4$ **b.** $-3 > -4$

3. Evaluate $9 + (-10) + 2$. (2.2) 1

4. Evaluate $-2.97 - 3.20 - (-0.02)$. (2.3) -6.15

5. Evaluate $(-6)(4)(-1)$. (2.5) 24

6. Evaluate $72 \div (-12)$. (2.7) -6

7. Evaluate $\frac{4-x}{-3}$, when $x = -1$. (2.7) $-\frac{5}{3}$

8. Evaluate $1 - 2x^2$, when $x = -2$. (2.5) -7

9. Simplify $-2x + 3 + 3x - 9$. (2.6) $x - 6$

10. Simplify $3 - 4(x - 2)$. (2.6) $11 - 4x$

11. Add the matrices. (2.4)
$$\begin{bmatrix} 3 & -7 \\ \frac{1}{2} & 6 \\ 0 & 2 \end{bmatrix} + \begin{bmatrix} -4 & 2 \\ -\frac{1}{2} & 4 \\ 5 & \frac{1}{4} \end{bmatrix} \quad \begin{bmatrix} -1 & -5 \\ 0 & 10 \\ 5 & 2\frac{1}{4} \end{bmatrix}$$

12. Subtract the matrices. (2.4)
$$\begin{bmatrix} 2 & -\frac{2}{3} & 4 \\ 5 & 16 & -7 \end{bmatrix} - \begin{bmatrix} 5 & \frac{1}{3} & -1 \\ -2 & 8 & 4 \end{bmatrix} \quad \begin{bmatrix} -3 & -1 & 5 \\ 7 & 8 & -11 \end{bmatrix}$$

In Exercises 13–18, decide whether the statement is true or false.

13. $4(x + 3) = 4(x) + 4(3)$ (2.6) True

14. $(2 - y)-7 = -7(2) + (-7)(y)$ (2.6) False

15. Zero is its own opposite. (2.1) True

16. Zero is its own reciprocal. (2.7) False

17. One is its own opposite. (2.1) False

18. One is its own reciprocal. (2.7) True

19. A bus is moving at a rate of 100 kilometers per hour. How far will it travel in $3\frac{1}{2}$ hours? 350 km

20. A company had a first-quarter profit of $2,189.70, a second-quarter profit of $1,527.11, a third-quarter loss of $502.18, and a fourth-quarter loss of $266.54. What was its profit for the year? $2,948.09

21. Suppose you visit Taiwan, and the currency exchange rate is 27 Taiwanese dollars per U.S. dollar. How many Taiwanese dollars will you receive for $300 (U.S.)? 8100

22. A first-grade class is going on a bus trip to an amusement park. The park requires groups to have one adult for every eight children. There are 3 teachers, 2 parents, and 36 children planning to go on the trip. Are there enough adults to meet the park's requirements? Yes

23. A 1-liter bottle of seltzer water sells for $0.99. A 1.75-liter bottle sells for $1.79. Which size is the better buy? 1-liter bottle

24. The table shows the distribution of students at Lincoln High School by grade level and by sex. Write this table as a matrix. How many male students are in the tenth grade at Lincoln? 323

Lincoln	9th	10th	11th	12th
Males	307	323	290	289
Females	331	300	270	312

See margin.

25. A parachutist is falling from a height of 1800 feet at a rate of 15 feet per second. Find the velocity and speed of the parachutist. When will the parachutist land on the ground?
-15 ft per second, 15 ft per second, in 120 seconds

In this chapter, techniques for solving linear equations will be developed. Emphasize that each skill that is learned can be used as the foundation for other equation-solving skills. In fact, to start with, only the mental-math techniques that students used to solve equations in Chapters 1 and 2 will be formalized.

Stress that one reason to learn how to solve linear equations is because many real-life situations can be modeled using them. The techniques of solving equations will provide students with the mathematical power to understand and interpret the world around them.

The Chapter Summary on page 167 provides you and the students with a synopsis of the chapter. It identifies key skills and concepts. You may want to have students look at the Chapter Summary as an overview before beginning the chapter.

CHAPTER 3

Solving Linear Equations

LESSONS

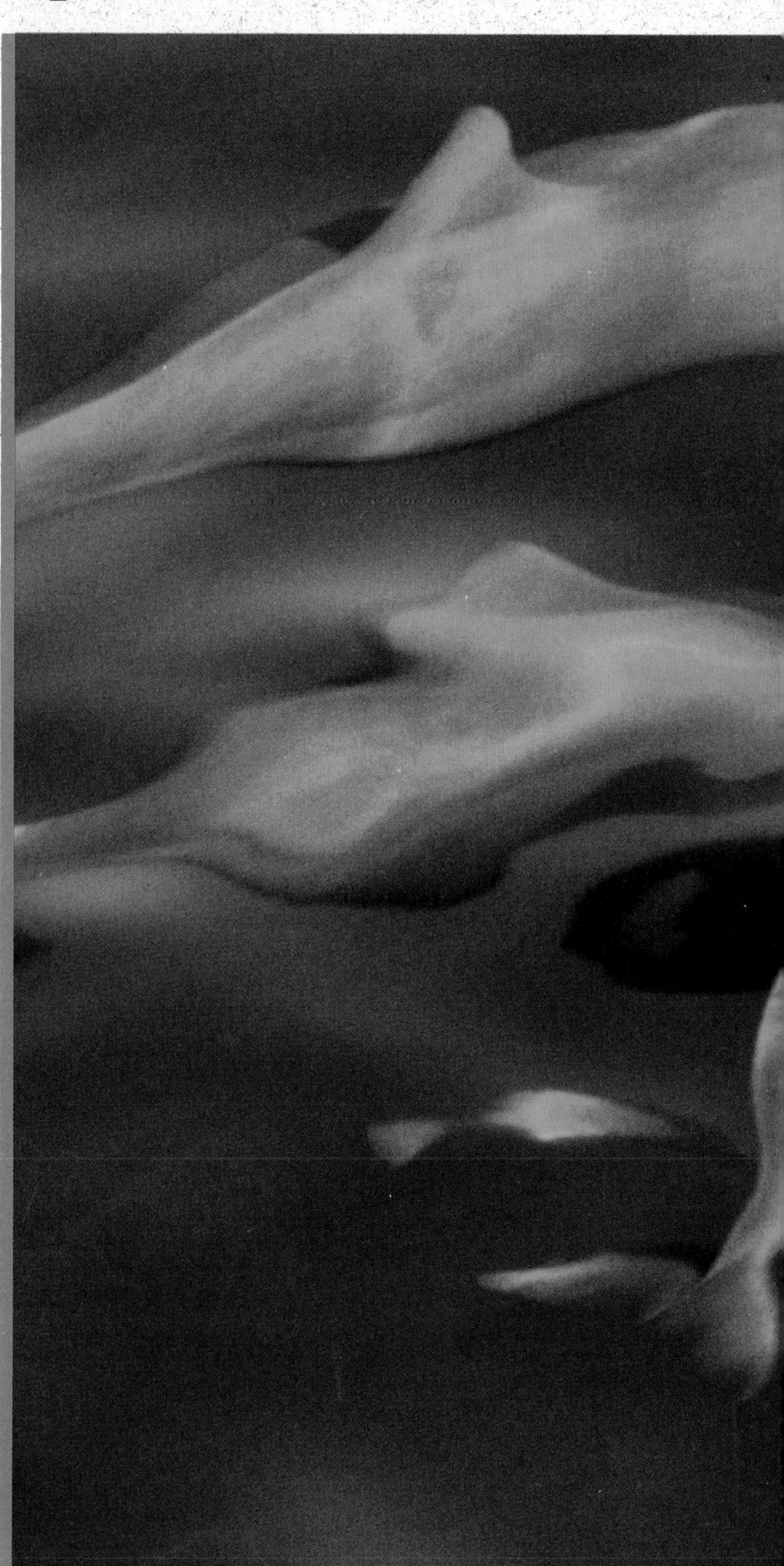

Dolphins are not only a beautiful example of graceful adaptation to ocean living, but they also seem to have a special fascination for humans.

Marine objects are not the only ones to experience drag while in motion. Cars and planes are examples of other objects that would benefit from a torpedo-shaped design.

Encourage students to use the encyclopedia to investigate the shapes of objects that facilitate faster motion. The shape of winning cars in the Indy 500 and the shape of passenger planes or military aircraft are good places to begin the investigation.

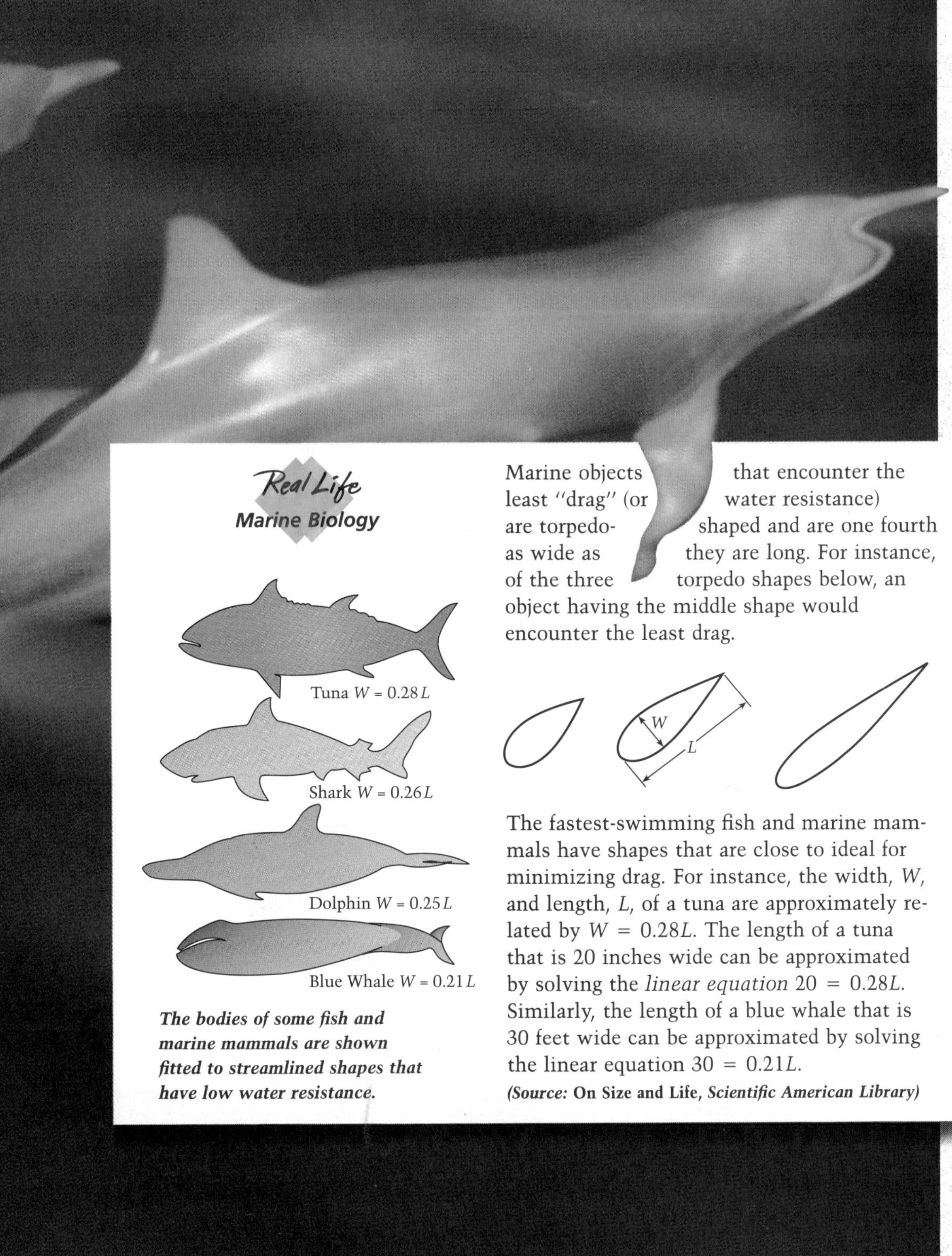

Real Life
Marine Biology

Marine objects that encounter the least "drag" (or water resistance) are torpedo-shaped and are one fourth as wide as they are long. For instance, of the three torpedo shapes below, an object having the middle shape would encounter the least drag.

The fastest-swimming fish and marine mammals have shapes that are close to ideal for minimizing drag. For instance, the width, W, and length, L, of a tuna are approximately related by $W = 0.28L$. The length of a tuna that is 20 inches wide can be approximated by solving the *linear equation* $20 = 0.28L$. Similarly, the length of a blue whale that is 30 feet wide can be approximated by solving the linear equation $30 = 0.21L$.

(Source:* On Size and Life, *Scientific American Library)

The bodies of some fish and marine mammals are shown fitted to streamlined shapes that have low water resistance.

Solving Linear Equations

PACING CHART

The daily Pacing Chart is meant to help you adjust your teaching pace. Students in the full course should finish the entire text by the end of the year. Students in the basic course may not complete the entire text in the school year. The Pacing Chart for each chapter contains suggestions for lessons that require more than one day and lessons that may be omitted for the basic course.

DAY	FULL COURSE	BASIC COURSE
1	3.1	3.1
2	3.2	3.2
3	3.3	3.3
4	3.4	3.4
5	3.4	3.4
6	Mid-Chapter Self-Test	Mid-Chapter Self-Test
7	3.5	3.5
8	3.6	3.6
9	3.7	3.6
10	Chapter Review	3.7
11	Chapter Review	Chapter Review
12	Chapter Test	Chapter Review
13	Cumulative Review	Chapter Test
14		Cumulative Review

CHAPTER ORGANIZATION

LESSON	PAGES	GOALS	MEETING THE NCTM STANDARDS
3.1	122-128	1. Solve equations systematically using addition, subtraction, division 2. Use reciprocals to solve equations	Problem Solving, Communication, Connections, Reasoning, Geometry
3.2	129-134	Use two or more transformations to solve an equation	Problem Solving, Communication, Connections
3.3	135-140	1. Collect variables on one side of an equation 2. Use algebraic models in real-life situations	Problem Solving, Communication, Connections, Reasoning
Mixed Review	140	Review of algebra and arithmetic	
3.4	141-146	Use a problem-solving plan for problems that fit a linear model	Problem Solving, Communication, Connections
Career Interview	146	Picture framer	Connections
Mid-Chapter Self-Test	147	Diagnose student weaknesses and remediate with correlated Reteaching worksheets	
3.5	148-153	1. Find exact and approximate solutions of equations with decimals 2. Solve problems that use decimal measurements	Problem Solving, Communication, Connections
3.6	154-159	Solve literal equations for a specified variable	Problem Solving, Communication, Connections, Geometry
Mixed Review	159	Review of algebra and arithmetic	
3.7	160-165	1. Use a coordinate plane to match points with ordered pairs of numbers 2. Use a scatter plot	Problem Solving, Communication, Connections, Statistics, Discrete Mathematics, Geometry
Using a Calculator	166	Draw scatter plots using a graphing calculator	Technology, Statistics
Chapter Summary	167	A restatement of what has been learned, why it has been learned, and how it fits into the structure of algebra	Connections, Structure
Chapter Review	168-170	Review of concepts and skills learned in the chapter	
Chapter Test	171	Diagnose student weaknesses and remediate with correlated Reteaching worksheets	
Cumulative Review	172-173	Review of concepts and skills from previous chapters	

LESSON RESOURCES

MEETING INDIVIDUAL NEEDS

RETEACHING For students who need to spend more time on basics:

If a mid-chapter self-test or chapter test indicates a deficiency, teachers can help students with the appropriate ***Reteaching Copymaster.***

PRACTICE For students who need more practice:

Additional exercises like those in the Pupil's Edition are provided for each lesson in ***Extra Practice Copymasters.***

ENRICHMENT For enriching and broadening students' experiences:

Problem of the Day copymasters in ***Teaching Tools*** provide a daily opportunity to use logical reasoning, looking for a pattern, writing an equation, and other routine and non-routine problem-solving strategies.

Math Log copymasters in ***Alternative Assessment*** provide opportunities to report on investigations, research, and open-ended problems.

Enriching activities with graphing and scientific calculators and computers are provided in ***Technology: Using Calculators and Computers.***

The ***Applications Handbook*** provides additional information about the cross-curriculum topics such as astronomy, chemistry, physics, sports, economics, genetics, and music that are integrated into the Pupil's Edition.

LESSON	3.1	3.2	3.3	3.4	3.5	3.6	3.7
PAGES	122-128	129-134	135-140	141-146	148-153	154-159	160-165
Teaching Tools							
Transparencies				✓			✓
Problem of the Day	✓	✓	✓	✓	✓	✓	✓
Warm-up Exercises	✓	✓	✓	✓	✓	✓	✓
Answer Masters	✓	✓	✓	✓	✓	✓	✓
Extra Practice Copymasters	✓	✓	✓	✓	✓	✓	✓
Reteaching Copymasters	Teacher-directed and independent activities tied to results on the Mid-Chapter Self-Tests and Chapter Tests						
Color Transparencies				✓	✓	✓	✓
Applications Handbook	Additional background information is supplied for many real-life applications.						
Technology Handbook	Calculator and computer worksheets are supplied for appropriate lessons.						
Complete Solutions Manual	✓	✓	✓	✓	✓	✓	✓
Alternative Assessment	Assess student's ability to reason, analyze, solve problems, and communicate using mathematical language.						
Formal Assessment	Mid-Chapter Self-Tests, Chapter Tests, Cumulative Tests, and Practice for College Entrance Tests						
Computer Test Bank	Customized tests can be created by choosing from over 2000 items.						

INSIGHTS

3.1 Solving Equations Using One Transformation

Many equations are too difficult to solve using mental math. A more systematic approach to equation solving gives more control.

Before complicated equations can be solved, students must understand what finding the solution of an equation means. They also need to know how to solve simple equations.

3.2 Solving Equations Using Two or More Transformations

The more systematic approach to solving equations is extended in this lesson to equations requiring more than one transformation. Emphasize that these equation-solving techniques will enhance students' ability to solve real-life problems using algebraic models.

3.3 Solving Equations with Variables on Both Sides

In this lesson, the systematic approach is extended to include solving equations with variables on both sides. Continue to emphasize that these equation-solving techniques enhance students' ability to solve real-life problems using algebraic models.

3.4 Linear Equations and Problem Solving

The systematic approach to solving equations will be put to use in this lesson. Students have seen how to construct algebraic models using the problem-solving plan found in Lesson 1.7: (a) write a verbal model, (b) assign labels, (c) write an algebraic model (an equation or inequality), (d) solve the algebraic model, and (e) answer the question and look back.

3.5 Solving Equations that Involve Decimals

Many real-life situations involving money, statistical data, measurements, and percents are modeled using decimal quantities. As a result, solutions to these algebraic models often involve round-off errors. This lesson focuses on exact and approximate solutions.

3.6 Literal Equations and Formulas

Most real-life relationships—for example, $I = Prt$, $d = rt$, or $A = lw$—depend upon more than one variable. In fact, relationships in business, geometry, chemistry, engineering, and physics frequently involve three or more variables. In practice, several variables in these algebraic models may represent constant values for a given application. The user needs to solve for the variable that has no initial replacement value. Thus, solving literal equations is a much-needed skill.

3.7 Exploring Data: Scatter Plots

"A picture is worth a thousand words." The visual representation of data is one of the most important mathematical tools available to students and is integrated throughout the text. The coordinate plane is also a bridge between algebra and geometry.

Problem of the Day

A brick balances exactly with $\frac{3}{4}$ of a pound and $\frac{3}{4}$ of a brick. What is the weight of a brick in pounds?

3 pounds

ORGANIZER

Warm-Up Exercises

1. Use mental math to solve these equations.
   a. $x + 3 = 7$ 4
   b. $x - 12 = -3$ 9
   c. $2x = 16$ 8
   d. $\frac{2}{3}x = 12$ 18
2. Describe the mental-math steps you used to solve for x in each of the equations in Exercise 1. Compare your solution approach with classmates'.

Lesson Resources

Teaching Tools
 Problem of the Day: 3.1
 Warm-up Exercises: 3.1
 Answer Masters: 3.1
Extra Practice: 3.1
Technology Handbook: p. 20

LESSON Notes

The Golden Rule of Solving Equations is easy to remember and essential to the equation-solving process: Do to one side of an equation what you do to the other.

It is important that students understand that the *solution* set of an equation or inequality is the collection of real numbers that make the equation or inequality a true statement when the number value is substituted for the variable in the equation or inequality.

3.1 Solving Equations Using One Transformation

What you should learn:

Goal 1 How to solve equations systematically using addition, subtraction, and division

Goal 2 How to use reciprocals to solve equations

Why you should learn it:

You can use techniques shown in this lesson to solve equations that are too difficult to solve with mental math.

Goal 1 Solving Equations Systematically

You have been using mental math to solve equations. Now you will study techniques for solving equations systematically. Although many equations in this lesson can be solved using mental math, you should make sure you also know how to use systematic techniques to find a solution.

Two equations are **equivalent** if they have the same solution set. For instance, the equations $x - 3 = 0$ and $x = 3$ are equivalent because both have the number 3 as their only solution. To *transform* an equation into an equivalent equation, it is helpful to think of the equation as having two sides that are "in balance." Simplifying one side does not change the balance, so nothing has to be done to the other side. Adding a (nonzero) number to one side does change the balance, so the same number has to be added to the other side. The following transformations keep the sides of an equation in balance.

Transformations That Produce Equivalent Equations

	Original Equation	Equivalent Equation
1. Add the same number to *both* sides.	$x - 3 = 5$	$x = 8$
2. Subtract the same number from *both* sides.	$x + 6 = 10$	$x = 4$
3. Multiply *both* sides by the same nonzero number.	$\frac{x}{2} = 3$	$x = 6$
4. Divide *both* sides by the same nonzero number.	$4x = 12$	$x = 3$
5. Interchange the two sides.	$7 = x$	$x = 7$

In solving an equation, your goal is to isolate the variable on one side by using **inverse operations.** Addition and subtraction are inverse operations. Multiplication and division are also inverse operations.

Example 1 Adding the Same Quantity to Both Sides

Solve $x - 5 = -13$.

Solution On the left side of the equation, 5 is subtracted from x. To isolate x, you need to apply the inverse operation by adding 5. Remember, however, to keep the balance you must add 5 to *both sides.*

$$x - 5 = -13 \quad \text{Rewrite original equation.}$$
$$x - 5 + 5 = -13 + 5 \quad \text{Add 5 to both sides.}$$
$$x = -8 \quad \text{Simplify.}$$

The solution is -8. You should check this solution by substituting -8 for x in the original equation.

Check $(-8) - 5 = -13$ ■

Each time you apply a transformation to an equation, you are writing a *solution step.* Notice that the solution steps are written "one below the other" with their equal signs aligned.

Example 2 Subtracting the Same Quantity from Both Sides

Solve $-8 = n + 4$.

Solution On the right side of the equation, 4 is added to n. You can isolate n by subtracting 4 from both sides of the equation.

$$-8 = n + 4 \quad \text{Rewrite original equation.}$$
$$-8 - 4 = n + 4 - 4 \quad \text{Subtract 4 from both sides.}$$
$$-12 = n \quad \text{Simplify.}$$

The solution is -12. Check this in the original equation. ■

Example 3 Dividing Both Sides by the Same Quantity

Solve $4x = 128$.

Solution On the left side of the equation, x is multiplied by 4. You can isolate x by dividing both sides by 4.

$$4x = 128 \quad \text{Rewrite original equation.}$$
$$\frac{4x}{4} = \frac{128}{4} \quad \text{Divide both sides by 4.}$$
$$x = 32 \quad \text{Simplify.}$$

The solution is 32. Check this in the original equation. ■

When mental math is used to solve equations or inequalities, steps that could be written explicitly and would be helpful in solving for the unknown variable are frequently omitted. For example, to solve $x - 3 = 5$, you could draw upon your knowledge of basic facts to decide that 8 minus 3 is 5, so $x = 8$. You could, however, see the equation as a statement that the variable x is 3 more than 5. In which case, $x = 5 + 3$. The same result can be obtained by using the Golden Rule of Equations, which gives rise to the following steps: $x - 3 + 3 = 5 + 3$, or $x = 5 + 3$, since $-3 + 3$ becomes 0.

A key concept in solving equations and inequalities is the isolation of the unknown. This is achieved by undoing the operation by which the unknown is affected (that is, performing the inverse operation).

Example 1

The solution to this example is an explanation of the steps used to obtain the solution of the equation, $x - 5 = -13$. The term, *solution* has two different meanings. First, it has a colloquial meaning of "explanation," which is used throughout the text to describe the process of solving examples. Second, it has a well-defined meaning of "a replacement value that makes a given equation a true statement." This second meaning must not be confused with the first.

Example 2

The solution is expressed as $-12 = n$. The conventional way of expressing this solution is $n = -12$, which is obtained by interchanging the sides of the equation. In other words, equality is symmetric, so if $a = b$, then $b = a$.

Examples 4–5

To solve $\frac{1}{5}x = 30$, you could multiply both sides of the equation by the reciprocal of the coefficient $\frac{1}{5}$, or 5. In that case, you get $5(\frac{1}{5}x) = 5(30)$ or $x = 150$.

Example 6

If the corresponding angles of two triangles are congruent (that is, have equal measure), then the two triangles are similar. Corresponding sides of similar triangles are proportional. Therefore, write this verbal model: the ratio of the length of *AB* to the length of *DE* is equal to the ratio of the length of *BC* to the length of *EF*.

Extra Examples

Here are additional examples similar to **Examples 1–6.**

1. Solve. Record your solution steps.
   a. $x - 3 = 15$ 18
   b. $x + 7 = -13$ −20
   c. $19 = m - 4$ 23
   d. $23 = 12 + m$ 11
   e. $8t = 24$ 3
   f. $26 = 6t$ $\frac{13}{3}$
   g. $\frac{2}{5}p = 4$ 10
   h. $7 = -\frac{1}{6}p$ −42
2. The two triangles below are similar. What is the length of the side *BC*? 8

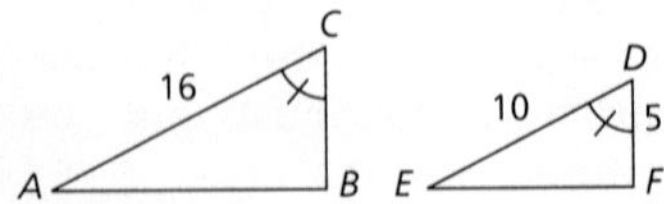

Goal 2 Using Reciprocals to Solve an Equation

When you are solving an equation that involves a fractional coefficient, such as $\frac{1}{5}x = 30$, you can isolate the variable by multiplying by the reciprocal of the fraction.

Example 4 *Multiplying Both Sides by a Reciprocal*

Solve $\frac{1}{5}x = 30$.

Solution On the left side of the equation, x is multiplied by the fraction $\frac{1}{5}$. You can isolate x by multiplying both sides of the equation by 5 (the reciprocal of $\frac{1}{5}$).

$\frac{1}{5}x = 30$ *Rewrite original equation.*

$5(\frac{1}{5}x) = 5(30)$ *Multiply both sides by 5.*

$x = 150$ *Simplify.*

The solution is 150. Check this in the original equation. ■

Example 5 *Multiplying Both Sides by a Reciprocal*

Solve $-\frac{2}{3}m = 10$.

Solution

$-\frac{2}{3}m = 10$ *Rewrite original equation.*

$(-\frac{3}{2})(-\frac{2}{3}m) = (-\frac{3}{2})(10)$ *Multiply both sides by $-\frac{3}{2}$.*

$m = -15$ *Simplify.*

The solution is -15. Check this in the original equation. ■

The equations in this lesson are called **linear equations** because each equation's variable occurs only to the first power and does not occur in a denominator. The equation $x^2 = 4$ is not linear because x occurs to the second power, and $6 = \frac{5}{x}$ is not linear because x is in the denominator. The transformations used to solve linear equations are based on rules of algebra called **properties of equality.**

Properties of Equality

1. If $a = b$, then $a + c = b + c$. Addition Property
2. If $a = b$, then $a - c = b - c$. Subtraction Property
3. If $a = b$, then $ca = cb$. Multiplication Property
4. If $a = b$ and $c \neq 0$, then $\frac{a}{c} = \frac{b}{c}$. Division Property

A common strategy for solving a real-life problem is to equate two ratios. Example 6 shows how this can be done using a property of similar triangles. Two triangles are **similar** if they have equal angles. The ratios of lengths of corresponding sides of similar triangles are equal.

Connections
Geometry

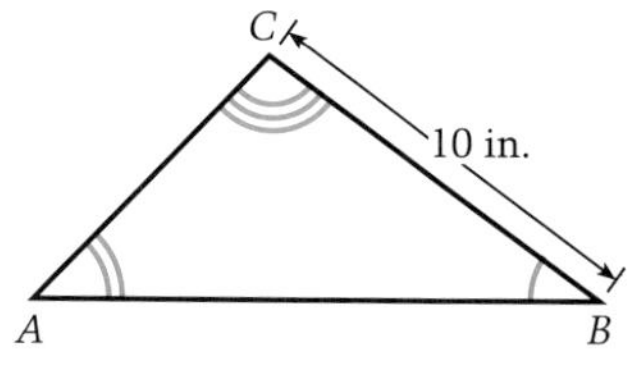

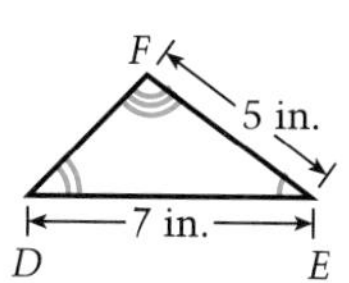

Example 6 *Similar Triangles*

The two triangles at the left are similar. What is the length of the side AB?

Solution Use the fact that the ratio of the length of AB to the length of DE is equal to the ratio of the length of BC to the length of EF.

Verbal Model $\dfrac{\text{Length of } AB}{\text{Length of } DE} = \dfrac{\text{Length of } BC}{\text{Length of } EF}$

Labels
Length of $AB = x$ (inches)
Length of $DE = 7$ (inches)
Length of $BC = 10$ (inches)
Length of $EF = 5$ (inches)

Equation
$\frac{x}{7} = \frac{10}{5}$ *Algebraic model*
$x = 7 \cdot \frac{10}{5}$ *Multiply both sides by 7.*
$x = 14$ *Simplify.*

The length of AB is 14 inches. ■

Communicating about ALGEBRA

SHARING IDEAS about the Lesson

Understand the Process Think about the transformations used to solve equations.

A. To solve $\frac{2}{3}x = 7$, you should multiply both sides of the equation by what number? $\frac{3}{2}$

B. Is 5 a solution of the equation $\frac{1}{7}x = \frac{5}{7}$? Explain why or why not. Does this equation have other solutions? Yes, because it checks; no

C. To solve $y - 5 = -13$, should you add 5 or 13 to both sides of the equation? Explain why. 5, to isolate the variable

D. Can you use mental math to solve $-\frac{3}{5}x = 81$? If not, multiply both sides of the equation by $-\frac{5}{3}$. Explain why this transformation gives the solution.

D. Transformation gives $x = -135$.

Check Understanding

1. Explain how the Golden Rule of Equations and the properties of equality are related. See Lesson Notes on pages 122–123.
2. Explain the two different meanings of *solution* found in the text. Which relates to solving equations? See Example 1 notes on page 123.
3. Why are systematic techniques needed to solve equations and inequalities? Should you abandon mental math to solve equations and inequalities? To solve more complicated equations and real-life algebraic models.

Communicating *about* ALGEBRA

Discuss how using mental math to solve linear equations is different from using solution steps. How important are the basic facts of addition and multiplication to using mental math?

Can multiplying by a reciprocal be performed in two steps? Explain?
Yes; if the reciprocal were a fraction such as the reciprocal of $\frac{2}{3}$, $\frac{3}{2}$, you can multiply each term by the numerator 3 and then divide the product by the denominator 2 .

EXERCISE Notes

ASSIGNMENT GUIDE

Basic/Average: Ex. 1–12, 13–51 odd, 55, 60–72 multiples of 3, 74, 75–79 odd

Above Average: Ex. 1–12, 14–24 even, 27–48 multiples of 3, 50–54 even, 56–57, 60–72 multiples of 3, 74, 75–80

Advanced: Ex. 1–12, 21–30, 43–48, 50–54, 56–58 even, 75–80

Selected Answers

Exercises 1–8, 9–77 odd

✪ More Difficult Exercises

Exercises 53-55, 57-58, 80, 83, 84

Guided Practice

▶ **Ex. 1** Ask students how solutions of equations are related to the solution set of an equation. Solutions are elements of the solution set.

▶ **Ex. 2–3** Ask students how many equivalent equations a given equation can have. An infinite number.

▶ **Ex. 6** MATH JOURNAL Have students record the properties of equalities in their math journals. See page 124 for a list of the properties.

Answer

5. Divide both sides by 2, add 5 to both sides, subtract 2 from both sides, multiply both sides by 4.

Common-Error ALERT!

In solving equations, students frequently forget to do to one side of the equation or inequality what they do to the other. Stress the Golden Rule of Equations (or Inequalities) and indicate that the properties of equality are simply a detailed version of the Golden Rule.

EXERCISES

Guided Practice

CRITICAL THINKING about the Lesson

1. **True or False?** Two equations are equivalent if they have the same solution set. True
2. **True or False?** Adding the same number to both sides of an equation produces an equivalent equation. True
3. **True or False?** Multiplying both sides of an equation by the same number produces an equivalent equation. False, the number must be nonzero
4. Which of the four basic operations (addition, subtraction, multiplication, and division) are inverses of each other? Addition and subtraction, multiplication and division
5. Describe the transformation you would use to solve each of the following equations. See margin.

   $2x = 6$ $x - 5 = 11$ $2 + n = 5$ $\frac{1}{4}s = 3$
6. State the four properties of equality. See page 124.

In Exercises 7 and 8, decide whether the equations are equivalent.

7. $x + 10 = 3$ and $x = -7$ Equivalent
8. $8 - x = 4$ and $x = -4$ Not equivalent

Independent Practice

In Exercises 9–12, state the inverse operation.

9. Add 31. Subtract 31
10. Subtract 27. Add 27
11. Multiply by -2. Divide by -2.
12. Divide by 5. Multiply by 5.

In Exercises 13–24, solve the equation.

13. $4 + t = 13$ 9
14. $m + 15 = -3$ -18
15. $w - 5 = 11$ 16
16. $x - (-3) = 4$ 1
17. $y + 7 = -5$ -12
18. $6 + n = 0$ -6
19. $14 = b + 4$ 10
20. $-3 = 7 + a$ -10
21. $|-3| + n = 0$ -3
22. $|-9| + z = -5$ -14
23. $22 = s + 7$ 15
24. $37 = r - (-8)$ 29

In Exercises 25–48, solve the equation.

25. $10x = 100$ 10
26. $-21m = 42$ -2
27. $16 = -2a$ -8
28. $15b = 5$ $\frac{1}{3}$
29. $256 = 16c$ 16
30. $-4m = -16$ 4
31. $400 = 25n$ 16
32. $330 = -15p$ -22
33. $\frac{1}{2}x = -40$ -80
34. $\frac{1}{3}z = 78$ 234
35. $0 = \frac{3}{5}t$ 0
36. $-x = 4$ -4
37. $-\frac{3}{4}L = 75$ -100
38. $-\frac{4}{5}c = 20$ -25
39. $\frac{1}{3}y = 3\frac{2}{3}$ 11
40. $\frac{3}{4}z = -5\frac{1}{2}$ $-7\frac{1}{3}$ or $-\frac{22}{3}$
41. $-\frac{2}{5}x = -4$ 10
42. $-\frac{2}{3}y = -4$ 6
43. $\frac{x}{4} = 2$ 8
44. $\frac{c}{7} = 15$ 105
45. $\frac{x}{2} = -3$ -6
46. $\frac{n}{-1} = -2$ 2
47. $\frac{t}{-2} = \frac{1}{2}$ -1
48. $\frac{m}{-4} = -\frac{3}{4}$ 3

49. *Checkbook Balance* You thought the balance in your checkbook was \$53, but when your bank statement arrived, you realized that you forgot to record a check. The bank says you have a balance of \$47. Let x represent the value of the check that you forgot to record. Which of the following is a correct model for the situation? Solve the correct equation.

a. $x + 47 = 53$ 6 **b.** $x - 47 = 53$

50. *Bonus Points* Suppose that a math test that you took had 100 regular points and 10 bonus points. You received a total score of 83, which included a 7-point bonus. Let x represent the score you would have had without the bonus. Which of the following is a correct model for this situation? Solve the correct equation.

a. $x + 83 = 100$ **b.** $x + 7 = 83$ 76

Similar Triangles **In Exercises 51 and 52, the two triangles are similar. Find the length of the side marked *x*.**

51. $2\frac{1}{7}$ ft

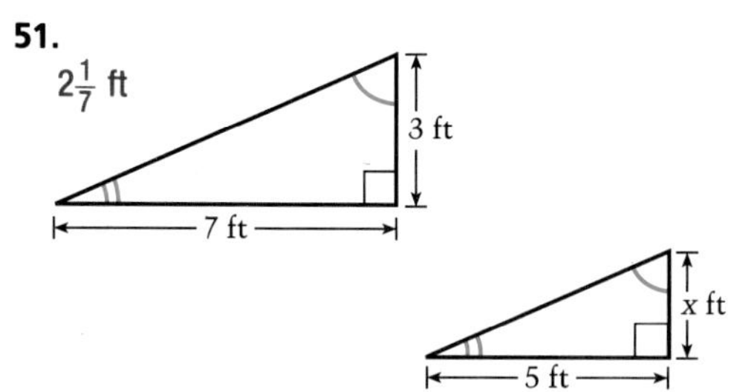

52. 7.5 cm

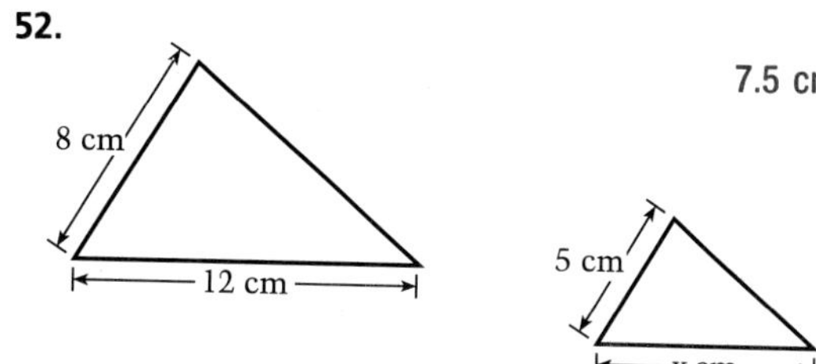

Balanced Beam **In Exercises 53 and 54, use the following information.**

For a uniform beam to balance on a fulcrum, as shown at the right, the ratio of the larger weight to the smaller weight must be equal to the ratio of the larger distance to the smaller distance.

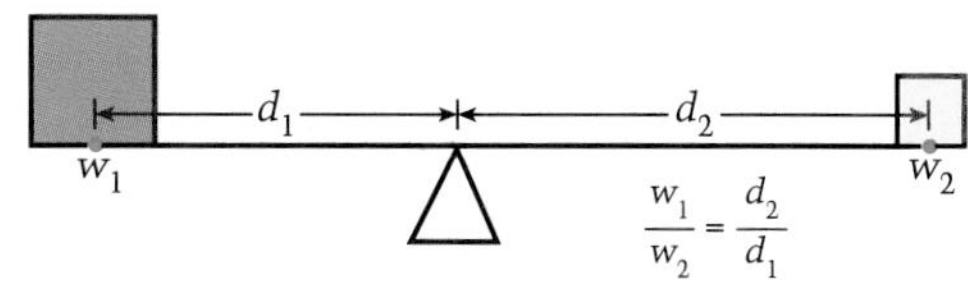

53. Two children, weighing 50 pounds and 70 pounds, are on a teeter-totter shown below. To balance, how far from the center should the 70-pound child sit? $4\frac{2}{7}$ ft

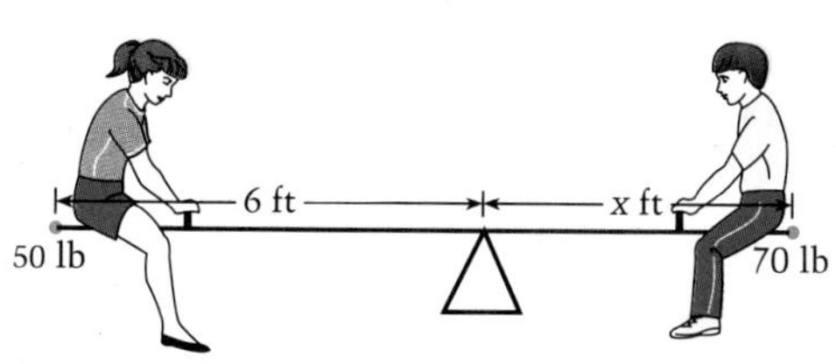

54. You want to lift a 300-pound block shown below. You weigh 100 pounds. How far from the fulcrum should you be? 6 ft

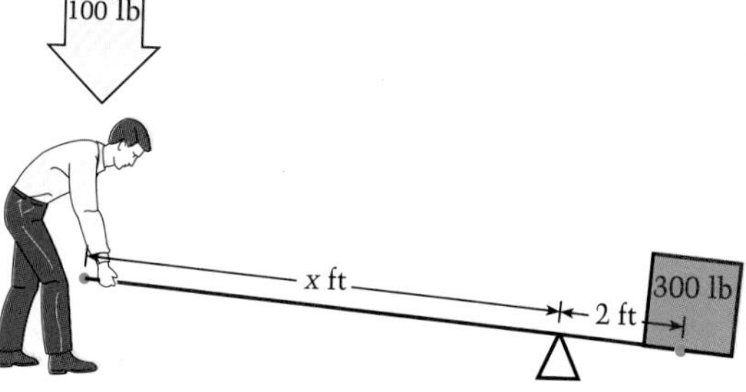

Independent Practice

▶ **Ex. 13–48** Remind students that the point of these exercises is to develop their equation-solving skills. Consequently, they should show their solution steps even when they can solve the exercise by using mental math.

▶ **Ex. 49–50, 55–58** **PROBLEM SOLVING** Recall the process that was used to express real-life situations as algebraic models: Write a verbal model, assign labels, and translate the verbal model into an equation or inequality.

Ask students if they would have given a different algebraic model in Exercises 49–50. What verbal models would they have used?

▶ **Ex. 51–52** **GEOMETRY** Recall that Example 6 of the lesson gives the relationship between two similar triangles.

▶ **Ex. 53–54** A *fulcrum* is the support on which a lever moves. In the diagram, the fulcrum is the triangle on which the seesaw rests.

Ex. 57 PROBLEM SOLVING Ask students to show the verbal model that describes the situation. One such verbal model is as follows.

New York day-care cost	= 190% •	Little Rock day-care costs

Ex. 81–84 Emphasize that expressions should be simplified before any solution steps are taken.

Enrichment Activities

WRITING Following the class discussion, students might record their personal responses in their math journals.

1. How are the terms *equal* and *equivalent* used to describe equations? In what ways are the terms alike? In what ways are the terms different? Equal equations have identical terms. Equivalent equations have identical solution sets.
2. Is it possible to solve any linear equation using only the addition and multiplication properties of equalities? Explain. Yes, subtraction is the inverse of addition, and division is the inverse of multiplication.
3. Equations such as $\frac{x}{5} = \frac{7}{10}$ are frequently solved using a technique called *cross multiplication*, in which you cross multiply 10 and x and 5 and 7 to get the equivalent equation, $10x = 5 \cdot 7$. Discuss why cross multiplication may be confusing to use when solving an equation such as $\frac{x}{5} + 2 = \frac{7}{10}$. Students often forget to multiply both the terms $\frac{x}{5}$ and 2 by 10.

55. *Baking* A recipe for holiday bread calls for 6 cups of flour and 2 cups of sugar. You have only $4\frac{1}{2}$ cups of flour. You want to reduce the amount of sugar by the same proportion. Let x represent the amount of sugar you should use. Which of the following is a correct model for this situation? Solve the correct equation.

a. $\frac{x}{2} = \frac{4\frac{1}{2}}{6}$ $1\frac{1}{2}$ **b.** $\frac{x}{2} = \frac{6}{4\frac{1}{2}}$

56. *Recycling Newspaper* You are asked to load a large pile of newspapers onto a truck. You decide to divide the pile into three smaller piles that each weigh 41 pounds. Let x represent the weight of the original pile. Which of the following is a correct model for this situation? Solve the correct equation.

a. $3x = 41$ **b.** $\frac{x}{3} = 41$ 123

57. *Day-Care Cost* In 1990, the average weekly cost of day care in New York City was 190% of the average weekly cost of day care in Little Rock, Arkansas. The average weekly cost in New York was \$95. Let x represent the Little Rock cost. Write an equation involving x and solve the equation. (*Source: Runzheimer International*) $1.90x = 95$ or $\frac{x}{95} = \frac{100}{190}$, 50

58. *Day-Care Cost* In 1990, the average weekly cost of day care in Mobile, Alabama, was $\frac{14}{29}$ of the average weekly cost of day care in Minneapolis, Minnesota. The average weekly cost in Minneapolis was \$87. Let x represent the Mobile cost. Write an equation involving x and solve the equation. $\frac{14}{29} \cdot 87 = x$, 42

Integrated Review

In Exercises 59–78, evaluate the expression.

59. $-8 - 4$ -12 **60.** $-3 + 7$ 4 **61.** $-9 + 5$ -4 **62.** $-5 + 9$ 4

63. $3(6 - 2)$ 12 **64.** $(-4 + 1)(-2)$ 6 **65.** $(7 - 11)4$ -16 **66.** $-5(8 + 2)$ -50

67. $(-12)(\frac{3}{2})$ -18 **68.** $(8)(-\frac{1}{2})$ -4 **69.** $(-5)(-11)$ 55 **70.** $(6)(-13)$ -78

71. $(-\frac{3}{2})(\frac{-2}{3})$ 1 **72.** $(\frac{4}{3})(-\frac{3}{4})$ -1 **73.** $(\frac{1}{2})(\frac{6}{4})$ $\frac{3}{4}$ **74.** $(\frac{2}{7})(\frac{7}{4})$ $\frac{1}{2}$

75. $-8 \div 2$ -4 **76.** $18 \div 6$ 3 **77.** $3.75 \div 0.25$ 15 **78.** $10.6 \div 0.4$ 26.5

Exploration and Extension

79. The perimeter of parallelogram $MATH$ below is 168. Find the lengths of its sides. 21 and 63

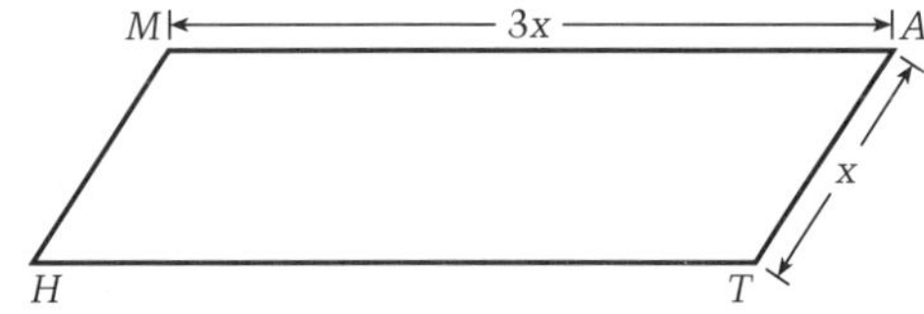

80. What is the greatest area of rectangle $QUAD$ below if x is chosen from $\{3, 5, 7\}$? 35, when $x = 5$

U
2x − 3
A
10 − x
Q
D

In Exercises 81–84, solve the equation.

81. $-(-x) = 15$ 15 **82.** $-(-3x) = -21$ -7 **83.** $-3 = x - (-7)$ -10 **84.** $-|-8| = 15 - x$ 23

3.2 Solving Equations Using Two or More Transformations

What you should learn:

Goal How to use two or more transformations to solve an equation

Why you should learn it:

You will need to use two or more transformations to solve most equations, such as formulas, that model real-life situations.

Goal Using Two or More Transformations

In Lesson 3.1, you solved equations using one transformation. Many equations, however, require two or more transformations. Here are some guidelines that can help you decide how to start. Once you have found a solution, check it in the original equation.

1. Simplify both sides of the equation (if needed).

2. Use inverse operations to isolate the variable.

Example 1 *Finding and Checking a Solution*

Solve $\frac{1}{3}x + 6 = -8$.

Solution Remember that your goal is to isolate the variable.

$\frac{1}{3}x + 6 = -8$	*Rewrite original equation.*
$\frac{1}{3}x + 6 - 6 = -8 - 6$	*Subtract 6 from both sides.*
$\frac{1}{3}x = -14$	*Simplify.*
$3\left(\frac{1}{3}x\right) = 3(-14)$	*Multiply both sides by 3.*
$x = -42$	*Simplify.*

Check Check that -42 satisfies the original equation.

$\frac{1}{3}(-42) + 6 \stackrel{?}{=} -8$	*Substitute -42 for x.*

Because both sides equal -8, the solution is correct. ■

Example 2 *Combining Like Terms First*

Solve $7x - 3x - 8 = 24$.

Solution

$7x - 3x - 8 = 24$	*Rewrite original equation.*
$4x - 8 = 24$	*Combine like terms $7x$ and $-3x$.*
$4x - 8 + 8 = 24 + 8$	*Add 8 to both sides.*
$4x = 32$	*Simplify.*
$\frac{4x}{4} = \frac{32}{4}$	*Divide both sides by 4.*
$x = 8$	*Simplify.*

The solution is 8. Check this in the original equation. ■

Problem of the Day

You can measure any whole number of grams up to 80 grams on a balance scale by using two identical sets of four weights.

a. What are the weights? The weights are 1g, 3 g, 9 g, and 27 g.

b. Look for a pattern in the weights. If you added the next logical weight to both sets, then up to how many grams could you balance? 242

ORGANIZER

Warm-Up Exercises

Provide the sequence you would follow to isolate the variable of each equation.

1. $x - 6 = 12$

2. $-13 = -5 - m$

3. $-7y = 42$ **4.** $\frac{4}{5}k = 20$

5. $2x + 3 = 23$

Note: Each sequence applies *to both sides.* 1. add 6; 2. add 5, divide by -1; 3. divide by -7; 4. multiply by $\frac{5}{4}$; 5. subtract 3, divide by 2

Lesson Resources

Teaching Tools
- Problem of the Day: 3.2
- Warm-up Exercises: 3.2
- Answer Masters: 3.2

Extra Practice: 3.2

Applications Handbook: p. 54

Technology Handbook: p. 22

LESSON Notes

Example 1

Another first step for this example is to clear all fractions. That is, multiply both sides of the equation by the least common denominator 3.

Examples 2–3, 5

Point out to students that when there is more than one term involving the variable x in the equation, they could first combine the like terms in x, then multiply, or multiply and then combine the x's. Before any variable can be isolated, it must be combined into one term.

Example 4

Compare the solution steps of this example with those of Examples 2, 3, and 5. Although quite similar, Example 4 can be solved very differently, as follows.

$$\begin{aligned} 18 - (x + 2) &= 21 \\ -(x + 2) &= 3 \\ x + 2 &= -3 \\ x &= -5 \end{aligned}$$

Because the variable x is only in one term a different sequence of solution steps was possible. Be alert to student variations.

An Alternate Approach
Using Flowcharts

Use flowcharts to model building and solving equations. These pictorial models help students recognize the sequence of steps needed to solve equations by showing the relationship between constructing equations and solving equations. See the example below.

You can use this flowchart to model the equation

$3(x - 6) + 5 = -10.$

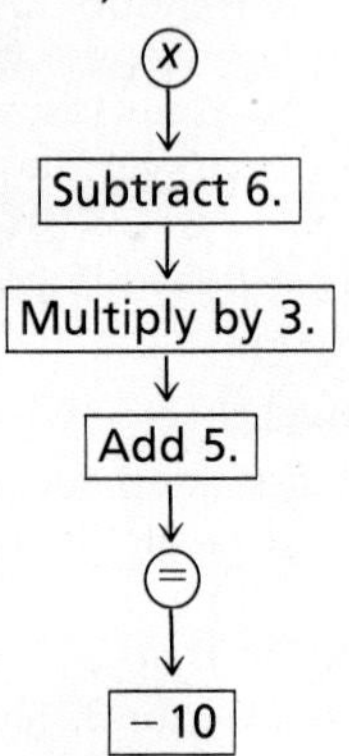

Some solution steps can be performed mentally. In Example 3, the solution on the left shows all steps. The format on the right allows you to indicate your solution process but shows less work.

Example 3 A Format for the "Expert" Equation Solver

Solve $5x + 3(x + 4) = 28$.

Solution

All Steps Shown

$$\begin{aligned} 5x + 3(x + 4) &= 28 \\ 5x + 3x + 12 &= 28 \\ 8x + 12 &= 28 \\ 8x + 12 - 12 &= 28 - 12 \\ 8x &= 16 \\ \frac{8x}{8} &= \frac{16}{8} \\ x &= 2 \end{aligned}$$

Steps Showing Process Only

$$\begin{aligned} 5x + 3(x + 4) &= 28 \\ 5x + 3x + 12 &= 28 \\ 8x + 12 &= 28 \\ 8x &= 16 && \textit{Subtract 12.} \\ x &= 2 && \textit{Divide by 8.} \end{aligned}$$

The solution is 2. Check this in the original equation. ■

Example 4 Showing the Process Only

Solve $18 - (x + 2) = 21$.

Solution

$$\begin{aligned} 18 - (x + 2) &= 21 && \textit{Rewrite original equation.} \\ 18 - x - 2 &= 21 && \textit{Opposite of a sum} \\ 16 - x &= 21 && \textit{Simplify.} \\ -x &= 5 && \textit{Subtract 16 from both sides.} \\ x &= -5 && \textit{Divide both sides by } -1. \end{aligned}$$

The solution is -5. Check this in the original equation. ■

Example 5 Showing the Process Only

Solve $-28 = 2(x + 3) - 5(x - 1)$.

Solution

$$\begin{aligned} -28 &= 2(x + 3) - 5(x - 1) && \textit{Rewrite original equation.} \\ -28 &= 2x + 6 - 5x + 5 && \textit{Distributive Property} \\ -28 &= -3x + 11 && \textit{Combine like terms.} \\ -39 &= -3x && \textit{Subtract 11 from both sides.} \\ 13 &= x && \textit{Divide both sides by } -3. \end{aligned}$$

The solution is 13. Check this in the original equation. ■

Example 6 Multiplying by a Reciprocal First

Solve $66 = -\frac{6}{5}(x + 3)$.

Solution

$$66 = -\frac{6}{5}(x + 3) \quad \text{Rewrite original equation.}$$
$$\left(-\frac{5}{6}\right)(66) = \left(-\frac{5}{6}\right)\left(-\frac{6}{5}\right)(x + 3) \quad \text{Multiply by reciprocal of } -\frac{6}{5}.$$
$$-55 = x + 3 \quad \text{Simplify.}$$
$$-58 = x \quad \text{Subtract 3 from both sides.}$$

The solution is -58. Check this in the original equation. ■

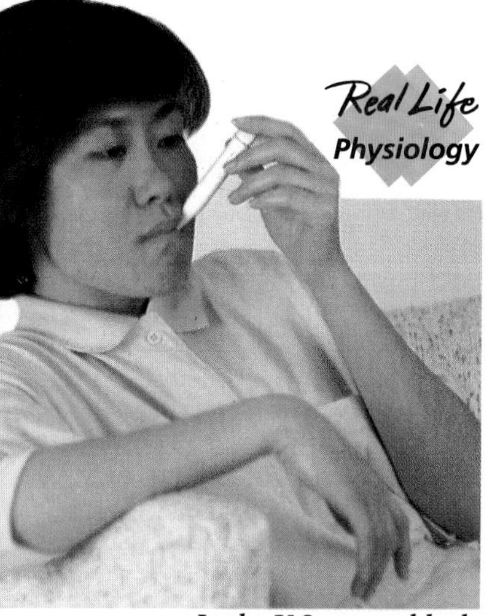

Real Life Physiology

In the U.S., normal body temperature is given as 98.6°F. Most other countries state normal body temperature in degrees Celsius.

Example 7 Normal Body Temperature

The Fahrenheit and Celsius scales are related by the formula $F = \frac{9}{5}C + 32$. What is normal body temperature on the Celsius scale? Does a temperature of 38.5°C indicate a fever?

Solution Begin by substituting the normal body temperature of 98.6° for F in the equation $F = \frac{9}{5}C + 32$. Then, solve the resulting equation for C.

$$F = \frac{9}{5}C + 32 \quad \text{Known formula}$$
$$98.6 = \frac{9}{5}C + 32 \quad \text{Substitute 98.6 for F.}$$
$$66.6 = \frac{9}{5}C \quad \text{Subtract 32 from both sides.}$$
$$\frac{5}{9}(66.6) = \frac{5}{9}\left(\frac{9}{5}C\right) \quad \text{Multiply by reciprocal of } \frac{9}{5}.$$
$$37 = C \quad \text{Simplify.}$$

Thus, on the Celsius scale, normal body temperature is 37°C. A person with a temperature of 38.5°C would have a fever. ■

Communicating about ALGEBRA

SHARING IDEAS about the Lesson

Understanding Errors Your friend has written the following solution steps, but the answers don't agree with those given in the back of the book. Is the book wrong or did your friend make mistakes? Correct any mistakes you find and explain what your friend did wrong.

A. $2(x - 3) = 5$
6 $2x - \not{3} = 5$
$2x = \not{8}$ 11
$x = \not{4}$ $\frac{11}{2}$

B. $5 - 3x = 10$
$-3x$ $\not{2x} = 10$ 5
$x = 5$ $-\frac{5}{3}$

See Additional Answers.

C. $\frac{1}{4}x - 2 = 7$
8 $x - \not{2} = 28$
$x = \not{30}$
36

Extra Examples

Here are additional examples similar to **Examples 1–6.**

Solve.

1. $4x - 9 = 15$ 6
2. $-17 = 3 + \frac{1}{2}x$ -40
3. $-13x + 8x = 23$ $-\frac{23}{5}$
4. $-2(x - 3) + 5x = 36$ 10
5. $37 = 3(2x - 4) + 5(x + 1)$ 4
6. $125 = \frac{5}{6}(x - 18)$ 168

An Alternate Approach
Building Equations

An excellent activity that assesses students' understanding of isolating the variable in solving equations is building equations. For example, given the equation $6 + (3x - 2) = 8$ and starting with x, what is the sequence of operations needed to construct this equation? The solution is to multiply x by 3, and then subtract 2 from the product. Now take the quantity and add 6 to it. This equals 8. (A change in direction in a flowchart, that is—bottom up—also allows you to go from solving equations to building equations.)

Observe that adding 6 at some other time in the sequence will result in a very different equation. Also note that the last building block in this activity is the first term of the equation that you undo, that the next-to-last building block is the second term that you undo, and so on.

Check Understanding

Build the following equations. Then, solve the equations and compare the building and solving processes.

a. $18 = 4 - 2x$ -7
b. $19 - 3(x + 4) = 7$ 0
c. $3x + \frac{1}{3}x = 27$ $\frac{81}{10}$

Communicating about ALGEBRA

COOPERATIVE LEARNING

Have students compare their solutions with classmates' in a cooperative learning group. Encourage them to explore different ways of approaching the same problems. For example, Part A can be solved in more than one way.

One way:

$$2(x-3)=5$$
$$2x-6=5$$
$$2x=11$$
$$x=\frac{11}{2} \text{ or } 5\frac{1}{2}$$

Another way:

$$2(x-3)=5$$
$$x-3=\frac{5}{2}$$
$$x=3+\frac{5}{2} \text{ or } 5\frac{1}{2} \text{ or } \frac{11}{2}$$

EXERCISE Notes

ASSIGNMENT GUIDE

Basic/Average: Ex.1–4, 5–49 odd, 53–57 odd

Above Average: Ex.1–4, 12–33 multiples of 3, 36–40 even, 41–45, 51–57 odd

Advanced: Ex.1–4, 12–33 multiples of 3, 35, 39–42, 51–58

Selected Answers

Exercises 1–4, 5–49 odd

More Difficult Exercises

Exercises 35–36, 38–39, 41–42, 53–58

Guided Practice

Ex. 4 Either a or b is an acceptable answer. However, because multiplying by the reciprocal results in a simpler computation ($\frac{3}{4}\cdot 16$ instead of distributing $\frac{4}{3}$ over the quantity $x+2$), using method b is more efficient.

EXERCISES

Guided Practice

CRITICAL THINKING about the Lesson

1. What is the first step you would use to solve $2x+1=9$? Subtract 1 from both sides.
2. Check whether -3 is a solution of $2x+4=-2$. It is.
3. Find the error(s) in the solution steps shown.

$$10x+8=-2$$
$$10x=6$$
$$x=\frac{3}{5}$$

8 was added to, instead of subtracted from, the right side.

4. For the equation $\frac{4}{3}(x+2)=16$, which of the following steps would you use first? Explain. Either, but using *b* is more efficient.
   a. Apply the Distributive Property to the left side.
   b. Multiply both sides of the equation by the reciprocal of $\frac{4}{3}$.

Independent Practice

In Exercises 5–10, check whether the number is a solution of the equation.

5. $x-10=16,\quad 6$ No
6. $\frac{1}{2}x-9=-2,\quad 14$ Yes
7. $\frac{2}{3}x+1=-5,\quad -9$ Yes
8. $4x-\frac{1}{2}=-\frac{3}{2},\quad -\frac{1}{2}$ No
9. $\frac{3x}{5}+1=7,\quad -10$ No
10. $-4-\frac{x}{9}=-8,\quad 36$ Yes

In Exercises 11–34, solve the equation.

11. $3x-7=23$ 10
12. $2x+5=9$ 2
13. $\frac{1}{4}x+2=3$ 4
14. $\frac{1}{2}x-1=-1$ 0
15. $6=14-2x$ 4
16. $9-\frac{2}{3}x=-1$ 15
17. $-4+\frac{4}{5}x=-6$ $-2\frac{1}{2}$ or $-\frac{5}{2}$
18. $3x+\frac{1}{2}=-1$ $-\frac{1}{2}$
19. $\frac{4x}{3}+5=-3$ -6
20. $16=2-\frac{2x}{5}$ -35
21. $4x-3x=9$ 9
22. $-6x+4x=2$ -1
23. $x+5x-5=1$ 1
24. $3x-7+x=5$ 3
25. $-2=\frac{2x}{3}-x$ 6
26. $-10=\frac{1}{2}x+x$ $-6\frac{2}{3}$ or $-\frac{20}{3}$
27. $3(x-2)=18$ 8
28. $12(2-x)=6$ $1\frac{1}{2}$ or $\frac{3}{2}$
29. $19=2(x+1)-x$ 17
30. $6=\frac{3}{2}x+\frac{1}{2}(x-4)$ 4
31. $3x+\frac{3}{2}(2x-1)=2$ $\frac{7}{12}$
32. $55x-3(9x+12)=-64$ -1
33. $\frac{9}{2}(x+3)=27$ 3
34. $\frac{4}{9}(2x-4)=48$ 56

35. *Hours of Labor* The bill (parts and labor) for the repair of an automobile was \$357. The cost for parts was \$285. The cost of labor was \$32 per hour. Let x represent the number of hours of labor. Write an equation involving x and solve the equation to find the number of hours of labor. $32x+285=357$, $2\frac{1}{4}$

36. *Band Fund-Raiser* Your school band sold stationery for \$12 a box. The profit, P, in dollars is given by $P=12x-3640$, where x is the number of boxes sold. If the band made a profit of \$2600, how many boxes of stationery were sold? 520

37. *Height of a Fountain* The upward velocity of the water in the stream of a particular fountain is given by the formula $v = -32t + 28$, where t is the number of seconds after the water leaves the fountain. While going upward, the water slows down until, at the top of the stream, the water has a velocity of zero. How long does it take each particle of water to reach the maximum height? $\frac{7}{8}$ second

38. *Using the Problem-Solving Plan* An office is making 6 copies of a report that has 500 pages. One photocopier can copy 300 pages in an hour. The other photocopier is faster and can copy 500 pages in an hour. To find the time it would take the two copiers together to complete the project, you could construct the following verbal model.

Rate for 1st machine	•	Time to complete	+	Rate for 2nd machine	•	Time to complete	=	Total pages

Assign labels to this model to form an equation. Then solve the equation.

39. *Time to Complete a Project* Two people are working on a project. The first can complete $\frac{1}{6}$ of the project in an hour, and the second can complete $\frac{1}{8}$ of the project in an hour. How many hours would it take each person working *separately* to complete the project? Find the number of hours it would take them *together* to complete the project by solving the following equation for t.

$\frac{1}{8}t + \frac{1}{6}t = 1$ 6 and 8, $3\frac{3}{7}$

40. *Temperature in Quito* What is the average temperature in Quito, Ecuador, in degrees Celsius? $\approx 13.9°$

Quito, Ecuador, lies in the Andes highlands where the weather is springlike all year, and the average temperature is 57°F.

Independent Practice

Ex. 11–34 A good way to diagnose students' equation-solving behavior is to tell students to begin each problem with an indication of the first solution step they took: combine, distribute, or multiply. This forces students to think before they undertake a sequence of solution steps and helps you identify possible false starts.

Ex. 35–36 PROBLEM SOLVING Remind students to begin with verbal models of the problems.

Answer

38. Rate for 1st machine = 300 (pages per hour)
Rate for 2nd machine = 500 (pages per hour)
Time to complete = t (hours)
Total number of pages = 6 • 500 (pages);
$300t + 500t = 6 \cdot 500$, $3\frac{3}{4}$

Ex. 39 PROBLEM SOLVING If $\frac{1}{6}$ of the project can be completed in 1 hour, then 6 hours are needed to complete the project. In particular, you use unit analysis, since you are given a rate, a number of hours, and the completion of the project, you can write $\frac{1}{6}\frac{\text{project}}{\text{hour}} \cdot 6 \text{ hours} = 1 \text{ project}$.
If two people work on the project, their combined efforts (the sum of their individual work) result in the completion of the project, thus giving the equation found in Exercise 39.

Ex. 40 In Example 7, the relationship between degrees Fahrenheit and Celsius is provided: $F = \frac{9}{5}C + 32$.

Enrichment Activities

These activities require students to go beyond lesson goals.

1. Solve these equations in more than one way. Explain the alternate solution steps.
   a. $120 = -\frac{4}{3}(2x + 9)$
   b. $34 - 5(x - 3) = 119$
   c. $\frac{5}{7}(3x - 14) - 4 = 31$
   a. $-\frac{99}{2}$, b. -14, c. 21

COOPERATIVE LEARNING Compare your solutions with your classmates'. Discuss the alternate solution steps.

✪ **41.** *Perimeter of a Rectangle* The length of a rectangle is 3 times its width, which means that the perimeter is given by

$$P = 2w + 2(3w),$$

where w is the width of the rectangle. The perimeter is 16 meters. What is the width? 2 m

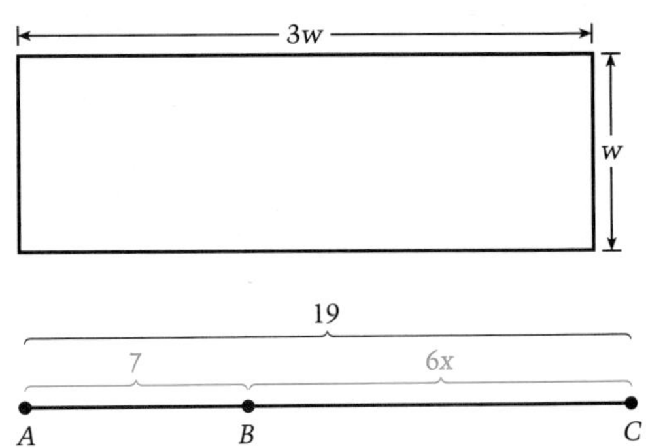

✪ **42.** *Distance on a Line* Solve for x in the figure at the right. 2

19
7
6x
A B C

Integrated Review

In Exercises 43–50, evaluate the expression.

43. $2x + 4x$, when $x = 4$ 24

44. $91y - 4 - 62y$, when $y = 2$ 54

45. $s^2 - 2s + 1$, when $s = 3$ 4

46. $42 - t^2$, when $t = 8$ -22

47. $16 + 9x - 12y - x$, when $x = 1$ and $y = 2$ 0

48. $3x - 4y + 9$, when $x = 4$ and $y = 4$ 5

49. $4.9s - 12.1 + 9.7t$, when $s = 3.2$ and $t = 1.7$ 20.07

50. $61.5 - 48.1x + 48.1y$, when $x = 2.2$ and $y = 0.3$ -29.89

Exploration and Extension

51. *College Entrance Exam Sample* If $x(p + 1) = m$, then $p =$ [?]

a. $m - 1$ **b.** m **c.** $\frac{m-1}{x}$ **d.** $m - x - 1$ **e.** $\frac{m}{x} - 1$

52. Are the solution steps at the right correct? If not, explain the error(s) and write a correct set of solution steps.
No, 12 was not subtracted from both sides.

$6x + 12 = 18$
$6x + 12 - 12 =$ ~~18~~ -12
$6x =$ ~~18~~ 6
$x =$ ~~3~~ 1

In Exercises 53–58, solve the equation two different ways. For each exercise, explain which method you prefer and why. See Additional Answers.

Sample: $\frac{1}{9}x + 1 = 4$

Method 1: Multiply first, subtract second.

$$\frac{1}{9}x + 1 = 4$$
$$x + 9 = 36$$
$$x = 27$$

Method 2: Subtract first, multiply second.

$$\frac{1}{9}x + 1 = 4$$
$$\frac{1}{9}x = 3$$
$$x = 27$$

✪ **53.** $\frac{2}{3}x + 1 = \frac{1}{3}$ -1

✪ **54.** $\frac{1}{2}x - \frac{1}{4} = \frac{3}{4}$ 2

✪ **55.** $\frac{1}{2}(3x - 7) = 4$ 5

✪ **56.** $\frac{6}{11}(x - 4) = -36$ -62

✪ **57.** $-56 = \frac{8}{9}(4 - x)$ 67

✪ **58.** $-9 = \frac{3}{7}(-2x + 5)$ 13

3.3 Solving Equations with Variables on Both Sides

What you should learn:

Goal 1 How to collect variables on one side of an equation

Goal 2 How to use algebraic models to answer questions about real-life situations

Why you should learn it:

You will need to write and solve equations with the variable on both sides for some real-life situations such as comparing costs.

Goal 1 Collecting Variables on One Side

Some equations have variables on both sides.

$73 - 5x = 7x - 11$, $7(2 - m) = 3m + 4$,
$-7a = -12a - 65$

The strategy for solving such equations is to *collect like variables* on the same side. Some people prefer to collect like variables on the left side, but we prefer the following guideline which avoids a negative coefficient.

- Collect variables on the side with the greater variable coefficient.

For instance, in the equation $x + 4 = 2x - 6$, the term $2x$ has a greater coefficient than x, so you can avoid a negative coefficient by collecting variables on the right.

Collect Variables on Right (Subtract x from both sides.)	Collect Variables on Left (Subtract $2x$ from both sides.)
$x + 4 = 2x - 6$	$x + 4 = 2x - 6$
$x + 4 - x = 2x - 6 - x$	$x + 4 - 2x = 2x - 6 - 2x$
$4 = x - 6$	$-x + 4 = -6$
$10 = x$	$-x = -10$
	$x = 10$

Example 1 *Solving an Equation with Variables on Both Sides*

Solve $7x + 19 = -2x + 55$.

Solution

$7x + 19 = -2x + 55$	*Rewrite original equation.*
$2x + 7x + 19 = 2x - 2x + 55$	*Add 2x to both sides.*
$9x + 19 = 55$	*Simplify.*
$9x + 19 - 19 = 55 - 19$	*Subtract 19 from both sides.*
$9x = 36$	*Simplify.*
$\frac{9x}{9} = \frac{36}{9}$	*Divide both sides by 9.*
$x = 4$	*Simplify.*

The solution is 4. Check this in the original equation. ■

Problem of the Day

3 cubes and 1 cone balance on a scale with 12 marbles. If 1 cone balances with 1 cube and 8 marbles, then how many marbles balance with 1 cone?
9

ORGANIZER

Warm-Up Exercises

1. Use the distributive property to multiply.
   a. $-3(4x + 2)$ $-12x - 6$
   b. $(-6x + 8)7$ $-42x + 56$
   c. $9(4 - 2x)$ $36 - 18x$
   d. $\frac{2}{3}(12 + 15x)$ $8 + 10x$
2. Find the additive inverse of each expression. Combine first, if necessary.
   a. 13 -13
   b. -17 17
   c. $3x$ $-3x$
   d. $-11x$ $11x$
   e. $-7x + 4x$ $3x$
   f. $17x - 6x$ $-11x$

Lesson Resources

Teaching Tools
Problem of the Day: 3.3
Warm-up Exercises: 3.3
Answer Masters: 3.3
Extra Practice: 3.3

LESSON Notes

Collecting the variable terms is another example of isolating the variable by first combining like terms.

The decision to collect variables on the side having the larger coefficient should be made after the expressions on either side of the equation have been rewritten as sums of terms. For example, $4 - 3x = 12 - 6x$ can be rewritten as $4 + (-3)x = 12 + (-6)x$, in which case -3 is the larger coefficient. Expert equation solvers can identify the larger coefficient mentally. You may now transform the equation by adding $6x$ to both sides, getting $4 + 3x = 12$.

Example 1

The larger coefficient is 7, so add $2x$ to both sides of the equation. Tell students that if you add $2x$ to the left of the left-hand side of the equation, it is customary to add $2x$ to the left of the right-hand side of the equation as well.

Examples 2–3

Students may need to write the subtractions as additions in order to decide which side of the equation has the greater variable term. For example, rewrite $80 - 6y = 4y$ as $80 + (-6)y = 4y$ in order to demonstrate that $4y$ is greater than $-6y$.

Example 4

To decide which membership plan is better, you can use the inequality $3x + 50 < 8x$ to determine when the cost of joining the video-game club is less than the cost of renting games from the video store.

Common-Error ALERT!

In **Example 3,** caution students about properly applying the distributive property. Remind them that the factor must be distributed to each of the terms in parentheses and that the product of a negative times a negative will be positive.

In the next example, the steps that can be done with mental math are not shown.

Example 2 *Showing the Process Only*

Solve $80 - 6y = 4y$.

Solution

$80 - 6y = 4y$	*Rewrite original equation.*
$80 = 10y$	*Add 6y to both sides.*
$8 = y$	*Divide both sides by 10.*

The solution is 8. Check this in the original equation. ■

The three equations in the next example are a little tricky. Be sure you can follow the steps that are shown.

Example 3 *Solving More Complicated Equations*

a.

$4(1 - x) + 3x = -2(x + 1)$	*Original equation*
$4 - 4x + 3x = -2x - 2$	*Distributive Property*
$4 - x = -2x - 2$	*Combine like terms.*
$4 + x = -2$	*Add 2x to both sides.*
$x = -6$	*Subtract 4 from both sides.*

The solution is -6. Check this in the original equation.

b.

$\frac{1}{4}(12x + 16) = 10 - 3(x - 2)$	*Original equation*
$3x + 4 = 10 - 3x + 6$	*Distributive Property*
$3x + 4 = 16 - 3x$	*Combine like terms.*
$6x + 4 = 16$	*Add 3x to both sides.*
$6x = 12$	*Subtract 4 from both sides.*
$x = 2$	*Divide both sides by 6.*

The solution is 2. Check this in the original equation.

c.

$7(2 - x) = 5x$	*Original equation*
$14 - 7x = 5x$	*Distributive Property*
$14 = 12x$	*Add 7x to both sides.*
$\frac{14}{12} = x$	*Divide both sides by 12.*
$\frac{7}{6} = x$	*Reduce fraction.*

The solution is $\frac{7}{6}$. Check this in the original equation. ■

Goal 2 Using Algebraic Models

At most clubs, members get the club's services more cheaply than nonmembers, but before joining, you should decide how much you will use these services.

Example 4 *Using the Problem-Solving Plan*

A local store charges \$8 to rent a video game for three days. You must be a member to rent from the store, but the membership is free. A video-game club in town charges only \$3 to rent a game for three days, but membership in the club is \$50 a year. Which membership is more economical?

Solution To decide which is more economical, find the number of rentals for which the two plans would cost the same.

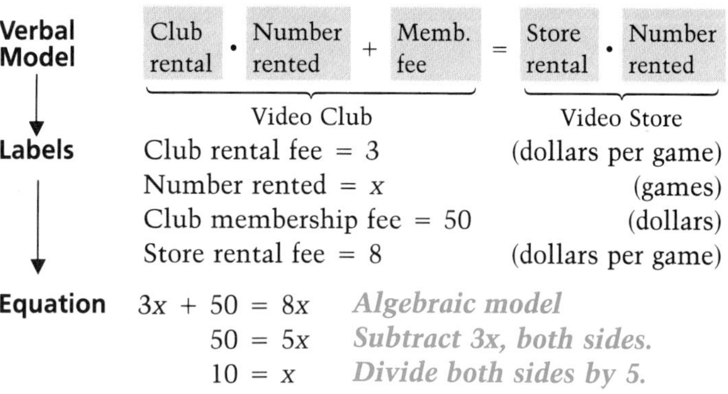

Verbal Model $\boxed{\text{Club rental}} \cdot \boxed{\text{Number rented}} + \boxed{\text{Memb. fee}} = \boxed{\text{Store rental}} \cdot \boxed{\text{Number rented}}$

Video Club — Video Store

Labels

Club rental fee $= 3$	(dollars per game)
Number rented $= x$	(games)
Club membership fee $= 50$	(dollars)
Store rental fee $= 8$	(dollars per game)

Equation

$3x + 50 = 8x$ *Algebraic model*

$50 = 5x$ *Subtract 3x, both sides.*

$10 = x$ *Divide both sides by 5.*

By 1990, Nintendo, Sega Genesis, Gameboy, and Atari had produced over 2000 video games.

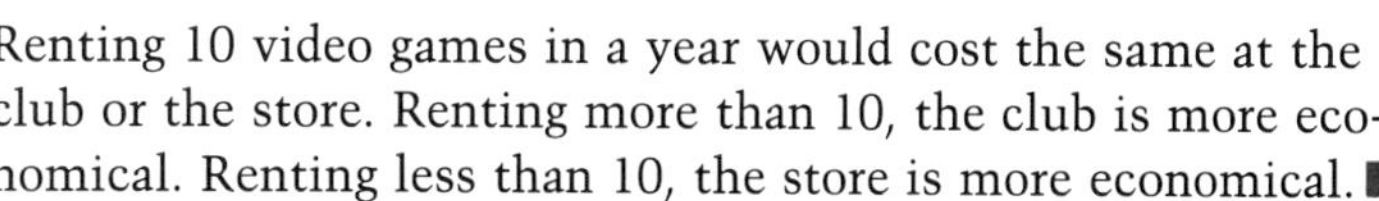

Renting 10 video games in a year would cost the same at the club or the store. Renting more than 10, the club is more economical. Renting less than 10, the store is more economical. ■

Communicating about ALGEBRA

SHARING IDEAS about the Lesson

Understanding the Process

A. Write the steps you use to solve $-3x + 5 = 9x - 19$. Beside each step, write an explanation of the step. Then show how to check your answer. A. See Additional Answers.

B. Which step would you do first to solve the equation $6y - (3y - 6) = -14 + 5y$? Explain your answer. Write $-(3y - 6)$ as $-3y + 6$ to combine like terms

Extra Examples

Here are additional examples similar to **Examples 1–3.**

1. Solve.

a. $9x - 8 = 4x + 12$ 4

b. $13 + 3x = -6x + 4$ -1

c. $13x + 26 = 2x + 62$ $\frac{36}{11}$

d. $27 - 8x = 31 - 16x$ $\frac{1}{2}$

2. Solve.

a. $3(5 + x) = -9$

b. $\frac{1}{3}(15x - 24) = 8 + 4(x - 3)$

c. $7x - 2(4 - 3x) = -3(2x + 1)$

d. $9 - 6(x + 3) = 4x - 8$

a. -8, b. 4, c. $\frac{5}{19}$, d. $-\frac{1}{10}$

Check Understanding

Build the following equations. Then, solve the equations and compare the building and solving processes. (See An Alternate Approach on page 131.)

a. $-8 + 13b = 15 - 3b$ $\frac{23}{16}$

b. $8(x - 4) = 3(5x - 2)$ $-\frac{26}{7}$

c. $(3m - 1)7 = 8m + 12$ $\frac{19}{13}$

***Communicating* about ALGEBRA**

EXTEND *Communicating*
USING FLOWCHARTS

A good way to explore students' understanding of the equation-solving process is to have them construct a flowchart for solving linear equations. (See Enrichment Activities on page 140.) Students should compare their flowcharts with classmates', discuss differences and similarities, and revise the flowcharts based upon those discussions. The completed flowcharts should be included in their math journals.

EXERCISE Notes

ASSIGNMENT GUIDE

Basic/Average: Ex.1–12, 15–36 multiples of 3, 37, 39–40, 45–59 odd

Above Average: Ex.1–12, 21–35 odd, 37–38, 41–42, 45–54 multiples of 3, 55–63 odd

Advanced: Ex.1–12, 31–40, 55–63

Selected Answers

Exercises 1–6, 7–53 odd

Use **Mixed Review** as needed.

More Difficult Exercises

Exercises 37, 39, 41–42, 61–63

Guided Practice

Use these as oral exercises for students to discuss in class prior to beginning Independent Practice.

Answers

1. False; the answer to Exercise 1 may vary. The approach used in the text is to collect the variables on the side with the larger variable coefficient. However, students who provide a rationale for the alternate approach should be supported in that decision.
2. True; subtract x from both sides, and $0 = x$ is the result.
3. Use the Distributive Property; $81 - 9x = 4x - 10$.
4. Add $9x$ to both sides; $81 = 13x - 10$.
5. Add 10 to both sides; $91 = 13x$.
6. $x = 7$; substitute 7 for x in the original equation and simplify.
7. Subtract x from both sides.
8. Add $4x$ to both sides.
9. Add 18 to both sides.
10. Add 16 to both sides.
11. Use the Distributive Property.
12. Use the Distributive Property.

EXERCISES

Guided Practice

CRITICAL THINKING about the Lesson See margin.

1. **True or False?** When solving an equation with variables on both sides, it is convenient to collect the variables on the side with the *smaller* variable coefficient. Explain.
2. **True or False?** The solution of $x = 2x$ is zero. Explain.

Exercises 3–6 refer to solving the equation $(9 - x)9 = 4x - 10$.

3. After rewriting the equation, what would be your first step? Perform this step.
4. Once you obtain $81 - 9x = 4x - 10$, should you subtract $4x$ from both sides or add $9x$ to both sides? Why? Perform this step.
5. Once you obtain $81 = 13x - 10$, what should the next step be? Perform this step.
6. Solve for x. Check your result in the original equation.

Independent Practice

In Exercises 7–12, describe the first step you would use to solve the equation. See margin.

7. $x + 2 = 3x - 1$
8. $-4x + 7 = 2x - 5$
9. $6y - 18 = 12$
10. $-16 + 4y = 11$
11. $7(1 - y) + 4y = -3(y + 2)$
12. $10(2y - 4) = -1(6 - 9y) + 3y$

In Exercises 13–36, solve the equation. Check your result.

13. $4x + 27 = 3x$ -27
14. $8y + 14 = 6y$ -7
15. $-2m = 16m - 9$ $\frac{1}{2}$
16. $7n = -35n - 6$ $-\frac{1}{7}$
17. $12c - 4 = 4c$ $\frac{1}{2}$
18. $-25d + 10 = 15d$ $\frac{1}{4}$
19. $5r + 6 = -14r - 13$ -1
20. $7s - 8 = 10s + 1$ -3
21. $12p - 7 = -3p + 8$ 1
22. $-6q + 8 = 4q - 12$ 2
23. $-7 + 4m = 6m - 5$ -1
24. $-9 + 13g = 11 - g$
25. $8 - 9t = 21t - 17$ $\frac{5}{6}$
26. $20 + 8r = -4 + 5r$ -8
27. $6(3 - x) = 3x$ 2
28. $-2(x - 5) = -x$ 10
29. $(-4 + y)10 = 2y$ 5
30. $(-6a - 2)4 = 16a$ $-\frac{1}{5}$
31. $9(b - 4) = 5(3b - 2)$
32. $-4(3 - n) = 11(4n - 3)$ $\frac{21}{40}$
33. $\frac{1}{2}(8n - 2) = 16 - 30n$ $\frac{1}{2}$
34. $\frac{1}{3}(42 - 18z) = 2(8 - 4z)$ 1
35. $\frac{1}{4}(100 + 36s) = 15 - 4s$ $-\frac{10}{13}$
36. $\frac{2}{3}(24t - 9) = 8t + 23$

37. *Bike Safety* Suppose you live near a park that has a bike trail you like to ride. The Park Department rents a bike with safety equipment for \$5 a day. If you provide your own safety equipment, the bike rental is \$3 a day. You could buy the equipment at a sports store for \$28. How many times must you use the trail to justify buying your own safety equipment? 15 or more

24. $1\frac{3}{7}$ or $\frac{10}{7}$
31. $-4\frac{1}{3}$ or $-\frac{13}{3}$
36. $3\frac{5}{8}$ or $\frac{29}{8}$

38. *Catching Up with Sis* Kate is always reminding her younger brother Tony that she is taller than he is. Kate is 63 inches tall and is growing at a rate of $\frac{1}{3}$ inch per year. Tony is 60 inches tall and is growing at the rate of $2\frac{1}{3}$ inches per year. To find how long it will take for Tony to catch up with Kate, assign labels to the following verbal model, write the algebraic model, and solve it. See margin.

Tony's height now	+	Tony's rate of growth	•	Number of years	=	Kate's height now	+	Kate's rate of growth	•	Number of years

Summer Swimming **In Exercises 39 and 40, use the following information.**

A new swimming pool is opening for 15 weeks during the summer. You can swim in the afternoon for \$3 or buy a membership for \$80 and pay only \$1 for the afternoon session. Let s equal the number of times you go swimming. You are trying to decide whether to buy a membership or buy daily passes.

39. Which equation is a correct model for this problem?

a. $s + 80 = 3s$ **b.** $3s + 80 = s$ **c.** $3s = s - 80$

40. Solve the correct equation in Exercise 39 and interpret the solution. 40; if you go swimming more than 40 times, you will benefit by buying a membership.

Remote-Control Cars **In Exercises 41 and 42, use the following information.**

The Fast Track company manufactures toy remote-control race cars, which it sells for \$18 each. The production cost for the company is \$2000 per day plus \$13 per race car. Let n be the number of race cars sold in a day. The company needs to know how many cars it must sell in a day to break even.

41. Which equation is a correct model for this problem?

a. $18n + 2000 = 13n$ **b.** $18n = 13n - 2000$ **c.** $18n = 13n + 2000$

42. Solve the correct equation in Exercise 41 and interpret its solution. 400; the Fast Track company must sell an average of 400 cars a day to break even.

Integrated Review

In Exercises 43–48, simplify the expression.

43. $x - 2(x - 1)$ $-x + 2$

44. $4 - 3(2x - 2)$ $10 - 6x$

45. $2(6 + 3x) - 8$ $4 + 6x$

46. $-4(2 + 5y) + 10$ $2 - 20y$

47. $7x[2 - (-5)] - 3x$ $46x$

48. $-4m[-2 - (-6)] - 11(m + 1)$ $-27m - 11$

In Exercises 49–54, decide whether the number is a solution of the equation.

49. $8(2 - x) = -4x$, 4 Yes

50. $-6(7 - a) = 4a$, 5 No

51. $\frac{1}{3}x + 10 = -12$, 7 No

52. $12 - \frac{1}{4}x = 32$, -80 Yes

53. $\frac{1}{2}x - 8 = \frac{1}{2}(x - 8)$, 18 No

54. $27 + 6x = 3x - 9$, -6 No

Independent Practice

EXTEND Ex. 7–36 Ask students to state the steps they would use to solve the equation.

Ex. 37–42 PROBLEM SOLVING Point out to students that the problem-solving model is used in these exercises. Ask students to state what the steps in the model are.

Ex. 37 Be sure students understand that the phrase *justify buying* is related to the break-even point.

Ex. 39 This decision can also be made using an inequality such as $3s > s + 80$, which determines when paying \$3 per day is greater than the cost of becoming a member.

Answer

38. Tony's height now = 60 (inches)
Tony's rate of growth = $2\frac{1}{3}$ (inches per year)
Kate's height now = 63 (inches)
Kate's rate of growth = $\frac{1}{3}$ (inches per year)
Number of years = y (years);
$60 + 2\frac{1}{3}y = 63 + \frac{1}{3}y$, $1\frac{1}{2}$ years

Answers

57. $-3y - 36 = 51$
$-3y = 87$
$y = -29$

58. $b \cdot 8 - 32 = -12$
$b \cdot 8 = 20$
$b = \frac{5}{2}$ or $2\frac{1}{2}$

59. $24y = 2y - 6$
$22y = -6$
$y = -\frac{3}{11}$

60. $10c = -30$
$c = -3$

▶ **Ex. 61–62** These questions foreshadow ideas in Chapter 7.

Enrichment Activities

These activities require students to go beyond lesson goals.

You can use this flowchart to model solving the equation $5x + 2 = 3x + 6$.

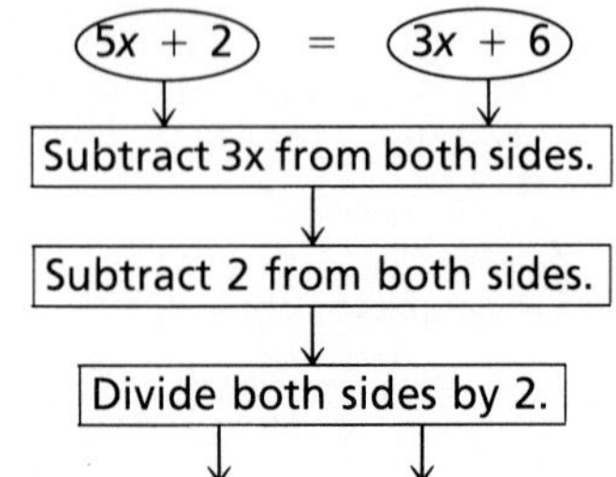

Use flowcharts to solve the following equations.

1. $\frac{2}{5}(25t - 30) = 6t - 17$ $-\frac{5}{4}$
2. $7 - 9t = 27t - 14$ $\frac{7}{12}$
3. $6(x + 3) = 8(6 - 4x)$ $\frac{15}{19}$

140

Exploration and Extension

In Exercises 55–60, find any mistakes in the solution steps. If there are none, write *correct*. If there is a mistake, find the correct solution. 57.–60. See margin.

55. $20x - 4 = 6$
$20x - 4 + 4 = 6 + 4$
$20x = 10$
$x = 2$ $x = \frac{1}{2}$

56. $-15 = 7x + 13$
$-15 - 13 = 7x + 13 - 13$
$\frac{-28}{7} = \frac{7x}{7}$
$x = -4$ Correct

57. $-3(y + 12) = 51$
$-3y + 12 = 51$
$-3y = 39$
$y = -13$

58. $(b - 4)8 = -12$
$b - 32 = -12$
$b = 20$

59. $8y(4 - 1) = 2y - 6$
$24y = 6y - 6$
$30y = -6$
$y = -\frac{1}{5}$

60. $(c + 6)7 = (4 - c)3$
$7c + 42 = 12 - 3c$
$10c = 30$
$c = 3$

In Exercises 61 and 62, try to solve the equation for *x*. Does the equation have no solution, exactly one solution, or many solutions?

✪ **61.** $3(4 - x) = -2x + 12 - x$ $0 = 0$, many solutions

✪ **62.** $8x - [4 - (-2)] = 8x$ $-6 = 0$, no solutions

✪ **63.** Solve for *a*, *b*, *c*, and *d*.

$$\begin{bmatrix} 2a - 3 & -4b + 2 \\ -17c + 7 & 3d - 19 \end{bmatrix} = \begin{bmatrix} -4a + 7 & 6b - 23 \\ 93c - 153 & 14 - 3(d - 1) \end{bmatrix}$$

$a = 1\frac{2}{3}$ or $\frac{5}{3}$, $b = 2\frac{1}{2}$ or $\frac{5}{2}$, $c = 1\frac{5}{11}$ or $\frac{16}{11}$, $d = 6$

Mixed REVIEW

1. Write 5% in fraction form. **(1.1)** $\frac{1}{20}$
2. Write 67% in decimal form. **(1.1)** 0.67
3. What is the quotient of 27 and 3? **(1.1)** 9
4. What is the product of $\frac{4}{3}$ and 9? **(1.1)** 12
5. Simplify $\frac{1}{3}(6 - 2x) - 4(1 + \frac{1}{3}x)$. **(2.3, 2.6)** $-2 - 2x$
6. Simplify $(3x \div \frac{1}{2}) \div 6$. **(1.2)** x
7. Use exponents to write $3 \cdot 3 \cdot x \cdot x \cdot x$. **(1.3)** $3^2 \cdot x^3$
8. What is the absolute value of -32? **(2.1)** 32
9. What is the reciprocal of $-\frac{2}{7}$? **(2.7)** $-\frac{7}{2}$
10. What is the opposite of 15.3? **(2.1)** -15.3
11. Are 3 and 4 solutions of $3x - 4 \le 6$? **(1.5)** 3 is, 4 is not.
12. Evaluate $\frac{1}{2}(y + 6) \cdot 4y$ when $y = 1$. **(1.2)** 14
13. Write an expression for the perimeter of a rectangle whose length is $3\frac{1}{2}$ times its width. **(1.2)** $w + 3\frac{1}{2}w + w + 3\frac{1}{2}w$ or $9w$
14. Insert grouping symbols so that $-9 + 4 \cdot x - 6$ has a value of -25 when $x = 11$. **(1.1, 1.2)** $(-9 + 4) \cdot (x - 6)$

In Exercises 15 and 16, translate into expressions, equations, or inequalities.

15. Two less than the sum of 3 times a number is less than 12. **(1.6)** $3x - 2 < 12$
16. A number plus the product of 5 and another number. **(1.6)** $x + 5y$

3.4 Linear Equations and Problem Solving

What you should learn:

Goal How to use a problem-solving plan for problems that fit linear models

Why you should learn it:

You can use linear models for many real-life situations such as comparing two quantities that are changing at different rates.

Goal Using a Problem-Solving Plan

In Lesson 1.7, you studied a general problem-solving plan.

Write a verbal model. → Assign labels. → Write an algebraic model. → Solve algebraic model. → Answer the question.

Remember that talking about a problem with another person is a good way to learn to form verbal models and to assign labels.

Example 1 The Gazelle and the Cheetah

A gazelle can run 73 feet per second for several minutes. A cheetah can run faster, but can only sustain its top speed for about 20 seconds. Gazelles seem to have an instinct for this difference because they will not run from a prowling cheetah until it enters their "safety zone." This is the distance the cheetah would need to run to overtake the gazelle in 20 seconds if both are running at top speed. How close should the gazelle let the cheetah come before it runs?

Solution During a 20-second run, the cheetah can travel (88)(20) feet, and the gazelle can travel (73)(20) feet. For the gazelle to outdistance the cheetah, it must start with a buffer distance.

Verbal Model: Gazelle's distance + Buffer distance = Cheetah's distance

Labels:
Gazelle's distance $= (73)(20)$ (ft)
Cheetah's distance $= (88)(20)$ (ft)
Buffer distance $= x$ (ft)

Equation:
$(73)(20) + x = (88)(20)$ *Algebraic model*
$1460 + x = 1760$ *Simplify.*
$x = 300$ *Subtract 1460 from both sides.*

Thus, the gazelle should not let the cheetah get closer than 300 feet (the length of a football field). ■

The long body and slender legs of the cheetah make it the fastest animal for running short distances. It can reach a speed of about 88 feet per second.

Problem of the Day

Write an expression for the whole numbers from 1 through 20 using *exactly* four fours. The expressions can include grouping symbols, addition, subtraction, multiplication, and division. Here are some examples.

$(4 + 4) \div (4 + 4) = 1$
$(4 \div 4) + (4 \div 4) = 2$
$(4 + 4 + 4) \div 4 = 3$

Sample answers:
$4 = 4 + [4 \cdot (4 - 4)]$,
$5 = [(4 \cdot 4) + 4] \div 4$,
$6 = 4 + [(4 + 4) \div 4]$,
$7 = \frac{44}{4} - 4$,
$8 = [(4 \cdot 4) - 4] - 4$,
$9 = (\frac{4}{4} + 4) + 4$,
$10 = \frac{44 - 4}{4}$, $11 = \frac{4}{.4} + \frac{4}{4}$,
$12 = \frac{44 + 4}{4}$, $13 = \frac{4 - .4}{.4} + 4$,
$14 = \frac{4(.4) + 4}{.4}$, $15 = \frac{44}{4} + 4$,
$16 = 4 + 4 + 4 + 4$,
$17 = 4 \cdot 4 + \frac{4}{4}$,
$18 = 4(4.4) + .4$,
$19 = \frac{(4 + 4) - .4}{.4}$,
$20 = 4(\frac{4}{4} + 4)$

ORGANIZER

Warm-Up Exercises

Solve.

1. $55 + 2x = (24)(15)$
2. $1479 - 35k = 2567 + 13k$
3. $\frac{2}{7} + 4x = 3\frac{3}{5}$

1. 152.5, 2. $-22\frac{2}{3}$, 3. $\frac{29}{35}$

Lesson Resources

Teaching Tools
Transparency: 3
Problem of the Day: 3.4
Warm-up Exercises: 3.4
Answer Masters: 3.4
Extra Practice: 3.4

LESSON Notes

Example 1

A gazelle can maintain its speed for several minutes. After 20 seconds, a cheetah's speed will start decreasing, thus allowing the buffer distance between the animals to increase.

Example 2

Another way to view the decrease in enrollment at Cleveland High is to express the rate as a negative value. The algebraic model would be similar: $-75x + 3150 = 60x + 2475$.

Example 3

A similar question can be asked about how long each picture should be, but more information would be needed.

Extra Examples

Here are additional examples similar to **Examples 1–3.**

1. Two joggers are running on a 10 kilometer course. One jogs at 8 kilometers per hour, and the other runs at 12 kilometers per hour.
   a. If the first jogger has a 3-kilometer head start, can the second jogger catch the first? Yes. Let t = time in hours; Solve $8t + 3 = 12t$; $t = \frac{3}{4}$ hour.
   b. How far will each have run when they are side by side? 9 km
2. You need to display four graphs on a sheet of poster board measuring 84 cm wide. If you decide to have a uniform border of 8 cm around the edge of the poster board and to space the graphs 4 cm apart, how wide should you make each graph in order to fit all four in a row on the poster board? Suggest that the students draw a picture. Sample equation: $4x + 12 = 84 - 16$; 14 cm wide

Example 2 ***High School Enrollments***

Cleveland High is in a city, and West Lake High is in one of the city's suburbs. Cleveland High's enrollment has been decreasing at an average rate of 75 students per year, whereas West Lake High's enrollment has been increasing at an average rate of 60 students per year. Cleveland High has 3150 students, and West Lake High has 2475. If enrollments continue to change at the same rates, when will the two schools have the same number of students?

Solution For each school, the change in enrollment is given by the rate per year times the number of years. Because Cleveland High's enrollment is decreasing, you subtract the change, and because West Lake High's enrollment is increasing, you add the change.

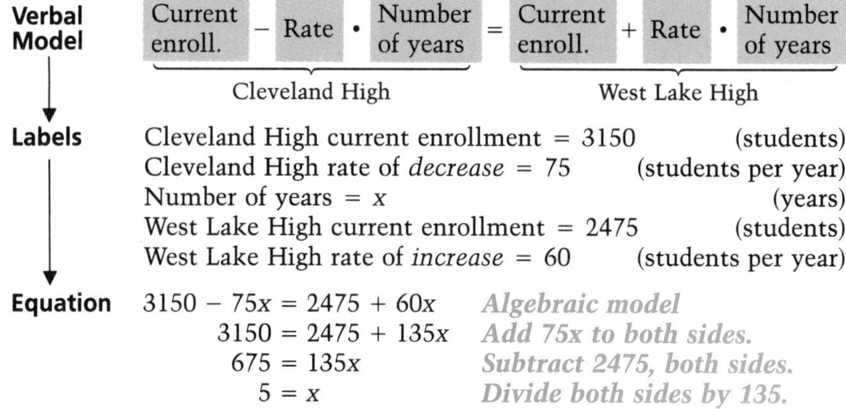

Verbal Model

[Current enroll.] − [Rate] • [Number of years] = [Current enroll.] + [Rate] • [Number of years]

(Cleveland High) (West Lake High)

Labels

Cleveland High current enrollment = 3150 (students)
Cleveland High rate of *decrease* = 75 (students per year)
Number of years = x (years)
West Lake High current enrollment = 2475 (students)
West Lake High rate of *increase* = 60 (students per year)

Equation

$3150 - 75x = 2475 + 60x$ *Algebraic model*
$3150 = 2475 + 135x$ *Add 75x to both sides.*
$675 = 135x$ *Subtract 2475, both sides.*
$5 = x$ *Divide both sides by 135.*

The schools will have the same enrollments in five years. ■

The enrollment information in Example 2 is a good candidate for a bar graph. Notice how the bar graph shown below helps you "see" the changing enrollments at the two schools.

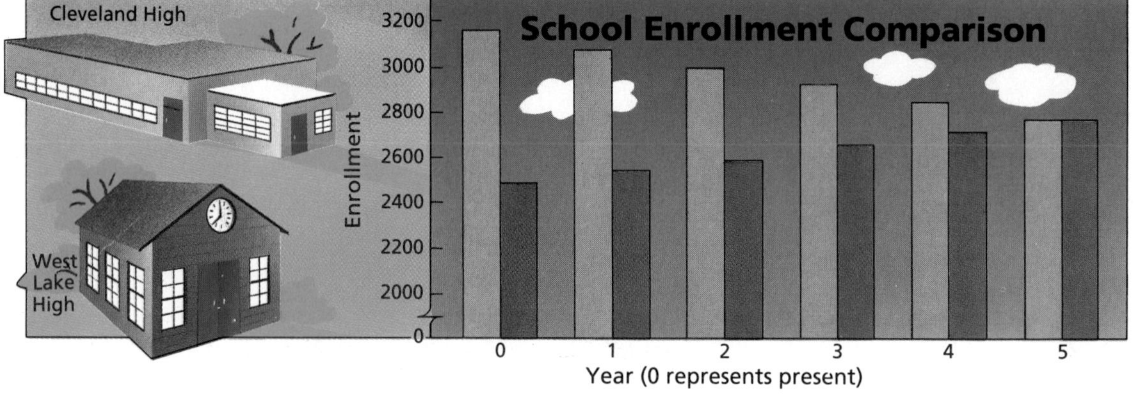

Example 3 *Designing a High School Yearbook*

The page of your school yearbook is $8\frac{1}{2}$ inches by 11 inches. The left and right margins are $\frac{3}{4}$ inch and $2\frac{7}{8}$ inches, respectively. The space between pictures is $\frac{3}{16}$ inch. How wide should you make each picture to fit three across the page?

Solution The total page width is made up of the widths of the left and right margins, 2 spaces, and 3 picture widths.

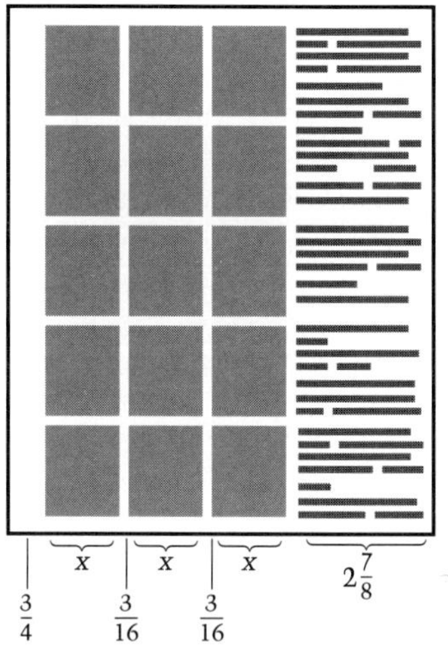

Verbal Model $\boxed{\text{Margin widths}} + 2 \cdot \boxed{\text{Picture space}} + 3 \cdot \boxed{\text{Picture width}} = \boxed{\text{Total width}}$

Labels

Margin width $= \frac{3}{4} + 2\frac{7}{8} = \frac{3}{4} + \frac{23}{8} = \frac{29}{8}$ (inches)

Picture-space width $= \frac{3}{16}$ (inches)

Picture width $= x$ (inches)

Total width $= 8\frac{1}{2}$ (inches)

Equation

$\frac{29}{8} + 2\left(\frac{3}{16}\right) + 3x = 8\frac{1}{2}$ *Algebraic model*

$\frac{29}{8} + \frac{3}{8} + 3x = \frac{17}{2}$ *Rewrite fractions.*

$\frac{32}{8} + 3x = \frac{17}{2}$ *Simplify $\frac{29}{8} + \frac{3}{8}$.*

$3x = \frac{17}{2} - \frac{32}{8}$ *Subtract $\frac{32}{8}$, both sides.*

$3x = \frac{17}{2} - \frac{8}{2}$ *Like denominators*

$3x = \frac{9}{2}$ *Simplify $\frac{17}{2} - \frac{8}{2}$.*

$x = \frac{3}{2}$ *Divide both sides by 3.*

Thus, each picture should be $1\frac{1}{2}$ inches wide. ■

Communicating about ALGEBRA

SHARING IDEAS about the Lesson

Extend the Problem Consider the page-design problem above.

A. The margins at the top and bottom of the page are $\frac{7}{8}$ inch each. The vertical space between pictures is $\frac{3}{16}$ inch. How high is each picture? $1\frac{7}{10}$

B. What is the width-to-height ratio of each photo? $\frac{15}{17}$

C. Is this the same ratio as a photo that is 8 inches by 10 inches? If not, will you have to "crop" (trim) photos that are reduced from 8-by-10 originals? No, Yes

Check Understanding

1. What are the key steps in the general problem-solving plan? See Lesson 1.7.
2. COOPERATIVE LEARNING Are any steps easier to complete when you work with others? Explain.
3. What are the verbal indicators that tell you the relationship among labels of a verbal model should be expressed as an inequality instead of as an equality? is less than, is greater than, at least, at most, etc.
4. Can information obtained from solving an inequality be obtained from solving a related equality? Explain. Yes; assign the proper inequality symbol to the solution of the equation.

Communicating about ALGEBRA

Require students to provide a verbal model of Part A. Discuss its similarities to the verbal model in Example 3.

EXTEND *Communicating*
WRITING
Ask students to create other questions that may be asked about this yearbook problem. For example, "What amount of page area do the photos cover? $38\frac{1}{4}$ in.^{2} About what percent of the page is not covered by photos? About 59.1%. What is the dimension of each photo? $1\frac{7}{10}$ in. by $1\frac{1}{2}$ in. What should the dimensions of each photo be if you want 24 photos to a page?" Sample answer: Assume there are 4 across and 6 down. Each photo would be $1\frac{5}{64}$ in. wide and $1\frac{37}{96}$ in. long.

EXERCISE Notes

ASSIGNMENT GUIDE

Basic/Average: Ex. 1–12, 18–25, 27–33 odd

Above Average: Ex. 1–8, 13–21, 26–32 even, 34–37

Advanced: Ex. 1–17, 22–25, 27–33 multiples of 3, 34–37

Selected Answers
Exercises 1–4, 5–33 odd

✪ More Difficult Exercises
Exercises 18–25, 34–37

Guided Practice

▶ **Ex. 1–4** COOPERATIVE LEARNING Begin these exercises by letting students work together in groups. Then go over student performance in the entire class.

Answers

1. Rate of 1st runner = 10 (km per hour)
   Time after 2nd runner starts = t(hours)
   Distance advantage of 1st runner = 7 (km)
   Rate of 2nd runner = 15 (km per hour)
5. (Rate of stray elephant) • (Time to catch up) = (Rate of herd) • (Time to catch up) + (Distance from elephant to herd)
6. Rate of stray elephant = 25 (mph); Time to catch up = $\frac{1}{12}$ (hour); Rate of herd = 10 (mph); Distance from elephant to herd = d (miles)

EXERCISES

Guided Practice

CRITICAL THINKING about the Lesson

Running a Race **In Exercises 1–4, use the following information.**

Two runners are running on a 21-kilometer course. The first runs at 10 kilometers per hour, and the second runs at 15 kilometers per hour. If the first runner is 7 kilometers past the starting line before the second runner starts, how far does each run before they are side by side? Use the following verbal model.

Rate of 1st runner	•	Time after 2nd runner starts	+	Distance advantage of 1st runner	=	Rate of 2nd runner	•	Time after 2nd runner starts

1. Assign labels to each part of the verbal model. Indicate the units of measure. See margin.
2. Write an algebraic equation. $10t + 7 = 15t$
3. Solve the equation. Interpret its solution. $t = 1\frac{2}{5}$ From the instant the 2nd runner started running, each runner ran for $1\frac{2}{5}$ hours.
4. How far does each run before they are side by side? 21 km

Independent Practice

Left Behind **In Exercises 5–8, use the following information and the caption to the photograph.**

An elephant herd is migrating to greener plains. The herd is moving at about 10 miles an hour. One elephant strays from the herd, stops, and is left behind. Then it senses danger and begins running to catch up with the herd. It takes the stray elephant 5 minutes to reach the others. How far had the herd traveled when the stray elephant became frightened? 5., 6. See margin.

5. Write a verbal model for this problem.
6. Assign labels to each part of the model. Indicate the units of measure.
7. Write an equation that represents the model. $25(\frac{1}{12}) = 10(\frac{1}{12}) + d$
8. Solve the equation and answer the question. $d = 1\frac{1}{4}$, $1\frac{1}{4}$ mi

An angry or a frightened elephant can run at a speed of about 25 miles per hour.

Population Growth **In Exercises 9–12, use the following information.**

From 1987 to 1988 in the United States, the population of the western region increased by 982,000 and that of the midwest region increased by 222,000. In 1988, the population of the western region was 50,679,000 and that of the midwest region was 58,878,000. If the populations continue to change at the same rates, when will the populations of the western region and the midwest region be the same?

9.–10. See Additional Answers.

9. Write a verbal model for this problem.

10. Assign labels to each part of the model.

11. Write an equation that represents the model. $50{,}679{,}000 + 982{,}000y = 58{,}878{,}000 + 222{,}000y$

12. Solve the equation. What year will the populations be the same? $y \approx 10.8$, 1999

Cover Design **In Exercises 13–17, use the following information.**

You want to include four photos on the cover of a program for the school play, two across the page. The cover is $6\frac{1}{2}$ inches wide, and the left and right margins are $\frac{3}{4}$ inch each. The space between the pictures is $\frac{1}{2}$ inch. How wide should you make the pictures?

13.–15. See Additional Answers.

13. Sketch a diagram of the cover.

14. Write a verbal model for this problem.

15. Assign labels to the model. Indicate the units of measure.

16. Write an equation that represents the model. $6\frac{1}{2} = 2 \cdot \frac{3}{4} + \frac{1}{2} + 2x$

17. Solve the equation and answer the question. $x = 2\frac{1}{4}$, $2\frac{1}{4}$ in.

Saving and Spending **In Exercises 18–21, use the following information.**

Currently, you have \$60 and your sister has \$135. You decide to save \$5 of your allowance each week, whereas your sister decides to spend her whole allowance plus \$10 each week. How long will it be before you have as much money as your sister?

18.–19. See Additional Answers.

✪ **18.** Write a verbal model for this problem.

✪ **19.** Assign labels to the verbal model.

✪ **20.** Write an equation that represents the model. $60 + 5w = 135 - 10w$

✪ **21.** Solve the equation and answer the question. $w = 5$, 5 weeks

Temperature Change **In Exercises 22–25, use the following information.**

Suppose you live in Greenville, South Carolina, where the temperature is 69°F and going up at a rate of 2°F an hour. You are talking on the phone to your friend who lives in Waterloo, Iowa, where the temperature is 84°F and going down at a rate of 3°F an hour. If the temperatures continue to change at the same rates, how long would you and your friend have to talk before they would be equal?

✪ **22.** Write a verbal model for this problem. See Additional Answers.

✪ **23.** Assign labels to the model. Indicate the units of measure. See Additional Answers.

✪ **24.** Write an equation that represents the model. $69 + 2h = 84 - 3h$

✪ **25.** Solve the equation and answer the question. $h = 3$, 3 hours

Independent Practice

▶ **Ex. 9–12** The rate of population change in the western region is an increase of 982,000 per year. The rate of change in the midwestern region is an increase of 222,000 per year.

▶ **Ex. 26–33** **PROBLEM SOLVING** Remind students that finding the units of measure is a good technique for checking solutions to problems.

Enrichment Activities

In Exercises 5–8, you used an equation to describe the situation. However, if the situation was changed to state that it took the stray elephant at *least* five minutes to reach the others, you could use the following inequality to find the *least* distance the herd traveled before the stray elephant caught up.

$$25\left(\frac{1}{12}\right) \geq 10\left(\frac{1}{12}\right) + d$$

1. Can you think of another way to rewrite the problem as an inequality? Answers may include restating the rates of the herd or using at most to find the greatest distance, etc.
2. For each of the problem sets in the Independent Practice sets, identify those situations that can be restated using an inequality. Reformulate such problems as inequalities and then solve them.

   All of the problem situations described in the Independent Practice could be changed to inequality situations in more than one way. Enourage students to share their changes with classmates.

Integrated Review

In Exercises 26–33, find the resulting unit of measure.

26. (miles per hour) • (hours) miles
27. (feet per minute) • (minutes) feet
28. (meters) • (meters) square meters
29. (liters) − (liters) liters
30. (hours per day) • (days) hours
31. (inches per foot) • (feet) inches
32. (centimeters) + (centimeters) centimeters
33. (hours) ÷ (days) hours per day

Exploration and Extension

The Rabbit and the Coyote **In Exercises 34–37, use the following information.**

As in the diagram, a rabbit is 30 feet from its burrow and can run 25 feet per second. A coyote that runs 50 feet per second spots the rabbit and starts running toward it.

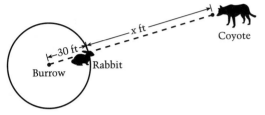

34. How long will it take the rabbit to get to its burrow? $1\frac{1}{5}$ seconds
35. Write an expression for the time it will take the coyote to get to the burrow. $\frac{30 + x}{50}$
36. Set the expressions from Exercises 34 and 35 equal to each other and solve for x. $x = 30$
37. Interpret your results. To get to the burrow at the same time as the rabbit, the coyote must be 30 ft farther from the burrow than the rabbit.

Career Interview

Picture Framer

Henry Leventhal has been a picture framer for 22 years. "I specialize in frames and mats that need artistic or special decorative skills."

Q: ***Does having a math background help you on the job?***
A: Yes. I use math every day. Adding, subtracting, dividing, and measuring in fractional amounts are all done before I cut the glass, mat board, paper, and frame. I also use math when I order materials. If I make a mistake, I have to pay for those materials!

Q: ***How much math do you need on the job?***
A: My math skills came from school, but I had no idea I'd be applying them the way I do now in my work. When I interview job applicants, the first thing I do is ask them to find one eighth or one sixteenth of an inch on a ruler. Half the people fail the interview right there. They can't read a ruler!

Q: ***What would you do if you were in high school again?***
A: I would learn more about how to use calculators and computers.

Mid-Chapter SELF-TEST

Take this test as you would take a test in class. The answers to the exercises are given in the back of the book.

In Exercises 1–12, solve the equation. (3.1, 3.2, 3.3) 7. $\frac{58}{15}$ or $3\frac{13}{15}$

1. $3m = 18$ 6

2. $\frac{1}{4}p = 7$ 28

3. $y - 6 = 7$ 13

4. $x + \frac{1}{2} = 3$ $\frac{5}{2}$ or $2\frac{1}{2}$

5. $3s + 2 = 17$ 5

6. $\frac{x}{4} + 10 = \frac{3}{4}$ -37 $\frac{1}{2}$

7. $16(x - 2) - (x - 3) = 29$

8. $\frac{1}{4}(y + 8) + 6 = 10$ 8

9. $12n + 2 - 3n = 5 - (n - 2)$ $\frac{1}{2}$

10. $y + 2 = 2y - 4$ 6

11. $\frac{1}{2}(x - 2) = 3 - x$ $\frac{8}{3}$ or $2\frac{2}{3}$

12. $3p - 2 = \frac{1}{3}(6p + 5)$ $\frac{11}{3}$ or $3\frac{2}{3}$

In Exercises 13–15, correct any mistakes in the solution steps. (3.3)

13.
$$-2(x + 3) = 3x + 1$$
$$-2x - 6 = 3x + 1$$
$$-2x = 3x + 7$$
$$x \not= 7$$
$0 = 5x + 7$; $-7 = 5x$; $-\frac{7}{5}$ or $-1\frac{2}{5} = x$

14.
$$20x - (3 + 5x) = 8x + 4$$
$$15x - 3 = 8x + 4$$
$$15x = 8x + 7$$
$$7x = 7$$
$$x = 1$$
Correct

15.
$$3\left(\tfrac{1}{2}y - 4\right) = 4(y + 2)$$
$$\tfrac{3}{2}y - 12 = 4y + 8$$
$$-12 = \tfrac{5}{2}y + 8$$
$$-20 = \tfrac{5}{2}y$$
$$-8 = y$$
Correct

16. You and three friends are eating a pizza with 12 pieces. Each person eats the same number of pieces. Let x represent the number of pieces each person eats. Which of the following equations is an algebraic model for the situation? **(3.4)**

a. $3x = 12$ **b.** $\frac{1}{3}x = 12$ **c.** $4x = 12$ **d.** $\frac{1}{4}x = 12$

17. You are mowing lawns for the summer. You charge \$5 per lawn and you have \$35 at the end of the day. Let x represent the number of lawns you mowed. Which *two* of the following equations are algebraic models for the situation? **(3.4)**

a. $5x = 35$ **b.** $\frac{1}{5}x = 35$ **c.** $x = \frac{1}{5}(35)$ **d.** $\frac{1}{35}x = 5$

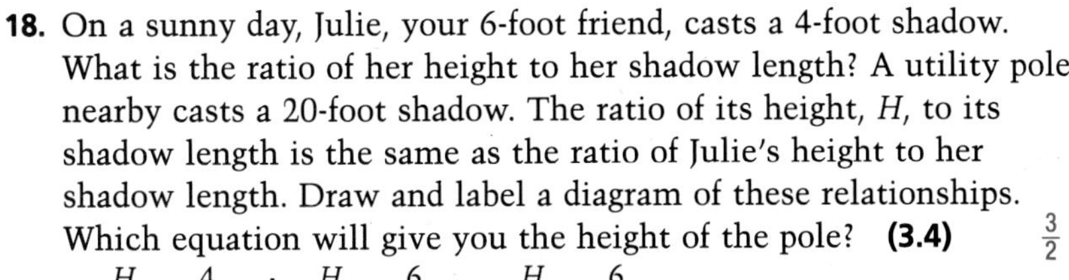

18. On a sunny day, Julie, your 6-foot friend, casts a 4-foot shadow. What is the ratio of her height to her shadow length? A utility pole nearby casts a 20-foot shadow. The ratio of its height, H, to its shadow length is the same as the ratio of Julie's height to her shadow length. Draw and label a diagram of these relationships. Which equation will give you the height of the pole? **(3.4)** $\frac{3}{2}$

a. $\frac{H}{20} = \frac{4}{6}$ **b.** $\frac{H}{20} = \frac{6}{4}$ **c.** $\frac{H}{20} = \frac{6}{3}$

19. What is the height of the pole in Exercise 18? 30 ft

20. The city park is sponsoring a pet show. The number of cats, birds, and hamsters equals the number of dogs and turtles. There are 2 birds, 2 turtles, and 5 hamsters. There are twice as many dogs as cats. Let n be the number of cats. Which of the following is an algebraic model for the situation? **(3.4)**

a. $2n + 2 + 5 = n + 2$ **b.** $n + 2n = 2 + 2 + 5$ **c.** $n + 2 + 5 = 2n + 2$

Problem of the Day

a. How does the sum of the first 100 even counting numbers compare with the sum of the first 100 odd counting numbers?

It is 100 greater.

b. How does the sum of the first 100 even whole numbers compare with the sum of the first 100 odd whole numbers?

It is 100 less.

ORGANIZER

Warm-Up Exercises

1. Compute the following. Round each answer to the nearest hundredth.
   a. $\frac{234}{72}$ 3.25
   b. $\frac{1298}{37}$ 35.08
   c. $23.461 + 8.085 - 17.48$ 14.07
   d. $\frac{5.37 - 34.70 + 2.38}{3.35}$ -8.04
2. Multiply each expression by 10 and simplify.
   a. $5.4 - 4.3 + 2.7$ 38
   b. $3.88 + 35.47 - 3.4$ 359.5
   c. $12.7 - 8.03 + 17.6$ 222.7
   d. $-3.9 + 22.356 - 4.7$ 137.56

Lesson Resources

Teaching Tools
- Problem of the Day: 3.5
- Warm-up Exercises: 3.5
- Answer Masters: 3.5

Extra Practice: 3.5
Color Transparency: 15

Lesson Investigation Answers

Number of people = 3
Each person's share = x
Cost of pizza = 12.89
Equation: $3x = 12.89$
$x = 4.30$

No. Three times $4.30 is $12.90, which is more than the price of the pizza.

3.5 Solving Equations That Involve Decimals

What you should learn:

Goal 1 How to find exact and approximate solutions of equations containing decimals

Goal 2 How to solve problems that use decimal measurements

Why you should learn it:

You will have to solve problems whose data are decimal amounts, such as with money, distances, and percents.

Goal 1 Solving Decimal Equations

All the equations in the first four lessons of this chapter involve integer or fraction coefficients and solutions. Techniques used to solve those equations can also be used to solve equations involving decimals. The difference, however, is that decimal solutions can involve *round-off error.*

LESSON INVESTIGATION

Investigating Round-Off Error

Partner Activity Three people want to share equally in paying for a pizza that costs $12.89. Use the verbal model and labels to write an equation that can be used to find each person's share. Then solve the equation. Can each person pay exactly the same amount? Explain. How is your answer related to round-off error?

Verbal Model Number of people · Each person's share = Cost of pizza

Labels
Number of people = ? (people)
Each person's share = ? (dollars per person)
Cost of pizza = ? (dollars)

Study Tip

When you check decimal answers that have been rounded, remember that the solution is not exact. So, after substituting, you should expect the two sides of the equation to be only approximately equal.

$$-38x - 39 = 118$$
$$-38(-4.13) - 39 \stackrel{?}{=} 118$$
$$-38(-4.13) \stackrel{?}{=} 157$$
$$156.94 \approx 157$$

Example 1 *Rounding for the Final Answer*

Solve $-38x - 39 = 118$. (Round to two decimal places.)

Solution

$-38x - 39 = 118$ *Rewrite original equation.*
$-38x = 157$ *Add 39 to both sides.*
$x = \frac{157}{-38}$ *Divide both sides by −38.*
$x \approx -4.13$ *Use a calculator to divide; round to two decimal places.*

The solution is approximately −4.13. ■

In Example 1, you did not have to use a decimal approximation until the last step. When the original equation involves decimals, you may have to round in intermediate steps, but you should keep as many decimal places as possible in those steps. For the final solution, you should normally write no more decimal places than appeared in the original equation.

Example 2 Original Equation Involves Decimals

Solve $3.57x - 37.40 = 0.23x + 8.32$.

Solution

$3.57x - 37.40 = 0.23x + 8.32$	*Rewrite original equation.*
$3.34x - 37.40 = 8.32$	*Subtract 0.23x from both sides.*
$3.34x = 45.72$	*Add 37.40 to both sides.*
$x = \frac{45.72}{3.34}$	*Divide both sides by 3.34.*
$x = 13.68862\ldots$	*Use calculator to divide.*
$x \approx 13.69$	*Round to two decimal places.*

The solution is approximately 13.69. Check this in the original equation. ■

The next example shows a technique for avoiding decimals in intermediate steps.

Example 3 Changing Decimal Coefficients to Integer Coefficients

Solve $4.5 - 7.2x = 3.4x - 49.5$.

Solution Since the coefficients and constant terms each have only one decimal place, we can "clear the equation of decimals" by multiplying both sides by 10.

$4.5 - 7.2x = 3.4x - 49.5$	*Rewrite original equation.*
$45 - 72x = 34x - 495$	*Multiply both sides by 10.*
$45 = 106x - 495$	*Add 72x to both sides.*
$540 = 106x$	*Add 495 to both sides.*
$\frac{540}{106} = x$	*Divide both sides by 106.*
$5.094339\ldots = x$	*Use calculator to divide.*
$5.1 \approx x$	*Round to one decimal place.*

The solution is approximately 5.1. Check this in the original equation. ■

LESSON Notes

Example 1

Note that the solution, $x \approx -4.13$, has been rounded down and is slightly less than the exact solution. Consequently, the check will result in a comparison that is slightly less than the value against which the check is made.

Example 2

TECHNOLOGY Encourage students to use calculators to perform the computations. If, however, computations are done by hand, remind students to align the decimal points before computing.

Example 3

Clearing the equations of decimals is a convenient step for students to know, particularly when computing by hand. Notice that in Example 2, you could have multiplied both sides of that equation by 100 to clear the equation of decimals.

Extra Examples

Here are additional examples similar to **Examples 1–3.**

Solve. (Round answers to two decimal places.)

a. $47x - 35 = 231$

b. $47 - 17x = 117$

c. $19.6x - 38.19 = 0.46x + 3.9$

d. $12.57 - 4.23x = -2.5x - 14.6$

e. $3.4 + 7.2x = 6.7x - 13.9$

a. 5.66, b. −4.12, c. 2.20, d. 15.71, e. −34.6

Example 4

Call students' attention to the need to round down to \$17.49, although it would be customary to round up to \$17.50. In real-life applications, always consider the reasonableness of the results. Also, note that this problem can be expressed as an inequality, such as $x + 0.05x \le 18.37$.

Check Understanding

1. Explain why one of the first steps in adding decimals is to align the decimal points. Or stated differently, in the language of fractions, what does aligning the decimal points achieve? Effectively, you are getting a common denominator for these decimal fractions.
2. Demonstrate how clearing an equation of decimals connects with clearing an equation of fractions.

Communicating about ALGEBRA

Ask students to identify other real-life situations in which cost consists of an initial fee plus an additional charge determined by a given rate. Express the relationships as both equations and inequalities. One example is a truck or car rental.

Goal 2 Solving Problems with Decimals

Real-life measurements often contain decimals such as a salad for \$2.45, a 1.3-mile jog, a body temperature of 98.6°F, a batting average of .315, and a sales tax of 7% = 0.07.

Real Life
Retail Purchase

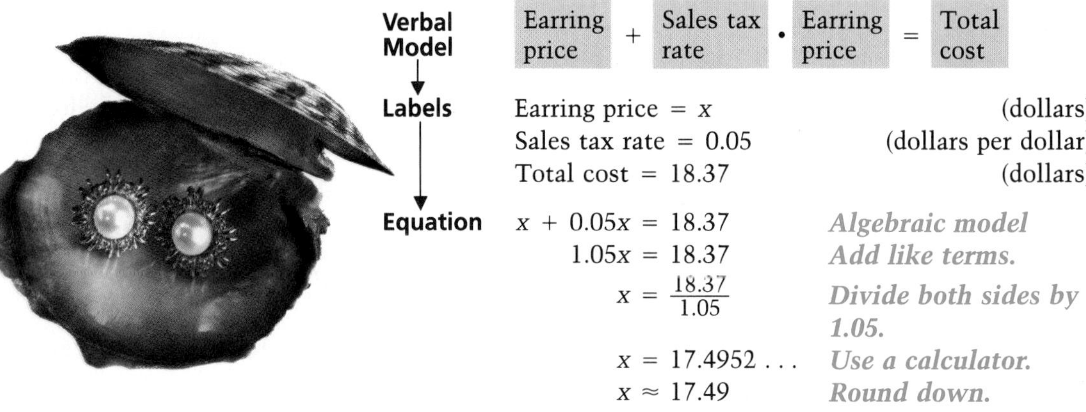

Example 4 *How Much Can They Cost?*

You are shopping for earrings. The sales tax is 5%. You have a total of \$18.37. What is your price limit for the earrings?

Solution

Verbal Model Earring price + Sales tax rate • Earring price = Total cost

Labels
Earring price = x (dollars)
Sales tax rate = 0.05 (dollars per dollar)
Total cost = 18.37 (dollars)

Equation

$x + 0.05x = 18.37$ *Algebraic model*

$1.05x = 18.37$ *Add like terms.*

$x = \frac{18.37}{1.05}$ *Divide both sides by 1.05.*

$x = 17.4952\ldots$ *Use a calculator.*

$x \approx 17.49$ *Round down.*

The answer was rounded *down* to \$17.49 rather than *up* to \$17.50. For a price of \$17.50, the sales tax is \$0.88 and you would be a penny short. ■

Communicating about ALGEBRA

SHARING IDEAS about the Lesson

Using the Problem-Solving Plan You have \$5.26 and want to call a town 105 miles away. How long can you talk?

Distance (in miles)	Cost for 1st minute	Cost for addl. minute
1–10	\$0.19	\$0.10
11–16	\$0.23	\$0.12
17–22	\$0.26	\$0.15
23–30	\$0.31	\$0.19
31–40	\$0.36	\$0.21
41–55	\$0.39	\$0.25
56–70	\$0.43	\$0.27
71–124	\$0.44	\$0.28
125–196	\$0.46	\$0.31
197–292	\$0.49	\$0.34
293–354	\$0.50	\$0.36

A. Assign labels to the following verbal model.

Verbal Model 1st min + Addl. min rate • Num. of addl. min = Total cost

Labels
Cost for 1st minute = ? 0.44 (dollars)
Rate for additional minutes = ? 0.28 (dollars per min)
Number of additional minutes = x (min)
Total cost = ? 5.26 (dollars)

B. Write and solve an equation for x, then answer the question.

B. $0.44 + 0.28x = 5.26$; $x \approx 17.21$, 18 minutes

EXERCISES

Guided Practice

CRITICAL THINKING about the Lesson

1. Give an example of round-off error. Examples vary; see page 148.
2. In 1990, the population of the United States was 249.6 million. During that year, Americans purchased 3.1 million basketballs. Which of the following is a better number to represent the number of Americans per new basketball in 1990? Explain. Nonwhole numbers are not meaningful; b is least objectionable.

   a. 80.516129 **b.** 80.5

3. How can you clear the following equation of decimals? Multiply both sides by 10.

   $0.2x - 1.4 = 1.1x + 3.2$

4. The solution of $7x = 3$, rounded to two decimal places, is 0.43. Which of the following is a better way to list the solution?

   a. $x = 0.43$ **b.** $x \approx 0.43$

Independent Practice

In Exercises 5–10, round the number to one decimal place. Use your calculator when appropriate.

5. -367.84159 -367.8
6. 83.7461 83.7
7. $1.0847 + 62.5583$ 63.6
8. $24.0321 - 21.8217$ 2.2
9. $3.21(4.56)$ 14.6
10. $4.57 \div 3.21$ 1.4

In Exercises 11–16, round the number to two decimal places.

11. 5.364 5.36
12. -2.495 -2.50
13. $-41.287 - 3.382$ -44.67
14. $11.051(3.467)$ 38.31
15. $-23.981(-4.598)$ 110.26
16. $15.953 \div 3.476$ 4.59

In Exercises 17–28, solve the equation. Round the result to two decimal places. Check your rounded solution. See page 148 for example of check.

17. $17x - 33 = 114$ 8.65
18. $-3x + 51 = 104$ -17.67
19. $-18 + 41a = 57$ 1.83
20. $31 = 44 - 12m$ 1.08
21. $25 = 14 - 10d$ -1.1
22. $99 = 100t + 56$ 0.43
23. $238 = 79z - 43$ 3.56
24. $28 - 68c = 241$ -3.13
25. $3(31 - 12x) = 82$ 0.31
26. $5(-8x + 15) = 49$ 0.65
27. $2(-5x + 4) = -x$ 0.89
28. $-(x - 3) = 2(3x + 1)$ 0.14

In Exercises 29–38, solve the equation. Round the result to two decimal places.

29. $15.74 - 2.36x = 18.66x - 12.23$ 1.33
30. $5.423 - 6.411x = 8.213x + 3.081$ 0.16
31. $2.7 - 3.6x = 8.4 + 23.7x$ -0.21
32. $5.3 + 9.2x = 7.4x - 8.8$ -7.83
33. $6(3.14 + 1.59x) = 12.29x - 4.37$ 8.44
34. $18(1.01 - 2.30x) = 4.93x + 6.22$ 0.26
35. $18.41x - 12.75 = (4.32x - 6.81)3$ -1.41
36. $25.79x - 18.24 = (7.77 - 13.91x)(-6)$ 0.49
37. $38.5x + 2.4 = -31.7 + 41.8x$ 10.33
38. $26.4x - 3.2 = 5.9x - 32.1$ -1.41

EXERCISE Notes

ASSIGNMENT GUIDE

Basic/Average: Ex. 1–4, 5–15 odd, 18–39 multiples of 3, 41–42, 47–48, 49–57 multiples of 3, 58–59

Above Average: Ex.1–4, 6–36 multiples of 3, 41–46, 57–59, 60–63

Advanced: Ex.1–4, 6–36 multiples of 3, 37–40, 43–48, 57–65

Selected Answers

Exercises 1–4, 5–59 odd

✪ More Difficult Exercises

Exercises 45–50, 62–65

Guided Practice

Ex. 1 GROUP ACTIVITY Discuss in class the meaning of *round-off error*. Refer to the lesson examples to help students understand.

Independent Practice

Ex. 17–38 You may want students to check their answers in the original equations. Point out that answers that are rounded down compare differently in the original equations than answers that are rounded up.

▶ **Ex. 39–40** Each of these exercises may be expressed as an inequality.

▶ **Ex. 43–50** COOPERATIVE LEARNING Encourage students to work together on these exercises.

Answers

45. (Rate of cold water) • (Time cold water is on) = (Rate of hot water) • (Time hot water is on)
46. Rate of cold water = 12.3 (liters per minute); Time cold water is on = t (minutes); Rate of hot water is 7.8 (liters per minute); Time hot water is on = $t + 2$ (minutes)

EXTEND Ex. 62–65 PROJECT Have students find the total number of households in your county and estimate the number of households with dogs and cats. They could check their dog estimate by finding the number of dog licenses issued this year countywide.

This is a good project for students. It gives them an opportunity to sharpen both their mathematics and research skills.

39. *Buying a Sweatshirt* You have \$32.14 to spend for a sweatshirt. The sales tax is 5%. What is the most the sweatshirt can cost? \$30.61

40. *Buying a Dinner* You have \$6.46 to spend for a dinner. There is no sales tax, but you want to leave a 15% tip. What is the most the dinner can cost? \$5.62

In Exercises 41 and 42, use the following information.

The cross-country team ran 10.3 kilometers in 42.5 minutes during their workout. Let r be the speed in kilometers per minute.

41. Which equation models this problem?

a. $42.5r = 10.3$ **b.** $10.3r = 42.5$

42. Solve for r and round the result to one decimal place. $r \approx 0.2$

In Exercises 43 and 44, refer to the information on page 121 on the "drag" of objects in the water.

43. Find the approximate length of a Greenland shark that is 23.5 inches wide. ≈ 90.4 in.

44. Find the approximate length of a dolphin that is 31.5 inches wide. 126 in.

Taking a Bath **In Exercises 45–48, use the following information.**

Suppose you are running water into a bathtub. You turn on the hot water first; it flows at a rate of 7.8 liters per minute. Two minutes later you turn on the cold water; it flows at a rate of 12.3 liters per minute. In how many minutes after the cold water is turned on will the hot and cold faucets have delivered the same amount of water?

45. Write a verbal model for this problem. See margin.

46. Assign labels to each part of the model. Indicate the units of measure. See margin.

47. Write an equation that represents your model. $12.3t = 7.8(t + 2)$

48. Solve the equation and answer the question. $t \approx 3.5$, ≈ 3.5

Fund-Raising **In Exercises 49 and 50, use the following information.**

To raise money, the Student Council of Daniels High School is selling magazine subscriptions. The Student Council will receive \$150 from the magazine publisher plus 38% of the subscription money.

49. How much must the council sell in subscriptions to raise \$300? \$394.74

50. How much must the council sell in subscriptions to raise \$500? \$921.05

Integrated Review

In Exercises 51–59, solve the equation.

51. $|-3| + n = 15$ 12

52. $2n + |-8| = 22$ 7

53. $5x + (-4)^2 = 21$ 1

54. $-(6)^2 + 19x = 15$ $\frac{51}{19}$ or $2\frac{13}{19}$

55. $3t - (-2t) + 14 = 6$

56. $7t + (-3t) - 8 = 16$ 6

57. $3(10y - 12) + 15y = 54$ 2

58. $x - 8 = 2 + 4x$ $-\frac{10}{3}$ or $-3\frac{1}{3}$

59. $12 + 2x = 8 - 5x$ $-\frac{4}{7}$

55. $-\frac{8}{5}$ or $-1\frac{3}{5}$

Exploration and Extension

60. Solve for a, b, c, and d. Round your results to two decimal places.

$$\begin{bmatrix} -2a & 3.1b \\ 0.57c & 3.2(d+48) \end{bmatrix} = \begin{bmatrix} 3.9 & -2.6b+5 \\ -8.6 & -d+1.7 \end{bmatrix}$$ $a = -1.95$, $b \approx 0.88$, $c \approx -15.09$, $d \approx -36.17$

61. *College Entrance Exam Sample* A 15-gallon mixture of 20% alcohol has 5 gallons of water added to it. The strength of the mixture, as a percent, is near which of the following?

a. 12.5 **b.** 13.33 **c.** 15.0 **d.** 16.66 **e.** 20.0

Pet Ownership **In Exercises 62–65, use the following information.**

Of all of the households in the United States, 30% have a dog, 22% have a cat, and 8% have some other pet. This map shows which counties in the Southwest have pet ownership rates above and below the average. Red counties have rates that are 20% or higher above the average. Yellow counties have rates that are up to 20% higher. Blue counties have rates that are lower than average. ***(Source: CACI Marketing Systems)***

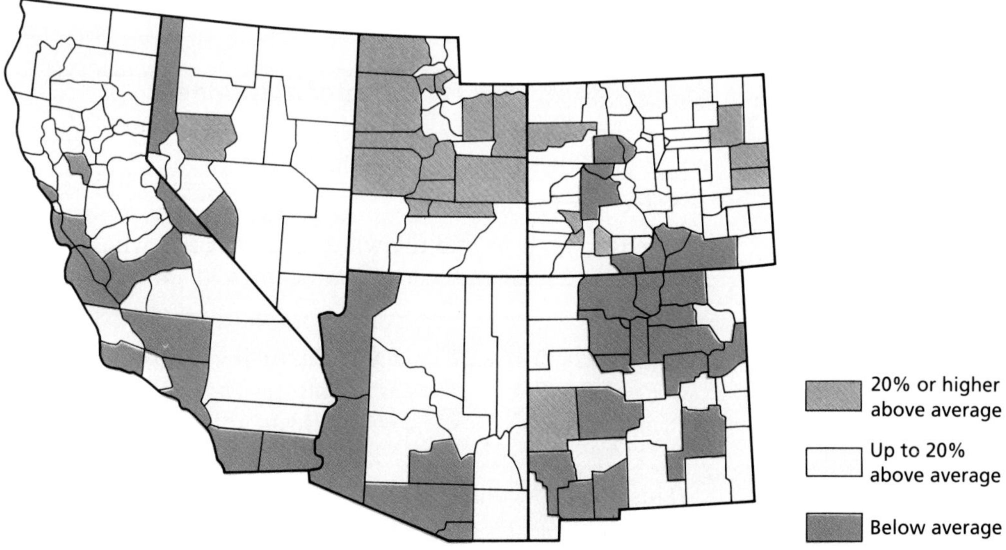

62. Consider a county with 56,000 households that is shaded red on the map. What is the least number of households in this county that owns a dog? 20,160

63. Consider a county with 112,000 households that is shaded blue on the map. What is the greatest number of households in this county that own a cat? 24,639

64. If a county that is shaded yellow on the map has 74,000 households that own cats, what is the least number of households that could be in the county? 280,304

65. If a county that is shaded red on the map has 23,000 households that own dogs, what is the least number of households that could be in the county? 38,333

Enrichment Activities

These activities require students to go beyond lesson goals.

1. Demonstrate how a terminating decimal number can be written as a common fraction. Use denominators of 10, 100, 1000, etc.
2. Can any decimal equation be rewritten as an equation involving only integer coefficients? Explain. No; only those with terminating decimals can be cleared of decimals.
3. Can any nonterminating decimal number be written as a common fraction? No; consider the decimal fraction, 0.1011011101111. . .

Problem of the Day

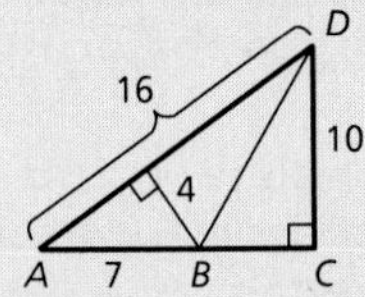

What's wrong with this diagram?

The area of triangle *ABD*, determined one way, is 32; determined another way, it is 35.

ORGANIZER

Warm-Up Exercises

1. Evaluate.
   a. $2x - 3y$ at $x = -2, y = 3$
   b. $P(1 + rt)$ at $P = 300, r = 0.06, t = 6$
   c. $\frac{1}{3}hs^2$ at $h = 7, s = 4$
   d. $2L + 2W$ at $L = 12, W = 5$

   a. -13, b. 408, c. $37\frac{1}{3}$, d. 34
2. Solve each equation for both x and y.
   a. $4y - 6x = 16$
   b. $x + 13y = 2.34$
   c. $-4.5x + 9y = 28$

   a. $x = \frac{4y - 16}{6}$, $y = \frac{16 + 6x}{4}$; b. $x = 2.34 - 13y$, $y = \frac{2.34 - x}{13}$; c. $x = \frac{9y - 28}{4.5}$, $y = \frac{28 + 4.5x}{9}$

Lesson Resources

Teaching Tools
- Problem of the Day: 3.6
- Warm-up Exercises: 3.6
- Answer Masters: 3.6

Extra Practice: 3.6
Color Transparency: 16
Technology Handbook: p. 27

3.6 Literal Equations and Formulas

What you should learn:

Goal How to solve literal equations, especially formulas, for a specified variable

Why you should learn it:

You can solve an equation such as $I = Prt$ for any of its variables, so you need to learn only one model, not many, for most real-life relationships.

Goal Solving Literal Equations

A **literal equation** is an equation that uses more than one letter as a variable. For instance, $3x + y = 4$ is a literal equation because it uses two variables, x and y. The word *literal* comes from the Latin word for "letter." You can solve a literal equation for any one of its "letters."

Solve for x.

$$3x + y = 4$$
$$3x = 4 - y$$
$$x = \frac{4 - y}{3}$$

Solve for y.

$$3x + y = 4$$
$$y = 4 - 3x$$

Connections **Geometry**

Example 1 ***The Perimeter of a Rectangle***

The perimeter, P, of a rectangle is given by $P = 2W + 2L$, where W is the width and L is the length.

a. Solve this formula for W.

b. Use the result to find the width of a rectangle whose length is 24 cm and whose perimeter is 70 cm.

L

W

Solution

a.

$P = 2W + 2L$	*Formula for perimeter*
$P - 2L = 2W$	*Subtract 2L from both sides.*
$\frac{P - 2L}{2} = W$	*Divide both sides by 2.*

This gives the width in terms of perimeter and length.

b. Substitute $L = 24$ and $P = 70$ into the formula for width.

$W = \frac{P - 2L}{2}$	*Formula for width*
$= \frac{70 - 2(24)}{2}$	*Substitute 70 for P and 24 for L.*
$= \frac{70 - 48}{2}$	*Simplify 2(24).*
$= \frac{22}{2}$	*Simplify 70 − 48.*
$= 11$	*Width*

Thus, the width is 11 cm. ■

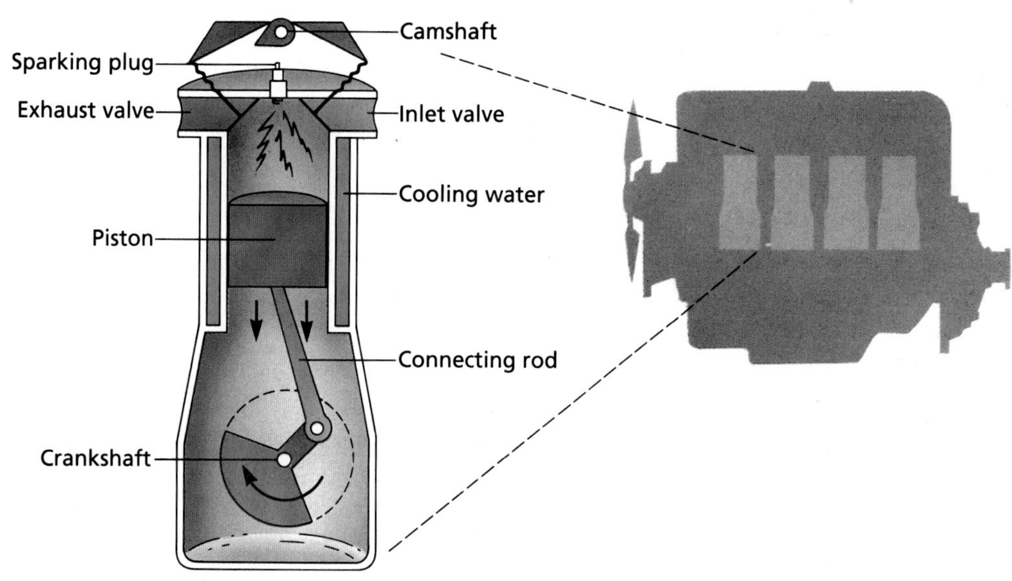

In a gasoline engine, power is produced by expanding gases that drive the pistons up and down. The pistons transfer the power to connecting rods, which turn the crankshaft.

The amount of horsepower produced by an engine whose pistons are turning a single crankshaft is

$$H = \left(\frac{1}{396{,}000}\right) PpArn,$$

where P is the average piston pressure, p is the piston stroke length, A is the piston area, r is the crankshaft revolutions per minute (RPM), and n is the number of pistons.

Real Life
Automotive Mechanics

Example 2 *Finding the Number of Pistons*

a. Solve the formula $H = \left(\frac{1}{396{,}000}\right) PpArn$ for n.

b. An automobile engine is producing 150 horsepower with an RPM of 2500. The average piston pressure is 120 pounds per square inch, the piston stroke length is 3.3 inches, and the piston area is 10 square inches. How many pistons does the engine have?

Solution

a.

$$H = \left(\frac{1}{396{,}000}\right) PpArn \quad \textit{Formula for horsepower}$$

$$396{,}000H = PpArn \quad \textit{Multiply both sides by 396,000.}$$

$$\frac{396{,}000H}{PpAr} = n \quad \textit{Divide both sides by factors } PpAr.$$

b. $H = 150$, $P = 120$, $p = 3.3$, $A = 10$, and $r = 2500$.

$$n = \frac{396{,}000H}{PpAr} = \frac{396{,}000(150)}{(120)(3.3)(10)(2500)} = \frac{5940}{990} = 6$$

Thus, the engine has 6 pistons. ■

LESSON Notes

Example 1

Some students will find it difficult to solve for W when there are two other variables in the equation. One technique that helps students solve literal equations asks them to replace the other variables with number values (which do not get combined further in the equation). For example, $P = 2W + 2L$ would become $10 = 2W + 2(4)$, where $P = 10$ and $L = 4$, the values having been picked arbitrarily. Students solve the resulting equation more easily to get

$$10 - 2(4) = 2W$$

$$\frac{10 - 2(4)}{2} = W.$$

Example 2

Caution students that the capital "P" and the lowercase "p" are different. A unit analysis of the measurements in this example reveals that horsepower is measured in foot-pounds per second.

Example 3

Ordinarily, $\frac{500}{1.065} \approx 469.4835$ would be rounded down to \$469.48. If you want to guarantee that at least \$500 will be earned, then you should round the amount up to \$469.49. To see the difference, compare the two evaluations.

$469.48(1 + 0.065(1)) \approx 499.9962$
$469.49(1 + 0.065(1)) \approx 500.0068$

Most financial institutions simply drop all decimal places beyond the hundredths place.

An Alternate Approach
Evaluating Formulas
In **Examples 1–3,** some students may evaluate the formulas before they solve them for the unknown. Students who recognize this approach should be commended. However, point out that there are times in which the literal constant may not be evaluated, making it necessary to know how to solve literal equations for any variable.

Extra Examples

Here are additional examples similar to **Examples 1–3.**

1. Solve $P = 2W + 2L$ for the length L. Use the result to find the length of a rectangle whose width is 36 and whose perimeter is 108. $L = \frac{P - 2W}{2}$, $L = 18$
2. Solve $A = \frac{1}{2}h(b_1 + b_2)$ for the height h. Use the result to find the height of a trapezoid with bases of length 14 and 8 inches and an area of 242 square inches. $h = \frac{2A}{(b_1 + b_2)}$, $h = 22$ inches
3. Solve for the interest rate r in $A = P + Prt$. Use the result to find the rate when the amount is 800, the principal is 680, and the time is 3.5 years. $r = \frac{A - P}{Pt}$, $r = 5\%$

Check Understanding

WRITING Explain, using examples, how the following properties help you to solve literal equations.

a. distributive property
b. commutative property
c. associative property
d. properties of equality

Communicating about ALGEBRA

A football field is 120 yards, or 360 feet, long. In Part B, you may wish to point out that the length of the base of the Great

The formula $I = Prt$ gives the interest earned after t years for a principal, P, at an annual interest rate, r. The balance after t years is the sum of the principal and the interest.

$$\underbrace{A}_{\text{Balance}} = \underbrace{P}_{\text{Principal}} + \underbrace{Prt}_{\text{Interest}}$$

Real Life **Finance**

Example 3 *Finding the Principal*

a. Solve the formula $A = P + Prt$ for P.
b. How much should you deposit in a savings account at 6.5% interest annually to have a balance of \$500 after one year?

Solution

a. The first step is to use the Distributive Property to write P as a factor.

$A = P + Prt$ *Formula for balance*
$A = P(1 + rt)$ *Distributive Property*
$\frac{A}{1 + rt} = P$ *Divide both sides by $1 + rt$.*

This formula gives the principal in terms of the balance, interest rate, and time.

b. Substitute $A = 500$, $r = 0.065$, and $t = 1$ into the formula for principal.

$P = \frac{A}{1 + rt}$ *Formula for principal*
$= \frac{500}{1 + 0.065(1)}$ *Substitute values for A, r, and t.*
$= \frac{500}{1.065}$ *Simplify.*
≈ 469.48 *Principal*

Thus, you should deposit \$469.48. ■

Communicating about ALGEBRA

SHARING IDEAS about the Lesson

Cooperative Learning

A. ***Work with a Partner*** The formula for the volume of a pyramid with a square base is $V = \frac{1}{3}hs^2$, where h is the height of the pyramid and s is the length of each side of the base. Solve this formula for h. $h = \frac{3V}{s^2}$

B. ***Apply*** The Great Pyramid in Egypt has a volume of about 90 million cubic feet and a square base with sides of about 750 feet. Use the formula for h to approximate the height of the Great Pyramid. 480 ft

EXERCISES

Guided Practice

CRITICAL THINKING about the Lesson

1. **True or False?** A literal equation is an equation that has a variable on both sides of the equation. False

2. Which of the following are literal equations? Explain how you know.
   a. $2x + 4 = 5 - 3x$ **b.** $a - 6 = 3b + 9$
   c. $4s + 9t = 16$ **d.** $3 = 27 - 24$

   Each uses more than one letter as a variable.

3. Solve the equation $3n + 4m = 9$ for m.
   $m = \frac{9 - 3n}{4}$

4. Solve the equation $I = Prt$ for r.
   $r = \frac{I}{Pt}$

Independent Practice

In Exercises 5–16, solve for the indicated variable.

5. *Area of a Triangle*
   Solve for h: $A = \frac{1}{2}bh$ $h = \frac{2A}{b}$

6. *Perimeter of a Rectangle*
   Solve for L: $P = 2L + 2W$ $L = \frac{P - 2W}{2}$

7. *Volume of a Rectangular Prism*
   Solve for L: $V = LWH$ $L = \frac{V}{WH}$

8. *Volume of a Circular Cylinder*
   Solve for h: $V = \pi r^2 h$ $h = \frac{V}{\pi r^2}$

9. *Markup*
   Solve for C: $S = C + rC$ $C = \frac{S}{1 + r}$

10. *Discount*
    Solve for L: $S = L - rL$ $L = \frac{S}{1 - r}$

11. *Investment at Simple Interest*
    Solve for r: $A = P + Prt$ $r = \frac{A - P}{Pt}$

12. *Investment at Compound Interest*
    Solve for P: $A = P(1 + r)^t$ $P = \frac{A}{(1 + r)^t}$

13. *Area of a Trapezoid*
    Solve for b_1: $A = \frac{1}{2}h(b_1 + b_2)$ $b_1 = \frac{2A}{h} - b_2$

✪ 14. *Sum of a Geometric Sequence*
    Solve for r: $S = \frac{rL - a}{r - 1}$ $r = \frac{S - a}{S - L}$

15. *Last Term of an Arithmetic Sequence*
    Solve for n: $L = a + (n - 1)d$ $n = \frac{L - a}{d} + 1$

✪ 16. *Sum of an Arithmetic Sequence*
    Solve for a: $S = \frac{n}{2}[2a + (n - 1)d]$

17. *Depth of a Water Trough* A water trough in the shape of a rectangular prism is 12 feet long and 3 feet wide. The trough has 9.4 cubic feet of water. How deep is the water in the trough? ≈ 0.26 ft

18. *Height of a Circular Cylinder* The volume of a circular cylinder is 48π cubic centimeters. The radius of the cylinder is 2 centimeters. What is the height of the cylinder? 12 cm

16. $a = \frac{S}{n} - (n - 1)\frac{d}{2}$

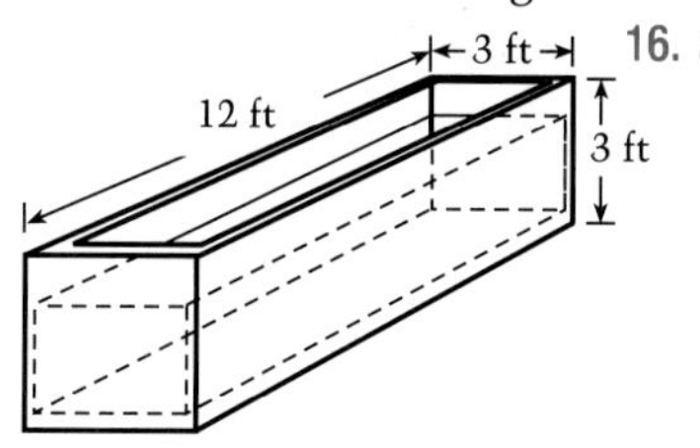

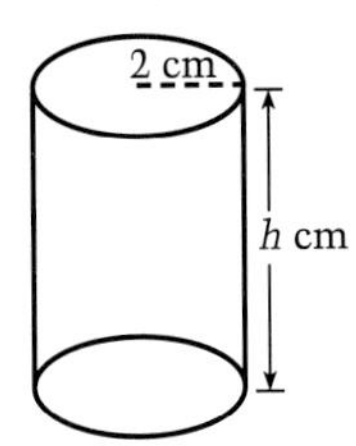

19. *Width of a Rectangle* A rectangle is 1.5 times as long as it is wide. The perimeter of the rectangle is 75 inches. Find the width of the rectangle. 15 in.

Pyramid at 750 feet is more than that of two football fields long.

EXTEND *Communicating*
COOPERATIVE LEARNING
Many of the formulas in **Exercises 5–16** are well-known to students. Have student pairs identify what the variables mean in each formula. They may need to use the encyclopedia or other subject texts to complete this assignment. Have student pairs share their results with the class.

EXERCISE Notes

ASSIGNMENT GUIDE

Basic/Average: Ex.1–4, 5–33 odd, 35–38

Above Average: Ex.1–4, 5–33 odd, 35–38

Advanced: Ex.1–4, 6–24 even, 25–33 odd, 35–39

Selected Answers
Exercises 1–4, 5–33 odd

Use **Mixed Review** as needed.

✪ **More Difficult Exercises**
Exercises 14, 16, 21–26, 35–39

Guided Practice

Use these exercises in class as oral exercises. Ask students to identify other examples of literal equations.

Independent Practice

▶ **Ex. 5–16** Discuss these formulas with your students. (See Communicating about Algebra activity.)

▶ **Ex. 17–18** Be sure students recognize the proper formulas to use in these exercises: $V = lwh$ or $V = \pi r^2 h$.

▶ **Ex. 21** Note that this formula assumes that the heel is square. Flooring manufacturers are much more concerned with the pressure exerted by women's high heels. Consider that the pressure exerted by a 150-pound woman on a heel that is 0.5-inch square is 720 pounds per square inch. That pressure is 16 times greater than the pressure exerted by Tim.

▶ **Ex. 23** Assume that the interest is compounded annually.

▶ **Ex. 35–38** TECHNOLOGY Students should use their calculators to complete the table.

▶ **Ex. 39** PROJECT Students should be encouraged to research the conditions surrounding this gift by Franklin, how the cities of Boston and Philadelphia have used their gift, and its current value. Let them compute a theoretical value (which can be compared to the actual value) by assuming that a fixed compound interest can be used. They can extend this problem into a project by determining the size of the interest rates of the times and the effects of changing rates on the value of such a gift.

Common-Error ALERT!

In **Exercises 23–26**, caution students about the differences between simple interest and compound interest. Unlike simple interest, which is determined using the original principal, compound interest adds the interest earned after one interest period to the principal at the beginning of that period. See Exercises 11 and 12 for the appropriate formulas.

20. *Length of Picture Frame* A rectangular picture frame has a perimeter of 3 feet. The width of the frame is 0.62 times its height. Find the height of the frame.

21. The pressure, P, exerted on the floor by a person's shoe heel depends on the weight, w, of the person and the width of the heel, H. The formula is given by

$$P \approx \frac{1.2w}{H^2},$$

where the weight is in pounds and the heel width is in inches. The heels on Tim's shoes are 2 inches wide, and each exerts a pressure of 45 pounds per square inch. Estimate Tim's weight. 150 pounds

22. *Physics* Newton's second law of motion states that force, F, is equal to the mass, m, times acceleration, a, or $F = ma$. A force of 20 newtons is acting on a mass of 10 kilograms. A newton is "one kilogram-meter per second per second." What is the acceleration? 2 meters per second per second

23. *Compound Interest* You earn 7.2% interest in one savings account and 5.3% interest in a second savings account. Your balance in *each* account at the end of two years is $324.75. How much money did you start with in each account? How much money did you earn from both accounts? $282.59, $292.88, $74.03

24. *Simple Interest* You deposit $400 in a savings account. After one year, your balance is $428. What was your annual interest rate? 7%

25. *Compound Interest* You have a balance of $179.20 in your savings account two years after the account was opened. Your only transaction was the initial deposit. The annual interest rate is 6%. What was your initial deposit? $159.49

26. *Simple Interest* You need $396.80 to buy a stereo. You deposit $320 into a savings account with 12% interest. How long will it be before you have enough money for the stereo? 2 years

20. ≈ 0.93 ft

Integrated Review

In Exercises 27–30, solve the equation.

27. $4t - 1 = 7$ 2

28. $s - 35 + 9 = 10 - s$ 18

29. $3(x - 1) = x + 9$ 6

30. $\frac{1}{2}\left(\frac{1}{2}a + 4\right) = 8 - \frac{3}{4}a$ 6

In Exercises 31–34, evaluate the expression.

31. $4V - 3H + 9$, when $V = \frac{1}{2}$ and $H = 4$ -1

32. $9x + 36 - 4y$, when $x = 2$ and $y = 14$ -2

33. $(1.2x - 1)^4 - 0.1$, when $x = 1.1$ ≈ -0.09

34. $\frac{1}{2}(x - y)^2$, when $x = 2$ and $y = 4$ 2

Exploration and Extension

Growing Money **In Exercises 35–38, use the following information.**

With an interest rate of $r = 8\% = 0.08$, the formula for an investment with compound interest, $A = P(1 + r)^t$, can be written as $A = P(1.08)^t$.

35. Solve the equation $A = P(1.08)^t$ for P. $P = \frac{A}{1.08^t}$

36. How much money must you deposit to have a balance of \$10,000 in one year? \$9259.26

37. How much must you deposit if you are willing to wait 5 years? \$6805.83

38. Continue this process by completing the table. You may want to use the y^x key on your calculator. From the table, how much would you have to deposit to earn a balance of \$10,000 if you were willing to wait 100 years? 179.50 years? \$4631.94, \$3152.42, \$2145.49, \$993.78, \$213.22, \$4.55, \$0.01

Number of Years, t	10	15	20	30	50	100	180
Deposit, P	?	?	?	?	?	?	?

39. *Benjamin Franklin* Benjamin Franklin died in 1790. In his will, he left \$5000 each to the cities of Boston and Philadelphia to be used for public works. Part was to be used after 100 years and the rest was to be used after 200 years. Estimate the amount of money that each city could have had available in 1990, if the funds were invested at 5% compound interest. Answers vary.

Mixed REVIEW

5. $3x^2 - 14x + 2$

1. Evaluate $(-37.2)(0.23) - 18.2 \div 4.9$. **(2.3, 2.5)** ≈ -12.27

2. Evaluate $\frac{3x}{4y} - \frac{1}{2}x$ when $x = \frac{1}{3}$ and $y = 2$. **(1.2)** $-\frac{1}{24}$

3. Find the sum of $\frac{1}{8}$ and $\frac{3}{4}$. **(1.1)** $\frac{7}{8}$

4. Find the difference of 3 and -7. **(2.3)** 10

5. Simplify $3x(x - 4) + 2(1 - x)$. **(2.6)**

6. Evaluate x^4 when $x = 3$. **(1.3)** 81

7. Simplify $-4 + 2x + (-12)$. **(2.2)** $2x - 16$

8. Evaluate $3 \div \left(\frac{4}{3} \cdot \frac{3}{8}\right) - 7$. **(1.1, 2.7)** -1

9. Solve $7x - 2x + 3 = 8$. **(2.6, 3.2)** 1

10. Solve $3x - 4 = 4$. **(3.2)** $\frac{8}{3}$ or $2\frac{2}{3}$

11. Solve $\frac{4x}{3} - 5 = 7$. **(3.2)** 9

12. Solve $6x + 29 = -4x - 1$. **(3.3)** -3

In Exercises 13–16, write the phrase or sentence in algebraic form. (1.6)

13. 24 is equal to 3 more than the product of 7 and a number. $24 = 7n + 3$

14. The product of 2 and a number, plus twice the product of 3 and another number is less than 20. $2x + 2 \cdot 3y < 20$

15. The sum of a number and 6, times the reciprocal of 4. $(x + 6)\frac{1}{4}$

16. The difference of 4 and a number equals the same number. $4 - x = x$

Enrichment Activity

CONNECTIONS This activity asks students about the formulas they use in their other subjects.

MATH JOURNAL Use your math journal to record a list of the most frequently used formulas in such subjects as biology, chemistry, economics, driver's education, etc.

Describe the formula in words and include an example of its use.

Problem of the Day

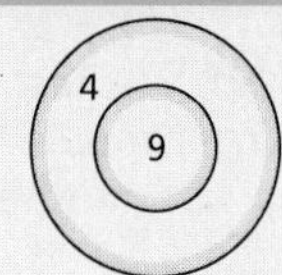

Here is a dartboard with only two regions worth 9 points and 4 points. If you could throw any number of darts, what is the largest positive score you cannot get?
23 points

ORGANIZER

Warm-Up Exercise

Draw a real number line. Then locate the following points: 13, 6, 0, $\frac{2}{3}$, $-3\frac{1}{2}$, 17.

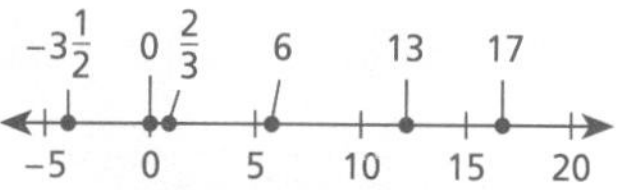

Lesson Resources

Teaching Tools
- Transparencies: 1, 2, 4
- Problem of the Day: 3.7
- Warm-up Exercises: 3.7
- Answer Masters: 3.7

Extra Practice: 3.7
Color Transparencies: 16, 17
Applications Handbook: p. 54
Technology Handbook: p. 31

LESSON Notes

Examples 1–3

Observe that scatter plots are finite sets of data. Scatter plots may sometimes reveal patterns that can be expressed by relationships between infinite sets of data.

Example 1

Have students find the pattern in the table (as *C* increases by 5, *F* increases by 9) in order to extend the table in both directions.

3.7 Exploring Data: Scatter Plots

What you should learn:

Goal 1 How to use a coordinate plane to match points with ordered pairs of numbers

Goal 2 How to use a scatter plot

Why you should learn it:

You can plot pairs of measurements in the coordinate plane as a visual model in your search for relationships among the measurements.

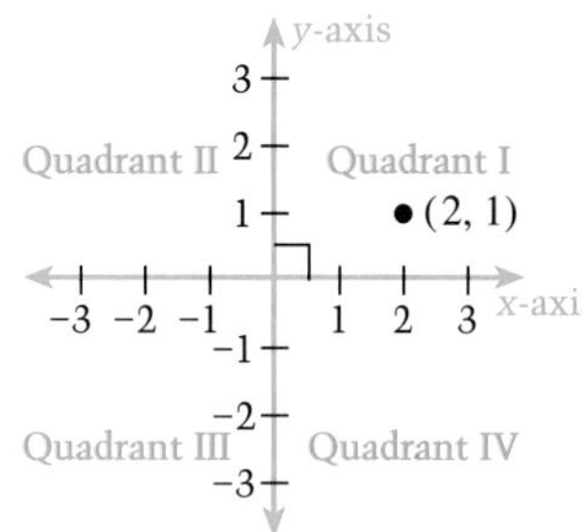

Temperature Scales

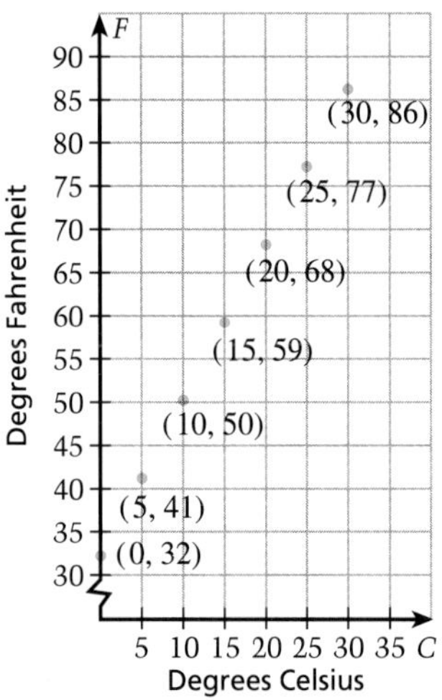

Goal 1 Using a Coordinate Plane

In this lesson, you will learn how to use a coordinate plane to visualize relationships between two variables.

A **coordinate plane** is formed by two real number lines intersecting at a right angle. The horizontal line is the **horizontal axis,** or ***x*-axis,** and the vertical line is the **vertical axis,** or ***y*-axis.** The point at which the two lines meet is the **origin,** and the axes divide the plane into four parts called **quadrants.**

Each point in a coordinate plane corresponds to an **ordered pair** of real numbers. The first number of an ordered pair is called its **horizontal coordinate,** or ***x*-coordinate** and the second number is called its **vertical coordinate,** or ***y*-coordinate.** For instance, the ordered pair (2, 1) has a horizontal coordinate of 2 and a vertical coordinate of 1.

When we *plot* the points that correspond to ordered pairs in a collection, we make a **scatter plot** of the collection. When we speak of "point (a, b)," it will be clear from its use whether we mean an ordered pair or a geometric figure.

Example 1 *Fahrenheit and Celsius Scales*

The table shows some temperatures on the Celsius scale and the corresponding temperatures on the Fahrenheit scale.

C	0	5	10	15	20	25	30
F	32	41	50	59	68	77	86

To make a scatter plot for this table, write each pair of *C*-values and *F*-values as an ordered pair, and plot the corresponding points on a coordinate plane.

(0, 32), (5, 41), (10, 50), (15, 59), (20, 68), (25, 77), (30, 86)

Each ordered pair has the form (C, F). Thus, the ordered pair (0, 32) tells you that F has a value of 32 when C has a value of 0. In the scatter plot at the left, you can see that as the Celsius temperature increases, the Fahrenheit temperature also increases. ■

Goal 2 Using a Scatter Plot

After the 1990 U.S. census, the 435 seats in the House of Representatives were redistributed among the 50 states (see table). State populations are given in thousands, and the rate is the number of people (in thousands) per representative.

	Pop.	Seats	Rate		Pop.	Seats	Rate		Pop.	Seats	Rate		Pop.	Seats	Rate
AK	552	1	552	IL	11,467	20	573	NC	6,658	12	555	SC	3,506	6	584
AL	4,063	7	580	IN	5,564	10	556	ND	641	1	641	SD	700	1	700
AR	3,678	6	613	KS	2,486	4	622	NE	1,585	3	528	TN	4,897	9	544
AZ	2,362	4	591	KY	3,699	6	617	NH	1,114	2	557	TX	17,060	30	569
CA	29,839	52	574	LA	4,238	7	605	NJ	7,749	13	596	UT	1,728	3	576
CO	3,308	6	551	MA	6,029	10	603	NM	1,522	3	507	VA	6,217	11	565
CT	3,296	6	549	MD	4,799	8	600	NV	1,206	2	603	VT	565	1	565
DE	669	1	669	ME	1,233	2	617	NY	18,045	31	582	WA	4,888	9	543
FL	13,003	23	565	MI	9,329	16	583	OH	10,887	19	573	WI	4,907	9	545
GA	6,508	11	592	MN	4,387	8	548	OK	3,158	6	526	WV	1,802	3	601
HI	1,115	2	558	MO	5,138	9	571	OR	2,854	5	571	WY	456	1	456
IA	2,787	5	557	MS	2,586	5	517	PA	11,925	21	568				
ID	1,012	2	506	MT	804	1	804	RI	1,006	2	503				

Example 2 *Number of Seats in the House*

In 1990, the population of the United States was 249,633,000. Ideally, how many people should each House seat represent? Make a scatter plot that depicts the number of seats, s, for a state and the number of people, p, represented by each seat.

Solution Each seat should ideally represent

$$\frac{\text{U.S. Population}}{\text{Seats in House}} = \frac{249,633,000}{435} \approx 574,000 \text{ people.}$$

From the scatter plot, you can see that states with only one representative vary greatly as to the number of people per representative. As the number of representatives increases, the population per representative gets closer to 574,000.

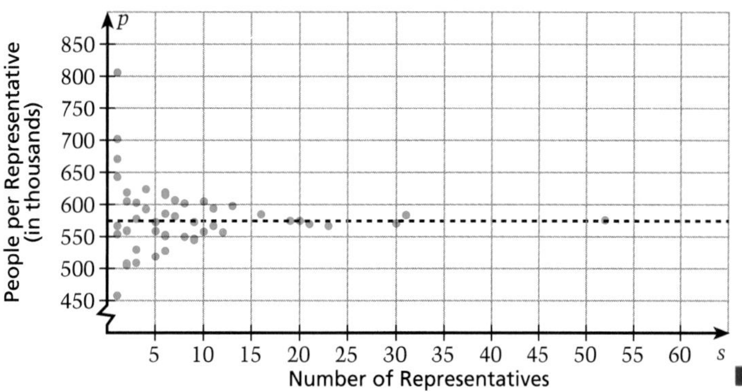

■

Example 2

Help students locate the dot on the scatter plot that corresponds to your state.

Example 3

Recall that in mathematics, the term *line* is used to mean a "straight line."

Extra Examples

Here are additional examples similar to Examples 1–3.

1. The table shows some cab fares (F, in dollars) for different distances (D, in miles) traveled. Make a scatter plot for this table, using the ordered pairs (D, F).

D	0.5	1.0	1.5	2.0	2.5	3.0
F	1.50	1.50	2.00	2.50	3.00	3.50

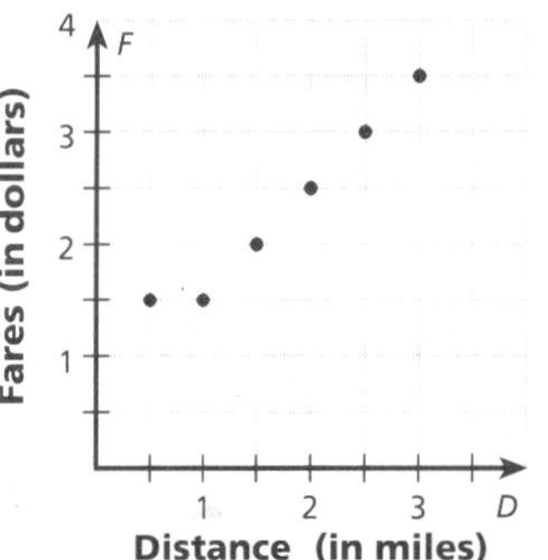

2. Make a scatter plot for the dates of entry for the original thirteen states of the United States.
CT (5), Jan. 9, 1788; DE (1), Dec. 7, 1787; GA (4), Jan. 2, 1788; MD (7), Apr. 28, 1788; MA (6), Feb. 6, 1788; NC (12), Nov. 21, 1789; NH (9), June 21, 1788; NJ (3), Dec. 18, 1787; NY (11), July 26, 1788; PA (2), Dec. 12, 1787; RI (13), May 29, 1790; SC (8), May 23, 1788; VA (10), June 25, 1788

Label the numbers 1–13 along the x-axis and the 30 months from Dec. 1787 through May 1790 along the y-axis.

Check Understanding

WRITING Following the class discussion, students might record their personal responses in their math journals.

1. How is a coordinate plane constructed from real number lines? The axes are made up of number lines.
2. Each point in a coordinate plane is an ordered pair of real numbers. Why is it important to order these pairs of numbers? Explain. To avoid ambiguity. The first number of an ordered pair names the horizontal coordinate; the second names the vertical coordinate.
3. How can scatter plots be used to interpret and understand information? See page 160.
4. If the order of the real numbers is maintained in each ordered pair of a relationship, does it matter which is given first in the scatter plot? For example, in Example 1 of the lesson, does it matter if you always plot (F, C) instead of always plotting (C, F)? The graph would indicate a different relationship. The relationship (F, C) indicates that the value of C depends on the value of F; the relationship (C, F) indicates that F depends on C.

Communicating *about* **ALGEBRA**

EXTEND *Communicating*
RESEARCH

Ask students to collect similar data for the states in their region of the country. Data may be obtained from newspapers (temperatures) and an atlas (latitudes). Ask students to provide a rationale for the selection of cities in their data sets.

Problem Solving
Find a Pattern

Example 3 *Archery Sales in United States*

The total amount A (in millions of dollars) spent on archery equipment in the United States from 1984 to 1994 is shown in the table. In the table, $t = 4$ corresponds to 1984. Draw a scatter plot for this data and describe the pattern.

Year, t	4	5	6	7	8	9	10	11	12	13	14
Amount, A	212	212	214	224	235	261	265	270	334	285	294

Solution The scatter plot for this data is shown below. From the scatter plot, you can see that from 1984 through 1994 the amount spent tended to increase each year, and that the pattern of points was approximately linear. Which year's sales is the most unusual?

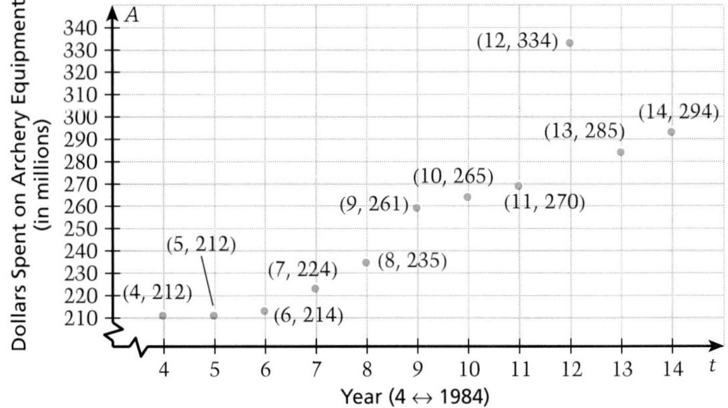

Communicating about ALGEBRA

SHARING IDEAS about the Lesson

Finding a Relationship The average June temperatures in several cities in the central part of the United States are shown in the table at the left. The table also shows the latitude degrees for the city.

City	Temp.	Lat.
Bismarck	64.3°	46.8°
Sioux Falls	68.4°	43.5°
Omaha	73.0°	41.2°
Kan. City	76.1°	39.1°
Wichita	73.6°	37.7°
Okla. City	77.0°	35.5°
Dallas	82.0°	33.2°
Houston	80.6°	29.8°

A. Construct a scatter plot for these data. See Additional Answers.

B. Describe the relationship between the average June temperatures and the latitude positions. As latitude increases, temperature decreases.

C. St. Louis has an average June temperature of 74.6°F. Use the scatter plot to estimate the latitude of St. Louis. $\approx$39.08

EXERCISES

Guided Practice

CRITICAL THINKING about the Lesson

1. Each point on a coordinate plane corresponds to which of the following?
 a. a real number
 b. an ordered pair of real numbers

2. In the ordered pair (3, 7), which number is the horizontal coordinate? Which is the vertical coordinate? 3, 7

3. Write the ordered pairs that correspond to the points on the coordinate plane shown at the right.

3. $A(2, 3)$, $B(-1, 4)$, $C(0, 0)$, $D(3, 0)$, $E(-2, -2)$, $F(2, -3)$

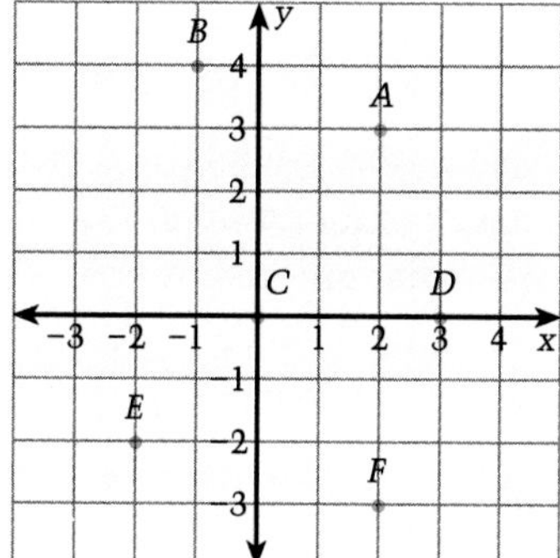

4. Does the point $(2, -3)$ lie above or below the horizontal (x) axis? Does it lie to the left or right of the vertical (y) axis? Below, right

5. Plot the ordered pairs $(-3, 4)$, $(0, 3)$, $(2, -1)$, $(4, 0)$, $(-3, -1)$, and $(2, 2)$ on a coordinate plane. See margin.

Independent Practice

In Exercises 6–8, write the ordered pairs that correspond to the points labeled *A*, *B*, *C*, *D*, *E*, and *F* in the coordinate plane. See below.

6.

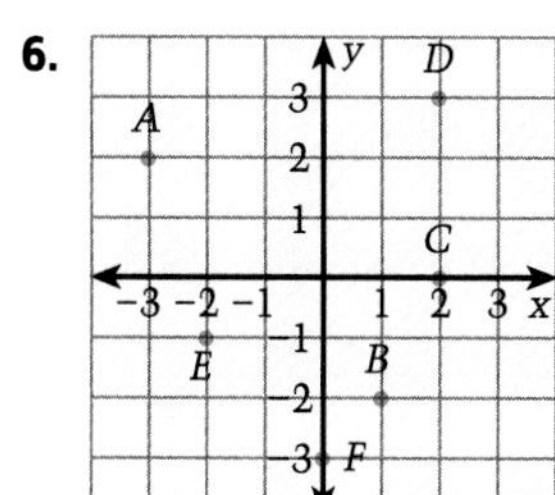

7.

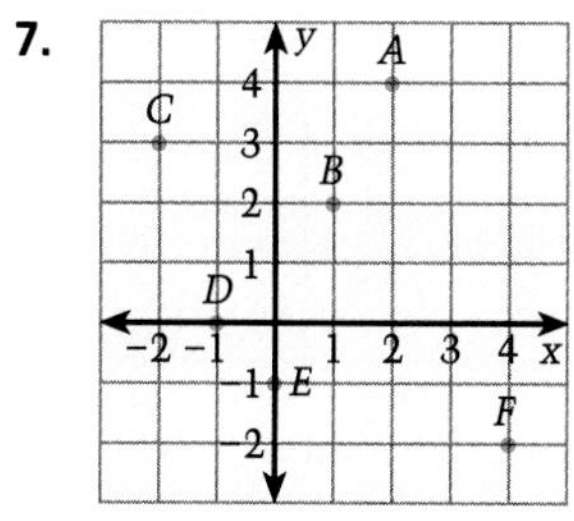

8.

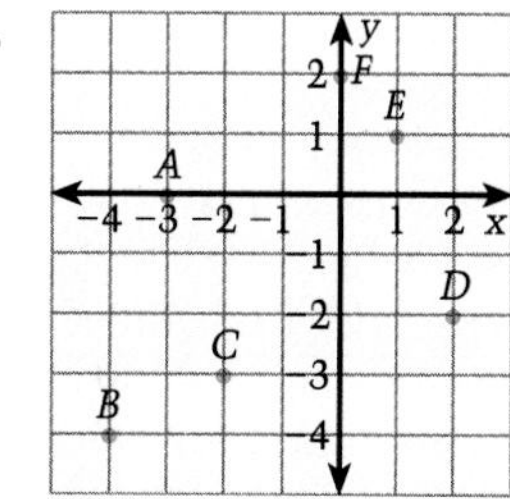

In Exercises 9–12, plot the ordered pairs on a coordinate plane. See Additional Answers.

9. $(-1, -4)$, $(0, 2)$, $(-3, 1)$, $(4, 0)$, $(-2, 1)$, $(3, 4)$

10. $(1, -4)$, $(-1, 0)$, $(2, 3)$, $(-4, -1)$, $(3, 2)$, $(1, 0)$

11. $(3, 1)$, $(0, -2)$, $(-1, -3)$, $(2, 2)$, $(-3, 0)$, $(4, 4)$

12. $(2, 0)$, $(0, 0)$, $(1, -1)$, $(-2, 0)$, $(-3, 3)$, $(-4, 4)$

6. $A(-3,2)$, $B(1, -2)$, $C(2,0)$ $D(2, 3)$, $E(-2, -1)$, $F(0, -3)$
7. $A(2, 4)$, $B(1,2)$, $C(-2,3)$, $D(-1,0)$, $E(0, -1)$, $F(4, -2)$
8. $A(-3, 0)$, $B(-4, -4)$, $C(-2, -3)$, $D(2, -2)$, $E(1, 1)$, $F(0, 2)$

EXERCISE Notes

ASSIGNMENT GUIDE

Basic/Average: Ex.1–5, 7–11 odd, 13–18, 23–24, 27–35 odd, 36–37

Above Average: Ex.1–5, 7–11 odd, 19–22, 25–26, 31–34, 36–37

Advanced: Ex.1–5, 19–22, 23–26, 31–34, 36–37

Selected Answers
Exercises 1–5, 7–31 odd

More Difficult Exercises
Exercises 20, 21, 24, 26, 37

Guided Practice

These exercises are good class-discussion problems about the coordinate plane.

Answer

5.

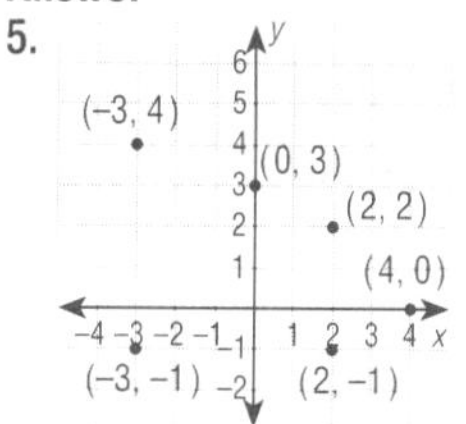

Common-Error ALERT!

When plotting points, students should pay close attention to which number values will be represented on the horizontal axis and which will be shown on the vertical axis. Stress with students that unless otherwise indicated, the horizontal axis is called the x-axis, and the vertical axis is called the y-axis.

Independent Practice

▶ **Ex. 13–18** It is important for students to be able to read information from the coordinate plane. Call attention to the labels of the horizontal and vertical axes. Point out that these axes are frequently named something besides the x-axis and the y-axis.

Suggest that students make up situations that could be described by the graphs. For example, you could use the graph in Exercises 13–15 to describe the cost of paperback books from 1988 to 1992.

▶ **Ex. 20–21** Be open to solutions that may be different from your own. Ask students to provide a rationale for any answer they offer. Accept reasonable solutions.

Answers

21. Some points might be considered as clustering around (38, 1.9); cities with similar January climates.
26. As the planet's distance from the sun increases, its temperature tends to decrease. Except for the average temperatures of Venus and Neptune, the average temperature of a planet actually does decrease as its distance from the sun increases. In general, planets that are far from the sun receive less heat energy than those that are near.

In Exercises 13–18, use the scatter plots to answer the questions.

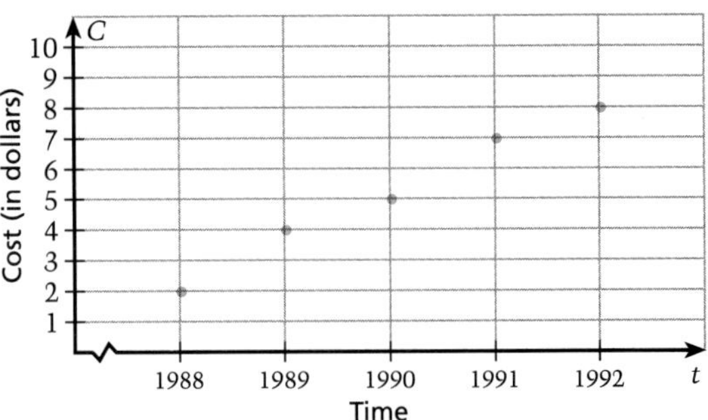

13. For the ordered pair (1990, 5), what is the value of t? What is the value of C? 1990, 5
14. What are the units of measure on the horizontal axis? What are the units of measure on the vertical axis? Years, dollars
15. Which of the following is true?
    a. C decreases as t increases.
    b. C is constant as t increases.
    c. C increases as t increases.

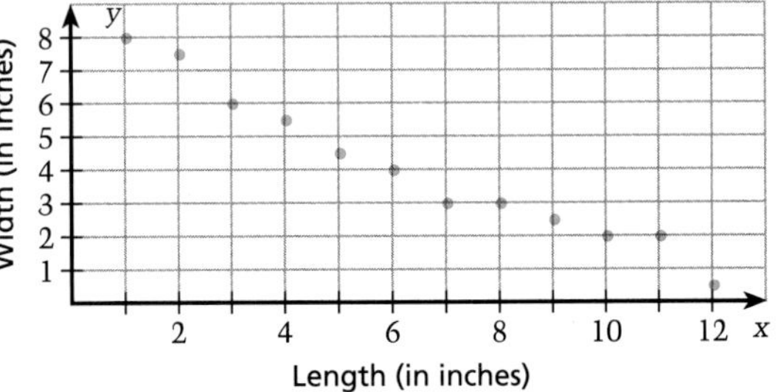

16. For the ordered pair (3, 6), what is the value of x? What is the value of y? 3, 6
17. What are the units of measure on the horizontal axis? What are the units of measure on the vertical axis? inches, inches
18. Which of the following is true?
    a. y decreases as x increases.
    b. y is constant as x increases.
    c. y increases as x increases.

Temperature and Rainfall **In Exercises 19–22, use the table below, which shows the average January temperature and average January rainfall for several cities.**

City	Temp.	Rainfall	City	Temp.	Rainfall
Baghdad, Iraq	50°F	0.9 in.	Mexico City, Mexico	54°F	0.2 in.
Beirut, Lebanon	57°F	7.5 in.	Montreal, Canada	14°F	3.8 in.
Berlin, Germany	31°F	1.9 in.	Moscow	15°F	1.5 in.
Buenos Aires, Argentina	74°F	3.1 in.	Nairobi, Kenya	66°F	1.5 in.
Calcutta, India	68°F	0.4 in.	Paris, France	37°F	1.5 in.
Colombo, Sri Lanka	79°F	3.5 in.	Reykjavik, Iceland	32°F	4.0 in.
Copenhagen, Denmark	33°F	1.6 in.	Rio de Janiero, Brazil	79°F	4.9 in.
Dublin, Ireland	41°F	2.7 in.	Rome, Italy	47°F	3.3 in.
Hong Kong	60°F	1.3 in.	Shanghai, China	40°F	1.9 in.
Jerusalem, Israel	48°F	5.1 in.	Taipei, Taiwan	60°F	3.8 in.
Los Angeles, United States	55°F	3.1 in.	Tokyo, Japan	38°F	1.9 in.
London, United Kingdom	40°F	2.0 in.	Wellington, New Zealand	63°F	3.2 in.

19. Construct a scatter plot that depicts the average January temperature and average January rainfall for the 24 cities. Use the horizontal axis to represent temperature. See Additional Answers.
20. Describe the relationship between average January temperatures and average rainfall around the world. There does not appear to be any relationship.
21. Do you notice any clusters of points on your scatter plot? If so, what do they represent?
22. What is the lowest average January temperature shown on your scatter plot? What is the highest? 14°F, 79°F

Snowmobile Sales **In Exercises 23 and 24, use the table showing the amount spent (in millions of dollars) each year on snowmobiles. In the table, $t = 4$ corresponds to 1984.** *(Source: National Sporting Goods Association)*

Year, t	4	5	6	7	8	9	10	11	12	13	14
Amount	140	162	177	188	273	301	322	362	376	495	530

23. Construct a scatter plot that shows the data. See Additional Answers.

24. Describe the sales pattern from 1984 to 1994. Sales generally increased.

Patterns in Our Solar System **In Exercises 25 and 26, use the table showing the average surface temperature of each planet and the average distance of each planet (in millions of kilometers) from the sun.**

Planet	Mercury	Venus	Earth	Mars	Jupiter	Saturn	Uranus	Neptune	Pluto
Distance	58	108	150	228	778	1430	2875	4504	5900
Temperature	127°C	462°C	−16°C	−63°C	−148°C	−178°C	−216°C	−214°C	−228°C

25. Construct a scatter plot that shows the data. See Additional Answers.

26. Describe the relationship between the average distance from the sun and the average surface temperature. As the distance increases, does the temperature tend to increase or decrease? Explain. See margin.

28. $y = \frac{1}{2}(4 - x)$ or $2 - \frac{1}{2}x$

30. $b = \frac{1}{3}(-16 - 2a)$

Integrated Review

In Exercises 27–32, solve for the indicated variable.

$a = \frac{1}{2}(-16 - 3b)$ or $-8 - \frac{3}{2}b$

27. $2y + x = 4,\ x$ $\quad x = 4 - 2y$

28. $2y + x = 4,\ y$

29. $3b - (-2a) + 9 = -7,\ a$

30. $3b - (-2a) + 9 = -7,\ b$

31. $5m + 3n + 4 = -2n,\ n$ $\quad n = -m - \frac{4}{5}$

32. $12t - 2(3s) = 5t,\ s$ $\quad s = \frac{7t}{6}$

In Exercises 33–35, solve the equation.

33. $3a - (-21) = 7a - 7$ $\quad 7$

34. $-18 - 4c = 5 + 9c$ $\quad -\frac{23}{13}$ or $-1\frac{10}{13}$

35. $5m - 3^2 = -4m - 18$ $\quad -1$

Exploration and Extension

36. *Favorite TV Shows* Write a list of ten television shows. Put your favorite ones on the list and some you don't like. Arrange the shows in alphabetical order and make two copies. On one copy, rank the shows from 1 to 10 (1 for the most liked and 10 for the least liked). Then ask a friend to rank the shows (without looking at your rankings). Answers depend upon preferences.

37. Make a scatter plot that relates your television likes and dislikes to your friend's likes and dislikes. What conclusions can you draw from the scatter plot? Preferences vary.

EXTEND Ex. 36–37 An illustration of a possible outcome is given here. Suppose for five TV shows—A, B, C, D, and E—the rankings are as follows.

	Rankings	
TV Shows	Yours	Your Friend's
A	3	1
B	1	2
C	5	4
D	2	3
E	4	5

Here is the resulting scatter plot.

Rankings of TV Shows

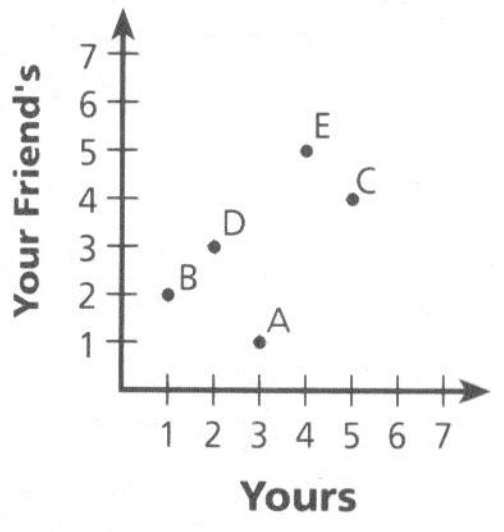

Enrichment Activities

Let students make measurements and compare the data using scatter plots.

1. Compare the length of your foot with the length of your forearm. Collect the data for your entire math class and represent them on a coordinate plane. Describe the pattern of points. The measurements should be about equal.
2. Compare the circumference of your wrist to the circumference of your neck. Collect the data for your entire math class and represent them on a coordinate plane. Describe the pattern of points. The circumference of your neck should be more than double the circumference of your wrist.

Answers

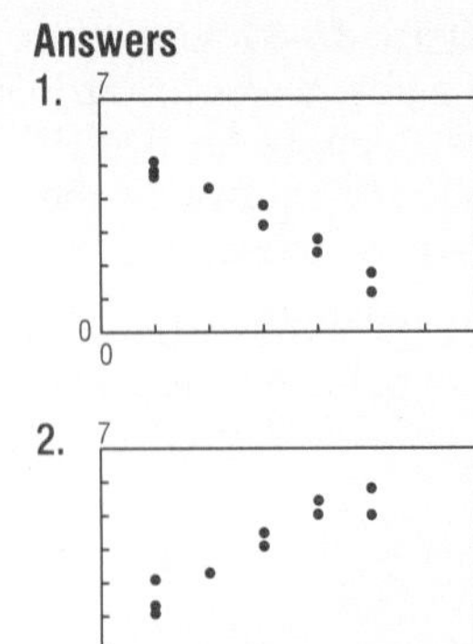

2\.

EXTEND Ex. 1–2 After completing the scatter plots, have students make up situations that could be described by the data points. For example, Exercise 1 could represent the tips earned by five different people. Exercise 2 could represent the weight losses of five different people one month at a health club. (Not everyone gets weighed in every time they go.)

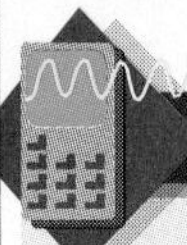

USING A GRAPHING CALCULATOR

A graphing calculator can be used to draw a scatter plot. Keystroke instructions for doing this on a TI-82, *Casio* fx-9700GE, *and Sharp* EL-9300C *are listed on page 736.*

The table shows the maximum time (in minutes) allowed for boys in the 1-mile run to qualify for the President's Physical Fitness Award.

Age (years)	6	7	8	9	10	11	12	13	14	15	16	17
Time (min)	10.25	9.37	8.80	8.52	7.95	7.53	7.18	6.83	6.43	6.33	6.13	6.10

You can write the data in the table as a set of ordered pairs.

(6, 10.25), (7, 9.37), (8, 8.80), (9, 8.52),
(10, 7.95), (11, 7.53), (12, 7.18), (13, 6.83),
(14, 6.43), (15, 6.33), (16, 6.13), (17, 6.10)

Drawing a scatter plot for this data helps you see patterns. What patterns can you see in the scatter plot at the right?

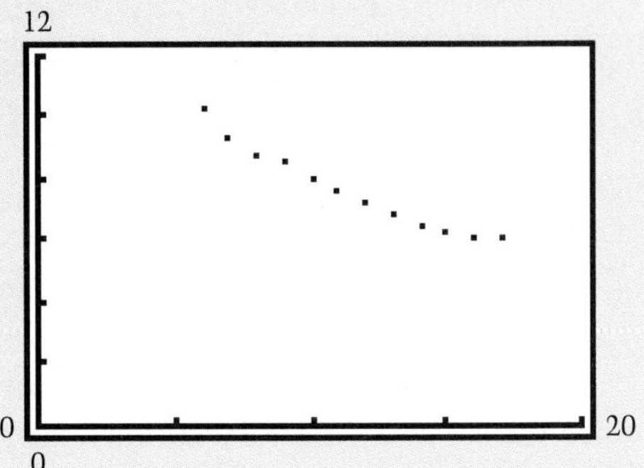

Focus on Functions

For this data, the maximum time depends on the age of the boy. Mathematically, you can describe this dependence by saying that the maximum time is a **function** of the age. Functions can be given by tables, graphs, equations, or verbal descriptions.

Throughout this text, you will study many functions that model real-life situations. Here are two examples.

Page 162 The amount A spent on archery equipment is a function of the year t. (The function is given by a table and by a graph.)

Page 215 The annual attendance A at National Hockey League games is a function of the year t. (The function is given by an equation.)

In Chapter 12, you will study some special notation and terminology that is used to describe functions.

Exercises

Use a graphing calculator or a computer to draw a scatter plot of the data. See margin.

1. (1, 5.1), (1, 4.8), (1, 4.7), (2, 4.3), (3, 3.8), (3, 3.2), (4, 2.8), (4, 2.4), (5, 1.8), (5, 1.2)

2. (1, 2.2), (1, 2.4), (1, 3.3), (2, 3.5), (3, 4.1), (3, 4.6), (4, 5.1), (4, 5.4), (5, 5.1), (5, 5.8)

3 Chapter Summary

What did you learn?

Skills

1. Solve a linear equation
   - using one transformation. **(3.1)**
   - using two or more transformations. **(3.2)**
   - with variables on both sides. **(3.3)**
   - involving decimals. **(3.5)**
2. Solve a literal equation
   - for one of its variables. **(3.6)**

Strategies

3. Create an algebraic model for a real-life situation
   - write a verbal model **(3.4)**
   - assign labels **(3.4)**
   - write a linear equation as an algebraic model **(3.4)**
   - use transformations to solve the equation **(3.4)**
   - answer questions about the real-life situation **(3.4)**

Exploring Data

4. Sketch a scatter plot of a data collection. **(3.7)**
5. Use a scatter plot to recognize patterns. **(3.7)**

Why did you learn it?

Linear models are the most commonly used algebraic models for real-life situations. By learning to write and solve linear models, you will be able to answer questions from a variety of fields: business, science, history, political science, engineering, and many more. Learning to create and solve linear models also prepares you for more complicated modeling that you will study later in this book.

How does it fit into the bigger picture of algebra?

In the first two chapters of this book, you used mental math to solve simple equations such as $x + 2 = 6$. But for solving complicated equations, such as $0.5(3x - 4) - 2(4x - 8) = 16$, you need other skills. But as messy as this equation might look, you saw that it really isn't that difficult, as long as you take it *one step at a time.* Remember that new mathematical skills are almost always built upon skills that you have already learned. The rules of algebra and the simplification techniques you learned in Chapter 2 form an essential part of the equation-solving techniques of Chapter 3.

ORGANIZER

TIME SCHEDULE: All levels, two days

The Chapter Summary helps students organize the main ideas of the chapter. In this chapter, the connection between modeling real-life problems and solving the resulting algebraic models was demonstrated.

Work with students to review the skills, strategies, and concepts of the chapter. The homework assignment can be used as the basis for the second day of review.

COOPERATIVE LEARNING

Encourage students to study together. Emphasize the importance of teaching a classmate how to perform a skill or how to recall a strategy. When students work together, everyone wins. The students receiving help get additional instruction, and the students giving help gain a deeper understanding of the skills and concepts involved.

Chapter SUMMARY

Review the chapter by asking students, "What did you learn? Why did you learn it? How does it fit in with the other algebra skills and concepts you have learned?"

Chapter REVIEW

ASSIGNMENT GUIDE

Basic/Average: Ex.1–21 odd, 25–28, 30–36

Above Average: Ex.1–21 odd, 25–28, 30–36

Advanced: Ex.1–21 odd, 25–27 odd, 29–36

✪ More Difficult Exercises

Exercises: 27, 28, 30, 34–36

▶ **Ex. 1–36** The lesson references in the Chapter Review direct students to the lesson in which each skill or strategy was first taught. Students should look back to this lesson if they are having difficulty attacking a set of exercises. Note also that the review exercises parallel the list of Skills, Strategies, and Exploring Data statements on page 167. Students may use this list to identify each skill or strategy, and perhaps check other chapters to see where this skill is used.

Chapter REVIEW

In Exercises 1–12, solve the equation. (3.1, 3.2, 3.3)

1. $12 = x - 2$ 14

2. $y + 4 = -6$ -10

3. $\frac{m}{8} = \frac{1}{4}$ 2

4. $7n = 28$ 4

5. $3p - (-4) = 13$ 3

6. $\frac{x}{6} + \frac{1}{3} = \frac{1}{2}$ 1

7. $\frac{1}{4}(y + 8) = 0$ -8

8. $16 = 5(1 - x)$ $-\frac{11}{5}$ or $-2\frac{1}{5}$

9. $9p + 7(p + 1) = 17 - 4p$ $\frac{1}{2}$

10. $\frac{1}{5}x - 3x + 2 = \frac{2}{5}x + 18$ -5

11. $\frac{2}{3}(x - 4) = 3(x + 6)$ $-\frac{62}{7}$ or $-8\frac{6}{7}$

12. $\frac{1}{4}x + 16 = 3(\frac{1}{4}x + \frac{4}{3})$ 24

In Exercises 13–18, solve the equation. Round your result to two decimal places. (3.5)

13. $5(x - 2) = -3(7 - 4x)$ 1.57

14. $\frac{4}{9}y = 6(y + 6)$ -6.48

15. $13.7t - 4.7 = 9.9 + 8.1t$ 2.61

16. $4.6(2a + 3) = 3.7a - 4$ -3.24

17. $-6(5.61x - 3.21) = 4.75$ 0.43

18. $7(81.74x - 0.31) - 6.21 = 6(94.28x + 21.11)$ 20.78

In Exercises 19–22, solve for the indicated variable. (3.6)

19. $4x - 3y = 2$, y $y = \frac{1}{3}(4x - 2)$

20. $4x - 3y = 2$, x $x = \frac{1}{4}(2 + 3y)$

21. $A = \frac{1}{2}bh$, b $b = \frac{2A}{h}$

22. $A = \frac{1}{2}h(b_1 + b_2)$, b_2 $b_2 = \frac{2A}{h} - b_1$

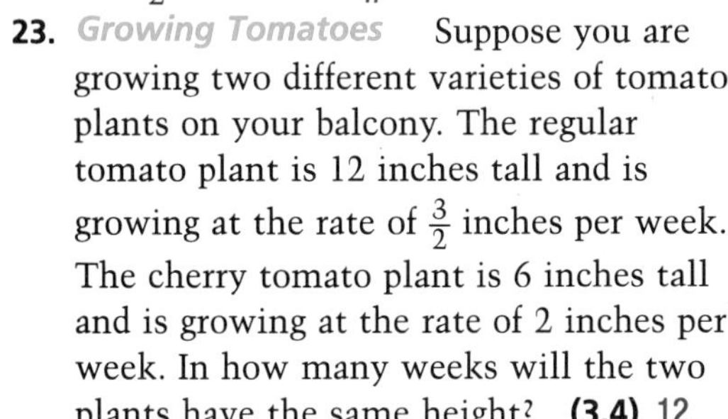

23. *Growing Tomatoes* Suppose you are growing two different varieties of tomato plants on your balcony. The regular tomato plant is 12 inches tall and is growing at the rate of $\frac{3}{2}$ inches per week. The cherry tomato plant is 6 inches tall and is growing at the rate of 2 inches per week. In how many weeks will the two plants have the same height? **(3.4)** 12

24. *Odometer Reading* You start a trip at 8:00 A.M. and the car's odometer reading is 65,660.1. At 4:00 P.M., when you reach your destination, the odometer reading is 66,035.8. What was your average speed? On the trip, you stopped for only an hour to have lunch and buy gas. What was your average *driving* speed? **(3.4)** ≈47.0 mph, ≈53.7 mph

25. *Baking Bread* You are baking raisin bread for a bake sale. For 2 loaves, the recipe calls for $2\frac{1}{2}$ cups of flour and $\frac{1}{3}$ cup of raisins. If you are making 5 loaves, how many cups of flour and raisins will you need? **(3.4)** Flour: $6\frac{1}{4}$, raisins: $\frac{5}{6}$

26. *Running a Race* Two runners left the starting line at the same time. The first runner crossed the finish line in 30 minutes and averaged 12 kilometers per hour. The second crossed the finish line 5 minutes later. Use the following verbal model to find the speed of the second runner. **(3.4)** 10.3 km per hour

Speed of 1st runner	•	Time for 1st runner	=	Speed of 2nd runner	•	Time for 2nd runner

✪ **27.** *Dwindling Rain Forest* In 1990, the area covered by rain forest was decreasing at about 78,000 square miles per year. Suppose that the desert areas were increasing at about 50,000 square miles per year. At those rates, how many years would pass before Earth's deserts cover twice the land area that the rain forests cover? **(3.4)** $148\frac{1}{2}$

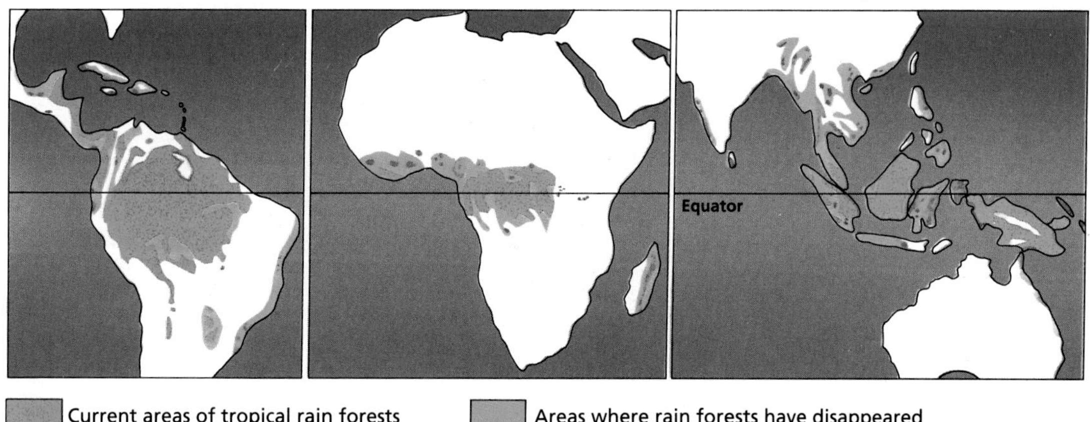

In 1990, there were about 35 million square miles of rain forest and 39.4 million square miles of desert. (Source: Smithsonian Institution)

✪ **28.** *Chemical Mixture* Your science teacher instructs you to add 10 milliliters of Solution B, which is 70% water and 30% citric acid, to 20 milliliters of Solution A, which is 80% water and 20% citric acid. How many milliliters of acid are in the combined mixture? What percent acid is the combined mixture? **(3.5)** 7 mL, $23\frac{1}{3}$%

29. *Chemical Mixture* You are still in the same science class (see Exercise 28), and your teacher says that she didn't want you to pour *all* of Solution B into Solution A—just enough to raise the acid percentage to 22%. How much of Solution B should you have poured into Solution A? Answer the question by assigning labels to the following verbal model and solving the resulting equation. **(3.5)** $x + 20$, 5 mL

Percent	Total amount of mixture	Percent	Amount of mixture used	Percent	Total amount of mixture
Acid in A		Acid in B		Acid in 22% mixture	

Percent acid in A = 0.2 (mL)
Total volume of A = 20 (mL)
Percent acid in B = 0.3 (mL)

Volume of B added = x (mL)
Percent acid in 22% solution = 0.22 (mL)
Total amount of 22% mixture = [?] (mL)

✪ **30.** *Buying a Book* A book that you have been wanting to buy has just been reduced to 65% of its original price. The sales tax is 7%. You pay the cashier \$25 and receive \$3.13 in change. Use the following model to find the original price, p, of the book. **(3.5)** \$31.45

[\$25] – ([Discounted price] + [Sales tax %] • [Discounted price]) = [\$3.13]

Ex. 35–36 Students may need help to solve these problems. For example, to solve Exercise 35, students first need to read the bar graph to find West Germany's total trade amount of 470 billion dollars. If x = the import amount then $470 - x$ = the export amount. Since there is a 60 billion dollar trade surplus, you solve the following equation for x.
$470 - x - x = 60$ ($x = 205$)
In Exercise 36, solve this equation for x.
$-(280 - x) = 4.5$ ($x = 142.25$)

Imports and Exports **In Exercises 31–36, use the scatter plot below and the stylized bar graph at the right. (3.7)**

The bar graph at the right shows the 48 countries that traded the most in 1987. Each crate represents $10 billion of trade. The total amount traded is the sum of the total value of exports and imports. A country that exports more than it imports has a trade surplus. A country that imports more than it exports has a trade deficit.

□ = $10bn

U.S.A
W. Germany
Japan
France
U.K.
Italy
Canada
Netherlands
Belgium
Switzerland
Hong Kong
Taiwan
S. Korea
Sweden
Spain
China
Singapore
Austria
Australia
U.S.S.R.
Denmark
Brazil
Norway
Saudi Arabia
Finland
S. Africa
Mexico
Indonesia
India
Malaysia
Turkey
Ireland
Thailand
Yugoslavia
Portugal
Israel
Iran
E. Germany
Iraq
Venezuela
U.A.E.
Algeria
Greece
New Zealand
Kuwait
Nigeria
Philippines
Argentina

1975, 1976–1988

31. During which year(s) from 1975 to 1988 did the United States have a trade surplus? During which year(s) did it have a trade deficit?

32. During which year(s) from 1975 to 1988 did the United States have the greatest trade deficit? 1986 and 1987

33. Use the scatter plot to estimate the total U.S. imports in 1987 and the total exports in 1987. Add the two values. Does your estimate agree with the number given by the bar chart at the right? (It should.) $405 billion, $255 billion, $660 billion, yes

✪ **34.** In 1987, Japan imported approximately $129 billion worth of merchandise. Approximate the amount that Japan exported in 1987. Did Japan have a trade surplus or a trade deficit?

✪ **35.** In 1987, West Germany had a trade surplus of approximately $60 billion. How much did it import? How much did it export?

✪ **36.** In 1987, France had a trade deficit of approximately $4.5 billion. How much did it import? How much did it export?

34. $221 billion, trade surplus 35. $205 billion, $265 billion

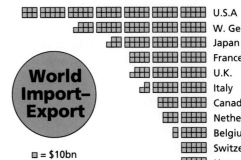

This scatter plot shows the total value of imports and exports (in billions of dollars) for the United States from 1975 to 1988.

U.S. Imports and Exports

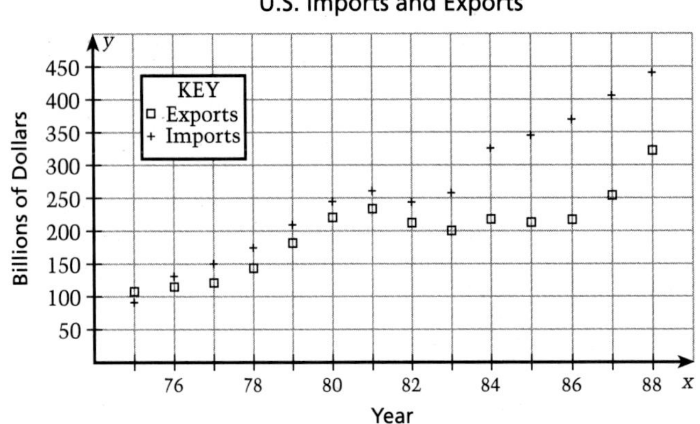

36. $142.25 billion, $137.75 billion

Chapter TEST

In Exercises 1–8, solve the equation. (3.1, 3.2, 3.3)

1. $3c = 36$ 12
2. $\frac{x}{4} = 20$ 80
3. $4y - 3(y + 8) = 12$ 36
4. $\frac{1}{4}m + 3 = 2$ -4
5. $6 + 9x - 2 = 7x + (-2x) + 8$ 1
6. $\frac{3x}{4} - 6 = x - 8$ 8
7. $\frac{3}{8}(p - 8) + \frac{1}{8}p = -\frac{1}{2}(p - 7)$ $\frac{13}{2}$ or $6\frac{1}{2}$
8. $(3 \div 4) - r(r - \frac{1}{2}) = r(3 - r)$ $\frac{3}{10}$

In Exercises 9–12, solve the equation. Round your result to two decimal places. (3.5)

9. $19x - 12 = 4(6 - 4x)$ 1.03
10. $\frac{18y}{3} - 7 = 7y$ -7
11. $27.3m - 45.1 = -13.2 + 6.8m$ 1.56
12. $7(8.65x - 37.61) = (28.45 + 3.75x)(-4)$ 1.98

In Exercises 13 and 14, solve for the indicated variable. (3.6)

13. $C = 2\pi r$, r $r = \frac{C}{2\pi}$
14. $S = 2\pi r^2 + 2\pi rh$, h $h = \frac{S - 2\pi r^2}{2\pi r}$

15. A rectangular picture frame has a perimeter of 44.2 centimeters. The width of the frame is seven tenths of the length. What are the dimensions of the frame? (3.4) 13 cm, 9.1 cm

16. You and your cousin are earning money for your family's vacation trip. Your cousin averages \$15 a week babysitting and receives one \$5 bonus. You average \$10 a week mowing lawns and \$8 a week running errands for neighbors. After working the same number of weeks, you end up with \$7 more than your cousin. How many weeks did you work? Use a verbal model, assign labels, and write an algebraic equation to answer the question. (3.4) 4

17. You deposited \$300 in a savings account. After one year, the balance was \$314.70. Use the formula $A = P + Prt$ to find the interest rate for your account. (3.6) 4.9%

18. Construct a scatter plot for the data in the table. Assign the average daily temperature to the horizontal axis and assign latitude to the vertical axis. Describe any patterns you see in the scatter plot. (3.7) See margin.

City	Average Temp.	Latitude	City	Average Temp.	Latitude
Albany, NY	47.3°F	42.39°N	Atlanta, GA	61.2°F	33.45°N
Atlantic City, NJ	53.1°F	39.27°N	Baltimore, MD	55.1°F	39.17°N
Boston, MA	51.5°F	42.21°N	Burlington, VT	44.1°F	44.28°N
Charlotte, NC	60.0°F	35.14°N	Columbia, SC	63.3°F	34.00°N
Concord, NH	45.3°F	43.12°N	Hartford, CT	45.3°F	41.46°N
Jacksonville, FL	68.0°F	30.20°N	Miami, FL	75.6°F	25.46°N
Mobile, AL	67.5°F	30.42°N	Norfolk, VA	59.5°F	38.40°N
New York, NY	54.5°F	40.43°N	Philadelphia, PA	54.3°F	39.47°N
Portland, ME	45.0°F	43.39°N	Providence, RI	50.3°F	41.50°N
Raleigh, NC	59.0°F	35.47°N	Richmond, VA	57.7°F	37.30°N

In general, the lower the latitude, the higher the temperature.

Answer

18.

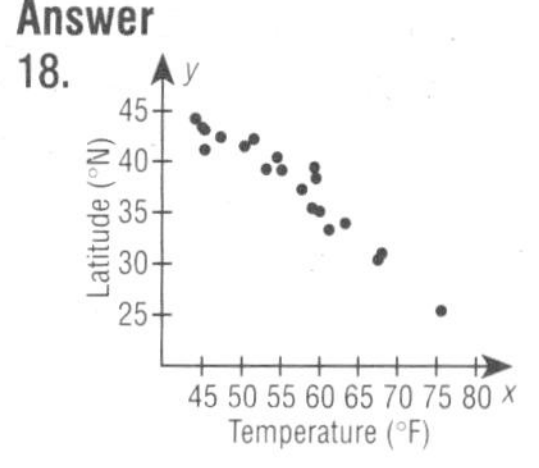

Cumulative REVIEW ■ *Chapters 1–3*

In Exercises 1–12, evaluate the expression.

1. $3 + 2(24 \div 6)$ 11
2. $4.9 \div (4 + 3) + 0.6$ 1.3
3. $[(13 - 8)^2 + 2] \div 3$ 9
4. $32 \div (-4) - (2 - 8)$ -2
5. $|3 - 10| - 3.7(4)$ -7.8
6. $(6.8)^2 - |11.1 - 2.2|$ 37.34
7. $(47 - 36.5)9 + 6.1$ 100.6
8. $(29 + 1) \div (-5)$ -6
9. $\frac{4(3 - 2)}{(6)(3)}$ $\frac{2}{9}$
10. $\frac{3^2 \cdot 4}{(42 - 21)}$ $\frac{12}{7}$
11. $\frac{3^2 - 8(6)}{7}$ $-\frac{39}{7}$
12. $\frac{8 + 2(-4)}{6(-6)}$ 0

In Exercises 13–18, solve the equation or inequality.

13. $4(x - 2) = 3^2 - x$ $\frac{17}{5}$
14. $\frac{1}{3}n + 3 = n - 2$ $\frac{15}{2}$
15. $9(2p + 1) - 3p > 4p - 6$ $p > -\frac{15}{11}$
16. $(q - 12)3 \le 5q + 2$ $q \ge -19$
17. $\frac{2y}{3} = \frac{8}{27}$ $\frac{4}{9}$
18. $\frac{m}{12} + \frac{5}{6} = \frac{5}{24}$ $-\frac{15}{2}$

In Exercises 19–22, solve the equation or inequality. Round your result to two decimal places.

19. $3.2(x - 0.1) = 4.2$ 1.41
20. $(19.8 - r)6.2 = 37.7r - 49.2$ 3.92
21. $18t(4.2 - 36.1) < -86.13$ $t > 0.15$
22. $41.2(2a - 0.6) > (0.25a + 3)99.9$ $a > 5.65$

In Exercises 23 and 24, perform the matrix operation.

23. $\begin{bmatrix} 47 & -3 & 22 \\ 6 & -17 & 9 \end{bmatrix} + \begin{bmatrix} -33 & 18 & 10 \\ 44 & -24 & -44 \end{bmatrix}$

23. $\begin{bmatrix} 14 & 15 & 32 \\ 50 & -41 & -35 \end{bmatrix}$

24. $\begin{bmatrix} 4 & 10 \\ -6 & 18 \\ 29 & -14 \end{bmatrix} - \begin{bmatrix} 56 & 2 \\ 4 & -11 \\ 29 & 7 \end{bmatrix}$ $\begin{bmatrix} -52 & 8 \\ -10 & 29 \\ 0 & -21 \end{bmatrix}$

In Exercises 25–30, translate the phrase or sentence into an expression, an equation, or an inequality.

25. The difference of three times a number and 27 $3n - 27$
26. The quotient of 65 and a number is less than 14. $\frac{65}{n} < 14$
27. The sum of 27 and a number, multiplied by $\frac{1}{2}$, is 37. $\frac{1}{2}(n + 27) = 37$
28. The square of a number, plus 18, is greater than or equal to 87. $n^2 + 18 \ge 87$
29. The absolute value of a number is less than the quotient of 36 and 5. $|n| < \frac{36}{5}$
30. The square of forty less than two times a number $(2n - 40)^2$

31. *Fund-Raiser* The members of the school band are selling wrapping paper to raise money to attend a band competition. For each order of wrapping paper, the band makes $1. The fifty band members need a total of $550 to pay for registration fees, transportation, and food. Each band member will pay $3 toward the cost. How many orders of wrapping paper must the band members sell to meet their goal? 400

32. *Maple Syrup* The table shows the retail price per gallon of maple syrup for 1976 through 1988. Construct a scatter plot of this information. Use the horizontal axis to represent the year and the vertical axis to represent the retail price. *(Source:* **The Vermont Almanac***)*

Year, t	1976	1977	1978	1979	1980	1981	1982
Price, p	\$11.20	\$12.00	\$12.93	\$14.26	\$16.64	\$18.60	\$18.39

Year, t	1983	1984	1985	1986	1987	1988
Price, p	\$17.95	\$18.08	\$19.05	\$23.45	\$28.10	\$33.00

See Additional Answers.

Cafeteria Tables **In Exercises 33 and 34, use the following information.**

The school cafeteria got rid of some of the old tables and bought new ones. The old tables were 2 feet wide and 6 feet long. The new tables were the same width but were 8 feet long.

33. What is the ratio of the area of a new table to the area of an old table? $\frac{4}{3}$

34. There are now 24 new tables in the cafeteria. If the cafeteria has the same table area as before, how many old tables were removed? 32

Seek-n-Find **In Exercises 35 and 36, use the following information.**

Your sister received a Seek-n-Find puzzle book for her birthday. It took her 15 minutes to complete the first puzzle containing 30 words.

35. What was the average number of words she found per minute? 2

36. The daily puzzle in the newspaper contained 18 words which your sister found in 8 minutes. Did she find words faster in the newspaper or the puzzle book? Newspaper

D	X	E	B	A	I	Z	W	K	M
O	E	L	O	N	E	C	H	B	O
G	B	E	A	R	K	A	A	A	U
C	G	P	R	O	K	M	L	T	S
M	E	H	O	R	S	E	E	F	E
O	H	A	S	E	A	L	T	O	X
M	O	N	K	E	Y	B	I	Y	X
L	G	T	I	G	E	R	B	O	W
Z	E	B	R	A	S	T	F	I	N
P	O	R	C	U	P	I	N	E	T

SEEK-N-FIND 20 Types of Mammals

37. *Shoe Sale* The shoes you were saving for are now on sale for \$17.56. The original price was \$21.95. What percent is the discount? 20%

38. *Geometry* The volume of a right circular cylinder with a radius of 1.5 inches is 42.4 cubic inches. What is the height of the cylinder? 6 inches

In this chapter, techniques for graphing linear and absolute value equations will be developed. Emphasis will be given to sketching "quick graphs" instead of plotting points. Connections between solutions of linear and absolute value equations and their graphs will be made.

Indicate to students that using graphs allows you to connect ideas in algebra with ideas in geometry. Another way of stating this is that graphing provides a way of "picturing" patterns and relationships.

Frequently, the sketch of a graph provides useful information about the real-life situation it models. More often, a comparison of graphs helps you decide among several options. Consequently, graphs are useful tools for understanding and managing the world around us.

The Chapter Summary on page 229 provides you and the students with a synopsis of the chapter. It identifies key skills and concepts. You may want to have students look at the Chapter Summary as an overview before beginning the chapter.

CHAPTER 4

Graphing Linear Equations

LESSONS

San Francisco traffic is a colorful reminder of the millions of automobiles in the United States.

You can use the graphs of these algebraic models to easily compare the registration rates of automobiles within and outside the United States. The straight line representing the registration of cars outside the United States is steeper than the line of registered cars in the United States. What interpretation can be assigned to this observation? In the same time period, the rate of registrations outside the United States is growing faster than inside the United States. What information is provided by the y-intercepts of each line? The number of registrations in 1970.

As students learn more about the graphs of linear equations, they can use their knowledge to interpret graphs to answer questions and solve problems.

Real Life

Automobile Statistics

Registered Automobiles

Outside the United States	United States

Automobiles (in millions): 0, 50, 100, 150, 200, 250, 300, 350 (y)

Year (0↔1970): 0 2 4 6 8 10 12 14 16 18 20 (t)

From 1970 to 1990, the number of registered automobiles in the United States increased by a rate of about 4 million automobiles per year. A model for this is $y = 4t + 110$, where y is the number of automobiles (in millions) and t is the year, with $t = 0$ corresponding to 1970. The graph of this model is shown in blue on the coordinate plane. It is a line whose y-intercept is 110 and whose slope is 4.

The black line in the coordinate plane is the graph of $y = 12t + 140$. It is a model for the number of registered automobiles in the world, excluding the United States.

(Source: Scientific American)

Graphing Linear Equations

PACING CHART

The daily Pacing Chart is meant to help you adjust your teaching pace. Students in the full course should finish the entire text by the end of the year. Students in the basic course may not complete the entire text in the school year. The Pacing Chart for each chapter contains suggestions for lessons that require more than one day and lessons that may be omitted for the basic course.

DAY	FULL COURSE	BASIC COURSE
1	4.1	4.1
2	4.2 & Using a Graphing Calculator	4.2 & Using a Graphing Calculator
3	4.3	4.3
4	4.4	4.4
5	Mid-Chapter Self-Test	Mid-Chapter Self-Test
6	4.5	4.5
7	4.6	4.6
8	4.7	4.7
9	4.7	4.7
10	4.7	4.7
11	4.8	4.8
12	4.8	4.8
13	4.8	4.8
14	Chapter Review	Chapter Review
15	Chapter Review	Chapter Review
16	Chapter Test	Chapter Test

CHAPTER ORGANIZATION

LESSON	PAGES	GOALS	MEETING THE NCTM STANDARDS
4.1	176-182	1. Graph horizontal and vertical lines 2. Use equations of horizontal and vertical lines in real-life settings	Problem Solving, Communication, Connections, Statistics, Geometry
4.2	183-188	1. Graph a linear equation from a table of values 2. Interpret graphs of linear equations	Problem Solving, Communication, Connections, Statistics, Reasoning, Geometry
Using a Calculator	189	Use a graphing calculator or computer to sketch graphs of linear equations	Technology, Geometry
4.3	190-195	1. Find the intercepts of the graph of a linear equation 2. Use intercepts to sketch a quick graph of a line	Problem Solving, Communication, Connections, Geometry
Mixed Review		Review of algebra and arithmetic	
Milestones		Calculating through the centuries	Connections
4.4	197-203	1. Find the slope of a line using two of its points 2. Interpret slope as constant or average rate of change	Problem Solving, Communication, Connections, Statistics, Geometry, Reasoning, Calculus
Mid-Chapter Self-Test	204	Diagnose student weaknesses and remediate with correlated Reteaching worksheets	
4.5	205-210	1. Find the slope and y-intercept from an equation 2. Use the slope-intercept form to sketch a line and solve problems	Problem Solving, Communication, Connections, Geometry, Reasoning
4.6	211-216	1. Use a graph as a quick check of a solution found algebraically 2. Approximate solutions of real-life problems by using a graph	Problem Solving, Communication, Connections, Reasoning, Geometry, Technology
Mixed Review	216	Review of algebra and arithmetic	
4.7	217-222	1. Graph an absolute value equation 2. Model a real-life situation using graphs of absolute value equations	Problem Solving, Communication, Reasoning, Connections, Technology, Geometry
4.8	223-228	1. Solve and check absolute value equations algebraically 2. Use a graph to check solutions of absolute value equations	Problem, Solving, Communication, Reasoning, Connections, Geometry
Chapter Summary	229	A restatement of what has been learned, why it has been learned, and how it fits into the structure of algebra	Structure, Connections
Chapter Review	230-232	Review of concepts and skills learned in the chapter	
Chapter Test	233	Diagnose student weaknesses and remediate with correlated Reteaching worksheets	

LESSON RESOURCES

MEETING INDIVIDUAL NEEDS

RETEACHING For students who need to spend more time on basics:

If a mid-chapter self-test or chapter test indicates a deficiency, teachers can help students with the appropriate ***Reteaching Copymaster.***

PRACTICE For students who need more practice:

Additional exercises like those in the Pupil's Edition are provided for each lesson in ***Extra Practice Copymasters.***

ENRICHMENT For enriching and broadening students' experiences:

Problem of the Day copymasters in ***Teaching Tools*** provide a daily opportunity to use logical reasoning, looking for a pattern, writing an equation, and other routine and non-routine problem-solving strategies.

Math Log copymasters in ***Alternative Assessment*** provide opportunities to report on investigations, research, and open-ended problems.

Enriching activities with graphing and scientific calculators and computers are provided in ***Technology: Using Calculators and Computers.***

The ***Applications Handbook*** provides additional information about the cross-curriculum topics such as astronomy, chemistry, physics, sports, economics, genetics, and music that are integrated into the Pupil's Edition.

LESSON	4.1	4.2	4.3	4.4	4.5	4.6	4.7	4.8
PAGES	176-182	183-188	190-195	197-203	205-210	211-216	217-222	223-228
Teaching Tools								
Transparencies	✓	✓	✓	✓	✓	✓	✓	✓
Problem of the Day	✓	✓	✓	✓	✓	✓	✓	✓
Warm-up Exercises	✓	✓	✓	✓	✓	✓	✓	✓
Answer Masters	✓	✓	✓	✓	✓	✓	✓	✓
Extra Practice Copymasters	✓	✓	✓	✓	✓	✓	✓	✓
Reteaching Copymasters	Teacher-directed and independent activities tied to results on the Mid-Chapter Self-Tests and Chapter Tests							
Color Transparencies	✓			✓			✓	✓
Applications Handbook	Additional background information is supplied for many real-life applications.							
Technology Handbook	Calculator and computer worksheets are supplied for appropriate lessons.							
Complete Solutions Manual	✓	✓	✓	✓	✓	✓	✓	✓
Alternative Assessment	Assess student's ability to reason, analyze, solve problems, and communicate using mathematical language.							
Formal Assessment	Mid-Chapter Self-Tests, Chapter Tests, Cumulative Tests, and Practice for College Entrance Tests							
Computer Test Bank	Customized tests can be created by choosing from over 2000 items.							

INSIGHTS

4.1 Graphing Linear Equations in One Variable

Many real-life relationships can be modeled by linear equations. For example, if the postage for a special-delivery package weighing less than 2 pounds is $7.65, then for packages weighing less than 2 pounds the relationship between the weight of a package (in pounds) and the cost of delivery (in dollars) can be modeled using a linear equation in one variable.

4.2 Graphing Linear Equations in Two Variables

Many real-life situations can be modeled by relationships between two variables, such as the cost, c, of producing x calculators, or the volume, v, of a cube with sides x cm long. Tell students that graphs of linear equations in two variables provide "pictures" or visual representations of relationships. Such models are often useful in understanding real-life problems.

4.3 Quick Graphs Using Intercepts

In Lesson 4.2, students saw that only two points were needed to determine a line. Thus, instead of constructing a table of values for several solutions of a linear equation, efforts can be reduced to identifying any two points on the line. This lesson shows students how to identify two points on the line easily: locate the x- and y-intercepts of the line.

4.4 The Slope of a Line

In this lesson, the concept of slope of a line is presented. Real-life examples of slope include the pitch of a roof, the grade of a highway, and the slant of a wall. The concept of slope is used to describe constant rates of change in economics, physics, and statistics, among other areas. Undoubtedly, students will encounter many real-life applications of slope. (See Exercises 43–50.)

4.5 Quick Graphs Using Slope-Intercept Form

In Lesson 4.2, the x- and y-intercepts were used to obtain a quick graph of a straight line. In this lesson, another quick-graph method that uses the y-intercept and the slope will be explored.

Quick graphs are a useful mental-math technique, even with the availability of graphing calculators, because knowing the steepness and y-intercept of a line often provides enough information to draw the graph.

4.6 Connections: Solutions and x-Intercepts

The link between geometry and algebra, the Cartesian plane (named for René Descartes), makes it possible for graphics calculators and computers to "solve" algebraic equations graphically using the "trace" feature. Consequently, it is helpful in real-life applications for students to understand the connection between solutions and x-intercepts.

4.7 Graphs of Absolute Value Equations

The distance from home to school is the same whether one is at home going to school, or at school returning home. This fundamental relationship of the distance between two locations has applications in physics, transportation, chemistry, and mathematics. The distance between two locations is described using absolute value, the focus of this lesson.

4.8 Solving Absolute Value Equations

Discuss the following with your students. Your doctor has prescribed an amount of medication for you. The tolerance (the amount by which the dosage may vary) for someone your age can be read from a chart. Because measuring the medication is not exact, what is the smallest amount of medication you should receive? What is the largest amount? The answers to such questions, which can have serious consequences, can be found by solving absolute value equations!

Problem of the Day

In each puzzle, each letter represents a different digit from 0 through 9. Find the digit each letter represents.

a. HAD × XD = ASH   b. SET × XS = THAT

a. H = 4, A = 9, D = 2, S = 8
b. S = 6, E = 8, T = 4, H = 1, A = 0

ORGANIZER

Warm-Up Exercises

1. Identify the y-coordinate in each ordered pair.
   a. $(-3, -6)$ b. $(-0.07, 3.5)$
   c. $(1.2, m)$ d. $(x + 2, -1)$
   a. -6, b. 3.5, c. m, d. -1
2. Plot these points on a coordinate plane. Describe any pattern observed.
   a. (1, 7.5), (2, 7.5), (3.5, 7.5), (4, 7.5), (12.5, 7.5), (15, 7.5)
   The points fall on the horizontal line $y = 7.5$.
   b. $(-3.3, 2)$, $(-3.3, -5)$, $(-3.3, 7.5)$, $(-3.3, 0)$, $(-3.3, -1)$, $(-3.3, 10)$
   The points fall on the vertical line $x = -3.3$.

Lesson Resources

Teaching Tools
- Transparency: 4
- Problem of the Day: 4.1
- Warm-up Exercises: 4.1
- Answer Masters: 4.1

Extra Practice: 4.1
Color Transparencies: 18, 19
Technology Handbook: p. 33

4.1 Graphing Linear Equations in One Variable

What you should learn:

Goal 1 How to graph horizontal and vertical lines in the coordinate plane

Goal 2 How to use equations of horizontal and vertical lines in real-life settings

Why you should learn it:

You can use horizontal and vertical lines as models of real-life situations, such as for approximating lines of latitude and longitude.

Goal 1 Horizontal and Vertical Lines

In this lesson, you will study relationships between two variables in which one of the variables always has the same value. For instance, in the following collection of ordered pairs, (x, y), the y-coordinates vary, but the x-coordinates are always 3.

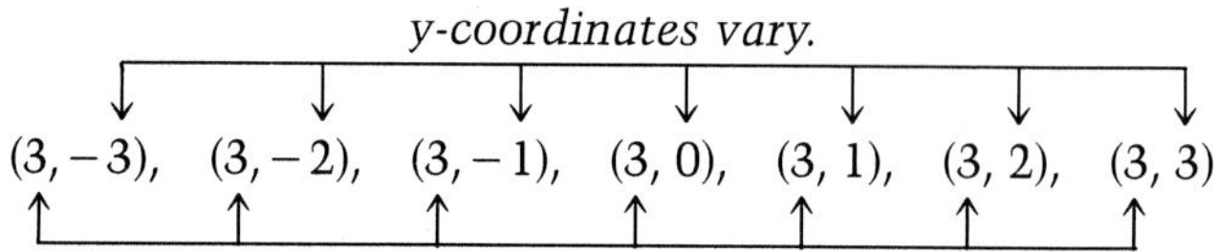

The equation that represents this relationship between x and y is $x = 3$. In the coordinate plane, the *graph* of $x = 3$ is the vertical line that crosses the x-axis at 3 because *all* the points of this line, and *no other* points, have x-coordinate 3.

In a similar way, we could show that the graph of the equation $y = 2$ is a horizontal line. On this line, the x-coordinates vary and the y-coordinates are all 2.

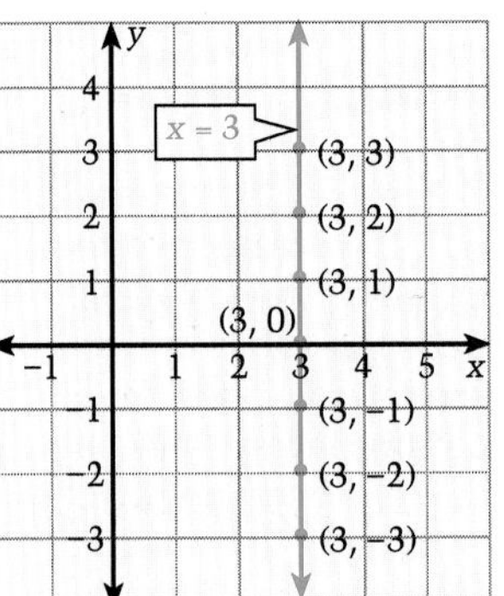

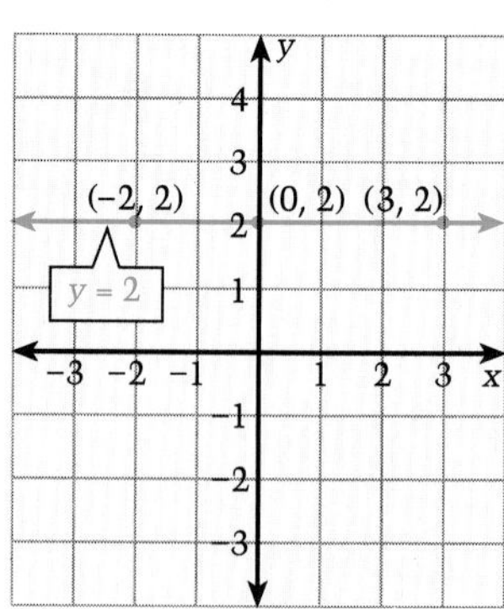

Equations of Horizontal and Vertical Lines

1. In the coordinate plane, the graph of $x = a$ is a vertical line.
2. In the coordinate plane, the graph of $y = b$ is a horizontal line.

Example 1 Sketching Horizontal and Vertical Lines

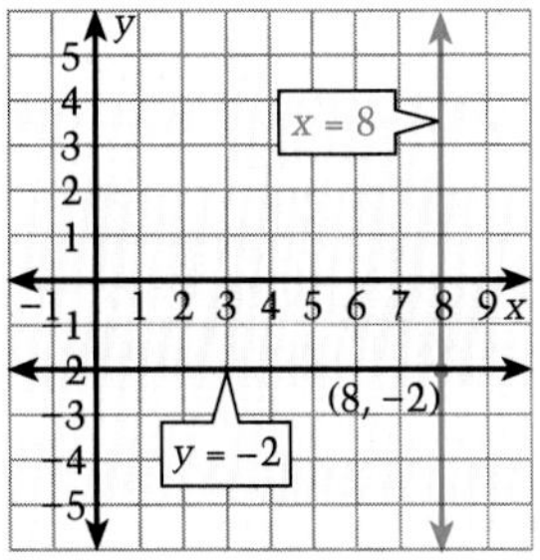

Sketch the graphs of $x = 8$ and $y = -2$. Find the point of intersection of the two graphs.

Solution The graph of $x = 8$ is a vertical line because the y-coordinates can have any value, and the x-coordinates must have a value of 8. The graph of $y = -2$ is a horizontal line because the x-coordinates can have any value, and the y-coordinates must have a value of -2.

Vertical line		Horizontal line	
x must be 8.	*y can have any value.*	*x can have any value.*	*y must be −2.*
$(8, y)$		$(x, -2)$	

The point of intersection of these two lines must have an x-coordinate of 8 and a y-coordinate of -2. Thus, the point of intersection is $(8, -2)$. ■

Example 2 Writing Equations for Horizontal and Vertical Lines

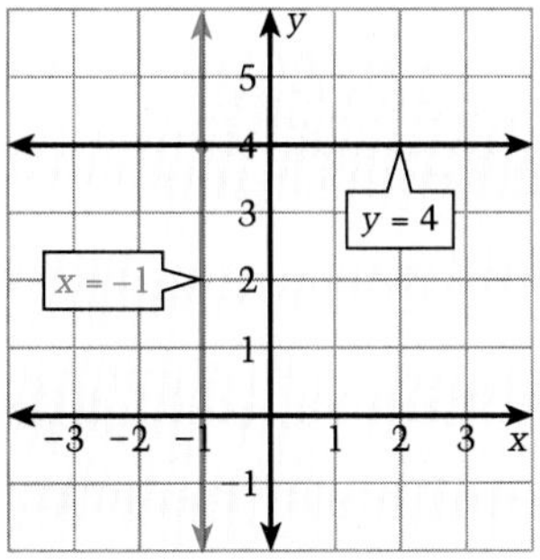

Write equations for the horizontal line and the vertical line that pass through the point $(-1, 4)$.

Solution Every point on the horizontal line through $(-1, 4)$ must have a y-coordinate of 4. Thus, the equation of the line is

$y = 4.$ *Horizontal line*

Every point on the vertical line through $(-1, 4)$ must have an x-coordinate of -1. Thus, the equation of the line is

$x = -1.$ *Vertical line* ■

Example 3 Describing an Activity to Match a Graph

The graph at the left shows three horizontal line segments. Describe a real-life situation that could be represented by this graph.

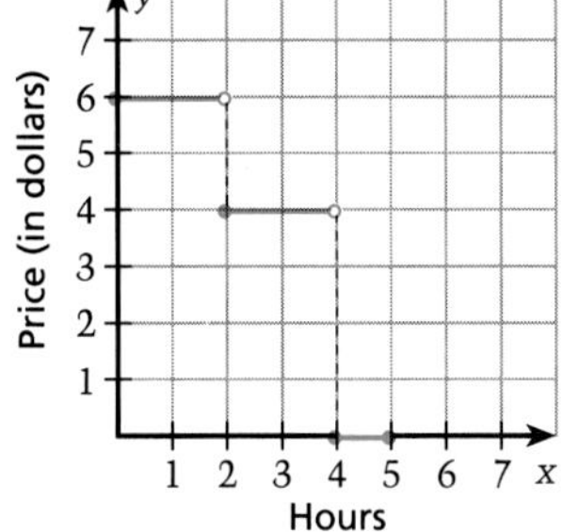

Solution Here is one possible solution. Kareem is running a garage sale. Among the many items for sale is an old sweatshirt that Kareem marked at \$6. After two hours, the sweatshirt has not been sold and Kareem marks the price down to \$4. After two more hours, he puts the sweatshirt in a box marked *free*. After one more hour, someone finally takes the sweatshirt. ■

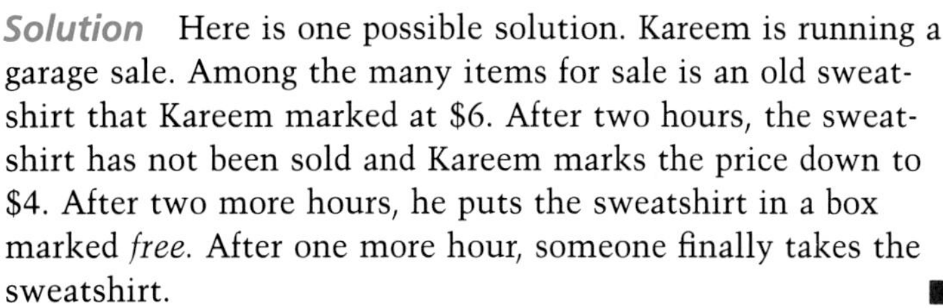

LESSON Notes

Linear equations in one variable represent vertical and horizontal lines in the coordinate plane. A linear equation in one variable, x, is any equation of the form $Ax = C$, where $A \neq 0$. The graph of $Ax = C$ is the set of all ordered pairs (x, y) for which x satisfies $Ax = C$. Similarly, $By = C$ is a linear equation in y, and its graph is the set of all ordered pairs (x, y) for which y satisfies $By = C$.

Example 1

The intersection is a point in the coordinate plane that satisfies both linear equations in one variable.

Example 3

Open dots in the graph of a relationship indicate that the point represented by the open dot is not a part of the graph. Closed dots indicate that the point represented by the closed dot is included in the graph of the relationship.

Example 4

A scatter plot is used to organize and interpret the data in the table.

Example 5

Remind students that the surface of the earth is spherical. Consequently, the borders of Colorado can only approximate a rectangle.

Common-Error ALERT!

In **Example 2,** many students confuse the line $y = 4$ with the point $y = 4$ on the real number line. Students need to be aware of the context in which $y = 4$ is presented to know whether it refers to the line in the plane or the point on the number line.

Extra Examples

Here are additional examples similar to **Examples 1–3.**

1. Sketch the graphs of $x = -4$ and $y = 6$. Find the point of intersection. The graph of $x = -4$ is a vertical line. The graph of $y = 6$ is a horizontal line. The point of intersection is $(-4, 6)$.
2. Sketch the line containing the points $(-3, 2)$, $(-3, -1)$, $(-3, -9)$, $(-3, 7)$, and $(-3, 0)$. Write an equation that models this line.
   $x = -3$

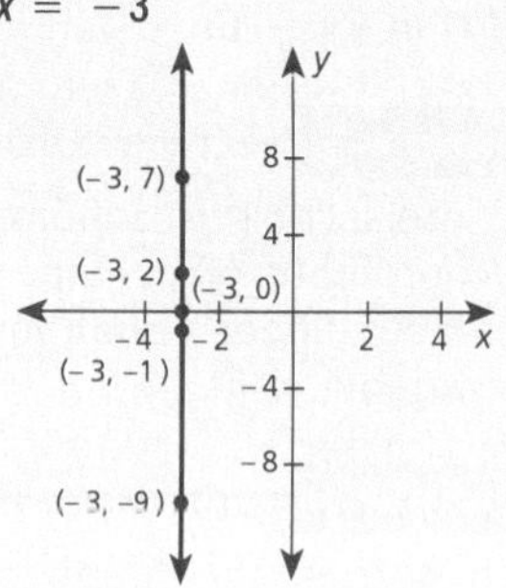

3. Describe a situation that this graph could model:

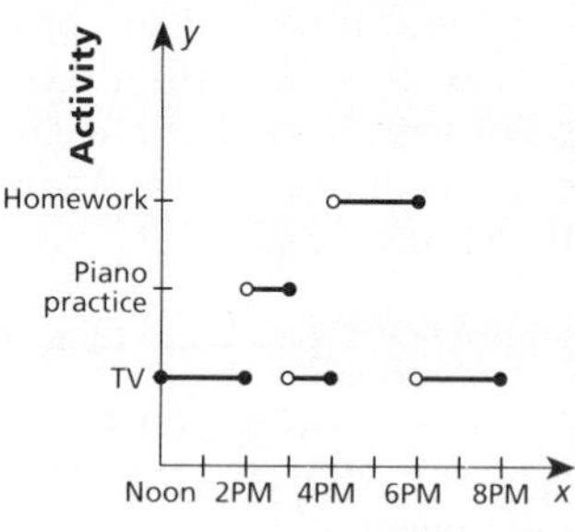

Sample answer: A student watched TV from noon to 2:00 PM, practiced piano until 3:00, watched TV until 4:00, did homework until 6:00, and then watched TV to 8:00 PM.

Check Understanding

1. Explain why the equation $x = a$ is a reasonable representation of the vertical line at $x = a$. x must equal a, while y can equal any number.
2. Where do the lines $x = 6$ and $y = 4$ intersect? Is the point, $(6, 4)$, a solution to each of the equations? Explain. At $(6, 4)$. Yes, the point $(6, 4)$ lies on the line $x = 6$ and it also lies on the line $y = 4$.

Goal 2 Using Horizontal and Vertical Lines

Real Life
Animal Studies

Example 4 *Alligator's Tail Length and Body Length*

The tail lengths and body lengths (in feet) of 24 alligators of different ages are shown in the table. (The ages range from 0 to 6 years.)

Tail	0.50	0.43	1.13	1.13	0.85	1.49	1.32	1.58
Body, B	0.44	0.40	1.05	1.04	0.80	1.37	1.23	1.47
Ratio, r	1.14	1.08	1.08	1.09	1.06	1.09	1.07	1.07
Tail	1.47	2.12	1.30	1.95	2.32	2.43	2.36	2.68
Body, B	1.36	1.95	1.22	1.80	2.14	2.25	2.17	2.47
Ratio, r	1.08	1.09	1.07	1.08	1.08	1.08	1.09	1.09
Tail	2.79	2.74	3.16	2.88	2.99	3.47	3.51	3.62
Body, B	2.57	2.54	2.91	2.66	2.76	3.19	3.22	3.35
Ratio, r	1.09	1.08	1.09	1.08	1.08	1.09	1.09	1.08

Sketch a graph that compares the body length, B, of each alligator with the ratio, r, of its tail length to its body length. What can you conclude from this graph?

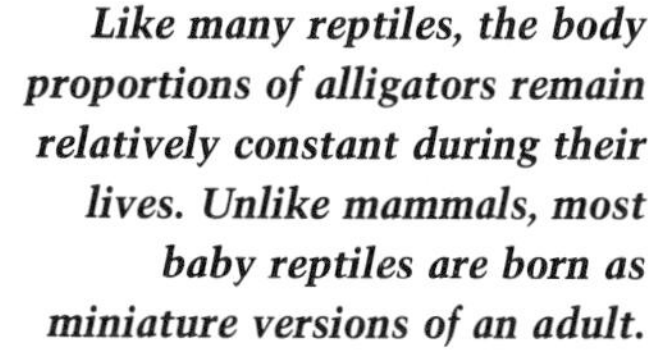
Like many reptiles, the body proportions of alligators remain relatively constant during their lives. Unlike mammals, most baby reptiles are born as miniature versions of an adult.

Alligator Dimensions

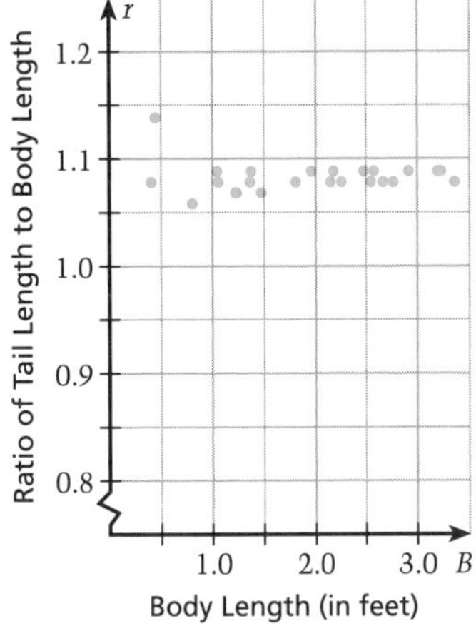

Solution Find an ordered pair for each alligator, with B as the first coordinate and r as the second coordinate.

Body Length, B Ratio, r
↓ ↓
$(0.44, 1.14), (0.40, 1.08), (1.05, 1.08), \ldots$

The graph of the 24 points is shown at the left. Notice that the points lie approximately on the horizontal line $r = 1.1$. Thus, you can conclude that a model for the approximate ratio of the tail length of an alligator to its body length is

$$r = 1.1.$$

■

Real Life **Mapmaking**

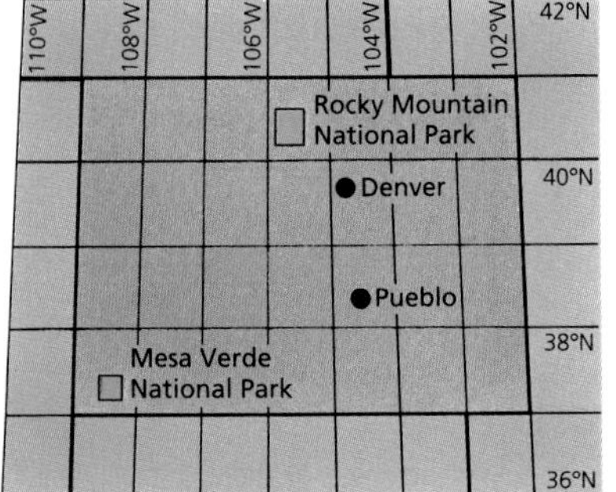

Example 5 ***Borders of Colorado***

On a map, the borders of Colorado approximate a rectangle. The north and south borders of Colorado are latitudes 41° N and 37° N, respectively. The west and east borders are longitudes 109.1° W and 102.1° W, respectively. Which is farther to the west: 109.1° W or 102.1° W? Write equations for the lines containing the four borders of this state.

Solution On maps, only nonnegative numbers name lines of latitude and longitude. Also, *N* and *S* mean "north" and "south" of the equator. *W* and *E* mean "west" and "east" of the prime meridian, which passes through Greenwich, England. Thus, the longitude line 109.1° W is farther west than the longitude line 102.1° W.

Latitude lines are also called *parallels* and longitude lines are also called *meridians.* Letting *P* represent the latitude and *M* represent the longitude, the four borders are as follows.

North Border	South Border	West Border	East Border
$P = 41$	$P = 37$	$M = 109.1$	$M = 102.1$
(Horizontal line)	(Horizontal line)	(Vertical line)	(Vertical line) ■

Communicating about ALGEBRA

SHARING IDEAS about the Lesson

Demonstrate Understanding Use grid paper and draw a large picture of the coordinate plane.

A. Label five points on the x-axis. See margin.

B. What do you notice about the second coordinate of each point? What does this observation imply about the equation for the x-axis? It is 0; it is $y = 0$.

C. Use a similar procedure to determine the equation of the y-axis. $x = 0$

D. What are the coordinates of the origin (the point at which the x-axis and y-axis intersect)? (0, 0)

E. What is the equation of the line passing through $(-4, -2)$, $(-4, 0)$, and $(-4, 2)$, $\underline{x = -4}$ or $y = -4$?

F. Explain why the equation $x = 1$ is a good way to describe the vertical line that intersects the x-axis at 1. See margin.

Communicating about ALGEBRA

Answers

A.

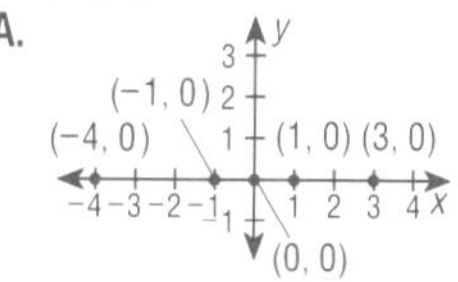

F. The x-coordinate is 1 for all points on the line $x = 1$.

EXTEND *Communicating*
STEP FUNCTIONS

A *step function* is the union of several linear relationships of the form $Ay = C$ for given values of x. Describe the step function below using linear equations in one variable. Have students complete the graph.

Let x = weight in pounds and y = charge in dollars. For $x \geq 0$ pounds and < 2 pounds, $y = \$7.65$; for $x \geq 2$ pounds and < 10 pounds, $y = \$7.95$; for $x \geq 10$ pounds, $y = \$8.55$.

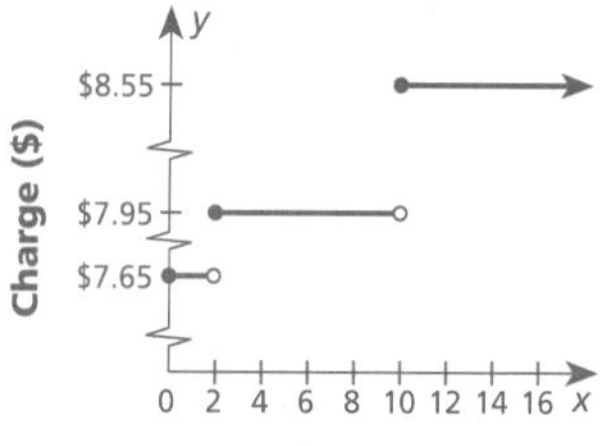

After students finish, have them suggest the real-life relationship that the graph could represent. Special-delivery postal rates

EXERCISE Notes

ASSIGNMENT GUIDE

Basic/Average: Ex. 1–22, 23–29 odd, 31–33, 45–51 multiples of 3, 53–54

Above Average: Ex. 1–22, 34–42, 44–52 even, 53–54

Advanced: Ex. 1–22, 31–42, 47–51 odd, 53–54

Selected Answers

Exercises 1–6, 7–51 odd

More Difficult Exercises

Exercises 37–42, 53–54

Guided Practice

Use these problems as oral exercises in class.

Answers

3. False. The graph of $x = 4$ consists of all points with x-coordinate 4. This is a vertical line.

5.

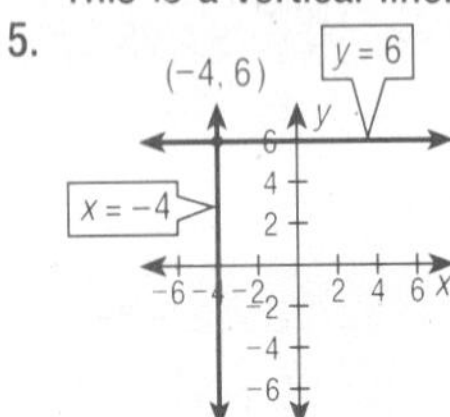

Independent Practice

▶ **Ex. 7–30** COOPERATIVE LEARNING These exercises should be discussed in class. Be attentive to errors resulting from the incorrect identification of *x*- and *y*-coordinates in ordered pairs, *(x, y)*.

EXERCISES

Guided Practice

CRITICAL THINKING about the Lesson

1. What is the x-coordinate in the ordered pair $(2, -1)$? 2
2. What is the y-coordinate in the ordered pair $(-5, -3)$? -3
3. **True or false?** The graph of the equation $x = 4$ is a horizontal line. Explain. See margin.
4. **True or false?** The graph of the equation $x = 4$ is a vertical line. True
5. At what point do the graphs of $x = -4$ and $y = 6$ intersect? Sketch a visual model (graph) to verify your answer. $(-4, 6)$ See margin.
6. Sketch the horizontal line and the vertical line that pass through the point $(3, -2)$. Write an equation for each. $y = -2$, $x = 3$

Independent Practice

In Exercises 7–10, match each equation with its graph.

7. $y = 2$ c
8. $x = -1$ a
9. $x = 2$ d
10. $y = -1$ b

a.

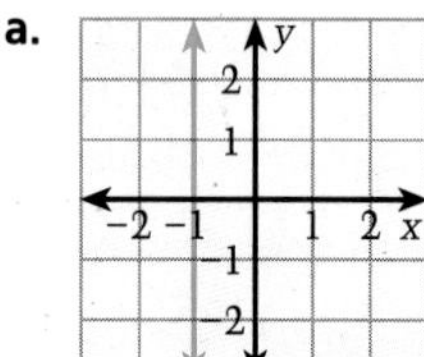

b.

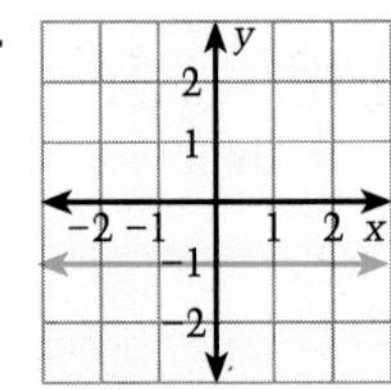

c.

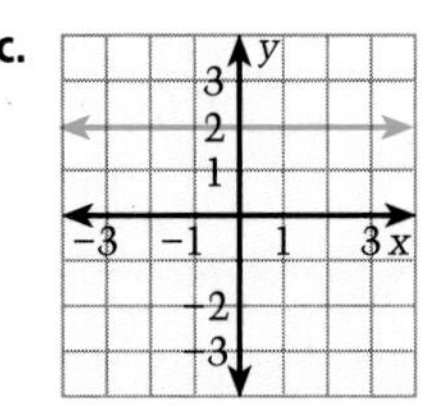

d.

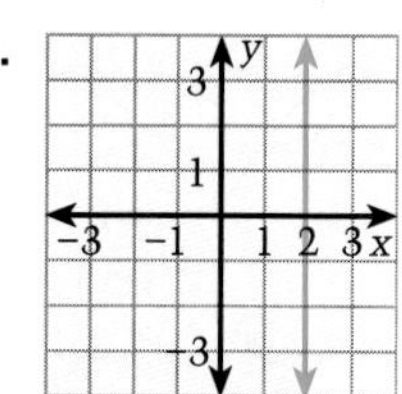

In Exercises 11–14, sketch the lines on the coordinate plane. Then find the point at which the two lines intersect. See Additional Answers.

11. $x = 6$, $y = -1$ $(6, -1)$
12. $x = 7$, $y = 3$ $(7, 3)$
13. $x = -4$, $y = -5$ $(-4, -5)$
14. $x = -8$, $y = 2$ $(-8, 2)$

In Exercises 15–18, decide whether the point lies on the line.

15. $(-3, 4)$, $x = -3$ Yes
16. $(-7, -4)$, $x = -4$ No
17. $(8, 10)$, $y = 8$ No
18. $(9, -5)$, $y = -5$ Yes

In Exercises 19–22, write an equation for the graph. 19. $x = 2$

19. A line, the x-coordinate of each point is 2.
20. A line, the y-coordinate of each point is 4. $y = 4$
21. A line, the y-coordinate of each point is -72. $y = -72$
22. A line, the x-coordinate of each point is -1. $x = -1$

In Exercises 23–30, write equations for the horizontal and vertical lines that pass through the point. 23. $y = -6$, $x = 10$

23. $(10, -6)$
24. $(8, -3)$ $y = -3$, $x = 8$
25. $(-21, -15)$ $y = -15$, $x = -21$
26. $(-4, -17)$ $y = -17$, $x = -4$
27. $(-5, 12)$ $y = 12$, $x = -5$
28. $(-9, 11)$ $y = 11$, $x = -9$
29. $(3, 0)$ $y = 0$, $x = 3$
30. $(0, -1)$ $y = -1$, $x = 0$

Television Rating **In Exercises 31–33, use the following information.**

Your family participates in a television rating system in which a monitor keeps track of the channels you watch. The results are plotted on the graph at the right.

31. Describe what happened from noon to midnight. See margin.

32. How many hours was the television on during this 12-hour period? 7

33. Which channel was on the most? Explain your reasoning. Channel 7; each of its segments is longer than any other segment.

Violin Family Ratios **In Exercises 34–36, use the following information.**

The violin family includes the bass, the cello, the viola, and the violin. The size of each instrument determines its tone. The violin produces the highest tones, while the bass produces the deepest (lowest) tones. The matrix shows different measurements, in inches, for different parts of each instrument.
See Additional Answers.

Violin Family	Bass	Cello	Viola	Violin
Total length	72	48	36	26
Body	48	32	24	17
Fingerboard	44	29	21	17
Neck	28	17	13	10

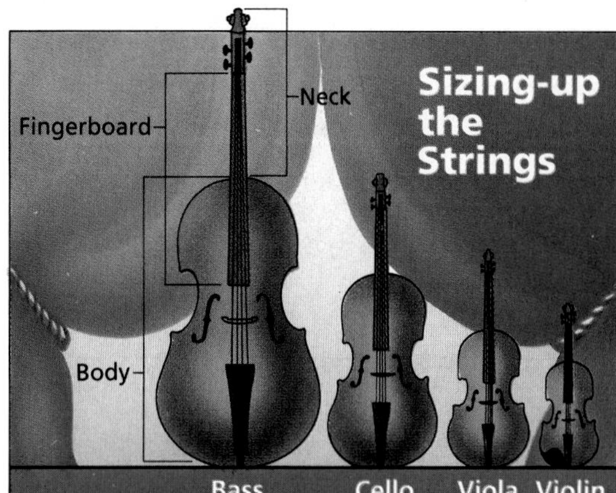

34. Find the ratios of the total lengths to the body lengths. Construct a graph that compares the ratios, y, with the total lengths, x. Describe the results.
1.5, 1.5, 1.5, ≈1.5

35. Find the ratios of the neck lengths to the fingerboard lengths. Construct a graph that compares the ratios, y, with the fingerboard lengths, x. Describe the results.
≈0.6, ≈0.6, ≈0.6, ≈0.6

36. Find and describe other similar ratios on the instruments.
Total length to fingerboard: ≈1.6, body to fingerboard: ≈1.1

Symphony Orchestras **In Exercises 37–39, use the following information.**

The number of symphony orchestras from 1980 to 1988 is shown in the table. ***(Source: American Symphony Orchestra League)***

	1980	1981	1982	1983	1984	1985	1986	1987	1988
College/Community	1311	1311	1304	1306	1317	1317	1298	1301	1253
Urban/Major	261	261	268	266	255	255	274	271	319
Total in United States	?	?	?	?	?	?	?	?	?

37. Complete the table. All entries are 1572.

38. Plot the data in the table on one coordinate plane. See Additional Answers.

39. Describe the results. See margin.

Answer

31. All times are P.M. The TV was off from noon until 1:00, from 2:00–4:00, from 6:30–8:00, and from 11:30 to midnight. The TV was on Channel 4 from 1:00–2:00; on Channel 6 from 6:00–6:30 and 11:00–11:30, on Channel 7 from 4:00–6:00 and 9:00–11:00; on Channel 8 from 8:00–9:00.

Ex. 34–36 HISTORICAL NOTE The violin was developed in the second half of the sixteenth century. The early renowned makers of the violin, such as the Amati family, their pupil Antonio Stradivari, and the Guarneri family, were all from Cremona, Italy.

Ex. 37–39 CONNECTIONS An orchestra may range in size from 60 to 100 performers and sometimes more. There are four instrument groups, usually in a ratio of 12:3:3:2, consisting of stringed instruments, woodwinds, brass (horns), and percussion instruments.

Answer

39. The points for the total number of orchestras lie on the horizontal line $y = 1572$. When college orchestras increased, urban orchestras decreased; when college orchestras decreased, urban orchestras increased.

▶ **Ex. 49–52** Remind students that values of one or more variables satisfy an equation if upon replacement of the variable with the values, the evaluation of the equation results in a true statement.

▶ **Ex. 53–54** GEOMETRY Be sure students understand that the transformed shapes are in the same orientation as the originals. So in Exercise 53, for example, the point $(-1, 1)$ corresponds to the point $(-3, -4)$ because $-3 + 2 = -1$ and $-4 + 5 = 1$.

Enrichment Activities

These activities require students to go beyond lesson goals.

GROUP ACTIVITY Have students take the following measurements of other students in class.

1. Measure the circumference of your wrist (W) and your neck (N).
2. Compute the ratio $R = \frac{N}{W}$ to two decimal places. The values usually range from about 1.9 through about 2.3.
3. Make a class list of ratios and wrist measurements. Use it to create a scatter plot of the ordered pairs, (W, R).
4. What pattern can you find in the graph of these points?
5. What other relationships can you find that can be modeled by a linear equation in one variable?

Recording the Temperature **In Exercises 40–42, use the following information.**

The temperature of Lake Michigan and the air temperature at O'Hare airport in Chicago are recorded at 2:00 P.M. during a week in October.

Date (x)	Oct. 17	Oct. 18	Oct. 19	Oct. 20	Oct. 21	Oct. 22	Oct. 23
Air Temperature (y)	62°F	58°F	52°F	51°F	59°F	61°F	60°F
Water Temperature (y)	53°F	53°F	53°F	53°F	53°F	53°F	53°F

40. Create two graphs, one for water temperature and one for air temperature. See Additional Answers.

41. For which graph do the points all have the same second coordinate? Water temperature

42. What does this tell you about the variability of air and water temperatures? Air temperature varies more than water temperature.

Integrated Review

In Exercises 43–48, solve the equation for y.

43. $2x + 5y = 10$ $\quad y = \frac{1}{5}(10 - 2x)$

44. $4y - 6x = 3$ $\quad y = \frac{1}{4}(3 + 6x)$

45. $2y - 5x = 14$ $\quad y = \frac{1}{2}(14 + 5x)$

46. $-6y - 8x = 22$ $\quad y = \frac{1}{3}(-11 - 4x)$

47. $3x - 7y = -2x + 19$ $\quad y = \frac{1}{7}(5x - 19)$

48. $9y + 10x = 21 + 18x$ $\quad y = \frac{1}{9}(21 + 8x)$

In Exercises 49–52, decide whether the values of x and y satisfy the equation.

49. $3x - y = 17$, $x = 0$ and $y = 8$ No

50. $2y - x = 21$, $x = 5$ and $y = 13$ Yes

51. $4y - 7x = -20$, $x = 8$ and $y = 9$ Yes

52. $5x + 2y = 30$, $x = 4$ and $y = 0$ No

Exploration and Extension

Computer Graphics **In Exercises 53 and 54, use the following information to find the coordinates of the vertices of the blue figure.**

Images on a computer screen can be represented by a collection of points in a coordinate plane. An object can be moved to the right or left by adding to or subtracting from the x-coordinate of each point. An object can be moved up or down by adding to or subtracting from the y-coordinate of each point.

53. 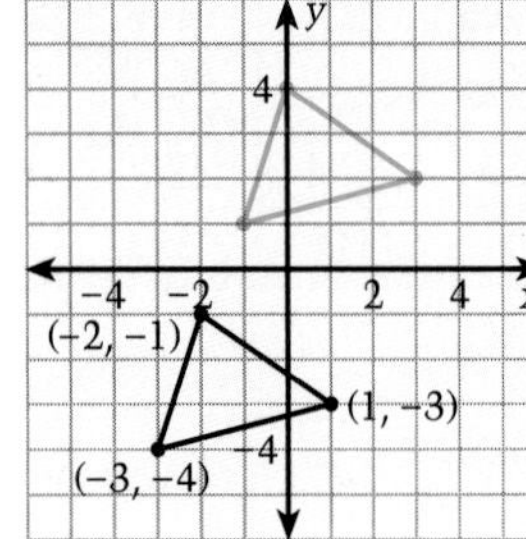

black triangle moves 2 units to the right and 5 units up to the blue triangle. $(-1, 1)$, $(0, 4)$, $(3, 2)$

54. 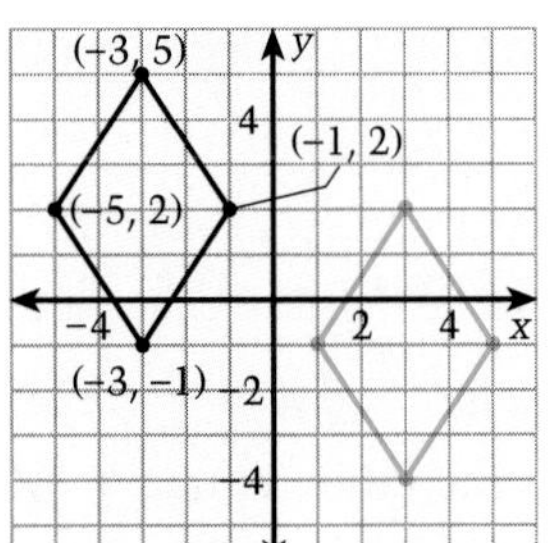

black diamond moves 6 units to the right and 3 units down to the blue diamond. $(1, -1)$, $(3, 2)$, $(5, -1)$, $(3, -4)$

4.2 Graphing Linear Equations in Two Variables

What you should learn:

Goal 1 How to graph a linear equation from a table of values

Goal 2 How to interpret graphs of linear equations

Why you should learn it:

You can use linear equations and their graphs as models for many real-life relationships, such as for sales figures over time.

Goal 1 Graphing a Linear Equation

In Lesson 4.1, you saw that the graph of an equation containing only one variable was either a horizontal or a vertical line. In this lesson, you will see that the graph of a **linear equation** with *two* variables, $Ax + By = C$, is also a line.

The **graph** of an equation involving two variables, x and y, is the collection of *all* points (x, y) that are **solutions** of the equation. For instance, since $(1, 3)$ is a solution of $y = 2x + 1$, you know that $(1, 3)$ is a point on the graph of $y = 2x + 1$. But $(1, 3)$ is only *one* point on the graph. How many points does the graph have in all?

The answer is that most graphs have far too many points to list. So how, then, could you ever sketch a graph? One way is to make a table of values, graph enough solutions to recognize a pattern, and then connect the points.

Problem Solving
Look for a pattern

Example 1 *Constructing a Table of Values and Finding a Pattern*

Construct a table of values for $y = 2x + 1$ and graph the corresponding solutions. What is the pattern?

Solution To construct a table of values, choose several x-values, substitute each x-value into the equation, and find the corresponding y-value.

Choose an x-value.	Substitute and find the y-value.	Find solution (x, y).
$x = -2$	$y = 2(-2) + 1 = -3$	$(-2, -3)$
$x = -1$	$y = 2(-1) + 1 = -1$	$(-1, -1)$
$x = 0$	$y = 2(0) + 1 = 1$	$(0, 1)$
$x = 1$	$y = 2(1) + 1 = 3$	$(1, 3)$
$x = 2$	$y = 2(2) + 1 = 5$	$(2, 5)$

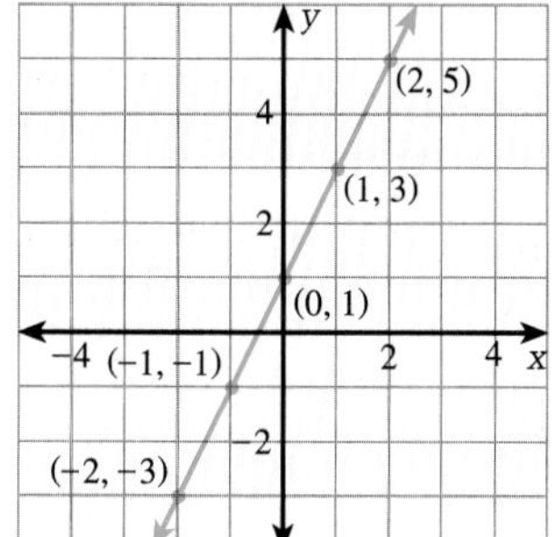

The five solutions are shown in the graph at the left. The pattern formed by the points is that they all lie on a line. The *entire* line, including the parts that are not shown, is the *graph* of $y = 2x + 1$. *Each* point on the line is a solution of the equation $y = 2x + 1$. ■

Problem of the Day

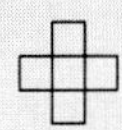

This pattern can be folded to make a topless box. Draw the other patterns of 5 squares that can be folded to make a topless box.

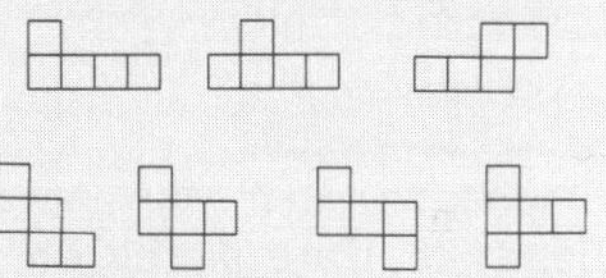

ORGANIZER

Warm-Up Exercises

Graph each set of ordered pairs in a coordinate plane and then connect them using a line. The ordered pairs are solutions of the linear equation given with each set.

1. $(2, 10), (-1, 4), (0, 6)$
   $y = 2x + 6$
2. $(0, -3), (1, -4), (-4, 1)$
   $y = -x - 3$
3. $(2, 4), (-2, -8), (0, -2)$
   $y = 3x - 2$
4. Evaluate $2x + 6$ at $x = 2$, -1, and 0. How are your answers related to the ordered pairs in Exercise 1? They are the same.
5. Evaluate $-x - 3$ at $x = 0$, 1, and -4. How are your answers related to the ordered pairs in Exercise 2? They are the same.
6. Evaluate $3x - 2$ at $x = -2$, 2, and 0. How are your answers related to the ordered pairs in Exercise 3? They are the same.

Lesson Resources

Teaching Tools
- Transparencies: 1, 2, 4
- Problem of the Day: 4.2
- Warm-up Exercises: 4.2
- Answer Masters: 4.2

Extra Practice: 4.2

Applications Handbook: p. 54

LESSON Notes

A linear equation in two variables is any equation of the form $Ax + By = C$, in which both A and B are not zero. (In Lesson 4.1, you worked with the special cases when either A or B was zero.)

Examples 1–3

Emphasize that the set of points that are placed on the coordinate plane can be used to identify a pattern that represents the linear equation. The pattern is a line. (Sets of points for other types of equations may result in patterns that are not straight lines.)

Example 4

Point out to students that the double arrow indicates "corresponds to." The year 1980 corresponds to an x-value 0. Using numerical values to represent the years is a convenient way to display the information.

Extra Examples

Here are additional examples similar to **Examples 1–3.**

1. Construct a table of values for $y = -2x + 4$ and graph the corresponding solutions. What is the pattern of points? Pattern: They all lie in a straight line.

x	−3	−2	−1	0	1	2	3
y	10	8	6	4	2	0	−2

2. Construct a table of values for the linear equation $3x + 4y = 12$ and graph the corresponding solutions. Begin by rewriting the linear equation by solving it for y. $y = -\frac{3}{4}x + 3$.

x	−8	−4	0	4	8
y	9	6	3	0	−3

3. Which of the following points is on the graph of $6x - 2y = 8$?
   a. (2, 1) b. (2, 2)
   c. (−1, −7) Points b and c

Before constructing a table of values, it is helpful to first solve for one variable in the equation, as shown in Example 2.

Example 2 *Rewriting before Constructing a Table*

Sketch the graph of $4x + 2y = 1$.

Solution Begin by solving the given equation for y.

$$4x + 2y = 1 \quad \textit{Original equation}$$
$$2y = -4x + 1 \quad \textit{Subtract 4x from both sides.}$$
$$y = -2x + \frac{1}{2} \quad \textit{Divide both sides by 2.}$$

Now choose several values of x and construct a table of values. For instance, when $x = 1$ the value of y is

$$y = -2(1) + \frac{1}{2} = -\frac{3}{2}.$$

Table of Values for $y = -2x + \frac{1}{2}$

Choose x.	−3	−2	−1	0	1	2	3
Evaluate y.	$\frac{13}{2}$	$\frac{9}{2}$	$\frac{5}{2}$	$\frac{1}{2}$	$-\frac{3}{2}$	$-\frac{7}{2}$	$-\frac{11}{2}$

With this table of values, you have found seven solutions.

$$\left(-3, \tfrac{13}{2}\right), \left(-2, \tfrac{9}{2}\right), \left(-1, \tfrac{5}{2}\right), \left(0, \tfrac{1}{2}\right), \left(1, -\tfrac{3}{2}\right), \left(2, -\tfrac{7}{2}\right), \left(3, -\tfrac{11}{2}\right)$$

By plotting the points, you can see that they all lie on a line. The line through the points is the graph of the equation. ■

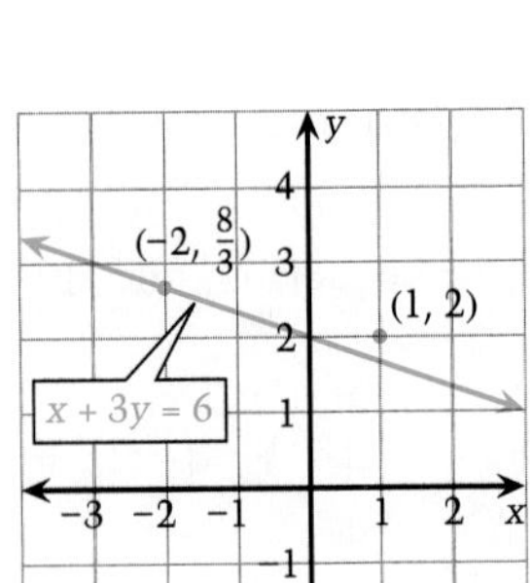

Example 3 *Verifying Solutions of an Equation*

Which of the points is on the graph of $x + 3y = 6$?

a. (1, 2) **b.** $\left(-2, \frac{8}{3}\right)$

Solution

a. The point (1, 2) is *not* on the graph of $x + 3y = 6$ because it is not a solution.

$$1 + 3(2) \stackrel{?}{=} 6 \quad \textit{Substitute 1 for x and 2 for y.}$$
$$7 \neq 6 \quad \textit{(1, 2) is not a solution.}$$

b. The point $\left(-2, \frac{8}{3}\right)$ is on the graph of $x + 3y = 6$ because it is a solution.

$$-2 + 3\left(\tfrac{8}{3}\right) \stackrel{?}{=} 6 \quad \textit{Substitute } -2 \textit{ for x and } \tfrac{8}{3} \textit{ for y.}$$
$$-2 + 8 \stackrel{?}{=} 6 \quad \textit{Simplify.}$$
$$6 = 6 \quad \left(-2, \tfrac{8}{3}\right) \textit{ is a solution.}$$
■

Goal 2 Interpreting Graphs of Linear Equations

The graph of a relationship between two variables can be used to give information about how the two variables are related.

Real Life Publishing

Example 4 *Interpreting Graphs*

The top graph shows the number of morning newspapers sold each weekday in the United States from 1978 to 1990 with $x = 0$ corresponding to 1980. The bottom graph shows the number of evening papers sold during the same years. What do these two graphs tell you about newspaper sales?
(*Source: Editor and Publisher Company*)

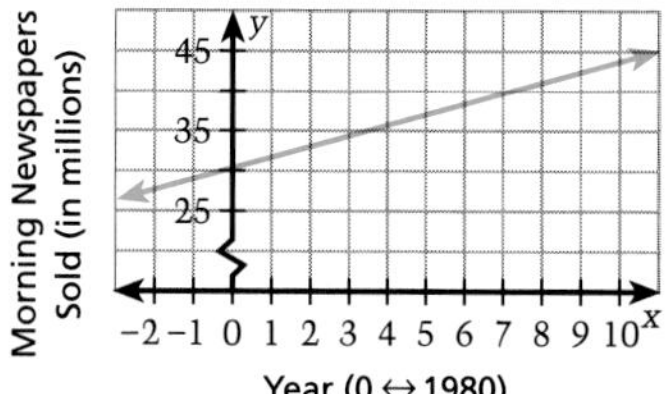

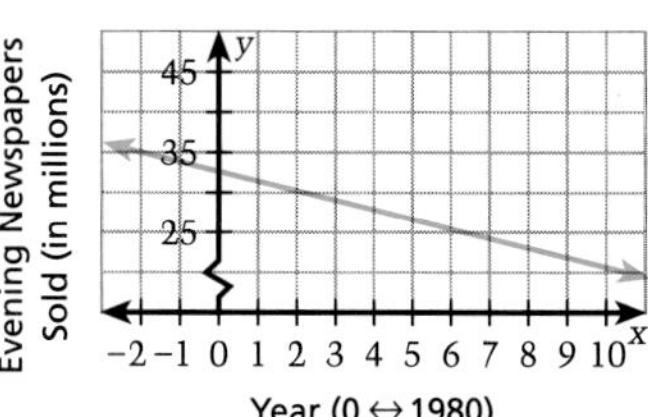

Solution The top graph shows that morning newspaper sales were *increasing* over time, from about 27 million in 1978 to about 43 million in 1990. The bottom graph shows that evening newspaper sales were *decreasing*, from about 35 million in 1978 to about 20 million in 1990. During the 13-year period, many Americans were switching from evening newspapers to morning newspapers. ■

Communicating about ALGEBRA

SHARING IDEAS about the Lesson

Drawing Graphs Remember, the graph of an equation is a visual model of the equation.

A. Sketch the graphs of $y = x$ and $y = -x$ on the same coordinate plane. What do you notice about the two graphs? They bisect the 4 quadrants of the plane.

B. Copy and complete the table for $y = -2x + 5$. Then sketch the graph of the equation. See Additional Answers.

x	-3	-2	-1	0	1	2	3
y	? 11	? 9	? 7	? 5	? 3	? 1	? −1

C. How many points are needed to determine a line? Justify your answer. See below.

D. To construct a table of values for $y = \frac{3}{4}x - 8$, which values of x would you choose? Why? See below.

C. 2; if two points are plotted and a line is drawn, the remaining points will be on the line.
D. Multiples of 4, so that y would be an integer.

Check Understanding

1. Sketch the graph of $2x + 7y = 14$. Explain each step of the process you used.

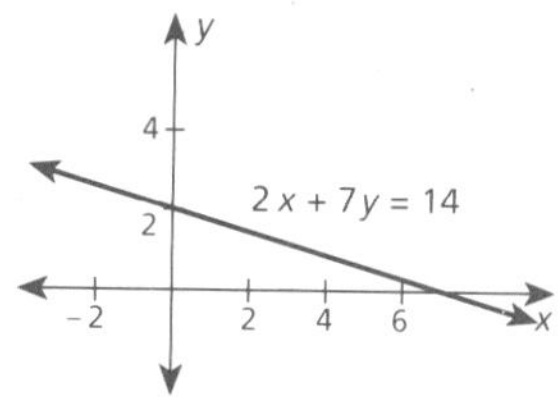

2. Use your completed graph to find the missing x- or y-coordinate for each of the ordered pairs given below.
   a. (0, ?) **b.** (?, 2) **c.** (?, $\frac{10}{7}$) **d.** (−7, ?)
   a. 2, b. 0, c. 2, d. 4

Communicating about ALGEBRA

In Part C, the discussion about the fact that only two points are needed to determine a straight line might suggest to students that only two points should be plotted when sketching graphs of lines. Emphasize that plotting several points and finding that all of them lie on a line provides a check of their computational work.

EXTEND *Communicating*
AN INTRODUCTION TO SLOPE

Have students examine the graphs of $y = x$ and $y = -x$ to determine the vertical and horizontal distances and directions needed to move from one point on the line to another point on the same line. For example, the vertical and horizontal distances and directions needed to move from the point (−2, 2) to the point (1, −1) on the graph of $y = -x$ are 3 down and 3 right. Have students compare several different points on each line. Ask students what they observe about this ratio.

$$\frac{\text{vertical distance}}{\text{horizontal distance}}$$

EXERCISE Notes

ASSIGNMENT GUIDE

Basic/Average: Ex. 1–8, 9–19 odd, 21–26, 29–31, 33–43 odd, 45–47 odd

Above Average: Ex. 1–8, 9–25 odd, 27–30, 32–34, 45–47

Advanced: Ex. 1–8, 9–23 odd, 25–27, 29–31, 45–48

Selected Answers
Exercises 1–4, 5–43 odd

✪ More Difficult Exercises
Exercises 27–28, 31–32, 47–48

Guided Practice

▶ **Ex. 1–4** MATH JOURNAL These make good oral exercises to complete in class. The definitions of a solution of a linear equation and the graph of an equation should be recorded in students' math journals.

Answer

4. $y = \frac{3}{2}x + \frac{5}{2}$; Solution: To isolate the y-term, add $6x$ to both sides. To get a y-coefficient of 1, multiply both sides by $\frac{1}{4}$.

Independent Practice

▶ **Ex. 5–8** Point out to students that the points that lie on a given line are solutions of the linear equation of the line.

An Alternate Approach
A Graphical Check
In Exercises 5–8, one way to verify that an ordered pair is a solution of an equation is to use an algebraic check as shown in Example 3. Another way is to graph the given equations and determine which of the points lie on the line.

EXTEND Ex. 21–24 WRITING Ask students to provide a written rationale for the matches that they make. Encourage students to discuss their selections with others.

EXERCISES

Guided Practice

CRITICAL THINKING about the Lesson

1. What is meant by a *solution* of $Ax + By = C$? A number pair that makes the equation true
2. Define *graph* of an equation. The collection of all points (x, y) that are solutions of the equation
3. Which of the following are solutions of $3x - y = 2$?
   a. $(3, 1)$ **b.** $(1, 2)$ **c.** $(-1, -5)$ **d.** $(0, -2)$
4. Solve $4y - 6x = 10$ for y. Explain each step of the solution. See margin.

Independent Practice

In Exercises 5–8, decide which of the two points lies on the line. Justify your choice algebraically and graphically. Justify algebraically by substituting coordinates into equations and simplifying.

5. $2x + 3y = 10$
a. $(2, 2)$
b. $(3, 1)$

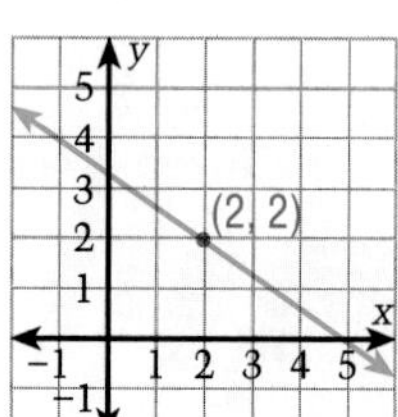

6. $2y - 4x = 6$
a. $(1, 4)$
b. $(-1, 1)$

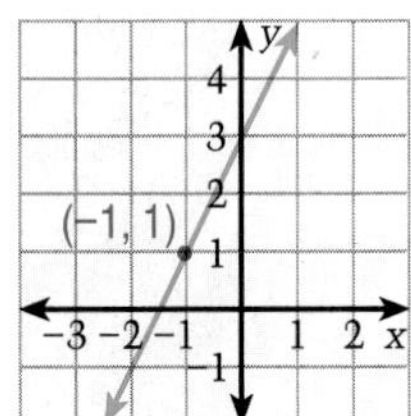

7. $6y - 3x = -9$
a. $(3, 0)$
b. $(2, -1)$

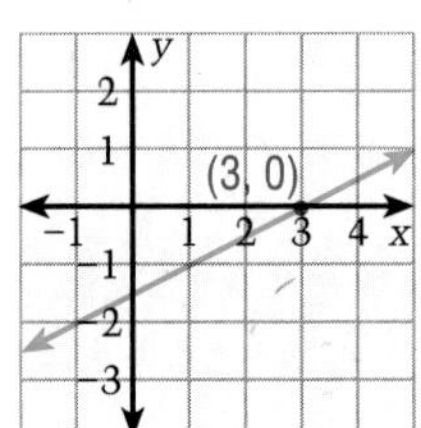

8. $-5x - 8y = 15$
a. $(-3, 0)$
b. $(-2, -1)$

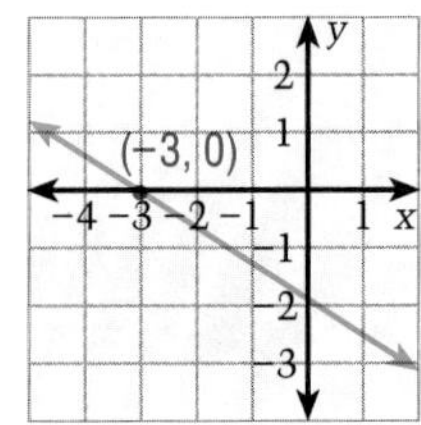

In Exercises 9–12, find several solutions of the equation.

9. $y = 2x - 4$ $(0, -4), (2, 0)$
10. $y = 5 - 3x$ $(0, 5), (2, -1)$
11. $y = 3(2x - 1)$ $(0, -3), (1, 3)$
12. $y = \frac{1}{2}(6 - 4x)$ $(0, 3), (1, 1)$

In Exercises 13–20, use a table of values to sketch the graph of the equation. See Additional Answers.

13. $y = -x + 2$
14. $y = -4x + 5$
15. $y = -(2 - x)$
16. $y = -2(x + 3)$
17. $y = \frac{1}{2}x + 4$
18. $y = -\frac{1}{2}x - 12$
19. $4x + 3y = 24$
20. $3x - 2y = -12$

In Exercises 21–24, match the equation with its graph.

21. $2x - y = 7$ b
22. $3x + y = 3$ d
23. $6x + 3y = 0$ a
24. $-2x + y = 5$ c

a.

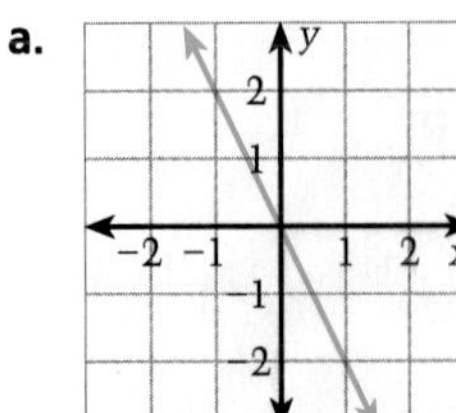

b.

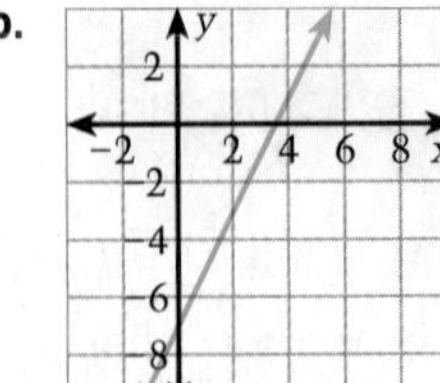

c.

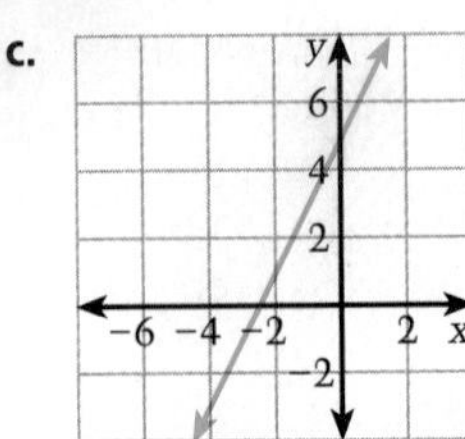

d.

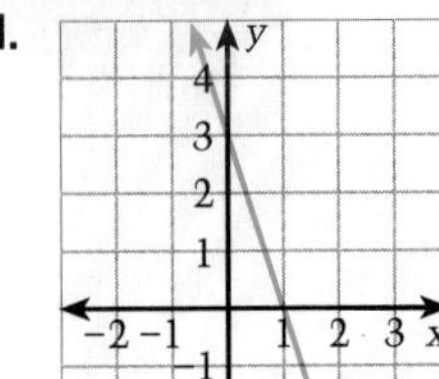

25. *What's the Temperature?* A surveyor on a trip to Antarctica sends back a report stating that the temperature has dropped to $-40°$. The surveyor's supervisor criticizes the report, saying that the surveyor should have said whether the temperature reading was in degrees Fahrenheit or Celsius. The surveyor replies that the scale wasn't necessary! Why? **$-40°C = -40°F$**

26. *Misleading Graphs* Graphs can help us visualize relationships between two variables, but they can also be misused to imply results that are not correct. The two graphs below represent the *same* data points. Which graph is misleading and why?

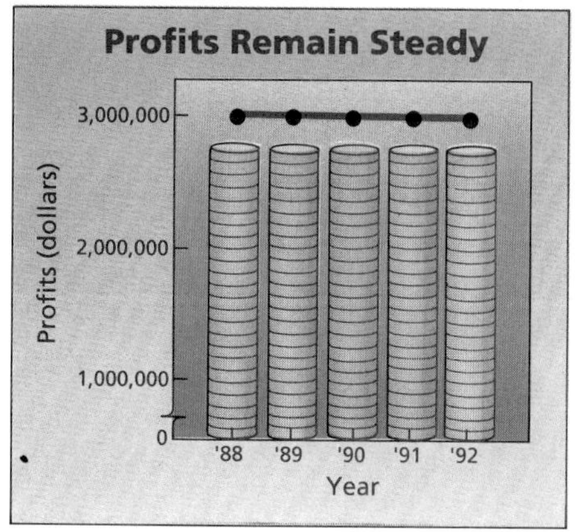

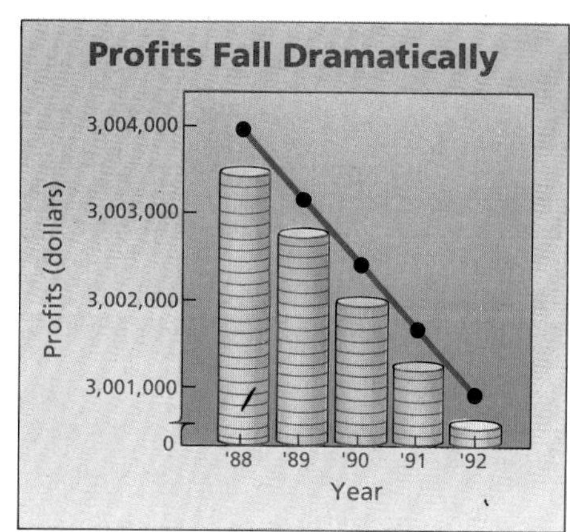

Graph on the right; a decrease of \$1000 per year per \$3 million or more is not dramatic.

27. *Flood Waters* A river has risen 6 feet above flood stage. Beginning at a time of $t = 0$, the water level drops at the rate of two inches per hour. The number of feet above flood stage, y, after t hours is given by $y = 6 - \frac{1}{6}t$. Sketch a graph of the equation over the 12-hour period from $t = 0$ to $t = 12$. **See margin.**

28. How long does it take the river to recede to flood stage? **36 hours**

In Exercises 29 and 30, match the description with its visual model. Explain.

29. Your friend starts driving from home at 55 miles per hour for 3 hours, where d is the distance from home. **a**

30. You start driving the 165 miles home at 55 miles per hour for 3 hours, where d is the distance from home. **b**

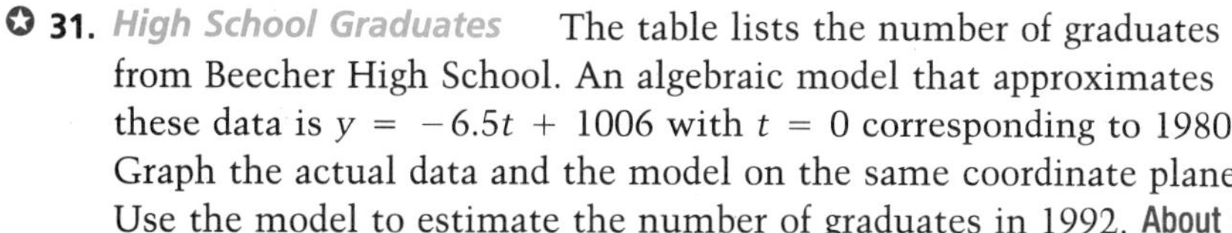

31. *High School Graduates* The table lists the number of graduates from Beecher High School. An algebraic model that approximates these data is $y = -6.5t + 1006$ with $t = 0$ corresponding to 1980. Graph the actual data and the model on the same coordinate plane. **See margin.** Use the model to estimate the number of graduates in 1992. **About 928**

Year	1983	1984	1985	1986	1987	1988	1989	1990	1991	1992
t	3	4	5	6	7	8	9	10	11	12
Graduates, y	984	987	962	968	953	952	946	944	931	?

Ex. 25 The formula that relates the Fahrenheit and Celsius temperature scales is $F = \frac{9}{5}C + 32$. Note that at $C = -40°$, $F = \frac{9}{5}(-40) + 32 = -40$. So, $-40°C = -40°F$.

EXTEND Ex. 26 MISLEADING GRAPHS Call students' attention to the values of the tick marks on the y-axis. Ask students to examine newspapers or magazines to see if they can find real-life data displayed in misleading graphs. Have them explain why the graphs they find may be misleading.

Ex. 27 Ask students to use their completed graphs to find the rate at which the water drops over the 12-hour period. Then look for a pattern. Using any two time periods, the rate is a constant drop of 1 foot in 6 hours or $-\frac{1}{6}$.

Answers

27.

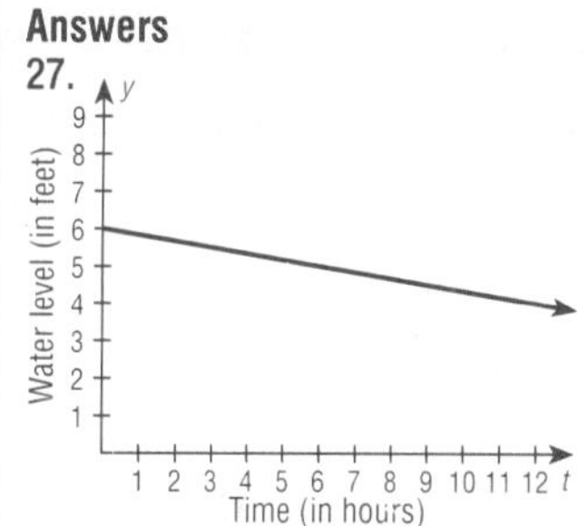

31.

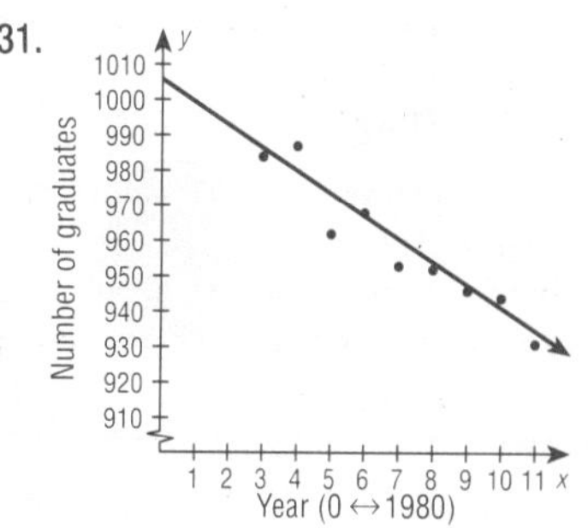

Ex. 45–46 Encourage students to look at Part A of Communicating about Algebra for similarities between exercises.

Answers

45. As x increases, y increases; $y = x - 4$
46. From -3 to 2, as x increases, y decreases; from 2 to 3, as x increases, y increases; $y = |x - 2|$ or $y = |2 - x|$
47. Any of the three vehicles is acceptable. Each vehicle increases its speed, travels at a constant rate, slows down, stops, then repeats that pattern.

Enrichment Activities

These activities require students to go beyond lesson goals.

1. Construct a table of values for $y = x^2$ using x values: $-5, -4, -3, -2, -1, 0, 1, 2, 3, 4, 5$. y-values: 25, 16, 9, 4, 1, 0, 1, 4, 9, 16, 25
2. Do the resulting ordered pairs represent solutions to the given equation? Explain. Yes, each ordered pair satisfies the equation.
3. Is the equation linear? What is the pattern of points? No, the graph is a *parabola*.

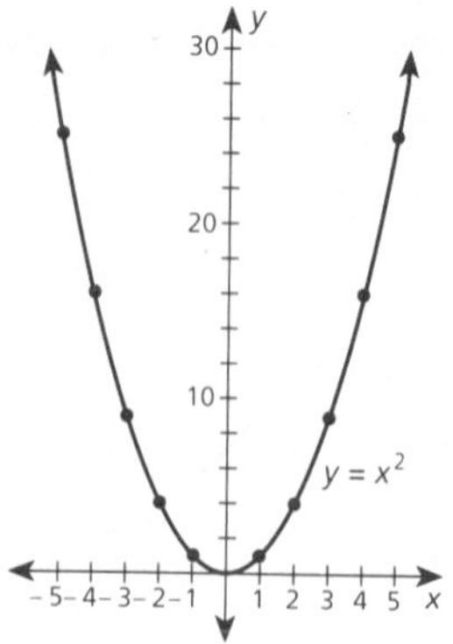

32. *Life Expectancy* The table gives the life expectancy in the United States for a child (at birth). A mathematical model for the life expectancy in years during this period is $y = 0.3t + 65.6$ with $t = 0$ corresponding to 1950. Graph the actual data points and the model on the same coordinate plane. Use the model to estimate the life expectancy in the year 2000. ***(Source: U.S. Bureau of Census)*** 80.6 See Additional Answers.

Year	1930	1940	1950	1960	1970	1980	1990	2000
t	-20	-10	0	10	20	30	40	50
Life Expectancy, y	59.7	62.9	68.2	69.7	70.8	73.7	75.6	?

Integrated Review

33.–36. See Additional Answers.

33. Sketch the graph of $x = 2$.

34. Sketch the graph of $y = -1$.

35. Sketch the graph of $y = 8$.

36. Sketch the graph of $x = -6$.

In Exercises 37–40, find the point of intersection of the two lines.

37. $x = -4,\ y = 7$ $(-4, 7)$

38. $x = 11,\ y = 5$ $(11, 5)$

39. $x = -10,\ y = -12$ $(-10, -12)$

40. $x = 6,\ y = -18$ $(6, -18)$

In Exercises 41–44, write equations for the horizontal and vertical lines that pass through the point.

41. $(6, -3)$ $x = 6, y = -3$

42. $(-12, -15)$ $x = -12, y = -15$

43. $(8, 21)$ $x = 8, y = 21$

44. $(-7, 9)$ $x = -7, y = 9$

Exploration and Extension

In Exercises 45 and 46, plot the points given in the table of values. See Additional Answers. **Describe the relationship between x and y. Then write an equation that represents the relationship.** See margin.

45.

x	-2	0	2	4	6	8	10
y	-6	-4	-2	0	2	4	6

46.

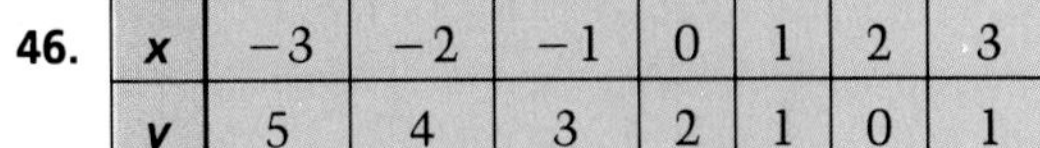

x	-3	-2	-1	0	1	2	3
y	5	4	3	2	1	0	1

47. Is the graph shown at the right more likely to represent a trip taken on a train, a bus, or an airplane? Explain your answer. See margin.

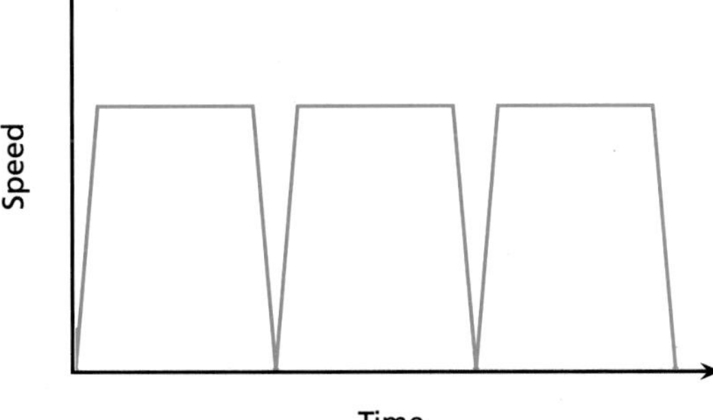

48. *Automobile Ownership* In 1990, the United States had $33\frac{1}{3}\%$ of all registered automobiles in the world. What percent did the United States have in 1980? (Use the information on page 175.) ≈36.6%

USING A GRAPHING CALCULATOR

A graphing calculator or computer can be used to sketch the graph of an equation. For instance, the screen at the right shows the graph of $x - 2y = 2$. *Keystroke instructions for sketching this graph on a* TI-82, *Casio* fx-9700GE, *and* Sharp EL-9300C *are listed on page 737.*

When using a graphing calculator to sketch a graph, there are three ideas you must remember.

- Solve the equation for y in terms of x.
- Describe the viewing rectangle by entering the least and greatest x and y values and the scale (units per mark). This is called *setting the Window or Range.*
- Use parentheses if you are unsure of the calculator's order of operations.

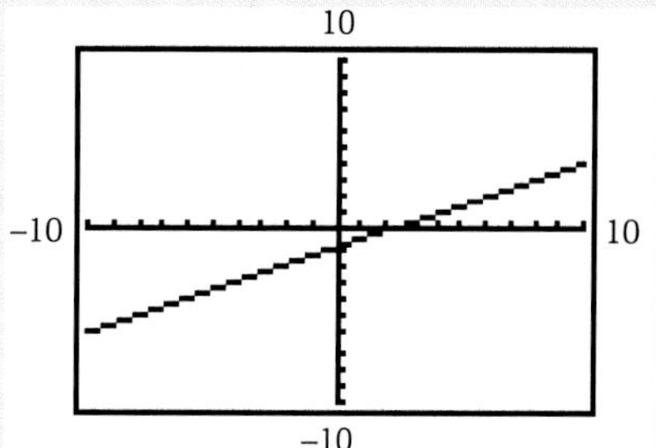

A standard viewing rectangle uses least x and y values of −10 and greatest x and y values of 10.

Be sure you see that the first step in using a graphing calculator to graph an equation in x and y is to solve the equation for y. When you do this, you are writing the equation in *function form* with y as a function of x.

$$x - 2y = 2 \qquad \textit{Original equation in x and y}$$

$$-2y = -x + 2$$

$$y = \tfrac{1}{2}x - 1 \qquad \textit{Function form}$$

Some equations, such as $y^2 = x$, do not represent y as a function of x. You cannot sketch the graph of this equation by entering this single equation into a function grapher. You *can* sketch the graph of $x^2 = y$ because this equation describes y as a function of x.

Exercises

In Exercises 1–8, use a graphing calculator to sketch the graphs. (For 1–4, use a standard viewing rectangle. For 5–8, use the indicated viewing rectangle.)

Odds: See margin.
Evens: See Additional Answers.

1. $y = -2x - 3$ **2.** $y = 2x + 2$ **3.** $x + 2y = -1$ **4.** $x - 3y = 3$

5. $y = x + 25$ **6.** $y = -x + 25$ **7.** $y = 0.1x$ **8.** $y = 100x + 2500$

5.	6.	7.	8.
RANGE	RANGE	RANGE	RANGE
Xmin = -10	Xmin = -10	Xmin = -10	Xmin = 0
Xmax = 10	Xmax = 10	Xmax = 10	Xmax = 100
Xscl = 1	Xscl = 1	Xscl = 1	Xscl = 10
Ymin = -5	Ymin = -5	Ymin = -1	Ymin = 0
Ymax = 35	Ymax = 35	Ymax = 1	Ymax = 15000
Yscl = 5	Yscl = 5	Yscl = 0.1	Yscl = 1000

Sample answers
Art

1.
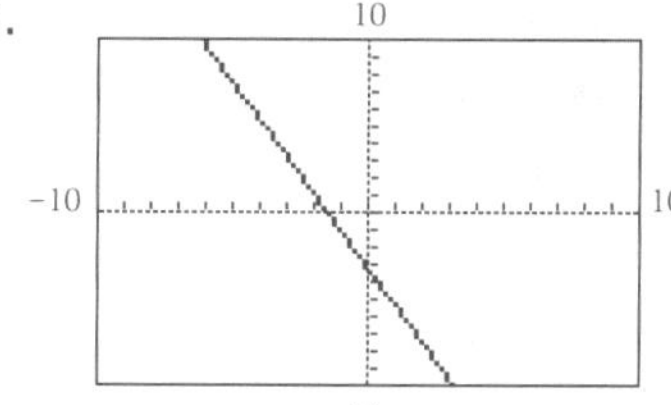

3.
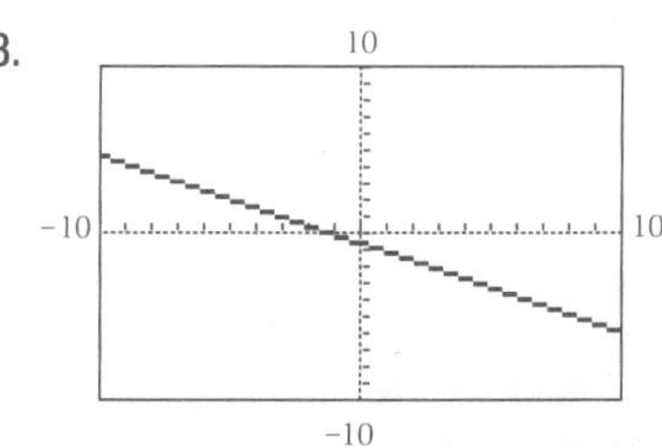

5.
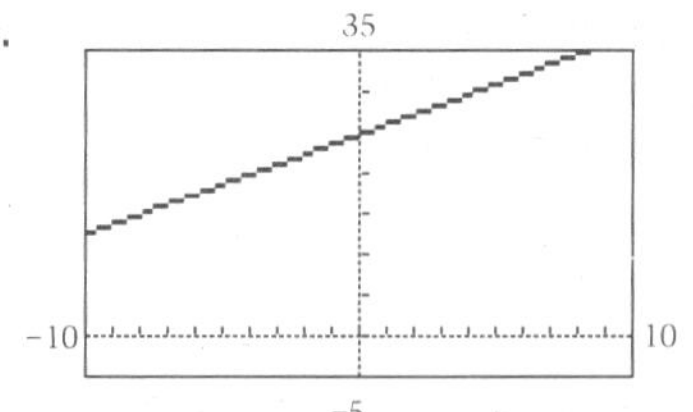

7.
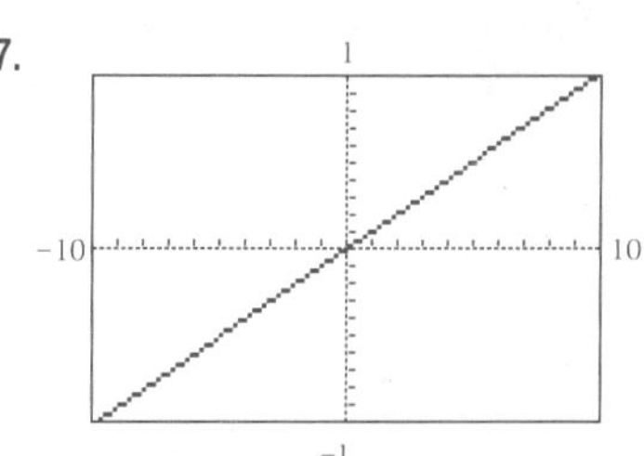

Problem of the Day

Alan, Bill, and Ezra each have different eye colors. One has gray eyes, one has hazel eyes, and one has blue eyes. The boy with gray eyes is taller than the boy with hazel eyes. Alan is taller than the boy with gray eyes. Bill is the shortest. What color are Ezra's eyes? Gray

ORGANIZER

Warm-Up Exercises

1. Solve the equations below by using mental math.
   a. $3x = 12$ — 4
   b. $-3y = 15$ — -5
   c. $6x + 3 = 24$ — 3.5
   d. $2y + 5 = -9$ — -7
2. Graph each pair of ordered pairs on a coordinate plane and connect the points.
   a. $(-3, 2)$ and $(4, 3)$
   b. $(0, 4)$ and $(3, 0)$
   c. $(-2, 0)$ and $(0, 5)$

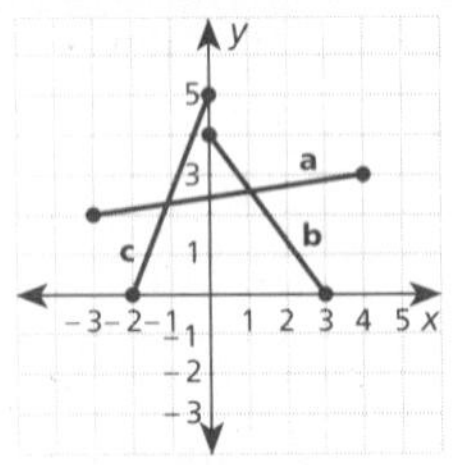

Lesson Resources

Teaching Tools
- Transparency: 4
- Problem of the Day: 4.3
- Warm-up Exercises: 4.3
- Answer Masters: 4.3

Extra Practice: 4.3

4.3 Quick Graphs Using Intercepts

What you should learn:

Goal 1 How to find the intercepts of the graph of a linear equation

Goal 2 How to use intercepts to sketch a quick graph of a line

Why you should learn it:

You can quickly sketch the graph of a linear equation by graphing its intercepts. The graph allows you to see relationships between the variables.

Goal 1 Finding the Intercepts of a Line

In mathematics, we like to find efficient ways to solve problems. For instance, in Lesson 4.2 you sketched the graph of an equation by creating a table of values, plotting the corresponding points, and finally drawing the graph through the points. For *linear* equations, the following two observations suggest a more efficient way to sketch the graph.

- The graph of any linear equation, $Ax + By = C$, is a line.
- Two points are all that are needed to determine a line.

Two points that are usually convenient to use are the points at which the line intersects the x-axis and the y-axis.

Intercepts of a Line

Consider the line given by $Ax + By = C$.

1. The x-intercept of the line is the value of x when $y = 0$. To find the x-intercept, let $y = 0$ and solve for x in $Ax = C$.
2. The y-intercept of the line is the value of y when $x = 0$. To find the y-intercept, let $x = 0$ and solve for y in $By = C$.

Example 1 *Finding Intercepts*

a. To find the x-intercept of $x + 5y = 10$, let y be zero.

$x + 5y = 10$	*Original equation*
$x + 5(0) = 10$	*Substitute 0 for y.*
$x = 10$	*Solve for x.*

The x-intercept is 10. It occurs at the point $(10, 0)$.

b. To find the y-intercept of $x + 5y = 10$, let x be zero.

$x + 5y = 10$	*Original equation*
$0 + 5y = 10$	*Substitute 0 for x.*
$y = 2$	*Solve for y.*

The y-intercept is 2. It occurs at the point $(0, 2)$. ■

Goal 2 Using Intercepts to Sketch a Quick Graph

The intercepts of a line provide a quick way to sketch the line.

Example 2 *Sketching a Quick Graph*

Sketch the graph of $3x + 4y = 12$.

Solution First, find the intercepts.

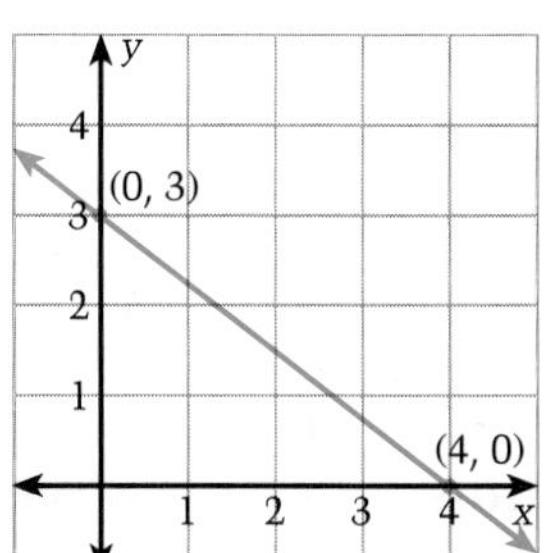

Quick Graph Using Intercepts

x-Intercept (Let $y = 0$.)	**y-Intercept (Let $x = 0$.)**
$3x + 4y = 12$	$3x + 4y = 12$
$3x + 4(0) = 12$	$3(0) + 4y = 12$
$x = 4$	$y = 3$
x-intercept is 4.	y-intercept is 3.

To sketch the graph, draw a coordinate plane, plot points for the intercepts, and draw a line through them. In the graph shown at the left, the x-intercept occurs at (4, 0), the point on the x-axis at which x is 4. The y-intercept occurs at (0, 3), the point on the y-axis at which y is 3. ■

When using the quick-graph method to sketch a line, you should find the intercepts *before* you draw the coordinate plane. This will enable you to use an appropriate scale on each axis.

Example 3 *Using Intercepts to Determine Appropriate Scales*

Sketch the line given by $y = 4x + 40$.

Solution First, find the intercepts.

Quick Graph Using Intercepts

x-Intercept (Let $y = 0$.)	**y-Intercept (Let $x = 0$.)**
$y = 4x + 40$	$y = 4x + 40$
$0 = 4x + 40$	$y = 4(0) + 40$
$-40 = 4x$	$y = 40$
$-10 = x$	
x-intercept is -10.	y-intercept is 40.

From the values of the intercepts, you can see that the coordinate plane should have a y-axis that includes 40 and an x-axis that includes -10. Plotting points for the intercepts at $(-10, 0)$ and $(0, 40)$, and drawing the line through them produce the graph shown at the left. ■

LESSON Notes

Example 1

Emphasize that the x- and y-intercepts are points on a coordinate plane. For example, the point (6, 0) lies on the x-axis and is the x-intercept of a line that contains it; the point (0, 4) lies on the y-axis and is the y-intercept of a line that similarly contains it. Point out to students that it is customary to speak of the x-intercept by its x-coordinate value (6) and the y-intercept by its y-coordinate value (4). You say, therefore, that the x-intercept is 6 and the y-intercept is 4.

Example 2

Stress to students that the x-intercept, 4, is the point (4, 0), and the y-intercept, 3, is the point (0, 3). Quick graphs will be used throughout their study of algebra, since they provide an efficient way to get a visual model of many algebraic relationships you will encounter.

EXTEND Example 3 GRAPHING This example shows that the scales (units per tick mark) that are selected for the coordinate plane often depend on the points you wish to graph.

Have students compare the slant of the lines obtained by graphing $y = 4x + 40$ with the indicated scale values for both the x- and y-axes.

a. Scale: 5 units per tick mark

b. Scale: 10 units per tick mark

c. Scale: 20 units per tick mark.

How is the visual display of the slant of the line (the slope) affected by the changes in scale values? The slant appears the same. However, if *different* scale values were used on the x-axis than on the y-axis, then the slant of the line would appear to be different.

How is the algebraic relationship between x and y affected? The algebraic relationship does not change.

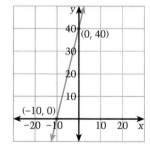

Example 4

PROBLEM SOLVING Emphasize the importance of beginning with a verbal model of the problem. Tell students to ask "what if" questions about the prices of adult and children's tickets to help formulate such a model. For example, "What if the adult tickets sold for \$5 and children's tickets sold for \$3? Then, if the same number of people come this year as last, you will raise 200(5) + 100(3) dollars, or \$1300." Although a guess may not solve the problem, it does indicate how the variables fit into the relationship and what quantities are fixed.

Extra Examples

Here are additional examples similar to **Examples 1–3.**

1. Find the x-intercept and y-intercept of each equation.
   a. $2x + 3y = 12$ 6, 4
   b. $-4x + 2y = 10$ −2.5, 5
   c. $x - 4y = 14$ 14, −3.5
2. Sketch the graph of each line using intercepts.
   a. $2x + 3y = 12$
   b. $3x - 5y = 15$
   c. $-4x + 6y = 18$

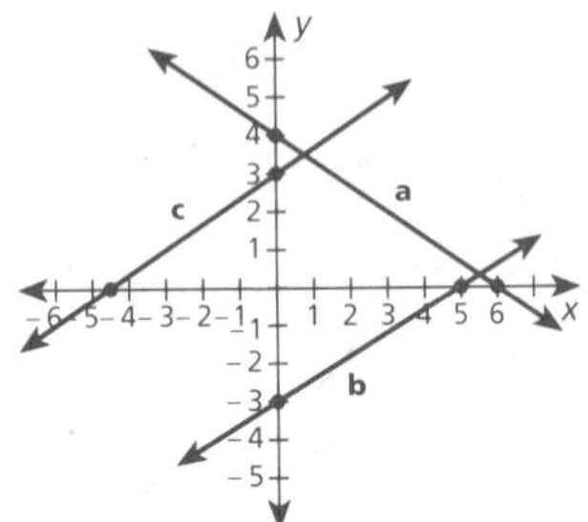

3. Sketch $y = 6x + 48$. Using intercepts, determine the appropriate scale of the coordinate plane. Intercept points: (−8, 0) and (0, 48). Discuss students' choice of scales.

Example 4 *Determining the Price of Tickets*

You are on the planning committee to organize a spaghetti dinner to raise funds for the zoo. Your goal is to sell \$1500 worth of tickets. Based on last year's attendance (200 adults and 100 children), how much should you charge for an adult's ticket and a child's ticket?

Solution

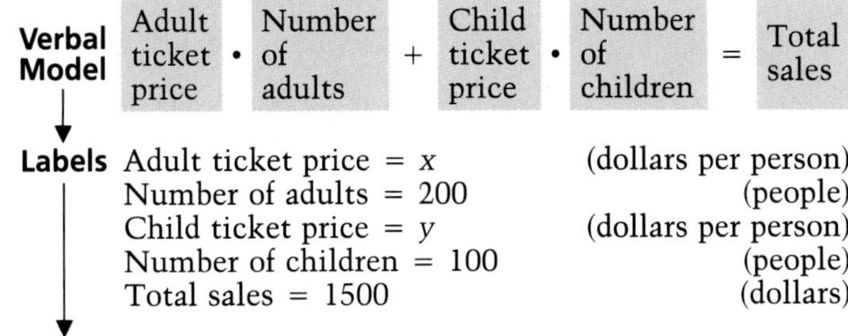

Verbal Model Adult ticket price • Number of adults + Child ticket price • Number of children = Total sales

Labels
Adult ticket price = x (dollars per person)
Number of adults = 200 (people)
Child ticket price = y (dollars per person)
Number of children = 100 (people)
Total sales = 1500 (dollars)

Equation $200x + 100y = 1500$ **Algebraic model**

This equation has many solutions. For instance, you could charge \$5 for everyone (5, 5), or you could charge \$6 per adult and \$3 per child (6, 3), or you could chargc \$4 per adult and \$7 per child (4, 7). To get a better idea of the options, use a quick graph, as shown at the left. ■

Communicating about ALGEBRA

SHARING IDEAS about the Lesson

Understand Limitations The quick-graph method using intercepts cannot be used for every line.

A. What is the x-intercept of the line given by $x = -19$? Does this line have a y-intercept? −19, no

B. What is the y-intercept of the horizontal line passing through (−34, 615)? Does this line have an x-intercept? 615, no

C. What is the y-intercept of the line given by $y = 2x$? What is the x-intercept of the line given by $y = 2x$? Are the intercepts the same point or different points? How did you decide? 0, 0; same, found the point where each occurs: (0, 0).

D. The quick-graph method using intercepts requires two intercepts that are different points. For which three types of lines is the quick-graph method using intercepts *not* applicable? Explain your reasoning. See below.

D. Horizontal lines, vertical lines, and lines that pass through the origin have only one point as an intercept instead of the two needed to do a quick graph.

EXERCISES

Guided Practice

CRITICAL THINKING about the Lesson

1. **True or False?** The graph of a linear equation is a line. True
2. How many points are needed to determine a line? Why? 2; only 1 line can be drawn between 2 points.
3. How do you find the x-intercept of the line $2x - y = 4$? Set $y = 0$ and solve for x.
4. How do you find the y-intercept of the line $2x - y = 4$? Set $x = 0$ and solve for y.
5. Is -8 the x-intercept or the y-intercept of the line $y = \frac{1}{2}x + 4$? x-intercept
6. Is the y-intercept of the line $y = 6x - 4$ equal to 6 or -4? Justify your answer. -4; $y = 6x - 4 = 6 \cdot 0 - 4 = -4$

Independent Practice

In Exercises 7–10, match the equation with its graph.

7. $y = 4x - 8$ c
8. $y = 4x - 4$ a
9. $y = 4x + 8$ d
10. $y = 4x + 4$ b

a.
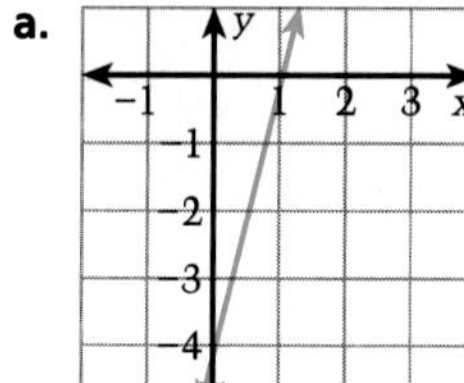

b.
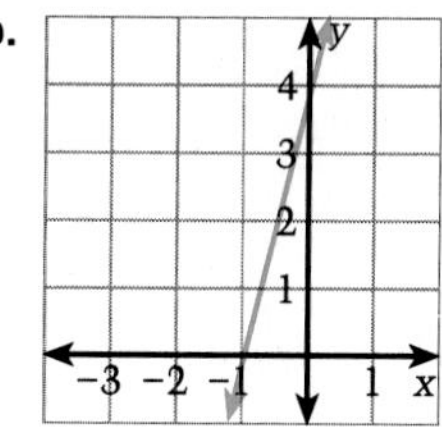

c.
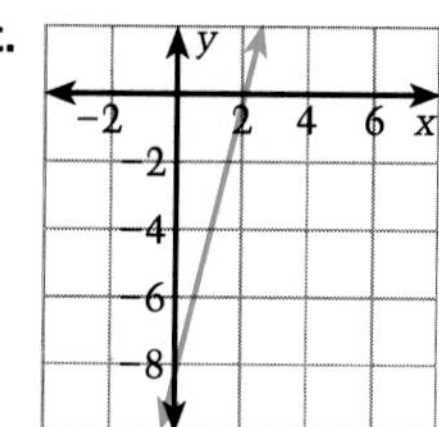

d.
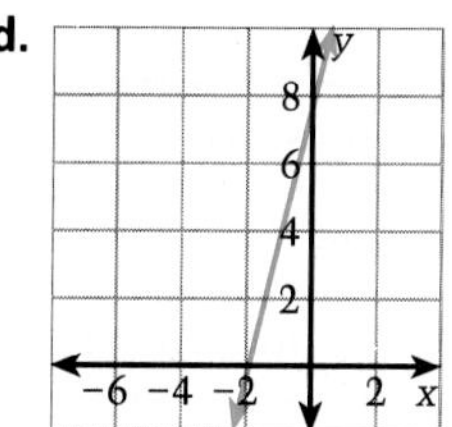

In Exercises 11–14, use the graph to find the x-intercept and y-intercept of the line.

11.
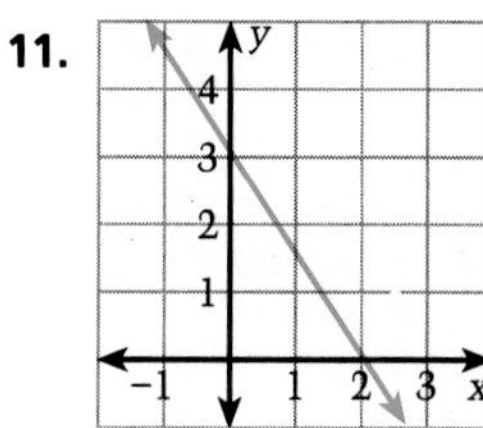
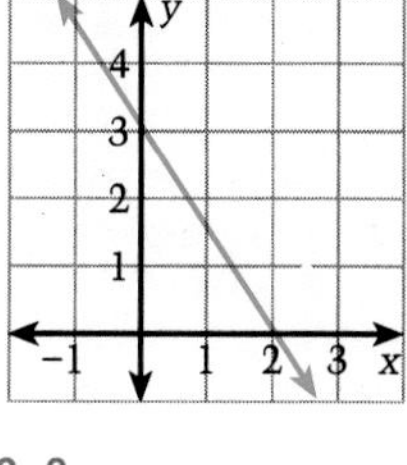

2, 3

12.
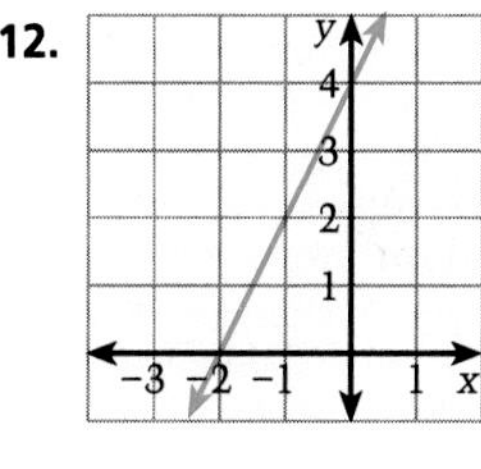

-2, 4

13.
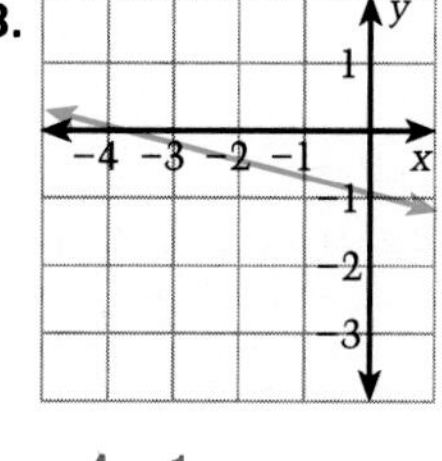

-4, -1

14.
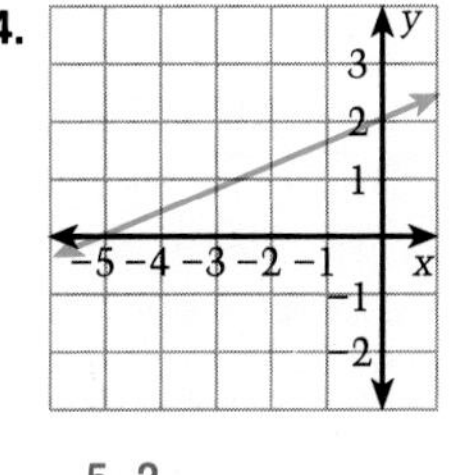

-5, 2

In Exercises 15–20, sketch the line with the given intercepts. See Additional Answers.

15. x-intercept: -3
 y-intercept: 7
16. x-intercept: 16
 y-intercept: 9
17. x-intercept: 9
 y-intercept: -8
18. x-intercept: -2
 y-intercept: -5
19. x-intercept: -11
 y-intercept: -7
20. x-intercept: -9
 y-intercept: 12

Check Understanding

1. What is the x-intercept of the line given by $Ax + By = C$? What is its y-intercept? $\frac{C}{A}$ and $\frac{C}{B}$
2. Explain why using intercepts to sketch linear equations is an efficient way to graph. Intercepts give you two points on a line, all that is necessary to determine a line.
3. What point on the coordinate plane corresponds to a y-intercept of -5? $(0, -5)$

Communicating about ALGEBRA

EXTEND *Communicating*
COOPERATIVE LEARNING
Have students work together to determine other examples of lines that cannot be sketched using the quick-graph approach. Then ask them to discuss whether lines that cannot be sketched using intercepts are, nevertheless, easy to draw. Can these lines be quickly drawn as well?

EXERCISE Notes

ASSIGNMENT GUIDE

Basic/Average: Ex. 1–20, 21–47 odd, 49–50, 53–56, 59–60

Above Average: Ex. 1–20, 21–48 multiples of 3, 49–52, 54–58 even, 59–60, 63–64

Advanced: Ex. 1–14, 15–48 multiples of 3, 49–52, 59–64

Use **Mixed Review** as needed.

Selected Answers
Exercises 1–6, 7–57 odd

✪ More Difficult Exercises
Exercises 51–52, 63–64

Guided Practice

▶ **Ex. 5–6** Students need to use critical thinking skills to determine x- and y-intercepts.

Independent Practice

EXTEND Ex. 7–10 WRITING Ask students to provide a written rationale for the matches that they make. Encourage students to discuss their selections with others. Ask students if they need to set x equal to 0 to find the y-intercepts for the linear equations.

▶ **Ex. 11–14** Caution students to be careful in identifying the x-axis and the y-axis.

▶ **Ex. 15–48** Remind students that the x- and y-intercepts are ordered pairs that can be graphed on the coordinate plane.

▶ **Ex. 49–52** PROBLEM SOLVING Encourage students to write verbal models of the relationships described in the exercises. Asking "what if" questions about the relationships will frequently make it easier to do this. In Exercise 49, what if she sold 100,000 paperback (pbk) books and 10,000 hardback (hbk) copies? Then she would receive 5% of (100,000) • (unit price of paperbacks) and 5% of (10,000) • (unit price of hardbacks). Or, using the information on the book cover at the right, 5% of (100,000 • 5 + 10,000 • 18). But, 0.05(100,000 • 5 + 10,000 • 18) is only $34,000. Alicia wants to earn $100,000. The following is one verbal model.

(0.05)(No. of Pbk • Pbk price + No. of Hbk • Hbk price) = 100,000

A linear model, with numbers of books in the thousands, is $y = -\frac{5}{18}x + \frac{1000}{9}$ with intercepts at (400, 0) and $\approx$(0, 111.11).

194

In Exercises 21–28, find the x-intercept of the line.

21. $x - 2y = 4$ 4
22. $x + 4y = -2$ −2
23. $2x - 3y = 6$ 3
24. $5x + 6y = 95$ 19
25. $-6x - 4y = 42$ −7
26. $9x - 4y = 54$ 6
27. $-x - 5y = 12$ −12
28. $-13x - y = 39$ −3

In Exercises 29–36, find the y-intercept of the line.

29. $y = 4x - 2$ −2
30. $y = -3x + 7$ 7
31. $y = 13x + 26$ 26
32. $y = 6x - 24$ −24
33. $3x - 4y = 16$ −4
34. $2x - 17y = -51$ 3
35. $-x + 8y = 40$ 5
36. $6x + 9y = -81$ −9

In Exercises 37–48, sketch the line. Label the x-intercept and y-intercept on the graph. See Additional Answers.

37. $y = 2 - x$
38. $y = x + 3$
39. $y = 3x + 9$
40. $y = -4 + 2x$
41. $3x + 5y = 15$
42. $4x + 3y = 24$
43. $-9x + 2y = 36$
44. $x - 6y = 36$
45. $5x - y = 35$
46. $10x + 7y = -140$
47. $y = -3x + 9$
48. $y = 10x + 50$

49. *Writing a Novel* Alicia Martínez has had her first novel accepted for publication. She signed contracts to receive a royalty rate of 5% of the retail price of each book. Alicia is now wondering how many paperback copies and how many hardback copies must be sold to earn her $100,000. Sketch a graph showing the possible sales combinations that could earn Alicia this amount. See Additional Answers.

50. Suppose Alicia's publisher sold 20,000 hardback copies. How many paperback copies must be sold for Alicia to realize her dream? Explain. 328,000; $0.05(18 \cdot 20{,}000 + 5x) = 100{,}000$

✪ **51.** *Saturday Matinee* A theater charges $4 per person before 6:00 P.M. and $7 per person after 6:00 P.M. The total ticket sales for Saturday was $11,228. Sketch a graph showing the possible number of people who attended the theater before and after 6:00 P.M.

✪ **52.** Suppose no one attended the theater before 6:00 P.M. How many people attended the theater after 6:00 P.M.? Explain.
1604; $7y = 11{,}228$

51. See Additional Answers.

Integrated Review

53. Is $(1, -3)$ a solution of $4x - 3y = -5$? No
54. Is $(-5, 4)$ a solution of $-5x - 4y = 9$? Yes

In Exercises 55–58, solve for y.

55. $6x + 2y = 18$ $y = 9 - 3x$
56. $4x - y = 7$ $y = 4x - 7$
57. $-9x - 3y = 2$ $y = -3x - \frac{2}{3}$
58. $5x + 20y = 10$ $y = \frac{1}{4}(2 - x)$

Exploration and Extension

Technology—Graphing Calculator **In Exercises 59 and 60, use the following information.**

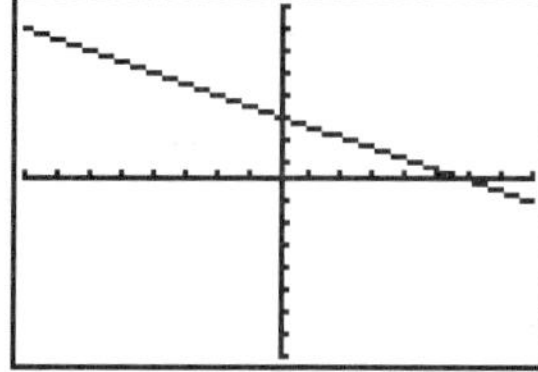

The graph of $30x + 60y = 1800$ was sketched with a graphing calculator. The calculator screen is shown at the right.

59. Find the x- and y-intercepts of the line. 60, 30

60. Use the result from Exercise 59 to determine the distance between adjacent tick marks on the x-axis and the distance between adjacent tick marks on the y-axis.
10 units, 10 units

Technology—Graphing Calculator **In Exercises 61 and 62, use the following information.**

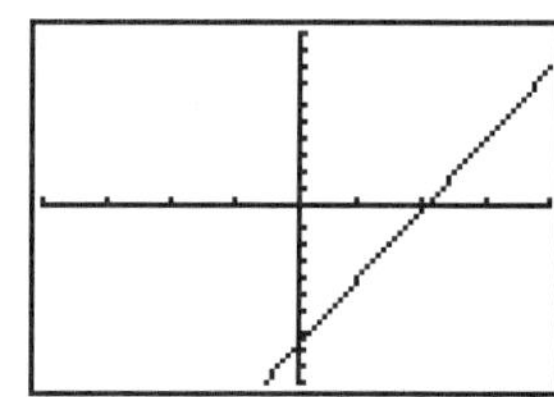

The graph of $64x - 16y = 256$ was sketched with a graphing calculator. The calculator screen is shown at the right.

61. Find the x and y-intercepts of the line. 4, -16

62. Use the result from Exercise 61 to determine the distance between adjacent tick marks on the x-axis and the distance between adjacent tick marks on the y-axis.
2 units, 2 units

Women's 200-Meter Dash **In Exercises 63 and 64, use the following information.**

The graph below shows the times for the women's 200-meter dash in the Summer Olympics from 1964 to 1996.

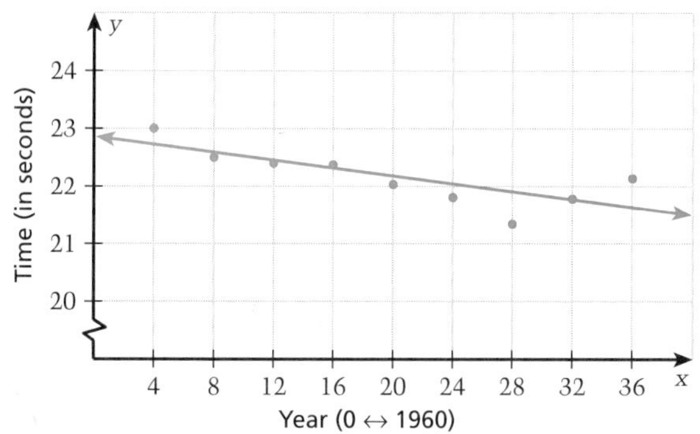

63. What are the coordinates of Florence Griffith Joyner's record time? (28, 21.34)

64. Do you think the line in the graph will continue to be a good model for the next 50 years? Explain.
No, there is some unknown limit between 21.34 and 0 that will take longer and longer to reach.

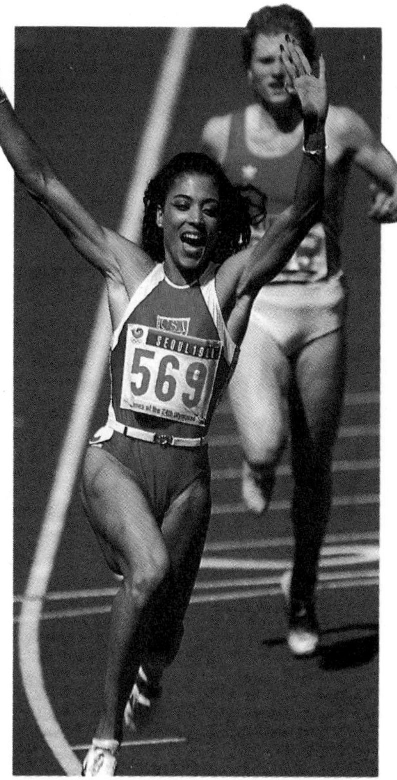

Florence Griffith Joyner of the United States set a record of 21.34 seconds in the women's 200-meter dash in the 1988 Summer Olympics in Seoul, South Korea.

Ex. 59–62 TECHNOLOGY
These exercises are similar to Example 3 in the lesson—using intercepts to determine the appropriate scale of a coordinate plane. If your students have graphics calculators, have them check the scale values they found by entering the information into the range of the corresponding linear equation.

Enrichment Activities

These activities require students to go beyond lesson goals.

1. Find the x- and y-intercepts of each line.

a. $y = 2x - 5$ $\quad \frac{5}{2}, -5$

b. $3y = 4 - 6x$ $\quad \frac{2}{3}, \frac{4}{3}$

c. $3y - 10x = 25$ $\quad -\frac{5}{2}, \frac{25}{3}$

d. $2x + 8y - 12 = 0$ $\quad 6, \frac{3}{2}$

e. $4x = 2 - 3y$ $\quad \frac{1}{2}, \frac{2}{3}$

WRITING Which of the linear equations above do you think are more suitable for finding intercepts and sketching quick graphs? Explain your answers.

Milestones

Language Arts: ***Library Skills***
Use your library's resources to answer these questions.

1. How old was Pascal when he built his adding and subtracting machine? nineteen years old
2. What calculating device did John Napier create? Napier's bones; used for multiplication.

Mixed REVIEW

9. $7x^2 - 4x - 3$
10. $-x - 2$
13. $\begin{bmatrix} -3.9 & 11.0 \\ 4.1 & 10.0 \end{bmatrix}$

1. Evaluate $3x + 7$ when $x = 0$. **(1.2)** 7
2. Evaluate $\frac{1}{3}6a(4 - b)$ when $a = 1$ and $b = 3$. **(1.2)** 2
3. Solve $4(x - 1) = 2y$ for y. **(3.6)** $y = 2(x - 1)$
4. Solve $\frac{10}{9}x = 3x + y$ for x. **(3.6)** $x = -\frac{9}{17}y$
5. What is the opposite of -3.2? **(2.1)** 3.2
6. What is the absolute value of $-3(-10)$? **(2.1)** 30
7. Evaluate $(3^2 - 1) \div 2^3$. **(1.3)** 1
8. Evaluate $4[(-5)^2 + (-10) - 7]$. **(2.5)** 32
9. Simplify $3(x^2 - 1) - 4(x - x^2)$. **(2.6)**
10. Simplify $x(x - 1) - (2 + x^2)$. **(2.6)**
11. Does $4 \cdot x + 3 \div 2 = 4 \cdot (x + 3) \div 2$? **(2.6)** No
12. Does $\frac{1}{5}(16x + 2) = (16x + 2) \div 5$? **(2.7)** Yes
13. Find $\begin{bmatrix} 0.4 & 9.2 \\ 6.3 & 7.8 \end{bmatrix} + \begin{bmatrix} -4.3 & 1.8 \\ -2.2 & 2.2 \end{bmatrix}$. **(2.4)**
14. Find $\begin{bmatrix} \frac{1}{4} & -\frac{3}{2} \\ \frac{3}{8} & \frac{7}{8} \end{bmatrix} - \begin{bmatrix} \frac{1}{8} & -\frac{3}{2} \\ -\frac{1}{8} & \frac{3}{4} \end{bmatrix}$. **(2.4)** $\begin{bmatrix} \frac{1}{8} & 0 \\ \frac{1}{2} & \frac{1}{8} \end{bmatrix}$
15. Is 0 a solution of $14x - 10 \geq 3x + 2$? **(1.5)** No
16. Is 3 a solution of $\frac{1}{2}(1 - x) < -2$? **(1.5)** No
17. Is -1 a solution of $4y^2 - 3 = 1$? **(1.5)** Yes
18. Is 5.5 a solution of $-3(a - 9.5) = -12$? **(1.5)** No
19. Is $(2, -1)$ a solution of $2x - y = 5$? **(4.2)** Yes
20. Is $(0, 4)$ a solution of $y = 4$? **(4.1)** Yes

Milestones

CALCULATING THROUGH THE CENTURIES

1550 1600 1650 1700 1750 1800 1850 1900 1950 2000

Pascal's Machine — Babbage's analytical engine — First electronic computer

Abacus (above). Pascal's "Machine Arithmetic" (below)

Next to fingers, the abacus is the oldest calculating device known. Different forms of the abacus were used by Egyptians, Greeks, Romans, Hindus, and Chinese. From a tray containing sand or dust in which to write, the abacus gradually evolved into a wooden frame with beads that slid on thin bamboo rods.

A knotted cord was used by various cultures to count days and inventory livestock. In part of India, an elaborate knotted cord system was used to take the census in 1872.

In 1642, Blaise Pascal built an adding and subtracting machine that carried tens mechanically. In the 1830's, Charles Babbage began work on his analytical engine, hoping it would do complex calculations from a series of instructions, but lacking the technology to make it work. All of these devices and ideas have led to today's calculators and super computers.

accept answers around 500 B.C.

Find out when the Chinese began using bamboo rods to compute.

4.4 The Slope of a Line

What you should learn:

Goal 1 How to find the slope of a line using two of its points

Goal 2 How to interpret slope as a constant rate of change or an average rate of change

Why you should learn it:

You can use slope to represent the rate of change of one quantity with respect to another, such as distance traveled compared to time used.

Goal 1 Finding the Slope of a Line

The *slope* of a nonvertical line is the number of units the line rises or falls for each unit of horizontal change from left to right.

The line shown below at the left rises 2 units for each unit of horizontal change from left to right. The slope of a line is represented by the letter m. Thus, this line has a slope of $m = 2$. Two points on a line are all that are needed to determine its slope.

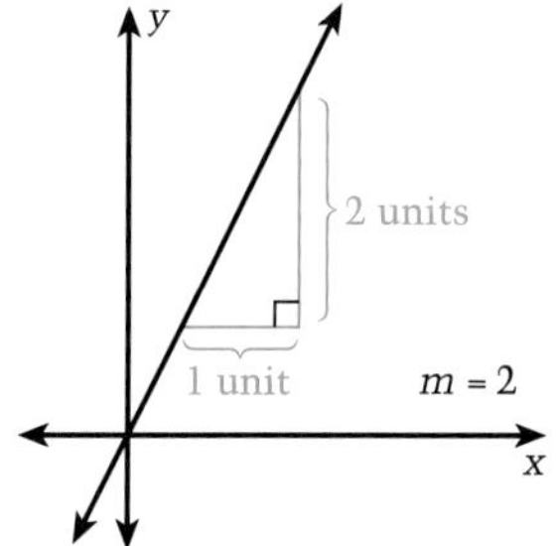

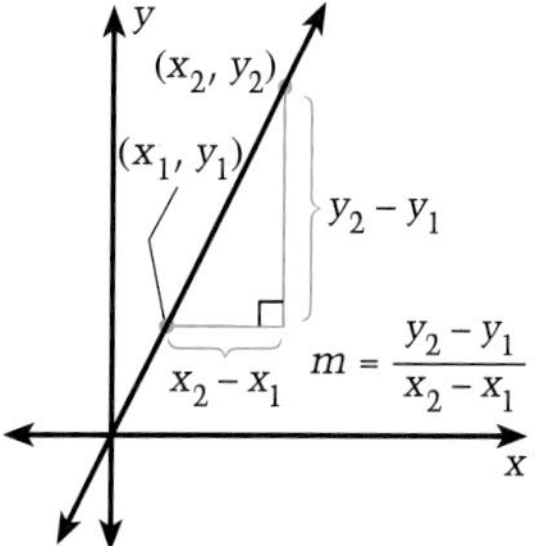

The Slope of a Line

The slope of the nonvertical line passing through the points (x_1, y_1) and (x_2, y_2) is

$$m = \frac{y_2 - y_1}{x_2 - x_1}.$$

The numerator is read as "y sub 2 minus y sub 1" and is called the *rise*. The denominator is read as "x sub 2 minus x sub 1" and is called the *run*.

When the formula for slope is used, the order of subtraction is important. Given two points on a line, you can label either as (x_1, y_1) and the other as (x_2, y_2). After doing this, however, you must form the numerator and the denominator using the *same* order of subtraction.

Subtraction order is the same. → $\frac{y_2 - y_1}{x_2 - x_1}$ Correct $\frac{y_2 - y_1}{x_1 - x_2}$ Incorrect ← *Subtraction order is different.*

Problem of the Day

A frog, a rabbit, and a kangaroo decide to go for a walk. The frog takes 2-foot hops, the rabbit 3-foot hops, and the kangaroo 7-foot hops. If they start their walk together, how far will they travel before all 3 have their feet on the ground at the same time? 42 feet

ORGANIZER

Warm-Up Exercises

1. Simplify the following ratios.
   a. $\frac{32}{48}$ — $\frac{2}{3}$
   b. $\frac{-15}{20}$ — $\frac{-3}{4}$
   c. $\frac{24}{-30}$ — $\frac{4}{-5}$
   d. $\frac{-13}{13}$ — -1
   e. $\frac{17-5}{-3+6}$ — 4
   f. $\frac{-7+7}{8}$ — 0
2. Evaluate for $x = -2, y = 3$. Express each result as a ratio.
   a. $\frac{5-y}{6-x}$ — $\frac{1}{4}$
   b. $\frac{y+11}{x-8}$ — $\frac{7}{-5}$
   c. $\frac{17-y}{14-x}$ — $\frac{7}{8}$
   d. $\frac{14-8}{16-x}$ — $\frac{1}{3}$
   e. $\frac{y-14}{x+9}$ — $\frac{-11}{7}$

Lesson Resources

Teaching Tools
  Problem of the Day: 4.4
  Warm-up Exercises: 4.4
  Answer Masters: 4.4
Extra Practice: 4.4
Color Transparency: 19, 20
Technology Handbook: p. 35

LESSON Notes

As you move on the x-axis of the coordinate plane from left to right, it is customary to speak of the rise or fall of a straight line.

The *slope* of a straight line is the ratio of the vertical change to the horizontal change as you move from one point (say P) on the line to another point (say Q) on the line. Using points P and Q, you can determine the slope of the line containing these points by using the formula for the slope of a line.

Examples 1–3

Remind students that the orientation that accounts for lines with positive slope, rising, and lines with negative slope, falling, is the perspective gained from moving on the x-axis from left to right.

Common-Error ALERT!

Students may not fully appreciate the use of subscripts to describe different coordinate values of points on a line. First, help students to read x_1, y_1, x_2, and y_2, as "x sub 1," "y sub 1," "x sub 2," and "y sub 2." Emphasize that subscripts are used to distinguish between different x- and y-coordinates. In other words, the subscripts provide a different way to name coordinates of the ordered pairs.

Another common error is the incorrect order in which the slope formula is applied. Caution students to carefully label the (x_2, y_2) point to be used, use y_2 as the first term of the numerator, and use x_2 as the first term of the denominator.

Example 1 A Line with a Positive Slope Rises

Find the slope of the line passing through $(-2, 0)$ and $(3, 1)$.

Solution Let $(x_1, y_1) = (-2, 0)$ and $(x_2, y_2) = (3, 1)$.

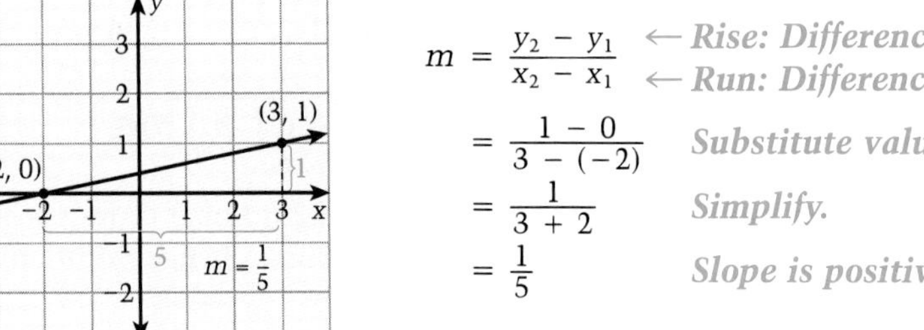

Positive slope, line rises.

$$m = \frac{y_2 - y_1}{x_2 - x_1}$$ ← Rise: Difference of y-values ← Run: Difference of x-values

$$= \frac{1 - 0}{3 - (-2)}$$ Substitute values.

$$= \frac{1}{3 + 2}$$ Simplify.

$$= \frac{1}{5}$$ Slope is positive.

■

Example 2 A Line with Slope Zero Is Horizontal

Find the slope of the line passing through $(-1, 2)$ and $(2, 2)$.

Solution Let $(x_1, y_1) = (-1, 2)$ and $(x_2, y_2) = (2, 2)$.

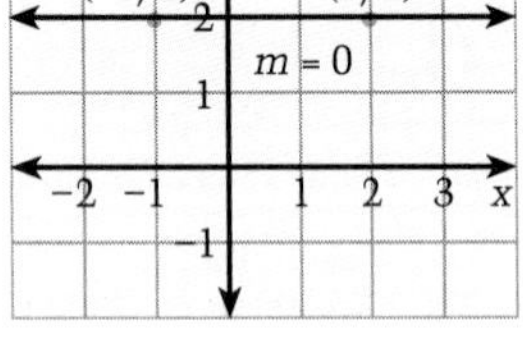

Slope zero, line is horizontal.

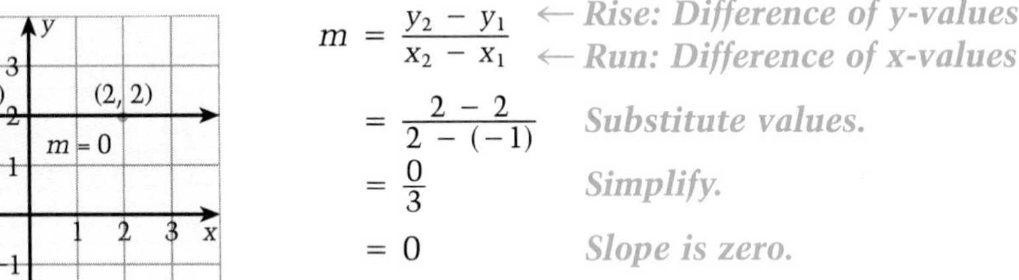

$$m = \frac{y_2 - y_1}{x_2 - x_1}$$ ← Rise: Difference of y-values ← Run: Difference of x-values

$$= \frac{2 - 2}{2 - (-1)}$$ Substitute values.

$$= \frac{0}{3}$$ Simplify.

$$= 0$$ Slope is zero.

■

Example 3 A Line with a Negative Slope Falls

Find the slope of the line passing through $(0, 0)$ and $(1, -1)$.

Solution Let $(x_1, y_1) = (0, 0)$ and $(x_2, y_2) = (1, -1)$.

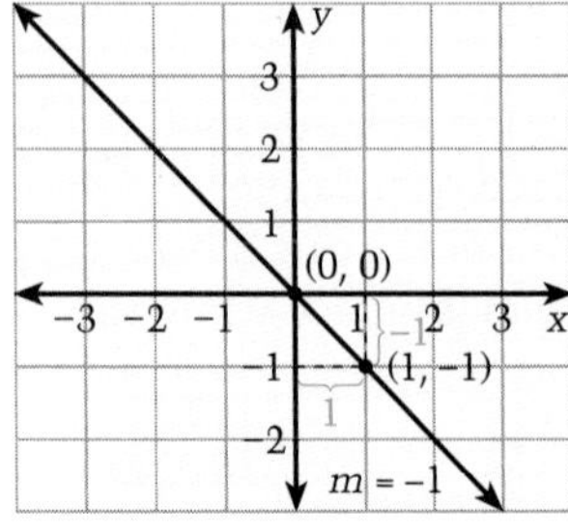

Negative slope, line falls.

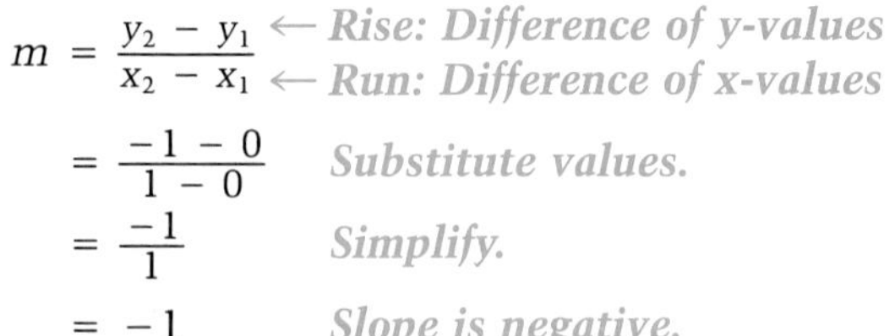

$$m = \frac{y_2 - y_1}{x_2 - x_1}$$ ← Rise: Difference of y-values ← Run: Difference of x-values

$$= \frac{-1 - 0}{1 - 0}$$ Substitute values.

$$= \frac{-1}{1}$$ Simplify.

$$= -1$$ Slope is negative.

■

Example 4 The Slope of a Vertical Line Is Not Defined

Apply the slope formula to the vertical line passing through (2, 4) and (2, 1).

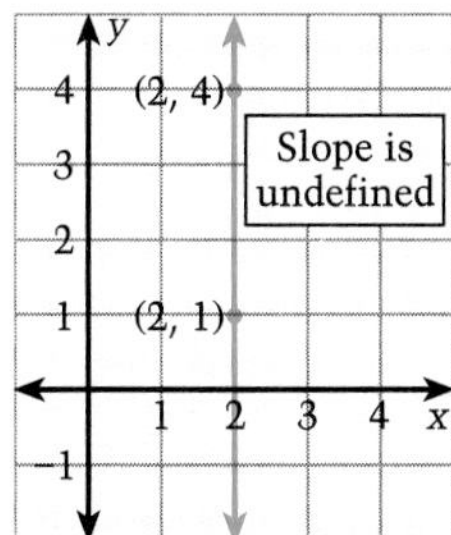

Vertical line

Solution Let $(x_1, y_1) = (2, 1)$ and $(x_2, y_2) = (2, 4)$

$$m = \frac{y_2 - y_1}{x_2 - x_1} \quad \begin{matrix} \leftarrow \text{Rise: Difference of } y\text{-values} \\ \leftarrow \text{Run: Difference of } x\text{-values} \end{matrix}$$

$$= \frac{4 - 1}{2 - 2} \quad \text{Substitute values.}$$

$$= \frac{3}{0} \quad \text{Division by 0 is undefined.}$$

Because division by 0 is undefined, the expression $\frac{3}{0}$ has no meaning. This is why the slope of a vertical line is undefined. (This will always happen with two points on a vertical line.) ■

Classification of Lines by Slope

1. A line with positive slope *rises* from left to right. $(m > 0)$
2. A line with negative slope *falls* from left to right. $(m < 0)$
3. A line with slope zero *is horizontal*. $(m = 0)$
4. A line with undefined slope *is vertical*. (m is undefined.)

Problem Solving
Make a Graph

The slope of a line not only tells you whether the line rises, falls, or is horizontal; it also tells you the steepness of the line. For two lines with positive slopes, the line with the greater slope is steeper. For two lines with negative slopes, the line with the slope of greater absolute value is steeper.

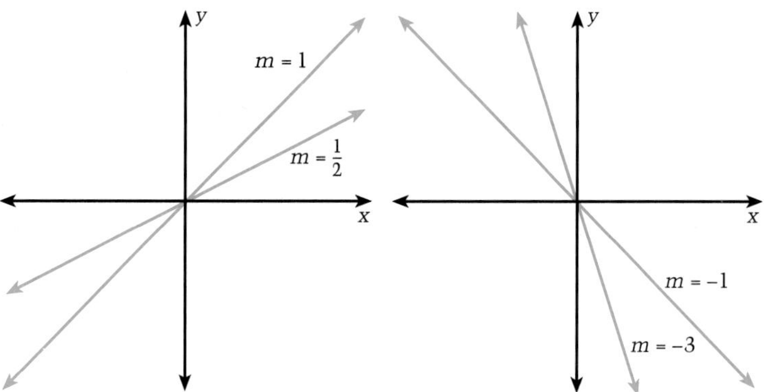

Lines with positive slopes *Lines with negative slopes*

An Alternate Approach
Using Graphics Calculators

An excellent way to demonstrate the effect of the magnitude of a given slope on the steepness of a line is to graph several lines in which each line has a slope greater than the one previously drawn on the same calculator display. For example, graph these four lines in the given order.

a. $y = x - 2$
b. $y = 3x - 2$
c. $y = 6x - 2$
d. $y = 12x - 2$

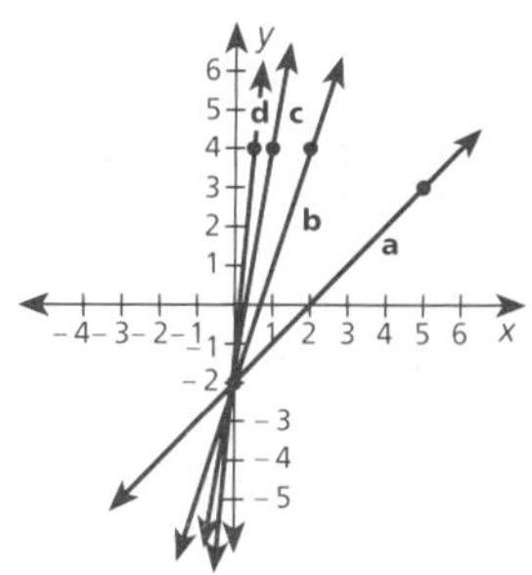

The dynamic display of the four lines drawn one after the other reveals the relationship: the greater the slope, the steeper the line. (You may wish to use this approach after you have talked about the slope-intercept form of a straight line.)

Common-Error ALERT!

In **Examples 2 and 4,** the fraction $\frac{0}{3}$ is well-defined ($\frac{0}{3} = 0$), while the fraction $\frac{3}{0}$ is not well-defined. The fraction $\frac{3}{0}$ is not well-defined because if $\frac{3}{0} = n$, then $0n = 3$, which cannot be. Emphasize the differences with students.

Example 5

As the steepness of a line, the ratio known as slope compares two number values representing changes in distances along the y- and x-axes. In this example, the changes are in two different types of quantities, distance and time. Thus, the graph shows "rates" instead of ratios.

Extra Examples

Here are additional examples similar to **Examples 1–4.**

Find the slope of the line passing through the pair of points.

1. (4, 6) and (−2, −8) $\frac{7}{3}$
2. (5, 9) and (13, 9) 0
3. (−3, 4) and (7, −5) $-\frac{9}{10}$
4. (−4, 8) and (−4, 12) undefined
5. (3, 7) and (5, 12) $\frac{5}{2}$

Check Understanding

Can you identify real-life uses of slope?

COOPERATIVE LEARNING

Compile a class list of jobs or situations in which slope is used.

Communicating about ALGEBRA

Plot the data in Parts A and B on different coordinate planes. Explain what the rise and run each represent for each. For example, in Part A the rise is distance in miles while the run is time in hours.

You might wish to have students discuss whether the *average* rate of change for each situation must have actually been obtained as the changes occurred. Sample answer: Because speed and temperature are continuous data, the average rate of change will actually occur.

Goal 2 Interpreting Slope as a Rate of Change

In real-life problems, slope is often used to describe a **constant rate of change** or an **average rate of change.** Units of measure such as miles per hour or dollars per year are usually used.

Real Life
Mountain Climbing

Using ropes, pitons (spikes), and snap links, a rock climber can scale a vertical cliff.

Example 5 *Slope as a Rate of Change*

Rebecca is climbing up a 500-foot cliff. By 1:00 P.M., she has climbed 125 feet up the cliff. By 4:00 P.M., she has reached an altitude of 290 feet. Find the average rate of change in Rebecca's altitude during the three hours. Give the rate in feet per hour.

Solution Let t represent the time and y represent Rebecca's altitude. Her two positions are represented by (t_1, y_1) and (t_2, y_2).

1st time ↓, 1st altitude ↓: $(t_1, y_1) = (1, 125)$

2nd time ↓, 2nd altitude ↓: $(t_2, y_2) = (4, 290)$

Use the formula for slope to find the average rate of change.

$$\text{Rate of change} = \frac{y_2 - y_1}{t_2 - t_1} \quad \begin{matrix}\leftarrow \textit{Difference of altitudes} \\ \leftarrow \textit{Difference of times}\end{matrix} \quad \left(\frac{\text{feet}}{\text{hours}}\right)$$

$$= \frac{290 - 125}{4 - 1} \quad \textit{Substitute values.} \quad \left(\frac{\text{feet}}{\text{hours}}\right)$$

$$= 55 \quad \textit{Simplify.} \quad \left(\frac{\text{ft}}{\text{hour}}\right)$$

If Rebecca had been climbing at the rate of 55 feet per hour during the entire three hours, the rate of change in her altitude would be *constant*. It is more likely, however, that her rate varied, and 55 feet per hour represents the *average* rate of change during the three hours. ■

Communicating about ALGEBRA

SHARING IDEAS about the Lesson

Using a Model Use the formula for slope.

A. At 3:00 P.M., a car leaves a city. By 5:00 P.M., it has traveled 90 miles. Find its average speed. 45 mph

B. At 7:00 P.M., the temperature is 63°F. At 10:00 P.M., it is 45°F. Find the average rate of change of the temperature.

−6°F per hour

EXERCISES

1. Did not use the same order of subtraction for numerator and denominator

Guided Practice

CRITICAL THINKING about the Lesson

1. A student found the slope of the line through (3, 2) and (1, 4) to be $m = \frac{4-2}{3-1}$. What did the student do wrong?
2. In the formula for slope, what is the numerator $y_2 - y_1$ called? Rise
3. In the formula for slope, what is the denominator $x_2 - x_1$ called? Run
4. Explain what happens when the formula for slope is applied to a vertical line. Denominator becomes zero, so slope is undefined.
5. What is the slope of a horizontal line? Zero
6. Which of the following lines has the *least* slope? Why? See margin.

a. 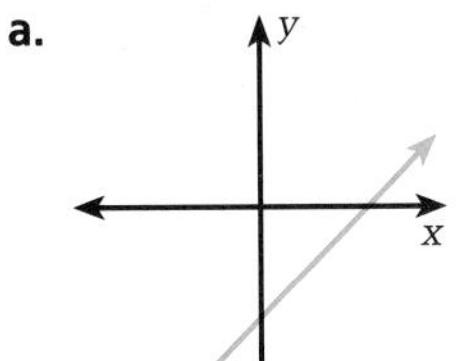

b. 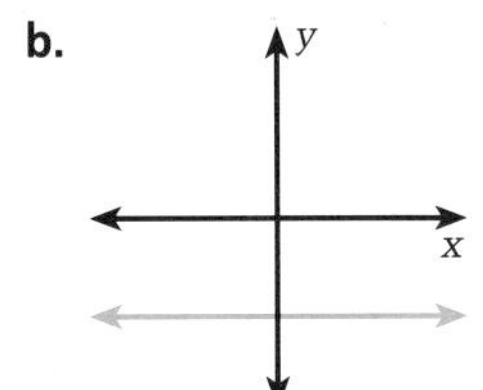

c. 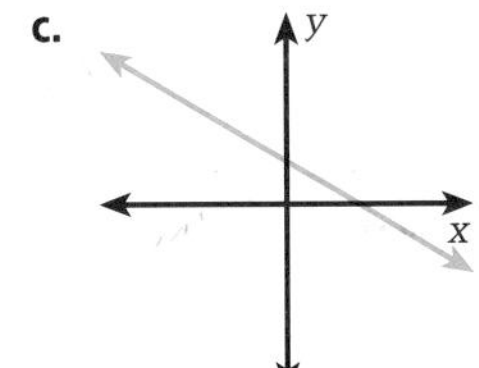

d. 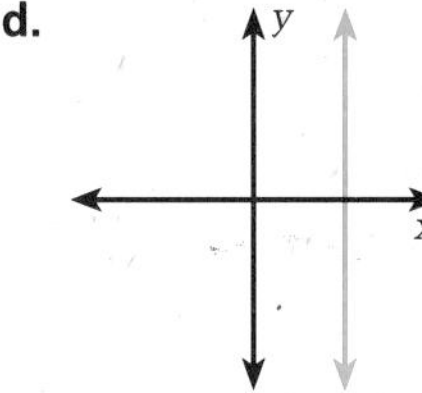

7. **True or False?** A line with positive slope rises from left to right. True
8. Give an example of the use of slope in a real-life problem. Answers vary; for example, miles per hour

Independent Practice

In Exercises 9–12, estimate the slope of each line.

9. 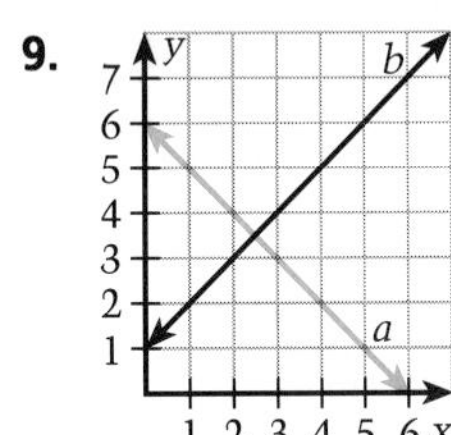
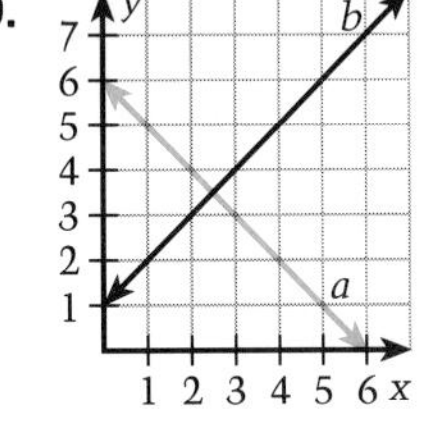

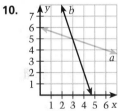

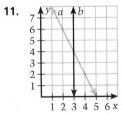

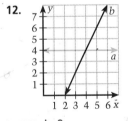

a: -1, b: 1

10. a: $-\frac{1}{3}$, b: -3

11. a: -2, b: undefined

12. a: zero, b: 2

In Exercises 13–16, sketch the line with the given slope through the given point. See Additional Answers.

✪ 13. (2, 1), $m = 2$
✪ 14. $(0, -2)$, $m = -\frac{1}{2}$
✪ 15. $(-1, -1)$, $m = -\frac{2}{3}$
✪ 16. $(-3, 1)$, $m = -1$

In Exercises 17–28, plot the points and find the slope of the line passing through the points. See Additional Answers.

17. (0, 0), (4, 5) $\frac{5}{4}$
18. (0, 0), $(-1, -3)$ 3
19. $(-3, -2)$, (1, 6) 2
20. $(0, -6)$, (8, 0) $\frac{3}{4}$
21. (0, 6), (8, 0) $-\frac{3}{4}$
22. (2, 4), $(4, -4)$ -4
23. $(-6, -1)$, $(-6, 4)$
24. $(0, -10)$, $(-4, 0)$
25. $(1, -2)$, $(-2, -2)$ 0
26. (3, 6), (3, 0) Undefined
27. (2, 2), $(-3, 5)$ $-\frac{3}{5}$
28. (4, 1), (6, 1) 0

23. Undefined
24. $-\frac{5}{2}$

EXERCISE Notes

ASSIGNMENT GUIDE

Basic/Average: Ex. 1–16, 17–37 odd, 43, 45, 47–50, 51–59 odd

Above Average: Ex. 1–16, 18–34 even, 35–42, 44, 46, 51–57 multiples of 3, 59–62

Advanced: Ex. 1–8, 9–36 multiples of 3, 39–42, 45, 46–50, 59–63

Selected Answers
Exercises 1–8, 9–57 odd

✪ More Difficult Exercises
Exercises 13–16, 37–38, 47–50, 59, 60, 63

Guided Practice

▶ **Ex. 6** Be careful to distinguish between the slant of a line and its numerical value, slope. Lines with negative slope may be said to have smaller slope than lines with positive slope, although the slant of the lines with negative slope may be steeper.

Answer

6. Consequently, line c has the *least* slope because the slope is negative.

Independent Practice

▶ **Ex. 13–16** These exercises should be discussed with students. It is important that students recognize that the slope provides a way (using rise over run) to move from one point on a line to another point on the same line.

▶ **Ex. 37–38** A good strategy is to plot and label the points. This will help students keep track of the pairs of points that need to be considered.

▶ **Ex. 39–41, 47–50** **PROBLEM SOLVING** Students are required to read from a graph or a table to get needed information. These are important skills for students to sharpen.

▶ **Ex. 39–41** The rate in these exercises is the average price per year.

▶ **Ex. 46** "Broadway" refers to the famous theater district in New York City, which began in 1895 with the Oscar Hammerstein play Olympia.

Enrichment Activities

These activities require students to go beyond lesson goals.

1. Points that lie on the same straight line are said to be collinear. Determine which points are collinear.
   a. (2, 1), (6, −7), (10, −15)
   b. (2, 18), (−6, −54), (8, 71)
   c. (4, −13), (−12, 34), (7, −23)
   d. (−6, 30), (−2, 14), (10, −34)
   e. (8, 21), (0, 4), (−7, −7)
   a. collinear, b. not collinear, c. not collinear, d. collinear, e. not collinear.
2. **WRITING** Examples of uses of slope in real life include pitching a roof and grading a highway. Investigate each situation and write a brief report on your findings.

In Exercises 29–34, determine whether the line through the points rises from left to right, falls from left to right, is horizontal, or is vertical.

29. (6, 9), (4, 3) Rises to right
30. (7, 4), (−1, 8) Falls to right
31. (5, 10), (5, −4) Vertical
32. (−3, 1), (9, 8) Rises to right
33. (1, 1), (4, −3) Falls to right
34. (−6, 7), (1, 7) Horizontal

In Exercises 35 and 36, find the rate of change between the two points. Give the unit of measure.

35. (2, 2), (9, 23) 3 inches per minute
x is measured in minutes.
y is measured in inches.

36. (3, 5), (11, 69) 8 dollars per year
x is measured in years.
y is measured in dollars.

In Exercises 37 and 38, use the following information to decide whether the points lie on the same line or form the vertices of a triangle.

Any two points on a line can be used to determine its slope. Any three points either lie on the same line or form the vertices of a triangle.

37. (−2, 1), (1, 2), (3, 3) Form the vertices of a triangle

38. (−2, −4), (2, −2), (6, 0) Lie on the same line

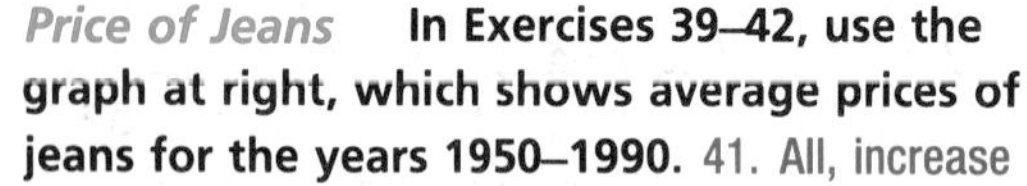

Price of Jeans **In Exercises 39–42, use the graph at right, which shows average prices of jeans for the years 1950–1990.** 41. All, increase

39. Estimate the average rate of change in price from 1950 to 1980. $\frac{1}{3}$ dollar per year

40. Estimate the average rate of change in price from 1980 to 1990. $\frac{6}{5}$ dollars per year

41. In which five-year periods is the rate of change positive? Does a positive rate of change mean a price increase or decrease?

42. Which five-year period had the largest price increase? 1985–1990

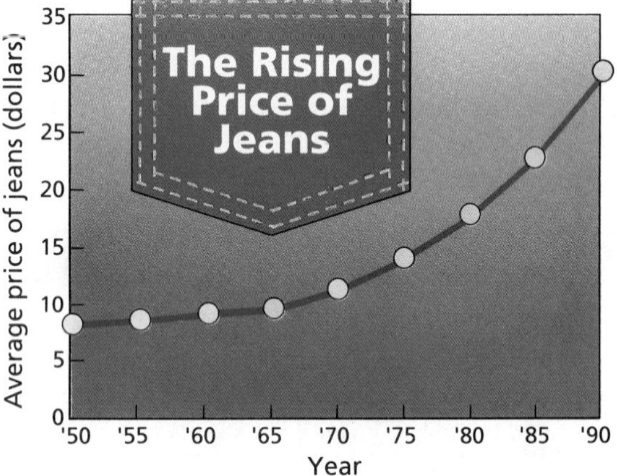

43. *Altitude of a Parachutist* At 12:05, a parachutist is 8000 feet above the ground. At 12:08, the parachutist is 5400 feet above the ground. Find the average rate of change in feet per minute. $-\frac{2600}{3}$ feet per minute

44. *Company Profits* In 1990, a company had a profit of $1,300,000. In 1992, the company had a profit of $1,200,000. Find the average rate of change of its profit. (Give the result in dollars per year.) −50,000 dollars per year

45. *Symphony Orchestras* In 1980, there were 1572 symphony orchestras in the United States. In 1988, there were 1572. Find the average rate of change in the number of symphony orchestras. (Give the result in symphony orchestras per year.) 0 symphony orchestras per year
(Source: American Symphony Orchestra League)

46. *Broadway Shows* In 1980, there were 67 new productions on Broadway. In 1987, there were 31 new productions on Broadway. Find the average rate of change in the number of new productions. (Give the result in new productions per year.)
(Source: Variety Magazine)
−5.14 new productions per year

Border Population Growth **In Exercises 47–50, use the following information.**

Seven major population regions lie along our border with Mexico. The table shows the populations of four of them.

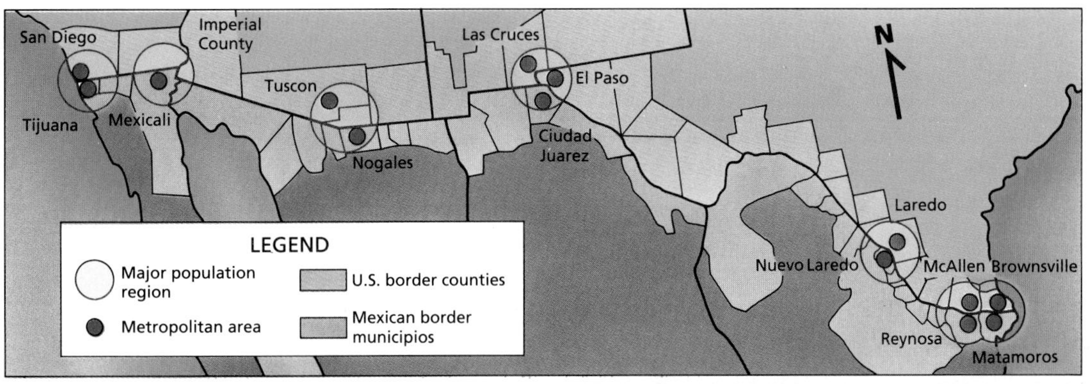

U.S.	San Diego	Tucson	El Paso	Laredo
1980	1,876,900	534,500	580,600	100,300
1990	2,461,100	655,000	737,800	132,200
Mexico	**Tijuana**	**Nogales**	**Juarez**	**N. Laredo**
1980	480,000	71,000	591,000	203,300
1990	742,600	107,100	727,700	217,900

47. 13,670 people per year 48. 3190 people per year

47. Find the average yearly rate of change in population for Juarez from 1980 to 1990.

48. Find the average yearly rate of change in population for Laredo from 1980 to 1990.

49. Which had a greater rate of change from 1980 to 1990: San Diego or Tijuana? San Diego

50. Which had a greater rate of change from 1980 to 1990: Tucson or Nogales? Tucson

Integrated Review

In Exercises 51–54, solve for *y*.

51. $9x - 2y = 14$ $y = -\frac{1}{2}(14 - 9x)$

52. $5x + 9y = 18$ $y = \frac{1}{9}(18 - 5x)$

53. $-6x + 3y = 18$ $y = 6 + 2x$

54. $-2x - 2y = 7$ $y = -\frac{1}{2}(7 + 2x)$

In Exercises 55–58, find the *x*-intercept and *y*-intercept of the line.

55. $3x - 12y = 16$ $\frac{16}{3}, -\frac{4}{3}$

56. $-2x + 4y = 16$ $-8, 4$

57. $-14x + 3y = -42$ $3, -14$

58. $x - 21y = 7$ $7, -\frac{1}{3}$

Exploration and Extension

59. Find x so that the line through (3, 4) and (x, 6) will have slope $-\frac{2}{5}$. $x = -2$

60. Find the slope of a line passing through (0, 0) and (5, –17). $-\frac{17}{5}$

61. Find the slope of the line $4x + 5y = 50$. Explain how you obtained the slope. See margin.

62. Find the slope of $y = -\frac{1}{2}x + 3$. $-\frac{1}{2}$

63. Find the y-intercept of each line graphed at the right. Describe an efficient way to find each y-intercept from the equation. Top to bottom: 2, 1, –1, –3; y-intercept is constant term in equation.

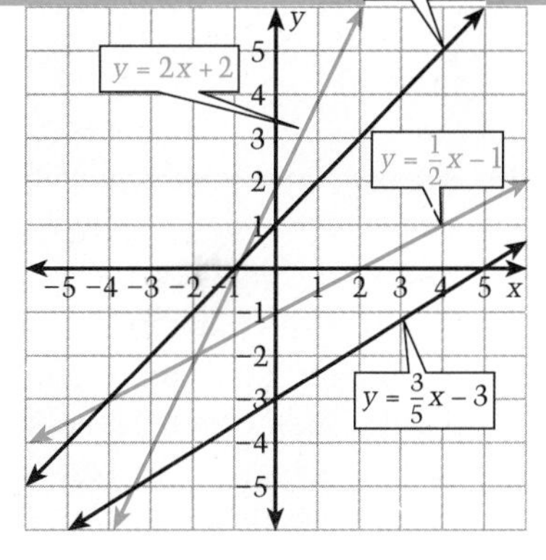

▶ **Ex. 47–50** The rate in these exercises is the average rate of change in population per year.

Answer

61. $-\frac{4}{5}$; Solve the equation for y. For each point, pick a value for the x-coordinate and evaluate the equation to determine the y-coordinate. Use the slope formula to determine the slope of the line.

▶ **Ex. 63** Ask students to identify two points on each line and determine the slope of each. Have them compare the slopes that they found and the algebraic statement that represents the line. Also ask students if they notice any patterns.

Answers

5.

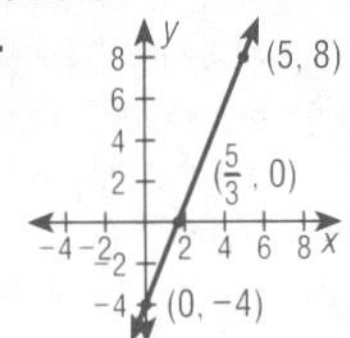

6.

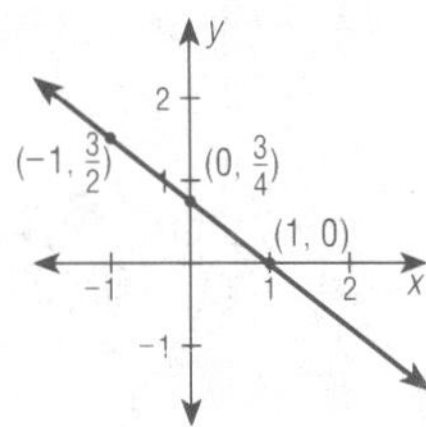

9.

10.

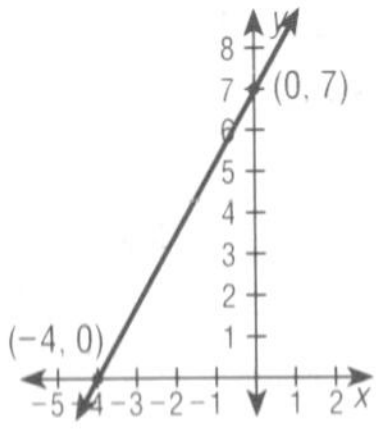

19.

Mid-Chapter SELF-TEST

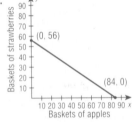

Take this test as you would take a test in class. The answers to the exercises are given at the back of the book.

1. Sketch the lines $x = -1$ and $y = 2$ on the same coordinate plane. What is the point at which the two lines intersect? **(4.1)** $(-1, 2)$ See Additional Answers.

2. Write equations for the horizontal line and the vertical line through $(4, -6)$. **(4. 1)** $y = -6$, $x = 4$

3. Solve for y: $6x + \frac{1}{3}y = 9$. **(4.2)** $y = 27 - 18x$

4. Solve for y: $x + \frac{3}{4}y = 4$. **(4.2)** $y = \frac{4}{3}(4 - x)$

In Exercises 5 and 6, solve the equation for y, construct a table of values, and plot the points in the table. (4.2) See margin.

5. $\frac{12}{5}x - y = 4$

6. $3x + 4y = 3$

7. Is $(3, -\frac{4}{3})$ a solution of $5x + 3y = 11$? Yes

8. Is $(3, 5)$ a solution of $\frac{9}{2}x - y = 20$? No

In Exercises 9 and 10, sketch the line and label the intercepts. (4.3) See margin.

9. $5x - 3y = 15$

10. $-7x + 4y = 28$

In Exercises 11–14, use the graph to determine your answer.

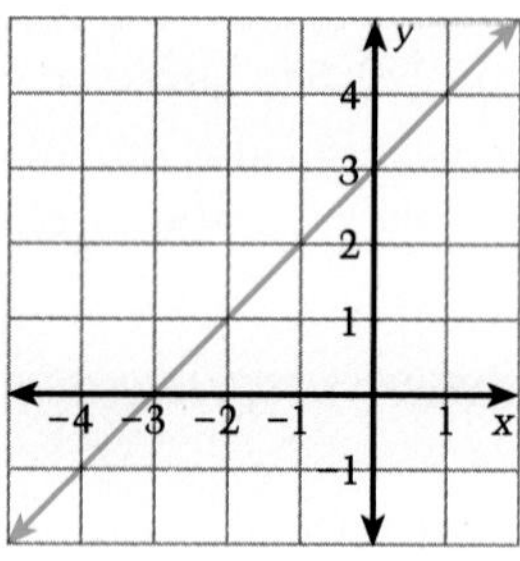

11. Which equation represents the line? **(4.2)**
a. $3x - 3y = 0$ **b.** $x + y = 3$ **c.** $y - x = 3$

12. Which is not a solution? **(4.2)**
a. $(-\frac{3}{2}, \frac{3}{2})$ **b.** $(\frac{3}{2}, -\frac{3}{2})$ **c.** $(1, 4)$

13. What is the x-intercept? **(4.3)** -3

14. What is the y-intercept? **(4.3)** 3

In Exercises 15 and 16, find the slope of the line through the two points. (4.4)

15. $(-1, 1), (20, 15)$ $\frac{2}{3}$

16. $(\frac{4}{3}, 2), (\frac{2}{3}, 6)$ -6

In Exercises 17 and 18, find the rate of change between the two points. Give the unit of measure. (4.4)

17. $(2.5, 3.75), (4.0, 5.95)$
x is in pounds. $1.47 per pound
y is in dollars.

18. $(\frac{1}{2}, 200), (2, 1000)$
x is in square miles.
y is in people. $\approx$533 people per square mile

19. A fruit stand is selling baskets of apples for $4 and baskets of strawberries for $6. The total sales for the two fruits on Tuesday was $336. Sketch a graph showing the possible numbers of baskets of each fruit sold. **(4.2)** See margin.

20. In January, Luis had a balance of $1140 in a savings account. In April, Luis had a balance of $1450 in the account. Find the average rate of change of the balance. (Give your result in dollars per month.) **(4.4)** $103.33 per month

4.5 Quick Graphs Using Slope-Intercept Form

What you should learn:

Goal 1 How to find the slope and y-intercept from an equation

Goal 2 How to use slope-intercept form to sketch a line and solve real-life problems

Why you should learn it:

You can write an equation of a line in slope-intercept form to make it easier to sketch and use.

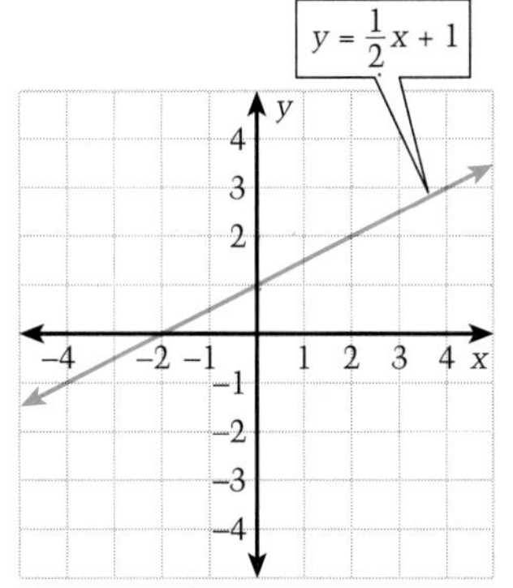

Goal 1 Finding the Slope and y-intercept

In Lesson 4.4, you were given two points on a line and you learned how to find the slope of the line. Now suppose you are given an equation of a line. How could you find its slope?

LESSON INVESTIGATION

Investigating the Slope of a Line

Partner Activity Draw the line $y = \frac{1}{2}x + 1$ on graph paper, as shown at the left. Copy and complete the list by finding five pairs of points on the line. (Choose some coordinates that are positive, some that are zero, and some that are negative.)

Pair 1	*Pair 2*	*Pair 3*	*Pair 4*	*Pair 5*
(?, ?)	(?, ?)	(?, ?)	(?, ?)	(?, ?)
(?, ?)	(?, ?)	(?, ?)	(?, ?)	(?, ?)

Use each pair of points to find the slope of the line. What do you notice? How is the slope related to the equation of the line?

In this investigation, you may have discovered that the slope of the line is the coefficient of x. By substituting 0 for x, you can also discover that the y-intercept is the constant term in the equation. This result suggests a quick way to find the slope and the y-intercept. Just write the equation in the form $y = mx + b$.

Slope-Intercept Form of the Equation of a Line

The linear equation

$$y = mx + b$$

is in slope-intercept form. The slope of the line is m. The y-intercept is b.

Problem of the Day

How many different 3-digit numbers can be formed using the digits 1, 2, and 3 once in each number? How many different 4-digit numbers can be formed using the digits 1, 2, 3, and 4 once in each number?
6; 24

ORGANIZER

Warm-Up Exercises

1. For each equation, find the value of y for which the ordered pair, $(0, y)$, is a solution.
   a. $3x - 6y = 9$ $\quad -\frac{3}{2}$
   b. $8x = 5y - 30$ $\quad 6$
   c. $2y = 7x - 11$ $\quad \frac{11}{2}$
   d. $y = \frac{2}{3}x + 8$ $\quad 8$
2. Solve each equation for y in terms of x.
   a. $3x - 6y = 9$ $\quad y = \frac{1}{2}x - \frac{3}{2}$
   b. $8x = 5y - 30$ $\quad y = \frac{8}{5}x + 6$
   c. $2y = 7x - 11$ $\quad y = \frac{7}{2}x - \frac{11}{2}$

Lesson Resources

Teaching Tools
- Problem of the Day: 4.5
- Warm-up Exercises: 4.5
- Answer Masters: 4.5

Extra Practice: 4.5
Technology Handbook: p. 37

Lesson Investigation Answers

Points will vary. Slope of the line is $\frac{1}{2}$. Slope is the same for all pairs of points. Slope of the line is the coefficient of x.

LESSON Notes

GOAL 1 Demonstrate for students how points on the line $y = \frac{1}{2}x + 1$ can be determined.

Example 1

Show students that the point (3, 3), at which you arrive, satisfies the equation $y = \frac{1}{3}x + 2$. This provides a check that you moved correctly from the y-intercept to another point on the line.

Example 2

You may wish to observe for students that any given line is parallel to itself! This occurs when different equations represent the same line.

Example 3

A "signature" in printing is a large sheet on which are printed a number of pages in multiples of four. This large sheet is then folded to page size and becomes one section of a book.

Extra Examples

Here are additional examples similar to **Examples 1–3.**

1. Identify three points on the line $y = \frac{3}{2}x - 3$ and verify that the slope of the line remains the same for each pair of points selected.
2. Sketch the graph of each line using its slope and y-intercept.
   a. $8x - 4y = 12$
   b. $5y + 3x = 8$
   a. $m = 2, b = -3$;
   b. $m = -\frac{3}{5}, b = \frac{8}{5}$
3. Determine which lines are parallel to each other. Sketch the set of parallel lines on a coordinate plane.
   a. $3x + 4y = 12$
   b. $-8y = 6x - 17$
   c. $4y = 3x + 8$
   d. $6y = -\frac{9}{2}x + 2$
   Lines a, b, and d are parallel.

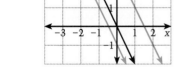

Goal 2 Using Slope-Intercept Form

In Lesson 4.3, you learned to sketch a line quickly using the x-intercept and y-intercept. The *slope* and *y-intercept* can also be used to sketch a quick graph.

Example 1 *Sketching a Quick Graph*

Sketch the graph of $-x + 3y = 6$.

Solution Begin by solving the equation for y.

$$-x + 3y = 6 \quad \textit{Original equation}$$
$$3y = x + 6 \quad \textit{Add x to both sides.}$$
$$y = \underbrace{\tfrac{1}{3}}_{m}x + \underbrace{2}_{b} \quad \textit{Divide both sides by 3.}$$

You can sketch the line as follows. Plot the y-intercept, (0, 2). Locate a second point by moving 3 units to the right and 1 unit up. Then draw the line through the two points.

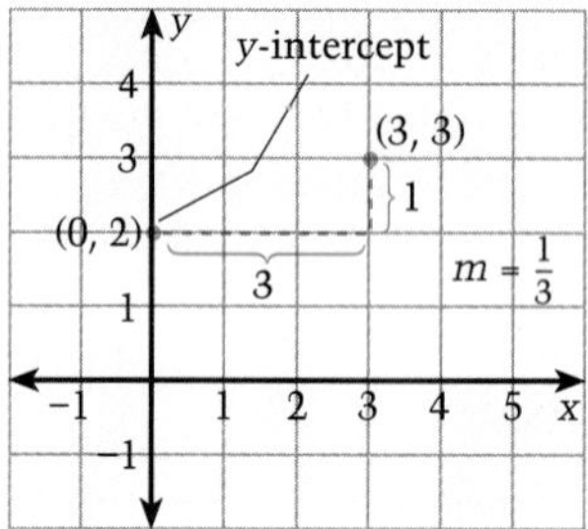

Plot the y-intercept. Then use slope to find a second point.

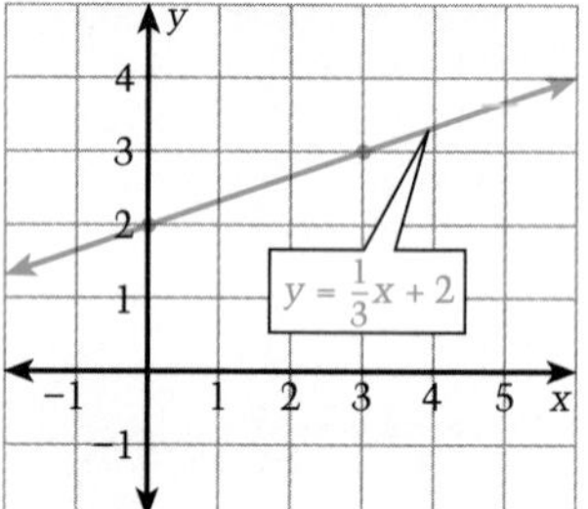

Draw the line that passes through the two points. ■

Two different lines that have the same slope are **parallel;** that is, they do not intersect.

Example 2 *Sketching a Family of Parallel Lines*

Sketch the following lines on the same coordinate plane.

a. $y = -2x + 3$ **b.** $y = -2x$ **c.** $y = -2x - 1$

Solution Each of the lines has a slope of $m = -2$. The y-intercepts of each are the values of b, as follows.

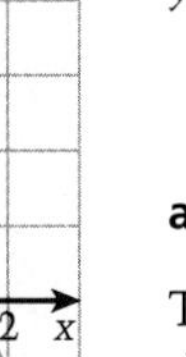

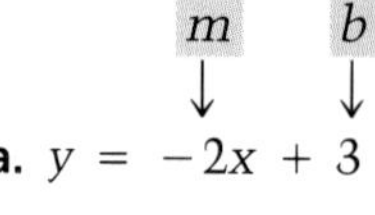

a. $y = -2x + 3$

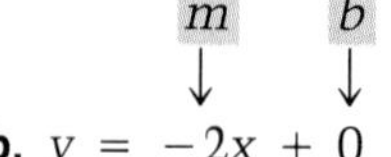

b. $y = -2x + 0$

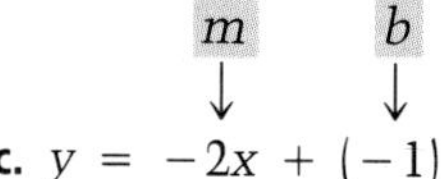

c. $y = -2x + (-1)$

To sketch each line, plot its y-intercept. Then move 1 unit to the right and 2 units *down* (the slope is negative), and plot a second point. Finally, draw a line through the two points. ■

These students are planning a creative writing project that will give them some valuable insights into the realities of publishing a small book.

Problem Solving
Make a Graph

Example 3 *Printing a Book*

Your creative writing class wants to have selections of its poems and prose printed and bound into a book. The book will have 128 pages (8 *signatures* of 16 pages each). The inside pages will be printed with black ink, but the cover may have a full-color picture. A printer gives you the following prices:

With full-color cover: Cost is \$750, plus \$2.00 per book.
With single-color cover: Cost is \$250, plus \$1.75 per book.

Sketch the graphs of these two costs. Let x be the number of books printed and let C be the cost of printing x books.

Cost of Printing Books in the U.S.

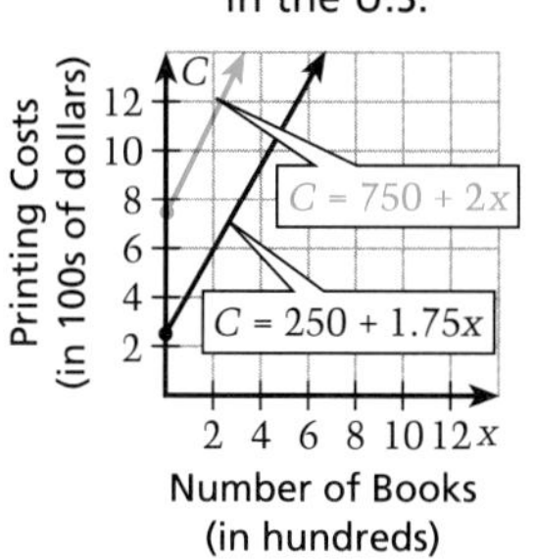

Number of Books (in hundreds)

Solution The two equations are as follows.

With full-color cover: $C = 750 + 2x$
With single-color cover: $C = 250 + 1.75x$

The graphs of both equations are lines. The first line has a C-intercept of 750 and a slope of 2. The second line has a C-intercept of 250 and a slope of 1.75. ■

Communicating about ALGEBRA

SHARING IDEAS about the Lesson

A. Cost that is independent of the number of books printed; cost per book printed

Decision Making Refer to the problem in Example 3.

A. What do the C-intercepts represent? What do the slopes represent?

B. Your class plans to order and sell 500 copies. With full-color covers, how much should you charge for each book to cover the printing cost? With single-color covers, how much should you charge? \$3.50, \$2.25

C. Suppose that your class could sell 750 copies. Could you afford to charge less per book? Explain your answer. See below.

C. Yes, the cost that is independent of the number of books printed would be spread among more books.

Check Understanding

1. Find the slope and y-intercept for each equation.
   a. $4x - 6y = 3$
   b. $11x = -5y + 14$
   a. $m = \frac{2}{3}, b = -\frac{1}{2}$;
   b. $m = -\frac{11}{5}, b = \frac{14}{5}$
2. Sketch the graphs of the line through the given point with the given slope.
   a. $m = -4, (-3, 6)$
   b. $m = \frac{2}{3}, (7, 1)$

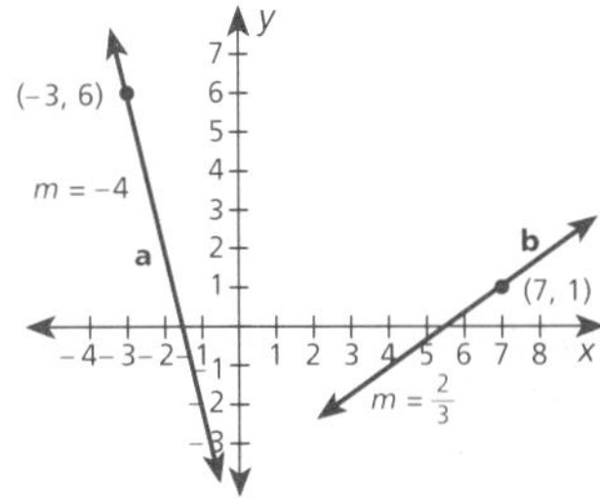

Communicating about ALGEBRA

Ask students to provide a rationale about whether the cost of printing the full-color version of the book will ever be the same as the cost of printing the single-color version.

Common-Error ALERT!

Remind students to label the two points selected for finding the slope of the line. This will help them avoid the mistake of mismatching the coordinates in the slope formula.

EXERCISE Notes

ASSIGNMENT GUIDE

Basic/Average: Ex. 1–4, 5–27 odd, 29–32, 33–43 odd, 45–54 multiples of 3, 58, 59

Above Average: Ex. 1–4, 6–27 multiples of 3, 29–32, 36, 39–40, 42–43, 45–54 multiples of 3, 57–62

Advanced: Ex. 1–4, 6–27 multiples of 3, 29–32, 37–40, 42–43, 57–62

Selected Answers

Exercises 1–4, 5–55 odd

✪ More Difficult Exercises

Exercises 37–40, 57–62

Guided Practice

▶ **Ex. 3** Ask students to create two different equations of a given line. Discuss the reasonableness of saying that these equations represent "lines" with the same slope.

Common-Error ALERT!

In **Exercises 15–16,** remind students that the coefficient of x in an equation of a straight line represents the slope of the line only when the equation is in slope-intercept form.

EXERCISES

1. Yes; slope of line does not change.

Guided Practice

CRITICAL THINKING about the Lesson

1. Can any pair of points on a line be used to calculate the slope of the line? Explain.

2. In the form $y = mx + b$, what does m represent? What does b represent? Slope, y-intercept

3. What word may be used to describe two different lines that have the same slope? Parallel

4. What is the slope of the line given by $3x + 2y = 1$? What is the y-intercept? $-\frac{3}{2}, \frac{1}{2}$

Independent Practice

In Exercises 5–10, solve for y. 22. $y = \frac{3}{2}x - 1$ 23. $y = \frac{3}{4}x + \frac{1}{2}$ 24. $y = -\frac{5}{3}x + \frac{1}{2}$

5. $5x + 7y = 0$ $y = -\frac{5}{7}x$ **6.** $2x - 4y = 0$ $y = \frac{1}{2}x$ **7.** $3x + 4y = 12$ $y = -\frac{3}{4}x + 3$

8. $-4x + 5y = 8$ $y = \frac{4}{5}x + \frac{8}{5}$ **9.** $y + 10 = 0$ $y = -10$ **10.** $y - 12 = 0$ $y = 12$

In Exercises 11–16, find the slope and the y-intercept of the line.

11. $y = -2x + 1$ $-2, 1$ **12.** $y = 3x - 6$ $3, -6$ **13.** $y = -4 + (-8x)$ $-8, -4$

14. $y = 4x - 20$ $4, -20$ **15.** $x - y = 3x + 4$ $-2, -4$ **16.** $2y - x = 7x - 9$ $4, -\frac{9}{2}$

In Exercises 17–28, write in slope-intercept form. Then sketch the line. See Additional Answers.

17. $2x - y - 3 = 0$ $y = 2x - 3$ **18.** $x - y + 2 = 0$ $y = x + 2$ **19.** $x + y = 0$ $y = -x$

20. $x - y = 0$ $y = x$ **21.** $x + 2y - 2 = 0$ $y = -\frac{1}{2}x + 1$ **22.** $3x - 2y - 2 = 0$

23. $3x - 4y + 2 = 0$ **24.** $10x + 6y - 3 = 0$ **25.** $y - 3 = 0$ $y = 3$

26. $y + 5 = 0$ $y = -5$ **27.** $2x + 3y - 4 = x + 5$ $y = -\frac{1}{3}x + 3$ **28.** $-x + 4y + 3 = 2x - 7$ $y = \frac{3}{4}x - \frac{5}{2}$

In Exercises 29–32, match the equation with its graph.

29. $y = \frac{1}{2}x + 1$ d **30.** $y = \frac{1}{2}x - 1$ c **31.** $y = x + 2$ a **32.** $y = -x - 1$ b

a. 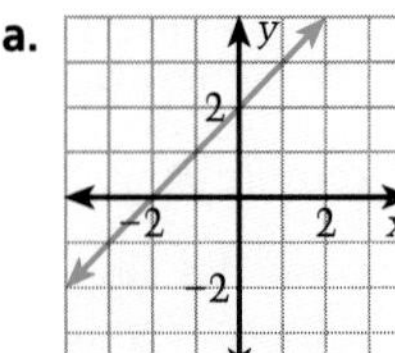

b. 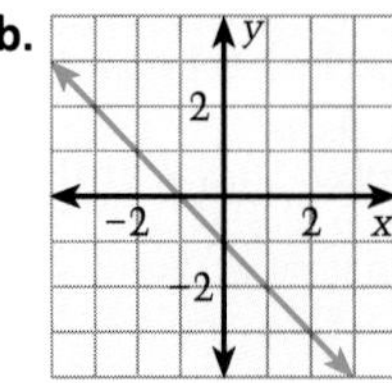

c. 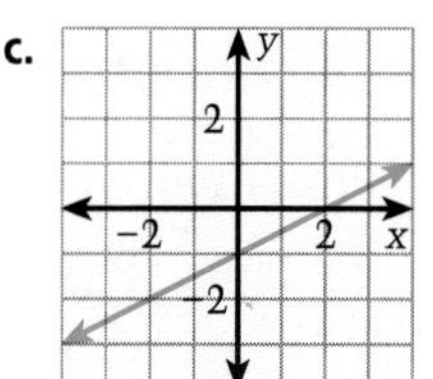

d. 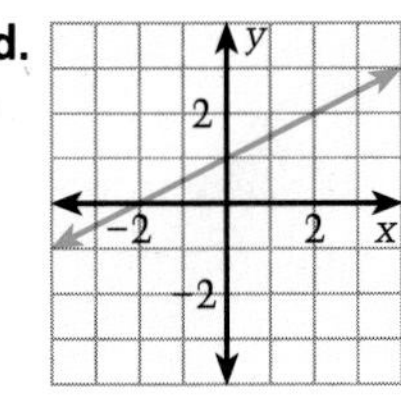

In Exercises 33–36, sketch the two lines on the same coordinate plane. See Additional Answers. **Find the slope and x- and y-intercepts of the lines.**

33. $y = -3x + 2$, $y = -3x - 2$

34. $y = -x + 6$, $y = -x + 10$ -1; (6, 0), (0, 6); (10, 0), (0, 10)

35. $y = 6x + 8$, $y = 6x - 2$ 6; $(-\frac{4}{3}, 0)$, (0, 8); $(\frac{1}{3}, 0)$, (0, −2)

36. $y = \frac{4}{3}x - 1$, $y = \frac{4}{3}x + 3$ $\frac{4}{3}$; $(\frac{3}{4}, 0)$, (0, −1); $(-\frac{9}{4}, 0)$, (0, 3)

33. −3; $(\frac{2}{3}, 0)$, (0, 2); $(-\frac{2}{3}, 0)$, (0, −2)

In Exercises 37–40, use the following information.

Two triangles are *similar* if their angles have the same measures. Corresponding sides of similar triangles have the same ratio. For instance, in the similar triangles at the right the ratio of AB to BC is equal to the ratio of DE to EF.

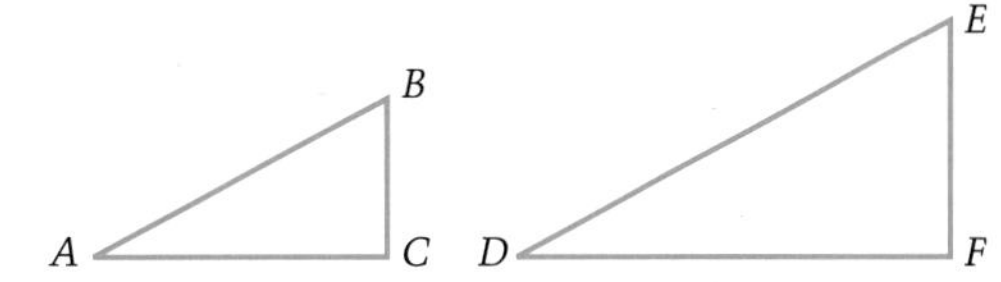

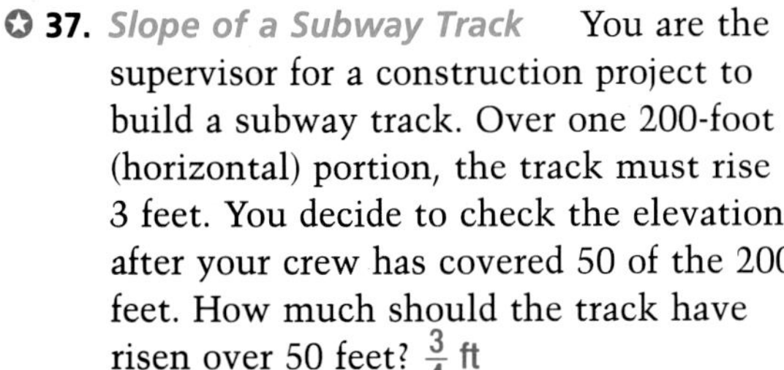

37. *Slope of a Subway Track* You are the supervisor for a construction project to build a subway track. Over one 200-foot (horizontal) portion, the track must rise 3 feet. You decide to check the elevation after your crew has covered 50 of the 200 feet. How much should the track have risen over 50 feet? $\frac{3}{4}$ ft

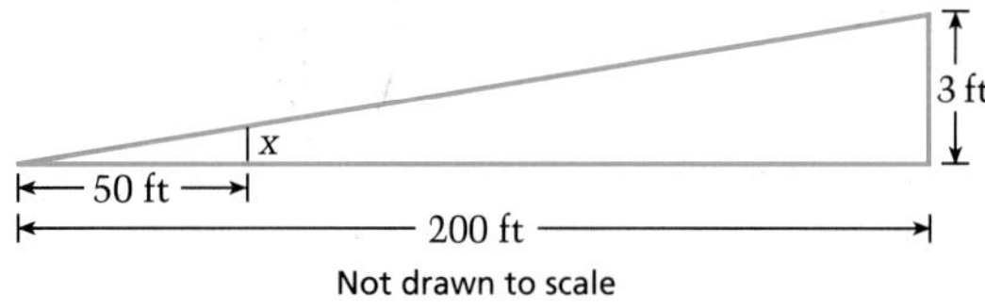

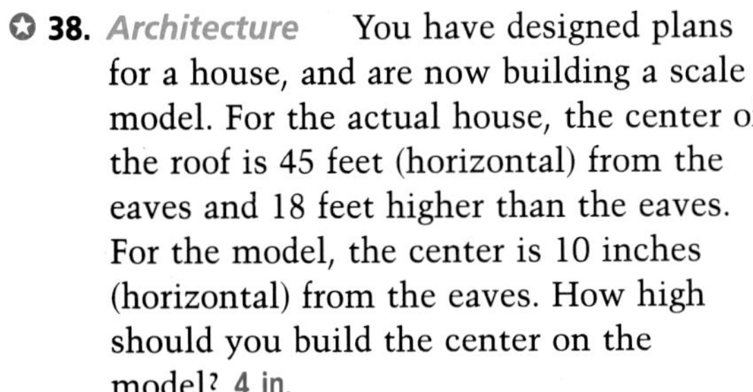

38. *Architecture* You have designed plans for a house, and are now building a scale model. For the actual house, the center of the roof is 45 feet (horizontal) from the eaves and 18 feet higher than the eaves. For the model, the center is 10 inches (horizontal) from the eaves. How high should you build the center on the model? 4 in.

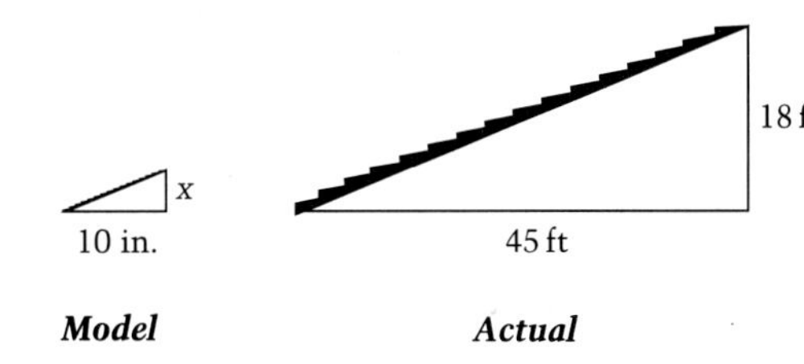

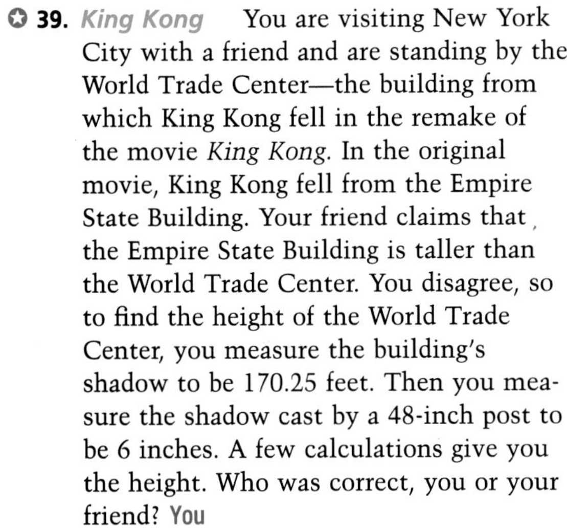

39. *King Kong* You are visiting New York City with a friend and are standing by the World Trade Center—the building from which King Kong fell in the remake of the movie *King Kong*. In the original movie, King Kong fell from the Empire State Building. Your friend claims that the Empire State Building is taller than the World Trade Center. You disagree, so to find the height of the World Trade Center, you measure the building's shadow to be 170.25 feet. Then you measure the shadow cast by a 48-inch post to be 6 inches. A few calculations give you the height. Who was correct, you or your friend? You

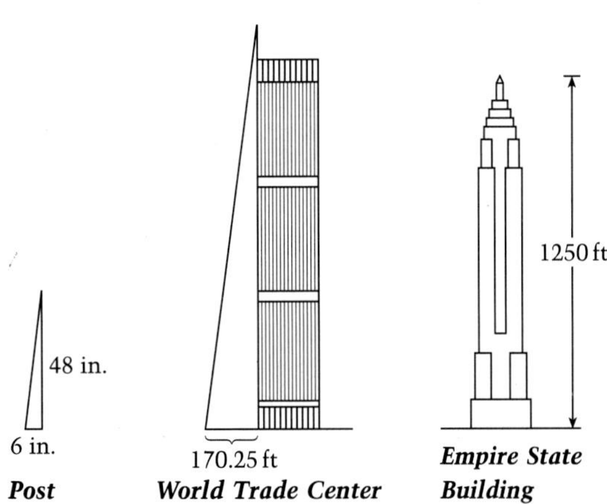

40. *Connecting Slope and Similar Triangles* Find the slope of the line passing through the hypotenuse of each triangle in Exercises 37–39. Explain how these "similar triangle" problems are related to the fact that any two points on a line can be used to calculate the slope of the line. See margin.

Independent Practice

▶ **Ex. 29–32** WRITING Ask students to provide a rationale for their matches.

▶ **Ex. 38** Eaves are the lower edge or edges of a roof, usually projecting beyond the sides of the building.

▶ **Ex. 40** COOPERATIVE LEARNING Discuss students' answers to this exercise in class. Through any two points on a line (neither vertical nor horizontal) a triangle is formed by a vertical line through one point and a horizontal line through the other. All such triangles are similar. So the $\frac{\text{rise}}{\text{run}}$ ratio is the same for any two points.

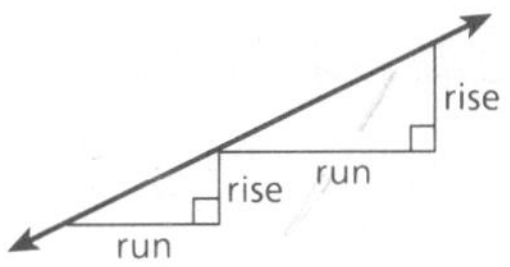

Enrichment Activities

WRITING Following the class discussion, students might record their personal responses in their math journals.

1. Have students write a narrative describing how a linear equation in two variables written in standard form, $Ax + By = C$, can be transformed to provide information which is then useful in graphing the equation of the line. Answers should include writing the equation in slope-intercept form.
2. Given the standard form of a linear equation in two variables, express the slope and y-intercept in terms of A, B, and C. $m = -\frac{A}{B}$, $b = \frac{C}{B}$

Ex. 43–44 PROBLEM SOLVING Ask students to identify the problem-solving strategies used to solve these problems. Stress the connections between the plan and uses of tables or graphs to solve the problems.

Answers

43.

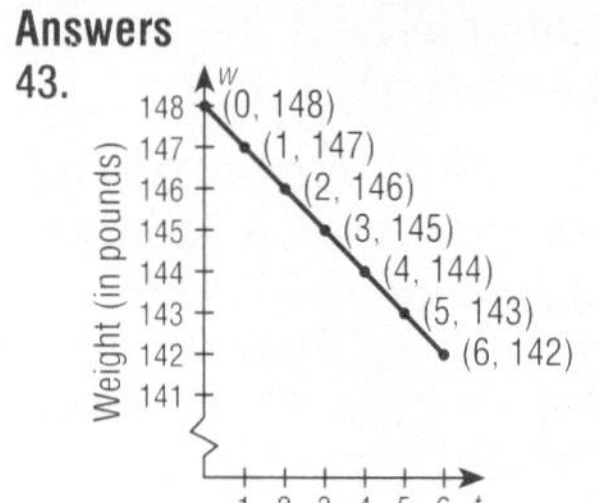

−1, rate of weight loss in pounds per week; 148, Mark's weight in pounds before he starts the weight-loss program.

44.

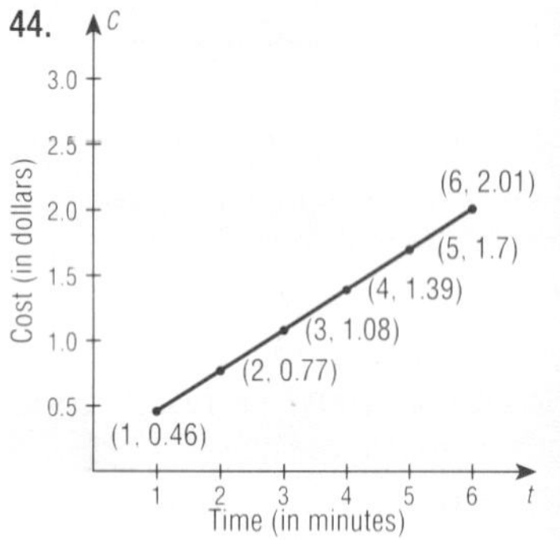

0.31, cost for each additional minute

Ex. 59–62 COOPERATIVE LEARNING These are useful exercises to go over in class.

41. Which line is parallel to $x + 3y = 2$?
a. $y = \frac{1}{3}x + 5$ **b.** $y = -\frac{1}{3}x + 4$

42. Which line is parallel to $2x - 4y = 6$?
a. $y = \frac{1}{2}x + 1$ **b.** $y = -\frac{1}{2}x - 1$

43. *Getting Ready for the Team* Huan decides to lose weight before wrestling team practice begins. He weighs 148 pounds now and loses one pound per week for six weeks. Let w represent Huan's weight and let t represent the time in weeks. Plot points for his weights at one-week intervals, then draw a line through the points. Find the slope of the line. What does it represent? Find the w-intercept of the line. What does it represent? **See margin.**

44. *Making a Phone Call* The cost of a telephone call between New York City and Philadelphia is \$0.46 for the first minute and \$0.31 for each additional minute. Let C represent the total cost of making a call that lasts for t minutes. Plot points for the costs of calls that last 1, 2, 3, 4, 5, and 6 minutes, then draw a line through the points. Find the slope of the line. What does it represent? **See margin.**

Integrated Review

In Exercises 45–48, evaluate the expression.

45. $\frac{-3-(-4)}{5-(-1)}$ $\frac{1}{6}$

46. $\frac{0-3}{9-3}$ $-\frac{1}{2}$

47. $\frac{-2-(-2)}{8-1}$ 0

48. $\frac{6-10}{0-(-4)}$ −1

In Exercises 49–52, find five solutions of the equation. Solutions vary.

49. $y = -2x + 4$ (0, 4), etc.

50. $y = 5x - 2$ (0, −2), etc.

51. $y = -\frac{5}{7}x$ (0, 0), etc.

52. $y = \frac{1}{2}x$ (0, 0), etc.

In Exercises 53–56, sketch the line that passes through (0, 2) with the given slope. See Additional Answers.

53. $m = 3$

54. $m = 0$

55. $m = -\frac{2}{3}$

56. $m = \frac{3}{4}$

Exploration and Extension

In Exercises 57 and 58, match the description with its visual model.

57. A person starts from home and rides a bike at a speed of 10 miles per hour. **B**

58. A person starts 25 miles from home and drives toward home at a speed of 25 miles per hour. **A** **59.–62. See Additional Answers.**

A.

y; 25, 20, 15, 10, 5; Miles; $\frac{1}{4}$ $\frac{1}{2}$ $\frac{3}{4}$ 1 x; Hours

B.

y; 40, 30, 20, 10; Miles; 1 2 3 4 x; Hours

59. Sketch the line that has a y-intercept of −4 and is parallel to $y = 2x + 3$.

60. Sketch the line that has a y-intercept of −8 and is parallel to $y = -\frac{1}{2}x + 5$.

61. Sketch the line that has a y-intercept of 2 and a slope of $-\frac{1}{8}$.

62. Sketch the line that has a y-intercept of 14 and a slope of 0.

4.6 Connections: Solutions and x-Intercepts

What you should learn:

Goal 1 How to use a graph as a quick check of a solution found algebraically

Goal 2 How to approximate solutions of real-life problems by using a graph

Why you should learn it:

You can use a graph to approximate related values by estimating coordinates of points on the graph.

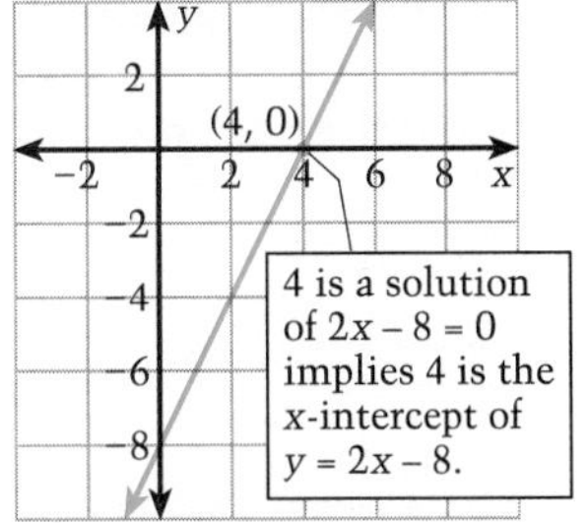

René Descartes (1596–1650)

Goal 1 Using a Graphic Check for a Solution

You have already solved linear equations in one variable. For instance, you solved the equation $2x - 8 = 0$ as follows.

$2x - 8 = 0$ *Original equation*

$2x = 8$ *Add 8 to both sides.*

$x = 4$ *Divide both sides by 2.*

In Chapter 3, we emphasized the importance of checking this solution in the original equation. *You should continue to do that.* You may, however, also check a solution graphically.

To see how to do this, consider the line $y = 2x - 8$. Recall that the x-intercept of this line is the value of x when $y = 0$. In other words, the x-intercept of the line $y = 2x - 8$ is the solution of the equation $2x - 8 = 0$.

One-Variable Equation	Two-Variable Equation
The *solution* of $2x - 8 = 0$ is 4.	The *x-intercept* of $y = 2x - 8$ is 4.

Thus, to check the solution of $2x - 8 = 0$ graphically, sketch the graph of $y = 2x - 8$ and observe the x-intercept.

Using a Graphic Check of a Solution

The solution of a linear equation involving one variable x can be checked graphically with the following steps.

1. Write the equation in the form $ax + b = 0$.
2. Sketch the graph of $y = ax + b$.
3. The solution of $ax + b = 0$ is the x-intercept of $y = ax + b$.

The coordinate-plane connection between algebra and geometry is one of the greatest discoveries ever made in mathematics. Before René Descartes introduced the coordinate plane in 1637, mathematicians had no easy way of "seeing" a solution of an algebraic equation. Because of Descartes's contribution, the coordinate plane is often called the **Cartesian Plane.**

Problem of the Day

Which element does not belong in the set? Why?

$A = \{18, 9, 20, 72\}$
$B = \{\frac{7}{5}, \frac{5}{6}, \frac{8}{7}, \frac{11}{10}\}$
$C = $ {circle, cone, cylinder, pyramid}

There may be more than one answer. for example,
A: 20 is not a multiple of 9, or 9 is a one-digit number
B: $\frac{5}{6} < 1$; or $\frac{7}{5}$ is the only fraction whose numerator and denominator do not differ by 1.
C: circle is 2-dimensional; pyramid is not round nor does it begin with a c.

ORGANIZER

Warm-Up Exercises

1. Solve for x.
   a. $3x + 23 = 12$
   b. $-13 - 5x = 19$
   c. $-4x + 16 = 0$
   d. $51 - 17x = 0$

   a. $-\frac{11}{3}$, b. $-\frac{32}{5}$, c. 4, d. 3
2. Solve for y in terms of x.
   a. $3y - 8x = 12$
   b. $16x - 4y = 54$

   a. $y = \frac{8}{3}x + 4$, b. $y = 4x - \frac{27}{2}$

Lesson Resources

Teaching Tools
  Problem of the Day: 4.6
  Warm-up Exercises: 4.6
  Answer Masters: 4.6
Extra Practice: 4.6
Technology Handbook: p. 40

LESSON Notes

Example 1

It is important to note here that a calculator is not required to make the check. You can use the quick-graph techniques to easily sketch the graph of the linear equation.

Example 2

TECHNOLOGY Using technology in this case eliminates tedious algebraic manipulations. Note, however, that students must still recognize that the x-intercept represents the solution of the linear relationship set equal to zero.

Extra Examples

Here are additional examples similar to **Examples 1–3.**

1. Solve each equation algebraically and graphically.
   - **a.** $8x - 5 = 13$
   - **b.** $43 + 7x = 34$
   - **c.** $4x + 14 = -6$

   a. $2\frac{1}{4}$, b. $-\frac{9}{7}$, c. -5
2. **TECHNOLOGY** Use a computer or a graphics calculator to solve each equation by rewriting the original equation set equal to zero. (Express answers using one decimal place.)
   - **a.** $6(3x - 15) = \frac{2}{3}(-2x + 10)$
   - **b.** $4x - 12 = \frac{1}{3}(51 - 17x)$
   - **c.** $23x - 34 = -72 + 17x$

   a. 5, b. 3, c. $x \approx -6.3$
3. **TECHNOLOGY** Use a computer or a graphics calculator to solve each equation by graphing the left and right sides of the equation separately and finding the point of intersection.
   - **a.** $64x - 32 = 128$
   - **b.** $-2(3x + 2) = 20$

   a. 2.5, b. -4

Example 1 *Using a Graphic Check of a Solution*

Solve the equation $3x + 1 = -8$. Check your solution algebraically and graphically.

Solution

$$\begin{aligned} 3x + 1 &= -8 && \textit{Rewrite original equation.} \\ 3x &= -9 && \textit{Subtract 1 from both sides.} \\ x &= -3 && \textit{Divide both sides by 3.} \end{aligned}$$

Algebraic Check Substitute -3 for x in the original equation.

$$\begin{aligned} 3(-3) + 1 &\stackrel{?}{=} -8 && \textit{Substitute } -3 \textit{ for } x. \\ -9 + 1 &\stackrel{?}{=} -8 && \textit{Simplify.} \\ -8 &= -8 && \textit{The solution checks.} \end{aligned}$$

Graphic Check Rewrite in the form $ax + b = 0$.

$$\begin{aligned} 3x + 1 &= -8 && \textit{Original equation} \\ 3x + 9 &= 0 && \textit{Add 8 to both sides.} \end{aligned}$$

x-intercept is −3

(−3, 0)

Now sketch the graph of $y = 3x + 9$, as shown at the left. Notice that the x-intercept is -3 (where $3x + 9 = 0$), which checks with the algebraic solution. ■

A graphic check can go wrong if you make a mistake when rewriting the original equation in the form $ax + b = 0$. Technology can be used to minimize this possibility, as shown in Example 2.

Connections **Technology**

Example 2 *Using a Graphing Calculator or Computer*

How could you use a graphing calculator to approximate the solution of $\frac{5}{6}(7x - 5) = 28 - \frac{3}{5}x$?

Solution There are several ways to do this. One nice way that requires a minimum amount of algebra is to rewrite the original equation as

$$\frac{5}{6}(7x - 5) - \left(28 - \frac{3}{5}x\right) = 0.$$

Then, without simplifying, use a graphing calculator or computer to sketch the graph of

$$y = \frac{5}{6}(7x - 5) - \left(28 - \frac{3}{5}x\right).$$

The result is shown at the left. Note that the x-intercept appears to be 5. (Each tick mark represents 1 unit.) You can verify this by substituting 5 for x in the original equation. ■

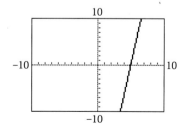

Goal 2 Approximating Solutions of Problems

In Examples 1 and 2, we rewrote each equation so that the right side was zero. Then we set y equal to the left side, sketched the graph, and used the x-intercept (where $y = 0$) as a graphic check or graphic approximation.

A slightly different approach is sometimes helpful. For equations of the form $ax + b = c$, sketch, on the same coordinate plane, the graphs of $y = ax + b$ and $y = c$. The solution is the x-coordinate of the point at which they cross.

Real Life
Medical Careers

Example 3 *Projecting the Number of Registered Nurses*

From 1970 to 1990, the number of registered nurses in the United States could be approximated by the linear model $N = 53{,}600t + 732{,}000$. In this model, t represents the year with $t = 0$ corresponding to 1970. According to this model, when will the United States have 2,250,000 registered nurses?
(Source: U.S. Department of Health and Human Services)

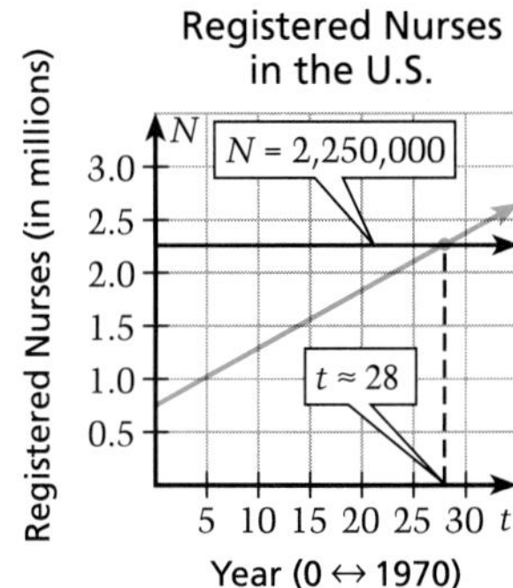

Solution Answer the question by solving the equation

$$53{,}600t + 732{,}000 = 2{,}250{,}000.$$

A way to approximate the answer is to sketch the graphs of

$$\underbrace{N = 53{,}600t + 732{,}000}_{\text{Left side of equation}} \quad \text{and} \quad \underbrace{N = 2{,}250{,}000}_{\text{Right side of equation}}$$

on the same coordinate plane. From the graph, you can see that there should be 2,250,000 nurses by 1998 or 1999. ■

Communicating about ALGEBRA

SHARING IDEAS about the Lesson

Graphic Check Match each solution with an x-intercept.

Solution	x-intercept
A. Solution of $2x = 10$ a	**a.** x-intercept of $y = 2x - 10$
B. Solution of $12x - 4 = 10x + 5$ c	**b.** x-intercept of $y = 3x - 2$
C. Solution of $2(x - 4) = 5(x - 2)$ b	**c.** x-intercept of $y = 2x - 9$

Check Understanding

Solve each equation algebraically and graphically. Then discuss the proposition: Obtaining algebraic solutions to linear equations is better than getting graphical solutions.

a. $7x - 4 = 24$ 4

b. $-3x - 7 = 38$ −15

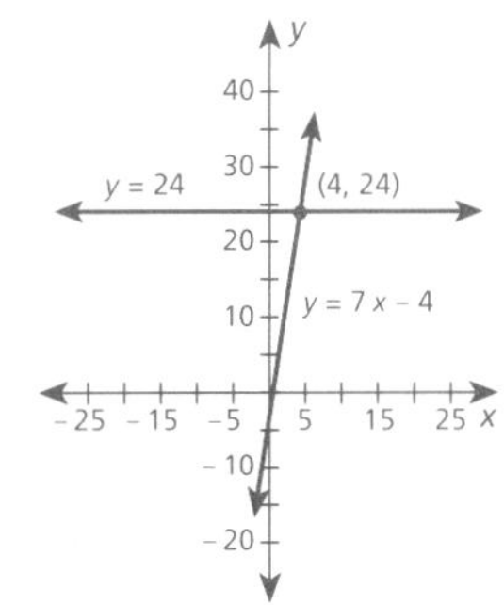

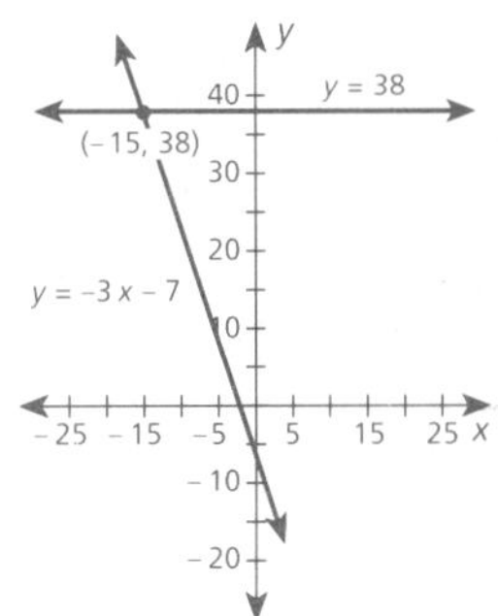

Solving equations algebraically will give an exact solution. Solving equations graphically will give a more visual representation. Neither method is better. Many real-life applications are difficult to solve algebraically and exact answers are not always required.

Communicating about ALGEBRA

COOPERATIVE LEARNING

These exercises should be presented in class after small groups of students have had an opportunity to discuss and complete each exercise. The value of these problems lies in students' recognition of the connection between the x-intercept and solutions to the linear equations.

EXERCISE Notes

ASSIGNMENT GUIDE

Basic/Average: Ex. 1–8, 9, 11, 15–24 multiples of 3, 25–31 odd, 33–42 multiples of 3, 43–46

Above Average: Ex. 1–8, 10, 12, 13–23 odd, 26–32 even, 42–46

Advanced: Ex. 1–8, 11–12, 19–24, 25–29 odd, 31–32, 43–46

Use **Mixed Review** as needed.

Selected Answers

Exercises 1–4, 5–43 odd

✪ More Difficult Exercises

Exercises 25–30, 45–46

Guided Practice

Discuss these exercises in class. Ask students to evaluate the techniques for finding solutions to linear equations indicating when one technique might be preferred over the other.

Independent Practice

▶ **Ex. 13–24** See Example 1 of the lesson.

▶ **Ex. 25–30** TECHNOLOGY A useful notation for indicating the range over which an equation is drawn on a graphics calculator or computer is as follows.

The interval on the x-axis is $(-10, 10)$ with ticks spaced 5 units apart. The interval on the y-axis is $(-15, 15)$ with ticks spaced 3 units apart.

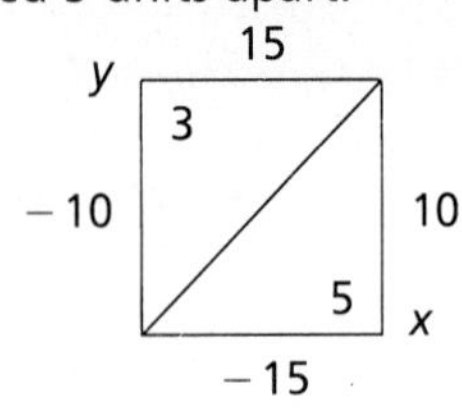

EXERCISES

Guided Practice

CRITICAL THINKING about the Lesson

In Exercises 1–3, use the equation $9x - 36 = 0$.

1. Solve the equation algebraically. $x = 4$
2. Write an equation whose graph you could use to check your solution. Sketch this graph. $y = 9x - 36$. See Additional Answers.
3. What is the x-intercept of the graph? 4
4. **True or False?** The solution of $ax + b = 0$ is the x-intercept of $y = ax + b$. Explain your reasoning. See below.

Independent Practice

In Exercises 5–8, match the equation with the line whose x-intercept is the solution of the equation.

5. $2x - 4 = 0$ d
6. $x + 6 = 0$ a
7. $x + 2 = 0$ c
8. $x - 6 = 0$ b

a. 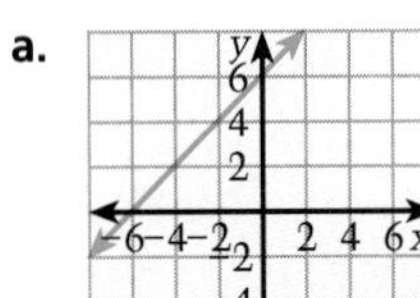

b. 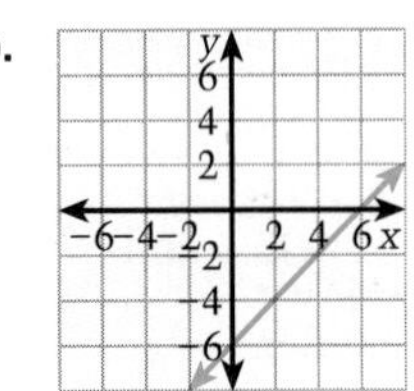

c. 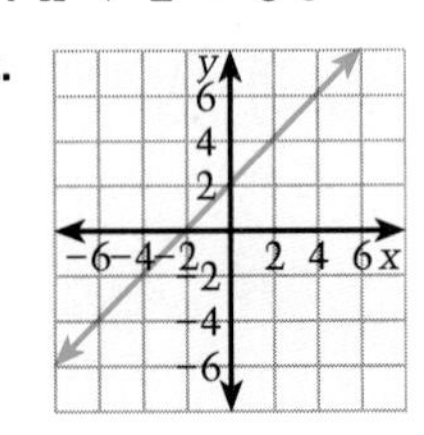

d. 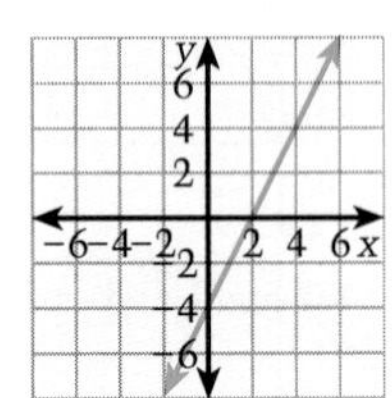

In Exercises 9–12, write an equation of a line whose x-intercept is the solution of the one-variable equation.

9. $9x + 4 = -4$ $y = 9x + 8$
10. $10x - 3 = 19$ $y = 10x - 22$
11. $5 - 3x = 16$ $y = -3x - 11$
12. $14 + 4x = 15$ $y = 4x - 1$

In Exercises 13–24, solve the equation. Check your solution algebraically and graphically.

13. $2x + 3 = 7$ 2
14. $3x - 2 = -5$ -1
15. $4x - 7 = -23$ -4
16. $x - 9 = -6$ 3
17. $5x + 17 = -13$ -6
18. $5x - 17 = 8$ 5
19. $4x + 3 = 5$ $\frac{1}{2}$
20. $5x + 7 = 9$ $\frac{2}{5}$
21. $9x - 6 = -10$ $-\frac{4}{9}$
22. $8x - 5 = 1$ $\frac{3}{4}$
23. $4x - 6 = 10$ 4
24. $-2x + 5 = 9$ -2

Technology **In Exercises 25–30, use a graphing calculator to find the solution. Check your solution algebraically in the original equation.**

✪ 25. $6(x + 3) = 2(x + 5)$ -2
✪ 26. $6(x + 2) = 5(x + 2)$ -2
✪ 27. $\frac{6}{5}(x + 2) = 5x + \frac{1}{2}$ $\frac{1}{2}$
✪ 28. $x - \frac{1}{3} = \frac{5}{4}(x - 6)$ $28\frac{2}{3}$
✪ 29. $\frac{3}{8}(x + 7) = \frac{1}{3}(x + 16)$ 65
✪ 30. $\frac{3}{2}(x - 10) = \frac{2}{3}(x - 12)$ $8\frac{2}{5}$

4. True. The x-intercept of $y = ax + b$ is the value of x when $y = 0$; this value of x is the solution of $ax + b = 0$.

31. *Attendance at Hockey Games* The total yearly attendance at hockey games for the National Hockey League between 1980 and 1990 is approximately given by

$$A = 1{,}100{,}000t + 10{,}534{,}000$$

where A represents the number of people who attended the games and t represents the year, with $t = 0$ corresponding to 1980. According to this model, in what year did 16,034,000 people attend the games? Solve algebraically and graphically. 1985. See Additional Answers

32. *Cost* The cost, y, of producing x units of a product is given by $y = 35x + 5000$. If the cost is \$7625, how many units are produced? Solve algebraically and graphically. 75. See Additional Answers.

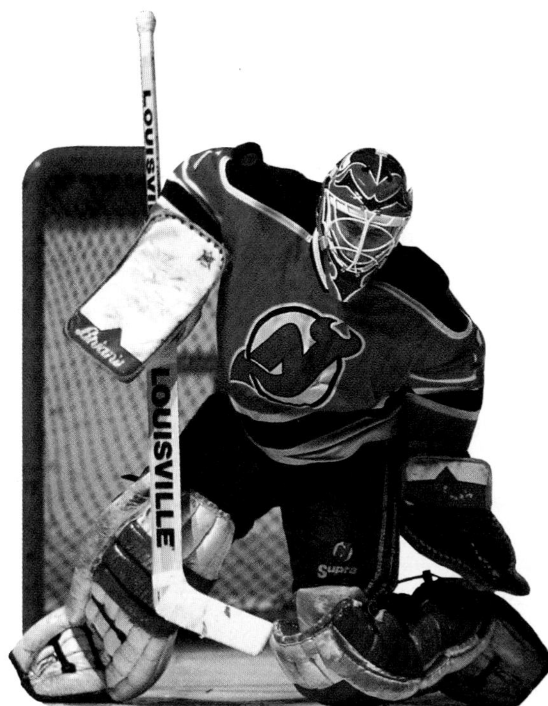

Integrated Review

In Exercises 33–40, evaluate the expression.

33. $|-4|$ 4
34. $-|6|$ −6
35. $4 - |-3|$ 1
36. $3|-5 + 2|$ 9
37. $|1 - 2| + 3$ 4
38. $16 - |-11 + 9|$ 14
39. $-2|-8 + 2| + 7$ −5
40. $-12 + |3 - 7|$ −8

In Exercises 41 and 42, complete the table.

41. $y = 4x - 1$ −13, −9, −5, −1, 3, 7, 11

x	−3	−2	−1	0	1	2	3
y	?	?	?	?	?	?	?

42. $y = -6x + 12$ 30, 24, 18, 12, 6, 0, −6

x	−3	−2	−1	0	1	2	3
y	?	?	?	?	?	?	?

43. *College Entrance Exam Sample* The lines a and b are parallel. The length of the line segment d between the lines
   a. steadily increases as it is moved along lines a and b toward the right.
   b. steadily decreases as it is moved toward the left.
   c. fluctuates in both directions.
   d. remains constant.
   e. None of the above

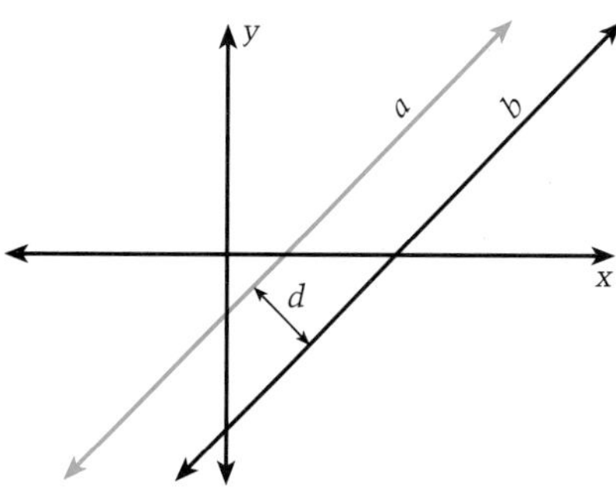

44. Sketch the line with a slope $-\frac{1}{2}$ that passes through the point (0, 9). See Additional Answers.

Ex. 25–27 Use this range.

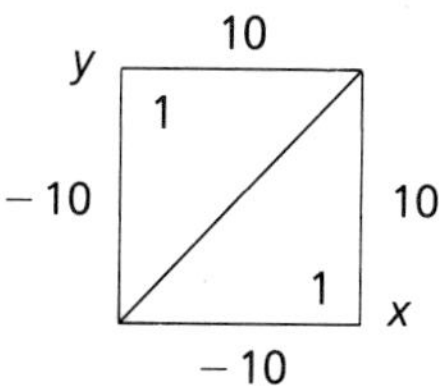

Solutions can easily be read from the display screen. Use the "trace" feature to identify the x-value where the line crosses the x-axis.

Ex. 28 Use this range.

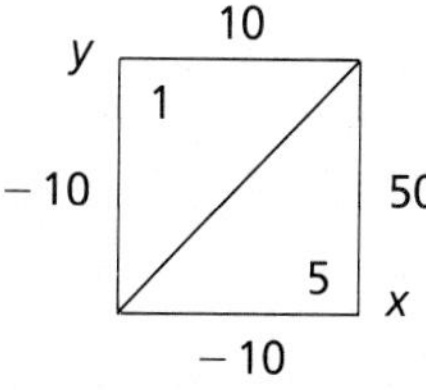

Use the trace feature and the "zoom-in" feature (about 2 or 3 times) to find the x-value, 28.6667 (probably $28\frac{2}{3}$).

Ex. 29 Use this range.

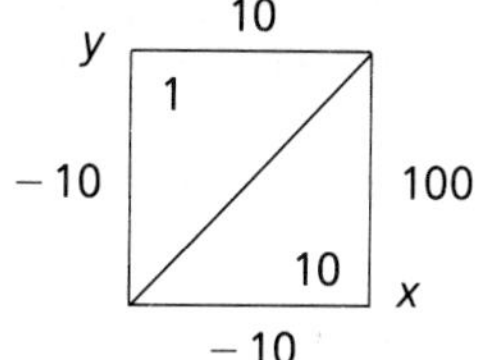

Use the trace feature and the zoom-in feature (about 2 or 3 times) to find the x-value, 65.

Ex. 30 Use this range.

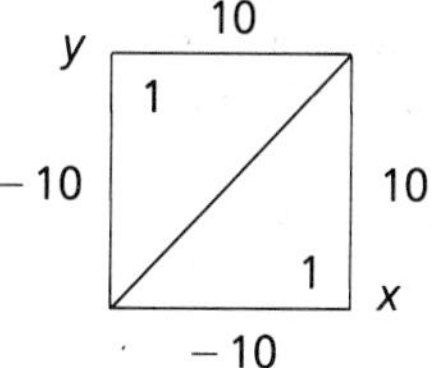

The solution appears to be 8.4.

Enrichment Activities

These activities require students to go beyond lesson goals.

In Example 3, you were introduced to a technique that used a graphics calculator or a computer to solve an equation. By graphing the left and right sides of the equation separately, you can identify the x-coordinate of the point of intersection as the solution.

1. Use the method to solve the equations given in Parts A–C on page 213. Require students to sketch the graphs and identify their solutions on the graphs.
   A. 5, B. $\frac{9}{2}$, C. $\frac{2}{3}$
2. Decide if the technique for solving the linear equation in Example 3 can be used to solve any equation. It can.
3. Why do you think this technique works? It works because of the properties of equations and of equalities.

Exploration and Extension

Technology—Graphing Calculator **In Exercises 45 and 46, use the following information.**

From 1980 to 1993, the number of dentists could be modeled as $D = 3470t + 139{,}170$, and the number of osteopaths could be modeled as $O = 1230t + 18{,}230$, where $t = 0$ corresponds to 1980.

45. Use a graphing calculator to estimate the year that the United States will have 205,000 dentists. Explain your reasoning. 1999

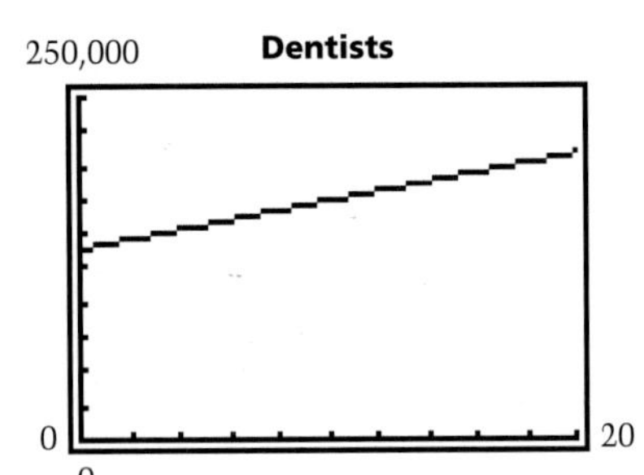

46. Use a graphing calculator to estimate the year that the United States will have 43,000 osteopaths. Explain your reasoning. 2000

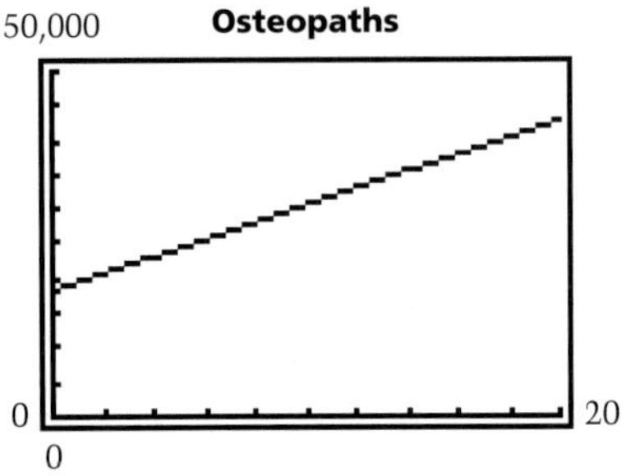

Mixed REVIEW

In Exercises 1–4, evaluate the expression.

1. $-6 - |-3|$ **(2.1)** -9 **2.** $6 - 2$ **(2.3)** 4 **3.** $|-7| - (-2)$ **(2.3)** 9 **4.** $4 - \left(-\frac{1}{3}\right)$ **(2.3)** $4\frac{1}{3}$

In Exercises 5 and 6, find the resulting unit of measure. (2.5)

5. (dollars per pound) • (pound) Dollars

6. (kilometers per second) • (second) Kilometers

7. Simplify $|-3| + 2(x - 2)$. **(2.6)** $2x - 1$

8. Simplify $2y - 3(y + 2)$. **(2.6)** $-y - 6$

9. Evaluate $3x^2 + x$ when $x = -4$. **(2.6)** 44

10. Evaluate $3(n - 1) + 2n$ when $n = 5$. **(2.6)** 22

11. Solve for a: $4a - 2b = 0$. **(3.6)** $a = \frac{b}{2}$

12. Solve for y: $-3(1 - y) + 4x = 1$. **(3.6)** $\frac{4 - 4x}{3}$

13. Is 5 a solution of $|x| + 2 = 7$? **(2.1)** Yes

14. Is -5 a solution of $|x| + 2 = 7$? **(2.1)** Yes

15. Evaluate $(22.66)(-41.20) \div 77.10$. **(2.7)** ≈ -12.11

16. Evaluate $\frac{1}{3}(x - 2)$ when $x = 3$. **(1.2)** $\frac{1}{3}$

In Exercises 17–20, translate the phrase or sentence into an algebraic expression, equation, or inequality. (1.6)

17. A number decreased by 6 is less than or equal to 10. $x - 6 \leq 10$

18. Six more than the number y is equal to the quotient of a number x and 3. $y + 6 = x \div 3$

19. The product of 3 and the opposite of 10, decreased by 1. $3(-10) - 1$

20. Twice the sum of x and y is equal to 20. $2(x + y) = 20$

4.7 Graphs of Absolute Value Equations

What you should learn:

Goal 1 How to graph an absolute value equation

Goal 2 How to model a real-life situation using graphs of absolute value equations

Why you should learn it:

You can model some real-life situations with absolute value equations.

Goal 1 Graphing Absolute Value Equations

The absolute value of a number can be defined as follows.

$$|x| = \begin{cases} x, & \text{if } x > 0 \\ 0, & \text{if } x = 0 \\ -x, & \text{if } x < 0 \end{cases}$$

Example

$|5| = 5$

$|0| = 0$

$|-3| = -(-3) = 3$

In this lesson, you will learn to sketch the graph of absolute value equations. To begin, let's look at the graph of $y = |x|$. By constructing a table of values and plotting points, you can see that the graph is V-shaped and opens up. The **vertex** of this graph is (0, 0).

x	$y = \|x\|$
−4	4
−3	3
−2	2
−1	1
0	0
1	1
2	2
3	3
4	4

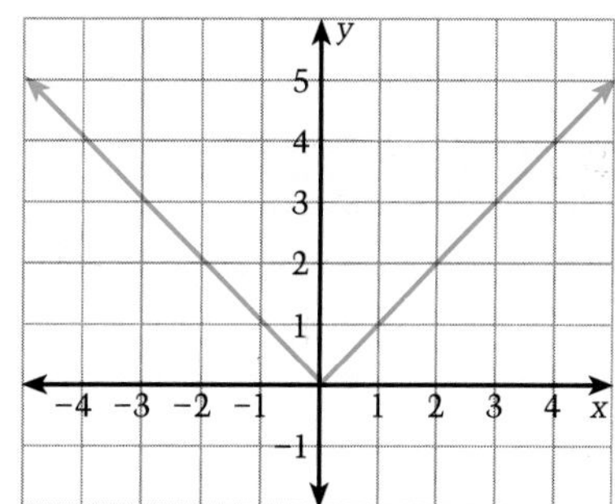

Other absolute value graphs are also V-shaped. Some open up and some open down. For those that open up, the vertex is the lowest point. For those that open down, the vertex is the highest point.

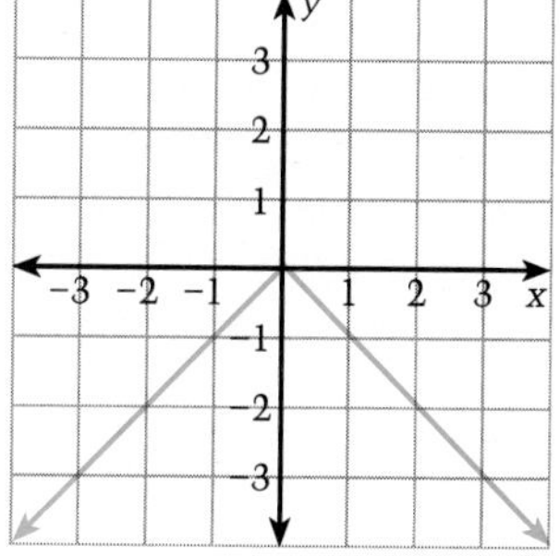

Graph of $y = -|x|$
Vertex at (0, 0), opens down.

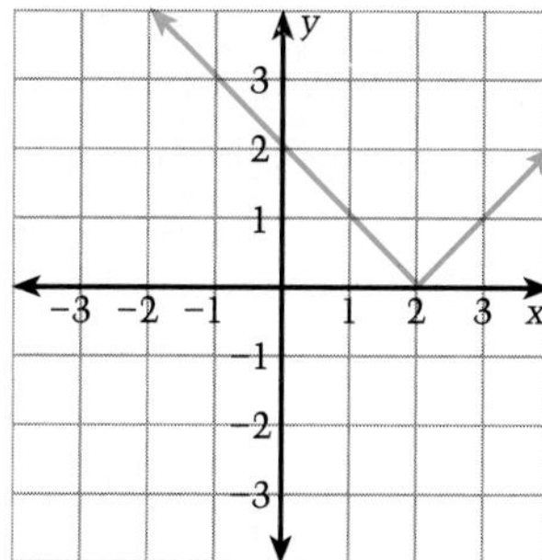

Graph of $y = |x - 2|$
Vertex at (2, 0), opens up.

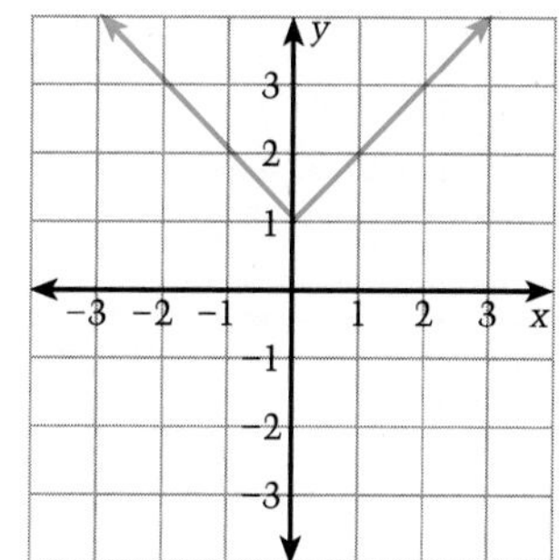

Graph of $y = |x| + 1$
Vertex at (0, 1), opens up.

Problem of the Day

If the area of the square is 4 cm², then what is the area of the triangle?

7 cm²

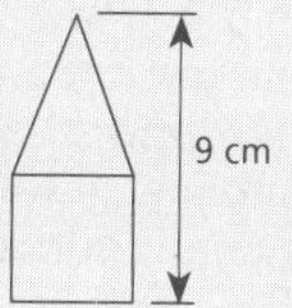

ORGANIZER

Warm-Up Exercises

1. Solve for x.
   a. $x - 4 = 0$
   b. $x + 12 = 0$
   c. $x + 2 = 0$
   a. 4, b. −12, c. −2
2. Complete the table for the equation $y = 2x + 3$.

x-value	y-value
−3	−3
−2	−1
−1	1
0	3
1	5
2	7
3	9

Lesson Resources

Teaching Tools
- Transparency: 10
- Problem of the Day: 4.7
- Warm-up Exercises: 4.7
- Answer Masters: 4.7

Extra Practice: 4.7

Color Transparencies: 20, 21

Technology Handbook: p. 42

LESSON Notes

GOAL 1 It is critical that students understand how the values of the absolute value table were determined. Spend time identifying how the table entries were obtained. This provides an opportunity for students to verify that they understand the definition.

Example 1

Students should know that the absolute value graph is V-shaped. Once the x-value of the vertex of the graph has been found, picking several x-coordinates on the left and right of the vertex and finding their corresponding y-values provides enough information to sketch the graph.

Example 3

CONNECTIONS Pool players put spin on the cue ball by striking it off-center with differing amounts of force. The cue ball may then leave the rail with a different angle than the one with which it hit the rail.

Point out that the motion of the cue ball in this example can be modeled by an absolute value equation.

Extra Examples

Here are additional examples similar to **Examples 1–2.**

Construct a table of values and sketch the following graphs. Label each vertex.

a. $y = \frac{2}{3}|x - 3| + 5$

b. $y = -5|x + 7| - 2$

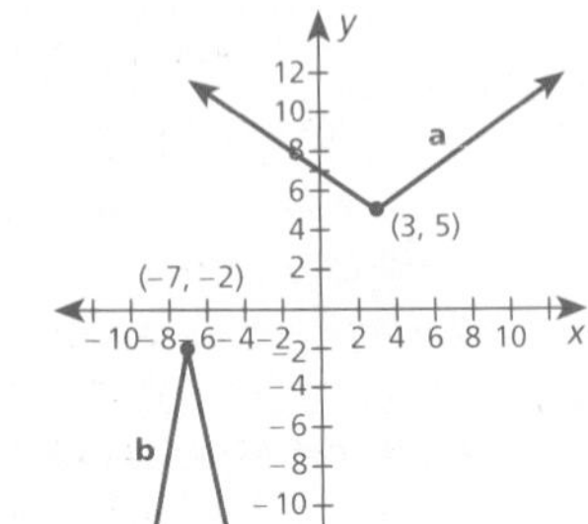

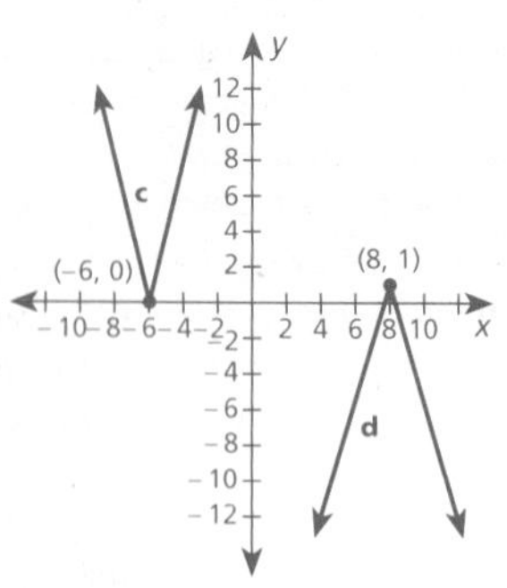

Here are some guidelines that can help you sketch the graph of an absolute value equation.

Sketching the Graph of an Absolute Value Equation

To sketch the graph of $y = a|x + b| + c$, use the following steps.

1. Find the x-coordinate of the vertex by finding the value of x for which $x + b = 0$.
2. Make a table of values using the x-coordinate of the vertex, some x-values to its left, and some to its right.
3. Plot the points given in the table and draw a V-shaped graph (opening up or down) through the points.

Example 1 *Graphing an Absolute Value Equation*

Sketch the graph of $y = \frac{1}{2}|x + 2| - 3$.

Solution

1. The x-coordinate of the vertex is -2 because that is the value of x for which $x + 2 = 0$.
2. Make a table, choosing x-values to the left and right of -2.

Vertex ↓

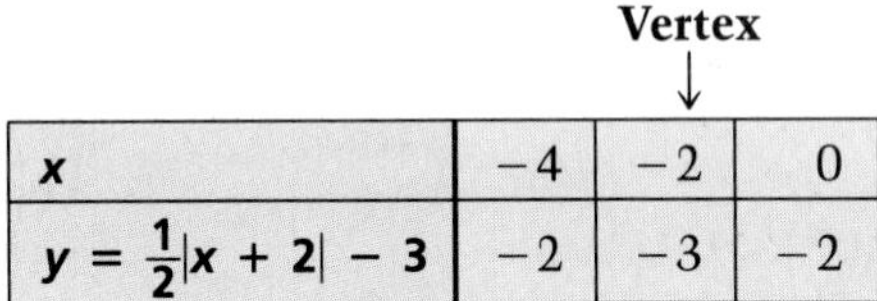

x	-4	-2	0
$y = \frac{1}{2}\lvert x + 2\rvert - 3$	-2	-3	-2

3. Plot the points. Notice that the vertex is at $(-2, -3)$. Then draw a V-shaped graph through the points. ■

Example 2 *Graphing an Absolute Value Equation*

Sketch the graph of $y = -2|x - 1| + 2$.

Solution

1. The x-coordinate of the vertex is 1 because that is the value of x for which $x - 1 = 0$.
2. Make a table, choosing x-values to the left and right of 1.

Vertex ↓

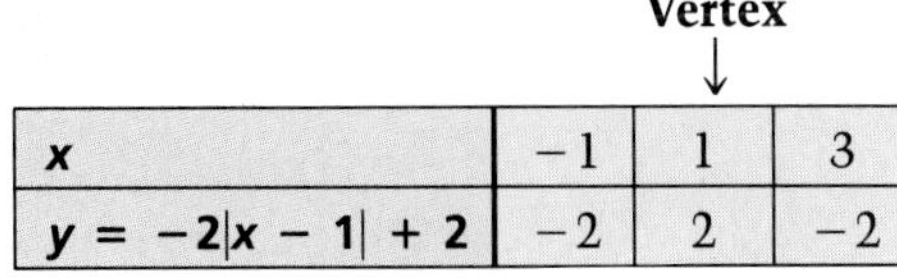

x	-1	1	3
$y = -2\lvert x - 1\rvert + 2$	-2	2	-2

3. Plot the points. Notice that the vertex is at $(1, 2)$. Then draw a V-shaped graph (opening down) through the points. ■

Goal 2 Modeling a Real-Life Situation

When a ray of light, a sound wave, or an object moving in a straight line hits a smooth surface, it is reflected in a V-shaped path. The angle formed by the incoming path and the surface is equal to the angle formed by the surface and the outgoing path.

Connections **Physics**

Example 3 *Finding the Path of a Billiard Ball*

Ana and Carlos are arguing about a pool shot. The cue ball cannot be moved, and each is trying to get the 10 ball into the corner pocket. Which one is right? (Assume that each is making a straight shot, with no spin on the ball.)

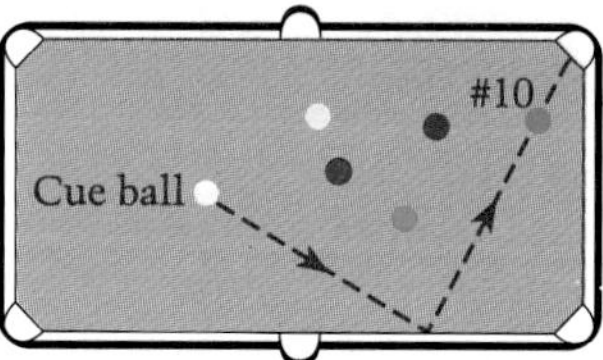

Ana says to imagine a line containing the corner pocket and the 10 ball. Then aim for the point at which the line intersects the opposite rail.

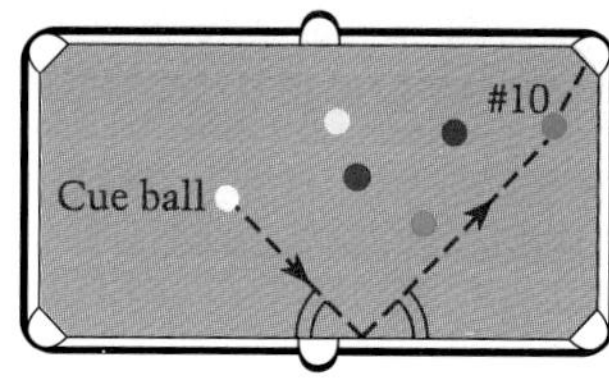

Carlos says to imagine equal angles between the path of the cue ball and the rail before and after the cue ball hits the rail. Then aim a hair past the point where the cue ball hits the rail to get a ricochet off the 10 ball.

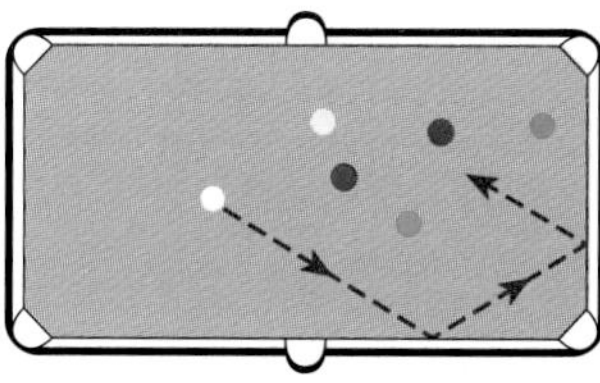

Solution Carlos is right. The rail of a billiard table is a reflecting surface, which means that a ball will leave the rail at the same angle it hits the rail. Ana's shot would travel the path shown at the left. ■

Communicating about ALGEBRA

Cooperative Learning

SHARING IDEAS about the Lesson

Work with a Partner and Compare Models Compare the graphs of the absolute value equations.

A. $y = |x|$ and $y = |x| - 1$ A.–C. See below. See Additional Answers for graphs.

B. $y = |x|$ and $y = |x - 2|$

C. $y = |x|$ and $y = -|x|$

D. Which has a sharper corner: $y = |x|$ or $y = 2|x|$? $y = 2|x|$

A. The graph of $y = |x| - 1$ is a vertical shift (a translation) 1 unit down of the graph of $y = |x|$.
B. The graph of $y = |x - 2|$ is a horizontal shift (a translation) 2 units to the right of the graph of $y = |x|$.
C. The graph of $y = -|x|$ is a reflection in the x-axis of the graph of $y = |x|$.

Check Understanding

1. Indicate how to find the absolute value of any real number x, positive, negative, or 0. See the definition on page 217.
2. How do you find the vertex of the graph of an absolute value equation? See step 1 on page 218. Is the vertex a real number or an ordered pair of real numbers? An ordered pair
3. How many points should be picked when constructing a table of values for the absolute value equation? Which points should be picked? Explain. See steps 2 and 3 on page 218.
4. Can the graph of the absolute value equation be something other than V-shaped? Explain. Yes, for example, here is a graph of $y = |x^2 - 1|$.

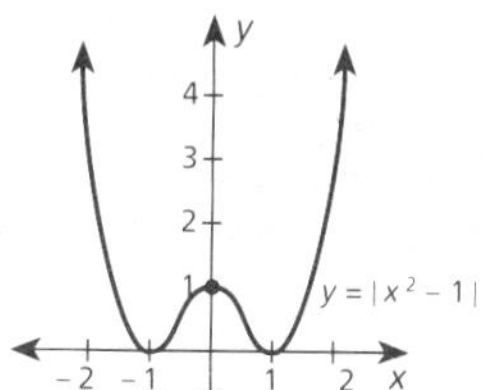

Communicating about ALGEBRA

A "sharper corner" describes a V-shaped graph that has a narrower opening than another.

In each exercise, the graph of the absolute value equation, $y = |x|$, is compared with the graph of another absolute value equation. Ask students to imagine that the graph of the second absolute value equation was formed from the first, the graph of $y = |x|$.

Students should compare, look for patterns from similar examples they create, and make conjectures about the effects of constants on the graph of the basic absolute value equation, $y = |x|$. (Note: Each exercise represents the different effects that constants can have on the equation.)

EXERCISE Notes

ASSIGNMENT GUIDE

Basic/Average: Ex. 1–4, 5–31 odd, 33–51 multiples of 3, 55–61 odd, 62

Above Average: Ex. 1–4, 5–31 odd, 33–51 multiples of 3, 52, 54, 55–63 odd

Advanced: Ex. 1–4, 5–31 odd, 33–51 multiples of 3, 52–54, 56–60 even, 62–63

Selected Answers
Exercises 1–4, 5–61 odd

More Difficult Exercises
Exercises 45–50, 52–53, 62–63

Guided Practice

Ex. 1–4 COOPERATIVE LEARNING Discuss these exercises with the class. Encourage and help students use the proper language in their explanations.

Independent Practice

Ex. 5–12, 25–28 It should be acceptable for students to construct tables and plot points to make the matches. If not used to make matches, the relationship between each graph and the graph of the absolute value equation $y = |x|$ should be identified.

COOPERATIVE LEARNING
These exercises make excellent oral problems for the class. Encourage students to describe the patterns they used as a basis for making decisions about matches.

EXERCISES

Guided Practice

CRITICAL THINKING about the Lesson

1. **True or False?** The vertex of the graph of $y = |x + b|$ is $(-b, 0)$. True
2. Is the vertex of the graph of $y = -|x|$ the lowest or highest point on the graph? Explain. Highest; the vertex is (0, 0) and $y \leq 0$.
3. Find the coordinates of the vertex of the graph of $y = \frac{1}{4}|x + 4| - 2$. $(-4, -2)$
4. Explain how to sketch the graph of $y = |x - 2| + 3$. See below.

Independent Practice

In Exercises 5–8, match the equation with its graph. Describe the relationship between each graph and the graph of $y = |x|$. See Additional Answers.

5. $y = |x| - 1$ c
6. $y = |x| + 1$ b
7. $y = |x| - 2$ d
8. $y = |x| + 2$ a

a. 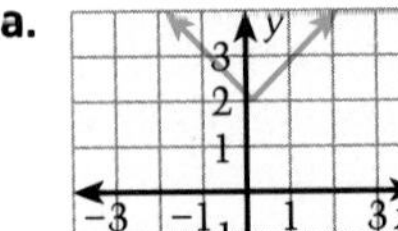

b. 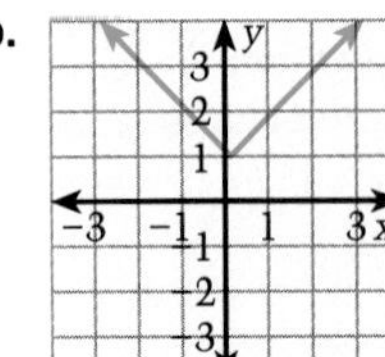

c. 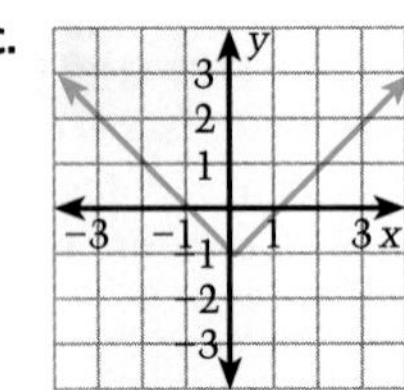

d. 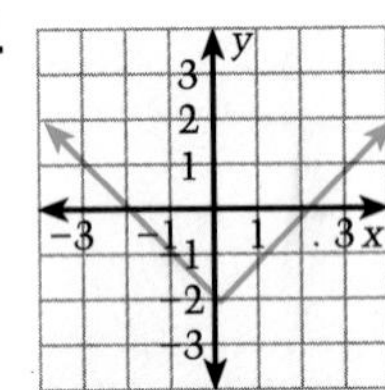

In Exercises 9–12, match the equation with its graph. Describe the relationship between each graph and the graph of $y = |x|$.

9. $y = |x - 1|$ b
10. $y = |x + 1|$ d
11. $y = |x - 2|$ c
12. $y = |x + 2|$ a

a. 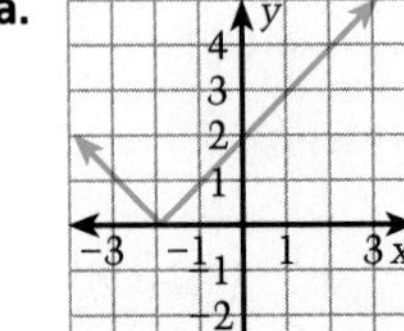

b. 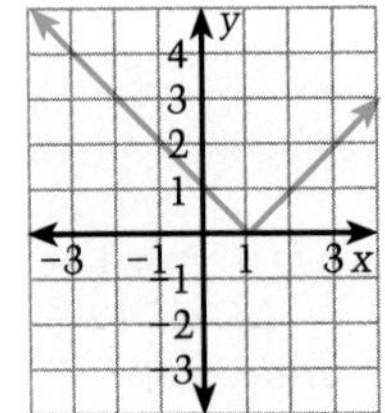

c. 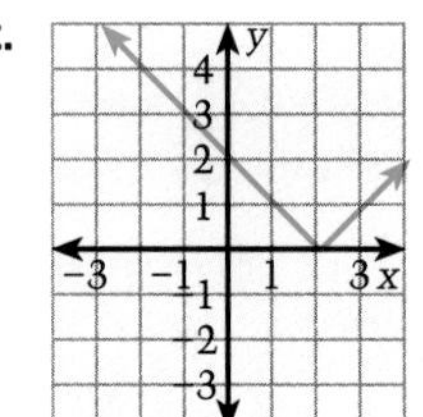

d. 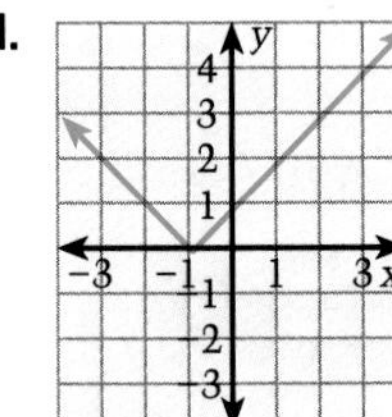

In Exercises 13–24, find the coordinates of the vertex of the graph.

13. $y = |x| - 6$ $(0, -6)$
14. $y = |x| + 5$ $(0, 5)$
15. $y = -|x| - 3$ $(0, -3)$
16. $y = |x + 6|$ $(-6, 0)$
17. $y = -2|x - 2|$ $(2, 0)$
18. $y = \frac{1}{2}|x - 3|$ $(3, 0)$
19. $y = |x + 3| - 2$ $(-3, -2)$
20. $y = |x - 3| + 1$ $(3, 1)$
21. $y = -|x - 9| + 10$ $(9, 10)$
22. $y = |x + 6| - 8$ $(-6, -8)$
23. $y = 2|x - 2| + 8$ $(2, 8)$
24. $y = -\frac{1}{2}|x - 3| + 5$ $(3, 5)$

4. Complete a table of values that contains the coordinates of the vertex and of several points to the right and left of the vertex. Then plot the points and draw the graph.

In Exercises 25–28, match the equation with its graph.

25. $y = |x - 1| + 2$ c **26.** $y = -|x - 2| - 1$ d **27.** $y = -|x + 1| - 2$ a **28.** $y = |x + 2| + 1$ b

a. 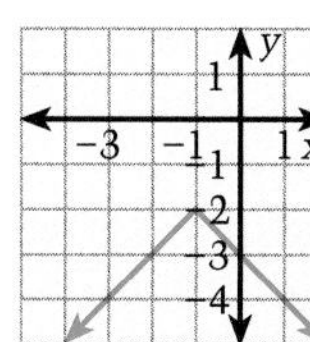

b. 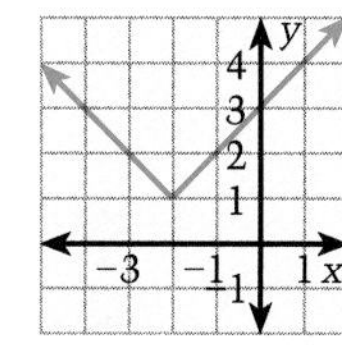

c.

d. 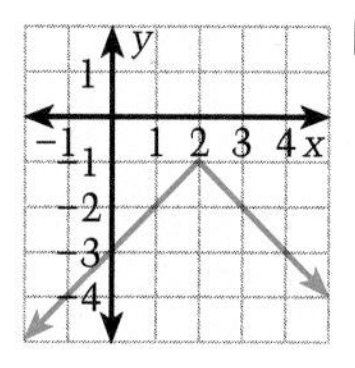

In Exercises 29–32, complete the table of values.

29. $y = -2|x - 1| + 3$ −3, −1, 1, 3, 1, −1, −3

x	−2	−1	0	1	2	3	4
y	?	?	?	?	?	?	?

30. $y = 3|x + 4| - 1$ 8, 5, 2, −1, 2, 5, 8

x	−7	−6	−5	−4	−3	−2	−1
y	?	?	?	?	?	?	?

31. $y = \frac{3}{2}|x + 3| - 4$ $\frac{1}{2}, -1, -\frac{5}{2}, -4, -\frac{5}{2}, -1, \frac{1}{2}$

x	−6	−5	−4	−3	−2	−1	0
y	?	?	?	?	?	?	?

32. $y = -\frac{3}{4}|x - 2| + 6$ $\frac{15}{4}, \frac{9}{2}, \frac{21}{4}, 6, \frac{21}{4}, \frac{9}{2}, \frac{15}{4}$

x	−1	0	1	2	3	4	5
y	?	?	?	?	?	?	?

In Exercises 33–50, sketch the graph of the equation. See Additional Answers.

33. $y = |x| + 3$ **34.** $y = |x| - 4$ **35.** $y = |x| - 3$ **36.** $y = |x| + 4$

37. $y = |x + 3|$ **38.** $y = |x - 5|$ **39.** $y = |x - 3|$ **40.** $y = |x + 4|$

41. $y = 2|x|$ **42.** $y = 3|x|$ **43.** $y = -2|x|$ **44.** $y = -\frac{1}{2}|x|$

✪ **45.** $y = |x + 3| - 2$ ✪ **46.** $y = -|x - 1| + 4$ ✪ **47.** $y = -2|x + 3| + 5$

✪ **48.** $y = \frac{1}{2}|x + 4| - 2$ ✪ **49.** $y = 3|x - 2| - 1$ ✪ **50.** $y = -3|x - 1| + 1$

51. True or False? The graphs of $y = -2|x - 3| + 4$ and $y = 2|x - 3| + 4$ have the same vertex. Explain. True, both have vertex (3, 4).

✪ **52.** *Rug Weaving* A coordinate plane has been drawn on the Navajo rug. Find absolute value equations for the four indicated graphs.

$y = |x + 2| + \frac{1}{2}$ $y = |x - 2| + \frac{1}{2}$

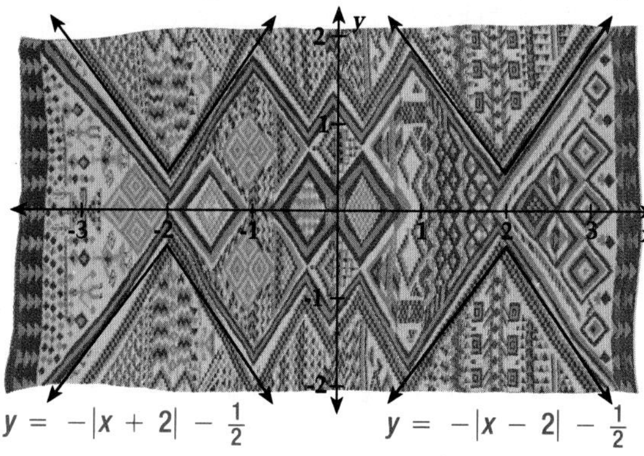

$y = -|x + 2| - \frac{1}{2}$ $y = -|x - 2| - \frac{1}{2}$

Wool used by the famous Navajo weavers is "homegrown," and dyed by hand.

9.–12. The graphs are all horizontal shifts (translations) of the graph of $y = |x|$.

9. 1 unit to the right **10.** 1 unit to the left **11.** 2 units to the right **12.** 2 units to the left

▶ **Ex. 13–24** Remind students that the vertex is an ordered pair.

▶ **Ex. 52** **MULTICULTURAL** Making computer-generated designs often depends on such algebraic relationships as absolute value equations. If you were interested in producing a computer image of a Navaho rug, knowing the absolute value equations as requested would be important.

EXTEND Ex. 52 You missed the end of the Wisconsin-Michigan football game and asked a friend who won. Your friend told you to sketch the graph of $y = |4|x| - 8|$. Who won the game? Sketch the graph of $y = |4|x| - 8|$ in two stages. First, graph the equation $y = 4|x| - 8$ using the techniques learned in this lesson. Next, reason that the negative y-values of the first graph will become positive (and the positive y-values will remain positive) when the absolute value is taken a second time. Consequently, Wisconsin won the game!

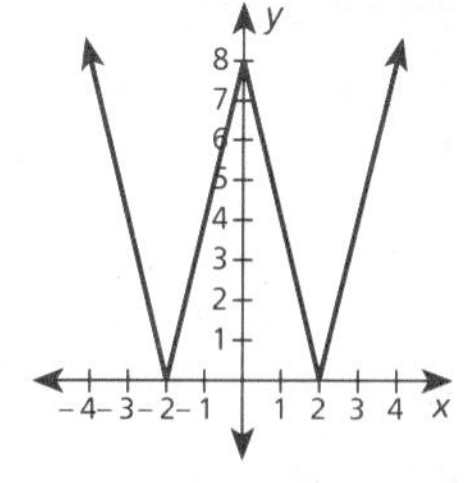

Enrichment Activities

Patterns can be used to sketch graphs of many algebraic relationships. Consider the following statement:

"The graph of the basic absolute value equation, $y = |x|$, can be stretched and shifted into the graph of any other absolute value equation."

TECHNOLOGY Graph each set of equations using a calculator or computer, if available. After you have graphed the five sets of equations below, discuss the graphs in light of the statement above.

1. a. $y = |x|$ b. $y = |x| - 4$, c. $y = |x| - 2$, d. $y = |x| - 1$, e. $y = |x| + 1$, f. $y = |x| + 2$, g. $y = |x| + 4$
2. a. $y = |x|$, b. $y = |x - 4|$, c. $y = |x - 2|$, d. $y = |x - 1|$, e. $y = |x + 1|$, f. $y = |x + 2|$, g. $y = |x| + 4|$
3. a. $y = |x|$, b. $y = -|x|$
4. a. $y = |x|$, b. $y = 2|x|$, c. $y = 4|x|$, d. $y = 6|x|$
5. a. $y = |x|$, b. $y = 0.90|x|$, c. $y = 0.75|x|$, d. $y = 0.50|x|$, e. $y = 0.25 x|$, f. $y = 0.10|x|$

1. The graph of the equation $y = |x| - b$ is a translation of the graph $y = |x|$ b units along the y-axis.
2. The graph of the equation $y = |x - a|$ is a translation of the graph $y = |x|$ a units along the x-axis.
3. The graph of $y = -|x|$ is the reflection of the graph $y = |x|$ about the x-axis.

4–5. The graph of $y = a|x|$ enlarges or shrinks the graph of $y = |x|$ by a factor of a. If $a > 1$, the graph narrows. If $a < 1$, the graph widens.

53. *Laser Show* You are setting up a laser show in the physical science lab. The laser is 40 feet from a corner of the lab and is aimed at a mirror on the adjoining wall. The mirror is the same height as the laser and is positioned as shown in the diagram. The laser beam travels along a path given by $y = \frac{4}{3}|x - 30|$. After the laser beam is reflected by the mirror, where will it hit the third wall?

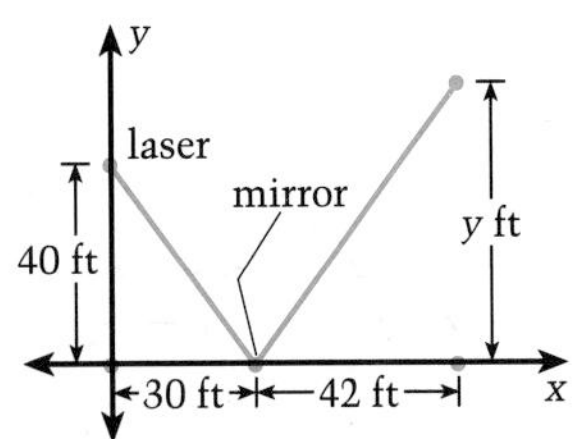

View from Above

56 ft from the other corner

Laser technology has come a long way since the first experiments with Light Amplification by Stimulated Emission of Radiation (L.A.S.E.R.).

Integrated Review

In Exercises 54–57, evaluate the expression.

54. $|-6 + 4| - 3$ −1
55. $-2|9 - 12| + 1$ −5
56. $-|-2 - (-1)| + 4$ 3
57. $|-5 - (-5)| - 2$ −2

In Exercises 58–61, evaluate the expression.

58. $|x - 2| - 5$, when $x = -2$ −1
59. $-|x + 3| - 4$, when $x = -5$ −6
60. $|-x + 1| + 3$, when $x = 3$ 5
61. $-\frac{1}{2}|x + 2| + 4$, when $x = -1$ $3\frac{1}{2}$

Exploration and Extension

Technology—Graphing Calculator **In Exercises 62 and 63, use the equation to determine the horizontal and vertical scale units for the graphing calculator screen.**

62. $y = \frac{3}{5}|2x - 34| - 20$

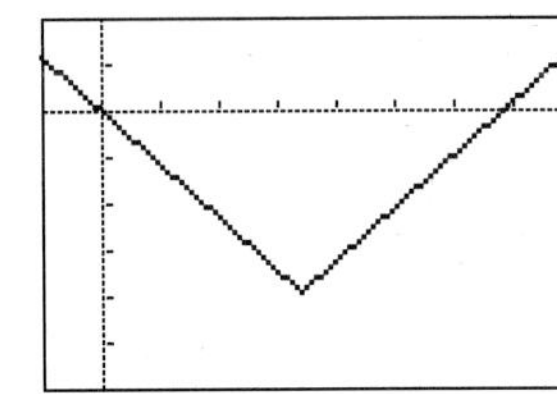

Each scale unit is 5.

63. $y = -\frac{1}{2}|x + 12| + 15$

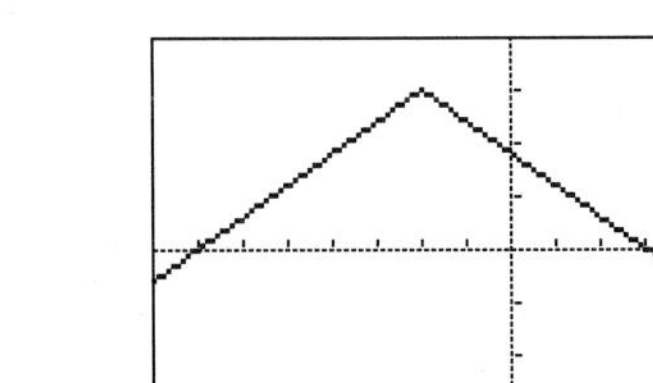

Horizontal scale unit is 6 and vertical scale unit is 5.

4.8 Solving Absolute Value Equations

What you should learn:

Goal 1 How to solve and check absolute value equations using algebra

Goal 2 How to use a graph to check solutions of absolute value equations

Why you should learn it:

You can use absolute value equations to model many real-life situations, such as representing minimum and maximum amounts.

Goal 1 Solving and Checking Algebraically

You already know how to use mental math to solve some absolute value equations. For instance, you know that the equation $|x| = 8$ has *two* solutions: 8 and -8.

Expression inside absolute value sign can be positive.	Expression inside absolute value sign can be negative.
8 is a solution of $\|x\| = 8$ because $\|8\| = 8$.	-8 is a solution of $\|x\| = 8$ because $\|-8\| = 8$.

To solve a more general absolute value equation, use the fact that the expression inside the absolute value sign can be either positive or negative.

Example 1 *Solving an Absolute Value Equation*

Solve $|x - 2| = 5$.

Solution

$\|x - 2\| = 5$	*Rewrite original equation.*
$x - 2 = 5$ or $x - 2 = -5$	*Expression can be 5 or -5.*
$x = 7$ or $x = -3$	*Add 2 to both sides.*

The equation has two solutions: 7 and -3. Check these solutions by substituting each into the original equation. ■

Example 2 *Isolating the Absolute Value Expression*

Solve $|2x - 7| - 5 = 4$.

Solution First, isolate the absolute-value portion on one side of the equation.

$\|2x - 7\| - 5 = 4$	*Rewrite original equation.*
$\|2x - 7\| = 9$	*Add 5 to both sides.*
$2x - 7 = 9$ or $2x - 7 = -9$	*Expression can be 9 or -9.*
$2x = 16$ or $2x = -2$	*Add 7 to both sides.*
$x = 8$ or $x = -1$	*Divide both sides by 2.*

The equation has two solutions: 8 and -1. Check these solutions by substituting each into the original equation. ■

Problem of the Day

Look for a pattern. What are the values of e and f?

$a - 34 = 11$
$b - 334 = 111$
$c - 3334 = 1111$
$d - 33{,}334 = 11{,}111$

$e - f = 11{,}111{,}111$

44,444,445 and 33,333,334

ORGANIZER

Warm-Up Exercises

1. Solve.
   a. $3x + 6 = 8$
   b. $11x - 35 = 9$
   c. $17 - 4x = 32$
   a. $\frac{2}{3}$, b. 4, c. $-3\frac{3}{4}$
2. Check each solution in Exercise 1 above graphically.
3. Use mental math to solve.
   a. $2x = 8$
   b. $x - 6 = 17$
   c. $|x| = 10$
   d. $|3x| = 9$
   a. 4, b. 23, c. 10 or -10, d. 3 or -3

Lesson Resources

Teaching Tools
- Transparencies: 4, 10
- Problem of the Day: 4.8
- Warm-up Exercises: 4.8
- Answer Masters: 4.8

Extra Practice: 4.8

Color Transparency: 21

Applications Handbook: p. 1, 2

LESSON Notes

GOAL 1 Point out to students that knowing $|x| = 3$ means x is 3 or -3 empowers them to solve absolute value equations. Ask students to solve several equations of the type, $|x| = k$, $(k \geq 0)$ before starting with Example 1.

Examples 1–5

Give students the opportunity to discuss how to find solutions to absolute value equations. Students may think it odd that equations that appear to be linear generate *two* solutions instead of one.

GOAL 2 Call attention to the similarity between using a graphic check for linear equations and using a graphic check for absolute value equations. This checking technique is useful in many equation–solving situations.

TECHNOLOGY Calculators and computers can be used to find graphic solutions to absolute value equations.

Extra Examples

Here are additional examples similar to **Examples 1–4.**

1. Solve.
   - a. $|x + 8| = 13$
   - b. $|7x - 11| = 34$
   - c. $|5x - 3| + 7 = 42$
   - d. $|6 - 3x| - 2 = 12$

   a. 5, −21; b. $6\frac{3}{7}, -3\frac{2}{7}$; c. $7\frac{3}{5}, -6\frac{2}{5}$; d. $-2\frac{2}{3}, 6\frac{2}{3}$
2. Check problems 1a and 1d above using a graphics check.

Goal 2 Using a Graphic Check

In Lesson 4.6, you learned to use a graphic check for solutions of linear equations. You can use the same technique to check solutions of absolute value equations.

- Rewrite the one-variable equation so that the right side is zero.
- Set y equal to the left side and sketch the graph of the resulting two-variable equation.
- The x-intercepts of the graph of the two-variable equation should match the solutions of the one-variable equation.

Example 3 Using a Graphic Check

Check the solutions of $|x - 2| = 5$ graphically.

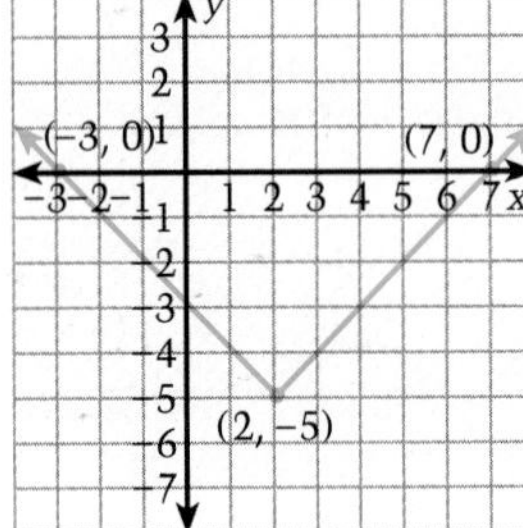

Solution In Example 1, the solutions were found to be 7 and −3. You can check these solutions graphically as follows.

$|x - 2| = 5$ *Original one-variable equation*

$|x - 2| - 5 = 0$ *Rewrite with zero on right side.*

Set y equal to the left side of the rewritten equation.

$y = |x - 2| - 5$ *Form two-variable equation.*

Using the techniques in Lesson 4.7, you know that the graph of this equation is V-shaped and has a vertex at $(2, -5)$. From the graph, you can see that the x-intercepts occur at 7 and −3, which match the solutions. ■

Example 4 Using a Graphic Check

Check the solutions of $|2x - 7| - 5 = 4$ graphically.

Solution In Example 2, the solutions were found to be 8 and −1. You can check these solutions graphically as follows.

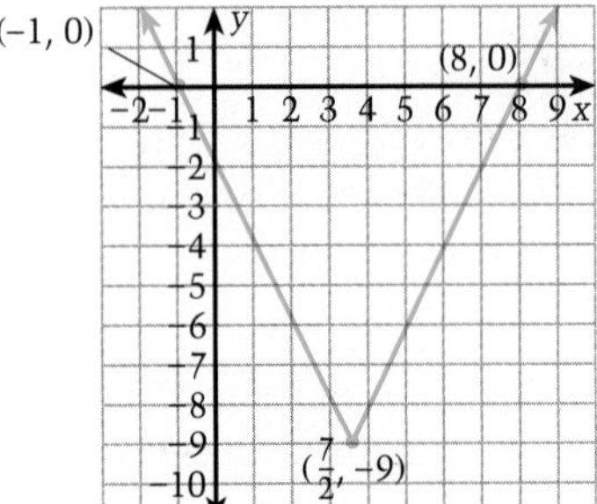

$|2x - 7| - 5 = 4$ *Original one-variable equation*

$|2x - 7| - 9 = 0$ *Rewrite with zero on right side.*

Set y equal to the left side of the rewritten equation.

$y = |2x - 7| - 9$ *Form two-variable equation.*

The graph of this equation is V-shaped, opens up, and has a vertex at $(\frac{7}{2}, -9)$. From the graph, you can see that the x-intercepts occur at 8 and −1, which match the solutions. ■

Real Life
Weights and Measures

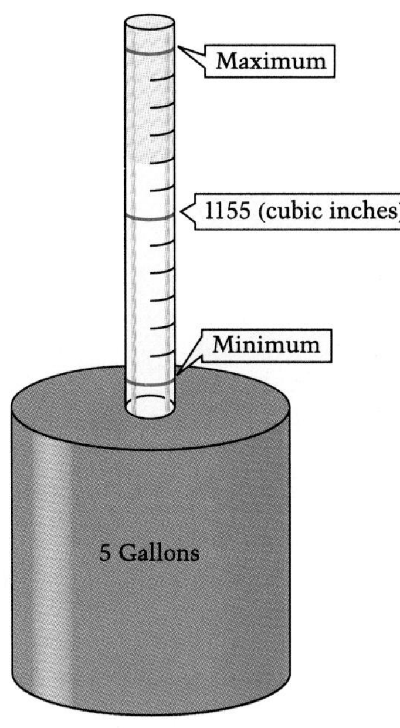

Example 5 *Accuracy of Measurement*

To obtain certification by the Bureau of Weights and Measures, a gasoline pump must pump within 6 cubic inches of 5 gallons (1155 cubic inches) when its gauge reads 5 gallons. The inspector pumps gasoline into a container like that shown at the left. The reading on the gas pump is 5 gallons. To be certified, the gas level in the container must fall within the minimum and maximum amounts on the scale. What are these minimum and maximum amounts? Find an absolute value equation that models this situation and has the minimum and maximum amounts as its solutions.

Solution Using the information given with the photo, the amount pumped must be within 6 cubic inches of 1155 cubic inches. Thus, the minimum and maximum amounts are:

Minimum

$1155 - 6 = 1149$ cubic inches

Maximum

$1155 + 6 = 1161$ cubic inches

These two numbers can be expressed as solutions of the equation $|x - 1155| = 6$. Do you see why? ■

Communicating about ALGEBRA

SHARING IDEAS about the Lesson

Interpret a Model Solve each equation. Use the solutions to rephrase the statement.

A. $|x - 160| = 1$ To pass certification, the scale must weigh within one ounce of 160 ounces. from 159 to 161 oz.

B. $|x - 189| = 86$ To be eligible for one of the 13 weight classes on the wrestling team, your weight must be within 86 pounds of 189 pounds. from 103 to 275 lb.

C. $|x - 30| = 10$ To obtain a bonus, the pizza delivery driver must deliver the pizza in 30 minutes, plus or minus 10 minutes. 20 to 40 minutes.

Check Understanding

1. Outline the steps for checking solutions to absolute value equations. Are the steps similar to those used with linear equations? The expression within the absolute value bars may be either positive or negative, thus giving two linear equations to solve and check.
2. Can the absolute value equation, $|2x + 3| = -4$ be solved. Explain. No, the absolute value of a number or expression is always positive.

Communicating *about* **ALGEBRA**

EXTEND *Communicating*
WRITING
Have students identify real-life situations in which upper and lower limits must be set. Then have them express these relationships using absolute value equations.

EXERCISE Notes

ASSIGNMENT GUIDE

Basic/Average: Ex. 1–33 odd, 38–39, 40–50 even, 52–53

Above Average: Ex. 1–4, 6–33 multiples of 3, 34–38, 41–51 odd, 52–53, 54, 56

Advanced: Ex. 1–4, 6–33 multiples of 3, 34–39, 41–51 odd, 52–57

Selected Answers
Exercises 1–4, 5–51 odd

More Difficult Exercises
Exercises 34–37, 52–57

Guided Practice

Ex. 4 Discuss this exercise orally with the class. Be sure that students understand how the coordinate plane is superimposed on the picture of the warship.

Independent Practice

Ex. 5–33 TECHNOLOGY Have students solve a collection of these exercises algebraically, using quick graphs, and using a graphics calculator or computer. Ask them to evaluate each approach for ease of use, accuracy, and time required.

Answers

14.

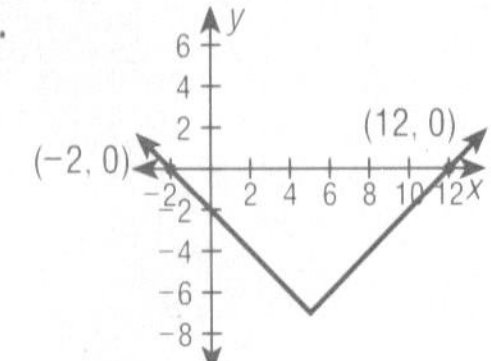

15.

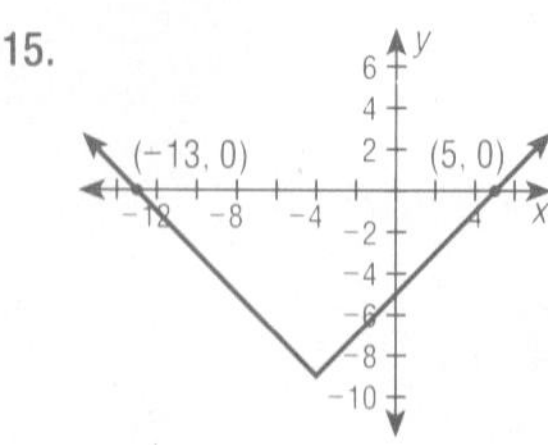

EXERCISES

Guided Practice

CRITICAL THINKING about the Lesson

1. Solve $|x + 1| = 2$. $1, -3$
2. What equation could you graph to check the solutions of $|x + 1| = 2$?
3. Which two equations would you solve to find the solutions of $|x + 1| = 32$?
4. If a coordinate plane is drawn through the galley at the right, the mainstays would lie on the graph of $y = 82.5 - 0.92|x|$, where x and y are measured in feet. How far above the deck does the mast rise? Find the x-coordinates of the points at which the mainstays are attached to the deck. 82.5 ft, 90 and -90

Around 650 B.C., the Greeks invented a warship called the trireme. It had three rows of oars. The photo shows a modern replica of a trireme.

2. $y = |x + 1| - 2$ 3. $x + 1 = 32$, $x + 1 = -32$

Independent Practice

In Exercises 5–10, solve the equation algebraically.

5. $|x + 4| = 3$ $-1, -7$
6. $|9 - x| = 4$ $5, 13$
7. $|6 - x| = 9$ $-3, 15$
8. $|x + 12| = 8$ $-4, -20$
9. $|2x + 6| = 14$ $4, -10$
10. $|3x - 4| = 7$ $\frac{11}{3}, -1$

In Exercises 11–13, rewrite the equation so that the term involving absolute value is isolated on one side.

11. $6 + |x + 1| = 9$ $|x + 1| = 3$
12. $|x - 17| + 9 = 1$ $|x - 17| = -8$
13. $7 - |4 - x| = 12$ $|4 - x| = -5$

14. Use a graphic check that -2 is a solution of $|x - 5| = 7$. See margin.
15. Use a graphic check that -13 is a solution of $|x + 4| = 9$. See margin.

In Exercises 16–24, solve the equation.

16. $|x + 1| = 7$ $6, -8$
17. $|x - 1| = 7$ $8, -6$
18. $|3x - 2| - 2 = 5$ $3, -\frac{5}{3}$
19. $|4x + 3| - 4 = 8$ $\frac{9}{4}, -\frac{15}{4}$
20. $3 + |-2x + 9| = 10$ $1, 8$
21. $6 + |-x - 5| = 9$
22. $|4 - 5x| - 8 = 15$ $-\frac{19}{5}, \frac{27}{5}$
23. $|7 - 3x| - 10 = 4$ $-\frac{7}{3}, 7$
24. $2|3x - 7| + 2 = 4$ $\frac{8}{3}, 2$

21. $-8, -2$

In Exercises 25–33, solve the equation. Use a graphic check.

25. $|x - 9| = 4$ $13, 5$
26. $|x + 7| = 16$ $9, -23$
27. $|2x - 4| + 6 = 9$ $\frac{7}{2}, \frac{1}{2}$
28. $4 + |x - 10| = 14$ $20, 0$
29. $10 + |3x + 1| = 24$ $\frac{13}{3}, -5$
30. $|2x + 9| - 15 = 36$
31. $2|x + \frac{1}{2}| - 1 = 9$ $\frac{9}{2}, -\frac{11}{2}$
32. $3|x + \frac{1}{6}| + 5 = 17$ $\frac{23}{6}, -\frac{25}{6}$
33. $3|4x + 9| - 2 = 6$ $-\frac{19}{12}, -\frac{35}{12}$

30. $21, -30$

A Tent on Ice **In Exercises 34–37, use the following information.**

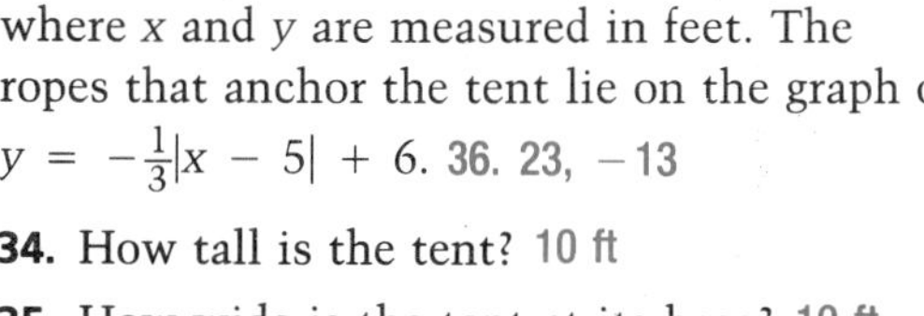

If a coordinate plane were drawn through the tent at the right, the edges of the tent would lie on the graph of $y = -2|x - 5| + 10$, where x and y are measured in feet. The ropes that anchor the tent lie on the graph of $y = -\frac{1}{3}|x - 5| + 6$. 36. 23, -13

34. How tall is the tent? 10 ft

35. How wide is the tent at its base? 10 ft

36. Find the x-coordinates of the points at which the stakes are secured to the ice.

37. How far are the stakes from the edge of the tent? 13 ft

38. *Earth's Orbit* Because Earth travels in an elliptical orbit around the sun, the distance between Earth and the sun varies. The average distance is 92,950,000 miles, but the distance can vary 1,550,000 miles from the average. Find an absolute value equation that models this situation and has the minimum and maximum distances between Earth and the sun as solutions. What are the minimum and maximum distances?

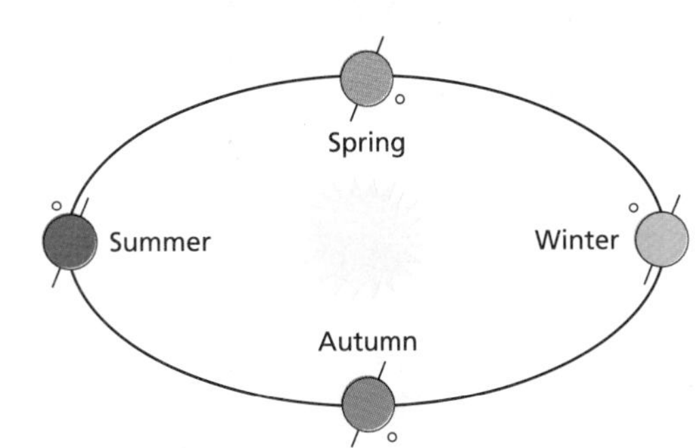

39. *Boxing Weights* A junior heavyweight boxer's weight must fall within 7 pounds of 183 pounds. Find an absolute value equation that models this situation and has the minimum and maximum weights for a junior heavyweight boxer as solutions. What are the minimum and maximum weights? $|x - 183| = 7$, 176 lb, 190 lb

38. $|x - 92,950,000| = 1,550,000$; 91,400,000 mi; 94,500,000 mi

Integrated Review

In Exercises 40–45, evaluate the expression.

40. $2|4| + 6$ 14

41. $3 - |8 - 4|$ -1

42. $3^2 - |-4|$ 5

43. $6^2 - 2|14 - 19|$ 26

44. $|3 - 9| \cdot 2^2 + 1$ 25

45. $-4|-5 + 1| - 12$ -28

46. Write an equation of the horizontal line passing through $(-2, 4)$. $y = 4$

47. Write an equation of the vertical line passing through $(5, -3)$. $x = 5$

48. Find the slope and y-intercept of $x + 2y = 18$. $-\frac{1}{2}$, 9

49. Find the slope and y-intercept of $6x - y = 3$. 6, -3

50. Sketch the line whose slope is $\frac{3}{2}$ and whose y-intercept is -5. See margin.

51. Sketch the line whose slope is -4 and whose y-intercept is 12. See margin.

Answers

50.

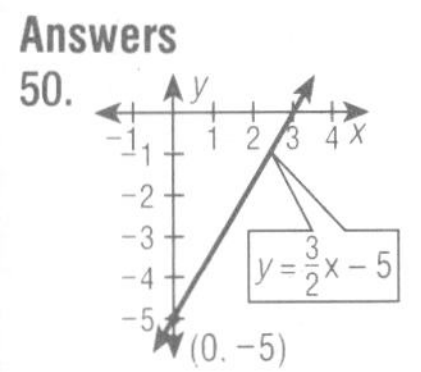

51.

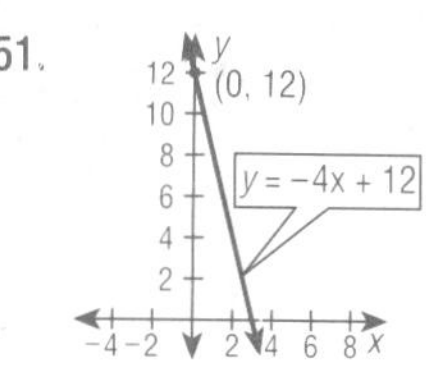

Enrichment Activities

WRITING Following the class discussion, students might record their personal responses in their math journals.

TECHNOLOGY Instead of graphically checking solutions to absolute value equations, use your graphics calculator or computer, if available, to solve them.

1. Discuss the similarities and differences between solving and checking absolute value equations.
2. COOPERATIVE LEARNING **Challenge!** Work in small groups. Identify the steps needed to solve absolute value equations using the graphing technologies. Outline a set of steps for your classmates to follow.

Ex. 52–53 From the vertex of the graph of an absolute value equation, you may move on the "V" shape toward the left or the right. Moving from the vertex toward the right is on the "right side of the graph." Similarly, the "left side of the absolute value graph" is moving toward the left from its vertex. The slopes for each side of the "V" are opposites.

Answers

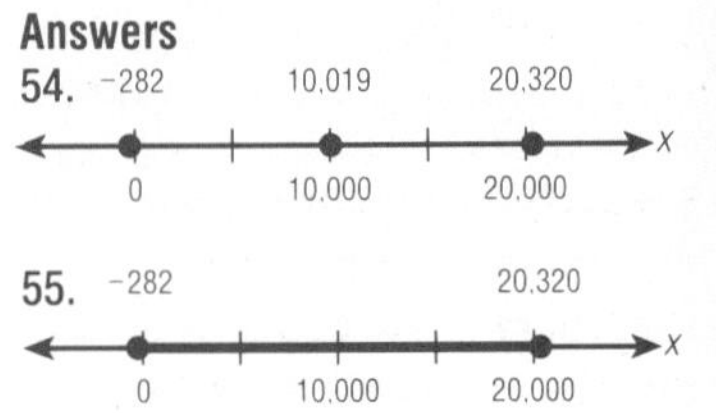

Exploration and Extension

In Exercises 52 and 53, find the slope of the lines that contain each side of the graph.

52. $y = \frac{2}{3}|x - 4| - 2$ $-\frac{2}{3}, \frac{2}{3}$

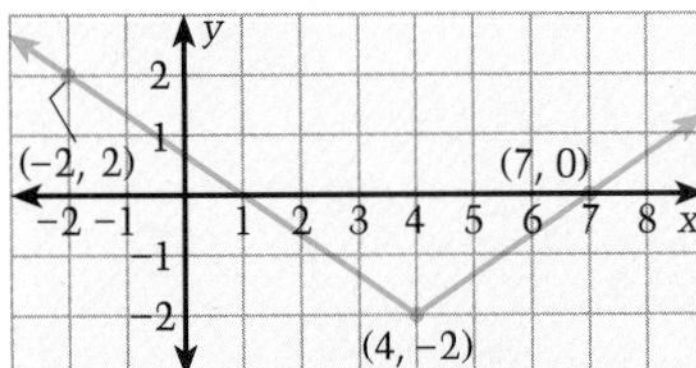

53. $y = -3|x + 10| + 4$ 3, −3

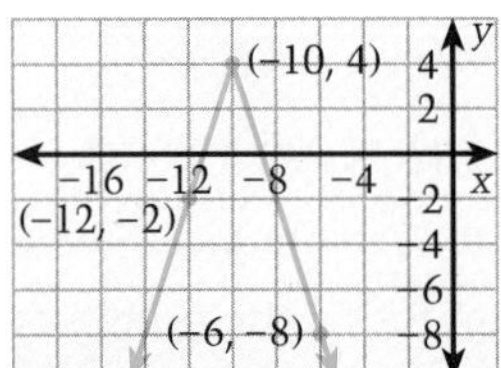

In Exercises 54–57, use the following information.

Sea level is considered to have an elevation of 0 feet. Elevation in North America ranges from below sea level in some valleys and canyons to far above sea level on mountain peaks.

The lowest elevation in North America is 282 feet below sea level near Badwater in Death Valley.

The highest elevation in North America is 20,320 feet above sea level at one of Mt. McKinley's peaks.

54. Plot the lowest and highest elevations in North America on a real number line. (Use a larger version of the number line shown at the right.) Find the point that lies midway between the highest and lowest points. **See margin; 10,019**

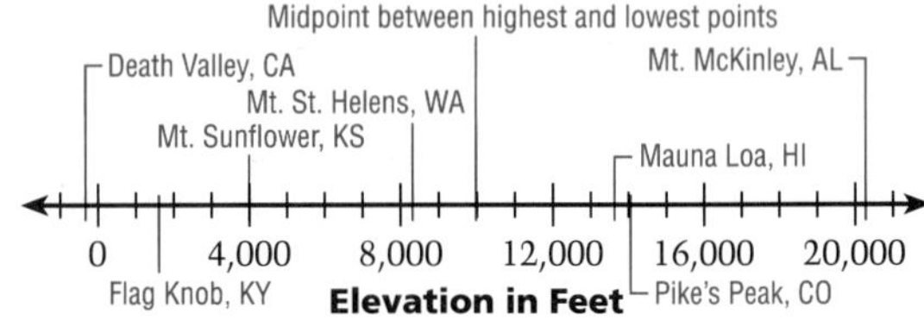

55. On your graph, shade the interval that represents all possible geographical elevations in North America. **See margin.**

56. Use the distance between the midpoint of the interval and each of the endpoints of the interval to write an *inequality* whose solutions represent all possible elevations in North America. $|x - 10{,}019| \leq 10{,}301$

57. Plot and label the following locations on your graph. **See above.**

Pikes Peak Colorado	Mount St. Helens Washington	Flag Knob Kentucky	Mount Sunflower Kansas	Mauna Loa Hawaii
14,100 feet	8365 feet	1658 feet	4039 feet	13,677 feet

4 Chapter Summary

What did you learn?

Skills

1. Sketch the graph of a linear equation in one variable. **(4.1)**
2. Write an equation of a horizontal or vertical line. **(4.1)**
3. Sketch the graph of a linear equation in two variables
   - using a table of values. **(4.2)**
   - using intercepts. **(4.3)**
   - using the slope-intercept form. **(4.5)**
4. Find the slope of a line
   - passing through two points. **(4.4)**
   - by writing its equation in slope-intercept form. **(4.5)**
5. Sketch the graph of an equation involving absolute value. **(4.7)**
6. Solve an equation involving absolute value. **(4.8)**

Strategies

7. Use a graph to check the solution(s) of
   - a linear equation. **(4.6)**
   - an equation involving absolute value. **(4.8)**
8. Interpret slope as a rate of change. **(4.4)**

Exploring Data

9. Create scatter plots and recognize linear patterns. **(4.1–4.6)**

Why did you learn it?

Linear models are the most commonly used algebraic models for real-life situations. By learning to sketch the graphs of linear models, you will be able to use them to help answer questions about the real world. Recognizing and using the algebra-geometry connection between equations and graphs is one of the most important skills you can acquire on your way to becoming a master problem solver.

How does it fit into the bigger picture of algebra?

If you were studying algebra 400 years ago, this chapter would not have been in your algebra text. At that time, algebra and geometry were separate branches of mathematics with few connections. Today, they are closely connected. You saw in this chapter that relationships between variables may be expressed in algebraic form as an equation, or in geometric form as a graph. For example, x and y could be related by a *linear equation* such as $y = x - 4$, whose graph is a line, or by an *absolute value equation* such as $y = |x - 4|$, whose graph is V-shaped. Drawing the graphs helps you understand the algebraic relationships.

ORGANIZER

TIME SCHEDULE: All levels, two days

The Chapter Summary helps students organize the main ideas of the chapter. In this chapter, many algebra skills needed for sketching graphs were developed. Graphs give us a useful way to interpret data from real-life situations. Throughout the chapter, graphs were used to categorize linear patterns, while scatter plots were used to organize data.

Work with students to review the skills, strategies, and concepts of the chapter. The homework assignment can be used as the basis for the second day of review.

COOPERATIVE LEARNING

Encourage students to study together. Emphasize the importance of teaching a classmate how to sketch a graph or how to use a strategy for graphing linear and absolute value equations. When students work together, everyone wins. The students receiving help get additional instruction, and the students giving help gain a deeper understanding of the skills and concepts involved.

Chapter SUMMARY

Review the chapter by asking students, "What did you learn? Why did you learn it? How does it fit with the other algebra skills and concepts you have learned?"

Chapter REVIEW

ASSIGNMENT GUIDE

Basic/Average: Ex. 1–39 odd, 43–46

Above Average: Ex.1–39 odd, 41–46

Advanced: Ex.1–39 odd, 41–50

✪ More Difficult Exercises
Exercises 42, 48–50

▶ **Ex. 1–50** The lesson references in the Chapter Review direct students to the lesson in which each skill or strategy was first taught. Students should look back to this lesson if they are having difficulty attacking a set of exercises. Note also that the review exercises parallel the list of Skills, Strategies, and Exploring Data statements on page 229. Students may use this list to identify each skill or strategy, and perhaps check other chapters to see where this skill is used.

Answers
Chapter Review

9.

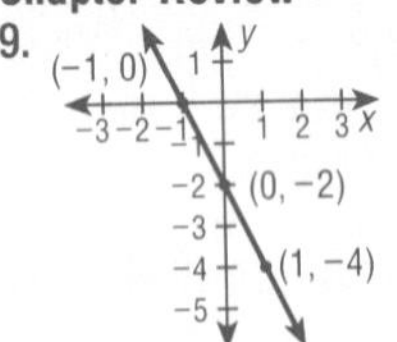

10.

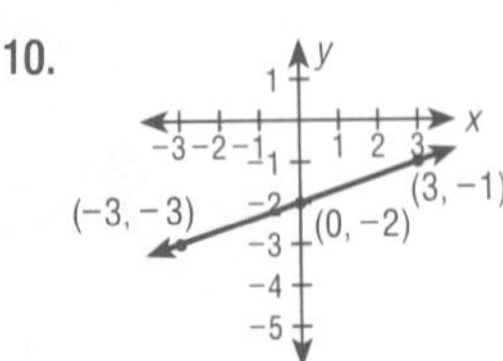

Chapter REVIEW

In Exercises 1–4, sketch the two lines on the same coordinate plane. Then find the point at which the two lines intersect. **(4.1)** See Additional Answers.

1. $x = 0, y = -3$ (0, −3)

2. $x = 7, y = -5$ (7, −5)

3. $x = -6, y = 4$ (−6, 4)

4. $x = -2, y = 0$ (−2, 0)

In Exercises 5–8, decide whether the point lies on the line. **(4.1)**

5. $(-6, -9), y = -9$ Yes

6. $(3, 4), y = 3$ No

7. $(-3, 8), x = 8$ No

8. $(5, -5), y = 5$ No

In Exercises 9–12, use a table of values to sketch the graph of the equation. **(4.2)** See margin.

9. $y = -(2x + 2)$

10. $y = \frac{1}{3}x - 2$

11. $y = 7 - \frac{1}{2}x$

12. $y = -4(x + 1)$

In Exercises 13–16, match the equation with its graph. **(4.3)**

13. $2x + 3y = 10$ c

14. $-x + 4y = 8$ b

15. $3x - 5y = 15$ d

16. $4x - 5y = -20$ a

a.

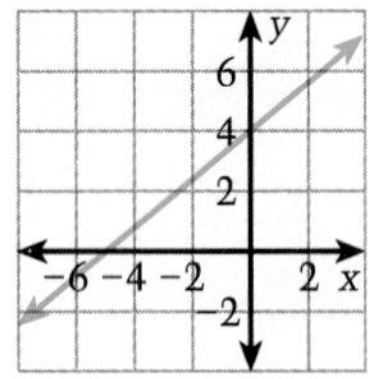

b.

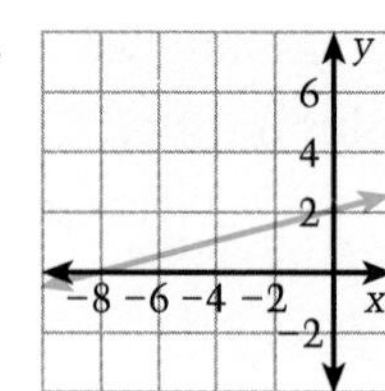

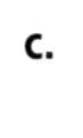

c.

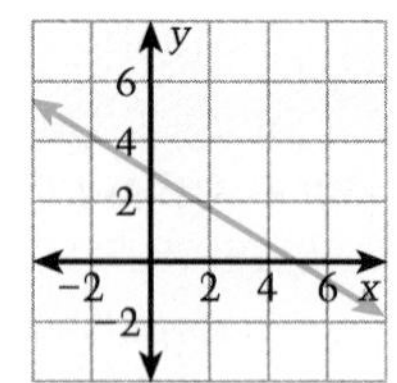

d.

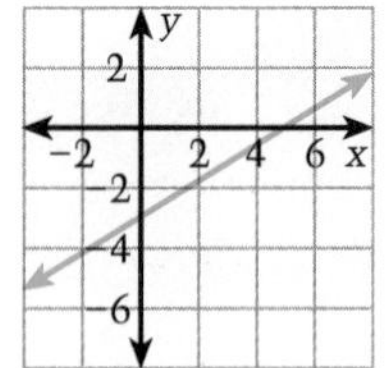

In Exercises 17–19, find the slope of the line passing through the points. **(4.4)**

17. $(2, -1), (3, 4)$ 5

18. $(0, 8), (-1, 2)$ 6

19. $(2, 4), (5, 0)$ $-\frac{4}{3}$

In Exercises 20–22, find the slope and *y*-intercept of the line. **(4.5)**

20. $x + 11y = 2$ $-\frac{1}{11}, \frac{2}{11}$

21. $x - 4y = 12$ $\frac{1}{4}, -3$

22. $-x + 6y = -24$ $\frac{1}{6}, -4$

In Exercises 23 and 24, sketch the line. Label the *x*-intercept and *y*-intercept. **(4.3)** See Additional Answers.

23. $x - 2y = 10$

24. $3x + 4y = -24$

In Exercises 25–28, solve the equation. Check your solution two ways. **(4.6)**

25. $3x - 6 = 0$ 2

26. $5x - 3 = 0$ $\frac{3}{5}$

27. $-x + 8 = 0$ 8

28. $-4x - 1 = 7$ −2

In Exercises 29–32, construct a table of values, sketch the graph, and label the vertex. **(4.7)** See margin.

29. $y = |x| - 4$

30. $y = 2|x + 3| - 1$

31. $y = -3|x + 1| + 5$

32. $y = \frac{1}{2}|x - 6| + 2$

In Exercises 33–36, match the equation with its graph. (4.7)

33. $y = |x| - 4$ d **34.** $y = |x - 1| + 1$ a **35.** $y = |x + 1| + 1$ b **36.** $y = |x + 1| - 1$ c

a. 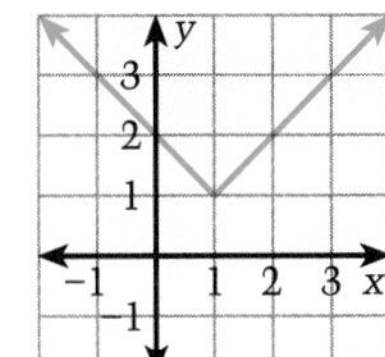
b. 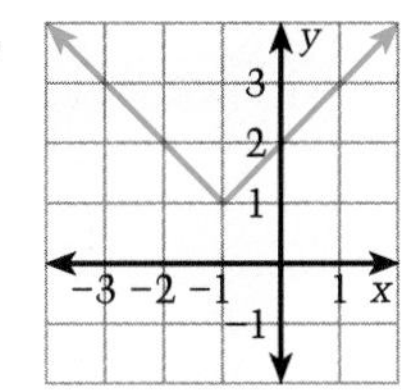
c. 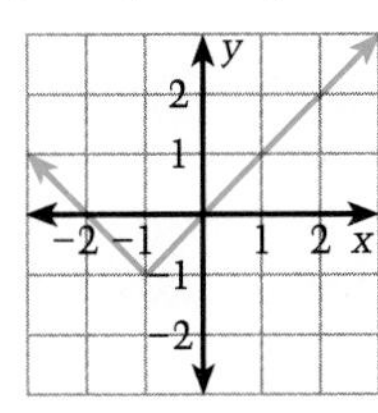
d. 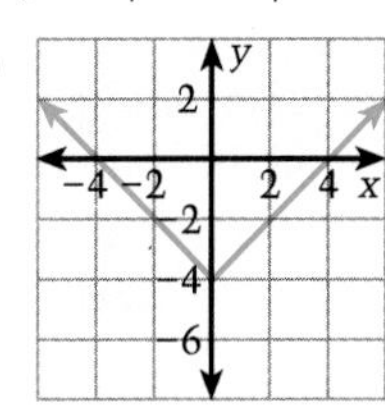

In Exercises 37–40, solve the equation. Check your solutions two ways. (4.8)

37. $4 + |x + 2| = 7$ 1, −5

38. $6 - 5|x - 1| = 1$ 2, 0

39. $7 - |2x + 6| = 4$ $-\frac{3}{2}, -\frac{9}{2}$

40. $|x + 8| - 6 = 2$ 0, −16

In Exercises 41–42, use the following information.

In 1990, a survey was taken to determine the ways in which American men and women measure success. Each respondent was allowed to list more than one measure. The results of the survey are shown below.
(Source: Roper Organization, New York)

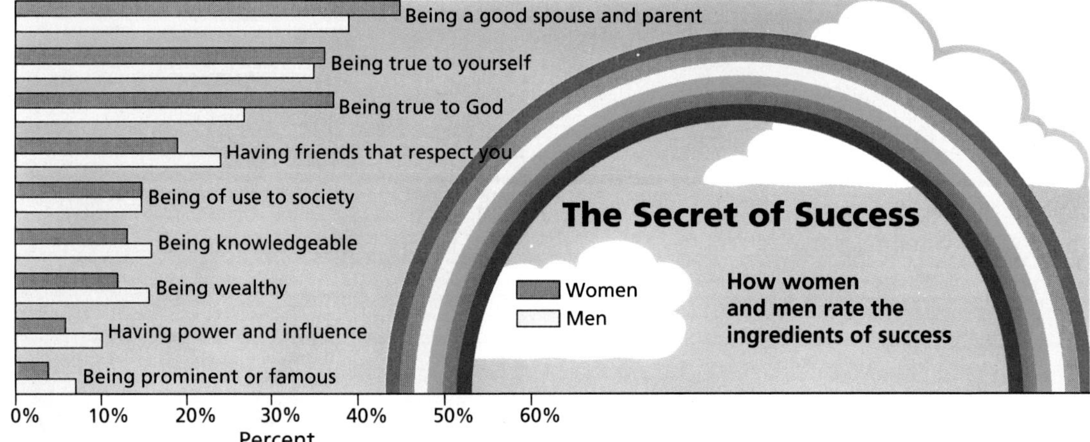

41. The graph at the right shows the responses of the first one hundred women surveyed, concerning one of the measures of success. After each group of 10 responses was tallied, the total number who had responded yes to this particular measure of success was plotted. Which measure of success was being tallied?

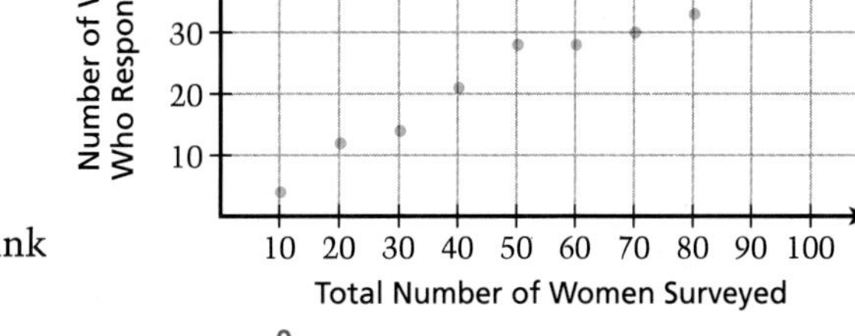

✪ **42.** What is the slope of the line that you think would best fit the points on the graph?

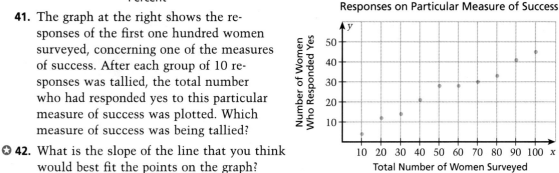

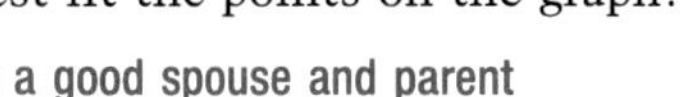
41. Being a good spouse and parent

Slopes vary; about $\frac{9}{20}$

11. 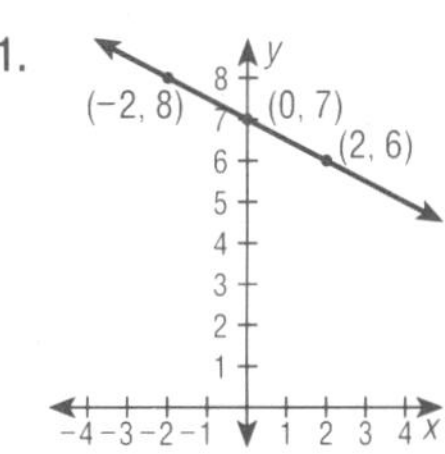

12. 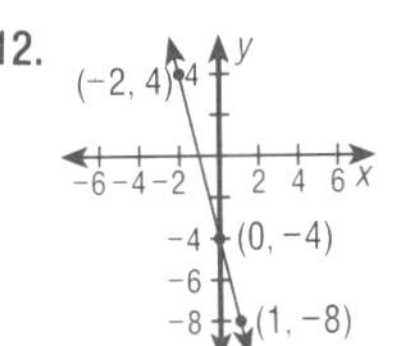

29. 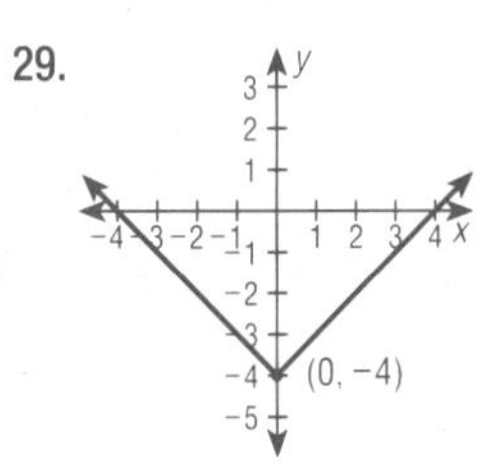

30. 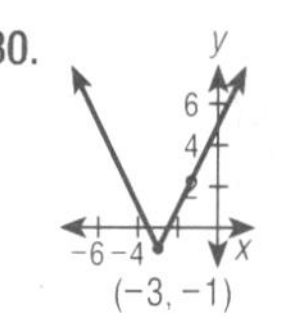

31. 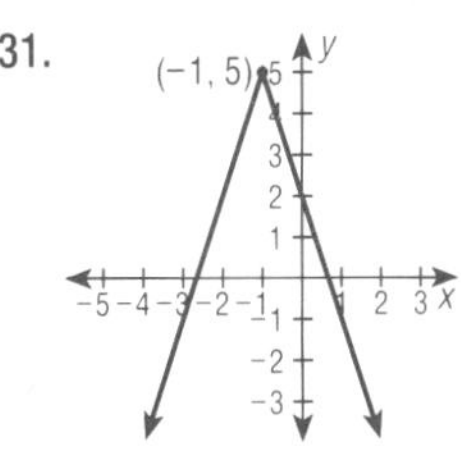

32. 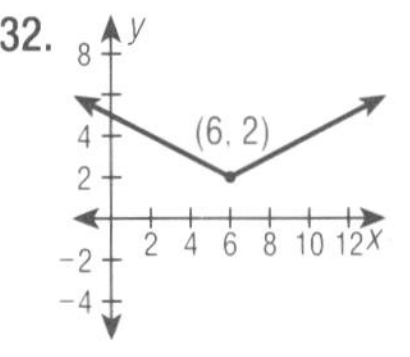

Chapter Test

Answers

1. 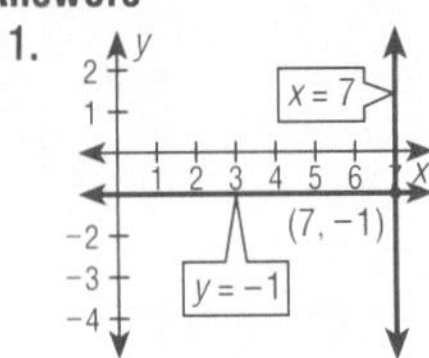

5. 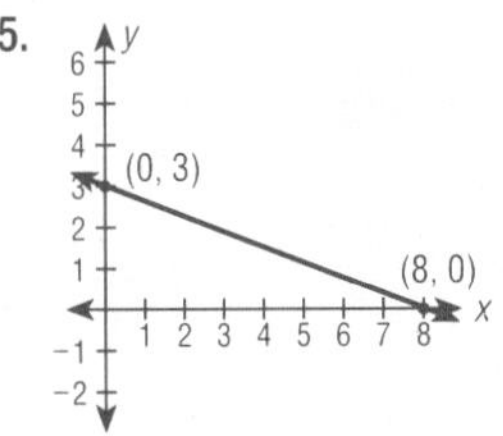

6. 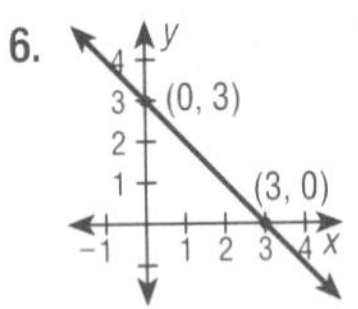

Recreational Boat Ownership **In Exercises 43–46, use the following information.**

The graph at the right shows the number of recreational boats, y (in millions), that were in use in the United States from 1983 to 1988. The numbers on the t-axis represent the year. ***(Source: National Marine Manufacturers Association)***

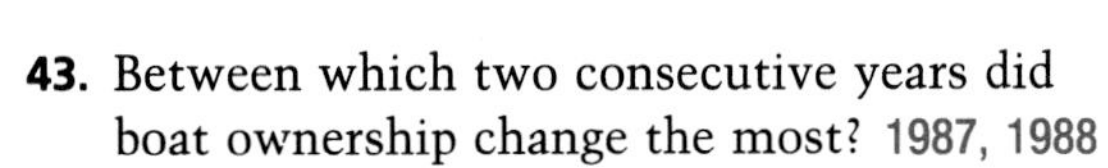

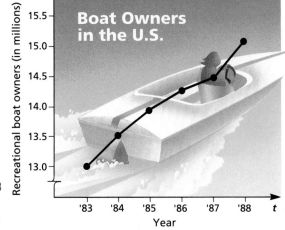

43. Between which two consecutive years did boat ownership change the most? 1987, 1988

44. Between which two consecutive years did boat ownership change the least? 1986, 1987

45. What was the average rate of change in boat ownership from 1983 to 1988? (List your result in millions of boats per year.) 0.4 millions of boats per year

46. What was the average rate of change in boat ownership from 1986 to 1988? (List your result in millions of boats per year.) 0.375 millions of boats per year

Sailboat Dimensions **In Exercises 47–50, use the following information.**

An xy-coordinate plane has been drawn through the sailboat at the right. The edges of the two largest sails lie approximately on the graph of the equation $y = -2.2|x| + 30$, where x and y are measured in feet.

47. How tall are the sails? 30 ft

48. Find the x-coordinates of the points that define the combined widths of the sails at their bases. What is the approximate combined base widths of the sails? 14, −14, 28 ft

49. What is the ratio of the height of the sails to the combined base widths of the sails? $\frac{15}{14}$

50. What are the slopes of the right and left edges of the two sails? −2.2, 2.2

Chapter TEST

1. Sketch the graphs of $x = 7$ and $y = -1$. Then find the point at which the two graphs intersect. **(4.1)** See margin; (7, −1)
2. Write the equation of the line passing through $(3, -1)$, $(3, 0)$, and $\left(3, -\frac{3}{4}\right)$. **(4.1)** $x = 3$
3. Which point is on the graph of $\frac{5}{3}x + y = 1$? **(4.2)**
   a. $\left(4, -\frac{17}{3}\right)$ b. $\left(2, \frac{10}{3}\right)$
4. Complete the table. **(4.2)**

x	−9	−6	−3	0	3	6	9
$y = \frac{1}{3}x - 2$	?	?	?	?	?	?	?
	−5	−4	−3	−2	−1	0	1

In Exercises 5 and 6, find the intercepts and sketch the line. (4.3) See margin.

5. $\frac{3}{8}x + y = 3$
6. $y = -x + 3$

In Exercises 7–10, find the slope of the line through the points. (4.4)

7. $(5, 6), (-9, 6)$ 0
8. $(4, 0), (1, -2)$ $\frac{2}{3}$
9. $\left(-3, -\frac{1}{2}\right), \left(-6, -\frac{1}{4}\right)$ $-\frac{1}{12}$
10. $(5, 70), (5, 76)$ Undefined

In Exercises 11–14, rewrite the equation in slope-intercept form. (4.5)

11. $3y + 4x = 7$ $y = -\frac{4}{3}x + \frac{7}{3}$
12. $3(y - 1) = x - 3$ $y = \frac{1}{3}x$
13. $\frac{1}{3}(y - 9) - (x - 1) = 6$ $y = 3x + 24$
14. $\frac{1}{4}x + 4y = 4$ $y = -\frac{1}{16}x + 1$

In Exercises 15 and 16, solve the equation. Check your solution(s) algebraically and graphically. (4.6)

15. $2x - 3 = 7$ 5
16. $2(x - 3) = -4$ 1

In Exercises 17 and 18, find the vertex and sketch the graph. (4.7) See margin.

17. $y = 2|x - 1| + 2$
18. $y = -3|x + 2| - 1$

In Exercises 19 and 20, solve the equation. (4.8)

19. $|3x + 6| - 7 = 3$ $\frac{4}{3}, -\frac{16}{3}$
20. $|x + 5| - 4 = 4$ 3, −13

21. A produce stand is selling two varieties of corn. Silver Queen sells for \$1.50 a dozen, and Butter and Sugar sells for \$1.25 a dozen. The total sales for one day was \$375. Let x be the number of dozen ears of Silver Queen sold and let y be the number of dozen ears of Butter and Sugar sold. Sketch a graph that shows the possible values of x and y. **(4.2)** See margin.
22. Water is in a liquid state if its temperature, t (in degrees Fahrenheit), satisfies the *inequality* $|t - 122| < 90$. Find the two solutions of the *equation* $|t - 122| = 90$. What do these solutions represent. **(4.8)** 212, 32; 212°F is the boiling point of water, 32°F is the freezing point.

17.
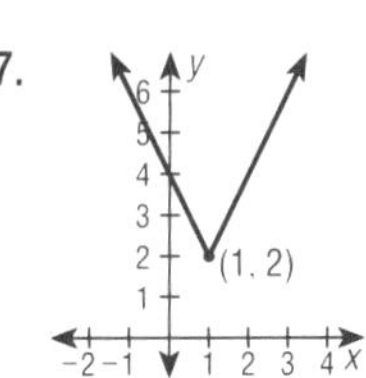

18.
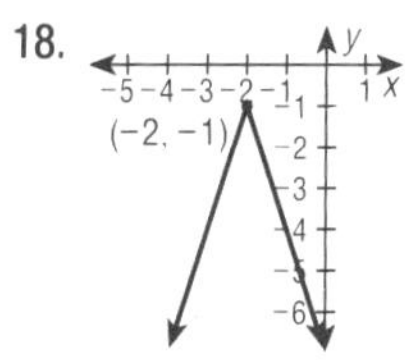

21.
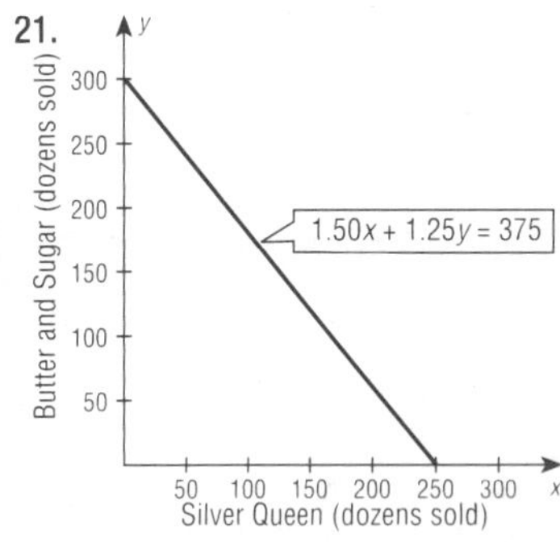

Students should understand that the techniques for writing linear equations are important algebraic skills because linear equations are often used to model real-life situations. In Lesson 5.4, an important application of writing linear equations—fitting a line to data—will be introduced. Encourage students to identify data from their own experiences that can be modeled by linear equations. In this way, students will experience applications of algebra that are meaningful to them.

The Chapter Summary on page 284 provides you and the students with a synopsis of the chapter. It identifies key skills, concepts, and vocabulary. You may want to have students look at the Chapter Summary as an overview before beginning the chapter.

CHAPTER 5

Writing Linear Equations

LESSONS

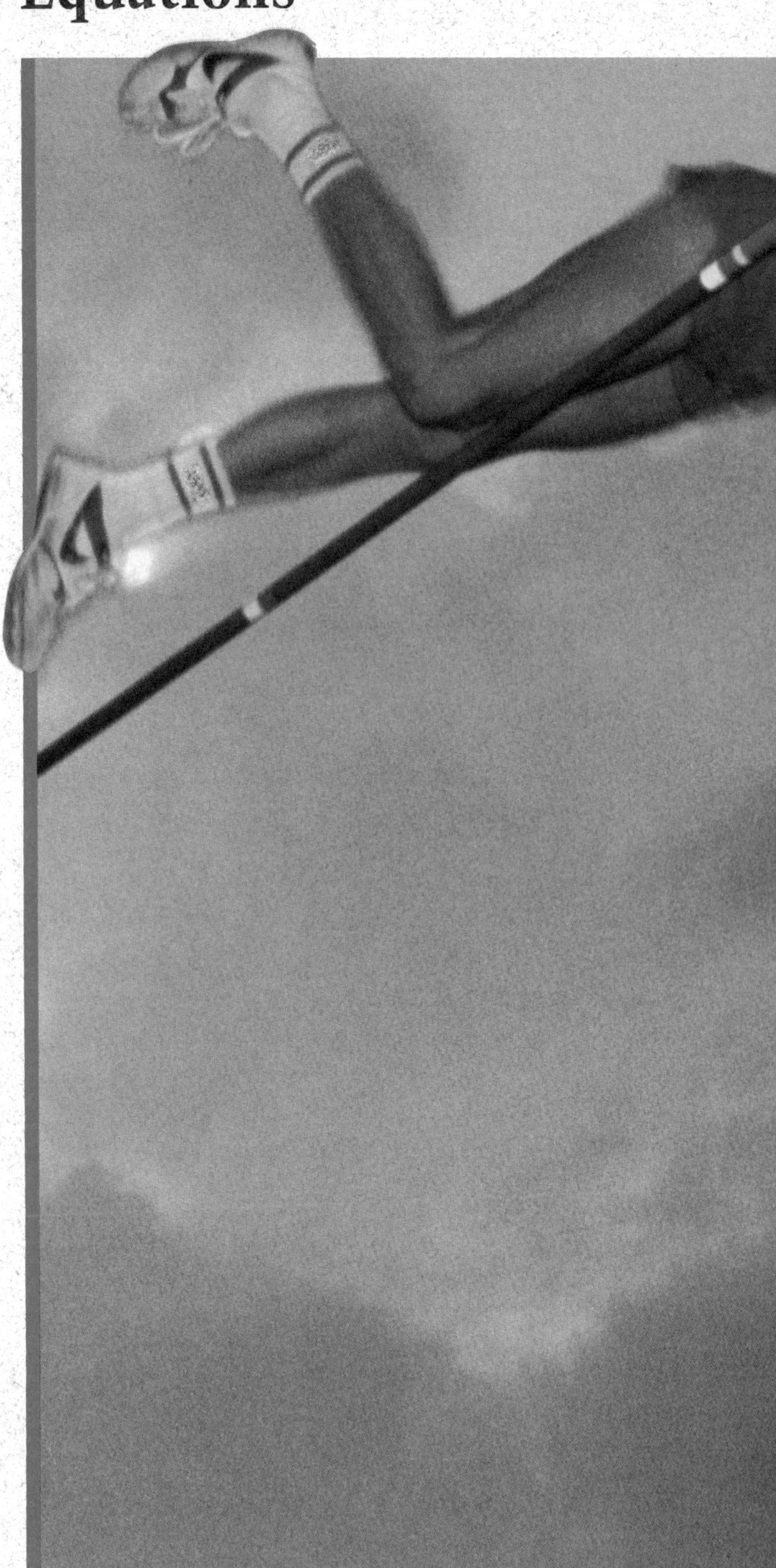

In Olympic competition, athletes push the parameters of physical achievement to higher and higher levels.

Real Life
Olympic Games

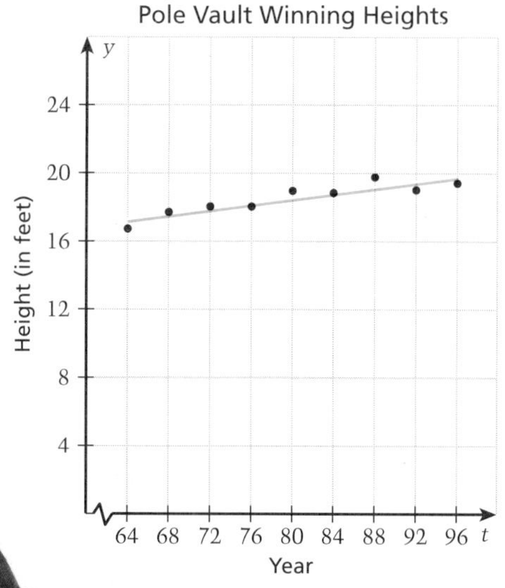

Pole Vault Winning Heights

Height (in feet)

Year

The points on this graph show the heights of the winning pole vault at the Olympic Games between 1964 and 1996. The winning heights can be approximated by the linear model $y = 0.08t + 12.17$. In this model, y is the height in feet and t represents the year, with $t = 64$ corresponding to 1964.

1964:	Fred Hansen (U.S.)	16.73 ft
1968:	Bob Seagren (U.S.)	17.71 ft
1972:	Wolfgang Nordwig (Germany)	18.04 ft
1976	Tadeusz Slusarski (Poland)	18.04 ft
1980:	Wladyslaw Kozakiewicz (Poland)	18.96 ft
1984:	Pierre Quinon (France)	18.85 ft
1988:	Sergei Bubka (U.S.S.R.)	19.77 ft
1992:	Maxim Tarassov (Unified Team)	19.04 ft
1996:	Jean Galfione (France)	19.42 ft

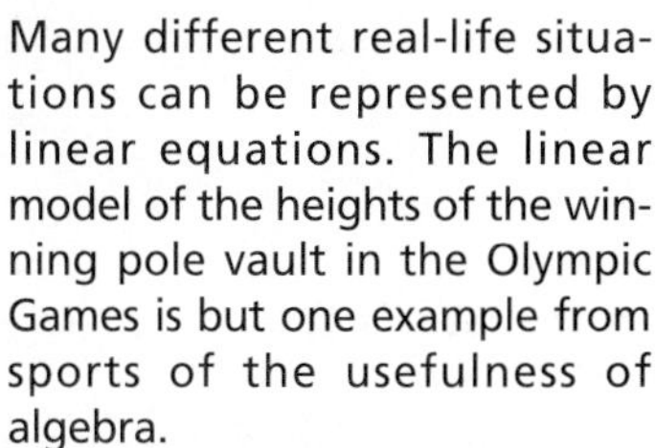

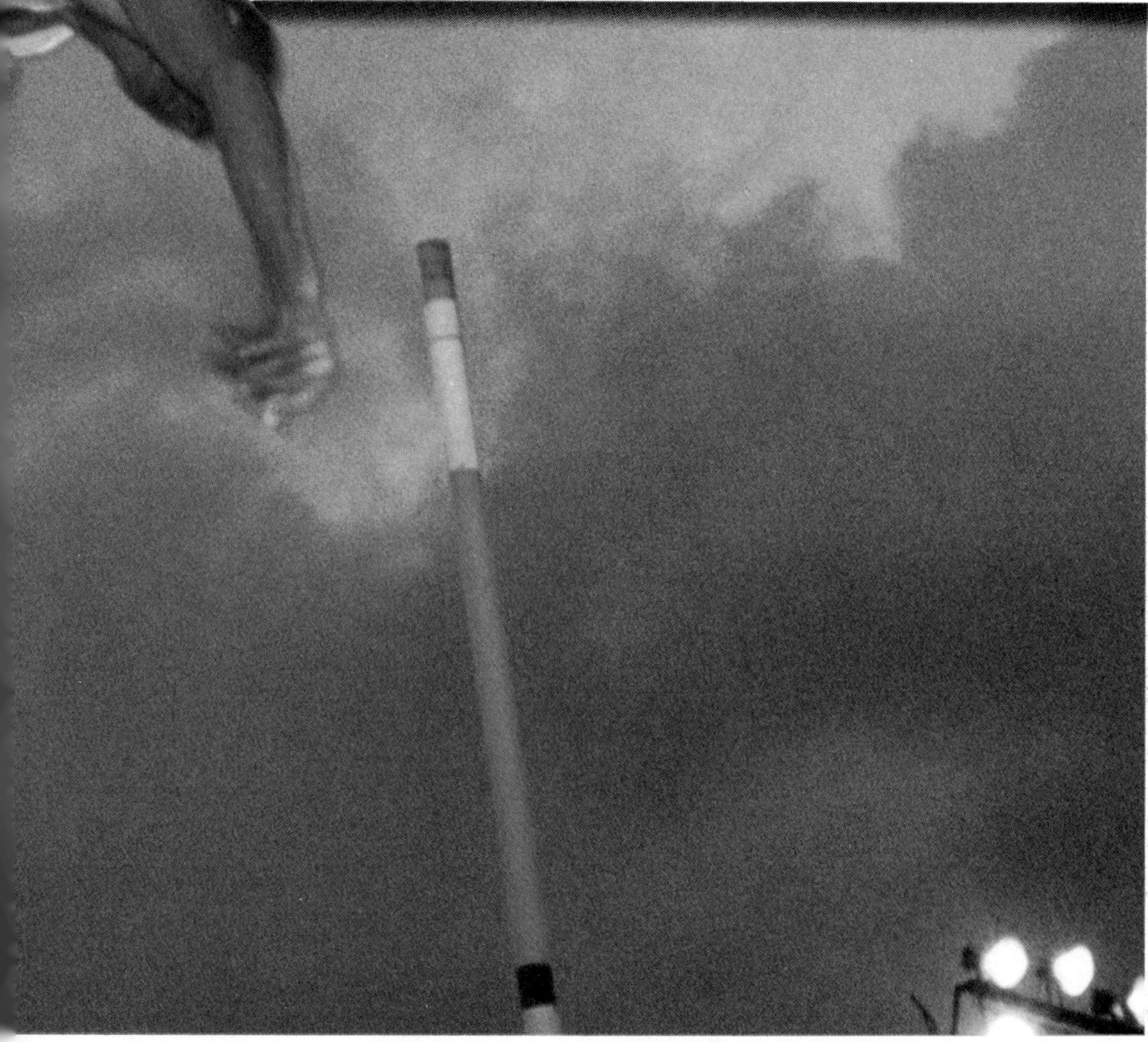

Many different real-life situations can be represented by linear equations. The linear model of the heights of the winning pole vault in the Olympic Games is but one example from sports of the usefulness of algebra.

1. Ask students, based upon the data in the graph, what they would expect the height of the winning pole vault to be in the 2000 Olympic Games to be held in Sydney, Australia. 20.17 feet
2. Discuss whether the winning height will continue to increase with successive Olympic Games. Discussion could involve factors affecting winning height: training methods, weather conditions, pole materials, limits to human abilities, etc.

Writing Linear Equations

PACING CHART

The daily Pacing Chart is meant to help you adjust your teaching pace. Students in the full course should finish the entire text by the end of the year. Students in the basic course may not complete the entire text in the school year. The Pacing Chart for each chapter contains suggestions for lessons that require more than one day and lessons that may be omitted for the basic course.

DAY	FULL COURSE	BASIC COURSE
1	5.1	5.1
2	5.2	5.2
3	5.3	5.2
4	5.4 & Using a Graphing Calculator	5.3
5	Mid-Chapter Self-Test	5.4
6	5.5	Using a Graphing Calculator
7	5.6	Mid-Chapter Self-Test
8	5.7	5.5
9	5.7	5.6
10	Chapter Review	5.7
11	Chapter Review	5.7
12	Chapter Test	Chapter Review
13		Chapter Review
14		Chapter Test

CHAPTER ORGANIZATION

LESSON	PAGES	GOALS	MEETING THE NCTM STANDARDS
5.1	236-241	1. Use the slope-intercept form to write an equation of a line 2. Model a real-life situation with a linear equation	Problem Solving, Communication, Statistics, Connections
5.2	242-247	1. Use the slope and any point on the line to write an equation of the line 2. Model a real-life situation with a linear equation	Problem Solving, Communication, Connections, Reasoning
Mixed Review	248	Review of algebra and arithmetic	
Career Interview	248	Outpatient counselor	Connections
5.3	249-254	1. Write an equation of a line given two points on the line 2. Model a real-life situation with a linear equation	Problem Solving, Communication, Connections
5.4	255-261	1. Find a linear equation that approximates a set of data points 2. Use scatter plots to determine positive, negative, or no correlation	Problem Solving, Communication, Statistics, Reasoning, Discrete Mathematics
Using a Calculator	262	Use a graphing calculator or computer to find the best-fitting line	Technology, Statistics
Mid-Chapter Self-Test	263	Diagnose student weaknesses and remediate with correlated Reteaching worksheets	
5.5	264-269	1. Transform a linear equation into standard form 2. Model a real-life situation using the standard form of a linear equation	Problem Solving, Communication, Connections, Statistics
5.6	270-276	1. Use the point-slope form to write a linear equation 2. Model a real-life situation using the point-slope form	Problem Solving, Communication, Connections, Reasoning, Geometry
Mixed Review	276	Review of algebra and arithmetic	
5.7	277-283	1. Create and use linear models to solve problems 2. Make accurate and easy-to-use linear models	Problem Solving, Communication, Connections, Statistics, Discrete Mathematics, Geometry
Chapter Summary	284	A restatement of what has been learned, why it has been learned, and how it fits into the structure of algebra	Structure, Connections
Chapter Review	285-288	Review of concepts and skills learned in the chapter	
Chapter Test	289	Diagnose student weaknesses and remediate with correlated Reteaching worksheets	

LESSON RESOURCES

MEETING INDIVIDUAL NEEDS

RETEACHING For students who need to spend more time on basics:

If a mid-chapter self-test or chapter test indicates a deficiency, teachers can help students with the appropriate ***Reteaching Copymaster.***

PRACTICE For students who need more practice:

Additional exercises like those in the Pupil's Edition are provided for each lesson in ***Extra Practice Copymasters.***

ENRICHMENT For enriching and broadening students' experiences:

Problem of the Day copymasters in ***Teaching Tools*** provide a daily opportunity to use logical reasoning, looking for a pattern, writing an equation, and other routine and non-routine problem-solving strategies.

Math Log copymasters in ***Alternative Assessment*** provide opportunities to report on investigations, research, and open-ended problems.

Enriching activities with graphing and scientific calculators and computers are provided in ***Technology: Using Calculators and Computers.***

The ***Applications Handbook*** provides additional information about the cross-curriculum topics such as astronomy, chemistry, physics, sports, economics, genetics, and music that are integrated into the Pupil's Edition.

LESSON	5.1	5.2	5.3	5.4	5.5	5.6	5.7
PAGES	236-241	242-247	249-254	255-261	264-269	270-276	277-283
Teaching Tools							
Transparencies	✓	✓	✓	✓	✓	✓	✓
Problem of the Day	✓	✓	✓	✓	✓	✓	✓
Warm-up Exercises	✓	✓	✓	✓	✓	✓	✓
Answer Masters	✓	✓	✓	✓	✓	✓	✓
Extra Practice Copymasters	✓	✓	✓	✓	✓	✓	✓
Reteaching Copymasters	Teacher-directed and independent activities tied to results on the Mid-Chapter Self-Tests and Chapter Tests						
Color Transparencies			✓	✓	✓	✓	✓
Applications Handbook	Additional background information is supplied for many real-life applications.						
Technology Handbook	Calculator and computer worksheets are supplied for appropriate lessons.						
Complete Solutions Manual	✓	✓	✓	✓	✓	✓	✓
Alternative Assessment	Assess student's ability to reason, analyze, solve problems, and communicate using mathematical language.						
Formal Assessment	Mid-Chapter Self-Tests, Chapter Tests, Cumulative Tests, and Practice for College Entrance Tests						
Computer Test Bank	Customized tests can be created by choosing from over 2000 items.						

INSIGHTS

5.1 Equations of Lines Using Slope-Intercept Form

Many real-life situations can be modeled by linear equations. For example, computing telephone bills, determining rental fees, and predicting water consumption are often linear. Perhaps the most common form of a linear equation is the slope-intercept form, which is studied in this lesson.

Students have an opportunity in this lesson to combine concepts and skills mastered in previous lessons.

5.2 Equations of Lines Given the Slope and a Point

When neither the slope nor the *y*-intercept is available, other information can be used to derive the equation of a line. This lesson presents a method of finding the equation of a line given the slope and *any* point on the line. Because the slope-intercept form of a line is so important, knowing different ways of obtaining this form of an equation of a line is essential.

5.3 Equations of Lines Given Two Points

Other information can also be used to derive the equation of a line. This lesson presents a method of finding the equation of a line given *any* two points on the line.

5.4 Exploring Data: Fitting a Line to Data

In previous lessons, it was frequently assumed that given relationships fit a linear model and could be represented by an equation of a line. In this lesson, data are organized to determine whether a line best describes the relationships under consideration.

The technique of finding a best-fitting line is used widely in education, physics, economics, and biology, among many other fields.

5.5 Standard Form of a Linear Equation

Perhaps the second most commonly used form of an equation of a line is the so-called "standard form," $Ax + By = C$, where A and B are constant coefficients and C is the constant term. This form is useful when the variables x and y represent quantities that need to be combined to produce fixed totals or amounts. This lesson focuses on transforming equations from slope-intercept form to standard form and vice-versa.

5.6 Point-Slope Form of the Equation of a Line

The slope-intercept form of a linear equation can be derived using several techniques described in previous lessons. This lesson teaches a technique that uses a distinctly different form of the equation for a line given a point on the line, namely, the point-slope form of a linear equation.

5.7 Problem Solving Using Linear Models

Algebra is important because it helps us interpret our world. This lesson illustrates more real-life problems that can be represented using a linear model. Students utilize their problem-solving skills and their understanding of when to employ the different forms of a linear equation.

Problem of the Day

Study these two diagrams representing balance scales that are in balance.

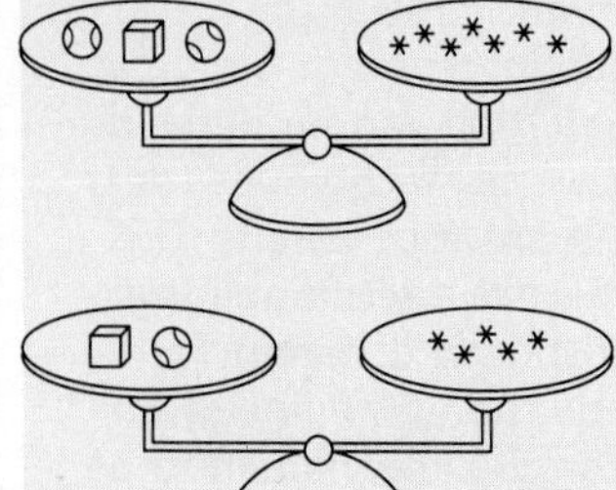

How many jacks balance one ball? 2 How many jacks balance one block? 3

ORGANIZER

Warm-Up Exercises

1. a. What is the y-intercept? The x-intercept? The value of y when $x = 0$, The value of x when $y = 0$
   b. What is the slope of a straight line? How is it determined? The $\frac{\text{rise}}{\text{run}}$ ratio
   c. How many points are needed to determine a straight line? 2
2. Find the slope of the line containing the given points.
   a. $(-3, -1)$ and $(2, 7)$ $\frac{8}{5}$
   b. $(5, 6)$ and $(-9, 1)$ $\frac{5}{14}$
   c. $(7, -4)$ and $(-8, -4)$ 0
3. Solve the equation for b. Then, evaluate b.
   a. $m = -\frac{1}{2}, x = 4, y = 3$
      $b = 5$
   b. $m = \frac{2}{3}, x = -5, y = 18$
      $b = \frac{64}{3}$
   c. $m = 5, x = 8, y = -20$
      $b = -60$

Lesson Resources

Teaching Tools
- Problem of the Day: 5.1
- Warm-up Exercises: 5.1
- Answer Masters: 5.1

Extra Practice: 5.1

5.1 Equations of Lines Using Slope-Intercept Form

What you should learn:

Goal 1 How to use the slope-intercept form to write an equation of a line

Goal 2 How to model a real-life situation with a linear equation

Why you should learn it:

You can use equations of lines to model many types of real-life situations such as cost problems (e.g., long-distance phoning), in which total cost is a flat fee plus a use fee that is based on a rate.

Goal 1 Using the Slope-Intercept Form

In Chapter 4, you learned to find the slope and y-intercept of a line whose equation is given. In this lesson, you will study the reverse process. That is, you will learn to write an equation of a line using its slope and y-intercept. To do this, you need to use the slope-intercept form of the equation of a line.

$y = mx + b$ *Slope-intercept form*

Remember that the slope of the line is m and the y-intercept is b.

Example 1 ***Writing an Equation of a Line***

Write an equation of the line whose slope is -2 and whose y-intercept is 5.

Solution From the given information about the line, you know that the slope, m, is -2 and the y-intercept, b, is 5. By substituting these values into the slope-intercept form, you obtain the following equation.

$y = mx + b$ *Slope-intercept form*
$y = -2x + 5$ *Substitute -2 for m and 5 for b.* ■

Example 2 ***Using a Graph to Write the Equation of a Line***

Write an equation of the line shown at the left.

Solution Since the line "intercepts" the y-axis at the point $(0, -4)$, the y-intercept, b, is -4. The line also passes through the point $(2, 0)$, which is four units up from, and two units to the right of the point $(0, -4)$. Therefore, the slope of the line is

$$m = \frac{\text{rise}}{\text{run}} = \frac{4}{2} = 2.$$

Knowing the slope and y-intercept, you can write the equation of the line.

$y = mx + b$ *Slope-intercept form*
$y = 2x - 4$ *Substitute 2 for m and 4 for b.* ■

Goal 2 Modeling a Real-Life Situation

In 1980, the population of California was approximately 23,668,000. During the next 10 years, the population increased by approximately 617,100 people per year. (Source: U.S. Census Bureau)

Example 3 *A Linear Model for the Population of California*

Write a linear model for the population of California between 1980 and 1990. Sketch a graph of the linear model.

Solution

Verbal Model: Population of California = 1980 population + Increase per year • Number of years from 1980

Labels:
Population of California = P (people)
1980 population = 23,668,000 (people)
Increase per year = 617,100 (people per year)
Number of years from 1980 = t (years)

Equation: $P = 23{,}668{,}000 + 617{,}100t$ **Linear model**

In slope-intercept form, this equation is

$$P = 617{,}100t + 23{,}668{,}000.$$

The graph of this model is shown below. Notice that the P-intercept represents the 1980 population, and the slope represents the rate of change in the population (in people per year).

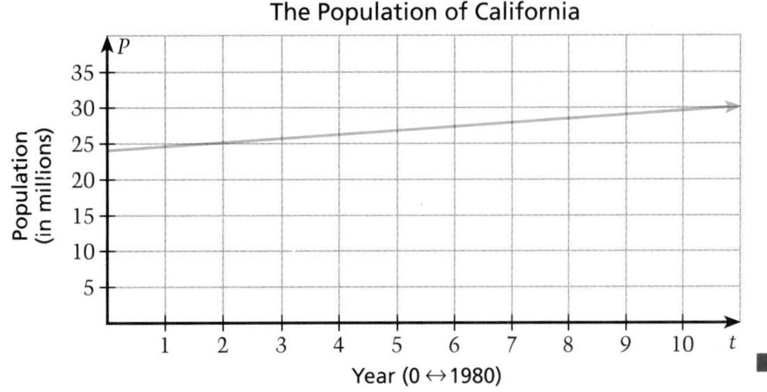

LESSON Notes

Examples 1–2 Stress the ease with which the equation of a line can be written, provided its slope and y-intercept are known.

Examples 3–4

Since the applications in this chapter involve writing linear equations, the general *algebraic* model described in the problem-solving plan has been replaced by a *linear* model.

Example 3

Emphasize that the slope of the linear model is the rate of change in the California population. You may wish to point out that a t-value of -3 represents the year 1977. Students should be told that rates of changes are typically real-life applications of the concept of slope.

Indicate to students that the linear model for the population of California is only one of several possible models.

Example 4

A linear model is used to describe the cost of long-distance calls. In practice, telephone companies charge the same for 8.5 minutes as they do for 9 minutes. Nevertheless, the linear model is a useful way to think of the cost of placing phone calls.

Extra Examples

Here are additional examples similar to **Examples 1–4.**

1. Write an equation of the line whose slope m and y-intercept b are given.
   a. $m = 8, b = -3$
   $y = 8x - 3$
   b. $m = -\frac{2}{3}, b = 6$
   $y = -\frac{2}{3}x + 6$

2. Write an equation for each line graphed below.

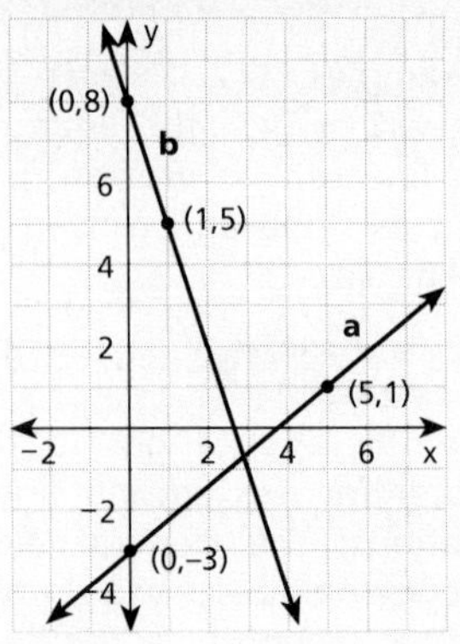

a. $y = \frac{4}{5}x - 3$, b. $y = -3x + 8$

3. The Tri-State car rental agency charges a flat fee of $25 plus $0.17 per mile to rent a subcompact car on weekends. Write a linear model for the cost in terms of the number of miles driven. Cost $= 0.17x + 25$

Check Understanding

1. If the rate of change for a real-life situation is given as 34,500 people per year, how should the x-axis and y-axis be labeled? x-axis: Time in years; y-axis: Population
2. Describe a real-life situation that can be modeled by a linear equation.
3. Suppose the Tri-State car rental agency charges a flat fee of $25 plus $0.17 per mile, with 100 free miles, to rent a subcompact car on weekends. Write a linear model for the cost in terms of the number of miles driven. $y = 0.17(x - 100) + 25$ for $x \geq 100$ miles; $y = 25$ for $x < 100$ miles.

Example 4 *Writing a Linear Model for Phone Charges*

Write a linear model for the cost of a long-distance call that costs $0.75 for the first minute and $0.50 for each additional minute.

Problem Solving
Verbal Model

Solution

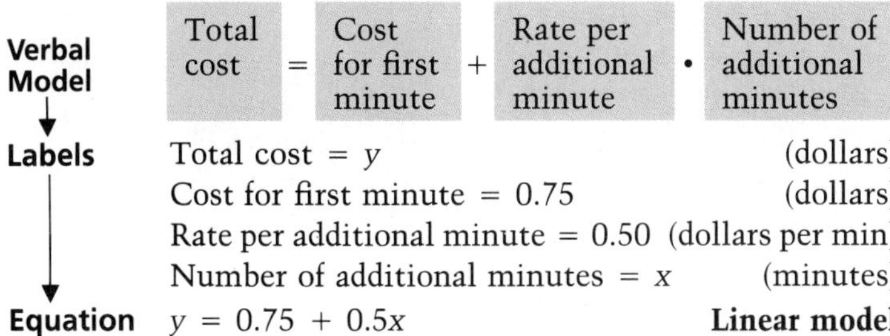

Verbal Model Total cost = Cost for first minute + Rate per additional minute • Number of additional minutes

Labels
Total cost $= y$ (dollars)
Cost for first minute $= 0.75$ (dollars)
Rate per additional minute $= 0.50$ (dollars per min)
Number of additional minutes $= x$ (minutes)

Equation $y = 0.75 + 0.5x$ **Linear model**

The slope of 0.5 is the cost per additional minute, and the y-intercept is the cost of the first minute. ■

This phone bill was generated by a computer, using the equation given in Example 4.

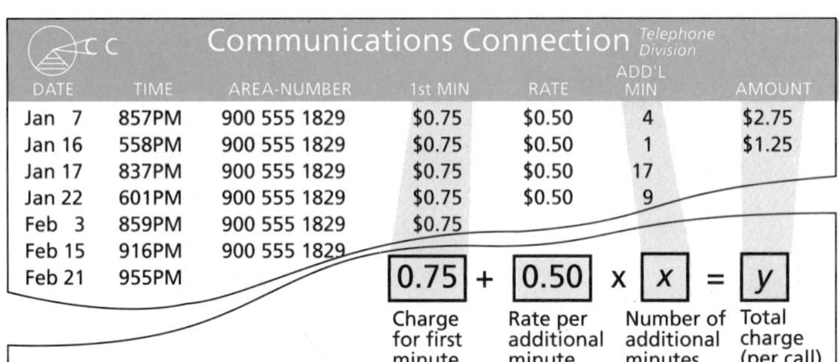

CC Communications Connection Telephone Division

DATE	TIME	AREA-NUMBER	1st MIN	RATE	ADD'L MIN	AMOUNT
Jan 7	857PM	900 555 1829	$0.75	$0.50	4	$2.75
Jan 16	558PM	900 555 1829	$0.75	$0.50	1	$1.25
Jan 17	837PM	900 555 1829	$0.75	$0.50	17	
Jan 22	601PM	900 555 1829	$0.75	$0.50	9	
Feb 3	859PM	900 555 1829	$0.75			
Feb 15	916PM	900 555 1829				
Feb 21	955PM					

0.75 + 0.50 x x = y

Charge for first minute; Rate per additional minute; Number of additional minutes; Total charge (per call)

Communicating about ALGEBRA

SHARING IDEAS about the Lesson

Compare with Linear Models The cost for calling a number in a city different from the one above is $1.25 for the first minute and $0.45 for each additional minute.

A. Write a linear model for the total cost of calling this second number. $y = 0.45x + 1.25$

B. Which would cost more: a seven-minute call to the first number (Example 4) or a seven-minute call to the second number?

C. Which would cost more: a 12-minute call to the first number or a 12-minute call to the second number?

EXERCISES

Guided Practice

CRITICAL THINKING about the Lesson

1. What is the name of the $y = mx + b$ form of an equation of a line? Slope-intercept form
2. Is 5 the x-intercept or y-intercept of the line $y = 3x + 5$? y-intercept
3. What is the slope and y-intercept of the line $y = -x$? $-1, 0$
4. Write an equation of the line whose slope is -7 and whose y-intercept is $-\frac{2}{3}$. $y = -7x - \frac{2}{3}$
5. Use the model given in Example 3 to estimate the population of California in 1994. 32,307,400

Renting a Bike **In Exercises 6–8, Ed's Rentals charges a flat fee of $11 plus $1.50 per hour to rent a motorbike.**

6. Write an equation for the cost, y, of renting the motorbike for x hours. $y = 1.50x + 11$
7. Sketch the graph of the cost equation. Label the y-intercept on the graph. See margin.
8. Use the equation to find the cost of renting the motorbike for 12 hours. $29

Independent Practice

In Exercises 9–14, write an equation of the line. 9. $y = 3x - 2$

9. The slope is 3; the y-intercept is -2.
10. The slope is 1; the y-intercept is 2. $y = x + 2$
11. The slope is -1; the y-intercept is 3.
12. The slope is -2; the y-intercept is 0.
13. The slope is $\frac{3}{2}$; the y-intercept is 3. $y = \frac{3}{2}x + 3$
14. The slope is $-\frac{1}{4}$; the y-intercept is 1. $y = -\frac{1}{4}x + 1$

In Exercises 15–18, write an equation of the line shown in the graph. See below.

15. 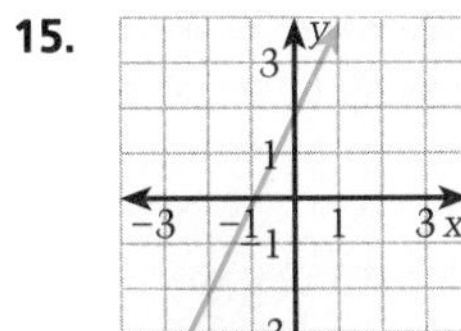

16. 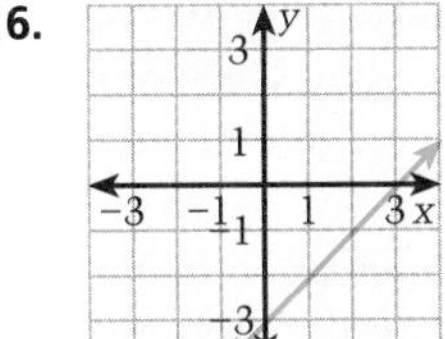

17. 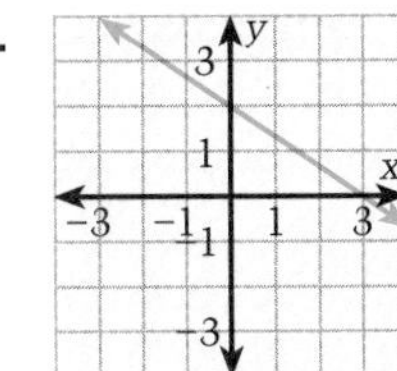

18. 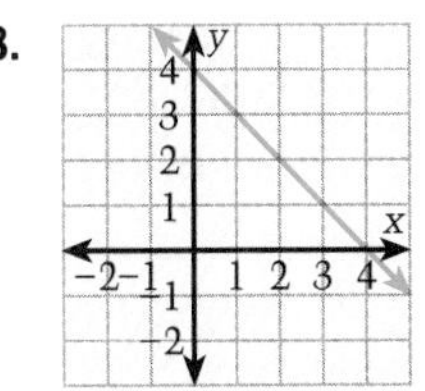

Parallel Lines **In Exercises 19 and 20, the lines are parallel. Write an equation for each.**

19. 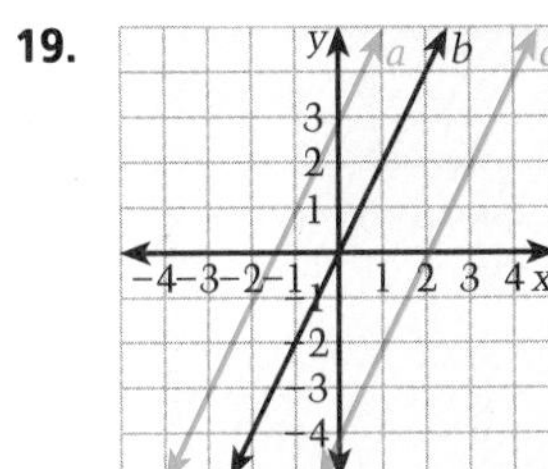

a: $y = 2x + 3$
b: $y = 2x$
c: $y = 2x - 4$

20. 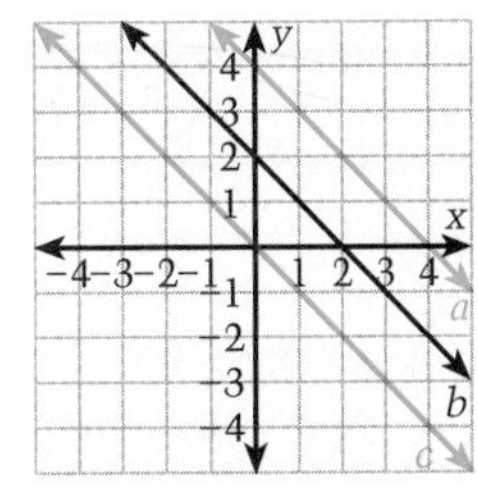

a: $y = -x + 4$
b: $y = -x + 2$
c: $y = -x$

11. $y = -x + 3$ 12. $y = -2x$ 15. $y = 2x + 2$ 16. $y = x - 3$
17. $y = -\frac{2}{3}x + 2$ 18. $y = -x + 4$

Communicating about ALGEBRA

In Example 4, it was noted that the telephone company would charge the same cost for 8.5 minutes as it would for 9 minutes.

a. Discuss what this means for the actual cost of phone calls compared to the linear model given in Example 4. For noninteger values of x, the actual cost of the call is greater than the model cost.

b. How should this latter method of billing telephone calls be graphed? Use a step graph like the one shown in Example 3 on page 177.

EXERCISE Notes

ASSIGNMENT GUIDE

Basic/Average: Ex. 1–8, 9–19 odd, 25–26, 31–41 odd, 47–48

Above Average: Ex. 1–8, 10–22 even, 24, 27–28, 31–45 odd, 47–48 odd, 49-52

Advanced: Ex. 1–8, 10–20 even, 25–28, 32–46 even, 47–52

Selected Answers
Exercises 1–8, 9–47 odd

✪ More Difficult Exercises
Exercises 23, 25, 29–30, 51–52

Guided Practice

▶ **Ex. 1–8** COOPERATIVE LEARNING A good way to reinforce the concepts and skills of this lesson is to have students complete the exercises in small cooperative learning groups.

Answer

7. 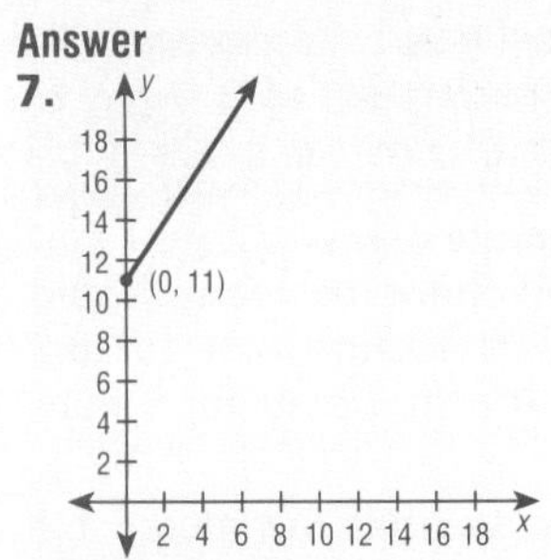

Independent Practice

▶ **Ex. 15–18** Caution students to use points on the graph that can be easily read when determining the slope of the line.

▶ **Ex. 19–20** You may wish to refer students to Example 2 in Lesson 4.5.

EXTEND Ex. 25 RESEARCH Use the 1980 census data for your state (obtained from an almanac or encyclopedia) and write a linear model of the population in your state based upon this information. Compare the 1990 census data with that predicted by your model.

Answer

27. $y = -50t + 200$, as the time increases the distance from home decreases.

▶ **Ex. 29–30** Students may have trouble going from the written data to the graphs on their own.

21. *Renting a Compact Car* The Star Car Rental Company charges a flat fee of $30 plus 25¢ per mile to rent a compact car. Write a linear model that gives the total cost in terms of the number of miles driven. $y = 0.25x + 30$

22. *Renting a Full-Size Car* The Maple Car Rental Company charges a flat fee of $45 plus 35¢ per mile to rent a full-size car. Write a linear model that gives the total cost in terms of the number of miles driven. $y = 0.35x + 45$

23. Use the equation you found in Exercise 21 to complete the table.

Miles (x)	25	50	75	100
Cost (y)	?	?	?	?

36.25, 42.50, 48.75, 55.00

24. Use the equation you found in Exercise 22 to complete the table.

Miles (x)	25	50	75	100
Cost (y)	?	?	?	?

53.75, 62.50, 71.25, 80.00

25. *Population of South Carolina* In 1980, the population of South Carolina was approximately 3,122,000. During the next ten years, the population increased by approximately 38,400 people per year. Write a linear model that gives the population, P, of South Carolina in terms of the year, t. Let $t = 0$ correspond to 1980. *(Source: U.S. Census Bureau)* $P = 38{,}400t + 3{,}122{,}000$

26. Use the equation you found in Exercise 25 to estimate the population of South Carolina in 1986. 3,352,400

27. *Traveling Home* At noon, you are 200 miles from home. You are traveling home at 50 miles per hour. Write an equation that gives your distance from home, y, in terms of the time, t. Let $t = 0$ correspond to noon. Why does the line given by this equation have a negative slope? See margin.

28. Use the equation you found in Exercise 27 to complete the table. 150, 100, 50, 0

Time (t)	1	2	3	4
Miles (y)	?	?	?	?

Water Conservation **In Exercises 29 and 30, use the following information.**

A standard toilet uses approximately 5 gallons of water each time it is flushed. Installing a water dam (a device costing about $10) reduces water usage by approximately 50%. Consider a city that has 100,000 households, each with toilets that are flushed an average of 15 times per day.

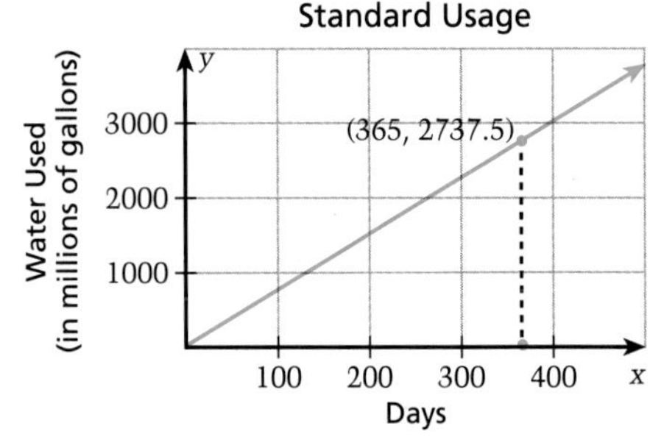

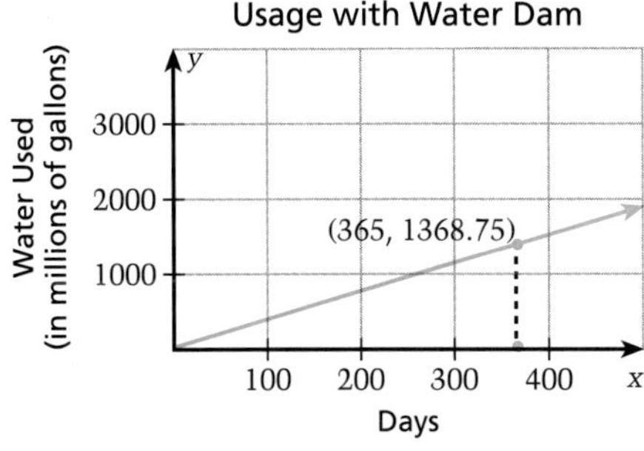

29. Write a linear model for the city's water usage without water dams. How much water is used in one year? $y = 7{,}500{,}000x$; 2,737,500,000 gallons per year

30. Write a linear model for the city's water usage with water dams. How much water was saved in one year? $y = 3{,}750{,}000x$; 1,368,750,000 gallons per year

Integrated Review

In Exercises 31–36, find the slope and the y-intercept of the line. Then sketch the line. See Additional Answers.

31. $y = -3x + 6$ $-3, 6$

32. $y = 2x - 5$ $2, -5$

33. $y + 2x = 2$ $-2, 2$

34. $-y + 3x = -5$ $3, 5$

35. $2y - 3x = 6$ $\frac{3}{2}, 3$

36. $2y + 4x = 6$ $-2, 3$

In Exercises 37–42, determine whether the line is horizontal or vertical. Then sketch the line. See Additional Answers.

37. $y = -2$ Horizontal

38. $x = 3$ Vertical

39. $x = 4$ Vertical

40. $y = -5$ Horizontal

41. $2y = 6$ Horizontal

42. $3x = -6$ Vertical

43. Can a horizontal line have an equation in slope-intercept form? Yes

44. Can a vertical line have an equation in slope-intercept form? No

45. Write an equation of the horizontal line whose y-intercept is 3. $y = 3$

46. Write an equation of the vertical line whose x-intercept is -4. $x = -4$

In Exercises 47 and 48, match the description with the linear model $y = 10$ or the linear model $y = 10x$.

47. A rental company charges \$10 per hour to rent a sailboard. (There is no flat fee.) $y = 10x$

48. A rental company charges a flat fee of \$10 to rent an electric drill. (There is no hourly charge.) $y = 10$

Exploration and Extension

49–50. Real-life problems vary.

In Exercises 49 and 50, write an equation of the line shown in the graph. Then describe a real-life problem that could be represented by the equation.

49.

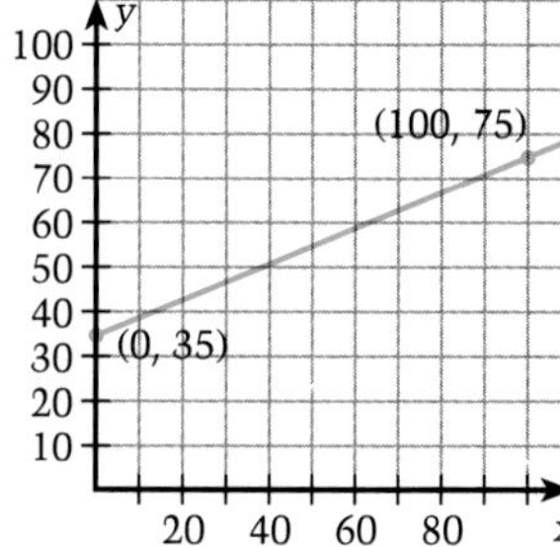

$y = 0.4x + 35$

50.

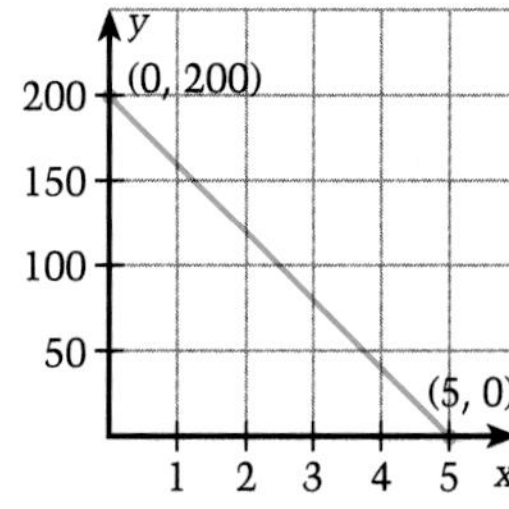

$y = -40x + 200$

51. Write an equation of the line that has the same slope as $y = -\frac{2}{5}x + 1$ and the same y-intercept as $y = 2x - 3$. Then sketch all three lines. $y = -\frac{2}{5}x - 3$

52. Write an equation of the line that has the same slope as $y = \frac{3}{2}x - 4$ and the same y-intercept as $y = -\frac{2}{3}x + 3$. Then sketch all three lines. $y = \frac{3}{2}x + 3$ See margin.

Ex. 51–52 Observe that the lines with the same slope are parallel lines. Ask students if parallel lines (in the same plane) always have the same slope.

Answers

51.

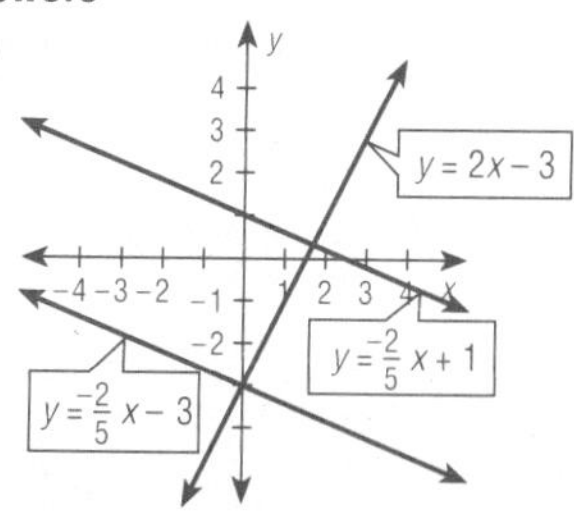

52.

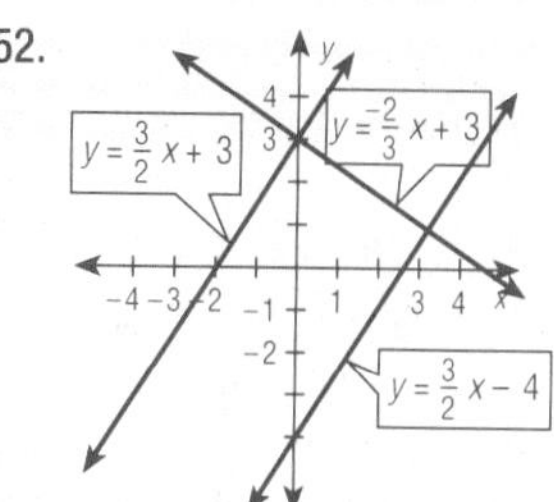

Enrichment Activities

These activities require students to go beyond lesson goals.

1. CONNECTIONS Use texts from other classes or the encyclopedia. Identify other models that could be used to describe the population of a state. Compare your models with the model in Example 3. See Chapters 8–10 for examples of exponential and quadratic population models.
2. Which model do you prefer? Why?

Problem of the Day

Study the pattern to help guess the values of e and f.

$9a = 81$ $\quad e = 999{,}999$

$99b = 9801$ $\quad f = 999{,}999$

$999c = 998{,}001$

$9999d = 99{,}980{,}001$

.
.
.

$e \cdot f = 999{,}998{,}000{,}001$

ORGANIZER

Warm-Up Exercises

1. Solve for b, given values for m and x.
   a. $4 = mx + b,\ m = 3,\ x = 3$
   b. $-8 = mx - b,\ m = -2,\ x = -11$
   c. $\frac{2}{3} = mx + b,\ m = 5,\ x = 2$

   a. $b = -5$, b. $b = 30$, c. $b = -\frac{28}{3}$
2. Solve for b, given values for m, x and y.
   a. $y = mx + b,\ m = 3,\ x = -2,\ y = 12$
   b. $y = mx + b,\ m = 9,\ x = 5,\ y = 1$
   c. $y = mx + b,\ m = \frac{3}{4},\ x = -8,\ y = -2$

   a. $b = 18$, b. $b = -44$, c. $b = 4$

Lesson Resources

Teaching Tools
  Problem of the Day: 5.2
  Warm-up Exercises: 5.2
  Answer Masters: 5.2
Extra Practice: 5.2

5.2 Equations of Lines Given the Slope and a Point

What you should learn:

Goal 1 How to use slope and any point on a line to write an equation of the line

Goal 2 How to model a real-life situation with a linear equation

Why you should learn it:

You can model a real-life situation with a linear equation when you are given the rate of change taking place (the slope) and any pair of related values (one point on the line).

Goal 1 Using Slope and a Point on a Line

In Lesson 5.1, you learned how to write an equation of a line when you are given the slope of the line and the y-intercept of the line. You used the slope-intercept form of the equation of a line

$y = mx + b.$ *Slope-intercept form*

In this lesson, the use of this form is extended by showing how it can help find an equation of a line when you are given the slope of the line and *any* point on the line. Here is an example.

Example 1 ***Writing an Equation of a Line***

Write an equation of the line that passes through the point $(6, -3)$ with a slope of -2.

Solution The first step is to use the given information to find the y-intercept of the line. Since the line has a slope, m, of -2 and passes through the point $(x, y) = (6, -3)$, you can substitute the values $m = -2$, $x = 6$, and $y = -3$ into the slope-intercept form and solve for b.

$y = mx + b$	*Slope-intercept form*
$-3 = (-2)(6) + b$	*Substitute -2 for m, 6 for x, and -3 for y.*
$-3 = -12 + b$	*Simplify.*
$9 = b$	*Solve for b.*

Thus, the y-intercept, b, is 9. Now you know both the slope and the y-intercept, and you can find an equation of the line using the slope-intercept form.

$y = mx + b$	*Slope-intercept form*
$y = -2x + 9$	*Substitute -2 for m and 9 for b.*

Graphic Check The graph of the line is shown at the left. Note that the line crosses the y-axis at the point $(0, 9)$ and passes through the given point $(6, -3)$. ■

y
(0, 9)
9
7
5
3
1
-1
-3
-5
-4 -2 2 4 6 8 x
(6, -3)

Writing an Equation of a Line Given Its Slope and a Point

To write an equation of a nonvertical line given its slope and a point on the line, use the following steps.

1. *Find the y-intercept:* Substitute the slope, m, and the coordinates of the given point (x, y) into the slope-intercept form, $y = mx + b$. Then solve for the y-intercept, b.
2. *Write an equation of the line:* Substitute the slope, m, and the y-intercept, b, into the slope-intercept form, $y = mx + b$.

Goal 2 Modeling a Real-Life Situation

Between 1980 and 1990, the number of vacation trips taken by Americans increased by about 15 million per year. In 1985, Americans went on 340 million vacation trips. (Source: U.S. Travel Data Center)

Example 2 A Linear Model for Vacation Travel

Find an equation that gives the number of vacation trips, y (in millions), in terms of the year, t. Let $t = 0$ correspond to 1980.

Solution Because the number of trips changed at a constant rate each year, you can model this problem using the linear equation $y = mt + b$. The constant rate is 15 million trips per year, so you know that the slope, m, is 15 (in millions). You also know an ordered pair, (5, 340), that represents a point on the line. (Why?)

$y = mt + b$	*Slope-intercept form*
$340 = (15)(5) + b$	*Substitute 15 for m, 5 for t, and 340 for y.*
$340 = 75 + b$	*Simplify.*
$265 = b$	*Solve for b.*

The slope-intercept form of the equation is $y = 15t + 265$. ■

LESSON Notes

Example 1

An important point to make with students is that in algebra the given information in a problem is not necessarily enough to complete the task. Sometimes you need to transform the given information in order to apply a formula or a technique you want to use. In this example, you are given the slope of the line and a point on the line. You use the information about a point on the line to get the y-intercept value. Since you now know both the slope and the y-intercept, you can write the equation of the line.

Example 2

This example illustrates how real data may be presented by finding the equation of a line when its slope and a point on the line are known.

Example 3

Point out to students that the linear model in this example does not explicitly take into account many factors that could affect the number of vacations that Americans actually take. (See Communicating about Algebra on page 244.)

Extra Examples

Here are additional examples similar to **Example 1.**

Write an equation of the line, given a point on the line and its slope.

1. point (8, 3) with slope 4
2. point (−5, 25) with slope −6
3. point (6, 7) with slope $\frac{2}{3}$

1. $y = 4x - 29$
2. $y = -6x - 5$
3. $y = \frac{2}{3}x + 3$

Check Understanding

1. Given a point P on a line with slope m, you can use the fact that $m = \frac{\text{rise}}{\text{run}}$ to graph the line. Explain. From the point P, move left or right by the magnitude of the run and move up or down by the magnitude of the rise to obtain a second point, Q, on the line. Joining points P and Q gives you a graph of the line.
2. Is the graphing technique described above sufficient to determine the equation of the line? Explain. Not really! The point at which the line crosses the y-axis cannot always be read exactly from the graph.

Communicating about ALGEBRA

Answer

D. No, extending the graph to the left would decrease the number of vacation trips to zero.

EXTEND *Communicating*
COOPERATIVE LEARNING

Ask student partners to identify, discuss, and write about the factors that could affect the linear model of vacation travel presented in Examples 2 and 3. Some factors are the condition of the economy, whether the country is at war or at peace, and the seasonal weather over long periods of time.

You can use the equation in Example 2 to predict the number of vacation trips that Americans will take in a given year.

Example 3 *Using a Linear Model to Predict Vacation Travel*

About how many vacation trips will Americans take in 1998?

Solution To answer this question, use the linear model from Example 2: $y = 15t + 265$. Since 1980 corresponds to $t = 0$, it follows that 1998 must correspond to $t = 18$.

$y = 15t + 265$	*Given model*
$y = 15(18) + 265$	*Substitute 18 for t.*
$y = 535$	*Trips in 1998*

Thus, you can estimate that Americans will take about 535 million vacation trips in 1998. The graph is shown below.

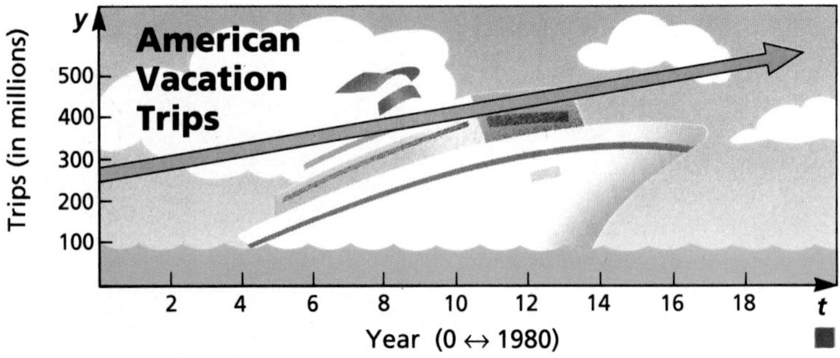

Communicating about ALGEBRA

SHARING IDEAS about the Lesson

Estimate Use the linear model in Example 3.

A. How many vacation trips were taken by Americans in 1990? 415 million

B. How many vacation trips will be taken by Americans in 1997? 520 million

C. During which year would there be about 400 million vacation trips? 1989

D. In Example 3, the model $y = 15t + 265$ was developed, based on travel data between 1980 and 1990. Do you think this model is valid for earlier years such as 1950 or 1960? Explain your reasoning. See margin.

EXERCISES

Guided Practice

CRITICAL THINKING about the Lesson

1. Describe the strategy used to find an equation of a line, given its slope and a point on the line. Sketch a flowchart for the steps. See margin.
2. When finding an equation of the line that passes through the point $(-2, 4)$ with a slope of $\frac{1}{3}$, begin by substituting values for m, x, and y into the slope-intercept form, $y = mx + b$. Which value should you substitute for m? Which value for x? Which value for y? What will these substitutions permit you to find? $\frac{1}{3}$, -2, 4, b

In Exercises 3–6, write an equation of the line that passes through the point and has the given slope. Give the equation in slope-intercept form.

3. $y = -2x + 5$
4. $y = 7$
5. $y = \frac{1}{2}x - 6$

3. $(1, 3)$, $m = -2$ 4. $(-6, 7)$, $m = 0$ 5. $(8, -2)$, $m = \frac{1}{2}$ 6. $(\frac{1}{2}, -\frac{3}{2})$, $m = 9$ $y = 9x - 6$

7. *Population of Naples* Between 1980 and 1990, the population of Naples, Florida, increased by approximately 6000 people per year. In 1985, the population of Naples was 116,000. Find an equation that gives the population, y, of Naples in terms of the year, x. Let $x = 0$ correspond to 1980. ***(Source: Naples Chamber of Commerce)*** $y = 6000x + 86{,}000$
8. Estimate the population of Naples in 1990. 146,000

Independent Practice

13. $y = -2x - 9$ 14. $y = 4x$
15. $y = \frac{1}{2}x - 4$ 16. $y = \frac{1}{3}x + 7$

In Exercises 9–20, write an equation of the line that passes through the point and has the given slope. Write the equation in slope-intercept form.

9. $(-3, 6)$, $m = 2$ $y = 2x + 12$
10. $(3, 2)$, $m = 1$ $y = x - 1$
11. $(4, -2)$, $m = -1$ $y = -x + 2$
12. $(3, 1)$, $m = -3$ $y = -3x + 10$
13. $(-2, -5)$, $m = -2$
14. $(1, 4)$, $m = 4$
15. $(4, -2)$, $m = \frac{1}{2}$
16. $(-6, 5)$, $m = \frac{1}{3}$
17. $(0, -1)$, $m = 3$ $y = 3x - 1$
18. $(0, 4)$, $m = 2$ $y = 2x + 4$
19. $(2, 5)$, $m = 0$ $y = 0 \cdot x + 5$
20. $(1, -3)$, $m = 0$ $y = 0 \cdot x - 3$

In Exercises 21–26, write the slope-intercept form of the equation of the line.

21. The line has a slope of $\frac{2}{3}$ and passes through the point $(-3, 4)$. $y = \frac{2}{3}x + 6$
22. The line has a slope of $-\frac{1}{4}$ and passes through the point $(8, 3)$. $y = -\frac{1}{4}x + 5$

23.
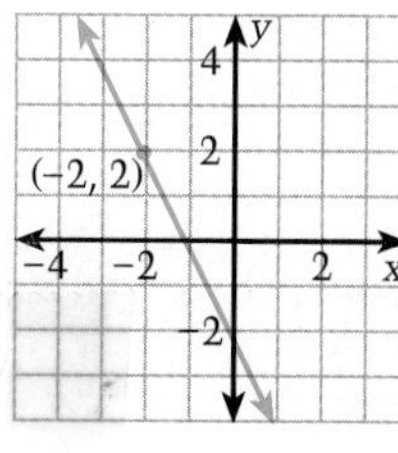

$y = -2x - 2$

24.
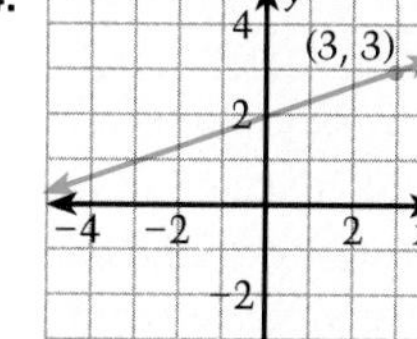

$y = \frac{1}{3}x + 2$

25.
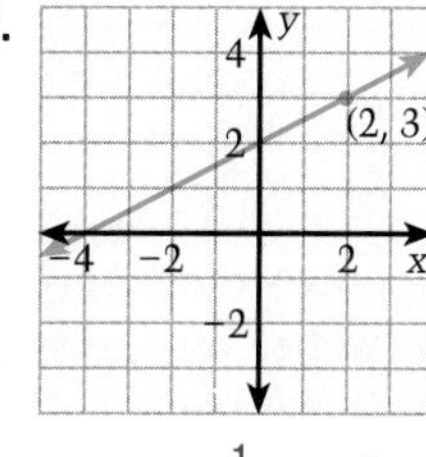

$y = \frac{1}{2}x + 2$

26.
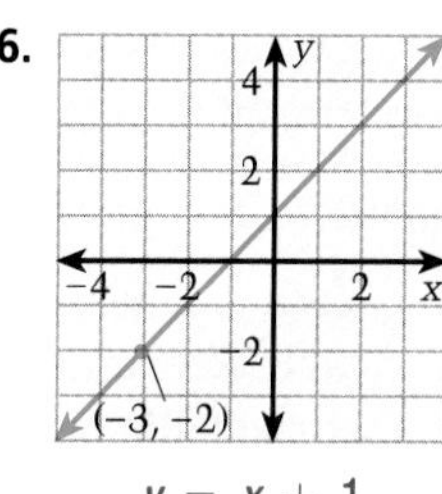

$y = x + 1$

EXERCISE Notes

ASSIGNMENT GUIDE

Basic/Average: Ex. 1–8, 9–27 odd, 29, 31, 35, 37–45 odd, 47–49

Above Average: Ex. 1–8, 9–27 odd, 31–35, 39–45 multiples of 3, 47–51

Advanced: Ex. 1–8, 9–19 odd, 20–32 even, 33–36, 39–45 odd, 47–51

Selected Answers
Exercises 1–8, 9–45 odd

Use **Mixed Review** as needed.

More Difficult Exercises
Exercises 33–34, 47–51

Guided Practice

Ex. 1 Ask students to compare flowcharts and to provide explanations for any differences that are found.

1. Sample answer:

Substitute the slope and the coordinates of the point into the slope-intercept form.

↓

Solve for the y-intercept.

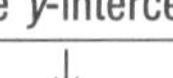

Substitute the slope and the y-intercept into the slope-intercept form.

Independent Practice

Ex. 9-20 You might wish to reserve the even-numbered exercises to use as additional practice at a later time.

Ex. 23-26 Caution students to identify carefully the x-axis and the y-axis in each exercise.

EXTEND Ex. 29–32

CONNECTIONS Each of the situations in these exercises has constraints limiting the increases that can be realized over time. Ask students to discuss what possible constraints might exist for each and what this means for the linear models used to describe the situations. For example, attendance at hockey games is limited by the size of the stadiums where games are played and the number of teams in the league. Cable subscriptions often reach a saturation point in a given area. Increases occur because more areas are included in the system. Greg will probably reach a plateau in his running. He cannot keep running forever. Christina will eventually run out of money.

▶ **Ex. 33–34** PROBLEM SOLVING Students might need help in interpreting the diagram and the given data.

Enrichment Activity

This activity requires students to go beyond lesson goals.

CONNECTIONS Identify situations that you, your family members, or their friends could describe using a linear model. See Exercises 29–32 for some ideas.

27. Find an equation of the line that crosses the x-axis at (3, 0) with a slope of $-\frac{1}{3}$. $27.\ y = -\frac{1}{3}x + 1$

28. Find an equation of the line that crosses the x-axis at (−2, 0) with a slope of 2. $y = 2x + 4$

29. *Hockey Attendance* Between 1980 and 1990, the attendance at National Hockey League games increased by about 1,100,000 people per year. In 1989, 20,434,000 people attended National Hockey League games. Find an equation that gives the attendance, y, at National Hockey League games in terms of the year, x. Let $x = 0$ correspond to 1980. *(Source: National Hockey League)* $y = 1{,}100{,}000x + 10{,}534{,}000$

30. *Cable Television Subscribers* Between 1980 and 1990, the number of cable television subscribers increased by about 3600 per year. In 1987, there were 41,100 cable television subscribers. Find an equation that gives the number, y, of cable television subscribers, in terms of the year, x. Let $x = 0$ correspond to 1980. *(Source: Television and Cable Factbook)* $y = 3600x + 15{,}900$

31. *Fitness Program* After 8 weeks on a fitness program, Greg can jog 30 miles a week. His average mileage gain has been 2 miles per week. Find an equation that gives Greg's mileage, m, in terms of the number, n, of weeks that he stays on the program. $m = 2n + 14$

32. *Savings Plan* Christina withdraws $7.50 per week from her savings account. After 12 weeks, the balance in her account is $452. Write an equation that gives the balance, y, of Christina's account in terms of the number of weeks, x. (Do not consider any interest earned by the account.) $y = -7.50x + 542$

Ancient Flour Mill **In Exercises 33 and 34, use the following information.**

The drawing at the right shows Roman flour mills on a hill near Barbégal in southern France. The mills are now in ruins, but during the fourth century A.D. this mill complex could grind enough flour to feed a population of 12,500 people. Water from the top of the hill flowed over 8 pairs of waterwheels, each powering grindstones. The angle of the hill is about 27°, which means that the slope of the hill is $\frac{1}{2}$. Each mill house was 15 feet by 8 feet. *(Source: Scientific American)*

33. Let the base of the left stairs represent the origin. Write a linear model for the line formed by the stairs. $y = \frac{1}{2}x$

34. Use the linear model to find the height of the stairs. 60 ft

35. *Phone Call* Suppose the cost of a telephone call between New York and Philadelphia is $0.46 for the first minute and $0.31 for each additional minute. Write an equation that gives the total cost of the call, C, in terms of the length of the call, t. $C = 0.46 + 0.31(t - 1)$

36. *Taxi Ride* Use the information given with the photo to find an equation that gives the total cost of a taxi ride, y, in terms of the number of miles, x.
$y = 2.50 + 1.30(x - 1)$
or $y = 1.30x + 1.20,\ x \geq 1$

The cost of a taxi ride in Detroit is $2.50 for the first mile plus $1.30 for each additional mile.

Integrated Review

In Exercises 37–40, solve the equation for b.

37. $-2 = \frac{3}{2}(-4) + b$ 4

38. $4 = \left(-\frac{2}{3}\right)(-6) + b$ 0

39. $\frac{3}{2} = \left(-\frac{1}{6}\right)(3) + b$ 2

40. $-\frac{2}{3} = \left(-\frac{1}{6}\right)(2) + b$ $-\frac{1}{3}$

In Exercises 41–44, sketch the line that passes through the given point and has the given slope. Use the graph to estimate the y-intercept. Then find the actual y-intercept. See margin.

41. $(2, 4),\ m = -\frac{1}{2}$ 5

42. $(4, 8),\ m = -2$ 16

43. $(3, 6),\ m = 2$ 0

44. $(4, 0),\ m = \frac{1}{2}$ -2

45. Find an equation of the line that passes through the point $(-1, 4)$ and is parallel to the line $y = -2x + 5$. $y = -2x + 2$

46. Find an equation of the line that passes through the point $(2, 6)$ and is parallel to the line $y = x - 3$. $y = x + 4$

Exploration and Extension

Fund-Raising **In Exercises 47–51, use the following information.**

The school math club decides to have a fund-raising project. The club rents the movie *The Life and Times of René Descartes* for $19.75, and charges an admission fee of $0.25 per person.

47. Find an equation that gives the profit, P, in terms of the number of people, n, who attend the movie. $P = 0.25n - 19.75$

48. Sketch the graph of the equation you found in Exercise 47. See margin.

49. Which values of n are not relevant to this linear model? (Hint: For instance, $\frac{1}{2}$ of a person could not attend.) Numbers that are not whole numbers

50. Is it possible that the club will not make a profit? How many people must attend the movie so that the club will get its $19.75 back? Yes, 79

51. After counting the receipts, the club members found that they earned a profit of $40.75. How many people attended the movie? 242

Answers

41.

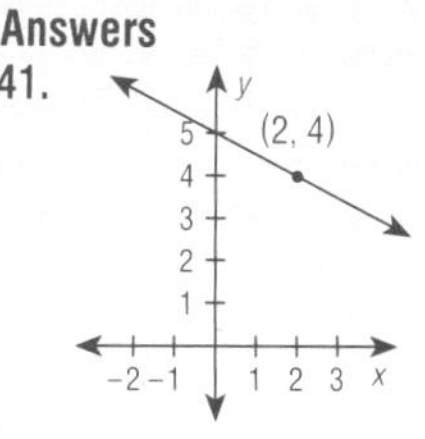

42.

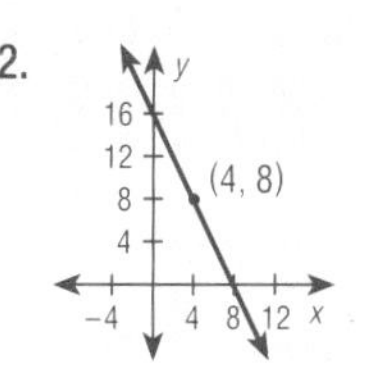

43.

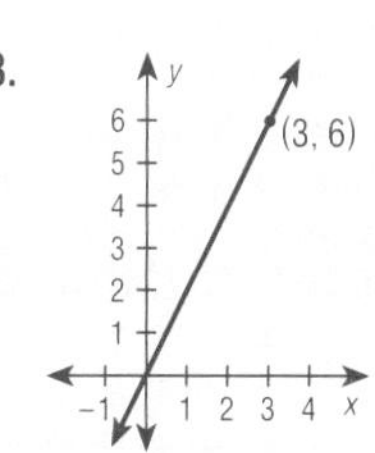

44.

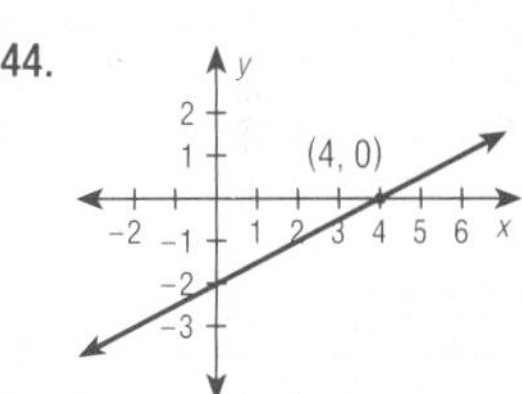

48.

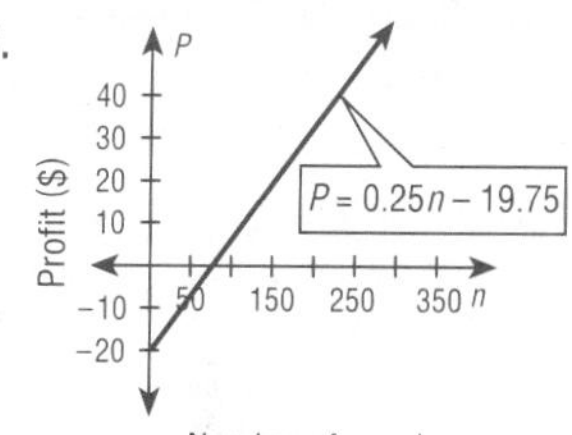

Answers

15.
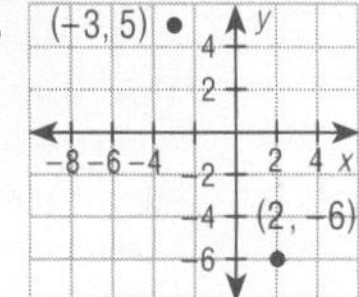

16.
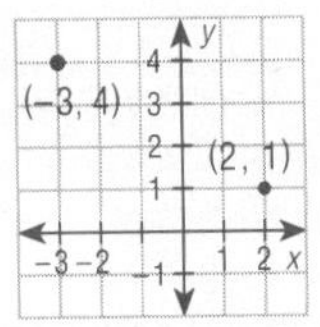

17.
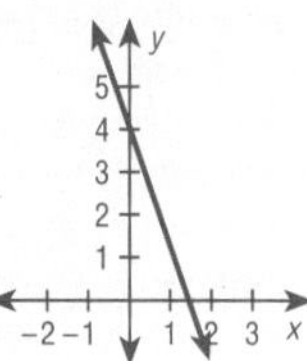

18.
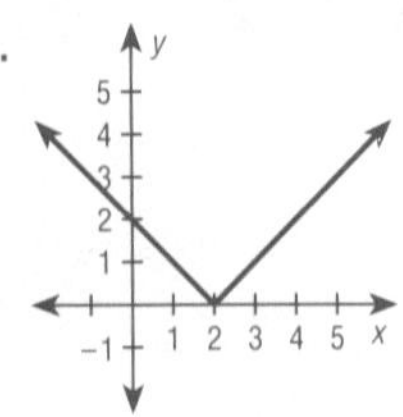

Mixed REVIEW

1. Which is larger, $-\frac{1}{3}$ or $-\frac{1}{2}$? **(2.1)** $-\frac{1}{3}$
2. Find the sum of $\frac{2}{3}$ and $\frac{3}{2}$. **(1.1)** $\frac{13}{6}$
3. Find the area of a 2-ft by 8-ft rectangle. 16 ft²
4. What is the average value of 3, 6, 8, and 13? $7\frac{1}{2}$
5. What number squared is 121? 11
6. What is 35% of 130? 45.5
7. Simplify $-(x - 4) + 2x$. **(2.6)** $x + 4$
8. Simplify $2(3x - 4) - 5(x - 4)$. **(2.6)** $x + 12$
9. Are the lines $2x = 3y$ and $3x = 2y$ parallel? **(4.5)** No
10. Are the lines $y = 2x$ and $y = -2x$ parallel? **(4.5)** No
11. Solve $2x - 4 = -x + 11$. **(3.3)** 5
12. Solve $3(x + 4) = -(x + 1) - 2$. **(3.3, 2.6)** $-3\frac{3}{4}$
13. Evaluate $-3x + 15$ when $x = -1$. **(2.5)** 18
14. Evaluate $|2x + 3|$ when $x = -2$. **(2.1, 2.5)** 1
15. Graph $(-3, 5)$ and $(2, -6)$.
16. Graph $(2, 1)$ and $(-3, 4)$. 15.–18. See margin.
17. Sketch the graph of $y = -3x + 4$. **(4.5)**
18. Sketch the graph of $y = |x - 2|$. **(4.7)**
19. Write an equation of the line with a slope of 2 and a y-intercept of 3. **(5.1)** $y = 2x + 3$
20. Write an equation of the line that passes through $(1, -3)$ and has a slope of -1. **(5.2)** $y = -x - 2$

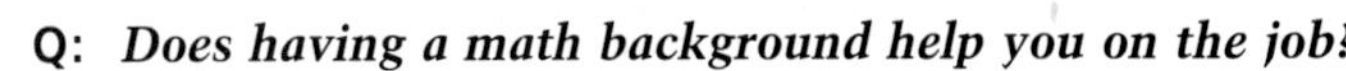

Counselor

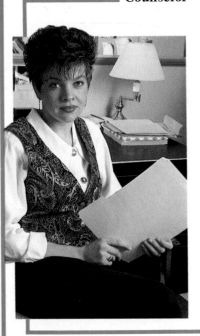

Anne Cavanaugh Sawan counsels outpatients in a community mental health center. She works with children, adolescents, and families experiencing difficulty. She acts as a resource to help them manage their problems.

Q: ***Does having a math background help you on the job?***

A: Mathematics provides the confirmation of research studies. I rely on mathematical information to interpret, validate, and express results of my work.

Q: ***How much of the math is done by computers? Does this mean that you do not need to understand math?***

A: Almost all research data are handled by computers; all formulas are input into programs to handle the data. However, I must still know how the formulas work, how to apply them, what type of a response will result, and how to judge the reasonableness of the results.

Q: ***What would you like to tell kids about math?***

A: Math is really fun, especially when you see the connections and relevance it has to your everyday life. Math is a subject you will rely on all your life. Don't be afraid of making mistakes; you can learn a lot from your errors.

5.3 Equations of Lines Given Two Points

What you should learn:

Goal 1 How to write an equation of a line given two points on the line

Goal 2 How to model a real-life situation with a linear equation

Why you should learn it:

You can model a real-life situation with a linear equation when you are given two pairs of related values (two points on the line) such as quantities sold in two different time periods.

Goal 1 Equation of a Line Given Two Points

So far in this chapter, you have been finding equations of lines for which you were given the slope. In this lesson, you will encounter problems in which you must first find the slope. To do this, you must use the formula given in Chapter 4. The formula states that the slope of the line passing through the points (x_1, y_1) and (x_2, y_2) is

$$m = \frac{\text{rise}}{\text{run}} = \frac{y_2 - y_1}{x_2 - x_1}. \quad \textit{Formula for slope, m}$$

Note how this formula is used in Example 1.

Example 1 Using the Slope Formula to Find the Equation of a Line

Find an equation for the line that passes through the points $(1, 6)$ and $(3, -4)$.

Solution The first step is to find the slope of the line. You let $(x_1, y_1) = (1, 6)$ and $(x_2, y_2) = (3, -4)$. Then you have

$$m = \frac{y_2 - y_1}{x_2 - x_1} = \frac{-4 - 6}{3 - 1} = \frac{-10}{2} = -5.$$

Now you know the slope and a point, and you can use the steps from Lesson 5.2 to find an equation of the line. You let $m = -5$, $x = 1$, and $y = 6$ and solve for b as follows.

$y = mx + b$	*Slope-intercept form*
$6 = (-5)(1) + b$	*Substitute −5 for m, 1 for x, and 6 for y.*
$6 = -5 + b$	*Simplify.*
$11 = b$	*Solve for b.*

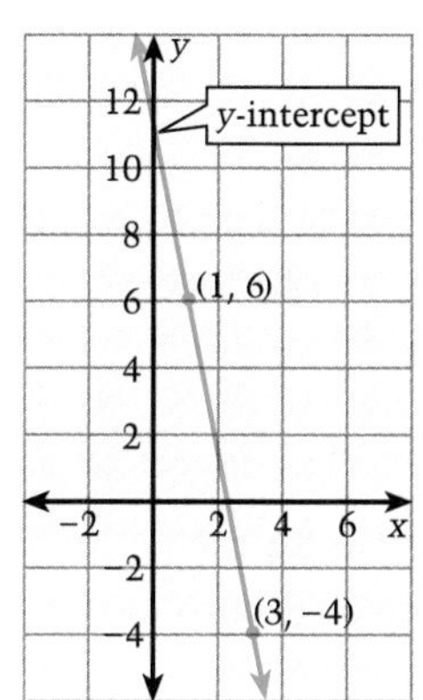

Thus, the y-intercept is $b = 11$, and the slope-intercept form of the equation of the line is $y = -5x + 11$.

Graphic Check Notice that the line has a y-intercept at 11. ■

In Example 1, we used the point $(1, 6)$ to find the y-intercept. Try using the other point, $(3, -4)$, to see that you obtain the same value for b.

Problem of the Day

Each of two inventions saves 40% on fuel. If you use both inventions, what percent of fuel is saved. 64%

ORGANIZER

Warm-Up Exercises

Find the slope of the line containing the two points P and Q. Then sketch the line.

a. $P(3, 5)$ and $Q(-7, 2)$
$m = \frac{3}{10}$

b. $P(6, -12)$ and $Q(9, 3)$
$m = 5$

c. $P(-13, 8)$ and $Q(4, -10)$
$m = -\frac{18}{17}$

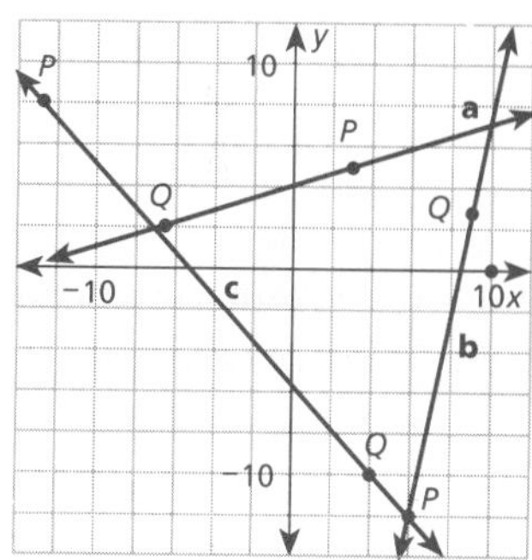

Lesson Resources

Teaching Tools
- Problem of the Day: 5.3
- Warm-up Exercises: 5.3
- Answer Masters: 5.3

Extra Practice: 5.3

Color Transparency: 24

Common-Error ALERT!

Caution students about improperly evaluating the $m = \frac{\text{rise}}{\text{run}}$ formula for the slope. A common mistake is placing x values into y variables. Another common error is combining x and y coordinates in either the numerator or denominator in the formula as in,
$m = \frac{y_2 - x_1}{x_2 - y_1}$.

LESSON Notes

Example 1

Point out to students that if (x_1, y_1) was $(3, -4)$, and (x_2, y_2) was $(1, 6)$, the slope m would still be -5.

Notice that the point (1, 6) can be used in computation of the slope m and in determining the slope-intercept form of the equation of the line containing the two given points.

Example 2

Encourage students to use a graphic check when writing the equation of a line, given two points.

Example 3

Emphasize that the actual number of pairs of jeans sold may not increase at a constant rate. However, the assumption of a constant rate permits writing a linear model to describe the sales.

Extra Examples

Here are additional examples similar to **Examples 1–2.**

1. Given the points P and Q, write the equation of the line containing the two points.
   a. $P(-1, 1)$ and $Q(0, 3)$
   b. $P(4, 4)$ and $Q(-2, -1)$
   c. $P(3, -4)$ and $Q(3, 8)$

   a. $y = 2x + 3$,
   b. $y = \frac{5}{6}x + \frac{2}{3}$,
   c. $x = 3$

Example 1 suggests the following strategy for writing an equation of a line given two points on the line.

Writing an Equation of a Line Given Two Points

To write an equation of a nonvertical line given two points on the line, use the following steps.

1. *Find the slope:* Substitute the coordinates of the two given points into the formula for slope, $m = \frac{(y_2 - y_1)}{(x_2 - x_1)}$.
2. *Find the y-intercept:* Substitute the slope m and the coordinates of one of the points into the slope-intercept form, $y = mx + b$. Then solve for the y-intercept b.
3. *Write an equation of the line:* Substitute the slope m and the y-intercept b into the slope-intercept form, $y = mx + b$.

The next example shows how you can use these guidelines to find an equation of a line.

Example 2 *Writing an Equation of a Line Given Two Points*

A line passes through the points $(-3, 2)$ and $(5, -2)$. Write its equation in slope-intercept form.

Solution The slope of the line is

$$m = \frac{y_2 - y_1}{x_2 - x_1} = \frac{-2 - 2}{5 - (-3)} = \frac{-4}{8} = -\frac{1}{2}.$$

To find the y-intercept, substitute $m = -\frac{1}{2}$, $x = -3$, and $y = 2$ into the slope-intercept form.

$y = mx + b$ *Slope-intercept form*

$2 = \left(-\frac{1}{2}\right)(-3) + b$ *Substitute $-\frac{1}{2}$ for m, -3 for x, and 2 for y.*

$2 = \frac{3}{2} + b$ *Simplify.*

$\frac{1}{2} = b$ *Solve for b.*

Using $b = \frac{1}{2}$ and $m = -\frac{1}{2}$, you can write the equation of the line as $y = -\frac{1}{2}x + \frac{1}{2}$. (Check to see that both of the given points lie on this line.)

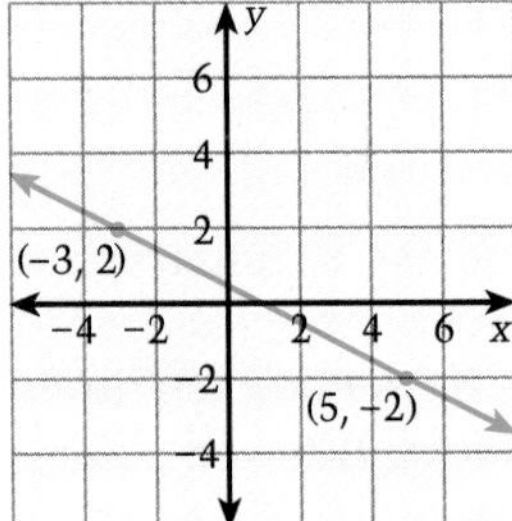

Graphic Check Plot the points $(-3, 2)$ and $(5, -2)$ and then draw the line through the two points as shown at the left. This line appears to have slope $-\frac{1}{2}$ and y-intercept $\frac{1}{2}$. ■

Goal 2 Modeling a Real-Life Situation

In Example 3, you can see how the guidelines for finding the equation of a line, given two points, can be used to create a linear model for a real-life problem.

Real Life
Retailing

A store at a shopping mall sold 4180 pairs of designer jeans in 1986 and 5710 pairs in 1992.

Example 3 ***A Linear Model for Sales of Jeans***

Write an equation that gives the number, y, of pairs of designer jeans sold in the store shown in terms of the year, x. (Let $x = 0$ correspond to 1980.)

Solution Assume that the number of jeans sold increased at a constant rate. This allows you to use a linear model. From the given information, you can find the following two points.

1986: (6, 4180) and 1992: (12, 5710)

The slope of the line passing through these two points is

$$m = \frac{y_2 - y_1}{x_2 - x_1} = \frac{5710 - 4180}{12 - 6} = 255.$$

Now, using $m = 255$, $x = 6$, and $y = 4180$, you can find the y-intercept of the line.

$y = mx + b$ *Slope-intercept form*
$4180 = (255)(6) + b$ *Substitute 255 for m, 6 for x, and 4180 for y.*
$4180 = 1530 + b$ *Simplify.*
$2650 = b$ *Solve for b.*

Thus, the equation is $y = 255x + 2650$. ■

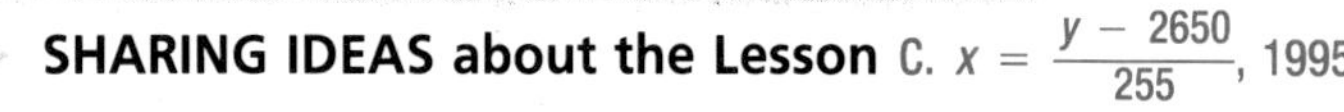

SHARING IDEAS about the Lesson C. $x = \frac{y - 2650}{255}$, 1995

Estimate and Predict In Example 3, we developed the linear model $y = 255x + 2650$ to represent the number of pairs of designer jeans sold by a store.

A. What does the slope of 255 represent? The store sold 255 more jeans per year.

B. Estimate how many pairs of jeans were sold in 1991. 5455

C. Solve the given linear model for x. Use the result to predict the year the store will sell 6475 pairs of jeans.

D. In 1990, the average price of jeans was \$35.95. Estimate how much money the store received for jeans in 1990.

\$186,940

2. Sketch the lines containing the points in Extra Example 1.

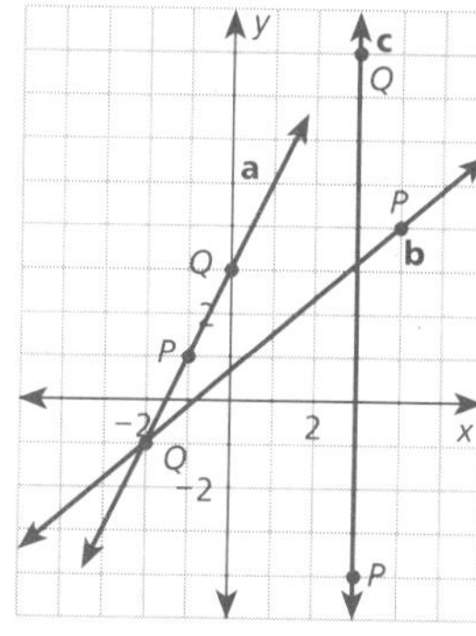

Check Understanding

1. Why is it important to have the slope-intercept form of a straight line? It enables one to easily visualize the graph.
2. What three sets of information can be used to determine the equation for a linear model? Slope and y-intercept; a point on the line and its slope; two points on the line
3. How does a graphic check help when finding the equation of a straight line? A graphic check provides a visual model of the algebraic model.

Communicating about ALGEBRA

Ask students to discuss how reasonable it is to assume that the number of pairs of jeans sold increases at a constant rate each year. Have students describe conditions under which a constant rate is likely and unlikely. What does the b value, 2650, represent? The number of jeans sold in 1980

EXERCISE Notes

ASSIGNMENT GUIDE

Basic/Average: Ex. 1–10, 11–25 odd, 27–28, 31–41 odd, 45–46

Above Average: Ex. 1–6, 8–26 even, 29–30, 32–46 even

Advanced: Ex. 1–6, 8–26 even, 27–30, 33–34, 36–45 multiples of 3, 46–47

Selected Answers
Exercises 1–6, 7–41 odd

More Difficult Exercises
Exercises 27–30, 43–47

Guided Practice

Ex. 1–6 COOPERATIVE LEARNING Have students work in small groups to complete these exercises.

Ex. 1 Have students compare flowcharts and resolve any differences that they find.

Answers

1. Sample answer:

Find the slope by substituting the coordinates of the two given points into the formula for slope.

↓

Find the y-intercept by substituting the slope and the coordinates of one of the points into the slope-intercept form.

↓

Substitute the slope and the y-intercept values into the slope-intercept form of the equation.

2. To write the equation of the line you need to know the slope and the coordinates of one point on the line.

EXERCISES

Guided Practice

CRITICAL THINKING about the Lesson

1. Describe the strategy used to find an equation of the line that passes through two given points. Sketch a flowchart that shows the steps. See margin.

2. When finding an equation of a line passing through two points, you should begin by finding the slope m. Why? See margin.

3. What is the slope of the line that passes through $(1, -1)$ and $(2, 3)$? 4

4. Write an equation of the line that passes through $(1, -1)$ and $(2, 3)$. $y = 4x - 5$

In Exercises 5 and 6, consider a car dealer who sold 450 new cars in 1986 and 600 new cars in 1992.

5. Write an equation that gives the number of cars sold, y, in terms of the year, x. Assume that sales increased at a constant rate, and let $x = 0$ correspond to 1980. $y = 25x + 300$

6. Use the model that you found in Exercise 5 to estimate the number of new cars sold by the dealer in 1990. 550

Independent Practice

In Exercises 7–10, write the slope-intercept form of the equation of the line.

7.
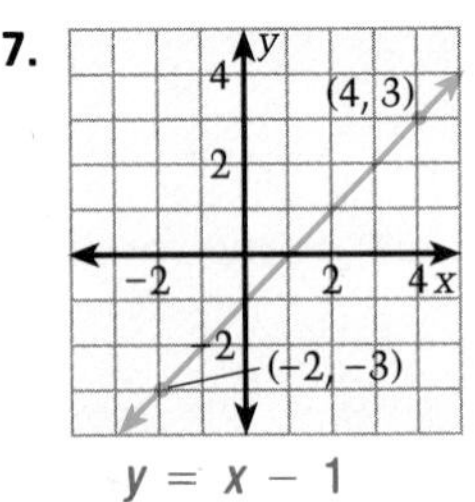

$y = x - 1$

8.
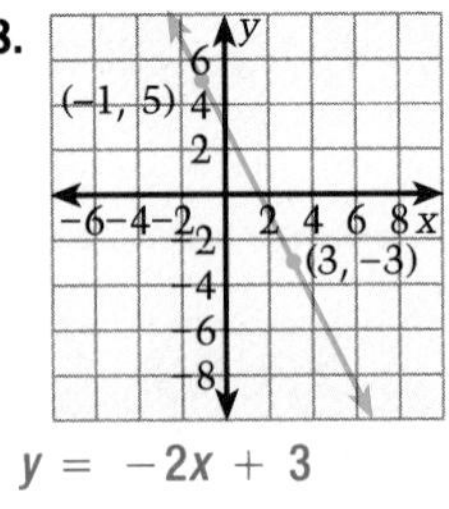

$y = -2x + 3$

9.
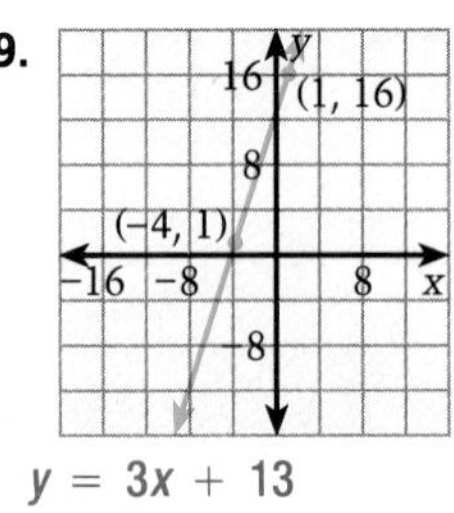

$y = 3x + 13$

10.
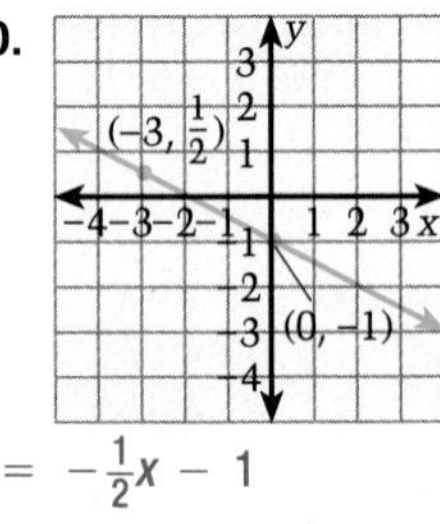

$y = -\frac{1}{2}x - 1$

In Exercises 11–22, write the slope-intercept form of the equation of the line that passes through the two points. See below.

11. $(-1, -1), (2, 8)$

12. $(1, 2), (4, -1)$

13. $(2, 0), (-4, -3)$ $y = \frac{1}{2}x - 1$

14. $(3, 1), (-3, 5)$ $y = -\frac{2}{3}x + 3$

15. $(1, -4), (-2, 8)$

16. $(0, -4), (3, 2)$

17. $(2, -5), (-1, 1)$

18. $(-2, -1), (4, 2)$ $y = \frac{1}{2}x$

19. $(1, 1), (4, 4)$ $y = x$

20. $(1, 2), (2, 4)$ $y = 2x$

21. $(1, 3), (3, 3)$ $y = 3$

22. $(-1, -2), (3, -2)$ $y = -2$

23. Sketch the line that passes through $(2, 6)$ and $(-4, 3)$. Write its equation in slope-intercept form. $y = \frac{1}{2}x + 5$. See margin.

24. Sketch the line that passes through $(3, -3)$ and $(-3, 1)$. Write its equation in slope-intercept form. $y = -\frac{2}{3}x - 1$. See margin.

25. Write an equation of the line whose x-intercept is -6 and whose y-intercept is -4. $y = -\frac{2}{3}x - 4$

26. Write an equation of the line whose x-intercept is -1 and whose y-intercept is 3. $y = 3x + 3$

11. $y = 3x + 2$ 12. $y = -x + 3$ 15. $y = -4x$ 16. $y = 2x - 4$ 17. $y = -2x - 1$

27. *Carpet Purchase* The Jacksons are planning to buy carpeting for several rooms in their house. They plan to use the same carpeting in each room. The family room requires 35 square yards of carpet and would cost \$665. The living room requires 28 square yards of carpet and would cost \$532. Write an equation that gives the cost, y, in terms of the total number of square yards of carpeting, x. $y = 19x$

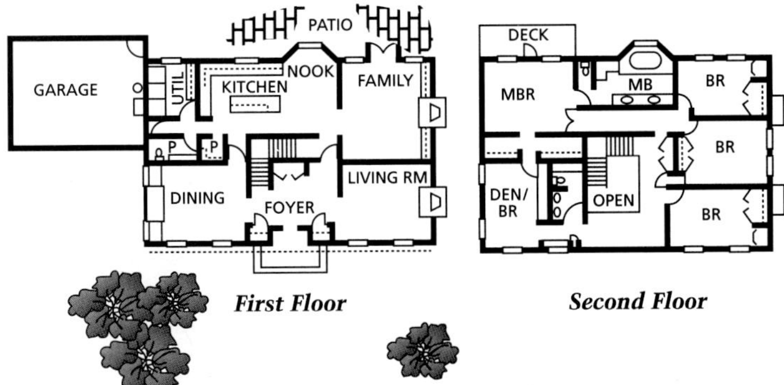

First Floor *Second Floor*

28. Use the equation found in Exercise 27 to complete the table.

Square yards (x)	15	26	41	63	104
Cost (y)	?	?	?	?	?

285, 494, 779, 1197, 1976

Chunnel **In Exercises 29 and 30, use the following information.**

The *Chunnel* is a tunnel built under the English Channel. It is one of the most ambitious engineering feats of the twentieth century.

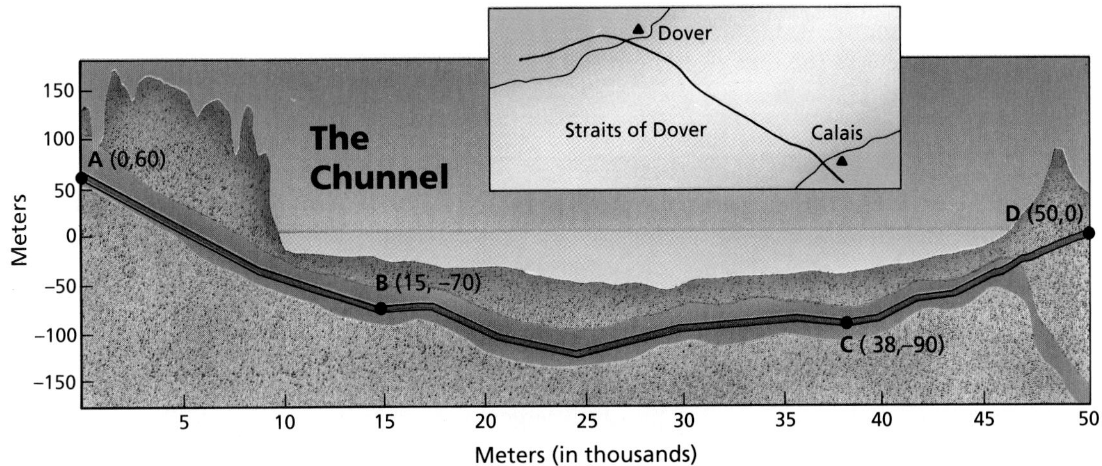

29. Use the diagram to write an equation of the line from point A to point B. What is the slope of this line? $y = -\frac{13}{1500}x + 60$, $-\frac{13}{1500}$

30. Use the diagram to write an equation of the line from point C to point D. What is the slope of this line? Is the Chunnel steeper on the French side or the English side? $y = \frac{3}{400}x - 375$, $\frac{3}{400}$, English side

Independent Practice

Ex. 7–10 Students must pay close attention to the points labeled on the given lines. Ask students how Exercise 10 differs from the others.

Answers

23.

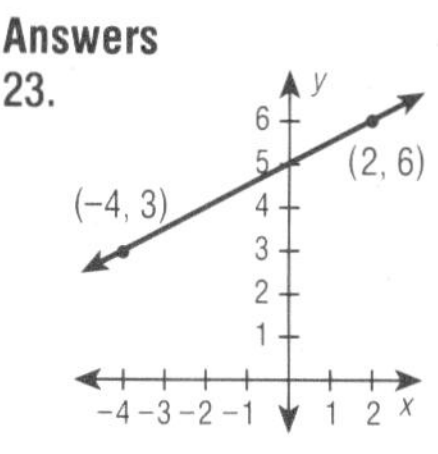

24.

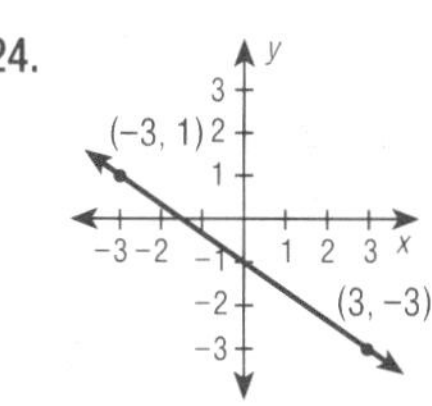

Ex. 25–26. Remind students that the x- and y-intercepts represent special points in the coordinate plane. Students should first identify the points representing the x- and y-intercepts.

Enrichment Activity

WRITING Following the class discussion, students might record their personal responses in their math journals.

Describe the techniques for finding the equation of a straight line. Use a flowchart to illustrate when and how the different techniques can be used. Student's flowcharts should reflect the steps shown on pages 236, 243, and 250.

Answers

35.

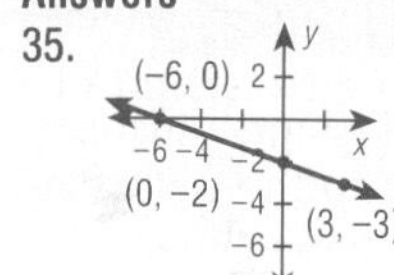

36.

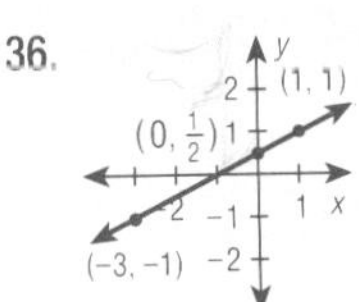

37.

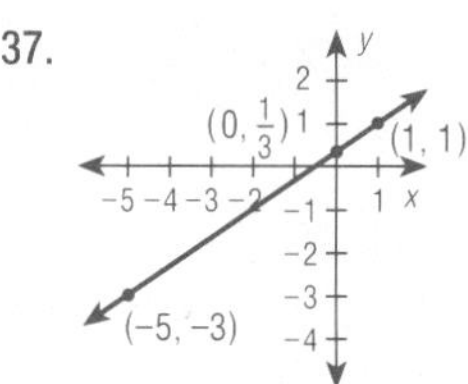

38.

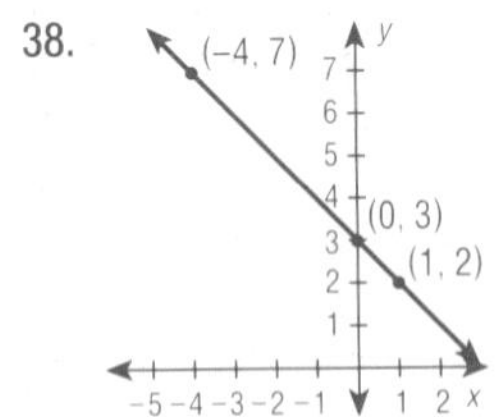

31. *Profit from Shoe Sales* A shoe store made a profit of \$14,510 in 1988 and a profit of \$21,260 in 1993. Write an equation that gives the profit, y, in terms of the year, x. Let $x = 0$ correspond to 1980 and assume that the profit followed a linear pattern. $y = 1350x + 3710$

32. *Bookstore Sales* A bookstore sold 576 copies of dictionaries in 1984 and 792 copies in 1992. Write an equation that gives the number, d, of dictionaries sold in terms of the year, t. Let $t = 0$ correspond to 1980 and assume that the sales followed a linear pattern. $d = 27t + 468$

33. *Newspaper Circulation* In 1975, Sunday newspapers in the United States had a circulation of 51,096,000. In 1985, the circulation had increased to 58,826,000. Write an equation that gives the Sunday circulation, y, in terms of the year, x. Let $x = 0$ correspond to 1970 and assume that circulation followed a linear pattern. ***(Source: American Newspaper Publishers Association)*** $y = 773{,}000x + 47{,}231{,}000$

34. *Population of Western United States* In 1950, the population of the western United States was 20,190,000. In 1980, the population was 43,200,000. Write an equation that gives the population, y, in terms of the year, x. Let $x = 0$ correspond to 1950 and assume that the population followed a linear pattern. ***(Source: U.S. Bureau of Census)*** $y = 767{,}000x + 20{,}190{,}000$

Integrated Review

In Exercises 35–38, sketch the line that passes through the given points. Use the graph to estimate the y-intercept of the line. Then find the actual y-intercept. See margin.

35. $(3, -3), (-6, 0)$ -2 **36.** $(1, 1), (-3, -1)$ $\frac{1}{2}$ **37.** $(1, 1), (-5, -3)$ $\frac{1}{3}$ **38.** $(1, 2), (-4, 7)$ 3

39. Find an equation of the line that passes through the point $(-4, 11)$ and has the same y-intercept as the line given by $y = -\frac{2}{3}x + 3$. $y = -2x + 3$

40. Find an equation of the line that passes through the point $(3, 7)$ and has the same y-intercept as the line given by $y = \frac{1}{2}x - 5$. $y = 4x - 5$

41. Find an equation of the line that passes through the point $(3, -4)$ and is parallel to the line passing through the points $(-4, -1)$ and $(2, 5)$. $y = x - 7$

42. Find an equation of the line that passes through the point $(4, -5)$ and is parallel to the line passing through the points $(2, 0)$ and $(-2, -2)$. $y = \frac{1}{2}x - 7$

Exploration and Extension

In Exercises 43–46, determine whether the given three points lie on the same line. If they do, find an equation of the line.

43. $(-3, -1), (0, 1), (12, 9)$ They do, $y = \frac{2}{3}x + 1$

44. $(4, -2), (-1, 2), (-8, 9)$ They do not.

45. $(-2, -1), (3, 2), (7, 5)$ They do not.

46. $(3, -3), (-1, 13), (1, 5)$ They do, $y = -4x + 9$

47. You are told that the points $(1.9, 7.11), (5, 13), (-3.3, -2.77)$, and $(-5, -6)$ lie on the same line. If you were asked to find the slope of the line, which two points would you choose and why? $(5, 13)$ and $(-5, -6)$, the slope is more easily determined using integers

5.4 Exploring Data: Fitting a Line to Data

What you should learn:

Goal 1 How to find a linear equation that approximates a set of data points

Goal 2 How to use scatter plots to determine positive correlation, negative correlation, or no correlation

Why you should learn it:

You can use the equation of a best-fitting line to investigate trends in data, such as sports performances, and to make predictions.

Goal 1 Fitting a Line to Data

In Lesson 5.3, you studied how to find an equation of the line that passes through *two* points. For this type of problem, there is always only one line—because two points determine exactly one line.

In this lesson, you will study problems that involve *several* data points. Usually there is no single line that passes through all the data points, so you try to find the line that best fits the data. This is called the **best-fitting line.** For instance, in the graph shown below, the line given by $y = 2x + 1$ is the best-fitting line for the data points.

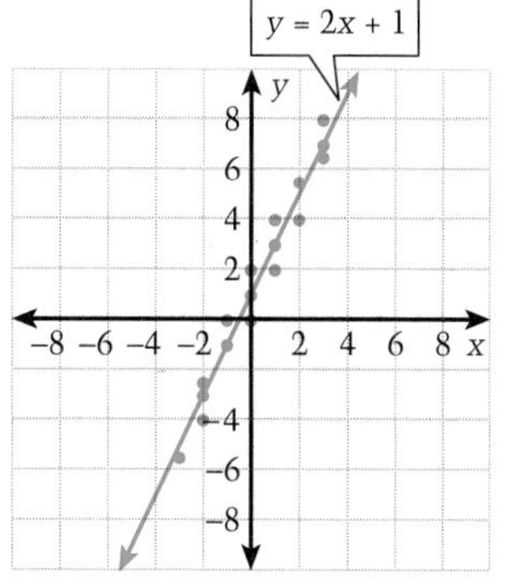

There are several ways to find the best-fitting line for a given set of data points whose graph suggests a linear relationship. One way is to use a graphing calculator (see page 262) or computer and formulas from a branch of mathematics called statistics. The statistical formula can *find* the best-fitting line. In this lesson, however, you will use a graphical approach, which *approximates* the best-fitting line.

LESSON INVESTIGATION

Investigating a Best-Fitting Line

Partner Activity With your partner, use the following steps to approximate a best-fitting line.

1. Carefully plot the following points on graph paper.

   (0, 3.3), (0, 3.9), (1, 4.2), (1, 4.5), (1, 4.8), (2, 4.7), (2, 5.1), (3, 4.9), (3, 5.6), (4, 6.1), (5, 6.4), (5, 7.1), (6, 6.8), (7, 7.5), (8, 7.8)

2. Use a ruler to sketch the line that you think best approximates the data points.
3. Locate two points on the line and approximate the x-coordinate and y-coordinate of each point. (These don't have to be two of the original points.)
4. Use the technique described in Lesson 5.3 to find an equation of the line that passes through the two points in Step 3.

Equations will vary.
$y = 0.5x + 3.8$

Problem of the Day

Each * represents a digit in this multiplication. Determine the digits.

```
   3 *          37
 × * *        × 19
 -----        ----
  3 * *       3 3 3
  * *         3 7
 -----        -----
  * * 3       7 0 3
```

ORGANIZER

Warm-Up Exercises

1. Plot the following points on a coordinate plane. (2, −1), (−14, −32), (3, 2), (6, 8), (−5, −20), (−8, −23), (10, 12), (16, 31), (13, 4), (−18, −37), (14, 26), (−1, 0)
2. Graph the line $y = 2x - 5$ on the coordinate plane used to plot the points in Exercise 1.

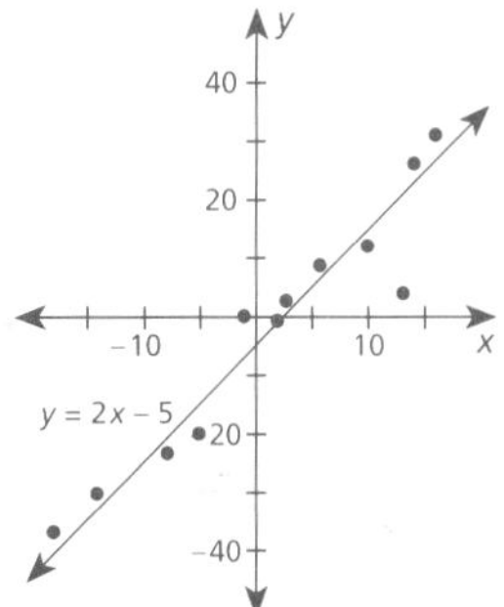

3. How many points are above the line $y = 2x - 5$? How many points are below the line? How many points are on the line? 7 points above, 4 points below, 1 point on the line.

Lesson Resources

Teaching Tools
- Transparencies: 1, 2, 4
- Problem of the Day: 5.4
- Warm-up Exercises: 5.4
- Answer Masters: 5.4

Extra Practice: 5.4

Color Transparencies: 24, 25

Technology Handbook: p. 45, 48, 51

LESSON Notes

GOAL 1 The best-fitting line for the data given on the coordinate plane passes through the point (0, 1). Because the point (1, 3) appears to be on the line, moving from (0, 1) to (1, 3) indicates that the rise is 2 for a run of 1—that is, the slope is 2. Consequently the equation of the best-fitting line is $y = 2x + 1$.

Example 1

Point out to students that the equations of the best-fitting lines may differ slightly from one person to the next.

Example 2

Note that the points (19, 20) and (26, 26) are not in the original set of data.

Common-Error ALERT!

Step 3 in Approximating the Best-Fitting Line indicates that the points on the line need not be in the set of original points. Caution students against assuming that all points of the line should be included in the original data. If the points selected to determine the line were not originally in the set of data, then these points should not be assumed to be part of the original data.

Connections
Statistics

Year	Winning Throw
1908	134.2 ft
1912	145.1 ft
1920	146.6 ft
1924	151.4 ft
1928	155.2 ft
1932	162.4 ft
1936	165.6 ft
1948	173.2 ft
1952	180.5 ft
1956	184.9 ft
1960	194.2 ft
1964	200.1 ft
1968	212.5 ft
1972	211.3 ft
1976	221.5 ft
1980	218.7 ft
1984	218.5 ft
1988	225.8 ft
1992	213.7 ft
1996	227.7 ft

Example 1 *The Olympic Games Discus Throw*

The winning Olympic Games discus throws from 1908 to 1996 are shown in the table. Approximate the best-fitting line for these throws. Let x represent the year, with $x = 8$ corresponding to 1908. Let y represent the winning throw.

Solution To begin, represent each of the winning throws by an ordered pair. For instance, the winning throw in 1908 is represented by (8, 134.2), and the winning throw in 1912 is represented by (12, 145.1). Next, plot the points given by the ordered pairs. Then sketch the line that best fits the points, as shown below.

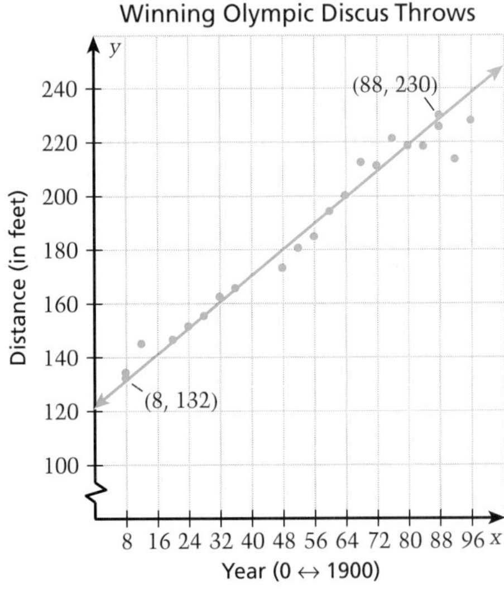

The next step is to find two points that lie on the best-fitting line. From the graph, you might choose the points (8, 132) and (88, 230). The slope of the line passing through these two points is

$$m = \frac{y_2 - y_1}{x_2 - x_1} = \frac{230 - 132}{88 - 8} = \frac{98}{80} = 1.225.$$

To find the y-intercept of the line, substitute the values $m = 1.225$, $x = 8$, and $y = 132$ into the slope-intercept form.

$y = mx + b$	*Slope-intercept form*
$132 = (1.225)(8) + b$	*Substitute 1.225 for m, 8 for x, 132 for y.*
$132 = 9.8 + b$	*Simplify.*
$122.2 = b$	*Solve for b.*

Thus, the y-intercept, b, is 122.2, and the slope-intercept form of the equation of the best-fitting line is approximately

$$y = 1.225x + 122.2.$$ ■

Forearm Length	Foot Length
22 cm	24 cm
20 cm	19 cm
24 cm	24 cm
21 cm	23 cm
25 cm	23 cm
18 cm	18 cm
20 cm	21 cm
23 cm	23 cm
24 cm	25 cm
20 cm	22 cm
19 cm	19 cm
25 cm	25 cm
23 cm	22 cm
22 cm	23 cm
24 cm	24 cm
20 cm	21 cm
18 cm	19 cm
24 cm	23 cm
24 cm	27 cm
21 cm	24 cm
22 cm	22 cm

Example 2 *Comparing Arm Length to Foot Length*

The data in the table contain the forearm lengths and foot lengths (without shoes) of 22 students in an algebra class.

Approximate the best-fitting line for these data. Let x represent the forearm length and let y represent the foot length.

Solution To begin, plot points for the data, as shown in the graph. (Note that only 19 points are plotted because three of the points are repeated.)

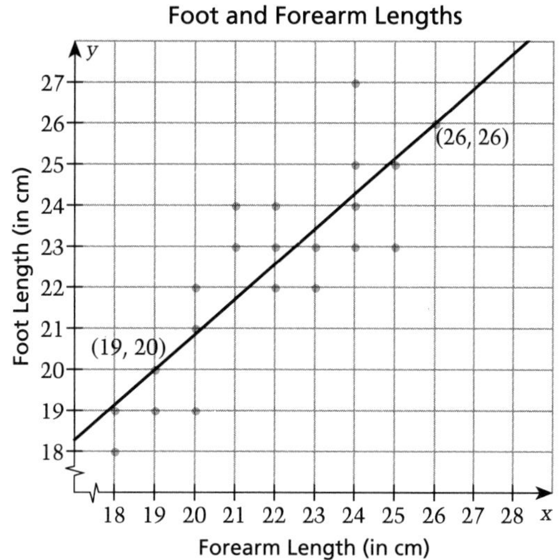

Next, sketch the line that appears to fit the points best and locate two points on the line: (19, 20) and (26, 26). The slope of the line passing through these two points is

$$m = \frac{y_2 - y_1}{x_2 - x_1} = \frac{26 - 20}{26 - 19} = \frac{6}{7} \approx 0.86.$$

To approximate the y-intercept of the line, substitute the values $m = 0.86$, $x = 19$, and $y = 20$ into the slope-intercept form.

$y = mx + b$	*Slope-intercept form*
$20 = (0.86)(19) + b$	*Substitute 0.86 for m, 19 for x, 20 for y.*
$20 \approx 16.3 + b$	*Simplify.*
$3.7 \approx b$	*Solve for b.*

Thus, the y-intercept, b, is approximately 3.7, and the slope-intercept form of the equation of the best-fitting line is approximately

$$y = 0.86x + 3.7.$$

■

Extra Example

Here is an additional example similar to **Examples 1–2**.

The winning Olympic times for the 100-meter run from 1928 to 1988 are shown in the table. Approximate the best-fitting line for these times. Let y represent the winning time and x the year ($x = 0$, corresponding to 1928).

100-Meter Run

Year	Winning time (in seconds)
1928	12.2
1932	11.9
1936	11.5
1948	11.9
1952	11.5
1956	11.5
1960	11.0
1964	11.4
1968	11.0
1972	11.07
1976	11.08
1980	11.6
1984	10.97
1988	10.54

The graph of the ordered pairs from the table and the sketch of a best-fitting line follows. The best-fitting line is $y = -0.02x + 12$. The points (0, 12) and (24, 11.5) are on the line of best fit. Because the selected line of best fit can differ from one student to another, the equation of the best-fitting line can vary.

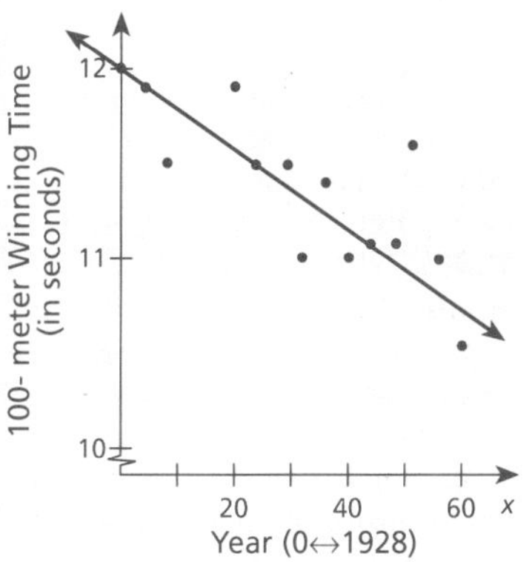

Check Understanding

1. Fitting a line to data is an application of writing an equation of a linear model. Explain. See page 255.
2. How is a line of best fit determined (that is, sketched) from a scatter plot of data? See page 255.
3. Can there be only one equation of the best-fitting line for a set of data? Explain. No, the lines of best fit are only approximations.

Communicating about ALGEBRA

Have the class compare their best-fitting lines by answering questions like the following. How many equations are alike? How many different equations were found?

Answers

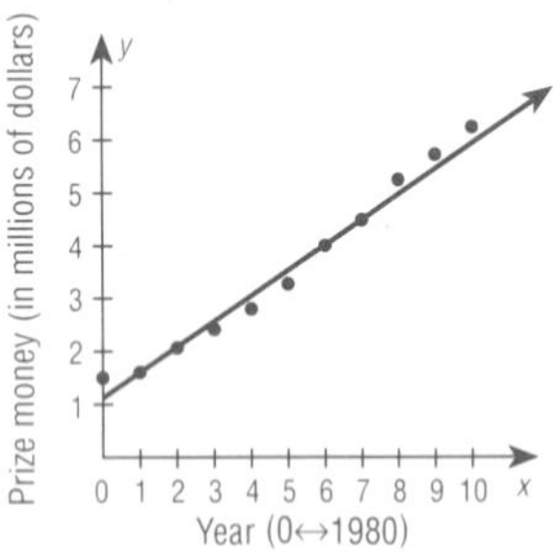

A, C

EXTEND *Communicating* ***RESEARCH***

As a class project, students could find out the actual total prize money for the Indianapolis 500 for each of the years 1991, 1992, and 1993. Compare the actual totals with those predicted by class models. Adjust the models by adding data from 1991 and compare them with outcomes of 1992 and 1993. Adjust the original models by including data from 1991 and 1992 and compare the results with the totals for 1993.

Goal 2 Determining the Correlation of *x* and *y*

Three scatter plots are shown below. In the first scatter plot, x and y have a **positive correlation,** which means that the points can be approximated by a line with *a positive slope.* In the second scatter plot, x and y have a **negative correlation,** which means that the points can be approximated by a line with a *negative slope.* In the third scatter plot, there is **no correlation** between x and y.

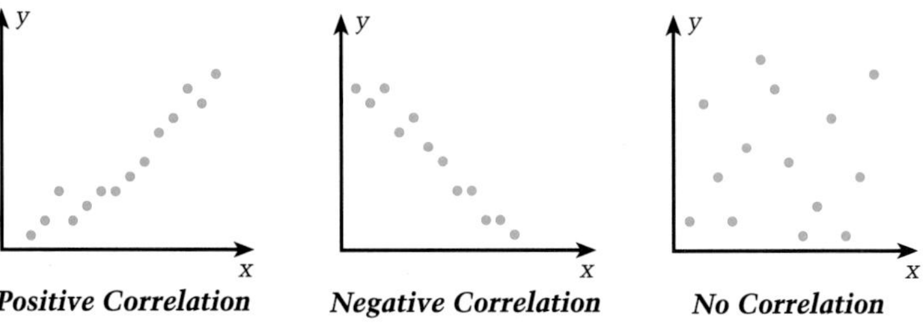

Communicating about ALGEBRA

SHARING IDEAS about the Lesson

Construct a Model The total prize money for the Indianapolis 500 from 1980 to 1990 is shown in the table. ***(Source: Indianapolis Motor Speedway)***

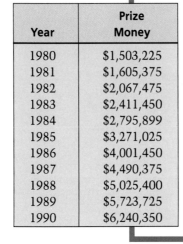

Year	Prize Money
1980	\$1,503,225
1981	\$1,605,375
1982	\$2,067,475
1983	\$2,411,450
1984	\$2,795,899
1985	\$3,271,025
1986	\$4,001,450
1987	\$4,490,375
1988	\$5,025,400
1989	\$5,723,725
1990	\$6,240,350

A. Sketch a scatter plot. Let x represent the year, with $x = 0$ corresponding to 1980, and let y represent the total prize money. A., C. See margin.

B. From your scatter plot, do x and y appear to have a positive correlation, a negative correlation, or no correlation? Positive

C. Sketch the best-fitting line.

D. Locate two points on the line. (Your two points need not be the same as the points chosen by others in the class.) (2, 2,067,475), (6, 4,001,450)

E. Use your two points to approximate an equation of the line. $y = 483{,}493.75x + 1{,}100{,}487.5$

F. Use the equation you found to predict the total amount of prize money for the Indianapolis 500 in 1996. \$8,836,387.50

EXERCISES

Guided Practice

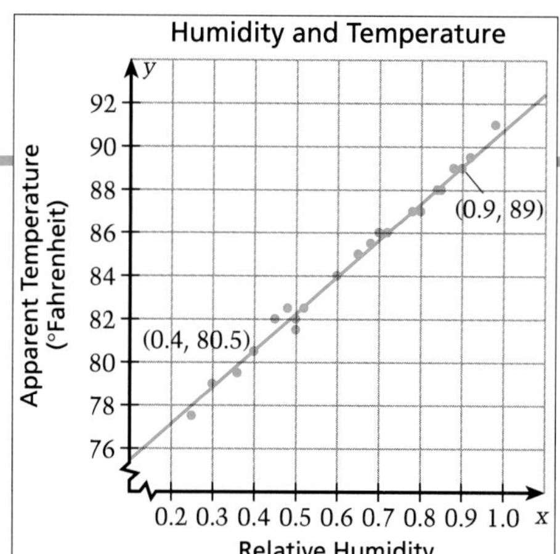

CRITICAL THINKING about the Lesson

In Exercises 1–6, use the data shown on the graph.

On humid days, people tend to feel warmer than the actual temperature. To illustrate this, several people were asked how warm they felt. In each case, the actual temperature was 80° Fahrenheit, but the relative humidity varied from 25% to 98%.

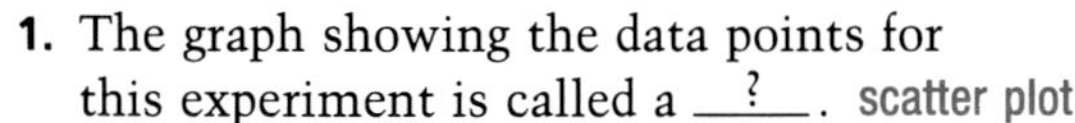

1. The graph showing the data points for this experiment is called a __?__. scatter plot

2. The line shown on the graph above is called the __?__-__?__ line. best-fitting

3. Use the two labeled points to find the slope of the line. 17

4. Find an equation of the line. $y = 17x + 73.7$

5. If the temperature is 80°F and the relative humidity is 0.80, how warm would it feel? 87.3°F

6. If the temperature is 80°F and it feels like 90°F, what is the relative humidity? 0.96

Independent Practice

In Exercises 7–10, decide whether x and y suggest a linear relationship.

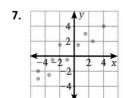

7. Suggest

8.

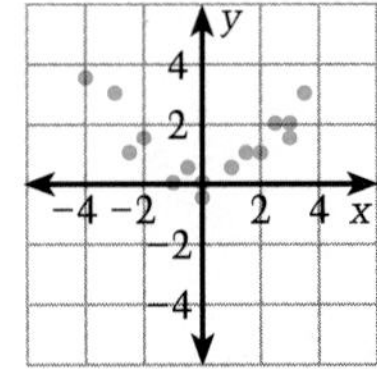

Do not suggest

9.

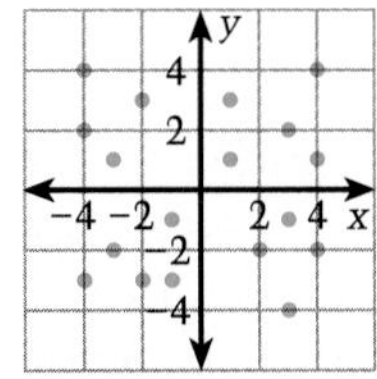

Do not suggest

10.

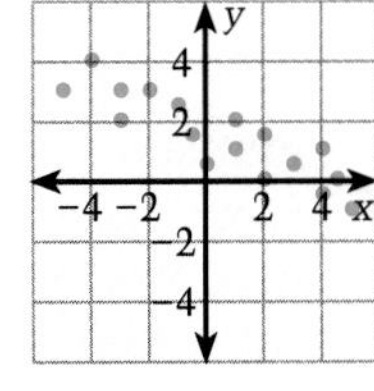

Suggest

In Exercises 11–13, approximate the best-fitting line for the scatter plot. State whether x and y have a positive or negative correlation. Best-fitting lines vary.

11.

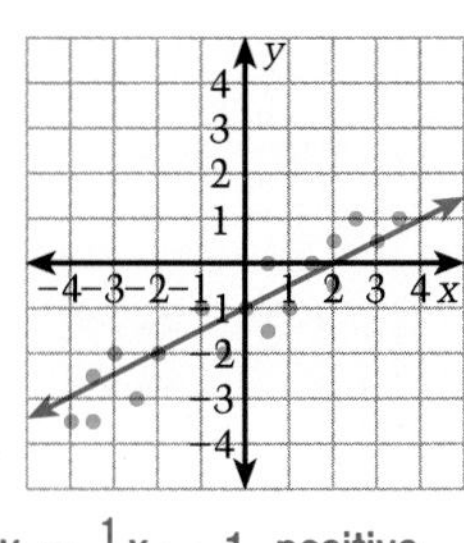

$y = \frac{1}{2}x - 1$, positive

12.

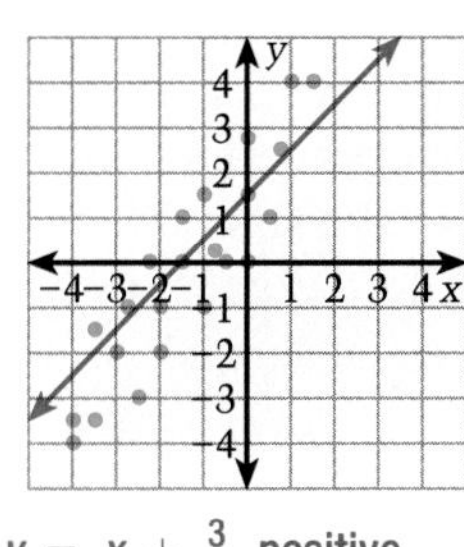

$y = x + \frac{3}{2}$, positive

13.

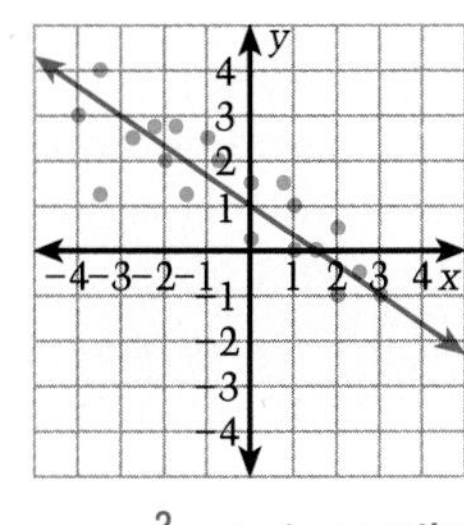

$y = -\frac{2}{3}x + 1$, negative

EXERCISE Notes

ASSIGNMENT GUIDE

Basic/Average: Ex. 1–6, 7–10, 11–27 odd, 28–29

Above Average: Ex. 1–6, 7–10, 12–26 even, 28–31

Advanced: Ex. 1–6, 7–10, 12–16 even, 17–19, 21–27 odd, 28–31

Selected Answers
Exercises 1–6, 7–27 odd

✪ More Difficult Exercises
Exercises 18–19, 28–31

Guided Practice

▶ **Ex. 1–6** Have students work these exercises individually. Then compare the collection of equations of the best-fitting line. Discuss why there should be no differences in the statement of the equation. Under what conditions are equations likely to vary? When the chosen pair of points, used to find the slope, varies.

Independent Practice

▶ **Ex. 7–13** These work well as oral exercises in which students discuss and provide rationales for their decisions.

Ex. 28–31 COOPERATIVE LEARNING Have students complete these exercises, working in small groups of two or three. Require each group to present a written argument for increasing the library's annual allocation of funds.

Enrichment Activities

COOPERATIVE LEARNING

These activities provide students with an opportunity to collect, organize, and analyze data obtained from the class.

1. Measure the circumference of your wrist and your neck. Organize the data in a table. Determine the best-fitting line for the data .
2. Identify other sets of data that can be collected and analyzed by using the technique of approximating a line of best fit. Students could compare arm span and height, foot length and height, etc., to determine whether there is any correlation.

Challenge! You might want to have students determine whether the measurements are a function of age or whether the results are different for males and females. They could repeat the experiment by measuring children and then adults to determine whether there are significant differences by age. Also, data can be separated into "males "and "females" to determine whether there are significant differences by sex .

In Exercises 14 and 15, construct a scatter plot for the given data. Then find an equation of the line that you think best represents the data. State whether x and y have a positive correlation or a negative correlation. See margin.

14.–18. Best-fitting lines vary.

14.

x	1.0	1.5	1.7	2.0	2.0	2.5	3.0	3.4	4.0	4.1	4.8	5.2
y	3.8	4.2	5.3	5.8	5.5	6.7	7.1	8.1	8.5	8.9	9.6	10.6

15.

x	3.0	3.5	3.7	4.0	4.0	4.5	5.0	5.4	6.0	6.1	6.8	7.2
y	10.1	9.7	8.6	8.1	8.4	7.4	6.8	5.8	5.6	5.2	4.3	3.5

16. *Football Salaries* The median base salary for national football league players from 1983 to 1993 is shown in the scatter plot at the right. In the graph y represents the salary and x represents the year, with $x = 3$ corresponding to 1983. Find an equation of the line that you think best fits the data. Then use the equation to approximate the median base salary in 1995. ***(Source: National Football League Players Association)***

$y = 24{,}000x + 8000$; \$368,000

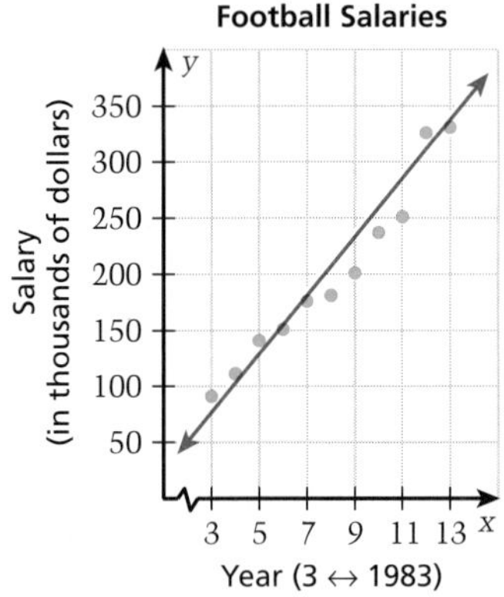

17. *Cricket Chirping* Crickets make a chirping sound by rubbing their legs together. The number of chirps that a cricket makes per minute is related to the temperature, as shown in the scatter plot at the right. In the graph, y represents the number of chirps per minute and x represents the temperature in degrees Fahrenheit. Find an equation of the line that you think best fits this data. Then use the equation to approximate the number of chirps per minute made by a cricket when the temperature is 85°.

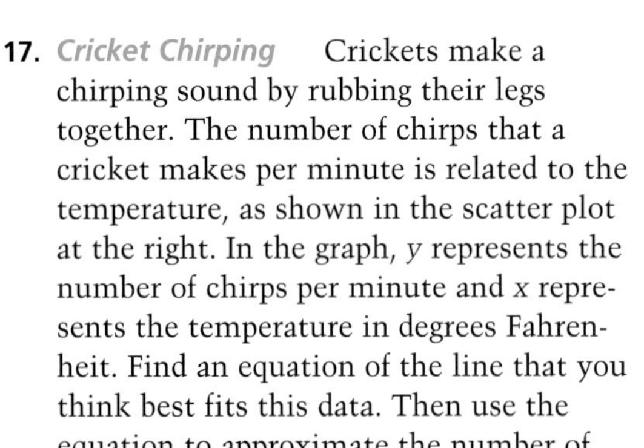

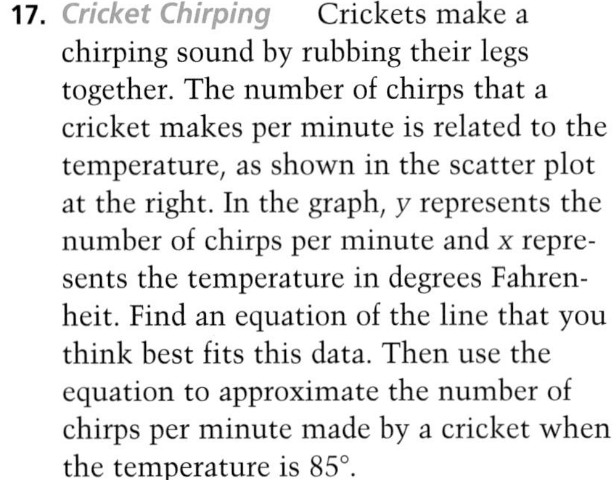

$y = 4x - 165$, 175

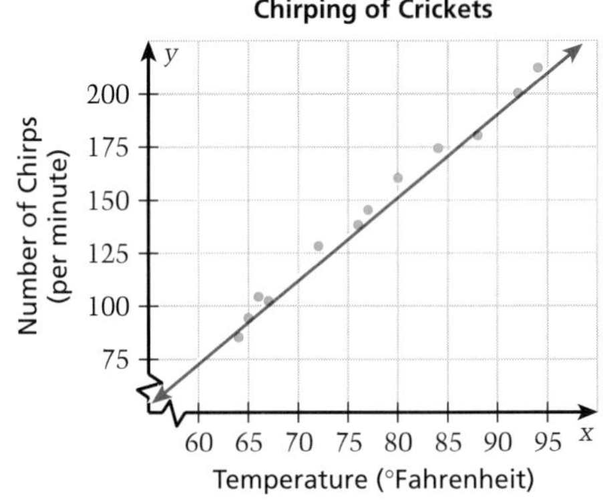

18. *American Workers* The number of Americans who were employed in nonmilitary jobs for selected years between 1982 and 1994 is given in the table at the right. Sketch a scatter plot for this data and find an equation of the line that best fits the data. Use your equation to approximate the number of American workers in 2010. ***(Source: U.S. Bureau of Labor Statistics)***

$y = 1{,}860{,}000t + 98{,}000{,}000$; 153,800,000. Graph in margin.

Year	Number of Workers
1982	99,526,000
1984	105,005,000
1986	109,597,000
1988	114,968,000
1990	117,914,000
1992	117,598,000
1994	123,060,000

19. *Average Temperatures* The average June temperatures and latitudes of several cities in the central United States are shown in the table. Sketch a scatter plot for these data and find an equation of the line that best fits the data. Use your equation to approximate the average June temperature for a city in the central United States with a latitude of 40°. *(Source: P.C. U.S.A.)*
$y = -x + 113$, 73°F. Graph in margin.

City	Lat.	Temp.
Bismarck	46.8°	64.3°F
Sioux Falls	43.5°	68.4°F
Omaha	41.2°	73.0°F
Kansas City	39.1°	76.1°F
Wichita	37.7°	73.6°F
Okla. City	35.5°	77.0°F
Dallas	33.2°	82.0°F
Houston	29.8°	80.6°F

Integrated Review

In Exercises 20–23, state whether the slope of the line passing through the two points is positive or negative.

20. (1, 4), (3, 5) Positive **21.** (2, 5), (4, 1) Negative **22.** (6, 1), (3, 8) Negative **23.** (5, 2), (4, 3) Negative

In Exercises 24–27, find an equation of the line that passes through the two given points.

24. (2, 5), (6, 4) $y = -\frac{1}{4}x + \frac{11}{2}$ **25.** (1, 4), (3, 7) $y = \frac{3}{2}x + \frac{5}{2}$ **26.** (3, 7), (7, 3) $y = -x + 10$ **27.** (2, 6), (5, 4) $y = -\frac{2}{3}x + \frac{22}{3}$

Exploration and Extension

Library Budgets **In Exercises 28–31, use the following information.**

A school library is allocated money, B, each year to buy new books (scatter plot at left). The average price, P, of a new book is shown in the scatter plot at the right. In each, $t = 0$ corresponds to 1980.

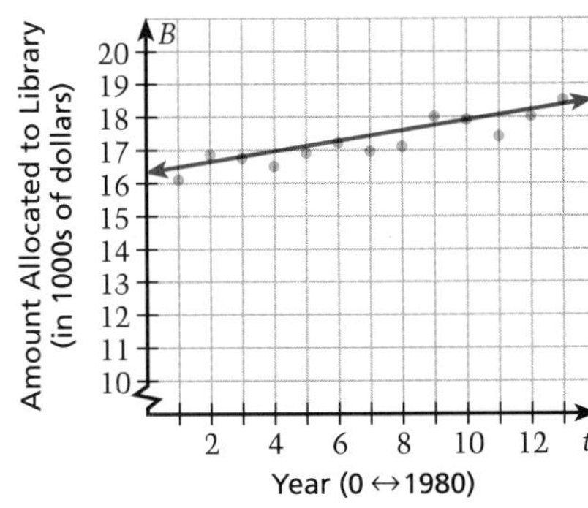

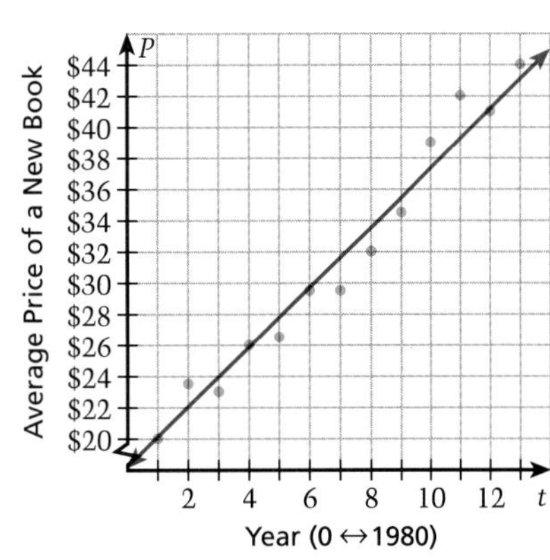

28. Find an equation of the line that you think best represents the amount of money given to the library for new books.

29. Find an equation of the line that you think best represents the average price of a new book. $P = 2t + 18$

30. What do the slopes of the two lines represent? See margin.

31. Suppose that you worked at the school library and wanted to ask the School Board to increase the library's annual allocation of money. How could you use the information to argue your case? See margin.

28. $B = 160t + 16{,}400$

Answers

14.
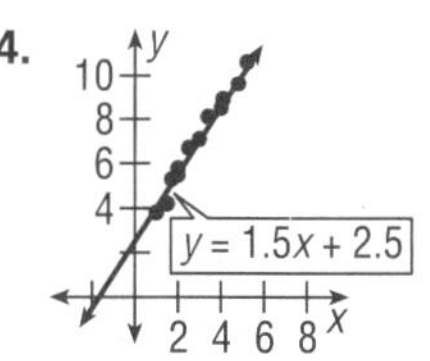

positive

15.
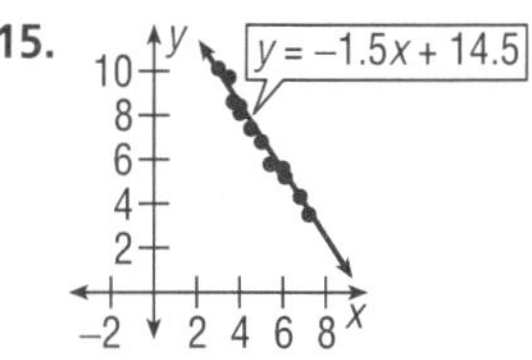

negative

18.
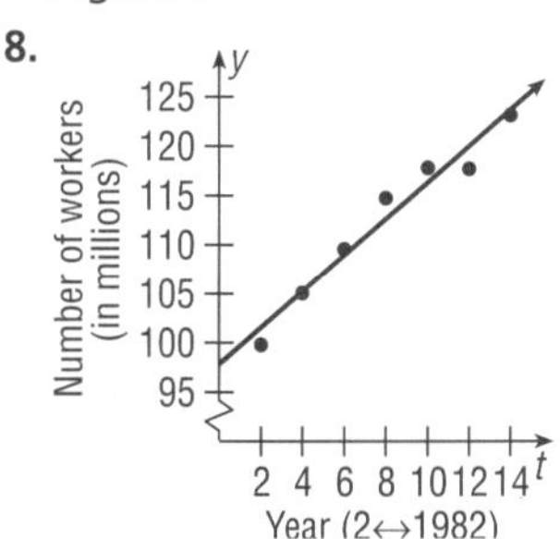

19.
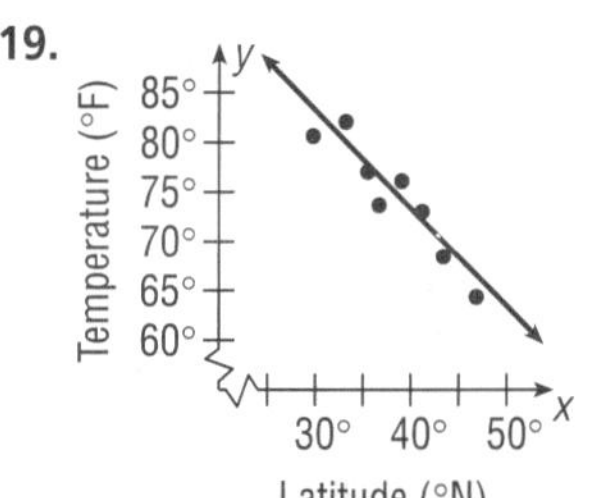

30. First slope: increase in the amount allocated to the library each year, \$160; Second slope: increase in the price of a new book each year, \$2

31. The rate of increase in allocated funds in 1990 was $\frac{160}{18,000}$ or 0.9%; the rate of increase in the cost of the books that same year was $\frac{2}{38}$, or 5.3%—six times the amount allocated.

EXTEND Ex. 1–2 TECHNOLOGY

Enter the data from Examples 1 and 2 of Lesson 5.4 on the graphing calculator. Compare the equations given in Examples 1 and 2 of Lesson 5.4 with the equations generated by the graphing calculator or computer. Are they the same or different? What might account for the differences? Students should understand that both the graphing calculator and the computer have a built-in series of mathematical instructions to find the line. The text suggests a more intuitive method for solving.

Example 1 of Lesson 5.4 suggests the equation $y = 1.225x + 122.2$. A graphing calculator suggests the equation $y = 1.10x + 126.77$, with a correlation coefficient of 0.98. A correlation coefficient of 1 indicates there is a perfect linear relationship between x and y. In Example 2, both the text and the graphing calculator suggest the equation $y = 0.86x + 3.7$, but the correlation factor is only 0.83.

Real-life Connection

Ask students what the word *measure* means in music. In music a measure is a unit of musical time containing an indicated number of beats.

Common-Error ALERT!

The linear model used by some graphing calculators is $y = a + bx$. Remind students that the a in this form of the equation represents the y-intercept and that the b represents the slope (not the y-intercept).

USING A GRAPHING CALCULATOR

In Lesson 5.4, you studied the graphical approach for approximating a best-fitting line. Another approach is to use a calculator or computer programmed to find the best-fitting line. Keystroke instructions for doing this on a TI-82, Casio fx-9700GE, and Sharp EL-9300C are listed on page 738.

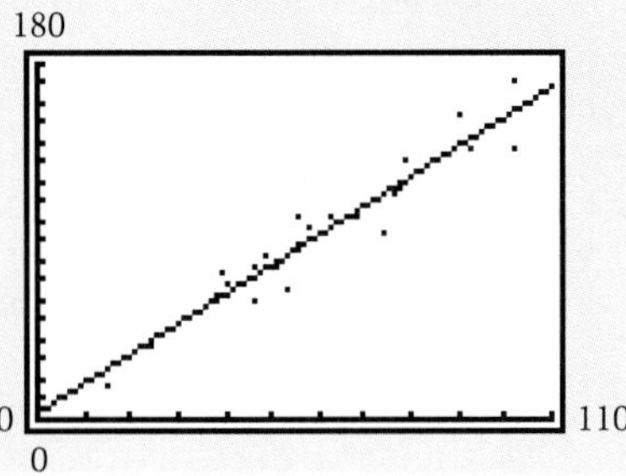

Mozart's Sonatas Sonatas can be naturally divided into two parts: the *exposition,* which introduces the musical theme, and the *development and recapitulation,* which develop and repeat the theme. The following data represent all of Mozart's (1756–1791) piano sonatas that have these two parts. The first coordinate x is the length of the exposition in measures. The second coordinate y is the length of the development and recapitulation in measures.

(38, 62), (28, 46), (56, 102), (56, 88), (24, 36), (77, 113), (40, 69), (46, 60), (15, 18), (39, 63), (53, 67), (14, 23), (102, 171), (51, 76), (58, 97), (49, 84), (39, 73), (58, 92), (31, 50), (74, 93), (46, 76), (28, 45), (78, 118), (79, 130), (63, 102), (68, 103), (93, 136), (90, 155), (102, 137)

This data was plotted on a graphing calculator, as shown at the left. Then, the graphing calculator was used to find the best-fitting line for the data. This line is $y = 1.50x + 3.24$. From this equation, how could you verbally describe the relationship between the two parts of Mozart's sonatas?

Exercises

In Exercises 1 and 2, find the best-fitting line for the points. Best-fitting lines vary.

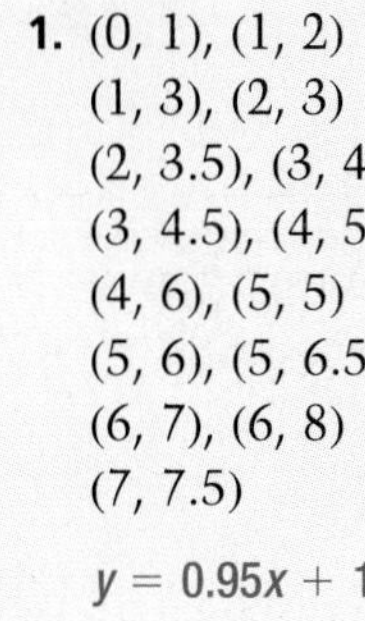

1. (0, 1), (1, 2)
(1, 3), (2, 3)
(2, 3.5), (3, 4)
(3, 4.5), (4, 5.5)
(4, 6), (5, 5)
(5, 6), (5, 6.5)
(6, 7), (6, 8)
(7, 7.5)

$y = 0.95x + 1.4$

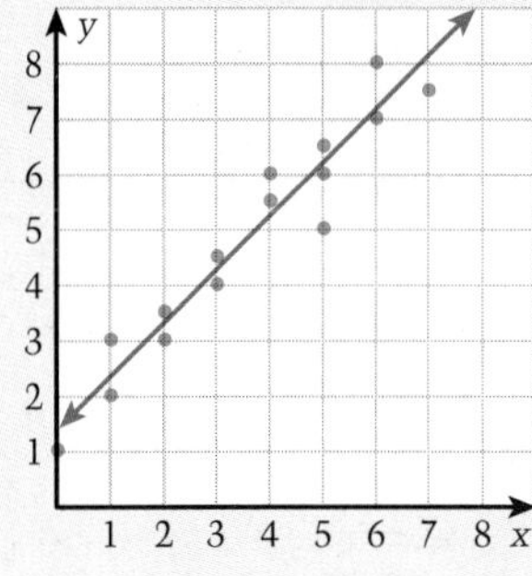

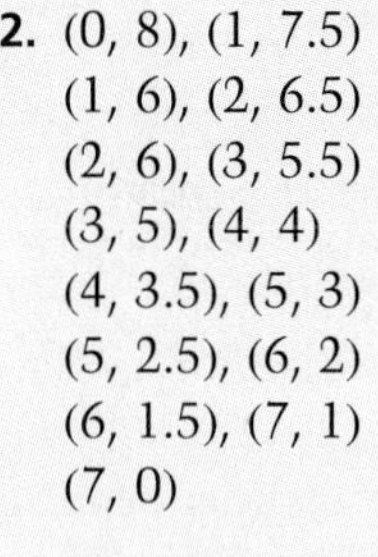

2. (0, 8), (1, 7.5)
(1, 6), (2, 6.5)
(2, 6), (3, 5.5)
(3, 5), (4, 4)
(4, 3.5), (5, 3)
(5, 2.5), (6, 2)
(6, 1.5), (7, 1)
(7, 0)

$y = -1.1x + 8.1$

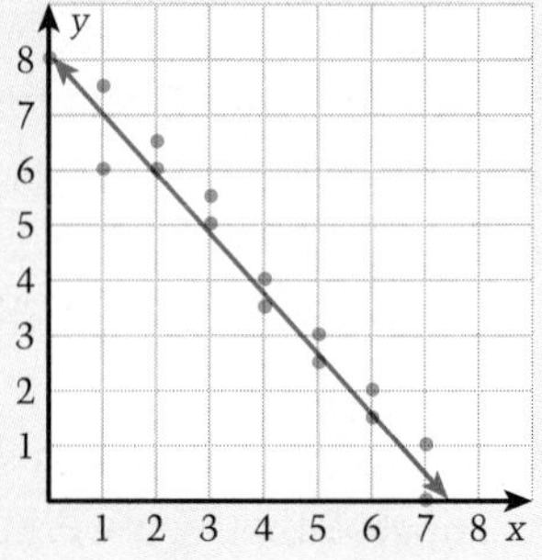

Mid-Chapter SELF-TEST

Take this test as you would take a test in class. The answers to the exercises are given in the back of the book.

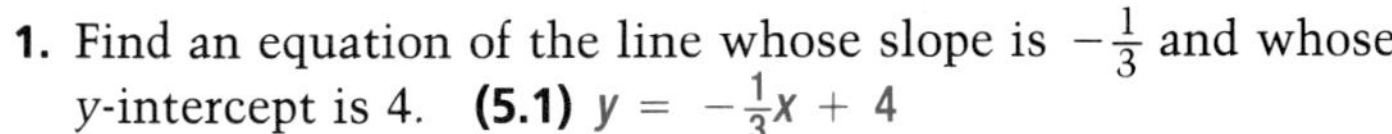

1. Find an equation of the line whose slope is $-\frac{1}{3}$ and whose y-intercept is 4. **(5.1)** $y = -\frac{1}{3}x + 4$

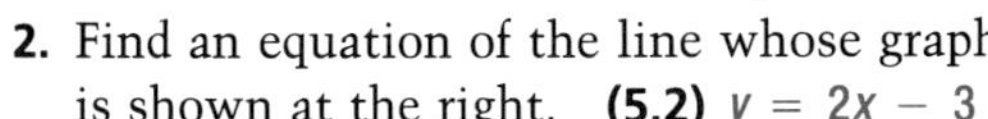

2. Find an equation of the line whose graph is shown at the right. **(5.2)** $y = 2x - 3$

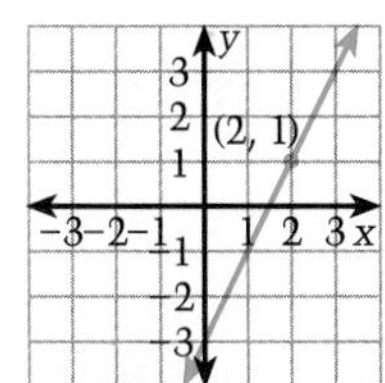

3. The Bargain Car Rental company charges a flat fee of \$35 plus \$0.40 per mile to rent a mid-size car. Write an equation that gives the total cost, C, in terms of the number of miles driven, x. **(5.1)** $C = 0.40x + 35$

4. Find an equation of the line that passes through $(-2, 4)$ with a slope of 3. **(5.2)** $y = 3x + 10$

5. Find an equation of the line that passes through $(3, -2)$ and is parallel to the line $y = -\frac{2}{3}x + 4$. **(5.2)** $y = -\frac{2}{3}x$

6. Find an equation of the horizontal line that passes through $(-4, 3)$. **(5.2)** $y = 3$

7. Between 1980 and 1990, the amount spent on advertising in the United States increased by approximately \$8.14 billion per year. In 1984, the total amount spent on advertising was \$88.1 billion. Find an equation that gives the total amount spent on advertising, A, in terms of the year, t. Let $t = 0$ correspond to 1980. **(5.2)** $A = 8.14t + 55.54$

8. Find an equation of the line that passes through $(-3, 5)$ and $(4, 1)$. **(5.3)** $y = -\frac{4}{7}x + \frac{23}{7}$

9. Find an equation of the line whose x-intercept is 5 and whose y-intercept is 4. **(5.3)** $y = -\frac{4}{5}x + 4$

10. The average weight of seven-year-old girls is 49 pounds. The average weight of eleven-year-old girls is 77 pounds. Use this information to find a linear model for the average weight, w, of girls who are x years old. Use the model to find the average weight of a ten-year-old girl. **(5.3)** $y = 7x$, 70 pounds

11. The table at the right shows the weight and wing area of several types of birds. Construct a scatter plot for these data and find an equation of the line that best fits the data. Let x represent the weight in grams. Let y represent the wing area in square centimeters. Use your equation to approximate the wing area of a bird whose weight is 400 grams. **(5.4)** See margin.

Bird	Weight	Wing Area
Sparrow	25 g	87 cm^2
Martin	47 g	186 cm^2
Blackbird	78 g	245 cm^2
Starling	93 g	190 cm^2
Dove	143 g	357 cm^2
Crow	607 g	1344 cm^2
Gull	840 g	2006 cm^2
Blue Heron	2090 g	4436 cm^2

Answer

11.

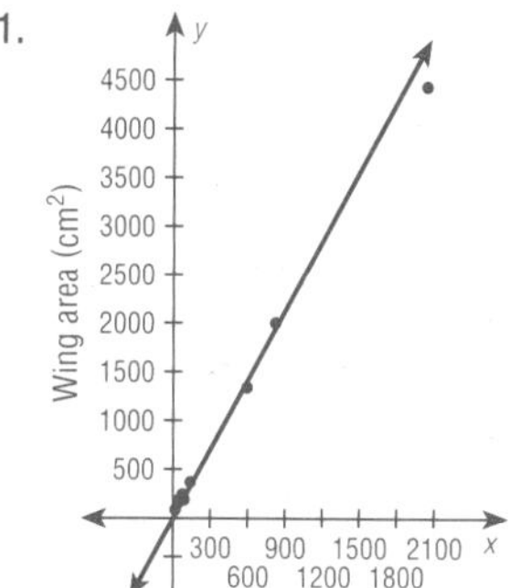

The best fitting line varies. $y = \frac{12}{5}x$; 960 cm^2

Problem of the Day

All faces of a cube are painted red. Then the cube is cut into 27 smaller cubes of the same size by two cuts in each of three directions through the large cube. How many of the little cubes have one, two, three, four, and no red faces?

6, 12, 8, 0, 1

ORGANIZER

Warm-Up Exercises

1. Solve each equation for y.
   a. $2x = 4y - 8$
   b. $12 - 3x = 5y$
   c. $9y - 6x + 18 = 0$

   a. $y = \frac{1}{2}x + 2$
   b. $y = \frac{-3}{5}x + \frac{12}{5}$
   c. $y = \frac{2}{3}x - 2$
2. Rewrite each equation, using integer coefficients.
   a. $\frac{2}{3}x - 2 = 7$
   b. $-9x + \frac{4}{5}y = 1$
   c. $2.3y - 0.25x = 0.86$

   a. $2x - 6 = 21$, b. $-45x + 4y = 5$, c. $230y - 25x = 86$

Lesson Resources

Teaching Tools
- Transparency: 4
- Problem of the Day: 5.5
- Warm-up Exercises: 5.5
- Answer Masters: 5.5

Extra Practice: 5.5

Color Transparency: 26

LESSON Notes

Example 1

Ask students if the equation $10x - 25y = 75$ is an equation equivalent to the original equation in this example. (It is.) Have students explain their responses.

5.5 Standard Form of a Linear Equation

What you should learn:

Goal 1 How to transform a linear equation into standard form

Goal 2 How to model a real-life situation using the standard form of a linear equation

Why you should learn it:

You can best model some real-life situations with a linear equation in standard form, such as when you must buy a combination of items with a fixed amount of money.

Goal 1 Transforming into Standard Form

Most of the linear equations in this chapter have been written in slope-intercept form.

$y = mx + b$ *Slope-intercept form*

Another commonly used form is

$Ax + By = C$ *Standard form*

which is called the **standard form** of the equation of a line. (This is also called the *general form* of the equation of a line.) Notice, in this form, that the variable terms are on the left side of the equation and the constant term is on the right side. One advantage of the standard form is that it can be used for *any* type of line, even a vertical line.

Example 1 *Transforming a Linear Equation into Standard Form*

Transform $y = \frac{2}{5}x - 3$ into standard form with integer coefficients.

Solution Note that this equation is given in slope-intercept form. To transform the equation into standard form, you need to isolate the variable terms on the left and the constant term on the right.

$y = \frac{2}{5}x - 3$ *Slope-intercept form*

$5y = 2x - 15$ *Multiply both sides by 5.*

$-2x + 5y = -15$ *Subtract 2x from both sides.*

A linear equation can have more than one standard form. For instance, if you multiplied both sides of the standard form listed above by -1, you would obtain $2x - 5y = 15$, which is also in standard form. ■

There really is no *best form* of a linear equation. For some purposes, such as graphing using intercepts, the standard form is better. For other purposes, such as mentally visualizing the graph, the slope-intercept form is better. So you need to be able to transform equations from one form to another.

Example 2 *Writing an Equation in Standard Form*

Rewrite $y = 0.55x + 1.35$ in standard form with integer coefficients.

Solution Since $0.55 = \frac{55}{100}$ and $1.35 = \frac{135}{100}$, you can clear the equation of decimals by multiplying both sides by 100.

$y = 0.55x + 1.35$	*Given equation*
$100y = 55x + 135$	*Multiply both sides by 100.*
$-55x + 100y = 135$	*Subtract 55x from both sides.*

This equation is in standard form with integer coefficients. However, you can simplify the equation further by dividing both sides of the equation by 5 to obtain $-11x + 20y = 27$. ■

Extra Examples

Here are additional examples similar to **Examples 1–3.**

1. Write each equation in standard form with integer coefficients.
   a. $y = \frac{3}{4}x + 2$
   b. $y = -3x - \frac{5}{8}$
   c. $y = -0.23x + 5.2$
   d. $1.5y = 3 - 1.11x$
   e. $\frac{2}{5}y = 3x - \frac{1}{3}$
   a. $-3x + 4y = 8$, b. $-24x - 8y = 5$ c. $23x + 100y = 520$, d. $111x + 150y = 300$, e. $-45x + 6y = -5$
2. You are given \$6.00 to buy bananas and apples. Bananas cost \$0.49 per pound, and apples cost \$0.34 per pound. Write a linear equation that represents the different amounts of fruit you can buy. $0.49x + 0.34y = 6$

Goal 2 Modeling a Real-Life Situation

The next example describes a real-life problem for which the standard form of a linear equation is a convenient model.

At the Kimchee Grocery Store, hamburger costs \$2 per pound and boned chicken costs \$3 per pound.

Problem Solving **Write an Equation**

Example 3 *A Linear Model for Buying Party Food*

You are in charge of buying the hamburger and boned chicken for a party. You have \$30 to spend, and are going to the Kimchee Grocery Store. Write an equation that represents the different amounts of hamburger and chicken that you can buy.

Solution

Verbal Model [Price per pound] · [Pounds of hamb.] + [Price per pound] · [Pounds of chicken] = [Total price]

Labels
Price of hamburger = 2 (dollars per pound)
Weight of hamburger = x (pounds)
Price of chicken = 3 (dollars per pound)
Weight of chicken = y (pounds)
Total price = 30 (dollars)

Equation $2x + 3y = 30$ **Linear model** ■

Example 4

Point out that obtaining the slope-intercept form of the equation makes it easier to sketch a graph.

Check Understanding

WRITING Write a paragraph that describes the advantages of using one representation over another. Cite examples of problems you have solved, where appropriate.

Communicating about **ALGEBRA**

Answers

C.

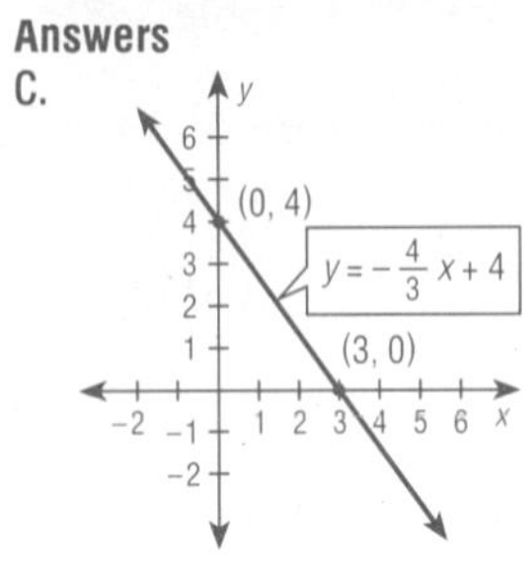

E. Any size greater than 3 cubic feet and less than 4 cubic feet

EXTEND *Communicating*
Suppose the garden-supply store can purchase topsoil for \$0.69 per pound and peat moss for \$0.99 per pound. How might this information affect the composition of the 12-pound packages of potting soil?

Since peat moss is more expensive than topsoil, the store could decide that the 12-pound bag should contain more topsoil.

266

Problem Solving
Make a Table

Example 4 *Interpreting a Linear Model*

Sketch the graph of the linear equation in Example 3. Then find several different amounts of hamburger and chicken that you can buy, and plot the corresponding points on the line.

Solution To sketch the graph, first write the equation in slope-intercept form.

$$2x + 3y = 30 \quad \textit{Standard form}$$
$$3y = -2x + 30$$
$$y = -\frac{2}{3}x + 10 \quad \textit{Slope-intercept form}$$

Party Shopping

Chicken (in pounds)

Hamburger (in pounds)

From this form, you see that the slope is $-\frac{2}{3}$ and the y-intercept is 10. Several different amounts of hamburger and chicken are shown in the table.

Hamburger, x	0 lb	3 lb	6 lb	9 lb	12 lb	15 lb
Chicken, y	10 lb	8 lb	6 lb	4 lb	2 lb	0 lb

The points corresponding to these different amounts are shown on the graph. Note that as the number of pounds of hamburger increases, the number of pounds of chicken decreases, and vice versa. ■

Communicating about ALGEBRA

Cooperative Learning

SHARING IDEAS about the Lesson

Work with a Partner and Use a Linear Model A garden supply store is making up 12-pound packages of potting soil. Each package contains topsoil that weighs 4 pounds per cubic foot and peat moss that weighs 3 pounds per cubic foot.

A. Write an equation in standard form that represents the different amounts of topsoil and peat moss that can be used in a 12-pound package. (Let x represent the number of cubic feet of topsoil and let y represent the number of cubic feet of peat moss.) $4x + 3y = 12$

B. Write the equation in slope-intercept form. $y = -\frac{4}{3}x + 4$

C. Sketch the line represented by the equation. See margin.

D. Find several different amounts of topsoil and peat moss that can be used to make a 12-pound package. Then plot the corresponding points on the line. $(\frac{3}{4}, 3)$, $(2, 1\frac{1}{3})$

E. What sizes, in cubic feet, are possible for the 12-pound packages of potting soil? See margin.

EXERCISES

Guided Practice

CRITICAL THINKING about the Lesson

1. What is the name of the $y = mx + b$ form of the equation of a line? Slope-intercept form

2. What is the name of the $Ax + By = C$ form of the equation of a line? Standard form

3. Write the equation $y = -\frac{1}{2}x + 5$ in standard form. $x + 2y = 10$

4. The equations $y = -\frac{2}{3}x + 3$ and $2x + 3y = 9$ are equivalent. Which form would you use for graphing the line represented by the equation? Which form would have been easier to write if you were creating a model for the different amounts of hamburger and chicken you could buy for \$9 at the Kimchee Grocery Store? $y = -\frac{2}{3}x + 3$, $2x + 3y = 9$

5. You have \$10 to buy tomatoes and avocados for a salad. Tomatoes cost \$1.25 per pound and avocados cost \$2 per pound. Write a linear equation that represents the different amounts of tomatoes, x, and avocados, y, that you could buy. $1.25x + 2y = 10$

6. Use the linear model in Exercise 5 to complete the table.

Tomatoes, x	0 lb	1.6 lb	4 lb	6.4 lb	8 lb
Avocados, y	?	?	?	?	?
	5 lb	4 lb	2.5 lb	1 lb	0 lb

Independent Practice

In Exercises 7–18, write the equation in standard form with integer coefficients.

7. $3x - y - 6 = 0$ $3x - y = 6$
8. $x + 2y - 4 = 0$ $x + 2y = 4$
9. $-4x + 3y + 18 = 0$ $4x - 3y = 18$
10. $2x - 3y - 12 = 0$ $2x - 3y = 12$
11. $x - 3 = 0$ $x = 3$
12. $y + 4 = 0$ $y = -4$
13. $y = -3x + 4$ $3x + y = 4$
14. $y = 2x - 7$ $-2x + y = -7$
15. $y = -\frac{1}{3}x - 2$
16. $y = -\frac{3}{4}x + \frac{5}{4}$ $3x + 4y = 5$
17. $y = -\frac{1}{8}x + \frac{3}{8}$ $x + 8y = 3$
18. $y = -0.4x + 1.2$

15. $x + 3y = -6$
18. $2x + 5y = 6$

In Exercises 19 and 20, write equations (in standard form) of the horizontal line and the vertical line that pass through the point.

19. $(-1, 3)$ $x = -1$, $y = 3$
20. $(2, -4)$ $x = 2$, $y = -4$

In Exercises 21–24, write an equation (in standard form) of the line that passes through the point and has the given slope.

21. $(-4, 3)$, $m = -1$ $x + y = -1$
22. $(0, 5)$, $m = 2$ $-2x + y = 5$
23. $(3, -1)$, $m = 0$ $y = -1$
24. $(-4, -2)$, $m = \frac{1}{2}$ $x - 2y = 0$

In Exercises 25–28, write an equation (in standard form) of the line that passes through the two points.

25. $(2, 4)$, $(5, 6)$ $-2x + 3y = 8$
26. $(-3, 3)$, $(6, 7)$ $-4x + 9y = 39$
27. $(-2, -1)$, $(2, -3)$ $x + 2y = -4$
28. $(-5, 4)$, $(2, -6)$ $10x + 7y = -22$

EXERCISE Notes

ASSIGNMENT GUIDE

Basic/Average: Ex. 1–6, 7–17 odd, 19–22, 25, 27, 29–32, 35–43 odd, 45–47

Above Average: Ex. 1–6, 10–18 even, 19–20, 23–24, 27–32, 36–44 even, 45–49

Advanced: Ex. 1–6, 9–18 multiples of 3, 19–27 odd, 31–34, 36–44 even, 45–49

Selected Answers

Exercises 1–6, 7–43 odd

More Difficult Exercises

Exercises 33–34, 43–44, 48, 49

Guided Practice

Ex. 1–6 COOPERATIVE LEARNING These exercises should be done in small groups and then discussed as a class. Ask students to explain their answers in Exercise 4.

Independent Practice

Ex. 7–18 Point out to students that the use of integer coefficients provides one way to write equations in standard form. You may wish to reserve the even-numbered exercises to use as additional practice at a later time.

Ex. 21–24 Use the techniques of Lesson 5.2 and this lesson to complete.

Ex. 25–28 Use the techniques of Lesson 5.3 and this lesson to complete.

Answers

30. 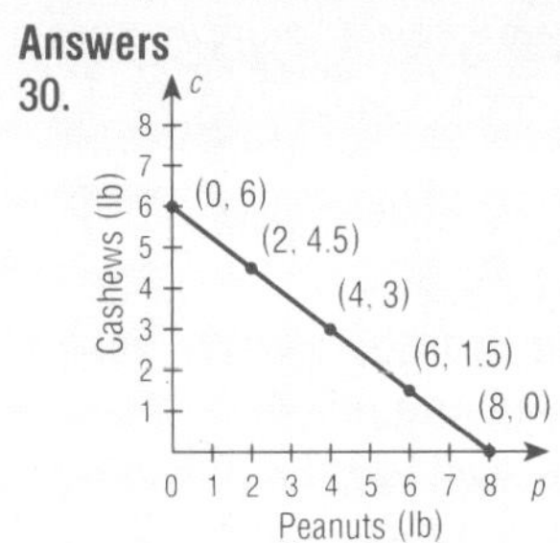

32. 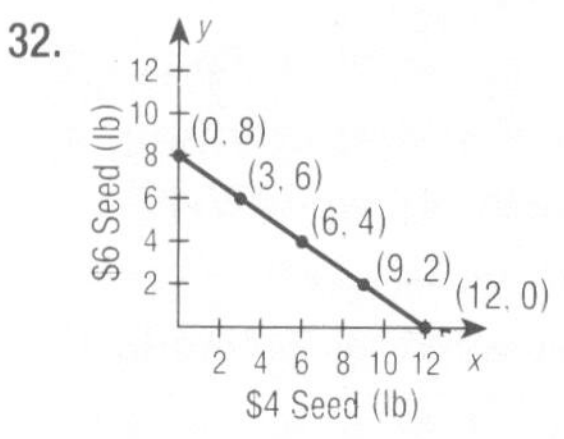

39. 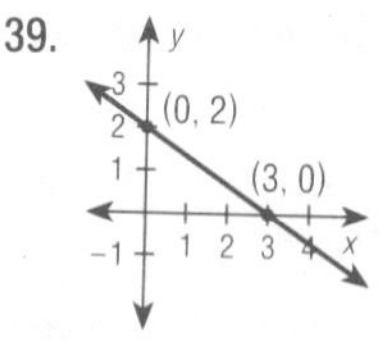

40. 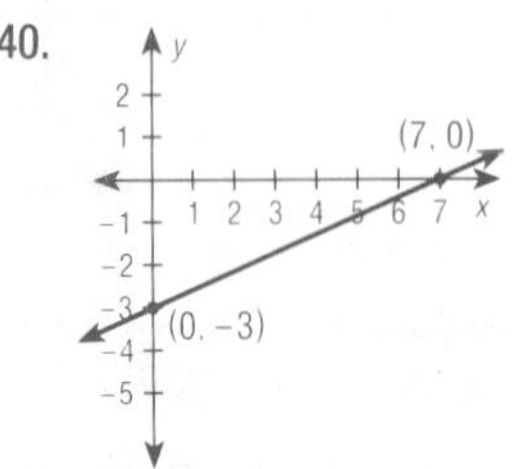

41. 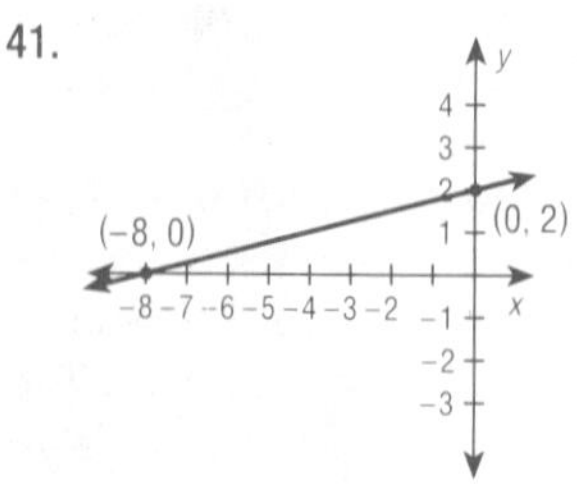

29. *Buying a Nut Mixture* You are buying $24 worth of peanuts and cashews for a party. The peanuts cost $3 dollars per pound and the cashews cost $4 per pound. Write an equation for the different amounts of peanuts, p, and cashews, c, that you could buy. $3p + 4c = 24$

30. Sketch the line representing the possible nut mixtures in Exercise 29. Then complete the table and label the points from the table on the graph. See margin.

Pounds of peanuts, p	0	2	4	6	8
Pounds of cashews, c	?	?	?	?	?

$6, 4\frac{1}{2}, 3, 1\frac{1}{2}, 0$

31. *Buying a Lawn Seed Mixture* You are buying $48 worth of lawn seed that consists of two types of seed. One type is a quick-growing rye grass that costs $4 per pound, and the other type is a higher-quality seed that costs $6 per pound. Write an equation that represents the different amounts of $4 seed, x, and $6 seed, y, that you could buy. $4x + 6y = 48$

32. Sketch the line representing the possible seed mixtures in Exercise 31. Then complete the table and label the points from the table on the graph. See margin.

Pounds of $4 seed, x	0	3	6	9	12
Pounds of $6 seed, y	?	?	?	?	?
	8	6	4	2	0

In Exercises 33 and 34, use the following information.

Color Printing In printing, different shades of color are often formed by "overprinting" two or more primary colors. In the color palette matrix at the right, the primary colors red and blue are overprinted to form different shades of purple. For instance, combining a 40% red screen with a 20% blue screen produces a more reddish purple than combining a 20% red screen with a 40% blue screen.

Percent Magenta (Red): 100%, 60%, 50%, 40%, 30%, 20%, 10%, 0%

Percent Cyan (Blue): 0%, 10%, 20%, 30%, 40%, 50%, 60%, 100%

33. You are designing an advertising piece and want the background to be overprinted with red and blue screens. You want the *combined* percentage of red and blue screens to be 60%. Write a linear equation that represents the different percentages of red, x, and blue, y, you could use. $x + y = 60$

34. Trace the red-blue color palette matrix at the right and identify the possible shades you could use for a combined percentage of 60%. How does the pattern formed by the possible shades relate to the line whose equation you found in Exercise 33? When x increases, y decreases; when x decreases, y increases.

Integrated Review

In Exercises 35–38, write the equation in slope-intercept form. Sketch the line. See Additional Answers.

35. $2x + y = 7$ $y = -2x + 7$

36. $x + 2y = 6$ $y = -\frac{1}{2}x + 3$

37. $-x + 3y = -6$ $y = \frac{1}{3}x - 2$

38. $6x - 2y = 1$ $y = 3x - \frac{1}{2}$

In Exercises 39–42, sketch the line. Label the *x*-intercept and *y*-intercept. See margin.

39. $2x + 3y = 6$

40. $3x - 7y = 21$

41. $-x + 4y = 8$

42. $-5x - 8y = 40$

43. *Temperature in Anchorage* The average monthly temperatures (in degrees Fahrenheit) in Anchorage, Alaska, for January through June are shown in the graph at the right. Approximate the best-fitting line for this data. Write your equation in standard form. Best-fitting lines vary; $-8x + y = 4$

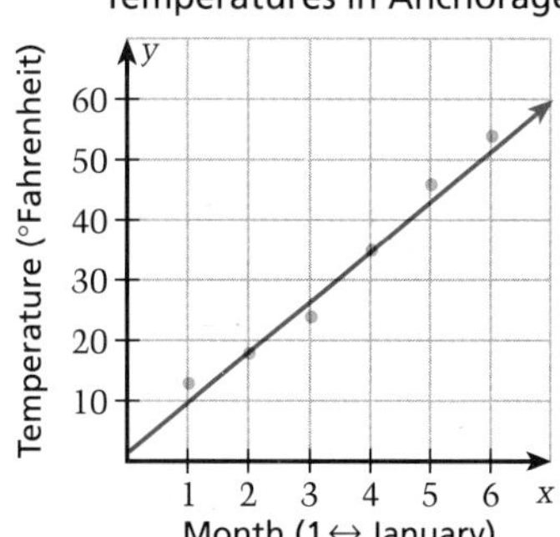

44. *Shoe Sales* The table shows sales (in billions of dollars) of athletic and sports footwear in the United States from 1980 to 1990. Sketch a scatter plot for the data. Let y represent the sales (in billions of dollars) and let x represent the year, with $x = 0$ corresponding to 1980. Then approximate the best-fitting line for the scatter plot. Write your equation in standard form. ***(Source: National Sporting Goods Association)*** See margin.

Year, *x*	1980	1981	1982	1983	1984	1985	1986	1987	1988	1989	1990
Sales, *y*	1.73	1.79	1.90	2.19	2.38	2.61	3.20	3.52	3.77	3.96	4.21

Best-fitting lines vary; $-x + 4y = 7$

Exploration and Extension

In Exercises 45–47, refer to the equation $2x + 7y = 14$.

45. What is the x-intercept of the graph? 7

46. What is the y-intercept of the graph? 2

47. Multiply each term in the equation by $\frac{1}{14}$. $\frac{1}{7}x + \frac{1}{2}y = 1$

48. For the equation in Exercise 47, find a relationship between the denominators on the left side of the equation and the x- and y-intercepts. See margin.

In Exercise 49, use the intercept form of the equation of a line. You can see from the equation that the *x*-intercept is *a* and the *y*-intercept is *b*.

$$\frac{x}{a} + \frac{y}{b} = 1$$

49. Use the intercept form to find an equation for the line whose x-intercept is 2 and whose y-intercept is 3. $3x + 2y = 6$

Answers

42.

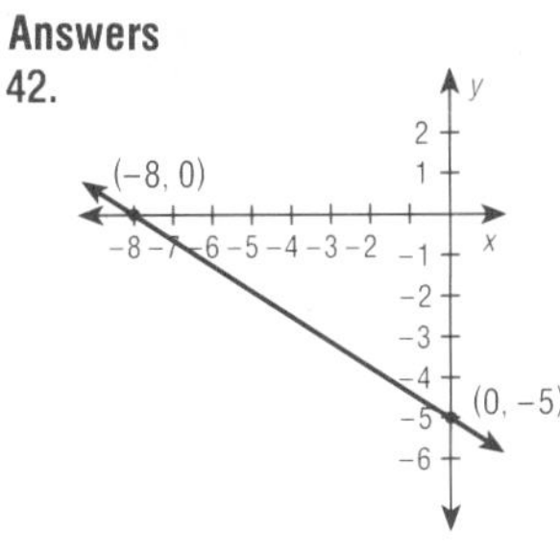

44.

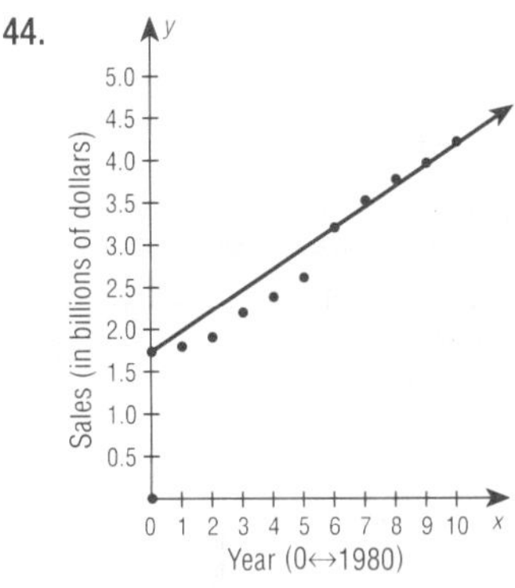

48. The denominator of the coefficient of x is the x-intercept; the denominator of the coefficient of y is the y-intercept.

Enrichment Activity

This activity requires students to go beyond lesson goals.

Exercise 49 introduces the intercept form of an equation: $\frac{x}{a} + \frac{y}{b} = 1$. Starting with the linear equation $Ax + By = C$, show how the intercept form of a linear equation can be derived by finding the x- and y-intercepts of $Ax + By = C$. The techniques are found in Chapter 4.

If $Ax + By = C$, then let $y = 0$ to find the x-intercept a and let $x = 0$ to find the y-intercept b. So, $a = \frac{C}{A}$ and $b = \frac{C}{B}$. For $C \neq 0$, $Ax + By = C$ is equivalent to $\frac{A}{C}x + \frac{B}{C}y = 1$. Substitute $\frac{1}{a}$ for $\frac{A}{C}$ and $\frac{1}{b}$ for $\frac{B}{C}$. So, $\frac{A}{C}x + \frac{B}{C}y = 1$ is equivalent to $(\frac{1}{a})x + (\frac{1}{b})y = 1$, or $\frac{x}{a} + \frac{y}{b} = 1$.

Problem of the Day

Count the triangles in these diagrams. 13; 118

ORGANIZER

Warm-Up Exercises

Find the slope m of a line, given two of its points.

a. $(-4, 0)$ and $(2, 7)$

b. $(3, -3)$ and $(6, 9)$

c. $(2, 8)$ and $(a, 5)$

d. $(-12, b)$ and $(-3, 4)$

a. $m = \frac{7}{6}$,

b. $m = 4$,

c. $m = \frac{-3}{a-2}$ or $\frac{3}{2-a}$,

d. $m = \frac{4-b}{9}$ or $\frac{b-4}{-9}$

Lesson Resources

Teaching Tools
- Problem of the Day: 5.6
- Warm-up Exercises: 5.6
- Answer Masters: 5.6

Extra Practice: 5.6

Color Transparency: 26

Common-Error ALERT!

Caution students about the differences between the points (x, y) and (x_1, y_1) when applying the point-slope form of an equation of a line. In practice, the point (x_1, y_1) is a known point on the line. Its x- and y-coordinates have number values assigned to them. The x- and y-coordinates of the point (x, y), on the other hand, remain as variables.

5.6 Point-Slope Form of the Equation of a Line

What you should learn

Goal 1 How to use the point-slope form to write an equation of a line

Goal 2 How to model a real-life situation using the point-slope form of a linear equation

Why you should learn it:

You can sometimes use the point-slope form as the most efficient way to find a linear model.

Goal 1 Using the Point-Slope Form

In Lesson 5.2, you studied one technique for using the slope of a line and a point on the line to write an equation of the line. In this lesson, you will study a different technique that involves a form called the **point-slope form** of an equation of a line.

To begin, consider the line containing the point $(x_1, y_1) = (2, 5)$ with a slope of $\frac{2}{3}$. Let (x, y) be any other point on the line. Because $(2, 5)$ and (x, y) are two points on the line, you know that the slope of the line is

$$m = \frac{\text{rise}}{\text{run}} = \frac{y-5}{x-2}.$$

Because the slope was given as $\frac{2}{3}$, you can write

$$\frac{y-5}{x-2} = \frac{2}{3}$$

$$y - 5 = \frac{2}{3}(x - 2).$$

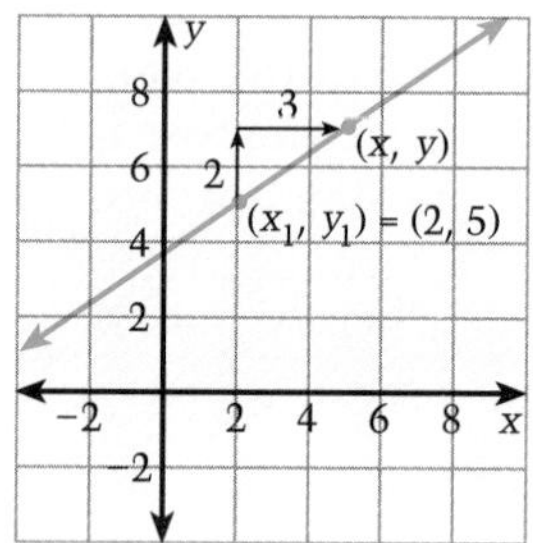

This equation is said to be in point-slope form

$$y - y_1 = m(x - x_1),$$

where $y_1 = 5$, $m = \frac{2}{3}$, and $x_1 = 2$.

Point-Slope Form of the Equation of a Line

Consider a nonvertical line that passes through the point (x_1, y_1) with a slope of m. The **point-slope form** of the equation of the line is

$$y - y_1 = m(x - x_1).$$

When using the point-slope form, be sure you see the difference between (x_1, y_1) and (x, y). The coordinates x_1 and y_1 represent the *given* point on the line, and the coordinates x and y represent *any other* point on the line.

Example 1 Using the Point-Slope Form to Write an Equation

Write an equation of the line shown at the left.

Solution Since the line passes through $(x_1, y_1) = (-3, 6)$ and $(x_2, y_2) = (1, -2)$, its slope is

$$m = \frac{y_2 - y_1}{x_2 - x_1} = \frac{-2 - 6}{1 - (-3)} = \frac{-8}{4} = -2.$$

To find an equation of the line, use the point-slope form and then simplify as follows:

$y - y_1 = m(x - x_1)$	*Point-slope form*
$y - 6 = -2(x - (-3))$	*Substitute -2 for m, -3 for x_1, 6 for y_1.*
$y - 6 = -2(x + 3)$	*Simplify.* ■

Remember that the point-slope form, $y - y_1 = m(x - x_1)$, has *two* minus signs. Be sure to account for these signs when the point (x_1, y_1) involves negative numbers.

Example 2 Changing from Point-Slope Form to Slope-Intercept Form

Write an equation of the line that passes through (2, 3) with a slope of $-\frac{1}{2}$.

Solution Using the point-slope form with $(x_1, y_1) = (2, 3)$ and $m = -\frac{1}{2}$, you can write the following.

$y - y_1 = m(x - x_1)$	*Point-slope form*
$y - 3 = -\frac{1}{2}(x - 2)$	*Substitute $-\frac{1}{2}$ for m, 2 for x_1, 3 for y_1.*

Once you have used the point-slope form to find an equation, you can simplify your result to slope-intercept form, as follows.

$y - 3 = -\frac{1}{2}(x - 2)$	*Point-slope form*
$y - 3 = -\frac{1}{2}x + 1$	*Distributive Property*
$y = -\frac{1}{2}x + 4$	*Slope-intercept form*

Graphic Check Try sketching the graph of the line that passes through the point (2, 3) with a slope of $-\frac{1}{2}$. From your graph, does it appear that the y-intercept is 4? ■

LESSON Notes

Examples 1–2

The point-slope form of an equation can be used to represent the line. It is more common, however, to transform the point-slope form into the slope-intercept form.

Extra Examples

Here are additional examples similar to **Examples 1–2.**

1. Write an equation of the line that passes through point P with slope m.
   - **a.** $P(-14, -8)$ and $m = \frac{4}{5}$
   - **b.** $P(-1, 3)$ and $m = -2$
   - **c.** $P(0, 7)$ and $m = 12$

   a. $y = \frac{4}{5}x + \frac{16}{5}$,
   b. $y = -2x + 1$,
   c. $y = 12x + 7$

2. Write an equation of the line that passes through the points P and Q.
   - **a.** $P(6, -8)$ and $Q(3, 4)$
   - **b.** $P(-2, -4)$ and $Q(9, 1)$
   - **c.** $P(0, -1)$ and $Q(2, 3)$

   a. $y = -4x + 16$,
   b. $y = \frac{5}{11}x - \frac{34}{11}$,
   c. $y = 2x - 1$

Common-Error ALERT!

In applying the point-slope form of an equation $y - y_1 = m(x - x_1)$, students frequently neglect to multiply both terms in the parentheses by the slope m. Caution students to apply the distributive property correctly.

You have now studied all of the commonly used forms of equations of lines. They are summarized in the following list.

Summary of Equations of Lines

Slope of a line through two points:	$m = \frac{y_2 - y_1}{x_2 - x_1}$
Vertical line (undefined slope):	$x = a$
Horizontal line (zero slope):	$y = b$
Slope-intercept form:	$y = mx + b$
Point-slope form:	$y - y_1 = m(x - x_1)$
Standard form:	$Ax + By = C$

Goal 2 Modeling a Real-Life Situation

The next example shows how to use a linear equation as a model for a parachute jump. A linear model is appropriate for that part of a parachute jump in which the parachute is open. This is why a parachute works. It creates such a large air resistance that the descent speed of the parachutist is constant (or nearly constant).

Descent speeds for parachutists vary. A speed of approximately 23 feet per second is common. That is the speed that would be attained by a person jumping (without a parachute) from a height of $8\frac{1}{2}$ feet. The parachutist must know proper landing techniques to avoid injury at such a landing speed.

Note that before the parachute opens, the parachutist's fall is an *acceleration,* which cannot be modeled with a linear equation.

Connections **Physical Science**

Example 3 *A Linear Model for Parachute Jumping*

At 12:15, a parachutist jumps from a plane and opens the parachute. Eight seconds later the parachutist is 2000 feet above the ground. After five more seconds, the parachutist's height is 1900 feet. Write a linear equation that gives the height, y (in feet), of the parachutist in terms of the time, t. Let t be measured in seconds, with $t = 0$ corresponding to 12:15. Find the height of the parachutist at 12:16.

Solution To find an equation for the model, use the two heights that are given in the problem.

Height at 12:15:*08* is *2000* feet: Point (8, 2000)
Height at 12:15:*13* is *1900* feet: Point (13, 1900)

The slope of the line that passes through these two points is

$$m = \frac{1900 - 2000}{13 - 8} = \frac{-100}{5} = -20.$$

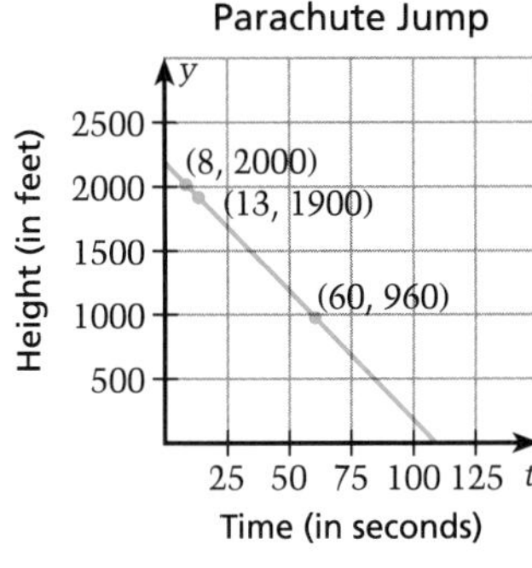

Now, using the point-slope form with $m = -20$ and $(t_1, y_1) = (8, 2000)$, you can write the equation of the line as follows:

$y - y_1 = m(t - t_1)$	*Point-slope form*
$y - 2000 = -20(t - 8)$	*Substitute -20 for m, 8 for t_1, 2000 for y_1.*
$y - 2000 = -20t + 160$	*Distributive Property*
$y = -20t + 2160$	*Slope-intercept form*

To find the height at 12:16, let $t = 60$ (60 seconds after 12:15). At that time, the height of the parachutist is

$$y = -20(60) + 2160 = 960 \text{ feet.}$$

■

Communicating about ALGEBRA

SHARING IDEAS about the Lesson

Interpret a Model Use Example 3 to answer the questions.

A. What does the slope of $m = -20$ represent in this problem? Why is this slope negative? See margin.

B. What does the y-intercept of $b = 2160$ represent? See margin.

C. What is the height of the parachutist at 12:15:30? 1560 ft

D. When will the parachutist land on the ground? 12:16:48

Check Understanding

Make a flowchart of the steps needed to write the equation of a line given two points on the line. Include all the possible paths that exist for all the different forms of a linear equation. Check students' charts.

Communicating about ALGEBRA

COOPERATIVE LEARNING

Students should be encouraged to answer the question and perform the experiment described below in a small-group setting.

Answers

A. The number of feet per second the parachutist is traveling; the parachutist is traveling downward.

B. The height at which the parachutist jumped

EXTEND *Communicating* ***AN EXPERIMENT***

1. Try making a small parachute using a handkerchief, some thread, and a small weight that is about as heavy as a pencil. Drop the parachute from a height of at least 8 feet and time its descent. Use your results to create a mathematical model for your "parachute jump". Models vary.

2. Repeat the experiment with different weights attached to the parachute. How do the resultant linear models compare to the original data? The heavier the weight, the faster the object falls.

EXERCISE Notes

ASSIGNMENT GUIDE

Basic/Average: Ex. 1–8, 9–17 odd, 20, 21–25 odd, 27–32.

Above Average: Ex. 1–8, 10–26 even, 27–32.

Advanced: Ex. 1–8, 10–26 even, 27–32.

Selected Answers
Exercises 1–8, 9–29 odd

Use **Mixed Review** as needed.

More Difficult Exercises
Exercises 20, 31–32

Guided Practice

Ex. 1–8 COOPERATIVE LEARNING Students should first complete each of these exercises. Then they should compare their work with others in a small-group setting.

Independent Practice

Ex. 15–18 The equation of the line is acceptable in either the point-slope form or the slope-intercept form.

Answers

9. $y = \frac{1}{3}x + \frac{8}{3}$
10. $y = -2x + 8$
11. $y = -x + 1$
12. $y = 2x + 7$
13. $y = \frac{1}{2}x - \frac{3}{2}$
14. $y = -\frac{1}{2}x - \frac{1}{2}$
15. $y + 2 = -\frac{4}{3}(x - 3)$ or $y - 2 = -\frac{4}{3}(x - 0)$
16. $y + 2 = -\frac{1}{3}(x - 2)$ or $y - 0 = -\frac{1}{3}(x + 4)$
17. $y - 0 = \frac{3}{5}(x - 5)$ or $y + 3 = \frac{3}{5}(x - 0)$
18. $y - 4 = \frac{4}{3}(x - 2)$ or $y - 0 = \frac{4}{3}(x + 1)$

EXERCISES

Guided Practice

CRITICAL THINKING about the Lesson

1. Name the $y - y_1 = m(x - x_1)$ form of the equation of a line. Point-slope form
2. A line passes through $(2, -3)$ with a slope of 2. For the form $y - y_1 = m(x - x_1)$, what is the value of m? What is the value of x_1? What is the value of y_1? 2, 2, −3
3. Write an equation of the line that passes through $(2, -3)$ with a slope of 2. $y = 2x - 7$
4. Write the equation $y - 4 = -\frac{1}{2}(x + 2)$ in slope-intercept form. $y = -\frac{1}{2}x + 3$
5. Which of the points lie on the line $y + 2 = -2(x - 3)$?
   a. $(-1, 6)$ **b.** $(2, 2)$

Mountain Climbing **In Exercises 6–8, use the following information.**

A mountain climber is scaling a 300-foot cliff. The climber starts at the bottom of the cliff at 12:00. At 12:30, the climber has moved 62 feet up the cliff. (Assume that the climber maintains a constant climbing rate.)

6. Write an equation that gives the height, y (in feet), of the climb in terms of the time, t (in hours). $y = 124t$
7. At what time will the mountain climber reach the top of the cliff? About 2:25 P.M.
8. What is the slope of the line in Exercise 6? What does this slope represent? 124, number of feet climbed per hour

Independent Practice

In Exercises 9–14, write an equation of the line that passes through the point and has the given slope. Then rewrite the equation in slope-intercept form. See margin.

9. $(1, 3)$, $m = \frac{1}{3}$
10. $(2, 4)$, $m = -2$
11. $(-1, 2)$, $m = -1$
12. $(-2, 3)$, $m = 2$
13. $(1, -1)$, $m = \frac{1}{2}$
14. $(3, -2)$, $m = -\frac{1}{2}$

In Exercises 15–18, write an equation of the line. See margin.

15.
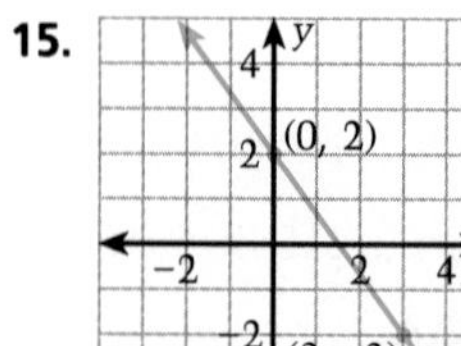

16.
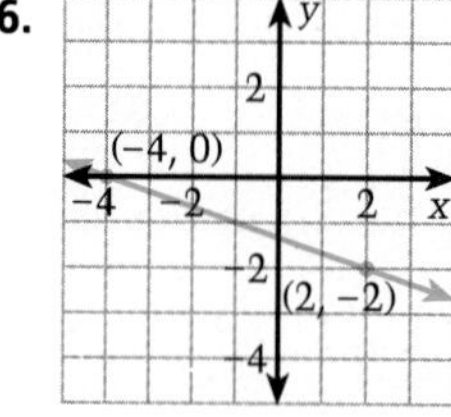

17.
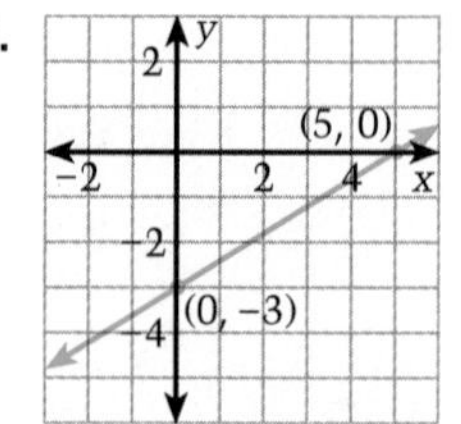

18.
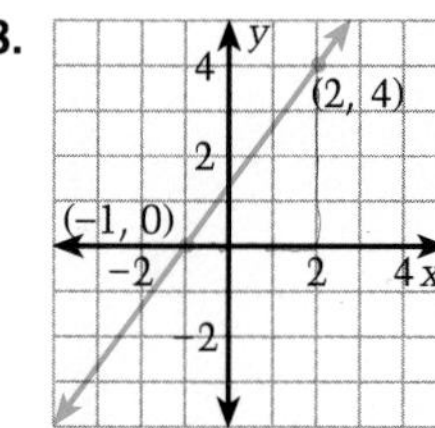

19. *Travel Time* You are flying from Montreal to Miami. You leave Montreal at 8:00 A.M. At 9:15 A.M., the pilot announces that you are flying over Washington, D.C. Use the map shown at the right to write a linear equation that gives the distance flown, y, in terms of the time, x. Let $x = 0$ correspond to 8:00 A.M. According to your equation, when will you reach Miami? (Assume that the plane is flying at a constant speed.)
$y = 639.2x$, 11:36 A.M.

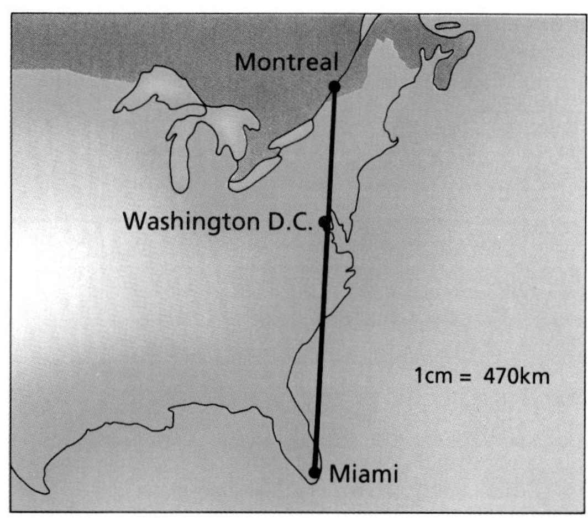

20. *Clue* The name of each of the six suspects is hidden on the graph. Write an equation of the line that contains each suspect's name. For example, the line that contains Mr. Green's name is $y = x + 11$. See margin.

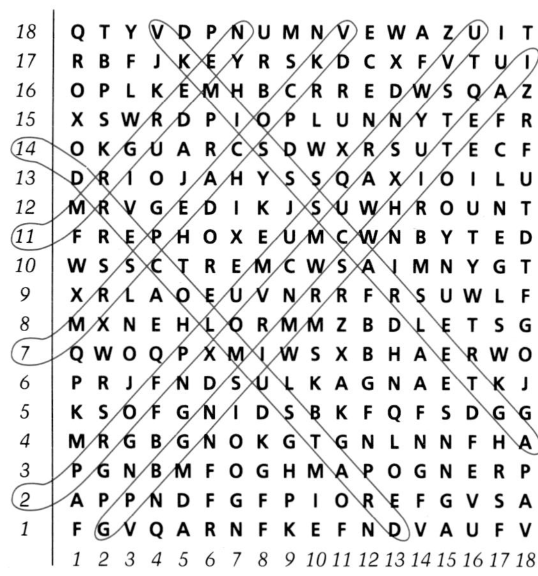

18 Q T Y V D P N U M N V E W A Z U I T
17 R B F J K E Y R S K D C X F V T U I
16 O P L K E M H B C R R E D W S Q A Z
15 X S W R D P I O P L U N N Y T E F R
14 O K G U A R C S D W X R S U T E C F
13 D R I O J A H Y S S Q A X I O I L U
12 M R V G E D I K J S U W H R O U N T
11 F R E P H O X E U M C W N B Y T E D
10 W S S C T R E M C W S A I M N Y G T
9 X R L A O E U V N R R F R S U W L F
8 M X N E H L O R M M Z B D L E T S G
7 Q W O Q P X M I W S X B H A E R W O
6 P R J F N D S U L K A G N A E T K J
5 K S O F G N I D S B K F Q F S D G G
4 M R G B G N O K G T G N L N N F H A
3 P G N B M F O G H M A P O G N E R P
2 A P P N D F G F P I O R E F G V S A
1 F G V Q A R N F K E F N D V A U F V
1 2 3 4 5 6 7 8 9 10 11 12 13 14 15 16 17 18

Integrated Review

In Exercises 21–26, determine which form would be best to use to find an equation of the line. Explain your answer. Then use the form to find an equation of the line. See margin.

21. The line has a slope of $\frac{2}{3}$ and passes through the point (0, 4).

22. The line passes through the points (1, 3) and (−2, 4).

23. The line has a slope of 2 and a y-intercept of −3.

24. The line passes through the points (1, 6) and (1, −5).

25. The line has a slope of 1 and passes through the point (1, −4).

26. The line has a slope of 0 and passes through the point (2, 5).

▶ **Ex. 20** Finding hidden words (or names) is a popular puzzle. (See Enrichment Activities that follows for ideas about using this puzzle to practice writing equations of linear models.)

Answers

20. Mrs. Peacock: $y = x + 7$; Prof. Plum: $y = x + 2$; Mrs. White: $y = x - 1$; Col. Mustard: $y = -x + 14$; Miss Scarlet: $y = -x + 22$
21. Slope-intercept form, because the slope and y-intercept are given; $y = \frac{2}{3}x + 4$
22. Point-slope form, because a point is given and the slope can be determined; $y - 3 = -\frac{1}{3}(x - 1)$
23. Slope-intercept form, because the slope and y-intercept are given; $y = 2x - 3$
24. Vertical line form, because the x-coordinates are the same; $x = 1$
25. Point-slope form, because the slope and a point are given; $y + 4 = (x - 1)$
26. Horizontal line form, because the line has a slope of 0; $y = 5$

Enrichment Activities

These activities require students to go beyond lesson goals.

1. **MATH JOURNAL** Summarize the different forms of linear equations and describe the conditions or situations for which one form is preferable over another.
2. Construct a hidden-word puzzle of algebra terms similar to the one in Exercise 20 of this lesson. Exchange your puzzle with a classmate and write equations for the lines containing the hidden words.

Answers

31. 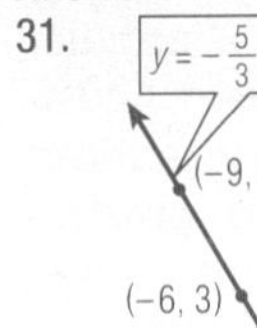

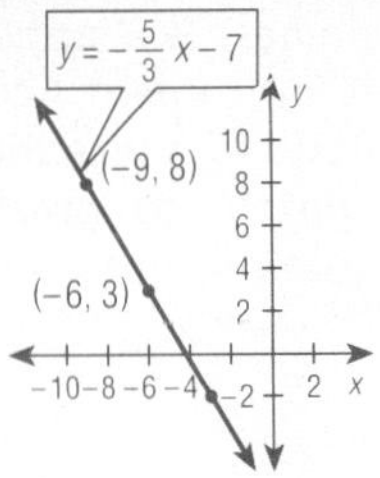

Integer coordinates fall 5 units up and 3 units to the left, or 5 units down and 3 units to the right, of the previous integer coordinates; $(-3, 2)$ and $(-9, 8)$

32. 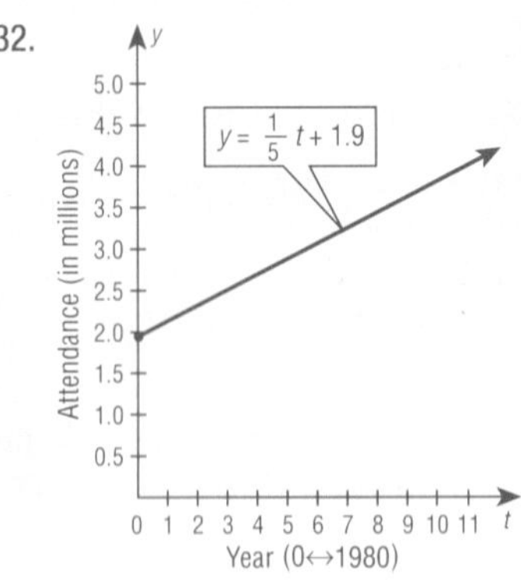

In Exercises 27–30, match the equation with its graph.

27. $y - 2 = -2(x - 1)$ b **28.** $y - 1 = \frac{1}{2}(x - 3)$ d **29.** $y = 2(x - 1)$ c **30.** $y = (-1)(x + 2)$ a

a. 

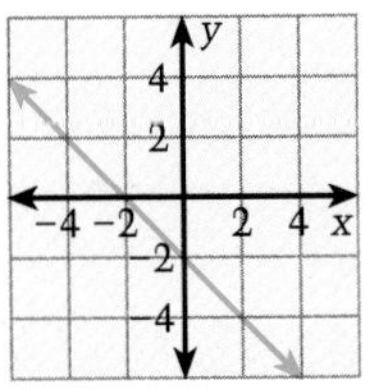

b. 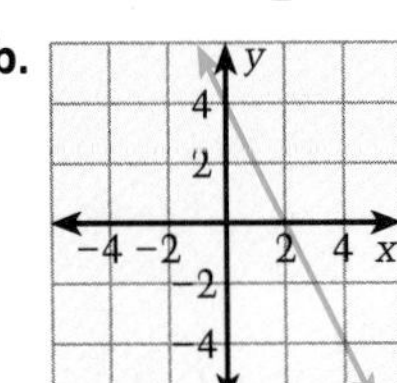

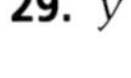

c. 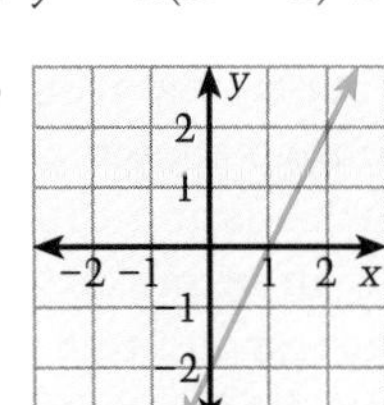

d. 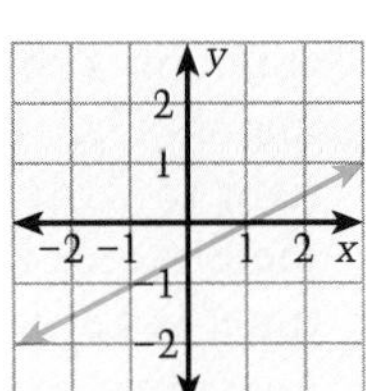

Exploration and Extension

31. Find an equation of the line passing through the point $(-6, 3)$ with a slope of $-\frac{5}{3}$. Sketch the line. Both coordinates of the point $(-6, 3)$ are integers. Describe the other points on the line that have integer coefficients. Of these, which two are closest to $(-6, 3)$?

32. *Basketball Attendance* Find an equation of the line that gives the N.C.A.A. women's basketball attendance, y (in millions), in terms of the year, t. Let $t = 0$ correspond to 1980. Sketch a graph of the line. Use the equation to estimate the attendance in 1995.

N.C.A.A. women's basketball attendance in 1985 and 1990 was 2.9 million and 3.9 million, respectively. (Source: N.C.A.A.)

31.–32. See margin.

Mixed REVIEW

1. Evaluate $\frac{1}{2}(x - \frac{1}{2})$ when $x = \frac{1}{4}$. **(2.6)** $-\frac{1}{8}$

2. Evaluate $\frac{3}{4}(3y - 2)$ when $y = \frac{1}{3}$. **(2.6)** $-\frac{3}{4}$

3. Solve $3a + 2b = 4$ for b. **(3.6)**

4. Solve $8m - 9n = 10$ for m. **(3.6)**

5. Is 7 a solution of $x(x - 3) < 21$? **(1.5, 2.6)** No

6. Is -6 a solution of $\frac{1}{4}(10 - x) \geq 4$? **(1.5, 2.6)** Yes

7. Are $3 \div 4 + 2$ and $3 \div (4 + 2)$ equivalent? **(1.4)** No

8. Write an equation of the line that passes through $(-2, 1)$ and $(-1, 3)$. **(5.3)**

9. Write an equation of the line that passes through $(2, 0)$ and $(-3, 4)$. **(5.3)**

10. Write an equation of the line that passes through $(5, 4)$ with a slope of 4. **(5.2)** $y - 4 = 4(x - 5)$

11. Write an equation of the line with a slope of 5 and a y-intercept of -3. **(5.1)**

3. $b = \frac{4 - 3a}{2}$ 4. $m = \frac{10 + 9n}{8}$

11. $y = 5x - 3$ 8. $y = 2x + 5$ 9. $y = -\frac{4}{5}x + \frac{8}{5}$

5.7 Problem Solving Using Linear Models

What you should learn:

Goal 1 How to create and use linear models to solve problems

Goal 2 How to make linear models that are accurate but simple to use

Why you should learn it:

You can use linear models efficiently and with good results only if they are simple and accurate.

Goal 1 Creating and Using Linear Models

There are two basic types of real-life problems that can be solved with linear models. One type of problem involves a constant rate of change. (See Example 1.) The other type of problem involves two variables, x and y, such that the sum $Ax + By$ is a constant. (See Example 2.) You have already studied examples of both types of problems. In this lesson, you will study several more.

As a salesclerk at a clothing store, Filipe receives a monthly base pay of \$800 plus a 5% commission on the amount that he sells.

Real Life
Retail Sales

Example 1 *A Linear Model for Monthly Pay*

Write a linear model that gives Filipe's total monthly pay in terms of the amount of clothes he sells.

Solution

Verbal Model: Total pay = Commission rate • Amount sold + Base pay

Labels:
Total pay $= y$ (dollars)
Commission rate $= 0.05$ (percent in decimal form)
Amount sold $= x$ (dollars)
Base pay $= 800$ (dollars)

Equation: $y = 0.05x + 800$ **Linear model**

The commission rate represents a constant rate of change in Filipe's total pay of \$0.05 per each dollar of clothes sold. For instance, if Filipe sold \$4000 worth of clothes, his total monthly pay would be $y = 0.05(4000) + 800 = \$1000$. ■

Problem of the Day

Study this diagram representing balance scales that are not in balance.

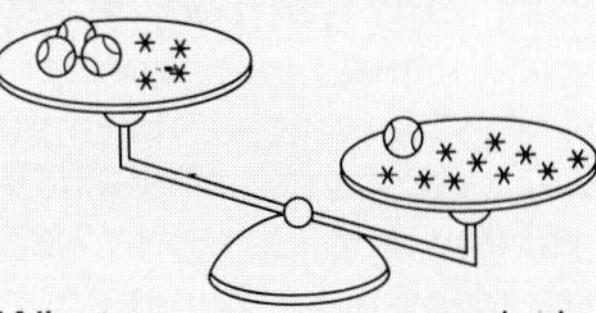

What can you conclude about the number of jacks that would balance a ball?
less than 3

ORGANIZER

Warm-Up Exercises

1. Order these steps in the problem-solving process.

   Steps
   - **a.** Write an algebraic model.
   - **b.** Answer the question.
   - **c.** Assign labels.
   - **d.** Write a verbal model.
   - **e.** Solve the algebraic model.

Order: d, c, a, e, b

2. Write the equation of the line containing the points $(-3, 6)$ and $(7, 4)$.
$y = \frac{-1}{5}x + \frac{27}{5}$

Lesson Resources

Teaching Tools
  Problem of the Day: 5.7
  Warm-up Exercises: 5.7
  Answer Masters: 5.7
Extra Practice: 5.7
Color Transparency: 27

LESSON Notes

Examples 1–3

Point out the two different types of problems represented in these examples. Example 1 is a constant-rate-of-change-problem, and Examples 2 and 3 represent a combination type problem in which the sum $Ax + By$ has a constant value.

Example 4

Point out that the table includes only data starting in 1984. This does not mean there were no personal computers in schools prior to 1984.

EXTEND Example 4

RESEARCH How accurate is the linear model in this example likely to be? What factors could cause there to be fewer computers in classroom use in 1992? Economic decline, limited funds for technology, school consolidations and closings, decline in student populations, etc. What factors could cause there to be more computers in classroom use? Lowering of the price, funded mandates from departments of education, etc.

Example 2 *Creating a Linear Model for Calorie Intake*

At Charles Drew High School, the main items on the lunch menu are often hamburgers and French fries. State law requires that school lunches contain at least 850 Calories. The hamburger bun has 80 Calories, a container of milk has 160 Calories, ground beef has 85 Calories per ounce, and French fries have 70 Calories per ounce. Write a linear equation that relates the number of ounces of ground beef, x, and the number of ounces of fries, y, that can be used to create a lunch consisting of 850 Calories.

Solution Since x ounces of ground beef have $85x$ Calories and y ounces of fries have $70y$ Calories, and 850-Calorie lunch consisting of a hamburger, fries, and milk must satisfy the following equation.

Ground beef + French fries + Hamburger bun + Milk = 850 Calories

$$85x + 70y + 80 + 160 = 850$$
$$85x + 70y = 610$$ ■

Mathematical modeling of real-life problems almost always has two phases. The first phase is to *create* the model, and the second phase is to *use* the model to answer questions about the real-life problem.

Example 3 *Using a Linear Model for Calorie Intake*

Use the linear model developed in Example 2 to find several ways that the school nutritionist could vary the amount of ground beef and French fries and still meet the state requirement of 850 Calories. Sketch a graph of your results.

Solution Before making up a table of values, it is helpful to rewrite the linear model in slope-intercept form.

$$85x + 70y = 610 \quad \textit{Standard form}$$
$$70y = -85x + 610 \quad \textit{Subtract 85x from both sides.}$$
$$y = -\frac{85}{70}x + \frac{610}{70} \quad \textit{Divide both sides by 70.}$$
$$y = -\frac{17}{14}x + \frac{61}{7} \quad \textit{Slope-intercept form}$$

Using this form, you can calculate several different amounts of ground beef and fries that could be used to create an 850-Calorie lunch.

Beef, x	3.10 oz	3.45 oz	3.80 oz	4.15 oz	4.50 oz
Fries, y	4.95 oz	4.525 oz	4.10 oz	3.675 oz	3.25 oz

■

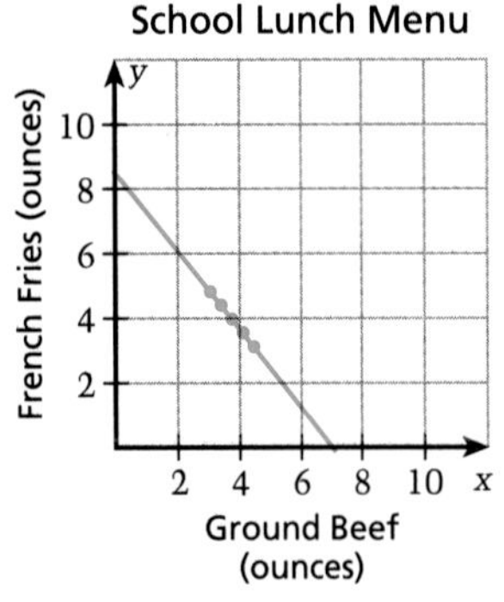

Making Models Accurate but Simple

Mathematical modeling has two basic goals: accuracy and simplicity. That is, you want the model to be accurate enough to be able to yield meaningful answers. On the other hand, you want the model to be simple enough to be usable.

Example 4 *A Mathematical Model for Classroom Computers*

The table shows the number of personal computers in use in American classrooms from kindergarten through high school between 1984 and 1988. ***(Source: Future Computing Datapro, Inc.)***

Year	1984	1985	1986	1987	1988
Computers	590,000	870,000	1,170,000	1,480,000	1,760,000

Estimate the number of computers in use in American classrooms in 1992.

Computers in U.S. Schools

Computers (in 1000s): 250, 500, 750, 1,000, 1,250, 1,500, 1,750, 2,000

Year (0 ↔ 1980): 1 2 3 4 5 6 7 8

Solution To begin, it is helpful to plot the given points so that you can get a picture of how computer use was increasing between 1984 and 1988. In the scatter plot shown at the left, x represents the year, with $x = 0$ corresponding to 1980, and y represents the number of computers. From the scatter plot, it appears that the points can be approximated by a linear model. So as you did in Lesson 5.5, you can sketch the line that appears to best fit the points. Using the points (4, 590,000) and (8, 1,760,000), you can find the slope of the line to be

$$m = \frac{1{,}760{,}000 - 590{,}000}{8 - 4} = \frac{1{,}170{,}000}{4} = 292{,}500.$$

Then find the equation of the line using the point-slope form.

$$y - y_1 = m(x - x_1) \quad \textit{Point-slope form}$$
$$y - 590{,}000 = 292{,}500(x - 4) \quad \textit{Substitute 292,500 for } m, \textit{ 4 for } x_1, \textit{ and 590,000 for } y_1.$$
$$y - 590{,}000 = 292{,}500x - 1{,}170{,}000 \quad \textit{Distributive Property}$$
$$y = 292{,}500x - 580{,}000 \quad \textit{Slope-intercept form}$$

Now, using this model, you approximate the number of classroom computers in use in 1992 ($x = 12$) to be

$$y = 292{,}500(12) - 580{,}000 = 2{,}930{,}000.$$

■

Extra Examples

Here are additional examples similar to **Examples 1–4.**

1. The cost of renting a lawn mower is $10 for the first hour and $5 for each additional hour. Write a linear model that gives the cost of renting a lawn mower in terms of the number of hours rented. $y = 5(x - 1) + 10$
2. In Exercise 1 , how much does it cost to rent a lawn mower for 3 hours? For 2.5 hours? cost = $20; cost = $17.50
3. The total income from the sale of raffle tickets was $201.25. If student tickets sold for $0.75 and adult tickets sold for $1.25, write a linear equation that describes the relationship between the number of student and adult tickets. $0.75s + 1.25a = 201.25$
4. Find several ways that the total income could have been attained in Exercise 3. Make a table of values.

Tickets	Number of Tickets				
Student	0	65	100	145	250
Adult	161	122	101	74	11

Example 5

The percent-error line of the table might have to be explained to the students. The error E is the difference between an actual value X and an approximated model value A. The percent of error is the ratio $\frac{E}{X}$ expressed as a percent.

Check Understanding

1. What are the different types of real-life problems that can be modeled with linear models?
2. Provide examples of how data is typically presented for each type of linear model identified in Check Understanding Exercise 1. Students could refer to any application presented in Chapter 5.

Communicating about ALGEBRA

Discuss the role of models in making decisions and identify some of the limitations of models. Refer to Examples 4 and 5 as needed.

EXTEND *Communicating*

Use the model to approximate the number of computers introduced to American classrooms in 1992. 292,500

Example 5 Comparing Model Results with Actual Results

In Example 4, the model $y = 292{,}500x - 580{,}000$ was developed to represent the number of computers used in American classrooms between 1984 and 1988. How accurately does this model fit the given data?

Solution Begin by using the model to find y-values for 1984 through 1988.

$$
\begin{aligned}
1984\ (x = 4)&: \quad y = 292{,}500(4) - 580{,}000 = 590{,}000 \\
1985\ (x = 5)&: \quad y = 292{,}500(5) - 580{,}000 = 882{,}500 \\
1986\ (x = 6)&: \quad y = 292{,}500(6) - 580{,}000 = 1{,}175{,}000 \\
1987\ (x = 7)&: \quad y = 292{,}500(7) - 580{,}000 = 1{,}467{,}500 \\
1988\ (x = 8)&: \quad y = 292{,}500(8) - 580{,}000 = 1{,}760{,}000
\end{aligned}
$$

The following table shows how the y-values given by the model compare to the actual y-values. The percent error is found by dividing the difference of each model y-value and actual y-value by the actual y-value.

Year	1984	1985	1986	1987	1988
Actual y-values	590,000	870,000	1,170,000	1,480,000	1,760,000
Model y-values	590,000	882,500	1,175,000	1,467,500	1,760,000
Percent error	0%	1.4%	0.4%	−0.8%	0%

■

Communicating about ALGEBRA

SHARING IDEAS about the Lesson

Use a Model Use the model developed in Example 4 to answer the questions.

A. Would you use this model to estimate the number of personal computers that were used in classrooms in 1980? Explain. No, the y-intercept is negative.

B. Based on the model, how many personal computers would you say were used in classrooms in 1990? 2,345,000

C. The model gives the total number of computers used in American classrooms. From the model, estimate how many computers were introduced to classrooms in each year. 292,500

EXERCISES

Guided Practice

CRITICAL THINKING about the Lesson

Monthly Pay **In Exercises 1–6, use the following information.**

Inés works at an electronics store in a shopping mall. Her base pay is $1000 a month. Inés also receives a commission of 2% of her sales total.

1. Write a linear model that gives Inés's total monthly pay, y, in terms of sales total, x. $y = 0.02x + 1000$
2. Use the model to find Inés's monthly pay for sales totaling $25,000. $1500
3. Sketch the graph of the linear model. See margin.
4. What is the slope of the line? What does the slope represent? 0.02, commission rate
5. What is the y-intercept of the line? What does the y-intercept represent? 1000, base pay
6. How much total sales must Inés have in order to earn $2000 in a month? $50,000

Independent Practice

Base Pay Plus Tips **In Exercises 7–12, use the following information.**

Tim is a waiter in a restaurant. His base pay is $750 per month. Tim also earns tips amounting to 15% of the value of the meals he serves.

7. Write a linear model that gives Tim's total monthly pay, y, in terms of the value, x, of the meals he serves. $y = 0.15x + 750$
8. Use the model to find Tim's monthly pay for serving $5000 worth of meals. $1500
9. Sketch the graph of the linear model. See margin.
10. What is the slope of the line? What does the slope represent? See margin.
11. What is the y-intercept of the line? What does the y-intercept represent? 750, base pay
12. How much must Tim serve in order to earn $2000 in a month? $8333.33

Infant Weight **In Exercises 13–16, use the following information.**

By the time an infant is one year old, she will weigh approximately three times the amount she weighed at birth.

13. Baby Carly weighed 7.5 pounds when she was born. Write a linear model that gives Carly's weight, y, in terms of her age, x, where x is measured in months. $y = \frac{5}{4}x + 7\frac{1}{2}$
14. Sketch the graph of the linear model. What does the slope represent? See margin.
15. What is the y-intercept of the line? What does the y-intercept represent? $7\frac{1}{2}$, weight at birth
16. How old will Carly be when she weighs 12.5 pounds? 4 months

EXERCISE Notes

ASSIGNMENT GUIDE

Basic/Average: Ex. 1–6, 7–12, 19–22, 23–26, 29–31

Above Average: Ex. 1–6, 7–16, 27–31, 32–35

Advanced: Ex. 1–6, 17–22, 27–35

Selected Answers
Exercises 1–6, 7–31 odd

More Difficult Exercises
Exercises 26–28, 32–35

Guided Practice

Ex. 1–6 GROUP ACTIVITY After students have completed Exercises 1–6, discuss their answers in class. Ask students to describe the type of linear model that this real-life problem represents (that is, in terms of a constant rate of change or a combination).

Answer

3.

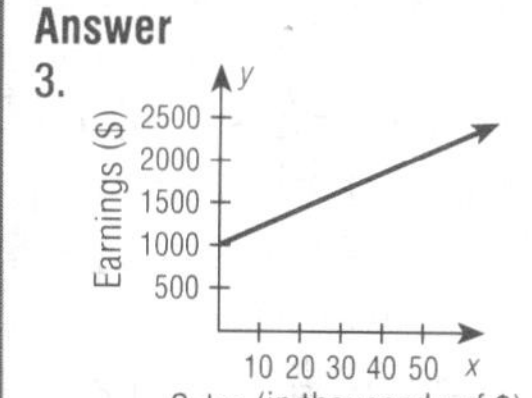

Independent Practice

Ex. 7–22 Ask students to identify the type of linear model represented in each of these exercises.

Answers

9.

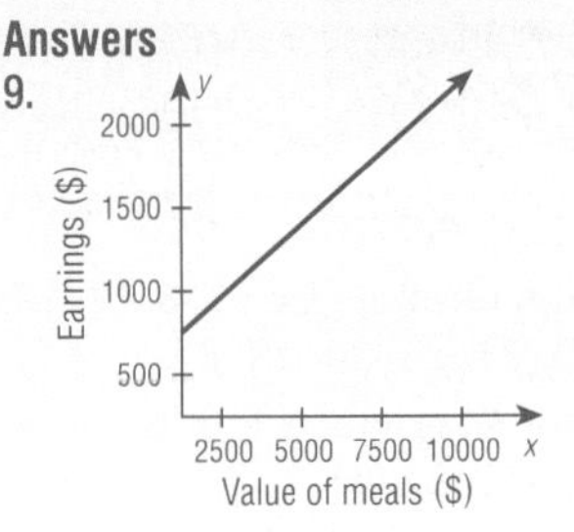

10. Slope: 0.15, Rate of earning based on value of the meals served

14.

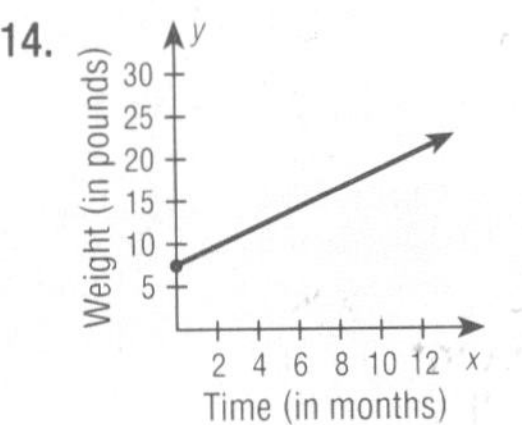

Amount of weight gained per month

▶ **Ex. 19–22** In basketball, a field goal is a score (of 2 or 3 points) that is made during the play of the game by tossing the ball through the basket. A free throw is given as a penalty (resulting from a foul) and is worth 1 point. Encourage students to share this information with their classmates.

Answers

22.

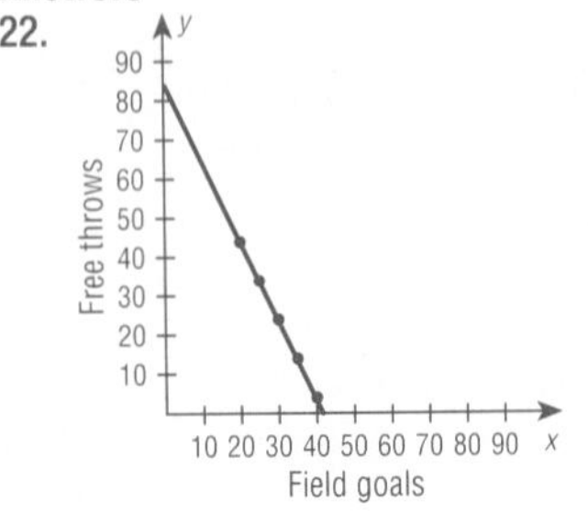

25–26.

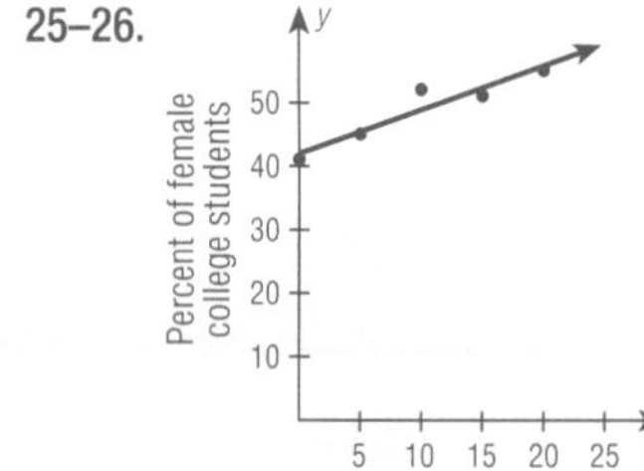

17. *Ticket Purchase* A school club visits a science museum. Student tickets cost \$4.50 each. Nonstudent tickets cost \$6.00 each. The club paid a total of \$288 for the tickets. 48 student tickets were purchased. How many nonstudent tickets were purchased? 12

18. *Savings Accounts* Jason has two savings accounts that pay 6% and 5.5% simple interest per year. Jason's interest for one year is \$40.50. If Jason has \$400 in the 6% account, how much does he have in the 5.5% account? \$300

Basketball Game **In Exercises 19–22, use the given information.**

A basketball team scored 84 points: 2 points for each field goal and 1 point for each free throw.

19. Write a linear equation that relates the number of field goals, x, with the number of free throws, y. $2x + y = 84$

20. Write the equation in slope-intercept form. $y = -2x + 84$

21. Use the linear model to complete the table. 44, 34, 24, 14, 4

Field goals, x	20	25	30	35	40
Free throws, y	?	?	?	?	?

22. Sketch the line and plot the points from the table. See margin.

Integrated Review

College Students **In Exercises 23–26, use the scatter plot, which shows the percent of male college students from 1970 to 1990.**

23. Write an equation of the best-fitting line for these data. $y = -\frac{2}{3}t + 58$

24. Complete the table, which compares the y-values from the best-fitting line with the actual y-values. 58%, 55%, 51%, 48%, 45%

Year, t	0	5	10	15	20
Actual, y	59%	55%	48%	49%	45%
Model, y	?	?	?	?	?

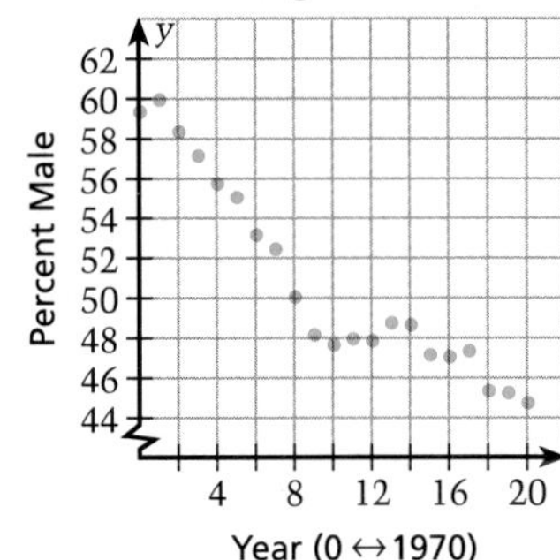

25. Construct a scatter plot that shows the percent of college students between 1970 and 1990 who were female. See margin.

26. Find an equation of the line that best fits the scatter plot found in Exercise 25. Describe a relationship between the equations for college males and college females. $y = \frac{2}{3}x + 42$, slopes are opposites and y-intercepts sum to 100

Restaurant Sales **In Exercises 27 and 28, use the scatter plot, which shows total sales for restaurants in the United States between 1980 and 1990.**

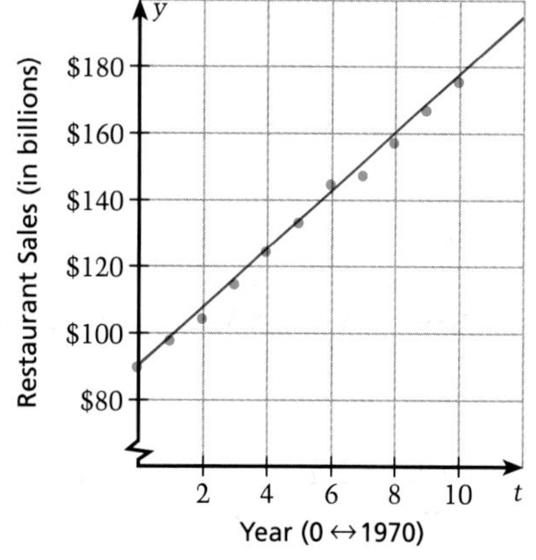

27. Write an equation of the best-fitting line.

28. Complete the table, which compares the y-values from the best-fitting line with the actual y-values. 27. $y = \frac{35}{4}x + 90$

Year, t	0	2	5	8	10
Actual, y	\$90	\$105	\$134	\$158	\$176
Model, y	?	?	?	?	?

90, 107.5, 133.75, 160, 177.5

In Exercises 29–31, match the description with its graph. In each case, tell what the slope of the line represents (as a rate of change).

29. A person is paying \$10 per week to a friend to repay a \$100 loan. b

30. An employee is paid \$12.50 per hour plus \$1.50 for each unit produced per hour. c

31. A sales representative receives \$20 per day for food, plus \$0.25 for each mile driven. a

a.

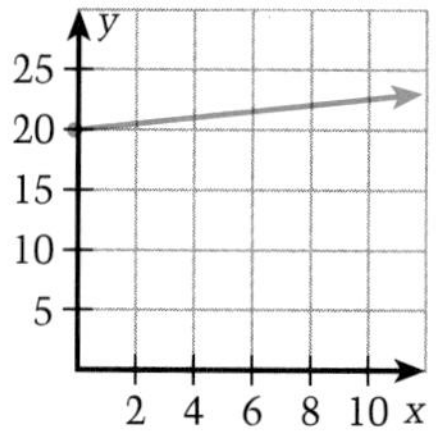

b.

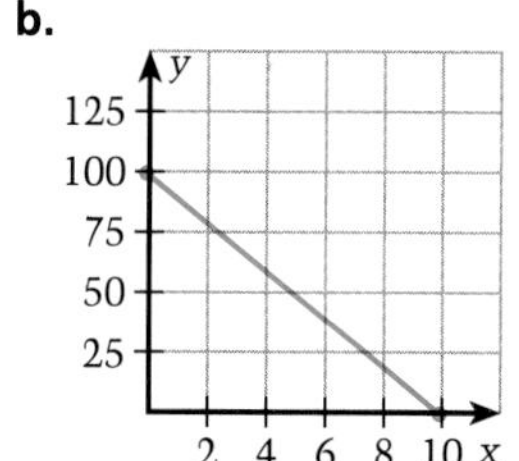

c.

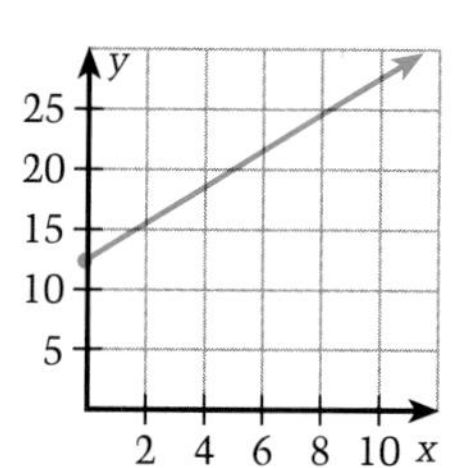

Exploration and Extension

Morning or Evening Paper? **In Exercises 32–35, use the following information.**

The total circulation of morning and evening newspapers in the United States in the 1980s can be approximated by the linear models

$y = 29.91 + 1.303t$ Morning newspaper circulation
$y = 32.27 - 1.216t$ Evening newspaper circulation

where y represents the circulation in millions and t represents the year, with $t = 0$ corresponding to 1980. ***(Source: Editor and Publisher Company)***

32. Sketch the graphs of both equations on the same coordinate plane. See margin.

33. What does the slope of each line represent? Which type of newspaper had an increasing circulation? Which had a decreasing circulation?

34. In 1983, were there more morning or evening papers sold? In 1989? Morning, morning

35. During which year were the two circulations the same? 1981

33. Rate of change of circulation per year, morning, evening

Ex. 29–31 The units for each scale have been omitted on purpose. Once students have matched each description with its graph, have them place the appropriate units on each of the x- and y-axes.

Answer

32.

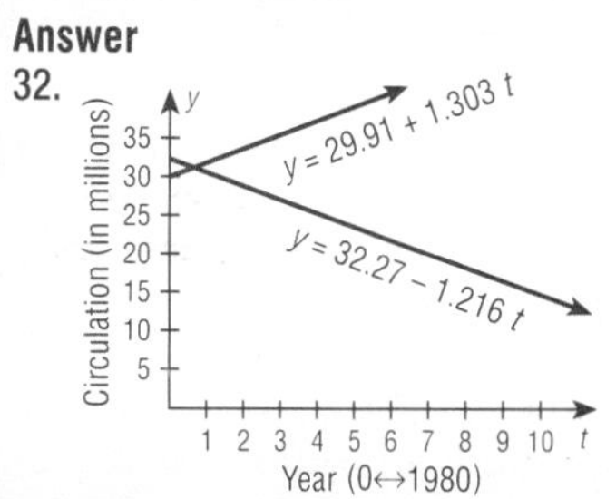

Ex. 35 Emphasize that the two circulations were the same at the point of intersection of the two lines.

Enrichment Activity

This activity requires students to go beyond lesson goals.

CONNECTIONS Identify real-life problems that you, your friends, or your family members have solved by using each of the two types of linear models discussed in this lesson.

ORGANIZER

TIME SCHEDULE: All levels, two days

The Chapter Summary helps students organize the main ideas of the chapter. In this chapter, techniques for writing an equation of a line using the slope-intercept form, the standard form, and the point-slope form were presented. Students were also instructed in using graphs to check their work. Applications of writing an equation of a line were made, including finding lines of best fit for a collection of data points.

Work with students to review the techniques and concepts of the chapter. It is suggested that the first day of review be a combination of you and students working problems from the Chapter Review exercises. The first day's homework assignment should be the basis for the second day of review.

COOPERATIVE LEARNING
Encourage students to study together. Emphasize the importance of teaching a classmate how to perform a skill or how to recall a strategy. When students work together, everyone wins. The students receiving help get additional instruction, and the students giving help gain a deeper understanding of the techniques of writing an equation of a line and the concepts involved.

Chapter SUMMARY

Review the chapter by asking students, "What did you learn? Why did you learn it? How does it fit with the other algebra skills and concepts you have learned?"

5 Chapter Summary

What did you learn?

Skills

1. Write an equation of a line using slope-intercept form
   - given its slope and its y-intercept. **(5.1)**
   - given its slope and one of its points. **(5.2)**
   - given two of its points. **(5.3)**
2. Write an equation of a line in standard form. **(5.5)**
3. Transform linear equations from one form to another. **(5.5)**
4. Write an equation of a line using point-slope form. **(5.6)**

Strategies

5. Create a linear model for a real-life situation
   - using the slope-intercept method. **(5.1–5.3)**
   - using the standard form model. **(5.5)**
   - using the point-slope method. **(5.6)**
6. Use a linear model to answer questions about real-life situations. **(5.5–5.7)**
7. Use graphing skills to provide graphic checks. **(5.1–5.7)**

Exploring Data

8. Find a best-fitting line for a collection of data points. **(5.4)**

Why did you learn it?

Many real-life situations have linear models. That is, the real-life situation can be represented by a linear equation. You can write the model in slope-intercept form if you know the constant rate of change (the slope) and an initial value (the y-intercept). Often, however, you are given the slope and a data point other than an initial value, or you are given two data points. In either case, you can still use the slope-intercept form by taking some additional steps. Sometimes the situation involves several scattered data points, so you try to find the linear model that best fits the data. A basic problem-solving strategy is to learn which form of a linear equation is best to use.

How does it fit into the bigger picture of algebra?

In this chapter, you studied techniques for writing equations of lines. Your basic strategy was to use the slope-intercept form that you had seen in Chapter 4. When an equation of a line is used to represent a real-life situation, it is called a **linear model.** In this chapter, you used the problem-solving strategies you studied in Chapter 1, the rules and simplification techniques you studied in Chapter 2, the systematic equation solving you studied in Chapter 3, and the graphing techniques you studied in Chapter 4.

Chapter REVIEW

In Exercises 1–4, write an equation of the line that has the given slope and y-intercept. (5.1)

1. Slope is 2; y-intercept is -6. $y = 2x - 6$

2. Slope is -1; y-intercept is 3. $y = -x + 3$

3. Slope is $-\frac{1}{3}$; y-intercept is -1. $y = -\frac{1}{3}x - 1$

4. Slope is $\frac{2}{5}$; y-intercept is -5. $y = \frac{2}{5}x - 5$

In Exercises 5–8, write an equation of the line. (5.1) See below.

5.

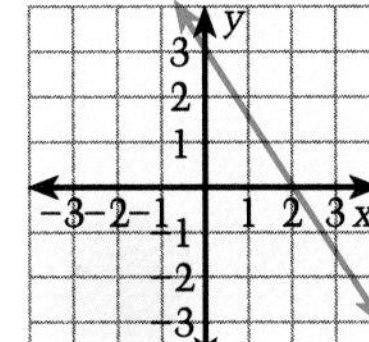

6.

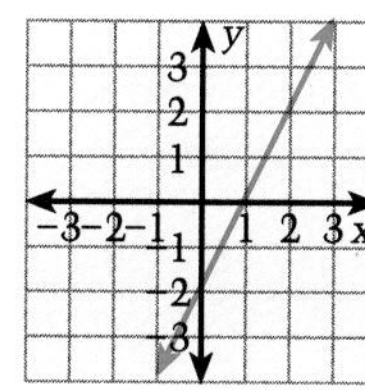

7.

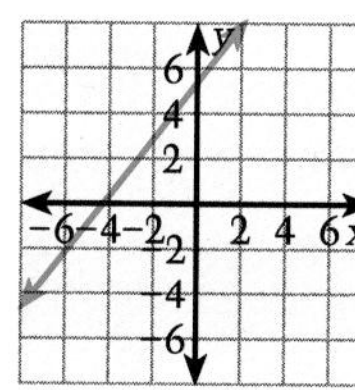

8.

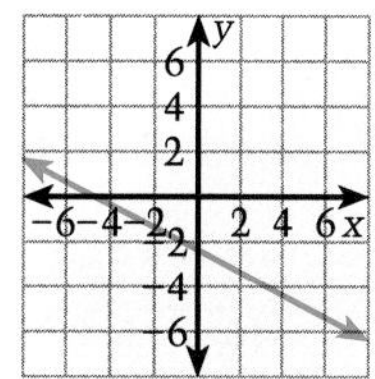

9. Find the slope and y-intercept of the line $y = -4x + 3$. Then sketch the line. **(5.1)** 9. -4, 3 See margin.

10. Find the slope and y-intercept of the line $y = x - 7$. Then sketch the line. **(5.1)** 1, -7 See margin.

✪ **11.** *Car Depreciation* A car was bought in 1990 for \$32,000. During the next five years, the value of the car depreciated by \$3000 per year. Write an equation that gives the value, y, of the car in terms of the year, x. Let $x = 0$ correspond to 1990. **(5.1)** $y = -3000x + 32{,}000$

✪ **12.** Use the equation found in Exercise 11 to estimate the value of the car in 1996. **(5.1)** \$14,000

In Exercises 13–20, write an equation of the line that passes through the point and has the given slope. Give the equation in slope-intercept form. (5.2)

13. $(6, -1)$, $m = 3$ $y = 3x - 19$

14. $(-9, 4)$, $m = 2$ $y = 2x + 22$

15. $(-3, 2)$, $m = -1$ $y = -x - 1$

16. $(-1, -6)$, $m = \frac{1}{3}$ $y = \frac{1}{3}x - \frac{17}{3}$

17. $(2, 4)$, $m = 0$ $y = 4$

18. $(2, -1)$, $m = -3$ $y = -3x + 5$

19. $(5, -3)$, $m = \frac{1}{5}$ $y = \frac{1}{5}x - 4$

20. $(2, -4)$, $m = \frac{1}{2}$ $y = \frac{1}{2}x - 5$

In Exercises 21–23, write the slope-intercept form of the equation of the line. (5.2) See below.

21.

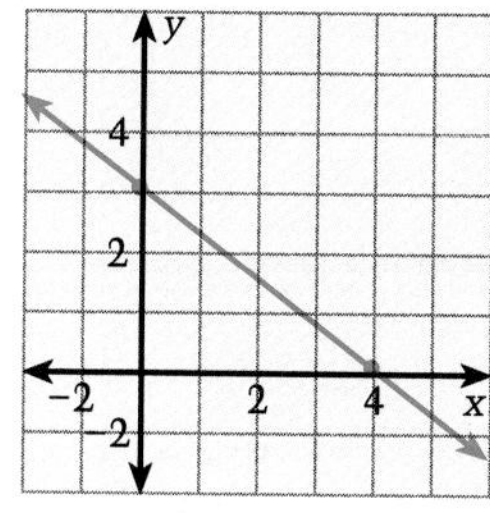

22.

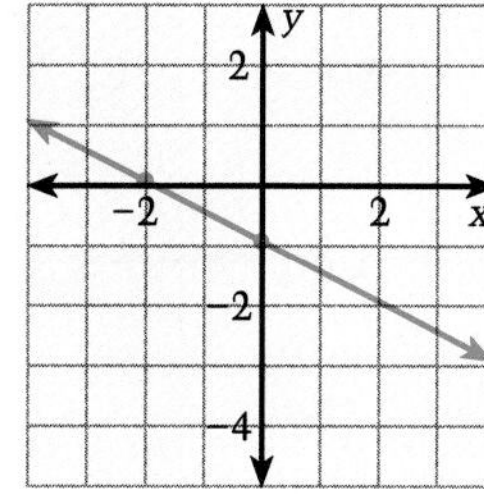

23.

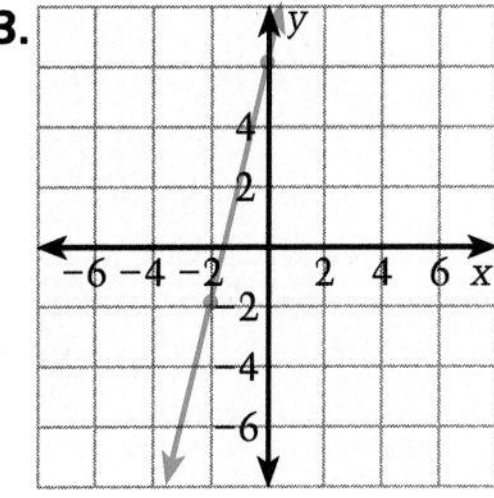

In Exercises 24 and 25, find an equation of the line passing through the point and parallel to the line. (5.2)

24. $(2, -6)$, $y = -x - 7$ $y = -x - 4$

25. $(-1, 2)$, $y = 3x + 9$ $y = 3x + 5$

5. $y = -\frac{3}{2}x + 3$ 6. $y = 2x - 2$ 7. $y = \frac{5}{4}x + 5$ 8. $y = -\frac{1}{2}x - \frac{5}{2}$

21. $y = -\frac{3}{4}x + 3$ 22. $y = -\frac{1}{2}x - 1$ 23. $y = 4x + 6$

Chapter SUMMARY

ASSIGNMENT GUIDE

Basic/Average: Ex.1–25 odd, 29–45 odd, 50–57 odd

Above Average: Ex.1–25 odd, 28–42 even, 49–50, 58–63

Advanced: Ex.1–27 odd, 30–33, 38–41, 45–50, 58–63

✪ More Difficult Exercises

Exercises 11, 12, 26–27, 48, 49, 58–63

▶ **Ex. 1–63** Since the Chapter Review exercises parallel the list of Skills, Strategies, and Exploring Data statements on page 284, tell students to use the Chapter Summary to recall the lesson in which a skill or concept was taught if they are uncertain about how to do a given exercise in the Chapter Review . This forces students to categorize the task to be performed and to locate information in the chapter about the task, both beneficial activities.

▶ **Ex. 21–23, 32–35** Stress the importance of reading points accurately from the graphs. Avoid selecting points that "look" as if they are on the line!

Answers

9.

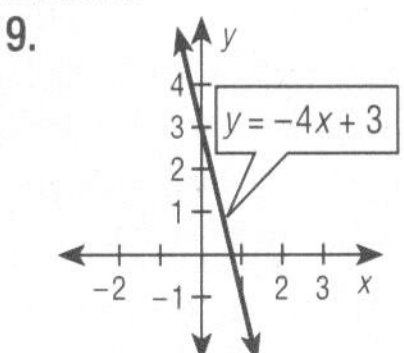

10.

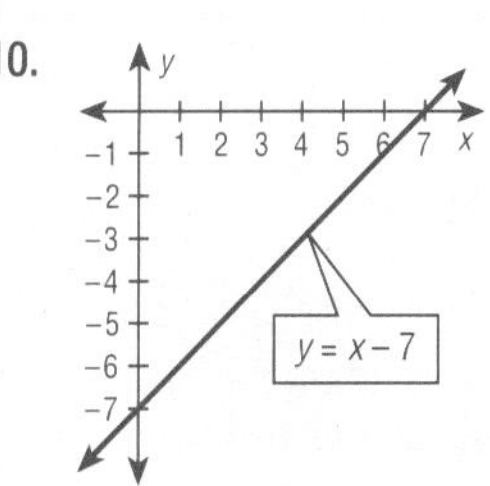

Answers

36.

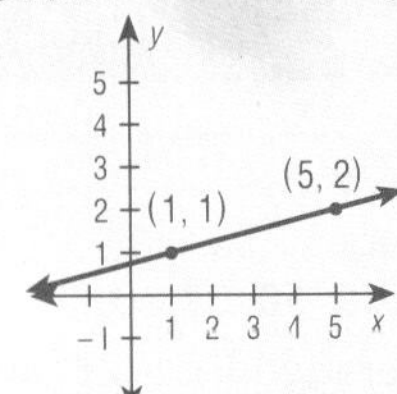

37.

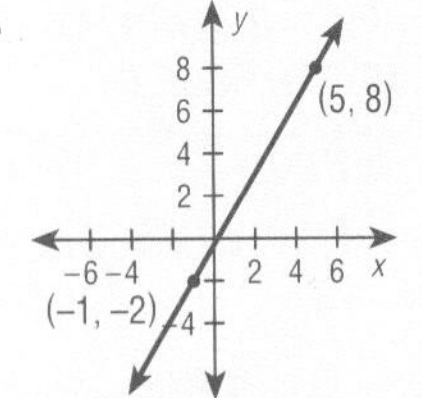

38.

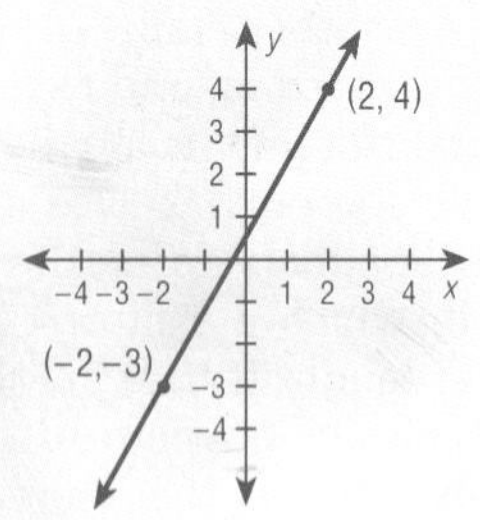

39.

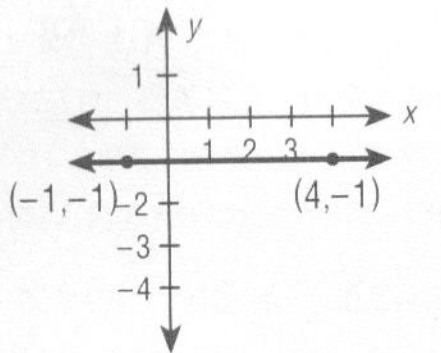

26. *Company Profits* Between 1980 and 1990, the annual profit for a company increased by about \$50,000 per year. In 1988, the company had an annual profit of \$1,300,000. Find an equation that gives the annual profit, P, in terms of the year, t. Let $t = 0$ correspond to 1980. **(5.2)**

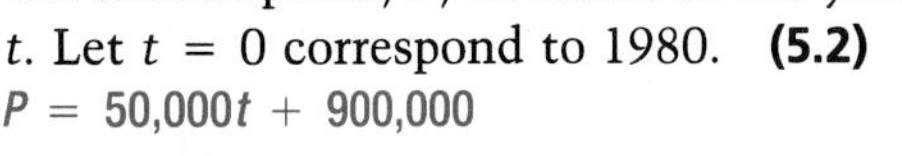

$P = 50{,}000t + 900{,}000$

27. *Advertising Expense* Advertising costs for the Save-More Store increase by about \$32,000 per year. In 1992, \$352,000 was spent on advertising. Find an equation that gives the amount spent on advertising, A, in terms of the year, t. Let $t = 0$ correspond to 1990. **(5.2)**
$A = 32{,}000t + 288{,}000$

In Exercises 28–35, write an equation of the line that passes through the given points. Give the equation in slope-intercept form. (5.3)

28. $(1, -6), (-3, 2)$ $y = -2x - 4$

29. $(1, 8), (-2, -1)$ $y = 3x + 5$

30. $(2, 5), (-4, 1)$ $y = \frac{2}{3}x + \frac{11}{3}$

31. $(3, -4), (-6, 2)$ $y = -\frac{2}{3}x - 2$

32.

33.
(−1, 3)
(4, −2)

34.

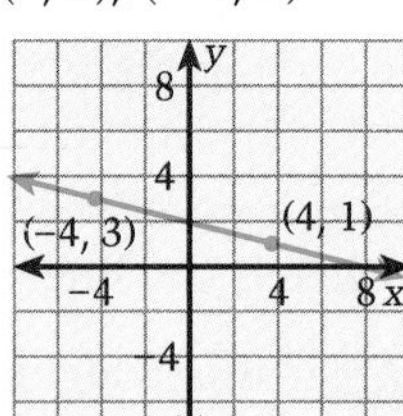

35.

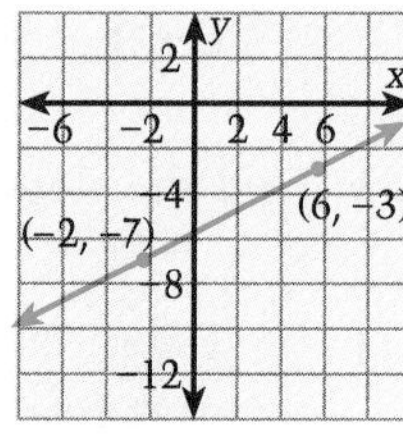

32. $y = \frac{1}{3}x - 2$

33. $y = -x + 2$

34. $y = -\frac{1}{4}x + 2$

35. $y = \frac{1}{2}x - 6$

In Exercises 36–39, sketch the line that passes through the points. Use the graph to approximate the *y*-intercept of the line. Then find the actual *y*-intercept. (5.3) See margin.

36. $(1, 1), (5, 2)$ $\frac{3}{4}$

37. $(-1, -2), (5, 8)$ $-\frac{1}{3}$

38. $(-2, -3), (2, 4)$ $\frac{1}{2}$

39. $(-1, -1), (4, -1)$ -1

In Exercises 40–43, find the equation of the horizontal line or the vertical line that passes through the point. (5.1)

40. Vertical line: $(2, -4)$ $x = 2$

41. Vertical line: $(-1, 8)$ $x = -1$

42. Horizontal line: $(8, -3)$ $y = -3$

43. Horizontal line: $(11, 6)$ $y = 6$

44. *U.S. Households with Televisions* Assume that the number of households with televisions increased at a constant rate from 1960 to 1985. Write an equation that gives the number, H, of households with televisions in terms of the year, t. Let $t = 0$ correspond to 1960. **(5.7)**
$H = 1{,}588{,}000t + 45{,}200{,}000$

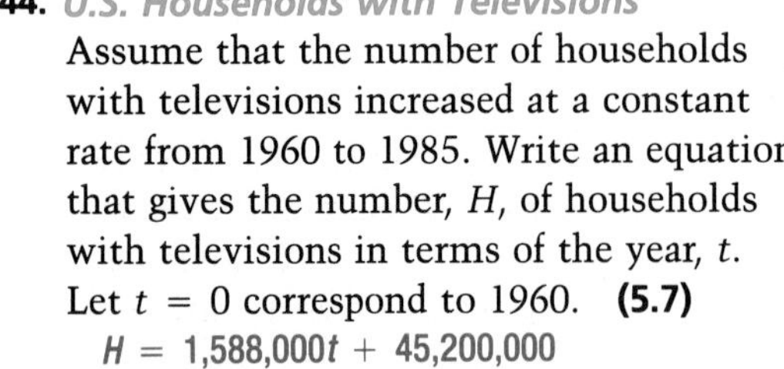

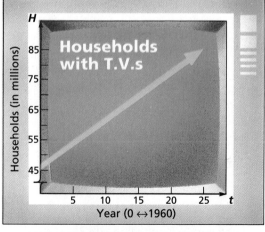

In 1960, 45,200,000 U.S. households had televisions. By 1985, the number had risen to 84,900,000. (Source: Television and Cable Factbook)

45. *Turner Broadcasting Systems* In 1985, the total revenue for Turner Broadcasting Systems was \$350 million. By 1987, the total revenue was \$650 million. Assume that the revenue increased at a constant rate. Write an equation that gives the revenue, R, in terms of the year, t. Let $t = 0$ correspond to 1980. ***(Source: Turner Broadcasting System, Inc.)*** **(5.7)** $R = 150t - 400$

In Exercises 46 and 47, sketch the line that you think best approximates the data shown in the scatter plot. Then find an equation of the line. (5.4)

46. 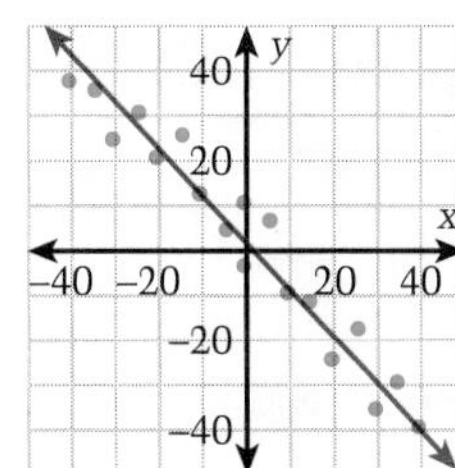

$y = -x + 1$

47. 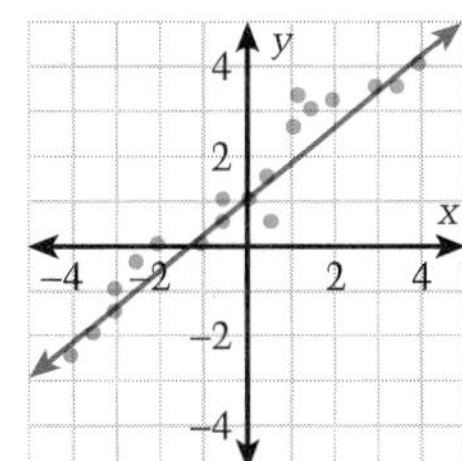

$y = \frac{4}{5}x + 1$

48. *World Population* The world population between 1986 and 1993 is given in the following table. Construct a scatter plot for this data and find an equation of the line that best fits the data. *(Source: U.S. Bureau of Labor Statistics)* **(5.7)** See margin.

Year	Population
1986	4,938,000,000
1987	5,024,000,000
1988	5,112,000,000
1989	5,202,000,000
1990	5,292,000,000
1991	5,385,000,000
1992	5,480,000,000
1993	5,572,000,000

49. *U.S. Voting* Construct a scatter plot showing the number of voters in presidential elections from 1964 through 1992. Then find an equation of the best-fitting line. *(Source: Committee for Study of American Electorate)* **(5.7)** See margin.

Year	U.S. Voters	President
1964	70,645,000	Johnson
1968	73,212,000	Nixon
1972	77,625,000	Nixon
1976	81,603,000	Carter
1980	86,497,000	Reagan
1984	92,653,000	Reagan
1988	91,610,000	Bush
1992	104,425,000	Clinton

50. Write $x + 9y = 7$ in slope-intercept form. **(5.5)** $y = -\frac{1}{9}x + \frac{7}{9}$

51. Write $3x + 5y = 5$ in slope-intercept form. **(5.5)** $y = -\frac{3}{5}x + 1$

52. Write $y = -\frac{1}{3}x + \frac{2}{3}$ in standard form. **(5.5)** $x + 3y = 2$

53. Write $y = \frac{3}{4}x + \frac{1}{2}$ in standard form. **(5.5)** $-3x + 4y = 2$

54. *Buying Party Dippers* Suppose that you buy \$10 worth of broccoli and cauliflower for party dippers. The broccoli costs \$2.00 per pound and the cauliflower costs \$1.25 per pound. Write an equation in standard form that represents the different amounts (in pounds) of broccoli, B, and cauliflower, C, that you could buy. **(5.7)**

$2B + 1.25C = 10$

55. *Racquetball or Tennis?* Spike and Renée have \$30 to spend at a health center. It costs \$10 an hour for the use of the racquetball court and \$5 an hour for the tennis court. Write an equation in standard form that represents the time they can spend on each court, with x representing the number of hours on the racquetball court and y the hours on the tennis court. **(5.7)** $10x + 5y = 30$

56. Use Exercise 54 to complete the table.

Broccoli, B (lb)	0	1	2	3	4	5
Cauliflower, C (lb)	?	?	?	?	?	?

8, 6.4, 4.8, 3.2, 1.6, 0

57. Use Exercise 55 to complete the table.

Racquetball, x (hours)	0	1	2	3
Tennis, y (hours)	?	?	?	?

6, 4, 2, 0

Answers

48. 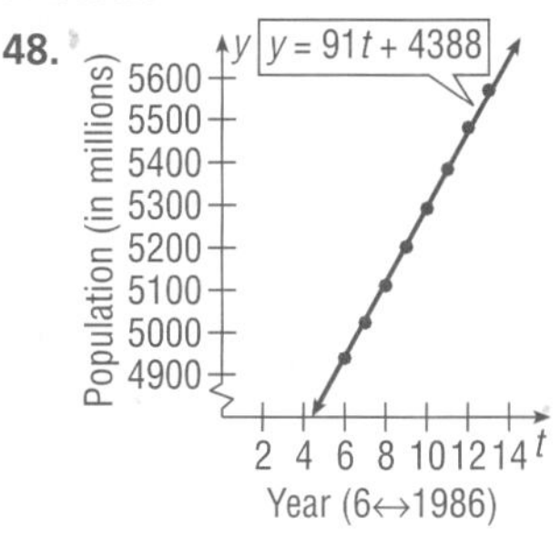

49. 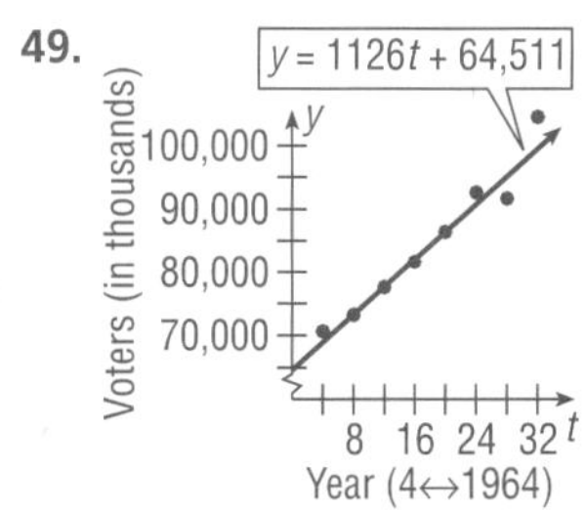

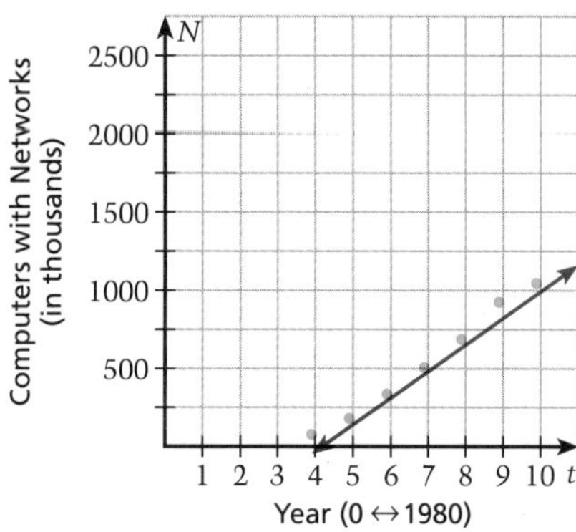

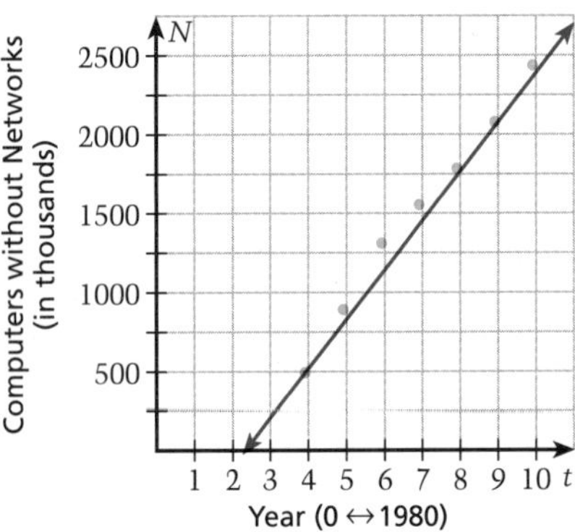

The number of office computers, both with and without networks, increased from 1984 to 1990. (Source: Gartner Group, Inc.)

58. *Office Computers with Networks* Find an equation of the best-fitting line for the scatter plot showing the number of office computers *with* networks from 1984 to 1990. Let N represent the number of computers and let t represent the year, with $t = 4$ corresponding to 1984. **(5.4)** $N = 170{,}000t - 680{,}000$

59. Use the equation found in Exercise 58 to complete the table, which compares the actual number of networked office computers with the number given by the model. **(5.4)**

Year, t	4	5	6	7	8	9	10
Actual, N	81,000	183,000	338,000	510,000	691,000	928,000	1,049,000
Model, N	?	?	?	?	?	?	?

0 170,000 340,000 510,000 680,000 850,000 1,020,000

60. *Office Computers without Networks* Find an equation of the best-fitting line for the scatter plot showing the number of office computers *without* networks from 1984 to 1990. **(5.4)** $N = 300{,}000t - 600{,}000$

61. Find the approximate rate of change in networked office computers from 1984 to 1990. **(5.4)** 170,000 per year

62. Find the approximate rate of change in nonnetworked office computers from 1984 to 1990. **(5.4)** 300,000 per year

63. Which type of office computer was increasing at a faster rate from 1984 to 1990: those with networks or those without networks? **(5.4)** Those without networks

Chapter TEST

1. Find an equation of the line whose slope is $\frac{1}{4}$ and whose y-intercept is -3. **(5.1)** $y = \frac{1}{4}x - 3$
2. Find an equation of the line that passes through the point $(-2, 3)$ and has a slope of -2. **(5.2)** $y = -2x - 1$
3. Sketch the line that passes through the points $(-4, 3)$ and $(2, -3)$. Then write the equation of the line in slope-intercept form. **(5.3)** $y = -x - 1$ See margin.
4. Construct a scatter plot for the following data. Then find an equation of the line that you think best represents these data. **(5.4)** See margin.

x	−6	−5	−3.5	−3	−2	−1	0	1	2	3
y	4	3	3	2	2	0	0.5	−1	−2.5	−3

5. Write an equation of the horizontal line that passes through the point (2, 5). **(5.1)** $y = 5$
6. Write an equation of the vertical line that passes through the point (2, 5). **(5.1)** $x = 2$
7. Rewrite the equation $-x + 5y = 20$ in slope-intercept form. **(5.2)** $y = \frac{1}{5}x + 4$
8. Sketch the line given by $-5x + 2y = 10$. Label the x- and y-intercepts. **(5.5)** See margin.
9. Rewrite the equation $y = \frac{2}{17}x + 3$ in standard form. **(5.5)** $-2x + 17y = 51$
10. Which of the following two lines are parallel to each other? **(5.5)**
    a. $-5x - 2y = 18$ **b.** $2x + 5y = 12$ **c.** $5x + 2y = 6$
11. Elizabeth goes to a store to buy $24 worth of flour and cooking oil for her restaurant. The flour costs $1.20 per bag, and the cooking oil costs $2.40 per quart. Write an equation that represents the different amounts of flour and cooking oil that Elizabeth can buy. **(5.1)** $1.20f + 2.40c = 24$
12. Use the point-slope form to write an equation of the line that passes through the point $(-3, 1)$ with a slope of $\frac{1}{3}$. **(5.6)** $y - 1 = \frac{1}{3}(x + 3)$
13. Bob is traveling home at a constant speed. After one-half hour, he is 77.5 miles from home and after one hour, he is 55 miles from home. Write a linear equation that gives the distance from home, y (in miles), in terms of the time x (in hours). How long will it take Bob to get home? **(5.7)** $y = -45x + 100$, 2.2 hours
14. A salesclerk for an appliance store receives monthly pay of $1250 plus a 4% commission on the appliances sold. Write a linear model that gives the total monthly pay, y, in terms of the amount of appliance sales, x. **(5.7)** $y = 0.04x + 1250$
15. Chris needs $150 to buy a new bike. She has already saved $30. Chris plans to earn the rest of the money by working at two part-time jobs. She earns $6 an hour at a grocery store and $4 an hour for baby-sitting. Write a linear equation that relates the number, x, of hours worked at the grocery store and the number, y, of hours baby-sitting that are necessary to earn enough money to buy the new bike. **(5.7)** $6x + 4y = 120$

Answers

3.

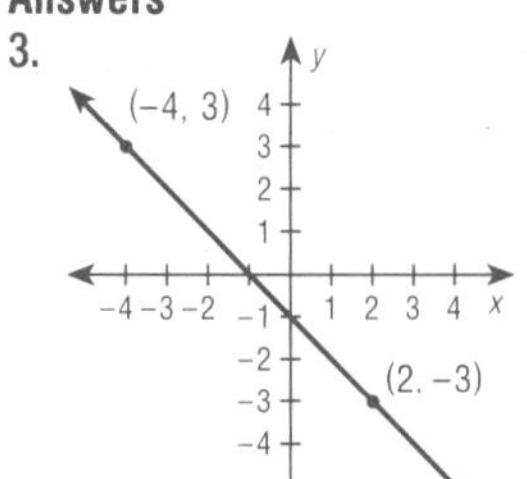

4.

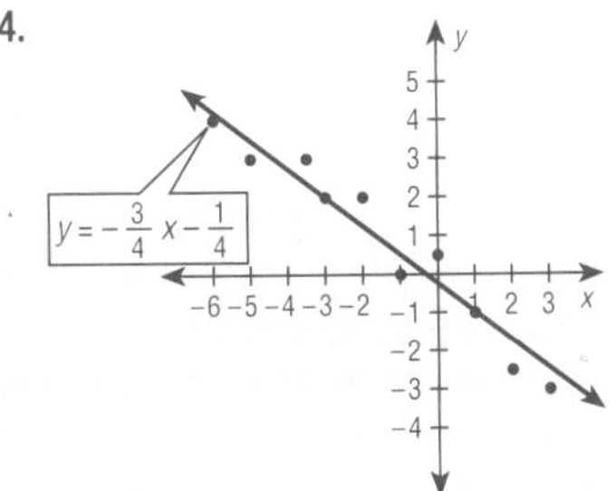

8.

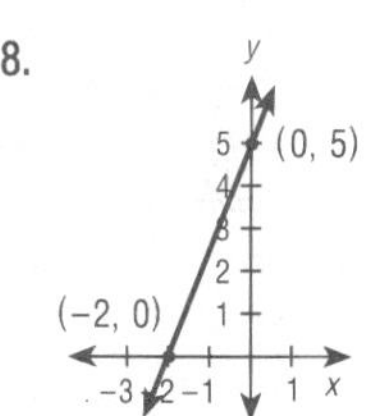

Solving and graphing linear inequalities are important algebraic skills because many real-life situations are modeled by inequalities. The connection between solving and graphing linear inequalities and solving and graphing linear equations should be pointed out throughout this chapter. The similarities will help students understand the skills and concepts associated with this chapter.

This chapter continues the Exploring Data strand. Encourage students to identify data from their own experiences that can be modeled by using time lines, picture graphs, and circle graphs.

The Chapter Summary on page 334 provides you and the students with a synopsis of the chapter. It identifies key skills, concepts, and vocabulary. You may want to have students look at the Chapter Summary as an overview before beginning the chapter.

CHAPTER 6

Solving and Graphing Linear Inequalities

LESSONS

The refraction of white light, whether in the rainbow or through a prism, creates many beautiful effects.

Real Life
Spectrometry

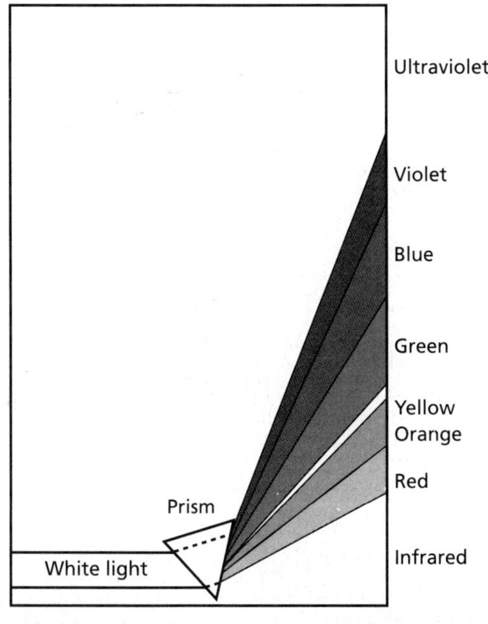

White light from the sun is composed of light with different wavelengths. When white light passes through a prism, the prism separates the light into bands called the *spectrum.* Infrared light has the longest wavelengths and is bent the least. Ultraviolet light has the shortest wavelengths and is bent the most. You cannot see ultraviolet or infrared light. The *visible spectrum* consists of the bands of light from red to violet.

The wavelengths for different colors in the spectrum are as follows.

Color	Wavelength, w (nanometers)
Ultraviolet	$w < 400$
Violet	$400 \le w < 424$
Blue	$424 \le w < 491$
Green	$491 \le w < 575$
Yellow	$575 \le w < 585$
Orange	$585 \le w < 647$
Red	$647 \le w < 700$
Infrared	$700 \le w$

Observe with students that the answer to the question *What wavelength of light produces the color green?* is the inequality $491 \le w < 575$, where w is measured in nanometers. A nanometer is equivalent to 10^{-9} meters. This is a practical unit of wavelength because the human eye can just about see the difference in color between two light sources that differ by one nanometer. (This accounts for the shades of green that you see.)

Many real-life situations in science, economics, education, and sports rely on a knowledge of inequalities. For instance, the sales tax of an item depends on its cost. Any cost above a certain amount and below another fixed amount is assessed the same sales tax. Or the performance on standardized tests is often measured in percentiles, which is a score indicating the percent of others taking the test who placed below the given percentile score. Ask students to identify other examples that use inequalities.

Solving and Graphing Linear Inequalities

PACING CHART

The daily Pacing Chart is meant to help you adjust your teaching pace. Students in the full course should finish the entire text by the end of the year. Students in the basic course may not complete the entire text in the school year. The Pacing Chart for each chapter contains suggestions for lessons that require more than one day and lessons that may be omitted for the basic course.

DAY	FULL COURSE	BASIC COURSE
1	6.1	6.1
2	6.2	6.2
3	6.2	6.2
4	6.2	6.2
5	6.3	6.3
6	6.3	6.3
7	6.3	6.3
8	6.4	6.4
9	6.4 & Using a Graphing Calculator	6.4
10	Mid-Chapter Self-Test	Using a Graphing Calculator
11	6.5	Mid-Chapter Self-Test
12	6.5	6.5
13	Chapter Review	6.5
14	Chapter Review	Chapter Review
15	Chapter Test	Chapter Review
16	Cumulative Review	Chapter Test
17		Cumulative Review

CHAPTER ORGANIZATION

LESSON	PAGES	GOALS	MEETING THE NCTM STANDARDS
6.1	292-297	1. Graph linear inequalities in one variable 2. Solve linear inequalities in one variable	Problem Solving, Communication, Connections, Structure, Reasoning, Geometry
6.2	298-303	Write and use a linear inequality as a model for a real-life situation	Problem Solving, Communication, Connections, Reasoning, Geometry
Mixed Review	304	Review of algebra and arithmetic	
Milestones	304	Origins of Ada	Connections
6.3	305-310	1. Solve and graph compound inequalities 2. Model a real-life situation with a compound inequality	Problem Solving, Communication, Connections, Reasoning, Geometry
Using a Calculator	311	Use a graphing calculator or computer to sketch the graph of an inequality	Technology, Geometry
Mid-Chapter Self-Test	312	Diagnose student weaknesses and remediate with correlated Reteaching worksheets	
6.4	313-319	1. Solve absolute value inequalities 2. Model a real-life situation with an absolute value inequality	Problem Solving, Communication, Connections, Reasoning, Geometry
Mixed Review	319	Review of algebra and arithmetic	
6.5	320-325	1. Graph a linear inequality in two variables 2. Model a real-life situation using a linear inequality in two variables	Problem Solving, Communication, Connections, Geometry, Reasoning
Using a Calculator	326	Use a graphing calculator to sketch the graph of an inequality in two variables	Technology, Geometry
6.6	327-333	Draw and interpret visual models such as time lines, picture graphs, and circle graphs	Problem Solving, Communication, Connections, Discrete Mathematics, Statistics, Reasoning
Chapter Summary	334	A restatement of what has been learned, why it has been learned, and how it fits into the structure of algebra	Structure, Connections
Chapter Review	335-338	Review of concepts and skills learned in the chapter	
Chapter Test	339	Diagnose student weaknesses and remediate with correlated Reteaching worksheets	
Cumulative Review	340-343	Review of concepts and skills from previous chapters	

LESSON RESOURCES

MEETING INDIVIDUAL NEEDS

RETEACHING For students who need to spend more time on basics:

If a mid-chapter self-test or chapter test indicates a deficiency, teachers can help students with the appropriate ***Reteaching Copymaster.***

PRACTICE For students who need more practice:

Additional exercises like those in the Pupil's Edition are provided for each lesson in ***Extra Practice Copymasters.***

ENRICHMENT For enriching and broadening students' experiences:

Problem of the Day copymasters in ***Teaching Tools*** provide a daily opportunity to use logical reasoning, looking for a pattern, writing an equation, and other routine and non-routine problem-solving strategies.

Math Log copymasters in ***Alternative Assessment*** provide opportunities to report on investigations, research, and open-ended problems.

Enriching activities with graphing and scientific calculators and computers are provided in ***Technology: Using Calculators and Computers.***

The ***Applications Handbook*** provides additional information about the cross-curriculum topics such as astronomy, chemistry, physics, sports, economics, genetics, and music that are integrated into the Pupil's Edition.

LESSON	6.1	6.2	6.3	6.4	6.5	6.6
PAGES	292-297	298-303	305-310	313-319	320-325	327-333
Teaching Tools						
Transparencies	✓	✓	✓	✓	✓	
Problem of the Day	✓	✓	✓	✓	✓	✓
Warm-up Exercises	✓	✓	✓	✓	✓	✓
Answer Masters	✓	✓	✓	✓	✓	✓
Extra Practice Copymasters	✓	✓	✓	✓	✓	✓
Reteaching Copymasters	Teacher-directed and independent activities tied to results on the Mid-Chapter Self-Tests and Chapter Tests					
Color Transparencies			✓	✓	✓	✓
Applications Handbook	Additional background information is supplied for many real-life applications.					
Technology Handbook	Calculator and computer worksheets are supplied for appropriate lessons.					
Complete Solutions Manual	✓	✓	✓	✓	✓	✓
Alternative Assessment	Assess student's ability to reason, analyze, solve problems, and communicate using mathematical language.					
Formal Assessment	Mid-Chapter Self-Tests, Chapter Tests, Cumulative Tests, and Practice for College Entrance Tests					
Computer Test Bank	Customized tests can be created by choosing from over 2000 items.					

INSIGHTS

6.1 Solving Inequalities in One Variable

Many real-life applications of algebra require determining whether one quantity is greater than or less than another. For example, to ride on a certain roller coaster, a rider must be at least 54 inches tall. A truck must not exceed the load limits of the bridges on its route. In each case, inequalities can be used to represent the relationships. In this lesson, the study of inequalities is begun by graphing and solving simple inequalities.

6.2 Problem Solving Using Inequalities

This lesson combines problem-solving approaches with the skills developed in Lesson 6.1 for solving inequalities. Because many relationships that can be expressed as equations may be better expressed as inequalities, it is important to consider problems similar to those found in Chapter 3. In so doing, problem-solving skills are strengthened as well.

6.3 Compound Inequalities

In the area of finance, compound inequalities occur frequently. Discuss the following with students. The income tax paid, t, depends on taxable income, x. The tax is determined by locating two given dollar amounts between which the amount of taxable income can be found. For instance, if taxable income is at least \$20,000 and less than \$39,000, the tax might be \$2,725. Written algebraically, if $20{,}000 \le x < 39{,}000$, then $t = 2{,}725$. Point out to students that $20{,}000 \le x < 39{,}000$ is a compound inequality.

6.4 Connections: Absolute Value and Inequalities

Students studied absolute value in Chapter 4. In this lesson, the connection between compound inequalities and absolute value provides students with an empowering perspective with which to view real-life situations. Having multiple ways in which to organize data increases the likelihood that problem situations can be fully understood and resolved.

6.5 Graphing Linear Inequalities in Two Variables

Many real-life applications in business, science, education, and manufacturing are represented by linear inequalities in two variables. Linear programming is one important example of how the algebraic concepts and skills of this lesson are used in everyday situations.

6.6 Exploring Data: Time Lines, Picture Graphs, and Circle Graphs

Newspapers use various types of graphs to capture a reader's interest and to illustrate some viewpoint. The more common graphs are time lines, picture graphs, bar graphs, and circle graphs. In order to be wise consumers and informed citizens, it is important to be able to interpret the data found in graphs.

Problem of the Day

The athletic boosters are going to have an annual awards banquet. The planning committee finds that, whether they seat 3 players at each table or 4, there is going to be one player left over. But, with 5 players at each table, all the players can be seated. If there will be between 30 and 100 players at the banquet, how many will be there? 85

ORGANIZER

Warm-Up Exercises

1. Graph each number on the same number line.
   a. -1.25 b. 3.5 c. 1.3
   d. 5 e. -4.5 f. -0.75

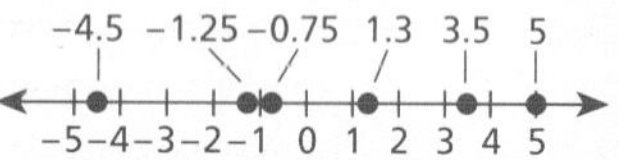

2. Graph each number on the same number line.
   a. -35 b. 20 c. -15
   d. 47 e. -5 f. 12

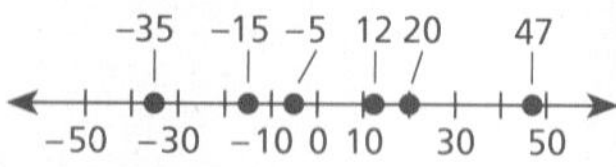

3. Solve each equation.
   a. $x + 5 = 8$
   b. $2y - 6 = 8$
   c. $\frac{1}{3}m = -5$
   d. $9 - 4x = 21$
   e. $5x + 7 = -2x$
   f. $3n = 7n - 24$

   a. $x = 3$, b. $y = 7$, c. $m = -15$, d. $x = -3$, e. $x = -1$, f. $n = 6$

Lesson Resources

Teaching Tools
- Transparency: 6
- Problem of the Day: 6.1
- Warm-up Exercises: 6.1
- Answer Masters: 6.1

Extra Practice: 6.1

6.1 Solving Inequalities in One Variable

What you should learn:

Goal 1 How to graph linear inequalities in one variable

Goal 2 How to solve linear inequalities in one variable

Why you should learn it:

You can model many real-life situations with linear inequalities, such as quantities that can be above or below zero (temperatures and profits, for example).

Goal 1 Graphing Linear Inequalities

The **graph** of a linear inequality in one variable is the graph on the real number line of all solutions of the inequality. There are four types of *simple inequalities.* (You will study compound inequalities in Lesson 6.3.)

Verbal Phrase	Inequality	Graph
All real numbers less than 3	$x < 3$	−4 −2 0 2 4
All real numbers greater than -2	$x > -2$	−3 −2 −1 0 1 2 3
All real numbers less than or equal to 1	$x \le 1$	−3 −2 −1 0 1 2 3
All real numbers greater than or equal to 0	$x \ge 0$	−3 −2 −1 0 1 2 3

Notice on the graphs that we use an open dot for $<$ or $>$ and a solid dot for $\le$ or $\ge$.

Real Life **Meteorology**

The July temperature in Mexico City is at least 12°C.

Example 1 *Sketching Graphs of Linear Inequalities*

a. The July temperature, T (in degrees Celsius), in Mexico City can be represented by the inequality $T \ge 12$. The graph is shown below.

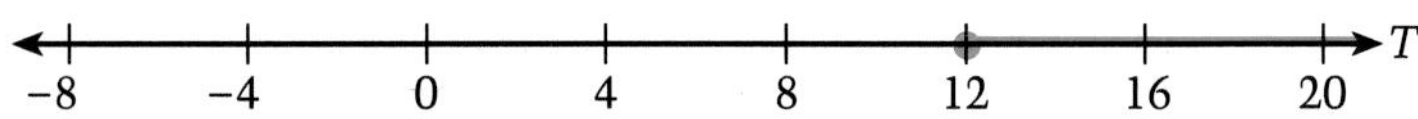

b. The highest elevation in Florida is 345 feet above sea level. If x is the elevation in feet at any location in Florida, then x must satisfy the inequality $x \le 345$. The graph is shown below.

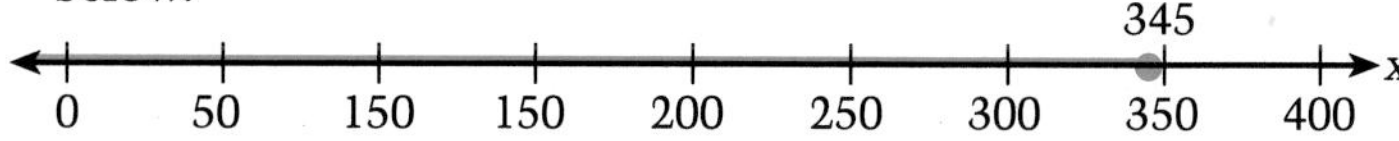

c. During 1992, the monthly company profits, p, were in the red. *In the red* means "negative." The inequality for this statement is $p < 0$. The graph is shown below.

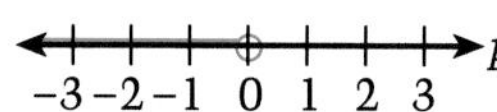

■

Goal 2 Solving Linear Inequalities in One Variable

Solving a linear inequality in one variable is much like solving a linear equation in one variable. The goal is to isolate the variable on one side using transformations that produce equivalent inequalities. The transformations are similar to those given in Lesson 3.1, *but* there are two important differences. In the following list, notice that when you multiply or divide both sides of an inequality by a negative number, you must *reverse* the inequality to maintain a true statement. For instance, to reverse $<$, replace it with $>$.

Transformations That Produce Equivalent Inequalities

	Original Inequality	Equivalent Inequality
1. Add the same number to *both* sides.	$x - 3 < 5$	$x < 8$
2. Subtract the same number from *both* sides.	$x + 6 \geq 10$	$x \geq 4$
3. Multiply *both* sides by the same *positive* number.	$\frac{1}{2}x > 3$	$x > 6$
4. Divide *both* sides by the same *positive* number.	$3x \leq 9$	$x \leq 3$
5. Multiply *both* sides by the same *negative* number and reverse the inequality.	$-x < 4$	$x > -4$
6. Divide *both* sides by the same *negative* number and reverse the inequality.	$-2x \leq 6$	$x \geq -3$

Example 2 *Solving a Linear Inequality (One Transformation)*

Solve $p + 5 \geq 3$.

Solution

$p + 5 \geq 3$ *Rewrite original inequality.*

$p + 5 - 5 \geq 3 - 5$ *Subtract 5 from both sides.*

$p \geq -2$ *Simplify.*

The solution is *all real numbers greater than or equal to* -2. The graph of this inequality is shown at the left. Check several numbers that are greater than or equal to -2 in the original inequality. ■

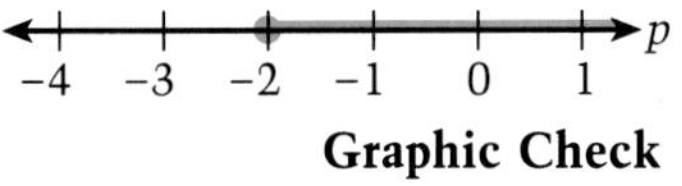

Graphic Check

LESSON Notes

GOAL 1 Emphasize that the graph of a linear inequality in one variable is a graph on the real number line and not on the coordinate plane.

GOAL 2 Remind students that equivalent equations (or inequalities) are equations (or inequalities) that have the same solution. That is, $2x + 3 = 4$ is equivalent to $2x = 1$ because $x = \frac{1}{2}$ is the solution to both equations.

Show the effect of subtracting the same number from both sides of an inequality as follows.

Original inequality	Equivalent inequality
$3 < 12$	$3 - 7 < 12 - 7$ or $-4 < 5$
$8 > -10$	$8 - 5 > -10 - 5$ or $3 > -15$

Show that the inequality sign must be reversed to maintain a true statement when both sides of an inequality are multiplied or divided by the same negative number, as follows.

Original inequality	Equivalent inequality
$3 < 12$	$3 \cdot -4 > 12 \cdot -4$ or $-12 > -48$
$8 > -10$	$8 \div -2 < -10 \div -2$ or $-4 < 5$

Common-Error ALERT!

In the discussion of the transformations that produce equivalent inequalities, students frequently forget to reverse the inequality sign when multiplying and dividing by a negative number. One way to help students understand the effect of the transformations on inequalities is to use real number values to illustrate the properties.

Examples 2–3

Observe with students that the solution steps to these inequalities are identical to the solution steps of the corresponding equations.

Examples 4–5

Remind students that multiplying or dividing by a negative number reverses the inequality sign in an inequality. As indicated in the solution following Example 5, the variable in an inequality can be isolated on either side of the inequality. Consequently you may choose to isolate the variable in such a way that multiplication and division by a negative number is never required.

Extra Examples

Here are additional examples similar to **Examples 1–5.**

1. Sketch each inequality on a number line.

a. $x \geq -3$ **b.** $t < 9$

c. $m \leq 25$ **d.** $k > -2$

a.

b.

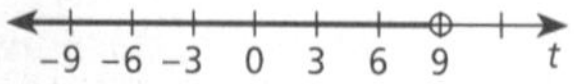

c.

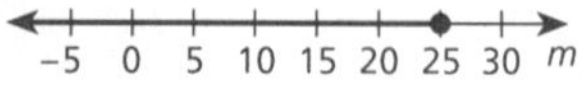

d.

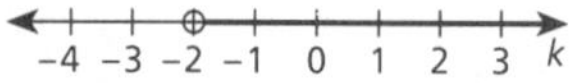

2. Solve each inequality.

a. $x - 8 < 15$

b. $4y + 3 > 7$

c. $13 - 7n \leq -8$

d. $3x > 11x + 4$

a. $x < 23$ b. $y > 1$

c. $n \geq 3$ d. $x < -\frac{1}{2}$

The solution shown in Example 2 required only one transformation. Often, finding the solution of a linear inequality requires two or more transformations.

Example 3 *Using More than One Transformation*

Solve $2y - 5 < 7$.

Solution

$2y - 5 < 7$	*Rewrite original inequality.*
$2y - 5 + 5 < 7 + 5$	*Add 5 to both sides.*
$2y < 12$	*Simplify.*
$\frac{2y}{2} < \frac{12}{2}$	*Divide both sides by 2.*
$y < 6$	*Simplify.*

The solution is *all real numbers less than* 6. The graph of this inequality is shown below. Check several numbers that are less than 6 in the original inequality. (*Also* try checking some numbers that are greater than or equal to 6 to see that they are *not* solutions of the original inequality.)

Graphic Check

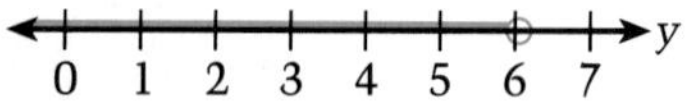

■

As you gain experience with solving linear inequalities, you may want to use the same "expert solver" format we discussed in Lesson 3.2. For instance, the solution in Example 3 could be streamlined as follows.

$2y - 5 < 7$	*Rewrite original inequality.*
$2y < 12$	*Add 5 to both sides.*
$y < 6$	*Divide both sides by 2.*

Example 4 *Reversing an Inequality*

Solve $5 - x > 4$.

Solution

$5 - x > 4$	*Rewrite original inequality.*
$5 - 5 - x > 4 - 5$	*Subtract 5 from both sides.*
$-x > -1$	*Simplify.*
$(-1)(-x) < (-1)(-1)$	*Multiply both sides by −1 and reverse inequality.*
$x < 1$	*Simplify.*

The solution is *all real numbers less than 1.* The graph of this inequality is shown below. Check several numbers that are less than 1 in the original inequality.

Graphic Check

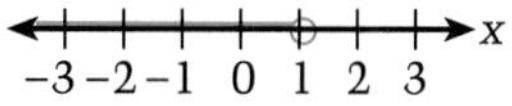

■

Example 5 *Solving an Inequality with Variables on Both Sides*

Solve $2x - 4 \le 4x - 1$.

Solution

$2x - 4 \le 4x - 1$	*Rewrite original inequality.*
$2x - 4 + 4 \le 4x - 1 + 4$	*Add 4 to both sides.*
$2x \le 4x + 3$	*Simplify.*
$2x - 4x \le 4x - 4x + 3$	*Subtract 4x from both sides.*
$-2x \le 3$	*Simplify.*
$\frac{-2x}{-2} \ge \frac{3}{-2}$	*Divide both sides by −2 and reverse inequality.*
$x \ge -\frac{3}{2}$	*Simplify.*

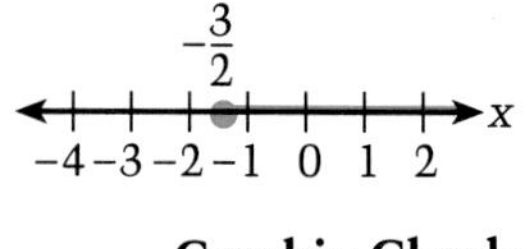

Graphic Check

The solution is *all real numbers greater than or equal to* $-\frac{3}{2}$. The graph of this inequality is shown at the left. Check several numbers that are greater than or equal to $-\frac{3}{2}$ in the original inequality. ■

In Examples 2, 3, 4, and 5, we have been writing the solution steps with the variable isolated on the left. Many people like to follow this convention, but it isn't required. Here's a streamlined format for Example 5.

$2x - 4 \le 4x - 1$	*Rewrite original inequality.*
$-4 \le 2x - 1$	*Subtract 2x from both sides.*
$-3 \le 2x$	*Add 1 to both sides.*
$-\frac{3}{2} \le x$	*Divide both sides by 2.*

Note that $-\frac{3}{2} \le x$ and $x \ge -\frac{3}{2}$ are equivalent inequalities.

Communicating about ALGEBRA

SHARING IDEAS about the Lesson

Explain Your Reasoning For each inequality, choose the inequality that is equivalent. Explain your reasoning. See margin.

Original Inequality	Which is equivalent?	
A. $x - 3 < 10$	**a.** $x < 7$	**b.** $x < 13$
B. $2x \le 8$	**a.** $x \le 4$	**b.** $x \le 6$
C. $-x \ge -2$	**a.** $x \ge 2$	**b.** $x \le 2$
D. $2 > x$	**a.** $x < 2$	**b.** $x > 2$

Check Understanding

1. **WRITING** Explain, in writing, the difference between the graph of the number 5 and the graph of the inequality, $x > 5$. Answers may include that the graph of the number 5 is one point on the number line while the graph of $x > 5$ is the infinite set of all the points that are greater than 5.
2. Identify the four types of simple inequalities. Give examples of each. See page 292.
3. Discuss the use of the six transformations that can be used to obtain equivalent inequalities. Provide numerical examples of each transformation. See page 293.

Communicating about ALGEBRA

COOPERATIVE LEARNING
Have students work in pairs to explore an alternate strategy for identifying equivalent inequalities. Instruct students to graph each pair of inequalities in the "Which is equivalent?" column, select number values from each graph, and decide which graph contains values that do not satisfy the original inequality. Encourage students to build strategies for testing number values, using the graph.

Answers

A. Add the same number to both sides.
B. Divide both sides by the same positive number.
C. Multiply or divide both sides by the same negative number and reverse the inequality sign.
D. Meaning of "<" and ">"

EXERCISE Notes

ASSIGNMENT GUIDE

Basic/Average: Ex. 1–8, 9–31 odd, 33–41, 45–46, 47–53 odd, 55

Above Average: Ex. 1–8, 10–32 even, 33–41, 42–46 even, 47–53 odd, 55–56

Advanced: Ex. 1–8, 9–30 multiples of 3, 33–41, 43–46, 49–53 odd, 55–56

Selected Answers
Exercises 1–8, 9–53 odd

✪ More Difficult Exercises
Exercises 55–56

Guided Practice

▶ **Ex. 1–8** COOPERATIVE LEARNING Have students complete these exercises in pairs. Provide answers using whole-group discussions.

EXTEND Ex. 6–7 Ask students to compare and contrast the solutions to these two exercises.

Answers:

3.

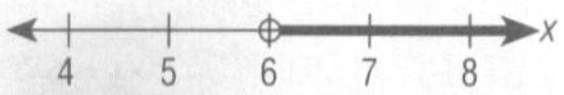

7. When you multiply a true inequality such as $3 > -3$ by -1, the inequality becomes false; to make the inequality true, you must reverse the inequality sign.

8.

Independent Practice

▶ **Ex. 15–32** Encourage students to graph the solutions of these exercises.

10, 12 15-31 odd
34-40 even

EXERCISES

Guided Practice

CRITICAL THINKING about the Lesson

1. Write the inequality for *all real numbers greater than or equal to 4.* $x \ge 4$
2. Write a verbal phrase that describes $x \le -2$. All real numbers less than or equal to negative two
3. Sketch the graph of $x > 6$. See margin.
4. Solve the inequality $5 - 2x \ge 3$. $x \le 1$
5. Is the number 4.78 a solution of $x < 5$? Is the number 6 a solution of $x < 6$? Explain. 5. Yes, no; $4.78 < 5$ and 6 is not < 6
6. Write an inequality that is equivalent to $2 < x$ but has x on the left side. $x > 2$
7. Explain why you must reverse the direction of the inequality symbol when multiplying by a negative number. See margin.
8. Sketch the graph of *all real numbers less than or equal to 4.* Then write an inequality that represents the graph. $x \le 4$ See margin.

Independent Practice

In Exercises 9–14, sketch a graph of the inequality. See Additional Answers.

9. $x > -4$ 10. $x < 15$ 11. $x \le 9$ 12. $x \ge -8$ 13. $2 \le x$ 14. $3 \ge -x$

In Exercises 15–32, solve the inequality.

15. $x + 5 < 7$ $x < 2$
16. $4 + x > -12$ $x > -16$
17. $-x - 8 > -17$ $x < 9$
18. $-3 + x < 19$ $x < 22$
19. $6 + x \le -8$ $x \le -14$
20. $x - 10 \ge -6$ $x \ge 4$
21. $2x + 7 \ge 4$ $x \ge -\frac{3}{2}$
22. $6 - 3x \le 15$ $x \ge -3$
23. $-4x - 2 \ge 10$ $x \le -3$
24. $2x + 5 \le -13$ $x \le -9$
25. $-3 \le 6x - 1$ $-\frac{1}{3} \le x$
26. $17 \ge 4x + 11$ $\frac{3}{2} \ge x$
27. $-x - 4 > 3x - 2$ $x < -\frac{1}{2}$
28. $3x + 5 < -7x - 9$ $x < -\frac{7}{5}$
29. $x + 3 \le 2(x - 4)$ $x \ge 11$
30. $2x + 10 \ge 7(x + 1)$ $x \le \frac{3}{5}$
31. $-x + 4 < -2(x - 8)$ $x < 12$
32. $-x + 6 > -(2x + 4)$ $x > -10$

In Exercises 33–41, solve the inequality. Then match its solution with one of the graphs.

33. $3x - 1 \le -7$ $x \le -2$, d
34. $10 \le x + 6$ $4 \le x$, a
35. $5x - 7 < 3x + 9$ $x < 8$, e
36. $12 - 2x > 10$ $x < 1$, c
37. $3x + 2 \le 14$ $x \le 4$, f
38. $-3 \ge 6x - 1$ $-\frac{1}{3} \ge x$, g
39. $-9x + 2 < 14$ $x > -\frac{4}{3}$, h
40. $2x - 6 < -x + 6$ $x < 4$, b
41. $2x \ge -x + 6$ $x \ge 2$, i

a.
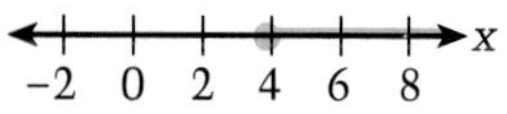

b.
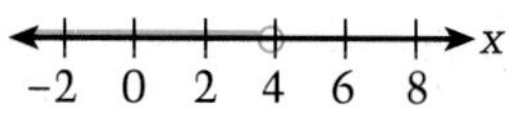

c.
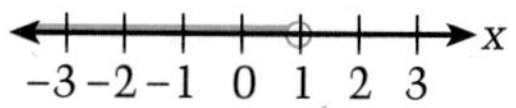

d.
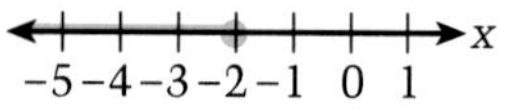

e.
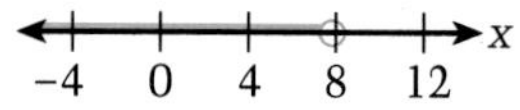

f.
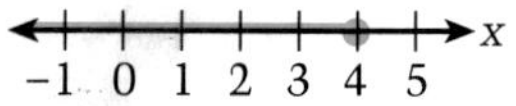

g.
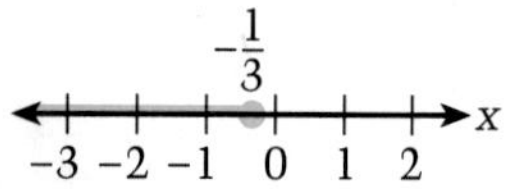

h.
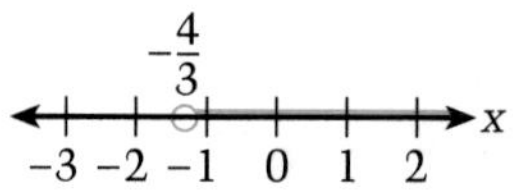

i.
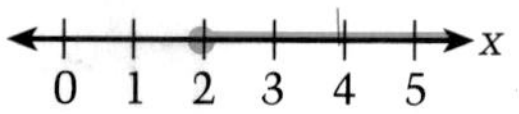

42. *Temperatures in Buffalo* In Buffalo, New York, January temperatures often do not exceed 0° Celsius. Write an inequality that describes the January temperatures, T, in Buffalo. Graph the inequality. $T \le 0$

43. *Elevations in California* The lowest elevation in California is 282 feet below sea level. Let E represent the elevation of any location in California. Write an inequality for E. Graph the inequality.

44. *Company Profits* During 1993, each monthly company profit, P, was at least \$1200. Write an inequality for this statement. Graph the inequality. $P \ge 1200$

45. *Viking Ships* The longest Viking ship that has been found is not quite 95 feet in length. Let V represent the lengths of known Viking ships. Write an inequality for V. Graph the inequality. $V < 95$

46. *Album Sales* Let d represent the number of copies sold to date of Whitney Houston's debut album. Write an inequality describing d. Graph the inequality. $d \ge 14{,}000{,}000$

43. $E \ge -282$

Whitney Houston's 1985 debut album sold 14 million copies in 2 years.

42.–46. See Additional Answers.

Integrated Review

In Exercises 47–50, solve the equation.

47. $|x + 2| = 8$ 6, −10 **48.** $|12 - x| = 6$ 6, 18 **49.** $|x - 3| = 9$ −6, 12 **50.** $|2x - 1| = 10$ $-\frac{9}{2}, \frac{11}{2}$

51. *Fitness Program* After 6 weeks, Marcia can continue an aerobics workout for 45 minutes. Her average weekly increase has been 5 minutes. Find an equation that gives Marcia's workout time, t, in terms of the number of weeks, w, that she is on the fitness program. $t = 5w + 15$

52. *Savings Account* Nate deposits \$6.25 per week in his savings account. After 15 weeks, the balance in his account is \$598. Write an equation that gives the balance, b, of Nate's account in terms of the number of weeks, w. Disregard any interest earned. $b = 6.25w + 504.25$

53. Find an equation of the line that contains the point $(2, -4)$ and has a slope of $\frac{3}{2}$. $y = \frac{3}{2}x - 7$

54. Find an equation of the line that contains the points $(-3, 6)$ and $(9, 1)$. $y = -\frac{5}{12}x + \frac{19}{4}$

Exploration and Extension

55. Use the graph of $y = \frac{2}{3}x - 2$ to solve the inequality $\frac{2}{3}x - 2 < 0$. Explain your reasoning. See margin.

56. Use the graph of $y = -2x + 4$ to solve the inequality $-2x + 4 > 0$. Explain your reasoning. See margin.

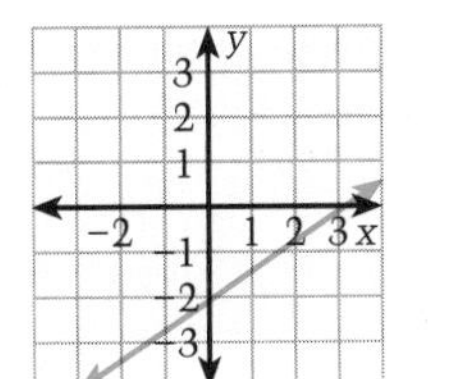

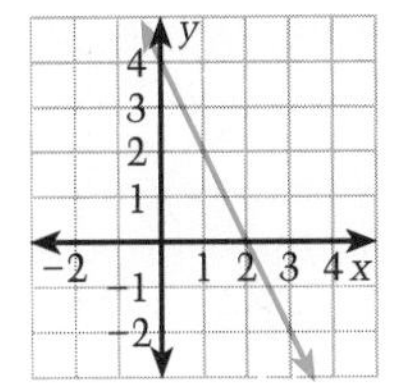

Ex. 55–56 COOPERATIVE LEARNING Students should work in small groups to complete these exercises. Encourage students to use the Guess and Check strategy to formulate hypotheses about the relationship between the x and y variables in each inequality.

Answers

55. $x < 3$. For values of $x < 3$, the value of y is less than zero. Therefore, any value of x that results in $y < 0$ (where the line is below the x-axis) is a solution to $\frac{2}{3}x - 2 < 0$.

56. $x < 2$. For values of $x < 2$, the value of y is greater than zero. Therefore, any value of x that results in $y > 0$ (where the line is above the x-axis) is a solution to $-2x + 4 > 0$.

Enrichment Activity

This activity requires students to go beyond lesson goals.

Identify the transformations that can be omitted from the list in this lesson and indicate why they may not be needed. (*Hint:* With only two transformations, given inequalities can be transformed into equivalent ones.) Since addition and subtraction and multiplication and division are inverse operations, you can rewrite the transformations list using only addition and multiplication.

Since you can express any subtraction as addition by using the additive inverse, then for all real numbers a, b, and c, if $a < b$, then $a + c < b + c$. Since you can express any division as multiplication using the multiplicative inverse, then for all real numbers a, b, and c, $c \ne 0$, if $a < b$, then $ac < bc$ when c is positive; and $ac > bc$ when c is negative. (Similar rules can be stated for the other inequality signs.)

Problem of the Day

These diagrams represent balance scales that are not in balance. What can you conclude about the number of jacks that would balance a ball?

Less than 5

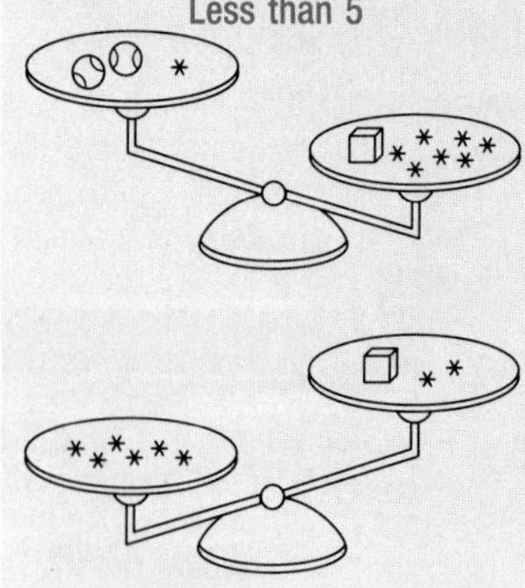

ORGANIZER

Warm-Up Exercises

1. Identify the general problem-solving strategy used in algebra. Write a verbal model, assign labels, write an algebraic model, solve the algebraic model, answer the original question.
2. Solve these inequalities.
   a. $3.4m - 0.2 \geq 64.4$
   b. $13t < 456 - 27t$
   c. $-0.23k > 317.5 - 1.5k$

   a. $m \geq 19$, b. $t < \frac{57}{5}$, c. $k > 250$

Lesson Resources

Teaching Tools
- Transparencies: 3, 4
- Problem of the Day: 6.2
- Warm-up Exercises: 6.2
- Answer Masters: 6.2

Extra Practice: 6.2

6.2 Problem Solving Using Inequalities

What you should learn:

Goal 1 How to write and use a linear inequality as a model for a real-life situation

Why you should learn it:

You can model many real-life situations with linear inequalities, such as in finding quantities that exceed given amounts (records in sports, popular trends, profits over expenses).

Real Life **Athletics**

Goal 1 Using Linear Inequalities as Models

Many of the real-life problems that you studied in Chapter 3 could have been worded so that the algebraic model was an inequality, rather than an equation. For instance, in Lesson 3.5, Example 4 described a situation in which you had \$18.37 to buy a pair of earrings. The example asked you to find the price limit for the earrings so that the money you had could pay the 5% sales tax. Using an inequality as a model, where x is the price of the earrings, we could have written

$x + 0.05x \leq 18.37.$ *Linear inequality as model*

The solution would then be $x \leq 17.49$, which says that the price of the earrings can be any amount less than or equal to \$17.49.

Example 1 *Writing a Linear Model*

Suzanne and Jorah both ran in a 2-kilometer race. Suzanne finished ahead of Jorah, and Jorah's time for the race was 8 minutes. Write an inequality that describes Suzanne's average speed for the race.

Solution

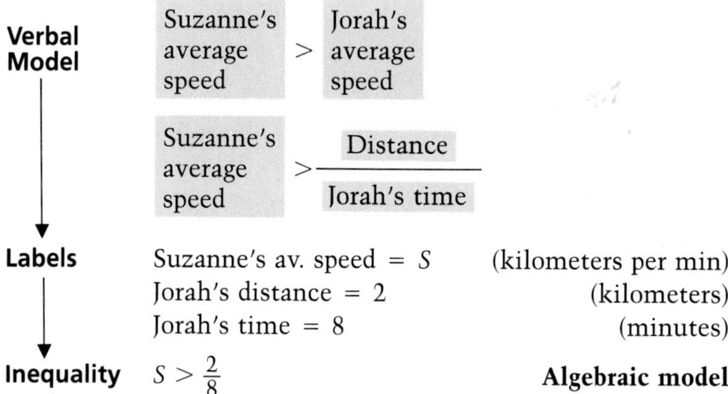

Verbal Model

[Suzanne's average speed] > [Jorah's average speed]

[Suzanne's average speed] > $\frac{\text{Distance}}{\text{Jorah's time}}$

Labels

Suzanne's av. speed $= S$ (kilometers per min)
Jorah's distance $= 2$ (kilometers)
Jorah's time $= 8$ (minutes)

Inequality $S > \frac{2}{8}$ **Algebraic model**

Thus, Suzanne's average speed was greater than $\frac{1}{4}$ kilometer per minute. ■

From 1970 to 1990, the average annual per person consumption of whole milk in the United States dropped from 103 quarts to 43 quarts. The average annual consumption of low-fat milk rose from 25 quarts to 60 quarts. (Source: U.S. Department of Agriculture)

Example 2 Linear Models for Milk Consumption

For which years did the consumption of low-fat milk by Americans exceed the consumption of whole milk? (Use a linear model.)

Solution Let W represent quarts of whole milk, L represent quarts of low-fat milk, and t represent the year, with $t = 0$ corresponding to 1970. Since whole milk consumption dropped from 103 quarts to 43 quarts in 20 years, the average rate of change was -3 quarts per year. The linear model is

$W = 103 - 3t.$ *Whole milk consumption*

Since low-fat milk consumption increased from 25 to 60 quarts per year, the average rate of change was 1.75 quarts per year. The linear model is

$L = 25 + 1.75t.$ *Low-fat milk consumption*

To find the years when low-fat milk consumption exceeded whole milk consumption, you can solve the following inequality.

Low-fat milk consumption > Whole milk consumption

$25 + 1.75t > 103 - 3t$	*Algebraic model*
$1.75t > 78 - 3t$	*Subtract 25 from both sides.*
$4.75t > 78$	*Add 3t to both sides.*
$t > 16.4$	*Divide both sides by 4.75.*

This suggests that Americans began drinking more low-fat milk than whole milk sometime during 1986. The graphs at the left show the linear models for both types of milk consumption. ■

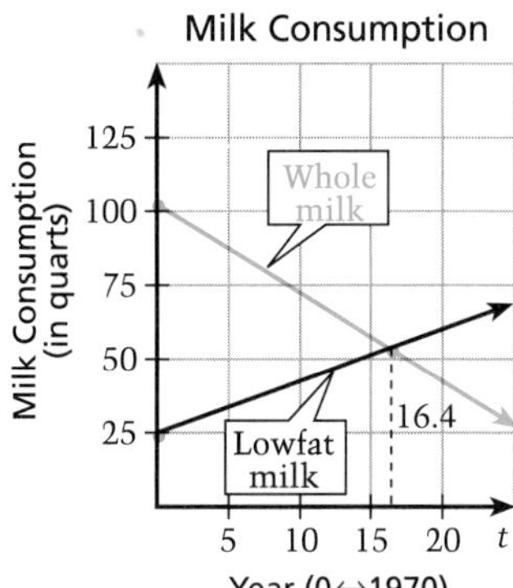

LESSON Notes

Example 1

Discuss the reasonableness of this solution with students. From the available information, you can conclude only that Suzanne ran faster than Jorah. You don't know how much faster she ran.

Example 2

Be sure that students understand what the solution 16.4 stands for. Since 0 corresponds to 1970, then 16.4 represents a time value in the year 1986. Observe also that the solution to this linear inequality can be found graphically. Ask students to state how the graphs were used to solve the inequality.

Extra Examples

Here are additional examples similar to **Examples 1–3.**

1. The Fast-Track Company manufactures toy remote-control race cars that sell for $18. The production cost for the company is $2000 per day plus $13 per race car. How many cars (*c*) must the company produce in a day in order to realize a profit? $c \geq 400$ cars
2. You have $32.14 to spend on a blouse. The sales tax is 6%. How much can the blouse cost (*c*) so you still have enough money to pay the sales tax? $c \leq \$30.32$

Check Understanding

1. Illustrate how both an inequality and an equation can often be used to solve the same problem. See page 298.
2. The width of a cardboard poster is 6 cm less than its length. The perimeter of the poster is at least 212 cm. Find the minimum integer dimensions for the poster. minimum dimensions: 56 by 50

Communicating about ALGEBRA

EXTEND *Communicating*
WRITING PROBLEMS
COOPERATIVE LEARNING

An important problem-solving skill is formulating problems. Have students work in small groups and construct another problem similar to the ones shown. Students could exchange problem sets with other classmates.

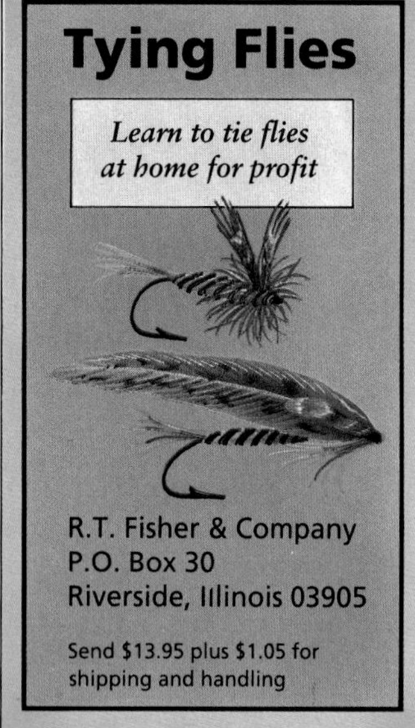

Tying Flies
Learn to tie flies at home for profit
R.T. Fisher & Company
P.O. Box 30
Riverside, Illinois 03905
Send $13.95 plus $1.05 for shipping and handling

Example 3 *Starting a Small Business*

You see an advertisement for instructions on how to tie flies for fly-fishing. The cost of materials for each fly is $0.15. You plan to sell each fly for $0.58, and you want to make a profit of at least $200. How many flies will you need to tie and sell?

Solution Let x represent the number of flies. Then your sales in dollars will be $0.58x$, and your total expenses will be $0.15x + 15$. Why?

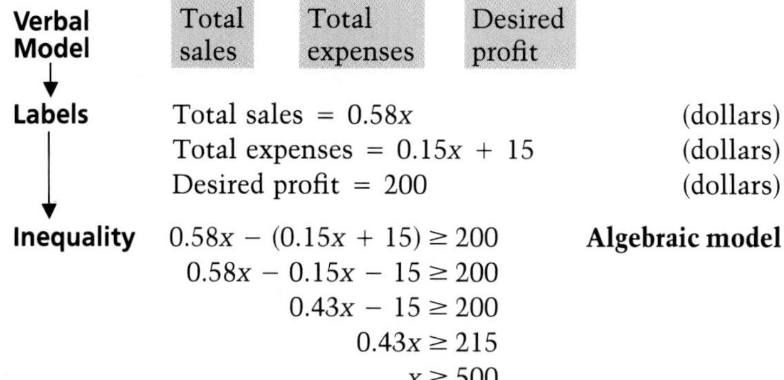

Verbal Model: Total sales − Total expenses ≥ Desired profit

Labels
Total sales $= 0.58x$ (dollars)
Total expenses $= 0.15x + 15$ (dollars)
Desired profit $= 200$ (dollars)

Inequality $0.58x - (0.15x + 15) \geq 200$ **Algebraic model**

$$0.58x - 0.15x - 15 \geq 200$$
$$0.43x - 15 \geq 200$$
$$0.43x \geq 215$$
$$x \geq 500$$

You need to tie and sell at least 500 flies. ■

Communicating about ALGEBRA

Cooperative Learning

SHARING IDEAS about the Lesson

Work with a Partner Use an inequality as a model for the following.

A. You want to run 2 miles in 12 minutes at the most. Write an inequality showing how fast you must run. Solve the inequality. $\frac{2}{r} \leq \frac{1}{5}$, $r \geq 10$

B. You want the area of a rectangle to be at least 100 square inches. The width of the rectangle is 10 inches. Write an inequality describing the length of the rectangle. Solve the inequality. $10L \geq 100$, $L \geq 10$

C. You want to start a business making wind chimes. Your equipment will cost $48. It costs you $3.50 to make each wind chime, and you sell each for $7.50. Write an inequality that shows the number you must make and sell to earn a profit of at least $300. Explain why you think this is the correct inequality. Then solve. See below.

C. $7.50x - (48 + 3.50x) \geq 300$; $7.50x$ is the income; $48 + 3.50x$ is the total cost, and the difference is the profit; $x \geq 87$

EXERCISES

Guided Practice

CRITICAL THINKING about the Lesson

Mail-Order Purchase **In Exercises 1–4, use the following information.**

You are placing an order for blank VCR tapes through the mail. Each tape costs \$3, including tax. Shipping and handling costs are \$2 for the entire order. If you cannot spend more than \$20, how many tapes can you order? 1., 2. See margin.

1. Write a verbal model for this inequality.
2. Assign labels to each part of the model.
3. Write an inequality. $3x + 2 \leq 20$
4. Solve the inequality. $x \leq 6$

The New River Gorge Bridge is 1700 feet long.

Independent Practice

5. *Mercury* Nearly 80% of the known elements are metals. The metallic element with the lowest melting point is Mercury. Write an inequality that describes the melting point, m (in degrees) Fahrenheit, of every other metallic element. $m > -37.8$°F

The melting point of mercury is −37.8°F.

6. *Steel Arch Bridge* The longest steel arch bridge is the New River Gorge Bridge near Fayetteville, West Virginia. Write an inequality that describes the length in feet, l, of every other steel arch bridge. $l < 1700$ ft

7. *Walking to School* Walking at 250 feet per minute, it takes you 10 minutes to walk from your home to school. Your uncle's home is closer to school than yours. Write an inequality for the distance your uncle lives from school. $x < 2500$ ft

8. *Walk, Don't Run* You enter a 3-mile-walk race and win. Your time is 30.2 minutes. Write an inequality about the average speed of the second-place walker. $x < 5.96$ mph

9. *Amusement Park* An amusement park charges \$5 for admission and \$0.80 for each ride. Suppose you go to the park with \$13. Write an inequality that represents the possible number of rides you can go on. What is your maximum number of rides? $5 + 0.8x \leq 13$, 10

10. *Pizza* You and your friends have a total of \$12 to spend on a pizza. A large pizza with cheese costs \$8 plus \$0.40 for each additional topping, tax included. Use an inequality to find the number of toppings you can afford. $0.40x + 8 \leq 12$, 10

EXERCISE Notes

ASSIGNMENT GUIDE

Basic/Average: Ex. 1–4, 5–11 odd, 15–27 odd, 29–32

Above Average: Ex. 1–4, 6–16 even, 23–27 odd, 29–32

Advanced: Ex. 1–4, 6–12 multiples of 3, 14–17, 20–26 even, 27–28, 29–32

Selected Answers
Exercises 1–4, 5–27 odd

Use **Mixed Review** as needed.

More Difficult Exercises
Exercises 13, 16–17, 29–32

Guided Practice

Ex. 1–4 After students have had an opportunity to solve the mail-order purchase problem, go over its solution with the entire class.

Answers

1. (cost per tape • number of tapes + shipping and handling) ≤ amount to spend
2. Cost per tape = 3 (dollars)
   Number of tapes ordered = x
   Shipping-and-handling cost = 2 (dollars)
   Amount to spend = 20 (dollars)

Independent Practice

Ex. 5 CHEMISTRY The melting point of any metal is that temperature at which the metal as a solid and the metal as a liquid can exist together, moving back and forth between states.

Ex. 16–17 GEOMETRY These exercises provide a connection to geometric concepts.

Ex. 27–28 NUMBER SENSE Students should be able to estimate the correct responses, based on the reasonableness of their answers. If students have difficulty, you might suggest they use other data as referents. For example, if students know that the equator (the circumference of the earth) is approximately 25,000 miles, then that would suggest that the answer of 4000 miles would be the most reasonable response.

Ex. 29–32 Note that B.C. is written following the numerical date and that A.D. precedes the date. B.C. stands for "before Christ," and A.D. stands for "Anno Domini," or "in the year of the Lord."

EXTEND Ex. 29-32 MULTICULTURAL The calendar used in the Americas, Europe, and parts of Africa and Asia has as its zero year the year Christ was born. Have students investigate the history of this standard calendar as well as the calendars of other cultures—for example, the Hebrew or Chinese calendars. Suggest that they then write this year's date using the calendar notation of the other cultures.

11. *Telephone Cost* The cost of a long-distance telephone call is \$0.48 for the first minute and \$0.22 for each additional minute. Use an inequality to find the longest call that would not cost more than \$4. $4 \geq 0.48 + 0.22(t - 1)$, 17 minutes

12. *Florida and New Jersey* From 1960 to 1990, the population in Florida rose from about 4,900,000 to about 13,000,000, and the population in New Jersey rose from about 6,067,000 to about 7,747,000. For which years did the population of Florida exceed the population of New Jersey? *(Source: U.S. Census Bureau)* 1966 to 1990

13. *Chicken or Pork?* For which years did the United States consumption of chicken exceed the consumption of pork? 1981 to 1990

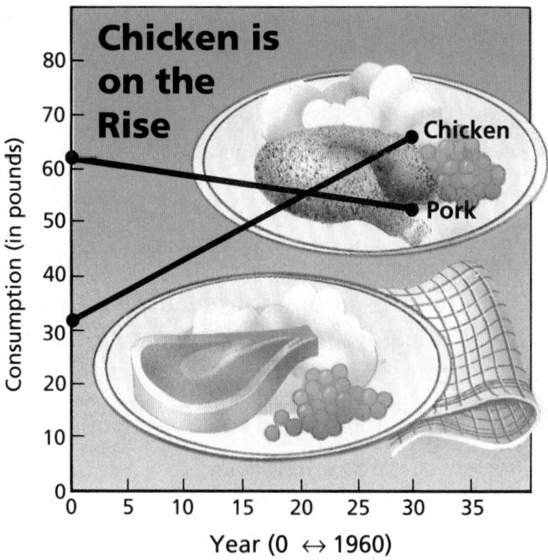

From 1960 to 1990, the average annual U.S. consumption of pork dropped from 62 pounds to 52 pounds. The average annual consumption of chicken rose from 31 pounds to 66 pounds. (Source: U.S. Department of Agriculture)

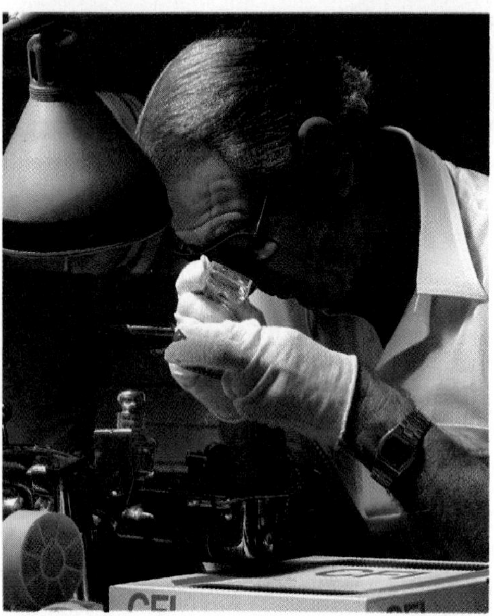

Editing film demands much concentration and expertise. When projected onto a screen, 35-millimeter film runs at 24 frames per second. A $2\frac{1}{2}$-hour movie has about 216,000 frames, after editing. The original film footage usually involves millions of frames.

14. *Animated Films* Suppose each frame in an animated feature film takes at least one hour to draw. Write an inequality that describes the number of hours it would take to draw the frames needed for a 90-minute animated feature film. $x \geq 1 \cdot 24 \cdot 60 \cdot 90$ or $x \geq 129{,}600$

15. *Work Force* Suppose that you have 36 artists with each working 40 hours a week for 45 weeks a year. How many years would it take your artists to draw the film described in Exercise 14? (The first animated feature-length film was *Snow White and the Seven Dwarfs,* 1937. It took Walt Disney Studio artists over three years to draw the thousands of frames needed to make the film.) 2

16. For the area of the rectangle to be at least 36 square meters, what can x be? $x \geq 4$ m

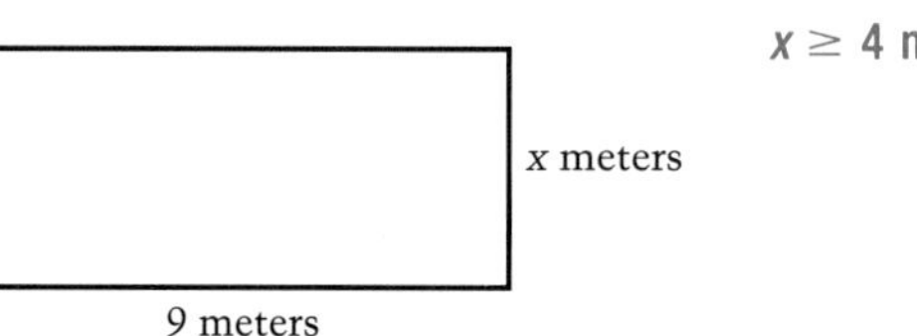

17. For the perimeter of the triangle to be no more than 65 feet, what can x be? $x \leq 18$ ft

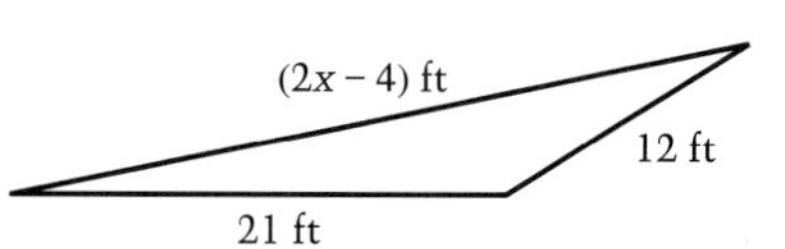

Integrated Review

In Exercises 18–25, solve the inequality.

18. $4x < -28$ $x < -7$ **19.** $6 \geq x + 1$ $x \leq 5$ **20.** $9y - 4 \geq 59$ $y \geq 7$ **21.** $13 + 5x < 8$ $x < -1$

22. $5 - 3x > 11$ $x < -2$ **23.** $-2x + 3 > 3x - 7$ $x < 2$ **24.** $4 + x \leq 16 - x$ $x \leq 6$ **25.** $-9x - 2 \leq 16$ $x \geq -2$

26. *College Entrance Exam Sample* If $4 - x > 5$, then

a. $x > 1$ **b.** $x > -1$ **c.** $x < 1$ **d.** $x < -1$ **e.** $x = -1$

27. *Powers of Ten* Which is the closest estimate of the radius of Earth?

a. 40 miles **b.** 400 miles
c. 4000 miles **d.** 40,000 miles

28. *Powers of Ten* Which is the closest estimate of the weight of a car?

a. 200 pounds **b.** 2000 pounds
c. 20,000 pounds **d.** 200,000 pounds

Exploration and Extension

In Exercises 29–32, match the description with the correct inequality. Let negative numbers represent B.C. years and positive numbers represent A.D. years.

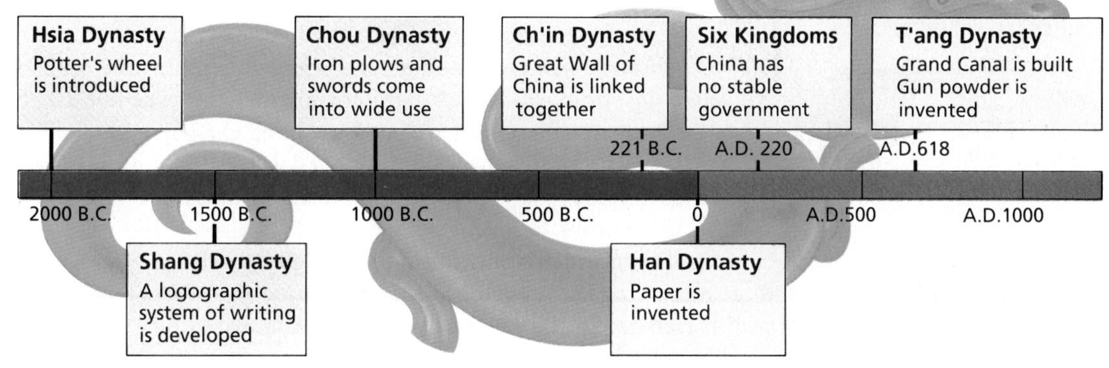

29. Years in which paper could be used b

30. Pre-Chou Dynasty years c

31. Years before the Great Wall was completed a

32. Years after gunpowder was invented d

a. $x < -221$ **b.** $x > 0$ **c.** $x < -1000$ **d.** $x > 618$

Enrichment Activities

These activities require students to go beyond lesson goals.

COOPERATIVE LEARNING
Students might work with partners to complete these activities. Then have student-pairs compare their responses with each other.

1. Starting with Examples 2 and 3, explore how graphs can be used to solve these linear inequalities. Hint: Graph the left and right sides of the inequality separately and observe the point of intersection.
2. After graphing several inequalities, make a conjecture about how to solve inequalities using a graphing approach. Graph both sides of the inequality in the coordinate plane. The x-coordinate of the point of intersection is the boundary point of the solution of the inequality. Test an x-value on either side of this value in order to determine the correct inequality sign in the solution.

Answers

Mixed Review

17.

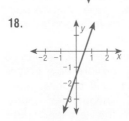

18.

Milestones

Language Arts: Critical Reading

1. What two programming languages are named in this reading? Name two other programming languages. Ada, Pascal; Answers may vary but may include: BASIC, COBOL, and FORTRAN.
2. Find out why Charles Babbage was unable to make his machine work. He was not able to find or create the parts he needed to operate his machine.

Mixed REVIEW

11. $-2x + 4$ 12. $3x - 2$ 15.–16. See Additional Answers. 17.–18. See margin.

1. Find the quotient of $\frac{3}{8}$ and 2. **(1.1)** $\frac{3}{16}$
2. Find the difference of $\frac{1}{3}$ and $\frac{1}{4}$. **(1.1)** $\frac{1}{12}$
3. Evaluate $-|x - \frac{2}{3}| + 2x$ when $x = \frac{1}{6}$. **(2.1)** $-\frac{1}{6}$
4. Evaluate $y^2 + y(4 - y)$ when $y = 3$. **(2.6)** 12
5. Solve for b: $3(a - b) = 7$. **(3.6)** $b = a - \frac{7}{3}$
6. Solve for y: $\frac{1}{2}(x - 2) + 3y = 6$. **(3.6)** $y = \frac{1}{6}(14 - x)$
7. In Exercise 5, evaluate b when $a = \frac{1}{3}$. **(1.2)** $b = -2$
8. In Exercise 6, evaluate y when $x = 0$. **(1.2)** $y = \frac{7}{3}$
9. Is 1 a solution for $5x - 3 < 2$? **(1.5)** No
10. Is 3 a solution for $x^2 + 2 \geq 11$? **(1.5)** Yes
11. Simplify $\frac{1}{4}(x - 2) + \frac{3}{4}(6 - 3x)$. **(2.6)**
12. Simplify $24 \div 6 + 3(x - 2)$. **(1.4, 2.6)**
13. Solve $|x - 2| + 3 = 5$. **(4.8)** 4, 0
14. Solve $|16 - x| - 30 = -4$. **(4.8)** -10, 42
15. Graph the points $(1, \frac{1}{2})$ and $(-1, -\frac{1}{2})$.
16. Graph the points $(-\frac{3}{4}, 6)$ and $(\frac{3}{4}, -6)$.
17. Sketch the graph of $|x + 2| - 3 = y$. **(4.7)**
18. Sketch the graph of $y = 3x - \frac{3}{2}$. **(4.7)**
19. Write an equation of the line that passes through $(-3, 0)$ and $(1, 2)$. **(5.3)** $y = \frac{1}{2}x + \frac{3}{2}$
20. Write an equation of the line that passes through $(-4, 3)$ with a slope of $-\frac{1}{3}$. **(5.6)** $y = -\frac{1}{3}x + \frac{5}{3}$

Milestones ADA

1800 — Augusta Ada Byron born
1850 — Translation and notes published
1900
1950
2000 — Programming language Ada developed

```
WITH IO;
PACKAGE BODY COUNT IS
USE IO;

BEGIN
  PUT ("BEGIN");
  NEW:=LINE;
```

Example of an Ada Program; Augusta Ada Byron

When Charles Babbage strove to make his mechanical computer a reality in the 1830's and 1840's, his assistant was Augusta Ada Byron, Countess of Lovelace. Despite social customs that discouraged women from the pursuit of mathematical knowledge, Countess Lovelace studied mathematics through discussions, correspondences, and self-teaching. In 1842, a paper describing Babbage's ideas for his machine was published in French. Countess Lovelace translated the paper and added to it notes of her own, including a description of a rudimentary computer program. Some consider Countess Lovelace to be the world's first programmer.

In honor of her contribution to computer science, a programming language developed for the United States Department of Defense in the 1980's was named Ada. Based on Pascal, Ada addressed the need for a single programming language that could be used in every branch of the Defense Department. Today Ada is used by numerous organizations in many different applications. **What do you think were some of the social customs that discouraged women from studying mathematics? Do similar customs exist today?**

6.3 Compound Inequalities

What you should learn:

Goal 1 How to solve and graph compound inequalities

Goal 2 How to model a real-life situation with a compound inequality

Why you should learn it:

Using compound inequalities, you can model many real-life situations, such as being above the timberline on a mountain *but* below the snow line.

Goal 1 Solving Compound Inequalities

In Lesson 6.1, you studied four types of simple inequalities. In this lesson, you will study **compound inequalities.** There are many types of compound inequalities. Here are two examples.

- All real numbers that are greater than or equal to zero *and* less than 4 can be represented by

  $0 \le x < 4.$

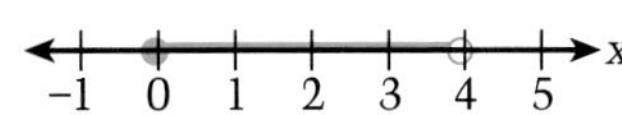

  This compound inequality is also written as $0 \le x$ *and* $x < 4$. The compound inequality $a < x < b$ is read as "x is **between** a and b." The compound inequality $a \le x \le b$ is read as "x is between a and b, *inclusive*."

- All real numbers that are less than -1 *or* greater than 2 can be represented by

  $x < -1$ *or* $x > 2$.

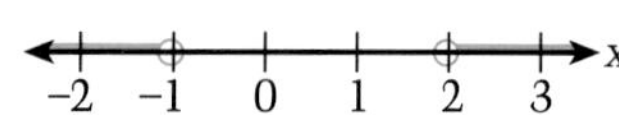

Real Life
Biology

Example 1 *Compound Inequalities in Real Life*

On Mount Rainier, in Washington, the timberline occurs at about five thousand feet. Alpine meadows occur between about five thousand and six thousand feet. Above about six thousand feet, the mountain is covered by glaciers and permanent snowfields. Write inequalities that help describe these three regions of Mount Rainier.

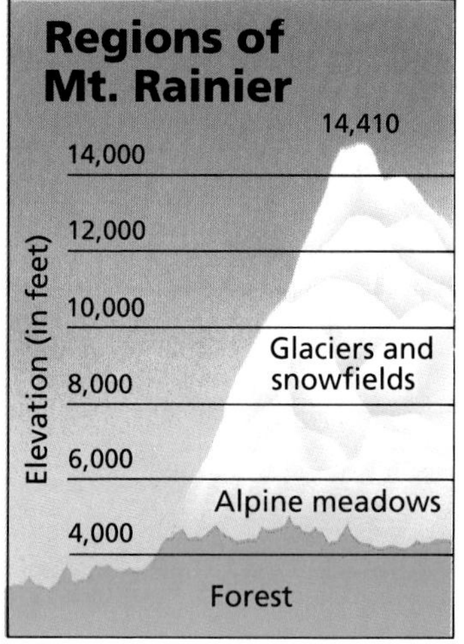

Solution

a. The timber has elevations y (in feet) with

$y \le 5000$ approximately.

b. The alpine meadows have elevations y (in feet) with

$5000 \le y \le 6000$ approximately.

c. The glaciers and permanent snowfields have elevations y (in feet) with

$6000 \le y \le 14{,}410$ approximately. ■

Problem of the Day

Two candles are the same length, but are made of different waxes. One will burn down in 4 hours, and the other in 5. If they are lighted at the same time, in how many hours will one be three times as long as the other? $3\frac{7}{11}$

ORGANIZER

Warm-Up Exercises

1. Solve.
   a. $3x > 45$ $x > 15$
   b. $-1.5m < 12$ $m > -8$
   c. $18 \le 34 - 3x$ $x \le \frac{16}{3}$
2. Graph the solution of each inequality above.

a.

b.

c.

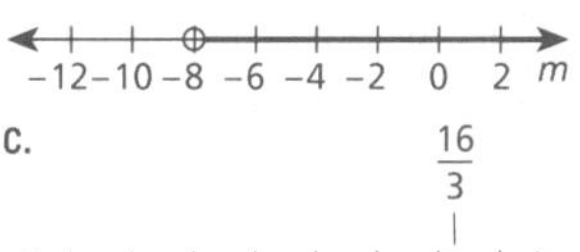

Lesson Resources

Teaching Tools
- Transparencies: 1, 2, 4, 6
- Problem of the Day: 6.3
- Warm-up Exercises: 6.3
- Answer Masters: 6.3

Extra Practice: 6.3

Color Transparency: 28

LESSON Notes

Example 1

Ask students if they can describe some aspect of their environment using compound inequalities. A starting point is to ask students what the words *or* and *and* mean in everyday language.

Example 2

Observe the use and meaning of the conjunction *and*. The solution values of x must, at the same time, be at least 2 and at most 6. So, x is between 2 and 6, inclusive.

An Alternate Approach
Solving Inequalities Using Sets

If students have difficulty using the solution technique suggested in Example 2, you might wish to introduce this alternate method. The compound inequality may be broken apart and each part solved separately.

In set terms, the solution of compound inequalities joined by *and* is the intersection of the solution sets of both inequalities. So to solve the compound inequality $2 \le 3x - 8 \le 10$, use the following technique.

$$2 \le 3x - 8 \quad \text{and} \quad 3x - 8 \le 10$$
$$10 \le 3x \qquad\qquad 3x \le 18$$
$$\frac{10}{3} \le x \quad \text{and} \quad x \le 6$$

Graph both inequalities on the number line. Since the inequalities are joined by *and*, look for the intersection.

$\frac{10}{3} \le x \le 6$

$(\frac{10}{3} \le x) \cap (x \le 6)$

So, $\frac{10}{3} \le x$ and $x \le 6$ is equivalent to $\frac{10}{3} \le x \le 6$.

Note: The solution of compound inequalities joined by *or* is the union of the solution sets of both graphs.

Example 3

Contrast the conjunction *or* with *and*. Here, the two inequalities must be solved separately. The solution is two sets of numbers. Remind students always to check their solutions in the original inequality.

Example 2 *Solving a Compound Inequality*

Solve $-2 \le 3x - 8 \le 10$.

Solution For this type of inequality, your goal is to isolate the variable x between the two inequality signs.

$-2 \le 3x - 8 \le 10$	*Rewrite original inequality.*
$6 \le 3x \le 18$	*Add 8 to each expression.*
$2 \le x \le 6$	*Divide each expression by 3.*

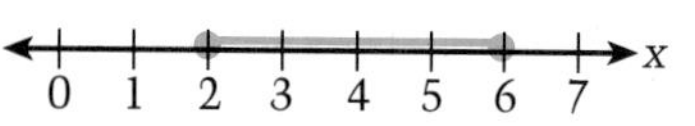

Graphic Check

Because x is between 2 and 6, inclusive, the solution is *all real numbers that are greater than or equal to 2 and less than or equal to 6.* The graph is shown at the left. Check several numbers between 2 and 6, inclusive, in the original inequality. ■

Example 3 *Solving a Compound Inequality*

Solve $3x + 1 < 4$ *or* $2x - 5 > 7$.

Solution Any solution of this compound inequality is a solution of *either* of the simple parts, so you must solve each part separately.

$3x + 1 < 4$	*Left part*	$2x - 5 > 7$	*Right part*
$3x < 3$	*Subtract 1.*	$2x > 12$	*Add 5.*
$x < 1$	*Divide by 3.*	$x > 6$	*Divide by 2.*

Graphic Check

The solution is *all real numbers that are less than 1 or greater than 6.* The graph is shown at the left. ■

When you multiply or divide by a negative number to solve a compound inequality, remember that you have to reverse *both* inequality signs.

Example 4 *Reversing Both Inequality Signs*

Solve $-2 < -2 - x < 1$.

Solution

$-2 < -2 - x < 1$	*Rewrite original inequality.*
$0 < -x < 3$	*Add 2 to each expression.*
$0 > x > -3$	*Multiply each expression by −1 and reverse both inequalities.*

To reflect the order of the number line, this compound inequality would usually be written as

$$-3 < x < 0.$$

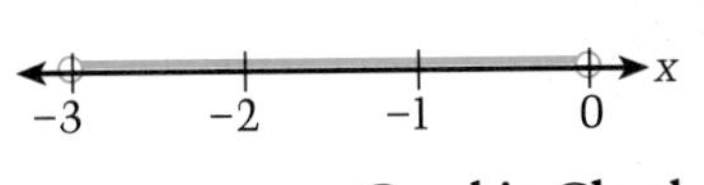

Graphic Check

The solution is *all real numbers that are greater than −3 and less than 0.* The graph is shown at the left. ■

A. Subtract the same number from the sides.
B. Divide the sides by the same positive number.

Goal 2 Modeling a Real-Life Situation

Problem Solving
Draw a Diagram

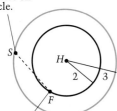

School is somewhere on this circle.

Friend's home is somewhere on this circle.

Example 5 *Describing Distances*

Suppose that you live three miles from school and your friend lives two miles from you. Write an inequality that describes the distances between the school and your friend's home.

Solution A good way to begin this problem is to draw a diagram like the one shown at the left. Point H represents your home. Since you live three miles from school, the school must be located at a point on the circle with a radius of 3 and a center at H. By similar reasoning, your friend's home must be located at some point on the smaller circle. The closest your friend could live to the school would be 1 mile. The farthest would be 5 miles. Thus, the distance, d, between your friend's home and the school can be described by the inequality

$$1 \le d \le 5.$$

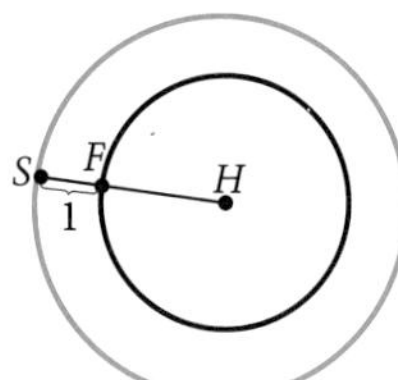

Smallest distance is 1 mile.

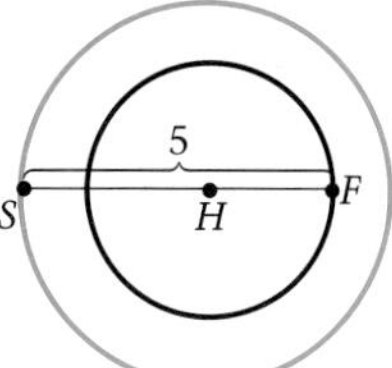

Greatest distance is 5 miles.

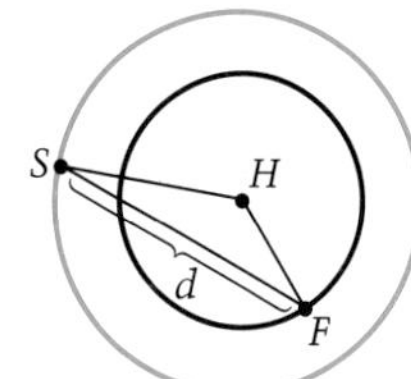

Any distance between 1 and 5 is possible. ■

Communicating about ALGEBRA

SHARING IDEAS about the Lesson

Explain Your Reasoning For each inequality, choose the inequality that is equivalent. Explain your reasoning.

Original Inequality	Which is equivalent?	
	See below.	
A. $4 < x + 2 < 10$	**a.** $2 < x < 8$	**b.** $6 < x < 12$
B. $-4 \le 2x \le 6$	**a.** $-6 \le x \le 4$	**b.** $-2 \le x \le 3$
C. $-3 \le -x \le -2$	**a.** $3 \le x \le 2$	**b.** $2 \le x \le 3$
D. $3 < 2x + 1 < 7$	**a.** $1 < x < 3$	**b.** $2 < x < 4$

C. Multiply (or divide) the sides by the same negative number and reverse the inequalities.
D. Subtract the same number from the sides and divide the sides by the same positive number.

Example 5

Stress the importance of drawing a diagram to the understanding of the problem situation.

Extra Examples

Here are additional examples similar to **Examples 2-4.**

1. Solve each compound inequality.
 a. $-9 < 6x + 3 \le 39$
 b. $8 + 2x < 6$ or $3x - 2 > 13$
 c. $17 < 5 - 3x < 29$
 a. $-2 < x \le 6$
 b. $x < -1$ or $x > 5$
 c. $-8 < x < -4$

2. Graph the solutions of each compound inequality in Exercise 1.

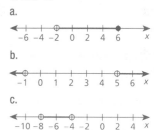

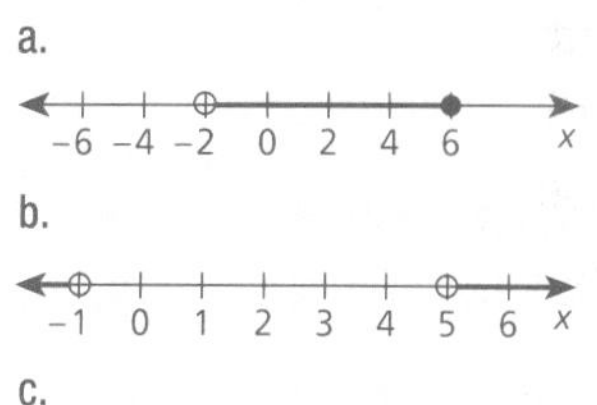

a.
b.
c.

Check Understanding

1. Identify real-life situations that can be described by compound inequalities. See Example 1 on page 305.

2. Explain the differences in solving compound inequalities that are joined by the conjunctions *and* and *or*. See Examples 2–4 on page 306.

3. How can solutions to compound inequalities be checked? Solutions can be checked graphically and algebraically. However, solutions must always be checked in the original compound inequality.

Communicating *about* ALGEBRA

COOPERATIVE LEARNING

Have students work in pairs to explore an alternate strategy for identifying equivalent compound inequalities. Each pair of inequalities in the "Which is equivalent?" column should be graphed. By selecting number values on each graph, decide which graph contains values that do not satisfy the original inequality. Encourage students to build their own strategies for selecting number values.

EXERCISE Notes

ASSIGNMENT GUIDE

Basic/Average: Ex. 1–6, 9–27 multiples of 3, 29–30, 33, 35, 37–43 odd, 45

Above Average: Ex. 1–6, 11–14, 17–20, 21–24, 25–31 odd, 34–35, 36–42 multiples of 3, 45–46

Advanced: Ex. 1–6, 11–13, 17–20, 23–24, 26–35, 42–48

Selected Answers

Exercises 1–6, 7–43 odd

✪ More Difficult Exercises

Exercises 33–34, 45–48

Guided Practice

▶ **Ex. 1–6** These exercises should be covered in whole-group discussions.

EXTEND Ex. 1–4 GROUP ACTIVITY Have students create their own collection of compound inequalities and graphs that other students would be asked to match.

Answers

5.

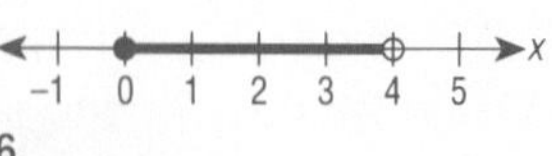

6.

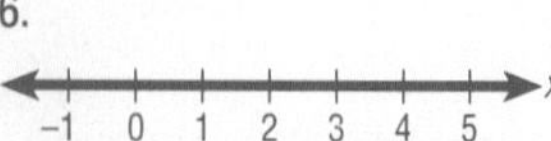

EXERCISES

Guided Practice

CRITICAL THINKING about the Lesson

In Exercises 1–4, match the inequality with its graph.

1. $-2 \le x < 5$ c

2. $-1 < x \le 4$ a

3. $x < -1$ or $x \ge 4$ d

4. $x \le -2$ or $x > 5$ b

a. (number line: −2 −1 0 1 2 3 4, x)

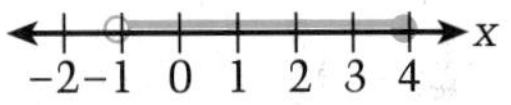

b. (number line: −4 −2 0 2 4 6, x)

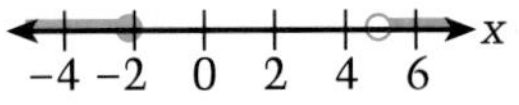

c. (number line: −4 −2 0 2 4 6, x)

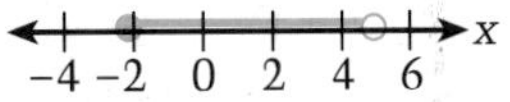

d. (number line: −2 −1 0 1 2 3 4, x)

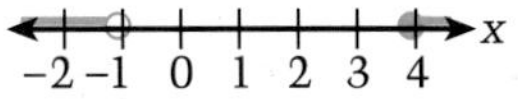

5. Solve the inequality $-2 \le 3x - 2 < 10$ and graph the solution. See margin. $0 \le x < 4$

6. Solve the inequality $2x - 3 < 5$ *or* $x - 4 \ge 0$ and graph the solution. See margin. $x < 4$ or $x \ge 4$ (all real numbers)

Independent Practice

In Exercises 7–14, solve the inequality.

7. $-6 < x - 4 \le 4$ $\;-2 < x \le 8$

8. $-5 < x - 6 < 4$ $\;1 < x < 10$

9. $-4 < 3 - x < 2$ $\;1 < x < 7$

10. $-10 \le 6 - 2x < 8$ $\;-1 < x \le 8$

11. $-10 \le -4x - 18 \le 30$ $\;-12 \le x \le -2$

12. $7 \le -3x + 4 \le 19$ $\;-5 \le x \le -1$

13. $6 - 2x > 20$ or $8 - x \le 0$ $\;x < -7$ or $x \ge 8$

14. $-3x - 7 \ge 8$ or $-2x - 11 \le -31$ $\;x \le -5$ or $x \ge 10$

In Exercises 15–20, solve the inequality and graph the solution. See Additional Answers.

15. $2 \le x < 5$ As is

16. $-3 < x \le -1$ As is

17. $-12 < 2x - 6 < 4$ $\;-3 < x < 5$

18. $-3 \le 6x - 1 < 3$ $\;-\frac{1}{3} \le x < \frac{2}{3}$

19. $-8 \le 1 - 3(x - 2) < 13$ $\;-2 < x \le 5$

20. $-19 < 3 - 4(x + 7) \le 5$ $\;-\frac{15}{2} \le x < -\frac{3}{2}$

In Exercises 21–24, write an inequality for the statement and draw its graph. See Additional Answers.

21. x is between 7 and 4. $4 < x < 7$

22. x is between -9 and 2. $-9 < x < 2$

23. x is less than 4 but is at least 0. $0 \le x < 4$

24. x is less than 2 but is at least -7. $-7 \le x < 2$

In Exercises 25–28, solve the inequality and draw its graph. Then graphically check whether the given x-value is a solution by graphing the x-value on the same number line. See margin.

	Inequality	x-Value		Inequality	x-Value
25.	$-2 < 5x - 7 < 13$ $\;1 < x < 4$	1 No	**26.**	$-4 < 8x - 20 \le 12$ $\;2 < x \le 4$	4 Yes
27.	$-3 < 3 - 4x < 3$ $\;0 < x < \frac{3}{2}$	$\frac{1}{2}$ Yes	**28.**	$7 \le -2x + 21 < 31$ $\;-5 < x \le 7$	0 Yes

29. *Stamp Values* Write a compound inequality that represents the different prices that a block of four "inverted" Curtiss Jenny stamps could have cost between 1918 and 1991. $0.96 \le x \le 550{,}000$

This block of Curtiss Jenny stamps was printed in 1918. It sold for $550,000 in 1991.

30. *Size of Antelopes* Antelopes come in all different sizes. The largest antelope, the derby eland, can weigh up to 2000 pounds. The springbok of South Africa, so-called from its ability to leap repeatedly 10 feet into the air when frightened, weighs a modest 75 pounds. The royal antelope, smallest of all, weighs as little as 7 pounds at adulthood. Write a compound inequality that represents the different weights of adult antelopes. $7 \le x \le 2000$

Dimensions of Our Galaxy **In Exercises 31 and 32, use the diagram of distances in our galaxy.**

31. Which compound inequality best describes the distance, d (in miles), between Earth and one of the other eight planets?

a. $10^7 < d < 10^{10}$ **b.** $10^8 < d < 10^{11}$

c. $10^9 < d < 10^{12}$

32. Which compound inequality best describes the distance, d (in miles), between Earth and a star (other than the sun) in our galaxy?

a. $10^{13} < d < 10^{18}$ **b.** $10^{13} < d < 10^{17}$

c. $10^{17} < d < 10^{18}$

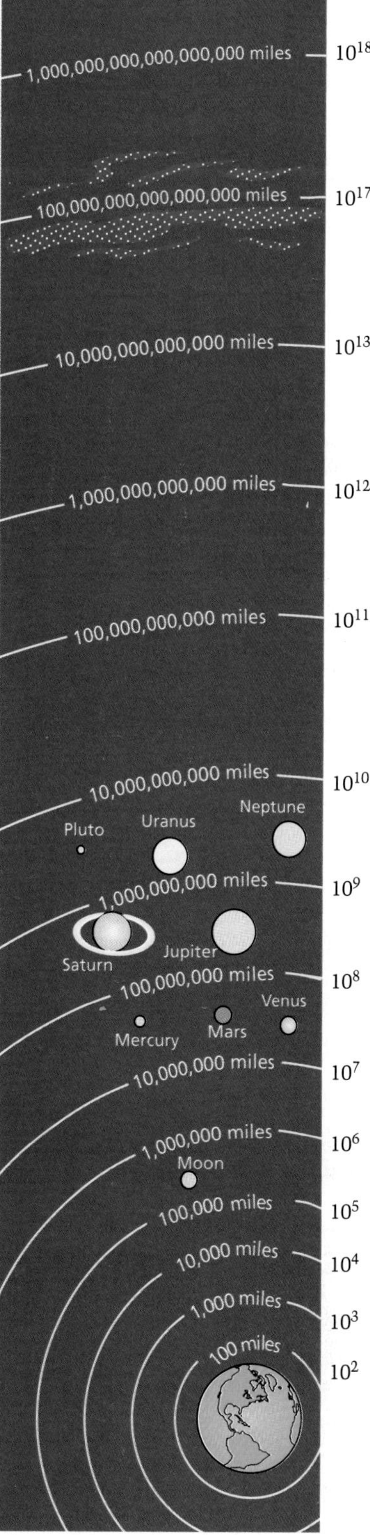

Independent Practice

▶ **Ex. 7–14** Encourage students to graph their solutions.

▶ **Ex. 21–24** COOPERATIVE LEARNING You may wish to use these exercises as oral exercises in class to be sure students can translate from words to inequalities.

Answers

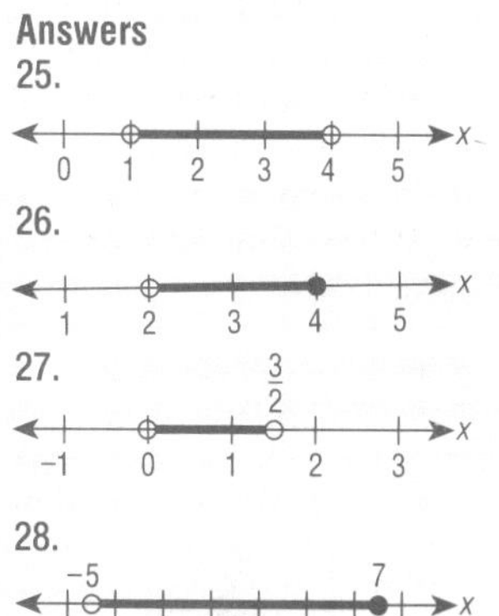

EXTEND Ex. 29 PROJECT An interesting project for students asks them to research the history of the Curtiss Jenny stamps as part of the bigger picture of what makes certain stamps or coins more valuable than others.

▶ **Ex. 31–32** Note that the scale used to measure distances in the galaxy is an exponential scale. Discuss why a linear scale is not useful.

EXTEND Ex. 31–32 HISTORY OF MATHEMATICS An interesting project in the history of mathematics is to have students investigate the uses of exponential and logarithmic scales and why they are needed.

▸ **Ex. 33–34 PROBLEM SOLVING** Emphasize the importance of using diagrams to understand problem situations. Refer students back to Example 5 on page 307.

▸ **Ex. 45–48** The connection to geometry should be pointed out to students. It should also be noted that an inequality is used to express an important relationship among the sides of any triangle—that is, the sum of the lengths of any two sides is greater than the length of the third side.

Enrichment Activities

WRITING Following the class discussion, students might record their personal responses in their math journals.

1. Summarize the techniques used to solve both simple and compound inequalities in your math journal. Include special cautions for solving inequalities where useful. Students should summarize the techniques presented in Examples 2–4, noting that the inequality sign is reversed when multiplying or dividing by a negative number.
2. Identify procedures for checking solutions to inequalities. Point out the similarities of solving inequalities to solving equations. Checking can be done algebraically by substituting values in the original inequalities or graphically by using the intersection or union of sets.

TECHNOLOGY The technique that the TI-82 graphing calculator uses to test the condition of whether -5 is less than $2x + 3$ is stated by the expression $y_1 = -5 < 2x + 3$. Point out to students that this syntax is unique to this graphing tool and is not common mathematical syntax.

Meet Me at the Y **In Exercises 33 and 34, use the following information.**

Jan lives in a mobile home 1 mile from the YWCA. Arlene lives $\frac{1}{2}$ mile from Jan. Jan walks from her home to the Y, where she meets Arlene, walks to Arlene's apartment, and then walks back home.

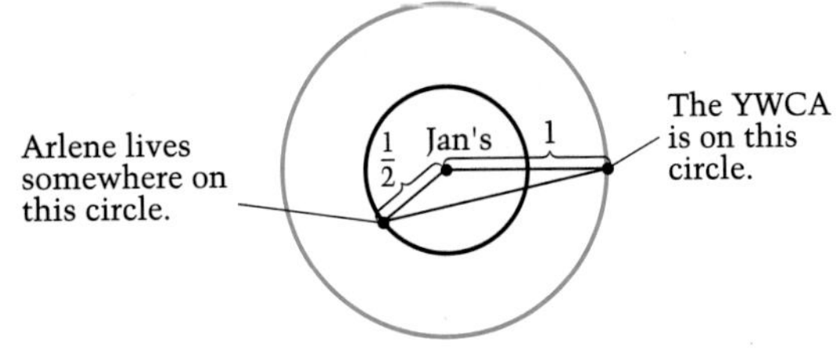

33. Write an inequality that describes the various distances that Jan could have walked. (Assume that each part of the trip followed a straight path.) $2 \le x \le 3$

34. Write an inequality that describes the various distances that Arlene could live from the YWCA. $\frac{1}{2} \le x \le 1\frac{1}{2}$

Integrated Review

35. *College Entrance Exam Sample* Which of the following is the graph of $x \ge -2$ *and* $x \le 3$? e

a. 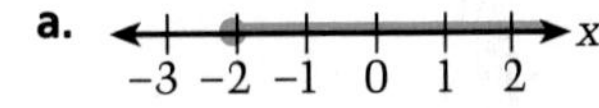

b. 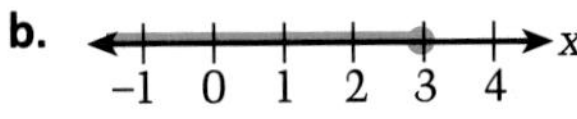

c. 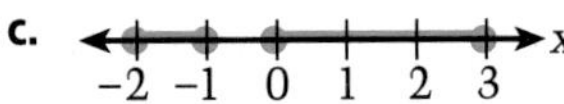

d. 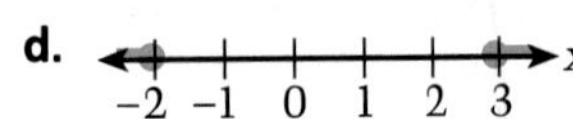

e. 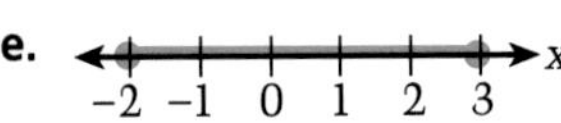

In Exercises 36–38, write the inequality for the graph.

36. 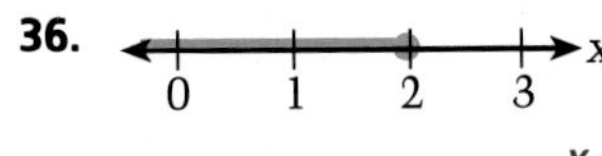

$x \le 2$

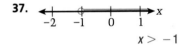

37. $x > -1$

38. 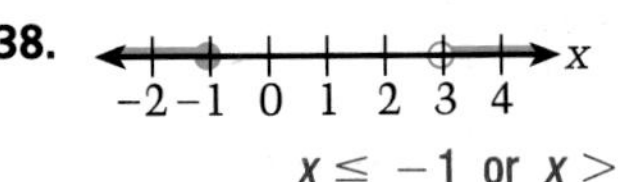

$x \le -1$ or $x > 3$

In Exercises 39–44, solve the equation.

39. $|x - 1| = 7$ 8, −6

40. $|x - 3| = 9$ 12, −6

41. $|2x + 1| = 11$ 5, −6

42. $|3x - 2| = 5$ $\frac{7}{3}, -1$

43. $17 = |5x - 2|$ $\frac{19}{5}, -3$

44. $24 = |6x + 8|$ $\frac{8}{3}, -\frac{16}{3}$

Exploration and Extension

Geometry **In Exercises 45–48, write a compound inequality that describes the possible lengths of the side of the triangle marked x. Use the fact that the sum of the lengths of any two sides of a triangle is greater than the length of the third side.**

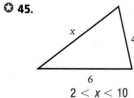

45. $2 < x < 10$

46. 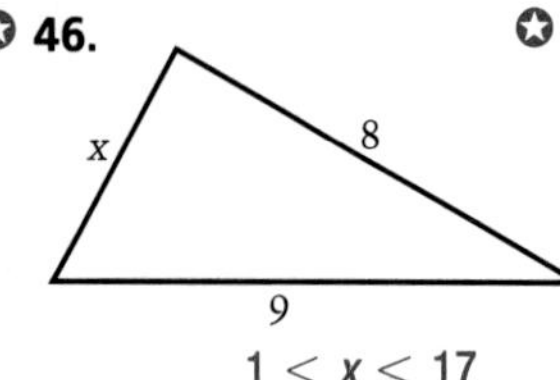

$1 < x < 17$

47. 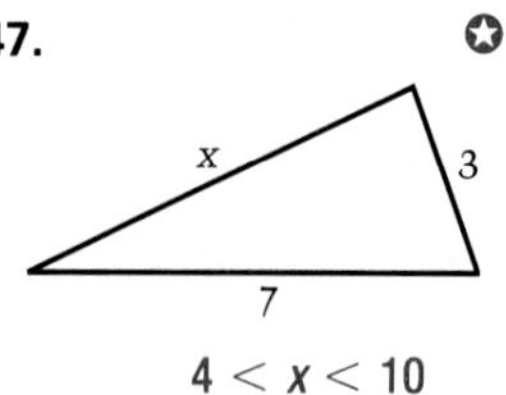

$4 < x < 10$

48. $6 < x < 16$

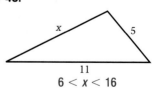

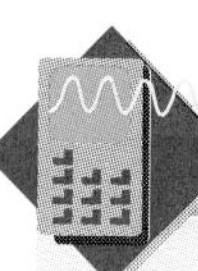

USING A GRAPHING CALCULATOR

Graphing calculators and computers can be used to sketch the the solution of an inequality in one variable. Keystroke instructions for doing this on a TI-82 are listed on page 739. Typically, the graph is not shown on the x-axis. Instead, it is shown as a horizontal line segment one unit above the x-axis.

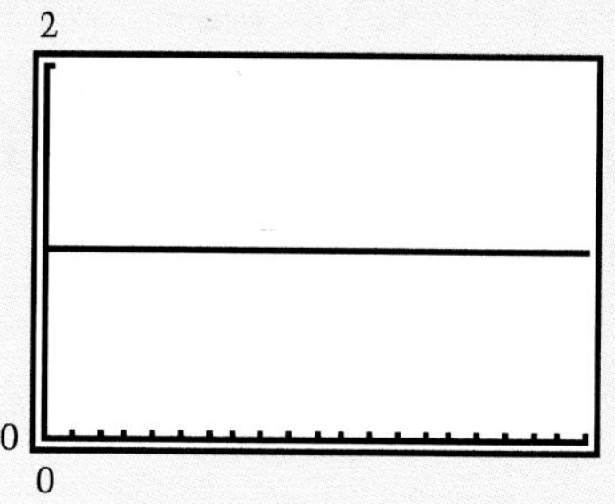

Graph of Inequality

Real Life Business

Sales Tax and Discount You are shopping for a wristwatch. You have \$30.00 to spend, and you have a coupon that will give you a discount of \$7.50. The sales tax rates for five states are shown at the left below. What retail prices will fit your budget? (The discount is applied before the sales tax is computed.)

Verbal Model $\left(\text{Price} - \text{Discount}\right) + \text{Sales tax} \le \$30$

Labels
Price $= x$ (dollars)
Discount $= 7.5$ (dollars)
Sales tax $=$ (tax rate)$(x - 7.5)$ (dollars)

Inequality $(x - 7.5) + (\text{tax rate})(x - 7.5) \le 30$

State	Sales Tax
Wyoming	3%
Michigan	4%
Massachusetts	5%
Texas	6%
Rhode Island	7%

If you were shopping in Texas, the tax rate would be 0.06. So, the inequality would be

$$(x - 7.5) + (0.06)(x - 7.5) \le 30.$$

From the graph of this inequality, you can estimate that you can buy a wristwatch that costs up to about \$36. An algebraic check shows that the actual solution is $x \le 35.80$. How would this answer change in the other four states whose sales tax rates are shown at the left?

Exercises

Use a graphing calculator or computer to sketch a graph of the inequality. State the solution shown by the graph. Then check the solution algebraically. See margin.

1. $-2x + 3 \ge 7$ $x \le -2$
2. $-3 < -x + 5$ $x < 8$
3. $-3x + 2 \le 4$ $x \ge -\frac{2}{3}$
4. $5x + 1 < 4x + 5$ $x < 4$
5. $-2x > -3x - 1$ $x > -1$
6. $-x - 2 \le 3x + 6$ $x \ge -2$

Answers

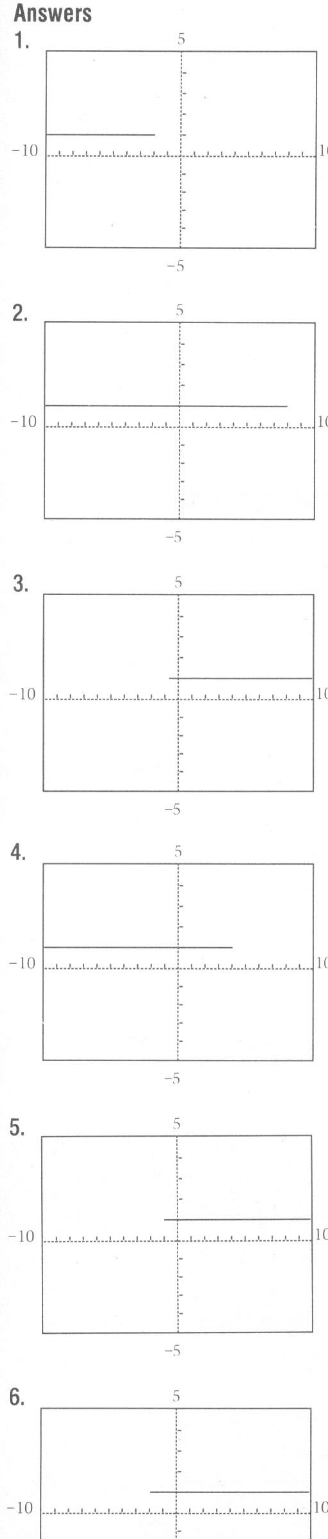

Answers

10.

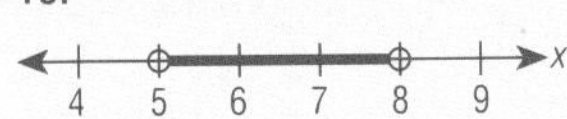

11.

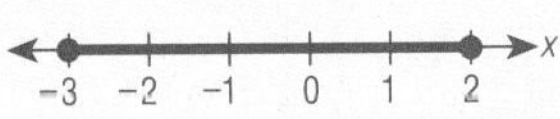

12.

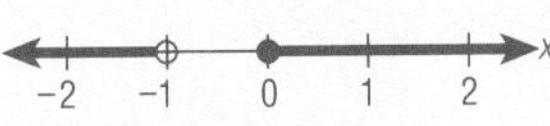

13.

−4 −3 −2 −1 0 1 2 3 x

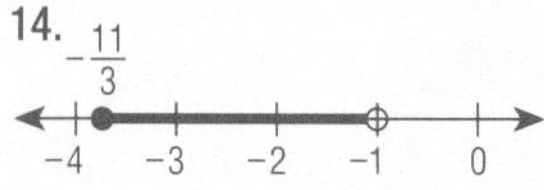

15.

4 5 6 7 8 x

16.

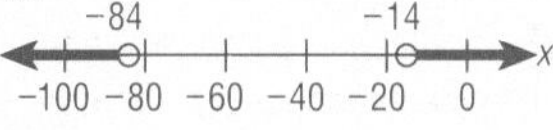

Common-Error ALERT!

In **Exercises 6–8,** students frequently forget to reverse the inequality sign when multiplying or dividing by a negative number. Point out that inequality signs must be reversed in these problems.

Mid-Chapter SELF-TEST

Take this test as you would take a test in class. The answers to the exercises are given in the back of the book.

In Exercises 1–3, solve the inequality. Then match it with one of the graphs. (6.1)

1. $x + 2 > 4$ c
2. $x - 7 \le 6$ a
3. $\frac{1}{4}x + 6 \ge 5$ b

a. 10 11 12 13 14 15 x

b. −5 −4 −3 −2 −1 0 x

c. 0 1 2 3 4 x

In Exercises 4–9, solve the inequality. (6.1)

4. $-3 \le x + 2 \le 5$ $-5 \le x \le 3$
5. $-7 < 3x + 5 < 6$ $-4 < x < \frac{1}{3}$
6. $6 < 10 - 4x \le 34$ $-6 \le x < 1$
7. $-20 \le -5x < -5$ $1 < x \le 4$
8. $-3x > 7$ or $3x \ge 7$ $x < -\frac{7}{3}$ or $x \ge \frac{7}{3}$
9. $\frac{3}{2}(x - 2) \le 6$ or $\frac{1}{5}(2x - 3) \ge 3$ $x \le 6$ or $x \ge 9$

In Exercises 10–13, write an inequality for the statement and draw its graph. (6.2) See margin.

10. x is more than 5 and less than 8. $5 < x < 8$
11. x is between -3 and 2, inclusive. $-3 \le x \le 2$
12. x is greater than or equal to 0 or x is less than -1. $x < -1$ or $x \ge 0$
13. x is at least -4 but no more than 3. $-4 \le x \le 3$

In Exercises 14–16, sketch a graph of the inequality. (6.3) See margin.

14. $-3 < -3(x + 2) \le 5$
15. $0 \le -2(4 - x) < 8$
16. $\frac{1}{14}x > -1$ or $\frac{-1}{14}x + 1 > 7$
17. The Dodge'em Cars at Fun Park require children to be at least 52 inches tall to drive the cars. Write an inequality that describes the required heights. **(6.2)** $x \ge 52$
18. The length of a new pencil before sharpening is $7\frac{1}{2}$ inches, including the eraser. Suppose you discard the pencil when it is $2\frac{1}{2}$ inches long. Write a compound inequality that describes the size of a usable pencil. **(6.3)** $2\frac{1}{2} < x < 7\frac{1}{2}$
19. Granola in the bulk-food section of the grocery store is \$1.96 per pound. You have \$3 to spend on granola and strawberry yogurt. Write an inequality that describes how many pounds of granola you can purchase if you pay \$0.69 for strawberry yogurt. Solve the inequality. **(6.2)** $1.96x + 0.69 \le 3.00$, $x \le 1.18$
20. Write a compound inequality that gives the possible lengths, x, of the side of the triangle. **(6.3)** $1 < x < 9$

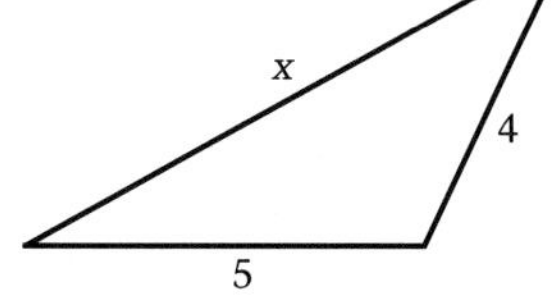

6.4 Connections: Absolute Value and Inequalities

What you should learn:

Goal 1 How to solve absolute value inequalities

Goal 2 How to model a real-life situation with an absolute value inequality

Why you should learn it:

You can use absolute value inequalities to model real-life situations about quantities within or outside of a number line interval, such as expenses being within 5% of budget.

Goal 1 Solving Absolute Value Inequalities

LESSON INVESTIGATION

Investigating Absolute Value Inequalities

Partner Activity Use a guess-and-check strategy to find values of x that satisfy each absolute value inequality.

a. $|x| < 2$ **b.** $|x + 2| \geq 1$ **c.** $|x - 3| \leq 2$

Sketch the solution set of each inequality on a number line. Then use a double inequality to describe the solution set.

In this investigation, you used a guess-and-check strategy to solve absolute value inequalities. The following properties give you another way to solve such inequalities.

Translating Absolute Value Inequalities

1. The inequality $|ax + b| < c$ is equivalent to

   $-c < ax + b < c.$

2. The inequality $|ax + b| > c$ is equivalent to

   $ax + b < -c$ or $ax + b > c.$

Study Tip

The first property at the right is true when either < or ≤ is used. The second property is true when either > or ≥ is used.

Example 1 *Solving an Absolute Value Inequality*

Solve $|x - 4| < 3$.

Solution

$\lvert x - 4\rvert < 3$	*Rewrite original inequality.*
$-3 < x - 4 < 3$	*Write equivalent compound inequality.*
$1 < x < 7$	*Add 4 to each expression.*

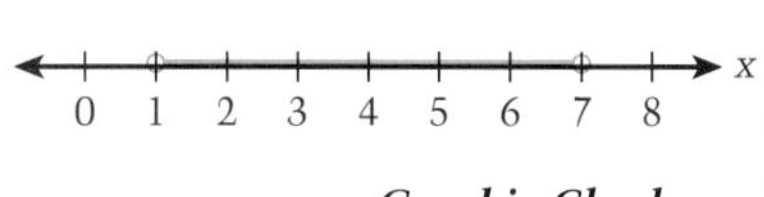

Graphic Check

The solution is *all real numbers that are greater than 1 and less than 7.* The graph is shown at the left. ■

Problem of the Day

Find the missing digits.

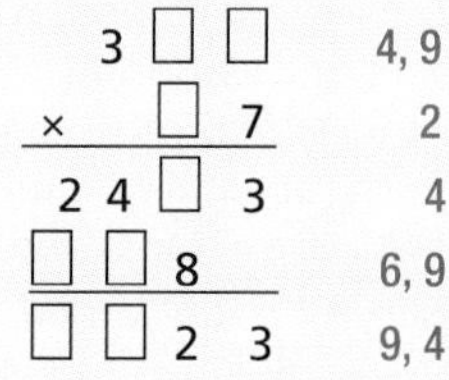

ORGANIZER

Warm-Up Exercises

1. Evaluate each expression, given that $x = 3$ and $y = -2$.
   a. $|2x - 9|$ 3
   b. $|y - x|$ 5
2. Sketch the graph of each absolute-value equation.
   a. $y = |x| - 4$
   b. $y = |2x - 3| + 1$

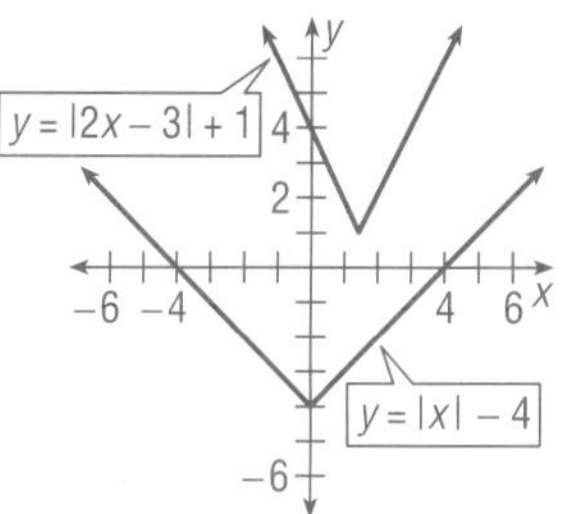

3. Solve.
   $|3x + 6| = 9$
   $x = 1, x = -5$

Lesson Resources

Teaching Tools
- Transparencies: 4, 6, 10
- Problem of the Day: 6.4
- Warm-up Exercises: 6.4
- Answer Masters: 6.4

Extra Practice: 6.4
Color Transparency: 29
Applications Handbook: p 36
Technology Handbook: p. 53

Lesson Investigation Answers

a. $-2 < x < 2$
b. $x \geq -1$ or $x \leq -3$
c. $1 \leq x \leq 5$

LESSON Notes

GOAL 1 Use a combination of numbers and variables to illustrate and explain the methods by which absolute value inequalities are translated into compound inequalities. For example, consider $|M| < 9$. Let students determine the values of M for which the inequality is true. Stress that M is a variable and may represent any real number value. As a result of discussing the possible number values of M, students must recognize that the solution, M, is between -9 and 9; or stated differently, M is greater than -9 but less than 9.
Point out that M may stand for any real-valued number or expression, so it follows that $|ax + b| < 9$ means that $ax + b$ is greater than -9 but less than 9. Written algebraically, $-9 < ax + b < 9$. Encourage students to delay making this substitution until they feel comfortable with it. Finally, replace the constant, 9, with a literal constant, c. Again, discuss as required. Using a gradual approach that builds onto the meaning of the expression makes students feel a greater sense of comfort about the method described.

Example 4

You may have to help students interpret the values in each inequality. For example, $|B - A| < 0.05B$ because $5790.18 < 6225.00$.

Example 2 *Solving an Absolute Value Inequality*

Solve $|x + 1| \geq 2$.

Solution This absolute value inequality is equivalent to the following compound inequality.

$x + 1 \leq -2$ *or* $x + 1 \geq 2$

You can solve this inequality using the procedure described in Lesson 6.3.

$x + 1 \leq -2$	*First inequality*	$x + 1 \geq 2$	*Second inequality*
$x \leq -3$	*Subtract 1.*	$x \geq 1$	*Subtract 1.*

The solution is *all real numbers that are less than or equal to* -3 *or greater than or equal to 1*. The graph is shown below.

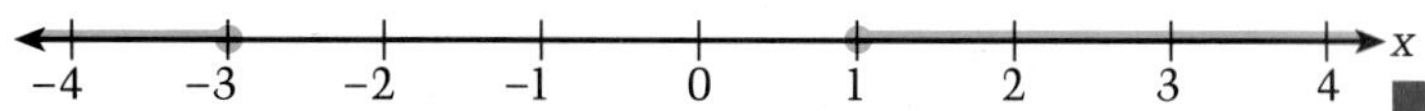

The next example shows how to write an absolute value inequality for a given graph.

Example 3 *Writing an Absolute Value Inequality*

Write an absolute value inequality to fit the graph shown below.

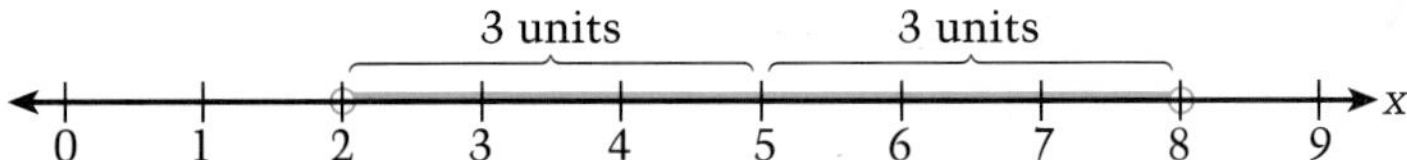

Solution To begin, write the compound inequality for the graph.

$2 < x < 8$

Then, to write the absolute value inequality, find the number that lies halfway between 2 and 8. This number is 5. (It lies 3 units from 2 and 3 units from 8.) Subtract 5 from each expression in the compound inequality.

$$2 - 5 < x - 5 < 8 - 5$$
$$-3 < x - 5 < 3$$

In this form, you can recognize the absolute value inequality to be

$|x - 5| < 3.$

For the inequality $|x - 5| < 3$, you can say that the range of solutions is from 3 less than 5 to 3 greater than 5.

Goal 2 Modeling a Real-Life Situation

This computer program compares actual expenses with budgeted expenses. It flags any expense that differs from its budget by 5% or more.

```
10 PRINT "BUDGET EXP. ACTUAL EXP."
20    FOR N=1 TO 2
30    READ B,A: PRINT USING "$###, ###.## ";B,A;
40    IF ABS(B-A)<.05*B THEN PRINT "APPROVED" ELSE PRINT
      "QUERIED"
50 NEXT
60 END
70 DATA 124500, 130290.18, 36500, 34383.29
```

Real Life
Accounting

Example 4 *Budget Variance*

You are the manager of the accounting department of a large retail store and receive the following *budgeted* and *actual* expenses from the marketing department. How will the computer respond?

Item	Budgeted Expense, B	Actual Expense, A
a. Salaries	\$124,500.00	\$130,290.18
b. Travel	\$36,500.00	\$34,383.29

Solution

a. For salaries, $|B - A| = 5790.18$ and $0.05B = 6225.00$. The inequality $|B - A| < 0.05B$ is satisfied. The computer will print "APPROVED."

b. For travel, $|B - A| = 2116.71$ and $0.05B = 1825.00$. The inequality $|B - A| < 0.05B$ is *not* satisfied. The computer will print "QUERIED." ■

Communicating about ALGEBRA

SHARING IDEAS about the Lesson

Translate between Models Write an absolute value inequality for each graph.

A. 

$|x - 1| < 4$

B. 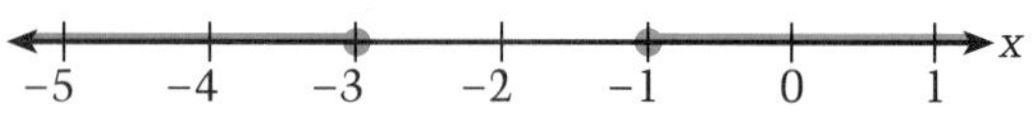

$|x + 2| \geq 1$

Extra Examples

Here are additional examples similar to **Examples 1–3.**

1. Solve.
   a. $|x + 9| > 13$ $\quad x < -22$ or $x > 4$
   b. $|3x - 15| < 12$ $\quad 1 < x < 9$
2. Graph the solutions.
   a. 

   b. 

3. Write an absolute-value inequality to fit each graph.
   a. 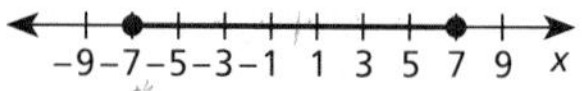

   b. 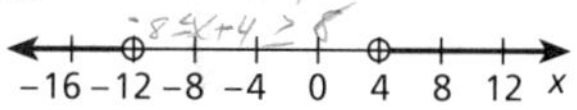

   a. $|x| \leq 7$
   b. $|x + 4| > 8$

Check Understanding

1. Explain why the expression $|x| < a$ represents real number values between $-a$ and a. The absolute value of a number is the distance between the number and the origin. It follows that the distance of a number $|x|$ less than a falls between the origin and the distance a on either side of the origin, or $|x|$ is between $-a$ and a.
2. Explain why the expression $|x| > b$ represents real number values that are not between $-b$ and b. Adapt the argument given above.
3. Describe the steps needed to translate an absolute value inequality into a compound inequality. See page 313.

Communicating about ALGEBRA

COOPERATIVE LEARNING
Have students work in small groups. Ask students to show the compound inequalities that lead to the absolute value inequalities. Students should be encouraged to create graphs for students who are not in their small group.

EXERCISE Notes

ASSIGNMENT GUIDE

Basic/Average: Ex. 1–12, 19–37 odd, 41–45 odd, 48–57 multiples of 3, 59–65 odd

Above Average: Ex. 1–12, 20–38 even, 42–46 even, 51–54, 59–66

Advanced: Ex. 1–12, 15–36 multiples of 3, 39–42, 45–57 multiples of 3, 59–66

Selected Answers
Exercises 1–6, 7–57 odd

Use **Mixed Review** as needed

✪ More Difficult Exercises
Exercises 41–44, 45, 59–66

Guided Practice

▶ **Ex. 1-6** GROUP ACTIVITY
Use the exercises as oral or small-group exercises with the class. They provide a quick overview of the concepts and skills taught in the lesson.

Answers
1. True; $|x + 1| < 3$ is equivalent to the compound inequality $-3 < x + 1 < 3$.
2. True; $|x| \geq 5$ is equivalent to the compound inequality $x \leq -5$ or $x \geq 5$.

Independent Practice

▶ **Ex. 7–18** Some students may wish to graph the solutions to these exercises.

316

EXERCISES

Guided Practice

CRITICAL THINKING about the Lesson 1., 2. See margin.

1. **True or False?** $|x + 1| < 3$ means that $x + 1$ is between -3 and 3. Explain.
2. **True or False?** $|x| \geq 5$ means that x is less than -5 or greater than 5. Explain.
3. Solve $|x - 1| < 4$. $-3 < x < 5$
4. Solve $|x + 3| > 9$. $x < -12$ or $x > 6$
5. Match the inequality $|x| \geq 6$ with its graph. a
6. Match the inequality $|x| < 8$ with its graph. b

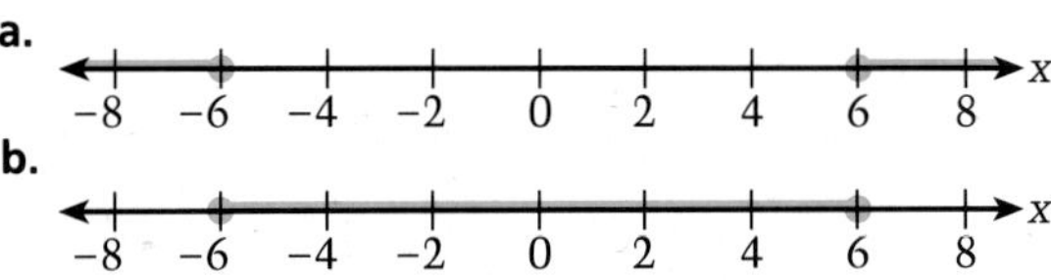

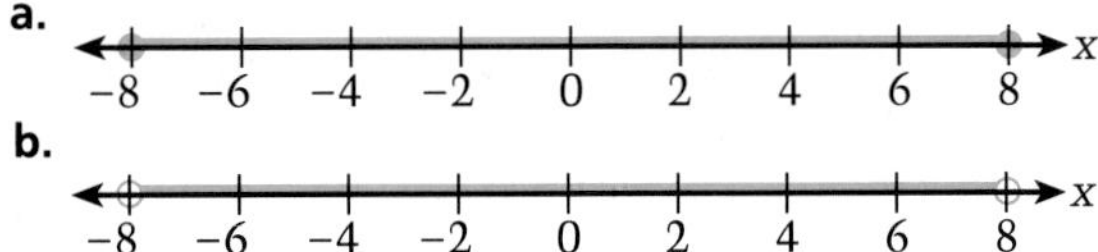

Independent Practice

In Exercises 7–12, match the absolute value inequality with its equivalent compound inequality.

7. $|x - 1| < 3$ d
8. $|x - \frac{1}{2}| \leq \frac{3}{2}$ e
9. $|x + 1| \leq 2$ b
10. $|x + 2| \geq 4$ c
11. $|x - \frac{5}{2}| > \frac{3}{2}$ f
12. $|x + 2| > 1$ a

a. $x < -3$ or $x > -1$
b. $-3 \leq x \leq 1$
c. $x \leq -6$ or $x \geq 2$
d. $-2 < x < 4$
e. $-1 \leq x \leq 2$
f. $x < 1$ or $x > 4$

In Exercises 13–18, solve the inequality.

13. $|x - 2| < 5$ $-3 < x < 7$
14. $|1 + x| \geq 4$ $x \leq -5$ or $x \geq 3$
15. $|2x - 1| > 3$ $x < -1$ or $x > 2$
16. $|4 + x| \leq 9$ $-13 \leq x \leq 5$
17. $|4 - x| \geq 2$ $x \geq 6$ or $x \leq 2$
18. $|-x + 1| < 1$ $0 < x < 2$

In Exercises 19–30, solve the inequality. Then sketch its graph. See margin.

19. $|x + 8| < 9$ $-17 < x < 1$
20. $|9 + x| \leq 7$ $-16 \leq x \leq -2$
21. $|4 - x| < 5$ $-1 < x < 9$
22. $|x + 12| < 36$ $-48 < x < 24$
23. $|x - 12| \geq 6$ $x \leq 6$ or $x \geq 18$
24. $|5 - x| > 18$
25. $|x + 1| > 17$ $x < -18$ or $x > 16$
26. $|x - 9| \geq 4$ $x \leq 5$ or $x \geq 13$
27. $|3x - 6| \geq 0$
28. $|10 - 4x| < 2$ $2 < x < 3$
29. $|1 + 2x| \leq 9$ $-5 \leq x \leq 4$
30. $|2x + 3| > 4$

24. $x > 23$ or $x < -13$ 27. $x \leq 2$ or $x \geq 2$ (all real numbers) 30. $x < \frac{-7}{2}$ or $x > \frac{1}{2}$

In Exercises 31–36, write an absolute value inequality to fit the graph.

31. 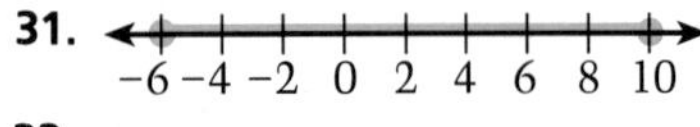 $|x - 2| \leq 8$

32. $|x + 1| < 5$

33. $|x + 4| < 2$

34. 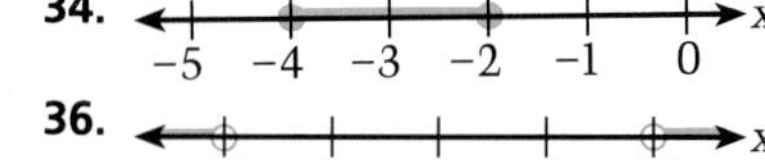 $|x + 3| \leq 1$

35. $|x| \geq 4$

36. $|x| > 10$

37. *Game Show* A contestant on a television show must guess within \$1000 of the actual price of a trip in order to win the trip. The actual price of the trip is \$14,000. Write an absolute value inequality that shows the range of possible guesses that will win the trip.

38. *Car Mileage* Your car averages 25 miles per gallon. The actual mileage for city driving varies from the average by 4 miles per gallon, depending on the traffic conditions. Write an absolute value inequality that shows the range for the mileage that your car gets in the city. $|x - 25| \le 4$

39. *Test Scores* Suppose the test scores in your class range from 60 to 100. Write an absolute value inequality describing test-score range. $|x - 80| \le 20$

37. $|x - 14{,}000| < 1000$

40. *400-Meter Run* The members of your track team can run 400 meters between 58 seconds and 70 seconds. Write an absolute value inequality describing the times for the runners. $|x - 64| < 6$

Fireworks **In Exercises 41–44, match the color of the star burst with one of the color ranges in the spectrum. (Use the wavelengths for various colors of the spectrum given on page 291.)**

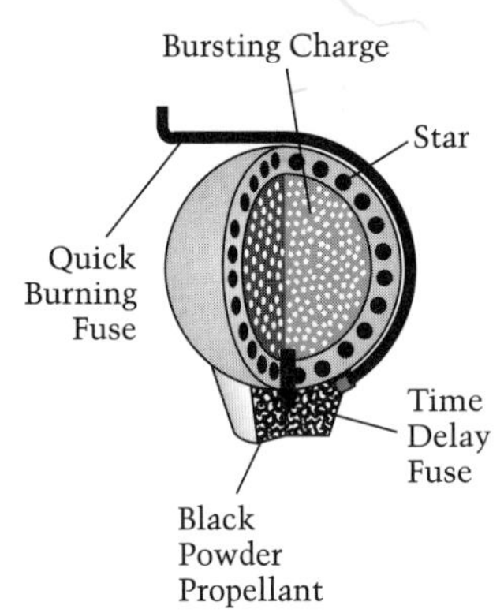

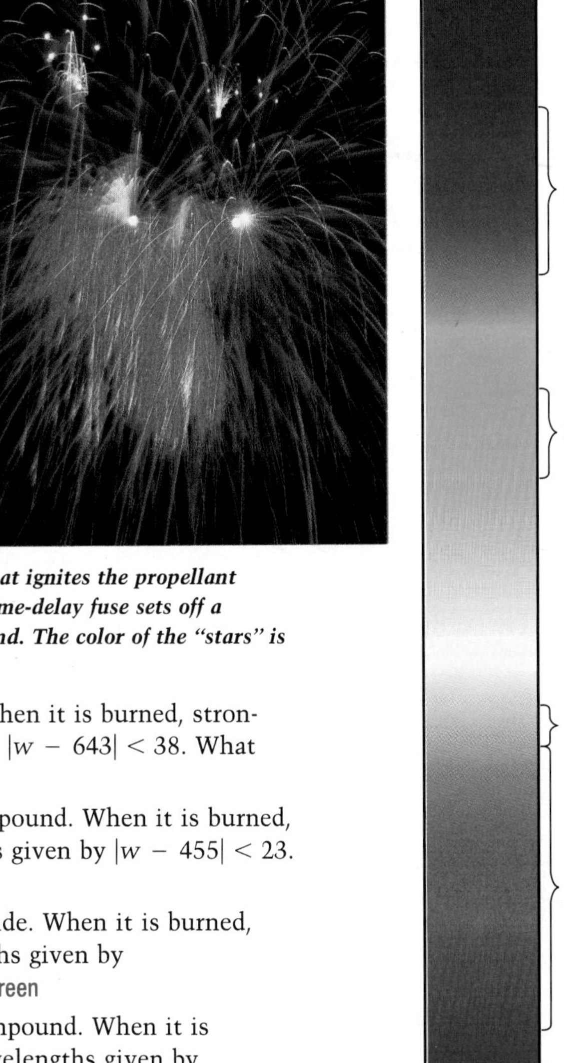

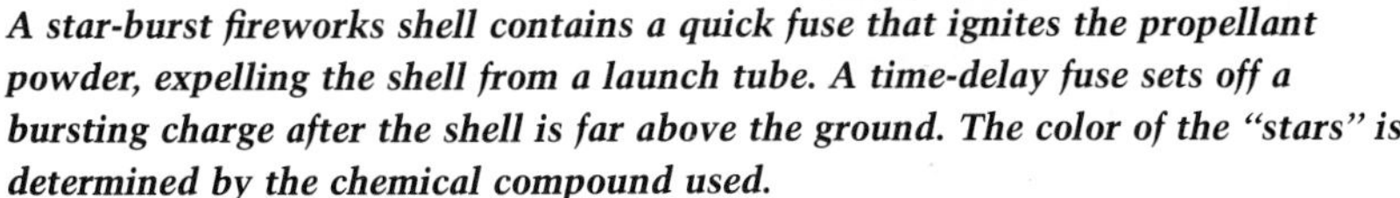
A star-burst fireworks shell contains a quick fuse that ignites the propellant powder, expelling the shell from a launch tube. A time-delay fuse sets off a bursting charge after the shell is far above the ground. The color of the "stars" is determined by the chemical compound used.

41. A fireworks star is made of strontium. When it is burned, strontium emits light at wavelengths given by $|w - 643| < 38$. What color is it? c–d, orange-red

42. A fireworks star is made of a copper compound. When it is burned, the compound emits light at wavelengths given by $|w - 455| < 23$. What color is it? a, blue

43. A fireworks star is made of barium chloride. When it is burned, barium chloride emits light at wavelengths given by $|w - 519.5| < 12.5$. What color is it? b, green

44. A fireworks star is made of a sodium compound. When it is burned, the compound emits light at wavelengths given by $|w - 600| < 5$. What color is it? c, orange

Ex. 13–30 You may wish to reserve some of these exercises to use as additional practice at a later time.

Ex. 31–36 Encourage students first to write the compound inequality by reading the graph and then to write the absolute value inequality.

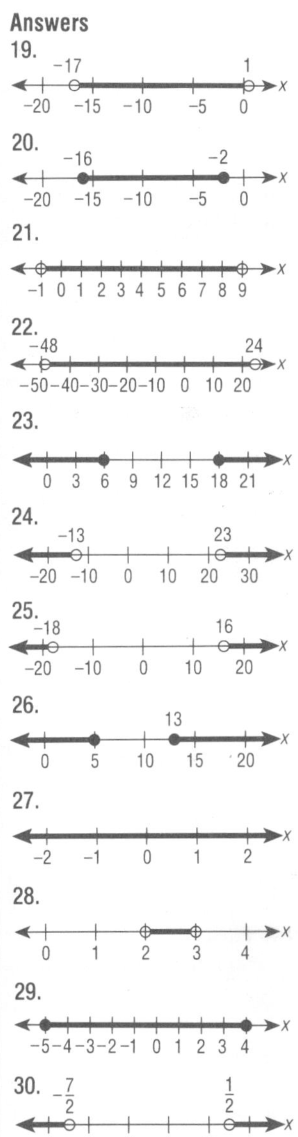

▶ **Ex. 59–60** Many people enjoy snorkeling in tropical waters because of all the beautiful colors they can see. Do the data in the exercises indicate that it is possible to witness a wide range of colors beneath the sea? No; most snorkelers would never see color more than a few feet beneath the surface.

Enrichment Activity

This activity requires students to go beyond lesson goals.

An important use of absolute value inequalities is found in addressing the problems of quality control. (See Example 4 in this lesson and Example 5 in Lesson 4.8.)

PROJECT Identify situations at school, on the job, or at play, that are subject to some type of quality control. Express the quality-control relationship as an absolute value inequality. For example, consider a drop of two letter grades on mathematics homework. This would alert the teacher to identify possible reasons for the change in performance.

Answers

53.

54.

55.

56.

57.

58.

45. *Grains of Sand* Sand is made up of loose pieces of rocks or minerals that are larger than silt but smaller than pebbles. Geologists define sand as grains that measure in diameter according to the graph below. Write an absolute value inequality describing the diameters of sand grains.

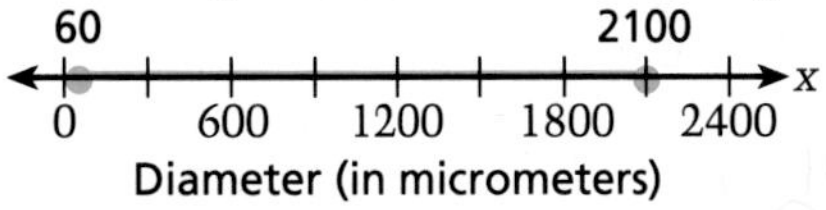

60 2100

0 600 1200 1800 2400

Diameter (in micrometers)

45. $|x - 1080| \leq 1020$

46. *It's Either Too Hot or Too Cold* Suppose you start the furnace whenever the temperature inside is 66°F or below and start the air conditioner whenever the temperature inside is 74°F or above. The graph below shows the temperatures at which you turn on the heat or air conditioner. Write an absolute value inequality that represents these temperatures. $|x - 70| \geq 4$

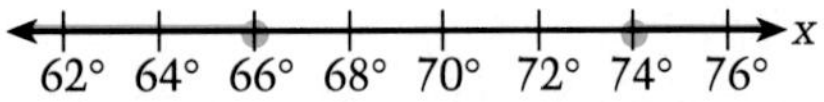

62° 64° 66° 68° 70° 72° 74° 76°

Integrated Review

In Exercises 47–52, solve the inequality.

47. $-6 < x - 4 < 3$ $-2 < x < 7$

48. $2 \leq 9 + x \leq 12$ $-7 \leq x \leq 3$

49. $-4 \leq -x + 9 \leq 6$ $3 \leq x \leq 13$

50. $-11 < -2x + 3 < 11$ $-4 < x < 7$

51. $-7 < 6x - 5 < 4$ $-\frac{1}{3} < x < \frac{3}{2}$

52. $2 \leq 8x - 14 \leq 58$ $2 \leq x \leq 9$

In Exercises 53–58, graph the inequality. See margin.

53. $3x + 7 > 1$

54. $4x + 5 \geq 21$

55. $2x + 4 \leq x + 3$

56. $4x + 7 < 3x + 3$

57. $-19 \leq 2x - 9 \leq -1$

58. $-4 < 5x - 9 < 26$

Exploration and Extension

Underwater Color **In Exercises 59 and 60, use the following information.**

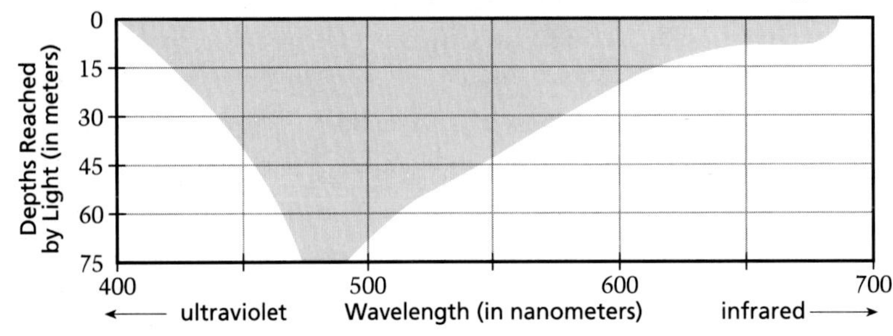

As light passes through seawater, most of the light with a long wavelength is absorbed. Much of the light with a short wavelength is also absorbed.

59. Write an absolute value inequality approximating the wavelengths of light that reach a depth of 30 meters in seawater. $|x - 510| \leq 70$

60. Write an absolute value inequality approximating the wavelengths of light that reach a depth of 60 meters in seawater. $|x - 495| \leq 20$

Biology **In Exercises 61–66, use the chart to write an absolute value inequality for the given sound frequency range.**

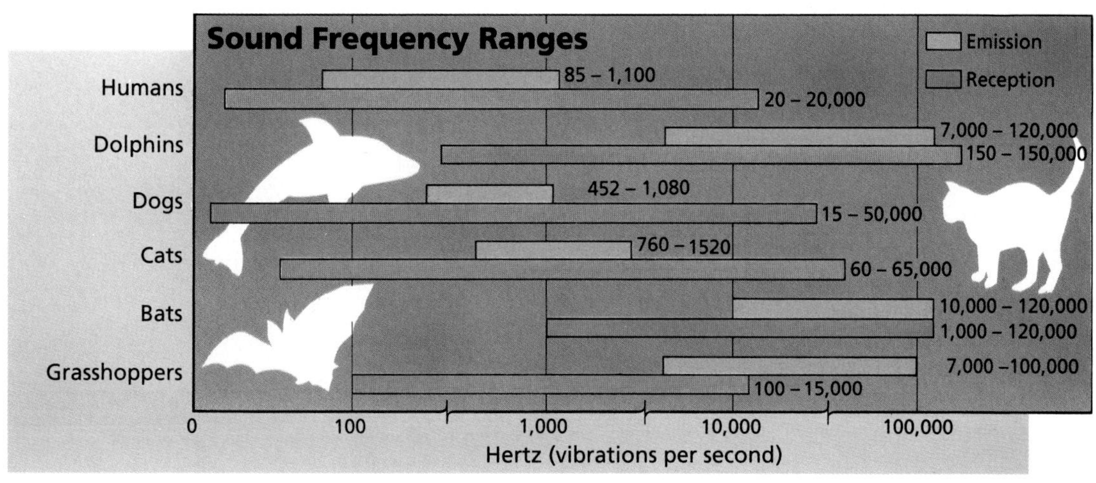

61. Frequencies people can emit $|x - 592.5| < 507.5$

62. Frequencies people can receive that bats emit $|x - 15{,}000| < 5000$

63. Frequencies that dogs can receive that people cannot receive

64. Frequencies that a cat can receive but cannot emit $|x - 410| < 350$ and $|x - 33{,}260| < 31{,}740$

65. Frequencies that a grasshopper can emit that people can receive $|x - 13{,}500| < 6500$

66. Frequencies that people can emit that dolphins cannot receive $|x - 117.5| < 32.5$

63. $|x - 17.5| < 2.5$ and $|x - 35{,}000| < 15{,}000$

Mixed REVIEW

4. 150 11. $5x - \frac{15}{2}$ 12. $6x - \frac{9}{4}y$
15., 16., 18. See margin.

1. Write 35% in decimal form. **(1.1)** 0.35
2. Use exponents to write $x \cdot x \cdot y \cdot y \cdot y$. **(1.3)** x^2y^3
3. What is 7% of \$3.40? **(2.7)**
4. What is two fifths of 375? **(2.7)**
5. What is the reciprocal of $-\frac{4}{3}$? $-\frac{3}{4}$
6. What is the opposite of $\frac{5}{6}$? **(2.1)** $-\frac{5}{6}$
7. Solve $3x + 4y = -20$ for y. **(3.6)** 21
8. Solve $\frac{1}{4}m - 2n = 3$ for n. **(3.6)** 5
9. Evaluate $x^2 + 2x - 3$ when $x = 4$. **(1.3)**
10. Evaluate $|3 - x| + |x|$ when $x = -1$. **(2.1)**
11. Simplify $3(x - 2) + \frac{1}{2}(4x - 3)$. **(2.6)**
12. Simplify $3(x - y) + \frac{3}{4}(4x + y)$. **(2.6)**
13. Solve $9x - 4 = \frac{1}{4}(3 - x)$. **(3.2)** $\frac{19}{37}$
14. Solve $\frac{1}{4}(y - 4) = (2y + 6)\frac{3}{8}$. **(3.2)** $-\frac{13}{2}$
15. Sketch the graph of $y = -3|x + 2| - 2$. **(4.7)**
16. Sketch the graph of $6x + 7y = 42$. **(4.3)**
17. Are the lines $y = \frac{3}{4}x$ and $y = \frac{3}{4}x - 2$ parallel? **(4.5)** Yes
18. Graph the inequality $\frac{1}{9}x - 1 \geq -x - \frac{1}{3}$. **(6.1)**
19. Find the slope and y-intercept of the line $y = -\frac{1}{2}x + 3$. **(4.5)** $-\frac{1}{2}$, 3
20. Find the x-intercept and y-intercept of the line $-3y + 2x = 6$. **(4.3)** 3, -2

7. $y = \frac{1}{4}(-20 - 3x)$ 8. $n = \frac{1}{2}(\frac{1}{4}m - 3)$

Answers

15.

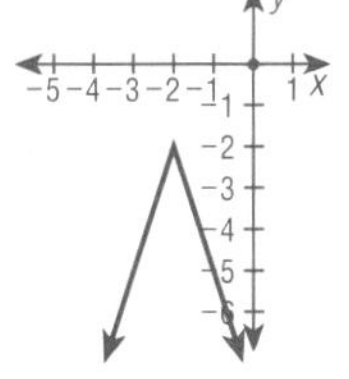

16.

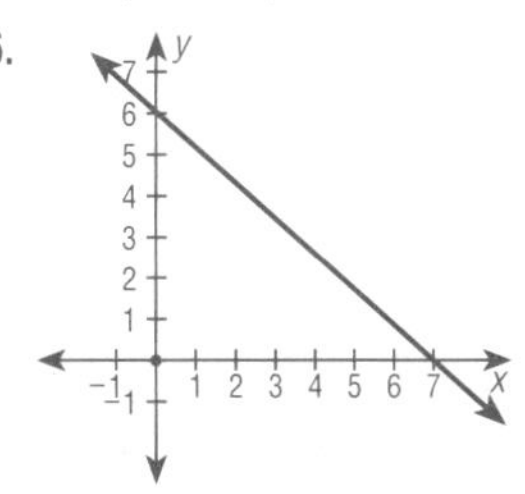

18. $\frac{3}{5}$; −1 0 1 2 3 x

Problem of the Day

25 billiard balls are placed inside a square wooden frame in 5 rows of 5 so that they are firmly held together. How many additional billiard balls are required to complete a pyramid on top of the 25-ball foundation layer?

30

ORGANIZER

Warm-Up Exercises

1. Check whether the value of x makes the given inequality a true statement.
   a. $x = -7$, $2x - 4 > 19$
   b. $x = 12$, $34 - 5x < -3$
   c. $x = 31$, $18 < -6x + 203$
   a. no, b. yes, c. no
2. Sketch the graph of the given equation.
   a. $y = 2x - 3$ **b.** $y = -4$

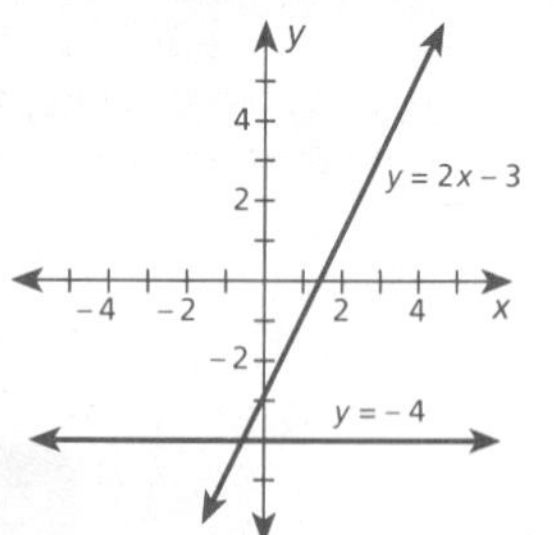

Lesson Resources

Teaching Tools
- Transparencies: 4, 11, 12
- Problem of the Day: 6.5
- Warm-up Exercises: 6.5
- Answer Masters: 6.5

Extra Practice: 6.5

Color Transparency: 30

Common-Error ALERT!

In **Example 1,** caution students to select test points that are clearly not on the line. The test point must lie in either half-plane. Never select a test point on the line.

6.5 Graphing Linear Inequalities in Two Variables

What you should learn:

Goal 1 How to graph a linear inequality in two variables

Goal 2 How to model a real-life situation using a linear inequality in two variables

Why you should learn it:

You can use linear inequalities to model real-life situations such as the possible gold content of ocean treasure.

Goal 1 Graphing Linear Inequalities

A **linear inequality** in x and y is an inequality that can be written in one of the following forms.

$$ax + by < c, \quad ax + by \le c, \quad ax + by > c, \quad ax + by \ge c$$

An ordered pair (a, b) is a **solution** of a linear inequality in x and y if the inequality is true when a and b are substituted for x and y, respectively. For instance, $(1, 3)$ is a solution of

$$4x - y < 2$$

because $4(1) - 3 < 2$ is a true statement.

Example 1 *Checking Solutions of Linear Inequalities*

Check whether the ordered pairs are solutions of $2x - 3y \ge -2$.

a. $(0, 0)$ **b.** $(0, 1)$ **c.** $(2, -1)$

Solution

	(x, y)	Substitute		Conclusion
a.	$(0, 0)$	$2(0) - 3(0)$	$= 0 \ge -2$	$(0, 0)$ is a solution.
b.	$(0, 1)$	$2(0) - 3(1)$	$= -3 < -2$	$(0, 1)$ is not a solution.
c.	$(2, -1)$	$2(2) - 3(-1)$	$= 7 \ge -2$	$(2, -1)$ is a solution. ■

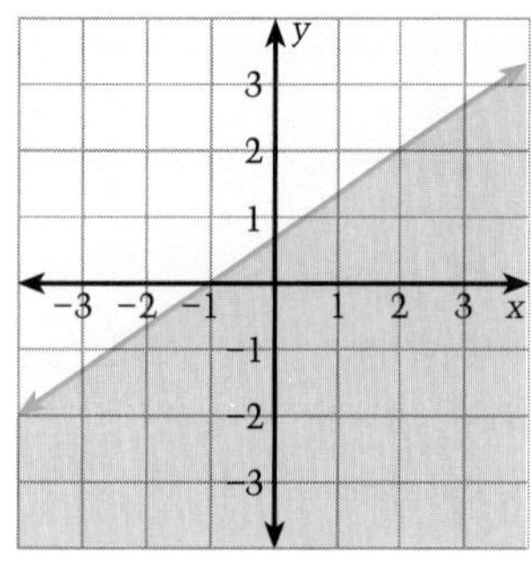

The graph of $2x - 3y \ge -2$ is shown at the left. Every point in the shaded region is a solution of the inequality, and every other point in the plane is not a solution.

Sketching the Graph of a Linear Inequality

1. Sketch the graph of the corresponding linear equation. (Use a *dashed* line for inequalities with < or > and a *solid* line for inequalities with ≤ or ≥.) This line separates the coordinate plane into two **half planes.**
2. Test a point in one of the half planes to find whether it is a solution of the inequality.
3. If the test point is a solution, shade its half plane. If not, shade the other half plane.

Example 2 *Sketching the Graph of a Linear Inequality*

Sketch the graphs of the linear inequalities.

a. $x < -3$ **b.** $y \leq 4$

Solution

a. First, sketch the vertical line given by

$x = -3$. *Use a dashed line.*

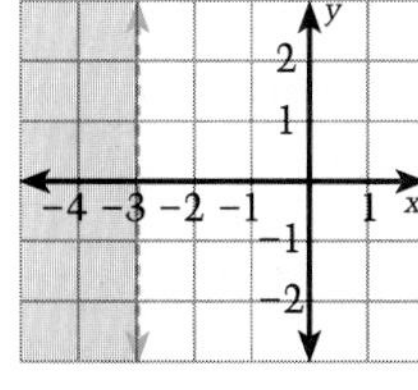

Next, test a point. The origin (0, 0) is *not* a solution and it lies to the right of the line. Therefore, the graph of $x < -3$ is all points to the left of the line $x = -3$.

b. First, sketch the horizontal line given by

$y = 4$. *Use a solid line.*

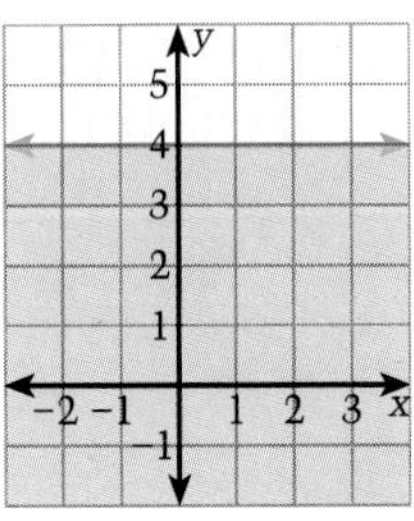

Next, test a point. The origin (0, 0) *is* a solution and it lies below the line. Therefore, the graph of $y \leq 4$ is all points on or below the line $y = 4$. ■

When sketching the graph of a line, remember that it is helpful to first write the line in slope-intercept form.

Example 3 *Writing in Slope-Intercept Form*

Sketch the graph of $x + y > 3$.

Solution The corresponding equation is $x + y = 3$. To sketch this line, first write the equation in slope-intercept form.

$y = -x + 3$ *Slope-intercept form*

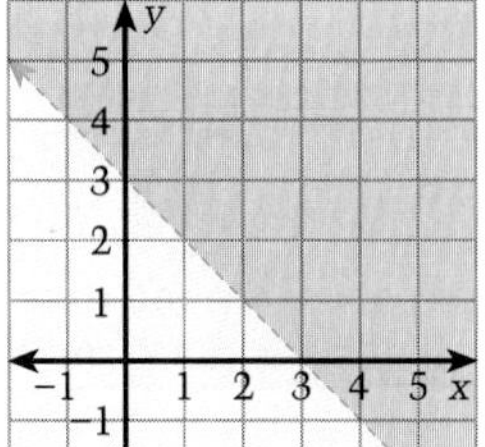

Then, sketch the line that has a slope of -1 and a y-intercept of 3. Use a dashed line. The origin (0, 0) is *not* a solution and it lies below the line. Therefore, the graph of $x + y > 3$ is all points above the line. (Try checking a point above the line. Any one you choose will satisfy the inequality.) ■

Notice that for each inequality, we used the origin as a test point. You can use any point that is not on the line. The origin is often convenient to use because it is easy to evaluate expressions in which 0 is substituted for each variable.

LESSON Notes

GOAL 1 Remind students that a solution of a linear equation, $ax + by = c$, is any ordered pair (m, n), for which the equation is true when m is substituted for x and n is substituted for y. This should help students understand what the solution of an inequality represents.

Examples 2–3

Encourage students to consider using several solution methods. Some students may wish to determine the truth of the inequality algebraically by substituting the x-y coordinate values. Other students may elect to sketch the graph of the corresponding linear equation and observe whether the point lies above, below, or on the line. This is a particularly effective strategy if students have a graphics calculator. Emphasize that the best test point is the origin, provided it does not lie on the line, because it is easy to evaluate expressions at $x = 0$, $y = 0$.

Example 4

Observe that the number of gold coins, x, must satisfy the inequality $x \geq 0$; and the number of silver coins, y, must satisfy the inequality $y \geq 0$. Consequently the shaded region that satisfies the conditions of this problem lies in the first quadrant of the coordinate plane. The solution is not a half-plane.

Extra Examples

Here are additional examples similar to **Examples 1–3.**

1. Check whether the given ordered pairs are solutions of $4x - 7y \geq 28$.

   a. $(-5, -7)$ **b.** $(-18, -3)$

   c. $(2, -2)$ **d.** $(0, -4)$

   a. solution, b. not a solution, c. not a solution, d. solution

2. Sketch the graphs of the given linear inequalities.

   a. $y < 6$ **b.** $6x + 5y \geq 30$

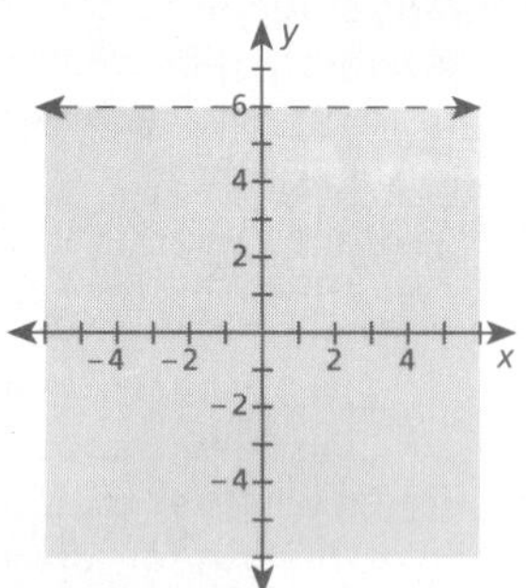

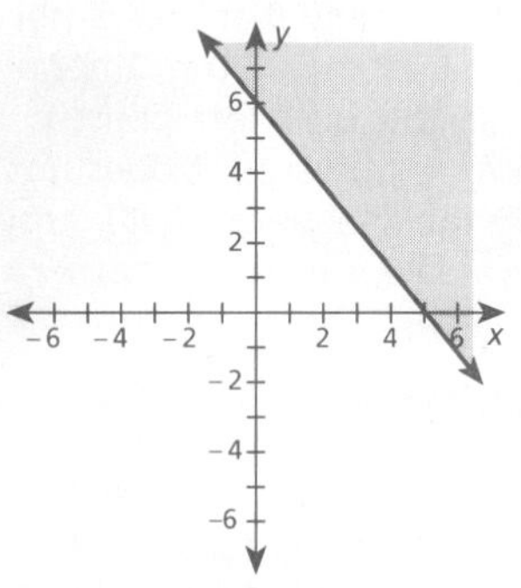

Check Understanding

1. What is a half-plane? In how many ways can a coordinate plane be divided into two half-planes? See Example 1 on page 320.
2. What is a test point? A point that lies on either side of the boundary line. Why is (0, 0) frequently chosen as a test point? Ease of computation
3. Is the solution to an inequality always a half-plane? (Consider Example 4 as you formulate your response.) No, the solution to a compound inequality could be a point or a line. An inequality such as $|a| < -3$ could have no solution.

Goal 2 Modeling a Real-Life Situation

In 1969, ocean divers found the remains of the Spanish ship **Girona** *that sank off the coast of Northern Ireland. Among other treasures, the divers found 405 gold coins and 756 silver coins.*

Real Life
Underwater Diving

Example 4 *Finding Ocean Treasure*

You are diving for precious coins. Your recovery basket contains about 50 pounds of material! If gold coins weigh about $\frac{1}{2}$ ounce each and silver coins weigh about $\frac{1}{4}$ ounce each, what are the various proportions of coins you may have?

Solution The basket contains 800 ounces of material.

Verbal Model

$$\boxed{\text{Weight per coin}} \cdot \boxed{\text{Num. gold coins}} + \boxed{\text{Weight per coin}} \cdot \boxed{\text{Num. silver coins}} \leq \boxed{\text{Weight in basket}}$$

Labels

Weight per gold coin $= \frac{1}{2}$	(oz. per coin)
Number of gold coins $= x$	(coins)
Weight per silver coin $= \frac{1}{4}$	(oz. per coin)
Number of silver coins $= y$	(coins)
Weight in basket $= 800$	(ounces)

Inequality $\frac{1}{2}x + \frac{1}{4}y \leq 800$ **Linear model**

All possible solutions are shown in the graph. ■

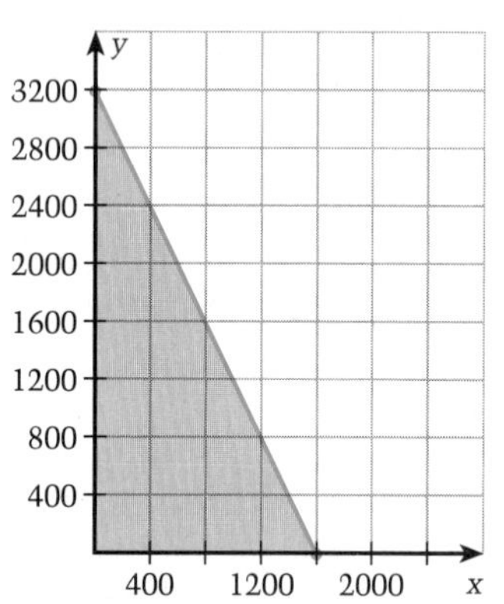

Communicating about ALGEBRA

A., B. See Additional Answers.

SHARING IDEAS about the Lesson

Make a Conjecture A conjecture is a statement that you think is true but that has not been proved true.

A. Sketch the graph of $y < -2x + 8$. Does the graph lie above or below the line $y = -2x + 8$?

B. Sketch the graph of $y > x + 1$. Does the graph lie above or below the line $y = x + 1$?

C. Make a conjecture based upon **A** and **B.** See below.

C. When y is less than the expression involving x, the graph is below the line; when y is greater than the expression involving x, the graph is above the line.

EXERCISES

Guided Practice

CRITICAL THINKING about the Lesson

1. What are the four forms of a linear inequality in x and y? $ax + by < c$, $ax + by \le c$, $ax + by > c$, $ax + by \ge c$

2. The graph of a linear inequality in x and y is a ____. half plane

In Exercises 3–5, are (0, 2) and (2, 1) solutions of the inequality?

3. $2x - 3y \le 4$ Both are.
4. $-x + 4y > 0$ Both are.
5. $y - \frac{2}{3}x \ge -2$ Both are.
6. Sketch the graph of $y - 2x \le 2$ and explain each step. See Additional Answers.

Independent Practice

In Exercises 7–10, is each ordered pair a solution of the inequality?

7. $3x - 2y < 2$; $(1, 3)$, $(2, 0)$ Only (1, 3) is.
8. $y - 2x > 5$; $(4, 13)$, $(8, 1)$ Neither is.
9. $5x + 4y \ge 6$; $(-2, 4)$, $(5, 5)$ Both are.
10. $5y + 8x \le 14$; $(-3, 8)$, $(7, -6)$ Neither is.

In Exercises 11–14, sketch the graph of the inequality. See Additional Answers.

11. $x \ge 5$
12. $x \le -3$
13. $y > -2$
14. $y < 8$

In Exercises 15–18, match the inequality with its graph.

15. $2x - y \le 1$ c
16. $2x - y < 1$ b
17. $2x - y \ge 1$ d
18. $2x - y > 1$ a

a. b. c. d.

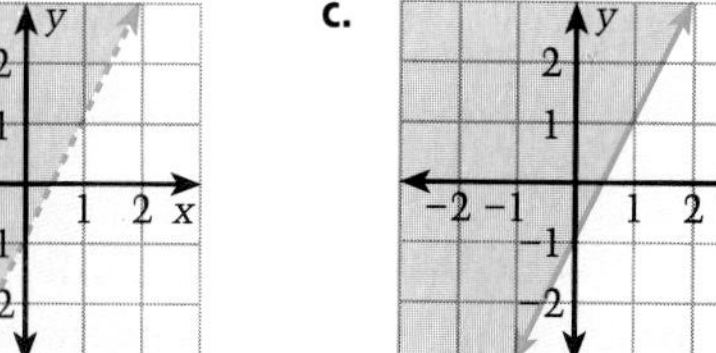

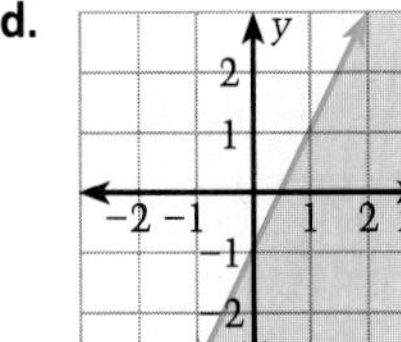

In Exercises 19–22, sketch the graph of the inequality. See Additional Answers.

19. $-x - y \le 4$
20. $2x - y \ge 6$
21. $3x + y > 9$
22. $y - 4x < 0$

In Exercises 23–26, write an inequality for the half plane.

23. 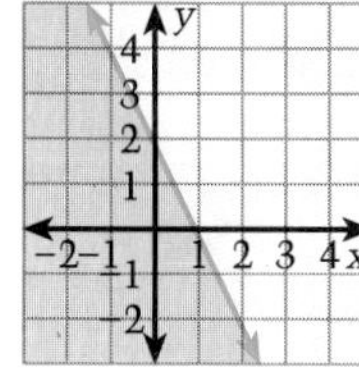

$y \le -2x + 2$

24. 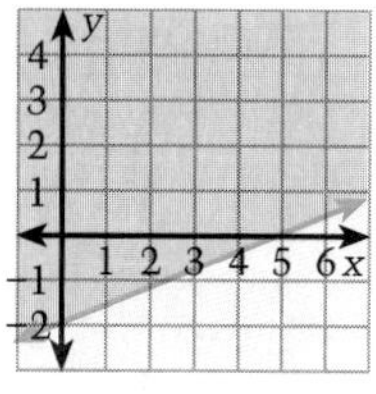

$y \ge \frac{2}{5}x - 2$

25. 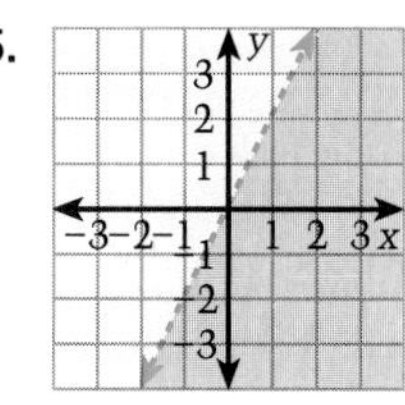

$y < 2x$

26. 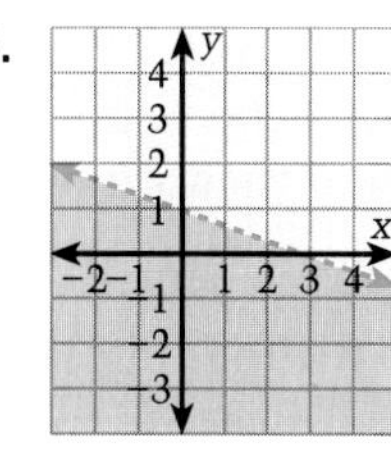

$y < -\frac{1}{3}x + 1$

Communicating about ALGEBRA

COOPERATIVE LEARNING

In Part C, have students work in small groups. Each group should come up with a conjecture about how one decides which half-plane should be shaded. Based upon the conjecture, identify a strategy for sketching the graph of any linear inequality in one of the four forms found at the beginning of this lesson.

EXERCISE Notes

ASSIGNMENT GUIDE

Basic/Average: Ex. 1–6, 7–10, 11–14, 15–21 odd, 23–25, 27–45 multiples of 3, 47

Above Average: Ex. 1–6, 8, 10, 11–14, 16–22 even, 24–25, 27–45 multiples of 3, 47–48

Advanced: Ex. 1–6, 7, 9, 11–14, 16–22 even, 24–26, 31–34, 47–48

Selected Answers

Exercises 1–6, 7–49 odd

More Difficult Exercises

Exercises 28–30, 51–52

Guided Practice

Ex. 3–5 Emphasize the importance of the concept of what a solution is. Note the similarities and differences between solutions to equations and solutions to inequalities.

Independent Practice

Ex. 27 GROUP ACTIVITY Before assigning the problem, you might have students reenact the experiment to see how many horses they noticed. Collect the data from the class and compare it to the data shown in Exercise 27.

EXTEND Ex. 27 Have students interpret the graph of the inequality. How many of the solutions in the graph are actually possible? (Assume that no one would see fractional parts of horses.) What are the coordinates of the correct answer? 28; (4, 1)

Answer

27.

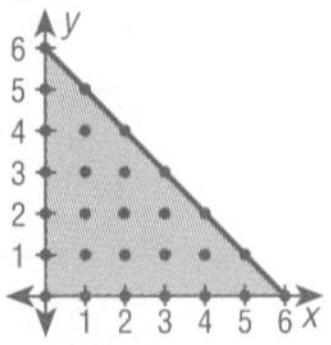

EXTEND Ex. 28 Have students interpret the graph of the inequality. Is (17, 14) a solution of the inequality? Explain. No, 17 + 14 = 31 and only 30 other students can ride the bus.

Answers

28.

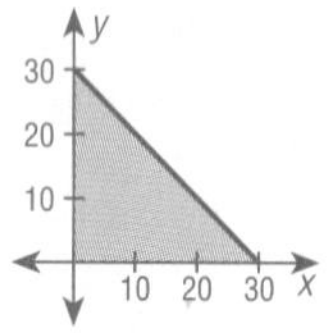

29.

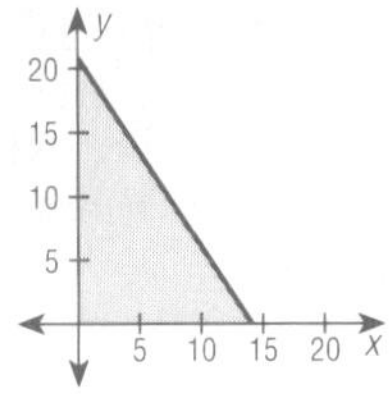

Horses in Hiding **In Exercise 27, use the following information.**

Camouflage experiment: Ask a person to stand 6 feet away. Before opening the book, tell the person that the picture you will show contains *at most* six horses. After letting the person look at the picture for 10 seconds, close the book and ask him or her to state the number of adult horses and the number of foals. Remind the person that the number of horses may be zero.

PINTOS by Bev Doolittle. © 1979 The Greenwich Workshop, Trumbull, Conn.

27. Let x represent the number of adult horses. Let y represent the number of foals. Write and graph an inequality that describes the numbers of adult horses and foals that the person could report seeing. $0 \le x + y \le 6$ See margin.

Billy Joe and Rosanne **In Exercise 28, use the following information.**

Billy Joe and Rosanne ride the same bus to school but have different homerooms. Thirty other students ride the same bus.

28. Let x represent the number of students other than Billy Joe on the bus from Billy Joe's homeroom. Let y represent the number other than Rosanne on the bus from Rosanne's homeroom. Write and graph an inequality that describes the different numbers of students on the bus who could be from Billy Joe's or Rosanne's homerooms. $x + y \le 30$ See margin.

Field Goals and Safeties **In Exercise 29, use the following information.**

Carver High School scored 42 points in its football victory over Central High School. In a football game, a field goal counts 3 points and a safety counts 2 points.

29. Let x represent the number of field goals scored. Let y represent the number of safeties scored. Write and graph an inequality that describes the different numbers of field goals and safeties that Carver could have scored. $3x + 2y \le 42$ See margin.

A Full Lot of Cars **In Exercise 30, use the following information.**

A car dealer sells several different car models. The dealer has \$1,200,000 worth of cars on the lot. A sports model costs \$20,000 and a sedan costs \$12,000.

 **30.** Let x represent the number of sports models on the lot. Let y represent the number of sedans. Write and graph an inequality in x and y that describes the different numbers of sports models and sedans on the lot. (Remember, none of the cars on the lot have to be sports models or sedans.) $20{,}000x + 12{,}000y \le 1{,}200{,}000$

Integrated Review

In Exercises 31–38, solve the inequality.

31. $3x - 1 > 7$ $x > \frac{8}{3}$

32. $6 - x \le 28$ $x \ge -22$

33. $14x + 5 \le 4x - 15$ $x \le -2$

34. $-3x - 1 > -2x + 19$ $x < -20$

35. $-5 < 3x + 4 \le 1$ $-3 < x \le -1$

36. $0 \le 9x - 63 < 18$ $7 \le x < 9$

37. $-10 \le 3 - 2x \le 9$ $-3 \le x \le \frac{13}{2}$

38. $-1 < 4x + 5 < 0$ $-\frac{3}{2} < x < -\frac{5}{4}$

In Exercises 39–42, determine whether the ordered pair is a solution of the inequality.

39. $5x - 2y < 14$; $(2, -2)$ Is not

40. $x - 3y \le 17$; $(-5, -4)$ Is

41. $-4x + y > 10$; $(-1, 14)$ Is

42. $3x + 9y \ge -3$; $(-4, 1)$ Is

In Exercises 43–50, sketch the graph of the line. See Additional Answers.

43. $x = -4$ **44.** $y = -9$ **45.** $y = 3$ **46.** $x = 7$

47. $x + y = 9$ **48.** $-2x + y = 11$ **49.** $3x - 2y = 12$ **50.** $x - 3y = 6$

Exploration and Extension

Geometry **In Exercises 51 and 52, find four different inequalities so that each point in the shaded region is a solution of each inequality *and* each point that is not in the shaded region is not the solution of at least one inequality.**

51.

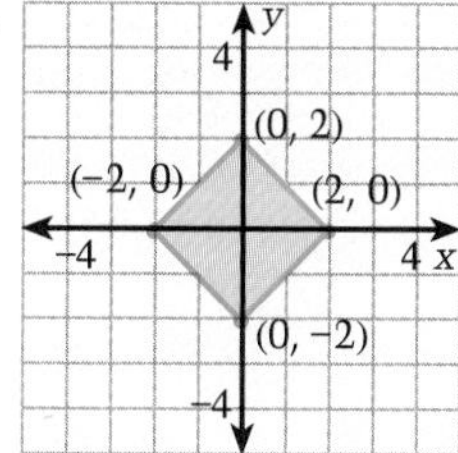

$y \le x + 2$, $y \le -x + 2$, $y \ge x - 2$, $y \ge -x - 2$

52.

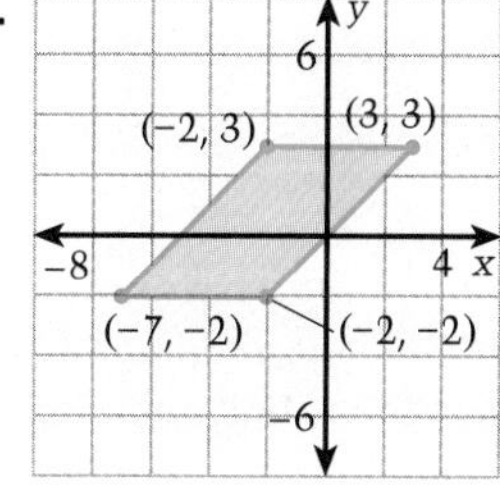

$y \le x + 5$, $y \le 3$, $y \ge x$, $y \ge -2$

EXTEND Ex. 51–52 GEOMETRY You can use the following activity to prepare students for these exercises.

Solve $x \ge -3$ and $x \le 3$ and $y \ge -3$ and $y \le 3$ by graphing the lines on the same coordinate plane. What do you notice? The solution defines a square region.

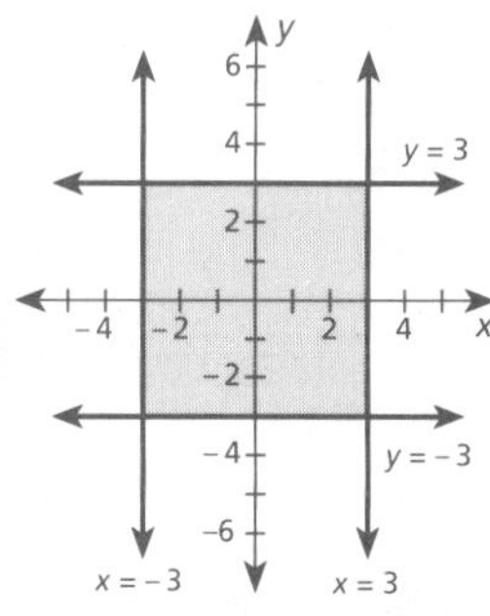

Enrichment Activities

These activities require students to go beyond lesson goals.

An Alternate Approach

The activities in Communicating about Algebra suggest the following alternate approach to graphing linear inequalities.

If the linear equation associated with a given linear inequality is in slope-intercept form, no test point is required to sketch a graph of the linear inequality.

WRITING Given that the above statement is true, how would you sketch the graph of any linear inequality. Demonstrate how the procedure works, using several examples. Graph the equation $y = mx + b$ by using the y-intercept point (0, b) and the slope m. If $y < mx + b$, then the solution is the half-plane below the line. If $y > mx + b$, then the solution is the half-plane above the line. If $y \le mx + b$ or if $y \ge mx + b$, then the equation of the line is part of the solution set.

TECHNOLOGY Refer to the user's manual of your graphing calculator for specific instructions on graphing inequalities.

Answers

1.

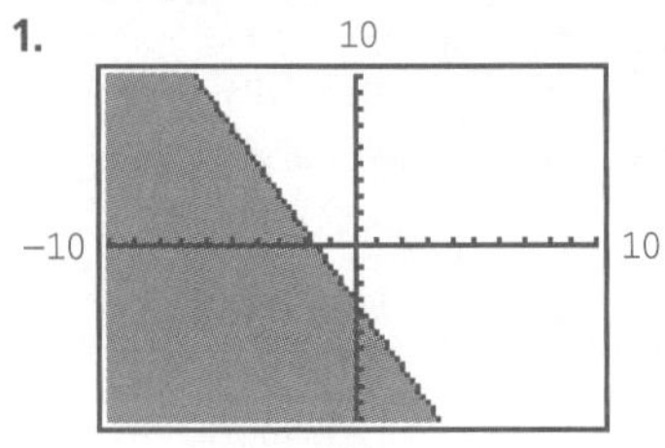

3.

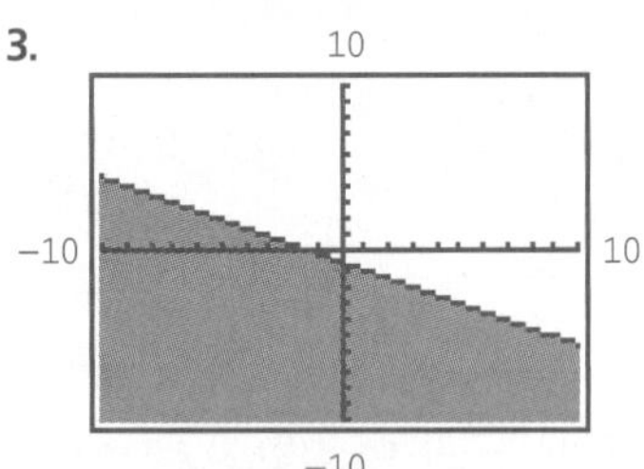

5.

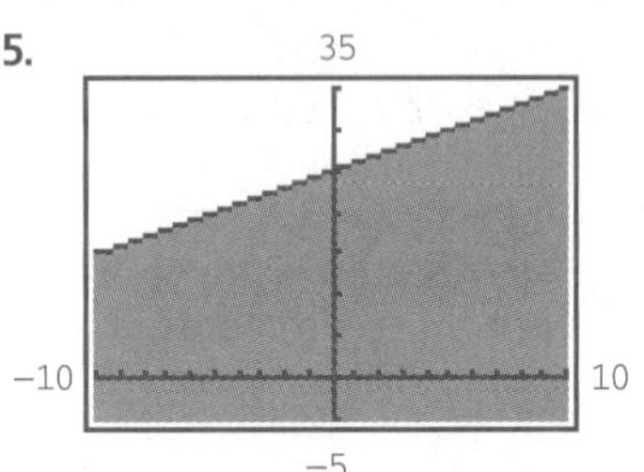

7.

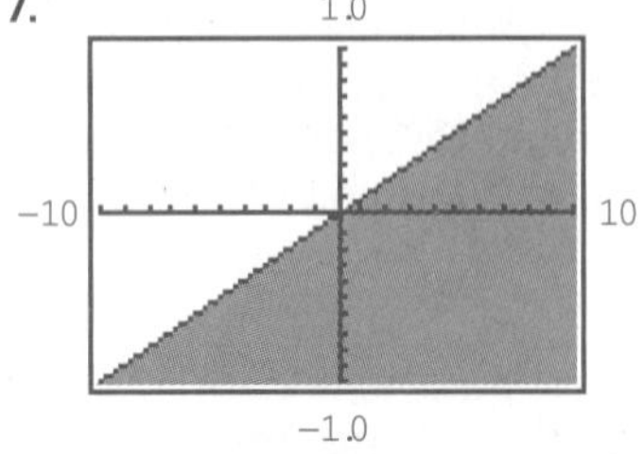

USING A GRAPHING CALCULATOR

Graphing calculators and computers can be used to sketch the solution of an inequality in x and y. Keystroke instructions for doing this on a TI-82 *and* Casio fx-9700GE, *are listed on page 740. As an example, consider the inequality*

$x - 2y \leq 6.$

The first step is to write the equivalent inequality that has y isolated on the left. For instance, $x - 2y \leq 6$ must first be written as

$y \geq \frac{1}{2}x - 3.$

The graph of this inequality is shown at the right.

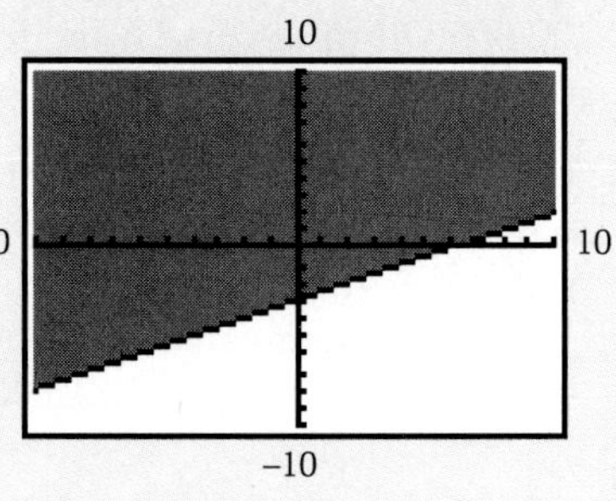

Graph of Inequality in x and y

Cooperative Learning

Partner Activity Graphs of two inequalities are shown below. With your partner, write an inequality to represent each graph. Then use a graphing calculator to check your answer.

a. **b.**

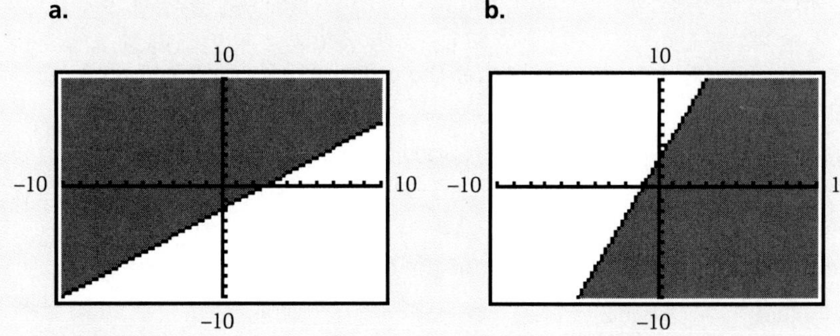

Exercises

In Exercises 1–8, use a graphing calculator to sketch the graph of the inequality. (For 1–4, use a standard viewing rectangle. For 5–8, use the indicated viewing rectangle.) Odds: See margin. Evens: See Additional Answers.

1. $y < -2x - 3$ **2.** $y > 2x + 2$ **3.** $x + 2y \leq -1$ **4.** $x - 3y \geq 3$

5. $y < x + 25$ **6.** $y > -x + 25$ **7.** $y \leq 0.1x$ **8.** $y \geq 100x + 2500$

	5.	6.	7.	8.
Range				
Xmin	-10	-10	-10	0
Xmax	10	10	10	100
Xscl	1	1	1	10
Ymin	-5	-5	-1	0
Ymax	35	35	1	15000
Yscl	5	5	0.1	1000

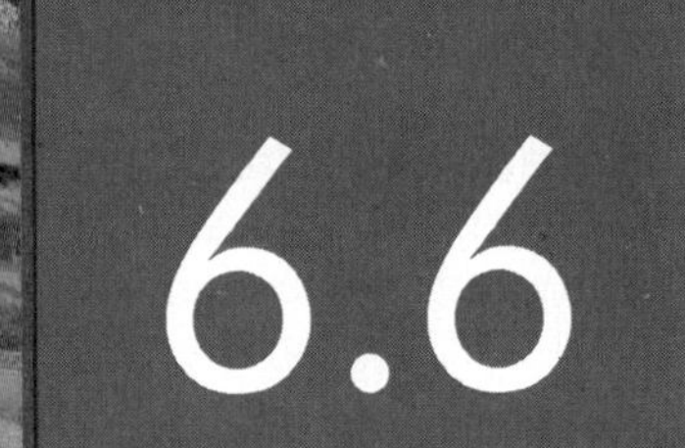

6.6 Exploring Data: Time Lines, Picture Graphs, and Circle Graphs

What you should learn:

Goal How to draw and interpret visual models, such as time lines, picture graphs, and circle graphs

Why you should learn it:

You can communicate effectively with visual models as well as with verbal models.

Goal Using Visual Models

The graphs used by many newspapers and magazines are meant to be eye-catching as well as informative. Each of the three types of graphs discussed in this lesson, *time-line graphs, picture graphs,* and *circle graphs,* is commonly used in presenting facts and figures in a somewhat informal way. Being able to interpret the information in such graphs is an important skill. Drawing these kinds of graphs gives you an opportunity to be creative and use your artistic ability.

A **time line** or **time-line graph** is often used to represent historical events. For instance, the time-line graph on page 303 shows the different dynasties that ruled China between 2000 B.C. and A.D. 1000.

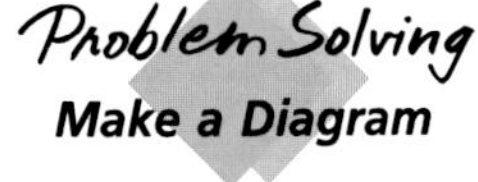

Example 1 ***The History of Automobiles***

Create a time line that shows the history of automobiles in the United States.

Solution This project could (and probably should) be very detailed. Here, however, we show a time line with only a few important events in the history of automobiles.

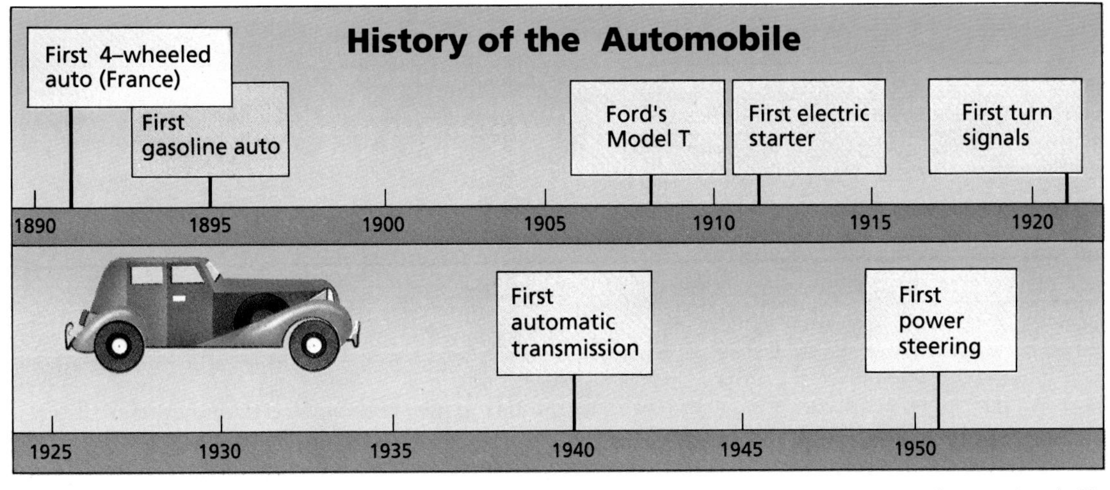

(Source: **Scientific American***)* ■

Problem of the Day

Study the pattern to help guess the values of d and e.

$$a - 6.7 = 2.2$$
$$b - 66.7 = 22.2$$
$$c - 666.7 = 222.2$$
$$\vdots$$
$$d - e = 222{,}222.2$$

$d = 888{,}888.9$
$e = 666{,}666.7$

ORGANIZER

Warm-Up Exercise

Create a graph on a number line that describes important events in your life, such as the following. Be as accurate as possible. Be creative in the way you represent your data. Discuss your design with your classmates to help you decide how to display the data.

Events

a. your birthdate
b. when you spoke your first word
c. when you first walked
d. when you started school
e. when you learned to ride a bike
f. when you met your best friend
g. when your brothers or sisters were born
h. when you first spent the night away from home without your parents

Answers will vary.

Lesson Resources

Teaching Tools
  Transparency: 5
  Problem of the Day: 6.6
  Warm-up Exercises: 6.6
  Answer Masters: 6.6
Extra Practice: 6.6
Color Transparencies: 30–34

LESSON Notes

Example 1

Point out the similarity of a time-line graph to a real number line. Time lines work well when sequential information must be displayed.

Example 2

Call students' attention to the legend in the top part of the graph. The legend indicates the value of each barrel displayed in the graph. Point out that the picture selected to convey information resembles barrels of oil but that the display is essentially a bar graph.

EXTEND Example 3 Ask students if they can think of another representation for the data. Construct another type of graph of the data.

Example 4

Emphasize that the data found in a circle graph should represent nonoverlapping (or disjointed) outcomes. For example, in this survey respondents should choose only one reason for rearranging their furniture. Another way to say this is that the total number of responses should not exceed 100 percent.

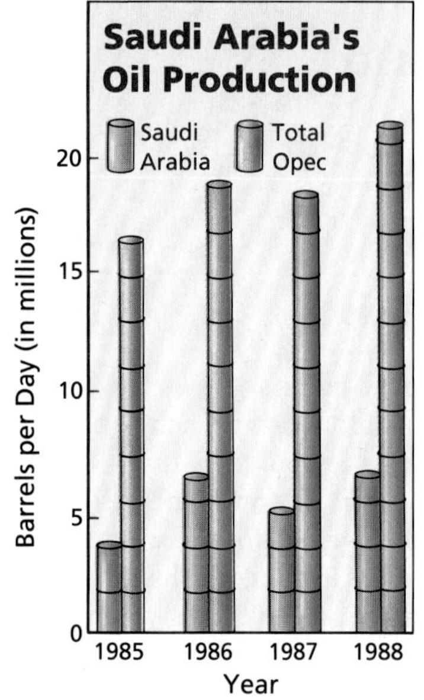

Picture graphs are usually stylized versions of bar graphs or graphs in the coordinate plane. The "pictures" in a picture graph provide a visual connection with the real-life quantities.

Example 2 *Oil Production by Saudi Arabia*

OPEC, the Organization of Petroleum Exporting Countries, was formed in 1960. In 1991, its members were Algeria, Ecuador, Indonesia, Iran, Iraq, Kuwait, Libya, Nigeria, Qatar, Saudi Arabia, United Arab Emirates, and Venezuela. Use the picture graph to estimate the ratio of Saudi Arabia's 1987 daily oil production to OPEC's 1987 daily oil production. ***(Source: OPEC)***

Solution From the graph, estimate that Saudi Arabia's daily production in 1987 was 5 million barrels and OPEC's was 18 million barrels. Thus, the ratio of Saudi Arabia's production to OPEC's was approximately

$$\frac{5}{18} \approx 28\%.$$

■

Example 3 *Raising a Litter of Kittens*

A mother cat was observed for six weeks after her kittens were born. In the graph on the left, each point of a line represents a time when the mother was *in* the box, while in the graph on the right, each point represents a time when all the kittens were *out of* the box. Notice that the first recorded information is at the top line of the graph, and that the vertical axis is then numbered from the top of the graph to the bottom, unlike most graphs you see. Describe the behavior of the mother and the kittens.

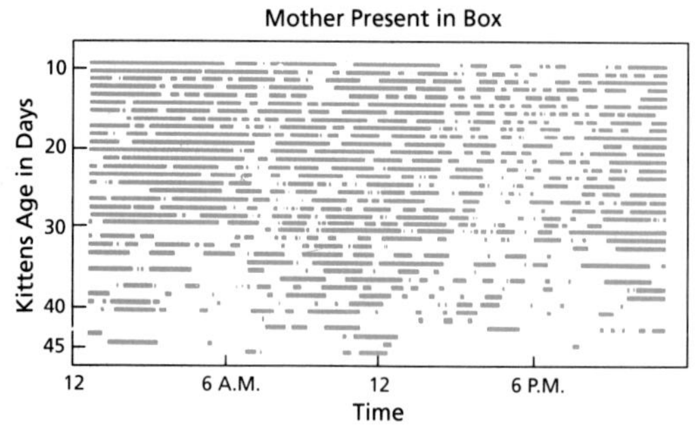

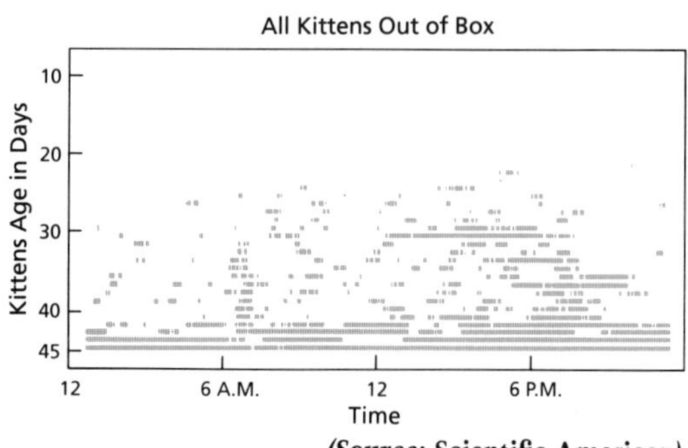

(Source:* Scientific American*)

Solution In the left graph, you can see that the older the kittens became, the more the mother left the box. In the right graph, you can see that the kittens stayed in their box until they were three weeks old. Then they began leaving for longer and longer times. ■

Circle graphs are used to show the relative sizes of different portions of a whole.

Example 4 *Rearranging Furniture*

Southwestern Bell Telephone Company surveyed 300 adults to find why they rearranged furniture in their homes. Of the 300, 108 said they were bored with the current arrangement, 57 said they moved to a new home, 48 said they redecorated, 45 said they purchased new furniture, and 42 had other reasons. Make a circle graph to show these results.

Solution To begin, you need to find the percentage that each type of response was of the total number of responses.

Bored	$\frac{108}{300} = 36\%$	Moved	$\frac{57}{300} = 19\%$
Redecorated	$\frac{48}{300} = 16\%$	New furniture	$\frac{45}{300} = 15\%$
Other	$\frac{42}{300} = 14\%$		

Next, you need to partition a circle into five sections corresponding to the five percentages. One way to do this is to divide the 360° "angle" at the center of the circle into 36% of 360° ($360 \cdot 0.36 \approx 130°$), 19% of 360° ($360 \cdot 0.19 \approx 68°$), etc. ■

Why People Rearrange Furniture

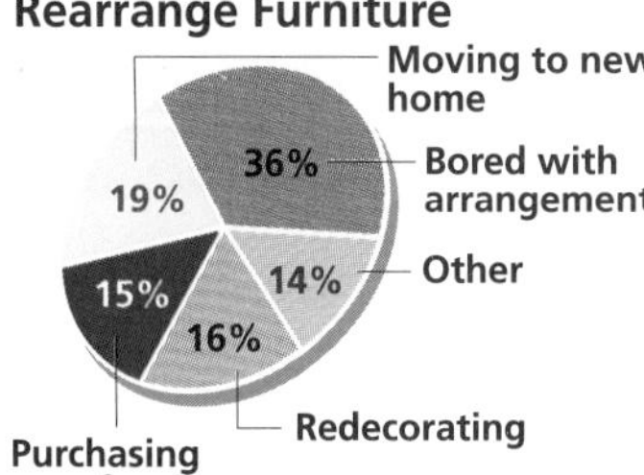

Communicating about ALGEBRA

SHARING IDEAS about the Lesson

Interpret a Graph In 1987, Japan had approximately 60 million people in its work force. The picture graph at the left shows the distribution of Japan's work force.

■ Manufacturing	■ Distribution
■ Services	■ Construction
■ Agriculture & Fishing	■ Other

Services

Distribution

Manufacturing

Other

Agriculture and Fishing

Construction

Japanese Workers

Each figure represents 2% of the workforce

9%

A. What percent of Japan's work force was in construction?

B. Approximate the number of manufacturing workers.

C. Approximate the number of service workers. **15 million**

D. How many more people worked in distribution than in agriculture and fishing? **7.2 million** **B. 14.4 million**

Extra Examples

Here are additional examples similar to **Examples 1–4.**

1. Construct a time-line graph that shows the history of your study of mathematics since the first grade.
2. Create a picture graph of the number of students in your class who watch TV for about one hour each day, less than one hour each day, and more than one hour each day.
3. Use a circle graph to show how many of the students in your class respond that vanilla, chocolate, or strawberry ice cream is their favorite ice-cream flavor.

Check Understanding

1. Name three different types of graphs that can be used to represent data. **For example, picture graphs, time lines, bar graphs, stacked bar graphs, line plots, circle graphs, frequency distributions, etc.**
2. Which graphs let you easily compute the total number of data points used to construct the graph? Explain. **Bar graphs and line plots because the heights of the bars (or *x*'s) represent the frequency for a given data point.**
3. Explain why graphs that represent the same data may look very different. How can such differences affect decisions that consumers make? **Graphs that use different scales can appear to be so different that they become misleading.**

Communicating about ALGEBRA

COOPERATIVE LEARNING
Encourage students to work with partners to create a circle graph and a standard bar graph of the information found here. Discuss the importance of color in conveying information in a graph. What other ways can different groups of data be distinguished in a graph?

Guided Practice

Ex. 1–4 GROUP ACTIVITY
Have students complete these exercises in small-group discussions. Reconvene students as a class to discuss their results.

ASSIGNMENT GUIDE

Basic/Average: Ex. 1–8, 11–15, 20–22, 25–39 odd, 40–41

Above Average: Ex. 1–10, 11, 16–19, 23–25, 26–39 multiples of 3, 40–42

Advanced: Ex. 1–4, 11–19, 22–23, 27–36 multiples of 3, 40–42

Selected Answers
Exercises 1–4, 5–37 odd

More Difficult Exercises
Exercises 16–19, 22–23, 40–42

Independent Practice

EXTEND Ex. 5–8 WRITING
Encourage students to select one of the inventors, research his biography in the library, and write about the inventor and the invention.

EXERCISES

Guided Practice

CRITICAL THINKING about the Lesson

1. Use Example 1. You are considering buying a 1932 Chevrolet. The owner of the car tells you that all the parts are original. You notice that the car has an automatic transmission. Is the owner telling the truth? No
2. Use Example 2. Estimate the ratio of Saudi Arabia's 1985 daily oil production to its 1988 daily oil production. $\frac{1}{2}$
3. Use Example 3. Did the kittens leave the box primarily during times the mother was in the box or primarily during times the mother was out of the box? Out of the box
4. Use Example 4. Suppose 500 people had been surveyed (instead of 300). Approximately how many people would you expect there to be in each category? Bored: 180, moved: 95, redecorated: 80, purchased: 75, other: 70

Independent Practice

Inventions **In Exercises 5–8 below, match the inventor with the time he lived.**

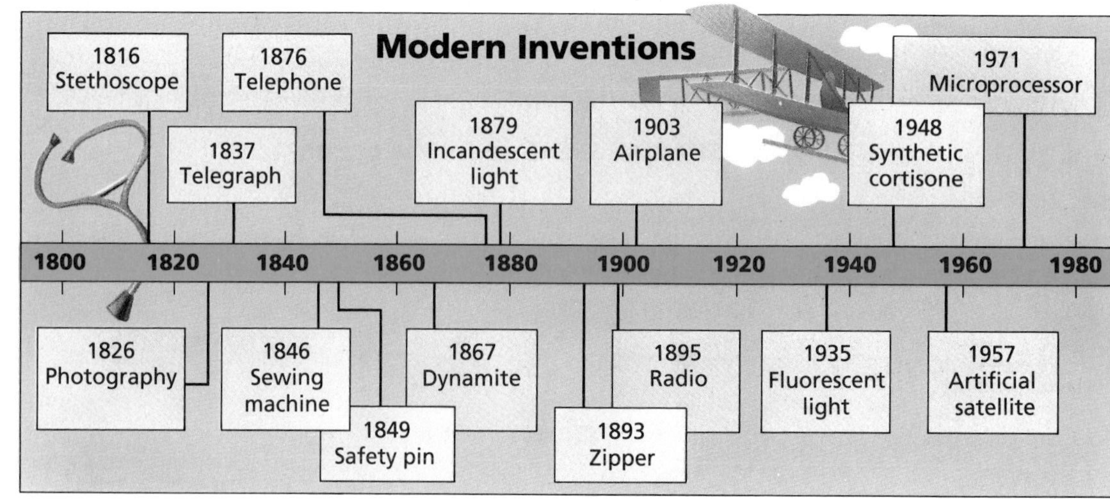

5. Guglielmo Marconi (radio) c
6. Alexander Graham Bell (telephone) d
7. Percy Julian (synthetic cortisone) b
8. Elias Howe (sewing machine) a

a. 1819–1867
b. 1899–1975
c. 1874–1937
d. 1847–1922

Activity Pattern **In Exercises 9 and 10, use the following time line.**

9. When did the woman have breakfast, lunch, and dinner?
10. How many hours did the woman sleep? 6

9. 7:00 A.M., 12 noon, 6:00 P.M.

A psychologist recorded a woman's activities during a 24-hour day.

11. *Schedule* Construct a time-line graph showing your schedule on an average day. Answer depends upon choice.

TV Rules **In Exercises 12–15, use the picture graph below, which compares the percent of parents in 1976 and 1990 who had strict rules for TV watching.** *(Source: "Roper Report," USA Today)*

12. In 1990, what percentage of parents had strict TV-watching rules for 13–17-year-old children? 45%
13. In 1976, what percentage of parents had strict TV-watching rules for 7–12-year-old children? 60%
14. Find the increase in percentage from 1976 to 1990 of parents who had strict TV-watching rules for 13–17-year-old children. 8%
15. Find the increase in percentage from 1976 to 1990 of parents who had strict TV-watching rules for children 6 and under. 13%

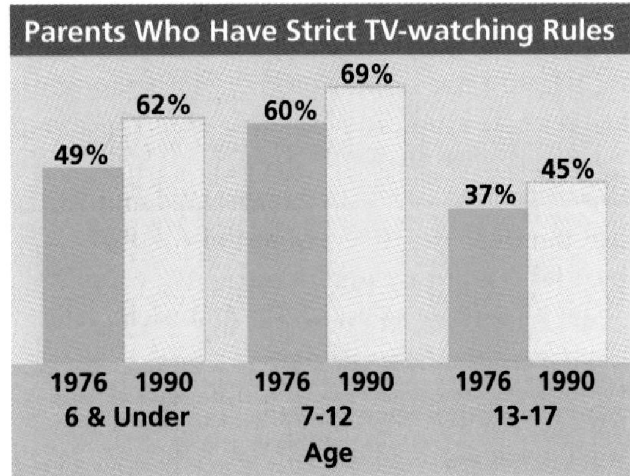

World Energy Sources **In Exercises 16–19, use the time-line graph, which shows the world's sources of energy from 1860 to 1985.** *(Source: Scientific American)* 18. 5 million barrels

16. In which decade did oil become the primary energy source? 1960's
17. In 1970, what was the second most common source of energy? Coal
18. Approximate the amount of natural gas used in 1920.
19. How much more primary electricity was used in 1970 than in 1960? ≈ 5 million barrels more

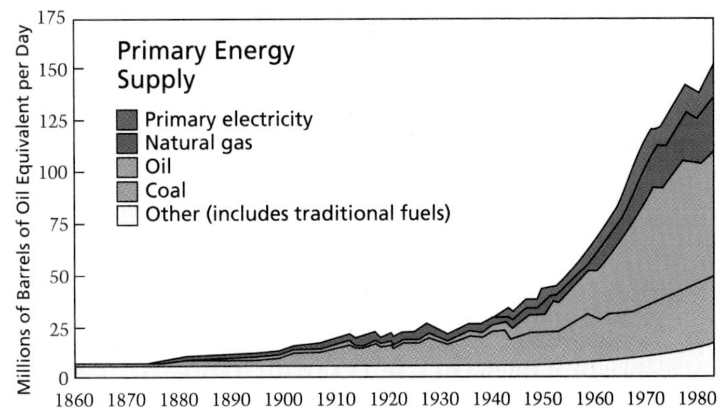

EXTEND Ex. 11 COOPERATIVE LEARNING Have students compare their timelines and state how they are alike and how they are different.

Ex. 16–19 Primary electricity is the electrical current produced by means of a generator which changes mechanical energy to electrical energy.

Answers

22.

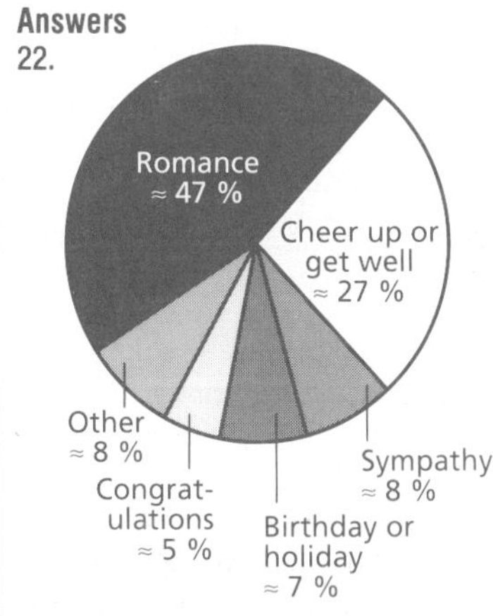

23.

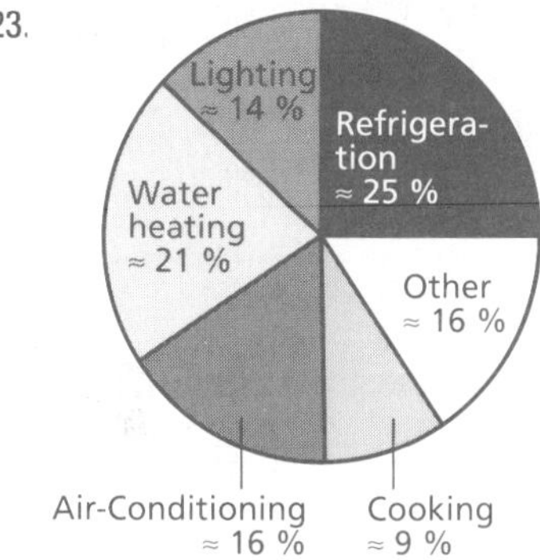

Enrichment Activities

These activities require students to go beyond lesson goals.

1. **RESEARCH**
   Newspapers and magazines contain lots of data. Select graphs from the print media that best illustrate the uses of the time-line graph, the picture graph, and the circle graph. Discuss your selections with other students before any final decisions are made.
2. **COOPERATIVE LEARNING**
   Work in a small-group setting. Each student in the group should locate examples of time lines, picture graphs, and circle graphs in the print media. Critique the sample graphs collected by the members of your group and choose the best example for each type of graph.
3. **WRITING**
   For each type of graph, write a paragraph that describes why the chosen graphs were the best examples and why the other examples were not. Students should look for examples of graphs that display differences of data accurately and graphs that are easy for readers to interpret.

What Size Bed? **In Exercises 20 and 21, use the circle graph showing the sizes of beds that Americans bought in 1990.**
(Source: International Sleep Products)

Bed Sizes People Buy
39% Twin
26% Full
24% Queen
Custom 5%
King 6%

20. What percentage of people bought either queen or full-size beds? 50%

21. What is the difference in the percentages of people who bought twin and full-size beds? 13%

22. *Giving Flowers* A survey of adults found that one third of American women gave flowers to a man during 1990. The reasons for giving the flowers were Romance (931), Cheer Him Up or Get Well (532), Sympathy (152), Birthday or Holiday (133), Congratulations (95), and Other (152). Construct a circle graph of this information. ***(Source: American Floral Marketing Council)***

23. *Electricity Consumption* The average amounts of electricity (in kilowatt-hours) used by a United States household in 1990 were as follows: Refrigeration (1810), Lighting (1000), Water Heating (1540), Air-Conditioning (1180), Cooking (635), Other (1180). Make a circle graph to show this information. ***(Source: Scientific American)***
22., 23. See margin.

What Color Bike? **In Exercises 24 and 25, use the circle graph showing the color of bicycles bought by Americans in 1989. During 1989, Americans spent $2,205,000 on bicycles.**
(Source: Bicycle Market Research Institute)

24. How much money was spent on blue bicycles in 1989? $507,150

25. How much more money was spent on black bicycles than on blue bicycles in 1989? $22,050

Integrated Review

26. *Test Points* You got 72% on a 150-point test. How many points did you get? 108

27. *Sales Tax* You paid $26.74 for a sweater. The price of the sweater is $24.99. What was the percent of sales tax? 7%

28. *Free Throws* A basketball forward makes 80% of free throws taken. If there were 240 free throws in a season, how many free throws were made? 192

29. *Completed Passes* A football team completes 56% of its passes. How many of 25 passes would you expect it to complete in a game? 14

30. *Baseball Card* You bought a baseball card for $3.50 and later sold it for $5.00. What percent profit did you make? 43%

In Exercises 31–36, graph the inequality. See Additional Answers.

31. $2x > 6$ **32.** $y \le 8$ **33.** $x + y \le 3$

34. $x - y > 5$ **35.** $x - 2y > 4$ **36.** $2x + y < 6$

37. *Estimation—Powers of Ten* Which best estimates the height of the Empire State Building?

a. 125 feet **b.** 1250 feet

c. 12,500 feet **d.** 125,000 feet

38. *Estimation—Powers of Ten* Which best estimates the annual salary of a manager of a fast-food restaurant?

a. $3000 **b.** $30,000

c. $300,000 **d.** $3,000,000

Exploration and Extension

United States History **In Exercises 39–42, use the time line showing the years in which the 26 amendments of the United States Constitution were passed.**

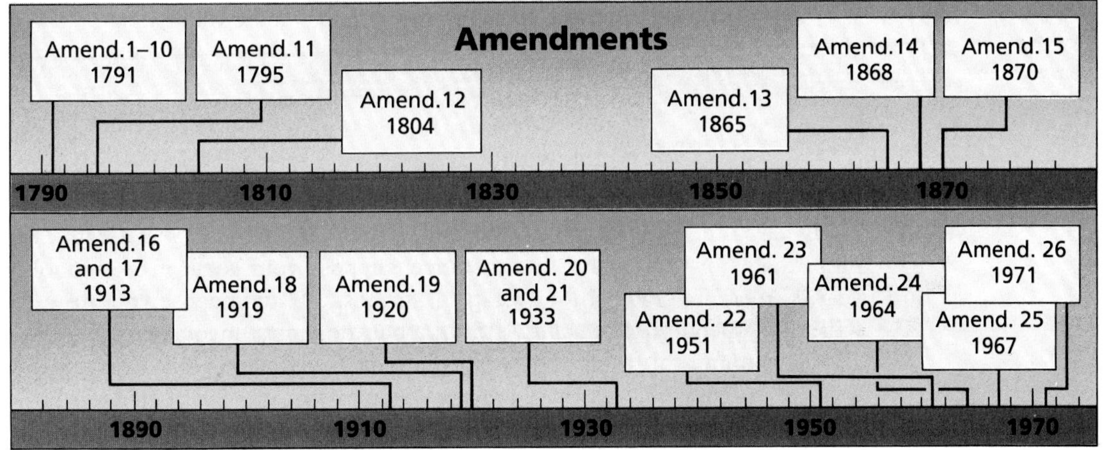

39. The Abolition of Slavery and the Civil Rights Amendments were passed three years apart, around the time of the Civil War. Which amendments are they? 13, 14

40. Two amendments were passed soon after World War I. The second of these gave women the right to vote. Which amendment is it? 19

41. Franklin Delano Roosevelt was elected President of the United States for four consecutive terms. The first amendment passed after his presidency limited presidents to two terms. Which amendment is it? 22

42. The two amendments, Prohibition of Liquor and Repeal of Prohibition, were passed 14 years apart. Which amendments are they? 18, 21

Ex. 37–38 Ask students to pay attention to the reasonableness of their answers. Students might consider using referrents to help them eliminate unreasonable choices when answering these questions. For example, in Exercise 37 students might consider that Mount Ranier at 14,000 feet (see page 305) would be considerably taller than the Empire State Building, so c and d are not reasonable estimates.

EXTEND Ex. 39–42 UNITED STATES HISTORY Encourage the class to investigate the amendments to the United States Constitution.

COOPERATIVE LEARNING It might be advantageous to students to divide the work load so that one or two students are required to report on any one amendment. The first ten amendments, the Bill of Rights, should be investigated as a unit.

ORGANIZER

TIME SCHEDULE: All levels, two days

The Chapter Summary helps students organize the main ideas of the chapter. In this chapter, students learned how to graph and solve simple and compound linear inequalities. The connection between absolute value and linear inequalities was also explored.

Work with students to review the skills, strategies, and concepts of the chapter. It is suggested that the first day of review be a combination of you and your students working problems together from the Chapter Review exercises. The first day's homework assignment can be used as the basis for the second day of review.

COOPERATIVE LEARNING

Encourage students to study together in small groups, perhaps with the same group members that are used for class activities. Emphasize the importance of teaching a classmate how to perform a skill or how to recall a strategy. When students work together, everyone wins. The students receiving help get additional instruction, and the students giving help gain a deeper understanding of the skills and concepts involved.

Chapter SUMMARY

Review the chapter by asking students, "What did you learn? Why did you learn it? How does it fit in with the other algebra skills and concepts you have learned?"

6

Chapter Summary

What did you learn?

Skills

1. Solve and graph an inequality in one variable.
   - Simple linear inequality **(6.1)**
   - Compound linear inequality **(6.3)**
   - Absolute value inequality **(6.4)**
2. Graph a linear inequality in two variables. **(6.5)**
3. Write an inequality to represent a graph.
   - Simple linear inequality in one variable **(6.1)**
   - Compound linear inequality in one variable **(6.3)**
   - Absolute value inequality in one variable **(6.4)**
   - Linear inequality in two variables **(6.5)**

Strategies

4. Use an inequality to model a real-life situation. **(6.1–6.5)**
5. Use an inequality to answer questions about real-life situations. **(6.1–6.5)**

Exploring Data

6. Create and interpret time lines, picture graphs, and circle graphs. **(6.6)**

Why did you learn it?

Many real-life situations have models that are inequalities. Think of how many verbal descriptions contain phrases such as "at most," "at least," "no more than," or "less than." For instance, suppose your goal for the school year is to earn a grade point average of *at least* 3.0, or suppose you are willing to spend *no more than* \$15 on a pair of shorts. Inequalities generally have many solutions. There are many possible grade point averages that are at least 3.0, and there are many prices that are less than \$15. When you are creating models of real-life situations, remember to look for phrases that indicate inequalities.

How does it fit into the bigger picture of algebra?

In this chapter, you studied techniques for solving and graphing linear inequalities. The first four lessons focused on linear inequalities in one variable and absolute value inequalities in one variable. In the fifth lesson, you learned that the graph of an inequality in two variables is a *half plane.* The sixth lesson described three types of visual models that can be used to represent data. When reviewing this chapter, look for connections between solving and graphing equations and solving and graphing inequalities.

Chapter REVIEW

In Exercises 1–4, graph the inequality on a number line. (6.1) See Additional Answers.

1. $x \geq -3$ **2.** $3 > x$ **3.** $-4 \geq x$ **4.** $x \leq 4$

In Exercises 5–20, solve the inequality. (6.1, 6.3, 6.4)

5. $3x + 5 \geq 4$ $x \geq -\frac{1}{3}$

6. $\frac{2}{3}(x - 6) \geq 0$ $x \geq 6$

7. $6 \geq \frac{1}{4}(3 + x)$ $x \leq 21$

8. $75x - 50 < 55x - 30$ $x < 1$

9. $-x + 5 < 2(x - 4)$ $x > \frac{13}{3}$

10. $(x - 6)4 > x + 30$ $x > 18$

11. $3 < 1 - 6x < 6$ $-\frac{5}{6} < x < -\frac{1}{3}$

12. $-53 < 37 - 5x < 42$ $-1 < x < 18$

13. $0 \leq \frac{1}{4}(x - 8) \leq 20$ $8 \leq x \leq 88$

14. $2(x + 3) < -4$ or $14x + (-3) > 4$

15. $-x + \frac{1}{2} \leq -\frac{3}{4}$ or $\frac{1}{4}(x + 2) < \frac{3}{4}$

16. $\frac{1}{3}(4 - x) > 6$ or $\frac{2}{5}(x + 7) \geq 4$

17. $|2x - 4| \geq 3$ $x \leq \frac{1}{2}$ or $x \geq \frac{7}{2}$

18. $|x - \frac{1}{2}| \leq 3$ $-\frac{5}{2} \leq x \leq \frac{7}{2}$

19. $|\frac{1}{3}x - \frac{2}{3}| > 6$ $x < -16$ or $x > 20$

20. $|\frac{25}{3} - 5x| < \frac{5}{3}$ $\frac{4}{3} < x < 2$

14. $x < -5$ or $x > \frac{1}{2}$

15. $x \geq \frac{5}{4}$ or $x < 1$

16. $x < -14$ or $x \geq 3$

In Exercises 21–26, solve the inequality. Then graph the solution on a number line. (6.1, 6.4) See Additional Answers.

21. $5 - \frac{1}{2}x \leq -3$ $x \geq 16$

22. $3(2x + 8) \geq 4(-2x - 4)$ $x \geq -\frac{20}{7}$

23. $-4 < 3x + 1 < 4$ $-\frac{5}{3} < x < 1$

24. $\frac{1}{5}(x + 5) < \frac{2}{3}$ or $\frac{1}{5}(x + 5) > \frac{5}{4}$

25. $|4x - 5| < 20$ $-\frac{15}{4} < x < \frac{25}{4}$

26. $|3 - \frac{1}{2}x| \geq 7$ $x \geq 20$ or $x \leq -8$

24. $x < -\frac{5}{3}$ or $x > \frac{5}{4}$

In Exercises 27–30, match the absolute value inequality with the equivalent compound inequality. (6.4)

27. $|x - 2| \leq 3$ b **28.** $|x + 1| \geq 2$ d **29.** $|x - 1| < 4$ a **30.** $|x + 2| > 1$ c

a. $-3 < x < 5$ **b.** $-1 \leq x \leq 5$ **c.** $x < -3$ or $x > -1$ **d.** $x \leq -3$ or $x \geq 1$

In Exercises 31–34, write an absolute value inequality to fit the graph. (6.4)

31. 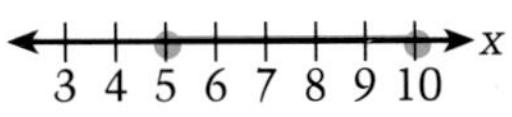

$|x - 7\frac{1}{2}| \leq 2\frac{1}{2}$

32. 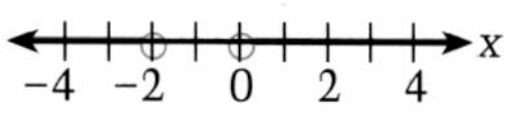

$|x + 1| > 1$

33. 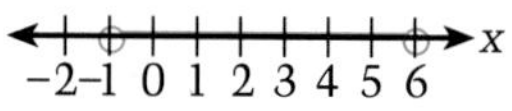

$|x - 2\frac{1}{2}| < 3\frac{1}{2}$

34. 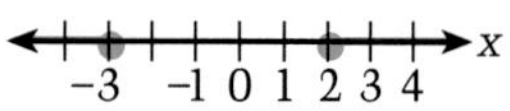

$|x + \frac{1}{2}| \geq 2\frac{1}{2}$

In Exercises 35–38, decide whether each ordered pair is a solution of the inequality. (6.5)

35. $\frac{1}{4}y - 2x + 3 \geq 4$; $(4, 0), (-4, -3)$ Only $(-4, -3)$ is.

36. $2(y - x) < 0$; $(0, 0), (2, -1)$ Only $(2, -1)$ is.

37. $\frac{1}{2}x + \frac{1}{3}y > -3$; $(-2, 1), (-8, -9)$ Only $(-2, 1)$ is.

38. $2x^2 + y \leq 14$; $(3, -2), (2, 7)$ Neither is.

In Exercises 39–42, sketch the graph of the inequality. (6.5) See margin.

39. $-2x + y > 4$ **40.** $-(\frac{1}{3}x + 3y) \leq 3$ **41.** $\frac{3}{8}(x + y + 8) \geq \frac{1}{4}$ **42.** $13x + y \geq 10x - 10$

Chapter REVIEW

ASSIGNMENT GUIDE

Basic/Average: Ex. 1–25 odd, 31–34, 35–45 odd, 47–50

Above Average: Ex. 1–25 odd, 27–30, 31–45 odd, 47–50

Advanced: Ex. 1–25 odd, 27–30, 35–45 odd, 49–54

More Difficult Exercises
Exercises 47–58

Ex. 1–58 Since the Chapter Review exercises parallel the list of Skills, Strategies, and Exploring Data statements on page 334, tell students to use the Chapter Summary to recall the lesson in which a skill or concept was presented if they are uncertain about how to do a given exercise in the Chapter Review. This forces the students to categorize the task to be performed and to locate relevant information in the chapter, both beneficial activities.

Answers

39. 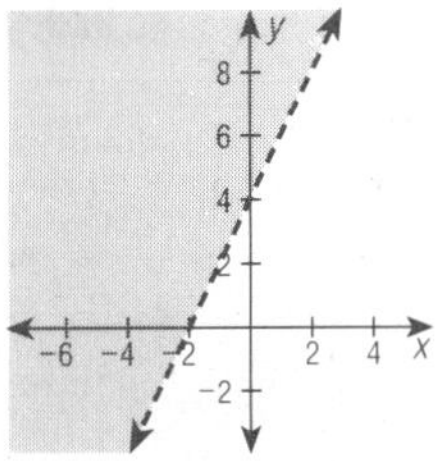

40. 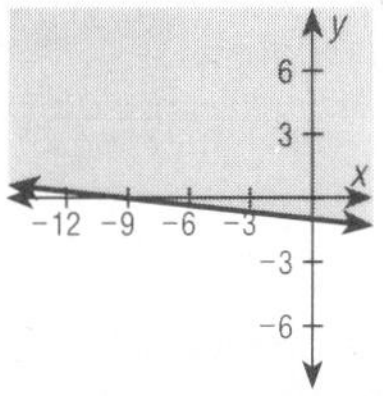

Answers

41.
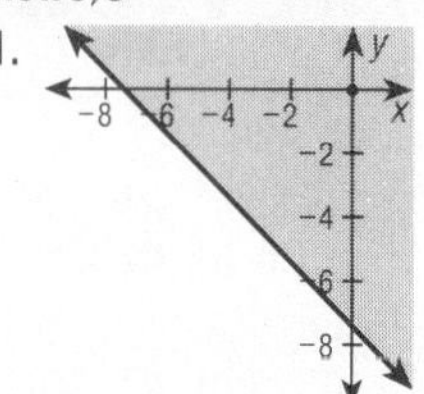

42.
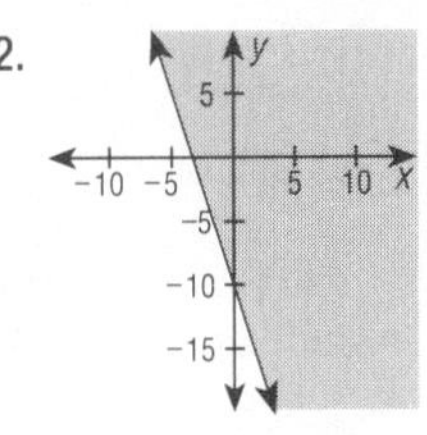

48.
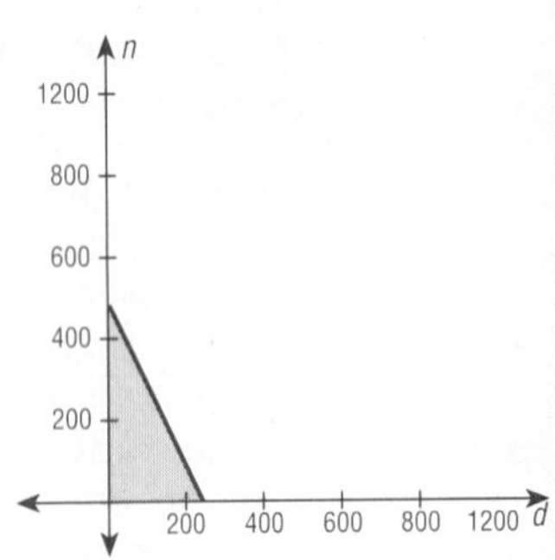

50.
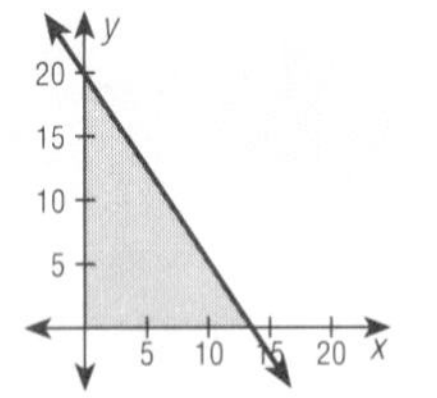

43. *Dinosaur Height* Let h represent the height of a dinosaur. Write an inequality that describes the range of heights of dinosaurs whose fossils have been found at Dinosaur National Park. $1 \le h \le 40$ or $12 \le h \le 480$

44. *Dinosaur Weight* Let w represent the weight of a dinosaur. Write an inequality that describes the range of weights of dinosaurs whose fossils have been found at Dinosaur National Park. $5 \le w \le 100{,}000$

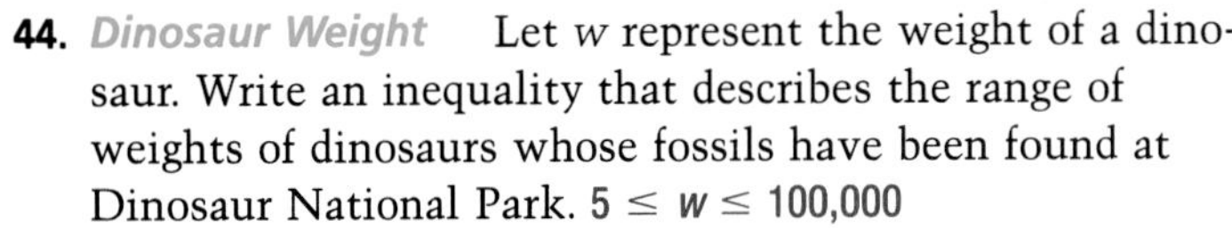

Dinosaur National Park, in northeastern Utah, has been one of the richest sources of dinosaur fossils in the world. Fossils of many different types of dinosaurs have been discovered at the park, ranging from Nanosaurus, about 12 inches tall and weighing about 5 pounds to Brachiosaurus, about 40 feet tall and weighing about 50 tons.

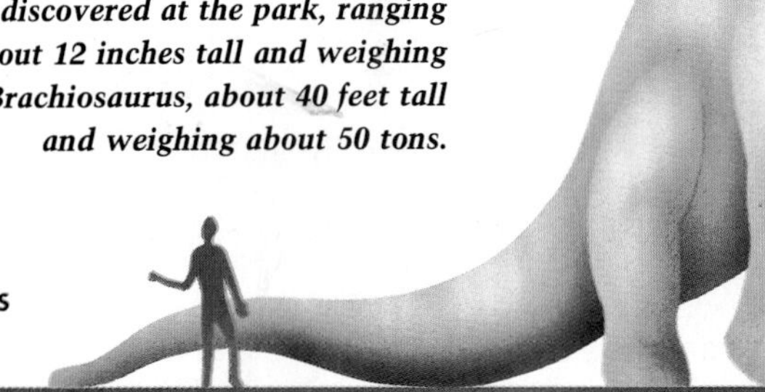

45. *Palindrome* A palindrome is a word (having 3 or more letters) that is spelled the same backward or forward. For instance, *pop, noon,* and *madam* are palindromes. In the English language, the longest palindrome is *tattarrattat.* Write an inequality that describes the number of letters, p, in a palindrome in the English language. (What does *tattarrattat* mean?) $3 \le p \le 12$

46. *Ocean Temperature* Let t represent the surface temperature of an ocean in degrees Fahrenheit. Write an inequality that describes the different surface temperatures of Earth's oceans. $28 \le t \le 86$

The highest surface temperature of an ocean is 86°F at different locations near the equator. The lowest temperature is 28°F at the North and South Poles.

Coin Collection **In Exercises 47 and 48, use the following information.**

Your grandfather has given you a sack of old coins that he had stored in his attic for several years. The sack is tied and has an old tag that reads, "$24.50 cash." You collect dimes and nickels, so you are anxious to see how many of each are in the sack. Let d represent the number of dimes and let n represent the number of nickels.

47. Write an inequality of d and n that describes the different numbers of dimes and nickels that could be in the sack. $0.10d + 0.05n \leq 24.50$

48. Sketch a graph of the inequality. Is (0, 35) a solution? If so, what does it represent? See margin. Yes, there are 0 dimes and 35 nickels.

Frozen Yogurt **In Exercises 49 and 50, use the following information.**

The softball team coach offers to buy frozen yogurt cones for each member of the team. Counting the coach, the team has 15 members. The coach has only $18 to spend on the cones. A large cone costs $1.35, and a small cone costs $0.90. Let x represent the number of large cones and let y represent the number of small cones.

49. Write an inequality that describes the different numbers of large and small cones the coach could buy. $1.35x + 0.09y \leq 18.00$

50. Sketch the graph of the inequality. Is (15, 0) a solution of the graph? Can the coach afford large cones for all the players? See margin. No, no

Entertainment **In Exercises 51–54, use the graph, which shows the average amount (in hundreds of dollars) spent by different age groups on entertainment in 1993.** ***(Source: U.S. Bureau of Labor Statistics)***

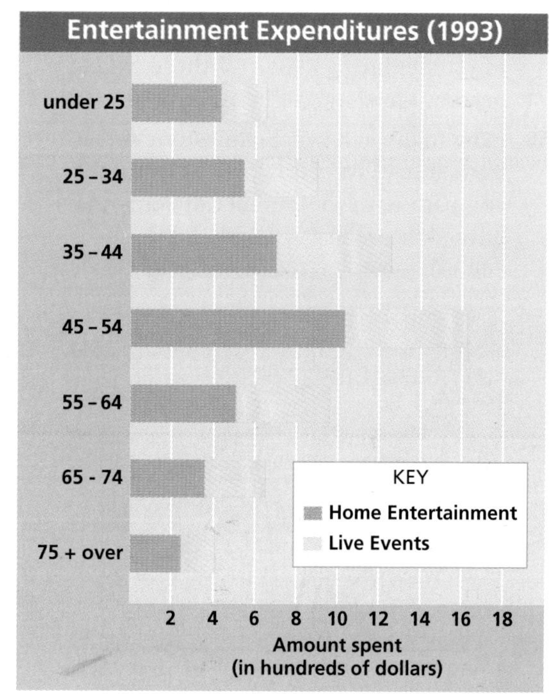

51. Which age groups spent approximately the same total amount on entertainment? Under 25 and 65–74

52. Which age groups spent nearly as much on live events as on home entertainment? 55–64 and 65–74

53. Estimate the average amount that a 28 year old spent on home entertainment. $\approx$ $550

54. Estimate how much more an 18 year old spent on home entertainment than on live events. $\approx$ $225

Answers

56. 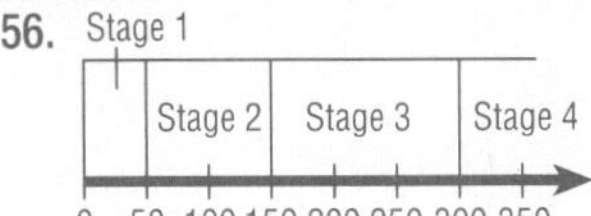

Chapter Test

1.–3. 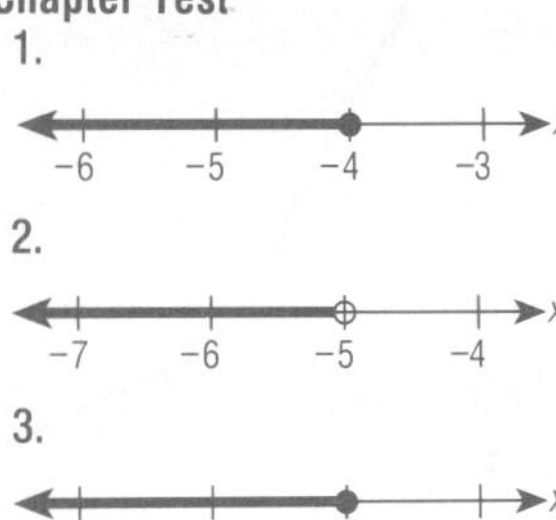

4.–5. 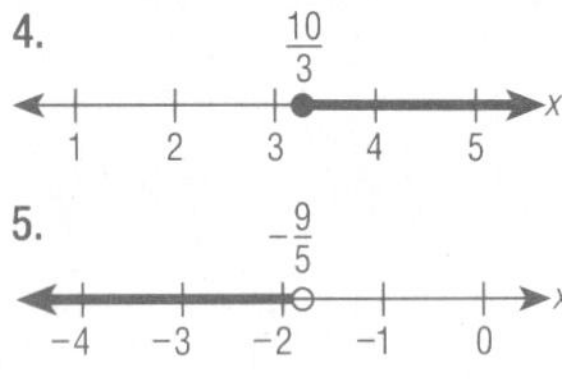

6. 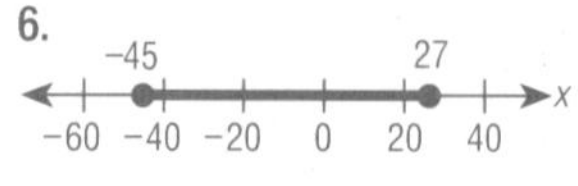

7. 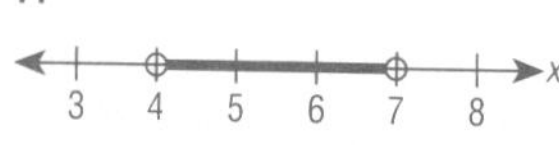

8. 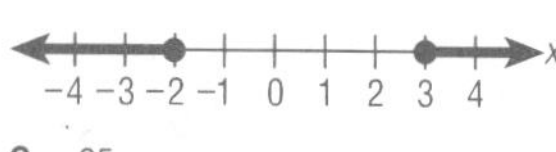

9.–10. 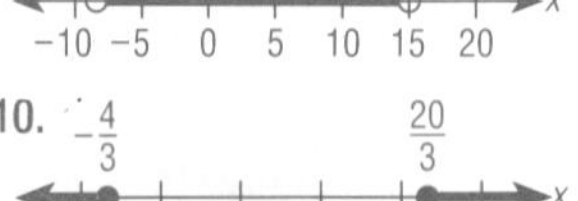

14. 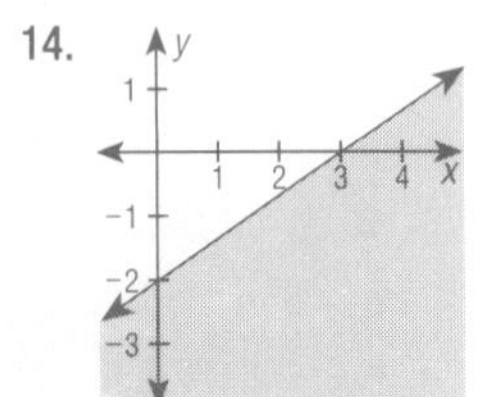

Forest Fire **In Exercises 55–58, use the information in the diagram showing ecological stages of a forest fire in Yellowstone National Park.**

Old-growth forest is struck by lightning. Forest burns. Most vegetation is destroyed. Lodgepole pines protect their seeds during the fire.

Stage 1. ***Lodgepole pine saplings grow to 15 or 20 feet. Many species of animals return to the forest.***

55. Solve the inequalities to find the years for each of the four stages of regrowth.

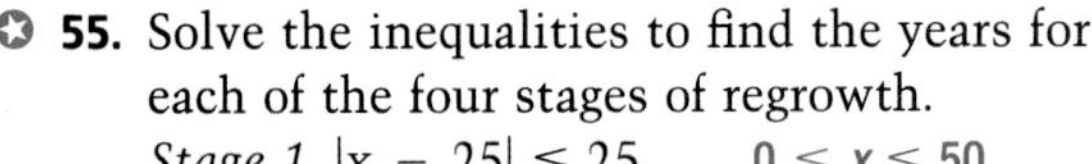

Stage 1 $|x - 25| \le 25$ $0 \le x \le 50$

Stage 2 $|x - 100| \le 50$ $50 \le x \le 150$

Stage 3 $|x - 225| \le 75$ $150 \le x \le 300$

Stage 4 $x - 300 \ge 0$ $x \ge 300$

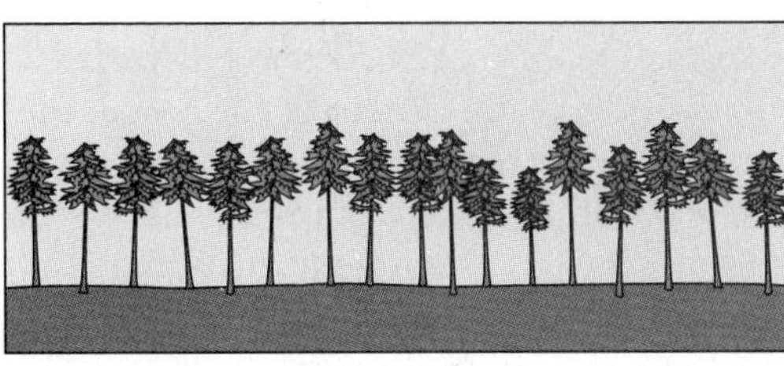

Stage 2. ***Pines reach heights up to 50 feet and form dense stands that block the sun from the forest floor.***

56. Use the solutions from Exercise 55 to create a time line showing the four stages of regrowth. See margin.

57. How many forest fires occurred in Yellowstone during 1988? Of these, what percent were started by lightning? ≈ 45, $\approx 87\%$

58. How many acres of Yellowstone forest burned in 1988? How does this number compare with the number of acres that burned between 1972 and 1987? 700,000; about 17,000 times as many

Stage 3. ***Lodgepole pines thin out. Second-generation trees appear. Increased sunlight aids vegetation growth on forest floor.***

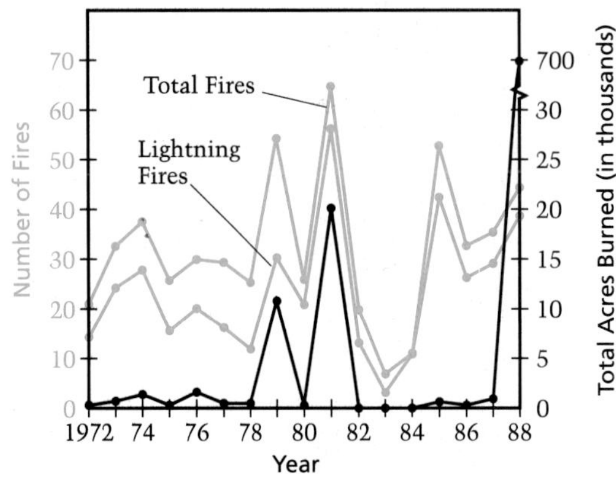

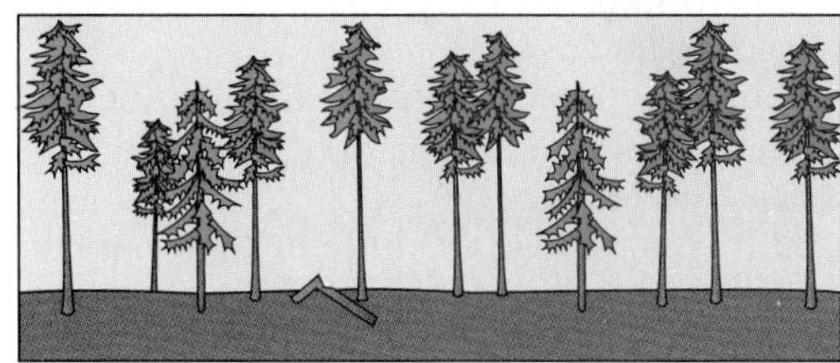

Stage 4. ***Original trees die. Dead branches and trees accumulate. The forest is again highly flammable.***

Chapter TEST

In Exercises 1–10, solve the inequality. Then sketch its graph. (6.1, 6.3, 6.4) See margin.

1. $\frac{3}{4}x + 6 \leq 3$ $x \leq -4$

2. $7 - x > 12$ $x < -5$

3. $4x + 3 \leq 3x - 1$ $x \leq -4$

4. $-(4 - x) \geq 2(3 - x)$ $x \geq \frac{10}{3}$

5. $-3x - 7 > 7x + 11$ $x < -\frac{9}{5}$

6. $-6 \leq \frac{1}{4}(3 - x) \leq 12$ $-45 \leq x \leq 27$

7. $8 < 3x - 4 < 17$ $4 < x < 7$

8. $2x + 1 \geq 7$ or $-3x - 4 \geq 2$

9. $|\frac{3}{5}x - 2| < 7$ $-\frac{25}{3} < x < 15$

10. $|8 - 3x| \geq 12$ $x \geq \frac{20}{3}$ or $x \leq -\frac{4}{3}$

In Exercises 11 and 12, write an absolute value inequality to fit the graph. (6.4) 8. $x \geq 3$ or $x \leq -2$

11. 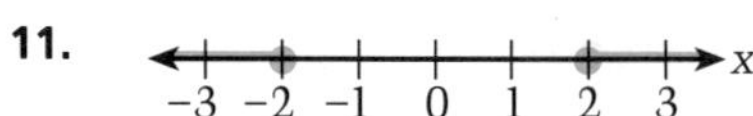 $|x| \geq 2$

12. 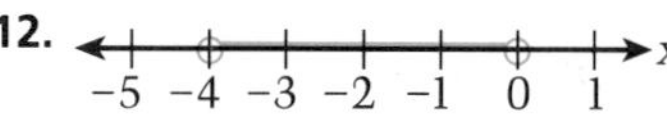 $|x + 2| < 2$

13. Which of the ordered pairs are solutions of $2x + 2y \geq 3$? **(6.5)**

a. $(0, 4)$ **b.** $(\frac{3}{4}, \frac{3}{4})$ **c.** $(-1, -\frac{1}{2})$

In Exercises 14–16, sketch the graph of the inequality. (6.5) See margin.

14. $-3y + 2x \geq 6$

15. $6y - 2x + 2 < 8$

16. $2x + 4y \leq 6$

17. In 1984, the United States produced 68.7 million tons of paper and paperboard. Other countries' production of paper that year was Japan, 21.3 million tons; Canada, 15.7 million tons; the Soviet Union, 10.5 million tons; and West Germany, 10.1 million tons. Make a circle graph to compare the amounts of paper produced by these five leading paper producers. **(6.6)** See margin.

18. The first step in producing paper is cutting the wood into chips $\frac{1}{2}$ inch to 1 inch long. Write an inequality to describe the size of the wood chips. **(6.3)** $\frac{1}{2} \leq x \leq 1$

19. Paper is made in a continuous roll that can be up to 33 feet wide with a paper-making machine called the Fourdrinier machine. Write an inequality describing the possible widths a paper-making machine can produce. (Assume that the smallest width is 12 feet.) **(6.3)**
$12 \leq x \leq 33$

20. Several friends go rafting on the Deschutes River in Oregon. The group rents six rafts and launches all the rafts at the same time. Write an inequality that describes the average rates (in miles per hour) for the six rafts. **(6.3)** $3 \leq x \leq 4$

The first raft to finish the 3-mile trip takes 45 minutes. The last raft to finish takes 60 minutes.

15.

16.

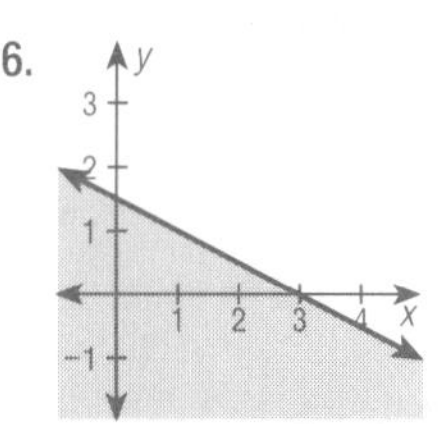

17.

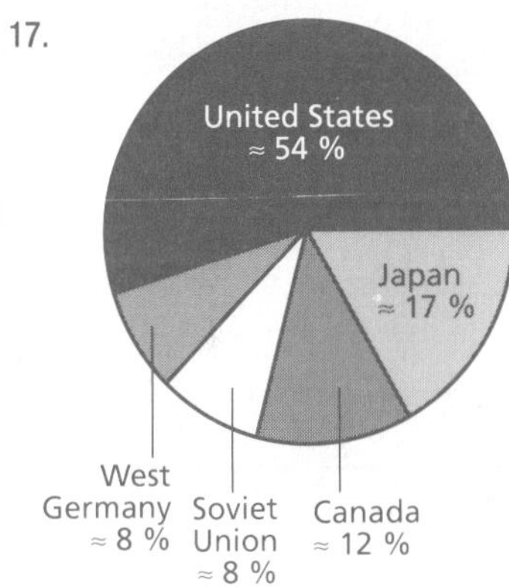

Hi mr. t -

Cumulative Review

5. 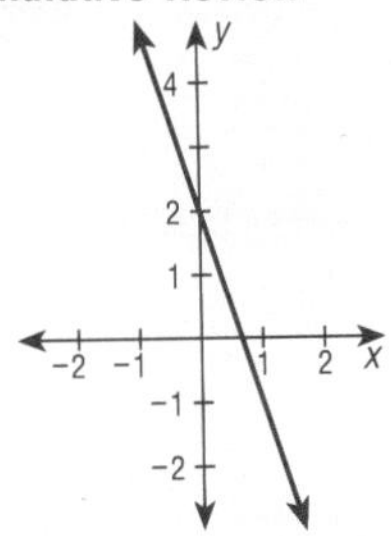

6. 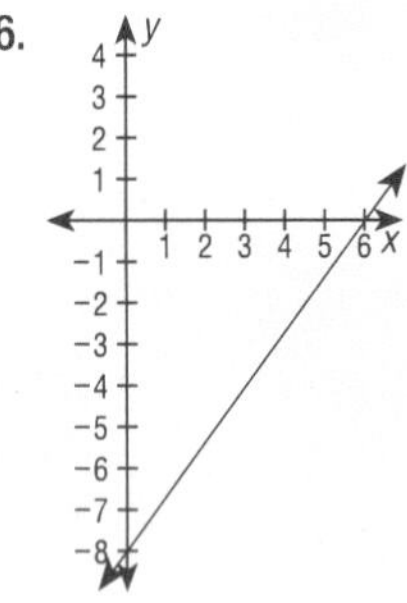

7. 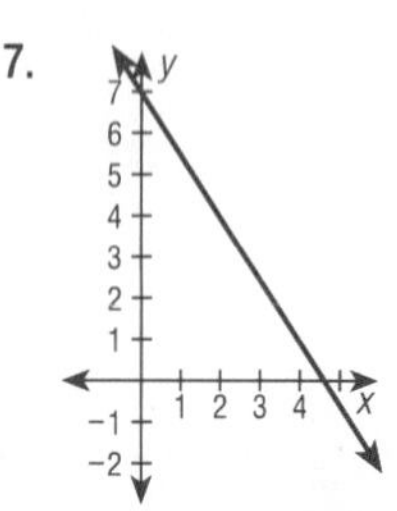

8. 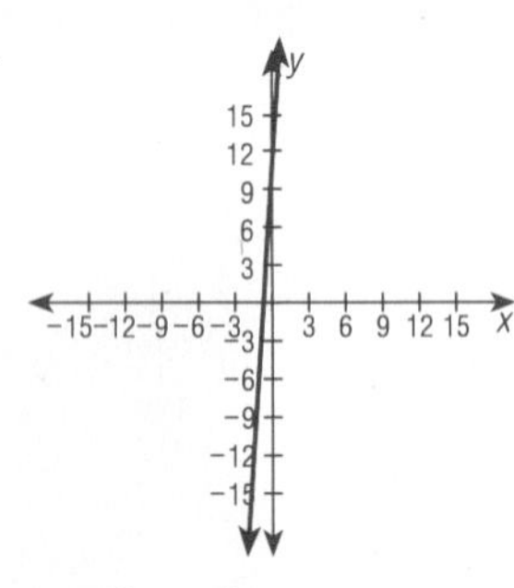

17. $y = 3x - 3$
18. $y = -\frac{1}{2}x + \frac{5}{2}$
19. $y = -5x - 9$
20. $y = \frac{7}{3}x + 3$

Cumulative REVIEW ■ Chapters 1–6

In Exercises 1–4, write equations of the horizontal line and the vertical line that pass through the point.

1. $(-3, 4)$ $y = 4; x = -3$
2. $\left(-2, -\frac{1}{2}\right)$ $y = -\frac{1}{2}; x = -2$
3. $(7, 6.2)$ $y = 6.2; x = 7$
4. $\left(0, -\frac{4}{3}\right)$ $y = -\frac{4}{3}; x = 0$

In Exercises 5–8, find the slope and y-intercept of the line. Sketch the graph of the line. See margin for graphs.

5. $y = -3x + 2$ $-3; 2$
6. $y = 2\left(\frac{2}{3}x - 4\right)$ $\frac{4}{3}; -8$
7. $3x + 2y = 14$ $-\frac{3}{2}; 7$
8. $\frac{1}{3}y - 6x = 4$ $18; 12$

In Exercises 9–12, find the slope of the line containing the points.

9. $(6, -2), (0, 5)$ $-\frac{7}{6}$
10. $(3, 7), (-6, 7)$ 0
11. $\left(\frac{1}{2}, \frac{3}{4}\right), (-3, 6)$ $-\frac{3}{2}$
12. $(5, 5), (20, 25)$ $\frac{4}{3}$

In Exercises 13–16, match the equation with its graph.

13. $y = 3x - 1$ c
14. $y = -\frac{1}{4}x + 2$ a
15. $y = \frac{1}{2}x + 3$ d
16. $y = -x - 2$ b

a. 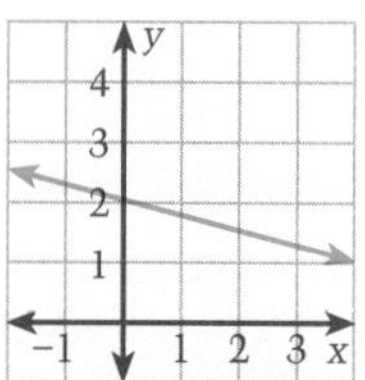

b. 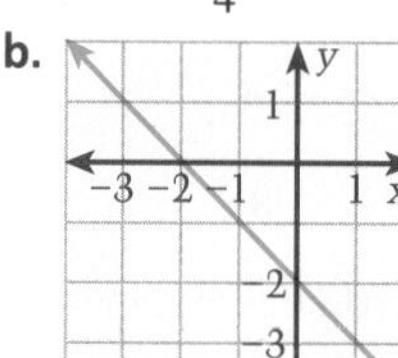

c. 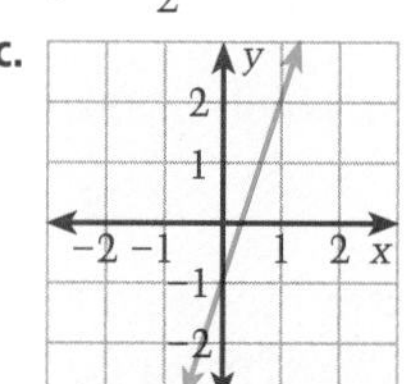

d. 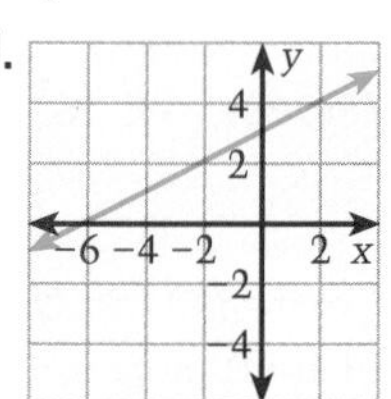

In Exercises 17–20, write an equation of the line that has the given slope and passes through the given point. See margin.

17. $m = 3, (4, 9)$
18. $m = -\frac{1}{2}, (5, 0)$
19. $m = -5, (-3, 6)$
20. $m = \frac{7}{3}, (0, 3)$

In Exercises 21–24, write an equation of the line that passes through the two points.

21. $(0, 6), (4, 9)$ $y = \frac{3}{4}x + 6$
22. $\left(\frac{2}{3}, -1\right), \left(\frac{4}{9}, 7\right)$ $y = -36x + 23$
23. $(4.2, -3.6), (7.0, 3.4)$ $y = 2.5x - 14.1$
24. $(56, 19), (37, -19)$ $y = 2x - 93$

In Exercises 25–28, complete the table. Then sketch the graph of the equation. See margin for graphs.

25. $y = -|x + 2| - 3$

x	−4	−3	−2	−1	0	1
y	?	?	?	?	?	?

$-5, -4, -3, -4, -5, -6$

26. $y = \frac{1}{3}|x - 4| + 6$

x	1	2	3	4	5	6	7
y	?	?	?	?	?	?	?

$7, 6\frac{2}{3}, 6\frac{1}{3}, 6, 6\frac{1}{3}, 6\frac{2}{3}, 7$

27. $y = -2|x + 2| - 2$

x	−4	−3	−2	−1	0	1
y	?	?	?	?	?	?

$-6, -4, -2, -4, -6, -8$

28. $y = |4x - 4| + 6$

x	−2	−1	0	1	2	3	4
y	?	?	?	?	?	?	?

$18, 14, 10, 6, 10, 14, 18$

In Exercises 29–32, find the vertex of the graph.

29. $y = |x + 3| - 6$ $(-3, -6)$

30. $y = |2x - 4| - 9$ $(2, -9)$

31. $y = \frac{1}{2}|x + 4| + 4$ $(-4, 4)$

32. $y = -3|x + 6| + 1$ $(-6, 1)$

In Exercises 33–36, match the inequality with its graph.

33. $x \le 3$ c

34. $-1 < x \le 2$ d

35. $x \ge -3$ a

36. $x < 1$ or $x > 2$ b

a.

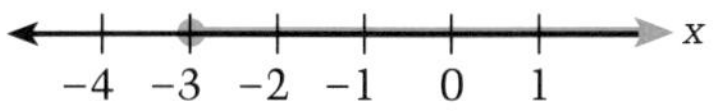

b.

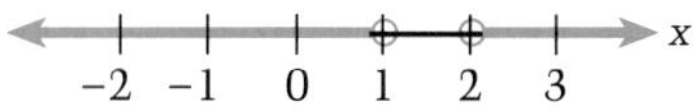

c.

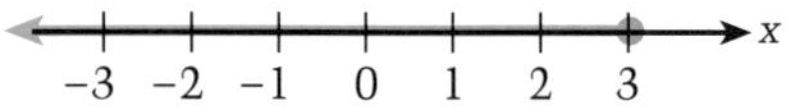

d.

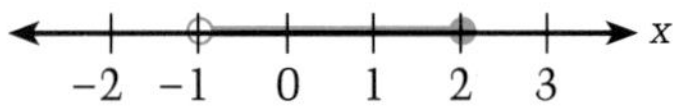

In Exercises 37–40, solve the inequality. Then match the inequality with its graph.

37. $|x - 2| < 4$ $-2 < x < 6$; c

38. $|4x - 6| \le 4$ $\frac{1}{2} \le x \le \frac{5}{2}$; d

39. $|\frac{1}{4}x + 2| > 1$ $x < -12$ or $x > -4$; a

40. $|x + \frac{3}{2}| \ge \frac{3}{2}$ $x \ge 0$ or $x \le -3$; b

a.

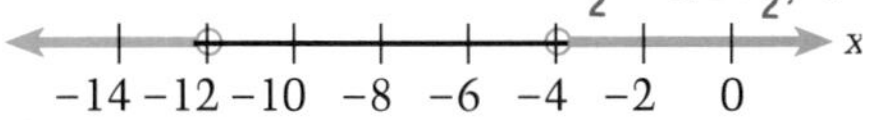

b.

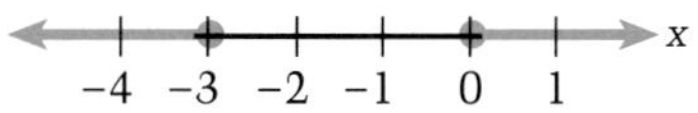

c.

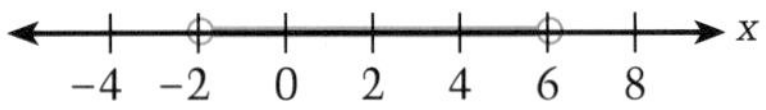

d.

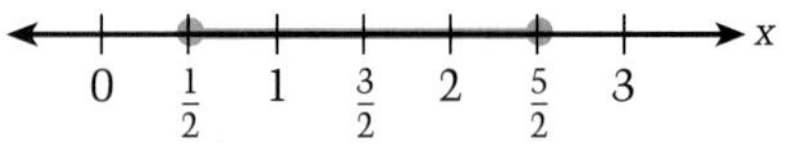

In Exercises 41–44, sketch the graph of the inequality. See margin.

41. $-2x + 3y \ge x + 9$

42. $\frac{1}{2}x + 2y > 4$

43. $-\frac{1}{2}(2x + y) < 1$

44. $-y \ge 3x - 1$

Popcorn and Peanuts **In Exercises 45 and 46, use the following information.**

At a baseball game you are selling popcorn for \$1.00 per box and peanuts for \$1.50 per bag. Total sales for the day are \$150. Let x be the number of boxes of popcorn sold and y the number of bags of peanuts sold.

45. Sketch a graph that shows the possible values of x and y. See margin.

46. If you sold 36 bags of peanuts, how many boxes of popcorn were sold? 96

Savings Plan **In Exercises 47–50, use the following information.**

At the beginning of the school year, you put \$20 in a savings account. During the school year, you plan to deposit \$3 a week in the account.

47. Write an equation that gives the account balance, y, in terms of the number of weeks, x. (Do not count interest.) $y = 3x + 20$

48. Sketch the graph of the equation. See margin.

49. What does the slope represent? An increase of \$3 for each week.

50. What does the y-intercept represent? The initial deposit

25.

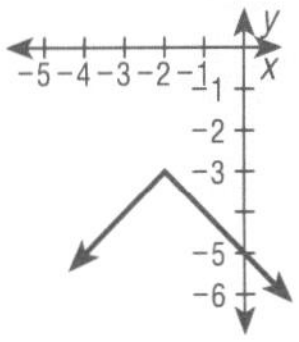

26.

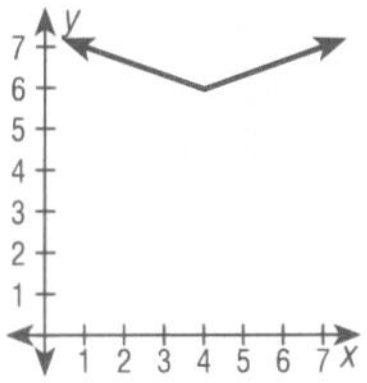

27.

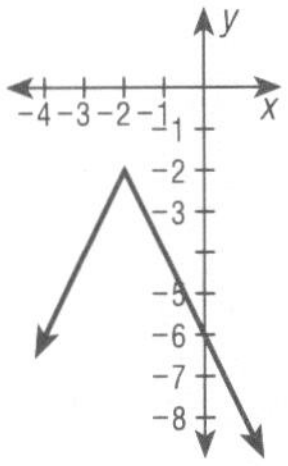

28.

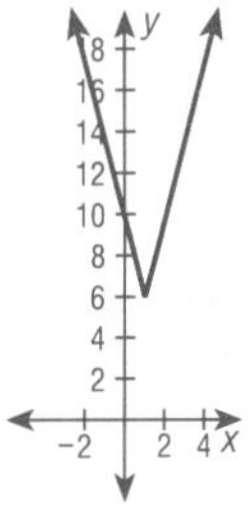

41.

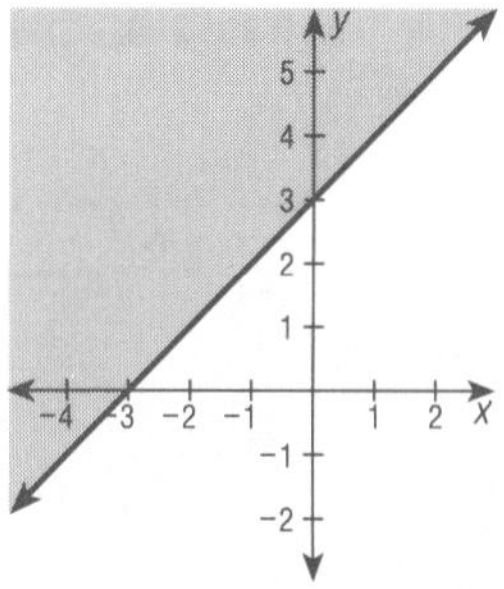

42.

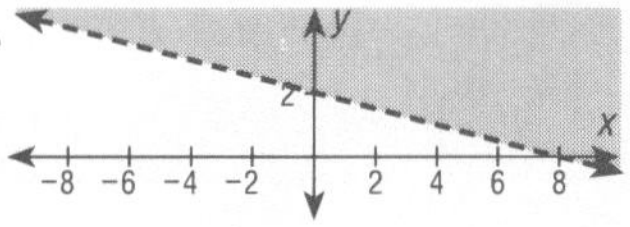

43.

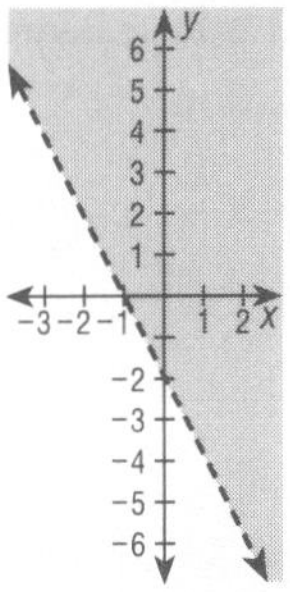

Answers

44.

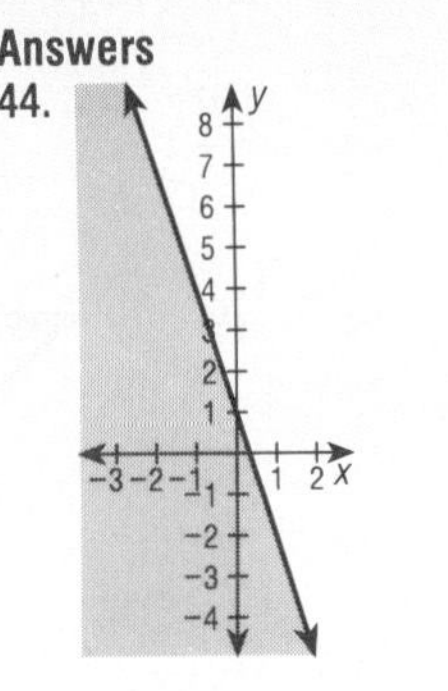

45.

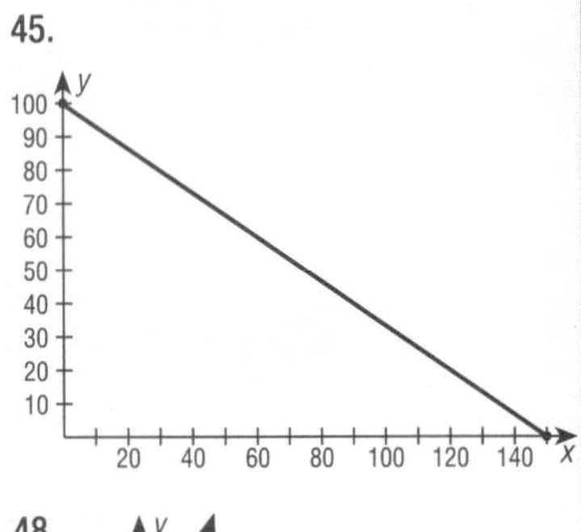

48.

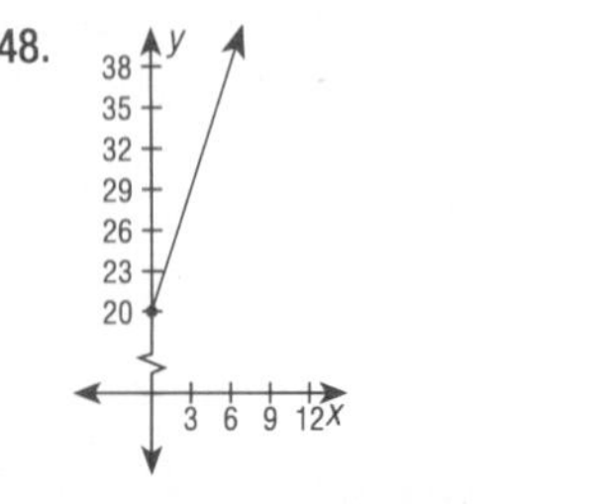

51.

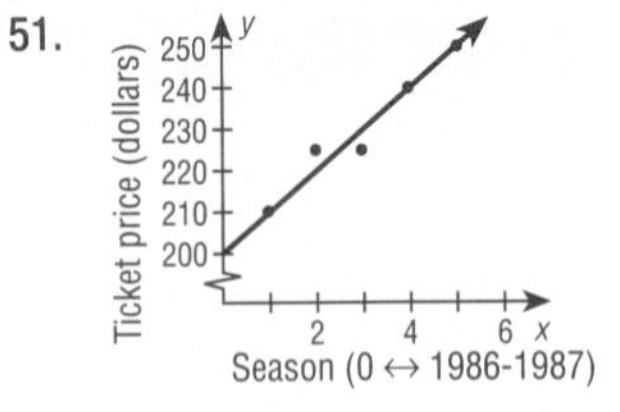

Season Tickets **In Exercises 51–54, use the following information.**

You want to save enough money to buy a season ticket for the local hockey team. Since ticket prices haven't been announced, you are not sure how much to save. Prices for a season ticket for the past six years are shown in the table.

Season	1986–87	1987–88	1988–89	1989–90	1990–91	1991–92
Price	\$200	\$210	\$225	\$225	\$240	\$250

51. Sketch a scatter plot for the data in the table. Let $x = 0$ represent the 1986–87 season. Let y represent the ticket price.

52. Sketch the best-fitting line. Find an equation for the line. $y = 10x + 200$

53. What does the slope represent? An average increase of \$10 per ticket each year

54. How much should you save to buy a season ticket? \$260

History of Computers **In Exercises 55–60, use the following information.**

In 5000 B.C., the abacus was invented by the Chinese. The following time line shows several other important dates in the history of calculating devices. Use the time line to match the inventor with the invention.

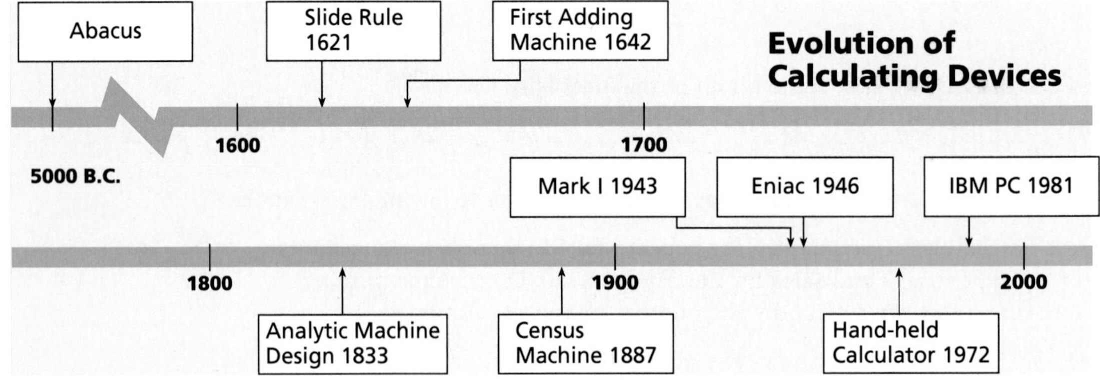

55. Blaise Pascal (1623–1662) a

56. Charles Babbage (1791–1871) b

57. William Oughtred (1574–1660) d

58. Herman Hollerith (1860–1929) c

a. First adding machine

b. Analytic machine

c. Census machine

d. Slide rule

59. *Slide Rule* The slide rule was used by engineers and scientists to make calculations before the widespread availability of hand-held calculators. How many years was the slide rule used in making calculations? 351

60. *UNIVAC I* In 1951, *UNIVAC I* became the first computer to be designed for business use. CBS News used it the following year to predict the winner of the presidential election. Who was elected President that year? Dwight Eisenhower

Jury Duty **In Exercises 61 and 62, use the following information.**

The circle graph at the right shows the results of a survey in which members of juries were asked why they had been willing to serve. ***(Source: The Defense Research Institute)***

Why Citizens Serve on Juries

Called/selected or picked 28%

Intrest in the Process/ wanted to 10%

Unable to get out of it 8%

Duty/responsibility as citizens 54%

61. A trial jury usually consists of 12 members. How many of the 12 jurors would you expect to be serving on the jury because they felt it was their responsibility?

62. Out of 75 people who have served on a jury, how many would you expect to have served because they were unable to get out of jury duty? 6

61. 6 or 7

Wild Geese **In Exercises 63–65, use the following information.**

There are more than 40 species of wild geese in the world. The adult birds typically weigh between $3\frac{1}{4}$ and $8\frac{3}{4}$ pounds. Geese are migratory birds, flying south for the winter and north for the summer. When migrating, the snow goose can fly 1700 miles in $2\frac{1}{2}$ days.

63. What is the average speed of a migrating snow goose? 680 miles per day

64. A duck's migration journey is usually shorter than a goose's. If a duck travels 975 miles in 24 hours, is it traveling faster or slower than a snow goose? Faster

65. Write a compound inequality for the weight of adult geese. $3\frac{1}{4} < W < 8\frac{3}{4}$

Flying Formation **In Exercises 66 and 67, use the following information.**

The graph at the right shows the flying pattern used by migrating geese. Some naturalists believe the flock follows a single lead bird during the entire migratory flight. Others believe that the lead bird changes and that the V-formation creates air currents that make flying easier for the rest of the birds.

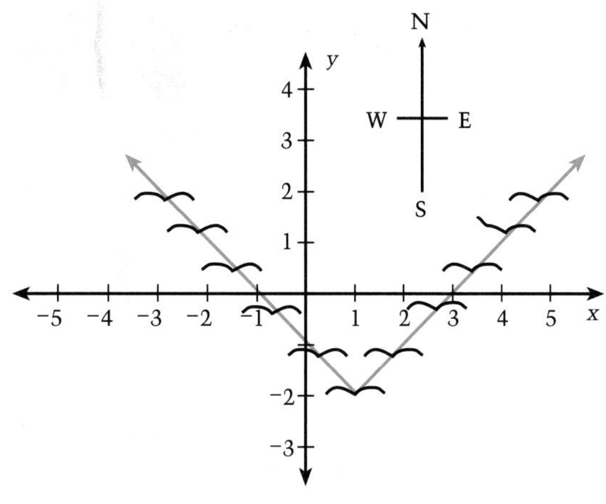

66. Write an absolute value equation that models the flying formation shown in the coordinate plane. $y = |x - 1| - 2$

67. The flying formation at the right is heading south. Write an absolute value equation of a flying formation heading north. $y = -|x - 1| - 2$

The techniques for solving systems of linear equations and inequalities empower students to handle many more types of real-life problem situations. It is important that students understand the connection between an algebraic solution to a system and its graphic solution. Emphasize that it is important to become proficient using each solution technique (graphing, substitution, linear combinations), and that it is equally important to be able to decide under which circumstances one of the techniques works best.

The Chapter Summary on page 393 provides you and the students with a synopsis of the chapter. It identifies key skills, concepts, and vocabulary. You may want to have students look at the Chapter Summary as an overview before beginning the chapter.

CHAPTER 7

Solving Systems of Linear Equations

LESSONS

According to the data analyzed on the opposite page, the slogan "Made in the U.S.A." may have special significance for this woman working in the garment industry.

The graph shows two lines that appear to have different slopes. The lines are the graphs of a system of two linear equations in two variables. In this chapter, students will learn how to solve systems of linear equations (and systems of linear inequalities).

Using the data in the graph, ask students to find the equations of the best-fitting lines. (Tell students to review Lesson 5.4 if necessary.) Sample answers: $4y = 69 + 3x$; $y = 4 + 2x$

Sketch the graph of the equation of each line and estimate when the value of import clothing will equal the value of clothing manufactured in the United States. Sometime in 1990

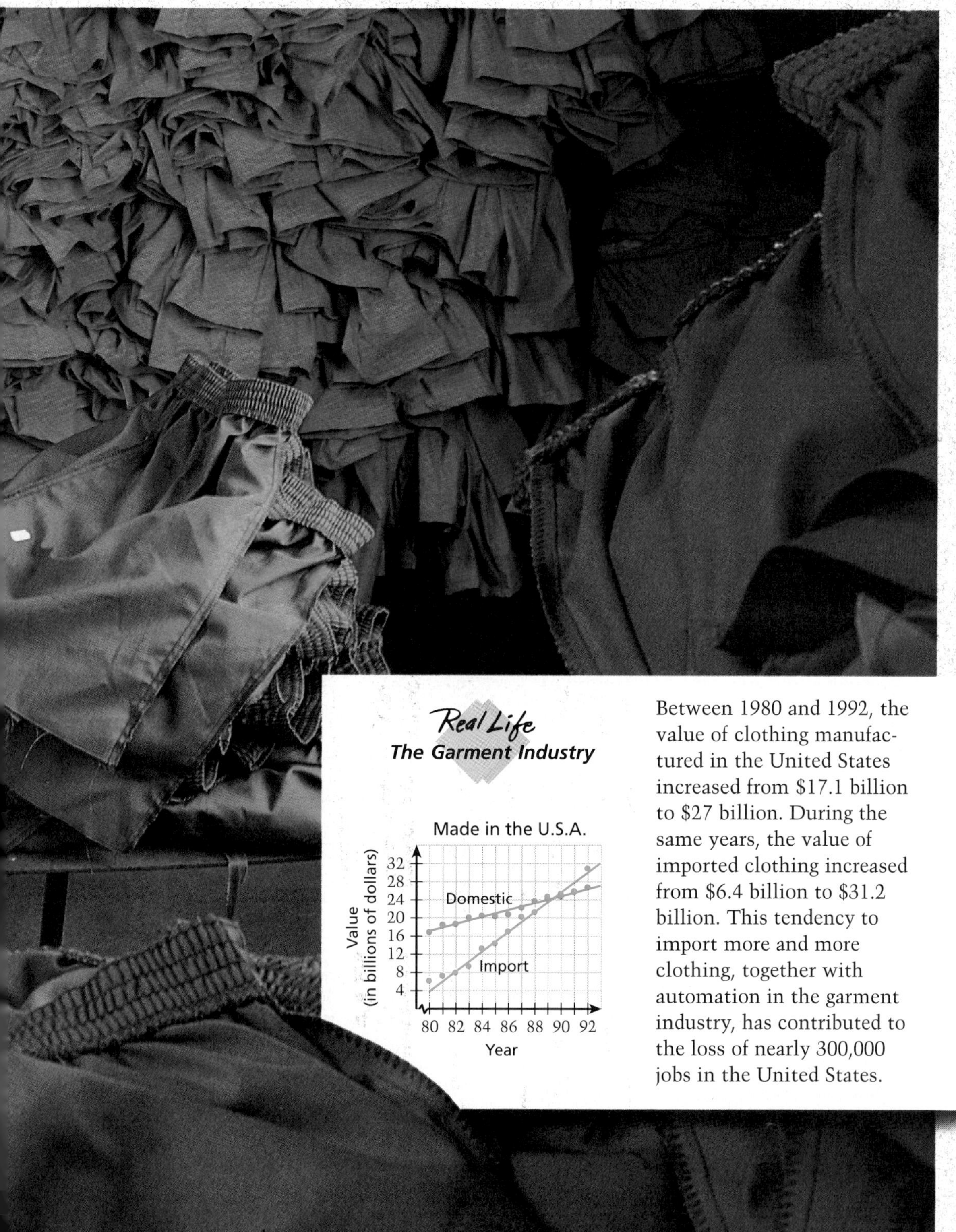

Between 1980 and 1992, the value of clothing manufactured in the United States increased from \$17.1 billion to \$27 billion. During the same years, the value of imported clothing increased from \$6.4 billion to \$31.2 billion. This tendency to import more and more clothing, together with automation in the garment industry, has contributed to the loss of nearly 300,000 jobs in the United States.

Solving Systems of Linear Equations

PACING CHART

The daily Pacing Chart is meant to help you adjust your teaching pace. Students in the full course should finish the entire text by the end of the year. Students in the basic course may not complete the entire text in the school year. The Pacing Chart for each chapter contains suggestions for lessons that require more than one day and lessons that may be omitted for the basic course.

DAY	FULL COURSE	BASIC COURSE
1	7.1	7.1
2	7.2	7.2
3	7.3	7.3
4	7.4	7.4
5	7.4 & Using a Graphing Calculator	7.4
6	Mid-Chapter Self-Test	Using a Graphing Calculator
7	7.5	Mid-Chapter Self-Test
8	7.6	7.5
9	7.7	7.6
10	7.7	7.7
11	Chapter Review	7.7
12	Chapter Review	Chapter Review
13	Chapter Test	Chapter Review
14		Chapter Test

CHAPTER ORGANIZATION

LESSON	PAGES	GOALS	MEETING THE NCTM STANDARDS
7.1	346-351	1. Solve a system of linear equations by graphing 2. Model a real-life situation using a system of linear equations	Problem Solving, Communication, Connections, Statistics, Discrete Mathematics, Geometry
Using a Calculator	352	Use a graphing calculator to graph a linear system	Technology, Geometry
7.2	353-358	1. Use substitution to solve a linear system 2. Model a real-life situation using a system of linear equations	Problem Solving, Communication, Connections, Reasoning, Discrete Mathematics, Geometry
Mixed Review	359	Review of algebra and arithmetic	
Career Interview	359	Pediatrician	Connections
7.3	360-366	1. Use linear combinations to solve a linear system 2. Model a real-life situation using a system of linear equations	Problem Solving, Communication, Connections, Discrete Mathematics
7.4	367-372	Write and use a linear system as a real-life model	Problem Solving, Communication, Connections, Discrete Mathematics, Geometry
Mid-Chapter Self-Test	373	Diagnose student weaknesses and remediate with correlated Reteaching worksheets	
7.5	374-380	1. Visualize the solution possibilities for linear systems 2. Identify a linear system that has many solutions	Problem Solving, Communication, Connections, Reasoning, Discrete Mathematics, Geometry
Mixed Review	380	Review of algebra and arithmetic	
7.6	381-386	1. Solve a system of linear inequalities by graphing 2. Model a real-life situation using a system of linear inequalities	Problem Solving, Communication, Connections, Discrete Mathematics, Geometry
7.7	387-392	1. Solve a linear programming problem 2. Model a real-life situation using linear programming	Problem Solving, Communication, Connections, Discrete Mathematics, Geometry
Chapter Summary	393	A restatement of what has been learned, why it has been learned, and how it fits into the structure of algebra	Structure, Connections
Chapter Review	394-396	Review of concepts and skills learned in the chapter	
Chapter Test	397	Diagnose student weaknesses and remediate with correlated Reteaching worksheets	

LESSON RESOURCES

MEETING INDIVIDUAL NEEDS

RETEACHING For students who need to spend more time on basics:

If a mid-chapter self-test or chapter test indicates a deficiency, teachers can help students with the appropriate ***Reteaching Copymaster.***

PRACTICE For students who need more practice:

Additional exercises like those in the Pupil's Edition are provided for each lesson in ***Extra Practice Copymasters.***

ENRICHMENT For enriching and broadening students' experiences:

Problem of the Day copymasters in ***Teaching Tools*** provide a daily opportunity to use logical reasoning, looking for a pattern, writing an equation, and other routine and non-routine problem-solving strategies.

Math Log copymasters in ***Alternative Assessment*** provide opportunities to report on investigations, research, and open-ended problems.

Enriching activities with graphing and scientific calculators and computers are provided in ***Technology: Using Calculators and Computers.***

The ***Applications Handbook*** provides additional information about the cross-curriculum topics such as astronomy, chemistry, physics, sports, economics, genetics, and music that are integrated into the Pupil's Edition.

LESSON	7.1	7.2	7.3	7.4	7.5	7.6	7.7
PAGES	346-351	353-358	360-366	367-372	374-380	381-386	387-392
Teaching Tools							
Transparencies	✓	✓	✓	✓	✓	✓	✓
Problem of the Day	✓	✓	✓	✓	✓	✓	✓
Warm-up Exercises	✓	✓	✓	✓	✓	✓	✓
Answer Masters	✓	✓	✓	✓	✓	✓	✓
Extra Practice Copymasters	✓	✓	✓	✓	✓	✓	✓
Reteaching Copymasters	Teacher-directed and independent activities tied to results on the Mid-Chapter Self-Tests and Chapter tests						
Color Transparencies	✓	✓	✓		✓		
Applications Handbook	Additional background information is supplied for many real-life applications.						
Technology Handbook	Calculator and computer worksheets are supplied for appropriate lessons.						
Complete Solutions Manual	✓	✓	✓	✓	✓	✓	✓
Alternative Assessment	Assess student's ability to reason, analyze, solve problems, and communicate using mathematical language.						
Formal Assessment	Mid-Chapter Self-Tests, Chapter Tests, Cumulative Tests, and Practice for College Entrance Tests						
Computer Test Bank	Customized tests can be created by choosing from over 2000 items.						

INSIGHTS

7.1 Solving Linear Equations by Graphing

Many more types of real-life situations will be encountered in this chapter. Finance, health care, energy, and chemistry are some of the areas with which students will be involved.

Solving a system of linear equations by graphing is a powerful way in which to solve the system and show the relationships among the variables at the same time. If a graphics calculator is available, this technique is quick and easy to use.

7.2 Solving Linear Systems by Substitution

Another technique for solving a linear system is substitution. Substitution is particularly useful when one or both linear equations of the system are expressed in slope-intercept form or when the coefficient of either unknown variable is 1.

Students must begin noting when a given technique for solving a linear system works well. Students gain mathematical power when they are capable of selecting from among different approaches to solutions.

7.3 Solving Linear Systems by Linear Combinations

The third technique for solving linear systems of equations is linear combinations. This method of solving is, in some ways, the most efficient of the three. Linear combinations may be the most powerful of the three because it readily handles linear systems of three or four variables.

Students can now select from among three methods to solve linear systems. It is necessary, therefore, that students develop an understanding of when to select a particular technique.

7.4 Problem Solving Using Linear Systems

In this lesson, students are asked to apply the three methods of solving linear systems to real-life problems. The key steps in solving verbal problems should be reviewed. Remind students that problem solving is a *process,* not an event. Encourage students to formulate problems for their classmates to solve. They will find this challenging and fun. It is an ideal way in which to reinforce the concepts of previous lessons.

7.5 Special Types of Linear Systems

Not every linear system of equations has exactly one solution. Students must recognize when linear systems have exactly one solution, no solution, or infinitely many solutions. Each type of linear system occurs in real-life situations.

7.6 Solving Systems of Linear Inequalities

Many types of real-life situations in areas such as finance, health care, and science can be represented by linear systems of inequalities. Solving such systems algebraically is not always adequate to make the solution understandable. Solving by graphing displays the relationships among the variables of the linear inequalities, giving greater insight into the solution.

7.7 Exploring Data: Linear Programming

Linear programming, a powerful tool for solving real-life problems, was developed during World War II to help in transporting men and supplies to the battlefields.

Students' work with solving systems of linear inequalities has prepared them for this type of exploration of data. Optimization problems occur in business and economics, engineering, and manufacturing. The skills students learn in this lesson can be readily applied.

Problem of the Day

Three women, each of them wearing a different color, were in the ski finals. Marie did not wear yellow. Eileen and the one in green had faster times than the one in yellow. The one in orange had the fastest time. One was named Jenny. What color did Jenny wear? Who was the fastest skier?

Yellow, Eileen

ORGANIZER

Warm-Up Exercises

1. Explain what it means for the ordered pair (a, b) to be a solution to the linear equation $Ax + By = C$. The statement $Ax + By = C$ is true when a and b are substituted for x and y, respectively.
2. Decide whether the ordered pairs are solutions to the linear equations.
   a. $3x - 8y = 4$, $(7, 2)$
   b. $-12x + 6y = 0$, $(5, 10)$
   c. $14x + 18y = -4$, $(10, -8)$
   a. not a solution, b. a solution, c. a solution
3. Sketch the graph of the linear equation.
   a. $3x - 2y = 12$
   b. $x + 3y = -9$

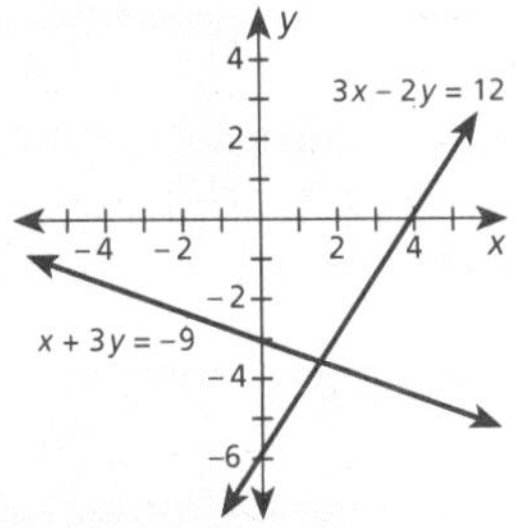

Lesson Resources

Teaching Tools
- Transparencies: 1, 2, 4
- Problem of the Day: 7.1
- Warm-up Exercises: 7.1
- Answer Masters: 7.1

Extra Practice: 7.1
Color Transparency: 36
Technology Handbook: p. 55

346

7.1 Solving Linear Systems by Graphing

What you should learn:

Goal 1 How to solve a system of linear equations by graphing

Goal 2 How to model a real-life situation using a system of linear equations

Why you should learn it:

You can model many real-life situations, such as multiple investments, with systems of linear equations. Graphing provides a visual model of situations and their solutions.

Goal 1 Graphing a Linear System

Until now, most of the problems in this text have involved just one equation in either one or two variables. Many real-life problems, however, involve two or more equations in two or more variables. For example, consider the following problem.

A total of \$9000 is invested in two funds paying 5% and 6% annual interest. The combined annual interest is \$510. How much of the \$9000 is invested in each fund?

Letting x be the amount in the 5% fund and y be the amount in the 6% fund, you can model the problem with *two* linear equations.

$$\begin{cases} x + y = 9000 & \text{Sum of two funds is \$9000.} \\ 0.05x + 0.06y = 510 & \text{Total interest is \$510.} \end{cases}$$

These two equations form a **system of linear equations** or simply a **linear system.** (We will solve this linear system in Example 3.)

A **solution** of a system of linear equations in two variables is an ordered pair (a, b) that satisfies each equation.

Example 1 *Checking a Solution Algebraically*

Check that $(2, -1)$ is a solution of the system.

$$\begin{cases} 3x + 2y = 4 & \text{Equation 1} \\ -x + 3y = -5 & \text{Equation 2} \end{cases}$$

Solution To check $(2, -1)$ as a solution, substitute 2 for x and -1 for y in each equation.

Equation 1: $3(2) + 2(-1) = 6 - 2 = 4$

Equation 2: $-(2) + 3(-1) = -2 - 3 = -5$ ■

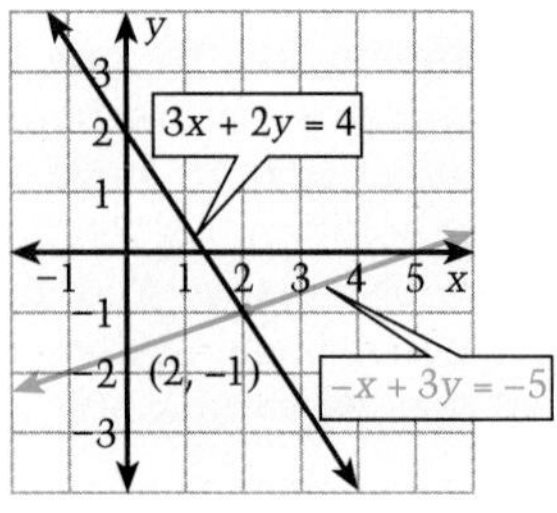

Graphic Check

Sketch a quick graph of each equation and find the point at which the two lines intersect. For instance, quick graphs at the left of the two linear equations of Example 1 suggest that they do indeed share $(2, -1)$ as a solution.

Example 2 *The Graph-and-Check Method*

Solve the system.

$$\begin{cases} x + y = -2 & \textit{Equation 1} \\ 2x - 3y = -9 & \textit{Equation 2} \end{cases}$$

Solution To sketch the two lines, first write the equations in slope-intercept form.

$$\begin{cases} y = -x - 2 & \textit{Slope: } -1, \textit{ y-intercept: } -2 \\ y = \frac{2}{3}x + 3 & \textit{Slope: } \frac{2}{3}, \textit{ y-intercept: } 3 \end{cases}$$

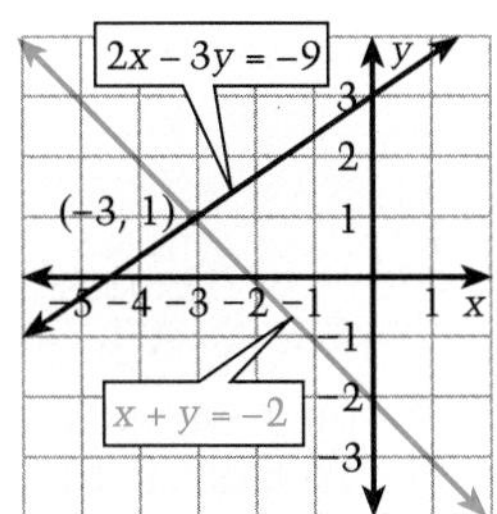

The two lines appear to intersect at $(-3, 1)$. To check this, substitute -3 for x and 1 for y in *each* equation in the original system.

Equation 1: $(-3) + (1) = -3 + 1 = -2$

Equation 2: $2(-3) - 3(1) = -6 - 3 = -9$ ■

Real Life Finance

Example 3 *Finding the Amounts Invested*

A total of \$9000 is invested in two funds paying 5% and 6% annual interest. The combined annual interest is \$510. How much of the \$9000 is invested in each fund?

Solution

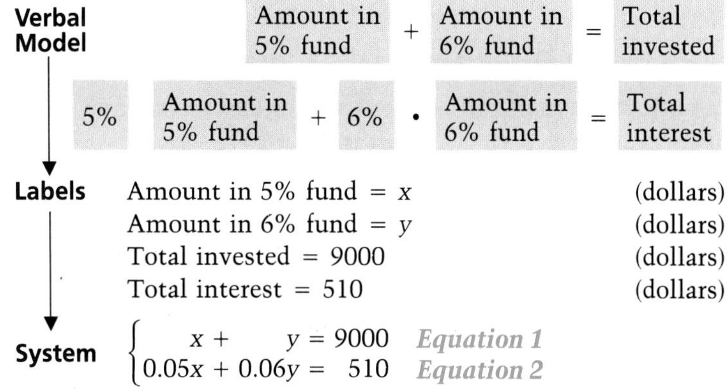

Verbal Model

[Amount in 5% fund] + [Amount in 6% fund] = [Total invested]

[5%] · [Amount in 5% fund] + [6%] · [Amount in 6% fund] = [Total interest]

Labels

Amount in 5% fund $= x$ (dollars)
Amount in 6% fund $= y$ (dollars)
Total invested $= 9000$ (dollars)
Total interest $= 510$ (dollars)

System

$$\begin{cases} x + y = 9000 & \textit{Equation 1} \\ 0.05x + 0.06y = 510 & \textit{Equation 2} \end{cases}$$

The first line has intercepts at (9000, 0) and (0, 9000). The second line has intercepts at (10,200, 0) and (0, 8500). The graphs of the two lines appear to intersect at (3000, 6000). This solution *checks* with the original statement of the problem.

$3000 + 6000 = 9000$ *Sum of funds is \$9000.*

$0.05(3000) + 0.06(6000) = 510$ *Total interest is \$510.* ■

y = 9000 − x
y = 8500 − 0.8333x
Dollars Invested at 6%
Dollars Invested at 5%

Graphic Check

LESSON Notes

Example 1

Emphasize that a solution to a linear system satisfies *both* linear equations in the system. Consequently, the ordered-pair solution lies on each line; or stated differently, the graphs of the lines in the system intersect at the ordered-pair solution.

Example 2

It is necessary to check solutions obtained by graphing. Frequently the ordered pairs obtained by graphing with paper and pencil are not as accurate as desired. Students must be encouraged to sketch their graphs with care to make them as precise as possible.

Example 3

There's a lot going on in this example. First, the verbal models have to be formulated. Take the time needed to be sure that students understand how the verbal models were created. Second, appropriate labels are needed. Third, the linear system is formed. Notice that the two linear equations are grouped together by the use of a set brace. Finally, a sketch of the linear equations of the system permits the identification of the ordered pair solution. Remember, the solution should always be checked in the original system.

Example 4

Observe that information is stored in a matrix format. Also note that the rate of increase in employees is obtained by subtracting the 1985 numbers from the 1990 numbers and then dividing by 5 years.

Extra Examples

Here are additional examples similar to **Examples 1–3.**

1. Decide whether the ordered pairs are solutions of the linear systems.

	Systems	Ordered pairs
a.	$\begin{cases} -4x + 3y = 12 \\ 3x - 4y = 5 \end{cases}$	(0, 4), (−9, −8)
b.	$\begin{cases} 15x + 3y = 30 \\ 5x + 6y = 10 \end{cases}$	(2, 0), (8, −5)
c.	$\begin{cases} x - 2y = -24 \\ 6x + 3y = 9 \end{cases}$	(−8, 16), (4, 11)

   a. (−9, −8) is a solution,
   b. (2, 0) is a solution,
   c. neither is a solution.

2. Solve the system.
$$\begin{cases} 4x - 7y = 21 \\ 11x + 3y = -9 \end{cases}$$
   The solution is (0, −3).

3. A total investment of \$11,000 was placed in two money market funds paying 6.6% and 7.5% annual interest. The combined yearly interest from the two accounts was \$766.50. How much of the \$11,000 was invested in each fund?
   The system is
$$\begin{cases} x + y = 11{,}000 \\ 0.066x + 0.075y = 766.50; \end{cases}$$
   the solution is \$6500 at 6.6%, \$4500 at 7.5%.

Check Understanding

1. How is a solution of a linear system similar to a solution of a linear equation? How is it different? See page 346.
2. An ordered pair (a, b) may be a solution to a linear equation in a linear system but not a solution of the linear system. Explain. To be a solution of a system of linear equations, an ordered pair must be a solution to both equations in the system.
3. Why is it necessary to check ordered pair solutions obtained by graphing? Graphing gives only an approximation of the solution. A check can indicate how accurate the approximation is and whether actual solutions are integer-valued.

Goal 2 Modeling a Real-Life Situation

Connections
Discrete Mathematics

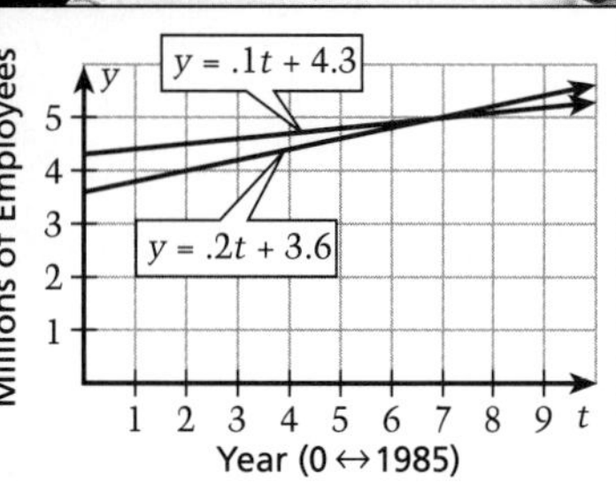

Example 4 *Health-Care Careers*

Use the matrix below to approximate the year in which the number of health-care workers in nonhospital jobs would equal the number in hospital jobs. (Assume that numbers in both types of employment fit linear models.)

Employees (in millions)	**1985**	**1990**
Hospital	4.3	4.8
Nonhospital	3.6	4.6

Solution Let t represent the year, with $t = 0$ corresponding to 1985. Let y represent the number of employees in millions. From 1985 to 1990, the number of workers in hospitals increased at a rate of 0.1 million workers per year. During that same time, the number of nonhospital health-care workers increased at a rate of 0.2 million workers per year. Using the slope-intercept form, you can write the following models.

$$\begin{cases} y = 0.1t + 4.3 & \textit{Health-care workers (hospital)} \\ y = 0.2t + 3.6 & \textit{Health-care workers (nonhospital)} \end{cases}$$

From the graph, you can see that the number of nonhospital workers would equal the number of hospital workers in 1991 or 1992. ***(Source: U.S. Bureau of Labor Statistics)***

■

Communicating about ALGEBRA

SHARING IDEAS about the Lesson

Use Models You should be able to use both the algebraic and the graphical models of a linear system.

A. Explain what it means to *solve* a system of linear equations. Find an ordered pair (a, b) that satisfies each equation.

B. Check whether (−3, 1) is a solution of the system. It is not.

$$\begin{cases} -4x + 5y = 17 & \textit{Equation 1} \\ 9x + y = -17 & \textit{Equation 2} \end{cases}$$

C. Use the graph-and-check method to solve the system.

$$\begin{cases} 2x - y = 8 & \textit{Equation 1} \\ -x + 2y = -1 & \textit{Equation 2} \end{cases}$$ (5, 2)

EXERCISES

Guided Practice

CRITICAL THINKING about the Lesson

In Exercises 1–3, use the system of linear equations at the right.

$$\begin{cases} 4x - 2y = 8 \\ -3x + 6y = 3 \end{cases}$$

1. Graph the system. See Additional Answers.
2. From the graph, which of the following appears to be a solution of the system: $(3, 2)$, $(3, -2)$, $(-3, 2)$, $(-3, -2)$?
3. Confirm your answer for Exercise 2 algebraically. Show, algebraically, why one of the pairs in Exercise 2 is not a solution of the system. See below.

In Exercises 4–6, graph and check to solve the system. See Additional Answers.

4. $\begin{cases} y = x + 1 \\ y = -x + 5 \end{cases}$ (2, 3)
5. $\begin{cases} y = -\frac{1}{2}x + 4 \\ y = 2x - 6 \end{cases}$ (4, 2)
6. $\begin{cases} 3x - y = 5 \\ -x + 2y = 0 \end{cases}$ (2, 1)

Independent Practice

In Exercises 7–12, decide whether the ordered pairs are solutions of the system.

7. $\begin{cases} 3x + 2y = 4 \\ -x + 3y = -5 \end{cases}$ $(2, -1)$, $(-3, 2)$ Only $(2, -1)$ is.
8. $\begin{cases} x + y = 6 \\ 2x - 5y = 10 \end{cases}$ $(3, 3)$, $(4, 2)$ Neither is.
9. $\begin{cases} x + y = -2 \\ 2x - 3y = -9 \end{cases}$ $(-2, 3)$, $(-3, 1)$ Only $(-3, 1)$ is.
10. $\begin{cases} 4x - 5y = 0 \\ 6x - 5y = 10 \end{cases}$ $(5, 4)$, $(10, 8)$ Only $(5, 4)$ is.
11. $\begin{cases} 2x + y = 4 \\ 4x + 3y = 9 \end{cases}$ $(\frac{5}{2}, 2)$, $(\frac{3}{2}, 1)$ Only $(\frac{3}{2}, 1)$ is.
12. $\begin{cases} 3x - 2y = -4 \\ -4x + 3y = 5 \end{cases}$ $(0, 2)$, $(-2, -1)$ Only $(-2, -1)$ is.

In Exercises 13–16, solve the linear system.

13. $\begin{cases} 2x + y = 4 \\ x - y = 2 \end{cases}$ (2, 0)
14. $\begin{cases} x + 3y = 2 \\ -x + 2y = 3 \end{cases}$ (−1, 1)
15. $\begin{cases} x - y = 0 \\ 3x - 2y = -1 \end{cases}$ (−1, −1)
16. $\begin{cases} 2x - y = 2 \\ 4x + 3y = 24 \end{cases}$ (3, 4)

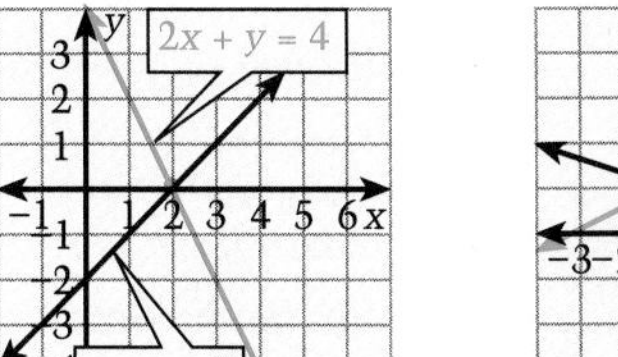

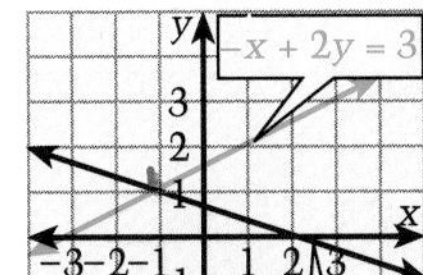

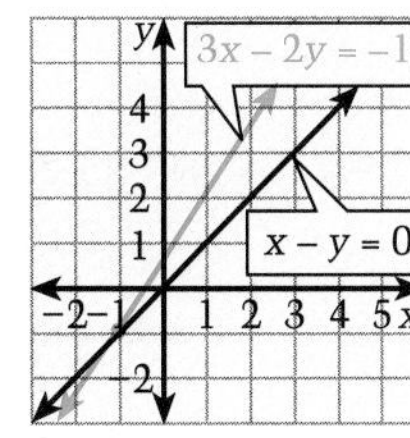

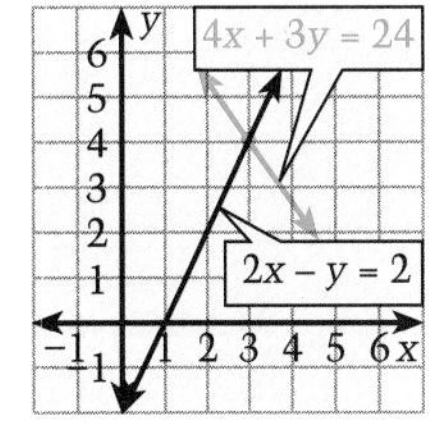

In Exercises 17–36, graph and check to solve the system. See Additional Answers.

17. $\begin{cases} y = -x + 3 \\ y = x + 1 \end{cases}$ (1, 2)
18. $\begin{cases} y = -x \\ y = x \end{cases}$ (0, 0)
19. $\begin{cases} y = 2x - 4 \\ y = -\frac{1}{2}x + 1 \end{cases}$ (2, 0)
20. $\begin{cases} y = -2x + 6 \\ y = 2x + 2 \end{cases}$ (1, 4)
21. $\begin{cases} 3x - 4y = 5 \\ x = 3 \end{cases}$ (3, 1)
22. $\begin{cases} 5x + 4y = 18 \\ y = 2 \end{cases}$ (2, 2)
23. $\begin{cases} 4x - 3y = -6 \\ 4x - 2y = 0 \end{cases}$ (3, 6)
24. $\begin{cases} 3x + 6y = 15 \\ -2x + 3y = -3 \end{cases}$ (3, 1)

3. $4(3) - 2(2) = 12 - 4 = 8$, $-3(3) + 6(2) = -9 + 12 = 3$; for example: $4(3) - 2(-2) = 12 + 4 = 16 \neq 8$, $-3(3) + 6(-2) = -9 - 12 = -21 \neq 3$

Communicating about ALGEBRA

EXTEND Communicating
WRITING

Ask students to write a few short paragraphs about solutions in algebra. That is, have them describe the various types of solutions that they have encountered in algebra to date. (These include solutions to linear equations, linear inequalities, and absolute-value equations.) Ask students to provide a definition of *solution* that represents each type of solution discussed.

EXERCISE Notes

ASSIGNMENT GUIDE

Basic/Average: Ex. 1–6, 7–29 odd, 41–47 odd, 53–57 odd, 61–62

Above Average: Ex. 1–6, 7–35 odd, 39–40, 41–59 odd, 61–62

Advanced: Ex. 1–6, 13–35 odd, 37–39, 49–59 odd, 61–62

Selected Answers
Exercises 1–6, 7–59 odd

More Difficult Exercises
Exercises 37–40, 61–62

Guided Practice

Ex. 1–6 COOPERATIVE LEARNING These provide an opportunity for an in-class partner activity. Student pairs can practice graphing and check solutions together. This allows them to be the "authority" on the information of the lesson.

Independent Practice

Ex. 7–12 Point out that the ordered pairs must be solutions to *both* equations to satisfy the system.

Ex. 17–36 Encourage students to use different graphing techniques from Chapters 4 and 5 for efficient graphing.

Ex. 37–38 Use Example 3 as a model for students to follow.

Ex. 39–40 Use Example 4 as a model for student assistance.

Answers

37. Amount in 5% fund = x (dollars)
Amount in $7\frac{1}{2}$% fund = y (dollars)
Total invested = 12,000 (dollars)
Total interest = 850 (dollars)
\$2000 at 5%; \$10,000 at $7\frac{1}{2}$%

39.

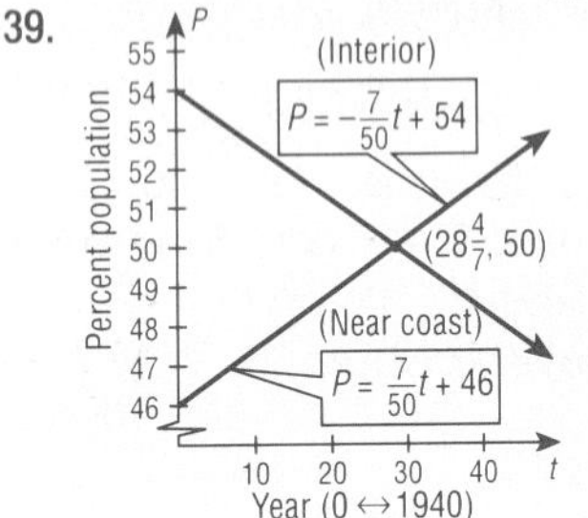

40.

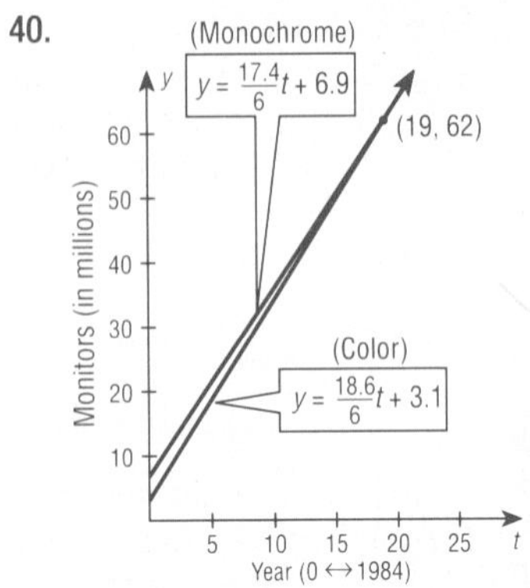

25. $\begin{cases} 2x - 3y = 8 \\ 4x + 3y = -2 \end{cases}$ **26.** $\begin{cases} -4x + 3y = 10 \\ 7x + y = 20 \end{cases}$ **27.** $\begin{cases} x - 8y = -40 \\ -5x + 8y = 8 \end{cases}$ **28.** $\begin{cases} -x + 5y = 6 \\ 2x - 6y = -4 \end{cases}$

29. $\begin{cases} 4x + 5y = 20 \\ \frac{5}{4}x + y = 4 \end{cases}$ **30.** $\begin{cases} -3x + 4y = -5 \\ 4x + 2y = -8 \end{cases}$ **31.** $\begin{cases} x + y = -1 \\ 2x - y = -8 \end{cases}$ **32.** $\begin{cases} -3x + y = 10 \\ -x + 2y = 0 \end{cases}$

33. $\begin{cases} x + 3y = 11 \\ -x + 3y = 7 \end{cases}$ **34.** $\begin{cases} 3x - y = -2 \\ x - 3y = 2 \end{cases}$ **35.** $\begin{cases} 2x + y = 5 \\ 3x - 5y = 1 \end{cases}$ **36.** $\begin{cases} x + 7y = -5 \\ 3x - 2y = 8 \end{cases}$

37. *Investment Problem* A total of \$12,000 is invested in two funds paying 5% and $7\frac{1}{2}$% annual interest. The combined annual interest is \$850. Copy the following verbal model. Assign labels. Then solve the resulting linear system to find how much of the \$12,000 is invested in each fund. **See margin.**

		Amount in 5% fund	+			Amount in $7\frac{1}{2}$% fund	=	Total invested
5%	•	Amount in 5% fund	+	$7\frac{1}{2}$%	•	Amount in $7\frac{1}{2}$% fund	=	Total interest

38. *Investment Problem* A total of \$25,000 is invested in two funds paying 5% and 6% annual interest. The combined annual interest is \$1400. How much of the \$25,000 is invested in each fund? **\$10,000 at 5%; \$15,000 at 6%**

39. *Living by the Coast* The matrix gives the percents of Americans who live in a county that is within 50 miles of a coastal shoreline (including the Great Lakes shoreline, but excluding Hawaii or Alaska) and who live further inland. For each location, use a linear model to represent the percent living there at time t, with $t = 0$ corresponding to 1940. Sketch the graphs of the resulting linear equations. From the graphs, estimate when the percent who lived near the coast equaled the percent who lived inland.
(Source: U.S. Bureau of Census) **1968 See margin for graphs.**

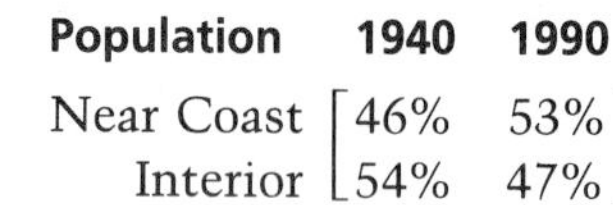

Population	1940	1990
Near Coast	46%	53%
Interior	54%	47%

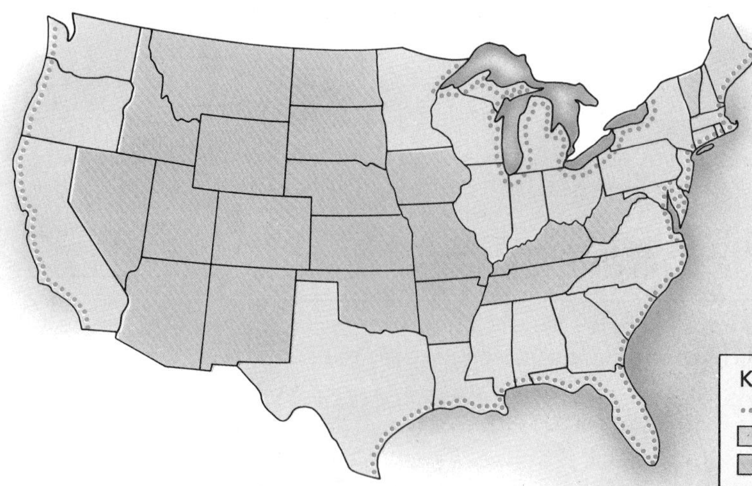

The coastal region of the 48 contiguous U.S. states contains 467,000 square miles, or 16% of the total land area. The population density in the coastal region is six times the population density in the interior.

25. (1, −2)
26. (2, 6)
27. (8, 6)
28. (4, 2)
29. (0, 4)
30. (−1, −2)
31. (−3, 2)
32. (−4, −2)
33. (2, 3)
34. (−1, −1)
35. (2, 1)
36. (2, −1)

40. *Color or Monochrome?* The matrix gives the number of monochrome and color computer monitors sold in the United States in 1984 and in 1990. For each type of monitor, use a linear model to represent the sales. (Let t represent the year with $t = 0$ corresponding to 1984.) Sketch the graphs of the resulting linear equations. From the graphs, estimate when the number of color monitors sold will equal the number of monochrome monitors sold. *(Source: DataPro, Inc.)* 2003 See margin for graphs.

Monitors (in millions)

	1984	1990
Monochrome	6.9	24.3
Color	3.1	21.7

Integrated Review

In Exercises 41–44, decide whether the ordered pair is a solution of the equation.

41. $y = 3x + 11$, $(3, 2)$ Is not

42. $y = 5x - 7$, $(3, 8)$ Is

43. $y = 8x - 4$, $(-1, 12)$ Is not

44. $y = 6x + 2$, $(2, -10)$ Is not

In Exercises 45–48, graph the equation. See Additional Answers.

45. $2x + y = 4$

46. $5x - 2y = 3$

47. $x + 4y = 8$

48. $-x - 3y = 12$

In Exercises 49–52, find an equation of the line passing through the two points.

49. $(-1, 3), (4, 8)$ $-x + y = 4$

50. $(2, 6), (5, 1)$ $5x + 3y = 28$

51. $(-2, -7), (5, 1)$ $-8x + 7y = -33$

52. $(3, -1), (2, 8)$ $9x + y = 26$

In Exercises 53–56, find the slope of the line.

53. $3x + 6y = 4$ $-\frac{1}{2}$

54. $7x - 4y = 10$ $\frac{7}{4}$

55. $2x - 8y = 12$ $\frac{1}{4}$

56. $-5x + 2y = 7$ $\frac{5}{2}$

In Exercises 57–60, solve for *y*. Then evaluate *y* for the given value of *x*.

57. $3x + y = 5;\ x = 2$ $y = -3x + 5,\ -1$

58. $-x + 4y = 7;\ x = 9$ $y = \frac{1}{4}x + \frac{7}{4},\ 4$

59. $-2x + 5y = 4;\ x = -3$ $y = \frac{2}{5}x + \frac{4}{5},\ -\frac{2}{5}$

60. $6x - 8y = 22;\ x = -1$ $y = \frac{3}{4}x - \frac{11}{4},\ -\frac{7}{2}$

Exploration and Extension

Red Meat or Poultry **In Exercises 61 and 62, use the following information.**

The table shows the average amounts of red meat and poultry eaten by Americans each year.

Year	1970	1975	1980	1985	1990
Red Meat	152 lb	139 lb	146 lb	141 lb	131 lb
Poultry	48 lb	50 lb	60 lb	68 lb	91 lb

61. red meat: $y = -\frac{3}{4}t + 150$; poultry : $y = 2t + 45$

61. Create scatter plots for the amounts of red meat and poultry eaten. Then approximate the best-fitting lines for the data. (Let t represent the year with $t = 0$ corresponding to 1970.) See Additional Answers for graphs.

62. Do you think that the average number of pounds of poultry eaten by Americans will ever equal the average number of pounds of red meat eaten? Explain. See below.

62. If the trends continue, they will be equal in about 2008.

Ex. 61–62 Students should plot the data in two separate colors. Draw one scatter plot for time and red meat, the other for time and poultry. The answer to Exercise 62 is found by asking for the first coordinate of the "solution" to the system.

Enrichment Activities

These activities require students to go beyond lesson goals.

TECHNOLOGY Refer to Example 4 on page 185 in Lesson 4.2. The graphs provided are related to the equations.

$$\begin{cases} y_1 = 29.91 + 1.303t \text{ (morning)} \\ y_2 = 32.27 - 1.216t \text{ (evening)} \end{cases}$$

Provide students with these equations and ask them to do the following.

1. Use a graphing calculator and the RANGE or WINDOW key to locate a similar viewing window, as provided in the example. x min: -2; x max: 10; x scl: 1; y min: 25; y max: 45; y scl: 5
2. Graph the equations. Use TRACE to approximate a solution to the system to two decimal places. Explain what this solution indicates. $x \approx 0.94$, $y \approx 31.11$. Therefore, by 1981 (< 1 year), evening sales = morning sales. At the time they were the same, there were about 31.11 million subscribers (the y-value).
3. In what year could you predict evening newspaper sales will no longer exist? For $y_2 = 0$, $32.27 - 1.216t = 0$, so $t = 26.5$ years. Therefore, at this rate, evening sales will be zero in the year 2006.

TECHNOLOGY Solving a linear system using a graphics calculator is quick and easy to do. Calculator technology makes the technique of solving by graphing as useful and powerful as the other two techniques (substitution and linear combination) to be discussed. By using the "zoom" feature of the calculator, ordered-pair solutions may be obtained with the degree of accuracy desired. It is still a good practice to check solutions algebraically, especially to determine whether observed intersections are integer-valued solutions.

Answers

1.

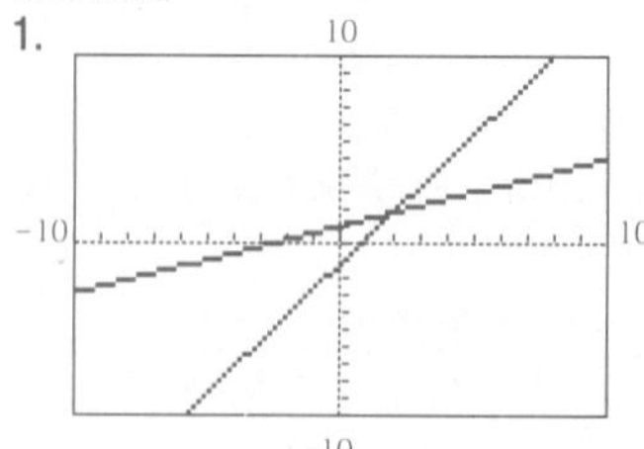

2.

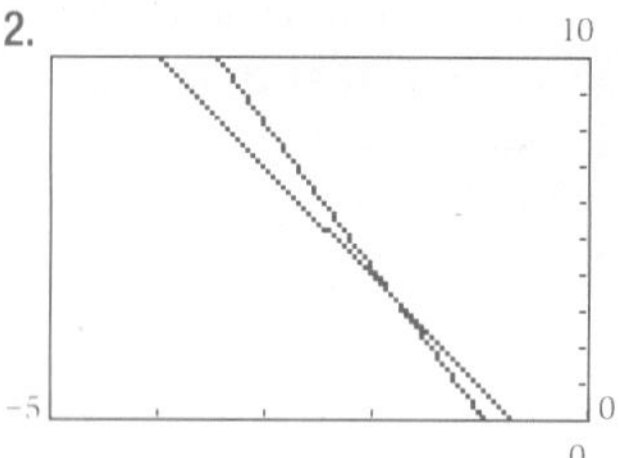

3.

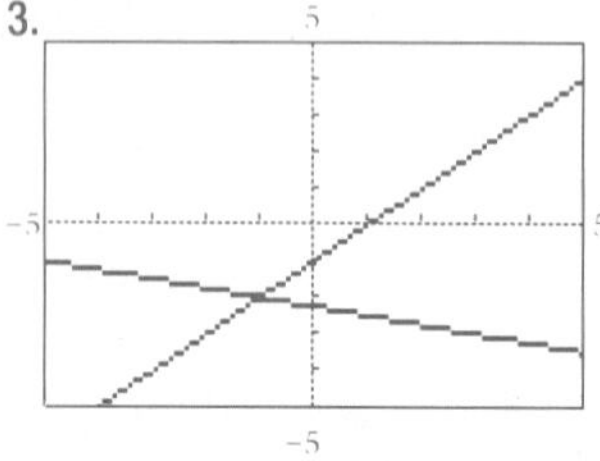

4.

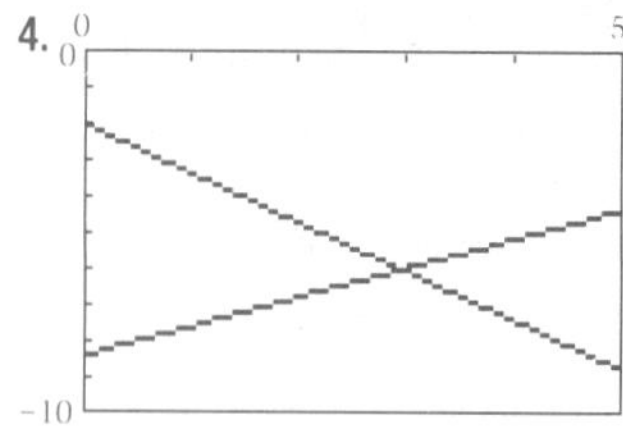

USING A GRAPHING CALCULATOR

A graphing calculator or computer can be used to graph a system of linear equations. Keystroke instructions for doing this on a TI-82, *Casio* fx-9700GE, *and Sharp* EL-9300C *are listed on page 741.*

Before you sketch the graph of a system of linear equations, you need to solve each linear equation for y in terms of x.

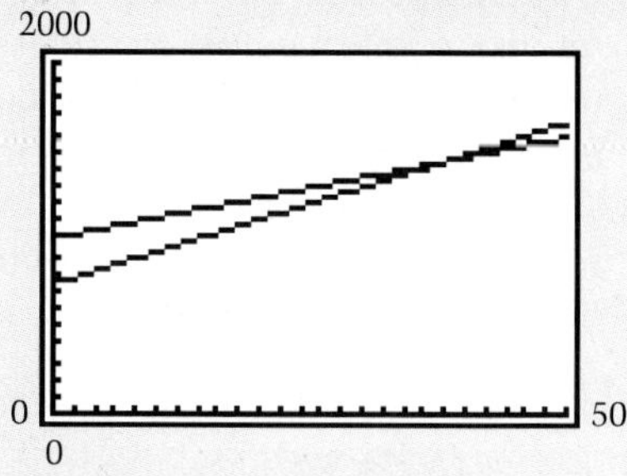

Graph of Linear System

Real Life Population

State Populations The 1970, 1980, and 1990 census population (in thousands) for Maine and New Hampshire are shown in the table. From these data, does it appear that the population of New Hampshire will exceed the population of Maine? If so, when?

Year	1970	1980	1990
Maine	994	1125	1228
New Hampshire	738	921	1109

One way to answer this question is to find the best-fitting line to describe each population.

$P = 11.7t + 998.7$ *Maine*

$P = 18.55t + 737.2$ *New Hampshire*

In these models, $t = 0$ represents the year 1970. When you graph these two lines on the same graphing calculator screen, you can see that they intersect near the year 2010. By substituting $t = 40$ for the year 2010, you can see that the population for Maine is predicted to be 1,467,000 and the population for New Hampshire is predicted to be 1,479,000.

Exercises

Use a graphing calculator to solve the linear system. (Use an appropriate viewing rectangle and check your result in each of the original equations.) See margin.

1. $\begin{cases} 2x - 5y = -6 \\ 3x - 2y = 2 \end{cases}$ (2, 2)

2. $\begin{cases} 3x + y = -2 \\ 4x + y = -4 \end{cases}$ (−2, 4)

3. $\begin{cases} -0.25x - y = 2.25 \\ -1.25x + 1.25y = -1.25 \end{cases}$ (−1, −2)

4. $\begin{cases} 0.8x + 0.6y = -1.2 \\ 1.25x - 1.5y = 12.75 \end{cases}$ (3, −6)

7.2 Solving Linear Systems by Substitution

What you should learn:

Goal 1 How to use substitution to solve a linear system

Goal 2 How to model a real-life situation using a system of linear equations

Why you should learn it:

You can model many real-life situations, such as cost comparisons, with linear systems. Some systems can be solved easily using substitution.

Goal 1 Substituting to Solve a Linear System

In this lesson, you will study an algebraic method—substitution—for solving a linear system. The basic steps in the method are as follows.

1. *Solve* one of the equations for one of its variables.
2. *Substitute* this expression into the other equation and solve for the other variable.
3. *Substitute* this value into the revised first equation and solve.
4. *Check* the solution pair in each of the original equations.

Example 1 *The Substitution Method*

Solve the linear system.

$$\begin{cases} -x + y = 1 & \text{Equation 1} \\ 2x + y = -2 & \text{Equation 2} \end{cases}$$

Solution Begin by solving Equation 1 for y.

$y = x + 1$ *Revised Equation 1*

Substitute this expression for y in Equation 2 and solve for x.

$$\begin{aligned} 2x + y &= -2 && \textit{Equation 2} \\ 2x + (x + 1) &= -2 && \textit{Substitute } x + 1 \textit{ for } y. \\ 3x + 1 &= -2 && \textit{Simplify.} \\ 3x &= -3 && \textit{Subtract 1 from both sides.} \\ x &= -1 && \textit{Solve for } x. \end{aligned}$$

You now know that the value of x in the solution is -1. To find the value of y, substitute the value of x into the revised Equation 1.

$$\begin{aligned} y &= x + 1 && \textit{Revised Equation 1} \\ y &= -1 + 1 && \textit{Substitute } -1 \textit{ for } x. \\ y &= 0 && \textit{Solve for } y. \end{aligned}$$

The solution is $(-1, 0)$. Check to see that it satisfies each of the original equations. ■

Problem of the Day

How can 8 quarts of water in an unmarked 8-quart pail be divided into two equal parts if the only other containers available are a 5-quart pail and a 3-quart pail, both with no markings?

8	5	3
8	0	0
3	5	0
3	2	3
6	2	0
6	0	2
1	5	2
1	4	3
4	4	0

or

8	5	3
8	0	0
5	0	3
5	3	0
2	3	3
2	5	1
7	0	1
7	1	0
4	1	3
4	4	0

ORGANIZER

Warm-Up Exercises

1. Solve each literal equation for the indicated variable.
   a. $2L + 2W = P$ W
   b. $\frac{1}{2}bh = A$ h
   c. $2x - y = 12$ y
   d. $4x + 8y = 16$ x
   a. $W = \frac{P-2L}{2}$, b. $h = \frac{2A}{b}$, c. $y = 2x - 12$, d. $y = -0.5x + 2$
2. Substitute $(2x + 1)$ for y and solve for x.
   a. $-3x + y = 4$
   b. $2y - x = 5$
   a. $x = -3$, b. $x = 1$

Lesson Resources

Teaching Tools
- Transparency: 4
- Problem of the Day: 7.2
- Warm-up Exercises: 7.2
- Answer Masters: 7.2

Extra Practice: 7.2

Color Transparency: 37

LESSON Notes

GOAL 1 MATH JOURNAL
The steps of the substitution method for solving linear systems are simpler to follow than it may appear! Demonstrate to students that the procedure is, indeed, straight-forward. Set a good example by labeling each step as you work. Encourage students to do the same. This is an effective way to help students memorize the sequence of steps, and it promotes writing in mathematics.

Ask students to write the steps in their math journals in a specially prepared section on solving linear systems.

Example 1

Remind students that it is important always to check the solution in the original equations.

Example 2

Ask students, "For which variable should we solve?" Discuss why it is better to solve for the variable x in the second equation. Remember, label each step as you work. (It may not be sufficient to say only what the next step is.) Labeling each step reinforces the sequence of steps for students.

Example 3

Point out to students that the same verbal model can be used to describe the total cost of either the conventional fixture or the retrofit fixture. (Note: A lumen is a unit of light related to the light produced by one candle.) Remind students to check the solution.

When using the substitution method, you will obtain the same solution (x, y) whether you solve for y first or x first. Thus, you should begin by solving for the variable that is more efficient to isolate. For instance, in the system

$$\begin{cases} 3x - 2y = 1 & \textit{Equation 1} \\ x + 4y = 3 & \textit{Equation 2} \end{cases}$$

it is easier to first solve the second equation for x. On the other hand, in the system

$$\begin{cases} 2x + y = 5 & \textit{Equation 1} \\ 3x - 2y = 11 & \textit{Equation 2} \end{cases}$$

it is easier to first solve the first equation for y.

If neither variable has a coefficient of 1 or -1, you can still use the substitution method. In such cases, however, the method discussed in Lesson 7.3 is usually more efficient.

Example 2 *The Substitution Method*

Solve the linear system.

$$\begin{cases} 2x + 2y = 3 & \textit{Equation 1} \\ x - 4y = -1 & \textit{Equation 2} \end{cases}$$

Solution Begin by solving Equation 2 for x.

$$x = 4y - 1 \qquad \textit{Revised Equation 2}$$

Substitute this expression for x in Equation 1 and solve for y.

$$\begin{aligned} 2x + 2y &= 3 && \textit{Equation 1} \\ 2(4y - 1) + 2y &= 3 && \textit{Substitute } 4y - 1 \textit{ for } x. \\ 10y - 2 &= 3 && \textit{Simplify.} \\ 10y &= 5 && \textit{Add 2 to both sides.} \\ y &= \tfrac{1}{2} && \textit{Solve for } y. \end{aligned}$$

The value of y in the solution is $\frac{1}{2}$. To find the value of x, substitute this value into the revised Equation 2.

$$\begin{aligned} x &= 4y - 1 && \textit{Revised Equation 2} \\ x &= 4\left(\tfrac{1}{2}\right) - 1 && \textit{Substitute } \tfrac{1}{2} \textit{ for } y. \\ x &= 1 && \textit{Solve for } x. \end{aligned}$$

The solution is $\left(1, \frac{1}{2}\right)$. Check to see that it satisfies each of the original equations. ■

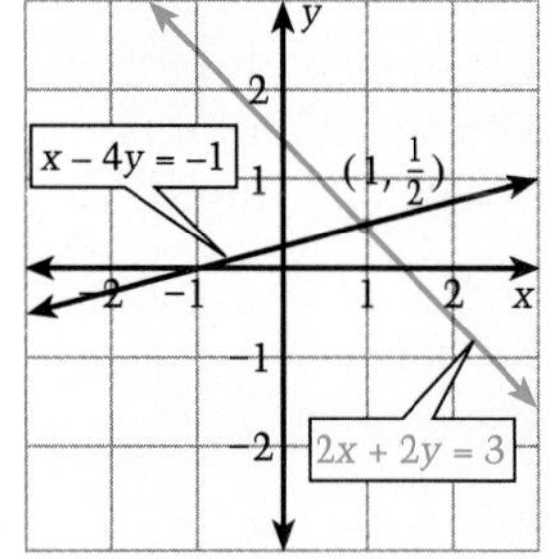

Graphic Check

When you use the substitution method, you can still use a graphic check. For instance, the graph at the left shows a graphic check for the linear system in Example 2.

Goal 2 Modeling a Real-Life Situation

A 4-bulb conventional fixture costs $50, plus $1 for each bulb. The fixture uses 175 watts.

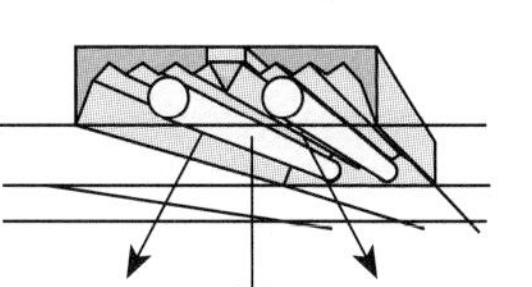

A 2-bulb retrofit fixture costs $120, plus $1 for each bulb. The fixture uses 70 watts.

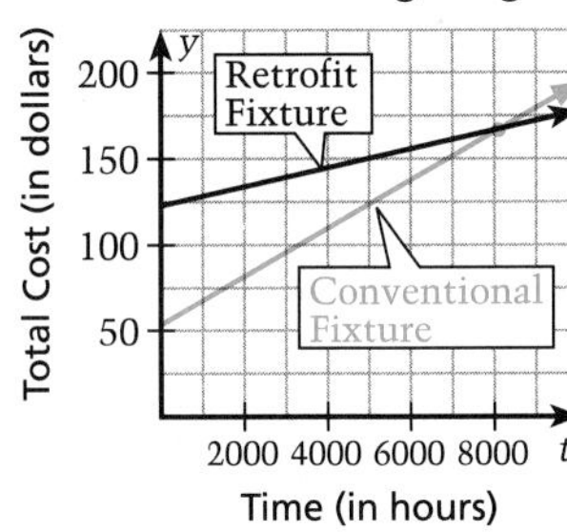

Example 3 *Comparing Costs*

You are building a convenience store and are considering two different types of fluorescent light fixtures: conventional and retrofit. A retrofit fixture is more expensive but uses less electricity. Electricity in your region costs $0.08 per kilowatt-hour. Show why a retrofit fixture can be considered more economical.

Solution The total cost of buying and operating each type of fixture is given by the following verbal model.

Total cost	=	Fixture & bulb cost	+	Cost per kilowatt-hour	•	Number of kilowatts	•	Number of hours

Let y represent the total cost (in dollars) of buying and operating a fixture for t hours. The verbal model gives two linear equations, one for a conventional fixture and one for a retrofit fixture.

$$\begin{cases} y = 54 + (0.08)(0.175)t & \text{Cost for conventional fixture} \\ y = 122 + (0.08)(0.070)t & \text{Cost for retrofit fixture} \end{cases}$$

The graphs suggest that at a certain time the retrofit fixture becomes more economical. Using substitution, the solution of the system is $t \approx 8100$ hours and $y \approx \$167$. This means it becomes more economical after working for about 8100 hours (about 1 year at 24 hours a day). ■

Communicating about ALGEBRA

Cooperative Learning

SHARING IDEAS about the Lesson

Work with a Partner and Extend Information Extend the ideas of Example 3 to the following.

A. The convenience store has 8 light fixtures. How much would the two types of fixtures cost to purchase and operate for 2 years? $2394.24, $1760.90

B. In 1990, there were 1.5 billion light fixtures in U.S. buildings. What are the potential energy savings per year for the U.S. by using more efficient light fixtures? 60%

Extra Examples

Here are additional examples similar to **Examples 1–2.**

Solve the linear system.

1. $\begin{cases} 7x + 3y = 43 \\ 2x - y = -10 \end{cases}$
2. $\begin{cases} -3x + 2y = 0 \\ x + 5y = 17 \end{cases}$
3. $\begin{cases} 2x + 9y = 40 \\ 4y = 2x + 38 \end{cases}$

1. (1, 12), 2. (2, 3), 3. (−7, 6)

Check Understanding

1. State the sequence of steps in the substitution method of solving a linear system. See page 353.
2. In the system $\begin{cases} x + y = 6 \\ 3x - y = -14 \end{cases}$ explain if it matters which variable is solved for first. Solve the system by finding x first, then by finding y. See the top of page 354; the ordered pair solution is (−2, 8).
3. Discuss: A solution to a linear system must be checked only when the system is solved graphically. False; a solution should always be checked in both equations of a linear system.

Communicating about ALGEBRA

EXTEND *Communicating*

How much could your school save in your mathematics classroom if more efficient lighting fixtures were adopted? (If your school uses efficient lighting fixtures, how much is it saving?) Estimate the savings for the entire school.

EXERCISE Notes

ASSIGNMENT GUIDE

Basic/Average: Ex. 1–4, 5–27 odd, 29, 35, 37–41 odd, 49, 53

Above Average: Ex. 1–4, 5–27 odd, 29–37 odd, 45–53 odd, 54

Advanced: Ex. 1–4, 9–27 odd, 30–36, 49, 51, 53, 54

Selected Answers
Exercises 1–4, 5–51 odd

Use **Mixed Review** as needed.

✪ More Difficult Exercises
Exercises 33, 34, 36, 54

Guided Practice

Answer

4. Algebraically: Substitute the solution into the original equations and simplify.
   Graphically: Graph the equations and see if their intersection agrees with the solution.

Independent Practice

▶ **Ex. 5–8** These exercises provide a visual meaning to the system that is being solved. Encourage students to use graphical checks whenever possible.

EXERCISES

Guided Practice

CRITICAL THINKING about the Lesson

In Exercises 1–4, use the following linear system.

$$\begin{cases} x + y = 9 & \textit{Equation 1} \\ 2x + 5y = 30 & \textit{Equation 2} \end{cases}$$

1. Solve Equation 1 for y. $y = -x + 9$
2. Substitute the expression for y into Equation 2 and solve for x. 5
3. Substitute the value of x into Equation 1 and solve for y. What is the solution of the linear system? 4, (5, 4)
4. Explain how you can check the solution algebraically and graphically. See margin.

Independent Practice

In Exercises 5–8, use substitution to solve the system. Use the graphs to check your solution.

5. $\begin{cases} x - y = 0 \\ x + y = 2 \end{cases}$ (1, 1)

6. $\begin{cases} x + y = 1 \\ 2x - y = 2 \end{cases}$ (1, 0)

7. $\begin{cases} 2x + y = 4 \\ -x + y = 1 \end{cases}$ (1, 2)

8. $\begin{cases} x - y = -5 \\ x + 2y = 4 \end{cases}$ (−2, 3)

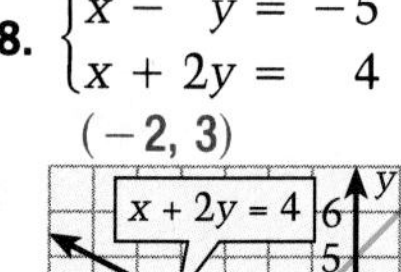

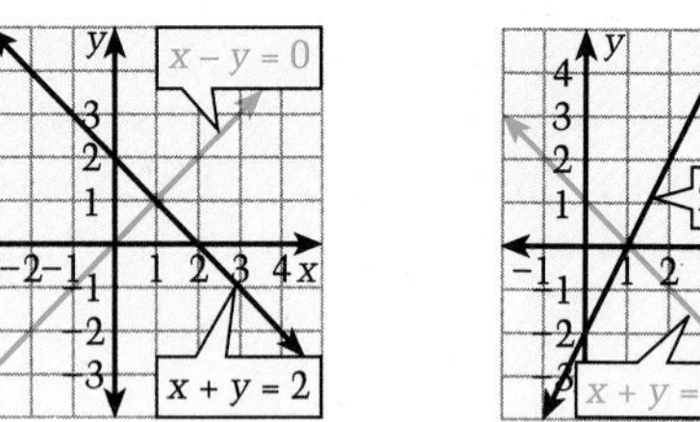

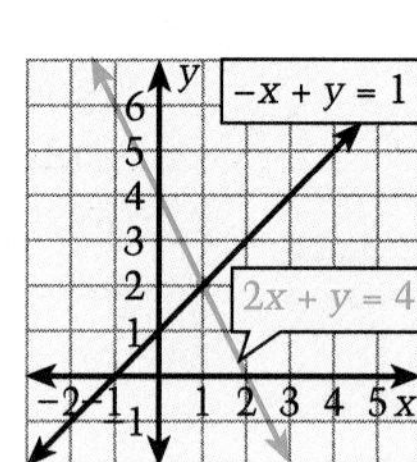

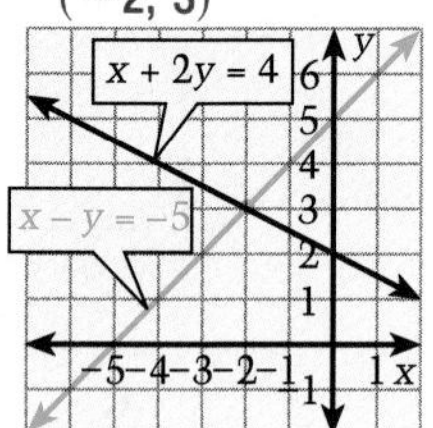

In Exercises 9–28, use substitution to solve the system.

9. $\begin{cases} y = x - 3 \\ 4x + y = 32 \end{cases}$ (7,4)

10. $\begin{cases} y = x + 4 \\ 3x + y = 16 \end{cases}$ (3,7)

11. $\begin{cases} 4x + 3y = 31 \\ y = 2x + 7 \end{cases}$ (1,9)

12. $\begin{cases} 4x + 5y = 48 \\ y = 3x + 2 \end{cases}$ (2,8)

13. $\begin{cases} 2x = 5 \\ x + y = 1 \end{cases}$ $(\frac{5}{2}, -\frac{3}{2})$

14. $\begin{cases} 3x - y = 0 \\ y = 6 \end{cases}$ (2,6)

15. $\begin{cases} x - y = 2 \\ 2x + y = 1 \end{cases}$ (1, −1)

16. $\begin{cases} x - 2y = -10 \\ 3x - y = 0 \end{cases}$ (2,6)

17. $\begin{cases} x - y = 0 \\ 2x + y = 0 \end{cases}$ (0,0)

18. $\begin{cases} x - 2y = 0 \\ 3x - y = 0 \end{cases}$ (0,0)

19. $\begin{cases} x - y = 0 \\ 5x - 3y = 10 \end{cases}$ (5,5)

20. $\begin{cases} x + 2y = 1 \\ 5x - 4y = -23 \end{cases}$ (−3,2)

21. $\begin{cases} 2x - y = -2 \\ 4x + y = 5 \end{cases}$ $(\frac{1}{2}, 3)$

22. $\begin{cases} -3x + 6y = 4 \\ 2x + y = 4 \end{cases}$ $(\frac{4}{3}, \frac{4}{3})$

23. $\begin{cases} \frac{1}{5}x + \frac{1}{2}y = 8 \\ x + y = 20 \end{cases}$ $(\frac{20}{3}, \frac{40}{3})$

24. $\begin{cases} \frac{1}{2}x + \frac{3}{4}y = 10 \\ \frac{3}{2}x - y = 4 \end{cases}$ (8,8)

25. $\begin{cases} -3x + y = 4 \\ -9x + 5y = 10 \end{cases}$ $(-\frac{5}{3}, -1)$

26. $\begin{cases} 5x + 3y = 11 \\ x - 5y = 5 \end{cases}$ $(\frac{5}{2}, -\frac{1}{2})$

27. $\begin{cases} x + 4y = 300 \\ x - 2y = 0 \end{cases}$ (100, 50)

28. $\begin{cases} 3x + y = 13 \\ 2x - 4y = 18 \end{cases}$ (5, −2)

29. *Buffet Dinner* The cost of a buffet dinner for a family of six was \$61.70 (\$11.95 adults, \$6.95 children). Assign labels to the verbal model below. Then solve the resulting linear system to find how many family members paid each price. **See margin.**

				Number of people over 12	+	Number of people 12 & under	=	Total number
Price per person over 12	•	Number of people over 12	+	Price per person 12 & under	•	Number of people 12 & under	=	Total price

30. *Football Tickets* You are selling tickets at a high school football game. Student tickets cost \$2 and general admission tickets cost \$3. You sell 1957 tickets and collect \$5035. How many of each type of ticket did you sell? **836 student, 1121 general**

31. *Comparing Costs* One car model costs \$12,000 and costs an average of \$0.10 per mile to maintain. Another car model costs \$14,000 and costs an average of \$0.08 per mile to maintain. If one of each model is driven the same number of miles, after how many miles would the total cost of one model be the same as the other? **100,000**

32. *Comparing Costs* A discount grocery store offers two types of memberships with annual membership fees of \$25 and \$100. With the \$25 membership, you get the "regular discount." With the \$100 membership, you get an additional 10% discount. For how many dollars worth of undiscounted groceries would the two memberships cost the same? **\$750**

Caught in the Doldrums **In Exercises 33 and 34, use the following information.**

On May 3, 1976, a team of adventurers, piloted by a Caroline Islands navigator, sailed from Hawaii on a traditional twin-hulled canoe called the *Hokule'a.* Sailing into northeast trade winds, the canoe maintained a course represented on the chart by $y = -\frac{3}{2}x - 215$. Sailing into southeast trade winds, the canoe maintained a course corresponding to $y = 7x + 1026$. At the point of intersection of these two paths, the canoe was caught in the "doldrums" and made little headway for 5 or 6 days.

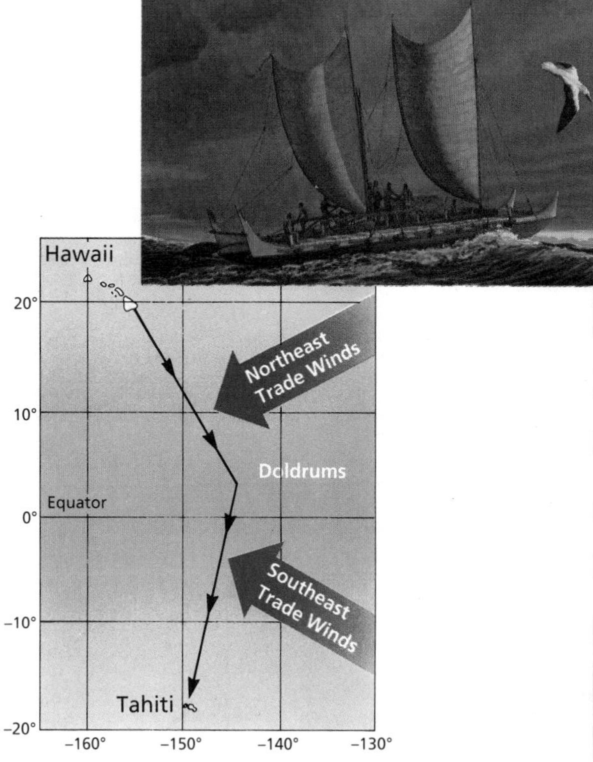

33. Find the coordinates of the doldrums. **(−146, 4)**

34. The equation of the line passing through Hawaii and Tahiti on the chart is $36x + 7y = -5483$. Find the coordinates of Hawaii and Tahiti. **(−156, 19), (−149, −17)**

Ex. 29–32 WRITING Remind students to interpret their solutions by providing a word sentence answer.

Ex. 33 TECHNOLOGY The coordinates of the doldrums are found by solving the system of equations. Students with graphics calculators may wish to use the pupil edition's art to help them set an appropriate viewing window to use a graphical method of finding the coordinates.

Answers

29. Number of over 12 = x
Number of 12 and under = y
Total number of people = 6
Total price = 61.70 (dollars)
4 of over 12, 2 of 12 and under

35. Meters uphill = x
Meters downhill = y
Rate uphill = 180 (meters per minute)
Rate downhill = 250 (meters per minute)
Total meters = 1557
Total time = 7.6 (minutes)
882 meters uphill, 675 meters downhill

Ex. 54 GEOMETRY You may wish to remind students that the area of a triangle is $\frac{1}{2}bh$.

Enrichment Activities

These activities require students to go beyond lesson goals.

1. Ask students to show graphically the different ways in which a linear equation and an absolute-value equation can intersect in a finite number of points. Line a has no points of intersection; line b has 1 point of intersection; line c has 2 points of intersection.

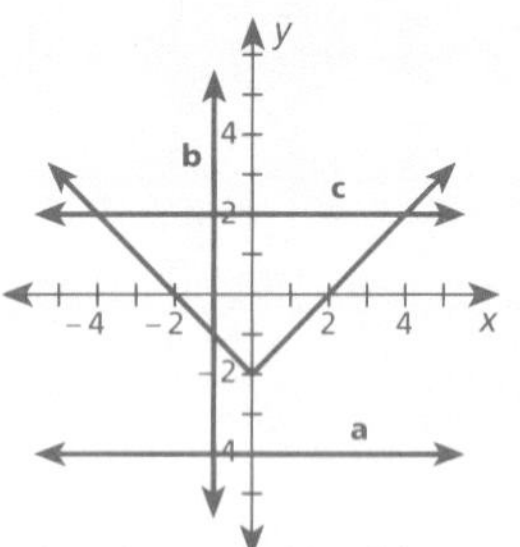

2. Solve by graphing:
$\begin{cases} y = |x + 1| \\ y = x + 2 \end{cases}$
Use quick-graph techniques and check your solutions.
$\left(-1\frac{1}{2}, \frac{1}{2}\right)$

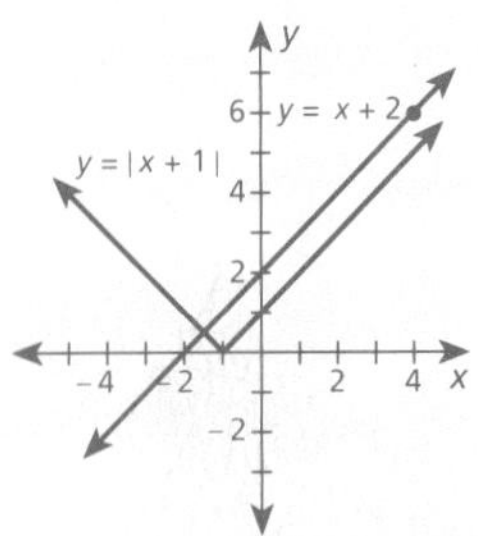

35. *Uphill and Downhill* Suppose you can run 250 meters per minute downhill and 180 meters per minute uphill. One day you run 1557 meters in 7.6 minutes always either uphill or downhill. Assign labels to the verbal model. Then solve the resulting linear system to find the number of meters you ran uphill and downhill.
See margin.

$$\text{Meters uphill} + \text{Meters downhill} = \text{Total meters}$$

$$\frac{\text{Meters uphill}}{\text{Rate uphill}} + \frac{\text{Meters downhill}}{\text{Rate downhill}} = \text{Total time}$$

36. *How Many of Each?* Your math teacher tells you that next week's test is worth 100 points and contains 29 problems. Each problem is worth either 5 points or 2 points. Because you are studying systems of linear equations, your teacher says that for extra credit you should be able to figure out how many problems of each value are on the test. How many of each value are there?
14 5-point problems, 15 2-point problems

Integrated Review

In Exercises 37–42, solve for *y*.

37. $2x + 3y = 8$ $y = -\frac{2}{3}x + \frac{8}{3}$

38. $x - 5y = 9$ $y = \frac{1}{5}x - \frac{9}{5}$

39. $9x - 2y = 24$ $y = \frac{9}{2}x - 12$

40. $2x - 2y = 19$ $y = x - \frac{19}{2}$

41. $3x - 4y = 14$ $y = \frac{3}{4}x - \frac{7}{2}$

42. $-6x + 8y = 36$ $y = \frac{3}{4}x + \frac{9}{2}$

In Exercises 43–48, simplify the expression.

43. $(2x + 3y) + (-2x + 5y)$ $8y$

44. $26 - (16y + 4) + (9y - x)$ $-x - 7y + 22$

45. $(x - 4y) - 14(3 - x) + y$ $15x - 3y - 42$

46. $(5y + 6x) - (4x - 3y)$ $2x + 8y$

47. $2x(4 + x) - (y - 5)$ $2x^2 + 8x - y + 5$

48. $9(y + 6) + (7x + y)(-3)$ $-21x + 6y + 54$

In Exercises 49–52, solve the linear system by graphing. See Additional Answers for graphs.

49. $\begin{cases} x - y = 1 \\ 2x + y = 8 \end{cases}$ (3, 2)

50. $\begin{cases} x - 2y = -10 \\ 4x + y = -4 \end{cases}$ (−2, 4)

51. $\begin{cases} x - 4y = 2 \\ 5x + y = -32 \end{cases}$ (−6, −2)

52. $\begin{cases} -x - y = 3 \\ 2x - y = 6 \end{cases}$ (1, −4)

Exploration and Extension

53. *Dimensions of a Rectangle* The area of the rectangle is y and the perimeter is $2y + 2$. What are the dimensions of the rectangle?
15 by 1

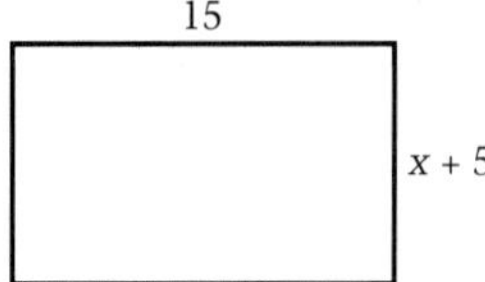

54. *Dimensions of a Triangle* The area of the triangle is $3x$ and the perimeter is $5x - 4$. Find the length of each side of the triangle.
5, 5, 6

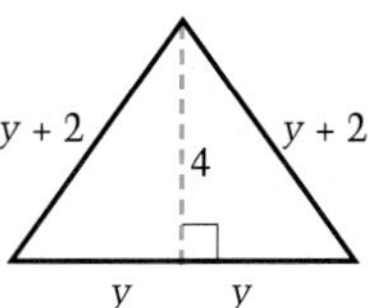

Mixed REVIEW

9. $x < 1\frac{3}{4}$ or $x > 2\frac{1}{4}$

1. Simplify $3(x - 2) + (-4)^2$. **(2.6)** $3x + 10$
2. Simplify $3(x - y) + 4x - y$. **(2.6)** $7x - 4y$
3. Evaluate $\frac{1}{3}(4 - x) + 2x$ when $x = \frac{1}{2}$. **(2.6)** $2\frac{1}{6}$
4. Evaluate $(|x| + 3)x + 2$ when $x = -3$. **(2.6)** -16
5. Solve $m(n + 1) + 2n = \frac{1}{2}$ for n. **(3.6)** $n = \frac{1 - 2m}{2m + 4}$
6. Solve $\frac{2}{3}r + 3s = 4$ for r. **(3.6)** $r = \frac{3}{2}(4 - 3s)$
7. Solve $3y + \frac{1}{2}(4 - y) = 3$. **(3.2)** $\frac{2}{5}$
8. Solve $|x + \frac{3}{4}| \le 3$. **(6.1)** $-3\frac{3}{4} \le x \le 2\frac{1}{4}$
9. Solve $|x - 2| - \frac{1}{4} > 0$. **(6.1)**
10. Solve $3(x - \frac{2}{3}) = \frac{1}{4}(2x + 3)$. **(3.2)** $\frac{11}{10}$
11. What is the product of $\frac{3}{4}$ and $\frac{8}{9}$? $\frac{2}{3}$
12. What is the quotient of $\frac{27}{5}$ and $\frac{9}{10}$? 6
13. Is 18 a solution for $\frac{7}{8}(x - 2) = 14$? **(1.2)** Yes
14. Is -3 a solution for $\frac{1}{4}|x| - x \ge 0$? **(2.1)** Yes

In Exercises 15–20, write an equation of the line that passes through the two points or that passes through the point and has the given slope. (5.2, 5.3)

18. $3x + 2y = -4$

15. $(-3, 2), (-4, 6)$ $4x + y = -10$
16. $(-4, 2), (-1, -3)$ $5x + 3y = -14$
17. $(-1, -1), (-3, -4)$ $-3x + 2y = 1$
18. $(0, -2)$, $m = -\frac{3}{2}$
19. $(4, -8)$, $m = \frac{1}{4}$ $-x + 4y = -36$
20. $(2, 5)$, $m = 0$ $y = 5$

Career Interview

Pediatrician

Bettye Kearse is a pediatrician. In her practice, Dr. Kearse oversees the proper growth and development of children, as well as provides acute care during emergencies.

Q: ***Did you always want to be a doctor?***

A: Well, I really thought I wanted to be a genetics researcher. While I found the results of research fascinating, I felt I needed a more people-oriented career. My father, grandfather, uncles, and husband are doctors, and found the practice of medicine personally fulfilling.

Q: ***What math classes did you take in high school?***

A: I took algebra, geometry, and trigonometry. I liked math, especially algebra. I hate to admit it, but I had no math courses in college.

Q: ***Does having a math background help you in your career?***

A: The main math application in pediatrics is in recommending medicinal dosages. Medicines for children are prescribed according to their weight. We use this rate: x mg per y kg of body mass.

Problem of the Day

Mr. Jordon, Mr. George, Mr. Ahearn, and Mr. Gomez are the principal, the custodian, the math teacher, and the librarian of a school, but not necessarily in that order. Mr. George is taller than the math teacher and the librarian. The principal lunches alone. Mr. Ahearn, the math teacher, and the librarian always lunch together. Mr. Gomez is older than the math teacher. Who has each position?

Mr. George: principal, Mr. Jordon: math teacher, Mr. Gomez: librarian, Mr. Ahearn: custodian

ORGANIZER

Warm-Up Exercises

1. By what constants would you multiply one or both expressions in each pair so that the sum of the products is zero?
   a. $-3y$, $12y$ b. $15x$, $3x$
   c. $4y$, $3y$ d. $-6x$, $8y$
   e. x, $-4x$

   a. multiply $-3y$ by 4;
   b. multiply $3x$ by -5;
   c. multiply $4y$ by 3, $3y$ by -4 or multiply $4y$ by -3, $3y$ by 4;
   d. multiply $-6x$ by 4, $8y$ by 3;
   e. multiply x by 4.
2. Solve for x.
   a. $3x = 18$
   b. $-2x = 32$
   c. $11x + 3(-4) = 10$
   d. $-2x - 2(\frac{1}{2}) = 7$

   a. 6, b. -16, c. 2, d. -4

Lesson Resources

Teaching Tools
- Problem of the Day: 7.3
- Warm-up Exercises: 7.3
- Answer Masters: 7.3

Extra Practice: 7.3
Color Transparency: 37
Applications Handbook: p. 3
Technology Handbook: p. 57

7.3 Solving Linear Systems by Linear Combinations

What you should learn:

Goal 1 How to use linear combinations to solve a linear system

Goal 2 How to model a real-life situation using a system of linear equations

Why you should learn it:

You can model many real-life situations, such as compositions of chemical mixtures and compounds, with linear systems. Some systems are solved most easily using linear combinations.

Goal 1 Using Linear Combinations

In this lesson, you will learn a third way to solve a linear system—the *linear combination method.* The basic steps in this method are as follows.

1. *Arrange* the equations with like terms in columns.
2. *Study* the coefficients of x (or y). *Multiply* one or both equations by an appropriate number to obtain new coefficients for x (or y) that are opposites.
3. *Add* the equations and solve for the remaining variable.
4. *Substitute* the value obtained in Step 3 into either of the original equations and solve for the other variable.
5. *Check* the solution in each of the original equations.

Example 1 *The Linear Combination Method*

Solve the linear system.

$$\begin{cases} 4x + 3y = 1 & \text{Equation 1} \\ 2x - 3y = 1 & \text{Equation 2} \end{cases}$$

Solution In this linear system, the coefficients for y are opposites. By adding the two equations, you obtain an equation that has only one variable, x.

$$\begin{array}{rcll} 4x + 3y &=& 1 & \text{Equation 1} \\ 2x - 3y &=& 1 & \text{Equation 2} \\ \hline 6x &=& 2 & \text{Sum of equations} \end{array}$$

Therefore, $x = \frac{1}{3}$. By substituting this value into Equation 1, you can solve for y.

$$\begin{array}{rcll} 4x + 3y &=& 1 & \text{Equation 1} \\ 4(\frac{1}{3}) + 3y &=& 1 & \text{Substitute } \frac{1}{3} \text{ for } x. \\ 3y &=& -\frac{1}{3} & \text{Subtract } \frac{4}{3} \text{ from both sides.} \\ y &=& -\frac{1}{9} & \text{Solve for } y. \end{array}$$

The solution is $\left(\frac{1}{3}, -\frac{1}{9}\right)$. Check this solution in each of the original equations. ■

To obtain coefficients (for one of the variables) that are opposites, you often need to multiply one or both of the equations by an appropriate number.

Example 2 *The Linear Combination Method*

Solve the linear system.

$$\begin{cases} 3x + 5y = 6 & \textit{Equation 1} \\ -4x + 2y = 5 & \textit{Equation 2} \end{cases}$$

Solution You can obtain coefficients of x that are opposite by multiplying the first equation by 4 and multiplying the second equation by 3. By adding the resulting equations, you obtain an equation that has only one variable, y.

$$\begin{array}{rcrl} 3x + 5y = 6 & \rightarrow & 12x + 20y = 24 & \textit{Multiply by 4.} \\ -4x + 2y = 5 & \rightarrow & -12x + 6y = 15 & \textit{Multiply by 3.} \\ \hline & & 26y = 39 & \textit{Sum of equations} \end{array}$$

Therefore, $y = \frac{3}{2}$. By substituting this value into Equation 2, you can solve for x.

$$\begin{array}{rl} -4x + 2y = 5 & \textit{Equation 2} \\ -4x + 2(\frac{3}{2}) = 5 & \textit{Substitute } \frac{3}{2} \textit{ for } y. \\ -4x + 3 = 5 & \textit{Simplify.} \\ -4x = 2 & \textit{Subtract 3 from both sides.} \\ x = -\frac{1}{2} & \textit{Solve for } x. \end{array}$$

The solution is $\left(-\frac{1}{2}, \frac{3}{2}\right)$. Check this solution in each of the original equations. ■

Guidelines for Solving a System of Linear Equations

To decide which method (graphing, substitution, or linear combinations) to use to solve a system of linear equations, consider the following.

1. The *graphing* method is useful for approximating a solution, checking a solution, and for providing a visual model of the problem.
2. To find an exact solution, use either *substitution* or *linear combinations.*
3. For linear systems in which one variable has a coefficient of 1 or −1, *substitution* may be more efficient.
4. In other cases, the *linear combinations* method is usually more efficient.

LESSON Notes

MATH JOURNAL The steps in the linear combinations method may seem complicated to some students. Carefully demonstrate and *label* each step of the sequence as you work examples. Ask students to write the steps in their math journals in a prepared section called solving linear systems.

Examples 1–2

Demonstrate how students should set up and manipulate linear systems when finding solutions. Label the steps and check the solution. Always model good system-solving techniques! Discuss the guidelines for solving linear systems. Ask students to explain why it is important to distinguish among the three methods of solutions.

Example 3

Ask students to express why the solution $G = 3$ and $S = 2$ is "bad news for the crown maker." The crown maker claimed the crown was all gold, so the results should have been $G = 5$, $S = 0$.

Extra Examples

Here are additional examples similar to **Examples 1–2.**

Solve the linear systems.

1. $\begin{cases} 8x + 2y = 16 \\ 5x - y = 28 \end{cases}$
2. $\begin{cases} 4x - 6y = -6 \\ 10x + 7y = -4 \end{cases}$

1. $(4, -8)$, 2. $(-\frac{3}{4}, \frac{1}{2})$

Check Understanding

1. List the steps in the linear combinations method of solving linear systems. See page 361.
2. For each linear system below, indicate which method of solving linear systems works best. Explain your reasoning.

   a. $\begin{cases} 3x + 8y = 13 \\ 4x - 4y = -12 \end{cases}$

   b. $\begin{cases} 7x + 8y = -24 \\ x - 5y = 15 \end{cases}$

   c. $\begin{cases} y = \frac{1}{3}x + 1 \\ y = -2x + \frac{9}{2} \end{cases}$

   a. linear combinations; no variable has a coefficient of 1

   b. substitution; x has a coefficient of 1

   c. substitution; y has a coefficient of 1
3. Solve the three systems in Exercise 2.

   a. $(-1, 2)$, b. $(0, -3)$, c. $\left(\frac{3}{2}, \frac{3}{2}\right)$

***Communicating* about ALGEBRA**

Your students may be interested in the following story about Archimedes (287–212 B.C.) who was a Greek mathematician famous for his ingenuity.

The king of Syracuse ordered a crown to be made of pure gold. When the crown was delivered, the king suspected that the crown was partly silver. He asked Archimedes to find a way of confirming his suspicion without destroying the crown. Later, at a public bath, Archimedes noticed that the water-level rose when he stepped into the bath. He suddenly realized how he could solve the problem, and became so excited that he ran home, stark naked, shouting "Eureka" ("I have found it!").

Goal 2 Modeling a Real-Life Situation

Ever since Archimedes jumped out of his tub, yelling "Eureka!", science students have known how to use Archimedes' fluid-displacement principle to relate the *weight* of an object to its *volume.* This relationship enabled Archimedes to prove that his king's "gold" crown was a forgery.

Example 3 *Exposing a Forgery*

A "gold" crown, suspected of containing some silver, was found to have a weight of 45 ounces, and a volume of 5 cubic inches. Gold weighs 11 ounces per cubic inch. Silver weighs 6 ounces per cubic inch. Is there silver mixed with the gold, and if so, how much?

Solution Using a system of equations, you can model the effect that silver would have on the volume and weight of the crown.

Verbal Model

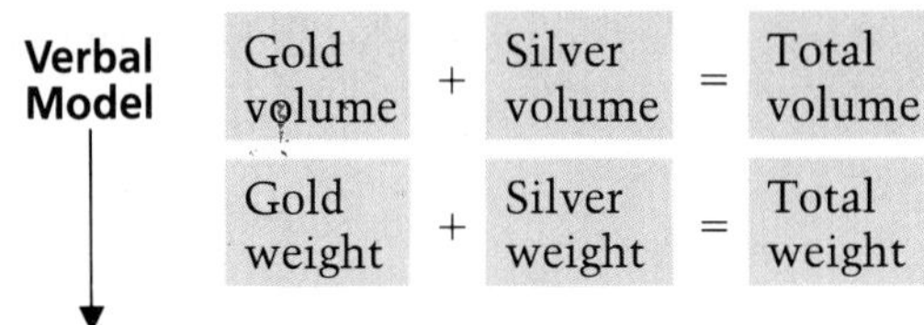

Gold volume + Silver volume = Total volume
Gold weight + Silver weight = Total weight

Labels

Volume of gold $= G$	(cubic inches)
Volume of silver $= S$	(cubic inches)
Total volume $= 5$	(cubic inches)
Weight of gold $= 11G$	(ounces)
Weight of silver $= 6S$	(ounces)
Total weight $= 45$	(ounces)

System

$$\begin{cases} G + S = 5 & \text{Total volume is 5 cubic inches.} \\ 11G + 6S = 45 & \text{Total weight is 45 ounces.} \end{cases}$$

You can solve the system by multiplying the first equation by -6 and adding the result to the second equation to give $5G = 15$. The solution is $G = 3$ in.^{3} and $S = 2$ in.^{3}. ■

Archimedes' Solution

1. Gold block balances crown.

2. Silver block balances crown.

3. Gold: low-water level.

4. Crown: mid-water level.

5. Silver: high-water level.

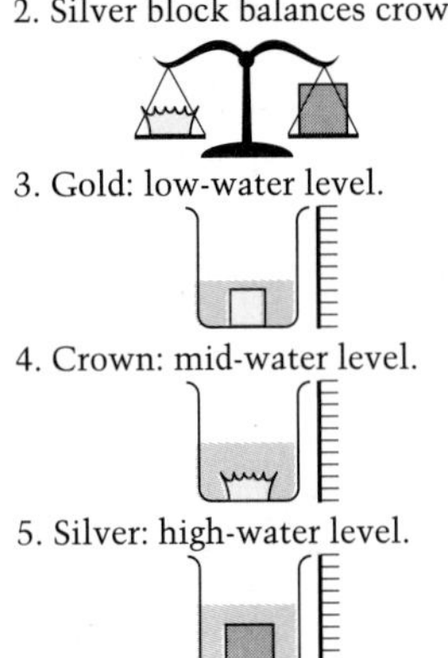

***Communicating* about ALGEBRA**

SHARING IDEAS about the Lesson

Relate Ideas How is the five-step diagram of Archimedes' solution related to the "system solution" shown in Example 3? See below.

Archimedes' solution compares the volume of the crown to the volumes of equal weights of gold and silver. The system solution compares the volume and weight of the crown with the volume and weight of a gold and silver compound.

EXERCISES

Guided Practice

CRITICAL THINKING about the Lesson

In Exercises 1–3, use linear combinations to solve the system.

1. $\begin{cases} 2x + 3y = 7 \\ -2x + 2y = -2 \end{cases}$ (2, 1)

2. $\begin{cases} 3x - 4y = 7 \\ 2x - y = 3 \end{cases}$ (1, −1)

3. $\begin{cases} 2x + 3y = 1 \\ 5x - 4y = 14 \end{cases}$ (2, −1)

In Exercises 4–6, solve the linear system by the method that you think is best. Explain why you chose that method. See margin.

4. $\begin{cases} 4x + 2y = 6 \\ -4x + 5y = 1 \end{cases}$ (1, 1)

5. $\begin{cases} y = -2x + 4 \\ y = \frac{1}{2}x - 1 \end{cases}$ (2, 0)

6. $\begin{cases} 4x + y = 9 \\ 7x - 8y = 6 \end{cases}$ (2, 1)

Independent Practice

In Exercises 7–10, use linear combinations to solve the system. Use the graphs to check your solution.

7. $\begin{cases} 2x + y = 4 \\ x - y = 2 \end{cases}$ (2, 0)

8. $\begin{cases} x + 3y = 2 \\ -x + 2y = 3 \end{cases}$ (−1, 1)

9. $\begin{cases} x - y = 0 \\ 3x - 2y = -1 \end{cases}$ (−1, −1)

10. 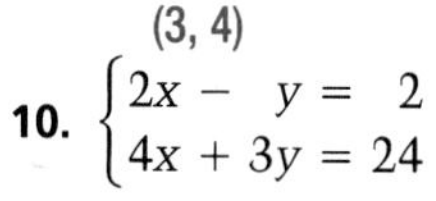 $\begin{cases} 2x - y = 2 \\ 4x + 3y = 24 \end{cases}$ (3, 4)

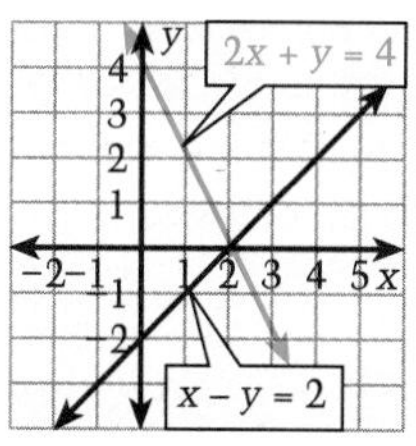

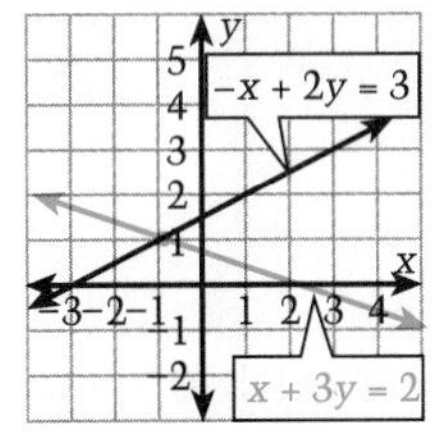

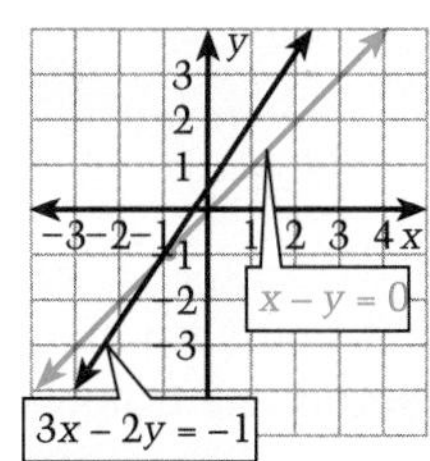

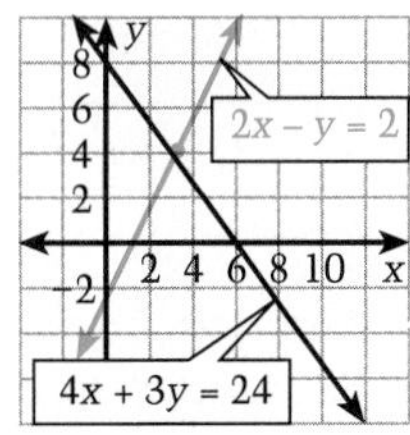

In Exercises 11–28, use linear combinations to solve the system.

11. $\begin{cases} x - y = 4 \\ x + y = 12 \end{cases}$ (8, 4)

12. $\begin{cases} -x + 2y = 12 \\ x + 6y = 20 \end{cases}$ (−4, 4)

13. $\begin{cases} 3x - 5y = 1 \\ 2x + 5y = 9 \end{cases}$ (2, 1)

14. $\begin{cases} x + 2y = 14 \\ x - 2y = 10 \end{cases}$ (12, 1)

15. $\begin{cases} x + 7y = 12 \\ 3x - 5y = 10 \end{cases}$ (5, 1)

16. $\begin{cases} 2x + 3y = 18 \\ 5x - y = 11 \end{cases}$

17. $\begin{cases} 5x + 2y = 7 \\ 3x - y = 13 \end{cases}$ (3, −4)

18. $\begin{cases} 4x + 3y = 8 \\ x - 2y = 13 \end{cases}$ (5, −4)

19. $\begin{cases} 3x + 2y = 10 \\ 2x + 5y = 3 \end{cases}$

20. $\begin{cases} 4x + 5y = 7 \\ 6x - 2y = -18 \end{cases}$ (−2, 3)

21. $\begin{cases} 6x - 5y = 3 \\ -12x + 8y = 5 \end{cases}$ $(-\frac{49}{12}, -\frac{11}{2})$

22. $\begin{cases} \frac{2}{3}x + \frac{1}{6}y = \frac{2}{3} \\ 3x - y = 12 \end{cases}$

23. $\begin{cases} 2u + v = 120 \\ u + 2v = 120 \end{cases}$ (40, 40)

24. $\begin{cases} 5u + 6v = 14 \\ 3u + 5v = 7 \end{cases}$ (4, −1)

25. $\begin{cases} 3a + 3b = 7 \\ 3a + 5b = 3 \end{cases}$ $(-1, -\frac{21}{13})$

26. $\begin{cases} 5a + 4b = 4 \\ 4a + 5b = \frac{31}{8} \end{cases}$ $(\frac{1}{2}, \frac{3}{8})$

27. $\begin{cases} 10m + 16n = 140 \\ 5m - 8n = 60 \end{cases}$ $(13, \frac{5}{8})$

28. $\begin{cases} 7m - 13n = 14 \\ 28m - 39n = 35 \end{cases}$

16. (3, 4) 19. (4, −1) 22. $(\frac{16}{7}, -\frac{36}{7})$ 25. $(\frac{13}{3}, -2)$

EXERCISE Notes

ASSIGNMENT GUIDE

Basic/Average: Ex. 1–6, 7–23 odd, 33, 39–47 odd, 51, 53

Above Average: Ex. 1–6, 7–27 odd, 29, 31, 33, 35–47 odd, 51, 53

Advanced: Ex. 1–6, 17–29 odd, 30, 32, 33–45 odd, 49–54

Selected Answers
Exercises 1–6, 7–49 odd

More Difficult Exercises
Exercises 30–34, 49–54

Guided Practice

Ex. 4–6 These exercises are designed to help students focus on the guidelines provided at the bottom of page 361. Choosing an efficient method is an important student focus.

Answers

4. Linear combination, because x-coefficients are opposites.
5. Substitution, because both equations are solved for y.
6. Substitution, because a y-coefficient is 1.

Independent Practice

Ex. 11–28 Encourage students to use a format similar to the one provided in Example 2 to ensure a neat and orderly presentation. You may wish to reserve some of these exercises to use as additional practice later.

Ex. 29 The focal length is the x-coordinate of the solution to the system of equations. It is also the x-intercept.

Ex. 30 Scout bees perform this dance by wiggling their bodies when they return to the hive. This dance tells worker bees where to go for food.

Ex. 31 Air speed is the speed the airplane would travel if there were no wind. Ground speed is the speed of the plane as measured from the ground.

Ex. 32 Students can use a verbal model similar to the model provided in Exercise 31. Be sure students understand that *upstream* means "against the current" and implies subtraction of the current speed from the water speed of the boat. Conversely, *downstream* means "traveling with the current" and implies addition of the speeds.

Ex. 33 This exercise is similar to Example 3 in the lesson.

Ex. 41–48 The reason for the method chosen is more important than the actual solution to the system.

29. *Focal Length of a Camera* When parallel rays of light pass through a convex lens, they are bent inward and meet at a *focus.* The distance from the center of the lens to the focus is called the *focal length.* The equations of the lines containing the two bent light rays in the camera are

$$\begin{cases} x + 3y = 1 \\ -x + 3y = -1 \end{cases}$$

$x + 3y = 1$, 1 inch

where x and y are measured in inches. Which of these equations is the upper ray? What is the focal length?

30. *Waggle Dance* Two bees have each found a food source. The first bee's path back to the hive follows the line $y = \frac{4}{5}x$. The second bee's path follows the line $y = -2x + 3$. Their paths cross at the hive. Find the coordinates of the hive. $(\frac{15}{14}, \frac{6}{7})$

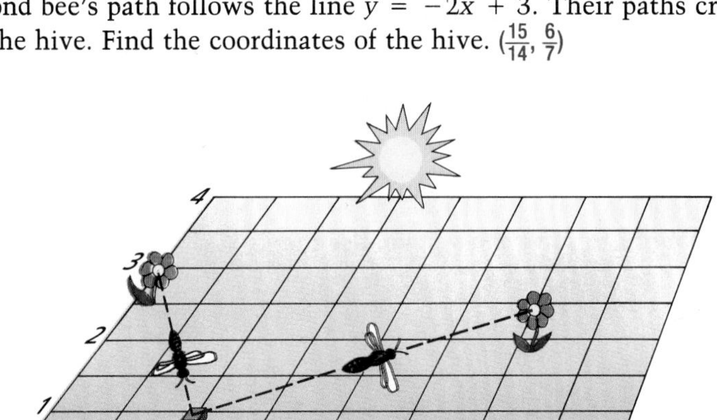

The waggle dance of a scout bee communicates the direction and distance of a food source. When the direction of the food source is the same as that of the sun, the waggle dance points straight up on the vertical honeycomb. When the source is 60° to the left of the sun, the waggle dance is oriented 60° to the left of vertical.

31. *Airplane Speed* It took 3 hours for an airplane, flying against a head wind, to travel the 900 miles from Birmingham, Alabama, to Duluth, Minnesota. On this portion of the trip, the "ground speed" of the plane was 300 miles per hour. On the return trip, the flight took only 2 hours, with a ground speed of 450 miles per hour. During both flights, the speed and direction of the wind and the airspeed of the plane were constant. Assign labels to the verbal model. Then solve the resulting equation to find the speed of the wind and the airspeed of the plane.

Airspeed − Wind speed = Ground speed against wind

Airspeed + Wind speed = Ground speed with wind

32. *Boat Speed* A motorboat went 10 miles upstream in 1 hour. The return trip took only 30 minutes. Assume that the "water speed" and the current speed were constant during both parts of the trip. Find the water speed of the boat and the current speed. 15 mph, 5 mph

33. *Gold Bracelet* A bracelet that is supposed to be 18-karat gold weighs 238 grams. The volume of the bracelet is 15 cubic centimeters. The bracelet is made of gold and copper. Gold weighs 19.3 grams per cubic centimeter, and copper weighs 9 grams per cubic centimeter. Is the bracelet really 18-karat gold? No

34. *Acid Mixture* You are making an acid solution in science class. Five liters of the 40% acid solution are obtained by mixing a 20% solution with a 60% solution. How much of each must you use? 2.5 liters of each

Jewelry is rarely made of 24-karat gold (which means 100% gold) because gold is a soft metal. Other metals such as copper are mixed with the gold to obtain greater strength. An 18-karat gold piece contains three-fourths gold (by weight) and one-fourth of some other metal.

Integrated Review

In Exercises 35–38, match the linear system with its correct graph. Then use linear combinations to solve the system.

35. $\begin{cases} 9x - 3y = -1 \\ 3x + 6y = -5 \end{cases}$ **36.** $\begin{cases} 5x + 3y = 18 \\ 2x - 7y = -1 \end{cases}$ **37.** $\begin{cases} x - 2y = 5 \\ 6x + 2y = 7 \end{cases}$ **38.** $\begin{cases} -6x + 4y = -12 \\ 3x - 4y = 8 \end{cases}$

a. 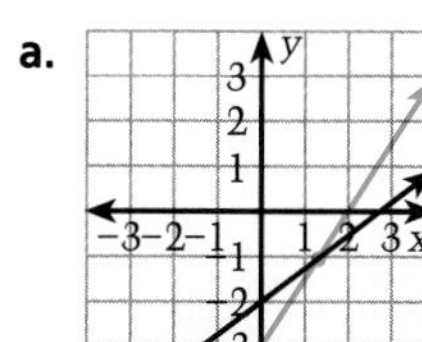

b. 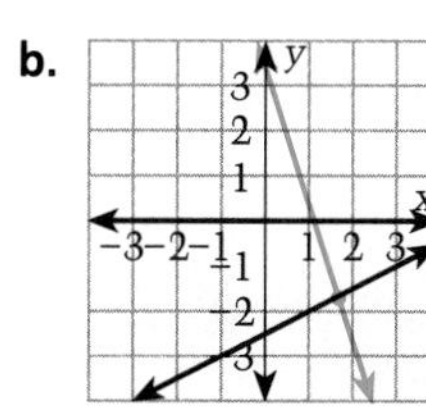

c. 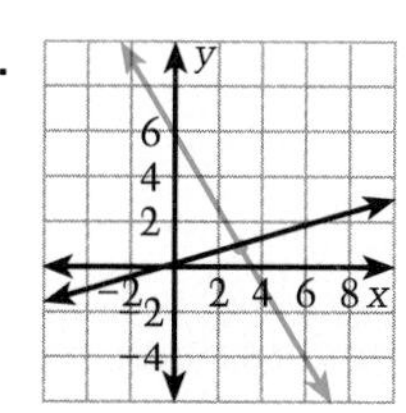

d. 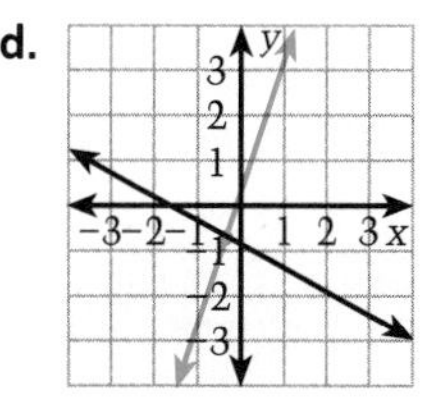

39. *College Entrance Exam Sample* Given the system of equations $3x + 2y = 4$ and $6x - 3y = 6$, what does y equal?

a. 14 **b.** $\frac{14}{6}$ **c.** 2 **d.** $\frac{11}{7}$ **e.** $\frac{2}{7}$

40. Check the solution found for the linear system in Exercise 39 by graphing. See Additional Answers.

35. d, $\left(-\frac{1}{3}, -\frac{2}{3}\right)$
36. c, (3, 1)
37. b, $\left(\frac{12}{7}, -\frac{23}{14}\right)$
38. a, $\left(\frac{4}{3}, -1\right)$

In Exercises 41–48, solve the linear system by the method you think is best. Explain why you chose that method. See margin.

41. $\begin{cases} 2x - 4y = 8 \\ x + 2y = 5 \end{cases}$ **42.** $\begin{cases} 2x - 3y = 6 \\ 4x + 9y = 8 \end{cases}$ **43.** $\begin{cases} 3x - 3y = 9 \\ 2x + 3y = 12 \end{cases}$ **44.** $\begin{cases} 5x + 2y = 10 \\ 3x - 6y = 12 \end{cases}$

45. $\begin{cases} 5x - y = 7 \\ x + 4y = 16 \end{cases}$ **46.** $\begin{cases} 6x + y = 8 \\ -4x + 2y = 9 \end{cases}$ **47.** $\begin{cases} -4x + 5y = 20 \\ 2x + 3y = 8 \end{cases}$ **48.** $\begin{cases} 7x - 5y = 18 \\ 8x + y = 24 \end{cases}$

Answers

31. Air speed = x (mph)
Wind speed = y (mph)
Ground speed against wind = 300 (mph)
Ground speed with wind = 450 (mph)
75 mph, 375 mph

41. $\left(\frac{9}{2}, \frac{1}{4}\right)$; substitution, because an x-coefficient is 1

42. $\left(\frac{13}{5}, -\frac{4}{15}\right)$; linear combination, because no coefficient is 1

43. $\left(\frac{21}{5}, \frac{6}{5}\right)$; linear combination, because no coefficient is 1

44. $\left(\frac{7}{3}, -\frac{5}{6}\right)$; linear combination, because no coefficient is 1

45. $\left(\frac{44}{21}, \frac{73}{21}\right)$; substitution, because an x-coefficient is 1

46. $\left(\frac{7}{16}, \frac{43}{8}\right)$; substitution, because a y-coefficient is 1

47. $\left(-\frac{10}{11}, \frac{36}{11}\right)$; linear combination, because no coefficient is 1

48. $\left(\frac{138}{47}, \frac{24}{47}\right)$; substitution, because a y-coefficient is 1

▶ **Ex. 51–52** These exercises check students' understanding of a solution to a system of equations being required to satisfy *both* equations.

Enrichment Activities

These activities require students to go beyond lesson goals.

Provide students with practice at solving systems "out of alignment" such as these.

1. $\begin{cases} 3x - 2y = 1, \\ 4y = 7 + 3x \end{cases}$
   (3, 4)
2. $\begin{cases} y - x = 4, \\ x - 2y = 7 \end{cases}$
   (−15, −11)
3. $\begin{cases} 2x + 5y = 6.35 \\ 4x = 13.20 - 12y \end{cases}$
   (2.55, 0.25)
4. Solve this system. What is the most efficient way to solve it?
   $\begin{cases} \frac{1}{3}x + \frac{2}{5}y = \frac{2}{3} \\ \frac{3}{4}x + 3y = \frac{4}{5} \end{cases}$
   $(\frac{12}{5}, -\frac{1}{3})$
   Multiply the first equation by 15 (the LCD of 3 and 5) and the second equation by 20 (the LCD of 4 and 5) before proceeding.

Winter and Summer Solstice **In Exercises 49 and 50, use the following information.**

Solstice is the time each year that the sun is at its northernmost or southernmost position. In the Northern Hemisphere, the northernmost position occurs on the summer solstice (the day with the most sunlight). The southernmost position occurs on the winter solstice (the day with the least sunlight).

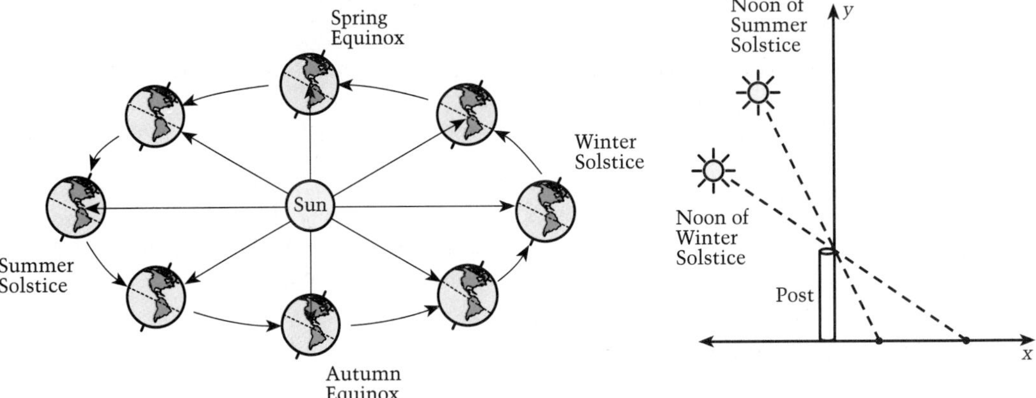

49. Match each solstice in the diagram on the right above with one of the equations in the system.

$\begin{cases} 2x + 3y = 30 \leftarrow \text{Winter} \\ 2x + y = 10 \leftarrow \text{Summer} \end{cases}$

(0, 10), the point of intersection, is the top of the post.

Solve this system and interpret your solution in the diagram. The units on the axes are measured in feet.

50. In the diagram, find the distance between the two points on the x-axis. Use the diagram to explain how ancient people could create calendars. 10 feet; see Additional Answers.

Exploration and Extension

In Exercises 51 and 52, complete the system with the constants that will ensure the given solution.

	System	Solution		System	Solution
51.	$\begin{cases} \frac{1}{2}x + 6y = \boxed{?} \\ 4x - \frac{1}{3}y = \boxed{?} \end{cases}$ 76, 28	(8, 12)	**52.**	$\begin{cases} -5x + \frac{6}{5}y = \boxed{?} \\ 3x - \frac{2}{3}y = \boxed{?} \end{cases}$ $-\frac{12}{5}, \frac{22}{15}$	$(\frac{3}{5}, \frac{1}{2})$

In Exercises 53 and 54, find the equations of the lines that pass through the given points. Then solve the resulting linear system.

53. Equation 1: (−2, −4), (6, 2)
Equation 2: (4, −2), (−1, 5)
$3x - 4y = 10, 7x + 5y = 18; (\frac{122}{43}, -\frac{16}{43})$

54. Equation 1: (−2, 3), (5, 2)
Equation 2: (−1, 3), (2, 4)
$x + 7y = 19, -x + 3y = 10; (-1.3, 2.9)$

7.4 Problem Solving Using Linear Systems

What you should learn:

Goal 1 How to write and use a linear system as a real-life model

Why you should learn it:

Many real-life situations, such as octane mixtures for gasoline, can be modeled with linear systems.

Goal 1 Using Linear Systems as Models

You have already modeled several situations with linear systems. In this lesson, you will practice using the linear-systems model to solve additional problems.

Remember that the key steps are writing a *verbal model*, assigning *labels*, writing an *algebraic model*, *solving* the algebraic model, and *answering* the question of the original problem. Remember that you have *three* methods for solving the system: graphing, substitution, and linear combinations.

Real Life
Petroleum Industry

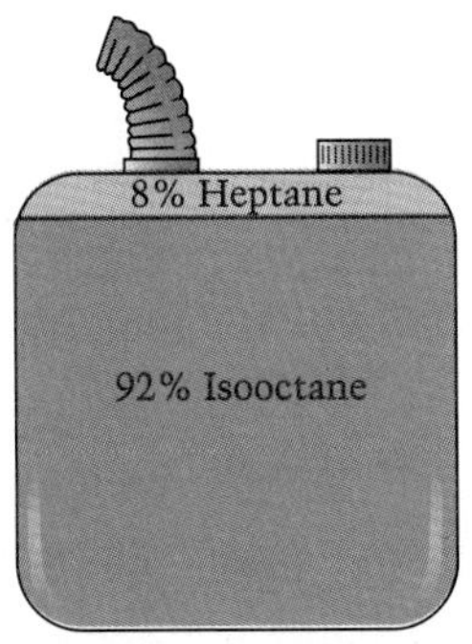

Octane rating is a measure of a gasoline's resistance to engine knock. Isooctane is 100-octane (knock resistant) and heptane is 0-octane (knock prone). A 92-octane gasoline has the same resistance to knocking as a mixture that is 92% isooctane and 8% heptane.

Example 1 *Octane Ratings*

You work in a gasoline refinery. One refining system produces 90-octane gasoline and another produces 96-octane gasoline. How much of each of these gasolines should you mix to produce 600 gallons of 92-octane gasoline?

Solution To begin, assume that 600 gallons of 92-octane gasoline contains $(0.92)(600) = 552$ gallons of isooctane.

Verbal Model

Volume of 90-octane + Volume of 96-octane = Volume of 92-octane

Isooctane in 90-octane + Isooctane in 96-octane = Isooctane in 92-octane

Labels

Volume of 90-octane gasoline $= x$ (gallons)
Volume of 96-octane gasoline $= y$ (gallons)
Volume of 92-octane gasoline $= 600$ (gallons)
Isooctane in 90-octane gasoline $= 0.90x$ (gallons)
Isooctane in 96-octane gasoline $= 0.96y$ (gallons)
Isooctane in 92-octane gasoline $= 552$ (gallons)

System

$$\begin{cases} x + y = 600 \\ 0.9x + 0.96y = 552 \end{cases}$$

Volume is 600 gallons.
Isooctane is 552 gallons.

Using the linear combinations method, you can multiply the first equation by -0.9 and add the result to the second equation. The solution is 400 gallons of 90-octane and 200 gallons of 96-octane. ■

Problem of the Day

Of the people who got off a train: 31 wore gloves, 20 carried an umbrella, 4 carried an umbrella and wore gloves but wore no hat, 9 carried an umbrella and wore a hat but wore no gloves, 12 wore gloves and a hat but carried no umbrella, 7 wore a hat but wore no gloves and carried no umbrella, 2 carried an umbrella but wore no gloves or hat, 3 wore no hat or gloves and carried no umbrella. How many people got off the train?

52

ORGANIZER

Warm-Up Exercises

1. Solve this system by graphing. $\begin{cases} y = 2x - 1 \\ y = \frac{x}{4} + \frac{5}{2} \end{cases}$
2. Solve this system by substitution. $\begin{cases} 7x + 2y = -11 \\ x - 5y = 9 \end{cases}$
3. Solve this system by linear combinations. $\begin{cases} 2x + 5y = 32 \\ 3x - 7y = -39 \end{cases}$

1. (2, 3), 2. (−1, −2), 3. (1, 6)

Lesson Resources

Teaching Tools
- Transparency: 3
- Problem of the Day: 7.4
- Warm-up Exercises: 7.4
- Answer Masters: 7.4

Extra Practice: 7.4
Applications Handbook: p. 54

LESSON Notes

Example 1

MOTIVATING THE APPROACH The mixture problem may be challenging for some students. Begin by asking, "What if all 600 gallons of gasoline were the 90-octane type? How many gallons of isooctane would there be?" 540 gallons "What if all 600 gallons were 96-octane?" 576 gallons "What if all 600 gallons were 92-octane?" 552 gallons

Ask students if either one of the 90-octane or 96-octane *alone* can produce the number of gallons of isooctane that you need. Let students reflect on the question until they realize that a mixture of the two will be needed—*and* that the 600 gallons will be made from the two types of gasoline (i.e., $x + y = 600$).

Next talk about the mixture in terms of how many isooctanes each type of gasoline will contribute for each gallon of gasoline in the mix (that is, $0.9x + 0.96y = 552$).

Encourage students to reflect on and talk about the relationships that make up the verbal model. Once the algebraic model is completed, ask students to decide on the method of solution and provide a rationale for the selection. Some may pick the substitution method because the coefficient of either x or y is 1. Talk about why the textbook may have used the linear combination method instead.

Example 3

Discuss why companies might offer annual salaries as described in this example.

In 1990, approximately $4.2 billion worth of athletic shoes were sold in the United States. Of this amount, 22.4% came from the sale of walking shoes, 21% from gym shoes and sneakers, 11.1% from jogging and running shoes, and 9.3% from tennis shoes.

Real Life
Inventory Control

Example 2 *The Cash Register Wasn't Working*

You are the manager of a shoe store. On Sunday morning, you are going over the sales receipts for the past week. They show that 240 pairs of walking shoes were sold. Style A sells for \$66.95, and Style B sells for \$84.95. The total receipts for the two types were \$17,652. The cash register was supposed to keep track of the number of each type sold. It malfunctioned. Can you find out how many of each type were sold?

Solution

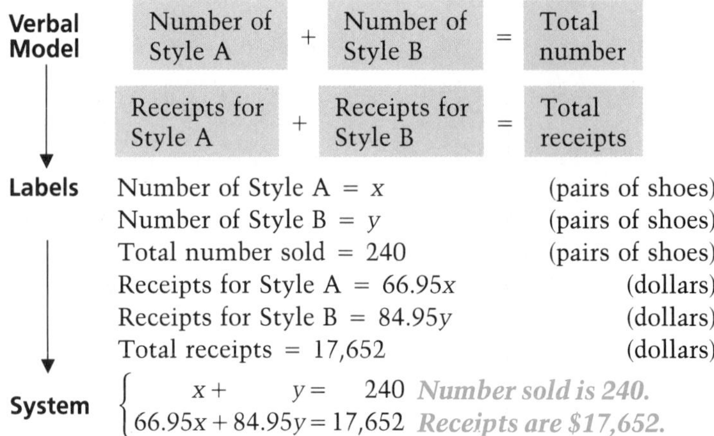

Verbal Model

Number of Style A + Number of Style B = Total number

Receipts for Style A + Receipts for Style B = Total receipts

Labels

Number of Style A $= x$ (pairs of shoes)
Number of Style B $= y$ (pairs of shoes)
Total number sold $= 240$ (pairs of shoes)
Receipts for Style A $= 66.95x$ (dollars)
Receipts for Style B $= 84.95y$ (dollars)
Total receipts $= 17,652$ (dollars)

System

$$\begin{cases} x + y = 240 & \text{Number sold is 240.} \\ 66.95x + 84.95y = 17,652 & \text{Receipts are \$17,652.} \end{cases}$$

Using the substitution method, you can solve the first equation for $x = 240 - y$ and substitute $240 - y$ into the second equation. After simplifying, you will obtain $18y = 1584$. The solution is 152 pairs of Style A and 88 pairs of Style B. ■

Example 3 *Choice of Two Jobs*

You are offered two different jobs selling dental supplies. One has an annual salary of \$20,000 plus a year-end bonus of 1% of your total sales. The other has a salary of \$15,000 plus a year-end bonus of 2% of your total sales. How much would you have to sell to earn the same amount in each job?

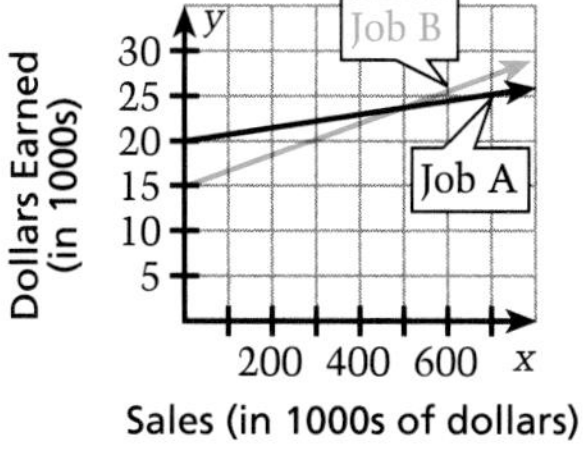

Solution

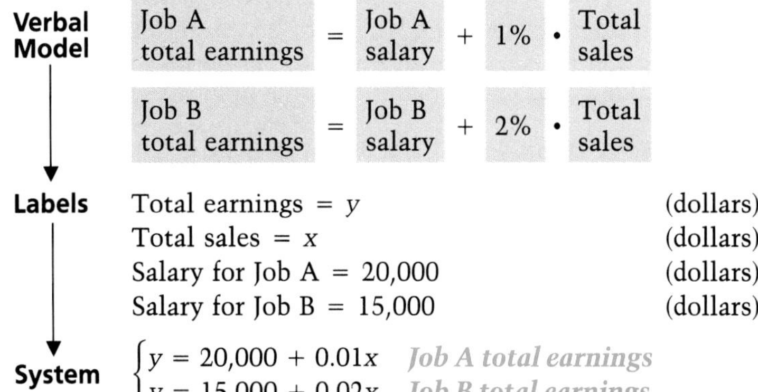

Verbal Model	Job A total earnings	=	Job A salary	+	1%	·	Total sales
	Job B total earnings	=	Job B salary	+	2%	·	Total sales

Labels
Total earnings = y (dollars)
Total sales = x (dollars)
Salary for Job A = 20,000 (dollars)
Salary for Job B = 15,000 (dollars)

System
$$\begin{cases} y = 20{,}000 + 0.01x & \textit{Job A total earnings} \\ y = 15{,}000 + 0.02x & \textit{Job B total earnings} \end{cases}$$

Problem Solving **Draw a Graph**

This system is a good candidate for solution by graphing or by substitution. You must sell about \$500,000 in dental supplies to earn the same amount in each job. If you think you can sell \$800,000 in supplies, which job would you take? ■

Communicating about ALGEBRA

SHARING IDEAS about the Lesson See Additional Answers for models.

Apply the Problem-Solving Plan Two airplanes fly in opposite directions from the same airport. The second plane flies 50 miles an hour faster than the first, but it starts one-half hour later. Two hours after the first plane started, the planes are 1825 miles apart. How fast is each flying?

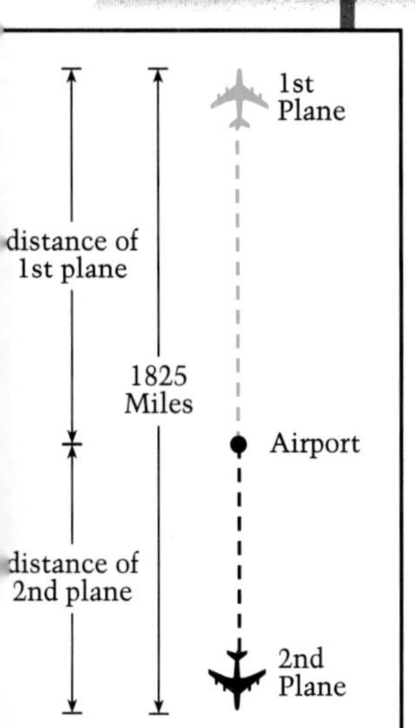

A. Write verbal models, one relating speeds, the other relating distances. Assign labels and write a linear system that models the problem. (Did each fly the same number of hours?) $s = r + 50$, $2r + 1.5s = 1825$; (No)

B. Solve the linear system using *all three* methods studied in this chapter. Which method seems best for this system? Explain $r = 500$, $s = 550$; substitution, 1st equation is already solved for s

Extra Examples

Here are additional examples similar to **Examples 1–3.**

1. **OCTANE RATINGS** In Example 1, what if you were required to use 87-octane gasoline and 96-octane in the mixture? How much of each of these gasolines should be mixed to produce 600 gallons of 92-octane

   The system is
   $$\begin{cases} x + y = 600 \\ 0.87x + 0.96y = 552. \end{cases}$$
   The ordered-pair solution is $(266\frac{2}{3}, 333\frac{1}{3})$.

2. **SELLING GLOVES** You work in the gloves section of a local department store. At the end of the day, you realize that you sold 20 pairs of gloves, but you sold only the plastic gloves at \$7.95 and the cloth gloves at \$11.50. None of the leather gloves at \$27.49 sold! Your total sales amounted to \$212.25. Although your cash register cannot keep track of the types of gloves that were sold, you want to know so you can push the most popular glove. How many of each type did you sell?

   The system is
   $$\begin{cases} x + y = 20 \\ 7.95x + 11.50y = 212.25. \end{cases}$$
   The solution is 5 pairs of plastic gloves, 15 pairs of cloth gloves, and of course no pairs of leather gloves.

Check Understanding

WRITING PROBLEMS An excellent way for students to determine how well they understand problem solving using linear systems is to formulate their own problem situations for their classmates to solve. Ask students to create one or two original problems for others to solve.

Communicating about ALGEBRA

COOPERATIVE LEARNING Encourage students to talk about the problem with classmates in a small group setting. Suggest that each member of the group verbalize an understanding of the problem. The group should try to determine what they agree on and disagree about. Verbalizing the situation will often suggest strategies for attacking the problem. Are any students using the formula, *distance* = *rate* × *time?*

EXERCISES

Guided Practice

CRITICAL THINKING about the Lesson

Price Per Gallon **In Exercises 1–4, use the following information.** 1., 2. See Additional Answers.

The total cost of 15 gallons of regular unleaded gasoline and 10 gallons of premium unleaded gasoline is \$35.50. Premium unleaded costs \$0.20 more per gallon than regular unleaded. What is the cost per gallon of each type of gasoline? 3. $15x + 10y = 35.50$, $y = x + 0.20$ 4. (1.34, 1.54); \$1.34, \$1.54

1. Write a verbal model for this problem.

2. Assign labels to the verbal model.

3. Use the labels to write a linear system.

4. Solve the system and answer the question.

Independent Practice

5. *Fuel Mixture* Five hundred gallons of 82-octane gasoline is obtained by mixing 80-octane gasoline with 86-octane gasoline. How much of each must be used? 80: $333\frac{1}{3}$ gal, 86: $166\frac{2}{3}$ gal

6. *How Many Karats?* A necklace that weighs 156.5 grams is made of gold and copper. The necklace has a volume of 12 cubic centimeters. What percent of the weight of the necklace is gold? Describe the percentage of gold in terms of karats. (Gold weighs 19.3 grams per cubic centimeter, and copper weighs 9 grams per cubic centimeter.) ≈ 58%, ≈ 14 karats

7. *Relay Race* The total time for a two-member team in a 5160-meter relay race is 16 minutes. The first runner on the team averages 300 meters per minute and the second runner averages 360 meters per minute. How many minutes did the first runner keep the baton before passing it to the second runner? Use the following verbal model to help answer the question. 10

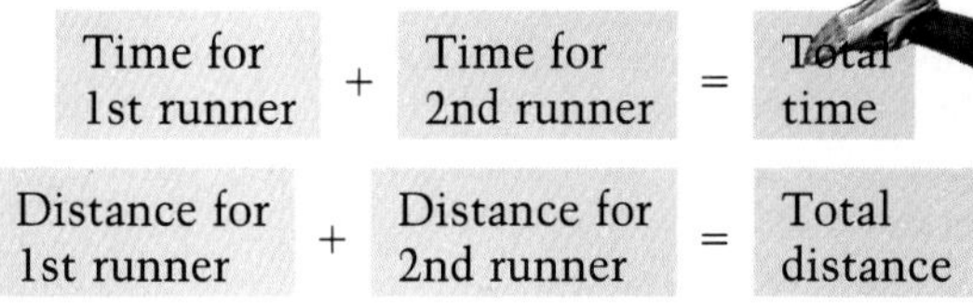

Time for 1st runner
+
Time for 2nd runner
=
Total time
Distance for 1st runner
+
Distance for 2nd runner
=
Total distance

8. *Average Speed* A van travels for two hours at an average speed of 40 miles per hour. For how long must the van then travel at an average speed of 55 miles per hour to raise the overall average speed to 45 miles per hour? What is the total time traveled? 1 hour, 3 hours

9. ***I Can't Believe I Ate All That*** You and your friend go to a Mexican restaurant. You order 2 tacos and 3 enchiladas, and your friend orders 3 tacos and 5 enchiladas! One bill was \$7.80 plus tax, and the other bill was \$12.70 plus tax. How much was each taco and each enchilada? **\$0.90, \$2.00**

10. ***Birthday Party*** You are buying decorations for a birthday party. The before-tax total for 3 rolls of crepe paper and 20 balloons is \$11.40. After you start decorating, you need more supplies. You buy 2 more rolls of crepe paper and 10 more balloons for \$7.20. How much did each roll of crepe paper and each balloon cost? **\$3.00, \$0.12**

11. ***Dimensions of a Rectangle*** The perimeter of the rectangle is 22 meters, and the perimeter of the triangle is 12 meters. Find the dimensions of the rectangle. **8 m by 3 m**

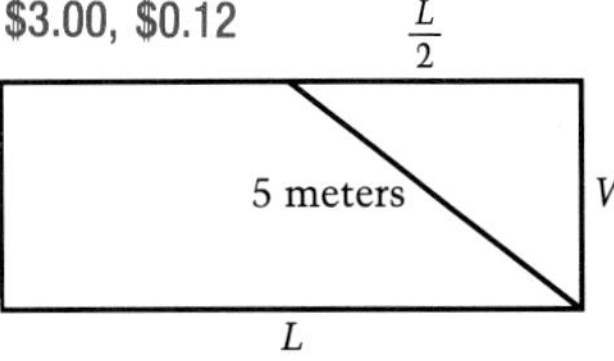

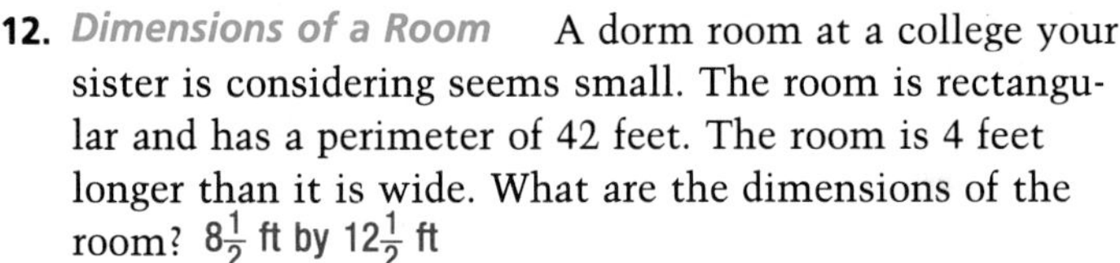

12. ***Dimensions of a Room*** A dorm room at a college your sister is considering seems small. The room is rectangular and has a perimeter of 42 feet. The room is 4 feet longer than it is wide. What are the dimensions of the room? **$8\frac{1}{2}$ ft by $12\frac{1}{2}$ ft**

✪ **13.** ***Finding Absolute Zero*** To calculate *absolute zero* (the complete absence of heat) a chemist uses the following procedure. First, the chemist takes several readings for the volume and temperature of three gases: CH_4 (methane), H_2O (water steam), and N_2O (nitrous oxide), as shown in the graph. Then the chemist finds a linear equation that relates the volume (in liters) and temperature (in degrees Celsius) of each gas.

$$\begin{cases} -2T + 302V = 546.4 & \textit{Methane} \\ -T + 228V = 273.2 & \textit{Water} \\ -3T + 1872V = 819.6 & \textit{Nitrous Oxide} \end{cases}$$

The measurements are taken only at temperatures for which the compounds are gases, as shown by the solid part of each line in the graph. However, if each line is extended, as shown by the dashed part, all three lines intersect at a common point, which is called *absolute zero.* What is absolute zero on the Celsius scale? **-273.2°C**

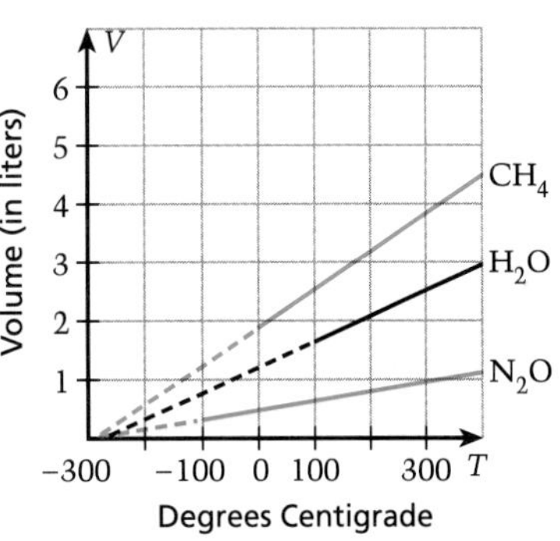

✪ **14.** ***Tree Growth*** You planted a 16-inch hemlock tree that grows at a rate of 4 inches per year. Then 5 years later you planted a 10-inch blue spruce tree that grows at a rate of 6 inches per year. In how many years after you planted the hemlock tree will the two trees be the same height? How tall will the trees be? **18, 88 in.**

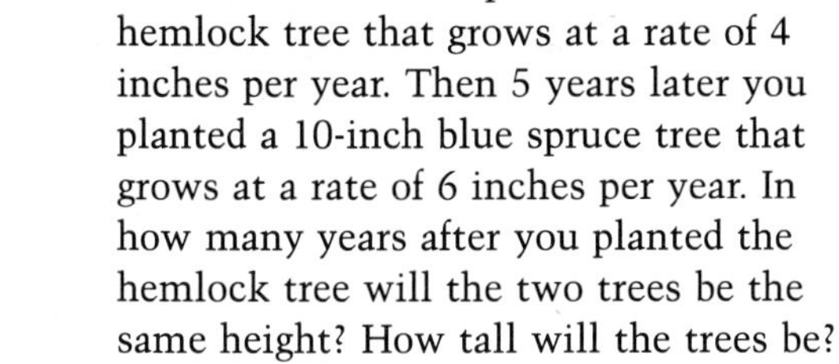

EXERCISE Notes

ASSIGNMENT GUIDE

Basic/Average: Ex. 1–4, 5–11 odd, 15–27 odd, 31, 32

Above Average: Ex. 1–4, 5–27 odd, 31, 32

Advanced: Ex. 1–4, 7–14, 28–33

Selected Answers
Exercises 1–4, 5–31 odd

✪ **More Difficult Exercises**
Exercises: 13, 14, 33

Guided Practice

▶ **Ex. 1–4** These exercises provide in-class practice of the problem-solving procedure modeled in Examples 1–3.

Independent Practice

▶ **Ex. 5** Tell students to refer to Example 1 for a similar model.

▶ **Ex. 13** Absolute zero is the lowest point on the Kelvin temperature scale. Ask students to solve any two of the three equations to compute the same *T*-value regardless of the pair chosen.

▶ **Ex. 14** Advise students to solve the linear models and then take into account the "5 years later" part of the problem.

Ex. 15–32 These exercises provide continued review and practice of necessary skills. Exercises 31–32 review students' understanding of the magnitude of measures.

Ex. 33 Discuss with students why the *y*-coordinate of *C* would represent the distance to the waterfall.

Enrichment Activities

These activities require students to go beyond lesson goals.

1. Solve this system using the graphing, substitution, and linear combination methods. Which method was most efficient? Why?

   $\begin{cases} y = \frac{2}{3}x + 5 \\ y = -\frac{1}{5}x + 2 \end{cases}$ $\left(-\frac{45}{13}, \frac{35}{13}\right)$

   Have students state their preference.

2. **JOB CHOICES** In Example 3, what if the annual salary offered by one dental supply company was \$20,000 plus a year-end bonus of 1% of your total sales, and the annual salary of the other company was \$15,000 plus a year-end bonus of 3% of your total sales? How much would you have to sell to earn the same amount in each job? Which job is more attractive now? Explain.

   The system is $\begin{cases} y = 20000 + 0.01x \\ y = 15000 + 0.03x. \end{cases}$ A graph of the system indicates that you must sell about \$250,000 worth of dental supplies to earn the same amount (\$22,500) in each job. Now the second job is more attractive because, compared to the situation in Example 3, you need sell only half as much as before for the second company.

Integrated Review

In Exercises 15–18, decide whether the ordered pair is a solution of the system.

15. $\begin{cases} 2x - 5y = 16 \\ -x + 4y = -11 \end{cases}$ $(3, -2)$ It is.

16. $\begin{cases} x + 5y = 26 \\ 3x - 6y = 33 \end{cases}$ $(1, 5)$ It is not.

17. $\begin{cases} 2x + 5y = -7 \\ -3x - 6y = 24 \end{cases}$ $(-6, 1)$ It is not.

18. $\begin{cases} 6x + 4y = 36 \\ 3x - 6y = -54 \end{cases}$ $(0, 9)$ It is.

In Exercises 19–30, solve the linear system by the method you think is best. Explain why you chose that method. See Additional Answers for methods. 27. $(14, -\frac{5}{4})$

19. $\begin{cases} -3x + 2y = 6 \\ y = 1 \end{cases}$ $(-\frac{4}{3}, 1)$

20. $\begin{cases} x + 3y = -3 \\ x = -3 \end{cases}$ $(-3, 0)$

21. $\begin{cases} 3x + y = 14 \\ -x + y = 2 \end{cases}$ $(3, 5)$

22. $\begin{cases} x + 5y = 45 \\ 2x - y = 2 \end{cases}$ $(5, 8)$

23. $\begin{cases} 9x - 2y = 4 \\ -5x + y = 1 \end{cases}$ $(-6, -29)$

24. $\begin{cases} 3x + 5y = 17 \\ \frac{1}{2}x + y = 3 \end{cases}$ $(4, 1)$

25. $\begin{cases} 2x - 7y = -27 \\ 6x + 5y = -3 \end{cases}$ $(-3, 3)$

26. $\begin{cases} 10x - 3y = 17 \\ -7x + y = 9 \end{cases}$ $(-4, -19)$

27. $\begin{cases} x + 4y = 9 \\ 3x + 8y = 32 \end{cases}$

28. $\begin{cases} 3x - 2y = 17 \\ -2x - 5y = 14 \end{cases}$ $(3, -4)$

29. $\begin{cases} 9x - 5y = 45 \\ -2x + 21y = -10 \end{cases}$ $(5, 0)$

30. $\begin{cases} 6x - y = -32 \\ 2x + 13y = 16 \end{cases}$ $(-5, 2)$

31. *Estimate within Reason* Which is the best estimate for the racing speed of a thoroughbred racehorse?
    a. 3.5 miles per hour b. 35 miles per hour
    c. 350 miles per hour d. 3500 miles per hour

32. *Estimate within Reason* Which is the best estimate for the population of the United States?
    a. 250,000 b. 2,500,000 c. 25,000,000 d. 250,000,000

Exploration and Extension

33. *Surveying* Two surveyors are trying to determine distances to a waterfall. They visualize a triangle formed by their positions, *A* and *B*, and the top of the waterfall, *C*. They also visualize a coordinate plane that contains the triangle. In this coordinate plane, point *A* has coordinates (1, 0) and point *B* has coordinates (1.5, 0), where *x* and *y* are measured in miles. The line containing *AC* is given by $24x - y = 24$, and the line containing *BC* is given by $6x + y = 9$. Find the distance between the line segment *AB* and the top of the waterfall. $2\frac{2}{5}$ mi

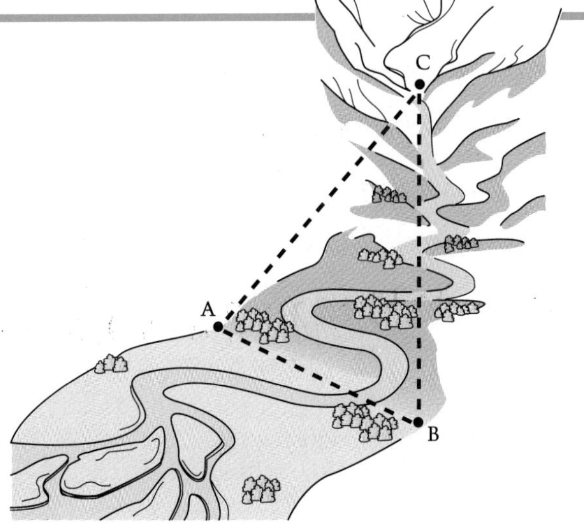

Mid-Chapter SELF-TEST

Take this test as you would take a test in class. The answers to the exercises are given in the back of the book.

In Exercises 1–6, determine which of the ordered pairs is a solution of the system. (7.1)

1. $\begin{cases} 4x - 4y = 4 \\ -2x + y = 2 \end{cases}$ $(-3, -4)$, $(2, 6)$

2. $\begin{cases} \frac{1}{2}x + 2y = \frac{27}{2} \\ 12x + y = -5 \end{cases}$ $(-1, 7)$, $(4, -5)$

3. $\begin{cases} 3x + 5y = 9 \\ -x + 2y = 8 \end{cases}$ $(-2, 3)$, $(4, 6)$

4. $\begin{cases} 3x + 2y = 11 \\ 11x + 8y = 35 \end{cases}$ $(1, 3)$, $(9, -8)$

5. $\begin{cases} -12x + y = 15 \\ 6x - 2y = -15 \end{cases}$ $(-\frac{3}{2}, -3)$, $(-\frac{5}{6}, 5)$

6. $\begin{cases} x + 7y = 4 \\ x + y = 1 \end{cases}$ $(\frac{1}{2}, \frac{1}{2})$, $(\frac{1}{4}, \frac{3}{4})$

In Exercises 7 and 8, solve the linear system by graphing. (7.1)

7. $\begin{cases} -2x + 3y = 6 \\ 2x + y = 10 \end{cases}$ (3, 4)

8. $\begin{cases} -2x + y = 2 \\ x - y = 1 \end{cases}$ $(-3, -4)$

In Exercises 9–14, use substitution to solve the system. (7.2)

9. $\begin{cases} x + 3y = 7 \\ 4x - 7y = -10 \end{cases}$ (1, 2)

10. $\begin{cases} -6x - 5y = 28 \\ x - 2y = 1 \end{cases}$ $(-3, -2)$

11. $\begin{cases} \frac{1}{2}x + \frac{3}{4}y = 9 \\ -2x + y = -4 \end{cases}$ (6, 8)

12. $\begin{cases} 4x + y = -1 \\ -5x - y = 0 \end{cases}$ $(1, -5)$

13. $\begin{cases} x - 6y = -19 \\ 3x - 2y = -9 \end{cases}$ $(-1, 3)$

14. $\begin{cases} 4x + y = 14 \\ 3x + 2y = 8 \end{cases}$ $(4, -2)$

In Exercises 15–20, use linear combinations to solve the system. (7.3)

15. $\begin{cases} -2x - 3y = 4 \\ 2x - 4y = 3 \end{cases}$ $(-\frac{1}{2}, -1)$

16. $\begin{cases} 5x + 7y = 35 \\ 4x - 7y = 1 \end{cases}$ $(4, \frac{15}{7})$

17. $\begin{cases} 4x + 3y = 1 \\ -2x + 9y = -4 \end{cases}$ $(\frac{1}{2}, -\frac{1}{3})$

18. $\begin{cases} 10x + 3y = 2 \\ 8x + 6y = 16 \end{cases}$ $(-1, 4)$

19. $\begin{cases} 3x - 5y = -4 \\ -9x + 7y = 8 \end{cases}$ $(-\frac{1}{2}, \frac{1}{2})$

20. $\begin{cases} 3x + 7y = -4 \\ -15x - 14y = 6 \end{cases}$ $(\frac{2}{9}, -\frac{2}{3})$

21. A total of \$8000 is invested in 2 funds paying 4% and 5% annual interest. The combined annual interest is \$350. How much of the \$8000 is invested in each fund? **(7.4)** \$5000 in 4%, \$3000 in 5%

22. You go to the video store to rent 5 movies for the weekend. Movies rent for \$2 and \$3. You spend \$13. How many \$2 movies did you rent? How many \$3 movies did you rent? **(7.4)** 2, 3

23. Your teacher is giving a test worth 150 points. There are 46 three- and five-point questions. How many of each are on the test? **(7.4)** 40 3-point, 6 5-point

24. You are in charge of ordering softballs for three different leagues. The Pony League uses an 11-inch softball priced at \$2.25. The Junior and Senior Leagues use a 12-inch softball priced at \$2.75. The invoice smeared in the rain. You can still read the totals, 80 softballs for \$210. How many of each size are there? **(7.4)** 20 11-inch, 60 12-inch

Answers

Mid-Chapter SELF-TEST

7.

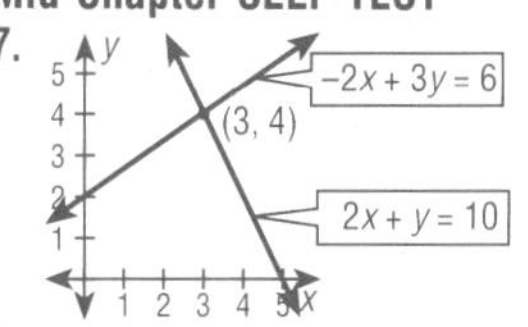

8.

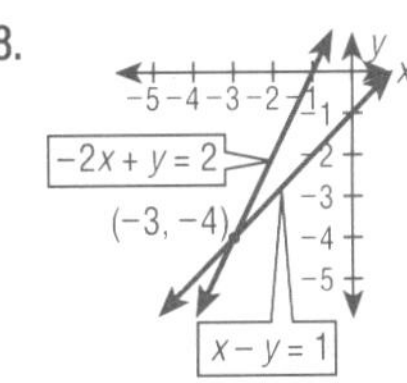

Problem of the Day

A farmer keeps track of his cows and chickens by counting legs and heads. If he counts 78 legs and 35 heads, how many cows and chickens does he have? 4 cows, 31 chickens

ORGANIZER

Warm-Up Exercises

1. Transform each equation into slope-intercept form and sketch its graph.
   a. $4x - 6y = 12$
   b. $3x + y = -8$
   c. $3x - 8y = 12$

   a. $y = \frac{2}{3}x - 2$;
   b. $y = -3x - 8$;
   c. $y = \frac{3}{8}x - \frac{3}{2}$

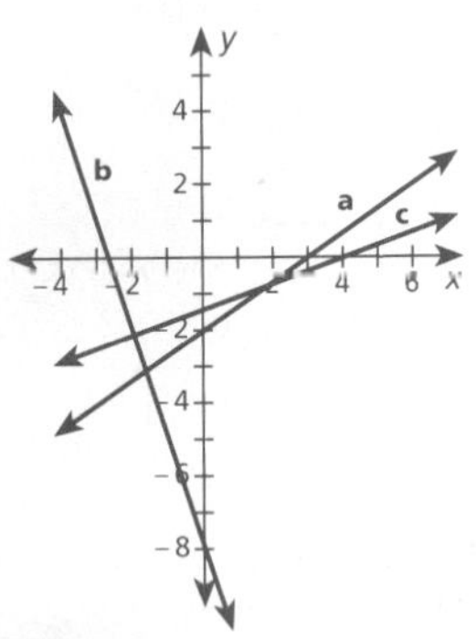

Lesson Resources

Teaching Tools
- Transparency: 4
- Problem of the Day: 7.5
- Warm-up Exercises: 7.5
- Answer Masters: 7.5

Extra Practice: 7.5

Color Transparency: 38

LESSON Notes

Example 1

The graphing method of solving the linear system lets students see the relationship between linear equations in a system with no solutions. It is

7.5 Special Types of Linear Systems

What you should learn:

Goal 1 How to visualize the solution possibilities for linear systems

Goal 2 How to identify a linear system as having many solutions

Why you should learn it:

You can interpret linear-system models for real-life situations as having no solution, one solution, and many solutions.

Goal 1 Visualizing the Solution Possibilities

In the first four lessons of this chapter, each linear system had exactly one solution. In the investigation below, you will discover that there are two other types of linear systems.

LESSON INVESTIGATION

Investigating Solution Possibilities

Partner Activity Use a graphing calculator to sketch the graphs of both equations in the system on the same screen. How are the two graphs in the system related?

a. $\begin{cases} 4x + 6y = 20 \\ 2x + 3y = 6 \end{cases}$ **b.** $\begin{cases} x - y = 4 \\ 3x - 3y = 12 \end{cases}$

In general, how can you recognize that a system has no solution? Many solutions?

a. The lines are parallel.
b. The lines are the same. A system has no solution if the lines are parallel. A system has many solutions if the lines are the same.

Example 1 *A Linear System with No Solution*

Show that the linear system has no solution.

$\begin{cases} 2x + y = 3 & \text{Equation 1} \\ 4x + 2y = 8 & \text{Equation 2} \end{cases}$

Solution You can use any of the three methods to show that the system has no solution. Using the graphing method, you could rewrite each equation in slope-intercept form.

$\begin{cases} y = -2x + 3 & \text{Revised Equation 1} \\ y = -2x + 4 & \text{Revised Equation 2} \end{cases}$

You can see that the lines have the same slope but different y-intercepts. Thus, the lines are parallel and nonintersecting, which means that the system has no solution. If you try using substitution or linear combinations to solve the system, you will obtain a meaningless "equation." For instance, multiplying Equation 1 by -2 and adding the result to Equation 2 produces $0 = 2$, which makes no sense. ■

Goal 2 Identifying a System with Many Solutions

Example 2 *A Linear System with Many Solutions*

Show that the linear system has many solutions.

$$\begin{cases} 2x + y = 3 & \textit{Equation 1} \\ 4x + 2y = 6 & \textit{Equation 2} \end{cases}$$

Solution You can use any of the three methods to show that the system has many solutions. Using the graphing method, rewrite each equation in slope-intercept form.

$$\begin{cases} y = -2x + 3 & \textit{Revised Equation 1} \\ y = -2x + 3 & \textit{Revised Equation 2} \end{cases}$$

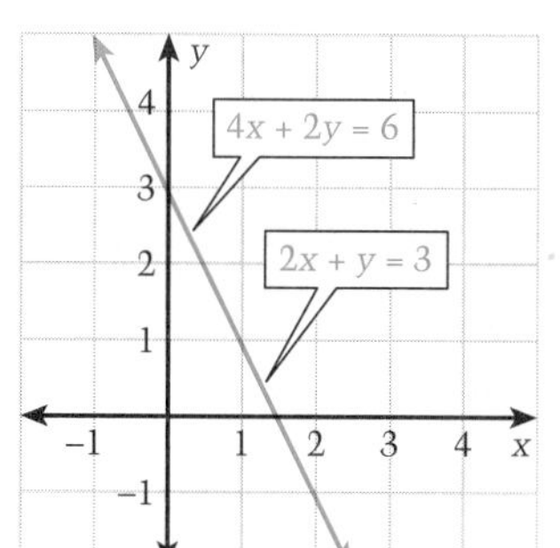

From these equations, you can see that each equation represents the same line. Thus, any point on the line is a solution. If you apply the substitution or linear combination method to the original equations, you will obtain an equation like $6 = 6$ or $0 = 0$, which is a signal that the system has many solutions. For instance, multiplying Equation 1 by -2 and adding the result to Equation 2 produces $0 = 0$. ■

Real Life
Fund Raising

Example 3 *Bake Sale*

At a 4-day bake sale, your club is selling cookies and brownies. Each day you try a different price. The results are shown in the matrix. Find linear models that relate the price to the numbers sold. From the models, will the number of cookies sold ever surpass the number of brownies sold?

Price in cents	20	25	30	35
Brownies	146	138	130	122
Cookies	133	125	117	109

Connections
Discrete Mathematics

Solution Let y represent the numbers of brownies or cookies sold. Let x represent the price in cents. The sale of each baked good decreased by 8 units with each 5 cent price increase. So, the sale decreased by 1.6 units per 1 cent increase.

$$\begin{cases} y = -1.6x + 178 & \textit{Brownies} \\ y = -1.6x + 165 & \textit{Cookies} \end{cases}$$

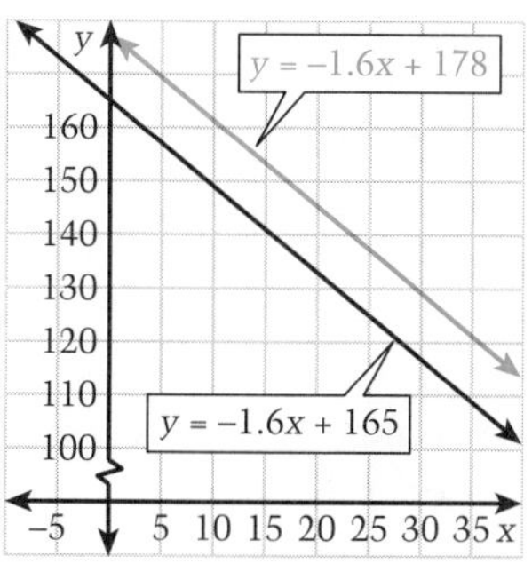

By graphing the system, you obtain a pair of parallel lines, as shown at the left. From the graph, you can see that if the sales trends remain the same and the cookie price is the *same* as the brownie price, then cookie sales will never surpass brownie sales. ■

important, however, that students recognize the algebraic indicator of a linear system with no solution, that is, an equation that makes no sense. For this reason, be sure to solve the system in this example by substitution or by linear combinations. Linear systems having no solution are frequently called "inconsistent" systems. Linear systems with exactly one solution are called "independent" systems.

Example 2

Graph the system, then solve by substitution or by linear combinations. The algebraic indicator of a linear system with infinitely many solutions is an equation that is always true. These systems are frequently called "dependent" systems.

An Alternate Approach
Technology

USING A COMPUTER PROGRAM You can use a BASIC program such as the following to solve a system of two linear equations in the form $Ax + By = C$ and $Dx + Ey = F$.

```
10 PRINT "ENTER A, B, C AND
   D, E, F";
20 INPUT A, B, C, D, E, F
30 M1 = -A / B
40 T1 = C / B
50 M2 = -D / E
60 T2 = F / E
70 IF M1 = M2 THEN 120
80 G = A * E - B * D
90 X = (C * E - B * F) / G
100 Y = (A * F - C * D) / G
110 PRINT "X = "X" Y = "Y:
    GOTO 10
120 IF T1 = T2 THEN PRINT "ONE
    LINE WITH SLOPE = "M1"
    AND Y-INTERCEPT = "T1:
    GOTO 10
130 PRINT "PARALLEL LINES
    WITH SLOPE = "M1" AND
    Y-INTERCEPTS "T1" AND "T2
140 END
```

Example 3

Observe that the constant change in the production of each type of paper is the same. Therefore, the slope of each line is the same. It follows that the lines of the linear equations in this system are parallel and that the linear system has no solution.

Example 4

GEOMETRY Point out the connection to geometry in the use of similar triangles.

Extra Examples

Here are additional examples similar to **Examples 1–3.**

Solve each linear system by graphing and by either substitution or linear combinations.

1. $\begin{cases} -4x + y = 2 \\ 12x - 3y = -6 \end{cases}$
2. $\begin{cases} -6x - 2y = -4 \\ 2y = 2x - 4 \end{cases}$

1. The lines are parallel, so the system has infinitely many solutions.
2. $(1, -1)$

Check Understanding

1. What is the algebraic indicator for a linear system with infinitely many solutions? Simplifying produces an equation that is always true, such as $0 = 0$ or $6 = 6$.
2. What is the algebraic indicator for a linear system with no solution? Simplifying produces an equation that is never true, such as, $1 = 0$ or $5 = 6$.
3. In light of this lesson, what are the advantages of solving linear systems by graphing? Graphing will indicate whether the system has no solution, one solution, or infinitely many solutions.

Real Life
Archaeology

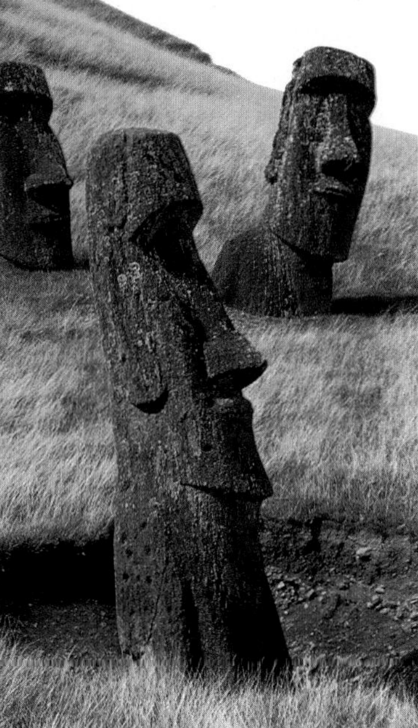

Easter Island, in the South Pacific, has more than 600 massive stone statues called moai. *The statues were carved hundreds of years ago, and some are over 40 feet high.*

Example 4 *Didn't We Already Know That?*

Paco and Diego, from Santiago, are visiting Chile's Easter Island 2300 miles to the west in the South Pacific. To measure the heights of two statues, Paco and Diego tried a technique they had learned in algebra. At 2:00 P.M., they measured the shadow lengths of the statues to be 18 feet and 27 feet. At 3:00 P.M., they measured the shadow lengths again and found them to be 20 feet and 30 feet. Later, however, they were unable to use their measurements to determine the heights of the statues. Why?

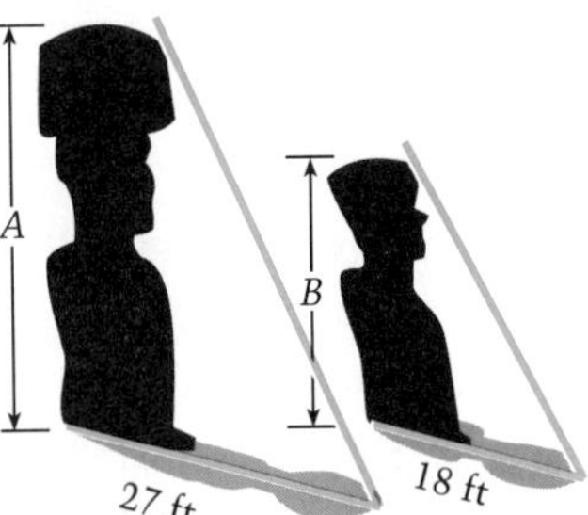

Shadow lengths at 2:00 P.M.

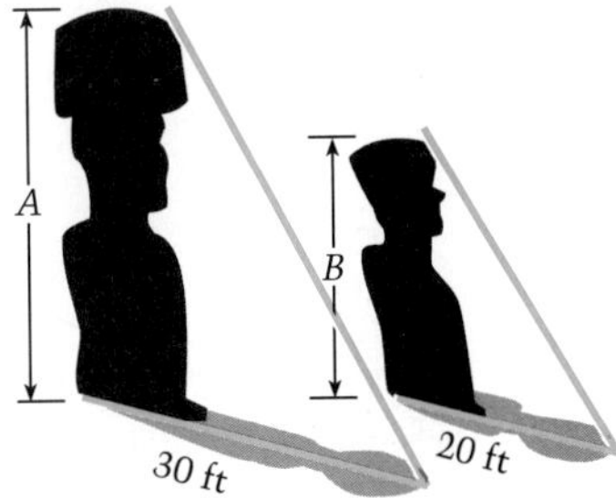

Shadow lengths at 3:00 P.M.

Solution Let A represent the height of the first statue and B represent the height of the second. Using the fact that the ratios of corresponding sides of similar triangles are equal, Paco and Diego wrote the following two equations.

$$\begin{cases} \frac{A}{B} = \frac{27}{18} & \text{2:00 P.M. measurements} \\ \frac{A}{B} = \frac{30}{20} & \text{3:00 P.M. measurements} \end{cases}$$

At this point, they got stuck—because both equations graph as the same line. All that Paco and Diego found from these measurements was that $A = \frac{3}{2}B$. In other words, they found that the height of the first statue is one and one-half times the height of the second. ■

Communicating about ALGEBRA

SHARING IDEAS about the Lesson

Plan Another Approach Help Paco and Diego find the heights of the statues.

What could Paco and Diego have done to find the heights of the two statues? (You may want to look at the exercises in Lesson 4.5.) See below.

Measure the height and shadow length of something else, then compare them to the height and shadow length of one of the statues.

EXERCISES

Guided Practice

CRITICAL THINKING about the Lesson

1. Describe the graphical model for a linear system that has no solution. 2 parallel lines
2. Describe the graphical model for a linear system that has many solutions. 1 line
3. Describe the graphical model for a linear system that has exactly one solution. 2 intersecting lines

In Exercises 4–6, give a graphical description of the system.

4. $\begin{cases} -4x + 2y = 12 \\ -6x + 3y = 8 \end{cases}$ 2 parallel lines

5. $\begin{cases} 3x - 5y = -14 \\ 2x + 6y = 8 \end{cases}$ 2 intersecting lines

6. $\begin{cases} -9x + 12y = 18 \\ 6x - 8y = 12 \end{cases}$ 1 line

Independent Practice

In Exercises 7–12, match the graphical model with one of the linear systems.

7. b

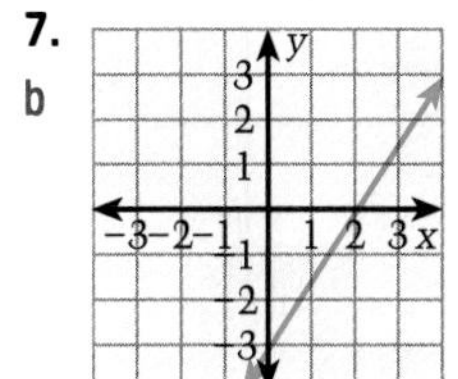

8. d

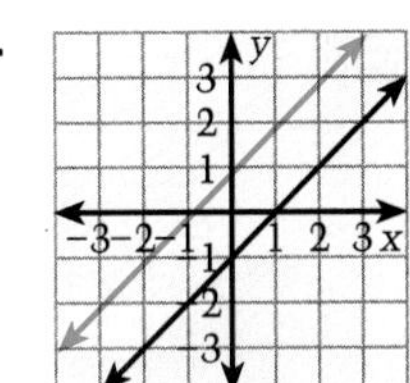

9. f

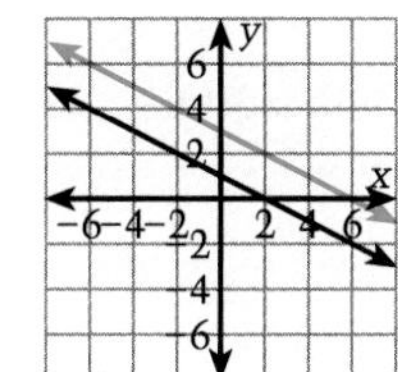

10. c

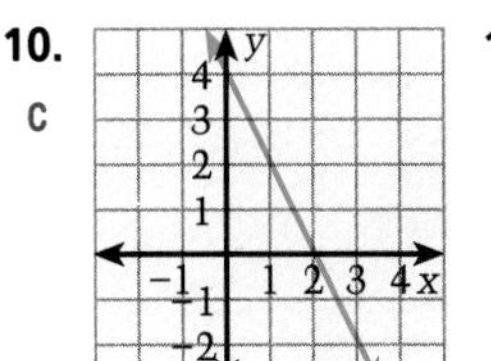

11. e

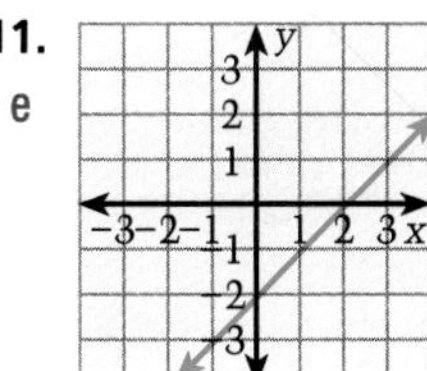

12. a

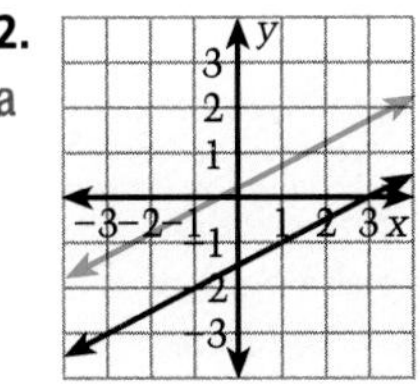

a. $\begin{cases} -2x + 4y = 1 \\ x - 2y = 3 \end{cases}$

b. $\begin{cases} 3x - 2y = 6 \\ -6x + 4y = -12 \end{cases}$

c. $\begin{cases} 2x + y = 4 \\ -4x - 2y = -8 \end{cases}$

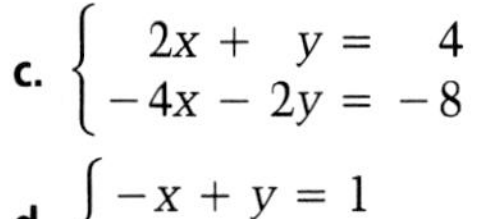

d. $\begin{cases} -x + y = 1 \\ x - y = 1 \end{cases}$

e. $\begin{cases} x - y = 2 \\ -3x + 3y = -6 \end{cases}$

f. $\begin{cases} x + 2y = 6 \\ x + 2y = 2 \end{cases}$

In Exercises 13–24, graph the system and describe its solution(s). See Additional Answers for graphs.

13. $\begin{cases} x + y = 8 \\ x + y = -1 \end{cases}$ No solution

14. $\begin{cases} 3x - 2y = 0 \\ 3x - 2y = -4 \end{cases}$ No solution

15. $\begin{cases} 3x - 2y = 3 \\ -6x + 4y = -6 \end{cases}$

16. $\begin{cases} x - 3y = 2 \\ -2x + 6y = 2 \end{cases}$ No solution

17. $\begin{cases} 3x + 2y = 2 \\ 6x + 4y = 14 \end{cases}$ No solution

18. $\begin{cases} -3x + 10y = 15 \\ 3x - 10y = -15 \end{cases}$

19. $\begin{cases} -x + 4y = -21 \\ 3x - 12y = -21 \end{cases}$ No solution

20. $\begin{cases} 2x - 6y = 5 \\ 3x - 9y = 2 \end{cases}$ No solution

21. $\begin{cases} 6x - 3y = 4 \\ -4x + 2y = -\frac{8}{3} \end{cases}$

22. $\begin{cases} 6x - 5y = 3 \\ -12x + 10y = 5 \end{cases}$ No solution

23. $\begin{cases} \frac{3}{4}x + \frac{1}{2}y = 10 \\ -\frac{3}{2}x - y = 4 \end{cases}$ No solution

24. $\begin{cases} \frac{2}{3}x + \frac{1}{6}y = \frac{2}{3} \\ 4x + y = 4 \end{cases}$ Many solutions

15. Many solutions 18. Many solutions 21. Many solutions

Communicating about ALGEBRA

Ask students to explain why the two relationships that Paco and Diego determined represent a dependent linear system. Since $A = \frac{3}{2}B$, the value of A depends entirely on the value of B.

EXERCISE Notes

ASSIGNMENT GUIDE

Basic/Average: Ex. 1–6, 7–12, 13–19 odd, 25–33 odd, 35–36, 48–50 odd

Above Average: Ex. 1–6, 7–12, 17–31 odd, 33–39, 44–52 odd

Advanced: Ex. 1–6, 7–12, 22–28, 29, 31, 35–39, 50–55

Selected Answers
Exercises 1–6, 7–51 odd

Use **Mixed Review** as needed.

More Difficult Exercises
Exercises 37–39, 52–55

Guided Practice

Ex. 1–6 These exercises provide practice in the visual connection of special-case systems. Use this as a summary check for the lesson.

Independent Practice

Ex. 7–12 These exercises need to be assigned as a group.

Ex. 25–28 These exercises provide students with practice in the inverse task of Exercises 13–24.

Sample answers:

25. $y = 4x - 2,\ 3y = 12x - 6$
26. $y = \frac{2}{3}x + 2,\ y = \frac{2}{3}x - 1$
27. $y = -\frac{1}{2}x + 4,\ y = -\frac{1}{2}x - 4$
28. $y = \frac{1}{3}x + 1,\ 3y = x + 3$

Ex. 29–32 Use these exercises to check whether the students understand the algebraic conditions necessary for the special-case situations.

Answers

29. Many: $n = 8$; no: $n \neq 8$
30. Many: $n = -6$; no: $n \neq -6$
31. Many: $n = -6$; no: $n \neq -6$
32. Many: $n = 20$; no: $n \neq 20$

Ex. 35–36 Remind students that $t = 0$ should correspond to 1960.

Answer

35. Male: $y = 89 + 1.1x$
    Female: $y = 91 + 1.3x$

In Exercises 25–28, find a linear system for the graphical model. If only one line is shown, find two different equations for the line. See margin.

25. 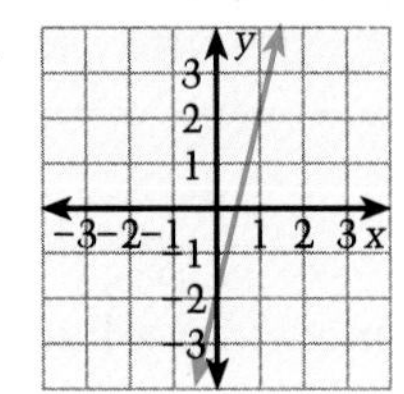

26. 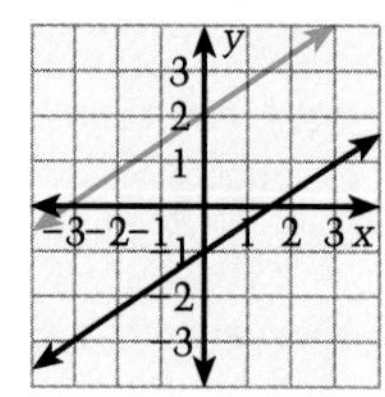

27. 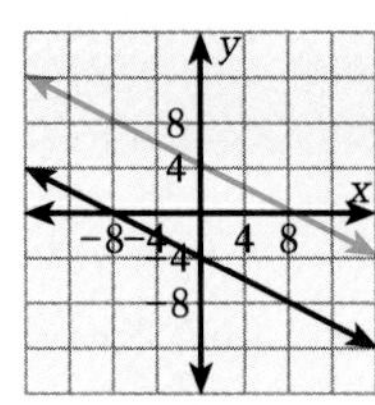

28. 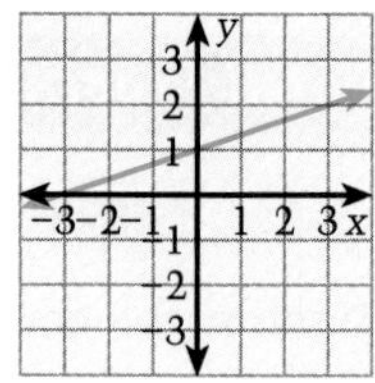

In Exercises 29–32, find one value of *n* so that the linear system has many solutions. Then find a second value of *n* so that the linear system has no solution. Graph both results. See margin.

29. $\begin{cases} x - y = 4 \\ 2x - 2y = n \end{cases}$ **30.** $\begin{cases} x - y = 2 \\ -3x + 3y = n \end{cases}$ **31.** $\begin{cases} 4x - 12y = n \\ -2x + 6y = 3 \end{cases}$ **32.** $\begin{cases} 9x + 6y = n \\ 1.8x + 1.2y = 4 \end{cases}$

33. *Perspective in Art* The drawing of a room at the right uses *perspective* to create the illusion of depth. Which of the lines that are supposed to be parallel (in three dimensions) are actually drawn parallel (in two dimensions)?

34. *Optical Illusion* If the lines that contain the line segments marked *AB* and *CD* were used to form a system of linear equations, would the system have exactly one solution, no solution, or many solutions? Explain. No solution, because the lines are parallel.

33. Horizontal and vertical lines

Which is longer? The line segment forming the front of the rug or the line segment forming the back of the room.

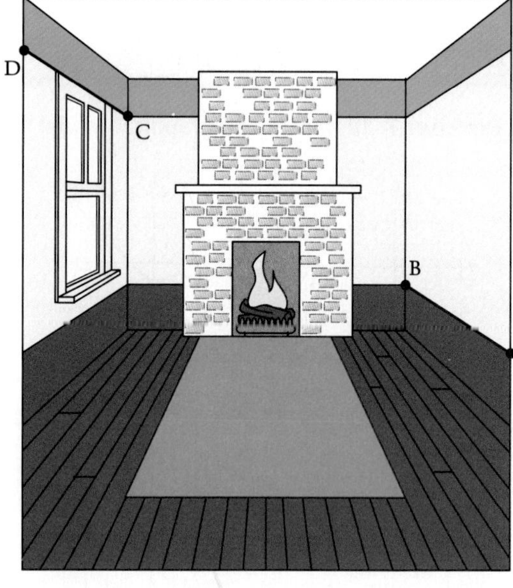

U.S. Population **In Exercises 35 and 36, use the following information.**

The male and female populations of the United States from 1960 to 1990 are shown in the matrix. ***(Source: U.S. Bureau of Census)***

Population (in millions)	Male	Female
1960	89	91
1970	100	105
1980	111	117
1990	122	128

35. Construct two scatter plots, one for the male population and one for the female population. Then find the line that best fits each scatter plot. See Additional Answers for graphs. See margin for equations.

36. Discuss the system of linear equations you found in Exercise 35. Are the two lines parallel? Do you think that the number of men in the United States will equal the number of women before the year 2000? Explain. No; no; number of women is increasing more rapidly than number of men

More Diversity in the 1990s

In Exercises 37 and 38, use the following information.

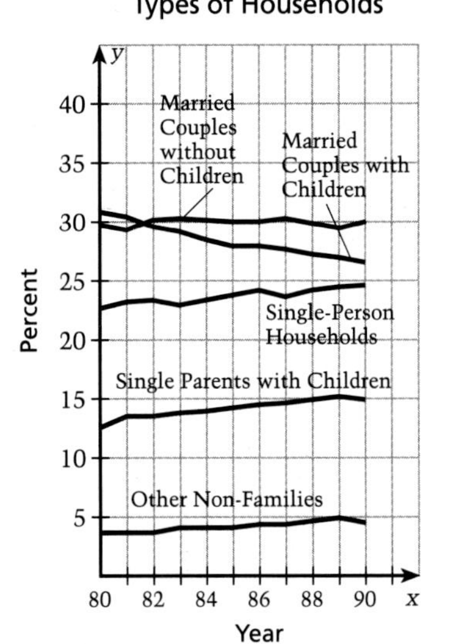

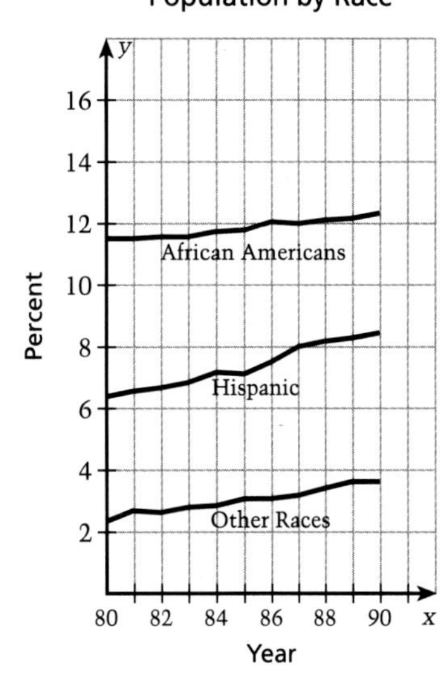

The graphs show the percentage changes in types of households in the United States and the percentage changes in racial or ethnic background of the U.S. population. (Source: U.S. Bureau of Census)

37. Which of the five graphs showing the percentages of different types of households do you think will intersect during the 1990s? Explain. See margin.

38. To make the second chart, the U.S. population was categorized into four groups: white, African-American, Hispanic, and other. Use the three given graphs to sketch the graph for the white population. See margin.

39. *A Number Trick* Write a linear equation for the description at the right. Let x represent the number that is chosen and let y represent the final result. To find which number a person would have to choose to obtain a final result of 1, solve the system consisting of the equation you wrote and the equation $y = 1$. Explain your result. See margin.

> *Choose any number.*
> *Add 10 to the number.*
> *Multiply the result by 2.*
> *Subtract 18 from the result.*
> *Multiply the result by one half.*
> *Subtract the original number.*

Integrated Review

In Exercises 40–51, solve the linear system by the method you think is best.

40. $\begin{cases} 2x - y = -4 \\ x + 2y = 3 \end{cases}$ $(-1, 2)$

41. $\begin{cases} x + y = 22 \\ x - y = 8 \end{cases}$ $(15, 7)$

42. $\begin{cases} y - 2x = 1 \\ y - 2x = -3 \end{cases}$ No solution

43. $\begin{cases} 3x + 2y = 18 \\ 5x - 2y = 14 \end{cases}$ $(4, 3)$

44. $\begin{cases} 2x + y = 4 \\ -2x + 3y = -12 \end{cases}$ $(3, -2)$

45. $\begin{cases} 2x - 5y = 3 \\ 4y - x = -3 \end{cases}$ $(-1, -1)$

46. $\begin{cases} \frac{1}{2}y = x + 3 \\ y = 2x + 6 \end{cases}$ 46. Any point on the line $y = 2x + 6$

47. $\begin{cases} x + y = 5 \\ x - y = 3 \end{cases}$ $(4, 1)$

48. $\begin{cases} 3x - y = 9 \\ 6x - 2y = 10 \end{cases}$ No solution

49. $\begin{cases} 3x + 2y = 10 \\ 4x - y = -1 \end{cases}$ $\left(\frac{8}{11}, \frac{43}{11}\right)$

50. $\begin{cases} 7 = 2x - y \\ 5 = x + y \end{cases}$ $(4, 1)$

51. $\begin{cases} x + y = 2 \\ x - 3y = 5 \end{cases}$ $\left(2\frac{3}{4}, -\frac{3}{4}\right)$

▶ **Ex. 37–38** These exercises develop student awareness of current social conditions.

Answers.

37. Married couples with children and single person households; the first type is decreasing to 25%, and the second type is increasing to 25%.

38.

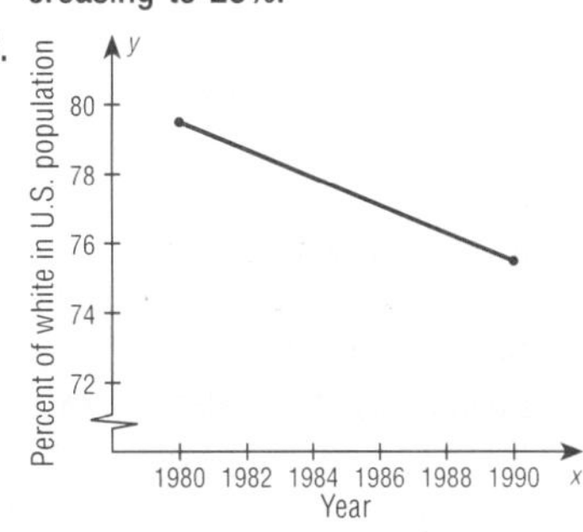

39. $y = [(x + 10)2 - 18]\frac{1}{2} - x$. The equations represent the same line, so there are many solutions; any number may be chosen.

▶ **Ex. 54–55** A quick check of slope will indicate whether the lines intersect.

EXTEND Ex. 54–55 TECHNOLOGY Have students use a graphics calculator or computer graphing program to locate the point of intersection.

Enrichment Activity

This activity requires students to go beyond lesson goals.

WRITING PROBLEMS Create your own "special case" linear system by using two linear models with the same rate of increase. For example, a population model for the growth rate of the number of men and women in the United States could be

$\begin{cases} y = 0.5x + 122 \text{ (men)} \\ y = 0.5x + 128 \text{ (women)} \end{cases}$

with $x = 0$ corresponding to 1990 and y representing the population in millions.

Exploration and Extension

In Exercises 52 and 53, correct the graphical model so that it matches the linear system.

52. $\begin{cases} 7x + 3y = 15 \\ \frac{7}{3}x + y = 5 \end{cases}$

Delete line on the left.

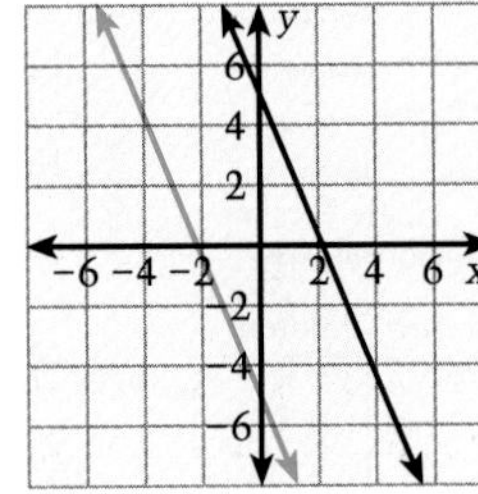

53. $\begin{cases} -x + 3y = 3 \\ 2x - 6y = 6 \end{cases}$

Insert line parallel to the given line with y-intercept 1.

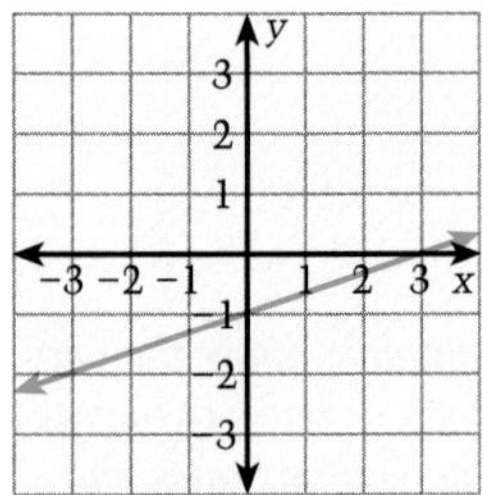

In Exercises 54 and 55, decide whether the lines are parallel. If they aren't, solve the system.

54. $\begin{cases} 200y - x = 200 \\ 199y - x = -198 \end{cases}$

Not parallel, (79400, 398)

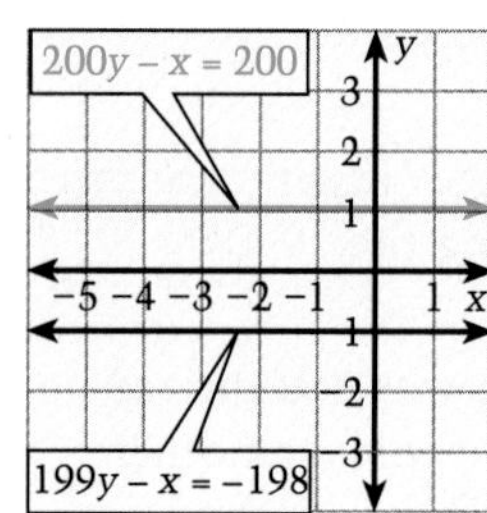

55. $\begin{cases} 25x - 24y = 0 \\ 13x - 12y = 120 \end{cases}$

Not parallel, (240, 250)

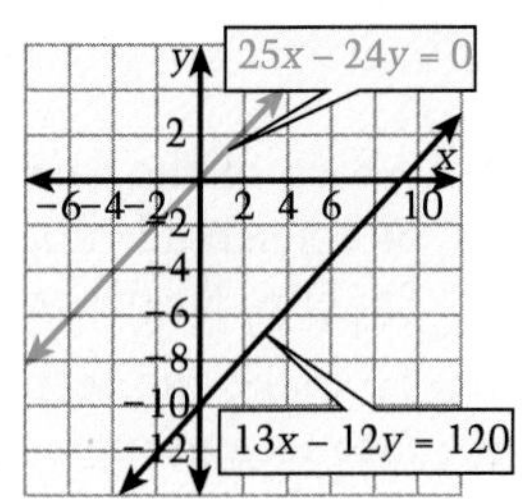

Mixed REVIEW

5. $b = \frac{1}{4}(6 - 3a)$ 6. $n = \frac{1}{4}(3m - 6)$ 7. Yes 8. Yes

19. $y = \frac{1}{4}x - 3$ 20. $-x + 9y = 14$ $-\frac{5}{4}$

1. Evaluate $3 + 6 \div 3 - 2$. **(1.4)** 3
2. Which number is larger, $-\frac{5}{4}$ or $-\frac{4}{3}$? **(2.1)**
3. Add. $\begin{bmatrix} -2 & 4 \\ 3 & 6 \end{bmatrix} + \begin{bmatrix} 6 & -2 \\ 3 & 8 \end{bmatrix}$ **(2.4)**
4. Add. $\begin{bmatrix} -3 & 7 \\ 4 & -5 \end{bmatrix} + \begin{bmatrix} 8 & 6 \\ -3 & -1 \end{bmatrix}$ **(2.4)**
5. Solve $3a + 4b - 6 = 0$ for b. **(3.6)**
6. Solve $3m - 2n = 2n + 6$ for n. **(3.6)**
7. Is $(-3, 2)$ a solution for $3x + 2y < -4$? **(6.5)**
8. Is $(4, 3)$ a solution for $\frac{1}{2}x - \frac{1}{6}y \geq 1$? **(6.5)**
9. Write $3x + \frac{1}{4}y = 6$ in slope-intercept form. **(5.1)** $y = -12x + 24$
10. Write $y = \frac{1}{4}(x + 4)$ in general form. **(5.5)** $-x + 4y = 4$
11. Is 6 a solution for $4a \div 3 = 8$? **(1.5)**
12. Is $\frac{7}{2}$ a solution for $3 \div b \cdot 7 < 6$? **(1.5)**
13. Find the difference of -3 and -6. **(2.3)** 3
14. Find the sum of $\frac{1}{3} + \frac{1}{6}$. **(2.2)** $\frac{1}{2}$
15. Evaluate $3x^2 + x - 7$ when $x = -1$. **(1.3)** -5
16. Evaluate $\frac{1}{4}(8 + |y|) + 6$ when $y = 4$. **(2.6)** 9
17. Solve $6x + \frac{1}{2}(x + 2) = 5$. **(3.2)** $\frac{8}{13}$
18. Solve $\frac{1}{3}(3 - y) - \frac{2}{3}y = 8$. **(3.2)** -7
19. Write an equation of the line passing through $(0, -3)$ with a slope of $\frac{1}{4}$. **(5.2)**
20. Write an equation of the line passing through $\left(-\frac{1}{2}, \frac{3}{2}\right)$ and $(4, 2)$. **(5.3)**

3. $\begin{bmatrix} 4 & 2 \\ 6 & 14 \end{bmatrix}$ 4. $\begin{bmatrix} 5 & 13 \\ 1 & -6 \end{bmatrix}$ 11. Yes 12. No

7.6 Solving Systems of Linear Inequalities

What you should learn:

Goal 1 How to solve a system of linear inequalities by graphing

Goal 2 How to model a real-life situation using a system of linear inequalities

Why you should learn it:

You can model many real-life situations, such as finding boundaries of regions, with systems of linear inequalities.

Goal 1 Solving Inequalities by Graphing

In Lesson 6.5, you studied linear inequalities in two variables. Remember from that lesson that the graph of a linear inequality in two variables is a half plane. The boundary line of the half plane is dashed if the inequality is $<$ or $>$ and solid if the inequality is $\le$ or $\ge$. Here are some examples.

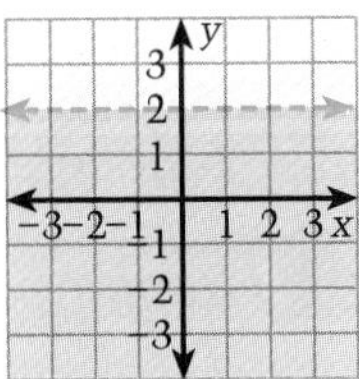

Graph of $y < 2$

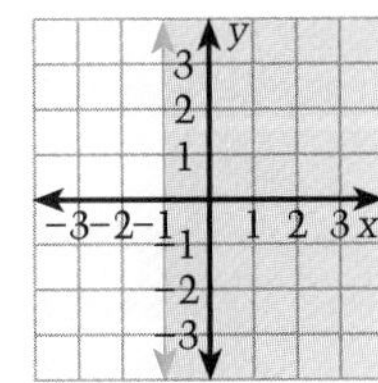

Graph of $x \ge -1$

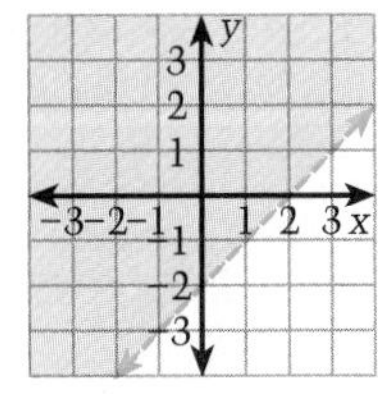

Graph of $y > x - 2$

Two or more linear inequalities form a system of linear inequalities. A solution is an ordered pair that is a solution of each inequality in the system. The graph of a system is the graph of *all* solutions of the system.

Example 1 *Graphing a System of Linear Inequalities*

Sketch the graph of the system of linear inequalities.

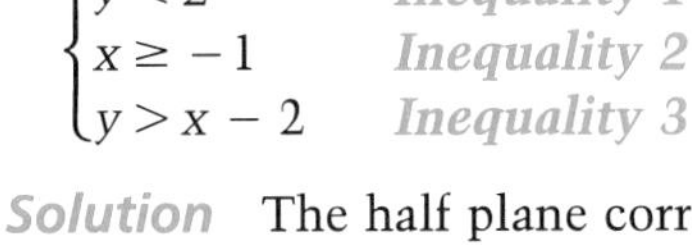

$$\begin{cases} y < 2 & \text{Inequality 1} \\ x \ge -1 & \text{Inequality 2} \\ y > x - 2 & \text{Inequality 3} \end{cases}$$

Solution The half plane corresponding to each inequality is shown above. The graph of the system is the *intersection* of the three half planes. For instance, the point (0, 0) is in the graph of the system because it is a solution of each inequality. The point (0, 3) is not in the graph of the system because it is not a solution of the first inequality. (It is a solution of the other two inequalities, but that isn't good enough to be a solution of the system.) ■

When sketching the graph of a system of linear inequalities, it is helpful to find each corner point (or *vertex*). For instance, the graph at the left has three corner points: $(-1, 2)$, $(-1, -3)$, and $(4, 2)$.

Problem of the Day

In 1977 Elson Darius married the sister of his widow. How did he do that?

He married the sister first, and later married the woman who eventually became his widow.

ORGANIZER

Warm-Up Exercises

1. What is a half-plane? Part of the coordinate plane that falls on one side of the graph of a line.
2. Decide if the given ordered pairs are solutions of the linear inequalities.
   a. $2x - y < 3$
      (3, 4), (−2, −6)
   b. $x > 3y$
      (0, 0), (9, 2)
   c. $2y \le -6x + 9$
      (−4, 8), (2, 11)

   a. Both are solutions, b. (9, 2) is a solution, c. (−4, 8) is a solution.
3. Sketch the graph of each linear inequality.
   a. $y \ge -3$
   b. $x < 7$
   c. $y > 2x - 3$

a. The graph is the region above the solid line $y = -3$.

b. The graph is the region to the left of the dashed line $x = 7$.

c.

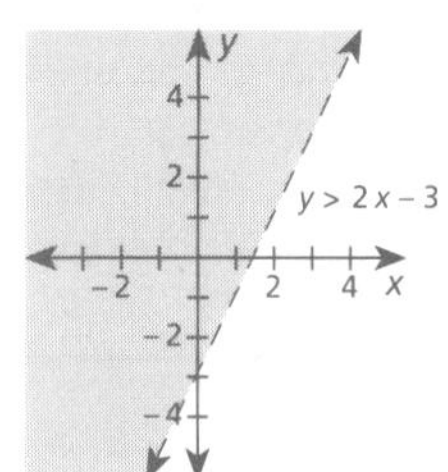

Lesson Resources

Teaching Tools
- Transparencies: 4, 11, 12
- Problem of the Day: 7.6
- Warm-up Exercises: 7.6
- Answer Masters: 7.6

Extra Practice: 7.6

LESSON Notes

Example 1

Point out to students that the linear system of inequalities consists of three inequalities. Emphasize that a solution of the system is any ordered pair that satisfies each linear inequality in the system. Consequently the solution of the system may not be a half-plane but, rather, the intersection of two or more half-planes.

A helpful practice when solving systems of linear inequalities by graphing is to use colored pencils to keep track of the half-planes that make up the system. Each half-plane is shaded with a different color. The intersection of all the different colors represents the solution of the system. Indicate that the intersections of corresponding linear equations of the system of linear inequalities determine the vertices of the system.

Examples 2–3

Observe that using a test point is not required to determine the half-planes in these examples because the inequalities are expressed in slope-intercept form. Ask students to verify this observation by using test points and formulating rules for locating the appropriate half-plane when linear inequalities are expressed in slope-intercept form.

Example 4

This is a good example to discuss in class. Ask how the linear inequalities were determined. Ask how the vertices can be found. Point out that this system consists of four linear inequalities.

Example 2 Graphing a System of Linear Inequalities

Sketch the graph of the system of linear inequalities.

$$\begin{cases} y < 4 & \text{Inequality 1} \\ y > 1 & \text{Inequality 2} \end{cases}$$

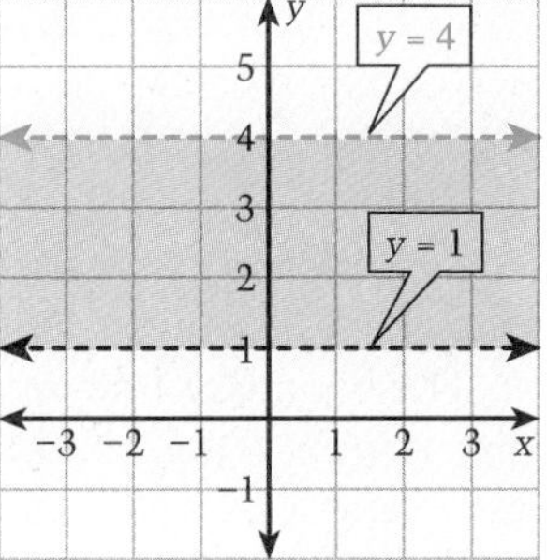

Solution The graph of the first inequality is the half plane *below* the horizontal line

$y = 4.$ *Upper boundary*

The graph of the second inequality is the half plane *above* the horizontal line

$y = 1.$ *Lower boundary*

The graph of the system is the horizontal band that lies *between* the two horizontal lines (where $y < 4$ and $y > 1$). ■

Example 3 Graphing a System of Linear Inequalities

Sketch the graph of the system of linear inequalities. Find a vertex of the graph.

$$\begin{cases} y \le x + 2 & \text{Inequality 1} \\ y \ge -\frac{1}{2}x + 5 & \text{Inequality 2} \end{cases}$$

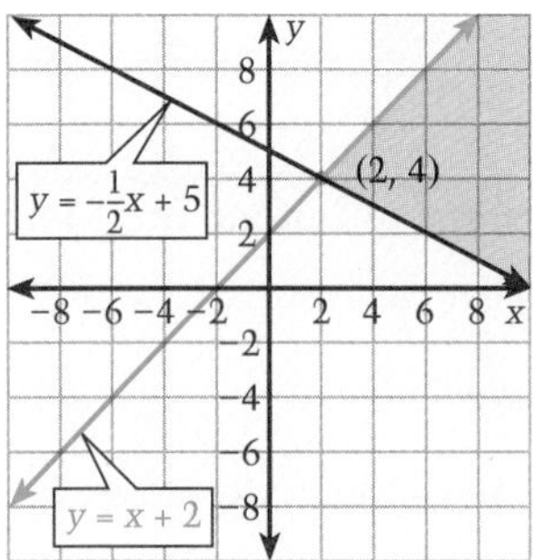

Solution The graph of the first inequality is the half plane *on and below* the line $y = x + 2$. The graph of the second is the half plane *on and above* the line $y = -\frac{1}{2}x + 5$. The graph of the system is the shaded region shown at the left. The vertex of the graph is the point of intersection of the two lines $y = x + 2$ and $y = -\frac{1}{2}x + 5$, which is (2, 4). ■

Here are some guidelines that may help you sketch the graph of a system of linear inequalities.

Graphing a System of Linear Inequalities

1. Sketch the line that corresponds to each inequality. (Use a dashed line for inequalities with < or > and a solid line for inequalities with ≤ or ≥.)
2. Lightly shade the half plane that is the graph of each linear inequality. (Colored pencils may help you distinguish the different half planes.)
3. The graph of the system is the intersection of the shaded half planes. (If you used colored pencils, it is the region that has been shaded with *every* color.)

Goal 2 Modeling a Real-Life Situation

Example 4 *Finding the Boundaries of a Region*

You are redecorating your bedroom with a scenic mural. One corner of the mural has to be cut away around the edges of a built-in bookcase. Use the bookcase dimensions shown (in feet) to find a system of inequalities that defines the region to be cut out.

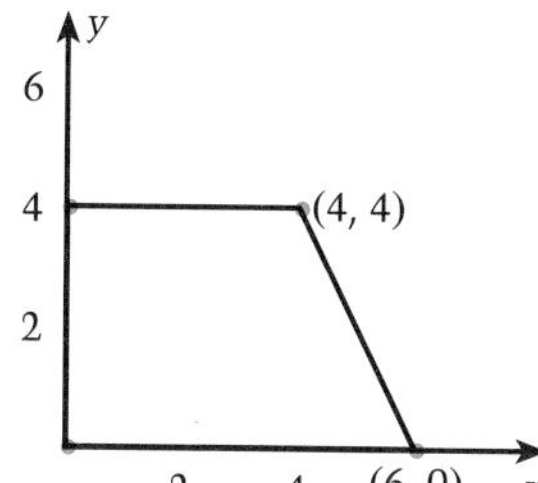

Solution First, define a coordinate plane that covers the entire mural, using the bottom and left edges as axes. Next, find the boundary for the diagonal line by writing an equation of the line passing through the points (4, 4) and (6, 0). The equation is $y = -2x + 12$. The system of linear inequalities that describes the region follows.

$$\begin{cases} y \leq 4 & \textit{On and below } y = 4 \\ y \geq 0 & \textit{On and above x-axis} \\ x \geq 0 & \textit{On and to right of y-axis} \\ y \leq -2x + 12 & \textit{On and below } y = -2x + 12 \end{cases}$$

■

Communicating about ALGEBRA

SHARING IDEAS about the Lesson

Use a Graph Describe or sketch the graph of the system.

A. Which quadrant forms the graph of $\begin{cases} x < 0 \\ y > 0 \end{cases}$? II

B. Sketch the graph of the system. See margin.

$$\begin{cases} y \geq 0 & \textit{Inequality 1} \\ x \geq 0 & \textit{Inequality 2} \\ y < -x + 4 & \textit{Inequality 3} \end{cases}$$

C. Sketch the graph of the system. See margin.

$$\begin{cases} y \geq 0 & \textit{Inequality 1} \\ x \geq 0 & \textit{Inequality 2} \\ y < -x + 4 & \textit{Inequality 3} \\ y > -x + 2 & \textit{Inequality 4} \end{cases}$$

Extra Examples

Here are additional examples similar to **Examples 1–3.**

1. Sketch the graph of the system of linear inequalities. Label the vertices of each graph.

   a. $\begin{cases} y \geq -4 \\ x < 2 \\ y \leq x + 2 \end{cases}$

   b. $\begin{cases} x \leq 6 \\ x \geq -2 \\ y \leq x + 2 \end{cases}$

   c. $\begin{cases} y > 2x - 3 \\ y < -x + 2 \end{cases}$

   Check student's graphs. For example, system a. produces the triangular region bordered by $y = -4$ (solid), $x = 2$ (dashed), and $y = x + 2$ (solid).

Check Understanding

1. Describe a solution of a system of linear inequalities. See page 381.
2. What procedure(s) can be employed to find the solution of a system of linear inequalities? (*Hint:* There are at least two approaches.) See pages 381–382.
3. How are the vertices of the graph of a system of linear inequalities determined? See page 382.

Communicating about ALGEBRA

Answers

B.

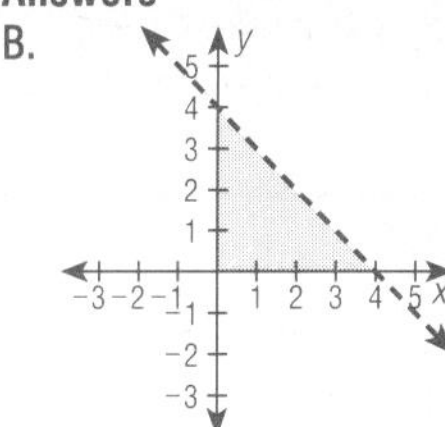

C.

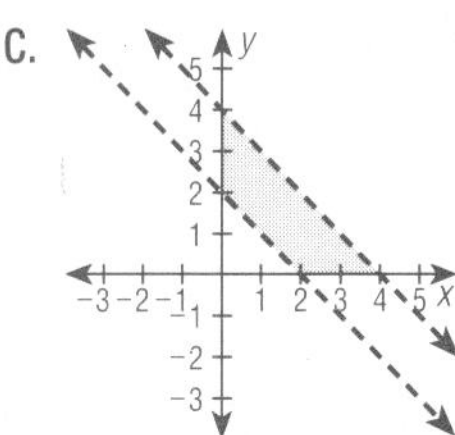

EXERCISE Notes

ASSIGNMENT GUIDE

Basic/Average: Ex. 1–6, 7–12, 13–27 odd, 35–38, 41–47 odd

Above Average: Ex. 1–6, 7–12, 13–27 odd, 35–39, 43, 47–49

Advanced: Ex. 1–6, 7–12, 13–31 odd, 33–39, 48, 49

Selected Answers
Exercises 1–6, 7–47 odd

✪ More Difficult Exercises
Exercises 29–32, 35–39, 48, 49

Guided Practice

▶ **Ex. 1–2** Refer students to Example 1.

▶ **Ex. 3–6** COOPERATIVE LEARNING These exercises can be used as a good board activity with pairs of students obtaining solutions together.

Independent Practice

▶ **Ex. 7–12** These exercises should be assigned as a group. Remind students to pay attention to solid versus dotted lines.

▶ **Ex. 16, 19, 20** It may be more efficient to sketch the graph for these exercises by eliminating the fractional coefficients first.

EXERCISES

Guided Practice

CRITICAL THINKING about the Lesson

1. **True or False?** A solution of a system of linear inequalities is an ordered pair that is a solution of any one of the inequalities in the system. Explain.
False; "any" should be "each."

2. What is a *vertex* of the graph of a system of linear inequalities?
Point of intersection of 2 lines in the system

In Exercises 3–6, decide whether the ordered pair is a solution of the system of linear inequalities. Justify your answer algebraically and graphically.

$$\begin{cases} x + y \le 2 \\ x - y > 0 \\ y \ge -2 \end{cases}$$

3. $(1, 0)$ It is. **4.** $(4, 3)$ It is not. **5.** $(1, -1)$ It is. **6.** $(-1, 2)$ It is not.

See Additional Answers for justifications.

Independent Practice

In Exercises 7–12, match the graph with one of the systems of linear inequalities.

7. c
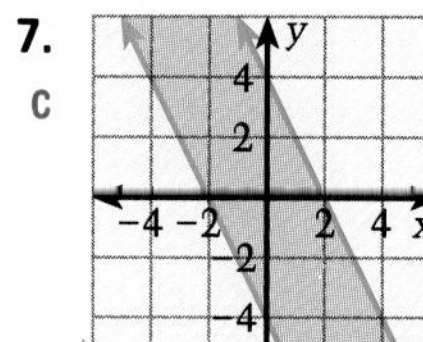

8. f
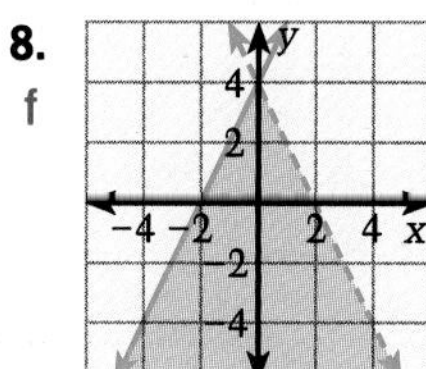

9. d
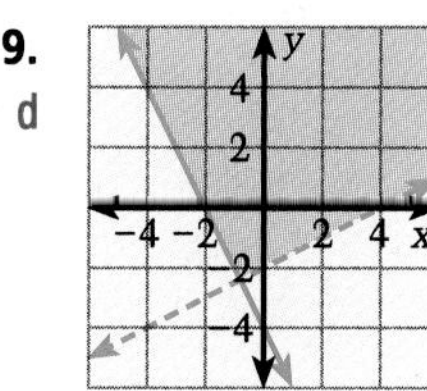

10. a
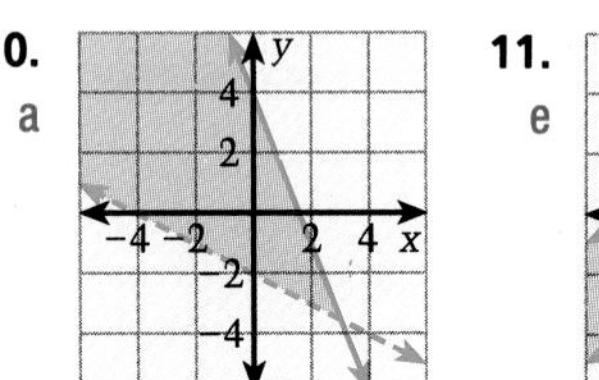

11. e
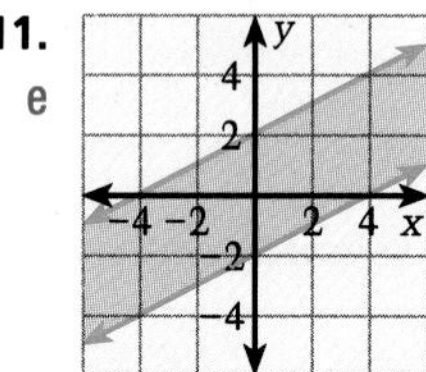

12. b
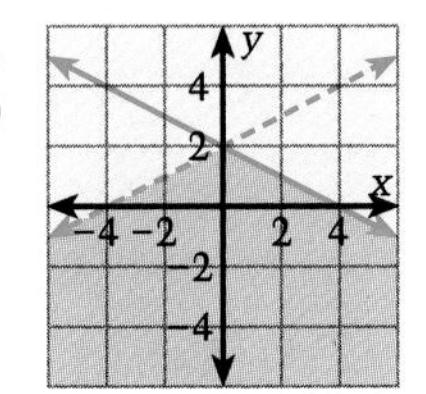

a. $\begin{cases} 2x + y \le 4 \\ x + 2y > -4 \end{cases}$

b. $\begin{cases} x + 2y \le 4 \\ -x + 2y < 4 \end{cases}$

c. $\begin{cases} 2x + y \le 4 \\ 2x + y \ge -4 \end{cases}$

d. $\begin{cases} 2x + y \ge -4 \\ x - 2y < 4 \end{cases}$

e. $\begin{cases} x - 2y \ge -4 \\ x - 2y \le 4 \end{cases}$

f. $\begin{cases} 2x + y < 4 \\ -2x + y \le 4 \end{cases}$

In Exercises 13–24, sketch the graph of the system of linear inequalities. See Additional Answers.

13. $\begin{cases} x \ge 0 \\ y \ge 0 \\ x \le 2 \\ y \le 4 \end{cases}$

14. $\begin{cases} x \ge -1 \\ y \ge -1 \\ x \le 1 \\ y \le 2 \end{cases}$

15. $\begin{cases} x + y \le 1 \\ -x + y \le 1 \\ y \ge 0 \end{cases}$

16. $\begin{cases} \frac{3}{2}x + y < 3 \\ x > 0 \\ y > 0 \end{cases}$

17. $\begin{cases} x + y \le 5 \\ x \ge 2 \\ y \ge 0 \end{cases}$

18. $\begin{cases} 2x + y \ge 2 \\ x \le 2 \\ y \le 1 \end{cases}$

19. $\begin{cases} -\frac{3}{2}x + y < 3 \\ \frac{1}{4}x + y > -\frac{1}{2} \\ 2x + y < 3 \end{cases}$

20. $\begin{cases} -\frac{1}{7}x + y < \frac{36}{7} \\ \frac{5}{2}x + y < \frac{5}{2} \\ \frac{6}{5}x + y > \frac{6}{5} \end{cases}$

21. $\begin{cases} x \geq 1 \\ x - 2y \leq 3 \\ 3x + 2y \geq 9 \\ x + y \leq 6 \end{cases}$

22. $\begin{cases} x + y < 10 \\ 2x + y > 10 \\ x - y < 2 \end{cases}$

23. $\begin{cases} x - 3y \geq 3 \\ x - 3y \leq 12 \end{cases}$

24. $\begin{cases} x + y < 3 \\ x + y > -1 \end{cases}$

See Additional Answers.

In Exercises 25–28, find the vertices of the graph of the system.

25. $\begin{cases} x + y \leq 12 \\ 3x - 4y \leq 15 \\ x \geq 0 \\ y \geq 0 \end{cases}$ (0, 12) (9, 3) (5, 0) (0, 0)

26. $\begin{cases} \frac{1}{2}x + y \leq 2 \\ x - y \leq 1 \\ x \geq 0 \\ y \geq 0 \end{cases}$ (2, 1) (0, 2) (1, 0) (0, 0)

27. $\begin{cases} 5x - 3y \geq -7 \\ 3x + y \leq 7 \\ x + 5y \leq -7 \end{cases}$ (1, 4) (3, −2) (−2, −1)

28. $\begin{cases} x + 2y \geq 0 \\ 5x - 2y \leq 0 \\ -x + y \leq 3 \end{cases}$ (0, 0) (−2, 1) (2, 5)

In Exercises 29–32, write a system of linear inequalities that defines the polygon. See margin.

	Polygon	Vertices		Polygon	Vertices
✪ **29.**	Rectangle	(2, 1), (5, 1), (5, 7), (2, 7)	✪ **30.**	Parallelogram	(0, 0), (4, 0), (1, 4), (5, 4)
✪ **31.**	Triangle	(0, 0), (5, 0), (2, 3)	✪ **32.**	Triangle	(−1, 0), (1, 0), (0, 1)

33. *Planning a Party* You are planning a party and want to have two types of hot snacks: stuffed mushrooms and cheese sticks. Each stuffed mushroom costs \$0.25 to make, and each cheese stick costs \$0.15 to make. You want to have enough so that each person can have at least two stuffed mushrooms and at least three cheese sticks. There are going to be at most 40 people at the party, and you do not want to spend more than \$45. Write a system of linear inequalities that shows the various numbers of stuffed mushrooms and cheese sticks that you could make. Graph your result.

33.–34. See Additional Answers for graphs.
33. $0.25x + 0.15y \leq 45$, $x \geq 80$, $y \geq 120$

34. *Work Schedule* Suppose you can work a total of no more than 20 hours a week at your two jobs. Your baby-sitting job pays \$3 an hour, and your cashier job pays \$5 an hour. You need to earn at least \$80 a week. Write a system of linear inequalities that shows the various numbers of hours you could work at each job. Graph your result. $x \geq 0$, $y \geq 0$, $x + y \leq 20$, $3x + 5y \geq 80$

An Herb Garden **In Exercises 35–38, use the picture of the herb garden.** See below.

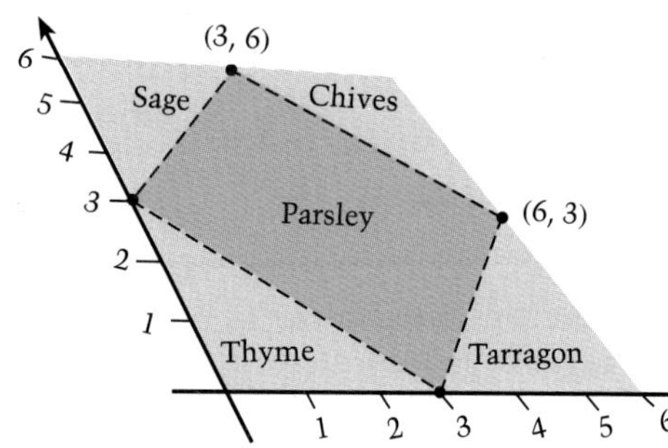

✪ **35.** Find a system of inequalities that defines the region containing thyme.

✪ **36.** Find a system of inequalities that defines the region containing sage.

✪ **37.** Find a system of inequalities that defines the region containing parsley.

✪ **38.** Find the area of each of the herb regions.

35. $x \geq 0$, $y \geq 0$, $y \leq -x + 3$

36. $x \geq 0$, $y \leq 6$, $y \geq x + 3$

37. $y \geq -x + 3$, $y \leq x + 3$, $y \leq -x + 9$, $y \geq x - 3$

38. Each corner area $= \frac{9}{2}$, center area $= 18$

▶ **Ex. 25–29** Refer students to Example 3 to remind them that the vertices of the graph give the points of intersection for each pair of lines.

▶ **Ex. 35–38** These exercises should be assigned as a group.

▶ **Ex. 29–32, 35–39, 48, 49** These exercises require students to generate the system from a given set of data.

Answers

29. $x \geq 2$, $y \geq 1$, $x \leq 5$, $y \leq 7$
30. $y \leq 4x$, $y \geq 0$, $y \geq 4x - 16$, $y \leq 4$
31. $y \geq 0$, $y \leq \frac{3}{2}x$, $y \leq -x + 5$
32. $y \geq 0$, $y \leq x + 1$, $y \leq -x + 1$

EXTEND Ex. 29–32

MAKING DESIGNS Have students create designs in the coordinate plane, which they then must describe by using systems of linear inequalities.

Have students create their individual designs (with vertices included if needed to determine the system), then challenge members of their cooperative group to write the system that describes it. For example, find the system that describes this design: My Tent.

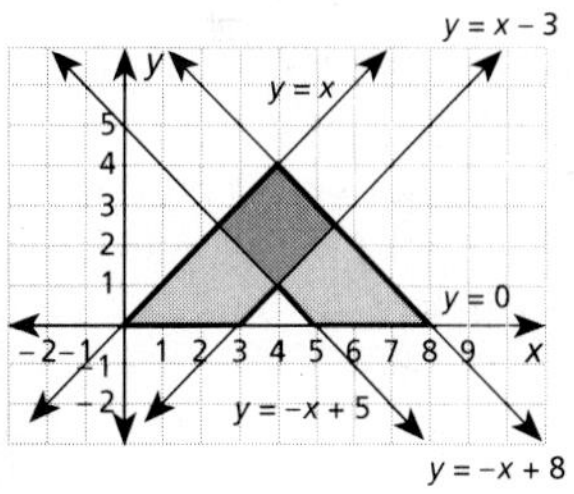

Answer

$\begin{cases} y \geq 0 \\ y \leq x \\ y \leq -x + 8 \\ y \geq x - 3 \\ y \geq -x + 5 \end{cases}$

▶ **Ex. 48, 49** GROUP ACTIVITY These exercises may be fun to do as an in-class activity.

Answer

49. No. Each inequality describes a half-plane; when you form one point of a star, you eliminate two other points of the star.

Enrichment Activities

These activities require students to go beyond lesson goals.

Use quick graphs to examine systems of absolute-value inequalities.

Solve:

1. $\begin{cases} y > |x + 6| \\ y \le x + 1 \end{cases}$ No solution

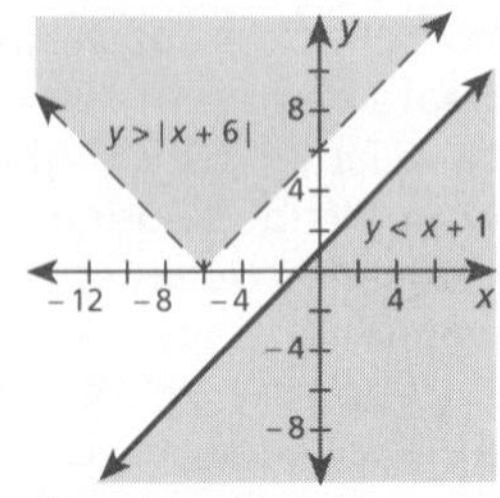

2. $\begin{cases} y \ge |x| + 4 \\ y \le 10 \end{cases}$

The region bordered by the equation $y = |x| + 4$ for $y \le 10$

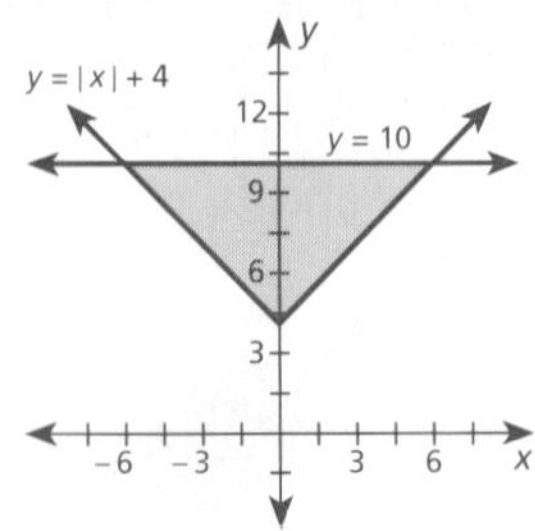

39. *Desktop Publishing* Suppose you are creating a "family-tree" book. You scan this photo into your computer. Then you use the "cut-and-paste" feature to electronically *crop* and insert the photo into a page. Use the coordinates below to write a system of linear inequalities defining the crop lines and the part of the photo containing the people.
$x \ge 0,\ x \le 19,\ y \ge 0,\ \frac{-2}{19}x + y \le 6$

(19, 8) •

(0, 6) •

(0, 0) •

(19, 0) •

School Kids with Teacher, 1924 ***The entire student body and faculty of a country school in Sanish (now New Town), North Dakota. (Note from one of the authors, Ron Larson: The fourth and fifth from the right are my mom and dad.)***

Integrated Review

In Exercises 40–43, sketch the line. See Additional Answers.

40. $2x - 6y = 18$ **41.** $x + 3y = -9$ **42.** $-3x + y = 21$ **43.** $7x + 2y = 28$

In Exercises 44–47, sketch the graph of the inequality. See Additional Answers.

44. $x \ge 2$ **45.** $y < 2 - x$ **46.** $2y - x \ge 4$ **47.** $4x - 2y \le 12$

48. $x \ge 1,\ x \le 19,\ y \ge 1,\ y \le 11$

Exploration and Extension

Complements of "Old Glory" **In Exercises 48 and 49, use the following drawing.**

If you stare at the white dot in the center of the flag for 30 seconds, then look at a blank sheet of white paper, the image of the United States flag should appear *in its correct colors!* This optical sensation stems from a property of human vision that no one quite understands. If the human eye stares at a color long enough, it begins to produce sensations of the complementary color. (Red is the complement of green, white is the complement of black, and blue is the complement of orange.)

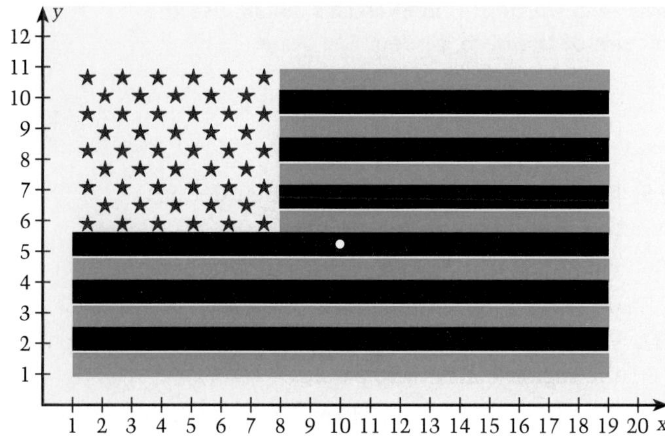

48. Write a system of linear inequalities that defines the region containing the flag.

49. Would it be possible to write a system of linear inequalities whose graph is one of the stars in the flag? Explain. See margin.

7.7 Exploring Data: Linear Programming

What you should learn:

Goal 1 How to solve a linear programming problem

Goal 2 How to model a real-life situation using linear programming

Why you should learn it:

With linear programming, you can model many real-life situations, such as finding ways to maximize profit and minimize cost.

Goal 1 Linear Programming Problems

Many real-life problems involve a process called *optimization,* which means finding the minimum or maximum value of some quantity. In this lesson, you will study one type of optimization process called **linear programming.**

A linear programming problem consists of a system of linear inequalities called **constraints** and an **objective quantity,** such as cost or profit, that can be minimized or maximized. To find the minimum or maximum value of the objective quantity, first find all the vertices of the graph of the constraint inequalities. Then evaluate the objective quantity *at each vertex.* The smallest value is the minimum, and the largest value is the maximum.

Example 1 *Solving a Linear Programming Problem*

Find the minimum value and maximum value of

$C = 3x + 2y$ *Objective quantity*

subject to the following constraints.

$$\begin{cases} x \geq 0 \\ y \geq 0 \\ x + y \leq 4 \end{cases}$$ *Constraints*

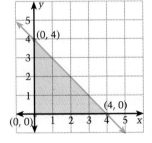

Solution The graph of the constraint inequalities is shown at the left. The three vertices are (0, 0), (4, 0) and (0, 4). To find the minimum and maximum values of C, evaluate $C = 3x + 2y$ at each of the three vertices.

At (0, 0): $C = 3(0) + 2(0) = 0$ *Minimum value of C*
At (4, 0): $C = 3(4) + 2(0) = 12$ *Maximum value of C*
At (0, 4): $C = 3(0) + 2(4) = 8$

The minimum value of C is 0. It occurs when $x = 0$ and $y = 0$. The maximum value is 12, when $x = 4$ and $y = 0$. ∎

It is important to try evaluating C at other points in the graph of the constraints. The value of C will be always greater than or equal to 0 and less than or equal to 12. For instance, at the point (2, 1) the value of C is 8.

Problem of the Day

The six faces of a cube are lettered A through F. Here are three views of the cube.

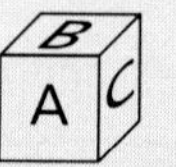

What letter is on the face opposite the A? The C? D, F

ORGANIZER

Warm-Up Exercises

1. Find the point of intersection of each pair of equations.
   a. $\begin{cases} y = 2x - 6 \\ y = -x + 3 \end{cases}$ (3, 0)
   b. $\begin{cases} y = 3 \\ y = 2x - 2 \end{cases}$ $(\frac{5}{2}, 3)$
   c. $\begin{cases} x - 3y = 8 \\ x = -1 \end{cases}$ $(-1, -3)$
2. Evaluate $y = 3x + 8$ at
   a. $x = 0$ $y = 8$
   b. $x = -3$ $y = -1$
   c. $x = 11$ $y = 41$

Lesson Resources

Teaching Tools
Transparencies: 4, 11, 12
Problem of the Day: 7.7
Warm-up Exercises: 7.7
Answer Masters: 7.7
Extra Practice: 7.7

LESSON Notes

Examples 1–3

Students may ask why the vertices are the only points of the solution of the system of constraint inequalities that must be used to optimize the objective quantity. The answer is beyond the scope of this text. It may, however, suffice to tell students that knowing how to use the vertices is the result of a basic theorem in the branch of discrete mathematics known as linear programming.

Example 4

This is an excellent example of the usefulness of linear programming in solving real-life problems. (See Communicating about Algebra on page 389.)

Extra Examples

Here are additional examples similar to **Examples 1–4.**

1. Find the minimum and maximum value of $D = 3x - 2y$, the objective quantity, subject to the following constraints.

$$\begin{cases} x \geq 0 \\ y \geq 0 \\ y \leq -0.5x + 10 \\ y \leq -x + 16 \end{cases}$$

Minimum value of D is 0 at $x = 0$ and $y = 0$.
Maximum value of D is 48 at $x = 16$ and $y = 0$.

2. Suppose you own a skateboard manufacturing company that makes a premium skateboard by using two very different processes. The hours of skilled labor and machine time per skateboard can be found in the matrix below. You can use up to 1796 hours of skilled labor and up to 3500 hours of machine time. How many skateboards should be made by each process to maximize your profits if skateboards made by Process A cost $35 each, and skateboards made by Process B cost $15 each?

	Process	
Assembly hours	**A**	**B**
Skilled labor	3	1
Machine time	1	3

The maximum profit is $24,580, obtained by making 236 skateboards with Process A and 1088 skateboards with Process B.

Example 2 Solving a Linear Programming Problem

Find the minimum value and maximum value of

$C = 4x + y$ *Objective quantity*

subject to the following constraints.

$$\begin{cases} x \geq 1 \\ x \leq 4 \\ y \geq 2 \\ y \leq 5 \end{cases}$$ *Constraints*

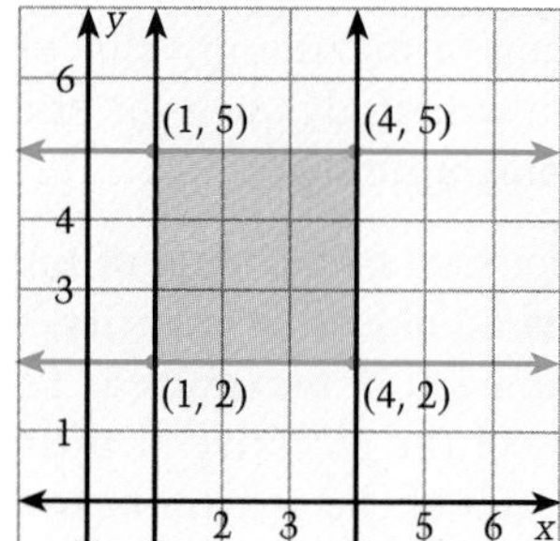

Solution The graph of the constraint inequalities is shown at the left. The four vertices are (1, 2), (1, 5), (4, 5) and (4, 2). To find the minimum and maximum values of C, evaluate $C = 4x + y$ at each of the four vertices.

At (1, 2): $C = 4(1) + (2) = 6$ *Minimum value of C*
At (1, 5): $C = 4(1) + (5) = 9$
At (4, 5): $C = 4(4) + (5) = 21$ *Maximum value of C*
At (4, 2): $C = 4(4) + (2) = 18$

The minimum value of C is 6. It occurs when $x = 1$ and $y = 2$. The maximum value of C is 21. It occurs when $x = 4$ and $y = 5$. ■

Example 3 Solving a Linear Programming Problem

Find the minimum value and maximum value of

$C = 5x + 4y$ *Objective quantity*

subject to the following constraints.

$$\begin{cases} x \geq 0 \\ y \geq 0 \\ x + 2y \leq 4 \\ x - y \leq 1 \end{cases}$$ *Constraints*

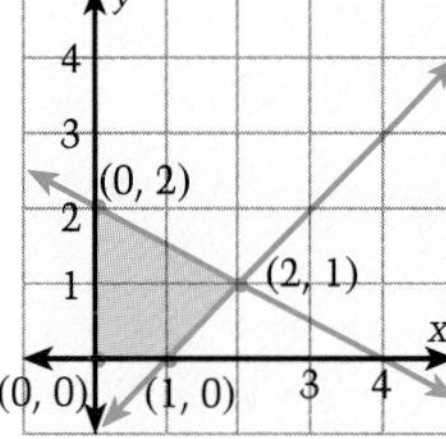

Solution The graph of the constraint inequalities is shown at the left. The four vertices are (0, 0), (1, 0), (2, 1) and (0, 2). To find the minimum and maximum value of C, evaluate $C = 5x + 4y$ at each of the four vertices.

At (0, 0): $C = 5(0) + 4(0) = 0$ *Minimum value of C*
At (1, 0): $C = 5(1) + 4(0) = 5$
At (2, 1): $C = 5(2) + 4(1) = 14$ *Maximum value of C*
At (0, 2): $C = 5(0) + 4(2) = 8$

The minimum value of C is 0. It occurs when $x = 0$ and $y = 0$. The maximum value of C is 14. It occurs when $x = 2$ and $y = 1$. ■

Goal 2 Modeling a Real-Life Situation

Connections
Discrete Mathematics

Example 4 *Finding the Maximum Profit*

You own a bicycle manufacturing plant and can assemble bicycles using two processes. The hours of unskilled labor, machine time, and skilled labor *per bicycle* are given in the matrix. You can use up to 4200 hours of unskilled labor and up to 2400 hours each of machine time and skilled labor. How many bicycles should you assemble by each process to obtain a maximum profit?

Assembly Hours	Process A	Process B
Unskilled labor	3	3
Machine time	1	2
Skilled labor	2	1

Process A earns a profit of $45 per bike, and Process B earns a profit of $50 per bike.

Solution Let a and b represent the number assembled with each process. Because you want a maximum profit, P, the objective quantity is

$P = 45a + 50b.$ *Profit factors for each process*

The constraints are as follows.

$$\begin{cases} 3a + 3b \le 4200 & \text{Unskilled labor: Up to 4200 hours} \\ a + 2b \le 2400 & \text{Machine time: Up to 2400 hours} \\ 2a + b \le 2400 & \text{Skilled labor: Up to 2400 hours} \\ a \ge 0, b \ge 0 & \text{Cannot produce negative amounts.} \end{cases}$$

The profits at vertices of the region are as follows.

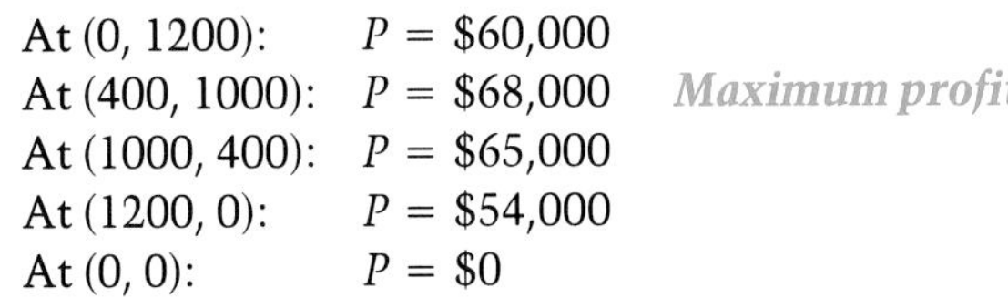

At (0, 1200): $P = \$60,000$
At (400, 1000): $P = \$68,000$ *Maximum profit*
At (1000, 400): $P = \$65,000$
At (1200, 0): $P = \$54,000$
At (0, 0): $P = \$0$

The maximum profit is obtained by making 400 bicycles with Process A and 1000 bicycles with Process B. ■

Communicating about ALGEBRA

SHARING IDEAS about the Lesson

Vary the Conditions How many bicycles should you produce by each process of Example 4 if the profit per bicycle is $30 for Process A, $65 for Process B? **0, 1200**

Check Understanding

1. How are the vertices of a linear programming problem used to solve the problem? See Examples 1–4.
2. What skills from solving systems of linear inequalities are needed to solve linear programming problems? Graphing skills, substitution skills
3. Why might the system of linear inequalities in a linear programming problem be called constraints? (*Hint:* Consider Example 4.) The linear inequalities can be thought of as boundaries.

Communicating about ALGEBRA

How many bicycles should you produce by each process of Example 4 if the profit per bicycle is as follows:

a. Process A: $50, Process B: $45
b. Process A: $65, Process B: $30

a. 1000, 400; b. 1200, 0

EXTEND *Communicating*

Ask students if there are activities (such as publishing the school's yearbook, completing student government projects, or performing tasks on their after-school/weekend jobs) in which maximizing or minimizing some objective quantity would make a real difference.

EXERCISE Notes

ASSIGNMENT GUIDE

Basic/Average: Ex. 1–6, 7–25 odd, 33, 35–41 odd

Above Average: Ex. 1–6, 7–13 odd, 15–25 odd, 26, 35, 43

Advanced: Ex. 1–6, 15–26, 37, 43–44

Selected Answers
Exercises 1–6, 7–41 odd

✪ More Difficult Exercises
Exercises 25–26, 43–44

Guided Practice

▶ **Ex. 1–6** Use these exercises as a homework-readiness check in class. They provide student reflection on critical information from the lesson.

Answer

4.

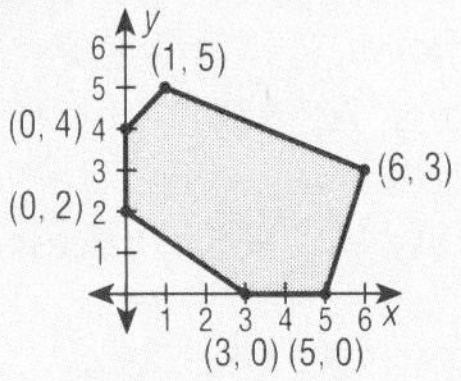

Independent Practice

▶ **Ex. 7–14** These exercises prepare students with the two skills necessary to do Exercises 15–22 correctly.

Answers

7.

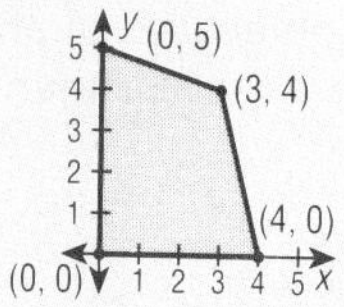

8.

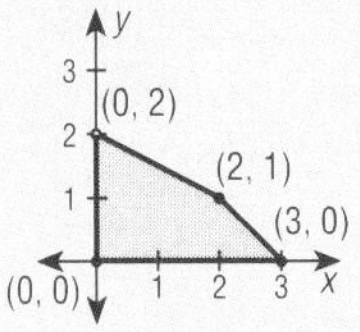

EXERCISES

Guided Practice

CRITICAL THINKING about the Lesson They define region of possible answers

1. How are constraints used to solve a linear programming problem?
2. How is the objective quantity used to solve a linear programming problem? It is evaluated at vertices formed by constraints.
3. What does the word *optimize* mean?
   Find the minimum or maximum value of some quantity.

In Exercises 4–6, use the objective quantity $C = 5x + 7y$ and the vertices (0, 2), (0, 4), (1, 5), (6, 3), (5, 0), and (3, 0).

4. Graph the region determined by the given vertices. See margin.
5. Find the minimum value of the objective quantity. $C = 14$
6. Find the maximum value of the objective quantity. $C = 51$

Independent Practice

In Exercises 7–10, sketch the graph of the constraints. Label the vertices of the graph. See margin.

7. $\begin{cases} x + 3y \le 15 \\ 4x + y \le 16 \\ x \ge 0 \\ y \ge 0 \end{cases}$

8. $\begin{cases} x + y \le 3 \\ x + 2y \le 4 \\ x \ge 0 \\ y \ge 0 \end{cases}$

9. $\begin{cases} 2x + 2y \le 10 \\ x + 2y \le 6 \\ x \ge 0 \\ y \ge 0 \end{cases}$

10. $\begin{cases} x + 2y \le 6 \\ 2x + y \le 6 \\ x \ge 0 \\ y \ge 0 \end{cases}$

In Exercises 11–14, find the minimum and maximum values of the objective quantity C.

11. $C = 3x + 2y$
Constraints:
$\begin{cases} x + y \le 7 \\ x \le 5 \\ y \le 5 \\ x \ge 0 \\ y \ge 0 \end{cases}$ $C = 0$, $C = 19$

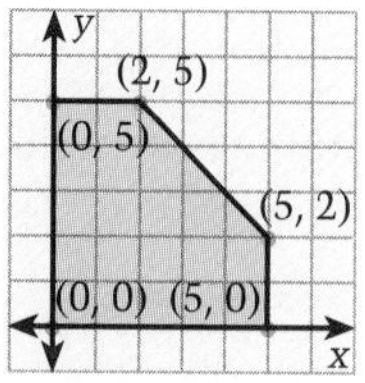

12. $C = 3x + 4y$
Constraints:
$\begin{cases} 3x + y \le 18 \\ x + y \le 8 \\ x + 4y \le 20 \\ x \ge 0 \\ y \ge 0 \end{cases}$ $C = 0$, $C = 28$

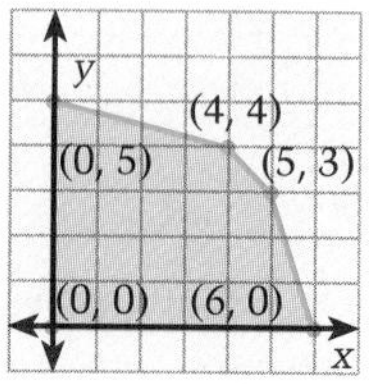

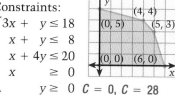

13. $C = 5x + 6y$
Constraints:
$\begin{cases} 2x + 3y \ge 6 \\ 4x + y \le 24 \\ x + 5y \le 25 \\ x \ge 0 \\ y \ge 0 \end{cases}$
$C = 12$, $C = 49$

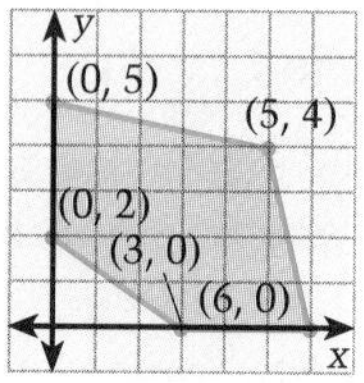

14. $C = 6x + 7y$
Constraints:
$\begin{cases} 6x + 5y \le 30 \\ 3x + 2y \ge 6 \\ x \ge 0 \\ y \ge 0 \end{cases}$
$C = 12$, $C = 42$

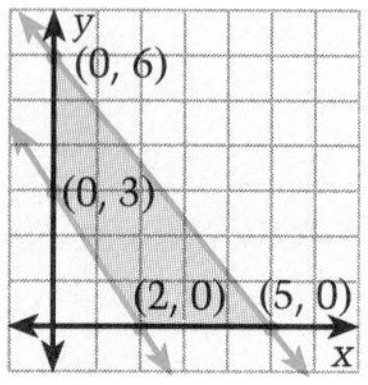

In Exercises 15–22, find the minimum and maximum values of the objective quantity *C*.

15. $C = x + 2y$

$C = 0,\ C = 12$

Constraints:
$$\begin{cases} x + y \le 6 \\ x \ge 0 \\ y \ge 0 \end{cases}$$

16. $C = 3x + y$

$C = 0,\ C = 6$

Constraints:
$$\begin{cases} 2x + y \le 4 \\ x \ge 0 \\ y \ge 0 \end{cases}$$

17. $C = 6x + 5y$

$C = 27,\ C = 50$

Constraints:
$$\begin{cases} x \le 5 \\ x \ge 2 \\ y \le 4 \\ y \ge 3 \end{cases}$$

18. $C = 7x + 2y$

$C = 25,\ C = 59$

Constraints:
$$\begin{cases} x \le 7 \\ x \ge 3 \\ y \le 5 \\ y \ge 2 \end{cases}$$

19. $C = 4x + y$

$C = 0,\ C = 24$

Constraints:
$$\begin{cases} x + 2y \le 8 \\ x + y \le 6 \\ x \ge 0 \\ y \ge 0 \end{cases}$$

20. $C = 2x + 3y$

$C = 0,\ C = 12$

Constraints:
$$\begin{cases} x + 3y \le 9 \\ 2x + y \le 8 \\ x \ge 0 \\ y \ge 0 \end{cases}$$

21. $C = 4x + 3y$

$C = 6,\ C = 29$

Constraints:
$$\begin{cases} 2x + 3y \ge 6 \\ 3x - 2y \le 9 \\ x + 5y \le 20 \\ x \ge 0 \\ y \ge 0 \end{cases}$$

22. $C = x + 6y$

$C = 3,\ C = 24$

Constraints:
$$\begin{cases} 2x + 3y \ge 6 \\ 3x - 2y \le 9 \\ x + 5y \le 20 \\ x \ge 0 \\ y \ge 0 \end{cases}$$

23. *Mixing Juices* You work at a store that sells bottled fruit juice. You are trying to decide how much of two different blends of apple and pineapple juice to prepare.

- Blend A: 30% apple juice, 70% pineapple juice, profit of \$0.60 per liter.
- Blend B: 60% apple juice, 40% pineapple juice, profit of \$0.50 per liter.
- Fruit juice in stock: 1200 liters of apple juice, 1400 liters of pineapple juice.

Use linear programming to find how much of each blend you should prepare to obtain a maximum profit. (Hint: How much apple juice is in a liter of Blend A?)

24. *Truck Farming* You have 100 acres of land to grow lettuce and peas. You want to decide how many acres of each crop to plant to make a maximum profit.

- Lettuce: Investment per acre is \$120, and income per acre is \$150.
- Peas: Investment per acre is \$200, and income per acre is \$260.
- Maximum amount you can invest: \$15,000

Use linear programming to find how many acres of each crop you should plant. (Hint: How much is the investment for x acres of lettuce?)

23. A: 1200 L, B: 1400 L, maximum profit: \$1420

24. Lettuce: 0 acres, peas: 75 acres, maximum profit: \$4500

✪ **25.** *Computer Monitors* You are the manager of a store that sells home computers. You are getting ready to order next month's stock and are trying to decide how many of each of two models of monitors to order to obtain a maximum profit.

- Model A: Your cost is \$250; your profit over cost is \$45.
- Model B: Your cost is \$400; your profit over cost is \$50.
- Your combined sales of Models A and B will not exceed 250 units.
- You do not want to order more than \$70,000 worth of the two models.

Use linear programming to find how many of each model to order.

A: 200, B: 50, maximum profit: \$11,500

Answers

9.

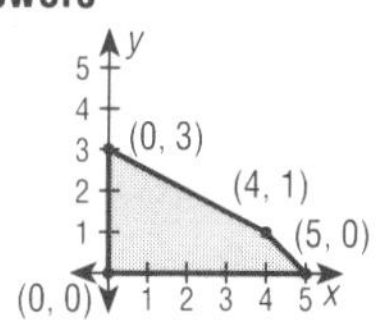

10.

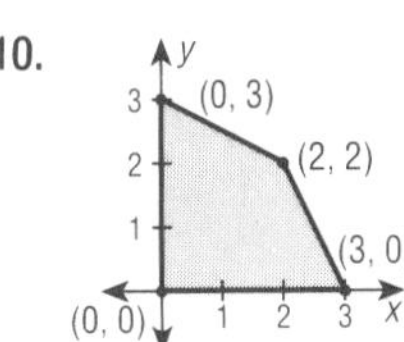

EXTEND Ex. 15–22
WRITING PROBLEMS Have students pick one or two of these exercises and write real-life applications that could be solved by the systems.

▶ **Ex. 23–26** These exercises are similar to the model provided in Example 4. Advise students on how to separate the quantity to be maximized or minimized from the constraint inequalities.

Answers

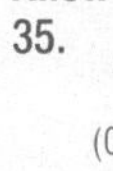

35.

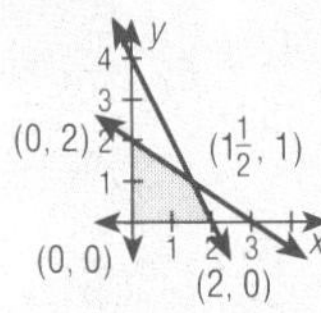

36.

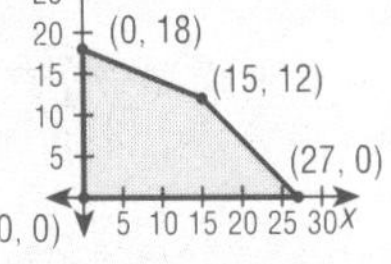

37.

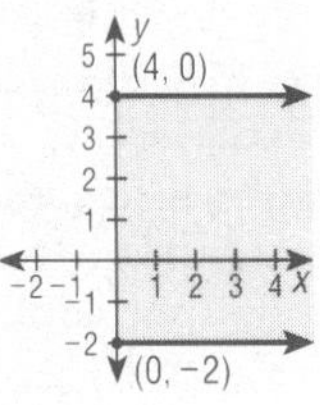

38.

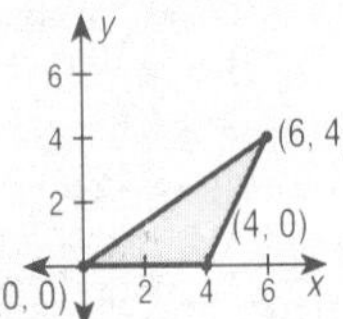

44. $x + 2y \le 4$, $x - y \le 1$, $x \ge 0$, $y \ge 0$; for example: $C = x + y$

Enrichment Activity

This activity requires students to go beyond lesson goals.

RESEARCH Use the school library to investigate actual applications of linear programming in industry, such as its use in airline flight schedules.

26. *Health Drink* You are hunting for a special health drink that you read about, but you can't find one with the recommended amounts of protein and carbohydrates. You find two drinks that have the right ingredients but the wrong proportions. You wonder how much of each to drink and still keep your cost down.

- Recommended Minimum Daily Amount: 3 cups for protein, 5 cups for carbohydrates
- Blend A: 25% protein; 75% carbohydrate; cost is \$0.25 per cup
- Blend B: 50% protein; 50% carbohydrate; cost is \$0.20 per cup

Use linear programming to find how many cups of each blend you could drink to meet your minimum daily requirement with a minimum cost. A: 4 cups, B: 4 cups, minimum cost: \$1.80

Integrated Review

In Exercises 27–34, sketch the graph of the inequality. See Additional Answers.

27. $x \le 4$ **28.** $y \le 3$ **29.** $3x > y$ **30.** $y \le 3 + x$

31. $5x + 3y \ge -15$ **32.** $6 - 2y < x$ **33.** $y + 3x > 6$ **34.** $y > 2x - 4$

In Exercises 35–38, sketch the graph of the system of linear inequalities. See margin.

35. $\begin{cases} 2x + y \le 4 \\ 2x + 3y \le 6 \\ x \ge 0 \\ y \ge 0 \end{cases}$ **36.** $\begin{cases} x + y \le 27 \\ 2x + 5y \le 90 \\ x \ge 0 \\ y \ge 0 \end{cases}$ **37.** $\begin{cases} y \ge -2 \\ y \le 4 \\ x \ge 0 \end{cases}$ **38.** $\begin{cases} 2x - 3y \ge 0 \\ 2x - y \le 8 \\ y \ge 0 \end{cases}$

In Exercises 39–42, solve the system of linear equations.

39. $\begin{cases} 3x - 2y = 18 \\ 4x + y = 13 \end{cases}$ (4, −3)

40. $\begin{cases} 2x + 5y = 26 \\ -x + 3y = 20 \end{cases}$ (−2, 6)

41. $\begin{cases} -5x + y = 9 \\ 5x + 6y = 19 \end{cases}$ (−1, 4)

42. $\begin{cases} 4x - 2y = 2 \\ 3x + 3y = 24 \end{cases}$ (3, 5)

Exploration and Extension

In Exercise 43, find the minimum value of the objective quantity *C*. Explain why the objective quantity does not have a maximum value. Constraints are unbounded for $x \ge 0$ and $y \ge 0$.

43. $C = 2x + 2y$

Constraints:

$$\begin{cases} x + 2y \ge 3 \\ 3x + 2y \ge 5 \\ x \ge 0 \\ y \ge 0 \end{cases}$$

$C = 4$

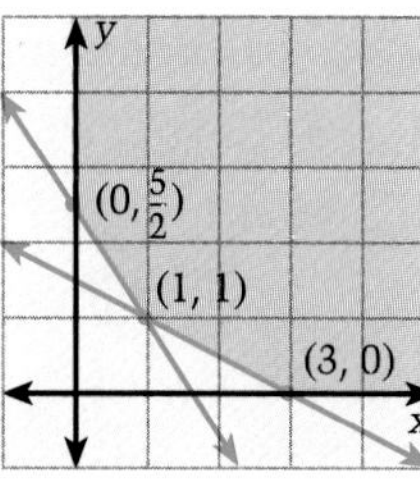

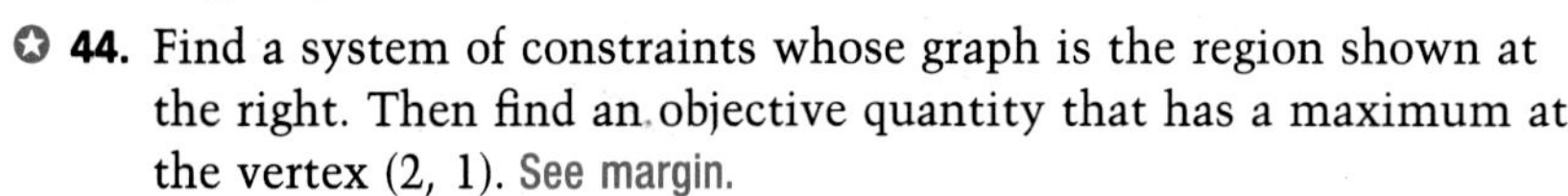

44. Find a system of constraints whose graph is the region shown at the right. Then find an objective quantity that has a maximum at the vertex (2, 1). See margin.

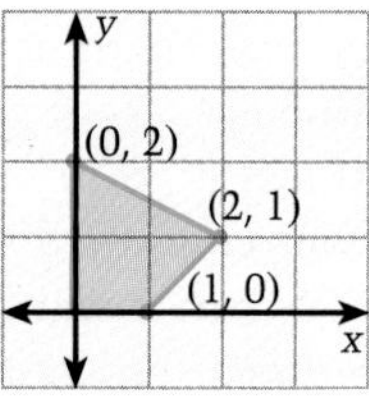

7

Chapter Summary

What did you learn?

Skills

1. Solve a system of linear equations
   - by graphing. **(7.1)**
   - by substitution. **(7.2)**
   - by linear combinations. **(7.3)**
2. Identify the number of solutions of a linear system. **(7.5)**
3. Solve a system of linear inequalities. **(7.6)**
4. Solve a linear programming problem. **(7.7)**

Strategies

5. Use a system of linear equations to model a real-life situation. **(7.1–7.5)**
6. Use a system of linear inequalities to model a real-life situation. **(7.6)**

Exploring Data

7. Use linear programming to solve a real-life problem. **(7.7)**

Why did you learn it?

Systems of linear equations and systems of linear inequalities occur often as models of real-life situations. One common application is to use a system of linear equations to compare two changing quantities and the rates at which they are changing. For instance, you might want to compare the number of a certain item manufactured in the United States in each of the past ten years with the number that were imported, and how those numbers changed over the ten years. Another common type of application is finding the parts of a mixture. For instance, you might want to find the number of adult tickets and the number of children's tickets sold for a performance. The information you gain from studying systems that model real situations will help you make decisions about those situations.

How does it fit into the bigger picture of algebra?

In this chapter, you studied *systems* of linear equations and linear inequalities. There are three basic methods for *solving* a system made up of two linear equations. You can graph both equations and find the point of intersection of the graphs. You can solve for a variable in one equation and substitute the resulting expression in the other equation. Or you can add a multiple of one equation to the other equation. One of your goals should be to become skilled at all three methods. Another goal—just as important, is to learn the advantages and disadvantages of each method. Using an appropriate method to solve a system of linear equations is part of becoming an efficient problem solver.

ORGANIZER

TIME SCHEDULE: All levels, two days

The Chapter Summary helps students organize the main ideas of the chapter. In this chapter, real situations were explored and questions about them were answered by using linear programming. The skills and concepts of the chapter reinforced the need to understand systems of linear inequalities.

Work with students to review the skills, strategies, and concepts of the chapter. It is suggested that the first day of review be a combination of you and the students working problems from the Chapter Review exercises. The first day's homework assignment should be the basis for the second day of review.

COOPERATIVE LEARNING

Encourage students to study together. Emphasize the importance of teaching a classmate how to perform a skill or how to conceptualize the ideas in the chapter. When students work together, everyone wins. The students receiving help get additional instruction, and the students giving help gain a deeper understanding of the skills and concepts involved.

Chapter SUMMARY

Review the chapter by asking students, "What did you learn? Why did you learn it? How does it fit with the other algebra skills and concepts you have learned?"

Chapter REVIEW

ASSIGNMENT GUIDE

Basic/Average: Ex. 3–39 multiples of 3, 41–42, 46–47.

Above Average: Ex. 3–39 multiples of 3, 43–46, 48

Advanced: Ex. 3–42 multiples of 3, 43–45, 48–49

✪ More Difficult Exercises

Exercises 43–45, 48, 49

▶ **Ex. 1–49** Since the Chapter Review exercises parallel the list of Skills, Strategies, and Exploring Data statements on page 393, tell students to use the Chapter Summary to identify the lesson in which a skill or concept was taught if they are uncertain about how to do a given exercise. This creates the need for students to categorize the task to be performed and to locate relevant information in the chapter about the task, both beneficial activities.

Answers

1.

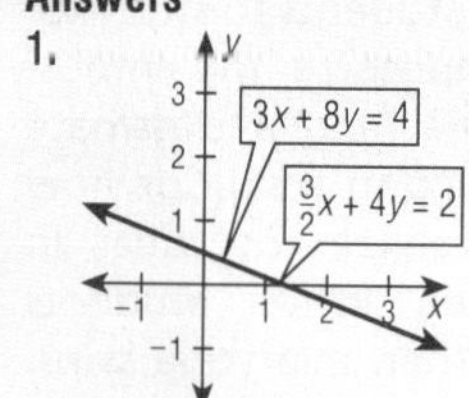

2.

3.

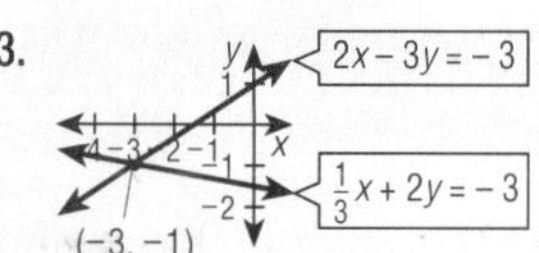

Chapter REVIEW

In Exercises 1–6, solve the linear system by graphing. (7.1, 7.5) See margin.

1. $\begin{cases} 3x + 8y = 4 \\ \frac{3}{2}x + 4y = 2 \end{cases}$ Any point on the line $3x + 8y = 4$

2. $\begin{cases} -x + y = 3 \\ x + y = 7 \end{cases}$ (2, 5)

3. $\begin{cases} 2x - 3y = -3 \\ \frac{1}{3}x + 2y = -3 \end{cases}$ (−3, −1)

4. $\begin{cases} 3x + 5y = 7 \\ 6x + 10y = 8 \end{cases}$ No solution

5. $\begin{cases} -2x - 3y = -8 \\ 6x + 2y = 24 \end{cases}$ (4, 0)

6. $\begin{cases} 5x - 2y = -17 \\ 3x + 6y = -3 \end{cases}$ (−3, 1)

In Exercises 7–18, use substitution to solve the linear system. (7.2, 7.5)

9. Any point on the line $x + 3y = 6$ 15. Any point on the line $5x - 2y = 4$

7. $\begin{cases} x + 3y = 0 \\ 4x - 2y = -14 \end{cases}$ (−3, 1)

8. $\begin{cases} 3x - 4y = 6 \\ -x + y = -1 \end{cases}$ (−2, −3)

9. $\begin{cases} \frac{1}{3}x + y = 2 \\ 2x + 6y = 12 \end{cases}$

10. $\begin{cases} 4x - 3y = -2 \\ 4x + y = 4 \end{cases}$ $(\frac{5}{8}, \frac{3}{2})$

11. $\begin{cases} x + \frac{3}{8}y = 1 \\ 8x + 3y = 4 \end{cases}$ No solution

12. $\begin{cases} -2x + y = 4 \\ 4x - 3y = -7 \end{cases}$

12. $(-\frac{5}{2}, -1)$

13. $\begin{cases} x - y = 14 \\ 7x - 15y = 50 \end{cases}$ (20, 6)

14. $\begin{cases} 3x + y = 11 \\ x + 2y = -3 \end{cases}$ (5, −4)

15. $\begin{cases} 5x - 2y = 4 \\ x - 0.4y = 0.8 \end{cases}$

16. $\begin{cases} -2x - 5y = 7 \\ 7x + y = -8 \end{cases}$ (−1, −1)

17. $\begin{cases} 0.25x - 1.25y = 10.25 \\ x - 5y = 20 \end{cases}$ No solution

18. $\begin{cases} x + \frac{1}{2}y = 2 \\ 6x + 4y = 15 \end{cases}$ $(\frac{1}{2}, 3)$

In Exercises 19–30, use linear combinations to solve the linear system. (7.3, 7.5) 24. Any point on the line $4x - 5y = 2$ 30. Any point on the line $5x - 8y = 4$

19. $\begin{cases} 8x + 3y = 15 \\ 5x - 2y = -10 \end{cases}$ (0, 5)

20. $\begin{cases} 3x + 5y = -16 \\ -2x + 6y = -36 \end{cases}$ (3, −5)

21. $\begin{cases} 7x - 3y = -1 \\ 4x + 6y = -16 \end{cases}$

21. (−1, −2)

22. $\begin{cases} 2x - 3y = 1 \\ -2x + 3y = 1 \end{cases}$ No solution

23. $\begin{cases} -4x - 5y = 7 \\ 3x + 10y = -24 \end{cases}$ (2, −3)

24. $\begin{cases} 4x - 5y = 2 \\ 600x - 750y = 300 \end{cases}$

25. $\begin{cases} -6x + 5y = 18 \\ 7x + 2y = 26 \end{cases}$ (2, 6)

26. $\begin{cases} 2x - 15y = -20 \\ -4x + 5y = 5 \end{cases}$ $(\frac{1}{2}, \frac{7}{5})$

27. $\begin{cases} 2x + y = 0 \\ x - 4y = 24 \end{cases}$ $(\frac{8}{3}, -\frac{16}{3})$

28. $\begin{cases} \frac{1}{2}x + \frac{1}{3}y = 8 \\ \frac{1}{4}x + \frac{1}{6}y = 12 \end{cases}$ No solution

29. $\begin{cases} \frac{1}{9}x + \frac{1}{3}y = 4 \\ -\frac{1}{3}x - 3y = -18 \end{cases}$ (27, 3)

30. $\begin{cases} \frac{5}{12}x - \frac{2}{3}y = \frac{1}{3} \\ \frac{1}{3}x - \frac{8}{15}y = \frac{4}{15} \end{cases}$

In Exercises 31–36, sketch the graph of the system of linear inequalities. (7.6) See margin.

31. $\begin{cases} x \geq -3 \\ y \geq 0 \\ y \leq 4 \end{cases}$

32. $\begin{cases} x < 5 \\ y > -2 \\ \frac{1}{2}x + y > -2 \end{cases}$

33. $\begin{cases} x \geq 0 \\ 3x - y \leq 7 \\ 2x + y \leq 2 \end{cases}$

34. $\begin{cases} x + y \leq 5 \\ x + y \geq -5 \\ x - y \leq 5 \\ x - y \geq -5 \end{cases}$

35. $\begin{cases} 3x + 2y \leq 12 \\ x + 2y \leq 8 \\ x \geq 0 \\ y \geq 0 \end{cases}$

36. $\begin{cases} \frac{2}{5}x + y \geq 4 \\ 3x + y \geq 6 \\ y \geq 0 \end{cases}$

In Exercises 37–39, find the maximum and minimum value of the objective quantity C. (7.7)

37. $C = 3x - 2y$

Constraints:

$$\begin{cases} x \geq 1 \\ x \leq 4 \\ y \geq -3 \\ y \leq 2 \end{cases}$$

$C = -1$, $C = 18$

38. $C = 4x + y$

Constraints:

$$\begin{cases} x \geq 0 \\ y \geq 0 \\ y \leq -2x + 6 \end{cases}$$

$C = 0$, $C = 12$

39. $C = x + 3y$

Constraints:

$$\begin{cases} x \geq 0 \\ y \geq 0 \\ -3x + 4y \leq 12 \\ 6x + y \leq 30 \end{cases}$$

$C = 0$, $C = 22$

In Exercises 40 and 41, use the following information. 40. (4, 3), (6, 0), (1, 0)

A triangle is formed by the lines $y = 0$, $x - y = 1$, and $3x + 2y = 18$.

40. Find the vertices of the triangle.

41. Find the area of the triangle. $7\frac{1}{2}$

42. *Common Nails* The manager of a hardware store orders the same number of 2-penny nails and 60-penny nails. The total weight of the order is 390 pounds. How many pounds of each did the manager order? 5 lb of 2-penny, 385 lb of 60-penny

Type of nail	2-Penny	60-Penny
Nails per pound	847	11

Southern and Northern Hemispheres **In Exercises 43 and 44, use the following information.**

The average daily high temperature in Auckland, New Zealand, decreases from January to May. The average daily high temperature in Mexico City, Mexico, increases during the same time of year.

$y = -3x + 84$, $y = 3x + 65$

Temperature Comparison

KEY: • Auckland • Mexico City

Auckland: (1, 81), (2, 78), (3, 75), (4, 72), (5, 69)

Mexico City: (1, 66.6), (2, 69.8), (3, 73), (4, 76.2), (5, 79.4)

Temperature (Fahrenheit)

Month (1 ↔ January)

43. Write equations for the average daily high temperatures in Auckland and Mexico City.

44. In which month is it most likely that the daily high temperatures for the two cities are closest together? Month 3 or March

45. *Concert Contract* You are a concert promoter and sign a contract with a rock group. In the contract, you guarantee an attendance of at least 12,500 and total ticket receipts of at least $225,000. The tickets to the concert will cost $15 and $25, and you agree to give the rock group $5 out of each $15 ticket and $7.50 out of each $25 ticket. Use linear programming to find the minimum amount of money you have guaranteed the rock group. (The graph of the constraints is shown at the right.) $71,875

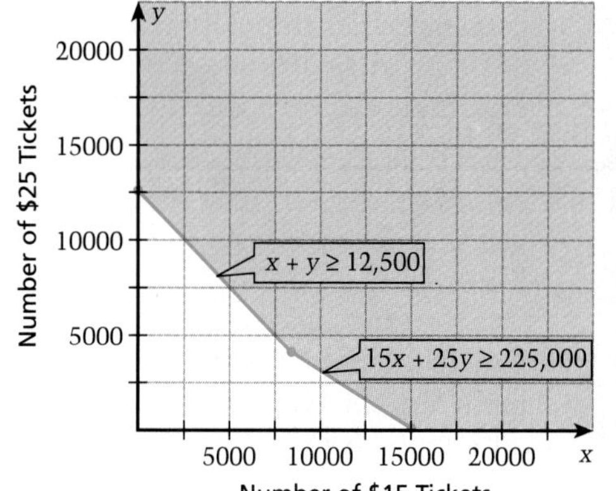

4.

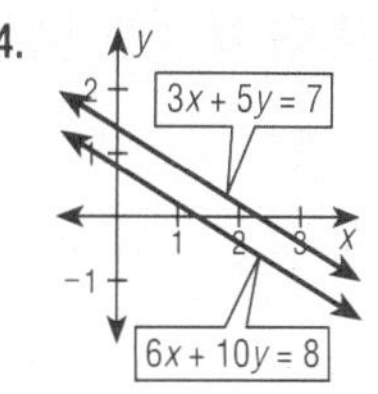

5.

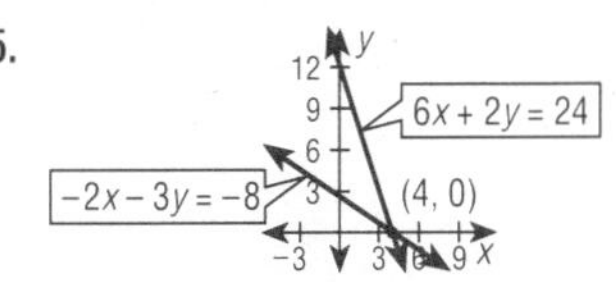

6.

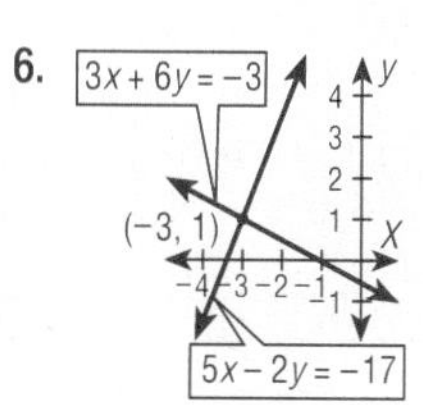

31.

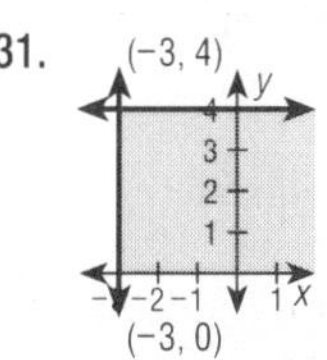

32.

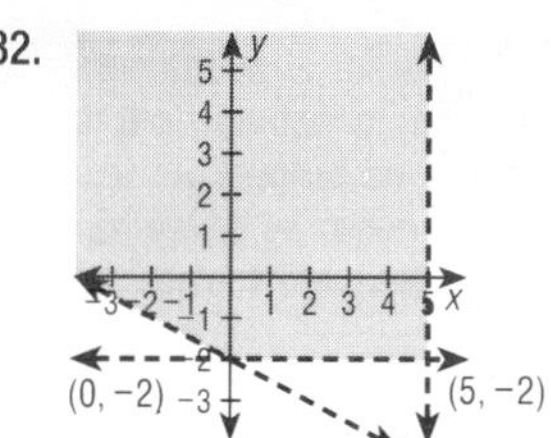

33.

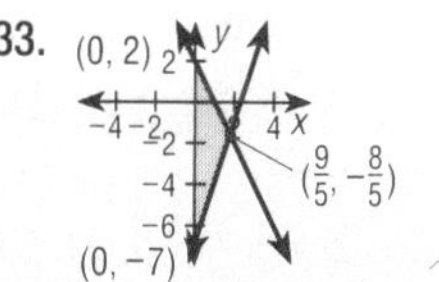

34.

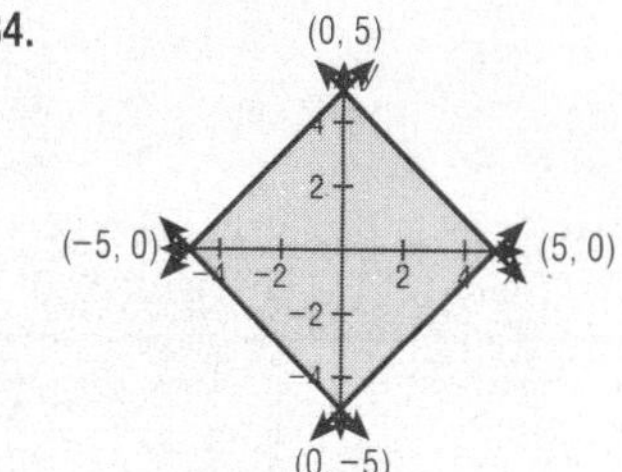

35.

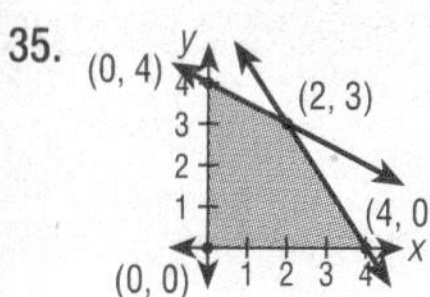

36.

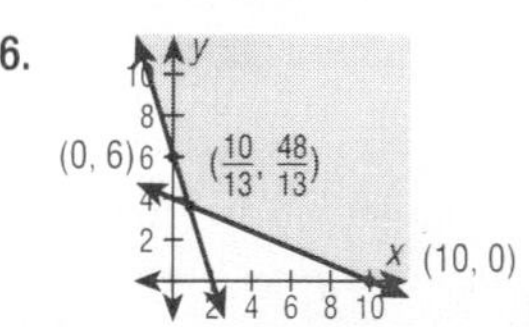

Answers
Chapter Test

1.

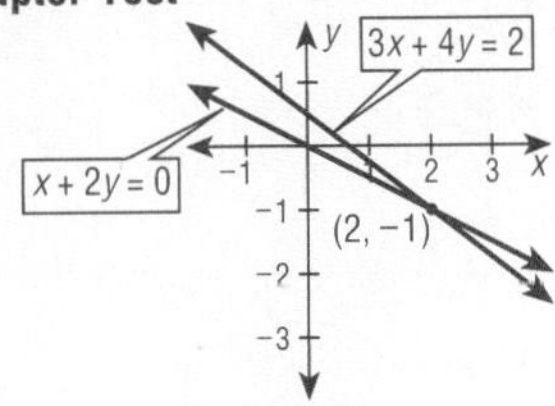

2.

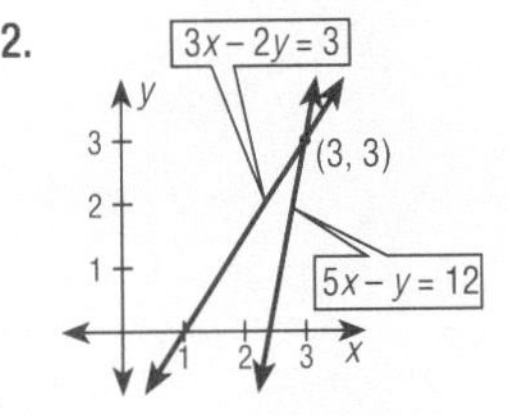

3.

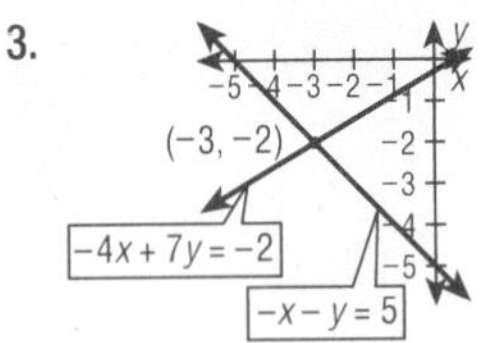

46. *Radio Frequency* One airplane is flying away from the VOR station. Its course is the line $3x - 2y = -2$. Another airplane is flying toward the VOR station on the line $x + y = 6$. Find the coordinates of the VOR station. **(2, 4)**

47. *On or Off Course?* The pilot of the first plane is alerted by the VOR equipment. The plane's coordinates are (10, 17). Is the plane off course or did the VOR equipment malfunction? **The plane is off course.**

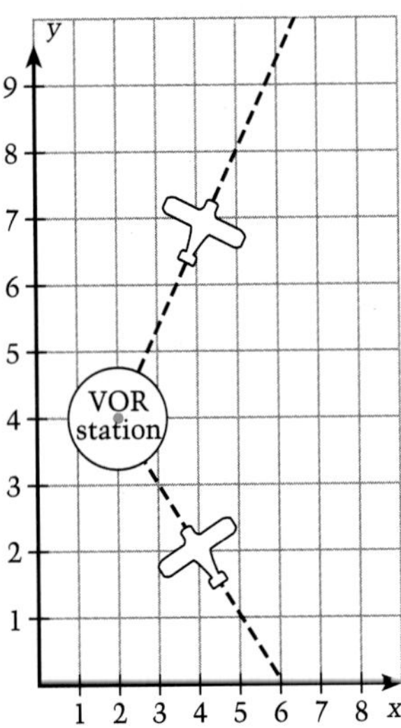

A VOR (very high frequency omnidirectional radio) station emits radio signals that enable a pilot to stay on course. If a plane veers off course, the VOR equipment alerts the pilot.

48. *Was That a Bird or a Plane?* A Beech P.35 leaves Dulles International Airport near Washington, D.C., and flies at 177 mph to Chicago's Midway Airport. The Beech flies at an altitude of 10,000 feet during the 600-mile trip. At the same time, a Douglas DC-9 leaves Midway Airport for Dulles International and flies at 561 mph at an altitude of 33,000 feet. Approximate the time that the DC-9 is over the Beech. **49 minutes**

49. *Pie-in-the-Sky?* Your friend Julie tells you that she flew from Reno, Nevada, to Salt Lake City, Utah, to watch the Utah Jazz play the Golden State Warriors. She flew in a Piper PA-31P Navajo at 266 mph. She claims to have made the 850-mile round–trip and watched the game in five hours. Is that possible? **Doubtful**

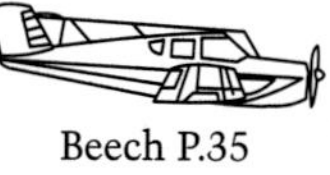
Beech P.35

Piper PA 31P

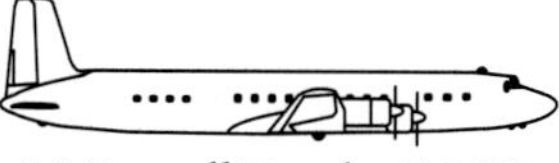
McDonnell Douglas DC-7C

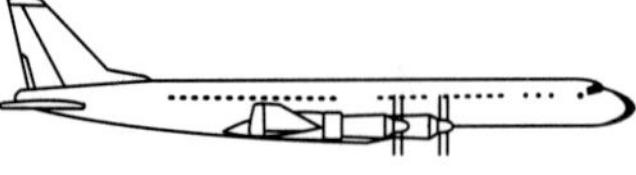
Tupolev TU-114

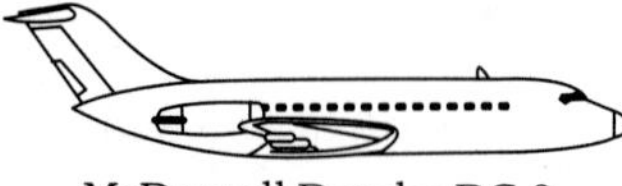
McDonnell Douglas DC-9

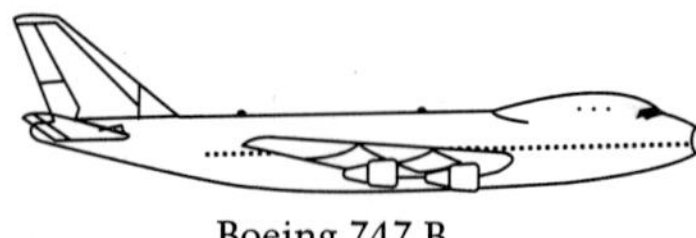
Boeing 747 B

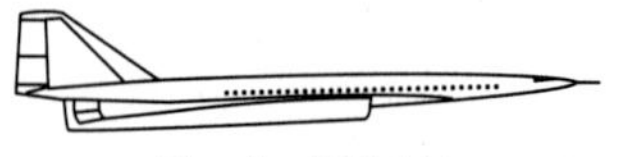
Tupolev TU-144

Chapter TEST

In Exercises 1–3, solve the linear system by graphing. **(7.1)** See margin.

1. $\begin{cases} x + 2y = 0 \\ 3x + 4y = 2 \end{cases}$ (2, −1)

2. $\begin{cases} 5x - y = 12 \\ 3x - 2y = 3 \end{cases}$ (3, 3)

3. $\begin{cases} -4x + 7y = -2 \\ -x - y = 5 \end{cases}$ (−3, −2)

In Exercises 4–9, use substitution to solve the linear system. **(7.2, 7.5)**

4. $\begin{cases} 2x - 5y = -12 \\ -4x + y = 6 \end{cases}$ (−1, 2)

5. $\begin{cases} -3x + 5y = 11 \\ x - 2y = -5 \end{cases}$ (3, 4)

6. $\begin{cases} -x + 3y = 4 \\ 2x - 6y = -8 \end{cases}$ Any point on the line $-x + 3y = 4$

7. $\begin{cases} 5x + y = 3 \\ 10x + 2y = 0 \end{cases}$ No solution

8. $\begin{cases} 6x + y = 12 \\ -4x - 2y = 0 \end{cases}$ (3, −6)

9. $\begin{cases} 7x + 4y = 5 \\ x - 6y = -19 \end{cases}$ (−1, 3)

In Exercises 10–15, use linear combinations to solve the linear system. **(7.3, 7.5)**

10. $\begin{cases} 8x - 4y = -11 \\ 6x + 2y = -2 \end{cases}$ $(-\frac{3}{4}, \frac{5}{4})$

11. $\begin{cases} -7x + 2y = -5 \\ 10x - 2y = 6 \end{cases}$ $(\frac{1}{3}, -\frac{4}{3})$

12. $\begin{cases} \frac{1}{4}x - \frac{3}{4}y = -2 \\ 2x - 6y = -16 \end{cases}$ Any point on the line $x - 3y = -8$

13. $\begin{cases} 6x + 7y = 5 \\ 4x - 2y = -10 \end{cases}$ $(-\frac{3}{2}, 2)$

14. $\begin{cases} 4x + 7y = -11 \\ 14x - 12y = -2 \end{cases}$ (−1, −1)

15. $\begin{cases} -x + \frac{1}{3}y = -6 \\ 3x - y = -16 \end{cases}$ No solution

In Exercises 16–18, graph the system of linear inequalities. **(7.6)** See margin.

16. $\begin{cases} y > \frac{3}{2}x + \frac{3}{2} \\ y < -\frac{1}{4}x - \frac{1}{2} \end{cases}$

17. $\begin{cases} x \geq 2 \\ y \leq 5 \\ y \geq x - 3 \end{cases}$

18. $\begin{cases} x > 1 \\ y \geq -1 \\ y > -2x + 5 \end{cases}$

In Exercises 19 and 20, find the minimum and maximum value of the objective quantity *C*. **(7.7)**

19. $C = 3x + 2y$ Constraints: $\begin{cases} x \leq 3 \\ x \geq 1 \\ y \leq 6 \\ y \geq 2 \end{cases}$

$C = 7$, $C = 21$

20. $C = 4x + 5y$ Constraints: $\begin{cases} 3x + y \leq 10 \\ x + y \leq 6 \\ x \geq 0 \\ y \geq 0 \end{cases}$

$C = 0$, $C = 30$

21. A local herb shop is producing two natural perfumes: *Gentle Rose* and *Rich Gardenia.* The owner of the shop asks you to help decide how much of each to produce. The owner, who has equipment that can make up to 3000 ounces of perfume, cannot afford to spend more than $9000. Use the table and linear programming to find how many bottles of each scent to produce to make a maximum profit. **(7.7)**
Rose: 600, gardenia: 1200, maximum profit: $8400

Scent	Gentle Rose	Rich Gardenia
Bottle size	2-ounce	1.5-ounce
Cost per bottle	$3	$6
Profit over cost (per bottle)	$4	$5

16.

17.
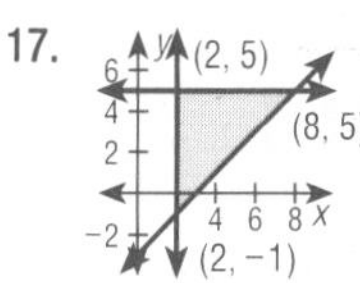

18.
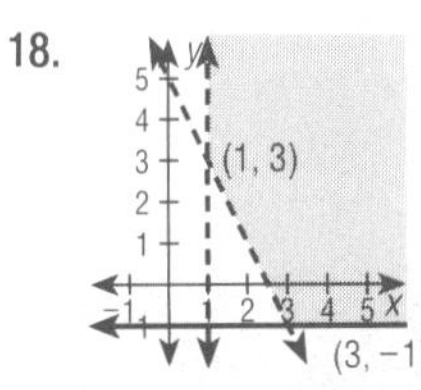

Real-life situations involving growth and decay can be modeled using exponential relationships. Population growth and inflation rates are just two of the types of topics that can be described by exponential models. Encourage students to use available calculators or computers to perform the more complicated computations. The rapid growth or decay of quantities found in exponential relationships makes scientific notation important. Scientific notation is used to represent very small or very large numbers. In calculator or computer displays, very small numbers or very large numbers may be automatically expressed in scientific notation.

COOPERATIVE LEARNING

Many of the applications in this chapter will appear difficult because of the very large and very small numbers used to express the algebraic models. Before assigning the application exercises for independent practice, you may find it beneficial to set aside some class time for students to work with partners and discuss possible calculator key sequences or other algebraic methods that can be used to solve the problems.

The Chapter Summary on page 445 provides you and the students with a synopsis of the chapter. It identifies key skills, concepts, and vocabulary. You may want to have students look at the Chapter Summary as an overview before beginning the chapter.

CHAPTER 8

Powers and Exponents

LESSONS

The cockpit displays shown here are actually part of a flight simulator that has been programmed, in this particular case, to represent for the pilot-in-training an approach to Hong Kong International Airport.

Real Life
Flight Training

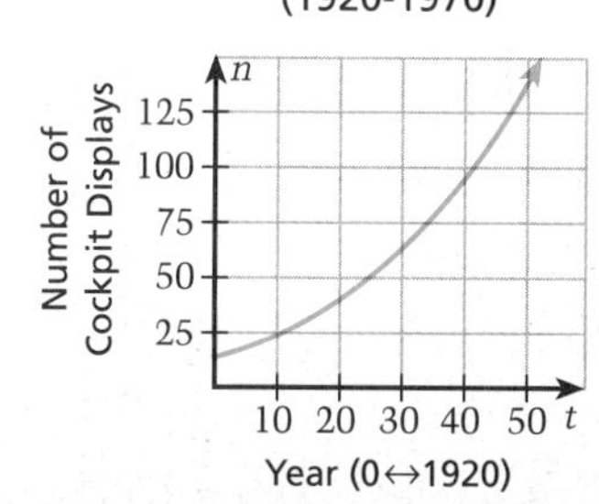

Between 1920 and 1970, the number, n, of cockpit displays in commercial aircraft increased about 4.5% per year. The French *Concorde,* built in 1970, had over 130 cockpit displays. A model that approximates this is $n = 15(1.045)^t$, where t is the year, with $t = 0$ corresponding to 1920. Concern for the amount of information that pilots can handle has reversed this trend. The *Airbus,* introduced in the late 1980's, has fewer than 100 cockpit displays.

(Source: Lockheed Corporation)

The study of powers and exponents is a first step toward modeling real-life situations by using other than linear relationships. Nonlinear relationships are frequently found in economics and in biological systems. Students will find it helpful to have a scientific or graphics calculator for exercises in this chapter.

According to the exponential model, $n = 15(1.045)^t$. Using this model, can you guess about how many cockpit displays could have been found in a commercial aircraft in 1920? In 1930? In 1940? Check by looking at the graph.
about 15, about 25, about 35

Powers and Exponents

PACING CHART

The daily Pacing Chart is meant to help you adjust your teaching pace. Students in the full course should finish the entire text by the end of the year. Students in the basic course may not complete the entire text in the school year. The Pacing Chart for each chapter contains suggestions for lessons that require more than one day and lessons that may be omitted for the basic course.

DAY	FULL COURSE	BASIC COURSE
1	8.1	8.1
2	8.2	8.2
3	8.3	8.2
4	8.4	8.3
5	Mid-Chapter Self-Test	8.4
6	8.5	Mid-Chapter Self-Test
7	8.6	8.5
8	8.6	8.6
9	8.6 & Using a Graphing Calculator	8.6
10	8.7	8.6 & Using a Graphing Calculator
11	8.7	8.7
12	8.7	8.7
13	Chapter Review	8.7
14	Chapter Review	Chapter Review
15	Chapter Test	Chapter Review
16		Chapter Test

CHAPTER ORGANIZATION

LESSON	PAGES	GOALS	MEETING THE NCTM STANDARDS
8.1	400-405	1. Use the multiplication properties of exponents to evaluate powers and simplify expressions 2. Use powers and the exponential change equation as models	Problem Solving, Communication, Connections, Structure, Geometry
8.2	406-411	1. Use negative and zero exponents in algebraic expressions 2. Use powers as models	Problem Solving, Communication, Connections, Geometry, Structure
Mixed Review	412	Review of algebra and arithmetic	
Milestones	412	Early computers	Connections
8.3	413-418	1. Use the division properties of exponents to evaluate powers and simplify expressions 2. Use powers as models	Problem Solving, Communication, Structure, Connections, Reasoning
8.4	419-424	1. Use scientific notation to express large and small numbers 2. Perform operations with numbers in scientific notation, with and without a calculator	Problem Solving, Communication, Connections, Technology
Mixed Review	424	Review of algebra and arithmetic	
Mid-Chapter Self-Test	425	Diagnose student weaknesses and remediate with correlated Reteaching worksheets	
8.5	426-431	Use scientific notation to solve real-life problems	Problem Solving, Communication, Connections, Reasoning
8.6	432-437	1. Use the compound interest formula 2. Use models for exponential growth to solve real-life problems	Problem Solving, Communication, Connections, Geometry
Using a Calculator	438	Use a calculator to find a best-fitting exponential growth or decay model	Technology, Statistics
8.7	439-444	Use models for exponential growth and decay to solve real-life problems	Problem Solving, Communication, Connections, Geometry
Chapter Summary	445	A restatement of what has been learned, why it has been learned, and how it fits into the structure of algebra	Structure, Connections
Chapter Review	446-448	Review of concepts and skills learned in the chapter	
Chapter Test	449	Diagnose student weaknesses and remediate with correlated Reteaching worksheets	

LESSON RESOURCES

MEETING INDIVIDUAL NEEDS

RETEACHING For students who need to spend more time on basics:

If a mid-chapter self-test or chapter test indicates a deficiency, teachers can help students with the appropriate ***Reteaching Copymaster.***

PRACTICE For students who need more practice:

Additional exercises like those in the Pupil's Edition are provided for each lesson in ***Extra Practice Copymasters.***

ENRICHMENT For enriching and broadening students' experiences:

Problem of the Day copymasters in ***Teaching Tools*** provide a daily opportunity to use logical reasoning, looking for a pattern, writing an equation, and other routine and non-routine problem-solving strategies.

Math Log copymasters in ***Alternative Assessment*** provide opportunities to report on investigations, research, and open-ended problems.

Enriching activities with graphing and scientific calculators and computers are provided in ***Technology: Using Calculators and Computers.***

The ***Applications Handbook*** provides additional information about the cross-curriculum topics such as astronomy, chemistry, physics, sports, economics, genetics, and music that are integrated into the Pupil's Edition.

LESSON	8.1	8.2	8.3	8.4	8.5	8.6	8.7
PAGES	400-405	406-411	413-418	419-424	426-431	432-437	439-444
Teaching Tools							
Transparencies						✓	✓
Problem of the Day	✓	✓	✓	✓	✓	✓	✓
Warm-up Exercises	✓	✓	✓	✓	✓	✓	✓
Answer Masters	✓	✓	✓	✓	✓	✓	✓
Extra Practice Copymasters	✓	✓	✓	✓	✓	✓	✓
Reteaching Copymasters	Teacher-directed and independent activities tied to results on the Mid-Chapter Self-Tests and Chapter Tests						
Color Transparencies	✓			✓	✓		
Applications Handbook	Additional background information is supplied for many real-life applications.						
Technology Handbook	Calculator and computer worksheets are supplied for appropriate lessons.						
Complete Solutions Manual	✓	✓	✓	✓	✓	✓	✓
Alternative Assessment	Assess student's ability to reason, analyze, solve problems, and communicate using mathematical language.						
Formal Assessment	Mid-Chapter Self-Tests, Chapter Tests, Cumulative Tests, and Practice for College Entrance Tests						
Computer Test Bank	Customized tests can be created by choosing from over 2000 items.						

INSIGHTS

8.1 Multiplication Properties of Exponents

Very large numbers, such as the distance from the earth to the sun (about 93,000,000 miles), and very small numbers, such as the charge of an electron of an atom (0.000 000 000 000 000 000 16 C; a coulomb, C, is a basic unit of electrical charge), often have many repeating factors. In the examples cited here, the repeating factor is 10. Such numbers can be expressed more easily using exponents.

8.2 Negative and Zero Exponents

In Lesson 8.1, the exponential equation, $y = C(a)^x$, was presented. For $0 < a < 1$, the exponential equation is used to model exponential decay. Frequently, the base a is a fraction. It is possible to extend the properties of exponents to powers of fractional bases. The extension results in negative and zero exponents.

8.3 Division Properties of Exponents

Frequently, real-life problems in science and business involve the quotient of powers. Although it is possible to use the properties of products of exponents, it is more efficient to develop division properties of exponents to compute the quotient of powers. It is important to expand students' understanding of exponents because it permits them to handle a wider range of problems more easily, both with and without a calculator.

8.4 Scientific Notation

The way number ideas are written can have an effect on the ease with which those ideas can be understood and recalled. Several written formats of numbers are frequently used in computation and communication. Students are already familiar with two of them: decimals and fractions. Scientific notation is a third format frequently used for both very large and very small numbers. Students should learn how to read, write, and compute numbers in scientific notation, especially since scientific calculators usually express very large or very small numbers in scientific notation automatically.

8.5 Problem Solving and Scientific Notation

There are many applications of scientific notation in real life. This lesson presents several of these applications. Remind students that mathematics in general and algebra in particular empower them to understand the world around them. Understanding algebraic concepts and mastering algebraic skills keep many career doors open to students.

8.6 Problem Solving: Compound Interest and Exponential Growth

Many applications in business and science use exponential relationships. Compound interest, an example of such an application, is used by millions of consumers every day. In real-life situations, calculators are typically used to complete any computations.

8.7 Exploring Data: Exponential Growth and Decay

Many applications in education, business, and science can be modeled by relationships that decrease by a given percent for each time interval. Memory loss over time is an example of such exponential decay. Calculators are useful in performing the required computations in such situations, as in exponential growth problems.

Problem of the Day

How can 6 drinking straws be placed so that each of them touches the other 5?

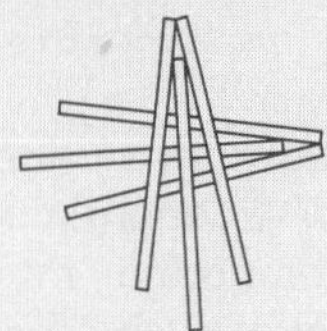

ORGANIZER

Warm-Up Exercises

1. Perform the indicated operations.
   a. 3^4 81
   b. $3^2 \cdot 2^3$ 72
   c. $2^5 \cdot 2^2$ 128
   d. $1^3 \cdot 1^4$ 1
2. a. How many factors make up the product in Exercise 1a? four factors of 3
   b. How many different factors are there for Exercise 1b? two, 3 and 2
   c. How many factors are there for Exercise 1c? seven factors of 2
   d. What is special about the product in Exercise 1d? The product is the same as the original base.
3. Rewrite each product using exponents.
   a. $3 \cdot 3 \cdot 3 \cdot 3 \cdot 3 \cdot 3 \cdot 3$ 3^7
   b. $7 \cdot 7 \cdot 7$ 7^3
   c. $m \cdot m \cdot m \cdot m \cdot m$ m^5
   d. $t \cdot t \cdot t \cdot t \cdot t \cdot t$ t^6

Lesson Resources

Teaching Tools
- Transparency: 5
- Problem of the Day: 8.1
- Warm-up Exercises: 8.1
- Answer Masters: 8.1

Extra Practice: 8.1
Color Transparencies: 39, 40

8.1 Multiplication Properties of Exponents

What you should learn:

Goal 1 How to use the multiplication properties of exponents to evaluate powers and simplify expressions

Goal 2 How to use powers and the exponential change equation as models in real-life settings

Why you should learn it:

You can use exponents in models of real-life situations that involve exponential change, and for repeated factors, such as in formulas for area and volume.

Goal 1 Multiplying with Exponents

Recall from Lesson 1.3 that exponents can be used to represent repeated multiplication. For instance, a^n represents the number that you obtain when a is used as a factor n times.

$$a^n = \underbrace{a \cdot a \cdot a \cdot \cdots \cdot a \cdot a}_{n \text{ factors of } a}$$

The number a is the **base,** and the number n is the **exponent.** The expression a^n is called a **power** and is read as "a to the nth power."

To multiply two powers that have the *same* base, you add exponents. To see why this is true, notice what happens when you multiply a^2 and a^3.

$$a^2 \cdot a^3 = \overbrace{\underbrace{a \cdot a}_{2 \text{ factors}} \cdot \underbrace{a \cdot a \cdot a}_{3 \text{ factors}}}^{5 \text{ factors}} = a^5 = a^{2+3}$$

Multiplication Properties of Exponents

Let a and b be numbers and let m and n be positive integers.

1. To multiply powers having the same base, add the exponents.
   $a^m \cdot a^n = a^{m+n}$ Product of Powers Property
2. To find a power of a power, multiply the exponents.
   $(a^m)^n = a^{m \cdot n}$ Power of a Power Property
3. To find a power of a product, find the power of each factor and multiply.
   $(a \cdot b)^m = a^m \cdot b^m$ Power of a Product Property

Example 1 *Using the Product of Powers Property*

a. $5^3 \cdot 5^6 = 5^{3+6} = 5^9$ **b.** $x^2 \cdot x^3 \cdot x^4 = x^{2+3+4} = x^9$ ■

Example 2 *Using the Power of a Power Property*

a. $(2^4)^3 = 2^{4 \cdot 3} = 2^{12}$

b. $(y^2)^4 = y^{2 \cdot 4} = y^8$

c. $[(a + 1)^2]^5 = (a + 1)^{2 \cdot 5} = (a + 1)^{10}$

d. $[(-3)^3]^2 = (-3)^{3 \cdot 2} = (-3)^6$ ■

Example 3 *Using the Power of a Product Property*

a. $(6 \cdot 5)^2 = 6^2 \cdot 5^2 = 36 \cdot 25 = 900$

b. $(2xy)^3 = (2 \cdot x \cdot y)^3 = 2^3 \cdot x^3 \cdot y^3 = 8x^3y^3$

c. $(-3a)^2 = (-3 \cdot a)^2 = (-3)^2 \cdot a^2 = 9a^2$

d. $-(3a)^2 = -(3 \cdot a)^2 = -(3^2) \cdot a^2 = -9a^2$ ■

Goal 2 Using Powers as Models in Real Life

The Great Pyramid at Giza was built about 2600 B.C. to house the burial chamber of King Khufu (called Cheops by the Greeks). About two million stone blocks, averaging 2.5 tons each, were used in the building. Deep within the pyramid is the Grand Gallery, a corridor measuring 153 feet long and 28 feet high, which is regarded as one of the marvels of ancient architecture.

Real Life Architecture

Example 4 *The Great Pyramid*

Some of the upper stones of the Great Pyramid of Cheops have long ago disappeared, but the original height was 481 feet. Each side of the square base measures 752 feet. How many cubic feet of space did the original pyramid occupy?

Solution The volume of a pyramid is one third the height times the area of the base. Thus, this pyramid has a volume of

$$\frac{1}{3}hB = \frac{1}{3}(481)(752)^2 \approx 91{,}000{,}000 \text{ cubic feet.}$$ ■

LESSON Notes

GOAL 1 Students understand these properties much better when number values are used to explain the statement of the properties.

For example,

Property 1:
$2^2 \cdot 2^3 = 2 \cdot 2 \cdot 2 \cdot 2 \cdot 2 = 2^5$

Property 2:
$(3^3)^2 = 3^3 \cdot 3^3 = 3^{3+3} = 3^{2 \cdot 3}$

Property 3:
$(2 \cdot 5)^3 = 2 \cdot 5 \cdot 2 \cdot 5 \cdot 2 \cdot 5 = 2 \cdot 2 \cdot 2 \cdot 5 \cdot 5 \cdot 5 = 2^3 \cdot 5^3$

Property 3 is frequently stated as "the power of a product is the product of the powers."

Examples 1–3

Be willing to expand each product into its basic factors if it helps students "see" the relationships between the actual multiplication and the use of exponents.

In Example 2, note that students frequently simplify powers by adding the exponents instead of multiplying them.

Example 4

The Great Pyramid is not the largest pyramid in the world. That distinction belongs to Quetzalcóatl, an Aztec pyramid near Mexico City. Its base is about 1,960,200 square feet. Its height is about 177 feet. Its volume has been estimated at 116,500,000 cubic feet.

Common-Error ALERT!

A common error in **Example 3** is "forgetting" to distribute the exponent 3 to the coefficient 2 in expressions such as $(2xy)^3$. Students recognize the error if the variables in parentheses are replaced by constants and the expression is evaluated in different ways.

Example 5

Use this situation to add meaning to the phrase, *growing exponentially*. Undoubtedly, students will be surprised at how fast the stack of paper grew. Emphasize that the effect of exponential growth can sometimes be very surprising.

Extra Examples

Here are additional examples similar to **Examples 1–3.**

1. Apply the product of powers property.
   a. $4^2 \cdot 4^5$ 4^7
   b. $p^3 \cdot p^4$ p^7
   c. $7^5 \cdot 7 \cdot 7^2$ 7^8
   d. $(-t)^2 \cdot (-t)^3$ $(-t)^5$
2. Apply the power of a power property.
   a. $(5^2)^4$ 5^8
   b. $(c^3)^6$ c^{18}
   c. $(y^5)^2$ y^{10}
   d. $[(k+2)^3]^2$ $(k+2)^6$
3. Apply the power of a product property.
   a. $(2 \cdot 3)^3$ 216
   b. $(-2xy)^3$ $-8x^3y^3$
   c. $-(4b)^2$ $-16b^2$
   d. $(xyz)^4$ $x^4y^4z^4$

Check Understanding

1. Simplify $(a^4)^3$. What property did you use? a^{12}; the power of a power property
2. Explain why the power of a product property is true. For example, $(2 \cdot 5)^3 = 2 \cdot 5 \cdot 2 \cdot 5 \cdot 2 \cdot 5 = 2 \cdot 2 \cdot 2 \cdot 5 \cdot 5 \cdot 5 = 2^3 \cdot 5^3$.
3. Evaluate -3^2. Note that the correct answer is -9. The accepted convention is that the value -3 is viewed as $-1 \cdot 3$, and consequently, the exponent governs only the base to its immediate left.

Real Life
Paper Production

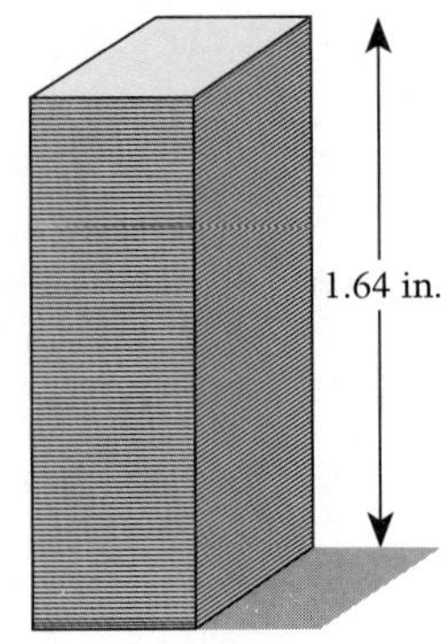

After doubling 9 times, the stack is about 1.64 inches tall.

Example 5 *Stacking Paper*

A piece of notebook paper is about 0.0032 of an inch thick. Start with one piece of paper and double the stack 25 times. How tall will the stack be in inches? In feet? In miles?

Solution After doubling the stack once, you will have 2 pages. After doubling twice, you will have 2^2. After doubling three times, you will have $2^2 \cdot 2$, or 2^3 pages. From this pattern, you can see that after doubling 25 times, you will have 2^{25} pages. To find the height, h, of this stack, multiply 0.0032 of an inch per page by 2^{25} pages.

$$h = 0.0032 \cdot 2^{25} \approx 107{,}374.2 \text{ inches} \quad \textit{Use a calculator.}$$
$$\approx 8{,}947.8 \text{ feet} \quad \textit{Divide by 12.}$$
$$\approx 1.7 \text{ miles} \quad \textit{Divide by 5280.}$$ ■

The equation $h = 0.0032 \cdot 2^{25}$ in Example 5 is an **exponential equation.** In the general form of an exponential equation $y = C(a)^x$, C is an initial amount, a is a "change" factor, and x is the number of times the change occurs. (Note that the variable x is an exponent.) If $a > 1$, the equation is a model for **exponential growth.** If $a < 1$, the equation is a model for **exponential decay.**

Communicating about ALGEBRA

SHARING IDEAS about the Lesson

Visualize Exponential Change Think about the expression 2^x.

A. Fill in the table of values. Then, for these values, sketch the graphs of $y = 2x$ and $y = 2^x$ on the same coordinate plane. See Additional Answers for graphs.

x	1	2	3
$2x$	?	?	?
2^x	?	?	?

2, 4, 6; 2, 4, 8

B. If you were offered a job that paid $2x$ dollars or 2^x dollars for x hours of work, which pay scale would you choose? Explain. 2^x; for $x > 2$, $y = 2^x$ pays more.

C. After working 8 hours, how much more could you make for working 2 additional hours if your hourly rate is 2^x dollars? $768

EXERCISES

Guided Practice

CRITICAL THINKING about the Lesson

1. Can x^8y^4 be simplified? Explain. No, the bases are not the same.
2. Simplify $(a^{10})^3$. What property did you use? a^{30}, Power of a Power
3. Is $a^5 \cdot a^3 = a^{15}$? Why or why not?
4. Is $(-3b)^4 = -12b^4$? Why or why not?
5. Simplify $a^3 \cdot a^4$. Confirm your result by letting $a = 2$ and evaluating the expression in both its original form and its simplified form. a^7
6. Use a calculator to evaluate $(1.06)^{11}$. Round your result to two decimal places. 1.90
7. In the general exponential equation $y = C(a)^x$, suppose that $a = 2$, and $x = 3$. Describe how y changes when x is increased by 1. y is doubled.
   3. No, $a^5 \cdot a^3 = a^{5+3} = a^8$
   4. No, $(-3b)^4 = (-3)^4b^4 = 81b^4$
8. Identify each of these equations as a model of exponential growth or of exponential decay.
   a. $y = 3^x$ **b.** $y = 0.5(3)^x$
   c. $y = (0.5)^x$ **d.** $y = 2(0.5)^x$
   growth, growth, decay, decay

Independent Practice

In Exercises 9–41, simplify, if possible. 23. $(-3xy)^6$ or $729x^6y^6$ 25. $3^2(-4)^4a^6$ or $2304a^6$ $(-9)^8$ or 43,046,721

9. $4^2 \cdot 4^3$ 4^5 or 1024
10. $6^5 \cdot 6^4$ 6^9 or 10,077,696
11. $[(-9)^2]^4$
12. $10^2 \cdot 10^9$ 10^{11}
13. $x \cdot x^5$ x^6
14. $(5^5)^4$ 5^{20}
15. $[(2x+3)^3]^2$ $(2x+3)^6$
16. $(2x)^3$ $8x^3$
17. $(3 \cdot 7)^4$ 21^4 or 194,481
18. $[(5+x)^3]^6$ $(5+x)^{18}$
19. $(-5a)^2$ $25a^2$
20. $(16 \cdot 2)^2$ 32^2 or 1024
21. $(4a)^2 \cdot a$ $16a^3$
22. $6^2 \cdot (6x^3)^2$ 6^4x^6 or $1296x^6$
23. $[(-3xy)^2]^3$
24. $(x \cdot x^2)^3 \cdot 3x$ $3x^{10}$
25. $(3a)^2 \cdot (-4a)^4$
26. $(9a^3)^2 \cdot (2a)^3$
27. $2x^3 \cdot (3x)^2$ $18x^5$
28. $3y^2 \cdot (2y)^3$ $24y^5$
29. $(-ab)(a^2b)^2$ $-a^5b^3$
30. $(-rs)(rs^3)^2$ $-r^3s^7$
31. $(-2xy)^3(-x^2)$ $8x^5y^3$
32. $(-3cd)^3(-d^2)$ $27c^3d^5$
33. $(4a^2)^3(\frac{1}{2}a^3)^2$ $16a^{12}$
34. $(8b^3)^2(\frac{1}{4}b^2)^2$ $4b^{10}$
35. $(-x)^5(-x)^2(-x)^3$ x^{10}
36. $(-y)^4(-y)^3(-y)^2$ $-y^9$
37. $(2t)^3(-t^2)$ $-8t^5$
38. $(-w^3)(3w^2)^2$ $-9w^7$
39. $(abc^2)^3(a^2b)^2$ $a^7b^5c^6$
40. $(r^2st^3)^2(s^4t)^3$ $r^4s^{14}t^9$
41. $(-3xy^2)^3(-2x^2y)^2$ $-108x^7y^8$

In Exercises 42–47, evaluate the expression when $a = 1$ and $b = 2$. 26. $2^39^2a^9$ or $648a^9$

42. $(a^4)^3$ 1
43. $b^3 \cdot b^4$ 128
44. $(a^2 \cdot b)^3$ 8
45. $(a^2b)^5$ 32
46. $(b^2 \cdot b^3) \cdot (b^2)^4$ 8192
47. $[(a+4)^2]^3 \cdot (a+4)$ 78,125

In Exercises 48–50, say which number is larger.

48. $(5 \cdot 7)^3$ or $5 \cdot 7^3$ $(5 \cdot 7)^3$
49. $5^4 \cdot 2^5$ or $(5 \cdot 2)^5$ $(5 \cdot 2)^5$
50. $(4^5 \cdot 4^{10})$ or 4^{50} 4^{50}

Communicating about ALGEBRA

EXTEND Communicating

Consider the sequence $\frac{1}{2}$, $\frac{1}{4}$, $\frac{1}{8}$, $\frac{1}{16}$, etc., obtained by "halving." It seems amazing, but by using halving and asking 20 yes-or-no questions, you could discover any entry word in a dictionary (or name in a phone book). Have a student select a word in a dictionary. Open to a page about halfway through the book. Ask "Does your word come before this page?" Depending on the answer, select another page about halfway through the remainder. By continuing this process, you can reach one page, then one word. This process works because the list is in alphabetical order, because $\frac{1}{2^{20}}$ is less than one millionth, and because most dictionaries define fewer than 1 million words.

EXERCISE Notes

ASSIGNMENT GUIDE

Basic/Average: Ex. 1–8, 9–48, multiples of 3, 55–77 odd

Above Average: Ex. 1–8, 21–77 odd, 78

Advanced: Ex. 1–8, 23–77 odd, 78–80

Selected Answers
Exercises 1–8, 9–77 odd

More Difficult Exercises
Exercises 53–54, 78–80

Guided Practice

Ex. 1–8 Give students time to write a sentence that explains their reasoning. These sentences often provide insight into a students' true understanding.

Ex. 4 Ask students to compare $(-3b)^4$ with $-(3b)^4$.

Ex. 22–41 These exercises require all three properties of exponents. Remember the order of operations. Evaluate within grouping symbols first, then apply exponents, then multiply.

Answers

53. $(a^m)^n = \underbrace{a^m \cdot a^m \cdot a^m \cdot \ldots}_{n \text{ times}}$
$= a^{nm}$
$= a^{mn}$

54. $(a \cdot b)^m =$
$\underbrace{a \cdot a \cdot a \cdot \ldots}_{m \text{ times}} \cdot \underbrace{b \cdot b \cdot b \cdot \ldots}_{m \text{ times}}$
$= a^m b^m$

Independent Practice

Ex. 55–60 TECHNOLOGY Be sure you help review the use of the power keys on the students' calculators.

Ex. 61–62 These are interesting problems that provide students with an interesting insight into history.

EXTEND Ex. 61–62 MULTICULTURAL Some students could research the Cahokia Indians and the Great Temple Mound and prepare a report for the class.

Common-Error ALERT!

Students often apply an exponent incorrectly. In **Exercise 78,** for example, since $t = 7$, students must be careful not to evaluate $[(3960) \cdot (1.7)]^7$.

In Exercises 51 and 52, solve the equation for *x*.

51. $4^2 \cdot 4^4 = 4^x$ 6

52. $(3^2)^9 = 3^x$ 18

In Exercises 53 and 54, give a convincing argument to explain why the property is true. See margin.

53. the Power of a Power

54. the Power of a Product

Technology **In Exercises 55–60, use a calculator to evaluate the expression. Round the results to two decimal places when appropriate.**

55. $(1.1 + 3.3)^3$ 85.18

56. $5.5^3 \cdot 5.5^4$ 152,243.52

57. $2.4^4 \cdot 2.4^2$ 191.10

58. $(4.0 + 3.9)^2$ 62.41

59. $(2.9^3)^5$ 8,629,188.75

60. $(9.1^2)^4$ 47,025,252.76

Native Americans in Cahokia **In Exercises 61 and 62, use the following information.**

Over 900 years ago, Native Americans lived in a city called Cahokia, in what is now Illinois. The city had over 30,000 inhabitants and contained several temple mounds. Some of the mounds were over 10 stories tall. The base of the Great Temple Mound covered 15 acres, 2 acres more than the base of the largest pyramid in Egypt.

Native American mounds were built as burial places and as platforms to hold temples and the houses of chiefs. Thousands of these mounds still stand in the U.S. and Canada.

61. The volume of the Great Temple Mound was $\frac{1}{3}\left(\frac{1}{2}\right)(800^3 - 600^3)$ cubic feet. Evaluate this volume. $49{,}333{,}333\frac{1}{3}$ ft³

62. A cubic foot of soil weighs about 90 pounds. How many pounds of soil had to be moved to construct the Great Temple Mound? 4,440,000,000

In Exercises 63–66, solve the equation $y = C(a)^x$ for the given values.

63. $C = 2$; $a = 1$; $x = 50$ 2

64. $C = 1$; $a = 2$; $x = 2$ 4

65. $C = 2$; $a = \frac{1}{2}$; $x = 3$ $\frac{1}{4}$

66. $C = 4$; $a = \frac{3}{4}$; $x = 2$ $\frac{9}{4}$

In Exercises 67–70, identify each of the equations as a model of exponential growth or of exponential decay.

67. $y = 500(1.03)^8$ growth

68. $y = 10(0.99)^{10}$ decay

69. $y = 1000(0.03)^{25}$ decay

70. $y = 0.75(1.01)^{50}$ growth

Integrated Review

In Exercises 71–74, evaluate the expression.

71. $a^2 + 2$ when $a = 3$ 11

72. $3x^2 - 5$ when $x = 5$ 70

73. $x^2 + y^3$ when $x = 1$ and $y = 4$ 65

74. $2a^4 - b^3$ when $a = 2$ and $b = 3$ 5

Geometry **In Exercises 75–77, use the following information.**

The opposite angles of a parallelogram have equal measures.

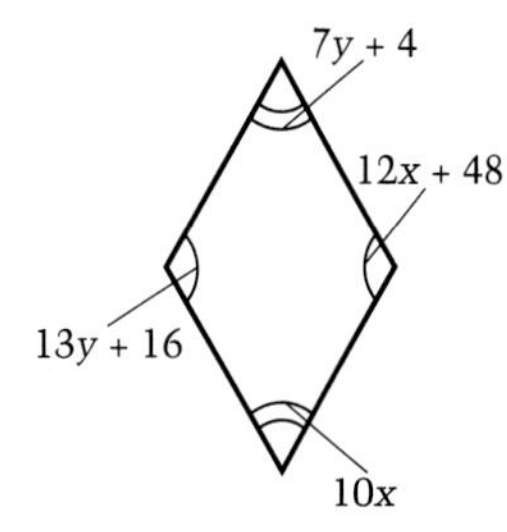

75. The figure shown is a parallelogram. Solve for x and y.

76. Find the measures of the angles of the parallelogram.

77. Show that the sum of the four angle measures is 360°.

60 + 120 + 60 + 120 = 360

75. $x = 6$, $y = 8$

76. 60°, 120°, 60°, 120°

Exploration and Extension

78. *Voice Messages* Between 1985 and 1990, the number, n, of businesses using voice messages increased about 70% per year. The number in year t can be modeled by

$$n = 3960 \cdot 1.7^t$$

where $t = 1$ corresponds to 1986. (3960 is how many thousands of businesses used voice messages in 1985.) If this pattern continued, estimate the number of businesses that used voice messages in 1992. ***(Source: Dataquest)*** 162,494

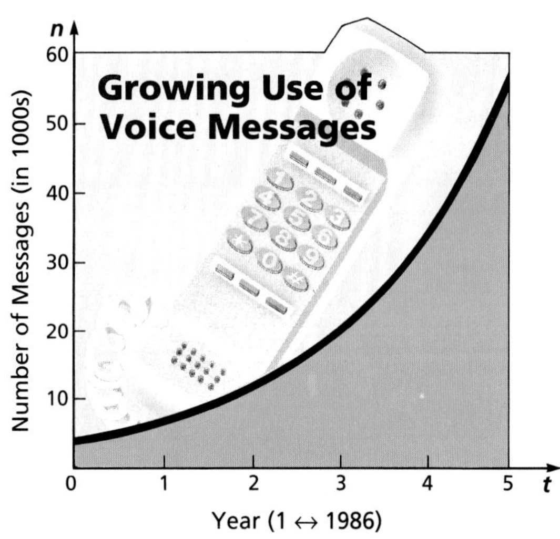

Profit **In Exercises 79 and 80, use the circle graph and the following information.**

Between 1981 and 1990, the annual profit, P, for the H. J. Heinz Company increased about 13% per year. The profit (in millions of dollars) in year t can be modeled by $P = 167.25 \cdot (1.13)^t$, where $t = 1$ corresponds to 1981. ***(Source: H. J. Heinz)***

79. Complete the table. If the pattern continues, what would you predict the profit to be in 1994? See margin.

Year, t	1981	1983	1985	1987	1990
Profit ($), P	?	?	?	?	?

The H.J. Heinz Company Division

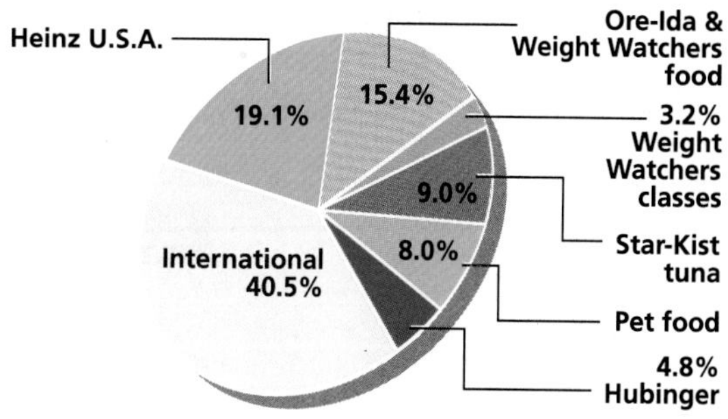

80. How much of the profit for 1990 was due to the Star-Kist Tuna division of H. J. Heinz? $51.10 million

EXTEND Ex. 78 Use the model to predict the number of businesses using voice messages in the year 2000, when $t = 15$. 11,335,195

Ex. 79–80 EXPLORING DATA These exercises should be assigned as a unit. They provide a connection between bar graphs, circle graphs, an algebraic equation, and real-life-data applications.

Answer

79. 188.99 million, 241.32 million, 308.15 million, 393.47 million, 567.74 million; $925.69 million

Enrichment Activities

These activities require students to go beyond lesson goals.

In a geometric sequence, each term can be computed from the previous term by multiplying by the same number. For example, in the sequence 2, 4, 8, 16, 32, . . ., each term is obtained by multiplying the previous term by 2. The next term after 32 is $32 \cdot 2 = 64$.

1. Find the next term in the geometric sequence: 1, 4, 16, 64 ? 256
2. Find the next term in the geometric sequence: −1, −6, −36, −216, ? −1296
3. Find the next term in the geometric sequence: −3, 6, −12, 24, ? −48
4. Find the next term in the geometric sequence: $\frac{1}{2}$, $\frac{1}{6}$, $\frac{1}{18}$, $\frac{1}{54}$, ? $\frac{1}{162}$
5. Evaluate $(-1)^m$ for $m = 1$, 2, 3, 4, 5, and 6. Look for a pattern. Use the pattern to predict $(-1)^{2001}$. −1

Problem of the Day

Study the pattern to find d and e.

$\frac{a}{11} = 11$

$\frac{b}{111} = 111$

$\frac{c}{1111} = 1111$

.
.
.

$\frac{d}{e} = 1{,}111{,}111$

$d = 1{,}234{,}567{,}654{,}321$
$e = 1{,}111{,}111$

ORGANIZER

Warm-Up Exercises

1. Simplify.
   a. 3^3 27 b. 5^3 125
2. Simplify.
   a. $\frac{9}{3^4}$ $\frac{1}{9}$ b. $\frac{15}{5^3}$ $\frac{3}{25}$

Lesson Resources

Teaching Tools
Transparency: 14
Problem of the Day: 8.2
Warm-up Exercises: 8.2
Answer Masters: 8.2
Extra Practice: 8.2
Technology Handbook: p. 59

Lesson Investigation Answers

4, 2, 1, $\frac{1}{2}$, $\frac{1}{4}$, $\frac{1}{8}$; $a^0 = 1$, $a^{-n} = \frac{1}{a^n}$

LESSON Notes

GOAL 1 Emphasize the importance of recognizing the pattern in explaining the meaning of negative and zero exponents. Use other patterns if necessary, such as $3^4 = 81$, $3^3 = 27$, $3^2 = 9$, $3^1 = 3$, $3^0 = 1$, $3^{-1} = \frac{1}{3}$, $3^{-2} = \frac{1}{9}$, $3^{-3} = \frac{1}{27}$.

8.2 Negative and Zero Exponents

What you should learn:

Goal 1 How to use negative and zero exponents in algebraic expressions

Goal 2 Using powers as models in real-life settings

Why you should learn it:

The study of negative and zero exponents helps you to interpret exponential models of real-life settings.

Goal 1 Using Negative and Zero Exponents

LESSON INVESTIGATION

■ Investigating Negative and Zero Exponents

Partner Activity Copy the table and discuss any patterns you observe. Use the patterns to complete the table.

Exponent, n	5	4	3	2	1	0	-1	-2	-3
Power, 2^n	32	16	8	?	?	?	?	?	?

With your partner, discuss a general procedure for evaluating expressions such as a^0 and a^{-n}.

Study Tip

Note in the definition at the right that a^0 is defined as 1 only if $a \neq 0$. The expression 0^0 is undefined.

Definition of Negative and Zero Exponents

Let a be a *nonzero* number and let n be a positive integer.

1. The expression a^{-n} is the reciprocal of a^n.

$$a^{-n} = \frac{1}{a^n}, \quad a \neq 0$$

2. A nonzero number to the zero power is 1.

$$a^0 = 1, \quad a \neq 0$$

Example 1 *Powers with Negative and Zero Exponents*

a. $2^{-2} = \frac{1}{2^2} = \frac{1}{4}$

b. $(-2)^0 = 1$

c. $(2x)^{-3} = \frac{1}{(2x)^3} = \frac{1}{2^3 \cdot x^3} = \frac{1}{8x^3}$

d. $3 \cdot 3^{-1} = 3 \cdot \frac{1}{3} = 1$ ■

Expressions with positive exponents are usually considered to be simpler than expressions with negative exponents.

Example 2 Simplifying Expressions That Have Negative Exponents

Simplify $2x^{-2}y^{-3}$ by rewriting with positive exponents.

Solution

$2x^{-2}y^{-3} = 2 \cdot \frac{1}{x^2} \cdot \frac{1}{y^3}$ *Definition of negative exponent*

$= \frac{2}{x^2y^3}$ *Multiply fractions.* ■

Informally you can think of rewriting expressions in positive exponent form as "moving factors" from a denominator to a numerator or vice versa.

$\frac{3x^{-2}}{4} = \frac{3}{4x^2}$ *Move from numerator to denominator.*

$\frac{1}{3x^{-4}} = \frac{x^4}{3}$ *Move from denominator to numerator.*

The next example shows how the multiplication properties of exponents can be used with negative or zero exponents.

Example 3 Using Multiplication Properties of Exponents

a. $3^{-2} \cdot 3^2 = 3^{-2+2} = 3^0 = 1$

b. $(4^{-2})^2 = 4^{-2 \cdot 2} = 4^{-4} = \frac{1}{4^4}$

c. $(5a)^{-2} = 5^{-2} \cdot a^{-2} = \frac{1}{5^2} \cdot \frac{1}{a^2} = \frac{1}{5^2a^2}$

d. $\frac{1}{a^{-n}} = (a^{-n})^{-1} = a^{-n \cdot (-1)} = a^n$ ■

Since you can now find a meaning for zero exponents and negative exponents, you can get a better idea of the shape of the graph of $y = 2^x$ in the next Example.

Example 4 Extending the Graph of $y = 2^x$

Sketch the graph of $y = 2^x$.

Solution First make a table of values that includes $x \leq 0$.

x	3	2	1	0	−1	−2	−3
2^x	8	4	2	$2^0 = 1$	$2^{-1} = \frac{1}{2}$	$2^{-2} = \frac{1}{4}$	$2^{-3} = \frac{1}{8}$

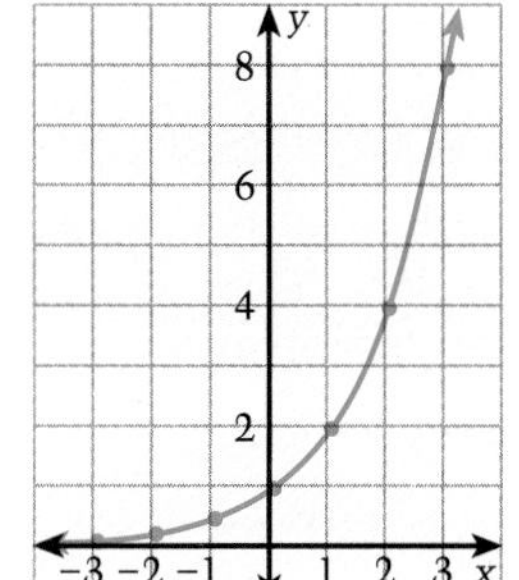

Now sketch the graph. Notice how the graph has a y-intercept of 1, and that it gets closer to the negative side of the x-axis as x values get less. ■

Example 3

Example 3*d* may be simplified differently, using the relationship $a^n = a^{-(-n)} = \frac{1}{a^{-n}}$. Stated differently, the sign of the exponent changes when the power is moved from the numerator to the denominator and vice versa. It follows that $\frac{1}{a^{-n}} = a^{-(-n)} = a^n$.

Extra Examples

Here are additional examples similar to **Examples 1–3.**

1. Evaluate.
   - **a.** $(-5)^{-3}$ **b.** $(3x)^{-2}$
   - **c.** $24 \cdot 4^{-3}$ **d.** -3^{-4}
   - a. $-\frac{1}{125}$ b. $\frac{1}{9x^2}$
   - c. $\frac{3}{8}$ d. $-\frac{1}{81}$
   - (Recall that the exponent governs only the base immediately to its left.)
2. Perform the indicated operations and simplify.
   - **a.** $3a^{-3}b^{-2}$ $\frac{3}{a^3b^2}$
   - **b.** $-4x^{-2}y^3$ $\frac{-4y^3}{x^2}$
   - **c.** $7^{-1} \cdot 7^{-2}$ $\frac{1}{343}$
   - **d.** $5^{-6} \cdot 5^3$ $\frac{1}{125}$
   - **e.** $\frac{1}{8^{-3}}$ 512
   - **f.** $(3^{-3})^2$ $\frac{1}{729}$

In **Example 2,** students frequently apply the exponent to a coefficient inappropriately. For example, in $2x^{-2}y^{-3}$ many students apply the exponent -2 to both the coefficient 2 and x. Caution students that the exponent governs only the base immediately to its lower left unless parentheses are used.

Example 4

TECHNOLOGY Students with graphics calculators might use them to graph $y = 2^x$.

Example 5

The exponent in this example represents values that are multiples of 25,000 years.

Check Understanding

1. What does it mean to simplify exponential expressions? See page 407.
2. Do the properties of positive exponents in Lesson 8.1 hold for negative and zero exponents? See Example 3, page 407.
3. A student was overheard saying, "To simplify an exponential expression involving negative exponents, move the power and change the sign of the exponent." Explain in greater detail what the student was saying. See Example 2. Simplifying with negative exponents is "moving factors" from a numerator to a denominator or vice versa.

Communicating about ALGEBRA

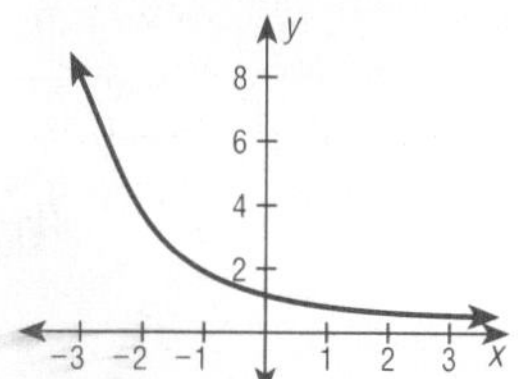

Compare this graph with $y = 2^x$. What do you notice? The graphs are mirror images of each other around the y-axis.

The worst (known) nuclear accident occurred in 1986 at the Chernobyl nuclear power plant. The amount of radioactive material released into the atmosphere was never revealed.

Goal 2 Using Powers as Models in Real Life

In Lesson 8.1, you studied some models of exponential growth. Example 5 models exponential decay. Note that the **half-life** of a radioactive substance is that period of time during which the initial mass of the substance is reduced by one-half. After two half-life periods have elapsed, the initial mass will have been reduced by one-quarter, and so on.

Example 5 *Modeling Exponential Decay*

Suppose that 10 grams of the plutonium isotope Pu-239 were released in the Chernobyl nuclear accident. (The half-life of Pu-239 is about 25,000 years.) The number of grams remaining after h half-life periods have elapsed is

$$W = 10\left(\tfrac{1}{2}\right)^h.$$

Sketch a graph that shows how much of the 10 grams will remain after 5 half-life periods.

Solution To begin, make a table of values showing the number of grams left after h half-life periods.

Half-life periods, h	0	1	2	3	4	5
Grams, W	10	5.00	2.50	1.25	0.625	0.3125

Half-Life of Plutonium-239

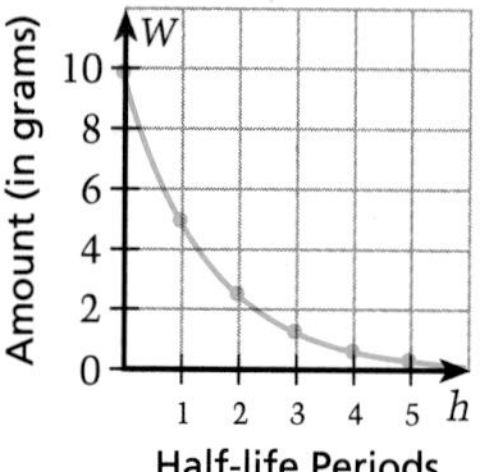

By plotting the points given in the table and connecting them with a smooth curve, you can obtain the graph shown at the left. The shape of the graph on the left is typical of an exponential decay model. ■

Communicating about ALGEBRA

SHARING IDEAS about the Lesson See margin.

Extend Your Thinking Graph $y = \left(\frac{1}{2}\right)^x$ for $-3 \le x \le 3$.

EXERCISES

1. True, the base determines whether the term is positive or negative.
3. 5 does not have a negative exponent.

Guided Practice

CRITICAL THINKING about the Lesson

1. **True or False?** If a is positive, a^{-n} is positive. Explain your reasoning.
2. Simplify $a^5 \cdot a^{-5}$. The result implies that a^5 and a^{-5} are [?] of each other. 1, reciprocals
3. Rewrite $5a^{-3}b^{-2}$ with positive exponents. Why does the 5 stay in the numerator?
4. Simplify $3c^{-5} \cdot 4c^4$. Can a simplified form have a negative exponent? $\frac{12}{c}$, no
5. If $a^0 = 1$ $(a \neq 0)$, what point do all graphs of the form $y = (a)^x$ have in common? Is this true for $y = 2(a)^x$? (0, 1); no

Independent Practice

In Exercises 6–17, rewrite the expression using positive exponents.

6. x^{-7} $\frac{1}{x^7}$
7. x^{-9} $\frac{1}{x^9}$
8. $5x^{-4}$ $\frac{5}{x^4}$
9. $3x^{-2}$ $\frac{3}{x^2}$
10. $\frac{1}{2x^{-3}}$ $\frac{x^3}{2}$
11. $\frac{1}{4x^{-5}}$ $\frac{x^5}{4}$
12. $x^{-2}y^3$ $\frac{y^3}{x^2}$
13. x^6y^{-7} 13. $\frac{x^6}{y^7}$
14. $3x^{-3}y^{-8}$ $\frac{3}{x^3y^8}$
15. $6x^{-2}y^{-4}$ $\frac{6}{x^2y^4}$
16. $\frac{1}{7x^{-4}y^{-1}}$ $\frac{x^4y}{7}$
17. $\frac{1}{2x^{-10}y^{12}}$ $\frac{x^{10}}{2y^{12}}$

In Exercises 18–29, evaluate the expression.

18. 3^{-2} $\frac{1}{9}$
19. 2^{-4} $\frac{1}{16}$
20. $-4^0 \cdot \frac{1}{2^{-2}}$ -4
21. $4^{-3} \cdot 4^2$ $\frac{1}{4}$
22. $6^3 \cdot 6^{-1}$ 36
23. $8^4 \cdot 8^{-4}$ 1
24. $7^{-9} \cdot 7^9$ 1
25. $(5^{-3})^2$ 25. $\frac{1}{15,625}$
26. $(-4^{-2})^{-1}$ -16
27. $-6 \cdot (-6)^{-1}$ 1
28. $5 \cdot 5^{-1}$ 1
29. $2^0 \cdot 3^{-3}$ $\frac{1}{27}$

In Exercises 30–41, rewrite the expression using positive exponents.

30. $(-3)^0x$ x
31. $(5y)^{-2}$ $\frac{1}{25y^2}$
32. $(-2x)^{-3}$ $-\frac{1}{8x^3}$
33. $(-4a)^0$ 1
34. $(-3x)^{-1} \cdot 2y$ $-\frac{2y}{3x}$
35. $(4xy)^{-2}$ $\frac{1}{16x^2y^2}$
36. $(3x)^{-1}$ $\frac{1}{3x}$
37. $(2a^{-3})^3$ 37. $\frac{8}{a^9}$
38. $\frac{4}{b^{-2}}$ $4b^2$
39. $\frac{5}{a^{-4}}$ $5a^4$
40. $\frac{1}{(4x)^{-3}}$ $64x^3$
41. $\frac{1}{(2y)^{-5}}$ $32y^5$

In Exercises 42–45, say if the graph of the function contains the point (0, 1).

42. $y = -3^x$ No
43. $y = 4^x$ Yes
44. $y = 3 \cdot 1^x$ No
45. $y = 50^x$ Yes

46. *Population of Missouri* Between 1970 and 1990, Missouri's population increased at the rate of 0.47% per year. The population, P, in year t is given by

$$P = 4{,}903{,}000 \cdot 1.0047^t.$$

where $t = 0$ corresponds to 1980. Find the population in 1970, 1980, and 1990. 4,678,406; 4,903,000; 5,138,376

47. *Population of Buffalo* Between 1970 and 1990, the population of Buffalo, New York, decreased at the rate of 0.82% per year. The population, P, in year t is given by 1,112,968; 1,025,000; 943,985

$$P = 1{,}025{,}000 \cdot 0.9918^t$$

where $t = 0$ corresponds to 1980. Find the population in 1970, 1980, and 1990.

EXERCISE Notes

ASSIGNMENT GUIDE

Basic/Average: Ex. 1–5, 7–45 odd, 46–47, 53–63 odd, 65–68

Above Average: Ex. 1–5, 7–45 odd, 46–49, 53–63 odd, 65–68

Advanced: Ex. 1–5, 19–45 odd, 46–52, 65–71

Selected Answers
Exercises 1–5, 7–63 odd

Use **Mixed Review** as needed.

More Difficult Exercises
Exercises 48–51, 69–71

Guided Practice

Ex. 1 The answer to this exercise can be found in the opening comments on page 406.

Ex. 3 This exercise provides a good check of students' understanding of the difference between $(5a)^{-2}$ and $5a^{-2}$.

Independent Practice

Ex. 18–41 Students can simplify in two ways. They can use the properties of exponents before simplifying, or they can convert the term with the negative exponent to a fraction and then simplify. For example,

Method 1:

$$5 \cdot 5^{-1} = 5^{1+(-1)} = 5^0 = 1$$

Method 2:

$$5 \cdot 5^{-1} = 5 \cdot \frac{1}{5} = \frac{5}{5} = 1$$

Ex. 46–47 Remind students how to convert a year such as 1980 to the expression $t = 10$. Ask them to explain the method.

Ex. 53–64 These exercises are a review from Lesson 8.1.

Ex. 65–68 These exercises are designed to visually help students to distinguish between linear and exponential models.

Answers

69. $a \approx 47.5, \approx 66.6, \approx 90.0, \approx 118.5$

Paleontology **In Exercises 48 and 49, use the following information.**

Sunlight produces radioactive carbon (carbon-14), which is absorbed by living plants and animals. Once the plant or animal dies, it stops absorbing carbon-14. Using carbon-14's half-life of 5700 years, scientists are able to approximate the age of plant or animal fossils by determining the percent of carbon-14 in the fossil.

A paleontologist uses a brush to clean soil deposits from an animal skull found in the Badlands, South Dakota.

48. Suppose the animal skull in the photograph contained 40 nanograms of carbon-14 when the animal was living. The amount (in nanograms) of carbon-14 in the skull h half-life periods after the animal died was

$$W = 40\left(\tfrac{1}{2}\right)^h.$$

Complete the table. 40, 20, 10

Half-life periods, *h*	0	1	2
Nanograms, *W*	?	?	?

49. Upon testing the animal skull, the paleontologist finds that it contains 15 nanograms of carbon-14. Use the table you completed in Exercise 48 to estimate the age of the skull in years. About 8000 years

Swiss Bank Account **In Exercises 50 and 51, use the following information.**

Swiss banks have the reputation of being safe and confidential. Because of the secrecy given to depositors, some accounts charge a "negative interest."

50. Suppose you deposited \$500,000 in a Swiss bank account that charged an interest of 2% per year. The balance in your account after t years is given as

$$A = 500{,}000 \cdot (0.98)^t.$$

500,000.00; 480,200.00; 461,184.08; 442,921.19; 425,381.51; 408,536.40

Complete the table.

Year, *t*	0	2	4	6	8	10
Dollars, *A*	?	?	?	?	?	?

51. In 1980, money was put into a Swiss bank account that charges 1% interest per year. The balance in the account is

$$A = 633{,}067.45 \cdot (0.99)^t$$

\$700,000.00; \$633,067.45, \$572,534.85

where $t = 0$ corresponds to 1990. How much was deposited in 1980? What is the 1990 balance? What will the balance be in 2000?

Common-Error ALERT!

In **Exercise 50,** warn students to treat expressions such as

$$500{,}000\ (0.98)^0$$

as 500,000(1), not just 1. The 500,000 represents the initial value of the equation at $t = 0$.

Integrated Review

52. *College Entrance Exam Sample* $p^8 \times q^4 \times p^4 \times q^8 = \boxed{?}$

a. $p^{12}q^{12}$ b. p^4q^4 c. $p^{32}q^{32}$ d. $p^{64}q^{64}$ e. $p^{16}q^{16}$

In Exercises 53–58, evaluate the expression.

53. 4^3 64
54. $(-2)^5$ -32
55. $(-\frac{2}{3})^2$ $\frac{4}{9}$
56. -8^2 -64
57. $25 - 3^2 \cdot 2$ 7
58. $(12 - 9)^3 \cdot 4 + 36$ 144

In Exercises 59–64, simplify the expression.

59. $a^4 \cdot a^6$ a^{10}
60. $x^2 \cdot y^4 \cdot x^3$ x^5y^4
61. $(-t^2)^3$ $-t^6$
62. $(4z)^2$ $16z^2$
63. $(m \cdot n)^5$ m^5n^5
64. $(3a \cdot 2b)^3$ $216a^3b^3$

Exploration and Extension

In Exercises 65–68, match the equation with the graph.

65. $y = 2x$ d
66. $y = 2^x$ a
67. $y = \frac{1}{2}x$ c
68. $y = (\frac{1}{2})^x$ b

a.

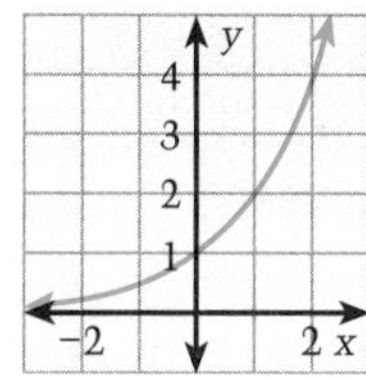

b.

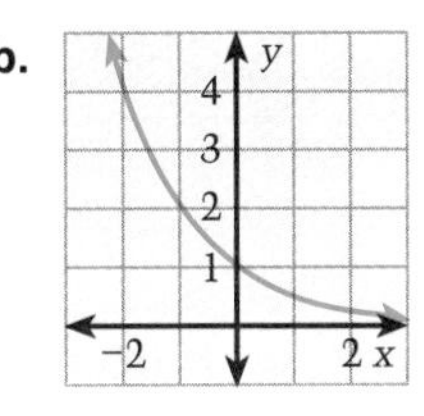

c.

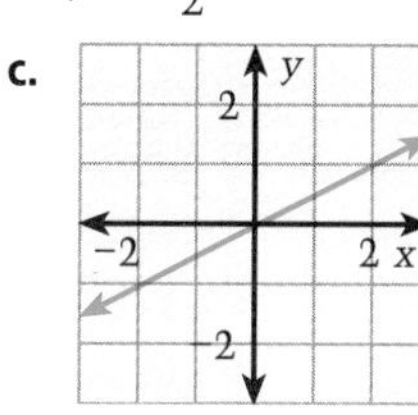

d.

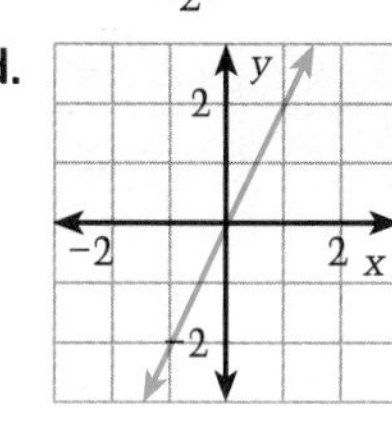

Elk Antlers **In Exercises 69–71, use the following information.**

The antler length, a, and the shoulder height, s, of an adult male elk are related (approximately) by the model

$$a = -40 + (32.5)(1.02)^s$$

where a and s are measured in inches.

69. Complete the table. See margin.

Shoulder height (in.), s	50	60	70	80
Antler length (in.), a	?	?	?	?

70. Estimate the antler length of the Irish elk shown in the photo at the right. 62.5 in.

71. Which is greater? An elk's shoulder height or its antler spread? (The spread is twice the length.) Antler spread

This is a skeleton of an Irish elk, now extinct. It had a shoulder height of about 58 inches.

Enrichment Activities

These activities require students to go beyond lesson goals.

COOPERATIVE LEARNING
Have students complete these activities in a partner setting.

1. Use your knowledge of the properties of exponents to solve these equations. Explain your reasoning.
   a. $5^x = 5$ 1
   b. $14^x = 1$ 0
   c. $6^{-x} = 6$ -1
   d. $2^x = \frac{1}{128}$ -7
   e. $(3^4)(3^x) = 3^{10}$ 6
   f. $\frac{1}{a^{10}} = a^x$ -10
   g. $(a^x)^{20} = a^{-40}$ -2
2. Sketch these graphs.
   a. $y = 3 \cdot 2^x$
   b. $y = \frac{1}{2} \cdot 5^x$

a.

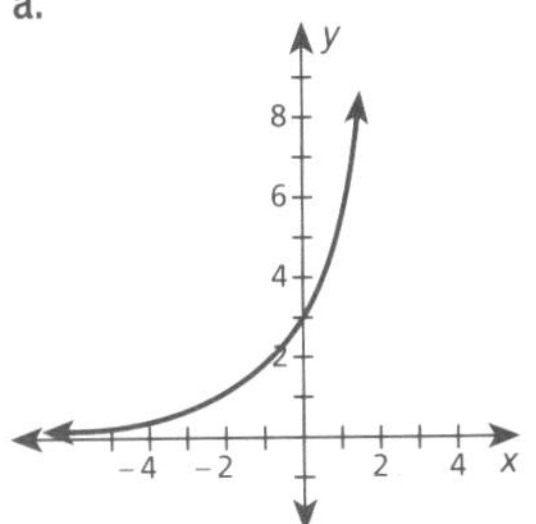

b.

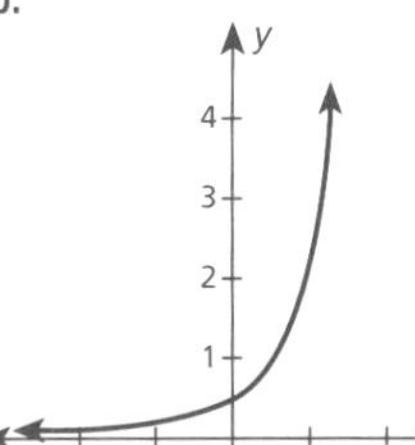

Milestones

Language Arts: WRITING

1. List five different ways in which computers are used in daily life. As cash registers, as automated teller systems, in cars, in appliances, as word processors, as telephone systems, etc.
2. Write a paragraph describing the role you see computers playing in society's future.

Mixed REVIEW

1. Find the difference of $\frac{3}{7}$ and $\frac{2}{5}$. **(1.1)** $\frac{1}{35}$
2. Find the sum of $2x$ and $3x$. **(2.6)** $5x$
3. What is 37% of 150? 55.5
4. What is $\frac{3}{4}$ of 64? 48
5. Solve $6a + 5b = 10$ for a. **(3.6)** $a = -\frac{5}{6}b + \frac{5}{3}$
6. Solve $45r - 9s = -27$ for r. **(3.6)** $r = \frac{1}{5}s - \frac{3}{5}$
7. Solve $x^2 - 3x + 12 = x(3 + x)$. **(3.3)** 2
8. Solve $\frac{1}{4}(8x + 2) = -\frac{1}{3}(4 - x)$. **(3.3)** $-\frac{11}{10}$
9. Write $3x + 2y = 6$ in slope-intercept form. **(5.1)** $y = -\frac{3}{2}x + 3$
10. Write $y = -\frac{1}{2}x + 2$ in standard form. **(5.5)** $x + 2y = 4$
11. Simplify $3x + 4(3 - x) + 2$. **(2.6)** $-x + 14$
12. Simplify $\frac{1}{2} - \frac{3}{4}(y - 12) + \frac{3}{2}$. **(2.6)** $-\frac{3}{4}y + 11$
13. Evaluate $3x^2 + 4x^3$ when $x = -1$. **(2.1)** -1
14. Evaluate $5p + 3p(4 - p)$ when $p = 3$. **(1.2)** 24
15. Is 3 a solution of $3x + 2 > 9$? **(1.5)** Yes
16. Is $\frac{1}{4}$ a solution of $4(x - 3) \leq -10$? **(1.5)** Yes
17. Is $(-2, 3)$ a solution of $3x + 6y \geq 12$? **(6.5)** Yes
18. Is $(4, -2)$ a solution of $\frac{1}{2}x - y \geq 3$? **(6.5)** Yes
19. Solve the system: $\begin{cases} 3x - 2y = -4 \\ -5x + y = 3 \end{cases}$ **(7.3)** $\left(-\frac{2}{7}, \frac{11}{7}\right)$
20. Solve the system: $\begin{cases} 6x - 2y = 3 \\ 4x + 2y = 6 \end{cases}$ **(7.3)** $\left(\frac{9}{10}, \frac{6}{5}\right)$

Milestones EARLY COMPUTERS

1800 · 1850 · 1900 · 1950

Babbage draws plans for computer · Mark I · ENIAC

ENIAC (Electronic Numerical Integrator and Calculator)

The idea of a computer was conceived by Charles Babbage in the 1830's. It took more than one hundred years to turn Babbage's idea into a working machine. In 1944, Howard Aiken completed the Mark I, an electromechanical machine programmed by punched cards. The first all-electronic computer, ENIAC, was completed in 1946. It was 100 feet long, 10 feet high, and 3 feet wide. The ENIAC was programmed by the setting of switches and wiring panels. It greatly increased calculation speed.

The first computers had to be programmed each time they were used. This was extremely time-consuming, and because of the number of steps involved, there were often mistakes made. In the 1940's, John von Neumann suggested that instructions could be stored electronically in the machines. This dramatically reduced the time required to operate the machines. Through technological advances such as the invention of the integrated circuit and its miniaturization onto a silicon chip, computers are now extraordinarily faster, smaller, and less expensive than the original machines.

Find out what each of the following people contributed to computer science: William Shockley, Grace Hopper, and John Mauchley.

8.3 Division Properties of Exponents

What you should learn:

Goal 1 How to use the division properties of exponents to evaluate powers and simplify expressions

Goal 2 How to use powers as models in real-life settings

Why you should learn it:

You can use exponents in models for many real-life situations, such as pollution standards that are expressed in "parts per million."

Goal 1 Dividing with Exponents

In Lesson 8.1, you learned that you can multiply powers that have the same base by adding exponents. To divide powers that have the same base, you subtract exponents. To see why this is true, let's look at the quotient of 2^5 and 2^3.

$$\frac{2^5}{2^3} = \frac{\overbrace{2 \cdot 2 \cdot 2 \cdot 2 \cdot 2}^{5 \text{ factors}}}{\underbrace{2 \cdot 2 \cdot 2}_{3 \text{ factors}}} = \underbrace{2 \cdot 2}_{2 \text{ factors}} = 2^2 = 2^{5-3}$$

This suggests one of the division properties of exponents.

Division Properties of Exponents

Let a and b be numbers and let m and n be integers.

1. To divide powers having the same base, subtract exponents.

   $\frac{a^m}{a^n} = a^{m-n}, \quad a \neq 0$ Quotient of Powers Property

2. To find a power of a quotient, find the power of the numerator and the power of the denominator and divide.

   $\left(\frac{a}{b}\right)^m = \frac{a^m}{b^m}, \quad b \neq 0$ Power of a Quotient Property

Example 1 *Using the Quotient of Powers Property*

a. $\frac{5^4}{5^3} = 5^{4-3} = 5^1 = 5$

b. $\frac{(-6)^3}{(-6)^3} = (-6)^{3-3} = (-6)^0 = 1$

c. $x^2 \cdot \frac{1}{x^3} = \frac{x^2}{x^3} = x^{2-3} = x^{-1} = \frac{1}{x}$

d. $\frac{8^2 \cdot 8}{8^5} = \frac{8^3}{8^5} = 8^{3-5} = 8^{-2} = \frac{1}{8^2}$ ■

Notice that the final steps of **c** and **d** above were to write the results as bases with positive exponents.

Problem of the Day

During our vacation it rained on 13 days. Sometimes it rained in the morning, sometimes it rained in the afternoon, but never at night. If it rained in the morning, the afternoon was clear; if it rained in the afternoon, the preceding morning was clear. We had 11 clear mornings, and 12 clear afternoons. How long was our vacation? 18 days

ORGANIZER

Warm-Up Exercises

1. Expand each power into a product of factors.
   a. 3^4 $3 \cdot 3 \cdot 3 \cdot 3$
   b. $(-3)^5$ $-3 \cdot -3 \cdot -3 \cdot -3 \cdot -3$
   c. 5^3 $5 \cdot 5 \cdot 5$
2. Simplify.
   a. $(-3m^2n^3)^2$ $9m^4n^6$
   b. $\frac{1}{x^{-5}}$ x^5
   c. $\frac{1}{y^{-3}z^{-6}}$ y^3z^6
   d. $x^5 \cdot x^{-3}$ x^2

Lesson Resources

Teaching Tools
Problem of the Day: 8.3
Warm-up Exercises: 8.3
Answer Masters: 8.3
Extra Practice: 8.3
Applications Handbook: p. 29–31

LESSON Notes

GOAL 1 Use number values to explain the statement of the properties. For example,

Property 1:

$$\frac{3^7}{3^4} = \frac{3 \cdot 3 \cdot 3 \cdot 3 \cdot 3 \cdot 3 \cdot 3}{3 \cdot 3 \cdot 3 \cdot 3} = 3 \cdot 3 \cdot 3 = 3^3 = 3^{7-4}$$

Property 2:

$$\left(\frac{5}{6}\right)^3 = \frac{5}{6} \cdot \frac{5}{6} \cdot \frac{5}{6} = \frac{5 \cdot 5 \cdot 5}{6 \cdot 6 \cdot 6} = \frac{5^3}{6^3}$$

Property 2 is sometimes stated as "the power of a quotient is the quotient of the powers."

Examples 1–3

Be prepared to expand each quotient into its basic factors if it helps students understand the relationships between the actual division and the use of exponents.

Example 4

Observe, if necessary, that 10^6 is one million. Check that students understand why the fish in this example has 50 percent more methylmercury than FDA standards allow.

Example 5

Check that students understand why the year 1940 is $t = -20$. Also note that the ratio of the number of shares traded in 1990 to the number of shares traded in 1940 can be entered directly into a graphics calculator, but it requires more than 30 keystrokes to do so! Many students will find that it is easier to form the power 1.09^{50} and then compute the power by using the calculator.

TECHNOLOGY The decimal display shown here is a truncated value. The TI-81 and Casio calculators both round up when they truncate. Other calculators just "drop" the remaining digits.

Example 2 *Using the Power of a Quotient Property*

a. $\left(\frac{2}{3}\right)^2 = \frac{2^2}{3^2} = \frac{4}{9}$

b. $\left(\frac{1}{x}\right)^3 = \frac{1^3}{x^3} = \frac{1}{x^3}$

c. $\left(-\frac{3}{2}\right)^3 = \left(\frac{-3}{2}\right)^3 = \frac{(-3)^3}{2^3} = \frac{-27}{8}$

d. $\left(\frac{4}{5}\right)^{-2} = \frac{4^{-2}}{5^{-2}} = \frac{5^2}{4^2} = \frac{25}{16}$ ■

Example 3 *Simplifying Expressions*

Simplify $\frac{2x^2y}{3x} \cdot \frac{9xy^2}{y^4}$.

Solution

$$\frac{2x^2y}{3x} \cdot \frac{9xy^2}{y^4} = \frac{(2x^2y)(9xy^2)}{(3x)(y^4)}$$ *Multiply fractions.*

$$= \frac{18x^3y^3}{3xy^4}$$ *Product of Powers Property*

$$= 6x^2y^{-1}$$ *Quotient of Powers Property*

$$= \frac{6x^2}{y}$$ *Write with positive exponents.* ■

Goal 2 Using Powers as Models in Real Life

Example 4 *Tolerable Amounts of Mercury*

Methylmercury is a constituent of industrial waste that, even in very small amounts, is harmful to humans and other animals. The U.S. Food and Drug Administration forbids the sale of fish having more than 1 part methylmercury per one million (10^6) parts of body weight. Suppose you are an FDA inspector and test a 2-kilogram fish that has 3 milligrams of methylmercury. Would this fish meet the FDA requirements? (One milligram is 10^{-3} grams, and one kilogram is 10^3 grams).

Solution The ratio of methylmercury to body weight is

$$\frac{3 \cdot 10^{-3} \text{ grams}}{2 \cdot 10^3 \text{ grams}} = \frac{1.5}{10^6} = 1.5 \text{ parts per million.}$$

This fish has 50% more methylmercury than FDA standards allow. ■

Toxic levels of methylmercury threaten bald eagle populations.

Although the next Example is not primarily concerned with exponential change, you may want to compare the graph of the stock listings with the graph on page 407.

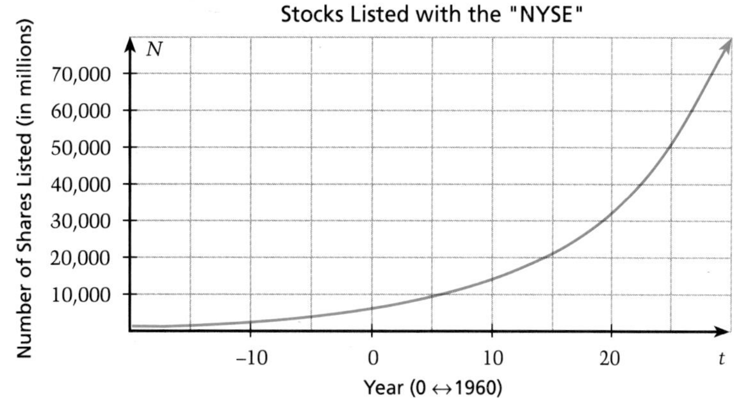

Real Life
Finance

Example 5 Shares Listed on the New York Stock Exchange

The number of shares (in millions) listed on the New York Stock Exchange from 1940 to 1990 can be modeled by

$$N = 6000 \cdot 1.09^t$$

where t is the year, with $t = 0$ corresponding to 1960. Find the ratio of shares listed in 1990 to the shares listed in 1940.

Solution For 1990, use $t = 30$ and for 1940, use $t = -20$.

$$\frac{\text{Number listed in 1990}}{\text{Number listed in 1940}} = \frac{6000 \cdot 1.09^{30}}{6000 \cdot 1.09^{-20}}$$
$$= 1.09^{30-(-20)}$$
$$= 1.09^{50} \qquad \textit{Use a calculator.}$$
$$\approx 74.4$$

There were 74.4 times as many listed in 1990 as in 1940. ■

Communicating about ALGEBRA

SHARING IDEAS about the Lesson

Evaluate a Process Which process for simplifying do you prefer, John's or Valerie's? Why? Simplify the expression, using steps different from these.

John's Solution

$$\left(\frac{a^{-2}}{b^5}\right)^{-3} = \frac{(a^{-2})^{-3}}{(b^5)^{-3}} = \frac{a^6}{b^{-15}} = a^6b^{15}$$

See margin.

Valerie's Solution

$$\left(\frac{a^{-2}}{b^5}\right)^{-3} = \left(\frac{1}{a^2b^5}\right)^{-3} = (a^2b^5)^3 = a^6b^{15}$$

Extra Examples

Here are additional examples similar to **Examples 1–3.**

1. Perform the indicated operation and simplify.
   a. $\frac{6^5}{6^3}$ 36
   b. $\frac{4 \cdot 4^3}{4^7}$ $\frac{1}{64}$
   c. $(\frac{2}{5})^6$ $\frac{64}{15{,}625}$
   d. $(\frac{3x}{4})^{-5}$ $\frac{1024}{243x^5}$
   e. $\frac{6x^4y^2}{4xy^5} \cdot \frac{8x^2y^6}{3x^3y}$ $4x^2y^2$

Check Understanding

1. Assuming the multiplication properties of exponents are true, justify Property 2.
   $(\frac{a}{b})^m = (a \cdot b^{-1})^m = a^m \cdot (b^{-1})^m = a^m \cdot b^{-m} = \frac{a^m}{b^m}$
2. Why must the denominator b of Property 2 be nonzero? Division by 0 is undefined.
3. Explain why there can be more than one way to evaluate a quotient of powers. Provide examples. Since division and multiplication are inverse operations, a quotient can always be written as a product. See Exercises 26–28 for examples.

Communicating about ALGEBRA

COOPERATIVE LEARNING

Have students work in a small-group setting to indicate the property that John and Valerie used to go from one step to the other. Then identify the properties used in your own solutions.

Sample answer:
$\left(\frac{a^{-2}}{b^5}\right)^{-3} = \frac{(a^{-2})^{-3}}{(b^5)^{-3}} = \frac{b^{15}}{a^{-6}} = a^6b^{15}$

EXERCISE Notes

ASSIGNMENT GUIDE

Basic/Average: Ex. 1–4, 5–27 odd, 31, 33, 35–47 odd, 53–56

Above Average: Ex. 1–4, 5–29 odd, 31–34, 45–48, 53–56

Advanced: Ex. 1–4, 15–29 odd, 30–34, 47–56

Selected Answers
Exercises 1–4, 5–47 odd

✪ More Difficult Exercises
Exercises 26–30, 32, 49–52

Guided Practice

▶ **Ex. 1–4** COOPERATIVE LEARNING These exercises provide a good in-class opportunity for students working with partners or in small groups to demonstrate an understanding of the essential skills necessary for success in the Independent Practice. Have students explain the answers to each other.

Independent Practice

▶ **Ex. 5–28** Students should refer back to Examples 1–3 when doing these exercises.

▶ **Ex. 29–30** Students should refer back to Example 4 if they need help with these problems.

EXERCISES

Guided Practice

CRITICAL THINKING about the Lesson

1. Can $\frac{x^{10}}{y^4}$ be simplified? Why or why not? No, the bases are not the same.

2. Does $\frac{x^{-4}}{x^{-5}}$ simplify as x or $\frac{1}{x}$? x

3. When you divide powers with the same base, do you add or subtract exponents? Subtract

4. What is the relationship between $\frac{x^4}{x^2}$ and $\frac{x^{-4}}{x^{-2}}$? Are they equivalent or are they reciprocals of each other? Explain. Reciprocals, their product is 1.

Independent Practice

In Exercises 5–16, evaluate the expression.

5. $\frac{6^6}{6^4}$ 36

6. $\frac{8^3}{8^1}$ 64

7. $\frac{(-4)^5}{(4)^5}$ -1

8. $\frac{(-3)^9}{(-3)^9}$ 1

9. $\frac{2^2}{2^{-3}}$ 32

10. $\frac{8^3 \cdot 8^2}{8^5}$ 1

11. $\frac{7^4 \cdot 7}{7^7}$ $\frac{1}{49}$

12. $\left(\frac{3}{4}\right)^2$ $\frac{9}{16}$

13. $\left(\frac{5}{3}\right)^3$ $\frac{125}{27}$

14. $\left(-\frac{2}{3}\right)^3$ $-\frac{8}{27}$

15. $\left(-\frac{4}{5}\right)^2$ $\frac{16}{25}$

16. $\left(\frac{9}{6}\right)^{-1}$ $\frac{2}{3}$

In Exercises 17–28, simplify the expression.

17. $\left(\frac{2}{x}\right)^4$ $\frac{16}{x^4}$

18. $\frac{x^4}{x^5}$ $\frac{1}{x}$

19. $\left(\frac{1}{x}\right)^6$ $\frac{1}{x^6}$

20. $x^3 \cdot \frac{1}{x^2}$ x

21. $x^7 \cdot \frac{1}{x^9}$ $\frac{1}{x^2}$

22. $\frac{3x^2y^2}{3xy} \cdot \frac{6xy^3}{3y}$ $2x^2y^3$

23. $\frac{4xy^3}{2y} \cdot \frac{5xy^{-3}}{x^2}$ $\frac{10}{y}$

24. $\frac{16x^3y}{-4xy^3} \cdot \frac{-2xy}{-x}$ $-\frac{8x^2}{y}$

25. $\frac{-9x^5y^7}{x^2y^3} \cdot \frac{(2xy)^2}{-6x^2y^2}$ $6x^3y^4$

✪ **26.** $\frac{6x^{-2}y^2}{xy^{-3}} \cdot \frac{(4x^2y)^{-2}}{xy^2}$ $\frac{3y}{8x^8}$

✪ **27.** $\frac{7x^{-1}y^3}{x^2y^{-2}} \cdot \frac{(3xy^2)^{-1}}{xy}$ $\frac{7y^2}{3x^5}$

✪ **28.** $\left(\frac{2xy^{-2}y^4}{3yx^{-1}}\right)^{-2} \cdot \left(\frac{4xy}{2x^{-1}y^3}\right)^2$ $\frac{9}{y^6}$

Mercury Levels **In Exercises 29 and 30, use the information from Example 4.**

✪ **29.** As the FDA inspector, you test a $4\frac{1}{2}$-kilogram fish and find that it has 4 milligrams of methylmercury. Does this fish meet FDA requirements? Yes

✪ **30.** A fish weighing 9 kilograms is found to contain 11 milligrams of methylmercury. As an FDA inspector, do you allow this fish to be sold? If not, how much would the fish have to weigh for 11 milligrams of methylmercury to be acceptable? No, 11 kg

31. *Toys "Я" Us* From 1980 to 1990, the sales for Toys "Я" Us increased by about the same percent each year. The sales, S (in millions of dollars), in year t can be modeled by

$$S = 640\left(\frac{5}{4}\right)^t$$

where $t = 0$ corresponds to 1980. Find the ratio of the 1989 sales to the 1984 sales. **≈3.05**

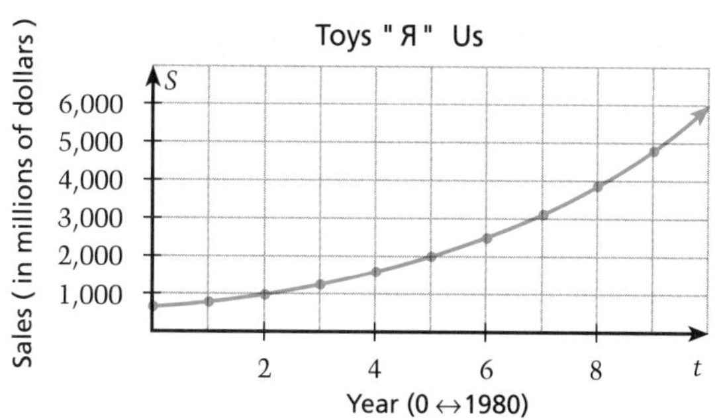

32. *Human Memory* Suppose that you memorized a list of 200 Spanish vocabulary words. Unfortunately, each week you forgot one fifth of the words that you knew the previous week. The number of words, S, you still remember after n weeks can be modeled by

$$S = 200\left(\frac{4}{5}\right)^n.$$

Complete the table showing the number of words you remembered after n weeks. How many weeks does it take to forget all but three words? Explain your answer.

Weeks, n	0	1	2	3	4	5	6	7	8
Words, S	?	?	?	?	?	?	?	?	?

200, 160, 128, ≈102, ≈82, ≈66, ≈52, ≈42, ≈34, ≈19 weeks, $S \approx 3$ when $n = 19$

33. *Olympic Rowing* Shells (boats) used in rowing competition usually have 1, 2, 4, or 8 rowers. Top speeds for racing shells in the Olympic 2000-meter races can be modeled by

$$s = 16.3(1.0285)^n$$

where s is the speed in kilometers per hour and n is the number of rowers. Use the model to estimate the ratio of the speed of an 8-rower shell to the speed of a 2-rower shell. **1.1837**

34. *Olympic Participation* From 1932 to 1988, the number of countries that participated in the summer Olympic Games increased by about 2.6% per calendar year. The number of participating countries, n, in year t can be modeled by

$$n = 83(1.026)^t$$ **4.210**

where $t = 0$ corresponds to 1960. (The model does not account for the 1976 or 1980 Olympics because of the boycotts in those years.) Use the model to estimate the ratio of 1988 participants to 1932 participants.

Ex. 31–34 TECHNOLOGY The ratio model of Example 5 can be used to complete these exercises. Scientific calculators provide students with a means of solving these real-life applications involving division of powers.

Ex. 41–48 You may wish to point out to students that they should simplify the expressions first in order to avoid evaluating cumbersome expressions.

Ex. 53–56 These exercises provide an interesting application of exponents. Even though Exercise 56 has no "right" answer, it gives students a chance to think about the problem in both an intuitive and mathematical manner.

Enrichment Activities

These activities require students to go beyond lesson goals.

1. RESEARCH Use the model from Exercise 34,
 $n = 83(1.026)^t$
 to predict the number of countries that take part in the 1992 Olympics. about 189 Use library resources to find the actual number.
2. Compare the predicted number with the actual number. How accurate was the model?
3. Can you think of some reasons why the model would probably be invalid for predicting the participants in the 1992 Olympics? The number of possible participants is very different in 1992 from what it was in 1988. World events since 1988 have changed the makeup of some countries. For example, in the 1988 Olympics, there were two Germanys. In the 1992 Olympics, there was only one Germany. In the 1988 Olympics, there was one Soviet Union.

Integrated Review

In Exercises 35–40, simplify the expression.

35. $(x^3)^5$ x^{15} **36.** $(a^5)^3$ a^{15} **37.** $x^5 \cdot x^3$ x^8 **38.** $n^3 \cdot n^{-5}$ $\frac{1}{n^2}$ **39.** $\frac{x^{-5}}{x^3}$ $\frac{1}{x^8}$ **40.** $\frac{b^3}{b^5}$ $\frac{1}{b^2}$

In Exercises 41–48, simplify the expression. Then evaluate it for the given values of the variables.

41. $x^3 \cdot x^2y;\ x = 2,\ y = 3$ x^5y, 96

42. $a^6b \cdot a^{-4}b^2;\ a = 10,\ b = -5$ a^2b^3; $-12{,}500$

43. $(2x)^2 \cdot (3y^2)^2;\ x = -1,\ y = 2$ $36x^2y^4$, 576

44. $(5n^3m)^4 \cdot (2nm)^2;\ n = 0,\ m = 2$

45. $(xy^4)^{-1} \cdot \left(\frac{1}{x^2y^3}\right)^{-2};\ x = 2,\ y = 5$ x^3y^2, 200

46. $\left(\frac{1}{3a^3b^2}\right)^{-4} \cdot (-3a^{10}b^9)^{-1};\ a = 4,\ b = 6$

47. $(10r^7s^{11})^0 \cdot \left(\frac{s^2}{rs^3}\right)^{-2};\ r = 12,\ s = -1$ r^2s^2, 144

48. $-\left(\frac{x^2y^3}{x^4y^2}\right) \cdot \left(\frac{2x^7y^4}{x^5y^6}\right)^{-2};\ x = 2,\ y = 3$ $-\frac{y^5}{4x^6}$, $-\frac{243}{256}$

44. $2500n^{14}m^6$, 0

Exploration and Extension

46. $-\frac{27a^2}{b}$, -72

In Exercises 49–52, simplify the expression.

49. $x^a \cdot x^a$ x^{2a} **50.** $2^{7x+1} \cdot 2^{3x+4}$ 2^{10x+5} **51.** $\frac{x^{a+2}}{x^{a-2}}$ x^4 **52.** $\frac{x^b}{x^{b+1}}$ $\frac{1}{x}$

Flapping Wings **In Exercises 53–56, use the following information.**

When in flight, small birds flap their wings more often than large birds. The number of times per second, n, that a bird flaps its wings during normal flight can be modeled by $nw = 6^3$, where w is the wing length in centimeters. *(Source:* **On Size and Life***)*

The bee hummingbird has a wing length of about 3.5 centimeters.

The California gull has a wing length of about 55 centimeters.

53. How many times per second does the hummingbird above flap its wings? ≈61.7

54. How many times per second does the California gull above flap its wings? ≈3.9

55. What is the ratio of the number of times a bee hummingbird flaps its wings in a second to the number of times a California gull flaps its wings in a second? ≈15.8

56. How would you explain the fact that large birds do not need to flap their wings as often in flight as small birds? See below.

56. Large wings offer more resistance to the air than small wings; so, a large-wing flap moves a large bird through the air more easily than a small-wing flap moves a small bird.

8.4 Scientific Notation

What you should learn:

Goal 1 How to use scientific notation to express large and small numbers

Goal 2 How to perform operations with numbers in scientific notation, with and without a calculator

Why you should learn it:

You can use scientific notation as a convenient way to write very small numbers, like those used in scientific measurements, or very large numbers, like those that give statistical information about the United States.

Goal 1 Using Scientific Notation

In this lesson, you will learn to write decimal numbers in scientific notation. Scientific notation is based on powers of ten.

Decimal form	Power of 10	Name
1,000,000,000	10^9	One billion
1,000,000	10^6	One million
1,000	10^3	One thousand
10	10^1	Ten
1	10^0	One
0.1	10^{-1}	One tenth
0.001	10^{-3}	One thousandth
0.000001	10^{-6}	One millionth

Scientific notation uses powers of ten to express decimal numbers. Here are two examples.

$1.2 \times 10^3 = 1.2(1000) = 1200$
$3.5 \times 10^{-1} = 3.5(0.1) = 0.35$

Numbers in scientific notation are written $c \times 10^n$, where c is a decimal number greater than or equal to 1 and less than 10.

Example 1 *Rewriting Scientific Notation in Decimal Form*

a. $1.345 \times 10^2 = 134.5$ *Move decimal point 2 places.*

b. $3.67 \times 10^4 = 36{,}700$ *Move decimal point 4 places.*

c. $7.8 \times 10^{-1} = 0.78$ *Move decimal point 1 place.*

d. $4.103 \times 10^{-5} = 0.00004103$ *Move decimal point 5 places.* ■

Example 2 *Rewriting Decimals Using Scientific Notation*

a. $2450 = 2.45 \times 10^3$ *Move decimal point 3 places.*

b. $1.78 = 1.78 \times 10^0$ *Move decimal point 0 places.*

c. $0.000722 = 7.22 \times 10^{-4}$ *Move decimal point 4 places.* ■

Problem of the Day

The Duo-County League has 8 football teams. If each team plays every other team exactly once, how many games are played? 28

ORGANIZER

Warm-Up Exercises

1. Express each number in decimal form.
   a. one million
   b. one billion
   c. one hundred thousand
   d. ten
   e. one ten-thousandth
   f. one billionth
   g. one hundredth

   a. 1,000,000
   b. 1,000,000,000
   c. 100,000
   d. 10 e. 0.0001
   f. 0.000 000 001 g. 0.01
2. Express each number in fraction form.
   a. one tenth
   b. one hundred-thousandth
   c. one thousandth
   d. one millionth

   a. $\frac{1}{10}$ b. $\frac{1}{100{,}000}$
   c. $\frac{1}{1000}$ d. $\frac{1}{1{,}000{,}000}$

Lesson Resources

Teaching Tools
Problem of the Day: 8.4
Warm-up Exercises: 8.4
Answer Masters: 8.4
Extra Practice: 8.4
Color Transparencies: 40, 41

LESSON Notes

Example 3

In Example 3a, point out the use of the commutative and associative properties in the step labeled "regroup." It may be necessary to indicate that $10.64 \times 10^7 = 1.064 \times 10 \times 10^7 = 1.064 \times 10^8$.

In Example 3b, show that $0.25 \times 10^3 = 2.5 \times 10^{-1} \times 10^3 = 2.5 \times 10^2$.

Example 4

TECHNOLOGY Graphics and scientific calculators allow users to perform computations in scientific notation mode. As a result, when in scientific notation mode, number operations entered in the decimal format will produce results given in scientific notation.

Extra Examples

Here are additional examples similar to **Examples 1–4.**

1. Rewrite these numbers in decimal form.
   a. 3.203×10^{-5} 0.00003203
   b. 2.42×10^7 24,200,000
2. Rewrite decimals in scientific notation.
   a. 3,902,000 3.902×10^6
   b. 0.000 000 0234 2.34×10^{-8}

Common-Error ALERT!

In **Examples 1–2,** a common mistake students make when expressing decimal numbers in scientific notation and when expressing numbers in scientific notation in decimal form is to assign the wrong sign to the exponent. Use the Alternate Approach suggested on page 423 to help those students.

Goal 2 Operations with Scientific Notation

To multiply, divide, or find powers of numbers in scientific notation, use the properties of exponents.

Example 3 *Computing with Scientific Notation*

a. $(1.4 \times 10^4)(7.6 \times 10^3)$

$= (1.4 \cdot 7.6) \times (10^4 \cdot 10^3)$ *Regroup.*

$= 10.64 \times (10^7)$ *Simplify.*

$= 1.064 \times 10^8$ *Scientific notation*

b. $\frac{1.2 \times 10^{-1}}{4.8 \times 10^{-4}}$

$= \frac{1.2}{4.8} \times \frac{10^{-1}}{10^{-4}}$ *Rewrite as a product of fractions.*

$= 0.25 \times 10^3$ *Simplify.*

$= 2.5 \times 10^2$ *Scientific notation* ■

Most scientific calculators automatically switch to scientific notation to display large (or small) numbers that exceed the display range. Try multiplying 98,900,000 times 500. If your calculator follows standard conventions, its display should be similar to [*4.945 E 10*]. This means that the result is 4.945×10^{10}.

To *enter* numbers in scientific notation, your calculator should have an exponential entry key labeled [*EE*], [*EXP*], or [*SCI*].

Example 4 *Using a Calculator with Scientific Notation*

Use a calculator to evaluate the following.

a. $78{,}000 \times 2{,}400{,}000{,}000$ **b.** $0.000000748 \div 500$

Solution

a. Since $78{,}000 = 7.8 \times 10^4$ and $2{,}400{,}000{,}000 = 2.4 \times 10^9$, you can use the following keystrokes.

7.8 [EE] 4 [×] 2.4 [EE] 9 [=] Display: [*1.872 E 14*]

Therefore, the product is 1.872×10^{14}.

b. Since $0.000000748 = 7.48 \times 10^{-7}$ and $500 = 5.0 \times 10^2$, you can use the following keystrokes.

7.48 [EE] 7 [+/−] [÷] 5.0 [EE] 2 [=] Display: [*1.496 E − 9*]

Therefore, the quotient is 1.496×10^{-9}. ■

Real Life **Retail Sales**

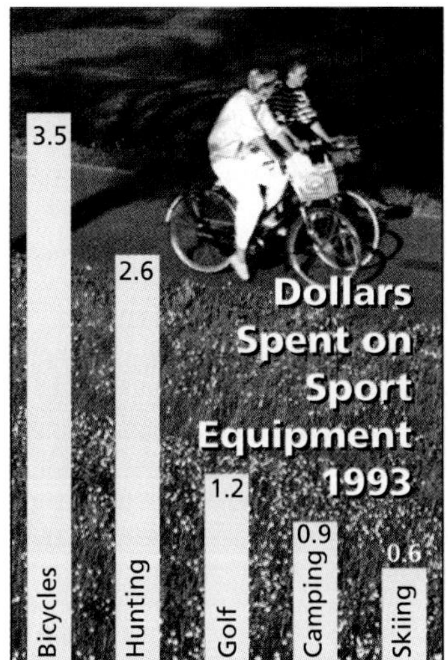

Example 5 *Average Amount Spent on Bicycles*

Find the average amount that an American spent on bicycles in 1993. (The U.S. population in 1993 was 258 million.)

Solution From the chart, you know that the total amount spent on bicycles in 1993 was \$3.5 billion.

$$\text{Average amount per person} = \frac{\text{Total amount}}{\text{Number of people}}$$
$$= \frac{3.5 \text{ billion}}{258 \text{ million}}$$
$$= \frac{3.5 \times 10^9}{2.58 \times 10^8}$$
$$\approx 1.36 \times 10^1$$
$$\approx 13.6$$

The average amount spent per person was about \$13.60. ■

In 1993, the USA produced 2.5 billion barrels of crude oil.

(Source: U.S. Energy Information Administration)

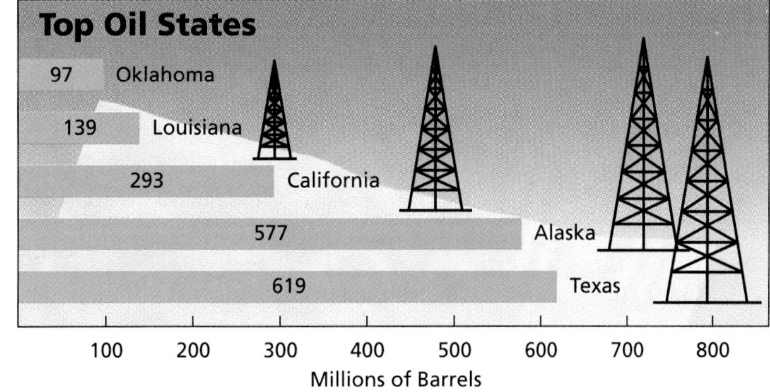

Communicating about ALGEBRA

Cooperative Learning

SHARING IDEAS about the Lesson

Work with a Partner and Use Notation Show how to use scientific notation to help solve the problem.

A. What percent of the crude oil produced in the United States in 1993 was produced in Louisiana? 5.56%

B. Evaluate successive powers of 5 with your calculator.

$5^6, 5^7, 5^8, 5^9, \ldots, 5^{17}$ See below.

List your results. For what powers of 5 does your calculator display scientific notation? 15 or greater

C. The mass of one proton is 1.67×10^{-24} gram. One atom of copper has 29 protons. Find the mass of the protons in 2.37×10^{17} atoms of copper. 1.147791×10^{-5} g

B. 15,625; 78,125; 390,625; 1,953,125; 9,765,625; 48,828,125; 244,140,625; 1,220,703,125; 6,103,515,625; 3.0517578×10^{10}; 1.5258789×10^{11}; 7.6293945×10^{11}

3. Perform the indicated operation. Express answers in scientific notation.

a. $(3.2 \times 10^6)(1.2 \times 10^4)$

b. $\frac{4.5 \times 10^{-7}}{9.0 \times 10^{-4}}$

c. $(3.45 \times 10^{-8})(1.3 \times 10^5)$

a. 3.84×10^{10}

b. 5.0×10^{-4}

c. 4.485×10^{-3}

Check Understanding

1. Why is scientific notation useful? See page 419.
2. Name at least three formats that real numbers may assume. (There are more than three formats in which real numbers may be written. Engineers, for example, use a format similar to scientific notation, in which exponents are always multiples of 3.) for example, whole number form, decimal form, fractional form, exponent form, scientific notation, etc.
3. Create a flowchart of the steps needed to rewrite any decimal number in scientific notation.
4. Create a flowchart of the steps needed to rewrite any number in scientific notation as a decimal number.

3.–4. Check students' charts.

Communicating about ALGEBRA

EXTEND *Communicating*

COOPERATIVE LEARNING

Have students work in small groups as a team. Ask each team to find newspaper or magazine reports in which scientific notation was used or should have been used. The three teams with the greatest variety of areas from which data was collected are the gold, silver, and bronze medal winners.

EXERCISE Notes

ASSIGNMENT GUIDE

Basic/Average: Ex. 1–4, 5–49 odd, 50, 53–54

Above Average: Ex. 1–4, 5–29 odd, 33–43 odd, 45–50, 52–54

Advanced: Ex. 1–4, 5–43 odd, 44–52, 55–56

Selected Answers
Exercises 1–4, 5–53 odd

Use **Mixed Review** as needed.

More Difficult Exercises
Exercises 38, 43, 44, 51–52, 55–56

Guided Practice

Ex. 1 Remind students that $1 \le c < 10$ for $c \times 10^n$ to be in scientific notation.

Ex. 2–4 Be sure to have students explain their answers.

Independent Practice

Ex. 21–26 Remind students they can regroup in order to simplify the multiplication.

Ex. 27–32 TECHNOLOGY Check students' understanding of the proper use of their calculators. The keystrokes shown in Example 4 may not be correct for every brand of calculator.

Answers

27. 4,984,000,000,000; 4.984×10^{12}
28. 129,000,000,000; 1.29×10^{11}
29. 3,098,100; 3.098×10^6
30. 29.67; 2.967×10^1
31. 0.000 000 000 000 000 000 000 11574317; $1.1574317 \times 10^{-22}$
32. 0. 000 000 000 000 091 125; 9.1125×10^{-14}

Ex. 33–48 GROUP ACTIVITY Encourage class discussions of these applications. The ability to write large and small numbers in scientific notation makes possible varied and interesting real-life applications.

EXERCISES

Guided Practice

CRITICAL THINKING about the Lesson

1. The following numbers are equal. Which one is in scientific notation?
   a. 912 b. <u>9.12×10^2</u>
2. To write 0.000032 in scientific notation, how many places must you move the decimal point? 5
3. Which is equal to 62,000, <u>6.2×10^4</u> or 6.2×10^{-4}?
4. What is one thousand times one millionth? Write your answer in scientific notation. 1×10^{-3}; one thousandth

Independent Practice

In Exercises 5–12, rewrite the scientific notation in decimal form.

5. 1.09×10^6 1,090,000
6. 2.345×10^8 234,500,000
7. 6.21×10^0 6.21
8. 9.4675×10^4 94,675
9. 8.52×10^{-3} 0.00852
10. 7.021×10^{-5} 0.00007021
11. 8.67×10^{-2} 0.0867
12. 4.73×10^0 4.73

In Exercises 13–20, rewrite the decimal in scientific notation.

13. 93,000,000 9.3×10^7
14. 900,000,000 9×10^8
15. 1,637,000,000 1.637×10^9
16. 67.8 6.78×10^1
17. 0.000435 4.35×10^{-4}
18. 0.008367 8.367×10^{-3}
19. 0.004392 4.392×10^{-3}
20. 0.0875 8.75×10^{-2}

In Exercises 21–26, evaluate the expression without a calculator. Write the result in decimal form.

21. $6 \times 10^{-2} \cdot 3 \times 10^4$ 1800
22. $5 \times 10^5 \cdot 5 \times 10^{-5}$ 25
23. $4 \times 10^4 \cdot 2 \times 10^{-1}$ 8000
24. $6 \times 10^{-3} \cdot 7 \times 10^{-4}$ 0.0000042
25. $9 \times 10^{-3} \cdot 4 \times 10^8$ 3,600,000
26. $8 \times 10^4 \cdot 10 \times 10^{-1}$ 80,000

Technology **In Exercises 27–32, use a calculator to evaluate the expression. Write the result in both decimal form and in scientific notation.** See margin.

27. $8{,}000{,}000 \cdot 623{,}000$
28. $3{,}000{,}000 \cdot 43{,}000$
29. $0.000345 \cdot 8{,}980{,}000{,}000$
30. $345{,}000 \cdot 0.000086$
31. $(3.28 \times 10^{-6})^4$
32. 0.000045^3

In Exercises 33–36, write the number in scientific notation.

33. *Carbon Atom* An atom of carbon has a mass of 0.00000000000000000000004 gram. 4×10^{-23}
34. *Population of the United States* In 1990, the United States population was about 250,000,000. 2.5×10^8
35. *Quarterback Salary* Jim Kelly, the quarterback for the Buffalo Bills football team, was paid $4,800,000 in 1990. 4.8×10^6
36. *Metric Conversion* One meter is equal to one thousandth of a kilometer. 1×10^{-3}

37. *A Flea and a Whale* A whale has a mass of 1.0×10^5 kilograms. A flea has a mass of 3.0×10^{-4} kilogram. What is the ratio of the mass of a whale to that of a flea? $\approx 3.3 \times 10^8$

38. *Population Increase* In 1990, Earth's population was increasing by about 8.5×10^7 people per year. Suppose you live in a city whose population increase includes 2000 newborn per year. What percent of Earth's population increase is your city contributing? $\approx 0.002\%$

39. *Size of the Universe* The universe is 2.1×10^{23} miles wide. Find the reciprocal of this distance. $\approx 4.8 \times 10^{-24}$

40. *Size of a Cell* Most cells are about 2.5×10^{-3} centimeter in diameter. Find the reciprocal of this distance. 4×10^2

41. *Atoms in the Sun* The volume of the sun is about 8.5×10^{31} cubic inches. There are about 1×10^{57} atoms in the sun. What is the average number of atoms per cubic inch in the sun? $\approx 1 \times 10^{25}$

42. *Sunlight* Light travels from the sun to Earth in approximately

$$\frac{9.3 \times 10^7}{1.1 \times 10^7} \text{ minutes.}$$

Write this time in decimal form. ≈ 8.5

43. *Distance to a Star* A light-year is approximately 5.88×10^{12} miles. Estimate the distance between Earth and Beta Andromedae in miles. $\approx 4.47 \times 10^{14}$ mi

44. *Mass of a Gold Atom* A proton and a neutron each weigh 1.67×10^{-24} gram. An electron weighs 9.11×10^{-28} gram. Find the mass of one atom of gold. $\approx 2.64 \times 10^{-22}$ g

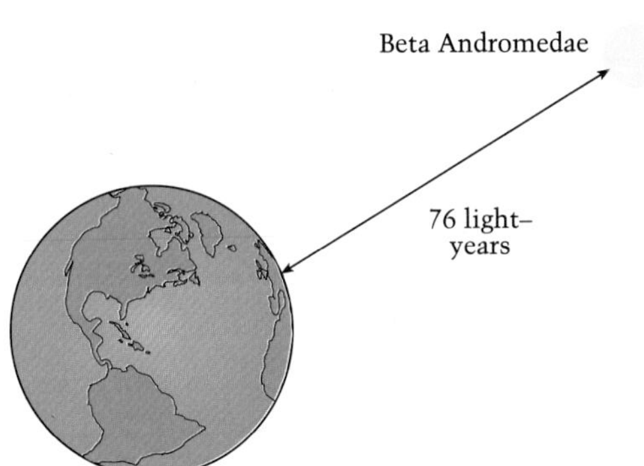

The star Beta Andromedae is approximately 76 light-years from Earth.

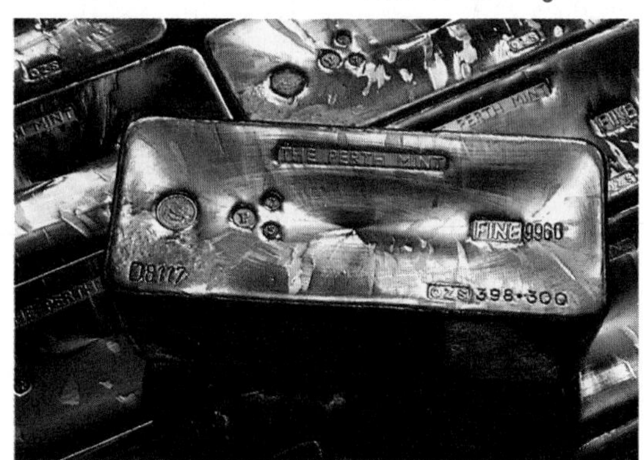

An atom of gold has 79 protons, 79 neutrons, and 79 electrons.

Daily Life in the United States **In Exercises 45–48, consider the population of the United States to be 250 million and consider a year to have 365 days.**

45. Approximately 1.6×10^{10} disposable diapers are thrown into the trash each year. How many are thrown each day? $\approx 4.4 \times 10^7$

46. Approximately 1.095×10^{12} gallons of sewage is dumped off the coasts of the United States yearly. How much is dumped each day? $\approx 3 \times 10^9$

47. Approximately 7.7×10^{10} dollars is given by individuals to charitable groups in a year. What is the average amount given per person? $\approx \$308$

48. Approximately 7.1×10^9 greeting cards are given yearly. What is the average number of greeting cards received per person? ≈ 28

Ex. 45–48 Be sure that students understand that 250 million is the same as 2.5×10^8.

An Alternate Approach
Rewriting Decimal Numbers

One way to keep track of the signs of exponents when expressing numbers in and out of scientific notation is to use a caret (‸) to mark the position behind the first significant nonzero digit in the number to be rewritten. For example, in 403,911 the first significant nonzero digit is 4; while in 0.000 039 03, the first significant nonzero digit is 3. (Note that the first four zeros following the decimal point in this number are significant zeros.)

To determine the sign (and magnitude) of the exponent once the caret has been placed, *count from the caret to the decimal point.* Moving to the left indicates that the sign is negative. Moving to the right indicates that the sign is positive. The magnitude, or size, of the exponent is the number of spaces from the caret to the decimal. For example,

Given: 134.5
Place the caret: 1‸34.5
Count: 2 spaces to the right to get 1‸3 4.5.
Exponent: +2 or 2
Scientific notation: 1.345×10^2

Given: 0.000 041 03
Place the caret: 0.000 04‸103
Count: 5 spaces to the left to get 0.0 0 0 0 4‸103
Exponent: −5
Scientific notation: 4.103×10^{-5}

Here is an example of rewriting scientific notation as a decimal number. Place the caret at the decimal point and move to the left or right from the caret as indicated by the exponent, annexing zeros as required.

Given: 7.22×10^{-4}

Place the caret: 7^22

Exponent: −4

Count: 4 spaces to the left, annexing three zeros and getting 0.0 0 0 7^22

Decimal number: 0.000 722

▶ **Ex. 51–52** These exercises are challenging. Some students may need extra help with them.

▶ **Ex. 56** To complete this exercise, students should refer to the conversion model provided in Exercise 55.

Enrichment Activity

This activity requires students to go beyond lesson goals.

RESEARCH Use newspapers, magazines, and an almanac to extend Exercises 45–48 and develop your own questions involving per person rates, per day rates, per year rates, etc. Some types of data you should look for are population data, athletic statistics and salaries, and space travel.

Integrated Review

49. Arrange the numbers in increasing order.

1st $(9.1 \times 10^4)^4$ 4th $(9.1 \times 10^2)^{10}$ 3rd $(9.1 \times 10^5)^4$ 2nd $(9.1 \times 10^{10})^2$

50. Which is greater, $\underline{\frac{1}{4 \times 10^5}}$ or $\frac{1}{4 \times 10^6}$?

In Exercises 51 and 52, find the slope of the line passing through the two points.

51. $(3.4 \times 10^4, 6.2 \times 10^5), (1.4 \times 10^4, 9.0 \times 10^5)$ −14

52. $(1.9 \times 10^9, 4.6 \times 10^2), (9.9 \times 10^9, 3.4 \times 10^2)$ -1.5×10^{-8}

In Exercises 53 and 54, use the chart.

53. About how much did people in the United States spend on recreational transportation in 1993? Write your answer in scientific notation. $\$1.6 \times 10^9$

54. About how much did people in the United States spend on snowmobiles in 1993? Write your answer in scientific notation. $\$4.8 \times 10^7$

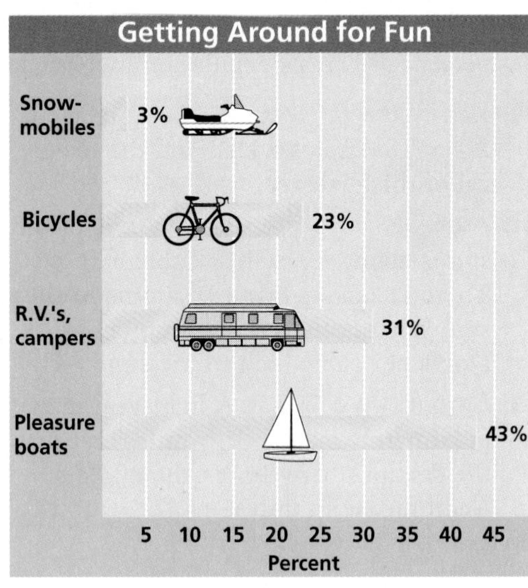

Total spent on recreational transportation in 1993 was about $1.6 billion.

Exploration and Extension

55. *Speed of Light* The speed of light is approximately 2.9979×10^{10} centimeters per second. Convert this speed to miles per second by evaluating the expression.

$$\frac{2.9979 \times 10^{10} \text{ cm}}{\text{sec}} \cdot \frac{1 \text{ in.}}{2.54 \text{ cm}} \cdot \frac{1 \text{ ft}}{12 \text{ in.}} \cdot \frac{1 \text{ mi}}{5280 \text{ ft}}$$

$\approx 1.86 \times 10^5$

56. *Speed of a Jet* The fastest speed attained by a jet aircraft is 1.93×10^5 feet per minute. What is this speed in feet per second? In miles per hour?

$\approx 3.22 \times 10^3$, $\approx 2.19 \times 10^3$

Mixed REVIEW

7. and 8. See margin.

1. Evaluate $(3 + 2) \div (27 - 2)$. **(1.1)** 0.2

2. Evaluate $12 \div 3 \cdot (4 - 1)$. **(1.1)** 12

3. Solve $|x + 2| = 3$. **(4.8)** −5, 1

4. Solve $-3|x + 2| = -9$. **(4.8)** −5, 1

5. Simplify $3x \cdot (3x)^2 \cdot x^4$. **(1.3)** $27x^7$

6. Simplify $p^2 \cdot q^3 \cdot p^2q^2$. **(1.3)** p^4q^5

7. Sketch the graph of $3|x - 2| = y$. **(4.7)**

8. Sketch the graph of $8x + 9y = 72$. **(4.3)**

9. What is the reciprocal of x^3? **(2.7)** x^{-3} or $\frac{1}{x^3}$

10. What is the opposite of $3y$? **(2.1)** $-3y$

11. Solve $3y + 2(3 - 6y) = 12$. **(3.2)** $-\frac{2}{3}$

12. Solve $(t - 2)\frac{1}{3} - 5t \geq 6$. **(6.1)** $t \leq -\frac{10}{7}$

13. Evaluate $3y + 2y^2$ when $y = -3$. **(2.5)** 9

14. Evaluate $\frac{1}{4}x \div (3 - x)$ when $x = \frac{1}{2}$. **(1.2)** $\frac{1}{20}$

Mid-Chapter SELF-TEST

Take this test as you would take a test in class. The answers to the exercises are given in the back of the book.

In Exercises 1–12, simplify the expression. (8.1, 8.2, 8.3)

1. $q^3 \cdot q^2$ q^5 **2.** $x^2 \cdot x^2 \cdot x^3$ x^7 **3.** $(n^5)^2$ n^{10}

4. $-(t^3)^2$ $-t^6$ **5.** $-3s^2 \cdot 8s^3$ $-24s^5$ **6.** $5x^{-5}$ $\frac{5}{x^5}$

7. $(-2b)^{-2}$ $\frac{1}{4b^2}$ **8.** $4c^0$ 4 **9.** $\frac{4x^2y}{2xy^2}$ $\frac{2x}{y}$

10. $\frac{p^{-3}}{p^2}$ $\frac{1}{p^5}$ **11.** $\frac{4q^4}{3q^{-3}}$ $\frac{4q^7}{3}$ **12.** $\frac{4w^2z^4}{5y^3z^2} \cdot \frac{10z^2}{2wy}$ $\frac{4wz^4}{y^4}$

In Exercises 13–18, evaluate the expression. (8.1, 8.2)

13. $6^2 \cdot 6^0$ 36 **14.** $4^3 \cdot 4^{-4}$ $\frac{1}{4}$ **15.** $\left(\frac{3}{4}\right)^2$ $\frac{9}{16}$

16. $\left(\frac{5}{2}\right)^{-2}$ $\frac{4}{25}$ **17.** $(0.3 + 2.5)^4$ 61.4656 **18.** $(3^2)^3$ 729

In Exercises 19 and 20, write the number in decimal form. (8.4)

19. 1.0079×10^7 10,079,000 **20.** 5.006×10^{-4} 0.0005006

In Exercises 21 and 22, write the number in scientific notation. (8.4)

21. 3,789,650 3.78965×10^6 **22.** 0.00000179 1.79×10^{-6}

In Exercises 23 and 24, evaluate the expression. (8.4)

23. $(6.2 \times 10^5) \cdot (1.1 \times 10^{-7})$ $\approx 6.8 \times 10^{-2}$ **24.** $(5.9 \times 10^{-2}) \div (4.7 \times 10^{-4})$ $\approx 1.3 \times 10^2$

25. The volume of a sphere can be found by the formula $V = \frac{4}{3}\pi r^3$, where $\pi \approx 3.14$ and r is the radius of the sphere. A spherical 9-inch balloon has a radius of $\frac{9}{2}$. How many cubic inches of air does it hold? **(8.1)** 381.51

26. If a water balloon has a diameter of 4 inches, how much water does it hold? ≈ 33.49 in.^{3}

27. Some radioisotopes have a short half-life. The half-life of C-11 is 20 minutes. If you start with 16 grams of C-11, the number of grams remaining after h half-life periods would be $W = 16\left(\frac{1}{2}\right)^h$. Complete the table and use the results to sketch the graph. **(8.2)**

Half-life periods, h	0	1	2	3	4
Grams, W	?	?	?	?	?

16, 8, 4, 2, 1
See margin.

28. In one year, 153,900,000,000 pieces of mail are sent via the U.S. Postal Service. Of these, 59,730,000,000 are direct-mail advertising. What percent of the total volume of mail is direct-mail advertising? Use scientific notation to determine your answer. **(8.4)** $\approx 38.81\%$

Answers
Mixed Review

7.
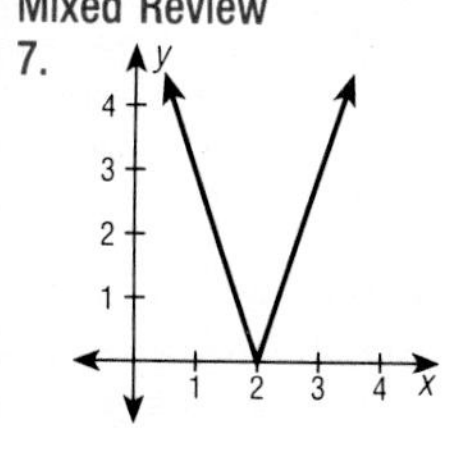

8.
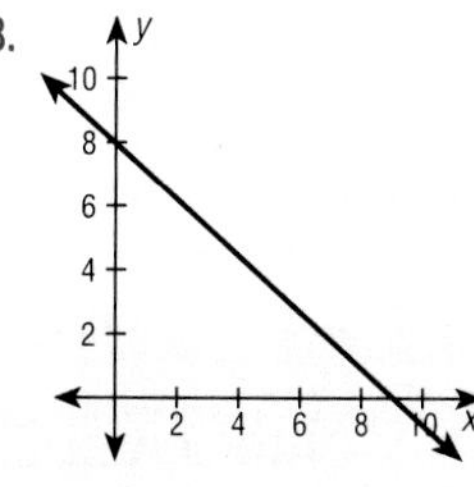

Mid-Chapter Test

27.
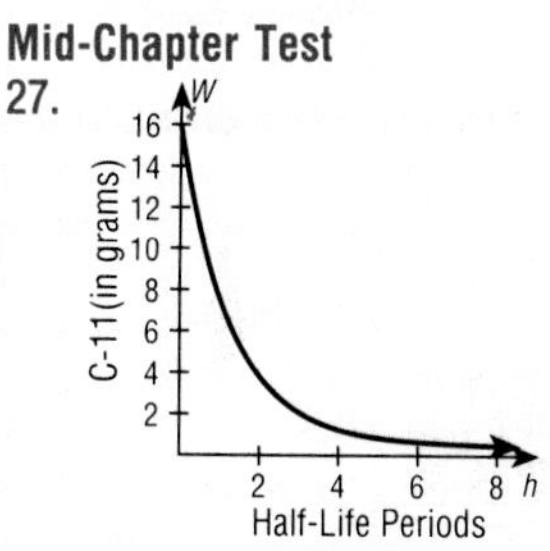

Problem of the Day

Yoko is willing to bet that the sum of the runs by the Saint Louis Cardinals during a season is greater than the product of the runs. Why is she likely to win such a bet?

Because a team is likely to be shut out in at least one game in a season, getting a score of 0. Then the product of all the scores will be 0.

ORGANIZER

Warm-Up Exercises

1. Perform the indicated operations.
   a. $(1.456 \times 10^{17})(4.327 \times 10^{23})$
   b. $(9.67 \times 10^{-45})(6.214 \times 10^{34})$
   c. $\frac{5.6 \times 10^{37}}{1.247 \times 10^{63}}$
   d. $\frac{4.88 \times 10^{13}}{2.40 \times 10^{27}}$

a. 6.300112×10^{40}
b. 6.008938×10^{-10}
c. $4.490777867 \times 10^{-26}$
d. 2.033333×10^{-14}

Lesson Resources

Teaching Tools
Transparency: 3
Problem of the Day: 8.5
Warm-up Exercises: 8.5
Answer Masters: 8.5
Extra Practice: 8.5
Color Transparency: 41
Applications Handbook: p. 9, 11, 12

LESSON Notes

EXTEND Examples 1–3
CRITICAL READING Ask students the following questions about the readings.

In **Example 1,** what is the ratio of the mass of the earth to the mass of the moon? about 81 to 1

In **Example 2,** what percent of Alaska was leased for gas and oil in 1969? about 0.12%

8.5 Problem Solving and Scientific Notation

What you should learn:

Goal How to use scientific notation to solve real-life problems requiring very large or very small numbers

Why you should learn it:

You can use scientific notation to help manage the manipulation of large and small numbers, such as in problems about space travel.

Goal Using Scientific Notation in Real Life

Remember the steps of the general problem-solving plan:

Write a verbal model. → Assign labels. → Write an algebraic model. → Solve the algebraic model. → Answer the question.

Example 1 *Landing on the Moon*

On July 20, 1969, the first humans set foot on the moon. The technology required was extremely complicated. The moon is a moving target. *Apollo 11* had to aim at a point far ahead in the moon's orbit. When *Apollo 11* left its temporary "parking orbit" around Earth, its speed was 24,300 miles per hour. After that, its speed decreased continuously until it reached a point at which Earth and the moon exerted equal gravitational pull. The balance of forces at this "equal-gravity" point can be modeled by

Moon
69 miles
Parking Orbit
2.446×10^4 mi
Equal Gravity Point
2.205×10^5 mi
Earth
118 miles
Parking Orbit

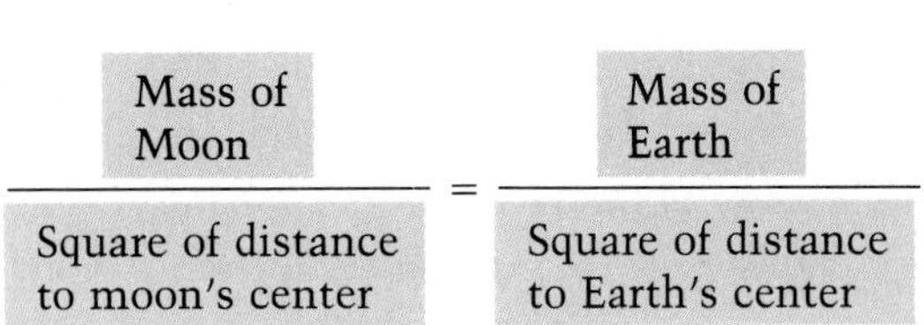

Mass of Moon
Square of distance to moon's center
=
Mass of Earth
Square of distance to Earth's center

The mass of Earth is 1.318×10^{25} pounds. The distance from the equal-gravity point to Earth's center is 2.205×10^5 miles. The distance to the moon's center is 2.446×10^4 miles. Find the mass of the moon.

Solution Let m be the mass of the moon. Substitute the mass of Earth and the two distances into the model and solve for m.

$$\frac{m}{(2.446 \times 10^4)^2} = \frac{1.318 \times 10^{25}}{(2.205 \times 10^5)^2}$$

$$m = \frac{(1.318 \times 10^{25})(2.446 \times 10^4)^2}{(2.205 \times 10^5)^2}$$

$$= \frac{1.318 \times 2.446^2 \times 10^{33}}{2.205^2 \times 10^{10}}$$

$$m \approx 1.622 \times 10^{23} \text{ pounds}$$

■

In 1969, Alaska auctioned gas and oil leases on 4.5 × 10⁵ acres for about $900 million, an average cost of about $2000 per acre.

Connections
U.S. History

Example 2 The Purchase of Alaska

In 1867, the United States purchased Alaska from Russia for $7.2 million. The total area of Alaska is about 3.78×10^8 acres. What was the price per acre?

Solution

$$\text{Price per acre} = \frac{\text{Total price}}{\text{Number of acres}} = \frac{7.2 \text{ million}}{3.78 \times 10^8}$$
$$= \frac{7.2 \times 10^6}{3.78 \times 10^8}$$
$$\approx 1.9 \times 10^{-2}$$
$$= 0.019 \text{ dollars per acre}$$

The price was about 2¢ per acre. ■

Real Life
Economics

Example 3 Comparing Prices

What is the ratio of the cost per acre of Alaskan land leased for gas and oil in 1969 to its purchase price in 1867?

Solution

$$\frac{\text{1969 cost per acre}}{\text{1867 cost per acre}} = \frac{2000}{0.02} = \frac{2.0 \times 10^3}{2.0 \times 10^{-2}} = 1.0 \times 10^5$$

Thus the land leased for about one hundred thousand times its purchase price. (The average cost of living in the United States increased by only three or four times from 1867 to 1969.) ■

In **Example 3,** what was the average yearly increase in the cost of living in the United States from 1867 to 1969? An average increase of between 3% and 4% each year Students may wish to check what the increase in the cost of living has been over the past several years.

Example 4

Observe that 1 meter is 1000 millimeters. Ask students to convert 1×10^{-4} meters into millimeters. Convert the diameter of a quarter into millimeters and compute the gear ratio, using millimeters. Before the computations are made, have students speculate about the gear ratio, given the change in measurement.

Extra Examples

Here are additional examples similar to **Examples 1–4.**

1. It is common in space travel to express distance in kilometers (km) and mass in kilograms (kg). If 1 mile is about 1.609 km and 1 pound is about 0.454 kg, what is the distance in kilometers from the "equal-gravity point" to (a) the earth's center and (b) the moon's center? What is the mass in kilograms of (c) the earth and (d) the moon?
   a. about 3.548×10^5
   b. about 3.936×10^4
   c. about 5.984×10^{24}
   d. about 7.364×10^{22}
2. Suppose that a microgear (see Example 4) is connected to a gear the size of a penny, which is about 1.9×10^{-2}m in diameter. What is the gear ratio of the penny to the microgear? about 190

3. The charge of one electron in coulombs (C), a unit of electrical charge, is about 1.60×10^{-19}C. How many electrons flow through the wire filament in a light bulb if the electrical charge in 1 millisecond is 50×10^{-6}C? about 3.125×10^{14} electrons

Check Understanding

1. Does the problem-solving model change for different real-life situations? Explain. No; see page 126.
2. Why is scientific notation used to solve problems of space travel? The numbers used in space travel would be too large and cumbersome to calculate.
3. Powers with negative exponents can be simplified by using techniques learned earlier in this chapter. In scientific notation, are powers of 10 with negative exponents similarly simplified? Explain.

Communicating about ALGEBRA

EXTEND *Communicating*

1. The total area of the United States is about 2.32×10^9 acres. What percent of the United States area is in Alaska? ≈ 16.3%

A Project

2. Have students research the area of your city or town and state. Then have them use the data to answer these questions: What percent of the total area of the United States is the area of your state? What percent is your city's area to that of the United States? What percent of your state's area is your city?

Engineers at the University of Wisconsin-Madison have created nickel microgears that have a diameter of only 1.0×10^{-4} meter. The holes in the gears are too small for a human hair to fit through.

$\leftarrow 2.4 \times 10^{-2}$ m $\rightarrow$

Example 4 *Gear Ratio*

Suppose that a microgear is connected to a gear the size of a quarter. What is the gear ratio of the quarter to the microgear?

Solution The way to get an exact answer would be to know the number of teeth on each gear. Without that, you can approximate the gear ratio by finding the ratio of the two circumferences. (The circumference of a circle is π times its diameter.)

$$\text{Gear ratio} \approx \frac{\text{Circumference of quarter}}{\text{Circumference of microgear}} = \frac{\pi(2.4 \times 10^{-2})}{\pi(1.0 \times 10^{-4})}$$

$$= 2.4 \times 10^2$$

$$= 240$$

The gear ratio is 240, which means that the microgear would turn 240 times each time the quarter turns once! ■

Communicating about ALGEBRA

SHARING IDEAS about the Lesson

Find Needed Information Information needed to solve this problem may be found elsewhere in this lesson.

The ratio of gravity on the surface of the moon to gravity on Earth's surface is given by

$$\frac{\text{Mass of moon}}{\text{Square of moon's radius}} \div \frac{\text{Mass of Earth}}{\text{Square of Earth's radius}}.$$

What is this ratio? (The radius of Earth is about 3960 miles. The radius of the moon is about 1080 miles.) ≈0.165

EXERCISES

1. Write a verbal model. Assign labels to the model. Write an algebraic model. Solve the algebraic model. Answer the question.

Guided Practice

CRITICAL THINKING about the Lesson

1. List the five steps of the general problem-solving plan.

Busy as a Bee **In Exercises 2–6, use the following information.** 2.–3. See Additional Answers

How many flowers must the 80,000 workers in a bee colony visit to provide the amount of honey the colony will consume in one year?

2. Write a verbal model for this problem.
3. Assign labels to the model.
4. Write an algebraic model for the problem.
5. Solve the algebraic model. See margin.
6. Answer the question. Give your answer in scientific notation.

6. between 1.0×10^{10} and 1.5×10^{10} flowers per year

Hard-working Bees

To make 2.2 pounds of honey, workers must visit 45–64 million flowers.

A large bee colony consumes 500 pounds of honey a year.

5. 1.0×10^{10} to 1.5×10^{10}

Independent Practice

7. *Sunlight on Jupiter* The distance between Jupiter and the sun is 7.8×10^8 kilometers. Light travels at a speed of about 3.0×10^5 kilometers per second. How long does it take light to travel from the sun to Jupiter? 2.6×10^3 seconds

8. *Earth and Jupiter* The distance between Earth and the sun is 1.5×10^8 kilometers. Write an inequality that describes possible distances, d, between Earth and Jupiter. $6.3 \times 10^8 \le d \le 9.3 \times 10^8$

Louisiana and Gadsden Purchases **In Exercises 9 and 10, use the following information.**

The Louisiana Purchase, in 1803, added 8.28×10^5 square miles to the United States. The cost of this land was \$15 million. The Gadsden Purchase, 1853, added 2.94×10^4 square miles and cost \$10 million. See margin.

9. Find the average cost of a square mile for each of these two purchases.
10. There are 640 acres in a square mile. Find the average cost of an acre for each of these two purchases.

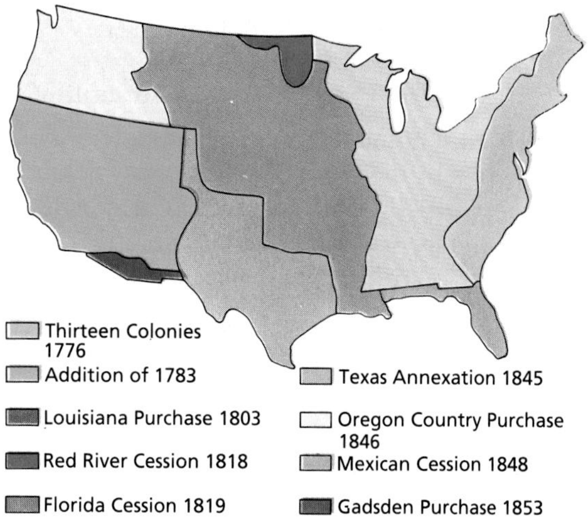

EXERCISE Notes

ASSIGNMENT GUIDE

Basic/Average: Ex. 1–6, 7–10, 13, 14, 16, 17, 19–31 odd, 33–34

Above Average: Ex. 1–6, 7–17, 19–31 odd, 33–35

Advanced: Ex. 1–6, 9–17, 19–31 odd, 33–36

Selected Answers
Exercises 1–6, 7–31 odd

More Difficult Exercises
Exercises 15, 33–36

Guided Practice

Ex. 1 Once again, review the problem-solving plan. Write a verbal model. Assign labels. Write an algebraic model. Solve the algebraic model. Answer the question. Check that the answer is reasonable.

Answer

4. Let n = the number of flowers to visit. Then,
$n = \frac{(45 \text{ "to" } 64) \times 10^6}{2.2} \times 500.$

Independent Practice

Ex. 7–17, 33–36 Students should use Examples 1–4 to remind them of the formats and models needed to solve these problems.

Ex. 9–10, 11–12 These exercises should be assigned in pairs.

Answers

9. Louisiana: ≈\$18.00; Gadsden: ≈\$340.00
10. Louisiana: \$0.028; Gadsden: \$0.531

EXTEND Ex. 11–14 CONNECTIONS You may wish to take advantage of the variety of connections outlined in these exercises to stimulate class discussions on the social issues and implications of these applications.

▶ **Ex. 12** This exercise requires students to add numbers in scientific notation. Remind them that numbers must be expressed to the same power of 10 before they are added or subtracted.

▶ **Ex. 16** *Voyager 2* was launched August 1977 as a space probe to study the distant planets of Jupiter and Saturn. It was not aimed at any one target, especially not at Sirius A.

EXTEND Ex. 16 Ask students to use their calculated percent to answer the following question. In September 1989, *Voyager 2* had been traveling in space for just over 12 years. At the same rate, how long would it take it to reach Sirius A? about 220,000 years

Manufacturers' Coupons **In Exercises 11 and 12, use the following information.**

The table shows the number of manufacturers' coupons distributed in the United States between 1985 and 1989. ***(Source: A. C. Nielson Company)***

Year	1985	1986	1987	1988	1989
Coupons distributed	1.999×10^{11}	2.252×10^{11}	2.383×10^{11}	2.474×10^{11}	2.676×10^{11}
Percent redeemed	3.25%	3.15%	3.02%	2.87%	2.65%
Number redeemed	?	?	?	?	?

11. Find the number of coupons *redeemed* in each year. $\approx 6.497 \times 10^9$, $\approx 7.094 \times 10^9$, $\approx 7.197 \times 10^9$, $\approx 7.100 \times 10^9$, $\approx 7.091 \times 10^9$

12. Suppose that the average coupon value was $0.25. What was the total value of unredeemed coupons over the five-year period? Write your answer in scientific notation. $\approx 2.859 \times 10^{11}$

13. *Cells of Mammals* Ribosomes are composed of different types of proteins and RNA (ribonucleic acid). A typical cell of a mammal contains about 10×10^7 ribosomes. The volume of a ribosome is about 2.5×10^{-17} cubic centimeter. What is the minimum volume of a typical mammal cell? 2.5×10^{-9} cm^3

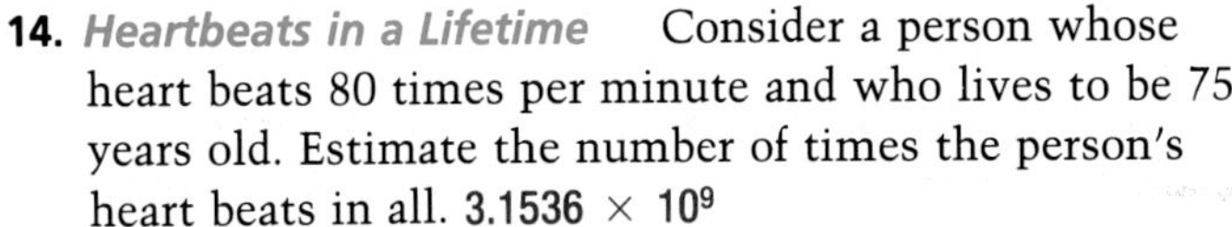

14. *Heartbeats in a Lifetime* Consider a person whose heart beats 80 times per minute and who lives to be 75 years old. Estimate the number of times the person's heart beats in all. 3.1536×10^9

15. *Batter Up!* A baseball pitcher can throw the ball to home plate in about 0.5 second. The distance between the pitcher's mound and home plate is 60.5 feet. This means that the ball is traveling at a rate of

$$\frac{121 \text{ feet}}{\text{second}} \cdot \frac{\text{meter}}{3.28 \text{ feet}} \cdot \frac{1000 \text{ millimeters}}{\text{meter}} \approx 36.9 \times 10^3 \text{ millimeters per second.}$$

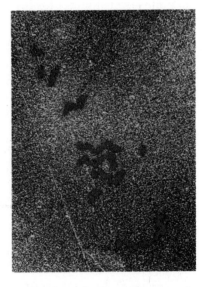

In this photo of a bacterium, the red areas are ribosomes attached to chains of RNA called polysomes.

To hit a home run, a batter has a leeway of only about 20 millimeters in the point of contact between bat and ball. What is the most that the batter's timing could be off and still result in a home run? (Hint: Find the time that it takes the ball to travel 20 millimeters.) $\approx 5.42 \times 10^{-4}$ second

16. *Travel to a Star* The star Sirius A is 8.265×10^{13} kilometers from Earth. The space probe *Voyager 2* was 4.494×10^9 kilometers from Earth in September 1989. Suppose that *Voyager 2* was headed for Sirius A. What percent of its journey was complete by September 1989? $\approx$0.005437%

17. *Population Density of the United States* The United States has a land area of 3.54×10^6 square miles. In 1990, the population of the United States was 2.5×10^8. What was the population density (people per square mile) of the United States in 1990? $\approx$71

Integrated Review

In Exercises 18–26, simplify the expression.

18. $(5x^4)(2x^0)$ $10x^4$

19. $(3y^2)(8y^3)$ $24y^5$

20. $\frac{2a^2}{a^3}$ $\frac{2}{a}$

21. $\frac{4b^3}{b^7}$ $\frac{4}{b^4}$

22. $(-3xy^4)(4xy^{-3})$ $-12x^2y$

23. $(6x^3y)(7x^{-1}y^8)$

24. $\frac{12(x^2y^4)^0}{-6x^{-2}y^2}$ $\frac{-2x^2}{y^2}$

25. $\left(\frac{5x^3}{y}\right)^{-2}$ $\frac{y^2}{25x^6}$

26. $\left(\frac{3}{4}\right)^2 \cdot \left(\frac{2}{3}\right)^4$ $\frac{1}{9}$

23. $42x^2y^9$

In Exercises 27 and 28, write the number in scientific notation.

27. 0.00000065 6.5×10^{-7}

28. 24,000,000,000 2.4×10^{10}

In Exercises 29 and 30, write the number in decimal form.

29. 2.65×10^{15} 2,650,000,000,000,000

30. 4.316×10^{-11} 0.00000000004316

In Exercises 31 and 32, evaluate the expression.

31. $4.2 \times 10^{-5} \cdot 6.1 \times 10^8$ 2.562×10^4

32. $5.04 \times 10^{12} \cdot 8.97 \times 10^{-7}$ 4.52088×10^6

Exploration and Extension

Lightning **In Exercises 33–36, use the following information.**

Approximately 44,000 thunderstorms and 8 million lightning flashes take place daily around the world. In the United States, lightning annually causes about 150 deaths and $20 million in property damage. It sets 10,000 forest fires a year, which destroy $30 million worth of timber.
(Source:* Scientific American*)

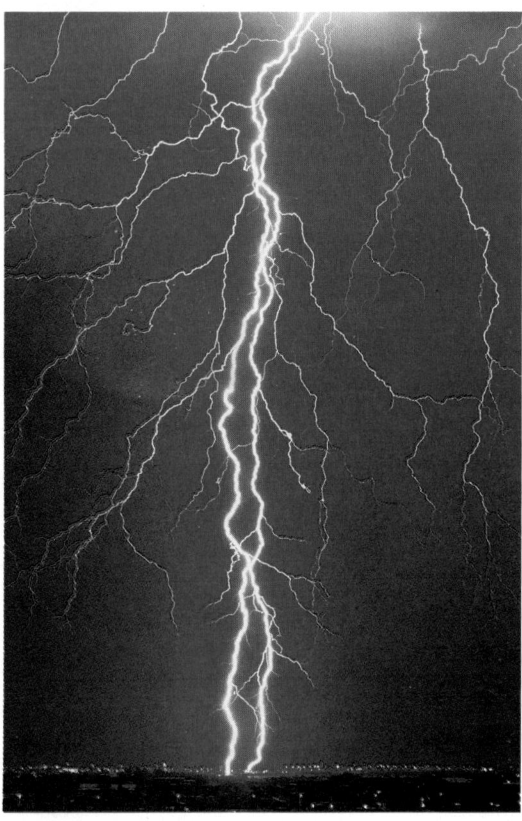

33. Approximate the number of lightning flashes that occur each year on Earth. 3×10^9

34. What is the average number of lightning flashes in a thunderstorm? ≈ 200

35. Sound travels at a speed of 1100 feet per second. Suppose you see a lightning flash and then count, "One-one-thousand, two-one-thousand, three-one-thousand, four-one-thousand, five-one-thousand" before you hear the thunder. About how far away was the lightning? 5500 ft

36. The length of a normal lightning flash varies between 300 feet and 4 miles. Write an inequality that describes the possible lengths (in feet) of a normal lightning flash. $300 \le x \le 21{,}120$

Enrichment Activity

This activity requires students to go beyond lesson goals.

WRITING PROBLEMS Look at **Exercises 7–17.** Use your school library or textbooks from other subjects to find out more about some of the applications mentioned in these exercises. Write some problems similar to the ones in the text for your classmates to solve. Include the solutions on the back.

Problem of the Day

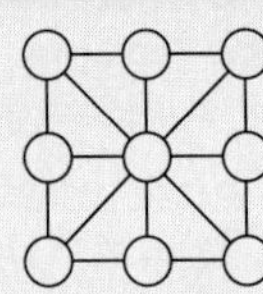

This pattern contains 8 rows of 3 coins each, including the diagonal rows. Revise the pattern to get 10 rows of 3.

ORGANIZER

Warm-Up Exercises

Use your calculator to compute to three decimal places.

1. 3^5 243
2. $(1.08)^{-6}$ 0.630
3. $(6 + 0.05)^3$ 221.445
4. $(-2 + 0.09)^4$ 13.309
5. $5(7)^2$ 245
6. $15(3.81)^3$ 829.595
7. $6.08(1.03)^6$ 7.260
8. $2.45(2.53)^5$ 253.962

Lesson Resources

Teaching Tools
- Transparencies: 1, 2, 3
- Problem of the Day: 8.6
- Warm-up Exercises: 8.6
- Answer Masters: 8.6

Extra Practice: 8.6

Technology Handbook: p. 62

LESSON Notes

GOAL 1 The letter A is used for "balance" because in the formula, A stands for the accumulated amount at the end of the term of the investment. Compound interest allows the investor to earn interest on interest acquired during previous

8.6 Problem Solving: Compound Interest and Exponential Growth

What you should learn:

Goal 1 How to use the compound interest formula

Goal 2 How to use models for exponential growth to solve real-life problems

Why you should learn it:

You can use an exponential growth model for situations that increase regularly by a fixed percent, such as the growth of an embryo.

Goal 1 Computing Compound Interest

Money can work for you if you place it in a savings or investment account. At an annual interest rate of r, a principal, P, earns *simple interest* of Pr in one year. Think of r as a decimal, and thus, 6% of P is $P(0.06)$. The balance, A_1, at the end of the first year is the sum of the principal and the interest.

$A_1 = P + Pr$

$A_1 = P(1 + r)$ *Balance after one year*

If the balance is left in the account for another year, the new balance, A_2, at the end of the second year would be

$A_2 = A_1(1 + r) = P(1 + r)^2.$ *Balance after two years*

This formula represents **compound interest.** Compound interest is paid on the original principal *and* on interest that becomes part of the account.

Compound Interest Formula

If a principal, P, earns an annual interest rate of r, then the balance after t years will be

$A = P(1 + r)^t.$

Real Life
Finance

Example 1 *Finding the Balance in an Account*

a. \$500 is deposited in an account that pays 6% annual interest compounded yearly. What is the balance after 10 years?

$A = P(1 + r)^t$ *Compound interest formula*

$= 500(1 + 0.06)^{10}$ *$P = 500, r = 0.06, t = 10$*

$= 500(1.06)^{10}$ *Use a calculator.*

$= \$895.42$

b. \$250 is deposited in an account that pays 5.5% interest compounded yearly. What is the balance after 8 years?

$A = P(1 + r)^t$ *Compound interest formula*

$= 250(1 + 0.055)^8$ *$P = 250, r = 0.055, t = 8$*

$= 250(1.055)^8$ *Use a calculator.*

$= \$383.67$ ■

Goal 2 Using Models for Exponential Growth

In Lesson 8.1, you learned that in the exponential model $y = C(a)^x$, if $a > 1$, the change is exponential **growth.** The compound interest formula is an example of an exponential growth model. An initial amount, C, is multiplied by a **growth factor** $(1 + r)$ each time period. At each stage, the amount increases by r percent. (The percentage increase, r, is positive and is written in decimal form.)

$C(1 + r)^1$ *Amount at time 1*
$C(1 + r)^2$ *Amount at time 2*

Example 2 *Cafeteria Workers*

In 1975, the average hourly wage of school cafeteria workers in the United States was \$2.53. Between 1975 and 1990, this increased by \$0.285 per year. During the same years, the cost of living in the United States increased by an average of 6.2% per year. Did the average hourly wage of school cafeteria workers keep up with inflation? ***(Source: National Survey of Salaries and Wages in Public Schools)***

Wages and Cost of Living

Average Hourly Wage; Year (0 ↔ 1975); Wages; Cost of Living

Solution Let W be the hourly wage in year t, with $t = 0$ corresponding to 1975. Because the average wage increased by the *same amount* each year, you can use a *linear* model.

$W = 0.285t + 2.53$ *Linear model for hourly wage*

Because the cost of living increased by the *same percent* each year, you can use an *exponential growth* model.

$C = 2.53(1.062)^t$ *Exponential growth model*

The table compares values of W and C for four different years.

Year, *t*	0 (1975)	5 (1980)	10 (1985)	15 (1990)
Hourly wage, *W*	\$2.53	\$3.96	\$5.38	\$6.81
Cost of living, *C*	\$2.53	\$3.42	\$4.62	\$6.24

You can use either the table or the graph to see that the hourly wage did keep up with inflation during the 15-year period. ■

In Example 2, the linear model grew more quickly than the exponential model over 15 years. Over long time periods, however, an exponential growth model will eventually overtake a linear model. For instance, in Example 2, suppose that the cost of living continues to increase at 6.2% per year. By the year 2000, cafeteria workers would need an hourly wage of \$11.38 to keep up. They couldn't do that by continuing to get annual raises of \$0.285 per hour. Why?

pay periods. Simple interest is paid only on the original investment, P. For t years, the balance of a simple interest investment is $A = P + Prt$. Notice that a simple interest investment is a linear relationship between the balance, A, and the time, t.

Example 2

Emphasize that an increase can be expressed by a percent change or by a constant increment. Percent changes are modeled by an exponential relationship; constant incremental changes are modeled by a linear relationship.

EXTEND Example 2 TECHNOLOGY If a computer plotter or a graphics calculator is available, ask students to graph each function of Example 2. Compare the graphs and find the value of each at $t = 20$. $W = 8.23$, $C \approx 8.43$

About when does the exponential function attain values that exceed those of the linear function? For $t = 19.1$, $W = 7.97$ and $C \approx 7.98$.

Extra Examples

Here are additional examples similar to **Examples 1–3.**

1. Suppose \$700 is deposited in an account that pays 7.5% annual interest compounded yearly. What is the balance after 8 years? \$1,248.43
2. Suppose in Example 2, the average hourly wage of school cafeteria workers increased by \$0.25 per year between 1975 and 1990. Would these wages have kept up with inflation over the same time period? yes

3. ***First Class Postage*** The cost of a first-class stamp in 1932 was 3 cents with an exponential growth rate of 3.87% per year. The cost of a first-class stamp in 1988 was 25 cents; in 1992 it was 29 cents. Compare the predicted costs of a first-class stamp to the actual costs, using the exponential model of 1932. How well does the model work? What would you expect the cost of a first-class stamp to be in the year 2000?
In both cases, the model predicted the actual cost to the nearest penny. In the year 2000, using the model, a first-class stamp would cost 40 cents.

Check Understanding

1. Explain how compound interest is different from simple interest. See page 432.
2. Explain why an increase by a percent results in a more rapid accumulation over time than a constant increment. See page 433.
3. About how long will it take \$100 to double in an account that pays 8.5% interest compounded yearly? about 9 years

Communicating about ALGEBRA

EXTEND *Communicating*
Show how the compound interest formula, $A = P(1 + r)^t$, can be derived. Using a table can be helpful.

Starting Balance: P

Interest	Ending Balance
Pr	$P + Pr = P(1 + r)$
$P(1 + r)r$	$P(1 + r) + P(1 + r)r = P(1 + r)(1 + r) = P(1 + r)^2$
$P(1 + r)^2r$	$P(1 + r)^2 + P(1 + r)^2r = P(1 + r)^2(1 + r) = P(1 + r)^3$
...	$P(1 + r)^{t-1}(1 + r) = P(1 + r)^t$

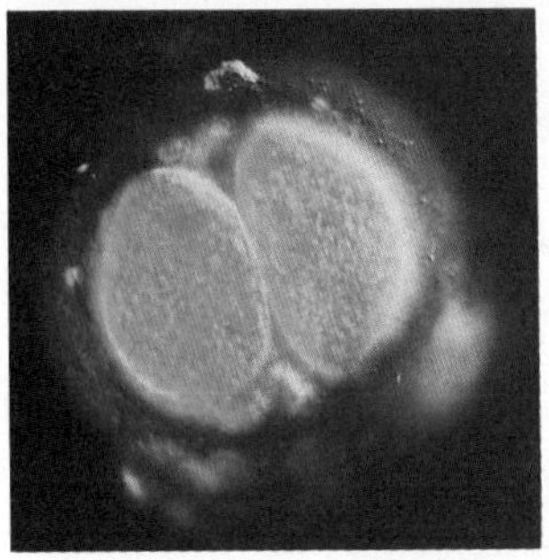
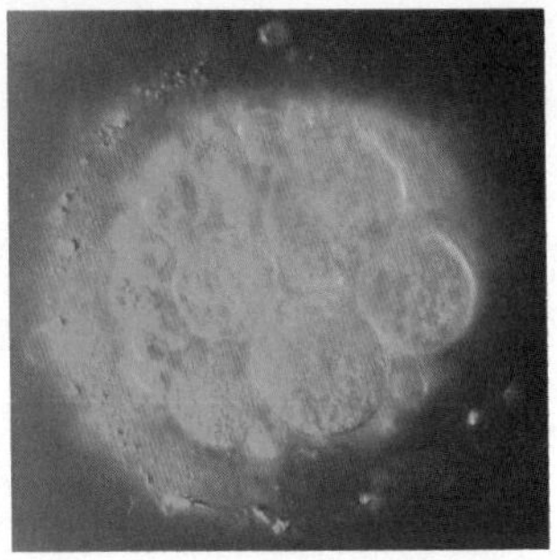

Through mitosis, a fertilized egg repeatedly doubles its number of cells. Cells soon begin to show new structural and biochemical characteristics, and within five weeks, the embryo shows recognizable features.

Connections **Biology**

Weight Gain by Human Embryos

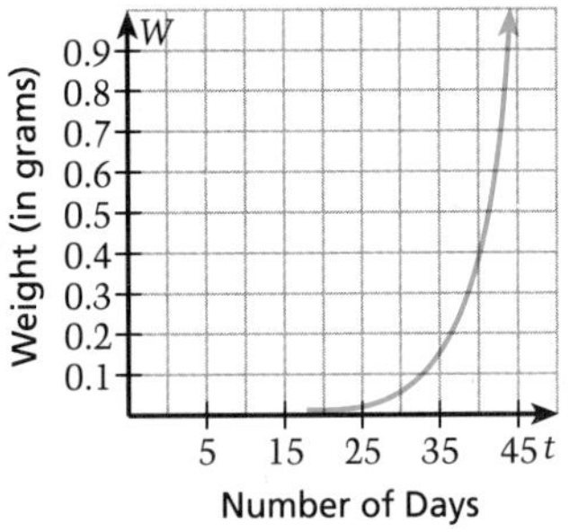

Example 3 Growth of a Human Embryo

Growth of a human embryo is approximately exponential, increasing at about 28.5% each day. A model for the weight of a human embryo is

$$W = 0.0125 \cdot 1.285^t$$

where W is the weight in milligrams and t is the age of the embryo in days. Sketch the graph of this model.

Solution Begin by making a table of values.

Age, t	0	7	14	21	28	35	42	45
Weight, W	0.0125	0.07	0.42	2.42	14.0	81.0	469	994

The graph of the model is shown at the left. ■

Communicating about ALGEBRA

SHARING IDEAS about the Lesson

Make a Decision Use tables and graphs to compare the two salary offers. Which is linear growth? Which is exponential growth? Which salary plan would you prefer? Explain. See below.

A. You are offered job A with a starting salary of \$24,000. Each year for 10 years you will be given a 4% raise. Exponential

B. You are offered job B with a starting salary of \$24,000. Each year for 10 years you will be given a \$1000 raise. Linear

Choose A; because after 3 years the increases of A are greater than the \$1000 increases of B, and because each increase of A after the first is greater than the previous increase.

EXERCISES

Guided Practice

CRITICAL THINKING about the Lesson

1. **True or False?** Any quantity that increases by the *same percent* each year can be represented by an exponential growth model. **True**

2. One thousand dollars is deposited in an account that pays interest at the rate of 5% compounded yearly. What is the balance after 10 years? **$1628.89**

3. In the exponential growth model $y = C(1 + r)^t$, what is the growth factor? $1 + r$

4. In the exponential growth model of Exercise 3, what does r represent? **% the amount increases**

Independent Practice

5. ***What Is the Balance?*** A principal of \$100 is deposited in an account that pays 7% interest compounded yearly. Find the balance after 5 years. **$140.26**

6. ***What Is the Balance?*** A principal of \$1200 is deposited in an account that pays 5% interest compounded yearly. Find the balance after 10 years. **$1954.67**

7. ***How Much Should I Deposit?*** How much must you deposit in an account that pays 6.5% interest compounded yearly to have a balance of \$600 after 8 years? **$362.54** $600 = x(1 + .065)^8$

8. ***How Much Should I Deposit?*** How much must you deposit in an account that pays 7% interest compounded yearly to have a balance of \$200 after 5 years? **$142.60**

In Exercises 9–12, match the graph with the description.

9. Deposit: \$200, **c**
   Annual rate: 6%

10. Deposit: \$200, **b**
    Annual rate: 9.5%

11. Deposit: \$550, **a**
    Annual rate: 8%

12. Deposit: \$550, **d**
    Annual rate: 4.5%

a.

b. 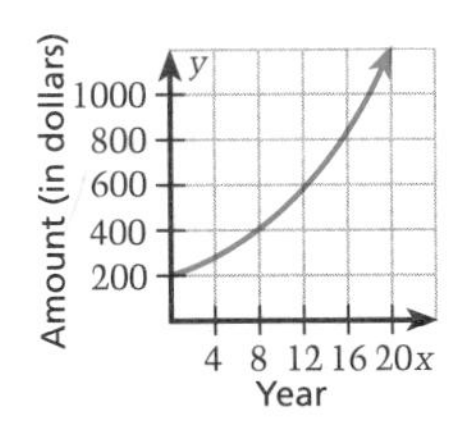

c. 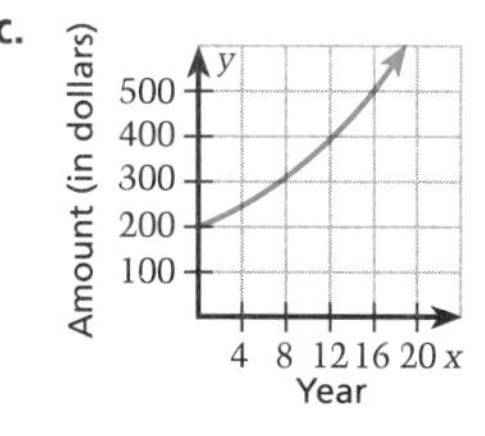

d. 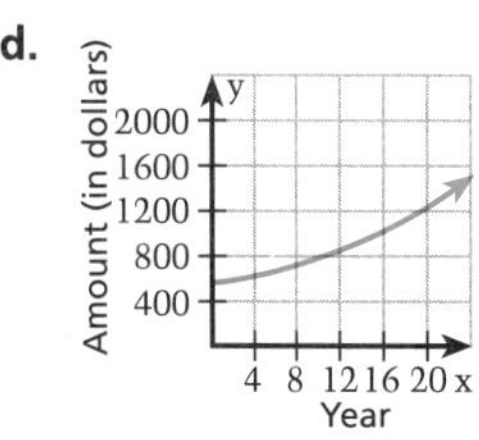

13. ***Population of Texas*** From 1960 to 1990, the population of Texas increased by about 1.94% per year. Write an exponential growth model that gives the population, P, in terms of the year, t. Let $t = 0$ represent 1960. $P = 9{,}580{,}000(1.0194)^t$

14. Estimate the population of Texas in 1970, in 1980, and in 1990. **11,600,000; 14,100,000; 17,000,000**

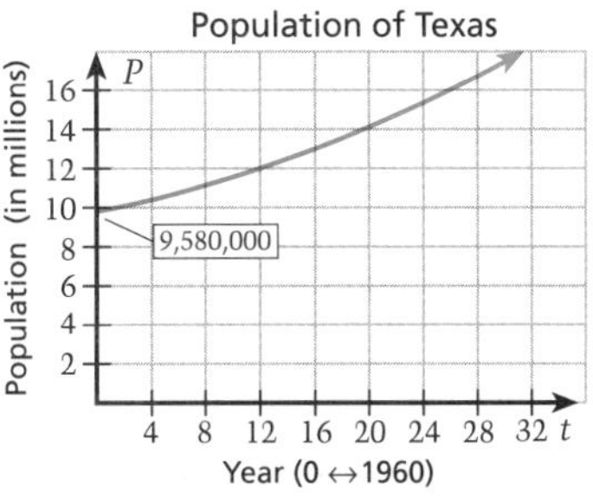

EXERCISE Notes

ASSIGNMENT GUIDE

Basic/Average: Ex. 1–4, 5, 7, 9–14, 17, 19, 27, 29, 31–33

Above Average: Ex. 1–4, 5–14, 16–20, 23–31 odd, 32–36

Advanced: Ex. 1–4, 9–20, 31–36

Selected Answers
Exercises 1–4, 5–31 odd

More Difficult Exercises
Exercises 19–20, 33–36

Guided Practice

EXTEND Ex. 1–4 Ask students to contrast the exponential growth model with a linear growth model. In linear growth, the increase is the same *amount* each year, not the same *percent.*

Independent Practice

TECHNOLOGY Note that students must use scientific calculators to complete many of the exercises in this lesson. Some students may need assistance in the use of their particular calculators. Students who have the same brand of calculators might work together to make the task of finding proper key sequences easier.

Ex. 9–12 Assign these exercises as a group. They facilitate the visual connection to exponential growth.

▶ **Ex. 15–16** Ask students to compare and contrast the linear and exponential growth models.

Answers

15., 16. Business in Exercise 16, because its profit is increasing at a faster rate.

▶ **Ex. 19–20** Refer students to Example 2 for a solution model if they have difficulty.

Answers

19. $y = 17.46t + 266$, $y = 266(1.062)^t$; no

20. $y = 0.36t + 3.75$, $y = 3.75(1.062)^t$; yes

EXTEND Ex. 19–20 To emphasize the phenomena of exponential growth, have students first estimate the value of the model in the year 2000, when $t = 25$. Then have them calculate the value and compare their estimates to the actual value.

Actual Values:

19. \$702.50, \$1196.73

20. \$12.75, \$16.87

Common-Error ALERT!

In **Exercises 13 and 17,** for example, students often forget to change the percent form to the decimal form before applying the formula. In Exercise 13, some students will write $y = 9{,}580{,}000(1 + 1.94)^t$ as the model instead of $y = 9{,}580{,}000(1 + 0.0194)^t$. Be sure to alert them that 1.94% is equivalent to 0.0194.

Managing a Business **In Exercises 15 and 16, match the graph with the description. Which business would you rather own? Explain.** See margin.

15. A business has a \$10,000 profit in 1980. The profit increases by \$2500 per year for 10 years. a

16. A business has a \$10,000 profit in 1980. The profit increases by 25% per year for 10 years. b

a.

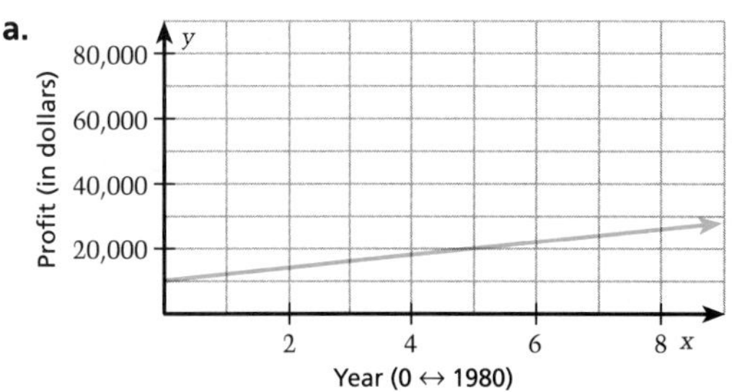

b.

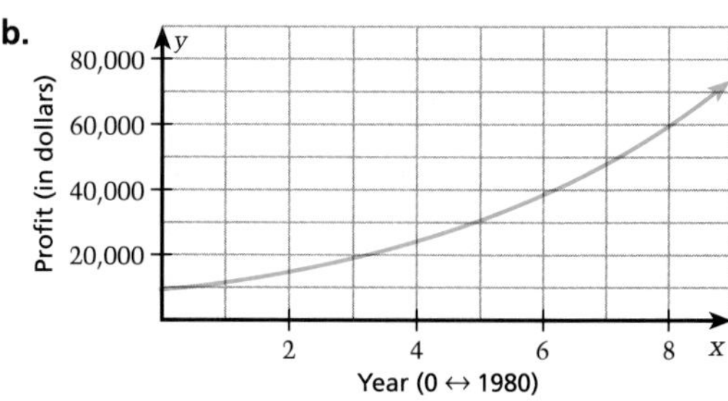

In Exercises 17 and 18, write an exponential growth equation.

17. *M.D.'s in the United States* From 1970 to 1990, the number of doctors of medicine in the United States increased by about 3.4% per year. $y = 334{,}000(1.034)^x$

18. *Furniture Sales* From 1970 to 1990, the total wholesale furniture sales in the United States increased by about 10% per year. $y = 17 \times 10^9(1.10)^x$

Medical Doctors in the United States

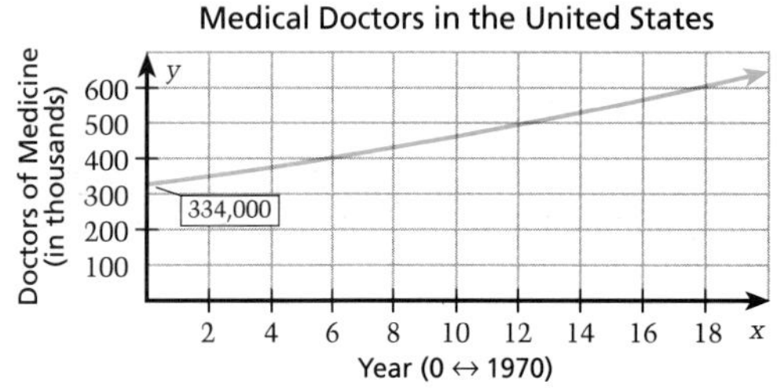

Furniture Sales in the United States

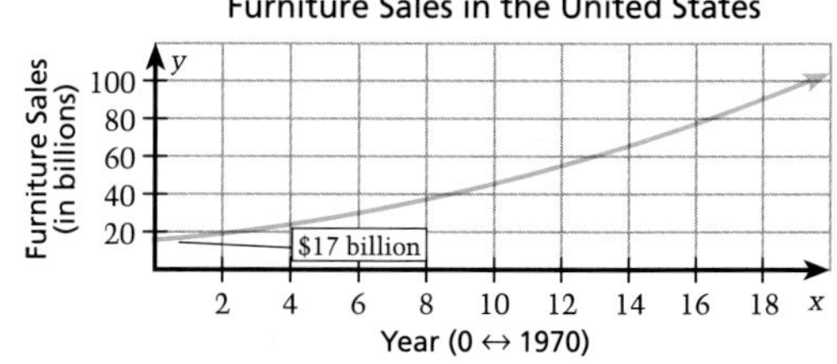

In Exercises 19 and 20, follow the instructions below.

Write a linear growth equation for the earnings. Then write an exponential growth equation showing what the earnings would have been in order to keep up with the inflation rate. (From 1975 to 1990, the inflation rate averaged 6.2% per year.) Did the workers' earnings keep up with inflation? See margin.

19. *Construction* From 1975 to 1990, the average weekly earnings of construction workers in the United States increased by \$17.46 per year.

20. *School Bus Drivers* From 1975 to 1990, the average hourly wage of school bus drivers in the United States increased by \$0.36 per year.

Average Weekly Earnings of Construction Workers

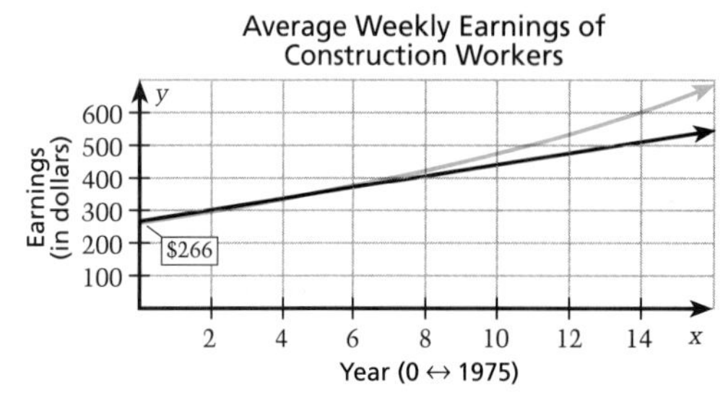

Average Hourly Wage of Bus Drivers

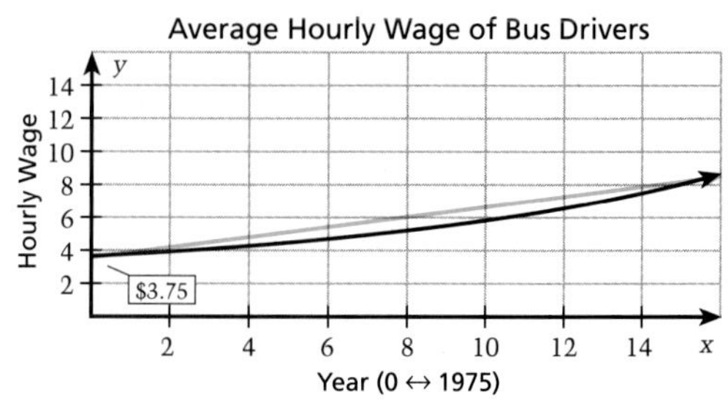

8.7 Exploring Data: Exponential Growth and Decay

What you should learn:

Goal How to use models for exponential growth and decay to solve real-life problems

Why you should learn it:

You can use an exponential decay model for situations that decrease regularly by a fixed percent, such as the dwindling number of farms in the United States.

Goal Using Exponential Models

In Lesson 8.6 you learned that compound interest is an example of *exponential growth,* which can be modeled with the general equation

$$y = C(1 + r)^t \quad \textit{Exponential growth}$$

where C is an initial amount, r is the percentage *increase* (expressed as a decimal), and t is the time period (in years).

Exponential decay occurs when the amount *decreases* by the same percent r for each time period. A model for exponential decay is

$$y = C(1 - r)^t \quad \textit{Exponential decay.}$$

Because the percentage r is greater than zero, the base $(1 - r)$ is less than 1.

Real Life **Agriculture**

Example 1 *Number of Farms in the United States*

Between 1960 and 1990, the number of farms in the United States decreased by about 2.1% each year. In 1960, there were 4 million farms. How many farms were there in 1990? ***(Source: U.S. Department of Agriculture)***

Solution Let F be the number of farms in year t, with $t = 0$ corresponding to 1960. Use the above model for exponential decay with $C = 4{,}000{,}000$ and $r = 0.021$.

$$\begin{aligned} F &= C(1 - r)^t && \textit{Exponential decay} \\ &= 4{,}000{,}000(1 - 0.021)^t && C = 4{,}000{,}000,\ r = 0.021 \\ &= 4{,}000{,}000(0.979)^t && \textit{Exponential decay model} \end{aligned}$$

To find the number of farms in 1990, substitute $t = 30$ into the model.

$$\begin{aligned} F &= 4{,}000{,}000(0.979)^{30} && \textit{Evaluate F when } t = 30. \\ &\approx 2{,}116{,}000 && \textit{Number of farms in 1990} \end{aligned}$$

There were about 2,116,000 farms in the United States in 1990. ■

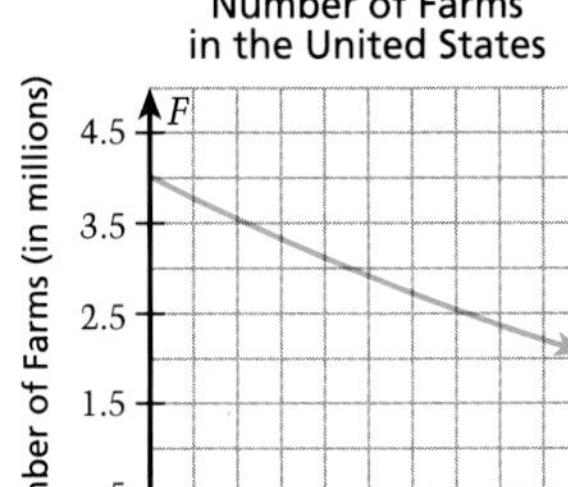

Problem of the Day

In a certain school, 4% of all those named John have a last name of Smith and 40% of all those with a last name of Smith are not named John. The number named both John and Smith plus the number named only John or only Smith is 80. How many are named John Smith? 3

ORGANIZER

Warm-Up Exercises

1. **TECHNOLOGY** Use your calculator to compute to three decimal places.
   - **a.** 0.3^3 0.027
   - **b.** $(0.08)^{-2}$ 156.250
   - **c.** $(1 - 0.05)^3$ 0.857
   - **d.** $(0.09)^4$ 6.561×10^{-5}
   - **e.** $5(0.27)^2$ 0.365
   - **f.** $1.5(0.81)^2$ 0.984
   - **g.** $6.08(1 - 0.03)^6$ 5.064
2. Create a table of values for integer values $-3 \le x \le 3$. What do you notice? Then sketch each graph.
   - **a.** $y = 2(0.5)^x$
   - **b.** $y = 0.25(0.1)^x$
   - a. $y =$ 16, 8, 4, 2, 1, 0.5, 0.25; each y is 0.5 of the previous one.
   - b. $y =$ 250, 25, 2.5, 0.25, 0.025, 0.0025, 0.00025; each y is 0.1 of the previous one.

Check students' graphs.

Lesson Resources

Teaching Tools
- Problem of the Day: 8.7
- Warm-up Exercises: 8.7
- Answer Masters: 8.7

Extra Practice: 8.7

LESSON Notes

Example 1

In 1990, there were 988 million acres of farmland. What was the average acreage per farm? about 467 acres

Example 2

Observe that the computer lost. When solving a problem that involves a large number of possible outcomes, the human mind can still exceed the analytical abilities of any modern-day computer.

Example 3

Point out to students that 0.45% in decimal form is 0.0045. Also observe that at $t = 0$ (high school graduation), the percent of high school algebra and geometry that a person remembers is 100%, or in decimal form, 1.

Extra Examples

Here are additional examples similar to **Examples 1–3.**

1. Between 1980 and 1990, the number of farms in the United States decreased by about 1.2% each year. In 1980, there were 2,437,000 farms. How many farms were there in 1990 according to these data? Compare this number with the number of farms obtained by the exponential model of Example 1. About 2,160,000 farms in 1990; this is about 44,000 more farms than indicated by the model in Example 1.
2. You buy a ten-speed bicycle for $150. Each year for three years, the bicycle loses 15% of its value. What is its value in three years? Its value is $92.12.

World Champion chess player Gary Kasparov is shown playing chess in 1989 with a computer using a program called* Deep Thought. *Kasparov won the match, despite the computer's ability to analyze ten moves ahead.

Example 2 *Computers and Chess*

A computer program for playing chess is rated on the number of future moves it can analyze. On a typical chess play, there are 38 possible positions. For subsequent moves, the number of possible positions, P, is given by $P = (38)^t$, where t is the number of moves. A computer program called *Deep Thought*, which was designed at Carnegie-Mellon University in Pittsburgh, can analyze 10 future moves. The computer can analyze 750,000 positions per second. Is it possible that *Deep Thought* analyzed all possible future positions for 10 moves into the future when it played Kasparov?

Solution To find the time required to analyze all possible positions, divide the number of positions by the number of positions that can be analyzed in a second.

$$\begin{aligned}\frac{\text{Number of positions}}{\text{Positions per second}} &= \frac{38^{10}}{750{,}000}\\ &= \frac{6.278 \times 10^{15}}{7.5 \times 10^{5}}\\ &\approx 8.37 \times 10^{9} \text{ seconds}\\ &= 2{,}325{,}000 \text{ hours}\end{aligned}$$

Thus, *Deep Thought* could not have analyzed all possible positions. In fact, the Carnegie-Mellon team found it was necessary to analyze only about 6 positions for each move. This meant that *Deep Thought* had to analyze only 6^{10} positions, which it could do in about 80 seconds.

(Source: **Scientific American***)*

Connections
Psychology

Example 3 *Human Memory Model*

Two faculty members at Ohio Wesleyan University conducted an experiment to determine how much high school algebra and geometry a person remembers after graduating from college. The experiment, which was conducted with 1743 volunteers, found that those who had studied math through calculus in college forgot only about 0.45% per year. Those who took no college math courses tended to forget about 2.38% per year. Show an algebraic model and a graphical model for the amount of high school algebra and geometry each group remembered over 50 years. ***(Source:* Science News*)***

Solution Let p be the percent (in decimal form) of high school algebra and geometry that a person remembers. Using exponential decay, you can write the following models.

$p = 1(1 - 0.0045)^t$ *Took college calculus.*
$= 0.9955^t$

$p = 1(1 - 0.0238)^t$ *Took no college math.*
$= 0.9762^t$

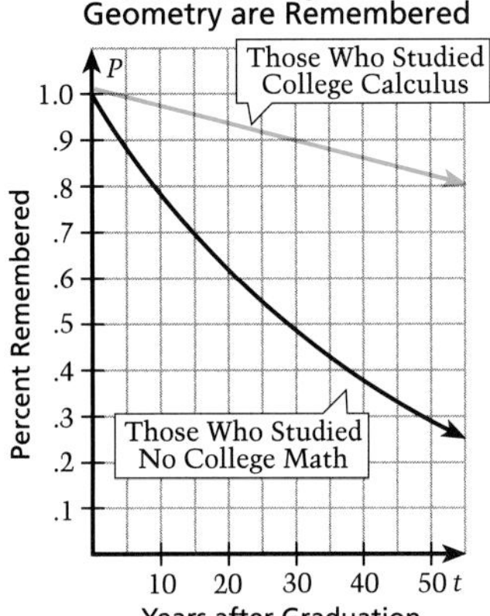

The table compares the percents remembered.

Years, t	0	10	20	30	40	50
Percent (calc), p	100%	95.6%	91.4%	87.3%	83.5%	79.8%
Percent (none), p	100%	78.6%	61.8%	48.5%	38.2%	30.0%

Using the table to sketch the graphs, you can see that forgetting only about 2% more per year results in a big difference after 50 years. ■

Communicating about ALGEBRA

SHARING IDEAS about the Lesson See margin for graphs.

Interpret a Model Use the algebraic or graphical form of a model for exponential growth and decay.

A. Sketch the graphs of $y = 0.8^x$ and $y = 0.9^x$ on the same coordinate plane. What can you conclude from the two graphs? Both decrease, but $y = 0.9^x$ decreases less rapidly.

B. An ocean sunfish, the mola mola, weighs about 1.5×10^{-5} pound when it hatches from an egg. By the time it reaches adulthood, it has doubled its weight 26 times! How much does an adult mola mola weigh? ≈1006.6 pounds

3. The population of Roanoke Rapids (NC) was 14,702 in 1970. From 1970 to 1980, the population declined by 0.84%. What was its population in 1980? What will its population be in 2000 if the population continues to decline at the same rate? In 1980, about 13,513 people; in 2000, about 11,415

Check Understanding

1. What is a general form of an exponential relationship? $y = c(1 + r)^t$ What are the two types of exponential relationships? growth and decay

2. Explain how the exponential decay model differs from the exponential growth model. See page 439; the decay model is $y = c(1 - r)^t$.

Communicating about ALGEBRA

Answer

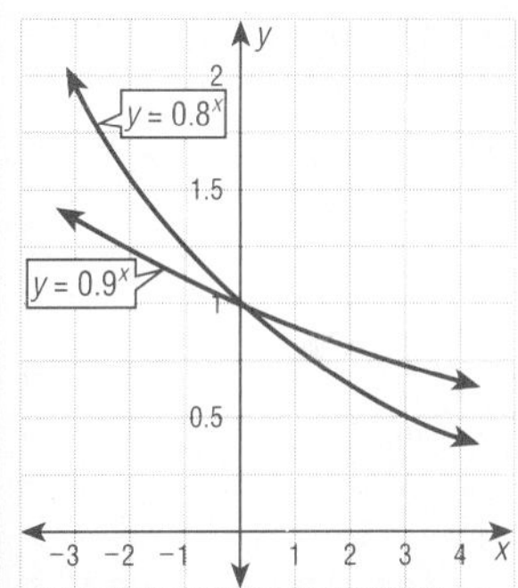

EXTEND *Communicating*

1. From 1980 to 1990, the population of Decatur, Illinois, decreased by about 1.14% per year. In 1980, the population of Decatur was 131,400. What was the population in 1990? 117,166

2. Have students track the population growth or decline in your community. Using these data, predict what the population in your community will be in the year 2000.

EXERCISE Notes

ASSIGNMENT GUIDE

Basic/Average: Ex. 1–4, 5–7, 10–17

Above Average: Ex. 1–4, 5–9, 10–17

Advanced: Ex. 1–4, 7–19

Selected Answers
Exercises 1–4, 5–15 odd

✪ More Difficult Exercises
Exercises 8–9, 18–19

Guided Practice

▶ **Ex. 1–4** These exercises concentrate on one of the important elements of exponential decay, that is, the base $(1 - r)$ of the power is less than 1.

▶ **Ex. 4** Help students translate a 12% decrease per year to the model $y = 10{,}000(0.88)^t$. Be sure they understand why 0.88 is used and not 0.12.

Independent Practice

TECHNOLOGY Some students may need assistance in the usage of their particular calculators. Students who have the same brand of calculators might work together to make the task of finding proper key sequences easier.

▶ **Ex. 5–9, 15–19 MATH JOURNAL** These exercises are all real-life applications of growth and decay. Students might be encouraged to enter some of these applications in a section of their math journals for future reference in Chapter 12.

Answer

7.

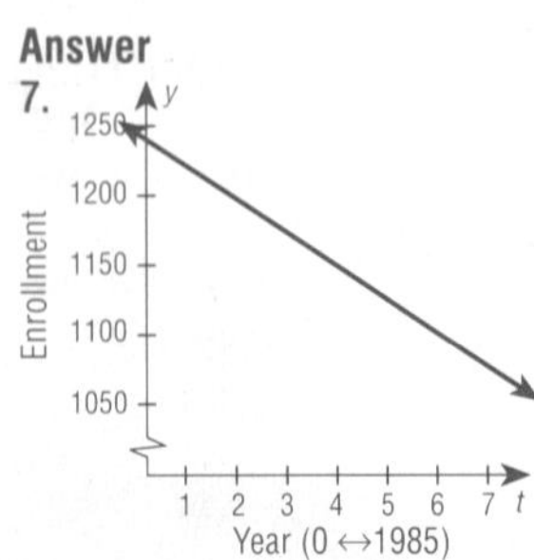

EXERCISES

Guided Practice

CRITICAL THINKING about the Lesson

1. Which of the following represent exponential decay? Explain. **Base of power < 1.**

   a. $y = 0.97^t$ **b.** $y = 1.02^t$ **c.** $y = \left(\frac{4}{3}\right)^t$ **d.** $y = \left(\frac{2}{3}\right)^t$

2. For the exponential decay model $y = 0.96^t$, by what percent does the amount decrease for each time period? **4%**

3. *Business Decline* A business earned \$75,000 in 1980. Then its earnings decreased by 2% each year for 10 years. Write an exponential decay model for the earnings, E, in year t. Let $t = 0$ represent 1980. $E = 75{,}000(0.98)^t$

4. *Trying to Be Perfect* In 1980, the company you own lost 10,000 hours of work time due to employee illness. Beginning that year, you introduced a health awareness program for company employees. Each year for 10 years the number of lost hours decreased by 12%. How many hours were lost to illness in 1990? If the number of lost hours continued to drop by 12% each year, would you live long enough to see it drop to zero? **≈2800, no**

Independent Practice

5. *I Thought the Test Was on Friday* You spend 3 hours Thursday night studying vocabulary words foı your French class. There are going to be 50 words on the test, and on Friday morning you know them all. When you get to school, however, you find out that the test is on Monday. Suppose that without further study, you forget 10% of the vocabulary words each day. How many will you remember for Monday's test? **≈36**

6. *Memory Course* You enroll in a weekend course that teaches you memorization techniques. The techniques improve your ability to remember vocabulary words. How many of 50 vocabulary words would you forget if, after completing the memory course, you forget 5% each day for three days? **≈7**

7. *Declining Enrollments* The Mesa School District had a declining student population from 1985 to 1992. The enrollment in 1985 was 1240. Each year for seven years, the enrollment decreased by 2%. Complete the table showing the enrollment for each year and sketch a graph of the results. **1240, ≈1215, ≈1191, ≈1167, ≈1144, ≈1121, ≈1098, ≈1076**

Year	1985	1986	1987	1988	1989	1990	1991	1992
Enrollment	?	?	?	?	?	?	?	?

See margin.

8. *Endangered Species* From 1980 to 1990, the number of elephants living in Africa decreased by about 6.8% per year. There were about 1,200,000 elephants living in Africa in 1980. Write an exponential decay model showing the number of elephants, E, in Africa in year t. Let $t = 0$ represent 1980. Use the model to estimate the number of elephants in Africa in 1990. $E = 1{,}200{,}000(0.932)^t$; $\approx 593{,}000$

(Source: African Wildlife Foundation, Kenya)

9. *Men and Women in the Work Force* From 1975 to 1990, the *difference* between the numbers of working men and working women in the United States was decreasing by about 3% per year. Write an exponential decay model for the difference, D, between the number of working men and working women in year t. Let $t = 0$ represent 1980. How many more men than women were in the work force in 1990? ***(Source: U.S. Bureau of Labor Statistics)*** $D = 18.8(0.97)^t$; ≈ 13.9 million

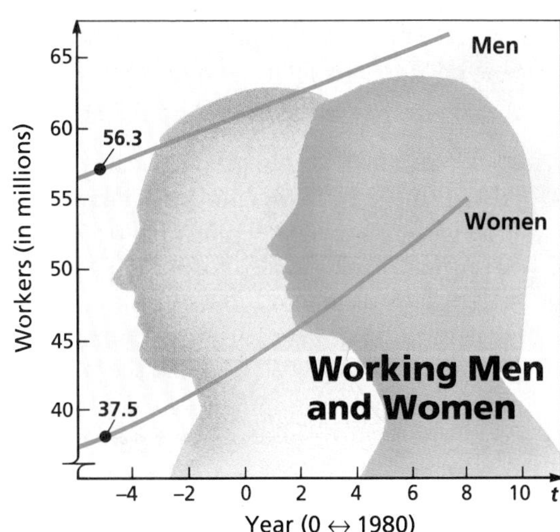

Integrated Review

In Exercises 10–13, match the equation with its graph.

10. $y = 5 - 3x$ c **11.** $y = 5(1.6)^x$ d **12.** $y = 5 + 3x$ a **13.** $y = 5(0.6)^x$ b

a. 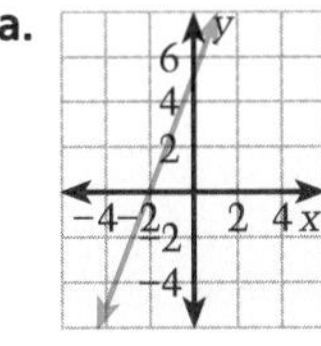
b. 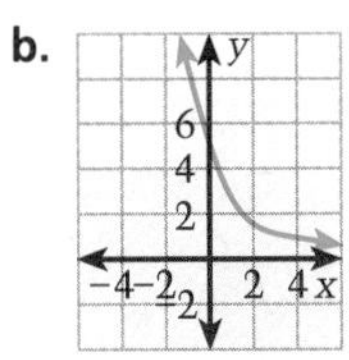
c. 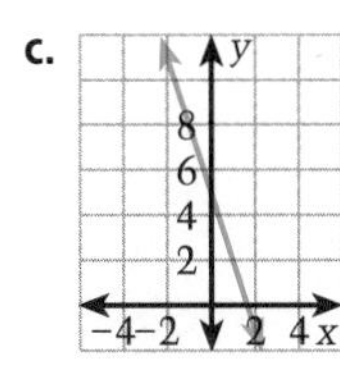
d. 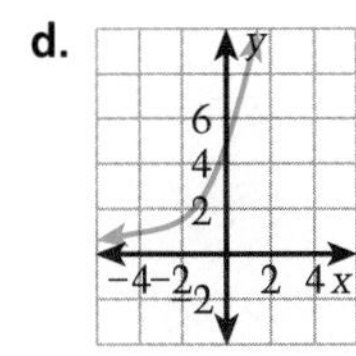

14. *Space Probes* On September 15, 1989, NASA calculated estimates for the distances (in miles) that different space probes had traveled from Earth. Place the probes in order, from the closest to Earth to the farthest from Earth.

6th **a.** *Pioneer 11:* 2.705×10^9 8th **b.** *Voyager 1:* 3.641×10^9 2nd **c.** *Pioneer 6:* 6.2×10^7

4th **d.** *Pioneer 12:* 1.0×10^8 1st **e.** *Magellan:* 4.0×10^7 7th **f.** *Voyager 2:* 2.793×10^9

9th **g.** *Pioneer 10:* 4.382×10^9 3rd **h.** *Voyager 7:* 7.8×10^7 5th **i.** *Pioneer 8:* 1.42×10^8

15. *Space Probes* *Voyager 1* was launched on September 5, 1977. *Pioneer 10* was launched on March 2, 1972. Which has had a greater average speed? Voyager 1

▶ **Ex. 9** Help students to identify the initial value by substituting the y-intercepts from the graph.

EXTEND Ex. 9 Ask students to predict the difference in the year 2000. Have students then calculate it and compare the predicted and calculated values. about 10,223 million

▶ **Ex. 14–15** These exercises should be assigned as paired exercises.

Enrichment Activities

These activities require students to go beyond lesson goals.

TECHNOLOGY Have students use their calculators to answer the following.

You can use the formula $P = \frac{R \cdot S}{1 - (1 + R)^{-n}}$ to determine the monthly payments, P, for borrowing S dollars at a rate of R percent for n months.

1. Suppose a store charges you 1% a month for 2 years to finance a \$1000 stereo system. What monthly payment can you expect?
   $P = \frac{0.01 \cdot 1000}{1 - (1 + 0.01)^{-24}} \approx \$47.07$
2. Use the monthly finance rates data from newspaper and TV advertisements to find the monthly payments on a favorite car (stereo, boat, etc.). Compare the monthly payments for 24 months with the monthly payments for 48 months. Compute how much more you pay in interest when you finance over a longer time.

Challenge! You may wish to have students compute the monthly mortgage payments for homes in your area for fixed mortgage rates over 20 years and over 30 years. They would probably be amazed at the differences in the amount of interest.

EXTEND Ex. 17 The given model is valid for larger mammals. Suppose it was valid for humans. Have students use the model to predict the size of a human baby at about 270 days gestation. about 40 kg, the same as for a cow

Ex. 18–19 These exercises provide an introduction to quadratic equations that will be studied in Chapter 9. Compare and contrast this model to the exponential growth model. Students should note that the amounts increase, but not as rapidly as with the exponential model.

Exploration and Extension

16. *Photocopy Reduction* You are using the photocopier to put a picture of a dollar bill on a graph you are making for your algebra class. The copier can only reduce to 72% of the original size. A dollar bill is 6.1 inches by 2.6 inches. How many times must you reduce the picture to obtain the size shown at the right? (It is legal to copy U.S. currency if the copy is black-and-white and smaller than $\frac{3}{4}$ actual size or greater than $\frac{3}{2}$ actual size.) 3 times to 72%, once to 90%

17. *Baby Mammals* The gestation time for a mammal is, in general, related to the birth weight of the mammal's baby. (The gestation time is the time between fertilization and birth.) Mammals that have small babies have short gestation times. Mammals that have large babies have long gestation times. A model that relates the gestation time, t (in days), to average birth weight, w (in kilograms), is $t = 142(1.016)^w$. Use the model to approximate the gestation time for the mammals listed in the table. Does the gestation time for humans (average birth weight $\approx$ 3 kilograms) fit this model? No

Mammal	Bear	Lion	Sheep	Deer	Seal	Cow	Giraffe	Elephant
Weight, w	0.5 kg	1.2 kg	4.1 kg	18 kg	33.5 kg	40 kg	68 kg	100 kg
Days, t	?	?	?	?	?	?	?	?
	143	145	152	189	242	268	418	694

Population of the United States **In Exercises 18 and 19, use the following information.**

The population of the United States is shown in the graph at the right. The population can be modeled by

$$P = 6{,}893{,}000 + 672{,}240t^2$$

where $t = 0, 1, 2, 3, \ldots$ represent the years 1800, 1810, 1820, 1830, ... This model is *not* an exponential growth model. It is a second-degree model of the type you will study in the next chapter.

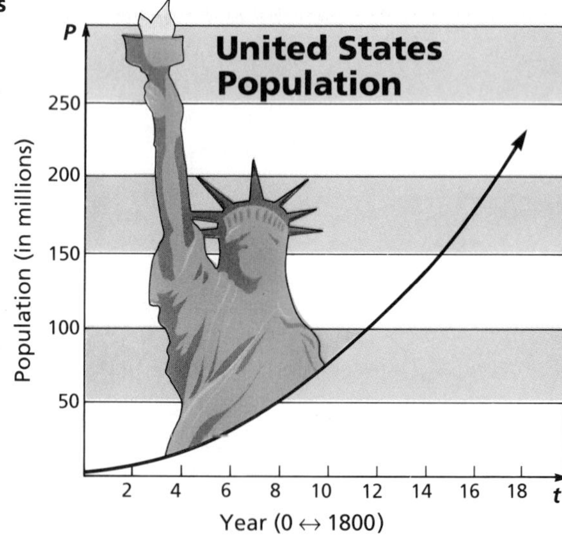

18. Use the model to estimate the United States population in 1990. 249,600,000

19. By what percent did the United States population increase from 1900 to 1930? By what percent did the population increase from 1960 to 1990? From these results, explain why the United States population has not increased according to exponential growth. ≈63%, ≈39%

8 Chapter Summary

What did you learn?

Skills

1. Evaluate powers
   - using multiplication properties of exponents. **(8.1)**
   - involving negative and zero exponents. **(8.2)**
   - using division properties of exponents. **(8.3)**
2. Convert numbers from decimal form to scientific notation. **(8.4)**
3. Convert numbers from scientific notation to decimal form. **(8.4)**
4. Perform operations with numbers in scientific notation. **(8.4)**
5. Use the formula for compound interest. **(8.6)**

Strategies

6. Use scientific notation in problem solving. **(8.5)**
7. Use exponential growth and decay to model real-life problems. **(8.6, 8.7)**

Exploring Data

8. Sketch the graphs of exponential growth and decay models. **(8.6, 8.7)**

Why did you learn it?

The linear models you studied in Chapters 3–7 are used very often in real-life situations. There are, however, many real-life problems that do not fit linear models. One of the most common types of nonlinear models is exponential growth and exponential decay. In this chapter, you studied many examples of both types of exponential models. For instance, the balance in a savings account can be modeled with exponential growth, and human memory can be modeled with exponential decay.

How does it fit into the bigger picture of algebra?

In this chapter, you studied properties of exponents, scientific notation, and situations such as growth and decay that can be modeled with exponents.

Using properties of exponents, you can multiply and divide powers, and find powers of products and quotients. Using scientific notation, you can perform operations with very large and very small numbers. And, using exponential growth and decay, you can model two more types of real-life situations.

The models you studied in Chapters 3–7 are *linear*—their graphs are lines. The exponential growth and decay models in this chapter are *nonlinear*—their graphs are curves.

ORGANIZER

TIME SCHEDULE: All levels, two days

The Chapter Summary helps students organize the main ideas of the chapter. In this chapter, real-life situations were modeled using exponential models of growth and decay. Scientific notation was used to represent very large and very small number values.

Work with students to review the skills, strategies, and concepts of the chapter. It is suggested that the first day of review be a combination of you and students working problems from the Chapter Review exercises. The first day's homework assignment can be used as the basis for the second day of review.

COOPERATIVE LEARNING

Encourage students to study together in small groups. Emphasize the importance of teaching a classmate how to perform a skill or how to recall a strategy. When students work together, everyone wins. The students receiving help get additional instruction, and the students giving help gain a deeper understanding of the skills and concepts involved.

Chapter SUMMARY

Review the chapter by asking students "What did you learn? Why did you learn it? How does it fit with the other algebra skills and concepts you have learned?"

Chapter REVIEW

ASSIGNMENT GUIDE

Basic/Average: Ex. 3–48 multiples of 3, 49–50, 53–56, 61–65 odd

Above Average: Ex. 3–48 multiples of 3, 51–52, 57–60, 61–67 odd

Advanced: Ex. 3–48 multiples of 3, 51–52, 57–60, 61–67 odd

More Difficult Exercises

Exercises 53–60, 65–68

EXERCISE Notes

Exercises 1–68 Since the Chapter Review exercises parallel the list of Skills, Strategies, and Exploring Data statements, tell students to use the Chapter Summary on page 445 to identify the lesson in which a skill, strategy, or concept was presented if they are uncertain about how to do a given exercise in the Chapter Review. This makes the students categorize each task in the exercise set and locate relevant information in the chapter about the task, both being beneficial activities.

Chapter REVIEW

In Exercises 1–24, simplify the expression. (8.1, 8.2, 8.3) 6. $108a^3p^2$ 21. $18pq^6$

1. $b^2 \cdot b^7$ b^9
2. $(p^3)^4$ p^{12}
3. $(a^2)^3 \cdot a^3$ a^9
4. $x^2 \cdot (xy)^2$ x^4y^2
5. $(4m)^2 \cdot m^3$ $16m^5$
6. $(3a)^3 \cdot (2p)^2$
7. $8^2 \cdot (xy)^2 \cdot 2x$ $128x^3y^2$
8. $w^3 \cdot (3w)^4$ $81w^7$
9. q^0 1
10. p^{-2} $\frac{1}{p^2}$
11. $(a^2b)^0$ 1
12. $(x^{-2}y^3)^{-2}$ $\frac{x^4}{y^6}$
13. $\frac{p^4}{p^2}$ p^2
14. $\frac{3b^2}{9b^5}$ $\frac{1}{3b^3}$
15. $\frac{(4x^2)^2}{4x^4}$ 4
16. $\frac{x^2 \cdot y^0 \cdot 3^2}{x^3 \cdot y^{-4}}$ $\frac{9y^4}{x}$
17. $\frac{(2p)^{-2} \cdot pq}{(p^2q)^4 \cdot p^0}$ $\frac{1}{4p^9q^3}$
18. $\left(\frac{m}{n}\right)^3 \cdot m^2n^{-4}$ $\frac{m^5}{n^7}$
19. $\left(\frac{3a^2}{b}\right)^3 \cdot \frac{2b^{-1}}{3^2 \cdot a}$ $\frac{6a^5}{b^4}$
20. $\left(\frac{w^{-2}}{z^4}\right)^{-3}$ w^6z^{12}
21. $\frac{(3pq)^2}{2p^{-2}} \cdot \frac{4p^{-1}q^4}{p^2}$
22. $\frac{5m^2n}{m^4} \cdot \left(\frac{m^{-3}}{n}\right)^{-2}$ $5m^4n^3$
23. $\frac{(3w)^4}{z^{-2}} \cdot (w^{-2}z)^{-2}$ $81w^8$
24. $\frac{(3x)^{-3}}{3y^3} \cdot \frac{x^{-2}y^2}{x^{-4}}$ $\frac{1}{81xy}$

In Exercises 25–30, write the number in decimal form. (8.4)

25. 6.667×10^{-3} 0.006667
26. 3.75×10^{-1} 0.375
27. 9.81×10^{-7} 0.000000981
28. 1.29×10^2 129
29. 7.68×10^5 768,000
30. 5.44×10^9 5,440,000,000

In Exercises 31–36, write the number in scientific notation. (8.4)

31. 0.0588 5.88×10^{-2}
32. 0.000769 7.69×10^{-4}
33. 0.0000000233 2.33×10^{-8}
34. 523,000,000 5.23×10^8
35. 7950 7.95×10^3
36. 1,234,000,000 1.234×10^9

In Exercises 37–48, evaluate the expression. (8.2, 8.3, 8.4)

37. $\frac{3^2}{3^5}$ $\frac{1}{27}$
38. $2^{-1} \cdot 2^5$ 16
39. $\left(\frac{2^{-1}}{3^2}\right)^{-2} \cdot \frac{1}{2 \cdot 3^3}$ 6
40. $(3^5 \cdot 5^2) \cdot 3^{-4}$ 75
41. $2^{-3} \cdot 4^2 \cdot 5^{-1} \cdot (3^2)^0$ $\frac{2}{5}$
42. $\frac{(37 \cdot (24)^2)^2}{37 \cdot (24)^4}$ 37
43. $\frac{2.4 \times 10^{-6}}{1.2 \times 10^{-12}}$ 2×10^6
44. $(7.6 \times 10^2) \cdot (2.0 \times 10^4)$ 15.2×10^6
45. $\frac{3.33 \times 10^4}{6.66 \times 10^{12}}$ 5×10^{-9}
46. $\frac{9.6 \times 10^{-4} \cdot 4.8 \times 10^6}{2.4 \times 10^8 \cdot 2.4 \times 10^{-2}}$ 8×10^{-4}
47. $\frac{3.6 \times 10^{-4}}{7.2 \times 10^6} \cdot 1.8 \times 10^{-2}$ 9×10^{-13}
48. $\frac{9.9 \times 10^6}{1.1 \times 10^{-4}}$ 9×10^{10}

49. *Retail Sales* In 1990, about 1.615×10^{12} dollars were spent in retail purchases in the United States. What was the average amount spent per day? What was the average amount spent per American in 1990? (Use a population of 250 million.) **(8.5)** $\approx\$4.425 \times 10^9$, $\approx\$6460$

50. Use the information given in Exercise 49 to find the average amount spent on retail purchases each day by an American in 1990. $17.70

Geometry **In Exercises 51 and 52, use the following information. (8.2)**

The edge of cube A is 5 meters. The edge of cube B is $\frac{5}{2}$ meters.

51. What is the volume of cube A? What is the volume of cube B? 125 m^3, $\frac{125}{8}m^3$

52. What is the ratio of the edge of cube A to the edge of cube B? What is the ratio of the volume of cube A to the volume of cube B? What do you notice about this number? 2, 8, $8 = 2^3$

Modes of Travel **In Exercises 53–56, use the following information. (8.5)**

In 1990, Americans traveled 417,800,000,000 miles by aircraft and 1,900,000,000,000 miles by motor vehicle.

53. How many more miles were traveled by motor vehicle than by aircraft? 1,482,200,000,000

54. What was the total traveled by motor vehicle and by aircraft? 2,317,800,000,000

55. What is the ratio of the miles traveled by aircraft to the miles traveled by motor vehicle? ≈0.220

56. Approximate the average number of miles traveled by motor vehicle per American in 1990. 7600

National Basketball Association (NBA) **In Exercises 57–60, use the following information and graph. (8.6, 8.7)**

The NBA ticket sales and television fees, T (in millions of dollars), from 1980 through 1990 can be modeled by $T = 60 + (50)(1.29)^t$, where $t = 0$ represents 1980. NBA sales, S (in millions of dollars), of items such as logo T-shirts and backboards can be modeled approximately by the equation $S = 20(1.48)^t$.

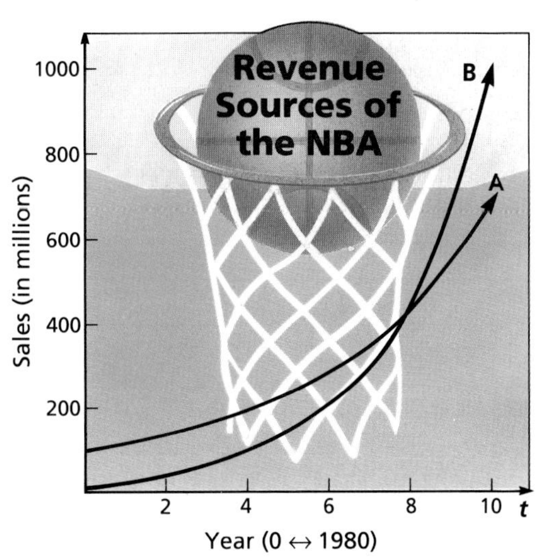

57. What was the annual percentage increase for sales of related items? 48%

58. Which model is represented by graph A? by graph B? T, S

59. For which years did sales of related items exceed ticket sales and television fees? 1988–1990

60. Use the graphs to estimate the NBA's total revenue from both sources in 1980; in 1990. Then use the models to estimate the total revenues. \$130 million; \$1700 million; \$130 million; \$1707 million

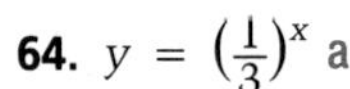

Answers

67. No, because the outer portion of A would then be visible beyond the circumference of B.

68. When the ratio of the distances from C is greater than the ratio of the diameters of the spheres, a total eclipse can occur; no.

In Exercises 61–64, match the equation with its graph.

61. $y = 3x$ b

62. $y = 3^x$ c

63. $y = \frac{1}{3}x$ d

64. $y = \left(\frac{1}{3}\right)^x$ a

a. 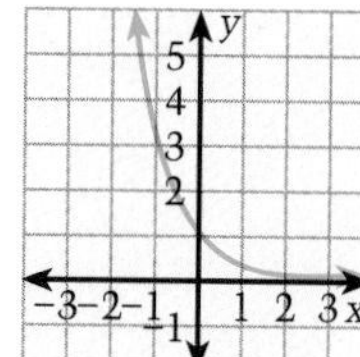

b. 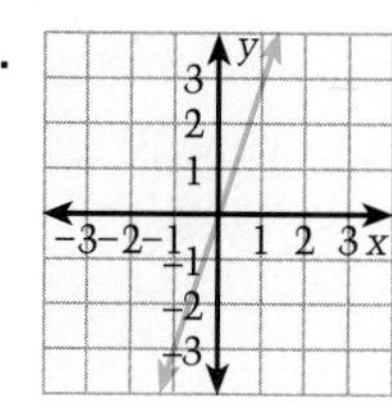

c. 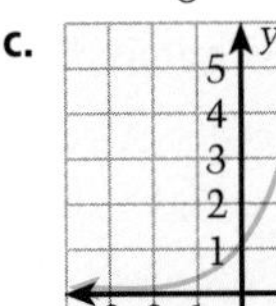

d. 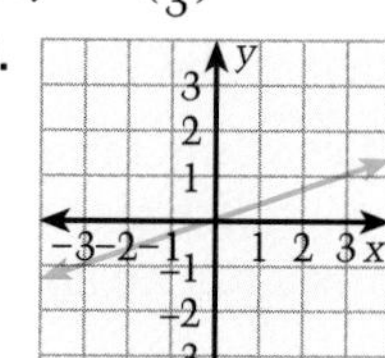

Total Eclipse of the Sun **In Exercises 65–68, use the following information.**

During a total solar eclipse, the moon completely covers Earth's view of the sun. The sun has a diameter of 8.64×10^5 miles, and the moon has a diameter of 2.16×10^3 miles. The distance between the sun and Earth varies between 9.15×10^7 miles and 9.45×10^7 miles. The distance between the moon's center and Earth's surface varies between 2.2×10^5 miles and 2.5×10^5 miles.

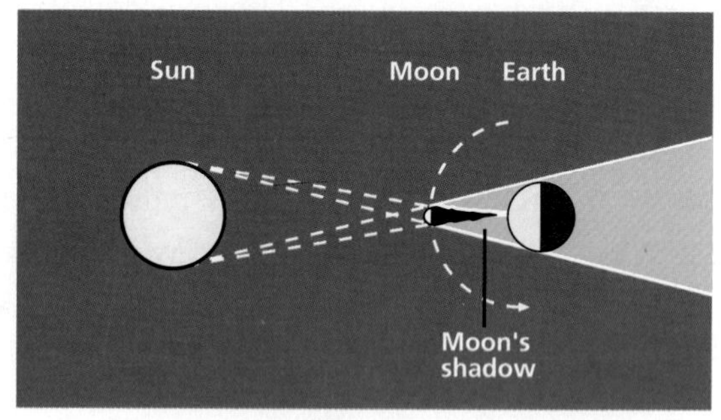

65. Find the ratio of the diameter of the sun to the diameter of the moon. 400

66. Complete the table by finding the ratios of the sun-to-Earth distances and moon-to-Earth distances.

Moon-to-Earth Distance (in miles)	Sun-to-Earth Distance (in miles)	
	Minimum 9.15×10^7	*Maximum* 9.45×10^7
Minimum 2.2×10^5	Ratio = ? ≈420	Ratio = ? ≈430
Maximum 2.5×10^5	Ratio = ? ≈370	Ratio = ? ≈380

67. The diagram at the right shows two spheres of diameters 15 inches and 3 inches. The ratio of these two diameters is the same as the ratio of the distance between C and the center of A and the distance between C and the center of B. If a person was farther than 4 inches away from B, could B totally block the person's view of A? Explain. See margin.

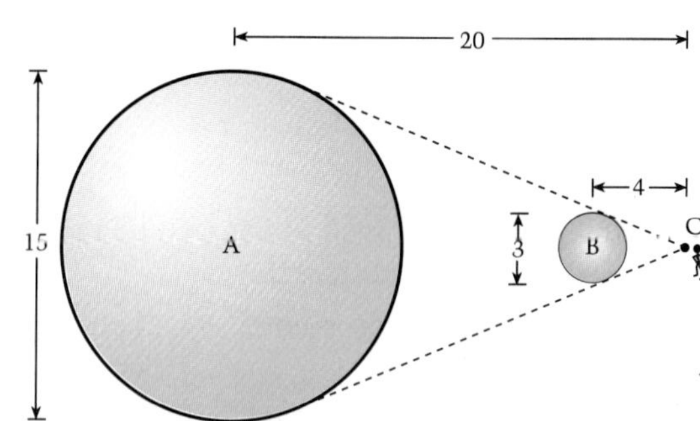

68. Use the results of Exercises 65–67 to explain why it is possible to have a total eclipse of the sun. Can this happen during the moon's *aphelion* (when the moon is farthest from Earth)? See margin.

Chapter TEST

In Exercises 1–12, simplify the expression. (8.1, 8.2, 8.3)

1. $a^2 \cdot a^5$ a^7
2. $x^0 \cdot x^3$ x^3
3. $x^{-4} \cdot x^2$ $\frac{1}{x^2}$
4. y^{-3} $\frac{1}{y^3}$
5. $(b^2)^8$ b^{16}
6. $(a^{-1})^2$ $\frac{1}{a^2}$
7. $\frac{q^3}{q^6}$ $\frac{1}{q^3}$
8. $(3a)^2(a^{-3})$ $\frac{9}{a}$
9. $(mn)^2 \cdot m^3$
10. $3x^4 \cdot 4x^{-2} \cdot x^3$ $12x^5$
11. $\left(\frac{m^3}{mn^2}\right)\left(\frac{n}{m}\right)^4$ $\frac{n^2}{m^2}$
12. $\frac{x^{-1}y^2}{xy} \cdot \frac{x^2y^3}{y^{-2}}$ y^6

9. m^5n^2

In Exercises 13–18, evaluate the expression. (8.1, 8.2, 8.3)

13. 3^4 81
14. 5^{-2} $\frac{1}{25}$
15. $(375^2)^0$ 1
16. $\frac{5 \cdot 2^5}{2^4}$ 10
17. $\left(\frac{3}{4}\right)^3 \cdot 4^2 \cdot 3^0$ $\frac{27}{4}$
18. $(6 \cdot 2)^3 \cdot 6^{-2}$ 48

In Exercises 19 and 20, write the number in decimal form. (8.4)

19. 3.65×10^6 3,650,000
20. 5.779×10^{-9} 0.000000005779

In Exercises 21 and 22, write the number in scientific notation. (8.4)

21. 2,440,000,000 2.44×10^9
22. 0.000000129 1.29×10^{-7}

23. In 1989, 1.2×10^5 United States companies spent 6.3×10^{10} dollars on research and development. What was the average amount per company spent on research and development? **(8.5)** $\approx 5.3 \times 10^5$

24. The world's largest pyramid was built by the Aztec Indians in Cholula, Mexico. It was 177 feet tall, and each side of its square base measured 1400 feet. What was the volume of this pyramid? (The volume of a pyramid is $\frac{1}{3}hb$, where h is the height and b is the area of the base.) **(8.2)**

25. Suppose that you inherited \$1000 when you were eight years old. You put \$500 in a savings account that earned 5% annually. You used the remainder of the money to buy stock in a new company. For the next 6 years, the value of the stock decreased 5% a year. Now that 6 years have passed, what is the balance of the savings account? How much is your stock worth? How does the sum of these two values compare to the value of your original inheritance? **(8.6)**
\$670.05, \$367.55, \$37.60 greater
24. 115,640,000 ft^3

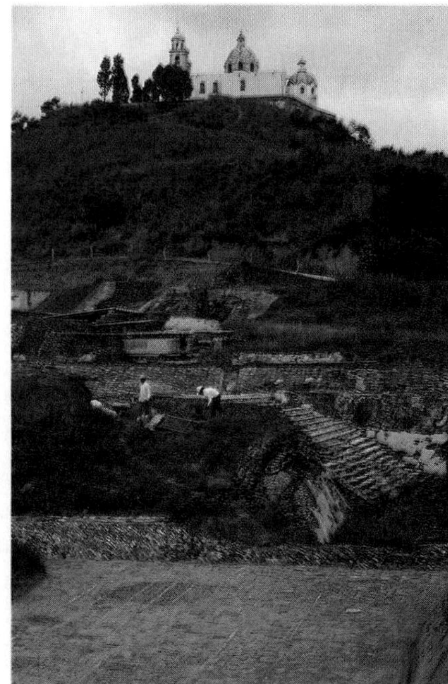

The Cholula Pyramid, in Mexico, is still largely buried underground. The base is estimated to cover nearly 45 acres.

Quadratic equations are used to model many real-life situations in science, economics, and engineering. Important tools for solving quadratic equations are the Pythagorean Theorem and the quadratic formula. As students continue to study mathematics, they will encounter many applications of quadratic equations, square roots, and the Pythagorean Theorem. In this chapter, students will have the opportunity to apply these and other algebraic tools.

COOPERATIVE LEARNING Many of the real-life applications in this chapter involve numbers that students might find difficult to calculate. You may wish to suggest that students work with partners to help each other apply the quadratic models and graphs to find the solutions to these exercises.

The Chapter Summary on page 498 provides you and the students with a synopsis of the chapter. It identifies key skills, concepts, and vocabulary. You may want to have students look at the Chapter Summary as an overview before beginning the chapter.

PHYSICS The opening of Chapter 9 contains many technical terms from physics. Students will want to know what they mean.

fission: the splitting of a nucleus of an atom of uranium or another heavy metal into two nearly equal fragments. A fission reaction was first accomplished in 1938 by German chemists and physicists Otto Hahn and Fritz Strassmann. In 1939, Austrian physicists Lise Meitner and her nephew Otto Frisch repeated the experiment and identified that what was accomplished was the first fission reaction.

CHAPTER 9

Quadratic Equations

LESSONS

In a nuclear reactor like this one, the controlled fission of a nuclear fuel such as uranium converts mass into energy.

Real Life

Energy Sources

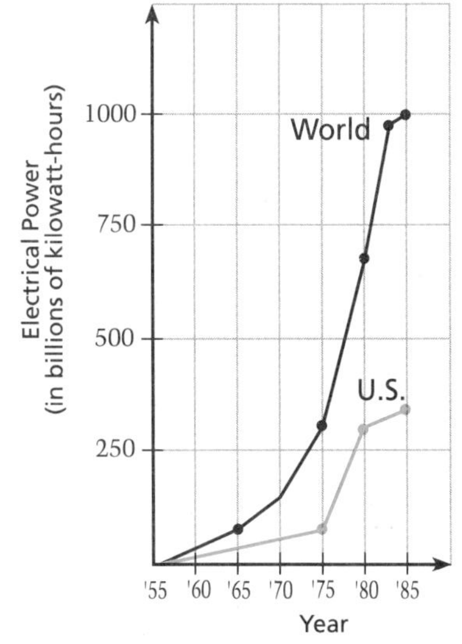

The kinetic (moving) energy of a mass m, traveling at a velocity, v, is $E = \frac{1}{2}mv^2$. Albert Einstein investigated derivatives of this quadratic equation. He suggested that when a mass m is transformed into energy, the total energy released is $E = mc^2$, where c is the speed of light. This equation, though not itself quadratic, is probably the most famous equation of modern times. Because the speed of light is so great, this equation tells us that even a small amount of mass can release an enormous amount of energy. For instance, the fission of 1 pound of uranium produces as much energy as the burning of 2,280,000 pounds of coal. The graph shows how much electrical power was produced by nuclear reactors during the first 30 years after their first use in 1956.

Sample Energy Values	Number of Joules
Energy released by one atom of U-235	1.80×10^{-11}
One Calorie	4.18×10^{0}
Energy equivalent of one gallon of gasoline	3.20×10^{7}
Energy in first atomic bomb	1.50×10^{14}
Energy used by the United States in 1990	3.00×10^{20}
Solar energy received by Earth each day	1.25×10^{22}
Total daily output of energy by the sun	1.00×10^{32}

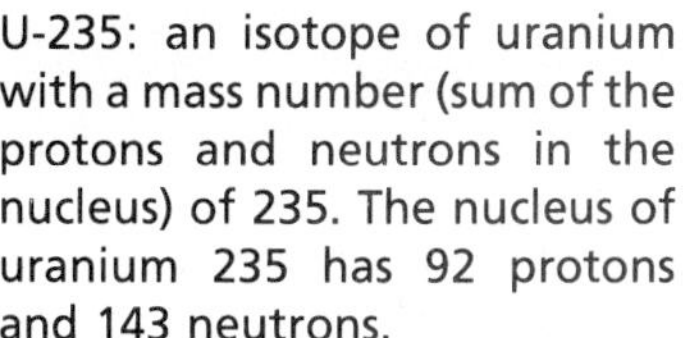

U-235: an isotope of uranium with a mass number (sum of the protons and neutrons in the nucleus) of 235. The nucleus of uranium 235 has 92 protons and 143 neutrons.

joule: the standard international unit of work and energy. Named after English physicist, James Joule.

BTU: British Thermal Unit, used to measure heat energy. One BTU is the amount of heat energy needed to raise the temperature of one pound of water one degree Fahrenheit.

calorie: the amount of heat energy needed to raise the temperature of one gram of water one degree Celsius.

TNT: trinitrotoluene, an explosive used in blasting.

Note that scientific notation is used to express energy levels in joules. Ask students the following types of questions.

1. How many times more energy is there for one calorie than for one electron volt? 4.18×10^{19}
2. How many times more energy is there for one btu than for one atom of U-235? 5.9×10^{13}

Quadratic Equations

PACING CHART

The daily Pacing Chart is meant to help you adjust your teaching pace. Students in the full course should finish the entire text by the end of the year. Students in the basic course may not complete the entire text in the school year. The Pacing Chart for each chapter contains suggestions for lessons that require more than one day and lessons that may be omitted for the basic course.

DAY	FULL COURSE	BASIC COURSE
1	9.1	9.1
2	9.2	9.2
3	9.3 & Using a Graphing Calculator	9.3
4	9.4	Using a Graphing Calculator
5	Mid-Chapter Self-Test	9.4
6	9.5	Mid-Chapter Self-Test
7	9.5	9.5
8	9.6	9.5
9	9.7	9.6
10	9.7	9.7
11	Chapter Review	9.7
12	Chapter Review	Chapter Review
13	Chapter Test	Chapter Review
14	Cumulative Review	Chapter Test
15		Cumulative Review

CHAPTER ORGANIZATION

LESSON	PAGES	GOALS	MEETING THE NCTM STANDARDS
9.1	452-457	1. Evaluate and approximate square roots 2. Use the Pythagorean theorem	Problem Solving, Communication, Connections, Technology
9.2	458-463	1. Solve a quadratic equation by finding square roots 2. Use quadratic models in real-life settings	Problem Solving, Communication, Connections, Geometry
Mixed Review	464	Review of algebra and arithmetic	
Career Interview	464	Manager of a printing company	Connections
9.3	465-470	1. Sketch the graph of a quadratic equation 2. Use quadratic models	Problem Solving, Communication, Connections, Geometry
Using a Calculator	471	Use the zoom feature of a graphing calculator to approximate solutions of an equation	Technology
9.4	472-477	1. Use the quadratic formula to solve a quadratic equation 2. Use quadratic models	Problem Solving, Communication, Connections
Mid-Chapter Self-Test	478	Diagnose student weaknesses and remediate with correlated Reteaching worksheets	
9.5	479-485	1. Find the number of solutions of a quadratic equation by using the discriminant 2. Use the discriminant in real-life models	Problem Solving, Communication, Connections
Mixed Review	485	Review of algebra and arithmetic	
9.6	486-491	1. Sketch the graph of a quadratic inequality 2. Use quadratic inequalities as real-life models	Problem Solving, Communication, Connections, Geometry
9.7	492-497	1. Choose a model that best fits a collection of data 2. Use models in real-life settings	Problem Solving, Communication, Connections, Geometry
Chapter Summary	498	A restatement of what has been learned, why it has been learned, and how it fits into the structure of algebra	Structure, Connections
Chapter Review	499-502	Review of concepts and skills learned in the chapter	
Chapter Test	503	Diagnose student weaknesses and remediate with correlated Reteaching worksheets	
Cumulative Review	504-505	Review of concepts and skills from previous chapters	

LESSON RESOURCES

MEETING INDIVIDUAL NEEDS

RETEACHING For students who need to spend more time on basics:

If a mid-chapter self-test or chapter test indicates a deficiency, teachers can help students with the appropriate ***Reteaching Copymaster.***

PRACTICE For students who need more practice:

Additional exercises like those in the Pupil's Edition are provided for each lesson in ***Extra Practice Copymasters.***

ENRICHMENT For enriching and broadening students' experiences:

Problem of the Day copymasters in ***Teaching Tools*** provide a daily opportunity to use logical reasoning, looking for a pattern, writing an equation, and other routine and non-routine problem-solving strategies.

Math Log copymasters in ***Alternative Assessment*** provide opportunities to report on investigations, research, and open-ended problems.

Enriching activities with graphing and scientific calculators and computers are provided in ***Technology: Using Calculators and Computers.***

The ***Applications Handbook*** provides additional information about the cross-curriculum topics such as astronomy, chemistry, physics, sports, economics, genetics, and music that are integrated into the Pupil's Edition.

LESSON	9.1	9.2	9.3	9.4	9.5	9.6	9.7
PAGES	452-457	458-463	465-470	472-477	479-485	486-491	492-497
Teaching Tools							
Transparencies			✓	✓	✓	✓	✓
Problem of the Day	✓	✓	✓	✓	✓	✓	✓
Warm-up Exercises	✓	✓	✓	✓	✓	✓	✓
Answer Masters	✓	✓	✓	✓	✓	✓	✓
Extra Practice Copymasters	✓	✓	✓	✓	✓	✓	✓
Reteaching Copymasters	Teacher-directed and independent activities tied to results on the Mid-Chapter Self-Tests and Chapter Tests						
Color Transparencies		✓	✓	✓	✓	✓	✓
Applications Handbook	Additional background information is supplied for many real-life applications.						
Technology Handbook	Calculator and computer worksheets are supplied for appropriate lessons.						
Complete Solutions Manual	✓	✓	✓	✓	✓	✓	✓
Alternative Assessment	Assess student's ability to reason, analyze, solve problems, and communicate using mathematical language.						
Formal Assessment	Mid-Chapter Self-Tests, Chapter Tests, Cumulative Tests, and Practice for College Entrance Tests						
Computer Test Bank	Customized tests can be created by choosing from over 2000 items.						

INSIGHTS

9.1 Square Roots and the Pythagorean Theorem

Square roots and the application of the Pythagorean Theorem occur frequently in economics, physics, and most branches of mathematics, including statistics and trigonometry. Consequently, there are a variety of reasons why students should be able to find square roots by estimation and computation, often using calculators.

9.2 Solving Quadratic Equations by Finding Square Roots

In Chapters 3 and 6, students learned how to solve linear equations and inequalities. This lesson is an introduction to solving quadratic equations. Solving quadratic equations and inequalities has many applications in science, engineering, and economics.

9.3 Graphs of Quadratic Equations

Quadratic equations in two variables, written in standard form, can be graphed on a coordinate plane. The graph is a visual model of the relationship between the two variables and is useful in determining solutions. That is, the graph of $y = ax^2 + bx + c$ can be used to find the solutions of $ax^2 + bx + c = 0$, $a \neq 0$. Because the graphics calculator is a powerful tool for graphing quadratic equations in two variables, it is important that students understand the types of information that are provided by the graphs of quadratic equations.

9.4 The Quadratic Formula

The quadratic formula is a powerful tool for solving quadratic equations. All real-valued solutions to any quadratic equation can be determined using this formula. The quadratic formula is especially efficient for noninteger values of a, b, and c, the coefficients of the quadratic equation written in standard form.

For students with access to a graphics calculator, it may be noted that such technological tools can be programmed to evaluate the quadratic formula for given coefficients, a, b, and c.

9.5 Problem Solving Using the Discriminant

The discriminant, $b^2 - 4ac$, provides information about the solutions of a quadratic equation in a simple and efficient way. Students may wish to employ this useful check prior to solving a quadratic equation. Checking helps to avoid needless computations.

9.6 Graphing Quadratic Inequalities

Economics, engineering, and physics are three real-life settings in which quadratic inequalities are used. The techniques employed in solving linear inequalities will be useful to students when solving quadratic inequalities.

9.7 Exploring Data: Comparing Models

Algebra helps us understand and interpret the world around us. In order to understand data that we encounter in real-life situations, we typically need to organize it and look for patterns. The algebraic models that have been studied provide the mathematical power with which to do this. In order to harness that power, students must recognize when to implement a given algebraic model.

In this lesson, students will have the opportunity to compare and contrast the different algebraic models for managing data.

Problem of the Day

When MacDonald was a young man and starting his famous farm, cows cost \$10 each, pigs cost \$3 each, and chickens cost 50¢ each. He bought 100 animals for \$100. If he bought at least one of each, how many chickens did he buy? 94

ORGANIZER

Warm-Up Exercises

1. Find values of x that make the following statements true.
   a. $x \cdot x = 16$ $x = 4, x = -4$
   b. $x \cdot x = 81$ $x = 9, x = -9$
   c. $x \cdot x = \frac{1}{9}$ $x = \frac{1}{3}, x = -\frac{1}{3}$
   d. $x \cdot x = 0.25$ $x = \pm 0.5$
   e. $x \cdot x = -4$ no solution
   f. $x \cdot x = 1.44$ $x = \pm 1.2$
2. TECHNOLOGY Use a calculator to compute to three decimal places by rounding and truncating.
   a. $4.03 \cdot \pi$
   b. $\pi^2 \cdot 2.56$
   c. $\frac{12}{17} \cdot \frac{5}{11}$
   d. $(\frac{8}{9} - 2.15) \cdot \pi$

	rounding	truncating
a.	12.661	12.660
b.	25.266	25.266
c.	0.321	0.320
d.	−3.962	−3.961

Lesson Resources

Teaching Tools
- Problem of the Day: 9.1
- Warm-up Exercises: 9.1
- Answer Masters: 9.1

Extra Practice: 9.1

9.1 Square Roots and the Pythagorean Theorem

What you should learn:

Goal 1 How to evaluate and approximate square roots

Goal 2 How to use the Pythagorean theorem

Why you should learn it:

You can use square roots to solve equations of the form $x^2 = a$, a form that frequently appears when you apply the Pythagorean theorem to right triangles.

Goal 1 Evaluating Square Roots

You know how to find the *square* of a number. The square of 3 is $3^2 = 9$. The square of -3 is also 9. Now you will study the reverse problem: finding a *square root* of a number.

Number	Positive $\sqrt{\ }$	Check	Negative $\sqrt{\ }$	Check
9	3	$(3)(3) = 9$	-3	$(-3)(-3) = 9$

Because the square of a number cannot be negative, we do not define the square root of a negative number.

Square Root of a Number

If $b^2 = a$, then b is a **square root** of a.

All positive real numbers have two square roots; one is the **positive** (or **principal**) **square root** and the other is the **negative square root.** Square roots are written with a **radical symbol** $\sqrt{\ }$. (The number inside a radical symbol is the **radicand.**)

$\sqrt{a}$ *Positive square root of a*

$-\sqrt{a}$ *Negative square root of a*

Example 1 ***Finding Square Roots of Numbers***

a. $\sqrt{81} = 9$ *Positive square root*

$-\sqrt{81} = -9$ *Negative square root*

b. $\sqrt{0} = 0$ *Zero has only one square root.*

c. $\sqrt{\frac{4}{9}} = \frac{2}{3}$ *Positive square root*

$-\sqrt{\frac{4}{9}} = -\frac{2}{3}$ *Negative square root*

d. $\sqrt{-4}$ is undefined.   *Negative numbers have no square root.* ■

The symbol $\pm$ (read as "plus or minus") is used to write *both* square roots of a positive number.

$\pm\sqrt{25} = \pm 5$ *Both square roots of 25, 5 and −5*

Numbers whose square roots are integers or quotients of integers are **perfect squares.** For example, 16 and 0.25, or $\frac{1}{4}$, are perfect squares. The square roots of numbers that are not perfect squares are **irrational numbers.** For example, the numbers $\sqrt{2}$, $\sqrt{3}$, $\sqrt{5}$, and $\sqrt{6}$ are all irrational. To approximate an irrational square root, you can use a calculator that has a *square root key* $\boxed{\sqrt{x}}$.

Example 2 *Evaluating Square Roots*

Evaluate the square root *exactly* if possible. Otherwise, approximate the square root and round the result to two decimal places.

a. $-\sqrt{121}$ **b.** $\sqrt{\frac{36}{25}}$ **c.** $\sqrt{0.09}$ **d.** $\sqrt{7}$

Connections Technology

Solution

a. $-\sqrt{121} = -11$ *$(-11)^2 = 121$*

b. $\sqrt{\frac{36}{25}} = \frac{6}{5}$ *$\left(\frac{6}{5}\right)^2 = \frac{36}{25}$*

c. $\sqrt{0.09} = 0.3$ *$(0.3)^2 = 0.09$*

d. $\sqrt{7} = 2.645751311\ldots$ *Use a calculator that has* $\boxed{\sqrt{x}}$.
≈ 2.65 *Round to 2 decimal places.* ■

Example 3 *Evaluating an Expression That Has a Square Root*

Evaluate $\sqrt{b^2 - 4ac}$ when $a = 1$, $b = -2$, and $c = -3$.

Solution Substitute the values of a, b, and c as follows.

$$\begin{aligned} \sqrt{b^2 - 4ac} &= \sqrt{(-2)^2 - 4(1)(-3)} \\ &= \sqrt{4 + 12} && \textit{Simplify.} \\ &= \sqrt{16} && \textit{Simplify.} \\ &= 4 && \textit{The positive square root of 16 is 4.} \end{aligned}$$ ■

Connections Technology

Example 4 *Evaluating an Expression with a Plus or Minus Symbol*

Evaluate $\frac{1 \pm 2\sqrt{3}}{4}$.

Solution This expression represents *two* numbers. Use a calculator as follows.

Keystrokes	Display	Rounded
(1 + 2 × 3 √x) ÷ 4 =	1.116025404	1.12
(1 − 2 × 3 √x) ÷ 4 =	−0.616025404	−0.62 ■

LESSON Notes

Example 1

Emphasize that the square root of a is the value b such that $b^2 = a$. Encourage students to check answers using this defining relationship. In particular, check by asking, If $\sqrt{81} = 9$, is $9 \cdot 9 = 81$?

Example 2

Finding the exact square root may also mean using the radical form to express the result when the square root is not a rational number. For example, $\sqrt{28} = \sqrt{4 \cdot 7} = \sqrt{4} \cdot \sqrt{7} = 2\sqrt{7}$ and $\sqrt{125} = \sqrt{25 \cdot 5} = \sqrt{25} \cdot \sqrt{5} = 5\sqrt{5}$. Note that, for a and b both positive, $\sqrt{a \cdot b} = \sqrt{a} \cdot \sqrt{b}$.

EXTEND Example 2 For more practice, have students try these.

a. $\sqrt{24}$ $2\sqrt{6}$ **d.** $\sqrt{175}$ $5\sqrt{7}$
b. $\sqrt{18}$ $3\sqrt{2}$ **e.** $\sqrt{44}$ $2\sqrt{11}$
c. $\sqrt{80}$ $4\sqrt{5}$ **f.** $\sqrt{54}$ $3\sqrt{6}$

Example 3

Students will soon learn that $b^2 - 4ac$ is a special expression called the "discriminant."

Example 4

TECHNOLOGY Different brands of calculators may differ in the displays that occur for the same computation.

EXTEND Example 5 Assuming that the catcher can throw a baseball about 75 miles per hour, and that the base runner can sprint about 8 yards per second, how much of a lead will the base runner need in order to beat the throw to second base from home plate? By the time the catcher releases the baseball, the runner needs a lead of about 62.2 feet!

Extra Examples

Here are additional examples similar to **Examples 1–4.**

1. Find these square roots. Round each decimal-valued number to two decimal places.
   a. $\sqrt{121}$ 11
   b. $-\sqrt{149}$ -12.21
   c. $\sqrt{\frac{36}{169}}$ $\frac{6}{13}$
   d. $-\sqrt{25}$ -5
   e. $\sqrt{0.0169}$ 0.13
   f. $\sqrt{0}$ 0
2. Evaluate $\sqrt{b^2 - 4ac}$ using the indicated values.
   a. $a = 3, b = 4, c = 1$ 2
   b. $a = 4, b = 5, c = 1$ 3
   c. $a = 1, b = 13, c = 12$ 11
3. TECHNOLOGY Use a calculator to evaluate each expression to two decimal places.
   a. $\frac{2 \pm 3\sqrt{5}}{4}$ 2.18, -1.18
   b. $\frac{1 \pm 5\sqrt{3}}{16}$ 0.60, -0.48

Check Understanding

1. Give examples of real-valued numbers whose square roots are integers. For example, 4, 144
2. Provide examples of real-valued numbers whose square roots are fractions. For example, $\frac{1}{4}$, $\frac{9}{25}$
3. Provide examples of real-valued numbers whose square roots are irrational. For example, 13, 24
4. Explain why each positive real number has two square roots. The product of two positive numbers is positive, and the product of two negative numbers is also positive. Thus, the squares of both positive and negative numbers are positive.

454

Goal 2 Using the Pythagorean Theorem

Connections **Geometry**

The word *theorem* means "a statement that can be proven to be true." The **Pythagorean theorem** states a relationship for the three sides of a right triangle. In a right triangle, the **hypotenuse** is the side opposite the right angle. The other two sides are the **legs** of the triangle. The theorem is named after the Greek mathematician Pythagoras (about 585–500 B.C.). Records of its use in northern Africa, however, predate Pythagoras.

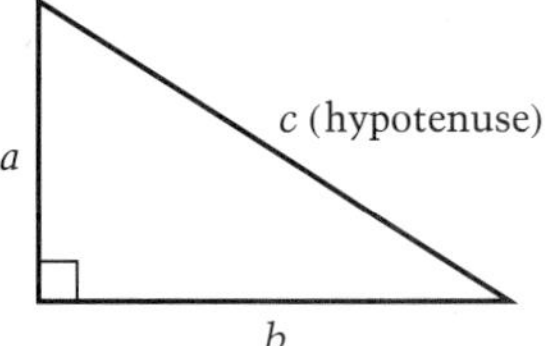

The Pythagorean Theorem

For any right triangle, the sum of the squares of the lengths of the legs, a and b, equals the square of the length of the hypotenuse, c.

$a^2 + b^2 = c^2$ *Pythagorean form*

Example 5 *Distance to Second Base*

Problem Solving **Draw a Diagram**

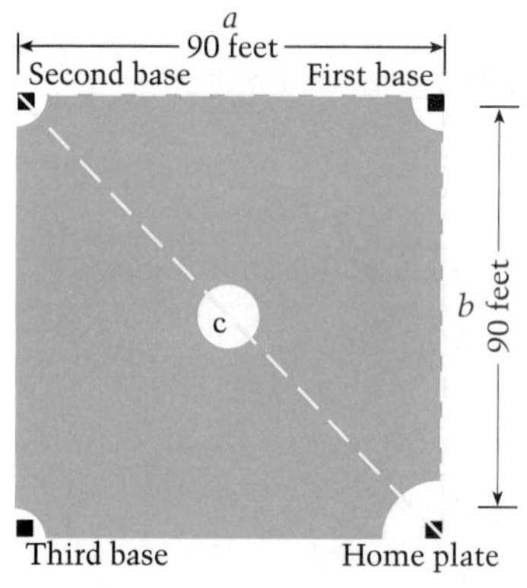

The length of each side of a square baseball diamond is 90 feet. What is the distance from home plate to second base?

Solution The diagonal from home plate to second base is the hypotenuse of a right triangle. Each leg is 90 feet long.

$$\begin{aligned} c^2 &= a^2 + b^2 && \textit{Pythagorean theorem} \\ &= 90^2 + 90^2 && \textit{Substitute 90 for a and b.} \\ &= 8100 + 8100 && \textit{Simplify.} \\ &= 16{,}200 && \textit{Simplify.} \end{aligned}$$

From the definition of a square root, c must be the positive square root of 16,200. (It wouldn't make sense to use the negative square root.) Thus, you can write

$$\begin{aligned} c &= \sqrt{16{,}200} && \textit{Use a calculator.} \\ &\approx 127.3 \text{ feet.} && \textit{Round to one decimal place.} \end{aligned}$$

The distance is 127.3 feet. ■

Communicating about ALGEBRA

SHARING IDEAS about the Lesson

Use Technology Use a calculator to evaluate each expression. (Round to two decimal places.) 0.56, -3.56

A. $2\sqrt{13}$ 7.21 **B.** $5 - 3\sqrt{2}$ 0.76 **C.** $\frac{-3 \pm \sqrt{(3)^2 - 4(1)(-2)}}{2}$

EXERCISES

Guided Practice

CRITICAL THINKING about the Lesson

1. **True or False?** If $b^2 = a$, then b is a square root of a. True
2. Which statements are true? Give an example of each true statement.
   a. A number can have no square root. True, -1
   b. A number can have only one square root. True, 0
   c. A number can have two different square roots. True, 25 has 5 and -5.
   d. A number can have more than two square roots. False
3. Evaluate $\sqrt{25}$, $-\sqrt{25}$, and $\pm\sqrt{25}$. 5, -5, ± 5
4. Find the numbers represented by $3 \pm \sqrt{(-3)^2 - 4(\frac{1}{2})(-8)}$. 8 and -2
5. Give three examples of perfect squares, one an integer, one a fraction, and one a decimal number. 9, $\frac{1}{4}$, 0.01
6. Is $\sqrt{2}$ irrational? Is $\frac{144}{99}$ a square root of 2? Is $\frac{7064}{4995}$ a square root of 2? Explain. Yes; no; no. An irrational number cannot be represented by a fraction

Independent Practice

In Exercises 7–14, find all square roots of the number. Check your results.

7. 64 8, -8
8. 144 12, -12
9. -9 None
10. 0 0
11. $\frac{4}{9}$ $\frac{2}{3}$, $-\frac{2}{3}$
12. $\frac{25}{16}$ $\frac{5}{4}$, $-\frac{5}{4}$
13. 0.16 0.4, -0.4
14. 0.25 0.5, -0.5

In Exercises 15–30, evaluate the expression. Give the exact value if possible. Otherwise, give an approximation to two decimal places. 22. 4.80 26. $-\frac{5}{10}$ or $-\frac{1}{2}$

15. $-\sqrt{256}$ -16
16. $\sqrt{49}$ 7
17. $\sqrt{11}$ 3.32
18. $\sqrt{121}$ 11
19. $\sqrt{100}$ 10
20. $-\sqrt{169}$ -13
21. $\sqrt{36}$ 6
22. $\sqrt{23}$
23. $\sqrt{42}$ 6.48
24. $\sqrt{0.75}$ 0.87
25. $\sqrt{0.04}$ 0.2
26. $-\sqrt{\frac{25}{100}}$
27. $\sqrt{\frac{81}{324}}$ $\frac{9}{18}$ or $\frac{1}{2}$
28. $\sqrt{6.25}$ 2.5
29. $-\sqrt{\frac{1}{64}}$ $-\frac{1}{8}$
30. $\sqrt{26}$ 5.10

In Exercises 31–34, evaluate $\sqrt{b^2 - 4ac}$ for the given values of a, b, and c.

31. $a = 4, b = 5, c = 1$ 3
32. $a = -2, b = 8, c = -8$ 0
33. $a = 3, b = -7, c = 6$ $\sqrt{-23}$ is undefined.
34. $a = 12, b = 13, c = 3$ 5

Technology **In Exercises 35–38, use a calculator to evaluate the expression. Round results to two decimal places. Use estimation to check results.**

35. $\frac{2 \pm 5\sqrt{6}}{2}$ 7.12, -5.12
36. $\frac{3 \pm 4\sqrt{5}}{4}$ 2.99, -1.49
37. $\frac{7 \pm 3\sqrt{2}}{-1}$ -11.24, -2.76
38. $\frac{5 \pm 6\sqrt{3}}{3}$ 5.13, -1.80

5. Explain why negative real numbers have no square roots. The product of a positive number and a negative number is negative. The square of a number cannot be negative, so a negative number has no real-number square root.

Communicating about ALGEBRA

EXTEND Communicating
Ask students to discuss how a calculator without a square root key can be used to determine the square root of any real-valued number. Once a procedure has been found, require students to show the process in a flowchart.

One way is to use the divide-and-average method. Another is to use a guess-and-check method.

EXERCISE Notes

ASSIGNMENT GUIDE

Basic/Average: Ex. 1–6, 7–37 odd, 41, 47, 49, 53, 57

Above Average: Ex. 1–6, 7–41 odd, 43, 44, 47–59 odd

Advanced: Ex. 1–6, 13–41 odd, 43–50, 61–64

Selected Answers
Exercises 1–6, 7–59 odd

More Difficult Exercises
Exercises 43, 44, 46, 61–64

Guided Practice

▶ **Ex. 1–6** COOPERATIVE LEARNING Have students complete these exercises in a partner setting and explain their reasoning to each other. If students have difficulty with Exercise 5, refer them back to Example 2.

Independent Practice

▶ **Ex. 13–30** TECHNOLOGY Check that students understand how to use the square-root keys on their calculators.

▶ **Ex. 31–34, 35–38** These exercises foreshadow the necessary numerical preparation for the quadratic formula. Refer to Example 4 in preparation for Exercises 35–38.

▶ **Ex. 45–46** Note that the numerical answer to both exercises is $\sqrt{3}$, but $\sqrt{3}$ in. $\neq \sqrt{3}$ in.^{2}.

▶ **Ex. 47–50** These exercises are applications of the Pythagorean Theorem.

Geometry **In Exercises 39–42, use the area, A, to find the length of the indicated side or radius. (Use $\pi \approx 3.14$.)**

39. Square: $s = \sqrt{A}$

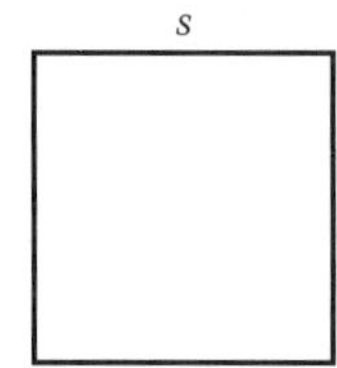

5.2 $A = 27.04$

40. Square: $s = \sqrt{A}$

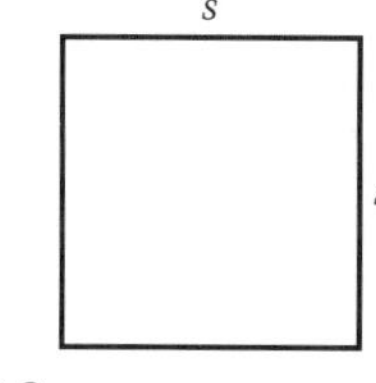

3.6 $A = 12.96$

41. Circle: $r = \sqrt{\frac{A}{\pi}}$

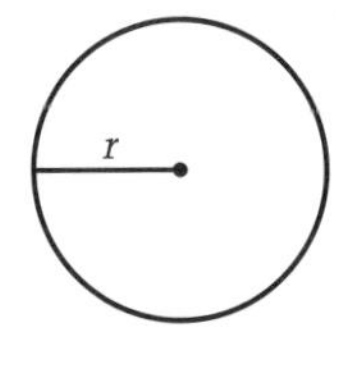

4 $A = 50.24$

42. Circle: $r = \sqrt{\frac{A}{\pi}}$

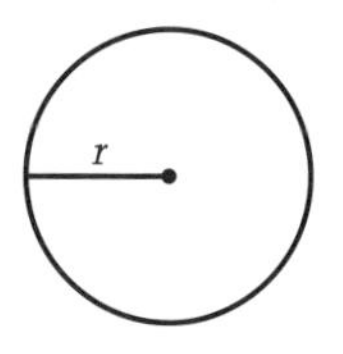

7 $A = 153.86$

Biosphere 2 **In Exercises 43 and 44, use the following information.**

Biosphere 2, in Arizona, is a large terrarium that covers 3 acres and reaches a height of 10 stories. The structure contains 7 biomes (miniature versions of ecological communities). These include a savanna, freshwater and saltwater marshes, a farm, a desert, and an ocean. To test the possibility of creating miniature "Earth environments" on other planets, 8 people were sealed in Biosphere 2 for 2 years.

✪ **43.** Biosphere 2 covers over 130,000 square feet. If its layout was in the shape of a single square, what would be its dimensions? $\approx$360 ft by $\approx$360 feet

✪ **44.** The "ocean" in Biosphere 2 has 110,000 cubic feet of sea water and is 25 feet deep. If a square swimming pool was 25 feet deep and held 110,000 cubic feet of water, how long would each side be? $\approx$66 ft

Geometry **In Exercises 45 and 46, use the formula $h = \frac{1}{2}\sqrt{3}s$ for the height, h, of an equilateral triangle whose sides are of length s.**

45. Find the height of an equilateral triangle with 2-inch sides. $\sqrt{3}$ in. or $\approx$1.73 in.

✪ **46.** Find the area of an equilateral triangle with 2-inch sides. $\sqrt{3}$ in.^{2} or $\approx$1.73 in.^{2}

2	2	2

47. *Length of a Ladder* A ladder is placed against the trunk of a cherry tree. It reaches a height of 10 feet on the tree. The bottom of the ladder extends 5 feet from the base of the tree. How long is the ladder? $\approx$11.2 ft

48. *Fallingwater* How long is the hanging staircase at Fallingwater from the bottom step to the top step? ≈13.9 ft

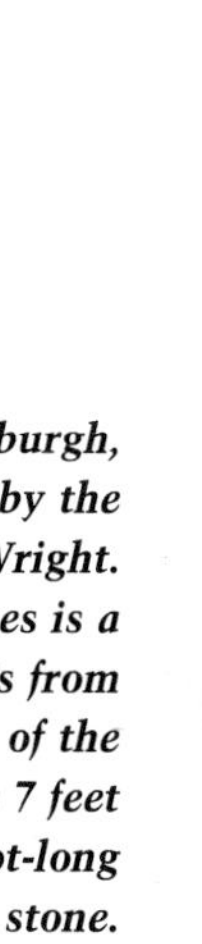

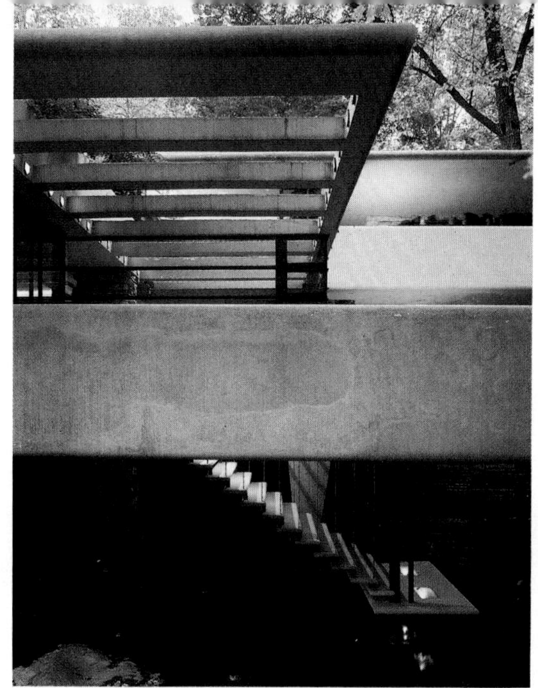

Fallingwater, near Pittsburgh, Pennsylvania, was designed by the American architect Frank Lloyd Wright. One of its many unique features is a hanging staircase that descends from the second floor to the stone steps of the "plunge pool." The staircase is 7 feet high and is cut from a 12-foot-long hole in the stone.

49. *Diagonal of a Football Field* A football field is 360 feet by 45 feet. How long is the walk from one corner diagonally to the opposite corner? ≈362.8 ft

50. *Repairing a Gate* The diagonal crossbar of an old wooden gate has rotted. The gate is rectangular, 3 feet by 4 feet. How long is the crossbar? 5 ft

Integrated Review

Geometry **What is the unit of measure for s?**

51. $s^2 = 64$ square miles miles

52. $s^2 = 200$ square inches inches

In Exercises 53–56, square the expression.

53. $-3a$ $9a^2$

54. $7b$ $49b^2$

55. $2x^2y$ $4x^4y^2$

56. $3xy^3$ $9x^2y^6$

In Exercises 57–60, use mental math to solve the equation.

57. $x^2 = 9$ $3, -3$

58. $y^2 = 16$ $4, -4$

59. $y^3 = 8$ 2

60. $-(x^2) = -36$ $6, -6$

Exploration and Extension

In Exercises 61–64, simplify the positive square root.

The positive square root of an integer that is the product of a *perfect square* and another integer can be simplified using the property $\sqrt{ab} = \sqrt{a} \cdot \sqrt{b}$. For example, 12 is the product of 4 and 3, which means that $\sqrt{12}$ can be written as $\sqrt{12} = \sqrt{4 \cdot 3} = \sqrt{4} \cdot \sqrt{3} = 2\sqrt{3}$.

61. $\sqrt{8}$ $2\sqrt{2}$

62. $\sqrt{18}$ $3\sqrt{2}$

63. $\sqrt{27}$ $3\sqrt{3}$

64. $\sqrt{75}$ $5\sqrt{3}$

Enrichment Activities

These activities require students to go beyond lesson goals.

GEOMETRY Students can use the following activities to find lengths that approximate $\sqrt{2}$ or $\sqrt{3}$. (If protractors are not available, have students draw the triangles on centimeter dot or grid paper.)

COOPERATIVE LEARNING

1. Work with a partner. You need a centimeter ruler and a protractor. Start in the lower left corner of the paper. Draw a right triangle ABC with legs 1 decimeter (10 centimeters) long. Use the Pythagorean Theorem. How long is hypotenuse AC? $\sqrt{2}$ dm
2. On the same diagram, draw right triangle ACD with leg $AC = \sqrt{2}$ dm and leg $CD = 1$ dm. How long is hypotenuse AD? $\sqrt{3}$ dm
3. Continue and draw the next four triangles. What would the hypotenuses of the next four triangles be? $\sqrt{4}$ dm $=$ 2 dm, $\sqrt{5}$ dm, $\sqrt{6}$ dm, $\sqrt{7}$ dm

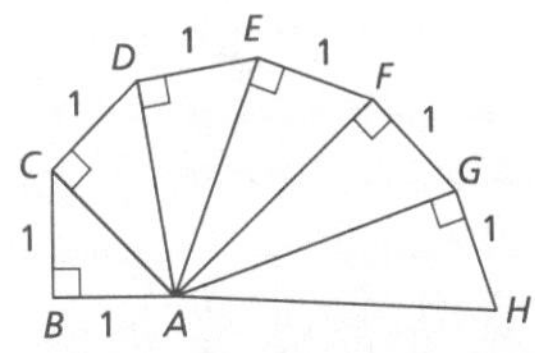

4. Use your ruler to measure the hypotenuses to the nearest millimeter. Use your measurements to give an approximation to the nearest tenth for $\sqrt{2}$, $\sqrt{3}$, $\sqrt{5}$, $\sqrt{6}$, and $\sqrt{7}$. 1.4, 1.7, 2.2, 2.4, 2.6

Problem of the Day

A water lily with a rigid stem extends one foot above the surface of the water. When pulled over, it disappears beneath the water at a distance 3 feet from where the stem originally entered the water. How deep is the pond? 4 ft

ORGANIZER

Warm-Up Exercises

1. Find each square root to three decimal places.
   a. 4 2 b. 0 0
   c. −1 no square root
   d. 16 4 e. 8 2.828
   f. 28 5.292
2. Evaluate each expression for the indicated values.
   a. $3x^2$, $x = 4$ and $x = -4$
   b. $2x^2 - 4x$, $x = 2$ and $x = -2$
   c. $x^2 + x - 3$, $x = 1$ and $x = -1$
   d. $-3x^2 + 4x - 5$, $x = 3$ and $x = -3$

   a. 48, 48 b. 0, 16
   c. −1, −3 d. −20, −44

Lesson Resources

Teaching Tools
- Transparency: 4
- Problem of the Day: 9.2
- Warm-up Exercises: 9.2
- Answer Masters: 9.2

Extra Practice: 9.2

Color Transparencies: 43, 44

Applications Handbook: p. 44, 45

Common-Error ALERT!

There are *two* solutions to the equation $x^2 = d$, $d > 0$. One is the positive square root of d and the other is the negative square root of d. Be sure that students give both solutions and not just the positive one.

458

Descending exponential order

9.2 Solving Quadratic Equations by Finding Square Roots

What you should learn:

Goal 1 How to solve a quadratic equation by finding square roots

Goal 2 How to use quadratic models in real-life settings

Why you should learn it:

You can model many situations, such as population growth in the United States, with quadratic models.

Goal 1 Solving a Quadratic Equation

A **quadratic equation** in x is an equation that can be written in the **standard form**

$$ax^2 + bx + c = 0, \qquad a \neq 0.$$

In this form, a is the **leading coefficient.**

In this lesson, you will learn how to find the real-number solutions of quadratic equations of the form $ax^2 + c = 0$.

Solving $x^2 = d$ by Finding Square Roots

1. If d is positive, then $x^2 = d$ has two solutions: $x = \pm\sqrt{d}$.
2. The equation $x^2 = 0$ has one solution: $x = 0$.
3. If d is negative, then $x^2 = d$ has no solution.

Example 1 *Solving Quadratic Equations*

a. The equation $x^2 = 4$ has two solutions: $x = \pm 2$.

b. The equation $x^2 = 5$ has two solutions: $x = \pm\sqrt{5}$.

c. The equation $x^2 = 0$ has one solution: $x = 0$.

d. The equation $x^2 = -1$ has no solution. ■

Example 2 *Transforming before Finding Square Roots*

Solve $3x^2 - 48 = 0$.

Solution

$3x^2 - 48 = 0$	*Rewrite original equation.*
$3x^2 = 48$	*Add 48 to both sides.*
$x^2 = 16$	*Divide both sides by 3.*
$x = \pm\sqrt{16}$	*Find square roots.*
$x = \pm 4$	*16 is a perfect square.*

The solutions are 4 and −4. Check these results in the original equation. ■

Goal 2 Using Quadratic Models in Real Life

In Exercise 22 of Lesson 8.7, you were asked to use a quadratic model to predict when the population of the United States would reach 300 million. In that lesson, you may have used a "guess and check" approach to solving the problem. Now, however, you can solve the problem by finding square roots.

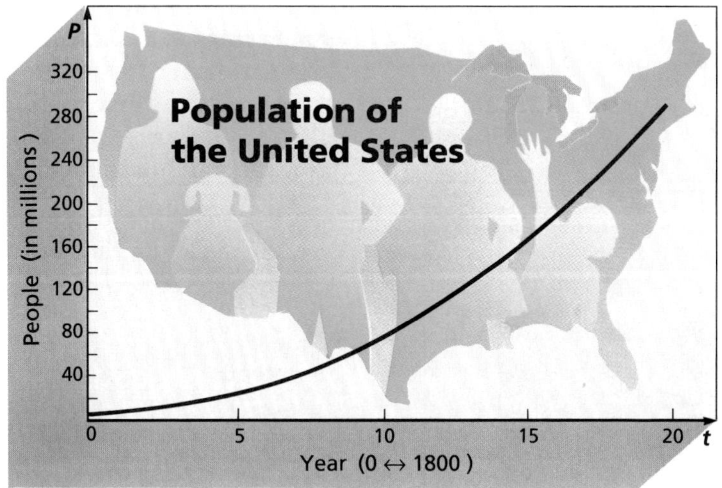

The U.S. population can be modeled by $P = 6{,}893{,}000 + 672{,}240t^2$, where $t = 0, 1, 2, \ldots,$ represents 1800, 1810, 1820, . . .

Real Life
Population Studies

Example 3 *Population of the United States*

Use the quadratic model given with the graph to estimate the year that the United States population will reach 300 million.

Solution

$6{,}893{,}000 + 672{,}240t^2 = P$	*Quadratic model*
$6{,}893{,}000 + 672{,}240t^2 = 300{,}000{,}000$	*Substitute 300 million for P.*
$672{,}240t^2 = 293{,}107{,}000$	*Subtract 6,893,000.*
$t^2 \approx 436$	*Divide by 672,240.*
$t \approx \sqrt{436}$	*Find positive square root.*
$t \approx 20.9$	*Use a calculator.*

The model predicts the United States population will reach 300 million in the year 2009.* (Note that we used the positive square root because a negative solution wouldn't make sense in this problem.) ■

*In 1990, the U.S. Census Bureau's predictions for 2009 varied between 263 million and 303 million. The differences come from different assumptions about birth, death, and immigration rates.

LESSON Notes

GOAL 1 Emphasize that the leading coefficient must be nonzero in the standard form of the quadratic equation. In this text, solutions of a quadratic equation are real-valued numbers.

Examples 1–2

Ask students to check the solutions.

Example 3

TECHNOLOGY A calculator was used to find the solution. Calculations should not be rounded until after the final computation. That is, find the square root of the value $t^2 = 436.0154112$ so that you get 20.88098204 and then round to 20.9.

Observe that time t is given in increments of 10 years. Accordingly, for $t = 0$ corresponding to 1800, $t = 3$ corresponds to 1830. Remember, $t = 20.9$ represents 209 years. So the population of the United States will reach 300 million in the year 1800 + 209 = 2009.

Example 4

The negative sign in the formula, $h = -16t^2 + s$, indicates that the object is falling (down). An equivalent formula uses *acceleration due to gravity* (the change in velocity of a falling body over time, $a = -32$ ft/s^2). That formula is $h = \frac{1}{2}at^2 + s$.

The formula provided assumes that any velocity is given in feet per second. Thus, it is important to express speeds, heights, and times in the appropriate units.

Extra Examples

Here are additional examples similar to **Examples 1–4.**

1. Solve.
   a. $x^2 = 36$
   b. $x^2 = 13$
   c. $4x^2 - 15 = 0$
   d. $-3x^2 + 40 = 0$
   e. $5x^2 + 125 = 0$
   a. $x = 6, x = -6$
   b. $x = 3.61, x = -3.61$
   c. $x = 1.94, x = -1.94$
   d. $x = 3.65, x = -3.65$
   e. no solution
2. Suppose the wrench in Example 4 were dropped from a height of 350 feet. How many seconds would it take to hit the ground? about $t = 4.68$ seconds

Check Understanding

1. Create a flowchart that shows how quadratic equations may be solved using square roots. Consider cases in which the quadratic equation must be transformed. Students can refer to Example 2 for ideas.
2. Why must the leading coefficient a of the quadratic equation in standard form be nonzero? If a were 0, then the x^2 term would be 0 and the equation would not be quadratic.
3. Do you think any quadratic equation can be solved by finding square roots? Explain. No, only those that can be written in the form $x^2 = d$.

Communicating about ALGEBRA

EXTEND *Communicating*

The formula $v = v_0 + at$ gives the velocity v of a body with initial velocity v_0 after time t seconds. The acceleration due to gravity, a, is -32 ft/s^2.

The distance formula $d = rt$ is used for problems in which the rate or speed is *constant.* When an object (with little air resistance) is dropped, its speed continually *increases.* The height, h (in feet), of such an object is given by the model

$$h = -16t^2 + s \quad \textit{Falling object model}$$

where s is the height in feet from which the object was dropped, and t is the number of seconds during which it has fallen. This model works for Earth's gravitational force.

Real Life
Construction Safety

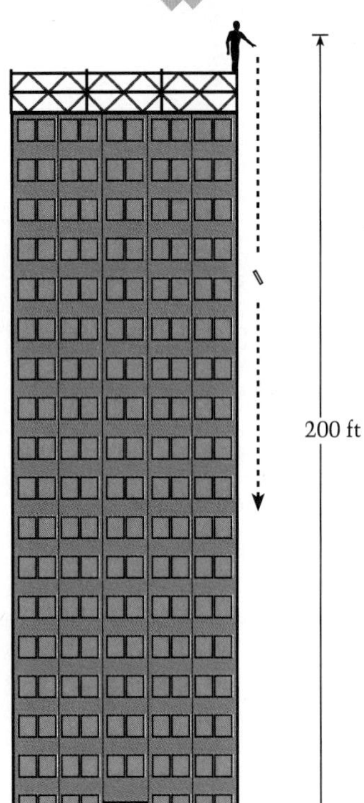

Example 4 *Watch Out Below!*

A construction worker on the top floor of a building accidentally drops a heavy wrench. How many seconds will it take to hit the ground? (Assume it drops from a height of 200 feet.)

Solution Because the wrench drops from a height of 200 feet, the model for the height of the wrench at time t is $h = -16t^2 + 200$. The table gives heights at different times.

Time, t in seconds	0.0	0.5	1.0	1.5	2.0	2.5	3.0	3.5
Height, h in feet	200	196	184	164	136	100	56	4

From the table, you can say that the wrench will take a little more than $3\frac{1}{2}$ seconds to hit the ground. Another way to solve the problem is to solve the quadratic equation for the time that gives a height of $h = 0$ feet.

$$\begin{aligned}
-16t^2 + 200 &= h && \textit{Falling object model} \\
-16t^2 + 200 &= 0 && \textit{Substitute 0 for h.} \\
200 &= 16t^2 && \textit{Add } 16t^2 \textit{ to both sides.} \\
12.5 &= t^2 && \textit{Divide both sides by 16.} \\
\sqrt{12.5} &= t && \textit{Find the positive square root.} \\
3.54 &\approx t && \textit{Use a calculator.}
\end{aligned}$$

The wrench will take about 3.54 seconds to hit the ground. ■

Communicating about ALGEBRA

SHARING IDEAS about the Lesson

Extend the Idea A free falling body accelerates at the rate of 32 feet per second every second.

In Example 4, find, or approximate, the speed of the wrench when it hit the ground. 113 feet per second

EXERCISES

Guided Practice

CRITICAL THINKING about the Lesson

1. Which of the following are quadratic equations?

   a. $-3x + 5 = 0$ b. $\underline{x^2 - 1 = 0}$ c. $x^2 - 3x^3 = 0$ d. $\underline{-3 + 4x + x^2 = 0}$

In Exercises 2–4, write in standard form and find the leading coefficient.

2. $-3x^2 + 5 = 0$ As is, -3
3. $\frac{1}{2}x^2 + 9x - 3 = 0$ As is, $\frac{1}{2}$
4. $-8x - x^2 + 4 = 0$ $-x^2 - 8x + 4 = 0$, -1

In Exercises 5–8, solve the equation. If there are no solutions, state the reason.

5. $x^2 = 17$ $\pm\sqrt{17}$
6. $x^2 = 0$ 0
7. $x^2 = -4$ No real solution
8. $x^2 = 6$ $\pm\sqrt{6}$

Independent Practice

In Exercises 9–20, solve the equation.

9. $x^2 = 9$ ± 3
10. $h^2 = 25$ ± 5
11. $6x^2 = 600$ ± 10
12. $\frac{1}{5}x^2 = 5$ ± 5
13. $3x^2 = 363$ ± 11
14. $2b^2 = 98$ ± 7
15. $t^2 + 2 = 11$ ± 3
16. $t^2 - 57 = 87$ 16. ± 12
17. $\frac{1}{2}x^2 - 1 = 7$ ± 4
18. $4y^2 + 7 = 8$ $\pm\frac{1}{2}$
19. $2s^2 - 5 = 27$ ± 4
20. $81x^2 - 5 = 20$ $\pm\frac{5}{9}$

Technology **In Exercises 21–28, use a calculator to solve the equation. Round the results to two decimal places.**

21. $3x^2 + 2 = 56$ ± 4.24
22. $7y^2 - 12 = 23$ ± 2.24
23. $2x^2 - 5 = 7$ ± 2.45
24. $\frac{2}{3}n^2 - 6 = 2$ ± 3.46
25. $\frac{1}{2}x^2 + 3 = 8$ ± 3.16
26. $4x^2 + 9 = 41$ ± 2.83
27. $6s^2 - 2 = 0$ ± 0.58
28. $5a^2 + 10 = 20$ ± 1.41

In Exercises 29–32, an object is dropped from a height h. How long does it take to reach the ground? (Assume there is no air resistance.)

29. $h = 64$ feet 2 seconds
30. $h = 144$ feet 3 seconds
31. $h = 500$ feet 31. ≈ 5.59 seconds
32. $h = 600$ feet 32. ≈ 6.12 seconds

✪ 33. *Geometry* The surface area of a cube is 150 square feet. Find the length of each edge. 5 ft

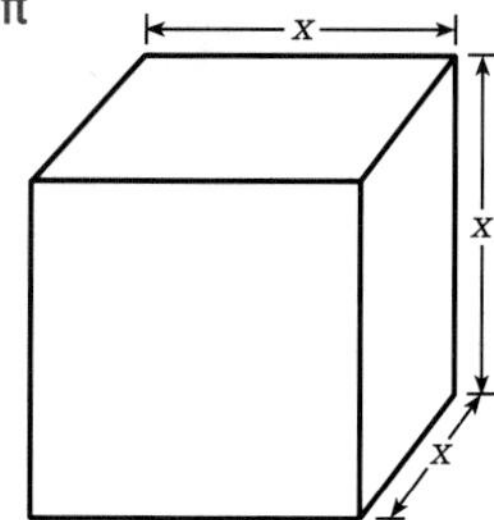

✪ 34. *Geometry* The surface area of a sphere is 80 square meters. Find the radius. (Use $\pi \approx 3.14$.) ≈ 2.52 m

$S = 4\pi r^2$

a. Assume the initial velocity of the wrench dropped in Example 4 was 0. At what speed did the wrench hit the ground? 113.28 ft/s or 77.24 mph

b. At what speed would the wrench hit the ground if the initial velocity of the wrench were 10 ft/s, now taking the wrench 3.23 seconds to hit the ground? 113.36 ft/s or 77.29 mph

EXERCISE Notes

ASSIGNMENT GUIDE

Basic/Average: Ex. 1–8, 9–25 odd, 33, 35–36, 39–40, 41–51 odd

Above Average: Ex. 1–8, 9–33 odd, 34–38, 39–49 odd, 51–52

Advanced: Ex.1–8, 17–33 odd, 34–38, 39–49 odd, 51–54

Selected Answers
Exercises 1–8, 9–49 odd

Use **Mixed Review** as needed.

✪ More Difficult Exercises
Exercises 33–34, 37–38, 53–54

Guided Practice

▶ **Ex. 1–8** These exercises help establish essential basic skills necessary for success in the lesson.

▶ **Ex. 2–4** There is more than one standard form for these exercises. Help students decide if the form they write is equivalent to another student's.

Independent Practice

▶ **Ex. 9–20** Refer students to Example 2 if they have difficulties with these exercises.

Ex. 21–28 Remind students to include both a positive and a negative answer.

Ex. 29–32 Refer students to the formula $h = -16t^2 + s$ as defined on page 460.

EXTEND Ex. 35 Ask students to use the model to predict the number of free school lunches in the year 2000 when $t = 18$. $L = -651{,}000{,}000$. The result is negative, indicating the model is valid only for the years 1970 through 1990.

Ex. 37–38 SCIENCE Share these exercises with the science teachers at your school.

Ex. 39–40 These exercises should be assigned as a pair. Remind students that s is the height from which the object was dropped.

EXTEND Ex. 39 What is the ratio of the gravity on the moon to the gravity on earth? about 0.16875

Answer

40. ≈ 0.56; no, air resistance would cause the feather to float to Earth instead of falling.

Ex. 41–44 These exercises are a good review of skills necessary for success in Lesson 9.4.

Answer

51. 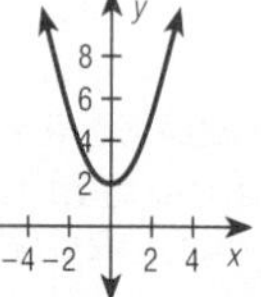

Ex. 53–54 First, students should find the choices for x.

35. *Free Lunches* The number, L (in millions), of free lunches served in public schools in the United States from 1970 to 1990 can be modeled by $L = -7.75t^2 + 1860$, where t is the year, with $t = 0$ corresponding to 1982. During which years were 1736 million free lunches served? During which years were 1364 million free lunches served?
(Source: U.S. Dept. of Agriculture)
1978 and 1986, 1974 and 1990

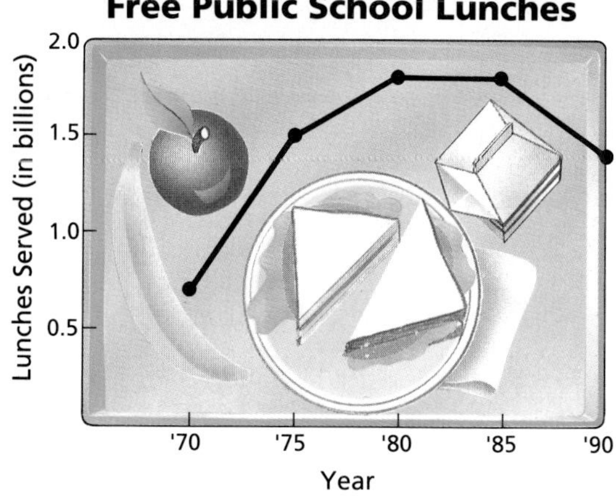

36. *Motion Picture Theaters* The number, M (in thousands), of motion picture theaters in the United States from 1970 to 1990 can be modeled by $M = 0.037t^2 + 10$, where t is the year, with $t = 0$ corresponding to 1980. Use this model to estimate the year in which there will be 19.5 thousand motion picture theaters in the United States. 1996

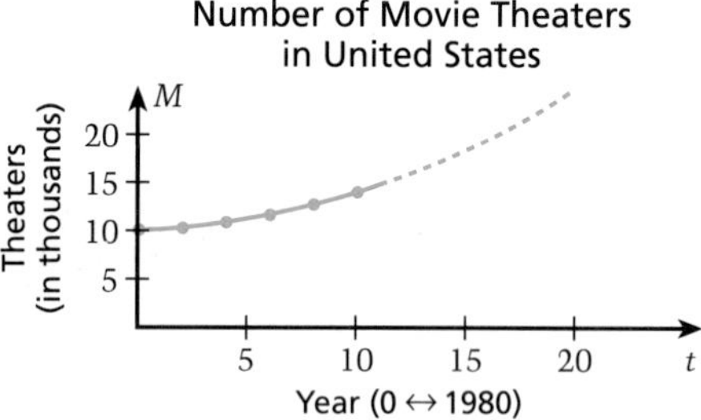

Hardness of Minerals **In Exercises 37 and 38, use the following information.**

Geologists use the Vickers scale to measure the hardness of minerals. The hardness, V, of a mineral can be determined by hitting the mineral with a wedge-shaped diamond and measuring the depth, d, of the indentation. A model that relates mineral hardness with the indentation depth in millimeters is $Vd^2 = 1.89$.

Fluorite: $V = 180$

Sulfur: $V = 40$

Wulfenite: $V = 85$

37. Find the depth of the indentation of each of the minerals shown above. 0.10 mm, 0.22 mm, 0.15 mm

38. Which mineral shown above is the hardest? Which is the softest? Why do you think the Vickers scale test uses a diamond to test the hardness of other minerals? Fluorite; sulfur; diamond is the hardest mineral known.

The Hammer and the Feather **In Exercises 39 and 40, use Example 4 on page 460 and the following information.**

In 1971, astronaut David Scott demonstrated that a feather and a hammer fall at the same rate on the moon because the moon has no atmosphere (and hence no air resistance). The height, h (in feet), of a falling object on the moon is given by

$$h = -\frac{27}{10}t^2 + s.$$

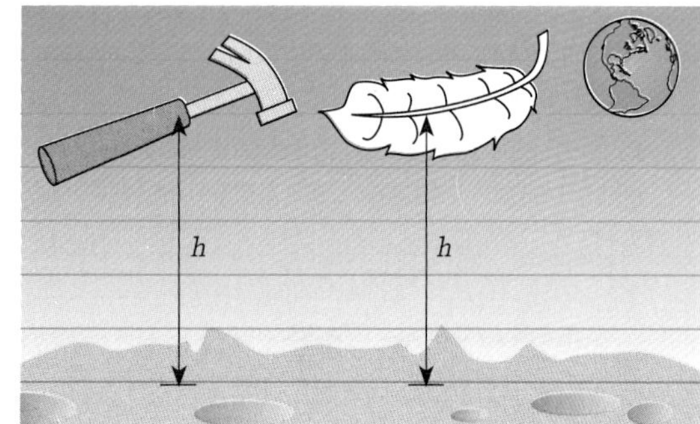

39. If a hammer and feather are dropped 5 feet from the surface of the moon, how long will it take for each to hit the surface? $\approx$1.36 seconds

40. On Earth, how long would it take the hammer to hit the ground? Would the feather hit the ground after the same amount of time? Explain. See margin.

Integrated Review

In Exercises 41–44, evaluate the expression.

41. $4x^2 + 9x - 14$ when $x = 3$ 49

42. $x^2 - 4x + 9$ when $x = 4$ 9

43. $-4ac$ when $a = 5$ and $c = 2$ -40

44. $-16t^2 + s$ when $s = 4$ and $t = 4$ -252

In Exercises 45–50, solve for x.

45. $2x - 9 = 7$ 8

46. $5x - 9 = 11$ 4

47. $\frac{1}{4}x + 6 = 8$ 8

48. $\frac{1}{3}x - 4 = -1$ 9

49. $\sqrt{x} - 1 = 5$ 36

50. $\sqrt{x} + 3 = 4$ 1

Exploration and Extension

51. Complete the table. Use the points (x, y) from the table to sketch the graph of $y = x^2 + 2$. See margin for graph.

11, 6, 3, 2, 3, 6, 11

x	−3	−2	−1	0	1	2	3
y	?	?	?	?	?	?	?

52. The points $(x_1, 7)$ and $(x_2, 10)$ are on the graph of the equation given in Exercise 51. What are the x-coordinates? $\pm\approx$2.24, $\pm\approx$2.83

In Exercises 53 and 54, there are two different ordered pairs (x, y) that are solutions of the system. Find the two ordered pairs.

53. $\begin{cases} -5 + 2x^2 = 27 \\ x + y = 4 \end{cases}$ (4, 0), (−4, 8)

54. $\begin{cases} \frac{1}{3}x^2 + 2 = 5 \\ 2x + y = 4 \end{cases}$ (3, −2), (−3, 10)

Enrichment Activities

These activities require students to go beyond lesson goals.

EXTEND Ex. 36 TECHNOLOGY Use a graphics calculator to graph the model $M = 0.037t^2 + 10$. Use the graph and the TRACE feature to answer the following questions.

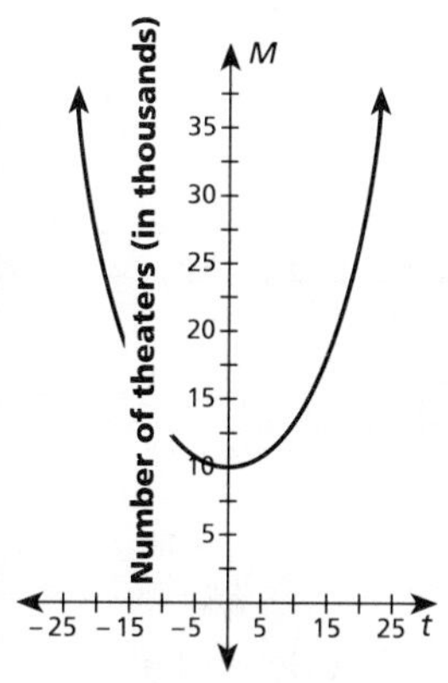

Year (0↔1980)

1. In what year did 15,500 theaters exist? sometime in 1992
2. How many theaters existed in 1970 ($t = -10$)? Explain your results. Is your answer reasonable? The model indicates the same number as for 1990: 13,700. No, the model is probably invalid for the years preceding 1980.
3. How many theaters are predicted for the year 2000? 24,800
4. WRITING Can you explain why the number of theaters increased after 1980? Popularity of movies, building of theater complexes at malls to attract more patrons, etc.

Mixed REVIEW

3. $p = \frac{2 + q}{1 + q}$ 4. $y = \frac{3}{4}x + 5$ 15. $2x^2 + 4x$

1. Write the reciprocal of a^4. **(2.7)** $\frac{1}{a^4}$
2. Write the reciprocal of 3. **(2.7)** $\frac{1}{3}$
3. Solve $p(1 + q) = 2 + q$ for p. **(3.6)**
4. Solve $3x - 4y = -20$ for y. **(3.6)**
5. Evaluate $3x^3 + 2x - 2$ when $x = -2$. **(1.3)** -30
6. Evaluate $3z^0 + 4z^{-1}$ when $z = \frac{1}{4}$. **(8.2)** 19
7. What is the slope of $y = 5x + 6$? **(4.4)** 5
8. What is the slope of $3y + 2 = 0$? **(4.4)** 0
9. Write 1,299,000,000 in scientific notation. **(8.4)** 1.299×10^9
10. Write 0.000000496 in scientific notation. **(8.4)** 4.96×10^{-7}
11. Is 2 a solution of $1 + 2x < 4x$? **(1.5)** Yes
12. Is 6 a solution of $3x - 6 = 10$? **(1.5)** No
13. Write $10x - 4y = -16$ in slope-intercept form. **(4.5)** $y = \frac{5}{2}x + 4$
14. Write $y = \frac{1}{3}x - 2$ in standard form. **(5.5)** $x - 3y = 6$
15. Simplify $3x(1 + x) - x^2 + x$. **(2.6)**
16. Simplify $(x^2y^{-3})^{-2} \cdot x^2y^{-4}$. **(8.1)** $\frac{y^2}{x^2}$
17. Solve $|x - 3| = 6$. **(4.8)** $-3, 9$
18. Solve $|2x + 4| \le 12$. **(4.8)** $-8 \le x \le 4$
19. Evaluate $2.2 \times 10^{-6} \cdot 3.6 \times 10^4$ **(8.4)**
20. Evaluate $7.6 \times 10^{-6} \div (2.5 \times 10^8)$. **(8.4)**

19. 7.92×10^{-2}
20. 3.04×10^{-14}

Career Interview

Printer

Jeffrey Wong is the printer at and manager of a printing company. He is involved in all aspects of running the business—from purchasing, to sales, to managing employees. He quotes estimates, sets up business agreements with customers, and processes orders.

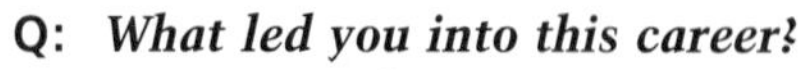

Q: ***What led you into this career?***
A: I was trained as an engineer. After working in that field for a number of years, I decided to take over the family printing business.

Q: ***Has new technology changed your job experiences?***
A: Many new printing machines and equipment have come on the market. Most have built-in functions that can take care of some computations we used to do by hand.

Q: ***Does this mean you don't need to understand math?***
A: No. All the functions on a new printer don't mean a thing if I don't know how they can be used, or how they impact on what's happening during a particular process.

Q: ***What would you like to tell kids about math?***
A: Often kids think that if they are good in English, then they don't need to worry about math, and vice versa. But we really need to do well in both. English is the vehicle to express your thoughts and feelings; math is the key to meeting everyday needs.

9.3 Graphs of Quadratic Equations

What you should learn:

Goal 1 How to sketch the graph of a quadratic equation

Goal 2 How to use quadratic models in real-life settings

Why you should learn it:

You can use visual models to help you see relationships between two variables, such as height reached and the distance traveled by a projectile.

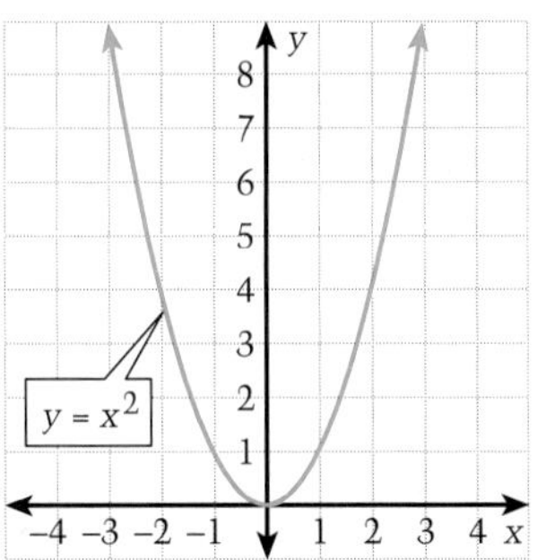

Goal 1 Sketching a Quadratic Equation

A **quadratic equation** in x and y is an equation that can be written in the **standard form**

$$y = ax^2 + bx + c, \qquad a \neq 0.$$

Every quadratic equation has a ∪-shaped graph called a **parabola.** For instance, the graph of $y = x^2$ is shown at the left.

On a parabola that opens up, the lowest point is the **vertex.** On a parabola that opens down, the highest point is the vertex. The vertical line passing through the vertex is the **axis of symmetry.** This line divides a parabola into two parts that are mirror images of each other.

LESSON INVESTIGATION

Investigating Graphs of Quadratic Equations

Partner Activity Use a graphing calculator to sketch several parabolas whose graphs are of the form

$$y = ax^2 + bx + c, \qquad a \neq 0.$$

Use a standard viewing rectangle and choose values of a, b, and c that lie between -2 and 2. With your partner, discuss which coefficient determines whether the parabola opens up or down. Does c affect the x-coordinate of the vertex?

Study Tip

To sketch a parabola by hand, first find the x-coordinate of the vertex. Then construct a table of values, using points to the left and right of the vertex. Finally, plot the points and connect them with a ∪-shaped graph.

Graph of a Quadratic Equation

The graph of $y = ax^2 + bx + c$ is a parabola

1. If a is positive, the parabola opens up.
2. If a is negative, the parabola opens down.
3. The vertex has an x-coordinate of $-\frac{b}{2a}$.
4. The axis of symmetry is the vertical line $x = -\frac{b}{2a}$.

Problem of the Day

Using only mathematical signs and without changing the positions of any of the figures, make 296 7 17 a valid equation:

$\sqrt{296 - 7} = 17$

ORGANIZER

Warm-Up Exercises

1. Evaluate each expression for $x = -3, -2, -1, 0, 1, 2,$ and 3. What do you notice?
   a. $x^2 + 3$ **b.** $-2x^2$
   a. 12, 7, 4, 3, 4, 7, 12
   b. $-18, -8, -2, 0, -2, -8, -18$
   The values are the same for x and for $-x$.
2. Evaluate the expression $-\frac{b}{2a}$ for values of a and b.
   a. $a = 3, b = -4$ $\quad\frac{2}{3}$
   b. $a = -1, b = 6$ $\quad 3$
   c. $a = 8, b = 5$ $\quad -\frac{5}{16}$

Lesson Resources

Teaching Tools
 Transparencies: 1, 2, 4, 13
 Problem of the Day: 9.3
 Warm-up Exercises: 9.3
 Answer Masters: 9.3
Extra Practice: 9.3
Color Transparencies: 45, 46
Applications Handbook: p. 46–48
Technology Handbook: p. 65

Lesson Investigation Answers

a determines if the parabola opens up or down. The coefficients are a and b.

Common-Error ALERT!

The coefficients of x^2 and x, and the constant term c, should be determined only after a given quadratic equation has been transformed into its standard form.

LESSON Notes

GOAL 1 The equation $y = ax^2 + bx + c$ in the two variables x and y may also be viewed as a function in x.

Examples 1–2

Observe with students that the y-values on either side of the y-coordinate of the vertex form a pattern. The x-intercepts of each graph can be identified in the table.

Example 3

Note that this may be a difficult example for students to follow because of the numbers involved in the model. Point out that the graph's x- and y-axes have different scales. You may also wish to observe that finding the x-value where the height becomes 0 is the same as solving the quadratic equation, $-0.01464x^2 + x + 5 = 0$.

Extra Examples

Here are additional examples similar to **Examples 1–3.**

1. Use a table to sketch each graph. Label the vertex and the x- and y-intercepts.
   a. $y = x^2 - x - 6$
   b. $y = 2x^2 + 2x - 40$
   c. $y = 2x^2 - 3x - 2$

   a. vertex: $(0.5, -6.25)$ intercepts: $(-2, 0)$, $(3, 0)$, $(0, -6)$
   b. vertex: $(-0.5, -40.5)$ intercepts: $(-5, 0)$, $(4, 0)$, $(0, -40)$
   c. vertex: $(0.75, -3.125)$ intercepts: $(-0.5, 0)$, $(2, 0)$, $(0, -2)$

   Check students' graphs.
2. In Example 3, what is the maximum height of the winning shot thrown if its path is given by the model $y = -0.0155x^2 + x + 5$, where x and y are measured in feet? The maximum, 21.13 ft, occurs at the vertex: (32.26, 21.13).

Example 1 *Sketching a Quadratic Equation*

Sketch the graph of $y = x^2 - 2x - 3$.

Solution In standard form $ax^2 + bx + c$, the coefficients are $a = 1$, $b = -2$, and $c = -3$. The x-coordinate of the vertex is

$$-\frac{b}{2a} = -\frac{-2}{2(1)} = 1. \quad \textit{x-coordinate of vertex}$$

Using x-values to the left and right of this x-value, construct a table of values.

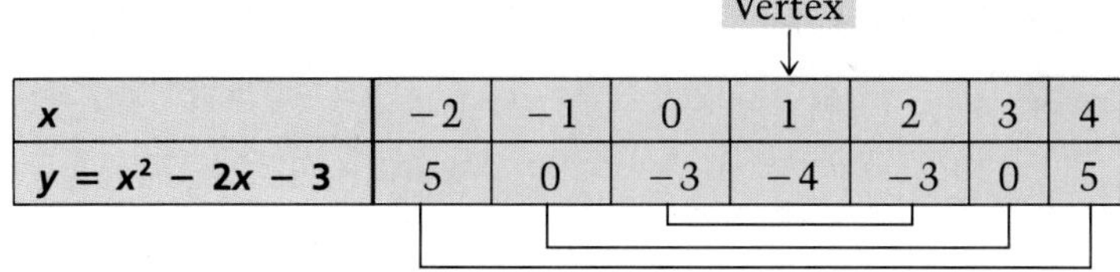

x	−2	−1	0	1 (Vertex)	2	3	4
$y = x^2 - 2x - 3$	5	0	−3	−4	−3	0	5

The vertex is $(1, -4)$ and the axis of symmetry is $x = 1$. Plot the points given in the table and connect them with a U-shaped curve that opens up. ■

In Example 1, notice that the graph has two x-intercepts, at $(-1, 0)$ and $(3, 0)$, and one y-intercept, at $(0, -3)$.

Example 2 *Sketching a Quadratic Equation*

Sketch the graph of $y = -\frac{1}{2}x^2 - x + 4$.

Solution For this quadratic equation, $a = -\frac{1}{2}$, $b = -1$, and $c = 4$. The x-coordinate of the vertex is

$$-\frac{b}{2a} = -\frac{-1}{2\left(-\frac{1}{2}\right)} = -1. \quad \textit{x-coordinate of vertex}$$

Using x-values to the left and right of this x-value, construct a table of values.

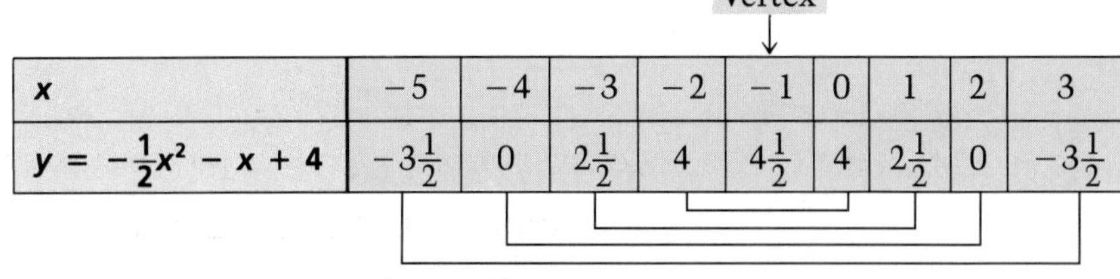

x	−5	−4	−3	−2	−1 (Vertex)	0	1	2	3
$y = -\frac{1}{2}x^2 - x + 4$	$-3\frac{1}{2}$	0	$2\frac{1}{2}$	4	$4\frac{1}{2}$	4	$2\frac{1}{2}$	0	$-3\frac{1}{2}$

The vertex is $\left(-1, 4\frac{1}{2}\right)$, and the axis of symmetry is $x = -1$. Plot the points given in the table and connect them with a U-shaped curve that opens down. ■

The graph in Example 2 has x-intercepts at $(-4, 0)$ and $(2, 0)$. The y-intercept occurs at $(0, 4)$.

Goal 2 Using Quadratic Models in Real Life

Natalya Lisovskaya of the Soviet Union won the shot put competition in the 1988 Olympics. The path of the winning throw is given by $y = -0.01464x^2 + x + 5$.

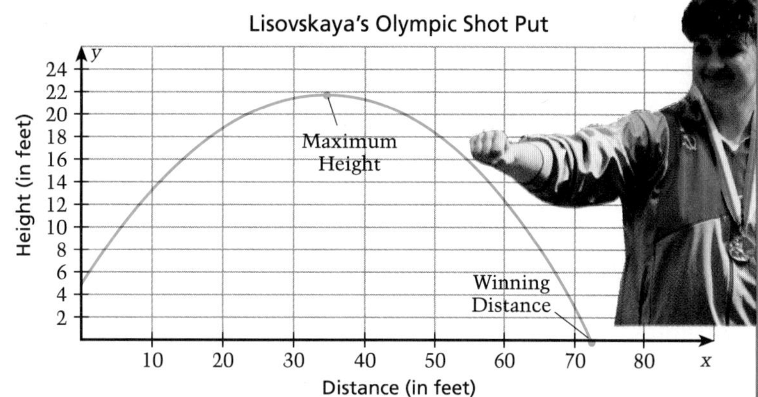

Real Life Sports

Example 3 *Length and Height of a Shot Put Throw*

What was the maximum height of the shot thrown by Lisovskaya? How long was her winning throw?

Solution The maximum height occurred at the vertex of the parabolic path. The x-coordinate of the vertex is

$$-\frac{b}{2a} = -\frac{1}{2(-0.01464)} \approx 34.15. \quad \textit{x-coordinate of vertex}$$

To find the height of the shot put at the vertex, substitute this x-value into the equation $y = -0.01464x^2 + x + 5$.

$$y = -0.01464(34.15)^2 + (34.15) + 5 \approx 22.1$$

The maximum height was about 22.1 feet. To find the length of the winning throw, you need to find the x-value that yields a y-value of 0. From the table, you can see that the shot hit the ground when x was a little less than 73 feet.

Distance, x	40 ft	50 ft	60 ft	70 ft	71 ft	72 ft	73 ft
Height, y	21.6 ft	18.4 ft	12.3 ft	3.3 ft	2.2 ft	1.1 ft	−0.02 ft

Communicating about ALGEBRA

SHARING IDEAS about the Lesson
See Additional Answers for graphs.

Use a Diagram Sketch the graphs of the quadratic equations on the same coordinate plane and describe the relationship among the graphs. See below.

A. $y = x^2 - 1$ $\quad y = x^2$ $\quad y = x^2 + 1$

B. $y = -x^2 - 4$ $\quad y = -x^2$ $\quad y = -x^2 + 4$

A. The axis of symmetry, $x = 0$, is the same; the graphs all open upward

B. The axis of symmetry, $x = 0$, is the same; the graphs all open downward

Check Understanding

1. How can the axis of symmetry and a table of y-values on the left of the vertex of a quadratic equation in x and y be used to complete the table and sketch the graph? The graph is symmetric about the axis of symmetry, so the y-values are the same on both sides of the axis of symmetry.
2. How may the x-intercepts of a quadratic equation in x and y be determined? Let $y = 0$ and solve the resulting equation.
3. Does every quadratic equation in x and y have a y-intercept? Yes, c Any x-intercepts? Explain. No, not all parabolas intersect the x-axis

Communicating about ALGEBRA

Ask students to make a conjecture about the effects of constants on the basic quadratic equation in two variables, $y = x^2$. Students should create examples to help them devise and test their conjectures. (Hint: Consider such equations as

1. $y = x^2 \pm k$, k real-valued,
2. $y = (x \pm h)^2$, h real-valued,
3. $y = ax^2$, $a \neq 0$, a real-valued,
4. combinations of 1–3.)

Given the general equation $y = a(x \pm h)^2 \pm k$, the value k translates the graph $y = x^2$ $|k|$ units up or down the y-axis; the value h translates the graph $|h|$ units left or right along the x-axis. The value a makes the graph wider if $|a| < 1$ and narrower if $|a| > 1$. If a is negative, the graph opens downward instead of upward.

EXERCISE Notes

ASSIGNMENT GUIDE

Basic/Average: Ex. 1–6, 7–33 odd, 41–45, 51 56

Above Average: Ex. 1–6, 11–29 odd, 39–45, 47, 51–56, 57–61

Advanced: Ex. 1–6, 11–29 odd, 37–48, 51–56, 61, 63–65

Selected Answers
Exercises 1–6, 7–59 odd

✪ More Difficult Exercises
Exercises 33–39, 42, 61–63

Guided Practice

▶ Ex. 1–6
It is suggested that you use these exercises as an in-class activity. Students' understanding of these exercises is critical for their success in the independent practice exercises.

Independent Practice

▶ Ex. 9–12 Remind students to solve these equations for y before determining the value of a in $y = ax^2 + bx + c$.

▶ Ex. 13–18 Remind students that the axis of symmetry is a vertical line whose equation is $x = -\frac{b}{2a}$.

EXTEND Ex. 13–18 Some students might be interested in a way in which to develop the axis of symmetry formula $x = -\frac{b}{2a}$. See Enrichment Activities, pages 470–471.

EXERCISES

Guided Practice

CRITICAL THINKING about the Lesson 4. True; $x = -\frac{b}{2a}$ is a vertical line

1. Write the equation $y = -3 + 4x - x^2$ in standard form. $y = -x^2 + 4x - 3$

2. The graph of a quadratic equation is called a [?]. parabola

3. How can you use a to decide whether the graph of $y = ax^2 + bx + c$ opens up or down? If $a > 0$, graph opens up; if $a < 0$, graph opens down.

4. True or False? The axis of symmetry of the graph of $y = ax^2 + bx + c$ is parallel to the y-axis (or *is* the y-axis). Explain.

5. Find the vertex of the graph of $y = 2x^2 + 4x - 2$. $(-1, -4)$

6. Find the axis of symmetry of the graph of $y = -3x^2 + 3x + 1$. $x = \frac{1}{2}$

Independent Practice

In Exercises 7–12, decide whether the graph of the equation opens up or down. Then find the coordinates of the vertex.

7. $y = 2x^2 + 4$ Up, $(0, 4)$
8. $y = -5x^2$ Down, $(0, 0)$
9. $y + 4x^2 = 0$ Down, $(0, 0)$
10. $y - 3x^2 = -2x$ Up, $(\frac{1}{3}, -\frac{1}{3})$
11. $y + 5x^2 = -x + 10$ Down, $(-\frac{1}{10}, \frac{201}{20})$
12. $y - 6x^2 = 3x + 12$ Up, $(-\frac{1}{4}, \frac{93}{8})$

In Exercises 13–18, find the coordinates of the vertex and the equation of the axis of symmetry. See below.

13. $y = 3x^2 + 2x + 4$
14. $y = 2x^2 + 3x + 6$
15. $y = -4x^2 - 4x + 8$
16. $y = 3x^2 - 9x - 12$
17. $y = 2x^2 + 7x - 21$
18. $y = -x^2 + 4x + 16$

In Exercises 19–36, sketch the graph of the equation. Label the vertex. See Additional Answers.

19. $y = x^2 + x + 2$
20. $y = -x^2 + 2x - 1$
21. $y = -2x^2 + 6x - 9$
22. $y = 2x^2 - 3x + 4$
23. $y = 6x^2 - 3x + 4$
24. $y = 5x^2 + 4x - 5$
25. $y = 4x^2 - x + 6$
26. $y = -3x^2 - x + 7$
27. $y = -5x^2 + 2x - 2$
28. $y = 6x^2 - 4x - 1$
29. $y = -3x^2 - 5x + 3$
30. $y = -2x^2 - 3x + 2$
31. $y = x^2 + 6x + 5$
32. $y = -4x^2 - 3x + 6$
✪ **33.** $y = -\frac{1}{2}x^2 - 3x + 4$
✪ **34.** $y = \frac{1}{3}x^2 + 3x - 2$
✪ **35.** $y = -2x^2 + \frac{1}{3}x - 1$
✪ **36.** $y = 3x^2 - \frac{1}{2}x + 4$

You've Got to Have the Right Angle **In Exercise 37, use the information given in Example 3.**

✪ **37.** Natalya Lisovskaya's winning throw in the shot put was at a 45° angle. If the shot had been thrown at a 40° angle or 50° angle, would it have gone farther? Explain. See margin.

Throw at 40° angle: $y = -0.0125x^2 + 0.84x + 5$
Throw at 50° angle: $y = -0.0177x^2 + 1.19x + 5$

13. $x = -\frac{1}{3}$; $(-\frac{1}{3}, \frac{11}{3})$ **14.** $x = -\frac{3}{4}$; $(-\frac{3}{4}, \frac{39}{8})$ **15.** $x = -\frac{1}{2}$; $(-\frac{1}{2}, 9)$
16. $x = \frac{3}{2}$; $(\frac{3}{2}, -\frac{75}{4})$ **17.** $x = -\frac{7}{4}$; $(-\frac{7}{4}, -\frac{217}{8})$ **18.** $x = 2$; $(2, 20)$

Water Sprinkler **In Exercises 38 and 39, use the following information.**

The paths of the water from a water sprinkler are shown for three different settings.

$35°: y = -0.06x^2 + 0.70x + 0.5$
$60°: y = -0.16x^2 + 1.73x + 0.5$
$75°: y = -0.60x^2 + 3.73x + 0.5$

Radii: ≈12 ft, ≈ 11 ft, ≈ 6 ft; heights: ≈2.5 ft, ≈5.2 ft, ≈6.3 ft

x and y are measured in feet.

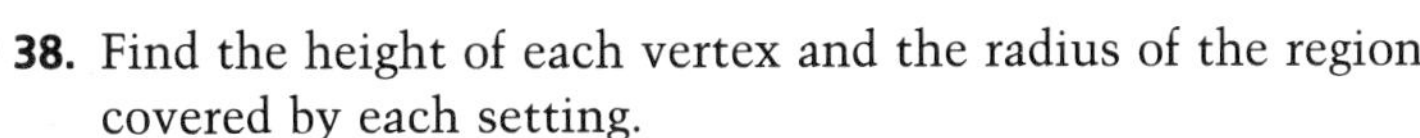
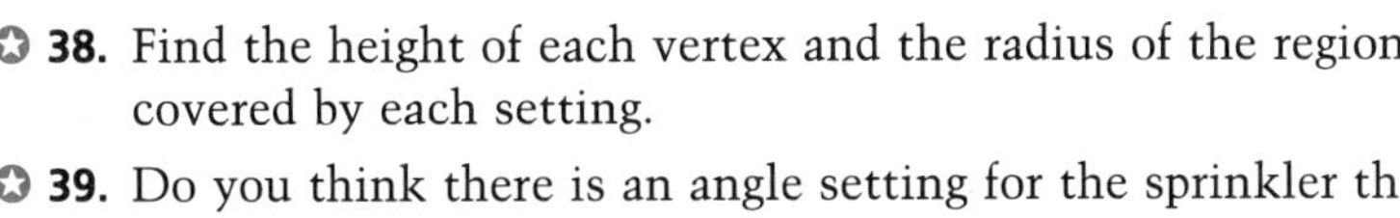
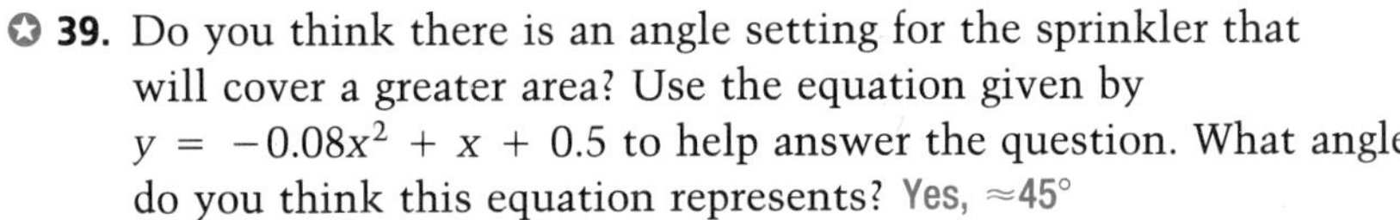

38. Find the height of each vertex and the radius of the region covered by each setting.

39. Do you think there is an angle setting for the sprinkler that will cover a greater area? Use the equation given by $y = -0.08x^2 + x + 0.5$ to help answer the question. What angle do you think this equation represents? Yes, ≈45°

World Production of Gold **In Exercises 40 and 41, use the following information.**

From 1970 to 1990, the annual world production, G, of gold in thousands of ounces, can be modeled by

$$G = 47{,}974 - 2446t + 167t^2$$

where t is the year, with $t = 0$ corresponding to 1970. The graph of this model is shown at the right. *(Source: U.S. Bureau of Mines)*

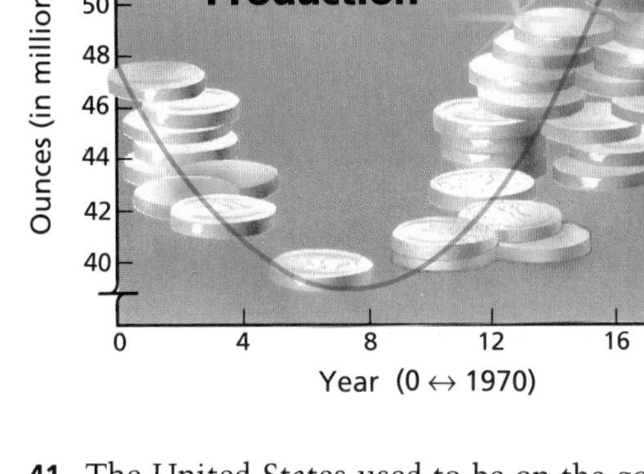

40. During which years between 1970 and 1990 was the world production of gold decreasing? During which years was the production increasing? How are these questions related to the vertex of the graph? 1970–1977; 1978–1990; at the vertex, decrease changes to an increase.

41. The United States used to be on the *gold standard,* which had the effect of keeping the price of gold artificially low. In 1971, the United States went off the gold standard. How did this affect the world production of gold? It decreased gold production.

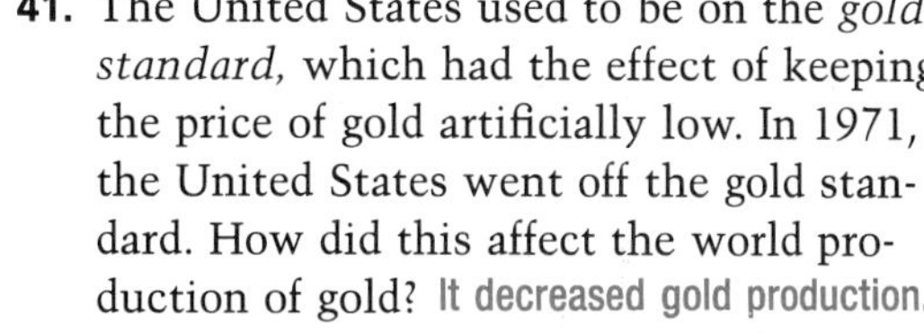

In Exercise 42, use the following information.

From 1950 to 1990, the average annual consumption, C, of cigarettes per American (18 or older) can be modeled by

$$C = 4024.5 + 51.4t - 3.1t^2$$

where t is the year, with $t = 0$ corresponding to 1960. The graph of the model is shown at the right. *(Source: U.S. Centers for Disease Control)*

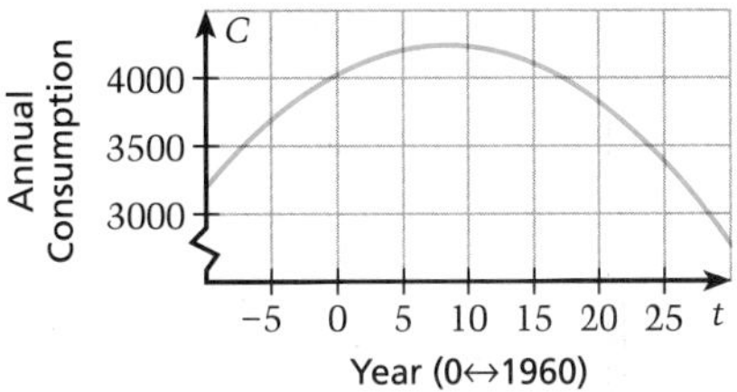

42. From 1966 on, all cigarette packages were required by law to carry a health warning. Did the warnings have any effect? Explain. Answers will vary. Yes, a decrease in consumption began in 1968.

Answers

37. No. At 72 ft, the 50° shot put is on the ground while the 45° shot put is in the air; at 72.8 ft the 40° shot put is on the ground while the 45° shot put is in the air.

▶ **Ex. 38–39, and 40–41** These exercises should be assigned in pairs.

▶ **Ex. 39** COOPERATIVE LEARNING If students need extra help with this problem, suggest they try to solve it in a partner setting. Having students help each other is beneficial for both partners.

▶ **Ex. 40–41** Note that the vertex is not at an integer year. It occurs at $t \approx 7.3$. Point out to students that the graph shows G in millions while the model gives G in thousands.

Ex. 49 and 53, 50 and 51, 52 and 54 These exercises are paired. Have students consider the graphs and note how they are alike and how they are different. For example, the paired graphs have the same y-intercepts and are reflections (mirror images) of each other.

Ex. 61–65 TECHNOLOGY Graphics calculators or computer graphing software would be very helpful for sketching these graphs.

Enrichment Activities

These activities require students to go beyond lesson goals.

1. **EXTEND** **Ex. 61–65** Graph the equations $y = x^2 + bx + 5$ when $b = -3, -2, -1, 0, 1, 2,$ and 3. As b increases in value, what happens to the parabola and its axis of symmetry? The graph's axis of symmetry shifts to the left. (Axes: $x = \frac{3}{2}$, $x = 1$, $x = \frac{1}{2}$, $x = 0$, $x = -\frac{1}{2}$, $x = -1$, $x = -\frac{3}{2}$.) The vertex of the graph shifts to the left upward, and then downward. (Vertices: $(\frac{3}{2}, \frac{11}{4})$, $(1, 4)$, $(\frac{1}{2}, \frac{19}{4})$, $(0, 5)$, $(-\frac{1}{2}, \frac{19}{4})$, $(-1, 4)$, $(-\frac{3}{2}, \frac{11}{4})$)
2. Connect the vertices of the parabolas. What do you notice? Can you write a model that would define the path for all the vertices for the family of parabolas $y = x^2 + bx + 5$? They define the parabola $y = -x^2 + 5$.
3. **Challenge!** The axis of symmetry $x = x_1$ divides the parabola $y = ax^2 + bx + c$, $a \neq 0$, into two parts that are mirror images of each other. For any horizontal line that intersects the graph at two points, those points are the

Integrated Review

In Exercises 43–48, solve the equation.

43. $16 + t^2 = 32$ ± 4

44. $s^2 - 40 = -4$ ± 6

45. $2m^2 - 30 = 20$ ± 5

46. $24 = 4n^2 - 12$ ± 3

47. $12 + 3a^2 = 2a^2 + 16$ ± 2

48. $6a^2 = 3a^2 + 108$ ± 6

In Exercises 49–54, match each equation with its graph.

49. $y = x^2 - 2x + 3$ b

50. $y = 2x + 3$ c

51. $y = 3 - 2x$ f

52. $y = 2^x + 2$ e

53. $y = -x^2 - 2x + 3$ d

54. $y = 2^{-x} + 2$ a

a.

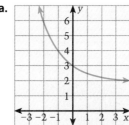

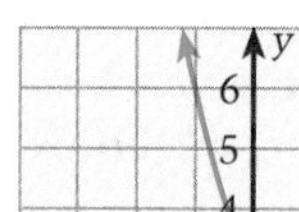

b.

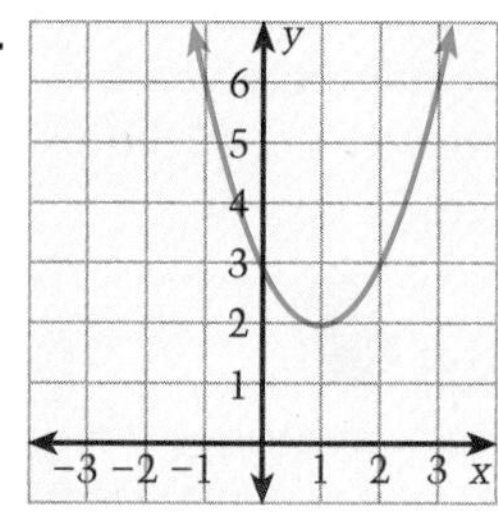

c.

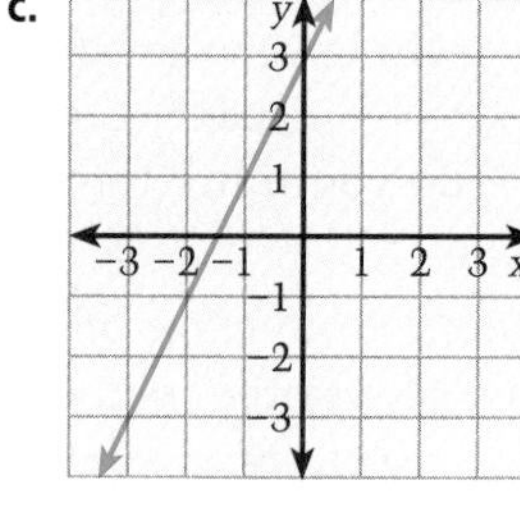

d.

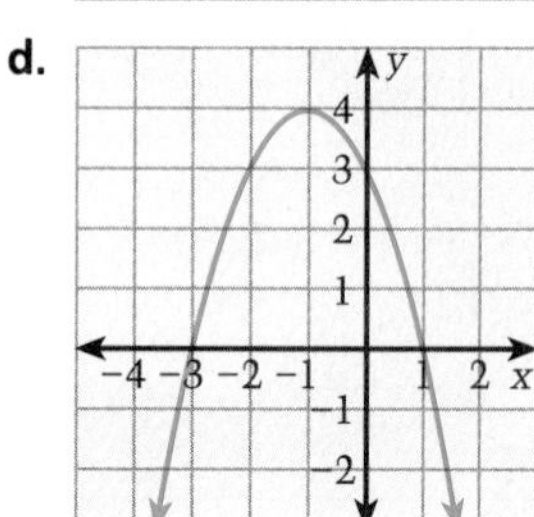

e.

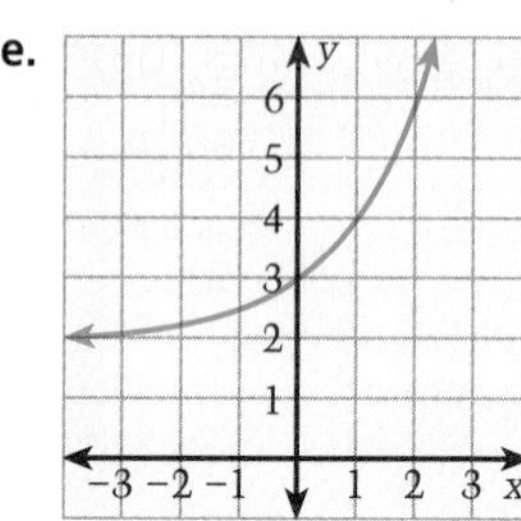

f.

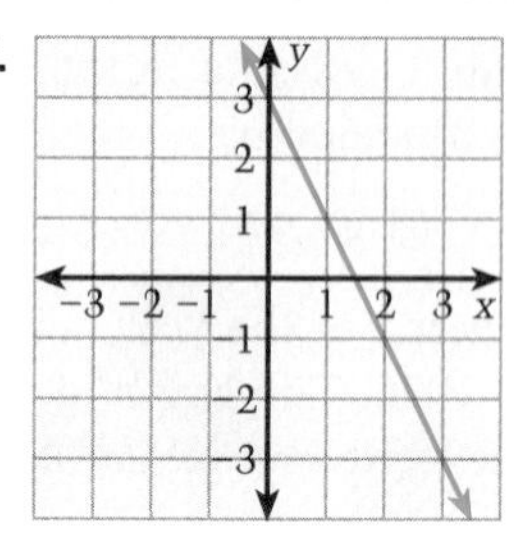

In Exercises 55–60, evaluate the expression. If necessary, use a calculator and round the result to two decimal places.

55. $-4ac$, $a = \frac{1}{2}$, $c = 12$ -24

56. $-4ac$, $a = -2$, $c = 7$ 56

57. $b^2 - 4ac$, $a = 6$, $b = 6$, $c = 1$ 12

58. $b^2 - 4ac$, $a = -5$, $b = 3$, $c = 1$ 29

59. $\sqrt{b^2 - 4ac}$, $a = 8$, $b = -10$, $c = 2$ 6

60. $\sqrt{b^2 - 4ac}$, $a = 6.2$, $b = 12.1$, $c = -4.3$ 15.91

Exploration and Extension

See Additional Answers for graphs.

In Exercises 61–63, sketch the graphs of all three equations on the same coordinate plane. Describe how the three graphs are related. See below.

61. $\begin{cases} y = x^2 - x + 1 \\ y = x^2 - x + 3 \\ y = x^2 - x - 2 \end{cases}$

62. $\begin{cases} y = x^2 + x + 1 \\ y = \frac{1}{2}x^2 + x + 1 \\ y = 2x^2 + x + 1 \end{cases}$

63. $\begin{cases} y = x^2 - 2x + 1 \\ y = x^2 - 4x + 4 \\ y = x^2 - 6x + 9 \end{cases}$

64. Use the result of Exercise 61 to describe how a change in the value of c changes the graph of $y = ax^2 + bx + c$.
The greater the value of c, the higher the graph.

65. Use the result of Exercise 62 to describe how a change in the value of a changes the graph of $y = ax^2 + bx + c$.
The greater the value of $|a|$, the narrower the graph.

61. Each graph has its vertex on $x = \frac{1}{2}$; each graph opens upward; the greater the constant term the higher up the graph is.

USING A GRAPHING CALCULATOR

You can use the zoom feature of a graphing calculator to approximate the solutions of an equation to any desired accuracy. The four graphs below show four steps in approximating the positive solution of

$x^2 - 3 = 0.$

Here, the approximating is $x \approx 1.732$. (The exact solution is $x = \sqrt{3}$.) By repeated zooming, you can obtain whatever accuracy you need.

1.

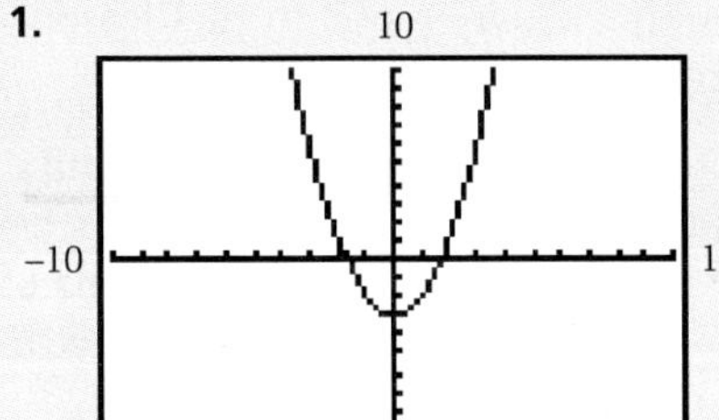

Sketch the graph of $y = x^2 - 3$.

2.

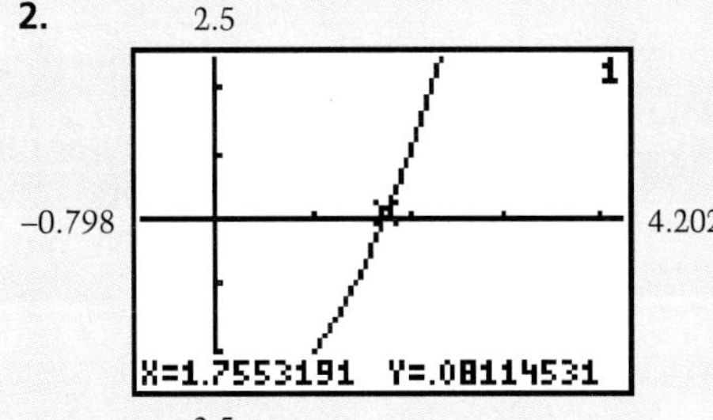

Zoom once to get a better view of the positive x-intercept. (The viewing rectangle will change automatically.)

3.

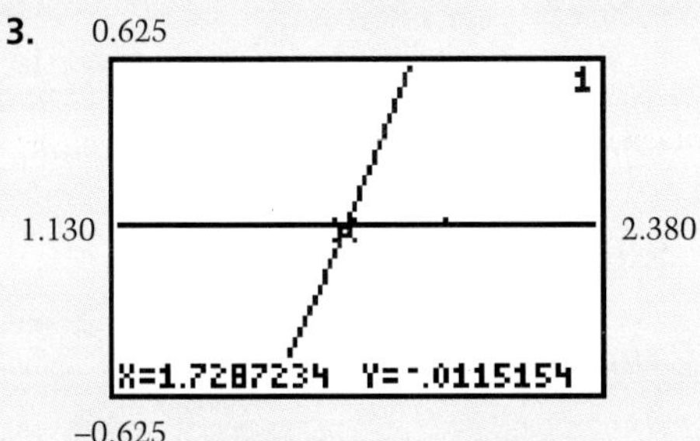

Zoom a second time to get an even better view. Use the cursor keys to determine that the x-intercept is about 1.73.

4.

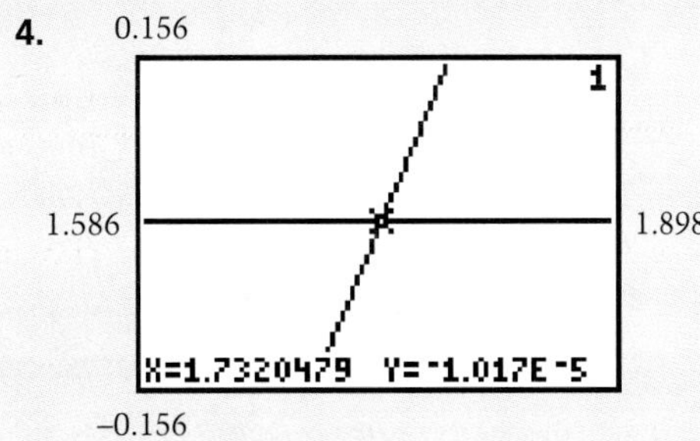

Zoom a third time. Use the trace feature to find the y-value that is closest to 0. The corresponding x-value, which is 1.732, is your approximation.

Exercises

Use a graphing calculator to approximate both solutions of the quadratic equation. Give each solution to two decimal places. 1. 0.26, 6.41 2. −0.77, 2.27 3. 0.21, 4.79

1. $3x^2 - 20x + 5 = 0$ **2.** $-4x^2 + 6x + 7 = 0$ **3.** $-x^2 + 5x - 1 = 0$

4. $0.1x^2 + 4.3x - 5.1 = 0$ 1.16, −44.16 **5.** $-1.4x^2 + 5.2x - 4.8 = 0$ 1.71, 2 **6.** $2.87x^2 - 9.43x - 4.53 = 0$ 3.71, −0.43

62. Each graph has a y-intercept of 1; each graph opens upward; the greater the leading coefficient, the narrower the graph is.

63. Each graph's vertex has a y-coordinate of 0; each graph opens upward

same distance d, $d > 0$, from the axis of symmetry. One point is $(x_1 - d, y_1)$. Its corresponding point across the axis of symmetry $x = x_1$ is $(x_1 + d, y_1)$. Use this information and the fact that $y = ax^2 + bx + c$ to show that the axis of symmetry is $x_1 = -\frac{b}{2a}$.

Sample answer:

Since $y = ax^2 + bx + c$,

$$y_1 = a(x_1 - d)^2 + b(x_1 - d) + c = a(x_1 + d)^2 + b(x_1 + d) + c.$$

Subtract c from both sides.

$$a(x_1 - d)^2 + b(x_1 - d) = a(x_1 + d)^2 + b(x_1 + d)$$

Combine similar terms on each side.

$$a(x_1 - d)^2 - a(x_1 + d)^2 = b(x_1 + d) - b(x_1 - d)$$

Expand.

$$a(x_1^2 - 2dx_1 + d^2) - a(x_1^2 + 2dx_1 + d^2) = bx_1 + bd - (bx_1 - bd)$$

Expand again.

$$ax_1^2 - 2adx_1 + ad^2 - ax_1^2 - 2adx_1 - ad^2 = bx_1 + bd - bx_1 + bd$$

Simplify.

$$-4adx_1 = 2bd$$

Since $a \neq 0$ and $d > 0$, you can divide both sides by $-4ad$.

$$x_1 = -\frac{b}{2a}$$

Therefore, the equation of the line of symmetry is $x = -\frac{b}{2a}$.

TECHNOLOGY Calculator technology allows you to solve quadratic equations to any desired degree of accuracy. Calculator functions such as "zoom" or "box," in addition to the window or range function, enable the user to get "closer" to any point on the graph. For instance, the vertex and the x-intercepts that correspond to the solutions of the quadratic equation, $ax^2 + bx + c = 0$, may be found.

Problem of the Day

A farmer has 600 plants arranged in rows so that each row has the same number of plants. He must take 5 plants from each row to make room for an irrigation ditch; but he can then make 6 more rows so that each row, both old and new, has the same number of plants. Find the original number of rows. 25

ORGANIZER

Warm-Up Exercises

1. Evaluate $b^2 - 4ac$ for the given values of a, b, and c.
   a. $a = 1, b = -6, c = 11$ -8
   b. $a = 1, b = -1, c = -6$ 25
   c. $a = 2, b = 2, c = -40$ 324
   d. $a = 2, b = -3, c = -2$ 25
2. Solve and check.
   a. $3x - 4 = 23$
   b. $-8 - 5x = 2x + 27$
   c. $2x^2 - 32 = 0$
   d. $81 - 9x^2 = 0$

   a. $x = 9$ b. $x = -5$
   c. $x = 4, x = -4$ d. $x = 3, x = -3$

Lesson Resources

Teaching Tools
 Transparencies: 4, 13
 Problem of the Day: 9.4
 Warm-up Exercises: 9.4
 Answer Masters: 9.4
Extra Practice: 9.4
Color Transparency: 46
Applications Handbook: p. 46–48
Technology Handbook: p. 68

9.4 The Quadratic Formula

What you should learn:

Goal 1 How to use the quadratic formula to solve a quadratic equation

Goal 2 How to use quadratic models in real-life settings

Why you should learn it:

You can model many situations, such as the motion of falling objects, with quadratic models.

Goal 1 Using the Quadratic Formula

In Lesson 9.2, you learned how to solve quadratic equations of the form $ax^2 + c = 0$ by finding square roots. In this lesson, you will learn how to solve *any* quadratic equation.

The formula for the solutions is called the **quadratic formula.** This is one of the oldest and most famous formulas in mathematics. No one knows who first discovered it, but versions of it were used as long ago as 2000 B.C. in Babylonia (now Iraq).

The Quadratic Formula

The solutions of the quadratic equation $ax^2 + bx + c = 0$ are

$$x = \frac{-b \pm \sqrt{b^2 - 4ac}}{2a}.$$

You may find it helpful to memorize this formula: *"The opposite of b, plus or minus the square root of b-squared minus 4ac, all divided by 2a."*

Example 1 *Using the Quadratic Formula*

Solve $x^2 + 5x - 14 = 0$.

Solution

$x^2 + 5x - 14 = 0$

$$x = \frac{-5 \pm \sqrt{5^2 - 4(1)(-14)}}{2(1)}$$ *$x = \frac{-b \pm \sqrt{b^2 - 4ac}}{2a}$; $a = 1, b = 5, c = -14$*

$$x = \frac{-5 \pm \sqrt{25 + 56}}{2}$$ *Simplify.*

$$x = \frac{-5 \pm \sqrt{81}}{2}$$ *Simplify.*

$$x = \frac{-5 \pm 9}{2}$$ *Solutions*

The equation has two solutions.

$$x = \frac{-5 + 9}{2} = 2 \quad \text{and} \quad x = \frac{-5 - 9}{2} = -7$$

Check these solutions in the original equation. ■

Remember that *before* you can apply the quadratic formula, you must first write the equation in standard form, $ax^2 + bx + c = 0$.

Example 2 *Writing in Standard Form*

Solve $2x^2 - 3x = 8$.

Solution

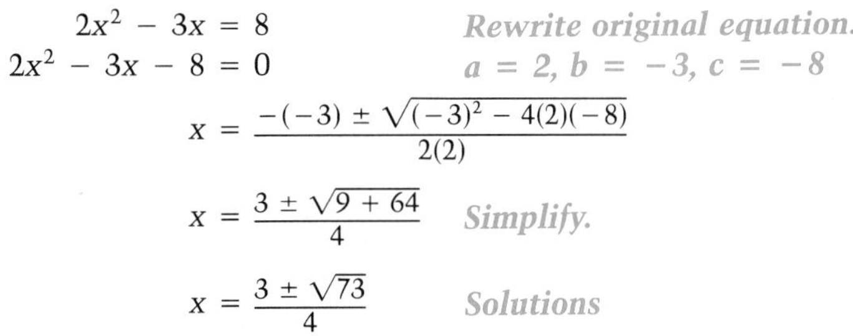

$$2x^2 - 3x = 8 \qquad \textit{Rewrite original equation.}$$

$$2x^2 - 3x - 8 = 0 \qquad a = 2, b = -3, c = -8$$

$$x = \frac{-(-3) \pm \sqrt{(-3)^2 - 4(2)(-8)}}{2(2)}$$

$$x = \frac{3 \pm \sqrt{9 + 64}}{4} \qquad \textit{Simplify.}$$

$$x = \frac{3 \pm \sqrt{73}}{4} \qquad \textit{Solutions}$$

The equation has two solutions.

$$x = \frac{3 + \sqrt{73}}{4} \approx 2.89 \quad \text{and} \quad x = \frac{3 - \sqrt{73}}{4} \approx -1.39$$

An algebraic check of solutions that have been rounded can be messy. Often, a graphic check is easier. To check the solutions in Example 2, sketch the graph of $y = 2x^2 - 3x - 8$. The two x-intercepts correspond to the two solutions.

Example 3 *Finding the x-Intercepts of a Graph*

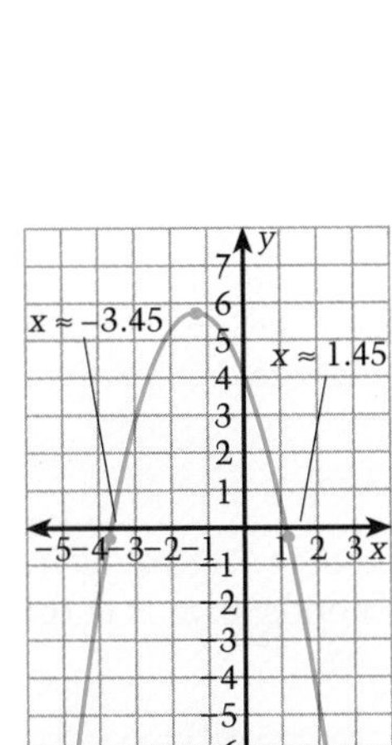

Find the x-intercepts of $y = -x^2 - 2x + 5$.

Solution The x-intercepts occur when $y = 0$.

$$0 = -x^2 - 2x + 5 \qquad \textit{Substitute 0 for y.}$$

$$x^2 + 2x - 5 = 0$$

$$x = \frac{-2 \pm \sqrt{2^2 - 4(1)(-5)}}{2(1)} \qquad x = \frac{-b \pm \sqrt{b^2 - 4ac}}{2a}$$

$$a = 1, b = 2, c = -5$$

$$x = \frac{-2 \pm \sqrt{4 + 20}}{2} \qquad \textit{Simplify.}$$

$$x = \frac{-2 \pm \sqrt{24}}{2} \qquad \textit{Solutions}$$

The x-intercepts of the graph of $y = -x^2 - 2x + 5$ are

$$x = \frac{-2 + \sqrt{24}}{2} \approx 1.45 \quad \text{and} \quad x = \frac{-2 - \sqrt{24}}{2} \approx -3.45.$$

Note that the x-intercepts are equidistant from the vertex of the parabola.

LESSON Notes

Examples 1–3

These examples make the connection between the solutions of the quadratic equation, $ax^2 + bx + c = 0$, and the x-intercepts of the graph of the quadratic equation in two variables, $y = ax^2 + bx + c$. Point out that the graph provides the means to check algebraic solutions easily, especially those involving decimal-valued solutions.

Example 3

The solutions may be simplified.

$$\frac{-2 \pm \sqrt{24}}{2} = \frac{-2 \pm \sqrt{4 \cdot 6}}{2}$$

$$= \frac{-2 \pm 2\sqrt{6}}{2} = \frac{-2}{2} \pm \frac{2\sqrt{6}}{2}$$

$$= -1 \pm \sqrt{6}$$

(See Lesson 9.1, Exercises 61–64 on page 457.)

Example 4

Observe that the vertical motion models are valid for motions expressed using the United States Customary System, not the metric system. In the International (Metric) System, acceleration due to gravity is -9.8 meters per second per second, or -9.8 m/s^2. The corresponding vertical motion models are $h = -4.9t^2 + s$ and $h = -4.9t^2 + vt + s$.

Common-Error ALERT!

Caution students to transform quadratic equations into standard form, if necessary, before assigning values to constants a, b, and c of the standard form.

Extra Examples

Here are additional examples similar to **Examples 1–4.**

1. Solve using the quadratic formula.
   a. $3x^2 + 8x - 3 = 0$ $\frac{1}{3}, -3$
   b. $x^2 - 4x = -2$ $2 \pm \sqrt{2}$
   c. $5x^2 + 12x = -4$ $-\frac{2}{5}, -2$
2. In Example 4, suppose your brother threw the rock straight up at a speed of 30 feet per second.
   a. How long would it take the rock to hit the water? about 16.8 seconds
   b. If the balloon were also rising at a rate of 5 feet per second when the rock was thrown, how long would it take the rock to hit the water? about 16.9 seconds

Check Understanding

1. Describe how the graph of a quadratic equation in two variables can be used to solve the corresponding quadratic equation in one variable. Its x-intercepts are the solutions.
2. What type of quadratic equations can be solved using the quadratic formula? A standard form quadratic, that is, $ax^2 + bx + c = 0$
3. In words, state the quadratic formula. See page 472.

Goal 2 Using Quadratic Models in Real Life

In Lesson 9.2, you studied the model for the height of a falling object that is *dropped*. For an object that is *thrown* down or up, the model changes to have an extra term. Problems involving these two models are *vertical motion* problems.

Models
$h = -16t^2 + s$ *Object is dropped.*
$h = -16t^2 + vt + s$ *Object is thrown.*

Labels
h = height (feet)
t = time in motion (seconds)
s = initial height (feet)
v = initial velocity (feet per second)

Remember that v is the velocity, not the speed.

Real Life
Hot Air Balloon

Example 4 When Does It Hit the Water?

Your brother is riding in a hot-air balloon over a lake. From an altitude of 4000 feet, he throws a rock straight down toward the water. When the rock leaves his hand, its speed is 30 feet per second. How long will it take the rock to hit the water?

Solution Because the rock is thrown down, its initial velocity is $v = -30$ feet per second. The initial height is $s = 4000$ feet. The rock will hit the water when height h is 0.

$h = -16t^2 - 30t + 4000$ *Vertical motion model*

$0 = -16t^2 - 30t + 4000$ *Substitute 0 for h.*

$16t^2 + 30t - 4000 = 0$ *a = 16, b = 30, c = −4000*

$$t = \frac{-30 \pm \sqrt{30^2 - 4(16)(-4000)}}{2(16)}$$

$$t = \frac{-30 \pm \sqrt{256{,}900}}{32}$$

$$t \approx 14.9 \quad \text{or} \quad -16.8$$

As a solution, −16.8 doesn't make sense in this case, so the rock will hit the water about 14.9 seconds after it is thrown. ■

Communicating about ALGEBRA

SHARING IDEAS about the Lesson

Extend the Idea In Example 4, the rock is thrown.

If the rock were *dropped*, when would it hit the water?

In ≈ 15.8 seconds

EXERCISES

Guided Practice

CRITICAL THINKING about the Lesson

1. **True or False?** The quadratic formula states that the solutions of the equation $ax^2 + bx + c = 0$ are "the opposite of b, plus or minus the square root of b minus $4ac$, all divided by $2a$." False
2. Describe the two models for vertical motion. See top of page 474.
3. State the values of a, b, and c from the standard form of the equation $5 = 6 + 9x - x^2$. $a = -1$, $b = 9$, $c = 1$
4. Solve $x^2 + x - 2 = 0$. $1, -2$
5. Sketch the graph of $y = x^2 + x - 2$ and label the x-intercepts. See Additional Answers.
6. Describe the relationship between the x-intercepts found in Exercise 5 and the solutions found in Exercise 4. They are the same.

Independent Practice

In Exercises 7–10, write the quadratic equation in standard form.

7. $-3x^2 + 5x = 9$ $\quad -3x^2 + 5x - 9 = 0$
8. $5 - 2x + x^2 = 0$ $\quad x^2 - 2x + 5 = 0$
9. $-4 + 3x + x^2 = 5$ $\quad x^2 + 3x - 9 = 0$
10. $9x - 7x^2 = 16$ $\quad -7x^2 + 9x - 16 = 0$

In Exercises 11–14, find the value of $b^2 - 4ac$ for the equation.

11. $2x^2 - 3x - 1 = 0$ $\quad 17$
12. $4x^2 + 4x + 1 = 0$ $\quad 0$
13. $3x^2 - 2x - 5 = 0$ $\quad 64$
14. $x^2 - 11x + 30 = 0$ $\quad 1$

In Exercises 15–20, use the quadratic formula to solve the equation.

15. $4x^2 - 13x + 3 = 0$ $\quad 3, \frac{1}{4}$
16. $3y^2 + 11y + 10 = 0$ $\quad -\frac{5}{3}, -2$
17. $2x^2 + 7x + 3 = 0$ $\quad -\frac{1}{2}, -3$
18. $x^2 - 6x + 7 = 0$ $\quad 3 + \sqrt{2} \approx 4.41,\ 3 - \sqrt{2} \approx 1.59$
19. $5y^2 + 2y - 2 = 0$
20. $2x^2 + 4x - 3 = 0$

19.–20. See below.

In Exercises 21–26, solve the quadratic equation by the most convenient method (finding square roots or the quadratic formula). Explain why you chose your method.

21. $6x^2 + 20x + 5 = 0$ $\quad \frac{-10 + \sqrt{70}}{6} \approx -0.27,\ \frac{-10 - \sqrt{70}}{6} \approx -3.06$
22. $t^2 = 27$ $\quad \sqrt{27} \approx 5.20,\ -\sqrt{27} \approx -5.20$
23. $x^2 - 625 = 0$ $\quad 25, -25$
24. $4u^2 - 49 = 0$ $\quad \frac{7}{2}, -\frac{7}{2}$
25. $-2x^2 + 6x + 1 = 0$ $\quad \frac{3 + \sqrt{11}}{2} \approx 3.16,\ \frac{3 - \sqrt{11}}{2} \approx -0.16$
26. $x^2 + 14x + 49 = 0$ $\quad -7$

In Exercises 27–32, find the x-intercepts of the graph of the equation.

27. $y = x^2 + 2x + 15$ $\quad$ None
28. $y = x^2 - 6x - 7$ $\quad 7, -1$
29. $y = x^2 + x - 20$ $\quad 4, -5$
30. $y = x^2 + 8x + 12$ $\quad -2, -6$
31. $y = x^2 + x - \frac{3}{4}$ $\quad \frac{1}{2}, -\frac{3}{2}$
32. $y = x^2 + \frac{7}{3}x - 2$ $\quad \frac{2}{3}, -3$

19. $\frac{-1 + \sqrt{11}}{5} \approx 0.46;\ \frac{-1 - \sqrt{11}}{5} \approx -0.86$ 20. $\frac{-2 + \sqrt{10}}{2} \approx 0.58;\ \frac{-2 - \sqrt{10}}{2} \approx -2.58$

Communicating about ALGEBRA

Under what conditions are the two vertical motion models identical? When $v = 0$

EXTEND Communicating

In Example 4, at what velocity would your brother have had to throw the rock for it to hit the water in about 10 seconds? -240 ft/s At what velocity would your brother have had to throw the rock for it to hit the water in about 20 seconds? 120 ft/s (Note the positive velocity; he would have to have thrown it up.)

EXERCISE Notes

ASSIGNMENT GUIDE

Basic/Average: Ex. 1–6, 7–31 odd, 39, 41–42, 43–48

Above Average: Ex. 1–6, 7–31 odd, 34, 36, 39–49 odd

Advanced: Ex. 1–6, 15–33 odd, 34–36, 39–40, 43–50

Selected Answers

Exercises 1–6, 7–47 odd

✪ More Difficult Exercises

Exercises 35–40, 49–50

Guided Practice

▶ **Ex. 3** There is more than one standard form for the equation.

▶ **Ex. 4–6** These exercises make a critical connection between the graph of a quadratic equation $y = ax^2 + bx + c$ and the solutions of the equation $0 = ax^2 + bx + c$. Be sure to emphasize that the solutions are the x-intercepts of the graph.

Independent Practice

Ex. 7–14, 15–20 These sets of exercises are related. Exercises 7–14, are preparatory skills for solving Exercises 15–20. In Exercises 15–20 students should round answers to the nearest hundredth.

Ex. 21–26 Remind students that in Exercises 22–24, since $b = 0$, using square roots is probably more efficient than using the quadratic formula. There is only one solution to Exercise 26 because $x^2 + 14x + 49$ equals $(x + 7)^2$, a perfect square trinomial. Students will study these quadratic equations in Chapter 10.

Ex. 27–32 TECHNOLOGY Students can either use the quadratic formula or the TRACE feature on a graphing calculator to find the solutions. You might suggest that students use one method to find the solutions and the other method to check their solutions.

Ex. 33–36 Students can use Example 4 as a solution model.

Answers

33. $\approx$ 8.1 seconds; yes, the rock would be traveling at about 90 miles per hour and could do a lot of damage to whatever it hit.

39.

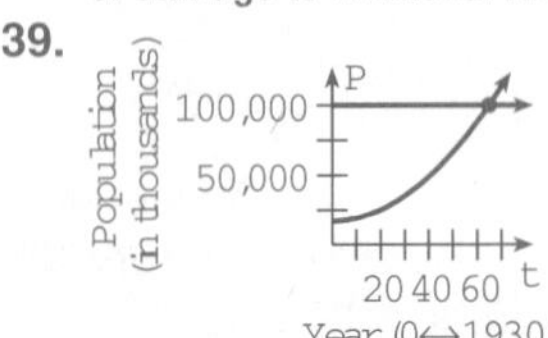

EXTEND Ex. 40 TECHNOLOGY The data in this exercise is presented in a bar graph. Ask students to graph the data points as a scatter plot to check the accuracy of the model $S = 27.4t^2 - 394.5t + 1831.3$. Have students predict the average salary of major league baseball players for 1992 and then check the accuracy of their predicted amount against data published by the Major League Baseball Players Association. $1,029,000

33. *Royal Gorge Bridge* If a rock was dropped from the Royal Gorge Bridge into the Arkansas River, how long would it take for the rock to hit the water? (Assume there is no air resistance.) Do you think there is a regulation *against* dropping objects off this bridge? Explain your answer. See margin.

34. *Royal Gorge Bridge* How much sooner would the rock reach the Arkansas River if it was thrown straight down with an initial speed of 30 feet per second? $\approx$0.9 seconds

The Royal Gorge Bridge near Canon City, Colorado, is the highest suspension bridge in the world. The bridge is 1053 feet above the Arkansas River.

35. *The Owl and the Mouse* An owl is circling a field at a height of 100 feet and sees a mouse. The owl folds its wings and begins to dive with an initial speed of 10 feet per second. Estimate the time the mouse has to escape. 2.2 seconds

36. *Play It Again in Metric* An owl is circling a field at a height of 30.5 meters and sees a mouse. The owl folds its wings and begins to dive with an initial speed of 3 meters per second. Estimate the time the mouse has to escape. (Use the vertical motion model $h = -4.9t^2 + vt + s$, where h is height in meters, v is initial velocity in meters per second, and s is initial height in meters.) 2.2 seconds

37. *Flying Time* The cities of Chicago, Atlanta, and Toronto approximate the vertices of a right triangle. How long would it take a plane flying at 500 miles per hour to fly from Atlanta to Toronto? 1.45 hours

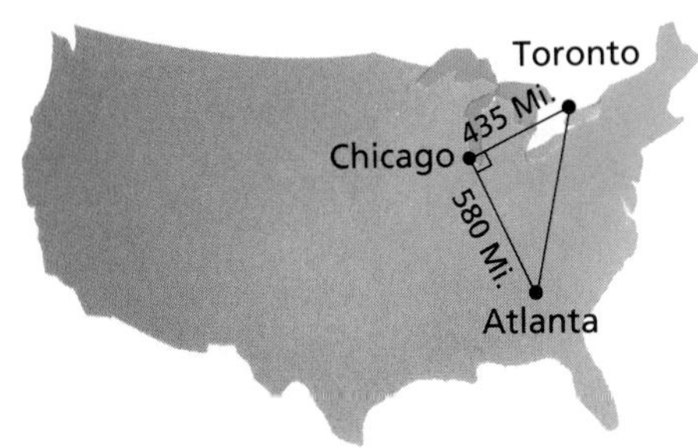

38. *Depth of a Submarine* The sonar of a Navy cruiser detects a submarine that is 3000 feet from the cruiser. The angle between the water surface and the submarine is 45°. How deep is the submarine? $\approx$2121 ft

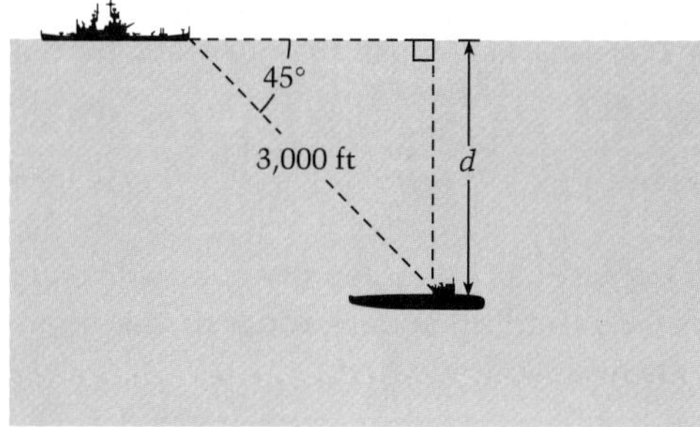

39. *Population of Mexico* The population, P (in thousands), of Mexico between 1930 and 1990 can be modeled by

$$P = 19.2t^2 + 31.6t + 16964.5$$

where t is the year, with $t = 0$ corresponding to 1930. Find the year in which the population of Mexico will reach 100,000,000. Demonstrate your result graphically. 1994 See margin.

40. *Baseball Salaries* The average salary S (in thousands) of major league baseball players between 1987 and 1992 can be modeled by

$$S = 27.4t^2 - 394.5t + 1831.3$$

where t is the year, with $t = 7$ corresponding to 1987. During which year was the average salary about $500,000? Does your algebraic solution agree with the graph?
(Source: Major League Baseball Players Association)
1989, yes

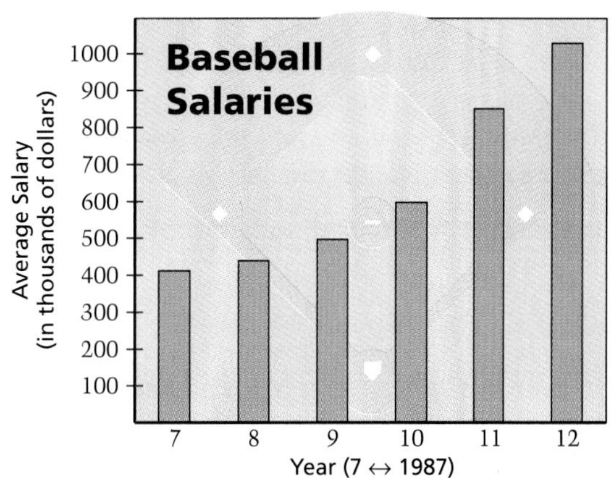

Integrated Review

In Exercises 41 and 42, evaluate the expression.

41. $\sqrt{16 - 4(3)(1)}$ 2

42. $\sqrt{36 - 4(2)(-4)}$ $\sqrt{68} \approx 8.25$

In Exercises 43–45, check whether the number is a solution of the equation.

43. $2x^2 - 4x + 9 = 9,\ 2$ Is

44. $3x^2 + 3x + 4 = 12,\ -2$ Is not

45. $x^2 - 9x + 7 = 11,\ 6$ Is not

In Exercises 46–48, sketch the graph of the equation. Label the x-intercepts on the graph. See Additional Answers.

46. $y = x^2 + 4x - 45$

47. $y = x^2 - x - 6$

48. $y = 2x^2 - 7x + 3$

Exploration and Extension

In Exercises 49 and 50, find the axis of symmetry of each graph and show that it lies midway between the two x-intercepts of the graph. Show how you could use this "two-part" form of the quadratic formula,

$$x = \frac{-b}{2a} \pm \frac{\sqrt{b^2 - 4ac}}{2a},$$

to find the distance between the axis of symmetry of a parabola and either of its x-intercepts. See margin.

49. $y = x^2 - 6x + 5$

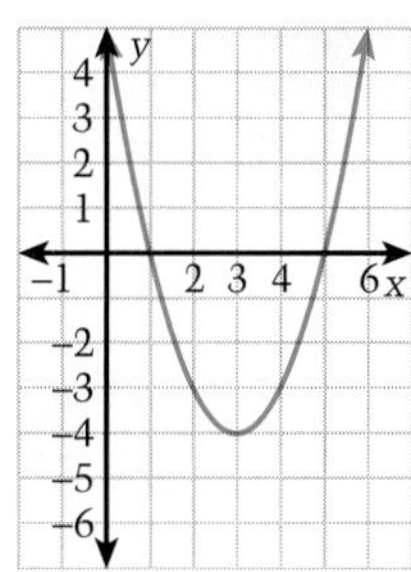

50. $y = x^2 + 3x - 18$

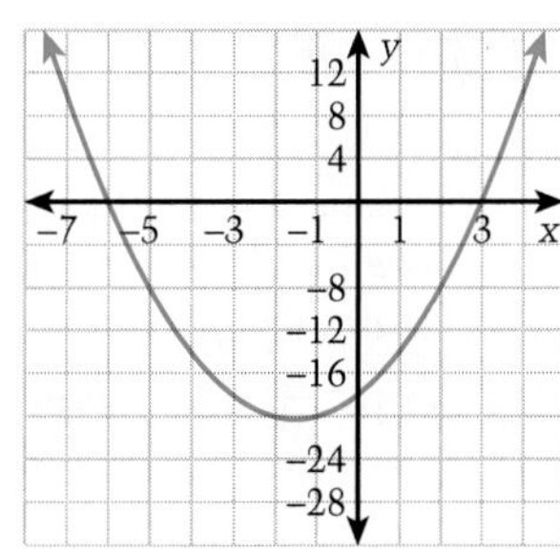

Ex. 49–50 Assign these exercises as a pair.

Answers

49. The axis $x = 3$ is 2 units from each of the x-intercepts, 1 and 5. The two-part form of the quadratic formula gives $x = 3 \pm 2$, where the "3" refers to the axis and the "$\pm$ 2" refers to the distance to each x-intercept.

50. The axis $x = -\frac{3}{2}$ is $\frac{9}{2}$ units from each of the x-intercepts, 3 and -6. The two-part form of the quadratic formula gives $x = -\frac{3}{2} \pm \frac{9}{2}$, where the "$-\frac{3}{2}$" refers to the axis and the "$\pm\frac{9}{2}$" refers to the distance to each x-intercept.

Enrichment Activities

These activities require students to go beyond lesson goals.

Graph the equation $y = 2x^2 - x + 1$.

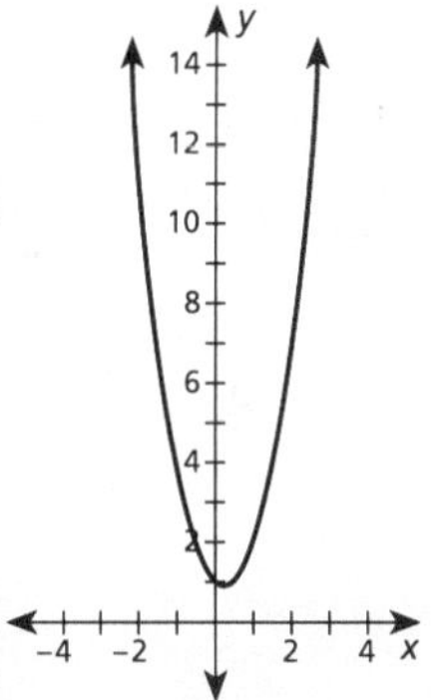

1. What do you notice about the graph? The graph does not cross the x-axis, so it has no x-intercepts.
2. Can you use the graph to solve the equation $0 = 2x^2 - x + 1$? No, the equation has no x-intercepts, so it has no real solutions.
3. How can you tell algebraically that the equation $0 = 2x^2 - x + 1$ has no real solutions? $b^2 - 4ac = -7$, a negative number

Answers

25. $\frac{3 + \sqrt{17}}{2} \approx 3.56,$
$\frac{3 - \sqrt{17}}{2} \approx -0.56$

26. $\frac{-5 + \sqrt{33}}{2} \approx 0.37,$
$\frac{-5 - \sqrt{33}}{2} \approx -5.37$

Mid-Chapter SELF-TEST

Take this test as you would take a test in class. The answers to the exercises are given in the back of the book.

In Exercises 1–6, find all square roots of the number. (9.1)

1. 64 ± 8
2. 121 ± 11
3. 0.0064 ± 0.08
4. $\frac{49}{100}$ $\pm \frac{7}{10}$
5. $\frac{1}{81}$ $\pm \frac{1}{9}$
6. 0.01 ± 0.1

In Exercises 7–12, evaluate the square root. Give the exact value, if possible. Otherwise, give an approximation to two decimal places. (9.1)

7. $\sqrt{4}$ 2
8. $\sqrt{1}$ 1
9. $\sqrt{8}$ 2.83
10. $\sqrt{2.25}$ 1.5
11. $\sqrt{0.1}$ 0.32
12. $\sqrt{50}$ 7.07

In Exercises 13–18, solve the equation. (9.2)

13. $\frac{1}{3}x^2 - 27 = 0$ ± 9
14. $25x^2 - 37 = 588$ ± 5
15. $7x^2 - 81 = 31$ ± 4
16. $\frac{3}{4}x^2 - 3 = 51$ $\pm\sqrt{72} \approx \pm 8.49$
17. $x^2 - \frac{1}{4} = 6$ $\pm \frac{5}{2}$
18. $x^2 + 37 = 199$ $\pm\sqrt{162} \approx \pm 12.73$

In Exercises 19–22, determine the vertex and *x*-intercepts of the graph. (9.3)

19. 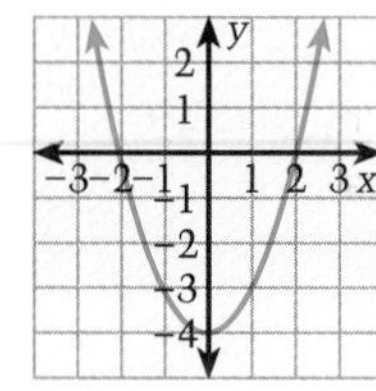
$(0, -4)$; 2, -2

20.  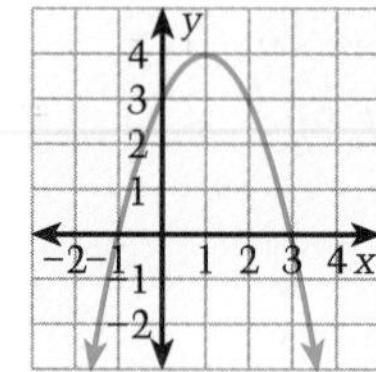
$(1, 4)$; -1, 3

21. 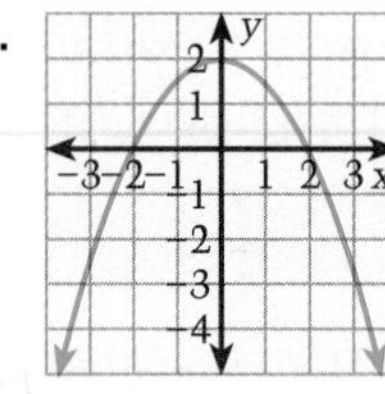
$(0, 2)$; 2, -2

22. 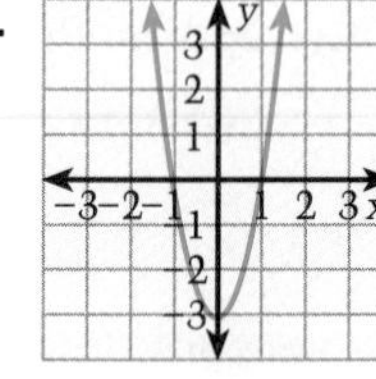
$(0, -3)$; 1, -1

In Exercises 23–26, solve the equation. (9.4)

23. $x^2 - x - 30 = 0$ 6, -5
24. $-x^2 - 8x - 16 = 0$ -4
25. $-x^2 + 3x + 2 = 0$ 25.–26. See margin.
26. $x^2 + 5x - 2 = 0$

27. *Geometry* The legs of a right triangle are 3 inches and 4 inches. What is the length of the hypotenuse? **(9.1)** 5 in.

28. *Geometry* The volume of a box with a square base and a height of 14 centimeters is 1400 cubic centimeters. What is the length of an edge of the base? **(9.2)** 10 cm

29. You are trying to build the world's tallest tower of interlocking toy blocks. (The record is $59\frac{1}{2}$ feet.) When your tower is 30 feet tall, the bag holding the loose blocks is knocked off the top of the tower. How many seconds did it take for the bag to hit the floor? **(9.2)** $\approx$1.37 seconds

30. If the bag of blocks in Exercise 29 is thrown down with an initial velocity of 10 feet per second, how many seconds does it take for it to hit the floor? **(9.4)** $\approx$1.09 seconds

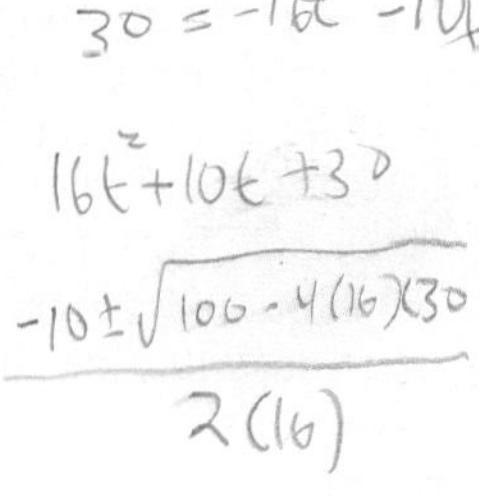

9.5 Problem Solving Using the Discriminant

What you should learn:

Goal 1 How to find the number of solutions of a quadratic equation by using the discriminant

Goal 2 How to use the discriminant in real-life models

Why you should learn it:

You can use the discriminant as a quick source of information about quadratic models.

Goal 1 Solutions of a Quadratic Equation

In the quadratic formula, the expression $b^2 - 4ac$ is the **discriminant.**

Discriminant

$$x = \frac{-b \pm \sqrt{b^2 - 4ac}}{2a}$$ *Quadratic formula*

The discriminant of a quadratic equation can be used to find the number of solutions of the quadratic equation.

The Number of Solutions of a Quadratic Equation

Consider the quadratic equation $ax^2 + bx + c = 0$.

1. If $b^2 - 4ac$ is positive, then the equation has two solutions.
2. If $b^2 - 4ac$ is zero, then the equation has one solution.
3. If $b^2 - 4ac$ is negative, then the equation has no solution.

Example 1 *Finding the Number of Solutions of a Quadratic Equation*

$ax^2 + bx + c = 0$	Discriminant ($b^2 - 4ac$)	Number of Solutions
a. $x^2 - 2x + 3 = 0$	$(-2)^2 - 4(1)(3) = -8$	None
b. $x^2 - 2x + 1 = 0$	$(-2)^2 - 4(1)(1) = 0$	One
c. $x^2 - 2x - 2 = 0$	$(-2)^2 - 4(1)(-2) = 12$	Two ■

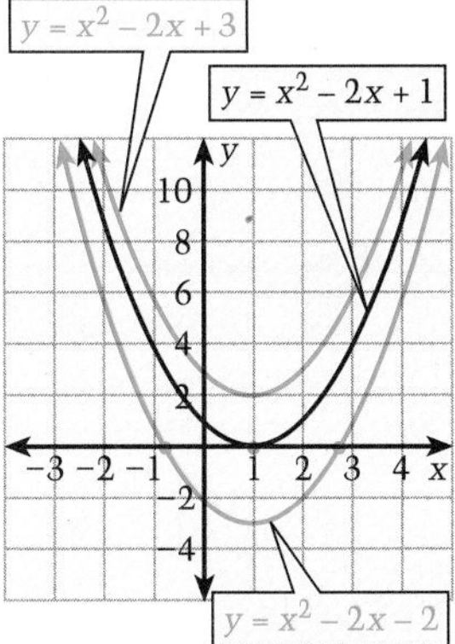

Graphic Check

In Example 1, note that the number of solutions can be changed by just changing the value of c. A graph can help you see why this occurs. By changing the value of c, you can move the graph of $y = x^2 - 2x + c$ up or down in the coordinate plane. If the graph is moved too high, it won't have an x-intercept and the equation $x^2 - 2x + c = 0$ won't have a solution.

$y = x^2 - 2x + 3$ *Graph above x-axis (no x-intercepts)*
$y = x^2 - 2x + 1$ *Graph touches x-axis (one x-intercept)*
$y = x^2 - 2x - 2$ *Graph crosses x-axis (two x-intercepts)*

Problem of the Day

All the blocks in this picture are identical. What letter is opposite the Y? The G? The W? B, R, P

ORGANIZER

Warm-Up Exercises

1. Compute $b^2 - 4ac$ for each quadratic equation.
   a. $x^2 - 5x = 9$ 61
   b. $-12x^2 + 6x - 8 = 0$ −348
   c. $8x^2 - 8x + 2 = 0$ 0
   d. $3x^2 + 18x = 16$ 516
2. Decide if each expression is positive, negative, or zero.
   a. $(-5)^2 - (4 \cdot -2 \cdot 6)$
   b. $(12)^2 - (4 \cdot -16 \cdot -10)$
   c. $(6)^2 - (4 \cdot 3 \cdot 3)$
   d. $(-13)^2 - (4 \cdot 15 \cdot 3)$

   a. positive b. negative c. zero d. negative

Lesson Resources

Teaching Tools
Transparencies: 3, 13
Problem of the Day: 9.5
Warm-up Exercises: 9.5
Answer Masters: 9.5
Extra Practice: 9.5
Color Transparencies: 47, 48, 49
Applications Handbook: p. 46–48
Technology Handbook: p. 70

LESSON Notes

Example 1

Emphasize that a solution is a real-valued solution of the quadratic equation. Help students make the connections between (a) "no real-valued solution and the graph of a quadratic equation with no x-intercepts," (b) "one real-valued solution and exactly one x-intercept," and (c) "two real-valued solutions and two x-intercepts." (Recall that the quadratic equation with no x-intercepts has imaginary solutions. Students will encounter imaginary numbers in Algebra 2.)

Example 2

Point out that when preparing to use the quadratic formula to solve real-life problems, evaluating the discriminant first can save time by helping you decide whether there are any real-valued solutions to consider. Frequently, in a life-threatening situation, people become stronger; so, there might be a different outcome! (See Extra Example 2 that follows.)

Example 3

Checking to see if a solution is possible does not slow down the computation associated with the quadratic formula. Students are simply asked to make a decision based on the discriminant before continuing the process.

Notice that in the graph, P is given in billions of dollars while in the model, P is given in millions of dollars.

Goal 2 Using the Discriminant in Real Life

Rick is a firefighter and is leaning out of a window on the eighth floor. He is trying to throw a grappling hook to a tenth-floor window that is 26 feet above him.

Connections **Physics**

Example 2 *Will the Grappling Hook Catch?*

Rick can throw the grappling hook with a maximum speed of 40 feet per second. Can he throw the grappling hook to the window above him?

Solution You don't know Rick's present height. In the vertical motion model, however, you can let his present height be $s = 0$. Use an initial velocity of $v = 40$ feet per second. To reach the window, Rick must be able to throw the grappling hook to a height of 26 feet.

$$h = -16t^2 + 40t + 0 \quad \textit{Vertical motion model}$$
$$26 = -16t^2 + 40t \quad \textit{Substitute 26 for h.}$$
$$16t^2 - 40t + 26 = 0 \quad \textit{Standard form}$$

Using $a = 16$, $b = -40$, and $c = 26$, the discriminant is

$$b^2 - 4ac = (-40)^2 - 4(16)(26) = -64.$$

Because the discriminant is negative, the equation $16t^2 - 40t + 26 = 0$ has no solution. This means that Rick cannot throw the grappling hook to the window above him. ■

If Rick could throw the hook so it travels just a little faster (41 feet per second) *or* if the window were just a little closer (25 feet), he could throw the hook to the window. Try to see why.

From 1960 to 1990, the total government payroll in the United States (local, state, and federal) can be modeled by $P = 35t^2 + 115t + 3410$, where P is in millions of dollars and $t = 0$ corresponds to 1960.

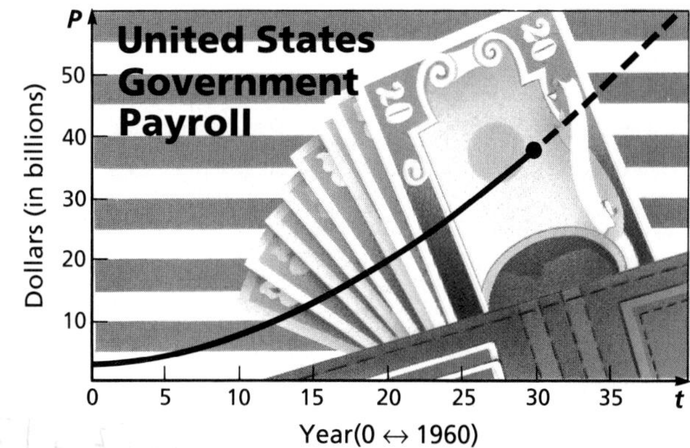

Real Life
Government

Example 3 *Predicting the Government Payroll*

Judging from the payroll pattern between 1960 and 1990, is it possible that the government payroll will reach 70 billion dollars a year?

Solution In the model for the payroll, P is given in millions of dollars. A payroll of 70 billion dollars would correspond to a P-value of 70,000. Why?

$$P = 35t^2 + 115t + 3410 \quad \text{Quadratic model}$$
$$70{,}000 = 35t^2 + 115t + 3410 \quad \text{Substitute 70,000 for } P.$$
$$0 = 35t^2 + 115t - 66{,}590 \quad \text{Standard form}$$

The discriminant of this equation is

$$b^2 - 4ac = 115^2 - 4(35)(-66{,}590) = 9{,}335{,}825.$$

Because the discriminant is positive, the equation has two solutions. Using the quadratic formula, you can find the solutions to be approximately -45 and 42. The solution 42 suggests that, based on the model, the government payroll will reach 70 billion dollars in 2002. ■

Communicating about ALGEBRA

SHARING IDEAS about the Lesson

Analyze Consider the equation $x^2 - 4x + c = 0$.

A. Find values of c so that the equation will have no solution, one solution, and two solutions. $c > 4$, $c = 4$, $c < 4$

B. Give a graphical interpretation of your answers to Part A. Graph has 0, 1, or 2 x-intercepts.

See Additional Answers for sample graphs.

Extra Examples

Here are additional examples similar to **Examples 1–3.**

1. Determine whether the given quadratic equations have no solutions, exactly one solution, or two solutions.
   a. $3x^2 + 4x - 5 = 0$
   b. $x^2 - 7x + 16 = 0$
   c. $x^2 + 10x = -25$
   a. two solutions
   b. no solution
   c. exactly one solution
2. Suppose that in Example 2, Rick could throw the grappling hook at an initial velocity of 50 feet per second, but that he needed to reach the 11th-floor window. Could Rick throw the grappling hook to the 11th-floor window? Assuming the 11th floor is 39 feet above Rick, Rick could reach the 11th floor!
3. From Example 3, is it possible that the government payroll will ever reach 80 billion dollars? yes By 2004? no

Check Understanding

Sketch typical graphs of the quadratic equations that correspond to the possible values of the discriminant. See Exercises 4–6 for samples.

Communicating about ALGEBRA

Ask students to discuss possible solutions of the quadratic equation $x^2 + 1 = 0$. Have them speculate about what a possible solution *might* be if there is, in fact, a solution to this simple quadratic equation! (The solution is $\sqrt{-1}$, which is not a real-valued number. Students will encounter "imaginary numbers" in Algebra 2.)

EXERCISE Notes

ASSIGNMENT GUIDE

Basic/Average: Ex. 1–6, 7–12, 15, 17, 18, 21–22, 23–25, 27–30

Above Average: Ex. 1–6, 7–15, 17,18, 21–22, 23–25, 27–34

Advanced: Ex. 1–6, 11, 13–18, 20–34

Selected Answers
Exercises 1–6, 7–29 odd

Use **Mixed Review** as needed.

More Difficult Exercises
Exercises 13, 14, 19, 20, 31–34

Guided Practice

Ex. 1–3 These exercises verify students' understanding of the discriminant.

Answer

2. If the discriminant is negative, there is no real-number solution. If the discriminant is 0, there is one solution. If the discriminant is positive, there are two solutions.

Ex. 4–6 Assign these exercises as a group.

Independent Practice

Ex. 7–12 Remind students to use the discriminant.

EXERCISES

Guided Practice

CRITICAL THINKING about the Lesson

1. Write the quadratic formula and circle the part that is called the discriminant. $x = \dfrac{-b \pm \sqrt{b^2 - 4ac}}{2a}$
2. Explain how the discriminant can be used to determine the number of solutions of $ax^2 + bx + c = 0$. See margin.
3. Find the discriminant of $3x^2 - 2x - 5 = 0$. How many solutions does this equation have? 64, 2

In Exercises 4–6, match the discriminant with the graph.

4. $b^2 - 4ac = 3$ b
5. $b^2 - 4ac = 0$ c
6. $b^2 - 4ac = -2$ a

a. 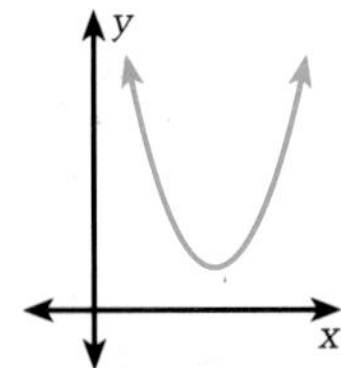

b. 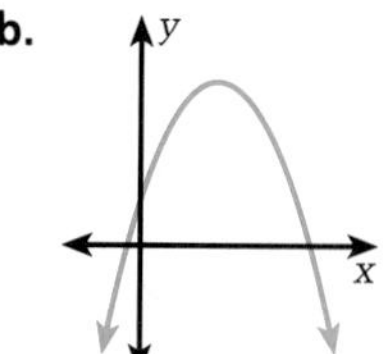

c. 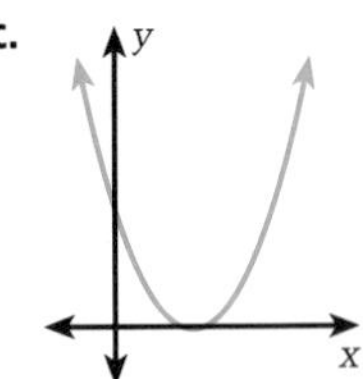

Independent Practice

In Exercises 7–12, decide how many solutions the equation has.

7. $2x^2 + 3x - 2 = 0$ 2
8. $x^2 - 2x + 4 = 0$ None
9. $-2x^2 + 4x - 2 = 0$ 1
10. $-\frac{1}{2}x^2 + x + 3 = 0$ 2
11. $5x^2 - 2x + 3 = 0$ None
12. $3x^2 - 6x + 3 = 0$ 1

13. *Spelunking* You and a friend are *spelunking* (exploring caves) in a section of the Onondaga Cave in Missouri. The two of you are standing beneath a ledge that is 15 feet high. Your friend can throw a grappling hook upward with an initial velocity of 30 feet per second. Will the grappling hook reach the ledge when it is thrown? Explain. No, the discriminant is negative.

14. *Spelunking, Part II* While you and your friend are attempting to reach the ledge, some other spelunkers join you. They see the trouble you are having and suggest that your friend stand on a foot-high rock to throw the hook. Would this help? Explain.
Yes, the discriminant would then be positive.

Onondaga Cave, in Leasburg, Missouri

15. *Geometry* Is it possible for a rectangle with a perimeter of 52 centimeters to have an area of 148.75 square centimeters? Explain. Yes, the discriminant is positive.

x

$26 - x$ $26 - x$

x

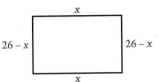

16. *Geometry* The area of the isosceles triangle is 192 square meters. What is its perimeter? 64 m

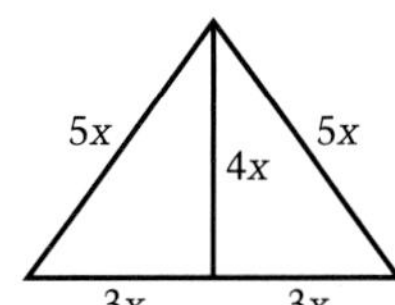

Dunking the Ball **In Exercises 17 and 18, use the vertical motion model:** $h = -16t^2 + vt + s$.

Chun and R.J. are playing basketball. Chun can jump with an initial velocity of 13 feet per second and needs to jump 2.6 feet to dunk the basketball. R.J. can jump with an initial velocity of 15 feet per second and needs to jump 3.6 feet to dunk the ball.

17. Can Chun dunk the ball? Can R.J.? Justify your answers. Yes, the discriminant is positive; no, the discriminant is negative.

18. Suppose that Chun can jump with an initial velocity of 12.5 feet per second and R.J. can jump with an initial velocity of 15.5 feet per second. How, if at all, would this change your answers to Exercise 17? Chun cannot dunk, R.J. can.

19. *Business as Usual?* For the past 30 years, profits of Specialty Aircraft have increased as shown in the graph. The profits can be modeled by

$$P = 1.4t^2 + 2.2t + 258,$$

where $t = 30$ represents the present year and P is in thousands of dollars. Now, you have been elected president of the company, and the board of directors expects you to increase annual profits by at least 10% per year for five years. Should you follow the previous president's strategies or do you need to make some changes? Explain. See margin.

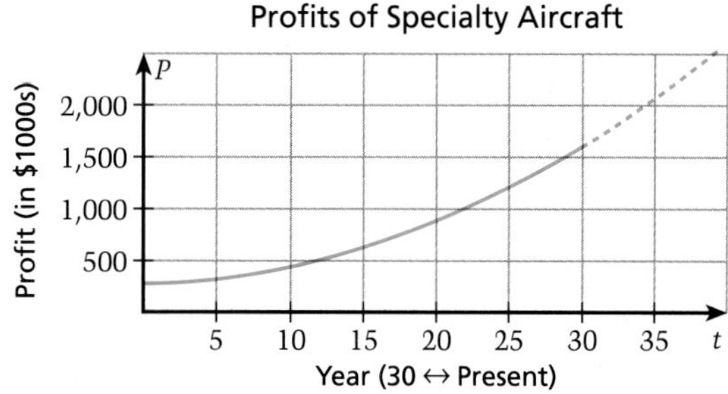

20. *Profit or Loss?* Your company's income and expenses for the past ten years can be modeled as follows.

$I = 2.3t^2 + 4.8t + 1049$ Income

$E = 6.5t^2 - 54.4t + 546$ Expenses

I and E are measured in thousands of dollars. Describe your company's profit record for the past ten years. How do the next ten years look? See margin.

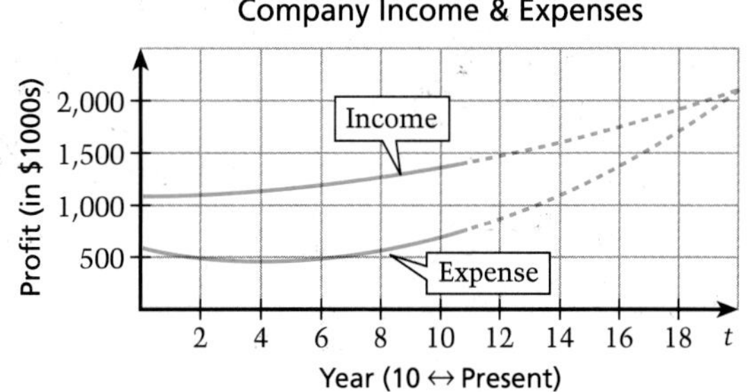

Ex. 13–14, 17–18 Assign these exercises in pairs. These problems are solved first in terms of time t even though they are not problems about time.

Answers

19. You need to make some changes. In 5 years, the profit from the model would be 2050 thousand; the board of directors expects a profit of 2551 thousand in 5 years.

20. During the first 7 of the 10 years, the profit increased yearly; during the last 3 of the 10 years, the profit decreased. During the next 10 years, the profit will decrease.

Ex. 21–22 Assign these exercises as a pair. Refer students back to Example 3 as a solution model.

Ex. 27–30 Assign these exercises as a group. Encourage students to use the quadratic formula.

Ex. 31–34 These exercises should be assigned as a long term assignment. Allow students several days to complete them. You may wish to set aside class time for students to share their responses and how they obtained them.

Enrichment Activities

These activities require students to go beyond lesson goals.

TECHNOLOGY The following computer program uses the discriminant to determine the number of real roots of the quadratic equation of the form $ax^2 + bx + c = 0$.

```
10 INPUT "ENTER THE VALUES OF A, B, AND C OF AX^2 + BX + C SEPARATED BY COMMAS"; A,B,C
20 LET D = B^2 -4*A*C
30 IF D < 0 THEN 110
40 LET X1 = (-B + SQR(D))/(2*A)
50 LET X2 = (-B - SQR(D))/(2*A)
60 IF D = 0 THEN 90
70 PRINT "THE TWO REAL ROOTS ARE ";X1;" AND ";X2;"."
80 GOTO 120
90 PRINT "THE ONE REAL ROOT IS ";X1;"."
100 GOTO 120
110 PRINT "THERE ARE NO REAL ROOTS."
120 END
```

484

21. *Public School Spending* For 1970 to 1990 in the United States, public school (K–12) spending can be modeled by

$$P = 195t^2 + 3480t + 41{,}540$$ 2001–2002

where P is in millions of dollars and $t = 0$ corresponds to 1970. Use the model to predict when annual spending will reach $350 billion.

(Source: U.S. National Center for Education)

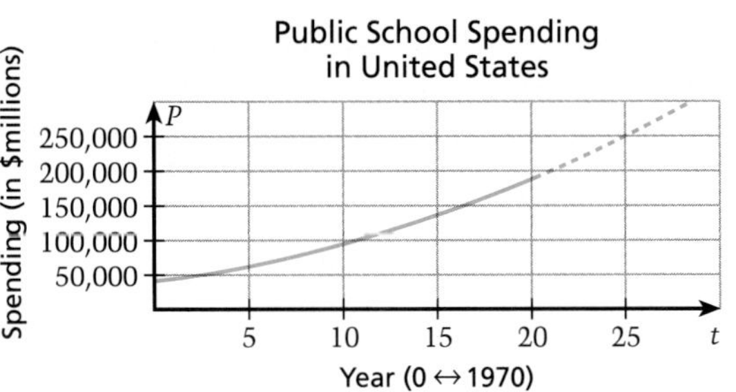

22. *Computers in the Classroom* For 1980 to 1990, the number of personal computers in grades K–12 (public and private) can be modeled by

$$C = 16t^2 + 96t$$

where C is in thousands of computers and $t = 0$ corresponds to 1980. Use the model to find when 5 million computers will be used in grades K–12. 1994–1995

(Source: Future Computing/Data Inc.)

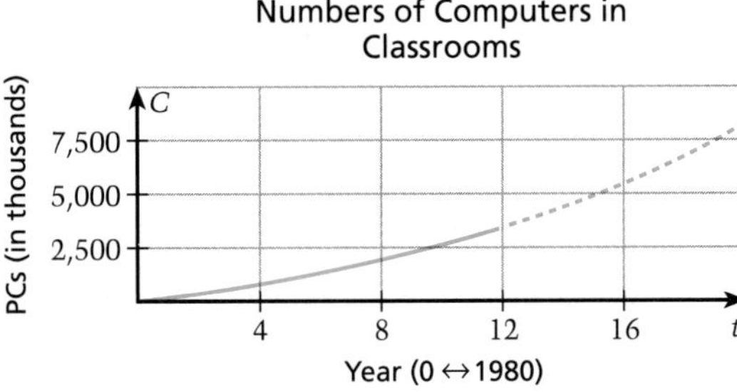

Integrated Review

23. *College Entrance Exam Sample* If $\frac{1}{x + 2} = \frac{1}{2}$, then $x =$ [?]

a. 2 **b.** 1 **c.** 0 **d.** −1 **e.** −2

24. *College Entrance Exam Sample* If $x = -3$, then $(x + 3)^2 - 3x =$ [?]

a. −6 **b.** −3 **c.** 0 **d.** 6 **e.** 9

25. *Estimation—Powers of Ten* Which is the best estimate of the time it takes an object to fall 100 feet?

a. 0.3 second **b.** 3 seconds **c.** 30 seconds **d.** 300 seconds

26. *Estimation—Powers of Ten* Which is the best estimate of distance between Providence, Rhode Island, and San Diego, California?

a. 250 miles **b.** 2500 miles **c.** 25,000 miles **d.** 250,000 miles

In Exercises 27–30, match the equation with its graph.

27. $y = 3x^2 - 2x + 1$ **28.** $y = \frac{3}{4}x^2 + 2x + 1$ **29.** $y = \frac{3}{4}x^2 - 2x + 1$ **30.** $y = 3x^2 + 2x + 1$

a. b **b.** c

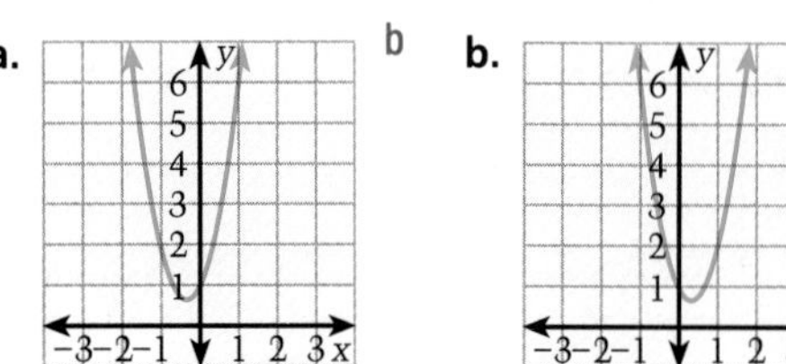

c. d

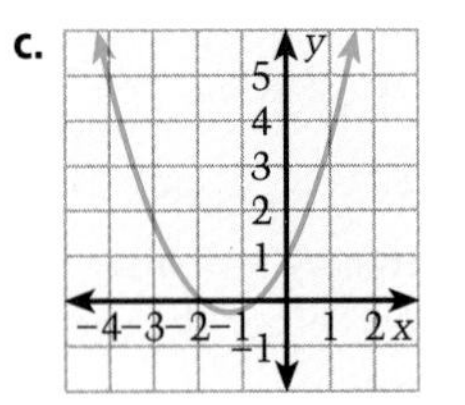

d. a

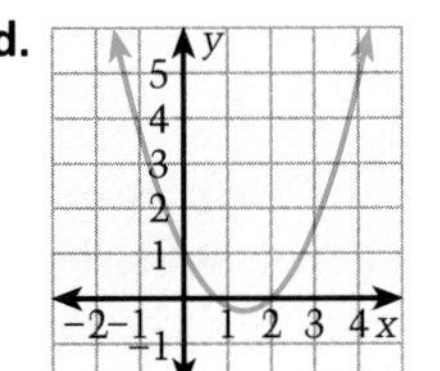

Exploration and Extension

Roller Coaster **In Exercises 31–34, use the following information.**

You are riding on a roller coaster whose path can be modeled by

$$y = 0.005x^2 - 2x + 200$$

where x and y are measured in feet. When $t = 0$ seconds, you are at the point (0, 200).

31. When $t = 4$ seconds, the y-coordinate of your location is 32. What is the x-coordinate of your location?

32. When $t = 4.23$ seconds, the y-coordinate of your location is 18. What is the x-coordinate of your location?

33. Use the Pythagorean theorem to approximate the length of the track between the two points $(x_1, 32)$ and $(x_2, 18)$.

34. What was your average speed between the two points?

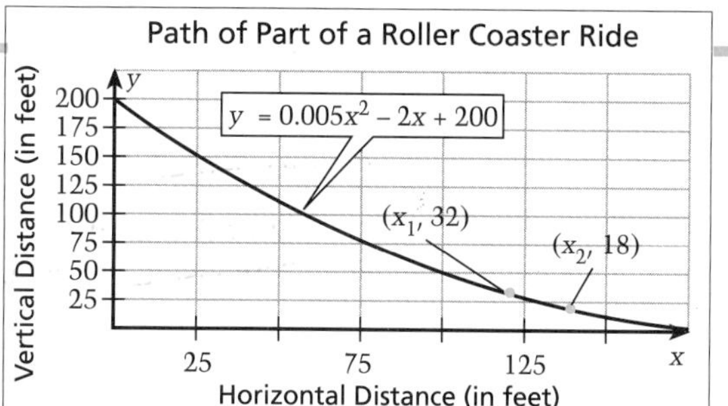

Roller coasters may include vertical drops of as much as 190 feet. At the bottom of such drops, the cars may reach a speed of about 72 miles per hour!

If you have access to a computer, enter the program and use it to determine the number of real roots of the equations. (If you do not have access to a computer, you may be able to program a graphics calculator.)

1. $3x^2 + 2x - 1 = 0$
2. $-7x^2 - 2.3x + 17 = 0$
3. $x^2 - 3.1x + 14 = 0$
4. $-0.03x^2 + 2.7x - 3.163 = 0$

1. THE TWO REAL ROOTS ARE 0.3333333 AND −1.
2. THE TWO REAL ROOTS ARE −1.7313087 AND 1.4027373.
3. THERE ARE NO REAL ROOTS.
4. THE TWO REAL ROOTS ARE 1.187140396 and 88.8128596.

Answers

7. 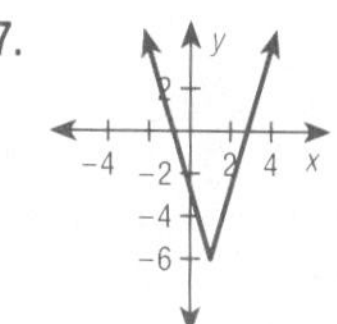

8. 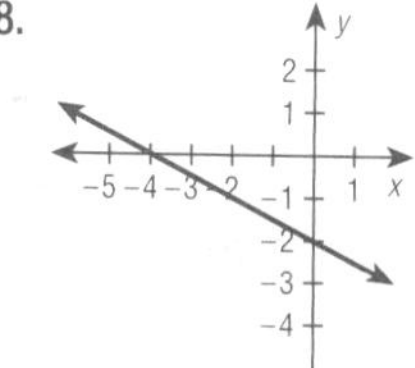

Mixed REVIEW

1. Find the quotient of x^2y and xy^2. **(1.3)**

2. Find the sum of $\frac{3}{7}$ and $\frac{1}{2}$. **(1.1)**

3. Evaluate $|x| - \sqrt{x}$ when $x = 9$. **(2.1, 9.1)**

4. Evaluate $|3y + 2| - \frac{1}{2}y$ when $y = -6$. **(2.1)**

5. What is the opposite of $-\frac{6}{7}$? **(2.1)**

6. What is the opposite of $-a^3$? **(2.1)**

7. Sketch the graph of $|3x - 3| - 6 = y$. **(4.7)**

8. Sketch the graph of $y = -\frac{1}{2}x - 2$. **(4.5)**

9. Evaluate $4[(3 \div 2) + \frac{1}{2}]$. **(1.1)**

10. Evaluate $(3x - 2)\frac{1}{2} + 2x$ when $x = 4$. **(1.4)**

11. Is (4, 6) a solution of $2x + 3y = 26$? **(1.5)**

12. Is (0, 3) a solution of $4x - 3y \le 5x + 1$? **(6.5)**

13. Solve $|x + \frac{1}{4}| - 8 < 0$. **(6.4)**

14. Solve $\frac{1}{3}(6x - 1) > 2$. **(6.1)**

In Exercises 15–20, write an equation for the indicated line. (5.2, 5.3)

15. Slope $m = -\frac{1}{2}$, through point $(3, -1)$

16. Slope $m = 5$, through point $(8, -6)$

17. Slope $m = 0$, through point $(-3, 6)$

18. Through two points: $(3, -5)$, $(3, 7)$

19. Through two points: (44, 28), (33, 17)

20. Through two points: $(-8, -3)$, $(-5, -5)$

Problem of the Day

Twenty-four tropical fish are on sale, some selling for \$6 each and the rest selling for \$10 each. Anyone buying all the cheaper fish would spend as much as anyone buying all the more expensive fish. How many cheaper fish are on sale? 15

ORGANIZER

Note: You might want to consider this lesson as optional for the Basic/Average level.

Warm-Up Exercises

1. Sketch the graphs.
   a. $y = 3x^2 - 4$
   b. $y = -x^2 + x + 6$

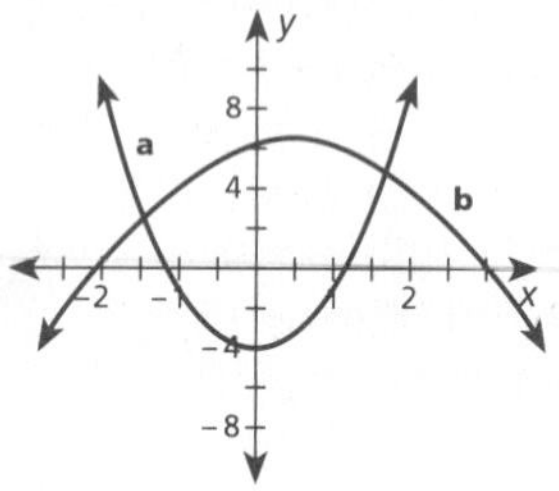

2. Sketch the solutions of the linear inequalities in two variables.
   a. $y > 12 - 4x$
   b. $2x - 3y \le 18$

a. Graph is region above dashed line with intercepts (0, 12) and (3, 0).

b.

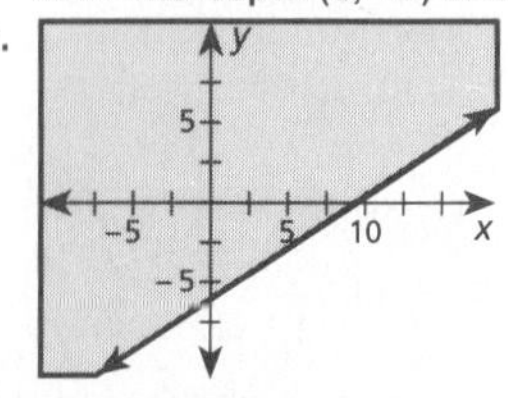

Lesson Resources

Teaching Tools
- Transparencies: 4, 11, 12, 13
- Problem of the Day: 9.6
- Warm-up Exercises: 9.6
- Answer Masters: 9.6

Extra Practice: 9.6

Color Transparency: 50

9.6 Graphing Quadratic Inequalities

What you should learn:

Goal 1 How to sketch the graph of a quadratic inequality

Goal 2 How to use quadratic inequalities as real-life models

Why you should learn it:

You can use quadratic inequalities to model many real-life situations, such as those in which the values of a quadratic expression cannot exceed a certain amount.

Goal 1 Sketching a Quadratic Inequality

In this lesson, you will study the following types of **quadratic inequalities.**

$y < ax^2 + bx + c \qquad y \le ax^2 + bx + c$

$y > ax^2 + bx + c \qquad y \ge ax^2 + bx + c$

The graph of any such inequality consists of the graph of all ordered pairs (x, y) that are solutions of the inequality. The steps used to sketch the graph of a quadratic inequality are similar to those used to sketch the graph of a linear inequality. (See Lesson 6.5.)

Sketching the Graph of a Quadratic Inequality

1. Sketch the graph of the parabola $y = ax^2 + bx + c$. (Use a *dashed* parabola for inequalities with $<$ or $>$ and a *solid* parabola for inequalities with $\le$ or $\ge$.)
2. Test a point inside the U-shape and one outside the U-shape.
3. Only one of the test points will be a solution. Shade the region that contains that test point.

Example 1 *Sketching the Graph of a Quadratic Inequality*

Sketch the graph of $y < 2x^2 - 3x + 4$.

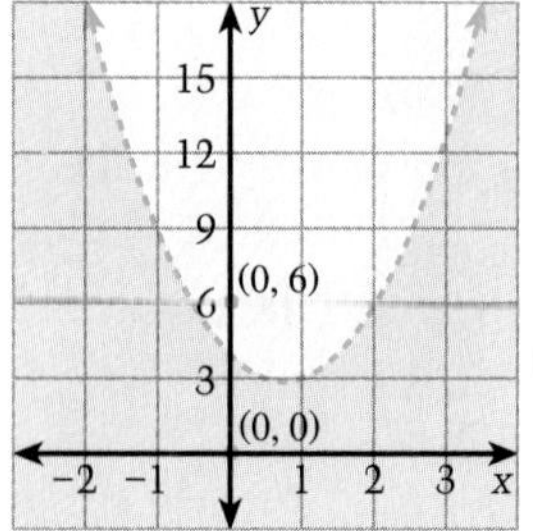

Solution The parabola $y = 2x^2 - 3x + 4$ has a vertex at $\left(\frac{3}{4}, \frac{23}{8}\right)$. (See graph at left.) Choose a point inside the U-shape, say (0, 6), and a point outside the U-shape, say (0, 0).

Point	Test the Point.	Conclusion
(0, 6)	$6 \overset{?}{<} 2(0)^2 - 3(0) + 4$	Because 6 is *not* less than 4, the ordered pair (0, 6) is not a solution.
(0, 0)	$0 \overset{?}{<} 2(0)^2 - 3(0) + 4$	Because 0 *is* less than 4, the ordered pair (0, 0) is a solution.

From this test, the graph must be the region outside the U-shape. ■

Goal 2 Quadratic Inequalities as Real-Life Models

The bridge over the Humber Estuary in England is the longest suspension bridge in the world. Its main suspension cables approximate the parabolic shape given by $y = 0.0001x^2$, where x and y are measured in feet. The vertex of the parabola is at the center of the bridge.

Example 2 ***The Humber Estuary Suspension Bridge***

The vertical cables that are suspended from the main cable lie in the region given by

$$y \le 0.0001x^2.$$

How high are the towers? Sketch a graph of the region containing the vertical cables between the towers.

Solution The distance between the two towers is 4626 feet. Because the vertex of the parabola $y = 0.0001x^2$ is at the center of the bridge, the x-coordinates of the tower tops must be ± 2313. Substituting either value into the equation gives a y-value of about 535. Thus, the towers are about 535 feet high. The graph of the parabola is shown below. The region that contains the vertical cables lies below the parabola and above the x-axis (the roadbed).

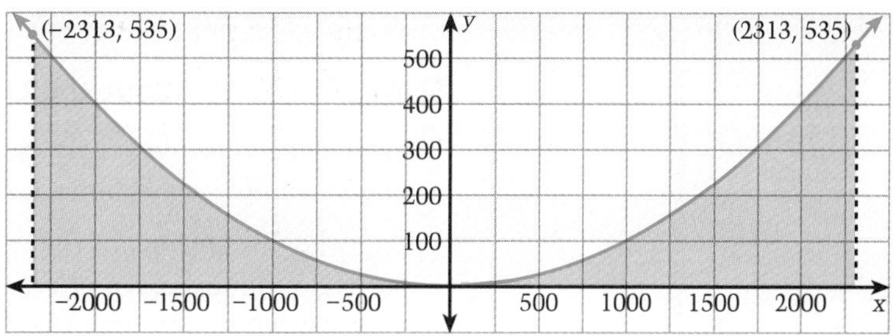

■

LESSON Notes

GOAL 1 Emphasize that the techniques used to solve linear inequalities are similar to those required here.

Example 1

Carefully go over how the test points are substituted into the original quadratic inequality.

Examples 2–3

Carefully explain the solutions presented in these examples. Encourage students to ask "what if" questions if needed to understand the relationships of the variables in each example.

Extra Examples

Here are additional examples similar to **Examples 1–3.**

1. Sketch the graphs of these quadratic inequalities.
   - **a.** $y \ge 3x^2 - 2x + 4$
   - **b.** $y < 0.0125x^2$

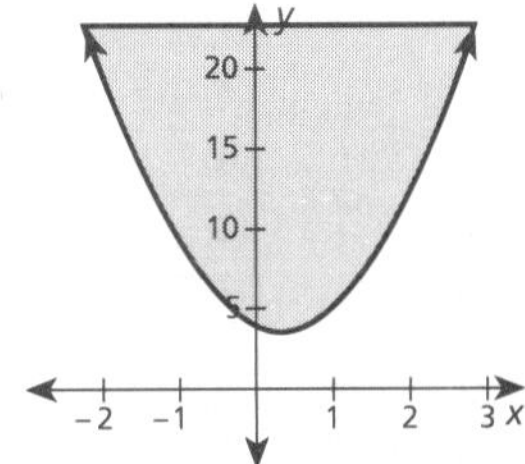

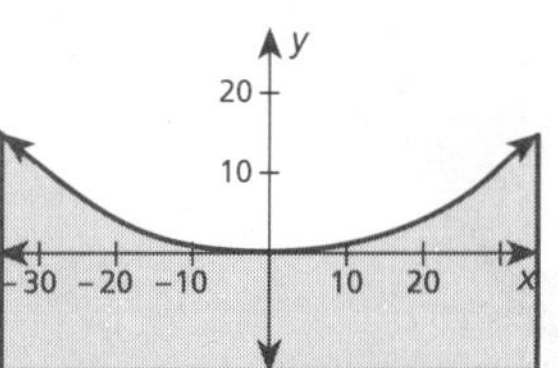

2. From the information given in Example 3, can you cover a bedroom door with dimensions 35 inches by 83 inches? Yes, when $x \approx 12.7$ or ≈ 14.3

Check Understanding

1. **MATH JOURNAL** Compare the techniques for solving linear and quadratic inequalities. How are they different? How are they alike? See Example 1 on page 486.
2. Include in your journal a definition of the solution of a quadratic inequality.

Communicating about ALGEBRA

Consider the sketch of the graph of the quadratic inequality $y \le 432x - 16x^2$ in Example 3. Explain why the area above the x-axis and below the parabola is the only region that should be shaded. The parabola represents the maximum dimensions, thus it forms one border. The dimensions are strictly positive, so the other border is $y = 0$.

Answer

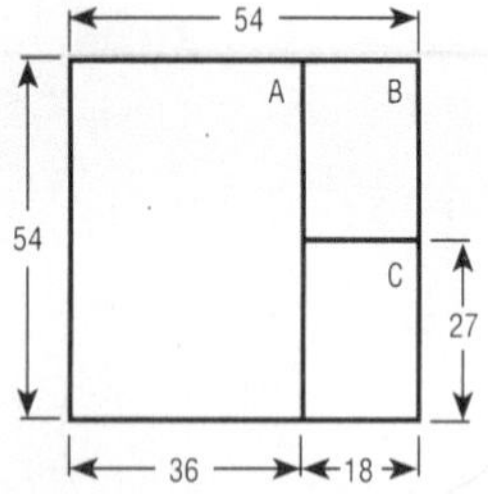

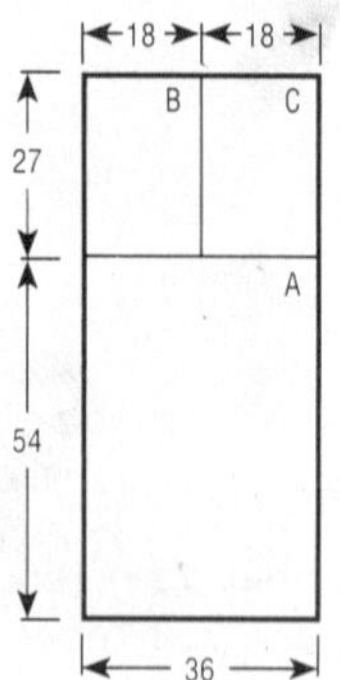

For a package to be accepted for mailing by the United States Postal Service, the sum of the length and girth of the package must not exceed 108 inches.

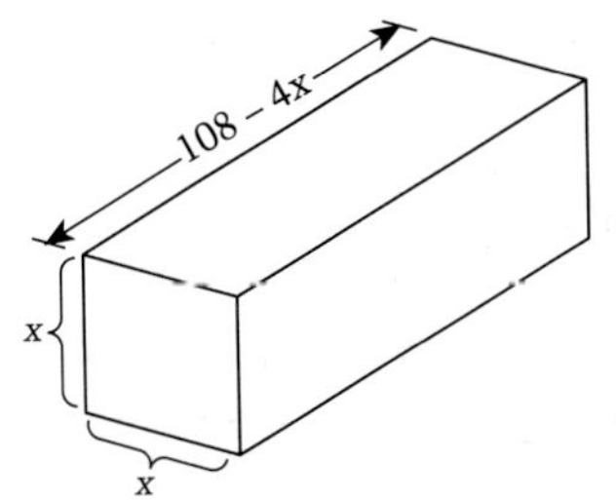

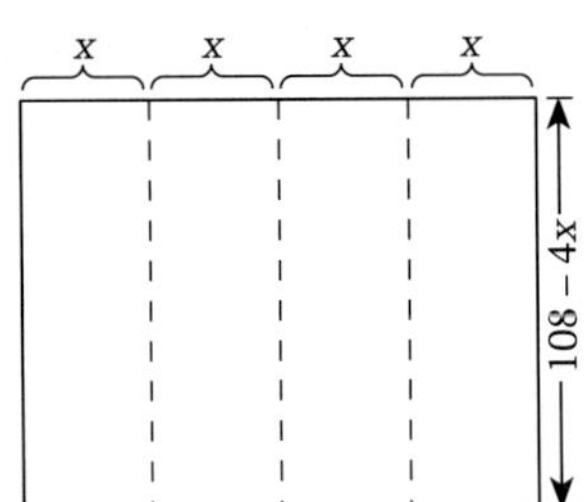

Connections
Geometry

Example 3 *Making a Poster*

The outside of a package has a pattern and pictures that you really like. You cut off and discard the square ends of the package. Your plan is to cut apart the sides to make hangings for your bedroom door. The door is 36 by 81 inches. Could there be enough cardboard to completely cover your door?

Solution Because the package was sent through the mail, you know that the sum of its length and girth (the distance around) is at most 108 inches. In the diagram, the girth of the box is $4x$. This means that the length is at most $108 - 4x$.

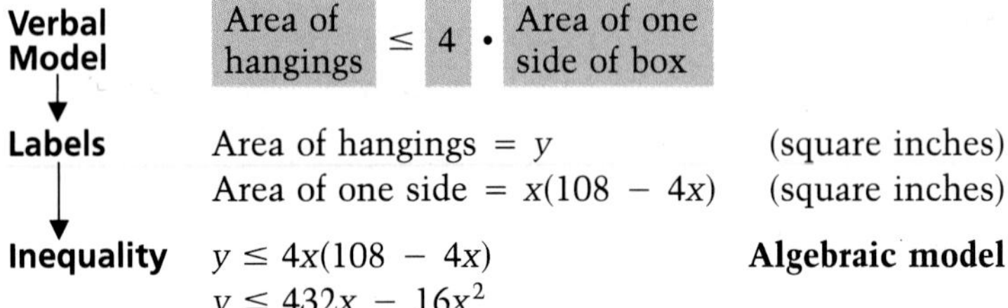

Verbal Model
Area of hangings ≤ 4 • Area of one side of box
Labels
Area of hangings = y (square inches)
Area of one side = x(108 – 4x) (square inches)
Inequality
y ≤ 4x(108 – 4x)
y ≤ 432x – 16x²
Algebraic model

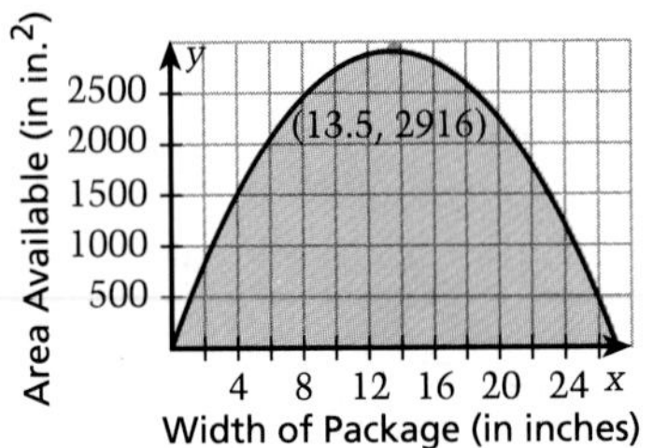

The graph of this inequality is shown at the left. Each point (x, y) in the shaded area represents a solution of the inequality, where y is the area of the cardboard that you intend to use for the hangings. The surface area of your door is $(36)(81) = 2916$ square inches, and the vertex of the parabola is $(13.5, 2916)$. If the ends of the package are 13.5 inches square, you have exactly enough to cover your door. ■

Communicating about ALGEBRA

SHARING IDEAS about the Lesson

Extend the Solution Only one shape of package will allow you to completely cover your door. (See Example 3.)

Find the dimensions of that package. Show how you could cut the unfolded sides of that package into only three rectangular pieces to cover the door. **13.5 in. by 13.5 in. by 54 in.**

EXERCISES

Guided Practice

CRITICAL THINKING about the Lesson 1., 2., 4. See margin.

1. Describe the steps used to sketch the graph of a quadratic inequality.
2. Sketch the graph of $y = x^2 + 3x + 2$.
3. Select a point inside the U-shape and a point outside the U-shape. Determine which of the points is a solution of the inequality $y \le x^2 + 3x + 2$.
   Points vary. (0, 0) is a solution.
4. Shade the region that contains the test point from Exercise 3 that satisfies the inequality.

Independent Practice

In Exercises 5–8, decide whether the point is a solution of the inequality.

5. $y \ge 2x^2 - x + 9$, (2, 10) Is not
6. $y > 4x^2 - 64x + 115$, (12, −60) Is
7. $y < x^2 + 6x + 12$, (−1, 6) Is
8. $y \ge x^2 - 7x + 9$, (−1, 16) Is not

In Exercises 9–14, match the inequality with its graph.

9. $y \ge -2x^2 - 2x + 1$ d
10. $y < -2x^2 + 4x + 3$ f
11. $y \ge 2x^2 + x + 1$ a
12. $y \ge 4x^2 - 6x + 5$ e
13. $y < 2x^2 - x - 3$ c
14. $y > -4x^2 + 6x - 5$ b

a. 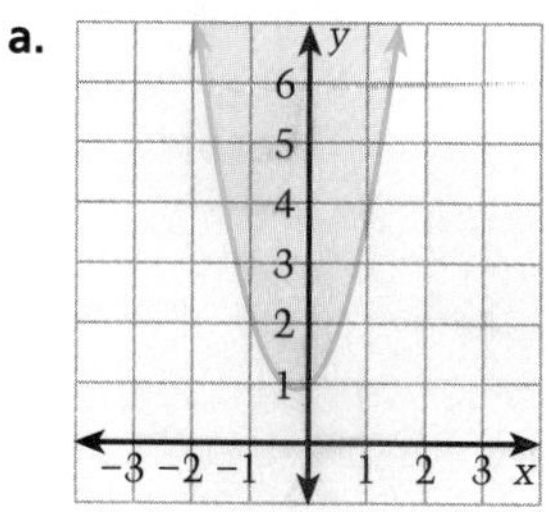

b. 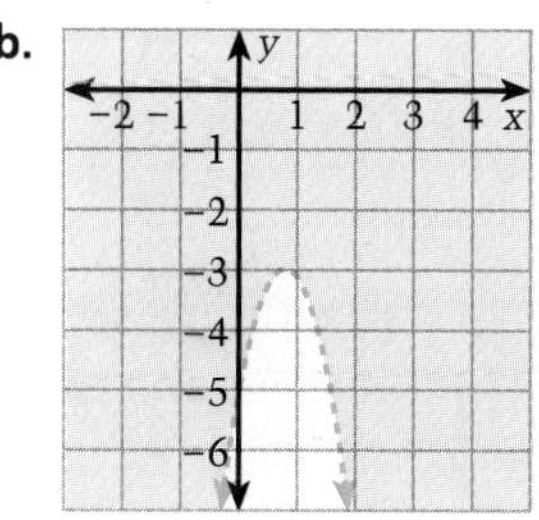

c. 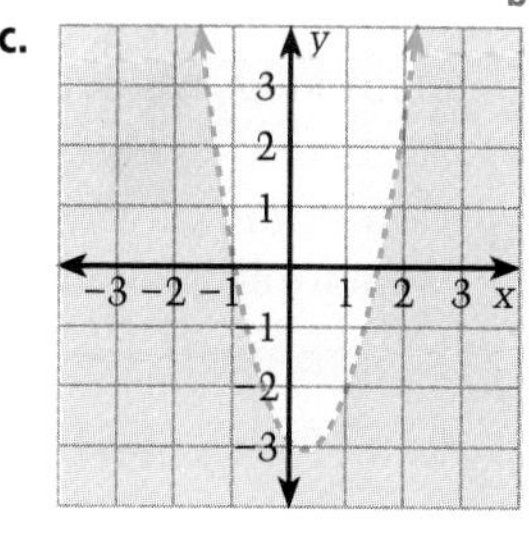

d. 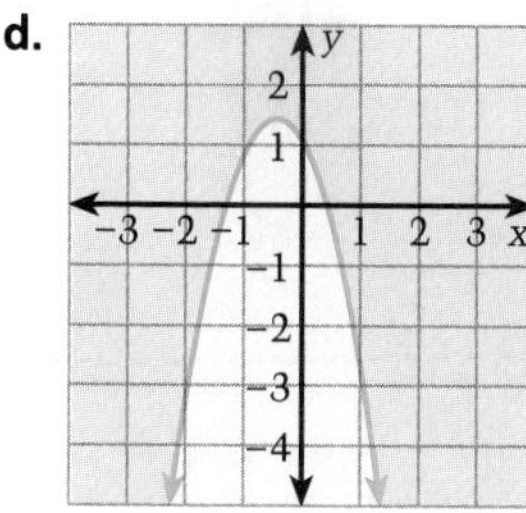

e. 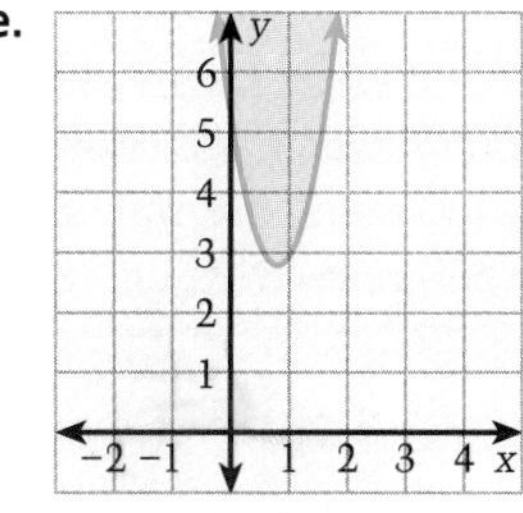

f. 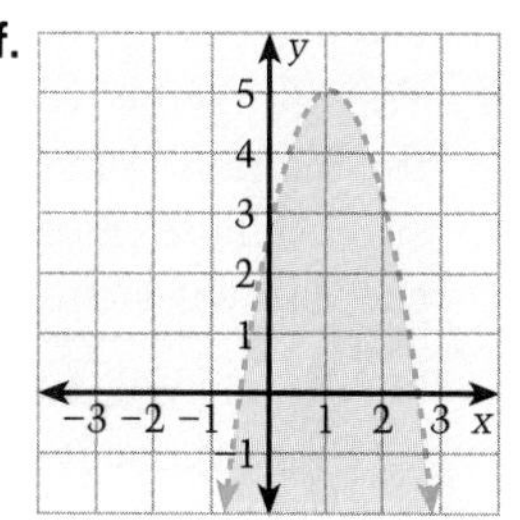

In Exercises 15–23, sketch the graph of the inequality. See Additional Answers.

15. $y < -x^2 - 4x - 3$
16. $y \ge x^2 - 3x + 1$
17. $y \ge 8x^2 - 3$
18. $y > -x^2 + 12x - 1$
19. $y \le 2x^2 + x + 6$
20. $y \ge 3x^2 + 9x - 1$
21. $y > -4x^2 - 8x - 4$
22. $y < -5x^2 - 3x + 6$
23. $y \le x^2 + 5x + 3$

EXERCISE Notes

ASSIGNMENT GUIDE

Basic/Average: Ex. 1–4, 5–7, 9–14, 15–23, 29–30, 31–41

Above Average: Ex. 1–4, 5–7, 9–14, 15–23 odd, 28–30, 31–43 odd

Advanced: Ex. 1–4, 9–17, 22–27, 29–30, 31–43 odd, 44

Selected Answers: Exercises 1–4, 5–41 odd

More Difficult Exercises
Exercises 24–27, 29–30, 43–44

Guided Practice

Ex. 1–4 GROUP ACTIVITY These exercises review the steps for graphing an inequality. They should be used as an in-class activity before assigning the independent practice.

Answers

1. Sketch the graph of the parabola. Test two points, one inside and one outside the U-shape. Then shade the region containing the point that is in the solution.

2., 4. 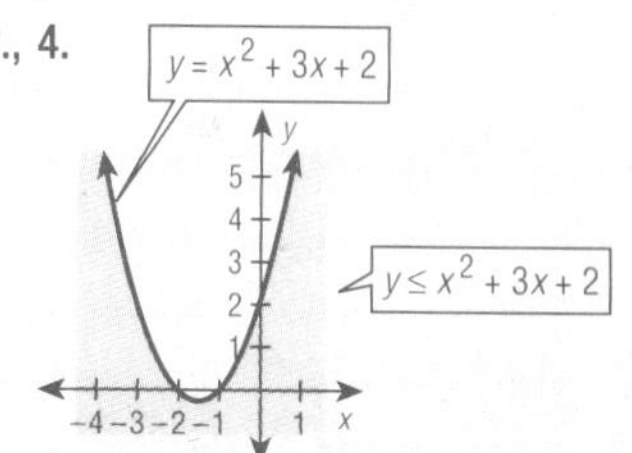

Independent Practice

Ex. 9–14 These exercises should be assigned as a group. Encourage students to use the quadratic formula to verify the value of the x-intercepts.

Ex. 15–23 TECHNOLOGY Have students use a graphics calculator to graph these equations.

▶ **Ex. 24–27** These exercises should be assigned as a group. Review the interpretation of the terminology for each graph before assigning them.

Notice that in the graph, y is given in thousands of dollars while in the model, y is given in dollars.

Answer

24. Above the graph for IF; if true, the diamond would be overpriced.

▶ **Ex. 29–30** These are paired exercises. Students need to realize that the population in Set A is less than the population in Set N. The 40 represents the 1990 population.

Answer

30. b; no; if true there would have been 506 million in the United States alone. The world, according to the model, had 496 million sets.

▶ **Ex. 43–44** TECHNOLOGY Students with graphics calculators can use the calculator's DRAW/ SHADE feature to complete these graphs.

Answers

43.

44.

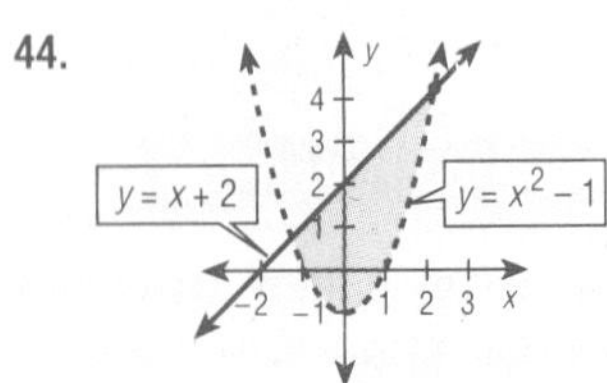

Diamonds, Diamonds, Diamonds! **In Exercises 24–27, use the following information.**

The price of a diamond is determined by its weight, cut, clarity, and flawlessness. Flawless diamonds with perfect clarity are very rare and very expensive. The equations and their graphs show the wholesale prices of a round-cut diamond with perfect clarity. The five highest grades of flawlessness vary from IF (flawless) to VS2. *(Source: Hook's Gallery)*

Flawlessness	Equation
IF	$y = 40 + 13{,}000x^2$
VVS1	$y = 31 + 10{,}160x^2$
VVS2	$y = 27 + 8670x^2$
VS1	$y = 20 + 6660x^2$
VS2	$y = 16 + 5150x^2$

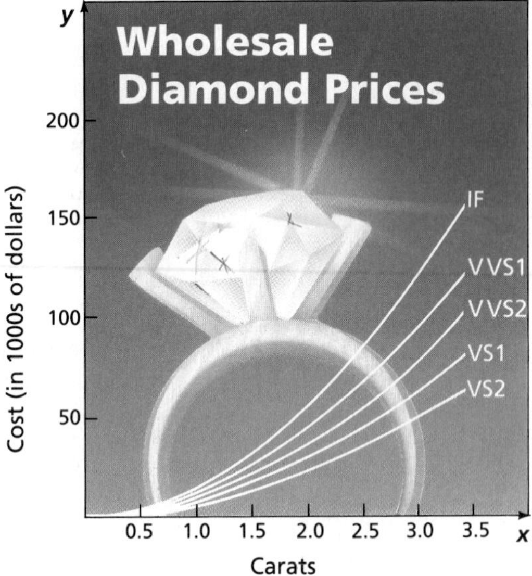

24. Your aunt tells you that she can get a two-carat diamond at the "wholesale price" of $70,000. She says the diamond is of the highest quality. Describe the region in the coordinate plane that contains the point (2, 70,000). What can you conclude about your aunt's statement? See margin.

25. The wholesale price of a $2\frac{1}{2}$-carat diamond with perfect clarity is $55,000. Assume that the price is a fair representation of the flawlessness of the diamond. Which region contains the point $(2\frac{1}{2}, 55{,}000)$?

a. $y > 31 + 10{,}160x^2$

b. $\underline{y < 31 + 10{,}160x^2}$

What grade is the diamond? VVS2

26. You are trying to decide how to cut a large rough diamond. Parts of the diamond have minor inclusions (foreign substances) and fissures (small cracks). Other parts, however, are flawless. Which of the following combinations has the greatest value? Explain.

a. One 3-carat VVS2 and one 1-carat VS1

b. One 2-carat IF and one 2-carat VS1

c. One $1\frac{1}{2}$-carat IF and two $1\frac{1}{4}$-carat VVS2

The diamonds in choice a have the greatest total value.

27. Many famous large diamonds have been found. The Jonker diamond, found in 1934, weighed 726 carats! The rough diamond was cut into 12 flawless stones of perfect clarity, the largest weighing 125 carats. Your uncle says he can get this latter diamond for you at a wholesale price! What price does he have in mind? ≈$200 million

28. *Light Reflector* A cross section of the parabolic light reflector at the right is described by the equation $y = \frac{1}{24}x^2 + 1$. The bulb inside the reflector is located at the point (0, 7). Light bounces off the reflector in parallel rays. This quality allows flashlights to direct light in a narrow beam. In which region is the bulb located?

a. $y > \frac{1}{24}x^2 + 1$ **b.** $y < \frac{1}{24}x^2 + 1$

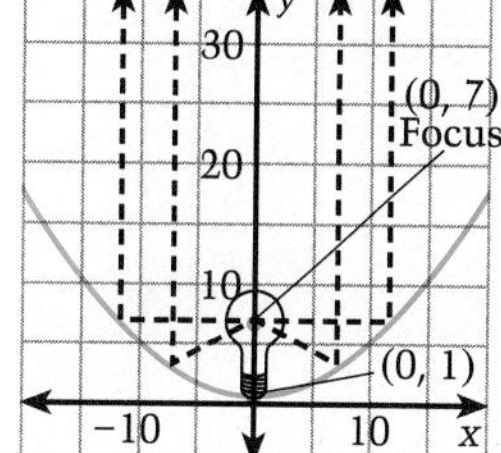

Glued to the Tube **In Exercises 29 and 30, use the following information.**

The number, N (in millions), of television sets in use in the world between 1950 and 1990 can be modeled by $N = \frac{1}{2}t^2 + 46$ where $t = 0$ corresponds to 1950.

29. Let A represent the number of television sets in North and South America in 1990. In which region is the point $(40, A)$ located? Explain.

a. $N < \frac{1}{2}t^2 + 46$ **b.** $N > \frac{1}{2}t^2 + 46$

The number in the Americas will be less than the number in the world.

30. In which region is the point (30, 500) located?

a. $N < \frac{1}{2}t^2 + 46$ **b.** $N > \frac{1}{2}t^2 + 46$

If someone told you that in 1980, Americans owned an average of 2.25 television sets each, would you believe them? Explain. (1980 U.S. Population: 225 million) See margin.

Integrated Review

In Exercises 31–34, solve the inequality.

31. $9x - 20 > 61$ $x > 9$

32. $2x + 9 < 3$ $x < -3$

33. $3x + 5 \le 17$ $x \le 4$

34. $\frac{1}{2}x - 2 \ge 6$ $x \ge 16$

In Exercises 35–38, sketch the graph of the inequality. See Additional Answers.

35. $y \ge x + 3$ **36.** $y < -x + 1$ **37.** $y > -2x - 4$ **38.** $y \ge 3x - 4$

In Exercises 39–42, sketch the graph of the equation. See Additional Answers.

39. $y = x^2 + 4x - 5$ **40.** $y = x^2 - 1$ **41.** $y = -\frac{1}{2}x^2 + \frac{3}{2}x - 4$ **42.** $y = -2x^2 - 5x + 3$

Exploration and Extension

In Exercises 43 and 44, shade the region whose points satisfy both inequalities.

43. $\begin{cases} y \le 3x^2 + 5 \\ y > 1 \end{cases}$

44. $\begin{cases} y > x^2 - 1 \\ y \le x + 2 \end{cases}$

Enrichment Activities

These activities require students to go beyond lesson goals.

1. Use a piece of grid paper. Draw all rectangles with integral sides that have a perimeter of 24. Use your drawings to complete this table.

Length	Width	Area
1	11	11
2	10	20
3	9	27
4	8	32
5	7	35
6	6	36
7	5	35
...	...	...
11	1	11

2. Make a scatter plot of length versus area.

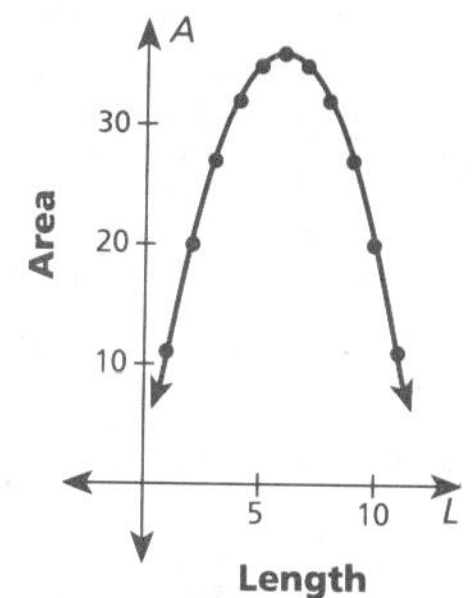

3. Since $A = L \cdot W$ and $P = 2L + 2W$, can you write the equation of a quadratic model that connects the points? (*Hint:* Solve for W in terms of P and L first.)

Since $P = 2L + 2W$, then $W = \frac{P}{2} - L$.

$A = L \cdot W = L(\frac{P}{2} - L)$.

Substitute 12 for $\frac{P}{2}$. Thus, $A = L(12 - L) = 12L - L^2$. The model is $A = 12L - L^2$.

Problem of the Day

What is the length of the diagonal of a 10-cm cube?

$\sqrt{300}$ cm, or about 17.32 cm

ORGANIZER

Warm-Up Exercises

1. Sketch the graph of each equation.

a. $y = 3x - 6$

b. $y = 0.5(1.4)^x$

c. $y = 2|x + 3| - 5$

d. $y = 2x^2 + 8x - 4$

a. line with intercepts (0, −6) and (2, 0).

b.

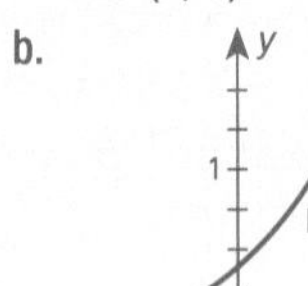

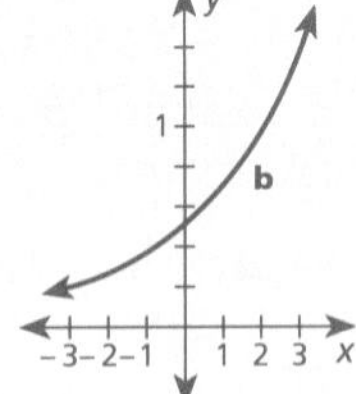

c.

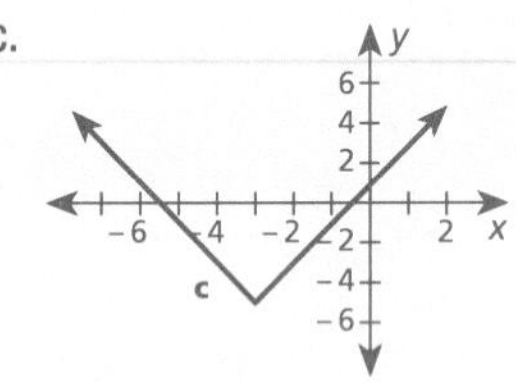

d. parabola with vertex (−2, −12).

x	−3	−2	−1	0	1	2
y	−10	−12	−10	−4	6	20

2. What is a scatter plot of data? How is it used to obtain a line of best fit? The graphs of points in a collection of data. By searching for a pattern among the points, a line that has about as many points above it as below it may be fitted.

Lesson Resources

Teaching Tools
- Transparencies: 4, 10, 13, 14
- Problem of the Day: 9.7
- Warm-up Exercises: 9.7
- Answer Masters: 9.7

Extra Practice: 9.7

Color Transparencies: 51, 52

Applications Handbook: p. 49–51

Technology Handbook: p. 76

9.7 Exploring Data: Comparing Models

What you should learn:

Goal 1 How to choose a model that best fits a collection of data

Goal 2 How to use models in real-life settings

Why you should learn it:

To be skilled at modeling, you must know characteristics of the different models and be able to recognize those characteristics in real-life settings.

Goal 1 Choosing a Model

You have now studied four basic types of algebraic models.

Type of Model	Equation	Graph
1. Linear	$y = ax + b$	Line
2. Absolute Value	$y = a\|x + b\| + c$	V-Shaped
3. Exponential	$y = C(1 \pm r)^x$	Exponential curve
4. Quadratic	$y = ax^2 + bx + c$	Parabola (U-Shaped)

This lesson will help you understand the differences between these basic types, and will help you choose the type of model that best fits a collection of data.

Example 1 *Choosing a Model*

Name the type of model that best fits each data collection.

a. $(-3, 4), \left(-2, \frac{7}{2}\right), (-1, 3), \left(0, \frac{5}{2}\right), (1, 2), \left(2, \frac{3}{2}\right), (3, 1)$

b. $(-3, 4), (-2, 3), (-1, 2), (0, 1), (1, 2), (2, 3), (3, 4)$

c. $(-3, 4), (-2, 2), (-1, 1), \left(0, \frac{1}{2}\right), \left(1, \frac{1}{4}\right), \left(2, \frac{1}{8}\right), \left(3, \frac{1}{16}\right)$

d. $(-3, 4), \left(-2, \frac{7}{3}\right), \left(-1, \frac{4}{3}\right), (0, 1), \left(1, \frac{4}{3}\right), \left(2, \frac{7}{3}\right), (3, 4)$

Solution Make scatter plots of the data. Then decide whether the points lie on a line, a V, an exponential curve, or a parabola.

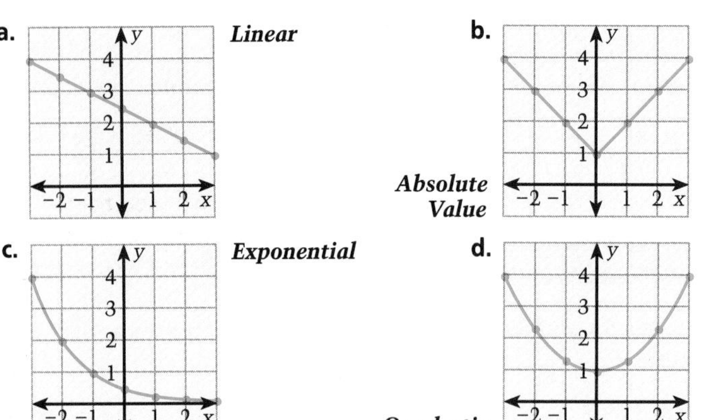

Goal 2 Using Models in Real-Life Settings

Time, t	Distance, d
0.25 s	0.61 in.
0.50 s	2.45 in.
0.75 s	5.50 in.
1.00 s	9.78 in.
1.25 s	15.28 in.
1.50 s	22.00 in.
1.75 s	29.95 in.
2.00 s	39.12 in.
2.25 s	49.51 in.
2.50 s	61.13 in.
2.75 s	73.96 in.
3.00 s	88.02 in.

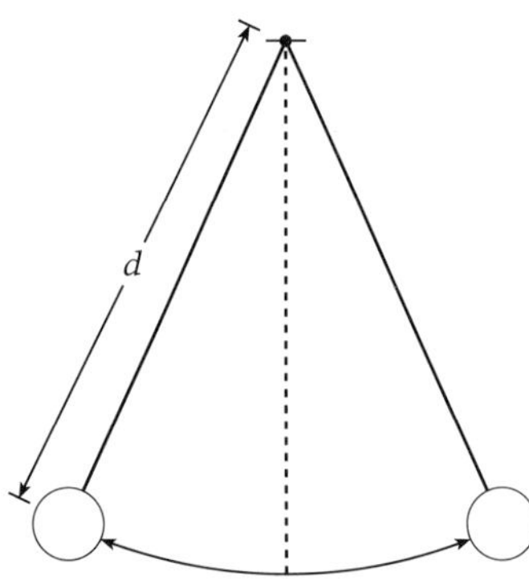

A simple pendulum can be constructed with a string and a weight. The distance, d (in inches), between the fixed point and the weight is related to the time, t (in seconds), it takes for the pendulum to swing through one full* period *(over and back).

Problem Solving
Make a Graph

Example 2 *Finding a Model*

Which type of model best fits the pendulum behavior described above? Find the model.

Solution Draw a scatter plot for the data, as shown at the left. From the scatter plot, you can see that the pattern is *not* linear or absolute value. If it fits any of the four basic types, it must be exponential or quadratic. If the pattern was exponential, then the pendulum lengths would have to increase by the same percentage (for equal increases in time). This, however, is not true, because $2.45 \div 0.61 \approx 4$, but $5.50 \div 2.45 \approx 2.24$ ($t = 0.25, 0.50, 0.75$).

Relationship between Period and Length of a Pendulum

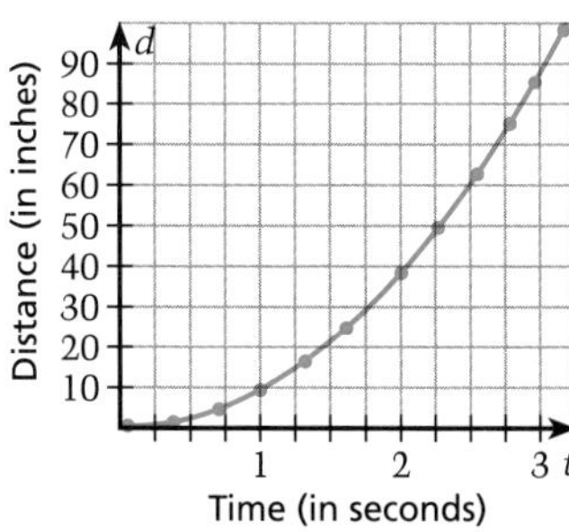

The only model left to try is a quadratic model. Let's try a simple one: $d = at^2$. To find a, substitute any known value of d and t.

$d = at^2$ *Simple quadratic model*

$9.78 = a(1^2)$ *Substitute $d = 9.78$ and $t = 1$.*

$9.78 = a$ *Solve for a.*

At this point, you do not know if the model $d = 9.78t^2$ works for any other values of d and t. However, after testing several values, you can see that the model fits the data. Therefore, the length of a pendulum is related to the period of the pendulum by the quadratic model

$$d = 9.78t^2.$$

The Dutch scientist Christiaan Huygens used this model when he patented the first pendulum clock in 1657. ■

LESSON Notes

Example 1

Observe that the patterns of scatter plots seldom fit either of the four possible models exactly. You must exercise care when deciding which model should be selected.

Examples 2–3

Point out that each model but one was considered and eliminated based upon characteristics of the model that were not met. This process of elimination is generally very effective. Be sure students recognize that the exponential model is not appropriate in Example 2 but is appropriate in Example 3, because of the results of the algebraic computations that were performed. See Lessons 8.6 and 8.7 for a review of exponential growth and decay models.

Extra Examples

Here are additional examples similar to **Examples 1–3.**

1. Name the model that best fits each collection of data.
   a. (−9, 26), (−5, 18), (−2, 12), (−1, 10), (1, 6), (4, 0), (6, 4), (7, 6), (10.5, 13)
   b. (−8.5, 20), (−8, 19), (−2, 7), (0, 3), (1, 1), (6, −9), (7.5, −12), (10, −17), (15, −27), (19.5, −36)
   c. (−4.11, 0.95), (−3.11, 1.42), (−2.1, 2.13), (−1.1, 3.2), (0.11, 5.23), (1.11, 7.84), (6.11, 59.55), (13.11, 1017.48)

   a. absolute value model
   b. linear model
   c. exponential model
2. Using the data in Exercise 1, find an algebraic model for Parts a–c.

   a. one model is $y = 2|x - 4|$
   b. one model is $y = 3 - 2x$
   c. one model is $y = 5(1.5)^x$.

Check Understanding

1. Make a flowchart describing the steps one takes to obtain a best-fitting line or curve for a collection of data. See Example 1.
2. How does one decide that a scatter plot is best described by an exponential relationship rather than a quadratic relationship? See Examples 2 and 3.

Communicating about ALGEBRA

Ask students to construct data that can be described by one of the four models. Once completed, other students should attempt to find the model that best fits the data.

Chamber, n	Volume, v
0	0.787
1	0.837
2	0.889
3	0.945
4	1.005
5	1.068
6	1.135
7	1.207
8	1.283
9	1.364
10	1.450
11	1.541

The volumes, v, of 12 consecutive chambers of a nautilus are shown in the table. The chambers are numbered from n = 0 to n = 11.

Problem Solving
Make a graph

Relationship between Position and Volume of Nautilus Chambers

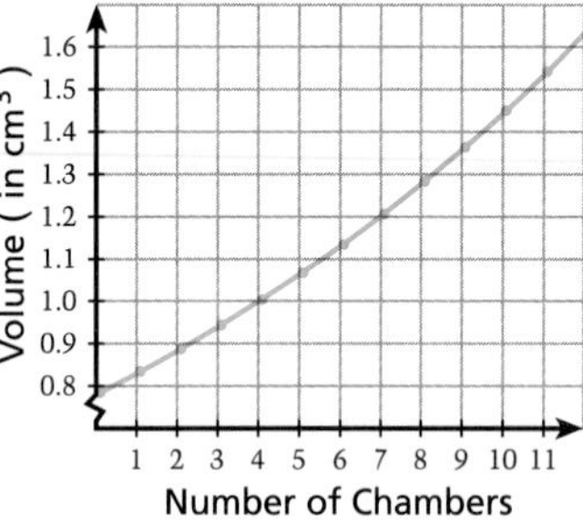

Example 3 Finding a Model

Find a model for the chambered nautilus data above.

Solution From the scatter plot, you can see that the model is not linear or absolute value. To see whether an exponential model fits, find the ratios of consecutive volumes.

$$\frac{0.837}{0.787} \approx 1.064 \qquad \frac{0.889}{0.837} \approx 1.062 \qquad \frac{0.945}{0.889} \approx 1.063$$

Each chamber's volume is about 6.3% larger than the previous chamber. So you can use the exponential model $v = C(1 + r)^n$. In this model, $r = 0.063$ and C is the volume when $n = 0$.

$$v = 0.787(1.063)^n$$

This model was first discovered by the French mathematician René Descartes. ■

Communicating about ALGEBRA

Cooperative Learning

SHARING IDEAS about the Lesson

Work with a Partner and Use a Table Find a model that relates the variables.

The area, A, of an *ellipse* whose radii are r and $r + 1$ is shown in the table. (This model occurred in the writings of Hypatia. She was a Greek mathematician who lived in Egypt.)

Radius, r	1	2	3	4	5	6
Area, A	2π	6π	12π	20π	30π	42π

$A = \pi(r^2 + r)$

EXERCISES

Guided Practice

CRITICAL THINKING about the Lesson

1. State the four types of algebraic models we have studied. Write an equation for each.
   1. See page 492 for examples.
2. Which model has a U-shaped graph? Quadratic
3. Which model has a V-shaped graph? Absolute value
4. Describe the graph of each model and sketch an example of each. See page 492 for a graph of each model.

Independent Practice

In Exercises 5–8, name the type of model suggested by the graph.

5. 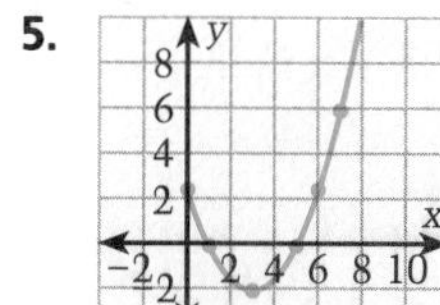
Quadratic

6. 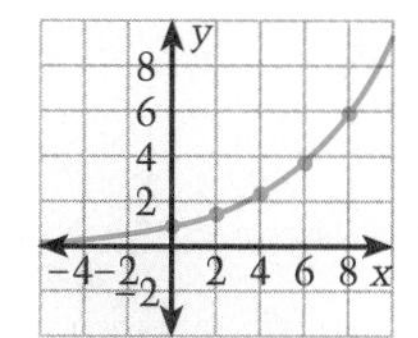
Exponential

7. 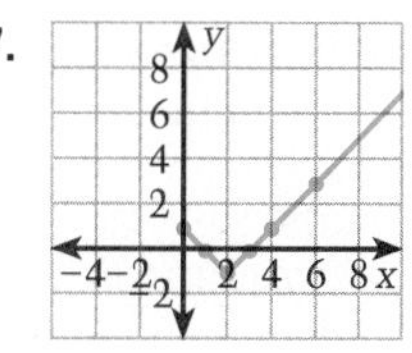
Absolute value

8. 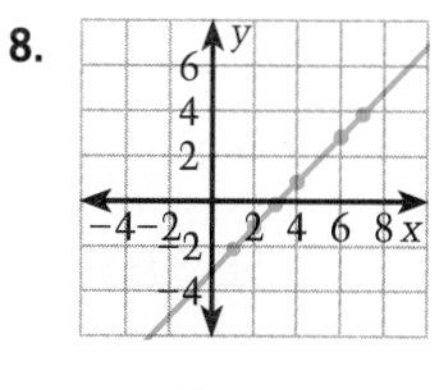
Linear

In Exercises 9–14, make a scatter plot of the data. Then name the type of model that best fits the data. See margin in page 497 for graphs.

9. $(-1, 16), (0, 4), (1, -2), (2, -2), (3, 4), (5, 34)$ quadratic
10. $(-3, 2), \left(-2, \frac{5}{2}\right), \left(-1, \frac{7}{2}\right), (0, 5), (1, 7), \left(2, \frac{19}{2}\right)$ exponential
11. $(-2, 2), \left(-1, \frac{5}{2}\right), (0, 3), \left(1, \frac{7}{2}\right), (2, 4), \left(3, \frac{9}{2}\right)$ linear
12. $(-4, 5), (-3, 3), (-2, 5), (-1, 7), (0, 9), (1, 11)$ absolute value
13. $(-2, -3), (-1, 4), (0, 9), (1, 12), (3, 12), (5, 4)$ quadratic
14. $(-5, 6), (-4, 3), (-2, -3), (-1, -6), (0, -9), (1, -12)$ linear

Cool, Cool Cotton **In Exercises 15 and 16, use the following information.**

The table gives the cotton consumption, C (in pounds), per American between 1950 and 1990. Let t represent the year, with $t = 0$ corresponding to 1960.

Year, t	−10	−5	0	5	10	15	20	25	30
Pounds of Cotton, C	30.6	27.6	24.6	21.6	18.6	15.6	12.6	13.2	16.2

✪ 15. Which of the following models best fits the data?
   a. $C = 24.6 - 0.6t$ b. $C = -1.8 + 0.6t$
   c. $C = 0.6|t - 22| + 11.4$

✪ 16. Describe the pattern of cotton consumption between 1950 and 1990. What caused this type of pattern? See margin.

EXERCISE Notes

ASSIGNMENT GUIDE

Basic/Average: Ex. 1–4, 5–14, 19–29 odd

Above Average: Ex. 1–4, 5–8, 9–21 odd, 23–29 odd, 31

Advanced: Ex. 1–4, 5–8, 15–18, 19–22, 23–29 odd, 31

Selected Answers
Exercises 1–4, 5–29 odd

✪ More Difficult Exercises
Exercises 15–18, 31

Guided Practice

▶ **Ex. 1–4** These exercises should remind students to make an intuitive visual connection for each model.

Independent Practice

▶ **Ex. 5–8** These exercises should be assigned as a group.

▶ **Ex. 15–16** TECHNOLOGY These exercises are paired.

Answer

16. Between 1950 and 1982, cotton consumption decreased. After 1982, consumption increased. The decrease in consumption was caused by the increase in man-made fibers. The increase in consumption was caused by the rediscovery of the comfort of cotton.

▶ **Ex. 18** This exercise could be done as an in-class model solution.

Answer

18. 61,400; 67,500; 75,000; 83,900; 94,200; 105,900; 119,000; 133,500; 149,400; 166,700; 185,400; 6100, 7500, 8900, 10300, 11700, 13100, 14500, 15900, 17300, 18700. Each difference is 1400.

EXTEND Ex. 18–21 RESEARCH Use library resources or almanacs to find examples of data similar to that found in Exercises 18–21 that can be defined by making use of linear, exponential, or quadratic models. Be sure to quote the sources of your data. Show the data on a scatter plot and approximate the model by writing a linear, exponential, or quadratic equation.

Ex. 31 Use the following to help students to discover the pattern.

$n = 4$; $1 + 2 + 3 + 4 = (1 + 4) + (2 + 3)$; 2 groups of 5

$n = 6$; $1 + 2 + 3 + 4 + 5 + 6 = (1 + 6) + (2 + 5) + (3 + 4)$; 3 groups of 7

$n = 8$; $1 + 2 + 3 + 4 + 5 + 6 + 7 + 8; = (1 + 8) + (2 + 7) + (3 + 6) + (4 + 5)$; 4 groups of 9

This pattern yields $\frac{n}{2}$ groups of $(n + 1)$ or $\frac{n}{2}(n + 1) = \frac{1}{2}n^2 + \frac{1}{2}n$.

Enrichment Activity

This activity requires students to go beyond lesson goals.

EXTEND Ex. 9–14 Use the graphs from Exercises 9–14 and write a model that describes the points. For some of the graphs, you may be able to use the STAT feature of the graphics calculator.

17. *Stretching a Spring* Different masses, M (in kilograms), are hung from a spring. The distances, d (in centimeters), that the spring stretches are recorded in the table. Which type of model best fits these data? Find the model. (This model was discovered by the English mathematician Robert Hooke in the late 1600's.) **Linear, $d = 2.6M$**

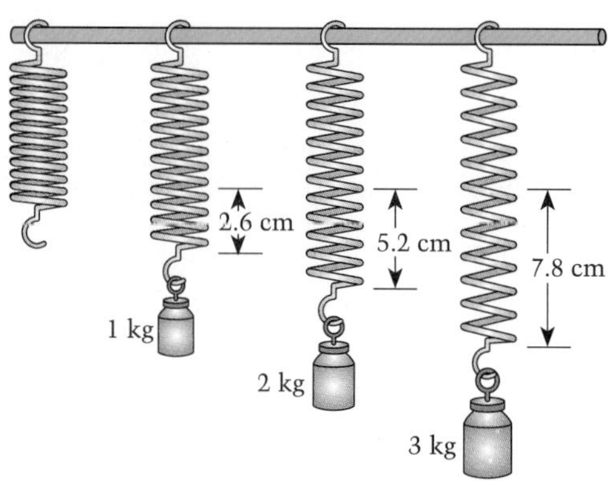

M	1	2	3	4	5	6	7
d	2.6	5.2	7.8	10.4	13	15.6	18.2

18. *Incompatible Computer Hardware* The estimated amount of money, A (in millions of dollars), that businesses spent on incompatible hardware between 1984 and 1994 can be modeled by

$A = 51{,}000 - 200t + 700t^2$,

where $t = 4$ corresponds to 1984. Complete the table. Then compute the difference of consecutive A-values.

(Amount in 1985) − (Amount in 1984)
(Amount in 1986) − (Amount in 1985)
(Amount in 1987) − (Amount in 1986)
⋮ See margin in page 495. ⋮

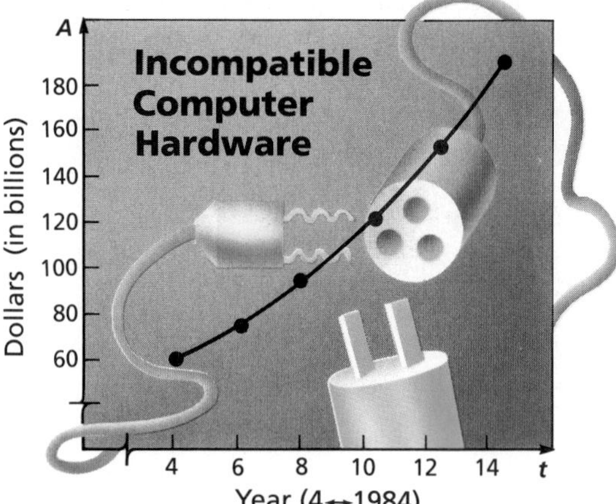

Describe the pattern given by these differences.

Year, t	4	5	6	7	8	9	10	11	12	13	14
Amount, A	?	?	?	?	?	?	?	?	?	?	?

19. *Population of the Philippines* In 1970, the population, P, in the Philippine Islands was 38,680,000. Since then, the average annual percent increase has been about 2.7%. Find a model that best fits the population of the Philippines for the years from 1970 on. **$P = 38{,}680{,}000(1.027)^t$**

20. *Grocery Sales* The sales of groceries, G (in billions of dollars), in the United States between 1970 and 1990 is given in the table. The year is represented by t, with $t = 0$ corresponding to 1970. Which type of model best fits this data? Find the model. **Linear, $G = 88 + 13.3t$**

Year, t	0	5	10	15	16	18	20
Grocery Sales, G	88	154.5	221	287.5	300.8	327.4	354

21. *Buy It through the Mail!* In 1970, the direct-mail purchases in the United States totaled \$2766 million. From 1970 to 1990, the purchases increased about 12% per year. Find a model for the total direct-mail expenditures, D (in millions), from 1970 to 1990. Let $t = 0$ represent 1970. **$D = 2{,}766{,}000{,}000(1.12)^t$**

22. *Surface Area of a Sphere* The surface area, S, of a sphere with radius r is shown in the table. Find a model that relates. (This model was discovered by mathematicians in early Greece.) $S = 4\pi r^2$

Radius, r	1	2	3	4	5
Area, S	4π	16π	36π	64π	100π

Integrated Review

In Exercises 23–25, find an equation of the line that passes through the two points.

23. $(3, -2)$, $(5, 4)$ $3x - y = 11$

24. $(-3, -9)$, $(5, 7)$ $2x - y = 3$

25. $(2, 3)$, $(-4, 6)$ $x + 2y = 8$

In Exercises 26–28, solve the equation.

26. $t^2 + t - 30 = 0$ $5, -6$

27. $\frac{1}{2}x^2 - 35 = 37$ ± 12

28. $3t^2 - 4t = 0$ $\frac{4}{3}, 0$

In Exercises 29 and 30, write an exponential model for the amount A in year t.

29. Initial value: 1000, 4% increase each year $A = 1000(1.04)^t$

30. Initial value: 160, 8% decrease each year $A = 160(0.92)^t$

Exploration and Extension

31. *Hey, Teacher, I Know!* Karl Gauss was one of the most famous mathematicians who ever lived. Read the following story.

One day in math class, during Gauss's early school life, Gauss's teacher wanted to give the students a project that would use up some of the class time. The teacher asked the students to find the sum of the integers from 1 to 100. After only a few seconds, Karl stood up and said, "The sum is 5050!"

To see how Karl may have found the total, consider the following:

$$1 = 1 \quad n = 1$$
$$1 + 2 = 3 \quad n = 2$$
$$1 + 2 + 3 = 6 \quad n = 3$$
$$1 + 2 + 3 + 4 = 10 \quad n = 4$$
$$1 + 2 + 3 + 4 + 5 = 15 \quad n = 5$$
$$1 + 2 + 3 + 4 + 5 + 6 = 21 \quad n = 6$$

$S = \frac{1}{2}n^2 + \frac{1}{2}n$; 1, 3, 6, 10, 15, 21, . . . are the numbers of dots used in making successive triangular arrays:

Let S be the sum of the first n integers. There is a quadratic model for S of the form $S = \boxed{?}n^2 + \boxed{?}n$. Can you find it? The numbers 1, 3, 6, 10, 15, and 21 were called triangular numbers by mathematicians in early Greece. Can you guess why?

9. $y = 3x^2 - 9x + 4$

10. $y = 5(1.4^x)$

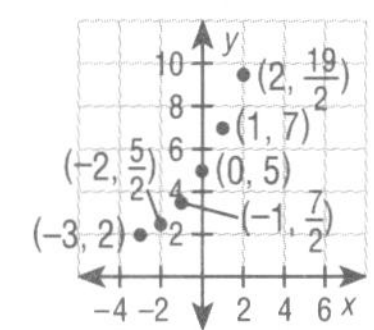

11. $y = \frac{1}{2}x + 3$

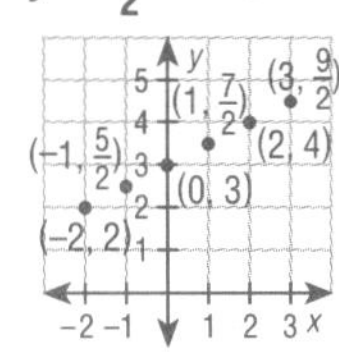

12. $y = 2|x + 3| + 3$

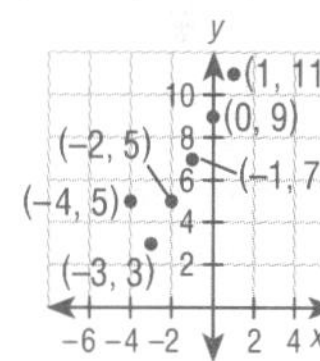

13. $y = -x^2 + 4x + 9$

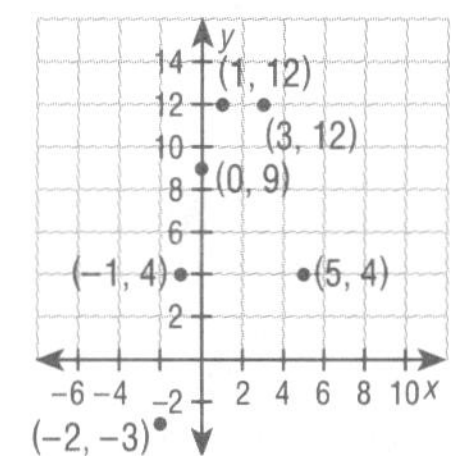

14. $y = -3x - 9$

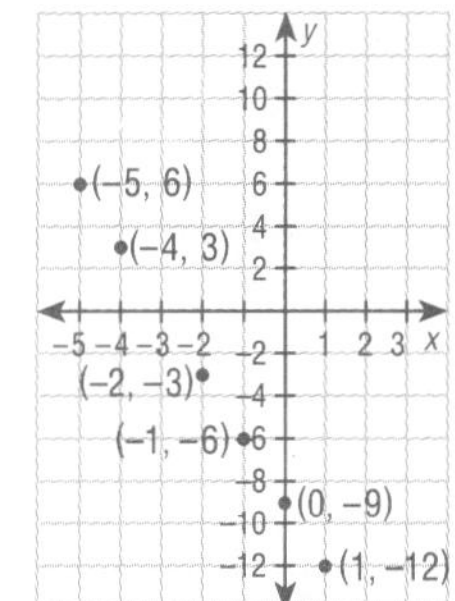

ORGANIZER

TIME SCHEDULE: All levels, two days

The Chapter Summary helps students organize the main ideas of the chapter. This chapter concluded by making the connection between modeling and solving real-life problems, and choosing appropriate algebraic models to interpret data. The chapter began with an introduction to the Pythagorean Theorem and how to solve quadratic equations using square roots. The quadratic formula was also used to solve quadratic equations and can be used with any quadratic equation to find its real-valued solutions.

Work with students to review the skills, strategies, and concepts of the chapter. It is suggested that on the first day of review you and your students work together to solve problems from the Chapter Review exercises. The first day's homework assignment can be used as the basis for the second day of review.

COOPERATIVE LEARNING
Encourage students to study together in small groups. Emphasize the importance of teaching a classmate how to perform a skill or how to recall a strategy. When students work together, everyone wins. The students receiving help get additional instruction, and the students giving help gain a deeper understanding of the skills and concepts involved.

Chapter SUMMARY

Review the chapter by asking students, "What did you learn? Why did you learn it? How does it fit in with the other algebra skills and concepts you have learned?"

9

Chapter Summary

What did you learn?

Skills

1. Evaluate square roots
   - exactly for some numbers. **(9.1)**
   - with a calculator for other numbers. **(9.1)**
2. Solve a quadratic equation
   - by finding square roots. **(9.2)**
   - by the quadratic formula. **(9.4)**
3. Find the number of solutions of a quadratic equation
   - by using the discriminant. **(9.5)**
   - by sketching a graph. **(9.3, 9.5)**
4. Sketch the graph of a quadratic equation. **(9.3)**
   - Find vertex and axis of symmetry of a parabola. **(9.3)**
5. Sketch the graph of a quadratic inequality. **(9.6)**

Strategies

6. Use quadratic models for real-life situations. **(9.1–9.7)**

Exploring Data

7. Decide which type of model best fits given data. **(9.7)**

Why did you learn it?

Real-life problems contain several different types of relations and patterns. For some, linear models fit well. For others, absolute value models or exponential models fit well. In this chapter, you have seen that many real-life situations can be modeled with quadratic equations. Some examples are vertical motion problems, Pythagorean theorem problems, parabolic path problems, and area problems.

How does it fit into the bigger picture of algebra?

Quadratic equations contain the square of a variable. This chapter discusses quadratic equations in one variable. For instance, $x^2 - 9 = 0$ is a quadratic equation in x. Quadratic equations in one variable can be solved using square roots. The chapter also discusses quadratic equations in two variables, such as $y = x^2 - 9$. The graph of this equation is a *parabola.* Try to remember the "algebra-geometry" connection between equations in one and two variables. For instance, the equation $x^2 - 9 = 0$ has two solutions which correspond to the two x-intercepts of the graph of $y = x^2 - 9$. The chapter introduced you to several types of real-life problems that can be modeled with quadratic equations. In the last lesson of the chapter, you compared quadratic models to the other types of algebraic models you have studied.

Chapter REVIEW

In Exercises 1–8, evaluate the expression. Give the exact value if possible. Otherwise give an approximation to two decimal places. **(9.1)**

1. $\sqrt{25}$ 5

2. $-\sqrt{0.06}$ -0.24

3. $\sqrt{47}$ 6.86

4. $\sqrt{-(-0.25)}$ 0.5

5. $\sqrt{32}$ 5.66

6. $\sqrt{30^2 - 4(6)(3)}$ 28.77

7. $\sqrt{\frac{4}{36}}$ $\frac{1}{3}$

8. $-\sqrt{\frac{4}{13}}$ -0.55

In Exercises 9–12, solve the equation using square roots. (9.2)

9. $w^2 - 144 = 0$ ± 12 **10.** $8y^2 = 968$ ± 11 **11.** $16t^2 - 81 = 0$ $\pm \frac{9}{4}$ **12.** $4x^2 - 19 = 6$ $\pm \frac{5}{2}$

In Exercises 13–16, match the equation with its graph. (9.3)

13. $y = x^2 - 9$ d **14.** $y = -x^2 + 25$ a **15.** $y = -x^2 + 3x - 2$ c **16.** $y = x^2 - x - 20$ b

a.

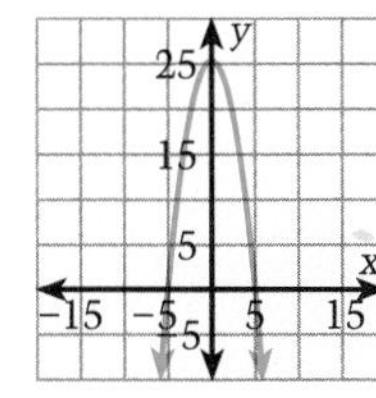

b.

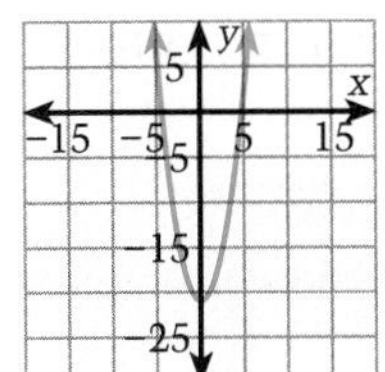

c.

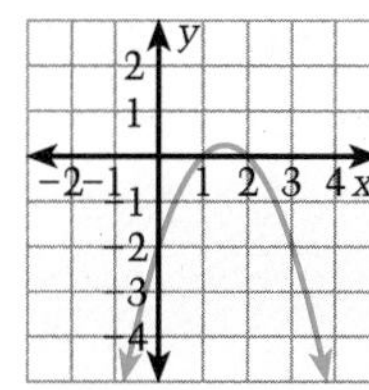

d.

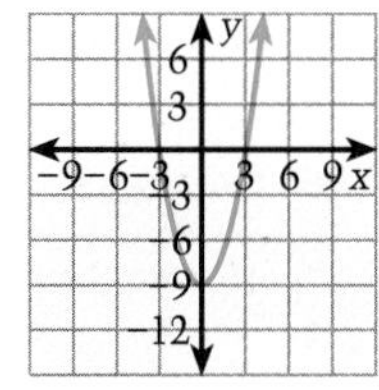

In Exercises 17–20, sketch the graph of the equation. Label the vertex. (9.3) See Additional Answers for graphs.

17. $-x^2 - 3x + 2 = y$ $\left(-\frac{3}{2}, \frac{17}{4}\right)$

18. $x^2 = y$ (0, 0)

19. $\frac{1}{4}x^2 + 2x - 4 = y$ $(-4, -8)$

20. $-6x^2 + 10x + 2 = y$ $\left(\frac{5}{6}, \frac{37}{6}\right)$

In Exercises 21–26, solve the equation. (9.4)

21. $x^2 - \frac{1}{4}x + \frac{1}{64} = 0$ $\frac{1}{8}$

22. $3x^2 - 4x + 1 = 0$ $1, \frac{1}{3}$

23. $-\frac{1}{4}x^2 + 3x - 9 = 0$ 6

24. $-x^2 + x - \frac{1}{8} = 0$

25. $-2x^2 + x + 6 = 0$ $-\frac{3}{2}, 2$

26. $-\frac{1}{4}x^2 + \frac{1}{4}x + \frac{1}{2} = 0$ $-1, 2$

24. $\frac{-1 + \sqrt{0.5}}{-2} \approx 0.15$, $\frac{-1 - \sqrt{0.5}}{-2} \approx 0.85$

In Exercises 27–30, sketch the graph of the inequality. (9.6) See margin.

27. $x^2 - 3 \geq y$

28. $\frac{1}{2}x^2 + 3x - 4 < y$

29. $-x^2 - 2x + 3 \leq y$

30. $-4x^2 + 2x + 5 > y$

In Exercises 31–34, construct a scatter plot of the data. Then decide which type of model best fits the data. (9.7) See margin.

31. $(-3, 4), (-2, 1), (-1, 0), (0, 1), (1, 4), (2, 9), (3, 16)$ Quadratic

32. $(-3, -7), (-2, -4), (-1, -1), (0, 2), (1, 5), (2, 8), (3, 11)$ Linear

33. $\left(-3, \frac{1}{8}\right), \left(-2, \frac{1}{4}\right), \left(-1, \frac{1}{2}\right), (0, 1), (1, 2), (2, 4), (3, 8)$ Exponential

34. $(-3, 0), (-2, -2), (-1, -4), (0, -2), (1, 0), (2, 2), (3, 4)$ Absolute value

ASSIGNMENT GUIDE

Basic/Average: Ex. 1–33 odd, 35, 38, 39, 43–48

Above Average: Ex. 1–33 odd, 37–42, 47–50

Advanced: Ex. 1–33 odd, 37–42, 47–50

More Difficult Exercises
Exercises 39–50

Ex. 1–50 Since the Chapter Review exercises parallel the list of Skills, Strategies, and Exploring Data statements on page 498, tell students to use the Chapter Summary to recall the lesson in which a skill or concept was presented if they are uncertain about how to do a given exercise. This forces the students to categorize the task to be performed and to locate relevant information in the chapter about the task, both beneficial activities.

Answers

27.

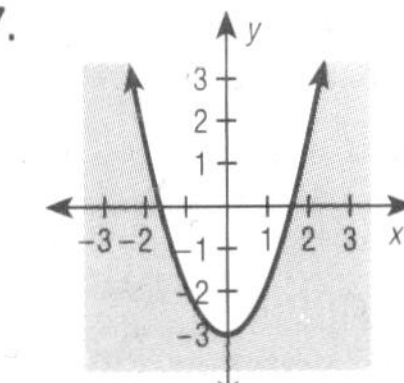

28.

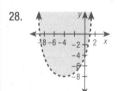

29.

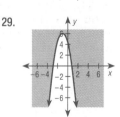

30. 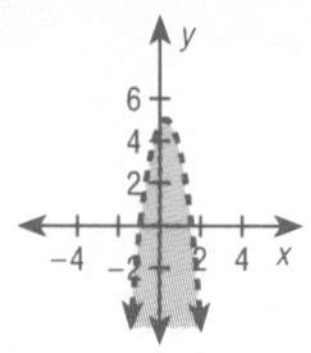

31. 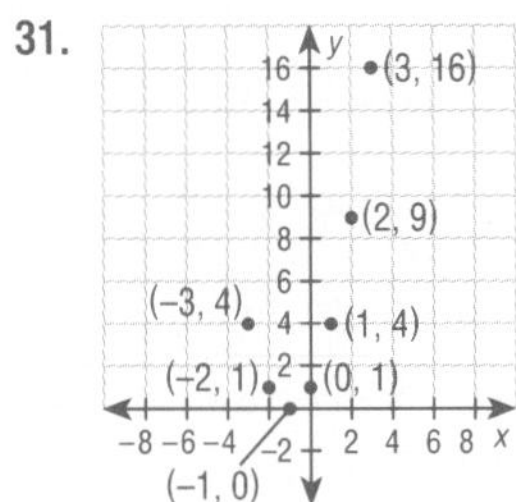

32. 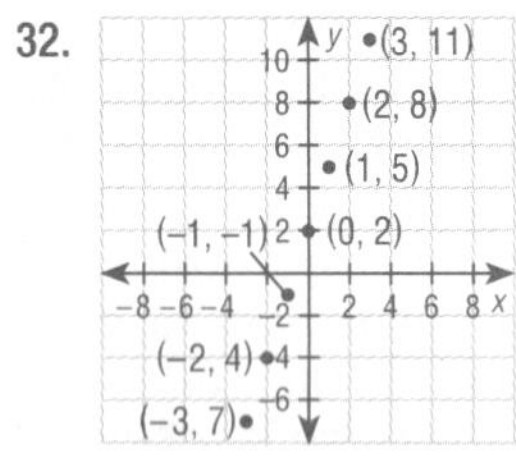

33. 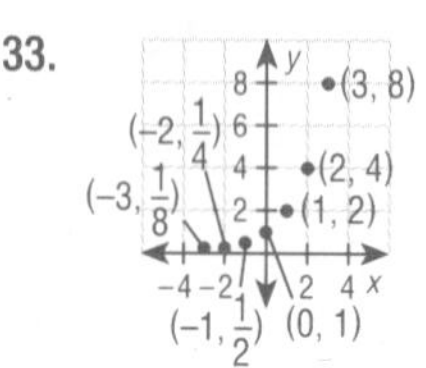

34. 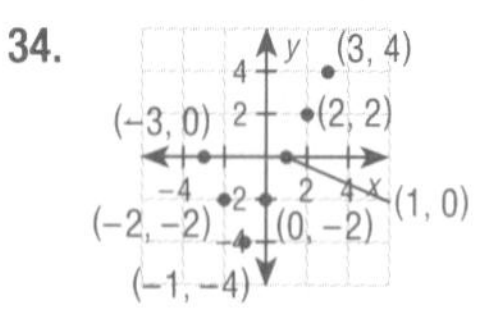

39. Yes; Connie can throw over the building, but Muriel can't.

40. 43.2; 41.8; 40; 37.8; 35.2; 32.2; 28.8; 25; 20.8; \$3.60 per dozen; any price increase results in a lower total sales.

42. By regulating the amount of nitrogen oxide that can be released into the air through setting standards and enforcing those standards through inspections and fines.

35. *Flower Garden* You are planting a triangular garden on the roof of your apartment building. To protect the flowers, you put up a small fence around the garden. How much fencing do you need?

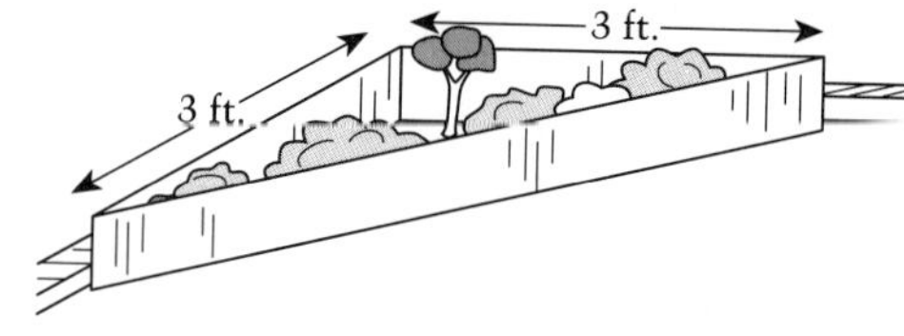

35. ≈10.24 ft

36. *Free Fall* On September 9, 1979, Kitty O'Neill dove 180 feet from a helicopter into a 30- by 60-foot air cushion for a TV film stunt. How long was O'Neill in the air? (Assume no air resistance and an initial velocity of zero.)
(Source:* Guinness Book of World Records*) ≈3.4 seconds

37. *Atom Smasher* The main instrument at the Fermi National Accelerator Laboratory is a particle accelerator, or atom smasher. In this circular tunnel, protons are accelerated to speeds that are close to the speed of light. Scientists direct a beam of protons at a target and study the results of the collisions. How far does a proton travel in one trip around the tunnel? (Circumference $= 2\pi r$.) 6.41 km

One of the world's largest accelerators lies underground at the Fermi National Accelerator Laboratory at Batavia, Illinois. The circular tunnel encloses an area of 3.27 square kilometers. (Area = πr^2.)

38. *Softball Teams* For 1965 to 1990, the number, S (in thousands), of youth softball teams in the United States can be modeled by

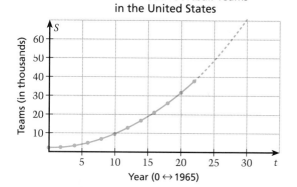

$$S = 0.075t^2 + 1.18$$

where t is the year, with $t = 0$ corresponding to 1965. Use the model to complete the table. Estimate the year for 60,000 teams.

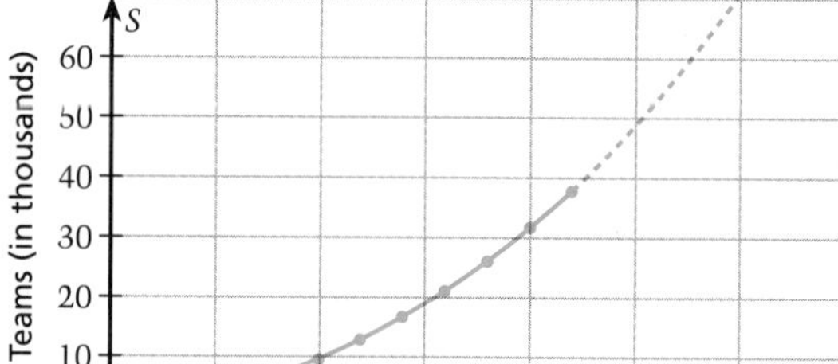

Year, t	0	5	10	15	20	25
Teams, S	?	?	?	?	?	?

1180; 3055; 8680; 18,055; 31,180; 48,055. 1993

39. *You Could Break a Window Doing That!* Muriel and Connie are playing a game of catch. They decide it would be fun to stand on different sides of a building and throw the ball back and forth. Muriel can throw the ball with a velocity of 14 meters per second. Connie can throw with a velocity of 16 meters per second. Each stands 10 meters from the building. The paths of the ball are:

Muriel: $y = -\frac{9.8}{14^2}x^2 + x + 2$

Connie: $y = -\frac{9.8}{16^2}x^2 + x + 2$

Might they have broken any windows? Explain. **See margin.**

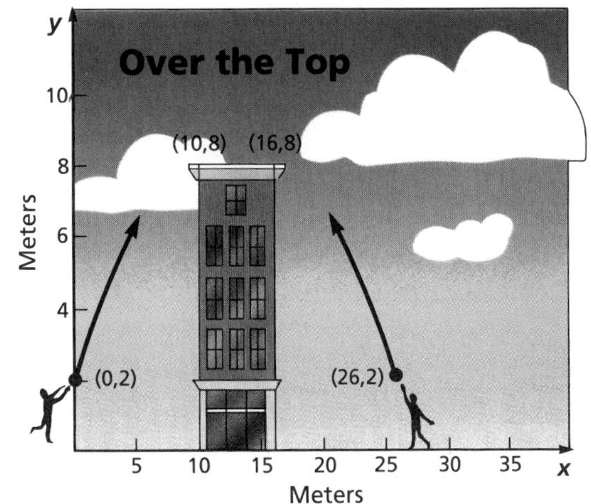

40. *Should Bud Charge More?* Bud sells about 12 dozen muffins a day at $3.60 per dozen. He estimates that he will sell one dozen fewer muffins a day for each $0.20 price increase per dozen.

Price	$3.60	$3.80	$4.00	$4.20	$4.40	$4.60	$4.80	$5.00	$5.20
Dozens Sold	12	11	10	9	8	7	6	5	4

The model that relates the total sales, M (in dollars), with the number of price increases, n, is

$$M = (12 - n)(3.60 + 0.2n) = -0.2n^2 - 1.2n + 43.2.$$

Find the total sales for $n = 0, 1, 2, \ldots, 8$. What should Bud charge to have the greatest sales? Explain. **See margin.**

Nitrogen Oxide **In Exercises 41 and 42, use the following information.**

For 1970 to 1990, the annual amount of nitrogen oxide, A (in thousands of tons), that was released into the air annually in the United States can be modeled by

$$A = 12{,}848 + 697t - 17t^2$$

where t is the year, with $t = 0$ corresponding to 1970. The graph is shown at the right.

(Source: U.S. Environmental Protection Agency)

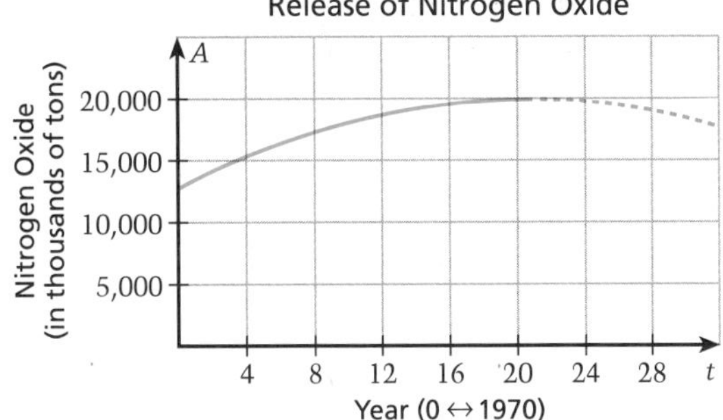

41. When did the United States begin to reduce the amount of nitrogen oxide it was releasing into the atmosphere? **1991**

42. The release of nitrogen oxide increases the production of ozone and contributes to acid rain. Nitrogen oxide is released by burning gasoline, coal, and oil. How do you think the United States has begun to reduce the amount of nitrogen oxide released into the air? **See margin.**

Ex. 43–46 Note that the graphs have two different scales because the Asia data is so much greater than the other continents.

Answers

Chapter Test

10. $\frac{-5+\sqrt{97}}{-4} \approx -1.21$, $\frac{-5-\sqrt{97}}{-4} \approx 3.71$

11. $\frac{-9+\sqrt{93}}{2} \approx 0.32$, $\frac{-9-\sqrt{93}}{2} \approx -9.32$

17. 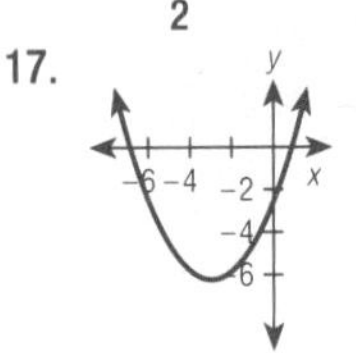

18. 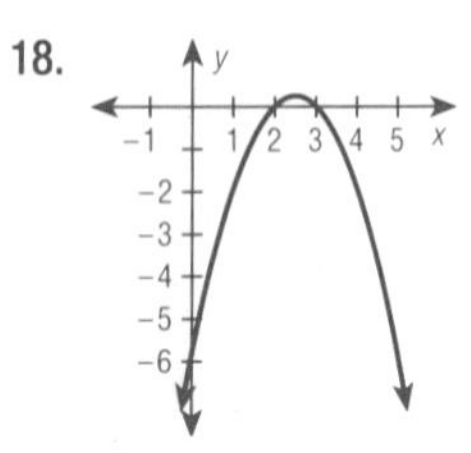

19. 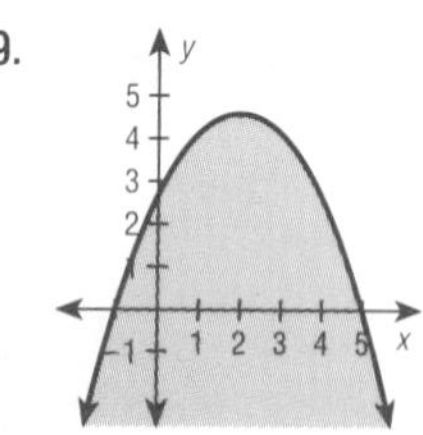

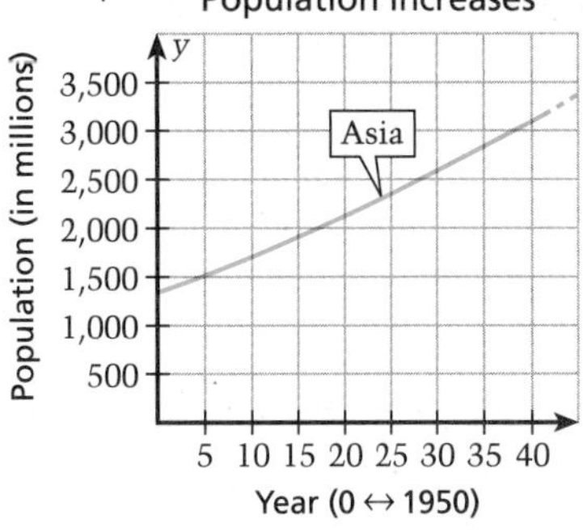

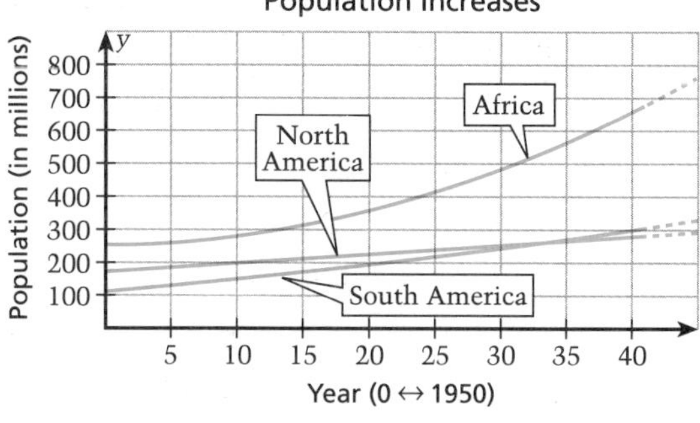

Earth's Populations **In Exercises 43–46, match the population model with one of the graphs above.** *(Source:* **1989 Demographic Year Book***)*

43. $y = 110 + 3.5t + 0.03t^2$ South America

44. $y = 0.23t^2 + 34.8t + 1330$ Asia

45. $y = 170 + 2.7t$ North America

46. $y = 0.25t^2 + 254$ Africa

47. According to the models, will the population of South America ever exceed the population of Africa? No

48. In what year did the population of Asia reach 2000 million? 1968

Breakfast Vendor in Beijing

Market in Ecuador

Market in Morocco

Continent	North America	Africa	South America	Asia (outside former U.S.S.R.)
Area (square kilometers)	21,525,000	30,305,000	17,819,000	27,582,000

49. In 1990, which continent above had the greatest population density? Which was the least densely populated? (The population density is the number of people per square kilometer.) Asia, North America

50. Estimate the population densities of Africa and South America in 2000. Compare the population densities of the two in 1990 and 2000. Which population density is increasing more rapidly? Africa's

Chapter TEST

In Exercises 1–6, evaluate the expression. Give the exact value if possible. Otherwise give an approximation to two decimal places. (9.1)

1. $-\sqrt{4}$ −2

2. $\sqrt{0.0081}$ 0.09

3. $\sqrt{\frac{4}{25}}$ $\frac{2}{5}$

4. $\sqrt{-9}$ No solution

5. $\sqrt{4^2 - 4(3)(-2)}$ 6.32

6. $\sqrt{8^2 - 4(2)(8)}$ 0

In Exercises 7–12, solve the equation. (9.2, 9.4) 10., 11. See margin.

7. $\frac{1}{2}x^2 - 8 = 0$ ± 4

8. $-3x^2 + 243 = 0$ ± 9

9. $x^2 - 8x + 7 = 0$ 7, 1

10. $-2x^2 + 5x + 9 = 0$

11. $x^2 + 9x - 3 = 0$

12. $x^2 + 3x + 3 = 0$ No solution

In Exercises 13–16, match the equation or inequality with its graph. (9.4)

13. $y = x^2 - 2x + 3$ c

14. $y = -x^2 + 3x - 4$ a

15. $y \geq x^2 + 4x - 5$ d

16. $y \leq -x^2 - 2x - 1$ b

a.

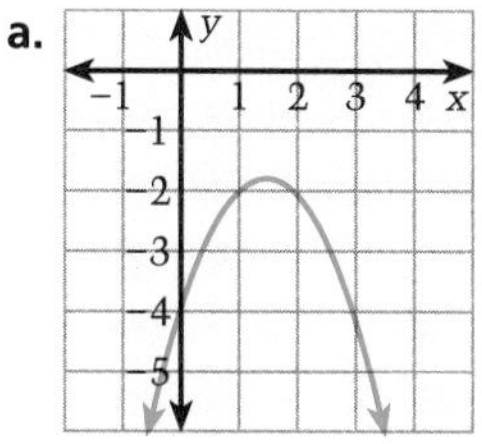

b.

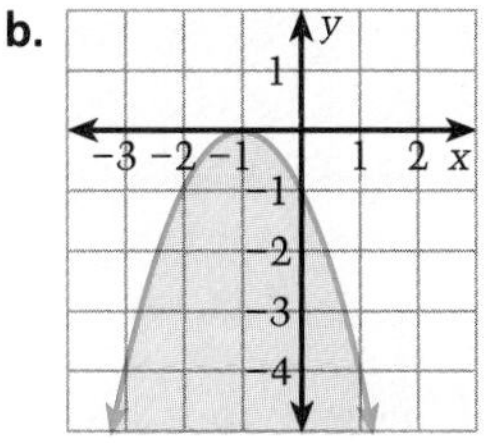

c.

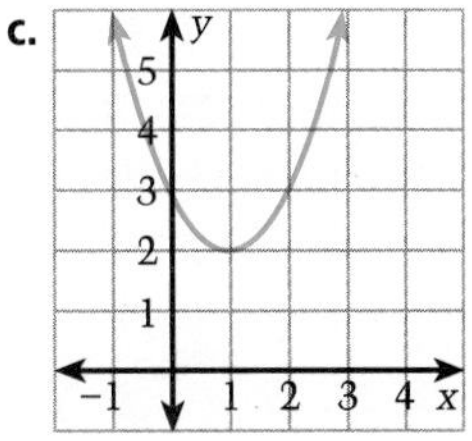

d.

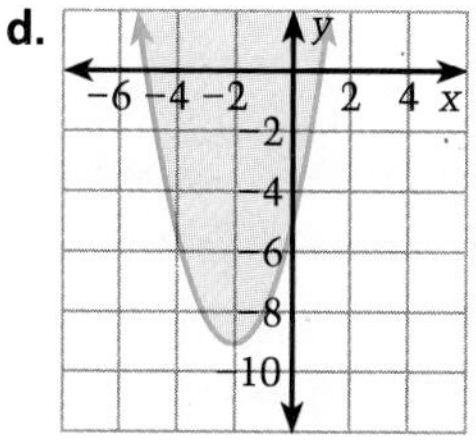

In Exercises 17–20, sketch the graph of the equation or inequality. (9.6) See margin.

17. $y = \frac{1}{3}x^2 + 2x - 3$

18. $y = -x^2 + 5x - 6$

19. $y \leq -\frac{1}{2}x^2 + 2x + \frac{5}{2}$

20. $y \leq x^2 + 7x + 6$

21. *Trying to Fit a Square Peg into a Round Hole* A round hole has a diameter of 2 inches. You have a square peg with sides of $1\frac{1}{2}$ inches. Will the square peg fit into the round hole? Explain. **(9.2)** See margin.

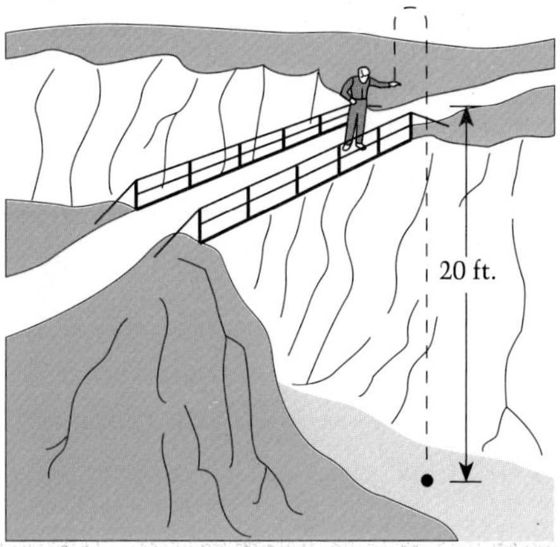

22. Henry is standing on a bridge over a creek. He releases a stone 20 feet from the water. How long will it take the stone to hit the water? **(9.2)** ≈1.12 seconds

23. Henry takes another stone and tosses it straight up with a velocity of 30 feet per second. How long will it take this stone to hit the water? **(9.4)** ≈2.40 seconds

24. If Henry could throw a stone straight up into the air with a velocity of 50 feet per second, could the stone reach a height of 60 feet above the creek? **(9.4)** No

20.

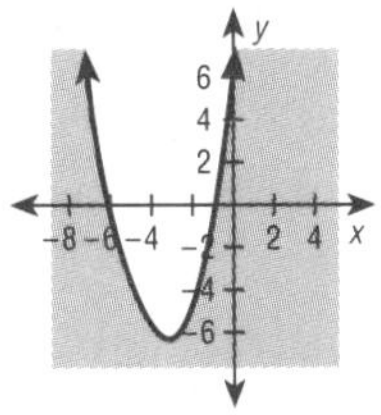

21. No, the diagonal of the square is 0.12 in. greater than the diameter of the hole.

Cumulative Review

11.
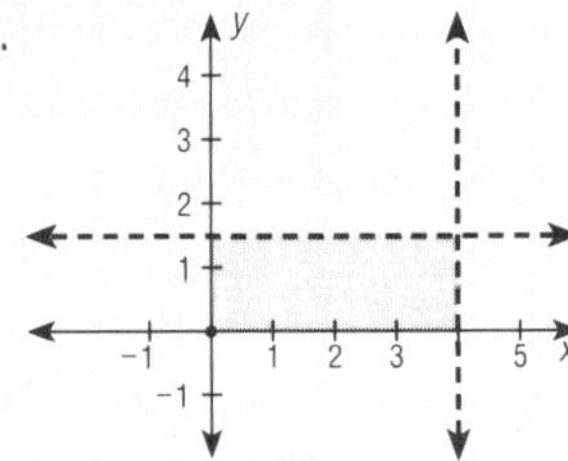

12.
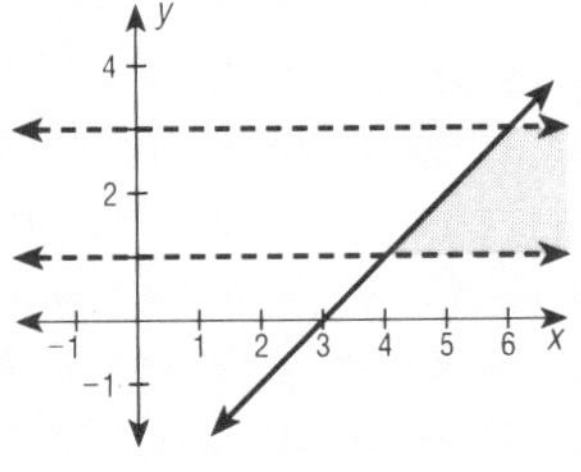

13.
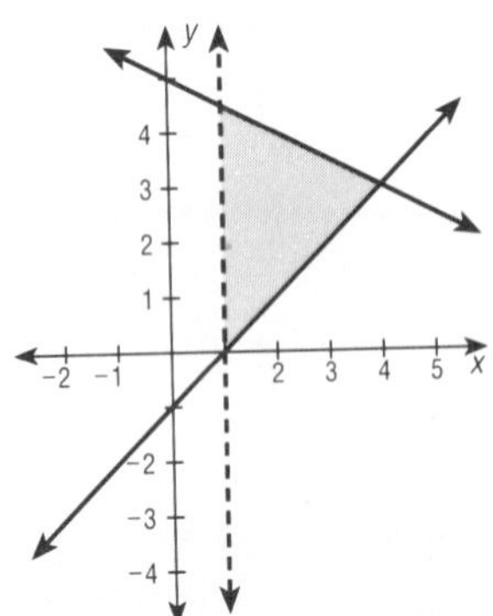

14.
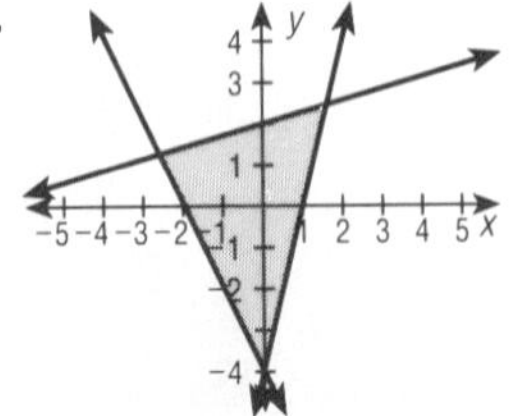

27.
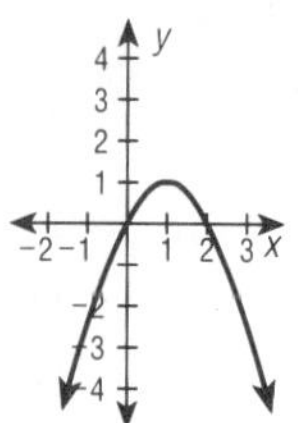

Cumulative REVIEW ▪ *Chapters* 7–9

In Exercises 1–6, solve the linear system.

1. $\begin{cases} x + y = 5 \\ 2x + y = 9 \end{cases}$ (4, 1)

2. $\begin{cases} \frac{1}{3}x - y = 4 \\ x + 3y = 0 \end{cases}$ (6, −2)

3. $\begin{cases} -2x + 16y = 10 \\ x - 4y = -4 \end{cases}$ $\left(-3, \frac{1}{4}\right)$

4. $\begin{cases} 3.2x + 1.1y = -19.3 \\ -32x + 4y = 148 \end{cases}$ (−5, −3)

5. $\begin{cases} \frac{1}{4}x - \frac{1}{2}y = -\frac{5}{2} \\ -4x + 3y = 5 \end{cases}$ (4, 7)

6. $\begin{cases} 1.4x + 2.1y = 17.99 \\ 2.8x - 4.2y = -11.99 \end{cases}$ (4.28, 5.71)

In Exercises 7–10, match the linear system with its graph.

7. $\begin{cases} 6x + 4y = 24 \\ \frac{1}{2}x + \frac{1}{3}y = 2 \end{cases}$ c

8. $\begin{cases} -\frac{1}{2}x + 3y = -2 \\ 4x - 24y = -10 \end{cases}$ d

9. $\begin{cases} -\frac{4}{3}x + \frac{1}{2}y = -4 \\ \frac{2}{3}x - \frac{1}{4}y = 2 \end{cases}$ b

10. $\begin{cases} x + 2y = -4 \\ 4.5x + 9y = 18 \end{cases}$ a

a.
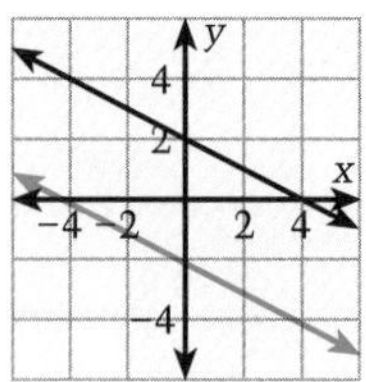

b.
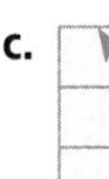

c.
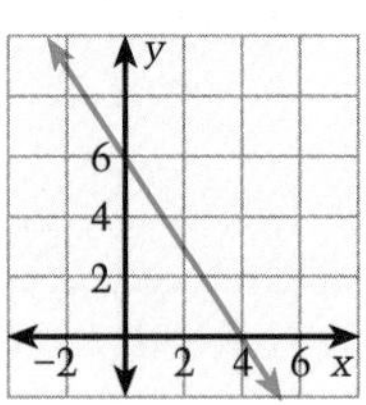

d.
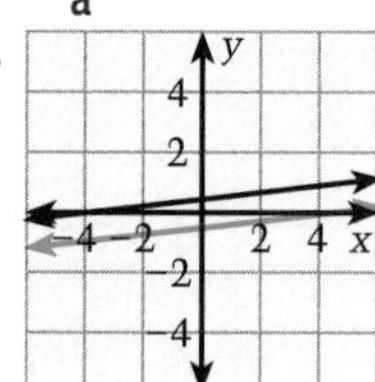

In Exercises 11–14, sketch the graph of the system of linear inequalities. See margin.

11. $\begin{cases} x \geq 0 \\ y \geq 0 \\ x < 4 \\ y < \frac{3}{2} \end{cases}$

12. $\begin{cases} x \geq 0 \\ y > 1 \\ y < 3 \\ x - y > 3 \end{cases}$

13. $\begin{cases} x > 1 \\ x - y \leq 1 \\ \frac{1}{2}x + y \leq 5 \end{cases}$

14. $\begin{cases} -\frac{1}{4}x + y \leq 2 \\ -4x + y \geq -4 \\ 2x + y \geq -4 \end{cases}$

In Exercises 15–18, simplify the expressions.

15. $\frac{(3w)^2}{2^2w} \cdot w^{-3}$ $\frac{9}{4w^2}$

16. $\frac{4ab^2}{2a^{-1}} \cdot (a^{-1}b)^4$ $\frac{2b^6}{a^2}$

17. $\frac{(37p)^0}{p^2} \cdot \frac{(q^2p)^4}{q^2p^2}$ q^6

18. $\frac{1}{4}r^2s^3t^{-1} \cdot 4r^{-3}t^4$ $\frac{s^3t^3}{r}$

In Exercises 19 and 20, evaluate the expressions. Write the results in scientific notation *and* in decimal form.

19. $(4.27 \times 10^3) \cdot (2.2 \times 10^2)$
9.394×10^5; 939400

20. $(3.6 \times 10^{-4}) \div (1.4 \times 10^{-3})$
2.57×10^{-1}; 0.257

In Exercises 21–26, solve the quadratic equation.

21. $3x^2 - 27 = 0$ 3, −3

22. $-\frac{1}{2}x^2 + 2 = 0$ 2, −2

23. $x^2 - 2x - 63 = 0$ 9, −7

24. $3x^2 + 11x + 6 = 0$ $-3, -\frac{2}{3}$

25. $x^2 + 3x - 9 = 0$

26. $5x^2 - 24x - 5 = 0$ $5, -\frac{1}{5}$

In Exercises 27–32, sketch the graph of the equation or inequality. See margin.

27. $y = -x^2 + 2x$

28. $y = x^2 - 3x + 2$

29. $y = 3x^2 + 6x + 1$

30. $y \geq \frac{1}{4}x^2 - 2x + 4$

31. $y > x^2 + 4x + 6$

32. $y \leq -2x^2 + 8x - 3$

25. $\frac{-3 + \sqrt{45}}{2}, \frac{-3 - \sqrt{45}}{2}$

Fashion Doll Clothes **In Exercises 33 and 34, use the following information.**

Jan is planning to start a small business, creating fashion doll clothes. She has designed two dresses: Style X (a daytime dress) and Style Y (an evening dress). The table below shows the cost of material, the number of hours of labor, and the profit for each style. The wages for employees is \$5 per hour. Jan can spend up to \$200 for the cost of material and up to \$900 for labor.

	X	Y
Cost of material (\$)	2	4
Number of hours of labor	2	3
Profit (\$)	6	10

33. Let x represent the number of dresses of Style X. Let y represent the number of dresses of Style Y. Use linear programming to find the number of dresses of each style that must be produced to obtain the maximum profit. 60 Style X and 20 Style Y dresses

34. In Exercise 33, what is the objective equation? What is the system of constraints? $P = 6x + 10y$; $2x + 4y \le 200$, $5(2x + 3y) \le 900$, $x \ge 0$, $y \ge 0$

Las Bolas Grandes **In Exercises 35 and 36, use the following information.**

The diameters of *Las Bolas Grandes* of Costa Rica range from a few inches to eight feet. Researchers believe that these granite spheres were carved 1600 years ago.

35. One of the great stone balls has a radius of four feet. What is its volume? Write the result in scientific notation and in decimal form. 2.68×10^2 ft^3; 268 ft^3

36. The ball in Exercise 35 weighs 16 tons. What is the weight in pounds? What is the density of the ball (in pounds per cubic foot)? 32,000 pounds; 119.4 pounds per cubic foot

37. *Compound Interest* Your savings account earns 4.8% interest, compounded annually. Another bank in town is offering 5.1% interest, compounded annually. The balance in your account is \$567. How much additional interest could you earn in 5 years by moving your account to the bank with the 5.1% interest? How much additional interest could you earn in 10 years? \$10.31, \$26.28

38. *Depreciation* A used car was purchased in 1992 for \$3700. If the value of the car depreciates 15% each year, what will the value be at the end of 1996? $\approx$\$1931

Model Rocket **In Exercises 39 and 40, use the following information.**

A model rocket is propelled straight up with an initial velocity of 300 feet per second.

39. How long will it take the rocket to reach an altitude of 1100 feet? 5 seconds

40. Will the rocket reach an altitude of 1500 feet? Explain.
No, it reaches its maximum height prior to 1500 feet

28.
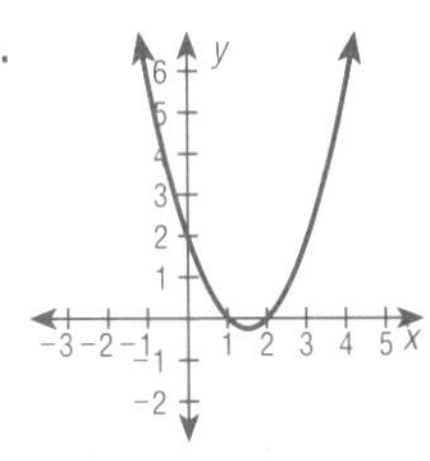

29.
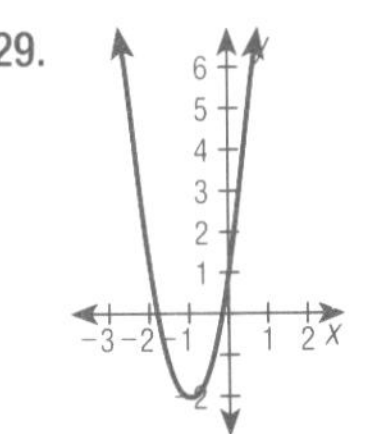

30.
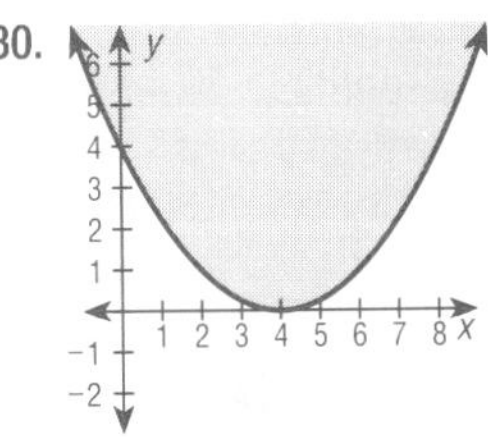

31.
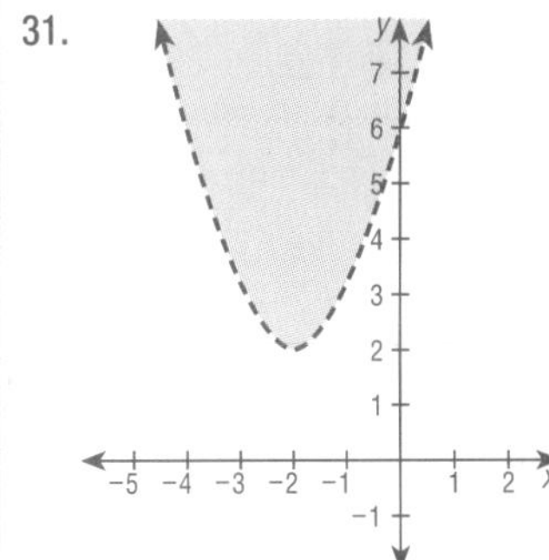

32.
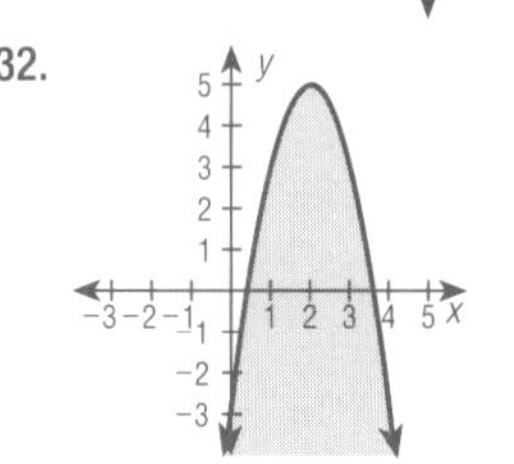

33.

Polynomials are an important class of algebraic expressions. Polynomial functions (which students will be introduced to in Chapter 12 and encounter more extensively in Algebra 2) have many applications in practically every branch of mathematics and science. In Lesson 10.7, a derivation of the quadratic formula will be possible and the connection between solving for the solution of a quadratic equation and the quadratic formula will be explained.

The Chapter Summary on page 555 provides you and the students with a synopsis of the chapter. It identifies key skills, concepts, and vocabulary. You may want to have students look at the Chapter Summary as an overview before beginning the chapter.

CHAPTER 10

Polynomials and Factoring

LESSONS

This brilliant pattern of colors is formed by ice crystals photographed in polarized light.

Real Life
Climatology

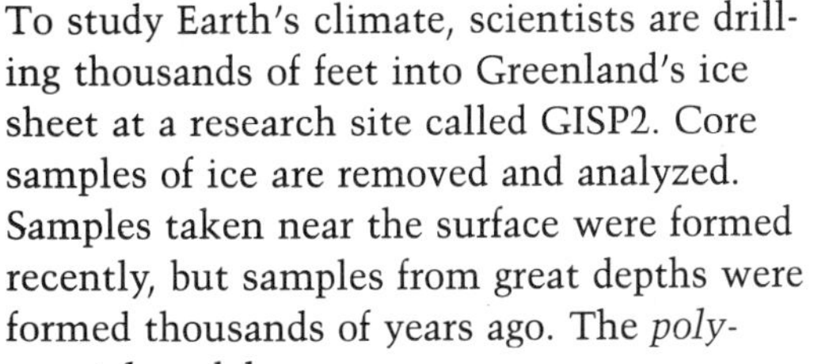

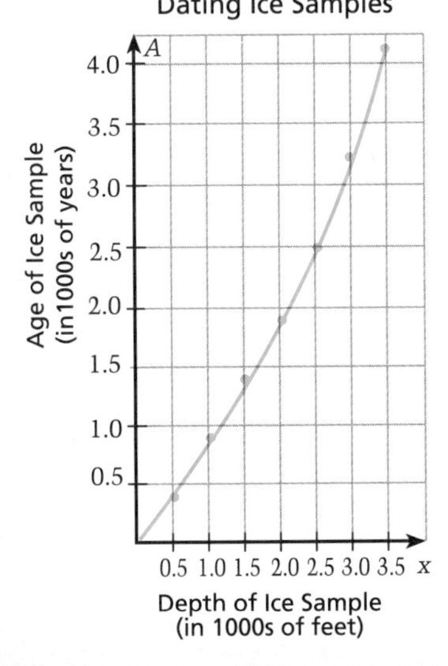

To study Earth's climate, scientists are drilling thousands of feet into Greenland's ice sheet at a research site called GISP2. Core samples of ice are removed and analyzed. Samples taken near the surface were formed recently, but samples from great depths were formed thousands of years ago. The *polynomial* model

$$A = \frac{9}{16}(x^3 - 7x^2 + 33x + 12)^2$$

relates the age of the ice, A (in years), with the depth of the core sample, x (in 1000's of feet). The ice crystals undergo changes in size, shape, and orientation with increasing depth. Researchers use polarized light to measure the changes in the ice crystals.

(Source: **Popular Science, 1991.*)***

Depth, x	0.5	1.0	1.5	2.0	2.5	3.0	3.5
Age, A	406	856	1357	1892	2478	3164	4028

Polynomial models are powerful tools for representing real-life phenomena. Climatologists use data such as the type described here to predict future weather patterns.

Ask students to verify the data in the table by evaluating the expression on the right-hand side of the polynomial model. Students may work individually if calculators are available. Otherwise, encourage students to share the computational work load among the members of their small cooperative groups.

Polynomials and Factoring

PACING CHART

The daily Pacing Chart is meant to help you adjust your teaching pace. Students in the full course should finish the entire text by the end of the year. Students in the basic course may not complete the entire text in the school year. The Pacing Chart for each chapter contains suggestions for lessons that require more than one day and lessons that may be omitted for the basic course.

DAY	FULL COURSE	BASIC COURSE
1	10.1	10.1
2	10.2	10.2
3	10.3	10.2
4	10.3	10.3
5	10.3	10.3
6	10.4	10.3
7	10.4	10.4
8	10.4	10.4
9	Mid-Chapter Self-Test	10.4
10	10.5	Mid-Chapter Self-Test
11	10.5	10.5
12	10.6 & Using a Graphing Calculator	10.5
13	10.7	10.6
14	Chapter Review	Using a Graphing Calculator
15	Chapter Review	10.7
16	Chapter Test	Chapter Review
17		Chapter Review
18		Chapter Test

CHAPTER ORGANIZATION

LESSON	PAGES	GOALS	MEETING THE NCTM STANDARDS
10.1	508-514	1. Add and subtract polynomials 2. Use polynomials as models	Problem Solving, Communication, Connections, Geometry
10.2	515-520	1. Multiply two polynomials using the Distributive Properties and the FOIL method 2. Use polynomial multiplication in real-life settings	Problem Solving, Communication, Connections, Geometry
Mixed Review	521	Review of algebra and arithmetic	
Milestones	521	Leonhard Euler	Connections
10.3	522-527	1. Use patterns for the product of a sum and difference and for the square of a binomial 2. Use special-product patterns in real-life models	Problem Solving, Communication, Reasoning, Connections, Geometry
10.4	528-533	1. Factor polynomials, including the difference of two squares and perfect square trinomials 2. Use factoring in real-life models	Problem Solving, Communication, Connections
Mid-Chapter Self-Test	534	Diagnose student weaknesses and remediate with correlated Reteaching worksheets	
10.5	535-540	1. Factor a quadratic trinomial or recognize that it cannot be factored 2. Use factoring in real-life models	Problem Solving, Communication, Connections, Structure, Geometry
10.6	541-547	1. Use factoring to solve a quadratic equation 2. Use factoring with real-life models	Problem Solving, Communication, Connections, Technology, Reasoning
Mixed Review	547	Review of algebra and arithmetic	
Using a Calculator	548	Use a graphing calculator to obtain a graphic interpretation of polynomial addition and subtraction	Technology, Geometry
10.7	549-554	Solve quadratic equations by completing the square	Problem Solving, Communication, Connections, Structure, Reasoning
Chapter Summary	555	A restatement of what has been learned, why it has been learned, and how it fits into the structure of algebra	Structure, Connections
Chapter Review	556-558	Review of concepts and skills learned in the chapter	
Chapter Test	559	Diagnose student weaknesses and remediate with correlated Reteaching worksheets	

LESSON RESOURCES

MEETING INDIVIDUAL NEEDS

RETEACHING For students who need to spend more time on basics:

If a mid-chapter self-test or chapter test indicates a deficiency, teachers can help students with the appropriate ***Reteaching Copymaster.***

PRACTICE For students who need more practice:

Additional exercises like those in the Pupil's Edition are provided for each lesson in ***Extra Practice Copymasters.***

ENRICHMENT For enriching and broadening students' experiences:

Problem of the Day copymasters in ***Teaching Tools*** provide a daily opportunity to use logical reasoning, looking for a pattern, writing an equation, and other routine and non-routine problem-solving strategies.

Math Log copymasters in ***Alternative Assessment*** provide opportunities to report on investigations, research, and open-ended problems.

Enriching activities with graphing and scientific calculators and computers are provided in ***Technology: Using Calculators and Computers.***

The ***Applications Handbook*** provides additional information about the cross-curriculum topics such as astronomy, chemistry, physics, sports, economics, genetics, and music that are integrated into the Pupil's Edition.

LESSON	10.1	10.2	10.3	10.4	10.5	10.6	10.7
PAGES	508-514	515-520	522-527	528-533	535-540	541-547	549-554
Teaching Tools							
Transparencies		✓	✓				
Problem of the Day	✓	✓	✓	✓	✓	✓	✓
Warm-up Exercises	✓	✓	✓	✓	✓	✓	✓
Answer Masters	✓	✓	✓	✓	✓	✓	✓
Extra Practice Copymasters	✓	✓	✓	✓	✓	✓	✓
Reteaching Copymasters	Teacher-directed and independent activities tied to results on the Mid-Chapter Self-Tests and Chapter Tests						
Color Transparencies	✓	✓		✓	✓		
Applications Handbook	Additional background information is supplied for many real-life applications.						
Technology Handbook	Calculator and computer worksheets are supplied for appropriate lessons.						
Complete Solutions Manual	✓	✓	✓	✓	✓	✓	✓
Alternative Assessment	Assess student's ability to reason, analyze, solve problems, and communicate using mathematical language.						
Formal Assessment	Mid-Chapter Self-Tests, Chapter Tests, Cumulative Tests, and Practice for College Entrance Tests						
Computer Test Bank	Customized tests can be created by choosing from over 2000 items.						

INSIGHTS

10.1 Adding and Subtracting Polynomials

Polynomials are basic expressions for which algebraic operations must be established. Begin with addition and subtraction (as with whole numbers many years ago). Once students understand how to combine polynomial expressions, solutions to many real-life problems will be within their grasp.

10.2 Multiplying Polynomials

Techniques for multiplying polynomials use the Distributive Property. Once the Distributive Property has been applied, it is usually necessary to simplify expressions using addition and subtraction, discussed in the previous lesson. Real-life problems in science, economics, and biology frequently require multiplication of polynomials.

10.3 Multiplying Polynomials: Two Special Cases

Some special products that occur in algebra should be memorized. The FOIL pattern in the previous lesson is an example of this. In this lesson, the connection between geometry and algebra is demonstrated and becomes a resource for understanding the product patterns.

10.4 Factoring: Special Products

An efficient way in which to solve many polynomial equations is by factoring, which is the "undoing" of the multiplication of polynomials. Undoing the Distributive Properties of Multiplication over Addition and Subtraction and reversing the special products of Lesson 10.3 suggest several strategies of factoring.

10.5 Factoring Quadratic Trinomials

Many real-life situations are modeled by quadratic trinomials. In some cases, the most efficient way in which to solve such polynomial equations is by factoring. Because factoring is the "undoing" of the multiplication of polynomials—typically by applying the Distributive Property—undoing the FOIL pattern is required.

10.6 Solving Quadratic Equations by Factoring

In this lesson, the connection is made between factoring polynomials and solving quadratic equations. Furthermore, students will be adding to their collection of quadratic equation-solving skills, which already include solving by square roots, by the quadratic formula, and by graphing. The collection of equation-solving techniques available to students continues to grow as students gain more and more mathematical power.

10.7 Solving Quadratic Equations by Completing the Square

The technique of completing the square is frequently used to rewrite quadratic functions (see Chapter 12) in order to facilitate the graphing and interpreting of quadratic relationships.

Problem of the Day

A dot pentagon with 2 dots per side contains 5 dots. How many dots would a pentagon with 35 dots per side contain? 170

ORGANIZER

Warm-Up Exercises

1. Find the coefficients of the given quadratic expression.
   a. $3x^2 - 5x + 8$
   b. $-2x^2 - 17$
   c. $-x^2$
   d. $x^2 + 8x$
   e. $23 - x + -3x^2$
   f. $-13 - x^2$

   a. 3, −5, 8 b. −2, 0, −17
   c. −1, 0, 0 d. 1, 8, 0
   e. −3, −1, 23 f. −1, 0, −13
2. In each group, determine which terms are alike.
   a. $-3x, -3, 4x, 5x^2, 3x^3$
   b. $6x^2, -x, -2xy, 11x^2, 6y^2$
   c. $16, 4x, 8y^2, -xy, y$

   a. $-3x$, $4x$ b. $6x^2$, $11x^2$
   c. none are alike

Lesson Resources

Teaching Tools
  Problem of the Day: 10.1
  Warm-up Exercises: 10.1
  Answer Masters: 10.1
Extra Practice: 10.1
Color Transparencies: 55, 56, 57

LESSON Notes

GOAL 1 Students frequently misunderstand the building blocks of a polynomial expression. Spend time making sure students understand the difference between a constant term and the general polynomial

10.1 Adding and Subtracting Polynomials

What you should learn:

Goal 1 How to add and subtract polynomials

Goal 2 How to use polynomials as models in real-life settings

Why you should learn it:

You can use polynomials as models for a majority of real-life situations such as the performance of an industry over an interval of time.

Goal 1 Adding and Subtracting Polynomials

An expression whose terms are of the form ax^k, where k is a non-negative integer, is a **polynomial in one variable** or simply a **polynomial.**

The integer k is the **degree** of ax^k. The term ax has a degree of one, and the **constant term** a has a degree of zero. The **degree of a polynomial** is the largest degree of its terms. Polynomials are usually written in **standard form,** which means that the terms are written in descending order, from the largest degree to the smallest degree.

The number a is the **coefficient** of the term ax^k. When a polynomial is written in standard form, the coefficient of its first term is the **leading coefficient** of the polynomial.

Example 1 *Identifying Coefficients of a Polynomial*

Identify the coefficients of $-4x^2 + x^3 + 3$.

Solution Write the polynomial in standard form. Account for each degree, even if you must include zero coefficients.

$$-4x^2 + x^3 + 3 = (1)x^3 + (-4)x^2 + (0)x + 3$$

The coefficients are 1, −4, 0, and 3. The leading coefficient is 1. ■

Polynomials that have only one term are **monomials.** Polynomials that have two terms are **binomials.** Polynomials that have three terms are **trinomials.**

Example 2 *Classifying Polynomials*

Polynomial	Degree	Classified by Degree	Classified by Terms
a. 6	0	Constant	Monomial
b. $-2x$	1	Linear	Monomial
c. $3x + 1$	1	Linear	Binomial
d. $-x^2 + 2x - 5$	2	Quadratic	Trinomial
e. $4x^3 - 8x$	3	Cubic	Binomial
f. $2x^4 - 7x^3 - 5x + 1$	4	Quartic	Polynomial

■

To add two polynomials, add the coefficients of like terms. You can use a horizontal or vertical format.

Example 3 *Adding Polynomials Horizontally*

Add $2x^2 + x - 5$ and $x^2 + x + 6$.

Solution

$$\begin{aligned}(2x^2 + x - 5) + (x^2 + x + 6) &= (2x^2 + x^2) + (x + x) + (-5 + 6)\\ &= 3x^2 + 2x + 1\end{aligned}$$

■

Example 4 *Using a Vertical Format to Add Polynomials*

Use a vertical format to find the sum.

$$(5x^3 + 2x^2 - x + 7) + (3x^2 - 4x + 7) + (-x^3 + 4x^2 - 8)$$

Solution Align the like terms of the polynomials.

$$\begin{array}{rrrr} 5x^3 + & 2x^2 - & x + & 7\\ & 3x^2 - & 4x + & 7\\ -x^3 + & 4x^2 & - & 8\\ \hline 4x^3 + & 9x^2 - & 5x + & 6\end{array}$$

■

To subtract a polynomial, you must subtract *each of its terms.*

Example 5 *Subtracting Polynomials Horizontally*

Subtract $2x^2 - x - 4$ from $3x^2 - 5x + 3$.

Solution

$$\begin{aligned}(3x^2 - 5x + 3) - (2x^2 - x - 4) &= 3x^2 - 5x + 3 - 2x^2 + x + 4\\ &= (3x^2 - 2x^2) + (-5x + x) + (3 + 4)\\ &= x^2 - 4x + 7\end{aligned}$$

■

One of the most common mistakes in algebra is to forget to subtract *each* term when subtracting one expression from another. Here is an example.

Wrong Signs
↓ ↓

$$(x^2 - 2x + 3) - (x^2 + 2x - 2) \neq x^2 - 2x + 3 - x^2 + 2x - 2$$

term, ax^k, a and k both nonzero. Make the connection between exponents in polynomial expressions and the degree of a polynomial. And, perhaps most importantly, provide students with practice rewriting polynomial expressions in standard form.

Examples 1–2

Observe that the leading coefficient is itself never a polynomial in x, and that the constant term is considered a coefficient.

An Alternate Approach
An Area Model Using Manipulatives

Paper cutouts like those shown (or algebra tiles) can be used to model linear and quadratic polynomials.

For example, $3x^2 + 2x + 4$ can be represented by this diagram.

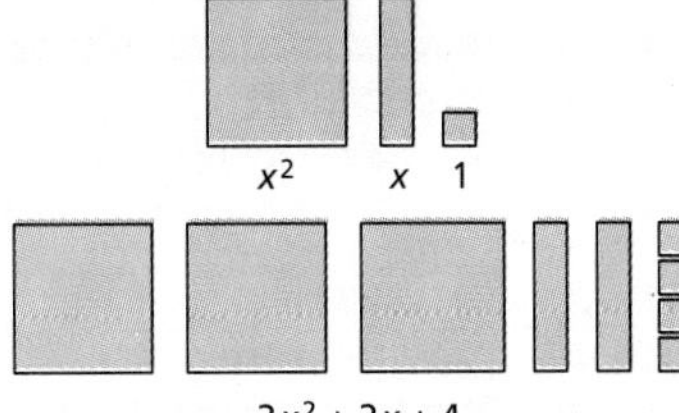

Cutouts can be used to add and subtract polynomials. Use grey cutouts to model negative polynomials. For instance, $-2x$ and $4x - 2x$ can be represented as shown.

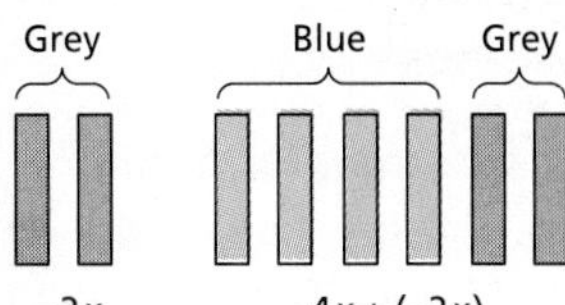

Examples 3–6

The key to adding and subtracting polynomial terms is recognizing when terms are alike. Remind students that terms in one variable are alike if the base and the exponent are the same, respectively. That is, x^3 and $-4x^3$ are alike because the base for each is x and the exponent for each is 3. Notice that the coefficients are different. All constant terms are alike.

Examples 3–4

The associative and commutative properties of addition permit expressions to be rearranged as illustrated in these examples. The horizontal and vertical formats should both be acceptable. It may be a good idea to insert missing terms using 0. For instance, in Example 4, the third expression could be rewritten as

$$-x^3 + 4x^2 + 0x - 8.$$

Examples 7–8

Have students reflect on the other algebraic models that have been used throughout the textbook. It should be observed that polynomial models were being used in a very natural way!

Extra Examples

Here are additional examples similar to **Examples 1–6.**

1. Add. Classify the results by degree and by terms, as in Example 2.
   a. $-4x^2 - 2x + 5$ and $6x^2 + 2x - 11$
   b. $9x^4 + 3x^2 - 24$ and $-9x^4 + 5x + 19$
   c. $6 - 4x + 17x^3$ and $-8x^3$ and $5x - 11x^3$

a. $2x^2 - 6$, quadratic, binomial
b. $3x^2 + 5x - 5$, quadratic, trinomial
c. $-2x^3 + x + 6$, cubic, trinomial

Example 6 *Using a Vertical Format to Subtract Polynomials*

Use a vertical format to find the difference.

$$(-2x^3 + 5x^2 - x + 8) - (-2x^3 + 3x - 4)$$

Solution To subtract each term, you *add its opposite.* One way is to multiply each "subtracted" term by -1 and then add.

$$\begin{array}{r} (-2x^3 + 5x^2 - x + 8) \\ -(-2x^3 \qquad\quad + 3x - 4) \\ \hline \end{array} \quad \rightarrow \quad \begin{array}{r} -2x^3 + 5x^2 - x + 8 \\ 2x^3 \qquad\quad - 3x + 4 \\ \hline 5x^2 - 4x + 12 \end{array}$$ ■

Goal 2 Using Polynomials in Real-Life Modeling

Real Life
Stained Glass

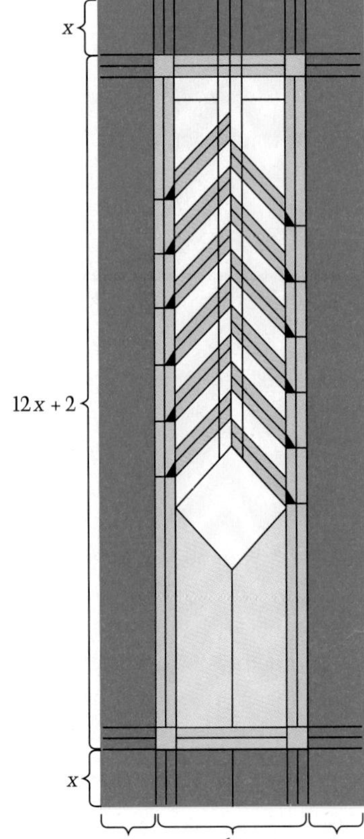

Example 7 *Finding the Area of a Border*

You are designing a stained-glass window. The border is made of red glass and is x inches wide. The interior uses other colors of glass. Write a polynomial expression for the amount of red glass you will use. Then use the expression to find how much red glass would be used for a 2-inch border and for a 3-inch border.

Solution

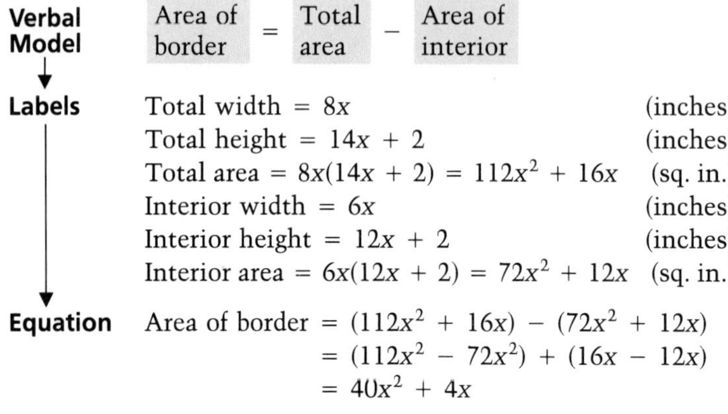

Verbal Model Area of border = Total area − Area of interior

Labels
Total width $= 8x$ (inches)
Total height $= 14x + 2$ (inches)
Total area $= 8x(14x + 2) = 112x^2 + 16x$ (sq. in.)
Interior width $= 6x$ (inches)
Interior height $= 12x + 2$ (inches)
Interior area $= 6x(12x + 2) = 72x^2 + 12x$ (sq. in.)

Equation
$$\begin{aligned} \text{Area of border} &= (112x^2 + 16x) - (72x^2 + 12x) \\ &= (112x^2 - 72x^2) + (16x - 12x) \\ &= 40x^2 + 4x \end{aligned}$$

If the border is 2 inches wide, then the area of red glass is

$$40(2^2) + 4(2) = 168 \text{ square inches.}$$

If the border is 3 inches wide, then the area of red glass is

$$40(3^2) + 4(3) = 372 \text{ square inches.}$$ ■

Lobster pots on a dock at West Tremont, Maine. Maine harvests more lobster than any other state.

Real Life
Fishing Industry

Example 8 Using Polynomial Models

The number of pounds of shrimp, S, and lobster, L, consumed by Americans from 1989 to 1993 can be modeled by

$S = 8.9t^2 - 166.9t + 1522.2$ *Shrimp (millions of pounds)*
$L = -3.4t + 121.4$ *Lobster (millions of pounds)*

where $t = 9$ represents 1989. Find a model that represents the total amount, T, of shrimp *and* lobster consumed from 1989 to 1993. Estimate the total amount consumed in 1994.

Annual Consumption of Lobster and Shrimp

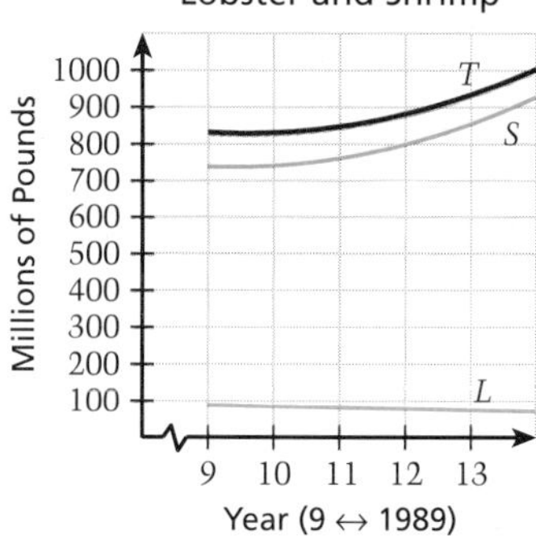

Solution By adding the two models, you can obtain the following model for the total consumption of shrimp and lobster.

$T = 8.9t^2 - 170.3 + 1643.6$ *Total (millions of pounds)*

Using this model, and substituting $t = 14$, you can estimate the 1994 consumption to be

$T = 8.9(14^2) - 170.3(14) + 1643.6 = 1003.8$ million pounds.

(Source: U.S. National Oceanic Atmospheric Administration) ■

Communicating about ALGEBRA

SHARING IDEAS about the Lesson

Use Terminology To "do" algebra, you must know how to use the language. See below.

A. The prefixed *poly, mono, bi,* and *tri* mean many, one, two, and three, respectively. Explain how these prefixes are used in *polygon, monopoly, binocular,* and *tricycle.*

B. Give an example of two polynomials of degree 3 whose sum is a polynomial of degree 2. $(x^3 + x^2) + (-x^3 - 1)$

A. *Polygon* refers to something with **many** sides; *monopoly* refers to ownership by **one**; *binocular* refers to something with **two** telescopes; *tricycle* refers to a vehicle with **three** wheels.

2. Subtract the second polynomial from the first. Classify the results by degree and by terms as in Example 2.

a. $8x^4 - 3x^2 - 11x - 3$ and $-13x^4 - 3x^2 + 2x - 17$
b. $9x^3 - 3x^2 - 1$ and $9x^4 + 5x^2 + 19$
c. $6x - 4x^2 + 17x^3$ and $-8x^3 + 5x^2 - 11x$

a. $21x^4 - 13x + 14$, quartic, trinomial
b. $-9x^4 + 9x^3 - 8x^2 - 20$, quartic, polynomial
c. $25x^3 - 9x^2 + 17x$, cubic, trinomial

Check Understanding

1. Explain why a constant term is a polynomial using the definition of a polynomial in one variable. The constant term has a degree of 0.
2. What is the leading coefficient of a polynomial expression? The coefficient of the term of highest degree Can the leading coefficient be zero? no
3. What is the standard form of a polynomial expression? Terms written in descending order

Communicating about ALGEBRA

EXTEND *Communicating*
WRITING

Language is critical to conveying both simple and complex ideas. Ask students to write a short essay describing the elements of any polynomial expression. Emphasis should be placed on communicating succinctly and accurately.

Consider also the effects of multiplying polynomial expressions with one another. Is it possible to multiply two monomials of degree 3 and get a product of degree 4? 5? 6? Explain. No, no, yes

EXERCISE Notes

ASSIGNMENT GUIDE

Basic/Average: Ex. 1–6, 7–31 odd, 35–38, 41–49 odd, 51, 52

Above Average: Ex. 1–6, 7–33 odd, 35–40, 41–49 odd, 51–53

Advanced: Ex. 1–6, 7–39 odd, 40, 41–49 odd, 51–53

Selected Answers
Exercises 1–6, 7–49 odd

More Difficult Exercises
Exercises 39–40, 53

Guided Practice

Ex.1–6 These exercises review essential terminology and tasks necessary for success in the lesson. Refer students to Examples 1 and 2 as a reference.

Common-Error ALERT!

In **Exercises 21–28,** students often forget to correctly apply the rule of subtraction, which states that you can subtract an expression by adding its opposite, that is, $a - b = a + (-b)$. But to obtain the opposite of a polynomial expression, you must remember to take the opposite of each term of the polynomial. The opposite of $5x^3 + 4x^2 - 19$ is $-5x^3 - 4x^2 + 19$.

EXERCISES

Guided Practice

CRITICAL THINKING about the Lesson

1. Describe a polynomial in one variable.
2. Name the terms of $-3x^3 - 2x^2 + 4x - 5$. $-3x^3, -2x^2, 4x, -5$
3. Name the coefficients in $-7x^3 + 12x - 31$. $-7, 0, 12, -31$
4. Write $15y - 6 + 10y^3 - 3y^2$ in standard form. $10y^3 - 3y^2 + 15y - 6$
5. What is the degree of $2x^2 - 4x^3 + 7$? 3
6. Subtract $(2x^2 - 4x + 1)$ from $(x^2 + 8)$. $-x^2 + 4x + 7$

1. See page 508.

Independent Practice

In Exercises 7–10, classify the polynomial by degree and by number of terms.

7. $-5x - 4$ Linear, binomial
8. -7 Constant, monomial
9. $16 - 4x + 3x^2 - x^4$ Quadratic, polynomial
10. $3x^2 + 6x + 1$ Quadratic, trinomial

In Exercises 11–16, add the polynomials. (Use a horizontal format.)

11. $x^2 - 3;\ 3x^2 + 5$ $4x^2 + 2$
12. $-3y + 2;\ y^2 + 3y + 2$ $y^2 + 4$
13. $2x^2 + 3x + 1;\ x^2 - 2x + 2$ $3x^2 + x + 3$
14. $2x^2 - x + 3;\ 3x^2 - 4x + 7$ $5x^2 - 5x + 10$
15. $12x^3 + 2x^2 - 4;\ 9x^2 + 3x - 8$ $12x^3 + 11x^2 + 3x - 12$
16. $-4x^3 - 2x^2 + x - 5;\ 2x^3 + 3x + 4$ $-2x^3 - 2x^2 + 4x - 1$

In Exercises 17–20, add the polynomials. (Use a vertical format.)

17. $2z - 8z^2 - 3;\ z^2 + 5z$ $-7z^2 + 7z - 3$
18. $6x^2 + 5;\ 3 - 2x^2$ $4x^2 + 8$
19. $5x^4 - 2x + 7;\ -3x^4 + 6x^2 - 5$ $2x^4 + 6x^2 - 2x + 2$
20. $4x^2 - 7x + 2;\ -x^2 + x - 2$ $3x^2 - 6x$

In Exercises 21–24, subtract the second polynomial from the first. (Use a horizontal format.)

21. $z^3 + z^2 + 1;\ z^2$ $z^3 + 1$
22. $10;\ u^2 + 5$ $-u^2 + 5$
23. $2x^2 + 3x - 4;\ x^2 + x - 1$ $x^2 + 2x - 3$
24. $3x^3 - 4x^2 + 3;\ x^3 + 3x^2 - x - 4$ $2x^3 - 7x^2 + x + 7$

In Exercises 25–28, subtract the second polynomial from the first. (Use a vertical format.)

25. $10x^3 + 15;\ 17x^3 - 4x + 5$ $-7x^3 + 4x + 10$
26. $y^2 + 3y^4;\ y^5 - y^4$ $-y^5 + 4y^4 + y^2$
27. $-2x^3 + 5x^2 - x + 8;\ -2x^3 + 3x - 4$ $5x^2 - 4x + 12$
28. $3x^2 + 7x - 6;\ 3x^2 + 7x$ -6

In Exercises 29–34, perform the indicated operations. Use either a horizontal or vertical format and explain why you chose the method you used. Reasons will vary.

29. $(6x - 5) - (8x + 15) + (3x - 4)$ $x - 24$
30. $(2x^2 + 1) + (x^2 - 2x + 1) - (2x^2 + 8)$ $x^2 - 2x - 6$
31. $-(x^3 - 2) + (4x^3 - 2x) - (2x^2 + 3)$
32. $-(5x^2 - 1) - (-3x^2 + 5) - (x^2 - x)$
33. $2(t^2 + 5) - 3(t^2 + 5) + 5(t^2 + 5)$ $4t^2 + 20$
34. $-10(u + 1) + 8(u - 1) - 3(u + 6)$ $-5u - 36$

31. $3x^3 - 2x^2 - 2x - 1$
32. $-3x^2 + x - 4$

Geometry **In Exercises 35 and 36, find the area of the shaded region.**

35.

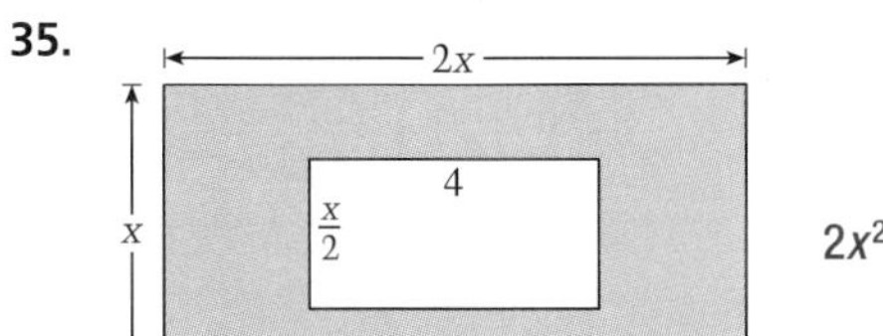

$2x^2 - 2x$

36.

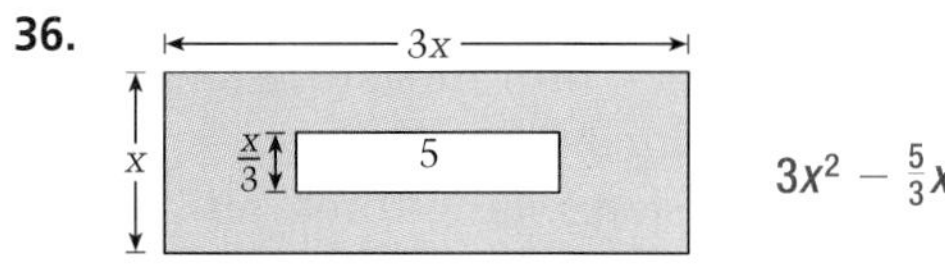

$3x^2 - \frac{5}{3}x$

37. *Chicken or Turkey?* For 1985 through 1993, the average numbers of pounds of chicken, C, and turkey, T, consumed per American can be modeled by

$C = -0.09t^2 + 2.26t + 31.07$

$T = -0.10t^2 + 2.34t + 0.30$

where $t = 5$ represents 1985. Find a model for the average number of pounds, P, of chicken *and* turkey consumed from 1985 to 1993. **(Source: U.S. Department of Agriculture)**

$P = -0.19t^2 + 4.60t + 31.37$

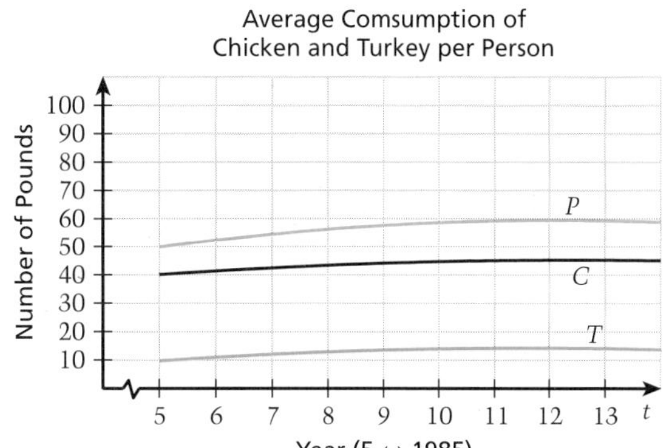

38. *Food Stores* For 1985 through 1993, the average numbers (in thousands) of grocery stores, G (including convenience stores), and specialty food stores, S, in the United States can be modeled by

$G = -0.12t^2 + 0.41t + 178.87$

$S = 0.004t^2 - 0.786t + 74.688$

where $t = 5$ represents 1985. Find a model that represents the total number of retail food stores, F, from 1985 to 1993. **(Source: National Restaurant Association)**

$F = -0.116t^2 - 0.376t + 253.558$

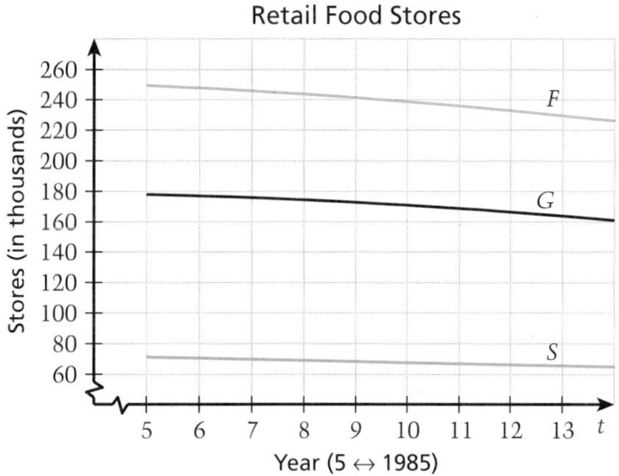

39. *American Men and Women* For 1950 through 1995, the total population, P, and male population, M, of the United States can be modeled by

$P = 2427.9t + 153{,}308$ (in 1000's)

$M = 1160.7t + 75{,}707$ (in 1000's)

where $t = 0$ represents 1950. Find a model that represents the female population, F, of the United States from 1950 to 1995. **(Source: U.S. Bureau of the Census)**

$F = 1267.2t + 77{,}601$

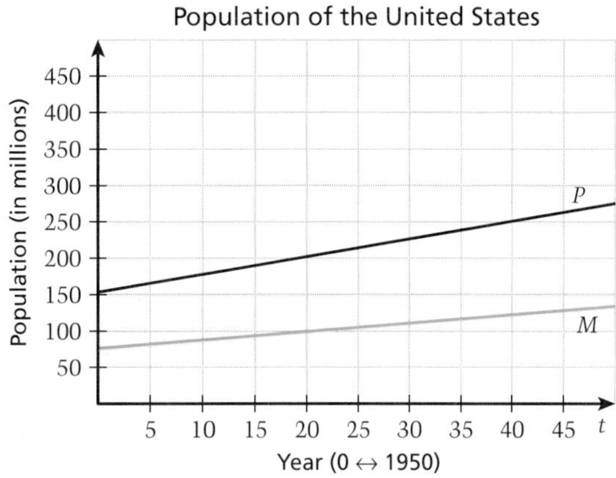

Independent Practice

Ex. 11–28 After completing these exercises, ask students which format they prefer.

Ex. 29–34 Be sure to ask students why they chose one method over the other.

Ex. 33–34 These exercises require use of the distributive property.

Ex. 35–36 In these exercises, students should assume that the figures are all rectangles.

Ex. 37–40 These exercises provide real-life data models for adding or subtracting polynomials. Students should find these exercises interesting discussion points the next day in class.

▶ **Ex. 45–48** Review basic matrix operations before assigning these exercises. The matrices provide another format for practicing polynomial operations.

Answers

45. $\begin{bmatrix} 2x^2 & -x \\ -7+9x & -3x^2+4 \end{bmatrix}$

46. $\begin{bmatrix} 3x^2+3 & -2x \\ 5x^2+3x & 3x-1 \end{bmatrix}$

47. $\begin{bmatrix} 20x-13 & 10-3x \\ x^3+7y^3 & 4y^2-11y+2 \end{bmatrix}$

48. $\begin{bmatrix} 7x^3+2x^2 & -x^2-4x \\ 2x-4 & 4-7x-x^2 \end{bmatrix}$

Enrichment Activities

This activity requires students to go beyond lesson goals.

Why are second-degree equations such as $-3x^2+2x-13=0$ called quadratic equations? Doesn't "quad" mean four? Look up the term in a dictionary or encyclopedia. Quadratic is derived from the Latin word "quadratus," which means squared.

✪ **40.** *U.S. Retail Sales* Retail stores are classified as durable-goods stores (car dealers, hardware stores, furniture stores, jewelers) and nondurable-goods stores (department stores, drugstores, restaurants, grocery stores, gas stations, general merchandise stores). For 1980 through 1990, the total sales for retail stores, R, and for durable-goods stores, D, in the United States can be modeled by

$$R = -0.21t^2 + 31.6t + 357.4 \quad \text{(in billions \$)}$$
$$D = 0.26t^2 + 2.9t + 34.3 \quad \text{(in billions \$)}$$

where $t = 0$ represents 1980. Find a model that represents the sales for nondurable-goods stores, N.
$N = -0.47t^2 + 28.7t + 323.1$

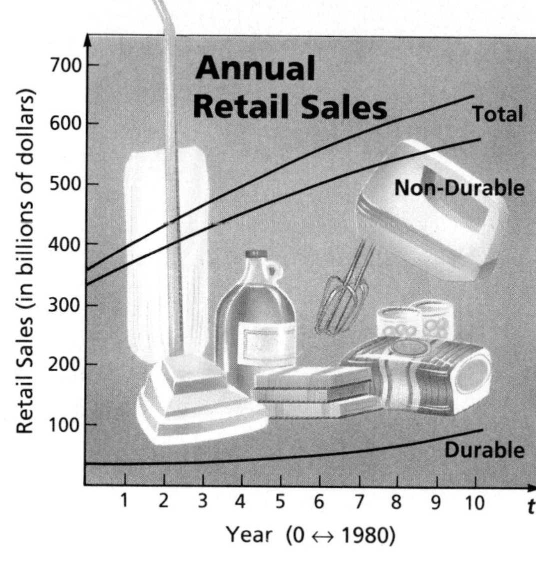

Integrated Review

In Exercises 41–44, simplify the expression.

41. $8y - 2x + 7x - 10y$ $5x - 2y$

42. $\frac{5}{6}x - \frac{2}{3}x + 8$ $\frac{1}{6}x + 8$

43. $\frac{1}{2}x + \frac{3}{2}y - 3y + 4x$ $\frac{9}{2}x - \frac{3}{2}y$

44. $-2y + 4x - 4x + 6y$ $4y$

Discrete Math **In Exercises 45–48, perform the indicated matrix operation.** See margin.

45. $\begin{bmatrix} x^2-4x & -3x \\ 7x & 2x^2+4 \end{bmatrix} + \begin{bmatrix} x^2+4x & 2x \\ -7+2x & -5x^2 \end{bmatrix}$

46. $\begin{bmatrix} 4 & -6x \\ 5x^2+x & 2x \end{bmatrix} + \begin{bmatrix} 3x^2-1 & 4x \\ 2x & x-1 \end{bmatrix}$

47. $\begin{bmatrix} 6x-5 & 10 \\ x^3 & 4y^2-5y \end{bmatrix} - \begin{bmatrix} 8-14x & 3x \\ -7y^3 & 6y-2 \end{bmatrix}$

48. $\begin{bmatrix} 7x^3 & x^2-3x \\ 2x-1 & 4-2x \end{bmatrix} - \begin{bmatrix} -2x^2 & 2x^2+x \\ 3 & x^2+5x \end{bmatrix}$

Geometry **In Exercises 49 and 50, find the value(s) for x.**

49. Area of shaded region is 260 square units. 5 units

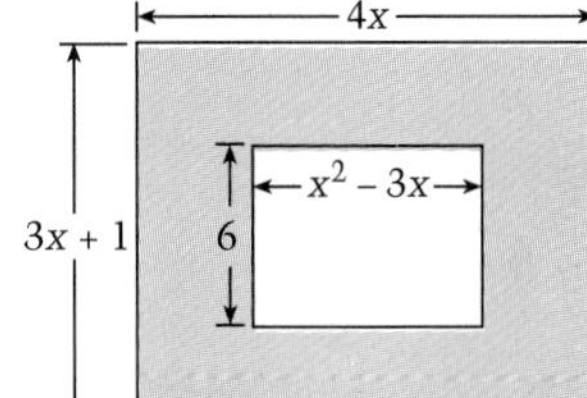

50. Area of shaded region is 84 square units. 2 units

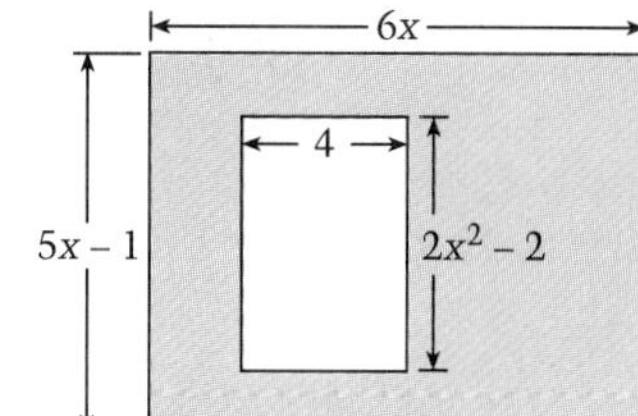

Exploration and Extension

51. What must you add to $4x^2 + x - 5$ to get $x^2 + 5x + 1$? $-3x^2 + 4x + 6$

52. What must you add to $x^2 - 7x + 3$ to get $-3x^2 + 2x - 1$? $-4x^2 + 9x - 4$

✪ **53.** What must you subtract from $3x^2 + 7x - 9$ to get $12x^2 - x + 6$? $-9x^2 + 8x - 15$

10.2 Multiplying Polynomials

What you should learn:

Goal 1 How to multiply two polynomials using the Distributive Properties and the FOIL method

Goal 2 How to use polynomial multiplication in real-life settings

Why you should learn it:

You can model many situations with polynomial expressions that have to be multiplied, such as a rate times a quantity.

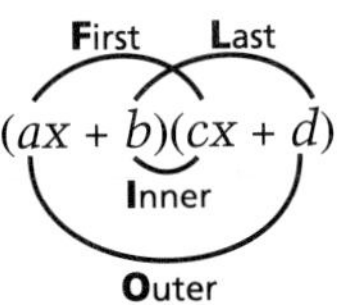

Goal 1 Multiplying Polynomials

You already know the simplest type of polynomial multiplication—multiplying a polynomial by a monomial.

$$\begin{aligned}(3x)(2x^2 - 5x + 3) &= (3x)(2x^2) - (3x)(5x) + (3x)(3)\\ &= 6x^3 - 15x^2 + 9x\end{aligned}$$

To multiply two binomials, you can use both left and right Distributive Properties.

$$\begin{aligned}(3x - 2)(2x + 7) &= 3x(2x + 7) - 2(2x + 7)\\ &= (3x)(2x) + (3x)(7) - (2)(2x) - 2(7)\\ &= 6x^2 + 21x - 4x - 14\end{aligned}$$

Product of First terms	Product of Outer terms	Product of Inner terms	Product of Last terms
$6x^2$	$21x$	$-4x$	-14

$$= 6x^2 + 17x - 14$$

With practice, you will be able to multiply two binomials in a single step using the **FOIL pattern.**

Example 1 *Multiplying Binomials (Distributive Property)*

Multiply: $(x - 3)$ by $(x + 2)$.

Solution

$(x + 2)(x - 3)$

$= x(x - 3) + 2(x - 3)$ *Right Distributive Property*

$= x^2 - 3x + 2x - 6$ *Left Distributive Property*

$= x^2 - x - 6$ *Combine like terms.* ■

Example 2 *Multiplying Binomials (FOIL Pattern)*

Multiply: $(3x + 4)(2x + 1)$.

Solution

$$\begin{aligned}(3x + 4)(2x + 1) &= \underbrace{6x^2}_{F} + \underbrace{3x}_{O} + \underbrace{8x}_{I} + \underbrace{4}_{L}\\ &= 6x^2 + 11x + 4\end{aligned}$$ ■

Problem of the Day

3 chickens and 1 duck sold for as much as 2 geese. 1 chicken, 2 ducks, and 3 geese sold for $25. If each sold for a whole number of dollars, what was the price of each?

Chicken: $2, duck: $4, goose: $5

ORGANIZER

Warm-Up Exercises

1. Multiply.
   a. $6x^4 \cdot 3x^2$ $18x^6$
   b. $-5x^3 \cdot 9x^5$ $-45x^8$
   c. $8x^2 \cdot -7x^4$ $-56x^6$
   d. $12x \cdot 3x^6$ $36x^7$
2. Perform the indicated operations.
   a. $(5x^2 - x + 23) + (-3x^3 + 6x^2 + 23)$
   b. $(19x^4 + 4x^2) - (8 + 5x^2 - x^3 + 19x^4)$
   c. $(34x^3 - 26x^2 + 8) - (14x^2 - 5x + 10)$

   a. $-3x^3 + 11x^2 - x + 46$
   b. $x^3 - x^2 - 8$
   c. $34x^3 - 40x^2 + 5x - 2$
3. Multiply using the distributive property.
   a. $12(3x + 7)$ $36x + 84$
   b. $(-18 - 23y)(-4x)$ $72x + 92xy$
   c. $-5(3x - 6y + 19)$ $-15x + 30y - 95$

Lesson Resources

Teaching Tools
- Transparency: 16
- Problem of the Day: 10.2
- Warm-up Exercises: 10.2
- Answer Masters: 10.2

Extra Practice: 10.2

Color Transparency: 58

LESSON Notes

Examples 1–2

The FOIL pattern uses the distributive property of multiplication over addition. Students can learn the pattern and multiply a binomial by a binomial using mental computations.

Examples 3–4

Most students find it easier to multiply polynomials of three or more terms using the vertical format. Whether the vertical or horizontal format is applied, it is a good practice to rewrite each polynomial being multiplied in standard form.

Example 5

Remind students that the area model is an interpretation of multiplying numbers. The same representation carries over into algebra. Consequently, paper cutouts or algebra tiles can be useful in enhancing the understanding of the multiplication of polynomials.

Example 6

Observe how the graph of the revenue relationship provides useful information. What algebraic steps can be taken to discover similar results? Using a guess-and-check strategy, compute different values of R from different values of t in order to see what value of t gives a maximum value of R.

Extra Examples

Here are additional examples similar to **Examples 1–5.**

1. Multiply by FOIL.
   - **a.** $(x - 5)(x - 3)$
   - **b.** $(2x + 4)(-7x + 8)$
   - **c.** $(5x - 6)(3x + 9)$
   - **d.** $(4x^2 - 4x)(3x + 2)$

   a. $x^2 - 8x + 15$
   b. $-14x^2 - 12x + 32$
   c. $15x^2 + 27x - 54$
   d. $12x^3 - 4x^2 - 8x$

To multiply two polynomials, remember that *each term of one polynomial must be multiplied by each term of the other polynomial.* This can be done using either a vertical or a horizontal format. In either case, it is best to begin by writing each polynomial in standard form.

Example 3 *Multiplying Polynomials (Vertical Format)*

Multiply: $(x - 2)(5 + 3x - x^2)$.

Solution Begin by aligning like terms in columns.

$$\begin{array}{rll} & -x^2 + 3x + 5 & \textit{Standard form} \\ \times & x - 2 & \textit{Standard form} \\ \hline & 2x^2 - 6x - 10 & \leftarrow -2(-x^2 + 3x + 5) \\ & -x^3 + 3x^2 + 5x \quad\quad & \leftarrow \quad x(-x^2 + 3x + 5) \\ \hline & -x^3 + 5x^2 - x - 10 & \textit{Add like terms.} \end{array}$$

■

Example 4 *Multiplying Polynomials (Horizontal Format)*

Multiply: $(4x^2 - 3x - 1)(2x - 5)$.

Solution

$$\begin{aligned} &(4x^2 - 3x - 1)(2x - 5) \\ &\quad = 4x^2(2x - 5) - 3x(2x - 5) - (1)(2x - 5) \\ &\quad = 8x^3 - 20x^2 - 6x^2 + 15x - 2x + 5 \\ &\quad = 8x^3 - 26x^2 + 13x + 5 \end{aligned}$$

■

Connections Geometry

Example 5 *An Area Model for Multiplying Polynomials*

Use an area model (or algebra tiles) to show that

$$(x + 2)(2x + 1) = 2x^2 + 5x + 2.$$

Solution Think of a rectangle whose sides have lengths $x + 2$ and $2x + 1$. The area of this rectangle is

$(x + 2)(2x + 1)$. *Area = (width)(length)*

Another way to find the area is to add the areas of the rectangular parts. There are two squares whose sides are x, five rectangles whose sides are x and 1, and two squares whose sides are 1. The total area of these nine rectangles is

$2x^2 + 5x + 2$. *Area = sum of rectangular areas*

Because each method must produce the same area, you can conclude that

$$(x + 2)(2x + 1) = 2x^2 + 5x + 2.$$

■

For many American dairy farmers, the 1980's were not good years. Although milk production per cow increased through computer management and breeding, the price of milk paid to dairy farmers decreased.

Goal 2 Polynomial Multiplication in Real Life

Example 6 *The Plight of the Dairy Farmer*

Dairies measure milk in 100-pound units. (A gallon of milk weighs about 8 pounds.) For 1980 through 1990, the average annual milk production, M, per dairy cow can be modeled by $M = 3t + 115$, where M is measured in hundreds of pounds and $t = 0$ represents 1980. The prices, p, paid to dairy farmers can be modeled by $p = -0.25t + 14.25$, where p is price in dollars per hundred pounds. Find a model for average annual revenue per cow. What can you conclude from the model?

Solution

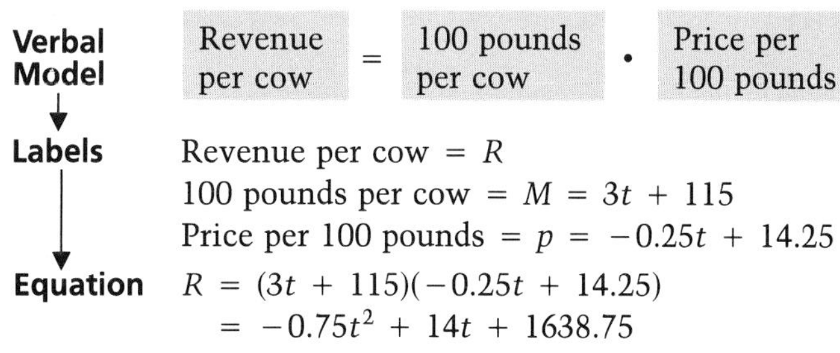

Verbal Model Revenue per cow = 100 pounds per cow • Price per 100 pounds

Labels Revenue per cow = R

100 pounds per cow = $M = 3t + 115$

Price per 100 pounds = $p = -0.25t + 14.25$

Equation $R = (3t + 115)(-0.25t + 14.25)$

$= -0.75t^2 + 14t + 1638.75$

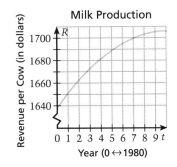

Milk Production

Revenue per Cow (in dollars): R — 1640, 1660, 1680, 1700

Year (0 ↔ 1980): 0 1 2 3 4 5 6 7 8 9 t

From the graph, you can see that the average revenue per cow peaked in about 1989. For many dairy farmers, the increased revenue was not sufficient to offset the increasing costs of machinery, fertilizers, and feed. ***(Source: U.S. Dept. of Agriculture)*** ■

See Additional Answers.

Communicating about ALGEBRA

Cooperative Learning

SHARING IDEAS about the Lesson

Work with a Partner and Make a Model Construct area models similar to Example 5 for each polynomial product.

A. $(x + 3)(x + 2)$ **B.** $(x + 1)(3x + 1)$

2. Multiply in both the vertical and horizontal formats.
   a. $(x + 3)(6 - 3x + 2x^2)$
   b. $(5x^2 + 11x - 8)(2x + 7)$
   a. $2x^3 + 3x^2 - 3x + 18$
   b. $10x^3 + 57x^2 + 61x - 56$
3. Use an area model to find the product.
   $(x + 3)(4x + 2)$.

$4x + 2$ (across), $x + 3$ (down)

x^2	x^2	x^2	x^2	x	x
x	x	x	x	1	1
x	x	x	x	1	1
x	x	x	x	1	1

$4x^2 + 14x + 6$

Check Understanding

1. What are the advantages and disadvantages of using the vertical and horizontal formats to multiply algebraic polynomials? See Lesson Notes for Examples 3 and 4.
2. The FOIL pattern provides a useful way to multiply two binomials. Can the product of two binomials be obtained if the order of the steps of FOIL are misspelled FIOL? Yes; because addition is commutative.
3. State the distributive properties of multiplication over addition and multiplication over subtraction. For all real numbers a, b, and c, $a(b + c) = ab + ac$ and $a(b - c) = ab - ac$.

Communicating about ALGEBRA

EXTEND *Communicating*

1. How could you use paper cutouts or algebra tiles or other physical models to represent the products,
   $(x + 3)(x^2 + 2)$ or
   $(2x + 3)(2x^2 + x + 1)$?
   Carefully explain how the products are created. The x^3 regions could be represented by an x-high stack of x^2-regions, or a cube.

2. Consider whether the FOIL pattern for multiplying two binomials can be extended to the product of a binomial and a trinomial. To two trinomials. Explain. The extension can be made as follows.
$(a + b)(c + d + e)$
$= (a + b)([c + d] + e) =$
$(a[c + d] + ae + b[c + d] + be$
$= ac + ad + ae + bc + bd + be$
The product of two trinomials is similar.

EXERCISE Notes

ASSIGNMENT GUIDE

Basic/Average: Ex. 1–4, 7–31 odd, 43–45, 47–61 odd

Above Average: Ex. 1–4, 9–35 odd, 38–43, 46, 48–63 multiples of 3, 64

Advanced: Ex. 1–4, 11–37 odd, 38–46, 63–65

Selected Answers
Exercises 1–4, 5–61 odd

Use **Mixed Review** as needed.

More Difficult Exercises
Exercise 46

Guided Practice

Ex. 1–4 These exercises help students with the connection between FOIL and the distributive property.

Independent Practice

Ex. 11–38 These exercises encourage students to use a variety of methods and formats for multiplying polynomials. The FOIL pattern provides a symbolic operation strategy, and the area model pattern provides visual operation support.

EXERCISES

Guided Practice

CRITICAL THINKING about the Lesson See below.

1. Show how the Distributive Properties can be used to multiply $(2x - 3)$ and $(x + 4)$.
2. Multiply: $(x + 1)(x^2 - x + 1)$. Explain your use of the Distributive Property. $x^3 + 1$
3. Multiply: $(x - 3)(2x + 5)$. $2x^2 - x - 15$
4. What does **FOIL** represent?

Independent Practice

In Exercises 5–10, multiply.

5. $(3x - 7)(-2x)$ $-6x^2 + 14x$
6. $3x^2(5x - x^3 + 2)$ $15x^3 - 3x^5 + 6x^2$
7. $(-x)(2x^2 - 3x)$ $-2x^3 + 3x^2$
8. $2x(3x^2 - 4x + 1)$ $6x^3 - 8x^2 + 2x$
9. $4x^2(5x^3 - 2x^2 + x)$ $20x^5 - 8x^4 + 4x^3$
10. $-x^2(6x^3 - 14x + 9)$ $-6x^5 + 14x^3 - 9x^2$

In Exercises 11–16, use the FOIL pattern to multiply.

11. $(3x - 2)(5x + 7)$ $15x^2 + 11x - 14$
12. $(3x + 5)(2x + 1)$ $6x^2 + 13x + 5$
13. $(x - 4)(x + 4)$ $x^2 - 16$
14. $(2x - 3)(x + 3)$ $2x^2 + 3x - 9$
15. $(x - 5)(2x + 10)$ $2x^2 - 50$
16. $(3x - 5)(2x + 1)$ $6x^2 - 7x - 5$

In Exercises 17–22, use an area model (or algebra tiles) to multiply. See Additional Answers.

17. $(x + 1)(x + 5)$ $x^2 + 6x + 5$
18. $(x + 2)(x + 6)$ $x^2 + 8x + 12$
19. $(x + 1)(x + 2)$ $x^2 + 3x + 2$
20. $(3x + 1)(2x + 2)$ $6x^2 + 8x + 2$
21. $(x + 2)(2x + 3)$ $2x^2 + 7x + 6$
22. $(2x + 1)(x + 3)$ $2x^2 + 7x + 3$

In Exercises 23–28, use the Distributive Property to multiply.

23. $(x - 3)(3x + 1)$ $3x^2 - 8x - 3$
24. $(2x + 1)(3x + 1)$ $6x^2 + 5x + 1$
25. $(3x^2 + x - 5)(2x - 1)$ $6x^3 - x^2 - 11x + 5$
26. $(2x^2 - 7x + 1)(4x + 3)$ $8x^3 - 22x^2 - 17x + 3$
27. $(x^2 + 9)(x^2 - x - 4)$ $x^4 - x^3 + 5x^2 - 9x - 36$
28. $(x + 3)(x^2 - 6x + 2)$ $x^3 - 3x^2 - 16x + 6$

In Exercises 29–37, multiply.

29. $(x + 3)(x - 4)$ $x^2 - x - 12$
30. $(2x - 1)(x + 9)$ $2x^2 + 17x - 9$
31. $(2x - 5)(x + 6)$ $2x^2 + 7x - 30$
32. $(3x - 4)(\frac{1}{3}x + 1)$ $x^2 + \frac{5}{3}x - 4$
33. $(x + \frac{6}{5})(4x - 5)$ $4x^2 - \frac{1}{5}x - 6$
34. $(x + \frac{1}{4})(x - \frac{5}{4})$
35. $(\frac{1}{2}x + 3)(\frac{1}{2}x - 2)$ $\frac{1}{4}x^2 + \frac{1}{2}x - 6$
36. $(-3x^2 + x - 1)(x + 3)$
37. $(x^2 + 4x - 9)(x - 4)$ $x^3 - 25x + 36$

36. $-3x^3 - 8x^2 + 2x - 3$

38. *Area of a Sail* The base of a triangular sail is x feet and its height is $\frac{1}{2}x + 5$ feet. Find an expression for the area, A, of the sail. $A = \frac{1}{4}x^2 + \frac{5}{2}x$

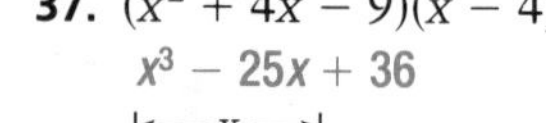

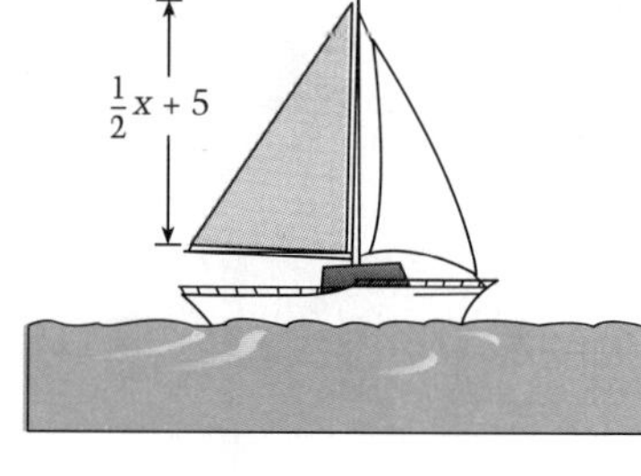

39. Use the expression in Exercise 38 to complete the table.

Base, x	5	6	7	8	9	10
Area, A	?	?	?	?	?	?

$18\frac{3}{4}$, 24, $29\frac{3}{4}$, 36, $42\frac{3}{4}$, 50

34. $x^2 - x - \frac{5}{16}$

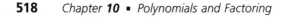

1. $(2x - 3)(x + 4) = 2x(x + 4) - 3(x + 4)$
$= 2x^2 + 8x - 3x - 12 = 2x^2 + 5x - 12$

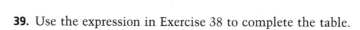

2. $(x^3 + 1)$, $x(x^2 - x + 1) + 1(x^2 - x + 1)$
$= x^3 - x^2 + x + x^2 - x + 1$

Crop Circles **In Exercises 40 and 41, use the following information.**

In recent years, people in Britain were puzzled by the overnight appearance of "crop circles." Many were surrounded by concentric rings. Those shown below appeared in a wheat field in 1987 about three miles from Stonehenge, England.

40. The area (in square feet) of the circular ring of grain shown in the upper right corner of the photo is given by

$A = \pi(4x + 28)(4x - 28)$.

Simplify this expression. Substitute $x = 9$ to obtain the area of this circular ring in square feet. $\pi(16x^2 - 784)$, 512π ft$^2 \approx 1608.5$ ft^2

41. If you doubled the value of x, would the value of A double? Explain. No; if $x = 18$, $A = 4400\pi \neq 2(512)\pi$

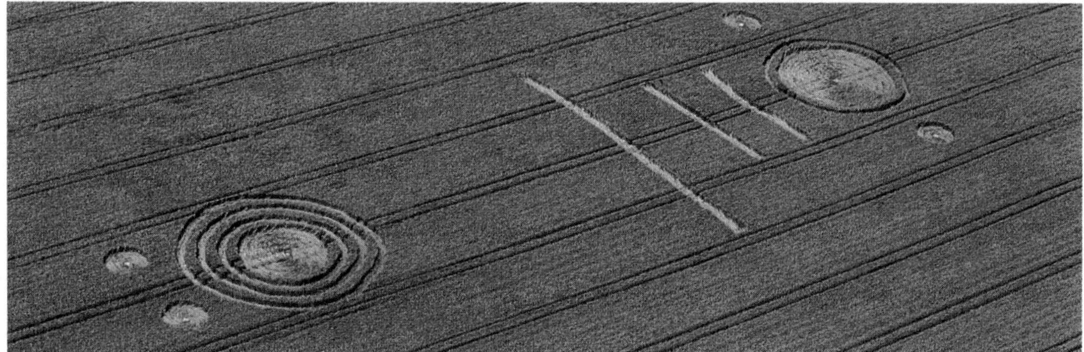

42. *FOIL Pattern* Find the area of the rectangle in two different ways: by multiplying length times width *and* by adding the areas of the smaller rectangles. Explain how this is related to the FOIL pattern for finding the product $(x + a)(x + b)$. See margin.

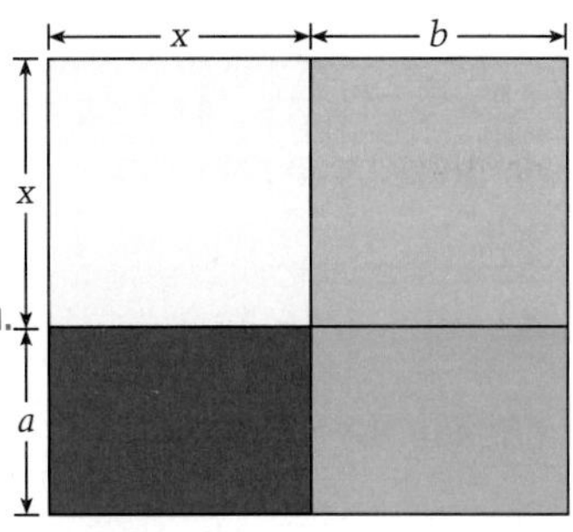

43. *College Entrance Exam Sample* $(x + 9)(x + 2) = \boxed{?}$ d

a. $x^2 + 18$ **b.** $11x$ **c.** $x^2 + 11$ **d.** $x^2 + 11x + 18$ **e.** $9(x + 2) + 2(x + 9)$

44. *Geometry* Find an expression for the area of the trapezoid. (Area $= \frac{1}{2}h(b_1 + b_2)$)

3x + 4

x + 1

5x + 7

$A = \frac{1}{2}(8x + 11)(x + 1) = 4x^2 + \frac{19}{2}x + \frac{11}{2}$

45. *Geometry* Find an expression for the area of the triangle.

6x – 5

6x – 5

x + 4

$A = \frac{1}{2}(6x - 5)(x + 4) = 3x^2 + \frac{19}{2}x - 10$

4. FOIL: Product of the First terms, product of the Outer terms, product of the Inner terms, product of the Last terms.

Ex. 40–41 A ring of grain can be thought of as concentric circles. The outside circle has radius R. The inside circle has radius r. So, the area of the ring is $\pi R^2 - \pi r^2 = \pi(R^2 - r^2) = \pi(R + r)(R - r)$ where $R = 4x$ and $r = 28$.

Ex. 42 This exercise provides visual support to the FOIL formula.

Answer

42. $x^2 + bx + ax + ab$. Multiplying each term of one side of the large rectangle by each term of the other side is the same as multiplying each term in the first binomial by each term in the second binomial.

Ex. 45 In this exercise, students should assume the diagram shows a triangle in a square.

▶ **Ex. 46** Note that $T = S \cdot P = (900t + 11{,}800)(228 + 2.2t)$.

Answer

46. $T = 1980t^2 + 231{,}160t + 2{,}690{,}400$

S: 11,800; 16,300; 20,800.
P: 228; 239; 250.
T: 2,690,400; 3,895,700; 5,200,000.

EXTEND Ex. 63 GROUP ACTIVITY Assign half the class to find the product by first multiplying $(x + 1)(2x - 3)$, then multiplying by $(x - 7)$. Ask the other half to first multiply $(2x - 3)(x - 7)$ and then multiply by $(x + 1)$. What do they notice?

Enrichment Activities

These activities require students to go beyond lesson goals.

Another method for multiplying two polynomials is an array. Consider the product of $(3x^2 - 3x + 1)(2x^2 - x + 7)$. To multiply, use the following steps.

A. Place the coefficients of each polynomial in an array as displayed.

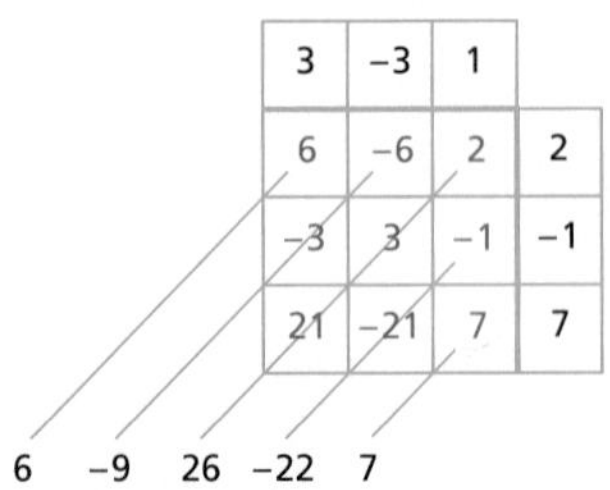

B. Complete the array with the products of each row and column.

C. Add diagonally as shown.

D. Transfer the sums to coefficients of the product. $6x^4 - 9x^3 + 26x^2 - 22x + 7$

1. Use this method to complete Exercises 26–28, 36–38, and 65.

46. *Gross Domestic Product (GDP)* For 1980 through 1990, each American's dollar share, S, of the GDP can be modeled by

$S = 900t + 11{,}800$ *dollar share per American*

where $t = 0$ represents 1980. The population, P (in millions), of the United States during the same time can be modeled by

$P = 228 + 2.2t.$ *number of Americans*

Find a model for the *total* Gross Domestic Product, T, from 1980 to 1990. Then use the three models to complete this table. See margin.

Year, t (1980 = 0)	0	5	10
Per capita GDP, S (dollars per American)	?	?	?
Population, P (millions of Americans)	?	?	?
Total GDP, T (millions of dollars)	?	?	?

(Source: Organization for Economic Cooperation and Development, Paris, France)

Integrated Review

In Exercises 47–50, rewrite the product in exponential form.

47. $(-3)(-3)(-3)(-3)$ $(-3)^4$
48. $(x)(x)(x)$ x^3
49. $(-2)(-2)(x)(x)(x)$ $(-2)^2x^3$
50. $(4)(4)(y)(y)$ 4^2y^2

In Exercises 51–54, rewrite the expression as repeated multiplication.

51. $\left(\frac{4}{5}\right)^4$ $\frac{4}{5} \cdot \frac{4}{5} \cdot \frac{4}{5} \cdot \frac{4}{5}$
52. $(4.5)^5$ $4.5 \cdot 4.5 \cdot 4.5 \cdot 4.5 \cdot 4.5$
53. $(x^2)^3$ $x^2 \cdot x^2 \cdot x^2$
54. $(y^2)^2$ $y^2 \cdot y^2$

In Exercises 55–58, simplify the expression.

55. $2(x - 4) + 5x$ $7x - 8$
56. $4(3 - y) + 2(y + 1)$ $-2y + 14$
57. $-3(z - 2) - (z - 6)$ $-4z + 12$
58. $(u - 2) - 3(2u + 1)$ $-5u - 5$

In Exercises 59–62, simplify the expression.

59. $\frac{x^4}{3x^{-3}}$ $\frac{1}{3}x^7$
60. $\frac{6x^{-1}}{2x}$ $\frac{3}{x^2}$
61. $\frac{5x^2y^0}{xy^3}$ $\frac{5x}{y^3}$
62. $\frac{x^3y^3}{x^0y}$ x^3y^2

Exploration and Extension

In Exercises 63 and 64, multiply.

63. $(x + 1)(2x - 3)(x - 7)$ $2x^3 - 15x^2 + 4x + 21$
64. $(3x + 7)(x + 4)(2x - 3)$ $6x^3 + 29x^2 - x - 84$

65. *Patterns* Multiply: $(x - 1)(x + 1)$, $(x - 1)(x^2 + x + 1)$, and $(x - 1)(x^3 + x^2 + x + 1)$. Find a pattern in your results. Use the pattern to guess the result of multiplying $(x - 1)$ and $(x^4 + x^3 + x^2 + x + 1)$. Then verify your guess by multiplying.
$x^2 - 1$, $x^3 - 1$, $x^4 - 1$, $x^5 - 1$

Mixed REVIEW

1. What is 22% of \$4,242.00? \$933.24
2. What is $\frac{1}{4}$ of $\frac{2}{3}$? $\frac{1}{6}$
3. What is the reciprocal of x^2y? **(2.7)** $\frac{1}{x^2y}$
4. What is the reciprocal of $-\frac{1}{2}$? **(2.7)** -2
5. Solve $6^{x+3} = (3 + 3)^2$. **(8.2)** -1
6. Solve $3x^2 - 2x - 1 = 0$. **(9.4)** $-\frac{1}{3}, 1$
7. Solve $x^2 + 4x + 4 = 0$. **(9.4)** -2
8. Solve $y^{x+3} = y^7$ for x. **(8.1)** 4
9. Evaluate $x^2 - 5x - 7$ when $x = -1$. **(1.3)** -1
10. Evaluate $|x + 2| + 3x$ when $x = -4$. **(2.1)** -10
11. Evaluate $(0.9)(3.2)^t$ when $t = 2$. **(1.3)** 9.216
12. Evaluate $(7.29)(6.2)^{-t}$ when $t = 1$. **(8.2)** ≈ 1.18
13. Is 3 a solution of $3x + 2^x \geq 15$? **(1.5)** Yes
14. Is -4 a solution of $-y + 7y^2 - 14 < 18$? **(1.5)** No
15. Is $(5, -3)$ a solution of $x + 2 = |-3y|$? **(2.1)** No
16. Is $(-4, -1)$ a solution of $2^x < (2^y)^2$? **(8.1)** Yes
17. Write 0.00794 in scientific notation. **(8.4)** 7.94×10^{-3}
18. Write 3.29×10^4 in decimal form. **(8.4)** 32,900
19. Find the slope of $2x - 5y = 10$. **(4.3)** $\frac{2}{5}$
20. Find the intercepts of $y = 3x - 2$. **(4.3)** $(\frac{2}{3}, 0), (0, -2)$

Milestones LEONHARD EULER (1707–1783)

1650 · 1700 · 1750 · 1800

Euler born — Euler at St. Petersburg Academy of Science — Euler heads Academy of Science, Berlin

When Leonhard Euler entered the University of Basel in Switzerland, he began studying for the clergy but, encouraged by one of his teachers, soon switched to mathematics. He received his master's degree at the age of sixteen. His arrival in Russia in 1727 coincided with the death of Czarina Catherine I and the beginning of a regime unfavorable to scientific pursuits. Later political change brought a more favorable climate for research, and he began teaching at the newly founded St. Petersburg Academy of Science.

Euler made significant contributions to the study of number theory, geometry, and calculus. A prolific writer, he published 866 papers on mathematical topics. He introduced the symbol i for imaginary numbers, e for the base of natural logarithms, and $f(\ )$ for functions. He was one of the first to adopt the use of the symbol π. He was also a pioneer in the field of topology. Although totally blind the last 17 years of his life, he continued his work in mathematics. An unassuming man, he said, "The path that I followed will be of some help perhaps."

Name a significant event in the history of the United States that occurred during Euler's lifetime.

Leonhard Euler

2. **WRITING** Write a paragraph comparing this method with the FOIL technique. Which do you prefer? Explain.

Milestones

Language Arts:
Critical Reading

You may need to use reference materials to find the answers to these questions.

1. What is topology? Topology is the study of geometric shapes and the properties of these shapes that are unchangeable under transformations such as stretching, shrinking, twisting, or other manipulations.
2. Find Euler's theorem for polyhedra and write the formula that relates the number of edges, e, the number of vertices, v, and the number of sides, s, of a polyhedron. $e + 2 = v + s$

Problem of the Day

Wigglesworth has a gold chain of seven links. An artist friend offers to sketch each of the seven members of Wigglesworth's family for one gold link per portrait, but wants to be paid for each portrait at its completion. What is the fewest number of cuts Wigglesworth needs to make in the chain?
One cut (in the 3rd link)

ORGANIZER

Warm-Up Exercises

1. Multiply using FOIL. Identify a pattern found in the products.
   a. $(x + 2)(x - 2)$
   b. $(x - 1)(x + 1)$
   c. $(x + 9)(x - 9)$
   a. $x^2 - 4$, b. $x^2 - 1$, c. $x^2 - 81$;
   One pattern is that there is no middle term; another is that the product is the difference of the squares of the terms in each binomial.
2. Multiply using FOIL.
   a. $(x - 3)(x - 3)$
   b. $(3x + 2)(3x + 2)$
   c. $(7 - 4x)(7 - 4x)$
   a. $x^2 - 6x + 9$
   b. $9x^2 + 12x + 4$
   c. $49 - 56x + 16x^2$

Lesson Resources

Teaching Tools
 Transparency: 16
 Problem of the Day: 10.3
 Warm-up Exercises: 10.3
 Answer Masters: 10.3
Extra Practice: 10.3
Applications Handbook:
 p. 13, 14

10.3 Multiplying Polynomials: Two Special Cases

What you should learn:

Goal 1 How to use special-product patterns for the product of a sum and difference and for the square of a binomial

Goal 2 How to use special-product patterns in real-life models

Why you should learn it:

You can multiply some polynomials very efficiently if they have a special form.

Goal 1 Using Special-Product Patterns

Some binomial products occur so frequently in algebra that it is worth remembering special patterns for them. For instance, the product $(x + 3)(x - 3)$ is called the *product of the sum and difference of two terms.* This special product has no "middle term."

$$(x + 3)(x - 3) = x^2 - 3x + 3x - 9 \quad \text{"Sum and difference"}$$
$$= x^2 - 9 \quad \text{No middle term}$$

Another common type of product is the *square of a binomial.* In its pattern, the middle term is always twice the product of the two terms in the binomial.

$$(2x + 5)^2 = (2x + 5)(2x + 5) \quad \text{Square of a binomial}$$
$$= 4x^2 + 10x + 10x + 25$$
$$= 4x^2 + 20x + 25 \quad \text{Middle term is } 2(2x)(5).$$

These two patterns are generalized in the following summary. When you use these special-product patterns, remember that a and b can be numbers, variables, or even variable expressions.

Special Products

Sum and Difference Pattern | **Example**

$(a + b)(a - b) = a^2 - b^2$ $\quad$ $(3x - 4)(3x + 4) = 9x^2 - 16$

Square of a Binomial Pattern | **Example**

$(a + b)^2 = a^2 + 2ab + b^2$ $\quad$ $(x + 4)^2 = x^2 + 2(x)(4) + 4^2 = x^2 + 8x + 16$

$(a - b)^2 = a^2 - 2ab + b^2$ $\quad$ $(x - 6)^2 = x^2 - 2(x)(6) + 36 = x^2 - 12x + 36$

You don't need to write the steps shown in color. They can be performed with mental math.

	a	b
a	a^2	ab
b	ab	b^2

Connections **Geometry**

The square-of-a-binomial pattern $(a + b)^2 = a^2 + 2ab + b^2$ can be verified with the area model shown at the left. The two rectangles with areas ab produce the middle term $2ab$.

Example 1 *The Product of the Sum and Difference of Two Terms*

Multiply: $(5t - 2)(5t + 2)$.

Solution This special product represents the sum and difference of two terms. The product has the form $(a - b)(a + b) = a^2 - b^2$.

$$\begin{aligned}(5t - 2)(5t + 2) &= (5t)^2 - 2^2\\ &= 25t^2 - 4\end{aligned}$$

■

Example 2 *Squaring a Binomial*

Multiply: $(2x - 7)^2$.

Solution This special product represents the square of a binomial. The product has the form $(a - b)^2 = a^2 - 2ab + b^2$. Note that the *middle* term of the product is twice the product of the two terms of the binomial.

$$\begin{aligned}(2x - 7)^2 &= (2x)^2 - 2(2x)(7) + 7^2\\ &= 4x^2 - 28x + 49\end{aligned}$$

■

Goal 2 Using Special Products in Real-Life Models

Connections **Geometry**

Example 3 *Finding an Area*

Find an expression for the area of the blue region shown at the left. Then evaluate the expression when x is equal to 2 inches; to 3 inches; and to 4 inches.

Solution

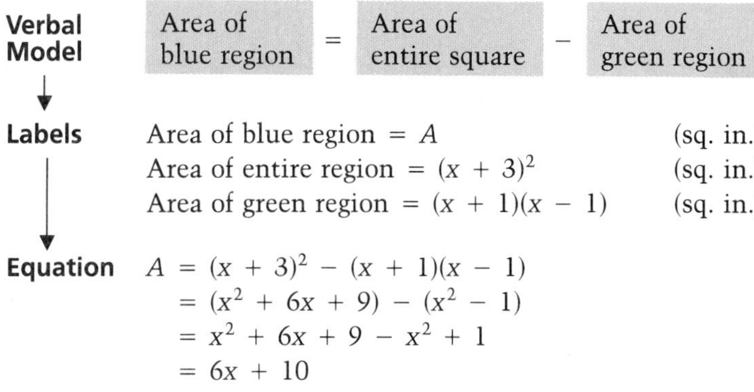

Verbal Model: Area of blue region = Area of entire square − Area of green region

Labels:
Area of blue region = A (sq. in.)
Area of entire region = $(x + 3)^2$ (sq. in.)
Area of green region = $(x + 1)(x - 1)$ (sq. in.)

Equation:

$$\begin{aligned}A &= (x + 3)^2 - (x + 1)(x - 1)\\ &= (x^2 + 6x + 9) - (x^2 - 1)\\ &= x^2 + 6x + 9 - x^2 + 1\\ &= 6x + 10\end{aligned}$$

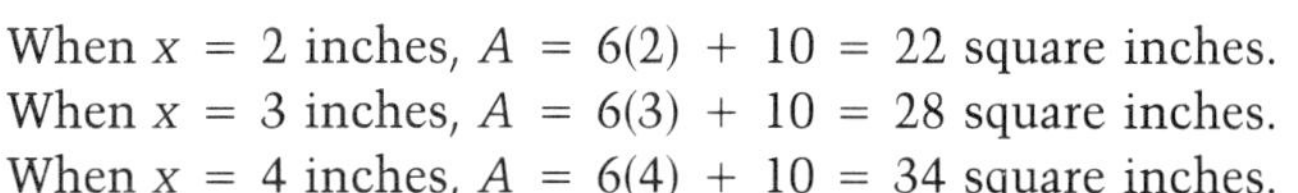

When $x = 2$ inches, $A = 6(2) + 10 = 22$ square inches.
When $x = 3$ inches, $A = 6(3) + 10 = 28$ square inches.
When $x = 4$ inches, $A = 6(4) + 10 = 34$ square inches. ■

LESSON Notes

Examples 1–2

One strategy for using the "sum and difference" and "square of a binomial" patterns is to write the general form of the patterns and then substitute into them. Accordingly,

$(a - b)(a + b) = a^2 - b^2$
$(5t - 2)(5t + 2) = (5t)^2 - (2)^2 = 25t^2 - 4.$

Students may find it helpful to see the area model of these patterns. For example, if blue is positive and grey is negative, then you can establish these rules.

a. Any time you multiply lengths with the same sign, the area of the product region is positive, or blue.

b. Any time you multiply lengths with different signs, the area of the product region is negative, or grey.

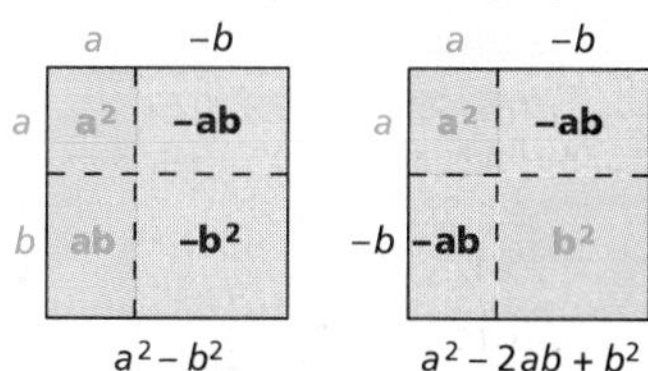

Observe that a has a different length than b.

Example 3

Have students explain how to determine the areas of the green and blue regions. Pay attention to the area of the green region; note that its width, $(x - 1)$, can be found by using information on both the left and right sides of the entire rectangular region.

Extra Examples

Here are additional examples similar to **Examples 1–3.**

1. Multiply.
   - **a.** $(x + 6)(x - 6)$
   - **b.** $(3x - 4)(3x + 4)$
   - **c.** $(7 - 5x)(7 - 5x)$
   - **d.** $(4x - 12)^2$
   - **e.** $(-6x + 2)^2$

   a. $x^2 - 36$ b. $9x^2 - 16$
   c. $49 - 70x + 25x^2$
   d. $16x^2 - 96x + 144$
   e. $36x^2 - 24x + 4$
2. Find an expression for the area of the unshaded region. Then evaluate the expression when x is equal to 5 cm.

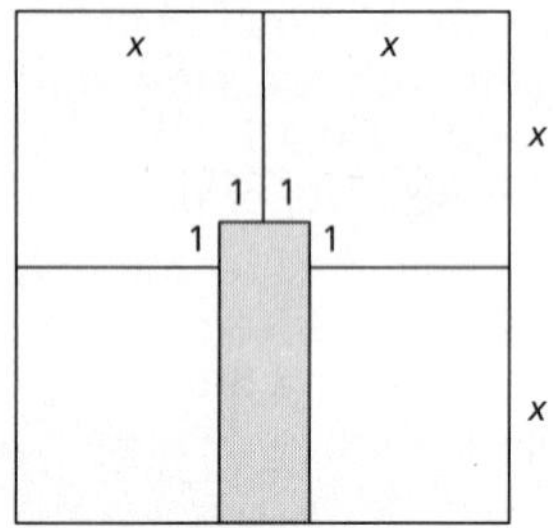

$4x^2 - 2x - 2$, 88 cm²

Example 4

Punnett squares are convenient ways to identify the genetic composition of offspring. They are named after the mathematician and geneticist Reginald Crundall Punnett, who devised them in the early 1900s. Encourage students to investigate Mendel's work on heredity involving dominant and recessive genes. (See Exercises 31 and 32, page 526.)

Check Understanding

1. **MATH JOURNAL** Write the general patterns of FOIL, the sum and difference of two terms, and the square of a binomial in your math journal. See pages 515 and 522.

A pink snapdragon has two genes—one white (W) and one red (R)—that determine its color. This Punnett square illustrates the possible results of crossing two pink snapdragons. Since each parent snapdragon passes along only one gene for color to an offspring (and pollination is at random), the snapdragon offspring will be approximately 25% red (RR), 50% pink (RW), and 25% white (WW).

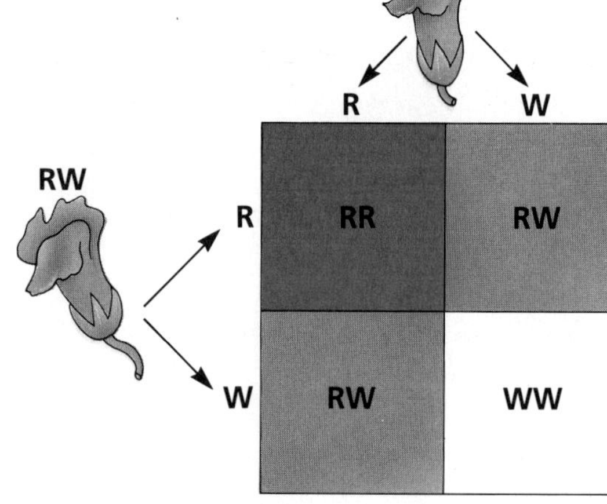

Connections **Biology**

Example 4 *Using Punnett Squares*

Show how the product of two binomials can be used to model the Punnett square shown above.

Solution Each of the two pink "parent" snapdragons have half red genes and half white genes. You can model the genetic makeup of each parent as $50\% R + 50\% W = 0.5R + 0.5W$. When the two parents are crossed, the genetic makeup of the offspring can be modeled by the product $(0.5R + 0.5W)^2$.

$$\begin{aligned}(0.5R + 0.5W)^2 &= (0.5R)^2 + 2(0.5R)(0.5W) + (0.5W)^2 \\ &= \underbrace{0.25R^2}_{\text{Red}} + \underbrace{0.5RW}_{\text{Pink}} + \underbrace{0.25W^2}_{\text{White}}\end{aligned}$$

Thus, 25% of the offspring should be red, 50% should be pink, and 25% should be white. ■

Communicating about ALGEBRA

SHARING IDEAS about the Lesson

Extend the Concept Is the trinomial the square of a binomial? Explain your reasoning.

A. $x^2 - 4x + 4$ Yes, $(x - 2)^2$

B. $2x^2 - 6x + 9$ No, the coefficient of x^2 is not a perfect square.

C. $4x^2 + 24x + 36$ Yes, $4(x + 3)^2$

EXERCISES

Guided Practice

CRITICAL THINKING about the Lesson

1. **True or False?** The product of $(a - b)$ and $(a - b)$ is $a^2 - b^2$. Explain. False. See margin.
2. Find the missing term: $(a + b)^2 = a^2 + \boxed{?} + b^2$. $2ab$
3. Write two expressions for the area of a square whose sides are each $x - 4$. $(x - 4)^2$, $x^2 - 8x + 16$
4. Give an example of each of the three types of special products in this lesson. $(x - 3)(x + 3) = x^2 - 9$; $(x + 3)^2 = x^2 + 6x + 9$; $(x - 3)^2 = x^2 - 6x + 9$

Independent Practice

In Exercises 5–10, use an area model (or algebra tiles) to write the square as a trinomial. See Additional Answers.

5. $(x + 2)^2$ $x^2 + 4x + 4$
6. $(x + 3)^2$ $x^2 + 6x + 9$
7. $(2n + 1)^2$ $4n^2 + 4n + 1$
8. $(3a + 2)^2$ $9a^2 + 12a + 4$
9. $(2x + 2)^2$ $4x^2 + 8x + 4$
10. $(3x + 1)^2$ $9x^2 + 6x + 1$

In Exercises 11–16, write the square as a trinomial.

11. $(n + 6)^2$ $n^2 + 12n + 36$
12. $(x + 4)^2$ $x^2 + 8x + 16$
13. $(2x + 1)^2$ $4x^2 + 4x + 1$
14. $(2m - 3)^2$ $4m^2 - 12m + 9$
15. $(3t - 2)^2$ $9t^2 - 12t + 4$
16. $(x - 9)^2$ $x^2 - 18x + 81$

In Exercises 17–22, multiply.

17. $(x + 5)(x - 5)$ $x^2 - 25$
18. $(x - 2)(x + 2)$ $x^2 - 4$
19. $(2x - 2)(2x + 2)$ $4x^2 - 4$
20. $(5x - 6)(5x + 6)$ $25x^2 - 36$
21. $(a + 2b)(a - 2b)$ $a^2 - 4b^2$
22. $(4x - 7y)(4x + 7y)$ $16x^2 - 49y^2$

In Exercises 23–28, write the square as a trinomial.

23. $(x + 6)^2$ $x^2 + 12x + 36$
24. $(x + 10)^2$ $x^2 + 20x + 100$
25. $(a - 2)^2$ $a^2 - 4a + 4$
26. $(2x - 5)^2$ $4x^2 - 20x + 25$
27. $(2x - 5y)^2$ $4x^2 - 20xy + 25y^2$
28. $(4s + 3t)^2$ $16s^2 + 24st + 9t^2$

Area Model (or Algebra Tiles) **In Exercises 29 and 30, write two different expressions for the area of the figure. Describe the special-product pattern that is represented.**

✪ 29.

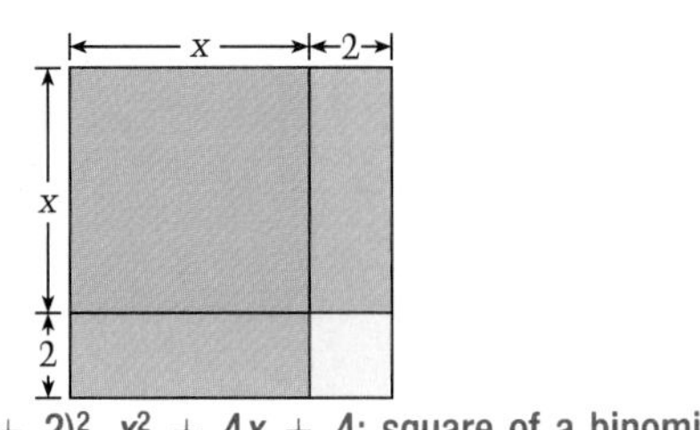

$(x + 2)^2$, $x^2 + 4x + 4$; square of a binomial

✪ 30. $4x^2 - 9$, $(2x + 3)(2x - 3)$; Product of a sum and difference of two terms

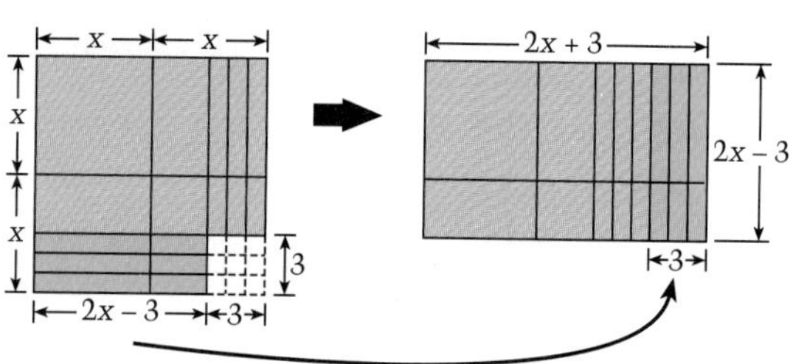

2. How is FOIL related to the sum and difference of two terms, and the square of a binomial?
3. Given the expression $(3x + y - 5)(3x + y + 5)$, explain how the pattern of the sum and difference of two terms can be used to find the product. $((3x + y) - 5)((3x + y) + 5) = (3x + y)^2 - 5^2 = 9x^2 + 6xy + y^2 - 25$

Communicating about ALGEBRA

EXTEND *Communicating*
Once decisions have been made about the trinomials A, B, and C, use the area model of polynomials to represent each trinomial that can be written as the square of a binomial.

EXERCISE Notes

ASSIGNMENT GUIDE

Basic/Average: Ex. 1–4, 5–27 odd, 31, 33, 36–44, 45–53 odd, 56

Above Average: Ex. 1–4, 5–27 odd, 29–34, 45–55 odd, 56

Advanced: Ex. 1–4, 9–27 odd, 29–38, 45–56

Selected Answers
Exercises 1–4, 5–53 odd

✪ More Difficult Exercises
Exercises 29–34, 55–56

Guided Practice

▶ **Ex. 1–4** Use these exercises in class to check for student understanding. Exercises 1–4 and Exercises 37–38 will focus students' attention on some common errors to avoid.

Answer

1. False $(a - b)(a - b) = a^2 - ab - ab + b^2 = a^2 - 2ab + b^2 \neq a^2 - b^2$

Independent Practice

▶ **Ex. 5–10** These exercises provide a visual model of the special-product process.

▶ **Ex. 23–28** Be sure that students understand that expressions like $(x + b)^2$ are not equal to $x^2 + b^2$.

▶ **Ex. 29–30** These exercises provide the visual model. Ask students to connect the model to the special-product formula.

▶ **Ex. 31–32** BIOLOGY Refer students to the use of Punnett squares in Example 4 on page 524.

EXTEND Ex. 34 Ask students to find the area of this shaded region.

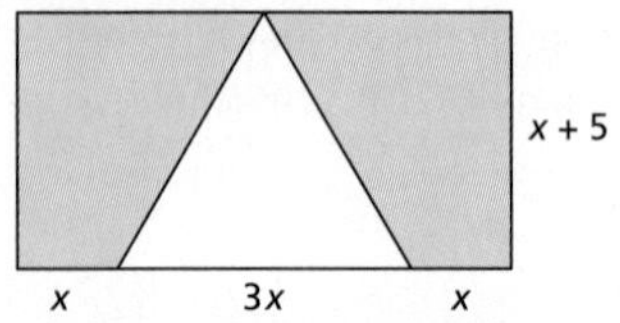

$\frac{7x^2 + 35x}{2}$

31. *Albino Tigers* In tigers, the normal color gene, C, is *dominant* and the albino color, A, is *recessive.* This means that tigers whose color genes are CC, CA, or AC will have normal coloring. A tiger whose color genes are AA will be an albino. The Punnett square at the right shows the possible results of crossing two tigers that have recessive albino genes. What percent of the offspring of two such tigers will be the normal color? What percent will be albino? Use the model

$$(0.5C + 0.5A)^2 = 0.25CC + 0.5CA + 0.25AA$$

to answer the question. 75%; 25%

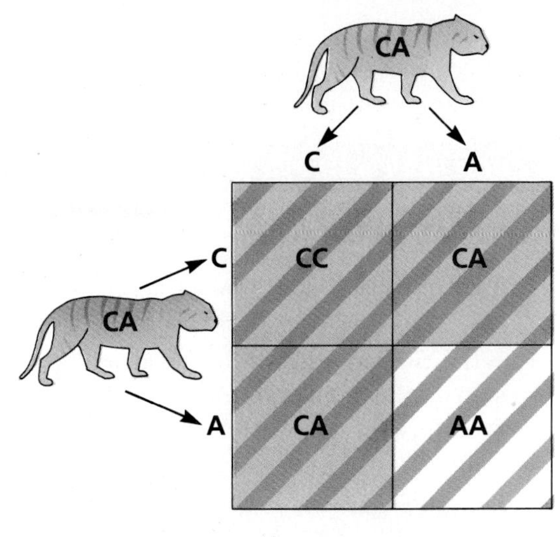

32. *Frizzle-Feathered Chickens* In chickens, neither the normal-feathered gene, F, nor the frizzle-feathered gene, f, is dominant. This means that chickens whose feather genes are FF will have normal feathers. Chickens with Ff or fF will have mildly frizzled feathers. Chickens with ff will have extremely frizzled feathers. The Punnett square at the right shows the possible results of crossing two chickens with mildly frizzled feathers. What percent of the offspring of two such chickens will have normal feathers? What percent will have mildly frizzled feathers? What percent will have extremely frizzled feathers? Use the model

$$(0.5F + 0.5f)^2 = 0.25FF + 0.5Ff + 0.25ff$$

to answer the question. 25%; 50%; 25%

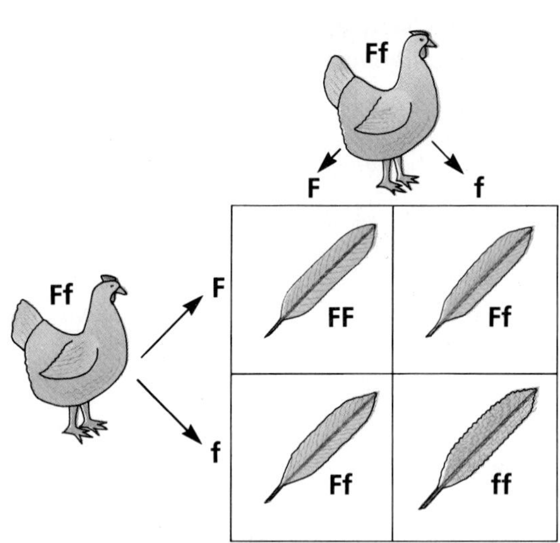

33. *Geometry* The ratio of the height and width of the smaller rectangle is equal to the ratio of the height and width of the larger rectangle. Find expressions for the perimeters and areas of both.

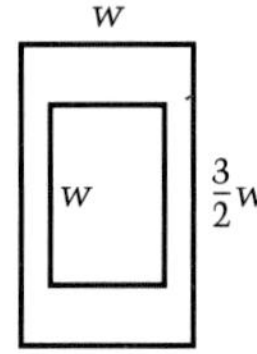

Large: $P = 5w$, $A = \frac{3}{2}w^2$. Small: $P = \frac{10}{3}w$, $A = \frac{2}{3}w^2$

34. *Geometry* Find the area of the rectangle and the area of the triangle.

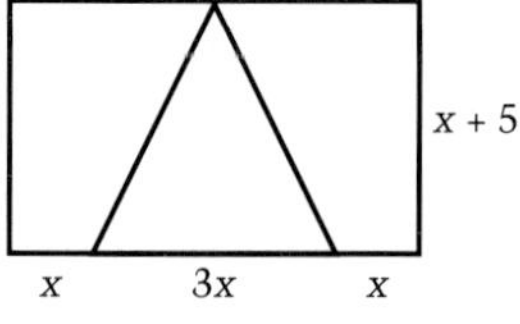

Triangle: $A = \frac{3}{2}x^2 + \frac{15}{2}x$. Rectangle: $A = 5x^2 + 25x$

35. *Compounded Interest* After two years, an investment of \$1000 compounded annually at an interest rate, r, will grow to the amount $1000(1 + r)^2$ in dollars. Write this product as a trinomial. $1000 + 2000r + 1000r^2$

36. *Compounded Interest* After two years, an investment of \$2000 compounded annually at an interest rate of 6% will have a dollar balance of

$$2000(1 + 0.06)^2 = 2000 + 2000(0.12) + 2000(0.0036).$$

Evaluate each side of this equation and compare your results. 2247.2, 2247.2; they are the same.

37. *Common Errors* Verify that $(a + b)^2 \neq a^2 + b^2$ by letting $a = 3$, $b = 4$, and evaluating both expressions. $49 \neq 25$

38. *Common Errors* Verify that $(a - b)^2 \neq a^2 - b^2$ by letting $a = 5$, $b = 4$, and evaluating both expressions. $1 \neq 9$

Integrated Review

In Exercises 39–44, evaluate the expression.

39. $(x + 5)^2$ when $x = 4$ 81
40. $(2x - 4)^2$ when $x = 1$ 4
41. $(-3x + 5)^2$ when $x = 0$ 25
42. $(x + 3)^2$ when $x = -6$ 9
43. $(\frac{1}{4}x + 9)^2$ when $x = -40$ 1
44. $(6x - 5)^2$ when $x = \frac{1}{3}$ 9

In Exercises 45–50, simplify the expression.

45. $\frac{x^2y}{xy^2}$ $\frac{x}{y}$
46. $\frac{4x^6y^8}{2y^2}$ $2x^6y^6$
47. $\frac{9x^5y^3}{3x^2y}$ $3x^3y^2$
48. $4x - (x + 5)(x - 5)$ $-x^2 + 4x + 25$
49. $(4 - x)^2 + 8$ $24 - 8x + x^2$
50. $(x + 2)^2 - x^2 + 9$ $4x + 13$

51. *College Entrance Exam Sample* If $x = -3$, then $(x + 3)^2 - 3x = \boxed{?}$ e

a. -6 b. -3 c. 0 d. 6 e. 9

52. *College Entrance Exam Sample* For what positive value of x does $\frac{4}{x} = \frac{x}{16}$? b

a. 2 b. 8 c. 20 d. 32 e. 64

53. *Estimation—Powers of Ten* Which of the following is the best estimate for the capacity of an automobile's fuel tank? b

a. 1.5 gallons b. 15 gallons
c. 150 gallons d. 1500 gallons

54. *Estimation—Powers of Ten* Which of the following is the best estimate for the number of fourteen-year-olds in the United States? d

a. 4×10^3 b. 4×10^4
c. 4×10^5 d. 4×10^6

Exploration and Extension

55. Find the product of $(a + b)(a + b)(a - b)$ by two methods.

a. First step: Find the product of the sum and difference of two terms. $a^3 + a^2b - ab^2 - b^3$

b. First step: Find the square of the binomial. (Compare your results.) $a^3 + a^2b - ab^2 - b^3$, results are the same

c. Which method do you prefer? Explain.

56. *Geometry* Use an area model to find the product: $(a + b + c)^2$. See margin for model. $a^2 + b^2 + c^2 + 2ab + 2ac + 2bc$

c. Probably method a, because it is easier to multiply two binomials than a binomial and a trinomial.

Ex. 55 It seems more efficient to first multiply the product $(a + b)(a - b)$. Ask students if they preferred doing this.

Answer

56.

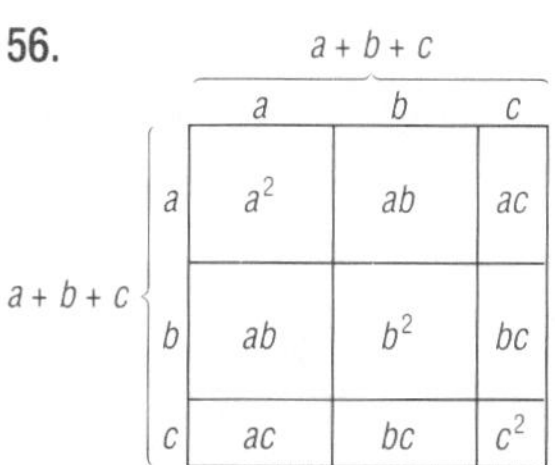

Enrichment Activities

ALTERNATIVE ASSESSMENT These activities require students to go beyond lesson goals.

This textbook has shown that mathematics has real-life connections to science, social studies, business, etc.

COOPERATIVE LEARNING Work with a partner. Choose a teacher of another subject in your school you would like to interview. Prepare interview questions on the connection between mathematics and the topics in the teacher's subject area. For example, you might wish to interview the biology teacher and ask for other examples of Punnett squares or other mathematical models used in the study of genetics.

2. Conduct your interview and prepare an oral report for your classmates.

Problem of the Day

When a shepherd counts his sheep by 2's, 3's, 4's, 5's, or 6's, there is 1 left over. When he counts them by 7, there are none left over. What is the smallest number he can have? 301

ORGANIZER

Warm-Up Exercises

1. Multiply.
   a. $5(2x + 3)$
   b. $-4(7 - x)$
   c. $(8x - 2)(8x + 2)$
   d. $(6x + 3)(6x + 3)$
   e. $(4x - 5)(4x - 5)$

   a. $10x + 15$
   b. $-28 + 4x$
   c. $64x^2 - 4$
   d. $36x^2 + 36x + 9$
   e. $16x^2 - 40x + 25$
2. Fill in the box to form a correct statement.
   a. $6x - 9 = 3(2x - \square)$
   b. $-8x + 12 = -4(\square - 3)$
   c. $15x^2 - 20x + 35 = \square(3x^2 - \square x + 7)$
   d. $x^2 - 25 = (x + 5)(\square - \square)$
   e. $16x^2 - 24x + 9 = (4x - \square)(\square - \square)$

   a. 3 b. $2x$ c. 5, 4
   d. x, 5 e. 3, $4x$, 3

Lesson Resources

Teaching Tools
- Problem of the Day: 10.4
- Warm-up Exercises: 10.4
- Answer Masters: 10.4

Extra Practice: 10.4

Color Transparency: 58

10.4 Factoring: Special Products

What you should learn:

Goal 1 How to factor polynomials that have a monomial factor, that are the difference of two squares, and that are perfect-square trinomials

Goal 2 How to use factoring in real-life models

Why you should learn it:

You can use factoring to help answer questions about polynomial models such as those used for geometric situations.

Goal 1 Factoring Polynomials

In this lesson and in the next lesson, you will learn strategies for factoring polynomials. **Factoring** is the *reverse* process of multiplying.

Part of factoring is being able to find the **greatest common factor** of two or more terms. For example, $2x$ is the greatest common factor of the terms of $(4x^2 + 6x)$. Writing this polynomial as the product $2x(2x + 3)$ illustrates one type of factoring. The factor $2x$ is a **monomial factor.** The Distributive Property is used to "factor out" a monomial factor.

Polynomial	Monomial Factor	Factored Form
$3x + 6$	3	$3(x + 2)$
$5x^2 - 15x$	$5x$	$5x(x - 3)$
$4x^2 + 6x + 8$	2	$2(2x^2 + 3x + 4)$

Each of the special product patterns that you studied in Lesson 10.3 can be used to factor polynomials.

Factoring Special Products

Difference of Two Squares Pattern	Example
$a^2 - b^2 = (a + b)(a - b)$	$9x^2 - 16 = (3x + 4)(3x - 4)$

Perfect Square Trinomial Pattern	Example
$a^2 + 2ab + b^2 = (a + b)^2$	$x^2 + 8x + 16 = (x + 4)^2$
$a^2 - 2ab + b^2 = (a - b)^2$	$x^2 - 12x + 36 = (x - 6)^2$

Note that *perfect square trinomials* come in two forms. The coefficient of the middle term can be positive or negative.

To recognize the **difference of two squares,** look for coefficients that are squares and for variables that are raised to the second power.

Polynomial	Difference of Squares	Factored Form
$x^2 - 4$	$x^2 - 2^2$	$(x + 2)(x - 2)$
$4x^2 - 25$	$(2x)^2 - 5^2$	$(2x + 5)(2x - 5)$
$25 - 49x^2$	$5^2 - (7x)^2$	$(5 + 7x)(5 - 7x)$

To recognize a **perfect square trinomial,** remember that the first and last terms must be positive and perfect squares, a^2 and b^2, and the middle term must be twice the product of a and b.

Example 1 *Factoring Perfect Square Trinomials*

a. $x^2 - 4x + 4 = x^2 - 2(2x) + 2^2$
$= (x - 2)^2$

b. $16y^2 + 24y + 9 = (4y)^2 + 2(4y)(3) + 3^2$
$= (4y + 3)^2$

c. $9t^2 - 30t + 25 = (3t)^2 - 2(3t)(5) + 5^2$
$= (3t - 5)^2$ ■

Before trying to apply one of the special product patterns for factoring, factor out any common monomial factor.

Example 2 *Removing a Common Monomial Factor First*

a. $3x^2 - 27 = 3(x^2 - 9)$
$= 3(x + 3)(x - 3)$

b. $3x^2 - 30x + 75 = 3(x^2 - 10x + 25)$
$= 3(x - 5)^2$

c. $16y^3 + 80y^2 + 100y = 4y(4y^2 + 20y + 25)$
$= 4y(2y + 5)^2$ ■

Goal 2 Using Factoring in Real-Life Models

Connections Geometry

Example 3 *Finding Dimensions*

You are shopping for a travel kennel. Your new puppy can just barely squeeze through a square opening that is $3\frac{1}{2}$ inches on each side. Are the grille openings in the kennel at the left too large to hold your puppy?

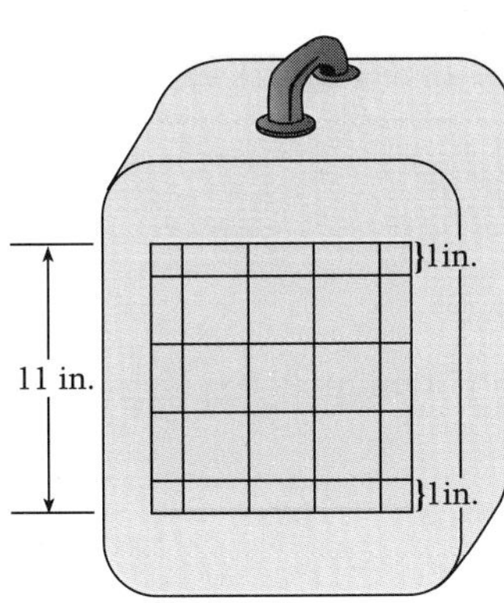

Solution Let each of the large grille openings be x inches on each side. The grille, then, forms 9 large squares with an area of x^2 each, 12 rectangles with an area of x each, and 4 small squares with an area of 1 each. By factoring,

$$9x^2 + 12x + 4 = (3x + 2)^2$$

you find that each side of the grille is $3x + 2$. By solving the equation $3x + 2 = 11$, you find x to be 3 inches. So your pup will be secure. ■

LESSON Notes

GOAL 1 When factoring special products, it may be helpful to show the connection between the given polynomial expression and the pattern to be used. For example, if factoring $16x^2 - 81$, write $16x^2 - 81 = (4x)^2 - (9)^2 = (4x + 9)(4x - 9)$. It may even be necessary to mark where the pattern is being used. Repeating the above example,

$$16x^2 - 81$$
$$\downarrow \quad \downarrow$$
$$= (4x)^2 - (9)^2$$
$$= (4x + 9)(4x - 9)$$

Examples 1–2

Encourage students to identify the values of a and b for each pattern used to factor the polynomials. For instance, in Example 1c, $a = 3t$ and $b = 5$.

Example 3

The problem has been expressed in terms of the quadratic relationship between the 11-inch by 11-inch square grille and the large grille openings that measure x inches by x inches. Using that relationship, the quadratic equation $9x^2 + 12x + 4 = 11^2$ can be formed.

Solving by square roots gives

$(3x + 2)^2 = 121$
$3x + 2 = 11$ (-11 has no
$3x = 9$ meaning)
$x = 3$

However, students may observe that one side of the square grille is 11 inches or $(3x + 2)$ inches. This linear relationship leads to the same results shown above. This approach to the problem should be accepted.

Example 4

This is an interesting and intriguing result that students will enjoy exploring.

Extra Examples

Here are additional examples similar to **Examples 1–4.**

1. Factor.
   - **a.** $9x^2 - 36$
   - **b.** $25x^2 - 30x + 9$
   - **c.** $-27x^3 - 18x^2 - 3x$
   - a. $9(x - 2)(x + 2)$
   - b. $(5x - 3)^2$
   - c. $-3x(3x + 1)^2$
2. Find the right-triangle triple for the positive integer 11. The triple is $a = 11$, $b = 60$, $c = 61$.

Check Understanding

1. Explain why a polynomial of the form $a^2 - 2ab + b^2$ is called a perfect square trinomial. The trinomial can be factored as the square of a binomial.
2. What role does the distributive property of multiplication over addition play in factoring polynomials? It is used to "factor out" common factors.
3. Explain the statement, "Factoring is the reverse process of multiplying." Factoring is dividing out common factors.

Communicating about ALGEBRA

Explain how one of the factoring patterns is used to determine right-triangle triples. Can a rationale be given for why this always works?

Answers

A. $(3t)^2 - 7^2 = (3t + 7)(3t - 7)$

B. $x^2 - 2(8x) + 8^2 = (x - 8)^2$

C. $(2y)^2 + 2(18y) + 9^2 = (2y + 9)^2$

D. $(11x)^2 - 10^2 = (11x + 10)(11x - 10)$

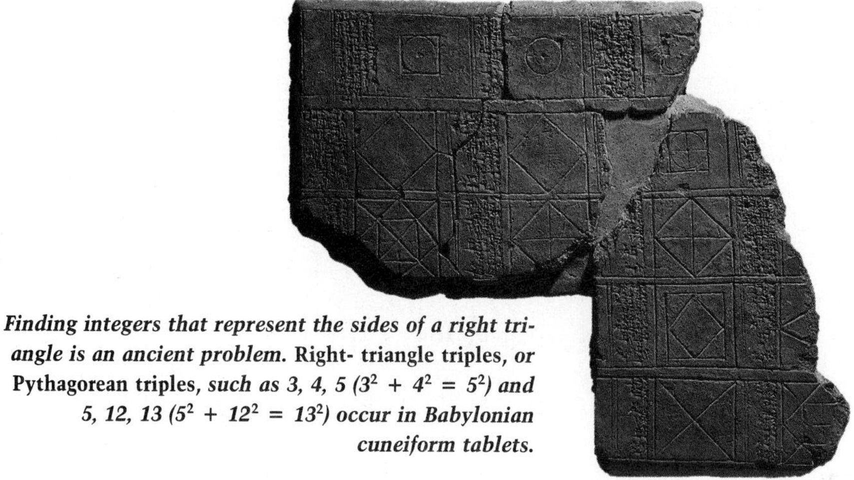

Finding integers that represent the sides of a right triangle is an ancient problem.* Right- triangle triples, *or* Pythagorean triples, *such as 3, 4, 5 ($3^2 + 4^2 = 5^2$) and 5, 12, 13 ($5^2 + 12^2 = 13^2$) occur in Babylonian cuneiform tablets.

Example 4 *Right-Triangle Triples*

Here is a technique for finding right-triangle triples.

Solution

1. Choose a positive integer and square it: $7^2 = 49$.
2. Factor into two odd factors or two even factors: $7^2 = (1)(49)$.
3. Write the product as the difference of two squares:

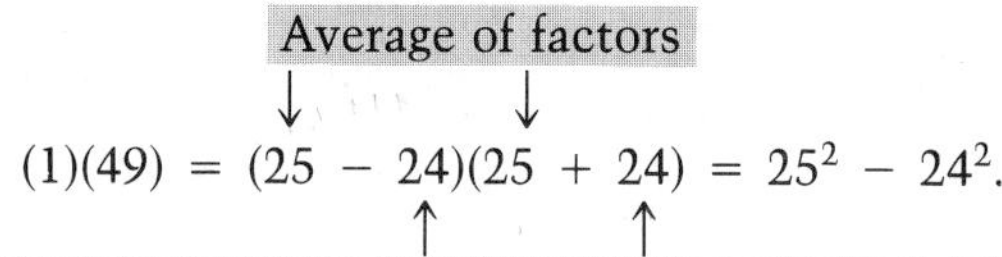

$$(1)(49) = (25 - 24)(25 + 24) = 25^2 - 24^2.$$

4. Write in Pythagorean form:

$$7^2 = 25^2 - 24^2$$
$$7^2 + 24^2 = 25^2 \qquad \textit{Pythagorean form, } a^2 + b^2 = c^2$$

The triple is $a = 7$, $b = 24$, and $c = 25$. ■

Communicating about ALGEBRA

SHARING IDEAS about the Lesson

Recognize Patterns Factor each polynomial. Explain your steps and reasoning. See margin.

A. $9t^2 - 49$ **B.** $x^2 - 16x + 64$

C. $4y^2 + 36y + 81$ **D.** $121x^2 - 100$

EXERCISES

Guided Practice

CRITICAL THINKING about the Lesson

1. Describe the relationship between multiplying polynomials and factoring a polynomial. Each is the reverse process of the other.
2. Factor out the greatest common monomial factor: $3x^3 - 6x^2 + 9$. $3(x^3 - 2x^2 + 3)$
3. Show how the Distributive Property can be used to factor $2x(2x - 3) + 3(2x - 3)$. 3., 4. See margin.
4. Give an example of each of the three special-product factoring patterns in the lesson.

Independent Practice

In Exercises 5–10, find the greatest common factor of the three terms.

5. $6x^5, 30x^4, 12x^3$ $6x^3$
6. $7x^3, 28x, 14x^4$ $7x$
7. $24x^3, 32x^2$ $8x^2$
8. $99x^6, 45x^3$ $9x^3$
9. $16x^2y, 84xy^2, 36x^2y^2$ $4xy$
10. $10xy^2, 25x^3y^2, 80x^2y$ $5xy$

In Exercises 11–19, factor out the greatest common monomial factor.

11. $2x^2 - 4$ $2(x^2 - 2)$
12. $3x + 6$ $3(x + 2)$
13. $4a - 12$ $4(a - 3)$
14. $14z^3 + 21$ $7(2z^3 + 3)$
15. $24x^2 - 18$ $6(4x^2 - 3)$
16. $-a^3 - 4a$ $-a(a^2 + 4)$
17. $21u^2 - 14u$ $7u(3u - 2)$
18. $36y^4 + 24y^2$ $12y^2(3y^2 + 2)$
19. $4x^2 - 8x + 8$ $4(x^2 - 2x + 2)$

In Exercises 20–28, factor the expression.

20. $x^2 - 64$ $(x + 8)(x - 8)$
21. $y^2 - 144$ $(y + 12)(y - 12)$
22. $2x^2 + 16x + 32$ $2(x + 4)^2$
23. $9x^2 - 30xy + 25y^2$ $(3x - 5y)^2$
24. $4y^2 + 20yz + 25z^2$ $(2y + 5z)^2$
25. $u^2 - \frac{1}{16}$ $(u + \frac{1}{4})(u - \frac{1}{4})$
26. $v^2 - \frac{9}{25}$ $(v + \frac{3}{5})(v - \frac{3}{5})$
27. $81 - (z + 5)^2$ $(4 - z)(14 + z)$
28. $3(x - 3)^2 - 12$ $3(x - 5)(x - 1)$

In Exercises 29 and 30, use the following information.

A manufacturer of television sets has modeled the revenue, R, for Model TXX to be

$$R = 800x - 0.05x^2 = xp$$

where x is the number of units sold, and p is the price of Model TXX. The graph of the model is shown at the right.

Revenue from Sale of Model TXX

Revenue (in millions of dollars): R, 1, 2, 3, 4

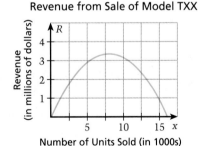

Number of Units Sold (in 1000s): 5, 10, 15, x

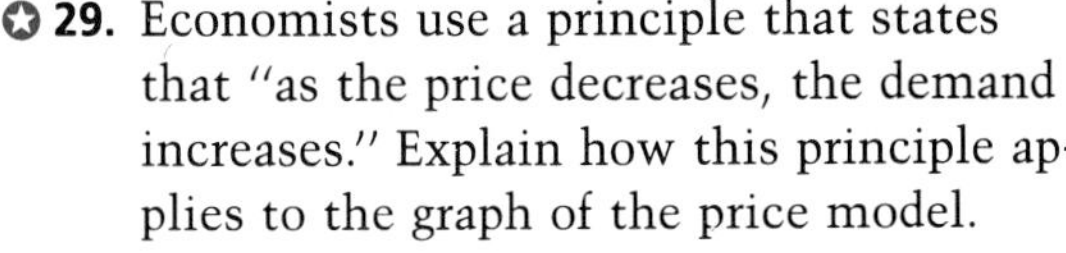

✪ 29. Economists use a principle that states that "as the price decreases, the demand increases." Explain how this principle applies to the graph of the price model.

✪ 30. If you were the manager of the marketing department, what price would you recommend for each television set? Explain your reasoning. 29., 30. See margin.

EXERCISE Notes

ASSIGNMENT GUIDE

Basic/Average: Ex. 1–4, 5–27 odd, 29–33, 37–49 odd, 51

Above Average: Ex. 1–4, 5–27 odd, 29–34, 36–38, 39–53 odd

Advanced: Ex. 1–4, 11–27 odd, 29–37, 45–51 odd, 52, 53

Selected Answers
Exercises 1–4, 5–51 odd

✪ More Difficult Exercises
Exercises 29–30, 34–37

Guided Practice

▶ **Ex. 1–4** GROUP ACTIVITY These exercises should be used for in-class discussion.

Answers

3. $2x(2x - 3) + 3(2x - 3) = (2x - 3)(2x + 3)$
4. $4x^2 - 25 = (2x - 5)(2x + 5)$
$x^2 + 14x + 49 = (x + 7)^2$
$x^2 - 16x + 64 = (x - 8)^2$

Independent Practice

▶ **Ex. 16** Caution students to watch the signs when factoring out $-a$.

▶ **Ex. 29–37** These exercises connect special-products factoring to quadratic models for real-life applications.

▶ **Ex. 29–30** These exercises should be assigned as a pair.

Answers

29. The graph of the price model shows that the lower the price is, the greater the number of TV sets sold.
30. \$400, R has a maximum value when $P = 400$.

Ex. 34, 36–37 These exercises are similar to Example 3 on page 529.

Ex. 36 Some students might question where the 60 feet in the diagram came from. It represents the total width of the street.

Ex. 44–49 These exercises can be simplified to review techniques for solving quadratic equations that were presented in Chapter 9.

Enrichment Activities

These activities require students to go beyond lesson goals.

GEOMETRY You can demonstrate the difference-of-two-squares pattern by using this visual model.

1. Draw a large square of length a on grid paper. Label its sides a. What is its area? a^2
2. Draw a smaller square of length b in the upper right-hand corner. Label its sides b. Label the sides of the remaining region in terms of a and b. What is the area of the small square? b^2

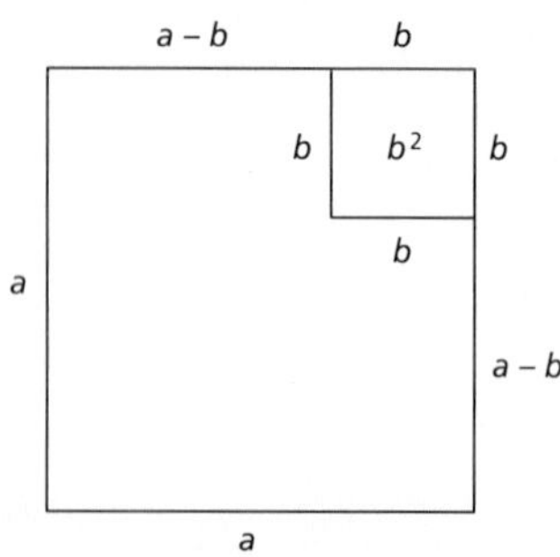

3. Cut out the smaller square. What is the area of the remaining region? $a^2 - b^2$

532

Right-Triangle Triples **In Exercises 31–33, use the technique described in Example 4 to find the indicated right-triangle triple. (Exercise 33 has two solutions.)** 33. $a = 8, b = 15, c = 17$ or $a = 8, b = 6, c = 10$

31. $4^2 + b^2 = c^2$ $a = 4, b = 3, c = 5$

32. $6^2 + b^2 = c^2$ $a = 6, b = 8, c = 10$

33. $8^2 + b^2 = c^2$

34. *A Jade Pi* The *jade pi* (pronounced "jade bee") at the right has an inside radius of 5 centimeters and an outside radius of 30 centimeters. Find the area of one of the flat surfaces of the pi. 2748.9 cm^2

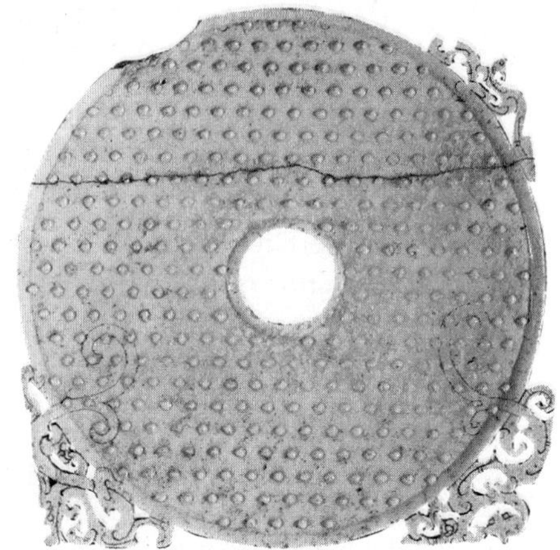

Early Chinese emperors used ring-shaped disks of jade in ceremonies to appeal to celestial spirits. The circular shape symbolized Heaven.

35. *Chemical Reaction* The rate of change of some chemical reactions can be modeled by

$$kQx - kx^2$$

where Q is the amount of the original substance, x is the amount of the substance formed, and k is a constant of proportionality. Factor this rate-of-change expression. $kx(Q - x)$

36. *Marching Band* Your band is asked to be in a local parade. It is allotted 3200 square feet in the parade and must stay at least x feet from each curb. Your director decides that rows will be 4 feet apart and each row will contain 8 band members. Can all of the 160 members of the band be in the parade? Yes

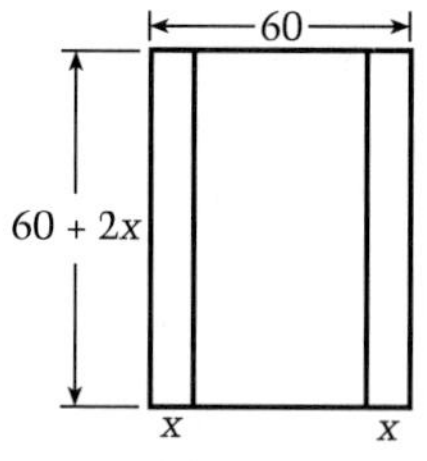

It takes musicianship, teamwork, and plenty of practice to maintain correct marching ranks, while still producing a well-balanced sound.

37. *Package from Grandma* A week before your birthday you receive this card from your grandmother, a retired math teacher. Will her package fit in your mailbox, which is 9 inches by 5 inches by 4 inches, or will it be returned to the post office, where you will have to pick it up later?
You will have to pick it up at the post office.

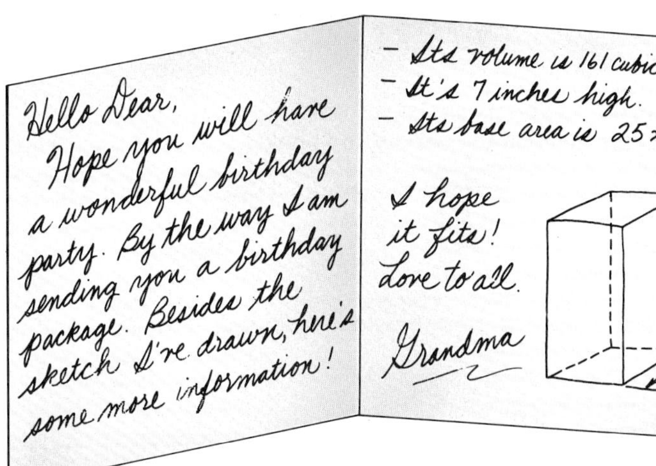

Hello Dear,
Hope you will have a wonderful birthday party. By the way I am sending you a birthday package. Besides the sketch I've drawn, here's some more information!
- Its volume is 161 cubic inches.
- It's 7 inches high.
- Its base area is $25x^2 - 1$.
I hope it fits!
Love to all.
Grandma
7 in.
5x + 1

Integrated Review

In Exercises 38–43, multiply.

38. $12x(2x - 3)$ $24x^2 - 36x$

39. $7y(4 - 3y)$ $28y - 21y^2$

40. $t(t^2 + 1) - t(t^2 - 1)$ $2t$

41. $2z(z + 5) - 7(z + 5)$ $2z^2 + 3z - 35$

42. $(11 - x)(11 + x)$ $121 - x^2$

43. $(6r + 5s)(6r - 5s)$ $36r^2 - 25s^2$

In Exercises 44–49, solve the equation.

44. $(x + 2)^2 - x^2 = 8$ 1

45. $(x - 1)^2 + 2x = 4$ $\pm\sqrt{3}$

46. $(x - 4)^2 + 8x = 32$ ± 4

47. $(x - 6)^2 - x^2 = 0$ 3

48. $(x + 3)(x - 3) + 5 = 0$ ± 2

49. $(x - 4)(x + 4) = -7$ ± 3

50. *Up, Up, and Away* The height (in feet) of a bottle rocket is modeled by

$$h = -16t^2 + 57t$$

where t is the time in seconds. Find the height of the rocket after 2 seconds. 50 ft

51. *Pythagorean Theorem* The lengths of the sides of a right triangle are $a = 5$, b, and $c = 13$. What is b? 12

Exploration and Extension

Geometry **In Exercises 52 and 53, find the perimeter of the rectangle and square. The lengths of the sides are obtained by factoring the expression for the area.**

52. Area $= x^2 - 49$ $4x$

53. Area $= x^2 + 6x + 9$ $4x + 12$

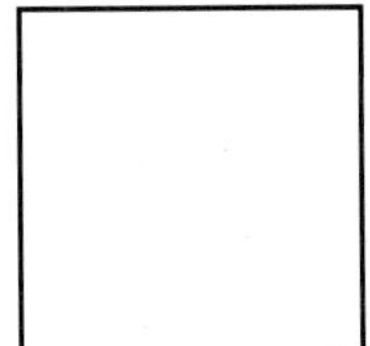

4. Draw the diagonal as shown and label the two regions A and B.

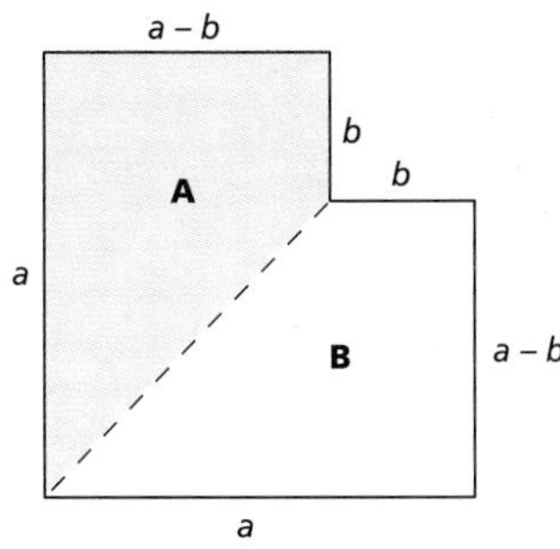

Cut along the diagonal and rearrange the pieces to form a rectangle. What is its area?
$(a + b)(a - b)$

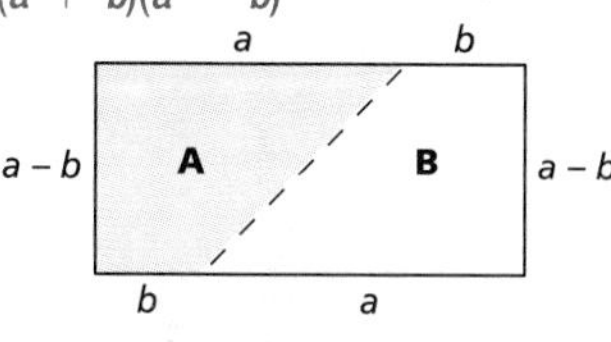

5. What can you conclude?
$a^2 - b^2 = (a + b)(a - b)$

Mid-Chapter SELF-TEST

Take this test as you would take a test in class. The answers to the exercises are given in the back of the book.

In Exercises 1 and 2, perform the indicated operation. Use a horizontal format. (10.1)

1. $(x^2 + 2x - 1) + (3x - 4x^2 + 2)$
$-3x^2 + 5x + 1$

2. $(-2x^2 + 4x - 5) - (3 - 7x + x^2)$
$-3x^2 + 11x - 8$

In Exercises 3 and 4, perform the indicated operation. Use a vertical format. (10.1)

3. $(3x^2 - x + 2) + (x^2 - 3x + 2)$
$4x^2 - 4x + 4$

4. $(7x^2 - 5x + 10) - (3x^2 + 2x + 5)$
$4x^2 - 7x + 5$

In Exercises 5–8, multiply. (10.2)

5. $(3x + 2)(3x + 5)$ $9x^2 + 21x + 10$

6. $(4x - 5)(2x + 10)$ $8x^2 + 30x - 50$

7. $(x + 1)(2x^2 - 3x + 2)$ $2x^3 - x^2 - x + 2$

8. $(2x + 1)(5x^2 + 7x - 3)$ $10x^3 + 19x^2 + x - 3$

In Exercises 9–12, multiply using the FOIL pattern. (10.2)

9. $(x + 2)(x - 3)$
$x^2 - x - 6$

10. $(x - 3)(x - 6)$
$x^2 - 9x + 18$

11. $(2x + 3)(4 - x)$
$-2x^2 + 5x + 12$

12. $(3 + x)(4 + x)$
$x^2 + 7x + 12$

In Exercises 13–16, use the special-product patterns to multiply. (10.3)

13. $(2x + 3)^2$
$4x^2 + 12x + 9$

14. $(4x - 5)^2$
$16x^2 - 40x + 25$

15. $(2x - 6)(2x + 6)$
$4x^2 - 36$

16. $(x - 7)(x + 7)$
$x^2 - 49$

In Exercises 17–22, factor the polynomial. (10.4)

17. $4x^2 - 49$ $(2x + 7)(2x - 7)$

18. $3x^2 - 108$ $3(x + 6)(x - 6)$

19. $x^2 - 2x + 1$ $(x - 1)^2$

20. $7x^2 + 42x + 63$ $7(x + 3)^2$

21. $9x^2 + 12x + 4$ $(3x + 2)^2$

22. $x^2 - 169$ $(x + 13)(x - 13)$

23. From 1970 to 1990, the number of pieces, F, of first-class U.S. mail can be modeled by

$F = 50{,}262 + 99.2t^2$ (in millions)

where $t = 0$ represents 1970. The number of pieces, K, of all other types of U.S. mail can be modeled by

$K = 33{,}045 + 140.7t^2$. (in millions)

Find a model for the number of pieces, M, of all types of U.S. mail. *(Source:* **Statistical Abstract,** *1990)* **(10.1)** $M = 83{,}307 + 239.9t^2$

24. Three storage cubes have sides of $(x + 1)$, $(x + 3)$, and $(x + 5)$. Write an expression for the total volume of the three cubes. **(10.2)**

24. $(x + 1)^3 + (x + 3)^3 + (x + 5)^3$

25. You are making a tick-tack-toe board for your cousin's birthday. Each square will have a side of x inches, and the board will have a 1-inch-wide border. Write a polynomial for the area of the board. **(10.3)**
$(3x + 2)^2$

26. You want to frame a 16-inch by 20-inch watercolor for your parents' anniversary. The frame is x inches wide on all four sides. Write a polynomial for the area of the framed picture. **(10.2)**
$(2x + 16)(2x + 20)$

10.5 Factoring Quadratic Trinomials

What you should learn:

Goal 1 How to factor a quadratic trinomial or recognize that it cannot be factored

Goal 2 How to use factoring in real-life models

Why you should learn it:

You can use factoring as an efficient step in solving some types of mathematical problems.

Goal 1 Factoring a Quadratic Trinomial

Let's look at quadratic trinomials of the form $x^2 + bx + c$. Try covering the factored forms in the left column. Can you find the factored form from the trinomial form?

Factored Form		F O I L		Trinomial Form
$(x + 2)(x + 4)$	=	$x^2 + 4x + 2x + 8$	=	$x^2 + 6x + 8$
$(x - 3)(x - 2)$	=	$x^2 - 2x - 3x + 6$	=	$x^2 - 5x + 6$
$(x - 1)(x + 4)$	=	$x^2 + 4x - x - 4$	=	$x^2 + 3x - 4$

To see how to factor this type of quadratic trinomial, consider the following.

$$(x + r)(x + s) = x^2 + \underbrace{(r + s)}_{\text{Sum of terms}}x + \underbrace{rs}_{\text{Product of terms}}$$

$$= x^2 + \boxed{b}\,x + \boxed{c}$$

Notice that you must find factors of c whose sum is b. If c is positive, the factors must have *like* signs. If c is negative, the factors must have *unlike* signs.

Example 1 *Factoring When c Is Positive*

Factor $x^2 + 3x + 2$.

Solution For this trinomial, $b = 3$ and $c = 2$.

$x^2 + 3x + 2 = (x + \boxed{?})(x + \boxed{?})$ *Think: You need factors of 2 whose sum is 3.*

$= (x + 1)(x + 2)$ *$2 = (1)(2), 1 + 2 = 3$* ■

Example 2 *Factoring When c Is Positive*

Factor $x^2 - 5x + 6$.

Solution For this trinomial, $b = -5$ and $c = 6$.

$x^2 - 5x + 6 = (x + \boxed{?})(x + \boxed{?})$ *Think: You need factors of 6 whose sum is −5.*

$= (x - 2)(x - 3)$ *$6 = (-2)(-3)$, $-2 - 3 = -5$* ■

Problem of the Day

A cube of wood 3 in. on each edge is to be cut into 1-inch cubes. After each cut, the pieces may be stacked in any way desired before making the next cut. What is the smallest number of cuts required?

Six: no matter how the pieces are stacked, six cuts are necessary for the central cube.

ORGANIZER

TIME SCHEDULE: All levels, two days

Warm-Up Exercises

1. Use the FOIL pattern to multiply.
   a. $(x + 3)(x - 4)$
   b. $(x - 5)(x - 6)$
   c. $(x + 9)(x + 1)$
   a. $x^2 - x - 12$
   b. $x^2 - 11x + 30$
   c. $x^2 + 10x + 9$
2. For each product in Exercise 1, find the sum and the product of their constant terms. Find a pattern relating the values with the answers found in Exercise 1.
   a. −1, −12, b. −11, 30, c. 10, 9; Each sum of constant terms is the middle term's coefficient of the trinomials and each product of constant terms is the constant term of the trinomials.

Lesson Resources

Teaching Tools
- Problem of the Day: 10.5
- Warm-up Exercises: 10.5
- Answer Masters: 10.5

Extra Practice: 10.5

Color Transparency: 59

LESSON Notes

You may wish to introduce factoring trinomials with the leading coefficient 1 on the first day (see Examples 1–3 with Exercises 1–2, 5–7, 11–20, 35, 47–48, 51–56) and factoring trinomials with the leading coefficient not 1 on the second day (see Examples 4–6 with Exercises 3–4, 8–10, 21–34, 36–46, 49–50, 57–60).

Examples 2–3

Many students may find it helpful to exhaust the list of possible factors for the boxes in expressions like

$x^2 - 5x + 6 = (x + \square)(x + \square)$.

Product: 6

$-1, -6$	$1, 6$
$-2, -3$	$2, 3$
$-3, -2 \leftarrow$ a repeat $\rightarrow$	$3, 2$
$-6, -1 \leftarrow$ a repeat $\rightarrow$	$6, 1$

From the list, students must identify pairs whose sum is -5. Observe that the second half of the list is not really a possibility because the sum must add to a negative number and this cannot occur with entries from the right-hand part of the table. Finally, have students enter the correct pair as $x^2 - 5x + 6 = (x + [-2])(x + [-3])$. Simplify to obtain $(x - 2)(x - 3)$.

In Example 3, the factors are $(x + [-4])(x + [2])$ or $(x - 4)(x + 2)$.

Example 4

Quadratic trinomials with leading coefficients different from 1 usually require a table to be constructed of possible factors for the leading coefficient and the constant term.

Example 3 *Factoring When c Is Negative*

Factor $x^2 - 2x - 8$.

Solution For this trinomial, $b = -2$ and $c = -8$.

$$
\begin{aligned}
x^2 - 2x - 8 \\
&= (x + \boxed{?})(x + \boxed{?}) && \text{Think: You need factors of } -8 \text{ whose sum is } -2. \\
&= (x + 2)(x - 4) && -8 = (2)(-4),\ 2 - 4 = -2
\end{aligned}
$$

■

To factor a trinomial whose leading coefficient is not 1, consider the following pattern.

Factors of a

$$ax^2 + bx + c = (\boxed{?}x + \boxed{?})(\boxed{?}x + \boxed{?})$$

Factors of c

The goal is to find a combination of factors of a and c so that the outer and inner products add to the middle term bx.

Example 4 *Factoring When Leading Coefficient Is Not 1*

Problem Solving Guess and Check

Factor $6x^2 + 7x - 5$.

Solution For this trinomial, $a = 6$, $b = 7$, and $c = -5$. Use a guess-and-check strategy to find the binomial factors.

Guess	Check
$(x + 1)(6x - 5)$	$= 6x^2 + x - 5$
$(x - 1)(6x + 5)$	$= 6x^2 - x - 5$
$(2x + 1)(3x - 5)$	$= 6x^2 - 7x - 5$
$(2x - 1)(3x + 5)$	$= 6x^2 + 7x - 5 \leftarrow$ *Correct factoring.* ■

It is important to remember that *many* quadratic trinomials cannot be factored with integer coefficients. A quadratic trinomial $ax^2 + bx + c$ will factor (using integer coefficients) only if the *discriminant* $b^2 - 4ac$ is a perfect square.

Example 5 *Using the Discriminant*

Show that $2x^2 + 3x - 6$ cannot be factored.

Solution For this trinomial, $a = 2$, $b = 3$, and $c = -6$. Since the discriminant

$$b^2 - 4ac = 3^2 - 4(2)(-6) = 57$$

is not a perfect square, the trinomial cannot be factored (using integer coefficients). ■

Goal 2 Using Factoring in Real-Life Models

For 1980 to 1990, the annual salary, S (in 1000s of dollars), paid by a manufacturer to its assembly-line workers is given by the quadratic model $S = 3t^2 + 130t + 1000$, where $t = 0$ represents 1980.

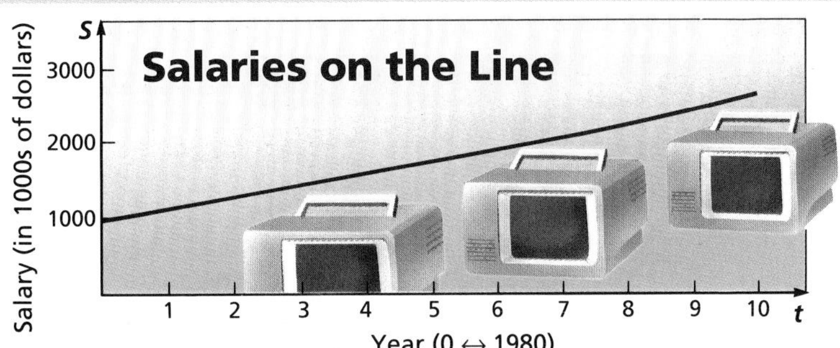

Example 6 Finding the Salary per Employee

In 1980, the above manufacturer had 100 assembly-line workers. Each year the number of workers increased by 3. Find a model for the workers' average annual salary from 1980 to 1990.

Problem Solving
Write an Equation

Solution Because the number of workers increased by 3 each year, the number of workers in year t is given by the linear model $3t + 100$.

Verbal Model Total salary = Number of workers × Average salary

Labels

Year	$= t$	(0 is 1980)
Total salary	$= 3t^2 + 130t + 1000$	($1000s)
Number of workers	$= 3t + 100$	(workers)
Average salary	$= W$	($1000s per worker)

Equation $3t^2 + 130t + 1000 = (3t + 100)W$

By factoring, you can see that

$$3t^2 + 130t + 1000 = (3t + 100)(t + 10)$$

which means the average salary was $W = t + 10$. From this model, you can see that the average salary was $10,000 in 1980. For the next 10 years, it increased by $1000 each year. ■

Communicating about ALGEBRA

SHARING IDEAS about the Lesson See Additional Answers.

Recognize Patterns Factor each polynomial. Explain your steps and reasoning.

A. $x^2 + 7x + 12$ **B.** $a^2 - 6a + 5$ **C.** $3n^2 + 4n - 4$

A. $(x + 4)(x + 3)$
B. $(a - 5)(a - 1)$
C. $(3n - 2)(n + 2)$

Leading Coefficient 6	Constant Term −5
1, 6	1, −5
2, 3	5, −1
−1, −6	
−2, −3	

Example 5

Note that the discriminant for the polynomial $\frac{1}{9}x^2 + x + 2$, $\frac{1}{9}$, is a perfect square. However, the factorization of the trinomial involves non-integer coefficients, $(\frac{1}{3}x + 2)(\frac{1}{3}x + 1)$.

Example 6

The average salary W may be obtained in the following way. Since $(3t + 100)(t + 10) = (3t + 100)W$, $(3t + 100)(t + 10) - (3t + 100)W = 0$. Factoring out $(3t + 100)$, $(3t + 100)[(t + 10) - W] = 0$. But either $(3t + 100) = 0$ (which makes no sense, why?) or, $t + 10 - W = 0$, which leads to $W = t + 10$. (See Lesson 10.6 for the Zero-Product Property.)

Extra Examples

Here are additional examples similar to **Examples 1–5.**

1. Factor.
  a. $x^2 + 2x - 24$
  b. $x^2 + 8x + 15$
  c. $x^2 + 14x + 48$
  d. $x^2 - 6x - 27$
  e. $6x^2 - 8x - 8$
  f. $24x^2 - 34x + 12$
  a. $(x - 4)(x + 6)$
  b. $(x + 3)(x + 5)$
  c. $(x + 6)(x + 8)$
  d. $(x + 3)(x - 9)$
  e. $(3x + 2)(2x - 4)$
  f. $(8x - 6)(3x - 2)$

2. Use the discriminant. Can the trinomials be factored with integer coefficients?
  a. $x^2 - 5x + 6$ yes
  b. $x^2 - 4x + 6$ no
  c. $2x^2 + 6x - 8$ yes
  d. $3x^2 + 8x - 6$ no

Check Understanding

1. List the patterns and strategies that can be used to factor polynomials.
2. How can a list of possible factors help in factoring quadratic trinomials? See Lesson Notes for Examples 2–4.
3. Can every polynomial be factored using integer coefficients? No Explain; provide examples. See Example 5.

Communicating about ALGEBRA

COOPERATIVE LEARNING Have students work in small groups to construct a flowchart of the patterns and strategies that may be used to factor polynomials. Compare all the flowcharts. How are they different? How are they the same? Can differences be resolved?

EXERCISE Notes

Assignment Guide

Day 1

Basic/Average: Ex. 1–2, 5–7, 11–19 odd, 47–48, 51–55 odd, 56

Above Average: Ex. 1–2, 5–7, 11–19 odd, 35, 47

Advanced: Ex. 1–2, 5, 7, 11–19 odd, 35, 47–48, 56

Day 2

Basic/Average: Ex. 3–4, 8–10, 21–23, 37–46

Above Average: Ex. 3–4, 9, 21–33, 37–46, 49–57 odd

Advanced: Ex. 3–4, 9, 21–33, 37–46, 49–50, 57–60

Selected Answers

Exercises 1–4, 5–55 odd

✪ More Difficult Exercises

Exercises 35–36, 49–50, 57–60

EXERCISES

16. $(x + 2)(x + 11)$ 19. $(y - 15)(y - 20)$
22. $(2x - 1)(3x + 4)$ 25. $(y - 12)(y - 4)$
28. $(1 + 7x)(5 - x)$

Guided Practice

CRITICAL THINKING about the Lesson

1. Factor $x^2 - 4x + 3$. When testing possible factorizations, why is it unnecessary to test $(x - 1)(x + 3)$ and $(x + 1)(x - 3)$?
2. Factor $x^2 + 2x - 3$. When testing possible factorizations, why is it unnecessary to test $(x - 1)(x - 3)$ and $(x + 1)(x + 3)$?
3. What is the discriminant of $ax^2 + bx + c$? $b^2 - 4ac$
4. If the discriminant of $ax^2 + bx + c$ is 35, can the trinomial be factored with integer coefficients? Explain. No. The discriminant must be the square of an integer.

1., 2. See margin.

Independent Practice

In Exercises 5–10, choose the correct factorization. (If neither is correct, find the correct factorization.)

5. $x^2 + x - 20$ a
   a. $(x - 4)(x + 5)$
   b. $(x + 4)(x - 5)$
6. $x^2 + 8x + 16$ b
   a. $(x + 2)(x + 8)$
   b. $(x + 4)(x + 4)$
7. $x^2 - 10x + 24$ a
   a. $(x - 6)(x - 4)$
   b. $(x - 12)(x + 2)$
8. $3x^2 - 7x - 6$ a
   a. $(x - 3)(3x + 2)$
   b. $(x + 3)(3x - 2)$
9. $6x^2 - 7x - 5$ b
   a. $(6x + 1)(x - 5)$
   b. $(2x + 1)(3x - 5)$
10. $2x^2 - 7x - 9$
   a. $(x - 1)(2x + 9)$
   b. $(2x - 1)(x + 9)$
   Neither, $(x + 1)(2x - 9)$

In Exercises 11–28, factor the trinomial.

11. $x^2 + 3x - 4$ $(x + 4)(x - 1)$
12. $x^2 - 5x + 6$ $(x - 2)(x - 3)$
13. $x^2 + 3x - 18$ $(x + 6)(x - 3)$
14. $y^2 - 16y - 36$ $(y - 18)(y + 2)$
15. $x^2 - 10x + 24$ $(x - 6)(x - 4)$
16. $x^2 + 13x + 22$
17. $x^2 + 15x + 50$ $(x + 10)(x + 5)$
18. $y^2 + 30y + 216$ $(y + 12)(y + 18)$
19. $y^2 - 35y + 300$
20. $t^2 - 4t - 21$ $(t - 7)(t + 3)$
21. $3x^2 + 8x + 5$ $(3x + 5)(x + 1)$
22. $6x^2 + 5x - 4$
23. $2x^2 - x - 21$ $(2x - 7)(x + 3)$
24. $3x^2 + 11x + 10$ $(3x + 5)(x + 2)$
25. $48 - 16y + y^2$
26. $32 + 12x + x^2$ $(x + 4)(x + 8)$
27. $2x^2 - x - 6$ $(2x + 3)(x - 2)$
28. $5 + 34x - 7x^2$

In Exercises 29–34, use the discriminant to decide whether the polynomial can be factored with integer coefficients. If it can be factored, then find the factors.

29. $12x^2 - 11x + 3$ Cannot
30. $2x^2 - 5x - 12$ $(2x + 3)(x - 4)$
31. $6x^2 - 10x + 4$ $2(3x - 2)(x - 1)$
32. $10x^2 - 9x + 6$ Cannot
33. $14x^2 - 19x - 40$ $(7x + 8)(2x - 5)$
34. $24x^2 + 3x - 11$ Cannot

✪ 35. *Geometry* The area of a rectangle is given by $A = x^2 + 4x - 5$. Find expressions for possible lengths and widths of the rectangle. $x + 5$, $x - 1$

✪ 36. *Geometry* The area of a circle is given by $A = \pi(4x^2 + 12x + 9)$. Find an expression for the radius of the circle. $2x + 3$

The Art of Africa **In Exercises 37–46, use the following information.**

The letters of the alphabet from A to Z (excluding W and X) are represented by nonzero integers from -12 to 12. Copy and complete the table by factoring the polynomials in Exercises 37–42. Then use the table to match the coded words in Exercises 43–46 with one of the pieces of African art.

Letter	A	B	C	D	E	F	G	H	I	J	K	L
Code Number	1	?	?	?	9	?	?	?	2	?	?	?
Letter	**M**	**N**	**O**	**P**	**Q**	**R**	**S**	**T**	**U**	**V**	**Y**	**Z**
Code Number	3	?	?	?	6	?	?	?	11	?	?	?

$-4, 8, -3, -8, 10, -7, -9, 5, -6$

$-10, 4, -1, -11, 7, -2, -12, 12, -5$

37. $8x^2 - 35x + 12 = (Ax + B)(Cx + D)$

38. $90x^2 - 143x + 56 = (Ex + F)(Gx + H)$

39. $10x^2 - 57x + 54 = (Ix + J)(Kx + L)$

40. $12x^2 - 43x + 10 = (Mx + N)(Ox + P)$

41. $42x^2 - 89x + 22 = (Qx + R)(Sx + T)$

42. $132x^2 - 199x + 60 = (Ux + V)(Yx + Z)$

43.

2	−12	4	−11	12		3	1	7	5

Ivory mask, a

44.

8	1	−11	−12	9	−3		7	−2	4	4	−6

Carved stool, b

45.

−1	1	−2	−2	9	−11	−10	9	−3		−12	9	7	7	9	−6

Patterned vessel, c

46.

3	1	7	1	2		−10	9	8	5	−6	1	8	9

Masai necklace, d

a. Ivory Coast

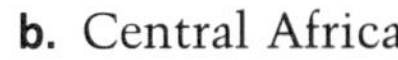

b. Central Africa

c. Central Africa

d. Kenya

Guided Practice

▶ **Ex. 1–4** Students may need to review using the discriminant from Chapter 9. In Exercise 4 remind students of the meaning of integer coefficients.

Answers

1. $(x - 1)(x - 3)$
   Since c is positive, its factors must have the same sign.
2. $(x + 3)(x - 1)$
   Since c is negative, its factors must have opposite signs.

Independent Practice

▶ **Ex. 25, 26, 28** Remind students to first rewrite the polynomials in descending order before applying the factoring patterns.

▶ **Ex. 31** The first step in factoring polynomials is to factor out common monomial factors.

▶ **Ex. 36** Remind students that $R^2 = 4x^2 + 12x + 9$.

▶ **Ex. 37–46** These exercises should be assigned as a group. Be sure that students understand the directions before assigning the exercises.

▶ **Ex. 49–50** Refer students to Example 6 on page 537.

▶ **Ex. 54–56** These exercises review using the distributive property with linear equations.

▶ **Ex. 57–60** Remind students what is meant by a linear factor.

Enrichment Activities

These activities require students to go beyond lesson goals.

1. Here is another factoring method that can be used to find the binomial factors of factorable trinomials of the form $ax^2 + bx + c$. It is sometimes called the "British method."

For example, here is a way to factor $6x^2 - x - 2$.

A. Find the magic number ac. The magic number is -12.

B. Write the magic number as the product of two factors whose sum is b. That is, factor -12 so that the sum of the factors is -1. The factors of -12 with a sum of -1 are 3 and -4.

C. Rewrite the trinomial using those factors as shown. $6x^2 - x - 2 = 6x^2 + 3x + (-4x) - 2$

D. Factor using the distributive properties.
$3x(2x + 1) + (-2)(2x + 1)$
$= (3x + (-2))(2x + 1)$
$= (3x - 2)(2x + 1)$

Do you think it would have made a difference if you had written $6x^2 - x - 2 = 6x^2 + (-4x) + 3x - 2$?
No, you would factor as $2x(3x + (-2)) + 3x - 2 = 2x(3x - 2) + 1(3x - 2) = (2x + 1)(3x - 2)$.

Try this method on the expressions in Exercises 37–42. Do you find it easier? Explain. This method calls for only one search for factors. This method is easier only when the coefficients are integers and when the "magic numbers" are numbers with recognizable factors.

2. MATH JOURNAL
Write about a general strategy that anyone can use in factoring trinomials. Include examples.
The strategies should include (a) writing the polynomials in descending order, (b) factoring out any common monomial factors first, (c) using the discriminant to decide if the expression is factorable and (d) use the guess, check, and revise method until the factors are found.

Algebra Tiles **In Exercises 47 and 48, factor the trinomial. Use an area model to illustrate your result. Use the following area model for $x^2 + 3x + 2 = (x + 1)(x + 2)$ as a sample.**

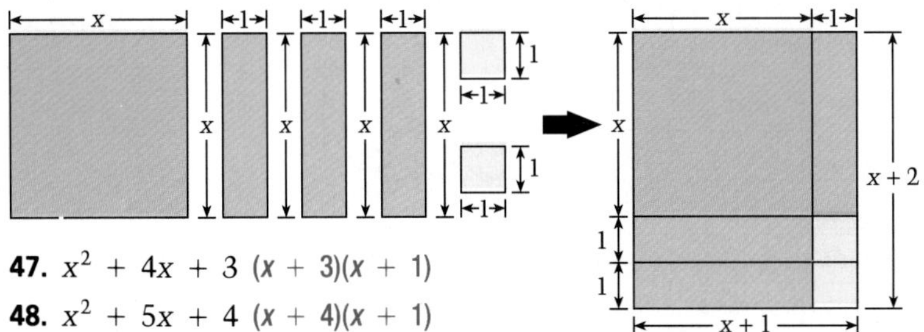

47. $x^2 + 4x + 3$ $(x + 3)(x + 1)$

48. $x^2 + 5x + 4$ $(x + 4)(x + 1)$

49. *Hot Dog Stand* You run a hot-dog stand for the adult softball league. Your revenue, R (in dollars), each week can be given by
$R = \frac{1}{100}(-40t^2 + 740t + 1200)$ $P = -\frac{1}{20}t + 1$, \$1.00, 95¢, 90¢, 85¢, 80¢, 75¢, 70¢, 65¢
where t represents the week with $t = 0$ for the first week. In the first week, you sold 12 hot dogs and each week after that the number of sales increased by 8. Find a model for the price of a hot dog over the 8-week season. Use the model to find the price of a hot dog during each week of the season.

50. *Summer Basketball* During the summer, the number, B, of people who played intramural basketball each week is given by the model

$B = t^2 + 11t + 30$ $A = t + 5$; 5, 6, 7, 8, 9, 10

where t represents the week, with $t = 0$ for the first week. In the first week, six teams were in the league. Each week, for five weeks, a new team joined the league. Find a model for the average number of members per team for the six-week season. Use the model to find the average number of players on each team during each week.

Integrated Review

In Exercises 51–53, multiply.

51. $-2y(y + 1)$ $-2y^2 - 2y$

52. $(x + 4)^2$ $x^2 + 8x + 16$

53. $(v - 1)(v - 6)$ $v^2 - 7v + 6$

In Exercises 54–56, solve the equation.

54. $3(x + 2) = 5$ $-\frac{1}{3}$

55. $-1(4x - 5) = x$ 1

56. $5(y - 3) = -1(y + 6)$ $\frac{3}{2}$

Exploration and Extension

In Exercises 57–60, factor the polynomial as the product of *linear* factors.

57. $2x^3 - 5x^2 - 3x$ $x(2x + 1)(x - 3)$

58. $48x^3 - 2x^2 - 20x$ $2x(3x - 2)(8x + 5)$

59. $x^4 - 5x^2 + 4$ $(x + 1)(x - 1)(x + 2)(x - 2)$

60. $x^4 - 13x^2 + 36$ $(x + 2)(x - 2)(x + 3)(x - 3)$

10.6 Solving Quadratic Equations by Factoring

What you should learn:

Goal 1 How to use factoring to solve a quadratic equation

Goal 2 How to use factoring with real-life models

Why you should learn it:

You can use factoring to help efficiently solve some quadratic equations.

Goal 1 Solving by Factoring

The **Zero-Product Property** states that if the product of two factors is zero, then one (or both) of the factors must be zero.

Zero-Product Property

If the product $ab = 0$, then $a = 0$ or $b = 0$.

This property connects factoring to solving equations. For instance, to solve the equation $(x - 2)(x + 3) = 0$, you can use the Zero-Product Property to conclude that either $x - 2 = 0$ or $x + 3 = 0$.

Set first factor equal to 0.	**Set second factor equal to 0.**
$x - 2 = 0$	$x + 3 = 0$
$x = 2$	$x = -3$

The equation $(x - 2)(x + 3) = 0$ has the solutions 2 and -3.

Problem Solving
Solve a Simpler Problem

Factoring and the Zero-Product Property allow you to solve a quadratic equation by converting it into two *linear* equations, which you already know how to solve. This is a common strategy of algebra—to break down a problem into simpler parts, each solved by previously learned methods.

Example 1 *Solving a Quadratic Equation by Factoring*

Solve $x^2 - x - 12 = 0$.

Solution

$$x^2 - x - 12 = 0$$

$$(x + 3)(x - 4) = 0 \quad \textit{Factor.}$$

$$x + 3 = 0 \text{ or } x - 4 = 0 \quad \textit{Zero-Product Property}$$

$$x = -3 \text{ or } x = 4 \quad \textit{Solve for x.}$$

The equation has two solutions: -3 and 4. Check these solutions in the original equation. ■

Problem of the Day

Veritas always speak the truth, and Prevars always lie. A stranger in their country meets a party of 3 and asks to which race each belongs. The first mumbles an answer which the stranger cannot understand, the second says "He said he was a Prevars," and the third says to the second, "You're a liar!" Of what race was the third?

Everyone, of either race, would say he was a Veritas, so the second lied, as the third said. So the third is a Veritas.

ORGANIZER

Warm-Up Exercises

1. Solve for x.
   a. $(x - 5) = 0$ 5
   b. $(x + 7) = 0$ -7
   c. $(2x + 1) = 0$ $-\frac{1}{2}$
   d. $(4x - 12) = 0$ 3
2. Factor.
   a. $x^2 - 11x + 24$
   b. $2x^2 + 3x - 5$
   c. $18 - 18x - 8x^2$
   d. $12x^3 - 21x^2 - 6x$

   a. $(x - 3)(x - 8)$
   b. $(2x + 5)(x - 1)$
   c. $(3 - 4x)(6 + 2x)$
   d. $3x(4x + 1)(x - 2)$

Lesson Resources

Teaching Tools
- Transparencies: 1, 2
- Problem of the Day: 10.6
- Warm-up Exercises: 10.6
- Answer Masters: 10.6

Extra Practice: 10.6

Technology Handbook: p. 78

LESSON Notes

Examples 1–3

TECHNOLOGY The connection between the solution of a quadratic equation and its graph should be made. Students may wish to use graphics calculators (or computers) to check their solutions by identifying where the graph of the equation written in standard form crosses the x-axis. In Examples 2 and 3, the solutions can be determined by identifying where the equations $y = 3x^2 + 5x$ and $y = 12$, and $y = x^2 - 6x + 11$ and $y = 2$ intersect, respectively.
In Example 3, the best way to dispel the conjecture that $ab = k$ implies $a = k$ or $b = k$ is to provide counterexamples. For instance, if $x(2x - 4) = 8$ implies that either $x = 8$ or $2x - 4 = 8$, it follows that for $x = 8$, $x(2x - 4) = 8$ becomes $8(2 \cdot 8 - 4) = 8$, or $8(12) = 8$, which is clearly untrue. Likewise, if $2x - 4 = 8$ then $x = 6$, and it follows that $x(2x - 4) = 8$ becomes $6(8) = 8$, again untrue.

Example 4

Not every algebraic solution is a reasonable solution to a real-life problem. Consequently, checking algebraic solutions for reasonableness is important. In this example, the solution -4 makes no sense as a height for a triangle.

Extra Examples

Here are additional examples similar to **Examples 1–4.**

1. Solve by factoring.
   a. $x^2 - 11x + 24 = 0$ 3, 8
   b. $2x^2 + 3x = 5$ $-\frac{5}{2}, 1$
   c. $0 = 18 - 18x - 8x^2$ $-3, \frac{3}{4}$
   d. $4x^2 = 7x + 2$ $-\frac{1}{4}, 2$
2. Suppose in Example 4, that the base of the triangular roadside sign is 3 times its

To use the Zero Product Property, the quadratic equation should be written in the standard form, $ax^2 + bx + c = 0$.

Example 2 Rewriting in Standard Form First

Solve $3x^2 + 5x = 12$.

Solution

$3x^2 + 5x = 12$	*Rewrite original equation.*
$3x^2 + 5x - 12 = 0$	*Write in standard form.*
$(3x - 4)(x + 3) = 0$	*Factor.*
$3x - 4 = 0$ or $x + 3 = 0$	*Set factors equal to 0.*
$x = \frac{4}{3}$ or $x = -3$	*Solve for x.*

The equation has two solutions: $\frac{4}{3}$ and -3. Check these solutions in the original equation. ■

In Examples 1 and 2, each equation had two solutions. Some quadratic equations have only one solution. This occurs when the quadratic is a perfect square trinomial.

Example 3 A Quadratic Equation with One Solution

Solve $x^2 - 6x + 11 = 2$.

Solution

$x^2 - 6x + 11 = 2$	*Rewrite original equation.*
$x^2 - 6x + 9 = 0$	*Write in standard form.*
$(x - 3)^2 = 0$	*Factor.*
$x - 3 = 0$	*Set factor equal to 0.*
$x = 3$	*Solve for x.*

The equation has only one solution: 3. Check this solution in the original equation. ■

Be sure you see that the Zero Product Property can only be applied to a product that is equal to *zero*. For instance, you *cannot* conclude from the equation $x(x - 1) = 6$ that $x = 6$ or $x - 1 = 6$. Instead, you must first write the equation in standard form and then factor the left side.

$$x(x - 1) = 6$$
$$x^2 - x = 6$$
$$x^2 - x - 6 = 0$$
$$(x - 3)(x + 2) = 0$$

From the factored form, the solutions are 3 and -2.

Goal 2 Using Factoring with Real-Life Models

Example 4 *Finding the Dimensions of a Sign*

A triangular sign has a base that is to be 4 feet less than twice its height. A local zoning ordinance restricts the area of signs to no more than 24 square feet. Find the base and height of the largest triangular sign that meets the zoning ordinance.

Problem Solving
Write an Equation

Solution Since the base is to be 4 feet less than twice the height, let h be the height and label the base as $2h - 4$.

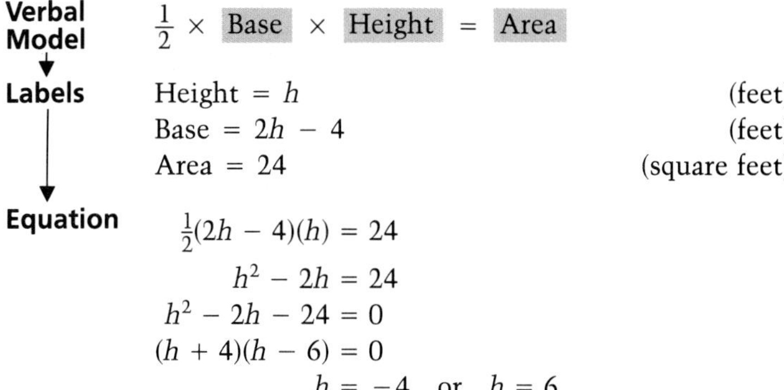

Verbal Model $\frac{1}{2} \times$ Base $\times$ Height $=$ Area

Labels Height $= h$ (feet)
Base $= 2h - 4$ (feet)
Area $= 24$ (square feet)

Equation

$$\tfrac{1}{2}(2h - 4)(h) = 24$$
$$h^2 - 2h = 24$$
$$h^2 - 2h - 24 = 0$$
$$(h + 4)(h - 6) = 0$$
$$h = -4 \quad \text{or} \quad h = 6$$

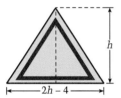

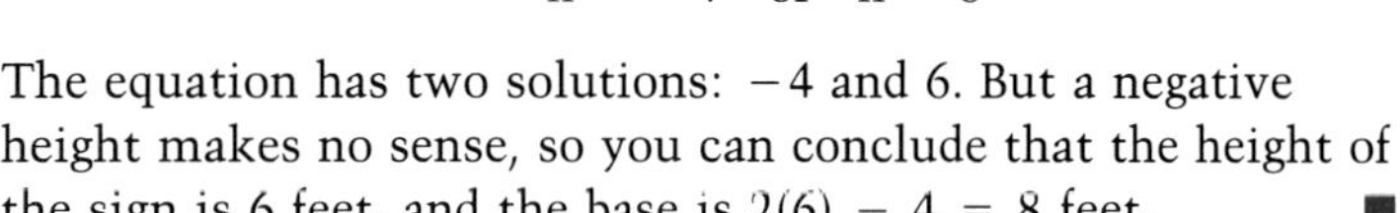

The equation has two solutions: -4 and 6. But a negative height makes no sense, so you can conclude that the height of the sign is 6 feet, and the base is $2(6) - 4 = 8$ feet. ■

Communicating about ALGEBRA

SHARING IDEAS about the Lesson See margin.

Choose a Method You have now studied three ways to solve quadratic equations: by finding square roots, by the quadratic formula, and by factoring. Solve each equation by the method you think is most efficient. Discuss the reasons for your choices and explain how the graphs can be used as a check of the solutions.

A. $x^2 - 3x + 2 = 0$ **B.** $x^2 - 2x - 1 = 0$ **C.** $x^2 - 3 = 0$

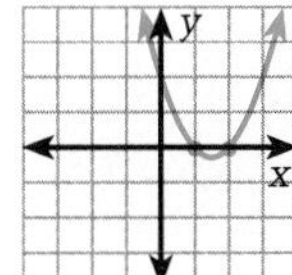

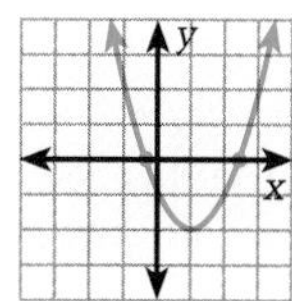

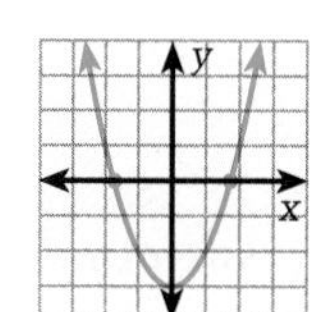

height. Find the base and height of the largest triangular sign that meets the zoning ordinance. height = 4 ft, base = 12 ft

Check Understanding

1. Why should quadratic equations be written in standard form before applying the Zero-Product Property? To find the factors a and b
2. How can the Zero-Product Property be used instead of the square root method for solving quadratic equations of the form $x^2 = k^2$. Rewrite as $x^2 - k^2$ and factor to $(x + k)(x - k)$.
3. What role should graphing play in solving quadratic equations? The x-intercepts of the graph are the real-number solutions of the quadratic equation if they exist.

Common-Error ALERT!

Students frequently misinterpret the Zero-Product Property to mean that if the product $ab = k$, for k any real number, then either $a = k$ or $b = k$. This seems reasonable to many students because they have been taught to solve quadratic equations of the form $ax^2 + c = 0$ using the square root method! In that method, if $x^2 = 16$, then $x = \pm\sqrt{16}$. Written differently, $ab = k$, where a and b happen to be equal. Of course, the conjecture that $ab = k$ implies $a = k$ or $b = k$ is incorrect, but not so farfetched.

Communicating about ALGEBRA

A fourth method of solving quadratic equations is by graphing. Discuss reasons why this approach should or should not be evaluated with the other three as an efficient method of solution. Explain the impact of real-life applications on the method of choice, including the graphing approach.

Answers

A. 1, 2; factoring, easily factored.

B. $1+\sqrt{2}\approx 2.41$, $1-\sqrt{2}\approx -0.41$; quadratic formula, not factorable.

C. $\sqrt{3}\approx 1.73$, $-\sqrt{3}\approx -1.73$; square roots, no x term.

A., B., C. x-intercepts are solutions.

EXERCISE Notes

ASSIGNMENT GUIDE

Basic/Average: Ex. 1–6, 7–15 odd, 17–24, 25–33 odd, 38–39, 41–51 odd, 59–65 odd

Above Average: Ex. 1–6, 11–15 odd, 17–24, 25–37 odd, 38–39, 42–63 multiples of 3, 66–68

Advanced: Ex. 1–6, 17–24, 25–35 odd, 36–39, 41–61 odd, 62–68

Selected Answers
Exercises 1–6, 7–61 odd

Use **Mixed Review** as needed.

More Difficult Exercises
Exercises 34, 35, 37, 66–68

EXERCISES

1. $a = 0$ or $b = 0$

Guided Practice

CRITICAL THINKING about the Lesson

1. Use the Zero-Product Property to complete the statement. If $ab = 0$, then [?].

2. Solve the equation: $(x-2)(x+1)=0$. 2, −1

3. Solve the equation: $3x^2+4x=0$. 0, $-\frac{4}{3}$

4. Which two numbers satisfy the statement, "The sum of a number and its square is zero."? 0, −1

5. **True or False?** If $(5x-1)(x+3)=1$, then $5x-1=1$ or $x+3=1$. Explain. False. The product of any number and its reciprocal is 1; so neither factor has to equal 1.

6. **True or False?** If $(x+3)(x-3)=0$, then $x+3=0$ or $x-3=0$. Explain True. A product cannot equal 0, unless one of the factors is 0.

Independent Practice

In Exercises 7–10, solve the equation.

7. $(x+1)(x+2)=0$ −1, −2

8. $(x-3)(x+7)=0$ 3, −7

9. $(x+3)(x+4)=0$ −3, −4

10. $(x+6)(x-5)=0$ −6, 5

In Exercises 11–16, solve the equation by factoring.

11. $x^2+5x-6=0$ −6, 1

12. $3x^2+11x-4=0$ $\frac{1}{3}$, −4

13. $2x^2+5x+3=0$ $-\frac{3}{2}$, −1

14. $6x^2+13x+5=0$ $-\frac{1}{2}$, $-\frac{5}{3}$

15. $3x^2+7x+2=0$ $-\frac{1}{3}$, −2

16. $12x^2-5x-3=0$ $\frac{3}{4}$, $-\frac{1}{3}$

In Exercises 17–24, match the equation with its solutions.

17. $x^2-5x+6=0$ f

18. $x^2+5x+6=0$ b

19. $x^2-7x+6=0$ g

20. $x^2+7x+6=0$ a

21. $x^2-5x-6=0$ d

22. $x^2+5x-6=0$ c

23. $x^2+x-6=0$ e

24. $x^2-x-6=0$ h

a. −1, −6 **b.** −2, −3 **c.** 1, −6 **d.** −1, 6

e. 2, −3 **f.** 2, 3 **g.** 1, 6 **h.** −2, 3

In Exercises 25–33, solve the equation by finding square roots, by the quadratic formula, or by factoring.

25. $x(x-9)=0$ 0, 9

26. $2y(y+6)=0$ 0, −6

27. $y^2-7y+6=-6$ 3, 4

28. $x^2-12=-3$ ±3

29. $x^2-8x=-16$ 4

30. $x^2+4x+7=3$ −2

31. $4x^2+2x=0$ 0, $-\frac{1}{2}$

32. $4y^2-18y=0$ 0, $\frac{9}{2}$

33. $x^2-12x+40=4$ 6

In Exercises 34 and 35, multiply both sides of the equation by an appropriate power of ten to obtain integer coefficients. Then solve by factoring.

34. $0.8x^2+3.2x+2.4=0$ −1, −3

35. $0.23x^2-0.54x+0.16=0$ $\frac{8}{23}$, 2

36. *How to Weigh an Elephant Seal* The rectangular scale in the photo is 8 feet longer than it is wide. The scale has an area of 33 square feet. What are the dimensions of the scale? 3 ft by 11 ft

37. *A California Raisin* During the breeding season, the male elephant seal in the photo lost 38% of his weight. His weight, w (in hundreds of pounds), at the beginning of the season was a solution of $w^2 - 38w + 361 = 0$. What did he weigh at the end of the season? 1178 pounds

To weigh this male elephant seal, a biologist at the University of California lured him to cross a special scale by pulling a model female seal named* Raisin. *(Source:* Discover, *1991)

The Taj Mahal **In Exercises 38 and 39, use the following information.**

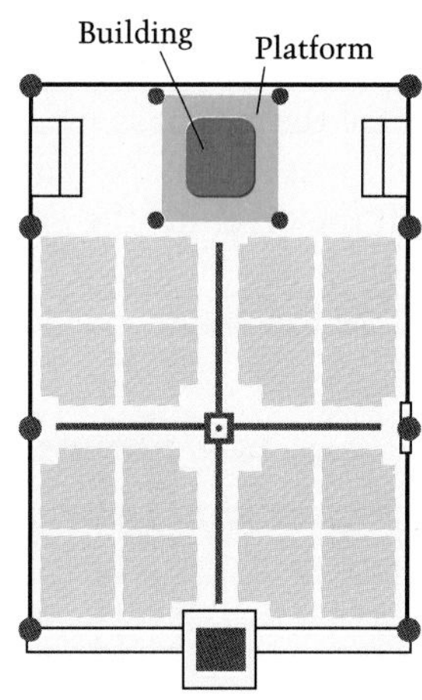

The Taj Mahal was built by the Indian ruler Shah Jahan as a tomb for his wife who died in 1629. The building is constructed of white marble and sits on a platform of red sandstone. Each corner of the platform has a* minaret *(tower) that is 133 feet high.

38. The platform is about 140 feet wider than the main building. The total area of the platform is about 102,400 square feet. Find the dimensions of the platform and the base of the building. (Assume each is a square.) 320 ft by 320 ft, 180 ft by 180 ft

39. Which of the following is the best estimate of the dimensions of the entire courtyard shown in the diagram?

a. 640 feet $\times$ 1280 feet **b.** 960 feet by 1280 feet **c.** 640 feet $\times$ 1600 feet

Guided Practice

Ex. 5–6 Use these exercises to determine whether students understand that the products must equal zero in order to apply the Zero-Product Property. Be careful to point out that you can find isolated cases where if $ab = k$, $a = k$ or $b = k$. Consider $(x - 1)(x - 2) = 2$, $x^2 - 3x + 2 = 2$, so $x = 0$ or $x = 3$. So, $(x - 1) = (3 - 1) = 2$ is true.

Independent Practice

Ex. 7–24 Have students practice factoring as a mental math technique for solving quadratic equations.

Ex. 25–33 Challenge students to choose the most efficient method for solving each equation. Help them to focus on why they chose a particular method. (See the Enrichment Activity on page 547.)

Ex. 36–39 Remind students to check the reasonableness of each solution. Be sure to assign Exercises 38 and 39 as a pair.

Answers

53. $\frac{7+\sqrt{13}}{2} \approx 5.30, \frac{7-\sqrt{13}}{2} \approx 1.70$

55. $\frac{11+\sqrt{409}}{6} \approx 5.20, \frac{11-\sqrt{409}}{6} \approx -1.54$

56. $\frac{8+\sqrt{124}}{5} \approx 3.83, \frac{8-\sqrt{124}}{5} \approx -0.63$

57. $\frac{17+\sqrt{73}}{18} \approx 1.42, \frac{17-\sqrt{73}}{18} \approx 0.47$

58. In 2 seconds. One of the factors must equal zero; the factor $(t-2)$ will give a positive answer.

59. In 10 seconds. One of the factors must equal zero; the factor $(t-10)$ will give a positive answer.

60.

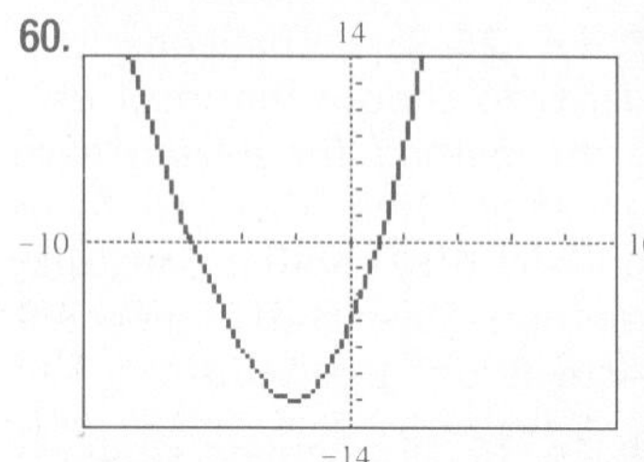

61.

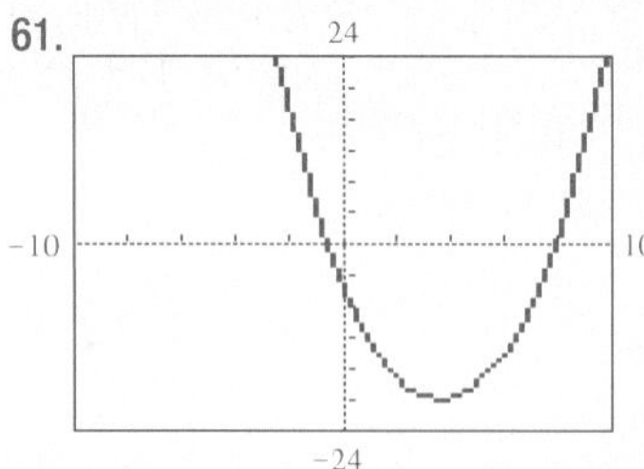

Ex. 62–65 Suggest to students that they might use the Zero Product Property to generate the equations. Note also that many equations are possible.

Ex. 68 Students can write and solve a system of equations in two variables. For example, let f be the number of weft threads and p be the number of warp threads; then $fp = 5000$. Since the pattern is more visible in the weft threads, the density of weft threads is twice that of the warp threads, or, $f = 2p$. So, $fp = (2p)p = 2p^2 = 5000$ and p is 50.

Integrated Review

In Exercises 40–45, multiply and simplify.

40. $(2x+6)(\frac{1}{2}x-3)$ $x^2-3x-18$

41. $(x+9)(4x-3)$ $4x^2+33x-27$

42. $36-(x+2)(x-3)$ $-x^2+x+42$

43. $3(x-\frac{1}{3})(3x+7)$ $9x^2+18x-7$

44. $4(x+3)(3x-1)+1$ $12x^2+32x-11$

45. $5+(2x-\frac{1}{2})(\frac{1}{2}x+1)$ $x^2+\frac{7}{4}x+\frac{9}{2}$

In Exercises 46–51, decide whether the number is a solution of the equation.

46. $x^2-3x+4=0;\ 2$ Is not

47. $x^2+5x+4=0;\ -1$ Is

48. $\frac{1}{3}x^2+2x-36=0;\ 6$ Is not

49. $2x^2-3x-4=0;\ 2$ Is not

50. $4x^2+3x-27=0;\ -3$ Is

51. $\frac{1}{2}x^2-x+8=0;\ 4$ Is not

In Exercises 52–57, use the quadratic formula to solve the equation. 53., 55.–57. See margin.

52. $x^2+4x+3=0$ $-1, -3$

53. $x^2-7x+9=0$

54. $2x^2-5x+3=0$ $\frac{3}{2}, 1$

55. $3x^2-11x-24=0$

56. $5x^2-16x-12=0$

57. $9x^2-17x+6=0$

58. *Height of a Diver* A diver jumps from a diving board that is 32 feet above the water. The height of the diver is given by

$$\text{Height} = -16(t-2)(t+1)$$

where the height is measured in feet, and the time, t, is measured in seconds. When will the diver hit the water? Can you see a quick way to find the answer? Explain.
58., 59. See margin.

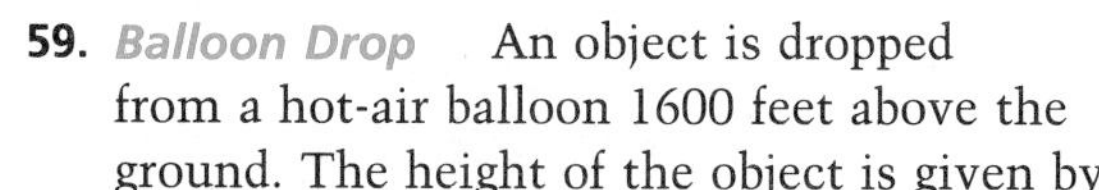

59. *Balloon Drop* An object is dropped from a hot-air balloon 1600 feet above the ground. The height of the object is given by

$$\text{Height} = -16(t-10)(t+10)$$

where the height is measured in feet, and the time, t, is measured in seconds. When will the object hit the ground? Can you see a quick way to find the answer? Explain.

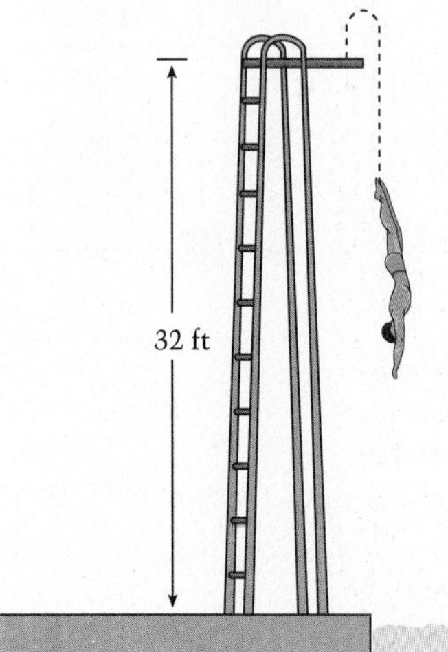

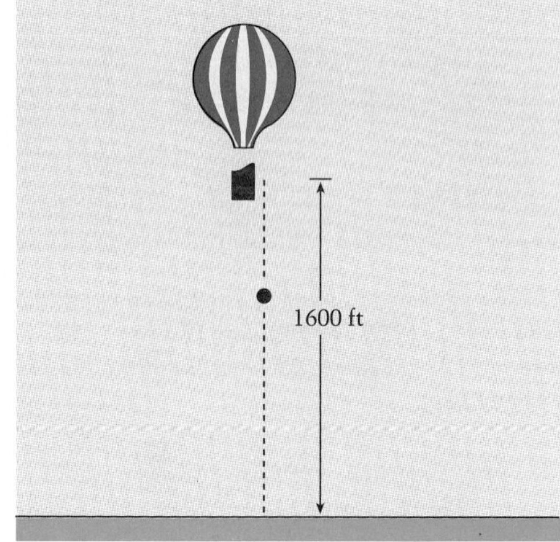

60. *Graphing Calculator* Sketch $y = x^2+5x-6$ and $y=(x+6)(x-1)$ on the same screen. What do you notice? They are the same graph.

61. *Graphing Calculator* Sketch $y = x^2-7x-8$ and $y=(x-8)(x+1)$ on the same screen. What do you notice? They are the same graph. See margin for graphs.

Exploration and Extension

62. $x^2 - x - 12 = 0$ 64. $x^2 + 7x + 10 = 0$

Tutoring a Friend **In Exercises 62–65, you are tutoring a friend and want to create some quadratic equations that can be solved by factoring. Find a quadratic equation that has the given solutions and explain the procedure you used to obtain the equation.**

62. 4 and -3

63. 5 and 5 $x^2 - 10x + 25 = 0$

64. -2 and -5

65. 0 and -1 $x^2 + x = 0$

66. Let a and b be real numbers such that $a \neq 0$. Find the solutions of $ax^2 + bx = 0$. $0, -\frac{b}{a}$

67. Let a be a nonzero real number. Find the solutions of $ax^2 - ax = 0$. 0, 1

68. *Ethiopian Weavers* In every square inch of the cotton fabric used for a shammas, the *warp* (lengthwise threads) intersects the *weft* (crosswise threads) 5000 times. The density (number of threads per inch) of the weft threads is twice that of the warp threads. How many weft threads are in each inch? How many warp threads are in each inch? 100, 50

These Ethiopian women are wearing a fine cotton fabric known as a shammas, *a national costume that is often edged with bright trimming.*

Mixed REVIEW

5. $m = \frac{-2n + 3}{4}$ 13. 8, -8 14. $x, -x$

1. Simplify $x^2y^{-1} \div 3x \cdot y^2$. **(8.3)** $\frac{xy}{3}$
2. Simplify $(p^2q^4 \div 2p) \cdot (4p^{-2} \div q^{-2})$. **(8.2)** $\frac{2q^6}{p}$
3. Find the quotient of $3x^2$ and $9x$. **(8.3)** $\frac{x}{3}$
4. Find the sum of $2p$ and $-4p$. **(2.6)** $-2p$
5. Solve $4m + 2n = 3$ for m. **(3.6)**
6. Solve $ab + a = 5$ for a. **(3.6)** $a = \frac{5}{b + 1}$
7. Write $2 \cdot x \cdot x \cdot x \cdot x$ in exponential form. **(1.3)** $2x^4$
8. Write 0.00012 in scientific notation. **(8.4)** 1.2×10^{-4}
9. Evaluate $(2 \times 10^{-3}) \cdot (4 \times 10^4)$. **(8.4)** 80
10. Evaluate $(3.2 \times 10^6) \div (1.6 \times 10^{-1})$. **(8.4)** 2×10^7
11. Solve $|x + 3| \leq 4$. **(4.8)** $-7 \leq x \leq 1$
12. Solve $2|3 - x| > 1$. **(4.8)** $x > \frac{7}{2}$ or $x < \frac{5}{2}$
13. What are the square roots of 64? **(9.1)**
14. What are the square roots of x^2? **(9.1)**
15. Sketch the graph of $2x + y = 6$. **(4.4)**
16. Sketch the graph of $2|x + 2| + 3 = 0$. **(4.7)**
17. Is -6 a solution of $2^{-x} < -7x$? **(8.2)** No
18. Is $(-1, -2)$ a solution of $y = -|2x|$? **(2.1)** Yes
19. Solve the system. $\begin{cases} x - 7y = -17 \\ 3x + y = -7 \end{cases}$ **(7.2, 7.3)** $(-3, 2)$
20. Solve the system. $\begin{cases} 4x + 3y = -1 \\ 5x - 6y = 28 \end{cases}$ **(7.2, 7.3)** $(2, -3)$

Enrichment Activities

These activities require students to go beyond lesson goals.

1. Choose a method for solving these quadratic equations. Give a reason for your choice. Do not solve.
   a. $-x^2 + 4 = 0$
   b. $x^2 + 4x + 4 = 0$
   c. $x^2 - 8x = 0$
   d. $13x^2 - 11x + 29 = 0$
   e. $0.237x^2 - 1.7x + 3.2 = 0$

   a. No linear term, use square roots
   b. Perfect square trinomial, factor
   c. Use the distributive property to factor; use Zero-Product Property
   d. Does not factor, use quadratic formula
   e. Too complicated for factoring, use quadratic formula
2. **EXTEND Exercises 25–33** Indicate the method that you used to solve the equations and why you chose that method. Compare your answers with your classmates' answers. Did you all agree?

Answers

Mixed Review

15.

16.

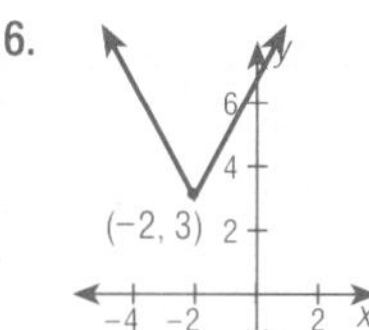

Answers

1.

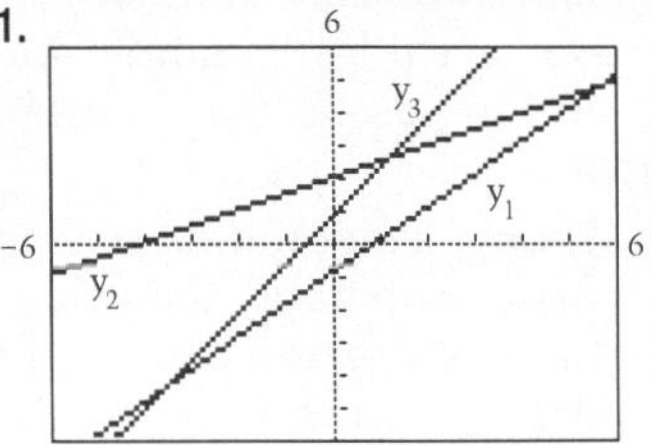

2.

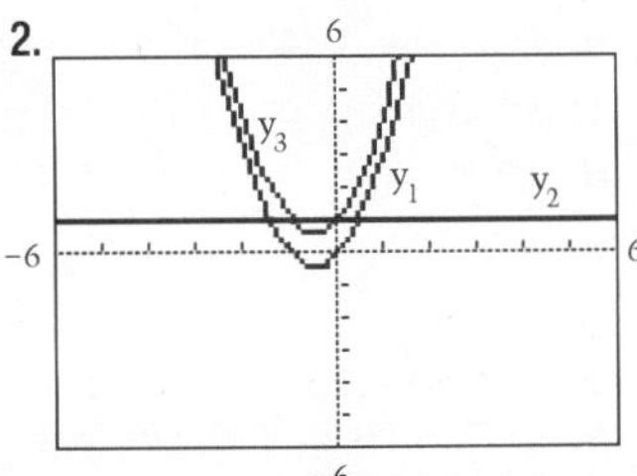

3.

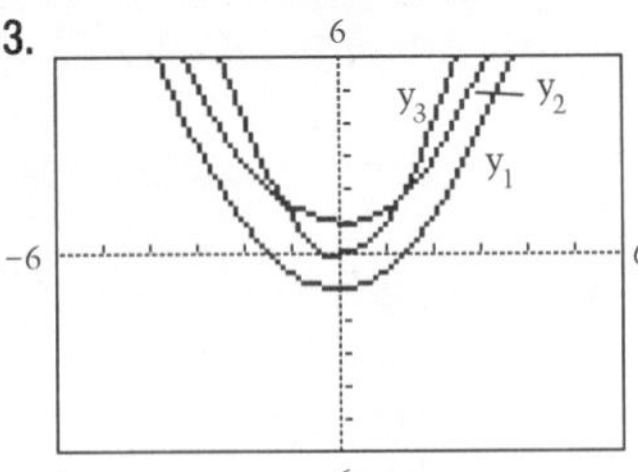

4.

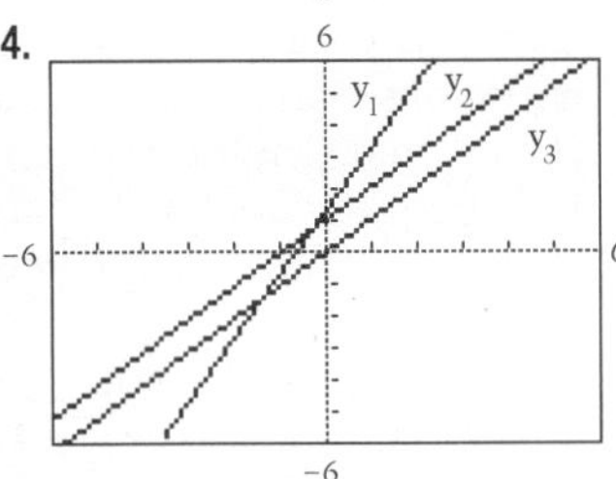

5.

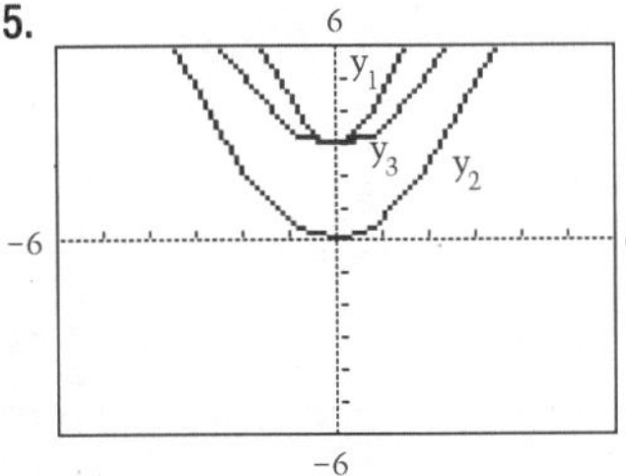

6.

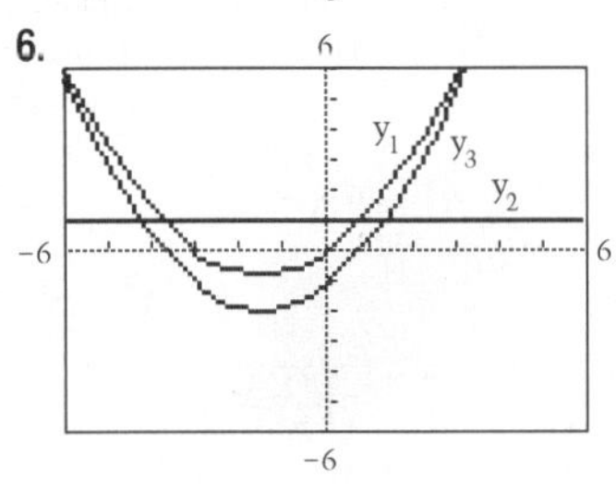

USING A GRAPHING CALCULATOR

Graphing calculators can be used to obtain a graphic interpretation of polynomial addition or subtraction. In the first graph below, notice that for any x*-value, the sum of* y_1 *and* y_2 *is equal to* y_3. *In the second graph, for any* x*-value, the difference of* y_1 *and* y_2 *is equal to* y_3. *Keystroke instructions for doing this on a* TI-82, *Casio* fx-9700GE, *and Sharp* EL-9300C *are listed on page 743.*

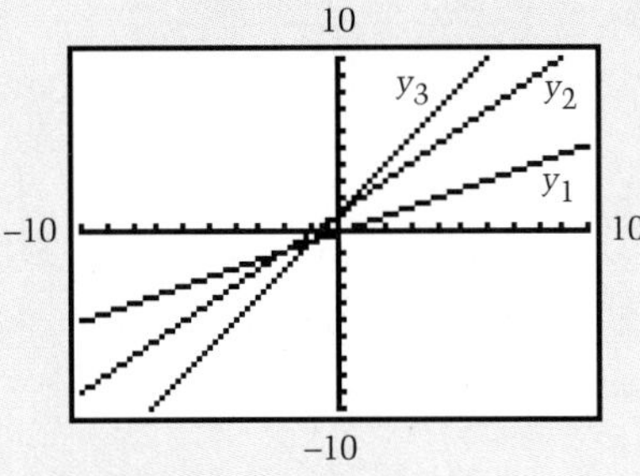

y_3 is the sum of y_1 and y_2.

First Graph

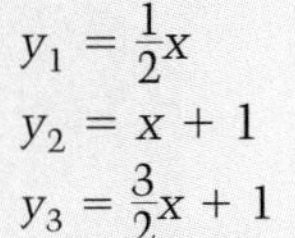

$y_1 = \frac{1}{2}x$

$y_2 = x + 1$

$y_3 = \frac{3}{2}x + 1$

Second Graph

$y_1 = \frac{1}{4}x^2 + 4$

$y_2 = 3$

$y_3 = \frac{1}{4}x^2 + 1$

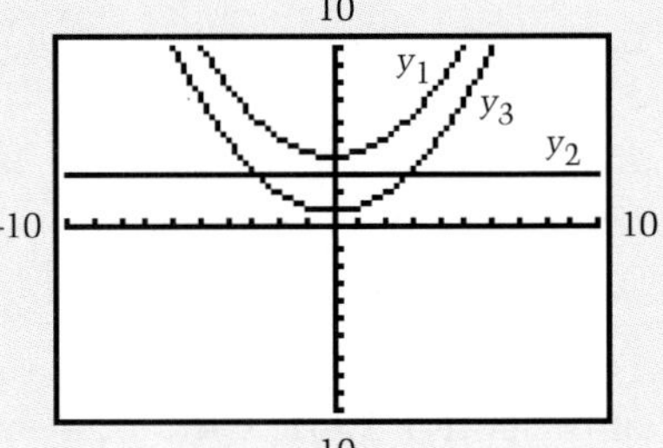

y_3 is the difference of y_1 and y_2.

Real Life Physics

Comparing Grain Consumptions From 1980 through 1994, the annual amount y (in pounds) of wheat and rice eaten by Americans can be modeled as follows.

$y_1 = 2t + 114$ *Wheat*

$y_2 = 0.75t + 8$ *Rice*

In the model $t = 0$ represents 1980. Use a graphing calculator to graph each model. Then, on the same screen, sketch the graph of the sum of the two models. What does the sum represent?
The total amount of wheat and rice eaten by Americans.

Exercises

In Exercises 1–3, let y_3 be the sum of y_1 and y_2. Sketch the graph of all three equations. See margin.

1. $y_1 = x - 1, y_2 = \frac{1}{2}x + 2$ **2.** $y_1 = x^2 + x, y_2 = 1$ **3.** $y_1 = \frac{1}{2}x^2 - 1, y_2 = \frac{1}{2}x^2 + 1$

In Exercises 4–6, let y_3 be the difference of y_1 and y_2. Sketch the graph of all three equations.

4. $y_1 = 2x + 1, y_2 = x + 1$ **5.** $y_1 = x^2 + 3, y_2 = \frac{1}{2}x^2$ **6.** $y_1 = \frac{1}{3}x^2 + x, y_2 = 1$

10.7 Solving Quadratic Equations by Completing the Square

What you should learn:

Goal How to solve quadratic equations by completing the square

Why you should learn it:

You can complete the square to help efficiently solve some quadratic equations.

Goal Solving by Completing the Square

In this lesson, you will learn to rewrite a quadratic equation in *completed-square form* by **completing the square.**

LESSON INVESTIGATION

■ Investigating Completed Square Form

Partner Activity Complete the following solution.

$x^2 + 8x + 5 = 0$	*Original equation*
$x^2 + 8x = -5$	*Subtract 5 from both sides.*
$x^2 + 8x + ? = -5 + ?$ (16, 16)	*Add ? to both sides.*
$(x + ?)^2 = ?$ (4, 11)	*Perfect square form*
$x = ?$ $-4 \pm \sqrt{11}$	*Solutions*

In the third step, how do you know what quantity to add to both sides? Check your conjecture by solving $x^2 - 6x + 5 = 0$ in two ways: once by factoring and once by completing the square.

Add the square of $\frac{1}{2}$ the coefficient of x.

Study Tip

In Example 1, notice in the third step that the quantity that is added to both sides is the square of one half the coefficient of x.

Example 1 *Completing the Square: Leading Coefficient is 1*

Solve $x^2 - 6x + 7 = 0$ by completing the square.

Solution

$x^2 - 6x + 7 = 0$	*Rewrite original equation.*
$x^2 - 6x = -7$	*Subtract 7 from both sides.*
$x^2 - 6x + (-3)^2 = -7 + 9$	*Add $(-3)^2$, or 9, to both sides.*
$(x - 3)^2 = 2$	*Binomial squared*
$x - 3 = \pm\sqrt{2}$	*Find square roots.*
$x = 3 \pm \sqrt{2}$	*Solve for x.*

The equation has two solutions: $3 + \sqrt{2}$ and $3 - \sqrt{2}$. ■

Problem of the Day

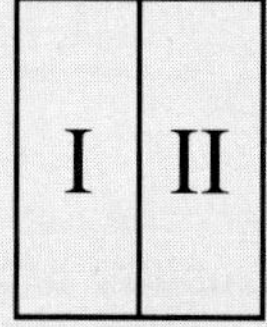

Volumes I and II of an encyclopedia set stand side by side on a book shelf. A book worm begins at page 1 of volume I, and burrows straight through to the last page of volume II. Each cover is $\frac{1}{8}$ inch thick, and each book without the covers is 2 inches thick. How far does the bookworm burrow? $\frac{1}{4}$ in.

ORGANIZER

Warm-Up Exercises

1. Perform the indicated operations and simplify.
   - **a.** $3 + 4^2$ 19
   - **b.** $-13 + (-3)^2$ -4
   - **c.** $-2 + (\frac{3}{2})^2$ $\frac{1}{4}$
   - **d.** $1 + -(\frac{1}{8})^2$ $\frac{63}{64}$
2. For each quadratic equation, find the constant that results from following the flowchart found below.
   - **a.** $3x^2 + 4x - 3 = 0$ 4
   - **b.** $5x^2 - 18x + 12 = 0$ 81
   - **c.** $19 - 11x + x^2$ $\frac{121}{4}$

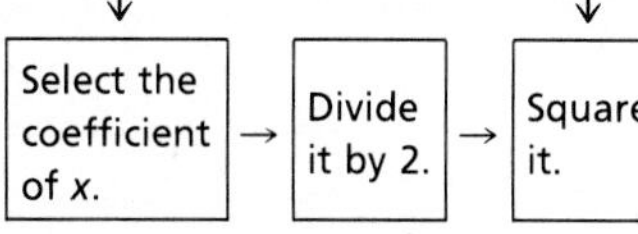

Lesson Resources

Teaching Tools
- Transparencies: 11, 12, 13
- Problem of the Day: 10.7
- Warm-up Exercises: 10.7
- Answer Masters: 10.7

Extra Practice: 10.7

LESSON Notes

Example 1

TECHNOLOGY Students may wish to check the two solutions using a computer or graphing calculator. On the TI-82, for instance, the following keystrokes allow students to enter the equation, $y = x^2 - 6x + 7$, and evaluate it at one of the solutions, $3 + \sqrt{2}$:

TI-82

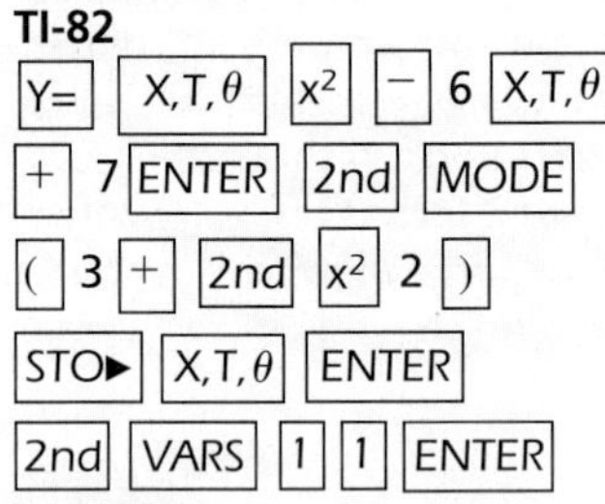

The result is 0. This shows that the equation evaluated at the solution equals zero, as expected. Of course, students may choose to check the solutions by graphing the equation and identifying where the graph crosses the x-axis.

Example 2

Emphasize that the quadratic equation must be divided by the leading coefficient before completing the square.

Example 3

Students may recognize this as an exponential growth problem with exponent 2.

If the leading coefficient of the quadratic is not 1, you should divide both sides of the equation by this coefficient *before* completing the square.

Example 2 Completing the Square: Leading Coefficient Is Not 1

Solve $2x^2 - x - 2 = 0$ by completing the square.

Solution

$2x^2 - x - 2 = 0$	*Rewrite original equation.*
$2x^2 - x = 2$	*Subtract −2 from both sides.*
$x^2 - \frac{1}{2}x = 1$	*Divide both sides by 2.*
$x^2 - \frac{1}{2}x + \left(-\frac{1}{4}\right)^2 = 1 + \frac{1}{16}$ (half of $-\frac{1}{2}$)^2	*Add $\left(-\frac{1}{4}\right)^2$, or $\frac{1}{16}$, to both sides.*
$\left(x - \frac{1}{4}\right)^2 = \frac{17}{16}$	*Binomial squared*
$x - \frac{1}{4} = \pm\frac{\sqrt{17}}{4}$	*Find square roots.*
$x = \frac{1}{4} \pm \frac{\sqrt{17}}{4}$	*Solve for x.*

The equation has two solutions, $x = \frac{1}{4} + \frac{\sqrt{17}}{4}$ and $x = \frac{1}{4} - \frac{\sqrt{17}}{4}$. As a check, you can use the quadratic formula on the original equation. ■

You have studied five different methods for solving quadratic equations.

Summary of Methods for Solving $ax^2 + bx + c = 0$

Method	Lesson	Comments
Finding Square Roots	9.2	Efficient way to solve $ax^2 + c = 0$.
Graphing	9.3	Can be used for *any* quadratic equation, but gives only approximate solutions.
Using Quadratic Formula	9.4	Can be used for *any* quadratic equation. Always gives exact solutions.
Factoring	10.5	Efficient way to solve equation *if* quadratic can be factored easily.
Completing the Square	10.6	Can be used for *any* quadratic equation, but is best suited for quadratics with $a = 1$ and b an even number.

A deposit of $6000 was put into a savings account that paid annual interest of r%, compounded yearly. After two years, the balance in the account was $6933.75.

Real Life Banking

Example 3 *Finding the Annual Interest Rate*

Find the annual interest rate for the above deposit.

Solution

Known Formula $A = P(1 + r)^t$

Labels

Number of years $= t = 2$		(years)
Principal (original deposit) $= P = 6000$		(dollars)
Balance $= A = 6933.75$		(dollars)
Annual interest rate $= r$		(decimal form)

Equation

$6933.75 = 6000(1 + r)^2$ *Quadratic model*

$1.156 = (1 + r)^2$ *Divide both sides by 6000.*

$1.075 = 1 + r$ *Take positive square roots.*

$0.075 = r$ *Subtract 1 from both sides.*

The annual interest rate is 7.5%. (Note that the only operation that was necessary to write this equation in completed square form was to divide both sides by 6000.) ■

Communicating about ALGEBRA

SHARING IDEAS about the Lesson

Choose a Method Solve each equation by the most efficient method. Explain the reasons for your choices. See below.

A. $x^2 - 6x + 8 = 0$ **B.** $x^2 - 6 = 0$ **C.** $x^2 - 6x = 0$

D. $x^2 - 6x - 8 = 0$ **E.** $2(x - 3)^2 = 12$ **F.** $2x^2 - 6x - 8 = 0$

A., C., F. Factoring, easy to factor. **A.**, 4, 2 **C.**, 0, 6 **F.**, 4 −1
B., E. Find square root, no x-term. **B.**, $\pm\sqrt{6}$ **E.**, $3 \pm \sqrt{6}$
D. Completing the square, $a = 1$ and b is even. $3 \pm \sqrt{17}$

Extra Examples

Here are additional examples similar to **Examples 1–3.**

1. Solve by completing the square.
   a. $4x^2 - 4x - 15 = 0$ $-\frac{3}{2}, \frac{5}{2}$
   b. $x^2 - 4x + 2 = 0$ $2 \pm \sqrt{2}$
   c. $x^2 + 5x - 8 = 0$ $-\frac{5}{2} \pm \frac{\sqrt{57}}{2}$
2. At what annual compound interest rate has money been invested if a principal of $2500 amounts to $3000 in 2 years? about 9.5%

Check Understanding

1. What is a perfect square trinomial? Give an example. Any trinomial of the form $a^2x^2 + 2ab + b^2$, or, $a^2x^2 - 2ab + b^2$
2. Which methods of solving a quadratic equation can always be used? Graphing, completing the square, using the quadratic formula
3. How is the quadratic formula related to completing the square? The quadratic formula is the result of applying the completing-the-square technique on the general, standard-form quadratic equation.

Communicating about ALGEBRA

Students may find it interesting to learn that the quadratic formula is the result of applying the completing-the-square technique on the general, standard-form quadratic equation. (See the Enrichment Activity for this lesson for a sample derivation.)

EXERCISE Notes

ASSIGNMENT GUIDE

Basic/Average: Ex. 1–4, 5–43 odd, 44, 51–63 multiples of 3, 67–70

Above Average: Ex. 1–4, 17–27 odd, 35–43 odd, 44–48, 49–65 odd, 71, 73, 75–76

Advanced: Ex.1–4, 17–27 odd, 35–43 odd, 44–48, 54, 60, 66, 71–77

Selected Answers
Exercises 1–4, 5–73 odd

More Difficult Exercises
Exercises 45–48, 75–77

Guided Practice

Ex. 1–4 Use these exercises as a class summary of the completing-the-square method. Refer students to the Summary Methods list on page 550.

Independent Practice

Answers

17. $\frac{1}{3} + \frac{\sqrt{28}}{3}, \frac{1}{3} - \frac{\sqrt{28}}{3}$

18. $-\frac{2}{5} + \frac{\sqrt{29}}{5}, -\frac{2}{5} - \frac{\sqrt{29}}{5}$

19. $-\frac{1}{2} + \frac{\sqrt{5}}{2}, -\frac{1}{2} - \frac{\sqrt{5}}{2}$

20. $\frac{1}{2} + \frac{\sqrt{5}}{2}, \frac{1}{2} - \frac{\sqrt{5}}{2}$

21. $-\frac{1}{2} + \frac{\sqrt{10}}{2}, -\frac{1}{2} - \frac{\sqrt{10}}{2}$

22. $4 + \frac{\sqrt{159}}{3}, 4 - \frac{\sqrt{159}}{3}$

26. $-\frac{3}{4} + \frac{\sqrt{41}}{4}, -\frac{3}{4} - \frac{\sqrt{41}}{4}$

27. $-1 + \sqrt{3}, -1 - \sqrt{3}$

29. $\frac{3}{2} + \frac{\sqrt{13}}{2}, \frac{3}{2} - \frac{\sqrt{13}}{2}$

34. $-\frac{5}{2} + \frac{\sqrt{17}}{2}, -\frac{5}{2} - \frac{\sqrt{17}}{2}$

36. $-1 + \sqrt{27}, -1 - \sqrt{27}$

37. $-\frac{5}{9} + \frac{\sqrt{244}}{18}, -\frac{5}{9} - \frac{\sqrt{244}}{18}$

EXERCISES

Guided Practice

CRITICAL THINKING about the Lesson

1. Which is a perfect square trinomial?
   a. $x^2 - 8x + 8$ **b.** $x^2 - 8x + 16$
   c. $x^2 - 8x + 64$

2. Solve $x^2 - 4x = 8$ by completing the square. Solve the same equation by the quadratic formula. Explain the difference in the results. $2 + \sqrt{12}, 2 - \sqrt{12}$; no difference

3. What term must be added to $x^2 + 6x$ to create a perfect square trinomial? 9

4. Name the five methods for solving a quadratic equation. See chart on page 550.

Independent Practice

In Exercises 5–10, find the term that must be added to the expression to create a perfect square trinomial.

5. $x^2 - 18x$ 81
6. $x^2 + 6x$ 9
7. $x^2 + 12x$ 36
8. $x^2 - 10x$ 25
9. $x^2 - 7x$ $\frac{49}{4}$
10. $x^2 - 5x$ $\frac{25}{4}$

In Exercises 11–28, solve the equation by completing the square. 16. $-\frac{1}{2}, -\frac{17}{2}$ 25. $-\frac{1}{3}, -1$

11. $x^2 + 10x - 11 = 0$ 1, −11
12. $x^2 + 14x - 15 = 0$ 1, −15
13. $y^2 - 24y + 63 = 0$ 21, 3
14. $y^2 - 8y + 12 = 0$ 6, 2
15. $t^2 + 3t - \frac{7}{4} = 0$ $\frac{1}{2}, -\frac{7}{2}$
16. $y^2 + 9y + \frac{17}{4} = 0$
17. $x^2 - \frac{2}{3}x - 3 = 0$
18. $x^2 + \frac{4}{5}x - 1 = 0$
19. $x^2 + x - 1 = 0$
20. $1 + x - x^2 = 0$
21. $4y^2 + 4y - 9 = 0$
22. $3x^2 - 24x - 5 = 0$
23. $2x^2 - 6x - 15 = 5$ 5, −2
24. $5x^2 - 20x - 20 = 5$ 5, −1
25. $3x^2 + 4x + 4 = 3$
26. $4x^2 + 6x - 6 = 2$
27. $x^2 + 2x = 2$
28. $x^2 - 2x = 2$ $1 + \sqrt{3}, 1 - \sqrt{3}$

17.–22., 26., 27. See margin.

In Exercises 29–43, use the most convenient method to solve the equation. Explain why you made your choice. 29., 34., 36., 37. See margin.

29. $x^2 - 3x - 1 = 0$
30. $4x^2 - 12 = 0$ $\sqrt{3}, -\sqrt{3}$
31. $y^2 + 6y - 24 = 0$ $-3 + \frac{\sqrt{132}}{2}, -3 - \frac{\sqrt{132}}{2}$
32. $4x^2 - 25 = 0$ $\frac{5}{2}, -\frac{5}{2}$
33. $x^2 + 7x + 10 = 0$ −5, −2
34. $u^2 + 5u + 2 = 0$
35. $3x^2 - 5x = 0$ 0, $\frac{5}{3}$
36. $y^2 + 2y - 26 = 0$
37. $9z^2 + 10z - 4 = 0$
38. $4x^2 + 4x + 1 = 0$ $-\frac{1}{2}$
39. $7x^2 - 14x = 0$ 0, 2
40. $4x^2 - 13x + 3 = 0$ 40. 3, $\frac{1}{4}$
41. $8x^2 - 10x + 3 = 0$ $\frac{3}{4}, \frac{1}{2}$
42. $7x^2 - 14 = 0$ $\sqrt{2}, -\sqrt{2}$
43. $y^2 + 20y + 10 = 0$ $-10 + \frac{\sqrt{360}}{2}, -10 - \frac{\sqrt{360}}{2}$

44. *Money in the Bank* At your seventh grade graduation, you and your twin sister each received \$200. You each deposited the money in savings accounts that compound interest annually. Two years later your sister's deposit has grown by \$28.98. Your account is in a different bank that pays an interest rate that is 1% more than your sister receives. What is your balance after two years? \$233.28

45. *Splash!* You and Jared are playing in the surf at the beach. Jared is 5 feet 4 inches tall and is standing at the point where the wave crests, as shown below. Did the wave go over his head? Explain. No. See margin.

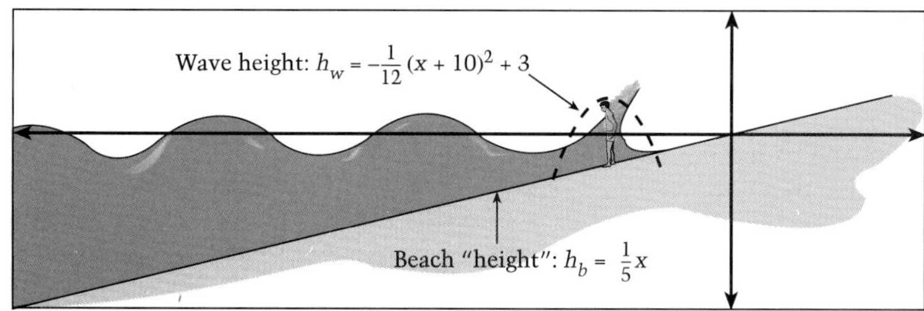

46. *Whale Watching* At the beach you see an 80-foot whale jump above water. The path followed by the whale is given by the model

$$h = -\frac{1}{400}x^2 + \frac{4}{5}x - 52$$

where h is the height (in feet) above the sea and x is the horizontal distance (in feet) traveled by the whale. Sketch a graph of this equation. For how many horizontal feet did the whale travel over the water before reaching its maximum height? 69.28 ft See margin.

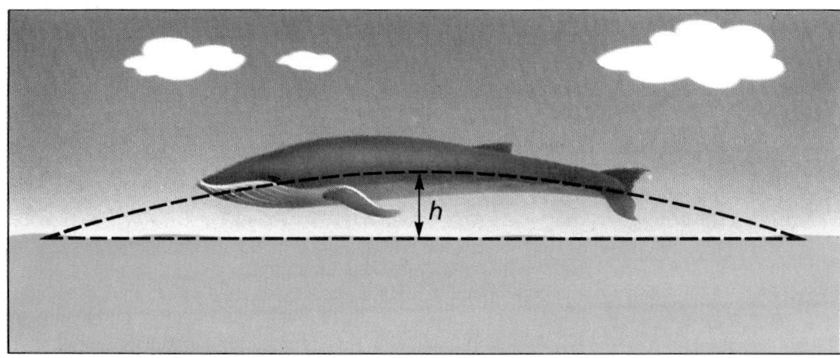

To attain a maximum thrust, a whale beats its tail down when a wave moves water upward (the front face of the wave) and beats its tail up when the wave moves water downward (the back face of the wave).

47. *Waterfall* The Vettisfoss waterfall falls over a vertical cliff. The path followed by the water as it flows over the Vettisfoss Falls can be modeled by

$$h = -\frac{1}{30}(x - 10)^2 + 900$$

where h is the height (in feet) above the lower river level, and x measures the horizontal distance (in feet) from the base of the cliff. How far from the base of the cliff does the water hit the river? $\approx$174 ft

48. How long does it take the water to hit the lower river? (Use a vertical motion model. See page 474.) 7.5 seconds

Vettisfoss Falls, in Norway, is one of the most famous waterfalls in the world. The distance between the lower river and the top of the falls is 900 feet.

Ex. 20–26 The leading coefficient is not 1. Refer students to Example 2.

Ex. 29–43 Encourage students to check solutions both graphically and algebraically.

Ex. 44 Refer students having difficulties to Example 3 on page 551.

Ex. 45 The wave crests where $h_1 = -\frac{1}{12}(x + 10)^2 + 3$ has its greatest value—the vertex; because this point must be $(-10, 3)$, the crest is 3 ft above sea level. Since $x = -10$, $h_2 = -2$; so, 2 feet of Jared is below sea level. Then 3 feet 4 inches of Jared is above sea level and the 3-foot crest does not go over his head.

Answer

45. No. The crest is 3 ft above sea level; Jared had 2 ft of his body below sea level and 3 ft 4 in. of his body above sea level.

46.

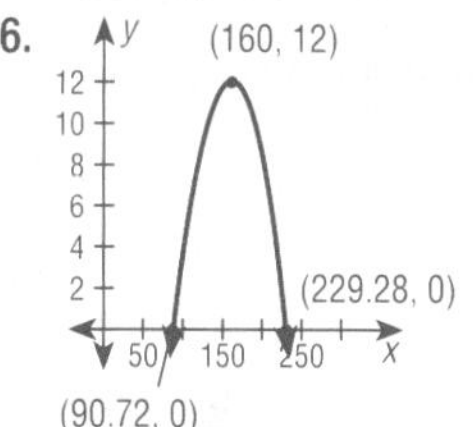

Ex. 47–48 These exercises should be assigned as a pair.

Ex. 49–70 These exercises review and connect techniques for solving quadratic equations from Chapter 9 with those in Chapter 10.

Ex. 55–60 Review with students the techniques for graphing quadratic equations.

Answer

After setting y equal to 0, factor the polynomial in x, set each factor to 0, and solve for x to obtain the x-intercepts.

Ex. 67–70 These exercises should be assigned as a group.

EXTEND Ex. 68 Students should understand that there are three tests they can use to determine whether an expression is factorable over the real numbers. For example, is the trinomial $3x^2 - 6x + 4$ factorable? (1) Look at graph *a*. What do you notice? There are no *x*-intercepts. The expression is not factorable. (2) What is the sign of the discriminant? Negative, so the expression is not factorable. (3) What happens when you try to complete the square? You obtain $-\frac{1}{3}$, a negative number.

Enrichment Activity

This activity requires students to go beyond lesson goals.

Derive the quadratic formula by completing the square of the general quadratic equation, $ax^2 + bx + c = 0$. Explain each step of the process.

One derivation follows.

$$ax^2 + bx + c = 0$$

Divide both sides by *a*.

$$x^2 + \frac{b}{a}x + \frac{c}{a} = 0$$

Subtract $\frac{c}{a}$ from both sides.

$$x^2 + \frac{b}{a}x = -\frac{c}{a}$$

Add $(\frac{b}{2a})^2$ to both sides.

$$x^2 + \frac{b}{a}x + (\frac{b}{2a})^2 = -\frac{c}{a} + (\frac{b}{2a})^2$$

Write as a binomial squared.

$$(x + (\frac{b}{2a}))^2 = -\frac{c}{a} + \frac{b^2}{4a^2}$$

Simplify.

$$(x + (\frac{b}{2a}))^2 = \frac{b^2 - 4ac}{4a^2}$$

Find the square root of both sides.

$$x + \frac{b}{2a} = \pm\sqrt{\frac{b^2 - 4ac}{4a^2}}$$

Solve for *x*.

$$x = -\frac{b}{2a} \pm \frac{\sqrt{b^2 - 4ac}}{2a}$$

$$x = \frac{-b \pm \sqrt{b^2 - 4ac}}{2a}$$

This is the quadratic formula.

Integrated Review

In Exercises 49–54, solve the equation.

49. $x^2 = 16$ ± 4 **50.** $x^2 + 3 = 7$ ± 2 **51.** $x^2 + 4 = 29$ ± 5

52. $\frac{1}{7}x^2 + 8 = 15$ ± 7 **53.** $2x^2 - 7 = 11$ ± 3 **54.** $3x^2 - 8 = 100$ ± 6

In Exercises 55–60, sketch the graph of the equation and find any *x*-intercepts. Explain how factoring could be used to find the *x*-intercepts. See Additional Answers for graphs.

55. $y = x^2 - 6x + 8$ (2, 0), (4, 0) **56.** $y = -2x^2 - 8x - 6$ (−3, 0), (−1, 0) **57.** $y = -x^2 - 2x - 1$ (−1, 0)

58. $y = x^2 + 8x + 16$ (−4, 0) **59.** $y = 2x^2 - x - 10$ $(\frac{5}{2}, 0)$, (−2, 0) **60.** $y = x^2 + 5x - 6$ (−6, 0), (1, 0)

In Exercises 61–66, factor the expression. See below.

61. $3x^2 - 15x - 18$ **62.** $12x^2 + 46x - 8$ **63.** $140y^2 + 340y + 120$

64. $-18y^2 + 156y + 54$ **65.** $12a^2 + 36a + 24$ **66.** $10x^2 + 15x + 5$

In Exercises 67–70, match the equation with its graph.

67. $y = -3x^2 + 6x - 1$ d **68.** $y = 3x^2 - 6x + 4$ a

69. $y = 3x^2 + 6x + 1$ b **70.** $y = -3x^2 - 6x - 4$ c

a. 

b. 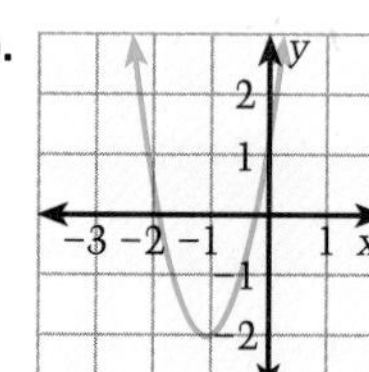

c. 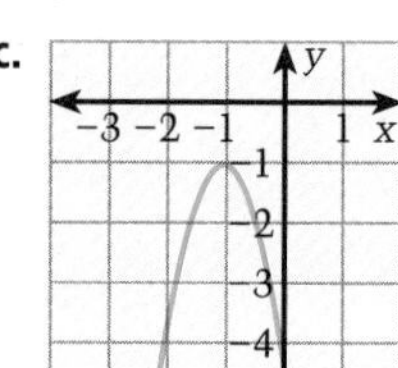

d. 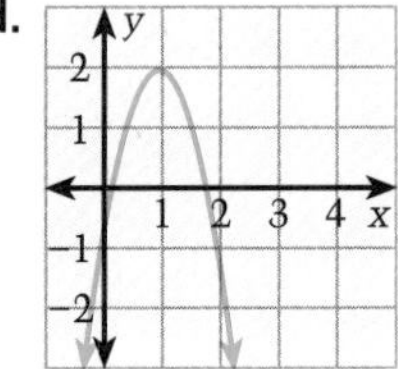

In Exercises 71–74, sketch the graph of the inequality. See Additional Answers.

71. $y \geq \frac{1}{2}x^2 - 4x + 6$ **72.** $y < \frac{1}{3}x^2 + 2x + 1$

73. $y > -x^2 + 8x - 16$ **74.** $y \leq -x^2 - 4x - 5$

Exploration and Extension

75. *Revenue* The revenue, R, for selling x units of a product is given by

$$R = x(50 - \tfrac{1}{2}x).$$

How many units must be sold to produce a revenue of \$1218? 58 or 42

76. *Revenue* The revenue, R, for selling x units of a product is given by

$$R = x(100 - \tfrac{1}{10}x).$$

How many units must be sold to produce a revenue of \$990? 990 or 10

77. *College Entrance Exam Sample* If x and y are positive integers, $x^2 + y^2 = 25$, and $x^2 - y^2 = 7$, then $y =$ [?].

a. 3 **b.** 4 **c.** 5 **d.** 9 **e.** 16

61. $3(x - 6)(x + 1)$ **62.** $2(6x - 1)(x + 4)$ **63.** $20(7y + 3)(y + 2)$ **64.** $-6(3y + 1)(y - 9)$

65. $12(a + 1)(a + 2)$ **66.** $5(2x + 1)(x + 1)$

10 Chapter Summary

What did you learn?

Skills

1. Classify a polynomial by degree and by number of terms. **(10.1)**
2. Add and subtract polynomials. **(10.1)**
3. Multiply polynomials
   - using the Distributive Property. **(10.2)**
   - using the FOIL pattern. **(10.2)**
   - using the sum and difference pattern. **(10.3)**
   - using the square of a binomial pattern. **(10.3)**
4. Factor a polynomial
   - using the difference of two squares pattern. **(10.4)**
   - using the perfect square pattern. **(10.4)**
   - using the guess-and-check strategy. **(10.5)**
5. Solve a quadratic equation
   - by factoring. **(10.6)**
   - by completing the square. **(10.7)**

Strategies

6. Use quadratic models in real-life problems. **(10.1–10.7)**

Exploring Data

7. Recognize sum, difference, and product models in real life. **(10.1–10.7)**

Why did you learn it?

Polynomials were among the first mathematical models to be used. By now, you can see why—they are relatively simple and yet they can model a great variety of real-life situations. In this chapter, you learned that some models are related to each other by addition, subtraction, or multiplication. For instance, a model for the total revenue, R, can be obtained by multiplying the models for the price per unit, p, and the number, x, of units sold. In other words, $R = xp$.

How does it fit into the bigger picture of algebra?

Polynomials are among the most commonly used models for real-life situations. You were already familiar with three types of polynomial models: $y = a$ (constant models), $y = ax + b$ (linear models), and $y = ax^2 + bx + c$ (quadratic models). In this chapter, you learned how to add and subtract polynomials. You also learned how to multiply polynomials and how to "undo" multiplication by a process called *factoring.* This chapter has many connections with the mathematics you studied in Chapter 9. For instance, in Lessons 10.6 and 10.7 you learned two additional methods—factoring and completing the square—for solving a quadratic equation.

ORGANIZER

TIME SCHEDULE: All levels, two days

The Chapter Summary helps students organize the main ideas of the chapter. In this chapter, concepts and computational skills associated with polynomials were developed. An important skill presented in this chapter was solving quadratic equations by several methods.

Work with students to review the skills, stategies, and concepts of the chapter. It is suggested that the first day of review be a combination of you and your students working problems form the Chapter Review exercises. The first day's homework assignment can be used as the basis for the second day of review.

COOPERATIVE LEARNING
Encourage students to work cooperatively. Emphasize the importance of teaching a classmate how to perform a skill or how to recall a strategy. When students work together, everyone wins. The students receiving help get additional instruction, and the students giving help gain a deeper understanding of the skills and concepts involved.

Chapter SUMMARY

Review the chapter by asking students, "What did you learn? Why did you learn it? How does it fit with the other algebra skills and concepts you have learned?"

Chapter REVIEW

ASSIGNMENT GUIDE

Basic/Average: Ex. 3–54 multiples of 3, 55–59 odd, 65–66

Above Average: Ex. 3–54 multiples of 3, 55–59 odd, 65–66

Advanced: Ex. 3–54 multiples of 3, 55–59 odd, 65–66

More Difficult Exercises

Exercises 59–66

Exercises 1–66 Since the Chapter Review exercises parallel the list of Skills, Strategies, and Exploring Data statements, tell students to use the Chapter Summary on page 555 to identify the lesson in which a skill, strategy, or concept was presented if they are uncertain about how to do a given exercise in the Chapter Review. This makes the students categorize each task in the exercise set and locate relevant information in the chapter about the task, both beneficial activities.

Answer

11. $3x^2 + x + 5$
12. $-x^3 - x^2 + 2x + 6$
15. $x^3 + 3x^2 - 4x + 10$

Chapter REVIEW

In Exercises 1–8, classify the polynomial by degree and by terms. (10.1)

1. $x^2 - 1$ Quadratic, binomial
2. $3x^2 + 2x - 2$ Quadratic, trinomial
3. 121 Constant, monomial
4. $4x$ Linear, monomial
5. $x^4 - x^2 + 2x + 3$ Quartic, polynomial
6. $49x - 2$ Linear, binomial
7. $8x^3 - 27$ Cubic, binomial
8. $2x^3 + 4x^2 - 5x + 6$ Cubic, polynomial

11. $3x^2 + x + 5$ 12. $-x^3 - x^2 + 2x + 6$ 13. $-3x^2 + 8x - 7$ 14. $4x^2 + x - 3$ 16. $x^3 + 2x^2 + 2x - 2$

In Exercises 9–20, perform the indicated operation. Use a horizontal format. (10.1, 10.2) See margin.

9. $(x + 2 - x^2) + (3x^2 + 4x + 5)$ $2x^2 + 5x + 7$
10. $(4x^3 + x^2 - 1) + (2 - x - x^2)$ $4x^3 - x + 1$
11. $(15 + 3x - x^2) + (4x^2 - 2x - 10)$
12. $(3x + 2 - x^2) + (4 - x - x^3)$
13. $(x^2 + 3x - 1) - (4x^2 - 5x + 6)$
14. $(x^2 + 9x + 2) - (5 + 8x - 3x^2)$
15. $(3x^2 - 2x + 4) - (-x^3 + 2x - 6)$ See margin.
16. $(x^3 + 5x^2 - 4x) - (3x^2 - 6x + 2)$
17. $(x - 5)(x - 10)$ $x^2 - 15x + 50$
18. $(2x + 2)(x + 4)$ $2x^2 + 10x + 8$
19. $(6 + x)(x^2 - 2x + 3)$ $x^3 + 4x^2 - 9x + 18$
20. $(7 - x)(3x^2 + 2x - 6)$ $-3x^3 + 19x^2 + 20x - 42$

26. $3x^2 + 2x - 21$ 27. $-5x^2 - 8x + 5$

In Exercises 21–32, perform the indicated operation. Use a vertical format. (10.1, 10.2) See margin.

21. $(6x^2 + 2x - 1) + (x^2 - 2)$ $7x^2 + 2x - 3$
22. $(x - 2) + (4x^2 - 7x + 5)$ $4x^2 - 6x + 3$
23. $(-x^2 - x + 2) + (x^2 + 2x - 4)$ $x - 2$
24. $(x^2 + 3x + 5) + (3x^2 - 4x + 6)$ $4x^2 - x + 11$
25. $(x^2 - 3) - (4x^2 - 3x + 2)$ $-3x^2 + 3x - 5$
26. $(x^2 + 3x - 7) - (-2x^2 + x + 14)$
27. $(x^2 - 4x + 2) - (6x^2 + 4x - 3)$
28. $(10x^2 + 3x - 4) - (5x^2 + 2x - 6)$ $5x^2 + x + 2$
29. $(x - 2)(3x^2 + 4x - 1)$ $3x^3 - 2x^2 - 9x + 2$
30. $(10 - x)(x^2 + x + 1)$ $-x^3 + 9x^2 + 9x + 10$
31. $(2x + 2)(4x^2 - 6x + 2)$ $8x^3 - 4x^2 - 8x + 4$
32. $(4 + 3x)(1 - 4x + 6x^2)$ $18x^3 + 12x^2 - 13x + 4$

In Exercises 33–36, multiply. (10.3)

33. $(x + 15)(x - 15)$ $x^2 - 225$
34. $(3x + 2)(3x - 2)$ $9x^2 - 4$
35. $(x + 2)^2$ $x^2 + 4x + 4$
36. $(5x - 6)^2$ $25x^2 - 60x + 36$

In Exercises 37–46, use the discriminant to determine whether the polynomial can be factored. If possible, factor the polynomial. (10.4, 10.5)

37. $x^2 - 2x - 15$ $(x - 5)(x + 3)$
38. $x^2 + 3x - 70$ $(x + 10)(x - 7)$
39. $x^2 - 64$ $(x - 8)(x + 8)$
40. $4x^2 + 25$ No factors
41. $x^2 - 8x + 8$ No factors
42. $9x^2 + 12x + 4$ $(3x + 2)^2$
43. $x^2 + 10x + 25$ $(x + 5)^2$
44. $x^2 - 8x + 16$ $(x - 4)^2$
45. $4x^2 - 32x + 60$ $4(x - 5)(x - 3)$
46. $3x^2 + 21x + 30$ $3(x + 2)(x + 5)$

In Exercises 47–54, solve the equation. (10.6) 50. 5, −2

47. $x^2 - 21x + 108 = 0$ 9, 12

48. $x^2 - 8x - 240 = 0$ 20, −12

49. $-x^2 + 30x - 200 = 0$ 10, 20

50. $-15x^2 + 45x + 150 = 0$

51. $36x^2 - 49 = 0$ $\pm\frac{7}{6}$

52. $x^2 + 26x + 169 = 0$ −13

53. $x^2 - 14x + 36 = 0$ $7 + \frac{\sqrt{52}}{2}, 7 - \frac{\sqrt{52}}{2}$

54. $x^2 + 10x - 3 = 0$ $-5 + \frac{\sqrt{112}}{2}, -5 - \frac{\sqrt{112}}{2}$

Right-Triangle Triple **In Exercises 55 and 56, find the right-triangle triple. (10.4)**

55. $11^2 + b^2 = c^2$ 11, 60, 61

11 c b

56. $17^2 + b^2 = c^2$ 17, 144, 145

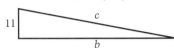

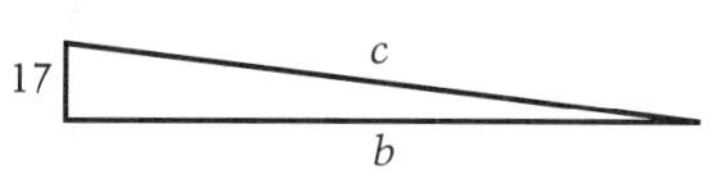

57. *Tossing a Ball* A ball is tossed into the air from a height of 10 feet with an initial velocity of 12 feet per second. Find the time, t (in seconds), for the object to reach the ground by solving the equation

$-16t^2 + 12t + 10 = 0.$ 1.25 seconds

58. *Summer Business* Your friend's weekly revenue, R (in dollars), from her tie-dye T-shirt business can be modeled by

$R = -2t^2 + 37t + 60$

where t represents the week of sales, with $t = 0$ for the first week. In the first week, 3 T-shirts were sold. After that, the sales increased by 2 T-shirts per week. Did the price of T-shirts remain constant during the 8-week summer season? Explain. No. See margin.

Huffing and Puffing **In Exercises 59 and 60, use the following information.**

Porcupine fish, members of the puffer fish family, range between 10 and 20 inches in length. When in danger, the body of the fish puffs up by taking in water or air. (The tail, which is about 3 inches long, does not puff up.)

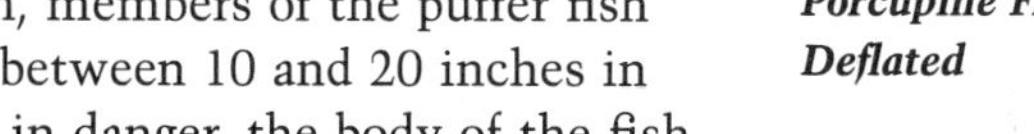

$\frac{\pi}{6}x^3 - \frac{3\pi}{2}x^2 + \frac{9\pi}{2}x - \frac{9\pi}{2}$

59. The volume of a "puffed-up" porcupine fish can be modeled by

$V = \left(\frac{1}{6}\right)\pi(x - 3)^3$

where x is the total length of the fish in inches. Write the right side of this equation in standard polynomial form.

60. Approximate the volume of an 11-inch porcupine fish. 268 in.^{3}

Porcupine Fish: Deflated

Inflated

Answer

58. No, the price of a t-shirt decreased by $1 each week.

Answer

63. No, it is related by a quadratic model since the x term is squared. The distance needed to stop increases quadratically with an increase in speed.

Stopping Distance **In Exercises 61–64, use the following information.**

The stopping distance of an automobile traveling at x miles per hour is the sum of the distance the car travels during the driver's reaction time *and* the distance the car travels after the brakes are applied. The distance, R, in feet, traveled during the driver's reaction time is approximately $R = 1.1x$. The distance, B, in feet, traveled after the brakes are applied is $B = 0.14x^2 - 4.43x + 58.4$ approximately.
(These models are based on normal road conditions.)

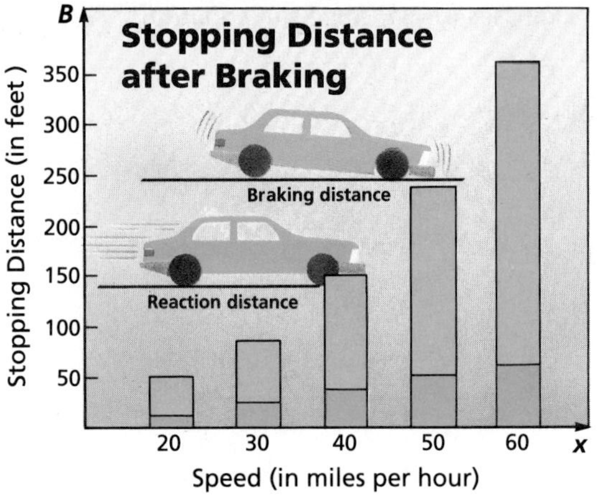

61. Find a model for the (total) stopping distance, S (in feet), of an automobile that is traveling x miles per hour. $S = 0.14x^2 - 3.33x + 58.4$

62. Estimate the total stopping distance for an automobile that is traveling 15 miles per hour; 30 miles per hour; 55 miles per hour. 39.95 ft, 84.5 ft, 298.75 ft

63. Are stopping distances and speeds related by a linear model? Explain. What does this tell you about the amount of distance you need to allow for stopping when traveling at various speeds? No. See margin.

64. The recommended safe *following distance* on highways under normal road conditions is given by $F = 1.8x$. How many feet should be between cars traveling at 55 miles per hour? 99 ft

Registered Cars **In Exercises 65 and 66, use the following information.**

From 1940 to 1990, the number, C, of registered cars in the United States can be modeled by $C = 20{,}500t^2 + 1{,}460{,}100t + 25{,}942{,}500$, where $t = 0$ represents 1940. ***(Source: Motor Vehicle Facts and Figures '90, Federal Highway Administration)***

65. During which year were 129,874,900 cars registered? 1984

66. According to this model, how many cars will be registered in 1995? 168,260,500

Chapter TEST

In Exercises 1–12, perform the indicated operations. **(10.1, 10.2, 10.3)** 4. $x^3 - 2x^2 + 7x - 4$

1. $(x^2 + 3x - 1) + (4x^2 + 2)$ $5x^2 + 3x + 1$

2. $(x^4 + 3x^2 + 2) + (2x^4 - 3x^2 + 6)$ $3x^4 + 8$

3. $(5x^2 - 2x + 1) - (7x + 10)$ $5x^2 - 9x - 9$

4. $(5x^3 + 2x - 4) - (4x^3 + 2x^2 - 5x)$

5. $(9x + 2)(9x - 2)$ $81x^2 - 4$

6. $(5x - 4)(5x + 4)$ $25x^2 - 16$

7. $(x - 14)^2$ $x^2 - 28x + 196$

8. $(3x + 5)^2$ $9x^2 + 30x + 25$

9. $(x + 2)(3x + 5)$ $3x^2 + 11x + 10$

10. $(2x - 1)(13x + 5)$ $26x^2 - 3x - 5$

11. $(x - 6)(4x^2 + 3x - 5)$ $4x^3 - 21x^2 - 23x + 30$

12. $(4x^3 - 6x + 7)(x + 1)$ $4x^4 + 4x^3 - 6x^2 + x + 7$

In Exercises 13–20, factor the expression. **(10.4, 10.5)** See margin.

13. $x^2 - 144$

14. $36x^2 - 25$

15. $x^2 - 12x + 36$

16. $x^2 + 10x + 25$

17. $3x^2 + 2x - 1$

18. $5x^2 - 3x - 2$

19. $x^3 + 2x^2 + x$

20. $2x^2 - 28x + 96$

In Exercises 21–26, solve the equation. **(10.6)**

21. $x^2 + 4x + 4 = 0$ -2

22. $x^2 - 7x + 6 = 0$ $1, 6$

23. $x^2 - 5x - 150 = 0$ $15, -10$

24. $x^2 + 6x - 91 = 0$ $7, -13$

25. $12x^2 + 15x + 3 = 0$ $-\frac{1}{4}, -1$

26. $4x^2 - 10x - 36 = 0$ $\frac{9}{2}, -2$

In Exercises 27–30, solve the equation by completing the square. **(10.7)**

27. $x^2 - 4x + 1 = 0$ $2 + \sqrt{3}, 2 - \sqrt{3}$

28. $x^2 + 6x - 9 = 0$ $-3 + \sqrt{18}, -3 - \sqrt{18}$

29. $x^2 + 20x + 3 = 0$ $-10 + \sqrt{97}, -10 - \sqrt{97}$

30. $x^2 - 2x - 5 = 0$ $1 + \sqrt{6}, 1 - \sqrt{6}$

31. Find a right-triangle triple such that $5^2 + b^2 = c^2$. **(10.4)** 5, 12, 13

32. The length of a bedroom is 3 feet less than twice its width. The area of the bedroom is 135 square feet. What are the dimensions of the room? **(10.5)** 9 ft by 15 ft

33. A deposit of $5000 was put into a savings account paying an annual interest of $r\%$, compounded yearly. After 2 years, the balance in the account was $5,644.53. What was the rate of interest? **(10.4)** $\approx 6.25\%$

34. The bed of a pond can be modeled by $25y = 2x^2 - 20x + 1$, where x and y are measured in meters and the x-axis matches the water level of the pond. What is the width of the pond? **(10.5)** ≈ 9.9 m

Answers

13. $(x - 12)(x + 12)$
14. $(6x - 5)(6x + 5)$
15. $(x - 6)^2$
16. $(x + 5)^2$
17. $(3x - 1)(x + 1)$
18. $(5x + 2)(x - 1)$
19. $x(x + 1)^2$
20. $2(x - 6)(x - 8)$

Many real-life situations can be described by using rational expressions. Numerous examples in physics, economics, and education will be shown in the lesson applications. Recognizing various types of problems that utilize rational expressions is a primary objective of this chapter. A secondary objective is the development of techniques that students can use effectively to solve problems.

The Chapter Summary on page 613 provides you and the students with a synopsis of the chapter. It identifies key skills, concepts, and vocabulary. You may want to have students look at the Chapter Summary as an overview before beginning the chapter.

CHAPTER 11

Using Proportions and Rational Equations

LESSONS

At the depth at which this photo was taken, the pressure exerted by the ocean is approximately 376 times the pressure at sea level.

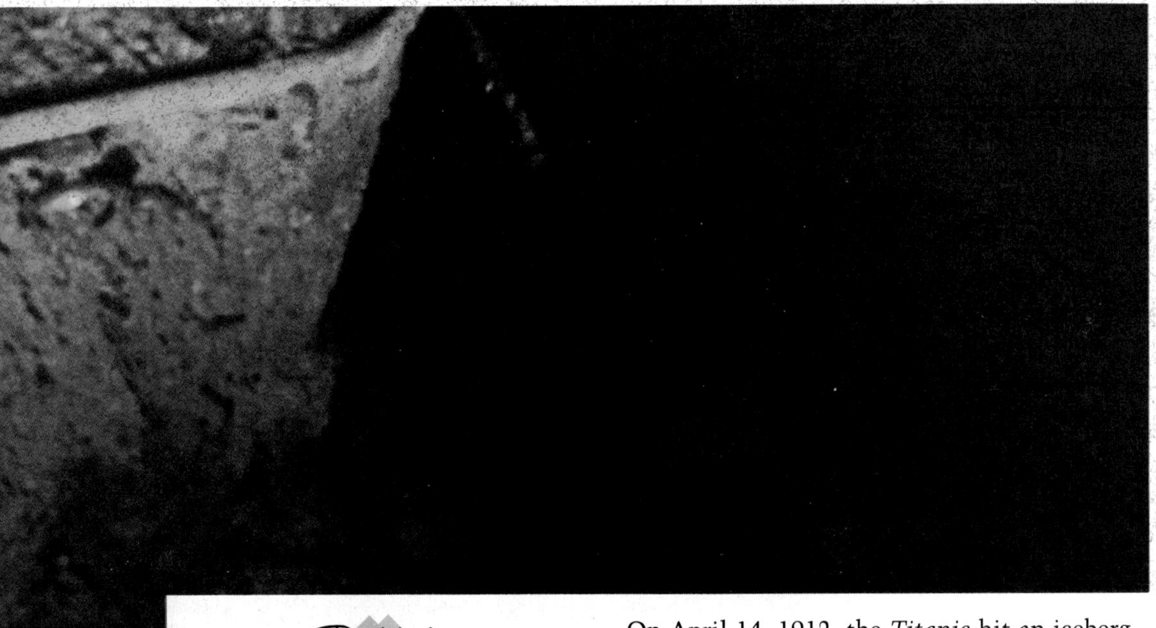

Real Life
Marine Exploration

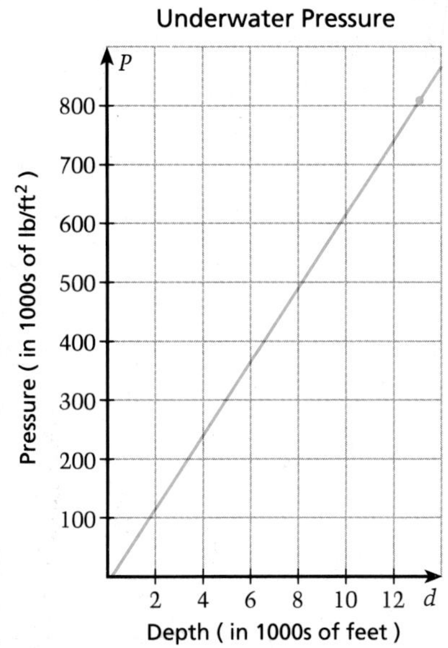

On April 14, 1912, the *Titanic* hit an iceberg and sank. The ship was considered unsinkable by its builders and owners, but it sank on its first Atlantic crossing. About 1522 of the 2227 passengers perished. For many years, the location of the *Titanic* was not known. In 1985, however, researchers found the wreckage about 500 miles southeast of Newfoundland. The wreckage was found at a depth of 13,000 feet.

Water pressure, P (in pounds per square foot), varies directly with the depth of the water, d, in feet. An equation that relates the two variables is

$$P = 62.4d.$$

The pressure at a depth of 13,000 feet is about 811,200 pounds per square foot. To take the photograph shown here, researchers used a special submarine called *Argo* that could bear such a tremendous pressure.

A number of mathematical questions about the sinking of the *Titanic* may be posed. For example, What percent of the passengers survived the disaster? about 32% survived. How many miles below the surface was the wreckage of the *Titanic*? 13,000 ft, almost 2.5 miles How much pressure per square inch is 811,200 pounds per square foot? about 5633 pounds per square inch

Pressure is defined as force per unit area. It is common to speak of pressure as so many "atmospheres," where one atmosphere (normal atmospheric pressure on Earth) is about 14.7 pounds per square inch (the pressure at a depth of 34 feet of water).

PACING CHART

The daily Pacing Chart is meant to help you adjust your teaching pace. Students in the full course should finish the entire text by the end of the year. Students in the basic course may not complete the entire text in the school year. The Pacing Chart for each chapter contains suggestions for lessons that require more than one day and lessons that may be omitted for the basic course.

DAY	FULL COURSE	BASIC COURSE
1	11.1	11.1
2	11.2	11.2
3	11.3	11.3
4	11.3	11.3
5	11.4	11.4
	Mid-Chapter Self-Test	Mid-Chapter Self-Test
6	11.5	11.5
7	11.5	11.5
8	11.5	11.5
9	11.6	11.6
10	11.6	11.6
11	11.6	11.6
12	11.7	Chapter Review
13	11.8	Chapter Review
14	Chapter Review	Chapter Test
15	Chapter Review	
16	Chapter Test	

CHAPTER ORGANIZATION

LESSON	PAGES	GOALS	MEETING THE NCTM STANDARDS
11.1	562-567	1. Solve proportions 2. Use proportions to solve real-life problems	Problem Solving, Communication, Connections
11.2	568-573	1. Solve percent problems 2. Use percents in real-life problems	Problem Solving, Communication, Reasoning, Connections
Mixed Review	574	Review of algebra and arithmetic	
Career Interview	574	Retailer	Connections
11.3	575-580	1. Use direct and inverse variation 2. Use direct and inverse variation in real-life settings	Problem Solving, Communication, Connections
11.4	581-586	1. Find the probability of an event 2. Use probability in real-life settings	Problem Solving, Communication, Connections, Discrete Mathematics, Geometry
Mid-Chapter Self-Test	587	Diagnose student weaknesses and remediate with correlated Reteaching worksheets	
11.5	588-593	1. Simplify a rational expression 2. Use rational expressions as real-life models	Problem Solving, Communication, Structure, Connections
Mixed Review	593	Review of algebra and arithmetic	
11.6	594-599	1. Multiply and divide rational expressions 2. Use rational expressions in real-life settings	Problem Solving, Communication, Reasoning, Connections
11.7	600-605	1. Divide a polynomial by a monomial or a binomial 2. Use polynomial division in real-life models	Problem Solving, Communication, Connections
Using a Calculator	606	Use a graphing calculator to find a range in which two graphs resemble each other	Technology, Geometry
11.8	607-612	1. Solve rational equations 2. Use rational equations in real-life settings	Problem Solving, Communication, Structure, Connections
Chapter Summary	613	A restatement of what has been learned, why it has been learned, and how it fits into the structure of algebra	Structure, Connections
Chapter Review	614-616	Review of concepts and skills learned in the chapter	
Chapter Test	617	Diagnose student weaknesses and remediate with correlated Reteaching worksheets	

LESSON RESOURCES

MEETING INDIVIDUAL NEEDS

RETEACHING For students who need to spend more time on basics:

If a mid-chapter self-test or chapter test indicates a deficiency, teachers can help students with the appropriate ***Reteaching Copymaster.***

PRACTICE For students who need more practice:

Additional exercises like those in the Pupil's Edition are provided for each lesson in ***Extra Practice Copymasters.***

ENRICHMENT For enriching and broadening students' experiences:

Problem of the Day copymasters in ***Teaching Tools*** provide a daily opportunity to use logical reasoning, looking for a pattern, writing an equation, and other routine and non-routine problem-solving strategies.

Math Log copymasters in ***Alternative Assessment*** provide opportunities to report on investigations, research, and open-ended problems.

Enriching activities with graphing and scientific calculators and computers are provided in ***Technology: Using Calculators and Computers.***

The ***Applications Handbook*** provides additional information about the cross-curriculum topics such as astronomy, chemistry, physics, sports, economics, genetics, and music that are integrated into the Pupil's Edition.

LESSON	11.1	11.2	11.3	11.4	11.5	11.6	11.7	11.8
PAGES	562-567	568-573	575-580	581-586	588-593	594-599	600-605	607-612
Teaching Tools								
Transparencies								
Problem of the Day	✓	✓	✓	✓	✓	✓	✓	✓
Warm-up Exercises		✓	✓		✓	✓	✓	✓
Answer Masters	✓	✓	✓	✓	✓	✓	✓	✓
Extra Practice Copymasters	✓	✓	✓	✓	✓	✓	✓	✓
Reteaching Copymasters	Teacher-directed and independent activities tied to results on the Mid-Chapter Self-Tests and Chapter Tests							
Color Transparencies	✓	✓	✓	✓	✓	✓	✓	✓
Applications Handbook	Additional background information is supplied for many real-life applications.							
Technology Handbook	Calculator and computer worksheets are supplied for appropriate lessons.							
Complete Solutions Manual	✓	✓	✓	✓	✓	✓	✓	✓
Alternative Assessment	Assess student's ability to reason, analyze, solve problems, and communicate using mathematical language.							
Formal Assessment	Mid-Chapter Self-Tests, Chapter Tests, Cumulative Tests, and Practice for College Entrance Tests							
Computer Test Bank	Customized tests can be created by choosing from over 2000 items.							

INSIGHTS

11.1 Problem Solving Using Ratios and Proportions

Many real-life problems involve proportions. For instance, finding the dimensions of similar objects and solving uniform motion problems frequently require proportions. Cross multiplying is a common approach to solving such equations. Students need to develop both an awareness of when to employ this technique and the skill with which to implement it.

11.2 Problem Solving Using Percents

Using percents is a practical skill. Students are likely to be familiar with the use of percents in shopping and in sports. However, percent problems also commonly occur in science, education, business, and statistics. Percent problems can be set up in at least two different ways. Encourage students to familiarize themselves with the different methods.

11.3 Direct and Inverse Variation

Many relationships in science and business are direct or inverse variations. For example, Ohm's Law states that the voltage in an electric circuit varies directly as the number of amperes of electric current in the circuit. The time required to complete a job varies inversely as the number of people working on the job. Students must learn to distinguish between the two types of variation in order to solve problems involving variation.

11.4 Exploring Data: Probability

Probability is an important branch of mathematics that utilizes many algebraic concepts and skills. Economic, political, and environmental decisions are often based upon information obtained from considerations of probability. Whenever students try to decide the likelihood that an event will occur they are using elements of probability.

11.5 Simplifying Rational Expressions

Rational expressions occur in many applications of physics, economics, and engineering. In this lesson, basic techniques for simplifying rational expressions are presented. These skills are required in order to perform the basic operations with rational expressions.

11.6 Multiplying and Dividing Rational Expressions

Rational expressions occur in many applications of physics, economics, and engineering. Techniques for simplifying rational expressions that were studied in Lesson 11.5 can be used here to perform the basic operations of multiplication and division of rational expressions.

11.7 Dividing Polynomials

To graph a rational expression with a divisor that is a monomial or binomial, it is convenient to have an alternative representation of the division of rational expressions. (See Chapter 12.) The "quotients" become sums of reduced fractions in which denominators are monomials or binomials. In this form, the behavior of graphs as x grows large can be determined algebraically.

11.8 Solving Rational Equations

Many relationships in physics, economics, and engineering can be modeled by rational equations. Techniques for solving rational equations rely on an understanding of the multiplication and division of rational expressions, and exploit the techniques that have been developed for solving linear and quadratic equations. Knowing how to solve rational equations increases the students' ability to handle a variety of real-life problem situations.

Problem of the Day

David, who had 15 oranges, and Alonzo, who had 9 oranges shared their oranges equally with Mary. Mary paid them with 24 dimes. David thought he should get 15 of the dimes and Alonzo should get 9. Alonzo thought he and David should each get 12 dimes, since they shared equally. Mary said that David should get 21 dimes and Alonzo only 3. What reasoning did she use?

When the 24 oranges were divided equally, each had 8. So David gave Mary 7, and Alonzo gave her 1. So they contributed in the ratio of 7 to 1, and should be paid in the same ratio.

ORGANIZER

Warm-Up Exercises

1. Solve for x.
   a. $2x = 4(x - 8) + 12$ 10
   b. $75 = 45x$ $\frac{5}{3}$
   c. $2x(x) = 2(6 - 5x)$ -6 or 1
2. Evaluate each fraction for given values of x.
   $\frac{24}{x}, \frac{x-3}{x}$
   a. $x = -3$ $-8, 2$
   b. $x = 4$ $6, \frac{1}{4}$
   c. $x = \frac{2}{3}$ $36, -\frac{7}{2}$
   d. $x = \frac{1}{4}$ $96, -11$

Lesson Resources

Teaching Tools
- Transparency: 3
- Problem of the Day: 11.1
- Warm-up Exercises: 11.1
- Answer Masters: 11.1

Extra Practice: 11.1

Color Transparency: 69

11.1 Problem Solving Using Ratios and Proportions

What you should learn:

Goal 1 How to solve proportions

Goal 2 How to use proportions to solve real-life problems

Why you should learn it:

You can solve many real-life problems using proportions, such as finding the size of an object from a scale model.

Goal 1 Solving Proportions

In Lesson 2.8, you learned that if two quantities, a and b, are measured in the *same* units, then their ratio is $\frac{a}{b}$. An equation that equates two ratios is a **proportion.** For instance, if the ratio $\frac{a}{b}$ is equal to the ratio $\frac{c}{d}$, then the proportion is the statement of equality

$\frac{a}{b} = \frac{c}{d}$. *a, b, c, and d are nonzero.*

This is read as "a is to b as c is to d." The numbers a and d are the **extremes** of the proportion. The numbers b and c are the **means** of the proportion.

Properties of Proportions

1. If two ratios are equal, their reciprocals are also equal.

   If $\frac{a}{b} = \frac{c}{d}$, then $\frac{b}{a} = \frac{d}{c}$. ***Reciprocal property***

2. The product of the extremes equals the product of the means.

   If $\frac{a}{b} = \frac{c}{d}$, then $ad = bc$. ***Cross-multiplying property***

Problems involving proportions often contain a variable. Solving for the variable is called *solving the proportion.*

Example 1 *Solving a Proportion*

Solve the proportion $\frac{3}{y} = \frac{5}{8}$.

Solution

$\frac{3}{y} = \frac{5}{8}$ *Rewrite original proportion.*

$\frac{y}{3} = \frac{8}{5}$ *Reciprocal property.*

$y = 3\left(\frac{8}{5}\right)$ *Multiply both sides by 3.*

$y = \frac{24}{5}$ *Simplify.*

The solution is $\frac{24}{5}$. Check this in the original proportion. ■

Example 2 *Solving a Proportion*

Solve the proportion $\frac{x}{x+4} = \frac{2}{x}$.

Solution

$$\frac{x}{x+4} = \frac{2}{x} \quad \textit{Original proportion}$$
$$x(x) = 2(x+4) \quad \textit{Cross multiply.}$$
$$x^2 = 2x + 8 \quad \textit{Simplify.}$$
$$x^2 - 2x - 8 = 0 \quad \textit{Write in standard form.}$$
$$(x-4)(x+2) = 0 \quad \textit{Factor.}$$
$$x = 4 \text{ or } x = -2 \quad \textit{Zero-Product Property}$$

The proportion has two solutions: 4 and −2. Check these solutions in the original proportion. ■

Goal 2 Using Proportions in Real-Life Problems

Mount Rushmore National Memorial is carved out of a granite cliff in the Black Hills of South Dakota. The carvings of George Washington, Thomas Jefferson, Theodore Roosevelt, and Abraham Lincoln are 80 times life size.

Real Life **Sculpture**

Example 3 *Estimating the Size of a Statue*

Estimate the height of the sculpture of George Washington's head. (Assume that an average distance from a man's chin to the top of his head is 9 inches or 0.75 ft.)

Solution Let x represent the height of the sculptured head. You can find x by solving the following proportion.

$$\frac{\text{Height of sculptured head}}{\text{Height of man's head}} = \frac{80}{1}$$

$$\frac{x}{0.75} = \frac{80}{1} \quad \frac{\text{feet}}{\text{feet}}$$

The solution is $x = 60$. The height of the sculptured head of Washington is about 60 feet. ■

LESSON Notes

Example 1

Stress that there is no reason to use the Cross-multiplying property in solving this proportion. In the third step, it is more direct to multiply both sides of the equation by 3.

Example 2

In this proportion, the Cross-multiplying property is needed. Observe that the final equation to be solved is a quadratic equation.

MATH JOURNAL Students should record examples of both properties in their math journals.

Example 3

The proportion expresses information in terms of feet. If students select the average length of a man's face to be 9 inches then the proportion $\frac{x}{9} = \frac{80}{1}$ can be formed by using inch ratios. From this proportion, $x =$ 720 inches or $\frac{720}{12}$ feet.

EXTEND Example 4 **RESEARCH** Students may want to investigate this archaeological finding in more detail. *National Geographic* and the encyclopedia are good places to begin.

Common-Error ALERT!

Students frequently identify a ratio as a proportion! Call their attention to the differences and the relationship between these two concepts.

Extra Examples

Here are additional examples similar to **Examples 1–3.**

1. Solve the proportions.
   a. $\frac{4}{x} = \frac{16}{3}$ $\frac{3}{4}$
   b. $\frac{10}{y} = \frac{5}{24}$ 48
   c. $\frac{3x}{4} = \frac{5}{6}$ $\frac{10}{9}$
   d. $\frac{2x}{3x-2} = \frac{4}{5}$ 4
   e. $\frac{x}{x+6} = \frac{3}{x}$ 6, −3
2. On a map, the distance from Memphis (TN) to Jackson (MS) is about $5\frac{1}{4}$ inches. If each inch represents about 38 miles, how far is it from Memphis to Jackson? about 200 miles

Check Understanding

1. What is the difference between a ratio and a proportion? See the top of page 562.
2. How is the Reciprocal property used to solve proportions? See Example 1.
3. For Examples 3 and 4, explain why the proportions $\frac{x}{80} = \frac{0.75}{1}$ and $\frac{x}{680} = \frac{316}{40}$ describe the given relationships equally as well as the original proportions. The proportions represent different relationships: sculpture-size measurements to life-size measurements and total numbers to excavation numbers.

Communicating about ALGEBRA

Have students explain why the Cross-multiplying property is not always efficient to use. Answers may include comments about the number of solution steps or the complexity of the computed expressions.

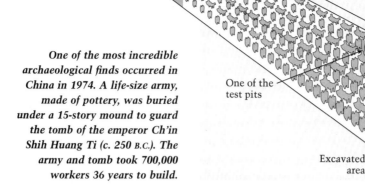

One of the most incredible archaeological finds occurred in China in 1974. A life-size army, made of pottery, was buried under a 15-story mound to guard the tomb of the emperor Ch'in Shih Huang Ti (c. 250 B.C.). The army and tomb took 700,000 workers 36 years to build.

10 ft
640 ft
One of the test pits
40 ft
10 ft
Excavated area

Real Life
Archaeology

Example 4 *Estimating the Size of an Army*

To estimate the size of the army, archaeologists excavated 3 test pits and a section at one end of the mound. They found 4 rows containing 320 soldiers in the first 10 feet of the section. (Four similar rows were at the other end of the mound.) In the next 40 feet, they found 316 soldiers, 24 horses, and 6 chariots, which they thought to be representative of the rest of the mound. How many soldiers, horses, and chariots did they estimate to be in the 680-foot portion of the mound between the eight "end rows"?

Solution Let x represent the total number of soldiers. You can find x by solving the following proportion.

$$\frac{\text{Total number of soldiers}}{\text{Number of soldiers found}} = \frac{\text{Total number of feet}}{\text{Number of feet excavated}}$$

$$\frac{x}{316} = \frac{680}{40}$$

The solution is $x = 5372$. (With the 640 soldiers at the ends, that makes 6012 soldiers.) In a similar way, you can use proportions to estimate that there were 408 horses and 102 chariots! *(Source:* **National Geographic***)* ■

Communicating about ALGEBRA

SHARING IDEAS about the Lesson

Explain the Process Solve each proportion. Explain your solution steps. See below.

A. $\frac{x}{4} = \frac{9}{2}$ 18 **B.** $\frac{5}{x} = \frac{4}{5}$ $\frac{25}{4}$ **C.** $\frac{x}{x+1} = \frac{10}{x+7}$ 5, −2

A. Multiply both sides by 4, simplify **B.** Reciprocal property, multiply both sides by 5, simplify
C. Cross multiply, simplify, write in standard form, factor, Zero-Product property

EXERCISES

Guided Practice

CRITICAL THINKING about the Lesson

1. In the proportion $\frac{a}{b} = \frac{c}{d}$, which numbers are the *extremes*? Which are the *means*? a, d b, c

2. Which of the following are equivalent to $\frac{a}{b} = \frac{c}{d}$? c
   a. $ac = bd$ b. $ba = dc$ c. $ad = bc$

3. Which of the following are equivalent to $\frac{a}{b} = \frac{c}{d}$? b
   a. $\frac{a}{b} = \frac{d}{c}$ b. $\frac{b}{a} = \frac{d}{c}$ c. $\frac{a}{d} = \frac{b}{c}$

In Exercises 4–6, solve the proportion.

4. $\frac{x}{4} = \frac{2}{5}$ $\frac{8}{5}$
5. $\frac{5}{x} = \frac{3}{2}$ $\frac{10}{3}$
6. $\frac{1}{x+1} = \frac{x}{2}$ $-2, 1$

Independent Practice

In Exercises 7–24, solve the proportion.

7. $\frac{2}{x} = \frac{3}{4}$ $\frac{8}{3}$
8. $\frac{3x}{4x-1} = \frac{1}{x}$ $\frac{1}{3}, 1$
9. $\frac{7x}{x+1} = \frac{5x}{2}$ $0, \frac{9}{5}$
10. $\frac{8x}{6} = \frac{2x}{x+3}$ $0, -\frac{3}{2}$
11. $\frac{6}{2y} = \frac{5}{8}$ $\frac{24}{5}$
12. $\frac{4}{5} = \frac{7}{y}$ $\frac{35}{4}$

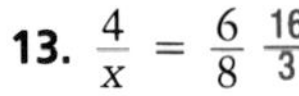

13. $\frac{4}{x} = \frac{6}{8}$ $\frac{16}{3}$
14. $\frac{2}{x} = \frac{6}{3}$ 1
15. $\frac{-3}{x} = \frac{5}{6}$ $-\frac{18}{5}$
16. $\frac{4}{x} = \frac{14}{-7}$ -2
17. $\frac{9}{2} = \frac{18}{y}$ 4
18. $\frac{10}{8} = \frac{5}{y}$ 4
19. $\frac{x}{5x+6} = \frac{1}{x}$ $6, -1$
20. $\frac{5}{x+1} = \frac{4x}{x}$ $\frac{1}{4}$
21. $\frac{6}{x+2} = \frac{x+1}{x}$ $1, 2$
22. $\frac{6}{19x} = \frac{-2}{x^2+2}$ $-\frac{1}{3}, -6$
23. $\frac{x+6}{x-2} = \frac{2}{x}$ -2
24. $\frac{x-3}{x} = \frac{2x}{-3}$ $\frac{3}{2}, -3$

25. *Planting Trees* In 1990, more than 2.1 billion trees were planted in the United States. The states that planted the most acres of trees were Georgia and Alabama. The ratio of Georgia's population to Alabama's population was about the same as the ratio of the numbers of trees planted in Georgia and Alabama, respectively. The 1990 population of Georgia was 6,508,000. What was the 1990 population of Alabama? ≈4,039,000
(Source: American Forest Resource Alliance)

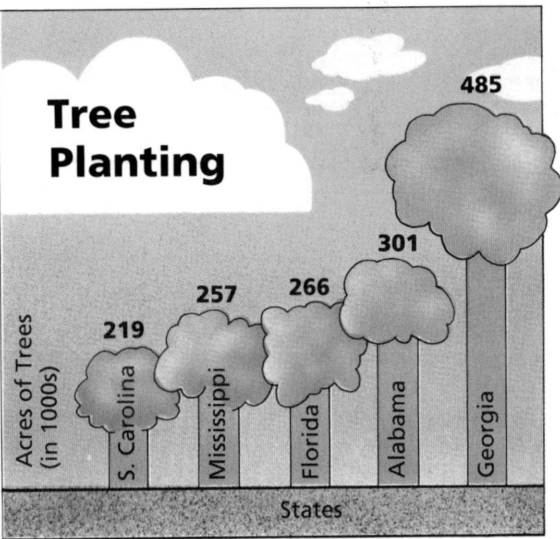

EXERCISE Notes

ASSIGNMENT GUIDE

Basic/Average: Ex. 1–6, 7–25 odd, 29, 30, 31–42, 49–50

Above Average: Ex. 1–6, 7–27 odd, 29–50

Advanced: Ex. 1–6, 7–27 odd, 28–30, 47–50

Selected Answers
Exercises 1–6, 7–47 odd

More Difficult Exercises
There are no more difficult exercises in this lesson.

Guided Practice

Ex. 4–6 Sometimes the quadratic formula is needed to solve the proportions. See Example 2 for a sample.

Independent Practice

Ex. 7–24 Remind students to always check their solutions in the original equations.

Ex. 25–30 These exercises provide students with examples of proportions from real-life data. Students who are having difficulty setting up the proportions should refer to Examples 3 and 4.

▶ **Ex. 31–46** These exercises review essential skills related to ratios.

▶ **Ex. 49–50** GROUP ACTIVITY These exercises should be completed as an in-class lab experiment.

EXTEND Ex. 49–50 AREA BY SAMPLING You can use a similar sampling technique to find the area of an irregularly shaped region. Enclose it in a region you know the area of, then use sampling to approximate the ratio of the shaded area to the total. For example, use sampling to find the ratio of the shaded area to the total area of this rectangle.

Students should work together with partners. By using a bent paper clip, they sample by taking turns closing their eyes and touching the point of the clip anywhere in the region. Students tally the number of times the point of the clip touches the shaded region and the total number of trials.

After each student pair completes 10 trials, combine the tallies for the entire class and approximate the ratio. Then use the ratio to find the area. For this sample, the ratio is $\frac{3}{5}$, or 60% of the region is shaded.

26. *Estimating Whale Populations* To estimate the number, W, of blue whales in a portion of the Indian Ocean, researchers marked 200 different whales. Months later the researchers checked 300 whales in the area and found only 15 that were marked. The researchers then assumed that the ratio of 300 to W was equal to the ratio of 15 to 200. What did they estimate the blue whale population to be? 4000

Monuments of Stone **In Exercises 27 and 28, use the following information.**

The Chief Crazy Horse Memorial is being carved from Thunderhead Mountain in South Dakota. When completed, it will be the world's largest stone sculpture: 641 feet wide and 563 feet high.

The Stone Mountain Monument in Georgia is a stone carving of Jefferson Davis, Robert E. Lee, and Stonewall Jackson. In 1990, it was the world's largest stone sculpture: 180 feet wide and 90 feet high.

27. The ratio of the sculpture of Chief Crazy Horse to actual size is 140 to 1. Estimate the diameter of the horse's nostril in the sculpture. Assume the diameter of a horse's nostril is 3 inches. 420 in. or 35 ft

28. The ratio of the Stone Mountain sculpture to actual size is 20 to 1. Estimate the length of a horse's head in the sculpture. Assume that the length of a horse's head is $2\frac{1}{2}$ feet. 50 ft

29. *Hoosiers vs. Boilermakers* The Purdue Boilermakers are playing the Indiana Hoosiers. Estimate the distance the Boilermakers have to travel (from West Lafayette to Indianapolis). between 60 to 70 mi

30. *Traveling on Interstate 95* You are driving on Interstate 95 from Philadelphia, Pennsylvania, to Baltimore, Maryland. Estimate the number of miles you will be traveling in Delaware. ≈20 mi

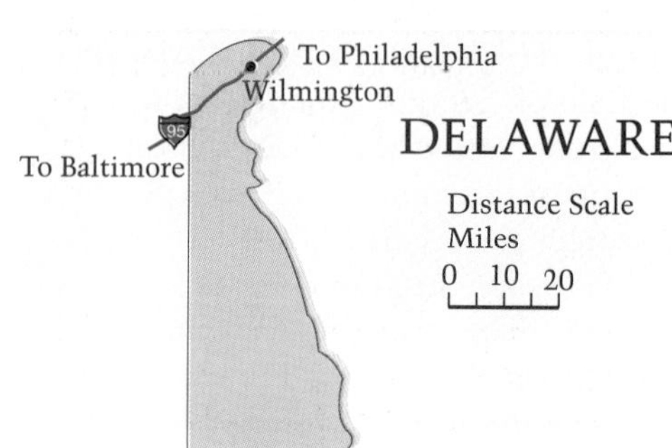

Integrated Review

In Exercises 31–34, find the reciprocal of the number.

31. $\frac{1}{3}$ 3 **32.** $\frac{4}{5}$ $\frac{5}{4}$ **33.** $\frac{-2}{7}$ $-\frac{7}{2}$ **34.** $\frac{7}{6}$ $\frac{6}{7}$

In Exercises 35–38, write the percent in decimal form.

35. 23% 0.23 **36.** 6.4% 0.064 **37.** 0.25% 0.0025 **38.** 3% 0.03

In Exercises 39–42, write the fraction as a percent.

39. $\frac{3}{4}$ 75% **40.** $\frac{2}{3}$ 66.7% **41.** $\frac{6}{2}$ 300% **42.** $\frac{1}{5}$ 20%

In Exercises 43–46, simplify the fraction.

43. $\frac{6x}{x}$ 6 **44.** $\frac{5xy}{15x^2}$ $\frac{y}{3x}$ **45.** $\frac{-2y^2}{14xy}$ $\frac{-y}{7x}$ **46.** $\frac{-3xy^3}{3x^3y}$ $-\frac{y^2}{x^2}$

47. *Estimation—Powers of Ten* Which is the best estimate for the length of an adult blue whale (the largest living mammal)?

a. 10 feet **b.** 100 feet **c.** 1000 feet **d.** 10,000 feet b

48. *Estimation—Powers of Ten* Which is the best estimate for the baking temperature of chocolate chip cookies?

a. 4°F **b.** 40°F **c.** 400°F **d.** 4000°F c

Exploration and Extension

49. *Buffon's Needle Experiment* The following experiment was devised by an 18th century French naturalist, Count Buffon.

Place a piece of ruled notebook paper on a flat surface. Cut a thin paper "needle" whose length is the distance between the lines on the paper. Drop the needle from a height of about one foot above the paper. If the needle touches a line, count the drop as a success. If it doesn't, count the drop as a failure. Repeat the experiment 30 or 40 times. Find the ratio of total drops to successes. It should be approximately equal to the ratio of π to 2. **Experimental results vary.**

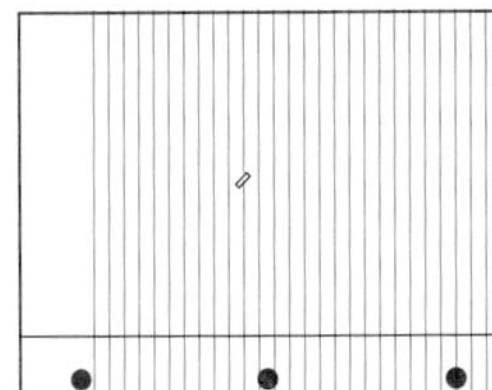

If the needle touches a line, count the drop as a success.

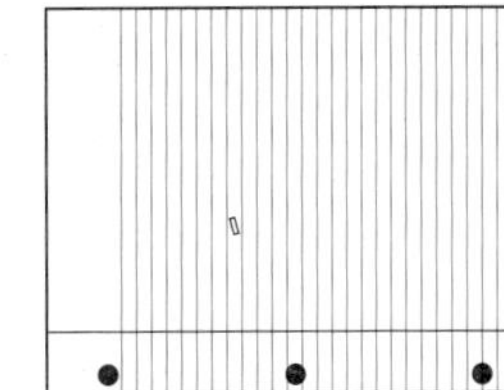

If the needle falls between two lines, count the drop as a failure.

50. If you performed this experiment 3195 times, how many successes would you need to obtain 3.1415929 as the value of π? 2034

Enrichment Activity

This activity requires students to go beyond lesson goals.

PROJECT Many newspapers and news magazines have articles that contain real-life data that can be used in ratios and proportions. Throughout this chapter, try to find situations in articles that can be expressed by using ratios, then write and solve problems using the data.

For example, in a recent article on the airline industry, it was stated, "In 1980, there were 31,665 pilots and 54,418 flight attendants. In 1990, the number of flight attendants had risen to 83,443." If you assume that the ratio of pilots to attendants is the same, then how many pilots were there in 1990? 48,554

Problem of the Day

Consider a square that just encloses a circle of radius x.

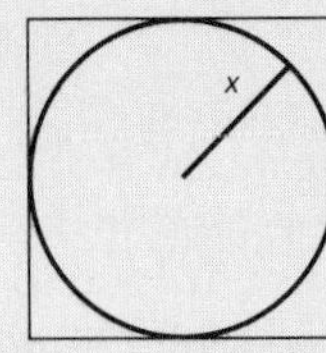

a. What is the ratio of the area of the square to the area of the circle? $\frac{4}{\pi}$

b. What is the ratio of the perimeter of the square to the circumference of the circle? $\frac{4}{\pi}$

ORGANIZER

Warm-Up Exercises

1. Solve the proportions.
   a. $\frac{4}{x} = \frac{6}{9}$ 6
   b. $\frac{2x}{12} = \frac{4}{16}$ $\frac{3}{2}$
   c. $\frac{7}{x+3} = \frac{21}{9}$ 0
2. Convert to decimal form.
   a. 24% 0.24
   b. 13% 0.13
   c. 0.12% 0.0012
   d. 436% 4.36
   e. 35.7% 0.357
   f. $\frac{45}{75}$ 0.6

Lesson Resources

Teaching Tools
- Transparency: 3
- Problem of the Day: 11.2
- Warm-up Exercises: 11.2
- Answer Masters: 11.2

Extra Practice: 11.2

Color Transparency: p. 61

11.2 Problem Solving Using Percents

What you should learn:

Goal 1 How to solve percent problems

Goal 2 How to use percents in real-life problems

Why you should learn it:

You can model and solve many real-life problems using percent, such as in presenting survey results.

Goal 1 Solving Percent Problems

Percent means *per hundred,* or *parts of 100.* For example, 30% means 30 parts of 100, which is equivalent to the fraction $\frac{30}{100}$ or the decimal 0.3.

When solving percent problems, we suggest converting percents to decimals *before performing arithmetic operations.*

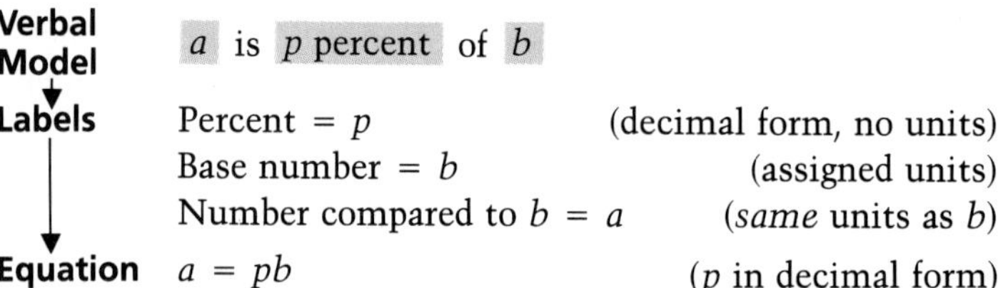

Verbal Model a is p percent of b

Labels
Percent $= p$ (decimal form, no units)
Base number $= b$ (assigned units)
Number compared to $b = a$ (*same* units as b)

Equation $a = pb$ (p in decimal form)

Example 1 Percent Problem: Unknown Number

What is 30% of 70 feet?

Solution

Labels
Unknown number $= a$ (feet)
Percent $= p = 0.3$ (decimal form)
Base number $= b = 70$ (feet)

Equation
$a = (0.3)(70)$ *a is p percent of b.*
$a = 21$ feet

Therefore, 21 feet is 30% of 70 feet. ■

Example 2 Percent Problem: Unknown Base Number

Fourteen dollars is 25% of what?

Solution

Labels
Number $= a = 14$ (dollars)
Percent $= p = 0.25$ (decimal form)
Unknown base number $= b$ (dollars)

Equation
$14 = 0.25b$ *a is p percent of b.*
$\frac{14}{0.25} = b$
$\$56 = b$

Therefore, $14 dollars is 25% of $56. ■

Example 3 *Percent Problem: Unknown Percent*

One hundred thirty-five miles is what percent of 27 miles?

Solution

Labels
Number $= a = 135$ (miles)
Percent $= p$ (decimal form)
Base number $= b = 27$ (miles)

Equation

$135 = p(27)$ *a is p percent of b.*

$\frac{135}{27} = p$

$5 = p$ *Decimal form*

$500\% = p$ *Percent form*

Therefore, 135 miles is 500% of 27 miles. ■

Goal 2 Using Percents in Real-Life Problems

Connections **Sociology**

Example 4 *Taking a Survey*

In 1991, a survey of American "teens" between the ages of twelve and nineteen had the following results.
(*Source: Simmons Market Research Bureau*)

Age Group	Number Surveyed	Number Who Date Regularly
12–14-year-olds	900	207
15–17-year-olds	990	594
18–19-year-olds	650	416

What percent of each age group surveyed said they dated regularly? What information about the survey would you need to know to conclude that the results are representative of *all* American teens?

Solution

Of the 12–14-year-olds, $\frac{207}{900}$, or 23%, said they dated regularly.

Of the 15–17-year-olds, $\frac{594}{990}$, or 60%, said they dated regularly.

Of the 18–19-year-olds, $\frac{416}{650}$, or 64%, said they dated regularly.

To know whether these percents are representative of *all* American teens, you would need to know that the teens used in the survey were chosen from many different backgrounds and many different parts of the country. For instance, if all 2530 teens who were surveyed were from the same city or county, then it would not be fair to form conclusions about a national population. ■

LESSON Notes

GOAL 1 The verbal model of percent problems, "a is p percent of b," indicates the relationship between the "base number b" (the number usually representing the total or a whole) and a part of the base number, the "number a."

An Alternate Approach
Rewriting Ratios as Percents
The ratio $\frac{a}{b}$ is the percent, p. But p can be rewritten as a ratio with 100 as the denominator. Consequently, percent equations may be written as proportions like those in Lesson 11.1. Offer students the option of setting up percent problems using either interpretation. See below for examples.

Examples 1–4

Many students will prefer the alternate approach to percent problems using proportions. In Examples 1–4, compare percent values with each other; then compare number values. For example, in Example 1, 30% is the ratio $\frac{30}{100}$. The number values may be compared after it is determined which number value represents the total or whole. Because you are looking for a part of (30% of) 70 feet, 70 represents the whole. It follows that the ratio is $\frac{n}{70}$ (what part of 70). The proportion becomes $\frac{30}{100} = \frac{n}{70}$, and $n = 70(\frac{30}{100})$ or $n = 21$.

For Examples 2 and 3,write the proportions $\frac{14}{n} = \frac{25}{100}$ and $\frac{135}{27} = \frac{p}{100}$. In Example 4, use the proportion $\frac{207}{900} = \frac{p}{100}$ to find that $p = 23\%$. Notice that the whole of a percent is always 100 and that n is used for number values while p is used for percent values.

Example 6

The markup includes overhead and profit. Overhead is the cost of doing business such as taxes, rents, maintenance, etc.

Extra Examples

Here are additional examples similar to **Examples 1–6.**

1. Solve.
   a. What is 40% of 85 yards?
   b. What is 65% of 120 dollars?
   c. Twenty-three dollars is 25% of what?
   d. Fifteen meters is 30% of what?
   e. Fifty-six pounds is what percent of 140 pounds?
   f. Seventy cars is what percent of 40 cars?

   a. 34 yards b. $78
   c. $92 d. 50 meters
   e. 40% f. 175%
2. In Example 4, what percent of 18–19-year-olds did not date regularly? 36%
3. In Example 6, suppose the wholesale cost of a pith helmet is $22.75 and the retail price is $38.95. What is the markup percent on the cost? about 71% markup

Check Understanding

1. Using the proportion model for percent problems, how is the base related to 100? The base is the whole or 100%.
2. Explain why the percent and proportion models are equivalent representations. Percents can always be expressed as ratios with 100 as the denominator.
3. There are three types of percent problems (three types of unknowns) using the percent model. Describe what they are. See pages 568–569.

Connections
Entomology

Example 5 *Butterflies and Moths*

Butterflies and moths comprise the insect order *Lepidoptera.* Entomologists (scientists who study insects) have classified about 20,000 species (types) of butterflies and about 140,000 species of moths. Lepidoptera make up 20% of all classified insect species. How many insect species have been classified?

Solution

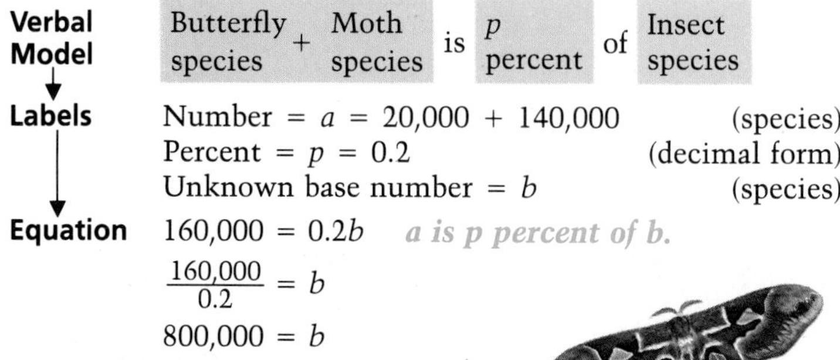

Verbal Model [Butterfly species] + [Moth species] is [p percent] of [Insect species]

Labels
Number $= a = 20{,}000 + 140{,}000$ (species)
Percent $= p = 0.2$ (decimal form)
Unknown base number $= b$ (species)

Equation
$160{,}000 = 0.2b$ *a is p percent of b.*

$\frac{160{,}000}{0.2} = b$

$800{,}000 = b$

There are about 800,000 classified species of insects. ■

Real Life
Retail Sales

Example 6 *Finding the Markup Percent*

The wholesale cost of a butterfly net is $18.50. The retail price is $32.95. What is the markup percent (on the cost)?

Solution

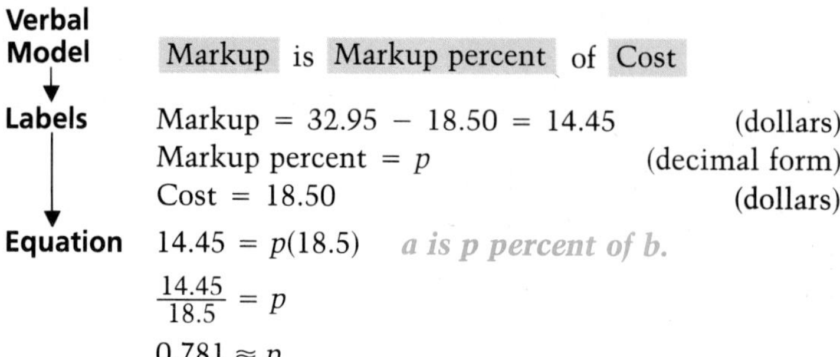

Verbal Model [Markup] is [Markup percent] of [Cost]

Labels
Markup $= 32.95 - 18.50 = 14.45$ (dollars)
Markup percent $= p$ (decimal form)
Cost $= 18.50$ (dollars)

Equation
$14.45 = p(18.5)$ *a is p percent of b.*

$\frac{14.45}{18.5} = p$

$0.781 \approx p$

The markup percent on the cost is about 78.1%. ■

Communicating about ALGEBRA

SHARING IDEAS about the Lesson

See below.

Derive the Meaning The Latin word *centum* means "one hundred." How is "100" related to *centimeter, cent,* and *centennial*? Name some other related "cent" words.

100 are in a meter, 100 are in a dollar, occurs every 100 years; century, centavo

EXERCISES

Guided Practice

CRITICAL THINKING about the Lesson

1. In the statement "16 is 10% of 160," what is the base number? Write an equation that represents this statement. 160, 16 = 0.10(160)
2. Thirty is 15% of what number? 200
3. What number is 26% of 450? 117
4. Nineteen is what percent of 10? 190%

Independent Practice

In Exercises 5–22, solve the percent problem.

5. What number is 23% of 90? 20.7
6. What number is 71% of 310? 220.1
7. 13 is what percent of 50? 26%
8. 48 is 60% of what number? 80
9. What distance is 6% of 400 meters? 24 m
10. How much is 32% of $625? $200
11. What distance is 21% of 580 miles? 121.8 mi
12. How many students is 86% of 950 students? 817 students
13. 35 feet is 50% of what length? 70 ft
14. 54 degrees is 45% of what angle measure? 120°
15. 3 dollars is 2% of what amount? 150 dollars
16. 9 grams is 12% of what weight? 75 grams
17. 11 people is what percent of 50 people? 22%
18. 13 years is what percent of 20 years? 65%
19. 24 pounds is what percent of 600 pounds? 4%
20. 486 dozen is what percent of 900 dozen? 20. 54%
21. 78 days is what percent of 39 weeks? 28.6%
22. 52 inches is what percent of 20 feet? 21.7%

Geometry **In Exercises 23–26, what percent of the region is shaded blue? What percent is shaded yellow?**

23. 13.3%, 86.7%

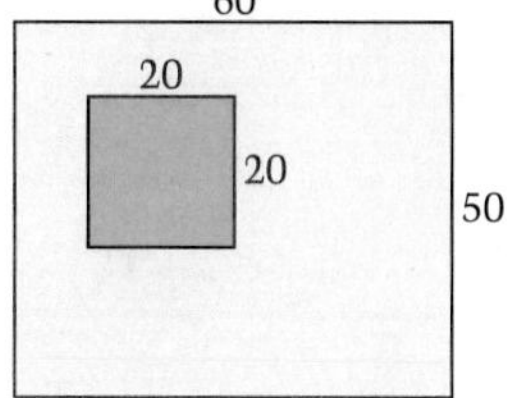

✪ 24. 40%, 60%

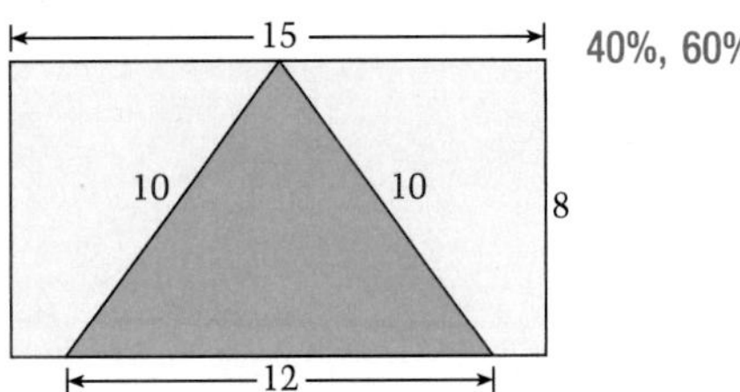

✪ 25. 25%, 75%

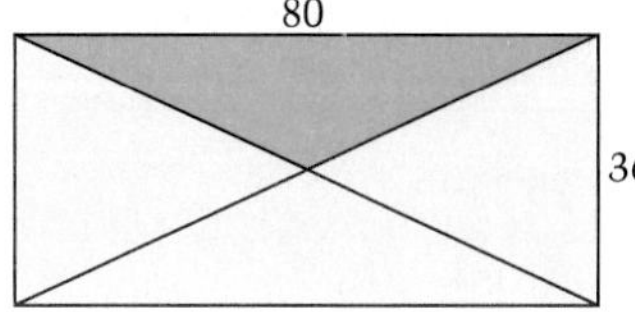

26. 29.4%, 70.6%

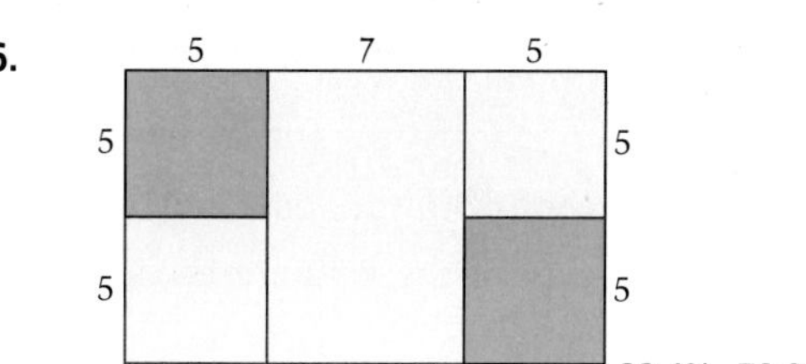

MATH JOURNAL Examples of the three types of percent problems should be included in students' math journals.

Communicating about ALGEBRA

COOPERATIVE LEARNING Make a class list of students' *cent* words and their definitions.

EXERCISE Notes

ASSIGNMENT GUIDE

Basic/Average: Ex. 1–4, 5–25 odd, 27–28, 31–33, 37–53 odd

Above Average: Ex. 1–4, 5–21 odd, 23–34, 43–53 odd

Advanced: Ex. 1–4, 11–25 odd, 27–36, 53–56

Selected Answers
Exercises 1–4, 5–53 odd

Use **Mixed Review** as needed.

✪ More Difficult Exercises
Exercises 24, 25, 30, 33, 34, 36, 55, 56

Guided Practice

▶ **Ex. 1–4** Students who have difficulty should review the verbal model presented before Example 1.

Independent Practice

▶ **Ex. 5–22** These exercises will ensure practice in each type of percent problem.

EXTEND Ex. 5–22 TECHNOLOGY Some calculators have percent keys. Have groups of students work together to determine how the percent keys work for different brands of calculators.

▶ **Ex. 23–26** These exercises provide a visual insight into the concept of percent using areas.

▶ **Ex. 33–34** These exercises can be used to check students' intuitive understanding of the meaning of percent.

▶ **Ex. 49–54** These exercises reinforce skills from Lesson 11.1.

▶ **Ex. 55–56 CRITICAL THINKING** These exercises are good questions to explore in class discussion.

Shutterbugs' Favorites **In Exercises 27 and 28, use the following information.**

One thousand ten amateur photographers were asked what their favorite subjects were. Each respondent was allowed to choose more than one subject.
(Source: Fuji Photo Film, Inc.)

Photographers' Favorite Subjects

Animals/pets 27%
Outdoor scenes 39%
Travel and vacations 48%
Children 49%
People they know 54%
Family celebrations 63%

27. How many of the respondents chose travel and vacations as their favorite subjects? ≈485

28. Suppose 1500 amateur photographers had been surveyed. How many would you expect to have chosen travel and vacations? 720

29. *Recycling Bottles and Jars* About 12% of the glass bottles and jars produced in the United States are recycled. In 1990, an average of 13,698,630 bottles and jars were recycled each day. Find the average number of bottles and jars that were produced each day. How many of these, on average, were *not* recycled? What percent is this of the number produced each day? 114,155,250; 100,456,620; 88%

✪ **30.** *Boots* Last year you bought a pair of designer boots for \$24.72. Your friend buys the same boots this year for \$37.08. Which statements are correct?

a. You paid $33\frac{1}{3}\%$ less than your friend.
b. Your friend paid 50% more than you did.
c. You paid $66\frac{2}{3}\%$ of what your friend paid. a, b, and c

Could You Lend Me a Day's Worth? **In Exercises 31 and 32, use the following information showing the average numbers of currency bills printed each day in 1991.** ***(Source: Bureau of Engraving and Printing)***

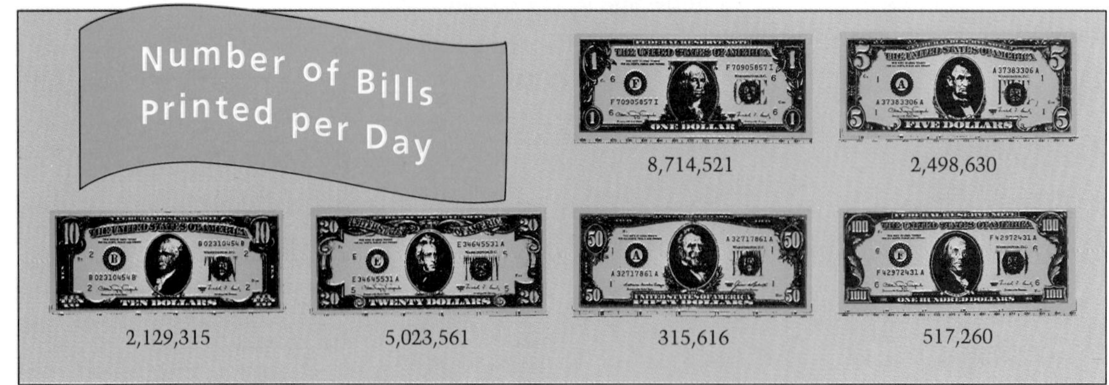

31. What percent of the six currency bills printed were \$20 bills? 26.2%

32. What percent of the value of the six types of bills printed was from \$20 bills? 47.7%

33. *Shampoo* You are shopping at a new store and buy a bottle of shampoo for $2.45. At the store where you usually shop, the price is $1.96. Which statements are correct?

a. You paid 25% more than your usual price.

b. You paid 125% more than your usual price.

c. You paid 125% of your usual price. a and c

34. *Automobile Expenses* Suppose that a person spent $748 on automobile gasoline in 1991. What would you estimate the total automobile expenses to have been?

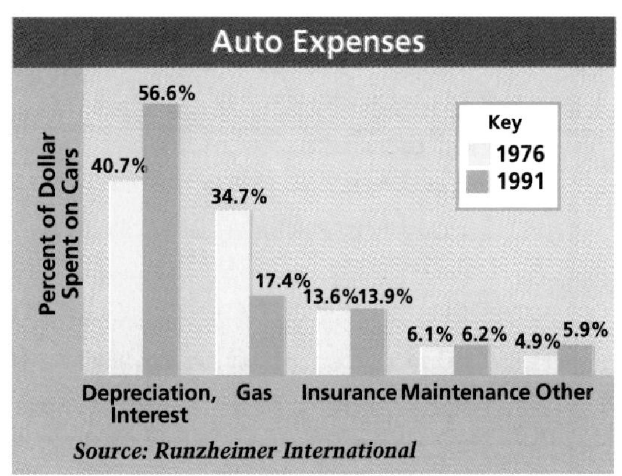

Source: Runzheimer International

35. *Sweater* A sweater you want is regularly priced at $19.95. The store is having a 20%-off sale. What is the sale price?

36. *Compact Disk Player* You buy a compact disk player on sale for $107.50. The sale price is 15% less than the original price. What was the original price? $126.47

34. ≈$4300 35. $15.96

Integrated Review

In Exercises 37–39, write the percent in decimal form.

37. 4% 0.04 **38.** 18% 0.18 **39.** 69% 0.69

In Exercises 40–42, write the decimal in percent form.

40. 1.34 134% **41.** 0.134 13.4% **42.** 0.0134 1.34%

In Exercises 43–48, solve the equation.

43. $6x = 3$ $\frac{1}{2}$ **44.** $5x = 2$ $\frac{2}{5}$ **45.** $108x = 100$ $\frac{25}{27}$

46. $96x = 50$ $\frac{25}{48}$ **47.** $16x = 9$ $\frac{9}{16}$ **48.** $110x = 9$ $\frac{9}{110}$

In Exercises 49–54, solve the proportion.

49. $\frac{2}{x} = \frac{x}{8}$ $4, -4$ **50.** $\frac{x}{3} = \frac{3}{x}$ $3, -3$ **51.** $\frac{x+1}{3} = \frac{1}{4x}$ $-\frac{3}{2}, \frac{1}{2}$

52. $\frac{3}{x-6} = \frac{1}{x}$ -3 **53.** $\frac{2}{2x-3} = \frac{x}{7}$ $\frac{7}{2}, -2$ **54.** $\frac{3x+1}{9} = \frac{10}{3x}$ $-\frac{10}{3}, 3$

Exploration and Extension

55. *Say That Again?* You earn 10% more money at your summer job than your sister does at hers. Does this mean that your sister earns 10% less than you? Explain your answer. No. Your sister earns $9\frac{1}{11}$% less.

56. *Tennis Shoes* You buy a pair of tennis shoes on sale for $48. The price is discounted 25% and the sales tax is 7%. Would it cost less to have the discount applied before the sales tax is added or after the sales tax is added? Explain. The cost would be the same, $38.52.

Enrichment Activity

This activity requires students to go beyond lesson goals.

Extend the project from Lesson 11.1 by finding examples of percents used in newspaper or magazine articles. For example, the following bar graph shows the salaries that some high-school students expect to earn after completing their education.

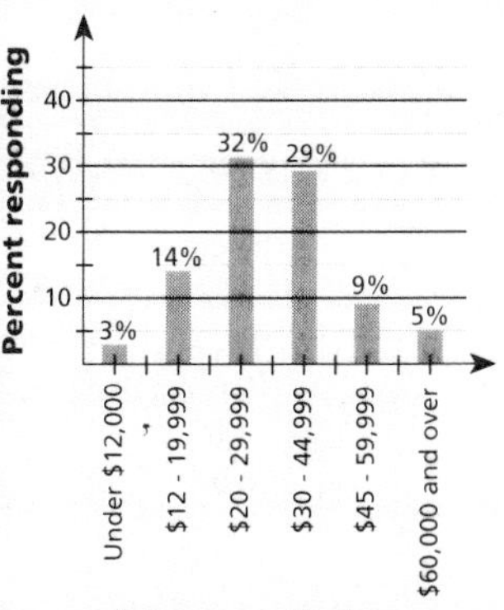

a. What percent of high-school students expect to earn from $20,000 up to $45,000? 61%

b. If 1879 high-school students were surveyed, how many more students said they expected to earn under $20,000 than $45,000 or more? about 56

c. Does the graph reflect the responses of all the 1879 students surveyed? Explain how you knew. No. The given percents total 92% so 8%, the responses of about 150 students, are not reflected in the graph.

Mixed REVIEW

9. 19.6608

1. Write $y - 3 = 2x + 6$ in standard form. **(5.5)** $-2x + y - 9 = 0$
2. Write $3x - 9 + 2x^2 = 0$ in standard form. **(9.2)** $2x^2 + 3x - 9 = 0$
3. Find the leading coefficient of $3x - 4x^2 + 9$. **(10.1)** -4
4. Solve the equation in Exercise 2. **(10.6)** $\frac{3}{2}, -3$
5. Simplify $(x - 1)(x + 2) + 3x - 5$. **(10.2)** $x^2 + 4x - 7$
6. Simplify $(xy^2 \div 3xy)(4x^{-2} \div 8x)$. **(8.3)** $\frac{y}{6x^3}$
7. Solve $8x + 2 = 3(4 - x)$. **(3.3)** $\frac{10}{11}$
8. Solve $8|x - 2| - 3 = 0$. **(4.8)** $\frac{19}{8}, \frac{13}{8}$
9. Evaluate $y = 3(1.6)^x$ when $x = 4$. **(8.1)**
10. Evaluate $|x + 3| + x^2$ when $x = -2$. **(2.3)** 5
11. Is 12 a solution of $m^2 - 3m \geq 110$? **(1.5)** No
12. Is $(4, -3)$ a solution of $y < -\frac{1}{4}x^2$? **(9.6)** No
13. Find the discriminant of $x^2 + 4x - 7$. **(9.5)** 44
14. Solve $x^2 + 4x - 7 = 0$. **(9.4)** $-2 + \frac{\sqrt{44}}{2}, -2 - \frac{\sqrt{44}}{2}$
15. Find the slope of the line $2x + 3y = 6$. **(4.4)** $-\frac{2}{3}$
16. Find the intercepts of the line $3x - y = 16$. **(4.3)** $(\frac{16}{3}, 0)(0, -16)$
17. Write $\frac{3}{8}$ in decimal form. **(11.2)** 0.375
18. Write 0.92 in percent form. **(11.2)** 92%
19. Evaluate $(3 \times 10^{-6})(6 \times 10^{10})$. **(8.4)** 1.8×10^5
20. Evaluate $(2.9 \times 10^{-3}) \div (4.6 \times 10^4)$. **(8.4)** 6.3×10^{-8}

Career Interview

Retailer

Melissa Durkee is a retailer specializing in Western wear, boots, Native American jewelry, and accessories.

Q: ***How did you get interested in retailing?***

A: I worked part-time as a salesperson through high school and most of college. From my experiences, I knew I wanted to and had the ability to work for myself. I wanted to be my own boss.

Q: ***How much math have you learned on the job?***

A: I haven't learned any new math skills per se. However, I am applying my understanding of math in different situations daily. The math skills and the knowledge I learned in school have sharpened my math business sense.

Q: ***What math classes did you take in high school?***

A: I took algebra and geometry.

Q: ***What would you like to tell kids about math?***

A: Almost every walk of life relies on some application of math to a certain degree. Math often ends up being the tool you'll rely on most in life.

11.3 Direct and Inverse Variation

What you should learn:

Goal 1 How to use direct variation and inverse variation

Goal 2 How to use direct variation and inverse variation in real-life settings

Why you should learn it:

You can solve many real-life problems using the direct variation (increase corresponds to increase) and inverse variation (increase corresponds to decrease) models.

Goal 1 Using Direct and Inverse Variation

Two variable quantities that have the same rate, or ratio, regardless of the values of the variables, have **direct variation.** For example, if you get paid \$5 per hour, then your total pay *varies directly* with the number of hours you work.

Total pay, p	\$10	\$20	\$30	\$40
Hours worked, t	2	4	6	8
Rate	$\frac{10}{2} = 5$	$\frac{20}{4} = 5$	$\frac{30}{6} = 5$	$\frac{40}{8} = 5$

The variables p and t are related by the equation $\frac{p}{t} = 5$ or $p = 5t$.

Models for Direct and Inverse Variation

1. The variables x and y **vary directly** if, for a constant k,
   $\frac{y}{x} = k$ or $y = kx$.
2. The variables x and y **vary inversely** if, for a constant k,
   $yx = k$ or $y = \frac{k}{x}$.

In both cases, the number k is the **constant of variation.**

Note that two quantities that vary inversely always have the same product.

Example 1 *Solving an Inverse Variation Problem*

The variables x and y vary inversely. When x is 3, y is 6. Find an equation that relates x and y.

Solution

$xy = k$ *Model for inverse variation*

$(3)(6) = k$ *Substitute 3 for x and 6 for y.*

$18 = k$ *Simplify.*

Thus, an equation that relates x and y is

$$xy = 18 \quad \text{or} \quad y = \frac{18}{x}.$$

■

Problem of the Day

A man brought 5 pieces of chain, each consisting of 3 links, to a shopkeeper and asked him to have them made into one continuous chain. The shopkeeper quoted him prices of \$1 per cut and \$1 per weld. What is the smallest total amount he could be charged? \$6

ORGANIZER

Warm-Up Exercises

1. Solve for the unknown.
   a. $45x = 75$ $\frac{5}{3}$
   b. $18y = 90$ 5
   c. $\frac{60}{x} = 48$ $\frac{5}{4}$
   d. $36 = \frac{78}{y}$ $\frac{13}{6}$
2. You are driving by car to see your aunt who lives 40 miles away. Complete the table showing how many hours the trip will take driving at the indicated speeds (or rates). Describe any patterns that you observe.

Distance, d	40	40	40	40	40
Speed, r	20	30	40	50	60
Time, t					

$t = 2, \frac{4}{3}, 1, \frac{4}{5}, \frac{2}{3}$; As the speed increases, the time decreases.

Lesson Resources

Teaching Tools
- Problem of the Day: 11.3
- Warm-up Exercises: 11.3
- Answer Masters: 11.3

Extra Practice: 11.3

Color Transparencies: 62, 63

LESSON Notes

GOAL 1 The table in Warm-Up Exercise 2 displays the inverse variation, $rt = 40$, or $t = \frac{40}{r}$.

Example 1

Observe that if two variables vary inversely, their product is constant.

Examples 2–3

Emphasize the connections to real-life situations. Ask if knowing the constant of variation is useful in solving similar problems.

Extra Examples

Here are additional examples similar to **Examples 1–3.**

1. Electric current, I, is measured in amperes. In a circuit, I varies directly as the voltage V. When 15 volts is applied, the current is 5 amperes. What is the current when 24 volts is applied? 8 amperes
2. The electric current, I, varies inversely as the resistance, R, of a conductor. (Resistance is measured in ohms.) If the current is 2 amperes when R is 60 ohms, what is the electric current when the resistance is 240 ohms? 0.5 amperes
3. The driving manual states that a car traveling 45 mph will require 150 feet to stop after the brakes have been applied. If D is the stopping distance of a car after the brakes have been applied and D varies directly as the square of the speed R, how far will a car travel before stopping if the brakes are applied when the car is traveling at a speed of 65 mph? about 313 ft

Goal 2 Direct and Inverse Variation in Real Life

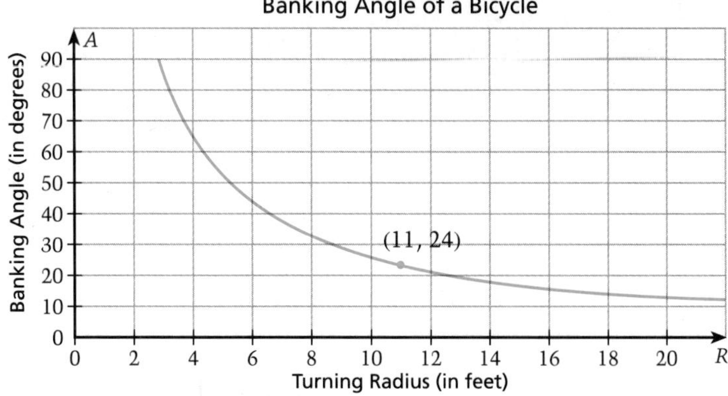

This graph shows the "banking angles" for a bicycle whose speed is 15 miles per hour. The banking angle and the radius of the turn vary inversely. As the radius of the turn gets smaller, the bicyclist must lean at greater angles to avoid falling over.

Real Life **Bicycling**

Bicycle race in San Rafael, California

Example 2 *Finding a Model for "Banking Angles"*

Find a model that relates the banking angle, A, with the turning radius, R, for a bicyclist whose speed is 15 miles per hour.

a. Use the model to find the banking angle for a turning radius of 8 feet.

b. Does the banking angle increase or decrease as the turning radius decreases?

Solution

a. From the graph, you can see that A is 24° when the turning radius, R, is 11 feet.

$A = \frac{k}{R}$ *Model for inverse variation*

$24 = \frac{k}{11}$ *Substitute 24 for A and 11 for R.*

$264 = k$ *Solve for k.*

The model is given by

$$A = \frac{264}{R}$$

where A is measured in degrees and R is measured in feet. When the turning radius is 8 feet, the banking angle is

$$A = \frac{264}{8} = 33°.$$

You can use the graph to confirm this result.

b. From the graph, you can see that the banking angle increases as the turning radius decreases. ■

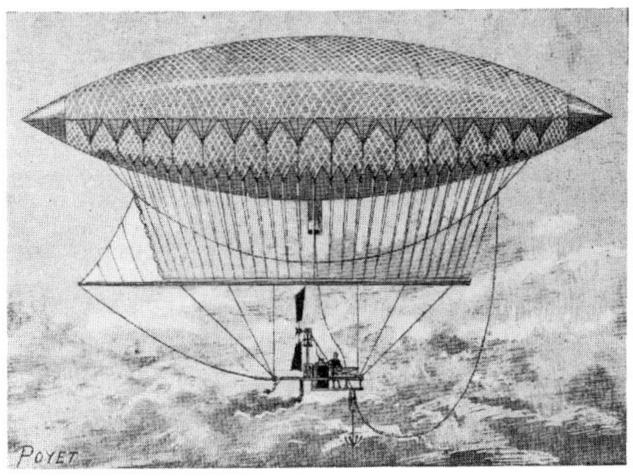

The first airship, built by Henri Giffard in 1852, had a volume of 8.83 × 10^4 cubic feet. The* Graf Zeppelin II*, 1933, was the largest (by volume) rigid airship ever built. Its volume was 7.06 × 10^6 cubic feet.

Real Life
Aeronautics

Example 3 *Finding a Model of Airship Weight*

The weight that an airship can support varies directly with its volume. Henri Giffard's airship could support a weight of 5650 pounds. Find a model that relates the volume, V (in cubic feet), to the weight, W (in pounds), that an airship filled with hydrogen can support. Use the model to find the weight that the *Graf Zeppelin II* could support. (Both airships were filled with hydrogen.)

Solution

$W = kV$	*Model for direct variation*
$5650 = k(88{,}300)$	*Substitute 5650 for W, 88,300 for V.*
$\frac{5650}{88{,}300} = k$	*Divide both sides by 88,300.*
$0.064 \approx k$	*Solve for k.*

The model is $W = 0.064V$. Since the *Graf Zeppelin II* had a volume of 7.06×10^6 cubic feet, it could support a weight of

$$W = 0.064(7{,}060{,}000) = 451{,}840 \text{ pounds.}$$

■

Communicating about ALGEBRA

SHARING IDEAS about the Lesson

Consider an Alternative Hydrogen is "lighter" than helium.

A. If Giffard's airship had been filled with helium, it could have only supported a weight of 5120 pounds. How much could the *Graf Zeppelin* have supported if it had been filled with helium? 409,480 pounds

B. Today's airships use helium rather than hydrogen. Since hydrogen can support more weight, why is helium used?

Hydrogen is flammable.

Check Understanding

1. Define direct variation and inverse variation. See the top of page 575.
2. Construct a table that displays a real-life example of direct variation. For example, the area of a rectangle whose width is 30 cm depends on its length.

Area	15	30	45	60	900
Length	0.5	1	1.5	2	30

3. Construct a table that displays a real-life example of inverse variation. For example, in a rectangle whose area is 30 cm^2, the length and width vary inversely.

Length	1	2	3	5	6	10	20
Width	30	15	10	6	5	3	1.5

Communicating about ALGEBRA

Students should investigate the Hindenburg disaster, which happened on May 6, 1937, for the answer to Part B.

EXERCISE Notes

ASSIGNMENT GUIDE

Basic/Average: Ex. 1–10, 11–27 odd, 28, 30, 32, 41–42

Above Average: Ex. 1–10, 11–29 odd, 30–32, 41–42

Advanced: Ex. 1–10, 11–29 odd, 30–32, 43

Selected Answers
Exercises 1–6, 7–39 odd

More Difficult Exercises
Exercises 29–32, 43

Guided Practice

Ex. 1–6 These exercises focus on the meanings of direct and inverse variation.

Independent Practice

Ex. 11–18 These exercises are designed to give students practice in a skill that is necessary for success in Exercises 19–32.

EXERCISES

Guided Practice

CRITICAL THINKING about the Lesson

1. Give a real-life example of two variable quantities that vary directly. See below.

2. Give a real-life example of two variable quantities that vary inversely.

In Exercises 3–6, classify each equation as direct or inverse variation, or neither.

3. $\frac{x}{y} = 33$ Direct

4. $y = 5x + 13$ Neither

5. $a = 10b$ Direct

6. $ab = 7$ Inverse

Independent Practice

In Exercises 7–10, state whether the variables have direct variation, inverse variation, or neither.

7. *Speed and Distance* You are riding your bike at an average speed of 12 miles per hour. The number of miles you ride, d, during t hours is given by $d = 12t$. Direct

8. *Mass and Volume* The mass, m, and volume, v, of a substance are related by the equation $2v = m$, where 2 is the density of the substance. Inverse

9. *Pieces Eaten and Pieces to Eat* The number of pieces, p, of pizza that Alicia ate for dinner and the number of pieces, q, she can eat for breakfast are related by $p = k + q$. Neither

10. *Waiting Time and People in Line* The time, t, spent waiting in line for the Super Looper roller coaster, and the number of people, n, in the line are related by the equation $t = kn$. Direct

In Exercises 11—14, the variables *x* and *y* vary directly. Given one pair of values for *x* and *y*, find an equation that relates the variables.

11. $x = 4,\ y = 8$
$y = 2x$

12. $x = 6,\ y = 24$
$y = 4x$

13. $x = 18,\ y = 4$
$y = \frac{2}{9}x$

14. $x = 22,\ y = 6$
$y = \frac{3}{11}x$

In Exercises 15–18, the variables *x* and *y* vary inversely. Given one pair of values for *x* and *y*, find an equation that relates the variables.

15. $x = 1,\ y = 4$
$xy = 4$

16. $x = 10,\ y = 3$
$xy = 30$

17. $x = \frac{1}{2},\ y = 7$
$xy = \frac{7}{2}$

18. $x = \frac{3}{4},\ y = 4$
$xy = 3$

In Exercises 19–22, find an equation that relates the two variables.

19. *Volume and Pressure* The volume, V, of a gas at a constant temperature varies inversely with the pressure, P. When the volume is 100 cubic inches, the pressure is 25 pounds per cubic inch. $VP = 2500$

20. *Time and Typing Speed* The time, t (in minutes), it takes for Roger to type his 2500-word term paper varies inversely with the number of words per minute, r, he can type. $tr = 2500$

1. Examples vary; distance and rate when time is constant, circumference and diameter of a circle
2. Examples vary; rate and time when distance is constant, distance from the fulcrum of a lever and weight.

21. *Pay and A's* Your great uncle sends you two hamburger gift certificates for each A you receive on your report card. The number, g, of gift certificates he sends varies directly with the number, n, of A's on your report card. $g = 2n$

22. *Percent and Correct Answers* The percent grade, p, on the exam varies directly with the number, n, of correct answers. (There are 20 questions each worth 5 points.) $p = 5n$

In Exercises 23 and 24, assume the variables vary directly.

23. If $x = 4$ when $y = 16$, find x when $y = 8$. 2

24. If $c = 6$ when $d = 3$, find c when $d = 5$. 10

In Exercises 25 and 26, assume the variables vary inversely. 25. 7

25. If $m = 2$ when $n = 7$, find m when $n = 2$.

26. If $q = \frac{1}{2}$ when $r = \frac{1}{3}$, find q when $r = 6$. $\frac{1}{36}$

27. *Pulse Rate* Your pulse rate, p, varies directly with the volume, b, of blood pumped from your heart each minute. Each time your heart beats, it pumps approximately 0.006 liter of blood. Find an equation that relates p and b. Take your pulse and find out how much blood your heart pumps per minute. $b = 0.006p$

28. *Weight* Weight depends on gravity. Neil Armstrong, the first man on the moon, weighed 360 pounds on Earth, including his heavy equipment, but only 60 pounds on the moon, with equipment. If the first woman in space, Valentina V. Tereshkova, had landed on the moon and weighed 54 pounds, with equipment, how much would she have weighed on Earth, with equipment? 324 pounds

29. *Mining Danger* One of the dangers of coal mining is the methane gas that can leak out of seams in the rock. Methane forms an explosive mixture with air at a concentration of 5 percent or greater. Suppose a steady leak of methane began in a coal mine so that the concentration of methane gas varied directly with time. Twelve minutes after the leak began, the concentration of methane in the air was 2 percent. If the leak continues at the same rate, when could an explosion occur? 30 minutes

30. *Snowshoes* When a person walks, the pressure, P, on each boot sole varies inversely with the area, A, of the sole. Denise is trudging through deep snow, wearing boots that have a sole area of 29 square inches each. The boot-sole pressure is 4 pounds per square inch. If Denise was wearing snowshoes, each with an area 11 times that of her boot soles, what would be the pressure on each snowshoe? The constant of variation in this problem is Denise's weight. How much does she weigh?
≈ 0.36 pounds per in.^{2}, 116 pounds

Snowshoeing on Mt. Osceola, N.H.

Ex. 27 GROUP ACTIVITY
Remind students that the direct variation $b = 0.006p$ is a linear model. To reinforce this concept, collect the answers from each student and graph the results. Students should observe that the points are part of a linear model.

▶ **Ex. 31–32** These exercises provide students with a visual model of an inverse variation. This model will be studied in Chapter 12 in a lesson entitled Rational Functions.

▶ **Ex. 41–43** These exercises review concepts from Lesson 11.2.

Enrichment Activities

These activities require students to go beyond lesson goals.

A good example of inverse variation is a "fixed area, variable perimeter" problem. For example, consider a rectangle with a fixed area of 12 units.

1. Make a table of values of possible lengths and widths.

Length	0.5	1	2	3	4	6	12
Width	24	12	6	4	3	2	1

2. Use the table to graph the length versus width.

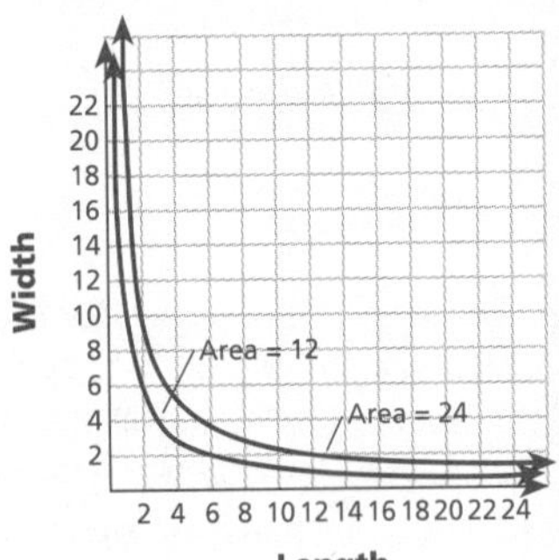

3. Repeat the activity for a fixed area of 24 square units. What do you notice about the graph?

Length	1	2	3	4	6	8	12	16	24
Width	24	12	8	6	4	3	2	1.5	1

The graph has a similar shape. Each y-coordinate is double the corresponding y-coordinate of the graph of $lw = 12$.

31. *Oil Spill* The graph shows the percent, p, of oil that remained in Chedabucto Bay, Nova Scotia, after an oil spill. The cleaning of the spill was left primarily to natural actions such as wave motion, evaporation, photochemical decomposition, and bacterial decomposition. After about a year, the percent that remained varied inversely with time. Find a model that relates p and t, where t is the number of years since the spill. Then use your model to find the amount of oil that remained $6\frac{1}{2}$ years after the spill. $pt = 114$, $\approx 17.5\%$

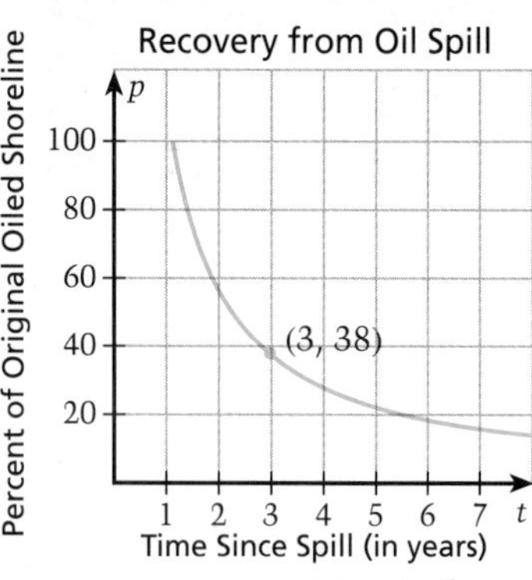

32. *Ocean Temperatures* The graph shows the temperature of the water in the north central Pacific Ocean. At depths greater than 900 meters, the water temperature varies inversely with the water depth. Find a model that relates the temperature, T, with the depth, d. What is the temperature at a depth of 4385 meters? $Td = 4{,}000$, $\approx 0.91°C$

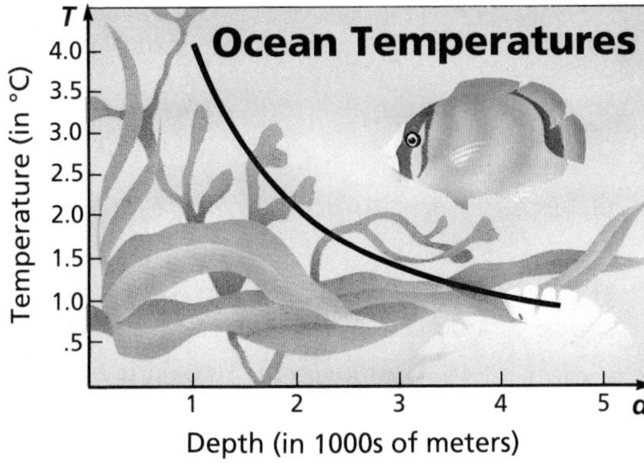

Integrated Review

In Exercises 33–36, solve the proportion.

33. $\frac{a}{20} = \frac{1}{5}$ 4

34. $\frac{3}{x} = \frac{4}{9}$ $\frac{27}{4}$

35. $\frac{3}{8} = \frac{3}{2d}$ 4

36. $\frac{5}{17} = \frac{r}{9}$ $\frac{45}{17}$

In Exercises 37–40, write the percent in decimal form.

37. 50% 0.5

38. 10% 0.1

39. 120% 1.20

40. 3% 0.03

Exploration and Extension

41. *Jacket Sale* A jacket is on sale for \$58. The regular price is \$70. What is the percent of discount? 17.14%

42. *Tennis Shoes* Your friend bought a pair of tennis shoes last week at the regular price. Today in the same store, you found the same kind of shoes at 20% off. You paid \$43.96. What did your friend pay? \$54.95

43. *Photocopies* Some photocopy machines can reduce and enlarge pictures. If a photocopier reduces a picture 75%, its new length, l, and width, w, vary directly with its original length, L, and width, W. Write two equations: one relating l and L and the other relating w and W. Then use the equations to find a model that relates the new area, a, with the original area, A. Is the area reduced by 75%? Explain. See below.

43. $l = 0.25L$, $w = 0.25W$, $a = 0.0625A$. No, it is reduced by $(100 - 6\frac{1}{4})\% = 93\frac{3}{4}\%$.

11.4 Exploring Data: Probability

What you should learn:

Goal 1 How to find the probability of an event

Goal 2 How to use probability in real-life settings

Why you should learn it:

You can use probability to anticipate results and to help you make decisions.

Goal 1 Finding the Probability of an Event

The **probability of an event** is a number between 0 and 1 that indicates the likelihood the event will occur. An event that is certain to occur has a probability of 1. An event that is certain to *not occur* has a probability of 0. An event that is **equally likely** to occur or not occur has a probability of $\frac{1}{2}$, or 0.5. For instance, the probability that a tossed coin will land heads up is 0.5.

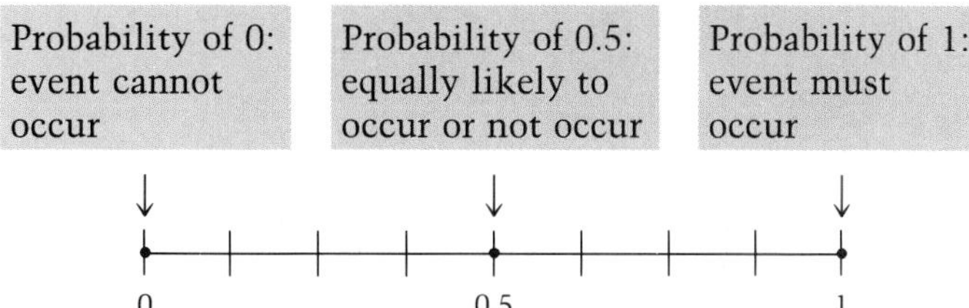

Finding the Probability of an Event

The **probability,** P, of an event is the ratio of successful outcomes, called **successes,** to all outcomes of the event, called **possibilities.**

$$P = \frac{\text{Number of successes}}{\text{Number of possibilities}}$$

Example 1 *Finding the Probability of an Event*

a. An algebra class has 17 boys and 16 girls. One student is chosen at random from the class. The probability, P, that the student is a girl is

$$P = \frac{\text{Number of girls}}{\text{Number of students}}$$
$$= \frac{16}{33} \approx 0.485, \quad \text{(a little less than one half).}$$

b. You are carrying seven cassette tapes in a backpack. On the way to school, you lean against a building and break one of the tapes. The probability that it is one of your two favorites is

$$P = \frac{\text{Number of favorite tapes}}{\text{Number of tapes}} = \frac{2}{7}.$$ ■

Problem of the Day

A group of more than 2 but less than 20 voters each paid the same number of dollars to attend a fund-raiser. The total contribution from the group was \$493. How many were in the group? 17

ORGANIZER

Warm-Up Exercises

1. Compute each ratio to three decimal places. Use a calculator if available.
   a. $\frac{14}{34}$ 0.412 **b.** $\frac{23}{50}$ 0.460
   c. $\frac{9}{24}$ 0.375 **d.** $\frac{14}{40}$ 0.350
2. **AN EXPERIMENT** Toss a coin 20 times. Record the number of heads and the number of tails that occur. Express these ratios as decimal numbers.

   $\frac{\text{number of heads}}{\text{20 tosses}}$

   $\frac{\text{number of tails}}{\text{20 tosses}}$

GROUP ACTIVITY Compare your outcomes with those of your classmates. Combine the outcomes of everyone in class.

a. What decimal numbers describe the number of heads and tails tossed in the class?

b. How do the combined class numbers compare with your results? **Answers will vary, but the combined-class outcomes should be nearly 0.50 for both heads and tails.**

Lesson Resources

Teaching Tools
  Problem of the Day: 11.4
  Warm-up Exercises: 11.4
  Answer Masters: 11.4
Extra Practice: 11.4
Color Transparencies: 63, 64
Technology Handbook: p. 81, 83

LESSON Notes

Example 1

A successful outcome is the occurrence of the desired event. Each outcome should be as likely to occur as any other. For instance, in Part b, it must be assumed that the tapes were placed in the backpack at random, that is, in no particular order.

Example 4

Point out the connection between the percent who said "yes" and the probability of saying "yes."

Extra Examples

Here are additional examples similar to **Examples 1–4.**

1. In Example 3, what is the probability that a person did not attend a jazz concert? $P = 0.9$
2. In Example 4, what is the probability that a high school student in 1986, picked at random, would express an interest in having lots of money? $P = 0.62$
3. Assume you have a standard deck of 52 playing cards that have been shuffled well.
   a. What is the probability that you will pick a 10 from the deck? about 0.08
   b. What is the probability that you will pick a face card (a Jack, Queen, or King)? about 0.23
   c. What is the probability that you will select a black card (a club or spade)? 0.5

Goal 2 Using Probability in Real-Life Settings

Real Life **Parachuting**

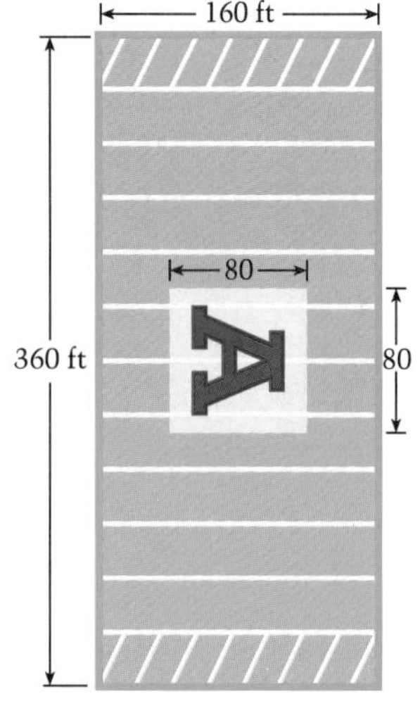

Example 2 *Finding the Probability of an Event*

You are on a committee to plan the halftime show for a football game. As part of the show, you hire a parachutist to land in the middle of the field. You want the parachutist to land in the center of the field on the square displaying the school emblem. The parachutist will guarantee landing on the field (otherwise the school will not be charged), but will not guarantee landing on the emblem. What is the probability that the parachutist will land on the emblem? (Assume that the landing on the field is sure to occur and that each spot on the field is an equally likely landing space.)

Solution The probability of landing on the emblem is

$$P = \frac{\text{Area of emblem}}{\text{Area of field}} = \frac{(80)(80)}{(160)(360)} = \frac{6400}{57{,}600} = \frac{1}{9}.$$

There is a 1 in 9 chance that the parachutist will land on the emblem. ■

Real Life **Performing Arts**

Mikhail Baryshnikov and Deirdre Carberry in a ballet performance.

Example 3 *Attendance at Performing Arts*

The table shows the results of a survey of 2300 Americans (18 and older). The numbers indicate how many of the 2300 attended at least one performance of the type listed during a year. *Based on this survey,* if an American (18 or older) is chosen at random, what is the probability that the person attended a ballet performance during the past year? What is the probability that the person attended a musical play?

(Source: U.S. National Endowment for the Arts)

Type of Performance	Jazz Concert	Classical Music	Opera	Musical Play	Other Play	Ballet
Number	230	299	69	391	276	92

Solution The probability that the person attended a ballet is

$$P = \frac{92}{2300} = 0.04. \qquad \textit{Probability of attending ballet}$$

The probability that the person attended a musical play is

$$P = \frac{391}{2300} = 0.17. \qquad \textit{Probability of attending musical}$$ ■

Percent of Americans Owning a Pet for Particular Reasons

Reason	Cat owners	Dog owners
Someone to play with	93%	90%
Companionship	84%	83%
Help children learn responsibility	78%	82%
Security	51%	79%

The bar graph at the right shows the results of a Gallup poll in which cat and dog owners were asked to state reasons why they owned their cats or dogs. For instance, security was a reason named by more dog owners than cat owners.

Example 4 *Using the Results of a Survey*

In a Gallup poll, a randomly selected dog owner was asked if he or she owned a dog for companionship. What is the probability that the person said yes? What is the probability that the person did not say yes?

Solution From the results of the Gallup poll, 83% of the dog owners said they owned their dogs for companionship. This means that 17% did not say they owned their dogs for companionship.

$$Y = \frac{83}{100} \quad \textit{Percent who said yes}$$
$$= 0.83 \quad \textit{Probability of saying yes}$$
$$N = \frac{17}{100} \quad \textit{Percent who did not say yes}$$
$$= 0.17 \quad \textit{Probability of not saying yes}$$

Note that the sum of the two probabilities is one. ■

Communicating *about* ALGEBRA

Cooperative Learning

SHARING IDEAS about the Lesson

Work with a Partner and Explain the Meaning See margin.

A. Explain what is meant by an 80% probability of rain.

B. Explain what is meant by a 50% probability of winning.

Check Understanding

1. Name three real-life situations in which probabilities are used. See applications in lesson.
2. Explain the concept of equally likely events. See top of page 581.
3. How are probabilities expressed? As ratios
4. How are probabilities related to fractions? To percents? Ratios can be expressed as fractions or as percents

COOPERATIVE LEARNING

As part of a cooperative group, identify other instances in the media (TV, magazines, newspapers, etc.) in which probabilities are used. Make a list of the different real-life situations found and categorize them in some logical manner.

Answers

A. Of all days on record with comparable meteorological conditions, rain occurred on 80% of the days.

B. Of all the times in a comparable situation, a person won 50% of the times.

EXERCISE Notes

ASSIGNMENT GUIDE

Basic/Average: Ex. 1–6, 7–11, 13–14, 17–19, 23–25

Above Average: Ex. 1–6, 7–14, 17–22, 23–25

Advanced: Ex. 1–6, 11–30

Selected Answers
Exercises 1–6, 7–27 odd

More Difficult Exercises
Exercises 12, 15–16, 29–30

Guided Practice

Ex. 1–3 Remind students that the probability of an event is a ratio from 0 through 1.

Ex. 4–6 It is easier for students when they know the format you expect the answers in, that is, as simplified fractions, as decimals, or as percents.

Independent Practice

Ex. 10–11 Students will need to multiply the total outcomes by the probability to obtain the required solutions. Exercise 11 provides a connection to language arts.

584

EXERCISES

Guided Practice

1. 0; The sun sets in the morning.
2. 1; The sun rises in the morning.

CRITICAL THINKING about the Lesson

1. What is the probability of an event that *cannot* occur? Give an example.
2. What is the probability of an event that *must* occur? Give an example.
3. How do you find the probability of an event?
   Find the ratio of successful outcomes to all outcomes.

Marbles in a Jar **In Exercises 4–6, consider a jar that contains 12 blue marbles, 8 green marbles, and 5 yellow marbles. Without looking at the colors, you reach into the jar and choose one marble.**

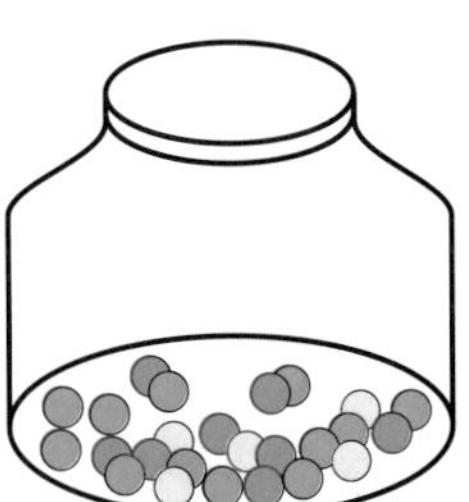

4. What is the probability that the marble is blue? 0.48
5. What is the probability that the marble is not blue? 0.52
6. What is the probability that the marble is not red? 1

Independent Practice

In Exercises 7–8, find the probability of an event given the number of successes and the number of possibilities.

7. Number of successes:12 0.1875
   Number of possibilities: 64
8. Number of successes: 2 0.2
   Number of possibilities: 10
9. *Archaeology* You are working on an archaeological site during the summer. One day you sift through 16 tables of soil and find artifacts in 3 of the tables. The next day your mother and stepfather visit and watch as you sift through one table of soil. What is the probability that they get to see you find an artifact? $\frac{3}{16}$
10. *Junior Achievement* Junior Achievement is an organization that teaches students how the American business system operates. The members manufacture and sell their own products. The probability of a member being a student is 0.95. How many of the 827,000 members are students? 785,650
11. *Emily Dickinson* Emily Dickinson (1830–1886) is one of America's most loved poets, yet during her lifetime, people rarely got the chance to read her poems. In fact, the probability that a particular poem was published during her lifetime is only 0.006. How many of her 1700 poems were published when she was living? From the Dickinson poem at the right, would you guess that the poems published while she was alive were done so with her permission or without her permission? 10, without

I'm Nobody! Who are you?
Are you—Nobody too?
Then there's a pair of us!
Don't tell! They'd advertise you know!

How dreary—to be—Somebody!
How public—like a Frog—
To tell one's name—the live long June—
To an admiring Bog!

12. *Baby Sea Turtles* A sea turtle buries 200 eggs in the sand. After the baby turtles hatch and dig to the surface, they scramble to make it to the relative safety of the ocean. Birds and other predators make the journey tough for the babies. Of the 176 babies that hatched, 125 were eaten by birds, and 39 were eaten by other predators. The rest made it to the ocean. Suppose that just after the eggs were laid, you marked one with an X. What is the probability that your egg hatched and the baby turtle made it to the ocean? $\frac{3}{50}$

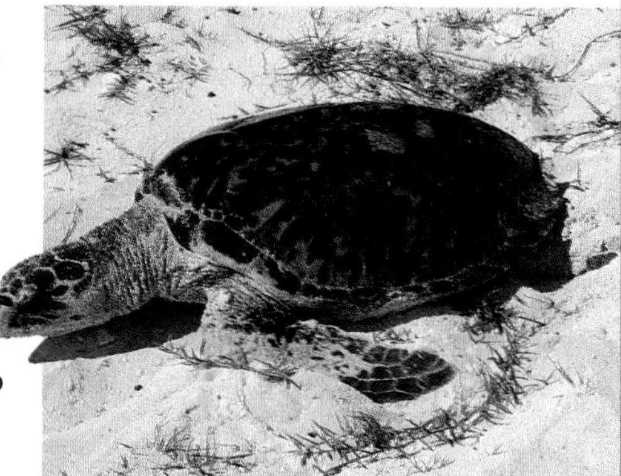

Meteorite! **In Exercises 13 and 14, use the following information.**

The Willamette meteorite was found in the Willamette Valley in Oregon in 1902. It is the largest found in the United States.

13. Earth contains 57,510,000 square miles of land and 139,440,000 square miles of water. What is the probability that a meteorite that hits Earth will fall onto land? $\frac{1917}{6565}$

14. What is the probability that a meteorite that hits Earth will fall into water? $\frac{4648}{6565}$

Bull's-Eye **In Exercises 15 and 16, use the following information.**

The bull's-eye of a standard dart board has a radius of about 1 inch. The inner circle has a radius of 5 inches, and the outer circle has a radius of 9 inches. Assume that when a dart is thrown at the board, the dart is equally likely to hit any point inside the outer circle.

15. What is the probability that a dart that hits the dart board lands on the bull's-eye? $\frac{1}{81}$

16. What is the probability that a dart that hits the dart board lands between the inner and outer rings? $\frac{56}{81}$

Farm Population **In Exercises 17 and 18, use the table, which shows the number of Americans in thousands who lived on farms in 1988.**

Age group	15 and under	16–19	20–24	25–44	45–64	65 and older
Number	1000	3951	421	1528	1305	697

17. Of all people who lived on a farm in 1988, one is chosen at random. What is the probability that the person was fifteen years old or younger in 1988? $\frac{500}{4451}$

18. In 1988, the United States population was 246,329,000. If a person was chosen at random from the population, what was the probability that the person lived on a farm? $\frac{8902}{246{,}329}$

Ex. 13 Students are usually surprised that so much of the surface of the earth is water.

Ex. 13–14, 15–16, 17–18, 19–20, and 21–22 These exercises should be assigned in pairs.

Ex. 15–16 Remind students that the area of a circle is equal to πr^2.

Ex. 17–20 TECHNOLOGY Students will need to use calculators to express these probabilities as decimals.

Enrichment Activities

These activities require students to go beyond lesson goals.

The Algebra I students of Lincoln High wanted to know which brand of orange juice Lincoln High students preferred. So they assigned students to the cafeteria entrances for each lunch period to ask other students to take part in a blind taste test. When the students combined their results, they found that they had sampled 296 different students. The results by grade level are shown in the table.

	9th	10th	11th	12th
Brand A	17	39	27	62
Brand B	28	42	38	43

1. a. If you were to pick one of the surveyed students at random, what is the probability that the student is a tenth-grader? $\frac{81}{296}$ or 27.4%
How does this percentage compare with the percent of students surveyed who were tenth-graders? It is the same.
b. Here is an example of conditional probability. Given that the random student is a tenth-grader, what is the probability that the student prefers Brand A? $\frac{39}{81}$ or $\approx$ 48.1%
c. There are 417 tenth-graders at Lincoln High School. Based on this sample, how many of them do you predict would prefer Brand A? about 201
d. Based on the sample, is there a clear favorite? No, the results are about equal. Brand B was chosen by about 51% compared to Brand A at 49%.

Immigrants **In Exercises 19 and 20, use the following information.**

The graph and table show the number of immigrants who have come to the United States during each decade since 1900.

Decade	1901–1910	1911–1920	1921–1930
Number	8,795,000	5,736,000	4,107,000
Decade	1931–1940	1941–1950	1951–1960
Number	528,000	1,035,000	2,515,000
Decade	1961–1970	1971–1980	1981–1990
Number	3,322,000	4,493,000	6,161,000

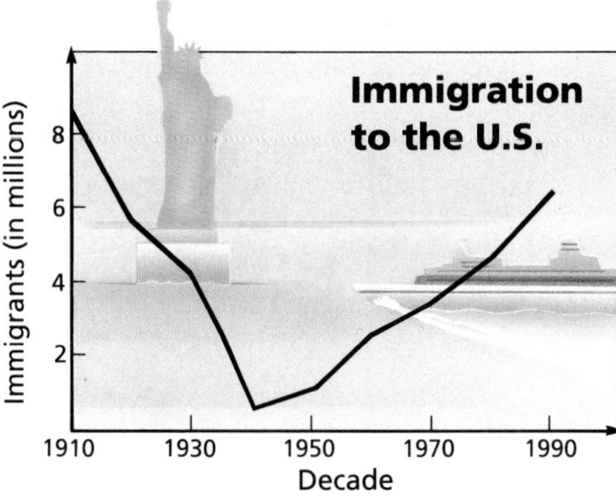

19. What is the probability that a person who immigrated to the United States between 1901 and 1990 came before 1941? $\frac{9583}{18,346}$

20. What is the probability that a person who immigrated to the United States between 1901 and 1990 came after 1940? $\frac{8763}{18,346}$

In Exercises 21 and 22, use the information given in Example 4 of this section.

21. In 1986, a randomly selected high school senior was asked if having a lot of money was important. What is the probability that the student said yes? $\frac{31}{50}$

22. In 1976, what was the probability that a randomly selected high school senior would say that he or she wanted 0 children? $\frac{3}{100}$

Integrated Review

In Exercises 23–25, simplify the expression.

23. $\frac{2x}{9} \div \frac{x}{3}$ $\frac{2}{3}$

24. $\frac{x^2 - 1}{x} \div \frac{x + 1}{3}$ $\frac{3(x-1)}{x}$

25. $\frac{32x + x^2}{2} \div \frac{x}{4}$ $2(32 + x)$

26. What is 90% of 50? 45

27. 11 is what percent of 185? 5.9%

28. 159 is 75% of what number? 212

Exploration and Extension

In Exercises 29 and 30, use the following information.

The *odds* of an event occurring is the ratio of the number of ways the event can occur to the number of ways the event cannot occur. In reduced form, odds of $\frac{a}{b}$ is read "*a* to *b*."

29. *Raffle Tickets* There are 600 tickets in a raffle. You buy 12 of them. What are the odds that the winning ticket is one of yours? $\frac{1}{49}$

30. *Collecting Insects* Your insect collection has several duplicates: 6 butterflies, 2 crickets, 8 grasshoppers, and 2 walkingsticks. Your cousin asks if he can have one and you agree. What are the odds that your cousin will choose a butterfly? (Assume that each insect is equally likely to be chosen.) $\frac{1}{2}$

Mid-Chapter SELF-TEST

Take this test as you would take a test in class. The answers to the exercises are given in the back of the book.

In Exercises 1–12, solve the proportion. (11.1)

1. $\frac{4}{x} = \frac{5}{6}$ $\frac{24}{5}$

2. $\frac{3}{7} = \frac{12}{x}$ 28

3. $\frac{a}{3} = \frac{5}{27}$ $\frac{5}{9}$

4. $\frac{a}{26} = \frac{5}{13}$ 10

5. $\frac{y}{27} = \frac{3}{y}$ 9, −9

6. $\frac{2y}{5} = \frac{40}{y}$ 10, −10

7. $\frac{x}{2x+1} = \frac{-1}{x}$ −1

8. $\frac{x+4}{x} = \frac{-6}{x-4}$ −8, 2

9. $\frac{x-3}{-3} = \frac{2x-3}{x-3}$ 0

10. $\frac{y+2}{5} = \frac{4}{y+3}$ −7, 2

11. $\frac{y+7}{6} = \frac{y+5}{3}$ −3

12. $\frac{5}{y-3} = \frac{6}{y+2}$ 28

13. How much is $6\frac{1}{2}\%$ of four dollars? **(11.2)** \$0.26

14. How many students is 75% of 1248 students? **(11.2)** 936

15. 24 feet is 80% of what length? **(11.2)** 30

16. 100 bottles is 26% of how many? **(11.2)** ≈385

17. 38 grams is what percent of 56 grams? **(11.2)** 67.9%

18. 17 meters is what percent of 68 meters? **(11.2)** 25%

19. Japan's citizens have four times as much money in their savings accounts as United States citizens have. Belgium's citizens have two times as much as United States citizens. Write a sentence that compares the average Japanese savings account with the average Belgian savings account. **(11.1)** $j = 2b$

20. A survey was conducted with 1540 junior and senior high school students. Of the 1540, 55% planned to attend college full-time, 26% planned to work and go to college part-time, 11% planned to go into the armed forces, 7% planned to work, and 1% were undecided. How many planned to attend college either part-time or full-time? How many planned to enter the armed forces? **(11.2)** 1247, 169

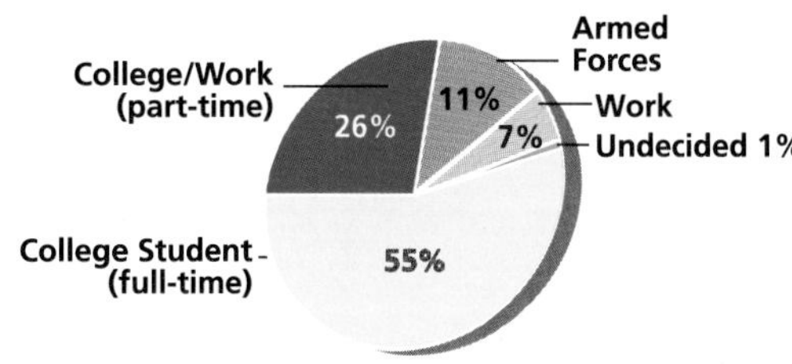

21. The variables a and b vary inversely. When a is 7, b is 35. Find an equation that relates a and b. If a is 5, what is b? **(11.3)** $ab = 245$, 49

22. You put \$5 into one of 6 pockets of your jeans, but you can't remember which pocket. What is the probability that you will find the money in the first pocket you check? After checking two pockets without success, what is the probability that the \$5 will be in the next pocket you check? **(11.4)** $\frac{1}{6}, \frac{1}{4}$

23. You have been mailed a discount catalog in which all regular prices have been discounted 20%. From this catalog you order \$60 worth of merchandise but you pay an additional \$5 for shipping and handling. What percent did you really save when you consider the \$5 charge for shipping and handling? **(11.2)** $13\frac{1}{3}\%$

2. COOPERATIVE LEARNING Work with a partner and design a similar survey that could be used in your school. Topics might include finding out students' preferences of juice or soft drink flavors, favorite classes, favorite colors, or favorite performing artists or television shows. Write some probability questions similar to those in this lesson that your survey could be used to answer.

3. WRITING Write a report on surveying as a sampling method. Include comments about how you ensured that the survey was random and unbiased, the difficulties you may have encountered in getting responses, and whether you would do anything different if you had to conduct another survey.

Problem of the Day

Lisa sold 50% of her books, then gave away 60% of those that were left, and then had 8 books left. How many did she have to begin with? 40

ORGANIZER

Warm-Up Exercises

1. Simplify.
   a. $\frac{45}{60}$ $\frac{3}{4}$
   b. $\frac{36}{252}$ $\frac{1}{7}$
   c. $\frac{360}{375}$ $\frac{24}{25}$
   d. $\frac{48}{54}$ $\frac{8}{9}$
2. Factor.
   a. $x^2 - 2x$ $x(x-2)$
   b. $x^2 - 3x - 10$ $(x+2)(x-5)$
   c. $6x^2 - 5x - 4$ $(2x+1)(3x-4)$
   d. $4x^2 + 9x - 9$ $(4x-3)(x+3)$
3. Evaluate.
   a. $\frac{3x-4}{x-3}$ at $x = 2$ -2
   b. $\frac{(x+8)(2x-5)}{x}$ at $x = -1$ 49
   c. $\frac{12(x^2 - 2x)}{x - 4x^2}$ at $x = 3$ $-\frac{12}{11}$

Lesson Resources

Teaching Tools
 Problem of the Day: 11.5
 Warm-up Exercises: 11.5
 Answer Masters: 11.5

Extra Practice: 11.5

Color Transparency: 64

LESSON Notes

GOAL 1 The domain of a rational expression must be determined for the original expression. Consequently, when a rational expression is reduced, its domain remains that of the original expression even if the simplified expression

11.5 Simplifying Rational Expressions

What you should learn:

Goal 1 How to simplify a rational expression

Goal 2 How to use rational expressions as real-life models

Why you should learn it:

You can use rational expressions as models of real-life situations such as in relating air pressure to altitude.

Goal 1 Simplifying a Rational Expression

A fraction whose numerator and denominator are polynomials is called a **rational expression.** Some examples are

$$\frac{3}{x+4}, \quad \frac{2x}{x^2-9}, \quad \text{and} \quad \frac{6}{x^2+1}.$$

The **domain** of a rational expression is the set of all real numbers *except* those for which the denominator is zero. For instance, the domain of the first expression is all real numbers except -4.

A rational expression is **simplified** or **in reduced form** if its numerator and denominator have no factors in common (other than ± 1). To reduce fractions, apply the following property. Be sure you see that this property allows you only to divide out factors, *not* terms.

Simplifying Fractions

Let a, b, and c be nonzero real numbers.

$\frac{a\not{c}}{b\not{c}} = \frac{a}{b}$ Divide out common factor of c.

Example 1 When and When Not to Divide

a. $\frac{\not{2} \cdot x}{\not{2}(x+5)} = \frac{x}{x+5}$ *You CAN divide out the common factor 2.*

b. $\frac{3+x}{3+2x}$ *You CANNOT divide out the common term 3.*

c. $\frac{\not{x}(x^2+6)}{\not{x} \cdot x} = \frac{x^2+6}{x}$ *You CAN divide out the common factor x.*

d. $\frac{x+4}{x}$ *You CANNOT divide out the common term x.* ■

Simplifying a rational expression usually requires two steps. First, factor the numerator and denominator. Then, divide out any *factors* that are common to both the numerator and denominator. Your success in simplifying a rational expression lies in your ability to factor its numerator and denominator.

Example 2 *Simplifying a Rational Expression*

Simplify $\frac{2x^2 - 6x}{6x^2}$.

Solution

$$\frac{2x^2 - 6x}{6x^2} = \frac{2x(x - 3)}{6x^2}$$ *Factor numerator and denominator.*

$$= \frac{\cancel{2x}(x - 3)}{\cancel{2x}(3x)}$$ *Divide out common factor 2x.*

$$= \frac{x - 3}{3x}$$ *Simplified form* ■

Goal 2 Using Rational Expressions in Real Life

Example 3 *Areas of Science Laboratories*

Rockville High School is building a new science wing. The architect has allocated spaces of x feet by x feet for both the biology and chemistry labs. At the request of the faculty, the principal agrees to increase the length of each lab by 5 feet and increase the width of the chemistry lab by 10 feet. Find an expression for the ratio of the area of the revised chemistry lab to the area of the revised biology lab. Evaluate the ratio when $x = 25$ feet and when $x = 30$ feet.

Problem Solving
Draw a Diagram

Solution

$$\text{Ratio} = \frac{\text{Area of Chemistry Lab}}{\text{Area of Biology Lab}}$$

$$= \frac{(x + 5)(x + 10)}{x(x + 5)}$$

$$= \frac{x + 10}{x}$$ *Divide out common factor.*

When $x = 25$ feet, the ratio is

$$\frac{25 + 10}{25} = \frac{35}{25} = 1.4$$

which means that the chemistry lab would be 40% larger than the biology lab. When $x = 30$ feet, the ratio is

$$\frac{30 + 10}{30} = \frac{40}{30} \approx 1.33$$

which means that the chemistry lab would be about 33% larger than the biology lab.

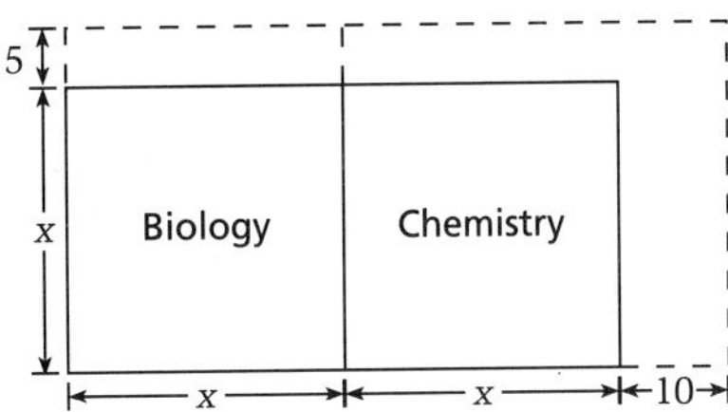

■

has a different domain. For example, the domain of $\frac{(2x + 3)(x - 3)}{x^2 - 9}$ is all real numbers except $x = 3$ and $x = -3$. The given rational expression can be reduced to $\frac{2x + 3}{x + 3}$ but its domain still remains all real numbers except $x = 3$ and $x = -3$.

Example 3

Point out that the diagram is helpful in setting up the relationships between the area of the chemistry and biology labs. The factor $(x + 5)$ in the ratio of the areas of the chemistry lab and the biology lab has been simplified like this: $\frac{\cancel{(x + 5)}(x + 10)}{x\cancel{(x + 5)}}$.

Example 4

The evaluation of P is simpler when the rational expression is simplified first. The difference in pressure between the inside of the cabin and the outside of a plane explains why objects inside the plane are pulled out when the hull of the plane is ruptured.

Common-Error ALERT!

In **Examples 1–2,** the Simplifying Fractions property is also known as the Cancelling property. Only factors that occur in both the numerator and denominator can be divided out or cancelled using this property. Stress that factors are multiplied (or divided). Terms (which are added or subtracted) cannot be cancelled.

Extra Examples

Here are additional examples similar to **Examples 1–4.**

1. Simplify.
   a. $\dfrac{8x^3 - 2x^2}{4x^2 - x}$ $2x$
   b. $\dfrac{3x^2 - 7x - 6}{2x^2 - 6x}$ $\dfrac{3x + 2}{2x}$
   c. $\dfrac{24x^2 - 3x^3}{15x^3}$ $\dfrac{8 - x}{5x}$
2. Given the following dimensions of two mathematics classrooms, find an expression for the ratio of the area of Classroom A to the area of Classroom B. Then evaluate the ratio when $x = 15$ feet.

Classroom A **Classroom B**

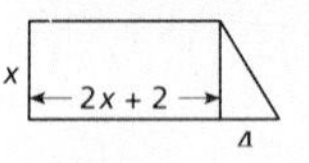

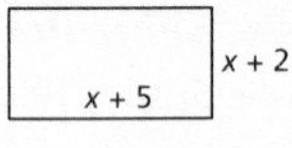

ratio $= \dfrac{2x}{x + 5}$; when $x = 15$, ratio $= 1.5$

Check Understanding

1. How is the domain of a rational expression changed if the rational expression can be reduced? It is unchanged.
2. How do terms and factors in algebraic expressions differ? See the Common-Error Alert on page 588.
3. Why should rational expressions be simplified before evaluating them? The evaluation is easier.

Connections Physics

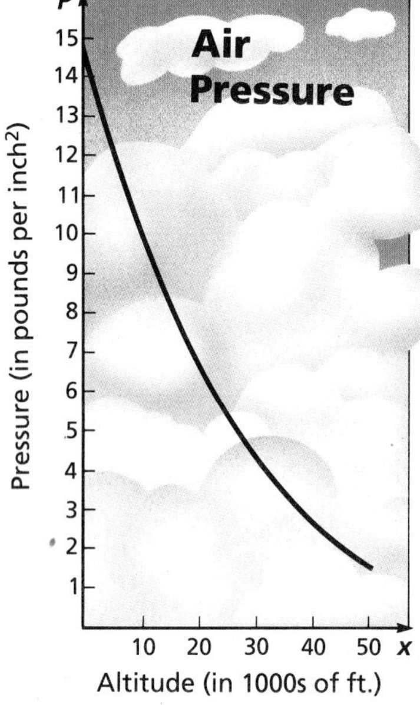

Example 4 *An Atmospheric Pressure Model*

The air pressure at sea level is about 14.7 pounds per square inch. As the altitude increases, the air pressure decreases. A model that relates air pressure to the altitude is

$$P = \frac{-9.05(x^2 - 65x)}{x^2 + 40x}$$

where P is measured in pounds per square inch and x is measured in 1000's of feet. (This model is valid in Earth's *troposphere,* the portion of the atmosphere closest to Earth.) Suppose you are in an airplane, flying at 40,000 feet. The pressure in the cabin is 14.7 pounds per square inch. What is the pressure outside?

Solution

$$P = \frac{-9.05(x^2 - 65x)}{x^2 + 40x} \quad \text{Air pressure model}$$

$$= \frac{-9.05x(x - 65)}{x(x + 40)} \quad \text{Factor.}$$

$$= \frac{-9.05\not{x}(x - 65)}{\not{x}(x + 40)} \quad \text{Divide out common factor.}$$

$$= \frac{-9.05(x - 65)}{x + 40} \quad \text{Simplified form}$$

Substitute 40 for x in the model.

$$P = \frac{-9.05(40 - 65)}{40 + 40}$$

$$\approx 2.8 \text{ pounds per square inch}$$

■

Communicating about ALGEBRA

SHARING IDEAS about the Lesson

Compare Two Expressions

A. Complete the table. 0, 0; 1, 1; 2, 2; 3, 3; 4, 4; $\frac{0}{0}$, 5; 6, 6

x-values	−2	−1	0	1	2	3	4
$\dfrac{x^2 - x - 6}{x - 3}$	?	?	?	?	?	?	?
$x + 2$	?	?	?	?	?	?	?

B. Write a paragraph describing the equivalence (or non-equivalence) of these two expressions. Support your argument with appropriate algebra from this lesson *and* a discussion of the domains of the two expressions. See below.

B. $\dfrac{x^2 - x - 6}{x - 3}$ is undefined at $x = 3$; the two expressions are equivalent for all values of x except $x = 3$.

EXERCISES

Guided Practice

CRITICAL THINKING about the Lesson

1. What is a rational expression? A fraction whose numerator and denominator are polynomials
2. Define the domain of a rational expression. What is the domain of $\frac{2}{x^2 - x - 2}$? The set of all real numbers except those for which the denominator is zero; all real numbers except 2 and -1.
3. Which is the correct reduced form of $\frac{2x + 6}{x^2 + 5x + 6}$?
   a. $\frac{2x}{x^2 + 5x}$ b. $\frac{2}{x + 5}$ c. $\frac{2}{x + 2}$ c
4. *True or False?* To simplify a rational expression, you can cancel common factors of the numerator and denominator. True

Independent Practice

In Exercises 5–10, find the domain of the rational expression.

5. $\frac{5}{x - 4}$ all real numbers except 4
6. $\frac{10}{x - 6}$ all real numbers except 6
7. $\frac{3}{x^2 - 1}$ all real numbers except 1 and -1
8. $\frac{z + 2}{z^2 - 4}$ all real numbers except 2 and -2
9. $\frac{x + 4}{x^2 + x - 6}$ all real numbers except 2 and -3
10. $\frac{x - 3}{x^2 + 5x - 6}$ all real numbers except 1 and -6

In Exercises 11–16, find the missing factor.

✪ 11. $\frac{5\boxed{?}}{6x^2} = \frac{5}{2x}$ $3x$
✪ 12. $\frac{3x\boxed{?}}{4x^2} = \frac{3x}{2}$ $2x^2$
✪ 13. $\frac{x(x + 2)}{2\boxed{?}} = \frac{x}{2}$ $(x + 2)$
✪ 14. $\frac{25x^2}{8\boxed{?}} = \frac{5x}{8}$ $5x$
✪ 15. $\frac{3x\boxed{?}}{x^2 - x - 6} = \frac{3x}{x - 3}$ $(x + 2)$
✪ 16. $\frac{(1 - z)\boxed{?}}{z^3 + z^2} = \frac{1 - z}{z^2}$ $(z + 1)$

In Exercises 17–28, simplify the expression.

17. $\frac{4x}{12}$ $\frac{x}{3}$
18. $\frac{18y}{36}$ $\frac{y}{2}$
19. $\frac{15x^2}{10x}$ $\frac{3x}{2}$
20. $\frac{18y^2}{60y^5}$ $\frac{3}{10y^3}$
21. $\frac{3x}{10x + x^2}$ $\frac{3}{10 + x}$
22. $\frac{x + 2x^2}{2x + 1}$ x
23. $\frac{y^2 - 16}{3y + 12}$ $\frac{y - 4}{3}$
24. $\frac{x^2 - 25z^2}{x - 5z}$ $(x + 5z)$
25. $\frac{3 - x}{x^2 - 5x + 6}$ $\frac{-1}{x - 2}$
26. $\frac{y^2 - 7y + 12}{y^2 + 3y - 18}$ $\frac{y - 4}{y + 6}$
27. $\frac{x^3 + 5x^2 + 6x}{x^2 - 4}$ $\frac{x(x + 3)}{x - 2}$
28. $\frac{t^3 - t}{t^3 + 5t^2 - 6t}$ $\frac{t + 1}{t + 6}$

29. *College Entrance Exam Sample* Which expression is equivalent to $\left(\frac{2x^2}{y}\right)^3$? e
   a. $\frac{8x^5}{3y}$ b. $\frac{6x^6}{y^3}$ c. $\frac{6x^5}{y^3}$ d. $\frac{8x^5}{y^3}$ e. $\frac{8x^6}{y^3}$
30. *College Entrance Exam Sample* Which has the same value as $\frac{P}{Q}$? d
   a. $\frac{P - 2}{Q - 2}$ b. $\frac{1 + P}{1 + Q}$ c. $\frac{P^2}{Q^2}$ d. $\frac{3P}{3Q}$ e. $\frac{P + 3}{Q + 3}$

Communicating about ALGEBRA

EXTEND *Communicating*
Set each expression in the table equal to y and graph the resulting equations. Discuss differences in the graphs and explain why the differences exist. Create a pair of algebraic equations whose domains are exactly the same except at two values of x. Construct a table of values for these equations and sketch their graphs. Discuss your results.

EXERCISE Notes

ASSIGNMENT GUIDE

Basic/Average: Ex. 1–4, 5–23 odd, 29, 30, 33–36, 37–47 odd

Above Average: Ex. 1–4, 7–27 odd, 29–36, 41–49 odd

Advanced: Ex. 1–4, 11–27 odd, 29–37, 41–50

Selected Answers
Exercises 1–4, 5–47 odd

Use **Mixed Review** as needed.

✪ More Difficult Exercises
Exercises 11–16, 31–32, 33–36, 50

Guided Practice

▶ **Ex. 1–4** Use these exercises in class to check the students' understanding of the lesson material preceding Example 1. The issue of domain will be further developed in Chapter 12.

Independent Practice

▶ **Ex. 5–10** Finding the domain of the expressions requires finding the zeros of the denominators.

Ex. 11–16 You might need to do one or two of these exercises in class so that students can see all the necessary solution steps.

Ex. 17–28 Refer students to Chapter 10 if they need to review factoring techniques. Students should note that sometimes common factors are binomials. Refer to Example 3 for an example of simplifying with binomial factors.

Ex. 31–32 These exercises should be assigned as a pair. The 0°F information in Exercise 32 permits the formula from Exercise 31 to be used. Ask students why they should substitute 15 for x to find the minimum wind chill.

Ex. 33–36 These exercises should be assigned as a group. Be sure students realize that $t = 0$ corresponds to 1980 in the equation.

Answer

33. $C = \dfrac{58 + 5t}{10 - 0.8t + 0.0592t^2}$.

Ex. 34–35 COOPERATIVE LEARNING Students might need to work together to develop calculator key sequences involving grouping symbols and memory keys that can be used to evaluate the rational expressions.

Ex. 49 COOPERATIVE LEARNING You may wish to use this exercise as an in-class paired activity.

Windchill Factor **In Exercises 31 and 32, use the following information.**

Meteorologists use the term *windchill* factor to estimate the temperature that a person feels when the wind is blowing. The faster the wind blows, the colder you feel. At 0° Fahrenheit, the windchill factor, w, is given by

$$w = \frac{7 - 16x - 23x^2}{(10 - 0.022x + 0.005x^2)(1 + x)}$$

where x is the wind speed in miles per hour.

At 0°F, the risk of frostbite is high when the wind is blowing at 15 (or more) miles per hour.

31. Simplify the expression for the windchill factor. $w = \dfrac{7 - 23x}{10 - 0.022x + 0.005x^2}$

32. You are driving and see a sign at a bank that gives the temperature as 0°F. On the radio, you hear that the risk of frostbite is high. What is the minimum that the windchill factor could be? ≈ -31.3°F

Mail-Order Catalogs **In Exercises 33–36, use the following information.**

The number of catalogs mailed, C (in billions), in the United States between 1980 and 1990 can be modeled by

$$C = \frac{116 + 68t + 5t^2}{(10 - 0.8t + 0.0592t^2)(2 + t)}$$

where $t = 0$ corresponds to 1980.

In 1990, there were about 94 million households in the United States.

33. Simplify the expression for the number of catalogs. See margin.

34. How many catalogs were mailed in 1990? ≈13.6 billion

35. What was the average number of catalogs received by an American household in 1990? 150

36. Sketch a bar graph showing the number of catalogs mailed in 1980, 1982, 1984, 1986, 1988, and 1990. See Additional Answers.

Integrated Review

In Exercises 37–42, simplify the expression.

37. $3x^2 \cdot x^4$ $3x^6$

38. $\frac{1}{4}x^3 \cdot x^{10}$ $\frac{1}{4}x^{13}$

39. $xy^3 \cdot y^2$ xy^5

40. $12xy^3 \cdot \frac{1}{2}x^4y$ $6x^5y^4$

41. $\dfrac{5xy^2}{15x^3y}$ $\dfrac{y}{3x^2}$

42. $\dfrac{a^2b^2}{5ab}$ $\dfrac{ab}{5}$

In Exercises 43–48, evaluate the expression.

43. $\frac{2x^2 - 1}{x^2 + 3}$, when $x = 1\frac{1}{4}$

44. $\frac{3x + 4}{x - 5}$, when $x = -2\frac{2}{7}$

45. $\frac{9 + x^2}{(x + 3)(x - 3)}$, when $x = 4$ $\frac{25}{7}$

46. $\frac{x + y^2}{x^2 - 1}$, when $x = 2$ and $y = -3$ $\frac{11}{3}$

47. $\frac{x^2 + y^2}{x^2 - y^2}$, when $x = -4$ and $y = 2$ $\frac{5}{3}$

48. $\frac{3 + x^2}{3 + y^2}$, when $x = 2$ and $y = 1$ $\frac{7}{4}$

Exploration and Extension

49. *Working as a Tutor* You are tutoring a friend in algebra. Your friend needs practice in simplifying expressions. Create some sample problems of the form

$$\frac{ax^2 + bx + c}{dx^2 + ex + f}$$

in which the numerator and denominator have a common factor. Describe the process you used to create numerators and denominators that have a common factor.

Write the product of two binomials as a trinomial in the numerator; then write the product of a third binomial times one of the previous two binomials as a trinomial in the denominator.

50. Solve the proportion.

$$\frac{x^2 + 5x + 6}{x^2 - 2x - 8} = \frac{x^2 - 4x - 5}{x^2 - 8x + 15}$$ $x = \frac{5}{3}$

Mixed REVIEW

1. Write 4,675,000,000 in scientific notation. **(8.4)** 4.675×10^9

2. Write 0.000649 in scientific notation. **(8.4)** 6.49×10^{-4}

3. Write 6.209×10^{-6} in decimal form. **(8.4)** 0.000006209

4. Write 4.255×10^4 in decimal form. **(8.4)** 42,550

5. What is the reciprocal of $-m^{-3}$? **(8.3)** $-m^3$

6. What is the opposite of $-m^{-3}$? **(2.1)** m^{-3}

7. What is 7.2% of \$69? \$4.97

8. What is $\frac{4}{9}$ of 36 inches? 16 in.

9. Solve $4|x - 3| \geq 8$. **(6.4)** $x \geq 5$ or $x \leq 1$

10. Solve $|3x + 2| < 6$. **(6.4)** $-\frac{8}{3} < x < \frac{4}{3}$

11. Simplify $(x^2y^{-3})^2 \cdot 4x^{-4}y^2 \cdot \frac{1}{8}x$. **(8.2)** $\frac{x}{2y^4}$

12. Simplify $3(x - 2) + x(-6 - x)$. **(2.6)** $-x^2 - 3x - 6$

13. Is (5, 6) a solution for $2^y < x^2 + x$? **(6.5)** No

14. Is (−2, 5) a solution for $y = 2x^2 - 1$? **(4.2)** No

15. Find the square roots of 81. **(9.1)** 9, −9

16. Solve $5x^2 = 720$. **(9.2)** ± 12

17. Evaluate $x^2 - 3\sqrt{x} + 3$ when $x = 4$. **(9.1)** 13

18. Evaluate $\sqrt{3x} + 2^x$ when $x = 3$. **(9.1)** 11

In Exercises 19 and 20, write an equation for the indicated line.

19. Slope is $\frac{3}{5}$, contains (4, −6) **(5.2)** $-3x + 5y + 42 = 0$

20. Contains (6, −2) and (27, −2) **(5.3)** $y = -2$

▶ **Ex. 50** Students should look for common factors before trying to solve the proportion.

Enrichment Activities

These activities require students to go beyond lesson goals.

1. Simplify. Are the expressions equivalent? Are the domains of the expressions equivalent? Explain.

a. $\frac{x - 3}{3 - x}$ and $-\frac{x + 3}{3 + x}$

b. $\frac{1}{x + 3}$ and $\frac{x - 3}{x^2 - 9}$

The paired expressions are equivalent except for their domains, which are not equivalent.

2. Ivanka and Tim were working together to simplify $\frac{x^2 - 4}{x - 2}$. Ivanka simplified as follows. $\frac{x^2}{x} = x$ and $\frac{-4}{-2} = 2$, so $\frac{x^2 - 4}{x - 2} = x + 2$

a. Is her answer correct? yes

b. Do you think her method of solution always works? no

c. Can you write an example where this method would not work? $\frac{x^2 + 4}{x - 2}$

Problem of the Day

Place 5 points on a sheet of paper so that no 3 are in a straight line. How many lines can be drawn connecting pairs of points? 10

ORGANIZER

Warm-Up Exercises

1. Perform the indicated operations.
   a. $\frac{15}{18} \cdot \frac{24}{25}$ $\quad \frac{4}{5}$
   b. $\frac{8x^3}{26xy} \cdot \frac{14y^4}{12x^2}$ $\quad \frac{14y^3}{39}$
   c. $\frac{32}{54} \div \frac{24}{25}$ $\quad \frac{50}{81}$
   d. $\frac{-3x^6}{20y} \div \frac{12xy^2}{15x^{-3}}$ $\quad \frac{-3x^2}{16y^3}$
2. Factor.
   a. $x^2 - 8x + 15$ $\quad (x - 3)(x - 5)$
   b. $8x^2 + 26x - 7$ $\quad (2x + 7)(4x - 1)$
   c. $x^3 + 3x^2 + 2x$ $\quad x(x + 1)(x + 2)$

Lesson Resources

Teaching Tools
- Problem of the Day: 11.6
- Warm-up Exercises: 11.6
- Answer Masters: 11.6

Extra Practice: 11.6
Color Transparency: 65

LESSON Notes

Examples 1–5

In each example, factors are cancelled before factored expressions are multiplied. This reduces the amount of work required to complete the computation. In Example 2, ask students to show how the factored form of the product was obtained.

11.6 Multiplying and Dividing Rational Expressions

What you should learn:

Goal 1 How to multiply and divide rational expressions

Goal 2 How to use rational expressions in real-life settings

Why you should learn it:

You can use rational expressions to model real-life situations, such as relating boiling points to altitudes.

Goal 1 Working with Rational Expressions

The rule for multiplying rational expressions is the same as the rule for multiplying numerical fractions. That is, *multiply numerators, multiply denominators, and write the new fraction in reduced form.*

$$\frac{a}{b} \cdot \frac{c}{d} = \frac{ac}{bd}$$

To divide one rational expression by another, multiply the first by the reciprocal of the second. Be careful not to confuse the operation. Invert the divisor expression only.

$$\frac{a}{b} \div \frac{c}{d} = \frac{a}{b} \cdot \frac{d}{c}$$

Example 1 *Multiplying Rational Expressions*

Multiply: $\frac{4x^3}{3x} \cdot \frac{-6x^2}{10x^4}$

Solution

$$\frac{4x^3}{3x} \cdot \frac{-6x^2}{10x^4} = \frac{-24x^5}{30x^5}$$ *Multiply numerators and denominators.*

$$= \frac{-4 \cdot \cancel{6} \cdot \cancel{x^5}}{5 \cdot \cancel{6} \cdot \cancel{x^5}}$$ *Factor, and divide out common factors.*

$$= -\frac{4}{5}$$ *Simplified form*

■

Example 2 *Multiplying Rational Expressions*

Multiply: $\frac{x}{5x^2 - 20x} \cdot \frac{x - 4}{2x^2 + x - 3}$

Solution

$$\frac{x}{5x^2 - 20x} \cdot \frac{x - 4}{2x^2 + x - 3}$$
$$= \frac{x(x - 4)}{(5x^2 - 20x)(2x^2 + x - 3)}$$
$$= \frac{x(x - 4)}{5x(x - 4)(x - 1)(2x + 3)}$$
$$= \frac{\cancel{x}\cancel{(x - 4)}}{5\cancel{x}\cancel{(x - 4)}(x - 1)(2x + 3)}$$
$$= \frac{1}{5(x - 1)(2x + 3)}$$

■

To multiply a rational expression by a polynomial, rewrite the polynomial as a fraction whose denominator is 1. This helps you to avoid confusion in simplifying the product.

Example 3 Multiplying a Rational Expression by a Polynomial

Multiply: $\frac{3x}{2x^2 - 9x + 10} \cdot (2x - 5)$

Solution

$$\begin{aligned}\frac{3x}{2x^2 - 9x + 10} \cdot (2x - 5) &= \frac{3x}{2x^2 - 9x + 10} \cdot \frac{2x - 5}{1} \\ &= \frac{3x(2x - 5)}{(2x - 5)(x - 2)} \\ &= \frac{3x\cancel{(2x - 5)}}{\cancel{(2x - 5)}(x - 2)} \\ &= \frac{3x}{x - 2}\end{aligned}$$

■

Example 4 Dividing Rational Expressions

Divide: $\frac{3x}{x + 2} \div \frac{x - 6}{x + 2}$

Solution

$$\frac{3x}{x + 2} \div \frac{x - 6}{x + 2} = \frac{3x}{x + 2} \cdot \frac{x + 2}{x - 6}$$ *Multiply by reciprocal.*

$$= \frac{(3x)(x + 2)}{(x + 2)(x - 6)}$$ *Multiply numerators and denominators.*

$$= \frac{(3x)\cancel{(x + 2)}}{\cancel{(x + 2)}(x - 6)}$$ *Divide out common factors.*

$$= \frac{3x}{x - 6}$$ *Simplified form* ■

Example 5 Dividing a Rational Expression by a Polynomial

Divide: $\frac{x^2 - 4}{2x^2} \div (x - 2)$

Solution

$$\frac{x^2 - 4}{2x^2} \div (x - 2) = \frac{x^2 - 4}{2x^2} \cdot \frac{1}{x - 2}$$ *Multiply by reciprocal.*

$$= \frac{(x + 2)(x - 2)}{2x^2(x - 2)}$$ *Multiply numerators and denominators.*

$$= \frac{(x + 2)\cancel{(x - 2)}}{2x^2\cancel{(x - 2)}}$$ *Divide out common factors.*

$$= \frac{x + 2}{2x^2}$$ *Simplified form* ■

Example 6

Observe that when water boils at a temperature lower than 212°F, cooking takes longer.

EXTEND Example 6 Have students use an almanac or atlas to find the elevations of certain world cities or mountains and then, by using the model in Example 6, determine the boiling time.

Extra Examples

Here are additional examples similar to **Examples 1–6.**

1. Perform the indicated operations.
   a. $\frac{4x^2}{2x^2 - 6x} \cdot \frac{x^2 - 9}{6x^2 - 18x}$
   b. $\frac{3x - 1}{2x^3 - 2x} \div \frac{x}{2x^3 - 2x^2}$
   c. $\frac{2x^2 - x - 3}{x^2 + x} \cdot \frac{x^2 + x}{2x + 3}$
   d. $\frac{x^2 + 4x + 4}{x + 3} \div (x + 2)$

   a. $\frac{x + 3}{3(x - 3)}$ b. $\frac{3x - 1}{x + 1}$
   c. $\frac{(2x - 3)(x + 1)}{2x + 3}$ d. $\frac{x + 2}{x + 3}$
2. Determine the elevation of your community and compute how long it should take to boil a medium-sized potato at your house. Answers will vary.

Check Understanding

1. How is multiplying rational expressions similar to multiplying numerical fractions? See top of page 594 and Example 1.
2. How is dividing rational expressions similar to dividing numerical fractions? See top of page 594 and Example 4.
3. At what point in the multiplication or division computation should rational expressions be reduced? Answers depend on personal preference. Students might reduce after each step to simplify the solution process.
4. Explain why it is important to know how to factor when multiplying or dividing rational expressions. It is easier to divide out common factors when the expressions are in factored form.

Communicating about ALGEBRA

Explain why division of rational expressions is equivalent to multiplication by the reciprocal of the divisor.

Answers

A. $2(x + 1)$; multiply numerators and denominators, divide out common factors

B. $\frac{x-3}{x}$; multiply by reciprocal, divide out common factors

C. $\frac{2}{x-2}$; multiply, divide out common factors

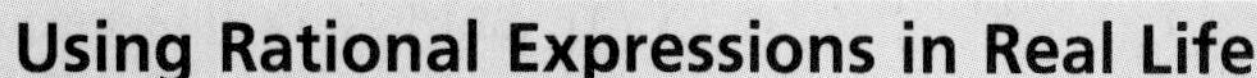

Goal 2 Using Rational Expressions in Real Life

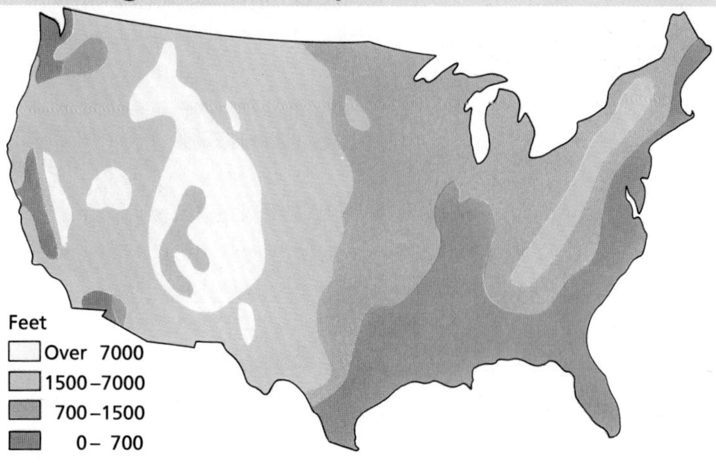

Elevations in the United States (excluding Alaska and Hawaii)

Real Life Cooking

Example 6 *How Long Does It Take to Boil a Potato?*

The boiling point of water is 212°F at sea level. At higher elevations, however, water boils at lower temperatures. A model for the boiling point, B, of water at x thousand feet is $B = 212 - 1.85x$. The time, T (in minutes), required to boil a medium-sized potato is given by the model

$$T = \frac{40(237 - B)}{B - 162}.$$

Write a model for the time, T, in terms of the elevation, x. Use the model to find the time it takes to boil a potato in Bozeman, Montana (4795 feet), Richwood, West Virginia (2194 feet), and Gary, Indiana (608 feet).

Solution Substitute $212 - 1.85x$ for B in the model.

$$T = \frac{40(237 - (212 - 1.85x))}{(212 - 1.85x) - 162} = \frac{40(25 + 1.85x)}{50 - 1.85x}$$

In Bozeman, $T = \frac{40(25 + 1.85(4.795))}{50 - 1.85(4.795)} \approx 33$ minutes.

In Richwood, $T = \frac{40(25 + 1.85(2.194))}{50 - 1.85(2.194)} \approx 25$ minutes.

In Gary, $T = \frac{40(25 + 1.85(0.608))}{50 - 1.85(0.608)} \approx 21$ minutes. ■

Communicating about ALGEBRA

SHARING IDEAS about the Lesson

Explain the Process Write each expression in simplified form. Explain your steps.

A. $\frac{x^2 - 1}{x} \cdot \frac{2x}{x - 1}$ **B.** $\frac{x^2 - 4x + 3}{2x} \div \frac{x - 1}{2}$ **C.** $\frac{2}{x^2 - 4} \cdot (x + 2)$

EXERCISES

Guided Practice

CRITICAL THINKING about the Lesson 1., 2. See top of page 594.

1. Explain the steps used to multiply two rational expressions.
2. Explain the steps used to divide two rational expressions.
3. Multiply $\frac{3x}{2x^2}$ and $-\frac{4x^2}{12x^3}$. Explain your steps. $\frac{-1}{2x^2}$, See Exercise 1
4. Divide $\frac{4x^2 - 25}{4x}$ by $(2x - 5)$. Explain your steps. $\frac{2x+5}{4x}$, See Exercise 2
5. What went wrong?

$$\frac{x^2-4}{5x} \div \frac{x+2}{x-2} = \frac{5x}{x^2-4} \bullet \frac{x+2}{x-2}$$
$$= \frac{5x}{(x+2)(x-2)} \bullet \frac{x+2}{x-2} = \frac{5x}{(x-2)^2}$$

6. What went wrong? 5., 6. See margin.

$$(2x+2) \bullet \frac{x^2-2x-3}{x+4} = \frac{2x+2}{1} \bullet \frac{x+4}{x^2-2x-3}$$
$$= \frac{2(x+1)}{1} \bullet \frac{x+4}{(x-3)(x+1)} = \frac{2(x+4)}{x-3}$$

Independent Practice

In Exercises 7–30, simplify the expression.

7. $\frac{6x}{5} \bullet \frac{1}{x}$ $\frac{6}{5}$
8. $\frac{8x^2}{3} \bullet \frac{9}{16x}$ $\frac{3x}{2}$
9. $\frac{3x^2}{2x} \bullet \frac{12x^2}{6x}$ $3x^2$
10. $\frac{12x^2}{6x} \bullet \frac{12x}{8x^2}$ 3
11. $\frac{25y^2}{8y} \bullet \frac{8y}{5y}$ $5y$
12. $\frac{11x^4}{3x} \bullet \frac{3x}{x^2}$ $11x^2$
13. $\frac{5-4x}{4} \bullet \frac{48}{10-8x}$ 6
14. $\frac{4x}{x^2-9} \bullet \frac{x-3}{8x^2+12x}$ $\frac{1}{(x+3)(2x+3)}$
15. $\frac{3x}{x^2-2x-24} \bullet \frac{x-6}{6x^2+9x}$ $\frac{1}{(x+4)(2x+3)}$
16. $\frac{5}{x-1} \bullet \frac{x-1}{25(x-2)}$ $\frac{1}{5(x-2)}$
17. $\frac{x^2-3x}{x^2-5x+6} \bullet \frac{(x-2)^2}{2x}$ $\frac{x-2}{2}$
18. $\frac{x+1}{x^3(3-x)} \bullet \frac{x(x-3)}{5}$ $-\frac{x+1}{5x^2}$
19. $\frac{x}{2x^2-x-3} \bullet (2x-3)$ $\frac{x}{x+1}$
20. $\frac{8}{2+3x} \bullet (8+12x)$ 32
21. $3(a+2) \bullet \frac{1}{3a+6}$ 1
22. $(4x^2+x-3) \bullet \frac{1}{(4x+3)(x-1)}$ See margin.
23. $\frac{x}{x+4} \div \frac{x+3}{x+4}$ $\frac{x}{x+3}$
24. $\frac{7x^2}{10} \div \frac{14x^3}{15}$ $\frac{3}{4x}$
25. $\frac{2(x+2)}{5(x-3)} \div \frac{4(x-2)}{5x-15}$ $\frac{x+2}{2(x-2)}$
26. $\frac{3x+12}{4x} \div \frac{x+4}{2x}$ $\frac{3}{2}$
27. $\frac{x^2-6x+8}{x^2-2x} \div (3x-12)$ $\frac{1}{3x}$
28. $\frac{5x^2-30x+45}{x+2} \div (5x-15)$ $\frac{x-3}{x+2}$
29. $\left(\frac{x^2}{5} \bullet \frac{x+2}{2}\right) \div \frac{x}{30}$ $3x(x+2)$
30. $\left(\frac{2u^2}{3} \bullet \frac{5}{u}\right) \div \frac{6u^2}{25}$ $\frac{125}{9u}$

EXERCISE Notes

ASSIGNMENT GUIDE

Basic/Average: Ex. 1–6, 7–25 odd, 31–34, 45–52, 55–56

Above Average: Ex. 1–6, 7–27 odd, 31–36, 43–55 odd, 57–60

Advanced: Ex. 1–6, 11–29 odd, 31–36, 49–57 odd, 59–63

Selected Answers
Exercises 1–6, 7–57 odd

More Difficult Exercises
Exercises 31–36, 59–63

Guided Practice

Ex. 1–6 Use these exercises in class to internalize the processes of multiplying and dividing rational expressions. Emphasize that factoring is usually the first step in multiplying and dividing.

Answers

5. When you divide, you multiply the first fraction by the reciprocal of the second fraction.
6. When you multiply, you do not find the reciprocal of the second fraction.

Independent Practice

Ex. 7–30 Remind students to factor before they multiply or divide. It is not always necessary to factor monomial expressions (Exercises 7–12) before multiplying and dividing.

Answer

22. $\frac{(4x-3)(x+1)}{(4x+3)(x-1)}$

▶ **Ex. 31–34** These exercises provide a visual model for rational expressions.

▶ **Ex. 35–36** These exercises should be assigned as a pair. Students are required to determine $S \div P$ and then evaluate the polynomial for $t = 0, 5, 10, 15$, and 20.

EXTEND Ex. 36 Have students use the model to predict the average sales for 1995 and 2000. Interpret the results. The values -0.83 and -0.07 are negative, indicating that the profit would be a loss. The results also indicate that the model is probably invalid for the years following 1990.

Answers

32.–33. Substitute two values for t into the ratio to determine which value of t gives a larger value to the ratio.

44. $\frac{2x}{x^2 + 4x + 4}$

53. All real numbers except 0 and $-\frac{3}{2}$

54. All real numbers except 3 and -3

55. All real numbers except 2 and 3

56. All real numbers except 4 and $-\frac{5}{2}$

▶ **Ex. 59–63** Students might need to review how to find restrictions on the domain.

Service Industry **In Exercises 31–34, use the following information.**

The total sales, S, of services in the United States from 1970 to 1990 can be modeled by

$$S = \frac{390(9 + t)}{30 - t} \qquad \text{(billion \$)}$$

where $t = 0$ represents 1970. The total sales of hotels, H, and auto repair services, A, during the same years can be modeled by

$$H = \frac{21(9 + t)}{32 - t} \qquad \text{(billion \$)}$$

$$A = \frac{27(8 + t)}{34 - t}. \qquad \text{(billion \$)}$$

(Source: U.S. Bureau of Economic Analysis)

Paramedical and other health services account for about 28% of the service industries in the U.S.

✪ **31.** Find the total sales given by each model in 1990. S: \$1131 billion; H: \$50.75 billion; A: \$54 billion

✪ **32.** Find a model for the ratio of hotel sales to the total service industry sales. Was this ratio increasing or decreasing from 1970 to 1990? Explain. $\frac{H}{S} = \frac{7(30 - t)}{130(32 - t)}$, decreasing. 32., 33. See margin.

✪ **33.** Find a model for the ratio of auto service sales to the total service industry sales. Was this ratio increasing or decreasing from 1970 to 1990? Explain. $\frac{A}{S} = \frac{9(8 + t)(30 - t)}{130(9 + t)(34 - t)}$, decreasing

✪ **34.** Match the models for S, H, and A with their graphs. a. A b. S c. H

a.

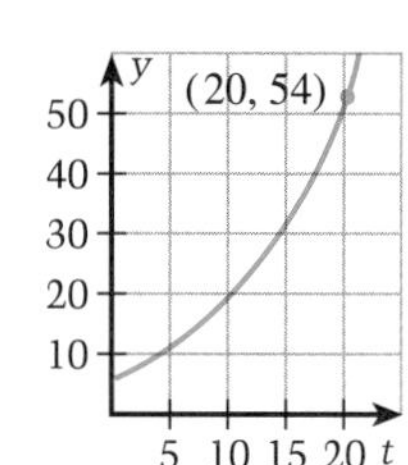

b.

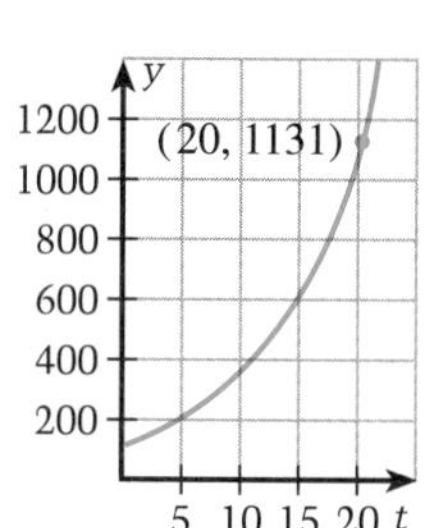

c.

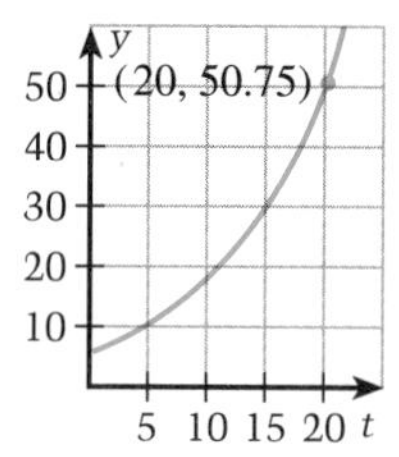

Printing and Copying Franchises **In Exercises 35 and 36, use the following information.**

In the United States from 1970 to 1990, the number, P, of franchised printing and copying businesses can be modeled by

$$P = \frac{10{,}000 + 5500t}{35 - t}$$

where $t = 0$ represents 1970.

✪ **35.** The total sales, S (in millions of dollars), of franchised printing and copying businesses from 1970 to 1990 can be modeled by

$$S = \frac{2200 + 400t}{24 - t}.$$

Find a model for the average sales per business. $\frac{S}{P} = \frac{2(11 + 2t)(35 - t)}{5(24 - t)(20 + 11t)}$

36. Use the model found in Exercise 35 to complete the table.

Year, *t*	0	5	10	15	20
Average Sales (in millions of dollars)	?	?	?	?	?

≈ 0.32, ≈ 0.18, ≈ 0.17, ≈ 0.20, ≈ 0.32

Integrated Review

In Exercises 37–44, simplify the expression.

37. $\frac{9}{15} \cdot \frac{3}{5}$ $\frac{9}{25}$

38. $\frac{4}{7} \div \frac{8}{5}$ $\frac{32}{35}$

39. $\frac{4x^2}{x}$ $4x$

40. $\frac{6x}{3x^3}$ $\frac{2}{x^2}$

41. $\frac{x^2 - 1}{(x + 1)(x + 3)}$ $\frac{x - 1}{x + 3}$

42. $\frac{x^2 - 4}{(x - 2)(x + 6)}$ $\frac{x + 2}{x + 6}$

43. $\frac{8x}{8x + 24}$ $\frac{x}{x + 3}$

44. $\frac{10x}{5x^2 + 20x + 20}$ See margin.

In Exercises 45–52, solve the equation.

45. $\frac{9}{6} = \frac{4}{x}$ $\frac{8}{3}$

46. $\frac{7}{3} = \frac{18}{y}$ $\frac{54}{7}$

47. $\frac{16}{2y} = \frac{3}{4}$ $\frac{32}{3}$

48. $\frac{8}{3x} = \frac{12}{18}$ 4

49. $\frac{3}{x + 2} = \frac{x}{5}$ $-5, 3$

50. $\frac{x - 1}{x + 1} = \frac{x - 2}{2}$ 0, 3

51. $\frac{x - 4}{9} = \frac{4}{x + 1}$ $8, -5$

52. $\frac{3}{x} = \frac{x + 11}{4}$ $1, -12$

In Exercises 53–56, find the domain of the expression. See margin.

53. $\frac{x - 3}{8x^2 + 12x}$

54. $\frac{4x}{x^2 - 9}$

55. $\frac{x^2 - 3x}{x^2 - 5x + 6}$

56. $\frac{x}{2x^2 - 3x - 20}$

Geometry **In Exercises 57 and 58, find the ratio of the area of the blue region to the total area of the figure.**

57. 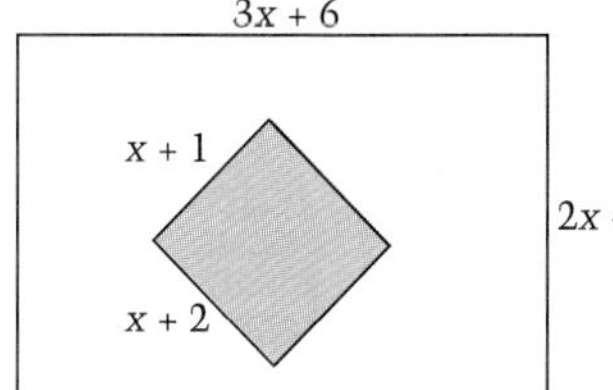

$\frac{1}{6}$

58. 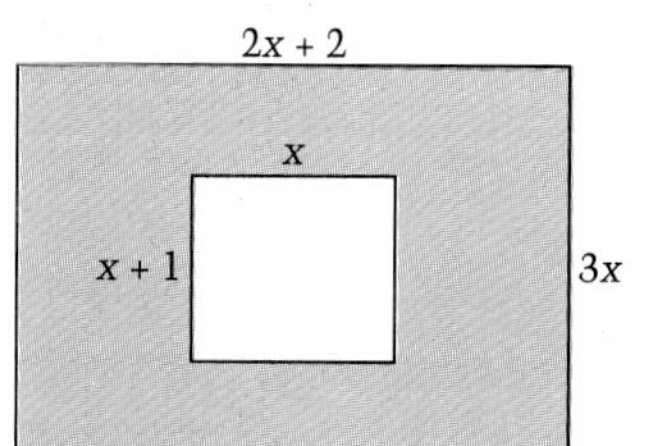

$\frac{5}{6}$

Exploration and Extension

In Exercises 59–62, decide whether the given value of the variable is in the domain of the rational expression.

59. $\frac{x - 2}{2x^2 - 2}$, $x = -1$ No

60. $\frac{3 - y}{y^2 - 5y + 6}$, $y = 2, 3$ No

61. $\frac{y^2 - 4}{y^3 + 3y^2 - 10y}$, $y = 0, 2, -5$ No

62. $\frac{x^2 - 7x}{x^3 - 8x^2 + 7x}$, $x = 0, -3, -2$ No, Yes, Yes

63. Find *two* different pairs of rational expressions (with first-degree numerators and first-degree denominators) whose product is

$$\frac{6x^2 - 39x + 18}{x^2 + 5x - 36}.$$

$\frac{3(2x - 1)}{x + 9}$ and $\frac{x - 6}{x - 4}$, $\frac{3(x - 6)}{x + 9}$ and $\frac{2x - 1}{x - 4}$

Enrichment Activities

These activities require students to go beyond lesson goals.

COOPERATIVE LEARNING

1. Work with a partner to graph the equations. What do you notice?

   a. $y = \frac{2}{x^2 - 4x + 4}$

   b. $y = \frac{4}{x^2 - x - 6}$

   a. As x approaches 2 from both the left and from the right, the graph of the equation rises sharply with y-values that are always positive.

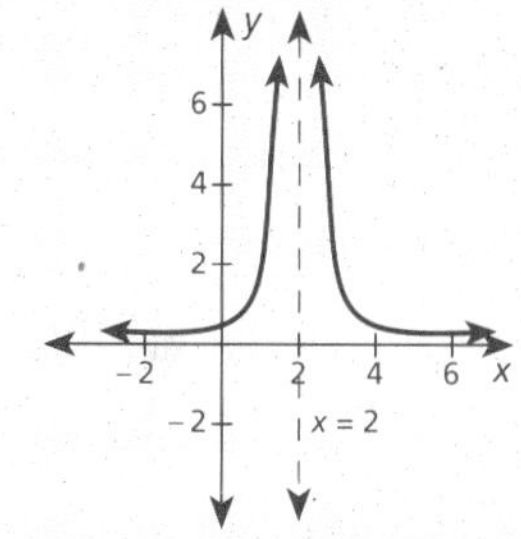

   b. As x approaches -2 from the left and approaches 3 from the right, the graph rises sharply with y-values that are always positive. For $-2 < x < 3$, the graph has a U-shaped branch that falls deeply with y-values that are always negative.

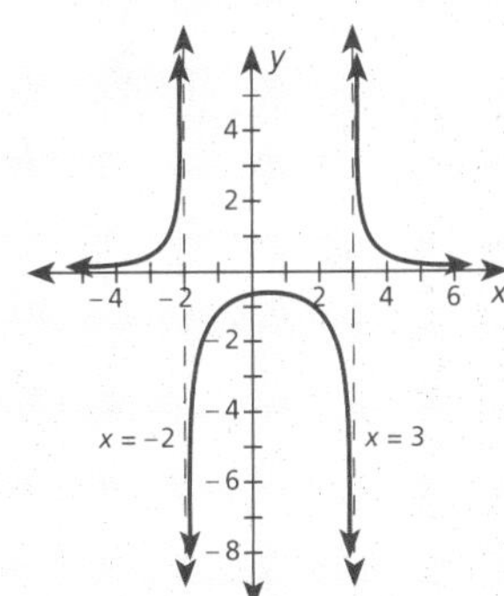

2. For each graph, describe what happens to the graph as the value of x becomes very large or very small.
   a. and b. The graphs of both equations approach but never reach the line $y = 0$.

Problem of the Day

87% of all A's are B's. 3% of all B's are A's. There are 13 A's that are not B's. How many B's are not A's? 2813

ORGANIZER

Warm-Up Exercises

1. **WRITING** Write a paragraph describing the division algorithm of whole numbers. If you wish, use the example $12\overline{)461}$.
Answers may vary.
2. Fill in the box and make each statement true.
   a. $x \cdot \square = 2x$ 2
   b. $3x \cdot \square = 12x^3$ $4x^2$
   c. $(x^2 + 2x) - (x^2 + x) = \square\, x$
   d. $x^3 - (x^3 - 4x^2) = \square$ $4x^2$

Lesson Resources

Teaching Tools
- Transparencies: 1, 2
- Problem of the Day: 11.7
- Warm-up Exercises: 11.7
- Answer Masters: 11.7

Extra Practice: 11.7
Color Transparency: 65

Lesson Investigation Answers

$\frac{x}{3} - \frac{1}{2x}$

a. $x - \frac{2}{x}$

b. $4x + 5$

11.7 Dividing Polynomials

What you should learn:

Goal 1 How to divide a polynomial by a monomial or a binomial

Goal 2 How to use polynomial division in real-life models

Why you should learn it:

You can use polynomial division to provide alternate forms for rational expressions to help you better understand the expressions.

Goal 1 Dividing a Polynomial

LESSON INVESTIGATION

Investigating Polynomial Division

Partner Activity With your partner, discuss how to complete the following division.

$$(2x^2 - 3) \div 6x = \frac{2x^2 - 3}{6x} = \frac{2x^2}{6x} - \frac{3}{6x}$$

Then use a similar procedure to perform the following divisions.

a. $(3x^2 - 6) \div 3x$ **b.** $(4x^3 + 5x^2) \div x^2$

Example 1 *Dividing a Polynomial by a Monomial*

Divide $12x^2 - 20x + 8$ by $4x$.

Solution

$$\begin{aligned}(12x^2 - 20x + 8) \div (4x) &= \frac{12x^2 - 20x + 8}{4x} \\ &= \frac{12x^2}{4x} - \frac{20x}{4x} + \frac{8}{4x} \\ &= \frac{(3x)(\cancel{4x})}{\cancel{4x}} - \frac{5(\cancel{4x})}{\cancel{4x}} + \frac{2(\cancel{4})}{\cancel{4}x} \\ &= 3x - 5 + \frac{2}{x}\end{aligned}$$

Study Tip
In the Lesson Investigation and in Example 1, notice that the first step in dividing a polynomial by a monomial is to divide each term of the polynomial by the monomial.

To divide a polynomial by a *binomial,* you can use the canceling technique of Lesson 11.6 when the numerator and denominator have common factors.

$$\frac{x^2 - 2x - 3}{x - 3} = \frac{(x + 1)\cancel{(x - 3)}}{\cancel{x - 3}} = x + 1$$

For polynomials that do not have common factors, you can use **polynomial long division.** To begin, you may need to review the process for long division in arithmetic.

Example 2 *The Long Division Algorithm*

Use the long division algorithm to divide 6584 by 28.

Solution

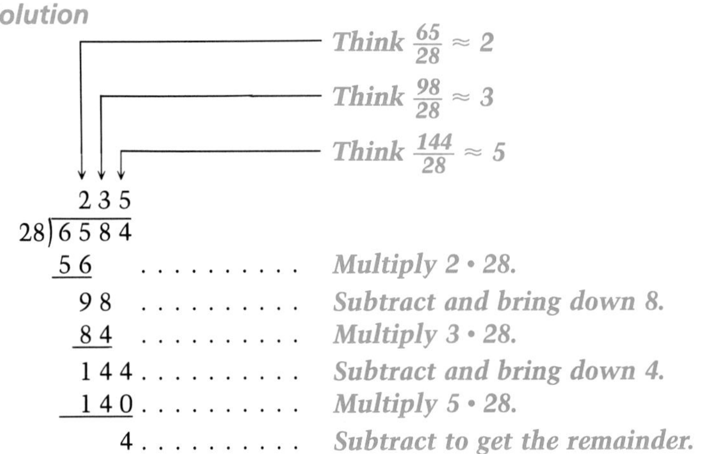

Think $\frac{65}{28} \approx 2$
Think $\frac{98}{28} \approx 3$
Think $\frac{144}{28} \approx 5$
235
28)6584
56 .... Multiply 2 • 28.
98 .... Subtract and bring down 8.
84 .... Multiply 3 • 28.
144 .... Subtract and bring down 4.
140 .... Multiply 5 • 28.
4 .... Subtract to get the remainder.

In this problem, 6584 is the **dividend,** 28 is the **divisor,** 235 is the **quotient,** and 4 is the **remainder.**

$$\underbrace{\frac{6584}{28}}_{\text{Divisor}}^{\text{Dividend}} = \overbrace{235}^{\text{Quotient}} + \underbrace{\frac{4}{28}}_{\text{Divisor}}^{\text{Remainder}} = 235 + \frac{1}{7}$$

■

Polynomial long division uses a process that is similar to the algorithm. Note: the divisor and dividend are in standard form.

Example 3 *Polynomial Long Division*

Divide $(x^2 + 2x + 4)$ by $(x - 1)$.

Solution

Think $\frac{x^2}{x} \approx x$

Think $\frac{3x}{x} \approx 3$

$$\begin{array}{r l}
x + 3 & \\
x - 1\overline{)x^2 + 2x + 4} & \\
\underline{x^2 - x} & \text{....... } \textit{Multiply } x(x - 1). \\
3x + 4 & \text{....... } \textit{Subtract and bring down 4.} \\
\underline{3x - 3} & \text{....... } \textit{Multiply } 3(x - 1). \\
7 & \text{....... } \textit{Subtract to get the remainder.}
\end{array}$$

The result is written as follows.

$$\underbrace{\frac{x^2 + 2x + 4}{x - 1}}_{\text{Divisor}}^{\text{Dividend}} = \overbrace{x + 3}^{\text{Quotient}} + \underbrace{\frac{7}{x - 1}}_{\text{Divisor}}^{\text{Remainder}}$$

■

LESSON Notes

Example 1

Contrast division by a monomial with division by a binomial (see Example 3). Although a long division procedure is appropriate when dividing by a monomial, it is simple just to think about dividing each term of the dividend by the monomial, as in this example.

Examples 2–3

Another way to express the result of dividing 6584 by 28 and getting a quotient of 235 with remainder 4 is the number sentence, 6584 = 28 • 235 + 4. In a similar manner, the result in Example 3 may be written as

$x^2 + 2x + 4 = (x - 1)(x + 3) + 7.$

The connection between long division of whole numbers and long division of polynomials should help students understand polynomial division.

Example 4

The additional expense is an increase from the 85% removal of pollutants to 90% removal. Be sure students recognize that the difference between these two costs is the answer to the question. Ask students if they think an additional 5% removal of pollutants is worth an additional cost of $400,000. Can students justify their responses? The removal cost is proportionally more expensive for the additional 5% than it was for the first 85%.

Extra Example

Here are additional examples similar to **Examples 1–4.**

1. Divide.
   a. $24x^2 - 32x$ by $8x$
   b. $(3x^2 + 4x - 14)$ by $(x + 1)$
   c. $(18x^3 - 3x^2 + 6)$ by $(2x - 3)$

   a. $3x - 4$
   b. $(3x + 1) + \frac{-15}{x + 1}$
   c. $9x^2 + 12x + 18 + \frac{60}{2x - 3}$
2. In Example 4, how much cheaper would it be to remove 80% of the pollutants than to remove 85%? $200,000 cheaper

Check Understanding

1. How are long division of whole numbers and long division of polynomials similar? See Examples 2 and 3.
2. Write two different number sentences for the following long division problem.
   $x - 1\overline{)x^2 - 3x + 1}$
   $\frac{x^2 - 3x + 1}{x - 1} = (x - 2) + \frac{-1}{x - 1}$
   or $x^2 - 3x + 1 =$
   $(x - 1)(x - 2) + (-1)$

Communicating about ALGEBRA

Discuss this activity in class to prepare students for Exercises 27–30.

Answer

$$\begin{array}{r} -120{,}000 + \frac{120{,}000}{-p+1} \\ -p + 1\overline{)120{,}000p} \\ \underline{120{,}000p - 120{,}000} \\ 120{,}000 \end{array}$$

Goal 2 Using Polynomial Division in Real Life

The cost of removing pollutants from smokestack emission does not vary directly with the percent removed. That is, if it costs C dollars to remove 25% of the pollutants, it would cost more than 2C dollars to remove 50% of the pollutants. As the percent of removed pollutants approaches 100%, the cost tends to become prohibitive.

Example 4 Cost-Benefit Model for Pollutants

A utility company burns coal to generate electricity. The cost, C (in dollars), of removing p percent of the smokestack pollutants can be modeled by

$$C = \frac{120{,}000p}{1 - p}.$$

As a member of the state legislature, you are considering a law that will require companies to remove 90% of their smokestack pollutants. The current law requires 85% removal. How much additional expense would the new law require of the utility company?

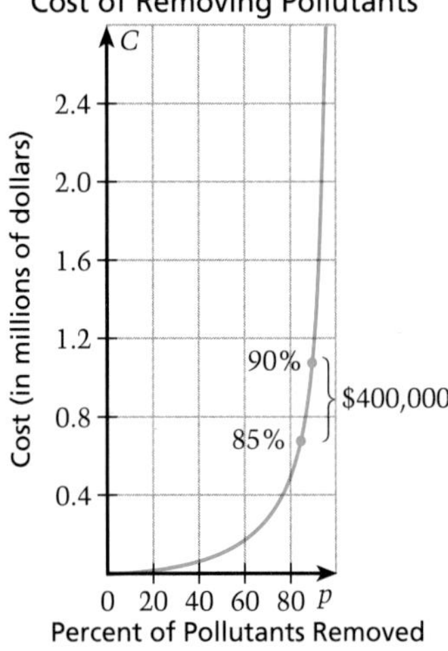

Solution Under the current law, the removal cost is

$$C = \frac{120{,}000(0.85)}{1 - 0.85} \approx \$680{,}000. \quad \text{85\% removal}$$

If the new law is passed, increasing the percentage removal to 90%, the cost to the utility company will be

$$C = \frac{120{,}000(0.9)}{1 - 0.9} \approx \$1{,}080{,}000. \quad \text{90\% removal}$$

The new law would require the company to spend $400,000 more to remove an additional 5% of the pollutants. ■

Communicating about ALGEBRA

SHARING IDEAS about the Lesson See margin.

Examine the Model Use long division to rewrite the model given in Example 4 as

$$C = \frac{120{,}000}{1 - p} - 120{,}000.$$

What happens to C when p gets closer and closer to 1? C gets larger.

EXERCISES

Guided Practice

CRITICAL THINKING about the Lesson

1. Match each number with its name: $\frac{5469}{42} = 130 + \frac{9}{42}$. a. 5469 b. 42 c. 130 d. 9

a. Dividend **b.** Divisor **c.** Quotient **d.** Remainder

2. Match each expression with its name: $\frac{x^2 + 2}{x - 3} = x + 3 + \frac{11}{x - 3}$. a. $x^2 + 2$ b. $x - 3$ c. $x + 3$ d. 11

a. Dividend **b.** Divisor **c.** Quotient **d.** Remainder

In Exercises 3–6, use long division.

3. Divide 365 by 52. $7 + \frac{1}{52}$

4. Divide 25,001 by 15. $1666 + \frac{11}{15}$

5. Divide $x^2 + 3x - 1$ by $x + 2$. $x + 1 - \frac{3}{x + 2}$

6. Divide $3x^3 - 4x - 1$ by $x + 1$. $3x^2 - 3x - 1$

Independent Practice

In Exercises 7 and 8, simplify the expression.

7. $\frac{10x^2}{2x(5x + 2)}$ $\frac{5x}{5x + 2}$

8. $\frac{33x^3(x + 1)}{27x^4(x + 2)}$ $\frac{11(x + 1)}{9x(x + 2)}$

In Exercises 9–16, divide.

9. $(6z + 10) \div 2$ $3z + 5$

10. $(9x + 10) \div 3$ $3x + \frac{10}{3}$

11. $(10z^2 + 4z - 12) \div 4$ $\frac{5}{2}z^2 + z - 3$

12. $(4u^2 + 8u - 24) \div 16$ $\frac{1}{4}u^2 + \frac{1}{2}u - \frac{3}{2}$

13. $(7x^3 - 2x^2) \div 14x$ $\frac{1}{2}x^2 - \frac{1}{7}x$

14. $(6a^2 + 7a) \div a$ $6a + 7$

15. $(m^4 + 2m^2 - 7) \div m$ $m^3 + 2m - \frac{7}{m}$

16. $(x^2 - 8) \div (-x)$ $-x + \frac{8}{x}$

In Exercises 17 and 18, determine the quotient, remainder, and divisor. (The long division is shown.)

17. $\frac{x^3 - 27}{x - 3} = \text{Quotient} + \frac{\text{Remainder}}{\text{Divisor}}$ Quotient: $x^2 + 3x + 9$; Remainder: 0; Divisor: $x - 3$

$$\begin{array}{r} x^2 + 3x + 9 \\ x - 3\overline{)x^3 + 0x^2 + 0x - 27} \\ \underline{x^3 - 3x^2} \qquad\qquad \\ 3x^2 + 0x \qquad \\ \underline{3x^2 - 9x} \qquad \\ 9x - 27 \\ \underline{9x - 27} \\ 0 \end{array}$$

18. $\frac{4x^3 + 5x}{2x - 1} = \text{Quotient} + \frac{\text{Remainder}}{\text{Divisor}}$ Quotient: $2x^2 + x + 3$; Remainder: 3; Divisor: $2x - 1$

$$\begin{array}{r} 2x^2 + x + 3 \\ 2x - 1\overline{)4x^3 + 0x^2 + 5x + 0} \\ \underline{4x^3 - 2x^2} \qquad\qquad \\ 2x^2 + 5x \qquad \\ \underline{2x^2 - x} \qquad \\ 6x + 0 \\ \underline{6x - 3} \\ 3 \end{array}$$

EXTEND *Communicating*
Use long division to rewrite $y = \frac{3x + 1}{x}$. Then make a table of values for the expression. Sketch the graph of the points in the table. Discuss the behavior of the graph as x grows large or small; as x approaches the value of 0.

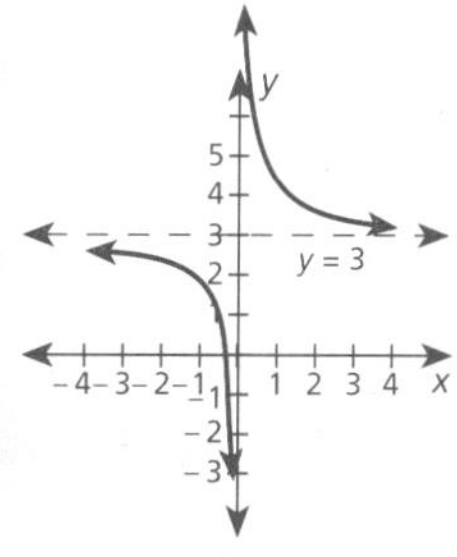

$y = 3 + \frac{1}{x}$; as x becomes very large or very small, the y-values approach the line $y = 3$. As x approaches 0, the y-values become very small or very large.

EXERCISE Notes

ASSIGNMENT GUIDE

Basic/Average: Ex. 1–6, 7–15 odd, 19–27 multiples of 3, 28, 31–41 odd, 43, 44

Above Average: Ex. 1–6, 7–25 odd, 29–30, 31–45 odd

Advanced: Ex. 1–6, 15–25 odd, 29–30, 31–43 odd, 45–47

Selected Answers
Exercises 1–6, 7–43 odd

More Difficult Exercises
Exercises 45–47

Guided Practice

Ex. 1–6 Have students compare and contrast polynomial long division and arithmetic long division.

Independent Practice

Ex. 19–26 Students can use a format similar to the one illustrated in Exercises 17–18.

Ex. 27–28 and 29–30 These exercises should be assigned in pairs.

Answers

21. $x + 10 + \frac{0}{x + 5}$

22. $y - 8 + \frac{0}{y + 2}$

25. $4x + 20 + \frac{30}{x - 2}$

26. $5y + 5 + \frac{7}{y - 1}$

28. More popular, because $\frac{C}{B}$ is increasing

30. Commercial sports revenue is not increasing as rapidly as amusement and recreation services revenue.

37.–42. Use long division, so you won't have to spend time looking for factors that may not exist.

In Exercises 19–26, divide. Write the result as Quotient + $\frac{\text{Remainder}}{\text{Divisor}}$. 21., 22., 25., 26. See margin.

19. $(x^2 - 8x + 15) \div (x - 3)$ $x - 5 + \frac{0}{x - 3}$

20. $(t^2 - 18t + 72) \div (t - 6)$ $t - 12 + \frac{0}{t - 6}$

21. $(x^2 + 15x + 50) \div (x + 5)$

22. $(y^2 - 6y - 16) \div (y + 2)$

23. $(x^2 - 4) \div (x - 3)$ $x + 3 + \frac{5}{x - 3}$

24. $(y^2 + 5y) \div (y + 1)$ $y + 4 - \frac{4}{y + 1}$

25. $(4x^2 + 12x - 10) \div (x - 2)$

26. $(5y^2 + 2) \div (y - 1)$

Canoes and Other Boats **In Exercises 27 and 28, use the following information.**

The number of canoes, C, in the United States from 1983 to 1993 can be modeled by

$$C = 90{,}000(t + 15)$$

where $t = 3$ represents 1983. The total number of recreational boats, B, can be modeled by

$$B = 360{,}000(t + 33).$$

27. Use long division to find a model for the ratio of canoes to boats. $\frac{C}{B} = \frac{1}{4} - \frac{9}{2t + 66}$

28. Use the model obtained in Exercise 27 to complete the table. Was canoeing becoming more or less popular compared to the total use of recreational boats? Explain.

Year, t	3	5	7	9	11	13
Ratio of C to B	?	?	?	?	?	?

0.125 ≈1.32 ≈0.138 ≈0.143 ≈0.148 ≈0.152 See margin.

Sports and Other Recreation **In Exercises 29 and 30, use the following information.**

From 1988 to 1993, amusement and recreation services revenue, A, can be modeled by

$$A = 3600(t + 2.8), \quad \text{(billions of dollars)}$$

where $t = 8$ represents 1988. Commercial sports revenue, S, can be modeled by

$$S = 800(t + 1.06), \quad \text{(billions of dollars).}$$

29. Use long division to find a model for the ratio of commercial sports revenue to the amusement and recreation services revenue. $\frac{S}{A} = \frac{2}{9} - \frac{3.48}{9t + 25.2}$

30. Use the model obtained in Exercise 29 to complete the table. How was commercial sports doing compared to amusement and recreation services? Explain your answer.

Year, t	8	9	10	11	12	13
Ratio of S to A	?	?	?	?	?	?

≈1.86 ≈0.189 ≈0.192 ≈0.194 ≈0.196 ≈0.198 See margin.

Integrated Review

In Exercises 31–36, simplify by factoring and dividing out common factors.

31. $\frac{2z^2 - 9z + 4}{z - 4}$ $2z - 1$

32. $\frac{2x^2 + 4x - 30}{2x - 6}$ $x + 5$

33. $\frac{6y + 2y^2}{2y}$ $3 + y$

34. $\frac{12p^2 - 15p}{3p^2}$ $\frac{4p - 5}{p}$

35. $\frac{2m^2 - 8m - 10}{m + 1}$ $2m - 10$

36. $\frac{2t^2 + 10t + 12}{4t + 12}$ $\frac{t + 2}{2}$

In Exercises 37–42, divide. Use long division *or* factoring and dividing out common factors. Explain why you chose the method you did. See margin.

37. $\frac{x^2 + 3x + 2}{x - 2}$ $x + 5 + \frac{12}{x - 2}$

38. $\frac{a^2 + 4a - 77}{a + 7}$ $a - 3 - \frac{56}{a + 7}$

39. $\frac{6y^2 + y - 2}{3y + 2}$ $2y - 1$

40. $\frac{3z^2 - 2z - 2}{3z - 2}$ $z - \frac{2}{3z - 2}$

41. $\frac{2b^2 + 21b + 10}{b - 10}$ $2b + 41 + \frac{420}{b - 10}$

42. $\frac{2c^2 + 19c - 10}{2c - 1}$ $c + 10$

43. *Estimation—Powers of Ten* Which is the best estimate of the ratio of Americans who are thirty-three years old or younger to *all* Americans?

a. 0.005 **b.** 0.05 **c.** 0.5 **d.** 5.0 c

44. *Estimation—Powers of Ten* Which is the best estimate of the ratio of Americans who are thirty-three years old or younger to those who are older than thirty-three?

a. 0.001 **b.** 0.01 **c.** 0.1 **d.** 1.0 d

45. 3, ≈18.5, ≈38.2, ≈58.2, ≈78.1, ≈98.1; −2, 18, 38, 58, 78, 98; 5, ≈0.45, ≈0.24, ≈0.16, ≈0.12, ≈0.10. 1st and 2nd increase, 3rd decreases

Exploration and Extension

In Exercises 45 and 46, use $\frac{x^2 + 6}{x + 2} = x - 2 + \frac{10}{x + 2}$.

45. Complete the table. What happens to the values of

$$\frac{x^2 + 6}{x + 2},\ (x - 2),\ \text{and}\ \frac{10}{x + 2}$$

as x gets larger and larger?

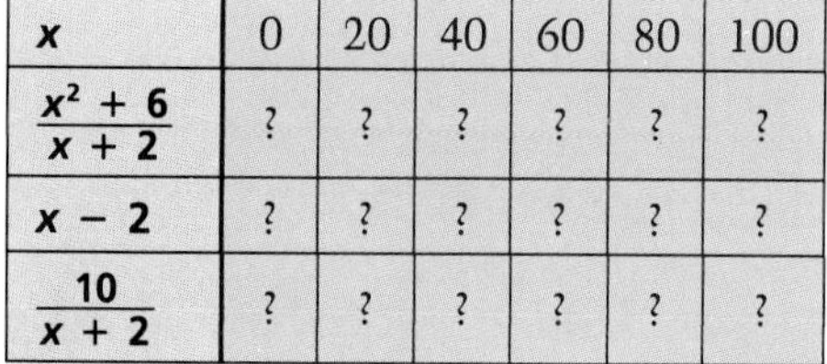

x	0	20	40	60	80	100
$\frac{x^2 + 6}{x + 2}$	?	?	?	?	?	?
$x - 2$	?	?	?	?	?	?
$\frac{10}{x + 2}$	?	?	?	?	?	?

46. Which of the graphs shown is the graph of

$$y = x - 2?$$

Which is the graph of

$$y = \frac{x^2 + 6}{x + 2}?$$ A. $y = \frac{x^2 + 6}{x + 2}$, B. $y = x - 2$

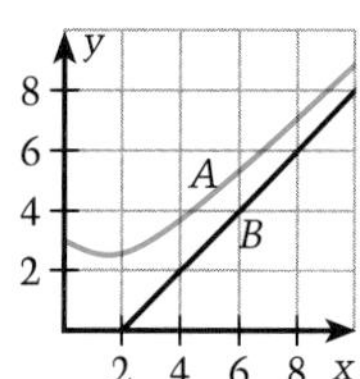

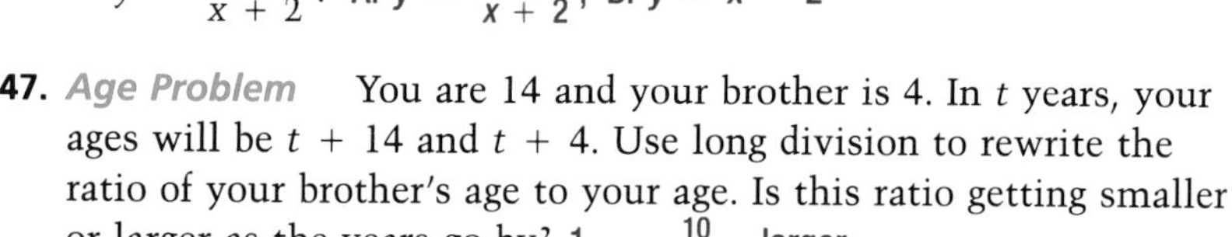

47. *Age Problem* You are 14 and your brother is 4. In t years, your ages will be $t + 14$ and $t + 4$. Use long division to rewrite the ratio of your brother's age to your age. Is this ratio getting smaller or larger as the years go by? $1 - \frac{10}{t + 14}$, larger

EXTEND Ex. 46 Ask students to use long division to divide $x^2 + 6$ by $x + 2$. Remind students that adding 0 to an expression does not change its value. So, you can set up the division using a zero placeholder: $x + 2\overline{)x^2 + 0x + 6} = x - 2 + \frac{10}{x + 2}$. What do you think happens to the expression $\frac{x^2 + 6}{x + 2}$ when x gets very large? The values approach the value of $x - 2$ since the value of the "remainder" fraction $\frac{10}{x + 2}$ approaches 0.

Enrichment Activities

These activities require students to go beyond lesson goals.

1. Use polynomial long division to simplify the following expressions. $\frac{x^3 - 27}{x - 3}$ and $\frac{x^3 + 8}{x + 2}$ $(x^2 + 3x + 9)$ and $(x^2 - 2x + 4)$
2. Remember, when there is no remainder, the dividend is the product of the divisor and the quotient. Use your answers to Activity 1 to generalize a factoring pattern for the sum and difference of two cubes that is similar to the factoring pattern for the difference of squares. $a^3 - b^3 = (a - b)(a^2 + ab + b^2)$ $a^3 + b^3 = (a + b)(a^2 - ab + b^2)$

TECHNOLOGY Ask students to monitor both vertical and horizontal asymptotes. Which does the graphing calculator draw? Which is not drawn? The horizontal asymptote $y = 2$ is drawn while the vertical asymptote $x = -5$ is not drawn.

Answers

1.

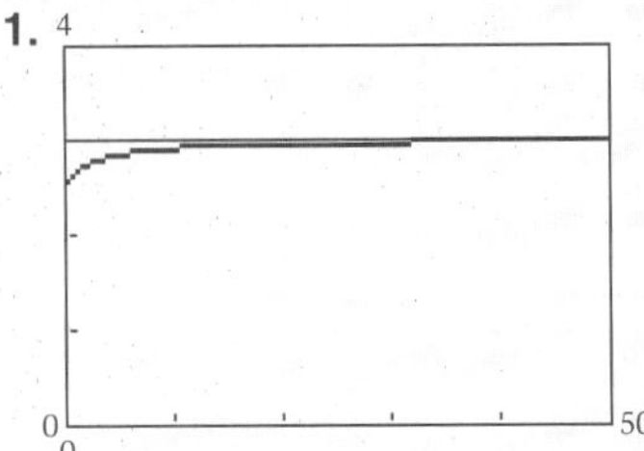

2.

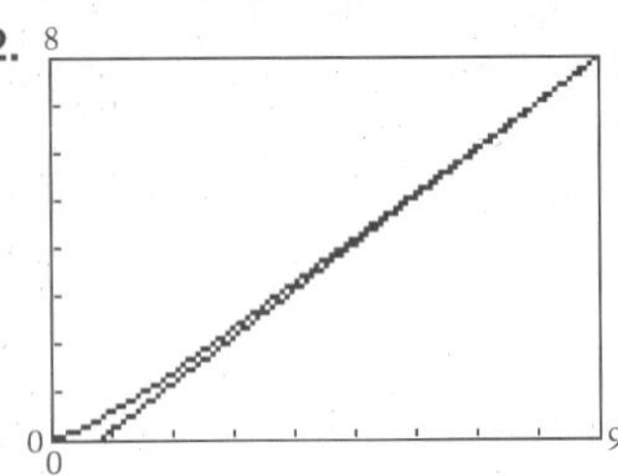

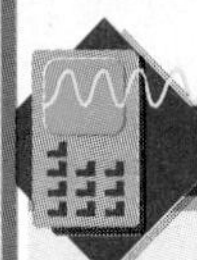

USING A GRAPHING CALCULATOR

Graphs of rational expressions often have ***asymptotes****— lines that resemble the graph in certain regions of the coordinate plane. For instance, consider the graphs of*

$$y_1 = \frac{2x}{x+5} \quad \text{and} \quad y_2 = 2.$$

Using long division, you can write

$$\underbrace{\frac{2x}{x+5}}_{y_1} = \underbrace{2}_{y_2} - \underbrace{\frac{10}{x+5}}_{\text{Remainder fraction}}.$$

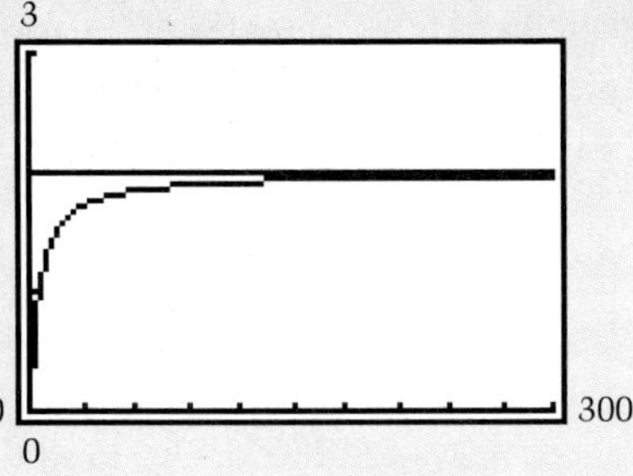

Graphs of y_1 and y_2

The value of the "remainder fraction" becomes very small as the values of x become large. This is why we say that the graph of y_1 resembles the line $y_2 = 2$ when x is large. The graphs of y_1 and y_2 are shown at the right. Keystroke instructions for graphing these equations on the *TI-82,* Casio *fx-9700GE,* and Sharp *EL-9300C* are listed on page 744.

Cooperative Learning

Partner Activity Read Example 4 on page 608. There you are shown how to set up an algebraic model that represents the average cost of a dozen muffins at your muffin shop. The solution shown on page 608 is obtained algebraically by solving a rational equation. With your partner, discuss how you could use a graphing calculator to graphically find how many dozens of muffins you must produce before the average cost per dozen drops to \$1.60. To do this, analyze the graph of

$$y = \frac{80{,}000 + 0.8x}{x}.$$

What is the asymptote for this graph? What does the asymptote represent?

Exercises

In Exercises 1 and 2, sketch the graphs of y_1 and y_2. Experiment to find a viewing rectangle in which the two graphs resemble each other.

1. $\underbrace{\frac{3x+10}{x+4}}_{y_1} = \underbrace{3}_{y_2} - \underbrace{\frac{2}{x+4}}_{\text{Remainder fraction}}$

The graphs resemble each other when $x > 20$.

2. $\underbrace{\frac{x^2}{x+1}}_{y_1} = \underbrace{x - 1}_{y_2} + \underbrace{\frac{1}{x+1}}_{\text{Remainder fraction}}$

The graphs resemble each other when $x > 6$.

11.8 Solving Rational Equations

What you should learn:

Goal 1 How to solve rational equations

Goal 2 How to use rational equations in real-life settings

Why you should learn it:

You can model many real-life problems with rational equations, such as comparing changing quantities.

Goal 1 Solving Rational Equations

In this lesson, you will learn how to solve equations that involve rational expressions. There are two basic techniques: multiplying by the least common denominator (LCD) and cross multiplying.

Multiplying by the least common denominator of each fraction in the equation will work for any rational equation. After multiplying by the least common denominator, simplify each term. The resulting equation will be a polynomial equation, which you can solve using standard techniques.

Example 1 *Multiplying by the LCD*

Solve $\frac{2}{x} + \frac{1}{3} = \frac{4}{x}$.

Solution

$\frac{2}{x} + \frac{1}{3} = \frac{4}{x}$	*LCD is 3x.*
$3x \cdot \frac{2}{x} + 3x \cdot \frac{1}{3} = 3x \cdot \frac{4}{x}$	*Multiply each term on both sides by 3x.*
$6 + x = 12$	*Simplify.*
$x = 6$	*Subtract 6 from both sides.*

The solution is 6. Check this in the original equation. ■

Example 2 *Multiplying by the LCD*

Solve $\frac{6}{x} + \frac{x}{2} = 4$.

Solution

$\frac{6}{x} + \frac{x}{2} = 4$	*LCD is 2x.*
$2x \cdot \frac{6}{x} + 2x \cdot \frac{x}{2} = 2x \cdot 4$	*Multiply each term on both sides by 2x.*
$12 + x^2 = 8x$	*Simplify.*
$x^2 - 8x + 12 = 0$	*Standard form*
$(x - 2)(x - 6) = 0$	*Factor.*
$x = 2$ or 6	*Zero Product property*

The solutions are 2 and 6. Check them in the original equation. ■

Problem of the Day

A group of students contains the-following: 18 who like math, 32 who like English, 25 who like language, 3 who like all 3 subjects, 8 who like both math and English, 16 who like both English and language, and 7 who like both math and language. Everyone likes at least one subject. How many are in the group? 47

ORGANIZER

Warm-Up Exercises

1. Solve for x.
   a. $\frac{12}{8} + \frac{2}{3} = x$ $\quad \frac{13}{6}$
   b. $\frac{25}{x} = \frac{40}{16}$ $\quad 10$
   c. $\frac{2}{3} + \frac{4}{5} - \frac{2}{8} = x$ $\quad \frac{73}{60}$
2. Solve for x.
   a. $x^2 - 2x - 35 = 0$ $\quad x = -5, x = 7$
   b. $3x - 14 = 5(x + 2)$ $\quad x = -12$
   c. $4x^2 + 12x = -9$ $\quad x = -\frac{3}{2}$

Lesson Resources

Teaching Tools
- Problem of the Day: 11.8
- Warm-up Exercises: 11.8
- Answer Masters: 11.8

Extra Practice: 11.8

Color Transparency: 66

Common-Error ALERT!

When multiplying both sides of an equation by the least common denominator, emphasize that *each* term of each side must be multiplied by the LCD.

LESSON Notes

Examples 1–3

Each example results in an equation that can be solved using techniques learned in previous lessons. Point out that it is important to check the solutions of a rational equation by using the original equation.

Example 4

Demonstrate how the average cost per muffin is computed. The initial investment of $80,000 is required to produce one dozen muffins. The cost of this dozen muffins is over $80,000! If 100 dozen muffins were produced, the cost would be merely $800.80 per dozen. If 1,000 dozen muffins were made, the cost would be $80.80 per dozen. Ask students for the verbal model of this series of computations.

It should be noted that the initial investment of $80,000 is unlikely to be spent entirely in unrecoverable costs. Many of the costs of starting the muffin shop are for items that have resale value such as cooking utensils and equipment.

Example 5

The future number of hits is the same as the future number of times at bat because the batter hits safely each time. (Hitting safely means that the batter gets on first, second, or third base, or hits a home run.) Point out that the new batting average is computed using 45 hits out of 150 times at bat. Have students justify the use of these number values in computing the batting average.

Cross multiplying (see Lesson 11.1) can be used only for equations in which each side is a single fraction.

Example 3 *Cross Multiplying*

Solve $\frac{5}{y+2} = \frac{y}{3}$.

Solution

$\frac{5}{y+2} = \frac{y}{3}$	*Rewrite original equation.*
$5(3) = y(y+2)$	*Cross multiply.*
$15 = y^2 + 2y$	*Simplify.*
$0 = y^2 + 2y - 15$	*Standard form*
$0 = (y+5)(y-3)$	*Factor.*
$y = -5 \text{ or } 3$	*Zero Product Property*

The solutions are -5 and 3. Check these in the original equation. ■

Goal 2 Using Rational Equations in Real Life

Real Life
Retail Business

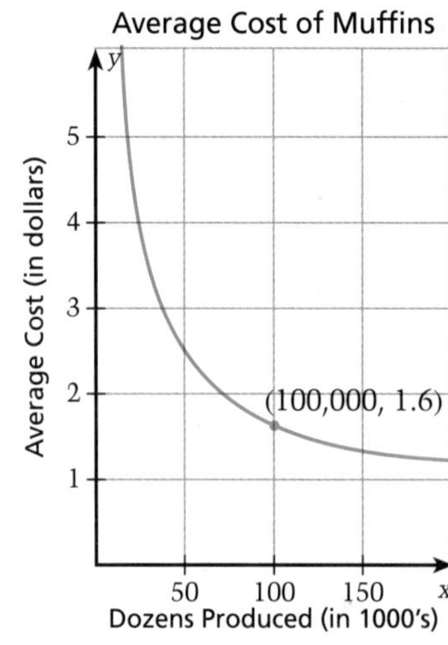

As your initial investment of $80,000 is distributed over more and more muffins, the average cost per dozen decreases.

Example 4 *Finding the Average Cost Per Unit*

You have invested $80,000 to start a muffin shop in a shopping mall. You can produce a dozen muffins for $0.80. How many dozens must you produce before your average cost per dozen (*including* your initial investment of $80,000) drops to $1.60?

Solution

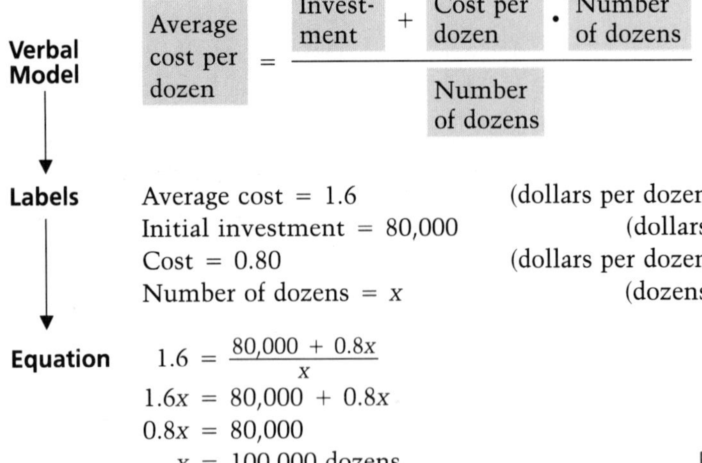

Verbal Model

$$\text{Average cost per dozen} = \frac{\text{Investment} + \text{Cost per dozen} \cdot \text{Number of dozens}}{\text{Number of dozens}}$$

Labels

Average cost = 1.6	(dollars per dozen)
Initial investment = 80,000	(dollars)
Cost = 0.80	(dollars per dozen)
Number of dozens = x	(dozens)

Equation

$$1.6 = \frac{80{,}000 + 0.8x}{x}$$
$$1.6x = 80{,}000 + 0.8x$$
$$0.8x = 80{,}000$$
$$x = 100{,}000 \text{ dozens}$$

■

Real Life
Baseball

Example 5 *Finding a Batting Average*

You have been officially at bat 140 times and have 35 hits. Your batting average is $\frac{35}{140} = .250$. How many consecutive hits must you get to increase your batting average to .300?

Solution If you hit safely in each of the next x times at bat, then x will be the number of future hits *and* the future times at bat.

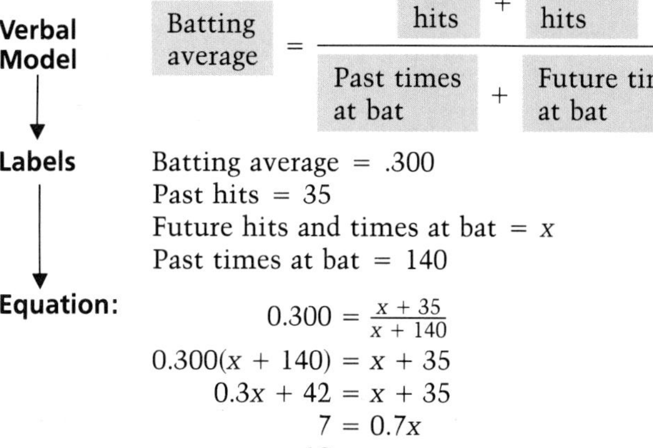

Verbal Model

$$\text{Batting average} = \frac{\text{Past hits} + \text{Future hits}}{\text{Past times at bat} + \text{Future times at bat}}$$

Labels

Batting average $= .300$
Past hits $= 35$
Future hits and times at bat $= x$
Past times at bat $= 140$

Equation:

$$0.300 = \frac{x + 35}{x + 140}$$
$$0.300(x + 140) = x + 35$$
$$0.3x + 42 = x + 35$$
$$7 = 0.7x$$
$$10 = x$$

You need to get a hit in each of your next 10 times at bat. Check this result by finding the batting average for 45 hits in 150 at bats. ■

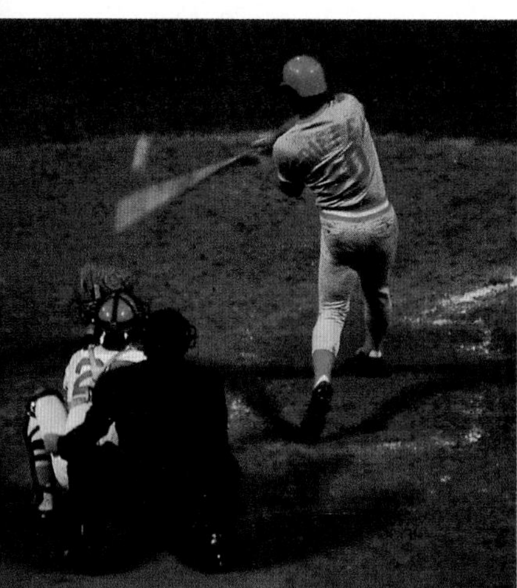

The batting average of a baseball player is the ratio of the number of hits to the number of official times at bat. In 1990, Daryl Strawberry (Mets) had a batting average of .225, and Barry Bonds (Pittsburgh Pirates) had a batting average of .248.

Communicating about ALGEBRA

SHARING IDEAS about the Lesson

In solving a rational equation, it is possible to obtain a "false solution." This is one reason it is important to check each solution in the original equation. Find the false solutions of the following.

A. $\frac{4}{x^2 - 2x} = \frac{4}{3x - 6}$ 2 **B.** $\frac{4}{x(x + 1)} = \frac{3}{x}$ 0

Extra Examples

Here are additional examples similar to **Examples 1–5.**

1. Solve.
   a. $\frac{12}{x} + \frac{2}{3} = \frac{6}{x}$ $x = -9$
   b. $\frac{x}{2} - \frac{8}{x} = 3$ $x = -2, x = 8$
   c. $\frac{12}{2x + 2} = \frac{x}{2}$ $x = -4, x = 3$
2. In Example 5, how many hits in a row are needed to raise your batting average up to 0.340? at least 20 consecutive hits

Check Understanding

1. When is it most appropriate to use the Cross-multiplying property in solving rational equations? When each side is a rational expression
2. When is it most appropriate to use the least common denominator in solving rational equations? When each side is composed of indicated additions or subtractions
3. Why is it necessary to be able to solve quadratic equations when solving rational equations? After multiplying, some terms may be quadratic.

Communicating about ALGEBRA

Determine the domain of each rational expression in Parts A and B. Compare the domains with the solution values. Use the comparisons to explain what "false solutions" are.

EXTEND *Communicating*
Identify two other rational equations that have false solutions. What patterns (if any) do you observe?

EXERCISE Notes

ASSIGNMENT GUIDE

Basic/Average: Ex. 1–6, 9–23 odd, 31–47 odd, 51–59 odd

Above Average: Ex. 1–6, 7–31 odd, 37–45 odd, 47–51, 57–63 odd

Advanced: Ex. 1–6, 15–45 odd, 46–51, 55–65 odd

Selected Answers
Exercises 1–6, 7–61 odd

More Difficult Exercises
Exercises 45–50, 63–65

Guided Practice

Ex. 1–6 These exercises help students review the critical elements of the solution techniques involving rational equations.

Independent Practice

Ex. 13–18 Point out to students the similarities between these exercises and proportions from Lesson 11.1.

Ex. 19–30 Review with students the ideas of finding least common denominators.

EXERCISES

Guided Practice

CRITICAL THINKING about the Lesson

1. What is the least common denominator for $\frac{1}{x}$, $\frac{x}{3}$, and $\frac{2}{3x}$? $3x$

In Exercises 2 and 3, solve $\frac{3}{x} = \frac{x}{12}$. Which method do you prefer? Why?

2. Solve the equation by cross multiplying. 6, −6.
3. Solve the equation by multiplying both sides by the least common denominator. 6, −6

In Exercises 4–6, decide which solution method is better and use that method to solve the equation.

4. $\frac{1}{x} + \frac{x}{3} = \frac{4}{3}$ 1, 3
5. $\frac{16}{x} = \frac{4}{2}$ 8
6. $\frac{x}{9} - \frac{1}{3} = \frac{6}{x}$ −6, 9

Independent Practice

In Exercises 7–12, find the greatest common factor.

7. $3, 12x, 6$ 3
8. $x, -4x, 3x^2$ x
9. $2x^2, 4x, -2$ 2
10. $3x^2, -6$ 3
11. $4x, 6x^2, 8x^2$ $2x$
12. $x^3, -5x, 3x^2$ x

In Exercises 13–18, solve the equation by cross multiplying.

13. $\frac{x}{3} = \frac{5}{2}$ $\frac{15}{2}$
14. $\frac{x}{10} = \frac{12}{5}$ 24
15. $\frac{3}{x} = \frac{9}{2(x+2)}$ 4
16. $\frac{5}{x+4} = \frac{5}{3(x+1)}$ $\frac{1}{2}$
17. $\frac{3}{x+5} = \frac{2}{x+1}$ 7
18. $\frac{7}{x+1} = \frac{5}{x-3}$ 13

In Exercises 19–30, solve the equation by multiplying both sides by the least common denominator.

19. $\frac{1}{2} + \frac{2}{x} = \frac{1}{x}$ −2
20. $\frac{1}{3} - \frac{2}{3x} = \frac{1}{x}$ 5
21. $\frac{x}{3} - \frac{1}{x} = \frac{2}{3}$ 3, −1
22. $\frac{x}{4} - \frac{5}{x} = \frac{1}{4}$ 5, −4
23. $\frac{25}{t} = 10 - t$ 5
24. $x + 4 = -\frac{4}{x}$ −2
25. $\frac{x}{x+4} = \frac{4}{x+4} + 2$ −12
26. $\frac{7}{3x-12} - \frac{1}{x-4} = \frac{2}{3}$ 6
27. $\frac{1}{x-3} + \frac{1}{x+3} = \frac{10}{x^2-9}$ 5
28. $\frac{1}{x-2} + 3 = \frac{16}{x^2+x-6}$ $-\frac{1}{3}$, −1
29. $5 + \frac{6}{x-3} = \frac{x+3}{x^2-9}$ 2
30. $\frac{2}{(x-2)^2} = 1 - \frac{1}{x-2}$ 4, 1

In Exercises 31–42, solve the equation (by the more convenient method).

31. $\frac{3x}{x+1} = \frac{2}{x-1}$ $-\frac{1}{3}$, 2
32. $\frac{20-x}{x} = x$ −5, 4
33. $\frac{4}{x} - \frac{x}{6} = \frac{5}{3}$ −12, 2
34. $x + \frac{1}{x} = \frac{5}{2}$ $\frac{1}{2}$, 2
35. $\frac{x+30}{x} = x$ 6, −5
36. $\frac{3}{x+5} = \frac{2}{x+1}$ 7

37. $\frac{18}{u} = \frac{u}{2}$ 6, −6

38. $\frac{4}{x+2} - \frac{1}{x} = \frac{1}{x}$ 2

39. $\frac{15}{x} - 4 = \frac{6}{x} + 3$ $\frac{9}{7}$

40. $\frac{5}{u+8} = \frac{1}{4}$ 12

41. $\frac{2}{x+3} + \frac{1}{x} = \frac{4}{3x}$ $\frac{3}{5}$

42. $\frac{10}{x+3} - \frac{3}{5} = \frac{10x+1}{3x+9}$ 2

43. Batting Average After 50 times at bat, a baseball player has a batting average of .160. How many consecutive hits must the player get to raise the batting average to .250? 6

44. Bowling Average After 12 games, your bowling average is 120. What must your average score be for the next 6 games to increase your overall average to 130? 150

45. The Karate Kid, Part 0 You are interested in attending a karate class at the Y. Membership at the Y is $36 per year. The charge per lesson is $3.50 for members. Write an expression that represents your average cost per lesson for n lessons. After completing the course, you figured that your average cost per lesson (including the membership fee) was $6.50. How many karate lessons did you take? $\frac{36 + 3.50n}{n}$, 12

46. Specialty Dog Food You have started a specialty dog food company. Your initial investment was $120,000. Each pound of dog food costs $0.50 to produce. Write an expression that represents your average cost per pound after producing x pounds of dog food. You are selling the dog food for $1 per pound. How many pounds must you produce to make your average cost per pound equal to your per-pound selling price? (This is called the *break-even point* for your business.)

By 1990, specialty dog foods and cat foods, produced mainly by small companies, accounted for $1.5 billion of the total $7.5 billion United States pet food market. (Source:* Forbes Magazine*)

46. $\frac{120{,}000 + 0.50x}{x}$, 240,000

47. Book Report You had 21 days to read a 420-page book for a book report. After finishing half of the book, you realize that you must read twice as many pages a day to finish on time. What was your average daily rate? What must your rate be to finish on time? Use the following model to answer the questions. 15 pages per day; 30 pages per day

$$\frac{\text{210 pages}}{\text{Previous rate}} + \frac{\text{210 pages}}{2 \cdot \text{Previous rate}} = \text{21 days}$$

48. Delivering Newspapers You have 40 minutes to deliver 90 newspapers in your neighborhood. After you have delivered half of the papers, you realize that you must begin to deliver them three times as quickly to finish on time. How many papers per minute were you delivering? How many papers per minute must you now deliver to finish on time? Use the following model to answer the questions. 1.5 papers per minute; 4.5 papers per minute

$$\frac{\text{45 newspapers}}{\text{Previous rate}} + \frac{\text{45 newspapers}}{3 \cdot \text{Previous rate}} = \text{40 minutes}$$

Ex. 25–30 Students have a choice of solution methods. They can combine terms on each side before finding the least common denominator, or they can multiply each term by the least common denominator before combining terms. You may wish to show examples of both solution methods in class.

Ex. 31–42 Refer students to Example 3 to help them recognize when it is appropriate to use the Cross-multiplying property.

Ex. 45–46 Refer students to Example 4 if they need help in setting up the equations in these exercises.

Answers

63.–65. When applying the Cross-multiplying property, you are not multiplying both sides of the equation by a 0 factor. Thus, in Exercise 63, cross multiplying leads to an extraneous solution of $x = 1$. However, multiplying both sides of the equation by the LCD does not lead to an extraneous solution. Checking your answer in the original equation will determine whether you obtain a false solution.

Enrichment Activites

These activities require students to go beyond lesson goals.

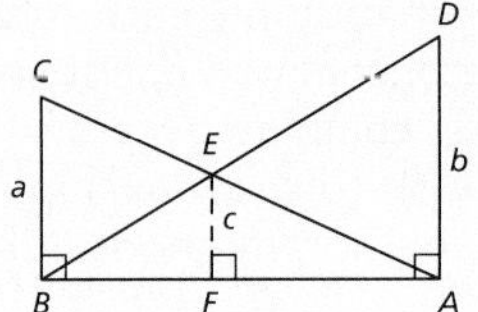

In this diagram, $\frac{1}{a} + \frac{1}{b} = \frac{1}{c}$, a, b, and $c > 0$.

1. Solve this equation for c in terms of a and b. $c = \frac{ab}{a+b}$
2. Triangles AEF and ACB are similar triangles as are triangles BFE and BAD. If $a = 5$, $b = 9$, and $BF = \frac{30}{7}$, find the dimensions of triangles ABC and ABD. $c = \frac{45}{14}$. Setting up proportions involving c leads to triangle ABC having dimensions 5, 12, and 13 and triangle ABD having dimensions 9, 12, and 15.
3. Do you think the relationship $\frac{1}{a} + \frac{1}{b} = \frac{1}{c}$ holds for other pairs of right triangles? Check your answer by drawing several diagrams on grid paper similar to the given one and measuring the values. It does.

612

49. *In Search of Pandora* Which of the following ratios can be used to find the depth, d, of the *Pandora* wreckage? (See page 561.) Explain your answer and find *Pandora's* depth.

a. The ratio of the pressure, P, to the depth, d, was 62.4.

b. The ratio of the *Titanic's* depth, D, to *Pandora's* depth, d, was 118.2.

b, $\frac{13{,}000}{d} = 118.2$, ≈ 110 ft

The Pandora sank on Australia's Great Barrier Reef in 1791. The ship was carrying some of the Bounty crew members who had been accused of mutiny. The wreckage was found in 1977.

50. *Finding the Bounty* The wreck of the *Bounty* was discovered n years before the wreck of the *Pandora*. Find the year in which the *Bounty* was discovered by solving 1922

$$\frac{5}{n+25} = \frac{3}{n-7}.$$

Integrated Review

In Exercises 51–53, decide whether the value of x is a solution of the equation.

51. $\frac{x}{5} - \frac{3}{x} = \frac{7}{10}$, $x = 6$ Yes

52. $\frac{3x}{5} + \frac{x^2}{2} = \frac{4}{5}$, $x = -2$ Yes

53. $\frac{5}{2x} - \frac{4}{x} = 3$, $x = 0$ No

In Exercises 54–56, solve the equation.

54. $(2x + 3)(x - 9) = 0$ $-\frac{3}{2}, 9$

55. $x^2 + x - 56 = 0$ $-8, 7$

56. $t^2 - 5t = 0$ $0, 5$

In Exercises 57–62, solve the proportion.

57. $\frac{3}{4} = \frac{2}{x}$ $\frac{8}{3}$

58. $\frac{5}{x} = \frac{1}{3}$ 15

59. $\frac{1}{x+1} = \frac{1}{6}$ 5

60. $\frac{1}{2x-3} = \frac{1}{9}$ 6

61. $\frac{1}{3-y} = -\frac{1}{10}$ 13

62. $\frac{5}{u+8} = \frac{1}{4}$ 12

Exploration and Extension

In Exercises 63–65, solve the equation by either method discussed in this lesson. Do you obtain "false solutions"? Explain. See margin on page 611.

63. $\frac{1}{x^2 - x} = \frac{1}{2x-2}$ 2

64. $\frac{1}{x(x+4)} = \frac{2}{x+4}$ $\frac{1}{2}$

65. $\frac{5}{x^2} = \frac{-1}{x}$ -5

11 Chapter Summary

What did you learn?

Skills

1. Solve a proportion. **(11.1)**
2. Solve a percent problem. **(11.2)**
3. Solve a problem involving
   - direct variation. **(11.3)**
   - inverse variation. **(11.3)**
   - probability. **(11.4)**
4. Simplify a rational expression. **(11.5)**
5. Multiply and divide rational expressions. **(11.6)**
6. Divide one polynomial by another polynomial. **(11.7)**
7. Solve a rational equation. **(11.8)**

Strategies

8. Use rational expressions in real-life problems. **(11.1–11.8)**

Exploring Data

9. Use data to find ratios, proportions, and probabilities. **(11.1–11.8)**

Why did you learn it?

Rational expressions occur frequently in real life as ratios, rates, percents, and probabilities. To solve problems involving rational expressions, you must be able to simplify rational expressions, multiply and divide rational expressions, and solve equations that involve rational expressions.

How does it fit into the bigger picture of algebra?

In this chapter, you studied *rational expressions*—fractions whose numerators and denominators are polynomials. The first four lessons in the chapter introduce several types of problems—ratios, proportions, percents, direct variation, inverse variation, and probability—that often involve rational expressions. Using ratio, proportion, and variation techniques in your approach to all kinds of real-life contexts should become almost an instinctive problem-solving tool.

The last four lessons describe techniques for simplifying rational expressions, multiplying and dividing rational expressions, and solving rational equations.

ORGANIZER

TIME SCHEDULE: All levels, two days

The Chapter Summary helps students organize the main ideas of the chapter. In this chapter, the basic operations of rational expressions were developed. These operations were then utilized to solve rational equations.

Work with students to review the skills, strategies, and concepts of the chapter. It is suggested that the first day of review be a combination of you and your students working problems from the Chapter Review exercises. The first day's homework assignment can be used as the basis for the second day of review.

COOPERATIVE LEARNING

Encourage students to work cooperatively. Emphasize the importance of teaching a classmate how to perform a skill or how to recall a strategy. When students work together, everyone wins. The students receiving help get additional instruction, and the students giving help gain a deeper understanding of the skills and concepts involved.

Chapter SUMMARY

Review the chapter by asking students, "What did you learn? Why did you learn it? How does it fit with the other algebra skills and concepts you have learned?"

Chapter REVIEW

ASSIGNMENT GUIDE

Basic/Average: Ex. 1–17 odd, 18, 21–54

Above Average: Ex. 1–17, 18, 21–36 multiples of 3, 39–53 odd, 59–60

Advanced: Ex. 1–17, 18, 21–36 multiples of 3, 39–53 odd, 59–60

More Difficult Exercises

Exercises 15–16, 55–60

Ex. 1–60 Since the Chapter Review exercises parallel the list of Skills, Strategies, and Exploring Data statements, tell students to use the Chapter Summary on page 613 to identify the lesson in which a skill, strategy, or concept was presented if they are uncertain about how to do a given exercise in the Chapter Review. This makes the students categorize each task in the exercise set and locate relevant information in the chapter about the task, both beneficial activities.

Chapter REVIEW

In Exercises 1–6, solve the proportions. (11.1)

1. $\frac{x}{2} = \frac{4}{7}$ $\frac{8}{7}$
2. $\frac{5}{x} = \frac{6}{11}$ $\frac{55}{6}$
3. $\frac{7}{10} = \frac{9+x}{x}$ -30
4. $\frac{8x}{4+x} = 2$ $\frac{4}{3}$
5. $\frac{x+3}{x} = \frac{4}{x-1}$ 3 or -1
6. $\frac{5}{(x+6)} = \frac{(x-6)}{x}$ 9 or -4

In Exercises 7–12, use the following information. (11.3)

The variables a and b vary directly. When a is 6, b is 19. The variables p and q vary inversely. When p is 2, q is 9.

7. Find an equation that relates a and b. $a = \frac{6}{19}b$
8. What is b when a is 3? $\frac{19}{2}$
9. What is a when b is 14? $\frac{84}{19}$
10. Find an equation that relates p and q. $pq = 18$
11. What is q when p is 30? $\frac{3}{5}$
12. What is p when q is 27? $\frac{2}{3}$
13. *Miniature Horses* The ratio of the weight of an adult miniature horse to the weight of a baby miniature horse is about 8. What does an adult weigh? **(11.1)** 192 lb
14. *Spiral Sculpture* Before creating his Spiral Jetty, Robert Smithson made a model using a ratio of 1 inch to 50 feet. How long was the coil in the model? **(11.1)** 30 in.

This baby miniature horse is standing on a man's arms. It weighs only 24 pounds.

This environmental sculpture in the Great Salt Lake, by Robert Smithson, is a coil of rocks 1500 feet long.

In Exercises 15 and 16, find the ratio of the area of the blue region to the entire area. (11.1)

15. $\frac{23}{24}$

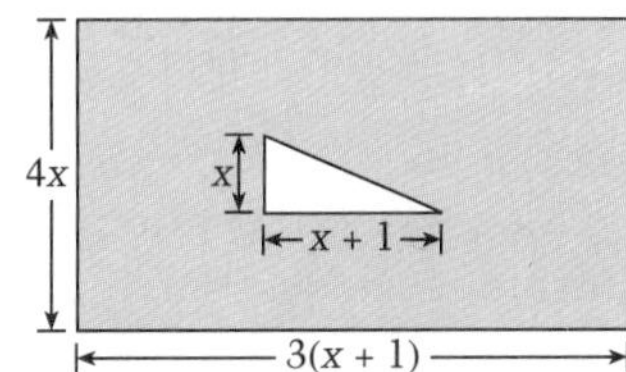

16. $\frac{1}{6}$

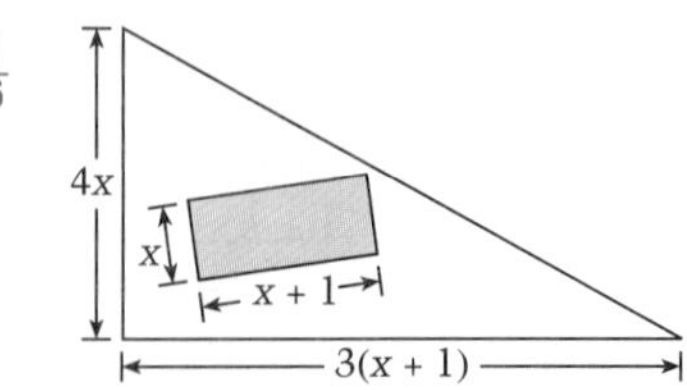

Three-Point Shots **In Exercises 17 and 18, use the following information. (11.4)**

The three-point shot was introduced to college basketball games in 1987. The graph at the right shows the average number of three-point shots attempted per game and the average number of three-point shots made.

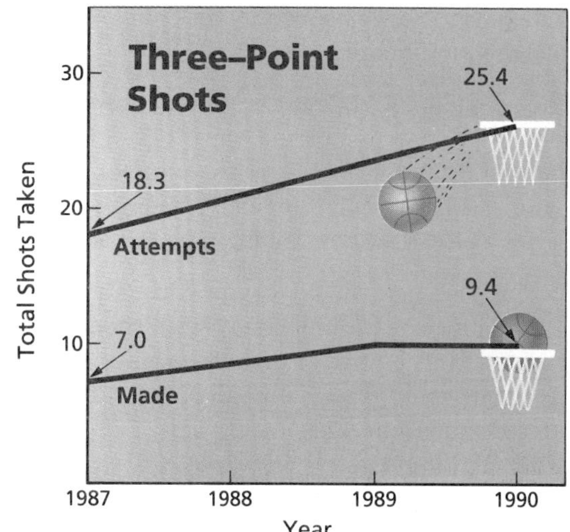

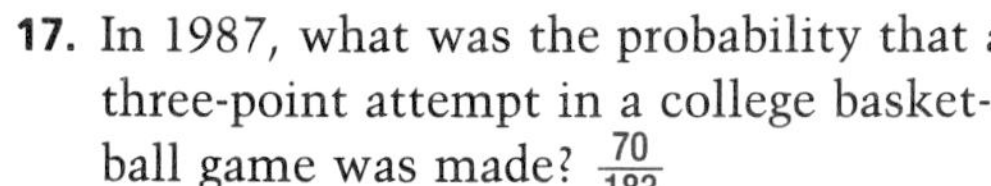

17. In 1987, what was the probability that a three-point attempt in a college basketball game was made? $\frac{70}{183}$

18. What was the probability in 1990? $\frac{47}{127}$

In Exercises 19–36, simplify the rational expressions. (11.5, 11.6)

19. $\frac{3x}{9x^2+3}$ $\frac{x}{3x^2+1}$

20. $\frac{4x^2-2x}{8x^2}$ $\frac{2x-1}{4x}$

21. $\frac{5x^2+10x-5}{25x+100}$

22. $\frac{6x^2}{12x^4+18x^2}$

23. $\frac{8x^2-4x}{16x^3}$ $\frac{2x-1}{4x^2}$

24. $\frac{7x^4}{49x^2+x^2}$ $\frac{7x^2}{50}$

25. $\frac{12x^2}{5x^3} \cdot \frac{25x^4}{3x}$ $20x^2$

26. $\frac{2x^3}{4x} \cdot \frac{3x^2}{12x^4}$ $\frac{1}{8}$

27. $\frac{14x}{x^3} \cdot \frac{2x^2}{7x^4}$ $\frac{4}{x^4}$

28. $\frac{2x}{x^2+2x+1} \cdot \frac{x+1}{2x^2-2x}$ $\frac{1}{(x+1)(x-1)}$

29. $\frac{3x^2-3}{x-1} \cdot \frac{4}{x+1}$ 12

30. $\frac{x^2-6x-7}{6x+30} \cdot \frac{2x^2-50}{2x-14}$ $\frac{(x+1)(x-5)}{6}$

31. $\frac{3x}{x+4} \div \frac{x^2}{x+4}$ $\frac{3}{x}$

32. $\frac{6x^3-6x}{3x^2} \div \frac{x+1}{2x^3}$ $4x^2(x-1)$

33. $\frac{x^2+5x+6}{x+3} \div \frac{x+5}{x^2+4x+3}$ $\frac{(x+1)(x+2)(x+3)}{x+5}$

34. $\frac{9x^3}{x^3-x^2} \div \frac{x-8}{x^2-9x+8}$ $9x$

35. $\frac{8x+4}{32x^2} \div \frac{2x^2+x}{x^3}$ $\frac{1}{8}$

36. $\frac{x^2+3x+2}{x^2+7x+12} \div \frac{x^2+5x+4}{x^2+5x+6}$ $\frac{(x+2)^2}{(x+4)^2}$

In Exercises 37–48, perform the indicated division. (11.7)

37. $(12x^2+2x-3) \div 2x$ $6x+1-\frac{3}{2x}$

38. $(8x^2-6x+9) \div 2x$ $4x-3+\frac{9}{2x}$

39. $(9x^2+3x+1) \div 3x$ $3x+1+\frac{1}{3x}$

40. $(4x^2+10x-3) \div 2x$ $2x+5-\frac{3}{2x}$

41. $(3x^2-15x-10) \div 3x$ $x-5-\frac{10}{3x}$

42. $(6x^2-36x+5) \div 6x$ $x-6+\frac{5}{6x}$

43. $(3x^2+2x-1) \div (x-1)$ $3x+5+\frac{4}{x-1}$

44. $(x^2-4x+2) \div (x+2)$ $x-6+\frac{14}{x+2}$

45. $(4x^2+9x+5) \div (2x-1)$ $2x+\frac{11}{2}+\frac{21}{2(2x-1)}$

46. $(6x^2-3x-4) \div (3x+6)$ $2x-5+\frac{26}{3x+6}$

47. $(5x^2+3x+6) \div (5x-2)$ $x+1+\frac{8}{5x-2}$

48. $(12x^2-8x-2) \div (4x+6)$ $3x-\frac{13}{2}+\frac{37}{4x+6}$

In Exercises 49–54, solve the equation. (11.8)

49. $\frac{3}{x}+\frac{2}{3}=\frac{5}{x}$ 3

50. $\frac{1}{4}+\frac{6}{x}=\frac{3}{x}$ -12

51. $\frac{x}{5}-\frac{6}{x}=\frac{1}{5}$ $6, -5$

52. $\frac{x}{4}+\frac{7}{4}=-\frac{3}{x}$ $-4, -3$

53. $\frac{x+2}{2}=\frac{4}{x}$ $-4, 2$

54. $\frac{x}{5}=\frac{7}{x-2}$ $7, -5$

▶ **Ex. 17–18** Have students who follow basketball explain the three-point shot to their classmates.

Answers.

21. $\frac{x^2+2x-1}{5(x+4)}$

22. $\frac{1}{2x^2+3}$

Comparing Apples and Oranges **In Exercises 55 and 56, use the following information. (11.3)**

The graph at the right shows the per capita consumption of apples and oranges in 1990, the average retail price, and the number of Calories in an average apple and orange. ***(Source: United Fresh Fruit and Vegetable Association)***

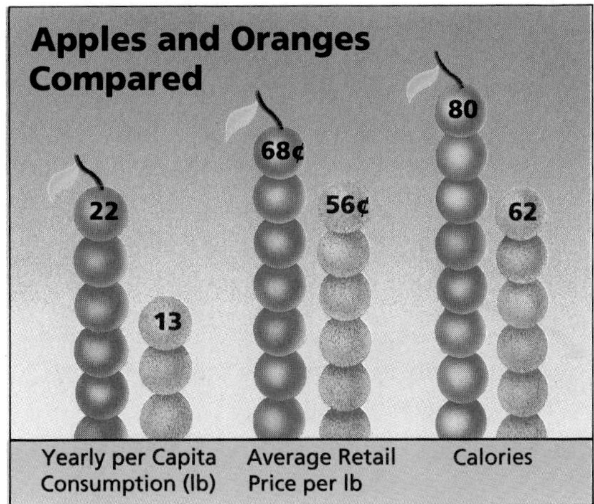

55. Suppose that the per capita consumption of oranges varies inversely with the retail price. If the price of oranges reaches 75¢, what would be the annual per capita consumption? 9.7 lb

56. The number of Calories consumed by eating apples varies directly with the number of apples eaten. If you consumed 558 Calories from apples in one week, how many apples did you eat? ≈7 apples

Apple Picking **In Exercises 57 and 58, use the following information. (11.2)**

The graph shows how the 231 million bushels of apples that were picked in 1990 were used. ***(Source: International Apple Institute)***

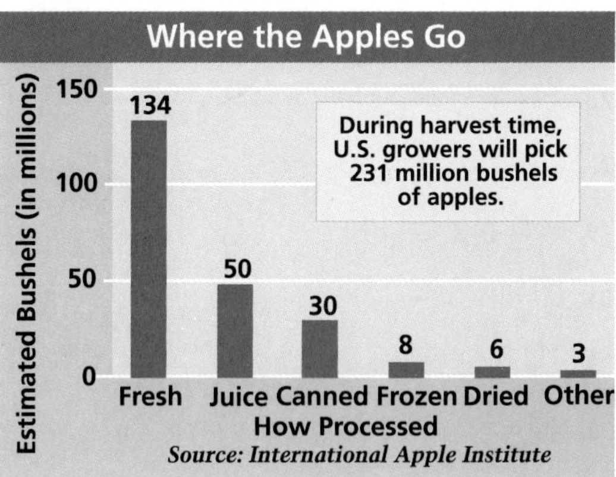

57. What percent of the apples were processed as dried apples? 2.6%

58. What was the probability that an apple that was picked would be processed for apple juice? $\frac{50}{231}$

Eating Ethnic **In Exercises 59 and 60, use the following information. (11.2)**

The graph at the right shows the results of a survey of 635 adults. Each person was asked about the types of ethnic restaurants at which they had eaten. ***(Source: National Restaurant Association)***

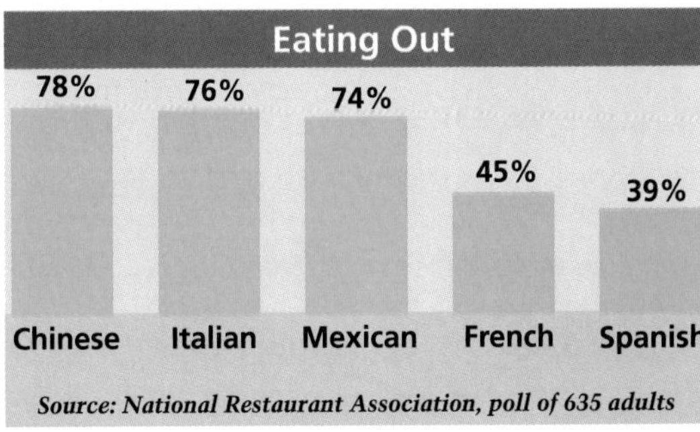

59. How many of the 635 polled have eaten in a Chinese restaurant? 495

60. Suppose that you took a survey about eating in ethnic restaurants and found that 336 people had eaten in a Mexican restaurant. If your results agreed with those shown, how many people did you survey? 454

Answers

19. $3x - 5 + \frac{9}{x + 2}$

20. $x + 1 + \frac{9}{x - 3}$

Chapter TEST

In Exercises 1–4, solve the proportion. **(11.1)** 5. 44 gal 7. 50m

1. $\frac{3}{x} = \frac{15}{4}$ $\frac{4}{5}$

2. $\frac{7}{3} = \frac{5}{y}$ $\frac{15}{7}$

3. $\frac{x + 2}{5} = \frac{-1}{x - 4}$ 3, −1

4. $\frac{x + 8}{x} = \frac{x + 2}{2}$ 4, −4

5. How much is 22% of 200 gallons? **(11.2)**

6. How many is 18% of 4700 people? **(11.2)** 846 people

7. 37 meters is 74% of what distance? **(11.2)**

8. 230 T-shirts is what percent of 500 T-shirts? **(11.2)** 46%

9. The variables x and y vary directly. When x is 3, y is 14. Find an equation that relates x and y. If x is 5, what is y? **(11.3)** $y = \frac{14}{3}x$; $\frac{70}{3}$

10. The variables m and n vary inversely. When m is 12, n is 42. Find an equation that relates m and n. If n is 2, what is m? **(11.3)** $mn = 504$; 252

11. Your cousin's basketball team has 15 members. During the game, each team member played the same amount of time. You walked into the game late. What is the probability that your cousin was playing when you walked in? (There are five team members playing at any given time.) **(11.4)** $\frac{1}{3}$

12. There are 14 plastic forks and 18 plastic spoons in a picnic basket. You grab one utensil. What is the probability that it is a fork? What is the probability that it is a spoon? **(11.4)** $\frac{7}{16}$; $\frac{9}{16}$

In Exercises 13–18, simplify the rational expressions. **(11.5, 11.6)** 19., 20. See margin.

13. $\frac{3x}{x^2 + 2x}$ $\frac{3}{x + 2}$

14. $\frac{x^2 + 2x + 1}{x^2 - 1}$ $\frac{x + 1}{x - 1}$

15. $\frac{4x^2}{10x^4} \cdot \frac{-3x^2}{6x}$ $\frac{-1}{5x}$

16. $\frac{x^3 + x^2}{x^2 - 9} \cdot \frac{x + 3}{4x^3 + x^2 - 3x}$ $\frac{x}{(x - 3)(4x - 3)}$

17. $\frac{x + 5}{3x^2} \div \frac{4x}{x + 5}$ $\frac{(x + 5)^2}{12x^3}$

18. $\frac{(x + 5)}{x^3 - x^2 - 6x} \div \frac{x^2 - 25}{x^2 + x - 12}$ $\frac{x + 4}{x(x + 2)(x - 5)}$

19. Divide $(3x^2 + x - 1)$ by $(x + 2)$. **(11.7)**

20. Divide $(x^2 - 2x + 6)$ by $(x - 3)$. **(11.7)**

In Exercises 21 and 22, solve the equation. **(11.8)**

21. $\frac{6}{x} + \frac{1}{2} = \frac{4}{x}$ −4

22. $\frac{x}{4} - \frac{21}{4x} = 5$ 21, −1

23. You have invested \$30,000 to start a bagel shop in town. You can produce bagels for \$1.20 a dozen. How many dozens must you produce before your average cost per dozen drops to \$1.80? **(11.8)** 50,000 dozen

24. You have taken four 100-point tests and your average is 88. What must you average on the remaining two 100-point tests to bring your overall average up to 92? **(11.8)** 100 points

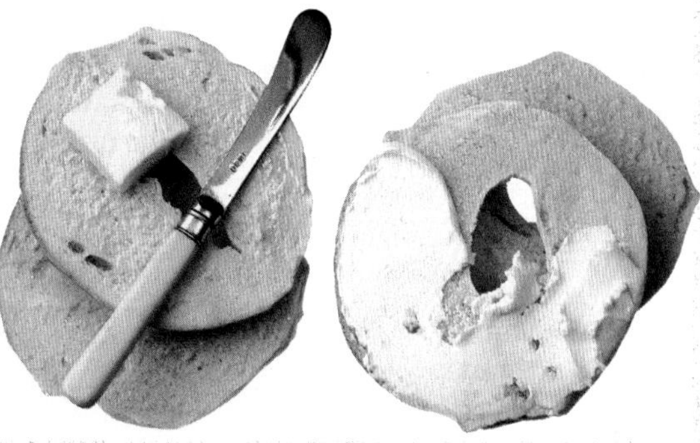

The table of values on the following page represents a special relationship called a function. The area of the n-sided regular polygon of fixed perimeter *depends* upon the value of n. For each value of n, there is only one area associated with it. We say that "the area of the n-sided regular polygon of fixed perimeter is a *function of n*." This chapter describes many of the important ideas about functions and how they are communicated in algebra.

The Chapter Summary on page 665 provides you and the students with a synopsis of the chapter. It identifies key skills, concepts, and vocabulary. You may want to have students look at the Chapter Summary as an overview before beginning the chapter.

CHAPTER 12

Functions

LESSONS

The Ballotini spheres shown here are tiny glass spheres that are used in Venetian glass mosaics.

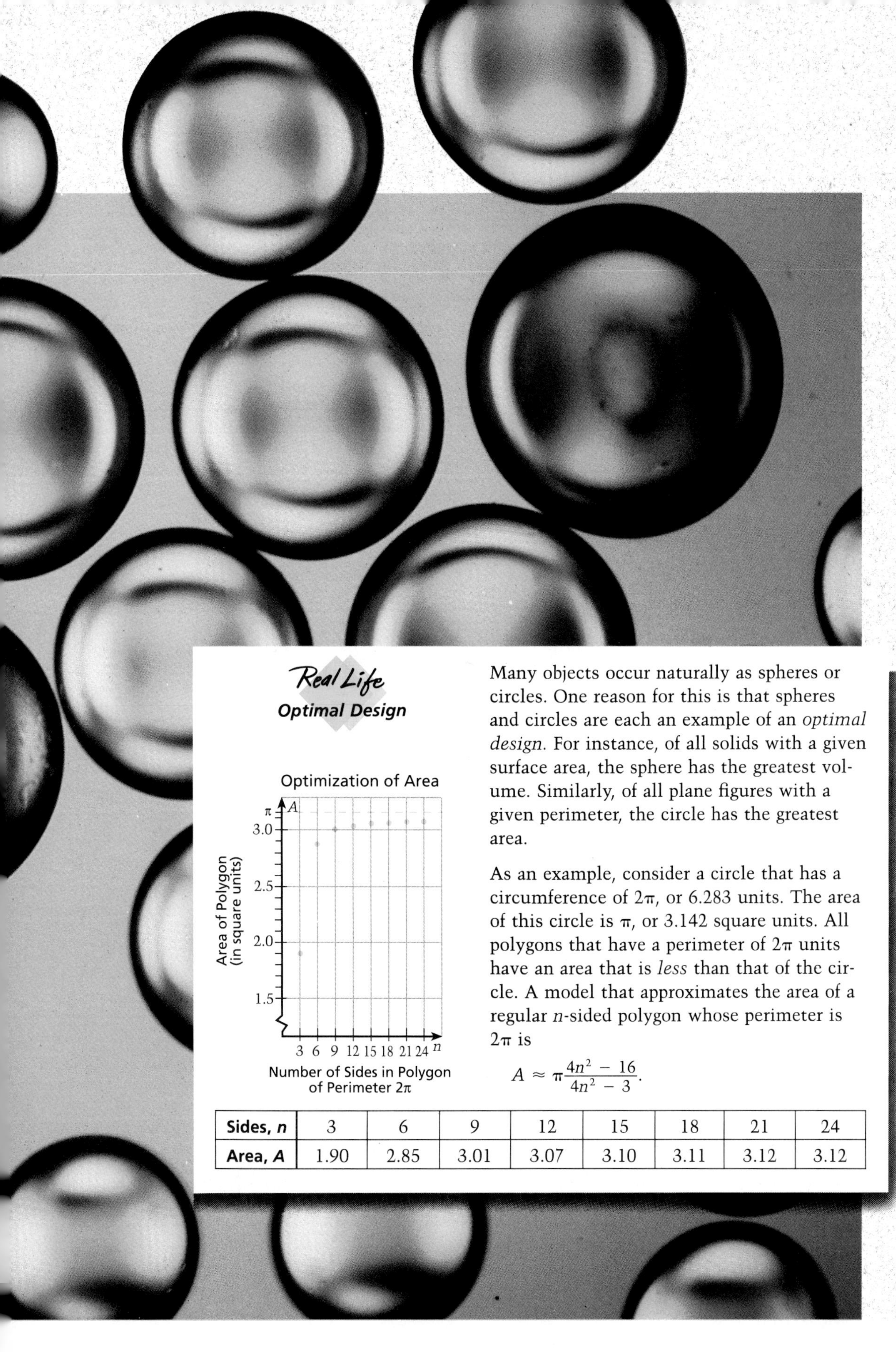

Many objects occur naturally as spheres or circles. One reason for this is that spheres and circles are each an example of an *optimal design.* For instance, of all solids with a given surface area, the sphere has the greatest volume. Similarly, of all plane figures with a given perimeter, the circle has the greatest area.

As an example, consider a circle that has a circumference of 2π, or 6.283 units. The area of this circle is π, or 3.142 square units. All polygons that have a perimeter of 2π units have an area that is *less* than that of the circle. A model that approximates the area of a regular n-sided polygon whose perimeter is 2π is

$$A \approx \pi\frac{4n^2 - 16}{4n^2 - 3}.$$

Sides, *n*	3	6	9	12	15	18	21	24
Area, *A*	1.90	2.85	3.01	3.07	3.10	3.11	3.12	3.12

Circles and spheres are, indeed, examples of optimal designs. Of all plane figures of a given area, a circle has the least perimeter, and of all solids of a given volume, a sphere has the smallest surface area.

Ask students to explain why a circle of circumference 2π has an area of π by having students use calculators to evaluate the given algebraic model for the area of a regular n-sided polygon of perimeter 2π. Ask students to provide a physical interpretation of successive entries in the table. As the number of sides increases, the polygon approaches the shape of a circle, and consequently, the area approaches the value of π.

Functions

PACING CHART

The daily Pacing Chart is meant to help you adjust your teaching pace. Students in the full course should finish the entire text by the end of the year. Students in the basic course may not complete the entire text in the school year. The Pacing Chart for each chapter contains suggestions for lessons that require more than one day and lessons that may be omitted for the basic course.

DAY	FULL COURSE	BASIC COURSE
1	12.1	12.1
2	12.2	12.2
3	12.3	12.2
4	12.4	12.3
5	Mid-Chapter Self-Test	12.4
6	12.5	Mid-Chapter Self-Test
7	12.6	12.5
8	12.7	12.6
9	Chapter Review	12.7
10	Chapter Review	Chapter Review
11	Chapter Test	Chapter Review
12	Cumulative Review	Chapter Test
13		Cumulative Review

CHAPTER ORGANIZATION

LESSON	PAGES	GOALS	MEETING THE NCTM STANDARDS
12.1	620-625	1. Identify functions and use function notation 2. Identify real-life relations that are functions	Problem Solving, Communication, Connections, Functions
12.2	626-631	1. Use linear functions to solve problems 2. Use linear functions to answer questions about real-life situations	Problem Solving, Communication, Connections, Functions
Mixed Review	632	Review of algebra and arithmetic	
Milestones	632	Unsolved problems in mathematics	Connections
12.3	633-638	1. Transform the graph of a function 2. Use exponential functions as models of real-life problems	Problem Solving, Communication, Connections, Functions, Geometry
12.4	639-644	1. Sketch the graph of a quadratic function 2. Use quadratic functions as models of real-life problems	Problem Solving, Communication, Reasoning, Connections, Functions, Geometry
Mid-Chapter Self-Test	645	Diagnose student weaknesses and remediate with correlated Reteaching worksheets	
12.5	646-651	1. Sketch the graph of a rational function 2. Use rational functions as models	Problem Solving, Communication, Connections, Functions, Geometry
Mixed Review	651	Review of algebra and arithmetic	
12.6	652-657	1. Construct a stem-and-leaf plot 2. Construct a box-and-whisker plot	Problem Solving, Communication, Statistics, Discrete Mathematics
12.7	658-663	1. Find the mean, median, and mode 2. Use measures of central tendency in real-life situations	Problem Solving, Communication, Connections, Statistics, Discrete Mathematics
Using a Calculator	664	Use a graphing calculator to construct a histogram	Technology, Statistics
Chapter Summary	665	A restatement of what has been learned, why it has been learned, and how it fits into the structure of algebra	Structure, Connections
Chapter Review	666-668	Review of concepts and skills learned in the chapter	
Chapter Test	669	Diagnose student weaknesses and remediate with correlated Reteaching worksheets	
Cumulative Review	670-673	Review of concepts and skills from previous chapters	

LESSON RESOURCES

MEETING INDIVIDUAL NEEDS

RETEACHING For students who need to spend more time on basics:

If a mid-chapter self-test or chapter test indicates a deficiency, teachers can help students with the appropriate ***Reteaching Copymaster.***

PRACTICE For students who need more practice:

Additional exercises like those in the Pupil's Edition are provided for each lesson in ***Extra Practice Copymasters.***

ENRICHMENT For enriching and broadening students' experiences:

Problem of the Day copymasters in ***Teaching Tools*** provide a daily opportunity to use logical reasoning, looking for a pattern, writing an equation, and other routine and non-routine problem-solving strategies.

Math Log copymasters in ***Alternative Assessment*** provide opportunities to report on investigations, research, and open-ended problems.

Enriching activities with graphing and scientific calculators and computers are provided in ***Technology: Using Calculators and Computers.***

The ***Applications Handbook*** provides additional information about the cross-curriculum topics such as astronomy, chemistry, physics, sports, economics, genetics, and music that are integrated into the Pupil's Edition.

LESSON	12.1	12.2	12.3	12.4	12.5	12.6	12.7
PAGES	620-625	626-631	633-638	639-644	646-651	652-657	658-663
Teaching Tools							
Transparencies	✓	✓		✓			
Problem of the Day	✓	✓	✓	✓	✓	✓	✓
Warm-up Exercises	✓	✓	✓	✓	✓	✓	✓
Answer Masters	✓	✓	✓	✓	✓	✓	✓
Extra Practice Copymasters	✓	✓	✓	✓	✓	✓	✓
Reteaching Copymasters	Teacher-directed and independent activities tied to results on the Mid-Chapter Self-Tests and Chapter Tests						
Color Transparencies	✓	✓	✓		✓	✓	
Applications Handbook	Additional background information is supplied for many real-life applications.						
Technology Handbook	Calculator and computer worksheets are supplied for appropriate lessons.						
Complete Solutions Manual	✓	✓	✓	✓	✓	✓	✓
Alternative Assessment	Assess student's ability to reason, analyze, solve problems, and communicate using mathematical language.						
Formal Assessment	Mid-Chapter Self-Tests, Chapter Tests, Cumulative Tests, and Practice for College Entrance Tests						
Computer Test Bank	Customized tests can be created by choosing from over 2000 items.						

12.1 Functions and Relations

Functions are a cornerstone in the mathematical structures that we use everyday. Understanding functions is central to the study of mathematics. This lesson discusses the ways in which functions are defined or described, and how they are evaluated.

12.2 Linear Functions

The connection between linear equations and linear functions is developed in this lesson. Students revisit many relationships that can now be expressed as linear functions. The lesson helps students see that mathematical ideas often have multiple representations.

12.3 Exponential Functions

This lesson investigates transformations of the graph of an exponential function, as a visual representation of adding and subtracting constants in the original function. It would be well to point out that transformations can be applied to all functions, not just to exponential functions. The next lesson will, in fact, illustrate this further. Transformations can be an exciting and enriching use of geometry (apart from being a justification of the technique of completing the square). On page 724, transformations of quadratic and exponential functions are summarized.

12.4 Quadratic Functions

The connection between quadratic models and quadratic functions is similar to that between linear models and functions, and exponential models and functions. Real-life quadratic relationships can now be expressed as quadratic functions. The technique of completing the square will be useful in understanding the effects of constants on the behavior of quadratic functions and in graphing them.

12.5 Rational Functions

Simple rational functions can be used to model real-life situations, including uniform motion problems, work problems, and problems involving the ratios of two quantities. This lesson is an introduction to some of the main ideas associated with rational functions. Graphs and asymptotes are important elements in the study of rational functions. As students encounter more complex functions, the role of asymptotes assumes greater importance.

12.6 Exploring Data: Stem-and-Leaf Plots, Box-and-Whisker Plots

The organization of data is an ever-increasing need in a technology-based society. Consumer, political, economic, and sports data are only a few of the data types that informed citizens use everyday. In this lesson, students see how to organize large sets of data for convenient use.

12.7 Exploring Data: Measures of Central Tendency

The mean, median, and mode are measures of central tendency that are commonly used to obtain or provide information about the "typical," "average," or "normal" individual, number, or behavior. Consumer, political, economic, and sports data can be interpreted differently, depending upon which of the three measures of central tendency is chosen by the student. It is important, therefore, to have a clear understanding of what these measures represent and the differences among them.

Problem of the Day

Some students had 91¢ in dimes and pennies. They divided the dimes evenly. They divided the pennies evenly. How many students shared the money? 7

ORGANIZER

Warm-Up Exercises

1. Evaluate each expression for the given values of x.
   a. $4x + 5$ at $x = 0, x = -1$
   b. $3x^2 - 4x + 1$ at $x = -2, x = 3$

   a. 5, 1; b. 21, 16

2. Match each set of ordered pairs with the equation for which the ordered pairs are solutions.

   Ordered Pairs
   a. $\{(-2, 4), (-1, 1), (0, 0), (1, 1), (2, 4)\}$
   b. $\{(-2, -4), (-1, -1), (0, 0), (1, -1), (2, -4)\}$
   c. $\{(-2, -8), (-1, -1), (0, 0), (1, 1), (2, 8)\}$

   Equations
   1. $y = -x^2$ 2. $y = x^3$
   3. $y = x^2$

   Matches are 1b, 2c, 3a.

Lesson Resources

Teaching Tools
Transparencies: 10, 13
Problem of the Day: 12.1
Warm-up Exercises: 12.1
Answer Masters: 12.1
Extra Practice: 12.1
Color Transparencies: 68, 69

Lesson Investigation Answers

a. Not a function. (1, 2) and (1, 3) have the same x-value.
b. Function.
Domain: $\{0, 1, 2, 3, 4, 5\}$
Range: $\{3\}$
c. Function.
Domain: $\{0, 1, 2, 3, 4, 5, 6\}$
Range: $\{0, 1, 2, 3\}$

12.1 Functions and Relations

What you should learn:

Goal 1 How to identify functions and use function notation

Goal 2 How to identify real-life relations that are functions

Why you should learn it:

Functions form a foundation for the study of advanced algebra and trigonometry.

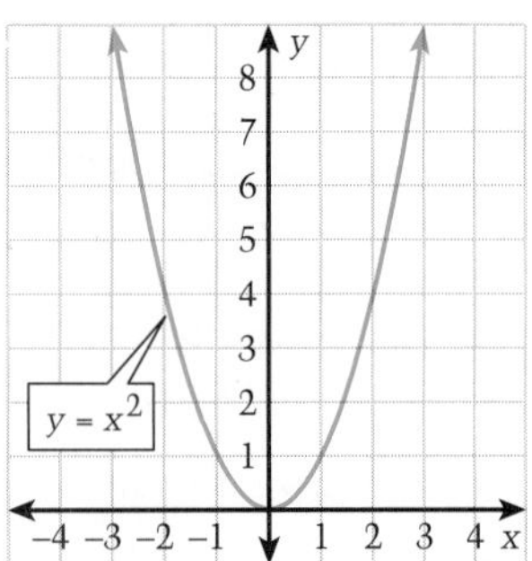

y is a function of x.

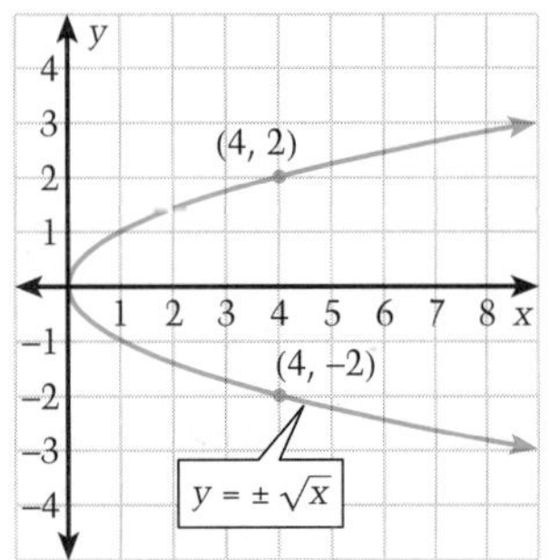

y is not a function of x.

Goal 1 Identifying Functions

A **relation** is a set of ordered pairs (x, y). A **function** of x is a relation in which no two ordered pairs have the same x-value. You can identify a function by checking its graph. The x-values make up the **domain** (or input) of the function. The y-values are the **range** (or output). There are many ways to define a function: by an equation, a table, a graph, a sentence, or a set of ordered pairs.

LESSON INVESTIGATION

Investigating Functions

Partner Activity Graph each relation and decide whether it is a function of x. Explain your reasoning. If the relation is a function, state its domain and range.

a. $\{(0, 1), (1, 2), (1, 3), (2, 4), (3, 5), (4, 5)\}$
b. $\{(0, 3), (1, 3), (2, 3), (3, 3), (4, 3), (5, 3)\}$
c. $\{(0, 3), (1, 2), (2, 1), (3, 0), (4, 1), (5, 2), (6, 3)\}$

Example 1 *Identifying Relations That Are Functions*

Let $A = \{a, b, c\}$ and let $B = \{1, 2, 3, 4, 5\}$. Which of the following are functions relating the set A to the set B?

a. $\{(a, 1), (b, 4), (c, 4)\}$
b. $\{(a, 1), (b, 2), (c, 3), (b, 4), (a, 5)\}$

c.

Input value	a	b	c	b	a
Output value	1	2	3	4	5

Solution

a. This set of ordered pairs *does* define a function relating A to B. No two ordered pairs have the same first coordinate.
b. This set of ordered pairs *does not* define a function relating A to B. Two of the pairs have the same first coordinate.
c. This is another way of representing the pairs in Part b. ■

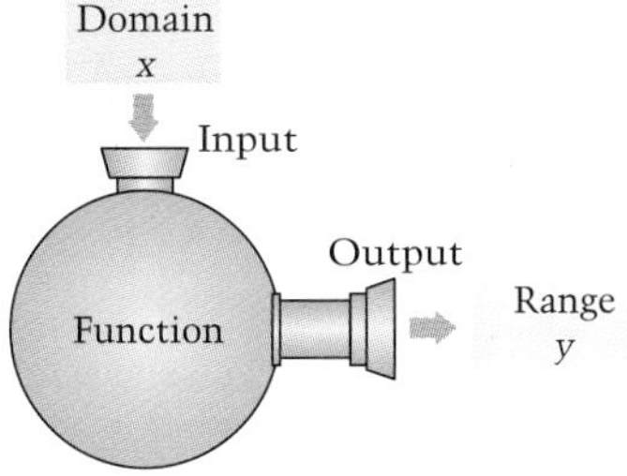

The terms *input* and *output* are used because a function can be thought of as a "machine" that changes numbers or other quantities. The input is put into the "function machine" and the function produces an output.

When a function is defined by an equation, it is convenient to name the function. For example, the set of ordered pairs (x, y) that satisfy the equation $y = 3x + 2$ form a function. By naming the function "f," you can use **function notation.**

***x-y* Notation**	**Function Notation**
(x, y) is a solution of $y = 3x + 2$.	$(x, f(x))$ is a solution of $f(x) = 3x + 2$.

The symbol $f(x)$ replaces y and is read as "the value of f at x" or simply as "f of x." When we evaluate $3x + 2$ for a particular value of x, we say we are evaluating f at x. Watch how this notation "works."

***x*-value**	**Value of *y*, or *f*(*x*)**
$x = 1$	$f(1) = 3(1) + 2 = 3 + 2 = 5$
$x = 2$	$f(2) = 3(2) + 2 = 6 + 2 = 8$

Up to this point in the text, we have been using parentheses to represent multiplication. The function notation $f(x)$ is a different use of parentheses. Remember that $f(x)$ means the value of f at x. It *does not* mean f times x.

Study Tip

In a function, x is the ***independent*** *variable and y is the* ***dependent*** *variable. In other words, if someone tells you the x-value, then you can find the y-value, because y depends on x. The reverse is not true. For instance, in Example 2 if someone tells you that the y-value is −3, then you cannot determine what the x-value is. Can you see why?*

Example 2 *Evaluating a Function*

Evaluate the function $f(x) = x^2 + 2x - 3$ at the given x-values.

a. $x = -2$ **b.** $x = 0$

Solution

a. To find the value of $f(x)$ when $x = -2$, substitute -2 for x in the equation that defines the function.

$f(x) = x^2 + 2x - 3$	***Equation for f***
$f(-2) = (-2)^2 + 2(-2) - 3$	***Substitute −2 for x.***
$= 4 - 4 - 3$	***Simplify.***
$= -3$	***Simplify.***

b. To find the value of $f(x)$ when $x = 0$, substitute 0 for x in the equation that defines the function.

$f(x) = x^2 + 2x - 3$	***Equation for f***
$f(0) = (0)^2 + 2(0) - 3$	***Substitute 0 for x.***
$= 0 + 0 - 3$	***Simplify.***

LESSON Notes

Examples 1–2

Another common definition of a function is "a rule of association for which each object in the domain corresponds to exactly one object in the range."

The domain of a function may be viewed as the set of objects for which the rule of association is defined, and the range of a function may be viewed as the set of objects generated by the application of the rule on the domain.

Example 3

In Part a, note that the coordinates of the ordered pairs do not have to be the same types of objects, nor do they have to be numbers. In Part b, the order of the coordinates in the ordered pairs makes a real difference! That is, the given relationship is not a function, but the ordered pairs consisting of (price, sales tax) would be. Ask students to explain. In Part c, ask students if it matters that the ordered pairs are in the given order in determining whether you have a function or not.

Extra Examples

Here are additional examples similar to **Examples 1 – 3.**

1. Let $A = \{t, u, v, w\}$ and $B = \{-2, -1, 0, 1, 2\}$. Which of the following are functions relating set A to set B?
   a. $\{(t, -2), (u, -2), (v, -2), (w, -2)\}$
   b. $\{(w, 0), (v, -1), (u, -2), (t, 1)\}$
   c. $\{(t, -1), (w, 2), (t, 2), (u, -1), (v, 0)\}$
   d. $\{(v, -2), (u, 2), (w, -1), (t, 0), (u, 1)\}$
   a., b. a function
   c., d. not a function

2. Evaluate each function at the given x-value.
   a. $f(x) = 3x - 5$ at $x = -3$
   b. $f(x) = \frac{4}{5}x + 12$ at $x = -20$
   c. $f(x) = 4x^2 - 9x - 3$ at $x = 2$
   d. $f(x) = 64 - \frac{1}{2}x^3$ at $x = 4$

   a. -14, b. -4, c. -5, d. 32
3. Which of the relations described below are functions? Explain.
   a. The area of a rectangular region is or is not a function of its length.
   b. The area of a triangular region with a base that is always 8 cm long is or is not a function of its height.
   c. The cost of a bag of apples, priced at $0.34 per pound, is or is not a function of its weight.
   d. The weight of an individual is or is not a function of his or her height.

   a. not a function because the width of the rectangle can vary
   b. a function
   c. a function
   d. not a function because people of the same height frequently have very different weights

Check Understanding

MATH JOURNAL Students might be encouraged to start a section on functions in their math journals, in which to enter their responses.

1. How are a relation and a function the same? How are they different? See the top of page 620.
2. List the ways in which a function can be defined or described. Give examples of each for the same function. As sets of ordered pairs, in words, using tables, using graphs, using function notation

Goal 2 Identifying Functions in Real Life

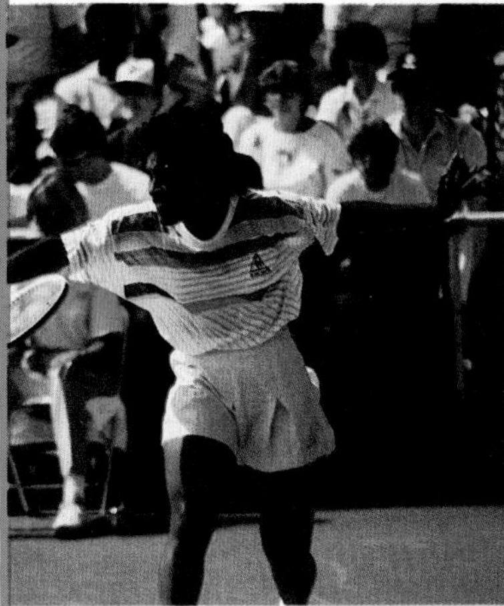

When (x, y) represents the ordered pairs of a function, we say that y (the second coordinate) *is a function* of x (the first coordinate).

Example 3 ***Identifying Relations That Are Functions***

a. The winner of the women's singles U.S. Open Tennis competition from 1987 to 1991 is a function of the year she won. (In no year were there cochampions.)

Year	1987	1988	1989	1990	1991
Winner	Martina Navratilova	Stefi Graf	Stefi Graf	Gabriela Sabatini	Monica Seles

b. In a state that has a sales tax of 8%, the price of a pair of sunglasses is *not* a function of the amount of sales tax. (A given amount of sales tax corresponds to several different prices.)

Sales Tax	$0.08	$0.08	$0.08	$0.08	$0.08	$0.08
Price	$0.94	$0.95	$0.96	$0.97	$0.98	$0.99

Sales Tax	$0.08	$0.08	$0.08	$0.08	$0.08	$0.08
Price	$1.00	$1.01	$1.02	$1.03	$1.04	$1.05

Connections **Geometry**

c. The volume of a sphere is a function of the radius of the sphere.

Radius, r	1	2	3	4
Volume, $\frac{4}{3}\pi r^3$	4.2	33.5	113.1	268.1

■

Communicating about ALGEBRA

SHARING IDEAS about the Lesson

A. Let $f(x) = 5x - 7$. Evaluate $f(3)$ and $f\left(\frac{2}{3}\right)$. 8, $-3\frac{2}{3}$

B. Let $f(x) = -x^2 + 2x + 4$. Evaluate $f(1)$ and $f\left(-\frac{1}{2}\right)$. 5, $2\frac{3}{4}$

C. Let $f(x) = \frac{1}{x}$. Evaluate $f(2)$ and $f\left(-\frac{3}{4}\right)$. $\frac{1}{2}$, $-\frac{4}{3}$

EXERCISES

Guided Practice

CRITICAL THINKING about the Lesson

In Exercises 1–3, state the domain of the function and find the indicated value of the function.

1. Find $f(4)$. 1, {1, 2, 3, 4}
$\{(1, 2), (2, 0), (3, 3), (4, 1)\}$

2. Find $f(2)$. 4, {1, 2, 3, 4}

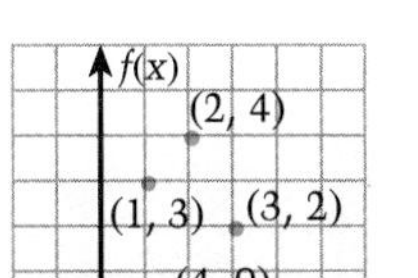

3. Find $f(3)$. 2, {1, 2, 3, 4}
$\{(1, 0), (2, 1), (3, 2), (4, 3)\}$

4. Evaluate $f(x) = 3x^2 + 2x - 1$ at $x = 1$ and at $x = 2$. Explain your steps.
4, 15; substitute 1 and 2 in equation for f and simplify.

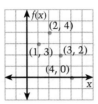

Independent Practice

In Exercises 5–8, decide whether the graph represents y as a function of x. Explain your reasoning. See below.

5. Yes

6.

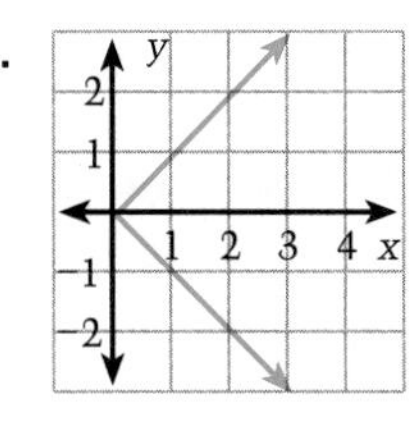

No

7.

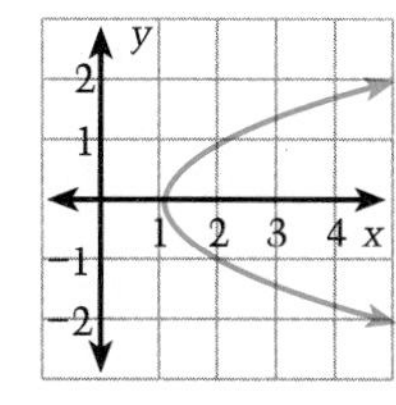

No

8.

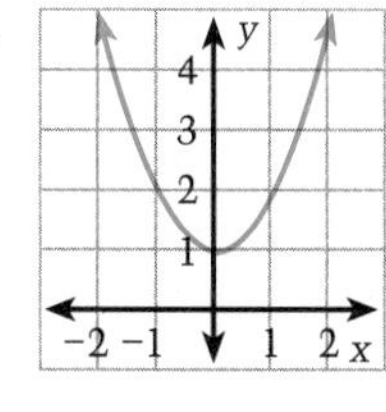

Yes

In Exercises 9–16, decide whether the information defines a function. If it does, state the domain of the function.

9. $\{(1, 5), (2, 4), (3, 3), (4, 2)\}$ Yes, {1, 2, 3, 4}

10. $\{(1, 5), (1, 4), (2, 3), (2, 2)\}$ No

11. $\{(1, 0), (2, 0), (3, 0), (4, 0)\}$ Yes, {1, 2, 3, 4}

12. $\{(0, 0), (1, 1), (2, 2), (3, 3)\}$ Yes, {0, 1, 2, 3}

13.

Input value	0	1	2	3	4
Output value	2	4	6	8	10

13. Yes, {0, 1, 2, 3, 4}

14. No

Input value	0	1	2	1	0
Output value	2	4	6	8	10

15.

Input value	1	3	5	3	1
Output value	1	2	3	4	5

No

16.

Input value	2	4	6	8	10
Output value	1	1	1	1	1

Yes, {2, 4, 6, 8, 10}

5., 8. No two ordered pairs have the same first coordinate.
6., 7. Two ordered pairs have the same first coordinate.

3. What is the connection between the solutions (x, y) of $y = 6x - 3$ and the function defined by $f(x) = 6x - 3$? The set of ordered pairs (x, y) is the same as the set of ordered pairs $(x, f(x))$.

4. Describe a real-life relationship that is a function.

Communicating about ALGEBRA

TECHNOLOGY Ask students to discuss how the functions can be evaluated using calculators. Have them describe or record the steps needed to evaluate the function $f(x) = 2x^2 - 4x + 3$. If students have graphics calculators, encourage them to explore how "storing" values in variables can be used to evaluate functions. On the TI-81 the key that stores values is STO> ; on the Casio fx-7700G the key is ⟶ .

EXERCISE Notes

ASSIGNMENT GUIDE

Basic/Average: Ex. 1–4, 5–8, 9–27 odd, 31–34, 43–46

Above Average: Ex. 1–4, 5–8, 9–33 odd, 34–48

Advanced: Ex. 1–4, 5–8, 9–31 odd, 32–48

Selected Answers
Exercises 1–4, 5–45 odd

✪ More Difficult Exercises
Exercises 29, 30, 37, 38, 47, 48

Guided Practice

Ex. 1–4 These exercises provide a final check for understanding and should be reviewed in class before assigning the Independent Practice. The notation is new to most students.

Independent Practice

▶ **Ex. 5–16** These exercises explore the three basic formats for presenting a function or relation, namely, as graphs, as sets, and as tables.

▶ **Ex. 17–26** Refer to Example 2 on page 621 for a solution model.

▶ **Ex. 31–34** Refer to Example 3 for a solution model.

EXTEND Ex. 31–34 Provide students with examples of relations such as the following and ask students to determine if they are functions.

a. Do the ordered pairs $(x, \pm\sqrt{x})$ define a function?

b. Do the ordered pairs $(x, |x|)$ define a function?

c. Do the ordered pairs (city, zip code) define a function?

d. Do the ordered pairs (telephone area code, city) define a function?

a. no b. yes c. no d. no

Answers

31., 33., 35., 36. No two ordered pairs have the same first coordinate.

32., 34. Two ordered pairs have the same first coordinate.

39. $y = \frac{3}{2}x - \frac{1}{2}, -3\frac{1}{2}$

40. $y = \frac{1}{3}x + \frac{5}{3}, 1$

41. $y = 2x^2 + 5, 13$

42. $y = 3x^2 - 2x + 10, 26$

In Exercises 17–26, evaluate the function at the given *x*-values.

	Function	x-values		Function	x-values
17.	$f(x) = 10x + 1$	$x = 0, x = 2$ 1, 21	18.	$f(x) = 8x - 2$	$x = -1, x = 1$
19.	$f(x) = x^2 + x - 1$	$x = 2, x = 3$ 5, 11	20.	$f(x) = 2x^2 - 2x + 1$	$x = 0, x = 2$ 1, 5
21.	$f(x) = 2x^2 - 3$	$x = -1, x = 4$	22.	$f(x) = 3x^2 + 5$	$x = 2, x = -4$
23.	$f(x) = 9x^2 - 10x - 1$	$x = -1, x = 1$	24.	$f(x) = 6x^2 + 2x - 4$	$x = -2, x = 3$
25.	$f(x) = x^2 - 10x + 10$	$x = -5, x = 5$	26.	$f(x) = 2x^2 + 5x - 20$	$x = 4, x = 5$

21. −1, 29 23. 18, −2 25. 85, −15 18. −10, 6 22. 17, 53 24. 16, 56 26. 32, 55

In Exercises 27 and 28, state the domain of the function. Find *f*(0) and *f*(2).

27.

Input values, x	0	2	4	6	8	10
Output values, $f(x)$	0	1	2	3	4	5

{0, 2, 4, 6, 8, 10}; 0, 1

28.

Input values, x	−3	−2	−1	0	1	2	3
Output values, $f(x)$	6	4	2	0	−2	−4	−6

{−3, −2, −1, 0, 1, 2, 3}; 0, −4

29. Find an equation that defines the function given in Exercise 27. $f(x) = \frac{1}{2}x$

30. Find an equation that defines the function given in Exercise 28. $f(x) = -2x$

31. *Pulse Rates of Mammals* The table gives typical pulse rates for six different mammals. Are the given mammals a function of the pulse rate? Why or why not?

Pulse rate, x	667	315	260	176	67	27
Type of mammal, y	Mouse	Rat	Rabbit	Cat	Human	Elephant

Yes. See margin.

32. *Symphony Orchestras* The number of college and community orchestras in the United States from 1980 to 1988 is given in the table. Is the year a function of the number of college and community orchestras? Why or why not?

Orchestras, x	1311	1311	1304	1306	1317	1317	1298	1301	1253
Year, y	1980	1981	1982	1983	1984	1985	1986	1987	1988

No. See margin.

Spring Semester Tests **In Exercises 33 and 34, the graph shows the average score on each test in an algebra class from January through June. Is the average score, *y*, a function of the month, *x*? Explain.** See margin.

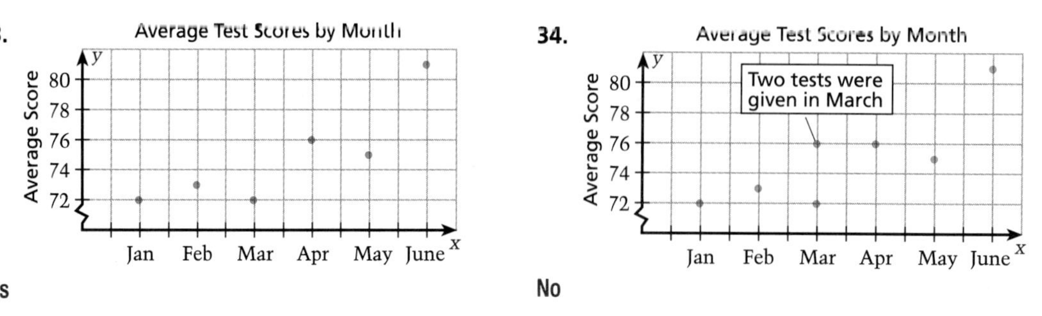

Yes No

High School and College Students **In Exercises 35–38, use the following information.**

The graph shows the projected number of students in the United States. **See margin.**
(Source: National Center for Education Statistics)

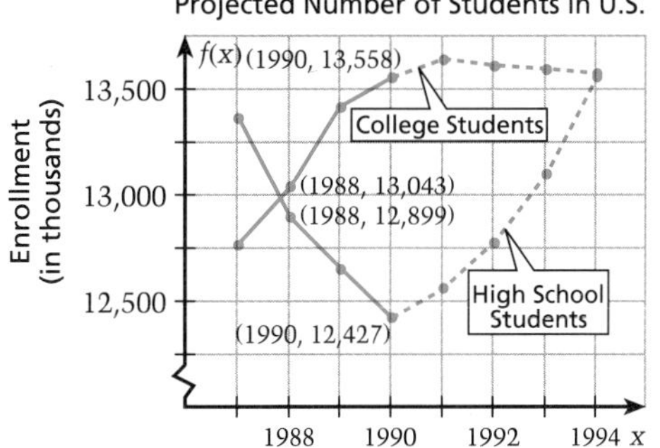

35. Is the projected high school enrollment a function of the year? Explain. **Yes**

36. Is the projected college enrollment a function of the year? **Yes**

37. Let $f(x)$ represent the projected number of high school students in year x. Find $f(1993)$. **≈13,100**

38. Let $f(x)$ represent the projected number of college students in year x. Find $f(1993)$. **≈13,600**

Integrated Review

In Exercises 39–42, solve the equation for *y*. Then evaluate *y* when $x = -2$. **See margin.**

39. $2y = 3x - 1$ **40.** $3y - 5 = x$ **41.** $y - 5 = 2x^2$ **42.** $y - 10 = 3x^2 - 2x$

In Exercises 43–46, match the equation with its graph.

43. $y = 2^x$ **c** **44.** $y = -x^2$ **d** **45.** $y = 2x + 1$ **b** **46.** $y = 2 - 2x$ **a**

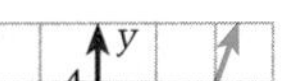

a.

b.

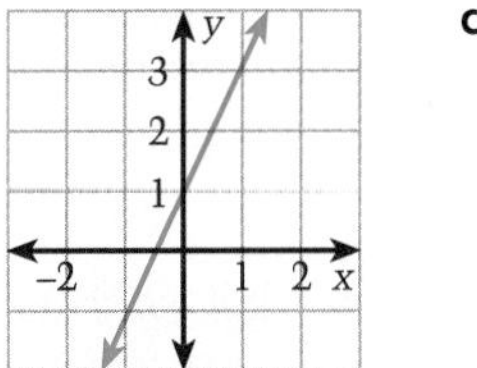

c.

d.

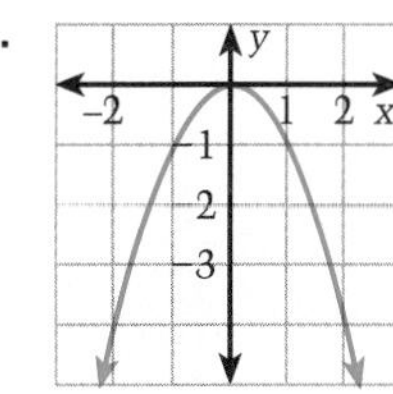

Exploration and Extension

Dog Years vs. Human Years **In Exercises 47 and 48, use the following information.**

The following *two* equations can be used to find the approximate human age, $f(x)$, that corresponds to a dog's age, x.

For $0.5 < x < 5.5$: $f(x) = 10\sqrt{3x - 1}$

For $5.5 \le x$: $f(x) = 25\sqrt{x - 3}$

(Source: Simon & Schuster's* Guide to Dogs*)

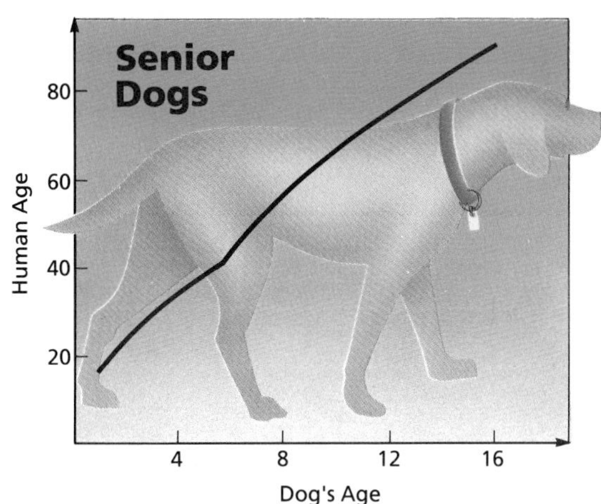

47. What is the human age of a one-year-old dog? **≈14.1**

48. What is the human age of a twelve-year-old dog? **75**

▶ Ex. 35–38 and 43–46 These exercises should be assigned as a group.

▶ Ex. 43–46 These exercises review the graphs of the basic equations or "functions" that were studied in previous chapters.

▶ Ex. 47–48 These exercises should be assigned as a pair.

Enrichment Activity

ALTERNATIVE ASSESSMENT This project promotes student creativity and understanding. You may wish to include the best examples as quiz items.

RESEARCH Find five examples of data similar to those given in Example 3 or in Exercises 31 and 32. Decide whether each relation defines a function or not. Share your results with your classmates.

Problem of the Day

The Lion lies on Mondays, Tuesdays, and Wednesdays, and tells the truth on other days. The Unicorn lies on Thursdays, Fridays, and Saturdays, and tells the truth on other days. In her adventures in Wonderland, Alice came across the two. The Lion said "Yesterday was one of my lying days." The Unicorn said "Yesterday was one of my lying days." What day of the week was it? Thursday

ORGANIZER

Warm-Up Exercises

1. Write the linear equation of the line that has slope m and y-intercept b.
   - **a.** $m = -3, b = 4$
   - **b.** $m = \frac{1}{2}, b = -7$
   - **c.** $m = 5, b = \frac{2}{3}$

   a. $y = -3x + 4$
   b. $y = \frac{1}{2}x - 7$
   c. $y = 5x + \frac{2}{3}$

2. Write the linear equation of the line that contains the given points.
   - **a.** (2, 0) and (3, −6)
   - **b.** (5, 1) and (−2, −8)
   - **c.** (−4, −4) and (6, 10)

   a. $y = -6x + 12$
   b. $9x - 7y = 38$ or $y = \frac{9}{7}x - \frac{38}{7}$
   c. $7x - 5y = -8$ or $y = \frac{7}{5}x + \frac{8}{5}$

Lesson Resources

Teaching Tools
- Transparency: 4
- Problem of the Day: 12.2
- Warm-up Exercises: 12.2
- Answer Masters: 12.2

Extra Practice: 12.2
Color Transparency: 70
Applications Handbook: p. 54

12.2 Linear Functions

What you should learn:

Goal 1 How to use linear functions to solve problems

Goal 2 How to use linear functions to answer questions about real-life situations

Why you should learn it:

You can use linear functions to model many real-life situations, such as those involving distance, rate, and time.

Goal 1 Using Linear Functions

In this and the next three lessons, you will review four of the basic types of models as *functions*.

- Linear function — Lesson 12.2
- Exponential function — Lesson 12.3
- Quadratic function — Lesson 12.4
- Rational function — Lesson 12.5

A **linear function** is a function defined by the equation

$f(x) = mx + b.$

The graph of a linear function is a *nonvertical* line whose slope is m and whose y-intercept is b.

Example 1 *An Equation for a Linear Function*

Find an equation of the linear function for which $f(2) = 5$ and $f(6) = 3$.

Solution Translate the information about the function f.

Function Notation	x-value and y-value	Point on graph
$f(2) = 5$	When $x = 2, y = 5$.	(2, 5)
$f(6) = 3$	When $x = 6, y = 3$.	(6, 3)

You need to find an equation for the line that passes through the points (2, 5) and (6, 3). The slope of the line is

$$m = \frac{y_2 - y_1}{x_2 - x_1} = \frac{3 - 5}{6 - 2} = -\frac{1}{2}.$$

Using the procedure you studied in Lesson 5.3, let $m = -\frac{1}{2}$, $x = 2$, and $y = 5$ and solve for b.

$y = mx + b$	*Slope-intercept form*
$5 = \left(-\frac{1}{2}\right)(2) + b$	$m = -\frac{1}{2}, x = 2, y = 5$
$5 = -1 + b$	*Simplify.*
$6 = b$	*Solve for b.*

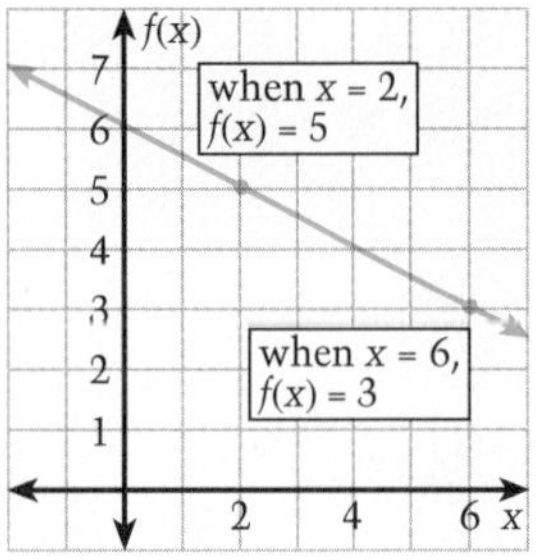

when $x = 2$, $f(x) = 5$

when $x = 6$, $f(x) = 3$

$f(x) = -\frac{1}{2}x + 6$

The y-intercept is $b = 6$, and the slope-intercept form of the line is $y = -\frac{1}{2}x + 6$. Using function notation, you can write

$f(x) = -\frac{1}{2}x + 6.$ ■

Goal 2 Using Linear Functions as Real-Life Models

Monarch butterflies migrate from the northern United States to Mexico. The trip of about 2400 miles takes about 30 days.

Monarch butterflies winter over in a mountainous region west of Mexico City. During their $4\frac{1}{2}$ month stay, they show little activity.

Real Life
Entomology

Example 2 A Linear Function to Model Distance

Find a linear function that models the distance traveled by a migrating monarch butterfly. Use the model to find the distance traveled in 20 days of migration.

Solution Because the butterflies travel about 2400 miles in 30 days, their average speed is 80 miles per day.

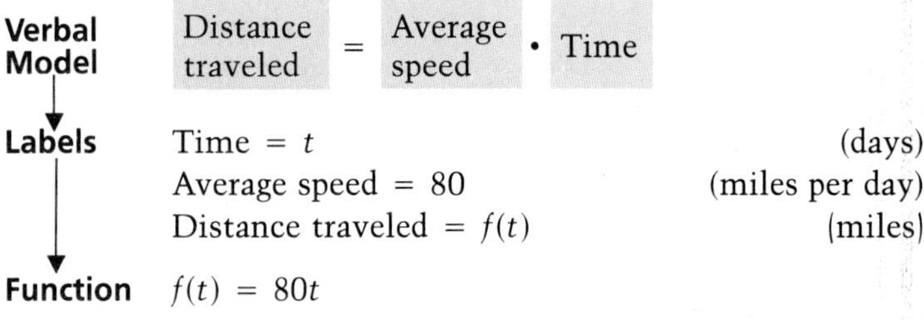

Verbal Model Distance traveled = Average speed · Time

Labels Time = t (days)
Average speed = 80 (miles per day)
Distance traveled = $f(t)$ (miles)

Function $f(t) = 80t$

After 20 days of migration, the distance traveled is

$f(t) = 80t$ *Linear function*
$f(20) = 80(20)$ *Substitute 20 for t.*
$= 1600$ miles *Simplify.*

Notice that it is not necessary to use x as the variable when writing function notation. Any letter can be used. ■

LESSON Notes

Example 1

The key to finding the equation of the linear function is the translation from functional notation to points on a graph. Stress the importance of the translation.

Example 2

The variable t is used to represent time. Similarly, the function notation d may be used instead of the f-notation to suggest distance. Consequently, the relationship would be expressed as $d(t) = 80t$.

Example 3

This example shows how algebra can be used to interpret real-life results. Point out to students that claims are frequently made in the press and in advertising that do not withstand algebraic scrutiny.

Extra Examples

Here are additional examples similar to **Examples 1–2.**

1. Find an equation of the linear function for which
   a. $f(3) = 7$ and $f(-2) = -3$
   b. $f(0) = -1$ and $f(5) = 8$
   a. $f(x) = 2x + 1$
   b. $f(x) = \frac{9}{5}x - 1$
2. Bill and Ted took 7 days to travel from Delaware to Oregon, a journey of about 2,730 miles. On average, they drove about 7.8 hours each day. Find a linear function that models the distance traveled by Bill and Ted. Use the model to find the distance traveled in 10 hours. Two models are possible: $f(t) = 390t$ where t is time in days, or $f(t) = 50t$ where t is time in hours. The second model lets us get the answer to the second question, 500 miles.

Check Understanding

MATH JOURNAL Students might be encouraged to enter their responses in the function section of their math journals.

1. What are different representations in algebra of a linear relationship? The expression $f(x) = mx + b$ or a graph of ordered pairs $(x, f(x))$
2. Describe the steps necessary to find the slope of a linear function. Find two ordered pairs of the function, $(x_1, f(x_1))$ and $(x_2, f(x_2))$. Find the ratio of the difference $f(x_2) - f(x_1)$ to the difference $x_2 - x_1$.
3. Write the formula for slope using function notation. $m = \dfrac{f(x_2) - f(x_1)}{x_2 - x_1}$

COOPERATIVE LEARNING

Encourage the class to collect the following data using students in the class in order to decide which may be linear relationships.

a. the circumferences of wrist and neck

b. the lengths of right foot and right forearm

c. the ages in months of individuals and their heights in inches

Answer

Answers vary. For example, height above sea level of a city and the average daily temperature of that city's outdoor air; someone's present salary for work and the number of past years that someone has worked at the same company.

Team	Salary	Percent	Team	Salary	Percent
Atlanta Braves	\$14.2	40.1%	Minnesota Twins	\$14.2	45.7%
Baltimore Orioles	\$ 8.1	47.2%	Montreal Expos	\$16.5	52.5%
Boston Red Sox	\$22.7	54.3%	New York Mets	\$22.2	56.2%
California Angels	\$21.9	49.4%	New York Yankees	\$20.6	41.4%
Chicago Cubs	\$13.8	47.5%	Oakland Athletics	\$22.3	63.6%
Chicago White Sox	\$11.1	58.0%	Philadelphia Phillies	\$14.2	47.5%
Cincinnati Reds	\$15.6	56.2%	Pittsburgh Pirates	\$15.5	58.6%
Cleveland Indians	\$15.4	47.5%	San Diego Padres	\$16.7	46.3%
Detroit Tigers	\$17.8	48.8%	San Francisco Giants	\$22.5	52.5%
Houston Astros	\$18.2	46.3%	Seattle Mariners	\$12.6	47.5%
Kansas City Royals	\$23.6	46.6%	St. Louis Cardinals	\$19.6	43.2%
Los Angeles Dodgers	\$20.4	53.1%	Texas Rangers	\$12.6	51.2%
Milwaukee Brewers	\$18.5	45.7%	Toronto Blue Jays	\$17.8	53.1%

The table shows the 1990 total team salaries (in millions) of the 26 baseball teams in the American and National Leagues and each team's percent of games won. (Source:* Inside Sports*)

Real Life
Baseball

Example 3 *Can Money Buy a Win?*

Construct a scatter plot for the data shown in the table.

a. Is the percent of games won by a team a *linear function* of the total salary paid to the team?

b. Can you find a *linear function that approximates* the relationship between the percent of games won by a team and the total salary paid to the team?

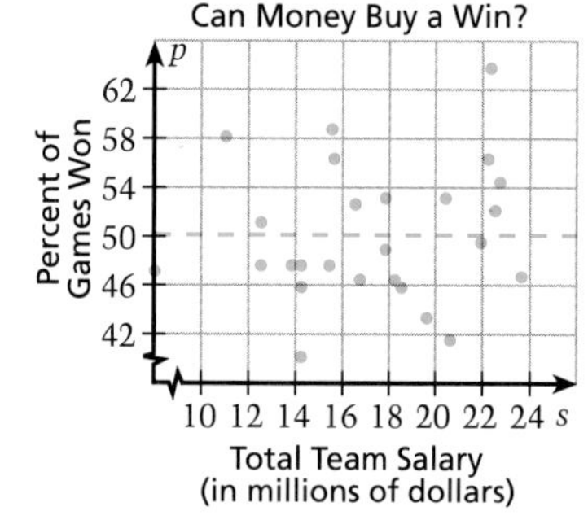

Solution Be sure you see that these questions are different.

a. From the scatter plot, you can see that the points do not all lie on a line. Therefore, the percent of games won by a team is *not* a linear function of the team's total salary.

b. The April 1991 issue of *Inside Sports* claimed that "a correlation exists between payroll and on-field performance," but we can't see it. (See page 258.) It is true that of the eight highest-paid teams, five won more than 50% of their games. Of the eight lowest-paid teams, only two won more than 50% of their games. But it would be hard to argue that a team can win a greater percentage of games by hiring more expensive players. What do you think? ■

Communicating about ALGEBRA

SHARING IDEAS about the Lesson

Use Resources Find two quantities whose relationship *can be* approximated by a linear function. Use a scatter plot and the approach discussed in Lesson 5.4. See margin.

EXERCISES

Guided Practice

CRITICAL THINKING about the Lesson 4. Use the points (1, 2), (3, −1), and the slope formula.

1. Which type of line *cannot* be the graph of a linear function? Explain. Vertical line
2. If $f(-3) = 5$, what point must be on the graph of f? Explain.
3. If the point (1, 5) is on the graph of f, what is the value of $f(1)$? Explain.
   5; when $x = 1$, $y = 5$
4. Given that $f(1) = 2$ and $f(3) = -1$, explain how to use this information to find the slope of the graph of f.

2. $(-3, 5)$; when $x = -3$, $y = 5$

Independent Practice

In Exercises 5–8, match the function with its graph.

5. $f(x) = 3x - 5$ c
6. $f(x) = 2x + 2$ a
7. $f(x) = -\frac{1}{2}x + 4$ d
8. $f(x) = -3x - 4$ b

a. 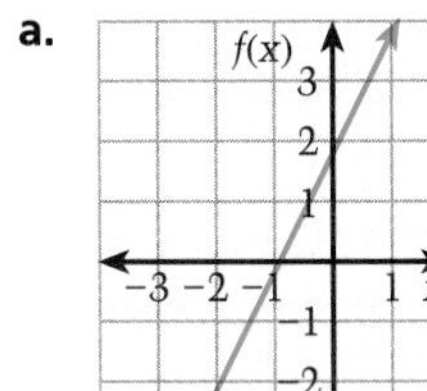

b. 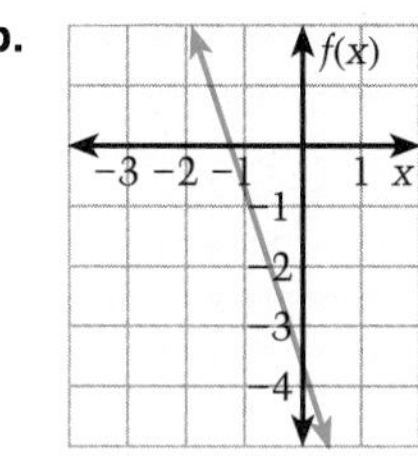

c. 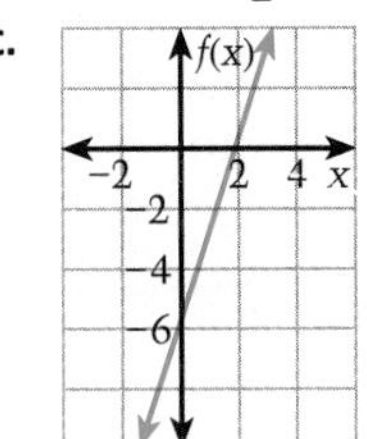

d. 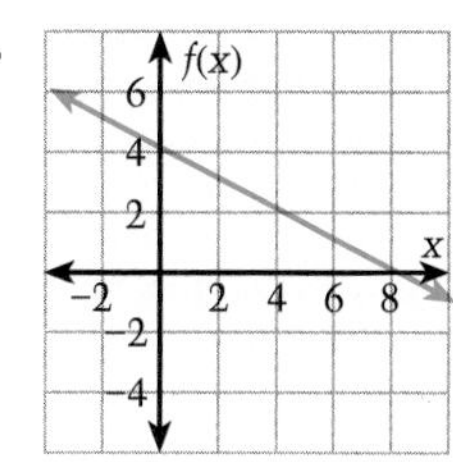

In Exercises 9–14, find the slope of the graph of the linear function *f*.

9. $f(2) = -3,\ f(-2) = 5$ -2
10. $f(0) = 1,\ f(1) = 0$ -1
11. $f(-3) = -9,\ f(3) = 4$ $\frac{13}{6}$
12. $f(6) = -1,\ f(3) = 8$ -3
13. $f(9) = -1,\ f(-1) = 2$ $-\frac{3}{10}$
14. $f(-1) = 1,\ f(1) = 6$ $\frac{5}{2}$

In Exercises 15–26, find an equation for the linear function *f*.

15. $f(0) = 1,\ f(2) = 1$
16. $f(2) = -2,\ f(-4) = 1$
17. $f(3) = -4,\ f(5) = 2$
18. $f(0) = 5,\ f(-6) = 3$
19. $f(-1) = 1,\ f(-2) = -4$
20. $f(4) = 6,\ f(9) = 1$
21. $f(0) = 4,\ f(-3) = 6$
22. $f(3) = -2,\ f(5) = -2$
23. $f(6) = 2,\ f(9) = -1$
24. $f(2) = 10,\ f(-4) = 16$
25. $f(-5) = 5,\ f(5) = -3$
26. $f(7) = 2,\ f(3) = -8$

In Exercises 27–32, sketch the graph of the linear function. See Additional Answers.

27. $f(x) = -2x + 4$
28. $f(x) = -x + 3$
29. $f(x) = \frac{1}{2}x + 1$
30. $f(x) = 2x - 3$
31. $f(x) = -\frac{1}{3}x + 2$
32. $f(x) = 3x - 8$

✪ 33. *The Goodyear Blimp* The Goodyear blimp traveled about 980 miles from Dover, Delaware, to Kansas City, Kansas, for a football game. It took the blimp 28 days to make the trip. Find a linear function that models the time taken by the blimp. Then use the model to estimate the time it would take the blimp to go from Kansas City to Indianapolis, Indiana (430 miles away).
$f(m) = \frac{1}{35}m$, $12\frac{2}{7}$ days, where m = distance

1. A vertical line, because all ordered pairs have the same first coordinate.

EXERCISE Notes

ASSIGNMENT GUIDE

Basic/Average: Ex. 1–4, 5–8, 9–21 odd, 27–33 odd, 36–42, 51, 53

Above Average: Ex. 1–4, 5–8, 11–23 odd, 27–35 odd, 36–42, 48–55 odd

Advanced: Ex. 1–4, 5–8, 13–33 odd, 34–42, 55

Selected Answers
Exercises 1–4, 5–53 odd

Use **Mixed Review** as needed.

✪ More Difficult Exercises
Exercises 33–38, 55

Guided Practice

▶ **Ex. 1–4** Some students may fail to translate the notation $f(-3) = 5$ to the notation $(-3, 5)$.

WRITING To observe students' writing abilities, have them provide written answers to the "Explain" part of the exercises.

Independent Practice

▶ **Ex. 5–8** These exercises should be assigned as a group.

▶ **Ex. 9–32** It might be appropriate to review linear models from Chapter 5.

Answers

15. $f(x) = 1$
16. $f(x) = -\frac{1}{2}x - 1$
17. $f(x) = 3x - 13$
18. $f(x) = \frac{1}{3}x + 5$
19. $f(x) = 5x + 6$
20. $f(x) = -x + 10$
21. $f(x) = -\frac{2}{3}x + 4$
22. $f(x) = -2$
23. $f(x) = -x + 8$
24. $f(x) = -x + 12$
25. $f(x) = -\frac{4}{5}x + 1$
26. $f(x) = \frac{5}{2}x - \frac{31}{2}$

▶ **Ex. 36–38 and 39–42** These exercises should be assigned as a group.

EXTEND Ex. 36–38 **TECHNOLOGY** Use the STAT feature of a graphics calculator to find the lines of best fit for the data.

Answers will vary. Sample models are given.

36. $f(t) = 500t + 4300$

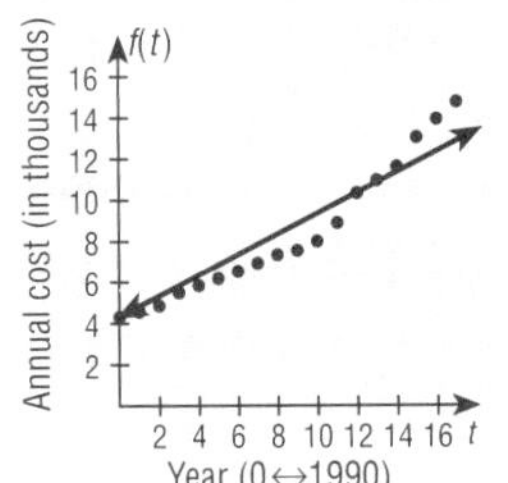

37.

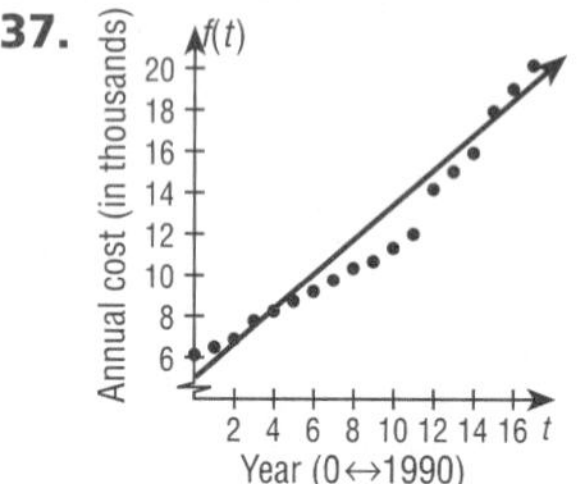

38.

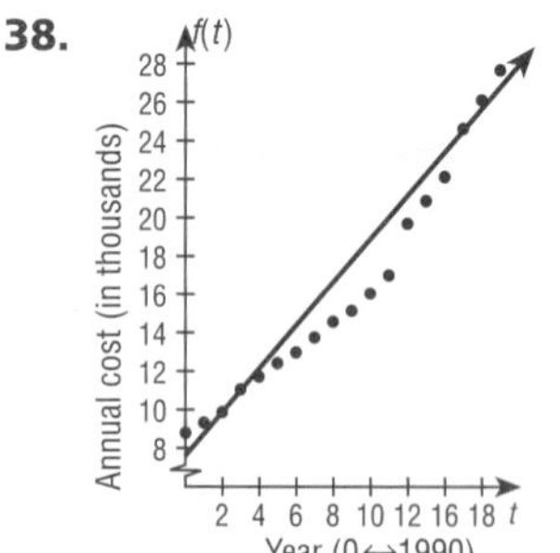

The calculator would probably give these models.

36. $f(t) = 3300 + 600t$

37. $f(t) = 4900 + 800t$

38. $f(t) = 7100 + 1100t$

34. *Go, Verne, Go!* On February 21, 1990, a garden snail named Verne set a world record at West Middle School in Plymouth, Michigan. Verne completed a 12.2-inch course in 133 seconds. Find a linear function that models the distance Verne traveled. (***Source:*** **Guinness Book of Records**) $f(t) = \frac{12.2}{133}t$

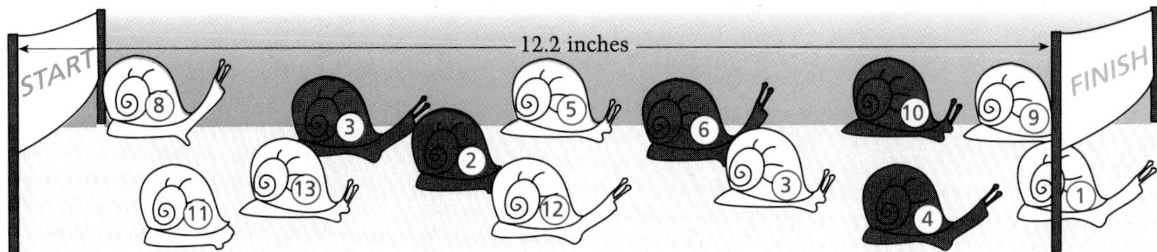

35. *Apparent Temperature* The temperature that a person feels depends on the actual temperature of the air and the humidity. The higher the humidity, the warmer the air feels. The graph at the right shows the *apparent temperature* when the actual temperature is 70°F. Find a linear function that models the apparent temperature. Use the model to find the apparent temperature when the actual temperature is 70°F and the humidity is 90%. $f(x) = \frac{1}{10}x + 63$; 72°F

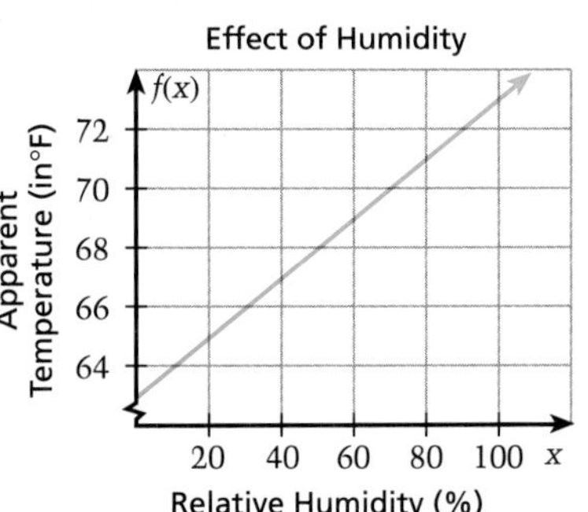

Cost of Raising Children **In Exercises 36–38, use the following information.**

The table at the right shows the projected average annual costs of raising a child that was born in 1990. In each income group, the cost increases steadily, then jumps. At what ages do the costs jump? Why do you think these jumps occur?
(***Source:*** **American Demographics Magazine**) See margin.

Year	Age	Low	Middle	High
1990	0	$4,330	$6,140	$8,770
1991	1	$4,590	$6,510	$9,300
1992	2	$4,870	$6,900	$9,850
1993	3	$5,510	$7,790	$11,030
1994	4	$5,850	$8,260	$11,690
1995	5	$6,200	$8,750	$12,390
1996	6	$6,550	$9,220	$12,950
1997	7	$6,950	$9,770	$13,730
1998	8	$7,360	$10,360	$14,550
1999	9	$7,570	$10,690	$15,120
2000	10	$8,020	$11,340	$16,030
2001	11	$8,500	$12,020	$16,990
2002	12	$10,360	$14,190	$19,680
2003	13	$10,980	$15,040	$20,860
2004	14	$11,640	$15,940	$22,110
2005	15	$13,160	$17,950	$24,610
2006	16	$13,950	$19,030	$26,090
2007	17	$14,780	$20,170	$27,650

36. Construct a scatter plot for the annual cost for a low-income family. Find a linear function that approximates the annual cost, $f(t)$. (Let $t = 0$ represent 1990.)

37. Construct a scatter plot for the annual cost for a middle-income family. Find a linear function that approximates the annual cost, $f(t)$. (Let $t = 0$ represent 1990.)

38. Construct a scatter plot for the annual cost for a high-income family. Find a linear function that approximates the annual cost, $f(t)$. (Let $t = 0$ represent 1990.)

In Exercises 39–42, construct a scatter plot. Then match the plot with the description.

39. Temperature, T (°K), and volume, V, of gas at constant pressure. b

V	T
10	200
14	280
15	300
17	340
19	380
20	400
22	440
25	500

40. Average annual salary, S, of chemical workers in year t. a

t	S
1972	8,731
1975	11,227
1977	13,839
1980	18,269
1982	20,836
1984	22,665
1985	23,345
1987	25,065

41. The density factor, d, of water at a temperature of T (°C). c

T	d
0	28.93
2	28.79
4	28.60
6	28.36
8	28.08
10	27.75
12	27.38
14	26.97

42. Atomic radius, r, of various atoms with a nuclear charge of N. d

N	r
3	1.52
9	1.36
11	1.86
13	1.43
14	1.17
18	1.54
19	2.31
37	2.44

a. Approximately linear, positive correlation.
b. Exactly linear.
c. Approximately linear, negative correlation.
d. Not linear.

Integrated Review

43. Evaluate $y = 6x + 9$ when $x = \frac{1}{2}$. 12

44. Evaluate $y = 3x - 1$ when $x = -4$. -13

45. Evaluate $y = 13 + x$ when $x = 0$. 13

46. Evaluate $y = -\frac{1}{3}x + 1$ when $x = 9$. -2

In Exercises 47–50, find the equation of the line that passes through the two points.

47. $(2, -5), (-4, 0)$ $y = -\frac{5}{6}x - \frac{10}{3}$

48. $(-1, 7), (0, 2)$ $y = -5x + 2$

49. $(0, 3), (6, 1)$ $y = -\frac{1}{3}x + 3$

50. $(5, 4), (-3, 1)$ $y = \frac{3}{8}x + \frac{17}{8}$

In Exercises 51–54, sketch the line. See margin.

51. $y = \frac{1}{2}x + 2$

52. $y = 3x$

53. $y = -\frac{3}{2}x - 3$

54. $y = -\frac{1}{4}x + 1$

Exploration and Extension

55. *NFL Players* The data in the table show the maximum number, n, of active players allowed on a NFL football team in year t, where $t = 0$ represents 1920. Do you think a linear function could be used to approximate these data? Explain your answer. No, the slope does not remain constant

t	5	6	8	11	13	15	17	18	19	23	27
n	24	25	30	33	28	33	34	35	32	33	35

t	29	30	31	33	38	44	46	50	52	54	58
n	36	38	36	37	40	47	43	45	48	49	45

▶ **Ex. 43–54** These exercises review content explored in both Chapter 4 and Chapter 5.

Answers

51.

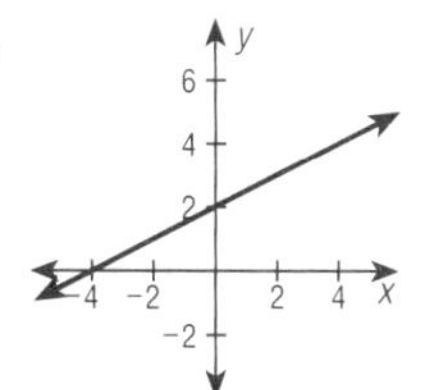

52.

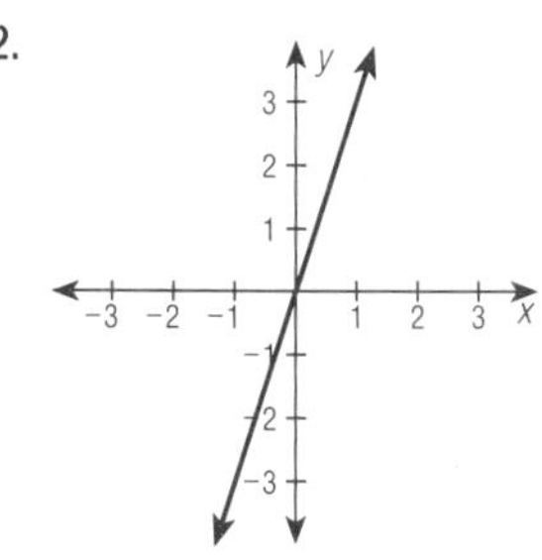

53.

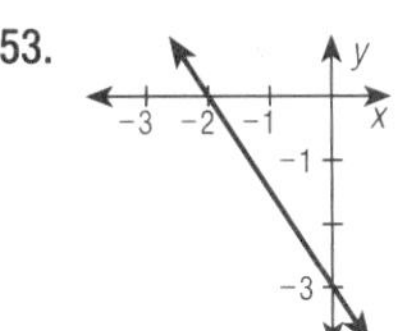

54.

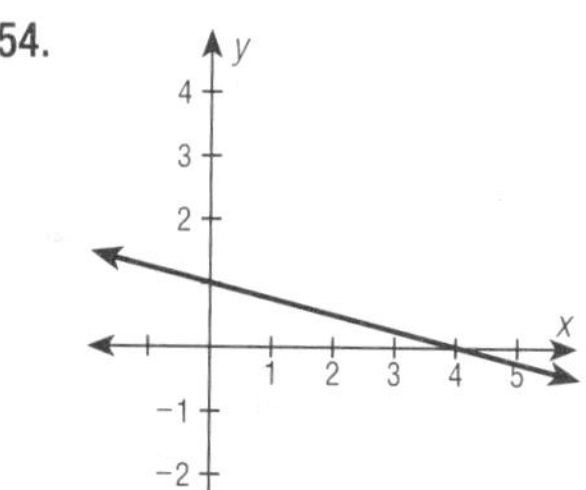

Enrichment Activity

ALTERNATIVE ASSESSMENT This project promotes student creativity and understanding.

RESEARCH Use an almanac or encyclopedia or other source of world records and demographic data. Find two examples of data sets you think might be related in a linear pattern. Decide whether the relations define a function or not. Plot the data in a scatter plot and explain why you think the data are related in a linear pattern. Share your results with your classmates.

Answers

1. 4.24×10^6
2. 6.2×10^{-7}
3. quadratic, trinomial
4. -4
5. 27.192
6. 0.078
7. $3x + 8 + \frac{10}{x - 2}$
8. $16x^4 - 4$
9. (3, 2)
10. (0, 3)
11. $\frac{3 + \sqrt{37}}{2}, \frac{3 - \sqrt{37}}{2}$
12. $-4 + \sqrt{11}, -4 - \sqrt{11}$
13. Yes
14. Yes
15. $-4, 1$
16. $3, -3$
17. $(-\frac{2}{3}, 0), (-4, 0)$
18. $(0, -3)$
19. 137.2
20. 22%

Milestones

Language Arts: *Critical Reading*

1. Find an even number greater than 4 that can be written as the sum of two odd prime numbers in two different ways. Answers may include 14 = 3 + 11 and 14 = 7 + 7; 16 = 3 + 13 and 16 = 5 + 11; or 20 = 3 + 17 and 20 = 7 + 13.
2. What are some of the practical applications of the solution to the four-color problem? Answers may include the fact that cartographers know they can use as few as four colors on any map they make.

Mixed REVIEW

1. Write 4,240,000 in scientific notation. **(8.4)**
2. Write 0.00000062 in scientific notation. **(8.4)**
3. Classify $3x^2 + 2x - 1$ by degree and terms. **(10.1)**
4. Find the leading coefficient of $2 + 3x - 4x^2$. **(10.1)**
5. Evaluate $y = 3 + 14(1.2)^t$ when $t = 3$. **(8.6)**
6. Evaluate $y = 3(6.2)^{-t}$ when $t = 2$. **(8.6)**
7. Divide $3x^2 + 2x - 6$ by $x - 2$. **(11.7)**
8. Multiply $4x^2 - 2$ by $4x^2 + 2$. **(10.2)**
9. Find the vertex of $-3|x - 3| + 2 = y$. **(4.7)**
10. Find the vertex of $y = 2x^2 + 3$. **(9.3)**
11. Solve $7 = -3x + x^2$ (quadratic formula). **(9.4)**
12. Solve $x^2 + 8x + 5 = 0$. (Complete the square.) **(10.7)**
13. Is $(-1, 1)$ a solution for $y \leq x^2 - x + 2$? **(6.1)**
14. Is (7, 2) a solution for $y + 2 = \frac{1}{3}(x + 5)$? **(4.2)**
15. Solve $-x^2 - 3x + 4 = 0$. **(10.6)**
16. Solve $3x^2 = 27$. **(9.2)**
17. Find the x-intercepts: $y = 3x^2 + 14x + 8$. **(9.3)**
18. Find the y-intercept: $3x + 2y = -6$. **(4.3)**
19. What is 28 percent of 490?
20. 99 is what percent of 450?

See margin for answers.

Milestones UNSOLVED PROBLEMS

1750 Goldbach's conjecture — 1850 Four-color problem posed — Four-color problem solved

1750 1800 1850 1900 1950 2000 2050

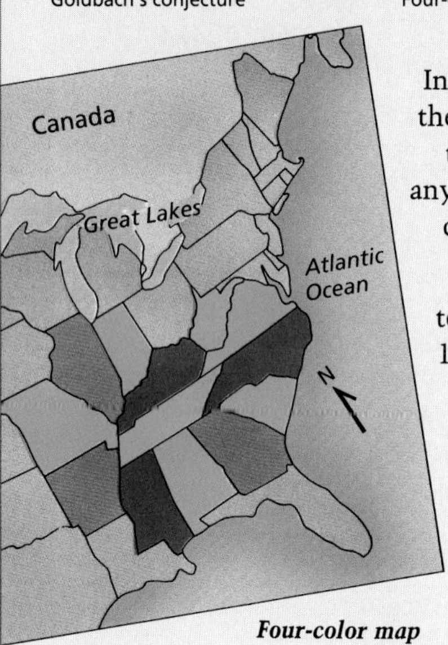

Four-color map

In the study of mathematics many questions are raised. Some, like the four-color problem posed in 1840, have been solved in this century. The problem was to prove that four colors could be used on any map so that no two regions sharing a border would be the same color. It was solved in 1976 through a lengthy computer analysis.

Some questions raised hundreds of years ago remain unanswered today. An example is a conjecture made by Christian Goldbach in 1742 that every even number greater than 4 can be written as the sum of two odd prime numbers and that every odd number greater than 7 can be written as the sum of three odd prime numbers. While no examples have been found to disprove this conjecture, no one has been able to prove that it is true.

Find an odd number greater than 7 that can be written as the sum of three odd prime numbers in three different ways.

12.3 Exponential Functions

What you should learn:

Goal 1 How to transform the graph of a function

Goal 2 How to use exponential functions as models of real-life problems

Why you should learn it:

You can model many real-life situations using exponential functions, such as college enrollments.

Goal 1 Transforming a Graph

In Chapter 8, you were introduced to graphs that model the exponential function.

Exponential Growth: $f(x) = 2^x$

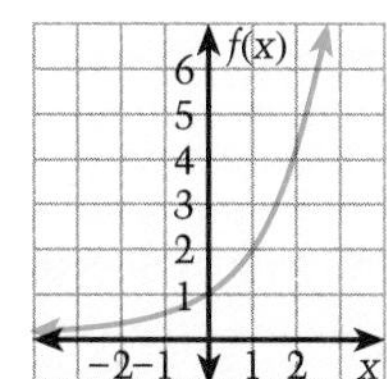

Exponential Decay: $f(x) = \left(\frac{1}{2}\right)^x$

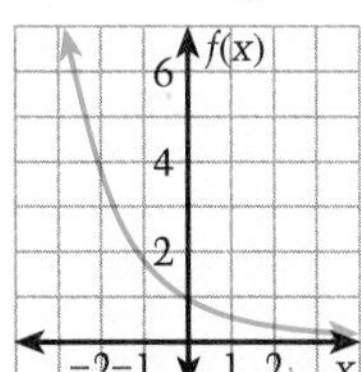

If the graph of a function is shifted about on the coordinate plane, it may then define a different function that is related in a simple way to the original function. Such a shift is one example of a **transformation.** In this lesson, you will learn to graph a function by transforming the graph of a simpler, related function.

In Example 1, note how letters other than f, such as g and h, are commonly used to denote functions.

Example 1 *Vertical Shifts*

a. To sketch the graph of

$$g(x) = 2^x + 3$$

shift the graph of $f(x) = 2^x$ *three units up.*

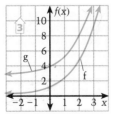

b. To sketch the graph of

$$h(x) = \left(\frac{1}{2}\right)^x - 2$$

shift the graph of $f(x) = \left(\frac{1}{2}\right)^x$ *two units down.*

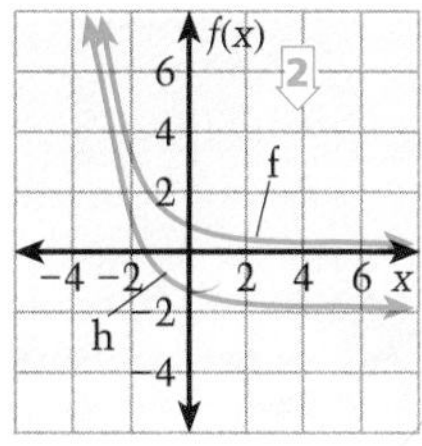

■

Problem of the Day

There are less than 100 apples in a basket. If they are counted by 3's, there are two left over; by 4's, there are three left over; and by 5's, there are four left over. How many apples are in the basket? 59

ORGANIZER

Warm-Up Exercises

1. Evaluate each exponential equation below at $x = -3, -2, -1, 0, 1, 2, 3$.
   a. $y = 3^x$ **b.** $y = 3^x + 1$
   c. $y = 3^x + 2$ **d.** $y = 3^x + 3$
   a. $y = 0.037, 0.111, 0.333, 1, 3, 9, 27$
   b. $y = 1.037, 1.111, 1.333, 2, 4, 10, 28$
   c. $y = 2.037, 2.111, 2.333, 3, 5, 11, 29$
   d. $y = 3.037, 3.111, 3.333, 4, 6, 12, 30$
2. Sketch the graph of each exponential equation in Exercise 1. Identify any patterns that you can among the four graphs. The graphs show that each equation has a graph that is identical in shape to the others. One pattern is that the graph of $y = 3^x$ can be transformed into the others by a vertical shift from one position to another as indicated by the constant term that is added.

Lesson Resources

Teaching Tools
- Transparencies: 4, 14
- Problem of the Day: 12.3
- Warm-up Exercises: 12.3
- Answer Masters: 12.3

Extra Practice: 12.3

Color Transparencies: 70, 71

Applications Handbook: p. 26–28, 58, 59

LESSON Notes

Example 1

Ask students to answer a series of "What if" questions about the effect of adding (or subtracting) a constant to (or from) an exponential function. For example, What if 4 is added to $y = 2^x$? How does the graph shift? What if 6 is added to $y = 2^x$? What if -4 is added? and so on.

Example 2

As in Example 1, ask students a series of questions about the effect of adding a constant to the variable in an exponential function. For instance, given $f(x) = 3^x$ and the point (1, 3) on the curve, can you define a function $g(x)$ so that the point (3, 3) would lie on the curve defined by 3^x? One such function is $g(x) = 3^{x-2}$. The y-coordinate 3 is now associated with the x-coordinate 3, which is 2 units to the right of the x-coordinate 1.

Example 3

Caution students that in $g(x) = -2^x$ the exponent x only governs the base 2.

Example 4

Point out to students that the broken lines in the sketches represent the asymptotes (or boundary lines) of the graphs. The function values here will never cross the asymptotes as t grows large.

Extra Examples

Here are additional examples similar to **Examples 1–3.**

Sketch the graphs.

a. $y = 3^x + 4$
b. $y = 3^x - 3$
c. $y = 3^{x+1}$
d. $y = 3^{x-3}$
e. $y = -3^x$

Example 2 Horizontal Shifts

a. To sketch the graph of

$$g(x) = 3^{x-2}$$

shift the graph of $f(x) = 3^x$ *two units to the right.*

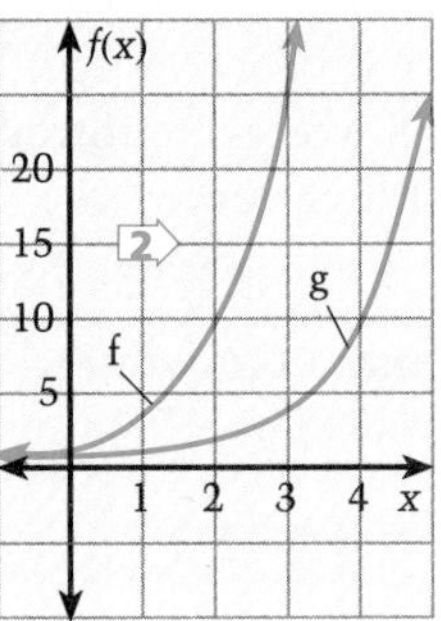

b. To sketch the graph of

$$h(x) = 3^{x+2}$$

shift the graph of $f(x) = 3^x$ *two units to the left.*

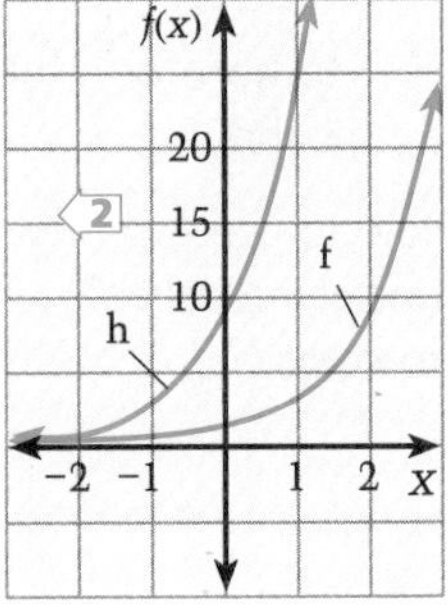

■

One graph is a **reflection** of another graph in the x-axis if the two graphs are mirror images of each other. (The mirror lies on the x-axis.)

Example 3 Reflections

a. To sketch the graph of

$$g(x) = -2^x$$

reflect the graph of $f(x) = 2^x$ in the x-axis.

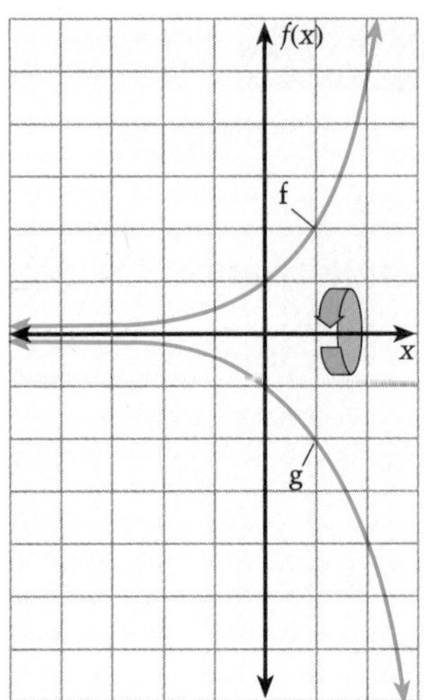

b. To sketch the graph of

$$h(x) = -\left(\frac{1}{2}\right)^x$$

reflect the graph of $f(x) = \left(\frac{1}{2}\right)^x$ in the x-axis.

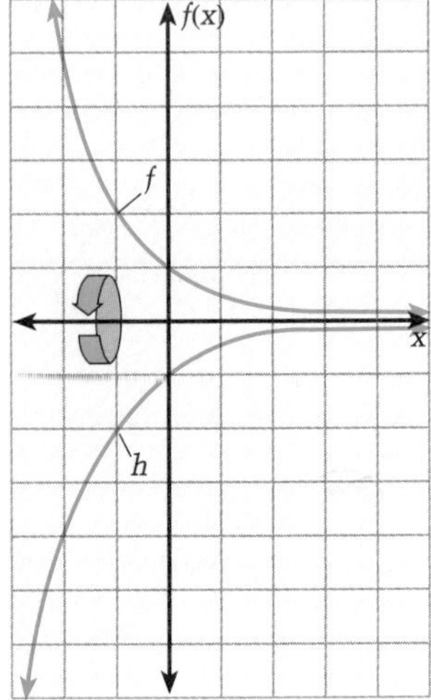

■

Lake LaCross, Mt. Steel, Olympic National Park

In 1975, 26% of the 9.73 million college students in the United States were enrolled in two-year colleges. By 1990, the percent of college students in two-year colleges had risen to about 38%.

Goal 2 Using Exponential Models in Real Life

Real Life **Education**

Example 4 ***Graphing an Exponential Function***

The percent of all college students in the United States from 1975 to 1990 who were enrolled in two-year colleges can be modeled by

$$f(t) = 38 - (0.8)^t$$

where $t = 0$ represents 1975. Find a model for the percent of college students who were enrolled in 4-year colleges. Then sketch the graphs of both functions.

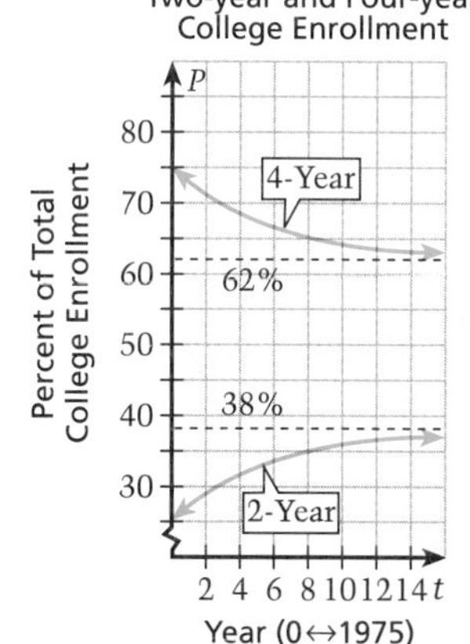

Solution

Verbal Model % in four-year colleges = 100% − % in two-year colleges

Labels % in two-year colleges $= 38 - (0.8)^t$ (%)
% in four-year colleges $= g(t)$ (%)

Equation $g(t) = 100 - (38 - (0.8)^t)$
$= 62 + 0.8^t$

The graphs of the two models are shown at the left. ■

Communicating about ALGEBRA

Cooperative Learning

SHARING IDEAS about the Lesson See Additional Answers.

Work with a Partner and Draw Visual Models Sketch the graphs of the functions. Which shift or reflection is useful for each?

A. $f(x) = 2^x - 4$ **B.** $f(x) = \left(\frac{1}{2}\right)^x + 2$ **C.** $f(x) = 2^{x+1}$
D. $f(x) = \left(\frac{1}{2}\right)^{x-2}$ **E.** $f(x) = -3^x$ **F.** $f(x) = -\left(\frac{1}{3}\right)^x$

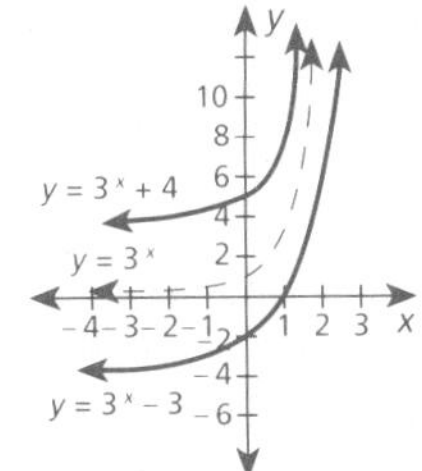

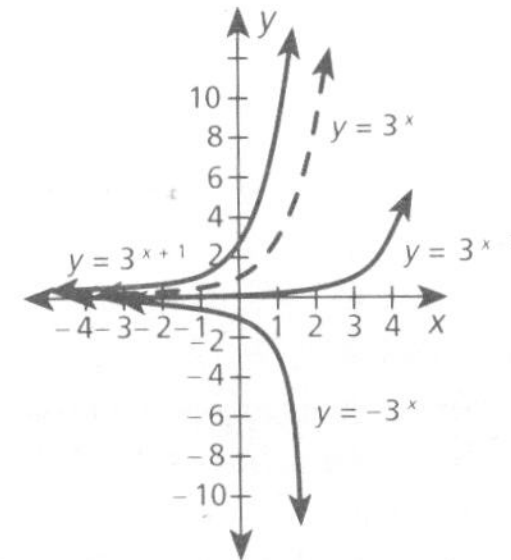

Check Understanding

1. Explain how the basic exponential function $f(x) = 4^x$ can be transformed by shifting it vertically down 5 units. What new function will describe this transformation? $g(x) = 4^x - 5$
2. What is the effect on the graph of multiplying an exponential function by −1? Reflection about the x-axis

***Communicating* about ALGEBRA**

EXTEND *Communicating*
Have students discuss the graphs of the following exponential functions.

$f(x) = (-1)^x$ and $g(x) = (-2)^x$

How are these graphs different from other exponential functions seen in this lesson? Ask students with graphics calculators to graph these functions. What happens? The functions are not defined for all x. Their graphs are a series of disconnected dots. The graphics calculator plots a few dots of $f(x) = (-1)^x$ and only one dot of $g(x) = (-2)^x$.

EXERCISE Notes

ASSIGNMENT GUIDE

Basic/Average: Ex. 1–6, 7–35 odd, 41, 43–46, 51–56

Above Average: Ex. 1–6, 11–35 odd, 40–48, 53–57

Advanced: Ex. 1–6, 13–35, 37–46, 53–57

Selected Answers
Exercises 1–6, 7–53 odd

More Difficult Exercises
Exercises 37–40, 55–57

Guided Practice

Ex. 3–6 These exercises might be graphed on the same coordinate axes in order to help students visualize the transformations of the basic function $f(x) = (\frac{1}{3})^x$. Be sure students understand that the difference between the graph of $f(x) = (\frac{1}{3})^x$ and $h(x) = 3^x$ is related to the value of the base of the exponent. The base $\frac{1}{3}$ is between 0 and 1; the base 3 is greater than 1.

Independent Practice

Ex. 7–18 Students should describe the basic relationship as a vertical shift, a horizontal shift, or an x-axis reflection.

Ex. 19–36 COOPERATIVE LEARNING You may have some students use quick-graph techniques to sketch these graphs. Other students might use graphics calculators. Students could compare results to verify the solutions.

Ex. 37–40 These exercises should be assigned as a group. Note that $\sqrt[12]{2}$ is read as "the twelfth root of two." In Exercise 37, $n = -48$ for the lowest note and $n = 39$ for the highest note. In Exercise 38, integer frequencies occur when n is a multiple of 12.

EXERCISES

Guided Practice

CRITICAL THINKING about the Lesson

In Exercises 1 and 2, does the function represent exponential growth or exponential decay? Explain your reasons.

1. $f(x) = (\frac{1}{4})^x$ Decay, $0 < \frac{1}{4} < 1$
2. $f(x) = 5^x$ Growth, $5 > 1$

In Exercises 3–6, state whether the graph of the function g is related to the graph of $f(x) = (\frac{1}{3})^x$ by a vertical shift, a horizontal shift, or a reflection. Use your answer as an aid to sketch the graph of g. See Additional Answers.

3. $g(x) = (\frac{1}{3})^{x-1}$ Horizontal shift
4. $g(x) = (\frac{1}{3})^x - 3$ Vertical shift
5. $g(x) = -(\frac{1}{3})^x$ Reflection
6. $g(x) = (\frac{1}{3})^x + 4$ Vertical shift

Independent Practice

In Exercises 7–18, describe the relationship between the graph of f and the graph of g. (You do not need to sketch the graphs of f and g to answer the question.)

7. $f(x) = (\frac{5}{3})^x$, $g(x) = -(\frac{5}{3})^x$ Reflection
8. $f(x) = 7^x$, $g(x) = 7^x + 7$ Vertical shift
9. $f(x) = 15^x$, $g(x) = 15^{x-3}$ Horizontal shift
10. $f(x) = (2.2)^x$, $g(x) = (2.2)^{x-9}$ Horizontal shift
11. $f(x) = 16^x$, $g(x) = 16^x - 1$ Vertical shift
12. $f(x) = (\frac{1}{5})^x$, $g(x) = -(\frac{1}{5})^x$ Reflection
13. $f(x) = 8^x$, $g(x) = 8^x - \frac{1}{3}$ Vertical shift
14. $f(x) = (\frac{2}{3})^x$, $g(x) = -(\frac{2}{3})^x$ Reflection
15. $f(x) = 2^x$, $g(x) = 2^x + 8$ Vertical shift
16. $f(x) = (\frac{1}{3})^x$, $g(x) = (\frac{1}{3})^{x+10}$ Horizontal shift
17. $f(x) = 7^x$, $g(x) = -7^x$ Reflection
18. $f(x) = 4^x$, $g(x) = 4^{x+2}$ Horizontal shift

In Exercises 19–24, use a vertical shift to sketch the graph of the function. See Additional Answers.

19. $f(x) = 2^x + 1$
20. $g(x) = 3^x - 2$
21. $h(x) = (\frac{1}{2})^x + \frac{1}{2}$
22. $h(x) = (0.6)^x - 4$
23. $f(x) = 2^x - 5$
24. $f(x) = (\frac{1}{3})^x + 10$

In Exercises 25–30, use a horizontal shift to sketch the graph of the function. See Additional Answers.

25. $f(x) = (\frac{1}{2})^{x-1}$
26. $f(x) = (\frac{1}{4})^{x+4}$
27. $h(x) = 2^{x+5}$
28. $g(x) = 1.5^{x-2}$
29. $h(x) = \frac{1}{3}(5)^{x-1}$
30. $g(x) = (\frac{1}{2})^{x-5}$

In Exercises 31–36, use a reflection to sketch the graph of the function. See Additional Answers.

31. $h(x) = -3^x$
32. $g(x) = -4^x$
33. $f(x) = -(0.25)^x$
34. $f(x) = -(0.75)^x$
35. $h(x) = -(\frac{1}{5})^x$
36. $g(x) = -(\frac{4}{3})^x$

Piano Notes **In Exercises 37–40, use the following information.**

Piano tuners use an *equal temperament* scale to tune pianos. The frequency, $f(n)$, of the nth note is given by the exponential function

$$f(n) = 440 \cdot (\sqrt[12]{2})^n \quad \text{(frequency in vibrations per second)}$$

where n represents the position of the note below or above the note called A-440. (The number $\sqrt[12]{2}$ is approximately 1.0594631, and $(\sqrt[12]{2})^{12} = 2$.)

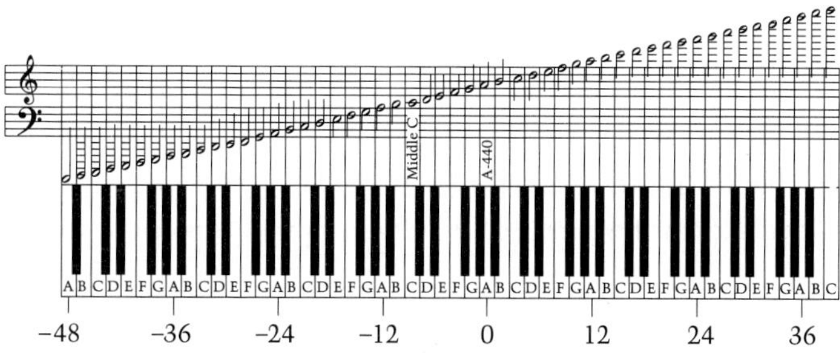

37. What are the frequencies of the lowest and highest notes on the piano? **27.5, ≈ 4186.0**

38. Which letter, A, B, C, D, E, F, or G, denotes notes with integer frequencies? **A, except the lowest A**

39. Two notes harmonize (sound pleasing to the human ear) if the ratio of their frequencies is an integer or a simple rational number. Find examples of notes whose frequencies have the following ratios. **See margin.**

Octave:	2	Fifth:	$\frac{3}{2}$	Fourth:	$\frac{4}{3}$		
Major third:	$\frac{5}{4}$	Minor third:	$\frac{6}{5}$	Major sixth:	$\frac{5}{3}$	Minor sixth:	$\frac{8}{5}$

40. Piano tuners tune piano keys by adjusting the tension on each string. How much tension should there be on the lowest note for a 5-foot string that weighs 0.00235 pounds per foot? The tension, T, weight per foot, W, length, L, and frequency, F, are related by $T = 4WF^2L^2$. **≈177.7 pounds per foot**

Snowflake **In Exercises 41 and 42, use the following information.**

In the snowflake sequence below, the snowflakes are numbered from 0 to 5. The perimeter, $f(n)$, of the nth snowflake is given by $f(n) = \left(\frac{4}{3}\right)^n$.

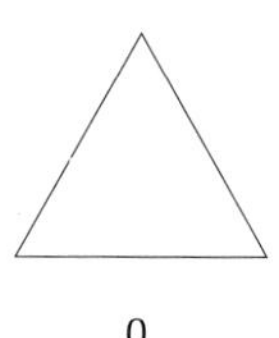
0

1

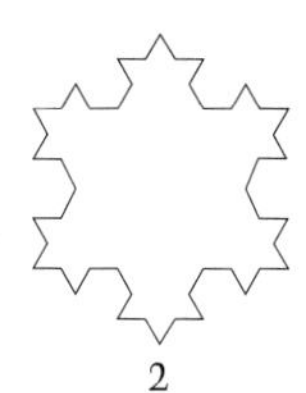
2

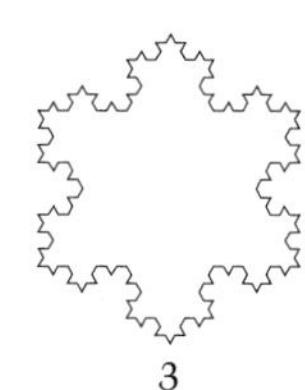
3

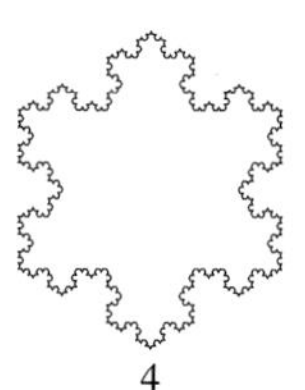
4

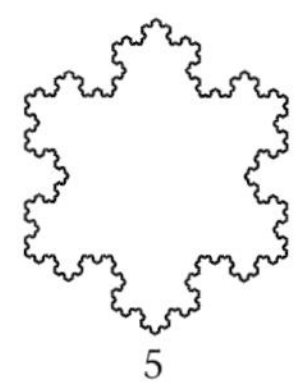
5

41. Find the perimeters of the 6 snowflakes. **See margin.**

42. If the pattern continues, what would be the perimeter of Snowflake 15? **$f(15) = \left(\frac{4}{3}\right)^{15} \approx 74.83$**

Answers

39. Answers vary.
Octave: $C(n = 3)$ to $C(n = -9)$
Fifth: $G(n = -2)$ to $C(n = -9)$
Fourth: $F(n = -4)$ to $C(n = -9)$
Major Third: $E(n = -5)$ to $C(n = -9)$
Minor Third: $G(n = -2)$ to $E(n = -5)$
Major Sixth: $A(n = 0)$ to $C(n = -9)$
Minor Sixth: $C(n = 3)$ to $E(n = -5)$

41. $f(0) = 1$
$f(1) = \frac{4}{3} \approx 1.33$
$f(2) = \left(\frac{4}{3}\right)^2 \approx 1.78$
$f(3) = \left(\frac{4}{3}\right)^3 \approx 2.37$
$f(4) = \left(\frac{4}{3}\right)^4 \approx 3.16$
$f(5) = \left(\frac{4}{3}\right)^5 \approx 4.21$

56.

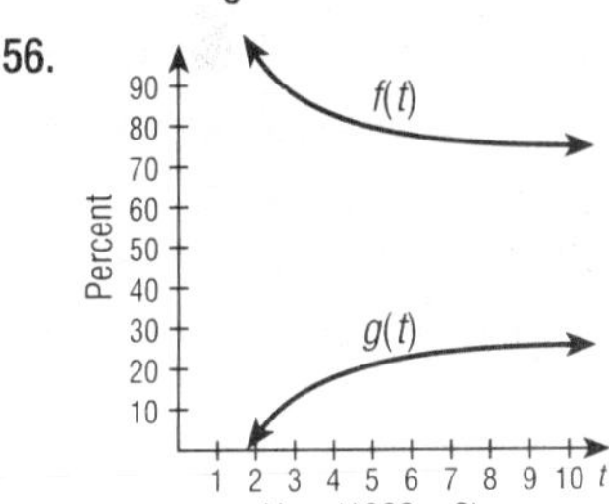

Yes, the percent that is sugar consumption is decreasing less and less each year.

57.

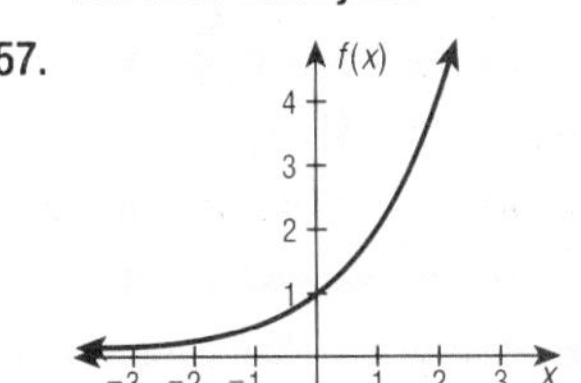

No, there is no solution; the solutions of $2^x = 0$ are the x-intercepts of $y = 2^x$.

▶ **Ex. 43–46** These exercises should be assigned as a group.

▶ **Ex. 47–52** TECHNOLOGY Encourage students to use calculators for these exercises.

▶ **Ex. 55–56** These exercises are similar to Example 4 modeled on page 635.

Enrichment Activities

These activities require students to go beyond lesson goals.

1. Can the effects of vertical and horizontal shifts and reflections be combined in one function? Describe a method of sketching these graphs.
   a. $g(x) = 2^{x-1} + 2$
   b. $h(x) = -3^{x+2} - 1$
   c. $j(x) = -4^x + 3$

   For example, suggest that $g(x)$ can be sketched by applying the effects of the horizontal and vertical shifts on the basic exponential function, $f(x) = 2^x$. That is, sketch the basic function $f(x)$ and then apply the effect of the horizontal shift to get a second sketch, followed by the effect of the vertical shift on the second sketch to arrive at the third sketch, a graph of $g(x)$.
2. WRITING
   The snowflake pattern in Exercises 41 and 42 is related to a pattern generated in the study of fractals. Explore the topic of "fractals" in the library and write a report to share with your classmates. In the 1970's, Benoit Mandelbrot, an IBM fellow, coined the term *fractal* to describe shapes in nature, such as clouds, riverbeds, and mountain ranges, that form irregular patterns.

Integrated Review

In Exercises 43–46, match the function with its graph.

43. $f(x) = 1.2^{x-1}$ d

44. $f(x) = -1.2^x$ c

45. $f(x) = 1.2^x$ a

46. $f(x) = 1.2^x - 1$ b

a.

b.

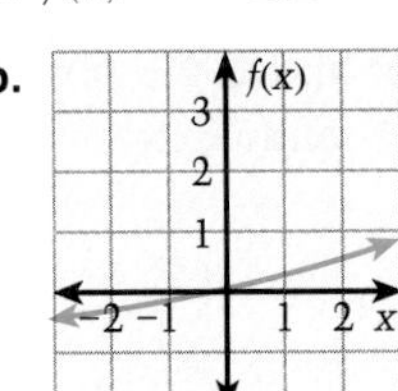

c.

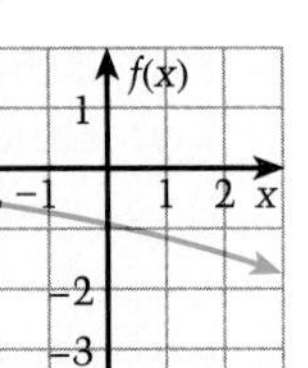

d.

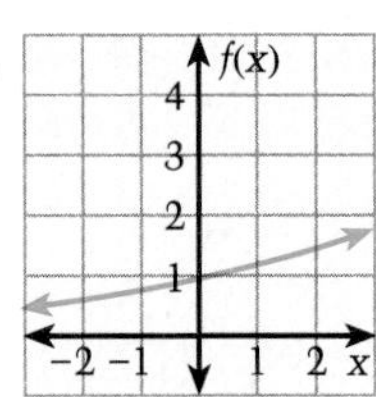

In Exercises 47–52, evaluate the function as indicated.

47. $f(x) = (0.05)^x$, $f(3) = \boxed{?}$ 1.25×10^{-4}

48. $f(x) = \frac{1}{7}(10)^x$, $f(5) = \boxed{?}$ 14285.71

49. $f(x) = 6^x - 12$, $f(2) = \boxed{?}$ 24

50. $f(x) = 21\left(\frac{1}{3}\right)^x$, $f(8) = \boxed{?}$ 0.003

51. $f(x) = \frac{1}{11}(2^x) + 4$, $f(5) = \boxed{?}$ 6.91

52. $f(x) = 6\left(\frac{1}{6}\right)^x$, $f(9) = \boxed{?}$ 5.95×10^{-7}

53. *College Entrance Exam Sample* If $x \neq 0$ and $3x^2 - 12x = 0$, what is x?

a. -9 **b.** -4 **c.** 3 **d.** 4 **e.** 9

54. *College Entrance Exam Sample* On a certain map, the distance of 100 miles is represented by 1 inch. What distance is represented by 2.4 inches on this map?

a. 200.4 mi. **b.** 204.0 mi. **c.** 225.0 mi. **d.** 233.3 mi. **e.** 240.0 mi.

Exploration and Extension

Sweet Tooth **In Exercises 55 and 56, use the following information.** 55. $g(t) = 25 - 37.25(0.615)^t$

For 1982 through 1990, the consumption of sugar in the United States, as a percent, $f(t)$, of the total consumption of sweeteners, can be modeled by

$$f(t) = 75 + 37.25(0.615)^t$$

where $t = 2$ represents 1982.

55. Find a model for the consumption of nonsugar sweeteners as a percent, $g(t)$, of the total consumption of sweeteners.

56. Sketch the graphs of f and g. Does the percent of sugar compared to the percent of nonsugar sweeteners appear to be stabilizing? Explain. See margin on page 637.

57. Sketch the graph of $f(x) = 2^x$. Does it have an x-intercept? What does this tell you about the number of solutions of the equation $2^x = 0$? Explain.

12.4 Quadratic Functions

What you should learn:

Goal 1 How to sketch the graph of a quadratic function

Goal 2 How to use quadratic functions as models of real-life problems

Why you should learn it:

You can model many real-life situations, such as maximizing the area of a figure, using quadratic functions.

Goal 1 Graphing a Quadratic Function

The graph of a quadratic function

$f(x) = ax^2 + bx + c$ *Standard form*

is a parabola. From Lesson 9.3, you know that the graph opens up if a is positive and opens down if a is negative.

In this lesson, you will learn how to sketch graphs of quadratic functions that are written in *completed square form.*

Completed Square Form

The completed square form of a quadratic function is

$f(x) = a(x - h)^2 + k.$ *Completed square form*

The vertex of the parabola is the point (h, k).

Example 1 *Sketching the Graph of a Quadratic Function*

a. The graph of

$$f(x) = (x - 2)^2 - 1$$

opens up ($a > 0$) and has its vertex at $(2, -1)$. Some additional points are given by the table.

x	0	1	2	3	4
f(x)	3	0	−1	0	3

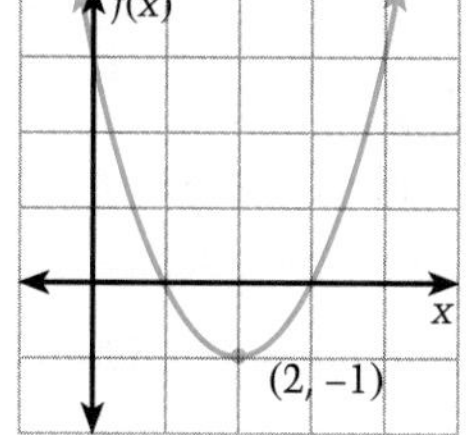

b. The graph of

$$f(x) = -(x + 1)^2 + 3$$

opens down ($a < 0$) and has its vertex at $(-1, 3)$. Some additional points are given by the table.

x	−3	−2	−1	0	1
f(x)	−1	2	3	2	−1

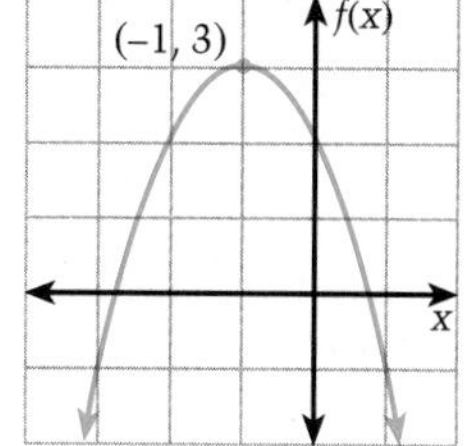

■

Problem of the Day

The lengths of the sides of a rectangle that is not a square are integers. The numerical values of the area and the perimeter are the same. How many units long is the longer side? 6

ORGANIZER

Warm-Up Exercises

1. Sketch the graph of the quadratic equation.

a. $y = x^2 + 2x - 4$

b. $y = -4x^2 + 6$

c. $y = (x - 3)^2 + x$

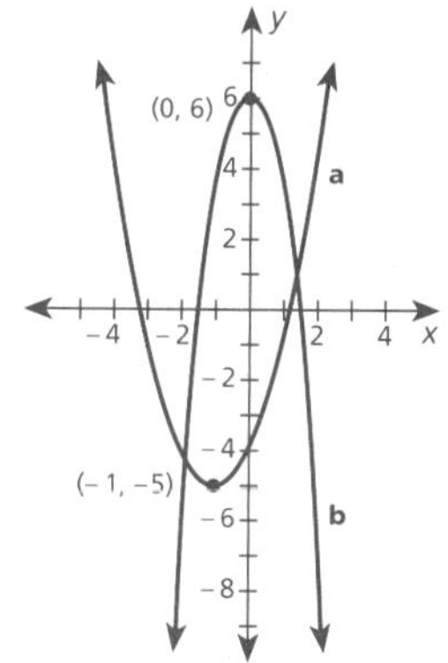

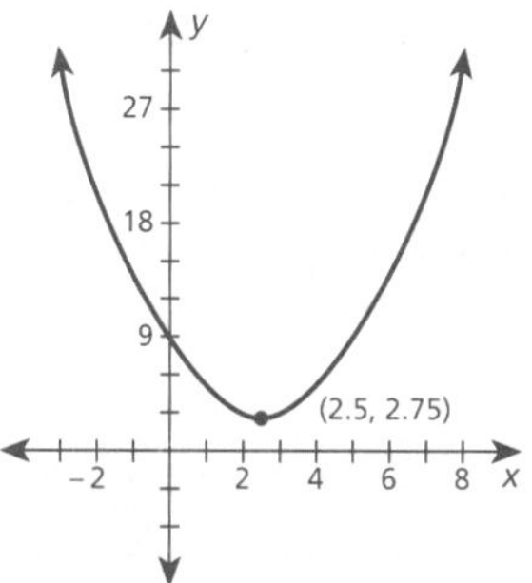

2. Complete the square.

a. $x^2 + 4x - 3 = 0$

b. $x^2 - 5x + 4 = 0$

a. $(x + 2)^2 = 7$

b. $(x - \frac{5}{2})^2 = \frac{9}{4}$

Lesson Resources

Teaching Tools
- Transparencies: 4, 13
- Problem of the Day: 12.4
- Warm-up Exercises: 12.4
- Answer Masters: 12.4

Extra Practice: 12.4

Technology Handbook: p. 84

LESSON Notes

GOAL 1 The function $f(x) = a(x - h)^2 + k$ is said to be in "completed square form" because it can be obtained by completing the square of any quadratic function in the form $f(x) = ax^2 + bx + c$. (See Examples 2 and 3.)

From Chapter 9, recall that the x-coordinate of the vertex of a quadratic equation of the form $y = ax^2 + bx + c$ is $-\frac{b}{2a}$. Students will need this relationship later to justify that the x-coordinate of the vertex for the quadratic function in completed square form is h.

Example 1

Remind students to select x-values on either side of the x-coordinate of the vertex when plotting points to sketch the graph of a quadratic function.

Common-Error ALERT!

The adding and subtracting of the square of half the coefficient of x in Example 3 is affected by the leading coefficient when the leading coefficient is different from 1. The product of the leading coefficient and the constant term that has been subtracted inside the parentheses must be obtained and combined with other constants located outside of the parentheses.

If a quadratic function is given in standard form, you can rewrite it in completed square form. When the leading coefficient is 1, you can complete the square by *adding and subtracting* the square of half the coefficient of x.

Example 2 Completing the Square: Leading Coefficient Is 1

Write the function $f(x) = x^2 - 6x + 5$ in completed square form. Then sketch its graph.

Solution

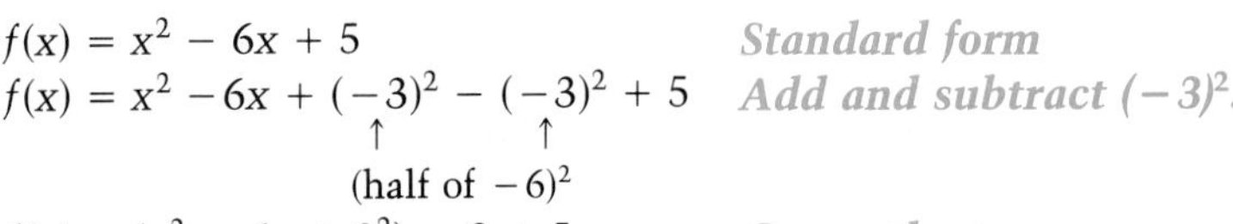

$f(x) = x^2 - 6x + 5$	*Standard form*
$f(x) = x^2 - 6x + (-3)^2 - (-3)^2 + 5$ (half of -6)^2	*Add and subtract $(-3)^2$.*
$f(x) = (x^2 - 6x + 3^2) - 9 + 5$	*Group the terms.*
$f(x) = (x - 3)^2 - 4$	*Completed square form*

From the completed square form, you can see that the vertex of the parabola is the point $(3, -4)$. See the graph at the left. ■

When the leading coefficient is not 1, you must first group the x^2-term and x-term. Then, factor the leading coefficient out of the grouping. Finally, add and subtract the square of half the coefficient of x *inside the grouping.*

Example 3 Completing the Square: Leading Coefficient Is Not 1

Write the function $f(x) = 2x^2 + 4x + 1$ in completed square form. Then sketch its graph.

Solution

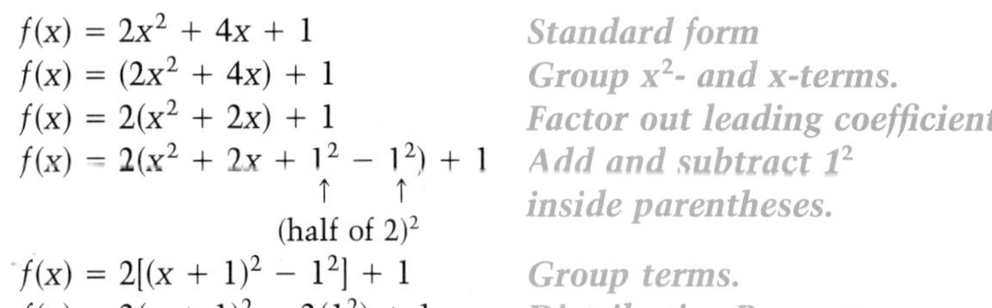

$f(x) = 2x^2 + 4x + 1$	*Standard form*
$f(x) = (2x^2 + 4x) + 1$	*Group x^2- and x-terms.*
$f(x) = 2(x^2 + 2x) + 1$	*Factor out leading coefficient.*
$f(x) = 2(x^2 + 2x + 1^2 - 1^2) + 1$ (half of 2)^2	*Add and subtract 1^2 inside parentheses.*
$f(x) = 2[(x + 1)^2 - 1^2] + 1$	*Group terms.*
$f(x) = 2(x + 1)^2 - 2(1^2) + 1$	*Distributive Property*
$f(x) = 2(x + 1)^2 - 1$	*Completed square form*

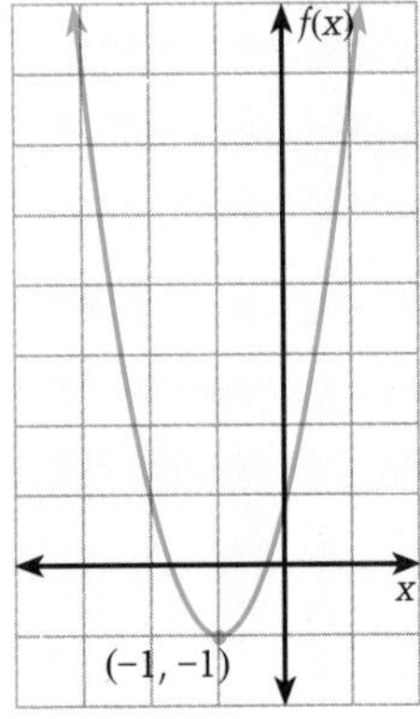

From the completed square form, you can see that the vertex of the parabola is $(-1, -1)$. The graph is shown at the left. ■

Goal 2 Using Quadratic Models in Real-Life

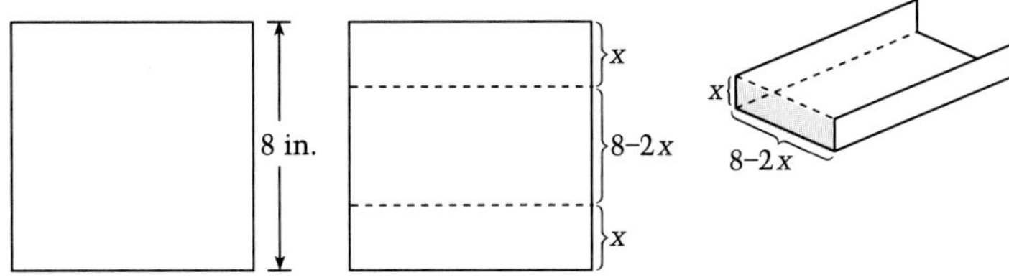

Start with a piece of paper that is 8 inches long. Fold both sides of the paper up by the same amount, x. The resulting "trough" has cross sections with area x(8 − 2x).

Connections
Geometry

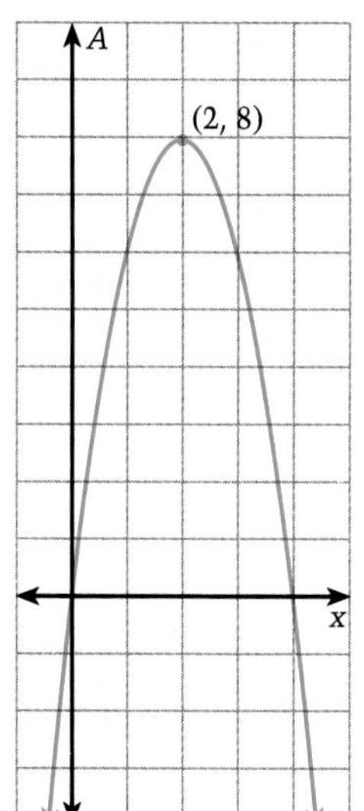

Example 4 *Finding the Maximum Area*

Where should the paper described above be folded to obtain the maximum cross-section area?

Solution

Verbal Model Area = Width × Height

Labels Height of cross section $= x$ (inches)
Width of cross section $= 8 - 2x$ (inches)
Area of cross section $= A$ (square inches)

Function
$A = (8 - 2x)(x)$
$A = -2x^2 + 8x$ *Standard form*
$A = -2(x - 2)^2 + 8$ *Completed square form*

The graph of this equation is a parabola that opens down and has its vertex at (2, 8) This tells you that the maximum cross-section area is 8 square inches and that you can obtain the maximum area by folding up 2 inches of paper on each end. ■

Communicating about ALGEBRA

SHARING IDEAS about the Lesson

See Additional Answers.

Use a Pattern Write each function in completed square form. Then identify the vertex and sketch the graph.

A. $f(x) = x^2 - 4x + 2$
B. $f(x) = -x^2 - 2x - 5$
C. $f(x) = 2x^2 - 4x + 7$
D. $f(x) = -3x^2 - 6x - 4$

Example 2

Caution students that the same value must be both added to and subtracted from the "definition side" of the function so that the relationship which defines the function remains unaffected.

Example 4

Remind students that the maximum or minimum value of a quadratic equation occurs at the vertex of the parabola associated with the equation.

The following steps can be used to transform the equation $A = (8 - 2x)(x)$ from standard form to completed square form.

Standard form
$A = -2x^2 + 8x$

Factor out the leading coefficient.
$A = -2(x^2 - 4x)$

Add zero written as 4 – 4.
$A = -2(x^2 - 4x + 4 - 4)$

Apply the distributive property.
$A = -2(x^2 - 4x + 4) + 8$

Completed square form
$A = -2(x - 2)^2 + 8$

Extra Examples

Here are additional examples similar to **Examples 1–3.**

Rewrite each function in completed square form and determine its vertex. Then sketch each graph.

a. $f(x) = x^2 - 6x - 3$
b. $g(x) = -x^2 + 4x$
c. $h(x) = x^2 - 2x + 7$
d. $j(x) = 4x^2 + 12x + 8$

a. $f(x) = (x - 3)^2 - 12$; vertex $(3, -12)$; graph opens upward
b. $g(x) = -(x - 2)^2 + 4$; vertex (2, 4); graph opens downward
c. $h(x) = (x - 1)^2 + 6$; vertex (1, 6); graph opens upward
d. $j(x) = 4(x + \frac{3}{2})^2 - 1$; vertex $(-\frac{3}{2}, -1)$; graph opens upward

Check Understanding

MATH JOURNAL Students might be encouraged to enter their responses in the function section of their math journals.

1. Describe the process of completing the square of a quadratic function when the leading coefficient is 1; when the leading coefficient is different from 1. See Examples 2 and 3.
2. How should points be selected for sketching the graph of a quadratic function? Find the vertex; select points on either side of it.
3. The leading coefficient of a quadratic function indicates whether its graph "turns up" or "turns down." Explain. The sign of the leading coefficient determines the direction of the graph.

Communicating about ALGEBRA

The standard form of a quadratic equation is $ax^2 + bx + c = 0$, while the standard form of a quadratic function is $f(x) = ax^2 + bx + c$. A quadratic equation can thus be thought of as a subset of its corresponding quadratic function. Have students discuss how quadratic functions and quadratic equations are the same and how they are different. In particular, they should identify differences in completing the square for quadratic functions versus quadratic equations. Students should recall that completing the square was a technique used to identify the solutions of quadratic equations.

EXERCISES

Guided Practice

CRITICAL THINKING about the Lesson 3., 4. See Additional Answers for graphs.

1. Write $f(x) = x^2 - 6x + 6$ in completed square form. $f(x) = (x - 3)^2 - 3$
2. What is the vertex of the graph of $f(x) = 3(x + 4)^2 - 3$? $(-4, -3)$
3. Sketch the graph of $f(x) = (x + 1)^2 + 1$. Explain why the vertex is the lowest point on the graph. See below.
4. Sketch the graph of $f(x) = -(x - 1)^2 - 1$. Explain why the vertex is the highest point on the graph. See below.

Independent Practice

In Exercises 5–10, find the vertex of the graph of the function. Is the vertex the lowest or highest point on the graph?

5. $f(x) = -2(x - 1)^2 + 5$ $(1, 5)$, highest
6. $f(x) = -(x - 3)^2 - 5$ $(3, -5)$, highest
7. $f(x) = 3(x - 3)^2 - 2$ $(3, -2)$, lowest
8. $f(x) = 4(x - 7)^2$ $(7, 0)$, lowest
9. $f(x) = (x + 2)^2 + 6$ $(-2, 6)$, lowest
10. $f(x) = 9(x + 5) + 9$ $(-5, 9)$, lowest

In Exercises 11–14, match the function with its graph.

11. $f(x) = (x - 4)^2 + 5$ c
12. $f(x) = -(x - 1)^2 + 3$ a
13. $f(x) = -(x + 3)^2 - 3$ b
14. $f(x) = (x + 2)^2 - 4$ d

a. b.

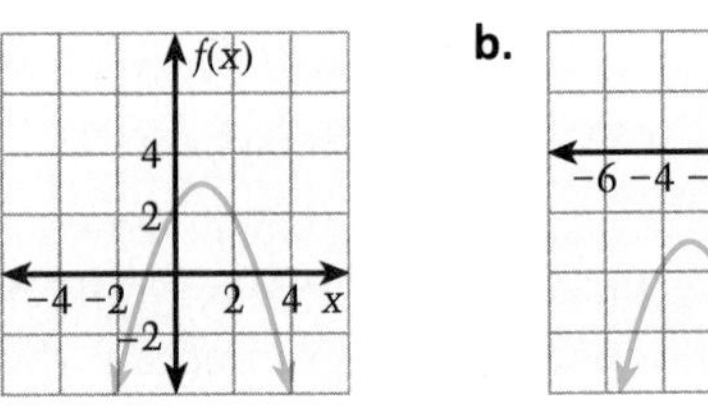

c.

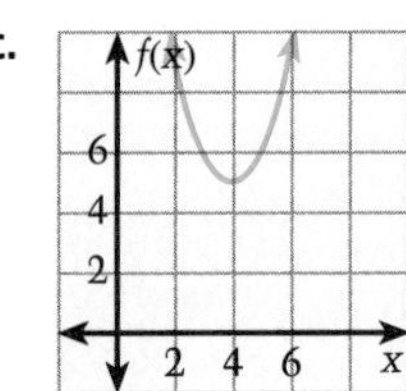

d.

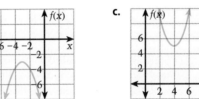

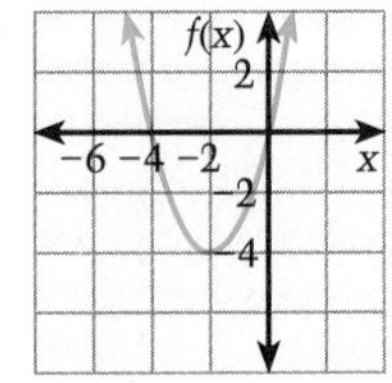

In Exercises 15–20, sketch the graph of the function. See Additional Answers.

15. $f(x) = (x - 1)^2 - 3$
16. $f(x) = -(x + 7)^2 + 2$
17. $f(x) = -(x + 2)^2 - 7$
18. $f(x) = -(x - 2)^2 + 6$
19. $f(x) = (x - 4)^2 + 2$
20. $f(x) = (x + 8)^2 - 2$

In Exercises 21–28, write the function in completed square form. Then find the vertex of its graph.

24. $f(x) = (x + 5)^2 + 8$, $(-5, 8)$
26. $f(x) = (x - 12)^2 + 7$, $(12, 7)$

21. $f(x) = x^2 - 18x + 79$ $f(x) = (x - 9)^2 - 2$, $(9, -2)$
22. $f(x) = x^2 + 8x + 11$ $f(x) = (x + 4)^2 - 5$, $(-4, -5)$
23. $f(x) = x^2 - 4x + 16$ $f(x) = (x - 2)^2 + 12$, $(2, 12)$
24. $f(x) = x^2 + 10x + 33$
25. $f(x) = x^2 + 16x + 62$
26. $f(x) = x^2 - 24x + 151$
27. $f(x) = -x^2 - 6x - 1$
28. $f(x) = -x^2 + 8x - 25$

$f(x) = -(x + 3)^2 + 8$, $(-3, 8)$ 25. $f(x) = (x + 8)^2 - 2$, $(-8, -2)$ $f(x) = -(x - 4)^2 - 9$, $(4, -9)$

3. The graph opens upward and the vertex is the point where the graph reverses direction from downward to upward.
4. The graph opens downward and the vertex is the point where the graph reverses direction from upward to downward.

In Exercises 29–36, write the function in completed square form. Then sketch its graph. See Additional Answers for graphs.

32. $f(x) = (x - 10)^2 - 150$
34. $f(x) = -3(x - 5)^2 - 6$

29. $f(x) = x^2 - 10x + 19$ $f(x) = (x - 5)^2 - 6$

30. $f(x) = x^2 - 8x + 25$ $f(x) = (x - 4)^2 + 9$

31. $f(x) = x^2 + 8x + 9$ $f(x) = (x + 4)^2 - 7$

32. $f(x) = x^2 - 20x - 50$

33. $f(x) = 2x^2 + 32x + 121$ $f(x) = 2(x + 8)^2 - 7$

34. $f(x) = -3x^2 + 30x - 81$

35. $f(x) = 3x^2 - 12x + 6$ $f(x) = 3(x - 2)^2 - 6$

36. $f(x) = 2x^2 - 4x + 5$ $f(x) = 2(x - 1)^2 + 3$

Cardboard Play Box **In Exercises 37 and 38, use the following information.**

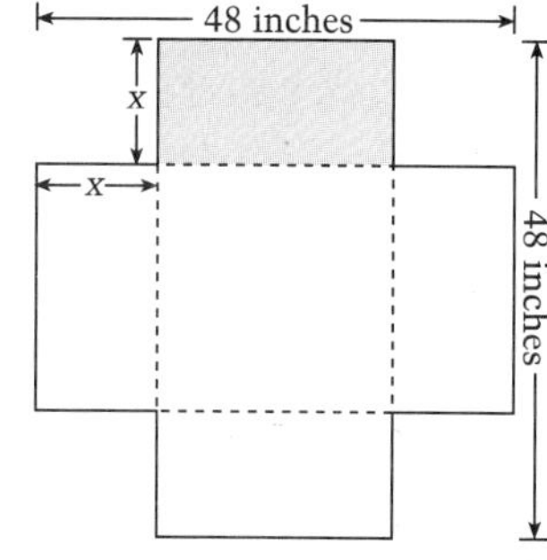

You are building a cardboard play box for a kitten from a 48-inch square piece of cardboard. You want to maximize the area of the side of the box. See Additional Answers.

37. Let $f(x)$ represent the area of the one side of the box. Write an equation for $f(x)$ in completed square form. Sketch its graph. $f(x) = -2(x - 12)^2 + 288$

38. How tall should you make the box? What are the dimensions of its base? 12 in., 24 in. by 24 in.

Human Cannonball **In Exercises 39 and 40, use the following information.**

A human cannonball is shot from a cannon and follows the path given by

$$h(x) = -\frac{1}{100}x^2 + x + 8$$

where $h(x)$ is the height in feet (above ground) and x is the horizontal distance traveled in feet.

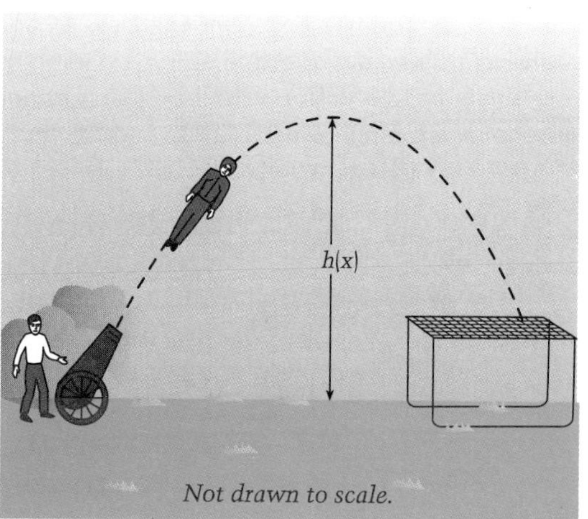

The human cannonball act originated in 1875 at the Circus d'Hiver in Paris, France. The record distance shot is 175 feet (1940, Ringling Brothers, Barnum & Bailey Circus).

39. Write the function in completed square form. Explain how this form can be used to find the maximum height of the human cannonball. See margin on page 644.

40. The human cannonball lands in a net that is 8 feet above the ground. How many feet does the human cannonball travel? 100 ft

Iron and Steel Production **In Exercises 41 and 42, use the following information.**

For 1950 through 1990, the United States iron and raw steel production, $f(t)$, as a percent of the world's production can be modeled by $f(t) = 1.7t^2 - 15.3t + 45.2$, where $t = 0$ represents 1950, $t = 1$ represents 1960, and so on. See Additional Answers for graph.

✪ **41.** Write the model in completed square form. Sketch its graph from $t = 0$ to $t = 7$. $f(t) = 1.7(t - 4.5)^2 + 10.775$

✪ **42.** The U.S. percent of world production has been decreasing. According to the model, when will this trend change? 1995

EXERCISE Notes

ASSIGNMENT GUIDE

Basic/Average: Ex. 1–4, 5–9 odd, 11–14, 15–33 odd, 39–40, 49–57 odd, 61–62

Above Average: Ex. 1–4, 7, 9, 11—14, 15–35 odd, 41–42, 49–57 odd, 61–63

Advanced: Ex. 1–4, 9, 11–14, 15–35 odd, 37–42, 49–57 odd, 61–63

Selected Answers

Exercises 1–4, 5–61 odd

✪ More Difficult Exercises

Exercises 41, 42, 63

Guided Practice

▶ **Ex. 1–4** You can use the completed square form as a quick-graph technique for quadratic functions. In Exercise 2, help students understand why -4 is used for the h-value and -3 is used for the k-value. Students will often write the incorrect sign. Recall that the vertex of the quadratic function $f(x) = ax^2 + bx + c$ can also be written as $(-\frac{b}{2a}, f(\frac{-b}{2a}))$.

Independent Practice

▶ **Ex. 5–20** These exercises are modeled by Example 1.

▶ **Ex. 11–14** These exercises should be assigned as a group.

▶ **Ex. 21–36** These exercises are modeled by Example 2 (leading coefficient 1) and Example 3 (leading coefficient not 1).

▶ **Ex. 37–38** These exercises are modeled by Example 4.

▶ **Ex. 39–40 and 41–42** These exercises should be assigned as pairs.

▶ **Ex. 39–42** TECHNOLOGY Students can verify their results by using the TRACE function of their graphics calculators.

Answer

39. $h(x) = -\frac{1}{100}(x - 50)^2 + 33$
The vertex is (50, 33), so the maximum height is 33 ft.

▶ **Ex. 49–54** Remind students of the alternative technique for finding the vertex that was presented in Lesson 9.3.

▶ **Ex. 63** This exercise connects the symbolic completed square form of a function with a visual representation.

Enrichment Activities

These activities require students to go beyond lesson goals.

1. Justify that the x-coordinate of the vertex of the quadratic function in completed square form is h. (*Hint*: Expand the completed square form and recall that the x-coordinate of the vertex of the quadratic function $f(x) = ax^2 + bx + c$ can also be written as $-\frac{b}{2a}$.)
Completed square form
$f(x) = a(x - h)^2 + k$
Expand.
$f(x) = a(x^2 - 2hx + h^2) + k$
$f(x) = ax^2 - 2ahx + ah^2 + k$
Rewrite subtractions as additions.
$f(x) = ax^2 + (-2ah)x + ah^2 + k$
Substitute $-2ah$ for b and simplify.
$-\frac{(-2ah)}{2a} = h$
Thus, the x-coordinate of the vertex is h.

Integrated Review

In Exercises 43–48, solve the equation by completing the square.

43. $x^2 + 6x - 7 = 0$ $-7, 1$

44. $x^2 - 2x - 35 = 0$ $-5, 7$

45. $x^2 - 4x - 9 = 0$ $2 + \sqrt{13}, 2 - \sqrt{13}$

46. $x^2 + 12x - 16 = 0$ $-6 + \sqrt{52}, -6 - \sqrt{52}$

47. $4x^2 - 16x - 5 = 0$ $2 + \frac{1}{2}\sqrt{21}, 2 - \frac{1}{2}\sqrt{21}$

48. $3x^2 + 24x - 131 = 0$ $-4 + \sqrt{\frac{179}{3}}, -4 - \sqrt{\frac{179}{3}}$

In Exercises 49–54, find the vertex of the graph in two ways: by using shifts from the origin, as in Lesson 9.3, and by the method presented in this lesson. Which do you prefer and why? See Additional Answers.

49. $y = x^2 - 4x + 11$ (2, 7)

50. $y = x^2 + 18x + 90$ (−9, 9)

51. $y = x^2 + 2x - 2$ (−1, −3)

52. $y = x^2 - 8x + 22$ (4, 6)

53. $y = 3x^2 - 72x + 428$ (12, −4)

54. $y = 2x^2 - 4x + 3$ (1, 1)

In Exercises 55–60, sketch the graph of the equation. Use any method and explain why you chose the method. See Additional Answers.

55. $y = x^2 + 2x - 2$

56. $y = x^2 - 14x + 53$

57. $y = x^2 + 10x + 18$

58. $y = x^2 + 2x - 9$

59. $y = 2x^2 - 4x + 18$

60. $y = 4x^2 + 24x + 33$

Homeowners and Renters **In Exercises 61 and 62, use the circle graphs to answer the question. (Single-person households have one or more single people with *no* children living at home. Female or male households are families with single parents *and* children living at home.)**

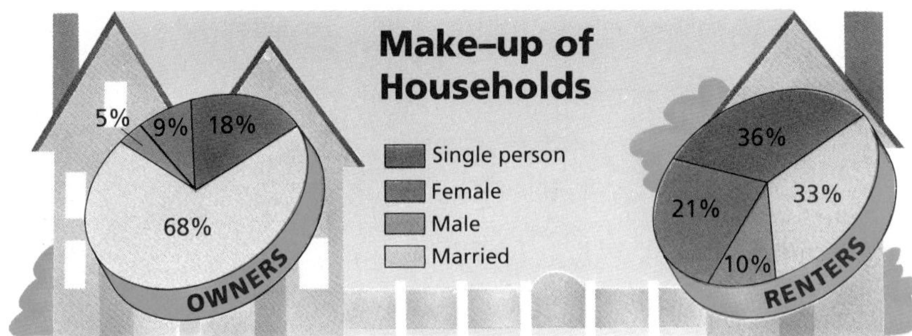

61. What is the probability that a randomly chosen homeowner is a female householder? $\frac{9}{100}$

62. What is the probability that a randomly chosen renter is a female householder? $\frac{21}{100}$

Exploration and Extension

63. Let a and b be positive numbers. Which two of the following functions have graphs that do not intersect the x-axis? Explain. b and c; b opens downward and the y of the vertex is negative, c opens upward and the y of the vertex is positive

a. $f(x) = -(x - a)^2 + b$

b. $f(x) = -(x - a)^2 - b$

c. $f(x) = (x - a)^2 + b$

d. $f(x) = (x - a)^2 - b$

Mid-Chapter SELF-TEST

Take this test as you would take a test in class. The answers to the exercises are given in the back of the book.

In Exercises 1 and 2, decide whether the table defines y as a function of x. (12.1)

1.

x	a	b	c	a	a
y	1	3	5	7	9

No

2.

x	g	h	i	j	k
y	9	3	0	4	5

Yes

In Exercises 3–8, evaluate the function at $x = -2$, $x = 0$, and $x = 2$. (12.1)

3. $f(x) = 3x + 2$ $-4, 2, 8$

4. $f(x) = -2x - 4$ $0, -4, -8$

5. $f(x) = x^2 - 1$ $3, -1, 3$

6. $f(x) = -3x^2 + 5$ $-7, 5, -7$

7. $f(x) = 4x^2 - 2x + 1$ $21, 1, 13$

8. $f(x) = -2x^2 + 4x + 6$ $-10, 6, 6$

In Exercises 9–11, find an equation for the linear function f. (12.2)

9. $f(0) = 3, f(3) = 9$ $f(x) = 2x + 3$

10. $f(-2) = 6, f(2) = -6$ $f(x) = -3x$

11. $f(-1) = 4, f(5) = 4$ $f(x) = 4$

In Exercises 12–14, how does the graph of g compare to the graph of $f(x) = 3^x$? (12.3)

12. $g(x) = 3^x + 2$ Vertical shift

13. $g(x) = 3^x - 4$ Vertical shift

14. $g(x) = 3^{x-2}$ Horizontal shift

In Exercises 15–17, write the function in completed square form. Then sketch its graph. (12.4) See Additional Answers.

15. $f(x) = x^2 - 4x + 9$

16. $f(x) = -x^2 - 2x - 4$

17. $f(x) = 2x^2 - 4x - 7$

18. The table shows the number of bowling establishments and the number of bowling lanes in use from 1984 to 1988 in the United States. Is the number of bowling lanes, L (in 1000's), a function of the number of bowling establishments, B? Yes

Year	1984	1985	1986	1987	1988
Establishments, *B*	8351	8275	8149	8031	7923
Lanes, *L*	155	155	153	151	150

19. While cartwheeling down a sand dune, a golden wheel spider can travel 5 feet per second. Find a linear function that gives the distance traveled as a function of the time. Use the function to find the distance traveled in 12 seconds. $f(x) = 5x$, 60 ft

20. A sheet of paper is 20 centimeters wide. When x centimeters of paper are folded up on each side, the area of the cross section is $A = x(20 - 2x)$ square centimeters. How many centimeters should be folded up to obtain a maximum cross-section area? 5

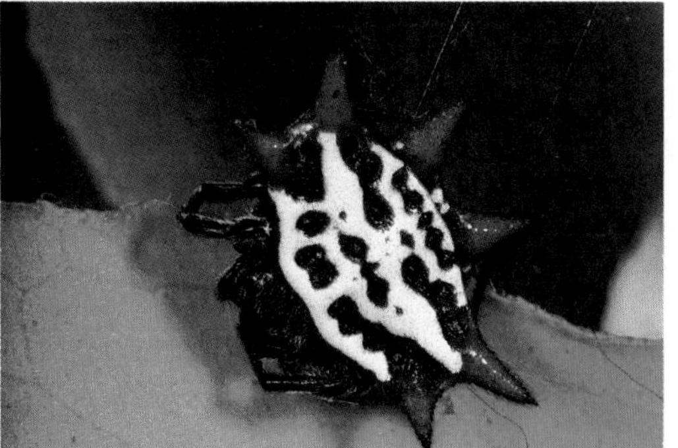

This is an orb weaver spider from Florida, which specializes in perfect webs.

The following activity provides students with an opportunity to discover a visual connection between the values of a, h, and k.

2. Write the equations for these graphs in completed-square form.

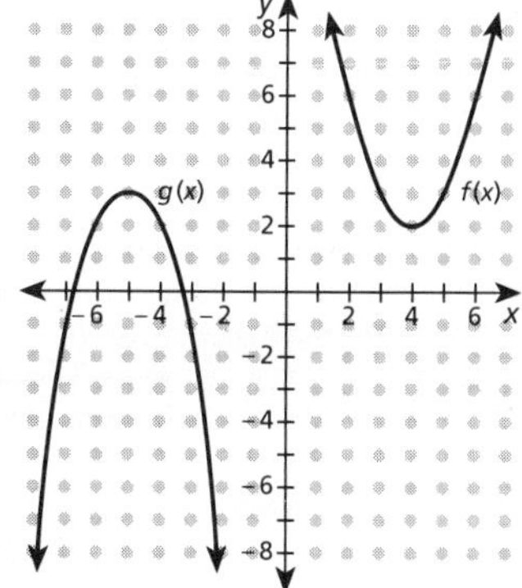

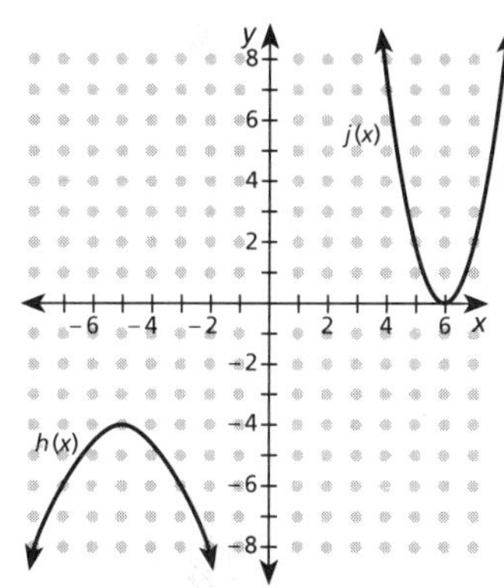

$f(x) = (x - 4)^2 + 2$;
$g(x) = -(x + 5)^2 + 3$;
$j(x) = 2(x - 6)^2$;
$h(x) = -\frac{1}{2}(x + 5)^2 - 4$

Problem of the Day

A prisoner was forced to serve 40 days under these conditions: for every day he worked breaking rocks he was to get \$20, but for every day he was idle he was to pay an \$8 fine. After 40 days he received \$380. How many days did he work, and how many days was he idle? Worked 25 days, was idle 15 days.

ORGANIZER

Warm-Up Exercises

1. Describe the behavior of the equation $y = \frac{1}{x}$ as $|x|$ grows large. Describe the behavior of y as $|x|$ grows small. As $|x|$ grows large the y-value gets closer to 0, but never assumes the value of 0. As $|x|$ grows small, $|y|$ gets large.
2. Describe the behavior of the equation $y = \frac{1}{x}$ as x approaches zero from the right and from the left. As x approaches 0 from the right, the y-value grows large; as x approaches 0 from the left, the y-value grows small.
3. Evaluate each equation at the given x-values.
   a. $y = \frac{1}{x}$ at $x = -2, 4, 0$
   b. $y = \frac{2}{x-3}$ at $x = -4, 3, 5$
   c. $y = \frac{4}{x+4}$ at $x = -4, 0, 4$

a. $\frac{1}{2}, \frac{1}{4}$, undefined b. $-\frac{2}{7}$, undefined, 1 c. undefined, 1, $\frac{1}{2}$

Lesson Resources

Teaching Tools
- Transparencies: 4, 15
- Problem of the Day: 12.5
- Warm-up Exercises: 12.5
- Answer Masters: 12.5

Extra Practice: 12.5

Color Transparency: 72

12.5 Rational Functions

What you should learn:

Goal 1 How to sketch the graph of a rational function

Goal 2 How to use rational functions as models of real-life problems

Why you should learn it:

You can use rational functions to model the change in the ratios of two quantities over a period of time.

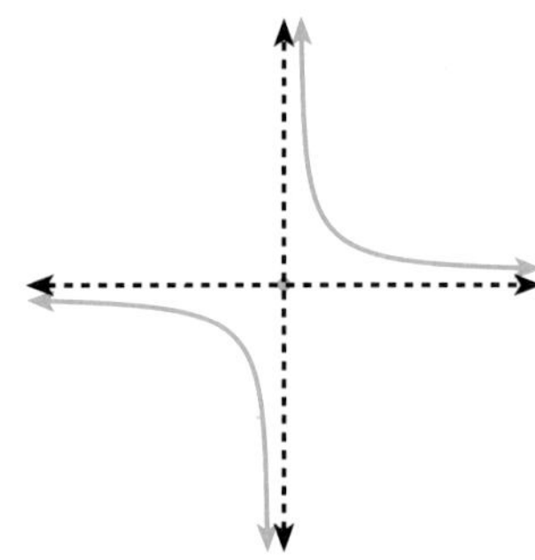

Hyperbola with asymptotes

Goal 1 Graphing a Rational Function

A **rational function** is a function of the form

$$f(x) = \frac{\text{polynomial}}{\text{polynomial}}.$$

In this lesson, you will study rational functions whose numerators and denominators are first-degree polynomials. Using long division, such a function can be written in the form

$$f(x) = \frac{a}{x-h} + k.$$

Rational Functions Whose Graphs Are Hyperbolas

The graph of the rational function

$$f(x) = \frac{a}{x-h} + k$$

is a **hyperbola** whose **center** is (h, k). The vertical and horizontal lines through the center are the **asymptotes** of the hyperbola.

The center and asymptotes of a hyperbola are used as sketching aids. They are *not* part of the graph.

Example 1 *Sketching the Graph of a Rational Function*

Sketch the graph of $f(x) = \frac{1}{x} + 2$.

Solution Think of the function as

$$f(x) = \frac{1}{x-0} + 2 = \frac{a}{x-h} + k. \quad f(x) = \frac{a}{x-h} + k$$

You can see that the center is (0, 2). The asymptotes can be drawn as dashed lines through the center. Construct a table of values to help locate several points on the graph. Choose a few x-values to one side of the x-coordinate of the center. (Because the hyperbola is symmetrical about the center, you need only plot points on one branch.)

x	0	0.25	0.5	1	2	4
$f(x) = \frac{1}{x} + 2$	Undefined	6	4	3	2.5	2.25

■

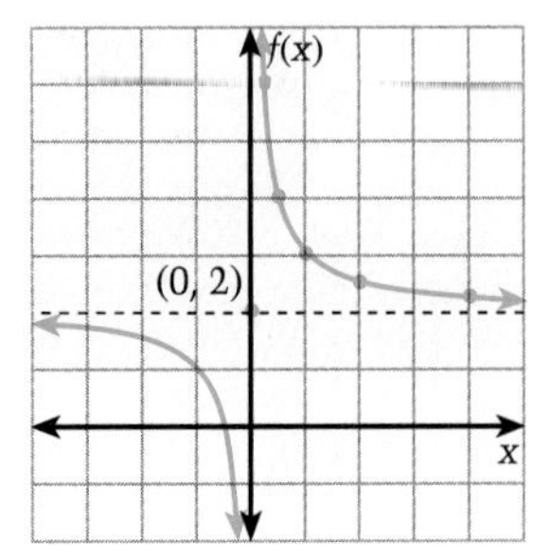

Example 2 *Graphing a Rational Function*

Sketch the graph of these rational functions.

a. $f(x) = \dfrac{-1}{x+2} + 1$ **b.** $f(x) = \dfrac{2}{x-3} - 2$

Solution

a. The graph of

$$f(x) = \frac{-1}{x+2} + 1$$

is a hyperbola with center at $(-2, 1)$.

x	−1.5	−1	0
f(x)	−1	0	0.5

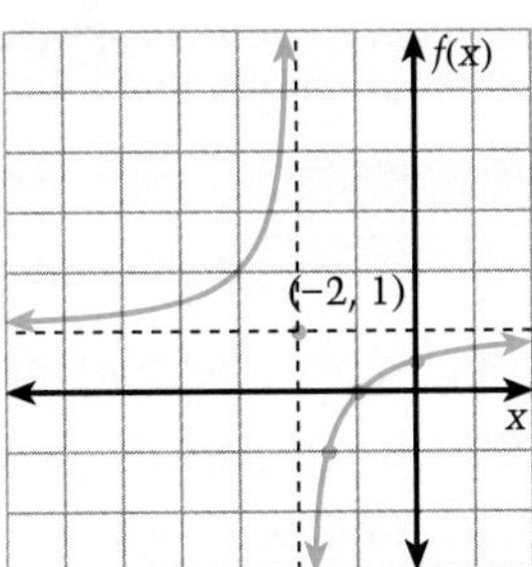

b. The graph of

$$f(x) = \frac{2}{x-3} - 2$$

is a hyperbola with center at $(3, -2)$.

x	3.5	4	5
f(x)	2	0	−1

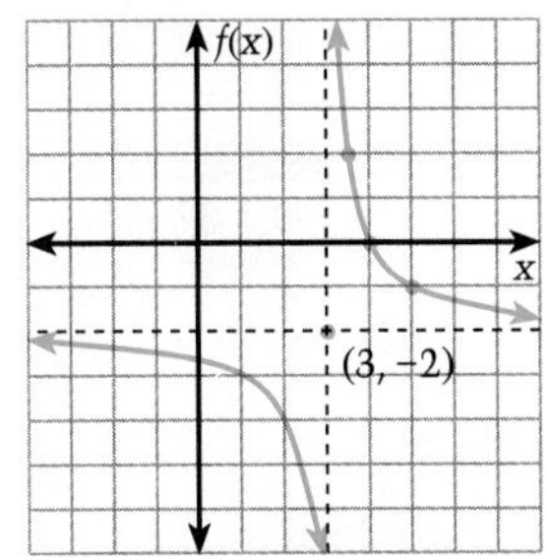

■

Polynomial long division can be used to write some rational functions in the form

$$f(x) = \frac{a}{x-h} + k.$$

Example 3 *Using Polynomial Long Division First*

Sketch the graph of $f(x) = \dfrac{2x+1}{x+2}$.

Solution Begin by dividing $2x + 1$ by $x + 2$.

$$\begin{array}{r} 2 \\ x+2\overline{)2x+1} \\ \underline{2x+4} \\ -3 \end{array}$$

From the form

$$f(x) = 2 - \frac{3}{x+2} \quad \text{or} \quad f(x) = \frac{-3}{x+2} + 2$$

you can see that the graph is a hyperbola with center at $(-2, 2)$. Some points on the hyperbola are shown in the table.

x	−1.5	−1	0	2
f(x)	−4	−1	0.5	1.25

■

LESSON Notes

GOAL 1 To make this lesson manageable, only linear polynomials are used to form rational functions. However, a rational function, in general, is the ratio of polynomials of any degree. For example,

$$f(x) = \frac{4x^5 - 3x^3 + 5x^2 - 2}{3x^4 + 13x^3 - x + 5}$$

is also a rational function.

Example 1

In the beginning, it is a good idea to ask students to locate points on both branches of the hyperbola. After they recognize the significance of the symmetry of the branches, then encourage them to locate points on only one branch.

Example 2

Show students that $x + 2$ can be written as $x - (-2)$ so $h = -2$ and $k = 1$.

Example 3

You may need to spend extra time on this example. Make sure students understand how to write the quotient.

Example 4

Observe that the sum of the *y*-intercepts for functions *A*, *I*, and *O* equals the *y*-intercept of the function *C*.

EXTEND Example 4 The ratio $\frac{A}{C}$ can be written in the form $\frac{A}{C} = \frac{a}{x-h} + k$. Divide $0.75t + 10.70$ into $0.34t + 6.75$. (Use a calculator, if available, to determine *h* and *k*.) The quotient is

$$\frac{1.8993}{0.75t + 10.70} + 0.4533,$$

or

$$\frac{\left(\frac{1.8993}{0.75}\right)}{t + 14.2666} + 0.4533$$

when written in the form $\frac{a}{x-h} + k$. So $\frac{A}{C} = \frac{2.5324}{t + 14.2666} + 0.4533$.

Compare the function $\frac{A}{C}$ and its graph presented here with the graph in the example.

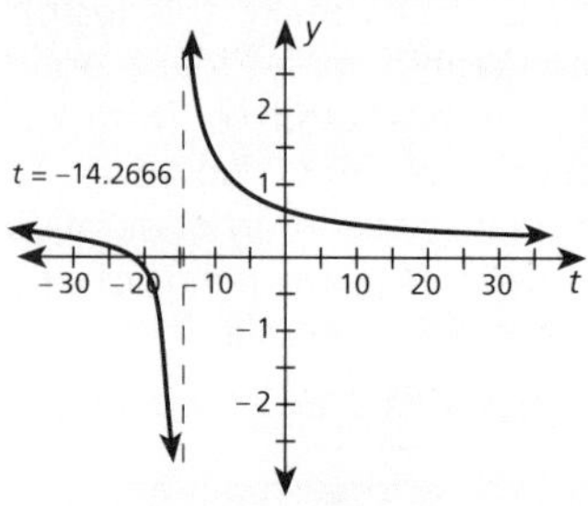

Extra Examples

Here are additional examples similar to **Examples 1–3.**

1. Identify the center of each rational function.
   a. $f(x) = \frac{3}{x} + 5$ (0, 5)
   b. $g(x) = -\frac{2}{x} + 1$ (0,1)
   c. $f(x) = \frac{3}{x+4} - 1$ (−4, −1)
   d. $h(x) = -\frac{1}{x-2} + 2$ (2, 2)
2. Sketch the graph of each rational function in Exercise 1 above. The graphs are all hyperbolas.
3. Rewrite these functions. Identify their centers.
   a. $g(x) = \frac{4x + 11}{x + 3}$
   b. $h(x) = \frac{-8x + 6}{x - 2}$

   a. $g(x) = \frac{-1}{x + 3} + 4$; (−3, 4)
   b. $h(x) = \frac{-10}{x - 2} - 8$; (2, −8)

Check Understanding

1. How do the center and asymptotes of a hyperbola help in sketching a graph of the rational function $f(x) = \frac{a}{x-h} + k$? See Example 1.
2. Why is it possible to plot points on only one branch of a hyperbola when sketching the graph of the rational function $f(x) = \frac{a}{x-h} + k$? The other branch is a reflection through the center (h, k).
3. What form does the general rational function take? $f(x) = \frac{\text{polynomial}}{\text{polynomial}}$

Goal 2 Using Rational Functions as Models

In 1970, Americans consumed a per capita average of about 11 pounds of cheese. By 1990, the average consumption had more than doubled.

Connections
Consumer Studies

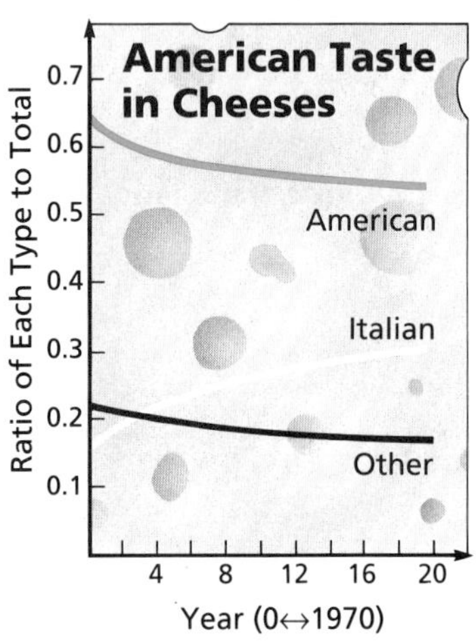

Example 4 Finding Ratios

For 1970 to 1990, the per capita consumption of cheese in the United States can be modeled as follows ($t = 0$ represents 1970).

Type of Cheese	Consumption Model
All types:	$C = 10.70 + 0.75t$ (pounds)
American (including cheddar):	$A = 6.75 + 0.34t$ (pounds)
Italian (including mozzarella):	$I = 1.60 + 0.31t$ (pounds)
Other (including cream cheese):	$O = 2.35 + 0.10t$ (pounds)

Use graphs to show how the ratio of each of the three consumptions to the total consumption was changing.
(Source: U.S. Department of Agriculture)

Solution The three ratios are as follows:

$$\frac{A}{C} = \frac{6.75 + 0.34t}{10.70 + 0.75t} \quad \frac{I}{C} = \frac{1.60 + 0.31t}{10.70 + 0.75t} \quad \frac{O}{C} = \frac{2.35 + 0.10t}{10.70 + 0.75t}$$

Notice that each graph is part of a branch of a hyperbola. ■

Communicating about ALGEBRA

SHARING IDEAS about the Lesson

See Additional Answers.

Sketch a Graph Sketch the graph of each function.

A. $f(x) = \frac{1}{x-1} + 1$ **B.** $f(x) = -\frac{2}{x+1} + 3$

C. $f(x) = \frac{3x+1}{x-2}$ **D.** $f(x) = \frac{3x-1}{x+1}$

EXERCISES

Guided Practice

CRITICAL THINKING about the Lesson

In Exercises 1–4, describe the center and asymptotes of the hyperbola. Explain the steps you used. See below.

1. $f(x) = \frac{4}{x+5} - 3$ $(-5, -3)$; $x = -5$, $y = -3$
2. $f(x) = \frac{-10}{x-10} + 4$ $(10, 4)$; $x = 10$, $y = 4$
3. $f(x) = \frac{-1}{x-3} + 16$ $(3, 16)$; $x = 3$, $y = 16$
4. $f(x) = \frac{3}{x+1} - 9$ $(-1, -9)$; $x = -1$, $y = -9$
5. Sketch the graph of $f(x) = \frac{4}{x}$.
6. Sketch the graph of $f(x) = \frac{1}{x+1} + 2$.

5., 6. See Additional Answers.

Independent Practice

In Exercises 7–12, find the center and asymptotes of the hyperbola.

7. $f(x) = \frac{1}{x-5} + 2$ $(5, 2)$, $x = 5$, $y = 2$
8. $f(x) = \frac{1}{x-3} - 8$ $(3, -8)$, $x = 3$, $y = -8$
9. $f(x) = \frac{2}{x-4} + 6$ $(4, 6)$, $x = 4$, $y = 6$
10. $f(x) = -\frac{3}{x+1} + 8$ $(-1, 8)$, $x = -1$, $y = 8$
11. $f(x) = -\frac{6}{x+9} - 7$ $(-9, -7)$, $x = -9$, $y = -7$
12. $f(x) = -\frac{3}{x+1} - 4$ $(-1, -4)$, $x = -1$, $y = -8$

In Exercises 13–18, match the equation with its graph.

13. $f(x) = \frac{1}{x-6} - 1$ c
14. $f(x) = -\frac{1}{x+7} - 1$ f
15. $f(x) = \frac{-2}{x+3} + 7$ d
16. $f(x) = \frac{2}{x-6} + 9$ a
17. $f(x) = \frac{1}{x-9} - 6$ b
18. $f(x) = \frac{1}{x-1} - 7$ e

a.
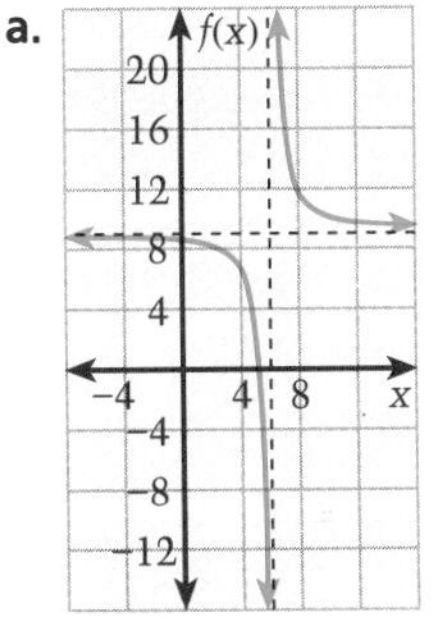

b.
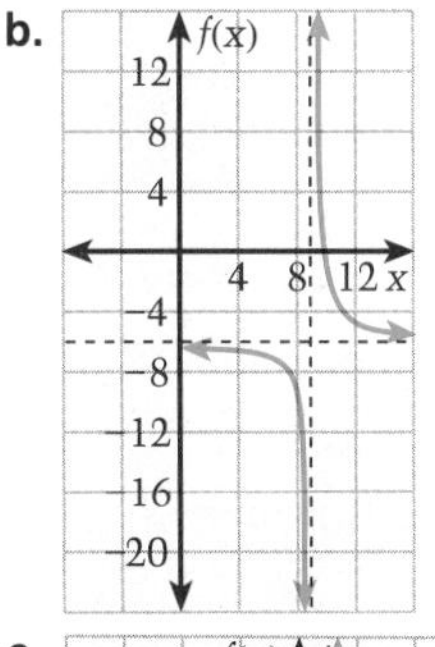

c.
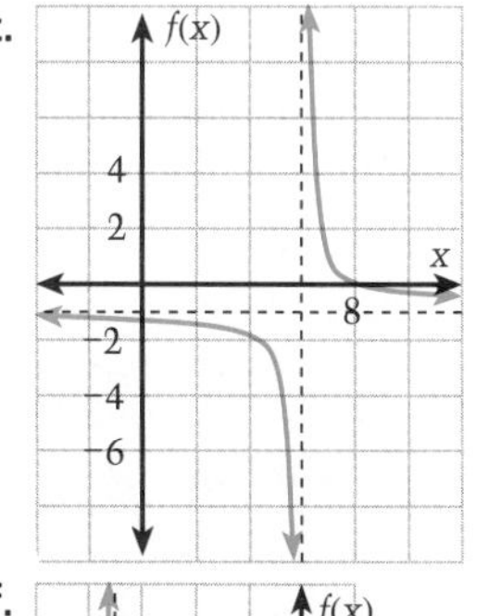

d.
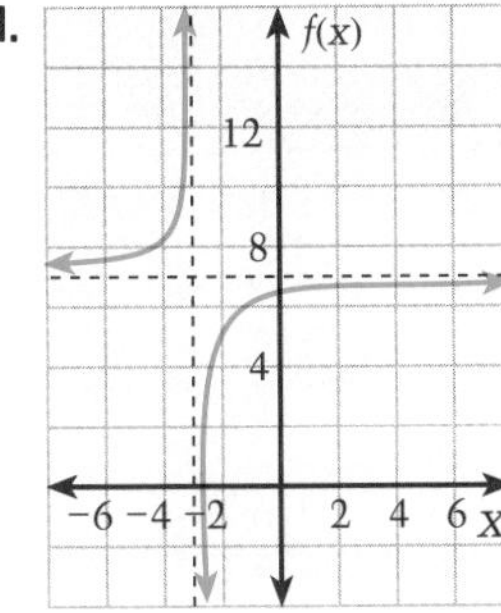

e.
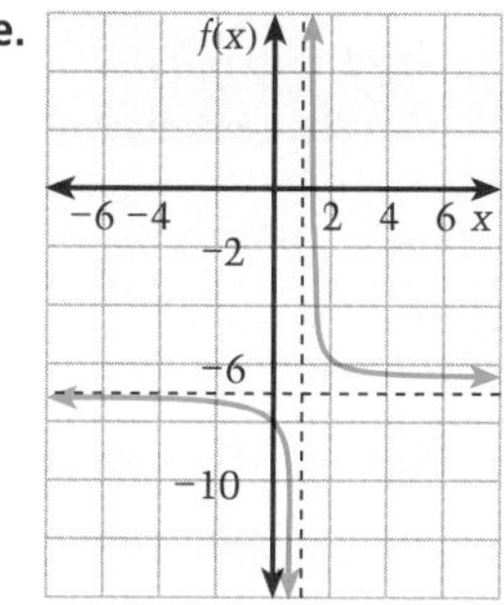

f.
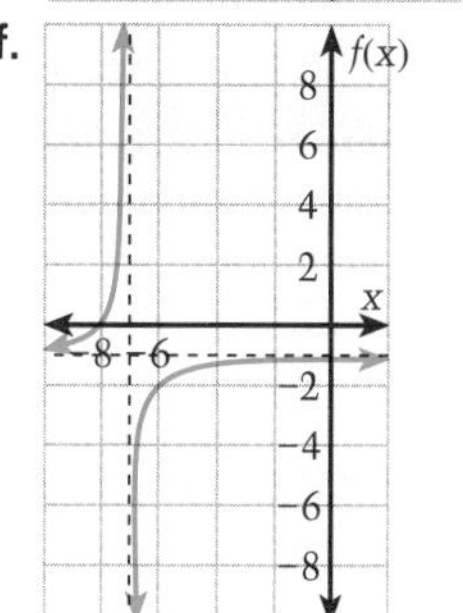

1.–4. $f(x) = \frac{a}{x-h} + k$ is a hyperbola whose center is (h, k) and whose asymptotes are the vertical and horizontal lines through the center.

Communicating about ALGEBRA

Ask students to investigate the effect of adding a constant k to $f(x) = \frac{1}{x}$ in terms of horizontal and vertical shifts in the basic function.

a. $g(x) = \frac{1}{x} - 5$

b. $g(x) = \frac{1}{x} + 3$

c. $g(x) = \frac{1}{x} - 1$

a.–c. All vertical shifts.

a. down 5

b. up 3

c. down 1

Next consider the effect of subtracting a constant h from the variable x in the basic function $f(x) = \frac{1}{x}$.

a. $h(x) = \frac{1}{x-5}$

b. $h(x) = \frac{1}{x+4}$

c. $h(x) = \frac{1}{x+6}$

a.–c. All horizontal shifts.

a. 5 right

b. 4 left

c. 6 left

EXERCISE Notes

ASSIGNMENT GUIDE

Basic/Average: Ex. 1–6, 7–11 odd, 13–18, 25–33 odd, 47–52

Above Average: Ex. 1–6, 7–11 odd, 13–18, 23–33 odd, 39–40, 47–52

Advanced: Ex. 1–6, 13–18, 19–35 odd, 37–40, 47–52

Selected Answers
Exercises 1–6, 7–49 odd

Use **Mixed Review** as needed.

More Difficult Exercises
Exercises 37–40, 51–52

Guided Practice

▶ **Ex. 1–6** These exercises help students make the connection between the h and k for rational functions and the h and k studied previously for exponential functions.

Independent Practice

▶ **Ex. 13–18** These exercises should be assigned as a group.

▶ **Ex. 23–24** These exercises are modeled by Example 2. Caution students regarding the negative sign.

▶ **Ex. 31–36** Students who have difficulty with long division could review Lesson 11.7 on Dividing Polynomials.

▶ **Ex. 37–38 and 39–40** Assign these exercises in pairs.

EXTEND Ex. 37 and 40
TECHNOLOGY For some students, you might wish to provide the graph before assigning the problems. Students with graphics calculators could plot several points and try to observe a pattern similar to the model in Example 4.

▶ **Ex. 47–50** These exercises review each of the four different types of functions studied thus far.

Answers

37.

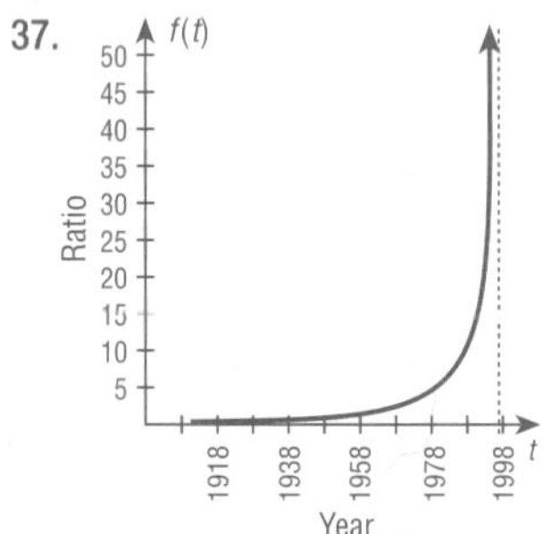

$t = 1997$; the year the lease was up

40.

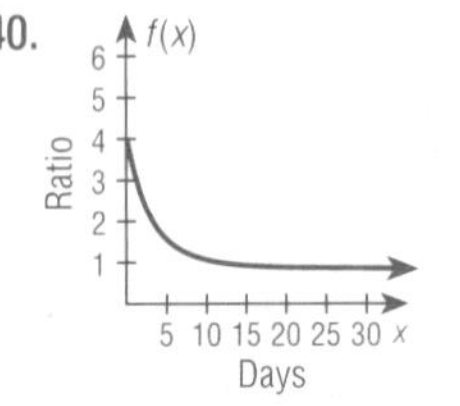

In Exercises 19–30, sketch the graph of the function. See Additional Answers.

19. $f(x) = \frac{1}{x+4} + 2$
20. $f(x) = \frac{1}{x+7} + 3$
21. $f(x) = \frac{3}{x-1} + 4$
22. $f(x) = \frac{2}{x-2} + 9$
23. $f(x) = -\frac{2}{x-5} - 4$
24. $f(x) = -\frac{1}{x+9} + 1$
25. $f(x) = \frac{2}{x+9} - 4$
26. $f(x) = -\frac{3}{x-3} - 2$
27. $f(x) = \frac{1}{x-7} + 4$
28. $f(x) = \frac{2}{x+8} - 8$
29. $f(x) = \frac{5}{x-5} - 3$
30. $f(x) = \frac{4}{x+4} + 5$

In Exercises 31–36, use polynomial division to help sketch the graph of the function. See Additional Answers.

31. $f(x) = \frac{-2x+11}{x-5}$
32. $f(x) = \frac{x+10}{x+8}$
33. $f(x) = \frac{-9x+16}{x-2}$
34. $f(x) = \frac{9x-6}{x-1}$
35. $f(x) = \frac{-5x+19}{x-3}$
36. $f(x) = -\frac{6x-50}{x+8}$

Hong Kong **In Exercises 37 and 38, use the following information.**

In 1898, Britain signed a lease with China. In the lease, Britain was allowed control of Hong Kong for a certain number of years. Let t represent any of the years between 1898 and 1997, the year the lease is up. The ratio, $f(t)$, of the years leased to the years yet to lease is given by

$$f(t) = \frac{t-1898}{1997-t}.$$

1997, the year the lease is up; see margin for graph.

✪ 37. Sketch the graph of the function f on the interval $1898 \leq t < 1997$. What is the vertical asymptote of the graph? In this problem, what does it represent?

✪ 38. During which year was $f(t)$ equal to 1? What was the significance of this year? 1947, half of the time of the lease

Hong Kong is known for its many shops. Over two million tourists visit Hong Kong each year.

Soaking Crabs **In Exercises 39 and 40, use the following information.**

You and your sister are visiting Christmas Island (near Australia) during the crab migration. The weather is hot, and you decide to set up a crab rescue station. At the rescue station, you soak the crabs in water, then release them. Your sister collects 5 crabs the first day and you collect 20. Each day after that your sister collects 2 more than the day before and you collect 1 more.

✪ 39. Create a model that represents the ratio, $f(x)$, of the number of crabs you collect each day to the number of crabs your sister collects. (Let $x = 0$ represent the first day.) $f(x) = \frac{20+x}{5+2x}$

✪ 40. Your rescue station is open for 30 days. Sketch a graph of the function f. See margin.

Integrated Review

In Exercises 41–43, simplify the quotient.

41. $\frac{x^2 - 2x - 3}{4} \div \frac{x - 3}{2}$ $\frac{x+1}{2}$

42. $\frac{x^2 + 3x - 10}{14} \div \frac{x + 5}{2}$ $\frac{x-2}{7}$

43. $\frac{1}{x^2 - 4} \div \frac{x + 2}{x - 2}$ $\frac{1}{(x+2)^2}$

In Exercises 44–46, solve the equation.

44. $\frac{1}{x^2} = \frac{1}{4}$ 2, −2

45. $\frac{x - 3}{4} = \frac{1}{x}$ 4, −1

46. $\frac{1}{x} - \frac{5}{x - 5} = 16$ $\frac{19}{8} + \frac{\sqrt{5456}}{32}, \frac{19}{8} - \frac{\sqrt{5456}}{32}$

In Exercises 47–50, match the equation with its graph.

47. $f(x) = \frac{2x - 3}{x + 1}$ b

48. $f(x) = x^2 - 4x + 5$ c

49. $f(x) = \frac{1}{3}x$ d

50. $f(x) = \left(\frac{5}{3}\right)^x - 1$ a

a.
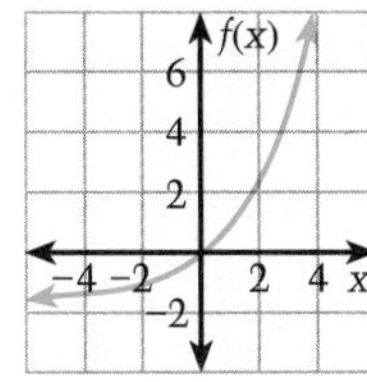
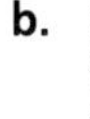
b.
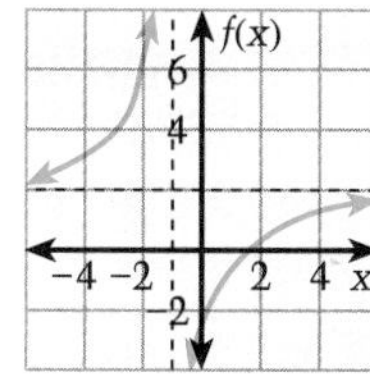

c.
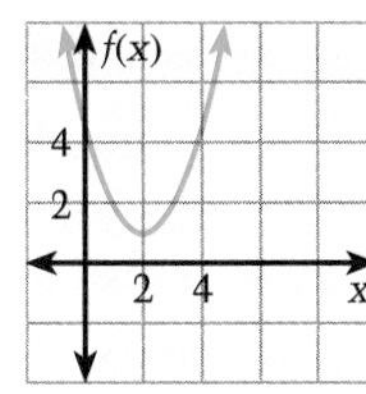
d.
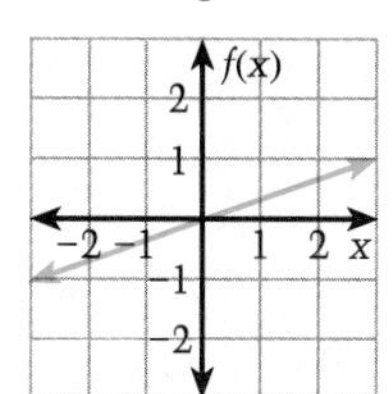

Exploration and Extension

Government Payroll **In Exercises 51 and 52, use the following information.**

For 1980 through 1990, the total government payroll (in millions of dollars), can be modeled by $T = 1836t + 1954$, where $t = 0$ represents 1980. The total payroll of the states in millions of dollars can be modeled by $S = 431.5t + 4189$.

51. Find a model that represents the ratio of the state payroll to the total government payroll. Then use a graphing calculator to sketch the graph of the model. See Additional Answers for graph. $f(t) = \frac{431.5t + 4189}{1836t + 1954}$

52. Estimate the horizontal asymptote of the graph. What does it represent? $y \approx \frac{1}{4}$, ratio of state payroll to federal payroll.

Mixed REVIEW

See Additional Answers.

1. Find the reciprocal of $2x^2y^{-1}$. **(8.2)**
2. Solve $\frac{1}{x} - \frac{2x}{3} = \frac{5}{3}$. **(11.8)**
3. Evaluate $3^n + 4n$ when $n = 3$. **(1.4)**
4. Solve $3|4 - m| \le 3$. **(6.4)**
5. Simplify $(x^2y^{-2})^{-1} \div 3(xy^4)^2$. **(8.2)**
6. What is 69% of 43?
7. Evaluate $(6.7 \times 10^{-3}) \div (2.0 \times 10^5)$. **(8.4)**
8. Find the vertex: $y = x^2 + 6x - 25$. **(9.3)**
9. Solve $512 = 8(1 + p)^2$. **(9.2)**
10. Solve the system: $\begin{cases} 2x - 3y = 4 \\ -2x - 4y = 10 \end{cases}$ **(7.3)**

EXTEND Ex. 52 In light of the functions discussed in Exercises 37 and 39, what does the horizontal asymptote represent? The eventual percent of the total government payroll that is the total payroll of the states. It is the ratio of the slopes of the two linear equations.

Enrichment Activities

These activities require students to go beyond lesson goals.

Here is the graph of the square root function. For all $x \ge 0$, $f(x) = \sqrt{x}$.

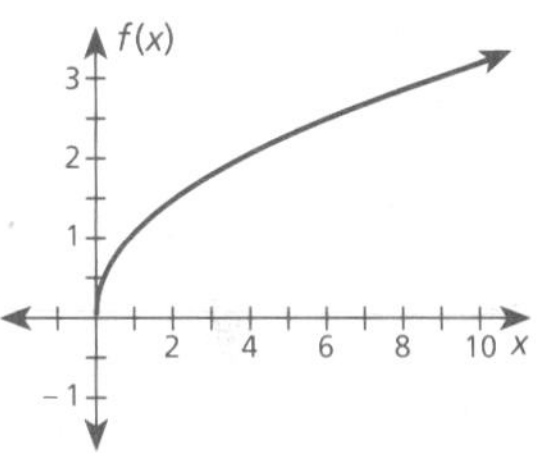

Extend your quick-graph techniques. What are the relationships between the graphs of the functions below and the graph of $f(x) = \sqrt{x}$? What are the domain and range of each function?

a. $f(x) = \sqrt{x - 2}$
b. $f(x) = \sqrt{x} - 2$
c. $f(x) = \sqrt{2 - x}$
d. $f(x) = -\sqrt{x - 2}$

a. Domain: $x \ge 2$; Range: $f(x) \ge 0$; horizontal shift of $f(x) = \sqrt{x}$ two units to the right

b. Domain: $x \ge 0$; Range: $f(x) \ge -2$; vertical shift of $f(x) = \sqrt{x}$ two units down

c. Domain: $x \le 2$; Range: $f(x) \ge 0$; a reflection of $f(x) = \sqrt{x - 2}$ about the line $x = 2$

d. Domain: $x \ge 2$; Range: $f(x) \le 0$; a reflection of $f(x) = \sqrt{x - 2}$ about the x-axis.

Problem of the Day

The sum, product, and quotient of two numbers are all equal. What are the numbers? $\frac{1}{2}$ and -1

ORGANIZER

Warm-Up Exercises

1. Find the average, or mean, of each set of numbers. Is each mean a member of its data set?
   a. 3, 4, 5, 3, 2, 1, 1, 2, 3, 6, 9, 3, 6, 8, 8, 8, 10, 2, 4, 1
   b. 1.3, 4.3, 3.2, 1.5, 2.3, 2.7, 2.3, 1.7, 3.0, 4.0, 3.3, 1.2
   c. 45, 34, 12, 56, 36, 41, 30, 10, 39, 38, 36, 36, 23, 32, 12, 58

   a. 4.45, b. 2.57, c. 33.625; None of the means is a member of its own set.
2. Rearrange the sets in Exercise 1 above in increasing order.
   a. 1, 1, 1, 2, 2, 2, 3, 3, 3, 3, 4, 4, 5, 6, 6, 8, 8, 8, 9, 10
   b. 1.2, 1.3, 1.5, 1.7, 2.3, 2.3, 2.7, 3.0, 3.2, 3.3, 4.0, 4.3
   c. 10, 12, 12, 23, 30, 32, 34, 36, 36, 36, 38, 39, 41, 45, 56, 58
3. a. What is 25% of 86?
   b. What is 75% of 125?
   c. What is 50% of 34?

   a. 21.5, b. 93.75, c. 17

Lesson Resources

Teaching Tools
- Transparency: 4
- Problem of the Day: 12.6
- Warm-up Exercises: 12.6
- Answer Masters: 12.6

Extra Practice: 12.6

Technology Handbook: p. 87

12.6 Exploring Data: Stem-and-Leaf Plots, Box-and-Whisker Plots

What you should learn:

Goal 1 How to construct a stem-and-leaf plot for data

Goal 2 How to construct a box-and-whisker plot for data

Why you should learn it:

Stem-and-leaf plots can help you organize large sets of data, such as the salaries of professional athletes.

Goal 1 Constructing Stem-and-Leaf Plots

A **stem-and-leaf plot** is a technique for ordering data in increasing order or decreasing order. In the following example, the *stem* consists of the tens digits for the data and the *leaves* consist of the units digits.

Unordered Data	Stem	Leaves	Ordered Data
52, 35, 26, 7, 31,	6	0 3 2 4	64, 63, 62, 60
31, 60, 14, 24, 54,	5	2 4 8 0	58, 54, 52, 50
15, 19, 24, 63, 58,	4	1 4 7 5 2	47, 45, 44, 42, 41
41, 23, 39, 8, 12,	3	5 1 1 9 4	39, 35, 34, 31, 31
18, 34, 62, 6, 44,	2	6 4 4 3 6 9 8	29, 28, 26, 26, 24, 24, 23
47, 45, 26, 29, 50,	1	4 5 9 2 8 7 1	19, 18, 17, 15, 14, 12, 11
17, 11, 64, 28, 42	0	7 8 6	8, 7, 6

Stem-and-Leaf Plot (Stem, Leaves)

Real Life Golf

Example 1 *Using a Stem-and-Leaf Plot*

The top 15 PGA Senior Tour golf winners for 1990 (through November 1) are shown in the table. Use a stem-and-leaf plot to order the winnings. *(Source:* **USA Today***)*

Winnings in 1000's	
George Archer	$640
Bob Charles	$460
Charles Coody	$670
Bruce Crampton	$420
Jim Dent	$640
Dale Douglass	$450
Al Gelberger	$310
Harold Henning	$340
Dave Hill	$320
Mike Hill	$680
Rives McBee	$430
Jack Nicklaus	$340
Gary Player	$490
Chi Chi Rodriguez	$630
Lee Trevino	$950

Solution To construct a stem-and-leaf plot, consider only the first two digits of each number.

9	5
6	4 7 4 8 3
4	6 2 5 3 9
3	1 4 2 4

Key: 9 | 5 represents $950,000.

Now, using the stem-and-leaf plot, order the winnings (in thousands) as follows.

$950	$680	$670	$640	$640
Trevino	M. Hill	Coody	Archer	Dent
$630	$490	$460	$450	$430
Rodriguez	Player	Charles	Douglass	McBee
$420	$340	$340	$320	$310
Crampton	Henning	Nicklaus	D. Hill	Gelberger

Goal 2 Constructing Box-and-Whisker Plots

You already know how to find the *average*, or **mean,** of n numbers: Add the numbers and divide the sum by n. It is one of the ways to represent the "middle" or "center" of a collection of numbers. (We will say more about measures of *central tendency* in the next lesson.) Another way to represent the middle of a collection of numbers is with a number called the *median.*

> **The Median of a Collection of Numbers**
>
> 1. The **median** of an odd collection of numbers, arranged *in order,* is the middle number.
> 2. The **median** of an even collection of numbers, arranged *in order,* is the average of the two middle numbers.

Example 2 *Finding the Median*

a. The median of 2, 4, 4, 5, $\underbrace{6}_{\text{Middle}}$, 8, 8, 9, 10 is 6.

b. The median of 1, 1, 3, 4, $\overbrace{4, 5}$, 6, 6, 7, 8 is $\frac{9}{2}$. ■

The median of an ordered collection of numbers roughly partitions the collection into two halves: those below the median and those above. The **first quartile** is the median of the lower half. The **second quartile** is the median of the entire collection. The **third quartile** is the median of the upper half. Quartiles partition an ordered collection into four **quarters.**

Example 3 *Finding the Quartiles*

The 15 golf winnings in Example 1 can be partitioned into quarters by the quartiles 340 (1st), 460 (2nd), and 640 (3rd).

1st Quarter:	310,	320,	340,	$x \le 340$
2nd Quarter:	420,	430,	450,	$340 \le x \le 460$
3rd Quarter:	490,	630,	640,	$460 \le x \le 640$
4th Quarter:	670,	680,	950	$640 \le x$

■

A **box-and-whisker plot** can be used to present a graphic summary of information shown, like that in Example 3.

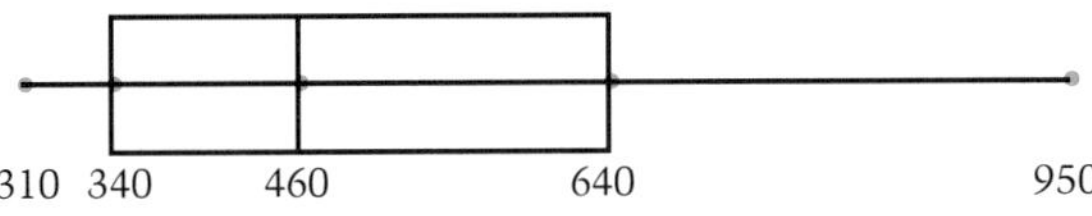

LESSON Notes

Example 1

An increasing order of the top 15 winners could be used if desired. Ask students to indicate why the stem-and-leaf representation of the data may be preferred over the list of winners in the table. (One reason is the ease with which information can be identified in the stem-and-leaf format.)

Example 2

Obtaining the median requires the set of numbers to be ordered. For Warm-Up Exercise 2, determine the median of each set.

a. $\frac{7}{2}$, b. $\frac{5}{2}$, c. 36

Example 3

Notice that since there are 15 entries, the quartiles appear in the data set. With other numbers of entries, some quartiles may not appear in the data set. Observe that the leftmost plot, 310, is the smallest value in the set, and the rightmost value, 950, is the largest number in the set. The other three values in the box-and-whisker plot are the quartiles, as discussed above.

Example 4

Here are other questions that can be asked related to this example.

a. What percent of 18–24-year-old males are shorter than Aaron? about 50%

b. What percent of 18–24-year-old females are shorter than Ann? about 75%

c. About how many 18–24-year-old males are less than 5′7″ in height? about 25%

d. About how many 18–24-year-old females are between 5′4″ and 5′6″ in height? about 25%

Extra Examples

Here are additional examples similar to **Examples 1–3.**

1. The 20 busiest United States airports in 1989 are shown in the table. Use a stem-and-leaf plot to order the number of passengers. (*Source: Air Transport Association of America*)

   Passengers arriving at and departing from U.S. airports in 1989 (in millions)

Airport	Number of Passengers
Atlanta	43.3
Boston	23.3
Chicago O'Hare	59.1
Dallas/Ft. Worth	47.6
Denver	27.6
Detroit	21.5
Honolulu	22.6
Houston	16.0
Las Vegas	16.7
Los Angeles	45.0
Miami	23.4
Minneapolis/St. Paul	19.4
New York (JFK)	30.3
New York (LGA)	23.2
Newark	20.9
Orlando	17.2
Phoenix	20.7
Pittsburgh	17.1
San Francisco	29.9
St. Louis	20.0

Key: 1|94 represents 19.4

5|91
4|76 50 33
3|03
2|99 76 34 33 32 26 15 09 07 00
1|94 72 71 67 60

2. The busiest airports given in Exercise 1 above can be partitioned using quartiles. Find the quartiles and construct a box-and-whisker plot of the data. 1st quartile = 19.7, median = 22.9, 3rd quartile = 30.1

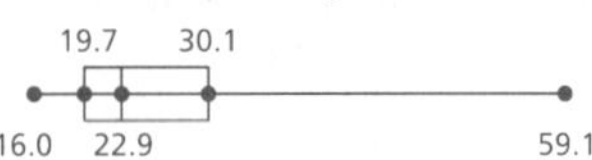

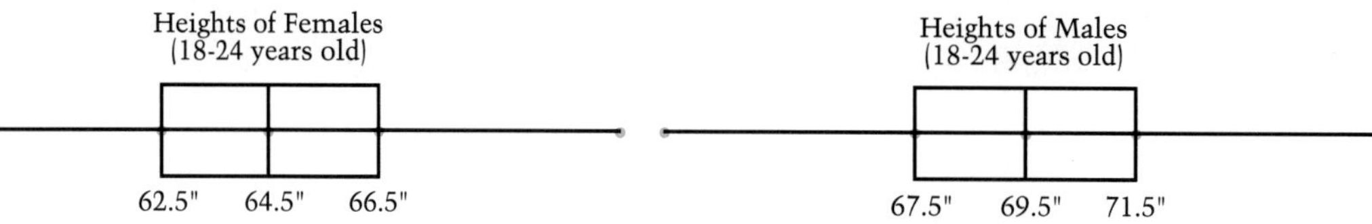

The box-and-whisker plots represent the heights of American females and males. (Source: U.S. Center for Health Statistics)

Real Life
Human Heights

Example 4 *Reading a Box-and-Whisker Plot*

Aaron and Ann are twins who will be starting college in the fall. Aaron is 5′10″ tall and Ann is 5′7″ tall.

a. What percent of the men at college are taller than Aaron?

b. What percent of the women at college are taller than Ann?

Solution

a. Aaron is 70 inches tall. For American males between 18 and 24 years of age, this is $\frac{1}{2}$ inch above the median. Less than 50% of the men are taller than Aaron.

b. Ann is 67 inches tall. This is $\frac{1}{2}$ inch above the third quartile. Less than 25% of the women are taller than Ann. ■

Communicating about ALGEBRA

SHARING IDEAS about the Lesson

The book *Life in America's Small Cities* rated 219 United States towns in several categories. The list shows the "diversions" ratings (on a scale from 0 to 100) for towns in North Carolina, South Carolina, and Georgia (diversions are shopping, dining, theaters, recreation). Make a stem-and-leaf plot *and* a box-and-whisker plot for these data. (High numbers indicate more diversions.) See Additional Answers.

Town	Rating	Town	Rating
Brunswick, GA	59	New Bern, NC	29
Greenwood, SC	30	Rocky Mount, NC	33
Hinesville, GA	22	Statesville, NC	30
Myrtle Beach, SC	75	Wilson, NC	36
Rome, GA	28	Greenville, NC	37
Shelby, NC	23	Hilton Head, SC	47
Valdosta, GA	43	Lumberton, NC	24
Goldsboro, NC	25	Orangeburg, SC	26
Havelock, NC	29	Sanford, NC	41
Kinston, NC	38	Sumter, SC	22

EXERCISES

Guided Practice

CRITICAL THINKING about the Lesson See top of page 652.

1. Why might you construct a stem-and-leaf plot for a data collection?
2. Describe the mean and median of a collection of numbers. Give an example in which they are the same. Give an example in which they are different. See top of page 653; same: 4, 5, 6; different: 1, 4, 10
3. Copy the data listing shown at the right. Label the first, second, third, and fourth *quarters* of the data collection. Label the first, second, and third *quartiles* of the data collection. Which quartile is equal to the median? Second
4. Construct a stem-and-leaf plot for the collection. Use your result to construct a box-and-whisker plot. See Additional Answers for box-and-whisker plot.

   55, 43, 12, 38, 49, 17, 32, 12, 39, 51, 44, 25, 46, 29, 20, 51, 24, 32, 13, 54

4. Stem	Leaves
5	1 1 4 5
4	3 4 6 9
3	2 2 8 9
2	0 4 5 9
1	2 2 3 7

[?] 1st [?] 2nd [?] 3rd

2, 2, 3, 4, 4, 5, 6, 6, 7, 8, 8, 9

[?] 1st [?] 2nd [?] 3rd [?] 4th

Independent Practice

In Exercises 5–8, find the median of the collection of numbers.

5. 16, 7, 25, 0, 3, 3, 21, 8 7.5
6. 150, 156, 160, 149, 150, 152 151
7. 1, 12, 15, 3, 11, 8, 12, 19, 4, 9, 10, 13, 7 10
8. 6.35, 4.40, 3.56, 3.56, 4.68 4.40

In Exercises 9 and 10, construct a stem-and-leaf plot for the data. Use the result to list the data in increasing order. See below.

9. 48, 10, 48, 25, 40, 42, 44, 23, 21, 13, 50, 17, 18, 19, 21, 35, 33, 25, 50, 13, 12, 46, 57, 41, 13, 59, 31, 48, 26, 15
10. 85, 61, 55, 78, 79, 86, 30, 76, 76, 87, 68, 82, 61, 84, 52, 33, 30, 68, 89, 70, 37, 42, 80

In Exercises 11–14, construct a box-and-whisker plot for the data. See Additional Answers.

11. 48, 60, 40, 68, 51, 47, 57, 41, 65, 61, 20, 65, 49, 34, 63, 53, 52, 35, 45, 35, 65, 65, 48, 36, 24, 53, 64, 48, 40, 22, 66, 66
12. 167, 191, 190, 154, 188, 174, 192, 166, 180, 155, 157, 161, 163, 172, 169, 167, 182, 184, 158, 160, 176, 192, 174, 161

13.
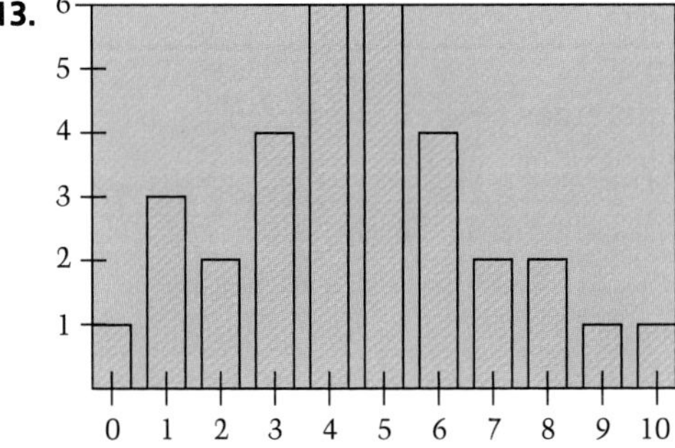

14.
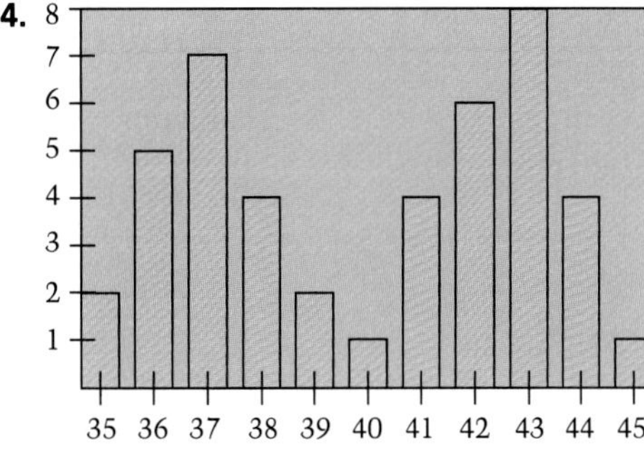

9. 10, 12, 13, 13, 13, 15, 17, 18, 19, 21, 21, 23, 25, 25, 26, 31, 33, 35, 40, 41, 42, 44, 46, 48, 48, 48, 50, 50, 57, 59
10. 30, 30, 33, 37, 42, 52, 55, 61, 61, 68, 68, 70, 76, 76, 78, 79, 80, 82, 84, 85, 86, 87, 89

Check Understanding

1. Define the median of a collection of data. The middle number
2. What are quartiles? How can they be used to organize data? See Example 2.
3. Compare the usefulness of an ordered stem-and-leaf plot with that of a box-and-whisker plot.

Communicating about ALGEBRA

Ask students to decide which plot (stem-and-leaf or box-and-whisker) of the data they prefer. They must provide a rationale for their selection.

EXERCISE Notes

ASSIGNMENT GUIDE

Basic/Average: Ex. 1–4, 5–13 odd, 19–26

Above Average: Ex. 1–4, 5–19 odd, 20–27

Advanced: Ex. 1–4, 5–19 odd, 21–27

Selected Answers
Exercises 1–4, 5–25 odd

More Difficult Exercises
Exercises 15–18, 21–27

Guided Practice

▶ **Ex. 1–4** Sometimes the first and third quartiles are referred to as the lower and upper quartiles, respectively.

Independent Practice

▶ **Ex. 5–18** Remind students to order the data first. In Exercises 13 and 14, remind students that the height of the bar graph represents the frequency values for a specific data point.

Ex. 23–27 These exercises connect previous techniques for organizing data, such as bar graphs, to the techniques presented in this lesson. Assign Exercises 23–26 as a group. Remind students to show a key that indicates the value of the stem-and-leaf plot. For example, in Exercise 23, the value 4|0 would represent 4.0, while in Exercise 24, the value 4|0 would represent 40.

Answers

15. Answers vary; 10, 11, 12, 13, 13, 14, 15, 16, 22, 23, 24, 25, 29, 32, 35, 38
16. Answers vary; 106, 112, 118, 124, 124, 132, 140, 148, 152, 155, 158, 161, 163, 173, 183, 193
17. Answers vary; 81, 84, 87, 90, 94, 94, 94, 94, 96, 96, 97, 97, 99, 104, 109, 114
18. Answers vary; 284, 290, 296, 302, 302, 309, 316, 323, 327, 345, 363, 381, 383, 390, 397, 404
19. 1–3, 1–24, 1–25, 2–7, 2–22, 3–17, 4–1, 4–14, 4–15, 4–30, 4–30, 5–10, 6–5, 6–13, 7–31, 8–21, 8–26, 9–12, 10–11, 10–17, 11–11, 12–9, 12–15, 12–28
20. 98, 99, 120, 121, 122, 126, 129, 133, 136, 139, 150, 155
21. 50.7, 34.7, 28.7, 26.3, 25.8, 15.2, 10.7, 9.7, 9.6, 8.5, 6.9, 5.4, 4.9, 4.5, 3.6, 3.6, 2.2, 1.6, 1.0
22. 13.2, 9.5, 6.8, 5.1, 3.9, 3.2, 2.9, 2.9, 2.5, 2.5, 1.9, 1.5, 1.5, 1.2, 1.2, 1.2, 0.8

Ex. 27 Generally, when there is a smaller number of data values, as in this exercise, a stem-and-leaf plot will order and display the data quickly. A box-and-whisker plot is more useful for larger data sets or when you are comparing within a data set.

In Exercises 15–18, create a collection of 16 numbers that could be represented by the box-and-whisker plot. (There are many correct answers.) See margin.

15.

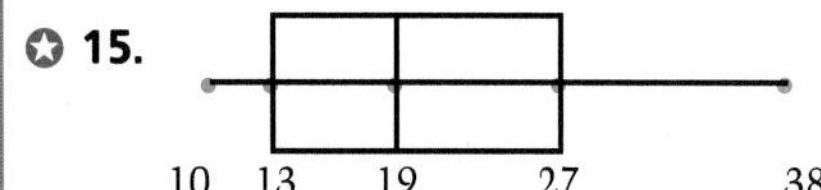

16.

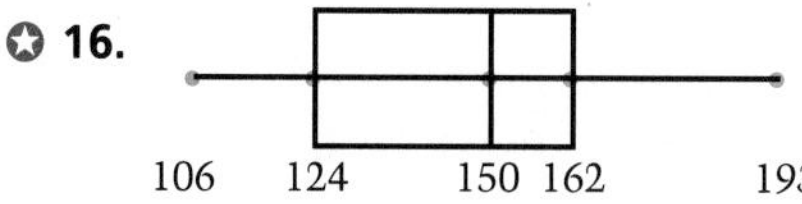

17.

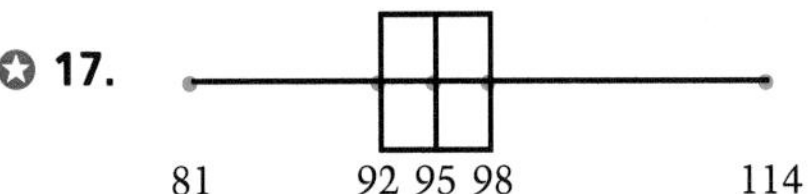

18.

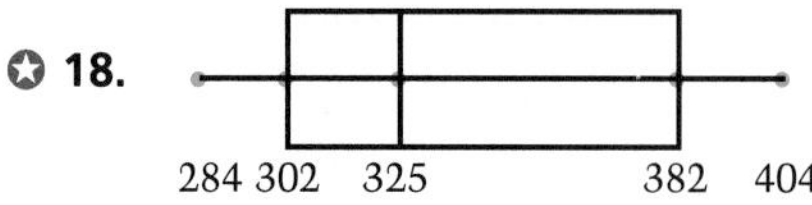

19. *Birthdays* Use a stem-and-leaf plot (months as stems, days as leaves) to write the following birthdays in order (earliest to latest). See margin.

10-11	4-14	7-31	12-28	4-15	4-30	2-22	8-21
1-24	9-12	1-3	2-7	4-30	10-17	6-5	1-25
5-10	11-11	12-9	4-1	8-26	12-15	3-17	6-13

20. *Food Drive* Twelve ninth-grade classes are having a food drive. Use a stem-and-leaf plot to order (from smallest to largest) the number of food items collected per class.

120, 98, 150, 129, 126, 122, 136, 121, 133, 139, 155, 99

Immigration **In Exercises 21 and 22, use the following information.**

In 1988, 264,500 people immigrated to the United States from Asia and 64,800 immigrated from Europe. The lists show the number of immigrants (in 1000's) from several Asian countries and several European countries. Use a stem-and-leaf plot to write the numbers in decreasing order. ***(Source: U.S. Immigration and Naturalization Service)*** See margin.

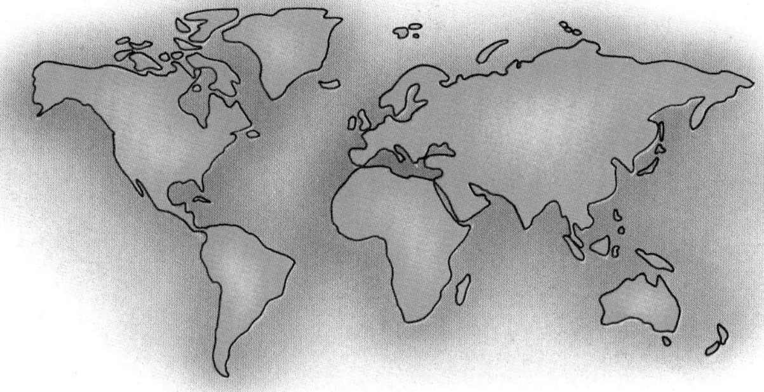

21.

Cambodia	9.6	China	28.7	Taiwan	9.7	Hong Kong	8.5
India	26.3	Iran	15.2	Iraq	1.0	Israel	3.6
Japan	3.6	Jordan	4.5	Korea	34.7	Laos	10.7
Lebanon	4.9	Pakistan	5.4	Philippines	50.7	Syria	2.2
Thailand	6.9	Turkey	1.6	Vietnam	25.8		

22.

Czechoslovakia	1.5	France	2.5	Germany	6.8	Greece	2.5
Hungary	1.2	Ireland	5.1	Italy	2.9	Netherlands	1.2
Poland	9.5	Portugal	3.2	Romania	3.9	Soviet Union	2.9
Spain	1.5	Sweden	1.2	Switzerland	0.8	United Kingdom	13.2
Yugoslavia	1.9						

Integrated Review

23. *Presidential Election* In the 1988 presidential election, 91,600,000 Americans voted. The following list shows the percent of this total in the 20 states with the greatest numbers voting. Use a stem-and-leaf plot to order the data in increasing order. Then use the results to construct *two* visual models of the data: a box-and-whisker plot and a bar graph. (Use intervals of width 2, beginning with $1 \le x < 3$.) Which visual model do you think helps to best present the data? Explain your reasons. See Additional Answers
(Source: Committee for the Study of the American Electorate)

California	10.8	Florida	4.7	Georgia	2.0	Illinois	5.0
Indiana	2.4	Maryland	1.9	Massachusetts	2.9	Michigan	4.0
Minnesota	2.3	Missouri	2.3	New Jersey	3.4	New York	7.1
North Carolina	2.3	Ohio	4.8	Pennsylvania	5.0	Tennessee	1.8
Texas	5.9	Virginia	2.4	Washington	2.0	Wisconsin	2.4

24. *Presidential Election* The following list shows the 20 states that had the greatest percent of their own populations vote in the 1988 presidential elections. Construct *two* visual models of the data: a box-and-whisker plot and a bar graph. (Use intervals of width 2, beginning with $40 \le x < 42$.) How would you explain the fact that this list is so different from the list given in Exercise 23?
(Source: Committee for the Study of the American Electorate) See Additional Answers

Colorado	42	Connecticut	45	Idaho	41	Iowa	43
Maine	46	Massachusetts	45	Minnesota	49	Missouri	41
Montana	45	Nebraska	41	New Hampshire	42	New Jersey	40
North Dakota	45	Ohio	40	Oregon	43	Rhode Island	41
South Dakota	44	Vermont	44	Washington	40	Wisconsin	45

25. Use the information given in Exercises 23 and 24 to find the 1988 population of Washington. 4,580,000

26. Use the information given in Exercises 23 and 24 to find the 1988 population of New Jersey. 7,786,000

Exploration and Extension

27. The heights, in meters, of the world's tallest skyscrapers are given in the list below. Choose one or more of the data-organization techniques discussed in this text to organize the data. Explain why you chose the method you used. Answers vary.

Allied Bank Plaza, Houston	302	Amoco, Chicago	346
Bank of Montreal Tower, Toronto	285	Bank of China, Hong Kong	305
Columbia Seafirst Center, Seattle	291	Citicorp Center, New York City	279
First Interstate World, Los Angeles	310	Empire State, New York City	381
NCNB Plaza, Dallas	281	John Hancock Center, Chicago	344
Scotia Plaza, Toronto	276	Overseas Union Bank, Singapore	280
Texas Commerce Tower, Houston	305	Sears Tower, Chicago	443
Two Prudential Plaza, Chicago	288	World Trade Center, New York City	411

Enrichment Activities

These activities require students to go beyond lesson goals.

The table shows the number of known moons for our solar system.

Planet	Number of Moons
Mercury	0
Venus	0
Earth	1
Mars	2
Jupiter	16
Saturn	23
Uranus	15
Neptune	2
Pluto	1

1. Use the data to construct a box-and-whisker plot. What do you notice? The data are skewed toward the lower end of the plot.

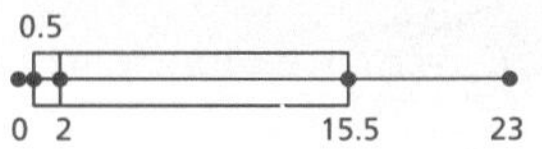

2. Construct a frequency bar graph using the same data.

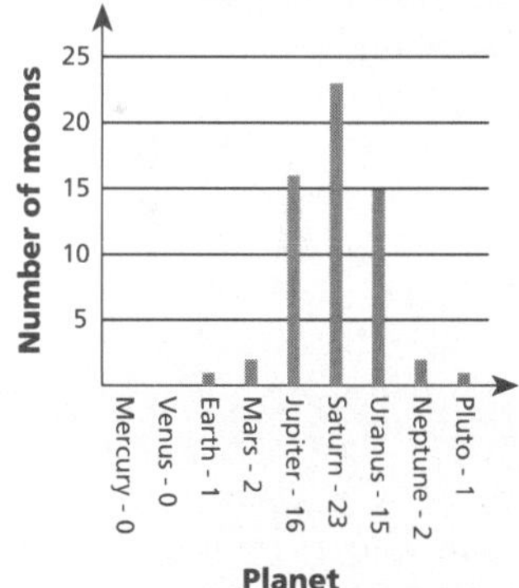

3. Write a paragraph describing which plot you feel shows the data in a more understandable way.

Problem of the Day

Each letter represents a different digit in this addition. Find the digits.

MADE	1692
+ MEAD	+ 1269
EDAM	2961

ORGANIZER

Warm-Up Exercises

1. Find the mean of each given set of numbers.
   a. 12, 10, 8, 3, 5, 5, 6, 4, 9, 10, 12, 6, 8, 7, 6, 6, 4, 5
   b. 2.3, 1.2, 4.0, 3.2, 4.2, 6.4, 4.5, 4.0, 7.3, 5.5, 6.8, 3.0, 4.8, 2.3

   a. 7, b. 4.25
2. Find the median of each set in Exercise 1. a. 6, b. 4.1
3. Identify the value(s) that occur most frequently in each set of Exercise 1.

   a. 6, b. a tie, 4.0 and 2.3

Lesson Resources

Teaching Tools
- Transparencies: 1, 2, 4
- Problem of the Day: 12.7
- Warm-up Exercises: 12.7
- Answer Masters: 12.7

Extra Practice: 12.7

Color Transparencies: 72, 73

Technology Handbook: p. 89

LESSON Notes

GOAL 1 Students may not understand why there is a need to find the mode and the median of data sets. Note that the mean is best used to describe continuous data sets such as scores, weights, and temperatures. The median is best used when the data are spread over a large range or skewed in one direction, such as in salary or

12.7 Exploring Data: Measures of Central Tendency

What you should learn:

Goal 1 How to find the mean, median, and mode of n numbers

Goal 2 How to use measures of central tendency to answer questions about real-life situations

Why you should learn it:

You can use the mean, median, and mode of a collection of data to represent a "typical" number in that collection.

Goal 1 Measuring Central Tendency

There are three commonly used measures of **central tendency** for a collection of numbers.

1. The **mean,** or **average,** of n numbers is the sum of the numbers divided by n.
2. The **median** of n numbers is the middle number when the numbers are written in order. (If n is even, the median is the average of the two middle numbers.)
3. The **mode** of n numbers is the number that occurs most frequently. If two numbers tie for most frequent occurrence, the collection has two modes and is called **bimodal.**

Many collections of numbers that are taken from real life have bar graphs that are *bell-shaped.* For such collections, the mean, median, and mode are approximately equal.

Example 1 *Central Tendency of a Bell-Shaped Distribution*

A tire manufacturer conducted a survey of 118 people who had purchased a particular type of tire. Each person was asked to report the number of miles driven before replacing tires. The responses are shown at the left. Construct a bar graph for these data. Find the mean, median, and mode.

Miles	Number of Responses
14,000	1
15,000	2
16,000	5
17,000	9
18,000	14
19,000	18
20,000	20
21,000	18
22,000	14
23,000	9
24,000	5
25,000	2
26,000	1

Solution The bar graph is bell-shaped. Most of the tires lasted between 18,000 and 22,000 miles. The median and mode are each 20,000. The mean is more difficult to calculate, but after some effort, you can find that it is also 20,000 miles.

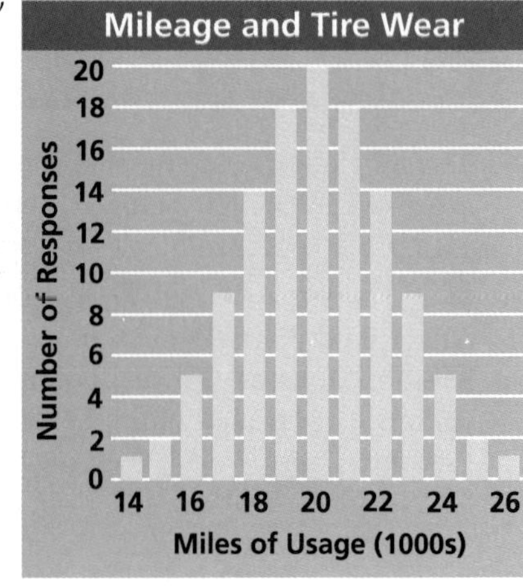

Mean: 20,000 miles
Median: 20,000 miles
Mode: 20,000 miles ■

Goal 2 Using Central Tendency in Real Life

Measures of central tendency are part of a field of mathematics called *statistics*. The goal of statistics is to organize large collections of data in ways that can be used to understand trends and make predictions.

Unfortunately, almost any data collection can be organized in several different ways—some of which are not at all representative of the data. Example 2 describes such an *abuse* of statistics.

Real Life
Education

Example 2 *Interpreting Measures of Central Tendency*

Elliot School, a small rural high school, had 28 graduates in 1992. Of the 28, 7 were 17 years old, 19 were 18 years old, 1 was 19 years old, and 1 was 80 years old! (A retired member of the community had decided to go back to high school to obtain a diploma.) The next fall a legislator from a neighboring school district wanted to have Elliot School closed. One reason given was that "the average age of a high school graduate at Elliot is 20 years." Was the legislator's mathematics correct? Was the choice of a measure of central tendency an honest representation of the typical age of an Elliot graduate?

Solution The legislator's mathematics was correct.

$$\text{Mean} = \frac{7(17) + 19(18) + 1(19) + 1(80)}{28} = \frac{560}{28} = 20$$

The choice of a measure of central tendency, however, was unfair to Elliot School. Both the median and the mode of the ages were 18 years.

In 1988, approximately 10% of the 20,000 public high schools in the United States had enrollments of less than 100 students.

■

population data. The mode is best used when the data clusters, as in students' ages or students' shoe sizes.

Example 1

Note that if each number value in the set occurs only once, there is no mode of the data set. Demonstrate (or ask students to demonstrate) how the average or mean of the data is computed. The total number of responses is the sum of all the entries in the Number of Responses column (the 118 people surveyed). The sum of all the miles driven by the 118 people is found by first multiplying the mileage in the table by its corresponding number of responses, then adding all the products. The sum of all these products (2,360,000 miles) is then divided by 118. (Observe that the mean (20,000 miles) times the number of respondents (118) is the total number of miles—2,360,000.)

Example 2

Point out to students that the improper use of statistics occurs frequently in the popular press. Have them identify (most notably from advertisements) instances of possible abuse. Have them report findings to the class.

Example 3

Usually, the line of the box-and-whisker plot should reflect the relative distances between the values displayed. Therefore, the distance between the first quartile (−50) and the median (−40) should be half as long as the distance between the first quartile and the smallest value in the data set (−70).

Extra Examples

Here are additional examples similar to **Examples 1–3.**

The table shows the responses of 90 people at a shopping mall who indicated what their individual annual incomes (in thousands of dollars) were.

1. Construct a bar graph of the data.
2. Find the mean, median, and mode of the data.
3. Construct a box-and-whisker plot of the data.
4. Can the data be used to indicate the average salary of the people who live in the community in which the shopping mall is located?

Annual Salary	Number of Responses
$10,000	1
$11,000	3
$12,000	2
$13,000	4
$14,000	2
$15,000	6
$16,000	8
$17,000	6
$18,000	9
$19,000	11
$20,000	7
$21,000	8
$22,000	6
$23,000	8
$24,000	5
$25,000	4

1. Check students' graphs
2. mean = $18,688.89; median = $19,000; mode = $19,000
3. the box-and-whisker plot (in thousands of dollars)

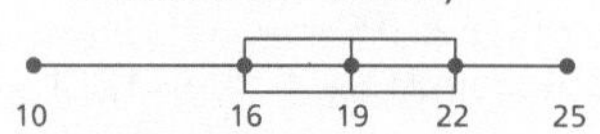

4. No; The people who shop at a mall do not necessarily live in the community in which the shopping mall is located.

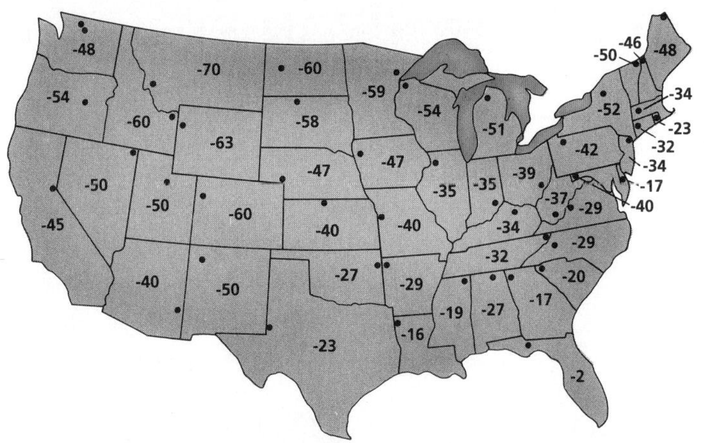

The map shows the lowest temperatures (°F) recorded in the 48 contiguous states.
(Source: National Oceanic and Atmospheric Administration)

Connections
Meteorology

Example 3 *Finding the Mean, Median, and Mode*

Find the mean, median, and mode of the temperatures shown in the map above. Sketch a box-and-whisker plot for the data.

Solution In increasing order, the low temperatures are:

$-70, -63, -60, -60, -60, -59, -58, -54, -54, -52,$
$-51, -50, -50, -50, -50, -48, -48, -47, -47, -46,$
$-45, -42, -40, -40, -40, -40, -39, -37, -35, -35,$
$-34, -34, -34, -32, -32, -29, -29, -29, -27, -27,$
$-23, -23, -20, -19, -17, -17, -16, -2$

The median is -40°F. There are two modes: -40°F and -50°F, each of which occurs four times. The mean is

$$\frac{-70 - 63 - 60 - 60 - \cdots - 16 - 2}{48} = \frac{-1914}{48} = -39.875°.$$

The box-and-whisker plot for these data is as follows.

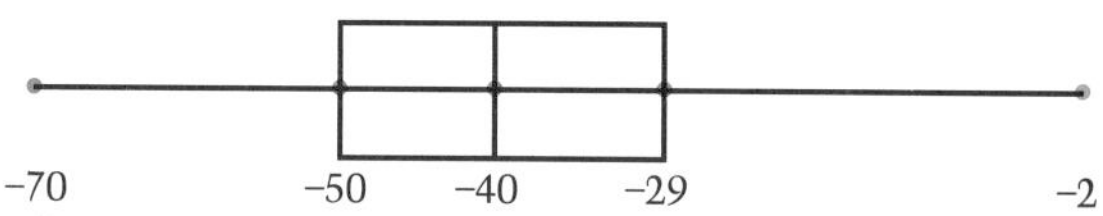

■

Communicating about ALGEBRA

SHARING IDEAS about the Lesson

Analyze Data At a family reunion, four brothers are comparing their weights. Which of the following have a mean of 180? A median of 180? A mode of 180?

A. 170, 171, 182, 197 **B.** 165, 178, 180, 197
C. 165, 178, 180, 180 **D.** 172, 180, 180, 188

Mean of 180: A, B, D; median of 180: D; mode of 180: C, D

EXERCISES

Guided Practice

CRITICAL THINKING about the Lesson

1. Define the mean, median, and mode of a collection of numbers. See page 658.
2. Find the mean, median, and mode of the following numbers:

   231, 235, 238, 226, 235, 239, 232 Mean: 233.7, median: 235, mode: 235
3. Give an example in which the mean of a collection of numbers is *not* representative of a typical number in the collection. Answers vary; a mean of 34 for the collection 1, 2, 99
4. The following bar graph and box-and-whisker plot represent the *same* data. Compare these two visual models. The graph shows the frequency of each number, the plot shows the median and quartiles.

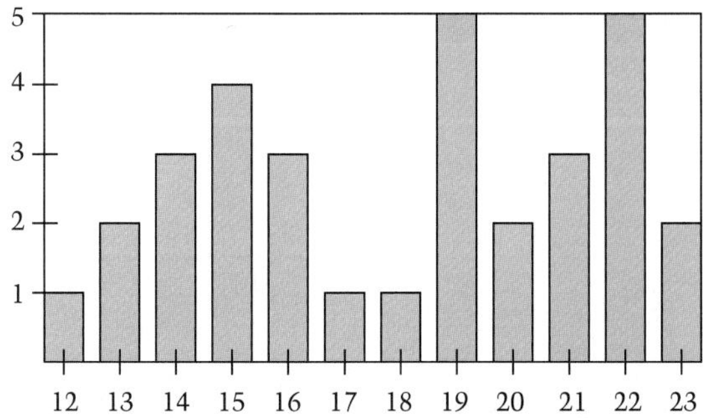

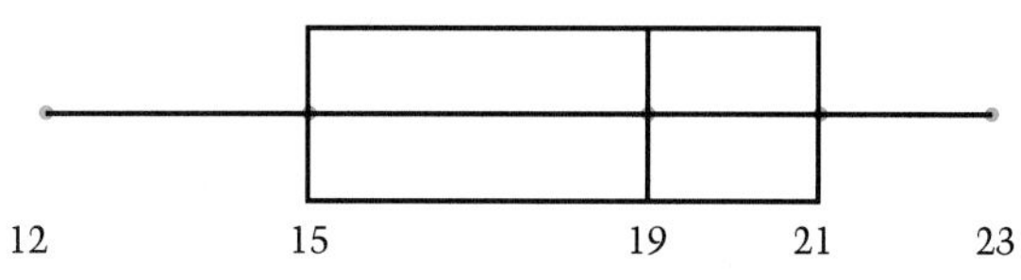

Independent Practice

5. *Hatching Eggs* You are incubating 55 eggs in science class. Every 20 minutes you record the percent humidity to see that it remains between 60% and 65%. Your friend tells you that the mean, median, and mode of your readings are equal. Do you agree? Explain. No; the mean and the median are 62.9, the mode is 62.8.

8:30 A.M.	62.5%	10:10 A.M.	63.1%	11:50 A.M.	60.9%	1:30 P.M.	63.9%
8:50 A.M.	62.8%	10:30 A.M.	63.0%	12:10 P.M.	62.6%	1:50 P.M.	62.8%
9:10 A.M.	61.3%	10:50 A.M.	62.8%	12:30 P.M.	63.1%	2:10 P.M.	64.3%
9:40 A.M.	61.0%	11:10 A.M.	61.7%	12:50 P.M.	64.2%	2:20 P.M.	64.5%
9:50 A.M.	63.2%	11:30 A.M.	61.7%	1:10 P.M.	64.0%	2:50 P.M.	64.6%

6. *Urban Lawyers* The number of lawyers in several large cities is shown in the table. Find the mean and median of these numbers. Are the mean and median equal? No

City	Boston	Chicago	Los Angeles	New York	San Francisco	Washington, D.C.
Lawyers	16,400	30,500	27,000	66,200	14,200	12,400

Check Understanding

1. Define the mean, median, and mode of a set of data. See the top of page 658.
2. Describe the conditions for which the mean, median, and mode of a set of data are equal. All data entries the same, or, a set of data like 1, 2, 3, 3, 4, 5
3. Explain how measures of central tendency can be used to misrepresent information.

Communicating about ALGEBRA

Ask students to create a data set that has (a) the same value for its mean and median, but a different value for its mode; (b) the same value for its median and mode, but a different value for its mean; and (c) the same value for its mean and mode, but a different value for its median.

a. −5, 5, 10, 20, 20
b. −5, 5, 10, 10, 20
c. −6, 0, 0, 1, 2, 3

EXERCISE Notes

ASSIGNMENT GUIDE

Basic/Average: Ex. 1–4, 5–19 odd, 20–24, 25

Above Average: Ex. 1–4, 5–11, 13–23 odd, 25

Advanced: Ex. 1–4, 5–12, 17–26

Selected Answers
Exercises 1–4, 5–23 odd

✪ More Difficult Exercises
Exercises 7, 8, 11

Guided Practice

▶ **Ex. 1–4** In Exercise 3, discuss the intuitive meaning of an "outlier," such as an 80-year-old graduate, to a set of data. How does this skew the mean?

Independent Practice

TECHNOLOGY Be sure students have access to scientific or graphics calculators before assigning the independent practice.

▶ **Ex. 6–10** These exercises represent real-life data models.

▶ **Ex. 7–8** COOPERATIVE LEARNING The box-and-whisker plots require time for students to complete them. They could be assigned as a group project.

Answers

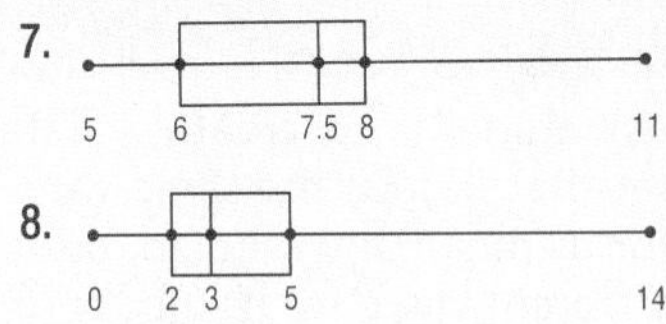

9. The mode is the smallest length and the mean is smaller than most of the lengths, so the median of 4.6 inches is the most representative.

10. The mode is the smallest number; the mean is smaller than 4 numbers and larger than 5 numbers; the median of 1696 is smaller than 4 numbers and larger than 4 numbers; so the median is most representative.

7. *Rain in Wichita* The bar graph shows the number of days of rain in Wichita, Kansas. Find the mean and median of the data. Construct a box-and-whisker plot for the data.

8. *All the Presidents' Children* The bar graph shows the number of children of American Presidents. Find the mean, median, and mode of the data. Construct a box-and-whisker plot for the data.

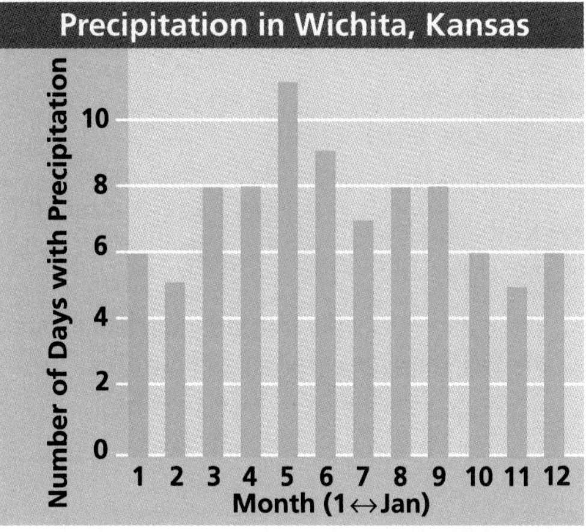

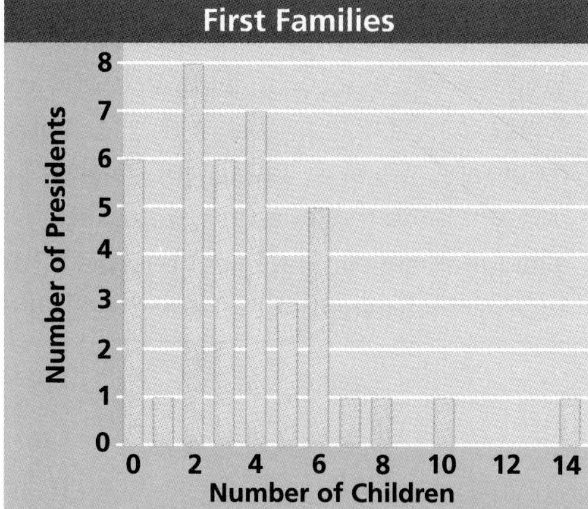

9. *Measuring Frogs* Your science class is collecting and measuring frogs of different ages. Which measure of central tendency is the most representative of a typical length of the 17 frogs the class has collected? Explain your reasoning.

5.4 in.	4.7 in.	0.2 in.	4.3 in.
6.3 in.	0.9 in.	5.1 in.	4.4 in.
4.0 in.	4.6 in.	6.1 in.	5.5 in.
0.2 in.	5.0 in.	0.5 in.	4.2 in.
6.2 in.			

See margin.

7. Mean: 7.25, median: 7.5.

8. Mean: 3.675, median: 3, mode: 2.

See margin for plots.

This is a blue poison dart frog from the savannah of Surinam.

10. *Southeastern Households* The number of households (in thousands) in each of the southeastern states (including the District of Columbia) is shown below. Which measure of central tendency do you think is most representative of the typical number of households in a south Atlantic state? Explain your reasoning. See margin.

Delaware	246	Georgia	2314	Florida	4922
Virginia	2228	West Virginia	708	Washington, D.C.	246
North Carolina	2444	Maryland	1696	South Carolina	1225

11. *Average Salary* A small town has an adult population of 100. One person is a multimillionaire who makes 5 million dollars a year. The mean annual salary in the town is \$74,750. Is this a fair measure of a typical salary in the town? What is the mean salary of the other 99 adults?

No; \$25,000

12. *Average Age* Create two different examples, each having 10 ages. Create one example so that the mean age is 14 and the median age is 16. Create the other example so that the median age is 14 and the mean age is 16. **Answers will vary. For example: 7, 7, 8, 10, 14, 18, 18, 19, 19, 20; 10, 12, 13, 13, 14, 14, 16, 20, 24, 24**

Integrated Review

13. 20 is what percent of 80? **25%**

14. What is 50% of 11.3? **5.65**

15. 7.3 is what percent of 9.5? **76.8%**

16. $\frac{1}{2}$ is 10% of what? **5**

17. What is 27% of 21.4? **5.778**

18. $\frac{6}{5}$ is 6% of what? **20**

19. *Estimation—Powers of Ten* Which is the best estimate of the number of calories in an orange?
    a. 6 **b.** 60
    c. 600 **d.** 6000 **b**

20. *Estimation—Powers of Ten* Which is the best estimate of the weight of an adult elephant?
    a. 100 pounds **b.** 1000 pounds
    c. 10,000 pounds **d.** 100,000 pounds **c**

21. *In-Line Skates* At five different stores, a pair of identical in-line skates costs \$119, \$149, \$139, \$100, and \$125. What is the mean cost? **\$126.40**

22. *Protective Gear* At five different stores, the cost of protective gear for in-line skating is \$90, \$89, \$100, \$97, and \$115. What is the mean cost? **\$98.20**

23. *Riding to School* You rode your bike to and from school each school day for two weeks and your daily riding times are given below. What was your mean riding time? **14.175 min**

    12.50 min 15.00 min 15.75 min 13.00 min 13.00 min
    14.50 min 14.75 min 14.75 min 14.00 min 14.50 min

24. *Practicing a Speech* You practiced a speech each school day for a week. What was your mean delivery time? **9.27 min**

    10.00 min 9.75 min 9.60 min 9.00 min 8.00 min

Exploration and Extension

Study Time **In Exercises 25 and 26, use the following information.**

In your math class, you were asked to record the number of hours you spend on homework each day for 4 weeks.

Study hours	0.0–0.5	0.5–1.0	1.0–1.5	1.5–2.0	2.0–2.5
Number of days	2	4	1	5	4

Study hours	2.5–3.0	3.0–3.5	3.5–4.0	4.0–4.5	4.5–5.0
Number of days	2	6	3	0	1

25. What measure of central tendency do you think best describes the number of hours you study on a typical day? Explain. **The median or the mean; they are the same: 2.0–2.5**

26. Construct a bar graph and a box-and-whisker plot for the data. Which visual model do you think is better? Explain your reasoning. **See answer to Exercise 23 on page 657, and see margin.**

▶ **Ex. 25–26** These exercises represent a cycle of data gathering, data organizing, and data interpretation. Remind students that the choice of mean or median and bar graph or box-and-whisker plot should be based on the most accurate representation of the data.

Answer

26.

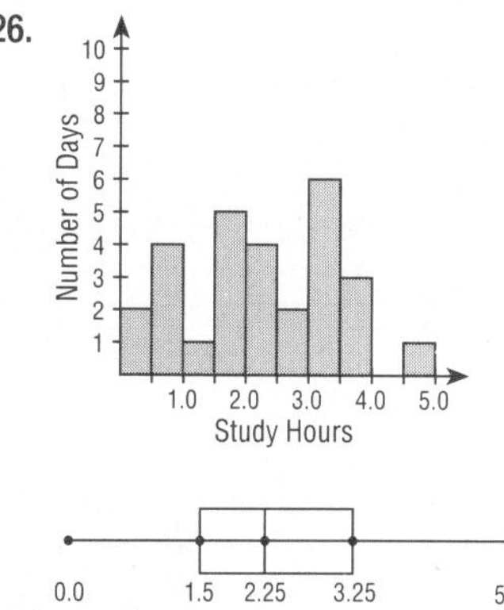

Enrichment Activities

These activities require students to go beyond lesson goals.

1. Collect data on something that interests you or on one of these topics.
    a. Number of hours students worked last week
    b. Number of hours students watched television last week
    c. Number of hours students spent on homework last week
    d. Weekly allowances of boys versus girls
    e. The ages of your parents or grandparents
2. State why you think the mean or the median better represents the typical number in the data set.
3. Construct a stem-and-leaf plot or a box-and-whisker plot for the data. Interpret the results.

TECHNOLOGY When entering data into a graphing calculator, the value of the y-coordinate may indicate the frequency of occurrence in the data set of the corresponding x-coordinate. A value, $k > 1$, for the y-coordinate means that the corresponding value of the x-coordinate has occurred k times in the set of data.

Answers

1. 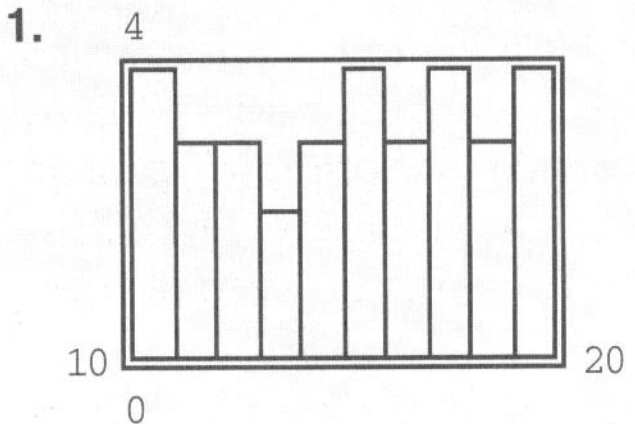

2. 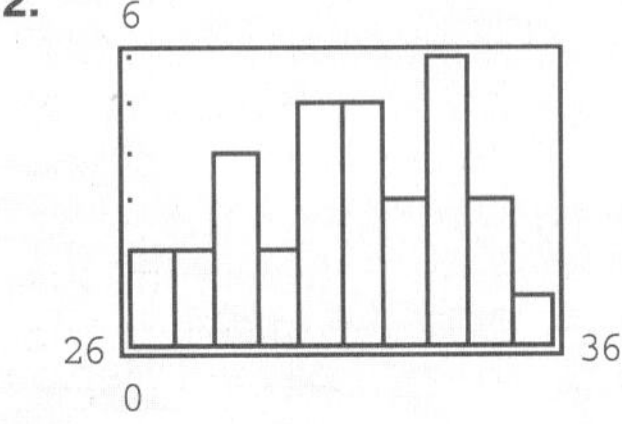

USING A GRAPHING CALCULATOR

Graphing calculators can be used to perform statistical computations such as drawing a bar graph or ***histogram*** *of a collection of numbers. Keystroke instructions for doing this on a* TI-82, *Casio* fx-9700GE, *and Sharp* EL-9300C *are listed on page 745.*

Cooperative Learning

Probability Experiment With a partner, toss four coins 50 times—one of you should toss the coins and the other should record the number of coins that land heads up on each toss. When we did, here are the results we obtained.

1, 2, 1, 0, 2, 3, 3, 4, 1, 2, 3, 2, 1, 2, 4, 2, 2,
0, 2, 1, 2, 2, 2, 1, 1, 2, 2, 2, 2, 4, 2, 3, 3, 0,
3, 1, 3, 2, 2, 2, 3, 1, 1, 2, 1, 2, 2, 3, 3, 1

A histogram for this data is shown below. Use a graphing calculator to draw a histogram for your data. Compare your result with that shown below. In each experiment (yours and ours) which outcome is most likely? From these experiments, estimate the probability that exactly two coins will land heads up when four coins are tossed.

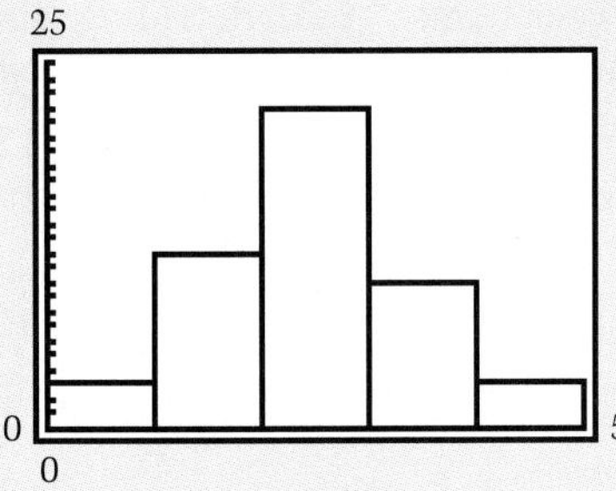

As an option to actually tossing the coins, you might want to use a computer to *simulate* the experiment. One way to do that is to use the following BASIC program.

```
10 RANDOMIZE
20 FOR I=1 TO 50
30 H=0
40 FOR J=1 TO 4
50 COIN=INT(2*RND)
60 IF COIN=1 THEN H=H+1
70 NEXT
80 PRINT, H "Heads"
90 NEXT
100 END
```

Exercises

In Exercises 1 and 2, construct a histogram for the data. See margin.

1. 10, 14, 11, 18, 14, 16, 17, 17, 10, 11, 19, 15, 13, 17, 11, 13, 19, 18, 12, 14, 19, 10, 12, 15, 17, 15, 15, 12, 18, 16, 16, 10, 19

2. 26, 29, 35, 34, 33, 30, 31, 32, 30, 30, 32, 27, 28, 28, 34, 33, 33, 28, 28, 31, 31, 32, 30, 31, 33, 27, 26, 34, 33, 33, 30, 29, 31

12 Chapter Summary

What did you learn?

Skills

1. Identify relations that are functions. **(12.1)**
2. Use function notation to evaluate a function. **(12.1)**
3. Find the domain of a function. **(12.1)**
4. Sketch the graph of
   - a linear function. **(12.2)**
   - a quadratic function. **(12.4)**
   - a rational function. **(12.5)**
5. Transform a graph. **(12.3)**

Strategies

6. Create models of real-life problems. **(12.2–12.5)**
7. Use functions to solve real-life problems. **(12.1–12.5)**

Exploring Data

8. Create stem-and-leaf plots for data. **(12.6)**
9. Create box-and-whisker plots for data. **(12.6)**
10. Use measures of central tendency to describe data. **(12.7)**

Why did you learn it?

Most of the models you studied in earlier chapters *are* functions—we just didn't use that name. This chapter gives you a chance for more practice with four basic types of models: linear, exponential, quadratic, and rational. The chapter also introduces you to transformations, and some function notation and terminology, which helps prepare you for other math courses.

How does it fit into the bigger picture of algebra?

This chapter explores functions. A *function* is a special type of relationship between two variables in which each *input* variable corresponds to exactly one *output* variable. For instance, the area of a circle is a function of its radius, and you can think of the radius of a circle as input and the area of a circle as output. If someone gives you the radius, r, of a circle, you know the area is $A = \pi r^2$.

In Lesson 12.1, you saw that functions can be described in several different ways. In the next four lessons, you studied functions that are described by equations, and how to transform the graph of a function. The last two lessons in the chapter described ways to organize data using stem-and-leaf plots, box-and-whisker plots, and measures of central tendency.

ORGANIZER

TIME SCHEDULE: All levels, two days

The Chapter Summary helps students organize the main ideas of the chapter. In this chapter, the connections between functions and previously-learned algebraic relationships were examined. The different ways in which to express functional relationships were also presented. Finally, techniques for organizing large sets of data were developed using measures of central tendency.

Work with students to review the skills, strategies, and concepts of the chapter. It is suggested that the first day of review be a combination of you and your students working together on problems from the Chapter Review exercises. The first day's homework assignment can be used as the basis for the second day of review.

COOPERATIVE LEARNING
Encourage students to study together. Emphasize the importance of "teaching" a classmate how to collect and organize data or how to recall a strategy for writing functions. When students work together, everyone wins. The students receiving help get additional instruction, and the students giving help gain a deeper understanding of the skills and concepts involved.

Chapter SUMMARY

Review the chapter by asking students, "What did you learn? Why did you learn it? How does it fit with the other algebra skills and concepts you have learned?"

ASSIGNMENT GUIDE

Basic/Average: Ex. 1–29 odd, 31–34, 35

Above Average: Ex. 1–29 odd, 31–34, 35–41

Advanced: Ex. 1–29 odd, 31–34, 35–41

✪ More Difficult Exercises

Exercises 35–41

EXERCISE Notes

▶ **Exercises 1–41** Since the Chapter Review exercises parallel the list of Skills, Strategies, and Exploring Data statements on page 665, tell students to use the Chapter Summary to recall the lesson in which a skill or strategy was first presented if they are uncertain about how to do a given exercise in the Chapter Review. This forces students to categorize the task to be performed and to locate relevant information in the chapter about the task, both beneficial activities.

Answers

5.

6.

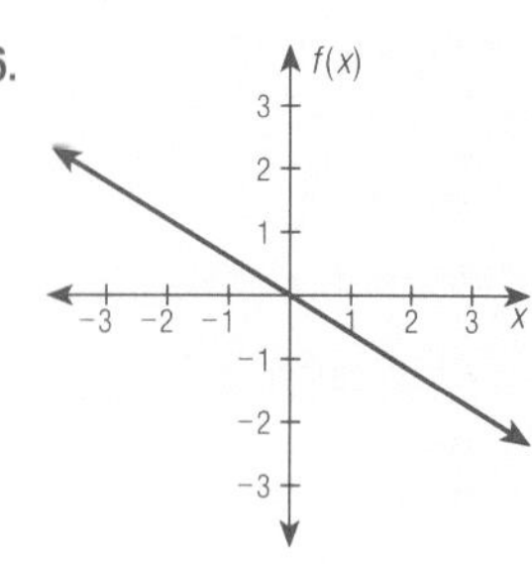

Chapter REVIEW

In Exercises 1–4, decide whether the set of ordered pairs defines a function. If it does, state the domain of the function. (12.1)

1. $\{(a, 0), (b, 3), (c, 3), (d, 4)\}$ Yes, $\{a, b, c, d\}$

2. $\{(m, 7), (n, 6), (p, 4), (p, 5)\}$ No

3. $\{(3, 8), (4, 8), (6, 8), (8, 8)\}$ Yes, $\{3, 4, 6, 8\}$

4. $\{(3, 5), (6, 5), (9, 8), (12, 9)\}$ Yes, $\{3, 6, 9, 12\}$

In Exercises 5–8, find an equation for the linear function, *f*. Then sketch its graph. (12.2) See margin.

5. $f(-3) = -6, f(2) = -1$ $f(x) = x - 3$

6. $f(0) = 0, f(5) = -3$ $f(x) = -\frac{3}{5}x$

7. $f(-3) = 2, f(1) = 6$ $f(x) = x + 5$

8. $f(3) = 3, f(7) = 2$ $f(x) = -\frac{1}{4}x + \frac{15}{4}$

In Exercises 9–12, match the function with its graph. (12.3)

9. $g(x) = 1.5^x$ b

10. $g(x) = 1.5^{x+2}$ a

11. $g(x) = 1.5^x + 3$ d

12. $g(x) = -1.5^x$ c

a.

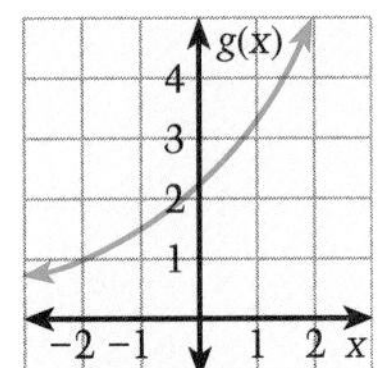

b.

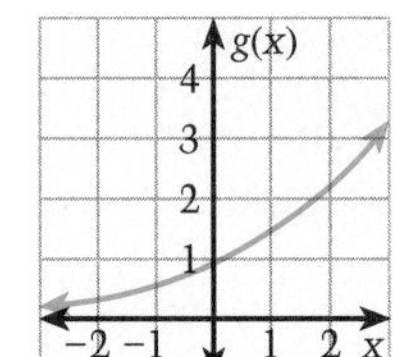

c.

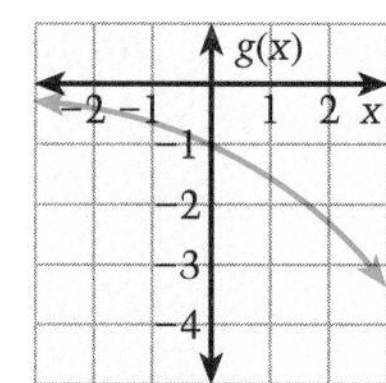

d.

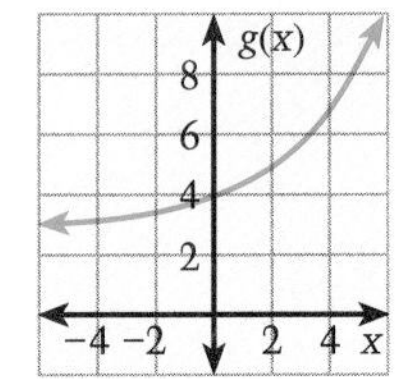

In Exercises 13–16, describe the relationship between the graph of *f* and the graph of *g*. (12.3)

13. $f(x) = \left(\frac{3}{4}\right)^x, g(x) = \left(\frac{3}{4}\right)^x - 2$ Shift 2 units down

14. $f(x) = 2^x, g(x) = 2^{x-5}$ Shift 5 units right

15. $f(x) = 3.9^x, g(x) = -3.9^x$ Reflection

16. $f(x) = 5^x, g(x) = 5^{x+2} - 5$ Shift 2 units left and 5 units down

In Exercises 17–20, write the function in completed square form. Then find the vertex of its graph. Does the graph open up or down? (12.4) See margin.

17. $f(x) = x^2 + 10x + 27$

18. $g(x) = -3x^2 + 42x - 151$

19. $h(x) = -\frac{1}{3}x^2 - \frac{4}{3}x - \frac{22}{3}$

20. $g(x) = \frac{1}{2}x^2 - 10x + 55$

In Exercises 21–24, match the function with its graph. (12.5)

21. $f(x) = \frac{1}{x+1} - 2$ c

22. $f(x) = \frac{1}{x-3} + 1$ b

23. $f(x) = \frac{2}{x+2} + 3$ d

24. $f(x) = \frac{-3}{x-1} - 2$ a

a.

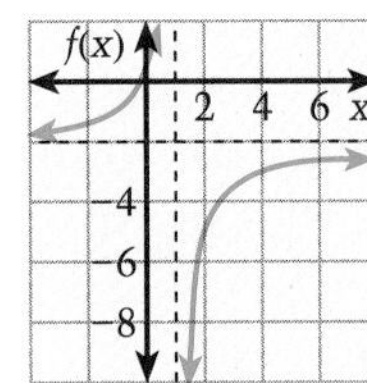

b.

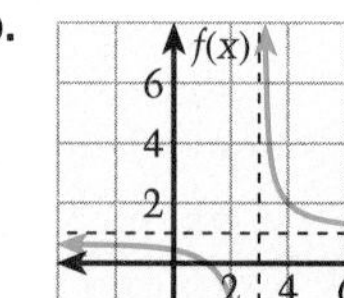

c.

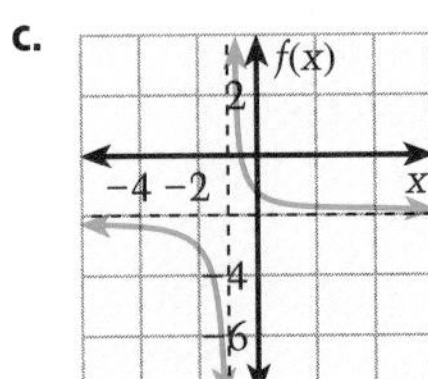

d.

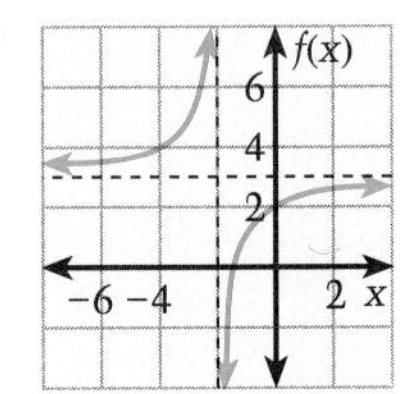

In Exercises 25–30, sketch the graph of the function. (12.5) See margin.

25. $f(x) = \frac{3x - 8}{x - 4}$ **26.** $f(x) = \frac{2x + 3}{x + 1}$ **27.** $f(x) = \frac{-5x + 12}{x - 2}$

28. $f(x) = \frac{-3x - 12}{x + 5}$ **29.** $f(x) = \frac{2x}{x - 2}$ **30.** $f(x) = \frac{-3x - 2}{x + 3}$

In Exercises 31–34, use the following collection of data points. (12.6, 12.7) See margin on page 668.

40, 60, 49, 43, 37, 69, 75, 67, 56, 70
44, 68, 45, 72, 70, 44, 36, 35, 68, 45
71, 64, 53, 46, 50, 49, 37, 47, 54, 72

Stem	Leaves
7	0 0 1 2 2 5
6	0 4 7 8 8 9
5	0 3 4 6
4	0 3 4 4 5 5 6 7 9 9
3	5 6 7 7

Mean: 54.5, median: 51.5, no mode

31. Construct a stem-and-leaf plot for the data.

32. Find the mean, median, and mode.

33. Construct a box-and-whisker plot for the data.

34. Construct a bar graph for the data using the following intervals. $35 \le x < 45$, $45 \le x < 55$, $55 \le x < 65$, $65 \le x < 75$

35. *The Sands of Time* Before the 1900's, sailors used a 30-second "hourglass" to determine the speed of a ship. A rope was tied to a log, and the log was thrown into the water to drift freely behind the ship. After 30 seconds, the rope was measured. Let x represent the length of the rope after 30 seconds. Write a linear function that represents the speed of the ship (in feet per second) in terms of x. Use the function to find the speed of a ship for which 100 feet of rope was let out. $f(x) = \frac{x}{30}$, $3\frac{1}{3}$ ft per second

36. *Life Span of Animals* Construct a box-and-whisker plot for the average life spans of the animals shown below. For which animals do the life spans fall between the second and third quartiles? See margin on page 668.

Bear	22	Cat	11
Chicken	7	Cow	10
Deer	12	Dog	11
Duck	10	Elephant	35
Fox	9	Goat	12
Groundhog	6	Guinea Pig	3
Hamster	2	Hippopotamus	30
Horse	22	Kangaroo	5
Lion	10	Monkey	13
Mouse	2	Parakeet	8
Pig	10	Pigeon	11
Rabbit	7	Rat	3
Sheep	12	Squirrel	8
Whale	45	Wolf	11

7.

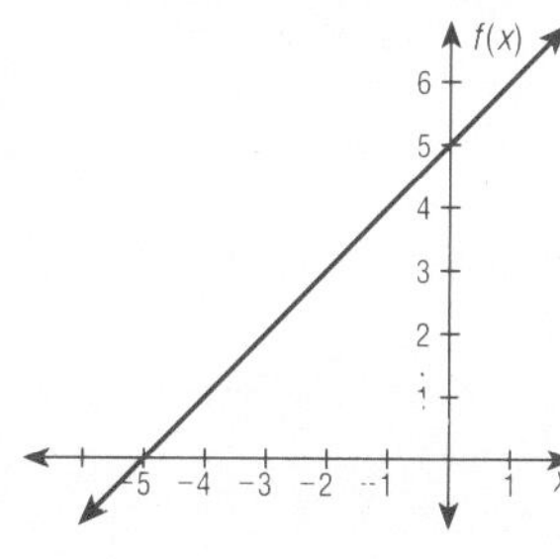

8.

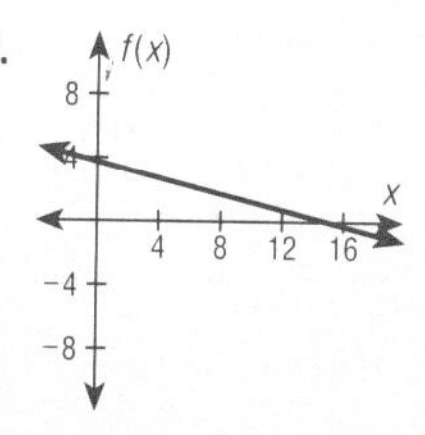

17. $f(x) = (x + 5)^2 + 2$, $(-5, 2)$; opens upward

18. $g(x) = -3(x - 7)^2 - 4$, $(7, -4)$; opens downward

19. $h(x) = -\frac{1}{3}(x + 2)^2 - 6$, $(-2, -6)$; opens downward

20. $g(x) = \frac{1}{2}(x - 10)^2 + 5$, $(10, 5)$; opens upward

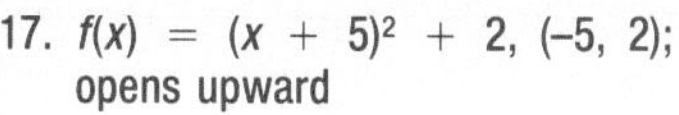

25.

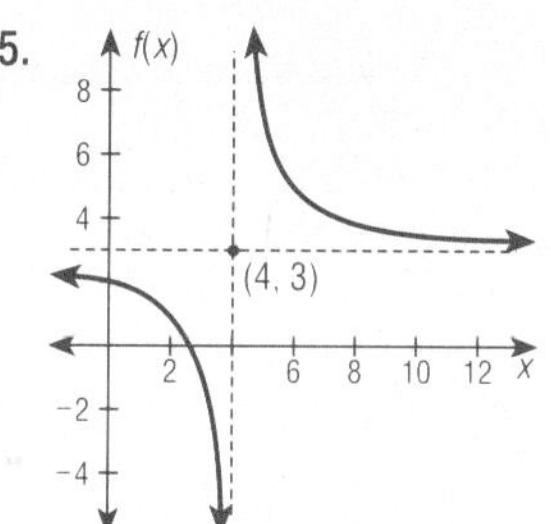

26.

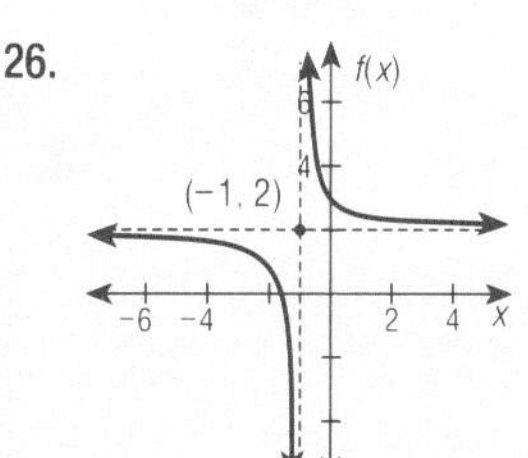

27.

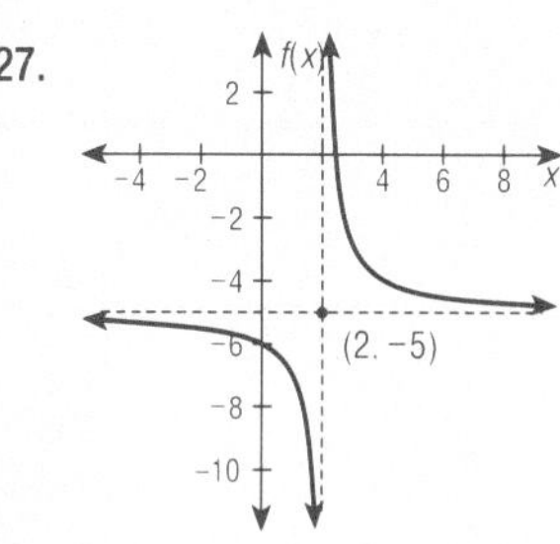

28.

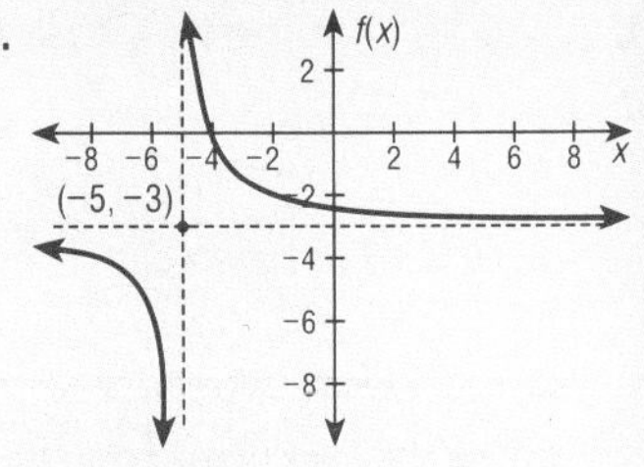

29.

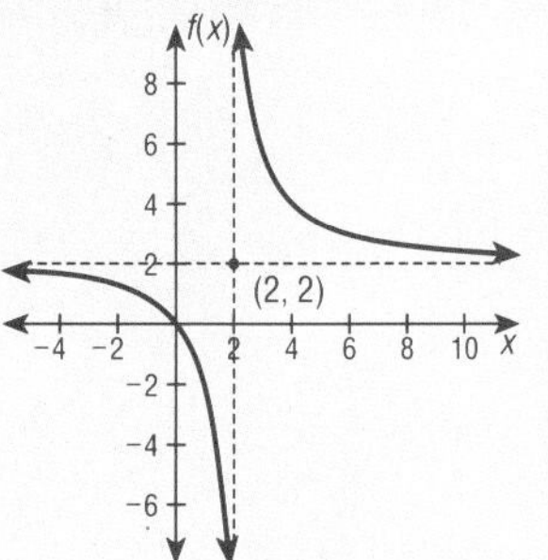

30.

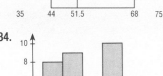

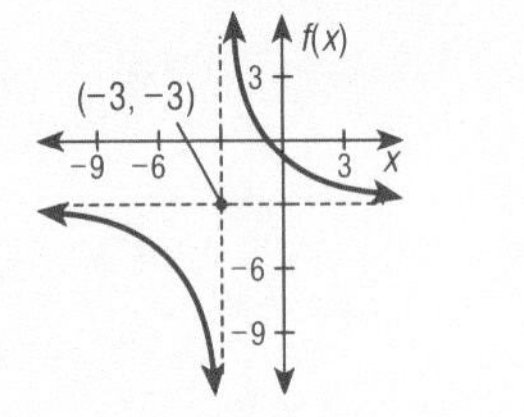

33. 35 44 51.5 68 75

34.

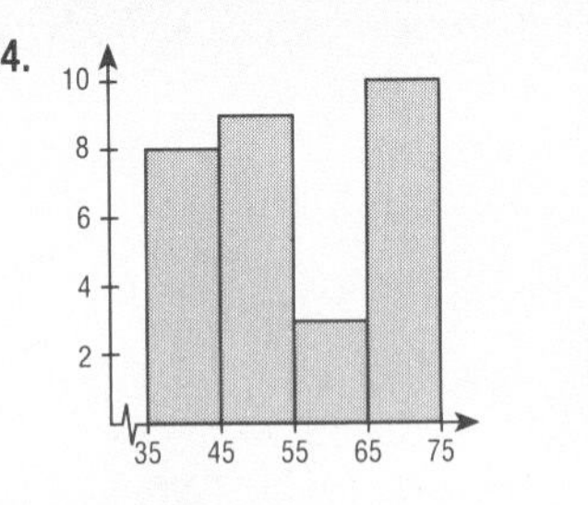

36. 2 7 10 12 45

Deer, Duck, Lion, Pig, Sheep, Cat, Cow, Dog, Goat, Pigeon, and Wolf

Wholesale Costs **In Exercises 37–41, use the following information.**

For 1980 through 1994, the total sales of U.S. wholesalers (in billions of dollars) are shown in the graphs. In each graph, a different type of function was used as a model. In each model, $t = 0$ represents 1980.
(Source: U.S. Bureau of the Census)

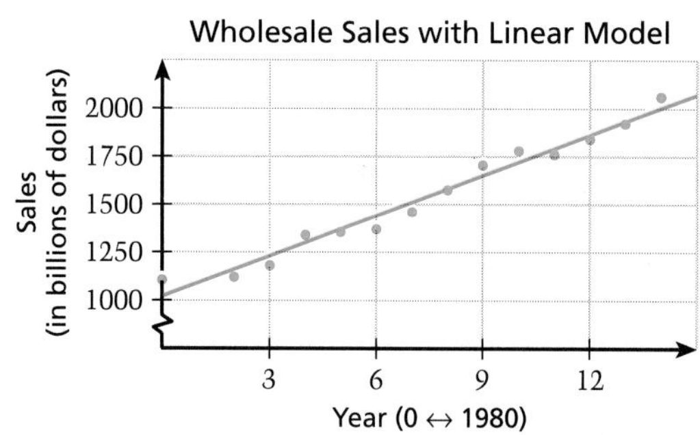

Model: $f(t) = 1040 + 70t$

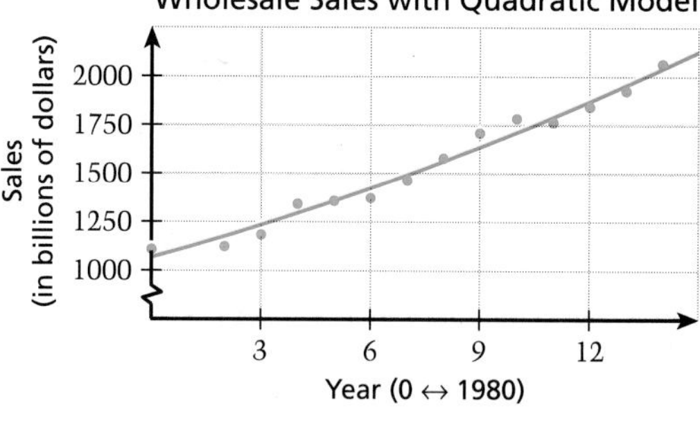

Model: $f(t) = 1.25t^2 + 51.9t + 1083$

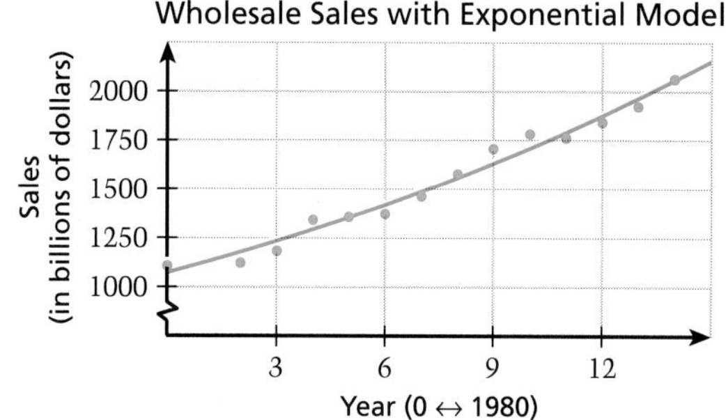

Model: $f(t) = 1090(1.047)^t$

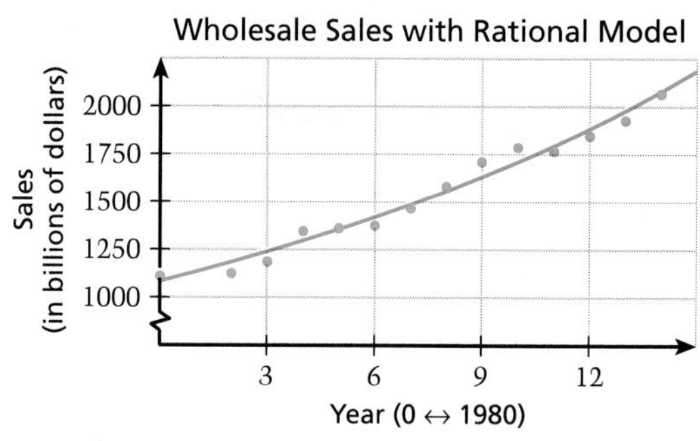

Model: $f(t) = \dfrac{22.85t + 1098.6}{-0.023t + 1}$

37. Use the linear model to predict the total wholesale sales in the year 2000. $2440 billion

38. Use the quadratic model to predict the total wholesale sales in the year 2000. $2621 billion

39. Use the exponential model to predict the total wholesale sales in the year 2000. $2731 billion

40. Use the rational model to predict the total wholesale sales in the year 2000. $2881 billion

41. Which of these models do you think best predicts the total wholesale sales in the year 2000? Explain your reasoning. Answers will vary.

Chapter TEST

In Exercises 1 and 2, use the table. (12.1)

Variable, x	a	b	c	d	e
Variable, y	2	3	2	1	5

1. Is y a function of x? Yes

2. Is x a function of y? No

3. Evaluate $f(x) = 17x - 25$ when $x = -3$ and $x = 2$. **(12.1)** -76, 9

4. Find an equation of the linear function such that $f(-2) = -4$ and $f(3) = 3$. **(12.2)** $f(x) = \frac{7}{5}x - \frac{6}{5}$

In Exercises 5–8, match the function with its graph. (12.3)

5. $f(x) = \left(\frac{1}{4}\right)^x$ b

6. $f(x) = \left(\frac{1}{4}\right)^x + 1$ d

7. $f(x) = \left(\frac{1}{4}\right)^{x-2}$ a

8. $f(x) = -\left(\frac{1}{4}\right)^x - 2$ c

a.

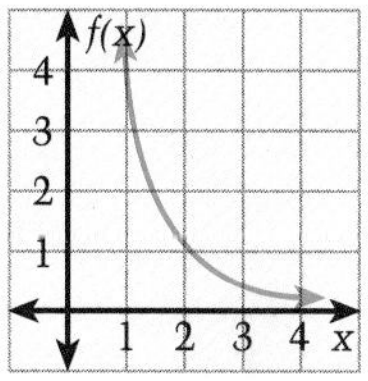

b.

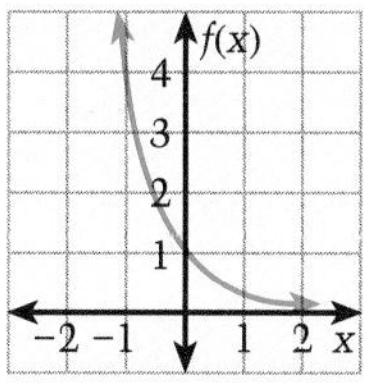

c.

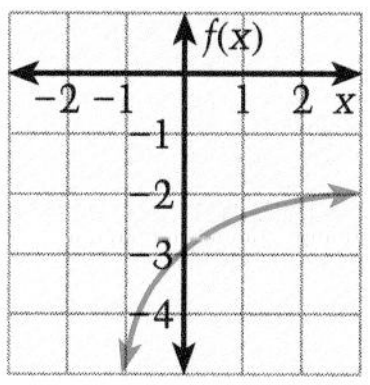

d.

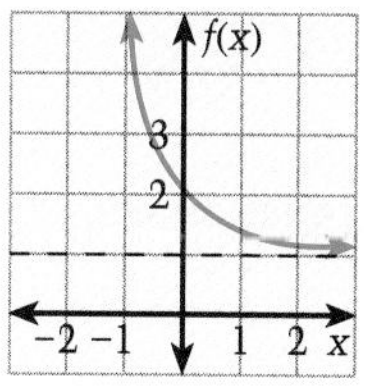

9. Write $f(x) = x^2 + 6x + 15$ in completed square form. Then sketch its graph. **(12.4)** $f(x) = (x + 3)^2 + 6$ See margin.

10. Sketch the graph of $f(x) = \dfrac{1}{x - 4} + 1$.

11. Sketch the graph of $f(x) = \dfrac{3x + 2}{x - 1}$. **(12.5)**

In Exercises 12–15, use the following data:
17, 7, 16, 8, 4, 9, 11, 3, 20, 8, 7, 9, 10, 14, 3, 9. (12.6, 12.7)

12.
2 | 0
1 | 0 1 4 6 7
0 | 3 3 4 7 7 8 8 9 9 9

12. Construct a stem-and-leaf plot for the data.

13. Find the quartiles of the collection. 7, 9, 12.5

14. Find the mean, median, and mode of the collection. Mean: 9.6875, median: 9, mode: 9

15. Construct a box-and-whisker plot for the data. See margin.

16. The human heart pumps about 2000 gallons of blood each day. Create a model that gives the total amount of blood, $f(x)$ in gallons, pumped in x days. Sketch a graph of the function. $f(x) = 2000x$ See margin.

17. Cissy's hair is 5 inches long and is growing at a rate of 4 inches per year. Julie's hair is 12 inches long and is growing at a rate of 3 inches per year. Write a model that gives the ratio of the length of Cissy's hair to the length of Julie's hair. (Let t represent the number of years and let $f(t)$ represent the ratio.) Sketch the graph of f over the interval $0 \le t \le 4$. $f(t) = \dfrac{5 + 4t}{12 + 3t}$ See margin.

18. The table gives the average number of days of rain in each month in Sioux Falls, South Dakota. Construct a bar graph for these data. Find the mean, median, and mode of the data. Mean: $7\frac{11}{12}$, median: $8\frac{1}{2}$, mode: 6 See margin.

Month	Jan	Feb	Mar	Apr	May	Jun	Jul	Aug	Sep	Oct	Nov	Dec
Days of Rain	6	6	9	9	10	11	9	9	8	6	6	6

Answers
Chapter Test

9.

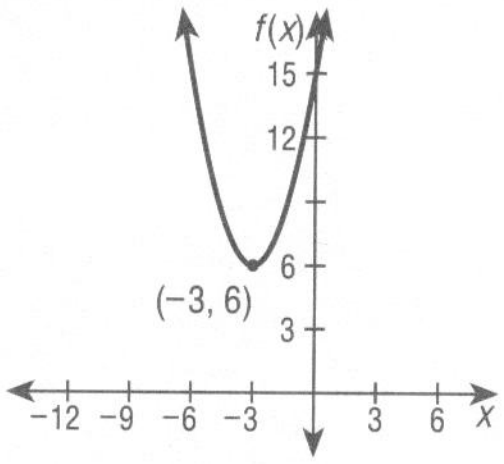

10.

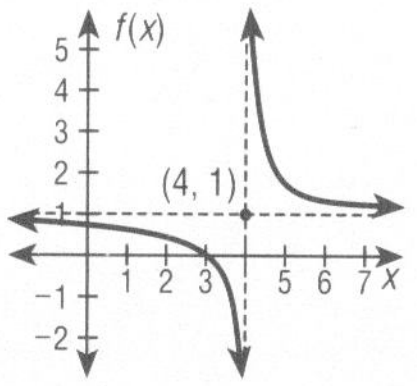

11.

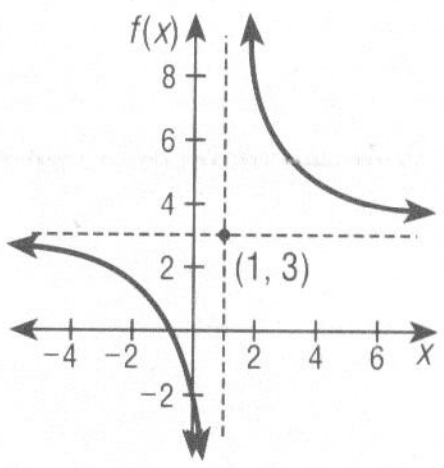

15.

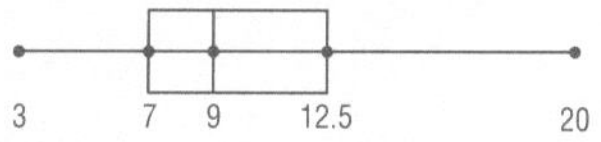

16.

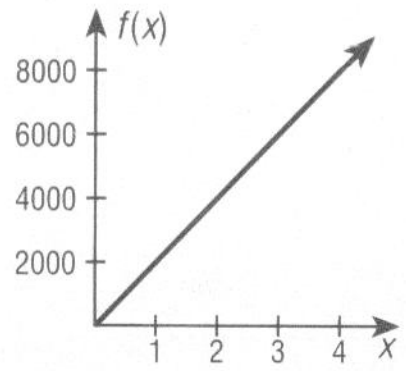

17.

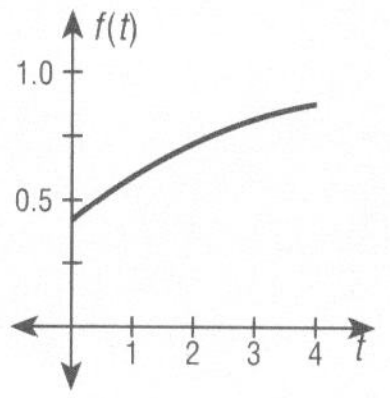

18.

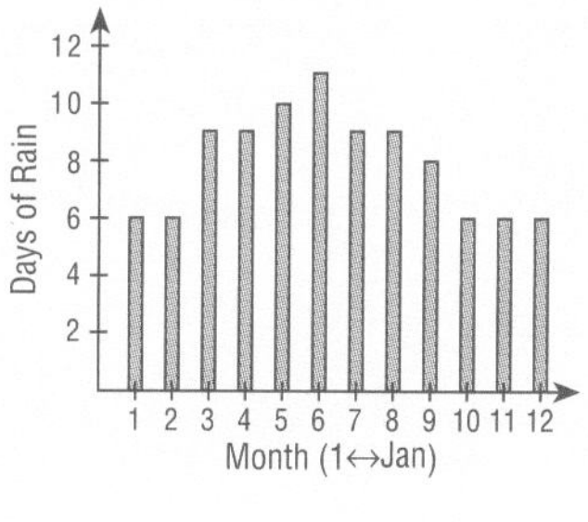

Cumulative Review Answers

7. $3x^3 + 9x^2 + 12x$
8. $2x^2 + 3x - 2$
9. $12x^3 + 20x^2 + 6x + 10$
10. $-\frac{1}{2}x^2 + 4x - 6$
11. $4x^2 - 16$
12. $4x^2 + 28x + 49$
19. $\frac{-3+\sqrt{29}}{2}, \frac{-3-\sqrt{29}}{2}$
20. $3+\sqrt{2}, 3-\sqrt{2}$
21. $2+\sqrt{17}, 2-\sqrt{17}$
22. $-4+\sqrt{17}, -4-\sqrt{17}$
23. $3+\frac{\sqrt{88}}{4}, 3-\frac{\sqrt{88}}{4}$
24. $4+\frac{\sqrt{504}}{6}, 4-\frac{\sqrt{504}}{6}$

Cumulative REVIEW ■ Chapters 7–12

In Exercises 1–6, classify the polynomial by term and by degree.

1. 436 monomial, 0
2. $7x^3 + 25x^2 + 36x - 1$ polynomial, 3
3. $\frac{1}{2}x^2 + 3x + 4$ trinomial, 2
4. $5x^4 - 1$ binomial, 4
5. $27x - 4$ binomial, 1
6. $-17x$ monomial, 1

In Exercises 7–12, perform the indicated operation. See margin.

7. $3x(x^2 + 3x + 4)$
8. $(x + 2)(2x - 1)$
9. $(4x^2 + 2)(3x + 5)$
10. $(\frac{1}{2}x - 3)(-x + 2)$
11. $(2x + 4)(2x - 4)$
12. $(2x + 7)^2$

In Exercises 13–18, use factoring to solve the equation.

13. $x^2 - 9x + 5 = -13$ 3, 6
14. $3x^2 + 8x - 4 = 12$ $-4, \frac{4}{3}$
15. $3x^2 + 18x = 0$ 0, −6
16. $-x^2 - 2x + 35 = 0$ −7, 5
17. $4x^2 - 37 = 12$ $-\frac{7}{2}, \frac{7}{2}$
18. $9x^2 + 30x + 24 = -1$ $-\frac{5}{3}$

In Exercises 19–24, solve the equation by any convenient method. See margin.

19. $x^2 + 3x - 5 = 0$
20. $x^2 - 6x + 7 = 0$
21. $x^2 - 4x - 13 = 0$
22. $x^2 + 8x - 1 = 0$
23. $2x^2 - 12x + 7 = 0$
24. $3x^2 - 24x + 6 = 0$

In Exercises 25–27, solve the proportion.

25. $\frac{3}{4m} = \frac{5}{2}$ $\frac{3}{10}$
26. $\frac{x+5}{3} = \frac{2}{x}$ −6, 1
27. $\frac{x+5}{x-8} = \frac{1}{3x}$ $-4, -\frac{2}{3}$

Geometry **In Exercises 28–30, what percent of the region is shaded blue?**

28. 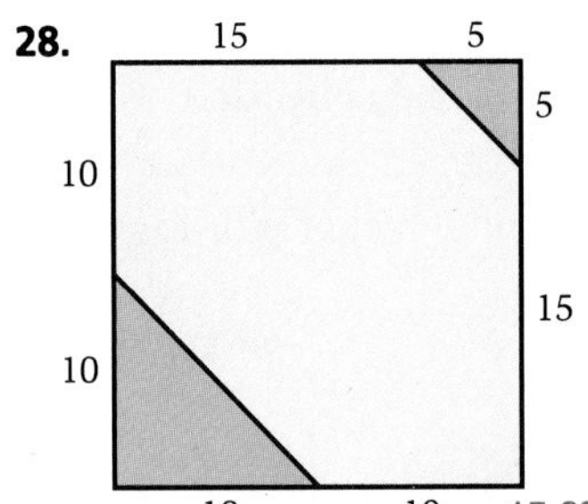

≈15.6%

29. 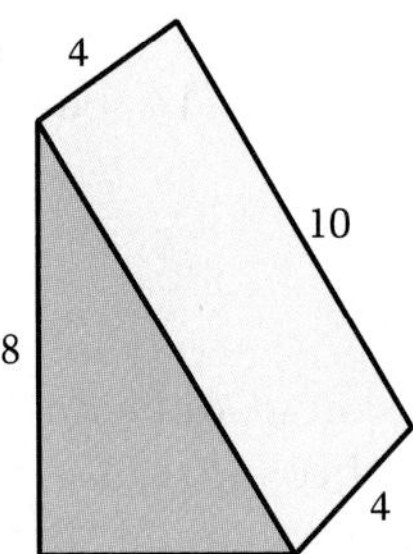

37.5%

30. 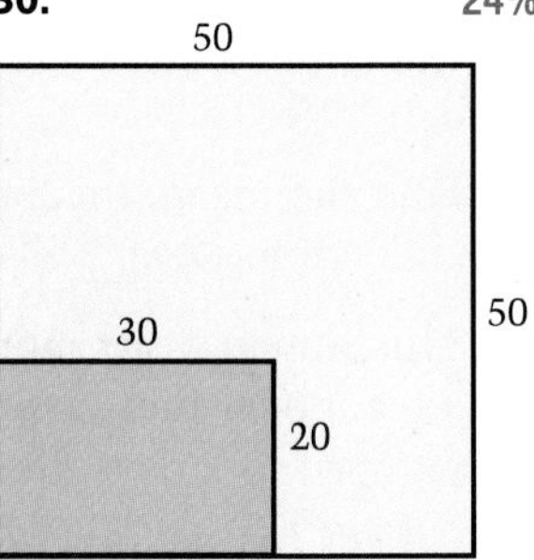

24%

31. What is 130% of 24 yards? 31.2 yards
32. 40 cm is what percent of 36 cm? 111.1%
33. \$120 is what percent of \$840? 14.3%
34. 4 days is what percent of 180 days? 2.2%

In Exercises 35–37, the variables *x* and *y* vary directly. Use the given values to find an equation that relates *x* and *y*.

35. $x = 3,\ y = 5$ $5x = 3y$
36. $x = \frac{1}{2},\ y = \frac{1}{3}$ $2x = 3y$
37. $x = 4.2,\ y = 3.1$ $3.1x = 4.2y$

In Exercises 38–40, the variables *x* and *y* vary inversely. Use the given values to find an equation that relates *x* and *y*.

38. $x = 12,\ y = 2$ $xy = 24$
39. $x = \frac{3}{2},\ y = 6$ $xy = 9$
40. $x = 6.3,\ y = 10.1$ $xy = 63.63$

Giving Book Reports **In Exercises 41–44, use the following information.**

There are 25 students in your English class. To determine the speaking order for presenting book reports, slips of paper numbered from 1 to 25 are placed in a box. Each student then draws a number to determine his or her speaking order.

41. What is the probability that the number you draw will be odd? $\frac{13}{25}$

42. What is the probability that the number you draw will be even? $\frac{12}{25}$

43. Five book reports will be given on each day. What is the probability that you will have to give your report on the first day? $\frac{1}{5}$

44. What is the probability that you will be the very last person to give his or her book report? $\frac{1}{25}$

In Exercises 45–56, simplify the expression. See margin.

45. $\frac{3y}{4} \cdot \frac{2}{6y^2}$ $\frac{1}{4y}$

46. $\frac{8x^3}{3} \cdot 5x^2$ $\frac{40x^5}{3}$

47. $\frac{5p^4}{3} \div \frac{3p^3}{2}$ $\frac{10p}{9}$

48. $\frac{(x-2)}{3x} \div \frac{4}{15x^2}$ $\frac{5x^2-10x}{4}$

49. $\frac{m}{m^2-4m+4} \cdot (m-2)$

50. $\frac{3n^2-2n-5}{n+1} \div (n+1)$

51. $(6z^2+4z) \div 2z$ $3z+2$

52. $(8x^2+4x-7) \div 4$

53. $(y^2+6y-2) \div (y-2)$

54. $(4m^3+m+6) \div (m+2)$

55. $\left(\frac{1}{3}z^3+z^2-7z\right) \div (z+3)$

56. $(27x^2+3x-2) \div (3x-1)$

In Exercises 57–62, solve the equation.

57. $\frac{5}{x+3} = \frac{6}{x+1}$ -13

58. $\frac{4}{2x+1} = \frac{1}{x}$ $\frac{1}{2}$

59. $\frac{2}{x} - \frac{1}{2} = \frac{5}{x}$ -6

60. $\frac{4}{x} + \frac{29}{30} = \frac{x}{5}$ $7\frac{1}{2}, -2\frac{2}{3}$

61. $\frac{4}{2x+1} = \frac{3}{x+1}$ $\frac{1}{2}$

62. $\frac{4x^2}{x-1} - \frac{4x}{x-1} = \frac{-3}{4x-4}$ $\frac{1}{4}, \frac{3}{4}$

In Exercises 63–68, evaluate the function at the given value of the variable.

63. $f(t) = 4t+2,\ t=4$ 18

64. $f(x) = 2^x+14,\ x=3$ 22

65. $f(x) = x^2+4x-3,\ x=-5$ 2

66. $f(t) = -3(t-7)^2,\ t=1$ -108

67. $f(x) = \frac{3}{x},\ x=6$ $\frac{1}{2}$

68. $f(t) = \frac{6}{t-2}+4,\ t=-10$ $3\frac{1}{2}$

In Exercises 69–72, match the function with its graph.

69. $f(x) = \frac{1}{2}(x+2)^2$ b

70. $f(x) = -\frac{4}{3}x+4$ c

71. $f(x) = \left(\frac{1}{2}\right)^x - 3$ d

72. $f(x) = \frac{1}{x-2}+3$ a

a.

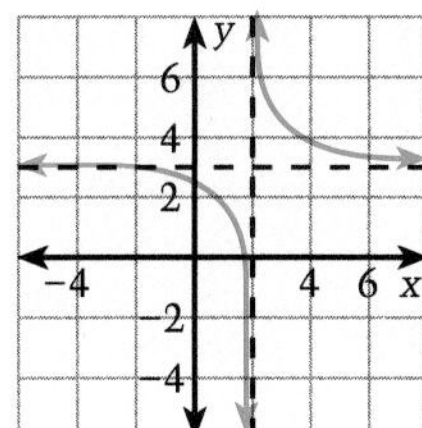

b.

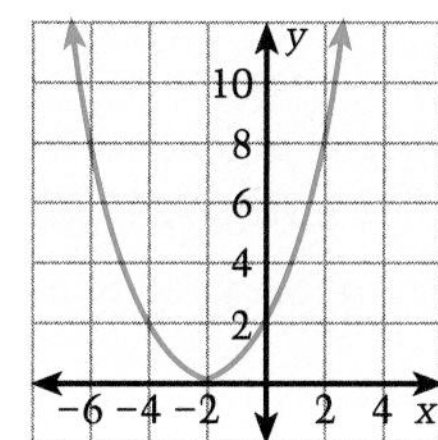

c.

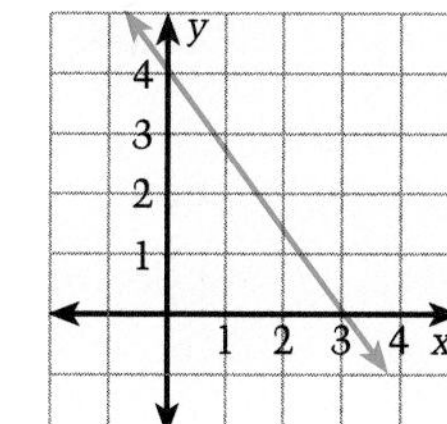

d.

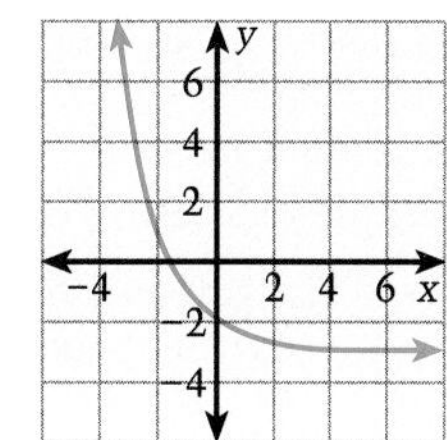

Answers

49. $\frac{m}{m-2}$

50. $\frac{3n-5}{n+1}$

52. $2x^2+x-\frac{7}{4}$

53. $y+8+\frac{14}{y-2}$

54. $4m^2-8m+17+\frac{-28}{m+2}$

55. $\frac{1}{3}z^2-7+\frac{21}{z+3}$

56. $9x+4+\frac{2}{3x-1}$

Answers

79.

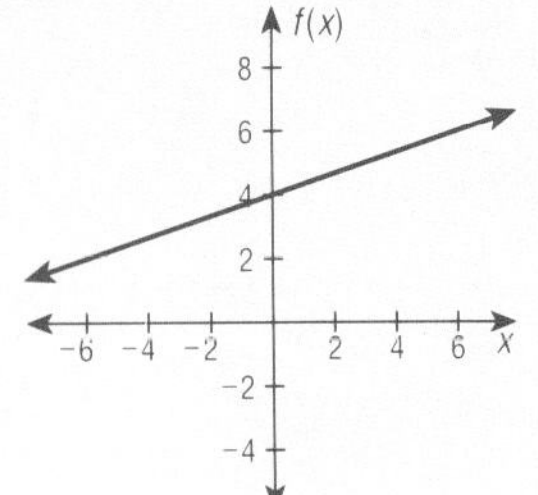

80.

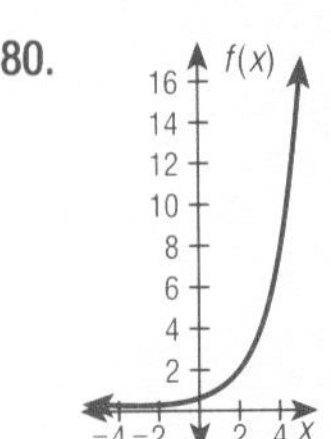

81.

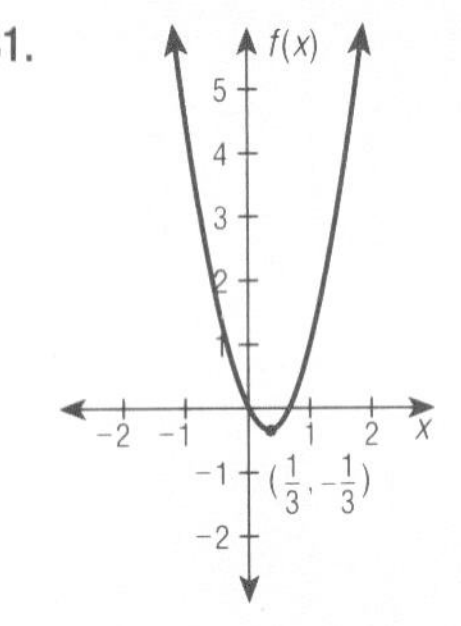

82.

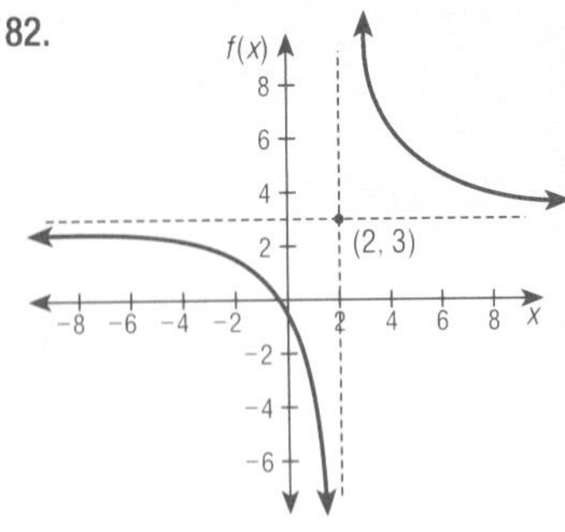

83. $y = \dfrac{54420 + 2900t}{31665 + 1545t}$

84. increasing; in 1980, the ratio $\dfrac{54420}{31665} \approx 1.72$; in 1990, the ratio $\dfrac{83420}{47115} \approx 1.77$

In Exercises 73 and 74, the two rectangles are proportional to each other. Find the length represented by *x*.

73.

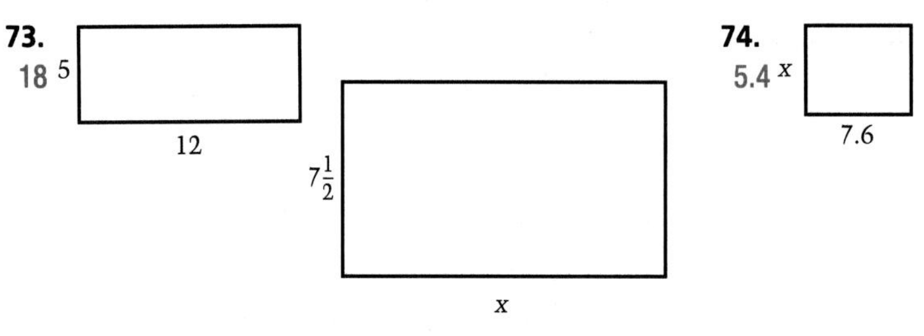

74.

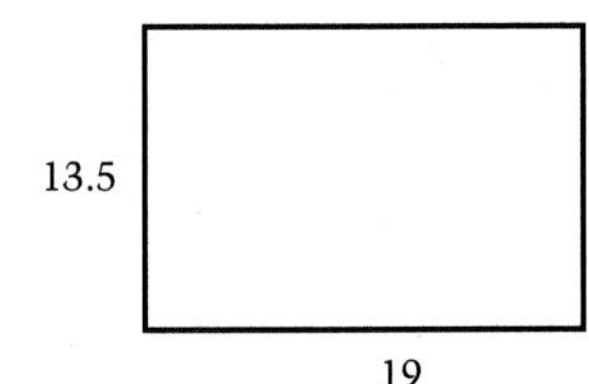

75. *Sweatshirt Sale* Sweatshirt City is having a "20%-off sale" on all college sweatshirts. The sale price is \$18.80. What was the original price? \$23.50

76. *Snack Recipe* A snack recipe calls for 5 cups of bite-sized cereal and $1\frac{1}{2}$ cups of pretzel sticks. Find an equation that relates the amount of cereal, c, to the amount of pretzel sticks, p. $\frac{3}{2}c = 5p$

Basketball Booster Club **In Exercises 77 and 78, use the following information.**

For a \$10 membership fee to the Basketball Booster Club, a parent of a high school student receives a \$1 discount off the \$3 price of a basketball ticket at each basketball game.

77. How many games must a parent attend to average \$2.50 per ticket? (Include the \$10 membership fee.) 20

78. After the basketball season, one member of the booster club said that she ended paying more per game than if she had not joined. What is the maximum number of games she attended? 9

In Exercises 79–82, sketch the graph of the function. See margin.

79. $f(x) = \frac{1}{3}x + 4$ **80.** $f(x) = 2^{x-1}$ **81.** $f(x) = 3x^2 - 2x$ **82.** $f(x) = \dfrac{3x + 1}{x - 2}$

Pilots and Flight Attendants **In Exercises 83 and 84, use the following information.**

In 1980, commercial airlines in the United States employed 31,665 pilots and 54,420 flight attendants. For the next 10 years these numbers increased by an average of 1545 pilots per year and 2900 flight attendants per year. ***(Source: Air Transport Association of America)*** See margin.

83. Create a model that gives the ratio of employed flight attendants to employed pilots for 1980 through 1990. (Let $t = 0$ represent 1980.)

84. Was the ratio of flight attendants to pilots increasing or decreasing from 1980 to 1990? Explain.

Snowfall **In Exercises 85–88, use the following information.**

The table shows the amount of measurable snow (in inches) that fell in Erie, Pennsylvania, during December 1988. (No snow fell on days not listed.)

Day	1	2	3	8	9	10	11	13
Snowfall	6.4	0.1	0.3	0.1	1.8	8.2	0.7	0.9

Day	15	16	17	18	25	27	28	29
Snowfall	0.2	0.6	2.9	1.1	0.3	0.2	4.0	0.4

85.

Stem	Leaves
0	1 1 2 2 3 3 4 6 7 9
1	1 8
2	9
4	0
6	4
8	2

85. Use a stem-and-leaf plot to order the recorded snowfall in increasing order.

86. What was the mean daily snowfall for the sixteen days of snow? 1.8 inches

87. What was the median daily snowfall for the sixteen days of snow? 0.65 inches

88. Construct a box-and-whisker plot for the data. What does this indicate about snowfall in Erie? See margin.

Snowblowing **In Exercises 89 and 90, use the following information.**

Suppose that you lived in Erie during December 1988. On December 1, to test your new snowblower, you cleared the sidewalks around the entire block shown in the diagram.

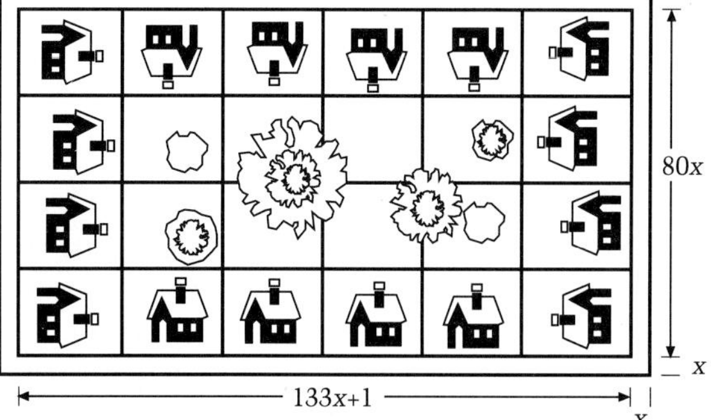

89. The sidewalk around the block is 3 feet wide. How many square feet of snow did you clear? 3876

90. How many *cubic feet* of snow did you clear? (Use the information given in the above table, and assume that all of the day's snow had fallen before you began clearing.) 2067.2

Snowblowing for Profit **In Exercises 91 and 92, use the following information.**

After seeing how well you cleared the sidewalks, three neighbors offered to pay you to snowblow their driveways. Each week after that, one more neighbor became a client. Your weekly earnings, R, during an eight-week period can be modeled by

$$R = 2t + 6$$

where $t = 0$ represents the first week.

91. How much did you charge to snowblow each driveway? $2 per week

92. How much did you earn during the eight weeks? $104

88. Box-and-whisker plot: 0.1, 0.25, 0.65, 2.35, 8.2

The usual daily snowfall in December is less than an inch; half the time the snowfall is less than 0.65 in., and three-fourths of the time the snowfall is less than 2.35 in.

As the *Voyager* example suggests, radicals are frequently used to express relationships found in physics, chemistry, economics, and mathematics, particularly in statistics. In this chapter, students will learn how to algebraically manipulate radicals and solve radical equations.

The Chapter Summary on page 716 provides you and the students with a synopsis of the chapter. It identifies key skills, concepts, and vocabulary. You may want to have students look at the Chapter Summary as an overview before beginning the chapter.

CHAPTER 13

Radicals and More Connections to Geometry

LESSONS

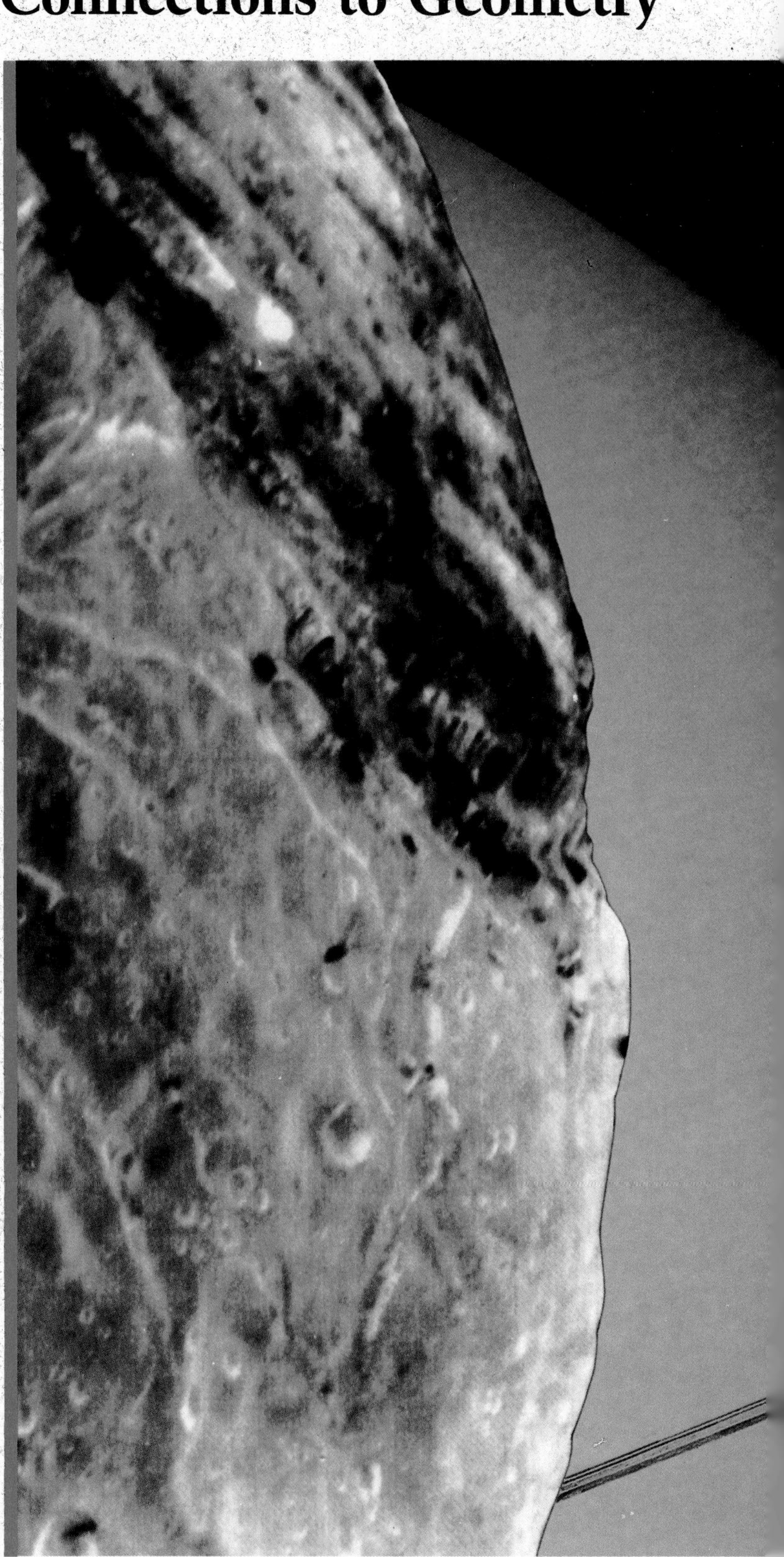

This montage of Voyager 2 images obtained in 1986 shows the blue-green planet Uranus overlaid with portion of a close-approach image of Miranda, Uranus' moon.

Real Life
The Voyager Program

The Moons of Uranus

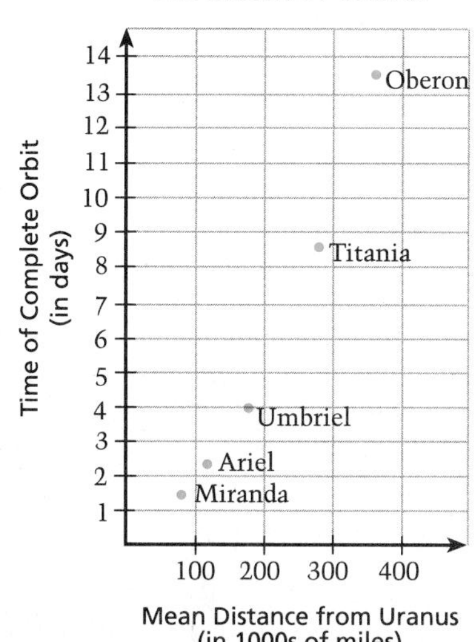

Voyager 1 and *Voyager 2* were launched by the United States in 1977 to observe Jupiter and Saturn. *Voyager 1* reached Jupiter in 1979 and Saturn in 1980. *Voyager 2* also passed by Jupiter and Saturn and then continued on to observe the outer planets. In 1986, *Voyager 2* transmitted stunning pictures of the rings of Uranus.

To escape Earth's gravitational pull, the *Voyagers* had to be propelled upward from Earth's surface with a velocity of at least

$$v = \sqrt{\frac{2Gm}{r}} \approx 11{,}186 \text{ (meters per second)}$$

where G is the universal gravitational constant, m is the mass of Earth, and r is Earth's radius.

For each of the five moons of Uranus, the graph relates the distance of the moon from Uranus with the time the moon takes to orbit the planet.

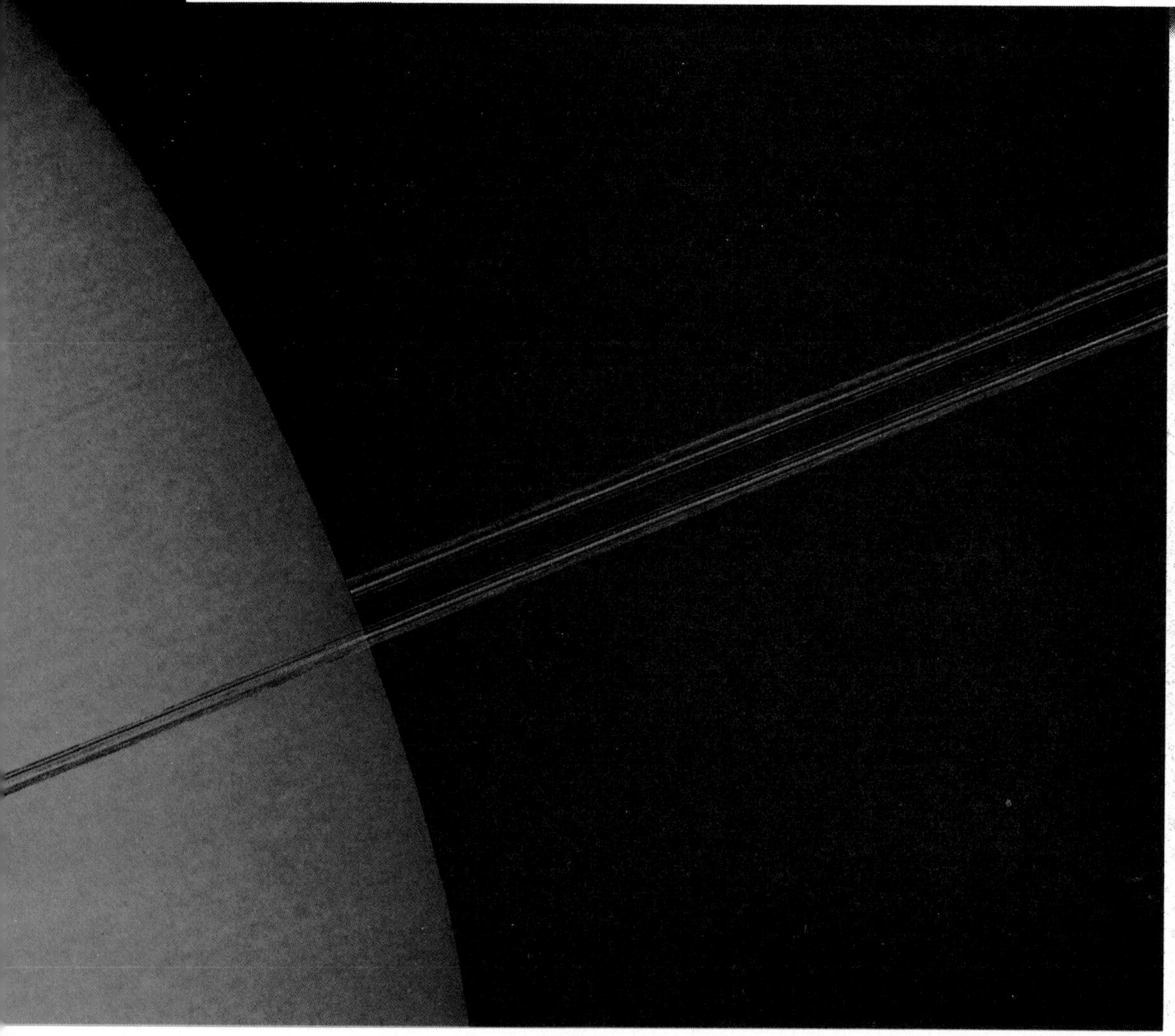

The universal gravitational constant, *G*, was predicted by Sir Isaac Newton in 1687, but the determination of its value, $G \approx 6.67 \times 10^{-11}\text{m}^3\text{kg}^{-1}\text{s}^{-2}$, was not made until 1798 by Henry Cavendish. Cavendish used the result to determine the mass and density of the earth.

Radicals and More Connections to Geometry

PACING CHART

The daily Pacing Chart is meant to help you adjust your teaching pace. Students in the full course should finish the entire text by the end of the year. Students in the basic course may not complete the entire text in the school year. The Pacing Chart for each chapter contains suggestions for lessons that require more than one day and lessons that may be omitted for the basic course.

DAY	FULL COURSE	BASIC COURSE
1	13.1	13.1
2	13.2	13.2
3	13.2	13.2
4	13.2	13.2
5	13.3	13.3
6	13.3	13.3
7	13.3	13.3
8	13.4 & Using a Graphing Calculator	Chapter Review
9	Mid-Chapter Self-Test	Chapter Review
10	13.5	Chapter Test
11	13.6	
12	Chapter Review	
13	Chapter Review	
14	Chapter Test	

CHAPTER ORGANIZATION

LESSON	PAGES	GOALS	MEETING THE NCTM STANDARDS
13.1	676-681	1. Find the distance between two points 2. Find the midpoint between two points	Problem Solving, Communication, Connections, Geometry
13.2	682-687	1. Simplify radicals by applying their properties 2. Use radicals in real-life situations	Problem Solving, Communication, Structure, Connections, Geometry
Mixed Review	688	Review of algebra and arithmetic	
Career Interview	688	Translation-service owner	Connections
13.3	689-694	1. Add and subtract radical expressions 2. Use radical operations in real-life problems	Problem Solving, Communication, Connections, Geometry
13.4	695-700	1. Solve radical equations 2. Use radical equations to solve real-life problems	Problem Solving, Communication, Connections, Geometry
Using a Calculator	701	Use a graphing calculator to explore the domain of a function	Technology
Mid-Chapter Self-Test	702	Diagnose student weaknesses and remediate with correlated Reteaching worksheets	
13.5	703-709	1. Use the tangent of an angle to find the lengths of the sides of a right triangle 2. Use the tangent to solve real-life problems	Problem Solving, Communication, Connections, Geometry, Trigonometry
Mixed Review	709	Review of algebra and arithmetic	
13.6	710-715	Use properties of algebra to prove a statement true and use counterexamples to prove a statement false	Problem Solving, Communication, Connections, Structure, Reasoning
Chapter Summary	716	A restatement of what has been learned, why it has been learned, and how it fits into the structure of algebra	Structure, Connections
Chapter Review	717-719	Review of concepts and skills learned in the chapter	
Chapter Test	720	Diagnose student weaknesses and remediate with correlated Reteaching worksheets	

LESSON RESOURCES

MEETING INDIVIDUAL NEEDS

RETEACHING For students who need to spend more time on basics:

If a mid-chapter self-test or chapter test indicates a deficiency, teachers can help students with the appropriate ***Reteaching Copymaster.***

PRACTICE For students who need more practice:

Additional exercises like those in the Pupil's Edition are provided for each lesson in ***Extra Practice Copymasters.***

ENRICHMENT For enriching and broadening students' experiences:

Problem of the Day copymasters in ***Teaching Tools*** provide a daily opportunity to use logical reasoning, looking for a pattern, writing an equation, and other routine and non-routine problem-solving strategies.

Math Log copymasters in ***Alternative Assessment*** provide opportunities to report on investigations, research, and open-ended problems.

Enriching activities with graphing and scientific calculators and computers are provided in ***Technology: Using Calculators and Computers.***

The ***Applications Handbook*** provides additional information about the cross-curriculum topics such as astronomy, chemistry, physics, sports, economics, genetics, and music that are integrated into the Pupil's Edition.

LESSON	13.1	13.2	13.3	13.4	13.5	13.6
PAGES	676-681	682-687	689-694	695-700	703-709	710-715
Teaching Tools						
Transparencies	✓					
Problem of the Day	✓	✓	✓	✓	✓	✓
Warm-up Exercises	✓	✓	✓	✓	✓	✓
Answer Masters	✓	✓	✓	✓	✓	✓
Extra Practice Copymasters	✓	✓	✓	✓	✓	✓
Reteaching Copymasters	Teacher-directed and independent activities tied to results on the Mid-Chapter Self-Tests and Chapter Tests					
Color Transparencies	✓	✓	✓			✓
Applications Handbook	Additional background information is supplied for many real-life applications.					
Technology Handbook	Calculator and computer worksheets are supplied for appropriate lessons.					
Complete Solutions Manual	✓	✓	✓	✓	✓	✓
Alternative Assessment	Assess student's ability to reason, analyze, solve problems, and communicate using mathematical language.					
Formal Assessment	Mid-Chapter Self-Tests, Chapter Tests, Cumulative Tests, and Practice for College Entrance Tests					
Computer Test Bank	Customized tests can be created by choosing from over 2000 items.					

13.1 The Distance Formula

The Distance Formula is a very useful tool, used routinely in the natural sciences and in engineering. It is essential in navigation and in any environment that uses coordinate geometry. Other mathematics courses, such as geometry, trigonometry, and calculus, will make use of the Distance Formula.

13.2 Simplifying Radicals

Real-valued radical expressions occur frequently in algebraic computations. Although radical expressions can be approximated by decimals, it is not always useful to do so. For example, suppose $\sqrt{3}$ is obtained in a computation and its decimal value, 1.732, is used. In a related but different computation, the decimal value 1.7320508 is obtained. Are the two quantities identical? We could easily determine the answer to that question if expressions involving radicals were written in exact form, that is, keeping the radical expressions intact.

13.3 Operations with Radicals

Many applications of radical expressions involve addition and subtraction of radicals. In this lesson, students use the Distributive Properties of Multiplication over Addition and over Subtraction to obtain sums and differences of radicals. Students will use the addition and subtraction techniques for radicals to check solutions of quadratic equations.

13.4 Solving Radical Equations

Many real-life situations are modeled by relationships involving radicals. It is necessary, therefore, to be able to solve equations containing radicals.

13.5 The Tangent of an Angle

Right triangles are used to model many real-life situations in engineering, construction, and the natural sciences. Ideas involving the tangent of an angle occur frequently in work involving right triangles. In addition, many algebraic situations make use of the relationships among the sides of a right triangle.

13.6 Formal Use of Properties of Algebra

The final lesson in the text is both a culmination of the preceding lessons and a bridge between algebra and geometry. The attention to proof identifies one of the primary strengths of our mathematical system and serves as a general introduction to deductive reasoning.

Problem of the Day

Barbara and Ramir raced each other. Barbara averaged 9 mph for the entire course, while Ramir averaged 8 mph for the first half of the course and 10 mph for the second half. Who won? Barbara

ORGANIZER

Warm-Up Exercises

1. Using a calculator, evaluate to two decimal places.
   a. $\sqrt{45}$ 6.71
   b. $\sqrt{4 + 23}$ 5.20
   c. $\sqrt{43 - 17}$ 5.10
   d. $\sqrt{-32}$ undefined
2. Determine the distance in both the vertical and horizontal directions from point A to point B.
   a. $A(3, 4)$, $B(-2, -5)$
   b. $A(-10, 7)$, $B(4, -12)$
   c. $A(6, 8)$, $B(9, 8)$

   Remember that distances are positive.

	Distances: Horizontal	Vertical
a.	5	9
b.	14	19
c.	3	0

3. Find the slopes of the lines containing the points in Warm-Up Exercise 2.
   a. $\frac{9}{5}$ b. $\frac{-19}{14}$ c. 0

Lesson Resources

Teaching Tools
- Transparency: 4
- Problem of the Day: 13.1
- Warm-up Exercises: 13.1
- Answer Masters: 13.1

Extra Practice: 13.1

Color Transparencies: 75, 76

13.1 The Distance Formula

What you should learn:

Goal 1 How to find the distance between two points

Goal 2 How to find the midpoint between two points

Why you should learn it:

On maps and other grids, you often need to find the distance or midpoint between two points not on the same grid line.

Goal 1 Finding the Distance Between Points

LESSON INVESTIGATION

Investigating Distance

Partner Activity Consider the two points $A(2, 1)$ and $B(6, 4)$ as shown in the graph at the left. Your goal is to find the distance between A and B.

a. Plot A and B on graph paper. Then draw a right triangle that has the line segment $\overline{AB}$ as its hypotenuse.

b. Label the coordinates of the vertex C.

c. Find the length of the legs of the right triangle. Label these lengths a and b.

d. Use the Pythagorean Theorem to find the length of the hypotenuse. It is the distance between A and B.

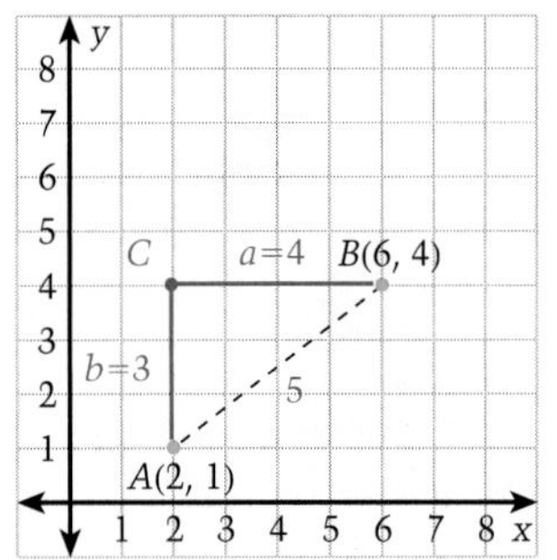

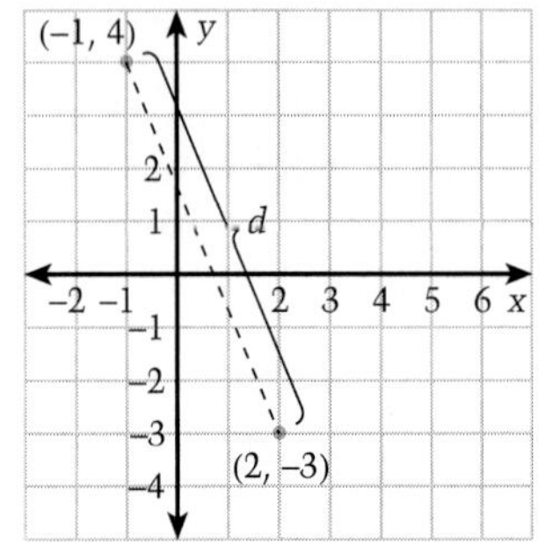

By performing this investigation with two general points (x_1, y_1) and (x_2, y_2), you can obtain the *Distance Formula.*

The Distance Formula

The distance, d, between the points (x_1, y_1) and (x_2, y_2) is

$$d = \sqrt{(x_2 - x_1)^2 + (y_2 - y_1)^2}.$$

Example 1 *Finding the Distance Between Points*

Find the distance between $(-1, 4)$ and $(2, -3)$.

Solution Let $(x_1, y_1) = (-1, 4)$ and $(x_2, y_2) = (2, -3)$.

$$\begin{aligned} d &= \sqrt{(2 - (-1))^2 + (-3 - 4)^2} \\ &= \sqrt{9 + 49} \\ &= \sqrt{58} \\ &\approx 7.62 \end{aligned}$$

The distance between the points is about 7.62 units. ■

So far in the book we have been using only one part of the Pythagorean theorem. *If a triangle is a right triangle,* then the lengths of its sides satisfy the equation $a^2 + b^2 = c^2$. It has another part that is also true. *If the lengths of the sides of a triangle satisfy the equation* $a^2 + b^2 = c^2$, then the triangle is a right triangle. (The second statement is the **converse** of the first statement.)

Connections
Geometry

Example 2 *Deciding Whether a Triangle Is a Right Triangle*

Show that the points (2, 1), (4, 0), and (5, 7) are vertices of a right triangle.

Solution Use the Distance Formula to find the lengths of the three sides.

$$d_1 = \sqrt{(5-2)^2 + (7-1)^2} = \sqrt{9+36} = \sqrt{45}$$
$$d_2 = \sqrt{(4-2)^2 + (0-1)^2} = \sqrt{4+1} = \sqrt{5}$$
$$d_3 = \sqrt{(5-4)^2 + (7-0)^2} = \sqrt{1+49} = \sqrt{50}$$

Because $d_1^2 + d_2^2 = 45 + 5 = 50 = d_3^2$, you can conclude that the triangle is a right triangle. ■

When you want to use the Distance Formula to find a distance in a real-life problem, the first step is to draw a diagram and assign coordinates to the points. This process is called *superimposing* a coordinate system on the diagram.

Example 3 *Finding the Distance of a Pass*

In a football game, a quarterback throws a pass from the 5 yard line, 10 yards from the sideline. The pass is caught by a wide receiver on the 45 yard line, 40 yards from the same sideline. How long was the pass?

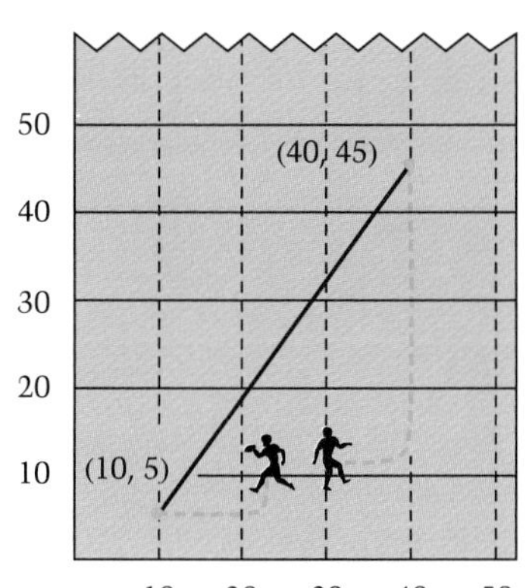

Solution You can begin by superimposing a coordinate system on the football field. The quarterback throws the ball from the point (10, 5). It is caught at the point (40, 45).

$$d = \sqrt{(40-10)^2 + (45-5)^2}$$
$$= \sqrt{900 + 1600}$$
$$= \sqrt{2500}$$
$$= 50 \text{ yards}$$

The pass was 50 yards long. ■

LESSON Notes

GOAL 1 The Distance Formula is an application of the Pythagorean Theorem. For points (x_1, y_1) and (x_2, y_2) in the diagram, the horizontal distance is $x_2 - x_1$ and the vertical distance is $y_2 - y_1$.

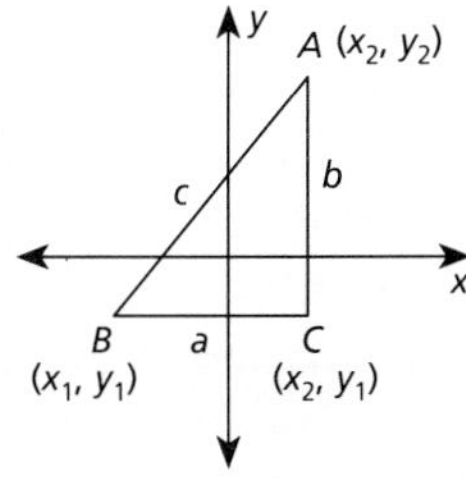

The line segments representing those distances can be used to form a right triangle, from which the length, c, of the third side can be found using the Pythagorean Theorem.

$c^2 = a^2 + b^2$, or, using d for distance,
$d^2 = (x_2 - x_1)^2 + (y_2 - y_1)^2$.

Take the square root of both sides of the equation to find d.

$$d = \sqrt{d^2}$$
$$= \sqrt{(x_2 - x_1)^2 + (y_2 - y_1)^2}$$

Examples 1–2

It may help students to see the sides of the triangle positioned on the coordinate plane.

Example 3

Although the pass goes on record as a 50-yard pass, the ball actually traveled a greater distance. Can your students explain why? The path of the football is most likely to be parabolic, and consequently, longer than the straight-line distance computed.

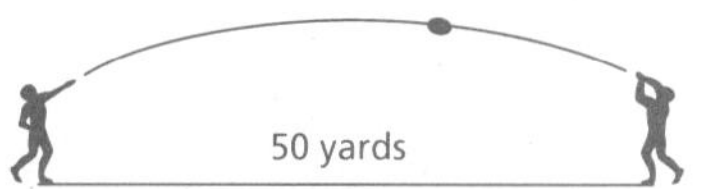

Example 4

SOCIAL STUDIES Students might like to investigate the history of the Pony Express. Who founded it? Were there any now-famous riders? Why did it end? etc.

Extra Examples

Here are additional examples similar to **Examples 1–4.**

1. Find the distance between points A and B.
   - **a.** $A(4, 5)$ and $B(-2, 6)$
   - **b.** $A(-14, -4)$ and $B(1, 8)$
   - **c.** $A(-2, 10)$ and $B(5, -3)$

   a. $\sqrt{37} \approx 6.08$
   b. $\sqrt{369} \approx 19.21$
   c. $\sqrt{218} \approx 14.76$
2. Are the given points vertices of a right triangle?
   - **a.** $A(9, 5)$, $B(-1, 1)$, and $C(1, -4)$
   - **b.** $A(-7, -5)$, $B(7, 1)$, and $C(-3, 5)$
   - **c.** $A(1, 4)$, $B(-3, -4)$, and $C(4, 9)$

   a. yes b. yes c. no
3. Find the midpoint between the points that are given in Exercise 1.

   a. $(1, 5.5)$
   b. $(-6.5, 2)$
   c. $(1.5, 3.5)$

Check Understanding

1. Explain the connection between the Distance Formula and the Pythagorean Theorem. See Lesson Notes for Goal 1.
2. Can the Distance Formula be used if the points A and B lie on a vertical line? yes
3. Explain how the Distance Formula can be used to determine if a given triangle is a right triangle. See Example 2.

Goal 2 Finding the Midpoint of Two Points

The coordinates of the midpoint between two points are the averages of the coordinates of the two points.

The Midpoint Formula

The **midpoint** between (x_1, y_1) and (x_2, y_2) is $\left(\frac{x_1 + x_2}{2}, \frac{y_1 + y_2}{2}\right)$.

The Pony Express was in operation from April 1860 to October 1861. Its riders were mostly young men, fourteen to eighteen years old, who rode about 70 miles each along the 2000-mile trip from St. Joseph, Missouri, to Sacramento, California. At each Pony Express station, riders were given a fresh horse. The entire trip took between 8 and 15 days, depending on the weather.

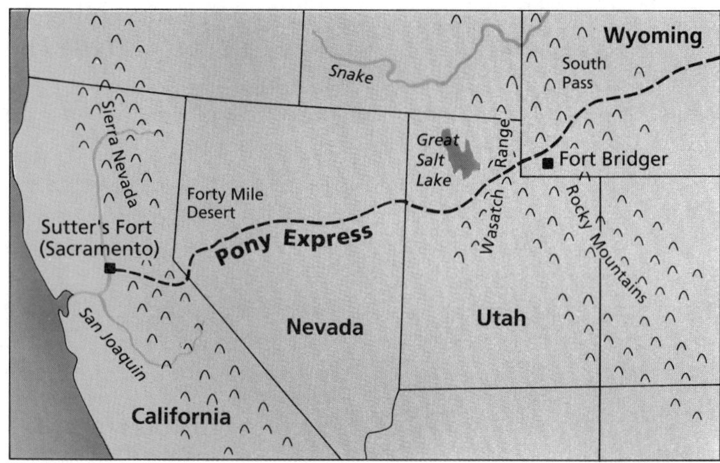

Real Life
Pony Express

Example 4 *Finding the Midpoint Between Points*

Pony Express stations were about 10 miles apart. The latitude-longitude coordinates of 2 former stations in western Nevada are (39.5 N, 117.8 W) and (39.3 N, 118.2 W). These two stations are about 20 miles apart. Estimate the coordinates of the station that was between them.

Solution Let (x_1, y_1) be $(39.5, 117.8)$ and (x_2, y_2) be $(39.3, 118.2)$. The approximate coordinates of the midpoint station would be

$$\text{midpoint} = \left(\frac{39.5 + 39.3}{2}, \frac{117.8 + 118.2}{2}\right) = (39.4, 118.0).$$ ■

Communicating about ALGEBRA

SHARING IDEAS about the Lesson

Understand a Formula Explain why either of the two points may be chosen to be (x_1, y_1) when using the Distance Formula or the Midpoint Formula. See margin.

EXERCISES

Guided Practice

CRITICAL THINKING about the Lesson

1. Plot the points (3, 6) and $(-3, -2)$ in a coordinate plane. Use the scale of the plane to estimate the distance between the two points. Then use the Distance Formula to find the exact distance between the points. 10
2. Find the midpoint of $(6, -1)$ and $(-4, 5)$. Check your result with a graph. (1, 2)
3. State the converse of the statement. "If a triangle is a right triangle, then the lengths of its sides satisfy the equation $a^2 + b^2 = c^2$." See margin.

Independent Practice

In Exercises 4–7, use the coordinate plane to estimate the distance between the two points. Then use the Distance Formula to find the distance between the points. Round the result, if necessary, to two decimal places. Estimates will vary.

4. (1, 5), $(-3, 1)$ 5.66 **5.** $(-2, 2)$, (2, 1) 4.12 **6.** $(-3, -2)$, (4, 1) 7.62 **7.** $(5, -2)$, $(-1, 1)$ 6.71

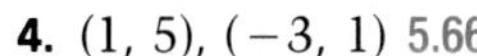
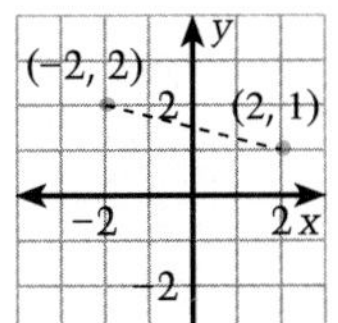

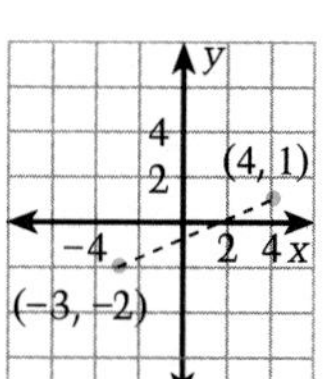

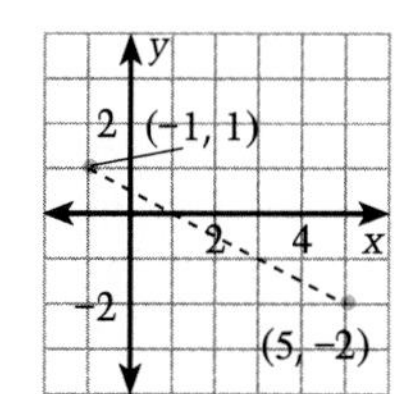

In Exercises 8–15, find the distance between the two points. Round the result, if necessary, to two decimal places. See margin.

8. $(1, 0), (9, -4)$ **9.** $(1, -7), (-2, 2)$ **10.** $(3, -1), (0, 3)$ **11.** $(1, -4), (-4, 6)$

12. $(6, 9), (-3, 1)$ **13.** $(-2, 2), (1, 5)$ **14.** $\left(\frac{1}{2}, \frac{1}{4}\right), \left(-\frac{1}{2}, \frac{5}{4}\right)$ **15.** $\left(\frac{1}{2}, \frac{3}{2}\right), (1, 1)$

In Exercises 16–19, decide whether the three points are the vertices of a right triangle.

16. $(4, 0), (2, 1), (-1, -5)$ Yes **17.** $(4, 5), (1, 0), (-1, 2)$ No

18. $(1, -3), (3, 2), (-2, 4)$ Yes **19.** $(-1, -1), (10, 7), (2, 18)$ Yes

In Exercises 20–25, find the midpoint between the two points.

20. $(-4, 4), (2, 0)$ $(-1, 2)$ **21.** $(0, 0), (0, 10)$ $(0, 5)$ **22.** $(2, 1), (14, 6)$ $(8, \frac{7}{2})$

23. $(-1, 0), (6, 2)$ $(\frac{5}{2}, 1)$ **24.** $(-2, 2), (3, -10)$ $(\frac{1}{2}, -4)$ **25.** $(1, 6), (4, 2)$ $(\frac{5}{2}, 4)$

Communicating *about* ALGEBRA

Answer

In the Distance Formula, squaring a positive difference gives the same result as squaring a negative difference. In the Midpoint Formula, addition of coordinates is commutative.

COOPERATIVE LEARNING
Encourage students to work together to complete the following task. Given any three points in a coordinate plane, use the Distance Formula to indicate whether or not they lie on the same straight line. Provide examples of this use of the Distance Formula for both cases.

EXERCISE Notes

Assignment Guide

Basic/Average: Ex. 1–3, 4–29 odd, 34–38, 39–49 odd

Above Average: Ex. 1–3, 4–31 odd, 32–36, 39–49 odd, 50

Advanced: Ex. 1–3, 9–31 odd, 32–36, 39–49 odd, 50

Selected Answers
Exercises 1–3, 4–47 odd

More Difficult Exercises
Exercises 26, 27, 31–33, 49, 50

Guided Practice

Ex. 1–3 COOPERATIVE LEARNING These exercises provide an opportunity for students to visually interpret the Distance and Midpoint Formulas. As a check for understanding, have students work together to draw, plot, or check the graphs.

Answer

3. If the lengths of the sides of a triangle satisfy the equation $a^2 + b^2 = c^2$, then the triangle is a right triangle.

Independent Practice

▶ **Ex. 4–7** Students could make scale drawings and measure the lengths of the line segments before applying the Distance Formula.

▶ **Ex. 8–19** Caution students that in Example 1, $\sqrt{9+49} = \sqrt{58}$ and not $\sqrt{9} + \sqrt{49}$.

Answers

8. 8.94 9. 9.49 10. 5
11. 11.18 12. 12.04 13. 4.24
14. 1.41 15. 0.71

▶ **Ex. 28–30** Caution students that each unit length represents an actual length of 95 miles. For example, Pierre, South Dakota is located at $(95 \cdot 0.2, 95 \cdot 4.3) = (19, 408.5)$.

Answers

32.

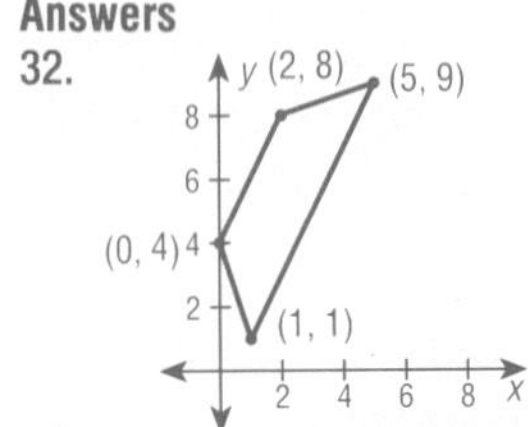

$m = \frac{8-4}{2-0} = 2$, $m = \frac{9-1}{5-1} = 2$;
$d = \sqrt{(2-5)^2 + (8-9)^2} = \sqrt{10}$,
$d = \sqrt{(1-0)^2 + (1-4)^2} = \sqrt{10}$

33.

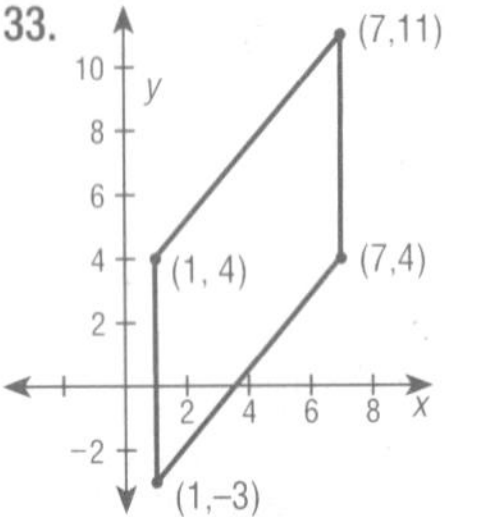

$d_1 = \sqrt{(7-1)^2 + (11-4)^2}$
$= \sqrt{85}$,
$d_1 = \sqrt{(7-1)^2 + (4-(-3))^2}$
$= \sqrt{85}$;
$d_2 = \sqrt{(7-7)^2 + (11-4)^2} = 7$,
$d_2 = \sqrt{(1-1)^2 + (4-(-3))^2} = 7$

Geometry **In Exercises 26 and 27, find the perimeter of the polygon.**

26.

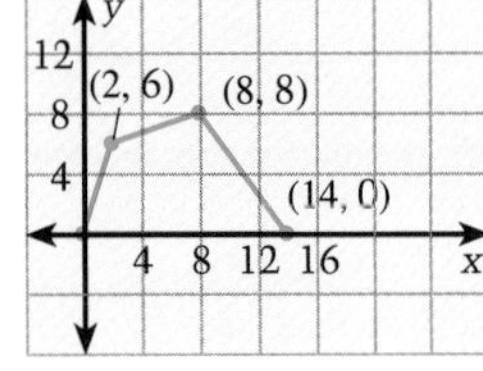

$24 + 2\sqrt{40}$

27.

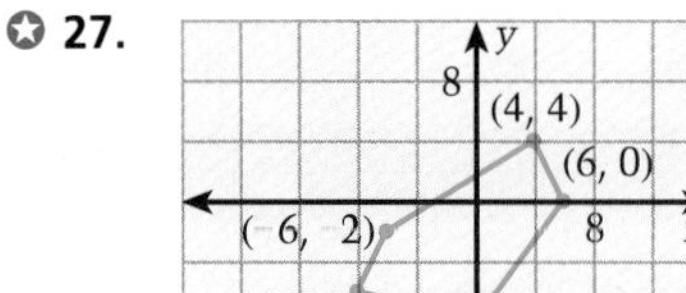

$10 + 2\sqrt{20} + \sqrt{68} + \sqrt{136}$

Distance on a Map **In Exercises 28–30, use the following information.**

Each square on the grid that is superimposed on the map represents a square that is 95 miles by 95 miles.

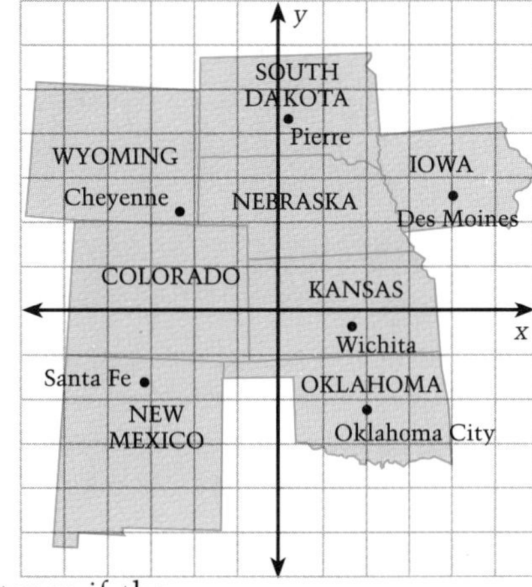

28. Use the map to estimate the distance between Pierre, South Dakota, and Santa Fe, New Mexico. ≈650 mi.

29. Use the map to estimate the distance between Wichita, Kansas, and Cheyenne, Wyoming. ≈460 mi.

30. Use the map to estimate the distance between Oklahoma City, Oklahoma, and Des Moines, Iowa. ≈470 mi.

31. *The Spider and the Fly* A spider and a fly are on opposite walls of a rectangular room. The spider asks the fly if it can come over and "visit" the fly. The fly believes that the shortest distance, walking on one of the room's surfaces, is 42 feet. (That is the distance if the spider walks straight up the wall, over the ceiling, and straight down the opposite wall.) So the fly agrees to the visit, *provided* the spider can find a path that is shorter than 42 feet. Use the "unfolded room" and the Distance Formula to explain the fly's fatal miscalculation. The shortest path is 40 ft.

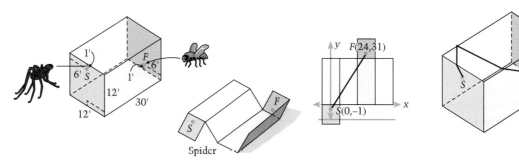

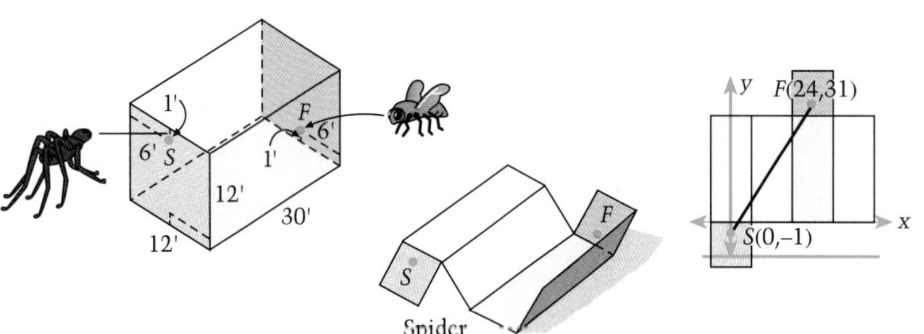

32. *Trapezoid* A trapezoid is **isosceles** if its two opposite nonparallel sides have the same length. Sketch the polygon whose vertices are (1, 1), (5, 9), (2, 8), and (0, 4). Show that it is a trapezoid by showing that two of the sides are parallel. Then use the Distance Formula to show that the trapezoid is isosceles. See margin.

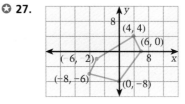

33. *Parallelogram* Sketch the parallelogram whose vertices are (1, 4), (7, 11), (1, −3), and (7, 4). Use the Distance Formula to show that lengths of opposite sides of the parallelogram are equal. See margin.

34. *Diagonals of a Parallelogram* Sketch the parallelogram whose vertices are (−1, 5), (−3, 5), (1, −3), and (3, −3). Find the midpoint of each diagonal. What do you notice about the midpoints of the diagonals of a parallelogram? Each midpoint is (0, 1); they are the same. See Additional Answers for the sketch.

Integrated Review

In Exercises 35–38, evaluate the expression.

35. $\sqrt{a^2 + b^2}$ when $a = 4$ and $b = 9$ $\sqrt{97}$

36. $c^2 - b^2$ when $c = 13$ and $b = 6$ 133

37. $(x - 2)^2 + (y + 3)^2$ when $x = 4$ and $y = -2$ 5

38. $\sqrt{x^2 + y^2}$ when $x = 1$ and $y = 6$ $\sqrt{37}$

Geometry **In Exercises 39–42, the lengths of two sides of a right triangle are given. Find the length of the other side.**

39. 24

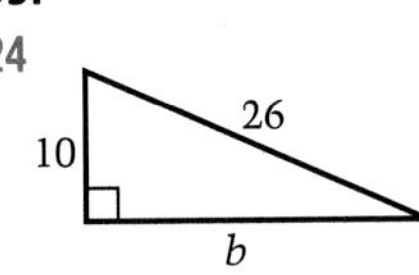

40. 10

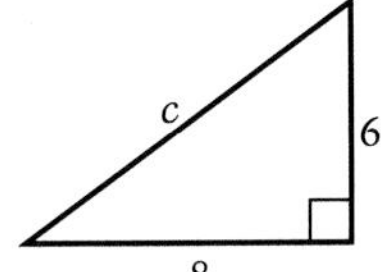

41. 60

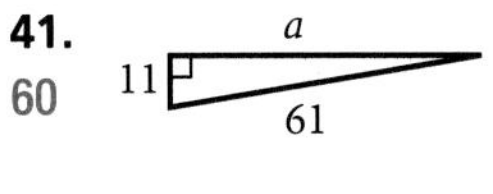

42. 25

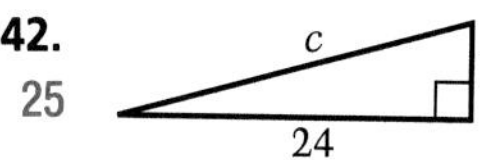

In Exercises 43–48, plot the points. Then find the equation of the line passing through the points.

43. (3, 1), (2, 11) $y = -10x + 31$

44. (−1, 6), (5, 3) $y = -\frac{1}{2}x + \frac{11}{2}$

45. (1, −2), (3, −2) $y = -2$

46. (9, 2), (−1, 0) $y = \frac{1}{5}x + \frac{1}{5}$

47. (2, 7), (6, −4) $y = -\frac{11}{4}x + \frac{25}{2}$

48. (8, 0), (8, −5) $x = 8$

Exploration and Extension

Hero's Formula **In Exercises 49 and 50, use Hero's Formula to find the area of the triangle. Hero's Formula states that the area of a triangle with sides of lengths *a*, *b*, and *c* is area = $\sqrt{s(s-a)(s-b)(s-c)}$, where $s = \frac{1}{2}(a + b + c)$.**

49. 96

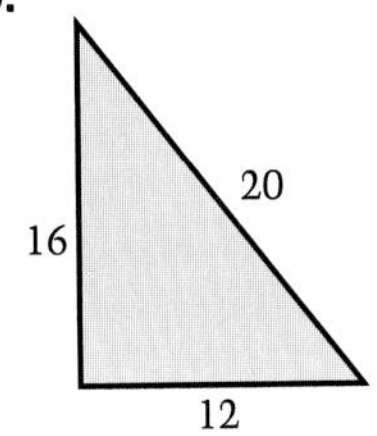

50. 252

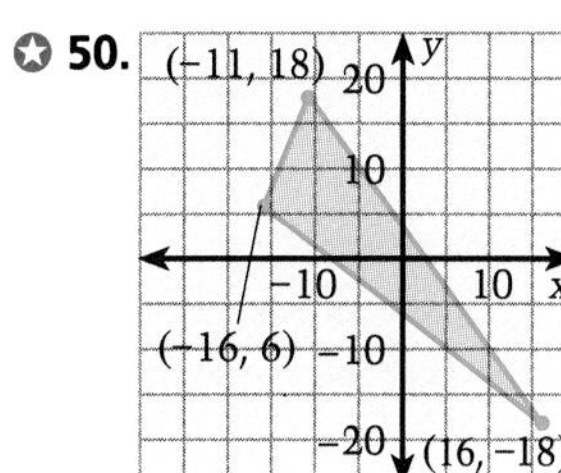

Ex. 49–50 The variable *s* in Hero's Formula refers to the "semi-perimeter," or half-perimeter.

Enrichment Activities

These activities require students to go beyond lesson goals.

1. One of your friends lives 3 miles north and 2 miles east of school. Another friend lives 5 miles south and 4 miles west of school. You live halfway between your two friends' homes. How far do you live from school? Sketch a diagram to show your results.
   $\sqrt{2}$ mi or 1.4 mi

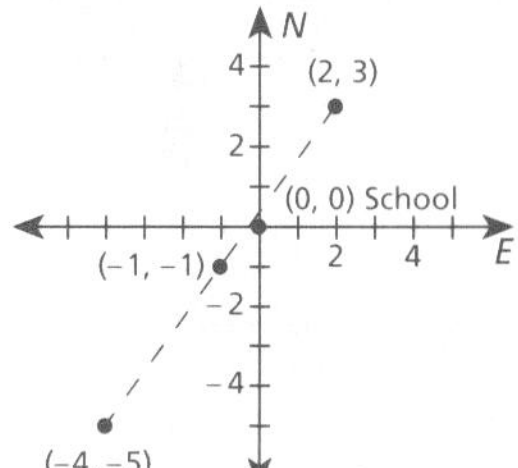

2. TECHNOLOGY You can use the following BASIC program to determine the distance between two points.

```
10 PRINT "ENTER
   COORDINATES OF POINT
   (X1, Y1)."
20 INPUT X1, Y1
30 PRINT "ENTER
   COORDINATES OF POINT
   (X2, Y2)."
40 INPUT X2, Y2
50 D = SQR((X2 − X1)*(X2 − X1)
   + (Y2 − Y1)*(Y2 − Y1))
60 PRINT "DISTANCE FROM
   (";X1;", ";Y1;") TO (";X2;",
   ";Y2;") IS ";D;" UNITS."
```

Use the program to verify your solutions for Exercises 8–15.

Problem of the Day

A farmer bought wire and posts to build a fence, planning to place the posts 8 feet apart. He found that he had been shortchanged 5 posts when he got to the farm, but that he could still build the fence if he placed the posts 10 feet apart. How many posts had the farmer intended to purchase? 26

ORGANIZER

Warm-Up Exercises

1. Evaluate.
   a. $\sqrt{36}$ 6
   b. $\sqrt{144}$ 12
   c. $\sqrt{121}$ 11
   d. $\sqrt{196}$ 14
   e. $\sqrt{81}$ 9
   f. $\sqrt{1}$ 1
2. In each case, $a^2 + b^2 = c^2$. Find the missing value a, b, or c.
   a. $a = 4, b = 5, c = ?$ $\sqrt{41}$
   b. $a = ?, b = 7, c = 8$ $\sqrt{15}$
   c. $a = 12, b = ?, c = 13$ 5
   d. $a = 2, b = 2, c = ?$ $\sqrt{8}$
3. Factor each number into its largest square factor and its corresponding nonsquare factor. For example, $12 = 4 \cdot 3$.
   a. 180 $36 \cdot 5$
   b. 448 $64 \cdot 7$
   c. 294 $49 \cdot 6$
   d. 2541 $121 \cdot 21$

Lesson Resources

Teaching Tools
  Problem of the Day: 13.2
  Warm-up Exercises: 13.2
  Answer Masters: 13.2
Extra Practice: 13.2
Color Transparencies: 76
Applications Handbook: p. 49–51

13.2 Simplifying Radicals

What you should learn:

Goal 1 How to simplify radicals by applying their properties

Goal 2 How to use radicals in real-life situations

Why you should learn it:

You can simplify radicals to make computations easier in many real-life problems, such as the period of a swinging pendulum.

Goal 1 Simplifying Radicals

An *expression with radicals* is in **simplest form** if the following are true.

1. No radicands (expressions under radical signs) have perfect square factors other than 1. $\sqrt{8} = 2\sqrt{2}$
2. No radicands contain fractions. $\sqrt{\frac{3}{4}} = \frac{1}{2}\sqrt{3}$
3. No radicals appear in the denominator of a fraction. $\frac{1}{\sqrt{4}} = \frac{1}{2}$

Properties of Radicals

Let a and b be positive numbers.

1. The square root of a product equals the product of the square roots of the factors.

   $\sqrt{ab} = \sqrt{a} \cdot \sqrt{b}$ *Product Property*

2. The square root of a quotient equals the quotient of the square roots of the numerator and denominator.

   $\sqrt{\frac{a}{b}} = \frac{\sqrt{a}}{\sqrt{b}}$ *Quotient Property*

Example 1 *Simplifying Radicals*

Simplify. **a.** $\sqrt{50}$ **b.** $\sqrt{\frac{3}{4}}$

Solution

a. $\sqrt{50} = \sqrt{25 \cdot 2}$ *Perfect square factor*
$= \sqrt{25} \cdot \sqrt{2}$ *Product Property*
$= 5\sqrt{2}$ *Simplest form*

b. $\sqrt{\frac{3}{4}} = \frac{\sqrt{3}}{\sqrt{4}}$ *Quotient Property*
$= \frac{\sqrt{3}}{2}$ *Simplest form*

The denominator in Example 1b could be rewritten without radicals because it was the square root of a perfect square. If the radical in a denominator is not the square root of a perfect square, then a different strategy is required.

Example 2 *Simplifying a Radical Expression*

Simplify $\frac{1}{\sqrt{2}}$.

Solution To simplify this expression, multiply the numerator and denominator by $\sqrt{2}$. (This is algebraically justified because it is equivalent to multiplying the original fraction by 1.)

$$\frac{1}{\sqrt{2}} = \frac{1}{\sqrt{2}} \cdot \frac{\sqrt{2}}{\sqrt{2}} \quad \textit{Multiply by } \frac{\sqrt{2}}{\sqrt{2}}, \textit{ or 1.}$$

$$= \frac{1 \cdot \sqrt{2}}{\sqrt{2} \cdot \sqrt{2}} \quad \textit{Multiply numerators and denominators.}$$

$$= \frac{\sqrt{2}}{\sqrt{2 \cdot 2}} \quad \textit{Product Property}$$

$$= \frac{\sqrt{2}}{2} \quad \textit{Simplest form}$$

■

Goal 2 Using Radicals in Real-Life Problems

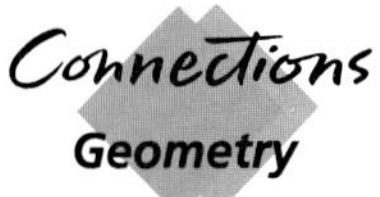

Example 3 *Finding the Area of a Rectangle*

Find the area of a rectangle whose width is $\sqrt{2}$ inches and whose length is $\sqrt{30}$ inches. Give the result in exact form and in decimal form.

Solution

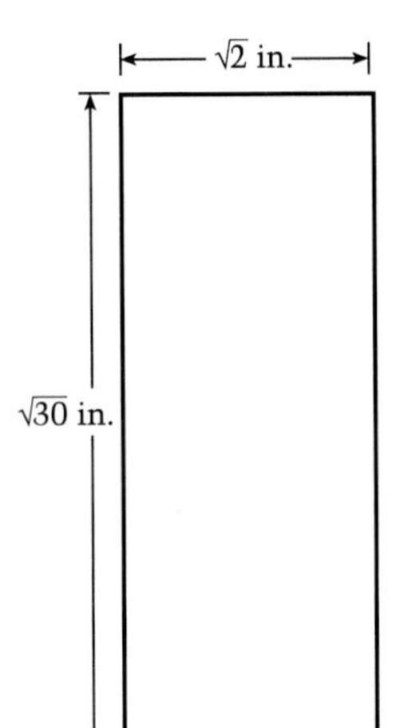

$$\text{Area} = \text{Length} \cdot \text{Width}$$

$$= \sqrt{30} \cdot \sqrt{2}$$

$$= \sqrt{30 \cdot 2} \quad \textit{Product Property}$$

$$= \sqrt{60} \quad \textit{Simplify.}$$

This expression is the exact area, but it is not in simplified form. The simplified form is

$$A = \sqrt{60} = \sqrt{4 \cdot 15} = \sqrt{4} \cdot \sqrt{15} = 2\sqrt{15} \text{ square inches.}$$

The decimal approximation of the area is

$$A = \sqrt{60} \approx 7.746 \text{ square inches.}$$

Try multiplying 2 by $\sqrt{15}$ on a calculator to see if you obtain the same approximation. ■

LESSON Notes

Example 1

For larger positive integers, such as 616 or 50050, use some of the properties of whole numbers to determine the prime factorization. Specifically, even numbers are divisible by the prime factor 2; a number that is divisible by 3 is composed of digits whose sum is divisible by 3; and numbers that end in 5 or 0 are divisible by 5.

For instance, 616 is even, so write $616 = 2 \cdot 308$. The factor 308 is even, so write $616 = 2 \cdot 2 \cdot 154 = 2 \cdot 2 \cdot 2 \cdot 77$. Since 7 divides 77, complete the factorization as

$$616 = 2 \cdot 2 \cdot 2 \cdot 7 \cdot 11.$$

Similarly, 50050 is even, so $50050 = 2 \cdot 25025$. Because 25025 ends in 5, it is divisible by 5 and $50050 = 2 \cdot 5 \cdot 5005$. Continuing,

$$50050 = 2 \cdot 5 \cdot 5 \cdot 1001.$$

Since the sum of the digits is 2, 1001 is not divisible by 3. The next prime to try is 7. So $50050 = 2 \cdot 5 \cdot 5 \cdot 7 \cdot 143$. Finally, factor 143 to obtain $50050 = 2 \cdot 5 \cdot 5 \cdot 7 \cdot 11 \cdot 13$.

Example 2

The technique used in this example is often called "rationalizing the denominator."

Example 3

The form that the answer in this example takes depends, in part, on the uses of the result. The exact form would be useful in recognizing like factors in more complex algebraic computations; a decimal approximation would be useful if measurements of physical space were involved.

Example 4

Ask students to speculate why different plants spiral in different directions.

In the square root spiral, point out that $r_0 = 1$. It may be necessary, for student understanding, to compute several more values of r_k:

$$r_3 = \sqrt{1^2 + (\sqrt{3})^2} = \sqrt{4}$$
$$r_4 = \sqrt{1^2 + (\sqrt{4})^2} = \sqrt{5}$$
$$r_5 = \sqrt{1^2 + (\sqrt{5})^2} = \sqrt{6}$$

Extra Examples

Here are additional examples similar to **Examples 1–3.**

1. Simplify the radical expressions.
   a. $\sqrt{75}$ $5\sqrt{3}$
   b. $\sqrt{180}$ $6\sqrt{5}$
   c. $\sqrt{\frac{49}{121}}$ $\frac{7}{11}$
   d. $\sqrt{\frac{12}{81}}$ $\frac{2\sqrt{3}}{9}$
2. Simplify the radical expressions.
   a. $\frac{3}{\sqrt{12}}$ $\frac{\sqrt{3}}{2}$
   b. $\frac{8}{\sqrt{18}}$ $\frac{4\sqrt{2}}{3}$
   c. $\sqrt{\frac{1}{8}}$ $\frac{\sqrt{2}}{4}$
   d. $\sqrt{\frac{3}{5}}$ $\frac{\sqrt{15}}{5}$
3. Find the area of each.
   a. A rectangle with width $\sqrt{3}$ and height $\sqrt{15}$
   b. A triangle with height $\sqrt{24}$ and base $\sqrt{6}$
   c. A square with height $\sqrt{8}$

   a. $3\sqrt{5}$ b. 6 c. 8

Check Understanding

1. State the Product and Quotient Properties of radicals. See page 682.
2. What is the exact form of a radical expression? An irrational number expressed in radical form.
3. Give examples of radical expressions that are not in simplest form. Justify your selections. Sample answers: $\sqrt{60}$ and $\sqrt{126}$.

Spirals occur in many forms in nature. Some are left-handed (spiraling counterclockwise) and some are right-handed (spiraling clockwise). Trumpet honeysuckle winds to the left and bindweed winds to the right. Some snails occur in both left- and right-handed versions.

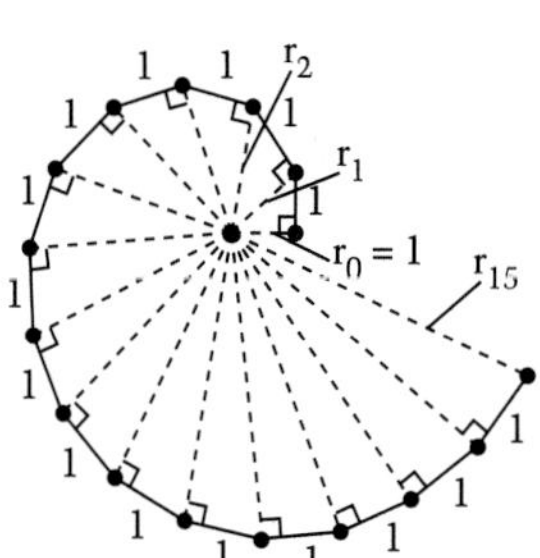

Connections
Geometry

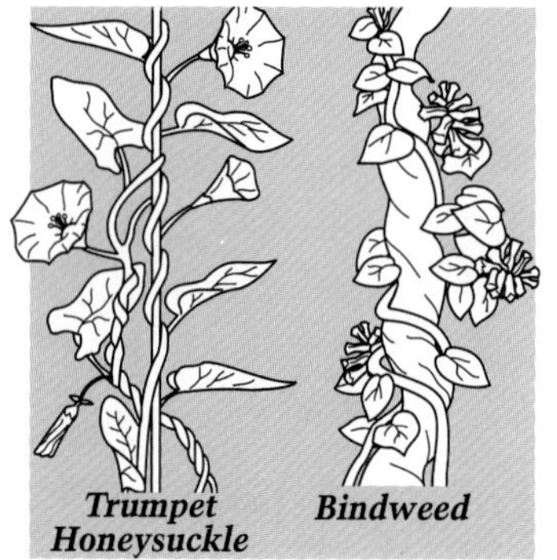
Trumpet Honeysuckle   Bindweed

Example 4 Using the Pythagorean Theorem

The *square root spiral* above is formed by a sequence of right triangles, each with a side whose length is 1. Let r_n be the length of the hypotenuse of the nth triangle. Find r_1, r_2, and r_{15}.

Solution Each leg of the first triangle is 1 unit. Use the Pythagorean theorem with $a = 1$, $b = 1$, and $c = r_1$.

$$r_1 = \sqrt{a^2 + b^2} = \sqrt{1^2 + 1^2} = \sqrt{2}$$

For the second triangle, the lengths of the legs are $a = 1$ and $b = \sqrt{2}$. Thus, the length of the hypotenuse is

$$r_2 = \sqrt{a^2 + b^2} = \sqrt{1^2 + (\sqrt{2})^2} = \sqrt{3}.$$

By continuing this procedure, you can find a pattern.

$r_0 = \sqrt{1}$	$r_1 = \sqrt{2}$	$r_2 = \sqrt{3}$	$r_3 = \sqrt{4}$
$r_4 = \sqrt{5}$	$r_5 = \sqrt{6}$	$r_6 = \sqrt{7}$	$r_7 = \sqrt{8}$
$r_8 = \sqrt{9}$	$r_9 = \sqrt{10}$	$r_{10} = \sqrt{11}$	$r_{11} = \sqrt{12}$
$r_{12} = \sqrt{13}$	$r_{13} = \sqrt{14}$	$r_{14} = \sqrt{15}$	$r_{15} = \sqrt{16}$ ■

Communicating about ALGEBRA

SHARING IDEAS about the Lesson See below.

Extend the Idea

A. Which of the r_n in Example 4 are simplified? Simplify those that are not.

B. Let A_n be the area of the nth triangle. Find a formula for A_n.

C. Find the ratio of A_8 to A_7. Give your answer in simplified form.

A. $r_1, r_2, r_4, r_5, r_6, r_9, r_{10}, r_{12}, r_{13}, r_{14}$; $r_0 = 1$, $r_3 = 2$, $r_7 = 2\sqrt{2}$, $r_8 = 3$, $r_{11} = 2\sqrt{3}$, $r_{15} = 4$

B. $A_n = \frac{1}{2}(1)(r_{n-1})$ **C.** $\frac{2\sqrt{14}}{7}$

EXERCISES

Guided Practice

CRITICAL THINKING about the Lesson

1. Describe the three conditions that must be true for a radical expression to be in simplest form. 1., 2., 3. See page 682.
2. Describe the Product Property of radicals in words. $\sqrt{28} = \sqrt{4 \cdot 7} = \sqrt{4} \cdot \sqrt{7} = 2\sqrt{7}$
3. Describe the Quotient Property of radicals in words.
4. Explain how the Product Property can be used to simplify $\sqrt{28}$.
5. Explain how the Product Property can be used to multiply $\sqrt{6}$ and $\sqrt{24}$. $\sqrt{6} \cdot \sqrt{24} = \sqrt{6 \cdot 24} = \sqrt{144} = 12$
6. Describe the steps you would use to simplify $\frac{1}{\sqrt{3}}$. $\frac{1}{\sqrt{3}} = \frac{1}{\sqrt{3}} \cdot \frac{\sqrt{3}}{\sqrt{3}} = \frac{\sqrt{3}}{\sqrt{3 \cdot 3}} = \frac{\sqrt{3}}{3}$

Independent Practice

In Exercises 7–22, simplify the radical expression.

7. $\sqrt{40}$ $2\sqrt{10}$
8. $\sqrt{18}$ $3\sqrt{2}$
9. $\sqrt{48}$ $4\sqrt{3}$
10. $\sqrt{75}$ $5\sqrt{3}$
11. $\sqrt{\frac{7}{9}}$ $\frac{\sqrt{7}}{3}$
12. $\sqrt{\frac{11}{16}}$ $\frac{\sqrt{11}}{4}$
13. $\sqrt{\frac{2}{50}}$ $\frac{1}{5}$
14. $\sqrt{\frac{20}{12}}$ $\frac{\sqrt{15}}{3}$
15. $\frac{1}{2}\sqrt{80}$ $2\sqrt{5}$
16. $\frac{1}{3}\sqrt{27}$ $\sqrt{3}$
17. $2\sqrt{\frac{5}{4}}$ $\sqrt{5}$
18. $18\sqrt{\frac{5}{81}}$ $2\sqrt{5}$
19. $\sqrt{\frac{1}{12}}$ $\frac{\sqrt{3}}{6}$
20. $\sqrt{\frac{4}{5}}$ $\frac{2\sqrt{5}}{5}$
21. $2\sqrt{\frac{1}{2}}$ $\sqrt{2}$
22. $3\sqrt{\frac{5}{6}}$ $\frac{\sqrt{30}}{2}$

In Exercises 23–34, perform the indicated operation. Simplify your result.

23. $\sqrt{5} \cdot \sqrt{15}$ $5\sqrt{3}$
24. $\sqrt{10} \cdot \sqrt{20}$ $10\sqrt{2}$
25. $\sqrt{2} \cdot \sqrt{6} \cdot \sqrt{3}$ 6
26. $\sqrt{2} \cdot \sqrt{3} \cdot \sqrt{5}$ $\sqrt{30}$
27. $(2\sqrt{13})^2$ 52
28. $(7\sqrt{3})^2$ 147
29. $(\frac{1}{2}\sqrt{8})^2$ 2
30. $(\frac{2}{3}\sqrt{3})^2$ $\frac{4}{3}$
31. $\frac{1}{\sqrt{18}}$ $\frac{\sqrt{2}}{6}$
32. $\frac{2\sqrt{5}}{\sqrt{4}}$ $\sqrt{5}$
33. $\frac{\sqrt{6}}{\sqrt{2}}$ $\sqrt{3}$
34. $\frac{6}{\sqrt{3}}$ $2\sqrt{3}$

Geometry **In Exercises 35–38, find the area of the figure. Give both the exact answer in simplified form and a decimal approximation rounded to two decimal places.**

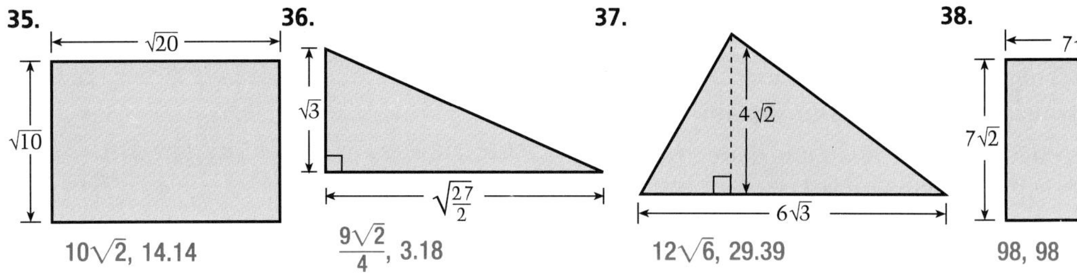

35. $10\sqrt{2}$, 14.14
36. $\frac{9\sqrt{2}}{4}$, 3.18
37. $12\sqrt{6}$, 29.39
38. 98, 98

Communicating about ALGEBRA

EXTEND Communicating

Suppose the constant length in the square root spiral were 2, instead of 1. If $r_0 = 2$, find r_1, r_2, and r_{15}. $2\sqrt{2}$, $2\sqrt{3}$, $2\sqrt{16} = 8$

Suppose the constant length and r_0 were each 3, instead of 1. Find r_1, r_2, and r_{15}. $3\sqrt{2}$, $3\sqrt{3}$, $3\sqrt{16} = 12$

What if the constant length in the square root spiral and r_0 are equal to the constant k? Find r_1, r_2, and r_{15}. $k\sqrt{2}$, $k\sqrt{3}$, $k\sqrt{16} = 4k$.

How would you express the length of the general term r_n? $k\sqrt{n+1}$

EXERCISE Notes

Assignment Guide

Basic/Average: Ex. 1–6, 7–35 odd, 41–42, 47–59 odd

Above Average: Ex. 1–6, 9–37 odd, 39–46, 47–59 odd

Advanced: Ex. 1–6, 11–39 odd, 40–47, 59–64

Selected Answers

Exercises 1–6, 7–61 odd

Use **Mixed Review** as needed.

More Difficult Exercises

Exercises 41–46, 63, 64

Guided Practice

Ex. 1–6 Remind students that the ability to simplify radical expressions will also be a useful skill in geometry.

Also, irrational numbers are expressed in this exact form using radical notation. Be sure students understand that $\sqrt{28}$ and $2\sqrt{7}$ are both exact representations. Expressing $\sqrt{28}$ as 5.3 is a rational approximation of an irrational number.

Independent Practice

▶ **Ex. 7–22** These exercises are modeled by Example 1.

▶ **Ex. 27–30** Remind students that $(\sqrt{10})^2 = 10$.

▶ **Ex. 35–40** These exercises provide a geometric application of operations with radicals.

▶ **Ex. 41–42** TECHNOLOGY Have students use calculators and check the numerical approximations of both the simplified and nonsimplified forms for accuracy. For example, in Exercise 41,

$$s = 3.1\sqrt{60} \approx 24.01$$
$$s = 6.2\sqrt{15} \approx 24.01$$

▶ **Ex. 43–46** Advise students that they might wish to first reduce the fraction and then express the radius as $\frac{\sqrt{12100}}{\sqrt{9\pi^2}} = \frac{110}{3\pi} \approx 11.67$ miles. This will provide the comparison value used for the exercises.

Answers

44.–46. When the distance between New Castle and a city is less than or equal to $\frac{110}{3\pi}$, then the city is in Delaware. If not, the city is in Pennsylvania.

Geometry **In Exercises 39 and 40, use the Pythagorean theorem to find x. Express the result as a simplified radical.**

39. $3\sqrt{2}$

40. $2\sqrt{15}$

Tsunami **In Exercises 41 and 42, use the following information.**

A *tsunami* (tsōō nä′ mē′), or tidal wave, is a destructive ocean wave that is caused by a large storm at sea or by an undersea earthquake. The speed, s (in meters per second), at which a tsunami moves is determined by the depth, d (in meters), of the ocean:

$$s = 3.1\sqrt{d}$$

41. Find the speed of a tsunami in a region of the ocean that is 60 meters deep. Write the result in simplified form. $6.2\sqrt{15}$

42. Find the speed of a tsunami in a region of the ocean that is 200 meters deep. Write the result in simplified form. $31\sqrt{2}$

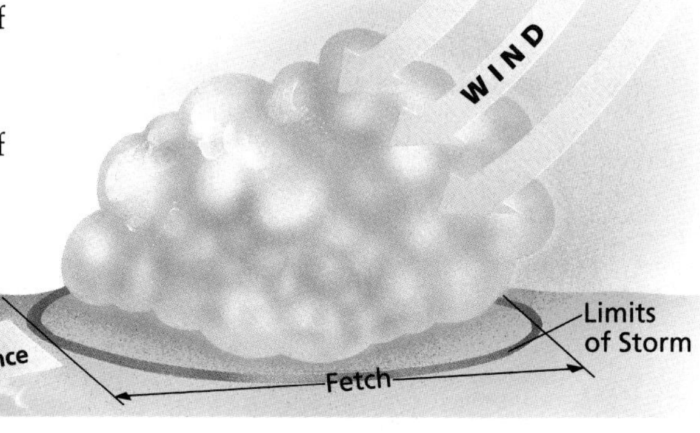

At the Top of Delaware **In Exercises 43–46, use the following information.**

The northern boundary of Delaware is the arc of a circle whose center is at New Castle, Delaware. The radius of the circle is

$$\sqrt{\frac{24{,}200}{18\pi^2}} \text{ miles.}$$

43. Simplify the expression for the radius. $\frac{110}{3\pi}$

44. On the coordinate plane at the right, the coordinates of Booths Corner are (1.5, 3.7). Is Booths Corner in Delaware or Pennsylvania? Explain. Pennsylvania

45. The coordinates of Yorklyn are (−1.6, 3.4). In which state is Yorklyn? Explain. Delaware

46. The coordinates of Naamans Gardens are (2.1, 3). In which state is Naamans Gardens? Explain. Delaware

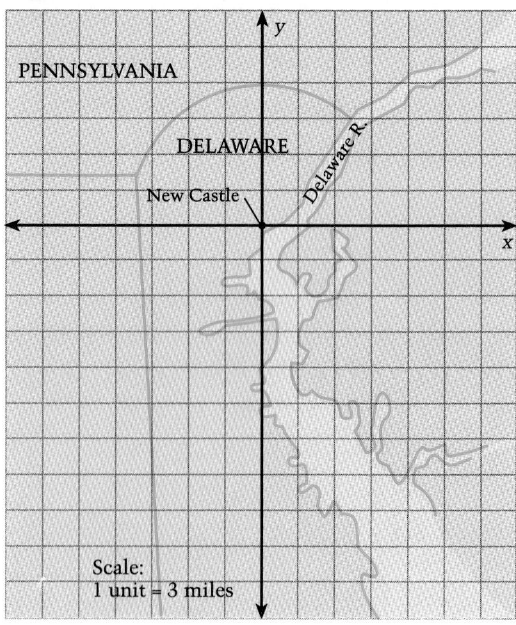

Integrated Review

In Exercises 47–52, evaluate the function at the given value of *x*. Write the result in simplified form.

47. $f(x) = x^2 - 7, x = \sqrt{6}$ −1

48. $f(x) = 2x^2 + 2, x = \sqrt{10}$ 22

49. $f(x) = \sqrt{2} \cdot \sqrt{x}, x = 8$ 4

50. $f(x) = \sqrt{3} \cdot \sqrt{x}, x = 12$ 6

51. $f(x) = x^2 - 5, x = 2\sqrt{5}$ 15

52. $f(x) = x^2 - 10, x = 4\sqrt{2}$ 22

In Exercises 53–58, use the quadratic formula to solve the equation. Write the result in simplified form. See margin.

53. $\frac{1}{2}x^2 + 6x - 6 = 0$

54. $\frac{1}{2}x^2 + 10x + 10 = 0$

55. $x^2 + 2x - 1 = 0$

56. $x^2 + 6x - 6 = 0$

57. $x^2 - 4x - 5 = 0$

58. $x^2 - 3x - 4 = 0$

Geometry **In Exercises 59 and 60, find the ratio of the area of the blue region to the area of the green region. Express the result in simplified form.**

59. $\frac{\sqrt{2}}{2}$

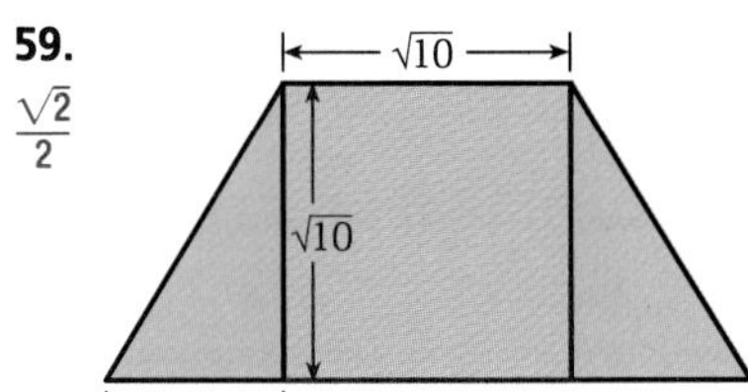

60. 3

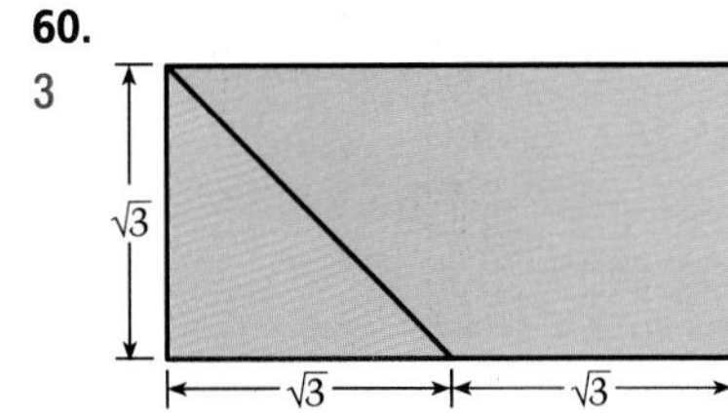

In Exercises 61 and 62, find the distance between the two points.

61. $(0, 9), (\sqrt{12}, 3)$ $4\sqrt{3}$

62. $(\sqrt{5}, 5), (0, 10)$ $\sqrt{30}$

Exploration and Extension

Pendulums and Resonance **In Exercises 63 and 64, use the following information.**

The period, T (in seconds), of a pendulum is the time it takes for the weights to swing back and forth. The period is related to the length, L (in feet), of the pendulum by

$$T = 2\pi\sqrt{\frac{L}{32}}.$$

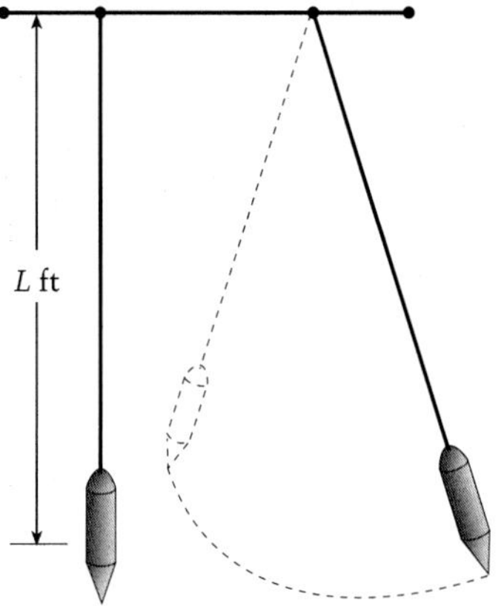

63. A string is pulled taut in a horizontal position. From the string, two pendulums are attached, each one the same weight and length. One is set in motion. Soon its partner will begin to swing with the *same* period. This is called *resonance.* Each pendulum is 1 foot long. What is their period? Write your result in simplified form. $\frac{\pi\sqrt{2}}{4}$

64. Try the experiment described in Exercise 63. Then change the length of one of the pendulums. When one is set in motion, does its partner begin to move? No

Ex. 47–52 and 59–60 These exercises integrate previous skills and radical expressions.

Answers

53. $-6 + 4\sqrt{3}, -6 - 4\sqrt{3}$
54. $-10 + 4\sqrt{5}, -10 - 4\sqrt{5}$
55. $-1 + \sqrt{2}, -1 - \sqrt{2}$
56. $-3 + \sqrt{15}, -3 - \sqrt{15}$
57. 5, −1
58. 4, −1

Ex. 63–64 GROUP ACTIVITY Have groups of students perform the experiments in class. The physics department could provide the necessary materials.

Enrichment Activities

These activities require students to go beyond lesson goals.

1. Use the Pythagorean Theorem. Find the altitude of an equilateral triangle with side length *s*. Use this value to write the formula for the area of an equilateral triangle.
   $A = \frac{\sqrt{3} \cdot s^2}{4}$
2. Use your formula to find the area of an equilateral triangle with side length $\sqrt{24}$ units.
   $6\sqrt{3}$ square units
3. Use your formula to find the area of an equilateral triangle with a perimeter of $\sqrt{48}$ units.
   $\frac{4\sqrt{3}}{3}$ square units
4. Use your formula to find the side length of an equilateral triangle with an area of $12\sqrt{3}$ square units.
   $4\sqrt{3}$ units

Answers

1. $(x + 7)(x + 1)$
2. $(-3x + 2)(x + 1)$
3. $\frac{3 \pm \sqrt{29}}{4}$
4. $\frac{-5 \pm \sqrt{17}}{2}$
5. 20
6. 300
7. 4.23×10^9
8. 2.6×10^{-3}
9. No
10. Yes
11. 

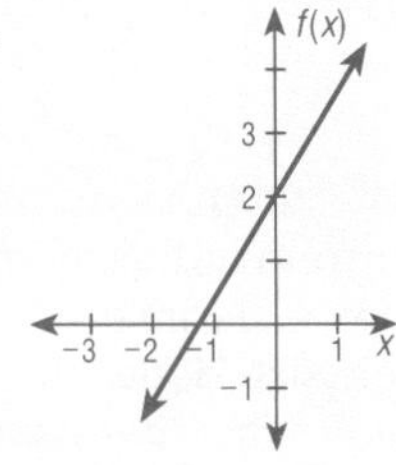

12. 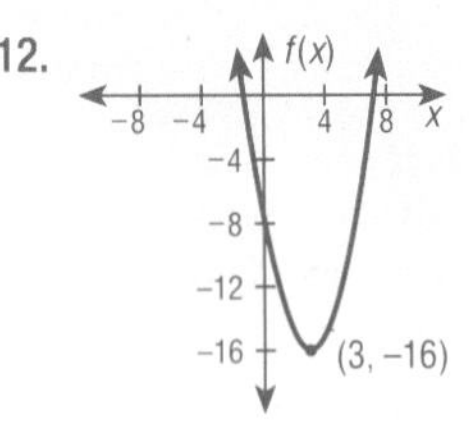

13. $x = 1$ or $x = 5$
14. $y = \frac{5}{3}$
15. $-3x^2 + 2$
16. $2x^{-2}y$
17. $3m^4$
18. $\frac{2z^6}{3x^2y^4}$
19. $\begin{bmatrix} 2b+3 & 7 \\ -2a+4 & 3b \end{bmatrix}$
20. $\begin{bmatrix} 4 & 3 \\ 8 & -9 \end{bmatrix}$

Mixed REVIEW

1. Factor. $x^2 + 8x + 7$. **(10.5)** See margin.
2. Factor. $-3x^2 - x - 2$. **(10.5)**
3. Solve. $4x^2 - 6x - 5 = 0$. **(9.4)**
4. Solve. $x^2 + 5x + 2 = 0$. **(9.4)**
5. Evaluate $f(x) = 3x + 2$ when $x = 6$. **(12.1)**
6. Evaluate $f(t) = 300(1.07)^t$ when $t = 0$. **(12.1)**
7. Write 4,230,000,000 in scientific notation. **(8.4)**
8. Write 0.0026 in scientific notation. **(8.4)**
9. Is (3, 4) a solution of $x^2 + y^2 = 5$? **(4.2)**
10. Is (2, 3) a solution of $y \geq 4 \cdot 3^{-x}$? **(8.2)**
11. Sketch the graph of $f(x) = \frac{3}{2}x + 2$. **(12.2)**
12. Sketch the graph of $f(x) = x^2 - 6x - 7$. **(12.4)**
13. Solve. $\frac{1}{x} + \frac{x}{5} = \frac{6}{5}$. **(11.8)**
14. Solve. $\frac{1}{y} + \frac{3}{5} = \frac{2}{y}$. **(11.8)**
15. Find the opposite of $3x^2 - 2$. **(2.1)**
16. Find the reciprocal of $\frac{1}{2}x^2y^{-1}$. **(8.3)**
17. Write $3 \cdot m \cdot m \cdot m \cdot m$ using exponents. **(1.3)**
18. Simplify $4x^2y^{-3}z^4 \div 6x^4yz^{-2}$. **(8.3)**
19. Add $\begin{bmatrix} 3 & 4 \\ -2a & b \end{bmatrix}$ and $\begin{bmatrix} 2b & 3 \\ 4 & 2b \end{bmatrix}$. **(2.4)**
20. Subtract $\begin{bmatrix} 3 & 6 \\ -2 & 5 \end{bmatrix}$ from $\begin{bmatrix} 7 & 9 \\ 6 & -4 \end{bmatrix}$. **(2.4)**

Career Interview

Translation Services

Jose Blanco is the owner and manager of a translation service specializing in English to Spanish translations. "As a small business I need to do my own accounting, payroll, and so on."

Q: ***Did you take math classes in high school?***

A: The math courses I took in school have turned out to be the most useful in my daily experiences. I rely on basic algebra and some geometry, which is especially useful when doing page designs.

Q: ***Does having a math background help you on the job?***

A: Definitely! I wouldn't be able to survive or be competitive without math skills to figure out or quantify projects properly. I often have to combine my basic math skills and my instincts. Being good with numbers is important.

Q: ***How much mental math and estimation do you use?***

A: Lots, and constantly. They are both necessary to make decisions! For example, when negotiating a bid, we don't always have much time; some decisions need to be made on the spot. I rely on estimation and mental math a lot.

Q: ***What would you like to tell kids who are in school?***

A: Having a strong number sense will be crucial to you.

13.3 Operations with Radicals

What you should learn:

Goal 1 How to add and subtract radical expressions

Goal 2 How to use radical operations in real-life problems

Why you should learn it:

You can use operations with radicals to solve many problems, such as finding the perimeters of figures on a grid.

Goal 1 Adding and Subtracting Radicals

Two radical expressions are **like radicals** if they have the same radicand. For instance, $\sqrt{2}$ and $3\sqrt{2}$ are like radicals. To add or subtract like radicals, add or subtract their coefficients.

Sum: $\sqrt{2} + 3\sqrt{2} = (1 + 3)\sqrt{2} = 4\sqrt{2}$

Difference: $\sqrt{2} - 3\sqrt{2} = (1 - 3)\sqrt{2} = -2\sqrt{2}$

Two radical expressions that are unlike may simplify to like radicals. (See Example 1b).

Example 1 *Adding and Subtracting Radicals*

a. $2\sqrt{3} + \sqrt{2} - 4\sqrt{3} = -2\sqrt{3} + \sqrt{2}$ *Add like radicals.*

b. $3\sqrt{2} - \sqrt{8} = 3\sqrt{2} - \sqrt{4 \cdot 2}$ *Perfect square factor*

$= 3\sqrt{2} - \sqrt{4} \cdot \sqrt{2}$ *Product Property*

$= 3\sqrt{2} - 2\sqrt{2}$ *Simplify.*

$= \sqrt{2}$ *Subtract like radicals.* ■

Example 2 *Multiplying Radicals*

a. $3\sqrt{2}(\sqrt{2} + 4\sqrt{6})$

$= (3\sqrt{2})(\sqrt{2}) + (3\sqrt{2})(4\sqrt{6})$ *Distributive Property*

$= 3(\sqrt{2} \cdot \sqrt{2}) + (3)(4)(\sqrt{2} \cdot \sqrt{6})$ *Regroup factors.*

$= 3\sqrt{4} + 12\sqrt{12}$ *Product Property*

$= 3\sqrt{4} + 12\sqrt{4 \cdot 3}$ *Perfect square factor*

$= 3\sqrt{4} + 12\sqrt{4} \cdot \sqrt{3}$ *Product Property*

$= 3(2) + 12(2)\sqrt{3}$ *Simplify.*

$= 6 + 24\sqrt{3}$ *Simplest form*

b. $(3 + 2\sqrt{2})^2$

$= 3^2 + 2(3)(2\sqrt{2}) + (2\sqrt{2})^2$ *Square of binomial*

$= 9 + 12\sqrt{2} + 2^2(\sqrt{2})^2$ *Simplify.*

$= 9 + 12\sqrt{2} + 4(2)$ *Definition of square root*

$= 17 + 12\sqrt{2}$ *Simplest form* ■

Problem of the Day

A man bought two different television sets, but later decided to sell them. He sold them each for \$600, making a 20% profit on one, and a 20% loss on the other. What was the percent profit or loss on the entire transaction? 4% loss

ORGANIZER

Warm-Up Exercises

1. Show the use of the Distributive Property in combining the following.
   a. $4x + 12x$
   b. $-3x^2y + 7x^2y$
   c. $8rs^3 - 13rs - 11rs^3$
   a. $4x + 12x = (4 + 12)x = (16)x = 16x$
   b. $-3x^2y + 7x^2y = (-3 + 7)x^2y = (4)x^2y = 4x^2y$
   c. $8rs^3 - 13rs - 11rs^3 = 8rs^3 - 11rs^3 - 13rs = (8 - 11)rs^3 - 13rs = -3rs^3 - 13rs$
2. Use the Product Property of Radicals to multiply.
   a. $\sqrt{5} \cdot \sqrt{15}$ $5\sqrt{3}$
   b. $\sqrt{6} \cdot \sqrt{12}$ $6\sqrt{2}$
   c. $\sqrt{14} \cdot \sqrt{21}$ $7\sqrt{6}$
3. Solve each quadratic equation using the Quadratic Formula.
   a. $x^2 - 5x + 6 = 0$ 2 or 3
   b. $2x^2 + 10x - 12 = 0$ -6 or 1

Lesson Resources

Teaching Tools
  Transparency: 4
  Problem of the Day: 13.3
  Warm-up Exercises: 13.3
  Answer Masters: 13.3
Extra Practice: 13.3
Color Transparency: 77

LESSON Notes

Example 1

Point out that the Distributive Property is used to factor out $\sqrt{3}$ in Example 1a, and $\sqrt{2}$ in Example 1b.

Examples 1–2

Emphasize the importance of simplifying radical expressions when adding and subtracting. Frequently, radical expressions that do not look alike may simplify into expressions that are alike.

Example 3

Note that solutions to quadratic equations can now be easily checked. Ask students to check the other solution, $2 - \sqrt{3}$.

EXTEND Example 4 An interesting relationship, Pick's Formula, can be used to find the area of the irregularly shaped lot. Refer to the grid that is superimposed on the lot. Count the number of intersections that are inside the boundary lines of the lot ($I = 26$). Count the number of intersections that occur on the boundary of the lot ($B = 10$). Pick's Formula states that the area of the lot depends on I and B, and is given by the formula, Area $= \frac{B}{2} + I - 1$. Therefore, the area of the lot is 30 square units. Because each unit is 15 feet, each square unit is 225 square feet, and the area is $30 \cdot 225$ ft^2 $=$ 6750 ft^2.

Example 5

COOPERATIVE LEARNING
Share the computations among the class. That is, form several groups and make them responsible for a subset of the computations. For example, if there are seven groups, let each group be responsible for three different computations.

Example 3 Checking Quadratic Formula Solutions

Solve $x^2 - 4x + 1 = 0$. Check the solutions.

Solution

$x^2 - 4x + 1 = 0$	*Rewrite original equation.*
$x = \dfrac{4 \pm \sqrt{4^2 - 4(1)(1)}}{2(1)}$	*Quadratic Formula*
$= \dfrac{4 \pm \sqrt{12}}{2}$	*Simplify.*
$= \dfrac{4 \pm 2\sqrt{3}}{2}$	*Product Property*
$= \dfrac{2(2 \pm \sqrt{3})}{2}$	*Factor numerator.*
$= 2 \pm \sqrt{3}$	*Cancel common factors.*

Check Here is the check for $2 + \sqrt{3}$.

$x^2 - 4x + 1 = 0$	*Rewrite original equation.*
$(2 + \sqrt{3})^2 - 4(2 + \sqrt{3}) + 1 \stackrel{?}{=} 0$	*Substitute $2 + \sqrt{3}$ for x.*
$4 + 4\sqrt{3} + 3 - 8 - 4\sqrt{3} + 1 \stackrel{?}{=} 0$	*Multiply.*
$0 = 0$	*Solution checks.*

■

Connections
Geometry

Example 4 Finding the Perimeter of a Polygon

You have purchased an irregularly shaped lot. The vertices of your lot have coordinates (0, 6), (2, 2), (4, 1), (7, 7), and (6, 9). Each unit represents 15 feet. Find the perimeter of your lot.

(6, 9)
(7, 7)
(0, 6)
(2, 2)
(4, 1)
y
x

Solution The lengths of the five sides are as follows.

$$d_1 = \sqrt{(2 - 0)^2 + (2 - 6)^2} = \sqrt{20} = 2\sqrt{5}$$
$$d_2 = \sqrt{(4 - 2)^2 + (1 - 2)^2} = \sqrt{5}$$
$$d_3 = \sqrt{(7 - 4)^2 + (7 - 1)^2} = \sqrt{45} = 3\sqrt{5}$$
$$d_4 = \sqrt{(6 - 7)^2 + (9 - 7)^2} = \sqrt{5}$$
$$d_5 = \sqrt{(0 - 6)^2 + (6 - 9)^2} = \sqrt{45} = 3\sqrt{5}$$

The perimeter, P, of the lot is

$$P = 2\sqrt{5} + \sqrt{5} + 3\sqrt{5} + \sqrt{5} + 3\sqrt{5}$$
$$= 10\sqrt{5} \text{ units.}$$

Because each unit is 15 feet, the perimeter is

$$P = 10\sqrt{5} \cdot 15$$
$$= 150\sqrt{5} \approx 335.4 \text{ feet.}$$

■

Goal 2 Using Radicals in Real-Life Problems

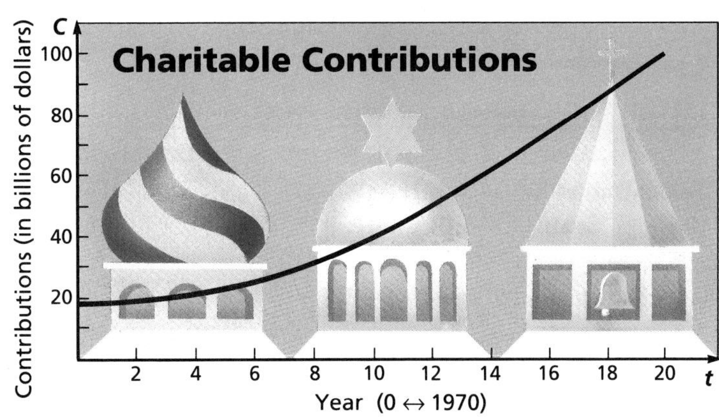

***Charitable contributions** by individuals account for about 83% of all charitable contributions in the United States. Of that, more than half is given to religious organizations.*

Real Life Philanthropy

Example 5 *Finding the Total Contributions*

For 1970 through 1990, the annual amount, C, given by individuals to religious organizations can be modeled by

$$C = 1.1\sqrt{310 + t^3} \qquad \textit{billions of dollars}$$

where $t = 0$ represents 1970. Estimate the total amount given during this 21-year period.

Solution Use a calculator to find the total.

$$\text{Total} = 1.1\sqrt{310 + 0^3} + 1.1\sqrt{310 + 1^3} + \cdots + 1.1\sqrt{310 + 20^3}$$
$$= 1.1(\sqrt{310} + \sqrt{311} + \cdots + \sqrt{8310})$$
$$\approx 992 \text{ billion dollars}$$

■

Communicating about ALGEBRA

SHARING IDEAS about the Lesson

Extend the Idea Use the following steps to find the total area of the lot described in Example 4.

A. Plot the vertices of the lot. See page 690.

$A = (0, 6)$, $B = (2, 2)$, $C = (4, 1)$, $D = (7, 7)$, $E = (6, 9)$

B. Show that $\angle BAE$, $\angle AED$, and $\angle BCD$ are right angles.

C. The area of the lot is the sum of the area of the trapezoid $ABDE$ and the triangle BCD. Compute the area of the lot. 6750 ft² B. See below.

Extra Examples

Here are additional examples similar to **Examples 1–4.**

1. Simplify the radical expressions.

a. $4\sqrt{3} - 5\sqrt{3}$ $\quad -\sqrt{3}$

b. $-6\sqrt{12} + \sqrt{75}$ $\quad -7\sqrt{3}$

2. Multiply and simplify.

a. $2\sqrt{3}(\sqrt{2} + 5\sqrt{6})$
$2\sqrt{6} + 30\sqrt{2}$

b. $(3\sqrt{8} - 5)^2$
$97 - 60\sqrt{2}$

3. Use the Quadratic Formula to solve $3x^2 - 9x + 3 = 0$.
$\frac{3 \pm \sqrt{5}}{2}$

4. Find the perimeter of the quadrilateral that is formed by the points $A(6, -2)$, $B(-6, -5)$, $C(-2, 3)$, and $D(7, 6)$. A sketch of the quadrilateral may help.

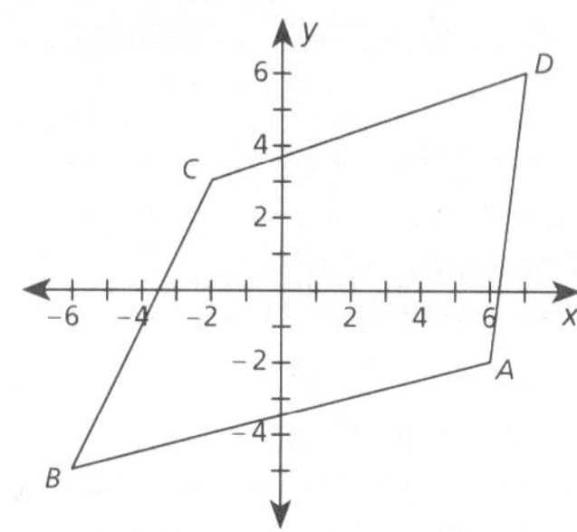

$P = (3\sqrt{17} + 4\sqrt{5} + 3\sqrt{10} + \sqrt{65})$ units, or approximately 38.86 units

Check Understanding

1. Explain how the Distributive Property is used in the addition and subtraction of radicals. See the top of page 689.

2. Consider the sums, $\sqrt{8} + \sqrt{2}$ and $\sqrt{15} + \sqrt{5}$. Which sum can be simplified into like radicals, and which cannot? Explain.
a, b; $\sqrt{8} + \sqrt{2} = 2\sqrt{2} + \sqrt{2} = 3\sqrt{2}$.

B. $(\sqrt{20})^2 + (\sqrt{45})^2 = (\sqrt{65})^2$, $(\sqrt{45})^2 + (\sqrt{5})^2 = (\sqrt{50})^2$, $(\sqrt{45})^2 + (\sqrt{5})^2 = (\sqrt{50})^2$

Communicating about ALGEBRA

Ask students to compare the results of Part C with Pick's Formula outlined in the extension of Example 4. Which technique is "better"? Explain.

EXERCISE Notes

Assignment Guide

Basic/Average: Ex. 1–6, 7–29 odd, 33–34, 41–46

Above Average: Ex. 1–6, 9–29 odd, 31, 33–36, 41–46

Advanced: Ex. 1–6, 11–29 odd, 31–36, 41–46

Selected Answers

Exercises 1–6, 7–43 odd

More Difficult Exercises

Exercises 31–34

Guided Practice

Ex. 6 Students may need additional practice in simplifying expressions like $(3 + \sqrt{6})^2$. Refer them to Example 2.

Independent Practice

Ex. 7–18 Remind students that one goal of this lesson is to develop a facility for operations with radical expressions without using a calculator.

EXERCISES

Guided Practice

CRITICAL THINKING about the Lesson

1. Give an example of two radical expressions that are like radicals. Give an example of two that are not. $3\sqrt{5}, \sqrt{5}; 2\sqrt{5}, \sqrt{2}$
2. Write $\sqrt{2}$ and $\sqrt{18}$ as like radicals. $\sqrt{2}, 3\sqrt{2}$

In Exercises 3–5, simplify the expression.

3. $3\sqrt{6} + \sqrt{24}$ $5\sqrt{6}$
4. $\sqrt{72} - 5\sqrt{2}$ $\sqrt{2}$
5. $\sqrt{3}(5\sqrt{3} - 2\sqrt{6})$ $15 - 6\sqrt{2}$
6. Is $3 + \sqrt{6}$ a solution of the equation $x^2 - 6x + 3 = 0$? Yes

Independent Practice

In Exercises 7–18, simplify the expression.

7. $4\sqrt{7} + 3\sqrt{7}$ $7\sqrt{7}$
8. $\sqrt{5} + 3\sqrt{5}$ $4\sqrt{5}$
9. $9\sqrt{3} - 12\sqrt{3}$ $-3\sqrt{3}$
10. $2\sqrt{6} - \sqrt{6}$ $\sqrt{6}$
11. $\sqrt{2} + \sqrt{18}$ $4\sqrt{2}$
12. $\sqrt{27} + \sqrt{3}$ $4\sqrt{3}$
13. $4\sqrt{5} + \sqrt{80} + \sqrt{20}$ $10\sqrt{5}$
14. $5\sqrt{5} + \sqrt{405} + \sqrt{5}$ $15\sqrt{5}$
15. $\sqrt{44} - 2\sqrt{11}$ 0
16. $\sqrt{72} - 9\sqrt{2}$ $-3\sqrt{2}$
17. $\sqrt{20} - \sqrt{45} + \sqrt{5}$ 0
18. $\sqrt{243} + \sqrt{75} - \sqrt{300}$ $4\sqrt{3}$

In Exercises 19–24, simplify the expression. 23. $-6 - 19\sqrt{2}$

19. $\sqrt{3}(3\sqrt{2} + \sqrt{3})$ $3\sqrt{6} + 3$
20. $\sqrt{6}(5\sqrt{2} + 6)$ $10\sqrt{3} + 6\sqrt{6}$
21. $(\sqrt{5} + 4)^2$ $21 + 8\sqrt{5}$
22. $(2\sqrt{3} - 5)^2$ $37 - 20\sqrt{3}$
23. $(\sqrt{2} + 4)(1 - 5\sqrt{2})$
24. $(\sqrt{7} + 3)(\sqrt{7} - 3)$ -2
25. *Geometry* Find the area and perimeter.

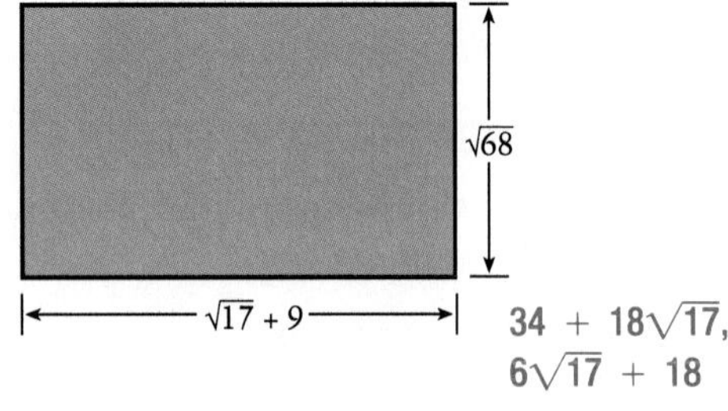

$34 + 18\sqrt{17}$, $6\sqrt{17} + 18$

26. *Geometry* Find the area. $45 + 20\sqrt{11}$

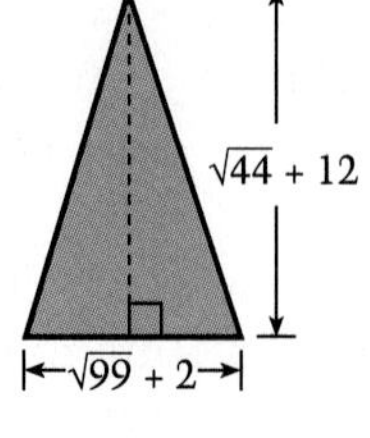

In Exercises 27 and 28, evaluate the function.

27. $f(x) = x^2 + 5x - 4$, when $x = \sqrt{10}$ $6 + 5\sqrt{10}$
28. $f(x) = x^2 - x - 6$, when $x = 2\sqrt{3}$ $6 - 2\sqrt{3}$

In Exercises 29 and 30, decide whether the *x*-value is a solution of the equation.

29. $x^2 - 4x - 15 = 0$, $x = 2 - \sqrt{19}$ Yes
30. $2x^2 + 16x - 21 = 0$, $x = -4 + \sqrt{106}$ No

31. *Swimming Pool* Find the perimeter of the swimming pool. (Each unit in the coordinate system corresponds to 2 feet.) The coordinates of the vertices are as follows. $20 + 24\sqrt{2}$ feet

$A = (0, 0)$	$B = (-2, 2)$
$C = (-2, 4)$	$D = (0, 6)$
$E = (2, 6)$	$F = (7, 1)$
$G = (7, -1)$	$H = (5, -1)$
$I = (3, 1)$	$J = (1, 1)$

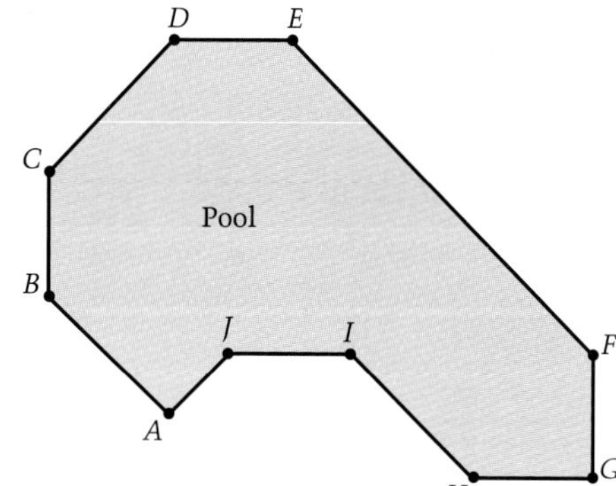

32. *The Troublesome T* The puzzle pieces below can be arranged to form the letter "T." Find the total perimeter of all four pieces. (Use the coordinate system to find the vertices of each piece: The piece at the lower left has coordinates (0, 0), (5, 0), (5, 3), and (0, 8).) Copy the pieces, cut out the copies, and form the letter "T." What do you notice about the perimeter of the "T"? Explain how to find the area of all four pieces. See margin.

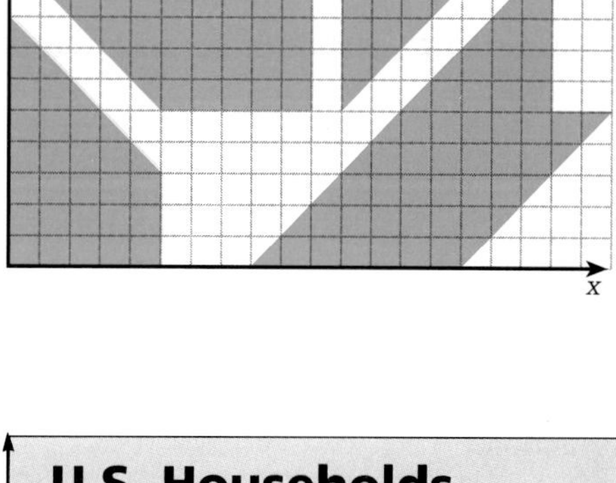
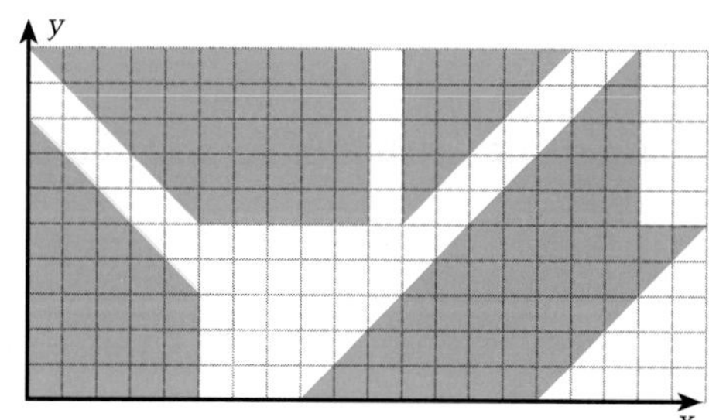

33. *U.S. Households* For 1970 through 1990, the number, H (in millions), of households in the United States can be modeled by

$$H = \sqrt{3978 + 241t}$$

where $t = 0$ represents 1970. How many households were there in 1970? How many households were there in 1971? How many households were there in 1972? Create a table that shows that the number of households was "increasing at a decreasing rate." See margin.

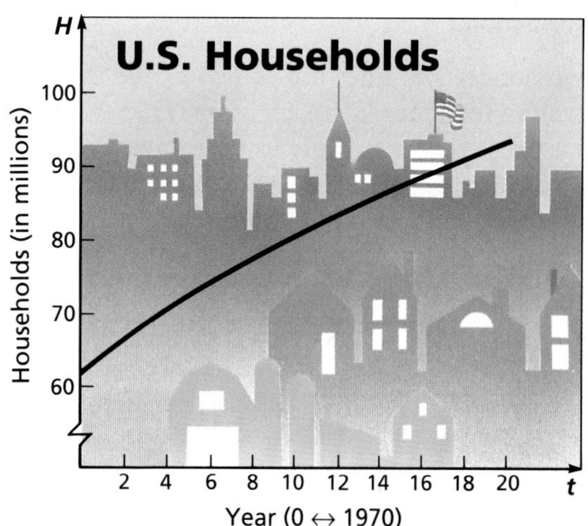

34. *Selling a Puzzle* The Spanish Club at school has its picture taken and sends the picture to a company that makes jigsaw puzzles. The club sells the puzzles to make money for a trip. The number, P, of puzzles sold each week for 20 weeks can be modeled by

$$P = 4\sqrt{6t - 3}$$

where $t = 1$ represents the first week. Use a table *and* graph to describe the sales pattern of the puzzles. If the club made a profit of \$4 on each puzzle, what is the total profit made in 20 weeks? \$2340.00

Answers

32. $60 + 30\sqrt{2}$, it contains no radicals (60), count the squares in the assembled T.

33.

Year	Households (in millions)	Increase from previous year
1970	≈63.07	
1971	≈64.95	1.88
1972	≈66.78	1.83
1973	≈68.56	1.78

34.

Week	Sold	Week	Sold
1	7	11	≈32
2	12	12	≈33
3	15	13	≈35
4	18	14	36
5	21	15	≈37
6	23	16	≈39
7	25	17	≈40
8	27	18	≈41
9	29	19	≈42
10	30	20	≈43

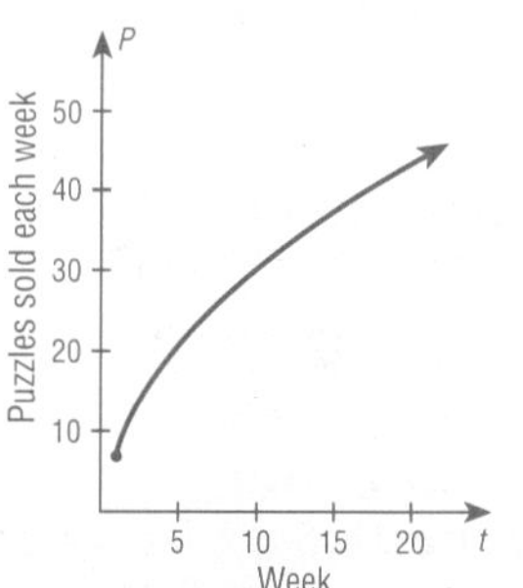

Ex. 44–46 GROUP ACTIVITY These exercises introduce the slopes of perpendicular lines, a concept that is a connection to geometry. You might want to use this exploration with all students as a class activity.

Answers

44. $(4\sqrt{5})^2 + (2\sqrt{5})^2 = 10^2$;
$m_1 = 2,\ m_2 = -\frac{1}{2}$,
$m_1 \cdot m_2 = -1$

45. $(\sqrt{10})^2 + (3\sqrt{10})^2 = 10^2$;
$m_1 = -3,\ m_2 = \frac{1}{3}$,
$m_1 \cdot m_2 = -1$

Enrichment Activities

These activities require students to go beyond lesson goals.

1. Remember this property of absolute value.
If $|x| < a$, then $-a < x < a$.
Solve these inequalities. Check your answers.
a. $|x - \sqrt{5}| < \sqrt{45}$
b. $|2x + \sqrt{3}| < \sqrt{27}$
a. $-2\sqrt{5} < x < 4\sqrt{5}$
b. $-2\sqrt{3} < x < \sqrt{3}$
2. COOPERATIVE LEARNING Project Work with a partner and design your own architectural plan for a house similar to that shown in Exercise 31. Superimpose your model on a grid and determine the perimeter and area of your plan.

Integrated Review

In Exercises 35–38, find the distance between the points. Write the result in simplified form.

35. $(8, \sqrt{5}), (5, 6\sqrt{5})$ $\sqrt{134}$

36. $(6\sqrt{2}, 5), (\sqrt{2}, 3)$ $3\sqrt{6}$

37. $(\sqrt{2}, 3), (-2\sqrt{2}, 2)$ $\sqrt{19}$

38. $(3, \sqrt{3}), (4, 4\sqrt{3})$ $2\sqrt{7}$

39. Is 4 a solution of $\sqrt{x+5} + \sqrt{x} = 5$? Yes

40. Is -2 a solution of $\sqrt{x+6} + \sqrt{14-x} = 20$? No

In Exercises 41 and 42, simplify the expression.

41. $\frac{\sqrt{243} \times 10^4}{\sqrt{27} \times 10^{-2}}$ 3×10^6

42. $\frac{\sqrt{112} \times 10^{-1}}{2\sqrt{7} \times 10^3}$ 2×10^{-4}

43. *College Entrance Exam Sample* The expression $\sqrt{a^2 + b^2}$ is equal to e

a. $a + b$ **b.** $a - b$ **c.** $(a + b)(a - b)$ **d.** $\sqrt{a^2} + \sqrt{b^2}$ **e.** none of these

Exploration and Extension

Perpendicular Lines **In Exercises 44 and 45, use the following information.**

Two nonvertical lines are perpendicular if their slopes, m_1 and m_2, are negative reciprocals of each other, or $m_1 \cdot m_2 = -1$. Show that the triangle is a right triangle in two ways. First, use the Distance Formula and the Pythagorean theorem. Second, show that two sides of the triangle are perpendicular. Which method do you prefer? Explain. See margin.

44. $(-3, -4), (1, 4), (5, 2)$

45. $(1, -7), (-5, 1), (-2, 2)$

46. *The Shortest Path to Shore* You are in a boat at the point $(-4, 7)$. The shoreline is given by the equation $y = \frac{5}{3}x$. The shortest distance between you and the shore is along the line that passes through the point $(-4, 7)$ and is perpendicular to the shoreline.

- Find the slope of the shortest path. $-\frac{3}{5}$
- Find an equation of the shortest path. $y = -0.6x + 4.6$
- Find the point of intersection of the shoreline and the shortest path. $\approx(2, 3.4)$
- Find the distance between the point of intersection and the point describing your location. ≈ 7

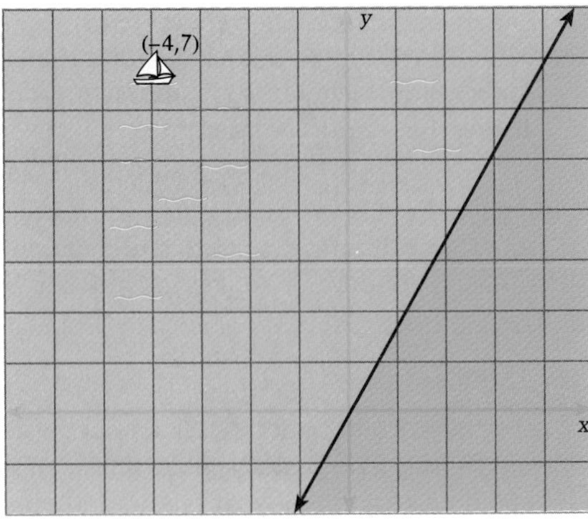

13.4

Solving Radical Equations

What you should learn:

Goal 1 How to solve radical equations

Goal 2 How to use radical equations to solve real-life problems

Why you should learn it:

You can use radicals to solve many real-life problems, such as compensating for inaccuracy of a balance.

Goal 1 Solving a Radical Equation

Solving an equation that contains radicals is somewhat like solving a rational equation—you try to rewrite the equation as a polynomial equation. Then you solve the polynomial equation using the standard procedures. The following property plays a key role.

Squaring Both Sides of an Equation

If $a = b$, then $a^2 = b^2$.

Squaring both sides of an equation often introduces **extraneous solutions** of $a^2 = b^2$ that are *not* solutions of $a = b$. So when you use this procedure, it is critical that you check each solution in the *original* equation.

Before squaring both sides of an equation, you should isolate the radical expression on one side of the equation.

Example 1 *Solving a Radical Equation*

Solve $\sqrt{x} - 8 = 0$.

Solution

$\sqrt{x} - 8 = 0$	*Rewrite original equation.*
$\sqrt{x} = 8$	*Add 8 to both sides.*
$(\sqrt{x})^2 = 8^2$	*Square both sides.*
$x = 64$	*Simplify.*

Check

$\sqrt{x} - 8 = 0$	*Rewrite original equation.*
$\sqrt{64} - 8 \stackrel{?}{=} 0$	*Substitute 64 for x.*
$8 - 8 = 0$	*Solution checks.*

The equation has 64 as its solution. ■

Try solving the equation $\sqrt{x} + 8 = 0$. You will obtain the same solution, 64. For this equation, however, 64 is extraneous. Try to see why.

Problem of the Day

The average weight of five freshmen was reported to be 60 kg. That was incorrect, because one freshman's weight of 80 kg had been miscopied. When correctly computed, the average weight was 70 kg. What had been written down instead of *80 kg*? 30 kg

ORGANIZER

Warm-Up Exercises

1. Solve for x.

a. $3x + 5 = 8$ — 1

b. $17 - 4x = 6x + 2$ — 1.5

c. $x^2 - 7x - 8 = 0$ — 8 or -1

d. $3x^2 + 4x = 4$ — -2 or $\frac{2}{3}$

2. Simplify.

a. $(\sqrt{x})^2$

b. $(\sqrt{3x})^2$

c. $(\sqrt{5 - 2x})^2$

d. $(\sqrt{7x - 3} + 2)^2$

a. x

b. $3x$

c. $5 - 2x$

d. $7x + 1 + 4\sqrt{7x - 3}$

Lesson Resources

Teaching Tools
- Transparencies: 1, 2
- Problem of the Day: 13.4
- Warm-up Exercises: 13.4
- Answer Masters: 13.4

Extra Practice: 13.4

Applications Handbook: p. 1–5, 55, 56

LESSON Notes

GOAL 1 Demonstrate the following property using numerical substitutions.

If $a = b$ then $a^2 = b^2$.

For example, since $5 = (3 + 2)$, $5^2 = (3 + 2)^2$, or $25 = 3^2 + 2(3)(2) + 2^2 = 9 + 12 + 4 = 25$.

Of course, the converse (if $a^2 = b^2$ then $a = b$) is not true. For example, for $a = 5$ and $b = -5$, $(5)^2 = (-5)^2$; that is, $25 = 25$, but $a = 5 \neq -5 = b$.

Example 1

If you square both sides of the equation in this example, you will get a trinomial whose middle term has a radical in it. But, if you isolate the radical on one side of the equation before squaring both sides, you will get a monomial (the radicand) that does not have a radical in it.

Squaring both sides of an equation is a nonequivalence transformation and may lead to extraneous roots. The equation $\sqrt{x} + 8 = 0$ becomes $\sqrt{x} = -8$. Squaring both sides gives $x = 64$. But checking this answer shows that 64 is an extraneous root because $\sqrt{64} + 8 \neq 0$.

Example 2

You may want to point out that the radicand, $2x - 1$, is assumed to be greater than or equal to zero. Ask students why the restriction is required. The restriction can be restated as $x \geq 0.5$.

Example 3

Since the square root of a real number must be nonnegative, and the square root in this equation equals x, $x = -1$ is not possible.

Example 2 Solving a Radical Equation

Solve $\sqrt{2x - 1} + 1 = 4$.

Solution

$\sqrt{2x-1} + 1 = 4$	*Rewrite original equation.*
$\sqrt{2x-1} = 3$	*Subtract 1 from both sides.*
$(\sqrt{2x-1})^2 = 3^2$	*Square both sides.*
$2x - 1 = 9$	*Simplify.*
$2x = 10$	*Add 1 to both sides.*
$x = 5$	*Divide both sides by 2.*

After checking $x = 5$ in the original equation, you can conclude that the equation has 5 as its solution. ■

Example 3 Solving a Radical Equation

Solve $x = \sqrt{x + 2}$.

Solution

$x = \sqrt{x+2}$	*Rewrite original equation.*
$x^2 = (\sqrt{x+2})^2$	*Square both sides.*
$x^2 = x + 2$	*Simplify.*
$x^2 - x - 2 = 0$	*Write in standard form.*
$(x-2)(x+1) = 0$	*Factor.*
$x = 2$ or $x = -1$	*Zero-Product Property*

Try checking *each* x-value in the original equation. You will find that $x = 2$ checks, but $x = -1$ *does not* check. Thus, the equation has 2 as its solution. ■

Example 4 Finding the Geometric Mean of Two Numbers

The **geometric mean** of a and b is $\sqrt{ab}$. If the geometric mean of a and 4 is 12, what is a?

Solution

$$\text{Geometric mean} = \sqrt{a \cdot 4}$$
$$12 = \sqrt{4a}$$
$$12^2 = (\sqrt{4a})^2$$
$$144 = 4a$$
$$36 = a$$

Check

$12 \stackrel{?}{=} \sqrt{(36)(4)}$

$12 \stackrel{?}{=} \sqrt{144}$

$12 = 12$ (*checks*)

The geometric mean of 36 and 4 is 12. ■

Goal 2 Solving Radical Equations in Real Life

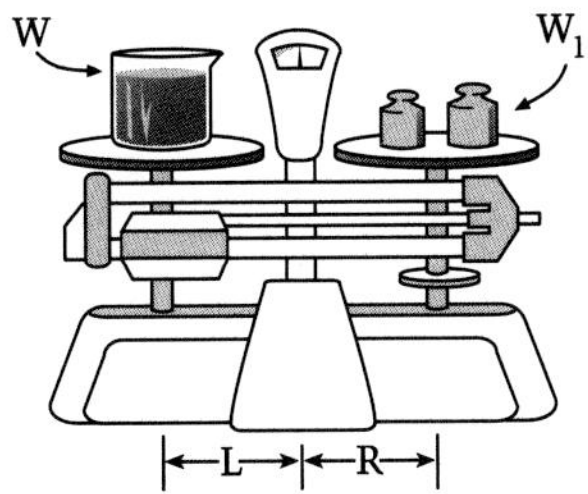

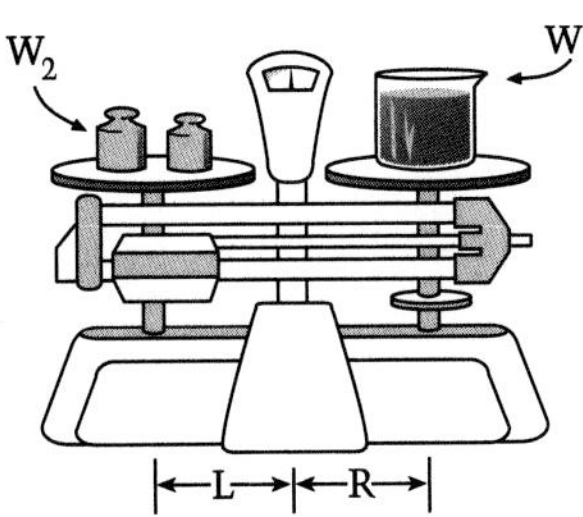

Balance scales can give inaccurate results if the lengths of the left and right arms, L and R, are not *exactly* the same. Here is a technique that scientists use to be sure of an accurate weight. Let W be the true weight of an object.

1. Place the object on the left side of the scale. Counterbalance the right side with a weight of W_1. By a property of levers, you can write $\frac{W}{W_1} = \frac{R}{L}$.
2. Place the object on the right side of the scale. Counterbalance the left side with a weight of W_2: $\frac{W_2}{W} = \frac{R}{L}$
3. Equate the two expressions for $\frac{R}{L}$ and solve for W. The solution is $W = \sqrt{W_1 W_2}$. Thus, the true weight is the geometric mean of the weights obtained by weighing the object on the left and right sides of the scale.

Connections
Physical Science

Example 5 *Finding an Accurate Weight*

In science class, you weigh a sulfur sample only once—on the left side of a scale: It weighs 92 grams. Your teacher says the actual weight is 95 grams. If 95 grams is correct, what weight would you have obtained on the right side of the scale?

Solution The actual weight is $W = 95$ grams. On the left side of the scale, you observed the weight to be $W_1 = 92$ grams. Let W_2 be the weight that would be observed on the right side of the scale.

$W = \sqrt{W_1 W_2}$ *W is the geometric mean of W_1 and W_2.*

$95 = \sqrt{92 W_2}$ *Substitute 95 for W and 92 for W_1.*

$9025 = 92 W_2$ *Square both sides of the equation.*

$98.1 \approx W_2$ *Divide both sides by 92.*

You would have obtained 98.1 grams on the right side. ■

Communicating about ALGEBRA

Cooperative Learning

SHARING IDEAS about the Lesson See Additional Answers.

Work with a Partner and Use Your Own Words Solve the equations. Explain your steps.

A. $\sqrt{x} - 5 = 0$ 25

B. $\sqrt{x} + 5 = 0$ No solution

C. $\sqrt{3x + 4} - 2 = 5$ 15

D. $x - 2 = \sqrt{8 - x}$ 4, −1

Example 4

Observe that the geometric sequence,

4, 12, 36, 108, 324, . . .

is characterized by the following pattern. The ratio of any two consecutive numbers in the sequence is constant; that is, $3 = \frac{12}{4} = \frac{36}{12} = \frac{108}{36} = \cdots$. Ask students to identify a pattern that might explain why the "geometric" mean of a and b is so named. Using the pattern, have them determine the geometric mean of 12 and 108; of 36 and 324. 36, 108

Example 5

It may be helpful to show students how W is derived from $\frac{W}{W_1} = \frac{R}{L} = \frac{W_2}{W}$, that is, from $\frac{W}{W_1} = \frac{W_2}{W}$. After cross multiplying, you get $W^2 = W_1 W_2$. Taking the square root of both sides gives $W = \sqrt{W_1 W_2}$.

Extra Examples

Here are additional examples similar to **Examples 1–5.**

1. Solve each radical equation.
   - **a.** $\sqrt{x} - 3 = 0$
   - **b.** $\sqrt{2x} + 4 = 0$
   - **c.** $\sqrt{4x + 1} = 7$
   - **d.** $\sqrt{24 - 5x} = x$
   - **e.** $\sqrt{2x - 1} = x - 8$

   a. 9 b. no solution
   c. 12 d. 3; −8 is extraneous
   e. 13; 5 is extraneous
2. If the geometric mean is $g = \sqrt{ab}$, find each missing value.
   - **a.** $g = 48, a = ?, b = 192$ 12
   - **b.** $g = 35, a = 5, b = ?$ 245
   - **c.** $a = 3, b = 363, g = ?$ 33
3. Find the actual weight of a sulfur compound if the sample weighed 19 grams on the left side of the scale and 16 grams on the right side of the scale. approximately 17.4 grams

Check Understanding

1. Explain why the converse of the property, "If $a = b$, then $a^2 = b^2$" does not hold. See Lesson Notes for Goal 1.
2. What are extraneous solutions? See Example 3.
3. Why must the values of x in a radical equation be restricted to certain real-number values? The radicand cannot be negative.

EXTEND *Communicating*

1. Explain how you would solve the following radical equations.
   a. $\sqrt{x+5} = \sqrt{2x-6}$
   b. $\sqrt{2x-4} = \sqrt{5-x} + 1$

   Square both sides first.

   a. 11

   b. 4; $\frac{20}{9}$ is extraneous.
2. When will the geometric mean and the arithmetic mean be the same?
   When $a > 0$, $b > 0$, and $a = b$

EXERCISE Notes

Assignment Guide

Basic/Average: Ex. 1–4, 5–27 odd, 33–37, 45–49 odd

Above Average: Ex. 1–4, 7–27 odd, 31–36, 39–49 odd

Advanced: Ex. 1–4, 9–29 odd, 31–37, 45–51 odd

Selected Answers

Exercises 1–4, 5–49 odd

More Difficult Exercises

Exercises 31–36, 51

EXERCISES

Guided Practice

CRITICAL THINKING about the Lesson

1. Describe a strategy for solving a radical equation. Then apply your strategy to solve $\sqrt{2x} - 10 = 0$. Isolate the radical, then square both sides, 50.
2. Is $x = 25$ a solution of $\sqrt{x} = -5$? Why or why not? No, it does not check.
3. One reason for checking a solution in the original equation is that you might have made an error in one of the steps of the solution. Describe another reason. You could have a false solution
4. The geometric mean of 12 and x is 6. What is x? 3

Independent Practice

12. No solution

In Exercises 5–16, solve the equation. (Some of the equations have no solution.)

5. $\sqrt{x} - 10 = 0$ 100
6. $\sqrt{x} - 1 = 0$ 1
7. $\sqrt{-x} - \frac{1}{2} = \frac{3}{2}$ -4
8. $\sqrt{3x} - 4 = 6$ $\frac{100}{3}$
9. $\sqrt{3x+2} + 2 = 3$ $-\frac{1}{3}$
10. $\sqrt{4-x} - 5 = 1$ -32
11. $4 = 6 - \sqrt{21x - 3}$ $\frac{1}{3}$
12. $10 = 17 + \sqrt{6x+7}$
13. $\sqrt{\frac{1}{2}x - 5} - 1 = 11$ 298
14. $\sqrt{\frac{1}{9}x + 1} - \frac{2}{3} = \frac{5}{3}$ 40
15. $-5 - \sqrt{10x - 2} = 5$ No solution
16. $6 - \sqrt{7x - 9} = 3$ $\frac{18}{7}$

In Exercises 17–24, solve the equation.

17. $x = \sqrt{20 - x}$ 4
18. $x = \sqrt{6 - x}$ 2
19. $\sqrt{-10x - 4} = 2x$ No solution
20. $x = \sqrt{100 - 15x}$ 5
21. $\sqrt{77 - 4x} = x$ 7
22. $2x = \sqrt{-13x - 10}$ No solution
23. $\frac{1}{2}x = \sqrt{x+3}$ 6
24. $x = \sqrt{4x + 32}$ 8

25. Find the geometric mean of 8 and 32. 16
26. Find the geometric mean of 4 and 32. $8\sqrt{2}$
27. The geometric mean of x and 5 is 15. What is x? 45
28. The geometric mean of 9 and x is 6. What is x? 4

Geometry **In Exercises 29 and 30, find the value of x.**

29. 26

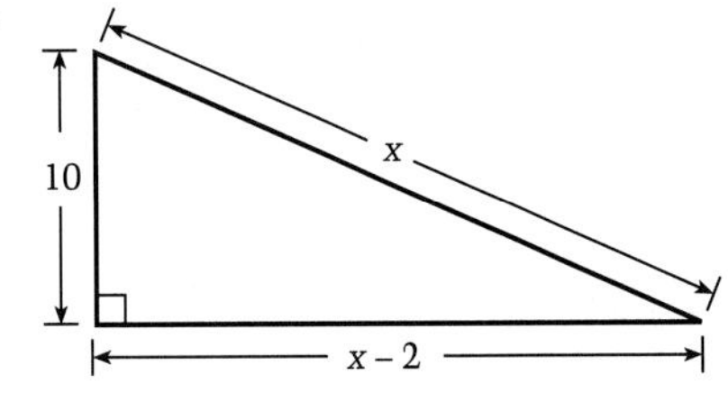

30. 7

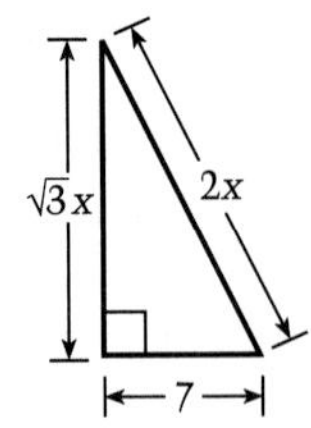

Spinning **In Exercises 31 and 32, use the following information.**

A ride at an amusement park spins in a circle of radius r, (in feet). The centrifugal force, F (in pounds), experienced by a passenger on the ride can be found by solving the equation

$$t = \sqrt{\frac{\pi^2 wr}{8F}}.$$

t is the number of seconds the ride takes to make one complete revolution and w is the weight, (in pounds), of the passenger.

31. A person who weighs 115 pounds is on a ride that is spinning at a rate of 10 seconds per revolution. The radius of the circular ride is 20 feet. How much centrifugal force does the person feel? $\approx$ 28.38 lb

32. You are spinning on ice skates with your arms outstretched. As you pull your arms in toward your body, you are decreasing the radius, r. What effect does this have on t? What effect does it have on the skater? (Assume that the centrifugal force, F, is constant.) t decreases, the skater spins at a faster rate.

The Speed of Sound **In Exercises 33 and 34, use the following information.**

The speed of sound near Earth's surface depends on the temperature. An equation that relates the speed, v (in meters per second), with the temperature, t (in degrees Celsius), is $v = 20\sqrt{t + 273}$.

33. Your friend is playing basketball 170 meters from you. You hear the sound of the ball hitting the backboard 0.5 seconds after seeing the ball hit the backboard. What is the temperature? 16°C

34. The temperature -273°C is called absolute zero. What is the speed of sound at this temperature? 0 m per second

Rifle Velocity **In Exercises 35 and 36, use the following information.**

Because of the air resistance, the velocity of a bullet decreases after the bullet is fired. For a 25–06 rifle, the velocity, v (in feet per second), of the bullet after t seconds is given by $v = \frac{3200}{t + 1}$. If the rifle is held horizontally, the distance, h (in feet), the bullet *falls* is related to the time by the equation $t = \frac{1}{4}\sqrt{h}$.

35. A 25–06 rifle is fired horizontally at a target. The bullet hits the target in $\frac{1}{8}$ second. How fast is the bullet moving when it hits the target?

36. How far did the bullet fall before hitting the target? $\frac{1}{4}$ foot

35. $2844\frac{4}{9}$ ft per second

The Winter Olympics Biathlon combines Nordic skiing with rifle shooting.

Guided Practice

Ex. 1–4 Refer students to Examples 1–4 for solution models. You can also relate the geometric mean, x, of two numbers ab as the solution of the proportion $\frac{a}{x} = \frac{x}{b}$, or $x^2 = ab$.

Independent Practice

Ex. 5–24 Remind students to check for extraneous solutions.

Ex. 31–36 TECHNOLOGY Students should use calculators to solve these applications.

Enrichment Activities

These activities require students to go beyond lesson goals.

An Alternate Approach
Checking by Graphing

In this lesson you learned an algebraic method of finding exact solutions to radical equations. You can also use a graphical method to approximate and verify solutions.

1. Solve $\sqrt{x - 1} = x - 3$ algebraically. Do not check your solutions. $x = 5$ or $x = 2$
2. Graph $y = \sqrt{x - 1}$ and $y = x - 3$ on the same coordinate plane. Where do they intersect? The point (5, 2).

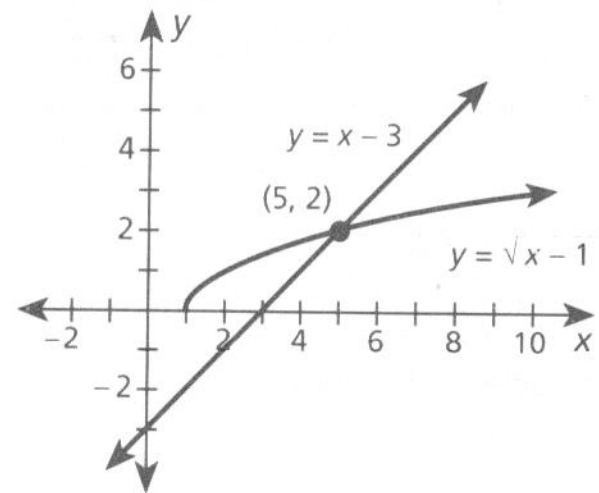

3. Which algebraic solution is extraneous? Why? $x = 2$; at $x = 2$ the two graphs do not intersect.

4. Use graphing to approximate the solutions of this radical equation: $5 - \sqrt{x} = \sqrt{x + 5}$.

$x = 4$

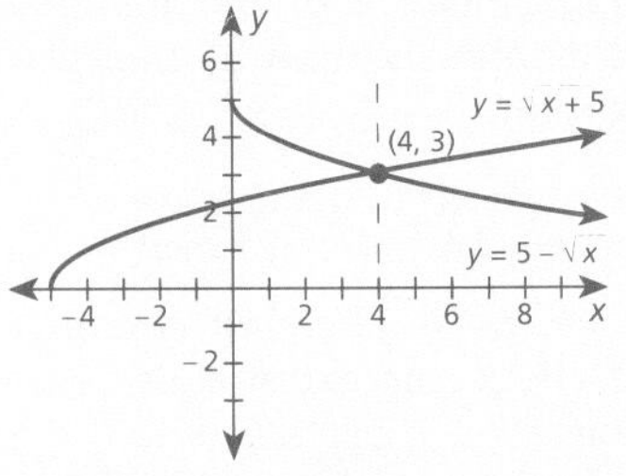

5. **WRITING** Write a paragraph on using graphing to solve and check radical equations.

Ex. 41–50 These exercises review the quadratic formula.

Ex. 51 Students should use a calculator and numerical substitutions to verify that if $x^3 = y^2$, then $y = \sqrt{x^3}$.

Answer

51. $y = \sqrt{x^3}$

	x^3	y^2
Mercury	≈0.058	≈0.058
Venus	≈0.378	≈0.378
Earth	1	1
Mars	≈3.533	≈3.538
Jupiter	≈140.852	≈140.683
Saturn	≈868.524	≈867.715

Graphing Calculator Answers

1.

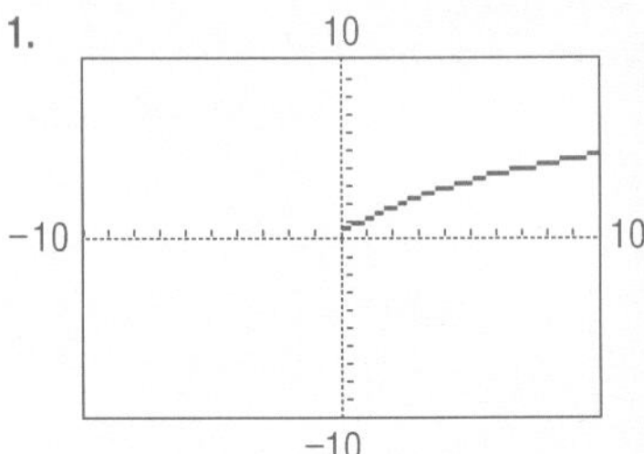

2.

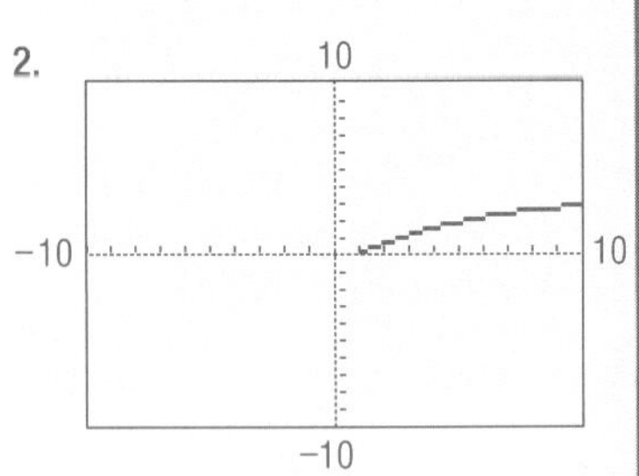

Integrated Review

In Exercises 37–40, are the given x-values solutions of the equation?

37. $x - \sqrt{9} = 0$, $x = \pm 3$ 3, yes; −3, no

38. $2x - \sqrt{16} = 0$, $x = \pm 2$ 2, yes; −2, no

39. $x = \sqrt{80 - 2x}$, $x = -10, x = 8$ 8, yes; −10, no

40. $x = \sqrt{12 - 11x}$, $x = -12, x = 1$ 1, yes; −12, no

In Exercises 41–46, write the solutions in simplified form.

41. $4x^2 + 22x + 30 = 0$ $-\frac{5}{2}, -3$

42. $3x^2 - 4x + 1 = 0$ $\frac{1}{3}, 1$

43. $x^2 - 3x - 9 = 0$ $\frac{3 + 3\sqrt{5}}{2}, \frac{3 - 3\sqrt{5}}{2}$

44. $x^2 + 14x - 12 = 0$ $-7 + \sqrt{61}, -7 - \sqrt{61}$

45. $x^2 - 14x + 33 = 0$ 11, 3

46. $10x^2 + \frac{5}{8}x - 2 = 0$ $\frac{-5 + 7\sqrt{105}}{160}, \frac{-5 - 7\sqrt{105}}{160}$

In Exercises 47–50, write the solutions in simplified form. (Multiply first by the least common denominator to "clear" the equation of fractions.)

47. $x^2 + \frac{5}{3}x + \frac{4}{9} = 0$ $-\frac{4}{3}, -\frac{1}{3}$

48. $\frac{1}{21}x^2 - \frac{3}{7}x - \frac{10}{21} = 0$ 10, −1

49. $\frac{1}{6}x^2 + \frac{2}{3}x + \frac{1}{3} = 0$ $-2 + \sqrt{2}, -2 - \sqrt{2}$

50. $\frac{1}{4}x^2 + \frac{5}{8}x - 2 = 0$ $\frac{-5 + \sqrt{153}}{4}, \frac{-5 - \sqrt{153}}{4}$

Exploration and Extension

51. *Orbits of the Sun* Johannes Kepler (1571–1630), discovered a relationship between the average distance of a planet from the sun and the time, or *period*, it takes the planet to orbit the sun. The following table shows the average distance, x (in astronomical units), and the period, y (in years), for the six planets that are closest to the sun. Complete the table. Then use the results to write y as a function of x. See margin.

	Mercury	Venus	Earth	Mars	Jupiter	Saturn
Average Distance, x	0.387	0.723	1	1.523	5.203	9.541
x^3	?	?	?	?	?	?
Period, y	0.241	0.615	1	1.881	11.861	29.457
y^2	?	?	?	?	?	?

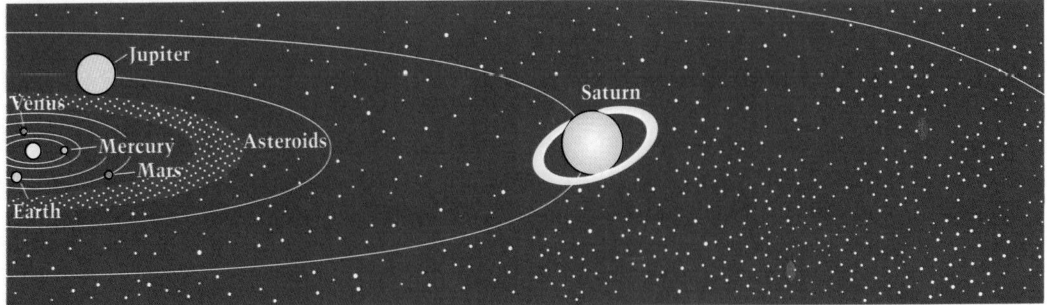

USING A GRAPHING CALCULATOR

The domain of the function $f(x) = \sqrt{x}$ *is the set of nonnegative numbers. The graph of this function is shown at the right. Notice that the graph begins at the origin and continues to the right. To the left of the origin, the graph does not exist because negative* x*-values are not in the domain of* f.

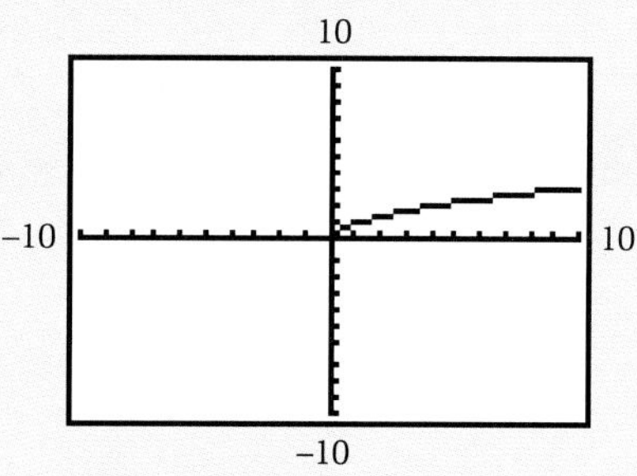

Graph of $f(x) = \sqrt{x}$

Keystroke instructions for sketching this graph on the *TI-82,* Casio *fx-9700GE,* and Sharp *EL-9300C* are listed on page 746.

Cooperative Learning

Writing a Radical Function With a partner, write radical functions of the form

$$f(x) = \sqrt{ax + b} \quad \text{or} \quad f(x) = \sqrt{ax - b}, \qquad b > 0$$

whose graphs resemble the graphs shown below.

a. $f(x) = \sqrt{x + 4}$

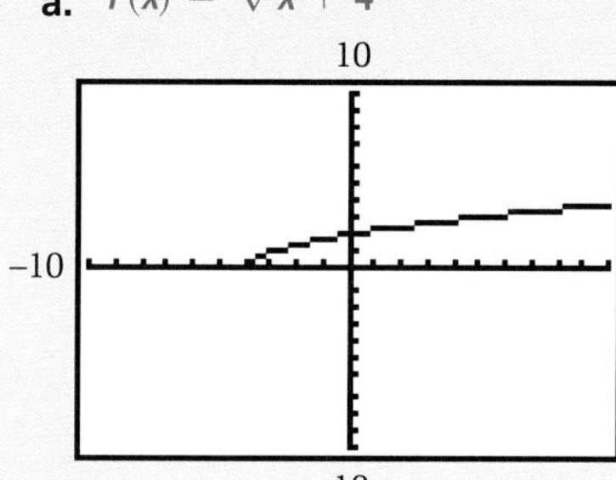

b. $f(x) = \sqrt{2x - 1}$

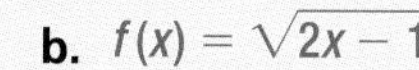

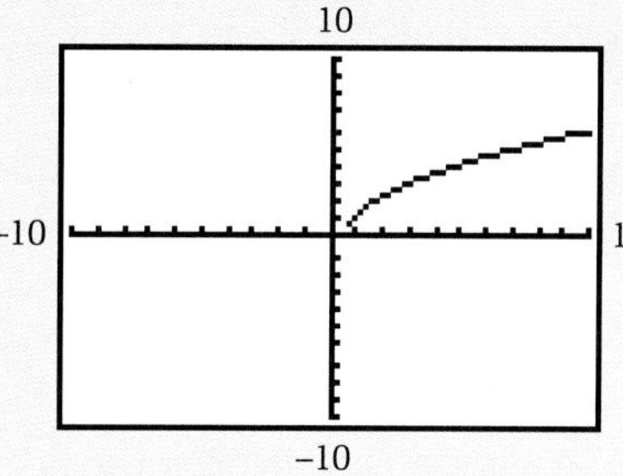

Use a graphing calculator to check that the graph of your functions match the graphs shown here.

Exercises

In Exercises 1–4, use a graphing calculator to sketch the graph of the function. From the graph, describe the domain of the function. See margin.

1. $f(x) = \sqrt{2x}$ $x \geq 0$

2. $f(x) = \sqrt{x - 1}$ $x \geq 1$

3. $f(x) = \sqrt{x + 2}$ $x \geq -2$

4. $f(x) = \sqrt{2x - 6}$ $x \geq 3$

In Exercises 5–8, use a graphing calculator to sketch the graphs of both functions. Explain how the graph of g **related to the graph of** f**.** See margin.

5. $f(x) = \sqrt{x}$, $g(x) = \sqrt{x - 3}$

6. $f(x) = \sqrt{x}$, $g(x) = \sqrt{x} - 3$

7. $f(x) = \sqrt{x}$, $g(x) = -\sqrt{x}$

8. $f(x) = \sqrt{x}$, $g(x) = \sqrt{x + 4}$

3.

4.

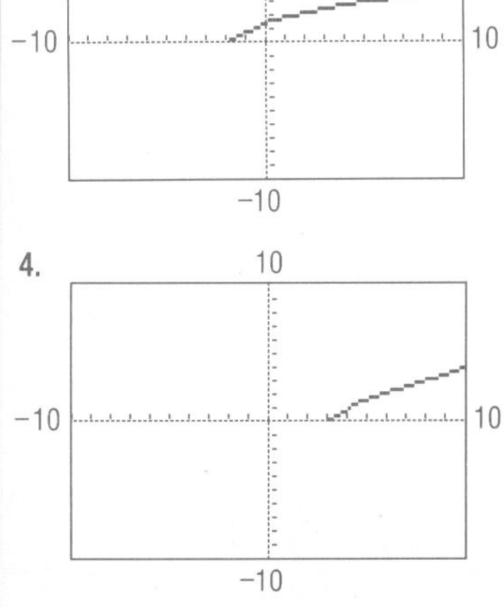

5.

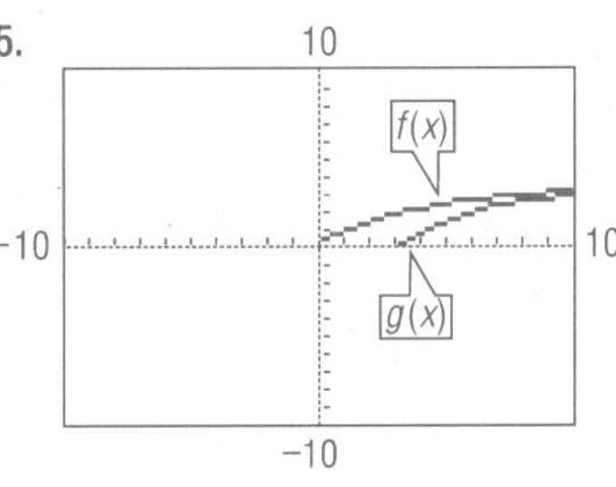

Horizontal shift 3 units right.

6.

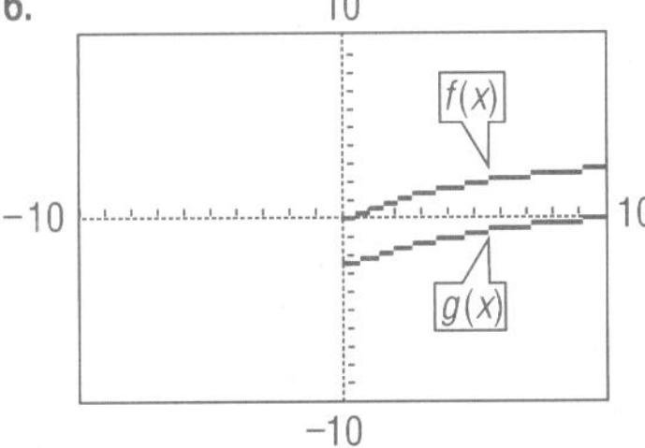

Vertical shift 3 units down.

7.

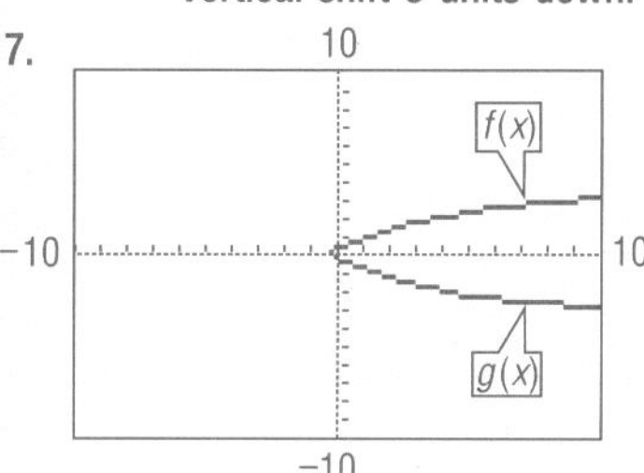

Reflection about the x-axis.

8.

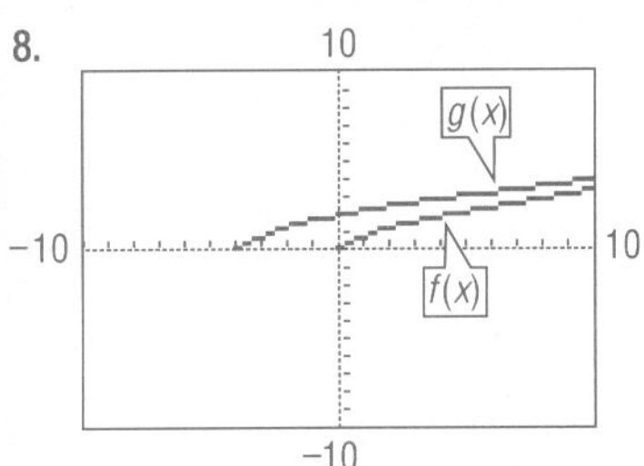

Horizontal shift 4 units left.

TECHNOLOGY Students should be careful to use parentheses when entering radical functions. For example, on a graphing calculator, the function $f(x) = \sqrt{x + 2}$ should be entered as

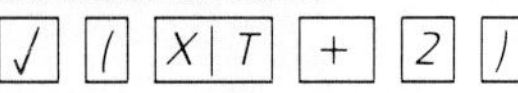

Answers

11. $6\sqrt{3} - \sqrt{2}$
12. $36 + 9\sqrt{2}$
13. $16 - 4\sqrt{5}$
14. $21 - 12\sqrt{3}$

Mid-Chapter SELF-TEST

In Exercises 1 and 2, find the distance between the two points. (13.1)

1. $(4, 5), (7, 2)$ $3\sqrt{2}$
2. $(1, -2), (-4, 8)$ $5\sqrt{5}$

In Exercises 3 and 4, decide whether the points are vertices of a right triangle. (13.1)

3. $(3, -2), (4, 6), (11, -1)$ No
4. $(1, 5), (2, 8), (-2, 6)$ Yes

In Exercises 5 and 6, find the midpoint between the two points. (13.1)

5. $(4, 2), (5, -6)$ $(\frac{9}{2}, -2)$
6. $(0, 10), (-12, 5)$ $(-6, \frac{15}{2})$

In Exercises 7–10, simplify the radical expression. (13.2)

7. $\sqrt{98}$ $7\sqrt{2}$
8. $\sqrt{432}$ $12\sqrt{3}$
9. $\sqrt{\frac{5}{36}}$ $\frac{\sqrt{5}}{6}$
10. $\sqrt{\frac{7}{20}}$ $\frac{\sqrt{35}}{10}$

In Exercises 11–14, perform the indicated operations. (13.3) See margin.

11. $5\sqrt{3} + \sqrt{3} - \sqrt{2}$
12. $3\sqrt{3}(4\sqrt{3} + \sqrt{6})$
13. $\sqrt{2}(8\sqrt{2} - 2\sqrt{10})$
14. $(2\sqrt{3} - 3)^2$

In Exercises 15–18, solve the equation. (13.4)

15. $\sqrt{x} - 3 = 0$ 9
16. $\sqrt{3x - 2} - 6 = 0$ $\frac{38}{3}$
17. $x = \sqrt{5x + 6}$ 6
18. $x = \sqrt{-3x + 4}$ 1

19. Find the geometric mean of 4 and 9. **(13.4)** 6

20. The geometric mean of a and 32 is 8. What is a? **(13.4)** 2

21. Find the perimeter of the polygon. **(13.3)** $10\sqrt{5}$

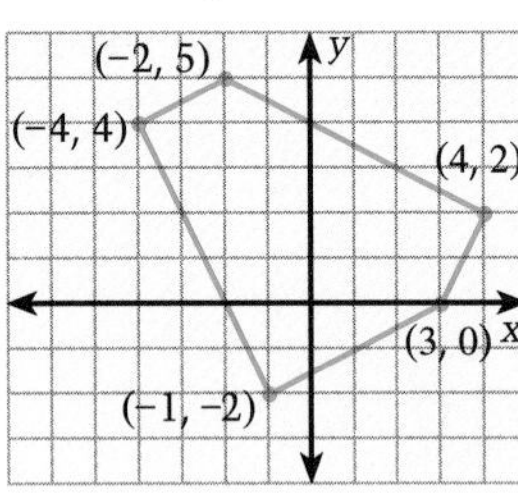

22. Find the area and perimeter of the polygon. **(13.3)**
$3\sqrt{2} + 4\sqrt{3}$,
$2\sqrt{3} + 4\sqrt{2} + 2\sqrt{6}$

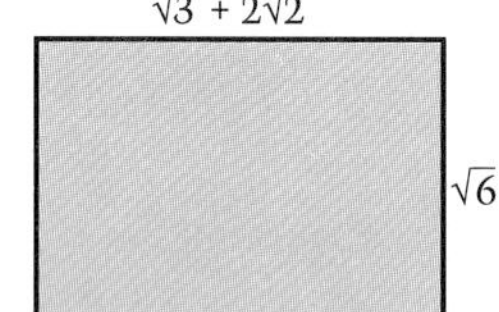

(−47, 90)

23. Two infielders position themselves so that second base is at their midpoint. Find the coordinates of second base. **(13.1)**

24. In the baseball stadium whose infield is shown at the right, any ball that is hit farther than 400 feet will be a home run. A batter hits a fly ball to the coordinates $(-245, -285)$. Is it a home run? (Each unit corresponds to 1 foot.) **(13.1)** Yes

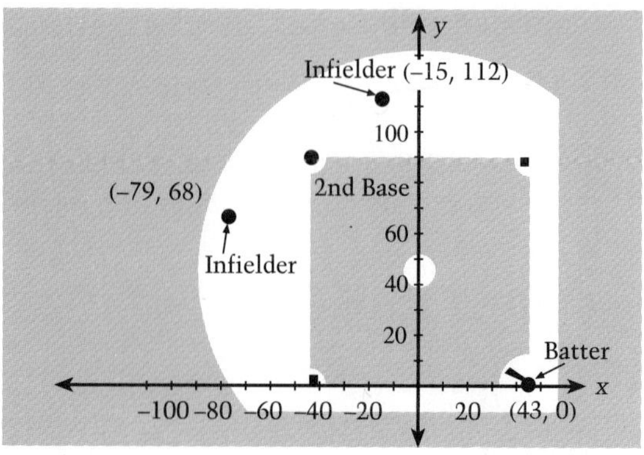

13.5 The Tangent of an Angle

What you should learn:

Goal 1 How to use the tangent of an angle to find the lengths of the sides of a right triangle

Goal 2 How to use the tangent of an angle to solve real-life problems

Why you should learn it:

You can use the tangent of an angle to solve many real-life problems, such as finding distances that cannot be measured directly.

Goal 1 Using the Tangent of an Angle

Angles are measured in degrees. An **acute angle** is an angle whose measure is greater than 0° and less than 90°. A right triangle has two acute angles and one right angle. The acute angles are denoted by A and B. The right angle is denoted by C. The measures of the three sides are denoted by the lowercase form of the same letters.

- The side BC is opposite angle A. Its length is a.
- The side AC is opposite angle B. Its length is b.
- The side AB is opposite angle C. Its length is c.

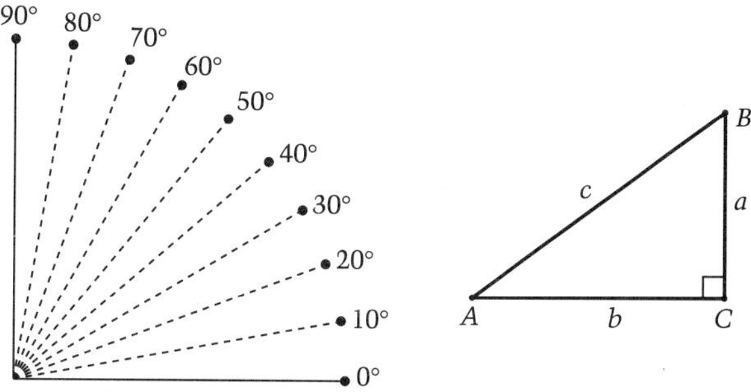

The Tangent of an Acute Angle

Let A and B be the acute angles of a right triangle. The **tangent** of A and the **tangent** of B are as follows:

$$\tan A = \frac{\text{length of side opposite } A}{\text{length of side adjacent to } A} = \frac{a}{b}$$

$$\tan B = \frac{\text{length of side opposite } B}{\text{length of side adjacent to } B} = \frac{b}{a}$$

Example 1 *Finding the Tangent of an Angle*

Find the tangent of A and the tangent of B for the triangle shown.

B, c, a = 5, A, b = 12, C

Solution $\tan A = \frac{a}{b} = \frac{5}{12}$

$\tan B = \frac{b}{a} = \frac{12}{5}$ ■

Problem of the Day

How many 4-digit numbers can you build with a 3, a 4, a 6, and an 8? How many of those are over 6000? 24, 12

ORGANIZER

Warm-Up Exercises

1. Use the Pythagorean Theorem to find the measure of the missing side.

a. $b = 12$

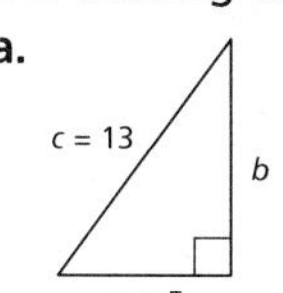

b. $c = 15$

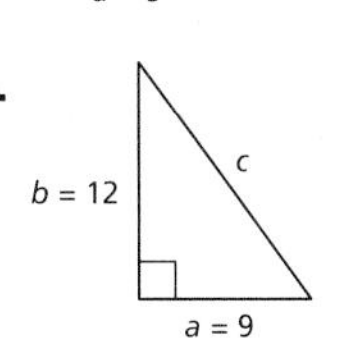

c. $a = 2$

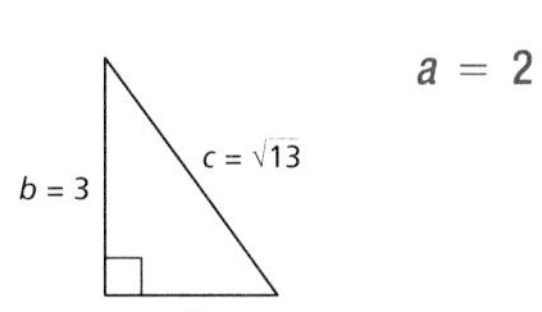

d. $a = 51$

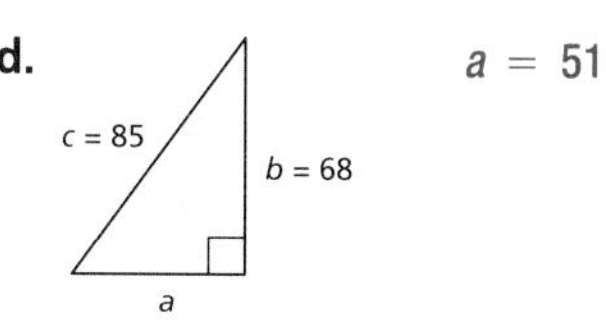

2. Solve each proportion.

a. $\frac{8}{7} = \frac{m}{35}$ $m = 40$

b. $\frac{3x}{5} = \frac{12}{45}$ $x = \frac{4}{9}$

c. $\frac{6}{5k} = \frac{8}{9}$ $k = \frac{27}{20}$

Lesson Resources

Teaching Tools
- Transparencies: 11, 12
- Problem of the Day: 13.5
- Warm-up Exercises: 13.5
- Answer Masters: 13.5

Extra Practice: 13.5

LESSON Notes

GOAL 1 The notation used to describe the right triangle ABC is used in both trigonometry and geometry. Although it is not required that the side opposite angle A be called side a, the practice is so common that nothing is gained by violating the convention.

Example 1

The third side, c, is called the hypotenuse. The length of c can be determined using the Pythagorean Theorem,

$$c = \sqrt{5^2 + 12^2} = 13.$$

Example 2

Point out that, for the given right triangle, $\tan A = \frac{12}{9} = \frac{4}{3}$. Notice that each right triangle below also has $\tan A = \frac{4}{3}$. Right triangles whose tan A values are equal are said to be similar triangles.

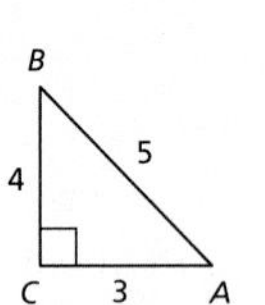

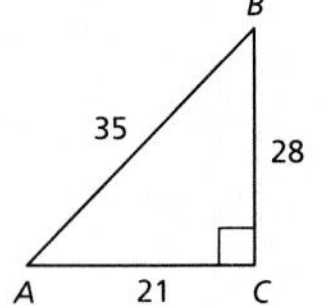

Example 3

TECHNOLOGY If using a graphics calculator, tan 30° is obtained by entering the following keystrokes:

[TAN] 30 (not 30 [TAN]).

The length of side AB is approximately 9.24.

Example 2 Finding the Lengths of the Sides of a Right Triangle

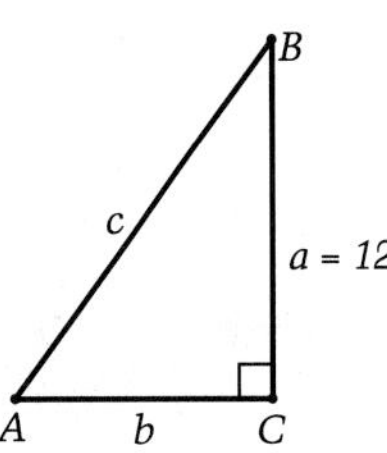

The tangent of angle A is $\frac{4}{3}$. The length of the side opposite A is 12. What are the lengths of the other two sides?

Solution

$\tan A = \frac{a}{b}$	*Definition of tangent of A*
$\frac{4}{3} = \frac{12}{b}$	*Substitute $\frac{4}{3}$ for tan A and 12 for a.*
$4b = 3(12)$	*Cross multiply.*
$b = 9$	*Divide both sides by 4.*

To find c, use the Pythagorean theorem.

$c^2 = a^2 + b^2$	*Pythagorean theorem*
$c^2 = 12^2 + 9^2$	*Substitute 12 for a and 9 for b.*
$c^2 = 225$	*Simplify.*
$c = \sqrt{225}$	*Find the positive square root.*
$c = 15$	*Simplify.*

The lengths of the other two sides are 9 and 15. ■

To find the tangent of an angle whose degree measure is given, you can use the tangent key, [TAN], on a scientific calculator.

Connections Technology

Example 3 Finding the Lengths of the Sides of a Right Triangle

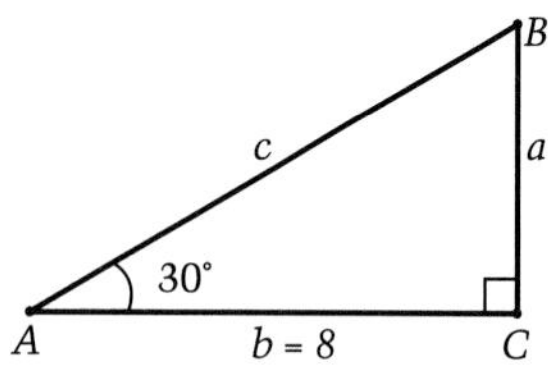

The measure of angle A is 30°, and b is 8. What is a?

Solution Use a scientific calculator to find the tangent of A. (First set the calculator to *degree mode.*)

Keystrokes	Display
30 [TAN]	0.577350

Thus, $\tan 30° \approx 0.57735$. You can solve for a as follows.

$\tan A = \frac{a}{b}$	*Definition of tangent of A*
$0.57735 \approx \frac{a}{8}$	*Substitute 0.57735 for tan A and 8 for b.*
$8(0.57735) \approx a$	*Multiply both sides by 8.*
$4.62 \approx a$	*Simplify.*

The length of a is about 4.62 units. ■

Goal 2 Using the Tangent in Real Life

The Grand Canyon is one of the most famous natural wonders in the United States. Its area is larger than Rhode Island. In some places, it is one mile deep.

Real Life Surveying

Example 4 *Finding the Depth of a Canyon*

Suppose that surveyors measure the angle between the horizontal and a point on the canyon floor to be 40°. Using a laser beam, they find the horizontal distance between their surveying equipment and a point above the canyon floor to be 5740 feet. What is the vertical distance to the canyon floor?

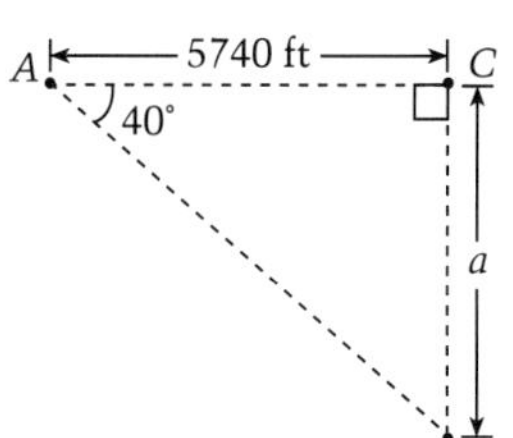

Solution Use a calculator to find $\tan 40° \approx 0.8391$.

$$\tan 40° = \frac{a}{5740}$$

$$0.8391 \approx \frac{a}{5740}$$

$$0.8391(5740) \approx a$$

$$4816.4 \approx a$$

The vertical distance is about 4816 feet. ■

Communicating about ALGEBRA

SHARING IDEAS about the Lesson

Extend the Idea Explain how to use the tangent to find the lengths of the sides. See margin.

A.

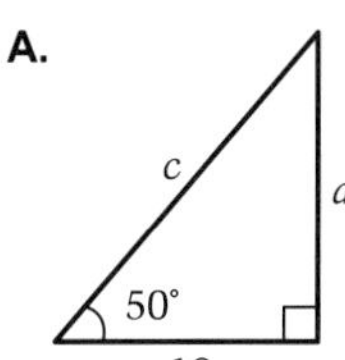

B.

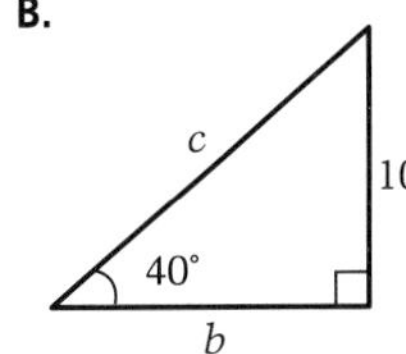

C.

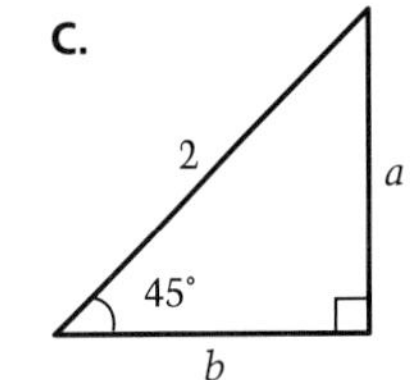

Example 4

Computations that involve intermediate calculations on a scientific or graphing calculator may be performed using the entire calculator display instead of an approximation of 2 to 4 decimal places. For example, when you enter:

[TAN] 40 [×] 5740 [ENTER]

on a graphing calculator, you can use the result, 4816.431883, in further calculations. The result is still an approximation.

Extra Examples

Here are additional examples similar to **Examples 1–4.**

1. Find the tangent of angles A and B.

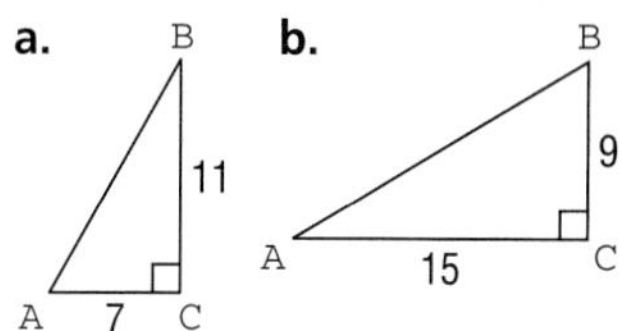

a. $\tan A = \frac{11}{7}$, $\tan B = \frac{7}{11}$

b. $\tan A = \frac{3}{5}$, $\tan B = \frac{5}{3}$

2. Find the lengths of the other two sides of the given right triangles.

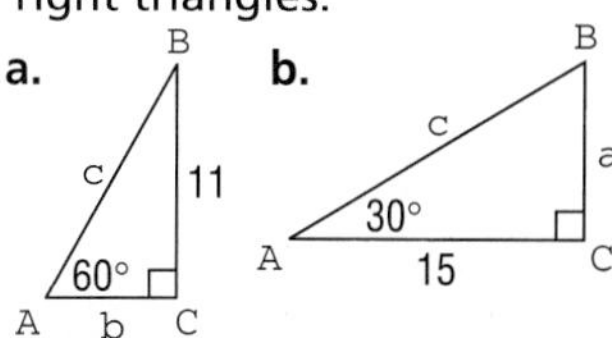

a. $b \approx 6.35$, $c \approx 12.70$
b. $a \approx 8.66$, $c \approx 17.32$

Check Understanding

1. Describe how right-triangle mathematics can be used to find distances that cannot be measured directly. See Examples 2 and 3.
2. Define the tangent of an acute angle. See page 703.
3. How is the Pythagorean Theorem used to find the lengths of sides of a right triangle? See Example 2.

Communicating about ALGEBRA

Answers

A. Solve $\tan 50° = \frac{a}{10}$ for a, substitute for a in $a^2 + 10^2 = c^2$, and solve for c.

B. Solve $\tan 40° = \frac{10}{b}$ for b, substitute for b in $10^2 + b^2 = c^2$, and solve for c.

C. Solve $a^2 + b^2 = 2^2$ for a in terms of b; substitute for a in $\tan 45° = \frac{a}{b}$ and solve for b; substitute for b in $a^2 + b^2 = 2^2$ and solve for a.

EXTEND *Communicating*
RESEARCH Ask students to investigate how right-triangle mathematics, which includes tangents of acute angles, is used in making topographical maps. What is the role of mathematics in building bridges and dams?

EXERCISE Notes

Assignment Guide

Basic/Average: Ex. 1–4, 5–15 odd, 19–27 odd, 33–34, 37–43

Above Average: Ex. 1–4, 5–27 odd, 28, 31–34, 38–45

Advanced: Ex. 1–4, 5–27 odd, 28–35, 41–45

Selected Answers
Exercises 1–4, 5–43 odd

Use **Mixed Review** as needed.

More Difficult Exercises
Exercises 17, 18, 27, 29–34, 44, 45

Guided Practice

Ex. 3 TECHNOLOGY This exercise is similar to Example 3. Students should use a calculator to evaluate tan 28. Students who have the same brand of calculator should help each other find the correct key sequence.

EXERCISES

Guided Practice

CRITICAL THINKING about the Lesson

1. What is an *acute* angle? In a right triangle, what is the sum of A and B? See top of page 703.

2. Find the tangent of the angle A. $\frac{14}{25}$

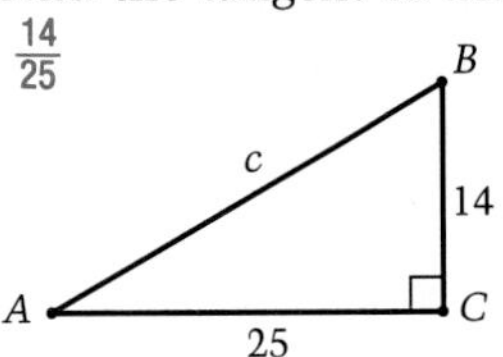

3. Find the lengths b and c. 15, 17

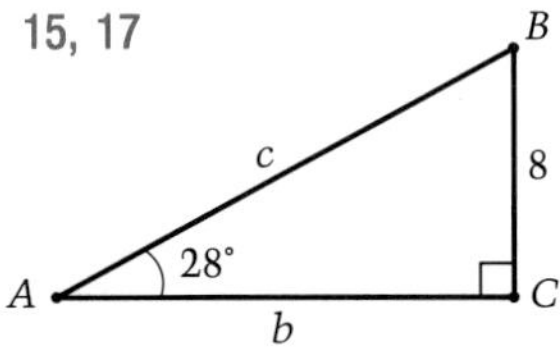

4. Use a calculator to evaluate tan 35°. Round the value to two decimal places. 0.70

Independent Practice

In Exercises 5–10, find tan *A* and tan *B*.

5. $\frac{4}{5}, \frac{5}{4}$

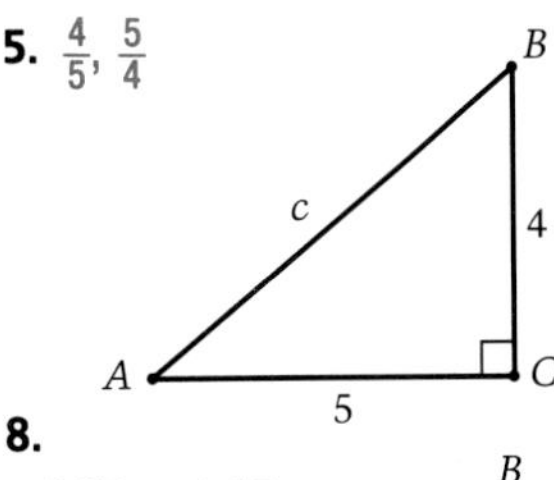

6. $\frac{3}{4}, \frac{4}{3}$

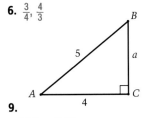

7. $\frac{\sqrt{3}}{3}, \sqrt{3}$

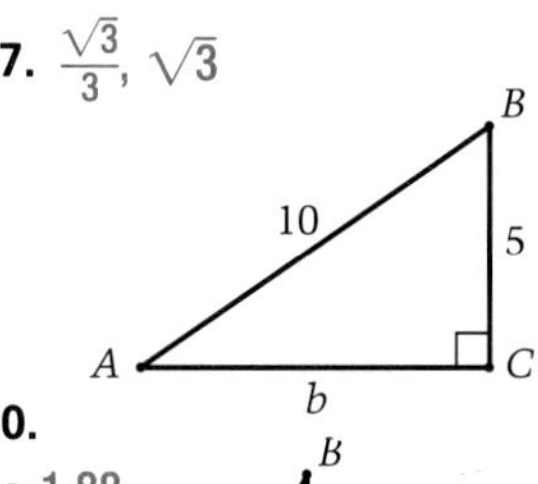

8. $\approx 0.87, \approx 1.15$

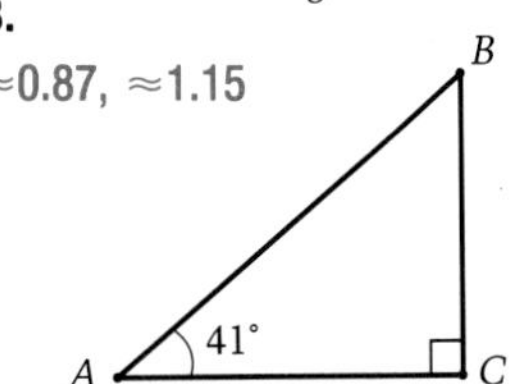

9. $\approx 0.36, \approx 2.75$

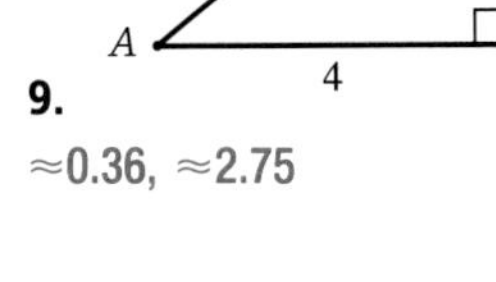
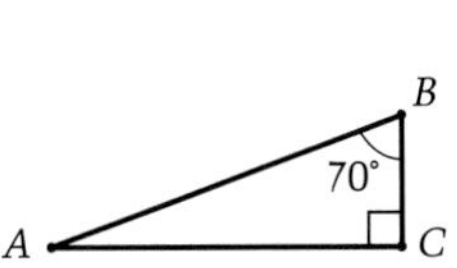

10. $\approx 1.88, \approx 0.53$

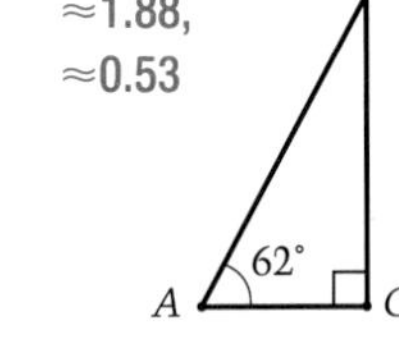

Technology **In Exercises 11–16, use a calculator to evaluate the expression. Round the value to two decimal places.**

11. tan 12° 0.21 **12.** tan 48° 1.11 **13.** tan 84° 9.51

14. tan 24.5° 0.46 **15.** tan 1.8° 0.03 **16.** tan 64.3° 2.08

17. The tangent of angle A in a right triangle is $\frac{2}{3}$. The length of the side opposite A is 8. What are the lengths of the other two sides? Sketch the triangle to scale and label the length of each side. 12, $4\sqrt{13}$ See margin.

18. The tangent of angle A in a right triangle is $\frac{7}{2}$. The length of the side adjacent to A is 3. What are the lengths of the other two sides? Sketch the triangle to scale and label the length of each side. 10.5, ≈ 10.9 See margin.

In Exercises 19–24, find the remaining side and angle measurements.

19.

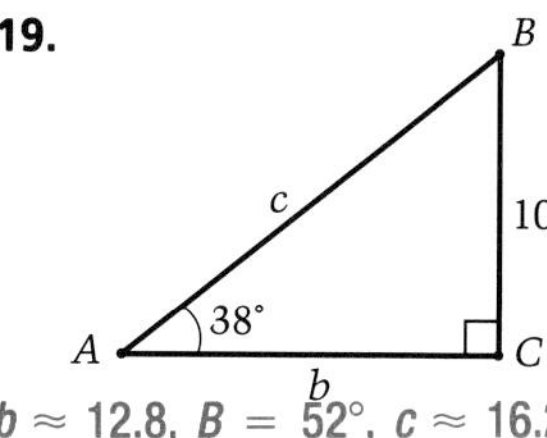

$b \approx 12.8$, $B = 52°$, $c \approx 16.2$

20.

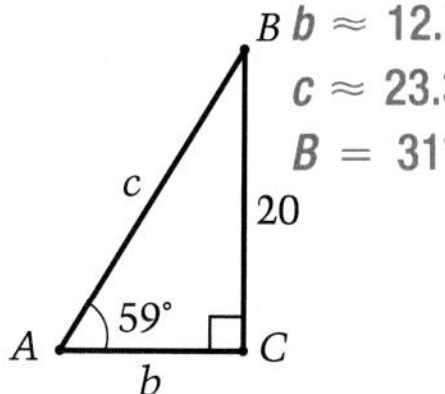

$b \approx 12.0$, $c \approx 23.3$, $B = 31°$

21.

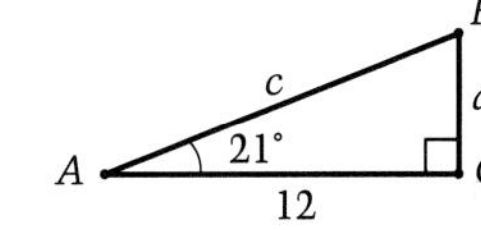

$a \approx 4.6$, $c \approx 12.9$, $B = 69°$

22.

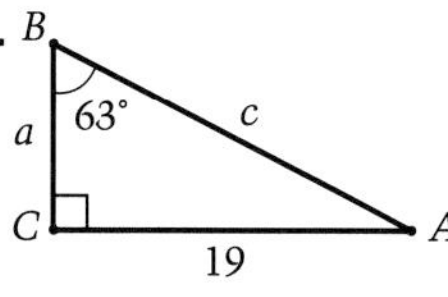

$a \approx 9.7$, $c \approx 21.3$, $A = 27°$

23.

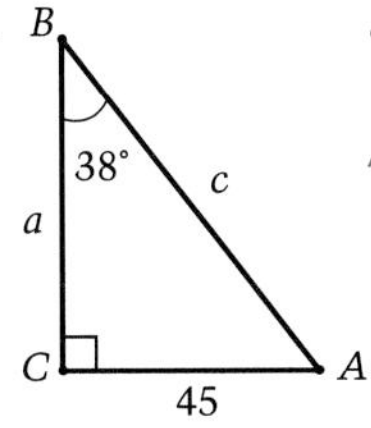

$a \approx 57.6$, $c \approx 73.1$, $A = 52°$

24.

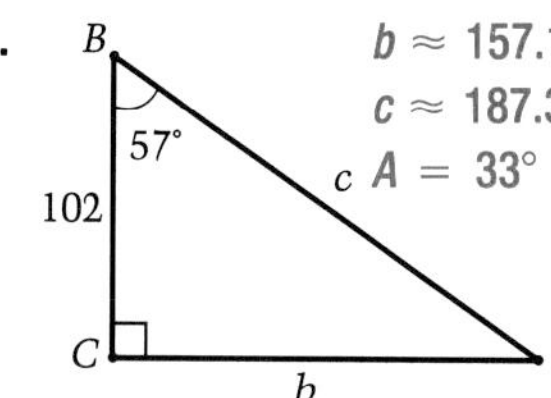

$b \approx 157.1$, $c \approx 187.3$, $A = 33°$

25. *45°-45°-90° Triangle* Use the triangle to find the exact value of the tangent of 45°. 1

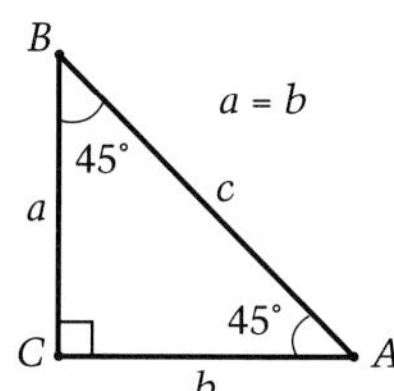

26. *30°-60°-90° Triangle* Use the triangle to find the exact values of tan 30° and tan 60°. $\frac{\sqrt{3}}{3}$, $\sqrt{3}$

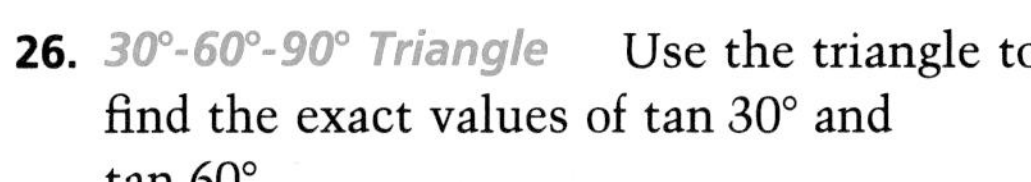

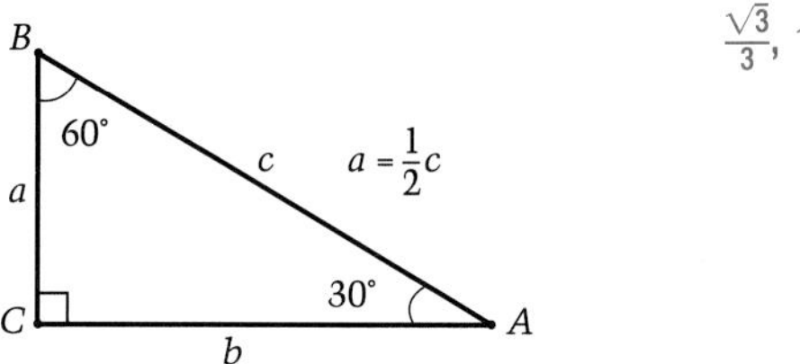

27. *Airplane Engine* The cylinders of the airplane engine form equal angles. The distance between the center of the engine and the tip of a cylinder is 28 inches. Find the distance between the tips of two consecutive cylinders. $\approx$24.3 in.

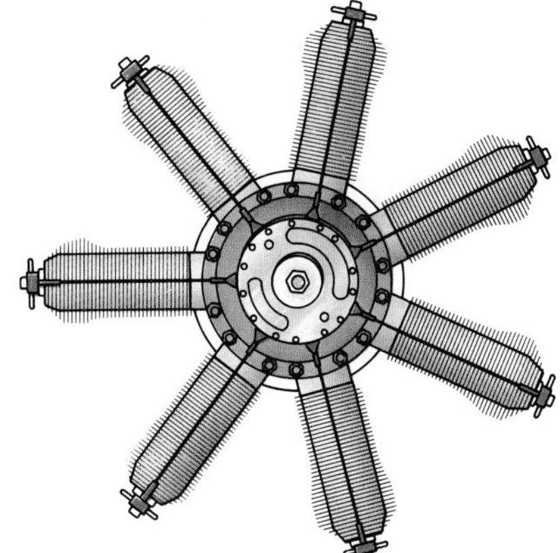

See margin for diagram.

28. *Air Force Rescue Helicopter* The blades of the helicopter meet at right angles and are all the same length. The distance between the tips of two consecutive blades is 32.5 feet. Find the measure of the angles in one of the triangles whose legs are consecutive blades. How long is each blade? 45°, 45°, 90°; 23.0 ft

Independent Practice

▶ **Ex. 5–10** Ask students if they notice a relationship between tan A and tan B. They are reciprocals.

▶ **Ex. 19–28** See Examples 2 and 3 for solution models. In Exercise 25, students should discover that in a 45°-45°-90° right triangle the legs are of equal length. They can then use this relationship to solve Exercise 28.

Answers

17.

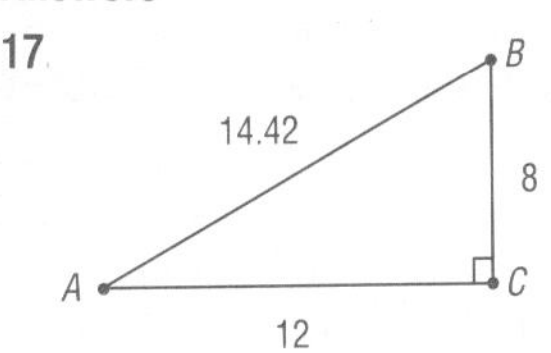

18.

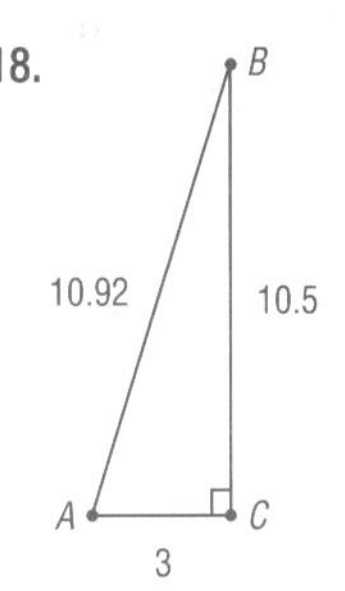

27.

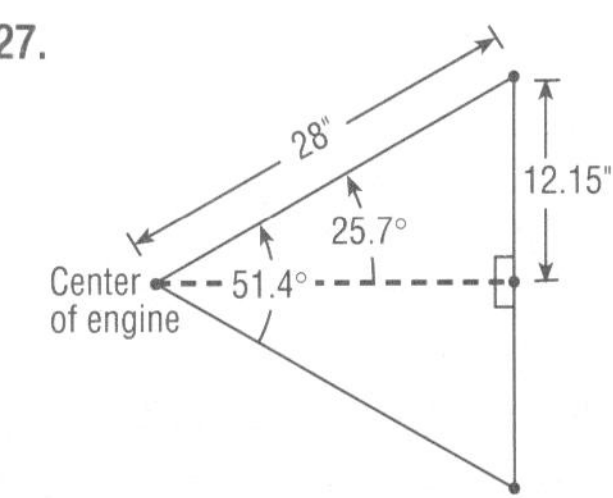

▶ **Ex. 31–34** Encourage students to draw a diagram as part of their solution.

▶ **Ex. 44–45** Note that $\tan A = m$, where m is the slope of the oblique line $y = mx + b$.

Enrichment Activities

These activities require students to go beyond lesson goals.

1. **TECHNOLOGY** Suppose you know the value of the tangent of an angle and want to find the measure of the angle. Scientific calculators have an inverse tangent key, $\tan^{-1}$, for use in finding the value of an angle given the value of its tangent. For example, here is right triangle *KRP*. What is the tangent of angle *P*? $\frac{12}{5}$

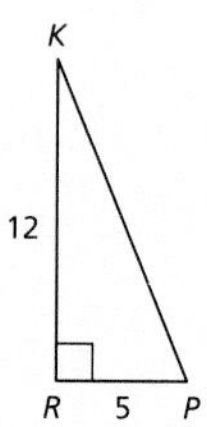

2. Check the manual for your calculator or try one of these calculator sequences to find that the approximate measure of angle *P* is 67.4°.

   12 [÷] 5 [INV] [TAN] [=] or
   12 [÷] 5 [2nd] [TAN] [=] or
   [2nd] [TAN] [(] 12 [÷] 5 [)] [ENTER]

3. Find the measure of angle A to the nearest degree. 37°

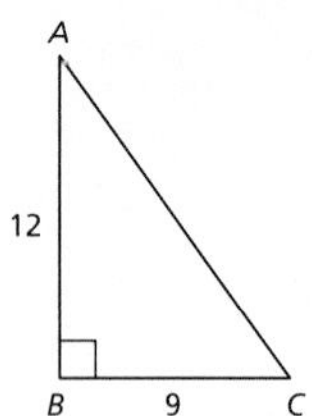

In Exercises 29 and 30, find the lengths *x* and *y*. Round your results to two decimal places.

29. $x \approx 7.28$, $y \approx 6.72$

30. $x \approx 3.53$, $y \approx 8.02$

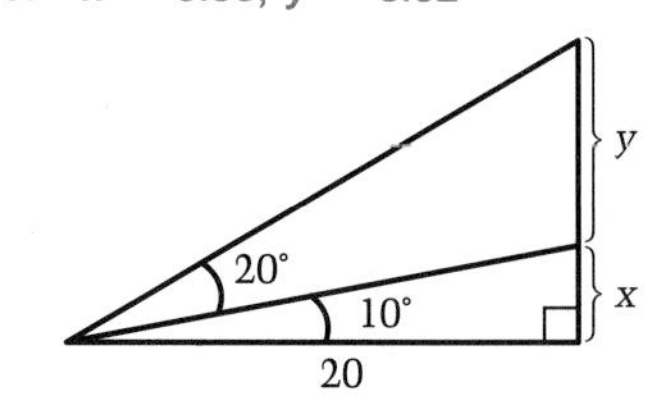

31. *Measuring an Arch* You are standing 75 feet from the base of a stone arch in Arches National Park, Utah. The angle formed by the ground and your line of sight to the top of the arch is 38°. How high is the arch? ≈58.6 ft

32. *Measuring an Arch* Use the result from Exercise 31 and the angle shown in the photo to find the thickness of the top of the stone arch. ≈17.0 ft

33. *Hagia Sophia* Imagine yourself standing 100 feet from the base of one of the four spires of the Hagia Sophia in Istanbul, Turkey—one of the most famous mosques in the world. The angle formed by the ground and your line of sight to the top of the spire is 65°. How tall is the spire? ≈214.5 ft

34. *Hagia Sophia* You are standing 120 feet from the same spire of the mosque. What is the tangent of the angle formed by the ground and your line of sight to the top of the spire? ≈1.79

Integrated Review

In Exercises 35–40, solve the equation.

35. $a^2 + 4^2 = 8^2$ $4\sqrt{3}, -4\sqrt{3}$

36. $6^2 + b^2 = 9^2$ $3\sqrt{5}, -3\sqrt{5}$

37. $4^2 + 5^2 = c^2$ $\sqrt{41}, -\sqrt{41}$

38. $\frac{2}{5} = \frac{a}{3}$ $\frac{6}{5}$

39. $\frac{7}{3} = \frac{4}{b}$ $\frac{12}{7}$

40. $\tan 20° = \frac{3}{b}$ ≈8.2

In Exercises 41 and 42, find the slope of the line that passes through the points. Explain how the slope of the line is related to the indicated angle. Slope is the tangent of the angle.

41. 1

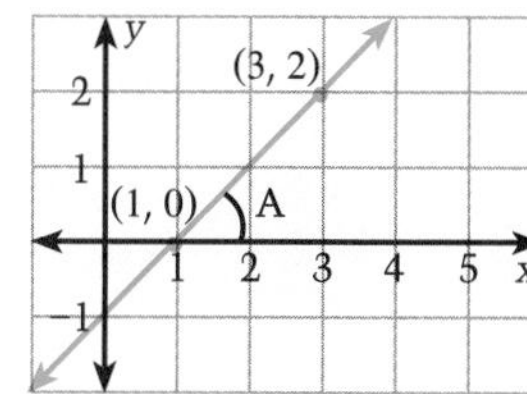

42. $\frac{2}{3}$

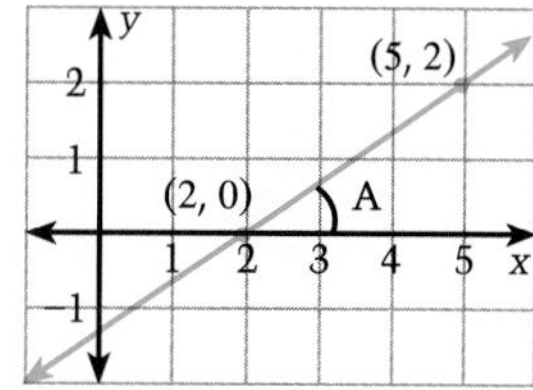

43. *College Entrance Exam Sample* In a right triangle, the ratio of the legs is 1:2. If the area of the triangle is 25 square units, what is the length of the hypotenuse? b

a. $\sqrt{5}$ **b.** $5\sqrt{5}$ **c.** $5\sqrt{3}$ **d.** $10\sqrt{3}$ **e.** $25\sqrt{5}$

Exploration and Extension

44. A line with a positive slope passes through the origin, making an angle of 60° with the positive x-axis. Write an equation of the line. $y = \sqrt{3}x$

45. A line with a negative slope passes through the origin. The angle between the line and the negative x-axis is 60°. Write an equation of the line. $y = -\sqrt{3}x$

Mixed REVIEW

See margin.

13.–18. See Additional Answers.

1. Simplify $2x(x + 3) - x(4 - x^2)$. **(2.6)**

2. Simplify $(3x \div 4x) + \frac{1}{4}$. **(2.7)**

3. Is 3 a solution of $x^2 - 2x + 4 < 5$? **(1.5)**

4. Is -4 a solution of $|3m + 2| + m > 5$? **(4.5)**

5. Solve $3p^2 - 6q = 0$ for p. **(3.6)**

6. Solve $3m + 3mn - 6x = 0$ for m. **(3.6)**

7. Evaluate $36x^2 + 2x - 6$ when $x = \frac{1}{4}$. **(1.4)**

8. Evaluate $3x + 9$ when $x = \frac{2}{9}$. **(1.4)**

9. Find the mean of 2, 3, 5, 7, and 8.

10. Find the median of 26, 42, 35, 16, and 81.

11. 216 is what percent of 360?

12. 212.5 is 25% of what number?

13. Sketch the graph of $-4x + 5y = -5$. **(4.3)**

14. Sketch the graph of $y = |3x + 2| - 4$. **(4.7)**

15. Sketch the graph of $f(x) = \left(\frac{1}{4}\right)^x + 2$. **(12.3)**

16. Sketch the graph of $f(x) = \frac{x + 2}{x + 1}$. **(12.5)**

17. Sketch the graph of $y < 3x + 2$. **(6.5)**

18. Sketch the graph of $y \geq x^2 - 4$. **(9.6)**

19. Solve the system $\begin{cases} x + 2y = -15 \\ 5x - 4y = 9 \end{cases}$ **(7.3)**

20. Solve the system $\begin{cases} 4x + 2y = 8 \\ 6x - 5y = 28 \end{cases}$ **(7.3)**

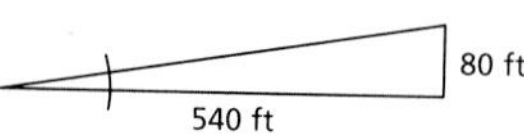

4. In Colorado, a portion of a mountain highway rises vertically 80 feet over a horizontal distance of 540 feet. Draw a diagram and determine the angle of elevation of the road to the nearest degree. 8°

Mixed Review Answers

1. $x^3 + 2x^2 + 2x$
2. 1
3. No
4. Yes
5. $\sqrt{2q}, -\sqrt{2q}$
6. $\frac{2x}{1 + n}$
7. $-\frac{13}{4}$
8. $\frac{29}{3}$
9. 5
10. 35
11. 60%
12. 850
13.–18. See Additional Answers.
19. $(-3, -6)$
20. $(3, -2)$

Problem of the Day

In numbering the pages of a book, a printer uses 522 digits. How many numbered pages does the book have? 210

ORGANIZER

Warm-Up Exercises

1. List the elements of A = the set of whole numbers and B = the set of integers.
   $A = \{0, 1, 2, 3, 4, 5, \ldots\}$
   $B = \{\ldots, -4, -3, -2, -1, 0, 1, 2, 3, 4, \ldots\}$
2. For sets A and B in Warm-Up Exercise 1, determine whether the sum of any two numbers in set A (or in set B) is also a number in the set.
   Yes, for sets A and B
3. For each set A and B in Warm-Up Exercise 1, determine whether the difference of any two numbers in a given set is also a number in the set.
   No for set A, yes for set B
4. Your friend said, "If an integer, a, is less than another integer, b, then ac is less than bc, if c is a third integer." What is your reply?
   The statement is not true. For example, consider the numbers $a = 5$, $b = 8$, and $c = -2$.

Lesson Resources

Teaching Tools
  Problem of the Day: 13.6
  Warm-up Exercises: 13.6
  Answer Masters: 13.6
Extra Practice: 13.6
Color Transparency: 78

13.6 Formal Use of Properties of Algebra

What you should learn:

Goal How to use properties of algebra to prove a statement true and to use counter-examples to prove a statement false

Why you should learn it:

You will use the properties of algebra more formally in higher-level mathematics courses.

GOAL Using Properties of Algebra

Mathematics evolved over the past few thousand years in many stages. *All* of the early stages were centered around the use of mathematics to answer questions about real life. This is much like the way we wrote *Algebra 1.* We centered the concepts around the real-life use of mathematics.

As mathematical development matured, people began to collect and categorize the different rules, formulas, and properties that had been discovered. This took place independently in many different parts of the world: Africa, Asia, Europe, North America, and South America. The mathematics that we use today is a combination of the work of literally thousands of people.

As the collections and catalogs of rules, formulas, and properties grew, some discrepancies were noticed—some of the rules didn't seem to fit with some of the other rules. This made people wonder which rules were "true," and they tried to see which could be **proved.** After some effort, however, they realized that you can't prove *every* rule—some have to be accepted "on faith." In mathematics, the rules that we accept to be true *without proof* are called **postulates** or **axioms.** Many of the rules in Chapter 2 fall into this category.

The Basic Axioms of Algebra

Let a, b, and c be any real numbers.

Name of Axiom	Addition Axioms	Multiplication Axioms
Closure	$a + b$ is a real number.	ab is a real number.
Commutative	$a + b = b + a$	$ab = ba$
Associative	$(a + b) + c = a + (b + c)$	$(ab)c = a(bc)$
Identity	$a + 0 = a$, $0 + a = a$	$a(1) = a$, $1(a) = a$
Inverse	$a + (-a) = 0$	$a\left(\frac{1}{a}\right) = 1$, $a \neq 0$
Axiom Relating Addition and Multiplication		
Distributive	$a(b + c) = ab + ac$	$(a + b)c = ac + bc$

Once a list of axioms has been accepted, you can add more rules, formulas, and properties to the collection. Some of the new concepts are **definitions.** You don't have to prove a definition, but you do need to check that it is consistent with previous definitions and axioms.

Other new concepts, called **theorems,** do have to be proved. For instance, it is possible (although it is not done here) to use the basic axioms to prove that $c(-b) = -cb$ for all real numbers b and c.

Example 1 *Proving a Theorem*

Use the definition of subtraction, $a - b = a + (-b)$ and prove that the following theorem is true for all real numbers a, b, and c.

$$c(a - b) = ca - cb$$

Solution

$$\begin{aligned} c(a - b) &= c(a + (-b)) && \textit{Definition of subtraction} \\ &= ca + c(-b) && \textit{Distributive Property} \\ &= ca + (-cb) && \textit{Theorem listed above} \\ &= ca - cb && \textit{Definition of subtraction} \end{aligned}$$

Proofs are not easy to write *or* to read. They take a lot of practice. Our reason for listing this proof is simply to point out that when you are proving a new formula, *every* step must be justified by an axiom, a definition, or a previously proved theorem. ■

Sometimes people propose new rules that *are not* consistent with the axioms, definitions, and theorems in the accepted list. To show that a proposed rule is *inconsistent* or *false,* you need to find only one counterexample.

Example 2 *Finding a Counterexample*

Assign values to a and b to show that the following "rule" is inconsistent: $a + (-b) = (-a) + b$ for all real numbers a and b.

Solution Almost any choices of a and b will show the inconsistency. For instance, let $a = 1$ and let $b = 2$. Then

$$a + (-b) = 1 + (-2) = -1 \quad \text{and} \quad (-a) + b = -1 + 2 = 1.$$

Because -1 and 1 are not the same number, you have shown one case in which $a + (-b)$ is not equal to $(-a) + b$. Therefore, the proposed rule must be false. (If it were true, it would have to work for any choices of a and b.) ■

LESSON Notes

GOAL 1 MATH JOURNAL Ask students to enter the formal definitions of postulate, axiom, definition, and theorem in their mathematics journals. They may wish to consult an unabridged dictionary or a mathematics dictionary if one is available.

Example 1

The theorem you are proving is the Distributive Property of Multiplication over Subtraction.

Note for students how the statement of the theorem involves subtraction. Furthermore, emphasize that the definition of subtraction can be read in both directions. That is, not only does "*a* minus *b* equal the sum of *a* and $-b$," but "the sum of *a* and $-b$ equals *a* minus *b*." Consequently, the definition of subtraction is used in two slightly different ways in the proof.

Finally, point out that you were told what definition to use to prove the theorem. Original proofs by mathematicians usually have no such hints!

Example 2

Warm-Up Exercise 4 can be solved again by finding a counterexample to the rule given by "your friend."

Example 3

Christian Goldbach was a Russian mathematician. In 1937, another Russian mathematician, I. M. Vinogradov, showed that every "sufficiently large" even number is the sum of at most four primes!

Extra Examples

Here are additional examples similar to **Examples 1–3.**

1. Prove that $(a + b) - b = a$ is true for all real numbers a and b.
   Solution:
   $(a + b) - b$
   $= (a + b) + (-b)$
   Definition of subtraction
   $= a + [b + (-b)]$
   Associative property
   $= a + [0]$
   Inverse property
   $= a$
   Identity property
2. Prove that the statement $-(a + b) = -a + b$ is true for all real numbers a and b.
   Solution: Close inspection will reveal that this is not a true statement. Assign the values 2 and 3 to a and b to verify the falseness of the statement. That is, $-(2 + 3) = -5$ is not equal to $-2 + 3 = 1$.
3. If possible, write the following even numbers as the sum of two primes.
   a. 204 **b.** 182
   a. $204 = 5 + 199$ is one possibility
   b. $182 = 3 + 179$ is one possibility

Check Understanding

1. What is the role of counterexamples in mathematical proof? See page 711.
2. What is the difference between an axiom and a theorem? Axioms are accepted without proof; theorems have to be proved.
3. If a "rule" has not been proven to be true, but no counterexamples have been found, should the rule be considered true?
   Not necessarily

Communicating about ALGEBRA

EXTEND *Communicating*
RESEARCH The Pythagorean Theorem is a famous result in algebra and geometry. Locate

Sometimes people have proposed rules and properties that they believe to be true but are unable to prove. For instance, Christian Goldbach (1690–1764) proposed the following rule, which is called *Goldbach's Conjecture.*

> *Every even number except 2 is equal to the sum of two prime numbers.*

No one has ever been able to prove this statement.

Example 3 *Can a Proof Consist of Examples Only?*

The following list shows that every even number between 4 and 26 is equal to the sum of two prime numbers. Does this prove that *every* even number greater than 2 is equal to the sum of two prime numbers? (Remember that a prime number is a number whose only factors are itself and 1. For instance, the numbers 2, 3, 5, 7, 11, 13, 17, 19, and 23 are prime.)

$4 = 2 + 2$	$6 = 3 + 3$	$8 = 3 + 5$	$10 = 3 + 7$
$12 = 5 + 7$	$14 = 3 + 11$	$16 = 3 + 13$	$18 = 5 + 13$
$20 = 3 + 17$	$22 = 3 + 19$	$24 = 5 + 19$	$26 = 3 + 23$

Solution This list of examples *does not* prove the conjecture. One counterexample is enough to show that the rule is false, but no amount of examples can prove that the rule is true for *every* even integer greater than 2. ■

Communicating about ALGEBRA

SHARING IDEAS about the Lesson

Explain the Consequences The famous French mathematician Pierre de Fermat (1601–1655) conjectured that the formula $2^{(2^n)} + 1$ produces a prime number for every nonnegative integer n. For instance,

$$2^{(2^0)} + 1 = 2^1 + 1 = 3 \qquad 2^{(2^1)} + 1 = 2^2 + 1 = 5$$
$$2^{(2^2)} + 1 = 2^4 + 1 = 17 \qquad 2^{(2^3)} + 1 = 2^8 + 1 = 257$$

are each prime. Later another famous French mathematician, Leonhard Euler (1707–1783), found that

$$2^{(2^5)} + 1 = 2^{32} + 1 = 4{,}294{,}967{,}297 = 641(6{,}700{,}417).$$

Explain the consequences to Fermat's conjecture.

Euler found a counterexample to Fermat's conjecture; so Fermat's conjecture was false.

EXERCISES

Guided Practice

CRITICAL THINKING about the Lesson See margin.

1. The **difference** of a and b is defined as the sum of a and the opposite of b. Write this definition symbolically. Does this definition have to be proved? Explain.

2. Prove the property $(a - b)c = ac - bc$. (Use only the basic axioms of algebra, the definition of subtraction, and the fact that $(-b)c = -(bc)$.)

3. A member of your algebra class proposes the following theorem. *The sum of the first n odd integers* is n^2. She gives four examples.

   $1 = 1^2$, $1 + 3 = 2^2$, $1 + 3 + 5 = 3^2$, and $1 + 3 + 5 + 7 = 4^2$

   Do the examples prove her conjecture? Explain. Do you think the conjecture is true?

4. *The Four-Color Problem* A famous theorem states that any map can be colored with four different colors so that no two countries that share a border have the same color. No matter how the map at the right is colored with *three* colors, at least two countries having a common border must have the same color. Does this map serve as a counterexample to the following proposal? Explain.

   Any map can be colored with three different colors so that no two countries that share a border have the same color.

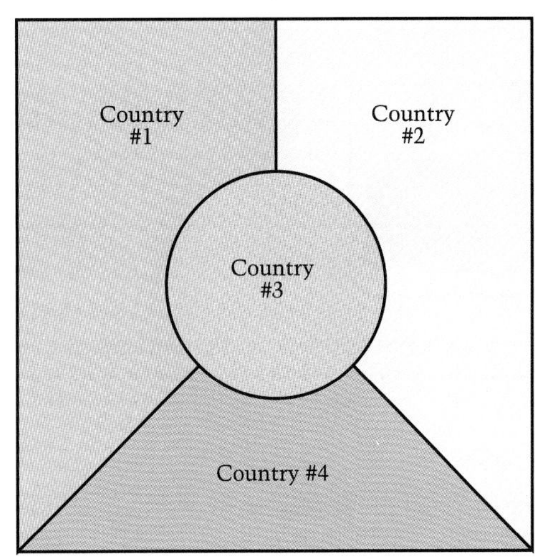

Independent Practice

Proving a Theorem **In Exercises 5 and 6, prove the property. (Use the basic axioms of algebra and the definition of subtraction given in Exercise 1.)** See Additional Answers.

5. If a and b are real numbers, then $a - b = -b + a$.
6. If a, b, and c are real numbers, then $(a - b) - c = a - (b + c)$.

Finding a Counterexample **In Exercises 7–10, find a counterexample to show that the statement is *not* true.** See margin of page 714.

7. If a and b are real numbers, then $(a + b)^2 = a^2 + b^2$.
8. If a and b are real numbers, then $(a - b)^2 = a^2 - b^2$.
9. If a, b, and c are real numbers such that $ab = ac$, then $b = c$.
10. If a, b, and c are nonzero real numbers, then $(a \div b) \div c = a \div (b \div c)$.

one or more of its many proofs and explain the steps of the proof to a classmate. You might also research history of mathematics books in the library to find some examples of proofs from ancient times and from other cultures.

EXERCISE Notes

Assignment Guide

Basic/Average: Ex. 1–4, 5–9 odd, 11–12, 15–21 odd

Above Average: Ex. 1–4, 5–11 odd, 12–14, 15–23 odd

Advanced: Ex. 1–4, 5–11 odd, 12–14, 15–23 odd

Selected Answers
Exercises 1–4, 5–21 odd

More Difficult Exercises
Exercises 13–14, 23

Guided Practice

Ex. 1–4 These exercises are concerned with the meaning of *proof.* A geometric argument for the proposed theorem in Exercise 3 will be considered in Exercise 13.

Answers

1. $a - b = a + (-b)$, No; a definition needs to be consistent with other definitions and with axioms, but is accepted as true rather than having to be proved true.
2. $(a - b)c$
   $= (a + (-b))c$
   Definition of Subtraction
   $= ac + (-b)c$
   Distributive Property
   $= ac + (-bc)$
   $= ac - bc$
   Definition of Subtraction
3. No; in order to prove her conjecture she would need to give every example (to avoid a counterexample), which would be impossible. Yes.
4. Yes; the map cannot be colored as proposed.

Independent Practice

▶ **Ex. 5–6** Help students to use an appropriate format as modeled in Example 1.

▶ **Ex. 7–10** Only one counterexample is necessary to disprove an assertion.

EXTEND Ex. 7–10 Are these statements true or false? Find a counterexample to show that the statement is false.

a. If a and b are real numbers such that $0 < a < b$ and n is an integer, then $a^n < b^n$. False, let $a = 3$, $b = 4$, and $n = -1$.

b. If a, b, and c are real numbers such that $ac < bc$, then $a < b$. False; let $a = 4$, $b = 2$, and $c = -1$.

▶ **Ex. 13** This exercise provides a visual model for the conjecture that the sum of the first n odd numbers is n^2.

Answers

7.–10. Counterexamples vary.

7. $(3 + 4)^2 \stackrel{?}{=} 3^3 + 4^2$
   $49 \neq 25$
8. $(3 - 4)^2 \stackrel{?}{=} 3^2 - 4^2$
   $1 \neq -7$
9. $0 \cdot 1 \stackrel{?}{=} 0 \cdot 2$
   $1 \neq 2$
10. $(12 \div 6) \div 2 \stackrel{?}{=} 12 \div (6 \div 2)$
    $1 \neq 4$
11. She drew a plan of the patio and colored it like a checkerboard.
12. Each rectangular tile must cover a red and a white square; the two that could not be covered were both red.

The Troublesome Tiles **In Exercises 11 and 12, use the following puzzle.**

Mr. Brown's patio is made from 40 square tiles. The tiles have deteriorated and he wants to cover them with a new set.

Betsy: What's the trouble Dad?
Mr.Brown: These blasted tiles won't fit. It's driving me nuts. I always end up with two squarcs I can't cover.

He chooses new tiles to match his lawn furniture. Unfortunately these tiles only come in rectangles, each of which covers two of his old tiles.
Storekeeper: How many of these do you want?
Mr.Brown: Well, I have to cover 40 squares, so I'll need 20, I guess.

Mr. Brown's daughter drew a plan of the patio and colored it like a checker board. Then she studied it for several minutes.

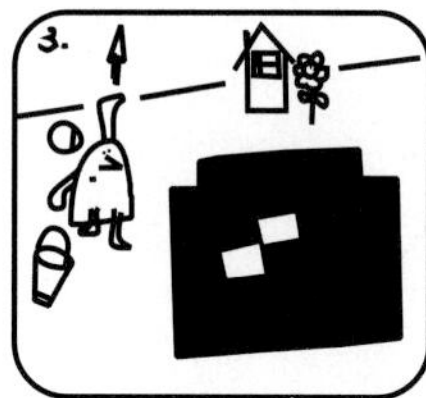

When Mr. Brown tried to cover his patio with the new tiles he became very frustrated. No matter how hard he tried, he couldn't make them fit.

Betsy: Aha! I see what the trouble is. It's obvious once you realize that each rectangular tile must cover a red square and a white square.
How does this help? Do you know what Betsy means?

(Source: From AHA! INSIGHT by Martin Gardner. Copyright © 1978 by Scientific American, Inc. Reprinted with permission of W.H. Freeman and Company.)

11. Explain how Betsy solved the problem. 11., 12., 13., 14. See margin.

12. How could Betsy prove to her father that the patio could not be covered with the 20 rectangular tiles *without* cutting one of the tiles in half?

✪ **13.** *Geometry* Explain how the following diagrams could be used to give a geometrical argument to support the theorem proposed in Exercise 3.

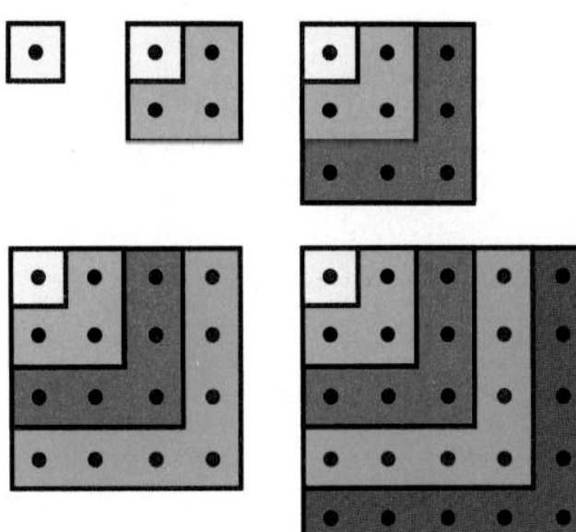

✪ **14.** Explain how the following diagram could be used to give a *geometrical* proof of the Pythagorean theorem.

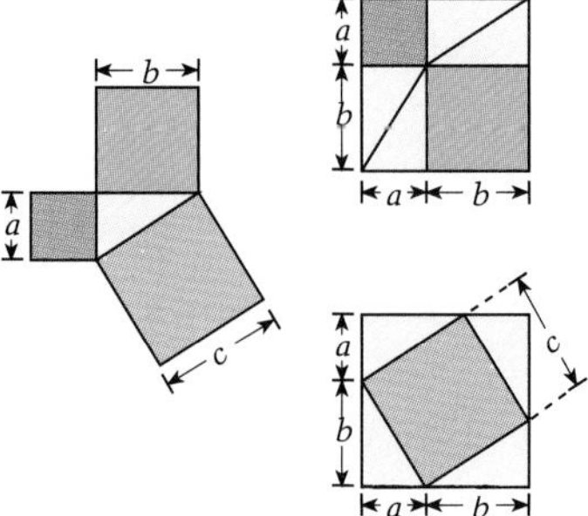

Integrated Review

In Exercises 15–18, solve the equation for a.

15. $ab - ac = 12$ $\frac{12}{b-c}$

16. $a^2 - 2a = 12$ $1 + \sqrt{13}, 1 - \sqrt{13}$

17. $ab - b^2 = 12$ $b + \frac{12}{b}$

18. $\frac{1}{a} + a = 12$ $6 + \sqrt{35}, 6 - \sqrt{35}$

19. *Diagonal of a Rectangle* A rectangular room is 12 feet wide and 14 feet long. How long is the diagonal line segment that connects two opposite ceiling corners of the room? $\approx$ 18.4 ft

20. *Height of a Tree* A rock is 20 feet from the base of a tree. The angle formed by the ground and the line segment connecting the rock to the top of the tree is 50°. How tall is the tree? $\approx$23.8 ft

21. *Estimation—Powers of Ten* Which is the best estimate of the distance between New Orleans, Louisiana, and Chicago, Illinois? b

a. 80 miles **b.** 800 miles **c.** 8000 miles **d.** 80,000 miles

22. *Estimation—Powers of Ten* Which is the best estimate of the weight of this book? b

a. 4 ounces **b.** 40 ounces **c.** 400 ounces **d.** 4000 ounces

Exploration and Extension

23. *Jumping Jim** Explain how you can "work backward" to solve the following maze. See margin.

Jumping Jim is about to begin his grand performance at the circus, but his jealous enemy, Dastardly Dan, has restrung all the trampolines. The number on each trampoline indicates how far Jim will move (horizontally or vertically, but *not* diagonally) when he bounces on the trampoline. Jim begins his routine by leaping onto the trampoline at the upper left. He must get to the Goal at the lower right, where he will take his bow. How can he get there?

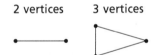

3	6	4	3	2	4	3
2	1	2	3	2	5	2
2	3	4	3	4	2	3
2	4	4	3	4	2	2
4	5	1	3	2	5	4
4	3	2	2	4	5	6
2	5	2	5	6	1	GOAL

* *Written by Robert Abbott,* Mad Mazes, *Copyright © 1990, Bob Adams, Inc., Publishers.*

13. Each additional *L*-shape contains the next odd number of dots; the dots can be lined up in a square form.

14. Add four triangles of height a and base b to $a^2 + b^2$ and you have an area equal to $(a + b)^2$. Add four triangles of height a and base b to c^2 and you have an area equal to $(a + b)^2$. Therefore, $a^2 + b^2 + 4(\frac{1}{2}ab) = c^2 + 4(\frac{1}{2}ab)$ and $a^2 + b^2 = c^2$.

23. Start at the goal and guess which squares to count horizontally and vertically until the number on a square matches the count; that square is the previous one. Continue until you reach the number 3 in the upper left-hand corner.

Enrichment Activity

This activity requires students to go beyond lesson goals.

In graph theory, a complete graph is one in which every pair of vertices is connected by part of the graph called an *edge*. The following graphs are complete.

2 vertices 3 vertices

4 vertices 5 vertices

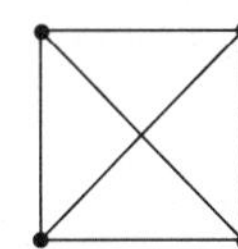

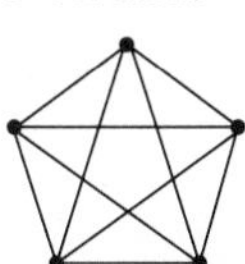

Count the number of edges. Look for a pattern and make a conjecture to complete the table.

Vertices	2	3	4	5	6	7	8	...	n
Edges	1	3	6	10	15	21	28	...	$\frac{n(n-1)}{2}$

ORGANIZER

TIME SCHEDULE: All levels, two days

The Chapter Summary helps students organize the main ideas of the chapter. In this chapter, concepts and skills involving radicals were presented, connections to geometry were identified, and a brief introduction to mathematical proof was made.

Work with students to review the skills, strategies, and concepts of the chapter. It is suggested that the first day of review be a combination of you and your students working problems from the Chapter Review Exercises together. The first day's homework assignment can be used as the basis for the second day of review.

COOPERATIVE LEARNING
Encourage students to study together. Emphasize the importance of "teaching" a classmate how to perform a skill or how to recall a strategy. When students work together, everyone wins. The students who receive help get additional instruction, and the students who give help gain a deeper understanding of the skills and concepts involved.

Chapter SUMMARY

Review the chapter by asking students, "What did you learn? Why did you learn it? How does it fit with the other algebra skills and concepts you have learned?"

13 Chapter Summary

What did you learn?

Skills

1. Find the distance between two points. **(13.1)**
2. Find the midpoint between two points. **(13.1)**
3. Simplify a radical expression. **(13.2)**
4. Multiply and divide radical expressions. **(13.2, 13.3)**
5. Add and subtract radical expressions. **(13.3)**
6. Solve a radical equation. **(13.4)**
7. Find and use the tangent of an angle. **(13.5)**

Strategies

8. Create models of real-life problems. **(13.1–13.5)**
9. Use properties of algebra formally. **(13.6)**

Why did you learn it?

In this chapter, you learned that the distance between two points in a coordinate plane can involve a radical expression. You also learned that many other real-life models involve radical expressions. Many of you will be going on to other courses in mathematics. Because of this, we included several topics in the chapter that are representative of other courses: geometry, trigonometry, and Algebra 2.

How does it fit into the bigger picture of algebra?

Radicals have many applications in geometry and in other fields. In this chapter, you learned to simplify radicals, multiply, divide, add, and subtract radicals, and solve radical equations.

The chapter also contains material that helps prepare you for other courses in mathematics. There are many applications involving geometry, an introduction to the tangent of an angle (which is part of a branch of mathematics called trigonometry), and an introduction to some of the formal properties of algebra (which is part of Algebra 2).

Chapter REVIEW

1. $\sqrt{10}$; $(\frac{-15}{2}, -\frac{1}{2})$

In Exercises 1–6, find the distance between the points. Then find the midpoint. (13.1)

1. $(-6, 0), (-9, -1)$

2. $(-3, 6), (1, 7)$ $\sqrt{17}$; $(-1, \frac{13}{2})$

3. $(0, 0), (-8, 9)$ $\sqrt{145}$; $(-4, \frac{9}{2})$

4. $(-3, -2), (1, 7)$ $\sqrt{97}$; $(-1, \frac{5}{2})$

5. $(-3, -3), (6, 7)$ $\sqrt{181}$; $(\frac{3}{2}, 2)$

6. $(-9, 17), (5, -7)$ $2\sqrt{193}$; $(-2, 5)$

In Exercises 7 and 8, show that the points are vertices of a right triangle. (13.1) See margin.

7. $(0, 0), (1, 2), (2, -1)$

8. $(-4, 2), (8, 5), (-2, -6)$

In Exercises 9–14, simplify the radical expression. (13.2)

9. $\sqrt{216}$ $6\sqrt{6}$

10. $\sqrt{175}$ $5\sqrt{7}$

11. $\sqrt{18}$ $3\sqrt{2}$

12. $\sqrt{\frac{4}{17}}$ $\frac{2\sqrt{17}}{17}$

13. $\sqrt{\frac{8}{9}}$ $\frac{2\sqrt{2}}{3}$

14. $\sqrt{\frac{25}{125}}$ $\frac{\sqrt{5}}{5}$

In Exercises 15–20, perform the indicated operations. Write the result in simplified form. (13.3)

15. $\sqrt{5} + 2\sqrt{5} - \sqrt{3}$ $3\sqrt{5} - \sqrt{3}$

16. $3\sqrt{3} - 6\sqrt{3} + \sqrt{2}$ $-3\sqrt{3} + \sqrt{2}$

17. $\sqrt{6}(2\sqrt{3} - 4\sqrt{2})$ $6\sqrt{2} - 8\sqrt{3}$

18. $\sqrt{2}(2\sqrt{2} - 6)$ $4 - 6\sqrt{2}$

19. $(3\sqrt{5} - \sqrt{10})^2$ $55 - 30\sqrt{2}$

20. $(2\sqrt{3} - 2)(2\sqrt{3} + 2)$ 8

In Exercises 21–26, solve the equation. (13.4)

21. $2\sqrt{x} - 4 = 0$ 4

22. $\frac{1}{3}\sqrt{x} + 2 = 8$ 324

23. $x = \sqrt{-4x - 4}$ No solution

24. $\sqrt{3x - 2} = x$ $2, 1$

25. $\sqrt{11x + 12} = x$ 12

26. $x = \sqrt{3x}$ $0, 3$

27. Find the geometric mean of 6 and 30. $6\sqrt{5}$

28. Find the geometric mean of 7 and 35. $7\sqrt{5}$

Geometry **In Exercises 29 and 30, find the perimeter of the polygon. (13.3)**

29.

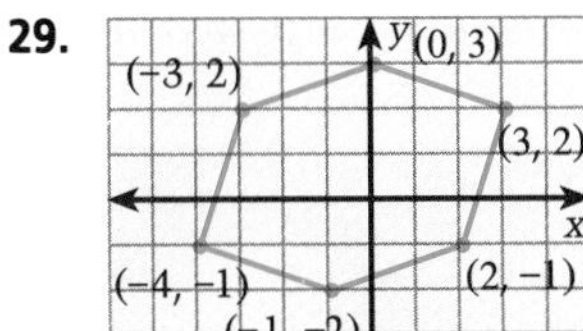

30.

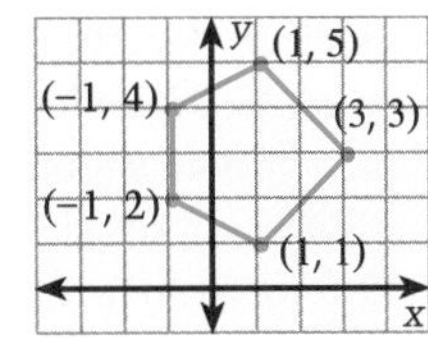

$2 + 4\sqrt{2} + 2\sqrt{5}$

Geometry **In Exercises 31 and 32, find the perimeter and the area of the rectangle. (13.3)**

31.

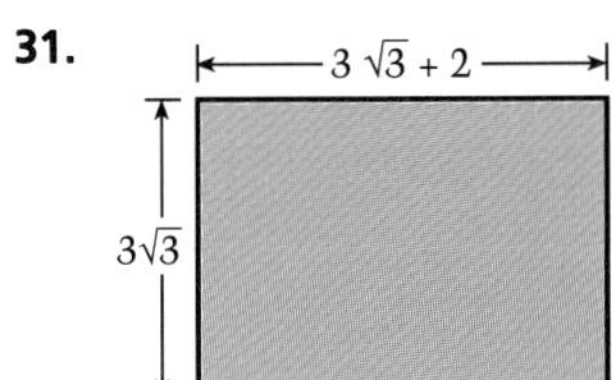

$12\sqrt{3} + 4$, $27 + 6\sqrt{3}$

32.

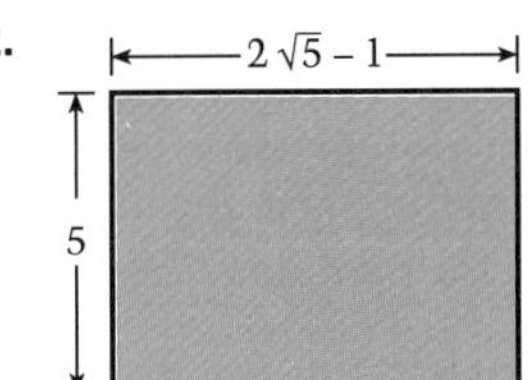

$4\sqrt{5} + 8$, $10\sqrt{5} - 5$

Chapter REVIEW

Assignment Guide

Basic/Average: Ex. 1–37 odd, 41–46

Above Average: Ex. 1–37 odd, 39, 41–46

Advanced: Ex. 1–37 odd, 39, 40, 41–48

More Difficult Exercises

Exercises 39–40, 45–48

Ex. 1–48 Since the Chapter Review Exercises parallel the list of Skills, Strategies, and Exploring Data statements, tell students to use the Chapter Summary on page 716 to identify the lesson in which a skill, strategy, or concept was presented if they are uncertain as to how to do a given exercise. This makes the students categorize each task in the exercise set and locate relevant information in the chapter about the task, both beneficial activities.

Answers

7. $(\sqrt{5})^2 + (\sqrt{5})^2 = (\sqrt{10})^2$

8. $(2\sqrt{17})^2 + (3\sqrt{17})^2 = (\sqrt{221})^2$

Answers

37. $1 + 3 = 4 = 2^2$
$3 + 6 = 9 = 3^2$
$6 + 10 = 16 = 4^2$
$10 + 15 = 25 = 5^2$

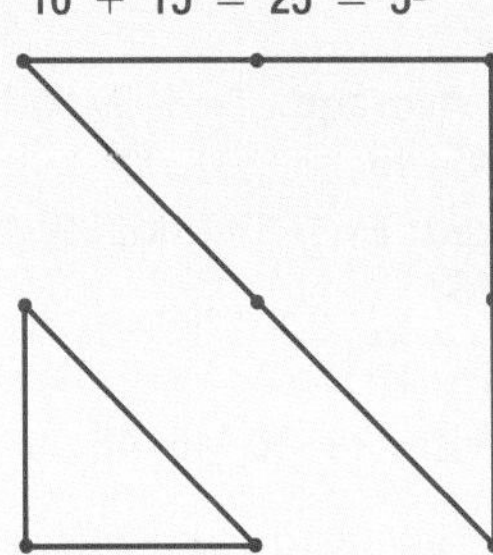

Geometry **In Exercises 33–36, find the remaining side and angle measurements.**

33. $a = \sqrt{3}$, $c = 2$, $B = 30°$

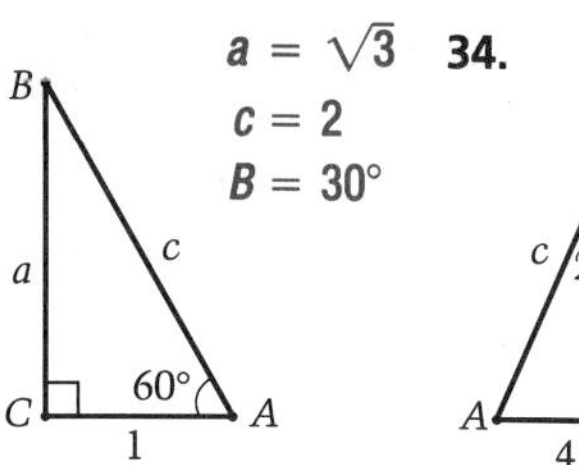

34. $a \approx 8.98$, $c \approx 9.83$, $A = 66°$

35. $b \approx 2.97$, $c \approx 7.60$, $B = 23°$

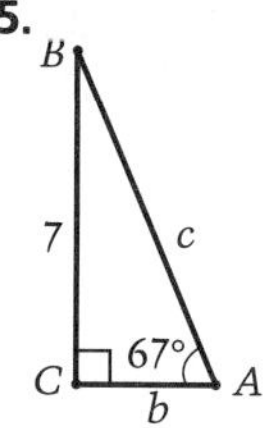

36. $c = 2\sqrt{13}$, $A \approx 56°$, $B \approx 34°$

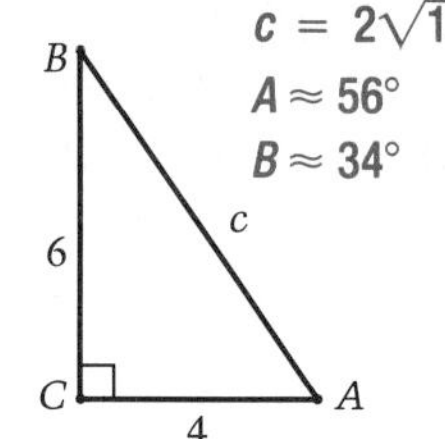

37. *Triangular Numbers* The numbers 1, 3, 6, 10, 15, . . . are **triangular numbers.** Each triangular number can be represented with a visual model, as shown below. Use the visual model to support the conjecture that the sum of two consecutive triangular numbers is a perfect square. See margin.

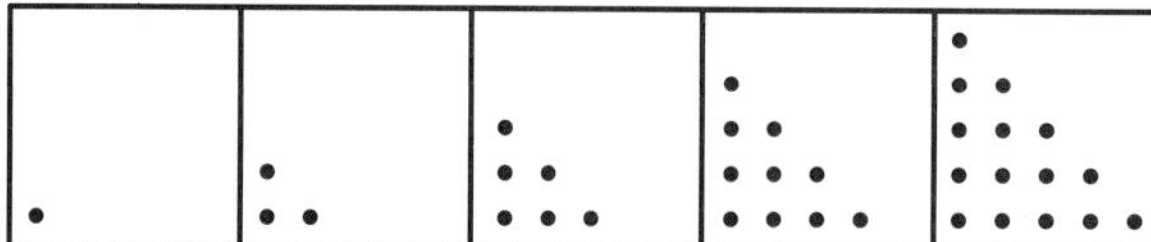

38. *The Song of Hiawatha* You are standing at a point that is 10 meters from the base of Minnehaha Falls. The angle formed by the ground and the line connecting the point with the top of the falls is 58°. How high is Minnehaha Falls? ≈16.0 m

In 1855, Henry Wadsworth Longfellow published the narrative poem "The Song of Hiawatha," based on the Native American folklore of a fifteenth-century Mohawk chief. The Minnehaha Falls mentioned in the poem are on Minnehaha Creek in Minneapolis, Minnesota.

Chemistry **In Exercises 39 and 40, use the following information.**

Your chemistry teacher asks you to weigh a magnesium sample as an exercise in accurate measurement. You are asked to weigh the sample once on each side of the scale and find the geometric mean of the two weights.

39. On the left side, the sample weighs 36 grams. On the right side, the sample weighs 40 grams. What is the true weight of the sample? ≈37.9 g

40. Using a scale *different* from the one used in Exercise 39, you weigh the sample on the left side and find the weight to be 37 grams. What would the sample weigh on the right side of this scale? ≈38.8 g

Sonar Fishing **In Exercises 41–44, use the following information.** 43. $a = 35$ 44. 35 ft

You are fishing in a boat and use a sonar device that detects schools of fish. The device indicates that a school of fish is 70 feet from your boat. The angle between the water's surface and the direction of the fish is 30°.

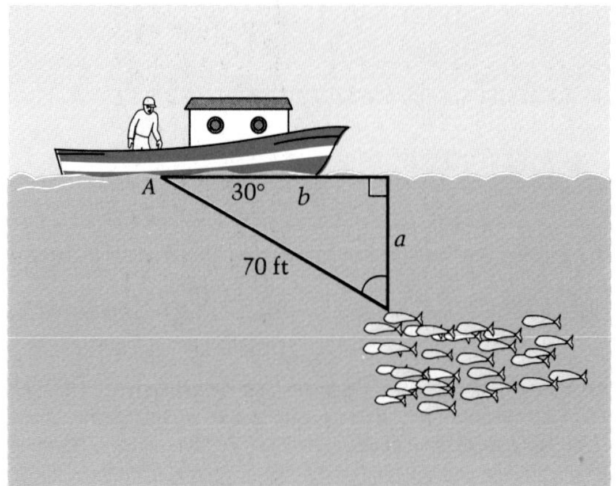

41. Write an equation that relates a and b with the acute angle A. $\tan 30° = \frac{a}{b}$

42. Solve the equation for b. $b = \frac{a}{\tan 30°}$

43. Substitute the expression for b into the equation $a^2 + b^2 = 70^2$, and solve for a.

44. How deep should you set your fishing lines to be at the same depth as the fish?

ZOTOK **In Exercises 45–48, use the following information.**

In the game of *ZOTOK*, there are three countries: South Zotok (purple), West Zotok (yellow), and East Zotok (orange). Green controls West Zotok and Pink controls East Zotok. Each unit on the x-axis and y-axis corresponds to 100 miles. A piece can move to any coordinate point (x, y) that is within the piece's range. See margin.

45. Can any of Green's pieces be hit by any of Pink's missiles? Explain.

46. Can any of Pink's pieces be hit by any of Green's missiles? Explain.

47. Which of Green's pieces can be hit by Pink's fighter planes in *two* moves?

48. Which of Pink's pieces can be hit by Green's fighter planes in *two* moves?

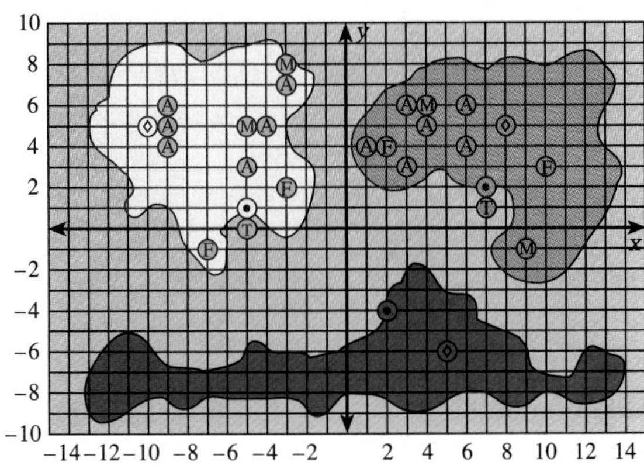

LEGEND

(A)	Armored Unit:	200 miles
(T)	Navy Transport:	300 miles
(M)	Missile:	700 miles
(F)	Fighter Plane:	500 miles
◇	Capital City	
⊙	Port City	

45. No; Green's piece (an armored unit at $(-3, 7)$) that is closest to one of Pink's missiles is ≈ 707 miles away from its location at $(4, 6)$.

46. Yes; Pink's armored unit at $(3, 6)$ is ≈ 632 miles from Green's missile at $(-3, 8)$; Pink's armored unit at $(1, 4)$ is ≈ 608 miles from Green's missile at $(-5, 5)$ and ≈ 566 miles from Green's missile at $(-3, 8)$.

47. $M(-3, 8)$, $M(-5, 5)$, $A(-3, 7)$, $A(-4, 5)$, $A(-5, 3)$, $F(-3, 2)$, $T(-5, 0)$, Port City $(-5, 1)$

48. $M(4, 6)$, $A(3, 6)$, $A(6, 6)$, $A(1, 4)$, $A(6, 4)$, $A(3, 3)$, $A(4, 5)$, $F(2, 4)$, Port City $(7, 2)$

Answers

1. $4^2 + 6^2 = (2\sqrt{13})^2$
2. $(2\sqrt{26})^2 + (4\sqrt{26})^2 = (2\sqrt{130})^2$
7. $3\sqrt{3} - \sqrt{2}$
8. $7\sqrt{6} + \sqrt{2}$
9. $30 - 90\sqrt{2}$
10. $81 + 8\sqrt{5}$
19. Counterexamples vary. $2 < 3$ but $\frac{2}{-1} > \frac{3}{-1}$

Chapter TEST

In Exercises 1 and 2, show that the three points are the vertices of a right triangle. (13.1) See margin.

1. $(-6, -4), (0, -4), (-6, 0)$ **2.** $(5, 2), (-5, 0), (1, 22)$

In Exercises 3–6, simplify the radicals. (13.2)

3. $\sqrt{500}$ $10\sqrt{5}$ **4.** $\sqrt{384}$ $8\sqrt{6}$ **5.** $\sqrt{\frac{1}{8}}$ $\frac{\sqrt{2}}{4}$ **6.** $\sqrt{\frac{25}{27}}$ $\frac{5\sqrt{3}}{9}$

In Exercises 7–10, perform the indicated operations. (13.3) See margin.

7. $4\sqrt{3} - \sqrt{2} - \sqrt{3}$ **8.** $6\sqrt{6} + \sqrt{6} + \sqrt{2}$ **9.** $3\sqrt{5}(2\sqrt{5} - 6\sqrt{10})$ **10.** $(4\sqrt{5} + 1)^2$

In Exercises 11—14, solve the equation. (13.4)

11. $\sqrt{y} + 6 = 10$ 16 **12.** $\sqrt{2m + 3} - 6 = 4$ $\frac{97}{2}$ **13.** $n = \sqrt{9n - 18}$ 6, 3 **14.** $p = \sqrt{-3p + 18}$ 3

15. Find the area and perimeter. **(13.3)** $6\sqrt{10} + 16$, $12\sqrt{2} + 6\sqrt{5}$

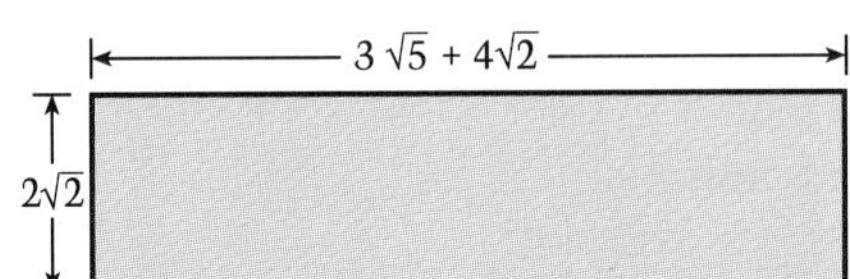

16. Find the perimeter of the parallelogram. **(13.3)** $8 + 4\sqrt{10}$

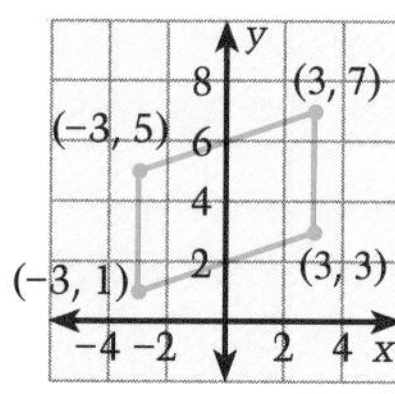

In Exercises 17 and 18, find the length of each side of the triangle. (13.5)

17. $b = 8, c = 10$

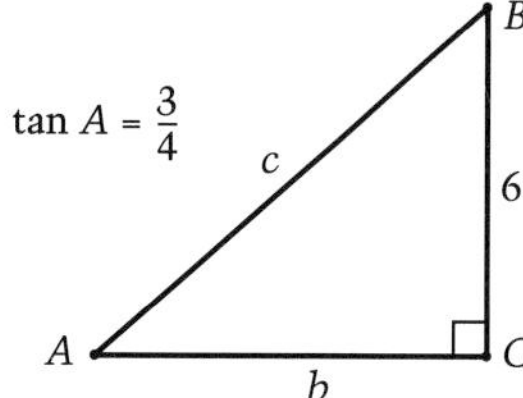

18. $a \approx 14.5$, $c \approx 24.7$

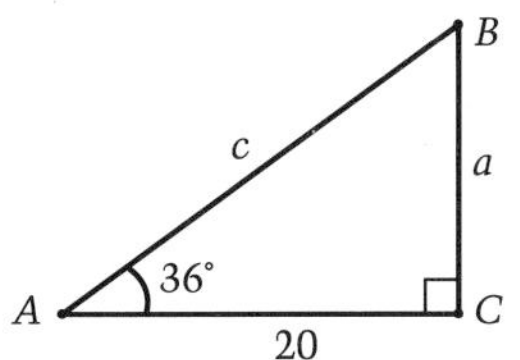

19. Find a counterexample to the statement: See margin.

If a, b, and c are numbers, $c \neq 0$, *and* $a < b$, *then* $\frac{a}{c} < \frac{b}{c}$.

20. The ratio of the length of a rectangular box to the width of the box is the same as the ratio of the width to the height. The length is 6 feet and the height is 3 feet. What is the width? $3\sqrt{2}$ ft

21. You are standing 134 feet from the Gateway Arch in St. Louis, Missouri. The angle formed by the ground and your line of sight to the top of the arch is 78°. What is the height of the Gateway Arch? ≈630 ft

End-of-Book CONTENTS

Table of Squares and Approximate Square Roots

n	n^2	$\sqrt{n}$	$\sqrt{10n}$	n	n^2	$\sqrt{n}$	$\sqrt{10n}$
1	1	1.000	3.162	**51**	2601	7.141	22.583
2	4	1.414	4.472	**52**	2704	7.211	22.804
3	9	1.732	5.477	**53**	2809	7.280	23.022
4	16	2.000	6.325	**54**	2916	7.348	23.238
5	25	2.236	7.071	**55**	3025	7.416	23.452
6	36	2.449	7.746	**56**	3136	7.483	23.664
7	49	2.646	8.367	**57**	3249	7.550	23.875
8	64	2.828	8.944	**58**	3364	7.616	24.083
9	81	3.000	9.487	**59**	3481	7.681	24.290
10	100	3.162	10.000	**60**	3600	7.746	24.495
11	121	3.317	10.488	**61**	3721	7.810	24.698
12	144	3.464	10.954	**62**	3844	7.874	24.900
13	169	3.606	11.402	**63**	3969	7.937	25.100
14	196	3.742	11.832	**64**	4096	8.000	25.298
15	225	3.873	12.247	**65**	4225	8.062	25.495
16	256	4.000	12.649	**66**	4356	8.124	25.690
17	289	4.123	13.038	**67**	4489	8.185	25.884
18	324	4.243	13.416	**68**	4624	8.246	26.077
19	361	4.359	13.784	**69**	4761	8.307	26.268
20	400	4.472	14.142	**70**	4900	8.367	26.458
21	441	4.583	14.491	**71**	5041	8.426	26.646
22	484	4.690	14.832	**72**	5184	8.485	26.833
23	529	4.796	15.166	**73**	5329	8.544	27.019
24	576	4.899	15.492	**74**	5476	8.602	27.203
25	625	5.000	15.811	**75**	5625	8.660	27.386
26	676	5.099	16.125	**76**	5776	8.718	27.568
27	729	5.196	16.432	**77**	5929	8.775	27.749
28	784	5.292	16.733	**78**	6084	8.832	27.928
29	841	5.385	17.029	**79**	6241	8.888	28.107
30	900	5.477	17.321	**80**	6400	8.944	28.284
31	961	5.568	17.607	**81**	6561	9.000	28.460
32	1024	5.657	17.889	**82**	6724	9.055	28.636
33	1089	5.745	18.166	**83**	6889	9.110	28.810
34	1156	5.831	18.439	**84**	7056	9.165	28.983
35	1225	5.916	18.708	**85**	7225	9.220	29.155
36	1296	6.000	18.974	**86**	7396	9.274	29.326
37	1369	6.083	19.235	**87**	7569	9.327	29.496
38	1444	6.164	19.494	**88**	7744	9.381	29.665
39	1521	6.245	19.748	**89**	7921	9.434	29.833
40	1600	6.325	20.000	**90**	8100	9.487	30.000
41	1681	6.403	20.248	**91**	8281	9.539	30.166
42	1764	6.481	20.494	**92**	8464	9.592	30.332
43	1849	6.557	20.736	**93**	8649	9.644	30.496
44	1936	6.633	20.976	**94**	8836	9.695	30.659
45	2025	6.708	21.213	**95**	9025	9.747	30.822
46	2116	6.782	21.448	**96**	9216	9.798	30.984
47	2209	6.856	21.679	**97**	9409	9.849	31.145
48	2304	6.928	21.909	**98**	9604	9.899	31.305
49	2401	7.000	22.136	**99**	9801	9.950	31.464
50	2500	7.071	22.361	**100**	10000	10.000	31.623

PROPERTIES

Basic Properties		**Addition**	**Multiplication**
	Commutative	$a+b=b+a$	$ab=ba$
	Associative	$(a+b)+c=a+(b+c)$	$(ab)c=a(bc)$
	Identity	$a+0=a,\ 0+a=a$	$a(1)=a,\ 1(a)=a$
	Inverse	$a+(-a)=0$	$a\cdot\frac{1}{a}=1,\ a\neq 0$
	Distributive	$a(b+c)=ab+ac$ or $(a+b)c=ac+bc$	
Properties of Equality	**Addition**	If $a=b$, then $a+c=b+c$.	
	Subtraction	If $a=b$, then $a-c=b-c$.	
	Multiplication	If $a=b$, then $ca=cb$.	
	Division	If $a=b$, and $c\neq 0$, then $\frac{a}{c}=\frac{b}{c}$.	
Properties of Exponents	**Product of Powers**	$a^m\cdot a^n=a^{m+n}$	
	Power of a Power	$(a^m)^n=a^{m\cdot n}$	
	Power of a Product	$(a\cdot b)^m=a^m\cdot b^m$	
	Quotient of Powers	$\frac{a^m}{a^n}=a^{m-n},\ a\neq 0$	
	Power of a Quotient	$(\frac{a}{b})^m=\frac{a^m}{b^m},\ b\neq 0$	
	Negative Exponent	$a^{-n}=\frac{1}{a^n},\ a\neq 0$	
	Zero Exponent	$a^0=1,\ a\neq 0$	
Properties of Radicals	**Product Property**	$\sqrt{ab}=\sqrt{a}\cdot\sqrt{b}$	**Quotient Property** $\sqrt{\frac{a}{b}}=\frac{\sqrt{a}}{\sqrt{b}}$
Properties of Proportions	**Reciprocal**	If $\frac{a}{b}=\frac{c}{d}$, then $\frac{b}{a}=\frac{d}{c}$.	
	Cross-multiplying	If $\frac{a}{b}=\frac{c}{d}$, then $ad=bc$.	
Properties of Matrices	**Equal Matrices**	If $\begin{bmatrix} a & b \\ c & d \end{bmatrix}=\begin{bmatrix} 2 & 4 \\ 7 & 5 \end{bmatrix}$ then $a=2, b=4, c=7$, and $d=5$.	
	Adding Matrices	$\begin{bmatrix} a & b \\ c & d \end{bmatrix}+\begin{bmatrix} 2 & 4 \\ 7 & 5 \end{bmatrix}=\begin{bmatrix} a+2 & b+4 \\ c+7 & d+5 \end{bmatrix}$	
Order of Operations	**1.** Parentheses **2.** Exponents **3.** Multiplication/Division	**4.** Addtion/Subtraction **5.** Left-to-Right	
Special Products and their Factors		**Product**	**Factors**
	Sum and Difference Pattern	$(a+b)(a-b)$ ⟷	a^2-b^2
	Square of a Binomial Pattern	$(a+b)^2$ ⟷	$a^2+2ab+b^2$
		$(a-b)^2$ ⟷	$a^2-2ab+b^2$
	FOIL The product $(a+b)(c+d)$ can be written: $a\cdot c$ + $a\cdot d$ + $b\cdot c$ + $b\cdot d$ First Outer Inner Last		

MODELS

A Problem-Solving Plan	**Ask** yourself what you need to know to solve the problem. Then **write a verbal model** that will give you what you need to know. **Assign labels** to each part of your verbal model. Use the labels to **write an algebraic model** based on your verbal model. **Solve** the algebraic model. **Answer** the original question. **Check** that your answer is reasonable.
Summary of Equations of Lines in Two Variables	**Slope-intercept Form** $y = mx + b$ **Point-slope Form** $y - y_1 = m(x - x_1)$ **Standard form** $Ax + By = C$ **Intercept Form** $\frac{x}{a} + \frac{y}{b} = 1$ **Slope of a Line through Two Points** $m = \frac{(y_2 - y_1)}{(x_2 - x_1)}$
Algebraic Models	**Linear** $f(x) = mx + b$ **Quadratic** $f(x) = ax^2 + bx + c$ or $f(x) = a(x - h)^2 + k$ **Exponential** $f(x) = C(a)^x$ **Rational** $f(x) = \frac{a}{x - h} + k$
Quick Graphs as Graphic Check of Solutions	The solution of a linear equation involving one variable, x, can be checked graphically with the following steps. **1.** Write the equation in the form $ax + b = 0$. **2.** Sketch the graph of $y = ax + b$. **3.** The solution of $ax + b = 0$ is the *x-intercept* of $y = ax + b$.
Approximating the Best-Fitting Line	To approximate the best-fitting line for a set of ordered pairs, use the following steps. **1. Sketch a Scatter Plot:** Sketch a scatter plot for the data. **2. Sketch the Line:** Sketch the line that appears to most clearly follow the pattern given by the points. There should be about as many points above the line as below the line. **3. Locate Two Points on the Line:** Locate two points on the line and approximate the x- and y-coordinates of each point. **4. Find an Equation of the Line:** Use the technique described in Lesson 5.3 to find an equation of the line that passes through the two points in step 3.
Models for Direct and Inverse Variation	**1.** The variables x and y vary directly if, for a constant k, $\frac{y}{x} = k$ or $y = kx$. **2.** The variables x and y vary inversely if, for a constant k, $yx = k$ or $y = \frac{k}{x}$. In both cases, the number k is the constant of variation.

TRANSFORMATIONS

Using Transformations to Solve Equations, Inequalities and Systems

Transformations that Produce Equivalent Equations

1. Add or subtract the same number to *both* sides.
2. Multiply or divide *both* sides by the same nonzero number.
3. Interchange the two sides.

Solving Equations Using Two or More Transformations

1. If the variable is on both sides of the equation, collect variables on the side with the greater variable coefficient.
2. Simplify both sides of the equation (if needed).
3. Use inverse operations to isolate the variable.
4. Check your solution in the original equation.

Transformations that Produce Equivalent Inequalities

1. Add or subtract the same number to *both* sides.
2. Multiply or divide *both* sides by the same *positive* number.
3. Multiply or divide *both* sides by the same *negative* number and reverse the inequality.

Solving a System of Linear Equations

To decide which method to use—graphing, or transformations (substitution or linear combinations)—consider the following.

1. The *graphing* method is useful for approximating a solution, checking a solution, and for providing a visual model of the problem.
2. To find an exact solution, use either *substitution* or *linear combinations.*
3. For linear systems in which one variable has a coefficient of 1 or -1, *substitution* may be more efficient.
4. In other cases, the *linear combinations* method is usually more efficient.

Using Transformations to Graph Related Functions

Quadratic Functions

When a quadratic function is written in the form $f(x)=(x-h)^2+k$ (by completing the square), the vertex of the parabola has coordinates (h, k).

- The function can be graphed by translating (shifting) the graph of $f(x)=x^2$ vertically by k units and horizontally by h units.
- If k is positive, the vertical shift is upward; if k is negative, the vertical shift is downward.
- If h is positive, the horizontal shift is to the right; if h is negative, the horizontal shift is to the left.
- The function $f(x)=-x^2$ can be graphed by reflecting the graph of $f(x)=x^2$ in the x-axis.

Exponential Functions

An exponential function can be written in the form $f(x)=a^{(x-h)}+k$.

- The function can be graphed by translating (shifting) the graph of $f(x)=a^x$ vertically by k units and horizontally by h units.
- If k is positive, the vertical shift is upward; if k is negative, the vertical shift is downward.
- If h is positive, the horizontal shift is to the right; if h is negative, the horizontal shift is to the left.
- The function $f(x)=-a^x$ can be graphed by reflecting the graph of $f(x)=a^x$ in the x-axis.

GLOSSARY

absolute value of a real number (64, 217) The distance between the origin and the point representing the real number.
For example: $|3| = 3$, $|-3| = 3$, or $|0| = 0$.

acute angle (699) An angle whose measure is greater than 0° and less than 90°.

addition property of equality (124) See Properties on pages 722-724.

additive identity (71) See Axioms on pages 722-724.

algebraic expression (8) A collection of numbers, variables, operations, and grouping symbols.

algebraic model (34) An algebraic expression or equation used to represent a real-life situation.

approximate (2) To express a numerical value to different degrees of accuracy. For example, $\frac{3}{7} = 0.42857$ can be approximated as 0.43.

associative axiom (706) See Axioms on pages 722-724.

asymptotes (606, 646) Lines that resemble the graph of a rational expression in certain regions of the coordinate plane.

axis of symmetry of a parabola (465) The vertical line passing through the vertex of a parabola and dividing the parabola into two parts that are mirror images of each other.

bar graph (50, 51) A graph that organizes a collection of data by using horizontal or vertical bars to display how many times each number occurs in the collection.

base of an exponent (14, 400) The number or variable that is used as a factor in repeated multiplication. For example, in the expression 4^6, 4 is the base.

best-fitting line (255) A line that best fits the data points on a scatter plot.

bimodal (658) A collection of numbers that has two modes.

binomial (508) A polynomial that has only two terms.

box-and-whisker plot (619, 652) A plot used to present a graphic summary of information. The quartiles and extreme values of a set of data displayed using a number line.

budget variance (315) The difference between budgeted expense and actual expense.

Cartesian plane (211) The *x*, *y* coordinate plane is often called the Cartesian plane because René Descartes first introduced it in 1637.

circle graph (327) A circle partitioned into sectors used to show the relative size of different portions of a whole.

closure axiom (706) See Axioms on pages 722-724.

coefficient (97, 508) In a term that is the product of a number and a variable, the number is the coefficient of the variable.

commutative axiom (706) See Axioms on pages 722-724.

completed-square form (549, 639) A quadratic equation in the form "perfect square = k", $k \geq 0$; a quadratic function in the form $f(x) = a(x - h)^2 + k$. The vertex of the parabola is the point (h, k).

completing the square (549) A process which can be used to solve any quadratic equation.

compound inequality (305) Two inequalities connected by *and* or *or*.

compound interest (432) Interest paid on the original principal and on interest that becomes part of the account.

conjecture (322) A statement that you think is true but that has not been proved true.

constant of variation (575) The constant k in the relationships of variation.

constant term (508) A monomial without a variable. For example, in $x^2 + 2x + 5$ the number 5 is the constant term.

constraints (387) The system of linear inequalities in a linear programming problem.

coordinate plane (160) A plane formed by two real number lines intersecting at a right angle.

counterexample (707) An example given to disprove a proposed axiom, or theorem.

cross-multiplying property (562) See Properties of Proportions on pages 722-724.

data (49) The facts, or numbers, that describe something.

definition (707) A statement or meaning of a word. A definition does not have to be proved but must be checked to be sure it is consistent with previous definitions and axioms.

degree of a monomial (508) The integer k is the degree of ax^k.

denominator (9) The divisor in a quotient expressed in fraction form. For example, 2 is the denominator of $\frac{5x}{2}$.

difference (3, 75, 709) The result obtained when numbers or expressions are subtracted. The difference of a and b is defined as $a - b$.

difference of two squares (522) See Special Products on pages 722-724.

direct variation (575) A function whose equation has the form $y = kx$ or $\frac{y}{x} = k$.

discriminant (479) The expression $b^2 - 4ac$, for a quadratic equation $ax^2 + bx + c = 0$.

distance formula (672) The distance, d, between any two points, (x_1, y_1) and (x_2, y_2), is given by $d = \sqrt{(x_2 - x_1)^2 + (y_2 - y_1)^2}$.

distributive property (95, 706) See Properties pages 722-724.

division property of equality (124) See Properties pages 722-724.

domain, or input, of a function (620) The x-values of a function $y = f(x)$.

domain of a rational expression (588) The set of all real numbers, except those for which the denominator is zero.

equal matrices (81) Matrices are equal if the entries in corresponding positions are equal.

equally–likely outcome (581) An event that is as likely to happen as not to happen.

equation (27) A statement formed when an equality symbol is placed between two equal expressions.

equivalent equations (122) Equations with the same solution set.

equivalent expressions (77) Expressions that have the same value for each number represented by the variable.

evaluate an algebraic expression (8) To find the value of an algebraic expression by replacing each variable in the expression by a number.

exponent (14, 400) The number or variable that represents the number of times the base is used as a factor. For example, in the expression 4^6, 6 is the exponent.

exponential decay (401) A situation in which the original amount is repeatedly multiplied by a change factor less than one.

exponential equation (402) An equation in which the variables appear as exponents.

exponential form of a power (14) The expression 4^6 is the exponential form of $4 \cdot 4 \cdot 4 \cdot 4 \cdot 4 \cdot 4$.

exponential growth (401) A situation in which the original amount is repeatedly multiplied by a change factor greater than one.

factors (9) The numbers and variables that are multiplied in an expression. For example, 4 and x are factors of $4x$.

FOIL pattern (515) A method used to multiply two binomials in a single step. Find the sum of the products of the **F**irst terms, **O**uter terms, **I**nner terms, and **L**ast terms.

formula (10) An algebraic expression that represents a relationship in real-life situations.

frequency distribution (51) A table that organizes a collection of data by counting how many times each number occurs in the collection.

function (619) A special type of relationship between two values in which each input value corresponds to exactly one output value. A relation in which no two ordered pairs (x, y) have the same x-value.

general form of a linear equation (264) A linear equation in the form $Ax + By = C$.

geometric mean (692) The geometric mean of a and b is $\sqrt{ab}$.

geometric sequence (405) A sequence in which each term after the first is the product of the preceding term and a constant, r. The constant is called the common ratio.

graph of an equation (183) A visual model of an equation.

greatest common factor (528) The greatest monomial factor that can be factored from the terms of a polynomial.

half-plane (320) In a plane, the region on either side of a boundary.

histogram (664) A bar graph that shows a frequency distribution.

horizontal shift of a graph (634) See Transformations on pages 722-724.

hyperbola (646) The graph of a rational function $f(x) = \frac{a}{(x - h)} + k$, whose center is (h, k).

hypotenuse (454) In a right triangle, the side opposite the right angle.

identity axiom (706) See Axioms on pages 722-724.

inequality (29) An open sentence formed when an inequality symbol is placed between two expressions.

integers (62) The set of numbers $\{..., -3, -2, -1, 0, 1, 2, 3, ...\}$.

intercept form of an equation of a line (269) See Equations on pages 722-724.

inverse axiom (706) See Axioms on pages 722-724.

inverse operations (122) Two operations that "undo" each other. For example, addition and subtraction are inverse operations. Multiplication and division are inverse operations.

inverse variation (575) A function defined by an equation of the form $xy = k$ or $y = \frac{k}{x}$.

irrational number (453) A real number that cannot be expressed as the quotient of two integers. For example, the square roots of numbers that are not perfect squares $\sqrt{2}$, $\sqrt{3}$, $\sqrt{5}$, and $\sqrt{6}$. When expressed as a decimal, it neither repeats nor terminates.

leading coefficient (458, 508) The coefficient of the first term in a polynomial written in standard form.

like terms (97) Terms in which the variable factors are the same.

line plot (51) A graph that organizes a collection of data by using an X each time a number occurs in the collection.

linear combination method (360) A method of solving systems of equations that involves adding multiples of the given equations.

linear function (626) A function defined by the equation $f(x) = mx + b$.

linear model (235) An equation of a line that is used to represent a real-life situation.

linear programming (387) A type of optimization process for finding the maximum or minimum value of an objective quantity by using a system of constraints.

literal equation (154) An equation which uses more than one letter as a variable. For example, $3x + y = 4$ is a literal equation.

matrix (81) A rectangular arrangement of numbers into rows and columns.

mean (658) The sum of a set of numbers divided by the number of numbers in the set.

measures of central tendency (619, 658) The mean, median, and mode of a distribution.

median (658) The middle number of an odd collection of numbers; the average of the two middle numbers in an even collection of numbers.

mental math (3) When operations are performed in one's head without other aids such as paper and pencil or calculator.

mode (658) The number that occurs most frequently in a collection of numbers. If two numbers tie for most frequent occurrence, the collection is said to be bimodal.

modeling (34) Writing algebraic expressions or equations to represent real-life situations.

monomial (508) A polynomial with only one term.

multiplication property of equality (124) See Properties on pages 722-724.

multiplicative identity (89) See Axioms on pages 722-724.

negative correlation (258) Data points on a scatter plot that approximate a line with a negative slope.

negative numbers (62) Numbers represented by points to the left of the origin on the real number line. Numbers less than 0.

numerator (9) The dividend in a quotient expressed in fraction form. For example, $5x$ is the numerator of $\frac{5x}{2}$.

numerical expression (4) A collection of numbers, operations, and grouping symbols.

objective quantity (387) In linear programming, the quantity or expression to be maximized or minimized.

odds (586) The ratio of the number of ways an event can occur to the number of ways the event cannot occur.

opposites (64) Two points on the real number line that are the same distance from the origin but on opposite sides of the origin. For example, 3 and -3 are opposites.

optimal design (618) The most desirable or satisfactory design.

optimization (387) Finding the minimum or maximum value of a quantity.

order of operations (20, 21) See Procedures on pages 722-724.

ordered pair (160) A pair of real numbers used to locate each point in a coordinate plane.

origin of the coordinate plane (160) The point in the coordinate plane at which the horizontal axis intersects the vertical axis. The point (0, 0).

origin of the real number line (62) The point that represents 0 on the real number line.

parabola (465) The graph of a quadratic equation.

percent (2, 258) Another way of expressing hundredths, or a number divided by 100, usually denoted by the symbol %.

perfect squares (453) Numbers whose square roots are integers or quotients of integers.

perfect square trinomial (536) A trinomial that factors into the square of a binomial.

perimeter of a polygon (12) The sum of the lengths of the sides of the polygon.

picture graph (327) Stylized versions of bar graphs.

point-slope form of the equation of a line (270, 272) See Equations on pages 722-724.

polynomial (508) The sum or difference of monomials. Monomials are also polynomials.

positive correlation (258) Data points on a scatter plot that approximate a line with a positive slope.

positive numbers (62) Numbers represented by the points to the right of the origin on the real number line. Numbers greater than 0.

postulate, or axiom (706) The rules in mathematics that are accepted to be true without proof.

power (14, 400) The result of repeated multiplication. For example, in the expression $4^2 = 16$, 16 is the second power of 4.

power of a power property (400) See Properties on pages 722-724.

power of a product property (400) See Properties on pages 722-724.

power of a quotient property (413) See Properties on pages 722-724.

probability (581) The ratio of successful outcomes to the total number of outcomes of the event.

probability of an event (581) A number between 0 and 1 that indicates the likelihood the event will occur.

problem–solving plan (46) See Procedures on pages 722-724.

product (3, 89) The result obtained when numbers or expressions are multiplied. The product of a and b is $a \cdot b$, or ab.

product property of radicals (678) See Properties on pages 722-724.

program (19) An ordered list of instructions for a calculator or computer.

properties of addition (69) See Properties on pages 722-724.

properties of multiplication (89) See Properties on pages 722-724.

properties of opposites (64) See Properties on pages 722-724.

proportion (562) An equation that equates two ratios. For example, $\frac{a}{b} = \frac{c}{d}$, where a, b, c, and d, are nonzero real numbers.

Pythagorean theorem (454) For any right triangle, the sum of the squares of the lengths of the legs, a and b, equals the square of the length of the hypotenuse, c. For example, $a^2 + b^2 = c^2$.

quadrant (160) In the coordinate plane, one of the four parts into which the axes divide the plane.

quadratic equation (458) An equation in x that can be written in the standard form $ax^2 + bx + c = 0, a \neq 0$.

quadratic formula (472) See Formulas on pages 722-724.

quartile (653) The values in a collection of numbers that separate the data into four equal parts.

quotient (3, 101) The result obtained when numbers or expressions are divided. The quotient of a and b is $\frac{a}{b}$, where a and b are real numbers, and $b \neq 0$.

quotient of powers property (413) See Properties on pages 722-724.

quotient property of radicals (678) See Properties on pages 722-724.

radicand (452, 678) The number or expression under a radical symbol.

range, or output, of a function (620) The y-values of a function $y = f(x)$.

rate of *a* to *b* (109) The relationship $\frac{a}{b}$ of two quantities a and b that are measured in different units.

rate of change (200) A relationship such as distance over time often described by using a slope.

ratio of *a* to *b* (109) The relationship $\frac{a}{b}$ of two quantities a and b that are measured in the same units.

rational equation (607) An equation that involves rational expressions with the variable occurring in the denominator(s).

rational expressions (561) Fractions whose numerators and denominators are polynomials.

rational numbers (109) Any number that can be written in the form $\frac{a}{b}$, where a and b are integers and $b \neq 0$.

real number line (62) A horizontal line that pictures real numbers as points.

real numbers (62) The set of numbers consisting of the positive numbers, the negative numbers, and zero.

reciprocal (101) If $\frac{a}{b}$ is a non-zero number, then its reciprocal is $\frac{b}{a}$.

reciprocal property (562) See Properties on pages 722-724.

reflection of a graph (634) See Transformations on pages 722-724.

relation (620) A set of ordered pairs (x, y).

round-off error (148) The error produced when a decimal result is rounded in order to provide a meaningful answer.

scatter plot (160) The plot, or graph, of the points that correspond to the ordered pairs in a collection.

scientific notation (419) Uses the powers of ten to express decimal numbers. Numbers in scientific notation are written $c \times 10^n$, where $1 \le c < 10$ and n is any integer.

set (62) A well-defined collection of objects.

similar triangles (125) Two triangles are similar if their corresponding angles are congruent.

simple interest (432) The amount paid or earned for the use of money for a unit of time.

simplest form of a radical expression (678) See Procedures on pages 722-724.

simplified, or reduced form, of a rational expression (588) See Procedures on pages 722-724.

slope of a line (197, 272) The ratio of the vertical change to the horizontal change; the ratio of the change in y to the corresponding change in x.

slope-intercept form (205, 272) See Procedures on pages 722-724.

solution of a system of equations (346) An ordered pair that is a solution of each of the equations in the system.

solution of a single-variable equation (27) A number that, when substituted for the variable in an equation, results in a true statement.

solving a proportion (562) Finding the value for the variable in a proportion.

solving an equation (28, 122) Finding all the solutions of an equation.

speed (65) The absolute value of the velocity of an object; expressed as the rate: distance over time.

square matrix (83) A matrix that has the same number of rows and columns.

square root (452) If $b^2 = a$, then b is a square root of a.

stacked bar graph (52) A bar graph that displays more than one collection of data simultaneously.

standard form of a polynomial (508) A polynomial in which the terms are written in descending order, from the greatest degree to the smallest degree.

standard form of the equation of a line (264) See Equations on pages 722-724.

statistics (659) A field of mathematics that organizes large collections of data in ways that can be used to understand trends and make predictions.

stem-and-leaf plot (619, 652) See Procedures on pages 722-724.

substitution method (353) A method of solving a system of equations in which one variable is written in terms of other variables in one equation and the resulting expression is substituted into the other equation.

subtraction property of equality (124) See Properties on pages 722-724.

subtraction rule (75) See Procedures on pages 722-724.

sum (3, 68) The result obtained when numbers or expressions are added. The sum of a and b is $a + b$.

system of equations (346) A set of equations in the same variables.

system of inequalities (381) A set of inequalities in the same variables.

tangent of angle *A* (699) See Formulas on pages 722-724.

terms (9, 75) In an expression such as $3x + 2$, $3x$ and 2 are called terms.

theorem (454, 707) A statement that can be proven to be true.

time-line graph (327) A graph used to represent events that occur in a given sequence in time.

transformations (633) See Transformations on pages 722-724.

trinomial (508) A polynomial that has only three terms.

unit analysis (91) A technique that can be used to decide the unit of measure assigned to a product.

variable (8) A letter that is used to represent one or more numbers.

velocity (65) The speed and direction in which an object is traveling.

verbal model (34) An expression that uses words to represent a real-life situation.

vertex of a parabola (465) The lowest point of a parabola that opens up and the highest point of a parabola that opens down.

vertex of convex polygons (linear programming) (381) The corner points of the graph of a system of constraints.

vertical shift of a graph (633) See Transformations on pages 722-724.

whole numbers (2) The set of numbers $\{0, 1, 2, 3, 4, \ldots\}$.

***x*-axis** (160) The horizontal axis in the coordinate plane.

***x*-coordinate** (160) The first number of an ordered pair, the horizontal coordinate.

***x*-intercept** (190) The value of x when $y = 0$.

***y*-axis** (160) The vertical axis in the coordinate plane.

***y*-coordinate** (160) The second number of an ordered pair, the vertical coordinate.

***y*-intercept** (190) The value of y when $x = 0$.

zero-product property (541) See Properties on pages 722-724.

TECHNOLOGY

Keystrokes for Other Graphing Calculators

Keystrokes for page 19

Graphing calculators can be instructed to evaluate an expression. The list of instructions is a ***program,*** *and each separate instruction is a* ***program step.*** *When the calculator performs the steps in a program, it is* executing *or* running *the program. Here is an example of a flowchart for a program that will evaluate the algebraic expression* $x^2 + 2x$ *for different values of the variable x.*

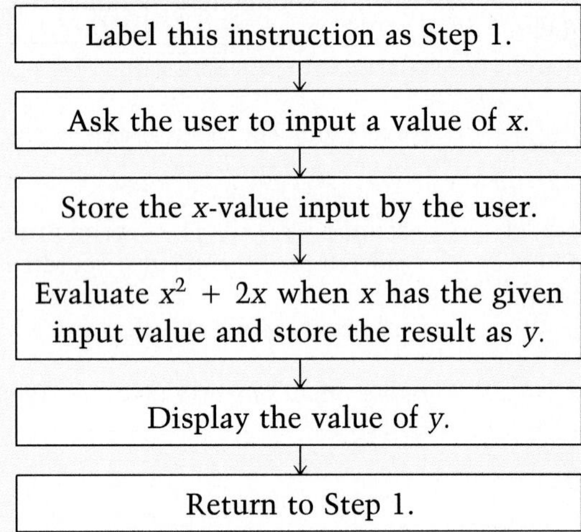

Label this instruction as Step 1.
Ask the user to input a value of x.
Store the x-value input by the user.
Evaluate $x^2 + 2x$ when x has the given input value and store the result as y.
Display the value of y.
Return to Step 1.

The following three versions of this program show steps that can be used with a *TI-82,* a Casio *fx-9700GE,* and a Sharp *EL-9300C.* When you run the program, enter the x-value when prompted, and the calculator will display the value of the expression $x^2 + 2x$.

TI-82

```
PROGRAM:EVALUATE
:Lbl 1
:Disp "ENTER X"
:Input X
:X^2+2X→Y
:Disp Y
:Goto 1
```

Casio *fx-9700GE*

```
EVALUATE
Lbl 1
"ENTER X"
?→X
X^2+2X→Y
Y
Goto 1
```

Sharp *EL-9300C*

```
evaluate
-------------------- REAL
Label 1
Print "enter X"
Input X
y-X^2+2X
Print y
Goto 1
```

Exercises

Use a graphing calculator to evaluate the following expressions when $x = 0$, $x = 2$, and $x = 5.36$. (Round each result to two decimal places.)

1. $x^2 + 2x$ **2.** $60 - 9x$ **3.** $0.5x + 10$ **4.** $\frac{1}{3}x$

5. x^2 **6.** $x^2 + 1$ **7.** $3x^2 + 8x$ **8.** $5x^2$

TECHNOLOGY
Keystrokes for Other Graphing Calculators

TECHNOLOGY APPENDIX

Keystrokes for page 87

Most graphing calculators can store, add, and subtract matrices. Here we show how to use a TI-82, *a Casio* fx-9700GE, *and a Sharp* EL-9300C *to add the matrices at the right.*

$$\begin{bmatrix} 2 & -1 & 0 \\ 3 & 2 & -1 \end{bmatrix} + \begin{bmatrix} 1 & 0 & -2 \\ 2 & -2 & -4 \end{bmatrix} = \begin{bmatrix} 3 & -1 & -2 \\ 5 & 0 & -5 \end{bmatrix}$$

TI-82

[MATRX]
Cursor to EDIT, [1]
2 [ENTER] (# of rows)
3 [ENTER] (# of cols)
Input 1st matrix in [A].
[A] 2×3
1,1=2 2,1=3
1,2=−1 2,2=2
1,3=0 2,3=−1
[MATRX]
Cursor to EDIT, [2]
2 [ENTER] (# of rows)
3 [ENTER] (# of cols)
Input 2nd matrix in [B].
[B] 2×3
1,1=1 2,1=2
1,2=0 2,2=−2
1,3=−2 2,3=−4
[2nd] [QUIT]
[MATRX] [1] + [MATRX] [2]
[ENTER]
Display will be
[[3 −1 −2]
[5 0 −5]]

Casio *fx-9700GE*

In MAIN MENU, enter 5
[F4] (LIST)
Cursor to Mat A.
[F2] (DIM) 2 [EXE] 3
[EXIT] [F1] (EDIT)
Input 1st matrix in [A].

A	1	2	3
1	2	−1	0
2	3	2	−1

[EXIT] Cursor to Mat B.
[F2] (DIM) 2 [EXE] 3
[EXIT] [F1] (EDIT)
Input 2nd matrix in [B].

B	1	2	3
1	1	0	−2
2	2	−2	−4

[EXIT] [EXIT]
[F1] (Mat) [ALPHA] [A] [+]
[F1] (Mat) [ALPHA] [B] [EXE]
Display will be

Ans	1	2	3
1	3	−1	−2
2	5	0	−5

Sharp *EL-9300C*

[⊞⊟⊠⊡] [MENU] [3] (MATRIX)
[MENU] Cursor to C DIM.
[ENTER] [ENTER]
2 [ENTER] 3 [ENTER]
Input 1st matrix in [A].
A[1,1]=2
A[2,1]=3
A[1,2]=−1
A[2,2]=2
A[1,3]=0
A[2,3]=−1
[MENU] [ENTER] (C DIM)
Cursor to matrix B. [ENTER]
2 [ENTER] 3 [ENTER]
Input 2nd matrix in [B].
B[1,1]=1
B[2,1]=2
B[1,2]=0
B[2,2]=−2
B[1,3]=−2
B[2,3]=−4
[QUIT]
[2nd F] [MAT] [A] [+]
[2nd F] [MAT] [B] [ENTER]
Display will be
Ans [1,1]=3
Ans [2,1]=5
Move cursor right to view other entries.

Exercises

Using a graphing calculator to find the sum and difference of the matrices.

1. $\begin{bmatrix} 1 & 2 & 5 \\ -2 & 0 & -3 \end{bmatrix}, \begin{bmatrix} 0 & -4 & -2 \\ 2 & 0 & 3 \end{bmatrix}$

2. $\begin{bmatrix} -3 & 4 & 2 \\ 6 & 1 & -4 \\ 0 & -2 & -5 \end{bmatrix}, \begin{bmatrix} 2 & 0 & -1 \\ -4 & 5 & 0 \\ 0 & -3 & 9 \end{bmatrix}$

TECHNOLOGY
Keystrokes for Other Graphing Calculators

Keystrokes for page 166

A graphing calculator can be used to draw a scatter plot. Here we show how to use a TI-82, *a Casio* fx-9700GE, *and a Sharp* EL-9300C *to draw a scatter plot for the following points.*

(1, 1.2), (1.5, 2.3), (2, 1.9), (2.5, 2.4), (3, 3.1)
(3.5, 3.5), (4, 3.8), (4.5, 4.7), (5, 4.9), (5.5, 5.7)

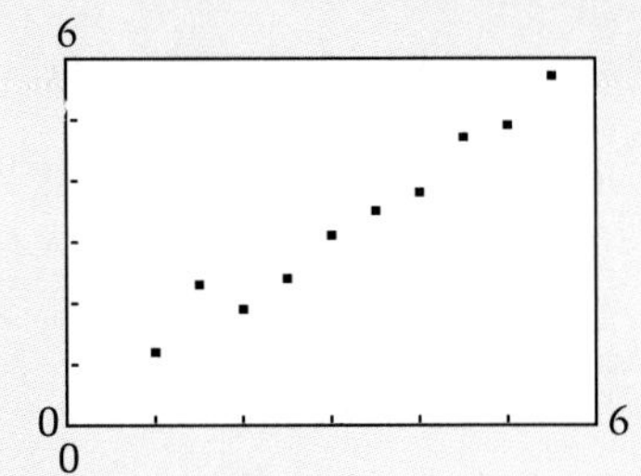

TI-82

WINDOW ▽ (Set window.)
Xmin = 0 Ymin = 0
Xmax = 6 Ymax = 6
Xscl = 1 Yscl = 1
Y= CLEAR
(Use arrow keys to clear other existing equations.)
STAT 4 (ClrList)
2nd L1 , 2nd L2
ENTER
STAT 1 (Edit)
Enter coordinates.

L1(1) = 1	L2(1) = 1.2
L1(2) = 1.5	L2(2) = 2.3
L1(3) = 2	L2(3) = 1.9
L1(4) = 2.5	L2(4) = 2.4
L1(5) = 3	L2(5) = 3.1
L1(6) = 3.5	L2(6) = 3.5
L1(7) = 4	L2(7) = 3.8
L1(8) = 4.5	L2(8) = 4.7
L1(9) = 5	L2(9) = 4.9
L1(10) = 5.5	L2(10) = 5.7

2nd STAT PLOT ENTER
Choose:
On, (scatter plot icon), Xlist:L1
Ylist:L2, Mark: □
GRAPH

Casio *fx-9700GE*

In MAIN MENU, enter 4.
Range (Set range.)
Xmin = 0 Ymin = 0
max = 6 max = 6
scale = 1 scale = 1
EXIT
SHIFT F5 (Cls) EXE
SHIFT CLR F2 EXE
SHIFT SET UP
Choose:
GRAPH TYPE :RECT
DRAW TYPE :PLOT
STAT DATA :STO
STAT GRAPH :DRAW
REG MODEL :LIN
M-DISP/COPY:M-DISP
EXIT F2 F3 F1
Enter coordinates.
1 F3 1.2 F1
1.5 F3 2.3 F1
2 F3 1.9 F1
2.5 F3 2.4 F1
3 F3 3.1 F1
3.5 F3 3.5 F1
4 F3 3.8 F1
4.5 F3 4.7 F1
5 F3 4.9 F1
5.5 F3 5.7 F1

Sharp *EL-9300C*

(menu key) MENU
D 2
ENTER 3
Enter coordinates.

X1 = 1	Y1 = 1.2
X2 = 1.5	Y2 = 2.3
X3 = 2	Y3 = 1.9
X4 = 2.5	Y4 = 2.4
X5 = 3	Y5 = 3.1
X6 = 3.5	Y6 = 3.5
X7 = 4	Y7 = 3.8
X8 = 4.5	Y8 = 4.7
X9 = 5	Y9 = 4.9
X10 = 5.5	Y10 = 5.7

2nd F (scatter plot key)
RANGE (Set range.)
Xdlt = 1
n = 6
Xmin = 0
Xmax = 6
Xscl = 1
Ymin = 0
Ymax = 6
Yscl = 1
QUIT
E ENTER

Exercises

Use a graphing calculator to draw a scatter plot of the data.

1. (1, 5.1), (1, 4.8), (1, 4.7), (2, 4.3), (3, 3.8), (3, 3.2), (4, 2.8), (4, 2.4), (5, 1.8), (5, 1.2)

2. (1, 2.2), (1, 2.4), (1, 3.3), (2, 3.5), (3, 4.1), (3, 4.6), (4, 5.1), (4, 5.4), (5, 5.1), (5, 5.8)

Keystrokes for page 189

A graphing calculator can be used to graph an equation. Here we show how to use a TI-82, *a Casio* fx-9700GE, *and a Sharp* EL-9300C *to sketch the graph of* $x - 2y = 2$.

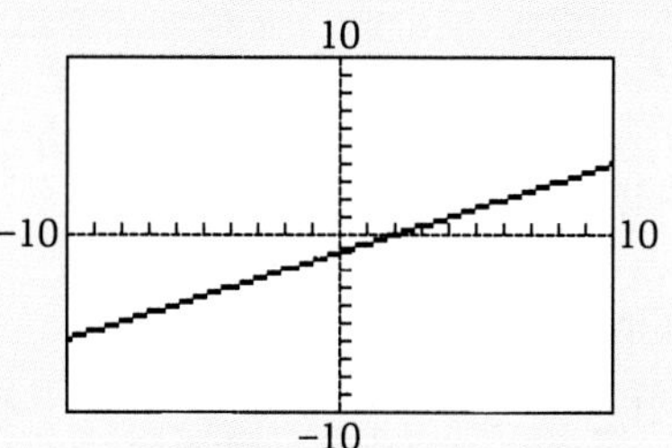

The first step is to solve for y: $y = \frac{1}{2}x - 1$

Using a graphing calculator is fairly easy, but there are four ideas you must remember.

- Solve the equation for y in terms of x.
- Set the range by entering the least and greatest x and y values and the scale (units per mark).
- Use parentheses if you are unsure of the calculator's order of operations.

TI-82

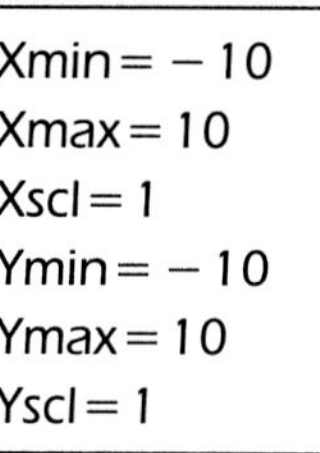

Xmin = −10
Xmax = 10
Xscl = 1
Ymin = −10
Ymax = 10
Yscl = 1

[WINDOW] [▽] (Set window as shown.)
[Y=] [CLEAR]
(Use arrow keys to clear other existing equations.)
[Y=] [(] 1 [÷] 2 [)] [X,T,θ] [−] 1
[GRAPH]
[CLEAR] (Clear screen.)

Sharp *EL-9300C*

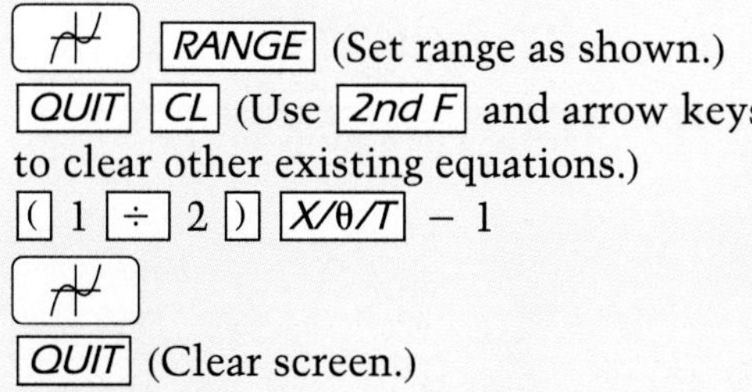

[graph key] [RANGE] (Set range as shown.)
[QUIT] [CL] (Use [2nd F] and arrow keys to clear other existing equations.)
[(] 1 [÷] 2 [)] [X/θ/T] − 1
[graph key]
[QUIT] (Clear screen.)

Casio *fx-9700GE*

In MAIN MENU, enter 1.
[Range] (Set range as shown.)
[EXIT] [SHIFT] [F5] (Cls) [EXE]
[Graph] [(] 1 [÷] 2 [)] [X,θ,T] [−] 1
[EXE]
[SHIFT] [F5] (Cls) [EXE] (Clear sketch.)

Exercises

In Exercises 1–8, use a graphing calculator to sketch the graphs. (For 1–4, use the range shown above. For 5–8, use the indicated range.)

1. $y = -2x - 3$

2. $y = 2x + 2$

3. $x + 2y = -1$

4. $x - 3y = 3$

5. $y = x + 25$
RANGE
Xmin = −10
Xmax = 10
Xscl = 1
Ymin = −5
Ymax = 35
Yscl = 5

6. $y = -x + 25$
RANGE
Xmin = −10
Xmax = 10
Xscl = 1
Ymin = −5
Ymax = 35
Yscl = 5

7. $y = 0.1x$
RANGE
Xmin = −10
Xmax = 10
Xscl = 1
Ymin = −1
Ymax = 1
Yscl = 1

8. $y = 100x + 2500$
RANGE
Xmin = 0
Xmax = 100
Xscl = 10
Ymin = 0
Ymax = 15000
Yscl = 1000

TECHNOLOGY
Keystrokes for Other Graphing Calculators

Keystrokes for page 262

In Lesson 5.4, you studied the graphical approach for approximating a best-fitting line. Another approach is to use a graphing calculator to find the best-fitting line.

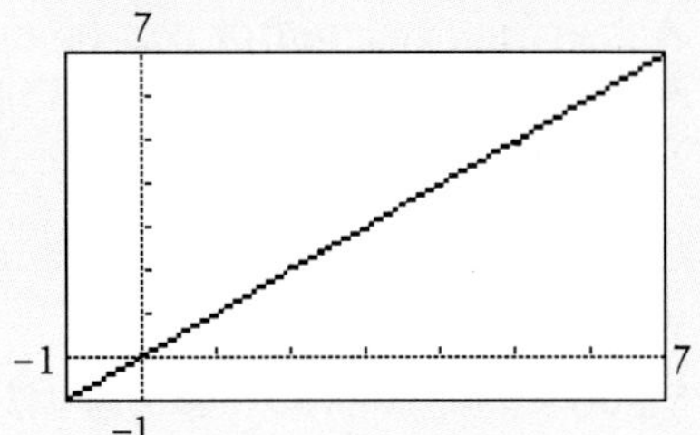

In Chapter 3, we showed how to use a graphing calculator to draw a scatter plot for the points listed below. To find the equation of the line that best fits these points, draw the scatter plot, then enter the given keystrokes. Each graphing calculator finds the coefficients a and b for the best-fitting line $y = ax + b$. Then each calculator sketches the line as shown at the right.

(1, 1.2), (1.5, 2.3), (2, 1.9), (2.5, 2.4), (3, 3.1)
(3.5, 3.5), (4, 3.8), (4.5, 4.7), (5, 4.9), (5.5, 5.7)

TI-82

Enter data.
(See p. 736.)
[STAT] Cursor to CALC.
[5] [LinReg(ax+b)]
[2nd] [L1] [,] [2nd] [L2]
[ENTER]
a=.936969697
b=.3048484848
[Y=] [CLEAR]
[VARS] [5] Cursor to EQ.
[7] (RegEQ)
[GRAPH]

Casio *fx-9700GE*

Enter data.
(See p. 736.)
[G↔T] [F6]
[F1] [EXE]
A=0.304848484848
[F2] [EXE]
B=0.93696969697
[Graph] [F1] [+] [F2] [X,θ,T]
[EXE]

Note: The line is in the form $y = a + bx$.

Sharp *EL-9300C*

Enter data.
(See p. 736.)
[2nd F] [ ] [ALPHA] [F] [1]
[MENU] [ENTER]
a=0.304848484
b=0.936969697

Note: The line is in the form $y = a + bx$.

Exercises

In Exercises 1 and 2, find the best-fitting line for the points. Then sketch the scatter plot and the line.

1. (0, 1), (1, 2)
(1, 3), (2, 3)
(2, 3.5), (3, 4)
(3, 4.5), (4, 5.5)
(4, 6), (5, 5)
(5, 6), (5, 6.5)
(6, 7), (6, 8)
(7, 7.5)

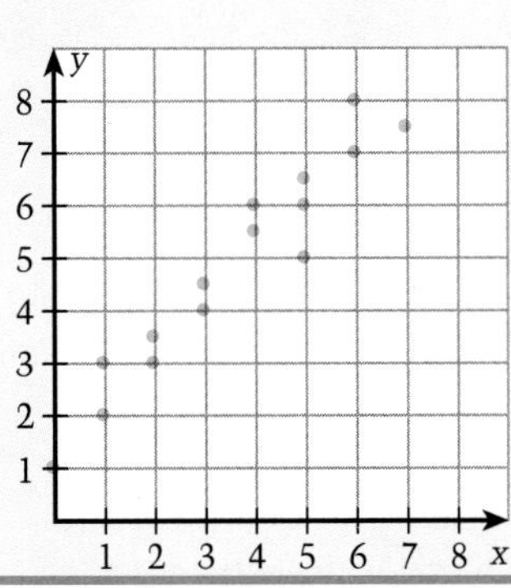

2. (0, 8), (1, 7.5)
(1, 6), (2, 6.5)
(2, 6), (3, 5.5)
(3, 5), (4, 4)
(4, 3.5), (5, 3)
(5, 2.5), (6, 2)
(6, 1.5), (7, 1)
(7, 0)

Keystrokes for page 311

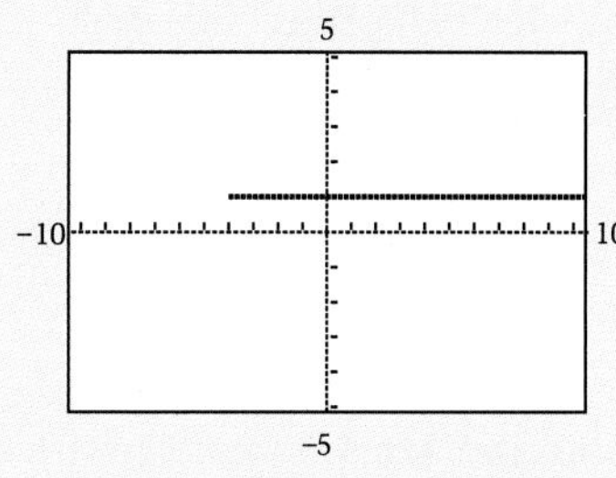

Some graphing calculators can be used to sketch the solution of an inequality in one variable. The following shows how to sketch a number-line graph using a Texas Instruments TI-82. *The inequality is*

$-5 < 2x + 3.$

The viewing window for the *TI-82* is shown at the right. The graph of the solution appears one unit above the horizontal number line. Thus, from the display, it appears that the solution of the inequality is

$x > -4.$

You can check this result algebraically.

TI-82

[MODE]
Cursor to DOT, [ENTER]
[WINDOW] [▽] (Set window.)
  Xmin = −10
  Xmax = 10
  Xscl = 1
  Ymin = −5
  Ymax = 5
  Yscl = 1
[Y=] [CLEAR] (Use arrow keys to clear other existing equations.)
[(−)] 5 [2nd] [TEST] [5] 2 [X,T,θ] [+] 3
  $Y_1 = -5 < 2X + 3$
[GRAPH]
[CLEAR] (Clear screen.)

Exercises

Use a graphing calculator to sketch the graph of the inequality. State the solution shown by the graph. Then check the solution algebraically.

1. $-2x + 3 \geq 7$ **2.** $-3 < -x + 5$ **3.** $-3x + 2 \leq 4$

4. $5x + 1 < 4x + 5$ **5.** $-2x > -3x - 1$ **6.** $-x - 2 \leq 3x + 6$

Keystrokes for Other Graphing Calculators

Keystrokes for page 326

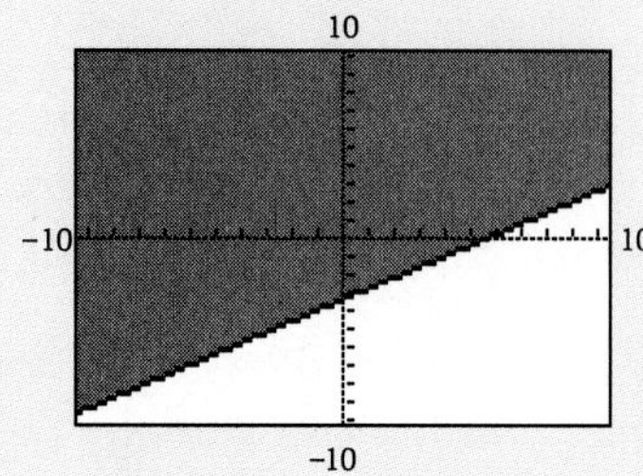

A graphing calculator can be used to sketch the graph of an inequality in x and y. Here we show how to use a Texas Instruments TI-82 *and a Casio* fx-9700GE *to sketch the graph of*

$$x - 2y \le 6.$$

The first step is to write the equivalent inequality that has y isolated on the left. For instance, $x - 2y \le 6$ must first be written as

$$y \ge \frac{1}{2}x - 3.$$

TI-82

[WINDOW] [▽] (Set window.)
Xmin = −10
Xmax = 10
Xscl = 1
Ymin = −10
Ymax = 10
Yscl = 1
[Y=] [CLEAR] (Use arrow keys to clear other existing equations.)
[2nd] [DRAW] [7] (Shade)
.5 [X,T,θ] [−] 3 [,] 10 [)]
Shade(.5X−3,10)
[ENTER]
[2nd] [DRAW] [1] (Clear sketch.)

Casio *fx-9700GE*

In MAIN MENU, enter 1.
[Range] (Set range.)
Xmin = −10
max = 10
scl = 1
Ymin = −10
max = 10
scl = 1
[EXIT]
[SHIFT] [SET UP] [F4] (INEQ) [EXIT]
[Graph] [F3] .5 [X,θ,T] [−] 3
Graph Y≥.5X−3
[EXE]
[SHIFT] [F5] [EXE] (Clear sketch.)

Exercises

In Exercises 1–8, use a graphing calculator to sketch the graph of the inequality. (For 1–4, use the range shown above. For 5–8, use the indicated range.)

1. $y < -2x - 3$ **2.** $y > 2x + 2$ **3.** $x + 2y \le -1$ **4.** $x - 3y \ge 3$

5. $y < x + 25$ **6.** $y > -x + 25$ **7.** $y \le 0.1x$ **8.** $y \ge 100x + 2500$

5.	6.	7.	8.
RANGE	RANGE	RANGE	RANGE
Xmin = −10	Xmin = −10	Xmin = −10	Xmin = 0
Xmax = 10	Xmax = 10	Xmax = 10	Xmax = 100
Xscl = 1	Xscl = 1	Xscl = 1	Xscl = 10
Ymin = −5	Ymin = −5	Ymin = −1	Ymin = 0
Ymax = 35	Ymax = 35	Ymax = 1	Ymax = 15000
Yscl = 5	Yscl = 5	Yscl = .1	Yscl = 1000

Keystrokes for page 352

A graphing calculator can be used to graph a system of linear equations. Here we show how to use a Texas Instruments TI-82, *a Casio* fx-9700GE, *and a Sharp* EL-9300C *to graph the following system.*

$$\begin{cases} 2x + 3y = 7 & \textbf{\textit{Equation 1}} \\ 4x - 5y = 3 & \textbf{\textit{Equation 2}} \end{cases}$$

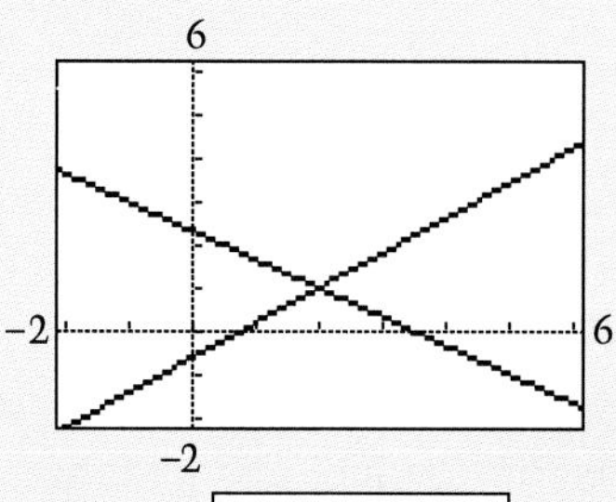

Xmin = −2
Xmax = 6
Xscl = 1
Ymin = −2
Ymax = 6
Yscl = 1

The display at the right shows the graph of the system. Note that the two lines appear to intersect at the point (2, 1). To check this, substitute 2 for x and 1 for y in each equation.

The first step is to solve each equation for y.

Equation 1: $y = -\frac{2}{3}x + \frac{7}{3}$ Equation 2: $y = \frac{4}{5}x - \frac{3}{5}$

TI-82

[WINDOW] [▽] (Set window as shown.)
[Y=] [(] [(−)] 2 [÷] 3 [)] [X,T,θ]
[+] [(] 7 [÷] 3 [)] [ENTER]
[(] 4 [÷] 5 [)] [X,T,θ]
[−] [(] 3 [÷] 5 [)]
[GRAPH]
[CLEAR] (Clear screen.)

Casio *fx-9700GE*

In MAIN MENU, enter 1.
[Range] (Set range as shown.)
[EXIT]
[SHIFT] [F5] (Cls) [EXE]
[SHIFT] [SET UP] [F1] (RECT) [EXIT]
[Graph] [(] [(−)] 2 [÷] 3 [)] [X,θ,T]
[+] [(] 7 [÷] 3 [)] [EXE]
[Graph] [(] 4 [÷] 5 [)] [X,θ,T]
[−] [(] 3 [÷] 5 [)] [EXE]
[SHIFT] [F5] [EXE] (Clear screen.)

Sharp *EL-9300C*

[graph key] [RANGE] (Set range as shown.)
[graph key] [MENU] [▷] [1]
[(] [(−)] 2 [÷] 3 [)] [X/θ/T]
[+] [(] 7 [÷] 3 [)] [ENTER]
[(] 4 [÷] 5 [)] [X/θ/T]
[−] [(] 3 [÷] 5 [)]
[graph key]
[QUIT] (Clear screen.)

Exercises

Use a graphing calculator to solve the linear system. (Use an appropriate range setting and check your result in each of the original equations.)

1. $\begin{cases} 2x - 5y = -6 \\ 3x - 2y = 2 \end{cases}$

2. $\begin{cases} 3x + y = -2 \\ 4x + y = -4 \end{cases}$

3. $\begin{cases} -0.25x - y = 2.25 \\ -1.25x + 1.25y = -1.25 \end{cases}$

4. $\begin{cases} 0.8x + 0.6y = -1.2 \\ 1.25x - 1.5y = 12.75 \end{cases}$

Keystrokes for Other Graphing Calculators

Keystrokes for page 438

The graphing calculator feature in Chapter 5 described how to use a graphing calculator to find a best-fitting line. Here we show how to use a graphing calculator to find a best-fitting exponential growth or decay model. The calculator will find two numbers, a and b, for the exponential growth model $y = a(b^x)$. The best-fitting exponential growth model for the following points is $y = 0.931(1.54)^x$. The graph of the model and the scatter plot of the points are shown.

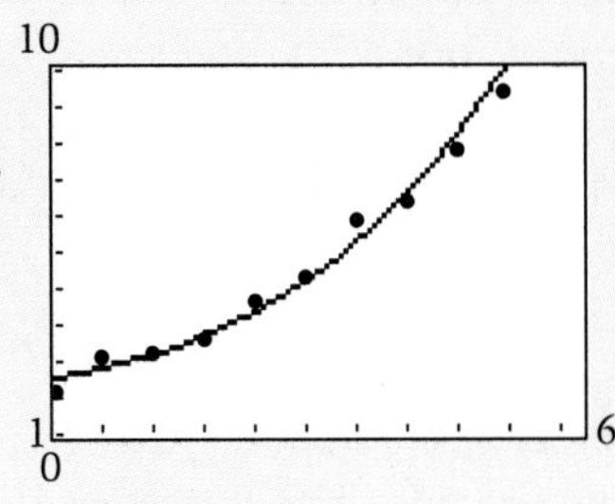

(1,1.2), (1.5, 2.1), (2, 2.2), (2.5, 2.6), (3, 3.7)
(3.5, 4.3), (4, 5.9), (4.5, 6.4), (5, 7.8), (5.5, 9.5)

TI-82

Enter data. (See p. 736.)

[STAT] [▷] [ALPHA] [A]
[2nd] [L1] [,] [2nd] [L2]
[ENTER]

y=a*b^x
a=.9311818813
b=1.542462151

[Y=] [CLEAR]

(Use arrow keys to clear other existing equations.)

[VARS] [5]

Cursor to EQ. [7]

[2nd] [STAT PLOT] [ENTER]

Choose:

On, [scatter plot icon], Xlist:L1
Ylist:L2, Mark: □

[GRAPH]

Casio *fx-9700GE*

In MAIN MENU, enter 4.

[SHIFT] [F5] [EXE]
[SHIFT] [CLR] [F2] [EXE]
[SHIFT] [SET UP]

Choose the following:

GRAPH TYPE :RECT
DRAW TYPE :PLOT
STAT DATA :STO
STAT GRAPH :DRAW
REG MODEL :EXP
M-DISP/COPY:M-DISP

[EXIT]

Enter data. (See p. 736.)

[G↔T] [F6] [F1] [EXE]

A=0.931181881305

[F2] [EXE]

B=0.433379939028

[Graph]
[F1] [SHIFT] [e^x] [(] [F2]
[X,θ,T] [)] [EXE]

Sharp *EL-9300C*

Enter data. (See p. 736.)

[2nd F] [stat icon]
[ALPHA] [E] [ENTER]
[2nd F] [stat icon]
[ALPHA] [F] [2]
[MENU] [ENTER]

a=0.931181881
b=0.433379939

Note: The Casio and Sharp calculators return an exponential function of the form $y = ae^{bx}$.

Exercises

Find the best-fitting exponential growth model for the points.

1. (0, 1), (1, 1.4), (2, 3), (3, 5), (4, 8) (5, 12), (6, 20), (7, 30), (8, 50), (9, 80)

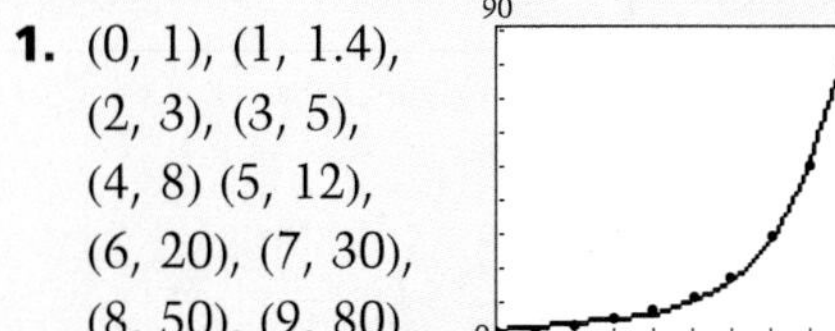

2. (0, 0.5), (1, 0.6), (2, 0.8), (3, 1.0), (4, 1.4) (5, 1.8), (6, 2.7), (7, 3.6), (8, 4.9), (9, 7.0)

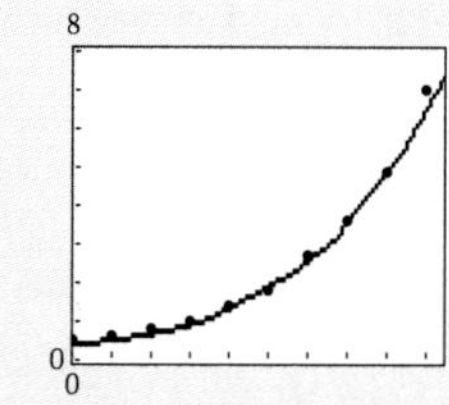

TECHNOLOGY
Keystrokes for Other Graphing Calculators

Keystrokes for page 548

Graphing calculators can be used to obtain a graphic interpretation of polynomial addition or subtraction. In the first graph below, notice that for any x-value, the sum of y_1 and y_2 is equal to y_3. In the second graph, for any x-value, the difference of y_1 and y_2 is equal to y_3.

$y_1 = \frac{1}{2}x,\quad y_2 = x + 1,\quad y_3 = \frac{3}{2}x + 1$

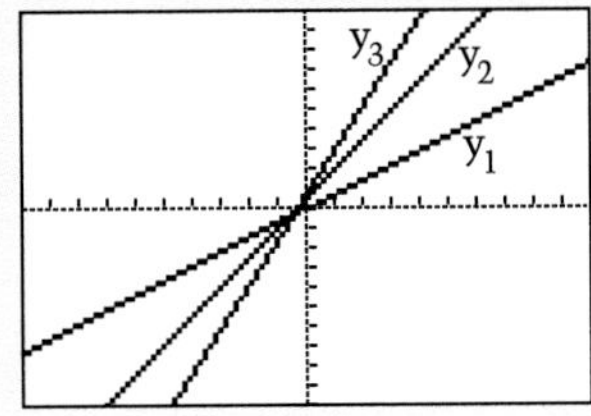

y_3 is the sum of y_1 and y_2.

$y_1 = \frac{1}{4}x^2 + 4,\quad y_2 = 3,\quad y_3 = \frac{1}{4}x^2 + 1$

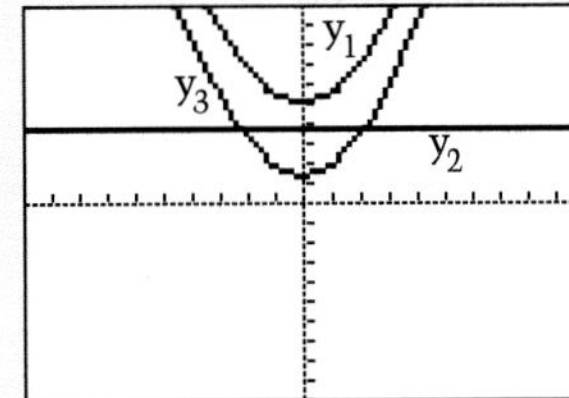

y_3 is the difference of y_1 and y_2.

Xmin = −10
Xmax = 10
Xscl = 1
Ymin = −10
Ymax = 10
Yscl = 1

TI-82

[WINDOW] [▽]
(Set window as shown.)
[Y=] [CLEAR]
(Use arrow keys to clear other existing equations.)
[Y=] .5 [X,T,θ] [ENTER]
[X,T,θ] [+] 1 [ENTER]
1.5 [X,T,θ] [+] 1
[GRAPH]
[CLEAR] (Clear screen.)

Casio *fx-9700GE*

In MAIN MENU, enter 1.
[Range] (Set range as shown.)
[EXIT]
[SHIFT] [F5] [EXE]
[SHIFT] [SET UP] [F1] [EXIT]
[Graph] .5 [X,θ,T] [EXE]
[Graph] [X, θ,T] [+] 1 [EXE]
[Graph] 1.5 [X,θ,T] [+] 1 [EXE]
[SHIFT] [F5] [EXE]
(Clear screen.)

Sharp *EL-9300C*

[graph key] [CL]
(Use [CL] and [2nd F] [△] to clear other existing equations.)
[RANGE]
(Set range as shown.)
[QUIT]
.5 [X/θ/T] [ENTER]
[X/θ/T] [+] 1 [ENTER]
1.5 [X/θ/T] [+] 1
[graph key] [QUIT] (Clear screen.)

Exercises

Let y_3 be the sum of y_1 and y_2. Sketch the graph of all three equations.

1. $y_1 = x - 1, y_2 = \frac{1}{2}x + 2$ **2.** $y_1 = x^2 + x, y_2 = 1$ **3.** $y_1 = \frac{1}{2}x^2 - 1, y_2 = \frac{1}{2}x^2 + 1$

Let y_3 be the difference of y_1 and y_2. Sketch the graph of all three equations.

4. $y_1 = 2x + 1, y_2 = x + 1$ **5.** $y_1 = x^2 + 3, y_2 = \frac{1}{2}x^2$ **6.** $y_1 = \frac{1}{3}x^2 + x, y_2 = 1$

TECHNOLOGY
Keystrokes for Other Graphing Calculators

Keystrokes for page 606

Graphs of rational expressions often have ***asymptotes****—lines that resemble the graph in certain regions of the coordinate plane. For instance, consider the graphs of*

$$y_1 = \frac{2x}{x+5} \quad \text{and} \quad y_2 = 2.$$

Using long division, you can write

$$\underbrace{\frac{2x}{x+5}}_{y_1} = \underbrace{2}_{y_2} - \underbrace{\frac{10}{x+5}}_{\text{Remainder fraction}}.$$

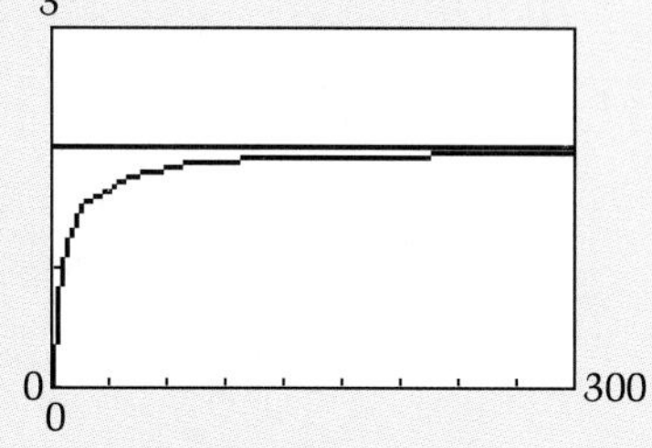

The value of the "remainder fraction" becomes very small as the values of x become large. This is why we say that the graph of y_1 resembles the line $y_2 = 2$ when x is large. A graphing calculator can show how close together two graphs can be.

Xmin = 0
Xmax = 300
Xscl = 30
Ymin = 0
Ymax = 3
Yscl = 1

TI-82

[WINDOW] [▽]
(Set window as shown.)
[Y=] [CLEAR] (Use arrow keys to clear other existing equations.)
2 [X,T,θ] [÷] [(] [X,T,θ] [+] 5 [)] [ENTER]
2 [ENTER] [GRAPH]
[CLEAR] (Clear screen.)

Casio *fx-9700GE*

In MAIN MENU, enter 1.
[Range] (Set range as shown.)
[EXIT] [SHIFT] [F5] [EXE]
[SHIFT] [SET UP] [F1] [EXIT]
[Graph]
2 [X,θ,T] [÷] [(] [X,θ,T] [+] 5 [)] [EXE]
[Graph] 2 [EXE]
[SHIFT] [F5] (Cls) [EXE] (Clear screen.)

Sharp *EL-9300C*

[graph key] [CL]
(Use [CL] and [2nd F] [△] to clear other existing functions.)
[RANGE] (Set range as shown.) [QUIT]

2 [X/θ/T] [÷] [(] [X/θ/T] [+] 5 [)] [ENTER]
2 [ENTER]
[graph key]
[QUIT] (Clear screen.)

Exercises

In Exercises 1 and 2, sketch the graphs of y_1 and y_2. Experiment to find a range setting in which the two graphs resemble each other.

1. $\underbrace{\frac{3x+10}{x+4}}_{y_1} = \underbrace{3}_{y_2} - \underbrace{\frac{2}{x+4}}_{\text{Remainder fraction}}$

2. $\underbrace{\frac{x^2}{x+1}}_{y_1} = \underbrace{x-1}_{y_2} + \underbrace{\frac{1}{x+1}}_{\text{Remainder fraction}}$

TECHNOLOGY

Keystrokes for Other Graphing Calculators

Keystrokes for page 664

Graphing calculators can be used to perform statistical computations such as drawing a bar graph or ***histogram*** *of a collection of numbers. The histogram at the right represents how many times each number appears in the following data.*

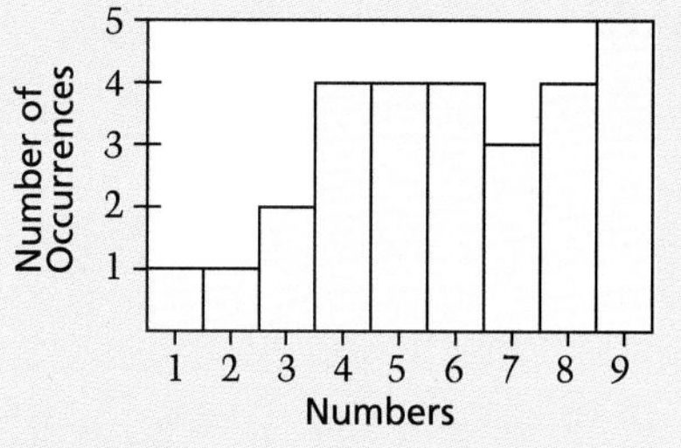

6, 1, 6, 4, 4, 7, 9, 4, 5, 7, 9, 5, 6, 2,
6, 9, 7, 5, 9, 3, 8, 5, 8, 4, 8, 3, 8, 9

Begin by setting the window or range as shown at the right.

Xmin = 1	Ymin = 0
Xmax = 10	Ymax = 6
Xscl = 1	Yscl = 1

TI-82

[Y=] [CLEAR] (Use arrow keys to clear other existing equations.)
[2nd] [Y-VARS] [5] [2] [ENTER]
[STAT] [4] [2nd] [L1] [,] [2nd] [L2] [ENTER]
[2nd] [STAT PLOT] [4] [ENTER]
[STAT] [ENTER] (Edit) Enter data.

L1(1) = 1	L2(1) = 1	L1(6) = 6	L2(6) = 4
L1(2) = 2	L2(2) = 1	L1(7) = 7	L2(7) = 3
L1(3) = 3	L2(3) = 2	L1(8) = 8	L2(8) = 4
L1(4) = 4	L2(4) = 4	L1(9) = 9	L2(9) = 5
L1(5) = 5	L2(5) = 4		

[2nd] [STAT PLOT] [ENTER] (Plot1)
Choose the following:
On, Type: , Xlist: L1, Freq: L2
[GRAPH]

Casio *fx-9700GE*

In MAIN MENU, enter 3.
[SHIFT] [CLR] [F2] (Scl) [EXE]
[SHIFT] [F5] (Cls) [EXE]
[SHIFT] [SET UP]

Choose: GRAPH TYPE :RECT
DRAW TYPE :PLOT
STAT DATA :STO
STAT GRAPH :DRAW
M-DISP/COPY :M-DISP

[EXIT] [F2] [F3] [F1]
[SHIFT] [Defm] 9 [EXE] Enter data.

1 [F3] 1 [F1]	6 [F3] 4 [F1]
2 [F3] 1 [F1]	7 [F3] 3 [F1]
3 [F3] 2 [F1]	8 [F3] 4 [F1]
4 [F3] 4 [F1]	9 [F3] 5 [F1]
5 [F3] 4 [F1]	

[Graph] [EXE]

Sharp *EL-9300C*

[MENU]
[ALPHA] [D] [2]
[ENTER] [2] Enter data.

X1 = 1	W1 = 1	X6 = 6	W6 = 4
X2 = 2	W2 = 1	X7 = 7	W7 = 3
X3 = 3	W3 = 2	X8 = 8	W8 = 4
X4 = 4	W4 = 4	X9 = 9	W9 = 5
X5 = 5	W5 = 4		

[2nd F] [ENTER]
(Note: Enter Xdlt = 1 and n = 10 in STAT RANGE.)

Exercises

In Exercises 1 and 2, construct a histogram for the data.

1. 10, 14, 11, 18, 14, 16, 17, 17, 10, 11, 19, 15, 13, 17, 11, 13, 19, 18, 12, 14, 19, 10, 12, 15, 17, 15, 15, 12, 18, 16, 16, 10, 19

2. 26, 29, 35, 34, 33, 30, 31, 21, 30, 30, 32, 27, 28, 28, 34, 33, 33, 28, 28, 31, 31, 32, 30, 31, 33, 27, 26, 34, 33, 33, 30, 29, 31

Keystrokes for page 701

The domain of the function $f(x) = \sqrt{x}$ is the set of nonnegative numbers. Some graphing utilities will display an error message when asked to sketch a graph over an interval that contains x-values that are not in the domain. The TI-82, *the Casio* fx-9700GE *and the Sharp* EL-9300C, *however, just ignore such values and draw the graph for values of x that are in the domain. Because of this feature, you can use any of these graphing calculators to explore the domain of a radical function. For instance, the display at the right shows the graph of $f(x) = \sqrt{x}$. Notice that the graph begins at the origin and continues to the right. To the left of the origin the graph does not exist because negative x-values are not in the domain of f.*

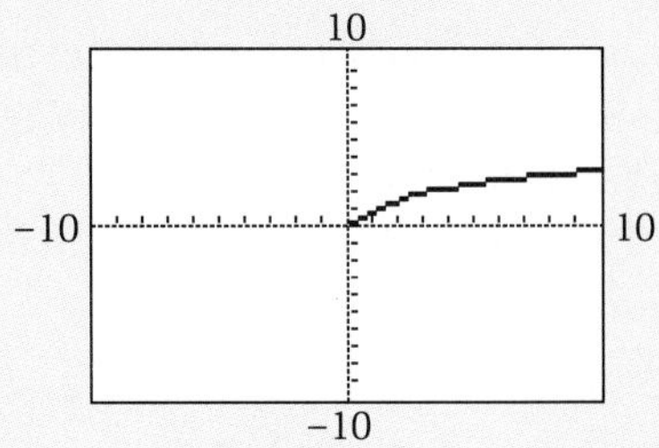

Graph of $f(x) = \sqrt{x}$

Before sketching the graph, set the window or range as shown at the right.

Xmin = −10
Xmax = 10
Xscl = 1
Ymin = −10
Ymax = 10
Yscl = 1

TI-82

[Y=] [CLEAR]
(Use arrow keys to clear other existing functions.)
[Y=] [2nd] [√] [X, T, θ]
[GRAPH]
[CLEAR] (Clear screen.)

Casio *fx-9700GE*

In MAIN MENU, enter 1.
[SHIFT] [F5] [EXE]
[Graph] [SHIFT] [√] [X,θ,T]
[EXE]
[SHIFT] [F5] [EXE]

Sharp *EL-9300C*

[graph key] [CL] (Use [2nd F] [△] to clear other existing functions.)
[MENU] [1]
[√] [X/θ/T]
[graph key]
[QUIT]

Exercises

In Exercises 1–4, use a graphing calculator to sketch the graph of the function. From the graph, describe the domain of the function.

1. $f(x) = \sqrt{2x}$ **2.** $f(x) = \sqrt{x - 1}$ **3.** $f(x) = \sqrt{x + 2}$ **4.** $f(x) = \sqrt{2x - 6}$

In Exercises 5–8, use a graphing calculator to sketch the graph of both functions. Explain how the graph of *g* is related to the graph of *f*.

5. $f(x) = \sqrt{x}$, $g(x) = \sqrt{x - 3}$ **6.** $f(x) \sqrt{x}$, $g(x) = \sqrt{x} - 3$

7. $f(x) = \sqrt{x}$, $g(x) = -\sqrt{x}$ **8.** $f(x) = \sqrt{x}$, $g(x) = \sqrt{x + 4}$

Use with Lesson 9.4, pp. 472–477

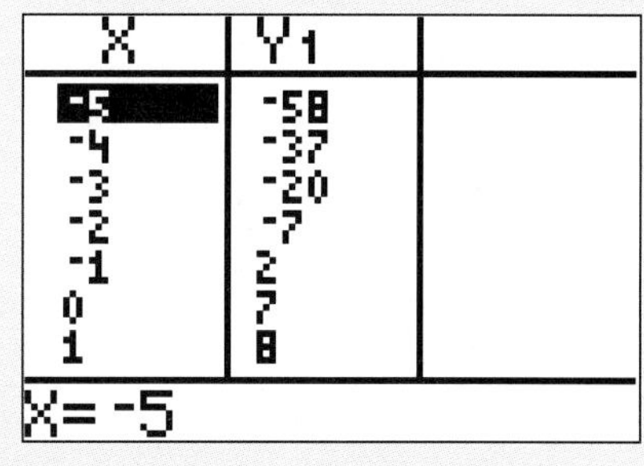

Some graphing calculators can be used to solve equations with tables. The following shows how to use the table features of a Texas Instruments TI-82 *and a Casio* fx-9700GE *to solve the equation*

$$-2x^2 + 3x + 7 = 0.$$

TI-82

[Y=] [CLEAR]
[(−)] 2 [X,T,θ] [x^2] [+] 3 [X,T,θ] [+] 7
[2nd] [TblSet]
TblMin = −5
△Tbl = 1
Indpnt: Auto
Depend: Auto
[2nd] [TABLE]

A possible table is shown above. Use [△] and [▽] to scroll through the table. A solution occurs when the y-value is 0. From the table, it appears that one solution occurs when x is between −2 and −1. Alter the table by changing TblMin to −2 and △Tbl to 0.1. Continue to alter TblMin and △Tbl until you can approximate the solution $x = -1.266$ (accurate to three decimal places).

Casio *fx-9700GE*

In MAIN MENU, enter 8.
[F1] [F1] [F1] (Clears any functions.)
[(−)] 2 [X,θ,T] [x^2] [+] 3 [X,θ,T] [+] 7
[F2] (RANGE)
Start : −5
End : 10
Pitch : 1
[F1]

A possible table is shown above. Use [△] and [▽] to scroll through the table. A solution occurs when the y-value is 0. From the table, it appears that one solution occurs when x is between −2 and −1. Alter the table by changing the Start to −2, End to −1, and Pitch to 0.1. (To do so, enter [F2] [F2].) Continue to alter the Start, End, and Pitch until you can approximate the solution $x = -1.266$ (accurate to three decimal places).

Exercises

1. Find the second solution to the equation $-2x^2 + 3x + 7 = 0$.

In Exercises 2–4, use a table to find the solutions to the equation. Each equation has two solutions. Your solution should be accurate to three decimal places.

2. $3x^2 + 4x - 2$

3. $-5x^2 - 7x + 10$

4. $0.5x^2 - 2x - 1$

ADDITIONAL ANSWERS

CHAPTER 1

Lesson 1.1

A. $[400 \div 8] - 3 = 47$, $[420 \div 7] - 3 = 57$, $[420 \div 6] - 3 = 67$

B. $0.4 \cdot 1 \cdot 20 = 8$, $0.5 \cdot 1 \cdot 17 = 8.5$, $0.5 \cdot 1 \cdot 18 = 9$

Lesson 1.5

33. 20: number of ft^2 needed for each station,
x: number of computers,
400: total number of ft^2 available.

34. 8: number of completed screens required to get 1 energy bar,
x: number of energy bars you will get,
96: number of completed screens.

35. 18: dollar cost of a tank of gas,
x: number of times you can fill the tank,
y: dollar amount left,
65: dollar amount available.

36. 5: dollar amount saved each week,
n: number of weeks you must save,
m: dollar amount left,
18.75: dollar cost of shirt.

Lesson 1.7

20. [Hours available for driving] • [Uncle's car speed] = [Greatest distance to and from grandmother's home]

21. Hours available for driving = 6
Uncle's car speed = 50 (mph)
Greatest distance to and from grandmother's home = x (miles)

25. [Number of weeks worked] • [Amount you save each week] = [Price of stereo with compact disk player]

26. Number of weeks worked = 12
Amount you save each week = m (dollars)
Price of stereo with compact disk player = 432 (dollars)

30. [Travel time] • [Rate of travel] = [Distance from robbery site to gas station]

31. Travel time = 18 (minutes)
Rate of travel = x (miles per minute)
Distance from robbery site to gas station = 12 (miles) or
Travel time = $\frac{18}{60} = \frac{3}{10}$ (hours)
Rate of travel = x (mph)
Distance from robbery site to gas station = 12 (miles)

Lesson 1.8

13.

Number	Tally	Frequency
0	|	1
1	|	1
2	||	2
3	||	2
4	||	2
5	||	2
6	|||	3
7	~~||||~~	5
8	||	2
9	|	1
10	||	2
11	|	1
12	||	2
13	|	1
14		0
15	|	1
16		0
17	|	1
18	|	1

16.

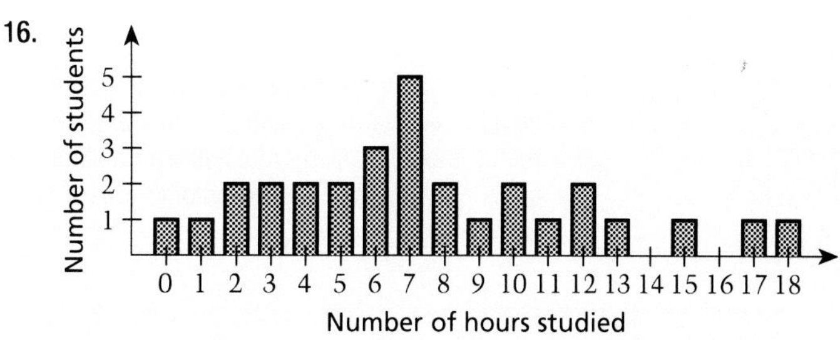

ADDITIONAL ANSWERS

Lesson 2.1

A. False; zero is neither positive nor negative.

B. True; the absolute value of a real number is a distance from zero, and distance is represented by a positive number.

C. False; if a is negative, then $-a$ is positive.

D. True; every negative number is to the left of every positive number on the number line.

E. True; one number has to be to the left of the other on the number line.

7.

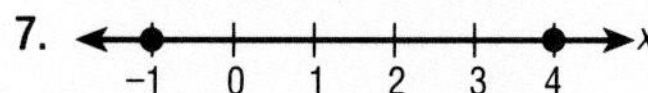

8.

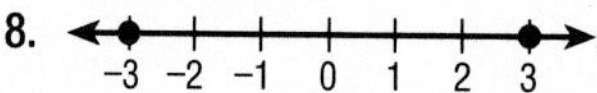

9.

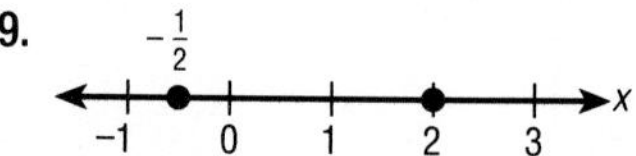

10.

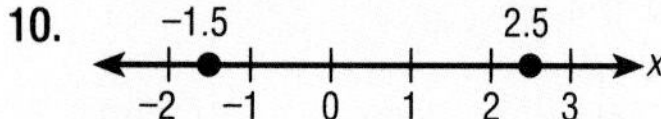

11.

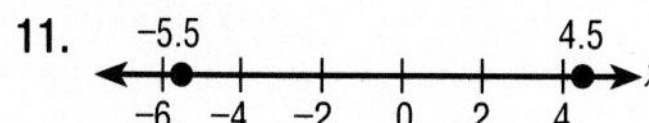

12.

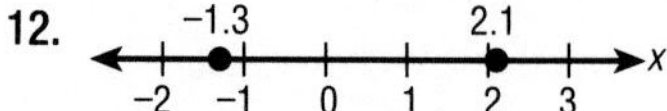

13.

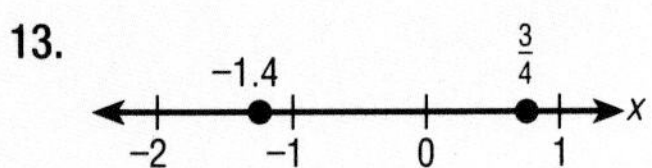

14.

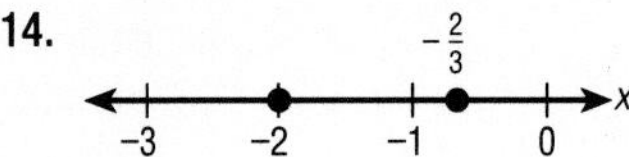

Lesson 2.2

A.

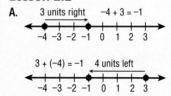

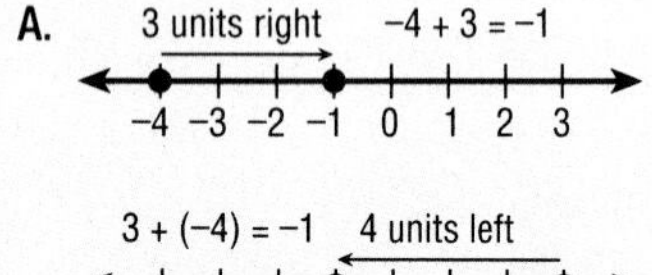

B. Start at -3, go right 2, go left 1; sum is -2.
Start at -3, go left 1, go right 2; sum is -2.
Start at 2, go left 3, go left 1: sum is -2.
Start at 2, go left 1, go left 3; sum is -2.
Start at -1, go right 2, go left 3; sum is -2.
Start at -1, go left 3, go right 2; sum is -2.

5.

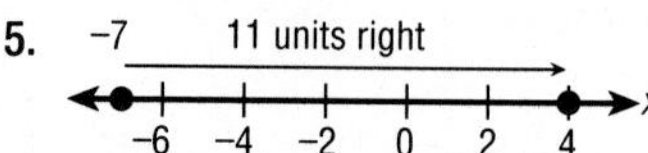

6.

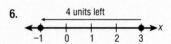

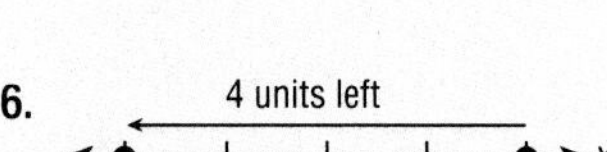

7.

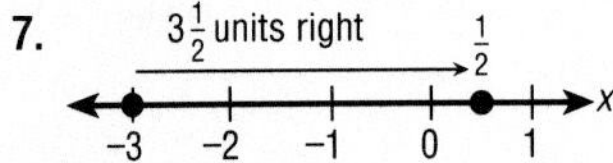

8.

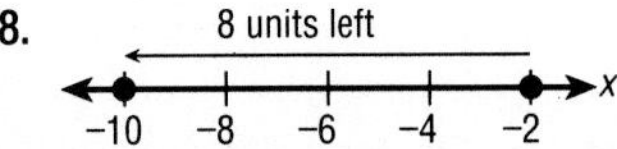

9.

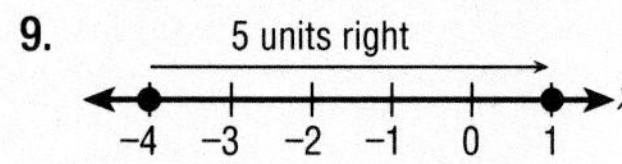

10.

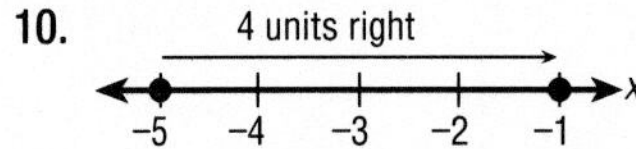

11.

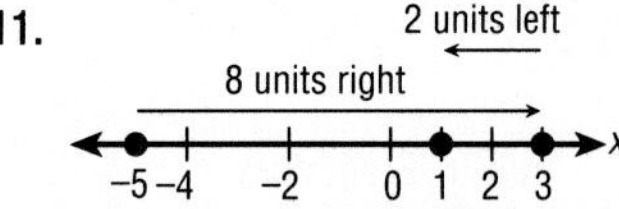

12.

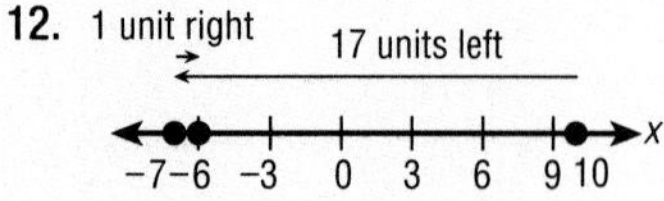

52. a and b; the sum of the five smaller areas equals the total area, and the total area minus the sum of the four known smaller areas equals the unknown smaller area.; 228

53. No. You can use one stroke (a hole in one) on the final hole to finish the round two over par, to win the round, and to tie overall. You can score two to finish the round three over par and to tie the round. Otherwise, your sister wins both the round and overall.

	Sister	You
1st round	+6	+8
2nd round	+4	+3
3rd round	+3	?

Lesson 2.3

63. 1 half note gets 2 beats, 2 quarter rests each get 1 beat, total: 4 beats

64. 6 eighth notes each get $\frac{1}{2}$ beat, 2 eighth rests each get $\frac{1}{2}$ beat, total: 4 beats

65. 1st measure—1 half note gets 2 beats, 1 quarter note gets 1 beat, total: 3 beats. 2nd measure—2 quarter notes each get 1 beat, 1 dotted eighth note gets $\frac{3}{4}$ beat, 1 sixteenth note gets $\frac{1}{4}$ beat, total: 3 beats.

66. 1 dotted eighth note gets $\frac{3}{4}$ beat, 1 sixteenth note gets $\frac{1}{4}$ beat, 2 quarter notes each get 1 beat, total: 3 beats.

Lesson 2.4

12. $\begin{bmatrix} 1 & 3 & 1 \\ 0 & 10 & 4 \\ -2 & -3 & 2 \end{bmatrix}$

13. $\begin{bmatrix} 1.3 & 4.7 & -13.8 \\ 17.5 & 3.2 & -9.2 \end{bmatrix}$

14. $\begin{bmatrix} 7.6 & 7.4 \\ -7.7 & 9.3 \\ 8.6 & -9.2 \end{bmatrix}$

24. 0.50, 0.50, 0.50, 0.75, 0.75, 0.75, 1.00, 1.00, 1.00

Wage-Rates ($)	0–1yr	2–3yr	4+yr
Grade level 4	8.50	9.50	10.00
Grade level 5	10.75	11.25	11.75
Grade level 6	12.50	13.00	13.50

33.

Incomes ($)	May	June
Store 1	98,000	81,500
Store 2	61,800	72,900

34.

Profits ($)	May	June
Store 1	16,500	10,500
Store 2	9,600	10,600

35.

Expenses ($)	May	June
Store 1	81,500	71,000
Store 2	52,200	62,300

Mid-Chapter Self-Test

25.

	Income ($)	Expenses ($)
Bowlathon	384	192
Car wash	150	10
Sub sale	400	200

yes, $32 over

Lesson 2.6

50. $0.9(180 - A)$, 54, 45, 36, 27, 18, 9, 0

Advantage of 90% handicap: Better for leagues with greater range of averages.

Disadvantage of 90% handicap: A low-average bowler has a greater likelihood for exceeding a low average by a large amount than a good bowler has for exceeding a high average by a large amount, thereby giving the low-average bowler an unfair advantage on occasion.

Advantage of 80% handicap: Better for leagues with a smaller range of averages and with the averages tending towards those of an average bowler (120–180).

Disadvantage of 80% handicap: Does not serve to "equalize" better bowlers with averages in the 180–200 range.

CHAPTER 3

Lesson 3.2

A. $2x - 6 = 5$
$2x = 11$
$x = \frac{11}{2}$

Should have multiplied $2 \times (-3)$.

B. $-3x = 5$
$x = -\frac{5}{3}$

Should not have subtracted 3 from 5 because 5 is not a coefficient of x.

C. $x - 8 = 28$
$x = 36$

Should have multiplied $4 \times (-2)$.

53.–58. First method is preferable because it eliminates the fraction.

53. $\frac{2}{3}x + 1 = \frac{1}{3}$
$2x + 3 = 1$
$2x = -2$
$x = -1$

$\frac{2}{3}x + 1 = \frac{1}{3}$
$\frac{2}{3}x = -\frac{2}{3}$
$x = -1$

54. $\frac{1}{2}x - \frac{1}{4} = \frac{3}{4}$
$2x - 1 = 3$
$2x = 4$
$x = 2$

$\frac{1}{2}x - \frac{1}{4} = \frac{3}{4}$
$\frac{1}{2}x = 1$
$x = 2$

55. $\frac{1}{2}(3x - 7) = 4$
$3x - 7 = 8$
$3x = 15$
$x = 5$

$\frac{1}{2}(3x - 7) = 4$
$\frac{3}{2}x - \frac{7}{2} = 4$
$\frac{3}{2}x = \frac{15}{2}$
$x = 5$

56. $\frac{6}{11}(x - 4) = -36$
$x - 4 = -66$
$x = -62$

$\frac{6}{11}(x - 4) = -36$
$\frac{6}{11}x - \frac{24}{11} = -36$
$\frac{6}{11}x = -\frac{372}{11}$
$x = -62$

Lesson 3.2 *(continued)*

57.
$$\begin{aligned} -56 &= \tfrac{8}{9}(4 - x) \\ -63 &= 4 - x \\ -67 &= -x \\ 67 &= x \end{aligned}$$

$$\begin{aligned} -56 &= \tfrac{8}{9}(4 - x) \\ -56 &= \tfrac{32}{9} - \tfrac{8}{9}x \\ -\tfrac{536}{9} &= -\tfrac{8}{9}x \\ 67 &= x \end{aligned}$$

58.
$$\begin{aligned} -9 &= \tfrac{3}{7}(-2x + 5) \\ -21 &= -2x + 5 \\ -26 &= -2x \\ 13 &= x \end{aligned}$$

$$\begin{aligned} -9 &= \tfrac{3}{7}(-2x + 5) \\ -9 &= -\tfrac{6}{7}x + \tfrac{15}{7} \\ -\tfrac{78}{7} &= -\tfrac{6}{7}x \\ 13 &= x \end{aligned}$$

Lesson 3.3

A.

$-3x+5 = 9x-19$	
$-3x+3x+5 = 9x+3x-19$	Add $3x$ to both sides.
$5 = 12x-19$	Simplify.
$5+19 = 12x-19+19$	Add 19 to both sides.
$24 = 12x$	Simplify.
$\frac{24}{12} = \frac{12x}{12}$	Divide both sides by 12.
$2 = x$	Simplify.

Lesson 3.4

9. [Population of western region in 1988] + [Rate of western increase] • [Time] = [Population of midwest region in 1988] + [Rate of midwest increase] • [Time]

10. Population of western region = 50,679,000 (people)
Rate of western increase = 982,000 (people per year)
Population of midwest region = 58,878,000 (people)
Rate of midwest increase = 222,000 (people per year)
Number of years = y

13.

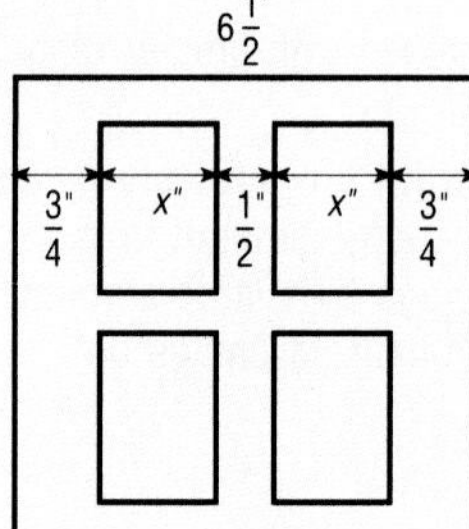

14. [Cover width] = 2 • [Margin width] + [Space between pictures] + 2 • [Picture width]

15. Cover width = $6\frac{1}{2}$ (inches)

Margin width = $\frac{3}{4}$ (inch)

Space between pictures = $\frac{1}{2}$ (inch)

Picture width = x (inches)

18. [Money you currently have] + [Money you save each week] • [Time] = [Money sister currently has] − [Extra money she spends each week] • [Time]

19. Money you currently have = 60 (dollars)
Money you save each week = 5 (dollars)
Money sister currently has = 135 (dollars)
Extra money she spends each week = 10 (dollars)
Time = w (weeks)

22. [Greenville temperature] + [Rate of temperature increase] • [Time] = [Waterloo temperature] − [Rate of temperature decrease] • [Time]

23. Greenville temperature = 69° (°F)
Rate of temperature increase = 2 (°F per hour)
Time = h (hours)
Waterloo temperature = 84 (°F)
Rate of temperature decrease = 3 (°F per hour)

Lesson 3.7

A.

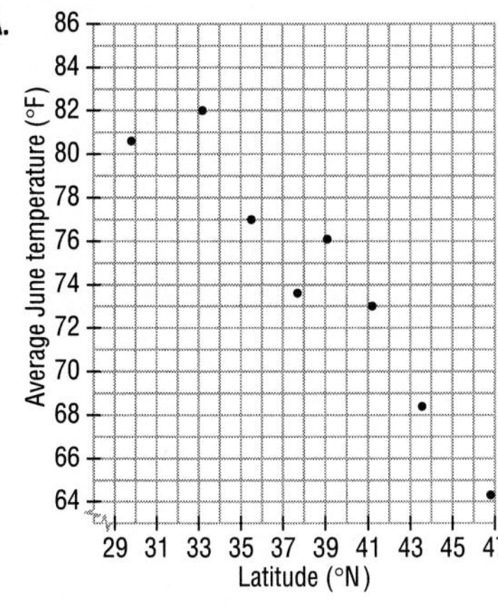

9.

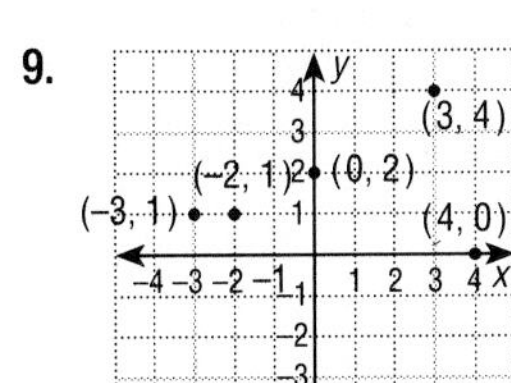

10.

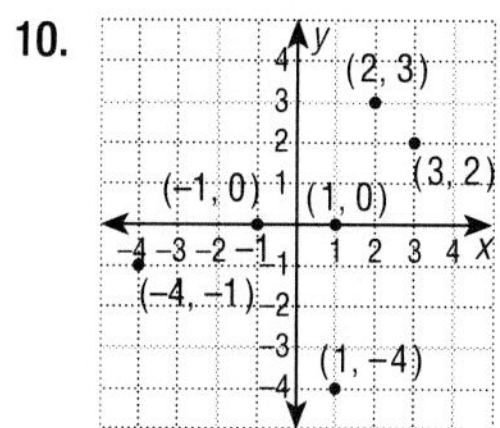

11.

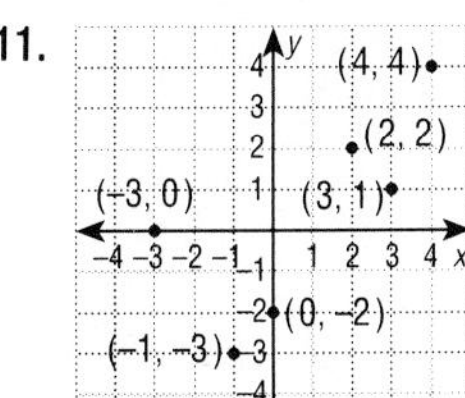

12.

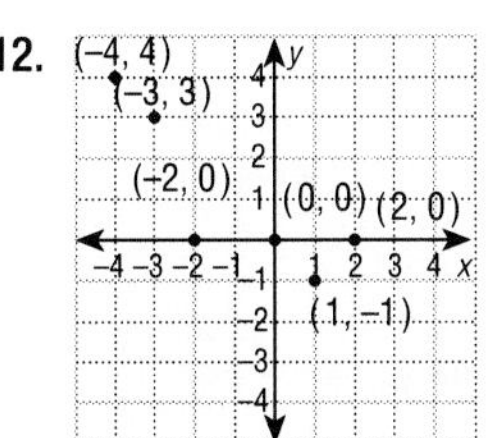

19.

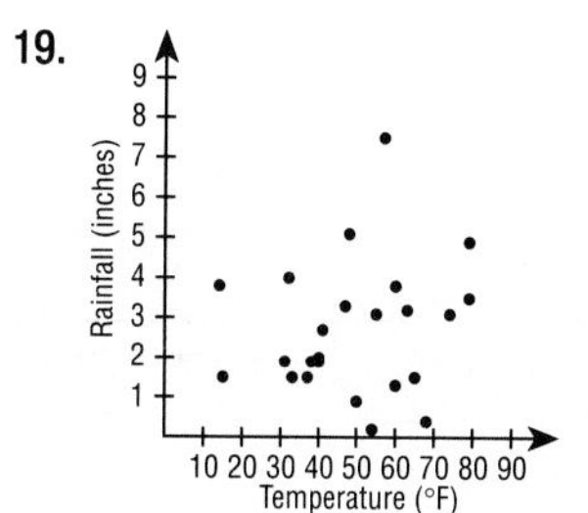

23.

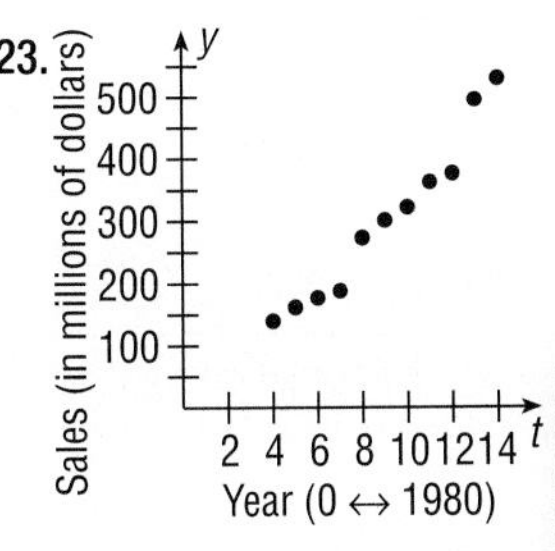

25.

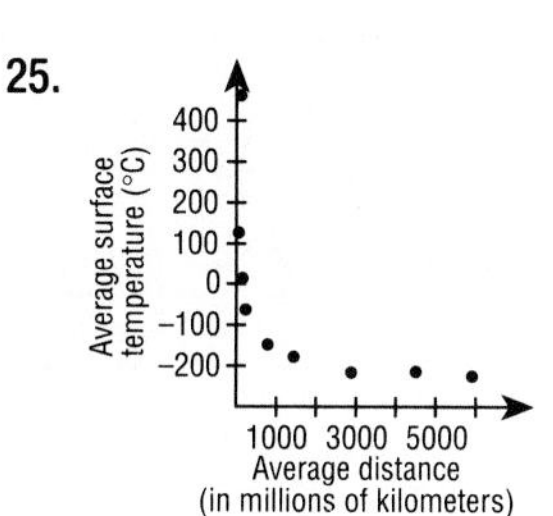

Cumulative Review

32.

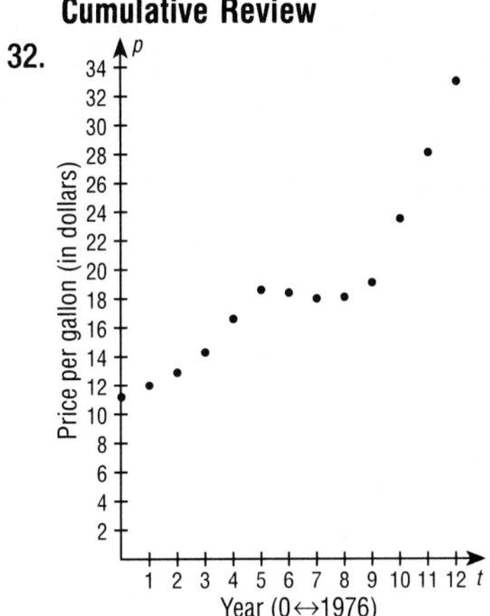

CHAPTER 4

Lesson 4.1

11.

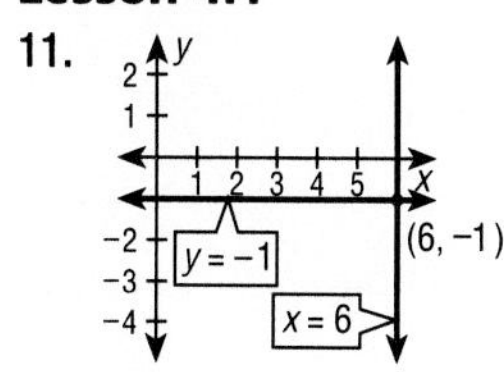

12.

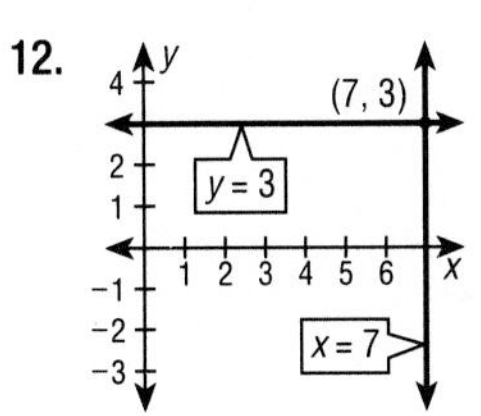

13.

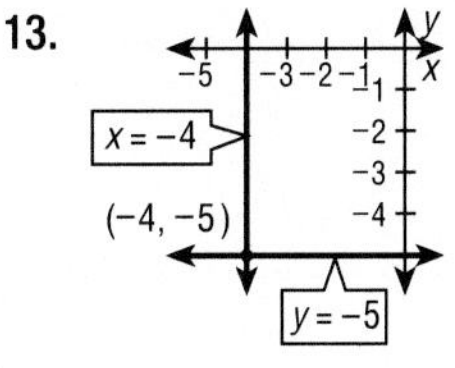

14.

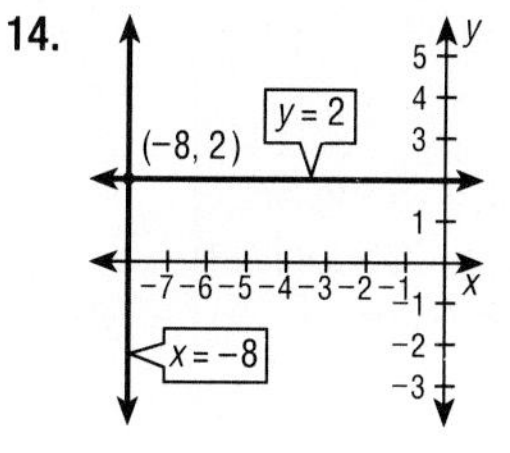

34.

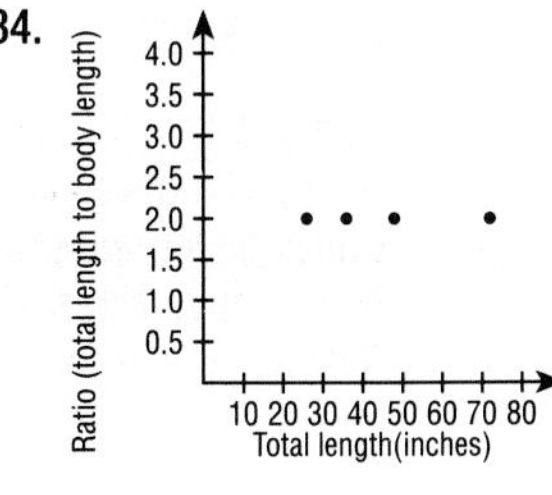

35.

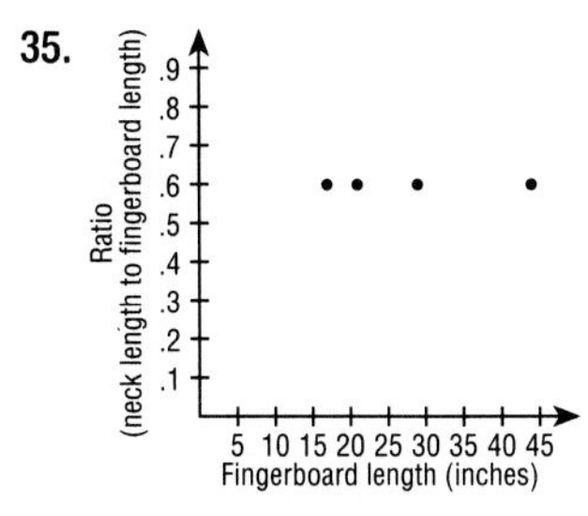

38.

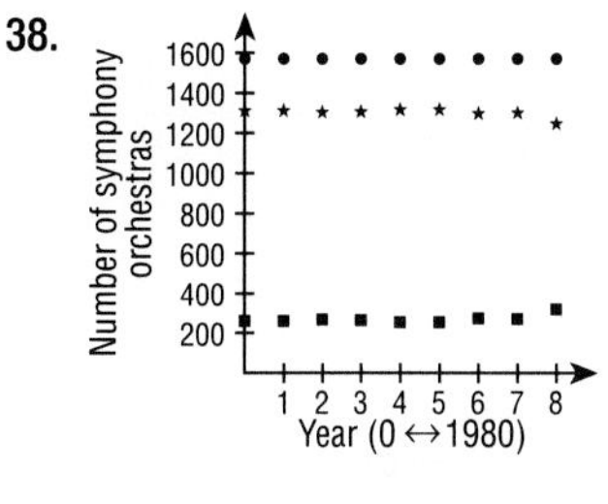

40. a.

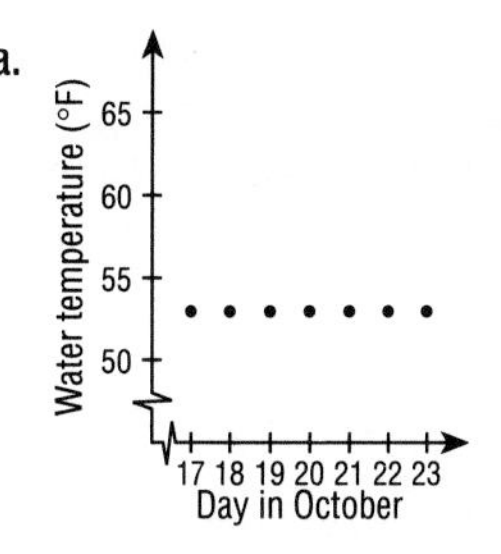

b.

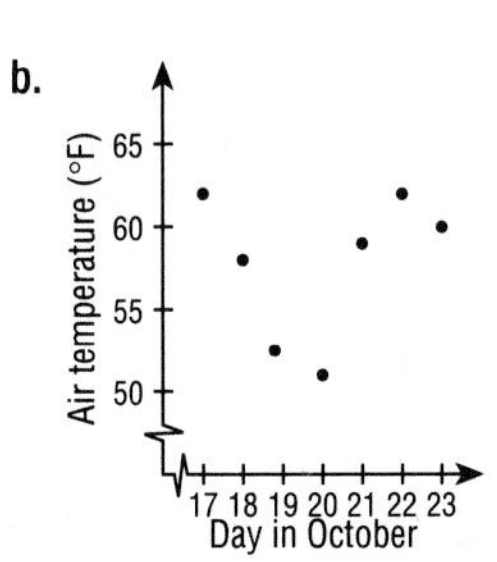

ADDITIONAL ANSWERS

Lesson 4.2

A. 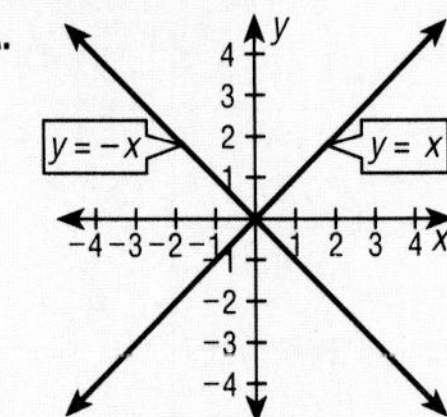

B. 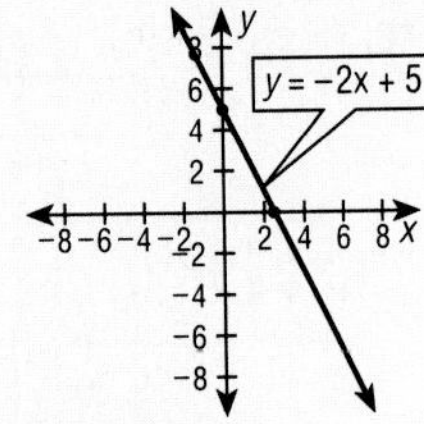

13. 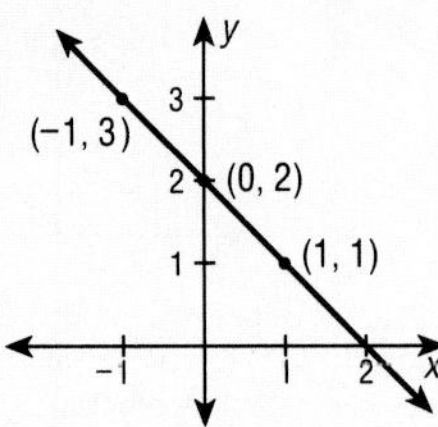

14. 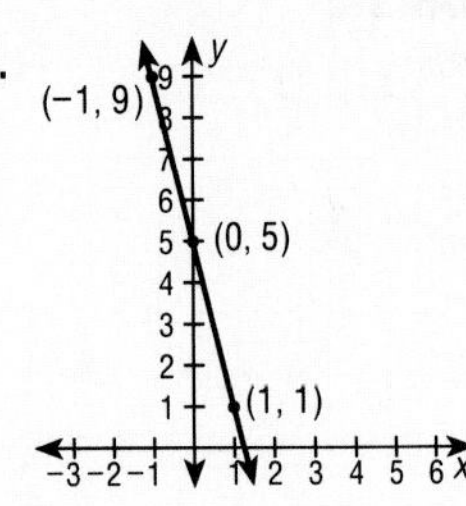

15. 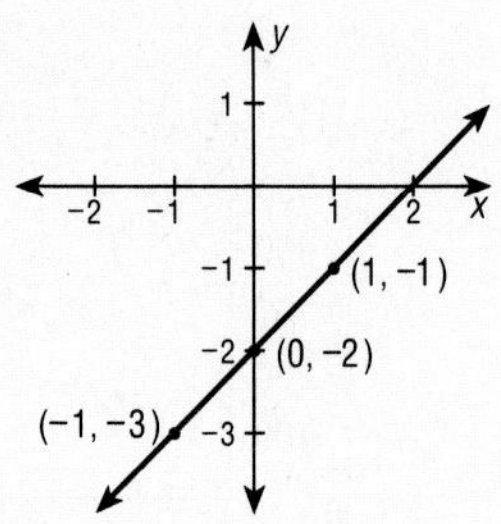

16. 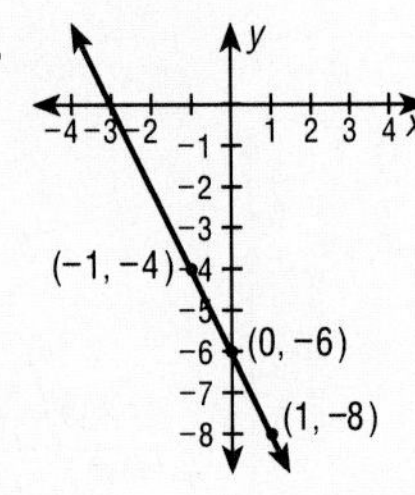

17. 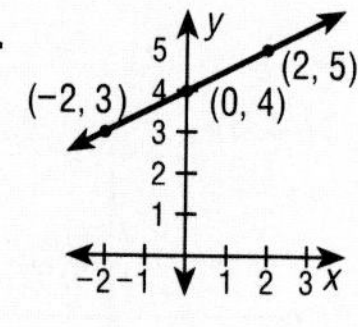

18. 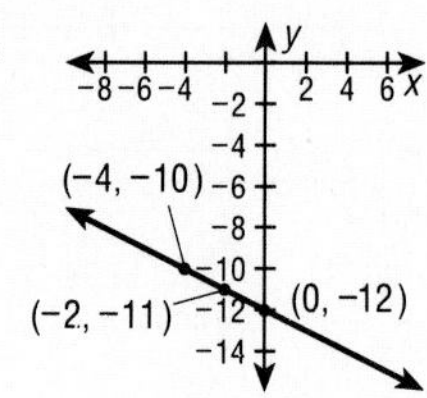

19. 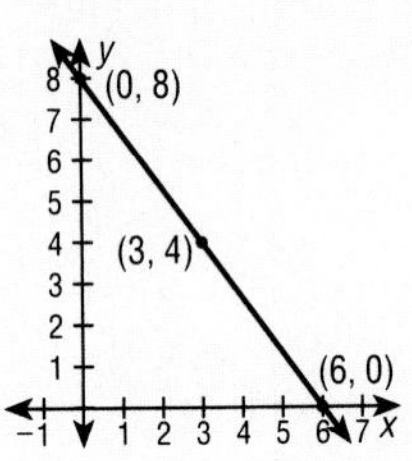

20. 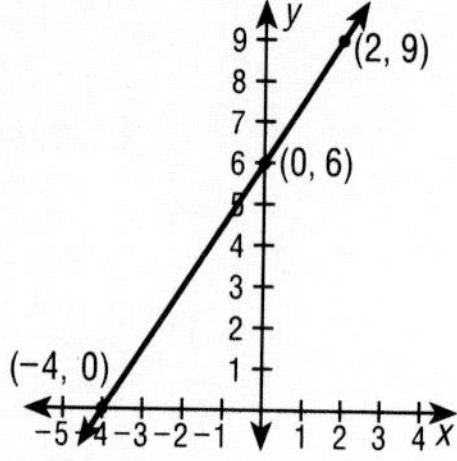

32. 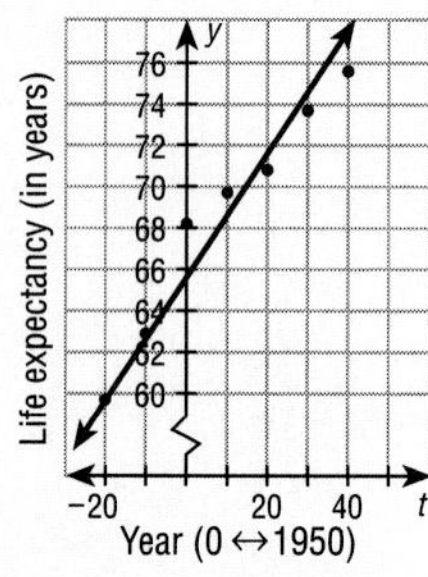

33. 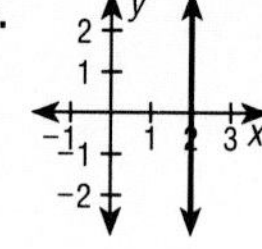

34. 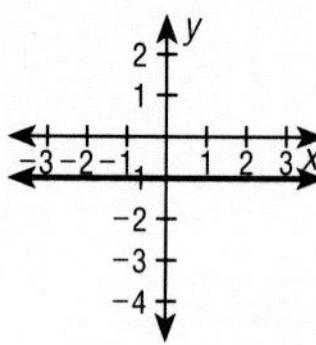

35. 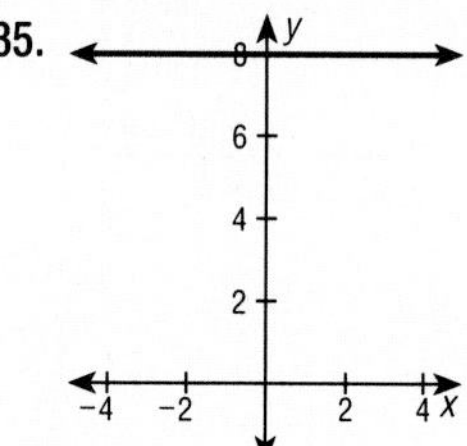

36. 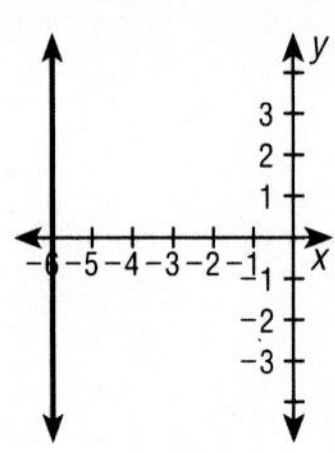

45. 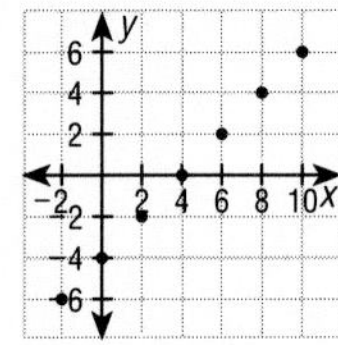

46. 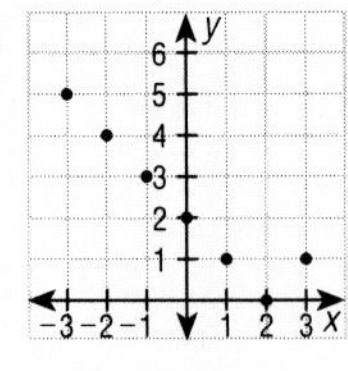

Using a Graphing Calculator

2. 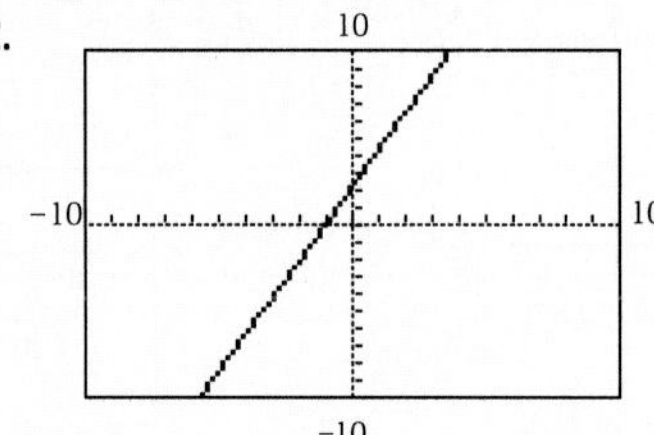

4. 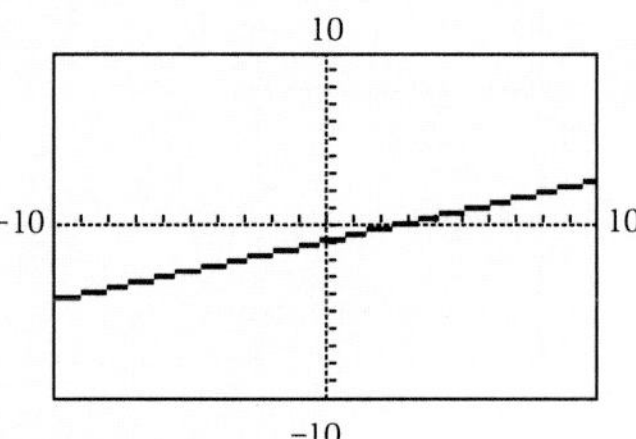

6. 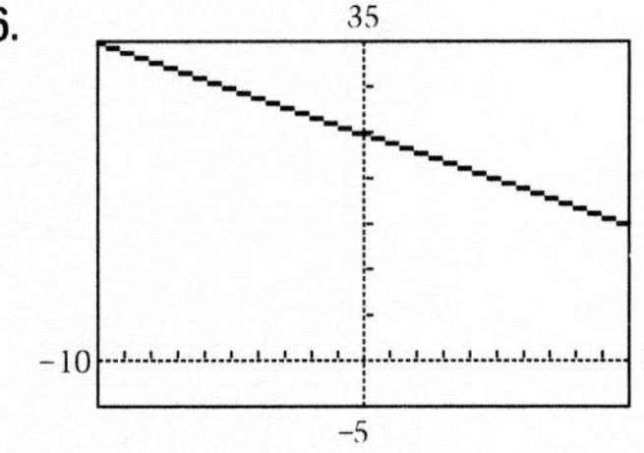

8. 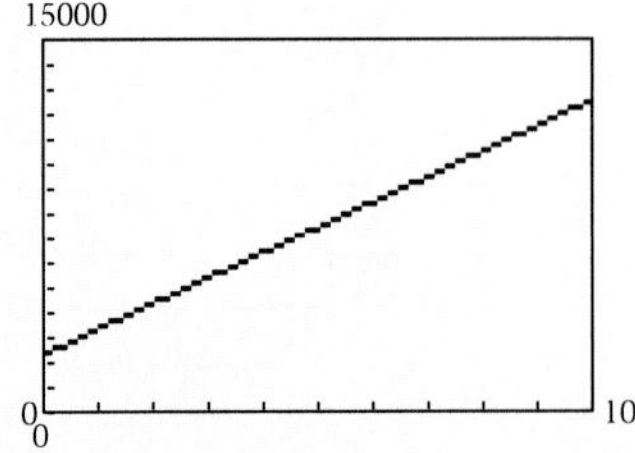

Lesson 4.3

15. 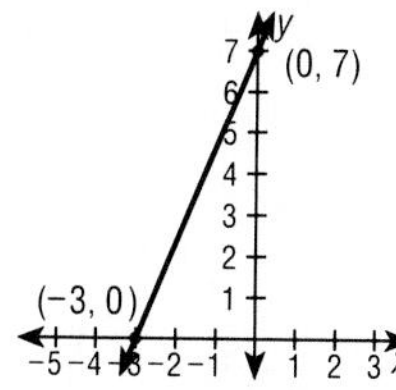

16. 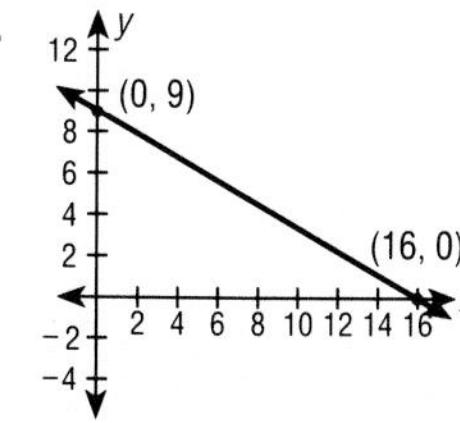

17. 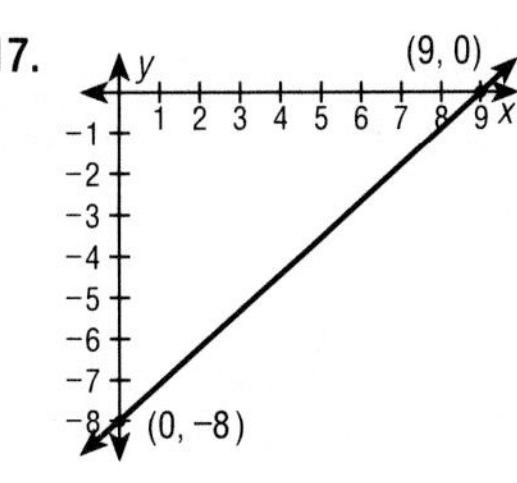

18. 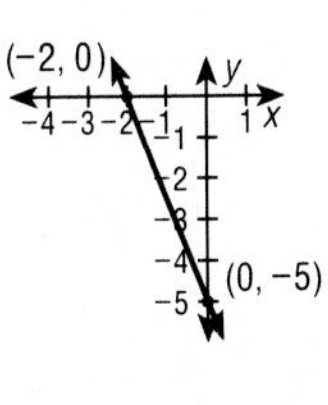

19. 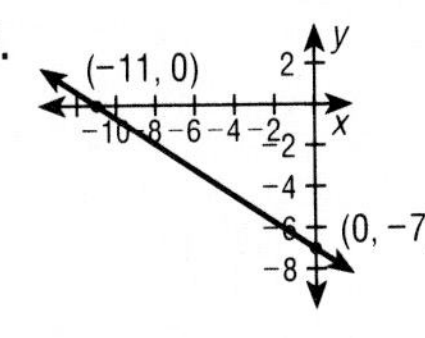

20. 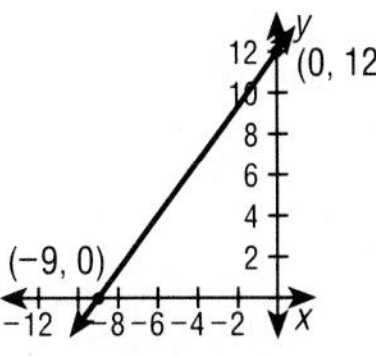

37. 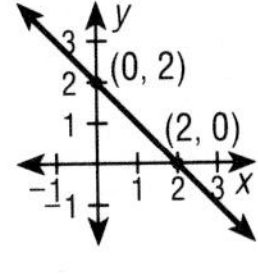

38. 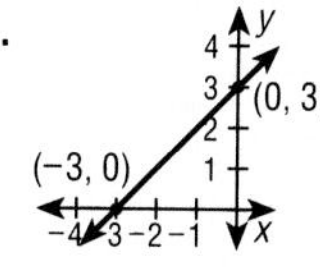

39. 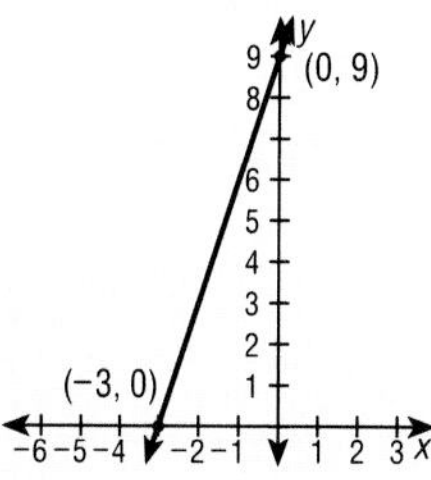

40. 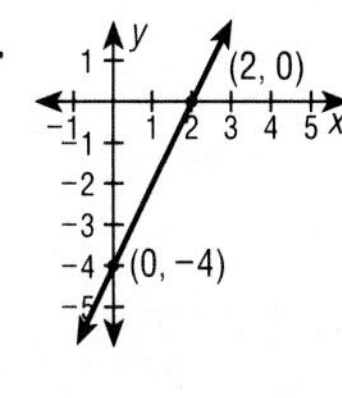

41. 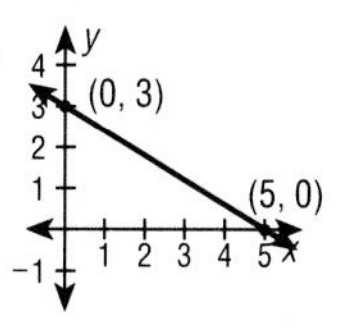

42. 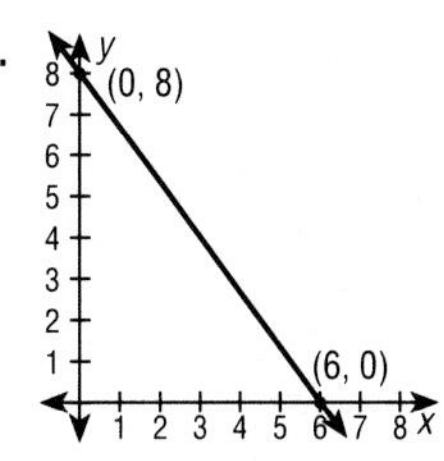

43. 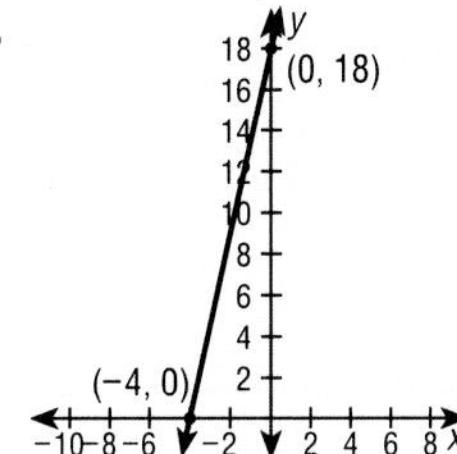

44. 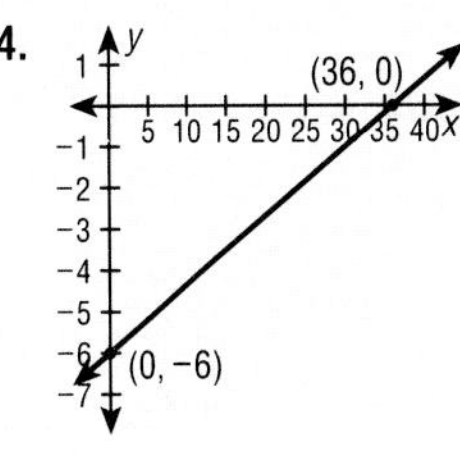

45. 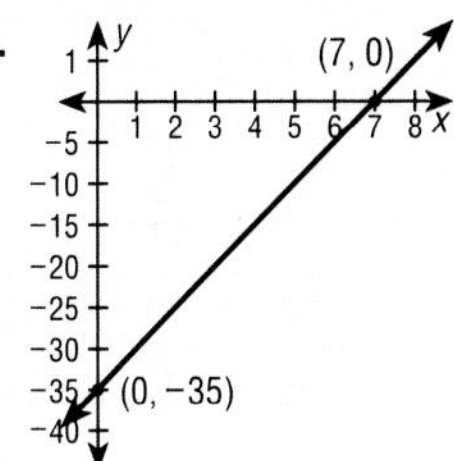

46. 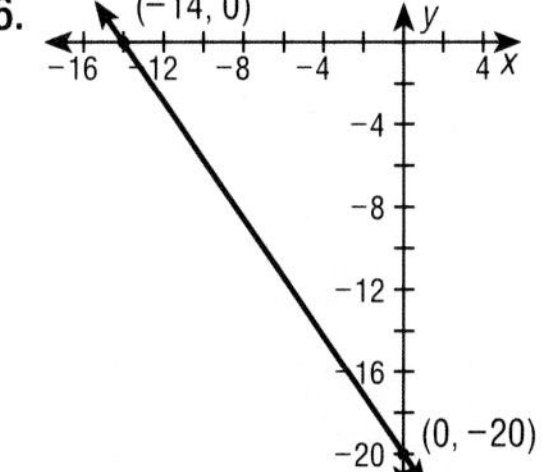

47. 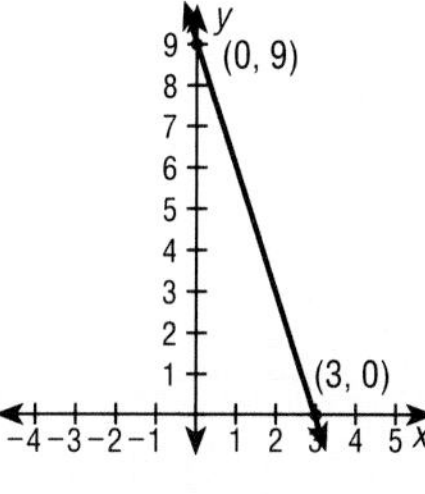

48. 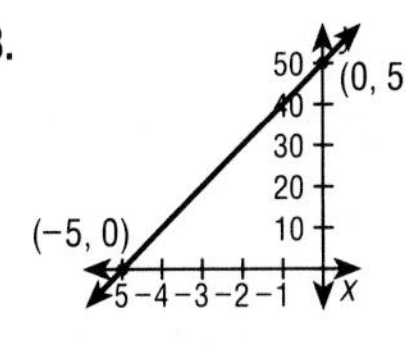

49. 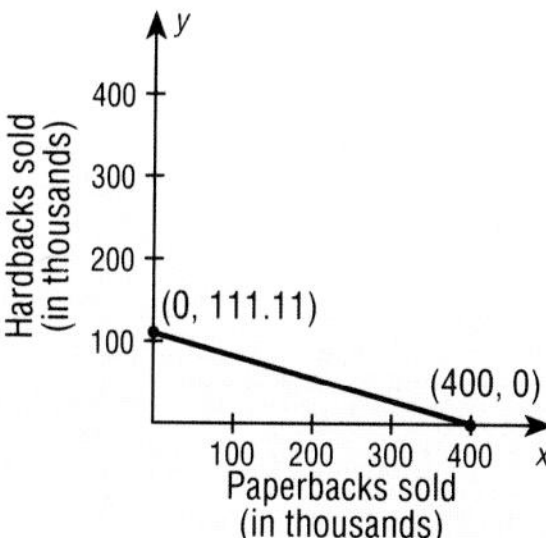

51. 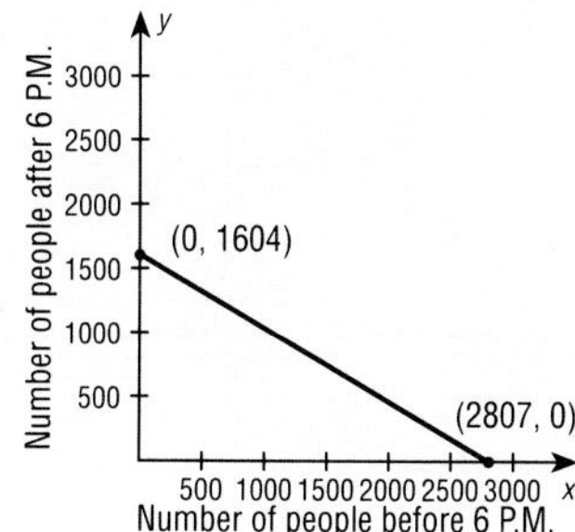

Lesson 4.4

13. 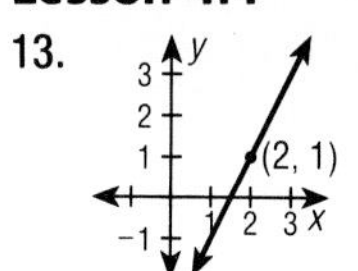

14. 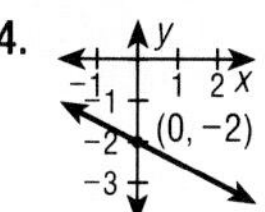

15. 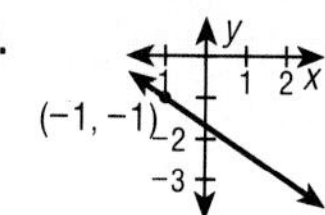

16. 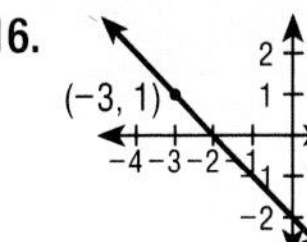

17. 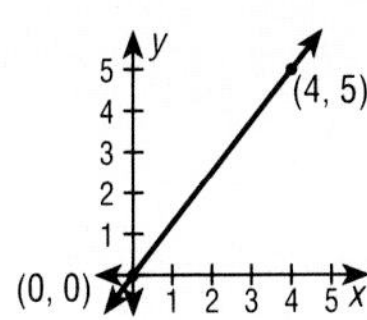

ADDITIONAL ANSWERS

Lesson 4.4 *(continued)*

18.

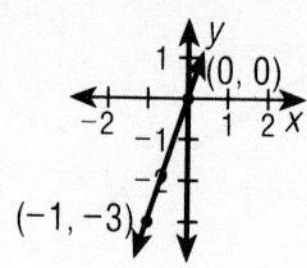

19.

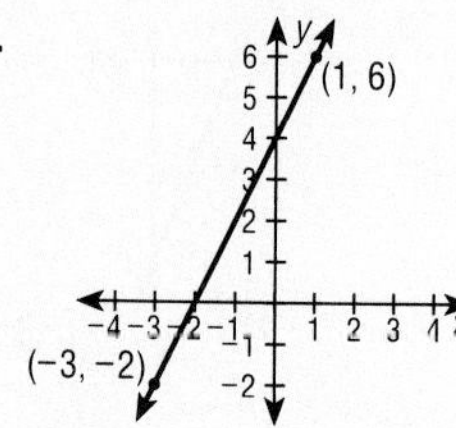

20.

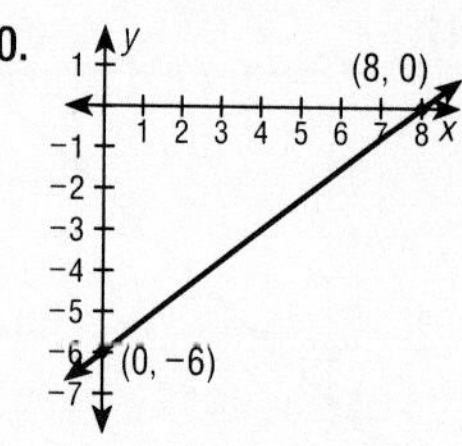

21.

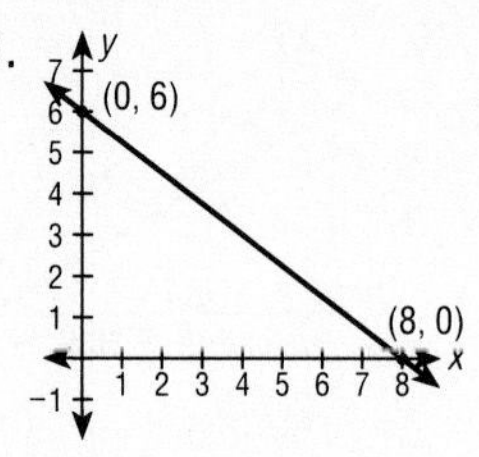

22.

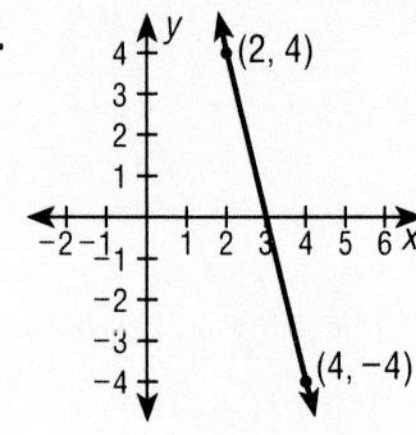

23.

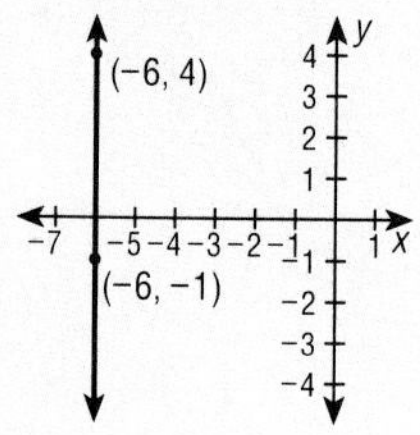

24.

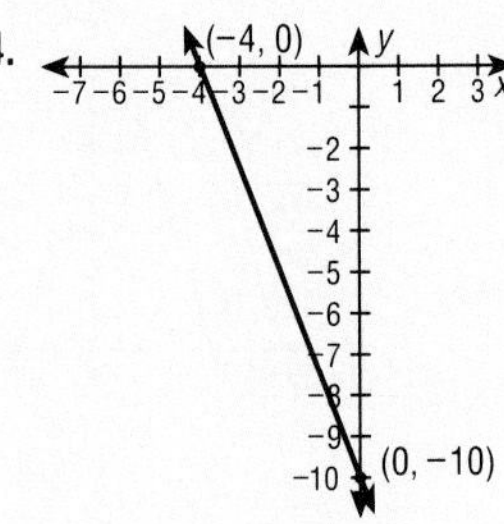

25.

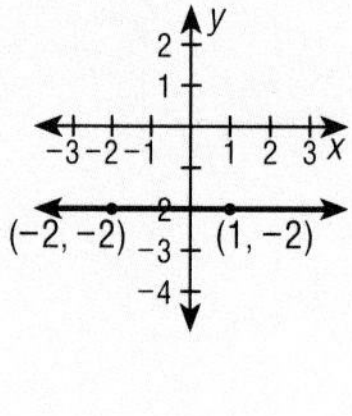

26.

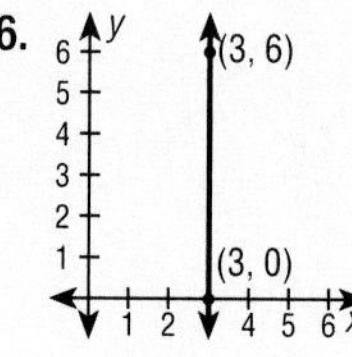

27.

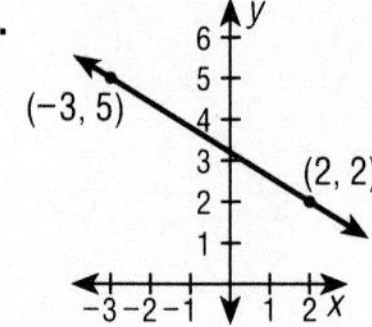

28.

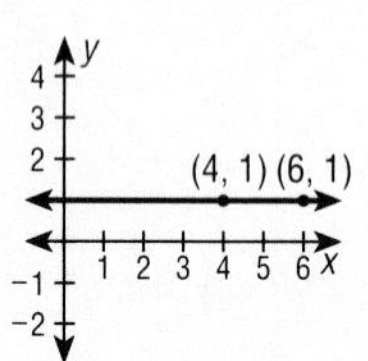

Mid-Chapter Self-Test

1.

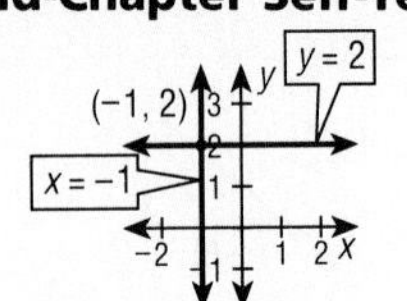

Lesson 4.5

17.

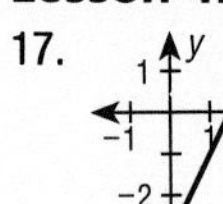

18.

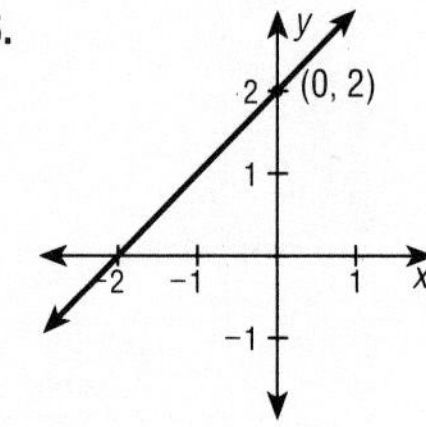

19.

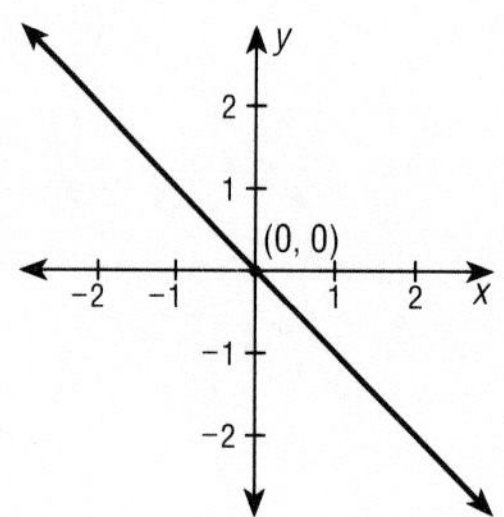

20.

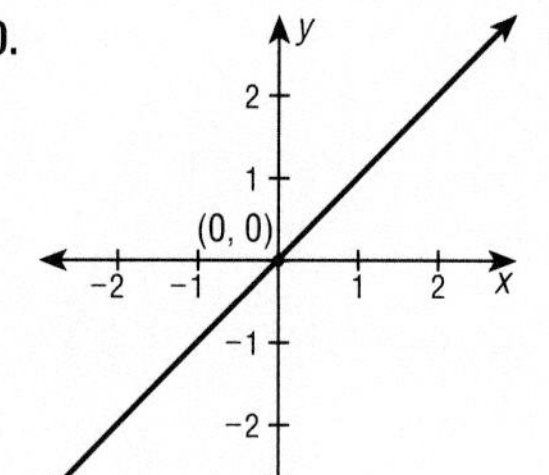

21.

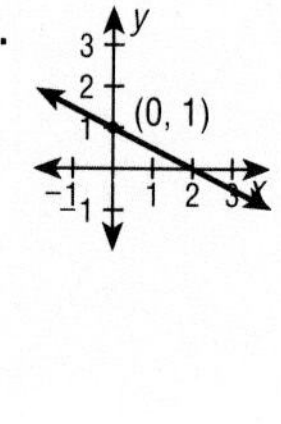

22.

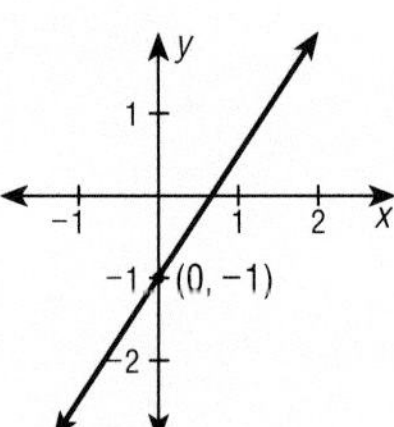

23.

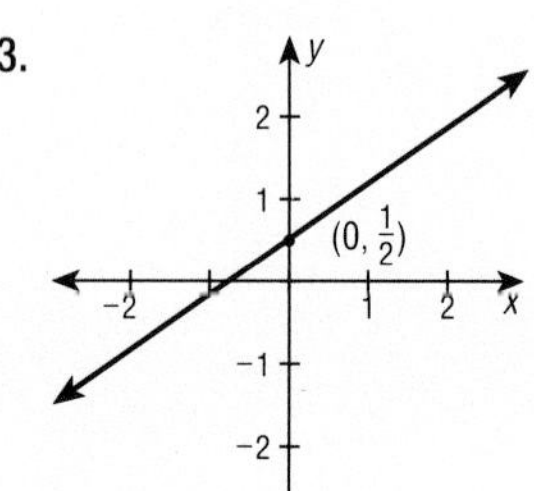

24.

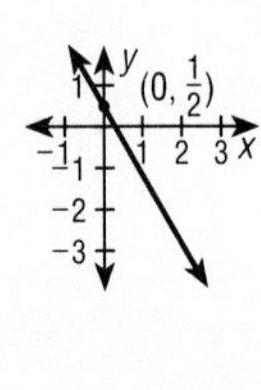

25.

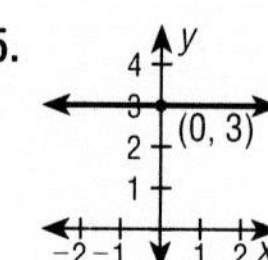

26.

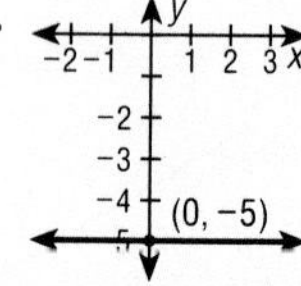

27.

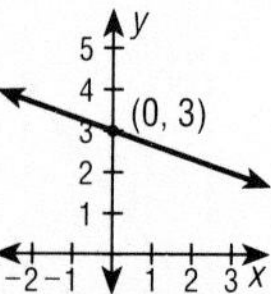

28.

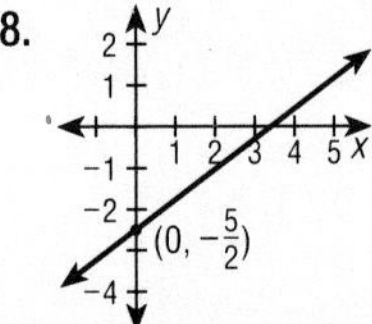

33.

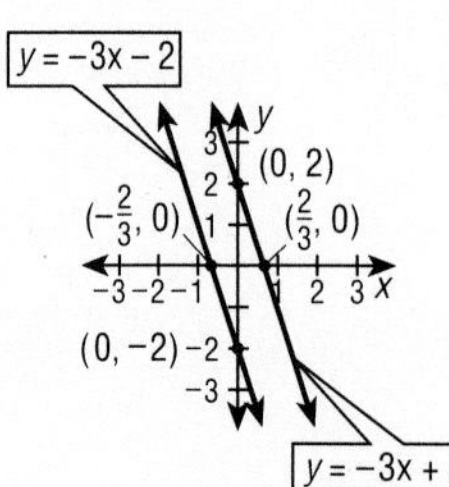

34.

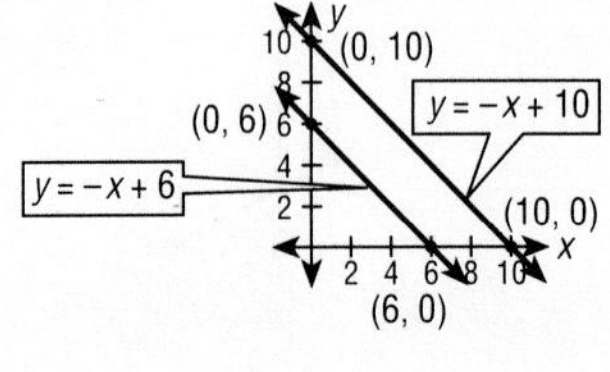

Lesson 4.5 *(continued)*

35. 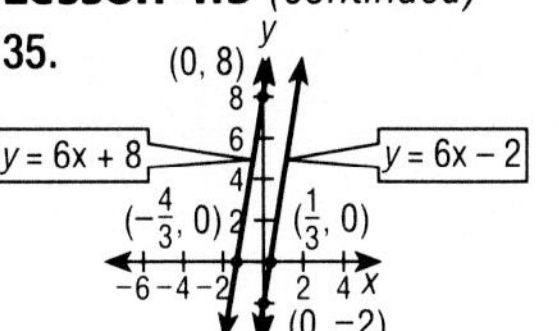

36. 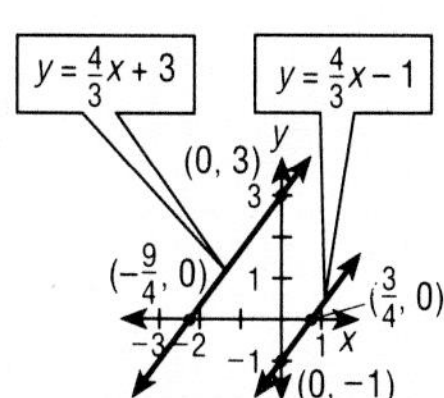

53. 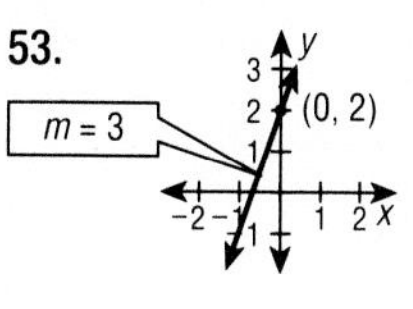

54. 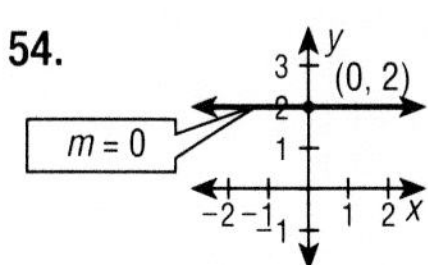

55. 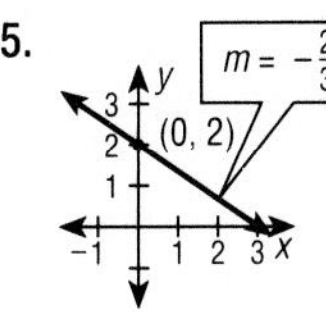

56. 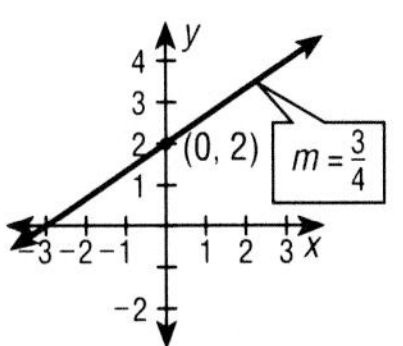

59. 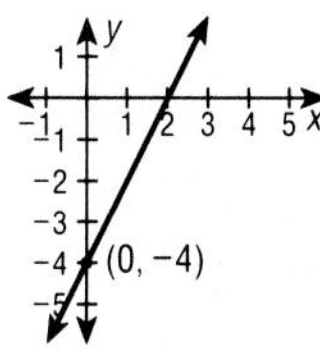

60. 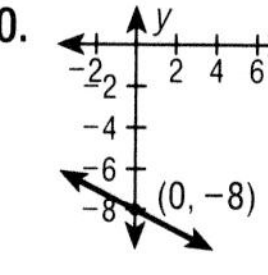

61. 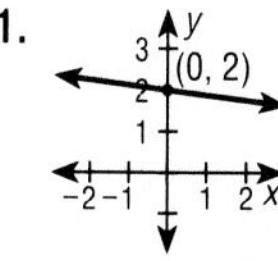

62. 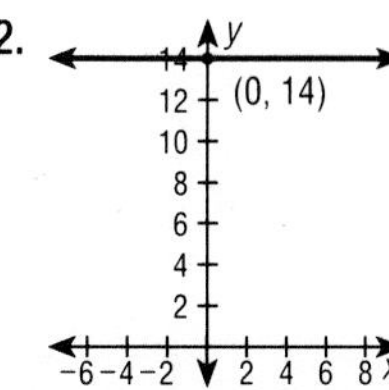

Lesson 4.6

2. 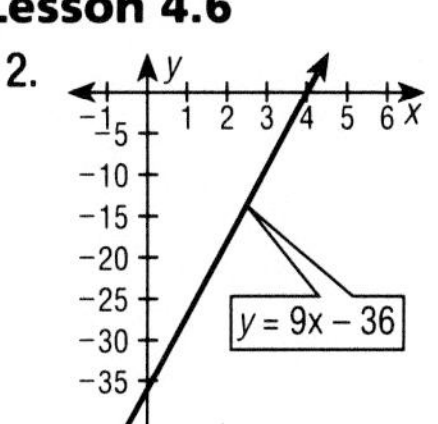

31. 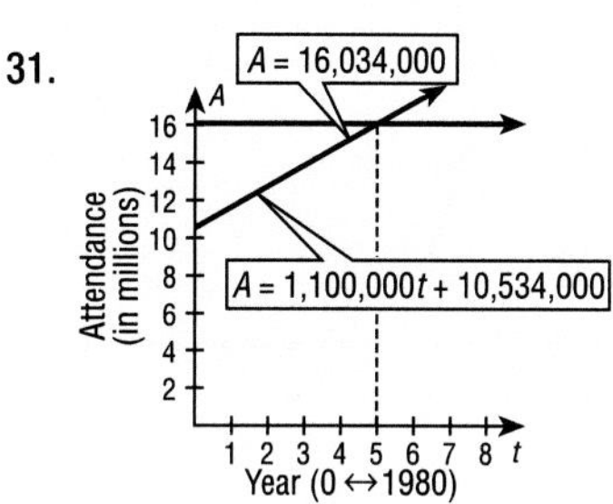

32. 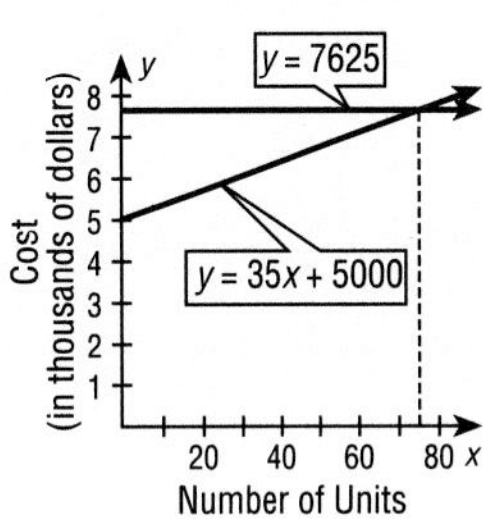

44. 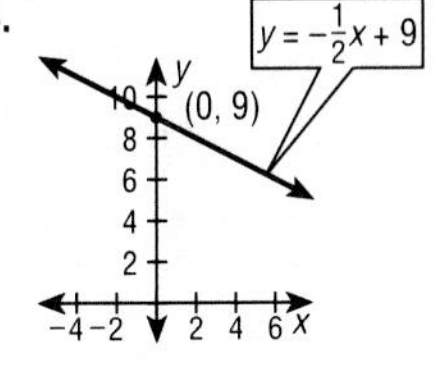

Lesson 4.7

A. 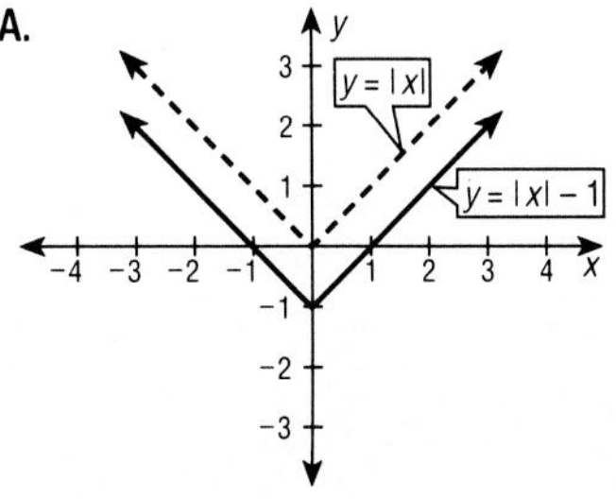

B. 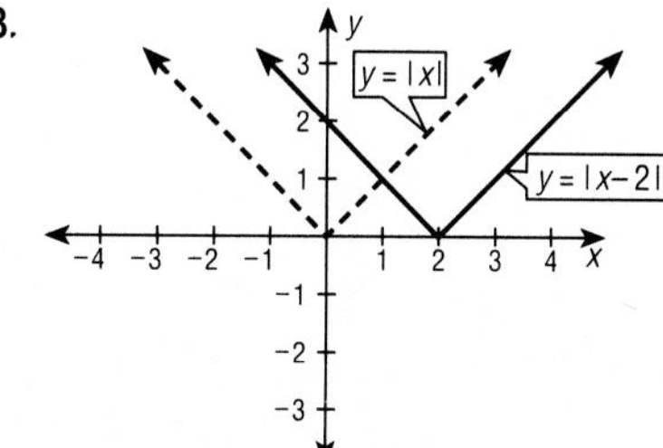

C. 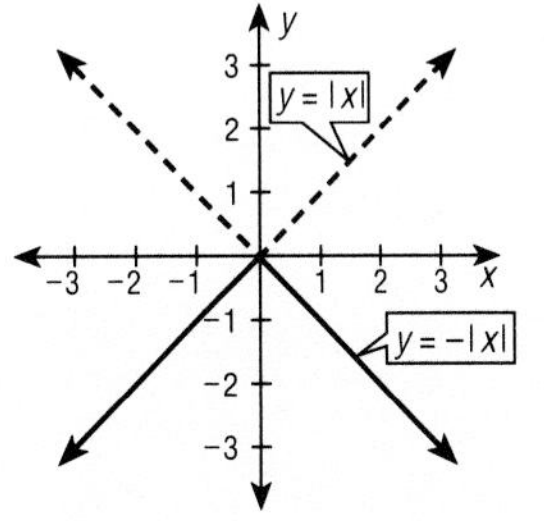

D. 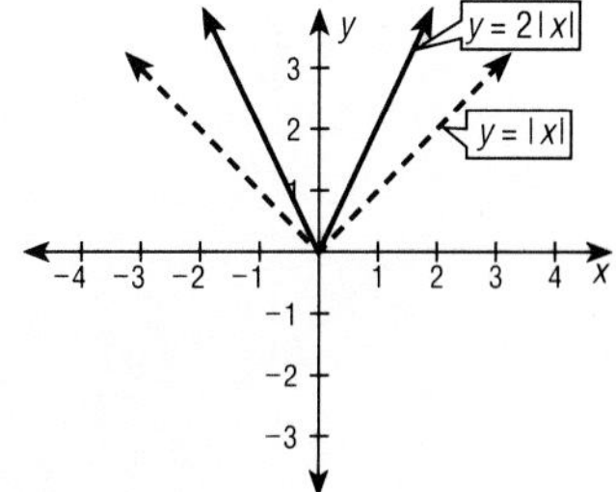

5. Graph is one unit below graph of $y = |x|$.

6. Graph is one unit above graph of $y = |x|$.

7. Graph is two units below graph of $y = |x|$.

8. Graph is two units above graph of $y = |x|$.

33. 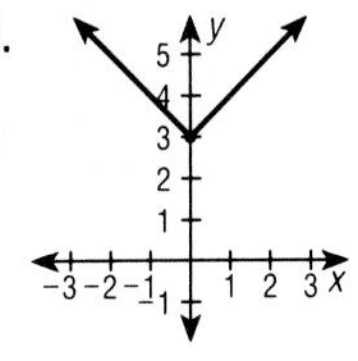

34. 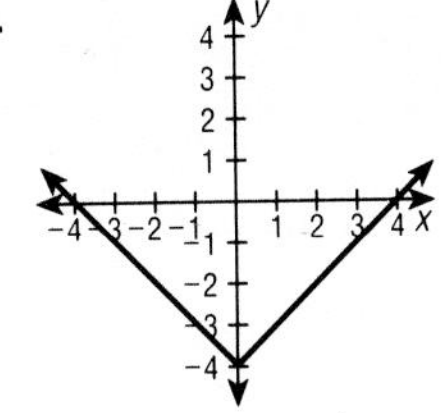

35. 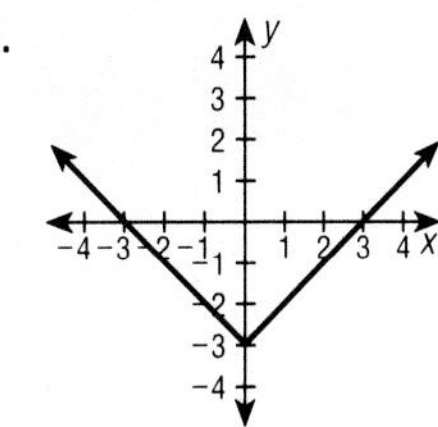

36. 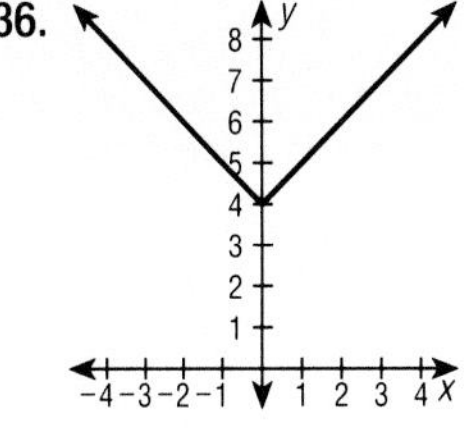

37. 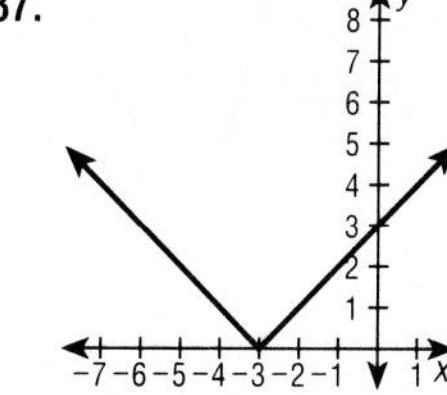

Lesson 4.7 *(continued)*

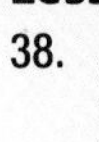

38. 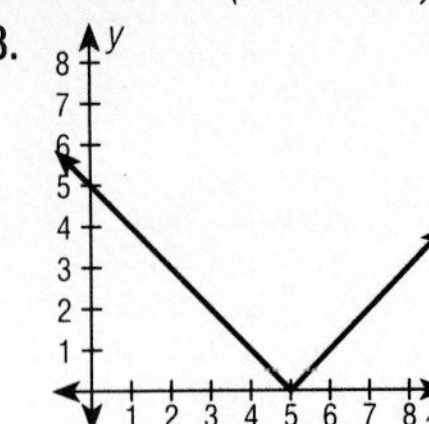

39. 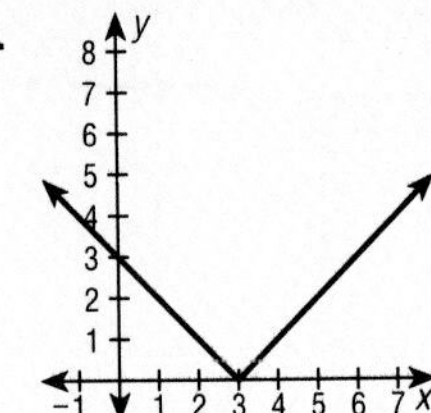

40. 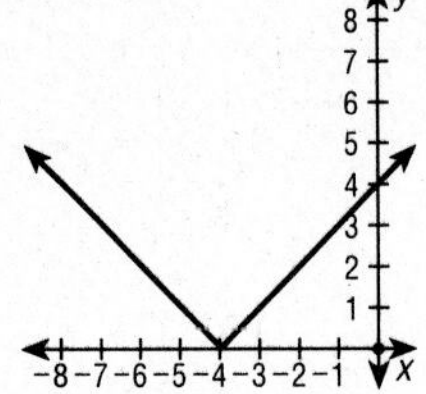

41. 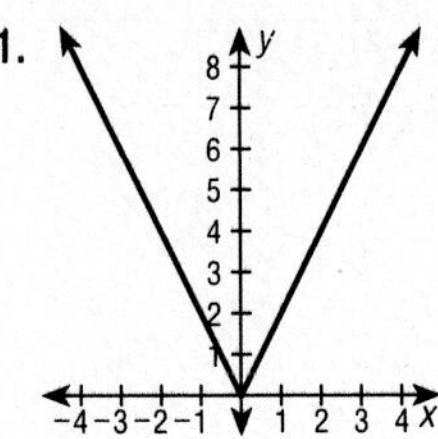

42. 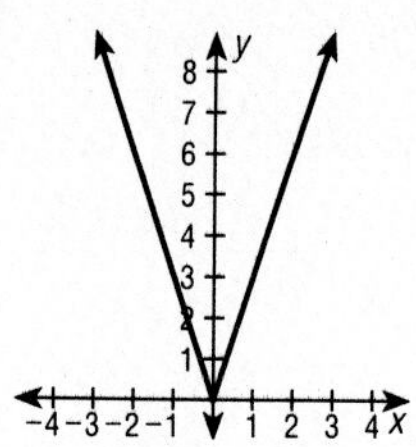

43. 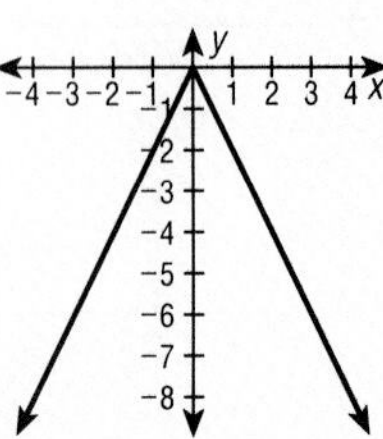

44. 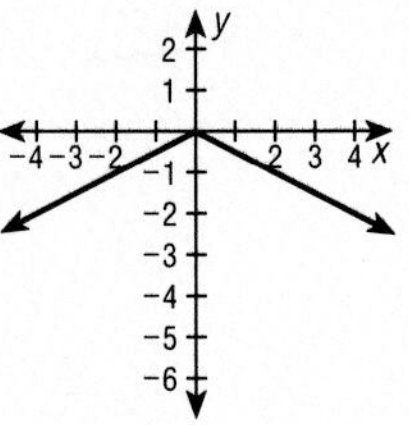

45. 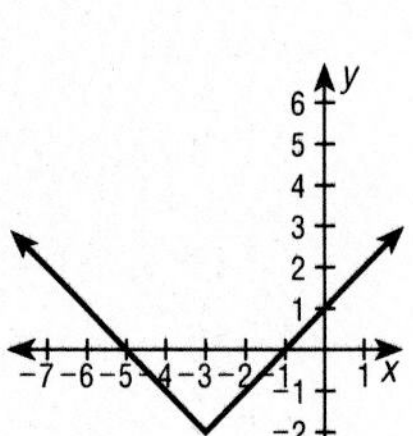

46. 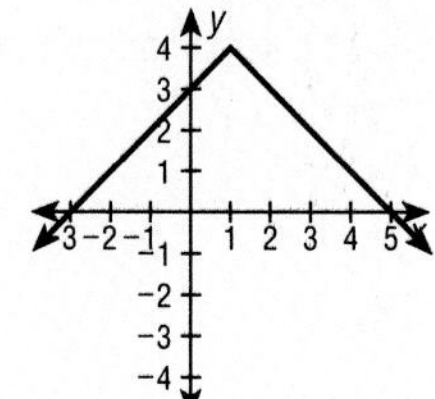

47. 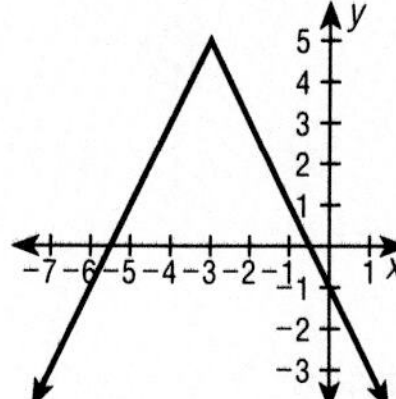

48. 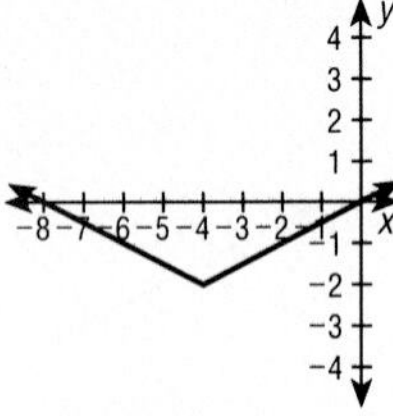

49. 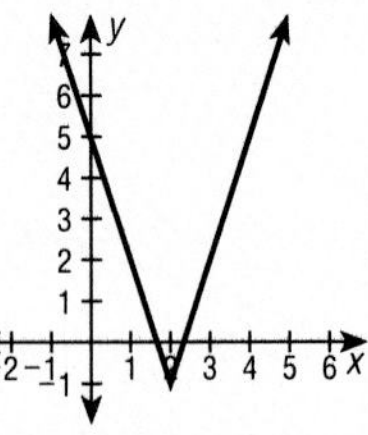

50. 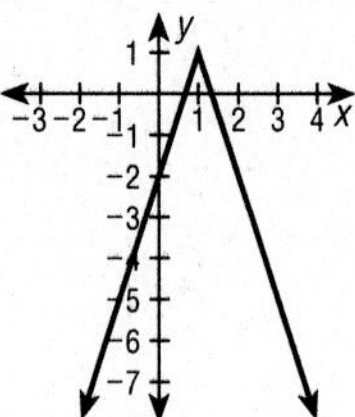

Chapter Review

1. 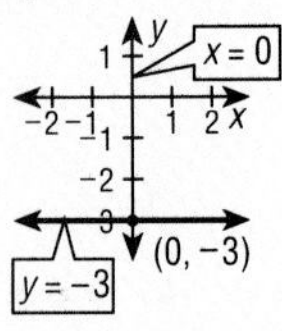

2. 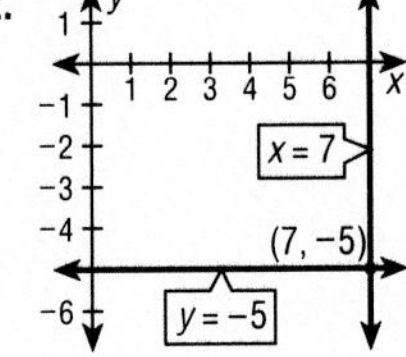

3. 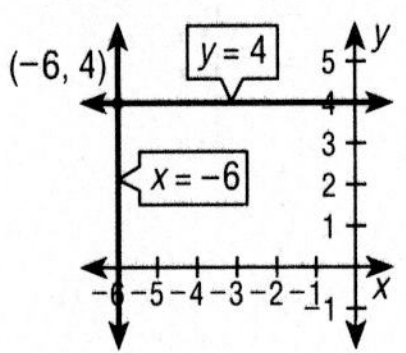

4. 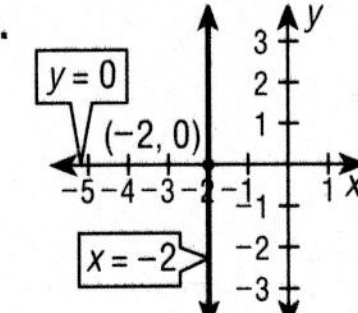

23. 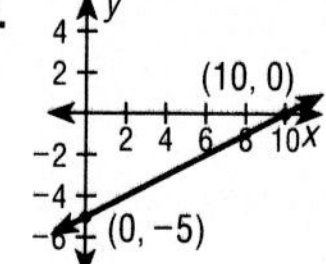

24. 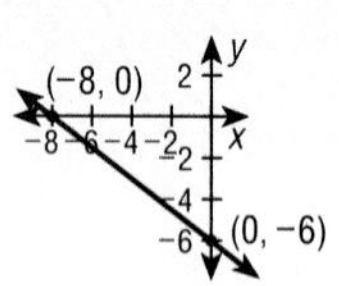

CHAPTER 5

Lesson 5.1

31. 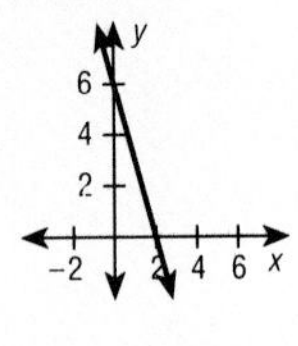

32. 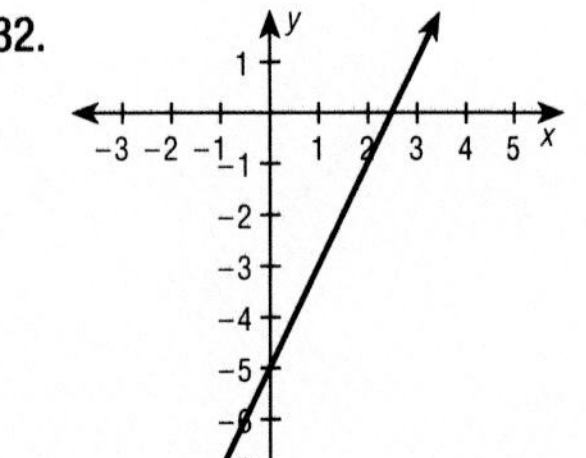

33. 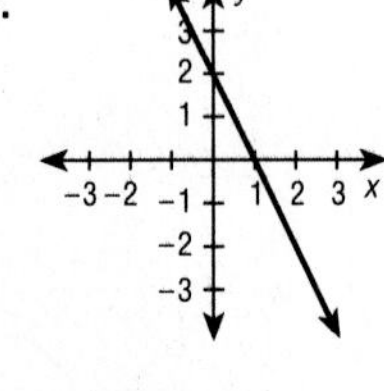

34. 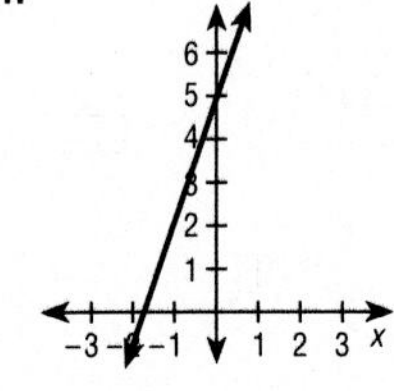

35. 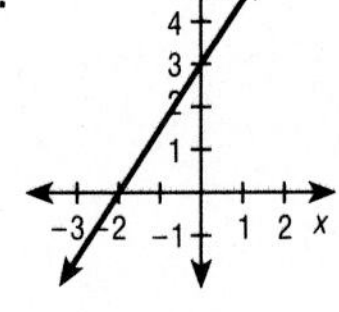

Lesson 5.1 *(continued)*

36. 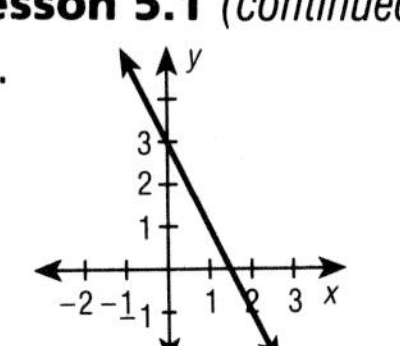

37. 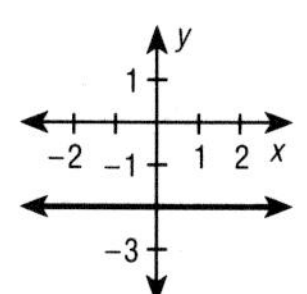

38. 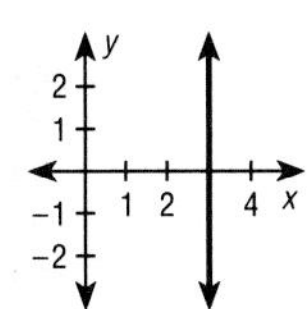

39. 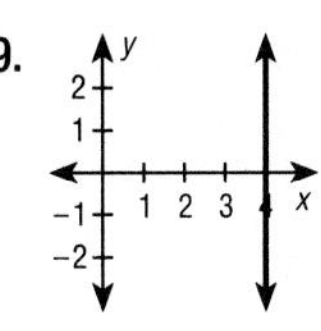

40. 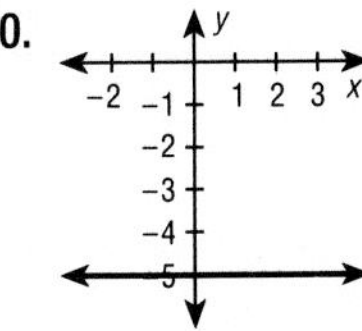

41. 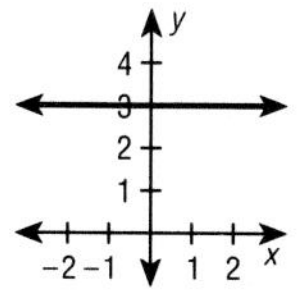

42. 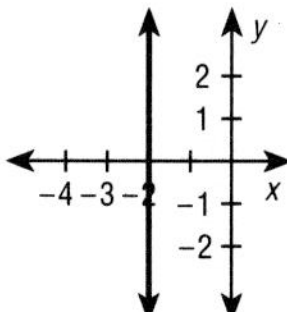

Lesson 5.5

35. 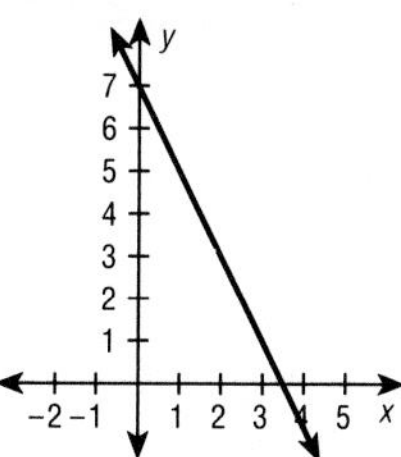

36. 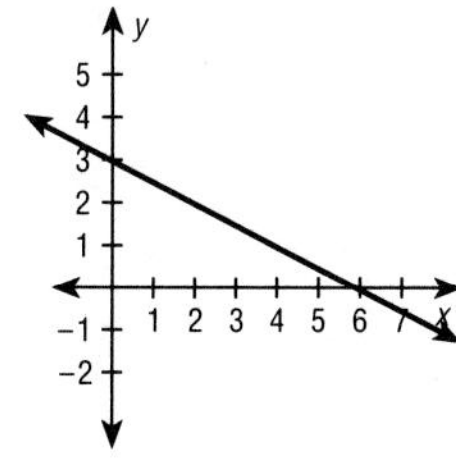

37. 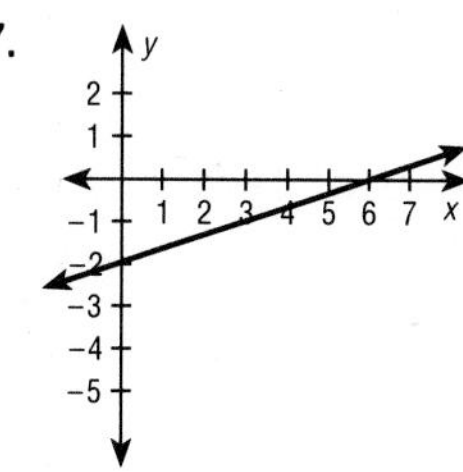

38. 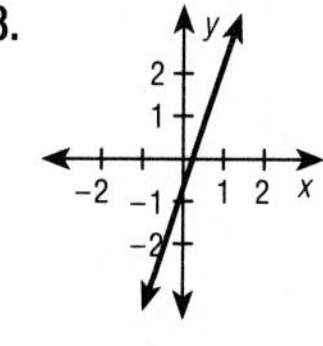

ADDITIONAL ANSWERS

CHAPTER 6

Lesson 6.1

9. −6 −5 −4 −3 −2 x

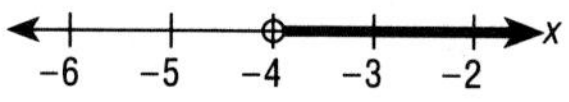

10. 13 14 15 16 17 x

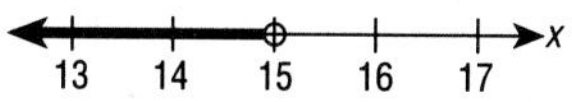

11. 7 8 9 10 11 x

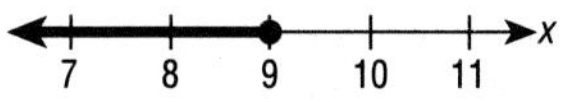

12. 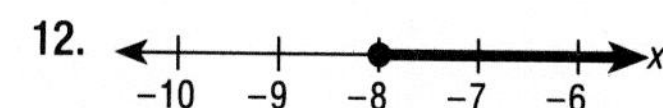

13. 0 1 2 3 4 x

14. 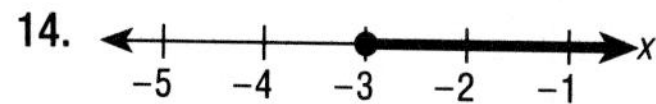

42. −2 −1 0 1 2 T

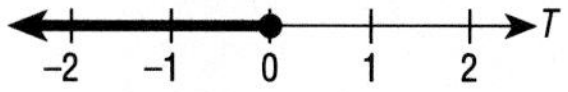

43. 

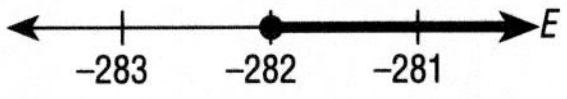

44. 1199 1200 1201 P

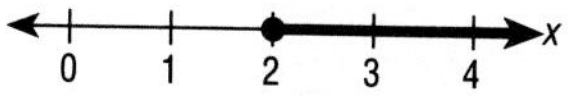

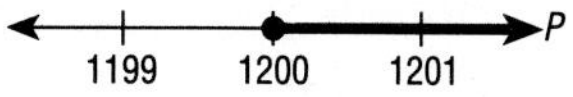

45. 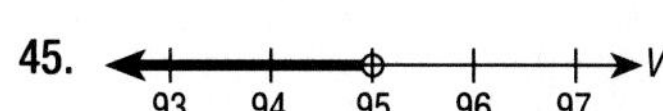

46. 13,000,000 14,000,000 15,000,000 d

Mixed Review

15. 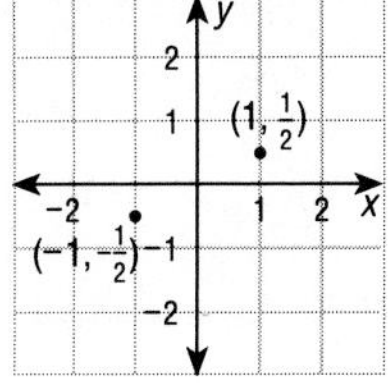

16. 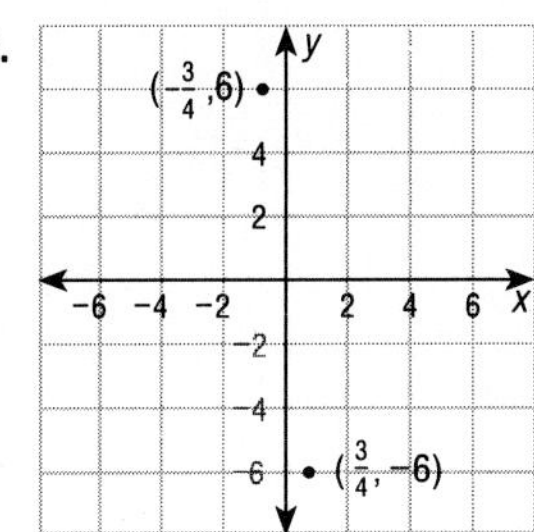

Lesson 6.3

15.

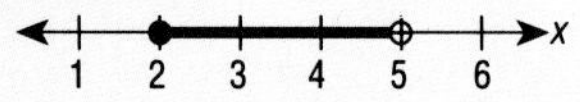

16.

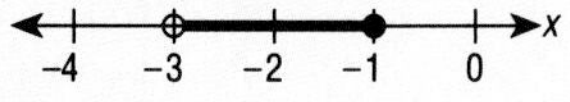

17.

18.

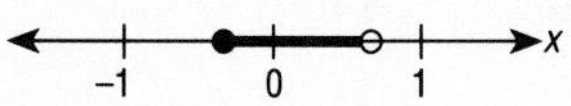

19.

20. 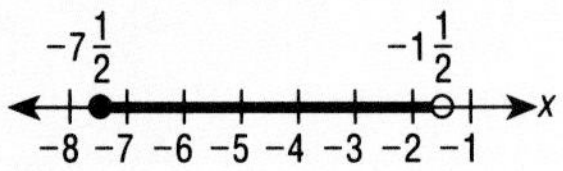 $-7\frac{1}{2}$ $-1\frac{1}{2}$

21.

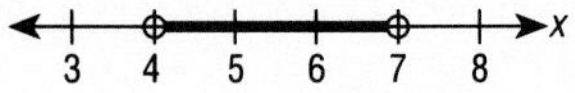

22. 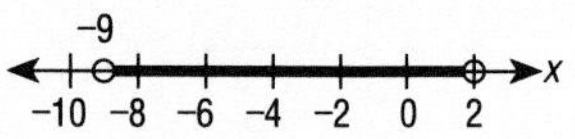 -9

23.

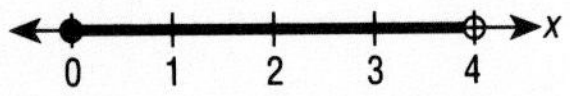

24. 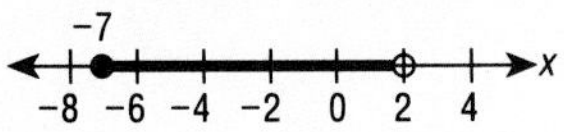 -7

Lesson 6.4

30. $-\frac{7}{2}$ $\frac{1}{2}$

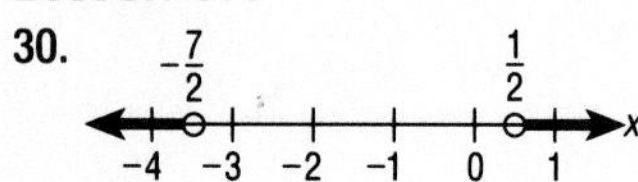

Lesson 6.5

A. The graph lies below the line.

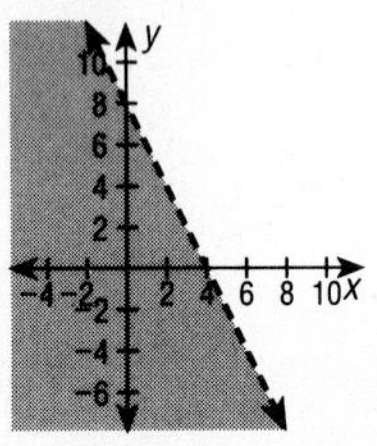

B. The graph lies above the line.

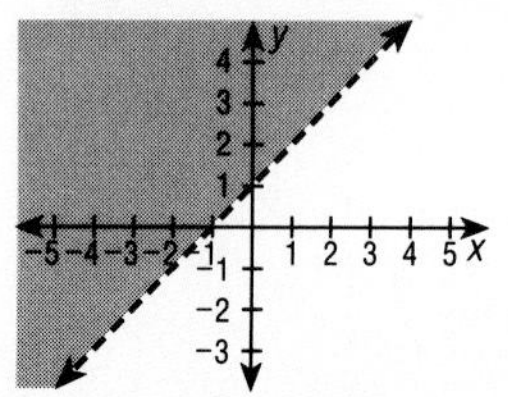

6. Write the inequality in slope-intercept form; sketch the line (here, solid) of the corresponding equation; test a point (here, the origin) to find whether it is a solution of the inequality; shade the correct half-plane.

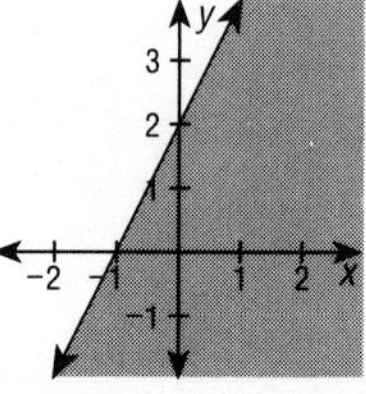

11.

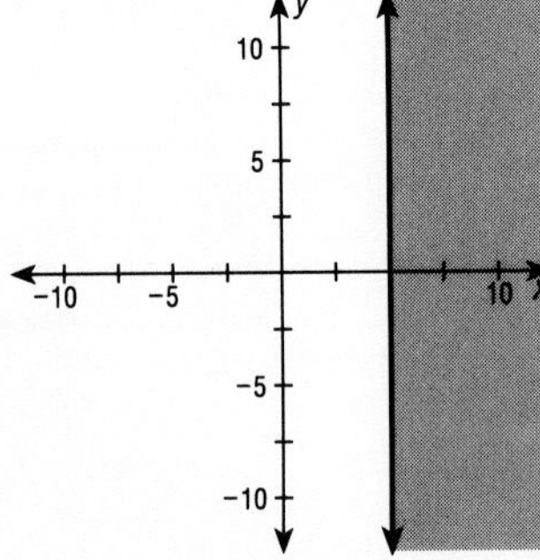

12.

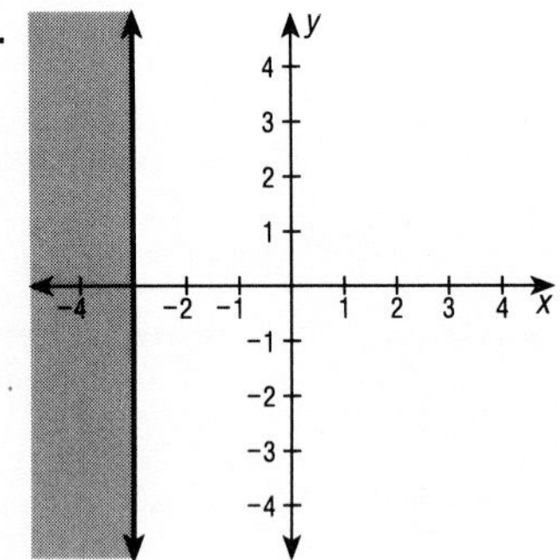

13.

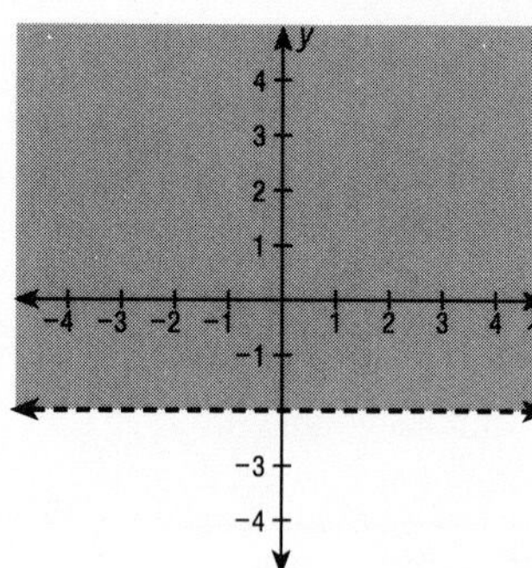

14.

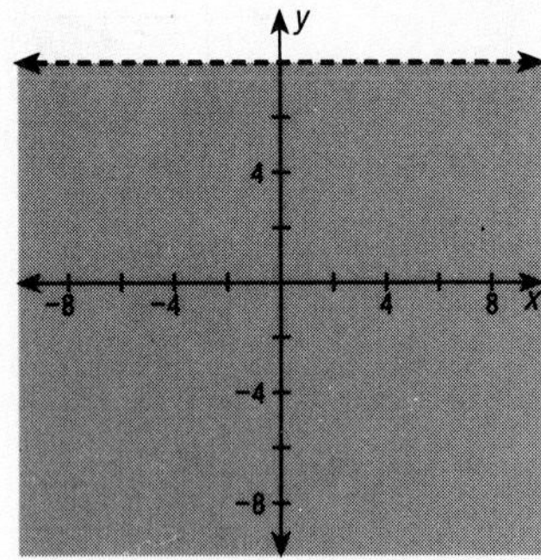

19.

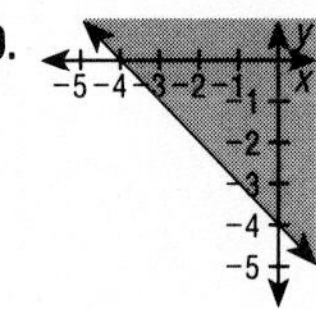

20.

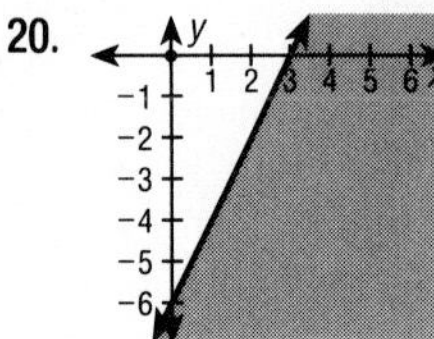

21.

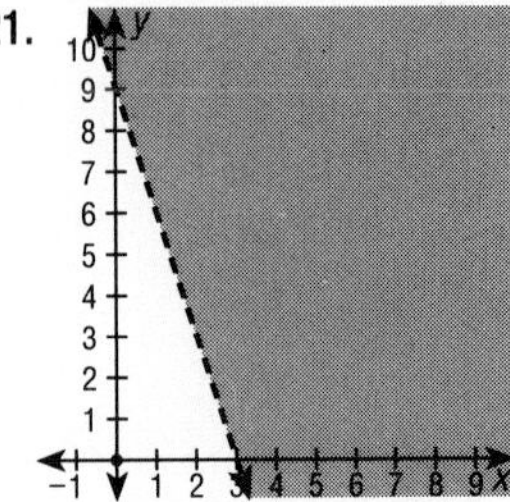

Lesson 6.5 *(continued)*

22.
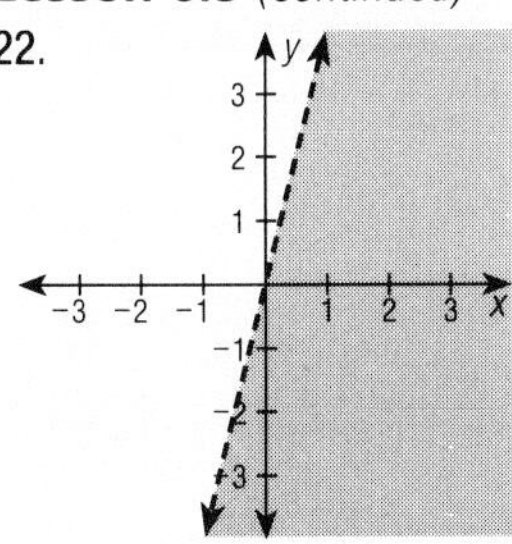

43.
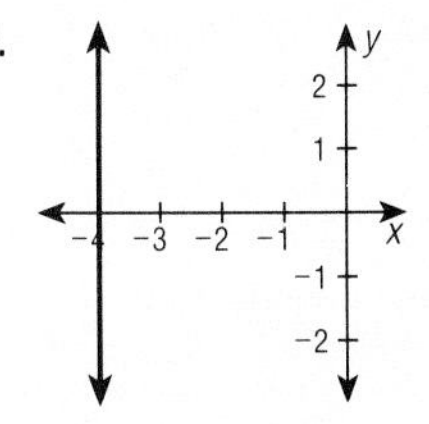

44.
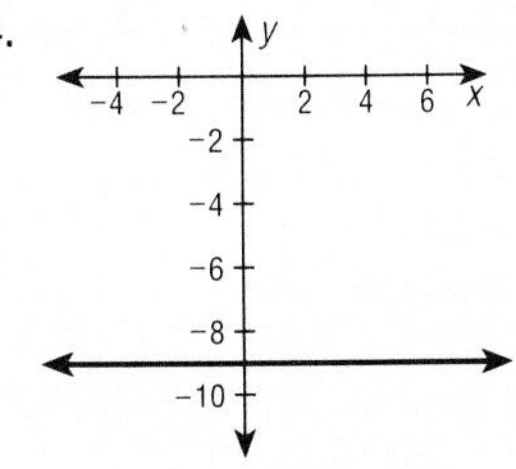

45.
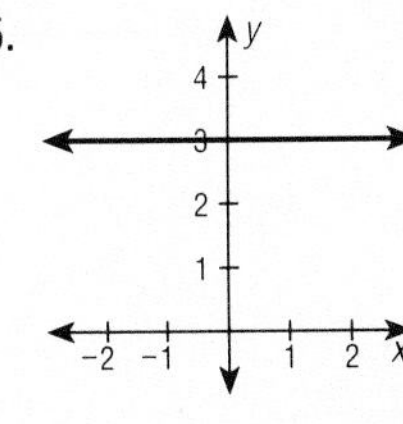

46.
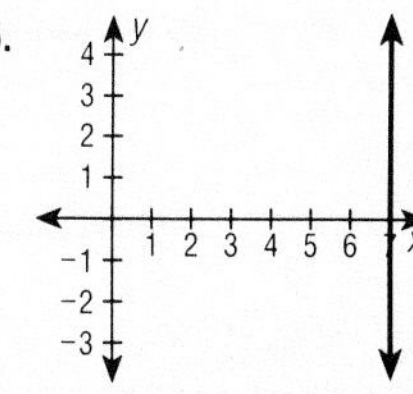

47.
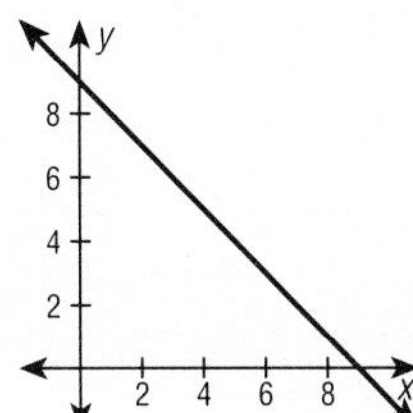

48.
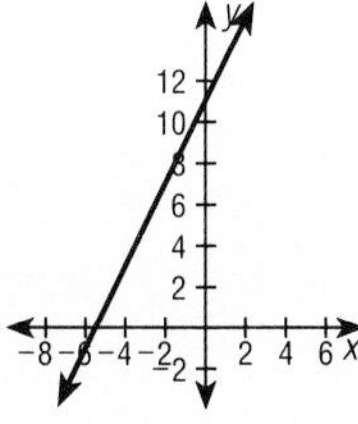

49.
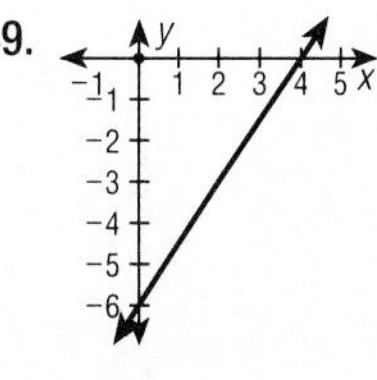

50.
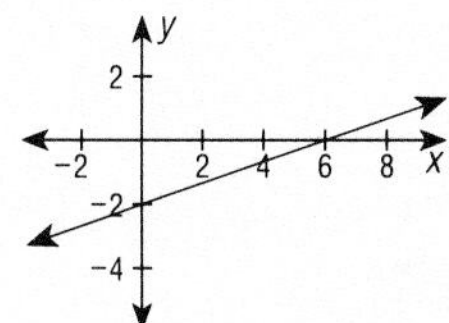

Using a Graphing Calculator

2.
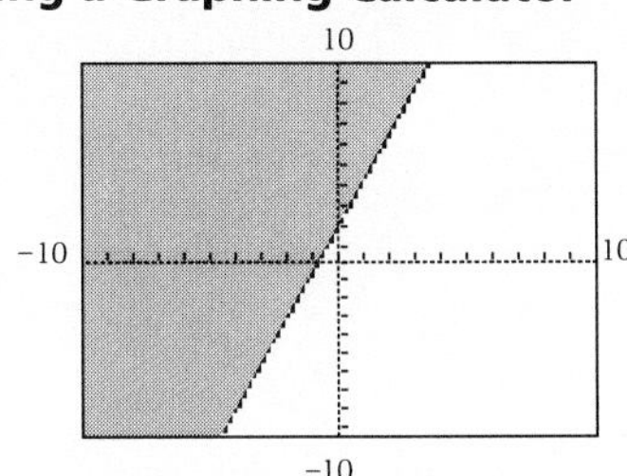

4.
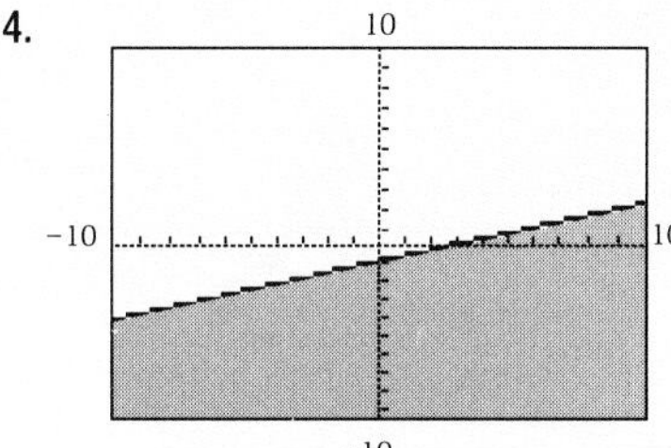

6.
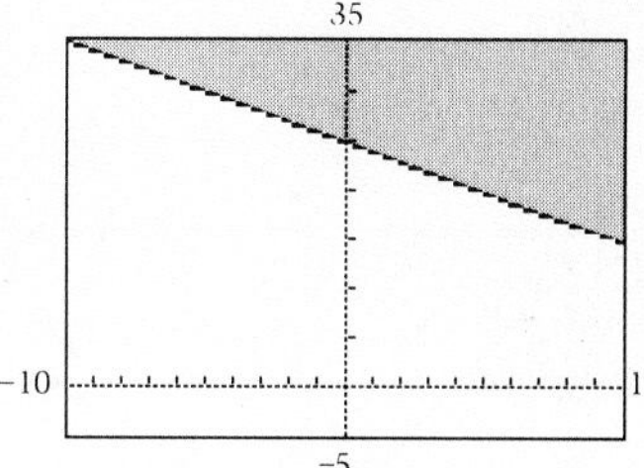

8.
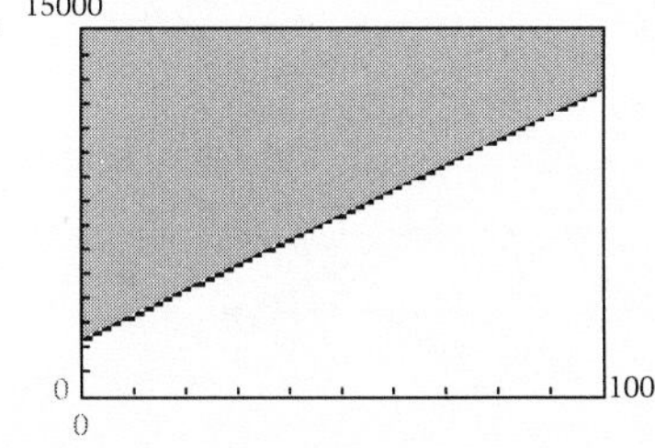

Lesson 6.6

31.
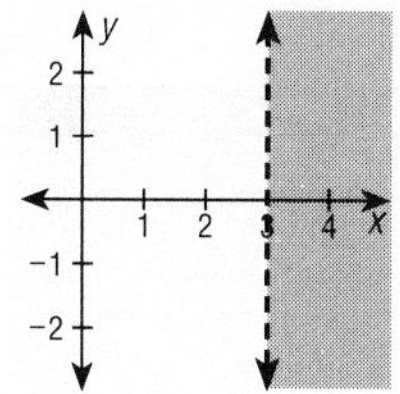

32.
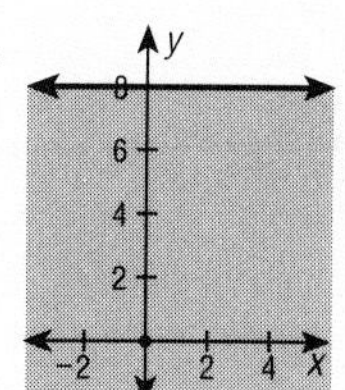

33.
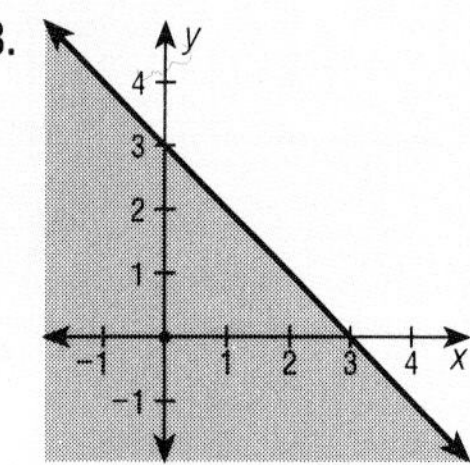

34.
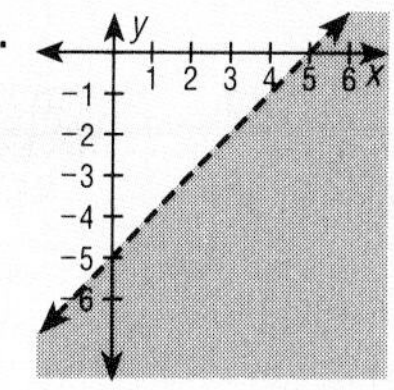

35.
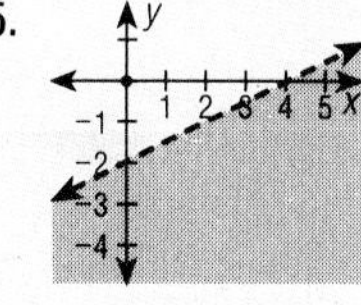

36.
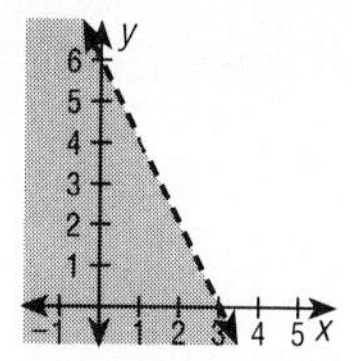

ADDITIONAL ANSWERS

Chapter Review

1. 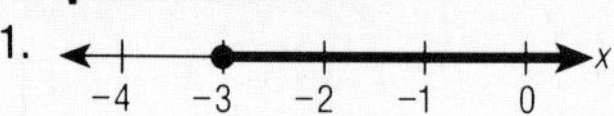

2. 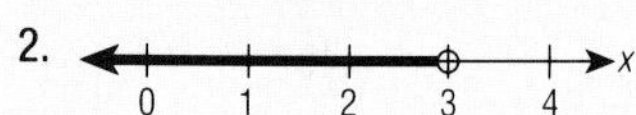

3. 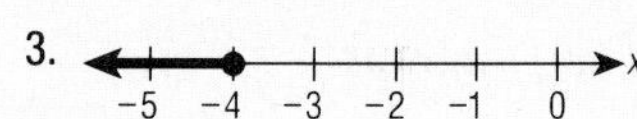

4. 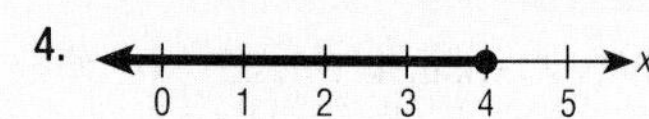

21.

22. 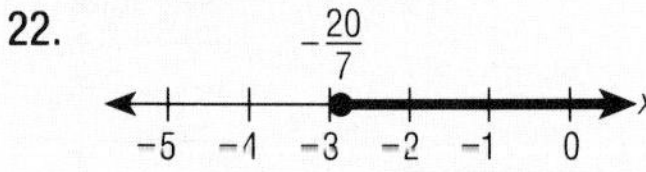

23.

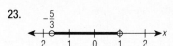

24. 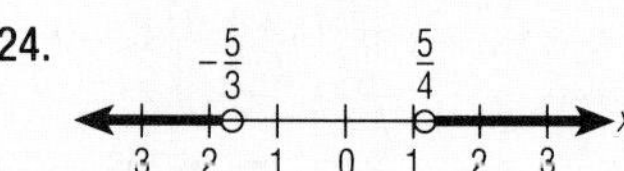

25.

26. 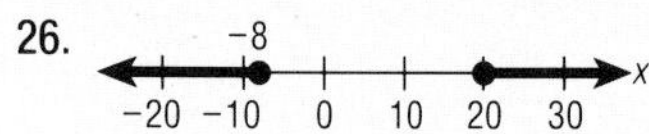

CHAPTER 7

Lesson 7.1

1. 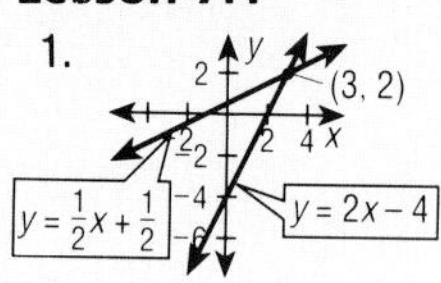

4. 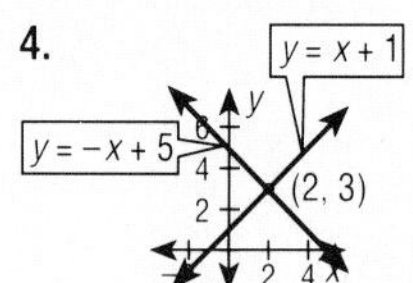

5. 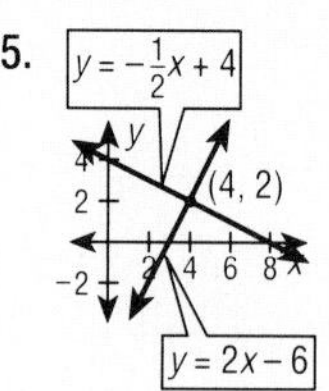

6. 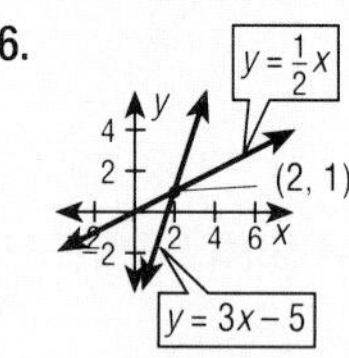

17. 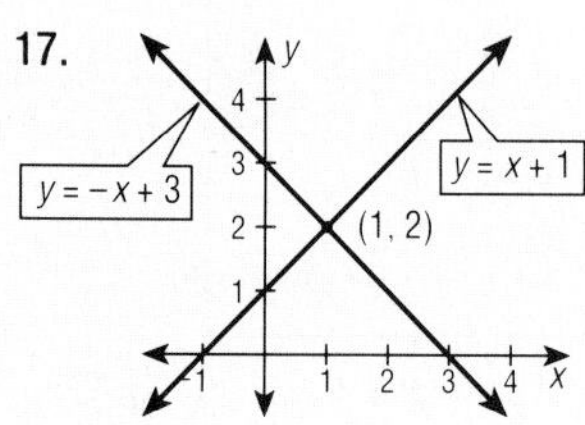

18. 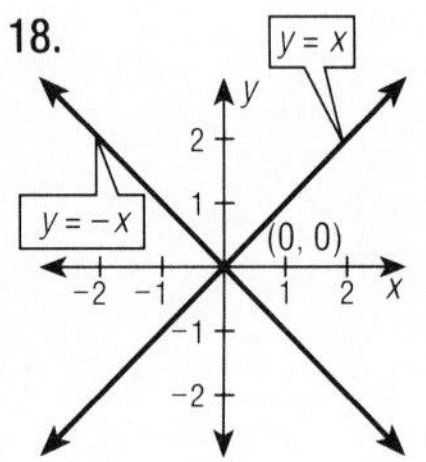

19. 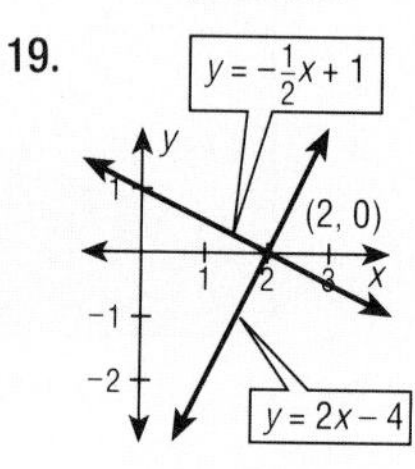

20. 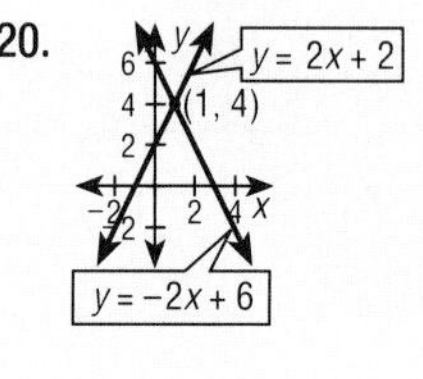

21. 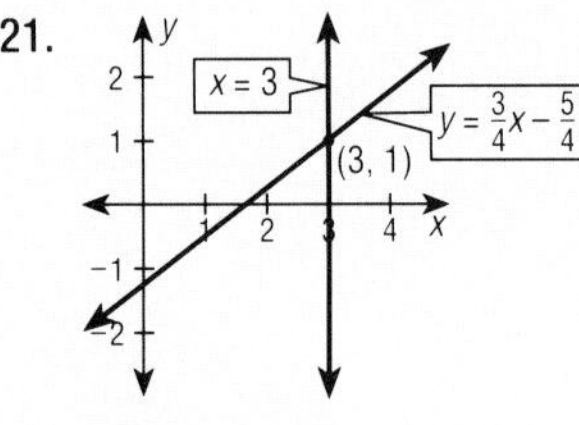

22. 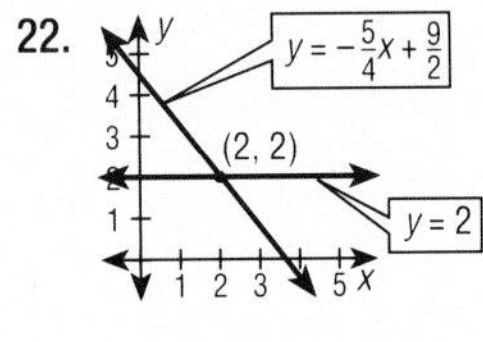

23. 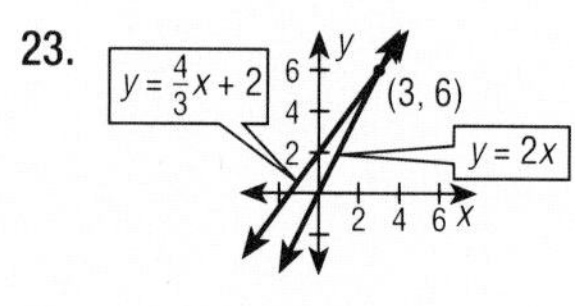

24. 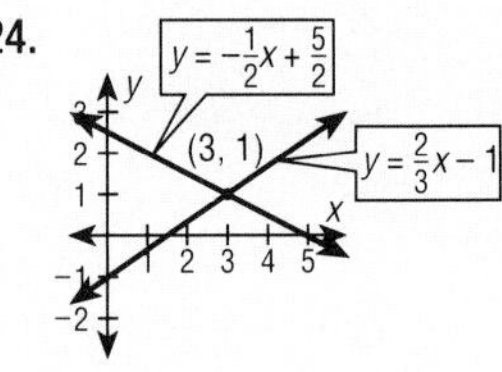

25. 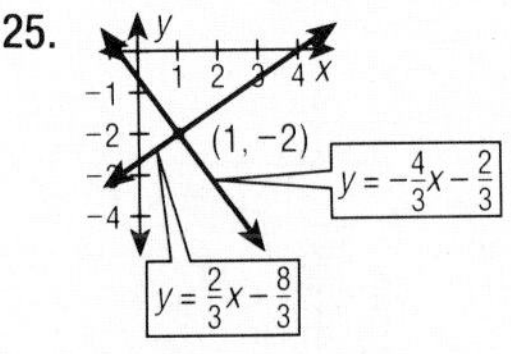

26. 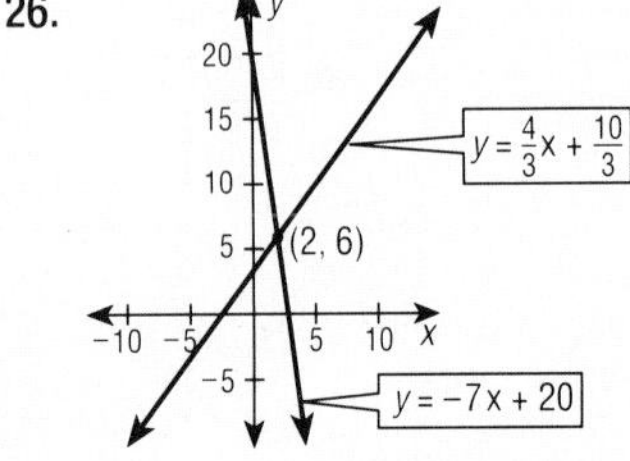

27. 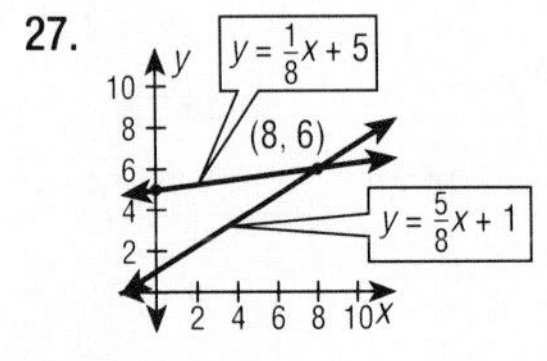

28. 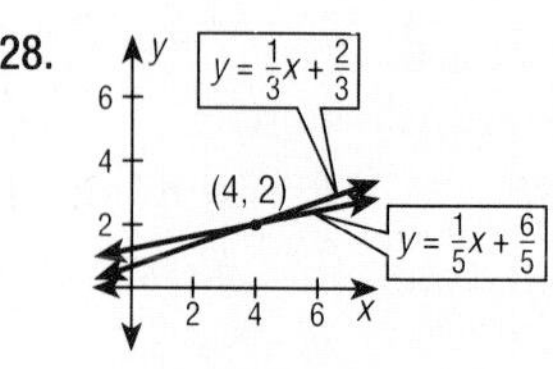

29. 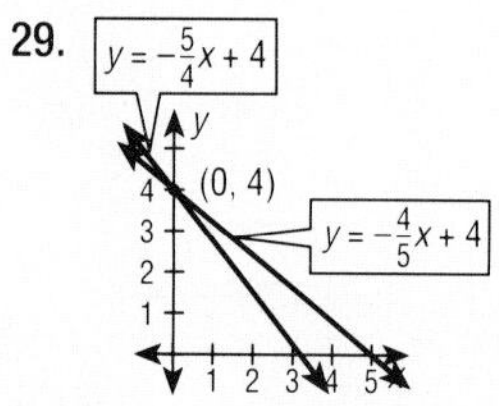

30. 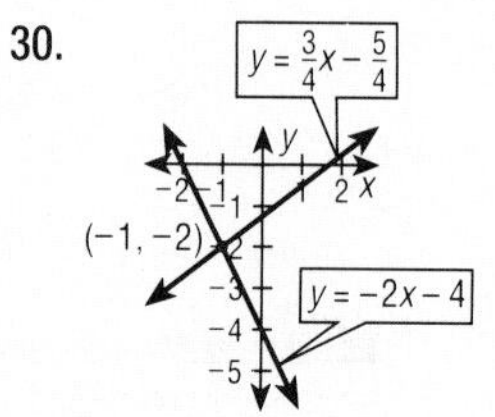

31. 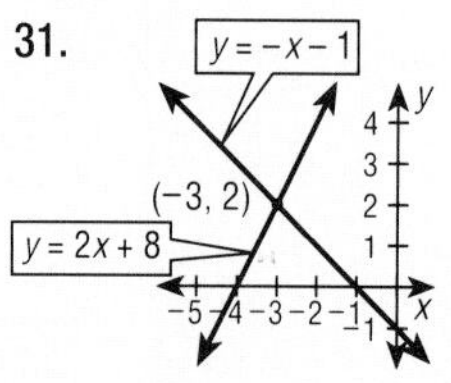

32. 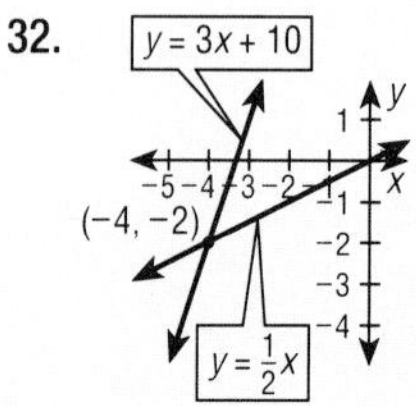

Lesson 7.1 *(continued)*

33.

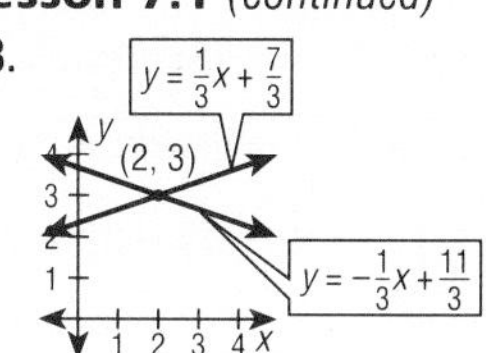

34.

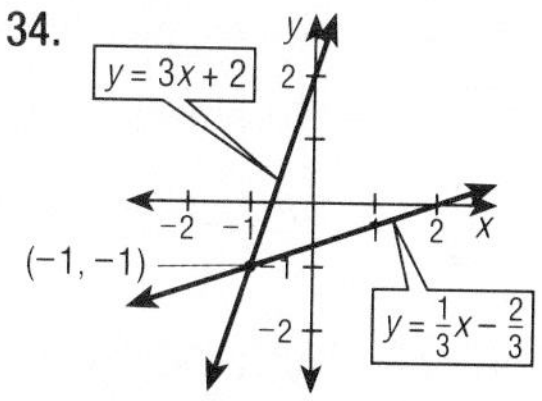

$y = 3x + 2$

(−1, −1)

$y = \frac{1}{3}x - \frac{2}{3}$

35.

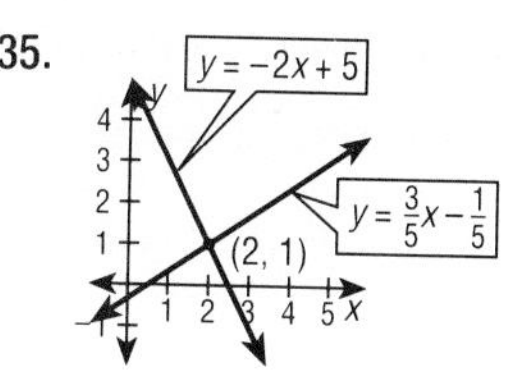

36.

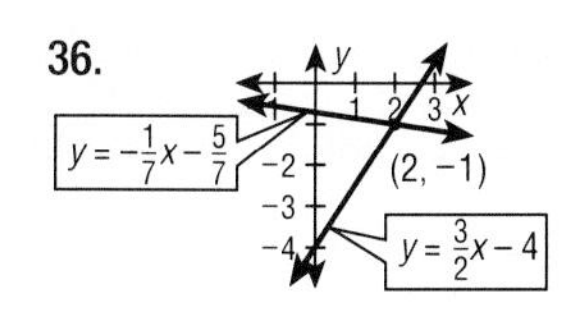

45.

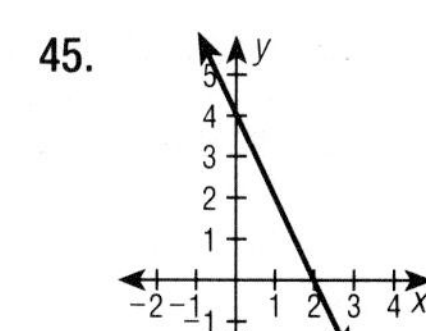

46.

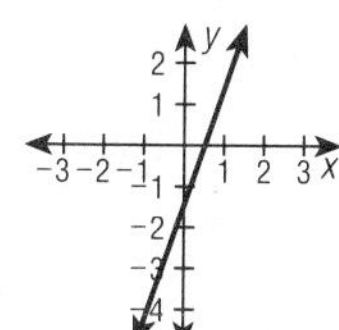

47.

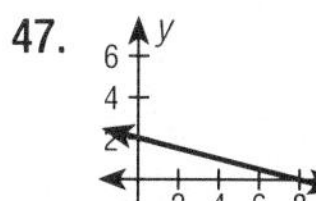

48.

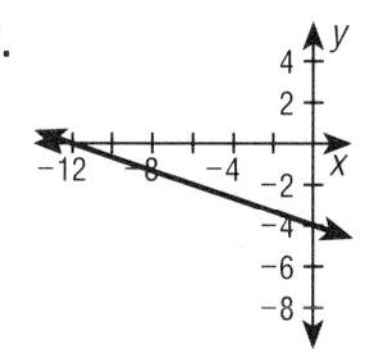

61.

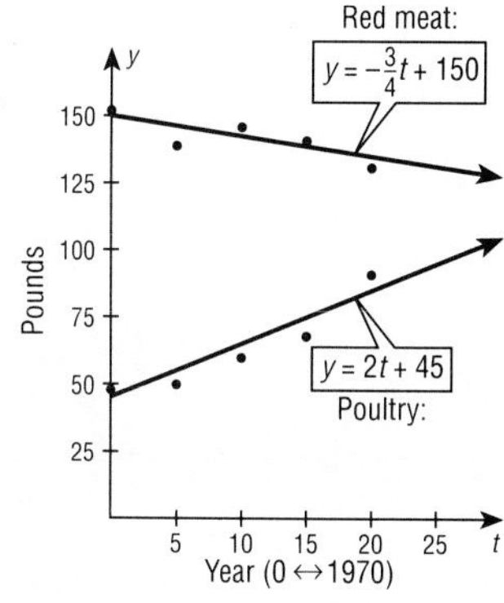

Lesson 7.2

49.

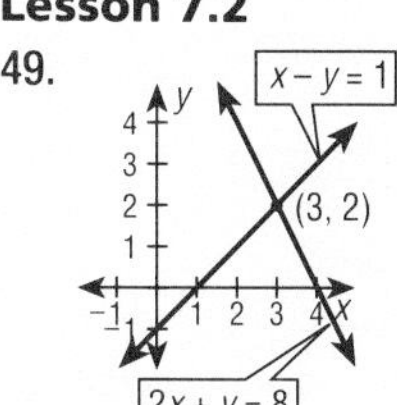

50.

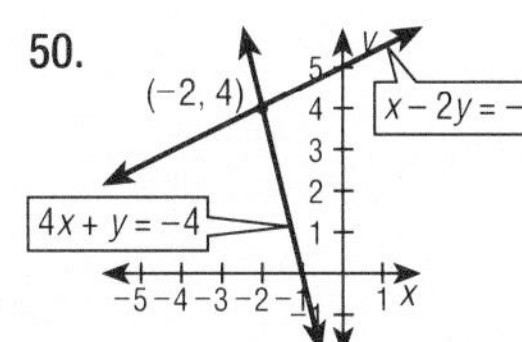

51.

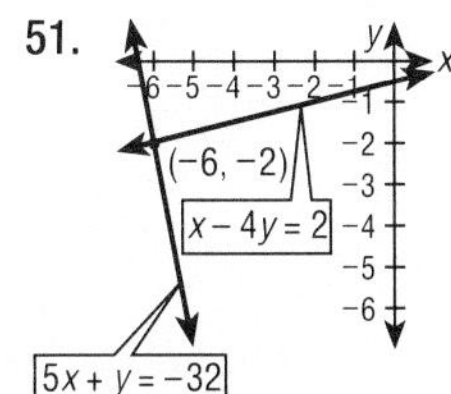

52.

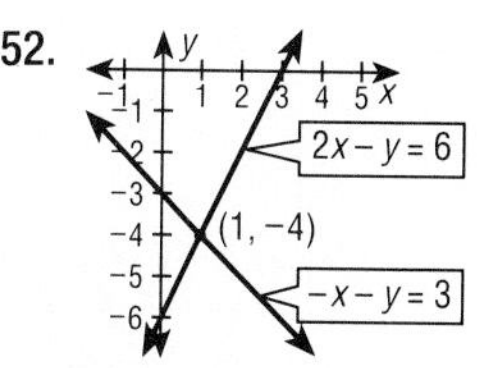

Lesson 7.3

40.

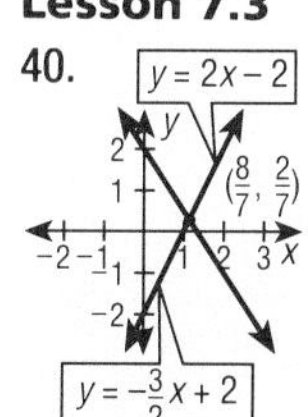

50. To get a calendar that records the seasons, divide the distance between the two points into four equal parts. The leftmost quarter represents summer, first going left and then going right; the rightmost quarter represents winter, first going right and then going left; the two middle quarters represent spring going left and fall going right.

Lesson 7.4

A. [Speed of 2nd plane] = [Speed of first plane] + 50 mph, 2 • [Speed of 1st plane] + 1.5 • [Speed of 2nd plane] = [Number of miles apart] ;

Speed of 1st plane = r (mph)
Speed of 2nd plane = s (mph)
Number of miles apart = 1825

1. [Number of gallons of regular unleaded] • [Cost per gallon of regular unleaded] + [Number of gallons of premium unleaded] • [Cost per gallon of premium unleaded] = [Total cost]

[Cost per gallon of regular unleaded] + 0.20 = [Cost of gallon of premium unleaded]

2. Number of gallons of regular unleaded = 15
Cost per gallon of regular unleaded = x (dollars)
Number of gallons of premium unleaded = 10
Cost per gallon of premium unleaded = y (dollars)
Total cost = 35.50 (dollars)

ADDITIONAL ANSWERS

Lesson 7.4 *(continued)*

19. Substitution, because y is given.

20. Substitution, because x is given.

21. Substitution, because a y-coefficient is 1.

22. Substitution, because an x-coefficient is 1.

23. Substitution, because a y-coefficient is 1.

24. Substitution, because a y-coefficient is 1.

25. Linear combinations, because no coefficient is 1.

26. Substitution, because a y-coefficient is 1.

27. Substitution, because an x-coefficient is 1.

28. Linear combinations, because no coefficient is 1.

29. Linear combinations, because no coefficient is 1.

30. Linear combinations, because no coefficient is 1.

Mid-Chapter Self-Test

7.

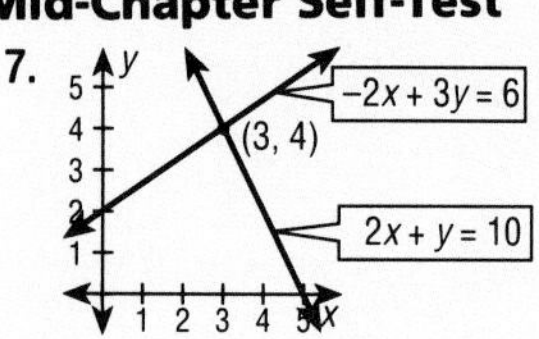

8.

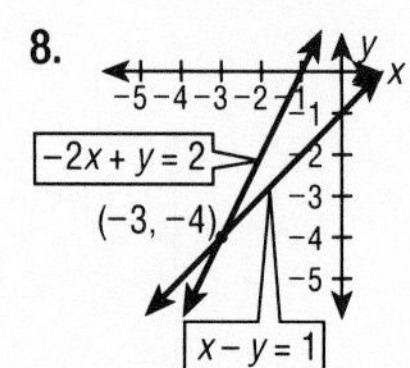

Lesson 7.5

13.

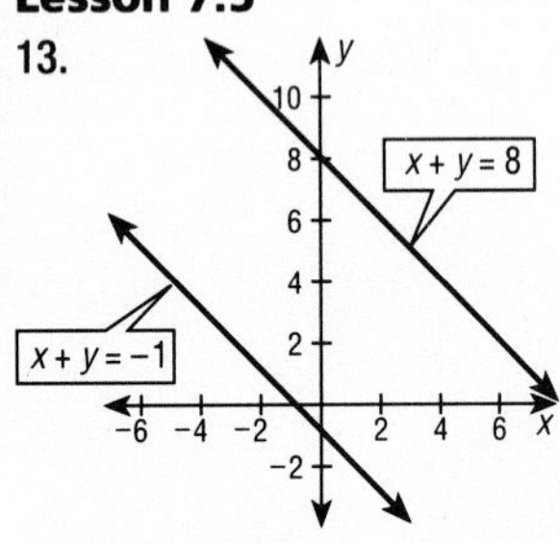

14.

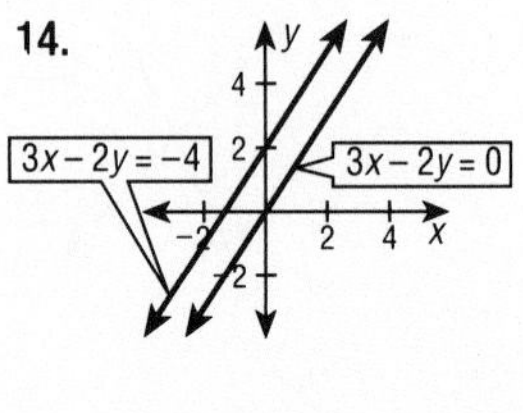

15.

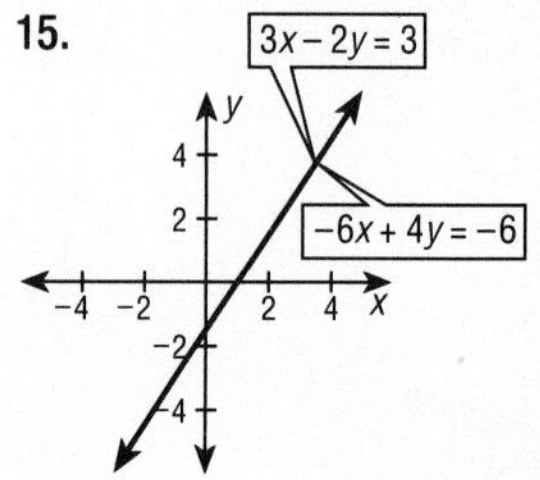

16.

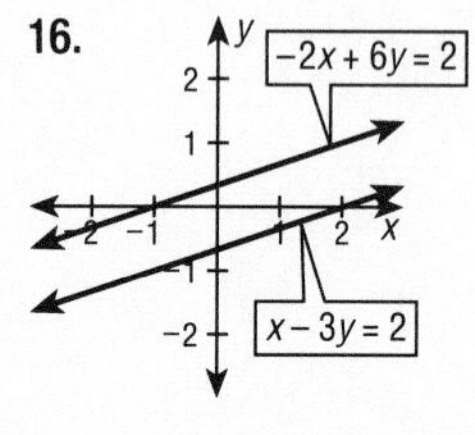

17.

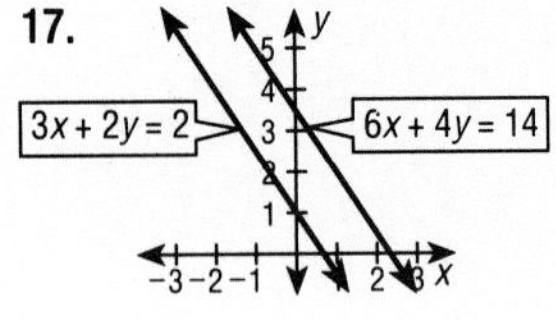

18.

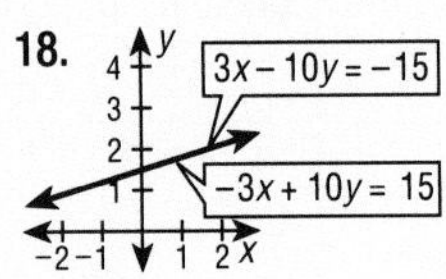

19.

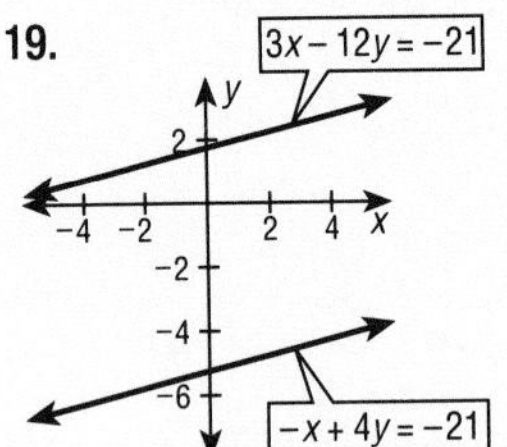

20.

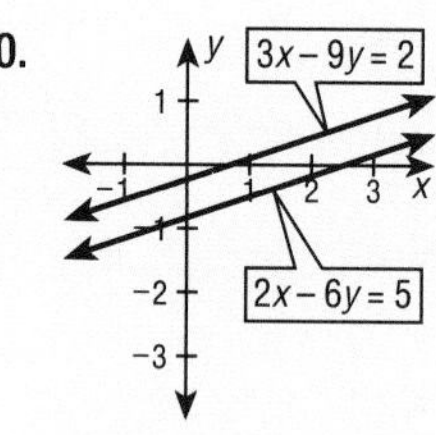

21.

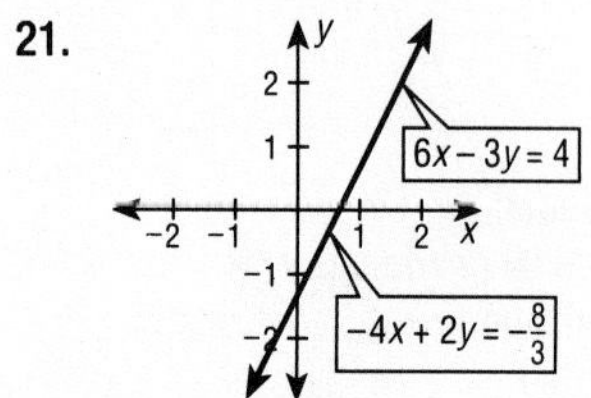

22.

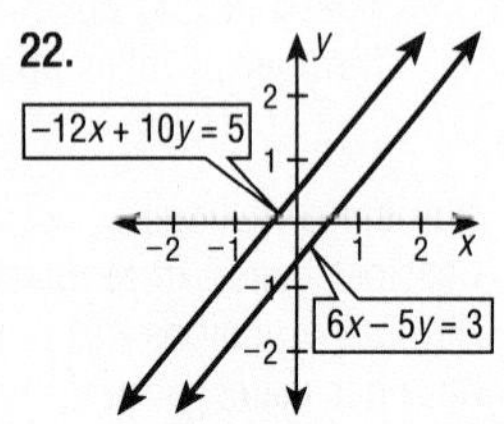

23.

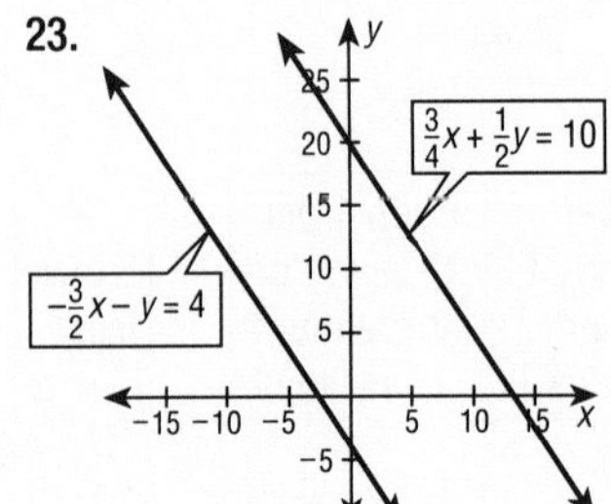

24.

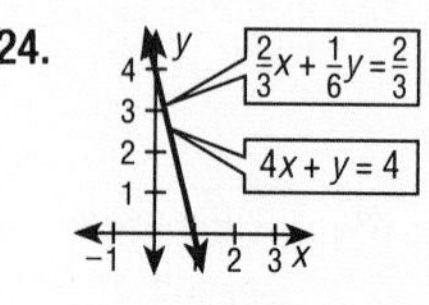

35.

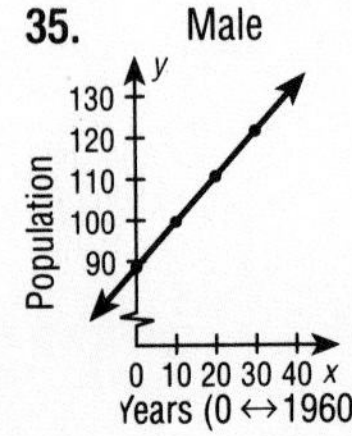

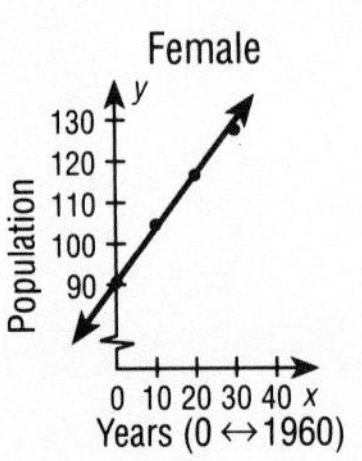

Lesson 7.6

3. $x + y \leq 2$, $1 + 0 \leq 2$;
   $x - y > 0$, $1 - 0 > 0$;
   $y \geq -2$, $0 \geq -2$

4. $x + y \leq 2$, $4 + 3 \leq 2$;
   false

5. $x + y \leq 2$, $1 + (-1) \leq 2$;
   $x - y > 0$, $1 - (-1) > 0$;
   $y \geq -2$, $-1 \geq -2$

6. $x - y > 0$, $-1 - 2 > 0$;
   false

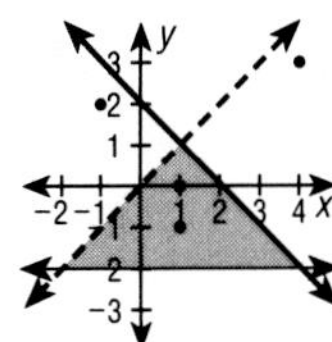

13. 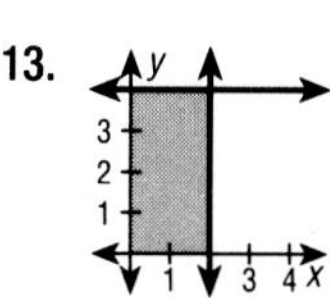

14. 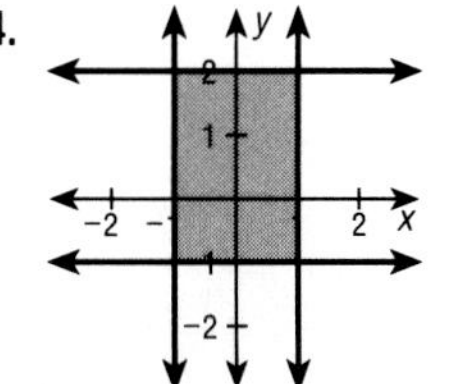

15. 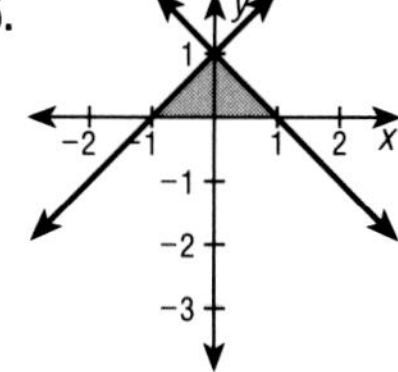

16. 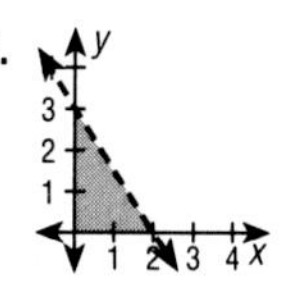

17. 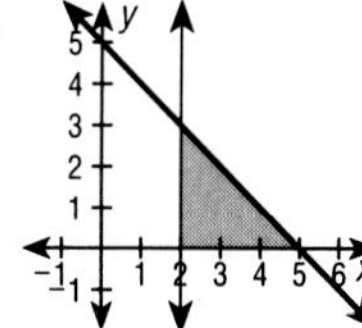

18. 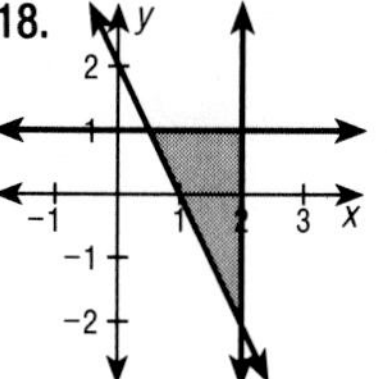

19. 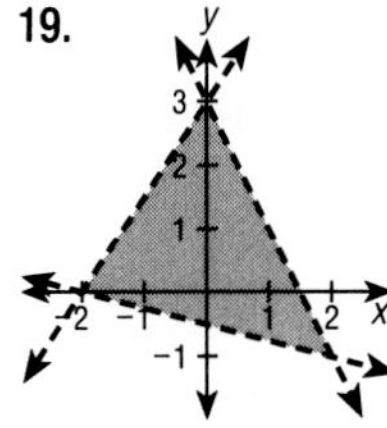

20. 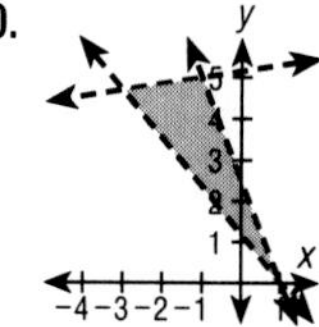

21. 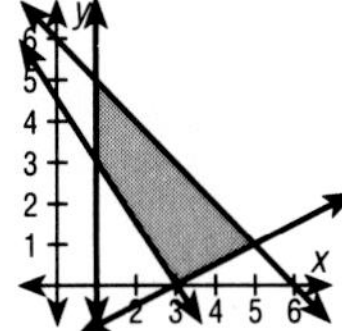

22. 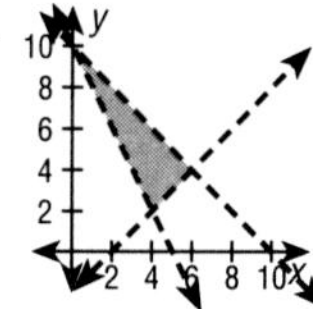

23. 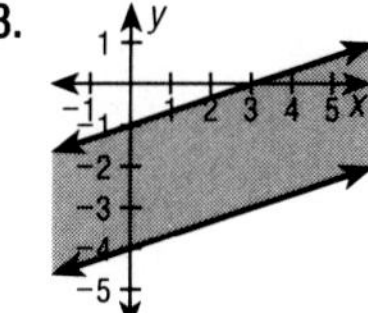

24. 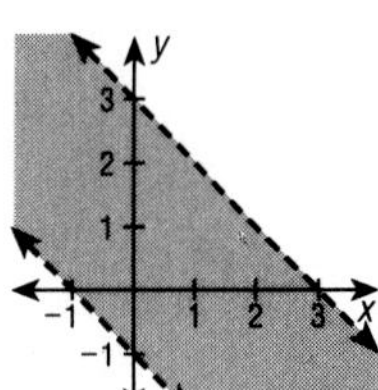

33. 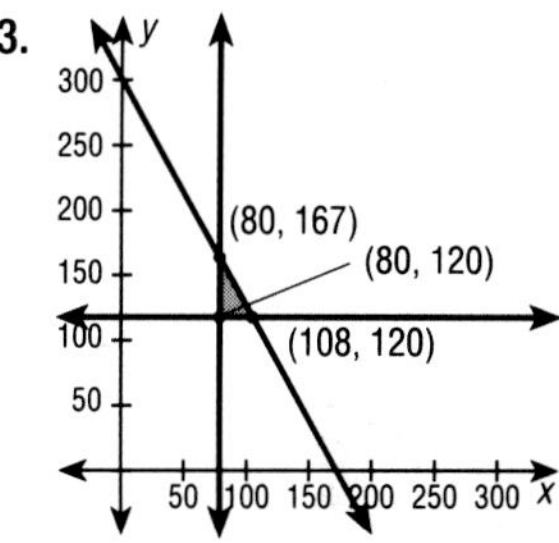

34. 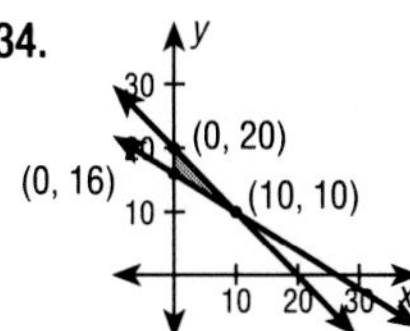

40. 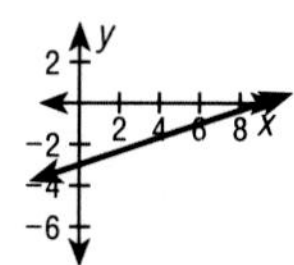

41. 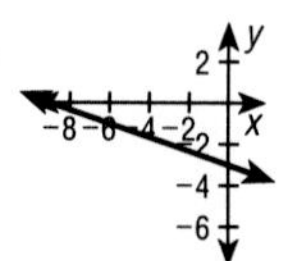

42. 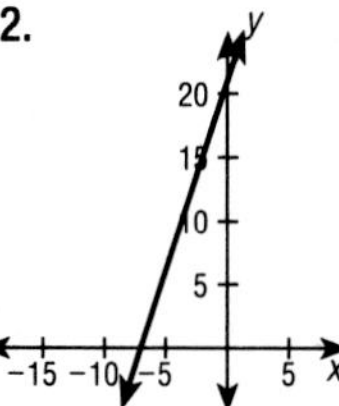

43. 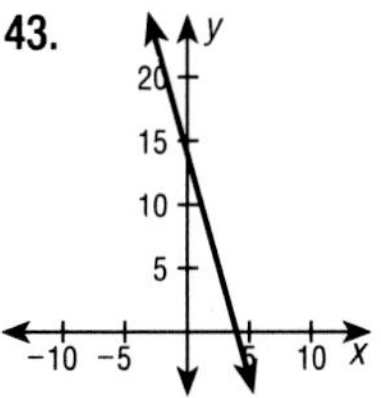

44. 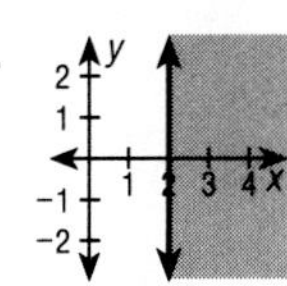

45. 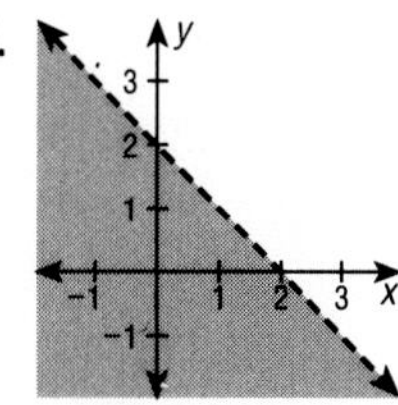

46. 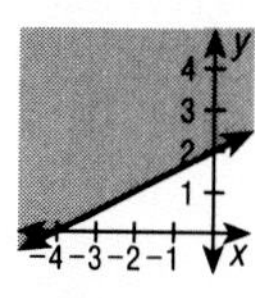

47. 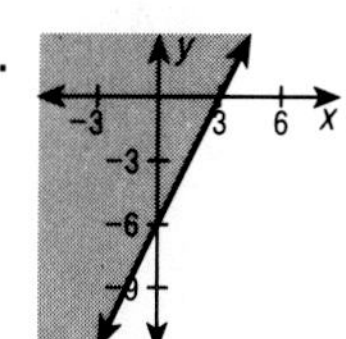

Lesson 7.7

27.

28. 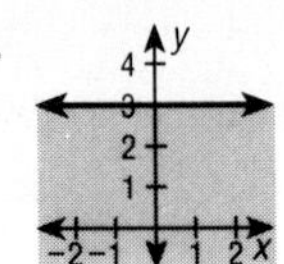

29. 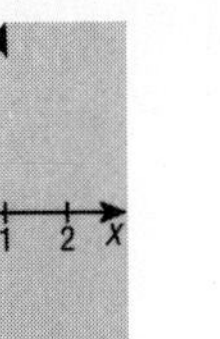

30. 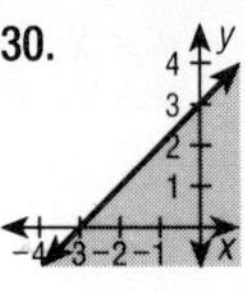

31. 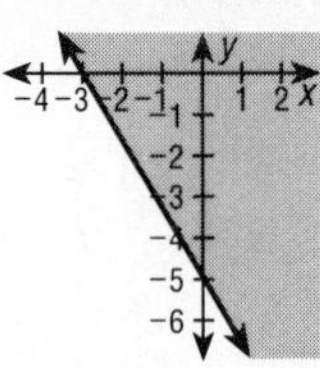

32. 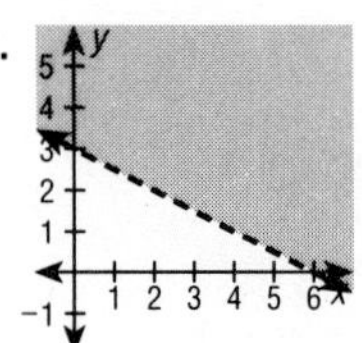

33. 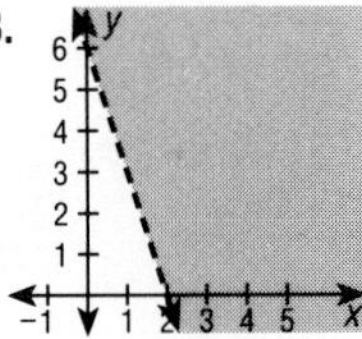

34.

CHAPTER 8

Lesson 8.1

A. 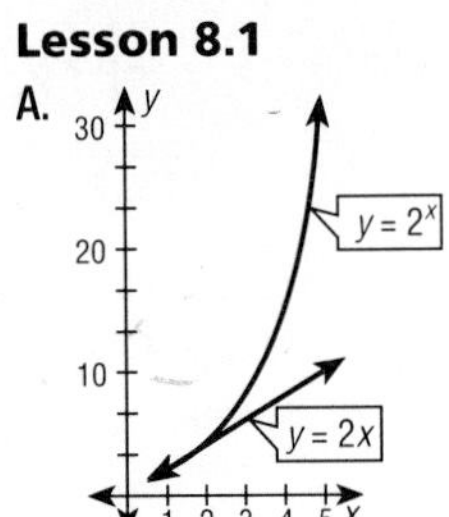

Lesson 8.5

2. 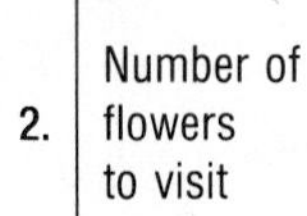

Number of flowers to visit	=	Number of flowers per pound of honey	•	Number of pounds of honey

3. Number of flowers to visit = n

Number of flowers per pound of honey $= \dfrac{(45 \text{ to } 64) \times 10^6}{2.2}$

Number of pounds of honey = 500

CHAPTER 9

Lesson 9.3

A. 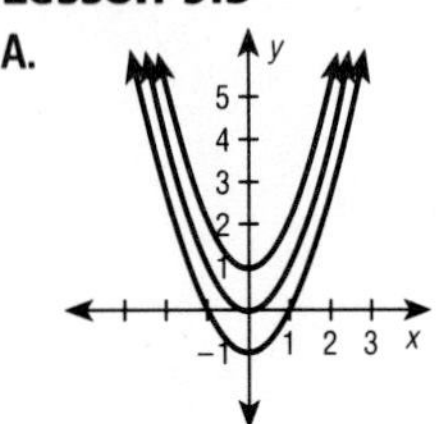

B. 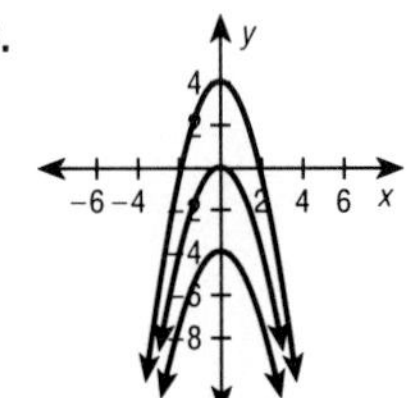

19. 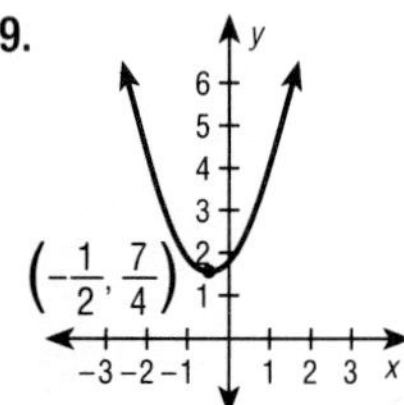

20. 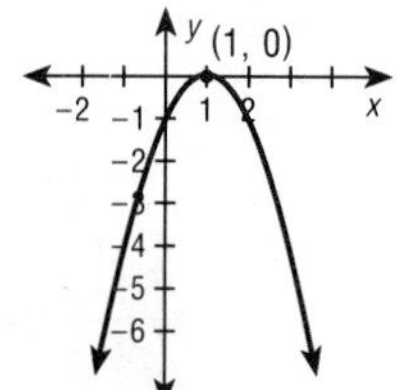

21. 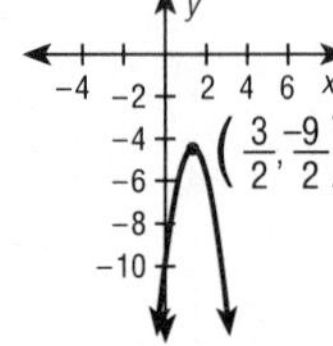

22. 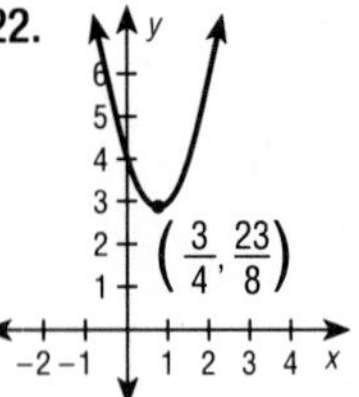

23. 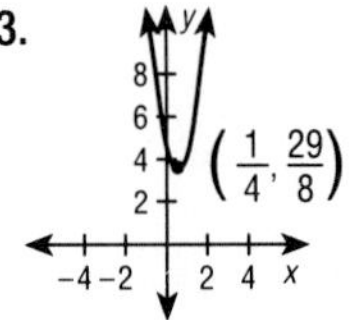

24. 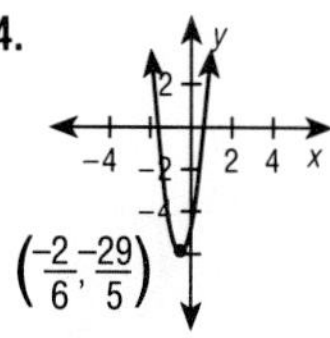

25. 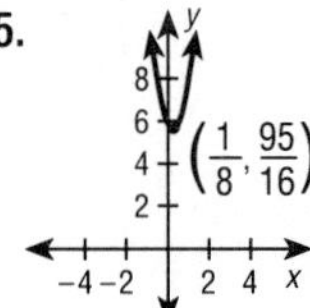

26. 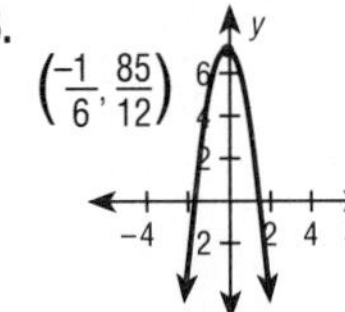

27. 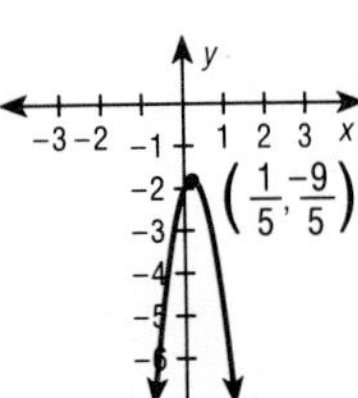

28. 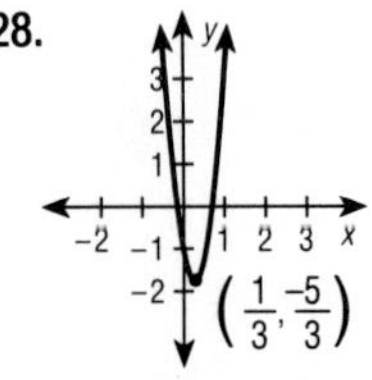

29. 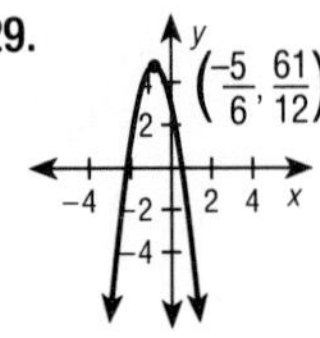

30. 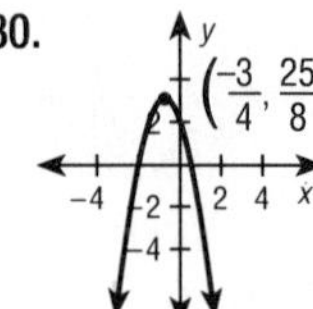

31. 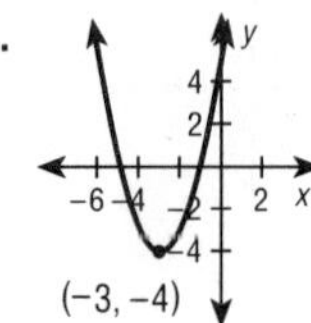

32. 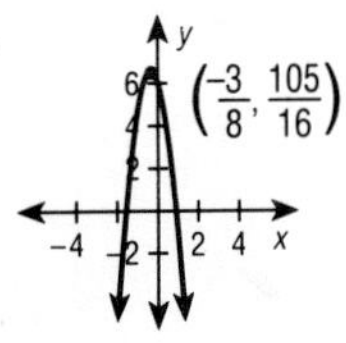

33. 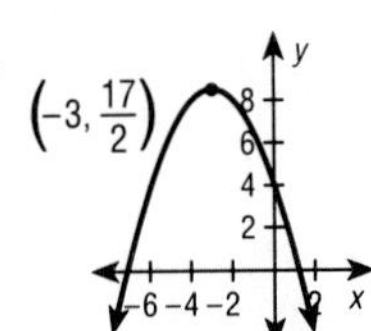

34. 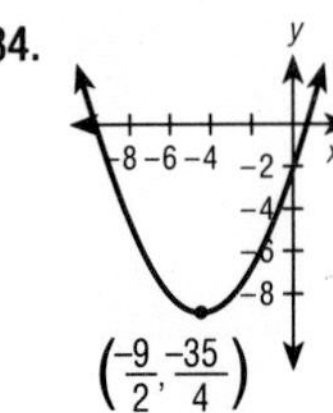

35. 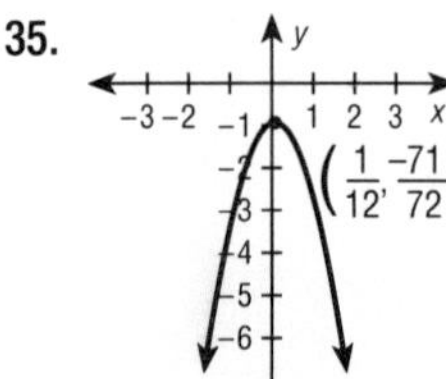

36. 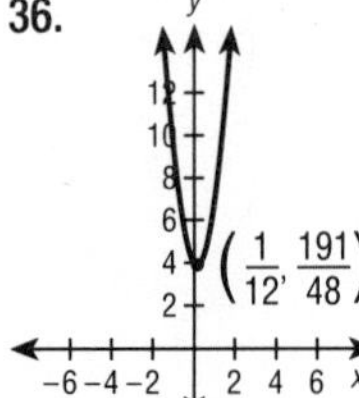

Lesson 9.3 *(continued)*

61. $y = x^2 - x + 3$, $y = x^2 - x + 1$, $y = x^2 - x - 2$

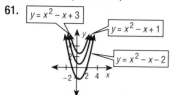

62.

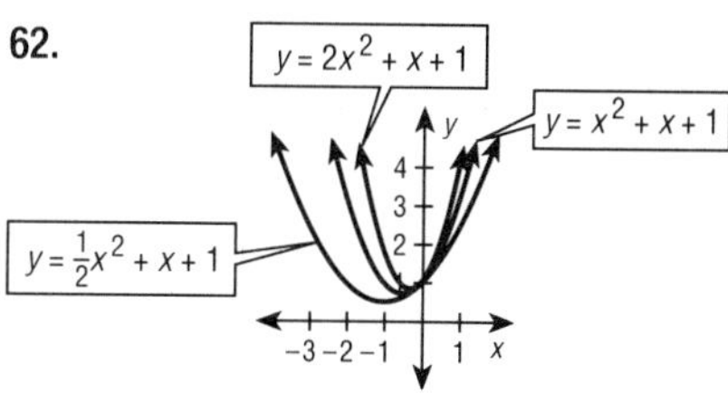

63.

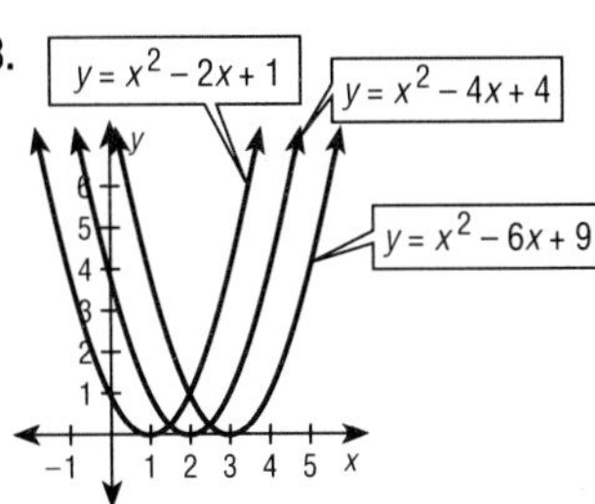

Lesson 9.4

5.

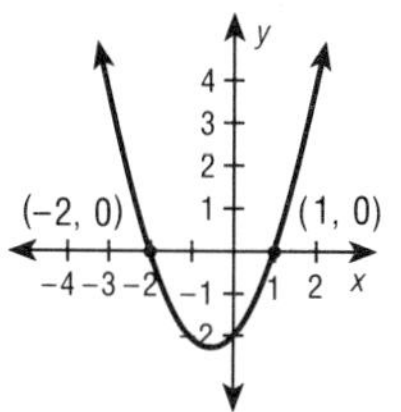

46.

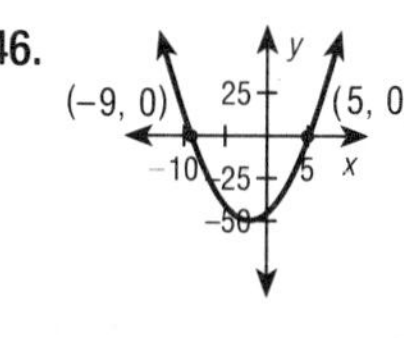

47.

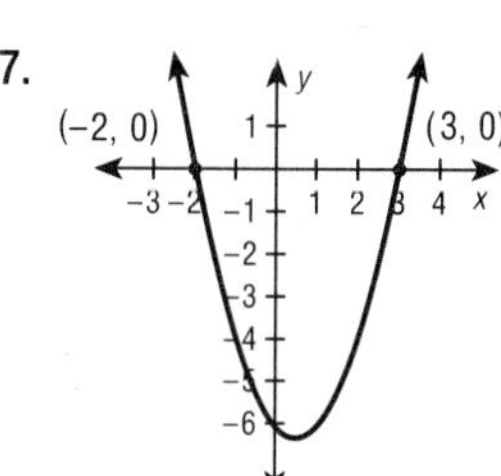

48.

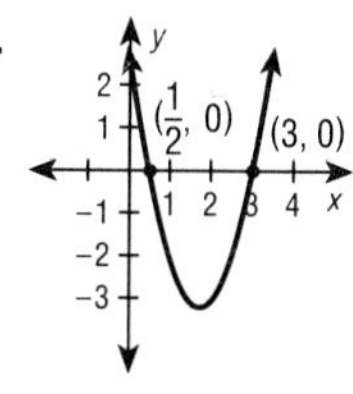

Lesson 9.5

B.

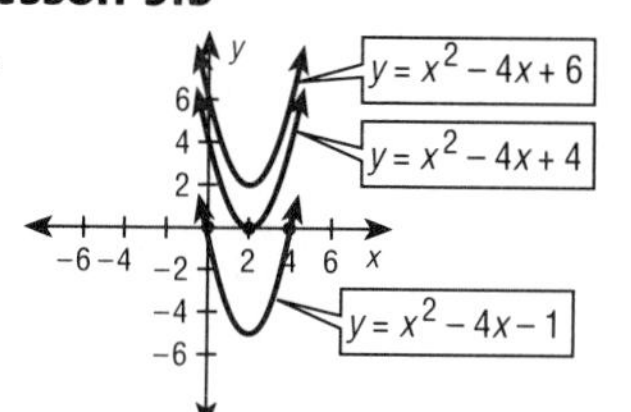

Lesson 9.6

15.

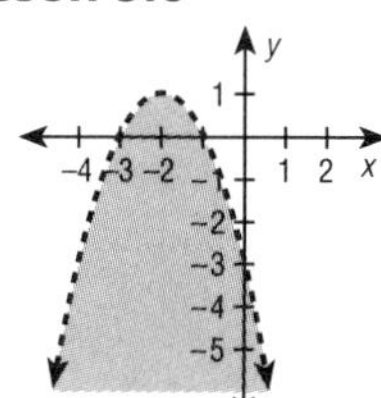

16.

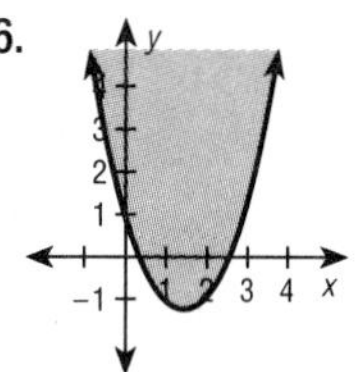

17.

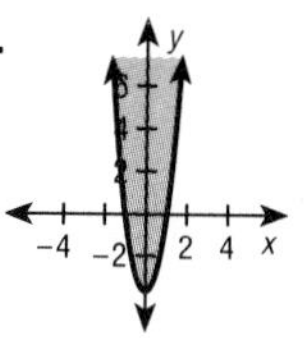

18.

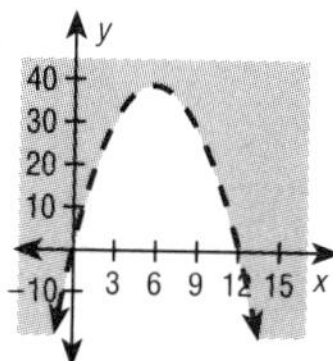

19.

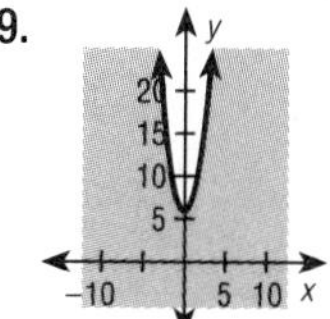

20.

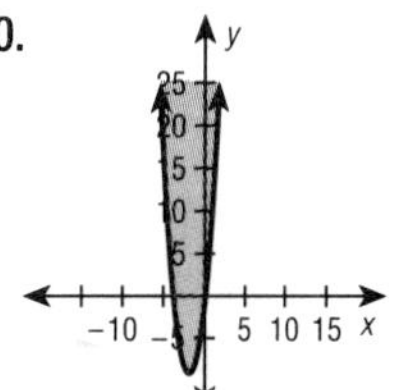

21.

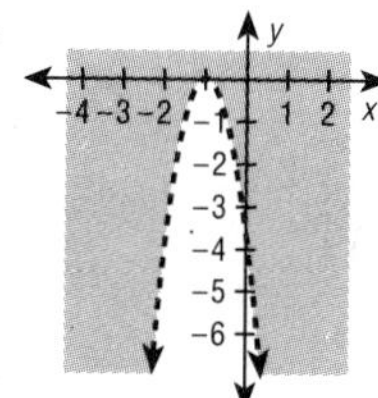

22.

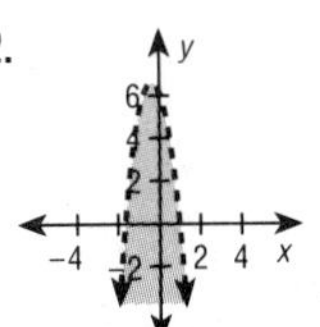

23.

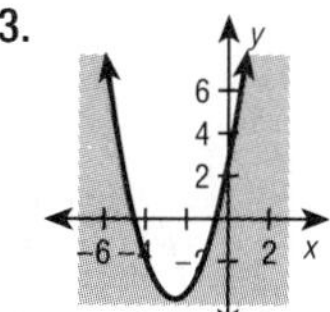

35.

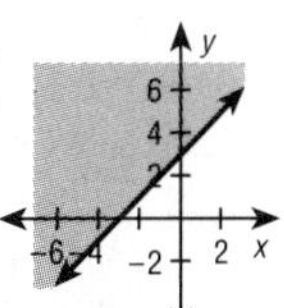

36.

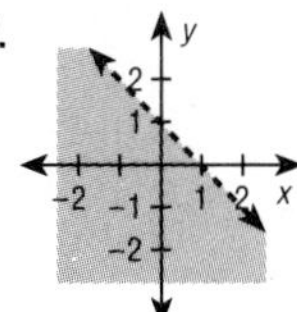

37.

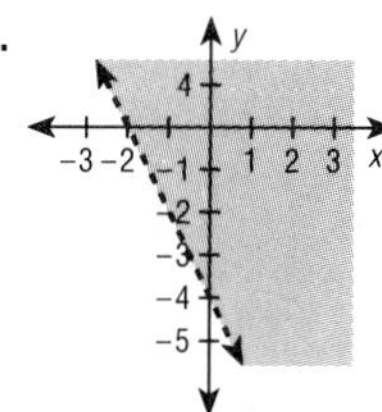

38.

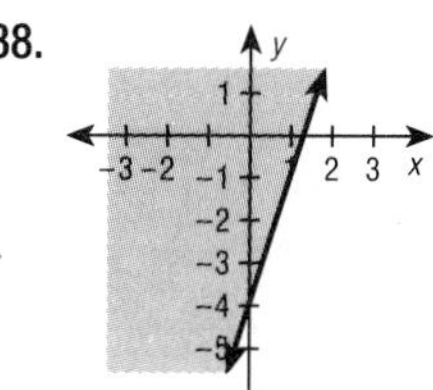

39.

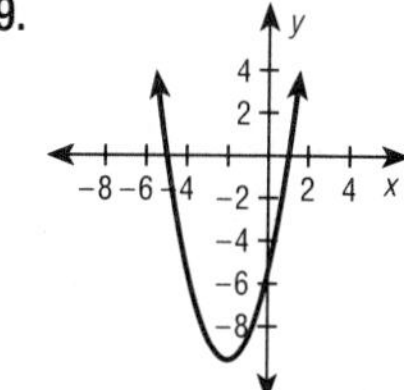

40.

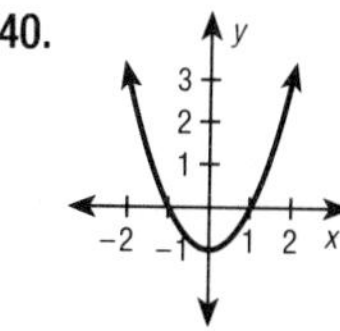

Lesson 9.6 *(continued)*

41.

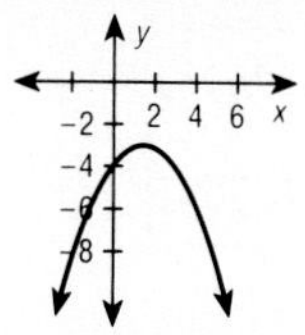

42.

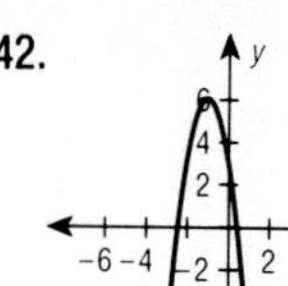

Chapter Review

17.

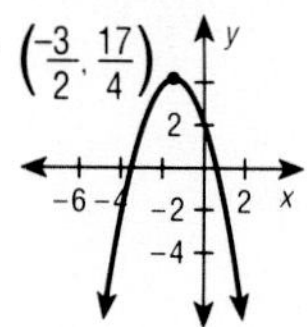

18.

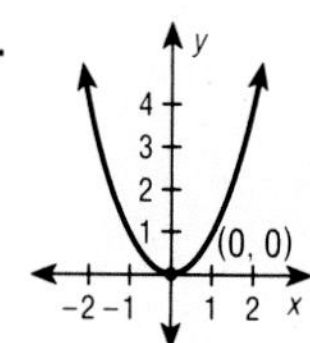

19.

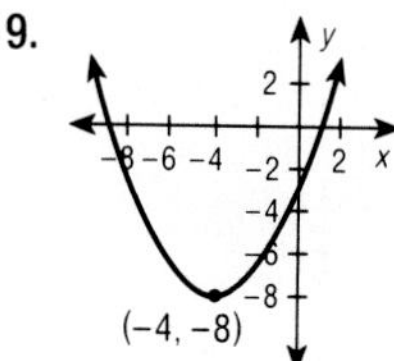

20.

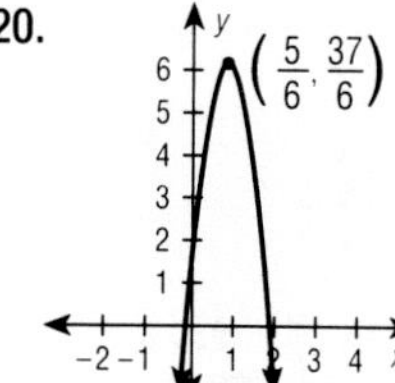

CHAPTER 10

Lesson 10.2

A. Top: $x + 3$; side: $x + 2$

	x	1	1	1
x	x^2	x	x	x
1	x	1	1	1
1	x	1	1	1

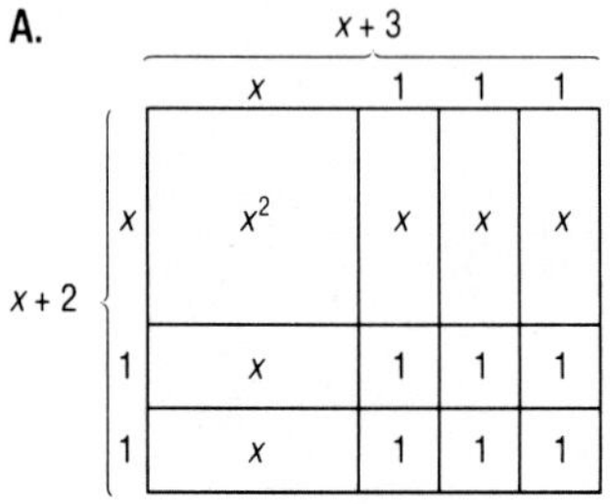

$x^2 + 5x + 6$

B. Top: $x + 1$; side: $3x + 1$

	x	1
x	x^2	x
x	x^2	x
x	x^2	x
1	x	1

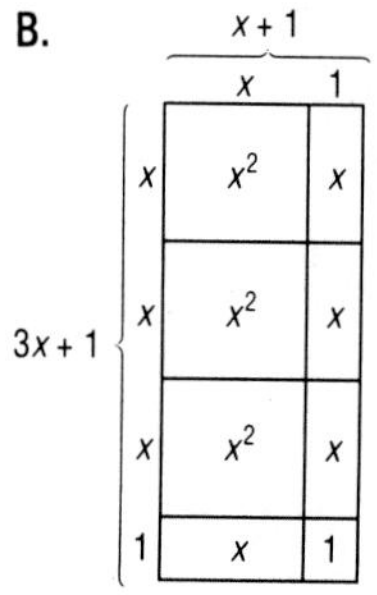

$3x^2 + 4x + 1$

17.

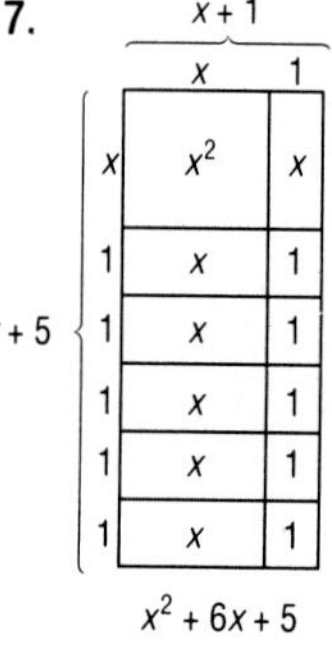

$x^2 + 6x + 5$

18.

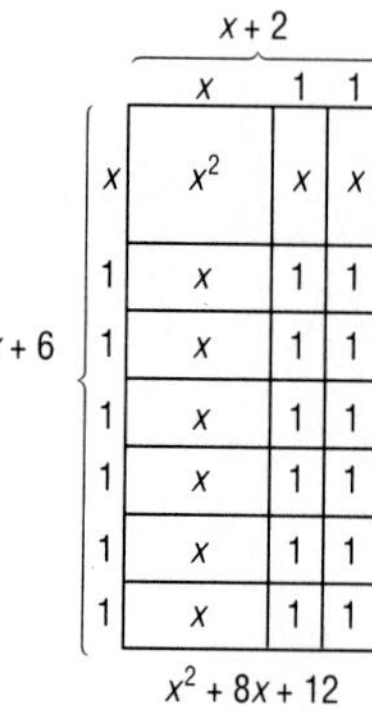

$x^2 + 8x + 12$

19.

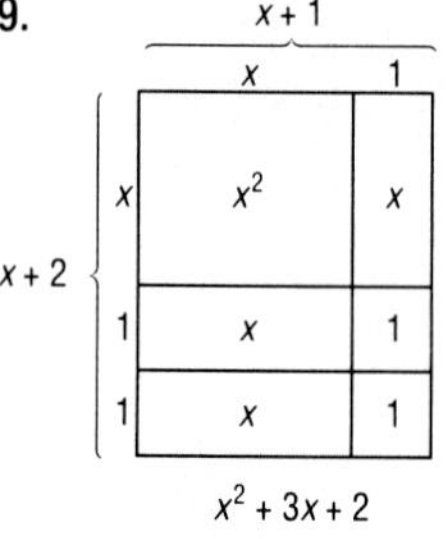

$x^2 + 3x + 2$

20. Top: $3x + 1$; side: $2x + 2$

	x	x	x	1
x	x^2	x^2	x^2	x
x	x^2	x^2	x^2	x
1	x	x	x	1
1	x	x	x	1

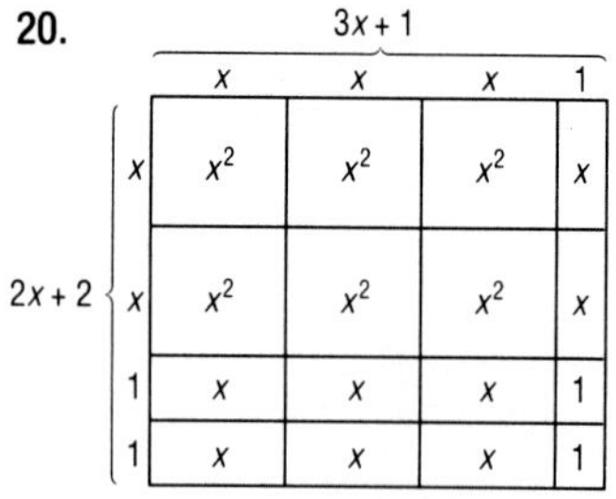

$6x^2 + 8x + 2$

21.

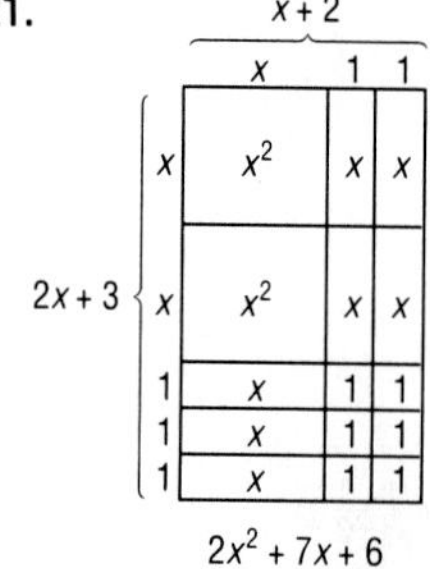

$2x^2 + 7x + 6$

22.

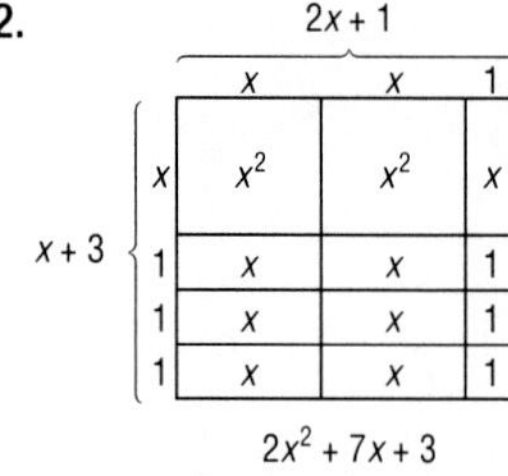

$2x^2 + 7x + 3$

Lesson 10.3

5. Top: $x + 2$; side: $x + 2$

	x	1	1
x	x^2	x	x
1	x	1	1
1	x	1	1

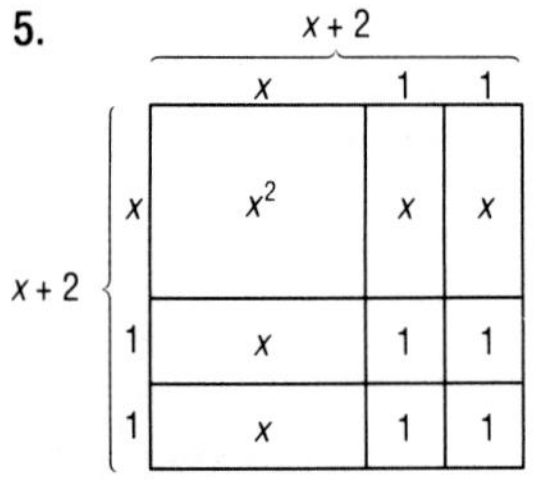

6. Top: $x + 3$; side: $x + 3$

	x	1	1	1
x	x^2	x	x	x
1	x	1	1	1
1	x	1	1	1
1	x	1	1	1

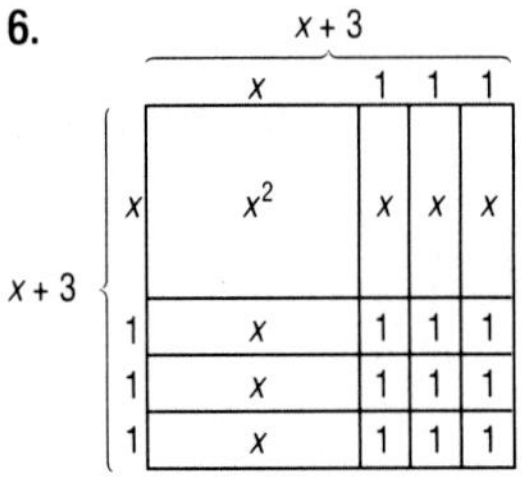

7. Top: $2n + 1$; side: $2n + 1$

	n	n	1
n	n^2	n^2	n
n	n^2	n^2	n
1	n	n	1

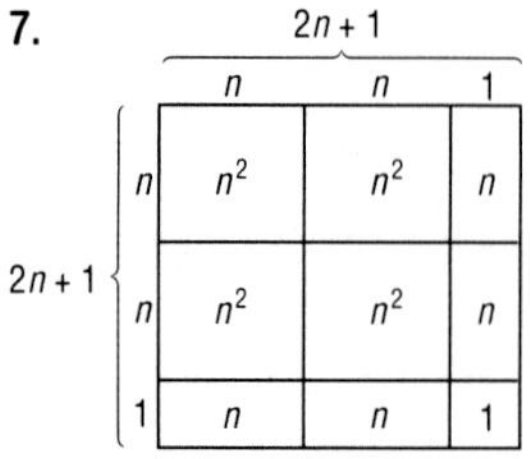

8. Top: $3a + 2$; side: $3a + 2$

	a	a	a	1	1
a	a^2	a^2	a^2	a	a
a	a^2	a^2	a^2	a	a
a	a^2	a^2	a^2	a	a
1	a	a	a	1	1
1	a	a	a	1	1

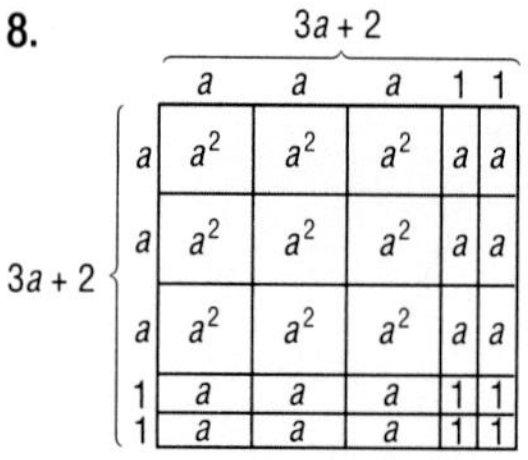

Lesson 10.3 *(continued)*

9. $2x + 2$ by $2x + 2$

	x	x	1	1
x	x^2	x^2	x	x
x	x^2	x^2	x	x
1	x	x	1	1
1	x	x	1	1

10. $3x + 1$ by $3x + 1$

	x	x	x	1
x	x^2	x^2	x^2	x
x	x^2	x^2	x^2	x
x	x^2	x^2	x^2	x
1	x	x	x	1

Lesson 10.5

A. $(x + 4)(x + 3)$; need two factors of 12 whose sum is 7: $4 \cdot 3 = 12$ and $4 + 3 = 7$

B. $(a - 5)(a - 1)$; need two factors of 5 whose sum is -6: $(-5)(-1) = 5$ and $-5 + (-1) = -6$

C. $(3n - 2)(n + 2)$; use a guess-and-check strategy involving $3n$ and n and pairs of factors of -4

Lesson 10.7

55. 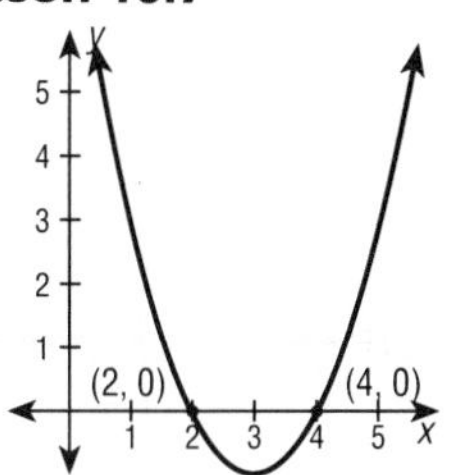

56. 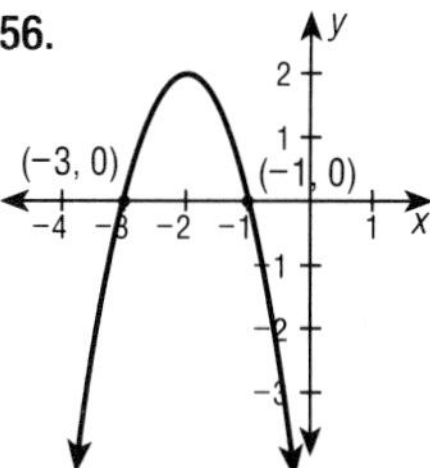

57. 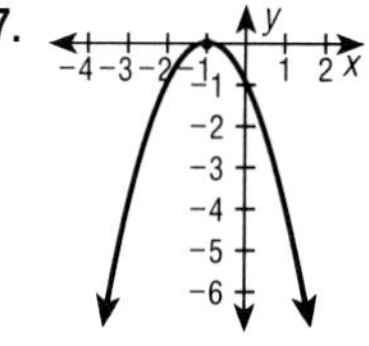

58. 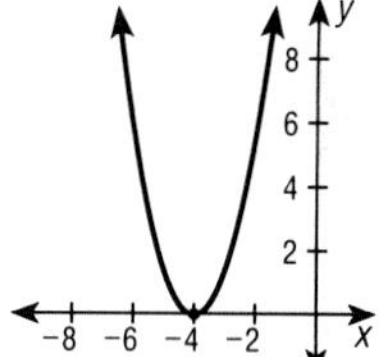

59. 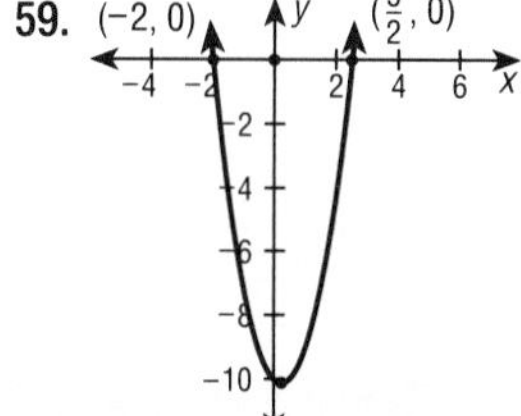

60. 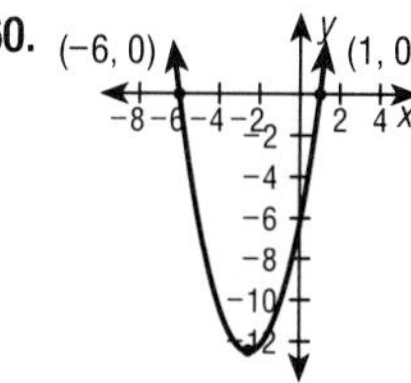

71. 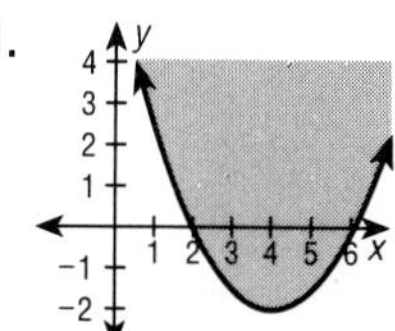

72. 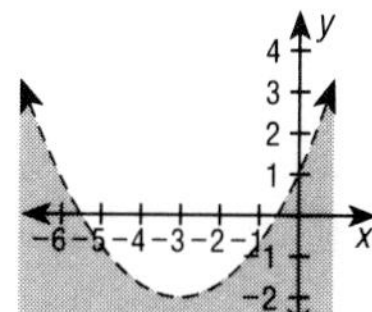

73. 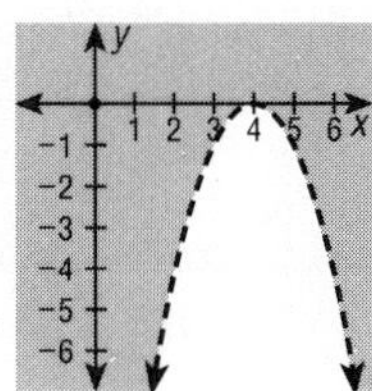

74. 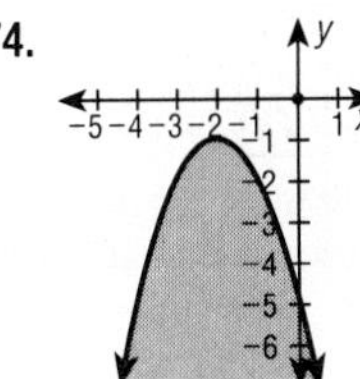

CHAPTER 11

Lesson 11.5

36. 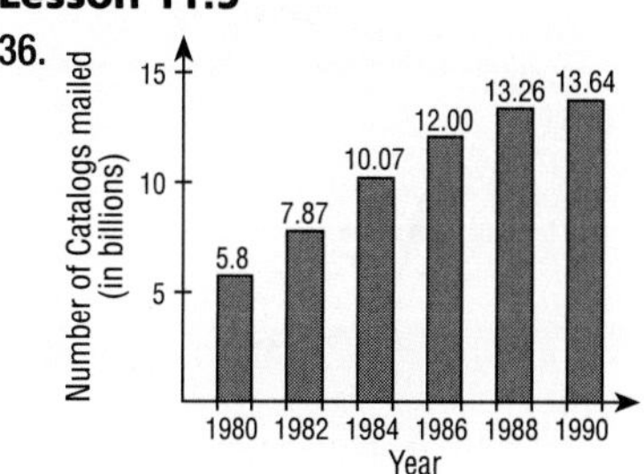

Lesson 12.2

27.
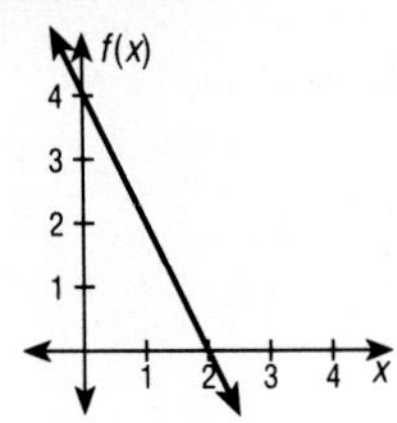

28.
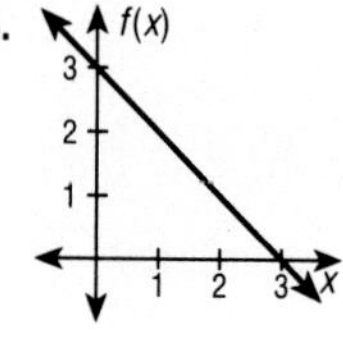

29.
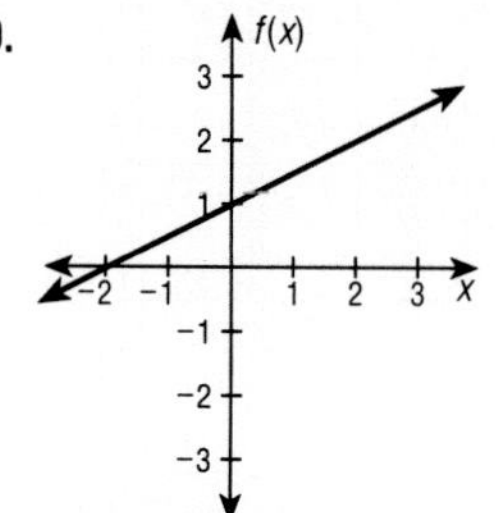

30.
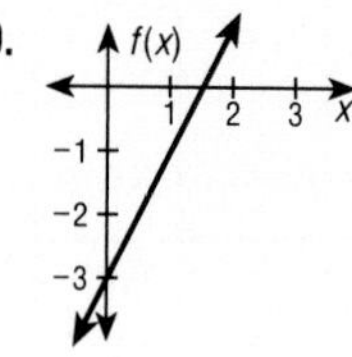

31.
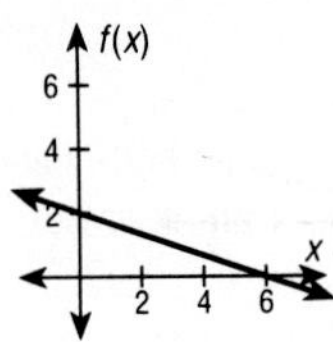

32.
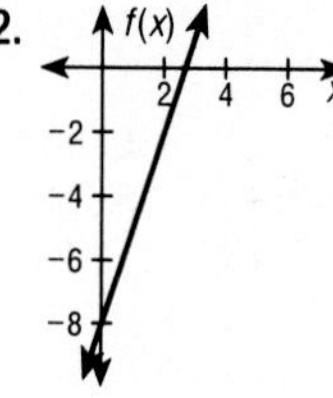

Lesson 12.3

A.

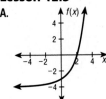

Shift the graph of $f(x) = 2^x$ four units down.

B.

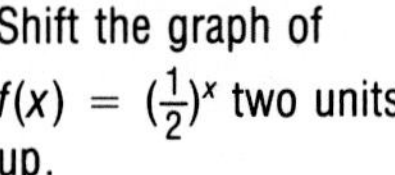

Shift the graph of $f(x) = (\frac{1}{2})^x$ two units up.

C.
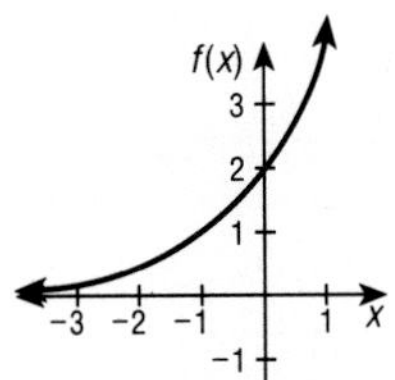

Shift the graph of $f(x) = 2^x$ one unit to the left.

D.
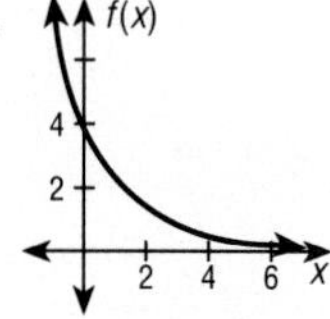
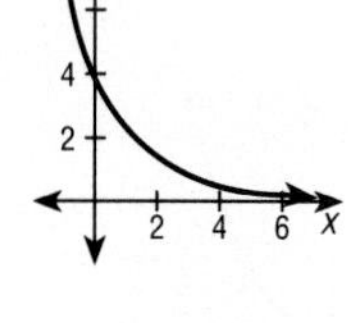

Shift the graph of $f(x) = (\frac{1}{2})^x$ two units to the right.

E.
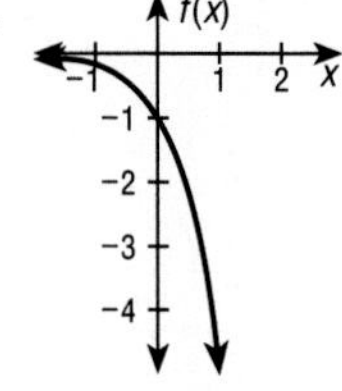

Reflect the graph of $f(x) = 3^x$ in the x-axis.

F.
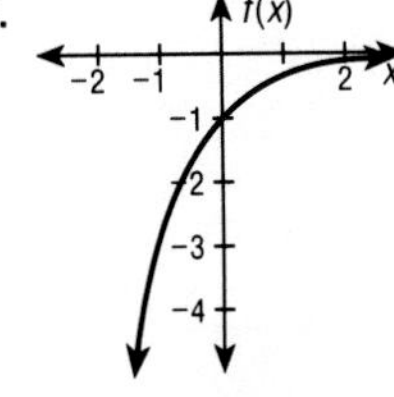

Reflect the graph of $f(x) = (\frac{1}{3})^x$ in the x-axis.

3.
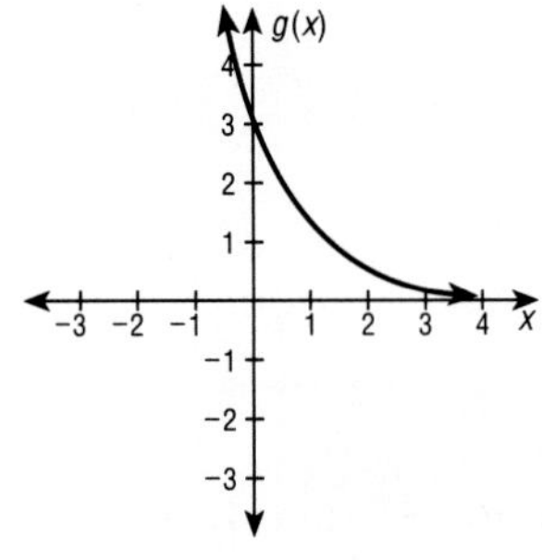

4.
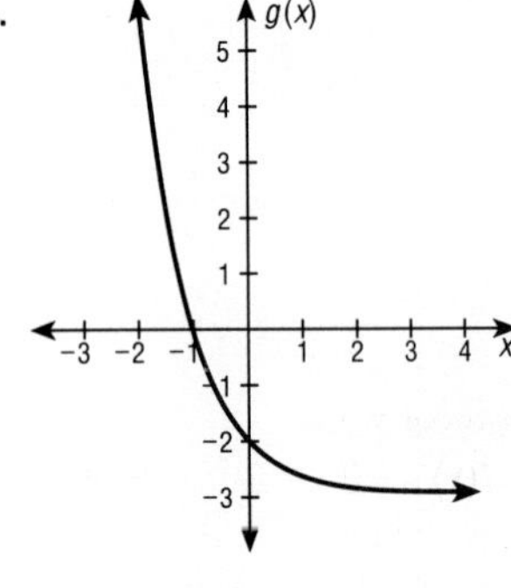

5.
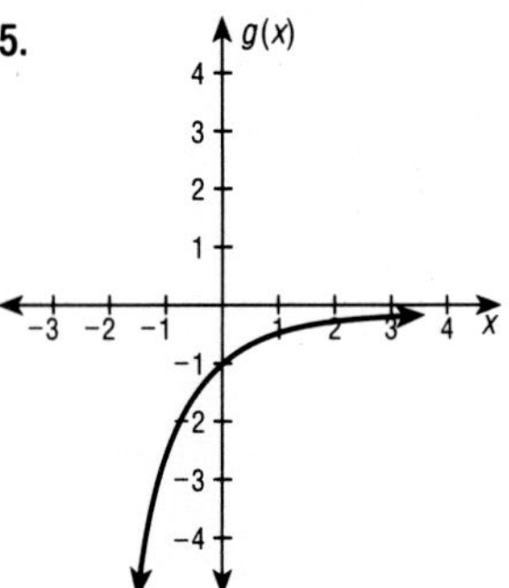

6.
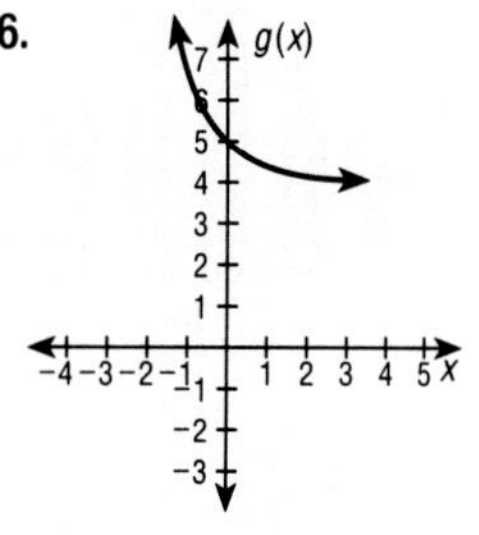

19.
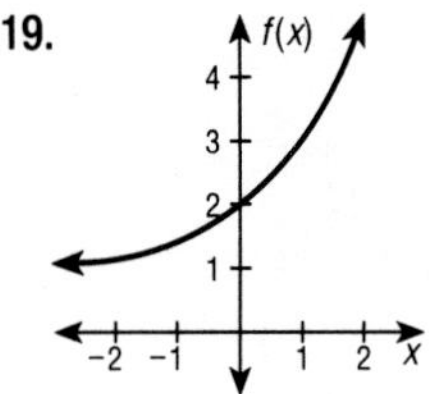

20.
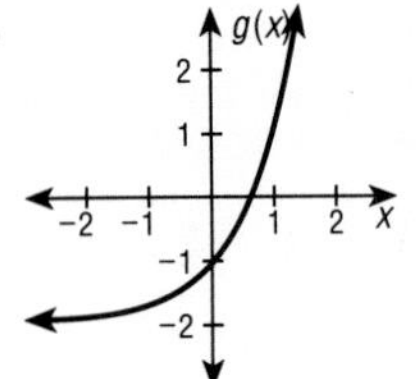

ADDITIONAL ANSWERS

Lesson 12.3 *(continued)*

21.

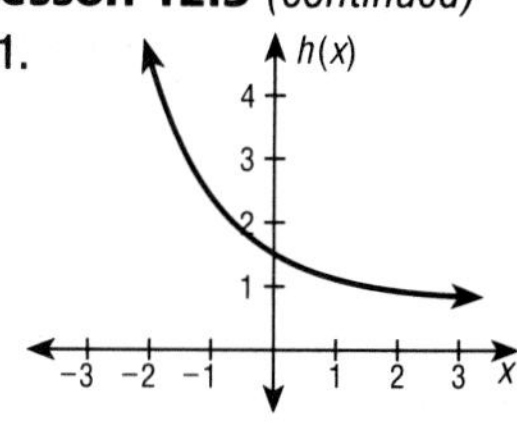

22.

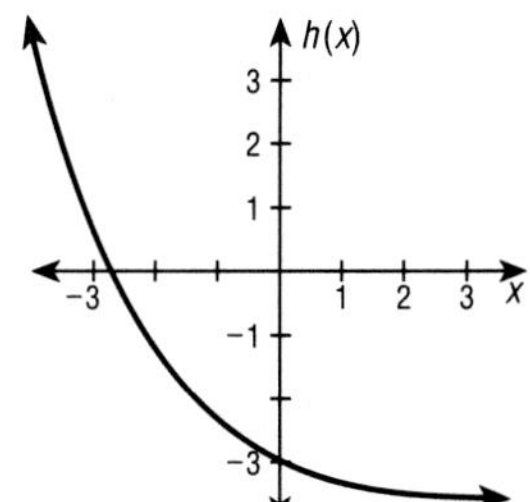

23.

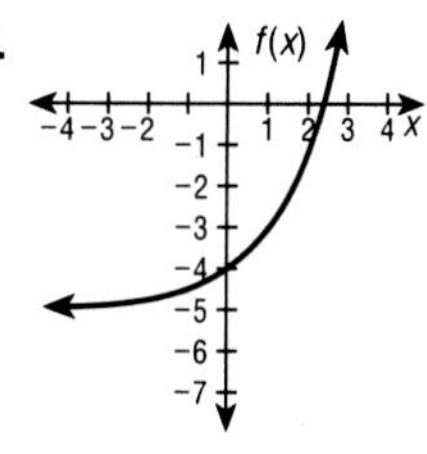

24.

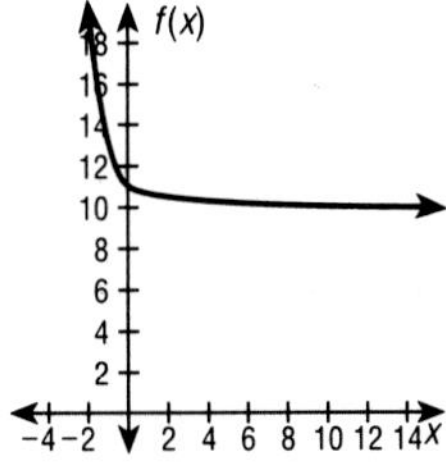

25.

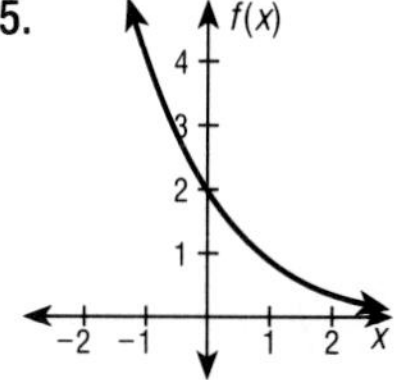

26.

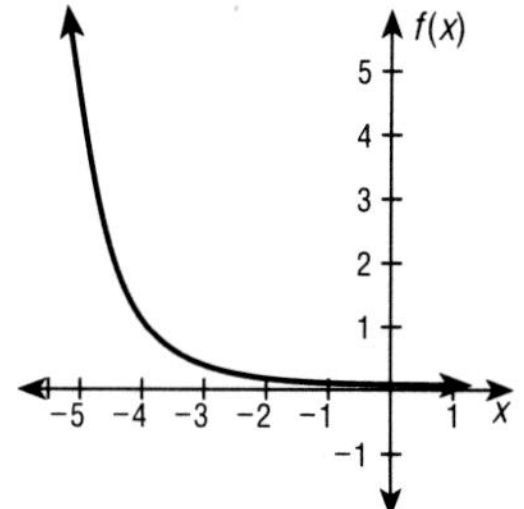

27.

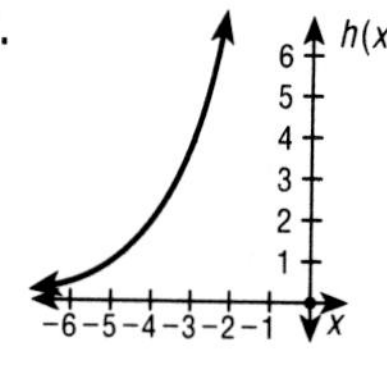

28.

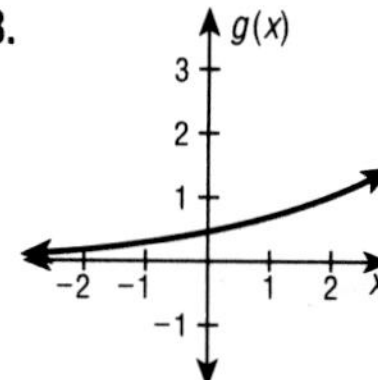

29.

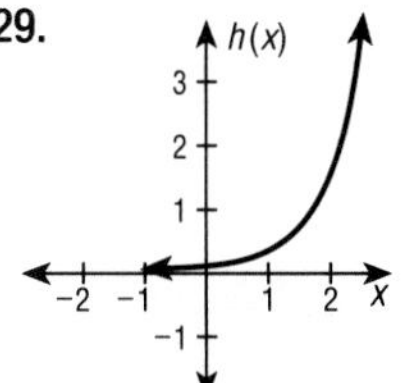

30.

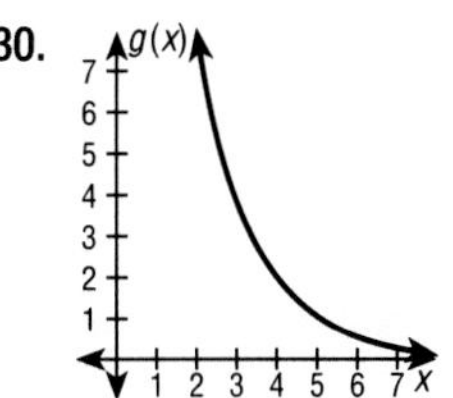

31.

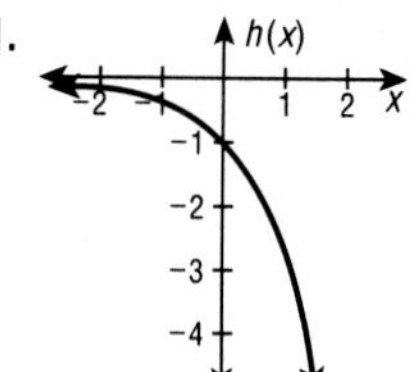

32.

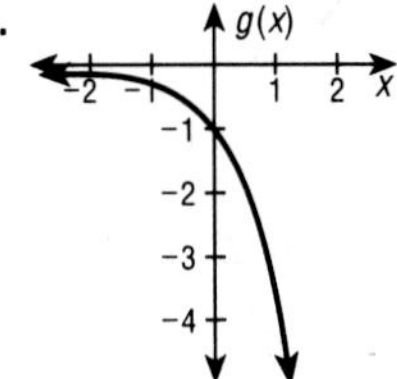

33.

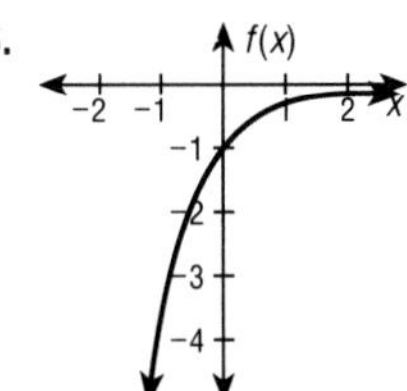

34.

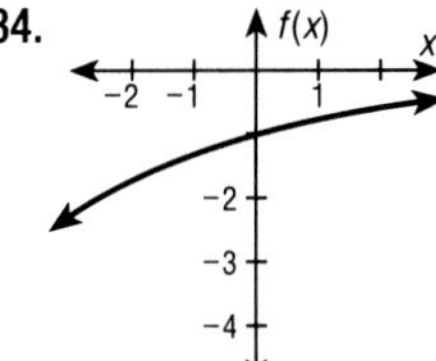

35.

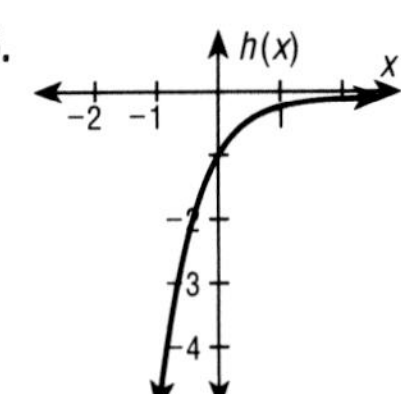

36.

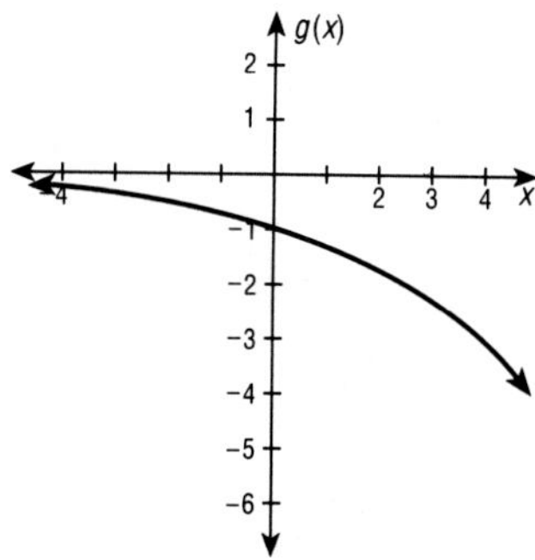

Lesson 12.4

A. $f(x) = (x - 2)^2 - 2$, vertex: $(2, -2)$

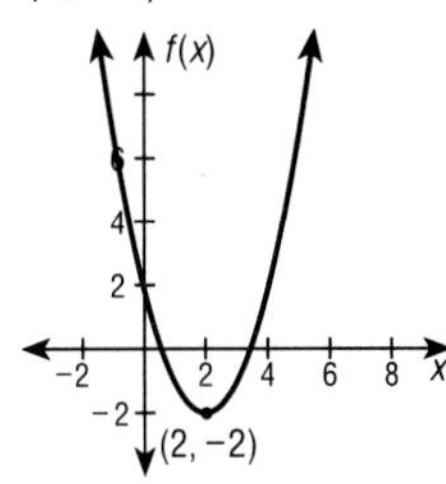

B. $f(x) = -(x + 1)^2 - 4$, vertex: $(-1, -4)$

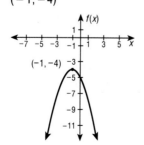

C. $f(x) = 2(x - 1)^2 + 5$, vertex: $(1, 5)$

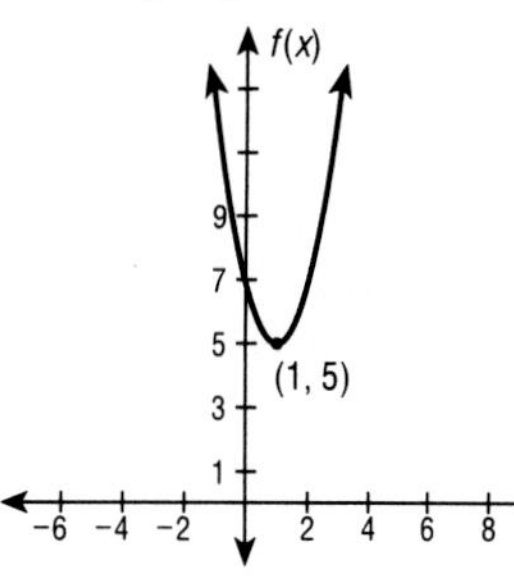

D. $f(x) = -3(x + 1)^2 - 1$, vertex: $(-1, -1)$

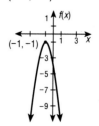

ADDITIONAL ANSWERS

Lesson 12.4 *(continued)*

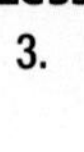

3.

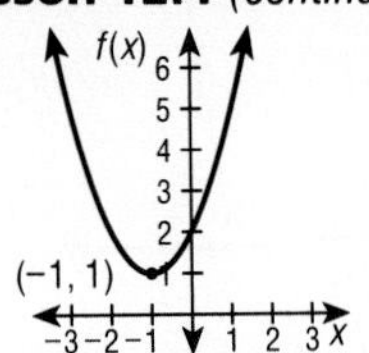

4.

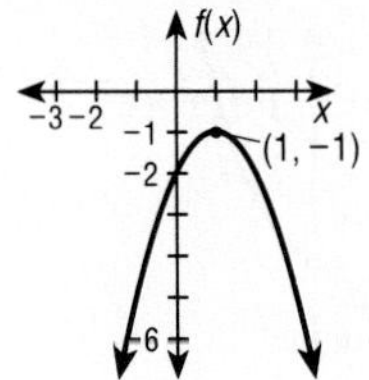

15.

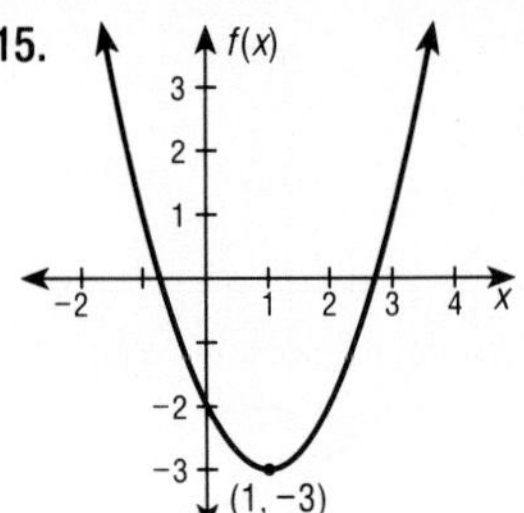

16.

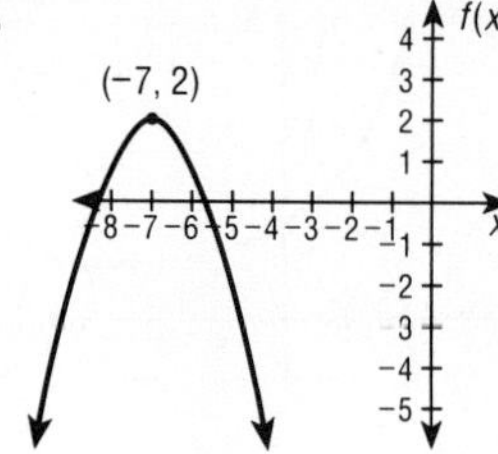

17.

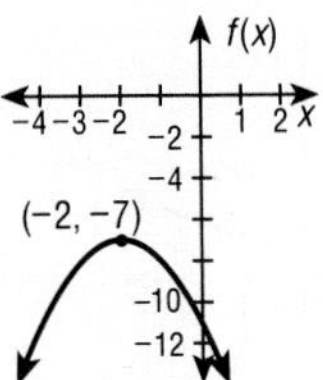

18.

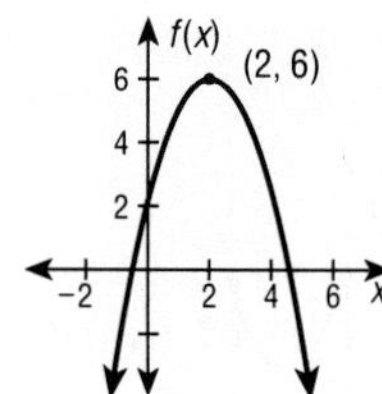

19.

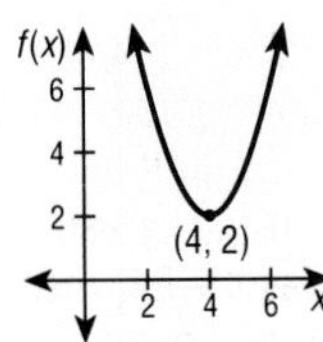

20.

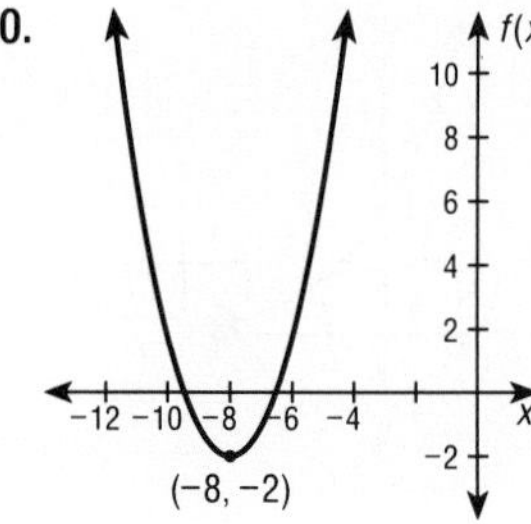

29.

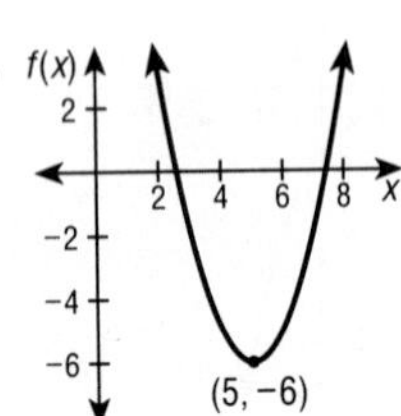

30.

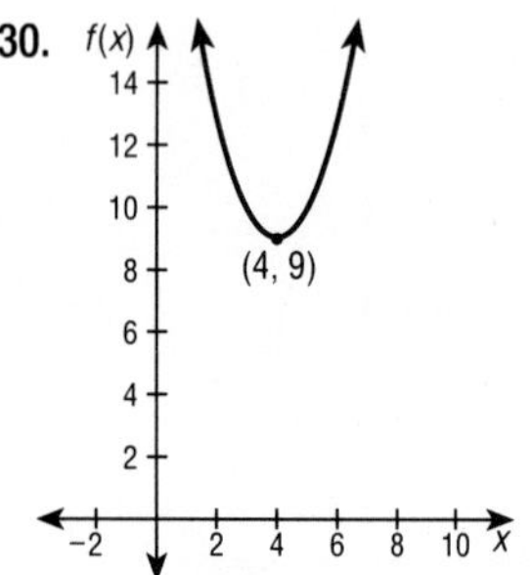

31.

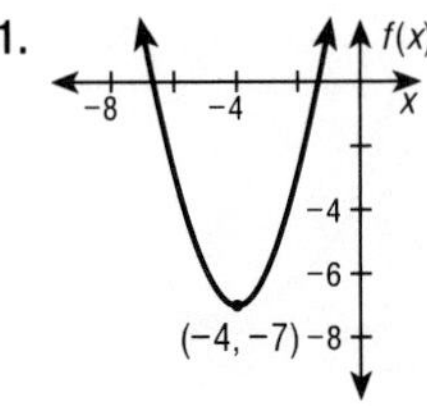

32.

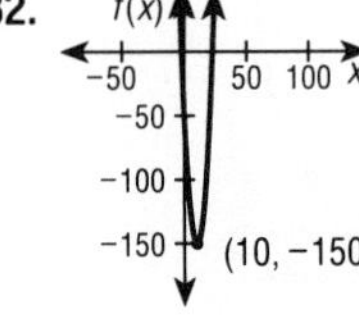

33.

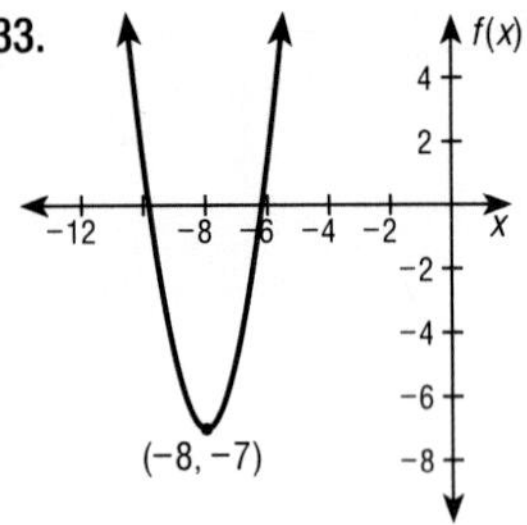

34.

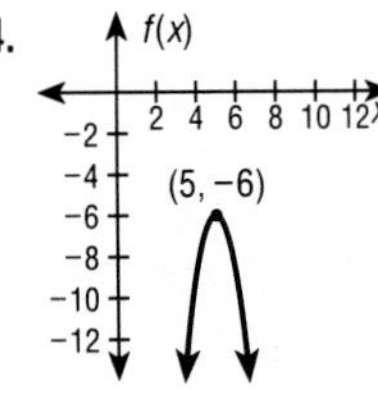

35.

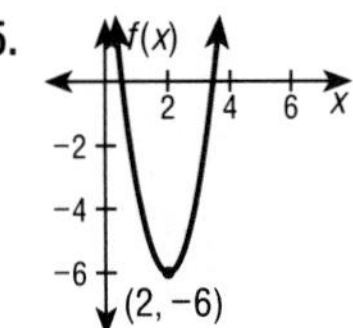

36.

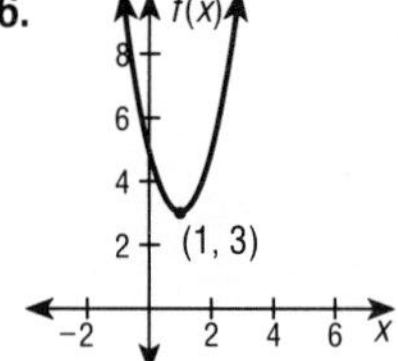

37.

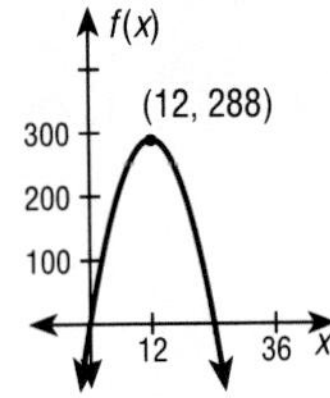

41.

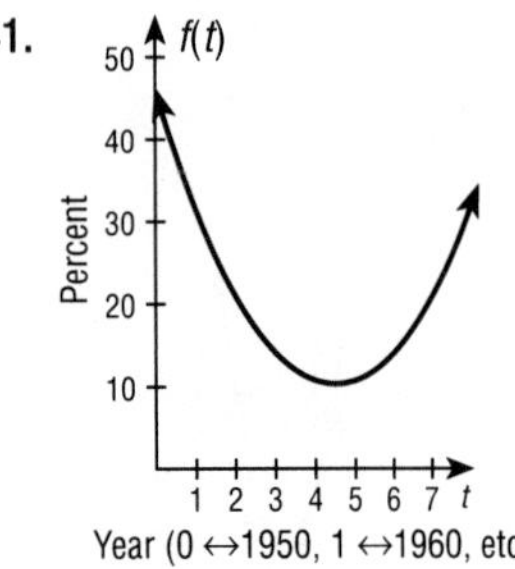

49.–54. The method in this lesson; each method enables you to easily find the *x*-coordinate of the vertex, but the method learned in this lesson enables you to find the *y*-coordinate of the vertex more easily.

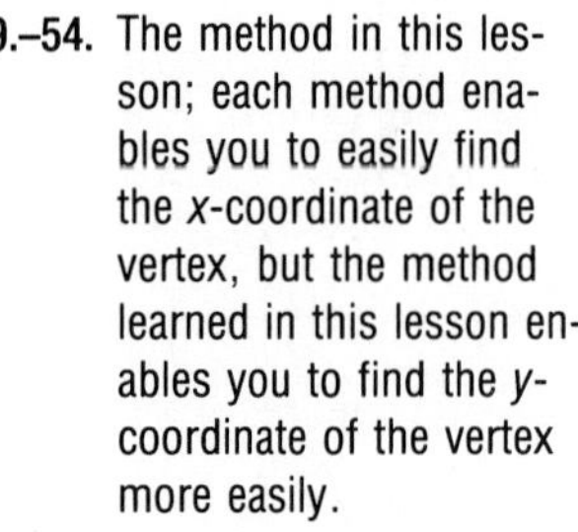

55.

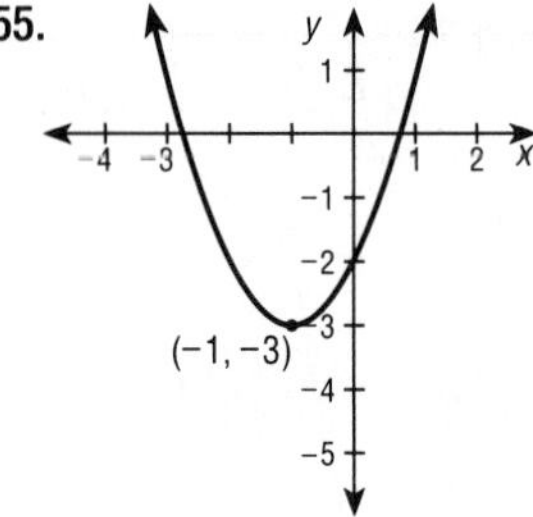

Lesson 12.4 *(continued)*

56.

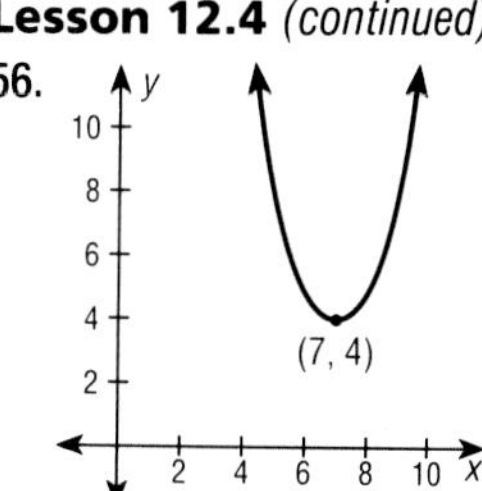

57.

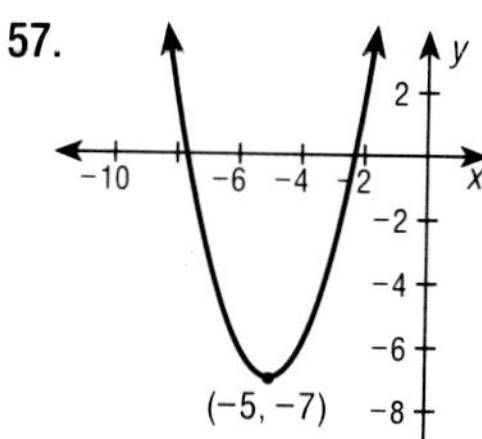

58.

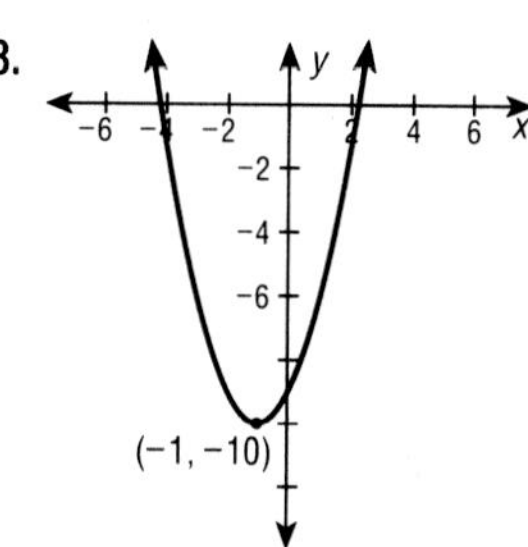

59.

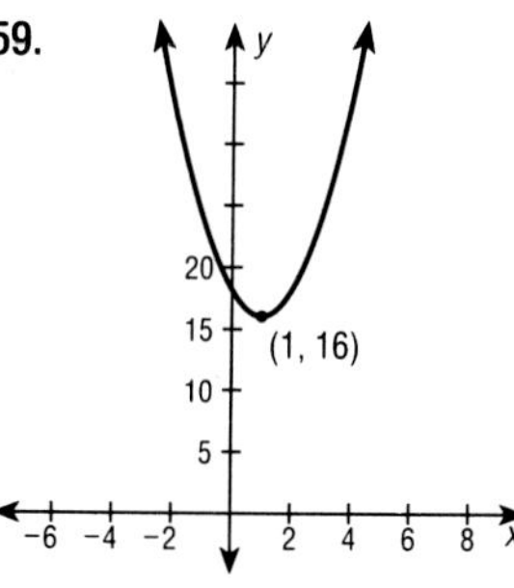

60.

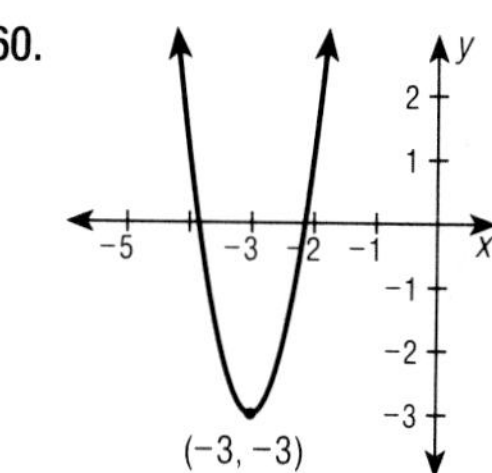

55.–60. Used the method in this lesson to find the vertex for the reason given in Exercises 49–54, then plot pairs of points that were the same horizontal distance from the vertex in order to easily draw a symmetric curve.

Mid-Chapter Self-Test

15. $f(x) = (x - 2)^2 + 5$

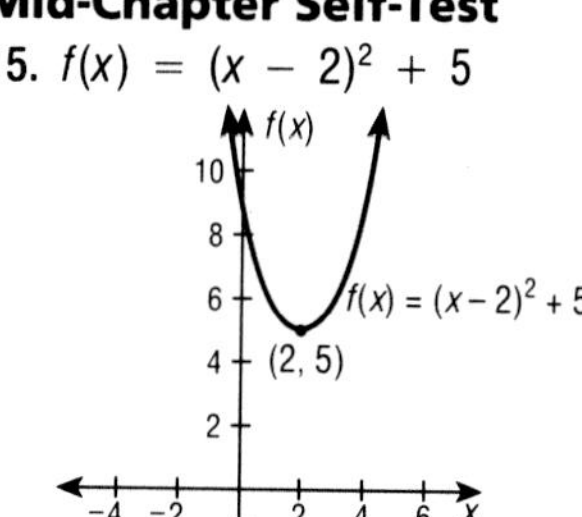

16. $f(x) = -(x + 1)^2 - 3$

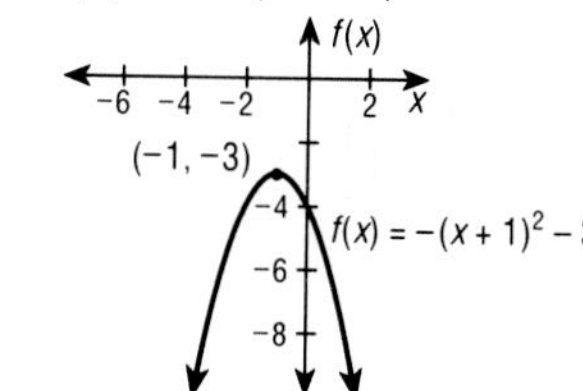

17. $f(x) = 2(x - 1)^2 - 9$

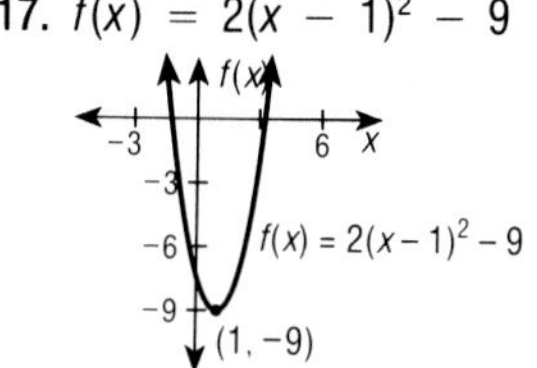

Lesson 12.5

A.

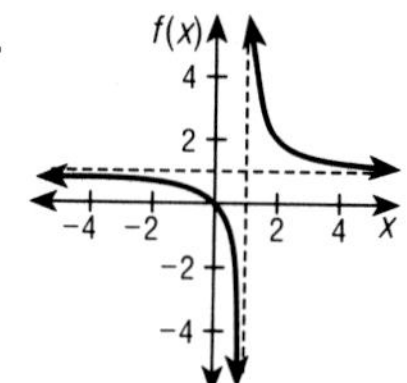

B.

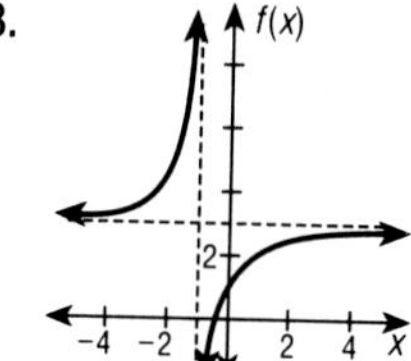

C.

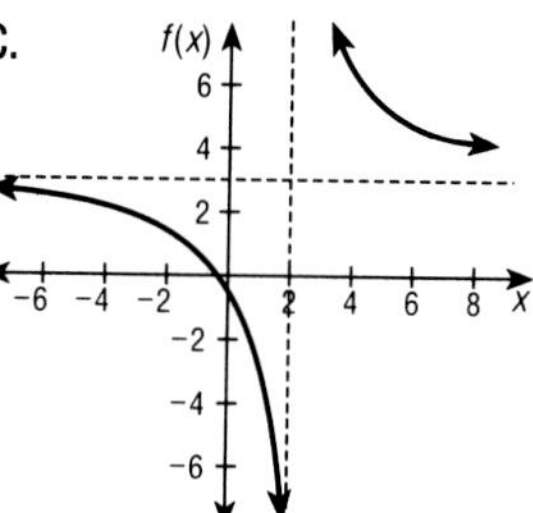

D.

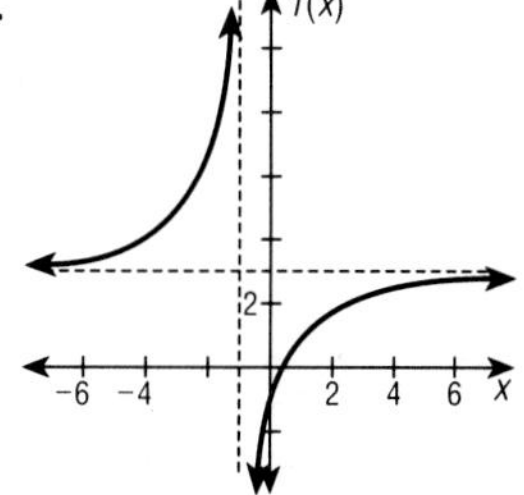

5.

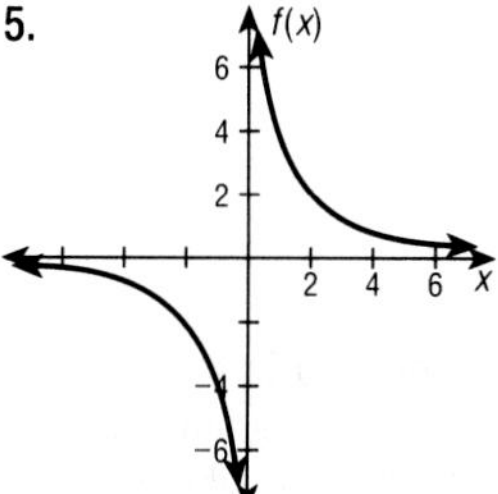

6.

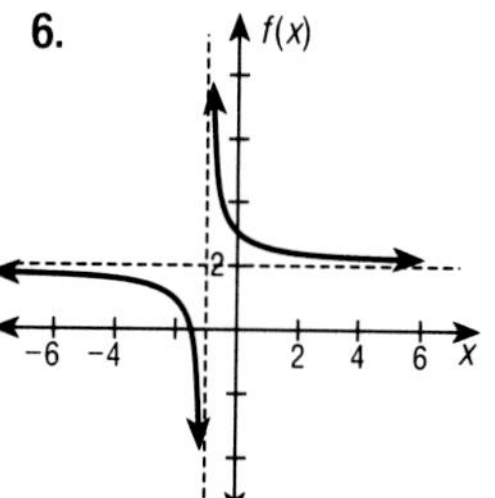

19.

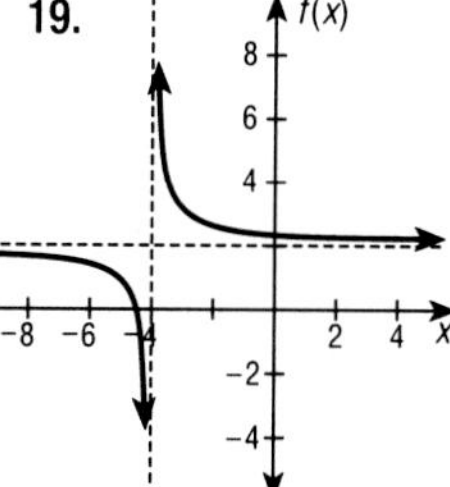

20.

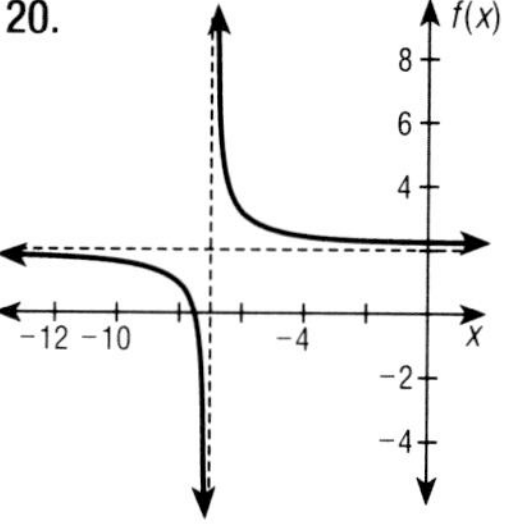

21.

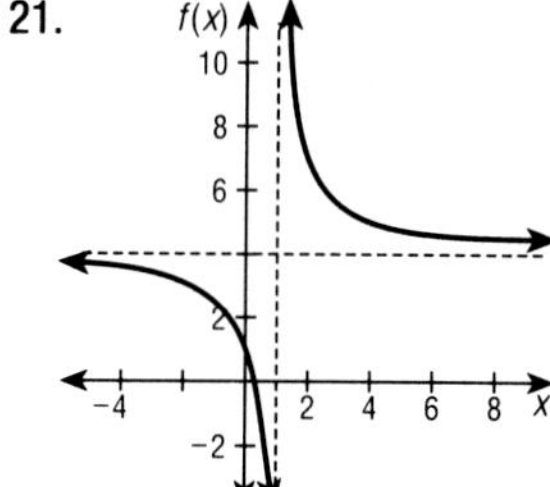

22.

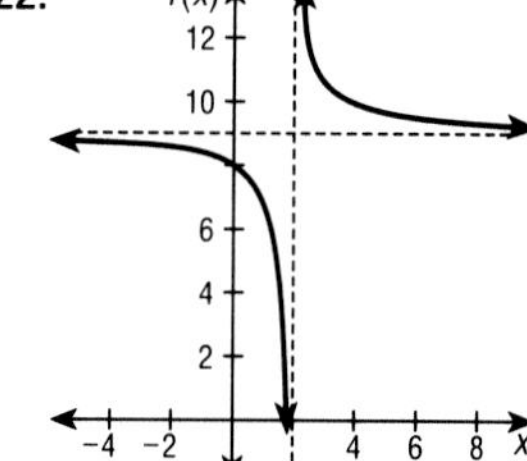

23.

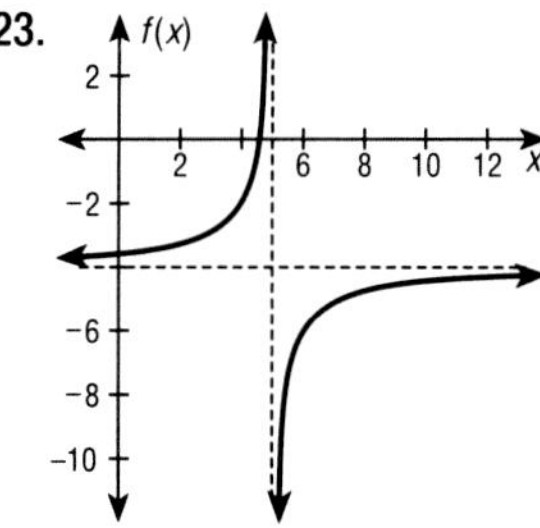

ADDITIONAL ANSWERS

Lesson 12.5 *(continued)*

24.
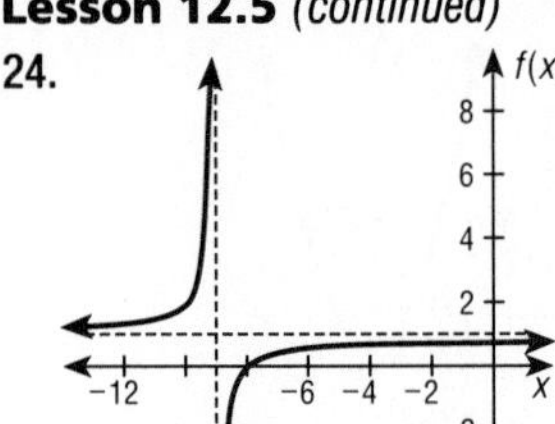

25.
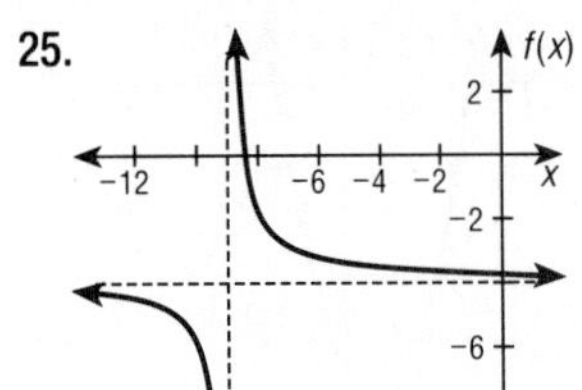

26.
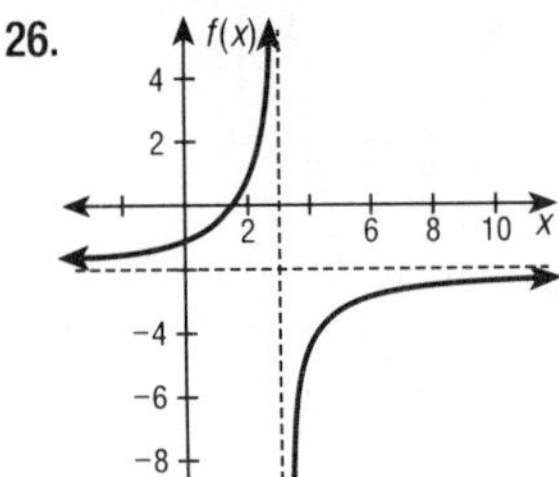

27.
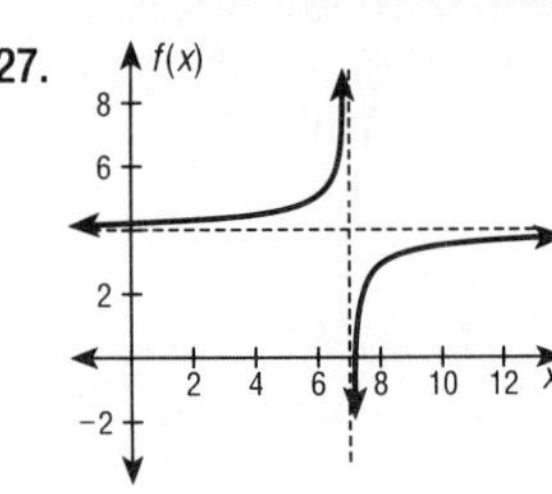

28.
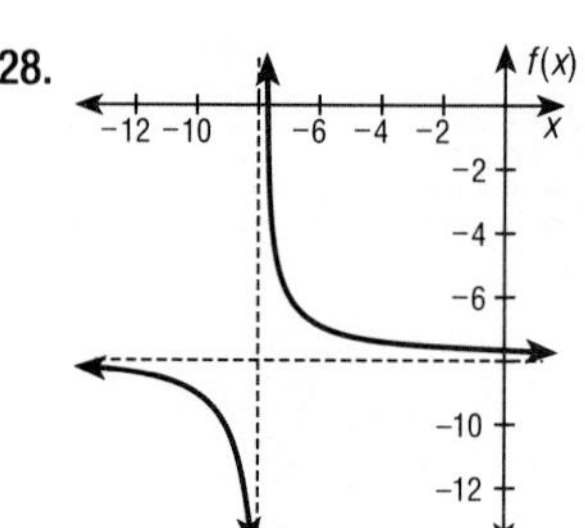

29.
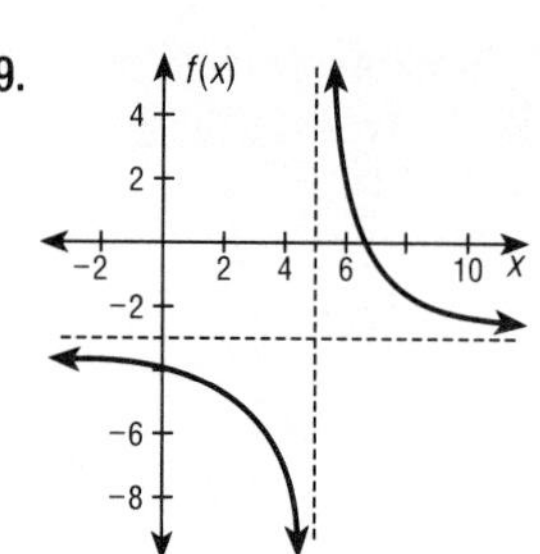

30.
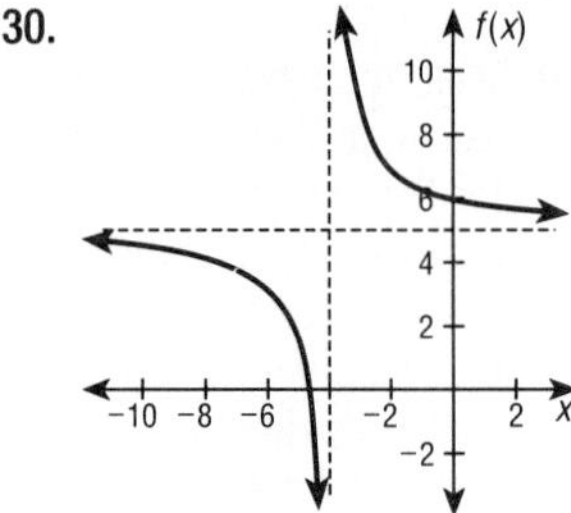

31.
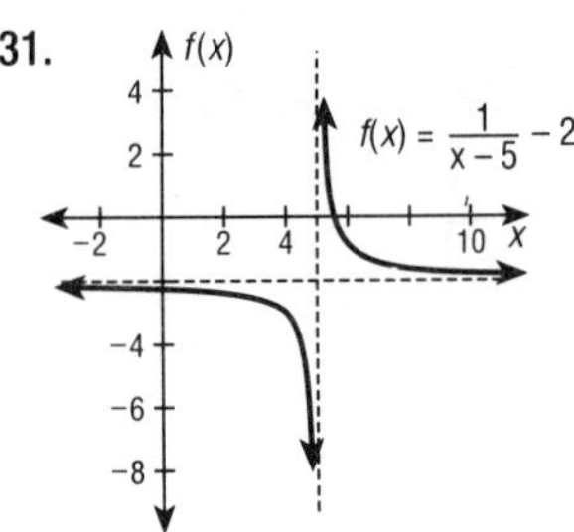

32.
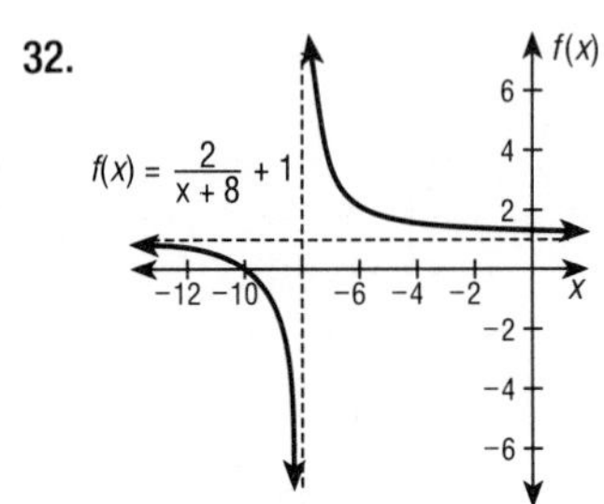

33.
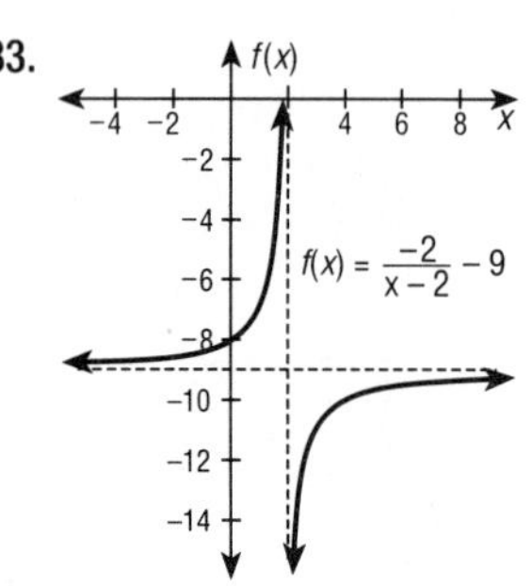

34.
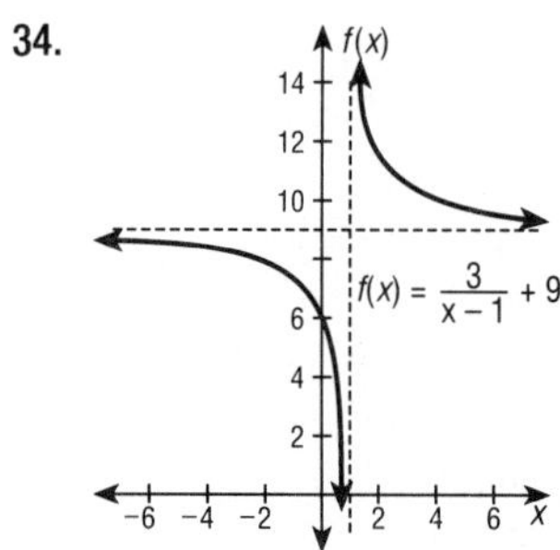

35.
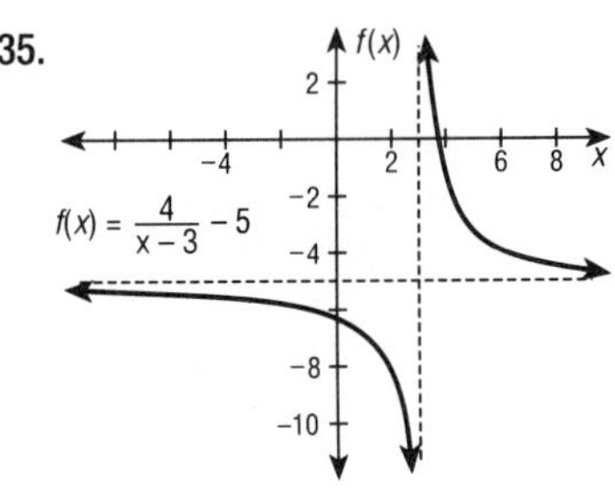

36.
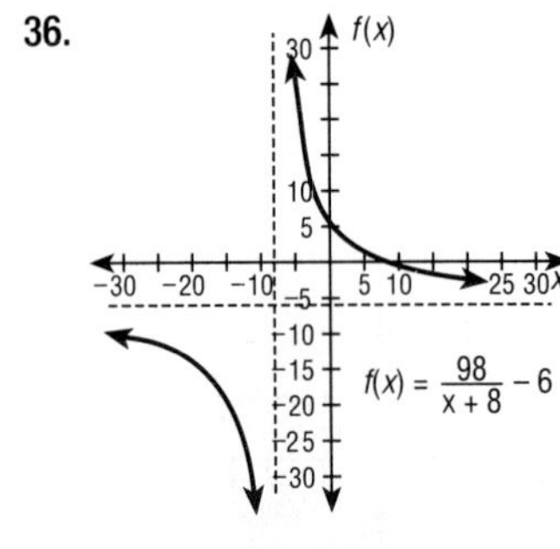

51.
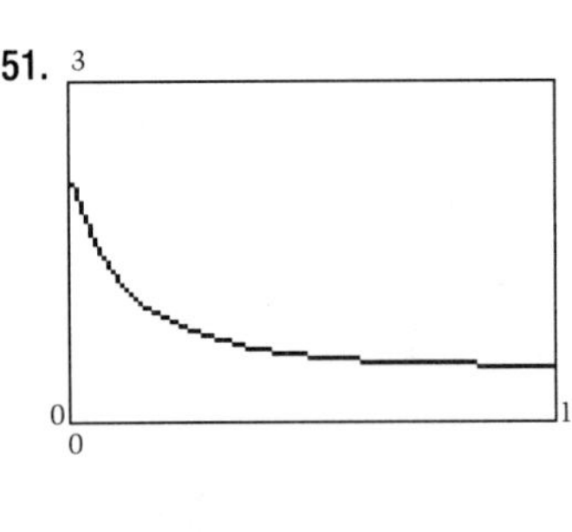

Mixed Review

1. $\frac{1}{2}x^{-2}y$
2. $-3, \frac{1}{2}$
3. 39
4. $3 \le m \le 5$
5. $\frac{1}{3x^4y^6}$
6. 29.67
7. 3.35×10^{-8}
8. $(-3, -34)$
9. 7, −9
10. $(-1, -2)$

Lesson 12.6

CAA

Stem	Leaf
7	5
5	9
4	1 3 7
3	0 0 3 6 7 8
2	2 2 3 4 5 6 8 9 9

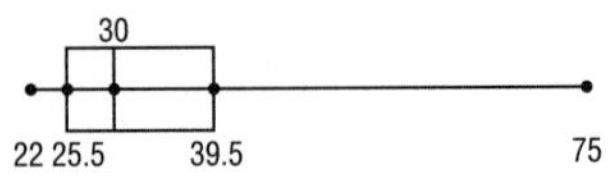

4.
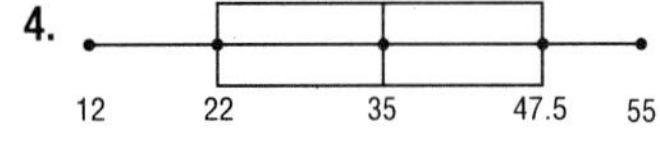

11.
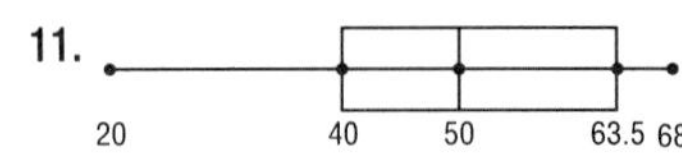

12.
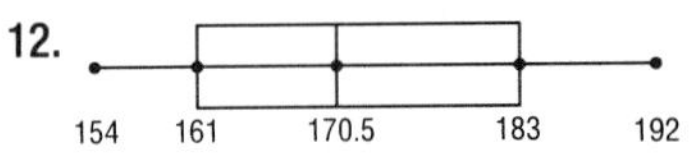

13.
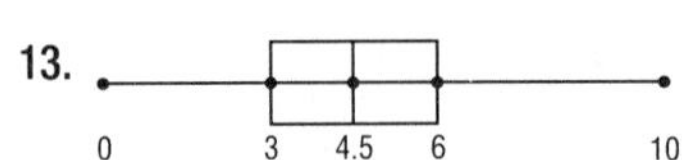

14.
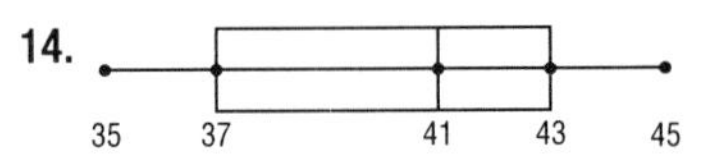

23. 1.8, 1.9, 2.0, 2.0, 2.3, 2.3, 2.3, 2.4, 2.4, 2.4, 2.9, 3.4, 4.0, 4.7, 4.8, 5.0, 5.0, 5.9, 7.1, 10.8

The box-and-whisker plot is better in showing how the data is clustered or spread out, while the bar graph is better in showing specific relationships among the data.

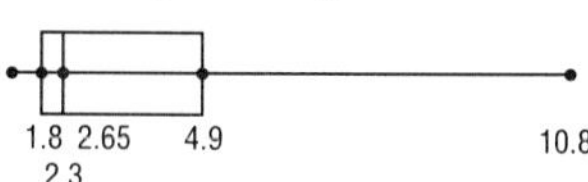

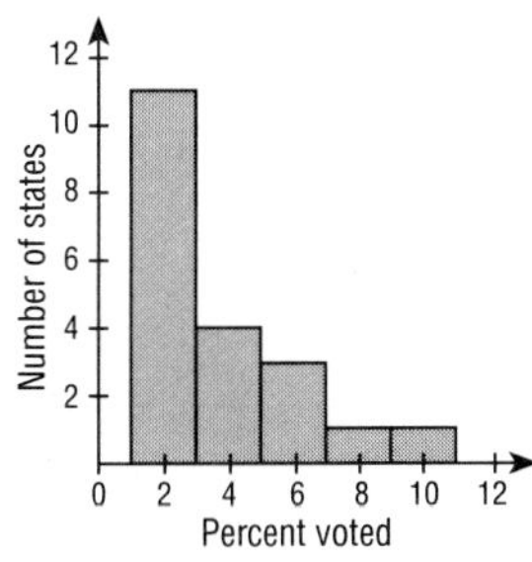

24. The "greatest number" and the "greatest percent" produce different data.

	State		
	A	B	C
Number voting	5	3	4
Population	10	4	6

For example, the state having the greatest number voting is State A, while the state having the greatest percent voting is State B.

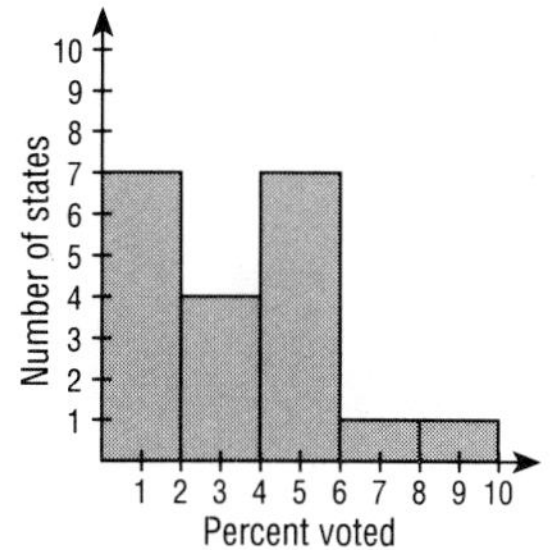

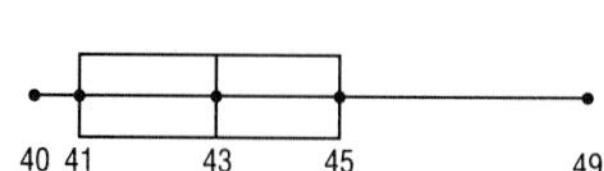

CHAPTER 13

Lesson 13.1

34.

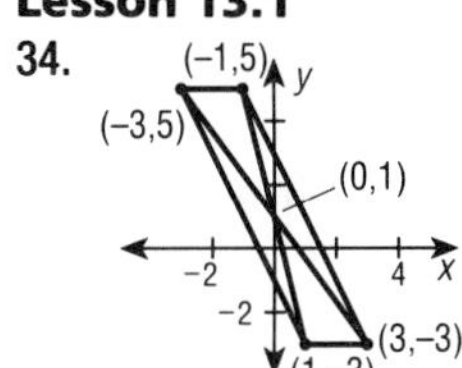

Lesson 13.4

A. Add 5 to both sides, square both sides.

B. Subtract 5 from both sides, square both sides, see that 25 is a false solution.

C. Add 2 to both sides, square both sides, subtract 4 from both sides, divide both sides by 3.

D. Square both sides, add $x - 8$ to both sides, factor, set each factor equal to zero, solve each factor for x.

Mixed Review

13.

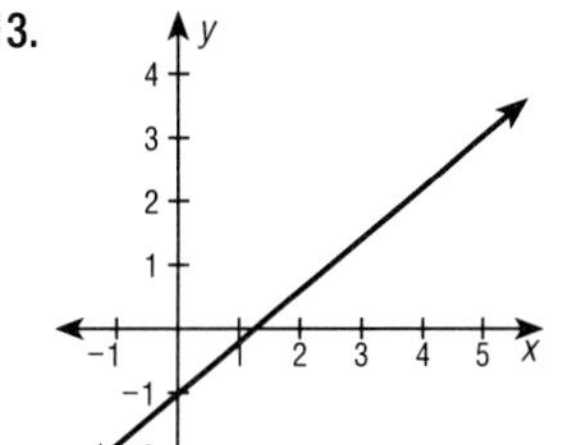

14.

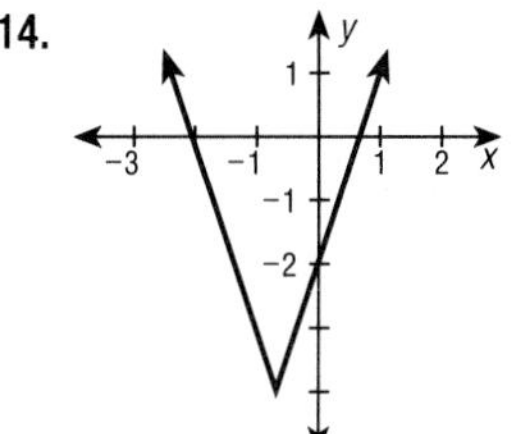

15.

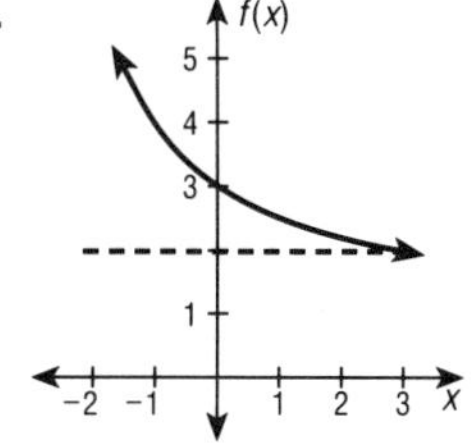

16.

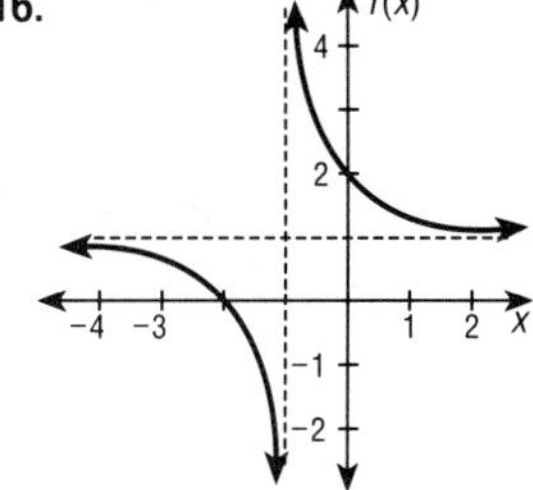

17.

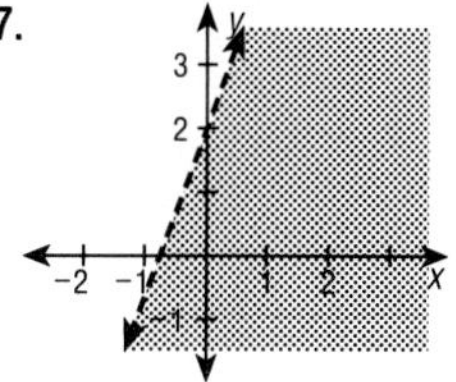

18.

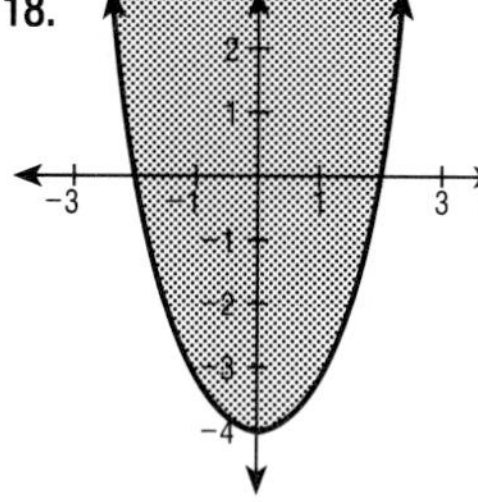

Lesson 13.6

5. $a-b=a+(-b)$ Def. of subtraction
$= -b+a$ Commutative axiom

6. $(a-b)-c=(a+(-b))+(-c)$ Def. of subtraction
$= a+((-b)+(-c))$ Associative axiom
$= a+((-1)b+(-1)c)$ Multiplication identity axiom
$= a+(-1)(b+c)$ Distributive axiom
$= a-(b+c)$ Multiplication identity axiom

A

B

C

D

INDEX

E

F

G

INDEX

N

INDEX

O

P

INDEX

Q

R

S

T

V

W

Z

Appreciation to the following art/photo production staff:

Leslie Concannon, Pam Daly, Irene Elios, Julie Fair, Martha Friedman, Susan Geer, Aimee Good, Carmen Johnson, Judy Kelly, Maureen Lauran, Mark MacKay, Helen McDermott, Penny Peters, Nina Whitney, Bonnie Yousefian.

ILLUSTRATION CREDITS

Calligraphy by **Jean Evans**
Illustration by **Pat Rossi and Associates**, p. 331 by **Cyndy Patrick**
Technical Illustration by **Tech-Graphics**

PHOTO CREDITS

vii: *t*, Michel Tcherevkoff (Image Bank); ***b***, Geraldine Prentice (Tony Stone Worldwide). viii: *t*, IKAN /Voigtmann (Peter Arnold, Inc.); *b*, Barrie Rokeach. ix: *t*, David Madison/Duomo; *b*, Peter Saloutos (The Stock Market). x: *t*, Murray and Associates (The Stock Market); *b*, Roger Ressmeyer (Starlight). xi: *t*, Y. Arthus Bertrand (Photo Researchers, Inc.); *b*, Alfred Pasieka (Peter Arnold, Inc.). xii: *t*, Woods Hole Oceanographic Institution; *b*, Manfred P. Kage (Peter Arnold, Inc.). xiii: NASA.

CHAPTER 1 0-1: Michel Tcherevkoff (The Image Bank). 6: The Arents Collections, The New York Public Library; Astor, Lenox and Tilden Foundations. 10: Alan Carey (The Image Works). 16: Dana Hyde (Photo Researchers, Inc.). 17: Mark Boulton (Photo Researchers, Inc.). 18: Stephan Johnson (Tony Stone Worldwide). 22: Richard Hutchings (Photo Researchers, Inc.). 25: Susan Doheny/©D.C. Heath. 26: Wesley Bocxe (/Photo Researchers, Inc.). 28: Bobbie Kingsley (Photo Researchers, Inc.). 31: Dion Ogust (The Image Works). 32: *b*, Zig Leszczynski (Animals Animals); *t*, John Garrett (Tony Stone Worldwide). 34: Will McIntyre (Photo Researchers, Inc.). 35: Jess Koppel (Tony Stone Worldwide). 38: Robert Tringali, Jr. (SportsChrome East/West, Inc.). 39: Bob Daemmrich Photography. 42: Kay Chernush (The Image Bank). 43: Mimi Forsyth (Monkmeyer Press). 43: Roger Dollarhide (Monkmeyer Press). 44: John Coletti (The Picture Cube). 45: Tony Stone Worldwide. 46: 1963 M.C. Escher/Cordon Art-Baarn-Holland. 48: 1955 M.C. Escher/Cordon Art-Baarn-Holland. 57: F.K. Schleicher © Vireo

CHAPTER 2 60-61: Geraldine Prentice (Tony Stone Worldwide). 65: Roger Ressmeyer (Starlight). 67: Tony Stone Worldwide. 70: Leonard Lessin (Peter Arnold, Inc.). 74: The British Museum (The Bridgeman Art Library, London). 76: John Neubauer (Monkmeyer Press). 79: Parker Brothers. 83: Giraudon (Art Resource). 86: Tony Freeman (PhotoEdit). 93: *b*, Frank Pedrick (The Image Works); *t*, Stephen Dalton (Animals Animals). 96: (SportsChrome East/West, Inc.). 105: Steve Maslowski (Photo Researchers, Inc.). 106: Tom & Pat Leeson (DRK). 108: Robert Frerck (Odyssey, Chicago). 116: Steve Kaufman (Peter Arnold, Inc.). 117: Scott Camazine (Photo Researchers, Inc.). 118: Don Griffin for H.A Perry Foundation.

CHAPTER 3 120-121: IKAN /Voigtmann (Peter Arnold, Inc.). 131: Sybil Shackman (Monkmeyer Press). 133: *b*, R.L. Kelly (Photo Researchers, Inc.); *t*, Larry Nicholson (Photo Researchers, Inc.). 137: Luis Castaneda (The Image Bank). 139: James H. Karales (Peter Arnold, Inc.). 141: Gunter Ziesler (Peter Arnold, Inc.). 144: Miriam Austerman (Animals Animals). 146: Susan Doheny/ © D.C. Heath. 147: Richard Hutchings (Photo Researchers, Inc.). 150: Roy Schneider (The Stock Market). 158: M.A.X. Mayer (The Stock Market).

CHAPTER 4 173-174: Barrie Rokeach. 178: F. Prenzel (Animals Animals). 192: John M. Burley (Photo Researchers, Inc.). 195: Lori Adamski-Peek (SportsChrome East /West, Inc.). 196: *t*, Culver Pictures; *b*,Vincent De Florio (Art Resource). 200: J.F. Causse (Tony Stone Worldwide). 207: Charles Gupton (Stock Boston). 211: North Wind Picture Archives. 215: Craig Milvin (SportsChrome East/West, Inc.). 221: *l*, Los Angeles County Museum of Natural History Foundation; *r*, Courtesy of the Southwest Museum, Los Angeles, Photo #22420. 222: Roger Ressmeyer (Starlight). 225: Maryland Department of Agriculture. 226: Julie Fair. 227: John Beatty (Tony Stone Worldwide). 228: *l*, Clyde H. Smith (/Peter Arnold, Inc.); *r*, Len Rue, Jr (Monkmeyer Press). 232: Lawrence Migdale

CHAPTER 5 234-235: David Madison/Duomo. 237: Spencer Grant (Stock Boston). 243: Paul Conklin (Monkmeyer Press). 247: James Foote (Photo Researchers, Inc.). 248: Susan Doheny/© D.C. Heath. 251: Bob Daemmrich (Stock Boston). 256: Mitchell B. Reibel (SportsChrome East/West, Inc.). 265: Brent Jones. 272: Pat Russo (SportsChrome East/West, Inc.). 275: Susan A. Anderson/© DC Heath. CLUE is a registered trademark of Waddingtons Games Ltd., England and is used with the kind permission of Waddingtons Games Ltd and their licensees Parker Brothers, Division of Tonka Corporation, Beverly, MA 01915, USA. 276: David Johnson (SportsChrome East/West,Inc.). 277: Peter Menzel (Stock Boston). 282: Frank Cezus (Tony Stone Worldwide). 288: Wayne Eastep (The Stock Market)

CHAPTER 6 290-291: Peter Saloutos (The Stock Market). 292: Robert Frerck (Odyssey, Chicago). 297: Raj Rama(LGI). 299: Larry Lefever (Grant Heilman Photography, Inc.). 301: *r*, Peter Menzel (Stock Boston); *l*, Fundemental Photographs. 302: Jay Freis (The Image Bank). 304: Culver Pictures. 309: Christie's, New York. 315: Steve Goldberg (Monkmeyer Press). 317: David Sutherland (Tony Stone Worldwide). 318: Bruce Iverson (Photomicrography). 322: Don Kincaid. 324: Pintos by Bev Doolittle © The Greenwich Workshop, Inc., Trumbull, CT . 336: Ralph Oberlander (Stock Boston). 339: Grant LeDuc (Monkmeyer Press)

CHAPTER 7 344-345: Murray and Associates (The Stock Market). 348: Richard Pasley (Stock Boston). 357: Herb Kawainui Kane. 359: Susan Doheny/© D.C. Heath. 362: North Wind Picture Archives. 365: Kentshire Gallery (Envision). 368: Steve Kaufman (Peter Arnold, Inc.). 370: Steven E. Sutton (Duomo). 376: George Holton (Photo Researchers, Inc.). 386: Courtesy of the author. 389: Charles Gupton (Tony Stone Worldwide).

CHAPTER 8 398-399: Roger Ressmeyer (Starlight). 402: David Sailors (The Stock Market). 404: Chahokia Mounds Historic Site. 408: Michael Springer (Gamma Liaison). 410: Jim Brandenburg (Minden Pictures). 411: Field Museum of Natural History (Neg. #GEO84564c), Chicago. 412: Bettmann Archives. 414: Tom & Pat Leeson (Photo Researchers,Inc.). 416: Eastcott/Momtiuk (Image Works). 417: David Madison (Duomo). 418: *r,* Charles Peake (Animals Animals); *l,* Robert A. Tyrrell (Animals Animals). 423: Cary Wolinsky (Stock Boston). 427: Nancy Simmerman (Tony Stone Worldwide). 428: *b,* Claudio Ferrer (Tony Stone Worldwide); *t,* H. Guckel, Center for Applied Microelectronics, University of Wisconsin. 430: Oscar Miller (Science Photo Library/Photo Reseachers, Inc.). 431: Ralph Wetmore (Tony Stone Worldwide) . 434: © Lennert Nilsson, *Being Born* (Bonnier Fakta). 437: Frans Lanting (Minden Pictures). 440: Louis Psihoyos (Matrix). 441: Bob Evans (Peter Arnold Inc.). 449: Barbara Cerva (DDB Stock Photo)

CHAPTER 9 450-451: Y. Arthus Bertrand (Photo Researchers, Inc.). 456: Peter Menzel. 457: Ezra Stoller (Esto). 462: Breck P. Kent. 464: Susan Doheny/© D. C. Heath. 467: Wide World Photo. 474: Grant LaDuc (Monkmeyer Press). 476: John Elk III (Stock Boston). 482: Frank Oberle (/Photographic Resources, Inc.). 485: Steve Goldberg (Monkmeyer Press). 487: Simon Jauncey (Tony Stone Worldwide). 490: R. Weldon (Gemological Institute of America). 494: Scott Camazine (Photo Researchers, Inc.). 497: Roger Dollarhide (Monkmeyer Press). 500: Fermilab Photography. 502: *c,* Hugh Rogers (Monkmeyer Press); *l,* Susan Van Etten (Tony Stone Worldwide); *r,* Frans Lanting (Minden Pictures).

CHAPTER 10 506-507: Alfred Pasieka (Peter Arnold, Inc.). 511: Robert Frerck (Tony Stone Worldwide). 517: Larry Lefever (Grant Heilman Photography, Inc.). 519: Spooner (Gamma Liaison). 521: Culver Pictures. 530: Michael Holford. 532: *b,* Mimi Forsyth (Monkmeyer Press); *t,* British Museum, London (Bridgeman Art Library). 537: Rick Brown (Stock Boston). 539: *bl,* Photograph by Eliot Elisofon, National Museum of African Art, Eliot Elisofon Archives, Smithsonian Institution; *br,* Photograph by Franko Khoury, National Museum of African Art, Eliot Elisofon Archives, Smithsonian Institution; *tl,* Photograph by Smithsonian Institution Staff, National Museum of African Art, Eliot Elisofon Archives, Smithsonian Institution; *tr,* Peabody Museum, Harvard University. Photograph by Hillel Burger. 545: *b,* Rameshwar Das (Monkmeyer Press); *t,* Chip Deutsch (Discover Magazine). 547: Ilene Perlman (Stock Boston). 551: Gabe Palmer (The Stock Market). 553: Paolo Koch (Photo Researchers, Inc.). 557: *b,* Zig Leszczynski (Animals Animals); *t,* W. Gregory Brown (Animals Animals). 558: Terry Farmer (Tony Stone Worldwide).

CHAPTER 11 560-561: Woods Hole Oceanographic Institution. 563: Don L. Crawford (Tony Stone Worldwide). 564: Joe Carini (The Image Works). 566: *l,* Mark C. Bernsau (The Image Works); *r,* Nancy Fix (Stock Boston). 569: Richard Hutchings (Photo Researchers, Inc.). 570: *t,* Pat & Tom Leeson (Photo Researchers, Inc.); *bl,* Ray Coleman (Photo Researchers, Inc.); *br,* John Serrao (Photo Researchers, Inc.). 574: Susan Doheny © D.C. Heath. 576: Frank Pedrick (The Image Works). 577: *l,* The Bettmann Archive; *r,* Culver Pictures. 579: Craig Blouin (New England Stock). 582: Martha Swope. 585: Jose Luis G. Grande (Photo Researchers, Inc.). 592: *t,* C.S. Gray (The Image Works); *b,* Mark Antman (The Image Works). 602: M. Siluk (The Image Works). 604: Philip & Karen Smith (Tony Stone Worldwide). 609: Carol Lee (Tony Stone Worldwide). 611: Ulrike Welsch (Photo Researchers, Inc.). 612. Jim Brandenburg (Minden Pictures). 614: *l,* Tom Nebbia; *r,* Gianfranco Gorgoni (Contact). 617: Steven Mark Needham (Envision)

CHAPTER 12 618-619: Manfred P. Kage (Peter Arnold, Inc.). 622: Nancy Ploeger (Monkmeyer Press). 627: Patti Murray (Animals Animals). 634: Del Mulkey (Photo Researchers, Inc.). 635: Bob Daemmrich (Tony Stone Worldwide). 638: Obremski (The Image Bank). 645: E.R. Degginger. 648: Steven Mark Needham (Envision). 650: Paul Conklin (Monkmeyer Press). 654: Mimi Forsyth (Monkmeyer Press). 659: Joe Devenney (The Image Bank). 661: Stacy Pick (Stock Boston). 662: Michael Dick (Animals Animals). 667 : *b,* Zig Leszczynski (Animals Animals); *t,* Peabody Museum of Salem. Photo by Mark Sexton

CHAPTER 13 670-671: NASA. 684: Susan Doheny © D.C.Heath. 695: Luis Castaneda (The Image Bank). 695: Steven E. Sutton (Duomo). 701: Peter Chapman (Stock Boston). 703: George Hall (Check Six). 704: Grant LaDuc (Monkmeyer Press). 706: Michael Holford. 714: Edna Douthat (Photo Researchers, Inc.). 716: Richard Pasley (Stock Boston)

COVER DESIGN

Linda Fishborne (Hemenway design).
Background Photo: (THE IMAGE BANK) Rob Atkins
Inset Photo: (COMSTOCK INC.) Lynn Eskenazi